1994 IEEE FIRST WORLD CONFERENCE ON PHOTOVOLTAIC ENERGY CONVERSION

CONFERENCE RECORD OF THE
TWENTY FOURTH IEEE PHOTOVOLTAIC SPECIALISTS CONFERENCE-1994

HILTON WAIKOLOA VILLAGE
WAIKOLOA, HAWAII
DECEMBER 5-9, 1994

VOLUME I

SPONSORED BY THE IEEE ELECTRON DEVICES SOCIETY (USA)

IEEE CATALOG NUMBER: 94CH3365-4

ISBN: 0-7803-1459-X Softbound
ISBN: 0-7803-1460-3 Casebound
ISBN: 0-7803-1461-1 Microfiche

LIBRARY OF CONGRESS: 93-61055

* *

Included in this Publication
24th IEEE PHOTOVOLTAIC SPECIALISTS CONFERENCE
ISSN: 0160-8371
LIBRARY OF CONGRESS: 84-64012

CHAIRPERSON'S MESSAGE

The First World Conference on Photovoltaic Energy Conversion represents a landmark event in our field. It is the first time a jointly planned meeting has ever been held by the three major R&D communities from the U.S., Japan, and Europe. Although it was not necessarily our goal, it also stands as the largest conference to-date dedicated exclusively to photovoltaics. Over 1200 scientists, engineers, government program managers and representatives from industry came together to share results from their research, development, applications experiences and efforts to commercialize photovoltaic technology with attendees from 43 countries around the globe. Though that fact in itself is not significant, what it may mean is. After over three decades of research and development, the reality of a broadly based, large-scale deployment of photovoltaic technology in a variety of commercial applications appears close at hand. This has to come as good news, not only to those of us intimately involved in the field, but to persons around the world who wait patiently for their own entry into the energy sufficiency that the industrialized nations have enjoyed for generations.

As you will discover when you examine these volumes, the conference contained an excellent mix of papers ranging from research reports on a variety of advanced devices to results of pilot commercialization projects. Special highlight sessions and workshops on the more practical problems relating to financing, installing and operating photovoltaic power generating systems were conducted in conjunction with the World Bank, the State of Hawaii, the U.S. Department of Energy and several private concerns as well. A strong area on space photovoltaic device development and system applications was also included.

Two special events were included with this conference. The first was a public exposition held on Sunday, Dec. 4, which provided an opportunity for local residents to gain firsthand knowledge of the potential of photovoltaic power systems, and to view the state-of-the art in commercial products displayed by the nearly 60 companies that filled the exhibits area to overflowing. The second was the school program. Although it is a regular feature of the U.S. IEEE PVSC, and usually involves schools from the immediate vicinity, this program involved all the high schools from the entire State of Hawaii, and included most of the junior high schools from the Island of Hawaii as well. In all, nearly 150 students participated in model solar car competitions and engineering and design projects. Their enthusiasm, energy and creativity are reassurance that staunch advocates for photovoltaic technology will abound in the generations to come.

This conference was organized by combining representatives from the executive committees of the IEEE Photovoltaic Specialists Conference in the U.S., the Photovoltaic Science and Engineering Conference from Japan, and the Photovoltaic Solar Energy Conference from the European Union into one Executive Committee. The dedication and support of each of those individuals is gratefully acknowledged. Over 750 abstracts were received from more than 50 countries, and were carefully reviewed by the program committee; their hard work and dedication to the task of making the content of the meeting as reflective as possible to the total spectrum of Interests of each of the three communities was spectacularly successful.

The success of this meeting has created agreement among the three communities to hold such a conference on a four year cycle. Responsibility for hosting it will revolve from one community to the next. It is my pleasure to announce that the Second World Conference on Photovoltaic Energy Conversion will be held in Vienna, Austria in the 1998-1999 time frame. The 2nd WCPEC will mark the first time that all three meetings will be held jointly: The 10th PVSEC, the 15th PSEC and the 27th IEEE PVSC will all be combined into one. Please mark your calendars now for what should be another history-making event.

Dennis J. Flood

EXECUTIVE COMMITTEE

General Chairperson
Dennis J. Flood
General Vice-Chairpersons
Masafumi Yamaguchi
Jurgen Schmid
Program Chairperson
John D. Meakin
Program Vice-Chairpersons
Makoto Konagai
Gerald Wrixon
International Co-Chairpersons
Allen M. Barnett
Yoshihiro Hamakawa
Leopoldo Guimaraes
Conference Administration
Americo F. Forestieri
Treasurer
Eldon C. Boes
Vice-Treasurers
Takashi Fuyuki
Peter Helm
Secretary
Ajeet Rohatgi
Vice-Secretarys
Hideyuki Takakura
Peter Helm
State of Hawaii Liaison
H. M. Hubbard
Industrial Liaison
David E. Carlson
Masayoshi Umeno
John Bonda
DOE Liaison
Lloyd Herwig
School Program
Lawrence Kazmerski

Publicity
Howard E. Pollard
Shoichi Nakano
H. Ehmann
Operations
Tim Coutts
Registration
Wendy Larsen
Developing Countries
Carlos Vargas-Aburto
Choochon Lee
A. K. Barua
Exhibits
Ronald M. Diamond
James Albeck
Yukinori Kuwano
Thomas Nordmann
Exposition
Richard Schwartz
Charles E. Backus
Posters
John Benner
Terunori Warabisako
Hans Schock
Tutorials
James Gee
Tadashi Saitoh
W. F. Van der Weg
Publication
Sheila Bailey
Hiroshi Sakai
Gerhard Willeke
Audit Committee
Cosmo Baraona
Social Program
Kirk Collier

PROGRAM COMMITTEE

Chairperson
John D. Meakin

Vice-Chairpersons
Makoto Konagai
Gerald Wrixon

Area Co-Chairs

1. Silicon Cells and Materials
Ted F. Ciszek
Martin A. Green
W. Wettling

2. II-IV and Related Materials and Devices
Bulent Basol
Tokio Nakada
Hans W. Schock

3. III-V and Multi-Junction Cells
Michael L. Timmons
Masafumi Yamaguchi
Antonio Luque

4. Amorphous Silicon
Rajiwa Arya
Hiroaki Okamoto
Emmanuel Fabre

5. Terrestrial Applications
Christina Jennings
Kosuke Kurokawa
Thomas Nordmann

6. Space Systems
Paul Stella
Sumio Matsuda
Klaus Bogus

7. Space Cells
Irving Weinberg
Susumu Yoshida
Carlo Flores

8. National Programs
Carlos Vargas-Aburto
Choochon Lee
E.V.R. Sastry

HONORARY EXECUTIVE COMMITTEE

Charles E. Backus

A.K. Barua

Werner H. Bloss

Klaus P. Bogus

Henry Brandhorst, Jr.

Martin A. Green

Yoshihiro Hamakawa

Robert Hill

Peter Iles

Joseph J. Loferski

Antonio Luque

Hiroyuki Matsunami

Wolfgang Palz

Morton Prince

Roger Van Overstraeten

Eugene L. Ralph

H. Scheer, M. P.

Kiyoshi Takahashi

Martin Wolf

Zou Xun

WILLIAM R. CHERRY COMMITTEE

Martin A. Green, Chairperson

Charles E. Backus

Allen M. Barnett

Cosmo R. Baraona

Henry W. Brandhorst, Jr.

David E. Carlson

Dennis J. Flood

Americo F. Forestieri

John Goldsmith

Peter A. Iles

Lawrence L. Kazmerski

Joseph J. Loferski

John D. Meakin

Eugene L. Ralph

Richard J. Schwartz

Richard L. Statler

Joseph Wise

Martin Wolf

INTERNATIONAL COMMITTEE

Victor Abramov
Zhores Alferov
Tapio Alvesalo
A. M. Andrade
Joseph Appelbaum
Hubert Aulich
Charles Backus
Allen Barnett
A. K. Barua
Paul Basore
Boris Berkovski
Werner Bloss
Eldon Boes
Reda Bottros
David E. Carlson
Ivan Chambouleyron
C. H. Chung
Emmanuel Fabre
Dennis J. Flood
Takashi Fuyuki
Marcello Garrozzo
Moustafa Ghannam
Adolf Goetzberger
Martin A. Green
Leopoldo Guimaraes
Yoshihiro Hamakawa
Robert Hill
Jorge Huacuz
Huey-Liang Hwang
Raj Jain
Patrick Jourde
Lawrence Kazmerski
Helmut Kiess
Makato Konagai
Kosuke Kurokawa
Yukinori Kuwano
Choochon Lee
P. Lund
Antonio Luque
Joachim Luther
Vassilios Makios
Santo Martinuzzi
Sumio Matsuda
Hiroyuki Matsunami
John D. Meakin
Pietro Menna
Robert Mertens
R. Murthy
Tokio Nakada
Hiroaki Okamoto
Nicolae Olariu
Pedro Paes
Wolfgang Palz
Ashok Parathasarathi
Alfredo Previ
Morton Prince
Dimitris Rapakoulias
Armin Rauber
Marcus Real
Ahmed Sadek
Tadashi Saitoh
Jurgen Schmid
Hans Schock
Richard J. Schwartz
Dipesh Shah
Ionel Solomon
Jinsoo Song
Jack L. Stone
Tuomo Suntola
Hideyuki Takakura
B. Topel
Masayoshi Umeno
Christian Verie
Hiroyuki Watanabe
Gerald Wrixon
Masafumi Yamaguchi
Susumu Yoshida
Peinuo Yu

WILLIAM R. CHERRY

The fourteenth IEEE Photovoltaic Specialists Conference organizing committee established an award to recognize persons who have made "outstanding contributions to the advancement of photovoltaic science and technology," and named it in memory of one of the founders of the photovoltaic field, William R. Cherry. The award is to be given at each Photovoltaic Specialists Conference. The selection is made by the William R. Cherry Committee which is composed of all past general chairmen of the specialists conferences (those who may be nominated for the award are not participants in the selection), and all past recipients of the award. Past recipients of this award are Dr. Paul Rappaport, January 1980; Dr. Joseph L. Loferski, May 1981; Prof. Martin Wolf, October 1982; Dr. Henry W. Brandhorst, May 1984; Mr. Eugene L. Ralph, October 1985; Dr. Charles E. Backus, May 1987; Dr. David E. Carlson, September 1988; Dr. Martin A. Green, May 1990; Mr. Peter A Iles, October 1991; Dr. Lawrence L. Kazmerski, May 1993.

PROFESSOR YOSHIHIRO HAMAKAWA

Professor Hamakawa holds the titles of Professor of the Department of Electrical Engineering, Director of the Institute for Extreme Materials, and Chair of the Semiconductor Laboratory in the Faculty of Engineering Science, Osaka, Japan. He has been with Osaka University since the completion of his MA Degree in 1958. He received the PhD in 1964. From 1965 until 1967 he was visiting Assistant Professor in the Electrical Engineering Department of the University of Illinois, and worked in the semiconductor Laboratory directed by Professors John Bardeen and Paul Handler. He has an extensive record of accomplishment in the fields of Photovoltaics, Semiconductor Physics, and Optoelectronics. He developed the technique of "modulation spectroscopy," now widely used for the characterization of the band structure of solids. He has conducted pioneering work on amorphous silicon multilayer tandem solar cells; demonstrated valence electron control in the production of amorphous silicon carbide films, and showed its applicability to successful fabrication of high efficiency, heterojunction solar cells and light emitting devices. He has authored or co-authored 32 books, 433 scientific research papers, and 135 invited talks.

In addition to these scientific activities, Professor Hamakawa has made many key contributions to the promotion of photovoltaic solar energy technology development around the world. He has been a member of the International Advisory Committee of the IEEE PVSC since 1978, and the EU-PSEC since 1981. He organized the Photovoltaic Science and Engineering Conference (PVSEC) in Japan in 1984, and was chairman of the first conference. He has been an active leader in the Industrial development of photovoltaic solar energy, and has been a Board Member for the Sunshine Project since 1973. He is now Chairman of the Solar Energy Division in the New Energy Section, Industrial Development Council, AIST, Ministry of International Trade and Industry of Japan.

TABLE OF CONTENTS

VOLUME I

203 CdTe and CIS Film Characterization and Device Modelling

Chairs: H.W. Schock
K. Sato

204 CIS Process and Substrate Effects

Chairs: J.R. Sites
V.K. Kapur

2P1 CIS and Related Alloys

Chair: S. Niki

2P2 Polycrystalline Thin Film Materials and Devices

Chair: N. Romeo

P4 Plenary Session Amorphous Silicon Technology

Chairs: M. Konagai and C. Wronski

401 Amorphous Silicon Cells and Modules-I

Chairs: H. Okamoto
J. Yang

402 Amorphous Silicon Materials-I

Chairs: J.M. Ruiz
V.L. Dalal

403 Amorphous Silicon Cells and Modules-II

Chairs: S. Tsuda
R. Arya

4P2 Amorphous Silicon-II

Chair: J. Kocka

P5 Plenary Session Terrestrial Photovoltaics

Chairs: A. Geotzberger
C. Jennings

501 Module and BOS Component Evaluation and Performance

Chairs: K. Sakuta
J. Bigger

502 Off Grid Systems

Chairs: J. Schmid
W. Bowe

503 Utility and Grid-Connected Systems

Chairs: S. Corsi
K. Kurokawa

504 PV On Buildings

Chairs: T. Nordmann
A. Rohatgi

505 System Economics, Market Trends and Issues

Chairs: K.M. Hynes and P.D. Maycock

506 Module and BOS Component Design and Manufacturing

Chairs: A. Kitamura and C.E. Witt

5P1 Module and BOS Component Evaluation and Performance

Chair: S. Krauter

5P2 PV on Buildings

Chair: R.W. Taylor

5P3 Utility and Grid Connected Systems

Chair: S. Corsi

5P4 Off Grid System Applications and Models

Chair: S.J. Durand

VOLUME II

P1 Plenary Session Silicon Cells and Materials

Chairs: W. Wettling
T. Ciszek

Silicon Material Growth and Characterization

Chairs: A. Ayer
J.P. Kalejs

102 Silicon Device and Process Modelling

Chairs: T. Warabisako
P.A. Basore

103 Single Crystal and High Efficiency Silicon Devices

Chairs: M.A. Green
J. Nijs

104 Multicrystalline and Thin-Layer Silicon Devices

Chairs: L.A. Verhoef
A. Blakers

105 Silicon Defect and Impurity Effects in Growth and Processing

Chairs: K. Kaneko
S. Narayanan

1P1 Silicon Crystal Growth Silicon Device Modeling

Chair: K.W. Mitchel

1P2 Crystalline Silicon Devices

Chair: R.A. Sinton

1P3 Impurities and Defects in Crystalline Silicon and Thin Layer Devices

Chair: R.B. Hall

P3 Plenary Session III–V and Multijunction Cells

Chairs: R.G. Little
M.L. Timmons

301 III–V Multijunction Devices

Chairs: Y.C.M. Yeh
K.A. Bertness

302 III–V Cell Growth and Characterization

Chairs: K. Bogus
J.A. Hutchby

3P1 III–V Single and Multijunction Cells-I

Chair: D. Friedman

3P2 III–V Single and Multijunction Cells-II

Chair: R.L. Statler

P6 Plenary Session Space Systems

Chairs: M. Tajima
P.M. Stella

601 Space Systems

Chairs: M. Tajima
Lee Goldhammer

6P1 Space Systems

Chair: T. Cross

P7 Plenary Session
Space Solar Cells

Chairs: S. Matsuda
G.A. Landis

701 Space Solar Cells-I

Chairs: C.R. Lewis
C. Flores

702 Space Solar Cells-II

Chairs: N.M. Pearsall
G. Summers

7P1 Space Solar Cells

Chair: D.C. Marvin

P8 Plenary Session National Programs

Chairs: A. Zervos
J.L. Stone

801 National Programs Economics and Technology

Chairs: H. Aulich
D.M. Allen

801 National Programs Status and Prospects

Chairs: C. Lee
C. Vargas-Aburto

8P1 National Programs

Chair: R.R. Hill

AUTHOR INDEX

ATTENDEES

(Name, Address, Phone #, Fax #)

ABE, TOMOHIRO
JAPAN QUALITY ASSURANCE
1-21-25-,KINUTA,SETAGAYA KU
TOKYO 157
JAPAN
03-3416-5560-Tel
03-3749-3505-Fax

ABERLE, ARMIN
INSTITUTE FOR SOLAR ENERGY RES
AM OHRBERG 1
D-31860
EMMERTHAL
GERMANY
5181-999-423-Tel
5181-999-400-Fax

ABOU-ELFOTOUH, FOUAD
NREL
1617 COLE BLVD
GOLDEN CO 80401
USA
303 384-6601-Tel
303 384-6490-Fax

ABROSIMOV, NIKOLAI
INSTITUTE OF CRYSTAL GROWTH
RUDOWE-CHAUSSEE 6, GEB 18 46
BERLIN 12489
GERMANY
49 030 6392 3016-Tel
49 030 6392 3003-Fax

ADANU, K. GODFREY
UNIVERSITY OF GHANA
LEGON
ACCRA
GHANA
233 21 777026-Fax

ADDIS, BILL
WASHINGTON STATE UNIVERSITY
100 SPROUT RD
RICHLAND WA 99352
USA
509-575-9242-Tel
509-375-5337-Fax

ADELMANN, PETER
STECA GMPH
MAMMOSTRABE 1
MEMMINGEN D-87700
GERMANY
08331 8558 0 Tel
08331 8558 11-Fax

ADLER, STEVE
INTERNATIONAL CONFERENCE MGMT.
7152 SW 47 STREET
MIAMI FL 33155
USA
305-661-5115-Tel
305-666-7134-Fax

AGARWAL, ARJUN
GLOCON, INC.
1 PRINCESS ST
PARSIPPANY NJ 07054
USA

AGREDANO, JAIME
IIE
MEXICO

AHM, PETER
PA ENERGY
SNOVDRUPPERVET 16
8340 MALLINS
DENMARK
45-8693-333-Tel
45-8693-3605-Fax

AHN, BYUNG TAE
KOREA ADV. INST. OF SCI. TECH.

AHRENKIEL, RICHARD
NREL
1617 CDE BLVD
GOLDEN CO 80401
USA
303-231-1770-Tel

AKAKA, SENATOR

AKASAKA, TETSUYA
TOKYO INSTITUTE OF TECHNOLOGY
4259 NAGATSUTA MIDORI
YOKOHAMA 227
JAPAN
045-924-5188-Tel
045-921-1089-Fax

AKHMAD, KHOLID
STUDENT OF OSAKA UNIVERSITY JA
1-3 MACHIKANEYAMA CHO
TOYONAKA 560
JAPAN
81-6-250-6315-Tel
81-6-250-6316-Fax

ALBECK, JAMES
HELIOKINETICS CORP
P.O. BOX 1040
AGOURA HILLS CA 91376
USA
818-991-7558-Tel

ALBRIGHT, SCOT
GOLDEN PHOTON, INC.
4545 MCINTYRE STREET
GOLDEN CO 80403
USA
303 271-7122-Tel
303 271-7410-Fax

ALDANA, FERNANDO
UNIVERSIDAD POLITECNICA MADRID
RAMIRO DE MAEZTU, 7
MADRID 28040
SPAIN
34-1-5343709-Tel
34-1-5352756-Fax

ALLEN, DOUGLAS
W. J. SCHAFER ASSOC., INC.
25000 GREAT NORTHERN CORP CTR
SUITE 300
NORTH OLMETED OH 44070
USA
216 779-3202-Tel
216 777-3972-Fax

ALTERMATT, PIETRO
UNIVERSITY OF NSW
CENTRE FOR PHOTOV. DEVICES
UNSW
SYDNEY 2052
AUSTRALIA

ALVAREZ, FERNANDO
UNICAMP
INST DE FISICA, UNICAMP
CAMPINAS, SAO PAU 13081
BRAZIL

ALVESALO, TAPIO
NESTE ADVANCED POWER SYSTEMS
P O BOX 3
FIN 02151 ESPOO
ESPOO FIN 02151
FINLAND
358-0-4501-5750-Tel
358-0-450-5744-Fax

ANDERSON, DENNIS
WORLD BANK
1818 H ST NW
WASHINTON DC 20035
USA
202 4731045-Tel
202 477 0545-Fax

ANDREEV, VYACHESLAV
IOFFE INSTITUTE
POLYTECHNICHESKAYA 26
ST PETERSBURG 194021
RUSSIA
812 247 9394-Tel
812 2471017-Fax

ANDRES, FLORDELIZA
PHILIPPINES DOE

APPELBAUM, JOSEPH
TEL AVIV UNIVERSITY
FACULTY OF ENGINEERING
TEL AVIV 69978
ISRAEL
972-3-640904-Tel
972-3-6407095-Fax

ARAMOTO, TETSUYA
MATSUSHITA BATTERY INDUSTRIAL
1 MATSUSHITACHO
MORIGUCHISHI
OSAKA 570
JAPAN
06-994-4531-Tel
06-994-7056-Fax

ARAUJO, JOSE CARLOS
DIGICON S.A.
DR 290 KM 75
GRAVATAI
RIO G. SUL 94000-970
BRAZIL
00 55 51 4893 33-Tel
55 51 489 11 10-Fax

ARCH, JOHN
TEXAS INSTRUMENTS, INC.

ARGAW, NEWAY
TAMPERE UNIVERSITY OF TECH
INSINOORINKATU 60A 26
TAMPERE 33720
FINLAND
358-31-3184007-Tel
358-31-3184007-Fax

ARIMOTO, SATOSHI
MITSUBISHI ELECTRIC CORP.
4-1 MIZUHARA
ITAMI 664
JAPAN
0727-84-7383-Tel
0727-80-2573-Fax

ARMSTRONG, JOSEPH
ITN ENERGY SYSTEMS
4155 E JEWELL AVE
DENVER CO 80222
USA

ARON, PAUL
295 YACHT CLUB DR
ROCKY RIVER OH 44116
USA
216-356-0873-Tel

ARRIVE, THIERRY
SOLER ENERGIE
PO BOX 4100
PAPEETE
TAHITI
689 43 02 00-Tel
689 43 46 00-Fax

ARYA, RAJEEWA
SOLAREX CORPORATION
826 NEWTOWN YARDLEY RD
NEWTOWN PA 18940
USA
215 860-0902-Tel
215 860-2986-Fax

ARYASOMAYAJULA, SUBRAHMANYAM
I.I.T. MADRAS, INDIA
DEPT OF PHYSICS
MADRAS 600036
INDIA

ASAOKA, MASAHISA
MITSUBISHI ELECTRIC CORP.
8-1-1 TSUKAGUCHIHONCHO
AMAGASAKISHI 661
JAPAN
06-497-8498-Tel
06-497-9360-Fax

ASAYAMA, NOBORU
NASU DENKI TEKKO CO., LTD.
YACHIYO FACTORY
YOSHIHASHI KOGYODAMICHI
YACHIYO-CITY, CHI 276
JAPAN

ASCHENBRENNER, PETER
ASTRO POWER, INC.
SOLAR PARK
NEWARK DE 19716-2000
USA
302-366-0400-Tel
302-368-6474-Fax

ASHIDA, YOSHINORI
MITSUI TOATSU CHEM INC.
1190 KASAMACHO SAKAE KU
YOKOHAMA
KANAGAWA 247
JAPAN
81 45 895 8141-Tel
81 45 895 8240-Fax

ATKINSON, DICK
FOSTER-MILLER, INC.
350 SECOND AVE
WALTHAM MA 02154-1196
USA
617 890-3200-Tel
617 890-3489-Fax

AUBREY, JOYSREE
LOS ALAMOS NATIONAL LABORATORY
MS B220
LOS ALAMOS NM 87545
USA
505 667-8949-Tel
505 665-4080-Fax

AULICH, HUBERT
SIEMENS SOLAR GMBH
FRANKFURTER RINGG 152
MUNCHEN 80807
GERMANY
49 89 3500 2618-Tel
49 89 3500 2840-Fax

AURIA, G.
EUROSOLARE S.P.A.
VIA A/ D ANDREA 6
NETTUNO(RM) 00048
ITALY
39-6-59883415-Tel
39-6-59883450-Fax

AZZAM, MONEER
ASE AMERICAS, INC.
4 SUBURBAN PARK DRIVE
BILLERICA MA 01821
USA
508-677-5900-Tel
508-663-2868-Fax

BACKUS, CHARLES
ARIZONA STATE UNIVERSITY
COLLEGE OF ENGINEERING
BOX 875506
TEMPE AZ 85287-5506
USA
602-965-2825-Tel
602-965-5518-Fax

BAE, SANG SOON
TAEDOK R&D CENTER
YUKONG LTD
140-1 WONCHON-DONG YUSUNG-GU
TAEJON 305-370
KOREA

BAE, SANGHOON
THE PENNSYLVANIA STATE UNIV
135 HAMMOND BLDG
UNIVERSITY PARK PA 16802
USA

BAHAJ, ABUBAKE SALZM
UNIVERSITY OF SOUTHAMPTON
SOLAR ENERGY CENTRE
SOUTHAMPTON 50171 GJ
UNITED KINGDOM
0703 592051-Tel
0703 593053-Fax

BAILEY, JEFF
U.C. BERKELEY
DEPT OF MATERIALS SCIENCE
BERKELEY CA 94720
USA

BAILEY, SHEILA
NASA LEWIS RESEARCH CENTER
MS 302-1
CLEVELAND OH 44135
USA
216-433-2228-Tel
216-433-6106-Fax

BAIRAMOV, BACKYSH
IOFFEE INSTITUTE
POLITEKHNICHESKAYA 26
ST PETERSBURG 194021
RUSSIA
7 812 247 2255-Tel
7 812 247 1017-Fax

BALDASARO, PAUL
KNOLLS ATOMIC LABORATORIES
1 RIVER RD
SCHENECTADY NY 12301
USA
518 395-6711-Tel
518 395-4422-Fax

BALL, TIMOTHY
APPLIED POWER CORP.
1210 HOMANY
LACEY CA 98503
USA
206-438-2110-Tel
206-438-2115-Fax

BALLANTYNE, RUSSELL
JX CRYSTALS, INC.
1105 12 AVE NW #A-2
ISSAQUAH WA 98027
USA
206 392 5237-Tel
206 392 7303-Fax

BANERJEE, ARINDAM
UNITED SOLAR SYSTEMS CORP.
1100 W MAPLE RD
TROY MI 48084
USA
810 362-4170-Tel
810 362-4442-Fax

BARAONA, COSMO
NASA LEWIS
21000 BROOKPARK RD
MS 301-5
CLEVELAND OH 44129
USA

BARKE, DHARHDAS
RES PHOTOVOLTAICS LTD
D 13A PHASE V JEEDIMETLA
HYDERABAD 500855
INDIA
91 40 897272-Tel
91 40 895807-Fax

BARNES, HARRY
EA TECHNOLOGY
CAPENHURST
CHESTER CH1-6ES
GREAT BRITAIN
051-339-4181-Tel
051-357-1581-Fax

BARNETT, ALLEN
ASTRO POWER, INC.
SOLAR PARK
NEWARK DE 19716-2000
USA
302-366-0400-Tel
302-368-6474-Fax

BARONE, FABIO
UNIV OF NSW AUSTRALIA
CTR FOR PHOTOVOLTAIC DEV
ELECTRICAL ENGINEERING
SYDNEY NSW 2052
AUSTRALIA

BARUA, ASOK KUMAR
INDIAN ASSOCIATION
FOR JADAVPUR CULT. OF SCIENCE
CALCUTTA
WEST BENGAL 700032
INDIA
91-33-4736612-Tel
91-33-4736612-Fax

BASOL, BULENT
INT'L. SOLAR ELECTRIC TECH.
8635 AVIATION BLVD.
INGLEWOOD CA 90301
USA
310-216-4427-Tel
310-216-2908-Fax

BASORE, PAUL
SANDIA NATIONAL LABORATORIES
MS 0752
P O BOX 5800
ALBUQUERQUE NM 87185
USA
505-844-3761-Tel
505-844-6541-Fax

BASS, ANDREW
BASS SOLAR TECHNOLOGY
516 E FRIAR TUCK
HOUSTON TX 77024
USA
713 688-4282-Tel
713 688-4516-Fax

BASS, FREDA
516 E FRIAR TUCK
HOUSTON TX 77024
USA
713 688-4516-Tel

BATHEY, BALA
EBARA SOLAR, INC.
811 ROUTE 51 SOUTH
LARGE PA 15025
USA
412-382-1269-Tel
412-382-1251-Fax

BAUMANN, ANGELIKA E.
NPAC
UNIVERSITY OF NORTHUMBRIA
ELLISON PLACE
NEW CASTLE UPON T NE1 8ST
UNITED KINGDOM
+44 91 227 4555-Tel
44 91 227 4561-Fax

BEAUCHAMP, WILLIAM T.
OCLI
2789 NORTH POINT PKWY
SANTA ROSA CA 95407
USA
707-525-7649-Tel
707-525-7018-Fax

BEAUCHENE, JAMES
OPTICAL RADIATION CORP
1300 OPTICAL DR
AZUSA CA 91702
USA
818 969 3344-Tel
818 969 3681-Fax

BEBERMEIER, HELMUT
DEUTSCHE AEROSPACE AG
MUNCHEN D-81663
GERMANY
89-607-25961-Tel
89-607-24202-Fax

BECK, NATALIE
IMT NEUCHATEL
BREQUET 2
UNIVERSITE DE NEUCHATEL
NEUCHATEL CH2000
SWITZERLAND
38 20 5452-Tel
38 25 4276-Fax

BEGAVIC, MIROSLAV
SCHOOL OF ECE, GA TECH
777 ATLANTIC DR. NW
ATLANTIC GA 30332-0250
USA
404-853-9826-Tel
404-853-9171-Fax

BEIER, JUTTA
WEIZMANN INSTITUTE OF SCIENCE
DEPT OF MATERIALS & INTERFACES
REHOVOT 76100
ISRAEL
972 8 34 3115-Tel
972 8 34 4139-Fax

BELL, RICHARD
24 AUSTIN ROAD
SUDBURY MA 01776
USA
508 443-8302-Tel

BENEKING, CLAUS
INSTITUTE OF THIN FILM AND ION
JULICH RESEARCH CENTRE
PO BOX 1913 D-52425
JULICH
GERMANY
49-2461-612616-Tel
49-2461-613335-Fax

BENEMANN, JOACHIM
FLACHGLAS SOLARTECHNIK
MUHLENGASSE 7
KOLN 50667
GERMANY
49 221 25 7311-Tel
49 258 1117-Fax

BENNER, JOHN
NREL
1617 COLE BLVD
GOLDEN CO 80401
USA
303 384-6496-Tel
303 384-6530-Fax

BENSON, MICHAEL
NORANDA TECHNOLOGY CENTRE
4950 RUE LEVY
ST LAURENT
QUEBEC H4R2P1
CANADA
514 856 6956-Tel
514 856 6925-Fax

BENTON, MAX
ABLE ENGINEERING COMPANY
93 CASTILIAN DRIVE
GOLETA CA 93117
USA
805 685-2262-Tel
805 685-1369-Fax

BERG, BILL
UMASS LOWELL
CENTER FOR SUSTAINABLE ENERGY
1 UNIVERSITY AVE
LOWELL MA 01851
USA

BERGMANN, RALF
MAX-PLANCK INSTITUTE
HEISENBERGSTRASSE 1
D-70569
STUTTGART
GERMANY
+49-711-689-1606-Tel
+49-711-689-1010-Fax

BERKOVSKI, BORIS
UNESCO
1 RUE MIOLLIS
PARIS 75015
FRANCE
33 1 45683901-Tel

BERRY, GENE
UNIVERSITY OF ILLINOIS
LAWRENCE LIVERMORE LABORATORY
1101 W SPRINGFIELD
URBANA IL 61801
USA
217 244 4123-Tel
217 244 1631-Fax

BERRY, WILLIAM
UNIVERSITY OF NOTRE DAME
DEPT OF ELECTRICAL ENGINEERING
NOTRE DAME IN 46556
USA
219-631-6102-Tel
219-631-4393-Fax

BERTNESS, KRIS
NREL
1617 COLE BLVD.
GOLDEN CO 80401-6559
USA
303-384-6559-Tel
303-384-6531-Fax

BETT, ANDREAS
FAUNNOFE JUSHTUT FUR SOL ENE
OLTMANNSSTR.22
FRIEBURY GE 79100
GERMANY
49-761-4588-8257-Tel
+49-761-4588-250-Fax

BETZ, GOTTFRIED
PHOENICON
BISMARCKSTRABE 58
BERLIN 10627
GERMANY
49 30 3421088-Tel
49 30 342 90 55-Fax

BETZ, GOTTFRIED
PHOENICON
BISMARCKSTRABE 58
BERLIN 10627
GERMANY
49 30 3421088-Tel
49 30 342 90 55-Fax

BHAKTHAVATSALAM, V.
IREDA
INDIA HABITAT CTR.,IHC COMPLEX
EAST CT,CORE-4 A, 1ST FLOOR
NEW DELHI 110 003
INDIA
4601344 460 1366-Tel
011 4602855-Fax

BHIMJIYANI, ANIL
BLUEBELL INTL LTD
12 SIDMOUTH RD
LONDON NW2 55X

BIGGER, JOHN E.
UTILITY PHOTOVOLTAIC GROUP
1800 M STREET NW
SUITE 300
WASHINGTON DC 20036
USA
202 857 0898-Tel
202 223 5537-Fax

BILGER, KEVIN M.
LOCKHEED MISSILES & SPACE CO.
1111 LOCKHEED MISSILES &SPACE
0/64-24 B/150
SUNNYVALE CA 94088
USA
408-742-1979-Tel
408-743-1813-Fax

BIRKMIRE, ROBERT
INSTITUTE OF ENERGY CONVERSION
UNIVERSITY OF DELAWARE
NEWARK DE 19716-3820
USA
302 831-6218-Tel
302 831-6226-Fax

BLAKERS, ANDREW
AUSTRALIAN NATIONAL UNIVERSITY
ENGEERING DEPARTMENT
CANBERRA 0200
AUSTRALIA
61-6-249-0506-Tel
61-6-249-0506-Fax

BLOSS, WERNER H.
INST.F. PHYSIDALISCHE ELEK
PFAFFENWALDRING 47
STUTTGART 70569
GERMANY
49-711-685-7140-Tel
49-711-685-7143-Fax

BLUNIER, STEFAN
AFIF TECHNOPARK, ETH ZURICH
PFINGSTNEIDSTR.30
ZURICH 8005
SWIZERLAND
41-1-4451473-Tel
41-1-4451499-Fax

BOBOBEE, EMMANUEL
UNIVERSITY OF SCIENCE TECH
AGRIC ENG DEPT U S T
KUMASI
ASH REG
GHANA
233-51-5354-Tel
233-51-3137-Fax

BOER, KARL WOLFGANG
UNIVERSITY OF DELAWARE
239 BUCKTOE HILLS RD
KENNETT SQUARE PA 19348
USA
610-444-3597-Tel
610-444-9472-Fax

BOES, ELDON
NAT'L RENEWABLE ENERGY LAB.
409 12TH STREET, S.W. SUITE710
WASHINGTON DC 20024
USA
202-484-1090-Tel
202-484-1096-Fax

BOGUS, KLAUS
EUROPEAN SPACE AGENCY
PO BOX 299
NL-2200 AG
NOORDWIJK
NETHERLANDS
31-1717-83973-Tel
31-1717-84994-Fax

BOHNKE, HEINZ
GTZ
7 V. KALAW ST, QUEZON CITY
TANGDANG SORA
MANILA
PHILLIPINES

BONDA, J.C.
EPIA
AVENEUE CHARLES QUINT 124
BOITE N 14 B 1080
BRUSSELS
32-02-4659162-Tel
30-02-4682430-Fax

BORELAND, MATT
CENTRE FOR PHOTOVOLTAIC SYSTE
UNIERSITY OF NSW
SYDNEY
NSW
AUSTRALIA
612-385-4060-Tel
612-662-4240-Fax

BORREGO, JOSE
RENSSELAER POLYTECHNIC INST
ECSE DEPT
110 8TH STREET
TROY NY 12180-3590
USA
518-371-7782-Tel
518-371-4659-Fax

BOSWELL, RANDY
USAF PHILLIPS LABORATORY
PL/VTPC
KIRKLAND AFB NM 87117
USA
505-846-5754-Tel
505-846-2885-Fax

BOTTENBERG, WILLIAM
DELAWARE SOLAR ELECTRIC POWE
6 WOODHELVEN CT
NEWARD DE 19711
USA
302 234 0676-Tel
202 234 9675-Fax

BOWDEN, STUART
UNIVERSITY OF NEW S. WHALES
SCHOOL OF ELECTRICAL ENGINEER
UNSW
SYDNEY 2052
AUSTRALIA
61 2 385 5246-Tel
61 2 662 4240-Fax

BOWER, WARD
SANDIA NATIONAL LABORATORIES
PV SYSTEMS APPLICATIONS
ALBUQUERQUE NM 87185-0753
USA
505 844-5206-Tel
505 844-6541-Fax

BOYEAUX, PIERRE
I.N.S.A. DE LYON BATIMENT 502
LAB. DE PHYSIQUE DE LA MATIERE
20, AVENUE ALBERT EINSTEIN
VILLEURBANNE CEDE 69621
FRANCE
33 72 43 8393-Tel
33 72 44 07 32-Fax

BRAGAGNOLO, JULIO
CONSULTANT
390 BRIAR LAND
NEWARK DE 19711
USA

BRAMMER, TORSTEN
AUSTRALIAN NATIONAL UNIVERSI
ENGINEERING DEPARTMENT
CANBERRA 0200
AUSTRALIA
06-249-3976-Tel
06-249-0506-Fax

BRAULT, JOHN
UNIVERSITY OF SAN FRANCISCO

BRAUN, GERALD

BREITENSTEIN, OTWIN
NTT LSI LABORATORIES
3-1, MORINOSATO WAKAMIYA
ATSUGI-SHI,
KANAGAWA PREF., 241-01
JAPAN
49-345-558250-Tel
81-462-40-2905-Fax

BRENNAN, PAUL
AEROSPACE CORPORATION M2-242
P.O. BOX 92957
LOS ANGELES CA 90009
USA
310-336-1983-Tel
310-336-5846-Fax

BRENNEMAN, BOB
7307 BROOKSIDE DR.
FREDERICK MD 21702
USA

BRIGANTI, ANTONIO
STUDIO BRIGANTI
VIA PIAVE 3
SARONNO VA 21047
ITALY
2-9620184-Tel
2-9620184-Fax

BRITT, JEFFREY
ENERGY PHOTOVOLTAICS, INC
PO BOX 7456
PRINCETON NJ 08543
USA
609-587-3000-Tel
609-587-5355-Fax

BRODERSEN, HEIKO
DEUTSCHE AEROSPACE
P O BOX 801168
MUNICH 81663
GERMANY
49 89 607 27458-Tel
49 89 607 21053-Fax

BROOKS, BILL
N C SOLAR CENTER
BOX 7401 NCSU
RALEIGH NC 27695
USA
919 515 7147-Tel
919 515 5778-Fax

BROWN, GAIL
AF WRIGHT LAB
WL/MLPO BLDG 651
3005 P. ST STE 6
WPAFB OH 45433
USA

BROWN, KEN
BP SOLAR AUSTRALIA
1/100 OLD PITTWATER ROAD
BROOKVALE
NSW 2100
AUSTRALIA
61 2 938 5111-Tel
61 2 939 1548-Fax

BROWN, STEVEN
BAKKEN HALE
P O BOX 38-4360
WAIKOLOA HI 96738
USA
808-325-3127-Tel
808-325-0928-Fax

BROWN, WILLIAM
UNIVERSITY OF ARKANSAS
3217 BELL
FAYETTEVILLE AR 72701
USA
501 575-6045-Tel
501 575-7967-Fax

BRUTON, TIM
BP SOLAR
PO BOX 191 CHERTSEY ROAD
MIDDLESEX
SUNBURY ON THAMES
ENGLAND
44 932 762181-Tel
44 932 762533-Fax

BUCHER, KLAUS
FRAUNHOTER INST.FOR SOLAR ENER
OLMANNISTR. 5
FREIBURG D79100
GERMANY
49 761 4588145-Tel
49 761 4588100-Fax

BULAWKA, ALEC
U.S. DEPARTMENT OF ENERGY
1000 INDEPENDENCE AVE SW
WASHINGTON DC 20585
USA
202-586-5633-Tel
202-586-5127-Fax

BULLEMER, BERNHARD
FED AF UNIVERSITY MUNICH
FAC.EE. INST PHYSICS
WERNER-HEISENBERG WEG 39
NEUBIBERG D-85579
GERMANY

BURNS, STEVE
HAWAII ELECTRIC LIGHT CO.

BYE, CHARLES
HONEYWELL INC.
11601 ROOSEVELT BLVD
ST PETERSBURG FL 33716-2202
USA
813 579-6703-Tel
813 579-6696-Fax

CABRAAL, ANIL
WORLD BANK
1818 H STREET NW
WASHINGTON DC 20433
USA
202-458-1538-Tel
202-522-3573-Fax

CALDEZON, GUSTAVO
INTER-AMERICAN DEVEL. BANK
1200 NEW YORK AVE. N.W.
WASHINGTON DC 20577
USA
202-623-1978-Tel
202-623-1786-Fax

CAMERON, CHRISTOPHER
SANDIA NATIONAL LABS
PO BOX 5800, MS 0753
ALBUQUERQUE NM 87185
USA
505 844 8161-Tel
505 844 6541-Fax

CAMPBELL, BRIAN
KROLLS ATOMIC POWER LAB.
BOX 1072
MAILSTOP D2-121
SCHENECTADY NY 12301-1072
USA
518 395-7438-Tel
518 395-4422-Fax

CAMPBELL, DONALD
NASA-LEWIS

CAMPBELL, PATRICK
UNIVERSITY OF NEW SOUTH WALES
SCHOOL OF ELECTRICAL ENGINEER.
CENTRE FOR P.V. DEV.
SYDNEY, NSW 2052
AUSTRALIA
02 385 5471-Tel
02 662 4240-Fax

CAREY, PAUL
LAWRENCE LIVERMORE NTL LAB
PO BOX 808, L-271
LIVERMORE CA 94550
USA

CARLIN, JOSEPH
KIPP & ZONEN
390 CENTRAL AVENUE
BOHEMIA NY 11716
USA
516-589-2885-Tel
516-589-2068-Fax

CARLSON, DAVID
SOLAREX
826 NEWTOWN YARDLEY RD.
NEWTOWN PA 18940
USA
215-860-0902-Tel
215-860-2986-Fax

CARPENTER, TOM
MOBILE SERVICES INTL
HCR 5201
KEAAU HI 96749
USA

CARTER, MICHAEL
UNIVERSITY OF NORTHUMBRIA
ELLISON PLACE
NEWCASTLE UPON TY NEI 8 ST
UNITED KINGDOM
44-91-227-4555-Tel
44-91-227-4561-Fax

CASTANER, LUIS
UPC
GRAN CAPITAN SLM
BARCELONA 08034
SPAIN
34-3-4016773-Tel
34-3-4016756-Fax

CASTELEYN, MARC
UNIVERSITY OF GENT, DEP ELIS
ST-PIETERSNIEVWSTR.41
GENT 9000
BELGIUM
+32-9-264-34-59-Tel
+32-9-264-35-94-Fax

CATALANO, ANTHONY
NREL
1617 COLE BLVD
GOLDEN CO 80401
USA
303-384-6446-Tel
303-384-6481-Fax

CAVICCHI, TERRY
SPECTROLAB INC.
12500 GLADSTONE AVE
SYLMAR CA 91342
USA
818-898-2867-Tel
818-361-5102-Fax

CESINI, GIANNI
DIP. ENERGETICA-UNIV OF ANCONA
VIA BRECCE BIANCHE
ANCONA
AN 60100
ITALY
+39 712204776-Tel
+39 712804239-Fax

CHALMERS, STEPHEN
S.M. CHALMERS
4044 E. WHITTON
PHOENIX AZ 85018
USA
602-955-4529-Tel
602-955-4529-Fax

CHAMBOULEYRON, IVAN
UNIVERSIDADE DE CAMPINAS
IFGN PO BOX 6165
CAMPINAS
S.P. 13083-970
BRAZIL
55 192 391232-Tel
55 192 393127-Fax

CHEN, ZHI ZHANG
GEORGIA TECH
ECE
ATLANTA GA 30312-0250
USA
404-853-9454-Tel
404-853-9882-Fax

CHERNIACK, MARK
TREXLER & ASSOC.,INC./SELF
1131SE RIVER FOKREST RD PRO
OAK GROVE OR 97267
USA
503-786-0559-Tel
503-781-9859-Fax

CHIANG, PENG-KUEN
SPECTROLAB INC.
12500 GLADSTONE AVE
SYLMAR CA 91342
USA
818-898-2885-Tel
818-361-5102-Fax

CHOI, YOUNG
ASTRO SPACE DIVISION
MS 59 P O BOX 800
KOREASAT PMO
PRINCETON NJ 08543
USA
609-490-4091-Tel
609-490-0475-Fax

CHOUDARY, U.V.
UVC CONSULTANTS
180 OAKCREEK ROAD
PO BOX 3427
HIGHTSTOWN NJ 08520
USA

CHUNG, CHOONG-HYUN
DEPT. OF PHYSICS YONSEI UNIV.
134 SHINCHON-DONG,
SUDAEMOON-KU
SEOUL 120-749
KOREA
82-2-361-2605-Tel
82-2-392-1592-Fax

CISZEK, TED
NREL
1617 COLE BLVD
GOLDEN CO 80401
USA
303 384-6569-Tel
303 384-6567-Fax

CLARK, ERIC
NASA LERC
21000 BROOKPARK RD
M/S 302-1
CLEVELAND OH 44135
USA

CLARK, MICHAEL
ATLANTIC PACIFIC TECH. INC.
44853 FREMONT BOULVARD
FREMONT CA 94538
USA
510-651-9995-Tel
510-651-9841-Fax

CLAVERIE, ANDRE
ADEME
500 ROUTE DES LUCIOLES
SOPHIA ANTIPOLIS 06560
FRANCE
33-93957913-Tel
33-93-653196-Fax

COLBOW, KONRAD
SIMON FRASER UNIVERSITY
BURNABY BC V5A1S6
CANADA
604 291 3162-Tel
604 291 3592-Fax

COLE, ERIC
GEORGE MASON UNIVERSITY
4400 UNIVERSITY DRIVE
FAIRFAX VA 22030
USA
703-993-1614-Tel
703-993-1601-Fax

COLLIER, R. KIRK
FLORIDA SOLAR ENERGY CENTER
300 STATE RD 401
CAPE CANAVERAL FL 32920
USA
407-783-0300 X185-Tel
407-783-2571-Fax

COLLIER, ROBERT
COLLIER ENGINEERING
109 TEQUESTA HARBOR DR.
MERRITT ISLAND FL 32952
USA
407-453-6030-Tel
407-7832571-Fax

COMBS, CHARLES
US NAVY
823BOAD
CHINA LAKE CA 93555
USA

COMPAAN, AL
UNIVERSITY OF TOLEDO
DEPT. OF PHYSICS ASTRON
TOLEDO OH 43606
USA
419-537-4787-Tel
419-537-2723-Fax

CONNOLLY, JAMES
IMPERIAL COLLEGE
SOLID STATE PHYSICS
PRINCE CONSORT ROAD
LONDON SW7 2BZ
UK

CONTRERAS, MIGUEL
NREL
1617 COLE BLVD
GOLDEN CO 80401
USA
303-384-6478-Tel
303-384-6430-Fax

COOK, CHRISTOPHER
MARYLAND PEOPLE'S COUNSEL
3110 NOVA SCOTIA ROAD
BEL AIR MD 21015
USA
410 767-8159-Tel
410 333-3616-Fax

COONEN, STEVEN
ATLANTIS ENERGY
14790 MOSSWOOD
GRASS VALLEY CA 96945
USA
916-274-2743-Tel
916-274-2564-Fax

COORS, JOHN
GOLDEN PHOTON, INC.
4545 MCINTYRE ST.
GOLDEN CO 80402
USA

CORIO, RON
ARRAY TECHNOLOGIES, INC
3402 STANFORD NE
ALBUQUERQUE NM 87107
USA

CORNISH, JOHN
MURDOCH UNIVERSITY
SOUTH STREET, MURDOCH
WA 6150
AUSTRALIA
09 360 2181-Tel
61 9 310 1711-Fax

CORSI, SANDRO
ENEL CENTRAL RICERCA DI AUTO
VIA A. VOLTA NO 1
20093 COLOGNO
MONZESE
ITALY
39 2 7224 56 3015-Tel

COSGROVE, MALCOLM
WORLD BANK
1818 H STREET NW
WASHINGTON DC 20433
USA
202-473-3121-Tel
202-522-3573-Fax

COTAL, HECTOR
NAVAL RESEARCH LABORATORY
4555 OVERLOOK AVE.
WASHINGTON DC 20375
USA
202-767-3938-Tel
202-404-8076-Fax

COUTTS, TIM
NREL
1617 COLE STREET
GOLDEN CO 80401
USA
303-384-6561-Tel
303-384-0430-Fax

CRAIG, MICKEY
WESTINGHOUSE ELECTRIC CORP
4400 ALAFAYA TRAIL
ORLANDO FL 32826-2399
USA
407 281 2444-Tel
407 281 5021-Fax

CRANE, RICHARD
SUNPOWER CORP
430 INDIO WAY
SUNNYVALE CA 94086
USA
408 991-0911-Tel
408 739-7713-Fax

CRAWFORD, ROBERT
ABLE ENGINEERING COMPANY
93 CASTILIAN DRIVE
GOLETA CA 93117
USA
805 685-2262-Tel
805 685-1369-Fax

CROSS, BRUCE
CARDIFF UNIVERSITY
BOX 925
CARDIFF CF21YF
UNITED KINGDOM
+44 1222 874314-Tel

CROSS, TREVOR
EEV LIMITED
106 WATERHOUSE LANE
CHELMSFORD
ESSEX CM1 2QU
ENGLAND
44 245 493493-Tel
44245 492492-Fax

CSIGE, EVA
ENERGY PHOTOVOLTAICS, INC.
PO BOX 7456
PRINCETON NJ 08543
USA
609-587-3000-Tel
609-587-5355-Fax

CUEVAS, ANDRES
AUSTRALIAN NATIONAL UNIVERSITY
DEPT OF ENGINEERING, FEIT
CANBERRA
ACT 0200
AUSTRALIA
616 249-3702-Tel
616 249-0506-Fax

CUOZZO, LUIGI
ATLANTIS
UNDENRAIN 4
BERN 3012
SWITZERLAND
004 1313003200-Tel
004 1313003210-Fax

CURRY, RICHARD
PV INSIDER'S REPORT
1011 WEST COLORADO BLVD.
DALLAS TX 75208
USA
214-942-5248-Tel
214-942-5248-Fax

CURTIS, HENRY
NASA
NASA/LEWIS 302-1
21000 BROOKPARK ROAD
CLEVELAND OH 44135
USA

CZANDERNA, ALVIN
NREL
1617 COLE BLVD.
GOLDEN CO 80401
USA
303-384-6460-Tel
303-384-6530 Fax

DAEY OUWENS, JEROEN
UTRECHT UNIVERSITY
PRINCETONPLEIN 5
UTRECHT 80000
NETHERLANDS
3130532904-Tel
3130543165-Fax

DAI, XIMING
UNIVERSITY OF NEW SOUTH WALES
C/O TED SZPITALAK
PHOTOVOLTAIC CENTRE, UNSW
SYDNEY, NSW 2052
AUSTRALIA
385 5246-Tel
662 4240-Fax

DAIELLO, BOB

DALAL, VIKRAM
IOWA STATE UNVERSITY
DEPT. OF EE +CPE
AMES IA 50011
USA

DANIELS, ERIC
SIEMENS SOLAR
4650 ADONR LN
CAMARILLO CA 93011
USA
805-388-6351-Tel
805-388-6395-Fax

DARKAZALLI, GHAZI
SPIRE CORPORATION
ONE PATRIOTS PARK
BEDFORD MA 01730-2396
USA
617 275 6000-Tel
617 275 7470-Fax

DASUKI, AGUS SALIM
BPP TEKNOLOGI

DAVIDSON, JOEL
SOLEC INTERNATIONAL, INC.
12533 CHADRON AVE.
HAWTHORNE CA 90250
USA
310-970-0065-Tel
310-970-1065-Fax

DAWSON, DAVID
CORNING, INC.
SP-FR-3-1
CORNING NY 14831
USA
607-974-3372-Tel
607-974-3675-Fax

DE ANDRADE, ADNEI M.
UNIVERSITY OFSAO PAULO/IEE
AV. PROF LUCIANO CUALBERTO
SAO PAULO SP 05508-900
BRAZIL
158 11 815 7216 Tel
55 112107750-Fax

DE BLASIO, RICHARD
NREL
1617 COLE BLVD
GOLDEN CO 80401
USA
303 231-1286-Tel
303 231-1199-Fax

DE HAAN, SJOERD
ECN RENEWABLE ENERGY
P O BOX 1
1755 2G PETTEN
NETHERLANDS
31 2246 4115-Tel
31 2246 3214-Fax

DE JONG, IRENE
OKE SERVICES
NIEUWSTRAAT 29
EINDHOVEN 5611DA
NETHERLANDS
31-40-445262-Tel
31-40-464133-Fax

DE NEUFUILLE, JOHN
VOLTAIX, INC.
197 MEISTER AVE
NORTH BRANCH NJ 08876
USA

DEANGELO, FRANK
NASA LEWIS
21000 BROOKPARK RD
MS 302-1
CLEVELAND OH 44135
USA

DEB, SATYEN
NREL
1617 COLE BLVD.
GOLDEN CO 80401-3393
USA
303-384-6405-Tel
303-384-6481-Fax

DEBOER, SCOTT
IOWA STATE UNIVERSITY
1925 SCHOLL RD
AMES IA 50011
USA
515-294-9794-Tel
515-294-95484-Fax

DEKAM, JAAP
FOKKER SPACE & SYSTEMS B.V.
C/O URBAIN TERMOTE
1199 NORTH FAIRFAX ST STE.500
ALEXANDRIA VA 22314
USA
011 31 71 245310-Tel
011 3171 245399-Fax

DEL BIAGGIO, WILLIAM
HERITAGE BANK
145 ALMADEN BLVD
SAN JOSE CA 95113
USA
408 947-6900-Tel
408 947-6910-Fax

DENDINGER, DAVID
SOLAREX CORP.
630 SOLAREX COURT
FREDERICK MD 21701
USA
301-698-4232-Tel

DENG, XUNMING
ENERGY CONVERSION DEVICES, INC
1675 WEST MAPLE ROAD
TROY MI 48084
USA
810 362-4780-Tel
810 362-0012-Fax

DENNIS, KEVIN
OMNION POWER ENGINEERING CORP.
P.O. BOX 879
EAST TROY WI 53120
USA
414-642-7200-Tel
414-642-7760-Fax

DESJARDINS, JOHN
UNIVERSITY OF CHICAGO
5735 SOUTH ELLIS AVE
CHICAGO IL 60637
USA
312-702-0218-Tel
312-702-0805-Fax

DHERE, NEELKANTH
FLORIDA SOLAR ENERGY CENTER
300 STATE ROAD 401
CAPE CANAVERAL FL 32920-4099
USA
407-783-0300X207-Tel
407-783-2571-Fax

DIAMOND, RON
IEEE
12500 GLADSTONE AVE.
SYLMAR CA 91342
USA
818-898-2802-Tel
818-361-5102-Fax

DIBBEN, MARK
US AIR FORCE
PL/XPS
3550 ABERDEEN AVE SE
KIRTLAND AFB NM 87117
USA
505-846-1634-Tel
505-846-4880-Fax

DIEDERICH, FRAMZ
FIZ, TECHNICAL UNIV OF AACHEN
AACHEN 52056
GERMANY
49 241 88947-0-Tel
49 241 8888284-Fax

DIETERINK, BEN
KIPP & ZONEN
390 CENTRAL AVENUE
BOHEMIA NY 11716
USA
516-589-2885-Tel
516-589-2068-Fax

DIETZ, DAVID PETER
P.B. TECHNIK
23 HOMER LANE
MENLO PARK CA 94025
USA
415 854 6801-Tel
415 854 3008-Fax

DINWOODIE, THOMAS
POWERLIGHT CORPORATION
2550 BENVENUE
BERKELEY CA 94704
USA
510 525 9495-Tel
510 525 7392-Fax

DISSELBREDE, VERA
F.F. LDA
SITIO DO PALHEIRINHO
ALJEZUR 8670
PORTUGAL
011-351-82-98745-Tel
011-351-82-98746-Fax

DOSHI, PARAG
GEORGIA TECH
ELECTRICAL & COMPUTER ENG
ATLANTA GA 30332
USA
404-894-3630-Tel
404-853-9882-Fax

DUCEY, ROCH
U.S. ARMY-CERL
P.O. BOX 9005
CHAMPAIGN IL 61826-9005
USA
217-398-5222-Tel
217-373-3430-Fax

DUPONT, FRANCOIS
S.P.I.R.E.
P. O. BOX 11530
MAHINA, TAHITI
FR. POLYNESIA
(689) 48 20 20-Tel
(689) 48 28 17-Fax

DURAN, JULIO CESAR
ATOMIC ENERGY COMMISSION
AV DEL LIBERTADOR 8250
BUENOS AIRES 1429
ARGENTINA
54 1 7547131-Tel
54 1 7547130-Fax

DURAND, STEVEN J
S W TECHNOLOGY DEV INST/NMSU
PO BOX 30001 DEPT 3 SOLAR
LAS CRUCES NM 88003
USA
505 646-4247-Tel
505 646-3841-Fax

DURBIN, STEVEN
FAMU/FSU COLLEGE OF ENGR.
TALLAHASSEE FL 32316-2175
USA
904-487-6481-Tel
904-487-6479-Fax

EASOZ, JOHN
EBARA SOLAR, INC.
811 RT. 51 SOUTH
LARGE PA 15025
USA
412-382-1255-Tel
412-382-1251-Fax

EBERSPACHER, CHRIS
UNISON
2700 W KELLY RD
NEWBURY PARK CA 91320
USA
805-499-7840-Tel
805-499-9502-Fax

EBRON, SONJA
SOLAR POWER & LIGHT

EDELMANN, ALBERT
ALTERNATIVE ENERGIE TECHNIK
INDUSTRIESTRASSE 12
66280
SULZBACH-NEUWEILE
GERMANY
49 6897 54337-Tel
49 6897 54359-Fax

EDELMANN, KONRAD
AET
AM KLOSTERHANG
AM KLOSTERHANG 66583
SPEISEN-ELVERSBER
GERMANY
49-6821-790068-Tel
49-6821-790503-Fax

EICKHOFF, DR. THOMAS
RESEARCH CENTRE JUELICH ISI-PV
P.O. BOX 1913
D-52425 JUELICH
GERMANY
49-2461-612615-Tel
49-2461-613735-Fax

EIDEN, GLENN
SUNTRACKER
302 US 30 E.
NEW HAVEN IN 46774
USA
219-749-4297-Tel

EIFFERT, PATRINA
OXFORD BROOKES UNIVERSITY
PRGS- SCHOOL OF ARCHITECTURE
GIPSY LANE-HEADINGTON
OXFORD OX3OBP
UNITED KINGDOM
865 483233-Tel
865 483298-Fax

EISENBERG, NAFTALI
J.C.T.
P O BOX 16031
JERUSALEM 91161
ISRAEL
972 2 751166-Tel
972 2 422075-Fax

EISGRUBER, INGRID
COLORADO STATE UNIVERSITY
PHYSICS DEPARTMENT
FORT COLLINS CO 80523
USA
303-491-1105-Tel
303-491-7947-Fax

ELGAMEL, HUSSAM ELDIN
IMEC
KAPELDREEF75
LEUVEN B-3001
BELGIUM
32 16 281555-Tel
32 16 281501-Fax

EMERY, KEITH
NREL
1617 COLE BLVD.
GOLDEN CO 80401
USA
303-384-6632-Tel
303-384-6490-Fax

ENDO, ELLCHI
ELECTROTECHNICAL LABORATORY
1-1-4 UMEZONO
TSUKUBA
IBARAKI 305
JAPAN
81-298-58-5826-Tel
81-298-58-5829-Fax

ENDO, KIYOSHI
TOKYO A&T UNIVERSITY
SAITO LAB

ERMER, JAMES
SPECTROLAB INC.
12500 GLADSTONE AVE
SYLMAR CA 91342
USA
818-898-2896-Tel
818-361-5102-Fax

ETHELVINA, ESCOBAR
C. TALAMANCA N 25
COL MIRAMONTE
SAN SALVADOR
EL SALVADOR C.A.

EVRARD, OLIVIER
IMEC
KAPELDREEF
LEUVEN 3001
BELGUIM
32-16-281-545-Tel
32-16-281-501-Fax

EYER, ACHIM
FHG-ISE
OLTMANNSSTR.5
FREIBURY GE 79100
GERMANY
+49-761-4588-261-Tel
+49-761-4588-250-Fax

FAHRENBRUCH, ALAN
STANFORD UNIVERSITY
DEPT. OF MATERIALS SCIENCE
STANFORD CA 94305-2205
USA
415-723-2951-Tel
415-725-4034-Fax

FAHRNER, WOLFGANG R.
UNIVERSITY OF HAGEN
P O BOX 940
HAGEN 58084
GERMANY
49 2331 987 4019-Tel
49 2331 987 321-Fax

FATEMI, NAVID
ESSENTIAL RESEARCH INC.
NASA LEWIS RESEARCH CENTER
21000 BROOKPARK RD MS 302 1
CLEVELAND OH 44135
USA

FATH, PETER
UNIVERSITY OF KONSTANZ
JAKOB-BURCKHARDT STR 22
KONSTANZ 88709
GERMANY

FAUR, MARIA
NASA LEWIS RES. CTR
21000 BROOKPARK ROAD
CLEVELAND OH 44135
USA
216 433-22933-Tel
216 443-6106-Fax

FAUR, MIRCEA
CLEVELAND STATE UNIVERSITY
1983 E 24 ST
CLEVELAND OH 44115
USA
216 687 3533-Tel
216 687 9280-Fax

FEREKIDES, CHRIS
UNIVERSITY OF SOUTH FLA
4202 B. FOWLER AVE
TAMPA FL 33620
USA
813 974-4818-Tel
813 974-5250-Fax

FIELD, HALDEN
NREL
1617 COLE BLVD.
GOLDEN CO 80401
USA
303-384-6685-Tel
303-684-6530-Fax

FIROR, KAY
BLUE MOUNTAIN ENERGY
59943 COMSTOCK RD
COVE OR 97824
USA
503 568 4473-Tel
503 568 4882-Fax

FITZGERALD, MARK
NREL
1617 COLE BLVD
GOLDEN CO 80401
USA
303-384-6570-Tel
303-384-6490-Fax

FITZGERALD, MATTHEW
APPLIED SOLAR ENERGY CORP.
15251 EAST DON JULIAN ROAD
CITY OF INDUSTRY CA 91745-1002
USA
818-968-6581-Tel
818-336-8694-Fax

FLAVIN, CHRISTOPHER
WORLDWATCH INSTITUTE

FLEDDERMANN, CHARLES
UNIVERSITY OF NEW MEXICO
EECE BUILDING
ALBUQUERQUE NW 87131
USA
505-277-5628-Tel
505-277-1439-Fax

FLOOD, DENNIS
LEWIS RESEARCH CENTER
MS 302-1
21000 BROOKPARK ROAD
CLEVELAND OH 44135
216-433-2303-Tel
216-433-6106-Fax

FLOOD, DENNIS M.
MOE CONSULTING

FLORES, CARLO
CISE S.P.A.
V.REGGIO EMILIA 39
SEGRATE 20090
ITALY
0039-2-216-72372-Tel
0039-2-21672620-Fax

FLOWERS, LAWRENCE
NREL
1617 COLE BLVD
GOLDEN CO 80439
USA
303-384-6910-Tel
303-384-6901-Fax

FONASH, STEPHEN
PENN STATE UNIVERSITY
ELECTRONIC MAT & PROC RESEARCH
111 EE WEST BLDG
UNIVERSITY PARK PA 16802
USA
814 865-4931-Tel
814 865-3018-Fax

FONG, KEN
HAWAIIAN ELECTRIC CO.

FORD, DAVID
ASTROPOWER
SOLAR PARK
NEWARK DE 19716-2000
USA
302-366-0400-Tel
302-368-6474-Fax

FOREST, HARVEY
SOLAREX
630 SOLAREX COURT
FREDERICK MD 21701
USA
301-698-4200-Tel
301-698-4201-Fax

FORESTIERI, MOE
MOE CONSULTING
74 BARBERRY DRIVE
BEREA OH 44017
USA
216 234 1574-Tel
216 433 6106-Fax

FORTUNATO, ELVIRA
CEMOP/UNINOVA
QUINTA DA TORRE
MONTE DE CAPARICA 2825
PORTUGAL
351.1.3500251-Tel
2941365-Fax

FOUQUET, JEAN-MICHEL
HELIOSTORE SYSTEMS
27 AVENUE PAUL VERLAINE
DOMAINE DE VILLARCEAU
F 77150
FRANCE

FRAAS, LEWIS
JX CRYSTALS INC
1105 12TH AVE NW #A-2
ISSAQUAH WA 98027
USA
206 392 5237-Tel
206 392 7303-Fax

FRANCIS, ROBERT
AEROSPACE CORP.
2350 E EL SEGUNDO BLVD
EL SEGUNDO CA 90245
USA
310-336-6272-Tel
310-336-5581-Fax

FRANK, ALLAN
SOLAR LETTER
9124 BRADFORD RD
SILVER SPRING MD 20901-4918
USA
301 565-2532-Tel
301 565-32298-Fax

FRANK, ALLAN
ALLAN L. FRANK ASSOCIATES
9124 BRADFORD RD.
SILVER SPRING MD 20901
USA

FRANTZIS, LISA
ARTHUR D. LITTLE.INC.
20 ACORN PARK
CAMBRIDGE MA 02140
USA
617-498-5688-Tel
617-4987007-Fax

FRASER, ALLISTER
ABLE ENGINEERING COMPANY
93 CASTILIAN DRIVE
GOLETA CA 93117
USA
805 685-2262-Tel
805 685-1369-Fax

FRELING, ROBERT
C/O DUBIN
156 N. WILLAMAN DRIVE
BEVERLY HILLS CA 90211
USA
202-234-7265-Tel

FREUNDLICH, ALEX
SPACE VACUUM EPITACY CENTER
UNIVERSITY OF HOUSTON
HOUSTON TX 77204-5507
USA
713 743 3621-Tel
713 747 7724-Fax

FRIEDLANDER, SANFORD
TRW
ONE SPACE PARK R4/1082
REDONDO BEACH CA 90278
USA
310-813-5967-Tel
310-813-6003-Fax

FRIEDMAN, DANIEL
NREL
1617 COLE BLVD
GOLDEN CO 80401
USA
303 384-6472-Tel
303 384-6531-Fax

FRIGERIO, HECTOR
SISTECNO SA
MIGUELETE
PO BOX 1561
MONTEVIDEO 11800
URUGUAY

FROST, JUDGE
R.A. ENERGY CO.
1949 E. 105TH STREET
CLEVELAND OH 44106
USA

FRYE, STUART L.
BECHTEL GROUP, INC
50 BEALE STREET
SAN FRANCISCO CA 94105
USA
415 768-1292-Tel
415 768-3580-Fax

FURUKUBO, STEVE
SOLEC INTERNATIONAL, INC.
12533 CHADRON AVE.
HAWTHORNE CA 90250
USA
310-970-0065-Tel
310-970-1065-Fax

FUYUKI, TAKASHI
KYOTO UNIVERSITY
YOSHIDAHONMACHI, SAKYO
KYOTO 606-01
JAPAN
81-75-753-5361-Tel
81-75-751-1576-Fax

GABOR, ANDREW
NREL
1617 COLE BLVD.
GOLDEN CO 80401
USA
303-384-6523-Tel
303-384-6130-Fax

GADDIPATI, RAMAYYA
KAPL, INC.
RIVER ROAD
SCHENECTADY NY 12302
USA

GADDY, EDWARD
NASA

GAFFNEY, KELLY
USAF PHILLIPS LABORATORY
PL VTPC
KIRTLAND AFB NM 87117
USA
505-846-2616-Tel
505-246-2885-Fax

GALICA, JAMES
SPRINGBORN MATERIALS SCIENCE
ONE SPRINGBORN CENTER
ENFIELD CT 06082
USA
203-749-8371-Tel
203-749-8234-Fax

GANGULY, GAUTAM
ETL
ELECTOTECHNICAL LAB
1-1-14 UMEZONO TSUKUBA
IBARAKI 305
JAPAN

GARBOUSHIAN, VAHAN
AMONIX, INC.
3425 FUJITA ST
TORRANCE CA 90505
USA
310-325-8091-Tel
310-325-0771-Fax

GARCIA, ALEXANDER
AEROSPACE CORPORATION
PO BOX 92957
LOS ANGELES CA 90009-2957
USA
310-336-8269-Tel
310-336-Fax

GARSTANG, COLLIN
UNITED SOLAR SYSTEMS CORP.
1100 W. MAPLE ROAD
TROY MI 48084
USA
1-800-843-3892-Tel
1-810-362-4442-Fax

GARVERICK, LINDA
ESSENTIAL RESEARCH
2460 FAIRMONT BLVD.
CLEVELAND OH 44106
USA

GAUCHAN, TEJ
CENTRE FOR RENEWABLE ENERGY
PO BOX 2963
KATHMANDU
NEPAL
011 977 1 227 041-Tel
011 977 1 226 369-Fax

GAY, CHARLES
UNISON
1399 SOUTHWIND CIRCLE
WESTLAKE VILLAGE CA 91361
USA
805 374-9992-Tel
805 373-6202-Fax

GAY, ROBERT
SIEMENS SOLAR INDUSTRIES
P.O. BOX 6032
CAMARILLO CA 93011-6032
USA
805-388-6507-Tel
805-388-6580-Fax

GEE, JAMES
SANDIA NATIONAL LABS
MS 0752
ALBUQUERQUE NM 87185-0752
USA
505-844-7812-Tel
505-844-6541-Fax

GERKEN, KEN
MORNINGSTAR CORP.
P.O. BOX 449
WASHINGTON CROSSI PA 18977
USA
215-321-8983-Tel
215-321-8987-Fax

GESSERT, TIMOTHY
NREL
ATTN: D. MORLAND, LIBRARY
1617 COLE BOULEVARD
GOLDEN CO 80401-3393
USA
303-384-6497-Tel

GHALI, FADIA
N.R.C. - E.R.I.
DOKKI-CAIRO
EGYPT

GHANNAM, MOUSTAFA
COLLEGE OF ENG. & PETROLEUM
P O BOX 5969,
SAFAT 13060
KUWAIT
965 4811188 7364-Tel
965 481 7451D-Fax

GIBART, PIERRE
CRHEA - CNRS
SOPHIA ANTIPOLIS
VALBONNE 06560
FRANCE
33 93 95 4227-Tel
33 93 95 83 61-Fax

GILES, GARY
SOLAR PLUS CANADA LTD
P O BOX 76
1755 SHAWNIGAN MILL BAY RD.
SHAWNIGAN LAKE BC VOR 2W0
CANADA
604 743-9796-Tel
604 743-9744-Fax

GILMORE, JIM
UNITED SOLAR SYSTEMS CORP.
5278 BASTGATE MALL
SAN DIEGO CA 92121
USA
619-625-2080-Tel
619-625-2083-Fax

GINLEY, DAVID
NREL
1617 COLE BLVD
GOLDEN CO 80401
USA

GIRAMBERK, YGAL
2841 MARIETTA CIRCLE
THOUSAND OAKS CA 91360
USA
805 493-4532-Tel
805 493-4532-Fax

GIRARD, JEAN-DENIS
S.P.I.R.E.
BP 11530
MAHINA
TAHITA
FR. POLYNESIA
(689) 48 20 20-Tel
(689) 48 28 17-Fax

GLEDHILL, KRISTEN
USAF PHILLIPS LABORATORY
PL VTPC
KIRTLAND NM 87117
USA
505-846-4899-Tel
505-846-2885-Fax

GODA, SHINJI
DAIDO HOXAN INC.
1007-58, IZUMISAWA
CHITOSE-SHI
HIKKAIDO 066
JAPAN
81-123-28-4035-Tel
81-123-28-3125-Fax

GODIN, RICHARD
BTU INTERNATIONAL
C/O JENNIFER BUTLER
23 ESQUIRE ROAD
NORTH BILLERICA MA 01862
USA
508-667-4111-Tel
508-667-9068-Fax

GOETHE, REINHARD
ASE GMBH
INDUSTRIESTRASSE 23-33
WEDEL D22880
GERMANY
49-4103-602626-Tel
49-4103-604704-Fax

GOETZBERGER, ADOLF
FRAUENHOFER INSTITUT FUR
SOLARE ENERGIESYSTEME
OLTMANNSSTR 5 D-79100
FREIBURG
GERMANY
07-6145880-Tel

GOLDENBERG, DR. BARBARA
HONEYWELL TECHNOLOGY CENTER
10701 LYNDALE AVE S
BLOOMINGTON MN 55420
USA
612-956-4099-Tel
612-956-4517-Fax

GOLDHAMMER, LEE
HUGHES SPACE/COMMUNICATIONS
2060 E. IMPERIAL HIGHWAY
EL SEGUNDO CA 90245
USA
310 416-3975-Tel
310 416-3933-Fax

GOMEZ GONZALEZ, AMANDA O
MEXICAN INST OF COMMUNICATIONS
AV DE LAS TELECOMUNICACIONES
S/N
MEXICO DF 09310
MEXICO
525 613 3126 531-Tel
525 613 3174-Fax

GOOD, BRIAN
NASA
LEWIS RESEARCH CENTER
CLEVELAND OH 44135
USA

GOODBODY, CHRISTOPHER
DEFENCE RESEARCH AGENCY
P234-BLDG
FARNBOROUGH GU146TD
UNITED KINGDOM
441252-394003-Tel
441252392669-Fax

GOODMAN, STANLEY
36 CONCORD DRIVE
NEW CITY NY 10956
USA
914-634-8640-Tel

GORDILLO, GERARDO
UNIV NATIONAL DE COLOMBIA
DEPT DE FISCA
SANTAFE DE BOGOTA
COLOMBIA
57 1 2690827-Tel
57-1-222-5716-Fax

GORDON, LEE
MORNINGSTAR CORP.
18 NORTH STATE STREET
NEWTOWN PA 18904
USA
215-504-9801-Tel
215-504-9802-Fax

GOULD, WAYNE
SOUTHERN CALIFORNIA EDISON CO.
2244 WALNUT GROVE AVENUE
ROOM 492
ROSEMEAD CA 91792
USA
818 302-9508-Tel
818 302-1102-Fax

GRAF, ERIC
TEXAS INSTRUMENTS INC.
P O BOX 655012 M535
DALLAS TX 75265
USA
214-995-0155-Tel
214-995-2337-Fax

GRAY, JEFFREY
PURDUE UNIVERSITY
1285 ELECTRICAL ENGINEERING
W LAFAYATTE IN 47907-1285
USA
317-494-6440-Tel
317-4946440-Fax

GREEN, MARTIN A.
UNSW ELECTRICAL ENGINEERING
P.O. BOX 1
KENSINGTON
NSW 2033
AUSTRALIA
61-2-385-4018-Tel
61-2-662-4240-Fax

GREEN, ROGER
INTEVAC
601 CALIFORNIA AVE
PALO ALTO CA 94304
USA

GREGORY, JENNIY
I. T. POWER LTD
THE WARREN, BRAMSHILL RD
EVERSLEY, HANTS RG27 OPR
UNITED KINGDOM
44 1734 730073-Tel
44 1734 730820-Fax

GRIFFIOEN, JAN H.
NV GEB ZUID-HOLLAND WEST/NVHOD
WOOSDUTNSEWG 913
P O BOX 60701
DEN-HAAG ZH 2906LS
NETHERLANDS
21 70 3132741-Tel
31 70 3132535-Fax

GRIFFITHS, TERENCE S.
PILKINGTON OPTRONICS, INC.
7550 CHAPMAN AVENUE
GARDEN GROVE CA 92641
USA
714-373-6061-Tel
714-373-6074-Fax

GRIMES, LAWRENCE
PHILLIPS LABORATORY
PL/WSAE KIRTLAND AFB
ALBUQUERQUE NM 87117
USA

GRIMMER, DERRICK
IOWA THIN FILM TECH. INC.
ISIS, ISU RESEARCH PARK
AMES IA 50010
USA
515-294-7005-Tel
515-294-9584-Fax

GROEHN, HANS-GUNTER
KFA
LEO BRANDT STR.
JULICH D-52425
GERMANY
02461616481-Tel
02461616695-Fax

GRONAU, MAGDALA
ESTSTV. 56
ESSEN D-45149
GERMANY

GROSSER, KAREN
MOE CONSULTING
16259 HUMMEL ROAD
BROOKPARK OH 44142
USA
216 267 6831-Tel

GRUNDHAUSER, DON
IDAHO POWER CO.
PO BOX 70
BOISE ID 83707
USA
208 388 2240-Tel
208 388 6910-Fax

GU, NING KE
HAI NAN EAST-ASIA PHOTOVOLTAIC
CO. LTD 18F ZHONGFANG BLDG
LONG KUN, NORTH RD
HAI KOU CITY, HAI 570005
R.O.C.
86 0898 8533473-Tel
86 0898 8533456-Fax

GUASTAVINO, FRANCIS
UNIVERSITY MONTPELLIAR II
C.E.M. BAT 21 CAS 075
MONTPELLIER 34095
FRANCE
33 67 143206-Tel
33 67 54 7134-Fax

GUED, LISA
SOLAR POWER & LIGHT

GUHA, SUBHENDU
UNITED SOLAR SYSTEMS
1100 W. MAPLE ROAD
TROY MI 48084
USA
810-362-3120-Tel
810-362-4442-Fax

GUIMARAES, LEOPOLDO
N. UNIVERSITY OF LISBON
2825 MONTE DE CAPARICA
QUINTA DA TORRE
ALMADA 2825
PORTUGAL
351-1-2954464-Tel
351-1-2957810-Fax

GUNARATNE, LALITH
POWER AND SUN PRIVATE LTD.

GUNN, JOHN
SCIENTIFIC ENGINEERING, INC.
5960 E SHIRLEY LANE
MONTGOMERY AL 36117
USA
205 271-0643-Tel
205 271-0818-Fax

GUNTHER, PETER
NMRC
LEE HALTINGS, PROSPECT ROW
CORK
IRELAND
35321 276871 4083-Tel
35321270271-Fax

GUPTA, DHARAM VIR
CENTRAL ELECTRONICS LTD
4 INDUSTRIAL AREA
SAHIBABAD UP 201010
INDIA
8771941 8771910-Tel
8771843 6482373-Fax

GUPTA, KEDAR
GT EQUIPMENT TECHNOLOGIES

HAARPAINTNER, GUIDO
IMPERIAL COLLEGE
PRINCE CONSORT ROAD
SOLID STATE PHYSICS
LONDON SW7 2BZ
UNITED KINGDOM
44 171 5947587-Tel
44 171 5947580-Fax

HAGEMAN, PAUL
UNIVERSITY OF NIJMEGEN
TOERNOOIVELD
NIJMEGEN 6525 ED
NETHERLANDS
080-653063-Tel
080-652314-Fax

HAGEMANN, INGO
TECHNICAL UNIVERSITY OF AACH
JOHANNITERSTRABE 21
D-52064 AACHEN
GERMANY
+49-241-30547-Tel
+49-241-30547-Fax

HALL, ROBERT
ASTRO POWER, INC
SOLAR PARK
NEWARK DE 19716-2000
USA
302 366-0400-Tel
302 368-6474-Fax

HALL, ROBYN
NORTHEAST PHOTOSCIENCES
18 FLAGG RD
HOLLIS NH 03049
USA

HALLE, LINDA
AEROSPACE CORPORATION
P O BOX 92957
LOS ANGELES CA 90009-2957
USA
310 336-4073-Tel
310 336-5846-Fax

HAMAKAWA, YOSHIHIRO
FACULTY OF ENGINEERING SCIENCE
OSAKA UNIVERSITY
1-3 MACHIKANE-YAMACHO
TOYONAKA, OSAKA 560
JAPAN
81-6-850-6315-Tel
81-6-850-6316-Fax

HAMARA, JAANA
NESTE
PL 3
FIN-02151
ESPOO
FINLAND
358-045-01550-Tel
358-0450-5744-Fax

HAMMOND, ROBERT
ARIZONA STATE UNIVERSITY
1971 E HUNTINGTON DRIVE
TEMPE AZ 85287 2848
USA
602 965 0377-Tel
602 838 2379-Fax

HAMSTRA, JUUL
FOKKER SPACE & SYSTEMS
NEWTEMWEG 1
LEIDEN
NETHERLANDS

HANAK, JOSEPH
240 RAPHAEL AVE, #9
AMES IA 50014
USA
515 292-8251-Tel
515 296-9910-Fax

HANCOCK, MARK
UNIVERSITY OF SOUTH WALES
SCHOOL OF ELECTRICAL ENGRNG.
UNIVERSITY OF NSW, SYDNEY
NSW 2052
AUSTRALIA
61 2 385 4061-Tel
61 2 385 5993-Fax

HANOKA, JACK
EVERGREEN SOLAR
107 YORK TERRACE
BROOKLINE MA 02146
USA
617 2775147-Tel

HANSEN, RICHARD
ENERSOL ASSOCIATES

HANSSON, PER-OVE
MAX PLANCK INSTITUT FKF
HEISENBERGSTR 1
STUTTGART D-70569
GERMANY
49-7116891454-Tel
49-7116891010-Fax

HANST, JUDY
POWER INFOR. CENTER (IAPG)
10700 PARKRIDGE ROAD, STE 250
RESTON VA 22091
USA
703 758 0531-Tel
703 758 0531-Fax

HARDINGHAM, CHRIS
EEV LTD
WATERHOUSE LANE
CHELMSFORD, ESSEX CMI 2QU
ENGLAND
0245 493493-Tel
0245 492492-Fax

HARNISCH, PETER
AKKU GESELLSCHAFT MBH
ALBERT SCHWEITZER STRABE 30
TAUBENHEIM 02689
GERMANY
0359-36-4377-Tel
0359-36-4397-Fax

HARR, MICHAEL
ANTEC GMBH
INDUSTRIESTRASSE 2-4
KELKHEIM/TS. D-65779
GERMANY
06195-973-0-Tel
06195-973-190-Fax

HARRINGTON, THOMAS
TRW SPACE & ELECTRONICS GROUP
1 SPACE PARK, R4/1206
REDONDO BEACH CA 90278
USA
310 812 1194-Tel
310 813 5777-Fax

HARRIS, DAVID
TEXAS INSTRUMENTS
7800 BANNER DR M/S 3922
MS 3922
DALLAS TX 75241
USA
214 917 1245-Tel
214 917 3953-Fax

HARTZELL, WILL
INTER ISLAND SOLAR SUPPLY
345 NIMITZ HWY.
HONOLULU HI 96817
USA
808-523-0711-Tel
808-536-5586-Fax

HASEGAWA, HIDEKI
HOKKAIDO UNIVERSITY

HASEGAWA, TAKASHI
SHOWA SOLAR ENERGY K.K.
TOKYO BLDG 7-3 MARUNAUCHI 2
CHOME
TOKYO 100
JAPAN
81-3-3215-9661-Tel
81-3-3284-1589-Fax

HASEGAWA, YASUYUKI
AIMS CORPORATION
ASAKAWA BLDG, 1-16-10
SHINBASHI, MINATO-KU
TOKYO 105
JAPAN
81-3-3501-1059-Tel
81-3-3501-1388-Fax

HASHIMOTO, EIJI
CRIEPI
2-11-1, IWATO-KITA,
KOMAE-SHI
TOKYO 201
JAPAN
81-3-3480-2111-Tel
81-3-34-80-3866-Fax

HASSETT, ROBERT
DEPT. OF ENERGY
WASHINGTON DC 20024
USA
202-586-8163-Tel
202-586-1628-Fax

HATANAKA, YUJI
DAIDO HOXAN INC.
1007-58,IZUMISAWA
CHITOSE-SHI
HOKKAIDO 066
JAPAN
81-123-28-4035-Tel
81-123-28-3125-Fax

HAUSER, CHARLES
HCT SHAPING
EN NOYER GIROD
ETOY 1163
SWITZERLAND
41-21-807-41069-Tel
41-21-807-3094-Fax

HAUSER, MARIE LAURE
HCT NOYER SHAPING 5455
EN NOYER GIROD
ETDY 1163
SWITZERLAND
41-21-807-4169--Tel
41-21-807-3094--Fax

HAYASHI, KATSUHIKO
KANEKA CORPORATION
1-1, HIEJTSUJI 2 CHOME
OTSU 520-01
JAPAN
0775-77-2184-Tel
0775-77-2125-Fax

HAYASHI, KIYOTAKA
NAGPYA INSTITUTE OF TECHNOLOGY
GOKISO CHO SHOWA KU
NAGOYA 466
JAPAN
81-52-732-2111-Tel
81-52-7325697-Fax

HE, DE YAN
TOKYO INSTITUTE OF TECHNOLOGY
4259 NAGATUSTA, MIDORI-KU
YOKOHAMA 227
JAPAN
045-924-5188-Tel
045-921-1089-Fax

HEARD, STEVE
POWER INFORMATION CTR (IAPG)
10700 PARKRIDGE RD, STE 250
RESTON VA 22091
USA
703 758 0531-Tel
703 758 0531-Fax

HEARTON, LENARD
ATSC
5150 KIMSCOTT COURT
ANNANDALE VA 22003
USA
703-323-0630-Tel
202-767-1407-Fax

HEGEDUS, STEVEN S.
INSTITUTE OF ENERGY CONVERSION
UNIVERSITY OF DELAWARE
NEWARK DE 19716-3820
USA
302-831-6253-Tel
302-831-8226-Fax

HEILSCHER, GERD
1ST ENERGIETECHNIK
STADTJAGERSTR.11
AUGSBRURG 86152
GERMANY
0049-821-156890-Tel
0049-821-513705-Fax

HEISER, GERNOT
UNIVERSITY OF NSW
SYDNEY
NSW 2052
AUSTRALIA
61 2 385 5156-Tel
61 2 385 5995-Fax

HELM, PETER
WIP
SYLVENSTEINSTR. 2
MUNICH
BAVARIA D-81369
GERMANY
49 89 7201235-Tel
49 89 7201240-Fax

HELMKE, CLAAS
JRC - ESTI
ATION, ESTI
FRC-ISPRA
TP450 ISPRA (VA) 21020
ITALY

HENDRICKSON, THOMAS
MANKATO STATE UNIVERSITY
MSU BOX 215
MANKATO MN 55044
USA
507-389-2477-Tel
507-389-1095-Fax

HERBST, WOLFGANG
UNIVERSITAT KAISERSLAUTERN
FACHBEREICH PHYSIK
ERWIN-SCHRODINGER STRABE
KAISERSLAUTERN D-67663
GERMANY
0631 205 4053-Tel
0631 205 2996-Fax

HERNANDEZ, LUIS
UNIVERSIDAD NACIONAL COLUMBIA
PHYSICS DEPT
BOGOTA
COLOMBIA
SOUTH AMERICA
57 1 269 0827-Tel
57 1 222 5716-Fax

HERTLEIN, HANSMARTIN P.
SOLAR ENERGY RESEARCH ASSOC
C/O DLR
D 51140 KOLN
GERMANY
02203 968 3625-Tel
02203 968 4740-Fax

HERWIG, LLOYD
U.S. DEPT OF ENERGY
EE-131
1000 INDEPENDENCE AVE S.W.
WASHINGTON DC 20585
USA
202 586-5538-Tel
202 586-5127-Fax

HESTER, STEVE
2497 MORELLO HEIGHTS CIRCLE
MARTINEZ CA 94553
USA
510-372-7531-Tel
510-372-7547-Fax

HIDE, ICHIRO
DAIDO HOXAN INC.
1007-58, IZUMISAWA
CHITOSE-SHI
HOKKAIDO 066
JAPAN
81-123-28-4035-Tel
81-123-28-3125-Fax

HIELSMAIR, HENRY
UNIV CALIFORNIA BERKELEY
484 HEARST MINING BLDG
BERKELEY CA 94320
USA
510 643 5304-Tel

HILL, ROBERT
UNIVERSITY OF NORTHUMBRIA
ELLISON PLACE
NEWCASTLE UPON TY NEI 8 ST
UNITED KINGDOM
44-91-227-4594-Tel
44-91-227-4561-Fax

HILL, ROGER
SANDIA NATIONAL LABORATORIES
ALBUQUERQUE NM 87185 0753
USA
505 844 6111-Tel
505 844 6541-Fax

HILL, STEVEN
MICROFABTEK, INC
590 LIPOA PARKWAY
MAIN RES TECH CENTER
KIHEI HI 96753
USA

HILLER, WERNER
DGSIESES
TECHNICAL UNIV CHEMNITZ
CHEMITZ
GERMANY
0491 0371 5313340-Tel
04910371 531 3376-Fax

HIRASAWA, TATSUO
KANDENKO CO. LTD
NO 4 8 33 SHIBAURA
MINATO KU
TOKYO 108
JAPAN
81 3 5476 3842-Tel
81 3 5476 3853-Fax

HIRASAWA, TERUHIKO
SHIN ETSU CHEMICAL CO, LTD
3-2-1 SAKATO
TAKATSU - KU
KAWASAKI KANAGAWA 213
JAPAN
044 819 3101-Tel
044 819 3103-Fax

HIRSCH, MICHELE
MA PL INSTITUTE F. FASTKE
HEISENBERH STR. 1
STUTTGART
GERMANY
99-711-689-1617-Tel
99-711-689-1010-Fax

HISHIKAWA, YOSHIHIRO
SANYO ELECTRIC CO., LTD.
1-18-13 HASHIRIDANI
HIRAKATA
OSAKA 573
JAPAN
81-720-41-1261-Tel
81-720-41-0386-Fax

HO, FRANK
APPLIED SOLAR ENERGY CORP.
15251 EAST DON JULIAN RD.
CITY OF INDUSTRY CA 91745-
USA
818-968-6581-Tel
818-336-8694-Fax

HOCKADAY, ROBERT
ENERGY RELATED DEVICES
3025 ARIZONA AVE
LOS ALAMOS NM 87544
USA
505-662-4807-Tel
505-662-4807-Fax

HODSON, FRANCES
NAT'L RENEWABLE ENERGY LAB
409 12TH STREET S.W. SUITE 710
WASHINGTON DC 20024
USA
202-484-1090-Tel
202-484-1096-Fax

HOFFMAN, BRIDGET
4996 HILDASUE TERR
FUEMONT CA 94555
USA

HOFFMAN, DAVID
IPRT/ IOWA STATE UNIVERSITY
112 OFFICE & LAB BLDG
AMES IA 50011
USA
515 294-6551-Tel
515 294-2361-Fax

HOFFMAN, RICHARD
ESSENTIAL RESEARCH
C/O NASA LERC
21000 BROOKPARK RD MS302-1
CLEVELAND OH 44135
USA

HOGG, DAVID
UNISEARCH LTD.
UNIVERSITY OF NEW SOUTH WALES
SYDNEY 2052
AUSTRALIA
612-385-5401-Tel
612-663-2949-Fax

HOHMEYER, OLAR H.
ZEW
P.O BOX 103443
MANNHEIM D-68034
GERMANY
49-0-621-1235-200-Tel
49-0-621-1235-226-Fax

HOLLEY, JR., WILLIAM H.
SPRINGBORN MATERIALS SCIENCE
1 SPRINGBORN CENTER
ENFIELD CT 06082
USA
203 749-8371-Tel
203 749-8234-Fax

HOMMA, TAKUYA
NEDO
SUNSHINE 60, 1-1
3 CHOME HIGASHI-IKEBUKURO
TOSHIMA-KU, TOKYO 170
JAPAN
81 3 3987 9422-Tel
81 3 5992 6440-Fax

HONG, SUNG MIN
SAMSUNG ELECTRONICS CO.
DAZHWA BLD 9TH FLOOR
SAMSUNG-DONG
KANGNAM-GU, SEOUL
KOREA

HONSBERG, CHRISTIANA
UNIVERSITY OF NEW SOUTH WALES
SCHOOL OF ELECTRICAL ENG.
SYDNEY, NSW 2052
AUSTRALIA
02 385 4018-Tel
02 662 4240-Fax

HORZEL, JORG
IMEC
KUPELDOLF 75
LEUVEN B-3001
BELGUIM
0032-16-281-555-Tel
0032-16-281-501-Fax

HOTEL TO AIRPORT,

HOU, JING-YA
PENN STATE UNIVERSITY
7711 BROUSE AVE.
PHILADELPHIA PA 19152
USA
215-333-2968-Tel

HOWLEY, WILLIAM
WINROCK INTERNATIONAL
1611 N KENT ST
SUITE 600
ARLINGTON VA 22209
USA
703 525 9430-Tel
703 243 1175-Fax

HSIEH, EDMUND
LAWRENCE LIVERMORE NAT'L LAB
P O BOX 808
L-482
LIVERMORE CA 94550
USA
510 422 7350-Tel

HUACUZ, JORGE
IIE

HULSTROM, ROLAND
NREL
1617 COLE BLVD
GOLDEN CO 80401
USA
303 384-6420-Tel
303 384-6490-Fax

HUND, THOMAS
SNL
P O BOX 5800
ALBUQUERQUE NM 87185-0753
USA
505-844-8627-Tel
505-844-6541-Fax

HUSSEIN, KHALID HASSAN
SAGA UNIVERSITY
1 HONJYO
SAYA 840
JAPAN
0952-24-5191-Tel
0952-29-4441-Fax

HUTCHBY, JAMES
RTI
1720 TROPICAL DRIVE
RALEIGH NC 27606
USA
919 541-6000-Tel

HYNES, KATHLEEN
UNIVERSITY OF NORTHUMBRIA
ELLISON PLACE
NEWCASTLE UPON TY NEI 8 ST
UNITED KINGDOM
44-91-227-4555-Tel
44-91-227-4561-Fax

ICHIJO, NORIMITSU
AIMS CORPORATION
ASAKAWA BLDG, 1-16-10,
SHINBASHI, MINATO-KU
TOKYO 105
JAPAN
81-3-3501-1059-Tel
81-3-3501-1388-Fax

ICHIKAWA, YUKIMI
FUJI ELECTRIC AND DEVELOP LTD
2-2-1 NAGASAKA
YOKOSUKA 240-01
JAPAN
468-57-6730-Tel
468-57-2791-Fax

IIZUKA, MASASHI
SUMITOMO SITIX CORP.
2-1-28 SHIBA MIRASO KU
TOKYO 105
JAPAN
81-3-3451-6442-Tel
81-3-3451-6040-Fax

IKKI, OSAMU
RESOURCES TOTAL SYSTEM CO ,LTD
2 7 11 SHINKAWA CHUOKU
TOKYO 104
JAPAN
81 3 3551 6345-Tel
81 3 3553 8954-Fax

ILES, PETER
APPLIED SOLAR ENERGY CORP.
15251 DON JULIAN RD
INDUSTRY CA 91745-1002
USA
818-968-6581-Tel
818-336-8694-Fax

IMAMURA, EI-ICHI
CENT. RES. INST. OF ELECT. POW
1-6-1 OHTE-MACHI
CHIYODA-KU
TOKYO 100
JAPAN
81-3-3201-6601-Tel
81-3-3287-2805-Fax

INGHAM, GEORGE W.
SOLAR SYSTEM ASSEMBLIES INC
198 PEMBERTON AUS
NORTH VANCOUVER BC V7P QR5
CANADA
604 983-2819-Tel
604 983-2869-Fax

ISHIDA, TAKEO
SANYO ELECTRIC
1-1 DAINICHI-HIGASHIMACHI
MORIGUCHI CITY
OSAKA 570
JAPAN

ISHIHARA, YOSHIYUKI
DOSHISHA UNIV.
TANABE-CHO
KYOTO 610-03
JAPAN
81-7746-5-6327-Tel
81-7746-5-6813-Fax

ISHIKAWA, NAOKI
DAIDO HOXAN INC.
1007-58,IZUMISAWA
CHITOSE-SHI
HOKKAIDO 066
JAPAN
81-123-28-4035-Tel
81-123-283125-Fax

ISHISAKA, YUICHI
DOSHISHA UNIVERSITY
TANABE
KYOTO 610-03
JAPAN
81-7746-5-6327-Tel
81-7746-5-6813-Fax

ITO, KENTARO
SHINSHU UNIVERSITY
FACULTY OF ENGINEERING
500 WAKASATO
NAGANO 380
JAPAN
+81-262-26-4101-Tel
+81-262-28-4295-Fax

ITO, TAKASHI
KUBOTA CO
STATIONED AT SHIGA FACTORY
2-1 TAKAMATSU, KOSEI-CHO
KOGA-CHO, SHIGA-K 520-32
JAPAN

IWAKATA, MIKE
CANON, INC.
ECOLOGY RESEARCH CENTER
4-1-1 KIZUGAWADAI, KIZU-CHO
KYOTO 619-02
JAPAN
81 7747 5 2111-Tel

IWATA, HIDEYUKI
TOYAMA PREFECTURAL UNIV.
5180 KUROKAWA,
KOSUGI-MACHI
TOYAMA 93903
JAPAN
81766-56-7500-Tel
81-766-56-6172-Fax

IZU, MASAT
ENERGY CONVERSION DEVICES INC.
1675 WEST MAPLE
TROY MI 480847
USA
810 280-1900-Tel
810 280-1456-Fax

JAAP DE KAM, MR.

JAEGERMANN, WOLFRAM
HAMN METNER INSTITUTE
GLIENICKER STR 100
BERLIN 14109
GERMANY
49-308062-2936-Tel
49-30862-2434-Fax

JAIN, R.K.
POLYPLEX CORPORATION LTD
75, AMRIT NAGAR
NEW DELHI 110 003
INDIA
91-11-4631761-Tel
91-11-4620729-Fax

JAIN, RAJ K.
NASA LEWIS RESEARCH CENTER
21000 BROOKPARK ROAD
MAIL STOP 302-1
CLEVELAND OH 44135
USA
216 433 2227-Tel
216 433 6106-Fax

JAMES, HILDA
INTERNATIONAL CONFERENCE MGMT
7152 SW 47 STREET
MIAMI FL 33183
USA
305-661-5115-Tel
305-666-7134-Fax

JAMES, LAWRENCE
JAMES ASSOCIATES
7329 MEADOW CT
BOULDER CO 80301
USA
303 530-9014-Tel
303 530-9014-Fax

JAMES, WILLIAM
USAF PHILLIPS LABORATORY/VTP
RALPH JAMES
3550 ABERDEAN AVE S.E.
KIRTLAND AFB NM 87117-5776
USA
505-846-4894-Tel
505-846-2885-Fax

JANG, JIN
KYUNG MEE UNIVERSITY
DONGDAEMOON-ICU
SEOUL KO 130-701
KOREA
82-2-961-0270-Tel
82-2-968-6924-Fax

JENKINS, PHILLIP
NYMA SETAR (NASA LERC)
NASA LEWIS RESEARCH CENTER
MS-302-1 21,000 BROOKPARK RD
CLEVELAND OH 44135
USA

JENNINGS, CHRISTINA
PACIFIC GAS & ELECTRIC
3400 CROW CANYON ROAD
SAN RAMON CA 94583
USA
510-866-5305-Tel
510-866-5447-Fax

JENNINGS, LAURENCE
PHOTRON, INC.
1220 BLOSSER LANE
WILLITS CA 95490
USA
707 459-3211-Tel
707 459-2165-Fax

JENSEN, TOM
STRATEGIES UNLIMITED
201 SAN ANTONIO CIR. SUITE 205
MOUNTAIN VIEW CA 94040
USA
415-941-3438-Tel
415-941-5120-Fax

JENSEN, VICTOR

JESTER, THERESA
SIEMENS SOLAR INDUSTRIES
P O BOX 6032
CAMARILLO CA 93010
USA
805 388 6500-Tel
805 388 6557-Fax

JESTER, THERESA
SIEMENS SOLAR INDUSTRIES
PO BOX 6032
CAMARILLO CA 93011
USA

JI, ILHWAN
SAMSUNG AIT

JIVACATE, CHAYA
EGAT
NONTHABURI
11000
THAILAND
662 4336317-Fax

JOACHIM, LUTHER
FRAUNHOFER INSTITUTE
FUR SOLAR ENERGIESYSTEME
FREIBURG
GERMANY
07 61 45 88 121-Tel
07 61 45 88 116-Fax

JOHANSON, GREG
SOLEC INTERNATIONAL
12533 CHARDON AVE
HAWTHORNE CA 90250
USA
310-970-0065-Tel
310-970-1065-Fax

JOHNSON, DARREN
USAF
PHILLIPS LAB
PL/VTPP
KIRKLAND AFB NM 87117
USA

JOHNSON, ROBERT O.
STRATEGIES UNLIMITED
201 SAN ANTONIO CIRCLE STE 20
MOUNTAIN VIEW CA 94040
USA
415 941-3438-Tel
415 941-5120-Fax

JOHNSON, STEVE
GOLDEN PHOTON, INC.
4545 MCINTYRE STREET
GOLDEN CO 80403
USA
303 271-7122-Tel
303 271-7410-Fax

JOHNSON, TIM
RICHMOND TECHNOLOGY INC.
12 FIELDWOOD CT.
SAN JOSE CA 95120
USA
408-268-5154-Tel
408-268-0794-Fax

JOHNSTON, STEVEN M.
TRACE ENGINEERING
0 McCONIGA RD
LYLE WA 98635
USA
509 365-2449-Tel
SAME-Fax

JONES, ALAN
BLE ENGINEERING COMPANY
3 CASTILIAN DRIVE
GOLETA CA 93117
USA
05 685-2262-Tel
05 685-1369-Fax

JONES, CARL
HOME BUILDERS INTERNATIONAL
55 EAST JEWELL AVENUE
TE 200
DENVER CO 80222
USA

JONES, DAVID
PILKINGTON SPACE TECHNOLOGY
UNIT 2 KIMMEL PARK
BODELWYDDAN RHYL LL18 5TY
UNITED KINGDOM
1-44-745-584584-Tel
1-44-745-584208-Fax

JONES, P ALAN
BLE ENGINEERING CO. INC
CASTILIAN DRIVE
GOLETA CA 93117
USA
5 685 2262-Tel
5 685 1369 Fax

NG WOOK, YUN
TRON, INC
3 IMSOODONG
MI
UNG BUK 730-350
REA

JORDAN, DAVID
SOLAR AUSTRALIA
00 OLD PITTWATER ROAD
BROOKVALE
W 2100
STRALIA
2-438-5111-Tel
2-939-1548-Fax

JOSEPH, SEBASTIAN
LABORATORIO DE ENERGIA
TEMIXCO 62580
MEXICO
73-250046-Tel
73-250044-Fax

JOURDE, PATRICK
GENEC
BAT 351
ST PAUL LEZ
DURANCE F 13108
FRANCE
33 42 25 66 67-Tel
33 42 25 73 73-Fax

JUDD, CULLY
INTER-ISLAND SOLAR SUPPLY
345 NIMITZ HWY
HONOLULU HI 96817
USA
808 523 0711-Tel
808 536 5886-Fax

KABAYA, MASAO
SOLAR SYSTEMS INSTITUTE, INC.
R #1103, NOA-SHIBADAIMON
1-4-4 SHIBADAIMON, MINATOKU
TOKYO 105
JAPAN
81 3 3434 1471-Tel
81 3 3434 3948-Fax

KADE, HERIBERT
SEO SOLAR SONNENENERGIE OPPACH
STR D FREUNDSCHAFT 8
OPPACH
BRO 02736
GERMANY
49 03 58 72 50-Tel
49 03 58 72 2337-Fax

KAGAN, MISHA
SRPE KVANT
129626, 3 MYTISHCHINSKAYA ST
16 MOSCOW
RUSSIA
2872959-Tel
2868922-Fax

KALES, JURIS
ASE AMERICAS
4 SUBURBAN PARK DRIVE
BILLERICA MA 01821
USA
508 667 5900 293-Tel
508 663 2868-Fax

KALMBACH, ERIC
LENBROOK ENERGY SYSTEMS
633 GRANITE COURT
PICKERING, ONT L1W 3K1
CANADA

KAMIMURA, KUNIO
SHARP CORPORATION
282-1 SHINJYO-CHO
KITAKATSURAGI-GUN
NARA 639-21
JAPAN
81 7456 51161-Tel
81 7456 28253-Fax

KAMISAKO, KOICHI
TOKYO UNIV OF AGR & TECH
2-24-16, KOGANEI
TOKYO 184
JAPAN

KANEKO, KYOJIRO
SUMITOMO SITIX CORP
1, HIGASHIHAMA-CHO
AMAGASAKI, HYOGO 660
JAPAN
06-411-1121-Tel
06-411-5928-Fax

KANG, TAE GU
RACER
KOREA EX BK CR SEV. B/D 3F
935-34 PANG BAE DONG SEOCHO KU
SEOUL 137-060
KOREA
82 2 525 3474-Tel
82 2 522 8094-Fax

KAPUR, JAGDISH
INFOS INDUSTRIES LIMITED
706-707 SURYA KIRAN
19 KASTURBA GANDHI MARG
NEW DELHI 110001
INDIA
3316196-98-Tel
91-11-3316331-Fax

KAPUR, VIJAY
INT'L. SOLAR ELECTRIC TECH.
8635 AVIATION BLVD.
INGLEWOOD CA 90301
USA
310-216-4427-Tel
310-216-2908-Fax

KARABANOV, SERGEY
GELION LTD.
55 NOVAYA STREET
RYAZAN 390000
RUSSIA
0912-93-01-43-Tel
0912-93-02-34-Fax

KARANIAN, LINDA
MARTIN MARIETTA
P O BOX 179 MS F 3300
DENVER CO 80201
USA
303 977 9140-Tel
303 977 4612-Fax

KARLINA, LUDMILA
IOFFE INSTITUTE
POLYTECHNICHEGKATA 26
ST PETERSBURG 194021
RUSSIA
7 812 247 9924-Tel
7 812 2471017-Fax

KARLSTETTER, BRIGITTE
DEUTSCHE AEROSPACE
P O BOX 801168
MUNICH 81663
GERMANY
49 89 607 27458-Tel
49 89 607-21059-Fax

KASAHARA, TADAO
MSK CORPORATION
ROOM 1011 TOKAN NO.2 CASTLE
5-12 NISHI-SHINJUKU 3-CHOME
TOKYO 160
JAPAN
03-3342-3838-Tel
03-3342-6534-Fax

KATIC, IVAN
SOLAR ENERGY LAB
P O BOX 141
TAASTRUP 2630
DENMARK
45 4350 4567-Tel
45 4350 7222-Fax

KATO, TOSHIMICHI
NAGOYA INST OF TECHNOLOGY
8-07
OYAMA-CYO
KARRYA, AICHI 448
AICHI

KATSU, TOMOJI
SHARP CORPORATION
282-1 SHINJYO-CHO
KITAKATSURAGI-GUN
NARA 639-21
JAPAN
81 7456 51161-Tel
81 7456 28253-Fax

KATSUMATA, HIROSHI
NEDO
SUNSHINE 60, 1-1,
3 CHOME HIGASHI-IKEBUKURO
TOSHIMA-KU, TOKYO 170
JAPAN
81-0-3-3984 9421-Tel
81-0-3-5992-6440-Fax

KATZ, DAVID
ALTERNATIVE ENERGY ENGINEERING
BOX 339
REDWAY CA 95560
USA
707 923 2277-Tel
707 923 3009-Fax

KAUFFMAN, BOB
PHOTOCOMM, INC.
BOX 14670
SCOTTSDALE AZ 85267-4670
USA
602 948 8003-Tel
602 483 6431-Fax

KAWABATA, AKIO
UNIVERSITY OF TSUKUBA
C/O PROF. KAWABE LAB
1-1-1 TENNODAI
TSUKUBA SCIENCE C IBARAKI305
JAPAN
81 298 53 6902-Tel
81 298 55 7440-Fax

KAWAMOTO, KEIICHI
FUJI ELECTRIC CORP R&D LTD.
5-20-14 SHIBA
MINATOKU
TOKYO 108
JAPAN
03-5476-2325-Tel
03-5476-0407-Fax

KAWAMURA, RITSUO
SUMITOMO SITEX CORP.
1 TOHAMACHO
AMAGASAKISHI
HYOGO 660
JAPAN
06-411-1121-Tel
06-411-1121-Fax

KAWASAKI, MASASHI
TOKYO INSTITUTE OF TECHNOLOGY
4259 NAGATSUTA MIDORI-KU
YOKOHAMA
KANAGAWA 226
JAPAN
81 45 924 5325-Tel
81 45 924 5362-Fax

KAWASAKI, OSAMU
NATIONAL SPACE DEV AGENCY
2-1-1 SENGEN
TSUKUBA, IBARAKI 305
JAPAN
81-298-52-2288-Tel
81-298-52-2438-Fax

KAYE, JOHN
UNIVERSITY OF N.S.W.
SCHOOL OF ELECTRIC ENGINEERING
SYDNEY 2052
AUSTRALIA
612 385-4892-Tel
612 385-5993-Fax

KAZMERSKI, LAWRENCE
NREL
1617 COLE BLVD
GOLDEN CO 80401
USA
303 275-3711-Tel
303 275-3701-Fax

KAZUTOSHI, ONOUE
FUJI ELECTRIC CO., LTD.
4-30-3, YOYOGI
SHIBUYA-KU
TOKYO 151
JAPAN
81 3 5388 7127-Tel
81 3 5388 7800-Fax

KEDOPAA, KINNO
NESTE
RALSSIRIE 7
1510 VANRAA
FINLAND
358-045-015750-Tel
358-0450-5744-Fax

KEEVERS, MARK
UNIVERSITY OF NEW SOUTH WALES
CENTRE FOR PHOTOVOLTAIC DEV.
SYDNEY, NSW 2052
AUSTRALIA
612 385 5246-Tel
61 2 662 4240-Fax

KENEDI, RON
PHOTOCOMM, INC.
BOX 14670
SCOTTSDALE AZ 85267-4670
USA
602 948 8003-Tel
602 483 6431-Fax

KERN, EDWARD C
ASCENSION TECHNOLOGY, INC
235 BEAR HILL RD
WALTHAM MA 02154
USA
617 890 8844-Tel
617 890 2050-Fax

KESSLER, JOHN
SOLAREX
826 NEWTOWN YARDLER ROAD
NEWTOWN PA 18940
USA
215-860-0902-Tel
215-860-2986-Fax

KEYES, JASON
JX CRYSTALS, INC.
1085 12TH AVE NW
SUITE D8
ISSAQUAH WA 98027
USA
206-392-5237-Tel
206-392-7303-Fax

KHATTAK, CHANDRA
CRYSTAL SYSTEMS, INC.
27 CONGRESS STREET
SALEM MA 01970
USA
508-745-0088-Tel
508-744-5059-Fax

KHEMTHONG, SCOTT
APPLIED SOLAR ENERGY CORP.
15251 EAST DON JULIAN RD.
CITY OF INDUSTRY CA 91745-10(
USA
818-658-6581-Tel
818-336-8694-Fax

KHOUZMAN, KAME
QUT, SCHOOL OF E & ES ENG
2 GEORGE ST
BRISBANE 4000
AUSTRALIA
617 864-2483-Tel
617 864-1516-Fax

KILMER, LOUIS
P.O BOX 92957
LOS ANGELES CA 40009-2957
USA
310-336-5321-Tel
310-336-5846-Fax

KIM, DONG SEOP
SAMSUND ADVANCED INT. OF TEC
440-600 SUWON PO BOX III
SUWON KYUNG KI 440-600
KOREA
82-391-280-9333-Tel
82-331-280-9308-Fax

KIM, DONGHWAN
COLORADO STATE OF MINES
DEPARTMENT OF PHYSICS
GOLDEN CO 80401
USA
303-273-3841-Tel
303-273-3919-Fax

KIM, SANG HO
RACER
KOREA EX. BK. CREDIT SERV.
B/D 3F, 935-34, PANG BAE DONG
SEO CHO KU, SEOUL 137-060
KOREA
82-2-525-3474-Tel
82 2 522 8094-Fax

KIMMAN, JACQUES
NOVEM
ST JACOBSSTRAAT 61
UTRECHT 3503RE
NETHERLANDS
31 30 363456-Tel
31 30 316491-Fax

KING, KENNETH
MULTI-CONTACT USA
22 SOUTH BROADWAY
SALEM NH 03079
USA

KING, RICHARD
SIEMENS SOLAR INDUSTRIES
4650 ADOHR LANE
P O BOX 6032
CAMARILLO CA 93001
USA
805-388-6263-Tel
805-388-6580-Fax

KINKAID, TOBY
SOLARDYNE CORPORATION
20 SOUTH MAIN ST.
GAINESVILLE FL 32601
USA

KINOSHITA, KAZUO
EBARA RESEARCH INSTITUTE
4 2 1 HONFUJISAWA
FUJISAWA CITY
KANAGAWA PREF 251
JAPAN
81 466 83 7260-Tel
81 3 3573 0685-Fax

KISS, ZOLTAN
ENERGY PHOTVOLTAICS, INC.
PO BOX 7456
PRINCETON NJ 08543
USA
609-587-3000-Tel
609 587 5355-Fax

KITAMURA, AKIO
KANSAI ELECTRIC POWER CO., INC
NAKOOJI 3-11-20
AMAGASAKI, HYOGO 661
JAPAN
81 6 494 9786-Tel
81 6 498 7662-Fax

KITCHEN, CEDRIC
PILKINGTON SPACE TECHNOLOGY
UNIT 2 KIMMEL PARK
BODELWYDDAN RHYL LL18 5TY
UNITED KINGDOM
011-44-745-584584-Tel
011-44-745-584208-Fax

KIYOTA, YOSHIHARU
SHARP CORPORATION
282-1 HAJIKAMI, SHINJO-CHO
KITAKATSURAGI-GUN
NARA 639-21
JAPAN
81 7456 5 1161-Tel
81 7456 5 1174-Fax

KJAER, THORVALD
DANFOSS A/S
ATTN. LAILA CHRISTIANSEN
E14S12
NORDBORG 6430
DENMARK
4574884255-Tel
4574884000-Fax

KNOBEL, ROLF
WACKER-CHEMITRONIC
BOX 1140
BURGHAUSEN 1140
GERMANY
08677 83 4502-Tel
08677 83 5824-Fax

KNOBLOCH, JOACHIM
FHG-ISE
OLTMANNSSTRABE 5
FREIBURG GE 79100
GERMANY
+49-761-4588-263-Tel
+49-761-4588-250-Fax

KNOESEN, DIRK
UNIVERSITY OF WESTERN CAPE
PO BOX X17
BELLVILLE 7535
SOUTH AFRICA
27 21959 3474-Fax

KOBAYASHI, HIKARU
OSAKA UNIVERSITY
FACULTY OF ENGINEERING SCIENCE
1-B MACHIKANEYAMA CHO TOYONAKA
OSAKA 560
JAPAN
81 6 850 6697-Tel
81 6 850 6699-Fax

KOBAYASHI, HIROMU
CRIEPI
2-11-1, IWATO-KITA
KAMAE-SHI
TOKYO 201
JAPAN
81-3-3480-2111-Tel
81-3-3480-3866-Fax

KOBAYASHI, HISAO
PHOTOVOLTAIC POWER GERNATION
TECHNOLOGY RESEARCH ASSC.
OTAWAMF BLDG 7F, 1-22-12,OTOWA
BUNKYO-KU 112
JAPAN
81-3-3946-2451-Tel
81-3-3946-6262-Fax

KOBAYASHI, KALVIN
COUNTY OF MAYI
250 S HIGH ST
WAILUKU HI 96793
USA
808 243 7735-Tel
808 243 7634-Fax

KOBAYASHI, SHIGEKI
TOYOTA CENTRAL R&D LABS., INC
41-1 YOKOMICHI, NAGAKUTE
NAGAKUTE-CHO
AICHI-KEN 480-11
JAPAN
81-561-62-6111-Tel
81-561-63-6298-Fax

KOBAYASHI, TETSUZO
SHOWA SOLAR ENERGY K.K.
TOKYO BLDG.7-3 HARUNOUCHI 2
CHOME
TOKYO 100
JAPAN
81-3-3215-9661-Tel
81-3-3284-1589-Fax

KOBAYASHI, TETSUZO
SHOWA SOLAR ENERGY
7-3 MARUNOUCHI 2-CHOME
CHIYODA-KU, TOKYO
JAPAN

KOCKA, JAN
INSTITUTE OF PHYSICS
CUKROVARNICKA 10
CUKROVARNICKA 10
CZECH REPULIC
CZECH REPUBLIC
42-2-24311137 449-Tel
42-2-3123184-Fax

KOCKOTT, DIETER
CONSULTANT
VOGELSBERGSTRASSE 27
HANAU STEINHIEM 63456
GERMANY
6181-659162-Tel
6181659175-Fax

KOENIGSTEIN, CHRISTIAN
VIENNA UNIVERSITY OF TECHNOLOG
C/O PROF A M SARGESON
R S OF C, AUSTRALIAN NATL UNIV
CANBERRA A C T 0200
AUSTRALIA
431-58801-Tel
61 6 2490750-Fax

KOGURE, NORIKO
PRESS

KOMOTO, KEIICHI
FUJI RESEARCH INSTITUTE CORP.
5-20-14, SHIBA
MIATO-KU
TOKYO 108
JAPAN
03-5476-2325-Tel
03-5476-0407-Fax

KONAGAI, MAKOTO
TOKYO INSTITUTE OF TECH
2-12-1 OHOKAYAMA, MEGURO-KU
TOKYO 152
JAPAN
81-3-5734-2554-Tel
81-3-5734-2897-Fax

KONDO, EIICHI
CANON, INC.
ECOLOGY RESEARCH CENTER, INC
4-1-1 KIZUGAWADAI, KIZU-CHO
KYOTO 619-02
JAPAN
81 7747 5 2111-Tel
81 7747 5 2030-Fax

KOUJITANI, KOUSAKU
TOYAMA PREFECTURAL UNIV.
5180 KUROKAWA,
KOSUGI-MACHI
TOYAMA 93903
JAPAN
81766-56-7500-Tel
81-766-56-6172-Fax

KRAFT, VICTOR
PRIMBUNKER CO LTD
KIROVA STREET 70/91
VIADIVOSTOK, 690068
RUSSIAN
423-2-260-435-Tel
423-2-260-435-Fax

KRAUTER, STEFAN
TECHNISCHE UNIV BERLIN
INST FOR ELECTRICAL MACHINES
EINSTEINUFER 11, SEC EM 4
BERLIN D-10587
GERMANY

KRAUTER, STEFAN
TECHNISCHE UNIV BERLIN
INST FOR ELECTRICAL MACHINES
EINSTEINUFER 11, SEC EM 4
BERLIN D-10587
GERMANY

KRAUTHAMER, STAN
JET PROPULSION LABORATORY
4800 OAK GROVE DRIVE
PASADENA CA 91109
USA
818-354-9686-Tel
818-393-4272-Fax

KRAUTWALD, REINHARD
DEUTSCHE AEROSPACE
P O BOX 801168
MUNICH 81663
GERMANY
49 89 607 27434-Tel
49 89 607 22661-Fax

KREUTZMANN, ANNE
SOLARENERGIE-FORDERVEREN
1150 306 STP
AACHEN NRW 52070
GERMANY
0048-241-511616-Tel
0048-241-535786-Fax

KRIEGEL, BERND
SILICON SENSOR
OSTENDSTR 1
BERLIN 12459
GERMANY
49 30 695 33901-Tel
49 30 695 33902-Fax

KROLL, WILLIAM
EMCORE CORPORATION
35 ELIZABETH AVE.
SOMERSET NJ 08873
USA
908-271-9090-Tel
908-271-8022-Fax

KROPOSKI, BENJAMIN
NREL
1617 COLE BOULEVARD
GOLDEN CO 80401
USA
303-384-6170-Tel
303-384-6490-Fax

KRUER, MARK
TRW
ONE SPACE PARK
REDONDO BEACH CA 90278
USA

KUBAT, KRISTINE
SUSTAINABLE DEVELOPMENT INST
GENERAL DELIVERY
KEAAU HI 96749
USA

KUBOTA, HIROSHI
KUMAMOTO UNIV
2-39-1 KUROKAMI
KUMAMOTO 860
JAPAN

KUHNS, DWIGHT
THE NEW WORLD POWER CORP.
5990 STONERIDGE DRIVE
SUITE 119
PLEASANTON CA 94588
USA
510-416-0600-Tel
510-416-0677-Fax

KUHNS, JOHN
THE NEW WORLD POWER CORP
558 LIME ROCK ROAD
LAREVILLE CT 06039
USA

KUKULKA, JERRY
SPECTROLAB INC.
12500 GLADSTONE AVE
SYLMAR CA 91342
USA
818 898-2837-Tel
818 361-5102-Fax

KUMEDA, MINORU
KANAZAWA UNIVERSITY
2-40-20- KODATSUNO
KANAZAWA 920
JAPAN
0762-34-4875-Tel
0762-34-4900-Fax

KUO, LEE-CHING
ITRI
BLDG 77 195 CHUNG HSING RD
SECTION 4 CHUTUNG
HSINCHU, TAIWAN
ROC

KURLAND, RICHARD
TRW SYSTEMS
4246 SESPE AVE
SHERMAN OAKS CA 91403
USA
818 783-4598-Tel
310 813-6003-Fax

KUROKAWA, KOSUKE
ELECTROTECHNICAL LABORATORY
1-1-4 UMEZONO
TSUKUBA-SHI
IBARAKI 305
JAPAN
+81-298-58-5824-Tel
+81-298-58-5829-Fax

KUROKAWA, SANAE

KURTZ, SARAH
NREL
1617 COLE BLVD
GOLDEN CO 80401
USA
303 384-6475-Tel
303 384-6531-Fax

KUSHIYA, KATSUMI
SHOWA SHELL SKEIYU K.K.
CENTRAL R&D LAB
123-1 SHIMOKAWAIRI ATSUGI CITY
KANAHAWA 243-02
JAPAN
81 462 45 1750-Tel
81 462 45 8761-Fax

KUWANO, YUKINORI
SANYO ELECTRIC CO. LTD.
1-18-13 HASHIRIDANI
HIRAKATA KA 573
JAPAN
+81-720-41-7833-Tel
+81-720-44-1378-Fax

KYOJIRO, KANEKO
SUMITOMO SITIX CORPORATION
1 TOHAMACHO
AMAGASAKISHI
HYOGO 660
JAPAN
06-411-1121-Tel
06-411-5928-Fax

KYUNG SUP, KIM
SILTRON
#283 IM 500 DONG
KUMI
K YUNG BUK 730-350
KOREA
82 546 468 0200-Tel
82 546 468 0283-Fax

LA ROCHE, GUENTHER
DEUTSHE AEROSPACE
POMMERNSTR. 10
OTTOBRUNN 85521
GERMANY
0049-89-60723078-Tel
0049-89-60724202-Fax

LAAKSO, MIKKO
MICRO CHEMISTRY
POB 45
ESPOO FIN 02151
FINLAND
358-0-450-5702-Tel
358-0-450-5700-Fax

LAALY, HESHMAT
ROOFING MATERIALS SCIENCE
9037 MONTE MAR DRIVE
LOS ANGELES CA 90035-4235
USA
310-559-6069-Tel
310-204-5642-Fax

LAM, YUEN MING
EVER STEP DEVELOPMENT LTD.
10/F , WIN WIN IND BLDG
18 LEE CHUNG STREET
CHAIWAN
HONG KONG
852 896 0341 Tel
852 897 2365-Fax

LAMBRIDES, MARK
WORLD BANK
1818 H STREET NW
WASHINGTON DC 20433
USA
202-473-3121-Tel
202-522-3573-Fax

LAMBRIDES, MARK
WORLD BANK-ASTAE
4300 KING STREET
ALEXANDRIA VA 23302
USA
703-998-3729-Tel
703-998-0887-Fax

LAMMASNIEMI, JOUKO
TAMPERE UNIV. OF TECHNOLOGY
DEPARTMENT OF PHYSICS
P.O. 692
TAMPERE FI 33101
FINLAND
-358-31-2162-607-Tel
+358-31-3162-600-Fax

LAMPERT, CARL
LAWRENCE BERKELEY LAB
UNIV OF CALIFORNIA
1 CYCLOTRON RD MS 62-203
BERKELEY CA 94720
USA

LAMPO, STEVEN
ASTRO POWER, INC.
SOLAR PARK
NEWARK DE 19716-2000
USA
302 366-0400-Tel
302 368-6474-Fax

LANDIS, GEOFFREY
NYMA, NASA LEWIS RESEARCH CTR
NASA LEWIS 302-1
21000 BROOK PARK RD
CLEVELAND OH 44142
USA

LANDSBERG, PETER
PROFESSOR
UNIVERSITY OF SOUTHAMPTON
UK 50G 5NH
UNITED KINGDOM
0703-556545-Tel
44 0703 593939-Fax

LANNING, BRUCE
MARTIN MARIETTA
P O BOX 179 M/SF3085
DENVER CO 80201
USA
303-971-9551-Tel
303-977-4612-Fax

LANTRATOV, VLADIMIR
IOFFE INSTITUTE
POLYTECHNICHESKAYA 26
ST PETERSBURG 194021
RUSSIA
7 812 247 9924-Tel
7 812 247 1017-Fax

LARRO, MIKE
SALES ENGINEER
2789 NORTHPOINT PKWY
SANTA ROSA CA 95703
USA
707-5257029-Tel

LARSEN, FINN

LARSEN, THOMAS

LARSEN, WENDY
NREL
1617 COLE BLVD
GOLDEN CO 80401
USA
303-384-6497-Tel
303-384-6481-Fax

LASICH, JOHN
SOLAR RESEARCH CORPORATION
6 LUTON LANE
HAWTHORN, VIC 3122
AUSTRALIA
61 3 819 9544-Tel
61 3 819 9063-Fax

LAUGIER, ANDRE
I.N.S.A. DE LYON BATIMENT 502
LAB. DE PHYSIQUE DE LA MATIERE
20 AVENUE ALBERT EINSTEIN
VILLEURBANNE CEDE
FRANCE
33 72 438393-Tel
33 7244 0732-Fax

LEDDY, JOHN
J.B. LEDDY INT'L. CONS.
1860 PARROTT DR
SAN MATEO CA 94402
USA
415 341 2298-Tel
415 341 3051-Fax

LEE, CHOOCHON
KOREA ADV. INST. OF SCIENCE
373-1 KUSONG-DONG, YUSONG-GU
TAAJEN 305-701
KOREA
82-42-869-25-Tel
82-42-869-2510-Fax

LEE, JONG KEUK
SAMSUNG ELECTRONICS
SUWON CITY
KYUNGKI DO 441 742
KOREA
82 331 200 4708-Tel
82 331 200 4722-Fax

LEE, SEUNGWON

LEE, SOO HONG
SAMSUNG
SANSUNG ADVANCED INST OF TECH
PO BOX 111
SUWON 440-600
KOREA

LEE, YOUN-JUNG
PURDUE UNIVERSITY
3812 BURBERRY DRIVE SOUTH #64
LAFAYETTE IN 47905
USA
317 449-8199-Tel
317-494-6440-Fax

LEISTIKO, OTTO
MICROELECTRONICS CENTER
TECH. UNIVERSITY OF DENMARK
LYNGBY DK2800
DENMARK
45 4593 1222 5712-Tel
45 4588 7762-Fax

LEMITI, MUSRAPHA
I.N.S.A. DE LYON BATIMENT 502
LAB. DE PHYSIQUE DE LA MATIERE
20, AVENUE ALBERT EINSTEIN
VILLEURBANNE CEDE 69621
FRANCE
33 7243 8393-Tel
33 72 44 07 32-Fax

LENG, GREGORY
ENERGY DIVER. RESEARCH LAB
1615 MONTEE STE JULIE CP 4800
VARENNES
QUEBEC J3X 156
CANADA
514 652 5154-Tel

LEONEDOVICH, VICTOR
PRIMBUNKER CO LTD
KIROVA STREET 70/91
VLADIVOSTOK 690068
RUSSIAN FED
423 2 260 435-Tel

LEPLEY, TOM
PHASOR ENERGY CO.
4202 E. EVANS DR.
PHOENIX AZ 85032
USA
602 788-7619-Tel
602 404-1765-Fax

LEUNG, SUNNY
CANADIAN COAST GUARD
344 SLATER STREET 8TH FLOOR
OTTOWA, ONTARIO K1A0N7
CANADA
613-998-1390-Tel
613-995-4700-Fax

LEVI, DEAN
NREL
1617 COLE BLVD
GOLDEN CO 80401
USA
303-384-6605-Tel
303-384-6490-Fax

LEWALD, NORBERT
KARLSRUHE UNIVERSITY
RENEWABLE ENERGY GROUP
KAISERSTR 12
KARLSRUHE 76128
GERMANY
00 49 721 691776-Fax

LEWIS, CAROL
JET PROPULSION LABORATORY
4800 OAK GROVE DRIVE
PASADENA CA 91109
USA
818 354-3767-Tel
818 393-4272-Fax

LEYDEN, THOMAS
ENERGY PHOTOVOLTAICS, INC.
PO BOX 7456
PRINCETON NJ 08543
USA
609-587-3000-Tel
608-587-5355-Fax

LIAN, YIN
NEW ENERGY FOR POWER

LILLINGTON, DAVID R.
SPECTROLAB INC.
12500 GLADSTONE AVE
SYLMAR CA 91342
USA
818-898-7515-Tel
818-361-5102-Fax

LIM, JOHN
REXOR
2240 E CEDAR ST.
ONTARIO CA 91761
USA
909 930-1433-Tel
909 930-1434-Fax

LIMSAWAT, WICHAI
EGAT
NONTHABURI
11000
THAILAND
662 4336317-Fax

LIN, XUAN YING
PHYSICS DEPARTMENT OF SHANTOU
SHANTOU UNIVERSITY
AUCHGDONG 515063
R.O.C.
754 8510505-Tel
754 851000032022-Fax

LINCOT, DANIEL
CNRL
ENSCP
11 RUE PIERRE ET MARIE CURIE
PARIS F75231
FRANCE
33143545385-Tel
331 44 27 67 50-Fax

LITTLE, ROGER
SPIRE CORPORATION
ONE PATRIOTS PARK
ATTN: CAROL EINARSSON
BEDFORD MA 01730-2396
USA
617-275-6000-Tel
617-275-7470-Fax

LOEFFLER, STEFFEN
UNIVERSITY OF HAGEN
P O BOX 940
HAGEN 58084
GERMANY
49 2331987 4012-Tel
49 2331 987 321-Fax

LOEFFLER, STEFFEN
UNIVERSITY OF HAGEN
P O BOX 940
HAGEN 58084
GERMANY
49 2331987 4012-Tel
49 2331 987 321-Fax

LOFERSKI, JOSEPH
BROWN UNIVERSITY OF PROVIDENCE
33 SLATER AVENUE
PROVIDENCE RI 02906
USA
401-863-2652-Tel
401-863-1157-Fax

LONG, RICHARD
GEORGIA INSTITUTE OF TECH.

LOOIS, GEERLING
ECOFYS
PO BOX 8408
UTRECHT NL 3503 RK
NETHERLANDS
31 030 913400-Tel
31 030 913401-Fax

LOVIS, TERRY
AIXTRON INC.
9150 S.W. PIONEER CT.
SUITE D-1
WILSONVILLE OR 97070
USA
503-682-4564-Tel
503-682-5673-Fax

LOWE, ROLAND
KENT STATE UNIVERSITY
NASA LEWIS RESEARCH CENTER
MS 302-1 21000 BROOKPARK RD
MIDDLEBURG HTS OH 44135
USA

LU, YIWEI
PENNSYLVANIA STATE UNIV
15S GRADUATE CIRCLE
COLLEGE STATION PA 16801
USA

LUQUE, ANTONIO
INSTITUTO DE ENERGIA SOLAR
E.T.S.I.T. -CIUDAD UNIV.
MADRID 28040
SPAIN
341 544-1060-Tel
341 544-6341-Fax

LUX-STEINER, MARTHA
UNIV OF KONSTANZ
FACULTY OF PHYSICS
SONNENBUHL
KONSTANZ D-78434
GERMANY

LYONS, JOHN
NASA GODDARD SPACE FLIGHT CTR
MAIL CODE 734.4
GREENBELT MD 20771
USA

MA, WEN
FACULTY OF ENGINEERING SCIENC
OSAKA UNIVERSITY
1-3 MACHIKANEYAMA-CHO
TOYONAK OSAKA 560
JAPAN
06-850-6317-Tel
06-850-6318-Fax

MACGILL, IAIN
CENTRE FOR PV DEVICES & SYSTEM
SCHOOL OF ELEC. ENGINEERING
UNIV OF N.S.W. 2052
AUSTRALIA
61 2 3854061-Tel
61 2 385 5993-Fax

MADAN, ARUN
MV SYSTEMS, INC
327 LAMB LANE
GOLDEN CO 80401
USA
303 526-9016-Tel
303 526-1408-Fax

MADSON, MICHAEL
THE MADISON LINE
PO BOX 338
CORTE MADERA CA 94976
USA
415 927 3600-Tel
415 927 4661-Fax

MAHONEY, ED
CONCORDE BATTERY CORP.
2009 SAN BERNARDINO RD
WEST COVINA CA 91790
USA
818 338-8549-Fax

MALDI, MICHAEL
OHIO UNIVERSITY
ALBANY ARMS APT'S 42
ALBANY OH 45710
USA

MANDELKORN, JOSEPH
JERUSALEM COLLEGE OF TECH.
P O BOX 16031
JERUSALEM 91160
ISRAEL
751145-Tel
02-422075-Fax

MANNINEN, LAURI
NAPS TECHNOLOGY LABRATORY
NESTE OY, P. O. BOX 310
PORVOO FIN 06101
FINLAND
358 9 155411-Tel
358 15 5417113-Fax

MARKVART, TOMAS
UNIVERSITY OF SOUTHAMPTION
HIGHFIELD
SOUTHAMPTON 509 5NH
UNITED KINGDOM
44 703 593783-Tel
44 703 593016-Fax

MARTINS, RODRIGO
CEMOP/UNINOVA
QUINTA DA TORRE
MONTE DE CAPARICA 2825
PORTUGAL
351.1.3500251-Tel
2941365-Fax

MARUYA, SHIGEHISA
SEISHIN TRADING CO. LTD.
2-1 SANNOMIYA CHO 1 CHOME
CHIO KU KOBE 650
JAPAN
078-391-5391-Tel
078 331 5735 Fax

MARVIN, DEAN
AEROSPACE CORPORATION
P O BOX 9045
ALBUQUERQUE NM 87119
USA
505 846-4980-Tel
505 846-1445-Fax

MASTAITIS, VICKI
INTERSTATE RENEWABLE ENERGY
2 EMPIRE STATE PLAZA STE 1801
ALBANY NY 12223-1252
USA
518 473-5577-Tel
518 473-8687-Fax

MATLAKOWSKI, GAELLE
DEPT OF ENGINEERING
ANU
CANBERRA 0200
AUSTRALIA
61-62-49-4884-Tel
61-62-49-0506-Fax

MATSON, RICHARD
NREL
1617 COLE BLVD
GOLDEN CO 80123
USA
303 275-3726-Tel
303 275-3701-Fax

MATSUBARA, HIDEKI
SUMITOMO ELECTRIC INDUSTRIES
1-1-1 KOYA-KIYA
ITAMI HYOGO 664
JAPAN
0727 72 4581-Tel
0727 72 2440-Fax

MATSUBARA, HIDEKI
SUMITOMO ELECTRIC INDUSTRIES
1-1-1, KOYA-KITA
ITAMI
HYOGO 664
JAPAN
0727-72-4581-Tel
0727-72-2440-Fax

MATSUDA, AKIHISA
ELECTROTECHNICAL LABORATORY
1-1-4 UMEZONO
TSUKUBA
IBARAKI 305
JAPAN
81-298-54-5252-Tel
81-298-58-5425-Fax

MATSUDA, SUMIO
NASDA
2-1-1 SENGEN
TSUKUBA, IBARAKI 305
JAPAN
81-298-52-2295-Tel
81-298-52-2438-Fax

MATSUDA, SUMIO
NASDA
2-1-1 SENGEN
TSUKUBA, IBARAKI 305
JAPAN
81-298-52-2295-Tel
81-298-52-2438-Fax

MATSUKUMA, KUNIHIRO
KUMAMOTO INST OF TECHNOLOGY
4-22-1 IKEDA
KUMAMATO 860
JAPAN

MATSUMOTO, YASUHIRO
CINVESTAV IPN
AV 1PN 2508 SN PEDRO ZACATENCO
D F 07360
MEXICO
752 06 77 3169-Tel
752 05 90-Fax

MATSUSE, MITSUTAKA
TOKYO INST. OF TEC.
4259 NAGATSUTA MIDORI WARD
YOKOHAMA
KANAGAWA 226
JAPAN
+81 45-924-5325-Tel
+81-45-924-5362-Fax

MAY, LEONARD
SOLAREX CORPORATION
630 SOLAREX CT.
FREDERICK MD 21701
USA
301-698-4338-Tel
301-698-4345-Fax

MAYCOCK, PAUL
PV ENERGY SYSTEMS
ROUTE 2 BOX 429
CATLETT VA 22019
USA
703-788-9626-Tel

MAYURI, IKKI
RESOURCES TOTAL SYSTEM
2-7-11 SHINKAWA
CHUOUKU
TOKYO 104
JAPAN

MC CARNEY, STEVE
PHOTOCOMM, INC.
BOX 14670
SCOTTSDALE AZ 85267-4670
USA
602 948 8003-Tel
602 483 6431-Fax

MC HUGO, SCOTT
UNIV OF CALIF AT BERKELEY
717 1/2 JOHNSON STREET
ALBANY CA 94706
USA
510 528 8626-Tel
510 642 2069-Fax

MC KENZIE, PETER
SIEMENS SOLAR
4650 ADOHR LN
CAMARILLO CA 93011
USA
805-388-6351-Tel
805-388-6395-Fax

MC MAHON, TOM
NREL
1617 COLE BLVD
GOLDEN CO 80401
USA
303-231-7039-Tel

MC NAUGHTON, WARREN
CORNICE ENGINEERING, INC.
POST OFFICE BOX 2542
DURANGO CO 81302
USA
303-247-9548-Tel
303-247-9577-Fax

MCCANDLESS, BRIAN
INSTITUTE OF ENERGY CONVERSION
UNIVERSITY OF DELAWARE
NEWARK DE 19716-3820
USA
302-831-6240-Tel
302-831-6226-Fax

MCDANAL, A.J.
ENTECH, INC.
PO BOX 612246
1015 ROYAL LANE
DALLAS/FT. WORTH TX 75261
USA

McNELIS, BERNARD
I.T. POWER LTD
THE WARREN, BRAMSHILL RD
EVERSLEY, HANTS RG27 OPR
UNITED KINGDOM
44 1734 730073-Tel
44 1734 730820-Fax

MEAKIN, DAVID
SOLAR ENERGY INDUSTRIES ASSOC.
11647 STONEVIEW SQ. 21C
RESTON VA 22091
USA
703 341 7915-Tel
202 383 2670-Fax

MEAKIN, JOHN D
UNIVERSITY OF DELAWARE
DEPT OF MECHANICAL ENGINEERING
126 SPENCER LABORATORY
NEWARK DE 19716-3140
USA
302 831 2421-Tel
302 831 3519-Fax

MEDER, STEPHEN
ARCHITECTURAL STUDIO
1918 UNIVERSITY AVE
HONOLULU HI 96822

MEDINA, ENRIQUE
UNION FENOSA
CAPITN HAYA 53
MADRID 28020
SPAIN
570-1773-Tel
570-2427-Fax

MEDRANO, ALBERTO
ATERSA MADRID
28045 MADRID
MADRID
SPAIN
34 1 474 72 11-Tel
34 1 474 74 67-Fax

MEDVEDKIN, GENNADIY
IOFFE INSTITUTE
26 POLYTECHNICHESKAYA
ST PETERSBURG 194021
RUSSIA
7 812 247 9173-Tel
7 812 247 1017-Fax

MEHLIN, MARTIN
TU MUNCHEN
LICHTENBERGSTR. 4
GARCHING BAVARIA D-85747
GERMANY
49 89 3209 3114-Tel
49 89 3209 3473-Fax

MEIER, DANIEL
EBARA SOLAR INC.
811 ROUTE 51 SOUTH
LARGE PA 15025
USA
412 382-1259-Tel
412 382-1251-Fax

MEIER, JOHANNES
IMT NEUCHATEL
BREQUEST 2
UNIVERSITE DE NEUCHATEL
NEUCHATEL CH2000
SWITZERLAND
038 20 5454-Tel
038 25 4276-Fax

MENEZES, LOUIS MANUEL
GOVERNMENT OF INDIA
MINISTRY OF NON CONV ENERGY SY
CGO COMPLEX, LODI RD
NEW DELHI
INDIA

MENEZES, SHALINI
INTERPHASES RESEARCH
722 RUSHING CREEK PLACE
THOUSAND OAKS CA 91360
USA
804 492-9814-Tel
805 492-9814-Fax

MERILLAT, MARSHA
REGISTRATION

MERTENS, ROBERT
IMEC
KAPELDREEF 75
LEUVEN 3001
BELGIUM
32-16-281280-Tel
32-16-281501-Fax

MERTIG, DIETER
11 EICHKAMP
WEDEL 22880
GERMANY
349-4103-88000-Fax

MESSENGER, SCOTT
SFA, INC.
UMBC PHYSICS DEPT
CP513
BALTIMORE MD 21228
USA

METSELAAR, WIM
DELFT UNIV. OF TECHNOLOGY
FAC. EL. ENG.,
MEKELWEG 4
NL-2628 CD DELFT
NETHERLANDS
31-15-785762-Tel
31-15-785691-Fax

MEYER, HANS
OMNION POWER ENGINEERING CORP.
P.O. BOX 879
EAST TROY WI 53120
USA
414-642-7200-Tel
414-642-7760-Fax

MEYER, KLAUS
FPM SPACE SENSOR GMBH
ST NICLAS SCHACHT 13
FREIBERG BRD D-09599
GERMANY

MEYERS, PETER
INSTITUTE OF ENERGY CONVERSION
UNIVERSITY OF DE
NEWARK DE 19716-3820
USA
302-831-3523-Tel
302-831-6226-Fax

MICHAEL, SHERIF
NAVAL POSTGRADUATE SCHOOL
CODE EC/MI
DEPT OF ELEC & COMPUTER ENG
MONTEREY CA 93943
USA

MINE, HISAICHI
FLEX INTERNATIONAL CO. LTD.
1084-2 HASSHOMACHI
HAMAMATSU 433
JAPAN
053-433-2731-Tel
053-433-2786-Fax

MINTAIROV, ALEXANDR
IOFFE INSTITUTE
POLYTECHNICHESKAYA 26
ST PETERSBURG 194021
RUSSIA
812 247 9924-Tel
812 247 1917-Fax

MISRA, AVINASH
POLYPLEX CORPORATION LTD.
75, AMRIT NAGAR
NEW DELHI 110003
INDIA
91-11-4631761-Tel
91-11-4620729-Fax

MISRA, MOHAN
ITN
14515 W BYERS PLACE
GOLDEN CO 80401
USA
303 279-1001-Tel
303 279-8179-Fax

MITCHELL, KIM
SIEMENS SOLAR
4650 ADOHR LANE
CAMRILLO CA 93010
USA
805-388-6571-Tel
805-388-6557-Fax

MIYAZAKI, MASAYUKI
CANON ECOLOGY R & D CENTER,D
4-1-1 KIZUGAWADAL, KIZU-CHO
KYOTO 619-02
JAPAN
81 7747 5 2100-Tel

MIZANY, ANOOSH
SOLAR DEPOT INC.
61 PAUL DR.
SAN REFAEL CA 94903
USA
415-499-1333-Tel
415-499-0316-Fax

MIZUHASHI, MAMORU
ASAHI GLASS CO
3-2-2 MARUNOUCHI
TIYODA-KU 100
TOKYO

MOCK, MARSHALL

MOEHLECKE, ADRIANO
INSTITUTO DE ENERGIA SOLAR
E.T.S.I.T. CIUDAD UNIVERSITY
MADRID 28040
SPAIN
341 544-1060-Tel
341 544-6341-Fax

MOESSLEIN, JOCHEN
ZSW/REGIOLUX
PHILIPP-REIS-STR.-15
HABFURT 97437
GERMANY
49 9521 5410-Tel
49 9521 64481-Fax

MOLLER, HANS JOACHIM
TU BAF, INST. F EXPERIM. PHYS
SILBERMANNSTRABE 1
FREIBERG 09596
GERMANY
49 3731 39 2896-Tel
49 3731 39 4314-Fax

MOR, SHALOM
IAI
ABARBANEL ST 18/31
RISHON LE ZION 75309
ISRAEL

MORALES-ACEVEDO, ARTURO
CINVESTAV-IPN
ELECTRICAL ENGRG DEPT
AV. IPN N. 2508 D.F.
MEXICO

MOREL, DON
UNIVERSITY OF SOUTH FLORIDA
DEPT EE
4202 E. FOWLER AVE
TAMPA FL 33620
USA
813 974-2508-Tel
813 974-5250-Fax

MORGADO, EDUARDO
INSTITUTO SUPERIOR TECNICO
AV ROVISCO PAIS
COMPLEXO 1
LISBOA 1000
PORTUGAL
351 1 352 4372-Fax

MORITA, KEIICHI
THE KUMAMOTO INST OF TECH
4-22-1 IKEDA
KUMAMOTO-SHI
KUMAMOTO-KEN 860
JAPAN

MORRIS, GRAEME
UNIVERSITY OF QUEENSLAND
DEPARTMENT OF CHEMISTRY
BRISBANE, Q'LD 4072
AUSTRALIA
011-61-7-365316-Tel
011-61-7-365-3628-Fax

MOSCHELLA, UMBERTO
ANIT
VIA PACINOTTI 20
GENOVA 16151
ITALY
39-10-5501-Tel
39-10 5504870-Fax

MOSHFEGHI, SHANN
SOLAR DEPOT INC.
61 PAUL DR
SAN RAFAEL CA 94903
USA
415-499-1333-Tel

MOSKOWITZ, PAUL D
BROOKHAVEN NATIONAL LABORTORY
BEAG, BLDG. 490D
UPTON NY 119735000
USA
516-282-2017 2-Tel
516-282-7867-Fax

MOWLES, THOMAS A.
SOLARWATT
POB 60638
PALO ALTO CA 94306
415-325-5191-Tel
415-437-5670-Fax

MRIG, LAXMI
NREL
1617 COLE BLVD
GOLDEN CO 80401
USA
303-231-7178-Tel
303-231-1381-Fax

MUEHL, WALLACE W.
CONSULTANT - NAVY CSS
PANAMA CITY FL 32448
USA

MUELLER, ROBERT
JET PROPULSION LABORATORY
4800 OAK GROVE DRIVE
PASADENA CA 91109
USA
818 354-4881-Tel
818 393-4272-Fax

MULAS DEL POZO, PABLO
IIE

MULLER, JEAN-CLAUDE
CNRS PHASE LAB
23 RUE DU LOESS, BP 20
STRASBOURG CEDEX 2
BAS-FMIN 67037
FRANCE
33 88 10 6330-Tel
33 88 10 6335-Fax

MUNECHIKA, KENNETH
NASA-AMES

MUNGENAST, JOHN
INTERTEC INTERNATIONAL INC
2472 EASTMAN AVE
BLDG 33-34
VENTURA CA 93003
USA

MURTHY, M.R.L.N.
SEMICON TECH
302 SAHARA APTS.
DEONAR FARM RD
DEONAR,BOMBAY 400088
INDIA
011-91-22-5563615-Tel
011-91-22-5560479-Fax

MURTY, SH. MSN
CENTRAL ELECTRONICS LTD
4 INDUSTRIAL AREA
SAHIBABAD UP 201010
INDIA
87719505-Tel
8771843-Fax

MUTA, ITSUYA
SAGA UNIVERSITY
1-HONJYO
SAGA 840
JAPAN
0952-24-5191-Tel
0952-29-4441-Fax

NAGAYOSHI, HIROSHI
TOKYO U-AGRI/TECH PASSIVATION
2-24-16 NAKAMACHI
KOGAMEI 184
JAPAN
81-423-64-3311-Tel
81-423-85-6729-Fax

NAGHEL, LIGIA
73-1273 AWAKEA ST
KILUA-KONA HI 96740

NAITO, NOBUO
TOKYO INSTITUTE OF TECH.
4259 NAGATSUTA MIDORI WARD
YUKOHAMA, KANAGAW 226
JAPAN
+81 45-924-5325-Tel
+81 45-924-5362-Fax

NAKADA, TOKIO
AOYAMA GAKUIN UNIVERSITY
TOKYO 157
JAPAN
81-3-5384-1111-Tel
81-3-5384-64103-Fax

NAKAJIMA, MASATO
KASHIMA KITA ELECTRIC CORP.
8-3-1 CHUO AMI-MACHI
INASHIKI-GUN 300-03
JAPAN
81-298-87-6734-Tel
81-298-87-7109-Fax

NAKAMURA, HIDESHI
SUMITOMO DENSETSU CO., LTD.
1 4, AWAZA 2 CHOME
NISHI-KU
OSAKA 550
JAPAN
81 6 537 3561-Tel
81 6 537 3569-Fax

NAKAMURA, KENJIRO
TOKYO INSTITUTE OF TECHNOLOGY
4259 NAGATSUDACHO, MIDORIKU
YOKOHAMASHI
KANAGAWA 227
JAPAN
045-924-5188-Tel
045-921-1089-Fax

NAKANO, SHIN-ICHI
DOW CHEMICAL
2030 DOW CENTER
MIDLAND MI 48674
USA
517 636 0588-Tel
517 636 9714-Fax

NAKANO, SHOICHI
SANYO ELECTRIC CO., LTD
1-18-13
HIRAKATA
OSAKA 573
JAPAN
81-720-41-7873-Tel
81-720-41-7891-Fax

NAKAO, TETSUYA
ADVANCED ENGINEERING SERV CO.
1-2-1 KASUGA
TSUBIBA, IBARAKI 305
JAPAN
81 298 52 2398-Tel
81 298 55 7596-Fax

NAKAYAMA, FUMIO
SHOWA SOLAR ENERGY K.K.
TOKYO BLDG. 7-3 MARUNOUCHI,2
CHOME
TOKYO 100
JAPAN
81-3-3215-9661-Tel
81-3-3284-1589-Fax

NAMJIL, ENEBISH
PV PRODUCTION & RES. ABE
COMPANY P.O. BOX 191
ULAANBAATAR 210620
MONGOLIA
4232-310-696-Tel
423-2-260-435-Fax

NARAYANAN, BRIG
CENTRAL ELECTRONICS LTD
4 INDUSTRIAL AREA
SAHIBABAD UP
INDIA
91 11 6413824-Tel
91 11 6422598-Fax

NARAYANAN, MOHAN
SOLAREX CORP.
630 SOLAREX CT
FREDERICK MD 21701
USA
301-698-4467-Tel
301-698-4201-Fax

NATH, PREM
UNITED SOLAR SYSTEMS CORP.
1100 W. MAPLE ROAD
TROY MI 48084
USA
810-362-4170-Tel
810-362-4442-Fax

NELSON, ART
NREL
1617 COLE BLVD
GOLDEN CO 80401
USA
303-384-6628-Tel
303-384-6490-Fax

NETELER, JOACHIM
RICHMOND TECH INC.
1897 E. COLTON AVE.
REDLANDS CA 92374
USA
909-389-2002-Tel
909-794-4003-Fax

NEUBUSER, BERND
INVO CONS.
KORKWITZER WEG 14
18311
RIBNITZ-DAMGARTEN
GERMANY
49 3821 5738-Tel
49 3821 5730-Fax

NICEWANDER, PATRICIA
OMNI ENGRG.
5931 STATE ROAD
WADSWORTH OH 44281
USA
216-239-2760-Tel
216-433-6106-Fax

NICOLAE, OLARIU
UNIV TARGOVISTE
CALEA DOMNEASCA 236
TARGOVISTE 0200
ROMANIA

NIELSEN, ERIK
SOLAR ELECTRIC ENG. HAWAII,INC
75-5812 NANI KAILUA PLACE
KAILUA-KONA HI 96740
USA
808-329-9193-Tel

NII, TETSURO
SHOWA SHELL SEKIYU K.K.
123-1 SHIMOKAWAIRI
ATSUGT CITY 243-02
JAPAN
81-462-45-1750-Tel
81-462-45-8761-Fax

NIJS, JOHAN
IMEC
KAPELDAEEF 75
LEUVEN 3001
BELGIUM
016-281284-Tel
016-281501-Fax

NIJS, JOHAN
IMEC
KAPELDAEEF 75
LEUVEN 3001
BELGIUM
016-281284-Tel
016-281501-Fax

NIKI, SHIGERU
ELECTROTECHNICAL LABORATORY
1-1-4 UMESONO
TSUKUBA
IBARAKI 305
JAPAN
81-298-58-5610-Tel
81-298-55-9664-Fax

NISHI, KAZUO
SEMICONDUCTOR ENERGY LAB CO.
398 HASE
ATSUGI
KANAGAWA 243
JAPAN
81 462 481131-Tel
81 462 702409-Fax

NISHITANI, MIKIHIKO
MATSUSHITA ELECTRIC INDUSTRIAL
3-4 HIKARIDAI, SEIKAMACHI
SAGARAGUN
KYOTO 519-02
JAPAN
07749-8-2518-Tel
07749-8-2585-Fax

NITTA, YOSHITERU
KANEKA CORPORATION
3-2-4, NAKANOSHIMA, KITA-KU
OSAKA 530
JAPAN
06-226-5032-Tel
06-226-5144-Fax

NOACK, MAX
IOWA STATE UNIVERSITY
MICROELECTRONICS RESEARCH CT
AMES IA 50011
USA
515 294-7732-Tel
515 294-9584-Fax

NOBRIGA, SHARI
MAUI ELECTRIC CO , LTD
210 W KAMEHAMEHA AVENUE
KAHULUI HI 96732
USA
808 871 8461-Tel
808 871 2350-Fax

NOGARET, ERIC
CENTRE ENERGETIQUE
RUE DAUNESSE BP 207
SOPHIA-ANTIPOLIS 06904
FRANCE
33 939-57599-Tel
33 939-57535-Fax

NONAKA, SAKUTARO
KINKI UNIVERSITY IN KYUSHU
11-6 KASHIWANOMORI
IIZUKASHI 820
JAPAN
0948-22-5655-Tel
0948-23-0536-Fax

NORDMANN, THOMAS
TNC CONSULTING AG
ASYLSTRASSE 84
MANNEDORF 8708
SWITZERLAND
41-1-921-13-14-Tel
41-1-921-13-77-Fax

NOUFI, ROMMEL
NREL
1617 COLE BLVD
GOLDEN CO 80401
USA

NOWAK, STEFAN
POLYGON, UNIVERSITY FRIBOURG
PEROLLES
FRIBOURG 1700
LIECHTENSTEIN
41 037 826 570-Tel
41 037 826 580-Fax

NOWLAN, MICHAEL
SPIRE CORPORATION
ONE PATRIOTS PARK
BEDFORD MA 01730-2396
USA
617-275-6000-Tel
617-275-7470-Fax

NUBILE, PAULO
BRAZIL

OBATA, JUN-ICHI
SYSTEM POINT
NUSASEINO
TOKYO
JAPAN
0422 55 5013-Tel

ODUNLANTI, TOLULOPE
BOX 5509
IKEJA
NIGERIA

OGIER, HUGUES
MAUI ELECTRIC CO. , LTD
210 W KAMEHAMEHA AVENUE
KAHULUI HI 96732
USA
808 871 8461-Tel
808 871 2350-Fax

OHTAKE, YASUTOSHI
TOKYO INSTITUTE OF TECHNOLOGY
2-12-1 OHOKAYAMA, MEGURO-KU
TOKYO 152
JAPAN
81-3-5734-2662-Tel
81-3-5734-2897-Fax

OKADA, YOSHITAKA
TSUKUBA UNIVERSITY
INSTITUTE OF MATERIALS SCIENCE
1-1-1 TENNODAI
TSUKUBA, IBARAKI 305
JAPAN
81 298 53 5293 Tel
81 298 55 7440-Fax

OKAMOTO, HIROAKI
FACULTY OF ENGINEERING SCIENCE
OSAKA UNIVERSITY
1-3 MACHIKANEYAMA-CHO
TOYONAKA, OSAKA 560
JAPAN
81-6-850-6317-Tel
81-6-850-6316-Fax

OKAMOTO, SATOSHI
FLEX INTERNATIONAL CO., LTD
282-1 FUTA, SHINJOCHO
KITAKATSUSHIKAGUN
NARA 639-21
JAPAN
07456-5-1161-Tel
07456-2-8254-Fax

OKEYA, EPHRAIM NKEM
MIPSSET/SPACE UNIV. (UNESCO)
149 HOSPITAL RD
ABA ABA
ABIA
NIGERIA

OLDENKAMP, HENK
OKE SERVICES
NIEUWSTRAAT 29
EINDHOVEN 5611 DA
NETHERLANDS
31-40-445262-Tel
31-40-464133-Fax

OLSEN, LARRY
WASHINGTON STATE UNIVERSITY
100 SPROUT ROAD
RICHLAND WA 99352
USA
509-375-9221-Tel
509-375-5337-Fax

OPPENHEIM, JERROLD
PACE LAW SCHOOL
91 THORNDIKE ST
BROOKLANE MA 02146-2516
USA
617 566-4474-Tel
617 566-0788-Fax

ORTABASI, UGUR
UNITED INNOVATIONS
1680 MEADOWGLEN LANE
ENCINITAS CA
USA
619 632 0041-Tel
619 632 6693 Fax

OSSENBRINK, HEINZ
EUROPEAN COMMISSION/ESTI
TP 450
ISPRA, VA 21020
ITALY
+39 332 789196-Tel
+39 332 789268-Fax

OSTERWALD, CARL
NREL
1617 COLE BLVD.
GOLDEN CO 80401-3393
USA
303-384-6630-Tel
303-384-6481-Fax

OTANI, KENJI
TOKYO A&T UNIVERSITY
SAITAN LAB

OTTO, THOMAS
CHEMITZ-ZWICKAU
FAKULTAT FOR ET AND IT -31105-
CHIMNITZ 09107
GERMANY
0049 371 531 3177-Tel
0049 371 531 3004-Fax

OVITT, SMITH
UNITED SOLAR SUSTEMS CORP.
5278 EASTGATE MALL
SAN DIEGO CA 92121
USA
619-625-2080-Tel
619-625-2083-Fax

OZKUL, AHMET
INTELSAT
3400 INTERNATIONAL DR.
WASHINGTON DC 20008
USA

PADULA, JIM
SOLAR ELECTRIC SPECIALTIES CO
96 N. MAIN ST
WILLITS CA 95490
USA

PAES, PEDRO
INETE
ESTRADA PASO DO LUMIAR
LISBON 1699
PORTUGAL
351-1-716-5141-Tel
351-1-716-0901-Fax

PANAYOTATOS, PAUL
RUTGERS UNIVERSITY
P.O. BOX 909
PISCATAWAY NJ 08855-0909
USA
908-445-3382-Tel
908-445-2820-Fax

PANUNTO, MICHAEL
UMBC

PAPULA, PETER
MARTIN MARIETTA ASTRO SPACE
399 LENAPE LANE
YARDLEY PA 19067
USA

PARK, KYUCHANG
KYUNG HEE UNIVERSITY
DONGDAEMOON-KU
SEOUL 130-701
KOREA

PARTAIN, LARRY
VARIAN ASSOCIATED
3075 HANSEN WAY
M/S K-402
PALO ALTO CA 94304 1025
USA
415 424 5645-Tel
415 855 8454-Fax

PASCH, GERD
DEUTCHLAND RADIO

PASQUA, VALTER
EUROSOLARE S.P.A.
VIA A/ D ANDREA 6
NETTUNO(RM) 00048
ITALY
39-6-59883415-Tel
39-6-59883450-Fax

PAULI, MATTHIAS
MAX-PLANCK INSTITUTE
HEISENBERGSTRASSE 1
D-70569 STUTTGART
STUTTGART
GERMANY
+49711-689-1617-Tel
+49-711-689-1602-Fax

PAULS, KAREN
2700 W KELLY RD
NEWBURY PARK CA 91320
USA
805-499-7840-Tel
805-499-9502-Fax

PAVELKA, TIBOR
SEMILAB RT.
FOTI UT 56
BUDAPEST H-1047
HUNGARY
36-1 169 4544-Tel
36-1 169-8284-Fax

PEARSALL, NICOLA
UNIVERSITY OF NORTHUMBRIA
ELLISON PLACE
NEWCASTLE UPON TY NEI 8 ST
UNITED KINGDOM
44-91-227-4595-Tel
44-91-227-4561-Fax

PEDROZA, ALEJANDRO
UNIV AUTONOMA DE PUEBLA
PUEBLA
PUEBLA
MEXICO

PEHNT, MARTIN
UNIVERSITY OF COLORADO
850 38 STREET
BOULDER CO 80303
USA
303 939-8940-Tel

PEPPER, JANIS
ENERTRON CONSULTANTS
418 BENVENUE AVE
LOS ALTOS CA 94024
USA
415 949-5719-Tel
415 948-3442-Fax

PERN, JOHN
NREL
1617 COLE BLVD
GOLDEN CO 80401
USA
303 384-6615-Tel
303 384-6490-Fax

PETERSON, TERRY
EPRI
3412 HILLVIEW AVE
PALO ALTO CA 94304
USA

PIAMONTE, EDGARDO
NATIONAL ELEC ADMINISTRATION
QUEZON CITY
PHILLIPINES

PIETRUSZKO, STANISLAW
WARSAW UNIV OF TECHNOLOGY
IMIO PW, KOSZYKOWA 75
WARSAW 00-662
POLAND

PIPREK, JOACHIM
UNIVERSITY OF DELAWARE
UNIVERSITY OF DELAWARE
NEWARK DE 19716-3106
USA
302-831-8048-Tel
302-831-2804-Fax

PISZCZOR, MICHAEL
NASA LEWIS
MS 302-1
21000 BROOKPARK ROAD
CLEVELAND OH 44135
USA

PIZZUTELLI, TERRIE
IEEE
12500 GLADSTONE AVE.
SYLMAR CA 91342
USA
818-898-2856-Tel
818-361-5102-Fax

PLIENINGER, RALF
MAX-PLANCK-INSTITUTE
HEISENBERGSTRABE 1
STUTTGART 70569
GERMANY
49 711 689-0-Tel
49 711 689-1010-Fax

POLLARD, HOWARD
SPACE SYSTEMS/LORAL

POPE, LEONARD
L & P ENTERPRISE SOLAR HEATING
P O BOX 305
LIHUE INDUSTRIAL CTR PHASE I
LIHUE HI 96766 0305
USA
808 246 9111-Tel
808 246 3450-Fax

POULEK, VLADISLAV
CZECH ACADEMY OF SCIENCE
NA SLOVANCE 2
PRAGUE
CZECH REPUBLIC CA180 40
CZECH REPUBLIC
+4226605 2308-Tel
+42 2 82 12 27-Fax

POUND, WILLIAM
LOCKHEED MISSILES & SPACE CO
1111 LOCKHEED WAY
B/101 O/24 O2
SUNNYVALE CA 94086
USA
408 743 1881-Tel
408 742 5553-Fax

POWELL, RICK
SOLAR CELLS, INC.
1702 N WESTWOOD AVE
TOLEDO OH 43607
USA
419-534-3377-Tel
419-534-2794-Fax

PRESSIE, NANCI

PRINCE, MORTON
DOE/RETIRED

RAEUBER, ARMIN
FLG - ISE
OLTMAWNNSSTRASSE 5
FREIBURG D-79100
GERMANY
+49-761-4588-177-Tel
+49-761-4588-250-Fax

RAGAY, PETER
T-4 EINHOVEN
P O BOX 513
EINDHOVEN 5600 MB
NETHERLANDS
31 40 47 48 65-Tel
31 40 45 3507-Fax

RAGHUNATHAN, KIDAMBI
IPTE, INC.
1 WOODLAKE DR.
PISCATAWAY NJ 08854
USA
908-463-1220-Tel
908-463-1221-Fax

RAJAKARUNANAYARE, YASANTHA
UNIV OF TOLEDO
5962 PEMBRIDGE DR
TOLEDO OH 43615

RAJAN, VARADA
SURYOVONICS/POLYENE

RALPH, EUGENE
APPLIED SOLAR ENERGY CORP.
15251 EAST DO JULIAN RD.
CITY OF INDUSTRY CA 91745-100.
USA
818-968-6581-Tel
818-336-8694-Fax

RAMADAS, VALDYANATHAN
ET & T CORPORATION
3001 REDHILL AVE
BLDG 5 SUITE 103
COSTA MESA CA 92626
USA
714 557 2703-Tel
714 545 2723-Fax

RAMAMURTHY, ERODE
RES PHOTOVOLTAICS LTD
D13A PHASE V JEEDINETLA
HYDERABAD 500855
INDIA
91 40 896676-Tel
91 40 895807-Fax

RAMOS, HORACIO
PORTUGAL TELECOM, SA
ESTRADA DA BELA VISTA
MONSANTO
LISBOA 1500
PORTUGAL
351 1 778 3013-Tel
351 1 778 0151-Fax

RAND, JAMES
ASTRO POWER INC
SOLAR PARK
NEWARK DE 19716 2000
USA
302 366 0400-Tel
302 368 6474-Fax

RAND, MARCUS
GREENPEACE INTERNATIONAL
GREENPEACE CANONBURY VILLAS
ISLLINGSTON
LONDON N129N
UK

RANNELS, JIM
US DOE
1000 INDEPENDENCE AVE
WASHINGTON DC 22585
USA

RAO, VENKATESWARA
IREDA
CORE 4-A, EAST COURT IHC
LODI ROAD, NEW DELHI
DELHI 110003
INDIA
4601344-Tel
4602855-Fax

RAPAKOULIAS, DIMITRIOS
UNIVERSITY OF PATRAS
DEPT OF CHEMICAL ENGINEERING
PATRA 26500
GREECE
30 61 493361-Tel
30 61 993361-Fax

RAU, UWE
UNIVERSITY BAYROUTH, EPII
PHYSICAL INSTITUTE
D-95440
BAGREUTH
GERMANY
+49 921-552608-Tel
+49-921-55261-Fax

RAVI, SUDHA RANI
RES PHOTOVOLTAICS LIMITED
D-13 A PHASE V IDA JEEDIMETLA
HYDERABAD
ANDHRA PRADESH 500 855
INDIA
91 040 896676-Tel
91 040 896879-Fax

REAL, MARKUS
ALPHA REAL LTD
FELDEGGSTR 89
ZURICH 8008
SWITZERLAND
41 383 0208-Tel
41 383 1895-Fax

REGAN, TOM
UNIV OF MASSACHUSETTS LOWELL
7 BEAVER BROOK RD
WESTFORD MA 01886
USA

REINDL, THOMAS
SIEMENS AG CORP RESEARCH & DEV
DEP ZFE BT EP 22
0H0/HAHN/RING 6
MUNICH 81139
GERMANY
+43/=89/636-41641-Tel
+43/+89/636-49164-Fax

REMOLINA, JOAQUIN
UNIV AUTOMATICA DE PUEBLA
AV POLITECNICO 2508
MEXICO
MEXICO D.F.

RENAUD, PHILIPPE
SPACE VACUUM EPITAXY CENTER
UNIVERSITY OF HOUSTON
HOUSTON TX 77204-5507
USA
713 743 3621-Tel
713 747 7724-Fax

REVER, BILL
SOLAREX CORPORATION
630 SOLAREX CT
FREDERICK MD 21701
USA
301 698 4208-Tel
301 698 4201-Fax

RIATSCH, J.
PMS ENERGIE A6
LINDHOFSTRASSE 13
HOENCHALTORF CH-8617
SWITZERLAND
41-1-994-9001-Tel
41-1-994-9005-Fax

RIATSCH, JON
ETH ZUERICH
ETH-ZENTRUM/ETL-H14
CH-8092 ZURICH 8092
SWITZERLAND

RIATSCH, JON
ETH ZUERICH
ETH-ZENTRUM/ETL-H14
CH-8092 ZURICH 8092
SWITZERLAND

RICAUD, ALAIN
CYTHELIA
3 PLACE DE LA GARE DE LOZERE
PALAISEAU 91120
FRANCE
33 1 60 12 42 92-Tel
33 1 60 12 42 32-Fax

RICCOBONO, JUANITA
NORTHEAST PHOTOSCIENCES
18 FLAGG RD
HOLLIS NH 03049

RICCOBONO, JUANITA
NORTHEAST PHOTOSCIENCES
18 FLAGG RD
HOLLIS NH 03049

RICO, RANDALL
P O BOX 5907
WOODLAND PARK CO 80866
USA
719-687-3968-Tel
719-687-8640-Fax

RIMMASCH, JORK
SIEMENS AG
DEP:ZFE T EP 2
DOMAGKSTR 11
MUNCHEN D 80807
GERMANY
49 89 3500 2431-Tel
49 89 3500 2422-Fax

RINGEL, STEVEN A.
OHIO STATE UNIVERSITY
205 DREESE LAB, 2015 NEIL AVE
COLUMBUS OH 43210-1272
USA
614-292-6904-Tel
614-292-7596-Fax

RJEVSKI, VIKTO
SPRE KBAHT
3 MYTISHENSKAY, 16
MOSCOW
RUSSIA
2872959-Tel
2868922-Fax

ROBBINS, CATHERINE
HAWAII ELECTRIC LIGHT CO.

ROBBINS, WILLIAM
ABLE ENGINEERING COMPANY
93 CASTILIAN DRIVE
GOLETE CA 93117
USA
805 685-2262-Tel
805 685-1369-Fax

ROBISON, GARY
SOLAR ASSEMBLY NETWORK
7061 FOX CIRCLE
PRESCOTT VALLEY AZ 86314
USA
602-722-0492-Tel
602-772-6121-Fax

ROCHELEAU, RICHARD E.
HAWAII NATURAL ENERGY INSTITUT
UNIVERSITY OF HAWAII 2540 DOLE
ST. HOLMES HALL 246
HONOLULU HI 96822
USA
808-956-2337-Tel
808-956-2336-Fax

ROCKETT, ANGUS
UNIV OF ILLINOIS
1-107 ESB
1101 W. SPRINGFIELD AVE
URBANA IL 61820
USA

RODRIQUES, ARUNA
SUNRAY HARVESTERS INDIA
8 LETCHFIELD, LEY HILL
CHESHAM
BUCKS HP5 3QU
UK

ROHATGI, AJEET
GEORGIA TECH
ELECTRICAL & COMPUTER ENG.
ATLANTA GA 30332-0250
USA
404-894-7692-Tel
404-853-9882-Fax

ROLAND, GEORGE
SIEMENS SOLAR
4650 ADOHR LAND
CAMARILLO CA 93108
USA

ROMEO, NICOLA
DEPTARTMENT OF FINANCE
VIALE DELLE SCIENZE
PARMA 43100
ITALY
0039/521-905257-Tel
0039 521 905223-Fax

ROS MARTINEZ, JUAN FRANCISCO
4 DOCTOR FLEMING 10
30500 MOLINA DE SEGURA
MURCIA
SPAIN
58 212822-Tel

ROSEY, RICHARD
EBARA SOLAR,INC.
811 ROUTE 51 SOUTH
LARGE PA 15025
USA
412-382-1254-Tel
412-382-1251-Fax

ROTHWARF, ALLEN
DREXEL UNIVERSITY
ECE DEPT
PHILADELPHIA PA 19104
USA
215 895-2262-Tel
215 895-1695-Fax

RUBEL, HERBERT
PHOTOTRONICS SOLARTECHNIK GBLT
HERMANNOOBERTH-STR.11
PUTZVUNN
PUTZBRUNN 85640
GERMANY
4989-607-30372-Tel
4989-607-30332-Fax

RUBY, DOUGLAS
SANDIA NATIONAL LABORATORIES
MS-0752, PO BOX 5800
ALBUQUERQUE NM 87185-0752
USA
505-844-0317-Tel
505-844-6541-Fax

RUCHH, MARTIN
UNIVERSTET STUTTGART
INST R PHYSIHALISHIE ELEK
PFAREENWALDRING 47
STUTTGART D-70569
GERMANY
49-711-685-7199-Tel
49-711-685-7143-Fax

RUDIN, ARTHUR
SUNWIZE SPECIALITY SYSTEMS INC
1151 FLATBUSH ROAD
KINGSTON NY 12401
USA
914 336 7700-Tel
914 336 7172-Fax

RUDISCHER, RAINER
ILK DRESDEN
BERTOLT-BRECHT-ALLEE 20
DRESDEN D-01309
GERMANY
49 351 4081700-Tel
49 351 4081725-Fax

RUF, FRITZ
OMNION POWER ENGINEERING CORP.
P O BOX 879
EAST TROY WI 53120
USA
414-642-7200-Tel
414-642-7760-Fax

RUIZ, JOSE M.
INSTITUTO DE ENERGIA SOLAR
E.T.S.I.T. - CIUDED UNIVERSITY
MADRID 28040
SPAIN
341 544-1060-Tel
341 544-6341-Fax

RUMYANTSEV, VALERI
IOFFE INSTITUTE
POLYTECHNICHESKAYA 26
ST PETERSBURG 194021
RUSSIA
812 247 99 33-Tel
812 247 10 17-Fax

RUPLEY, KATE
ROCKWELL INTERNATIONAL
6633 CANOGA AVE
CANOGA PARK CA 91309
USA

RUPP, STEFAN
AET
INDUSTRIESTR 17
SPIESEN ELVERSBER 66853
GERMANY

RUSSELL, T.W. FRASER
INSTITUTE OF ENERGY CONVERSION
UNIV. OF DE. WYOMING RD.
NEWARK DE 19716
USA
302-831-6220-Tel
302-831-6226-Fax

RUZANSKY, SHAYA
ISRAEL AIRCRAFT IND.
HASNUNIT 10
BEER YAACOV 70300
ISRAEL

RUZINSKY, MICHAL
SLOVAK TECHNICAL UNIVERSITY
FEI- ILKOVICOVA 3
BRATISLAVA SK-81219
SLOVAKIA
42 7 351724-Tel
42 7 720415-Fax

SACCHI, GINA
IEEE
445 HOES LANE
PISCATAWAY NJ 08855
USA
908-562-3875-Tel

SAIDA, TAKAHIRO
STANLEY ELECTRIC CO., LTD
PART 21-306 MEZON TOYOHAKA
2-22-17 MRYAYAMO CHO
TOYONAKA CITY OSA 560
JAPAN
06 846 3304-Tel

SAIFEE, SHABBAR
V.I. ENERGY OFFICE
OLD CUSTOMS HOUSE
FREDERIKSTED
USVI 00823
USA
809-772-2616-Tel
809-772-0063-Fax

SAITO, SEITOKU
TDK CORPORATION
1424-196 SAKU-CITY
NAGANO 385
JAPAN

SAITOH, TADASHI
TOKYO A&T UNIVERSITY
2 24 16 NAKAMOSHI
KOGANEI
TOKYO
JAPAN
81 423 88 7108-Tel
81 423 82 9055-Fax

SAKAMOTO, KAZUO
UNIVERSITY OF TSUKUBA
C/O PROF. KAWABE LABORATORY
1-1-1 TENNODAI
TSUKUBA SCIENCE C IBARAKI30
JAPAN
81 298 53 6902-Tel
81 298 55 7440-Fax

SAKATA, SUEO
MITSUBISHI ELECTRIC CORP
1-1, TSUKAGUCHI
HONMACHI 8 CHOME
AMAGASAKI 661
JAPAN
81 6 497 8497-Tel
81 6 497 9360-Fax

SAKUTA, KOICHI
ELECTROTECHNICAL LAB.
1-1-4 UMEZONO
TSUKUBA-SHI
IBARAKI 305
JAPAN
81-298-58-5740-Tel
81-298-58-5743-Fax

SALAMI, JALAL
EBARA
811 ROUTE 51 SOUTH
LARGE PA 15025
USA
414 382-1265-Tel
412 382-1251-Fax

SALGADO, JOSE E.
SOLARTEC S.A.
DEMARIA 4446
BUENOS AIRES 1425
ARGENTINA
54 1 307 9229-Tel
54 1 307 7858-Fax

SAMAK, I.
ENGCOTEC
KRONPRINZSTR. 12
P O BOX 101262
D70011 STUTTGART
GERMANY
0711 29 93 61-Tel
0711 29 93 66-Fax

SAMARAS, JOHN
JX CRYSTALS INC.
1105 12TH AVE NW #A-2
ISSAQUAH WA 98027
USA
206 392 5237-Tel
206 392 7303-Fax

SANA, PEYMAN
GEORGIA TECH
243 APT C
PEACHTREE HILLS AVE
ATLANTA GA 30305
USA

SANDWISCH, DAN
SOLAR CELLS, INC.
1702 N WESTWOOD AVE
TOLEDO OH 43507
USA
419-534-3377-Tel
419-534-2974-Fax

SANJURJO, ANGEL
SRI INTERNATIONAL
333 RAVENSWOOD AVENUE
MENLO PARK CA 94025
USA
415 859 5215-Tel
415 859 2111-Fax

SANNOMIYA, HITOSHI
FLEX INTERNATIONAL CO., LTD.
282-1 TSUBO, SHINJOCHO
KITAKATSURAGIGUN
NARA 639-21
JAPAN
07456-2-1161-Tel
07456-2-8254-Fax

SANO, HIROYUKI
STANLEY ELECTRIC CO., LTD.
1-3-1 EDA EDA-NISHI
AOBA-KU
YOKOHAMA 225
JAPAN
81 45 911 111-Tel
81 45 911 0089-Fax

SARDI, LUIGI
ANIT
VIA PACINOTTI, 20
GENDA
GE 16151
ITALY
39 10 550 4870-Tel
39 10 550 4842-Fax

SASALA, RICHARD
SOLAR CELLS, INC.
1702 N WESTWOOD AVE
TOLEDO OH 43607
USA
419-534-3377-Tel
419-534-2794-Fax

SATO, GLENN
KAUAI ENERGY EXTENSION SERV.
4280 B RICE STREET
LIHUE HI 96766
USA
808 241 6390-Tel
808 241 6399-Fax

SATO, KANJI
KASHIMA KITA ELECTRIC CORP.
8-3-1 CHUO AMI-MACHI
INASHIKI-GUN,IBAR 3002-03
JAPAN
81-298-87-6734-Tel
81-298-87-7109-Fax

SATO, KATSUAKI
TOKYO UNIVERSITY OF A & T
KOGANEI
TOKYO 184
JAPAN
81 423 88 7120-Tel
81 423 87 8151-Fax

SCHAEFER, JOHN
JCS CONSULTANT
P O BOX 60785
PALO ALTO CA 94306
USA
415 3229337-Tel
415 322 9337-Fax

SCHAEFFER, LORETTA
WORLD BANK
1818 H STREET NW
WASHINGTON DC 20433
USA
202-458-1434-Tel
202-522-3573-Fax

SCHEER, HERMANN
EUROSOLAR
PLITTERSDORFER STR 103
BONN D 53173
GERMANY
49 228 362373-Tel
49 228 361279-Fax

SCHEER, ROLAND
HAHN MEITNER INSTITUTE
GLIENICKER STR 100
BERLIN 14109
GERMANY
49-308062-2789-Tel
49-308062-2434-Fax

SCHEIMAN, DAVE
NYMA SETAR
NASA LERC MS 302-1
21000 BROOKPARK RD
CLEVELAND OH 44135
USA

SCHERRING, C.
DEUTSCHE AEROSPACE DASA
EGERLANDERSTR 19
TAUFKIRCHEN 82024
GERMANY
49 89 6072 2500-Tel
49 89 6072 4202-Fax

SCHMID, DIETER
UNIVERSITAET STUTTGART
PFAREENWALDRING 47
STUTTGART D-70569
GERMANY
49-711-685-7199-Tel
49-711-685-7143-Fax

SCHMID, FREDERICK
CRYSTAL SYSTEMS, INC.
27 CONGRESS STREET
SALEM MA 01970
USA
508-745-0088-Tel
508-744-5059-Fax

SCHMID, HEIDRUN

SCHMID, JUERGEN
KARLSRUHE UNIVERSITY
KARLSRUHE 76125
GERMANY
00 49 721 691776 Fax

SCHMIDT, ELLEN
GREENPEACE
1436 U-STREET NW
WASHINGTON DC 20009
USA

SCHMIDT, HERBERT
FRAUNHOFER INSTITUTE FUG-ISE
OLTUANNSSTRASSE 5
FREIBURG D-79100
GERMANY
49 7614588226/216-Tel
49 7614588 217-Fax

SCHMIT, RUSTY
TEXAS INSTRUMENTS
13502
NORTH CENTRALS,M/S 35
DALLAS TX 75243
USA
214-995-7961-Tel
214-995-2337-Fax

SCHOCK, HANS WERNER
UNIVERSITAET STUTTGART
INST F PHYSIKALISCHE
ELEKTRONIK, PFAFFENWALDRING 47
D-70569 STUTTGART
GERMANY
49 (711) 685-7180-Tel
49 (711) 685-7143-Fax

SCHOFTHALER, MARTIN
MAZ- PLANCK-INSTITUTE
HISENBERGGSTR.1
STUTTGURT GE 70565
GERMANY
49-711-689-1622-Tel
49-711-689-1010-Fax

SCHRODER, WINFRIED
INSTITUTE OF CRYSTAL GROWTH
D-RUDOWER CHAUSSEE 6
BERLIN D-12489
GERMANY
49 030 6392 3003-Tel
49 030 6392 3000-Fax

SCHROEDER, BERND
UNIVERSITAT KAISERSLAUTERN
P O BOX 3049
KAISERSLAUTERN 67
GERMANY
49 631-2052377-Tel
49-631-2052996-Fax

SCHROEDER, MONIKA

SCHROPP, RUUD E.I.
UTEDHT UNIVERSITY
P.O. BOX 80.000
UTRECHT 3508 TA
NETHERLANDS
31-30-533170-Tel
31-30-543165-Fax

SCHUBERT, MARKUS
UNIVERSITAET STUTTGART
PFAFFENWALDRING 47
D 70569 STUTTGART
GERMANY
49 (711)685-7181-Tel
49 (711)685-7143-Fax

SCHUBERT, ROBERT
196 N MAIN
OBERLIN OH 44074
USA

SCHUBERT, W.KENT
SANDIA NATIONAL LABORATORIES
MS-0752, POST OFFICE BOX5800
ALBUQUERQUE NM 87185-0752
USA
505-844-5935-Tel
505-844-6541-Fax

SCHULLER, MICHAEL
USAF
PL/VTPP
KIRTLAND AFB NM 87117
USA

SCHWARTZ, RICHARD
PURDUE UNIVERSITY
1285 ELECTRICAL ENGINEERING
WEST LAFAYETTE IN 47907
USA
317-494-3539-Tel
317-494-6440-Fax

SCHWARZ, BJORN
DEUTSCHE AEROSPACE
P.O. BOX 801168
MUNICH 81663
GERMANY
49-88-607-22661-Tel
49-89-607-29765-Fax

SCHWIRTLICH, INGO
BAYER AG
RHEINUFERSTR.7-9
KREFELD GE 47809
GERMANY
49-2151-88-7168-Tel
49-2151-88-7880-Fax

SCOFIELD, DR. JOHN
OBERLIN COLLEGE
PHYSICS DEPT
OBERLIN OH 44074
USA
216-775-8333-Tel
216-775-2254-Fax

SCUTO, MICHELE
CONPHOEBUS
ZONA INDUSTRIALE
PASSO MARTINO C.P. PIANO DARCI
CATANIA 95030
ITALY

SEBASTIAN, JOSEPH
SOLAR ENERGY LAB-UNAM
LAB DE ENERGIA SOLAR
IIM-UNAM
TEMIXCO, MORELOS 62580
MEXICO

SEIFERT, RICHARD
UNIV. OF ALASKA FAIRBANKS
P O BOX 756180
FAIRBANKS AK 99775-6180
USA
907 474-7201-Tel
907 474-5139-Fax

SEVERANCE, PAUL
PHILLIPS LABORATORY
29 RANDOLPH ROAD
HANSCOM AFB MA 01731
USA
617-377-3992-Tel
617-377-2491-Fax

SEVERON, BURKHARD
K.H. STEVERNAGEL LICHTTECHNIK
GERAVER STR. 56A
GMBH
MORFELDEN-WALLDOR D-64546
GERMANY
49 6705 7888-Tel
49 6705 3085-Fax

SHACKELFORD, PETER
RENEWABLE ENERGY SERVICES
P O BOX 278
PAAUILO HI 96776
USA
808-775-0852-Tel
808-328-1285-Fax

SHACKELFORD, PETER
RENEWABLE ENERGY SERVICES

SHAH, ARVIND
IMT UNIVERSITY OF NEUCHATEL
BREQUET 2
NEUCHATEL 2000
SWITZERLAND
41 38 205465-Tel
41 38 254276-Fax

SHAH, DIPESH
BP SOLAR INTERNATIONAL
CHERTSEY ROAD
MIDDX
SUNBURY ON THAMES TW167XA
ENGLAND
44 932 76 2225-Tel
44 932 76 2323-Fax

SHAHRYAQ, ISHAQ
SOLCE INTL INC
1132 TELLEM DR
PACIFIC PALISADES CA 90272
USA

SHAHRYAR, ISHAQ
SOLEC INTERNATIONAL, INC.
12533 CHADRON AVE.
HAWTHORNE CA 90250
USA
310-970-0065-Tel
310-970-1065-Fax

SHANKS, HOWARD
IOWA STATE UNIVERSITY
MICROELECTRONICS RESEARCH CNT
AMES IA 50011
USA
515-294-0206-Tel
515-294-9584-Fax

SHARMIT, KHALED
MARTIN MARIETTA
P O BOX 172 M/S'8700
DENVER CO 80201
USA
303 971-2222-Tel

SHARPS, PAUL
RESEARCH TRIANGLE INSTITUTE
P.O. BOX 12194
RESEARCH TRIANGLE NC 27709
USA
919-541-8859-Tel
919-541-6515-Fax

SHAW, JR., ROBERT W.
ARETE VENTURES, INC.
6110 EXECUTIVE BLVD.
#1040
ROCKVILLE MD 20852
USA
301-881-2555-Tel
301-770-2877-Fax

SHELDON, PETER
NREL
1617 COLE BLVD
GOLDEN CO 80401
USA

SHEPPARD, CHARLES
1-45 BREEZE HILL
LAKE WALES FL 33853
USA
813-696-3369-Tel

SHI, ZHENGRONG
UNIVERSITY OF NEW SOUTH WALE
CENTRE OF PHOTOVOLTAIC DEV.
SYDNEY, NSW 2052
AUSTRALIA
612 385 4951-Tel
612 662 4240-Fax

SHIBATA, AKIO
EBARA SOLAR INC
811 RT. 51 SOUTH
LARGE PA 15025
USA
412 382-1267-Tel
412 382-1251-Fax

SHIBATA, AKIRA
MITSUBISHI CHEMICAL CORP.
5Z MARUNOUCHI Z-CHOME
CHIYODA-KU TO 100
JAPAN
81-33283-5252-Tel
81-9-328-3-5828-Fax

SHIBUSAWA, HIDEO
AIMS CORPORATION
ASAKAWA BLDG, 1-16-10,
SHINBASHI
TOKYO 105
JAPAN
81-3-3501-1059-Tel
81-3-35011388-Fax

SHIBUYA, NORIAKI
PVTEC
1-22-12 OTOWA, BUNKYO-KU
TOKYO 112
JAPAN
81-3-3946-2451-Tel
81-3-3946-6262-Fax

SHIMIZU, HIDENORI
JAPAN QUALITY ASSURANCE ORG.
AKASAKA 2-17
MINATOKU
TOKYO 107
JAPAN
81 3 3584 9068-Tel
81 3 3584 9002-Fax

SHIMIZU, ISAMU
FLEX INTERNATIONAL CO., LTD.
4259 NAGATSUDA, MIDORIKU
YOKOHAMASHI
KANAGAWA 227
JAPAN
045-924-5424-Tel
045-921-1089-Fax

SHINJO, FUMIHIRO
NEDO
SUNSHINE 60, 1-1
3 CHOME HIGASHI IKEBUKURO
TOSHIMA-KU, TOKYO 170
JAPAN
81 0 3 3987 9422-Tel
81 0 3 5992 6440-Fax

SHINOHARA, HIROHUMI
TOSHIBA CORPORATION
NEW ENERGY TECHNOLOGY DEPT
116 UCHISAIWAI CHO CHIYODA KU
TOKYO 100
JAPAN
813 3597 2342-Tel
813 5512 8113-Fax

SHUGAR, DANIEL
ADVANCED PV SYSTEMS
150 TAHAMA COURT
SAN BRUNO CA 94066
USA
415-871-2275-Tel
415-871-2279-Fax

SIAMCHAI, PAVAN
TOKYO INSTITUTE OF TECHNOLOGY
2-12-1 OHOKAYAMA, MEGURO-KU
TOKYO 152
JAPAN
81-3-5734-2661-Tel
81-3-5734-2897-Fax

SIMBURGER, EDWARD J.
AEROSPACE CORPORATION
PO BOX 92957 MS, MS-688
LOS ANGELES CA 90009
USA
310-336-7126-Tel
310-336-5076-Fax

SINGH, JAWINDER
MINISTRY OF ENERGY
TELECOMMUNICATIONS & POSTS
LG FLOOR, WISMA DAMANSARA
KUALA LUMPUR 50668
MALAYSIA

SINKE, WIM
ECN RENEWABLE ENERGY
P O BOX 1
PETTEN 17557G
NETHERLANDS
31 2246 4058-Tel
31 2246 3214-Fax

SINKKONEN, JUHA
HILSINKI UNIV. OF TECHNOLOGY
OTAKAARI 7A
ESPOO 02150
FINLAND
358-0-451-2320-Tel
358-0-460-224-Fax

SINTON, RONALD
SINTON CONSULTING
4820 LA FIESTA PLACE
SAN JOSE CA 95129
USA
408-243-8819-Tel
408-243-8820-Fax

SITES, JAMES
COLORADO STATE UNIVERSITY
PHYSICS DEPARTMENT
COLLINS CO 80523
USA
303-491-6206-Tel
303-491-7947-Fax

SKINNER, DOUG
ADVANCED PV SYSTEMS
195 CLARKSVILLE ROAD
LAWRENCEVILLE NJ 08543
USA

SLATER, RUSS
SOLTEK
4493 BOUNDARY ROAD
VAN B.C. V5R 2N3
CANADA

SLOMINSKI, LARRY
UNITED SOLAR SYSTEMS CORP.
5278 EASTGATE MALL
SAN DIEGO CA 92121
USA
619-625-2080-Tel
619-625-2083-Fax

SMALL, KIM
EAST WEST CENTER
1617 COLE BLVD
GOLDEN CO 80439
USA
303-384-6910-Tel
303-384-6901-Fax

SMEKENS, GUY
ENE
AV. VAN DER MEERSCHEN
188
BRUSSELS 1050
BELGIUM
32 2 771 1328-Tel

SMEKENS, RICHARD
E.N.E.
AVENUE VAN DER MEERSCHEN 188
BRUSSELS 1150
BELGIUM
32-2-771-13-28-Tel
32 3 771 1328 Fax

SMILANICH, NICHOLAS
EVEREADY BATTERY CO.
USA

SMITH, BRIAN
SPECTROLAB
12500 GLADSTONE AVE
SYLMAR CA 91342
USA
818-898-2844-Tel
818-361-5102-Fax

SMITH, GARYL
DEPARTMENT OF DEFENSE
CODE 823000D
CHINA LAKE CA 93555
USA

SMOLE, FRANC
FACULTY OF ELECT. & COMP. ENG
TRZASKA 25
LJUBLJANA 61000
SLOVENIA
386 61 1768330-Tel
386 61 264 990-Fax

SMOLE, JOZE

SOGA, TETSUO
NAGOYA INSTITUTE OF TECHNOLOGY
GOKISO CHO SHOWA KU
NAGOYA 466
JAPAN
81-52-732-2111-Tel
81-52-732-9241-Fax

SOMMERFELD, ARNO
TECHNICAL UNIV. OF AACHEN
AACHEN 52056
GERMANY
49 241 88947 0-Tel
49 241 8888284-Fax

SONG, JINSOO
KOREA INST. OF ENERGY RESEARCH

SORENSEN, BENT
RUE
HJORTEVAENGET 18
ALLEROD DK 3450
DENMARK
45 4817 7540-Tel
45 4817 7540-Fax

SPENCE, BRIAN
ASTRO AEROSPACE
6384 VIA REAL
CARPINTERIA CA 93013
USA
805-684-6641-Tel
805-684-3372-Fax

SPROUL, ALISTAIR
UNIVERSITY OF NEW SOUTH WALES
SCHOOL OF ELECTRICAL ENG.
SYDNEY, NSW 2052
AUSTRALIA
61 2 385 4060-Tel
61 2 662 4240-Fax

SRINIVAS, K.S.
BHEL
MYSORE ROAD
BANGALORE 560026
INDIA
011-91-80-6610359-Tel
011-91-80-6610534-Fax

SRIVATSAN, R
RES PHOTOVOLTAICS LIMITED
D 13 A PHASE V IDA JEEDIMELTLA
HYDERABAD
ANDHRA PRADESE 500 855
INDIA
91 040 896676-Tel
91 040 896879-Fax

STAEBLER, DAVID
NREL
1617 COLE BLVD
GOLDEN CO 80401
USA
303-384-6456-Tel
303-384-6531-Fax

STAM, W
NETHERLANDS ENERGY RESEARCH
P O BOX 1
PETTEN 1755ZG
NETHERLANDS
+31 2246 4058-Tel
+31 2246 3214-Fax

STANBERY, B.J.
UNIVERSITY OF FLORIDA
119 CHEMICAL ENGINEERING BLDG
GAINESVILLE FL 32611-6005
USA
904-392-0882-Tel
904-392-9513-Fax

STARRS, THOMAS
ENERGY & RESOURCES GROUP
403 94TH AVE. S.E.
BELLEVUE WA 98004
USA
206 454-4570-Tel

STATLER, RICHARD
S.T.A. INC.
5606 SPRUCE DRIVE
CLINTON MD 20735
USA
301-297-9528-Tel
202-404-8076-Fax

STEINBACH, INGO
ACCESSE. V.
INTZESTR S
AACHEN 52072
GERMANY
49-241-806727-Tel
49-241-38578-Fax

STEINKE, OLAFF
TECHNISCHE UNIVERSTAT DRESDEN
INST. F. FESTKORPERELEKTRONIK
MOMMSENSTR. 13
DRESDEN SACHSEN 01062
GERMANY
011 493514634296-Tel
011493514632320-Fax

STELLA, PAUL
JPL, CALIF. INST OF TECH.
MS 303/308
4800 OAK GROVE DR.
PASADENA CA 91109
USA
818 354-6308-Tel
818 393-4272-Fax

STEPHAN, ULF
TECHNISCHE UNIVERSITAT DRESDEN
INST. F.FESTKORPERELEKTRONIK
MOMMSENSTR. 13
DRESDEN SACHSEN 01062
GERMANY
011 493514634296-Tel
011 493514632320-Fax

STEPHENS, S
THE COTTAGE, PAVENHAM ROAD
FELMERSHAM, BEDFO MK43 7EX
ENGLAND

STERK, STEFFEN
FHG-ISE
OLTMANNSSTSTRABE 5
FRIEBURG 79100
GERMANY
+49-761-4588-0-Tel
+49-761-4588-100-Fax

STERN, MICHAEL J.
UTILITY POWER GROUP
9410 DE SOTO AVE
CHATSWORTH CA 91311
USA
818 700-1995-Tel
818 700-2518-Fax

STIMSON, MARK
SIEMENS SOLAR INDUSTRIES
4650 ADOHR LANE
CAMARILLO CA 93012
USA
805-388-6289-Tel
805-388-6395-Fax

STOCKS, MATTHEW
AUSTRALIAN NATIONAL UNIVER
ENGINEERING DEPT.
CANBERRA 0200
AUSTRALIA
61-6-249-4914-Tel
61-6-249-0506-Fax

STOKES, ROBERT
NREL
1617 COLE BLVD
GOLDEN CO 80401
USA
303 275-3025-Tel
303 275-3097-Fax

STOLT, LARS
UPPSALA UNIVERSITY
DEPT OF TECHNOLOGY
PO BOX 534, S 751 21 UPPSALA
SWEDEN
46 18 18 30 39-Tel
46 18 55 50 95-Fax

STONE, JACK
NREL
1617 COLE BLVD
GOLDEN CO 80401
USA
303-384-6470-Tel
303-384-6481-Fax

STONE, KENNETH
MC DONNELL DOUGLAS
5301 BOLSA
HUNTINGTON BCH CA 92647
USA
714-896-3311-Tel
714-896-1244-Fax

STORELLI, JADE MARIE

STORELLI, VITTORIO
ISES
VIA CRISPI 72
NAPOLI 80121
ITALY
081-66-9325-Tel
011-398-140-7854-Fax

STORY, WALTER
23347 GILMORE STREET
WEST HILLS CA 91304
USA
818-887-7916-Tel

STRONG, STEVEN
SOLAR DESIGN ASSOCIATES
P O BOX 242
HARVARD MA 01451
USA
508 456-6855-Tel
508 456-3030-Fax

STUMM, PETRA
OHIO UNIVERSITY
PHYSICS DEPARTMENT
ATHENS OH 45710
USA

SUGIUE, KOJI
MINISTRY OF INT'L. TRADE
1-3-1, KASUMIGASEKI
CHIYODA-KU
TOKYO 100
JAPAN
81-3-3501-9279-Tel
81-3-3501-7926-Fax

SULIMA, OLEG
IOFFEE INSTITUTE
POLYTECHNICHESKAYA 26
ST PETERSBURG 194021
RUSSIA
812 247 9924-Tel
812 247 1017-Fax

SULLIVAN, ROBERT
MORNINGSTAR CORP.
18 NORTH STATE ST.
NEWTOWN PA 18940
USA
215-504-9801-Tel
215-504-9802-Fax

SUMMERS, GEOFFREY
NAVAL RESEARCH LAB

SUNTHARALINGAM, VYSHI
PENNSYLVANIA STATE UNIVERSITY

SUNTOLA, TUOMO
MICROCHEMISTRY
P O BOX 45
KEILARANTA 6
ESPOO 02151
FINLAND
358 0 450 5701-Tel
358 0 450 5700-Fax

SUREK, THOMAS
NREL
1617 COLE BLVD.
GOLDEN CO 80401
USA
303-384-6471-Tel
303-384-6530-Fax

SUZUKI, MAMORU
JAPAN QUALITY ASSURANCE ORGAN
21-25, KINUTA 1-CHOME SETAGAYA
KU
TOKYO 157
JAPAN
03-3416-5560-Tel
03-3749-3505-Fax

SVELTO, FRANCESCO
ASI
VIALE REGINA MARGHERITA 202
ROMA 00138
ITALY
39 68567356-Tel
3968367267-Fax

SWANEPOEL, RYNO
RAND AFRIKAANS UNIVERSITY
AUCKLAND PARK
JOHANNESBURG
SOUTH AFRICA
2711 489 2326-Tel
27 11 489 3207-Fax

SWANSON, RICHARD
SUNPOWER CORP.
430 INDIO WAY
SUNNYVALE CA 94086
USA
408 991 0908-Tel
408 739 7713-Fax

SWENSON, RON
ECO SYSTEMS
147 SOUTH RIVER ST SUITE 307
SANTA CRUZ CA 95061
USA

SWISHER, RON
OREGON INSTITUTE OF TECHNOLOGY
KLAMATH FALLS OR 97601
USA
503 883-1440-Tel

SYNHORST, ANNELIESE

SYNHORST, JOHN
METRO STATE COLLEGE- DENVER
73 WARD COURT
LAKEWOOD CO 80228
USA
303-556-2504-Tel
303-556-2159-Fax

SZOT, YOZEF
UNIVERSITY OF SYDNEY
17-32 MEADOW CRESCENT
MEADOWBANK NSW 2114
AUSTRALIA
61 2 809 2845-Tel
61 2 807 3567-Fax

SZPITALAK, TED
UNIVERSITY OF NEW SOUTH WALES
PHOTOVOLTAIC CENTRE, UNSW
SYDNEY, NSW 2052
AUSTRALIA
61 2 666 4902-Tel
61 2 666 4079-Fax

TAJIMA, MICHIO
ISAS
3-1-1 YOSHINODAI
SAGAMIHARA
KANAGAWA 229
JAPAN
0427-51-3911 2712-Tel
0427-59-4238-Fax

TAKAHASHI, KENJI
PVTEC
1-22-12 OTOWA,BUNKYO-KU
TOKYO 112
JAPAN
81-3-3946-2451-Tel
81-3-3946-6262-Fax

TAKAHASHI, MASAHIDE
SHIKOKU RESEARCH INSTITUTE INC
2109-8, YASHIMA-NISHIMACHI
TAKAMATSU-SHI
KAGAWA 761-01
JAPAN
81-878-43-8111-Tel
81-878-70-0046-Fax

TAKAKURA, HIDEYUKI
TOYAMA PREFECTURAL UNIVERSITY
5180 KUROKAWA, KOSUGI-MACHI
IMIZUGUN
TOYAMA 93903
JAPAN
81766567500 EX452-Tel
81-766-56-8021-Fax

TAKAMOTO, TATSUYA
JAPAN ENERGY CORPORATION
3-17-35 NIIZOMINAMI
TODASHI 335
JAPAN
048-433-2061-Tel
048-442-8535-Fax

TAKATA, HAJIME
MATSUSHITA BATTERY

TAKATO, HIDETAKA
ELECTROTECHNICAL LABORATORY
1-1-4, UMEZONO,
TSUKUBA-SHI
IBARAKI 305
JAPAN
81-298-58-5494-Tel
81-298-54-5497-Fax

TAKAYAMA, MICHIHIRO
KYOCERA CORPORATION
10-1 KAWAI, GAMO-CHO
GAMO-GUN
SHIGA 529-15
JAPAN

TAKESHITA, HIROSHI
SHOWA SHELL SEKIYU K.K.
123-1 SHIMOKAWAIRI, ATSUGI SHI
KANAGAWA KEN 243-02
JAPAN
0462-45-6216-Tel
0462-45-9343-Fax

TAKESHITA, TOSHIHIDE
TECHNORA INC.
FUKOKU BLDG,
2-2-2 UCHISAIWAI-CHO
CHIYODA-KU, TOKYO
JAPAN
03-3508-2280-Tel
03-3508-7578-Fax

TAKIGAWA, KIYOSHI
CRIEPI
2-11-1 IWATOKITA
KOMAESHI
TOKYO 201
JAPAN
03-3480-2111-Tel
03-3480-3886-Fax

TALBOTT, JONATHAN
GT EQUIPMENT TECHNOLOGIES INC.
472 AMHERST ST.
NASHUA NH 03063
USA
603-883-5200-Tel
603-595-6993-Fax

TAN, SIMON
SIEMENS SOLAR
4650 ADOHR LN
CAMILLO CA 93011
USA
805-388-6351-Tel
805-388-6395-Fax

TANG, YINGHUI
UNIVERSITY OF NEW SOUTH WALES
C/O MR. TED SZPITALAK
PHOTOVOLTAIC CENTRE, UNSW
SYDNEY, NSW 2052
AUSTRALIA
385 5246-Tel
662 4240-Fax

TAO, GUOQIAO
DELFT UNIVERSITY OF TECHNOLOGY
MEKELWEG 4
DELFT 2628 CD
NETHERLANDS
31 15 786779-Tel
31 15 785691-Fax

TARRANT, DALE
SIEMENS SOLAR INDUSTRIES
4650 ADOHR LANE P.O. BOX 6032
CAMARILLO CA 93010
USA
805-388-6328-Tel
805-388-6580-Fax

TATRO, MARJORIE
PV SYSTEM COMPONENTS
MS 0752
P O BOX 5800
ALBUQUERQUE NM 87185 0752
USA
505 844 3154-Tel
505 844 6541-Fax

TAYLOR, ROGER
NREL
1617 COLE BLVD
GOLDEN CO 80401
USA
303 384-6432-Tel
303 384-6490-Fax

TAYLOR, STEVE
CNRS
1 WINDLE GROVE, WINDLE
ST. HELENS
MERSEYSIDE WA106HN
ENGLAND

TEFERTILLER, BEN
THE DOW CHEMICAL CO.
2030 DOW CENTER
MIDLAND MI 48674
USA
517-636-0185-Tel
517-636-8127-Fax

TER HEIDE, FERDINAND
MASTERVOLT
LEMELERBERGTWEG 23
AMSTERDAM 1101 AJ
NETHERLANDS
31 20 6970736-Tel
31 20 6971006-Fax

TERMOTE, URBAIN
FOKKER SPACE & SYSTEMS
1199 NORTH FAIRFAX STREET
SUITE 500
ALEXANDRIA VA 22314
USA
703 838 0678-Tel
703 683 5476-Fax

THEENHAUS, ROLF
MEMBER OF THE BOARD OF DIREC.
FORSCHUNGSZENTRUM JULICH
GMBH
JULICH 52425
GERMANY
02461-61-4671-Tel
02461-61-2910-Fax

THIESSEN, KLAUS
GOS
RUDOWER CHAUSSEE 5
BERLIN D-12489
GERMANY
49-30-63924752-Tel
49-30-63924751-Fax

THOMAS, DENNIS
EAGLE- PICHER INDUSTRIES
P.O. BOX 737
QUAPAW OK 74363
USA
918-673-1650-Tel
918-673-2121-Fax

THOMAS, HOLLY
NREL
1617 COLE BLVD
GOLDEN CO 80401
USA
303 384-6400-Tel
303 384-6490-Fax

THOMAS, RAYE
NEWSUN TECHNOLOGIES LTD
5 FREDERICK PLACE
OTTAWA, ONT K15 3G1
CANADA

THONGSATHITYA, AMNUAY
DEPT OF ENERGY DEV & PROMOTION
OFFICE OF ENERGY STUDY RES
RAMA I RD
YODSE BANGKOK 10330
THAILAND

THORP, DAVID
P V CENTRE, UNSW
SYDNEY 2052
AUSTRALIA
61 2 385 5246-Tel
61 2 662 4240-Fax

TIMMONS, MICHAEL
RESEARCH TRIANGLE INSTITUTE
P O BOX 12194
RESEARCH TRIANGLE NC 27709
USA
919-541-5930-Tel
919-541-6515-Fax

TOMLINSON, ROBERT
UNIVERSITY OF SALFORD
DEPT. OF ELECTRONIC & ELECTRIC
ENGINEERING
SALFORD M54WT
UNITED KINGDOM

TOMORI, KUNIAKI
PVTEC
1-22-12 OTOWA, BUNKYO-KU
TOKYO 112
JAPAN
81-3-39462451-Tel
81-3-39466262-Fax

TOPEL, BRUNO
HELIODINAMICA
RODOVIA RAPOSO TAVARES
KM-41 PO BOX 111 VARGEM GRANDE
PAULISTA, SAO PAU 06730-970
BRAZIL
55-11-790-0888-Tel
55-11-790-1280-Fax

TOSHIO, MUROTANI
MITSUBISHI ELECTRIC CORP.
4-1 MIZUTANI
ITAMI
OSAKA 664
JAPAN
0727-84-7383-Tel
0727-80-2573-Fax

TOWNSEND, TIM
ENDECON
P.O. BOX 354
DAVIS CA 95616
USA
916-753-0795-Tel
916-753-8469-Fax

TREFNY, JOHN
COLORADO STATE OF MINES
DEPARTMENT OF PHYSICS
GOLDEN CO 80401
USA
303-273-3833-Tel
303-273-3919-Fax

TRIEBE, JOHN
AMOCO TECHNOLOGY/SOLAREX
55 SHUMAN BLVD
NAPERVILLE IL 60563
USA
708-717-2451-Tel
708-717-0024-Fax

TSUDA, IZUMI
ELECTORTECHNICAL LABORATOR
1-1-4 UMEBONO
TSUKUBA-SHI
IBARAKI 305
JAPAN
81-298-58-5827-Tel
81-298-58-5829-Fax

TSUDA, SHINYA
SANYO ELECTRIC CO., LTD
1-18-13
HIRAKATA
OSAKA 573
JAPAN
81-720-41-1261-Tel
81-720-41-0386-Fax

TSUJI, NAOTO
UNIVERSITY OF TOKYO
7-3-1 HONGO
BUNKYO 113
JAPAN
03-3812-2111--Tel
03-5684-8402-Fax

TSUO, SIMON
NREL
SERF-W115 1617 COLE BLVD
GOLDEN CO 80401-3039
USA
303 384-6433-Tel
303 384-6531-Fax

TUTTLE, JOHN
NREL
1617 COLE BLVD
GOLDEN CO 80401
USA
303-384-6534-Tel
303-384-6430-Fax

UCHIDA, YASUNORI
SHARP CORPORATION
282-1, SHINJYO-CHO
KITAKATSURAGI-GUN
NARA 639-21
JAPAN
81 7456 5 1161-Tel
81 7456 5 1174-Fax

UDAYAKUMAR, MR.N.
UDHAYA SEMICONDUCTORS LTD.
NO.1 PERA NAIDU STREET PEELAME
COIMBATORE TAMILN 641004
INDIA
91-422-572814-Tel
91-422-572675-Fax

UEDA, AKIRA
DOSHISHA UNIV.
TANABE
KYOTO 610-03
JAPAN
+81-7746-5-6327-Tel
+81 7746 5 6813 Fax

UH, DIETER
ATLANTIS GMBH
BERLIN 10969
GERMANY

ULLAL, HARIN
NREL
1617 COLE BLVD
GOLDEN CO 80401
USA

UMENO, MASAYOSHI
NAGAYA INSTITUTE TECHNOLOGY
SHOWA KU
NAGOYA 466
JAPAN
81-52-732-2111-Tel
81-52-732-2761-Fax

UMENO, MASAYOSHI
NAGOYA INSTITUTE OF TECHNOLOGY
GOKISO CHO SHOWA KU
NAGOYA 466
JAPAN
81-52-732-2111-Tel
81-52-732-5697-Fax

UMEO, YOSHIYUKI
1-33-27 MINAMINAKABURI
HIRAKATA
OSAKA 573
JAPAN
0720-33-0577-Tel
0720-33-0577-Fax

URBANO, J. ANTONIO
CINVESTAV-IPN
AV IPN 2508 SN PEDRO ZACATENCO
D.F. 07360
MEXICO
752 06 77 EXT3175-Tel
752 0590-Fax

USHER, ERIC
CANMET
1615 MONTEE STE-JULIE
PO BOX 4800
VARENNES PQ J3X 1S6
CANADA
514 652-5572-Tel
514 652-5177-Fax

USHIROKAWA, AKIO
TOKYO ENGINEERING UNIV
1401-1 KATAKURACHO
HACHIOJISHI
TOKYO 192
JAPAN
0426-37-2111-Tel
0426-37-2118-Fax

UTOMO, ENDRO
DGEED INDONESIA

VAN DEN BOS, KEES
KIPP & ZONEN
390 CENTRAL AVENUE
BOHEMIA NY 11716
USA
516-539-2885-Tel
516-589-2068-Fax

VAN DER WEIDEN, THEO C.J.
ECOFYS
PO BOX 8408
UTRECHT NL 3503RK
NETHERLANDS
31 030 913400-Tel
31 030 913401-Fax

VAN OVERSTRAETE, ROGER
IMEC
KAPELDREEF 75
LEUVEN 3001
BELGIUM
+3216281372-Tel
+321627150-Fax

VARADI, PETER F.
SOLAREX
4620 N PARK AVE
#1606W
CHEVY CHASE MD 20815
USA
301 657-9110-Tel
301 657-9110-Fax

VARGAS ABURTO, CARLOS
KENT STATE UNIVERSITY
SCHOOL OF TECHNOLOGY
KENT OH 44242
USA
216-672-2770-Tel
216-672-2894-Fax

VASANTH, KARTNIK
PRINCETON UNIVERSITY
B-4D5, E-QUAD
PRINCETON UNIVERSITY
PRINCETON NJ 08544
USA
609-258-5902-Tel
609-258-6279-Fax

VASCOTTO, GIAN
ONTARIO HYDRO TECHNOLOGY
800 KIPLING AVE
TORONTO ON M8Z 554
CANADA
416 207 5826-Tel
416 207 5558-Fax

VAZ, CELIO COSTA
INST. NACIONAL DE PESQ. ESPACI
AV. DOS ASTRONAUTAS, 1758
SAO JOSE DOS CAMP 12227070
BRAZIL
123-41-8977-Tel
123218743-Fax

VENDURA, GEORGE
TRW
2761 ST. ALBANS DRIVE
LOS ALAMITOS CA 90720
USA
310-812-8485-Tel
310-812-4978-Fax

VENKATASUBRAMANIAN, RAMA
RESEARCH TRIANGLE INSTITUTE
3040 CORNWALLIS ROAD
RESEARCH TRIANGLE NC 27709
USA
919 541-6889-Tel
919 541 6515 Fax

VERHOEF, LEENDERT
R&S RENEWABLE ENERGY SYSTEMS
P O BOX 3049
HELMOND NB 5700JC
NETHERLANDS
31 4920 23335-Tel
31 4920 49665 Fax

VERIE, CHRISTIAN
CONSULTANT
SOPHIA ANTIPOLIS
RUE BERNARD GREGORY
VALBONNE 06600
FRANCE
33 93 95 42 11-Tel
33 93 95 83 61-Fax

VERLINDEN, PIERRE
SUNPOWER CORP.
430 INDIO WAY
SUNNYVALE CA 94086
USA
408 991-0910-Tel
408 739-7713-Fax

VERNON, STANLEY
SPIRE CORPORATION
ONE PATRIOTS PARK
BEDFORD MA 01730 2396
USA
617 275 6000-Tel
617 275 74770-Fax

VESEY, ANDREW M.
NIAGARA MOHAWK POWER CORP.
300 ERIE BOUEVARD WEST
SYRACUSE NY 13202
USA
315-428-6800-Tel
315-428-5566-Fax

VEYAN, PHILIPPE
TOTAL ENERGIE
7 CH DU PLATEAU
69570
DARDILLY
FRANCE
33 78 474455-Tel
33 78 649100-Fax

VIEIRA, MANUELA
CEMOP/UNINOVA
QUINTA DA TORRE
MONTE DE CAPARICA 2825
PORTUGAL
351.1.3500251-Tel
2941365-Fax

VILELA, MAURO
SPACE VACUUM EPITAXY CENTER
UNIVERSITY OF HOUSTON
HOUSTON TX 77024-5507
USA
713 743 3621-Tel
713 747 7724-Fax

VIOTTO, MICHAEL
ISET
KONIGSTAR 5G
KESSEL D-36113
GERMANY
0049 5617294 219-Tel
0049 561 7294 100-Fax

WADA, TAKAHIRO
MATSUSHITA ELECTRIC INDUSTRIAL
3-4 HIKARIDAI
KYOTO 619-02
JAPAN
07749-8-2518-Tel
07749-82585-Fax

WAGNER, FRANZ
F.F. LDa
SITIO DO PALHEIRINHO
ALJEZUR 8670
PORTUGAL
011-351-82-98745-Tel
011-351-82-98746-Fax

WAGNER, HERIBERT
ISI-PV
FORSCHUNGSZENTRUM JUELICH
JUELICH 52425
GERMANY
49-2461-61-6507-Tel
49-2461-61-3735-Fax

WAGNER, SIEGFRIED
TU CHEMNITZ-ZWICKAU
FAKULTAT FUE ET UND IT -31105-
CHEMNITZ 09107
GERMANY
0049 371 531 3009-Tel
0049 371 531 3004-Fax

WAKAMATSU, SEIJI
PHOTOVOLTAIC POWER GENERATION
TECHNOLOGY RESEARCH ASSOC.
OTOWA BLDG 7F
OTOWA, TOKYO 1-22-12
JAPAN
81-3-3946-2451-Tel
81-3-3946-662-Fax

WAKISAKA, KENICHIRO
SANYO ELECTRIC CO., LTD.
1-18-13- HASHIRIDANI
HIRAKATA, OSAKA 573
JAPAN
81 720-41-1265-Tel
81 720-41-0386-Fax

WALD, FRITZ
ASE AMERICAS INC.
4 SUBURBAN PARK DRIVE
BILLERICA MA 01821-3890
USA
508-667-5900 x286-Tel
508-663-2868-Fax

WALLACE, RICHARD
STATE UNIV OF NY AT BUFFALO
226 SPRINGVILLE AVE
AMHERST NY 14226
USA

WALLACE, WILLIAM
NREL
1617 COLE BLVD
GOLDEN CO 80228
USA
303-384-6476-Tel
303-384-6490-Fax

WALLERSTEIN, DAVID
SIEMENS SOLAR INDUSTRIES
4650 ADHR LN.
CAMARILLO CA 93011
USA
805-388-6365-Tel
805-388-6395-Fax

WALPERT, MARIANNE
SOLAR ENGINEERING APPL. CORP.
3500 THOMAS RD
SUITE E
SANTA CLARA CA 95054
USA
408-986-9231-Tel
408-986-9233-Fax

WALT, ROBB
INTEGRATED POWER CORP
7524 STANDISH PLACE
ROCKVILLE MD 20855
USA

WALTERS, ROBERT
NAVAL RESEARCH LAB
CODE 6615 NAVAL RESEARCH LAB
4555 OVERLOOK AVE SE
WASHINGTON DC 20375
USA

WANG, AIHUA
UNIVERSITY OF NEW SOUTH WALES
CENTER FOR PV DEV.
SYDNEY, NSW 5246
AUSTRALIA
61 2 385 5246-Tel
61 2 662 4240-Fax

WANG, ANHUA
GANSU GNERE PV CO.

WANG, XIN
HAI NAN ASIA PHOTOVOLTAIC IND
CO. LTD 18F ZHONG FANG BLDG
LONG KUN NORTH RD HAI KOU CITY
HAI NAN 570005
CHINA
86 0898 8533473-Tel
86 0898 853 3456-Fax

WANLASS, MARK
NREL
1617 COLE BLVD
GOLDEN CO 80401
USA
303 384-6532-Tel
303 384-6532-Fax

WARABISAKO, TERUNORI
HITACHI RESEARCH LAB.
HITACHI LTD.
1-280 MIGASHI-KOIGAKUBO
KOKUBUNJI, TOKYO
JAPAN
+81-423-23-1111-Tel
+81-423-27-7761-Fax

WARTA, WILHELM
FHG-ISE
OLTMANNSSTR. 5
FREIBURG D-79100
GERMANY
+49-761-4588792-Tel
+49-761-4588-100-Fax

WATSON, LEE
WRIGHT RESEARCH FOUNDATION
5040 LYNNWOOD CENTER RD NE
BAINBRIDGE IS WA 98110
USA

WATT, MURIEL
CENTRE FOR PY DEVICES & SYS
SCHOOL OF ELECTRICAL ENGIN.
UNIV OF NEW SOUTH WALES
NSW 2052
AUSTRALIA
61 2 452 1408-Tel
61 2 662 4240-Fax

WEBB, JOHN
NREL
1617 COLE BLVD
GOLDEN CO 80401
USA
303-384-6703-Tel
303-384-6481-Fax

WEBER, EICKE
UNIVERSITY OF CALIFORNIA
DEPT OF MATERIALS SCIENCE
BERKELEY CA 94720
USA
510 642 0205-Tel
510 642 2069-Fax

WEBER, KLAUS
AUSTRALIAN NATIONAL UNIVERSITY
DEPT OF ENGINEERING
CANBERRA 0200
AUSTRALIA
61-62-249-4585-Tel
61-62-249-0506-Fax

WEI, GUANG-PU
SHANGHAI UNIV OF SCIENCE & TEC
210800
SHANGHAI

WEINGART, GEROME
NREL
1617 COLE BLVD
GOLDEN CO 80439
USA
303-384-6910-Tel
303-384-6901-Fax

WEISSMAN, JANE
PHOTOVOLTAKS FOR UTILITIES
15 HAYDN ST
BOSTON MA 02131-4013
USA
617 323-7377-Tel
617 325-6738-Fax

WELIK, GEORGE
LOCKHEED
1483 BELLEVILLE WAY
SUNNYVALE CA 94087
USA
408 756 1026-Tel
408 756 8636-Fax

WELLMAN, DON
NAVY LIAISON UNIT
NAVLU PO BOX 500
PEARL HARBOR HI 96860
USA

WELSER, ROGER
YALE UNIVERSITY
691 ORANGE ST APT.#503
NEW HAVEN CT 06511
USA
203 432 4240-Tel
203 432 7769-Fax

WENAS, WILSON
TOKYO INSTITUTE OF TECH
2-12-1 OHOKAYAMA, MEGURO-KU
TOKYO 152
JAPAN
81-3-5734-2661-Tel
81-3-5734-2897-Fax

WENGER, HOWARD
PACIFIC ENERGY GROUP
32 VALLA CT
WALNUT CREEK CA 94596
USA
510 937-1750-Tel
510 937-0385-Fax

WENHAM, STUART
UNIVERSITY OF NEW SOUTH WALES
ANZAC RD. KENSINGTON
NSW 2033
AUSTRALIA
612 385 5171-Tel
612 662 4240-Fax

WERNER, JURGEN H.
MAX PLANCK INSTITUTE
HEISENBERGSTR 1
STUTTGART 70506
GERMANY
49-714-689-1606-Tel
49-714-689-1602-Fax

WERNER, MARTINA
MAX PLANCK INSTITUTE
WEINBERG 2
HALLE 06120
GERMANY
49 345 5582 814-Tel
49 345 5511 223-Fax

WEST, RICHARD
UTILITY POWER GROUP
9410-G DE SOTO AVE
CHATSWORTH CA 91311
USA
818 700-1995-Tel
818 700-2518-Fax

WETTLING, WOLFRAM
FHG- ISE
OLTMANNSSTRABE 5
FREIBURG 79100
GERMANY
+49-761-4588-266-Tel
+49-761-4588-250-Fax

WHISNANT, RICHARD
RESEARCH TRIANGLE PARK
P.O.BOX 12144
RESEARCH TRIANGLE NC 27511
USA
919-541-5825-Tel
919-541-6515-Fax

WHITAKER, CHARLES
ENDECON ENGINEERING
3160 CROW CANYON RD
SUITE 210
SAN RAMON CA 94583
USA
510 867 3360-Tel
510 867 3362-Fax

WHYTE, DOUGLAS
SOUTHERN CAL EDISON
2244 WALNUT GROVE AVE
ROSEMEAD CA 91770
USA

WIETING, ROBERT
SIEMENS SOLAR INDUSTRIES
P.O. BOX 6032
CAMARILLO CA 93012
USA
805-445-2803-Tel
805-388-6580-Fax

WIJERATNA, D.G.D.C.
CEYLON ELECTRICITY BOARD

WIJESOORIYA, PRIYANTHA
SOLANKA ASSOCIATES
8 VISAKA ROAD,
COLOMBO 4, SKI LA
SRI LANKA
94-1-582876-Tel
94-1-590950-Fax

WILBANKS, WILLIAM
CENTER FOR HIGH TECHNOLOGY
ALBUQUERQUE NM 87131
USA
505-277-1410-Tel
505-277-6433-Fax

WILES, JOHN C.
S.W. TECHNOLOGY DEV. INST/NMSU
PO BOX 30001/DEPT 3 SOLAR
LAS CRUCES NM 88003-0001
USA
505-646-6105-Tel
505-646-3841-Fax

WILL, JURGEN
ATLANTIS GMSH

WILLEKE, GERHARD
UNIVERSITY OF KONSTANZ
P.O. BOX 5560
KONSTANZ D-78934
GERMANY
49-7531-882079-Tel
49-7531-883895-Fax

WILLEMEN, JOOST
DELFT UNIVERSITY OF TECHNOLOGY
P.O. BOX 5031
DELFT 2600GA
NETHERLANDS
31-15-78-2486-Tel
31-15-78-5691-Fax

WILLENBERG, RICH
EXHIBIT PRODUCTIONS INC.
10800 FORD ROAD
DEARBORN MI 48126
USA
313-582-9250-Tel
313-582-1398-Fax

WILLIAMS, NEVILLE
SOLAR ELECTRIC LIGHT FUND
1739 CONNECTICUT AVE NW
WASHINGTON DC 20009
USA
202 234-7265-Tel

WILLISON, MICHAEL
SUN ASSISTED PTY LTD
GPO BOX 725
SYDNEY, NSW 2001
AUSTRALIA
61-2-560-8353-Tel
61-2-560-8406-Fax

WILLS, ROBERT
SKYLINE ENGINEERING
1 COLBORN RD
TEMPE NH 03084
USA

WILT, DAVID
NASA LEWIS RESEARCH CENTER
21000 BROOKPARK RD
MS 302-1
CLEVELAND OH 44135
USA

WITT, ED
NREL
1617 COLE BLVD
GOLDEN CO 80401
USA
303 384-6402-Tel
303 384-6481-Fax

WOHLGEMUTH, JOHN
SOLAREX CORP.
630 SOLAREX CT.
FREDERICK MD 21701
USA
301-698-4375-Tel
301-698-4201-Fax

WOJTCZUK, STEVEN
SPIRE CORPORATION
ONE PATRIOTS PARK
BEDFORD MA 01730-2396
USA
617-275-6000-Tel
617-275-7470-Fax

WOLF, MARTIN
UNIVERSITY OF PENNSYLVANIA
179 BROOKSTONE DRIVE
PRINCETON NJ 08540
USA
609 921-6870-Tel

WOLLWEBER, JURGEN
INST. FOR CRYSTAL GROWTH
RUDOWER CHAUSSEE 6
BERLIN D12489
GERMANY
030 6392-3013-Tel
030 6392 3003-Fax

WOLVERTON, ROBERT
USAF PHILLIPS LABORATORY
PL VTPC
KIRTLAND AFB NM 87117
USA
505-846-5754-Tel
505-846-2885-Fax

WONG, DAVID
BTU INTERNATIONAL
C/O JENNIFER BUTLER
23 ESQUIRE ROAD
NORTH BILLERICA MA 01862
USA
508-667-4111-Tel
508-667-9068-Fax

WONG, JOSEPH
SOLAREX
FLT 38,11/F BLK D
WAH LOK IND. CENTRE
31-41 SHAN MEI ST NT FOTAN
HONG KONG
852-6056419-Tel
852-6010623-Fax

WOOD, RICHARD F.
OAK RIDGE NATIONAL LAB
SOLID STATE DIVISION
OAK RIDGE TN 37831
USA

WOOLVERTON, RICHARD
MAUI ELECTRIC CO., LTD.
210 W KAMEHAMEHA AVENUE
KAHULUI HI 96732
USA
808 871 8461-Tel
808 871 2350-Fax

WORAVECH, PAIRACH
DEPT OF ENERGY DEV & PROMOTION
OFFICE OF ENERGY STUDY RES
RAMA I RD
YODSE BANGKOK 10330
THAILAND

WRIXON, GERARD
NMRC
LEE MALTINGS, PROSPECT ROW
CORK
IRELAND
+353-21-277375-Tel
+353-21-277375-Fax

WROBLEWSKI, KIM
SANDIALABS
UNIVERSITY OF NEW MEXICO
7127 PROSPECT PLACE NE
ALBUQUERQUE NM 87110
USA
505-881-3881-Tel
505-842-5251-Fax

WRONSKI, CHRISTOPHER
PENN STATE UNIVERSITY
213 EE WEST
UNIVERSITY PARK PA 16802
USA
814-865-0930-Tel
814-865-7065-Fax

WYRSCH, NICOLAS
IMT UNIVERSITE OF NEUCHATEL
BREQUET 2
NEUCHATEL CH2000
SWITZERLAND
47 38 205453-Tel
47 38 254276-Fax

XIONG, YI-MING
TOKYO UNIV. OF AGRICULTURE
DIVISION OF ELECTRONIC ENGR.
KOGANEI
TOKYO 184
JAPAN
0423-88-7108-Tel
0423-85-9055-Fax

YAJIMA, NOBUTSUGU
PHOTOVOLTAIC POWER GENERATION
TECHNOLOGY RESEARCH ASSOC.
OTOWA MF BLDG 7F,1-22-12,OTOWA
BUNKYO-KU TOKYO
JAPAN
81-3-39-46-2451-Tel
81-3-3946-6262-Fax

YAMADA, AKIRA
TOKYO INSTITUTE OF TECH
2-12-1 O-OKAYAMA, MEGURO-KU
TOKYO 152
JAPAN
81-3-5734-2698-Tel
81-3-5734-2897-Fax

YAMADA, TORU
SHIN ETSU CHEMICAL CO. LTD.
3-2-1 SAKATO, TAKATSA-KU
KAWASAKI
KANAGAWA 213
JAPAN
044 819 3101-Tel
044 819 3103-Fax

YAMAGUCHI, MASAFUMI
TOYOTA TECH. INSTITUTE
2-12-1 HISAKATA, TEMPAKU
NAGOYA 468
JAPAN
1-502-802-1111-Tel
1-502-802-6069-Fax

YAMAGUCHI, MISAKO
TOKYO UNIV. OF AGRI TECH.
2-24-16 NAKAMACHI
KOGANEI 184
JAPAN
81-423-64-3311-Tel
81-423-85-6729-Fax

YAMAGUCHI, SHUICHI
KANDENKO CO., LTD
4-8-33 SHIBAURA
MINATO-KU
TOKYO 108
JAPAN

YAMAGUCHI, TOSHIYUKI
WAKAYAMA COLLEGE OF TECHNOI
77 NOSHIMA, NADA-CHO
GOBO
WAKAYAMA 644
JAPAN
81 738 29 2301-Tel
81 738 29 2574-Fax

YAMAJI, TETSUO
PVTEC
1-22-12 OTOWA BUNKYO-KU
TOKYO
JAPAN
81 3 3946 2451-Tel
81 3 3946 6262-Fax

YAMAMOTO, AKIO
FUKUI UNIVERSITY
3-9-1 BUNKYO
FUKUI 910
JAPAN
81-776-27-8566-Tel
81-776-27-8749-Fax

YAMAMOTO, KENJI
CENTRAL RESEARCH LABORATOR
KANEKA CORPORATION 2-80
1 CHOME, YOSHIDA-CHO, HYOGO-
YOSHIDA-CHO,HYOGO KOBE, H
652 JAPAN
078-652-4055-Tel
078-652-4081-Fax

YAMAMOTO, YOSHIHIRO
SHARP CORPORATION
282-1 HAJIKAMI
SHINJO-CHO, KITAKATURAGI-GUN
NARA PREF 639-21
JAPAN
81 7456 5 1161-Tel
81 7456 2 8254-Fax

YAMAMOTO, YUKIO
FUKUI COLLEGE FO TECHNOLOGY
GESHI-CHO
SABAE 916
JAPAN
81-778-62-1112-Tel
81-778-62-1108-Fax

YAMANO, YOSHIYA
JAPAN STORAGE BATTERY CO LTD
NISHINOSHO KISSHOIN
MINAMI-KO
KYOTO
JAPAN

YAMAWAKI, TAKEHARU
KANEKA CORPORATION
1-1 HIEITSUJI 2 CHOME,
OTSU
SHIGA 520-01
JAPAN
0775-77-2180-Tel
0775-77-2125-Fax

YANG, JEFFREY
UNITED SOLAR SYSTEMSCORP
1100 WEST MAPLE RD
TROY MI 48084
USA
810-362-4170-Tel
810-362-4442-Fax

YATER, JANE
NASA LEWIS RESEARCH CENTER
21000 BROOKPARK ROAD,MS 302-1
CLEVELAND OH 44135
USA
216-4336106-Tel
216-433-6106-Fax

YAZAWA, YOSHIAKI
HITACHI RES LAB., HITACHI, LTD
7-1-1, OMIKA-CHO
HITACHI-SHI
IBARAKI 319-12
JAPAN
81 294 52 5111-Tel
81-294-52-7618-Fax

YEH, MILTON
APPLIED SOLAR ENERGY CORP.
15251 EAST DON JULIAN RD.
CITY OF INDUSTRY CA 91745-1002
USA
818 658 6581 Tel
818-336-8694-Fax

YENAMANDRA, RAJU
SIEMENS SOLAR
4650 ADONR SOLAR
CAMARILLO CA 93011
USA
805-388-6351-Tel
805-388-6395-Fax

YIN, LIAN
MINISTRY OF ELECTRIC POWER
NO 137, FUGOU STREET
BEIJING 100031
CHINA

YIP, LAP SUM
MCGILL UNIVERSITY
ELECTRICAL ENGRG DEPT
3480 UNIVERSITY STREET
MONTREAL, QUEBEC H3A 2A7
CANADA

YOCUM, RICK
SOLAR CELLS, INC.
1702 N WESTWOOD AVE
TOLEDO OH 43607
USA
419-534-3377-Tel
419-534-2794-Fax

YOO, SEI WOONG
HANKUK GLASS
DONGCHUN-DONG
NAM-GU
INCHON 402-130
KOREA

YOON, SEWANG
AMONIX, INC.
3425 FUJITA ST.
TORRANCE CA 90505
USA
310-325-8091-Tel
310-325-0771-Fax

YOSHIDA, AKIRA
TOYOHASHI UNIVERSITY OF TECH.
TEMPAKU
TOYOHASHI 441
JAPAN
81 532 47 0111-Tel
81 532 48 3422-Fax

YOSHIDA, TAKASHI
CORPORATE RESEARCH AND DEVELOP
2-2-1 NAGASAKA
YOKOSUKA 240-01
JAPAN
468-57-6730-Tel
468-57-2791-Fax

YOSHINO, MASAHIRO
FLEX INTERNATIONAL CO, LTD
200 YOSHIDA
KUROBE SHI 938
JAPAN
0765-54-8333-Tel
0765-54-8338-Fax

YOSHIOKA, KAZUYA
TOKYO A & T UNIVERSITY
2-24-16 NAKAMACHI
KOGANEISHI
TOKYO 184
JAPAN
0423-88-7108-Tel
0423-85-9055-Fax

YU, GWONJONG
KOREA INSTITUTE OF ENERGY RES.
71-2 JANG DONG
YUSONG-KU
TAEJON 305 343
KOREA
82 42 860 3423-Tel
82 42 860 3132-Fax

YU, PEI NUO
TIANJIN INST. OF POWER SOURCES
TIANJIN 300381, P.O. BOX 277
TIANJIN 300381
CHINA
86-022-3382851-Tel
86-022-3383783-Fax

YUKAWA, MOTONOBU
FLEX INTERNATIONAL CO., LTD.
1084-2 HASSHOMACHI
HAMATSUSHI
SHIZUOKA 433
JAPAN
053-433-2731-Tel
053-433-2786-Fax

ZAAIMAN, WIM
EUROPEAN COMMISSION / ESTI
TP 450
ISPRA VA 21020
ITALY
+39 332 785750-Tel
+32 332 785750-Fax

ZAININGER, HENRY W.
ZAININGER ENGINEERING CO. INC.
1590 OAKLAND RD, STE B-211
SAN JOSE CA 95131
USA
408 452-1840-Tel
408 452-1838-Fax

ZANESCO, IZETE
INSTITUTO DE ENERGIA SOLAR
E.T.S.I.T. - CIUDAD UNIVERSITY
MADRID 28040
SPAIN
341 544-1060-Tel
341 544-6341-Fax

ZEMMRICH, DIETER
SPECTROLAB, INC.
12500 GLADSTONE AVE
SYLMAR CA 91342
USA
818-898-2845-Tel
818-365-7771-Fax

ZERVOS, ARTHOUROS
EUROPEAN COMMISSION
200, RUE DE LA LOI
BRUSSELS B-109
BELGIUM

ZHAO, JIANHUA
UNIVERSITY OF NEW SOUTH WALES
PV CENTRE, UNSW
SYDNEY, NSW 2052
AUSTRALIA
61 2 385 5246-Tel
61 2 662 4240-Fax

ZHAO, YUWEN
BEIJING SOLAR ENERGY
3 HUA YUAN RD
HAIDIAN DISTRICT
BEIJING 100083
CHINA

ZHOU, TEDDY
SOLAR CELLS, INC.
1702 N. WESTWOOD AVE.
TOLEDO OH 43607
USA
419-534-3377-Tel

ZOLPER, JOHN
SANDIA NATIONAL LABORATORIES
P O BOX 5800
ALBUQUERQUE NM 87185
USA
505-844-8571-Tel
505-844-8985-Fax

ZOTT, STEFAN
TU DRESDEN
MOMMSENSTRASSE 13
DRESDEN, SACHSEN 01062
GERMANY
49 351 463 2464-Tel
49 351 463 7065-Fax

ZWEIGART, SIEGWUUD
UNI STUTTGART
PFAFFER WALDRING 47
STUTTGART D-70569
GERMANY
49-711168517200-Tel
49-711168517143-Fax

FABRICATION OF A NEW TYPE OF SYNTHETIC SEMICONDUCTOR IN SPACE

Yoshihiro HAMAKAWA
Faculty of Engineering Science, Osaka University
Toyonaka, Osaka 560, Japan

ABSTRACT

A review is given on utilization of space environment for fabrication of a new type of synthetic semiconductor. Firstly, significancies of space environment for material processing are enumerated. It has been shown that oxigen-free space with high vacuum environment and microgravity would be quite useful for semiconductor material preparations and device fabrication processing. Secondly, as an example of material processing, fabrication of a ternary Si-As-Te amorphous semiconductors in the Spacelab J is introduced together with essential advantages of this material system in views of both basic physics and technology as a new type of synthetic semiconductor. Some basic technical data on electrical, optical and optoelectronic properties, energy gap and also valency electron controllabilities are demonstrated. In a final part, a large area thin film growth in space and its application to a bioproductive photovoltaic power generation is proposed and discussed as a new step of space power station .

INTRODUCTION

According to a NASA report, more than 4500 satellites has been launched off for their various missions; communication, broadcasting, weather watch, geographic resources and military purposes. Space development would be one of the most brilliant commemorable results established by the science and technology of the 20th century's mankind.

In view of semiconductor material processing, the space environment provides various charming points, such as microgravity, high vacuum with oxigen-free space, abundant solar energy which can be converted to thermal energy and electric power etc. On the other hand, the amorphous Si-As-Te chalcogenide system is also an interesting semiconductor. It has a very wide vitreous region in the Gibb's Triangle, and a wide range of energy gap controllability by regulation of the atmoic fraction. However, there exists a substantial barrier impending the fabrication of this atomic control due to large differences in the weight density, melting point and vapor pressure of the individual atomic elements.

A series of systematic experiments has been carried out in the terrestrial environment since 1974 [1] and also within the TT-500 Rocket experiment in 1980 [2]. On the basis of these data, the Spacelab J experiments have been accomplished in the FMPT (First Material Processing Test) project in 1992.

In this paper, some significancies of space environment for the functional material processing are pointed out and discussed. Then, fabrication of Si-As-Te chalcogenide semiconductor in FMPT Spacelab J is introduced, and a series of technical data on the basic characterizations on this material system are presented. Finally, a concept of large scale photovoltaic generation with full use of significancies in space environment is proposed and discussed.

UTILIZATION OF SPACE ENVIRONMENT FOR MATERIAL PROCESSING

According to Dr.M.Mori who was a Payload Specialist of the space shuttle "Endeavor" [3], there exist a lot of interesting experiences in both physics and physiological view points. Apart from these, in a view of material preparation technology, space environment has many varieties of significant points. The significances are enumerated as follows, that is:

(1) ***High Vacuum & Oxigen Free Environment***
As it is reported, elsewhere, the vacuum level at the geo-synchronized orbit is 10^{-12} Torr. If we utilize the *wake* effect it becomes higher than 10^{-14} Torr. One can imagine, for example, although the MBE facility in the terrestrial laboratory is very expensive, however, more than 70% of the apparatus cost is beeing paid only to produce high quality vacuum environment. At the same time, we could obtain much higher quality vacuum environment in space in free of charge. A high vacuum more than 10^{-14} Torr also means an oxigen and nitrogen free space which might be a useful environment for purification of metals and semiconductors.

(2) ***Microgravity***
In general, the production of high-quality crystals and/or thin films requires an extremely stable environment. In view of this point, microgravity environment less than 10^{-4} **g** in space provides a suitable condition for material processing.

(3) ***An Abundant Energy Resource***
Abundant solar energy, easliy converted into electricity by photovoltaics or high temperature by photothermal conversion, could be obtained in space. Because of the high vacuum environment, obtaining extremely high temperatures might be fairly easy.

(4) ***Availability of Wide Area and Huge Space***
As it has been proved by astronomical observ ations, extremely wide solar photovoltaic arrays and big factories could be built in space, and operatre free of charge.

CH3365-4/94/0000-0001 $4.00 © 1994 IEEE

First WCPEC; Dec. 5–9, 1994; Hawaii

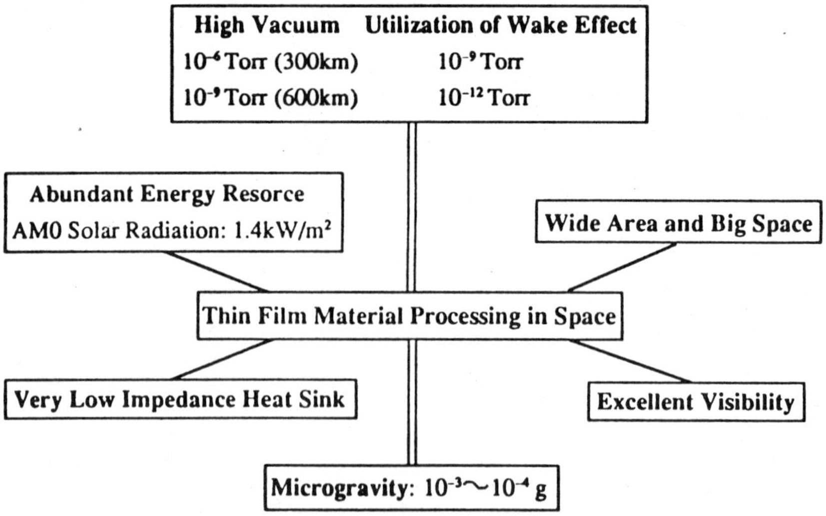

Figure 1. *Significances of space environment and thin film material processing*

(5) ***A Very Low Impedance Heat Sink***
Related to the item (4), the shadow part of the solar radiation acts as an infinite black body radiator which is corresponding to the zero impedance heat sink.

Figure 1 illustrates significancies of space environment mentioned above. Whith full use of these significances, material processing of the MOCVD, MBE, graphoepitaxy growth of multilayered thin films, and also plasma CVD would be easily accomplished in space. Thus it could be possible to producce multilayeredr functional devices and also new kinds of macroelectronics apparatus and facilities.

Apart from the industrial interest, less than 10^{-14} Torr would be also available in space and all semiconductor materials could be produced carbon-free and oxygen-free. That might be of great interest in basic physics and material characterization sciences.

As a NASDA-NASA joint project, the Japanese material processing experiments in space: FMPT (First Material Processing Test) has been carried out by the space shuttle "Endeavor" in a one-week program from September 13 to 20, 1992.

The FMPT consists of 34 experiments - ranging from crystallization superconducting materials to monitoring the health of the NSDA´s payload specialist. This complement of experiments allows the maximum number of Japanese scientists to participate in space-based research, gather data on mission operations, and demonstrate the effective use of space environment for research. Spacelab J was also the first time a NASDA astronaut has flown in space.

For NASDA, nine experiments expand knowledge of microgravity and life sciences gathered on previous Spacelab missions and will allow preparations for Space Station Freedom. Testing technology and procedures on this mission will serve as a precursor to Freedom´s system and will demonstrate the cooperative working relationship developed with one of our Space Station partners.

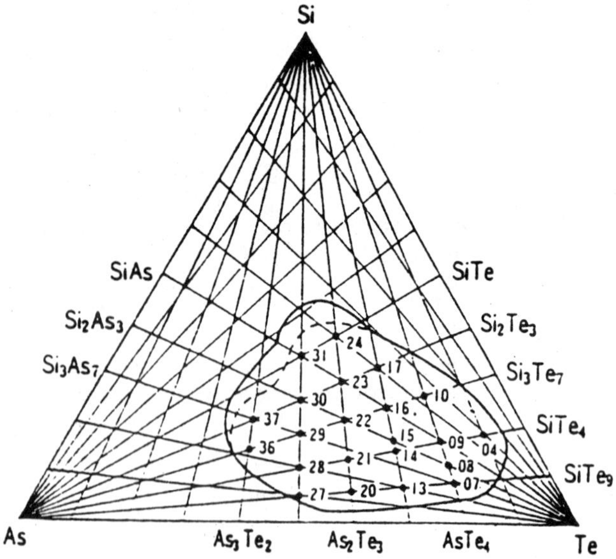

Figure 2. *Synthesized glasses and glass-forming region in Si-As-Te system*

A NEW TYPE OF SYNTHETIC SEMICONDUCTOR: Si-As-Te

Ternary chalcogenide Si-As-Te amorphous material system has various interesting points in views of both basic physics and technological applications. Since the Si-As-Te system consists of mixed combinations of IV-III-II hedral bonding, it has a large structural flexibility to construct a high possibility of random network, which yields a very large glass formation region with a wide physical constant controllability as shown in Figure 2. For example, the energy gap can be controlled continuously in the range from 0.6eV to 2.5eV, which covers the energy gaps of conventional crystalline semiconductors; Ge (0.66eV), Si (1.0eV), GaAs (1.43eV) and GaP (2.25eV).

Figure 3 shows the temperature dependence of the d.c. conductivity of Si-As-Te material system, while dashed line shows that of crystalline silicon as a reference. As can be seen in the change of the slope of 1/T plots, the energy gap E_g can be controlled with controlling fraction. Another experimental evidence on this energy gap

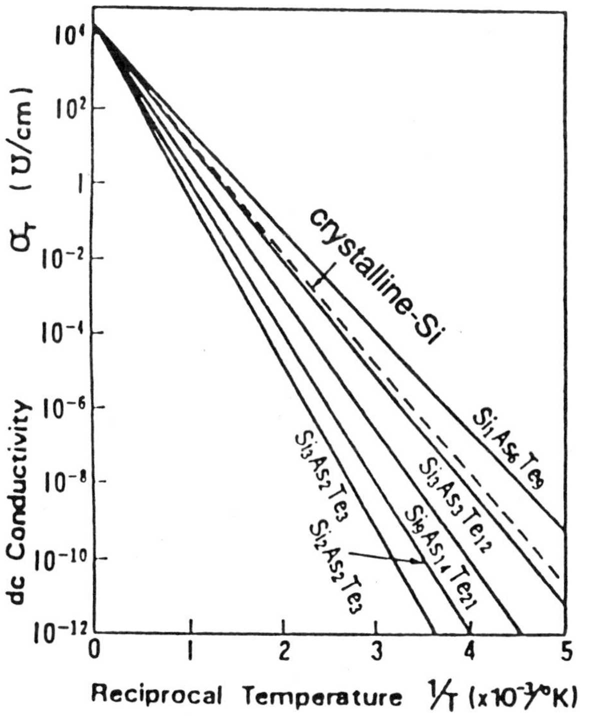

Figure 3. Temperature dependence of d.c. conductivity of terrestrial mode amorphous Si-As-Te semiconductors

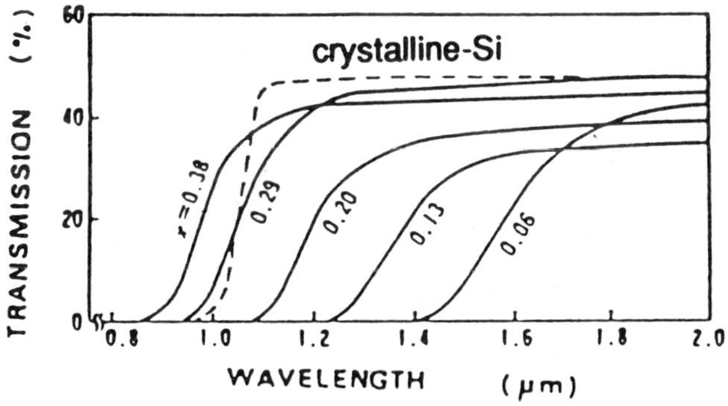

Figure 4. *Optical transmission spectra in the vicinity of fundamental absorption edge ao several Si-As-Te glasses 350 μm thick at room temperature*

controllability has also been identified from optical absorption spectra as shown in Figure 4. Figure 5 shows a summary of atomic and electric properties of the material system. An interesting feature of the summarized data is that both equi-contures of the specific gravity, glass transition temperature and energy gaps show similar shape of patterns.

This indicates that the Si-As-Te amorphous semiconductor would be a suitable system to explore the compositional dependence of the atomic and electronic propertis in the random bonding network in solids.

On the other hand, from an applied physics point of view, this material system could be applied to multi-layered heterostructure devices and also optoelectronic functional elements in a wide spectral region from near-infrared to visible light region. Therefore it can be called a new type of synthetic semiconductor. In addition, because of the amorphous network, the material can be deposited non-epitaxially on any inexpensive substrates such as glass, ceramics and metals. These advantages might be fairly well matching with the technological requirements for good massproducibility with low cost device fabrication processes in the future.

In spite of these significant advantages in the Si-As-Te amorphous system, there exists a substatial barrier impeding the wide utilization of this material, that is, an essential difficulty in the material fabrication due to large differences in the weight density and melting point of individual elements for alloying composition as shown in Table 1. Homogeneous materials which would have good electronic properties can never be formed in the presence of the terrestrial gravity.

Table 1. *Physical constants of Si, As and Te elements*

	Si	As	Te
melting point (°C)	1412	817	449
weight density (g/cm³)	2.33	5.72	6.24

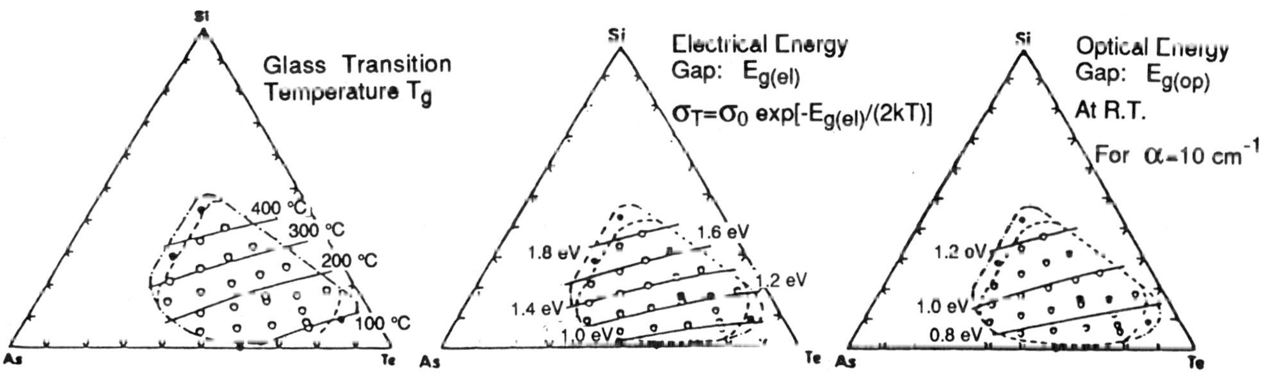

Figure 5. *Compositions of synthesized glasses and glass-forming region in the Si-As-Te system. Cirles express those quenched by cooled air, dots by cooled water.*

Focusing on these advantages in this material system, a series of preliminary experiments has been made in the terrestrial environment since 1974 [2], and also the TT-500 Rocket experiment in 1980 [3]. On the basis of these data, a proposal has been made to join the FMPT program.

MATERIAL PREPARATION FOR SPACELAB J

Five different specimens having compositions $Si_x(As_2Te_3)_{1-x}$ in which x is varied from 0.25 to 0.75 were selected and prepared on the basis of the terrestrial experimental data. For the alloy materials of these compositions, a systematic experimental data have been accumulated on structural, electrical and optical properties. These data, terrestrial and rocket experimental, will serve as the reference for those of materials processed in space. In order to examine the possibility of valency electron controllability, a trial has also been made separately for the effect of impurity doping with Ni element of several atomic % on $S_9As_{14}Te_{21}$ sample, for which the electronic properties have been most intensively investigated in this research.

The starting Si-As-Te material were prepared in the following way. Each of Si, As and Te raw materials (and optional Ni) of purity better than 99.999% was ground down to 100 mesh powder, weighed with an accuracy of 0.1 mg to a total amount of about 50 g and sealed up into a fused-quartz ampoule at a pressure below 10^{-5} Torr. The mixtures sealed in the ampoules of a diameter 17 mm were heated at 1200 °C continuously for 6-8 hours by the RF heating method. Self-agitating pinch effects in the molten zone by the electro-magnetic force ensures a good reaction of mixtures in a relatively short time. After that, a rapid air-quenching was accomplished by quickly removing the ampoule from the furnace, and 17 mm diameter Si-As-Te ingots were obtained.

The atomic fractions of constituent Si, As and Te elements are listed in Table 2. The pre-synthesation procedure here adopted is thought to be essential to avoid anticipated break-down of the material sealing due to a rapid increase in vapor pressure, particularly of As, occurring upon an accidental extreme overheatings.

The processed 100 mesh powder was sealed (total amount 1g) two-fold with fused quartz ampoules as shown in Figure 6 (a), and encapsulated in the Ta

Table 2. *Atomic fraction of Si-As-Te (Ni) amorphous semiconductors for FMPT experiments*

FMPT No.	LAB No.	Atomic Fraction			
		Si	As	Te	Ni
1	20	1	6	9	0
2	21	3	8	12	0
3	22	9	14	21	0
4	23	2	2	3	0
5	24	3	2	3	0
6	22*	9	14	21	3.6

*Ni 5wt.%, 7.6at.% doped

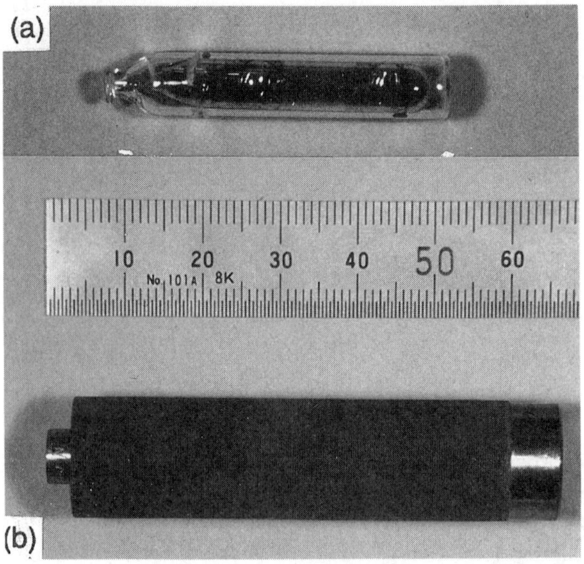

Figure 6. *Appearance of two-fold quartz ampoule (a) and Ta cartridge (b) for FMPT*

universal cartridge shown in Figure 6 (b) for the CHF (Continuous Heating Furnace) apparatus shown in Figure 7. Particular attention has been paid to the completeness of the quartz ampoule vacuum sealing, since constituent elements As and Te are of highly toxicity. Even if only a small fraction of these elements escapes from the vacuum crucible into the Spacelab., it tends to immediately exceed estimated SMAC (Space-lab. Maximum Allowable Content) values. In the course of pre-flight experiments, various sealing materials have been examined from the view points of mechanical strength and chemical stability. Carbon may be one promising candidate, however, it would be difficult to make complete sealing for vapor. Although Ta, which is used for the universal cartridge, has a sufficient mechanical strength under the As vapor pressure even at 1350 °C, it has been turned out to be easily eroded by the molten Si-As-Te.

EXPERIMENTAL RESULTS

All the scheduled material processings were performed safely and effectively during the allotted 6 hours; from 21:49, 13th to 3:45, 14th, September, 1992. The processed samples, which reached us at the end of November, 1992, were removed from the Ta cartridge.

The Si-As-Te samples were sliced into disks of thickness 0.2-0.3 mm, of which the surfaces were polished with 0.3 micron diamond powder and washed well in alcohol. These disk samples were employed for the measurements of optical properties as well as structure analysis. For the electrical measurements, Al coplaner electrodes of a 0.5 mm spacing were formed by vacuum evaporation.

Figures 8 shows plots of the absorption coefficient spectra of five undoped $Si_x(As_2Te_3)_{1-x}$ samples. The absorption coefficient of amorphous semiconductors near the fundamental edge are usually represented by [4]

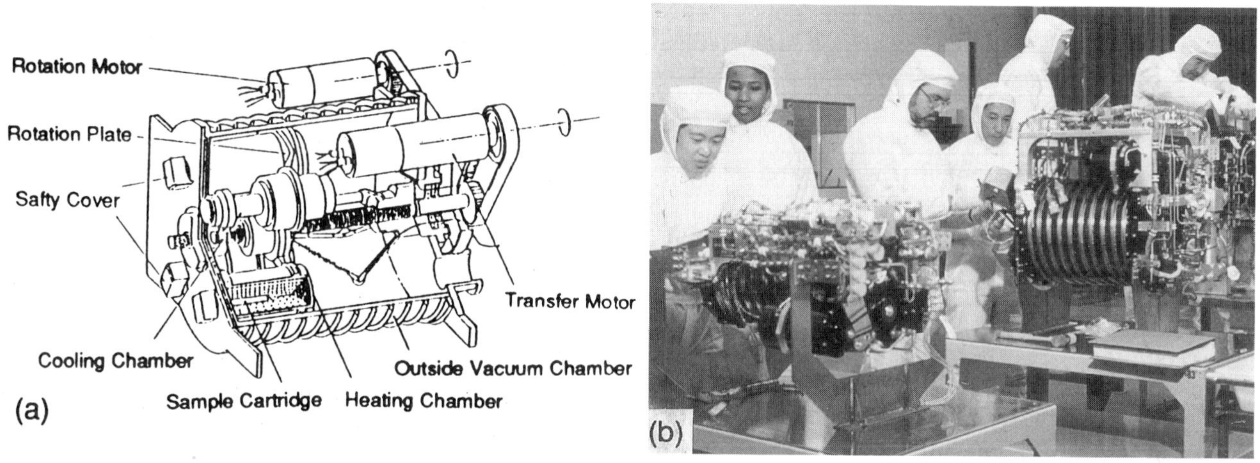

Figure 7. *Schematic diagram of CHF construction used for Si-As-Te amorphous semiconductor processing (a) and appearance of CHF (b)*

$$\alpha h\omega \propto (h\omega - E_o)^2 \qquad (1)$$

for high absorption region,

$$\alpha \propto \exp[\, h\omega/E_u] \qquad (2)$$

for low absorption region,
where E_0 defines the optical energy gap, and E_U denotes the Urbach energy.

The electrical conductivity of FMPT materials is shown in Figure 9 in the form of the Arrenius plot, comparing with that of terrestrial counterparts in Figure 7(b). Although he range of measurement temperature is limited, a clear activated conduction is confirmed for both

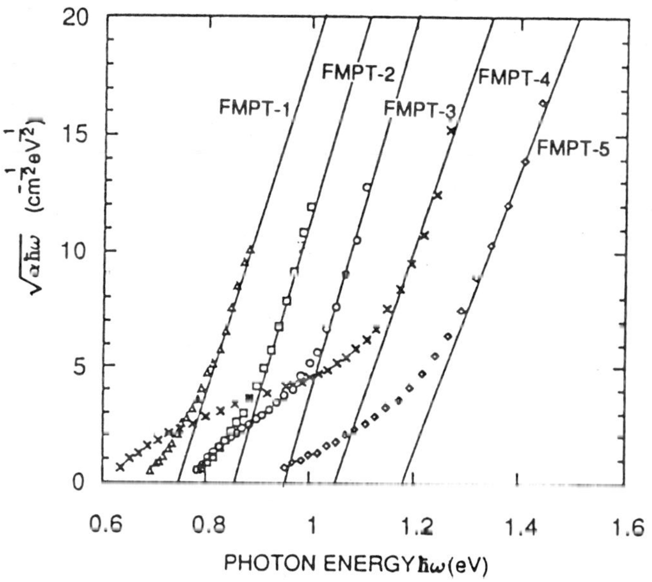

Figure 8. *Otptical absorption spectra of FMPT produced samples which shows a continuous energy gap controllability with the atomic contents*

the materials, which indicates that the band conduction of thermally activated carriers dominates the electrical conduction. The electrical conductivity of semiconductors is generally expressed by [4]

$$\sigma_{dc}(T) = \sigma_o \gamma \exp[- \Delta\varepsilon_\sigma/k_B T] \qquad (3)$$

where σ_0 and $\Delta\varepsilon_\sigma$ define the conductivity-prefactor and conductivity activation energy, while γ denotes correction factor due to temperature dependences of the energy band gap, Fermi level as well as transport energy. It has been shown from the detailed analysis an this data by the carrier transport prefactor [4-5] that upon assuming material insensitive g factor, the conductivity-prefactor s_0 should be larger in the FMPT materials. Since the conductivity-prefactor s_0 is determined primarily by the carrier mean free path [4], the result

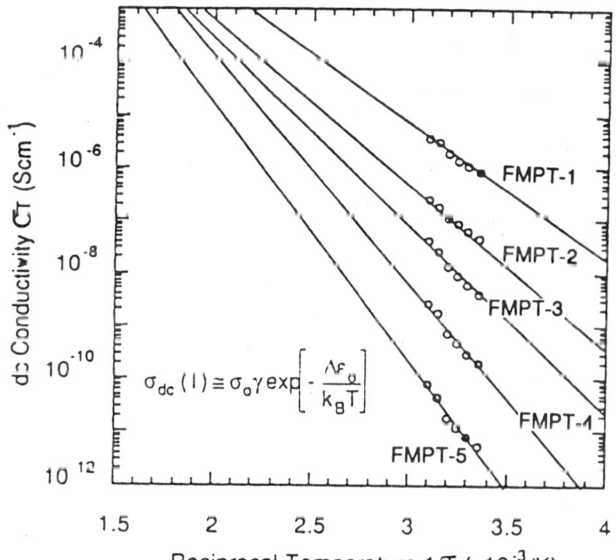

Figure 9. *Temperature dependence of d.c. conductivity*

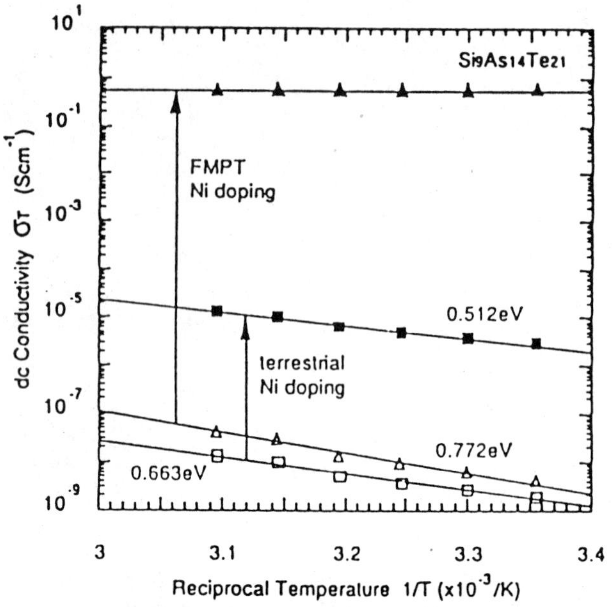

Figure 10. *Temperature dependence of dc electrical conductivity for Ni doped (4.6 at.%) FMPT and terrestrial materials.*

would lead us to conclude an extended carrier mean free path and thereby a greater mobility in FMPT materials.

Demonstrated in Figyre 10 are the effects of Ni doping on the electrical conductivity of $Si_9As_{14}Te_{21}$. A dramatic enhancement of the electrical conductivity, more than 7 orders of magnitude, is found for the FMPT material, whit an increase of only 3 orders on terrestrial material. Also found is almost zero activation energy for the doped FMPT material, which contrasts highly with a minor (25%) decrease in the activation energy upon doping in the terrestrial material. In both cases, a change in the conduction type from p-type (hole in undoped material) to n-type (electron in Ni doped material) is confirmed by means of thermoelectric power measurement. The great extent of the effect of doping on electrical conductivity, has never been observed for melt-quenched chalcogenide glass systems.

This high doping effiency in the FMPT sample implies either a lower native defect density or a greater doping efficiency of Ni additives.More detailed analyses and theoretical considerations on this item are discussed in our previous paper [8].

All the experimental results suggest the realization of homogeneous amorphous network structure with a minimized disorder by the material processing in a microgravity environment. The material processing in space would contribute largely to the physics of disordered materials as well as to the development of new materials for semiconductor electronics.

A new trial of the effect of high pressure and other advanced data on high electric field effect etc. are now in progress, and will be reported in the near future.

BIO-PRODUCTIVE SOLAR PHOTOVOLTAIC GENERATION IN SPACE

The development of a new technology for thin film solar cells might be one of the most direct approaches to the cost reduction of photovoltaic systems from the point of view of material saving and automated assembly in cell production. Various kinds of semiconductor thin film preparation technology such as chemical vapor deposition, liquid phase epitaxy, plasma decomposition, Molecular Beam Epitaxy (MBE) and the cluster ion beam process have now been intensively investigated for various semiconductor on low cost substrate materials by many researchers. Among these technologies, MBE is one of the best crystal growth techniques for semiconductor thin films; it provides in particular precise controllability of both the stoichiometry of the compound semiconductor and the thickness of the grown layer. By utilizing these advantages, even laser diodes of the AlGaAs/GaAs double heterostructure have been fabricated by MBE technology [9]. Fortunately the space around a space satellite power station (SSPS) having a geostationary Earth orbit has a vacuum level of 10^{-14} Torr, and the corresponding mean free path of atoms is estimated to be more than (7-10) x 10^5 km.This means that it might be possible to fabricate a solar cell even as large as 1 km^2 using a giant MBE system [10].

For the epitaxial growth of semiconductors, a single-crystal substrate is generally required. This is an enormous technological problem for the fabrication of large-area solar cells. One possible solution is the

Figure 11. *Schematic illustration of a fabrication method of large area substrate for MBGE technology*

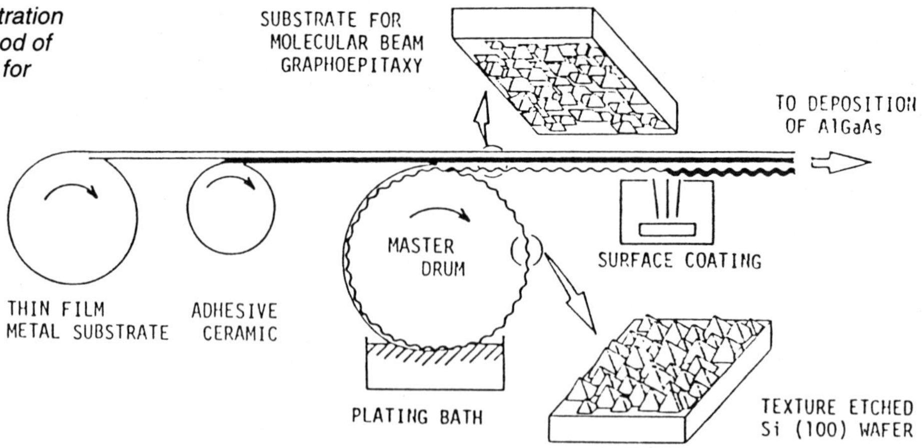

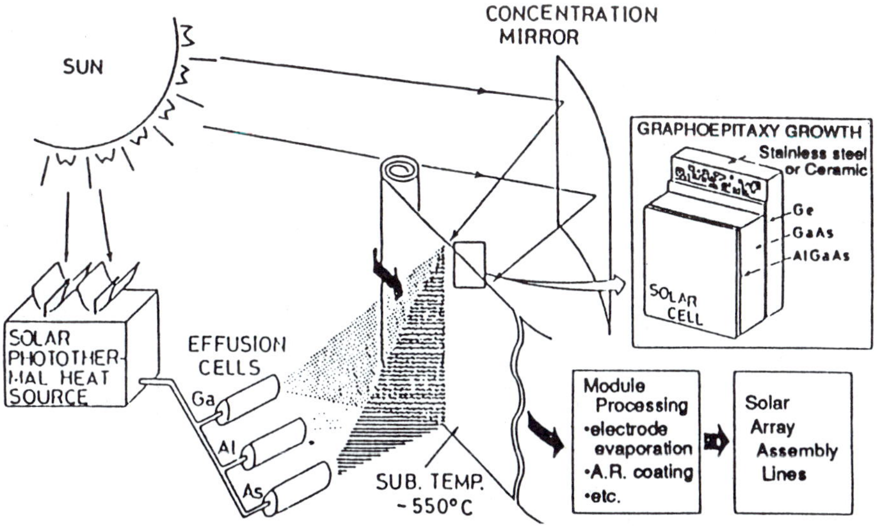

Figure 12. *Concept of large scale molecular beam graphoepitaxy system for low cost solar cells in space*

application of Grapho-Epitaxy technology [11]. The microgroove pattern on the substrate surface defines the crystal orientation of the growth layer. The principle of single-crystal growth is based on the Bridgeman method of crystal growth. For large-scale substrates, this technology can be applied to the molecular beam epitaxial growth technique and combined with an annealing process such as laser annealing. Figure 11 shows a schematic illustration of the fabrication method and the structure of the substrate. The fabrication technique of this substrate will be reported elsewhere [12]. A stainless steel film substrate having a replica pattern of micro pyramids from the textured surface of silicon might be useful for a very large solar cell. Figure 12 is a schematic diagram of the large-scale MBGE system for low cost solar cells in space. The proposed system might be applicable to AlGaAs/GaAs solar cell because of a GaAs film only a few micrometer thick is sufficient to absorb the solar energy and it also has an excellent efficiency.

The advantages of the proposed system would be as follows.

(i) The fabrication of *solar cells of very large area (even 1 km^2)* is easily attained.

(ii) *A very high fabrication yield* is expected because the substrate is free from contamination in the highvacuum environment (this is particularly important for thin film devices)

(iii) *A long life and a high reliability* should be achieved because there is no moisture or oxygen in space; the only possibility is radiation damage.

(iv) There should be *no pollution problem*. In terrestrial fabrication, the handling of GaAs, CdS and CdTe necessitates apparatus to protect these materials from pollution and this is very expensive.

Figure 13 shows a result of cost analysis of an SSPS, comparing the prototype 5000MW station by P. E. Glaser [13] and that with present technology[14]. As can be seen from this figure, the largest contributing factor to the costs is the space transportation fee which corresponds to 43.5%; the second largest is the cost of the solar cells which is 23.7%. Thus the cost of the solar cells and the transportation fee make up more than 67% of the total system cost. This means that the fabrication of a very large thin film solar cell by the MBGE technique might contribute greatly to reducing the total cost of the SSPS. For example, from a direct comparison of the power-to-mass ratio of an AlGaAs/GaAs solar cell (2.4mm thick substrate) with that of the very thin silicon solar cell (50mm thick substrate) developed by the OAST, NASA for use in space, it can be seen that the AlGaAs/GaAs MBGE cell is more than one order of magnitude less expensive.

Apart from the detailed feasibility analyses which must of course be even more specific in near future, one unique point of the proposed system is the multiplication of the electric power by the continuous solar energy supply as long as the ratio of the system operational life to the energy pay-back time is greater than unity. Solar photovoltaic conversion systems provide significant quantities of electric energy from an inexhaustible resource which is free of charge and also have nomechanically moving parts. Moreover, solar cells are made up of semiconductor modules which can be mass produced very well like transistors or integrated circuits. Therefore, if the operational life of the MBGE thin film cell exceeds its energy pay-back time in space, any size of small-scale system can be used initially and the system can be expanded gradually by the positive feedback of electric energy. The philosophy of the concept is similar to that of biological growth in living organisms, and the energy pay-back time of AlGaAs/GaAs MBGE thin film cells is estimated to be less than 2 years (which does not include the substrates and assembly of the array system). However, these peripheral costs might not exceed the solar cell cost. In contrast with this, the operational life of this system is assumed to be more than 15-20 years. In the silicon OAST cell, an operational life of 30 years was assumed [15]. Therefore the proposed system might be flexible enough to be used as an accelerating component in the SSPS system. Figure 14 is a block diagram illustration of the "bio-productive" electric power generation system proposed here. The system should work as an eternal energy resource for mankind in the future.

The major advantages of the proposed system are summarized as follows.

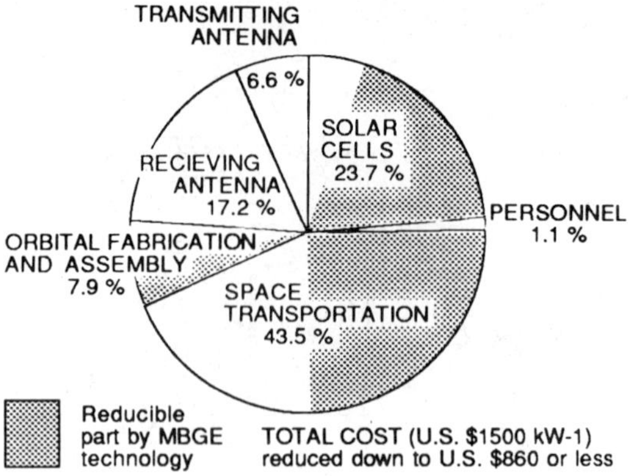

Figure 13. *Cost analysis of an SSPS for a 5000 MW prototype system and improvement by MBGE technology*

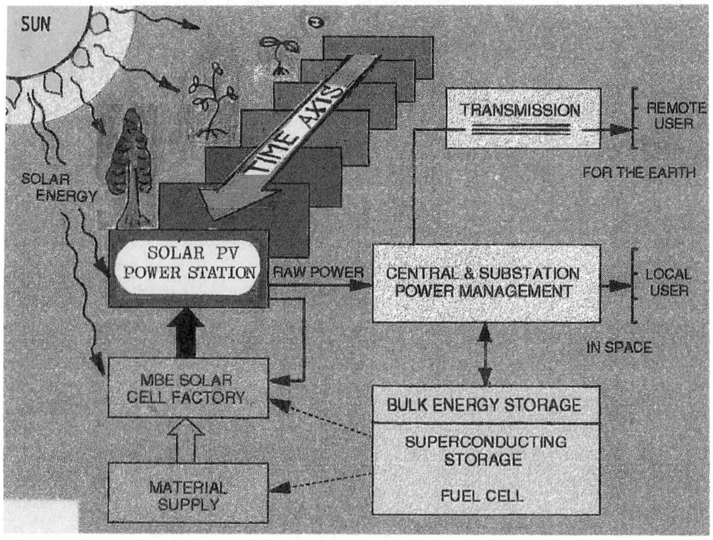

Figure 14. *Concept of a solar "bio-productive" electric power generation system*

(1) *Solar "bio-productive" power generation:* the energy pay-back time is less than 2 years which is much less than the life of the system.

(2) *No critical feasible power scale:* any size of power station can be used initially and the power scale can be expanded gradually, because of the semiconductor modular unit which can be mass produced well and because of the large scale merit.

(3) *High power-to-mass ratio:* a value of around 20W/g or more is achieved for the AlGaAs/GaAs cell as against 1 W/g for the silicon OAST cell.

(4) *Savings in semiconductor material and low cost:* a very large area and a thin film (2.5mm for the AlGaAs cell as against 50-100 mm for the silicon OAST cell) means lower costs.

(5) *Reduction in space transportation and assembly expenses:* thin films and automated solar cell fabrication in space with good mass reproducibility also reduce costs.

(6) *No surface contamination and freedom from pollution problems:* because of the ultrahigh vacuum in space, the materials are free of surface passivation.

In the future, more detailed assessments of the proposed system should be undertaken in the fields of physics and engineering and also the global-scale politics of the system should be considered. However, the system possibly one of the most promising eternal energy resources for future civilization in the world.

ACKNOWLEDGEMENTS

Author expresses his sincere thanks to the flight crews of Spacelab-J., particularly Payload Specialist Dr. M. Mori for their tremendous efforts for successing the FMPT. The authors are also much obliged to experimental supporting group in NASDA, Mr. A. Kubozono, President of JAMSS and Ishikawajima Harima Heavy Industries Co., Ltd. (IHI) for their asisstance of experimental apparatus building and set-up for Spacelab.

To perform the presented work, many of my colleagues, Prof. H. Takakura at Toyama Prefectural University, Prof. H. Okamoto, Dr. K. Hattori, Mr. C. Sada, Mr. W. S. Kolahi, Mr. M. Hambach at Osaka University, have been joined. Author wishes to express his sincere thanks for their enthusiastic assistance.

REFERENCES

[1] M. Nunoshita, H.Arai, Y. Hamakawa and T. Fujimoto: Proc. 10th International Congress on Glass, Kyoto Japan (1974) 7-37.
[2] Y. Hamakawa: *Ceramics*, **22** (1987) 277 [in Japanese]
[3] Dr. M. Mori's Conversation at Summary Report on Science & Results of Fuwatto '92 Space Experiment, NASDA, Dec. 6-7, 1993.
[4] H. Okamoto: "*Amorphous Silicon"* ed. by K. Tanaka (Ohm-Sha, 1992), Chapt.4, pp.101-188.
[5] H. Okamoto, K. Hattori and Y. Hamakawa: *J. Non-Cryst. Solids*, **137&138** (1991) 627.
[6] H. Okamoto and Y. Hamakawa: *J. Non-Cryst. Solids*, **33** (1979) 225.
[7] N.Rysava & L.Tichy J.: *mat. Sci.*, **27** (1992) 5209.
[8] Y.Hamakawa, W.Shams-Kolahi, K.Hattori, C.Sada and H.Okamoto: *Journal of Japan Association of Microgravity,* to be published.
[9] A. Y. Cho, R. W. Dixon, H. C. Casey, Jr., and R. L. Hartman: *Appl. Phys. Lett.*, **28** (1976) 501.
[10] *"Prometheus in Space"* edited by Y.Hamakawa, A survey report of the Research Comitte on Functional New Material, Japan Utilization Promotion Center, Tokyo (1993) March.
[11] M. W. Geis, D. C. Flanders and H. I. Smith: *Appl. Phys. Lett.*, **35** (1979) 71.
[12] Y. Hamakawa and H. Takakura: *Acta Astronautica*, **14** (1986) 439.
[13] P. E. Glaser: Proc. Joint U.S.A.-Canada Int. Solar Energy Society Conf., Winnipeg, 1976, 1.
[14] Y. Hamakawa: Proc. 31th Congr. International Astronautical Federation, Tokyo, 1980, G-309.
[15] Comparative assessment of orbital and terrestrial central power system, *JPL Program Review*, 1976.

EUROPEAN STRATEGIES IN PHOTOVOLTAICS

Wolfgang Palz
Renewable Energies Program Manager
European Commission
200, Wetstraat, 1049 Brussels, Belgium

ABSTRACT

The Council of Ministers of the European Union has just approved the new energy program for the period 1995-98. The budget allocated to R&D on photovoltaics within that program is estimated at 40 to 50 millions ECU. The European Commission is entrusted with the implementation of that program.

The program is oriented along 2 lines : one is the technological development of photovoltaic cells, including upscaling of manufacturing processes and systems. A new sector is added, called integration. It concerns the promotion of photovoltaics in the building sector, municipalities and regions. Particular emphasis will be given to the Power for the World concept.

INTRODUCTION

The European photovoltaic programs are drafted and executed by the European Commission in Brussels. The Commission is the executive body of the European Union. It proposes all decisions for adoption by the Council of Ministers and the European Parliament. The programs, once decided, are implemented in the 12 European Union member countries (from the beginning of 1995 : 15 member countries).

The European Union program is executed through cost-shared contracts between the Commission and transnational networks of industries, universities and research centers. The Commission's own Joint Research Center, namely its center at Ispra in Italy, with its extensive PV test facilities will also qualify to participate in the next program; this had not been possible in the past. The European Union member country governments have a leading role in the implementation of the program through the representation in a central program committee in Brussels which oversees program outline and selection of contracts.

Participation of entities in non-European Union member countries in the photovoltaic and other programs has been agreed for almost all West European countries (European Economic Area countries) and is being negotiated with Switzerland and Israel. Participation without financial contribution is also possible for entities from Eastern Europe including Russia and for Canada and the 3rd World Countries. In these cases, separate funding arrangements are possible for certain countries. The current photovoltaic program already includes a certain number of contracts with entities in Eastern Europe (the PECO countries) on the subject of research and development in photovoltaics.

PARTICULARITIES OF THE EUROPEAN PHOTOVOLTAIC PROGRAM

The European program is not the only source of public financing in photovoltaics in Europe. Some member countries have their own national programs, the biggest of them being the German one. But regions and cities are also improving their financing in this field. The particularity of the European Union program is twofold: firstly by nature it is broadly international. It is a program of excellence involving, through widely publicized calls for proposals, the best European competences and activities. Also, it is a common link between all national programs and activities in the European Union member countries and for PV research and development joined national and European funding are possible for the same project. Secondly, it is an integral part of an overall renewable energy research, development and demonstration strategy, which in turn is linked to the political rationale of security of European energy supply and improvement of the environment.

BUDGETS

Current photovoltaic contracts are part of the JOULE II energy research and development program 1992-94. Most of these contracts were concluded in 1992 and run until the end of 1995. This year, there was an additional funding as part of a so-called APAS. Table 1 gives an overview.

The photovoltaic budget is currently 22% of the renewable energy budget. Originally, that renewable energy budget for the JOULE II period 1992-94 was approximately 57 million ECUs (1 ECU : European Currency Unit = 1.28 US$ in November 1994). However, through late additions, that JOULE II budget was increased to 141 million ECUs. This

First WCPEC; Dec. 5–9, 1994; Hawaii

was due to political recognition by the European Union that renewable energies deserve much greater attention in an overall energy and environment policy.

Photovoltaic R&D Projects

	Number of contracts	EU Contribution (Mio ECU)	Number of participants
Joule II (1992)	34	20.4	158
Joule II (1993)	12	4.1	61
APAS (1994)	14	5.3	71
PECO (93 - 94)	11	1.4	27
Total	**71**	**31.2**	**317**

Table 1. Photovoltaic R&D budgets 1992-94

A new JOULE-THERMIE program and budget, extending until the end of 1998, was finally adopted by the Council of Ministers in November 1994. The renewable energy research and development budget for contracts (excluding administration and other expenses) for 1995-98 is about 220 million ECU. This is a four fold increase with respect to the original renewable energy budget of JOULE II. However, thanks to the additions which were included previously in JOULE II, the actual increase between the funding period 1992-94 and that of 1995-98 will only be 50%. The Photovoltaic budget within the renewable energy program 1995-98 is presently not yet defined but it is expected to fall also in the range of 20 to 25% of the total.

There is a separate budget for demonstration which will be 134 million ECU for 1995-98. This is approximately the same amount as in the previous period without any real increase. The photovoltaic part will also be in the range of 20 to 25% here.

Strategies

The photovoltaic Research and Development program will be implemented in the frame of two calls for proposals in the energy sector. One on December 15th, 1994 and another in 1996. The program, as decided by the Council follows two strategic lines. Firstly, the techno-economic development of technological options and secondly the development of **Market Integration** as part of the establishment of **Road Maps** for the implementation of renewable energies in Europe's and the world's future energy systems.

An outline of the technological development targets in the program is given in Table 2.

As in the past, the European program will give equal importance to solar cells and systems	
Cells	- All forms of silicon - CIS and CdTe - Scaling up of manufacturing processes (up to 500 MW p.a.) - Module technology
Systems	- Emphasis on standardized small scale PV appliances and systems - Phasing out of systems above 100 Kw - Renewal of concentration systems

Table 2. Technology development

In the solar cell area, main emphasis will be on mono- and polycrystalline silicon; for thin films the three options CIS, a-Si and CdTe will be pursued. The development objective is cost reduction through technical improvements. A new item in this program is the development of manufacturing processes. Taking into account the target of the "Power for the World" program for wide dissemination of photovoltaics in the Third World, production lines of up to 500 MW are considered in this context. Some effort will be done to reduce cost of assembling and encapsulation of solar cells which is currently excessively high.

For systems, the emphasis will be on small standardized photovoltaic appliances and stand alone systems. Grid connected systems will be considered mainly for building applications while large arrays above 100 kW will be no more a priority. A new effort - though modest -will be started on concentration systems.

A new strategic item is integration. An outline is given on Table 3. The renewable energy program 1995-98 has a new program sector on this subject, which includes photovoltaics together with other forms of solar energy applications, biomass, wind and water.

- PV as part of renewable energy integration scenarios and plans for Europe, its countries, regions, cities and towns, utilities, professions, authorities etc.
- PV integration in urban development and buildings
- Power for the World

Table 3. Development of integration

The objective is to develop scenarios for future large scale integration in Europe as a whole, its member countries, the regions and municipalities. The main operator of energy systems, such as utilities, professions and responsible authorities will be involved in the program. The "Power for the World" concept [1] is part of this program sector.

Table 4 gives a summary of the various initiatives for integration which have been taken in the frame of the previous and current renewable energy programs of the European Commission. They will also play a role in the implementation of the next program. A number of professional networks targeted towards the furthering of renewable energies as a whole have been set up. They group in each category the leaders of some relevant professions throughout Europe. All groups together and the European industry in renewable energy and their associations are forming the European Solar Council also called Club of Paris, which was created in 1993.

All the networks are currently already being funded for establishing more detailed policies for renewable energy technology integration in the various economic sectors and society at large.

- The **European Solar Council** - Club of Paris, chaired by W. Palz, in
  cooperation with EUROSOLAR, the political association chaired by Dr.Scheer, MP, and the renewable energy forum of Madrid chaired by Dr. Robles Picquer, MEP

- **Professional networks** (members of the European Solar Council)

  EUREC-Agency, chaired by Baron van Overstraeten, Belgium, representing Europe's leading research centers in renewable energies

  READ, the architects network, chaired by Sir Norman Foster, UK, with the participation of R. Piano, R. Rogers, Th. Herzog and N. Kaiser

  EURE, the utilities network, chaired by Dr. H. Hoffmann of RWE in Germany

  CERE, the cities network, chaired by Prof. W. Leonhardt, Energy Minister in the Saarland, Germany

  EURENET, the new network for renewable energies in regions, chaired by Dr. Scheer, MP, Bonn

- The **European PV conferences** under the European Commission's own
  steering

Table 4. Instruments for Integration and Promotion of Renewable Energies

Conclusions

Efforts for photovoltaics as part of a broad renewable energy strategy in Europe are currently being increased. In comparison to the equivalent previous period, the budget of the European Union will be increased by 50% over the next 4 years. Also the scope of work will go much beyond the strict technology development and will embrace new targeted involvement of the leading European professions. **Power for the World** will be a cornerstone of the future European R&D efforts on photovoltaics.

References

[1] *Int. J. Solar Energy*, 1994, Vol. 14, pp. 231-237

"European PV Industry - The Future Beckons."

Mr. D.J. Shah
MD & CEO BP Solar International &
Chairman of the European Photovoltaic Industry Association (EPIA)

Good Morning, ladies and gentleman. I expect many of you will recall that brilliant childhood book, "Alice's Adventures in Wonderland", in which the Red Queen introduces a hungry Alice to the concept of "jam tomorrow". And, of course, tomorrow never comes. "Jam tomorrow" could very well have been the motto of the solar industry over the last two decades. A golden future of growth and of industry viability seems always to be tantalisingly just around the corner. "If only $/W$_p$ costs could be reduced." "If only Governments would spend more on research and demonstration programmes". "If only the utility companies would make a major investment in photovoltaics". "If....."

So you could be forgiven for an element of cynicism if I were to paint a rosy picture of the future to you. And yet, there are reasons to believe that we are indeed poised on the brink of rapid commercial growth, leading to broadly-based industry viability. And these have less to do with ambitious cost reduction targets, however seductive, than with the mechanisms of getting the product to market and the wider ambitions of society for alternative and cleaner forms of energy.

What role has the European industry to play in helping to bring about this better future? As, in looking at where we hope to go, it is usually helpful to take stock of where we actually <u>are</u>, I will begin, with a brief look at the European industry today.

Slide 1 : PV in Europe - "The state of the union".

If I may borrow from President Roosevelt, "the state of the PV industry within the European Union is good". Much of the developed world is slowly emerging from the worst recession it has experienced since the 1930s. Europe has been far from immune from the effects of the recession and yet its PV industry is emerging in robust health. Capacity utilisation within the industry is now above 75%. Rationalisation, notably in Germany, has produced stronger and better resourced companies. Over the last four years, despite the recession, European manufacturers have increased their share of the global market from 22% to 25%.

Slide 2 : Regional Shipment Growth 1990-1994.

Why has the European PV industry emerged relatively unscathed? In part it is due to the strength of the national programmes adopted by the member states.

Slide 3 : Europe - National Programmes.

Some of these are summarised here.
And these are supplemented by the policies of the European Union (EU).

Slide 4 : European Union PV Programmes.

The next EU four-year programme (the so called "4th Framework"), which runs from 1995-1998, will contain both research and development programmes and further "Thermie" Energy Demonstration projects. Previous Themie PV projects include the roof top power projects, under construction at Amsterdam in Holland and at the Gelsenkirchen Science Park in Germany, and also the 40kW Building Integration project that has just been completed at Newcastle University in the UK.

Slide 5 : Newcastle University.

In addition to the 4th Framework programme, the EU has developed a comprehensive programme ("Altener"), which calls for doubling the contribution from renewable energy sources, by the year 2005. This includes 500 kW of installed PV energy generating capacity by 2005 - an enormous increase over current levels of PV-generated power.

It would be foolish, of course to assume that either the individual National or EU programmes will be implemented in full. Budgetary constraints, alternative priorities and the difficulty of implementing such ambitious programmes within the planned time-scales will ensure that achievement falls short of plan. Nonetheless, such plans are an important demonstration of intent and form a useful pointer to the future.

Let me return to the question I asked at the outset - what role can the Europeans play in the future development of a large viable photovoltaics industry? For the Developed World, photovoltaics have a unique advantage over other forms of renewable energy

Slide 6 :PV - the only renewable capable of mass application in an urban environment.

Let us look now at some of the market sectors in which Europe can play a significant role in developing the future.

Slide 7 : Europe - Key PV Initiatives.

Europe can fairly claim to have established world leadership in the integration of photovoltaic power into the building infrastructure. The inclination of the sun in

CH3365-4/94/0000-0012 $4.00 © 1994 IEEE

northern latitudes and the shortage of land for central PV generating plants provides an incentive to turn south facing sides of buildings into sites for PV electricity generation. Many of the most exciting developments may be found in Switzerland and Germany.

Slide 8 : 10 kW roof integrated, Geneva Switzerland.

Slide 9 : amorphous silicon window, Head Office of Stawag electricity utility ,Aachen, Germany.

Slide10 : 18 kW sunshade ,Kirchberg Switzerland.

For architectural use, the visual appearance of the solar panels will be extremely important and the colour possibilities that can be achieved will become of increasing
importance. These coloured modules have been developed under the EU "Eurofacade" programme.

Slide 11 : BP Solar coloured Saturn modules.

Architects are excited by the possibilities conferred by the steel blue and in particular the gold cells.

Roof top power, installed on either commercial or industrial premises, is another application offering huge potential for the future,

Slide 12 : - Model of Gelsenkirchen science park

or for domestic use

Slide 13 : - 3 kW Saturn array (+ solar thermal) on house, Switzerland.

The spread of rate based incentives, begun in Switzerland at Burgdorf in 1991, offers a significant pointer to the future development of PV applications in Europe. The concept of rewarding private individuals for installing PV systems, through a premium rate for each kilowatt-hour of solar power returned to the grid, has now been adopted by five German cities. A number of others are actively considering the concept. Key advantages of the concept include :

- breaking the dependence of the PV industry on Government funding (which is always subject to over-riding factors, such as budget constraints, changes in Government etc..)
- the creation of long term, sustainable, markets through consistent funding.
- the establishment of a climate of confidence. This will encourage manufacturers to increase production capacity, creating a "virtuous circle" of cost reduction, through process automation, economies of scale etc.

In southern Europe, particularly where land space is more readily available, we will see the further development of central power stations, of the type pioneered by ENEL of Italy. Recently, they opened a giant 3 MW station at Serre, some 50 miles south-east of Naples.

Among the more unusual applications of central power is this 24 kW array outside the Swiss parliament building in Berne. This is used to provide power for a funicular railway.

All these applications have been in the developed world. We should not neglect the needs of those less fortunate than ourselves. We hope that the EU will be able to sustain, and indeed increase, the aid programmes that have allowed systems such as this microwave repeater station for a telephone system in Sierra Leone.

Slide 14 : - Microwave repeater station, Sierra Leone.

This shows one of the 15 arrays.

We have seen some of the market sectors which will come into play in building the European and indeed the global PV business of the future. What else should we be doing to ensure that a viable industry is more than a mere pipe dream?

Slide 15 : - Europe - The Future. What should we be doing?

Firstly, we should encourage the use of PV as part of a policy to reduce environmental damage resulting from energy use. This could include regulations that promote energy efficiency in buildings and encourage PV generating capacity to be integrated into their structures. Today, it is calculated that the installation of PV on the roof and south facing walls of an office building could generate some 10%-15% of its energy needs. Within a decade, this could rise to as much as 40%, due not only to advances in PV technology, but also to increased sophistication in its use and improved building design, leading to much lower energy requirements.
Legislation should be extended throughout Europe, requiring the electricity distribution companies to permit PV generating capacity to be connected to the grid via two-way metering. The wide-spread use of rate based incentives, of the type pioneered at local level in Switzerland and Germany, should be encouraged.
Freeing Governments from the burden of funding large procurement programmes would allow them to focus on funding R&D into future technologies, on demonstration projects and to increase Overseas aid programmes, allowing the benefits conferred by PV to be enjoyed more widely.
Finally, education. The solar world is small (if not always cosy!). We tend to assume that all are as familiar as we are with the technology and benefits of photovoltaics. They are not. If PV is to take a significant share of the *renewable* energy - and still more the *total* energy - market, it will need to achieve a much higher profile. Architects, designers, local authorities and the general public will all have to be made aware of the

possibilities conferred by PV power. In this, we all have a role to play.

Which brings me neatly to the last point that I wish to make today.

Slide 16: - Europe - The Future - The Global Village.

We do not live in isolation in Europe. Improved transport and the ever increasing sophistication in communications is shrinking the world. National boundaries are becoming of decreasing significance. And naturally, this is reflected in our own industry.
The rationalisation of the photovoltaic industry in Europe, to which I alluded earlier, has not been entirely introspective. Its effects have been felt more widely. Overseas investment is rendering traditional concepts of national ownership and identity less meaningful. Photovoltaic companies are becoming increasingly multi-national and indeed multi-cultural.
Moreover, I suspect that there is probably no business serving such a diverse customer base as the photovoltaic industry. From Washington to the Andes, from Africa to the Pacific, photovoltaics are bringing a quiet revolution to the quality of life.

And this is appropriate. After all, the resource is universal.

In the European PV industry we reach out in recognition of the global village. We ask you to work with us in building a better future for our children.

It is our belief that, with the co-operation of the manufacturers, the EU, National and local governments, we can improve the quality of life for *many* people and the quality of the environment for *everybody*. However, we are by no means adverse to a little help from above. Perhaps this solar powered church will help us with our endeavours!

Slide 17: - PV Powered Church, Steckborn Switzerland.

END

THIN FILM PHOTOVOLTAICS - STRATEGY OF EUREC AGENCY

W.H. Bloss
Institut für Physikalische Elektronik, Universität Stuttgart
Pfaffenwaldring 47, D-70569 Stuttgart, Germany

ABSTRACT

European activities in the field of thin film photovoltaics are coordinated in a network by Eurec Agency. Main emphasis lies in the development of an appropriate production technology of CIS and CdTe based photovoltaic modules in an industrial scale. These efforts are supported by a research program on relevant materials, structures and processes for thin film photovoltaics. Substantial progress has been achieved during the last years which opens new perspectives for future trends. Joint efforts in research and development based on CIS are coordinated by the network EUROCIS. A screening program on natural minerals with relevance to photovoltaic performance provides the basis for further strategic steps.

"Eurec Agency" (European Renewable Energy Centres Agency), an interesting group presently consisting of 30 research centres, is aiming at increased and coordinated research, development and implementation of renewable energies. It supports the European Commission in its efforts to promote joint activities in this field. Eurec Agency presents one of 4 supporting groups which form an interacting network with "CERE" (Communities of Europe for Renewable Energies), "EURE" (European Utilities for Renewable Energy), and "READ" (Renewable Energies in Architecture and Design).

In the frame of this network increasing cooperation in joint projects on different aspects of utilization has been stimulated and coordinated.Special emphasis has been put on close links between research institutes and industry including utilities and users.

Besides providing a basis for international and concerted cooperation Eurec Agency pursues the goal to increase exchange of information and scientific staff between the member countries. Eurec Agency moreover aims to organize a platform for dialogue and cooperation with non-EC-countries and to support and to cooperate with developing countries in the frame of EU programmes.

Different tasks and strategies have been identified in the Eurec Agency Programme: 1.) Integrated Systems; 2.) Photovoltaics; 3.) Wind Energy; 4.) Biomass; 5.) Buildings and Urbanization. Other relevant items such as hydropower, wave and tidal energy and others have been included. In the field of photovoltaics the activities of Eurec Agency have been concentrated to major progammes like Monochess and Multichess Concepts of high efficiency silicon solar cells and EUROCIS, chalcopyrite based thin film solar cells, upscaling for submodule production and fundamental studies for improved materials and devices.

In this paper thin film programmes carried out and coordinated by the network of Eurec Agency or in close relation to the network are reviewed. In this connection also EUROCAD, the CdTe Thin Film Solar Cell Project, which has proceeded towards industrial production, will be discussed and compared.

Major emphasis has been laid on research and development work on solar cells and modules based on $Cu(InGa)(SeS)_2$ compounds aiming at highly efficient and cost effective modules.

1.)	Eurec Agency	Technical coordiation
2.)	Zentrum fuer Sonnenenergie und Wasserstoff-Forschung (ZSW), Stuttgart, Germany	7x7 cm^2 and 30x30 cm^2 module development
3.)	Universitaet Stuttgart, Institut fuer physikalische Elektronik (IPE), Stuttgart	New and modified absorber materials, window deposition, alternative buffer layers
4.)	Royal Institute of Technology (KTH), Stockholm, Sweden	Advanced coevaporation for high efficiency devices, upscaling from 5x5 cm^2 to 30x30 cm^2
5.)	Ecole Nationale Supérieure de Chimie (ENSCP), Paris, France	Electrodeposition of CIS, upscaling of CBD (CdS) to 30x30 cm^2
6.)	University of Northumbria (UNUMB), Newcastle, United Kingdom	CIS deposition by Stacked Elemental Layers (SEL), laser scribing

 First WCPEC; Dec. 5–9, 1994; Hawaii

7.)	Universiteit Gent (UG), Gent, Belgium	Screenprinting of CIS, selenization by ICSVT, patterning by photolithography
8.)	ENEA, Portici, Italy	Sputtering of Cu/In precursors, selenization, laser scribing
9.)	Instituto de Energias Renovables (CIEMAT), Madrid, Spain	Electrodeposition of CIS, alternative buffer layers by CBD, upscaling 30x30 cm^2
10.)	Universidad Complutense de Madrid (UCM), Madrid, Spain	RF sputtering of $Cu(InGa)Se_2$ and $CuIn_3Se_5$ cells, impedance spectroscopy
11.)	Microchemistry (MC), Espoo, Finland	Atomic layer epitaxy (ALE) of ZnO at 30x30 cm^2 for window layers
12.)	Centre d'Electronique de Montpellier (CEM), Montpellier, France	CIS deposition by MOCVD
13.)	Arbeitsgemeinschaft fuer industrielle Forschung (AFIF), Zürich, Switzerland	Epitaxial films on Si substrates

Table 1. Groups of the EUROCIS Project

The research institutes participating in the EUROCIS programme are listed in Table 1, where the main contributions are indicated.

Complementary work is carried out with associate members from Central and Eastern Europe:

Tallin Technical University, Estland	Photoluminescence, spray deposition
Warsaw University of Technology, Poland	Defect studies, crystal growth
Bulgarian Academy of Sciences, Sofia	Electrodeposition
New Energy Sources Laboratory, Bucharest	Sputtering selenization

Table 2. Central European Groups

The research studies undertaken by EUROCIS network are devoted to material research of chalcopyrite based compounds and to photovoltaic structures. Since no final decision of materials and structures for large-scale production has been made so far, all potential materials have to be taken into consideration and all alternative low cost processes for production have to be studied. The comparison of different growth mechanisms should provide the basis for optimizing the deposition technology for high quality films based on the alloy system $Cu(In,Ga)(Se,S)_2$.

Module production of the size 30x30 cm^2 is envisaged as an intermediate step for transfer to large-scale production and all contributions of the institutions participating are to be combined for optimized paths for pilot production. The updated results achieved in the frame of EUROCIS are listed in Table 3.

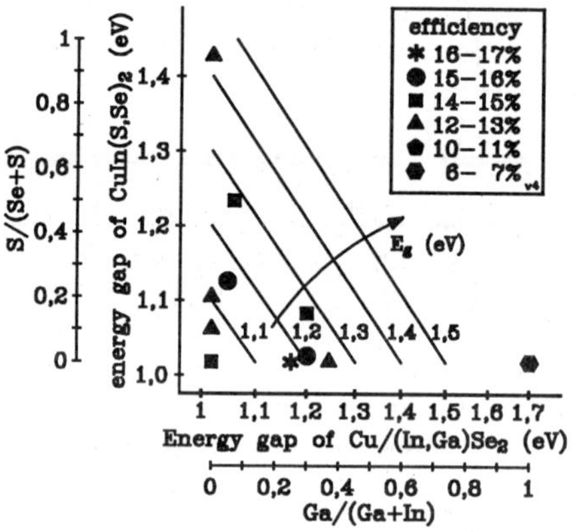

Fig. 1. Chalcopyrite Material System

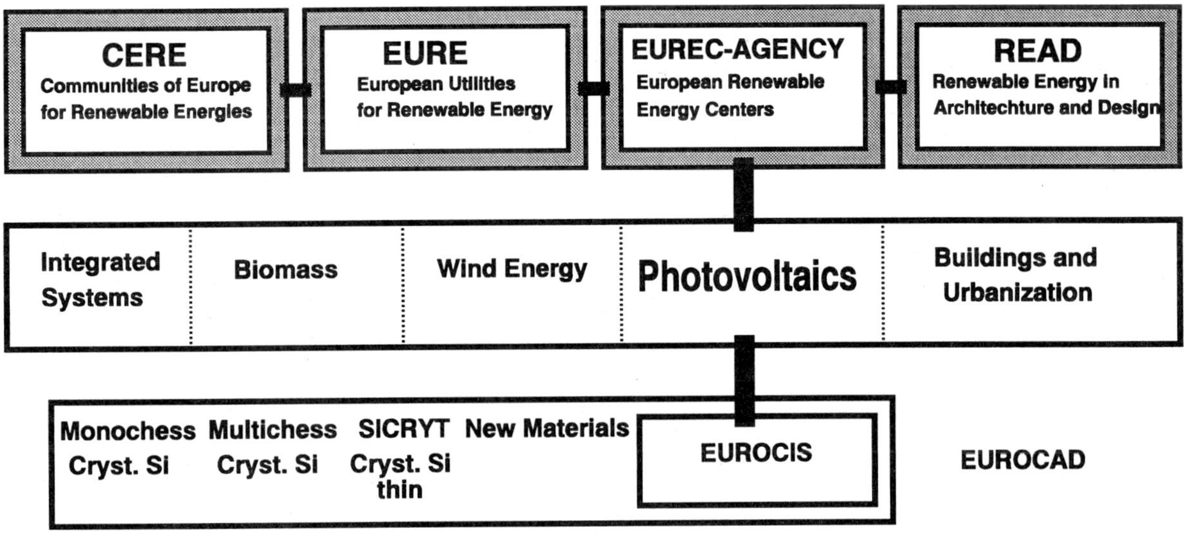

Fig. 2. Structure of the European Organisations

The cell types listed indicate the potentials in materials and composition. Coevaporation of the elements still leads to the highest performance of cells and modules.

Efficiencies of laboratory type modules in the size of 5x5 cm^2 (KTH) and 7x7 cm^2 (ZSW) exceed 10%.

Solar Cell Type	E_g (eV)	V_{oc} [mV]	j_{sc} [mA/cm^2]	FF	Effi. [%]	Area [cm^2]	Lab.
MgF_2/ZnO/CdS/ Cu(Ga,In)Se_2	1.17	641	35.8	0.73	16.9	0.33	KTH IPE
MgF_2/ZnO/CdS/ $CuInSe_2$	1.02	519	41.2	0.75	16.1	0.35	KTH IPE
MgF_2/ZnO/CdS/ CuIn(S,Se)$_2$	1.12	613	33.5	0.74	15.2	0.15	IPE
MgF_2/ZnO/CdS/ CuInSSe	1.27	667	28.8	0.74	14.2	0.08	IPE
MgF_2ZnO/CdS/ Cu(In,Ga)Se_2	1.19	651	31.0	0.70	15.0	0.25	IPE
ZnO/CdS/$CuInS_2$	1.45	715	23.7	0.71	12.0	0.25	IPE
Cu(In,Ga)Se_2 Submodule 5x5cm^2		2880	31.4	0.69	12.4	14.5	KTH
Cu(In,Ga)Se_2 Submodule 7x7cm^2		5508	30.74	0.69	10.2	37.8	IPE ZSW

Table 3. Updated Results achieved in the EUROCIS groups (100mW/cm2 Illumination at 25 C)

The main goals of EUROCIS activities are:
- upscaling to 30x30 cm^2
- optimization of deposition processes
- replacement of CdS by nontoxic buffer layers
- optimization of material composition and structure.

The flow diagram of a pilot production plant as a joint project with industry is presented in Fig. 3.

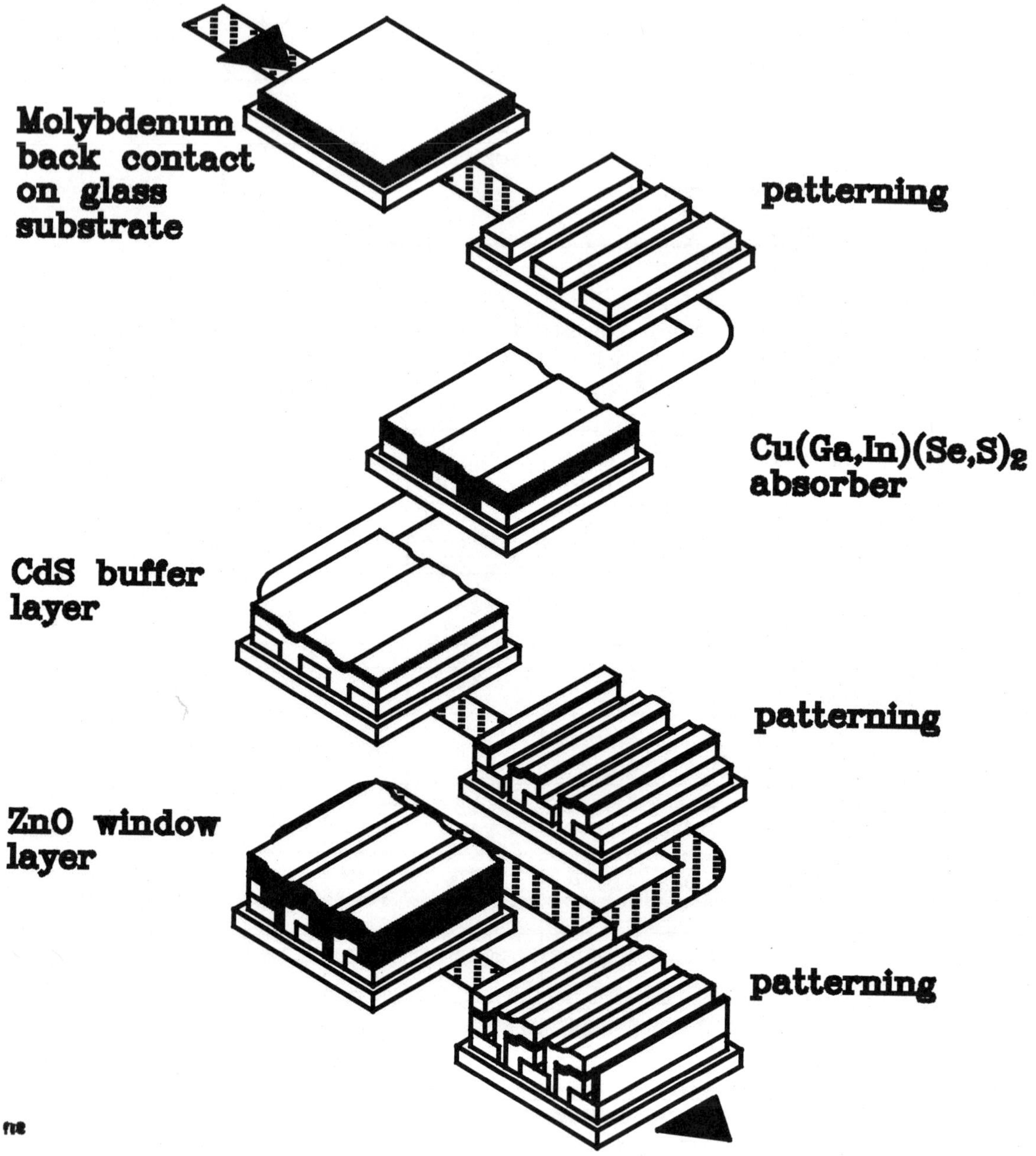

Fig. 3. Flow Diagram

A study group is accompanying this work to analyze the technologies for large-scale module production in respect with material, costs and ecological aspects, at different degrees of upscaling capacities. As a target for the year 2000 module costs of 1 ECU/W_p at a production capacity of 10 MW/a have been identified.

One of the most remarkable advancements in the European activities is BP Solar's programme on Apollo thin film CdTe photovoltaic cells. This industrial programme is beyond the frame of Eurec Agency which is based on research institutions primarily; however, the programme of BP Solar is linked to a number of research institutions, mainly universities, in a joint European project EUROCAD. Even though there is no direct organizational link with Eurec Agency there exists an intense information exchange and interaction with Eurec Agency members and the EUROCAD group.

The achievements of BP Solar lead to an efficiency of 8.4% for a module consisting of 6 parallel connected plates with an aperture 89x51 cm^2. The efficiency distribution resulting from a small production line is rather narrow between about 8 and 9% and indicates the potential to apply low cost deposition processes - electrodeposition of CdTe and chemical bath deposition of CdS - and low cost glass substrates .

EUROCAD (ANTEC Applied New Technologies) has demonstrated efficiencies of 12% on low cost glass substrates and high deposition rates (10 μm/min) with Close Spaced Sublimation. Based on this deposition process 10x10 cm^2 modules with 9 to 10% efficiencies have been realized which will be upscaled to 20x30 cm^2 modules early in 1995 to the final product size of 50x100cm^2.

A low cost potential of 1 ECU/W_p at a production capacity of 100.000 m^2/a is deduced from the underlying processes and the material and energy inputs. $Cu(In,Ga)(Se,S)_2$ compounds and CdTe in its advanced state of Apollo thin films of BP Solar are considered as a the most promising options for polycrystalline photovoltaic modules in the European scene. Besides, research on amorphous a-Si:H and a-Si alloy films in European research institutions and module manufacturing at Phototronics and Neste Advanced Power Systems are continued.

Research work is carried out in European institutes on WSe_2, WS_2, and $MoSe_2$ and their use in solar cells and photoelectrochemical cells as well as on FeS_2 and $FeSi_2$. The status of these programmmes is basically oriented to materials and structures and has not yielded so far perspectives for industrial manufactures.

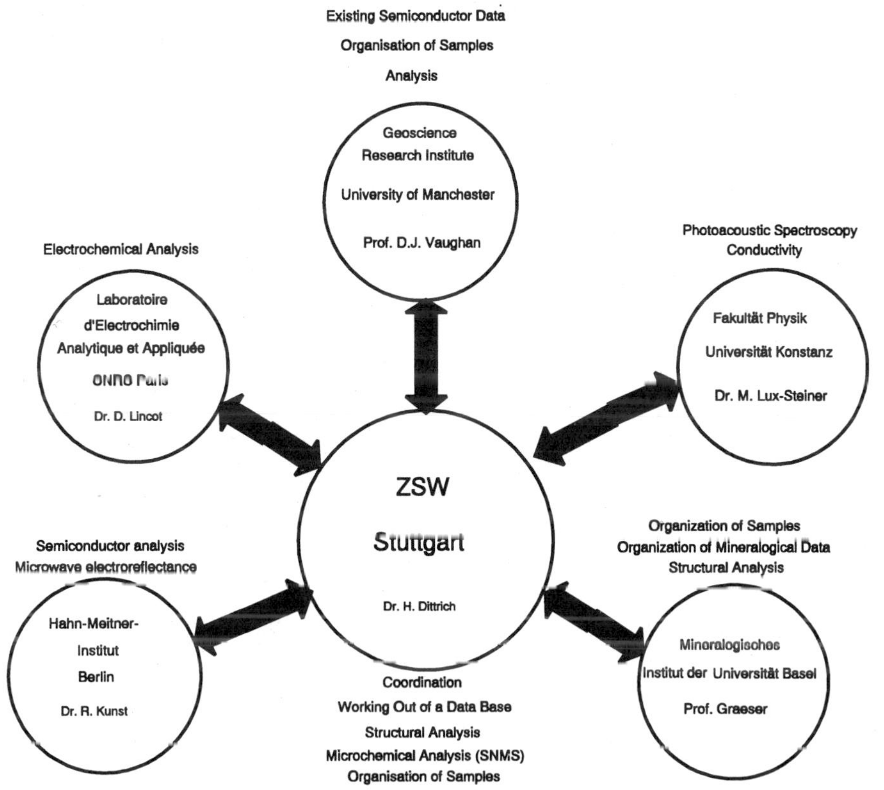

Fig. 4. Structure of the Network

The process which has been achieved with thin film polycrystalline semiconductor compounds has stimulated a systematic search for photovoltaic materials which can be derived from mineralogy hereby indicating the natural availability and stability. Thousands of mainly sulfidic ore minerals are screened in the frame of a Eurec

Results from photoacoustic spectroscopy:

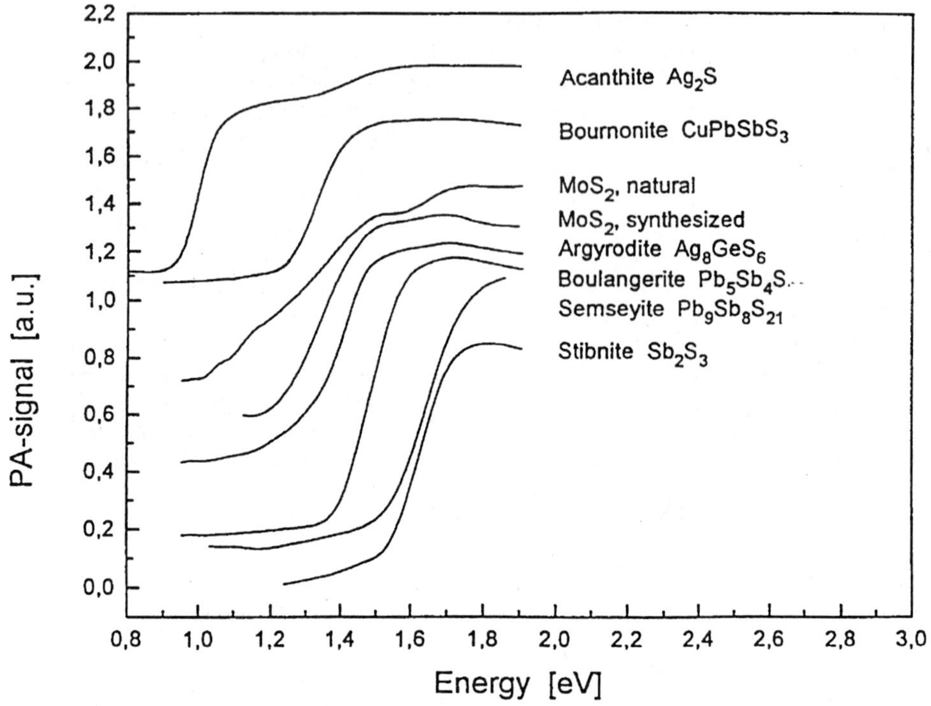

Fig.5. Results of Photoacoustic Measurements

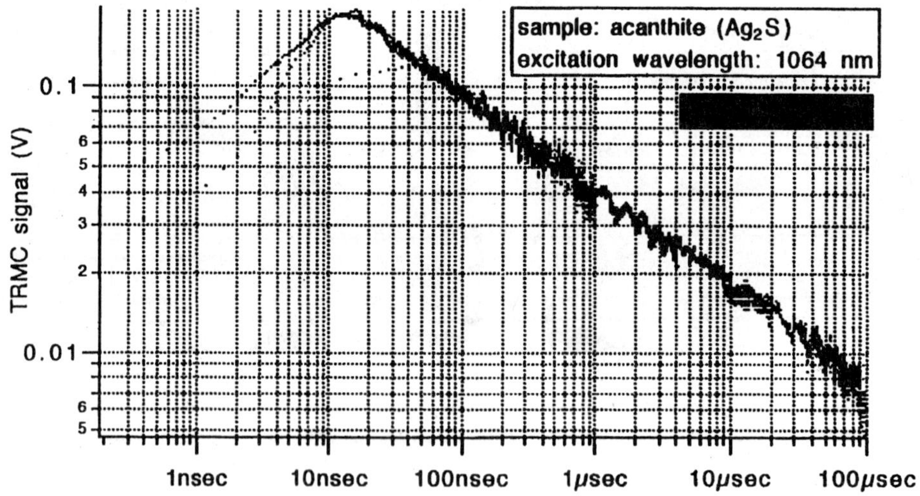

Fig.6. Microwave Reflectance Signal

Agency Programme "New Photovoltaic Materials from Systematic Mineralogy" in view of potential applications in photovoltaics.

Screening procedures include x-ray powder diffraction, microchemical analysis, photoacoustic spectroscopy, electrochemical characterization, and - contactless - electrical measurements (time resolved microwave reflectance). The structure of the network, contributions and coordination are demonstrated in Fig. 4.

As an example results of photoacoustic measurements obtained from different natural minerals are illustrated in Fig.5. which indicates bandgaps in the relevant region between about 1.0 and 1.7 eV and in Fig.6. the microwave reflectance signal as a response of a light pulse which leads to photocarrier lifetimes of about 15 μsec.

At this stage of the systematic screening process it has been shown that the quaternary fahlore group

$Me^{1+}{}_{10}Me^{2+}{}_{2}Me^{3+}{}_{4}Ch^{2-}{}_{13}$

with Me^{1+}Cu,Ag

Me^{2+}Cu,Fe,Zn,Mn,Cd

Me^{3+}Sb,As,Bi

Ch^{2-},S,Se

could be a source of photovoltaic relevant materials and that naturally grown crystals show a high degree of purity from the semiconductor point of view (e.g. Ag_2S and $CuPbSbS_3$).

This procedure allows a rapid screening of natural minerals in view of their structural, optical and electronic properties for potential use in photovoltaic structures and realized in thin films by presently applied deposition processes.

Conclusions

Eurec Agency has given to research and development of thin film photovoltaics as well as its transfer and implementation to industrial production an appropriate weight and emphasis. It has supported a comprehensive research program on materials, structures and processes related to the production of CIS based cells and modules, including a systematic screening of new materials.

«SWITZERLAND'S APPROACH TO PHOTOVOLTAIC APPLICATIONS»

Thomas Nordmann, Programme Director:
Swiss Federal PV-Promotion Programme
c/o TNC Consulting AG, Asylstrasse 84/P.O. Box 152
CH-8708 Männedorf / Switzerland
Tel: .41 1 921 13 14 Fax: .41 1 921 13 77

ABSTRACT

«ENERGY 2000» The Swiss national programme

- Policy target: 50 MWp (by the year 2000)
- Existing capacity of grid connected plants: 4.4 MWp Dec. 1993, 5.2 MWp Dec. 1994 (estimate)
- Strong commitment to integration in buildings
- Rate based subsidies in certain towns (Burgdorf, Lyss and Geneva)
- Small public budget for PV

The contribution of the Swiss Federal Office of Energy is:

  - 3 million sfr/a for the promotion programme and P&D projects
  - 3 million sfr/a for R&D projects

- The total funding in the field of PV in Switzerland was about sfr. 23 million in 1993 for R&D and P&D, of which 15 million from public and about 8 million from private funds.

1. THE NATIONAL PROGRAMME «ENERGY 2000»

1.1. Introduction

The adoption of a new article in the Swiss Constitution relating to energy and a 10 year moratorium on nuclear energy by Swiss voters on 23rd September 1990 had political consequences. The Swiss government (Federal Council), supported by the ruling parties, launched a 10 year national programme in November 1990, known as «Energy 2000» [1,2,3,4].

The concerted efforts of the entire Swiss national economy are intended to achieve ambitious objectives with respect to energy:

- To stabilise the total energy consumption of fossil fuels (oil, petrol, gas, coal), and the quantity of CO^2 emitted, by the year 2000 at or below the level pertaining in 1990, and thereafter reduced.
- To slow down the increase rate in consumption of electricity during the decade followed by stabilising demand from 2000 onwards.
- To increase the contribution level of renewable energy sources by an additional 0.5% to electricity and 3% of the heating energy production, by the year 2000.
- To increase hydroelectric power production by 5%, and the power rating of existing nuclear power stations by 10%, by the year 2000.

The national «Energy 2000» program sets ambitious objectives for the use of solar energy. The objective for thermal use of the sun's power is 1.5 million m^2 (see Fig. 2), and for photovoltaics the objective is 50 MW, (see Fig. 3).

By the turn of the millennium, photovoltaic grid-connected installations amounting to a total of 50 MWp should be installed and brought into operation within the context of the Swiss national programme «Energy 2000». The local and regional utilities are supporting this ambitious objective by reimbursing the marginal costs of the energy supplied and by additional accompanying measures.

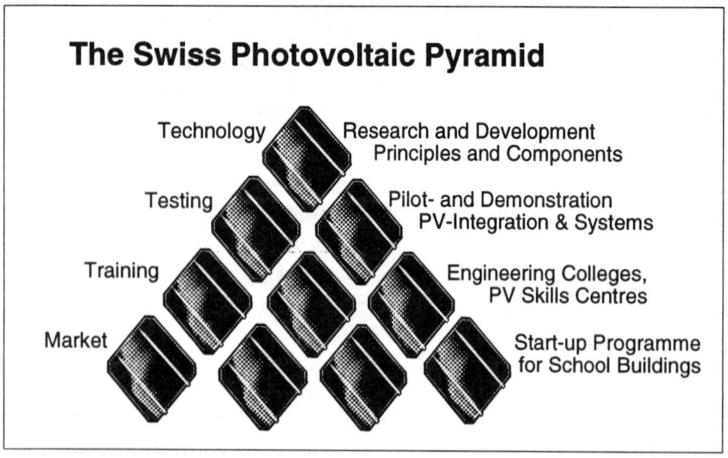

Due to limited land resources, the main emphasis for the use of photovoltaics in Switzerland is placed on integration with buildings. Over the last five years, a number of technically interesting pilot and demonstration projects have been set up.

One of the main thrusts of Switzerland's promotional activities in the field of photovoltaics has been in the area of Schools and the educational sector. In 1992 Switzerland launched a promotional program for public schools. By the end of 1995, it is anticipated that there will be about 1 MWp of grid connected PV-plants operational in the public schooling sector in Switzerland. Further development of the market will not be possible with public sector finance alone (see Fig. 4). The emphasis must be on partnership between producers and consumers. At present, promotional models are being tested by some utilities.

1.2 «Energy 2000» Divisions

To achieve its objectives, the implementation Programme «Energy 2000» is directed specifically at the market. The Programme consists of eight divisions.

- **Motor Fuels**
- **Renewables**
- **The State**, aimed at energy managers in federal, regional (cantons) and local (communities) authorities.
- **Residential Buildings**, aimed, for example, at house owners, promoters, commercial users and tradesmen.
- **Industry**, aimed at industrial firms and organisations.
- **Trades**, aimed at energy-intensive concerns, large vending areas and the catering trades.
- **Services**, aimed at designers, promoters and operators of service buildings, as well as buyers and users of office and household equipment and entertainment media.
- **Hospitals**, aimed at hospital services.

In the case of new renewable energy sources the 3 main points of emphasis are:

- **Solar Energy** (thermal solar, passive solar, solar drying, photovoltaic)
- **Ambient Heat** (heat pumps)
- **Bio Energy** (wood)

2. SOLAR ENERGY TARGETS IN «ENERGY 2000»

2.1 Thermal Solar Energy Objectives of «Energy 2000»

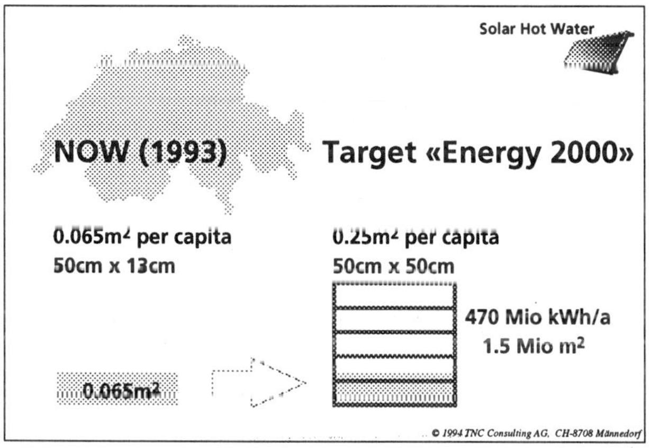

Fig. 2. Graphical representation of the **Thermal Solar Objectives** for «Energy 2000». The intention is to increase the area from 0.3 million. m² (54 GWh (th) /a) to 1.5 million m² (470 GWh (th) /a) by the turn of the millennium.

2.2 Photovoltaic Objectives of «Energy 2000»

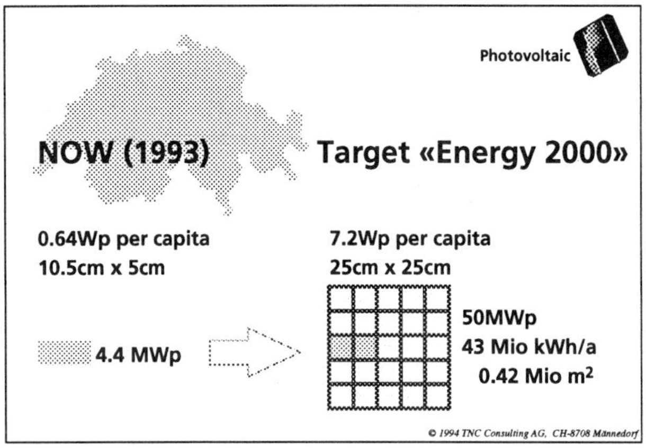

Fig. 3. Graphical representation of the **Photovoltaic Objectives** for «Energy 2000». The intention is to increase output from 4.4 MWp (3.9 GWh (el) /a) to 50 MWp (43 GWh (el) /a) by the turn of the millennium.

2.3 Development of the Market for Solar Power in Switzerland

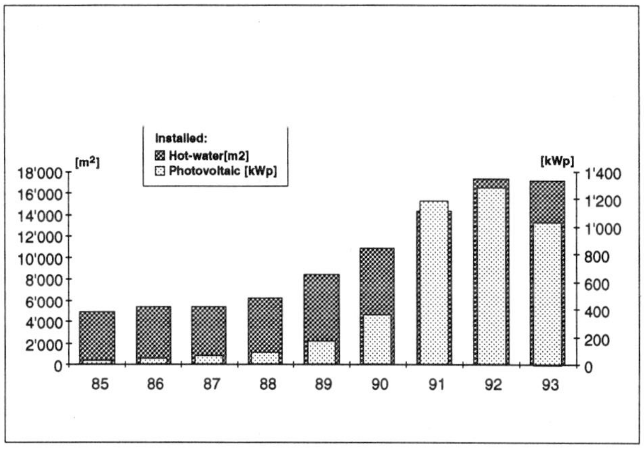

Fig. 4. Diagram showing the growth of the market for solar power in Switzerland (new thermal collectors and new PV modules installed annually in m² or kWp). Survey carried out by the Swiss Solar Industry Association (SOFAS) [5].

3. WHY PV IN BUILDINGS FOR SWITZERLAND?

Switzerland is a densely populated country and has only limited land resources available (see Fig. 5). The use of photovoltaic in the energy economy must not fail because of the land area that this source requires. A strategy must therefore be put forward in which photovoltaic applications only take up areas that are already used. Figure 5 and 6 show the distribution of land use in Switzerland. The area of land that is required for road traffic is surprisingly large in relation to that used for buildings (see Fig. 6) [6].

The aims of the national «Energy 2000» programme, though ambitious from the photovoltaic point of view, are still insignificant as far as land use is concerned. Figure 7 shows the level of photovoltaic usage per head of population achieved by the end of 1993. If the «Energy 2000» target is achieved, the additional land use will be minimal (see Fig. 8).

Comparison of selected EU Member States or the average of the European Union shows that national ratios in Europe are similar. The United States, on the other hand, has far more land available and so there is naturally less motivation for photovoltaic integration in buildings (see Fig. 9).

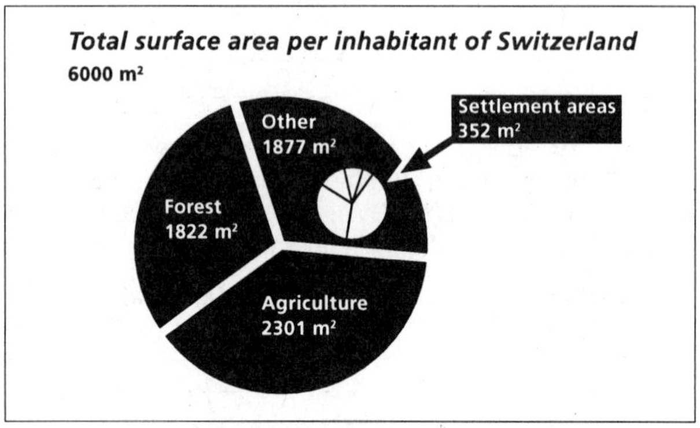

Fig. 5 Total land resources in Switzerland.

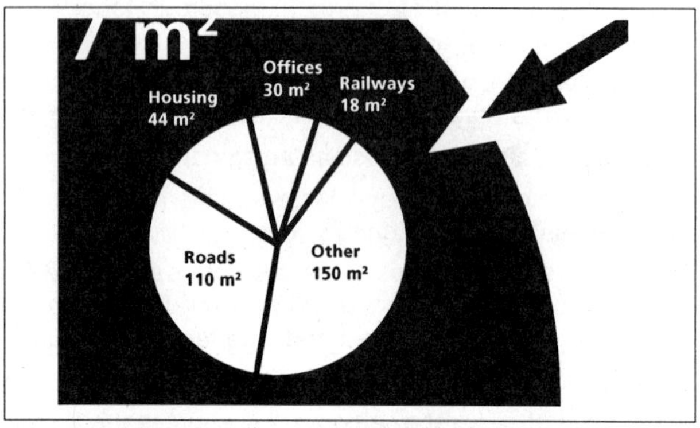

Fig. 6 Breakdown of land resources.

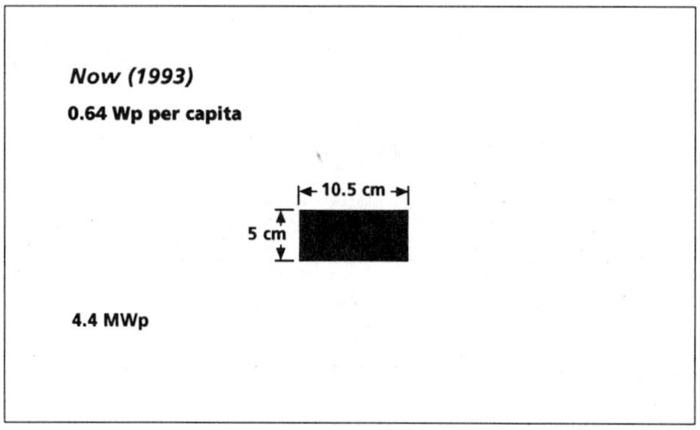

Fig. 7 PV in Switzerland now (1993).

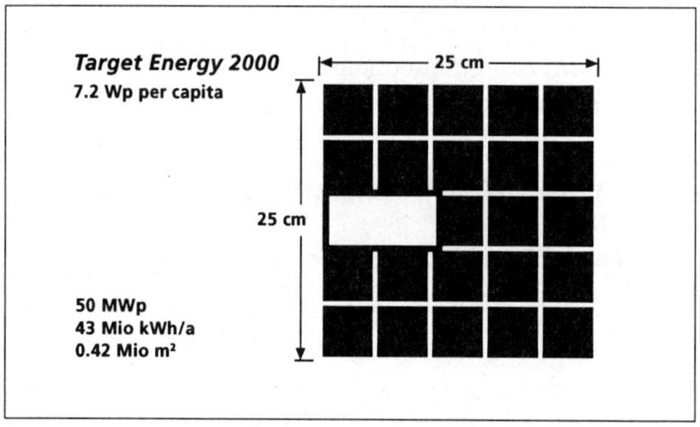

Fig. 8 ENERGY 2000 PV target.

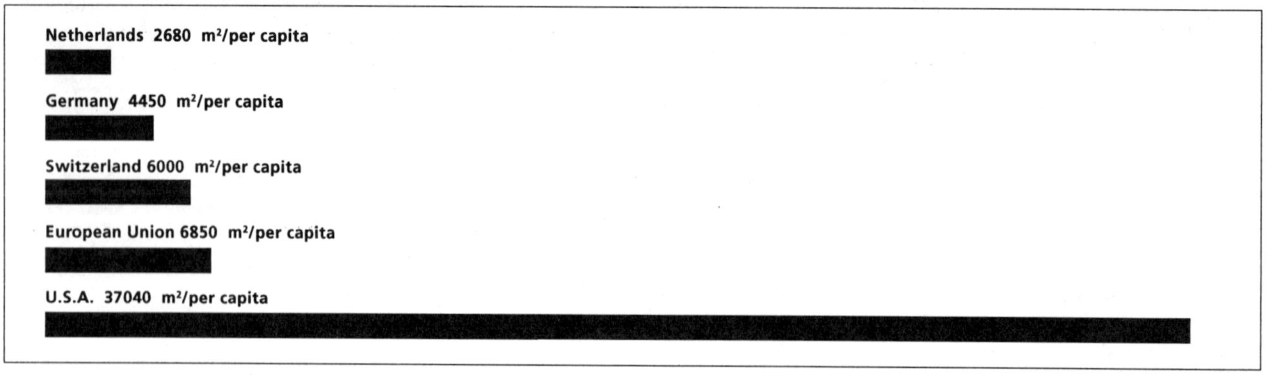

Fig. 9 Comparison of land resources.

4. SUBSTITUTION WITH INTEGRATION OF PV

The Swiss photovoltaic industry has tested a number of products and applications for specific integration into buildings. To ensure that they have reached a stage where they are ready to be marketed practical trials on pilot and demonstration projects have been carried out [7].

The Solar Tile

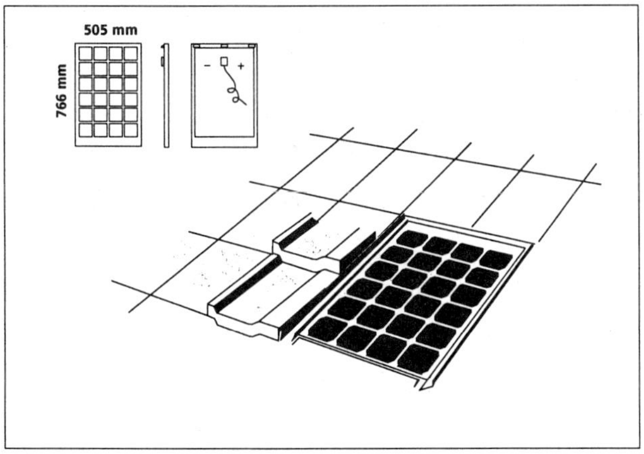

Fig. 10. The Swiss photovoltaic tile replaces four standard roof tiles with a photovoltaic module which can be plugged into its neighbour. Insulated plugs are used so that they can be laid by a roofer.

The Solar Skylight

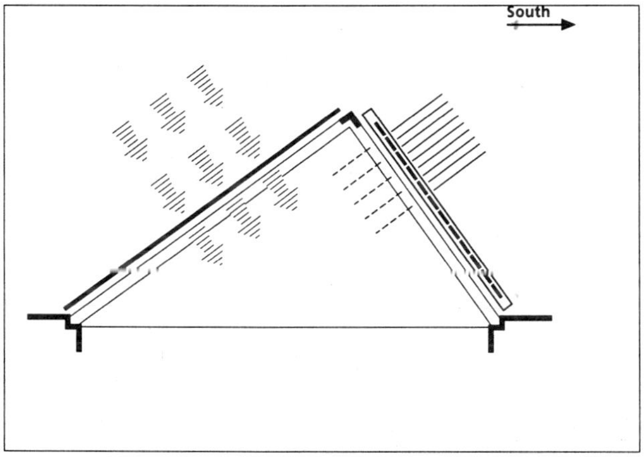

Fig 11 The standard use of saw tooth skylight to obtain maximum daylight an d maximum shading in large buildings can be practically complemented by photovoltaic modules on the south-facing side.

Shadow-Voltaic

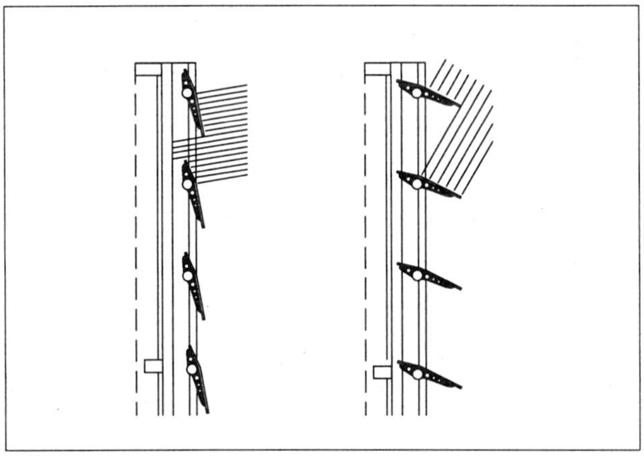

Fig. 12. Shadow-voltaic is a technical solution which advantageously combines photovoltaics with exterior shading devices added to give protection from the sun. Adjustable fines throw the maximum possible shadow and at the same time provide maximum exposure of the strips of photovoltaic modules attached to them.

The AC-Module

Fig. 13. A number of technical developments in Switzerland are aimed at integrating a miniaturised inverter directly in the photovoltaic module's plug connector. This largely does away with the expensive direct current wiring that is responsible for a large proportion of the balance-of-system costs. It will then be easier for electricians, whose experience is with alternating current, to use photovoltaic technology. Even a plug type connector between the alternating current module and the mains wall socket is conceivable. The electrical safety devices of the mini inverter must in one way or another ensure that the inverter switches off when the mains voltage is not present.

5. A PV-TOUR THROUGH SWITZERLAND

A large proportion of the five megawatt capacity that will be in operation in Switzerland by the end of 1994 consists of pilot installations and demonstration plants which set new standards, in particular the integration into buildings. All these projects are monitored so that important conclusions can be drawn and information obtained on the functionality, operational safety and electrical output [8].

PV-Parking Lot 1992

Fig. 14. PV integration by European PV specialists on show at the IEA Solar Heating and Cooling, Task XVI demonstration site at Swiss Federal Institute of Technology (EPFL) in Lausanne. The collection of demonstration models is constantly being extended and is monitored and evaluated by Laboratory for Solar Energy (LESO) using a variety of measurement techniques.

PV Meets the Matterhorn 11.5 kWp (1993)

Fig. 15. A photovoltaic installation in a high-profile location. Hundreds of thousands of visitors to the Matterhorn can now view a photovoltaic system too, integrated into a stone roof.

Alpine PV-Dairy 13.3 kWp (1991)

Fig. 16. This installation provides testing conditions for roof-mounted large modules, situated some 1650 meters above sea level. The equipment has to withstand snow, cold temperatures and high winds.

PV Solar Dome 13.3 kWp (1993)

Fig. 17. A special shadow-voltaic installation. Movable PV panels provide optimum lighting in the building whilst allowing the cells to be arranged in the most appropriate way.

6. PROMOTIONAL STRATEGIES FOR THE MARKET DEVELOPMENT OF PHOTOVOLTAICS

Technological and economic development will be promoted: i.e. achievement of lower specific prices (francs per watt) and the improvement of PV module efficiencies for the same specific price (francs per watt). This can have a beneficial effect on the balance-of-system costs of overall systems, among other things.

Promotion of experience of application: integration of PV in buildings where specialists at all levels are trained, erection of representative pilot and demonstration plants throughout the country. This will stimulate discussion and debate about photovoltaics by citizens and politicians.

6.1 Why put photovoltaic installations on engineering colleges and electrical training college buildings?

In these colleges, of which there are about 100, the future electrical specialists of Switzerland are trained. The confrontation of teaching staff and apprentices with this new technology produces a snowball effect:

- training college → training staff → apprentices → industrial training staff → industry specialists.

A visible «market share» of photovoltaic action has already been achieved in Swiss education with a relatively low expenditure of 3 million Swiss Francs a year (see Fig. 18). By the end of 1996 approximately 1 MWp of pilot and demonstration projects will be in place on school and college buildings (see Fig. 19). Tens of thousands of students, school pupils and apprentices will have been positively confronted with the potential of using photovoltaics.

6.2 Why put PV installations on school buildings?

To achieve a maximum promotional effect, the main thrust of photovoltaic demonstrations should be directed at the schools.

Like the introduction of computers, photovoltaics should also progress through the education system. Young people are very open to acceptance of new, environmentally friendly energy techniques. They are our future political decision-makers and also the energy and power consumers of the future.

- Long-term and lasting informative and promotional effect, promotion of photovoltaic applications on school buildings.
- Reduction of the national contribution to promotion through the use of photovoltaics in state day schools. The government, the cantons and the community all contribute financially.
- School buildings represent a public duty for the education of young people.
- School buildings are a highly effective means of spreading the word, thanks to the chains:
- school → pupils → parents → citizens or
- school → teachers → authorities → citizens.
- Every community has schools of its own, thus everyone is approached and involved.
- Grid connected photovoltaic plants are not a commercial proposition yet, but they represent a good and lasting investment in the education of our young people.

Why PV in Swiss School Buildings? — The Present

PV-Schools to 94	Total CH School Build.	School Buildings, Photovoltaic Projects: Installations	Implemented	Total Power
Engineering colleges (HTL's) «PV skills centres»	20	6	30%	115 kWp
Electrical engineering colleges «PV training centres»	100	4	4%	22 kWp
Start-up programme for state schools «PV demonstration install.»	3'018	45	1%	233 kWp
Total installed power (31-10-94)				370 kWp
Annual production				~ 315 MWh

Fig. 18 PV in schools, now (1994).

Why PV in Swiss School Buildings? — The Future

PV-Schools to 94 + planned	Total CH School Build.	School Buildings, Photovoltaic Projects: Installations	Implemented	Total Power
Engineering colleges (HTL's) «PV skills centres»	20	14	70%	140 kWp
Electrical engineering colleges «PV training centres»	100	20	20%	249 kWp
Start-up programme for state schools «PV demonstration install.»	3'018	90	3%	554 kWp
Total future power (31-10-94 + planned)				943 kWp
Annual production				~ 800 MWh

Fig. 19 PV in schools, future plans.

6.3 Future market development of PV in Switzerland

The current restrictions on public spending mean that funding for the «Energy 2000» target cannot be expected to come from taxation revenue. The aim is, through the electricity companies, to get electricity consumers financially involved in market development. Various models for financing photovoltaic installations are currently being tested in the market using the bottom-up approach. These are contributions to investment (sfr/kWp) or contributions to operating production (sfr/kWh) that will be supplied to the national grid. The second method of remuneration links the electrical yield of a plant with the amount of support but does not involve a large expenditure on administration. Under the current legislation electricity companies are obliged to purchase photovoltaic power at marginal rates (approximately sfr 0.16/kWh). This is not sufficient, however, to cover the true commercial cost of producing this electricity, which is between 1 and 2 sfr per kWh (see Fig. 20 and 24). The additional requirement can be raised by increasing the purchase price to meet the true costs of production (see Fig. 20 and 23). Another model allows consumers to order photovoltaic power from the electricity company.

The electricity company undertakes to supply the amount of photovoltaic power ordered (in kWh) using its own plant, built at its own cost and risk (see Fig. 21).

A constitutional initiative for a "Solar Centime", which admittedly is only due to be approved towards the end of the century, may ensure the future development of photovoltaics in Switzerland. Consumer acceptance of a small rise in the price of electricity, of the order of less than 1%, to be put towards photovoltaics, is very high, according to representative surveys of the Swiss population. At the end of 1993 over 80% of the Swiss people questioned were in favour (see Fig. 24).

There is now a sensible work/task division between technological and market development. Technological development continues to be the responsibility of the state, the technical universities and the photovoltaic industry. Market development can be guaranteed between the consumers and producers (the power sector) using an appropriate incentive model.

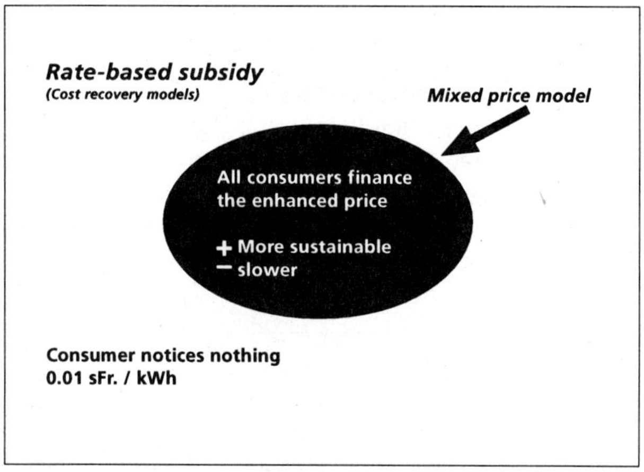

Fig. 20. Mixed price cost recovery model

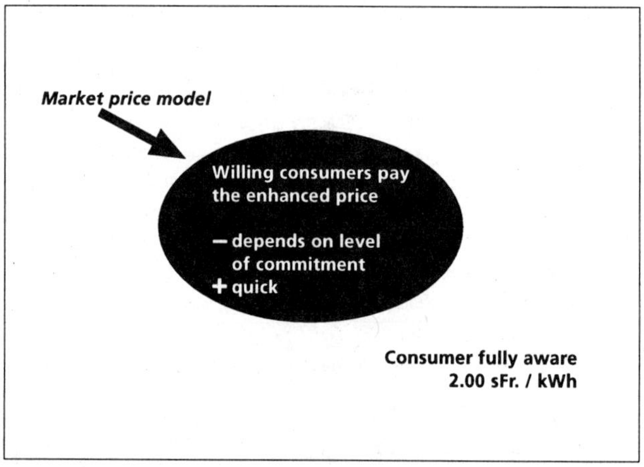

Fig. 21. Market price cost recovery model

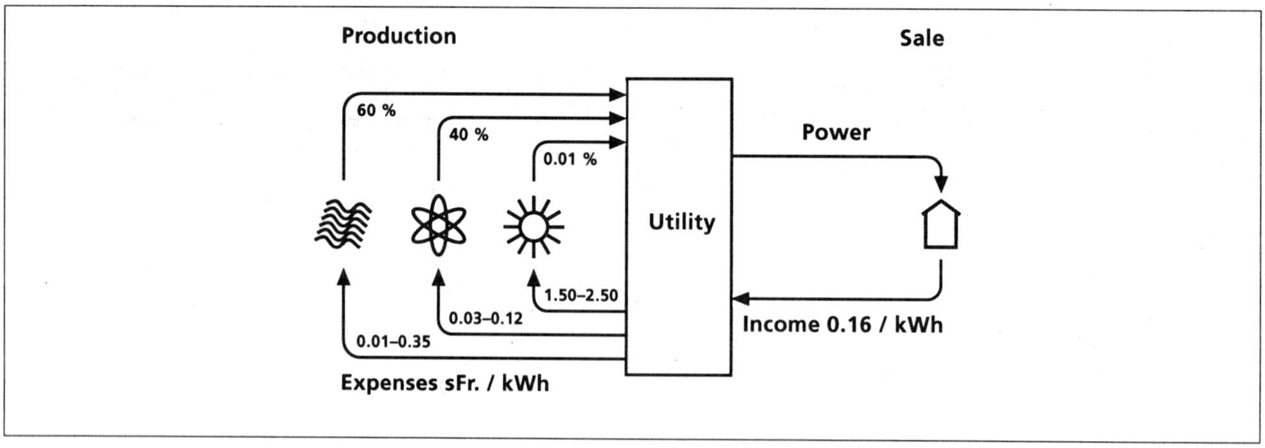

Fig. 22. The flow of energy and money for the rate-based subsidy in the mixed price cost recovery model

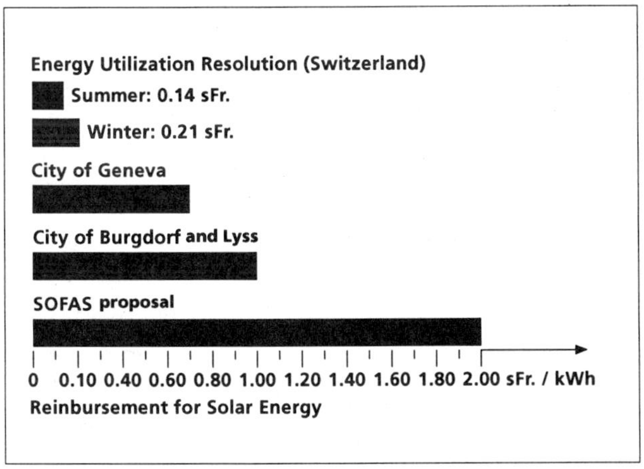

Fig. 23. Payment for PV-Energy in the mixed price model

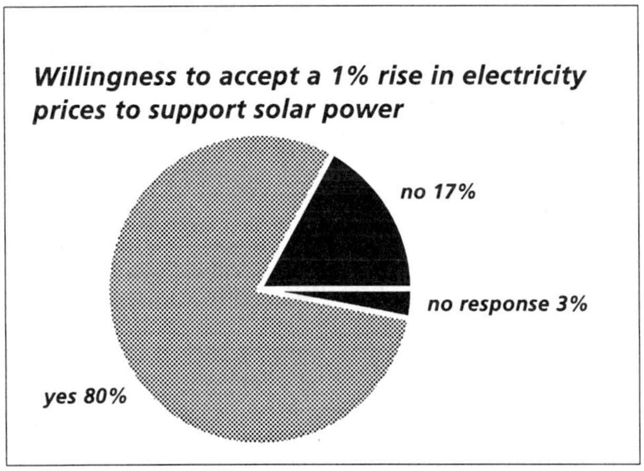

Fig. 24. The goodwill of the Swiss citizens for the mixed price model (Survey 1994).

6.4 Preliminary ideas for photovoltaic supply with rate-based subsidy

- Cost-covering and cost-effective reimbursement from solar energy for the operating costs instead of capital investment
- Fixed-term contracts between operators of photovoltaic plants and electricity supply stations, covering:
- Purchase of solar energy by the utilities at predetermined prices which cover costs and are profitable
- Apportionment of the costs incurred in purchasing solar power sales income
- Periodic adjustment of solar power price scales as investment costs fall

References

[1] Swiss Federal Office of Energy; " The Action Programme Energy 2000" (Summary in English, 14 pgs.) Order No.: 805.071, EDMZ, Berne, Switzerland.

[2] CADDET, Energy Efficiency Newsletter; "The Swiss Approach to Energy Efficiency", Special Issue, Sept. 1994, CADDET, PO. Box 17, 6130 AA Sittard, The Netherlands.

[3] Th. Nordmann, "Summary of The National Photovoltaic Transfer Programme" (in German); *VSE/SEV Bulletin Vol. 16/1991*; PO. Box 6140, CH-8023 Zurich, Switzerland.

[4] Federal Office for Economic Affairs; "Impuls" (German/French); Impuls Programme Journal; CH-3003 Berne, Switzerland.

[5] SOFAS; "Annual market survey"; Solar Industry Association of Switzerland; PO. Box, CH-8050 Zurich, Switzerland.

[6] O. Humm, P. Toggweiler; "Photovoltaics in Architecture"; Swiss Federal Office of Energy; IBSN 3-7643-2891-6; Birkhäuser, 1993, Basel, Switzerland.

[7] Swiss Federal Office of Energy; "Overview & Annual Reports, R&D Projects", Polygon, April 1994, Fribourg, Switzerland.

[8] The Swiss Federal Promotion Programme for PV, c/o TNC Consulting AG; "PV Through Switzerland, Volume I & II"; Documentation about outstanding examples of P&D Photovoltaic grid connected installations in Switzerland; Photographer: Martin Peer, Kontrast, Zurich, Commissioned by the Swiss Federal Office of Energy, © 1994; available at ENET, PO. Box 142, CH-3000 Berne 6, Switzerland.

PRESENT STATUS AND FUTURE PROSPECTS OF ELECTRO-MAGNETIC CASTING FOR SILICON SOLAR CELLS

Kyojiro Kaneko, Ritsuo Kawamura and Teruoki Misawa
Research & Development Center, Sumitomo SiTiX Corporation
Higashihama-cho 1, Amagasaki, Hyogo, 660, Japan

Introduction

The Development research of Electro-Magnetic Casting (EMC) for silicon crystal manufacturing technology has been carried out for years with the purpose of providing low cost multicrystalline silicon substrate for solar cells. The EMC technology is a new concept, in which electromagnetic force is utilized to suspend molten metal without contact to crucible wall for melting and solidification of silicon material. The past and future development researches are presented in Table 1. The first fundamental research for casting was begun in 1987 with a small round cross-sectioned ingot of 5 cm diameter, and the ingot size was gradually expanded with a square cross-sectioned ingot shape. At present, the research has been carried out for the development of casting technique with an ingot size of 22 x 22 cm ^{2} cross section, and the furnace construction for producing a 35 x 35 cm ^{2} cross sectioned ingot has been begun.

Table 1. Past, present and future development researches of Electro-Magnetic Casting

Ingot size [cm]	Melt weight [kg]	Ingot Weight [kg]	Research year
5.0 ϕ	0.3	1	1987~88
8.8×8.8	1.8	6	1988~90
11.8×11.8	3.6	30	1989~90
17.0×17.0	7.0	100	1990~92
22.0×22.0	12	200	1991~
33×33	20	500	1994~

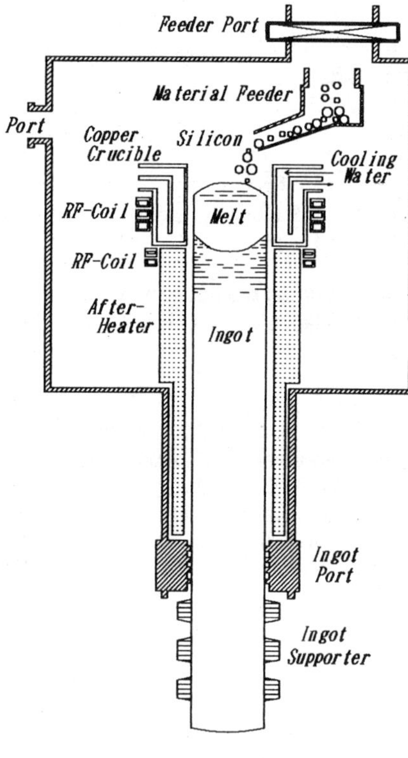

Fig. 1 Schematic illustration of silicon EMC furnace for solar cells

Solar cell conversion efficiencies using EMC ingot crystals are ranging 13 ~ 14 % at the present, and the quality of EMC material reaches within that of conventional mold casting material. By the improvements of higher casting speed, higher material quality and larger ingot size at the EMC technology, it is expected that a new casting technique for lower cost ingot production will be realized.

Features of EMC technology

As a new promising method for manufacturing low cost and high quality silicon substrates, the EMC method differs from the conventional mold-using casting methods in some points. The conventional casting methods give rise to higher production cost due to a burden of consumables for the crucible and mold equipment, and impose restrictions on productivity from the intermittent furnace operation. The EMC has the following features and advantages.

1)less contamination melting is possible by contactless suspension of melt,
2)less consumables cost is attainable by non-consumable crucible, and
3)high productivity is realized by high speed casting of larger ingot.

First WCPEC; Dec. 5–9, 1994; Hawaii

Silicon EMC furnace

One of advantageous points of EMC to the conventional mold using casting is that the EMC method can perform both melting and solidification simultaneously in an induction melting crucible. The schematic furnace structure is illustrated in Fig. 1, showing the continuous casting operation of melting and solidification. Silicon materials are fed from the open top of the water-cooled copper crucible to melt by the induction heating. The silicon melt in the crucible is suspended contactlessly to the crucible wall by an electro-magnetic pinching force. The silicon melt is only contacting with the solid-liquid interface of silicon ingot. By increasing the casting speed of ingot, the latent heat of solidification and the heat flow from the melt to the solid at the solid-liquid interface increase so that the shape of the interface tends to be deep and unstable.

Electro-magnetic phenomenon of induction melting is governed by the following well-known equation to suspend the melt.

$$F = [J \times B],$$

where, F is electromagnetic force, J is electric current density, and B is magnetic flux density. The electric current in the surface layer of melt generates Joule heat to raise the temperature of charged melt. Meanwhile, the electric current interacts with the magnetic field in the surface layer to exert the magnetic pressure, which is known as Lorentz force. The magnetic pressure works in the normal direction to the molten surface to pinch the charged melt.

Fig. 2 EMC ingot of 170 cm length with $22\times22cm^2$ cross section

Table 2. Chemical and physical properties of EMC silicon materials

Chemical Impurities		
Carbon	$5 \sim 10 \times 10^{16}$	atoms/cm^3
Oxygen	$1 \sim 5 \times 10^{16}$	atoms/cm^3
Chromium	< 0.06	ppbw
Copper	< 2	ppbw
Iron	≒ 0.2	ppbw
Manganese	< 0.05	ppbw
Nickel	≒ 0.2	ppbw
Titanium	< 0.5	ppbw
Doping Level	P-Type 0.5～1.5 Ω.cm	
Etch Pit Density	$5\sim20\times10^5$ / cm^2	
Diffusion Length	50～120 μm	
Life Time	2 ～ 3 μsec	

An after-heater system is equipped under the cold crucible open bottom to control the solidification process. Another induction heating is used for the temperature control of after-heater system.

The casting furnace chamber is evacuated by a vacuum pump to exchange air for argon of one atmospheric pressure. Silicon materials are continuously fed after a stable silicon melt is formed, and the solidification of

Fig. 3 $10\times10cm^2$ solar cell wafer cut from EMC ingot (a quarter of cross section)

ingot starts simultaneously. At the end of casting, the top of ingot is hot-top-heated to avoid a cracking by the volume expansion of silicon at the solidification. In Fig. 2, an ingot of 22×22 cm^2 cross section cast by EMC is shown.

Crystalline Properties of EMC Ingot

The physical and electrical properties of cast silicon multicrystals are determined both by the concentrations of included chemical impurities and by various kinds of induced crystalline defects. Chemical impurities and physical properties including the etch pit density, minority carrier diffusion length and life time were measured on EMC silicon crystals, and are shown in Table 2. The silicon material which was used for the casting was the solar grade. As one of characteristic features of EMC ingot, the low concentrations of carbon and oxygen are revealed in the table. Etch pit densities are rather high as $5 \sim 20 \times 10^5$ / cm^2 , but minority carrier diffusion lengths in average and life times range $80 \sim 90$ μm and $2 \sim 3$ μsec. A 10×10 cm^2 quarter of ingot cross section is shown in Fig. 3, in which small grain sized crystals are distributed at the corner and the side of ingot but grain sizes more than 1 mm diameter in average are grown at the inside of ingot.

Solar cell conversion efficiencies of EMC ingots were investigated along the whole length of the casting direction. The results are shown in Fig. 4. The length of cast ingot was 170 cm with the 22×22 cm^2 cross section, and the casting speed was operated at 2.0 mm/min. The relative conversion efficiency, that is, the percentage of conversion efficiency of EM cast ingot relative to the conversion efficiency of silicon single crystal, is plotted in the vertical axis of the figure. It can be seen that the conversion efficiency of EMC ingot is unchanged along the whole length, and about 84 % of conversion efficiency is attained relative to the conversion efficiency of single crystal. Conversion efficiencies at

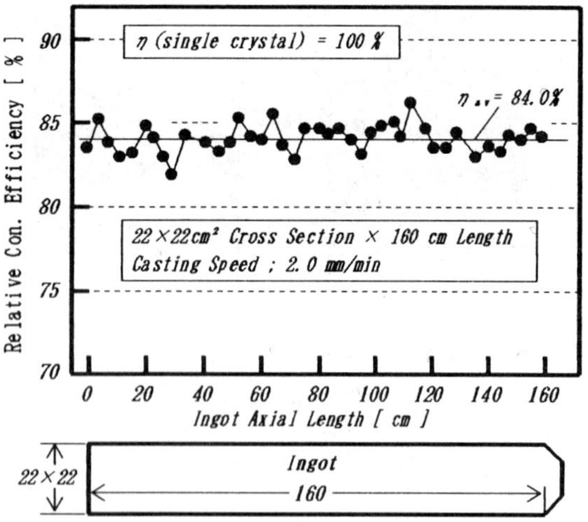

Fig. 4 Solar cell conversion efficiencies of EMC ingot in the axial direction

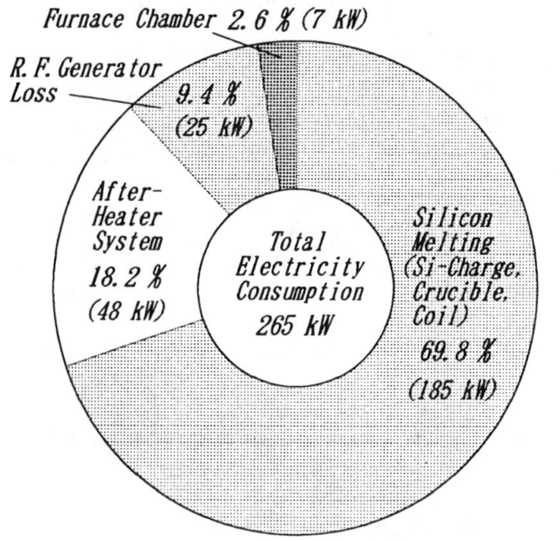

Fig. 5 Electric power consumptions of EMC for 22×22 cm² cross-sectioned ingot

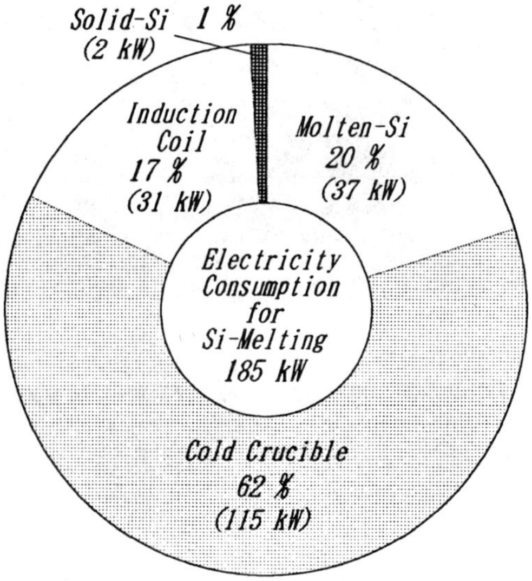

Fig. 6 Electric power consumption of silicon melting at cold crucible system for EMC

different casting speeds are also measured in the range from 1.0 to 3.0 mm/min. It resulted in that the conversion efficiency was unchanged with the casting speed.

Electric Power Consumption of EMC

Energy used for EMC is supplied by electric energy. At the technique of cold crucible induction melting, the surface of high temperature molten metal is so opposed to the cold surface of water-cooled crucible wall that heat loss from the melt is most disadvantageous for melting. The demerit of this technique at the large expense of electric power consumption must be compensated by other merits of the technique, namely, high speed casting and less consumption of consumables.

The electric power consumed at EMC for the 22×22 cm^2 cross sectioned ingot was monitored, and the energy distribution was determined by the measurement of temperature rise at the cooling water system. The result is shown in Fig. 5. At the EMC operation of 22 ×22 cm^2 cross sectioned ingot, an electric energy of 265 kWh was consumed for one hour, and 69.8 % (185 kWh) out of it was distributed to the silicon melting system, 18.2 % (48 kWh) was to the after-heater system, and the rest of it was dissipated to the radio-frequency generator system (9.4 %, 25 kWh) and the furnace chamber (2.6 %, 7 kWh) as the heat loss. The detail of the energy consumption at silicon melting system was calculated by the computer simulation using a three-dimensional finite element method. Fig. 6 is the computational result, and shows Joule heat generation at each element of the cold crucible melting system. Nearly two-thirds (62 %) of total electric energy were dissipated at the cold crucible as a useless heat generation. Only a 37 kWh (20 %) out of 185 kWh was utilized for the silicon melt heating. An energy loss of 17 % was also generated at the induction coil.

Cost Comparison of EMC and CZ methods

Crystalline silicon substrates for solar cells are provided from two kind of crystals, namely, single and multi crystals. The silicon solar cell conversion efficiency of single crystal is superior to that of multicrystal. Meanwhile the multicrystal has a potential for low cost production. The manufacturing of single crystal is done by the Czochralski pulling, and the multicrystal is manufactured by the casting method.

The cost analysis is compared between the single crystal production of 6 inches diameter ingot by Czochralski pulling and the multicrystal production of 33 ×33 cm^2 square cross sectioned ingot by EMC. The direct production cost and the depreciation cost are shown in Table 3. The direct production cost is converted to 100 for the present technology level of Czochralski pulling with an annual production of 50 tons, and other comparative numbers are listed in the table. A high speed Czochralski pulling is postulated as the doubled pulling speed (annual production of 100 tons) comparing to the present technology. For the EMC technology, the cost is estimated at the annual production of 100 tons. The direct production cost of the high speed Czochralski pulling can be reduced to 78 % compared with the present technology, meanwhile it can be reduced to 49 % by the EMC technology. The depreciation costs amount to 17 in EMC, and to 22 and 11 in Czochralski pullings.

Table 3. Comparison of direct cost and depreciation cost between CZ pulling and EM-Casting

(Direct Cost; CZ pulling at Present = 100)

	CZ (6 in.-φ)		EMC (33×33cm²)
	Present	High-Speed(x2)	(in 1996)
Production Scale [Ton/Y]	50	100	100
Si-Material	20	20	20
Consumables	44	39	7
Electricity	16	9	11
Labour	20	10	11
Direct Cost	100	78	49
Depreciation Cost	22	11	17

Future Task of EMC Technology

Multicrystalline silicon ingot was successfully produced by the EMC technique in the ingot size of 22 ×22cm^2 cross section with 170 cm length. However, future tasks are still necessary for the promotion of this technology to provide lower cost solar cells, namely,

1) expansion of ingot size to 35 ×35 cm^2 square cross section, and
2) improvement of electrical properties up to 150 μm minority carrier diffusion length.

This work was a part of NEDO new sunshine program at MITI of the Japanese Government.

RECENT ADVANCES OF THIN FILM SOLAR CELLS AND THEIR TECHNOLOGIES

Yoshihiro HAMAKAWA
Faculty of Engineering Science, Osaka University
Toyonaka, Osaka 560, Japan

ABSTRACT

The current state of the art in amorphous silicon (a-Si:H) and thin film solar cell R & D efforts and their photovoltaic system applications are reviewed. Firstly, progress in a-Si alloy production technologies is overviewed with their significance as the champion material for a low cost solar cells. Secondly, some new approaches and key technologies to improve solar cell efficiency with stabilized performance by new amorphous silicon alloys and polycrystalline thin film for a-Si basis stacked solar cells are demonstrated. Recent aspect of some other new material devices such as CIS and CdS/CdTe are also introduced. In the final part of paper, progress of the conversion efficiency in various types of amorphous silicon solar cells are surveyed and summarized.

INTRODUCTION

Developing clean energy resources as alternative to oil has become one of the most important tasks assigned to modern science and technology. The reason for this strong motivation is to stop air pollution resulting from the mass consumption of fossil fuels and to keep the ecological cycles of the biosystems on the earth. Views of future energy envision that the energy structure in the 21st century will be characterized as a "Best Mix Age" of different energy forms. Among a wide variety of renewable energy projects in progress, photovoltaics is the most promising one as a future energy technology. It is pollution-free and abundantly available everywhere in the world, even in space, and also can operate by diffused light.

A big barrier impeding the expansion of the large-scale power source application in the photovoltaic system was the high price of solar cell module, which was more than $5/Wp(¥600/Wp) in 1992. Therefore, the cost reduction of the solar cell is of prime importance. To achieve this objective, tremendous R & D efforts have been made in a wide variety of technical fields from solar cell material, cell structure, and mass production processes to photovoltaic systems over the past ten years. As the result, about an order of magnitude cost reduction has been achieved in the period of 1980. However, another one fifth of cost reduction should be required to levelize the present conventional electricity price according to the recent survey report [1]. To enhance the utilization of large scale merit, some new strategies such as PV roof construction materials, building solar tiles, PV electricity purchasing system have been initiated in the recent few years.

In this paper, present status of the a-Si solar cell R & D efforts with some new trials on the polycrystalline thin film are reviewed in the first part. Then, some new approaches and key technologies to improve cell performances in progress are demonstrated. Finally, recent topics in the application and industrialization fields are reviewed and discussed.

Table 1 Significance of a-Si alloy as a solar cell material

A Physical Properties	B Fabrication Technology
a-1) Excellent photoconductivities with high absorption coefficient for sun light *	b-1) Wide area thin film * # (Plasma CVD, ECR CVD, etc.)
a-2) Low dark conductivity	b-2) Low temperature deposition * # $100°C < T_s < 380°C$
a-3) High structural sensitivity @ (Valency electron controllability)	b-3) p-n control can be accomplished only by mixture gas regulations #
a-4) Wide range of energy gap controllability @ $1.0\ eV < \varepsilon_{gopt} < 3.6\ eV$ a-SiGe - a-Si - a-SiC	b-4) Can be deposited on any in expensive substrates
a-5) Easy to make heterojunction # (Relatively low interface states)	b-5) Easy to apply integration technology # (HJ, super lattice, tandem cells)
a-6) Mechanically strong (because of amorphous network)	b-6) Low cost and good massproduceability #

First WCPEC; Dec. 5–9, 1994; Hawaii

PROGRESS IN a-Si ALLOY TECHNOLOGIES

As has been reported elsewhere, a-Si alloys can be deposited onto any inexpensive substrates with low temperature less than 300°C. Moreover, it is possible to form very wide area thin films due to a vapor phase growth onto non-crystalline substrates. Another noticeable property of these a-Si alloys is existence of valency controllability by doping of substitutional impurity atoms with mixing of impurity gasses such as B_2H_6 and PH_3, etc. Therefore, in view of production sequence, this material system has an excellent massproduceability with large scale merit. Significances of a-Si alloys as a solar cell materials are summarized in Table 1.

As the preparation technology for a-SiC alloy, the plasma CVD is now widely utilized everywhere. While ECR CVD and Ion-beam CVD (IB-CVD) are intensively investigated in a recent few years. The unique advantage of the ECR CVD is that the growing surface receives almost no bombardment damages by electrons and/or other heavy species soft landing having an energy of several tens of eV [2]. This effect might result not only in prevention of weak bonds from being introduced into the network but also suppression of the diffusion of long-lifetime radical species due to the raised surface temperature. It is expected that films with dense network and low defect density are formed.

For the deposition of a-SiC and μc-SiC, hydrogen is used as an ECR plasma excitation gas, and a mixture of SiH_4, CH_4 and B_2H_6 or PH_3 are usually employed as a reaction gas for the growth of p and n type SiC:H. Details of the preparation conditions are reported in reference [3]. Since the operation pressure is in the range of 10^{-3} and 10^{-4} Torr, the lifetime of chemically active hydrogen radicals is quite long, so that a large amount of hydrogen radicals will reach the growing surface and play an important role in determining the properties of growing films. Therefore, the dependence of the material properties on the hydrogen dilution ratio in the reaction gas has been investigated. Figure 1 shows the dependence of the optical energy gap and dark conductivity of the samples on the H_2 dilution ratio. As the ratio increases, the optical gap (E_o) and also the dark conductivity (σ_d) of both p- and n-type films increase.

As can be seen from Fig. 1, there are two main factors which determine the optical energy gap; one is the composition ratio of Si:C:H corresponding to the source gas ratio CH_4/SiH_4, and the other is the grade of H_2 dilution which might be related to the details of the network structure. The film properties are strongly dependent not only on the substrate temperature and microwave power but also on the ratio of hydrogen to reaction gases $H_2/(CH_4+SiH_4)$ during deposition. Although the optical energy gap increases with the flow rate of CH_4, the effect is not as remarkable as the dependence of hydrogen dilution. Hydrogen dilution has the effect of reducing the hydrogen content in the film, and also of enhancing the degree of microcrystallinity.

The formation of Si and SiC microcrystallites is confirmed by Raman spectra as mentioned in the original work [2]. The Raman spectrum of the films prepared at microwave powers higher than 250Watts exhibits distinct structures at around 520 and 740 cm^{-1}, which correspond to TO phonon modes of crystalline Si and SiC clusters.

The conductivity of μc-SiC:H can also be controlled by adjusting the flow ratio of dopant gas to host reaction gas. It has been shown in the valency electron control studies [3]. The total doping efficiency in p μc-SiC:H is

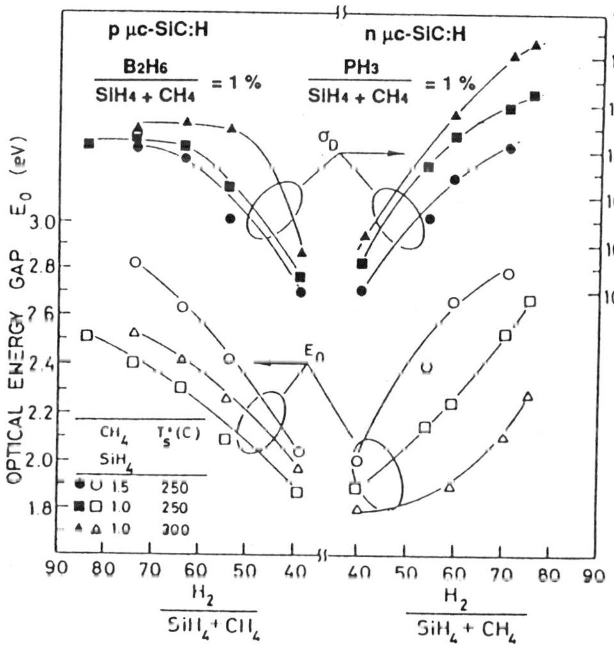

Fig. 1 Dependence of the optical energy gap and dark conductivity of p- and n-type μc-SiC:H on the hydrogen dilution ratio in the reaction gas in ECR CVD.

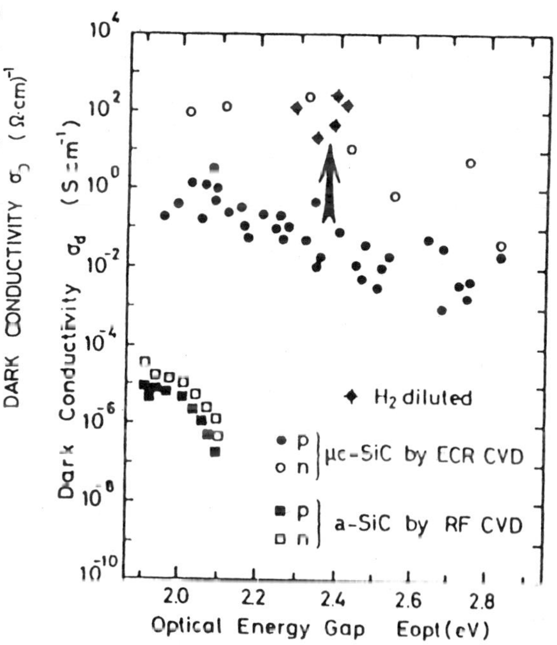

Fig. 2 Relationship between dark conductivity and optical energy gap of amorphous and microcrystalline SiC:H prepared by RF and ECR plasma CVD.

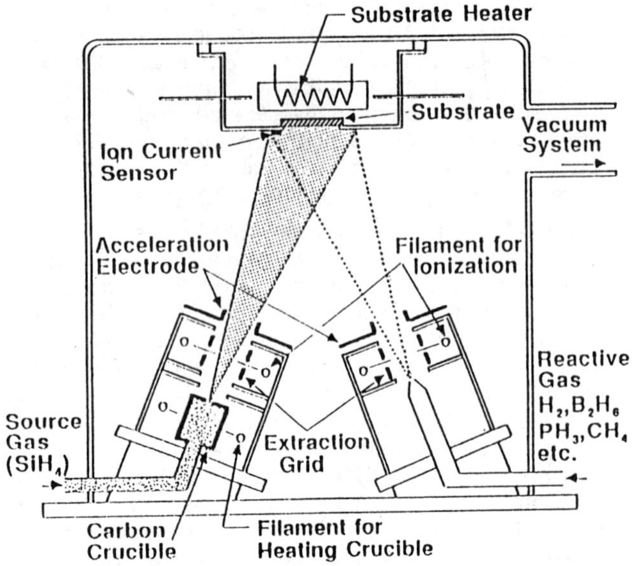

Fig. 3 A schematic representation of IBD (Ion Beam Deposition) system.

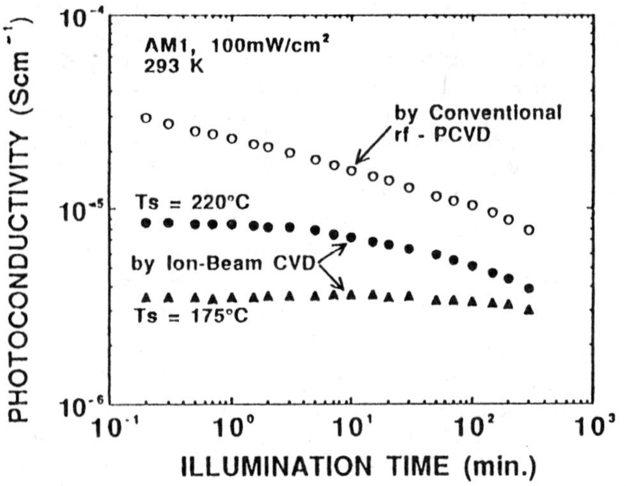

Fig. 4 Change in the photoconductivity with illumination time in IBD produced and PCVD produced a-Si:H films.

higher than that in p a-SiC:H by several orders of magnitude. Figure 2 summarizes the relation between the dark conductivity and the optical energy gap of p- and n-type a-SiC:H prepared by conventional RF plasma CVD and p- and n-type μc-SiC:H prepared by ECR plasma CVD. As the optical energy gap increases, the dark conductivity of the films prepared by the RF plasma CVD rapidly decreases, while that of films prepared by the ECR plasma CVD remains higher than $10^{-3}Scm^{-1}$ even when the optical energy gap exceeds 2.5eV. In the figure, the star marked highly H_2 diluted p-type μc-SiC:H films has more than $3x10^{19}$ cm^{-3} hole concentrations with $24cm^2/Vs$ mobility which are identified by Hall measurements [2].

Another topics in the material preparation technology field is the IBD (Ion Beam Deposition) method [4]. A high stability film against the light-induced degradation with high deposition rates is still important key issue in the field. To conquer these problems, tremendous R & D efforts have been in progress by the wide varieties of approaches, e.g.; the Chemical Annealing Treatment (CAT) [5], Intense Xenon Light Pulse Assisted Plasma CVD [6], IBD and so on. Among these, IBD method is a unique challenge in views of a wide range controllability on both hydrogen content and decomposed species ion energy.

Figure 3 shows a schematic illustration of the IBD system. A typical a-Si film growth condition in this system is: the ion acceleration voltage V_{acc}=100-300V, ionic current I_e, with voltage V_e are 0.8A at 400V, and substrate temperature T_S=100–300°C, while chamber base pressure is 10^{-4} Torr or less.

As the preliminary experiment, a systematic investigations on the undoped a-Si film deposition has been made at a series of deposition parameters. The

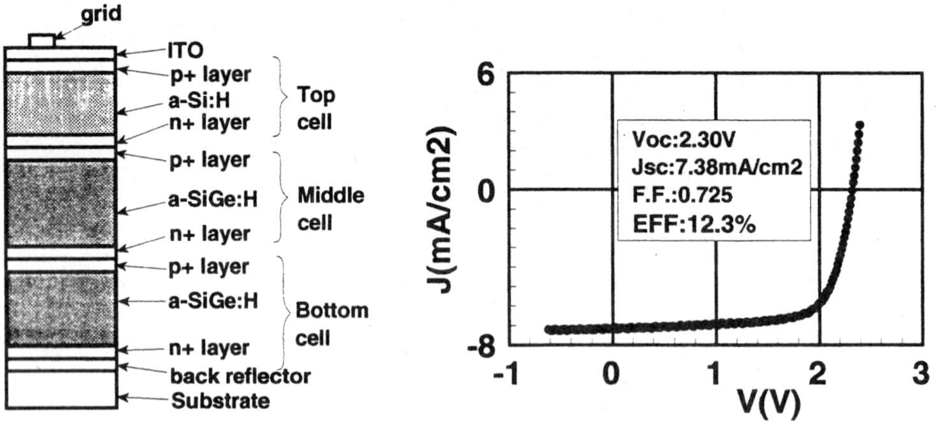

Fig. 5 Structure and V-I characteristics of triple stacked cell fabricated by the highest deposition rates of a-Si:H (7.5Å/s)/a-SiGe:H(100Å/s)/a-SiGe:H(100Å/s) (presented by Canon Corp.).

result shows a considerably good film quality having 10^4–10^6 photo- to dark-conductivity ratio σ_{ph}/σ_d with optical energy gaps of 1.7–1.8eV. A noticeable feature of IBD produced film is a better stability against light exposure. Figure 4 shows a comparison of changes in the photoconductivity σ_{ph} with AM1 light illumination for IBD produced film and conventional plasma CVD produced film [4].

An increase of deposition rate is also another important key issue for the a-Si basis solar cells. Because it rises up the through-put in the mass production line which might be directly connected to the reduction of module cost. Recently, Cannon group has announced the highest deposition rate of 100Å/s for the amorphous silicon germanium alloy (a-SiGe) film [5] in their roll-to-roll process. Figure 5 shows the construction of three stacked solar cell (a) and its output V-I characteristics having the initial efficiency of 12.3% (b). As a summary of this section, Fig. 6 shows plots of recent results of dark σ_d and photoconductivity σ_{ph} of a-Si alloys with their optical energy gap.

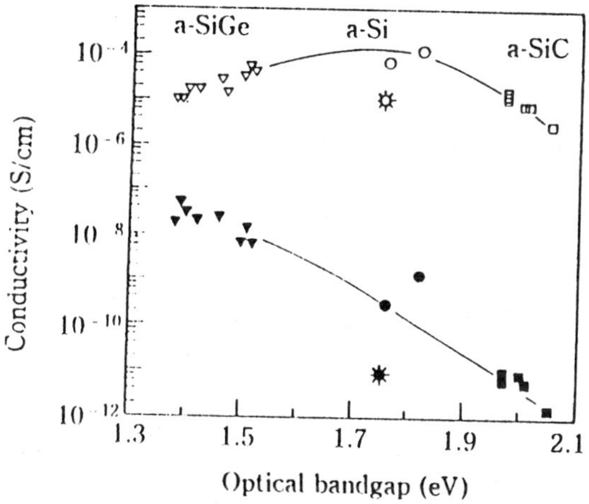

Fig. 6 Dark and photo-conductivity of a-SiGe:H (triangles), a-Si:H (circles), a-SiC:H (squares) versus optical gap. Star marks ☼ and ✸ show the property of IBD produced film.

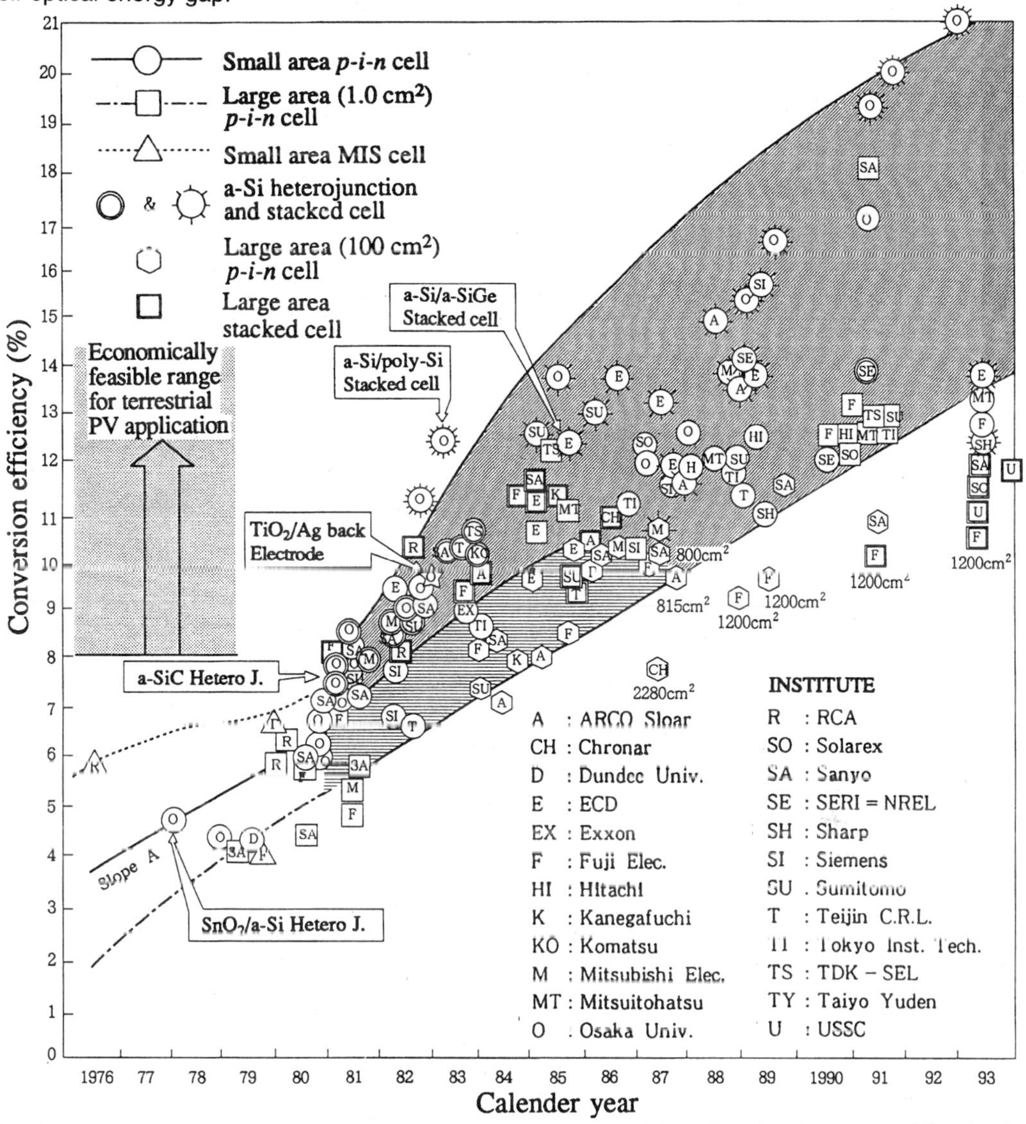

Fig. 7 Progress of a-Si solar cell efficiencies for various types of junction structures, as of September 1994.

R&D EFFORTS ON PV PERFORMANCE IMPROVEMENTS

A wide variety of R & D efforts for high-efficiency with low cost approaches have been in progress on each process of the PV conversion steps to develop. Among these, a remarkable advances has been seen in the technology of the wide gap windows heterojunction, graded band profiling, doping, superlattice, BSF (Back Surface Field Effect) treatment and also the stacked junctions with new materials such as a-SiC, μc-SiC alloys. As the results, amorphous silicon solar cell efficiency is improving day by day as shown in Fig. 7. As can be seen from the figure, a steep slope change is seen with the inventions of a-Si alloys such as a-SiC in 1981, a-SiGe in 1985, and also stacked junction solar cells. According to a histogram of several thousands modules massproduction data [6], the In-Line module efficiency is being reached to 9%, and several institutes reported more than 10% efficiency for the 100 cm^2 or more area as the top data in a recent few years. While, the laboratory phase efficiencies are more than 13% with a-SiC/a-Si heterojunction on the single junction cells as shown in Fig. 8 for example [7].

As it has been reported elsewhere [8] that the amount of percentage degradation increase with increasing i-layer thickness. The reason is that the volume recombination of the photo-generated carriers become relatively large with decreasing the lowest electric field in the i-layer [8]. For the purpose of suppressing this effect, the tandem type solar cell has been recommended as a more reliable a-Si solar cell [8-9]. Along with this concept, wide varieties of experimental trials are in progress on the multi-bandgap stacked solar cells [7-16]. For example, on large area (1200 cm^2) a-Si/a-Si stacked solar cells, the efficiency of 10% is reported by Fuji Electric Co. Ltd. as shown in Fig. 9 [12].

Recently, a concept of the band profiling design has been initiated as an optimum design of the ambipolar carrier transport in i-layer of the multi-bandgap junction. As an example of the band profiling study, a result on the a-SiC/a-Si/a-SiGe triple band gap tandem solar cell is shown in Fig. 10 [17]. As can be seen from the figure, an initial conversion efficiency of 12.4 % has been achieved.

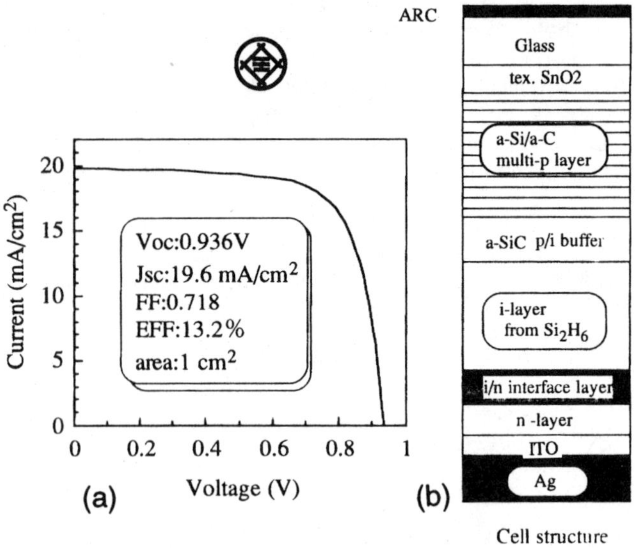

Fig. 8 I-V characteristics of a-Si single junction solar cell (a) and its junction structure (b) (presented by Mitsui Toatsu Chem. Inc.).

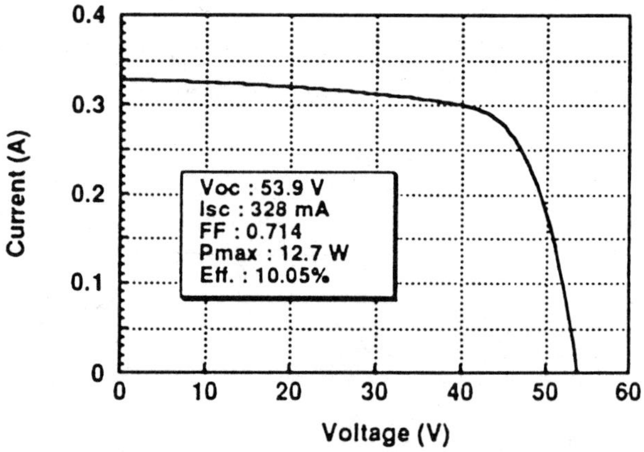

Fig. 9 Output characteristics of a-Si//a-Si double junction stacked submodule having 30x40 cm^2 area (presented by Fuji Electric Co. Ltd.).

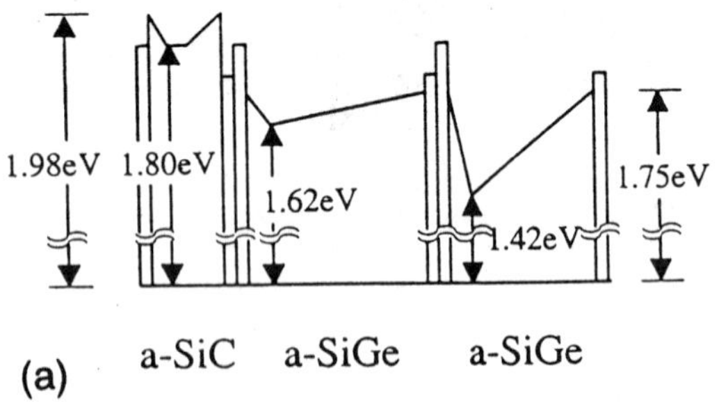

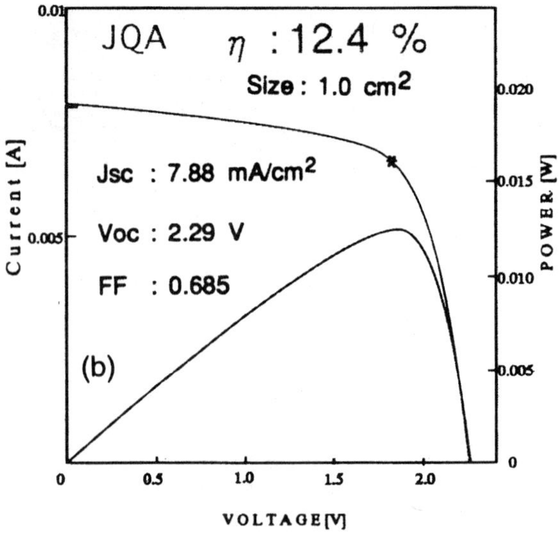

Fig. 10 An example of band profiling study on triple junction stacked cell (a) and its V-I characteristics (b) having the initial efficiency of 12.4% (presented by Sharp Corp.).

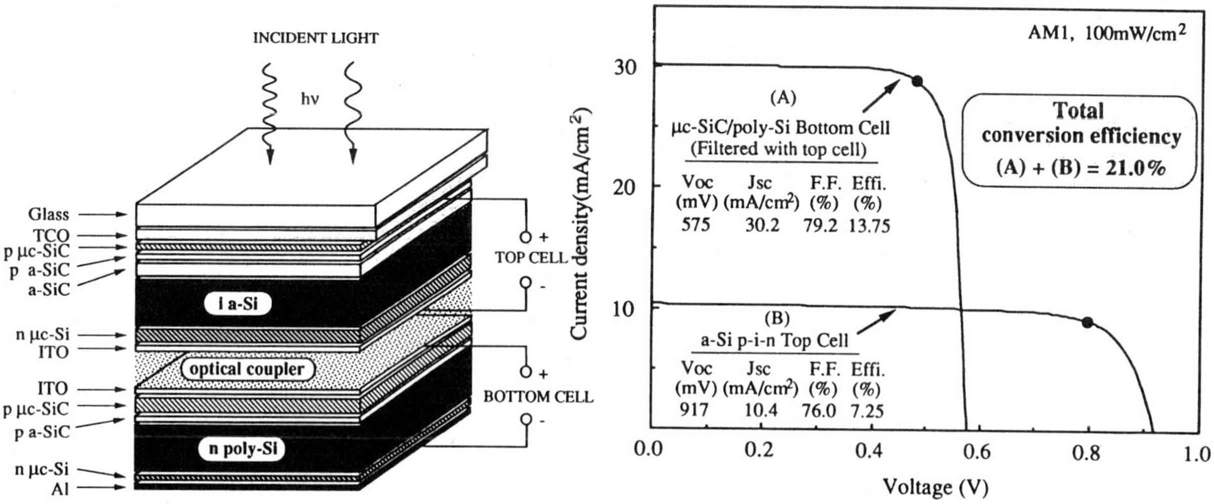

Fig. 11 Cell structure (a) and V-I characteristics of a-Si//poly-Si 4 terminal stacked solar cell by Osaka Univ.

It has been shown in the early stage of the stacked junction solar cell in 1987 [11], the best combination of stacked junction material with a-Si would be poly-Si thin film in viewpoints of theoretical efficiency, stability and cost [15]. Then, a series of systematic investigations has been initiated elsewhere. As a result, on the two-terminal a-Si solar cell stacked with poly-Si [14], a conversion efficiency of 15.04% with V_{OC}=1.478 V, J_{SC}=16.17 mA/cm^2 and FF=63% has been obtained under AM1 illumination. Quite recently, by the same combination on four-terminal cell, 21.0% efficiency has been obtained by Ma Wen et al. [16]. Table 2 shows the summarized results of a-Si solar cell performance reported as of May, 1993, on single junction solar cells of R&D phase (a), submodules (b) and a-Si basis stacked junction solar cells (c).

TOPICS IN THE POLYCRYSTALLINE THIN FILM SOLAR CELLS

As a candidate of the most promised bottom cell stacked with the a-Si basis tandem solar cell discussed in the former section, a considerable attention has been recently gathered on the polycrystalline silicon thin film. To realize this material combination so called "Honeymoon Cell" [11], a number of new technological approach to fabricated polycrystalline thin film grown or deposited on inexpensive substrates has been going on elsewhere. For example; "Molecular Beam Graphoepitaxial Growth (MBEG)" [11], "Solid Phase Crystallization (SPC) Method" [18], "Zone Melting Recrystallization (ZMR) Method" [19] and "Plasma Spray of Silicon Growth (PSSG)" [20].

There are no space to explain details of those technologies. Some final results are introduced as following. Figure 12 shows the junction structure and V-I characteristics of SPC grown n-type poly-Si basis solar cell. The poly-Si film was grown from 10μm thick GD a-Si film deposited on a textured substrate with a high substrate temperature of 500–600°C. The n-type film has an electron Hall mobility of 623cm^2/Vs with the grain size around 6μm. As can be seen from the figure, 8.5% efficiency with I_{sc}=24.2A/cm^2 and V_{oc}=0.537V obtained was an excellent performance in considering the active layer thickness of only 10μm poly-Si.

Figure 13 shows a schematic illustration of ZMR

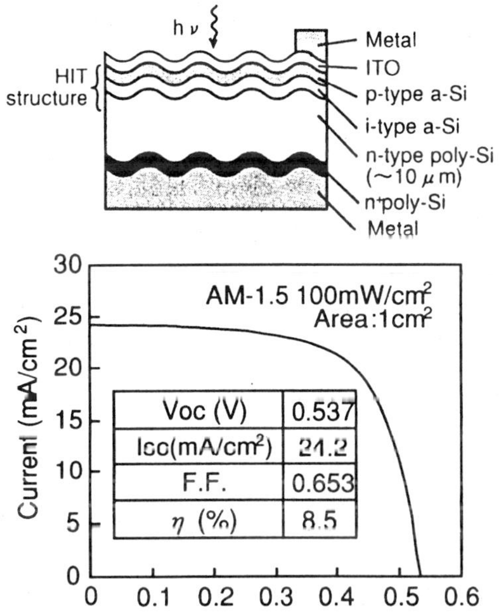

Fig. 12 Junction structure (a) and V-I characteristics of SPC produced poly-Si solar cell (b) (presented by Sanyo Electric Co. Ltd.).

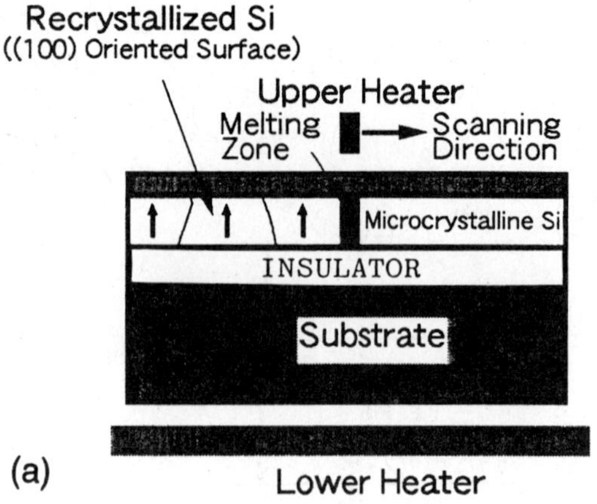

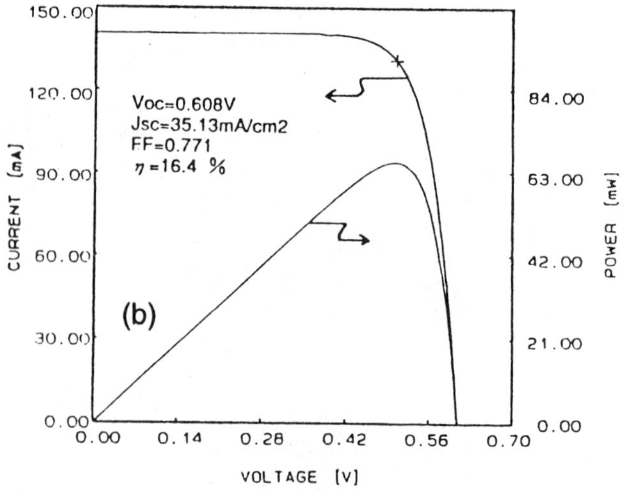

Fig. 13 A schematic illustration of ZMR technology (a) and output characteristics of ZMR grown poly-Si solar cell (b)

method and V-I characteristics of ZMR poly-Si solar cell. As it has been seen in the figure (a), 3–5μm microcrystalline silicon (μc-Si) is deposited by LP-CVD on the SiO_2 coated silicon substrate, then SiO_2 coating is made on the deposited μc-Si. After that, ZMR processing is carried out by a traveling copper heater to perform a kind of zone melting grain growth. As a final step, 50–60μm active layer has been formed by the thermal CVD method. As a top data, 16.4% efficiency has been obtained so far with 2x2 cm^2 cell area.

In the way of R&D efforts for the honeymoon cell technology, an interesting byproduct has born. That is, a- or μc-Si (SiC) / poly-Si heterojunction solar cell as shown in Fig. 14 [21]. Significances of this kind of cell are, 1) a very simple fabrication processing, 2) low cost, 3) more than 17% efficiency easily obtained. Recently, Sanyo group [22] has reported 21% efficiency with HIT (Heterojunction with Intrinsic Thin-layer) structure junction. If the low cost poly-Si thin-film production in progress might become a realistic technology, this kind of a-Si/poly-Si heterojunction solar cell will be one of the most promised champion in the PV technology. Recently CIS and CdS/CdTe solar cell technologies have also been remarkably in progress [23]. Because of no area in the paper, some top records are summarized in Table 2.

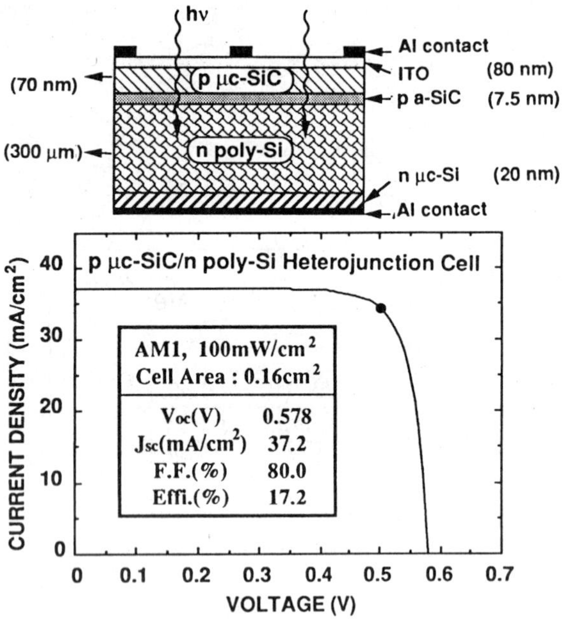

Fig. 14 μc-SiC/poly-Si heterojunction solar cell and its output characteristics (presented by Osaka Univ.).

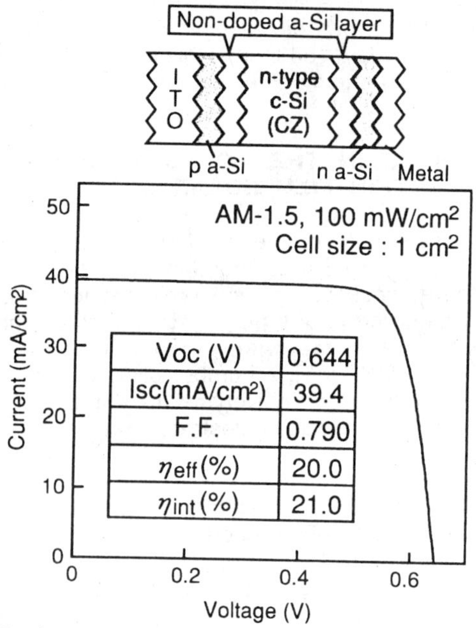

Fig. 15 Junction structure and I-V curve of an a-Si/c-Si heterojunction solar cell (presented by Sanyo).

Table 2. Summary of some top efficiency records in thin film solar cells and modules (As of Sept.,1994)

		Voc (V)	Jsc (mA/cm^2)	F.F. (%)	η (%)	A (cm^2)	Institute	Remarks	
a-Si Single Junction Small Area Cells More Than 12% Efficiency		0.967	17.7	70.3	**12.0**	0.033	Osaka Univ.	p μc-SiC (ECR)	[87]
		0.891	19.13	70.0	**12.0**	1.0	Solarex		[89]
		0.927	18.4	70.5	**12.0**	1.0	Fuji-Elect.	BF3 Pulse CVD	[91]
		0.90	18.9	72.0	**12.3**	0.09	TIT	δ-doped p layer	[90]
		0.923	18.4	72.5	**12.3**	1.0	Fuji-Elect.	Pulse CVD; p-a-SiC:H	[92]
		0.899	18.8	74.0	**12.5**	1.0	Fuji-Elect.	p a-SiO:H/a-Si:H/n a-Si:H	[93]
		0.885	19.13	74.7	**12.65**	1.0	SEL	Reverse Bias Annealing	[91]
		0.887	19.4	74.1	**12.7**	1.0	Sanyo	Superchamber	[92]
		0.909	19.8	73.3	**13.2**	1.0	Mitsui-Toatsu	a-C/a-Si ML/a-SiC/a-Si	[92]
a-Si Sub-Modules (M) More Than 10%		12.53/M	130.1/A	73.5	**12.0**	10x10	Sanyo	TCO improvement	[93]
		2.409/M	611.6/A	68.6	**10.1**	10x10	Mitsubishi	a-Si/a-Si/a-SiGe	[92]
		53.9/M	328/A	71.4	**10.05**	30x40	Fuji-Elect.	a-Si/a-Si, a-SiO p layer	[93]
a-Si Basis Stacked Solar Cells	2 T	1.75	8.16	71.2	**10.2**	1.0	Solarex	a-SiC/a-Si	[93]
		1.80	9.03	74.1	**12.0**	1.0	Fuji	a-Si/a-Si	[93]
		2.29	7.90	68.5	**12.4**	1.0	Sharp	a-SiC/a-SiGe/a-SiGe	[92]
		2.32	7.30	73.0	**12.4**		Sumitomo	a-Si/a-Si/a-SiGe	[92]
		2.55	7.66	70.1	**13.7**	1.0	ECD/Sovonics	a-Si/a-Si /a-SiGe	[92]
		1.48	16.2	63.0	**15.0**	0.033	Osaka Univ.	a-Si/c-Si	[91]
		2.4	6568/A	67.5	**11.8**	905.9	Canon/USSC	a-Si/a-SiGe/a-SiGe	[93]
	4 T	0.871	16.4	72.0	**10.3**		ARCO	a-Si//CuInSe2	[91]
		0.432	17.4	68.0	**+ 5.3 = 15.6**				
		0.917	10.4	76.0	**7.25**	0.033	Osaka Univ.	a-Si//poly-Si	[92]
		0.575	30.2	79.2	**+ 13.75 = 21.0**				
Poly-crystalline Thin Film Solar Cells		0.66	31.5	78.7	**16.4**	0.4	NREL	MgF2/ZnO/CdS/CIS	[94]
		0.84	25.1	74.5	**15.8**	1.0	Univ.S.F.	MgF2/SnO2/CdS/CdTe	[92]
		0.69	7488/A	68.0	**12.7**	69	Siemens Solar	ZnO/CdZnS/CIGS	[94]

References

[1] K. Nakagawa: *Japan 21st,* Vol. 39, No. 9 (1994).

[2] Y. Hamakawa, Y. Matsumoto, G. Hirata and H. Okamoto: *Mat. Res. Soc. Symp. Proc.* **164** (1990) 291.

[3] Y. Hamakawa and H. Okamoto: *Mat. Res. Soc. Symp. Proc.* **242** (1992) 651.

[4] Y. Hamakawa, K. Hattori and H. Okamato: *Report of a-Si Solar Cell Contractors Meeting, Sunshine Project, MITI,* May 8-10 (1990).

[5] H. Shirai, D. Das, J. Hanna and I. Shimizu: *Tech. Digest of PVSEC-5,* Kyoto (1990) 59.

[6] Y. Tawada and H. Yamagishi: *Report of a-Si Solar Cell Contractors Meeting, Sunshine Project, MITI,* May 8-10 (1990).

[7] H. Tanaka, N. Ishiguro, Y. Ashida and N. Fukuda: *Proc. 23rd IEEE PVSC,* (1993) 811.

[8] Y. Nakata, H. Sannomiya, K. Nomoto and T. Tsuji: *Optoelectronics –Devices and Technologies–,* Vol. 5, No. 2 (1990) 209.

[9] Y. Hamakawa: *Current Topics in Photovoltaics,* ed. T. J. Coutts and J. D. Meakin (Academic Press, New York, 1985) pp. 111-167.

[10] For example, S. Tsuda, H.Tatsui, T. Matsuyama, H. Haku, K. Watanabe, Y. Nakashima, S. Nakano and Y. Kuwano: *Proc.2nd Int. Photovoltaic Science and Engineering Conf.,* Beijing (1986) 409.

[11] Y. Hamakawa, Y. Matsumoto, Xu Chong-Yang, M. Okuyama, H. Takakura and H. Okamoto: *Mat. Res. Soc. Symp. Proc.* **70** (1986) 481.

[12] Y. Ichikawa, S. Fujikake, T. Katayama, S. Saito, H. Ota, T. Yoshida, T. Ihara and H. Sakai: *Proc. 23rd IEEE PVSC* (1993) 27.

[13] R. R. Arya, L. Yang, M. Benett and A. Wilczynski: *ibid.* (1993) 790.

[14] A. Banerjee, E. Chen, R. Clough and S. Guha: *ibid.* (1993) 795.

[15] Y. Hamakawa: *Solar Energy Materials* ***23*** (1991) 139.

[16] W. Ma, T. Horiuchi, C. C. Lim, M. Yoshimi, H. Okamoto and Y. Hamakawa: *Proc. 23rd IEEE PVSC* (1993) 338.

[17] M. Ito, T. Tomita, Y. Takeda and Y. Yamamoto: to be publ. in *Proc. 1st WCPEC,* Hawaii (1994).

[18] T. Matsuyama, T. Baba, T. Takahama, K. Wakisaka and S. Tsuda: *Optoelectronics –Devices and Technologies–,* Vol. 9, No. 3 (1994) 391.

[19] A. Takami, S. Arimoto, H. Morikawa, S. Hamamoto, T. Ishihara, H. Kumabe and T. Murotani: *Proc. 12th EC-PVSEC,* Amsterdam (1994) 59.

[20] K. Kumagai: *Report of 31st Solar Energy Tech. Promotion Committee, New Sunshine Project HQ, MITI,* (June, 1994) 83.

[21] W. Ma , T. Horiuchi, M. Yoshimi, K. Hattori, H. Okamoto and Y. Hamakawa: *Proc. PVSEC-6,* New Delhi (1992) 463.

[22] K. Kumagai: *Report of 31st Solar Energy Tech. Promotion Committee, New Sunshine Project HQ, MITI,* (June, 1994) 87.

[23] L. L. Kasmerski: *Proc. 3rd WREC,* Part I (1994) 252.

Japanese Space Solar Cell Activities
- GaAs and Si -

Sumio Matsuda, Yasunari Yamamoto, Osamu Kawasaki

Parts and Material Technology Laboratory,
Space Subsystems and Technology Department,
Office of Research and Development
National Space Development Agency of Japan

2-1-1, Sengen, Tsukuba, Ibaraki 305, JAPAN

ABSTRACT

This paper describes Japanese Space solar cell activities using about GaAs and Si materials. First, we review GaAs solar cell performance, design, fabrication and flight data of Japanese domestic communications satellite-3 (CS-3). We introduce Japanes next satellite employing GaAs solar cell employed satellite in Japan, which is called COMETS for Communications and Broadcasting Engineering Test Satellite. Second, we review Si solar cell performance, design, fabrication and so on. And we discuss a high efficiency Si solar cell which is 100μm Non Reflective Surface (NRS) / Local Back Surface Field (LBSF) and average efficiency of 18.0% at AM0 conditions. Third, we analyze flight data of Solar Cell Monitor (SCM) on Engineering Test Satellite-V (ETS-V). Through collecting the SCM datas, we could observe degradation phenomenon, due to solar flares which occurred in October 1989.

Introduction

Solar cells have been a dependable power source for use in space for the last 30 years, and have served as the primary power source for satellites. The performance of Si solar cells has increased from 10% air mass zero (AM0) solar energy conversion efficiency in the early 1960s to almost 15% on today's spacecraft. However, as larger satellites with greater output and longer life are developed, and their power supplies will require solar cells with higher efficiency, lighter weight and longer life to increase the specific power (W/Kg) and the areal power density (W/m^2) of the solar array. Reduction of solar array mass is especially important for high earth altitude or geosynchronous orbit missions due to launch vehicle cost and a restricted launch vehicle capability to boost payloads into high earth orbit. In addition, reducing solar array mass and/or increase in specific power enables a greater allocation for more sensors, transponders, or additional payload on existing spacecraft. Furthermore, higher areal power densities are important for low-altitude earth orbit due to the environmental drag on the surface of the solar array, causing altitude adjustments with an auxiliary reactor control gas thruster subsystem decreasing the fuel supply and mission life.

Gallium arsenide (GaAs) solar cells have been expected to supplant Si solar cells because of their high efficiency and high radiation tolerance[1]-[4]. GaAs solar cells were adopted for main solar arrays to satisfy the requirements of CS-3, which were successfully launched in 1988 and have been generating almost the expected power.

Meanwhile , thin (50μm to 100μm) silicon solar cells have the advantages of lighter weight and higher radiation resistance than thicker (≧ 200μm) solar cells. Since the development of the 50μm and 100μm thin silicon solar cells in 1983, they have been used for many Japanese satellite programs.

NASDA (National Space Development Agency of Japan) has launched and had put it into a geostationary orbit at longitude 150 degrees east . This spacecraft has the Technical Data Acquisition equipment (TEDA) to obtain technical data for developing spacecrafts. The TEDA consists of 8 monitors which are Dosemeter (DOM), Discharge Monitor (DIM), Potential Meter (POM), Integrated Circuit Monitor (ICM), Solar Cell Monitor (SCM), Paddle Vibration Monitor (PVM), Thermal Control Material Degradation Monitor (TDM) and RAM Soft-error Monitor (RSM).

The status and trends in development of several kinds of GaAs and Si solar cells for space use are shown in Table 1. Table 2(a) and (b) show solar cell application of NASDA and The Institute of Space and Astronautical Science (ISAS). Figure 1 shows the improvement in space solar cell energy conversion efficiency in Japan.

CH3365-4/94/0000-0042 $4.00 © 1994 IEEE First WCPEC; Dec. 5-9, 1994; Hawaii

GaAs Solar Cell

GaAs solar cell has anAlGaAs/GaAs heretofore structure, which consisting of p-AlGaAs/p-GaAs/n-GaAs buffer layer /n-GaAs substrate. This type of heretofore structure can be produced by VSTC-LEP[5]. The radiation resistance of p-AlGaAs/p-GaAs/n-GaAs heretofore structure solar cells is known to vary greatly with the thickness of the p-GaAs layer (i.e., the pn junction depth Xj), so Xj is controlled to 0.5±0.1μm[6],[7].

The p-GaAs layer thickness is less than 0.2μm to reduce the light absorption in the short-wavelength region. The CVD Si_3N_4 film with excellent chemical resistance and moisture resistance was used as an anti-reflective (AR) coating [8]. The p and n electrode are metallized with evaporated silver of 0.4μm or greater in thickness, which is suitable for either soldered or welded interconnections. The cell dimension for CS-3 are 2cm × 2cm square and 280μm in thickness. The top on the cell was covered with a 150μm-thick coverglass to prevent the low energy proton damage.

CS-3 a and b were launched by Japanese H-1 rocket in February and September in 1988 from NASDA's Tanegashima Space Center and have been generating almost the expected power. GaAs solar cells were adopted for the main solar array to satisfy the CS-3 power requirement.

The single mount method [9] was applied for solar array assembly to make the maximum use of the solar array area.

Interconnectors are made of silver-plated KOVAR plate (40μm thickness). Silver-plated KOVAR plate is punched and shaped to form the interconnector.

KOVAR was selected for the interconnector because its coefficient of thermal expansion is the same as that of GaAs crystal. This similarity of coefficient of thermal expansions and the out-of-plane stress relief loop in two direction reduce thermal induced stress of solar cells, and increase the interconnector fatigue life to 10^3cycles (about 11.3 years on GEO).

To investigate the radiation degradation differences between GaAs solar cells and Si solar cells, the annual generated powers of CS-3a, ETS-V and BS-2b were compared. ETS-V was launched on August 1987, and Broadcasting Satellite-2b (BS-2b) was launched on February 1986.

The solar arrays of ETS-V and BS-2b use silicon solar cells. Their annual power degradation from 1987 to 1988 when there were no large solar flares was about 3 to 4 percent. However their degradation was 8 to 10 percent from 1988 to 1989 because many solar flares occurred in 1989.

In contrast, the CS-3a solar array, which uses GaAs solar cells, showed only a 2 percent degradation from 1988 to 1989, which includes a large solar flare on October in 1989.

As mentioned above, the annual radiation degradation rate of the CS-3 solar array was one fourth or one fifth that of ETS-V and BS-2b. Therefore, if the shielding of the cylindrical substrate is considered in comparing the three satellites, the flight data shows that the radiation hardness of GaAs solar cells was for better than that of Silicon solar cells because the degradation from the backside of the panel was generally less than that from the front side.

The Communications and Broadcasting Engineering Test Satellite (COMETS) is a research and development satellite for developing new technologies in communications and broadcasting fields. These include inter-orbit communications, advanced satellite broadcasting, advanced mobile satellite communications, and upgrading large geostationary satellites. It will also be used to conduct experiments and verify the above subjects. COMETS is scheduled to be launched in the winter of 1997 from Tanegashima Space Center.

COMETS mission equipment includes inter-orbit communications equipment developed by NASDA, advanced mobile satellite communications equipment developed by NASDA, advanced mobile satellite communications equipment developed by the Communications Research Laboratory (CRL) of the Ministry of Posts and Telecommunications, and advanced satellite broadcasting equipment jointly developed by both organizations. An upgraded version of the 2 ton class, three-axis stabilized geostationary satellite bus developed for ETS- VI is used as the satellite bus to carry the mission equipment. Technologies will be developed for an electric power subsystem employing nickel-hydrogen batteries and light-weight panels with highly effective solar cells, integrated propulsion subsystem with a bipropellant pressurized apogee engine and gas blow down jets, and high-accuracy attitude control subsystem to control the pointing angles of large antennas.

COMETS has a 6kW class, fold-out, retractable solar array. Manufacturing and integration of the subsystem engineering models is currently in progress, and the development tests are scheduled to start in September 1993.

There are 31,080 GaAs solar cells in the main array (518/panel × 30 panels × 2 wings) and 3,936 (492/panel × 4 panels × 2 wings) in the charge array.

Thin diodes with high reliability have been developed.

The diodes will be used for blocking diodes or bypass diodes on the solar cell array for space use. In view of the high reliability parts, we improved the insulation of the conventional thin diode to avoid short failures.

The thin diode consists of a diode chip made from 100 μm thick N-type silicon, a silver connector and an optical solar reflector. The optical solar reflector was used to decrease the optical temperature of the diode and to shield the diode from sun light and particle radiation.

The diode chip has a planar type pn-junction, and its front surface is covered with a silicon oxide layer and polyimide to prevent shorts between the diode edge and the connector. The connector is coated by an insulation layer except in welding and stress relief areas. The insulators formed on both connector and diode chips are fail-safe to prevent shorts between the connector and diode chip.

There are a total of 1,112 thin diodes (556/wing × 2 wings).

Si Solar Cell

Solar cells for space use must satisfy various requirements which differ from those for the terrestrial solar cells. They are required to produce high electrical outputs at the operating temperature in space and at the end of life of the satellites. To have a long life and low operating temperature in space, they must have high efficiency as well as also high radiation resistance and low solar absorptance (α s).

We have investigated the feasibility of improving these characteristics of the thin solar cells for about 2 years. The experimental results of the 70μm thick solar cells were published[10]. The basic design of the solar cell with nonreflective surface (NRS), passivated rear surface and locally P^+ diffused BSF layer was confirmed to be very effective for improving characteristics of the space solar cells.

Four inch diameter, CZ, (100) oriented, 200μm thick, chemically polished silicon wafers were used to fabricaes 2cm × 2cm thin solar cells. Before surface texturing, silicon wafers were etched to the target thickness in a hot sodium hydroxide solution. The fine gridlines on the surface were formed by the photolithography technique. The outline of fabrication process is outlined below.

1. Thinning etch
2. Texturing etch (Formation of NRS)
3. Passivation of rear surface
4. p and n diffusion
5. Contact formation
6. A. R. C. formation
7. Cutting to 2cm × 2cm

In general, power dissipation under a reverse bias condition in a shadowed portion of illuminated solar arrays is well known as a hot spot phenomenon and discussed in various papers [11]. This phenomenon may become the cause of a module failure with power output loss. To protect this failure, modules are designed to have adequate intervals and a bypass diode is connected in parallel to each separated module. But this method has the problems that a manufacturing cost of arrays increases and it restricts the satellite design. In this paper, we propose new silicon space solar cells with IBF (Integrated Bypass Function) which enable prevention of a failure caused by a reverse biasing and present the result of some preliminary experiments performed to these cells.

SCM of TEDA on ETS-V

ETS-V is a three-axis stabilized geostatioary satellite weighing about 550kg. It was launched using the first three-stage H-I rocket. Its purpose is to establish new domestic technology for geostationary three-axis stabilized satellites. Effectiveness of the technologies has been proved by long-term operations on orbit.

ETS-V also carries mobile communications equipment, and communications experiments will be conducted with ships and aircraft.

In the sensor part of SCM, 24 solar cells are attached to Aluminum panels[12],[13].

Cells differ in:

- Cell material: Gallium Arsenide (GaAs) and Silicon (Si)
- Cell thickness: 50μm and 200μm (Si), 110μm and 280μm (GaAs)
- Surface: flat and Non-Reflective Surface (NRS)
- Cell structure: Back Surface Field and Reflector (BSFR), Back Surface Reflector(BSR)
- Growth method of epitaxial layer: Liquid Phase Epitaxy (LPE) and Metal Organic Chemical Vapor Deposition (MOCVD)
- Cover glass thickness: 0 (no cover), 50μm, 100μm, 150μm
- Surface electrode: Bar and Connector Integrated solar Cell (CIC)

In SCM the generated currents and voltages are measured under ten load conditions for each cell. Then the degradation of current-voltage characteristics can be obtained in space radiation environment.

Figure 2 shows the flight data of three cells, GaAs LPE, Si BSFR 200μm and Si BSFR 50μm cells, after correction. The abscissa indicates the percentage of normalized Isc, which is also called remaining factor (RF) of Isc. On 25 May 1994 the RFs of GaAs was 84%, that of Si BSFR 200μm ,82% and that of Si BSFR 50μm, 88%. These data also show degradation due to a solar flare on 19 December 1989.

The results obtained are shown in Table 3. A solar flare which occurred on 19 October 1989 degraded Si BSFR 50μm cells 0.7%, 200μm cells 2.3% and GaAs cells 1.3%. This degradation of Si BSFR cells is almost equal to one year's degradation. Solar flare degradation of GaAs cells almost equals a half year degradation.

The comparison of these data can be summarized as follows;

(1) Si thin cells (BSFR 50μm, BSR 100μm) best resist degradation due to Earth-trapped electrons and the solar flare.

(2) The degradations of Si thick cells (BSFR 200μm, BSR 200μm, Black 200μm) and GaAs cells (LPE, MOCVD) due to electrons are almost equal.

(3) GaAs cells (LPE, MOCVD) resist solar flares better than Si thick cells (BSFR 200μm, BSFR 200μm, Black 200μm).

(4) These results suggests that the most critical factor of radiation hardness is not so much the structure as the thickness of Si cells.

(5) No differences were observed between the type of cover glasses.

(6) If cover glass thickness is changed from 50μm to 200μm, the degradation remains almost constant for Si thick cells.

(7) The manufacturing process and the thickness are not critical for GaAs cells.

(8) The normal and soldered types have almost equal hardness, but the CIC type is harder than others.

Figure 3 shows the accumulated value in the Dosimeter (DOM) proton channel. The large solar flares from August to October in 1989 rapidly increased the values in this Figure. About 510,000 proton were counted. We converted proton count numbers into fluence of 10MeV proton from outside of satellite, and found it to be 7.1×10^9(proton/cm^2) in the normal incident condition.

Data from the space environmental monitor (SEM) on Geostationary Operational Environmental Satellite (GOES) 7 will also be used[15]. The spectrum of protons from the solar flare is converted into the equivalent fluence of 10MeV protons using damage-equivalence functions of the Si cell. The result was 3.1×10^{10}(protons/cm^2).

Conclusion

We reviewed Japanese programs using GaAs and Si solar cells as well as GaAs solar cell flight data of the CS-3 program and SCM on ETS-V. GaAs solar cells have high resistance to the for space radiation environment including large solar flares. We will continue to develop SCM for NASDA satellites, because SCM data is very useful for analyzing trends of satellite power and the space environment when combined with the DOM sensor.

GaAs solar cells are said to have two major disadvantages compared to Si solar cells. One is that a GaAs solar cell is heavier than a Si solar cell, the other is cost.

In order to solve these problem, GaAs on Ge and GaAs on Si solar cells have been developed. GaAs on Ge solar cells are now being put to practical use. We strongly expect GaAs on Si solar cells will be useful in the future because they are lighter and cost less.

The 100μm NRS/LBSF Si solar cells for space use were developed and their efficiency is accomplished to 18% AM0.

We proposed new silicon space solar cells with IBF which prevent a failure caused by shadowing.

NASDA will continue to develop new GaAs and Si solar cells for future space applications.

Acknowledgments

The authors wish to thank Dr. M Tajima of ISAS and to Mr. T. Nakao and Mr. Y. Wako of Advanced Engineering Services, Dr. S. Yashida of Mitsubisi Electric Corporation and Mr. T. Saga of Sharp Corporation.

References

[1] J. M. Woodall and H.J. Hovel: "High-efficiency Ga1-xAlx-GaAs solar cells", Appl. Phys. Letter, Vol.21 No.8 p.379 (1972)

[2] R.G. Knecktli, G.S. Kamath, J. Even and R.y. Loo: "Development of Space-Qualified GaAs Solar Cells", Proc. 2nd European Symposium, "Photovoltaic Generators in Space", p.121 (1980)

[3] Yoshida S., Mitsui K., Oda T. and Shirahata K.: "High efficiency AlGaAs/GaAs solar cells with high resistance to radiation damage", Proc. 15th IEEE Photovoltaic Specialists Conference p.239(1981)

[4] S. Yoshida, N. Imura, M. Goto and S. Matuda: "Availability of AlGaAs/GaAs Solar cells in Space", Proc. 4th European Symposium, "Photovoltaic Generators In Space", p.65 (1984)

[5] Yoshida S., Mitsui K., Kato M., Yukimoto Y. and Matsuda S.: "Liquid phase epitaxy technology of large area AlGaAs/GaAs wafers of GaAs solar cells for space applications", Rec. 17th IEEE Photovoltaic Specialists Conference, p.42 (1984)

[6] S. Yoshida et al.: "High Efficiency Al/GaAs/GaAs Solar Cells with High Resistance to Radiation Damage" Proc.15Th IEEE Photovoltaic specialists Conf., p.1384 (1980)

[7] S. Matsuda et al.: "Manufacturing status of GaAs solar cells for Space Applications" Proc. 5th Europ. Symp. Photovoltaic Generators in Space p.31 (1986)
[8] K. Mitsui et al.: "A High Quality AR coating for AlGaAs/GaAs Solar Cells" Proc. 17th IEEE Photovoltaic Specialists Conf. p.106 (1984)
[9] Takata N., Kurakata H., Matsuda S., Okuno T., Yoshida S., Matusumoto H., Goto M., Ohkubo M. and Ohomura M.: "Space Power GaAs Solar Cell -Main Power Generation for CS-3-", Proc. 21st IEEE Photovoltaic Specialists Conference, p.1219(1990)
[10] T. Katsu et al.: "High Efficiency Silicon Solar Cells for Space Use" 18th International Symposium on Space Technology and Science (ISTS), Kagoshima, Japan, May 18, 1992
[11] H. S. Rauschenbach: "Solar Cell Array Desing Handbook" Litton Educational Publishing,Inc. p.74, 1980
[12] Uesugi M., Kuboyama S., Tamura T., Matsuda S. and Goka T.: "Flight Data of Solar Cell Monitor and RAM Soft-Error Monitor on Engineering Test Satellite-V(ETS-V)", Proc. The International Workshop on Radiation Effects of Semiconductor Devices for Space Application (1992)
[13] Uesugi M., Tamura T. and Matsuda S.: "Flight data of Solar Cell Monitor on Engineering Test Satellite-V (ETS-V)" Proc. of Solar Terrestrial Predictions Workshop-IV Vol. 2 p.346 (1992)
[14] Siedle, A. H. and R. Freeman: "RADIATION EFFECTS ENGINEERING HANDBOOK", ESA-CR(p)-1067, (1987)
[15] Marvin, D. C. and D. j. Gorney: "Solar Proton Events of 1989: Effects on Spacecraft Solar Arrays" J.SPACECRAFT, 28, No.6, (Nov.-Dec.,1991)

Table 1 Status and Trends in GaAs Solar Cells for Space Applications (GaAs & Si)

Solar Cell		Currrent	Goal and Trend
Si	In production efficiency Size /Thickness	13% to 15% 4cm×6cm/50μm	High power to weight ratio Low cost Large Area/Light Weight
	In development efficiency	17% (2cm×2cm)	High Efficieny(20%)
GaAs	In production efficiency Size /Thickness	18% to 20% 2cm×4cm/200μm	Light weight concentrator
GaAs on Si	In production efficiency Size /Thickness	18% 2cm×2cm/200μm	High efficiency Light weight
GaAs on Ge	In production efficiency Size /Thickness	18.6% 2cm×2cm/200μm	High efficiency Light weight

Table 2(a) Solar Cell Applications of NASDA Satellites

	ETS-V	CS-3a CS-3b	GMS-4	MOS-1b	BS-3a BS-3b	JERS-1	ETS-VI	ADEOS	COMETS
Launch Date	27.Aug.87	19.Feb.88 16.Sep.88	6.Sep.89	7.Feb.90	28.Aug.91 25.Feb.91	11.Feb.92	28.Aug.94	Winter 96	Winter 97
Satellite Shape	Box	Cylindrical	Cylindrical	Box	Box	Box	Box	Box	Box
Satellite Weight	550kg	550kg	325kg	740kg	550kg	1340kg	2000kg	3500kg	2000kg
Satellite Life	1.5y	7y	5y	2y	7y	2y	10y	3y	3y
Satellite Power (EOL)	845W	628W	265W	540W	1482W	2053W	4100W	4500W	5400W
Solar Cell Type	Si BSFR	GaAs	Si BSFR	Si BSF	Si BSR	Si BSFR	Si BSFR	Si BSFR	GaAs
Solar Cell Size (cm)	2×4	2×2	2×2	2×4	2×4	2×4	2×4	2×4	2×4
Solar Cell Thickness	280μm	280μm	200μm	280μm	200μm	50μm	50μm	100μm	200μm
Coverglass	CMX150	CMS150	OCLI150	CMS150	OCLI150	AS100	AS100	AS100	AS100

Table 2(b) Solar Cell Applications of ISAS Satelites

	GINGA	AKEBONO	HITEN	YOHKOH	GEOTAIL	ASCA	SFU	MUSES-B	LUNAR-A
Launchd date	5.Feb.87	22.Feb.89	24.Jan.90	30.Aug.91	24.Jul.92	20.Feb.93	Winter 95	Summer 96	Summer 97
Satellite Shape	Box	Box	Cylindrical	Box	Cylindrical	Box	Box	Box	Cylindrical
Satellite Weight	411Kg	295Kg	140Kg	420Kg	981Kg	420Kg	3850Kg	800Kg	515Kg
Satellite Life	1yr (4.75yr*)	1yr**	1yr (3.2yr*)	1yr**	3.8yr**	3yr*	0.5yr	1yr	1yr
Satellite Power (EOL)	490W	270W	115W	506W	338W	492W	2880W	700W	310W
Solar Cell Type	Si BSFR	Si BSFR	Si BSFR	Si BSFR	Si BSFR	Si BSFR	Si BSFR	Si BSFR	Si BSFR
Solar Cell Size (cm)	2×4 2×2	2×4	2×4	2×4	2×4 2×2	2×4 2×2	2×4	2×4	2×4 2×2
Thickness	200μm	100μm	50μm	50μm	100μm	50μm	100μm	50μm	75μm
Coverglass (Thickness)	CMS (150μm)	BDX (500μm)	BDX (100μm)	BDX (100μm)	BDX (100μm)	BDX (100μm)	BDX (100μm)	BDX (100μm)	BRR (OCLI) (100μm)

*Actual life

**Still under operating

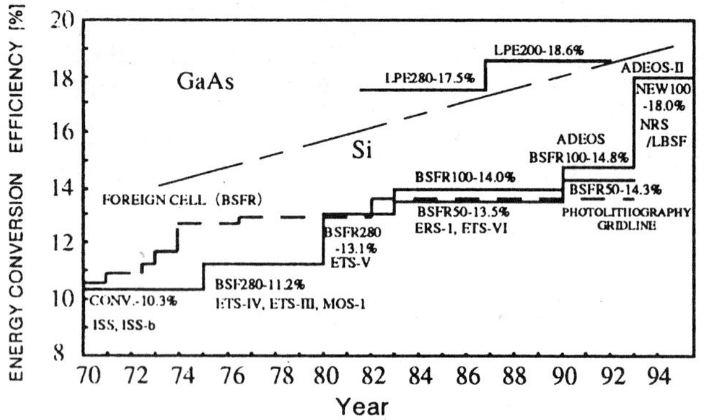

Figure 1 Immprovemt of Space Solar Cell Energy Conversion Efficiency

Table 3 Summary of degradation in short circuit current

No.	Material	Process	Thickness	Cover glass	Degradation Rate	Solar flare	Aug. 27, 1987	May. 25, 1984	Note
1	GaAs	LPE	280	None	Can't be acquired.				
2	GaAs	LPE	280	CMX100	-2.3	-1.3	96.9	83.7	
3	GaAs	LPE	280	CMS150	-2.2	-1.1	98.0	85.6	
4	GaAs	LPE	280	None	Can't be acquired.				
5	GaAs	LPE	280	CMX100	-2.3	-1.3	97.3	83.8	Soldered
6	GaAs	LPE	280	CMS150	-2.1	-1.3	96.1	83.7	Soldered
7	Si	BSFR	200	CMX100	-2.3	-2.2	97.7	81.6	
8	Si	BSFR	200	AS100	-2.0	-2.3	98.2	83.3	
9	Si	BSFR	200	AS50	-2.6	-3.1	98.1	N/A	**
10	Si	BSFR	50	AS100	-1.4	-0.6	97.9	88.7	
11	Si	BSFR	50	AS50	-1.6	-0.7	97.7	87.6	
12	Si	BSR	200	CMX100	-2.2	-1.2	97.8	82.7	
13	Si	BSR	200	AS100	-2.0	-1.3	98.8	84.0	
14	Si	BSR	200	AS50	-1.7	-2.0	99.1	78.1	
15	Si	BSR	100	AS100	-1.4	-0.6	99.6	90.6	
16	Si	BSR	100	AS50	-1.6	-0.7	99.0	89.3	
17	Si	Black	200	CMX100	-1.8	-1.8	98.4	85.9	
18	Si	Black	200	AS100	-1.6	-1.6	98.5	87.7	
19	GaAs	MOCVD	110	CMX50	-2.4	-1.1	93.3	79.4	
20	GaAs	MOCVD	110	AS50	-2.7	-1.0	93.7	78.1	
21	GaAs	MOCVD	110	CMX50	-2.2	-1.2	95.6	81.0	
22	GaAs	MOCVD	110	AS50	-2.5	-0.3	95.0	81.1	
23	Si	BSFR	200	CMX100	-1.5	-2.1	97.8	86.3	CIC
24	Si	BSFR	200	AS100	-1.4	-2.2	98.1	86.5	CIC

* Cells No.1-6 and 19-20 were manufactured by Mitsubishi Electric Co.;
the others were manufactured by Sharp Co.

** These value are calculated using the data collected before June 12, 1991, when the failure was observed.

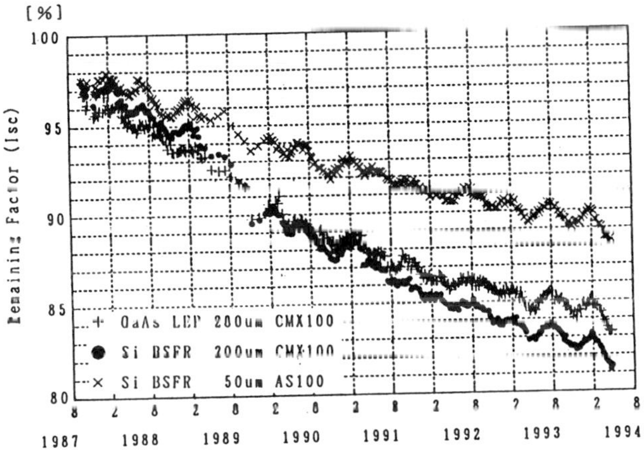

Figure 2 Flight Data of Three Cells

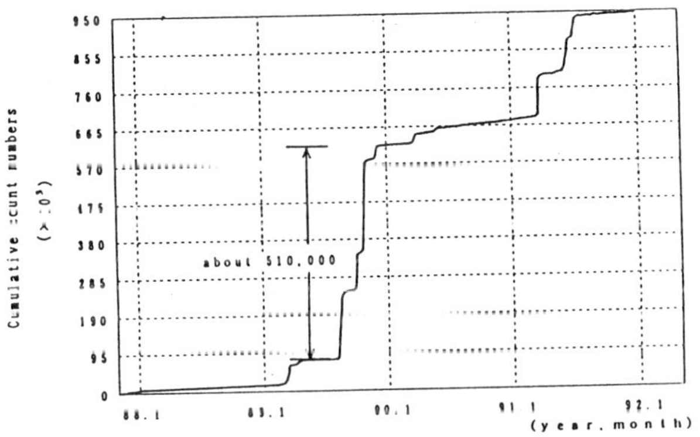

Figure 3 Cumnlative Counts in DOM Proton Channnel

JAPANESE ACTIVITIES FOR INTRODUCING RESIDENTIAL PV SYSTEMS AS A NATIONAL ENERGY SUPPLY

Kosuke Kurokawa
Electrotechnical Laboratory

ABSTRACT

The paper summarizes Japanese activities concerning photovoltaic systems, specially for the residential use as a national energy supply. This year 1994 is just the 20th anniversary of the Sunshine Project, which started in July 1974. In the Project the utility-connected, residential applications mounted on roofs have been its major target from an early stage of R&D. Recently, it can be considered that technologies for the target have been fundamentally established and a series of new activities have been being introduced to promote the commercialization and diffusion of PV systems. To review those activities, several condensed tables are presented, *i.e.*, R&D history of residential applications, recent trends in regulation and code improvements, and new institutional activities to spread PV systems. Possible R&D items from now on are also reviewed and listed, which are necessary to support these effort.

Table 1 Progresses of R&D relating to residential applications in Japan

Topics	Period	Outline	Remarks
The Sunshine Project started.	July 1974	Fundamental R&D of Photovoltaic cells were included in Solar Energy Program.	
R&D of PV 'system' fundamentals started.	FY1978 | FY1979	Fundamental R&D of Photovoltaic system including 'utility connected systems'.	contractor: CRIEPI
NEDO was established.	Oct. 1980	NEDO is responsible for alternative energy development by the Government law.	mainly financed by the special account budget from electricity consumption tax.
R&D of PV system for private house	FY1980 | FY1984	· 3kW roof-integrated; bipolar transistor PWM inverter · experiments without/with reversal power flow · experiments without/with battery	contractor: Fuji Electric Co.
R&D of PV system for multi-family house	FY1980 | FY1984	· 20kW roof-top array; GTO thyristor inverter · switch-over type utility backup	contractor: Sharp
actual size grid connection experiments	FY1980 | FY1985	· 2kW×2, 3kW×4, 5kW×1 · line- and self-commutated inverters · utility connection; interference among plural systems	Akagi Test Facility, CRIEPI
inverter development for residential use	FY1985 | FY1995	· 3kW FET PWM inverter(phase 1) · 3kW IGBT HF-isolation inverter (phase 2) · 3kW IGBT transf.-less power conditioner(phase 3)	contractor: Toshiba
roof-integrated module development for residential use	FY1985 | FY1989	· glass roof tile with amorphous Si type · slate roof tile with poly-Si type · roof structure model demonstrated	contractor: Sanyo
roof-mounted array installation for residential use	FY1989 | FY1992	· mechanical strength design rule · 2.4kW/house stand-off structure roof arrays · 5,000yen/m^2 installation cost by 2 men/half day	contractor: Kandenko
PV air conditioners for private home	FY1990 | 1992	· development and evaluation of PV air conditioner as an electric appliance	as a part of R&D of BOS evaluation system contractor: CRIEPI
utility grid operation with a number of PV systems	FY1986 | FY1991	· 2kW×100 houses + 100/200V-6kV actual-scale distribution line; various types of inverters · autonomous, individual control and protection	Rokko Island Test Facility contractor: Kansai Electric Power Co. and CRIEPI
PV modules integrated into building construction materials	FY1993 | FY1996	· exchangeable shingle; · prefabricated roof panel; · heat insulated roof panel; · glass curtain wall; · metal curtain wall; · flexible roofing sheet	PVTEC/Sanyo-Kubota; Matsushita Electric Works-Matsushita Battery; Kaneka; Daido Hoxan-Kajima-Showa Shell; YKK-Kyocera; Fuji Electric-Misawa Homes
IEA Task V started as a part of PV Power System R&D Program.	Spring, 1993	· grid integration of roof-mounted PV power systems · operational experience, guidelines, component evaluation experiment	6 tasks are planned in parallel.

First WCPEC; Dec. 5–9, 1994; Hawaii

Table 2 Recent Systems R&D by NEDO

● R&D of BOS technologies	
– compact, package power conditioner	transformerless, 3kW,1 ϕ3W, $\eta \geq$ 95%, ¥50/W (3,000 units/mon.) 1.5kW/m^2,
– vanadium redox flow battery	74% η, 20yrs life, 14¥/Wh
– hydrogen battery	100Ah(10hR), 6,000 cycles, ¥70/Wh (200MWh/yr)
– modules integrated into building construction materials	exchangeable shingle; prefabricated roof panel; heat insulated roof panel; glass curtain wall; metal curtain wall; flexible roofing sheet
– protection for utility interconnection	statistical simulation; fault probability
– array installation variety	installation methods for rail load, traffic road, riverbed, etc.
● R&D of system evaluation	
– test for inverters and batteries	Mt. Akagi Facility, CRIEPI; standardization for JIS & IEC
– evaluation for various applications	Hamamatsu Facility, JQA; system sizing rules
● Feasibility study of advanced applications	PVTEC Study Committees
● RD&D of PV systems	
– DG hybrid system for remote island	Miyako Is., PV:750kW, bat.:3,000kWh
● International cooperation RD&D	
– pumping for highland conditions	Nepal, (p) PV:4kW, inverter:1.5kVA × 3 pumps; (a) PV:40kW, 2 pumps
– mobil systems for nomadic tribe	Mongolia, PV:204W × ≈200 units
– battery charge station	Thailand, (p) PV:4.3kW to charge 16 batteries; (a) PV:40kW for 500 homes
– village electrification for the tropics	Malaysia, (p) PV:10kW for 10 homes; (a) PV:100kW
● Survey of meteorological data	national irradiation database for PV system sizing, world irradiation database

INTRODUCTION

This year 1994 is just the 20th anniversary of the Sunshine Project, which started in July 1974. The Project originally planned to develop alternative energy resources by R&D efforts, including solar energy, geothermal energy, coal gasification/liquefaction, and hydrogen energy system. Although solar thermal house and power generation were more focused in the Solar Energy Program at the start of the Project, the fundamental studies of photovoltaic cells were also included. Several years after that, system R&D also began.

Japan is almost entirely covered by the electric utilities. It was a quite natural course to think about utility connection for PV systems if the large amount of their use is considered. Therefore, in the Project "the utility-connected, residential applications mounted on roofs" have been chosen as its major target from an early stage of R&D.

Recently, the R&D phase can be considered to have reached the state "fundamentally" established, including system composing technology, BOS such as array construction, inverter hardware, its internal controls, utility interconnection control and protection. According to these R&D fruits, a series of new activities have been being introduced to create social acceptance for the recent 2 or 3 years. Among all, the most drastic changes were "utility-interconnection guidelines for PV systems with

Table 3 Case study of capital and running costs assuming 100MW/Y annual production scale

	FY2000 target		early 2000's	[reference] Case with technology established in FY1992
	Case I	Case II	Case III	
total system cost	≈320¥/W	≈260¥/W	≈220¥/W	≈450¥/W
component cost PV module BOS cost	≈210¥/W (≈15%) ≈110¥/W	≈170¥/W (≈10%) ≈90¥/W	≈150¥/W (≈12%) ≈70¥/W	≈330¥/W (≈15%) ≈120¥/W
assumptions	system utilization factor ≈ 12%, ≈100MW/Y production scale			
electricity generation cost	≈30¥/kWh	≈25¥/kWh	≈20¥/kWh	≈40¥/kWh

[Note] a) 3kW approx. utility-interactive systems with bi-directional power flow without battery are considered for all cases.
b) "FY2000 target" and "early 2000's" are assuming a situation when and where R&D starting from FY1993 will be completed and their outcomes will be applied.†
c) "Case with technology established in FY1992" is assuming a situation when and where the outcomes of R&D completed in FY1992 will be applied.†
d) "Case I" and "Case with technology established in FY1992" assume the utilization of polycrystalline silicon technology.
e) "Case II" assumes the utilization of thin-film modules unified with building materials; e.g., Amorphous Si, CdTe.
f) "Case III" assumes the utilization of thin-film modules unified with building materials; e.g., Amorphous Si, CdTe, $CuInSe_2$.
g) The cell efficiency of "technology established in FY1992" corresponds to 17% for poly crystalline cells and 10–12% for thin-film cells such as amorphous Si and CdTe. This level is considered to be practically satisfactory at the moment and may still be improved until 2000. However in this case study, the cost-down efforts is more emphasized than the requirement for efficiency improvement.
† Generally speaking, it may need several years to establish manufacturing lines after R&D is completed.

Table 4 Future direction of PV systems R&D

R&D Target	● System study backing-up introduction and diffusion program ● Possibility increase of dissemination / area expansion ● "Compact, Light-weight and Low cost" → End-less, Common target !!
Improve-ment R&D	● Total system - efficiency improvement - slimming of equipment size and capacity - value improvement - system sizing and evaluation technology ● BOS Simplification - efficiency improvement & cost reduction - simplified, packaged and standardized equipments - unified materials of roof/wall array - standardized evaluation and testing ● Diversification of utilization form - creation of additional value - high-performance, small-volume batteries for remote application, high rate, etc. - total system optimized for community use - peak value evaluation for utility grid - social cost study - measurement of solar energy area distribution - modular PV system
Long Term R&D	● Advanced concept for global utilization - extensive diversification of utilization form - large scale bulk system - urban city scale optimization - location on the sea or desert - plantation system to control desertification - space satellite power generation - renewable energy network conception - advanced seeds, e.g., electrolytic PV cell, photo-regenerative fuel cell, bio-energy conversion, unified PV conversion and energy storage, etc.

Tentative conclusions discussed at PV Workshop, April 25-28, organized by PVTEC

power reversal" by the Government in 1993 and "one-to-one basis electricity sell-back" by the utilities in 1992.

In spite of these trends, the author thinks that there exist still certain technical problems left behind to improve the economical features of PV systems and to create a real solar PV network in a world scale.

Long History of Residential R&D

In FY1974 the Sunshine Project started including the R&D of Photovoltaics. At first no program concerning 'system' was accommodated. In 1977 the Sunshine Headquarters started to make R&D plan for the utility-connected systems. The introduction of such concepts to PV technology was quite early as compared with other countries. The reason is that Japan is entirely covered by the utility networks. Under these circumstances, the utility connected systems, especially the residential systems were thought to be promising because of land restrictions, which means no enough space and high price. If PV arrays are installed on the roofs of residential houses, they can supply a large amount of energy according to the following assumptions.

The total number of private houses in 1987 was 42 million in Japan [1]. If 22% of those houses can accommodate a 3kW PV system for each, it makes 27.6GW potential. In addition, 580 thousands multi-family houses give another 8.13GW assuming that a half of building area can be covered by 10% efficiency modules. Therefore, residential application can potentially accept 35.73GW PV systems [2]. If they produce energy with the capacity factor of 12%, it corresponds to 37.6TWh/year. This amount can not be neglected, considering that Japanese gross final electricity consumption was 758.4TWh in 1990 [3]. It means that approximately 5% of the gross electricity can be delivered by PV residences on the realistic basis.

For the past 15 years Japan has made extensive R&D progresses for the residential applications and their utility interconnection as shown in Table 1. Especially, it can be noted that a series of utility interconnection control and protection R&D's gave us a lot of fruits for the future expansion of the residential systems.

RECENT R&D AND NEW WAVE

Table 2 shows the recent system R&D program. More than a half of them are still closely related to the

Table 5 Relaxation of regulations and codes relating to PV systems utilization in Japan

Item	Established in	Outline	Remarks
revision of ministerial ordinance prescribing technical standards of electrical facilities relating to electric utility industry law	April 1990	addition of definitions and rules of PV systems, etc. to the standard	· withstand voltage specification · installation method of PV modules · PV power systems without a full-time supervisor
	April 1990	revision of enforcement regulations (construction plan approval, qualified engineer assignment, etc.)	· no legal formalities for construction plan lower than 100kW · prior notification of plan for higher than or equal to 100kW and lower than 500kW · no assignment of the qualified engineer for a system less than 500kW
guidelines of technical requirements of utility interconnection	June 1990 | April 1993	description of protective devices	· with/without reversal power flow connected with EHV transmission line and HV exclusive line · with/without reversal power flow with low/high voltage general line · without power flow reversal with spot network
revision of ministerial ordinance under electric appliances regulatory law	Oct. 1991	addition of electric air cooler and dehumidifier having PV modules	
surplus energy purchase by electric utilities	April 1992	Reversal power flow is accepted by the utility.	purchased at the same level price to selling one

Table 6 Recent activities for promoting the introduction and diffusion of PV systems in Japan

Activity	Outline	Remarks
tax relaxation for investment to energy demand and supply structure improvement	· tax reduction for private industry investment to facilities relating to energy foundation · PV systems and PV air conditioners are included in the scope. · tax credit of income tax or corporation tax corresponding to 7% of installed facility and its special depreciation	The existing investment tax reduction system was expanded to this in FY1992.
special treatment of assessment standard of municipal property tax for local energy utilization facilities	· Local energy supplies such as PV systems, solar thermal heating cooling systems are in the scope · reduction of assessment standard to 5/6 for 3 years	
subsidy for model project of power generation by local energy resource	· Local energy supplies such as PV systems, wind energy conversion are covered. · 50% subsidy for project expenses to regional public organization and private enterprise	
subsidy for the promotion of power generating industries adopting local energy system	· Local energy supplies such as PV systems, wind energy conversion are covered. · grant for paying a fixed rate of interest to the financial institution which fiances regional public organizations and private enterprises at a low rate	MITI/ New Energy Foundation(NEF)
○field test of PV systems for public facilities, etc.	· acquisition and evaluation of operational data of public facilities, etc. · 2/3 subsidy for PV system facility cost, installation fee and data acquisition	MITI/NEDO
○monitor plan of private PV houses (700 roof plan)	· 1/2 subsidy for residential PV system installation cost · monitoring report submission monthly	MITI/NEF

residential applications. The cost target for PV R&D is summarized in Table 3. To accomplish the target, extensive R&D is necessary including the device field as well as 'system side' as shown in Table 4.

In recent years as a result of 20 year R&D efforts, the government has been getting more positive to employ a new policy for the introduction of photovoltaics as shown in Tables 5 and 6. The former table lists the relaxations of regulations and codes. The latter describes institutional activities to spread PV systems. The present structure of R&D and other activities is shown in Fig. 1.

Moreover, the Government has been setting up "New Energy Introduction General Policies" for Cabinet determination. A scenario as shown in Table 7 is expected to be described in this Policies. PV systems are assumed to be 300,000¥/kW mainly with 10% market of newly built houses in FY2000, which corresponds to 60,000 houses/year. In FY2010 the installation cost should be reduced to 200,000¥/kW by 20% share of newly built houses as well as other business and industrial applications. To realize the scenario, necessary action will be taken by the Government.

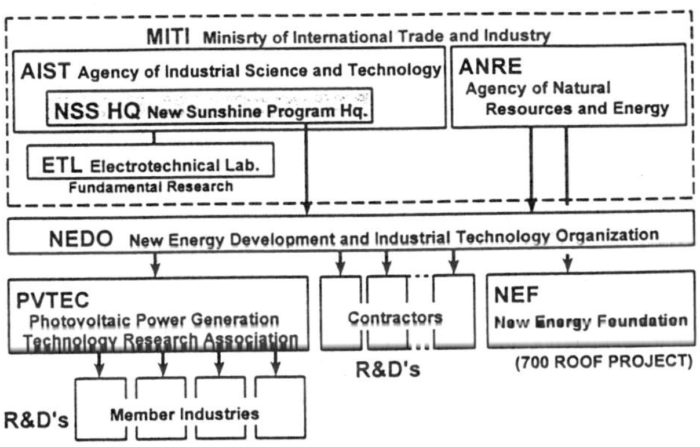

Fig. 1 Structure of PV R&D and its related activities in Japan

Table 7 New energy introduction scenario

New Energy	FY2000	FY2010
PV systems	400MW	4,600MW
solar systems	2,200,000 cases	5,000,000 cases
solar boiler	4,000,000 cases	3,700,000 cases
wind energy	20MW	150MW
clean car	488,000 cars	2,440,000 cars
co-generation	14,520MW	19,120MW
unused resources	268MℓOE	719MℓOE
fuel cells	200MW	2,200MW
generation by waste	2,000MW	4,000MW

These values were tentatively released by MITI, June 1994.

CONCLUSIONS

The author wishes the successful determination of the New Government Policies and expects that necessary and sufficient actions will be taken for PV technologies to take off toward their future prosperity from now on. However, he can also point out that this residential application in Japan is quite the small entrance of the global diffusion of PV system for the future human beings and that further fundamental research works are very much needed for that.

REFERENCES

[1] Ministry of Construction, "Statistical Manual of Construction", 1989.
[2] MITI, "PV Interim Review Report", 1992.
[3] "Energy Statistics of OECD Countries, 1989-1990", OECD,1992

POLYCRYSTALLINE THIN-FILM TECHNOLOGIES: STATUS AND PROSPECTS

A. Catalano
National Renewable Energy Laboratory
Golden, CO 80401

ABSTRACT

The recent progress towards high-efficiency thin-film polycrystalline solar cells is examined in this paper together with a brief history of their development. We focus on the three materials of contemporary interest for such devices: polycrystalline Si, $CuInSe_2$ and its alloys, and CdTe. Although there has often been an implicit assumption that thin-film devices required a compromise of lower efficiency than their single-crystal, or bulk polycrystalline competitors, we now see, for the first time, that polycrystalline thin-film solar cells can rival these devices. In dispelling this myth, we examine the elements that have contributed to the progress in each area.

INTRODUCTION

Although polycrystalline thin-film (PTF) solar cells have been recognized as presenting the opportunity for low-cost solar electricity, most have regarded the technology as implying a sacrifice in performance in order to realize that goal. Recently, however, results obtained for small-area cells and large-area modules have been at such high performance levels that they shatter this concept. We have come to realize that PTF solar cells have the potential to meet or exceed the performance of traditional commercial crystalline Si technology without sacrificing the potential for low cost. Even more encouraging, new products based on the PTF technologies are now entering the marketplace.

In this paper we briefly review the history of PTF solar cells and examine the recent progress that has led to the current state of the art in the leading material systems. We will examine the latest performance results for PTF devices and discuss the areas of work needed to realize their potential.

HISTORICAL OVERVIEW

Like crystalline Si solar cells, one of the first PTF material systems, Cu_2S/CdS, was seriously pursued as a result of interest in its use for space-based power [1]. Then, as now, there has been a need for lightweight, low-cost solar cells to satisfy commercial and military satellite requirements.The advent of the world energy crisis in the early 1970s, however, caused a skyrocketing interest in terrestrial photovoltaic (PV) applications, and with it "traditional" single-crystal Si and III-V devices, as well as new materials that promised the low cost necessary to produce electricity competitive with conventional fossil fuels. This national need led to increased funding for research, and quickly produced a number of promising candidate thin-film materials. An examination of the Proceedings of the 11th IEEE Photovoltaic Specialists Conference (1975), shows five candidate materials being pursued for application in thin polycrystalline film solar cells. These materials are thin film Si, Cu_2O, CdTe, $CuInSe_2$, and Cu_2S. [2-6] It is interesting to note that, with the exception of Cu_2S and Cu_2O, which are no longer generally considered viable candidates for PTF solar cells, these materials, identified as the most promising near the start of the major work in PV, continue to dominate the technology today, nearly 20 years later.

The funding for PV research in the United States increased rapidly throughout the period from 1975 to 1980, reaching a peak of $150M in 1980. This led to programs that sought superior semiconductor materials for use in thin-film solar cells that would reduce the cost and improve the performance of products. International programs, particularly in Japan and Europe, also stimulated considerable funding and research. In 1980 when the U.S. funding for PV peaked, the IEEE PV Specialists Conference proceedings contained papers on a wide range of new PTF materials for use in solar cells. This was a direct result of funding for exploratory research. Materials reported on at the conference included all those mentioned above, plus quaternary chalcopyrite materials, $CuGaSe_2$, Zn_3P_2, polycrystalline thin films of GaAs and InP, InSe, CdSe, and organic PV devices. Work on the new amorphous silicon (a-Si:H) PV solar cells was also presented. The subsequent drop in federal funding for PV, combined with either the technical failure or perceived lack of promise during the relatively short investment period, resulted in nearly all the new materials being casualties of the funding cuts. The promising results obtained with CIS and CdTe led to an increasing number of papers in these areas, as well as a-Si:H. Conversely, the inability to solve the stability problems with Cu_2S led to the abandonment of work in that area.

To illustrate these points, Figure 1 plots the number of papers published in the IEEE proceedings of the PV Specialists Conference in each of the years 1975, 1980, 1985, and 1990, for the PTF materials, a-Si:H, and terrestrial-based crystalline Si. Unlike crystalline Si, where the number of papers published dominates all others and follows the overall level of funding, the few successful PTF materials, with relatively few publications, continued to grow in number as did those for a-Si:H. Interestingly, the number of a-Si:H papers exceeded those for terrestrial crystalline Si in 1985 and was on a par with crystalline silicon in 1990. The number of papers in CIS and CdTe continued to grow slowly throughout this period, although, as is obvious from Figure 1, they remained at far lower levels than either crystalline Si or a-Si:H.

The three PTF materials for which interest remains strong today are $CuInSe_2$ (and related alloys), CdTe, and polycrystalline Si. We shall briefly review progress in each of these three material systems.

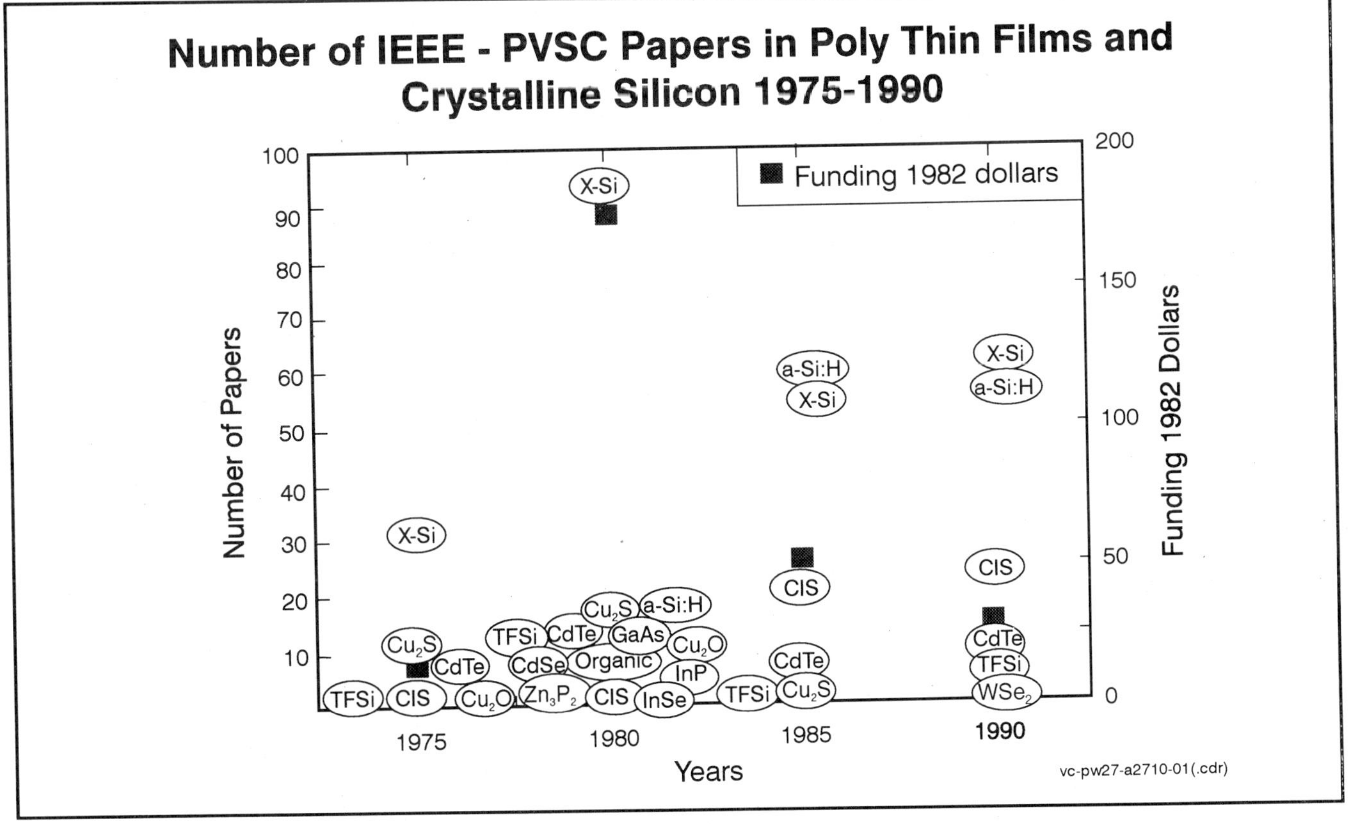

Figure 1. The number of papers published in a variety of polycrystalline thin film technologies by year in the Proc. IEEE PV Spec. Conf. for the years 1975, 1980, 1985, 1990. Also shown for comparison is the U.S. DOE Program level in current dollars, and the number of papers published in terrestrial crystalline Si and a-Si:H.

COPPER INDIUM DISELENIDE

Interest in $CuInSe_2$ dates back to the work of Wagner, et al., [7] who demonstrated a 12% conversion efficiency using single crystals. The extraordinarily high absorption coefficient of the material make thin- film devices practical, although the rather low optical band gap of 1.05 eV is usually considered too low for optimum conversion efficiencies. Thin-film solar cells, at 6.6% conversion efficiency, were prepared by Kazmerski et al., [8] by simultaneous evaporation from $CuInSe_2$ and Se sources. Interest in $CuInSe_2$ for thin-film solar cells increased markedly with the results reported by researchers at Boeing, who demonstrated 10.6% efficient cells in 1982 [9] prepared by the physical evaporation of the three elements onto heated, Mo-coated substrates. Key elements to achieving the high conversion efficiency appear to be the use of a $Cd_{1-x}Zn_xS$, n-type heterojunction partner for the normally p-type compound [9] and the realization that the growth of a Cu-rich layer early in the deposition process was the key to high quality material [10]. The use of the wider band gap heterojunction partner enhanced both the open-circuit voltage and photo-current of the device over previous structures that simply employed CdS. Although the use of a Cu-rich nucleating layer appears to have been motivated primarily by considerations of morphology, the need to avoid the formation of Cu nodules, and the control of film conductivity, the development appears to have presaged a more complete understanding of the material that is only now evolving. The Boeing group also recognized the importance of increasing the optical band gap of the absorber layer as the preferred method of increasing efficiency and introduced the use of alloys in which Ga is substituted for In. The use of $Cu(In,Ga)Se_2$ as the absorber layer resulted in a then-record total-area conversion efficiency of 14.6%.

The preparation of $CuInSe_2$ thin-film devices has been accomplished, with varying success, by many sometimes subtly different methods. The two principle methods, as given by the success in preparing high-efficiency devices are physical vapor deposition (PVD) and selenization of the deposited metals. Among the many other methods that have been less successful in yielding high-efficiency devices are electrodeposition [11], sputtering [12], spray deposition [13], direct evaporation [14], and screen printing [15]. Although it may be argued that sufficient study has not permitted development of these methods, it is also possible that the complex nature of the growth process required for high-performance devices is not easily achieved by methods other than PVD and selenization. Table 1 summarizes the methods and the performance achieved in devices.

The electrical properties of $CuInSe_2$ have consistently been shown to vary strongly with overall composition of the films, ranging from low-resistivity p-type to high-resistivity intrinsic or n-type conductivity. One relationship involves the relative amounts of the cation metals Cu and In. Films with greater than 25 at.% Cu are mixed phase and strongly p-type, whereas Cu-deficient films (<25%) are generally high resistivity intrinsic or n-type. A second critical relationship involves the relative occupancy of the anion and cation sublattice

Method	Efficiency, %	Organization	Comments
PVD $Cu(In,Ga)Se_2$	14.6	Boeing	From elements
Selenization	14.1	ARCO Solar	Method Unknown
Reactive Sputtering	6	Univ. Illinois	
Electrodeposition	~8	McGill	Single Bath
Electrodeposition	11	ISET	
Selenization	4–5	Parma	$CuInSe_2$ Target
Sputtering	6.1	ARCO	Cu_2Se-In_2Se_3
Sputtering	8.5	Univ. Illinois	Hybrid PVD
Sputtering Spray	2–3	Univ. Science & Tech., France	
Spray	3.1	Indian Institute of Tech.	
Screen Printing	1–3	Univ. Simon Bolivar, Matsushita	

Table 1. Methods used to prepare $CuInSe_2$ solar cells and their conversion efficiencies

by structure elements (Cu, In, Se, vacancy). This is referred to as valence stoichiometry. The deviation of the ideal valence stoichiometry ΔZ, is given by

$$\Delta Z = \{2[Se]/[(Cu)+3(In)]\} - 1.$$

It is likely that positive values of ΔZ result in p-type films via the presence of the metal vacancies (acceptors), while negative values imply Se vacancies, resulting in intrinsic, or slightly n-type, films.

That such electrical behavior exists is less controversial than the specific crystallographic phase(s) and composition that give rise to these observations, and their importance to device performance. Adding additional complexity is the kinetics of growth that require that the given composition desired be "approached" or synthesized in a manner that ensures device quality films. Further difficulty is added by the unwieldy nature of working with only thin films and the challenges posed by characterizing both the films' bulk and surface composition and discerning their crystallographic structure. Because the understanding of these important factors has been obtained primarily for films prepared by the physical vapor deposition (PVD) process, the discussion that follows may be assumed to be based on this process.

Figure 2 gives the ternary-phase field containing $CuInSe_2$. Note that the tie-line connecting Cu_2Se and In_2Se_3, encompassing $CuInSe_2$, does not require deviations from the normal valency of the intermediate compositions. Deviations normal to the tie line, however, result in an excess or deficiency in the number of bonding electrons. Tuttle et al., [16] have suggested that either discrete or phase-mixed films with an overall In-rich composition are required to fabricate the best devices. One such alternative composition that is found at the surface of finished absorbers is that of the $CuIn_3Se_5$ (1-3-5) phase, often referred to as the ordered vacancy compound (OVC) [17]. These materials, which appear to be intrinsic to slightly n-type, may be key to forming a "pseudo-homojunction" within the largely $CuInSe_2$ film. Schmid et al., [18] have provided definitive evidence for the 1-3-5 or OVC material and characterized it as having a 1.3-eV band gap. Direct evidence of the $CuInSe_2$/OVC junction was given by electron beam induced current (EBIC) measurements and photo emission measurements. The buried nature of the junction raises the idea that CdS may be eliminated as the pseudo-heterojunction partner. Kessler et al., [19] have shown that the surface treatment of the $CuInSe_2$ plays an important role in determning the open-circuit voltage of the junction. Recently, 14.9% (active area) devices have been prepared using $In_x(OH,S)_y$ as the heterojunction partner by giving the proper chemical treatment to the surface of the semiconductopr prior to the deposition of the transparent oxide.

As described above, relatively small changes in surface composition appear to be responsible for junction formation in the $CuInSe_2$ devices. Understanding how such junctions form has only recently been clarified and is based on the rather complex phase diagram of the Cu-Se and Cu_2Se-In_2Se_3 pseudo binary systems. At the early stages of growth, a Cu-rich flux is utilized to create a mixed-phase $CuInSe_2$+Cu_xSe

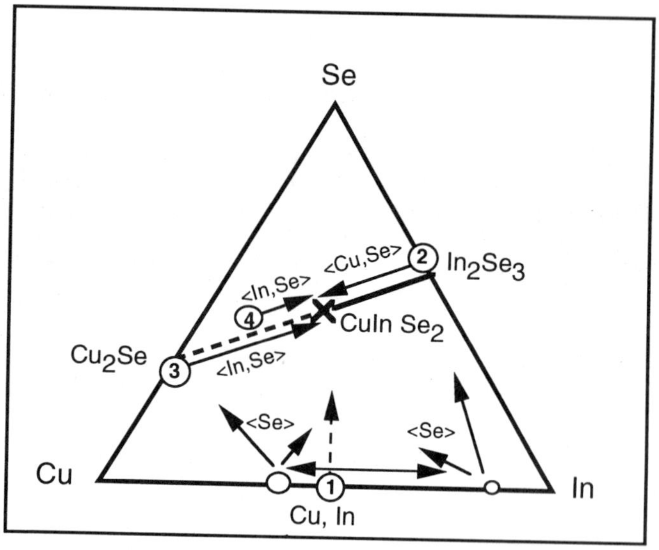

Figure 2. Ternary-phase field for Cu-In-Se

precursor. A key element of the process appears to be the formation of a liquid Cu_xSe phase within which the growth of the $CuInSe_2$ columnar grains occurs. This mixture, when cooled, has a characteristic large-grained structure, which has poor PV properties, and is referred to as the large-grained precursor or LGP. In order to convert the unwanted Cu_xSe to useful $CuInSe_2$, the vapor phase is adjusted to In-rich or In-Se so that the exess Cu_xSe is consumed. The $CuInSe_2$ growth occurs epitaxially onto the existing nuclei. Once the Cu_xSe has been consumed, the In-Se delivered to the surface forms In_ySe compounds that eventually convert the surface to the desired composition and structure of the OVC $CuIn_3Se_5$ phase. Any excess In_ySe compounds that do not react immediately are volatized. Evidence for the wide process tolerance of the OVC was noted by Schmid et al, [18] who noted almost no lateral variation in the composition of the films despite the non-uniformity in the flux; they concluded that the loss of excess In and Se as In_2Se was responsible for their observations. Figure 3 summarizes the model for the growth of $CuInSe_2$ films.

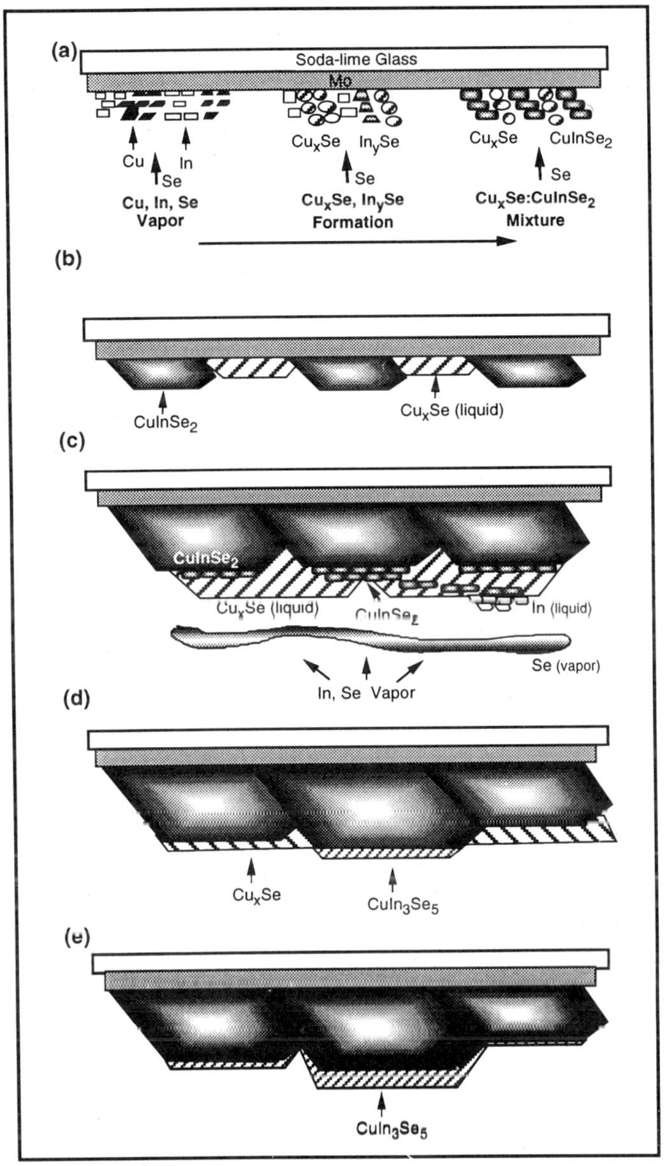

Figure 3. Model for the process of growth of device-quality CuInSe2 films and the OVC buried pseudohomojunction

Despite the success in fabricating devices with conversion efficiencies greater than 10% with $CuInSe_2$-based devices, it has been recognized that the low 1.04-eV band gap limited the open-circuit potential and overall efficiency. Improvements in efficiency beyond those outlined earlier have come from increasing the optical band gap of the absorber layer. Both S and Ga have been alloyed with $CuInSe_2$ to increase the optical gap and conversion efficiency. The highest efficiency devices, measured at 16.8% total area, reported at this conference, are graded to create a wide band gap, Ga-rich layer near the semiconductor-metal rear contact. Other impressive results have been reported by Hedstrom et al [20] who have reported high active-area efficiencies. Those devices are prepared in a three-step process in which the graded $(InGa)_2Se_3$ layer is first deposited on Mo-coated soda-lime glass at about 300° C followed by CuSe deposition at high temperature (approximately 560° C) for the $Cu(In,Ga)Se_2$ layer. In the final step, the film is driven Cu-poor to form what is presumed to be the alloy-equivalent of the OVC layer by exposing it to a flux of In+Ga+Se at 560°C followed by a cool-down in Se vapor. Table 2 summarizes the performance of cells by various groups.

Despite the fact that the highest thin-film solar cell efficiencies have been obtained with $CuInSe_2$ alloy-based materials, there is currently no organization manufacturing commercial products. Nonetheless, impressive results have been obtained for large-area modules, and prototype production runs. Siemens Solar Industries (SSI) has demonstrated 11.4% conversion efficiency on 0.37-m^2 (4-ft^2) modules. High manufacturing yields have also been reported in pilot production runs. Although no intrinsic degradation mechanism has been identified for $CuInSe_2$, SSI has reported that failures of modules occur as a result of defects in the film or substrate. While such problems would appear to be amenable to a technical solution, the problem underscores the need for careful testing and understanding of failure mechanisms in any emerging product required to have a 20+ year lifetime outdoors.

CADMIUM TELLURIDE

CdTe is an attractive material for solar cell applications because of the nearly ideal direct optical gap near 1.45 eV that makes thin-film devices possible. The ability to make both n- and p-type films is an obvious advantage, although the difficulty in fabricating shallow homojunctions combined with the parasitic surface recombination losses, is challenging in the preparation of high-efficiency devices.

The first thin film CdTe solar cells were prepared by Cusano [21] via chemical transport using $CdCl_2$ to deposit the n-type film. $Cu_{2-x}Te$ was used as the p-type component of the solar cell, which, though unstable, achieved efficiencies close to 6%. Schottky-barrier solar cells have been fabricated that avoid the difficult requirement of fabricating a contact to the p-type material [22], but they have achieved only low conversion efficiencies. The use of p-type CdTe/CdS heterojunction deposited onto SnO_2-coated glass was first reported by Andirovich [23] who, while obtaining only 1%-efficient devices, devised the device structure commonly used today. Bonnet and Rabenhorst [24] reported the first CdTe/CdS devices with an efficiency of about 5%, prepared by high temperature (500°–700°C) vapor transport onto Mo foil substrates.

Organization	Method	Total-Area Efficiency, %	Active-Area Efficiency, %	Comments
NREL	PVD	16.8	17.7*	$Cu(In,Ga)Se_2$
NREL	PVD	13.2	—	$Cu(In,Ga)Se_2/CdS$
ISET	Selenization	—	—	$Cu(In,Ga)Se_2/CdS$
EuroCIS	PVD	—	16.9	$Cu(In,Ga)Se_2$
Siemens	Selenization	—	16.2	Ga, S Grading
EPV	Selenization	12.5	—	$CuInSe_2$
Solarex	PVD	14	—	$CuInSe_2$
Stuttgart	PVD	—	10.5	$CuInSe_2/ZnO$
Stuttgart	PVD	—	10.4	$CuInS_2/In(O,H,S)$
Stuttgart	PVD	—	14.9	$Cu(In,Ga)Se_2/$ $In(O,H,S)$
* Calculated for 5% grid blockage				

Table 2. Conversion efficiencies of CuInSe2-based solar cells prepared by various groups

One of the more attractive features of CdTe is the large number of methods by which device-quality films may be prepared. Film formation methods currently in use for preparing high-efficiency devices include close-spaced vapor transport (CSVT), atomic layer epitaxy (ALE), screen printing, electrodepositon, PVD, and metallorganic chemical vapor deposition (MOCVD). Table 3 lists the highest conversion efficiencies obtained by each of the methods.

Despite the low temperature methods available for deposition, high-temperature treatments, frequently employing $CdCl_2$ as a flux, are necessary to prepare large, low-defect-level crystallites. Even the high-temperature methods such as CSVT, which yield large-sized grains, often require a $CdCl_2$ treatment to reduce inter- and intra-grain recombination and achieve adequate transport properties. High temperature treatments may also play a role in tailoring the p-n junction and improving performance. It is likely that some inter-diffusion at the heterojunction is desirable, perhaps because an alloyed junction is formed reducing interface states. However, difficulties in forming the alloy may be why wider band-gap, n-type heterojunction partner materials such as ZnSe or $Cd_{1-x}Zn_xS$ often fail to achieve higher photovoltages. An exception is the 0.870 volt open-circuit voltage obtained in the 12.5% cells reported by Seol and Im [25]. Mitchell has also shown that CdS may not be necessary in the 9.7% cells that use only the SnO_2 window layer as the n-type contact [26].

Deposition Process	Efficiency %	V_{oc}, mV	Organization	Comments
CSS	15.8	843	Univ. S. Florida	Solution CdS
CSS	11.6	797	Battelle Europe	
CSS	12.0	833	NREL	
ALE	14.0	804	Microchemistry	
Spray	12.7	799	Photon Energy	
Electrodeposition	12.7	807	BP Solar	Solution CdS
Electrodeposition	11.2	767	AMETEK	
PVD	11.0	789	Inst. Energy Conversion	
MOCVD	9.9	812	Univ. S. Florida	
Screen Printing	12.5	870	KAIST	CdZnS
Screen Printing	11.3	797	Matsushita	

Table 3. Performance of CdTe solar cells prepared by various means.

One of the most vexing problems associated with the p-type absorber layer is the formation of a low-resistance, stable contact. One factor that appears to be important is the formation of a Te-rich surface layer, which appears to lower the barrier at the semiconductor/metal interface for Te thicknesses greater than about 5 nm. Etching is the preferred method of forming the Te layer. Low-resistance contacts can be fabricated by depositing ZnTe, which is easily doped p-type. For example, 11.2% devices have been fabricated on electrodeposited CdTe using evaporated, Cu-doped ZnTe by Meyers [26]. Alternately, conducting pastes containing Ag, Cu, or Hg have been used to achieve low-resistance contacts by forming interlayers of Ag_2Te, Cu_2Te, or $Hg_xCd_{1-x}Te$, respectively.

The large number of successful deposition methods that have given rise to 10%+ performance have enabled a number of potential manufacturers to develop prototype products with relative ease, and to quickly obtain high performance in the range of 8% conversion efficiency. Table 4 lists the performance of module prototypes developed by several organizations. Like the other thin-film candidates, monolithic interconnections can be made to form modules. As with CIS, no intrinsic degradation mechanism has been identified, although Cu-alloy rear contacts can exhibit the instabilities associated with the non-stoichiometric CuTe. Moisture sensitivity has also been implicated in the instability of modules, but the mechanism may simply be corrosion of the vulnerable semiconductor and metallic elements.

THIN FILM POLYCRYSTALLINE SILICON

Crystalline Si, with its indirect band gap at 1.1 eV, generally uses slabs of the material that are in the range of 200-μm or more thick in commercial products. Among the first suggestions was that of Redfield [28], saying that much thinner crystalline Si could be used while still achieving the short-circuit current densities required for high-efficiency devices. Redfield showed, via calculations as well as experiment, that optical confinement could make up for, at least in part, what nature had omitted by making crystalline Si, an indirect band gap material. Those early devices, measuring 2.5-μm thick, were, however, single-crystal in nature, but demonstrated that high performance was indeed possible in a relatively thin film. Later work by Spitzer showed theoretical evidence that thin Si films could achieve high efficiency by using both optical confinement and minority carrier "mirrors" [29]. For the purposes of this discussion of crystalline Si films, the term "thin film" is stretched to include films thinner than 100 μm, but in fact may translate to higher costs than "true" thin films such as those discussed above that require only 1-2 μm films to obtain high performance.

Another approach that has been used to prepare thin-film silicon devices is deposition via the melt onto a conducting or insulating substrate. The Honeywell group, for example, prepared devices greater than 10%-efficient using 250 μm films on a mullite substrate [30]. Truly thin polycrystalline cells were fabricated by Feldman et al, [31] who prepared Si films 22 μm thick on sapphire or alumina substrates made conducting by using TiB_2 films. These devices exhibited efficiencies as high as 3.1 %. Chu et al reported efficiencies of 9.75% using 25 μm films, but these were prepared on metallurgical-grade Si substrates [32]. Barnett et al., reported the preparation of Si films on steel substrates via solution growth that yielded 9.5 % conversion efficiencies without antireflection coatings. The thickness of the silicon on these devices was not reported [33]. Other Si-film devices have approached 16% on 1-cm² devices, but again, the film thickness is not described. Single-crystal thin-film devices, prepared by liquid phase epitaxy (LPE) on heavily doped substrates, have demonstrated the proof of concept that thin Si cells can indeed achieve high efficiency. For example, Kolodinski et al., [34] have employed Si films only 16 μm thick to yield conversion efficiencies of 14.7%. Those devices were fabricated by preparing epitaxial films on heavily doped single-crystal substrates.

More recently, Sanyo reported the preparation of 8.5% polycrystalline thin-film solar cells fabricated on a metal sub-

Group	Material	Area, cm²	Efficiency, %	Power, Watts
Siemens Solar**	$CuInSe_2$	3830	11.2	43.1
Siemens Solar	$CuInSe_2$	3859	10.3*	39.7*
Siemens Solar	$CuInSe_2$	3261	10.1*	35.7*
Siemens Solar	$CuInSe_2$	938	11.1*	10.4*
ISET	$CuInSe_2$	845	6.9*	5.8*
EPV	$CuInSe_2$	791	7.2	5.7
EPV	$CuInSe_2$	722	6.0*	4.3*
Golden Photon	CdTe	832	8.1*	6.8*
BP Solar***	CdTe	706	10.1	7.1
Golden Photon	CdTe	3528	7.7*	27.2*
Solar Cells, Inc.	CdTe	6862	8.1*	55.4*

* NREL Measurements; All aperture-area efficiency ** Unencapsulated *** Non-monolithic integration

Table 4. Performance of Polycrystalline Thin-Film Photovoltaic Modules

Organization	Silicon Thickness, microns	Efficiency, %	Comments
RCA, Redfield	2.5	few percent	Optical confinement, single crystal
Honeywell, Feldman, et al	250 22	10% 3.1%	mulite substrate TiB_2 contact on Alumina or saphire
SMU, Chu et al	25	9.75%	metallurgical Si substrate
Kolodinski	16	14.7%	single crystal
Sanyo	10.27	8.5	poly on metal substrate
Kaneka	1.5–2.5	3.1	Plasma CVD, LBL growth

Table 5. Selected thin-film silicon solar cells

strate. [35] These devices, utilizing a-Si:H as the starting material, are recrystallized (solid phase crystallization) at 600°C to yield a polycrystalline film with a minority-carrier diffusion length of 24 μm, but have an active layer that is only 10-μm thick. The overall device thickness is 10.27 μm. Kaneka Corp. has reported 3.1% devices for polycrystalline Si film solar cells only 1.5–2.5-μm thick that were deposited using a modified plasma-enhanced CVD process referred to as "layer-by-layer" growth.[36] This information is summarized in Table 5.

Green [37] has also conceptually described a multilayered polycrystalline Si solar cell that would comprise vertically stacked and interconnected thin p-n junctions. The stacking of the extremely thin devices is intended to increase the optical path length leading to the necessary high currents, while the thin junctions are intended to take advantage of the admittedly poor diffusion length likely to result from such films. Such a concept may be able to take advantage of the materials science technology being developed for Si films, by those interested primarily in flat panel display applications. This concept is ambitious and will require aggressive development of the material and device processes.

AREAS FOR FURTHER DEVELOPMENT

Despite the substantial progress over the last 15 years, and the especially remarkable progress in CdTe and $CuInSe_2$ in the last 5 years, much remains to be done. As described above, modules of $CuInSe_2$ appear to be sensitive to defects and may lead to serious performance losses. It is imperative that module reliability issues be understood and eliminated, so that customers' favorable view of the reliability of crystalline Si products be shared by their thin-film cousins as well.

More basic technical issues remain to be solved. Our basic understanding of the mechanism of film growth for $CuInSe_2$ needs further work to be generally accepted; clearly, more must be done to extend the model in a detailed fashion to the alloys, both Ga- and S- substituted. One might speculate that the more rapid commercialization of CdTe has been due not to chance but the many choices available for the deposition of high-quality films of that material. Hence, it may be important to the commercialization of $CuInSe_2$ to develop straightforward alternate means of depositing the device-quality alloy films. It is unknown if the present process, which requires high-temperature substrates and sources, will allow low-cost, large-area manufacturing. Alternative methods, such as sputtering, spray deposition, and electrodeposition should be pursued with the goal of obtaining the same high-performance devices, keeping in mind the mechanism of film formation of evaporated films. A less serious problem—the use of a CdS n-type heterojunction partner—also needs to be addressed to simplify processing. Clearly, long- term research subjects such as superstrate devices and the use of flexible substrates will broaden the range of $CuInSe_2$ -based products.

CdTe devices, in contrast, have different problems that should be addressed. The film treatment required to produce device-quality films, now accomplished via heating the films with a $CdCl_2$ flux, is an annoying, costly step in the process that should be eliminated or combined with the film formation. The development of more refined processes may be one avenue of approach. Similarly vexing is the limited choice of contact materials to the p-type film; highly reflective, low-resistance materials and processes would be be an asset for high performance devices.

All the thin film materials are likely to face their own unique reliability issues. It will be of the utmost importance to identify failure mechanisms and devise successful solutions in order for these materials to realize their full potential in the marketplace.

ACKNOWLEDGEMENTS

The author is indebted to many people who contributed to this manuscript: to professors Bloss and Hamakawa for communicating the latest developments in Asia and Europe for inclusion in the talk; to the staff at NREL for their suggestions and providing many sources of information, especially Rommel Noufi, John Tuttle, Miguel Contreras, Harin Ullal, Ken Zweibel, and Pete Sheldon. Thanks are especially due to Wendy Larsen and Mark Fitzgerald for help in preparing the manuscript and presentation. This work was supported under the U.S. Department of Energy contract DE-AC36-83CH10093.

REFERENCES

[1] R. C. Reynolds, "Brief History of the Development of the CdS Cell," *Proc. 18th IEEE PV Spec. Conf.*, p.13 (1985)
[2] D. Redfield, "Enhanced PV Performance of Thin Silicon Films by Multiple Light Passes," *Proc. 11th IEEE PV Spec. Conf.*, p.431, (1975)
[3] L. C. Olsen and R. Bohara, "Experimental and Theoretical Studies of Cu_2O Schottky Barrier Solar Cells," *Proc. 11th IEEE PV Spec. Conf.*, p. 381(1975)
[4] A. L. Fahrenbruch, F. Buch, K. Mitchell, and R. H. Bube, "II-VI Photovoltaic Heterojunctons for Solar Energy Conversion," *Proc. 11th IEEE PV Spec. Conf.*, p. 490 (1975)
[5] J. L. Shay, S. Wagner, K. Bachmann, E. Beuhler, and H. M. Kasper, "Preparation and Properties of InP/CdS and CuInSe2/CdS Solar Cells," *Proc. 11th IEEE PV Spec. Conf.*, p. 503(1975)
[6] G. H. Hewig, F. Pfisterer, and W. H. Bloss, "Optical Properties and Theoretical Model of the Cu_xS-CdS Heterojunction," *Proc. 11th IEEE PV Spec. Conf.*, p. 441(1975)
[7] S. Wagner, J. L. Shay, P. Migliorato and H. M. Kasper, *APL*, 25, p. 434 (1974)
[8] L.L. Kazmerski, P.J. Iveland, P.R. White and R. B. Cooper, *Proc. 13th IEEE PV Spec. Conf.*, Washington, DC, p. 184 (1978)
[9] R. A. Mickelsen and W.S. Chen, *Proc 16th IEEE PV Spec. Conf.*, pp. 781-5 (1982)
[10] R. A. Mickelsen and W.S. Chen, *Proc. 15th IEEE PV Spec.Conf.*, pp. 800-4 (1981)
[11] R. N. Bhattacharya and K. R. Rajeshwar, "Electrodeposition of CuInX (X=Se,Te) Thin Films," *Solar Cells*, 16, p. 237-43(1986)
[12] N. Romeo, A. Bosio, V. Canevari, and L. Zanotti, "$CuInSe_2$/CdS Thin Film Solar Cells by RF Sputtering," *7th E.C. PV Solar Energy Conference*, p. 656-61 (1986)
[13] R. Ram, P. R. Thangaraj, A. K. Sharma,and O.P. Agnihotri, " Thin Film CdZnS/$CuInSe_2$ Solar Cells by Spray Pyrolysis," *India Bull. Mater. Sci.*,8(13), p. 279-84 (1985).
[14] N. Romeo, V. Canevari, G. Sberveglieri,A. Bosio,, and L. Zanotti, "Growth of Large-grained $CuInSe_2$ Films by Flash Evaporation and Sputtering," *Solar Cells*, 16, pp. 155-64 (1986)
[15] F. J. Garcia, and M. S. Tomar, "A Screen Printed CdS/$CuInSe_2$ Solar Cell," *J. Appl. Phys. Suppl.* (Japan) pp. 535-8. *Proceedings of the 14th Conference (1982 International) on Solid State Devices*, 24-26 Aug. 1982, Tokyo, Japan
[16] J. R. Tuttle, D.S. Albin and R. Noufi, "Thoughts on the Microstructure of Polycrystalline Thin Film CuInSe2 and its Impact on Material and Device Performance," *Solar Cells*, 30 pp. 21-38 (1991).
[17] J. R. Tuttle, M. Contreras, M.H. Bode, D. Niles, D.S. Albin, R. Matson, A.M. Gabor, A. Tennant, and R. Noufi, "The Structure, Chemistry, and Growth Mechanisms of Thin-Film $CuInSe_2$ Grown from a Mixed-Phase Precursor," *J. Appl. Phys.* 77 (1), p. 1-9 (1995)
[18] D. Schmid, M. Ruckh, F. Grunwald and H. W. Schock, *J. Appl. Phys.* 73, p. 2902 (1993).
[19] J. Kessler, M. Ruckh, D. Hariskos, U. Ruhle, R. Menner, and H. W. Schock, "Interface Engineering Between $CuInSe_2$ and ZnO," *Proc. 23rd IEEE PV Spec. Conf.*, p. 447 (1993).
[20] J. Hedstrom, H. Ohlsen, M. Bodegard, A. Kylner, L. Stolt, D. Hariskos, M. Ruckh, H. W. Schock, "ZnO/CdS/Cu(In,Ga)Se_2 Thin Film Solar Cells With Improved Performance," *Proc. 23rd IEEE PV Spec. Conf.*, p. 364 (1993).
[21] D. A. Cusano, *Sol. St. Elec.*, 6, 217 (1963)
[22] R. W. Aldrich, and R. E. Halstead, *Proc. 6th IEEE PV Spec. Conf.*, p. 188 (1967)
[23] E. I. Andirovich, Y. M. Yuabov, and G. R. Yagudaev, *Sov. Phys. Semicond.*, 3, p 61 (1969)
[24] D. Bonnet, "The CdTe Thin Film Solar Cell-An Overview," *Int. J. Solar Energy*, 12, pp.1-14 (1992)
[25] Y. Seol and H. B. Im, "Photovoltaic Properties of Cd1-x ZnxS/CdTe Solar Cells," *Proc 9th European PV Solar Energy Conf.*
[26] P. V. Meyers, "Advances in CdTe n-i-p Photovoltaics," *Solar Cells* 27,p.91-98 (1989)
[27] M. Spitzer, J. Shewchun, E. S. Vera, J. J. Loferski, "Ultra High Efficiency Thin Silicon p-n Junction Solar Cells Using Reflecting Surfaces," *Proc. 14th IEEE PV Spec. Conf.*, p. 375 (1980)
[28] J. D. Heaps, S. B. Schuldt, B. L. Grung, J. D. Zook and C. D. Butter, "Continuous Coating of SIlicon on Ceramic," *Proc. 14th IEEE PV Spec. Conf.*, p. 39 (1980).
[29] C. Feldman, N. A. Blum and F. G. Satkiewicz, "Vacuum Deposited POlycrystalline Silicon Solar Cells For Terrestrial Use," *Proc. 14th IEEE PV Spec. Conf.*, p. 375 (1980).
[30] T. L. Chu, S. S. Chu, E. D. Stokes, *Solar Energy Materials*, 2, p. 256 (1979/80)
[31] A. Barnett, R. B. Hall, D. A. Fardig and J. S. Culik, "Silicon-Film Solar Cells on Steel Substrates," *Proc 18th IEEE PV Spec. Conf .*, p. 1094 (1985)
[32] A. M. Barnett, S. R. Collins, W. R. Bottenberg, R. B. Hall, C. L. Kendall, J. C. Checchi, S. M. Lampo, W. Bloothoofd, D. H. Ford, J. A. Rand, A. T. Kendall, "Large Area Silicon-Film(TM) Solar Cells for Power Applications," *Tech. Digest of the International PVSEC-7*, Nagoya, Japan, p.245 (1993)
[33] S. Kolodinski, J. H. Werner, H. Rau, J. K. Arch, F. Bauser, "Thin Film Silicon Solar Cells From Liquid Phase Epitaxy," *11th E. C. PV Solar Energy Conference Proceeedings*, p.53, 1992.
[34] T.Matsuyama, T. Baba, T. Takahama, S. Tsuda, and S. Nakano "Polycrystalline Si Thin Film Solar Cell Prepared by Solid Phase Crystallization (SPC) Method," *Proc Intl. PVSEC-7*, p. 447 (1993)
[35] I. Shimizu, "Fabrication of poly-Si Thin Film in Non-Equilibrium Conditions," *Proc. of the 7th "Sunshine" Workshop on Thin Film Solar Cells,* 29-30 Nov. 1993, Shinjuku, Tokyo, Japan p. 75 (1993); Also Y. Hamakawa, Private Communication.
[36] Martin Green, Private Communication

SEQUENTIAL DEPOSITION OF $Cu(In,Ga)(S,Se)_2$

S. Zweigart, Th. Walter, Ch. Köble, S. M. Sun, U. Rühle, and H.W. Schock
Institut für Physikalische Elektronik, Universität Stuttgart
Pfaffenwaldring 47, D-70569 Stuttgart, Germany

ABSTRACT

Several sequential deposition processes for $Cu(In,Ga)(S,Se)_2$ leading to highly efficient thin film solar cells are described. Secondary phases and interdiffusion determine the growth mechanism of a process consisting of $InSe_x$ and Cu deposition at a low substrate temperature followed by a high temperature selenization step. Complex phase transitions are involved in the formation of $CuInS_2$ by this *inverted process*. The presence of a secondary CuS phase leads to a complete recrystallization of the film. Alloys of $CuInSe_2$ with $CuInS_2$ can be formed with the inverted process. Devices based on these absorber layers exhibit efficiencies up to nearly 16% for $Cu(In,Ga)Se_2$ and nearly 12% for $CuInS_2$.

INTRODUCTION

Deposition of $CuInSe_2$ or related alloys is a crucial step for large scale commercialization of this technology. Even though very high efficiencies for small area devices and large area modules with an output power exceeding 40 W have been realized [1,2,3], easily scalable processes with wide process windows are still being developed. Successful approaches are selenization/sulfurization of metallic precursors or coevaporation of the elements. The first method involves a number of reactions and complex interdiffusion of intermediate phases which affect the controllability of the final film quality. The second method leads to defined film properties and easily allows control of the alloy composition. However, an important prerequisite is the precise control of the fluxes of the single elements. A compromise between these two extreme cases is the sequential deposition of precursors which contain chalcogenides (S,Se) of the group I (Cu) or the group III (In,Ga) elements. It is possible to start either with a Cu precursor [4] or to deposit first the In,Ga chalcogenides. In both cases only a defined amount of elemental Cu needs to be added; either through deposition or in the vapour phase. The inverted process involves less mass transfer than selenization and its sequential nature makes it easier to control. In this contribution approaches for sequential deposition are described and reaction schemes during the process are analyzed. Since this method differs from the "conventional" approaches, also with respect to film morphology, the results of cell performance are compared to cells made by coevaporation. The specific segregation mechanism also opens new ways for the fabrication of $CuInS_2$ based cells.

DEPOSITION STEPS AND FILM FORMATION

Fig.1 illustrates the process sequences which have been investigated. A common feature is the use of $(In,Ga)(S,Se)_x$ precursors in a sequential process. In subsequent steps copper and the chalcogen are added in order to form $Cu(In,Ga)(S,Se)_2$. The main difference between the two process sequences shown in Fig.1 is the substrate temperature during the deposition of copper. In the first approach (process (a) in Fig.1) $InSe_x$ is deposited from a In_2Se_3 source at a substrate temperature of 200 °C followed by the evaporation of copper at the same low substrate temperature. The substrate temperature is increased up to 600 °C only during the last stage of the process, when selenium is evaporated onto the stack consisting of $InSe_x$ and Cu. It should be noted that this is a

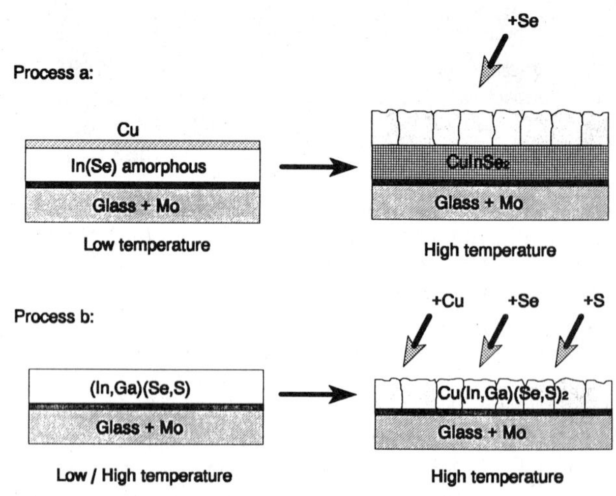

Fig. 1 Principle of the low and high temperature sequential processes

CH3365-4/94/0000-0060 $4.00 © 1994 IEEE First WCPEC; Dec. 5–9, 1994; Hawaii

Fig. 2 SEM micrographs of $InSe_x$ precursors

true sequential process which allows the control of the composition by simple techniques. Additionally a high substrate temperature is only required during the selenization step. Therefore, other deposition methods such as sputtering could also be applied.

In the second approach (process (b) in Fig.1), $(In,Ga)(S,Se)_x$ precursors are deposited by coevaporation of the elements at either low substrate temperatures of about 250 °C or at elevated temperatures around 600 °C. In the subsequent stage of the process copper, sulfur and selenium are diffused into the $(In,Ga)(S,Se)_2$ precursor at a substrate temperature of about 600 °C.

The samples were characterized by means of energy dispersive X-ray spectroscopy (EDX), secondary electron microscopy (SEM), X-ray diffraction (XRD), ultraviolet photoelectron spectroscopy (UPS), X-ray photoelectron spectroscopy (XPS) and metallurgical techniques. The latter characterization technique consists of imbedding the films in a resin, abrading and polishing them under an angle of 10 ° prior to SEM and optical microscope investigations. This is useful to study phase segregations.

In the following, three aspects of the processes described above will be presented and discussed. First a description of the film formation of $InSe_x$ precursors will be given. In the second chapter the growth of $CuInSe_2$ in the process consisting of the deposition of $InSe_x$ and copper at a low substrate temperature (process (a) in Fig.1) will be discussed. In a third part the film formation of $CuInS_2$ in a process which involves the indiffusion of copper and sulfur into In_2S_3 precursors at high substrate temperatures (process (b) in Fig.1) will be described. This process is referred to as *inverted process* because, contrary to the traditional Cu-rich/Cu-poor sequence of the Boeing bilayer recipe, this process starts with an extreme Cu-poor precursor, i.e. $(In,Ga)(S,Se)_x$.

Precursor film formation

The structure and morphology of the $(In,Ga)(S,Se)_2$ precursors depends strongly on the source material and of course on the substrate temperature. A layered semiconductor exhibiting the structure of In_6Se_7 is obtained (Fig.2) [5] in the deposition with In_2Se_3 as source material followed by a recrystallization step at substrate temperatures above 620 °C. For coevaporated precursors the composition is always close to $(In,Ga)_2(S,Se)_3$. Precursors consisting of $In_2(S,Se)_3$ can be recrystallized at substrate temperatures around 600 °C leading to large crystallites. As an example the morphology of a recrystallized In_2Se_3 thin film is shown in Fig.2 . When small amounts of Gallium are added to the precursor layers this recrystallization is no longer possible at substrate temperatures below 650°C.

Process (a): Low temperature deposition of In(Se)/Cu and selenization

The growth mechanism involved in the process (a) (Fig.1) was analyzed using the temperature profile given in Fig.3. The substrate temperature was increased by several steps up to about 600°C during the selenization. The duration of the 600°C step was varied between a few minutes and two hours. Another variable is the Se partial pressure which can be controlled by the temperature of the selenium source. Fig.4 shows typical cross sections of the final film for different compositions. Two layers with different morphologies are visible for In- rich films. We will mainly focus on the growth of the top part of this film in this chapter.

XRD measurements of the In(Se) precursor show no defined reflection beside the Mo- peaks, which indicates an amorphous structure of the In(Se) layer. A small increase in the background around 2θ=35° indicates a scattering effect from the near range ordering or extremely small nuclei [6]. Under our process conditions the In(Se)

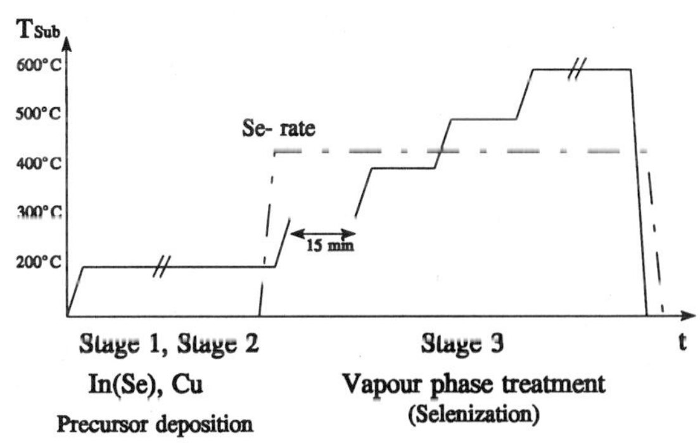

Fig. 3 Temperature Profile for the analysis of the low temperature sequential process.

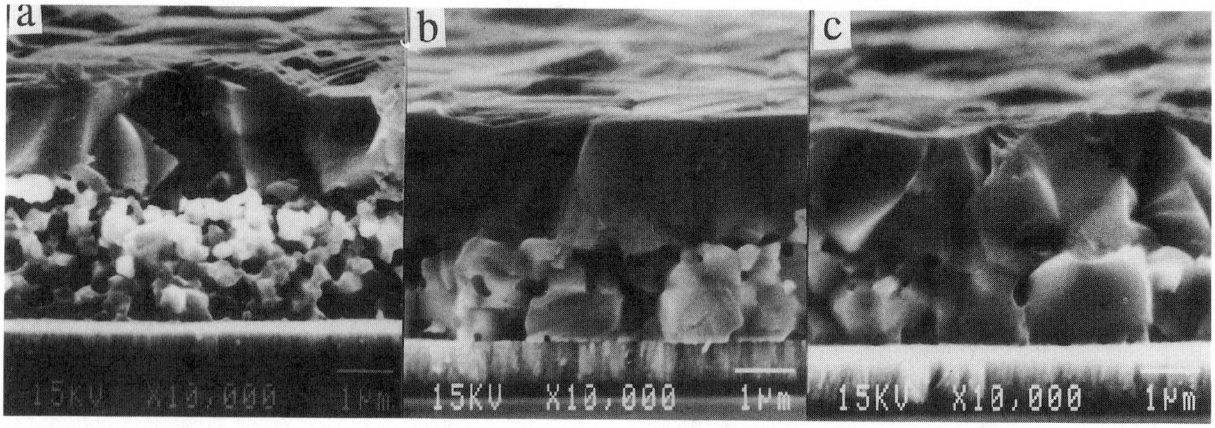

Fig. 4 Cross sections of the final film for different compositions: a) Cu/In ⟨ 1; b) Cu/In ≈ 1; c) Cu/In ⟩ 1

layer has the composition : ~51% In and ~49% Se. The situation has changed after the deposition of copper . If the Se background pressure in the deposition chamber is low enough, the Cu (111) and (200) can be seen by XRD [6]. The plot shown in Fig.6a represents the situation after the deposition of In(Se) and Cu. Most of the detected peaks can be related to $CuSe_2$ (JCPDS: 19-400) and Cu_2Se (JCPDS: 27-1131). Some additional peaks may be caused by an In(Se) phase due to a catalytic crystallization during the evaporation of Cu. The Cu peaks are also visible in high resolution measurement. The intensity of the $CuSe_2$ peaks decreases during the selenization as shown in Fig.6. These peaks are no longer detectable after the 300°C step. The intensity of the peaks which belong to Cu_2Se, $Cu_{2-x}Se$ (JCPDS: 6-0680) and In_2Se_3 (JCPDS: 20-0494) increase when rising $T_{Sub.}$ up to 400°C. For substrate temperatures above 500°C only $CuInSe_2$ (JCPDS: 40-1487) and InSe (JCPDS: 42-0919) peaks can be seen.

The In rich films show for all substrate temperatures a two layer structure (Fig.4, Fig.5). When the films are imbedded in a resin it is possible to separate the two layers mechanically. From XRD measurements on the separated layers it can be seen that the InSe phase exists only in the bottom layer. After the final selenization step only $CuInSe_2$ peaks are visible. The very well defined splitting of the (204/220) and (116/312) peaks indicates high crystalline quality in the top layer.

These results can be compared to EDX measurements performed on the samples after the different selenization steps. The sample after the 400°C step (Fig.5a) shows two

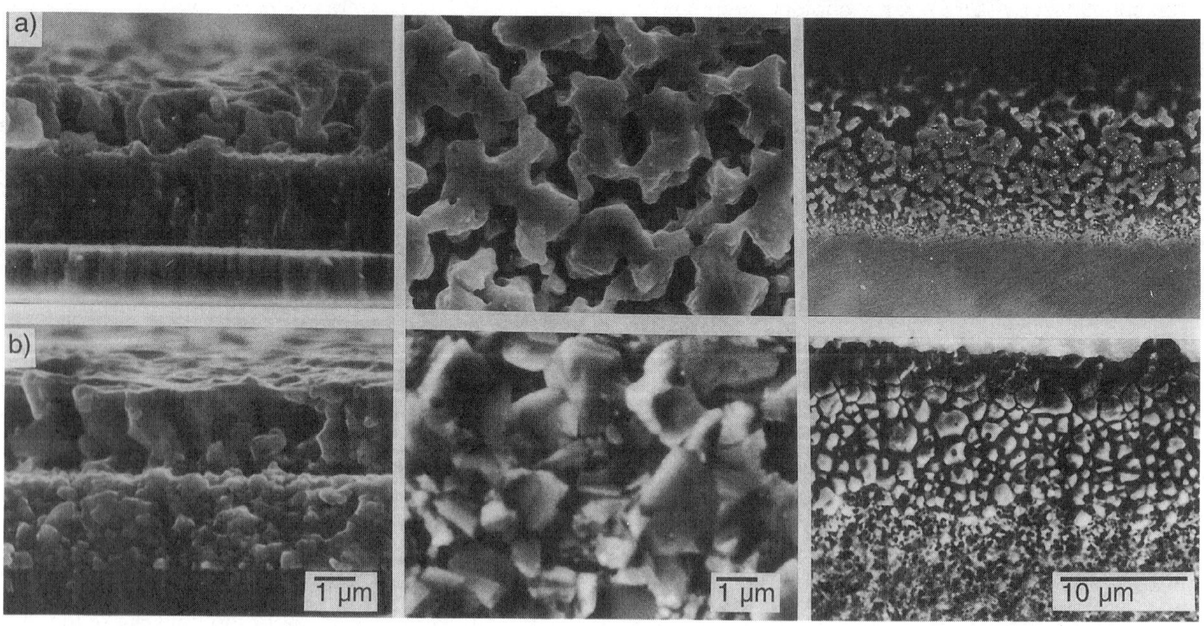

Fig. 5 SEM micrographs of the cross sections, surfaces and polished samples after the 400°C (a), and 600°C (b) selenization step

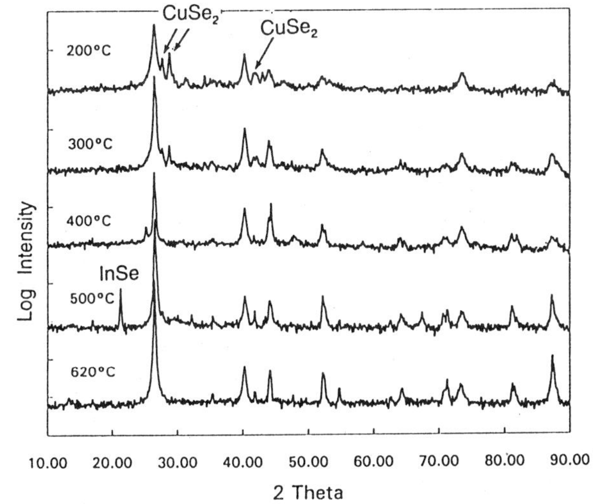

Fig. 6 XRD- diagrams of the precursor (a),the intermediate states during selenization (b: 300°C, c:400°C, d: 500°C) and the final film (d)

different phases. The measurements indicate that one phase (bright particles) has the composition of $Cu_{2-x}Se$, whereas the other contains approximately 20% indium. After the 500°C step only a very small amount of the $Cu_{2-x}Se$ phase is still present. The final film shows only one phase ($CuInSe_2$). The bevelled section (Fig.5b) exhibits a very densely packed top layer and relatively small crystallites together with some holes in the bottom layer. A schematic model is shown in Fig.7. In the temperature range between 200°C and 400°C the Cu reacts to $Cu_{2-x}Se$. A dendritic growth of the $Cu_{2-x}Se$ is clearly seen in the metallurgical studies. The amount of the $CuInSe_2$ phase is very low at this temperature. This changes at about 450°C. The $Cu_{2-x}Se$ phase is than transformed into $CuInSe_2$. The growth of the $CuInSe_2$ can be explained by the reaction path shown in Fig 7. Such a chain reaction is very fast. Additionally the diffusion of indium to the reaction front is quite easy because it diffuses through a defect structure ($(Cu,In)_{2-x}Se$). A diffusion along a lot of defects is very fast. Additionally the Se- sublattice is the same in $Cu_{2-x}Se$, $(Cu,In)_{2-x}Se$ and $CuInSe_2$ and therefore the activation energy for the recrystallization is very low. The sticking of Se is very high because of the reactive Cu at the surface. $Cu_{2-x}Se$ was also detected with surface analyses measurements (UPS, XPS). It is clearly seen that the morphology is influenced by the presence or absence of a Cu-rich growth front. In the bottom layer there is an indium rich growth mechanism and therefore the crystallites are very small. A recrystallization of the bottom part of the film is only possible for integral Cu-rich films with an excess of $Cu_{2-x}Se$ in the top layer once all indium has been consumed.

Initial experiments show that a recrystallization of the whole film is possible without a two layer structure and

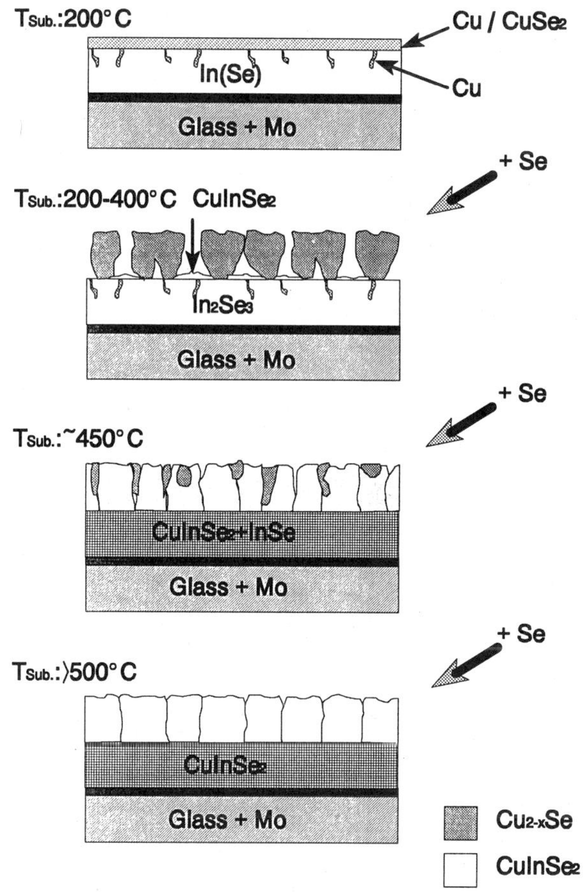

Fig. 7 Model of the growth mechanism

with a crystallite size of a few µm when the selenization is performed under a high selenium partial pressure. This selenization process was performed by isothermal close spaced vapor transport as presented in reference [7].

XRD measurements of the top layer show a strong 112 orientation (Intensity of (112) / Intensity of (204, 220) ~800) whereas the bottom layer exhibit no orientation. It seems that a strong orientation occurs only in films which where grown with excess Cu.

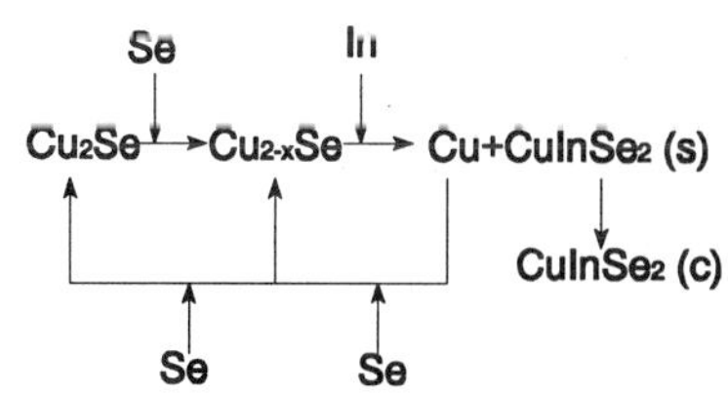

Fig. 8 Reaction path for the formation of $CuInSe_2$ in the top layer

Process (b): Indiffusion of Cu & S,Se into $(In,Ga)_2(S,Se)_3$ precursors at a high substrate temperature

The process sequence (b) of Fig.1 has already been investigated [1,8] for the $Cu(In,Ga)Se_2$ system. In this contribution the material system is extended to contain sulfur as well. $Cu(In,Ga)Se_2$ absorber layers were also deposited by the inverted process as a comparison.

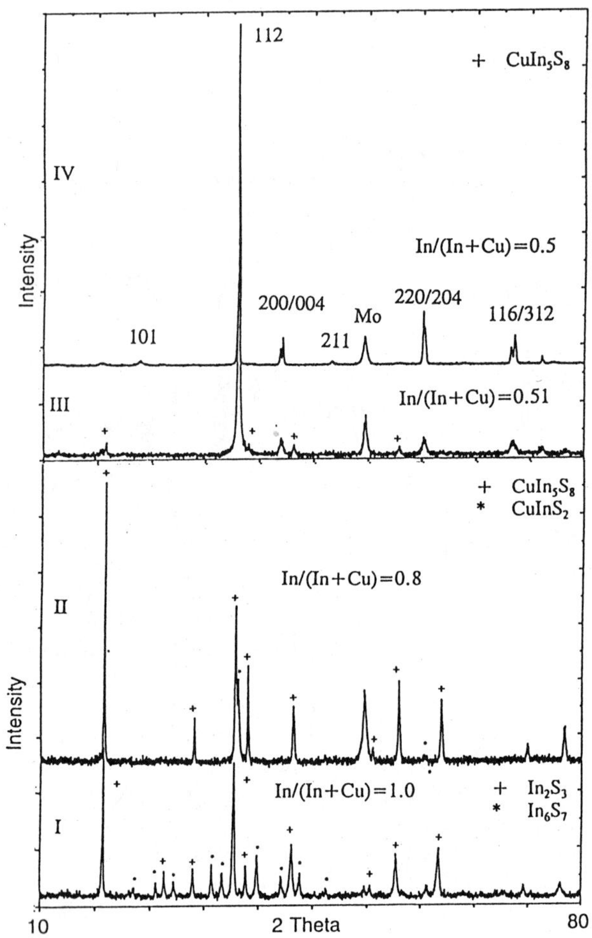

Fig. 9 Diffusion of Cu+S into InS_x: XRD diagrams at different stages of the process

The film formation of $CuInS_2$ involving the diffusion of copper and a chalcogen into a indiumchalcogenide precursor, the film formation of $CuInS_2$ will be discussed in the following. According to the pseudobinary phase diagram of $CuInS_2$ [9] only one distinct phase with a composition of $CuIn_5S_8$ exists between $CuInS_2$ and In_2S_3. $CuIn_5S_8$ crystallizes in a direct spinel structure with In atoms on octahedrally coordinated sites. The structure of In_2S_3 is a defect spinel and therefore closely related to the spinel structure of $CuIn_5S_8$. Thus it can be expected that during the indiffusion of copper and sulfur the transition from In_2S_3 to $CuIn_5S_8$ can occur easily by the diffusion of copper atoms onto metal sites which are not occupied in the defect spinel structure of In_2S_3. However, the transition from the spinel structure to the chalcopyrite structure of $CuInS_2$ involves severe structural changes, despite a preserved sulfur sublattice. In atoms must move from the octahedrally coordinated sites to tetrahedrally coordinated sites.

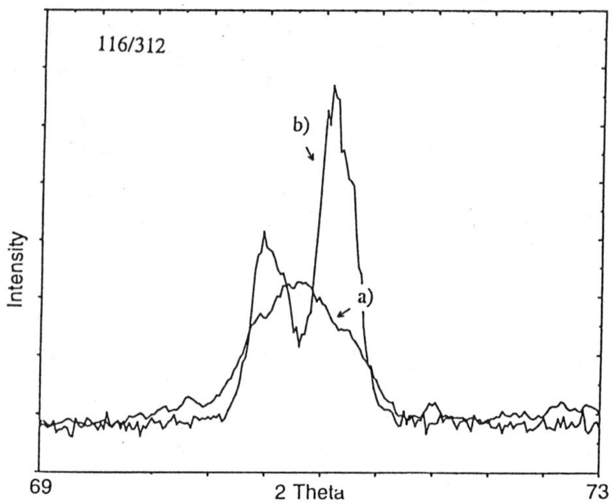

Fig. 10 Recrystallization under excess of CuS:
a) slightly Cu- poor $CuInS_2$
b) Cu- rich $CuInS_2$

In order to study phases and phase transformations involved in the film formation, the indiffusion of copper and sulfur into In_2S_3 precursors was interrupted at different stages of the process. XRD analysis of In_xS films which were deposited at T_{Sub}=250 °C and subsequently annealed in a sulfur atmosphere at T_{Sub}=600 °C reveals primarily In_2S_3 with additional peaks belonging to In_6S_7 (Fig.9 I). After the indiffusion of copper and sulfur (In:Cu=4:1) (Fig.9 II) $CuIn_5S_8$ is the predominant phase in the films. A XRD diagram of a thin film which is only slight-

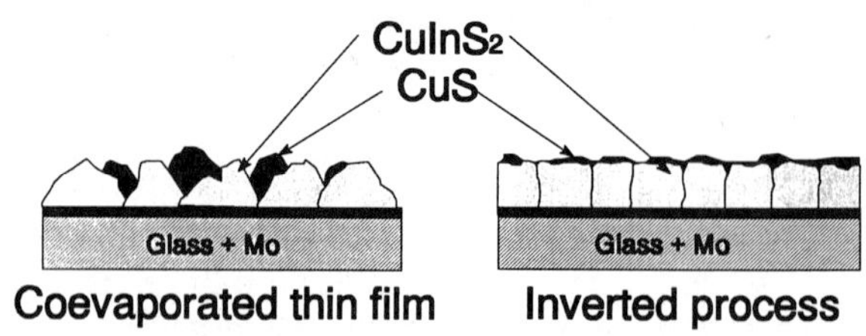

Fig. 11 Comparison of coevaporation and inverted process: segregation of CuS on Cu- rich $CuInS_2$ thin films

ly on the In_2S_3 rich side of $CuInS_2$ is shown in Fig.9 III. The main peaks belong to the chalcopyrite structure. However, small amounts of $CuIn_5S_8$ are still present. Additional indiffusion of copper and sulfur results in the segregation of CuS on the film surface as detected by grazing incidence XRD measurements. This secondary phase can be easily removed by a quick KCN treatment. The XRD diagram of the resulting film is shown in Fig.9 IV. Only peaks which belong to the chalcopyrite phase are visible. A close inspection of the 116/312 peak reveals a complete recrystallization of the film during the transition from slightly Cu-poor to Cu-rich material (Fig.10). The slightly Cu-poor film exhibits no traces of the tetragonal splitting of these two peaks whereas the Cu-rich film shows a clearly resolved peak splitting. This implies a complete reorganization of the crystallites by the existence of CuS as secondary phase. As a conclusion of these XRD studies on $CuInS_2$ thin films grown by the inverted process it can be stated that the phases and the corresponding film compositions agree with the pseudobinary phase diagram along the Cu_2S-In_2S_3 tie line on the In_2S_3 rich side of $CuInS_2$. Cu-rich films exhibit a CuS phase predominantly on the film surface. During the formation and segregation of this CuS phase the crystallites undergo a significant recrystallization process.

It is well known from coevaporated $CuInS_2$ thin films that Cu-rich material with a subsequent removal of the secondary CuS phase is required for efficient devices. The failure of Cu-poor thin films has been attributed to a very low carrier concentration and therefore to the formation of an unsufficiently high barrier in the heterojunction [10]. This Cu-rich step is also necessary for the inverted process. However, there is a significant difference in the segregation behavior of the CuS phase as illustrated in Fig.11. In case of coevaporated thin films there is a tendency for CuS to form island-like structures on the film surface and also to segregate in grain boundaries. For the inverted process the coverage of the surface with CuS is more homogeneous and there is less segregation at the grain boundaries. This is due to the fact that the CuS phase appears only in the final stage of the process and not continuously during the whole deposition as in case of coevaporated thin films. This implies consequences for heterojunctions after the removal of the CuS which will be pointed out below.

ALLOY FORMATION

Alloying of $CuInSe_2$ with $CuInS_2$ and or $CuGaSe_2$ opens the possibility to adjust the optical and electrical film properties. However, it is not clear a priori if these penternary materials can be formed by the inverted process. Initial experiments on the alloy formation in $CuIn(S,Se)_2$ thin films grown by the inverted process have been performed. Copper, selenium and sulfur were diffused in $In_2(S,Se)_3$ precursors with the same S/Se-ratio in the vapor phase during the deposition of the precursor and during the indiffusion of copper, sulfur and selenium. In Fig.12 the (112)-peak

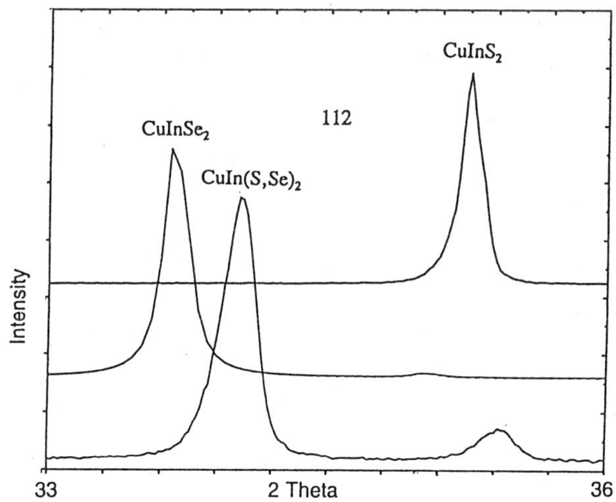

Fig. 12 XRD diagram: $CuInSe_2$, $CuInS_2$, $CuIn(S,Se)_2$

(XRD) of such a thin film is shown together with the (112)-peak of $CuInSe_2$ and $CuInS_2$. The peak profile of the alloyed sample is symmetric and the lattice spacing shifted indicating a homogeneous incorporation of sulfur and selenium. It can be concluded that a homogeneous $CuIn(S,Se)_2$ alloy was formed.

DEVICES

Thin film solar cells were completed by the deposition of a thin buffer layer and a RF-sputtered ZnO window layer. The buffer layer consisted in most cases of CdS or, in order to avoid Cd containing materials, of In(OH,S); both of which were deposited in a chemical bath [11]. In Table I the performance of devices based on $Cu(In,Ga)(S,Se)_2$ deposited by the inverted process is summarized. The $CuInSe_2$ based device results from process (a) in Fig.1. The absorber layers of the other devices were deposited according to process (b) in Fig.1. In case of the $CuInS_2$ based devices the indiffusion of Cu and S was extended to a stage resulting in the segregation of CuS on the film surface which was subsequently removed by a chemical treat-

Table 1 Device Performance

Absorber material	V_{OC}(mV)	i_{SC}(mA/cm^2)	ff (%)	η(%)
$CuInSe_2$	493	38.7	73	13.9
$CuInS_2$	705	23.3	72	11.8
$CuInS_2$ & Cd-free buffer layer	693	21.9	68	10.3
$Cu(In,Ga)Se_2$	620	32.5	77	15.5
$Cu(In,Ga)Se_2$ & Cd-free buffer layer	606	30.8	76	14.2

ment. For the $Cu(In,Ga)Se_2$ based devices no Cu-rich stage was reached during the process. The current densities and therefore efficiencies refer to active areas (grid coverage about 8%).

In general these devices are characterized by high open circuit voltages and fill factors up to 77%. Additionally, these solar cells exhibit a remarkable tolerance in the device performance with respect to the composition, i.e. the (In+Ga)/(Cu+In+Ga) ratio. Comparing the two different buffer layers, the V_{oc} of devices with a Cd-free buffer layer is in general about 20-30 mV higher than with a CdS buffer. The fill factor reaches similar values (up to 76%), however with lower j_{sc}. This is caused by a reduced red response despite the better blue response due to the higher bandgap of the buffer layer. An explanation for these observations is given in [11]. An efficiency exceeding 10% could be obtained for a device free of cadmium and selenium.

COMPARISON OF TECHNOLOGIES

We would now like to compare devices based on $Cu(In,Ga)(S,Se)_2$ grown by coevaporation and by the inverted process. These two processes should differ significantly because in the former case a layer by layer growth is intended whereas in the latter case diffusion processes

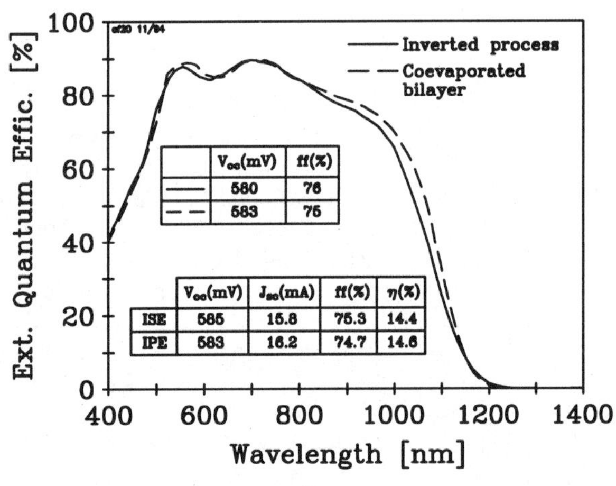

Fig. 13 Comparison of coevaporation and inverted process: spectral quantum efficiency and device performancece of $Cu(In,Ga)Se_2$ based solar cells

are responsible for the film formation. In Fig.13 the spectral quantum efficiencies of devices with similar bandgaps based on $Cu(In,Ga)Se_2$ grown by the inverted process and by the modified bilayer approach, i.e. by coevaporation, are shown, indicating no significant difference in current collection. Both open circuit voltage and fill factor exhibit similar values for both processes. Despite this similarity of the device performance the morphologies differ significantly as shown in Fig.14. This implies that the mechanism which limits and determines the device performance is relatively

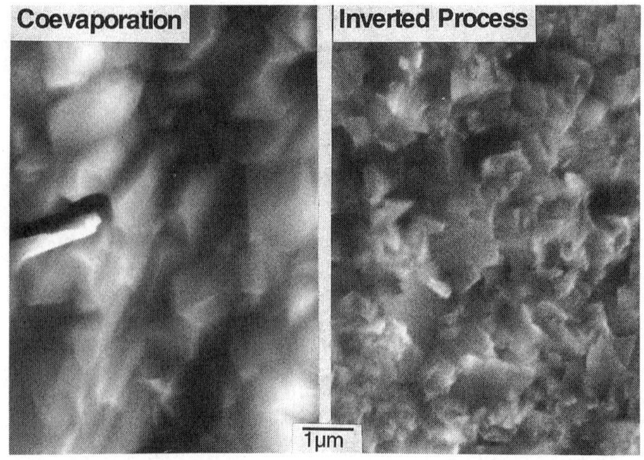

Fig. 14 Comparison of coevaporation and inverted process: SEM micrographs of $Cu(In,Ga)Se_2$ absorber layer

independent of the morphology of the absorber film. In order to check the calibration of our jV setup reference measurements (total area) were performed on the device grown by coevaporation at the ISE in Freiburg/Germany. A comparison of the device parameters is included in Fig.13 demonstrating the accuracy and reliability of our own measurements. In Fig.15 the open circuit voltage of devices grown by the inverted process and by the coevaporation method are shown together with the corresponding bandgaps as deduced from quantum efficiency measurements. Devices are included with a Cd-free buffer layer. Additionally Fig.15 contains results from experiments where sulfur was added during the diffusion of copper and selenium into the $(In,Ga)_2Se_3$ precursors. These results demonstrate that the increase of the open circuit voltage is equivalent to the increase of the bandgap due to higher Ga/(In+Ga)- or S/(S+Se)-ratios for both types of devices.

An evident advantage of the inverted process

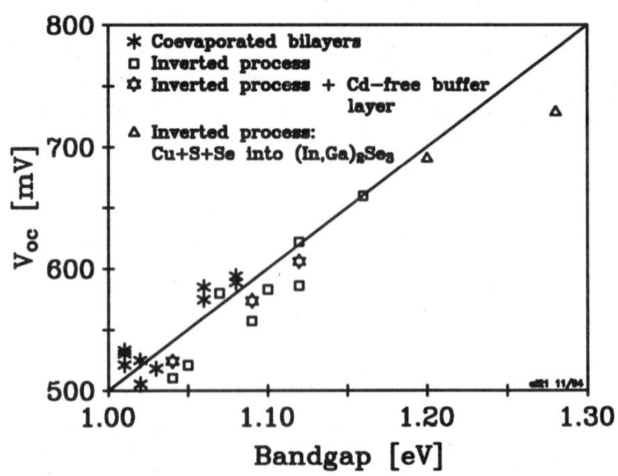

Fig. 15 Comparison of coevaporation and inverted process: V_{OC} vs. bandgap of $Cu(In,Ga)(S,Se)_2$ based heterojunctions

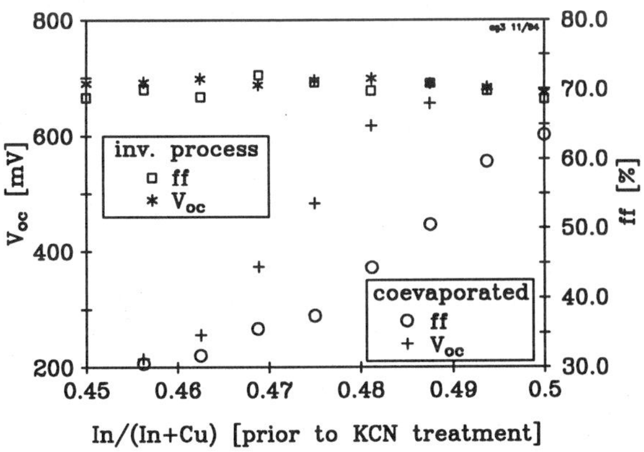

Fig. 16 Comparison of coevaporation and inverted process: V_{OC} and ff vs. composition prior KCN treatment of $CuInS_2$ based devices

versus coevaporated thin films can be demonstrated for $CuInS_2$ based devices. As mentioned above the absorber layers must be grown under an excess of copper followed by a KCN treatment in order to remove the segregated CuS. As illustrated in Fig.11 coevaporated thin films tend to form CuS islands, especially when a high excess of copper during the film growth is present, which results in shunting paths after the chemical treatment. Films grown by the inverted process are more dense and exhibit a more complete coverage of the surface with CuS thus reducing these shunting paths. The impact on the device performance is illustrated in Fig.16 where the fill factor and the open circuit voltage of devices based on $CuInS_2$ grown by coevaporation and by the inverted process versus the In/(In+Cu) ratio prior to the chemical treatment are compared. For the composition close to stoichiometry the device performance is similar. However, for coevaporated thin films fill factor and open circuit voltage decrease for higher copper contents due to shunts as deduced from jV-measurements. In case of the inverted process these shunting paths are reduced resulting in a device performance which is almost independent of the initial composition of the $CuInS_2$ absorber layer.

CONCLUSIONS

The experiments described above show different possibilities to obtain efficient $Cu(In,Ga)(S,Se)_2$ based devices using the inverted process. The processes lead to completely different morphologies also as compared to coevaporated chalcopyrite thin films. However, there are only minor differences in the device parameters and performance. Therefore, a process has to be qualified according to its advantages and disadvantages regarding upscaling, reproducibility and tolerance with respect to deviations from stoichiometry. We described a process for highly efficient $CuInSe_2$ based devices which requires a high substrate temperature only during the final selenization step. This certainly reduces problems related to high substrate temperatures during the metal deposition. Also a process for the deposition of $CuInS_2$ was proposed which has a greatly enlarged tolerance with respect to the initial composition of the thin film, since the secondary phase segregates differently. The device efficiency ranged between 15 and 16% for $Cu(In,Ga)Se_2$ based solar cells. Efficiencies up to nearly 12% could be obtained for $CuInS_2$ based heterojunctions. Highly efficient devices based on $Cu(In,Ga)Se_2$ and $CuInS_2$ could be demonstrated with a Cd-free buffer layer.

Acknowledgements

The authors are thankful to H. Kuschke of the Max- Plank Institute für Metallforschung in Stuttgart and H. Dittrich from the Center for Energy and Hydrogen Research for the XRD measurements and to Marc Casteleyn (University of Gent) for the ICSVT experiments.
This work was sponsored by the German Ministry for Science and Technology (BMFT), contract 0328059E and the Commission of the European Communities, contract JOUR2-CT92-0141.

REFERENCES

[1] A.Gabor, J.Tuttle, D.Albin, A.Tennant, M.Contreras, and R.Noufi, AIP Conf. Proc., 306 (1993) 59
[2] J.Hedström, H.Ohlsen, M.Bodegard, A.Kylner, L.Stolt, D.Hariskos, M.Ruckh, and H.W.Schock, 23rd IEEE PVSC, Louisville, (1993) 364
[3] D.Tarrant and J.Ermer, 23.IEEE PVSC, Louisville, (1993) 372
[4] J.Kessler and H.W.Schock, 11th E.C. Photovoltaic Solar Energy Conf., Montreux, (1992) 838
[5] J.Kessler, presented at the MRS spring meeting, San Francisco, (1993)
[6] S. Zweigart, D. Schmid, J. Kessler, H. Dittrich, H. W. Schock, accepted for publication in J. of Crystal Growth
[7] M. Casteleyn, M. Burgelman, B. Depuydt, A. Niemegeers, I. Clemminck, this conference
[8] J. Kessler, D. Schmid, S. Zweigart, H. Dittrich, and H.W.Schock, 12th E.U. Photovoltaic Solar Energy Conf., Amsterdam, (1994) 684
[9] J.J.M.Binsma, L.J.Giling, and J.Bloem, J.Crystal Growth, 50 (1980) 429
[10] T.Walter, R.Menner, Ch.Köble, and H.W.Schock, 12th E.U. Photovoltaic Solar Energy Conf., Amsterdam, (1994) 1755
[11] D.Hariskos, M.Ruckh, U.Rühle, T.Walter, and H.W.Schock, this conference

HIGH EFFICIENCY $Cu(In,Ga)Se_2$-BASED SOLAR CELLS: PROCESSING OF NOVEL ABSORBER STRUCTURES

Miguel A. Contreras, John Tuttle, Andrew Gabor, Andrew Tennant, Kannan Ramanathan, Sally Asher, Amy Franz, James Keane, L. Wang, John Scofield and, Rommel Noufi.
National Renewable Energy Laboratory, Golden, CO 80401

ABSTRACT

Our effort towards the attainment of high performance devices has yielded several devices with total-area conversion efficiencies above 16%, the highest measuring 16.8% under standard reporting conditions (ASTM E892-87, Global 1000 W/m^2). The first attempts to translate this development to larger areas resulted in an efficiency of 12.5% for a 16.8-cm^2 monolithically interconnected submodule test structure, and 15.3% for a 4.85-cm^2 single cell. Achievement of a 17.2% device efficiency fabricated for operation under concentration (22-sun) is also reported. All high efficiency devices reported here are made from graded bandgap absorbers. Bandgap grading is achieved by compositional Ga/(In+Ga) profiling as a function of depth. The fabrication schemes to achieve the graded absorbers, the window materials and contacting will be described.

INTRODUCTION

In this paper, we report on device optimizations that have led to important efficiency achievements. We present two different absorber fabrication approaches, which we believe represent promising and viable paths to large-area manufacturing of $CuInSe_2$ (CIS)-based modules. These processes are (1) coevaporation and (2) compound formation from Se containing precursors.

Regardless of the absorber fabrication process, all high efficiency $Cu(In,Ga)Se_2$ (CIGS) materials appear smooth and specular, with a morphology characterized by well-faceted and columnar grains. They always fall within a $0.86 \leq Cu/(In+Ga) \leq 0.96$ compositional range—as determined by electron probe microanalysis (EPMA) and include compositional gradients in Ga/(In+Ga) as a function of depth. The graded Ga incorporation results in changes mostly in the conduction band (CB), hence, producing absorbers with graded bandgaps. X-ray diffraction (XRD) spectra show well-defined chalcopyrite structures without the presence of secondary phases (i.e., Cu_2Se, In_2Se_3). Capacitance-voltage (C-V) measurements also reveal graded doping profiles.

DEVICE PROCESSING

All devices reported in this contribution are fabricated with the layered structure shown schematically in Fig. 1. The base substrate of preference is soda-lime glass (SLG), but we have used other materials as well. Among them 7059 Corning glass, Al_2O_3, and metallic foils such as Mo, Ni, and stainless steel. For terrestrial applications SLG has yielded the best results from a device performance point of view. In terms of lightweight flexible substrate—for potential space applications—Mo foils about 2.5x10^{-3} cm (0.001 in) thick have resulted in encouraging device performance results (efficiencies >10%). The most significant limitation for SLG substrates is its inability to withstand high processing temperatures (>500°C). In some instances of our high efficiency absorber fabrication processes, samples are warped because of this thermally induced stress on the glass. We have partially addressed the problem by going to thicker SLG substrates (2-3 mm), but this situation could be more optimally solved by either lowering substrate temperature or finding other suitable and low-cost substrates.

Common to all 1-sun devices reported here, is the e-beam evaporated top-contact grid made of 500-Å Ni and 3-μm of Al. This top contact grid has about 4% shadowing loss. Details for the design of concentrator devices based on CIGS absorbers can be found in ref. [1]. A description of the other components in the CIS-based solar cell is presented below.

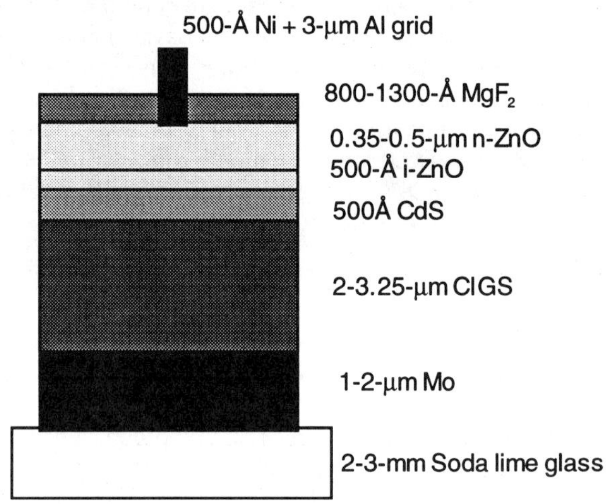

Fig. 1. Schematics of CIGS based solar cells

Mo Back Contact

The Mo back contact was fabricated by dc magnetron sputtering at two different Ar pressures resulting in a 1μm bilayer. The first 0.1-μm is deposited at a pressure of 10 mTorr and exhibits good adhesion and a resistivity of about 60 μΩ-cm. The second layer (~0.9-μm) is deposited at 1 mTorr; it shows poor adhesion (to the SLG) and resistivity of 10 μΩ-cm. The resulting bilayer adheres very well to the SLG and retains the lower resistivity. Details of the deposition process and characteristics of the Mo bilayer can be found in ref. [2] of these proceedings.

ZnO and CdS window processing

The ~500-Å CdS layer deposited by a chemical bath deposition (CBD) technique has been described elsewhere[3].

Perhaps the most significant improvement on our window layers has come from an overall better quality ZnO. The ZnO is a bilayer structure composed of a thin (~500-Å) high resistivity i-ZnO film capped with a high conductivity n-ZnO layer (3500-5500 Å). The high resistivity layer is deposited by RF magnetron sputtering from an intrinsic ZnO target in a mixture of Ar and O_2 at a sputtering pressure of 10 mTorr with no intentional heating. The resistivity of the thin intrinsic layer is highly dependent on O_2 partial pressure during sputtering. The best results have been obtained with a highly dilute gas mixture with a concentration of less than 1% O_2. The doped ZnO layer is sputtered from a 2 wt% Al_2O_3-doped ZnO target with neither O_2 nor intentional heating. Typical sheet resistance for the 500-Å intrinsic layer, 3500-Å doped layer, and 4000-Å bilayer are 10^7, 25, and 15 Ω/sq., respectively. Characteristic transmission and reflection data of the doped layer ZnO window component are shown in Fig. 2.

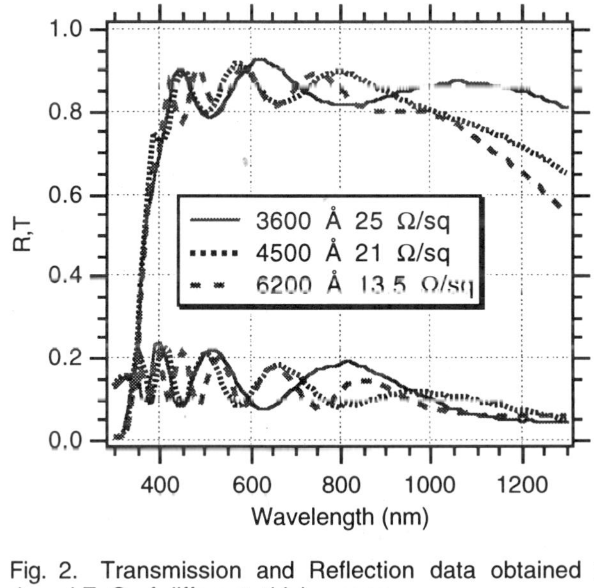

Fig. 2. Transmission and Reflection data obtained for doped ZnO of different thicknesses

Hall measurements performed on a 6300-Å thick n-ZnO sample, characterize the film with a 600-800 μΩ-cm resistivity, an electron density of 5.1-7.5 x10^{20} cm^{-3}, and a mobility of 14-15 cm^2/V-s. The measurement was done using a 2 Tesla magnetic field.

MgF_2 antireflection (AR) coating

The effect of the AR coating can be described as an effective optical coupling that allows more photons to reach the diode's junction with a net enhancement of current generation from the device. This coating is deposited by e-beam evaporation using optical-grade MgF_2 granules. Optimum thickness of this layer depends on the underlying ZnO properties and thickness. In our case—and for the ZnO quality we showed above—best results have been achieved with a MgF_2 thickness of 800-1200 Å. The gain in short circuit current is usually 4%-8% with the corresponding enhancement in conversion efficiency.

ABSORBER FABRICATION

Use of evaporation from elemental sources has proved to be a very flexible and powerful technique to explore different reaction pathways to compound formation and to fabricate specific graded band-gap absorbers in the CIGS system. Our focus on absorber fabrication for the past year covers two main approaches: (i) coevaporation, and (ii) CIGS formation from Se-containing precursors.

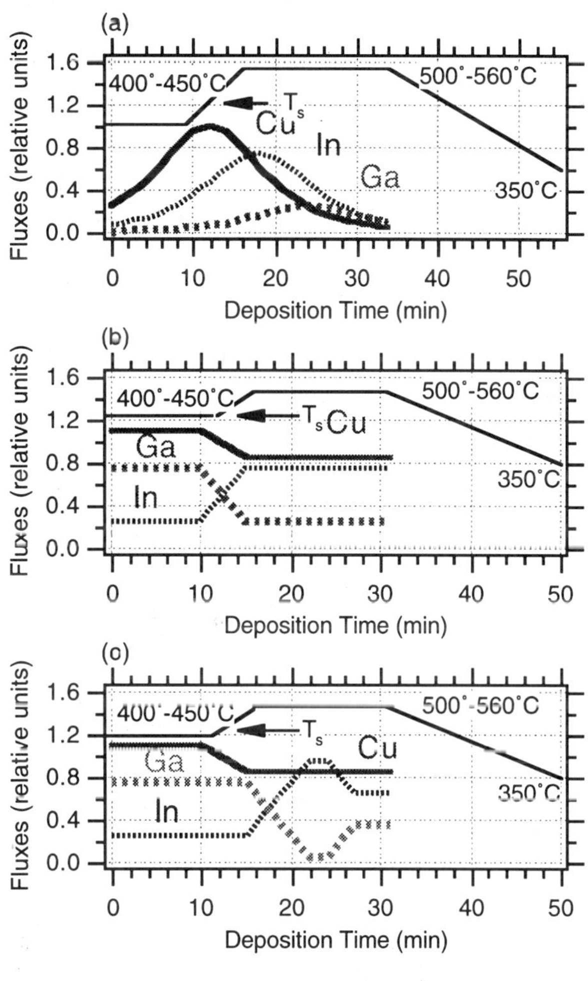

Figure 3. Deposition schemes for: (a) in-line emulation, (b) profiles for a normal grading, and (c) double grading

Coevaporation

This type of processing, enhanced by a computer interface for evaporant species deposition control, has allowed us to emulate potential large-scale in-line evaporation processes [4,5]. Furthermore, with the precise control of the metal fluxes during deposition—via preprogrammed profiles—we have been able to fabricate absorber materials that intentionally incorporate a compositional gradient in Ga/(Ga+In) as a function of depth.

Fig. 3 shows three coevaporation schemes with which we have experimented: (a) emulation of in-line coevaporation, (b) profiles for an absorber with normal band-gap grading, and (c) profiles for an absorber with double band-gap grading. The terms "normal" and "double grading" have been borrowed from amorphous Si (a-Si) technology where these type of graded band-gap structures have been previously implemented [6]. Common to all three coevaporation processes is a Se vapor selenization step done while cooling from the high substrate temperature (500˚-560˚C) to 350˚C in 20 min.

The in-line coevaporation approach emulates a continuous evaporation process in which a substrate travels at a constant speed over the sources that evaporate at a constant rate. CIGS-based graded band-gap materials have been shown to enhance some device parameters [7] as well as improve adhesion [8] to the Mo/SLG substrate. Graded structures with normal- and double-graded profiles are discussed later in this paper in terms of a simple but highly idealized one-dimensional solar cell model.

CIGS formation from Se-containing precursors

First attempts to compound formation by this kind of reaction pathway were done in two stages. The first stage, done at a low substrate temperature (250˚C<Ts<300˚C), forms an $(In_{1-x}Ga_x)_2Se_3(s)$ precursor layer (0<x<0.30). This Se-containing precursor is then subjected —in a second stage—to a Cu and Se flux, along with a higher substrate temperature (500˚C<Ts<560˚C) to form the CIGS compound. We note that, in both stages the Se flux was three times that of the metals. The compound formation can be represented by the following generic equations:

$$(In,Ga)(l)+Se(g) \longrightarrow (In_{1-x}Ga_x)_2Se_3(s)$$
$$(In_{1-x}Ga_x)_2Se_3(s) + Cu(s) + Se(g) \longrightarrow Cu(In,Ga)Se_2(s)$$

XRD data for a film obtained from the first-stage processing described above clearly reveals the crystallographic structure of an $(In_{1-x}Ga_x)_2Se_3(s)$ phase (see Fig. 4). Morphology and film evolution during Cu incorporation have been described in ref. [9,10].

The driving idea in this type of processing is to simplify manufacturing. It is conceivable to deposit the $(In_{1-x}Ga_x)_2Se_3$ precursor layer and the subsequent Cu-Se at low substrate temperatures by well-developed sputtering techniques, for example. A high-substrate temperature selenization step (via Se vapor or H_2Se) will then be required for compound formation and recrystallization.

This two-stage approach proved useful in fabricating thin films with near stoichiometric composition but with a rather small grain morphology and rough surfaces. A rough surface morphology may affect the device performance adversely because it can lead to poor metallurgical contact at the window/absorber interface. Furthermore, the added area at the junction will increase the forward current. Best devices fabricated by this two stage approach resulted in efficiencies <10%.

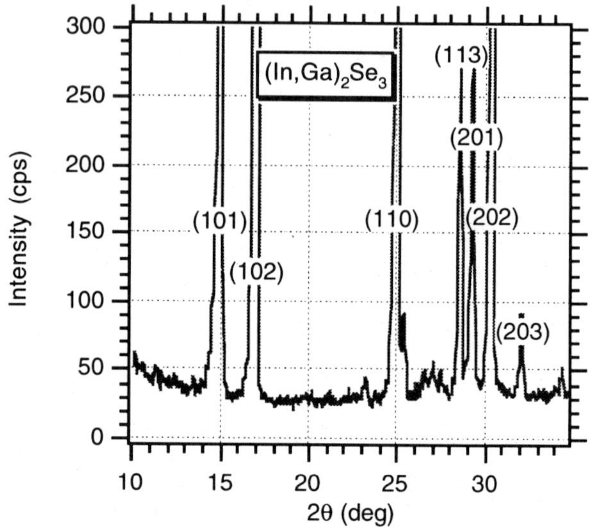

Fig. 4. XRD spectra of the $(In_{1-x}Ga_x)_2Se_3$ precursor film.

To improve bulk and surface morphology, we designed a 3-stage approach [11,12] in which a Cu-rich regime is reached at some point during Cu-Se deposition (second stage). We have previously reported on a growth model for CIS where we invoked the existence of a Cu_xSe liquid-phase as a fluxing agent that aids in grain growth [13]. This liquid-phase-assisted film growth yields a large and columnar grain morphology (see ref. [5,9,14]). Based on this premise, we decided to incorporate about 80%-90% of the In and Ga during the first stage, and the remaining 20%-10% after the completion of the second stage (see Fig. 5). With this scheme, the Cu-rich film at the end of the second stage consists of large-grain mixed phase of $Cu(In,Ga)Se_2$ and $Cu_{2-x}Se$.

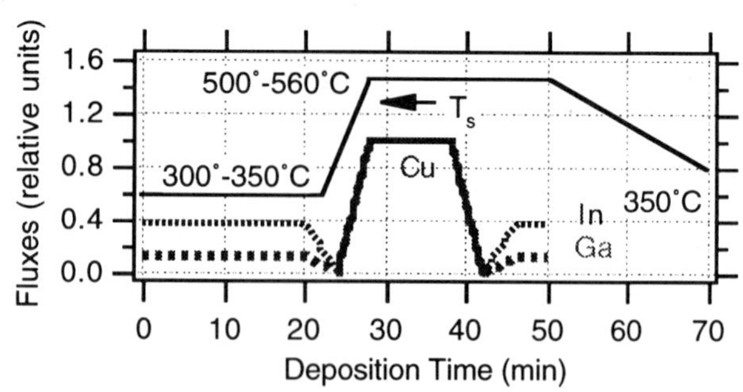

Fig. 5. Flux profiles for 3-stage processing

Adding the final 20%-10% of In and Ga at the end (third stage) is expected to aid in the formation of a smooth surface and to facilitate the formation of a Cu-poor defect chalcopyrite that has previously been shown to exist at the surface of CIS films [15]. The process also includes a final Se vapor treatment while cooling from 560°C to 350°C in 20 min.

We note that the 3-stage processing results in absorbers with a graded bandgap structure with double profiling in Ga (see next section). The incorporation of graded Ga and In as a function of depth has been reported elsewhere (see ref. [11,12]).

Fig. 6 shows the representative morphology for a high-efficiency absorber made by the 3-stage processing approach. There are two aspects to this morphology: (1) columnar grains that facilitate current transport across the thickness of the absorber, and (2) a rather densely packed microstructure free of pinholes and microcracks.

Fig. 6. SEM of film grown by 3-stage processing

XRD analysis (see Fig. 7) of absorbers grown by this 3-stage processing reveals a well-defined chalcopyrite structure with a strong (112) orientation and free of secondary phases (e.g., In_2Se_3, $Cu_{2-x}Se$)

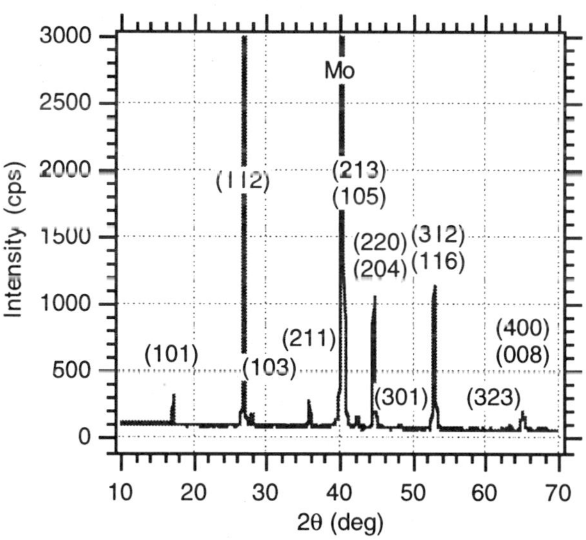

Fig. 7. XRD spectra of CIGS absorber

GRADED BAND-GAP STRUCTURES

Exchange of In atoms by Ga atoms in a base CIS matrix results in band-gap enhancement by raising mostly the conduction band (CB) edge [16]. In some cases, we have intentionally profiled Ga through the bulk of the film and, in others, compositional gradients were a result of the different diffusion mechanisms of In and Ga [10] during film growth (3-stage process). Graded Ga incorporation as a function of depth has implications on the shape of energy bands (conduction and valence) within the absorber structure. There are two types of band-gap grading we have focused upon. These are (a) normal profiling and (b) double profiling. These graded band-gap structures have been previously modeled and applied to a-Si:Ge alloys (see ref. [6]), with significantly enhanced device performance as compared to single band-gap absorbers. In the following paragraphs, we discuss both structures and their influence on device performance.

Normal grading structure

An absorber with normal profiling (nomenclature after ref. 6) is one where the band-gap is increased from front to back (see Fig. 8). The natural tendency of an electron excited to the CB will be to "roll down" the potential edge. The force exerted on such an electron comes from the additional quasi-electric field (due to the potential difference in the CB), and it can be seen intuitively from Fig. 8. The force acting on electrons in the CB (in p-type absorbers) is foreseen to "push" the electrons toward the space charge region and to reduce recombination at the back contact. Hence, the probability of these minority carriers contributing to current generation can be enhanced.

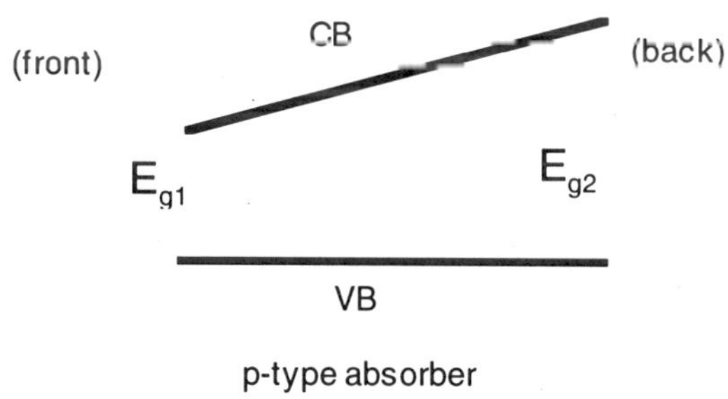

Fig. 8. One-dimensional band diagram of absorber incorporating normal profiling

To utilize the normal profiling concept in CIGS materials, we engineer a higher Ga content toward the back of the CIGS absorber (as shown in Fig. 3b). Because of the arguments presented above, this will result in an increase in the CB edge and the attainment of an effective force field repelling minority carriers from the back contact.

We have fabricated several absorbers that incorporate a higher Ga content toward the back of the device. Evidence of the graded Ga concentration has been verified using XRD and depth profiling techniques such as Auger Electron Spectroscopy (AES) (see Fig. 9) and Secondary Ion Mass Spectrometry (SIMS). Fig. 9a shows the XRD spectra of the (112) peak for one of the normal profiling absorbers (sample S423). The XRD data reveal clearly the existence of at least two different CIGS compositions. Figure 9b is the Auger depth profile of the same absorber. Included in the figure are the EPMA results for accelerating voltages of 10 and 20 kV. The depth profile clearly shows a significant variation in Ga and In through the depth of the film. The lower Ga content found at the surface provides the lower band-gap composition. The higher Ga content found towards the back gives the wider band-gap composition. The compositional depth profile is complimentary to XRD data and serves as a qualitative verification of the normal profiling structure.

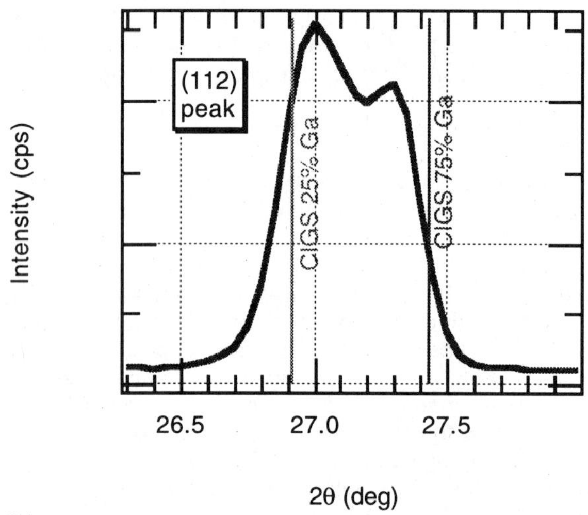

(b)

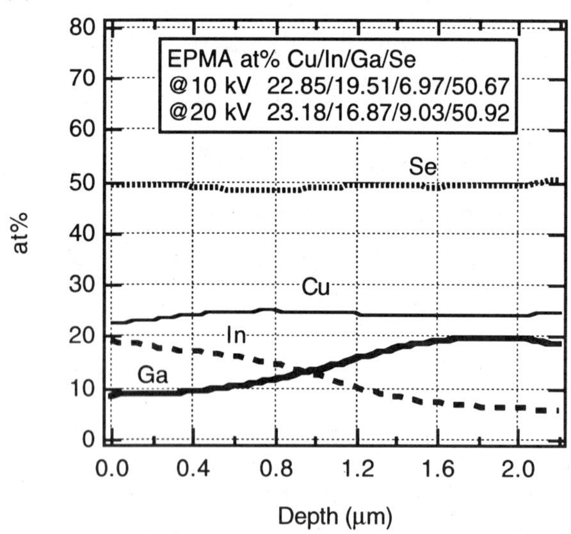

Fig. 9. (a) (112) peak XRD spectra of normal profiling absorber (sample S423), and (b) Auger depth profile for the same sample

Absorber with Double Profiling

The double profiling structure incorporates normal profiling plus an inverse profiling (reverse of a normal profiling) in the region adjacent to the surface of the film (see Fig. 10). The first absorbers incorporating this type of structure were fabricated by the coevaporation process shown in Fig. 3(c), but also have been attained by the 3-stage scheme shown in Fig. 5. This double profiling structure has been theoretically modeled in a-Si devices [in Ref. 6]. The authors conclude that a double-profiled structure with a front graded a-Si layer extended for about 10%-20% of the total device thickness should give improved performance over simple unidirectional profiling.

A particular case of this double profiling is a notch structure. The latter can be thought of as a low band-gap material sandwiched between two wider band-gap materials. A theoretical analysis and numerical computer simulations of this and other graded structures based on CIGS has been reported by Dhingra *et al.* [17]. Among all the different simulations carried out by these authors, the notch structure yields the highest projected device performance.

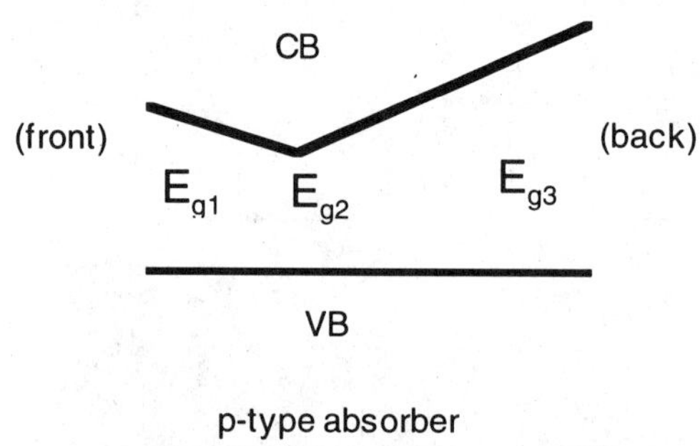

Figure 10. One-dimensional band diagram of absorber incorporating double profiling

Referring to Fig. 10, we anticipate the double profiling structure to absorb photons with energies higher than E_{g1}; furthermore, due to a decreasing energy gap as a function of depth, we can expect some of the photons with energies as low as E_{g2} to be absorbed ($E_{g1}>E_{g2}$). Hence, the structure has the potential for optimized photon absorption in a specific photon energy range.

The two energy gaps at the front of the double profiling structure (E_{g1} and E_{g2} in Fig. 10), could be engineered to match certain bands of the terrestrial solar spectrum (AM1.5) in order to capture more efficiently from the blue and red spectral regions. The improved quantum efficiency in such a device will translate into enhanced current generation (increase in J_{sc}). An additional enhancement of current generation could also come from the force field effect on electrons created by the increasing CB edge found towards the back (similar to the normal profiling structure discussed above).

Fig. 11 shows the SIMS depth profile for an absorber incorporating double profiling (sample C266) fabricated using the 3-stage process. The SIMS profile is calibrated with a set of standards with known compositions (as determined by EPMA). This method results in quantitative accuracy of ±1 at% for Ga and In, and ±3 at% for Cu and Se. The analysis clearly reveals the qualitative aspect of the double profiling structure. Atomic composition (at%) of the same film (C266) as obtained from EPMA are:

Cu/In/Ga/Se=23.25/18.83/7.73/50.20 at 10 kV and 22.82/19.67/6.96/50.56 at 20 kV.

The model adopted in ref. [17] points out an optimum notch position as close as possible to the front surface. The depth profile in Fig. 11 meets this condition to some degree. There are, however, questions regarding the optimum width and depth of the notch, variables that we continue to investigate along with abrupt transitions between different compositions.

DEVICE PERFORMANCE

Table 1 is a summary of I-V measurements for representative small-area devices. All I-V parameters reported in the table are for total-area measurements under standard reporting conditions of 1000 Wm^{-2} irradiance and a cell temperature of 25°C (ASTM E892). The table includes the "effective" bandgap value as obtained from cut-off extrapolation of the quantum efficiency measurements. In the table, we also include results on a 22-sun concentrator device. Table 2 is a summary of I-V results for larger area devices.

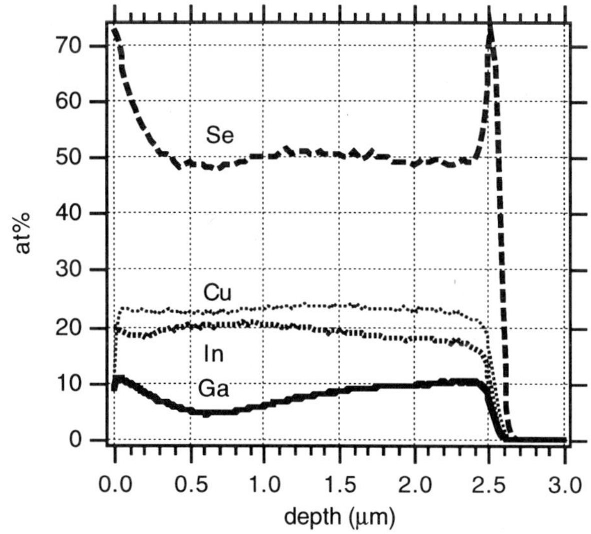

Fig. 11. SIMS depth profile of absorber incorporating double profiling (sample C266)

sample	type of grading	area (cm^2)	V_{oc} (mV)	J_{sc} (mA/cm^2)	FF (%)	Total Area Efficiency (%)	E_g (eV)
M1201	$CuInSe_2$ (no Ga)	0.395	484	36.29	75.10	13.2	0.95
S456	normal on Mo foil	0.444	496	32.93	62.70	10.2	1.07
S319	normal	0.251	**687**	23.73	74.78	12.2	1.17
S423	normal	0.422	664	29.24	78.04	15.1	1.10
C253	double	0.432	611	**34.13**	76.00	15.8	1.03
C257	double	0.458	623	32.82	75.36	15.4	1.07
C266	double	0.418	660	31.53	**78.73**	16.4	1.08
C362	double	0.437	652	33.18	77.44	**16.8**	1.09
C364	22-sun concent.	0.074	0.767	49.68	75.2	17.2	-
C371-14	double	1.025	678	31.96	75.80	16.4	1.11

Table 1. I-V parameters obtained under standard reporting conditions of 25°C and irradiance of 1000 W/m^2 (ASTM E892). Concentrator cell is a grided device with 15% shadowing.

comments	area (cm^2)	V_{OC} (V)	I_{SC} (mA)	FF (%)	Efficiency (%)
submod. 6 cells	16.8	3.43	92.1	66.3	12.5
single cell	6.64	0.644	198.5	72.21	13.9
single cell	4.85	0.657	150.7	74.75	15.3

Table 2. I-V parameters for larger area devices. Submodule structures incorporate a monolithic interconnection between cells.

From Table 1, we note that devices incorporating a double profiling absorber preserve the high V_{oc} values attained in the normal profiling devices, moreover, they yield higher short-circuit currents for comparable effective band-gaps. This situation can be explained from the enhanced spectral response of the double profiling structures. Fig. 12 shows the normalized quantum efficiency (QE) of selected normal and double profiling devices. The figure reveals that normal profiling devices do not collect efficiently in the long wavelengths. The double profiling devices on the other hand, not only have an improved response in the long wavelengths but in the short wavelengths as well. However, we like to emphasize that we have not explored extensively other normal profiling schemes which might improve the red response. The short wavelength cutoff is common to all devices and is only limited by the transmission properties of the interface window material (CdS).

To calculate carrier concentration from capacitance data (in polycrystalline materials) we have followed the approach described in ref. [18]. Carrier concentrations in the CIGS graded structures are determined mostly by the film compositions resulting from the growth conditions described above.

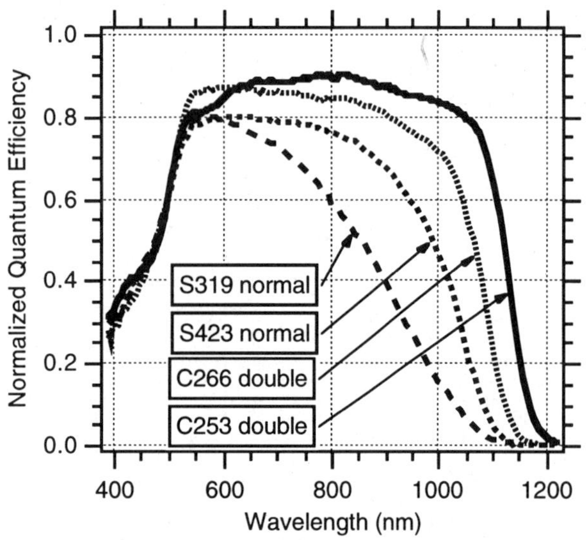

Fig. 12. Normalized QE for selected devices incorporating absorbers with normal and double profiling

Capacitance measurements of all graded band-gap devices reveal a graded carrier concentration profile. Fig. 13 shows the carrier concentration as a function of distance from the space charge edge for both graded structures (normal and double profiling).

FINAL REMARKS

The results presented above are very encouraging. For absorber band gaps used in this study, device parameters exist for greater than 18% efficiency. Improvement in collection from the short wavelength region (400-500nm) may account for additional 1-2 mA/cm^2. Demonstration of laboratory cell efficiency above 18% facilitates the effort of attaining 15% flat plat modules.

Even though the results described above may favor one deposition scheme over another, or one device structure compared to another, we caution that more investigations into the variables affecting the different device structures and other deposition schemes are needed before a final conclusion is made. We expect, in the near future, to expand on this theme. We also anticipate further improvement in device performance with further optimization of the window materials, and additional improvement in the absorber by the incorporation of sulfur.

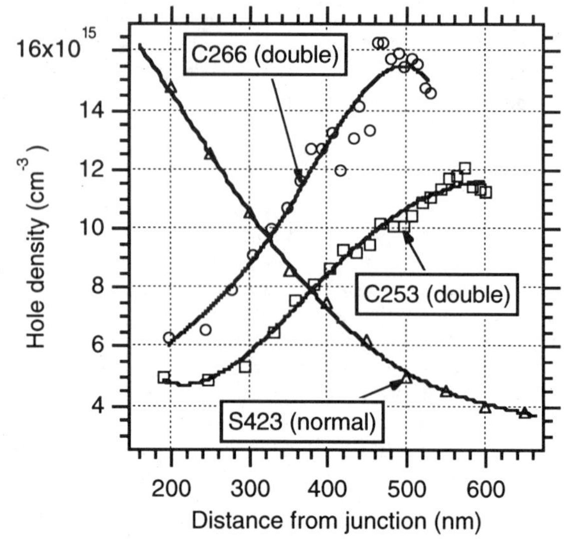

Figure 13. Carrier concentration as a function of distance from the junction

ACKNOWLEDGMENTS

The author would like to thank James Dolan for technical assistance, Alice Mason and Rick Matson for composition analysis and SEM, James Sites for assistance with capacitance measurements, and Siemens Solar Industries for assistance with sub-module fabrication and Mo substrates. This work was performed at NREL under Contract No. DE-AC36-83CH10093 to the U.S. Dept. of Energy.

REFERENCES

[1] J. Tuttle, M.A. Contreras, J.S. Ward, A.M. Gabor, K.R. Ramanathan, A.L. Tennant, L. Wang, J. Keane, and R. Noufi. These proceedings.

[2] John Scofield, S. Asher, D. S. Albin, J. Tuttle, M. Contreras, A. Tennant, D. Niles, R. Reedy, and R. Noufi. These proceedings.

[3] J. Kessler, K.O. Velthaus, M. Ruckh, R. Laichinger, and H.W. Schock. *6th International PVSEC.* New Delhi, India, 1992, pp. 1005-1010.

[4] L. Stolt, J. Hedström, J. Kessler, M. Ruckh, K. Velthaus, and H. Schock. *Appl. Phys. Lett.* 62 (6), 8 February 1993, pp. 597-599.

[5] M. Contreras, J. Tuttle, D. Albin, A. Tennant, and R. Noufi. *Proceedings of theTwenty-Third IEEE PVSC*, 1993, pp. 486-490.

[6] A. Pawlikiewicz and S. Guha. *IEEE Transactions on Electronic Devices*, Vol. 37, No. 2, February 1990, pp. 403-409.

[7] M. Contreras, J. Tuttle, D. Du, Y. Qu, A. Swartzlander, A. Tennant, and R. Noufi. *Appl. Phys. Lett.* **63** (13), 27 September 1993, pp. 1824-1826.

[8] C. Jensen, D. Tarrant, J. Ermer and G. Pollock. *Twenty-Third IEEE PVSC*, 1993, pp. 577-580

[9] A. M. Gabor, J.R. Tuttle, D.S. Albin, R. Matson, A. Franz, D.W. Niles, M.A. Contreras, A.M. Hermann and, R. Noufi. Mat. Res. Soc. Symp. Proc., Vol. 343, 1994, pp. 143-148.

[10] A.M. Gabor, J.R. Tuttle, A. Schwaryzlander, A.L. Tennant, M.A. Contreras, and R. Noufi. These proceedings

[11] M. Contreras, A.M. Gabor, A. Tennant, S. Asher, J. Tuttle, and R. Noufi. *Progress in Photovoltaics*, Vol. 2, pp. 287-292, (John Wiley &Sons,1994).

[12] A. Gabor, J.R. Truttle, M. Contreras, D.S. Albin, A. Franz, D.W. Niles, and R. Noufi. *Proceedings from the 12th European Photovolaic Solar Energy Conference* (H.S. Stephens and Assoc., UK), Amsterdam, The Netherlands, 11-15 April 1994, pp. 939-943.

[13] J. Tuttle, M. Contreras, A. Tennant, D. Albin, and R. Noufi. *Twenty-Third IEEE PVSC*, 1993, pp.415-421.

[14] A.M. Gabor, J.R. Tuttle, D. S. Albin, A. L. Tennant, M.A. Contreras, and R. Noufi. AIP Conference Proceedings 306, (AIP Press, New York), 1994, pp. 59-66.

[15] D. Schmid, M. Ruckh, F. Grunwald, and H.W. Schock. J. Appl. Phys. 73 (6), 15 March 1993, pp. 2902-2909.

[16] S. Wei and A. Zunger (internal communication). Data obtained from theoretical calculations based on local density formalism as implemented by Linear Augmented Plane Waves Method (S. Wei and A. Zunger, *Physical Review Letters*, Vol. 59, No. 1, 6 July 1987, p. 144).

[17] A. Dhingra and A. Rothwarf. *Twenty-Third IEEE PVSC*, 1993 pp. 475-480.

[18] P. Mauk, H. Tavakolian, and J. Sites. *IEEE Transactions On Electron Devices*, vol. 37, No. 2, February 1990, pp. 422.

Processing Issues for Thin-Film CdTe Cells and Modules

Robert W. Birkmire and Peter V. Meyers
Institute of Energy Conversion, University of Delaware

United States Department of Energy
University Center of Excellence
for Photovoltaic Research and Education
(National Renewable Energy Laboratory)
Newark, Delaware 19716 USA

ABSTRACT

Among emerging thin-film PV technologies, CdTe-based modules have displayed perhaps the most flexibility with respect to their method of manufacture. Device efficiencies have reached >15% and there appears to be no inherent device instability. Nonetheless, there remain serious technical challenges which must be met if CdTe modules are to be manufactured at a commercially attractive level. This paper identifies and discusses these issues.

INTRODUCTION

Thin-film CdTe-based photovoltaic technology continues to attract serious attention within the photovoltaic community. Solar cells with an area of 1 cm^2 have been made with efficiencies over 15% and modules have been made with efficiencies over 10%. In addition, CdTe solar cells with efficiencies over 11% have been made by a variety of deposition methods and several of these methods have been adapted to module manufacturing. Table 1 summarizes the performance of CdTe solar cells and modules fabricated by a variety of methods and lists the research groups involved. In this paper the status of CdTe technology is discussed with emphasis on key technical issues relating to device fabrication and commercialization.

MATERIALS AND DEVICE ISSUES

All high efficiency CdTe/CdS solar cells have a superstrate device configuration. This is shown schematically in Figure 1 along with the processing steps required to fabricate the device. In this section, we discuss several material and device issues associated with the fabrication of CdTe cells with respect to developing a manufacturing process for modules.

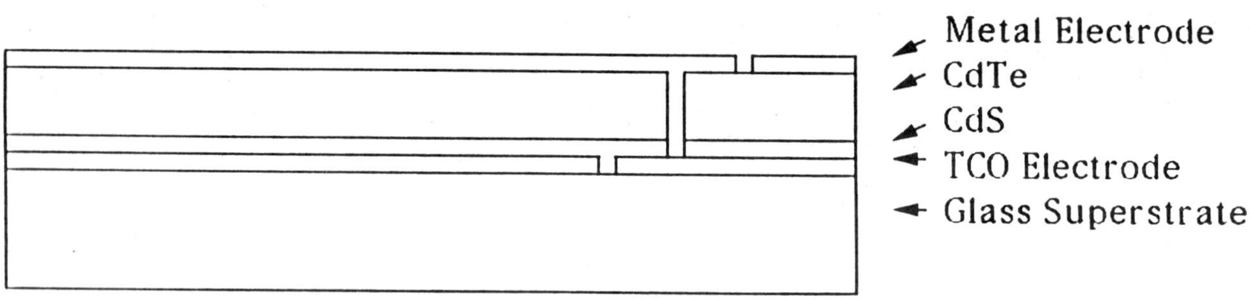

1. Glass superstrate - soda lime or 7059
2. TCO - by pyrolysis or CVD8888
3. CdS deposition
4. Heat treatment
5. CdTe deposition
6. Chloride heat treatment
7. Surface preparation
8. Contact deposition

Figure 1. Generic CdTe Solar Cell Structure and Fabrication Sequence

First WCPEC; Dec. 5–9, 1994; Hawaii

Table 1. Leading CdTe Deposition Technologies [1,2]

CdTe Deposition Process	Academic Interest	Industrial Interest	Efficiency/Area
Compound Sublimation and Vapor Transport (CSS)	USF [1], NREL, IEC, Stanford	SCI, Antec [3]	15.8% / 1.05 cm^2 [4] 8.4% / 7200 cm^2 [5]
Electrodeposition	CSM, Queensland	BP Solar, A.D. Plating	14.2% / 0.02 cm^2
Atomic Layer Epitaxy		Microchemistry	14.0% / 0.12 cm^2
Screen printing	KAIST, U. Gent [7]	Matsushita	12.8% / 0.78 cm^2 8.1% / 1200 cm^2
Spraying		Golden Photon	12.7% / 0.3 cm^2 8.1% / 832 cm^2
MOCVD	GIT, USF, U. Durham, RPI [8]		11.9% / 0.08 cm^2 [9]
PVD - vacuum evaporation	IEC, U. Parma, New Castle-on-Tyne, Stanford		11.8% / 0.3 cm^2 [10]
Laser Driven PVD	U. Toledo		10.5% / 0.1 cm^2 [11]
MBE	NREL [12], Swiss Fed. Inst. Tech, GIT		10.5% / 0.08 cm^2 [13]
Sputtering	NREL, U. Toledo		10.4% / 0.1 cm^2 [11]
APCVD	USF [14]	Batelle [15], SCI [16]	9.9% / 1 cm^2 [14]

CdS Window Layer and J_{SC}

Based on the standard AM1.5 Global spectrum, the maximum J_{SC} of a CdTe solar cell is about 30.8 mA/cm^2 [17]. Approximately 7.1 mA/cm^2 of that current is due to photons with energy above the CdS absorption edge. Thus absorption losses in the CdS layer can have a significant effect on J_{SC}.

The highest efficiency CdTe/CdS solar cells are distinguished by their high short circuit current >25 mA/cm^2. This has been achieved using thin CdS, ~70 nm, deposited from an aqueous solution, by a chemical bath deposition (CBD)[1,18]. The use of the thin CdS is often accompanied by reduced yield of devices which is, in turn, attributed to pinholes or other defects in the CdS. The use of thin CdS is also complicated by a further reduction in the thickness of the CdS which occurs during CdTe film growth or post deposition $CdCl_2$ heat treatment [19,20].

From a manufacturing perspective, CBD film growth of CdS has certain disadvantages :

1) the deposition rate of CBD films, ~10 Å/s, is relatively slow;
2) the process is complicated since the best cell results are achieved only when the CdS film is heat treated in H2 [1] or with $CdCl_2$ [19];
3) the CdS grows heterogeneously on the glass/TCO superstrate and vessel walls as well as homogeneously forming particulates in the solution which can subsequently adhere to the superstrate; and
4) the materials utilization of precursor chemicals is low requiring recycling or disposal of a hazardous waste.

Thus, there is a clear need to develop alternative methods to deposit pinhole free thin CdS or to identify a suitable wider bandgap heterojunction/window layer if the high solar cell efficiencies achieved in the laboratory are to be translated to a manufacturing environment.

$CdCl_2$ Heat Treatment

In fabricating high efficiency CdTe/CdS solar cells, there is a post deposition heat treatment in the presence of $CdCl_2$ or a chloride is present during the growth of the CdTe layer. The $CdCl_2$ treatment improves the overall performance of the solar cells and appears to improve the yield and uniformity of devices.

Typically, the $CdCl_2$ heat treatment consists of: 1) coating the surface of the CdTe/CdS with $CdCl_2$ from an alcohol solution containing $CdCl_2$, [21], by laser ablation [22] or by vacuum evaporation [23], or by performing the heat treatment in a $CdCl_2$ vapor [24]; 2) annealing the coated CdTe/CdS at 400°C for 10 to 30 min in air or in an oxygen containing atmosphere; and 3) the residual $CdCl_2$ is removed by a water rinse or mild etch.
The effect of the $CdCl_2$ heat treatment on the structural and optical properties of evaporated CdTe/CdS has

been documented. The heat treatment promotes grain growth of both the CdS and CdTe films. Figure 2 is a schematic representation based on TEM results showing the film [25] structure before and after heat treatment. In the as-deposited state, the CdTe grows pseudo-epitaxially on the CdS most likely creating a strained interface due to the ~10% lattice mismatch. After heat treatment there is no longer a correlation between the CdS and CdTe grain structures, and the grain size of the CdTe has increased from 200 nm to greater than 1μm. Spot EDS of the TEM samples and shifts in the lattice parameter of the CdTe measured by XRD have shown that S has diffused into the CdTe layer creating a CdTeS layer. The incorporation of S into the CdTe reduces the optical band gap as can be seen in Figure 3. This shift in band gap is also seen in spectral response measurements on completed devices[19]. Additionally, there is diffusion of Te into the CdS, reducing the optical band gap or 'softening' the CdS absorption edge. Based on Auger/SIMS measurements, there are oxygen and chlorine incorporated into the CdTe/CdS structure. Rohatgi [9] has suggested that the Cl acts as a deep level trap in the CdTe. The effects of the O and Cl on the properties of the CdTe/CdS are not well understood.

The $CdCl_2$ heat treatment or use of $CdCl_2$ during film growth is a complex, yet critical process in fabricating CdTe/CdS solar cells. In fact, it could be argued that it is the $CdCl_2$ heat treatment which enables the use of the wide variety of CdTe deposition procedures. A quantitative understanding of the roles of $CdCl_2$, O and Cl on the fabrication and operation of CdTe solar cells is needed to effectively design manufacturing facilities for CdTe modules.

Back Contact

The most difficult to process and most vulnerable component of CdTe solar cells is a stable low-resistance back contact [3,26]. There are many procedures - some proprietary - for producing a back contact to CdTe, but most researchers admit that fabrication of the low resistance contact is more art than science. Most procedures to form a back contact include 1) an etch or surface preparation step; 2) application of a film containing Cu, Hg, Pb, or Au [27-30]; and 3) a subsequent heat treatment above 150°C. The contact fabrication process can be interpreted as: 1) the etch produces a Te rich surface resulting in a highly doped or degenerate p^+ interfacial layer; 2) the incorporation of a p-type dopant into the contact material may have two functions: formation of a degenerate Te compound with the dopant and modification of the 'bulk' CdTe properties by diffusion of the dopant into the CdTe making it more p-type; and 3) the heat treatment promotes reaction and diffusion of the contact materials.

Based on this interpretation, it is not surprising that most contact procedures depend on the specific CdTe deposition process since film parameters such as thickness and grain size vary with fabrication procedures. The empirical nature of the CdTe contact fabrication process makes it difficult to define a suitable procedure for contacting CdTe. A clear understanding of the interfacial chemistry, including an understanding of the mechanisms of dopant diffusion in a polycrystalline film, is needed in order to develop a procedure for producing a robust contact suitable for modules.

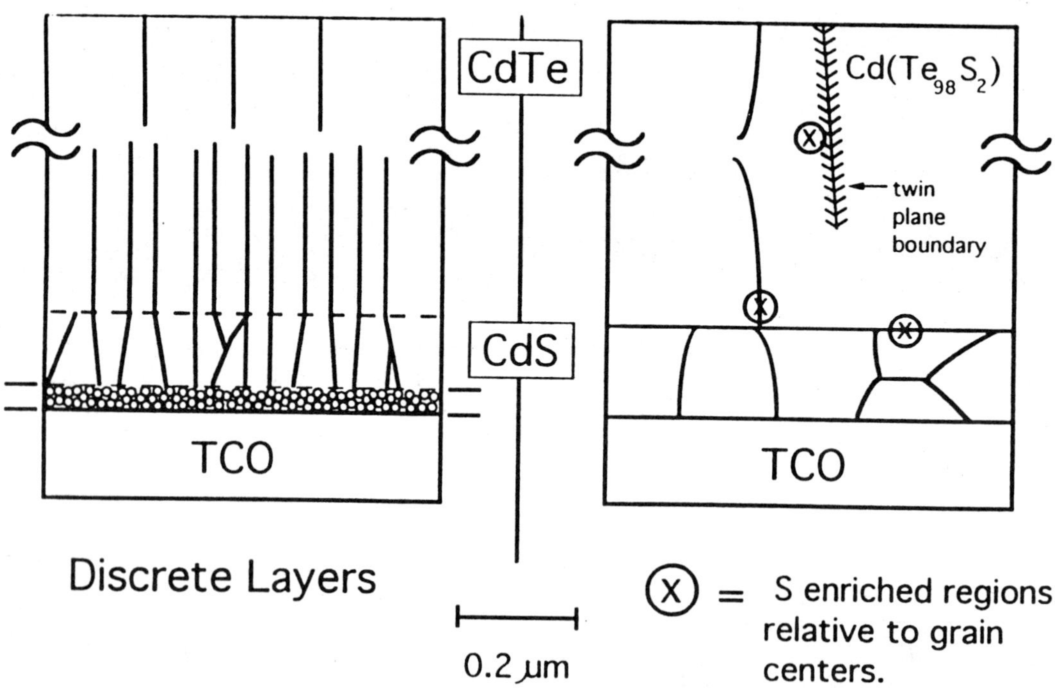

Figure 2. Schematic representation of a CdTe/CdS structure before and after $CdCl_2$ heat treatment

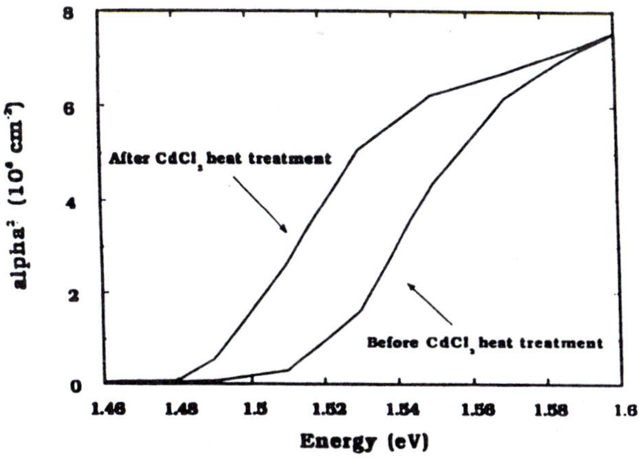

Figure 3. Change in the optical band edge before and after $CdCl_2$ heat treatment.

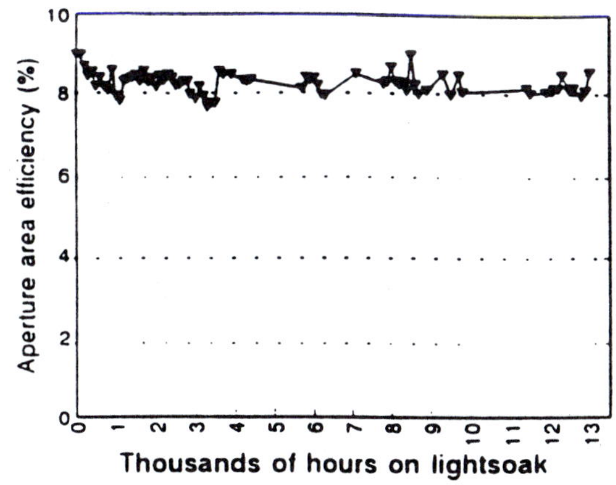

Figure 4. Efficiency of a CdTe module fabricated by BP Solar during external testing in Spain [37].

COMMERCIALZATION ISSUES

In this section, technical issues regarding commercialization of CdTe technology are discussed.

Stability

Reports in the literature attest to the inherent stability of CdTe/CdS PV devices [31-35]. Figure 4 shows the performance of a CdTe module during 12,000 hours of exposure. Nonetheless, at this time no thin film technology has established the field track record for stability and reliability that single crystal and large-grained polycrystalline silicon PV has demonstrated. The reported results of stable CdTe-based PV modules should be viewed as "existence proofs" and do not eliminate the need for more rigorous testing.

The principal concerns regarding CdTe device stability center on the low- resistance back contact. Degradation of the contact is manifested by increases in series resistance or by the emergence of a second diode in the current voltage characteristics. There are several plausible explanations for these effects including oxidation of the metal at the CdTe-metal interface and migration of species which alter the chemistry or electrical properties within the interfacial region. At this point it is not clear that all CdTe devices are susceptible to this form of degradation or if there are other degradation mechanisms which could contribute to reduced power conversion over time. In addition there may be durability issues associated with encapsulation. However, it is important for the CdTe PV community to analyze stability issues, to build upon the existence proofs of stable CdTe modules, and to develop an in-depth understanding of the fundamental issues affecting stability.

There are two major goals of such an analysis:

1) the stability of specific device structures should be demonstrated to enable researchers to optimize device structures and processing to maximize stability.
2) procedures for accelerated life testing must be established to shorten the time required to assess stability from years to months or even days. The way to establish a credible accelerated testing procedure is to determine the physical mechanisms responsible for degradation, establish their dependence on stress factors such as temperature, voltage bias, humidity or light level, and then to extrapolate the results obtained at elevated stress levels to those levels anticipated in normal use.

Production Technology Issues

One of the greatest attractions of CdTe PV technology for the design engineer is the freedom that it provides with respect to its method of semiconductor deposition. The evidence to date suggests that, if one is willing to include a post deposition $CdCl_2$ heat treatment step, any process which will deposit essentially stoichiometric CdTe can be adapted to produce 10% efficient cells. Thus prospective PV module manufacturers have wide latitude to adapt existing high throughput deposition technologies to the deposition of CdTe/CdS films. Production technology must be established to achieve large area and high throughput. In this case, large area is defined as one square meter and high throughput is defined as one square meter per minute. At 8% efficiency this is equivalent to about 10 MW/yr. per 8 hr shift.)

At present there are industrial activities directed toward commercialization of CdTe devices deposited by electrodeposition, screen printing, spraying, and CSVT. However this does not exhaust the list of attractive options.

Potential alternatives include sputtering, vacuum evaporation and atmospheric pressure chemical vapor deposition (APCVD). Sputtering is widely used for coating millions of square feet per year of architectural glass [36]. Vacuum evaporation has not only been used commercially for coating large area architectural glass and automobile sunroofs, but has also been shown to be suitable for deposition of thin film CdS for PV [37]. APCVD is used to apply TCO's to glass while it is still on the float line at the rate of >20 square meters per minute[38].

At this point there are many attractive options but no clear cut leader. Important parameters in selecting a deposition technology might include throughput, equipment cost, operating cost, source material cost, source material utilization efficiency, process yield, environmental impact, equipment up-time (or cycle time), equipment reliability, equipment lifetime, and compatibility with other process steps. An analysis of the pros and cons of the various procedures could be a useful tool in quantifying the relative merits of each and might serve as a tool to focus the CdTe PV development effort.

Environmental Concerns

One technical and perceptual barrier to the widespread utilization of CdTe PV technology is related to the toxicity of the semiconductor materials and specifically to cadmium. There is a legitimate concern that at the end of their useful life, CdTe modules might be disposed of in locations and under conditions in which cadmium might enter the environment. A particular concern is the water supply.

The EPA defines waste to be toxic based upon the results of the Toxicity Characteristic Leaching Procedure, TCLP [39]. With respect to cadmium in PV modules, this means that spent modules are defined to be toxic if 1 ppm of Cd can be leached from crushed modules under conditions that are intended to mimic those which might be encountered in a landfill. Preliminary results are mixed [40,41], but they suggest that CdTe PV modules may have difficulty meeting this standard. Future standards may be even more stringent. Although it is by no means certain that defective or discarded CdTe PV modules will be regarded as regulated wastes, it is prudent for the CdTe PV community to prepare for that contingency. Concerns of potential manufacturers, distributors and users would be greatly allayed by the demonstration of technology by which CdTe-based PV modules can be disposed of in a manner which captures Cd-bearing wastes, promotes recycling, and is cost-effective. The most widely suggested solution is to use the spent modules as flux in smelting operations in which the cadmium and tellurium would not affect the toxicity of the byproducts of existing operations. While this solution satisfies existing requirements, it has the shortcoming of relying on the continued existence and good will of the smelter operator.

Given these environmental concerns and their potential to significantly impact the commercial viability of CdTe PV technology, it seems prudent to demonstrate that responsible means of disposing of CdTe PV modules are available. The technology is straight -forward [41, 42] and new science is not required. In addition to providing additional options to existing industrial operations, a well-documented, carefully controlled, laboratory scale demonstration would address the concerns of environmentalists and potential investors in existing or future CdTe PV enterprises. Conversely, if the proposed procedures are not cost effective, then the industry is best served by taking this information into consideration as soon as possible.

SUMMARY

CdTe continues to be a strong candidate in the competition to set the standard for the next generation of thin- film PV technology. It has some very strong points in that it has high demonstrated efficiency, no inherent problems with stability, and provides wide flexibility in terms of its method of manufacture. Nonetheless, significant technical issues remain unresolved. These issues, which are all being actively investigated, include establishing improved process control over the CdS window layer thickness and the $CdCl_2$ heat treatment, controlling the formation and stability of the low resistance back contact, and demonstration of cost-effective and environmentally responsible methods for dealing with the environmental issues surrounding cadmium. There appear to be no "show stoppers" to the commercialization of CdTe PV. It is a technology well suited to high throughput manufacturing and is well poised to play a significant role in the future of PV.

ACKNOWLEDGEMENT

We wish to thank William Shafarman from the Institute of Energy Conversion. This work was supported by National Renewable Energy Laboratory's subcontract #X-AV-13170-01.

REFERENCES

1. Bonnet, D., "The CdTe Thin Film Solar Cell - An Overview", *Int. J. Solar Energy*, **12**, pp 1-14 (1992).

2. Zweibel, K., "Toward Low Cost CdTe PV", *Int. J. Solar Energy*, pp 285-292 (1992).

3. Bonnet, D., B. Henrichs and H. Richter "Some Phenomena in CdTe/CdS Thin Film Solar Cells Made by Close-Spaced Sublimation", *Int. J. Solar Energy*, **12**, pp 133-136 (1992).

4. Ferekides, C., J. Britt, Y. Ma, and L. Killian, "High Efficiency CdTe Solar Cells by Close Spaced Sublimation", *Proc. 23rd IEEE PVSC*, pp 389-393 (1993).

5. Sandwisch, D., "Development of CdTe Module Manufacturing" *Proc. 1st WCPEC (this conference)*, IEEE, (1994).

6. Barker, J., S.P. Binns, D.R. Johnson, R.J. Marshall, S. Oktik, M.E. Ozsan, N.H. Patterson, S.J. Ransome, S. Roberts, N. Sadeghi, J. Sherborne, A.K. Turner and J.M. Woodcock, "Electrodeposited CdTe for Thin Film Solar Cells", *Int. J. Solar Energy*, **12**, pp 79 - 94 (1992).

7. Clemminck, I., M. Burgelman, N. Casteleyn and B. Depuydt, "Screen Printed and Sintered CdTe-CdS Solar Cells", *Int. J. Solar Energy*, **12**, pp 67-78 (1992).

8. Ghandi, S.K., N.R. Tsakar, and I.B. Bhat, *Appl. Phys. Lett.* , **50**, pp 900-902 (1987).

9. Rohatgi, A., "A Study of Efficiency Limiting Defects in Polycrystalline CdTe/CdS Solar Cells", *Int. J. Solar Energy*, **12**, pp 37-49 (1992).

10. McCandless, B.E., Y. Qu, and R.W. Birkmire, "A Treatment to Allow Contacting CdTe with Different Conductors", *Proc. 1st WCPEC (this conference)*, IEEE, (1994).

11. Compaan, A.D., C.N. Tabory, Y. Li, Z. Feng and A Fisher, "CdS/CdTe Solar Cells by RF Sputtering and by Laser Physical Vapor Deposition", *Proc. 23rd IEEE PVSC*, pp 394-399 (1993).

12. Nelson, A.L., F. Hasoon and D. Levi, "Processing and Characterization of Large-Grain Thin-Film CdTe", *J. Vac. Sci. Tech. A* **12(5)**, pp 2803-2807 (1994).

13. Rohatgi, A., Georgia Institute of Technology, personal communication.

14. Chu, T.L., "Thin Film Cadmium Telluride Solar Cells by Two Chemical Vapor Deposition Techniques", *Solar Cells*, **23**, pp 31-48 (1988).

15. Bonnet, D. and H. Rabenhorst, "New Results on the Development of a Thin-Film p-CdTe-n-CdS Heterojunction Solar Cell", *Proc. 9th IEEE PVSC*, pp 129-131 (1972).

16. Nolan, J.F., and P.V. Meyers, "Fabrication of Stable Large Area Thin Film CdTe Photovoltaic Modules", *Annual Tech. Status Rept. #2, NREL Subcontract ZR-1-11059 Solar Cells*, (1993).

17. ASTM E892 AM1.5 global Spectrum at 1000 W/m2, K. Emery, private communication.

18. Chu, T.L., S.S. Chu, J. Britt, G. Chen, C. Ferekides, N. Schultz, C. Wang and C.Q. Wu, "High Efficiency Thin Film Cadmium Telluride Solar Cells", *Proc. 11th Eur PVSEC*, pp 988-990(1992).

19. McCandless, B.E., and S.S. Hegedus, "Influence of CdS Window Layers on Thin Film CdS/CdTe Solar Cell Performance", *Proc. IEEE 21st PVSC*, pp 967-972 (1991).

20. Ozsan, M.E. and D.R. Johnson, "Analysis of Interface Structures in CdS/CdTe Solar Cells", *Proc. 12th Eur. PVSEC*, pp 1600-1603 (1994).

21. Meyers, P.V., C.H. Liu, and M. Doty, "Method of Making Photovoltaic Cell with Chloride Dip", US Pat 4,873,198 (1989).

22. Compaan, A., and A. Bhat, " Laser-Driven Physical Vapor Deposition for Thin-Film CdTe Solar Cells", *Int. J. Solar Energy*, pp 155-169 (1992).

23. Romeo, N., A. Bosio and V. Canevari, "Large Crystalline Grain CdTe Thin Films for Photovoltaic Application", *Int. J. Solar Energy*, **12**, pp 183-186 (1992).

24. Birkmire, R.W., J.E. Phillips, W.A. Buchanan, S.S. Hegedus, B.E. McCandless, W.N. Shafarman, and T.A. Yokimcus, "Processing & Modeling Issues for Thin Film Solar Cell Devices", *Annual Report, NREL Subcontract X-AV-13170-01*, (1994).

25. Al-Jassim, M.M., F.S. Hasoon, K.M. Jones, B.M. Keyes, R.J. Matson and H.R. Moutinho, "The Morphology, Microstructure, and Luminescent Properties of CdS/CdTe Thin Film Solar Cells", *Proc. 23rd IEEE PVSC*, pp. 459-465, (1993).

26. Ikegami, S., "CdS/CdTe Solar Cells by the Screen-Printing-Sintering Technique: Fabrication, Photovoltaic Properties and Applications", *Solar Cells*, **23**, pp 89-105 (1988).

27. Tyan, Y.S., and E.A. Perez-Albuerne, "Efficient Thin-film CdS/CdTe Solar Cells"*Proc. 16th IEEE PVSC*, pp794-800(1982).

28. Chu, T.L. and S.S. Chu, "High Efficiency Thin Film CdS/CdTe Solar Cells", *Int. J. Solar Energy*, **12**, pp 121-132 (1992).

29. Mitchell, K.W., C. Eberspacher, F. Cohen, J. Avery, G. Duran, and W. Bottenberg, "Progress Towards High Efficiency, Thin Film CdTe Solar Cells", *Proc. 18th IEEE PVSC*, pp 1359-1364 (1985).

30. Meyers, P.V. "Polycrystalline CdS/CdTe/ZnTe n-i-p Solar Cell", *Proc. 7th Eur. PVSEC*, Kluwer Acad. Pub., Dordrecht, The Netherlands, pp 1211-1213 (1986).

31. Ikegami, S. and A. Nakano, :Effect of Oxygen on Photovoltaic Properties of Screen Printed CdS/CdTe Solar Cell and Module", *Int. J. Solar Energy*, **12**, pp 53-65 (1992).

32. Meyers, P.V., "Polycrystalline Thin Film Cadmium Telluride n-i-p Solar Cells", *SERI Subcontract ZL-7-06031-2, Final Report*, (1990).

33. Woodcock, J.M., M.E. Ozsan, A.K. Turner, D.W. Cunningham, D.R. Johnson, R.J. Marshall, N.B. Mason, S. Oktik, N.H. Patterson, S.J. Ransome, S. Roberts, N. Sadeghi, H.J.M. Sherborne, D. Sivapathasundaram,

and I.A. Walls, "Thin Film CdTe Photovoltaic Cells", *Proc. 12th Eur. PVSEC,* pp948-950, (1994).

34. Albright, S.P., R.C. Chamberlin, B. Ackerman, and J.F. Jordan, "Performance Measurement Irregularities on CdS/CdTe Devices and Modules", *Int. J. Solar Energy,* **12**, pp 109-120 (1992).

35. Tyan, Y-S., "Topics on Thin Film CdS/CdTe Solar Cells", *Solar Cells,* **23**, pp19-29 (1988).

36. Product literature, Leybold Technologies Inc., 120 Post Road, Enfield, CT.

37. Baron, B.N., S.C. Jackson, and TWF Russell, "Fundamental Engineering Analysis of CdTe/CdS Photovoltaic Processing", *Final Subcontract Report, Subcontract No. XL-4-03146-1* (1986).

38. Nikodem, R.B., "New Generation Solar Control Coated Products", *J. Vac. Sci. Technol.* **A 10(4)**, pp 1884-1891 (1992).

39. EPA Regulations for Identifying Hazardous Waste, 40 CFR 261, Appendix II, (1993).

40. "Recycling of Cadmium and Selenium from Photovoltaic Module and Manufacturing Wastes: A Workshop Report", Moskowitz, P. D. and K. Zweibel, eds., *Proc. of Conf. on Recycling of Cd and Se from PV Modules and Manufacturing Wastes*, Golden, CO, Mar. 11-12, BNL 47787, (1992).

41. Patterson, MS, AK Turner, M Sadeghi and RJ Marshall, "Health, Safety and Environmental Aspects of the Production and Use of CdTe Thin Film Photovoltaic Modules", *12th Eur PVSEC,* 11-15 April 1994, pp 950-952., (1994).

42. Doty, M. and P. Meyers, "Safety Advantages of a CdTe based PV Module Plant", *AIP Conf. Proc.* **166**, New York, pp 10-18, (1988).

BAND-GAP ENGINEERING IN $Cu(In,Ga)Se_2$ THIN FILMS GROWN FROM $(In,Ga)_2Se_3$ PRECURSORS

Andrew M. Gabor, John. R. Tuttle, Amy Schwartzlander, Andrew. L. Tennant, Miguel. A. Contreras, and Rommel Noufi
National Renewable Energy Laboratory, 1617 Cole Blvd., Golden, CO 80401

ABSTRACT

A three-stage process starting with the deposition of $(In,Ga)_2Se_3$ precursor films has been successful in the fabrication of graded band-gap $Cu(In,Ga)Se_2$ thin films. In this work we examine (1) the reaction of Cu+Se with $(In,Ga)_2Se_3$, which leads to a spontaneous grading in the Ga content as a function of depth through the film, and (2) modification of the Ga content in the surface region of the film through a final deposition of In+Ga+Se. We show how band-gap grading can be enhanced by the formation of non-uniform precursors, how counterdiffusion limits the degree of grading possible in the surface region, and how the Cu_xSe secondary phase acts to homogenize the film composition.

INTRODUCTION

Photovoltaic solar cells based on $Cu(In,Ga)(Se,S)_2$ have recently gained status as the most efficient type of thin film solar cells. The National Renewable Energy Laboratory has been successful in growing $CuIn_{1-x}Ga_xSe_2$ (CIGSe) thin films through the conversion of $(In,Ga)_2Se_3$ precursor layers [1], and a variation of this recipe is now responsible for a solar cell with a 16.8% total-area efficiency [2]. The reaction of Cu and Se with an $(In,Ga)_2Se_3$ precursor can result in the spontaneous formation of a graded $CuIn_{1-x}Ga_xSe_2$ layer where *x* varies through the depth of the film. Such a grading in the quantity [Ga]/([In]+[Ga]) corresponds to a grading of the band gap ($CuInSe_2$ = 1.04 eV, $CuGaSe_2$ = 1.67 eV). Manipulation of the recipe also allows us to deliberately engineer different graded band-gap structures.

Band-gap grading of the absorber has been recommended for several reasons. Increasing the band gap toward the back of the device can establish a quasielectric field [3] where no electrostatic field exists. Such a field can aid minority-carrier transport toward the junction. It can also reduce recombination of minority carriers at the back contact and thus decrease J_o.

Grading the absorber in the opposite direction by increasing the band gap toward the junction has been recommended in order to boost V_{oc} by decreasing the forward current recombination rate for a given forward bias [4,5]. Lower band-gap material deeper within the absorber can still absorb the lower energy photons to retain high values of J_{sc}. The assumptions here are that the recombination rate is dependent on the band gap, that the density of gap states and carrier concentrations do not increase with the higher band-gap alloy compositions, and that the dominant forward current mechanism is indeed recombination in the depletion region. This type of grading could conceivably present a conduction-band barrier to electron collection. However, experiments with $Al_xGa_{1-x}As$ Schottky-barrier solar cells have shown that a raising of the conduction band level near the junction yields improved device performance as long as the gradient occurs within the depletion region [6]. Using the device simulation program ADEPT developed at Purdue University [7], we show in Fig. 1 how the ZnO/CdS/CIGSe device possesses a barrier only for the case of the gradient occurring outside the high electric-field region.

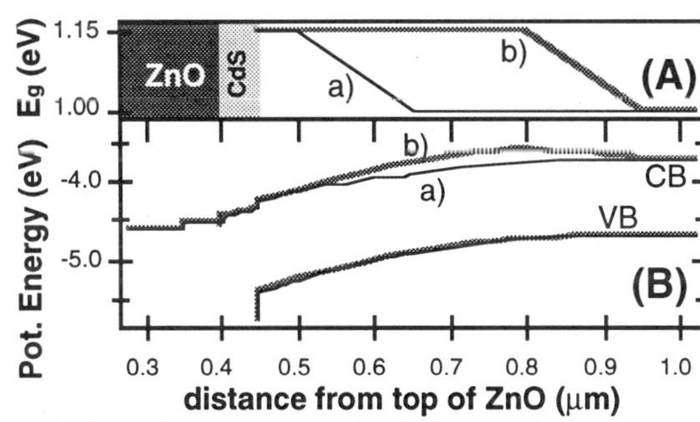

Fig. 1. (A) The band gap of CIGSe vs. depth as used in (B) equilibrium simulations for CIGSe devices with the gradient occurring (a) inside; and (b) outside the high electric-field region.

The band gap at the surface of the absorber will also determine the conduction band offset with the CdS. Since calculations have shown the electron affinity of CIGSe to decrease with increasing Ga contents [8], increasing the Ga content will move the conduction band of the CIGSe upward relative to the CdS. Such an increase could aid the device by a reduction in any conduction-band barrier [9,10] to photogenerated electron transport at the interface and by possibly increasing any barrier to forward current thermionic emission into the CIGSe. However, the increase could also hurt the device by reducing any surface inversion in the CIGSe [7], which is thought to protect the device from recombination at the CIGSe/CdS interface. Finally, the surface composition is likely to change with Ga content. An n-type $CuIn_3Se_5$ composition has been reported at the surface of bulk $CuInSe_2$ [11], but any surface phase for Ga-containing material may have different composition and properties.

Given that bulk and junction properties of the absorber may be enhanced by band-gap grading, it is worthwhile to understand how to engineer graded structures. In this work we develop a growth model for the conversion of $(In,Ga)_2Se_3$, which explains the spontaneous grading of the films, and we examine how variations to the recipe can yield different film morphologies and graded structures. We also examine how moderate variation of Ga content in the surface region affects device performance.

EXPERIMENTAL

The films were deposited by physical vapor deposition from resistively heated elemental sources within a bell-jar system. The substrates were soda-lime glass with a back contact of sputtered Mo. The physical separation of the sources allowed the establishment of a [Cu]/([In]+[Ga]) gradient across the substrate so that different compositions could be studied from a single deposition. A time-temperature profile of the basic 3-stage recipe is shown in Fig. 2. In the first stage, In, Ga, and Se are coevaporated at T_S ~260°C to form the $(In,Ga)_2Se_3$ precursor. The substrate is then ramped to ~560°C within a flux of Se, at which point Cu and Se are coevaporated with the Cu flux < 3 Å/s. Enough Cu is added to bring the overall composition Cu-rich ([Cu] > ([In]+[Ga])) or near to Cu-rich. The average film composition at the end of this second stage is typically such that $0.97 \leq [Cu]/([In]+[Ga]) \leq 1.08$. Then, in a third stage, additional In, Ga, and Se are codeposited to bring the overall composition back into the Cu-poor regime. Typically, the amount of In and Ga in the third stage is around 10% of the total in the first and third stages. The samples are then cooled down within a flux of Se to ~350°C. The total film thickness is around 2.7 μm. In some cases, we interrupted the recipe before completion in order to probe intermediate states of the films and understand the film evolution.

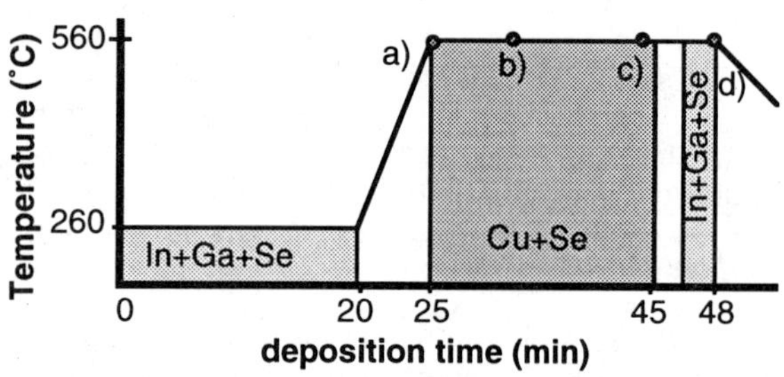

Fig. 2 A time-temperature profile of the 3-stage recipe.

Figure 3 shows SEM micrographs of the cross sections of two films corresponding to points (c) and (d) in Fig. 2. Film (c) was formed by stopping the deposition during the second stage before the film became Cu rich, while film (d) went through all three stages such that it became Cu-rich at the end of the second stage and then reverted to Cu-poor during the third stage. The films have similar percentages of Cu, yet the second film has larger grains.

We probed the composition through the films by performing Auger electron spectroscopy (AES) as function of sputter depth. The quantity [Cu]/([In]+[Ga]) was found to be fairly uniform, as might be expected from the high diffusion rate of Cu in CISe [12]. The quantity [Ga]/([In]+[Ga]) varied as function of film depth. Figure 4 shows [Ga]/([In]+[Ga]) as a function of film depth for four different films as indicated in Fig. 2: (a) the $(In,Ga)_2Se_3$ precursor after being ramped up to 560°C; (b) part way through the second stage such that [Cu]/([In]+[Ga]) ~ 0.33; (c) near the end of the second stage such that [Cu]/([In]+[Ga]) ~ 0.92; and (d) a finished film after completion of the third stage with [Cu]/([In]+[Ga]) ~ 0.92. The precursor has fairly uniform [Ga]/([In]+[Ga]) as a function of film depth, but with addition of Cu and Se, the film becomes progressively more depleted of Ga in the surface region. Depending on the percentage of Ga in the flux during the third stage, the top few thousand Ås of a finished film may have a higher percentage of Ga than the Ga-depleted region below it.

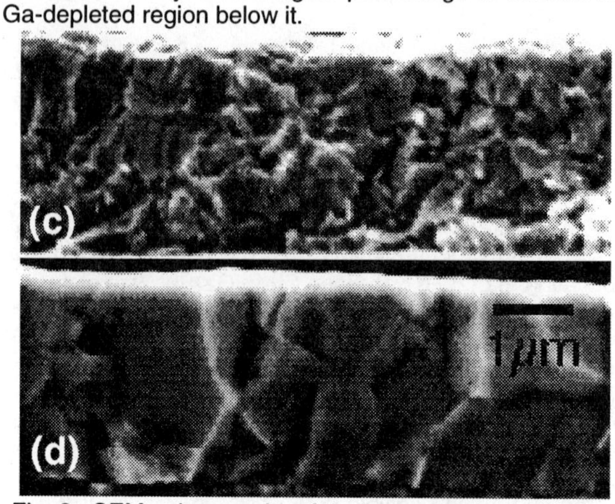

Fig. 3. SEM micrographs of the cross section of CIGSe films corresponding to points (c) and d) from Fig. 2.

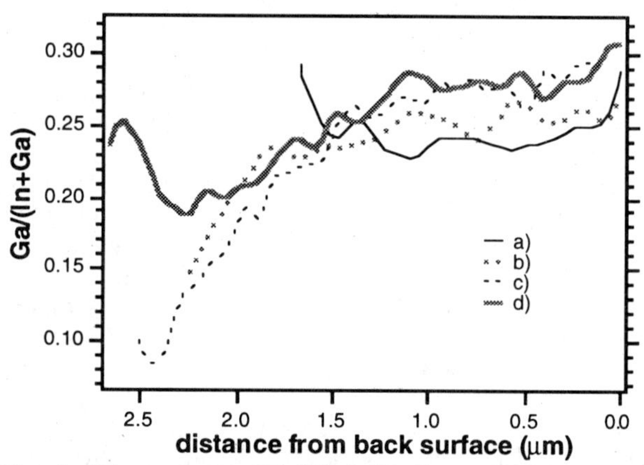

Fig. 4. The quantity [Ga]/([In]+[Ga]) vs. film depth as determined by AES for four films: (a) the $(In,Ga)_2Se_3$ precursor; (b) part way through the second stage; (c) near the end of the second stage; (d) at the end of the third stage.

Films were also made from non-uniform precursors. By depositing first In+Se and then Ga+Se, we formed a precursor with high Ga in the front. After ramping the sample to 560°C, we cooled the sample and removed part of it for analysis. We then placed the rest of the sample back in the bell jar for addition of enough Cu+Se at 560°C such that the sample was converted to Cu-poor CIGSe. As is seen in the AES profile of the annealed precursor and of the converted film in Fig. 5, even though the precursor has high Ga at the surface, the converted film again has little Ga at the surface.

In a similar experiment, we also made a nonuniform precursor by depositing first Ga+Se and then In+Se. Enough Cu+Se was added such that half of the sample became Cu-rich and half remained Cu-poor due to the compositional gradient across the film in the substrate plane. Figure 6 shows AES depth profiles of the annealed precursor and of Cu-poor and Cu-rich spots on the converted piece. The transition region between high and low Ga regions of the Cu-poor spot is considerably sharper than is the case for the Cu-rich spot.

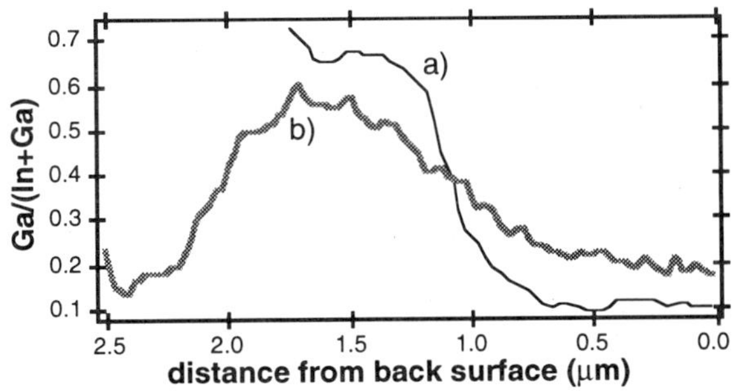

Fig. 5. The quantity [Ga]/([In]+[Ga]) vs. film depth as determined by AES for (a) a precursor with high Ga in the front; and (b) a Cu-poor CIGSe film made from the precursor.

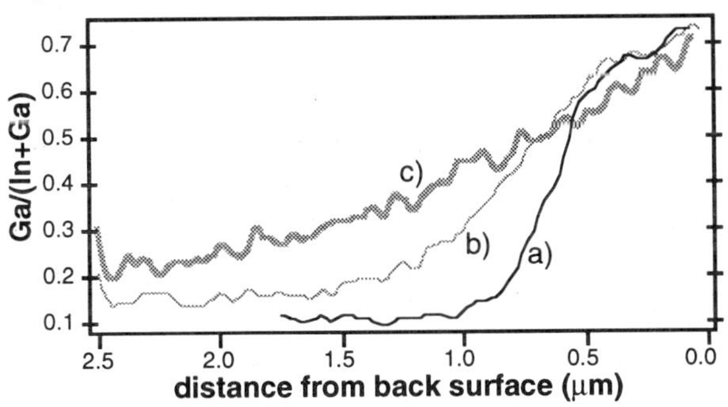

Fig. 6. The quantity [Ga]/([In]+[Ga]) vs. film depth as determined by AES for (a) a precursor with high Ga in the back; (b) Cu-poor and (c) Cu-rich spots on the CIGSe film made from the precursor.

Table 1. Parameters for cells made from absorbers with varying percentages of Ga in the third stage.

Ga/(In+Ga) in 3rd stage flux (%)	Voc (mV)	Jsc (mA/cm^2)	FF (%)	η (%)
0.40	638	30.9	76.7	15.1
0.34	642	31.9	73.7	15.1
0.31	645	30.9	71.9	14.4
0.27	632	30.0	75.3	14.3
0.22	630	30.3	74.5	14.2

In a final series of depositions, we kept the first and second stages of the process constant and varied the percentage of Ga in the third stage from $0.22 \leq [Ga]/([In]+[Ga]) \leq 0.40$ to examine the tolerance of the process to variations in composition of the surface region. A uniform precursor with [Ga]/([In]+[Ga])~0.31 was used in these experiments. We fabricated cells from this series of films by depositing solution-grown CdS, ion-beam-sputtered ZnO, and top contacts (no antireflective coating). Cell efficiencies (total-area, non-official) were quite tolerant to this third stage variation, as is seen in Table 1. The spectral response cutoff of the cells in the infrared varied systematically with the third stage Ga contents, with ~25 meV difference between the highest and lowest Ga cases (see Fig. 7).

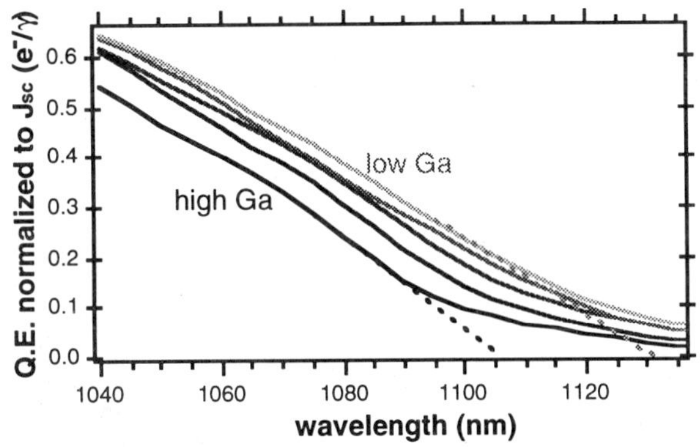

Fig. 7. The quantum efficiency of the five cells in Table 1 made from absorbers with varying percentages of Ga in the third stage.

DISCUSSION

In the conversion of $(In,Ga)_2Se_3$ to CIGSe, the film thickness increases, and this additional thickness of film is depleted of Ga relative to the precursor. A growth model consistent with this phenomenon and with defect adamantine theory [13] goes as follows. The reaction of enough Cu and Se with the precursor to bring the composition to [Cu]/([In]+[Ga])~0.3 restructures the precursor into a defect diamond-like lattice, as X-ray diffraction spectroscopy has shown [14]. This lattice is similar to the chalcopyrite lattice of $Cu(In,Ga)Se_2$ except for In and Ga antisite defects (In_{Cu} and Ga_{Cu}) and Cu vacancies (V_{Cu}). Film compositions are most stable if they fall on the Cu_2Se - $(In,Ga)_2Se_3$ tie line, as such compositions fill the valence shell of the Se anions [13]. In order for compositions to remain on the tie line, further diffusion of Cu into the defect lattice must be accompanied by outdiffusion of some In or Ga. For every three atoms of indiffusing Cu, one atom of In or Ga must diffuse out. This outdiffusing In and Ga can react with additional Cu and Se delivered to the film to form new unit cells at the surface. These new cells contribute to a growth in the film thickness. However, because In has a larger atomic radius than does Ga, the strain energy of the film is lowered by preferential outdiffusion of In relative to Ga (see ref. [15] for similar behavior in III-V materials). In this way the additional film thickness becomes depleted of Ga relative to the back. If Se is not coevaporated with the Cu [16], then In will likely leave the film as $In_2Se(g)$, and the film will not increase in thickness.

If enough Cu is added during the second stage, the film will become Cu-rich, with the excess Cu existing in a secondary Cu_xSe phase. At sufficiently high temperatures, the Cu_xSe may exist as a liquid phase [17]. If the CIGSe grains are soluble in the Cu_xSe, then the rate of grain growth can be greatly increased through liquid-phase sintering [18]. The kinetics associated with the diffusion due to concentration gradients, as described by Fick's law, are also increased with the existence of the Cu_xSe at grain boundaries. Thus the gradient in [Ga]/([In]+[Ga]) decreases when a film becomes Cu-rich, as was shown Fig. 6. Similarly, alloying of $CuInSe_2$ and

$CuGaSe_2$ layers grown within the Cu-rich regime has been previously demonstrated [19,20]. The homogenizing potential of the Cu_xSe makes it difficult to engineer sharp gradients in [Ga]/([In]+[Ga]) if a film enters the Cu-rich regime. Alternatively, undesired gradients can be decreased by bringing a film into the Cu-rich regime. For example, selenization of a Cu-In-Ga film also results in a nonuniform CIGSe film with most of the Ga present at the back of the film [21]. Ga could be alloyed into the bulk of the film by making the film Cu-rich.

After completion of the second stage, the film should exist as large stoichiometric CIGSe grains, with Cu_xSe present at the grain surfaces. The In+Ga+Se delivered in the third stage reacts with the Cu_xSe to form additional CIGSe until the Cu_xSe is consumed. At that point, a counterdiffusion process must occur that is opposite to that which was present during the second stage. Now Cu must diffuse out of the CIGSe grains to react with new In+Ga+Se while some In and Ga will diffuse into the bulk grains to bring them to more Cu-poor compositions.

In adding In+Ga+Se to the film, interdiffusion prevents the formation of a sharp step in Ga content, as is clear from Fig. 4. The cutoff in the spectral response of Fig. 7 corresponds to the lowest band-gap region of the film, which is the Ga-depleted surface formed after the second stage deposition. After the third stage, this low band-gap region varies in [Ga]/([In]+[Ga]) by ~0.04 between the highest and lowest Ga cases, as determined from the variation in the cutoff. Thus, the greater the Ga content in the third stage, the more the indiffusion of Ga raises the band gap of this region. The relative insensitivity of the device efficiencies to the Ga content in the third stage (see Table 1) may be attractive from a manufacturing perspective, as the Ga and In fluxes in the third stage may vary somewhat from run to run or across the substrate. Also noteworthy is the fact that the cell with highest Ga content in the third stage did not have the highest V_{oc}. Sharper surface gradients could conceivably be formed by codeposition of Cu along with the In+Ga+Se, as cation counterdiffusion with the underlying material would then likely decrease. However, such an approach has not yielded efficient devices in the past [4,5].

CONCLUSIONS

The engineering of graded band-gap structures in CIGSe by the 3-stage process must take into account two competing mechanisms: (1) the greater tendency of In over Ga to diffuse to the surface during growth, and (2) the homogenizing force described by Fick's law, which washes out concentration gradients. Larger gradients in [Ga]/([In]+[Ga]) than spontaneously occur by mechanism (1) may be established by forming non-uniform precursors. Complete separation of the precursor deposition into delivery of Ga+Se and then In+Se has yielded a band-gap gradient in the converted film of ~0.4 eV. However, the potential for engineering a sharp increase in the band gap by delivering an In+Ga+Se flux to a CIGSe film with low Ga at the surface, is limited due to indiffusion of Ga. In addition, it is not yet clear that larger gradients than those we have obtained are desirable in the surface region. The kinetics of mechanism (2) have been shown to increase with the presence of the secondary Cu_xSe phase, and the deliberate formation of Cu_xSe by any growth recipe may be an effective way to reduce undesired compositional gradients. Finally, only a few minute soak in the Cu-rich regime at 560°C is necessary to induce significant recrystallization of a small-grain film.

ACKNOWLEDGMENTS

We thank J. Dolan, J. Keane, R. Matson, L. Wang, and K. Ramanathan for technical assistance, A. Zunger and J. Phillips for helpful discussions, and Siemens Solar Industries for Mo-coated glass substrates. A. Gabor thanks Associated Western Universities for fellowship support. This work was performed at NREL under Contract No. DE-AC36-83H10093 to the U.S. DOE.

REFERENCES

[1] A. M. Gabor, J. R. Tuttle, D. S. Albin, M. A. Contreras, R. Noufi and A. M. Hermann, *Appl. Phys. Lett.* **65**, 1994 p. 198.
[2] J. R. Tuttle, M. A. Contreras, J. S. Ward, A. M. Gabor, K. R. Ramanathan, A. L. Tennant, L. Wang, J. Keane and R. Noufi, *1st WCPEC*, 1994, (these proceedings).
[3] H. Kroemer, *RCA Review* **18**, 1957, p. 332.
[4] J. R. Tuttle, M. Ruth, D. Albin, A. Mason and R. Noufi, *20th IEEE PVSC,* 1988, p. 1525.
[5] R. W. Birkmire, W. N. Shafarman and R. D. Varrin, *20th IEEE PVSC*, 1990, p. 550.
[6] H. T. Yang, Y. D. Shen, D. Edwall, D. L. Miller and J. S. J. Harris, *IEEE Trans. Elect. Dev.* **ED-27**, 1980, p. 851.
[7] R. J. Schwartz, J. L. Gray and Y. J. Lee, *22nd IEEE PVSEC*, 1991, p. 920.
[8] S. H. Wei and A. Zunger (to be published).
[9] D. W. Niles et. al. (to be published).
[10] S. H. Wei and A. Zunger, *Appl. Phys. Lett.* **63**, 1993, p. 2549.
[11] D. Schmid, M. Ruckh, F. Grunwald and H. W. Schock, *J. Appl. Phys.* **73**, 1993, p. 2902.
[12] G. Dagan, T. F. Ciszek and D. Cahen, *J. Phys. Chem.* **96**, 1992, p.11009.
[13] E. Parthe, *7th Internat.Conf. on Ternary and Multinary Compounds*, 1986 (MRS), p. 3.
[14] A. M. Gabor, J. R. Tuttle, D. S. Albin, R. Matson, A. Franz, D. W. Niles, M. A. Contreras, A. M. Hermann and R. Noufi, *Mat. Res. Soc. Symp. Proc.* **343**, 1994, p. 143.
[15] Z. Lin, F. Xu and J. H. Weaver, *Phys. Rev. B* **36**, 1987, p. 5777.
[16] J. Kessler, D. Schmid, S. Zweigart, H. Dittrich and H. W. Schock, *12th European PVSEC*, 1994, p. 648.
[17] J. R. Tuttle, M. Contreras, A. Tennant, D. Albin and R. Noufi, *23rd IEEE PVSC*, 1993, p. 415.
[18] D. S. Albin, G. D. Mooney, A. Duda, J. Tuttle, R. Matson and R. Noufi, *Solar Cells* **30**, 1991, p. 47.
[19] T. Walter and H. W. Schock, *Thin Solid Films* **224**, 1993, p. 74.
[20] J. R. Tuttle, D. S. Albin, A. Tennant, A. M. Gabor, M. Contreras and R. Noufi, *Solar Energy Materials and Solar Cells* **35**, 1993, p. 193.
[21] D. S. Albin, J. R. Tuttle, M. Contreras, A. M. Gabor, A. Mason, R. Noufi and P. Singh, submitted to *Solar Energy Materials and Solar Cells* .

DEVELOPMENT OF HIGH-EFFICIENCY $CuIn_xGa_{1-x}Se_2$ THIN-FILM SOLAR CELLS BY SELENIZATION WITH ELEMENTAL Se VAPOR IN VACUUM

Katsumi KUSHIYA*, Akira SHIMIZU, Koki SAITO°, Akira YAMADA and Makoto KONAGAI

Department of Electrical and Electronic Engineering, Tokyo Institute of Technology
2-12-1 O-okayama, Meguro-ku, Tokyo, 152, JAPAN
*Showa Shell Sekiyu K.K. Central R&D lab
°Department of Electronics and Information Science, The Nishi-Tokyo University

ABSTRACT

On the basis of the establishment of the two-stage method developed in this study which is directly linked to the control of [redacted] the Se beam-flux intensity during the precursor preparation stage, achievements in the development of CIGS thin-film absorbers are reported with remarkably high efficiency of 14.9 % in the band gap of about 1.3 eV. Furthermore, it is, for the first time, demonstrated that the two-stage method developed in this study is applicable to the formation of a graded band-gap structure by adjusting the Se beam-flux intensity during the precursor preparation stage to over 1.5×10^{-5} Torr, which leads to rather high efficiency of 10.1 %. These results demonstrate 1) a significantly high capability of the two-stage method developed in this study to prepare device-quality CIGS thin-film absorbers and 2) the control of Se is the most important factor to fabricate a sufficiently high-quality CIGS thin-film absorber in this method.

INTRODUCTION

With the impressive achievements in polycrystalline Cu-III-VI_2 thin-film solar cells, the research activities have been focused on the development of a fabrication process of large-area thin-film modules. The two-stage method, which consists of " a precursor preparation stage, followed by a selenization stage ", is considered to be promising to resolve this key issue. Various types of two-stage methods have already been developed which enabled high efficiencies of over 13 % [1,2].

In this study, achievements obtained by paying more attention to the Se amounts to optimize the two-stage method of CIGS thin-film absorbers, in which they are fabricated by the selenization of a Se-contained, coevaporated precursor layer with elemental Se vapor in vacuum, are reported. Sufficiently high efficiencies over 14% with CIGS thin-film solar cells will demonstrate a favorable capability of the two-stage method developed in this study.

EXPERIMENTAL

The substrate-temperature program for the fabrication process in this study is shown in Fig. 1. The precursor layer was prepared by a coevaporation method with four constituent elements (Cu, In, Ga and Se) on a Mo sputter-coated glass substrate at the substrate temperatures of 150 to 200°C for 30 min. The selenization stage was performed at the substrate temperature of 500°C for 60 min supplying elemental Se vapor with a Se beam flux intensity of about 1×10^{-5} Torr in vacuum continuously including the 5-min ramp-up stage from the end of the precursor preparation stage, which led to the formation of a 2- to 3-μm thick CIGS thin-film absorber.

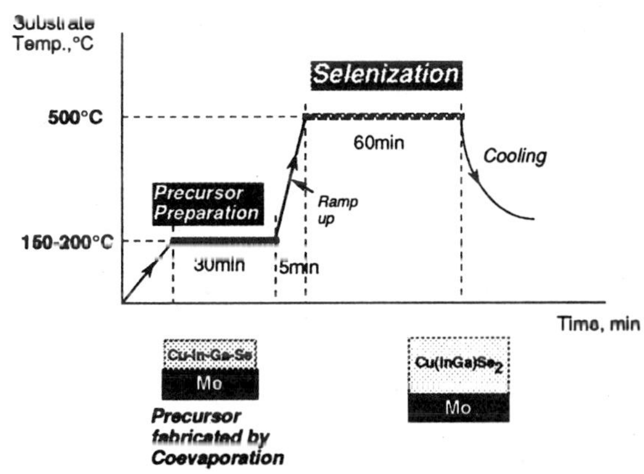

Fig. 1 Substrate-temperature program for the fabrication of CIGS thin film absorbers.

In the case of the fabrication of CIGS thin-film solar cells, a CdS buffer layer less than 20 nm thick was deposited on the CIGS thin-film absorber by a chemical bath deposition method. Then, after annealing in air at about 200°C for 30 min, a textured ZnO window layer

CH3365-4/94/0000-0087 $4.00 © 1994 IEEE

about 2 μm thick was deposited on the CdS buffer layer by a metal-organic chemical vapor deposition technique. In reducing a series-resistance component, Al top metal-grid was deposited on the ZnO window layer by a vacuum evaporation method.

RESULTS AND DISCUSSION

Optimization of the two-stage method for CIGS thin-film absorbers

Before starting the fabrication of the CIGS thin-film absorbers by the two-stage method, the optimization of the fabrication of $CuInSe_2$ (CIS) thin-film absorbers was performed. The optimized conditions are summarized as follows [6]:

- In the precursor preparation stage, the optimized precursor layer fabricated by coevaporation at the substrate temperatures of about 150°C was obtained with Se content of 30 to 40 at.% in order to prevent a rapid volume expansion which caused a peeling at the interface of Mo/CIS and In loss or In aggregation during the selenization stage which was correlated to the reevaporation of In in the form of volatile In_2Se and, hence, led to a serious shunt.
- In the selenization stage, device-quality CIS thin-film absorbers were fabricated by adjusting the Se beam-flux intensity to about 1 x 10^{-5} Torr at the substrate temperature of 500°C including the 5-min ramp-up stage.
- It was demonstrated that the control of Se was directly linked to the control of the Cu/In ratio. Therefore, both the Se content in the precursor layer and the Se beam-flux intensity during the selenization stage was the most important factor to fabricate a device-quality CIS thin-film absorber. Based upon these findings, 12.3 % efficiency was achieved with CIS thin-film solar cells.

The same analogy as obtained for the CIS thin film absorbers in the two-stage method as mentioned above was applied to the fabrication of CIGS thin-film absorbers. As a good evidence that the two-stage method developed in this study was successfully applicable to prepare device-quality CIGS thin-film absorbers, AES depth profiles indicate a uniform distribution of Ga in the CIGS thin-film absorber as shown in Fig. 2 (a). In order to obtain the device-quality CIGS thin-film absorbers, the Se content of about 50 at.% in the precursor layer was required in this study. The Se beam-flux intensity in the precursor preparation stage to fabricate such precursor layers was in the range of 0.7 to 1.0 x 10^{-5} Torr. The surface morphology and cross section of the precursor layer fabricated by the above condition is shown in Figs. 3 (a) and (b), respectively. As seen in these figures, the surface is covered partially with round grains on the top of the densely-packed, columnar, fine-grain ground which Se content is about 50 at.% and contains relatively less In and more Se than the round grains on the top.

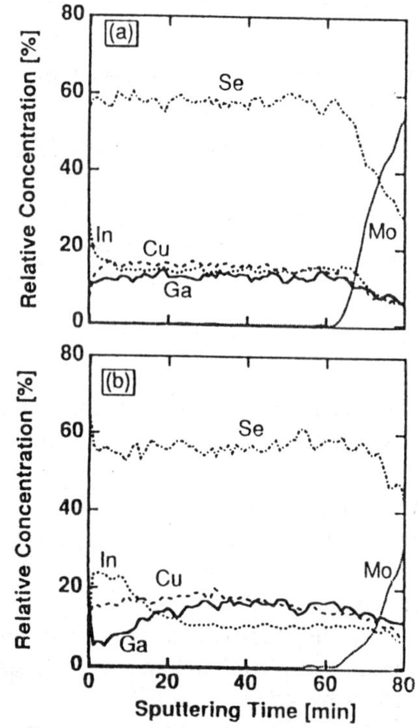

Fig. 2 AES depth profiles of CIGS thin-film absorber as a function of Se beam-flux intensity during the precursor preparation stage: (a) 0.8 x10^{-5} Torr, (b) 1.5 x 10^{-5} Torr.

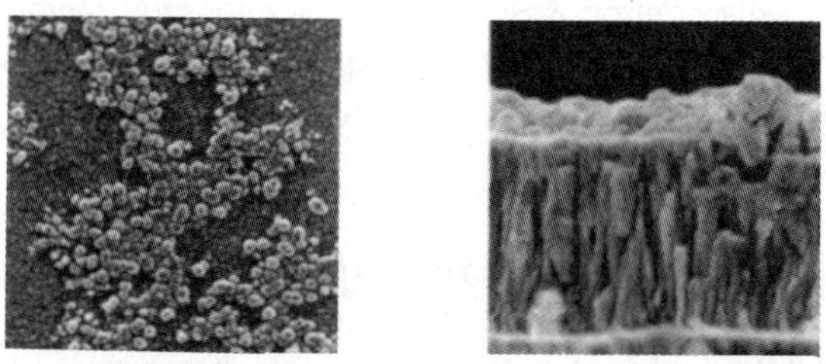

(a) Surface (b) Cross section

6 μm 1.5 μm

Fig. 3 Surface morphology and cross section of the precursor layer fabricated with Se beam-flux intensity of 0.7 x 10^{-5} Torr.

The cross section of the CIGS thin-film absorber shown in Fig. 2 (a) is shown in Fig. 4 (a), which shows the CIGS thin-film absorber with densely-packed, columnar grains.

The most striking result in this study was this fabrication method enabled the spontaneous formation of a graded band-gap structure in the CIGS thin-film absorbers as shown in Fig. 2 (b). This graded band-gap structure was formed when the Se beam-flux intensity during the precursor preparation stage was increased to over 1.5 x 10^{-5} Torr. The surface morphology of the precursor layers, which led to the graded band-gap structure, was basically the same as that shown in Fig. 3 (a), although the growth of porous round-grains was enhanced in the Se beam-flux intensity and some of them grew down into the columnar, fine grains in a conical form. Comparing Figs. 4 (a) and (b) to Fig. 3 (b), it is found that CIGS thin-film absorbers have grown in a similar form to their precursor layers.

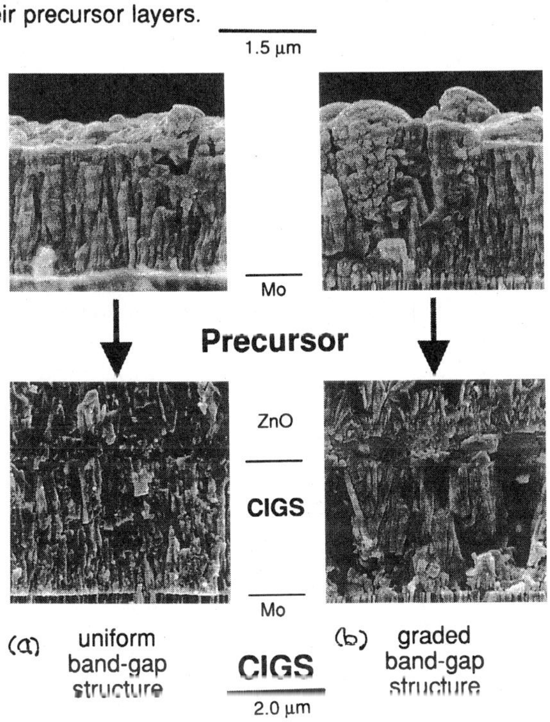

Fig. 4 Cross section of the CIGS thin-film absorbers with different band-gap structure: (a) uniform band-gap, (b) graded band-gap structures.

• As seen in Fig. 4 (b), big chunks were observed on the top of the columnar, fine-grain structure. This chunk seems to be formed from the porous round grains in the precursor layer as shown in Fig. 3 (b). The formation of this chunk may be correlated to the graded band-gap structure, since this chunk was not observed in the uniform band-gap structure shown in Fig. 4 (a) and some of them grew down into the columnar, fine-grain structure in the same form as the precursor layer. As a result of phase separation probably related to the formation reactions of volatile In_2Se and stable InSe phases on the surface, the round grains may be formed. Further investigations are necessary to verify the formation processes of the graded band-gap structure in this study.

Performance of CIGS thin-film solar cells

The performance of CIGS thin-film solar cells is summarized in Fig. 5.As seen in this figure, both the increase in V_{oc} and the decrease in J_{sc} are consistent with the increase in the band gap. The currently best efficiency in this study is 14.9 % which is achieved using the CIGS thin-film absorber in the band gap of about 1.3 eV (Ga content of about 40 at.% in the precursor layer) as shown in Fig. 6.

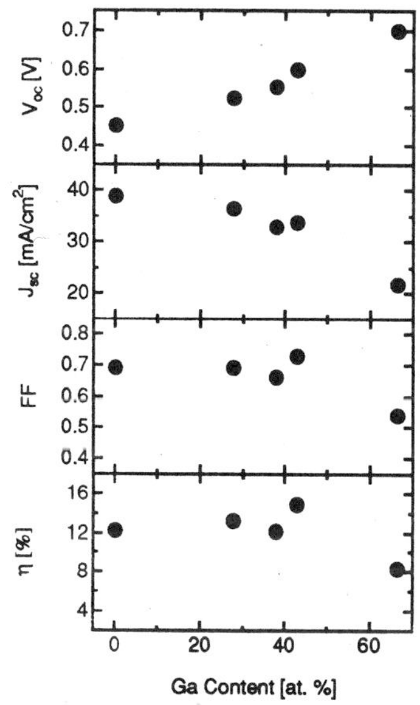

Fig. 5 Performance of CIGS thin-film solar cells in this study.

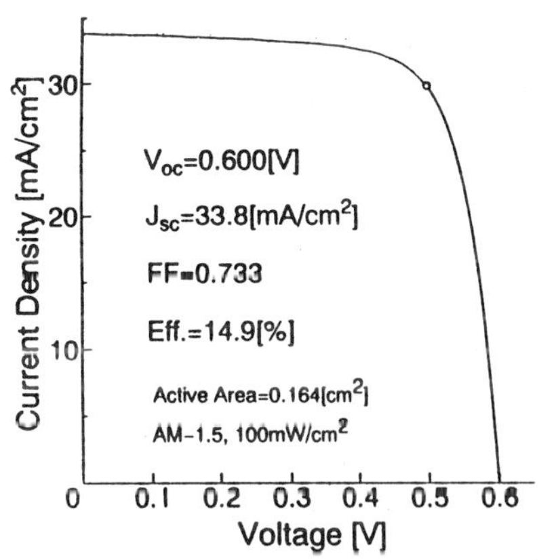

Fig 6 I-V characteristic of the currently best CIGS thin-film solar cell.

Spectral response measurements of CIGS thin-film solar cells as a function of Ga content are shown in Fig. 7. As seen in Fig. 7, the long-wavelength absorption edge of CIGS thin-film absorbers shifts towards the shorter-wavelength side increasing the fractional Ga content in the CIGS thin-film absorbers.

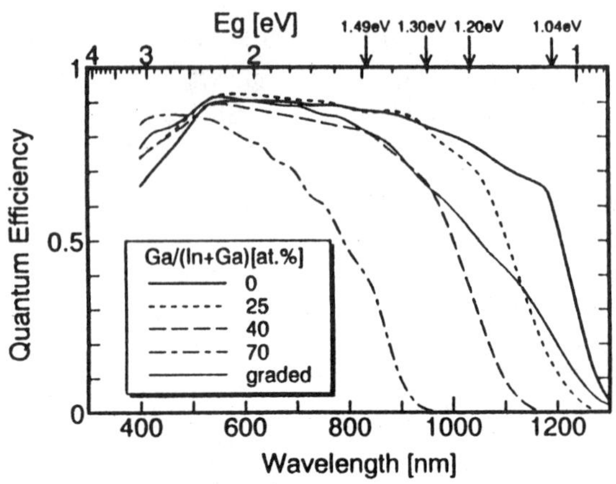

Fig. 7 Spectral response measurements of CIGS thin-film solar cells in this study.

Furthermore, the CIGS thin-film absorbers fabricated using the precursor layer with Se content of about 50 at.% by this method in general show a tendency of slight tailing in the longer-wavelength range. This may be correlated to the presence and amounts of round grains in the precursor layer, which contain relatively more In and less Se than the columnar, fine-grain ground as mentioned above. The CIGS formation reactions which progress from the surface may also be another factor to dominate the quality of CIGS thin-film absorbers, since the surface is the only place Se is incorporated into the CIGS phase during the selenization stage.

In the CIGS thin-film solar cell with Ga content of about 70 at.% in the precursor layer, the quality of the absorber is indicated to be worse. This may be correlated to the experimental fact that the substrate temperature at the selenization stage is fixed at 500°C to avoid the warping of soda-lime glass substrates. It may be indicated that this substrate temperature is too low to enhance the Ga diffusion and fabricate higher-quality CIGS thin-film absorbers in the high Ga contents of over 50 at.%. An increase in the substrate temperature should be favorable to attain the formation of more uniform, higher-quality CIGS thin-film absorbers through the interdiffusion of each constituent element, phase mixing or grain growth.

Other specific remarks seen in Fig. 5 are described as:

- The efficiencies over 12 % were attained using the CIGS thin-film absorbers in the band-gap range from 1.04 eV for CIS to about 1.4 eV in this study.
- 10.1 % efficiency was achieved using the CIGS thin- film absorber with Ga content of about 40 at.% in the precursor layer, which showed the graded band-gap structure formed spontaneously by the two-stage method. This structure may be applicable as one of the options to fabricate device-quality CIGS thin-film absorbers, although the approach to improve the efficiency with this structure requires to establish the optimum combination of Se and Ga contents in the precursor layer and the substrate temperatures at the selenization stage.

CONCLUSIONS

On the basis of the establishment of the two-stage method developed in this study which was directly linked to the control of Se in the precursor preparation stage, achievements in the development of CIGS thin-film absorbers were reported with remarkably high efficiency of 14.9 % in the band gap of about 1.3 eV.

It was, for the first time, demonstrated in this study that the two-stage method was applicable to the formation of a graded band-gap structure by adjusting the Se beam-flux intensity to over 1.5×10^{-5} Torr during the precursor preparation stage, which led to rather high efficiency of 10.1 %.

These results have clearly demonstrated the significantly high capability of the two-stage method developed in this study to prepare device-quality CIGS thin-film absorbers.

Based upon the above achievements, further improvement of efficiency is strongly expected with CIGS thin-film absorbers fabricated by the two-stage method developed in this study.

ACKNOWLEDGEMENT

This study was supported in part by NEDO as a part of the New Sunshine Program under Ministry of International Trade and industry.

REFERENCES

[1] D. Tarrant and J. Ermer, Proc. 23rd IEEE PVSC (1993) p.372-378.

[2] J. Kessler, D. Schmid, S. Zweigart, H. Dittrich and H.W. Schock, Proc. 12th EC PVSEC (1994) p.648-652.

[3] K.Kushiya, Akira Shimizu, Akira Yamada and M. Konagai (to be published in Jpn. J. Appl. Phys. 34 (1995))

A NOVEL CADMIUM FREE BUFFER LAYER FOR $Cu(In,Ga)Se_2$ BASED SOLAR CELLS

Dimitri Hariskos, Martin Ruckh, Ulfert Rühle, T. Walter and Hans Werner Schock
Institut für Physikalische Elektronik, Universität Stuttgart
Pfaffenwaldring 47, D-70569 Stuttgart, Germany

ABSTRACT

Solar cells based on $Cu(In,Ga)Se_2$ were prepared replacing the 'standard buffer layer' CdS with a $In_x(OH,S)_y$ thin film. The film is deposited in a chemical bath (CBD) process using an aqueous solution containing $InCl_3$ and thioacetamide. X-ray photoemission spectroscopy measurements were performed in order to characterize the growth kinetics and the chemical composition. The influence of different concentrations of $InCl_3$ and thioacetamide in the solution on the electrical properties of the solar cells was studied by measuring the j-V characteristics and the spectral quantum efficiencies. Capacitance-voltage (C-V) measurements indicate that the high V_{OC} values of devices with the novel buffer layer are correlated with narrower space charge widths and higher effective carrier concentrations in the absorber materials. The achieved conversion efficiency of approximately 15% using the cadmium free $In_x(OH,S)_y$ buffer demonstrates the potential of this process as an alternative to the standard chemical bath deposition of CdS.

INTRODUCTION

Thin film solar cells using $Cu(In,Ga)Se_2$ absorber layers have recently surpassed the 16% efficiency level [1,2]. They are promising candidates for large scale production due to the low material consumption and preparation techiques appropriate for large areas. The highest conversion efficiencies of $Cu(In,Ga)Se_2$ based solar cells have been achieved with a chemical bath deposited (CBD) CdS buffer layer prior to rf sputtered ZnO. The motivation of searching for alternatives for this 'standard' buffer layer is not only to eliminate the toxic cadmium but also to improve the light transmision in the blue wavelength region by using a material with a wider bandgap compared to CdS. Several attempts have been made to replace the CBD-CdS with other materials in a chemical bath process [3,4], but in all cases the standard recipe lead to superior device characteristics, mainly in fill factor (FF) and open circuit voltage (V_{oc}).

In this work we developed a chemical bath process for the deposition of a buffer layer containing In, S and O which leads to electrical device properties of the solar cells comparable to the results using the standard CBD-CdS buffer layer.

CHEMISTRY

The deposition takes place in an aqueous $InCl_3$ and thioacetamide (CH_3CSNH_2) solution at temperatures up to 70°C. The reaction paths during the deposition are shown schematically in Fig.1.

Thioacetamide is known to decompose in aqueous solutions at higher temperatures to ammoniaacetate and H_2S. For In^{3+}(aq) two reaction paths are possible.

1. The hydrolysis to $In(OH)_3$ is forced by heating aqueous In^{3+} solutions in the presence of ammoniumacetate.
2. In^{3+}(aq) can react with H_2S to In_2S_3.

Dipping a substrate into the bath, the composition of the growing film depends on the concentrations of $InCl_3$ and thioacetamide, temperature and reaction time.

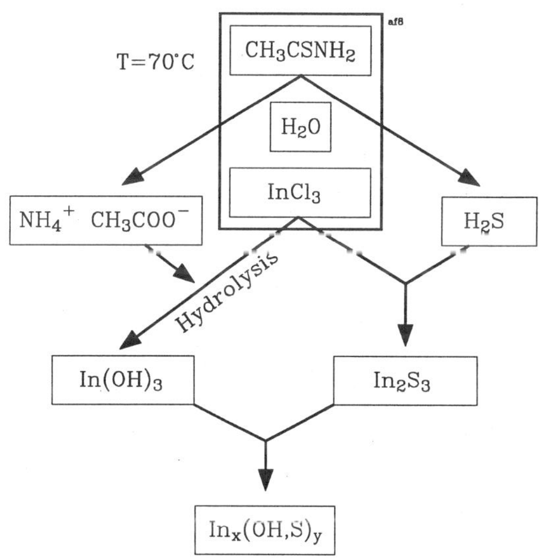

Fig. 1. Schematic representation of the different possible reaction paths leading to the formation of a $In_x(OH,S)_y$ compound

SURFACE ANALYSIS

X-ray photoemission spectroscopy (XPS) measurements were performed to characterize different stages of the film growth. Thus it was possible to

CH3365-4/94/0000-0091 $4.00 © 1994 IEEE

investigate the influence of the chemical bath on the absorber surface. We compared a Cu(In,Ga)Se_2 film exposed to air for several days and rinsed with H_2O with absorbers treated for different times in the chemical bath. After some seconds in the cold bath we observe a precleaning effect similar to that in the standard CBD-CdS process [5]. If native oxides are present at the surface only Se-O is removed completely while In-O and Ga-O species remain partially at the surface. The evolution of the film growth was studied after 30", 1', 2.5', 5' and 20' deposition time in a bath with the concentrations [$InCl_3$]= 5 mmol/l, [CH_3CSNH_2]= 0.15 mol/l at T= 70°C. For these experiments we used freshly

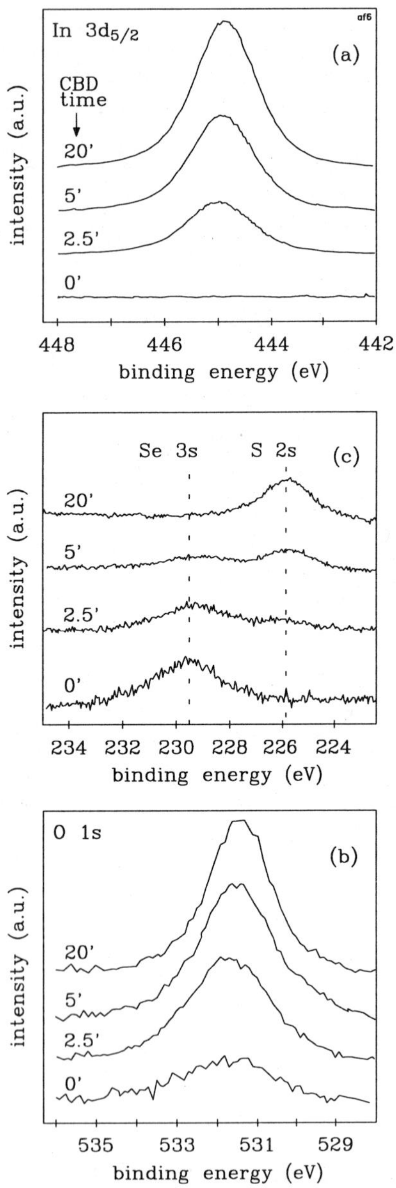

Fig. 2. Photoemission signals of the involved elements during the formation of the $In_x(OH,S)_y$ film on a $CuGaSe_2$ surface. The evolution of In, S and O signals is parallel. The Se substrate signal dissappears completely after 20 min.

prepared single layer $CuInSe_2$ and $CuGaSe_2$ substrates with a Cu-poor composition. The surfaces were carefully rinsed with deionized H_2O after the deposition to remove unreacted parts of the solution.

There is no indication of a film growth in the first minute under the chosen conditions. In Fig. 2 the surfaces of $CuGaSe_2$ samples after different deposition times are compared with a surface which was only rinsed with H_2O. Here $CuGaSe_2$ samples are shown because in this case it is possible to observe the evolution of the In signal without interference from the absorber layer. The film growth is comparable to that on $CuInSe_2$ surfaces. Between 1' and 2.5' a species containing In, S and O begins to grow at the surface. This can be seen in the increase of the corresponding photoelectron signals. There is no indication of a change in the composition of the deposited film during the growth. After 20 minutes the absence of any absorber layer signals (i.e. Se) proves, that a dense film thicker than the information depth of XPS (≈100 Å) covers the absorber surface.

A detailed investigation of the chemical state of the film components indicates the formation of an In-O-S alloy. With XPS there is no way to directly prove the presence of hydrogen in the compound, but from a chemical point of view an In-OH formation is favoured in aqueous solutions. The well-defined XPS signal of the indium (MNN) Auger structure (Fig. 3) cannot be explained by a mixture of $In(OH)_3$ and In_2S_3, which were characterized as pure films in a previous work [4]. An alloy would be characterized by a broader structure and the dip in the structure at about 404.5 eV would be smeared out by In($M_4N_{45}N_{45}$) electrons originating from $In(OH)_3$. The Auger parameter α=851.6 eV is between the values for $In(OH)_3$ (α=850.8 eV) and In_2S_3 (α=852.2 eV) which can be found in the literature [6] and support this indication.

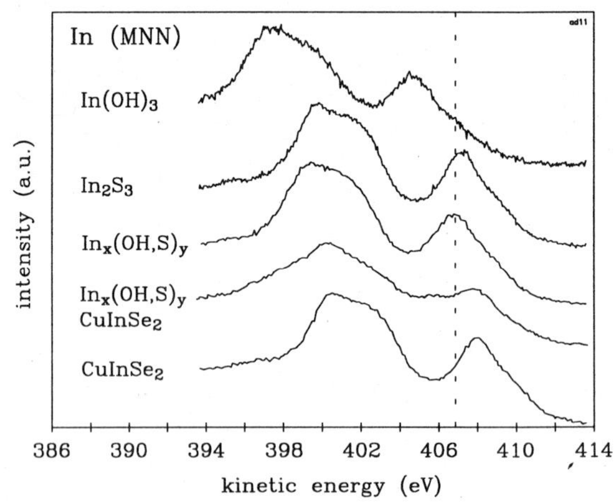

Fig. 3. XPS signal of the indium (MNN) Auger structure of different compounds. The signal of $In_x(OH,S)_y$ can not be explained by a mixture of $In(OH)_3$ and In_2S_3

DEVICE PROPERTIES

The absorbers for studying the influences of parameter variations of the chemical bath were prepared in a thermal evaporation process analogous to the one presented by Hedström et al. [1]. The Ga/(In+Ga) ratio was varied between 10 and 20%. The use of a concentric coaxial source guaranteed a homogeneous film composition over a 7x7 cm^2 subatrate area. Each parameter variation was carried out on absorbers from one preparation run.

In view of solar cell device optimization the concentrations of $InCl_3$ and thioacetamide in the solution were varied. $Cu(In,Ga)Se_2$ absorbers were diped in the bath, heated to 70°C and kept there for different times. The main influences of this variations are changes of the film thickness and film composition.

Variation of the thioacetamide concentration at 5 mmol/l $InCl_3$ indicates (see Fig. 4), that superior conversion efficiencies can be expected for higher thioacetamide concentrations . A high amount of ammoniumacetate as a stabilizing H^+ buffer or hydrogensulfide as the source for sulfurization of In^{3+} ions could influence the composition of the deposited layer and the growth kinetics.

Variation of the $InCl_3$ concentration between 5 and 20 mmol/l does not influence the device performance in a systematic way. We observed a slightly increased film growth for higher $InCl_3$ concentrations.

Variation of the deposition time reflects changes in the film thickness. Using the conditions given in Fig. 4 we got the highest efficiencies for 20 min deposition time.

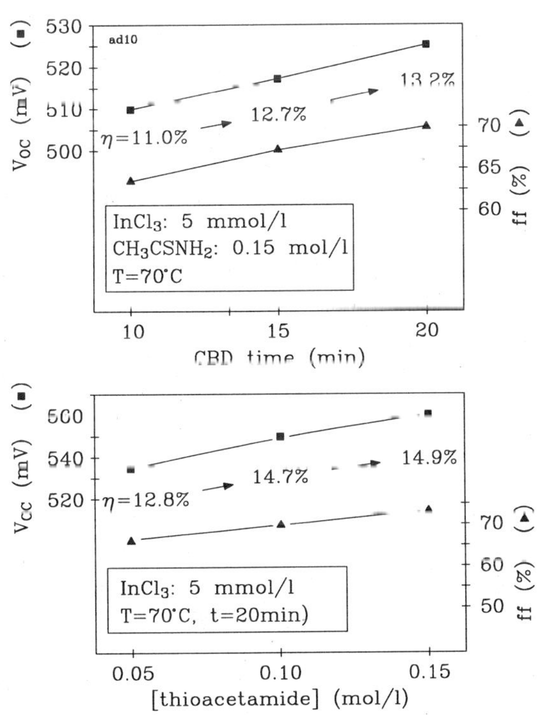

Fig. 4. Influence of chemical bath concentrations on the electrical properties of $Cu(In,Ga)Se_2$ solar cells

The electrical properties of solar cells prepared under optimized conditions were studied by measuring the j-V characteristics and the spectral quantum efficiencies. Comparing the device performance of solar cells using a standard CBD-CdS buffer layer with the cadmium free CBD-$In_x(OH,S)_y$ layer we find an improved V_{oc}, comparable values for FF and sometimes slightly reduced j_{sc} (Fig. 5). The conversion efficiencies (active area with MgF AR- coating) are comparable to the best standard cells and independent of the substrate. The spectral quantum efficiency measurement shows a significant gain below 540 nm, the absorption edge of CdS, due to the higher band gap of $In_x(OH,S)_y$. Nevertheless, the integrated current over the AM 1.5 solar spectrum does not exceed the values for the standard devices with CdS buffer layer. All devices prepared with the new CBD process exhibit reduced red response. The effect is most significant for samples prepared on float glass substrates.

This observation has important consequences. Changes in the spectral quantum efficiency in the red wavelenth region are due to changes in the absorber properties. The deposition process includes not only the growth of a buffer layer with a good electronic and geometric match to the absorber material, but also modifies the electrical properties of the absorber. One explanation for the reduced red response is the possible influence of a reduced space charge width.

C-V measurements show significant differences in the space charge width of samples depending on the substrate and the buffer layer (see Tab. 1).

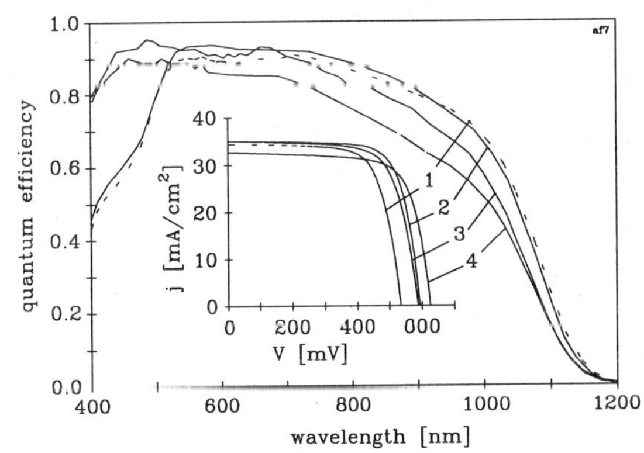

Nr.	Buffer layer	Glass substrate	V_{OC} [mV]	ff [%]	jsc [mA/cm²]	η [%]
1	CdS	Corning 7059	500	73	34.8	13.3
2	CdS	float	595	73	34.9	15.2
3	$In_x(OH,S)_y$	Corning 7059	590	72	34.9	14.8
4	$In_x(OH,S)_y$	float	630	73	32.4	14.9

Fig. 5. j-V curves, spectral quantum efficiency and electrical device properties of $Cu(In,Ga)Se_2$ based solar cells (Ga/(Ga+In)=18%). The novel $In_x(OH,S)_y$ buffer layer is compared to the standard CdS buffer for cells prepared on different substrates

buffer layer	substrate	space charge width (nm)
CdS	Corning 7059	~1000
CdS	Float	300-500
In(OH,S)	Corning 7059	~200
In(OH,S)	Float	~150

Tab. 1. space charge width of various $Cu(In,Ga)Se_2$ solar cells as calculated from C-V measurements

The space charge width d depends on the effective carrier concentration N_A. If the dominant recombination is in the space charge region [7] (neglecting the influence of tunneling mechanisms and surface recombination) an increased acceptor concentration N_A in the absorber material leads to a smaller space charge width, less recombination occurs and V_{OC} is higher compared to a reference absorber with an acceptor base concentration N_{A0}.

$$d \propto \frac{1}{\sqrt{N_A}} \quad \text{and} \quad \Delta V_{OC} = \frac{2kT}{q} \ln \sqrt{\frac{N_A}{N_{A0}}}$$

The observations are compatible with the model of an increased effective acceptor concentration in the absorber due to the chemical bath process under investigation. Comparing the space charge width of the devices on soda lime glass with CdS and In(OH,S) buffer layer an increase of the carrier concentration by a factor of 4-9 for the Cd-free buffer layer can be deduced. This corresponds according to the formula above to an increase of V_{oc} of about 35-50 mV. The correlation between weak red response, reduced space charge width and increased V_{OC} in $Cu(In,Ga)Se_2$ solar cells is known from studies concerning substrate influences on device properties [8]. When the carrier collection is supported by the electric field in the space charge region an increase in the acceptor concentration reduces this space charge width. Therefore the effective collection width is reduced resulting in a worse red response. In the case of the chemical bath process, it is thermodynamically improbable that the procedure changes the bulk properties of the absorber through material diffusion. The consequence is that charged surface states at the grain boundaries of the polycrystalline $Cu(In,Ga)Se_2$ can be influenced by a chemical bath process.

The positive effect on V_{OC} can be demonstrated clearly for samples prepared on Corning 7059 glass where the substrate influence can be neglected. The V_{OC} for devices with the novel buffer layer is significantly higher compared to cells with the standard CdS buffer.

CONCLUSIONS

A novel Cd-free CBD process for buffer layers in $Cu(In,Ga)Se_2$ solar cells has been developed. The achieved conversion efficiencies of nearly 15% clearly demonstrates the potential of this process as an alternative to the standard CBD-CdS deposition.

The film was characterized by XPS and found to contain In, O and S, possibly a $In_x(OH,S)_y$ compound, which grows in 20 minutes to a dense film with a thickness of at least 10 nm.

The observation of a narrow space charge region, the high values for V_{OC} and the reduced red response support the hypothesis that the effective acceptor concentration in the absorber is higher in devices using the novel buffer layer. We favour the model that charged surface states at the grain boundaries of the polycrystalline $Cu(In,Ga)Se_2$ can be influenced by the chemical bath process. A neutralization of donor-like surface states like Se vacancies or the formation of acceptor-like surface states could be the reason for the observed properties.

This model of an increased effective acceptor concentration caused by modifications of charged surface states could also shine a light on the reason for the success of the CBD CdS process compared to a evaporated CdS window layer (high V_{OC} and reduced red response). Systematic studies to control this influence are underway.

ACKNOWLEDGMENTS

This work has been supported by the Bundesministerium für Forschung und Technologie (BMFT) contract 032 8059E and by the CEC contract No. JOU2-CT92-0141. T. Magorian is gratefully acknowledged for reviewing.

[1] J.Hedström, H.Ohlsén, M.Bodegård, A.Kylner, L.Stolt, D. Hariskos, M.Ruckh, H.W.Schock, Proc. 23rd Photov. Spec. Conf., Louisville (1993)

[2] A.Gabor, J. Tuttle, M. Contreras, D. Albin, A. Franz, D. Niles, and R. Noufi, 12th EC PVSEC Amsterdam (1994) 939

[3] J.Kessler, M.Ruckh, D.Hariskos, U.Rühle, R. Menner, H.W.Schock, Proc. 23rd Photov. Spec. Conf., Louisville (1993)

[4] K.O.Velthaus, J.Kessler, M.Ruckh, D.Hariskos, D.Schmid, H.W.Schock, Proc. 11th E.C. Photov. Solar Energy Conf., Montreux (1992)

[5] J. Kessler, K.O.Velthaus, M.Ruckh, R.Laichinger, H.W.Schock, D.Lincot, R.Ortega, 6th PVSEC, New Delhi (1992)

[6] C.D. Wagner, in *Practical Surface Analysis*, 2nd ed., edited by D.Briggs and M. P. Seah (Wiley, Chistester, NY, 1990), Vol.1 p. 614

[7] T. Walter, R. Menner, Ch. Köbele and H.W. Schock, Proc. 12th E.C. Photov. Solar Energy Conf. (1994) 1755

[8] M. Ruckh, D. Schmid, M.Kaiser, R. Schäffler, T. Walter and H.W. Schock, this conference

SUPERSTRATE-TYPE $CuInSe_2$ SOLAR CELLS WITH CHEMICALLY DEPOSITED CdS WINDOW LAYERS

Tokio Nakada, Nobukata Okano, Yoshiyuki Tanaka,
Hiroyuki Fukuda and Akio Kunioka

Department of Electrical Engineering and Electronics, Aoyama Gakuin University,
Setagaya-ku, Tokyo 157, Japan

ABSTRACT

Chemically deposited CdS was used for window layers of superstrate-type CIS solar cells. The CBD-CdS layers had no pinholes and had higher optical transmission at short wavelengths than evaporated CdS layers, which led to the increase in Jsc. Relatively high substrate temperature of 450 ℃ was used for CIS deposition, resulting in high Voc due to improved crystallinity of CIS films. However, the interdiffusion and alloy formation at CIS/CdS deteriorated cell performance. The best cell yielded an efficiency of 8.1 % with Voc = 440 mV, Jsc= 34 mA/cm^2 and FF= 0.54 using near-stoichiometric (Cu/In = 0.91-0.98) CIS films.

INTRODUCTION

The superstrate type solar cells with metal contact/ CIS/CdS/TCO/glass structure have several advantages such as no requirement of surface cover glass for encapsulation and interconnection between unit cells in the manufacturing of modules. However, the best device efficiencies reported so far were in the 6 % range with CIS films grown on thermally evaporated CdS layers [1-5]. The main reason for the low efficiency is the poor crystallinity of CIS films since the substrate temperatures below 350 ℃ was used in order to avoid undesirable interdiffusion at the CIS/CdS interface. However, we have considered that the most important point for achieving an efficient devise is to improve the crystallographic quality of absorber layers.

We thus tried to deposit CIS films at high substrate temperatures above 400 ℃ , and further replaced thermally evaporated CdS layers with chemically deposited CdS layers which possibly offer different junction behaviors. In this paper, the interdiffusion behavior at CIS/CdS interface and the crystallographic properties of CIS films are described. Light J-V characteristics of superstrate-type solar cells with Au/CIS/CdS/ZnO:Al/glass structure are discussed in relation to the Cu/In ratio and deposition conditions of the CIS films .

EXPERIMENTAL

The superstrate-type CIS solar cells with Au/CIS /CdS/ZnO:Al /glass structure were fabricated as follows. The 2-µm-thick ZnO:Al films were first deposited on soda-lime glass substrates at 500-550 ℃ by DC magnetron sputtering with 2 wt%Al doped-ZnO ceramic target [6]. The ZnO:Al films exhibited sheet resistance of 2Ω/sq and optical transmission of more than 85% at 600 nm. The CdS layers were then deposited on ZnO:Al-coated soda-lime glass substrates by chemical bath deposition (CBD) with CdI_2(1.4 x 10^{-3} M)-thiourea(0.14 M)-ammonia(1 M) aqueous solution at 25-60℃ for 4 min [7]. Approximately 40 nm-thick CdS layer was obtained by one deposition step. For the window layers, 300 nm thick films were deposited by repeating the same CBD process. The CdS layers were also deposited by thermal evaporation at 150-200℃ for comparison with CBD-CdS layers. Following that the 2-2.5 µm thick CIS films were deposited onto CdS/ZnO:Al/soda-lime glass substrates by three-source coevaporation method at 350-500 ℃, in which Cu-rich layer was first deposited on CdS layers prior to In-rich layer. The fluxes of Cu, In and Se were measured by quartz thickness monitor. The Se/metal flux ratio was maintained at more than three times that of total metal flux. Finally, 3.5 x 3 mm^2 Au back contact film was deposited on the CIS films.

RESULTS AND DISCUSSION

CIS absorber layers

The SEM photographs of near-stoichiometric CIS films (Cu/In=0.98) deposited on CBD-CdS/ZnO:Al/glass

CH3365-4/94/0000-0095 $4.00 © 1994 IEEE First WCPEC; Dec. 5–9, 1994; Hawaii

substrates at different substrate temperatures are shown in Fig. 1. The films deposited at 450 ℃ showed triangular large grains, whereas the film deposited at 350 ℃ consisted of small grains, showing poor crystallinity of the films. This result suggests that the efficient solar cells can not be fabricated with CIS films deposited at low substrate temperature of 350 ℃.

Figure 2 shows the X-ray diffraction patterns of CIS films with different Cu/In ratios deposited on CBD-CdS/ZnO:Al layers at 450 ℃. A secondary phase, $CuCd_2InSe_4$, was observed in Cu-rich and near-stoichiometric films. The formation of this alloy may originate from the diffusion of Cd into CIS layers. On the other hand no alloy formation was observed in In-rich film with Cu/In ratio below 0.8, so that Cu diffusion into CdS layers from In-rich CIS film was suppressed.

CdS window layers

The SEM and EPMA analyses revealed that thermally evaporated CdS layers had columnar structure with small pinholes. Therefore, relatively thick layers of 0.8-3.0 μm were required for the best cells [1-3]. On the other hand, chemically deposited layers were uniform and dense structure, therefore thinner layers than thermally evaporated layers can be used as window layers. The optical transmission curves of 0.3-0.5 μm thick CBD-CdS and 0.5 μm thick evaporated CdS layers on glass substrates are shown in Fig. 3. Two points are immediately obvious in this figure. The CBD-CdS layers exhibited transmission curve spread towards short wavelength region than thermally evaporated layers, which led to the increase in Jsc as mentioned later. Another feature in chemically deposited layers is a few interference fringes in the transmission curve. These facts suggest that the CBD-CdS layers by our process are not pure CdS.

We also investigated the composition of these layers by XPS. As predicted a large amount of oxygen (O1s) was observed in CBD-CdS layers in the form of $Cd(OH)_2$ and/or CdO which were determined by the chemical shift of binding energy. On the contrary, no oxygen was observed in thermally evaporated layers. Furthermore, the oxygen was observed in CBD-CdS layers grown with other Cd sources such as $CdSO_4$, $CdCl_2$ and $Cd(CH_3COO)_2$ although it depended on ammonia concentration. The influence of Cd oxide or hydroxide on the cell performance is not fully understood at present.

Interdiffusion at CIS/CdS interface

At high substrate temperatures, very complex interdiffusion at CIS/CdS interface probably occurs. For instance, the secondary phase $CuCd_2InSe_4$ was formed by heavy doping of Cd as identified from XRD patterns. We also confirmed the increase in donor concentration by Cd doping into CIS matrix.

Fig. 1. The SEM photographs of near-stoichiometric CIS films (Cu/In=0.98) deposited on CBD-CdS /ZnO:Al/glass substrates at (a) 350 ℃ and (b) 450 ℃.

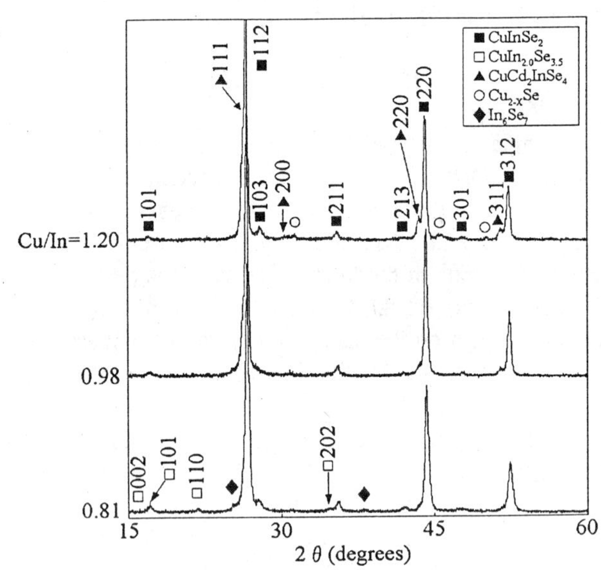

Fig. 2. XRD patterns of CIS films with different Cu/In ratios deposited on CBD-CdS/ZnO:Al layers at 450 ℃.

The diffusion behavior of CIS/CdS/ZnO:Al/glass multilayers was checked by AES depth profile as shown in Fig. 4. The 2μm thick CIS film with Cu/In ratio of 0.98 was deposited at 450 ℃ on 0.3 μm thick CBD-CdS

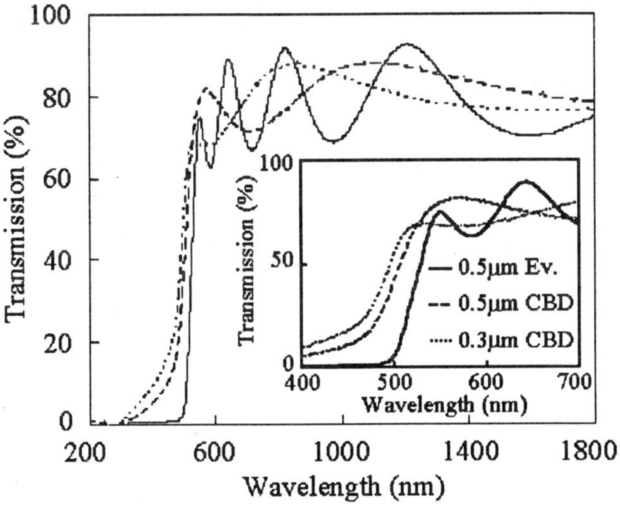

Fig. 3. Optical transmission curves of 0.3-0.5μm thick CBD-CdS and 0.5μm thick evaporated CdS layers on glass substrates.

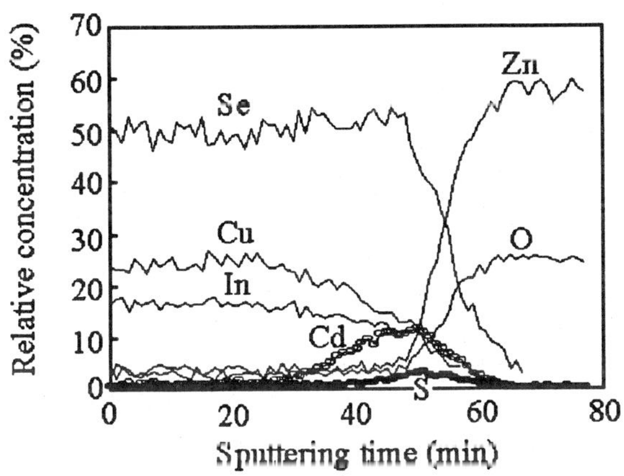

Fig. 4. AES depth profiles of CIS/CBD-CdS/ZnO:Al layers.

layer. The depth profile revealed that Cd diffused easily into CIS layer than sulfur. This result is consistent with XRD data shown in Fig. 2. The relative concentrations of Cu, In, Se, Cd and S reached zero at the same sputtering time, suggesting that the intermixing of two layers occurred completely. However we could not detect any compounds except CIS, $CuIn_2Se_{3.5}$, $Cu_{2-x}Se$ and $CuCd_2InSe_4$ from XRD patterns. The interdiffusion behavior was exactly the same as thermally evaporated CdS layers, within accuracy of AES and XRD, although CBD-CdS layers contained oxide and/or hydroxide.

CIS Cells

Figure 5 shows J-V characteristics of Au/CIS/CdS ZnO:Al/glass cells with CIS films(Cu/In=0.90) deposited at different substrate temperatures. The cell performances are summarized in Table 1. The best cell yielded an efficiency of 8.1 % with Voc = 440 mV, Jsc=34 mA/cm^2

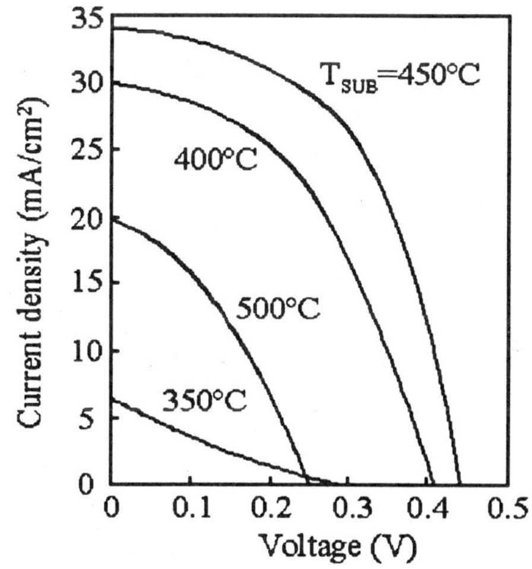

Fig. 5. J-V characteristics of superstrate cells with CIS films deposited at various substrate temperatures on 0.3μm-thick CBD-CdS layers.

Table 1. Performance of superstrate devices with CIS films deposited at various substrate temperatures.

T_{SUB}(℃)	V_{OC}(mV)	J_{SC}(mA/cm^2)	FF	η (%)
500	250	20.1	0.36	1.8
450	440	34.0	0.54	8.1
400	410	30.1	0.44	5.4
350	280	6.6	0.21	0.39

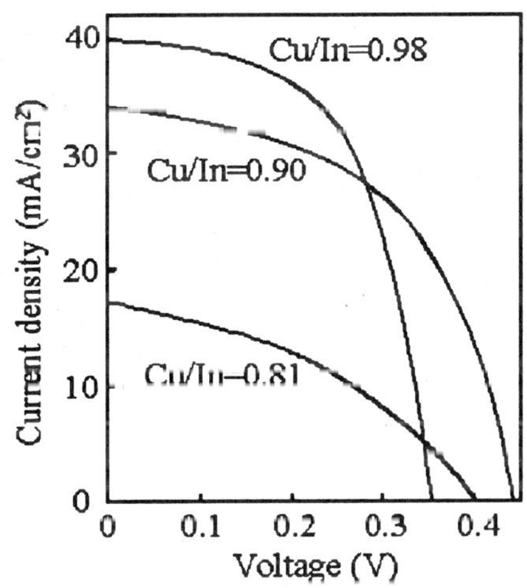

Fig. 6. J-V characteristics of superstrate CIS cells with different Cu/In ratios.

Table 2. Performance of superstrate devices using CIS films with different Cu/In ratios.

Cu/In	V_{OC}(mV)	J_{SC}(mA/cm^2)	FF	η (%)
0.98	350	39.8	0.58	8.1
0.90	440	34.0	0.54	8.1
0.81	400	17.5	0.38	2.7

and FF=0.54 using near-stoichiometric CIS films deposited at 450℃, 100 ℃ higher than that previously reported in superstrate devices. The highest Voc obtained with these cells was 500 mV for 3 μm thick CIS films. We speculate the high Voc may be due to the decrease in grain boundaries by using high substrate temperatures. Some of the devices fabricated under these conditions showed no light-dark crossover in the J-V characteristics. When the In-rich CIS layers were first deposited on CdS layers, the devices shorted presumably because of the formation of low resistivity In-doped CdS. The best CIS cells were obtained when Cu-rich layers were first deposited on CdS layers prior to In-rich layers. This can be explained as In diffusion into CdS layer being suppressed by Cu-rich layer. However heavy doping of Cu decreased the cell efficiency because of the formation of high resistivity CdS:Cu so that we optimized the Cu and In flux profile.

The effect of the Cu/In ratio on the device performance was also investigated for CIS films deposited at 450 ℃. The results are shown in Fig. 6 and Table 2. The best cells were obtained with Cu/In ratios between 0.90 and 0.98, which is similar to that of substrate type solar cells. For the ratios above unity the devices shorted. On the other hand, for the ratios below 0.8 cell efficiency decreased due to the increase in resistivity and the coexistence of $CuIn_2Se_{3.5}$. The optimized Cu/In ratio is quite different from that previously reported ratios of about 0.5-0.7 [3].

Figure 7 shows the spectral response curves of two CIS devices in which 0.3 μm thick CBD-CdS layer and thermally evaporated 0.6 μm thick CdS were used as window layers. For the latter cell, best efficiency was 5.4% with Voc=362 mV, Jsc= 27.7 mA/cm^2, FF= 0.537. The photoresponse for the cell using CBD-CdS spread towards short wavelength region. This is attributed to the reduction in optical absorption loss at short wavelengths due to the change in CdS layer by the interdiffusion. At long wavelengths the photoresponse increased slightly, which is probably attributed to the decrease in Cu diffusion into CdS layer. Our CBD-CdS layers included a large amount of $Cd(OH)_2$ and/or CdO, therefore their diffusion properties may be different from pure CdS layers. It is noted from the response curves that one of the ways to improve efficiency of superstrate CIS cell is to eliminate Cu diffusion into CdS layers.

ACKNOWLEDGMENTS

The authors would like to thank T. Shimatoh, N. Karasawa and K. Migita for their sample preparation. This work was partly supported by the Photovoltaic Power Generation Technology Research Association.

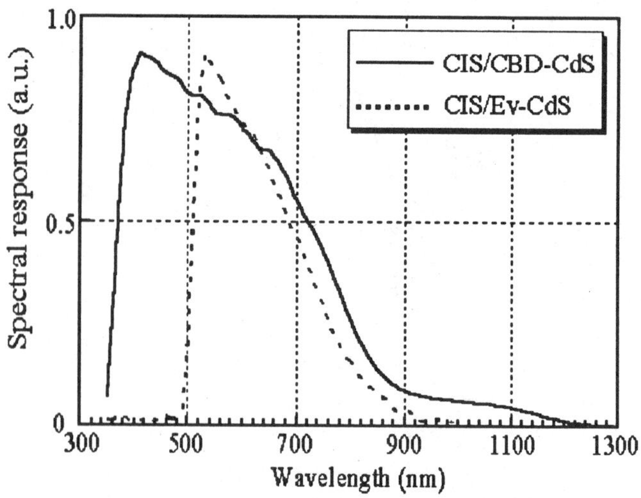

Fig. 7. Spectral response curves of superstrate cells with CIS/CBD-CdS and CIS/Ev-CdS structures.

REFERENCES

[1] T. Negami, M. Nishitani, M. Ikeda and T. Wada, " Preparation of $CuInSe_2$ Films on Large Grain CdS Films for Superstrate-type Solar Cells", Solar Energy Materials and Solar Cells, 35, 1994, pp. 215-222.

[2] T. Negami, M. Nishitani, T. Wada and T. Hirao, " Preparation of $CuInSe_2$ Films on CdS Films for Photovoltaic Applications" Proc. 11th European Photovoltaic Solar Energy Conf., 1992, pp. 783-786.

[3] T. Yoshida and R. W. Birkmire, " Fabrication of $CuInSe_2$ Solar Cells in a Superstrate Configuration", Proc. 11th European Photovoltaic Solar Energy Conf., 1992, pp. 811-814.

[4] R. W. Birkmire, R. B. Hall, J. E. Phillips, " Material Requirements for High Efficiency $CuInSe_2$/ CdS Solar Cells", Proc. 17th IEEE Photovoltaic Specialists Conf. , pp. 882-886.

[5] R. W. Birkmire, W. N. Shafarman and R. D. Varrin Jr., " Options for Fabrication and Design of $CuInSe_2$ Based Solar Cells" Proc. 21st IEEE Photovoltaic Specialists Conf. , pp. 550-560.

[6] T. Nakada, N. Murakami and A. Kunioka," Transparent Conducting Al-, AlB_{12}, and B-Doped ZnO Films for Solar Cells by DC magnetron Sputtering", Proc. 12th European Photovoltaic Solar Energy Conf. , 1994, PP. 1507-1510.

[7] J. Kessler, K. O. Velthaus, M. Ruckh, R. Laichinger, H. W. Schock, D. Lincot, R. Ortega and J. Vedel, " Chemical Bath Deposition of CdS on $CuInSe_2$, Etching Effect and Growth Kinetics", Tech. Dig. 6th Photovoltaic Science and Engineering Conf., 1992, pp. 1005-1009.

THE EFFECTS OF CdS PROCESSING AND GLASS SUBSTRATES ON THE PERFORMANCE OF CdTe SOLAR CELLS

C. S. Ferekides, K. Dugan, V. Ceekala, J. Killian, D. Oman,
R. Swaminathan, and D. L. Morel
Department of Electrical Engineering
University of South Florida
4202 E. Fowler Ave.
Tampa, FL 33620

ABSTRACT

Cadmium Sulfide films prepared by rf sputtering and close spaced sublimation (CSS) have been used for the fabrication of CdTe/CdS thin film solar cells on borosilicate glass substrates. The CdTe layer was prepared by CSS at high processing temperatures (600°C). CdS films prepared by the chemical bath deposition process (CBD) were deposited on tin oxide coated soda lime glass substrates. For these devices the CSS CdTe films were prepared at low substrate temperatures (<550°C). Devices prepared at low processing temperatures (CdTe-CSS/CdS-CBD) on soda lime glass substrates exhibited efficiencies in excess of 13% as measured under AM 1.5 conditions at the National Renewable Energy Laboratory.

INTRODUCTION

Thin film CdTe/CdS solar cells have in recent years demonstrated remarkable increases in photovoltaic conversion efficiencies[1]. Electrodeposition, rf sputtering, close spaced sublimation, evaporation, and chemical bath deposition are among the most widely used processes for the fabrication of CdTe/CdS solar cells. Certain processing steps such as a post-deposition heat treatment of the CdTe/CdS structure prior to back contact formation, and a chemical surface treatment of the CdTe surface to facilitate contact formation are common to almost all fabrication methods. However, significant processing differences exist leading to variation in bulk semiconductor and junction properties. Developing an effective manufacturing process for module fabrication requires that easily scalable, high thruput techniques with the potential to yield high efficiency devices are selected. Understanding the complex issues of junction formation/properties and how these are influenced by processing is also crucial to successfully converting laboratory processes into manufacturing ones. This paper reports on recent progress toward achieving high efficiency CdTe/CdS solar cells using manufacturing friendly processes. The CSS process is used for the deposition of the CdTe layers. The high deposition rates achievable by CSS make it a good candidate for manufacturing purposes. To date this process has been used successfully at relatively high temperatures (600°C) to prepare CdTe solar cells with efficiencies over 15%[2]. However, these devices were fabricated on expensive borosilicate glass substrates. The deposition temperature of the CSS process has been significantly lowered and solar cells have been fabricated on soda lime glass substrates. RF sputtering and CSS have been used for the deposition of CdS films as alternatives to the CBD CdS. Both these processes offer manufacturing advantages over the CBD which yields large amounts of liquid waste. CdTe/CdS solar cells using rf sputtered and CSS CdS films have been prepared and optimized on borosilicate glass substrates.

EXPERIMENTAL

Thin film CdTe/CdS solar cells of the superstrate configuration have been prepared on SnO_2 coated glass substrates. Devices utilizing rf sputtered and CSS CdS films were primarily fabricated on borosilicate glass substrates, with the CSS CdTe layer prepared at high substrate temperatures (600°C). The CSS of the CdTe has been previously described[2]. Tables I & II provide summaries of the deposition conditions used for the rf sputtered and CSS CdS films.

Solar cells prepared on SnO_2 coated soda lime glass substrates utilized a CdS layer prepared by the CBD process[3], and a CSS CdTe layer prepared at low substrate temperatures (460 - 550°C). In all cases the CdS films were annealed in the presence of H_2 in the temperature range of 350 - 450°C. After the deposition of the CdTe the structures were annealed in the

CH3365-4/94/0000-0099 $4.00

First WCPEC; Dec. 5-9, 1994; Hawaii

presence of $CdCl_2$. A back contact was formed by applying a layer of doped graphite after chemically treating the surface of the CdTe.

Material	CdS 5N purity
RF Power (watts)	100 -150
Substrate Temperature (°C)	100 - 450
P(Ar) (mtorr)	5
Substrate-Source Distance (in)	3 - 6

Table I. Process conditions used for the deposition of the rf sputtered CdS films.

Source Temperature (°C)	450 - 700
Substrate Temperature (°C)	250 - 600
Pressure (torr)	1 - 400
Source-Substrate Spacing (cm)	0.2 - 1
Ambient	He, H_2, O_2

Table II. Process conditions used for the deposition of the CSS CdS films.

RESULTS AND DISCUSSION

RF Sputtered CdS Films

The structural, and electrical properties of the rf sputtered CdS films prepared on glass and SnO_2/glass substrates have been studied as a function of the process parameters and a heat treatment in the presence of H_2. The grain size of the CdS layer is an important parameter influencing the performance of CdTe/CdS solar cells. Small grains may enhance the formation of an interdiffused layer at the CdTe/CdS interface which influences the junction's transport properties. The grain size of the rf sputtered CdS films was found to be in the range of 0.10 - 0.2μm and it increased with increasing substrate temperature. The films exhibited preferential orientation along the (002) axis. The intensity of the (002) peak was also found to increase with increasing temperature. The dark electrical resistivity was measured in the lateral direction using four collinear contacts. It was found to be higher than 10^6 Ω-cm independent of the substrate temperature.

Annealing of the CdS films in the presence of H_2 has been found to influence film properties and solar cell performance[2,4]. Figure 1 shows the mobility of the CdS films as a function of the annealing and deposition temperatures. In general, the mobility was found to be about 1 cm^2/V-s for films annealed at low temperatures (<400°C), and increased to 9 - 11 cm^2/V-s for films annealed at higher temperatures. This behavior was virtually independent of the deposition temperature.

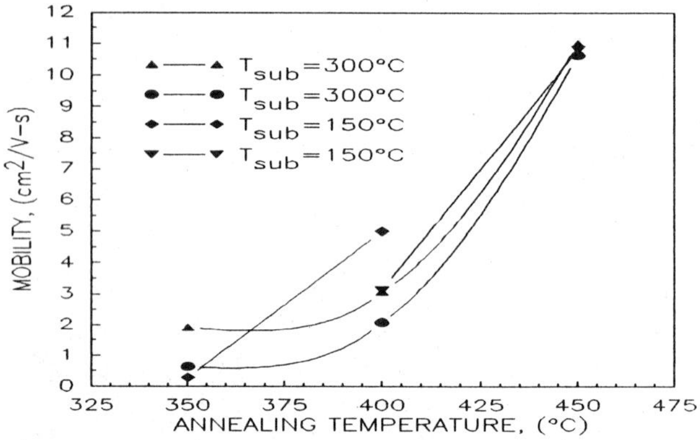

Fig. 1. The effect of H_2 annealing on the mobility of the rf sputtered CdS films.

Annealing the rf sputtered CdS films in H_2 improved the performance of solar cells, in particular the open-circuit voltage (V_{oc}) and spectral response (SR). Typically, the V_{oc} increased by about 30 - 50 mV when the CdS films were annealed in H_2 as shown in table III. The best results were obtained for annealing temperatures around 400 °C.

Deposition T (°C)	Annealing T (°C)	V_{oc} (mV)
150	400	850
150	450	820
300	450	830
300	None	800

Table III. The effect of H_2 annealing on the open-circuit voltage.

Figure 2 shows the SR of solar cells fabricated using rf sputtered CdS annealed in H_2 at different temperatures. The roll off in the SR at 600 nm is due to the formation of a $CdS_{1-x}Te_x$ layer. Increasing the annealing temperature enhances the SR in the 500 - 600 nm region suggesting that the higher annealing temperatures limit the extend of interdiffusion at the CdTe/CdS interface. The roll off in the SR at 600 nm has also been been observed for devices with CSS and CBD CdS, but the decrease in the SR in this region (500 - 600 nm) was considerably less. This behavior seems to explain the fact that high V_{oc}'s (>800mV) were obtained only when the thickness of the rf sputtered CdS films was in excess of 2000 Å. When thinner films were used excessive interdiffusion lead to lower V_{oc}'s. Another unique feature associated with the rf sputtered CdS films is the excessive crossover between the light and the dark I-V curves as shown in fig. 3. Based on cuurent data the crossover mechanism is not believed to be critical to device performance[4]. The current transport mechanism and junction behavior have been investigated and will be the topic of a future publication.

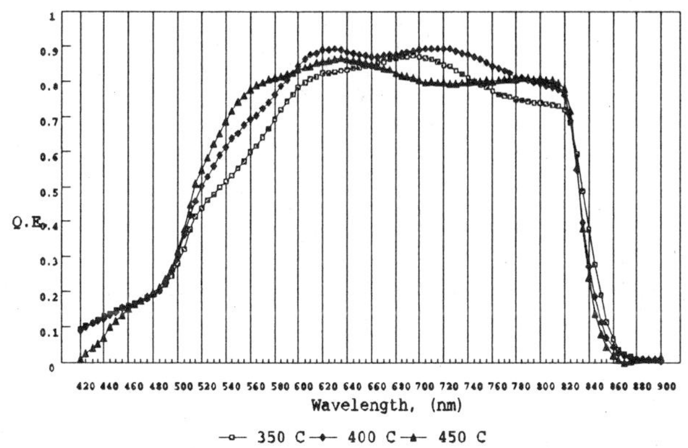

Fig. 2. The SR of solar cells prepared with rf sputtered CdS.

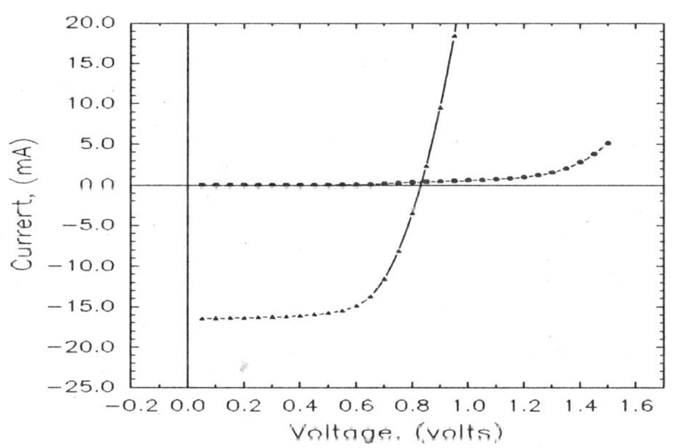

Fig. 3. The crossover mechanism in a CdTe/CdS(rf) solar cell.

The best solar cell prepared using rf sputtered CdS exhibited a conversion efficiency of 11.6% (V_{oc} = 843 mV, J_{sc} = 20.6 mA/cm^2, FF = 67%). The efficiencies of these devices are currently limited by the rather low FF's and J_{sc}'s.

CSS CdS Films

Several attempts to deposit CSS CdS on glass substrates resulted in no or minimal deposits due to the surface properties of the glass, and therefore CSS CdS films were almost exclusively deposited on SnO_2 coated glass. From a limited number of depositions carried out on glass substrates the resistivity of the CSS CdS films was found to be in the order of 10^6 Ω-cm. Unlike the CBD and rf sputtered CdS, the grain size of CSS CdS films could be varied over a wider range up to several μm depending on the deposition conditions. However, large grain films were typically avoided since these often contained pinholes when prepared at small thicknesses. The CSS CdS grain size was also affected by the ambient. Typically, inert ambients yielded high deposition rates and large grain sizes. When O_2 was added to the ambient the deposition rate decreased and denser films consisting of smaller grains were obtained.

The presence of O_2 during the deposition of the CSS CdS also seemed to affect solar cell performance. When the CdS films were deposited at small thicknesses, 700 - 800 Å, those prepared in inert ambient yielded V_{oc}'s up to 50 mV lower than those prepared in O_2. However, as the film thickness was increased, some of the voltage, up to 30 mV, was recovered.

Figures 4 and 5 show the SR of two devices prepared with CSS CdS films of different thicknesses. Both CdS layers were prepared in the presence of O_2. It appears that the thickness of the CdS layer has an effect on the SR in the 500 - 600 nm region. For the thin CdS device a roll off at about 600 nm is present indicating the formation of the $CdS_{1-x}Te_x$ mixed crystal layer. The thick CdS device shows a sharp cut off at 510 nm corresponding to the CdS energy gap. This could be due to the fact that smaller grain size enhances the formation of $CdS_{1-x}Te_x$. Both devices exhibited V_{oc}'s in excess of 810 mV. Table IV lists some of the most significant solar cell results obtained to date using CSS CdS. Open-circuit voltages in excess of 850 mV, and ff's as high as 74% have been achieved.

Sample #	214A	21812B	2187A
Eff., (%)	12.93	11.7	12.52
V_{oc}, (mV)	833	852	838
J_{sc}, (mA/cm^2)	21	19.5	21.33
FF	74.08	70.6	70

Table IV. Solar cell parameters for devices prepared with CSS CdS.

Solar Cells on Soda Lime Substrates

One of the most important tasks undertaken is to lower the temperature of the CSS process and fabricate high efficiency CdTe/CdS solar cells on soda lime glass

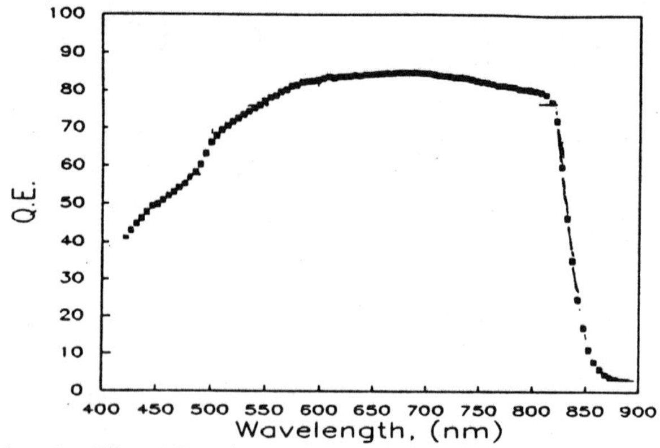

Fig. 4. The SR of a solar cell fabricated with a thin CSS CdS layer.

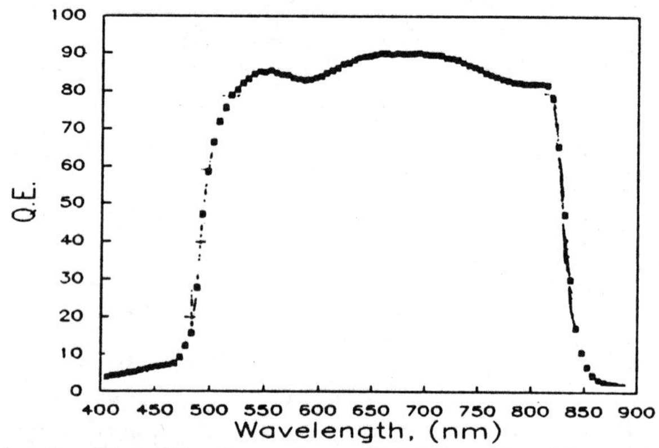

Fig. 5. The SR of a solar cell fabricated with a thick CSS CdS layer.

substrates. The high substrate temperature used during the CSS process is believed to be one of the important processing parameters that lead to efficiencies in excess of 15%. Early work indicated that solar cells prepared at low temperatures (<550 °C) exhibited low V_{oc}'s as a result of increased shunting[2]. Lowering the CSS substrate temperature also resulted in reduced deposition rates and increased deposition times. The total pressure, source-substrate spacing, and source temperature were adjusted in order to increase the deposition rates to over 1 μm/min. The increase in deposition rate also produced denser CdTe films. Figure 6 shows the SEM micrographs of CdTe films prepared at a substrate temperature of 500 °C. Both devices exhibited V_{oc}'s in excess of 820 mV. After optimizing solar cell performance on borosilicate glass substrates, a variety of soda lime glass substrates were incorporated in the process. The various soda lime glass substrates used presented different problems. Defects formed frequently leading to film blistering. In the worst cases the CdTe films peeled off completely. Further improvements in processing including lowering both the CdTe deposition temperature as well as the heat treatment temperature resulted in considerable improvements in device performance. Table V lists the

Fig. 6. SEM micrographs of CdTe deposited at 500°C.

most significant results obtained to date for CdTe solar cells prepared on soda lime glass substrates. The best device exhibited a conversion efficiency of 13.5% under AM1.5 as measured at the National Renewable Energy Laboratory.

Sample #	213A3	41B21	41B11
Eff., (%)	13.2	13.3	13.5
V_{oc}, (mV)	851	851	850
J_{sc}, (mA/cm^2)	20.6	21.5	21.45
FF	74.9	72.6	73.8

Table V. Solar cell parameters for devices prepared with CSS CdS.

SUMMARY

Considerable advances toward improving the efficiencies of CdTe/CdS solar cells prepared using manufacturing friendly techniques have been made. A conversion efficiency of 13.5% under AM1.5 has been achieved for a CdTe/CdS solar cell prepared on a soda lime glass substrate. Acknowledgements: This work was conducted under the sponsorship of NREL.

REFERENCES

[1] K. Zweibel, H. Ullal, B. G. Roedern, R. Noufi, T Coutts, and M. M. Al-Jassim, Conference Proceedings of the 23rd IEEE PVSC, 1993, p. 379.

[2] C. Ferekides, J. Britt, Y. Ma, and L. Killian, Conference Proceedings of the 23rd IEEE PVSC, 1993, p. 389.

[3] T. L. Chu, S. S. Chu, C. Ferekides, C. Q. Wu, J. Britt, and C. Wang, Conference Proceddings of the 22nd IEEE PVSC, 1991, p. 952.

[4] R. Sundharsanan and A. Rohatgi, Conference Proceedings of the 21st IEEE PVSC, 1990, p. 504.

[5] D. Oman, C. S. Ferekides, and D. L. Morel, to be published.

VAPOR CHLORIDE TREATMENT OF POLYCRYSTALLINE CdTe/CdS FILMS

T.X. Zhou, N. Reiter, R.C. Powell, R. Sasala, P.V. Meyers*
Solar Cells, Inc., Toledo, Ohio 43607

ABSTRACT

Heat treatment with $CdCl_2$ near 400 °C has been known to consistently improve the electronic properties of thin-film CdTe. The treatment process usually involves application and drying of the $CdCl_2$ solution, annealing and residue removal. The conventionally treated films often appear non-uniform with insoluble Cl-containing residues. In order to make the process more suitable to manufacturing alternative methods have been investigated. $CdCl_2$ vapor, hydrogen chloride and chlorine gases have been applied to treat CdTe/CdS films. The vapor $CdCl_2$ and hydrogen chloride treated CdTe surface are found to be uniform with little Cl-related residues and to have electonic properties similar to the conventionally treat films. Total area efficiency of 6.9% has been demonstrated on a $CdCl_2$ vapor treated 7200 cm^2 module.

INTRODUCTION

Heat treatment near 400°C with $CdCl_2$ has been routinely used to significantly improve electronic characteristics of polycrystalline CdTe/CdS films. The $CdCl_2$ treatment usually results in grain growth in small grain films prepared with a number of methods such as electrochemical deposition or vacuum evaporation [1]. In the case of high-temperature vapor deposition such as the close-spaced sublimation (CSS) method and its variations, the as-deposited average grain size is large (~1 μm). Nonetheless the $CdCl_2$ heat treatment is still necessary in order to achieve consistently good device characteristics, i.e., V_{oc} > 750 mV and FF > 60%. There is evidence that in large grain films the treatment promotes grain boundary re-growth and small grain elimination. In either case $CdCl_2$ is believed to act as a flux agent that increases the atomic mobility of CdTe at annealing temperatures. The conventional $CdCl_2$ treatment (hereafter referred to as solution treatment) usually involves the following steps: 1) Application of $CdCl_2$ in methonal solution; 2) Heat treatment in air at about 400 °C for 30 to 60 minutes; 3) Removal of the $CdCl_2$ residue from the treated film. $CdCl_2$ has also been applied by laser-driven physical vapor deposition[2]. Incorporation of these processes may cause substantial delay in mass production. This work reports the investigation at Solar Cells, Inc. (SCI) of alternative methods of post-deposition chloride treatment potentially better suited to a manufacturing process. SEM studies of surface morphology changes and photovoltaic device performance are used to evaluate the effectiveness of the various chloride treatment techniques. All PV devices reported in this work have been fabricated using SCI's standard process with the exception of the chloride treatment.

EXPERIMENTS AND DISCUSSION

Immediate post-deposition treatment by vapor $CdCl_2$

It is known that $CdCl_2$ preserves its linear molecular structure in the vapor phase and exhibits a high vapor pressure as shown in Fig. 1. Thus at 400 °C a significant amount of $CdCl_2$ exist near the CdTe surface in the conventional $CdCl_2$ annealing step. The interaction of $CdCl_2$ vapor with a heated CdTe film may be sufficient to affect the polycrystalline grain structures in a manner similar to the conventional $CdCl_2$ treatment if the adsorbed $CdCl_2$ molecules have a sufficiently long dwell time on the film surface.

Experiments of $CdCl_2$ vapor treatment in air were performed by simultaneously heating the CdTe/CdS film in close proximity to a try of $CdCl_2$ powder. The treatment was conducted in room air for 5 to 30 minutes. The treated films appeared non-uniform and were covered by condensed $CdCl_2$ crystallites. Despite the poor appearance, 1.1 cm^2 cells with greater than 8% efficiencies have been made. This approach is similar to one discussed extensively by Birkmire et al. [3].

The $CdCl_2$ vapor treatment on 100 cm^2 CdTe film was experimented in SCI's multistage CSS deposition system described in a previous publication[4]. The CdS and CdTe were deposited in N_2 ambient at 1 to 10 Torr at superstrate temperature near 600 °C. Immediately following deposition the chamber was back-filled with N_2 to pressures at 1 to 200 Torr while allowing the film to cool to about 400 °C. The film was then inserted into a CSS zone loaded with anhydrous $CdCl_2$ for the treatment. The upper and lower heaters of the $CdCl_2$ zone were set between 420 and 450 °C to maintain a sufficient $CdCl_2$

*Current Address: Institute of Energy Conversion, University of Delaware, Newark, DE 19716

$4.00 © 1994 IEEE
First WCPEC; Dec. 5-9, 1994; Hawaii

vapor pressure. The treatment lasted between 5 to 20 minutes. If the treated film is allowed to cool quickly, fine grains of $CdCl_2$ in the form of a dark powder remain on the surface and can be easily rinsed off with de-ionized water. It is more convenient, however, to remove the $CdCl_2$ residue by evaporation in the preheat zone under modest vacuum for several minutes.

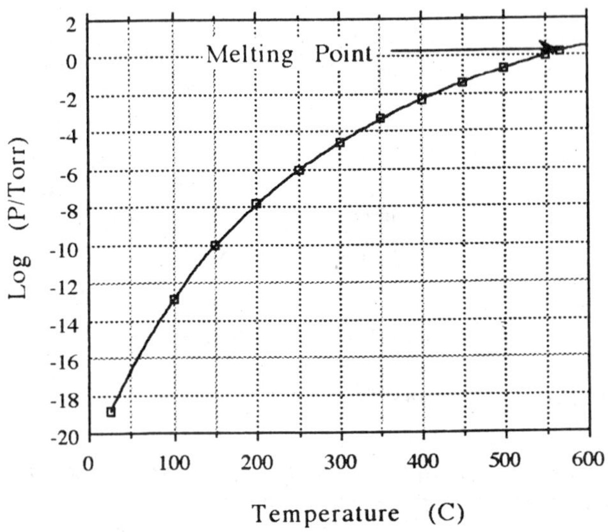

Fig. 1 $CdCl_2$ vapor pressure.

The $CdCl_2$-vapor treated (hereafter referred to as "vapor treated") CdTe films appear different from the conventionally solution treated ones. The vapor treated films appear very uniform whereas the solution treated films usually exhibit swirl marks formed during the drying of the solvent. SEM studies show subtle but definite difference between the vapor and solution treated samples (see Fig.2). In general the vapor treatment leaves smoother compact grains and lower chlorine residue (undetectable by EDS). On the solution treated surface, however, certain fine structures are often found to be Cl-containing. It is important to note that the conventional $CdCl_2$ treatment is usually conducted in oxygen and moisture containing environment. Several chlorine and oxygen related compounds on the CdTe surface after $CdCl_2$ heat treatment in room air have been identified by X-ray analysis [5]. It has been observed that a CdTe film exposed to room air prior to the vapor treatment would appear similar to a conventionally treated film in terms of surface inhomogeneity. This suggests that surface oxides may have an latent effect which develops during the vapor $CdCl_2$ treatment. Clearly the immediate post-deposition treatment significantly reduces the possibility of contamination on as-deposit films.

Table 1 lists typical vapor treatment parameters along with small-area cell performance measured under simulated AM1.5 illumination. The listed average efficiency is obtained from six 1.1 cm^2 cells. It has been found that treatment time longer than 5 minutes and ambient pressure higher than 10 Torr tend to result in best and most consistent device performance.

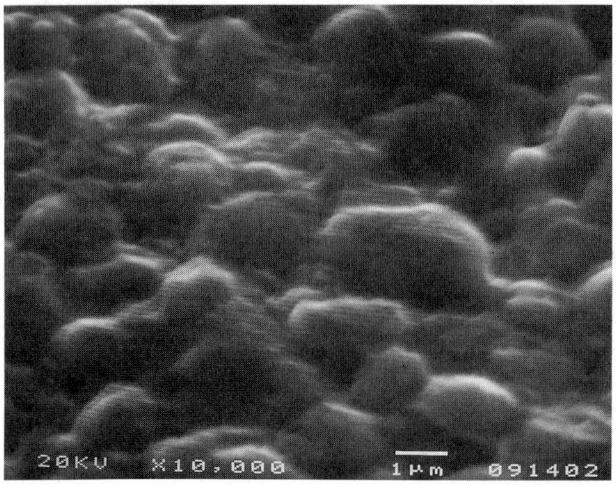

(a)

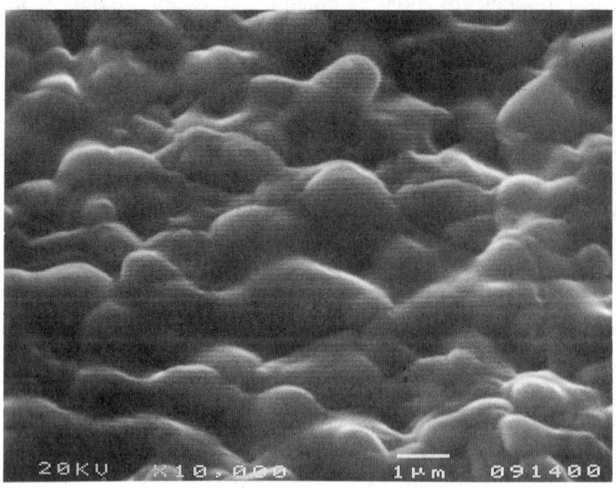

(b)

Fig. 2 SEM photograph of a) solution $CdCl_2$ treated, b) vapor $CdCl_2$ treated CdTe surface.

Table 1. Representative data of vapor-$CdCl_2$ treatment experiments conducted immediately following the CSS deposition of CdTe.

Sample	T(heater) (°C)	T(sample) (°C)	P (Torr)	t (min.)	Avg. Eff (%)
N2017	440/450	396	200	10	10.2
N2018	440/450	397	200	5	10.0
N2059	400/440	377	10	4	9.3
N2033*	440/440	398	200	10	10.7
N2031	440/440	392	1	2	8.4
N2043	400/460	385	1	1	8.3

* CdTe deposition was in air at 10 Torr.

Vapor $CdCl_2$ treatment of 7200-cm^2 films

The vapor $CdCl_2$ treatment has also been applied to 7200 cm^2 CdTe films in the large-area deposition system described in Ref. 6 . Due to equipment limitations the CdS/CdTe depositions and the vapor treatment were not conducted in a single vacuum cycle. To perform the vapor treatment the semi-cylindrical boats normally used for CdS and CdTe source materials were loaded with powder $CdCl_2$. A pre-deposited CdS/CdTe film, re-heated to ~400 °C, was slowly oscillated under the $CdCl_2$ boats. The treatment was conducted at different N_2 ambient pressures for several minutes. Three 7200 cm^2 modules with total area efficiencies between 6.7% and 6.9% have been made with this process (See Table 2). In this case the ambient pressure did not seem to be a critical parameter. It must be pointed out that the effective anneal time may be ~3 times longer than indicated in the table due to a $CdCl_2$ dwell time longer than the period of oscillation.

Table 2. Vapor $CdCl_2$ treated 7200 cm^2 modules

Sample	T(heater) (C)	T(sample) (C)	P (Torr)	t (min.)	Total Area Eff.
B9418	450/430	400	200	3.5	6.8%
B9419	450/430	400	10	2	6.9%
B9421	450/420	400	1	1.1	6.7%

Vapor treatment by other chlorides

In both conventional and vapor treatments $CdCl_2$ is applied to the film surface. However, the amount of $CdCl_2$ needed for an effective treatment is low, usually no more than 2 molar percent of the CdTe. It is also possible to form a $CdCl_2$ layer by reaction of chlorine (Cl_2) or certain chloride vapors with the CdTe surface. It has been observed that even at room temperature HCl vapor renders the CdTe surface hydrophobic and electrically more conductive. Although an aqueous HCl solution does not etch CdTe, a certain amount of Cd leaching is believed to occur, resulting in a slightly Te-rich surface. The first experiments to investigate the interaction of HCl and CdTe were performed in a simple apparatus involving a Pyrex tube and a heater. A flow of N_2 was established in the tube until the sample temperature reached 400 °C. The gas flow was then diverted through a bubbler containing a concentrated HCl solution. The HCl:N_2 flow was maintained at about 500 sccm for < 5 minutes. The sample was then allowed to cool in N_2 ambient. After the reaction, the film was covered with powdered $CdCl_2$ residue similar to that on the $CdCl_2$ vapor treated films. The small cell (0.25 cm^2) average efficiency was 8.4% with V_{oc} greater than 750 mV. These devices are considerably better than those made on untreated films.

Production-scale HCl treatment experiments were performed in the large-area deposition system. A pre-deposited 7200 cm^2 CdTe/CdS film was heated to 400 °C in N_2. A flow of premixed HCl:N_2 was bled into the chamber through a shower head gas distributor approximately 1 cm above the film surface. The film was oscillated with respect to the gas distributor which was held at a temperature slightly higher than the film temperature. The effective time that a particular section of film was directly in the stream of the incoming HCl flow was < 10 minutes. As shown in Fig. 3 the treated film appeared to be homogeneous with little or no Cl residue detectable by EDS. Small cells with efficiencies above 10% have been repeatedly produced from HCl-treated CdTe 7200 cm^2 films. Satisfactory spatial uniformity of treatment has been indicated by an average 48-cm^2 aperture-area efficiency of 8.4% (± 0.4%) over 59 minimodules randomly selected from a single 7200 cm^2 film.

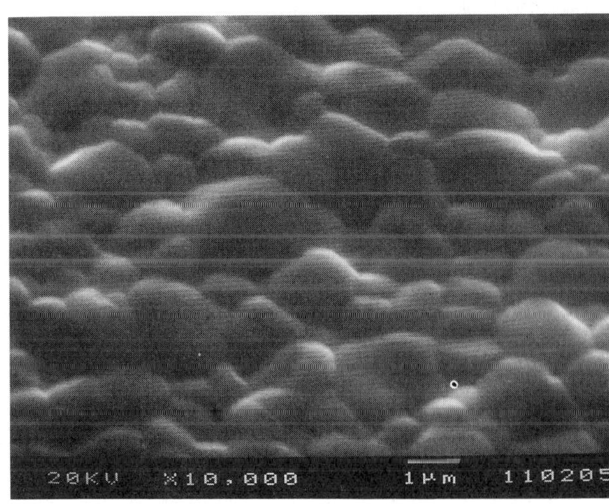

Fig. 3 SEM micrograph of CdTe surface after a heat treatment in HCl:N_2.

A dilute mixture of Cl_2 with N_2 has also been used to treat the CdTe/CdS film in a similar manner. The Cl_2-

treated film appeared non-uniform. Isolated area of films, however, have been made into cells with efficiency in the range of 9 - 10%. Apparently a considerable amount of $CdCl_2$ was formed during the treatment. This was expected from thermodynamic calculations which predict a stronger reaction of CdTe with Cl_2 than with HCl (elemental halogens and their solutions are known to etch CdTe at room temperature). While reaction control may be difficult, chlorine remains a viable candidate as a processing gas for the post-deposition treatment of CdTe. Further experiments have been planned to optimize chlorine or chloride gas treatment to fulfill the standard production requirements such as yield, throughput and safety.

Capacitance-Voltage (CV) and Quantum Efficiency (QE) have been measured on samples processed with the various chloride treatments with all other process steps being identical. For cells with similar efficiencies and I-V characteristics, there are no significant differences in the CV or QE curves for the solution, vapor or HCl treated samples. From the slope of the CV curves, the carrier density is determined to be in the low 10^{14} cm^{-3} range throughout the majority of the CdTe and increases slightly near the back contact. Photoluminesence (PL) measurements have been performed on representative samples from each treatment and show only slight differences. The films are all good quality and have carrier lifetimes of 1-3 ns.

Since PL, CV and QE measure average values for a cell, any microscopic differences that may be present are not detectable. There is, however, evidence from accelerated tests which suggests that there may be some critical microscopic parameters that are different between the solution treated and the vapor or HCl treated samples that might significantly increase the stability. Further work along these lines is needed to solidify this speculation.

SUMMARY

Post deposition heat treatment of CdTe films with $CdCl_2$ vapor has been demonstrated equivalent to the conventional solution $CdCl_2$ treatment in its enhancement of electronic properties of the film. 7200 cm^2 modules have been made on vapor treated films with efficiencies comparable to the conventionally treated films. Chlorine and hydrogen chloride have been effectively applied to form $CdCl_2$ on the CdTe surface during the heat treatment to improve the electronic properties. The vapor treatment generally results in a more uniform surface with less surface contamination and impurities. Effects of more uniform and cleaner film surface on long term device stability need be further studied. Optimization of the alternative chloride treatments is planned to better serve the purpose of low-cost high-throughput manufacturing of CdTe PV modules.

ACKNOWLEDGEMENTS

The authors gratefully acknowledge Mike Steele for his technical assistance. The assistance and support of the entire SCI team are sincerely appreciated. This work has been partially supported by the NREL subcontract ZR-1-11059-1.

REFERENCES

1 B. Baron et al, "Polycrystalline Thin Film Materials and Devices", Annual Report to SERI under subcontract No. XN-0-10023-1, 1991.

2 A. Compaan, *AIP Conference Proc.* Vol. **268**, 255 (1992).

3. R. Birkmire et al, "Polycrystalline Thin Film Materials and Devices", Final Report to NREL under subcontract No. XN-0-10023-1. 1993.

4. P.V. Meyers, *AIP Conference Proc.* Vol. **268**, 250 (1992).

5. B. Baron et al, "Polycrystalline Thin Film Materials and Devices", Annual Report to SERI under subcontract No. XN-0-10023-1, 1991.

6 . J. Nolan, *23rd IEEE PVSC Proc.*, (1993) pp. 37-41.

A TREATMENT TO ALLOW CONTACTING CdTe WITH DIFFERENT CONDUCTORS

B.E. McCandless, Y. Qu, and R. W. Birkmire

Institute of Energy Conversion
University of Delaware

United States Department of Energy
University Center of Excellence
For Photovoltaics Research and Education
(National Renewable Energy Laboratory)
Newark, Delaware 19716 USA

ABSTRACT

A new method is presented for forming low resistance contacts to the CdTe surface of CdTe/CdS thin-film solar cells. After $CdCl_2$ processing, the CdTe surface is coated with a thin layer of copper which is heated and then reacted. The heat treatment diffuses Cu into CdTe, doping it p-type while the reaction removes elemental Cu and produces a conductive surface which is easily contacted using different materials. This process allows great flexibility in selection of a back contact material. High efficiency devices having fill factors up to 77% were fabricated. The process is applicable to CdTe/CdS devices having CdTe deposited by different techniques and is compatible with module fabrication technologies.

INTRODUCTION

CdTe polycrystalline thin-film solar cells are promising for terrestrial application because of high conversion efficiency and potentially low cost. Cells with efficiency higher than 11.0% have been produced using a variety of deposition techniques such as close-spaced sublimation, electrodeposition, physical evaporation, and spray pyrolysis. In many of the reported devices the maximum power is limited in part by the series resistance which limits fill factor [1]. Achieving low device series resistance requires controlling the conductivity of the CdTe layer and forming a non-rectifying contact. Formation of low resistance contacts to CdTe has been facilitated either by direct formation of a p^+ layer such as CdTe:Te [2], Cu_2Te [3], ZnTe:Cu [4], or HgTe [5] or by processes in which the contact materials such as Cu/Au layers [6,7,8,9] or carbon paste containing Hg or Cu salts [10,11,12] are applied to a suitably prepared surface and are heated. The basis for these approaches has been extensively reviewed [13,14,15]. Devices with contacts containing elemental Cu, such as Cu/Au, have achieved fill factors over 70% but are not always stable, and the mechanism for instability may be linked to chemical diffusion processes that are quite active at room temperature [16,17]. Furthermore, fabrication processes using Cu/Au contacts couple the CdTe doping with contact formation [18] which restricts the latitude for independent control over CdTe and contact properties. Devices with paste contacts containing Hg compounds have achieved fill factors up to 75% [10] but may suffer from issues related to large scale commercialization such as toxicity and monolithic integration for module fabrication. We have developed a process, hereafter referred to as the diffused Cu process, that is compatible with module fabrication, separates the doping and contacting steps, leaves no elemental Cu in the finished device, is not sensitive to CdTe deposition method, and has demonstrated devices with high V_{oc} and fill factors exceeding 75%.

EXPERIMENTAL

All of the devices were fabricated in a superstrate configuration on glass coated with a transparent conductive oxide (TCO). Contact development was carried out on CdTe/CdS films deposited by physical vapor deposition (PVD) and were compared to devices made by the baseline process with Cu/Au contacts. The general applicability of the contacting process to devices made by other methods was tested using CdTe/CdS films fabricated at other laboratories by: elemental vapor deposition (EVD) at Solar Cells, Incorporated; close-spaced sublimation (CSS) at the University of South Florida; and metal organic chemical vapor deposition (MOCVD) at Georgia Institute of Technology. The CdTe and CdS film thicknesses ranged from 2-6 μm and 0.15-0.25 μm, respectively. After deposition the CdTe/CdS/TCO structures were coated with $CdCl_2$ and heat treated at ~400°C as described elsewhere [6].

The diffused Cu contact process consists of deposition of a Cu layer on the CdTe surface, heat treatment of the entire structure, etching, and application of the desired current-carrying contact material. Cu was deposited by electron beam evaporation to a thickness of 5 nm to 20 nm at a rate of 0.2 nm/sec at a base pressure of 1×10^{-6} torr. Heat treatment was carried out in a tube furnace at 150°C to 200°C for 15 to 60 minutes with flowing argon to prevent oxidation of the Cu layer. Next the

CH3365-4/94/0000-0107 $4.00 © 1994 IEEE

First WCPEC; Dec. 5–9, 1994; Hawaii

structure was immersed in 0.004 mol % bromine-methanol solution (Br_2:CH_3OH) for 5-10 sec to remove metallic Cu. The action of the Br_2:CH_3OH on metallic Cu layers was verified by reaction of 10 nm thick Cu on Corning 7059 glass in 0.004 mol % Br_2:CH_3OH. The Cu films completely dissolved in 1-3 sec. The etch rate of CdTe in this strength solution is 0.3 μm/min, determined by removal time of 2-4 μm thick CdTe films on CdS/ITO/glass, and is estimated to remove 20-40 nm of CdTe during the etch step.

After rinsing in methanol the samples were contacted with the different current carrying conductors such as Ni, Cr, Pt, Mo, carbon paste, and indium-tin oxide (ITO). No subsequent treatments were necessary to optimize devices. Cell areas up to 1 cm^2 were prepared by scribing through the entire structure down to the TCO. Indium solder was applied to the TCO to provide a low resistance contact point to the emitter side of the device.

Devices were evaluated by current-voltage, spectral response, and capacitance-voltage analysis techniques and were compared to devices fabricated with Cu/Au contacts as described elsewhere [18]. A set of samples was fabricated with dot contacts on PVD and EVD samples, evaporated through an aperture mask, to leave the Cu-treated surface exposed for XPS and SIMS analysis to determine both surface composition and Cu concentration in CdTe.

RESULTS

Devices

The current-voltage (J-V) parameters obtained using the diffused Cu process and different conductors and CdTe deposition techniques are shown in Table 1. For the devices made with PVD CdTe, the performance of Cu/Au contact is shown for comparison. The diffused Cu contact yields similar performance to that obtained with the baseline Cu/Au contact for PVD devices and to that obtained by the contacts of the respective groups.

Table 1. J-V parameters of CdTe/CdS devices made with the diffused Cu contact process using the conductors and CdTe deposition techniques shown.

CdTe Method (Group)	Conduct-tor	Voc (mV)	Jsc (mA/ cm^2)	FF (%)	Eff (%)
PVD	Cu/Au	790	20.1	69	11.0
PVD	Cr	811	20.7	66	11.1
(IEC)	Mo	821	22.4	63	11.6
	Ni	775	21.3	71	11.7
	Pt	795	19.6	75	11.7
	C	830	19.4	73	11.8
EVD	Ni	804	18.0	77	11.0
(SCI)	C	828	18.1	74	11.1
CSS	Ni	792	22.4	73	12.9
(USF)	C	864	22.4	73	14.1
CVD (GIT)	C	757	20.6	61	9.5

Figure 1 is a light J-V plot of a 1 cm^2 device with EVD CdTe deposited at Solar Cells, Inc. and processed at IEC with the diffused Cu contact and carbon. The performance of this device is the same as the best devices made by Solar Cells, Inc.. Other devices, with Ni conductors, have been measured with fill factors as high as 77%, V_{OC} > 800 mV and resistance at V_{OC} of only 2 Ω-cm^2. These fill factors are as high as can expected from CdTe devices with A-factors of 2.

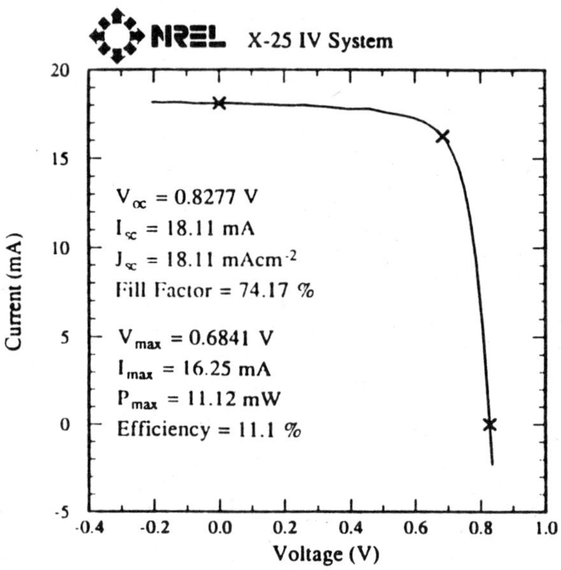

Figure 1. J-V curve of a CdTe/CdS device with EVD CdTe (Solar Cells, Inc.) using the Cu diffused process.

No systematic dependence of device parameters or efficiency with contact material was found, indicating no influence of contact material work function. This together with the low resistances measured suggests that the etch produces a conductive surface on the CdTe which acts as the primary contact to CdTe; its presence allows for great flexibility in choice of contact material. A sample made with indium-tin oxide conductor demonstrates that truly novel device designs, such as transparent cells, can utilize this contacting approach.

Capacitance-voltage measurements of devices made with the diffused Cu process were compared to those of devices made with Cu/Au contacts. Measurements were made at 100 kHz from -2 to 0 volts. $1/C^2$ versus voltage plots for a device with Cu/Au contact and with the diffused Cu process and a chrome conductor are shown in Figure 2. The device with Cu/Au contact exhibits non-linear behavior of $1/C^2$ versus voltage, indicating either a changing carrier concentration or the presence of a second capacitor in series. The linear portion of the curve represents an acceptor concentration of 2.4×10^{15} cm^{-3}. The device made with the diffused Cu process exhibits linear behavior over the same voltage range, indicating uniform carrier concentration of 2.4×10^{15} cm^{-3} near the junction.

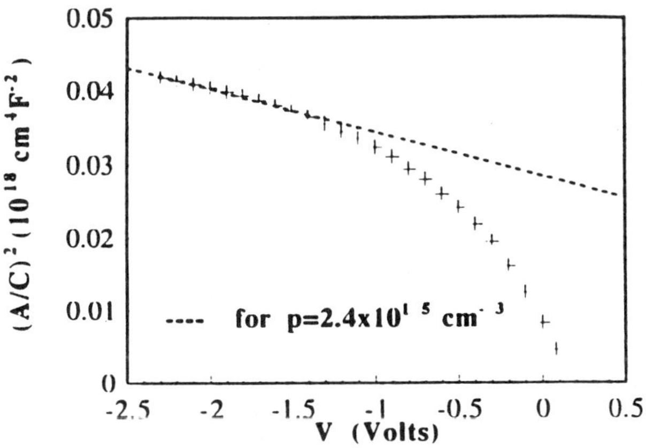

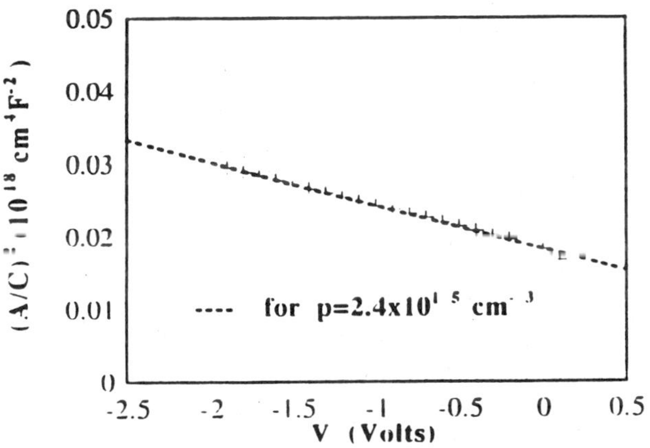

Figure 2. $1/C^2$ versus voltage curves for CdTe/CdS devices made with Cu/Au contact (top) and the diffused Cu process with chrome (bottom).

The sensitivity of this process to its individual steps was tested with respect to Cu layer thickness, heat treatment temperature and time, and etch time. No difference in device performance was obtained using 5 nm, 10 nm, and 20 nm of Cu, leading to the target thickness of 10 nm used for all other evaluations. Cu in excess of 50 , however, leads to shunting behavior. Devices were fabricated with no Cu using Ni, Au, and carbon paste contacts on PVD and EVD films that were Br_2:CH_3OH etched after $CdCl_2$ treatment. In all cases, the J-V characteristic exhibited low V_{OC} and severe double diode behavior in the forward bias portion of the light J-V curves. The low V_{OC} is attributed to low doping concentration in the CdTe while the non-ohmic forward bias characteristic arises from rectification between CdTe and the back contact materials.

It was necessary to optimize the treatment temperature and time for the CdTe thickness employed. In particular, insufficient treatment temperature and time yields low V_{OC} and suggests that the Cu plays a role in controlling the V_{OC} by modifying the CdTe conductivity. For 2 µm thick CdTe, 150°C for 60 minutes was found to be sufficient, while 5-7 µm thick films required 180°C for 30 minutes. This result is qualitatively similar to that found for devices with Cu/Au [19] and ZnTe:Cu [20] contacts, i.e., optimum Voc is obtained when sufficient Cu is diffused into CdTe.

The diffused Cu process was not found to be very sensitive to etch time in the range of 5 sec to 1 minute, yielding linear J-V behavior in forward bias for etch times up to 1 minute, which removed ~0.3 µm of CdTe. Thus, the etch time of 5 sec was selected on the basis of removal of the Cu metal layer.

Chemical Interpretation

EDS measurements of device structures after etching do not reveal the presence of Cu or excess surface Te, within the respective detection limits, which implies that any residual surface layer that might be produced during etching is less than 20 nm in thickness. XPS revealed a quantity of Cu near the detection limit [21]. SIMS analysis to determine copper distribution is in progress.

We speculate that the reaction with Br_2:CH_3OH forms a conductive layer on the CdTe consisting of either elemental tellurium or copper telluride of composition Cu_xTe, where x lies between 1 and 2. The reaction to form elemental Te is:

$$CdTe + Br_2 - Te + CdBr_2$$

The reaction to form Cu_xTe may occur in a single step as:

$$CdTe + xCu + Br_2 = Cu_xTe + CdBr_2$$

or via three step reaction as:

$$2Cu + Br_2 = 2CuBr$$

$$CdTe + Br_2 = CdBr_2 + Te$$

$$xCuBr + Te = Cu_xTe + (x/2)Br_2$$

In either case, the reaction product $CdBr_2$ is soluble in methanol and is expected to remain dissolved given the very low quantity of Cu and Cd reacted. The presence of a beneficial surface layer was tested by reacting the surface further prior to contact deposition. Samples were reacted in 0.2 M KCN at 50°C for 10 sec after the Br_2:CH_3OH etch and were contacted with carbon paste dots. KCN solution dissolves copper compounds of the type Cu_2S, Cu_2Se, and Cu_2Te as well as elemental tellurium while not reacting strongly with CdTe nor with Cu. The devices with KCN etched surfaces exhibited a strong double diode characteristic in forward bias, indicative of a blocking contact between the carbon paste and the Cu-doped CdTe suggesting that the beneficial layer formed by Br_2:CH_3OH was removed by the KCN treatment.

CONCLUSIONS

A contacting process has been developed that is compatible with module fabrication and which has been demonstrated on devices having CdTe made by several methods. The process discretely separates Cu diffusion doping of the CdTe layer from the formation of a conductive surface, permitting independent control of each step. Devices fabricated with the diffused Cu process contain no elemental Cu layer in the finished device which may reduce chemically-induced instability. The process is applicable to CdTe devices made by different methods, and CdTe devices have been demonstrated in which the device performance is not limited by series resistance. V_{oc} and bulk series resistance, which are related to doping of the CdTe layer, can be controlled by the treatment time. Capacitance measurements of optimized devices show that the CdTe layer is uniformly doped near the junction. Fill factors not limited by contact resistance can be achieved by using a suitable etchant which may promote Te or Cu_xTe formation. With this contact and doping approach, the choice of contact materials can be based on manufacturing issues rather than on CdTe requirements.

ACKNOWLEDGEMENTS

The authors would like to acknowledge James E. Phillips for device analysis and many helpful discussions. Evaluation of the process on CdTe devices made by other methods was made possible by Teddy Zhou of Solar Cells Incorporated who provided EVD samples, Chris Ferekides of the University of South Florida who provided CSS samples, and Sachin Kamra and Ajeet Rohatgi of Georgia Institute of Technology who provided MOCVD samples. XPS and SIMS analyses were performed by David Niles and Sally Asher, respectively, both of the National Renewable Energy Laboratory in Golden, Colorado. This work was sponsored by the U.S. Department of Energy's National Renewable Energy Laboratory under subcontract number X-AV-13170-01.

REFERENCES

1. R. A. Sasala, X. X. Liu and J. R. Sites, *Int. J. Solar Energy*, 12 (1992) p. 17.
2. L.F. Szabo and W.J. Biter, U.S. Patent 4,735,662 (1988).
3. N. Nakayama, H. Matsumoto, K. Yamaguchi, S. Ikegami, and Y. Hioki, *Japan. J. Appl. Phys*, 15 (1976) 11, p. 2281.
4. A. Mondal, B. E. McCandless and R. W. Birkmire, *Solar Cells*, 26 (1992) p. 181.
5. T. L. Chu, S. S. Chu, K. D. Han and M. Mantravadi, *Proc. IEEE PVSC*,(1988) p. 1422.
6. R. W. Birkmire, B. E. McCandless and S. S. Hegedus, I*nt. J. Solar Energy*, 12 (1992) p.145.
7. J. Skarp, Y. Koskinen, S. Lindfors, A. Rautianen, and T. Suntola, *Proc. 10th European PVSC* (1991).
8. G. C. Morris, P. G. Tanner, and A. Tottszer, Proc. *21st IEEE PVSC*, (1990) p. 575.
9. A. Rohatgi, I*nt'l Jour. Solar Energy*, 12 (1992) p. 37.
10. J. Britt and C. Ferekides, *Appl. Phys. Lett. 62* (22) (1993) p. 2851.
11. J. W. Park, B. T. Ahn, and H. B. Im, *J. Electrochem. Soc.*, 139 (1991) 11, p. 3351.
12. D.R. Johnson, European Patent Application 92310170.3 (1992).
13. T. C. Anthony, A. L. Fahrenbruch, and R. H. Bube, *J. Elect. Materials*, 11 (1982) p. 89.
14. E. Janik and R. Triboulet, *J. Phys. D: Appl. Phys.*, 16 (1983) pp. 2333-2340.
15. J. P. Ponpon, *Solid-State Electronics*, 28, No 7 (1985), pp. 689-706.
16. C.-A. Chang, *Appl. Phys. Lett.* 55(1989) 26, 2754.
17.W.-Y. Lin, C. Wei, and K. Rajeshwar, *J. Appl., Phys.* 76 (1994) 7, p. 4145.
18. B. E. McCandless and R. W. Birkmire, *Solar Cells*, 31 (1991) p. 527.
19. Annual Report to National Renewable Energy Lab, Subcontract No. X-AV-13170-01 (1994) 90.
20. Final Report to National Renewable Energy Laboratory, Subcontract No. XM-0-18110-2 (1992) p. 32.
21. David Niles, National Renewable Energy Laboratory (personal communication).

RF SPUTTERED CdS/CdTe SOLAR CELLS: EFFECTS OF MAGNETIC FIELD, RF POWER, TARGET MORPHOLOGY, AND SUBSTRATE TEMPERATURE

A.D. Compaan, M. Shao, C.N. Tabory, Z. Feng, A. Fischer, I. Matulionis, and R.G. Bohn
Department of Physics and Astronomy, The University of Toledo
Toledo, OH 43606

ABSTRACT

In this paper we present a study of the influence of substrate temperature, rf power, target erosion, and magnetic field configuration on RF sputtering of CdTe and CdS. These sputtering parameters are shown to affect deposition rate, film morphology, photoluminescence efficiency, and cell performance. The magnetic field shape and strength affects the charged particle flux on the growing film and appears to have a strong influence on the final cell performance.

BACKGROUND

We have shown that rf planar magnetron sputtering of both CdS and CdTe can produce CdS/CdTe solar cells on soda-lime glass with efficiencies above 10% [1]. However, the process of rf magnetron sputtering, especially for semiconductor film growth, is not yet well understood. Film growth rate, morphology, and electrical quality depend on a number of parameters, some of which we have studied in this work. We present here the results of growth rate vs. substrate temperature, target condition, and rf power. In addition we present some results indicating a strong dependence of film quality on the magnetic field structure of the magnetron gun.

The results reported here were obtained with two different 50 mm diameter planar magnetron sputtering guns. One gun had an approximately balanced magnetic field [2] and the other had a strongly unbalanced magnetic field [3]. Substrates of borosilicate glass (Corning 7059 or NEG 0A2) or SnO_2-coated soda lime glass (LOF 8 or 10 Ω/square) were held at a distance of 70 mm and heated radiatively with a tantalum wire heater. Glass temperature was checked by ruby fluorescence [4] and by optical interference fringe counting [5]. A sketch of the system showing the magnetic field distributions of the two guns is shown in Fig. 1.

OPTICAL EMISSION SPECTROSCOPY AND PLASMA PROPERTIES

The strength and configuration of the magnetic fields of the two sputtering guns has a significant effect on the electron confinement and therefore also on the degree of excitation and ionization in the plasma at some distance from the sputter guns. This is illustrated by the plot of emission intensity of the neutral Cd emission at 508.6 nm shown in Fig. 1, Panel c. In addition to the optical emission measurements, other characteristics of the plasmas have been measured with a Langmuir probe placed 1 cm from the substrate on the center line of the substrate-target system. The probe consisted of a 0.13 mm diameter tantalum wire with 5 mm exposed beyond the end of a glass capillary sheath. A summary of the plasma characteristics is given in Table 1. Note that the unbalanced magnetron has higher electron density and temperature, higher ion flux onto the substrate, and somewhat higher V_p-V_{sb}, the acceleration potential of ions across the substrate sheath for the case of a substrate at floating (self-bias) potential. In addition to the higher electron flux, note that the unbalanced gun yields a higher ratio of positive-ion-to-neutral Cd+S flux. The data presented below show that these differences in plasma characteristics also affect the quality of the as-grown CdS and CdTe films.

SUBSTRATE TEMPERATURE DEPENDENCE

Increasing substrate temperature provides increased mobility for surface atoms and generally improves film quality. However, higher temperature also increases the sublimation of atoms from the film surface and slows the growth rate. The net growth rate is a balance between the incoming flux and the sublimation flux. Fig. 2 shows the rate for CdTe film growth for a fresh two-inch diameter CdTe sputtering target with a planar surface and for a deeply eroded target with an erosion track approximately 1/4 inch deep [4]. The curves in Fig. 2 are fits of the data to the functional form [6]

$$\text{growth rate} = A - BT^2\sinh(h\nu/2kT)\exp(-e\Phi/kT), \qquad (1)$$

where, for 60 W of rf power into the gun and 18 mTorr of Ar pressure, A = 26 and 48 nm/min for the fresh and deeply eroded targets respectively, B = 4600 nm/(min-K^2), ν = CdTe optical phonon frequency = 4.8 x 10^{12} Hz, T is the substrate temperature, and Φ = 0.73 eV. The difference between the fresh (planar) target and the old (eroded) target we believe to be due to the increased surface area of the target directly under the magnetron annulus and the oblique incident angles of Ar^+ sputter ions in this region.

CH3365-4/94/0000-0111 $4.00 © 1994 IEEE

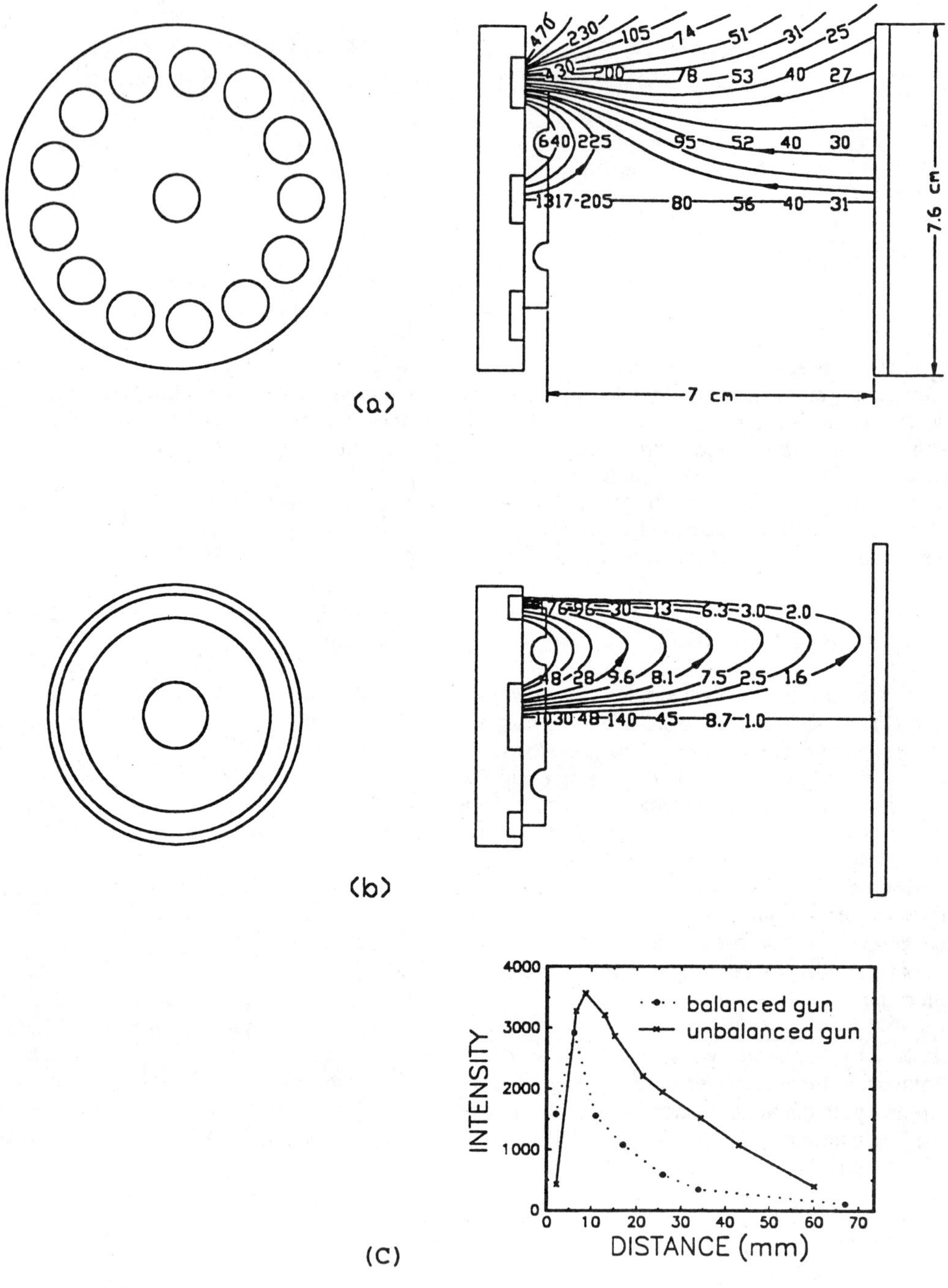

Fig. 1. Magnetic field structure of (a) unbalanced and (b) balanced field planar magnetron sputter guns used in this work. Left side shows top view of sputter gun head with central magnet and a ring of magnets for the unbalanced gun (a) and an annular magnet for the balanced gun (b). Right side shows side views of sputter gun - substrate arrangements. Magnetic field strengths, measured with a Bell model 615 gaussmeter, are shown in Gauss. Panel (c) shows the optical emission signal, obtained with a PAR OMA-II vidicon and 1/4 meter Aries spectrometer, for the CdI line at 508.6 nm at 18 mTorr and 60 W of rf power.

Table 1. Unbalanced vs. Balanced Magnetron Characteristics (P_{RF} = 75 W, p_{Ar} = 18 mT, CdS Target)	UNBAL.	BALANCED
substr. self bias V_{sb}	-2.2 V	+8 V
plasma potential, V_p	+19 V	+26 V
substr. sheath, V_p-V_{sb}	+21 V	+18 V
ion flux, F_I ($cm^{-2}sec^{-1}$)	$1.8x10^{15}$	$1.0x10^{15}$
deposition flux, F_{Cd+S} ($cm^{-2}sec^{-1}$)	$2.4x10^{15}$	$2.6x10^{15}$
F_I/F_{Cd+S}	0.75	0.4
electron temp., T_e	5.3 eV	4.0 eV
electron density, n_e (cm^3)	$7.6x10^{10}$	$3.9x10^{10}$

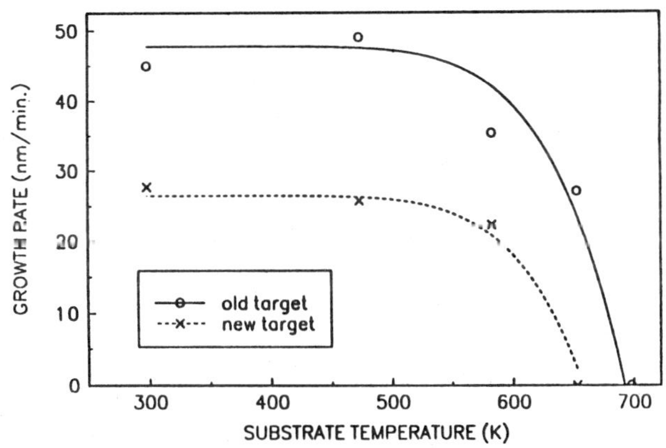

Fig. 2. Growth rate of CdTe films vs. substrate temperature for p=18 mTorr, P_{rf}=60W, unbalanced gun. Curves according to Eq. 1.

RF POWER DEPENDENCE

The growth rate of CdTe vs. rf power for a substrate temperature of 310 C and argon pressure of 18 mT is shown in Fig. 3 for the two target conditions. We have directly measured the negative D.C. self-bias potential which develops between the sputter target and ground (chamber walls). Thus we find that the sputtering rate is approximately linear in the self-bias potential above a threshold power of about 35 W corresponding to ~80 V self-bias potential. At 60 W and 100 W of rf power, the self-bias potentials are ~130 V and ~170 V respectively.

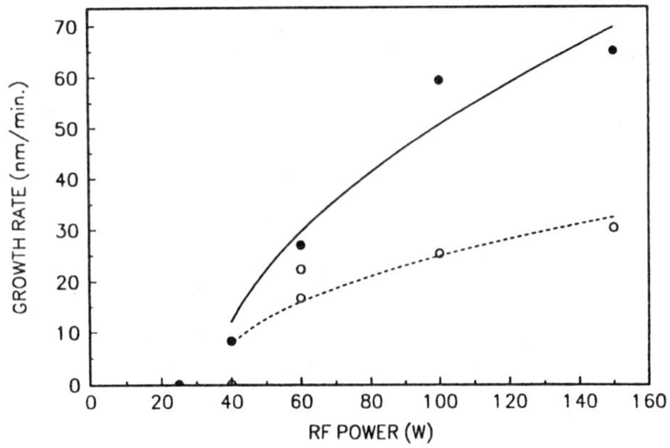

Fig. 3. Growth rate of CdTe films vs. rf power for T_s=310 C and p=18mTorr; •/o=fresh/eroded target; unbalanced gun. Curves show a square root dependence on rf power above a threshold of 35 W.

MAGNETIC FIELD DEPENDENCE

Plasma properties

In the planar magnetron sputtering gun, the presence of a static magnetic field is used to modify the electron trajectories to enhance the ionization near the target, thus increasing deposition rates and permitting lower sputter gas pressures. In addition, the magnetic field may alter the electron flux incident on the growing film. (The magnetic field of the magnetron gun has a strong influence on the electron trajectories but little effect on the ion motions.) In many sputtering applications, mechanical properties such as adhesion and hardness or optical properties are the critical parameters of the deposited films. [7]. Generally the electrical properties of sputtered films have not been studied extensively, particularly for the case of semiconductors. However, in the use of sputtering for the deposition of photovoltaic films, optimizing the electrical properties is critical for achieving highest cell performance. Sputtering is characterized by significant densities of ions and electrons (~5 x $10^{10}/cm^3$ in the background of ~4 x $10^{14}/cm^3$ of neutrals at 18 mT); however, their influence on the electrical properties of the film is not well understood.

In order to examine the effect of the different plasma characteristics on the film quality, we compared the electrical and optical properties of individual films of CdS and CdTe grown with the two sputter guns having the magnetic field characteristics shown in Fig. 1.

Photoluminescence

Fig. 4 shows a comparison between the photoluminescence (PL) observed from CdS films grown with these two sputtering guns but with otherwise identical growth parameters. The growth rates for the two guns were nearly the same. The CdS film grown with the unbalanced gun exhibits approximately 1000 times stronger photoluminescence than the film grown with the balanced magnetron. Films grown with the unbalanced gun exhibited strong near-band-edge PL at 500 nm indicating relatively few deep traps, whereas films grown with the balanced gun show very weak PL from a band of radiative centers relatively deep in the gap (~250 meV). The weakness of the PL indicates a high density of nonradiative recombination centers. We believe that the improved quality of films grown with the unbalanced magnetron is due to the enhanced flux of low energy electrons, ions, and/or excited atoms onto the substrate during growth with the unbalanced magnetron.

Studies of CdTe prepared with the two different sputter guns indicate effects on the PL similar to those described above for the CdS films.

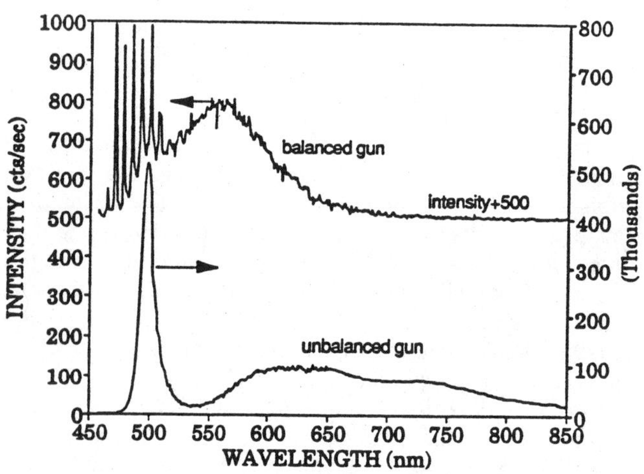

Fig. 4. PL at 77K from two CdS films grown with balanced and unbalanced magnetrons; T_s = 385C, p = 18 mTorr, laser excitation at 458 nm. Sharp line structure in upper trace is Raman scattering. This is masked in the lower trace by the very strong PL.

Cell performance

We have prepared solar cells using combinations of these two types sputter guns and find preliminary indications that cell performance is seriously degraded when the CdTe layer is deposited with the balanced magnetron gun. The cell performance is less seriously affected when the CdS layer is grown with the balanced gun. These differences are consistent with the poor PL observed in as-grown films from the balanced gun. In typical CdS/CdTe cells, electron beam induced current (EBIC) measurements [1] show that most of the carrier collection arises from the CdTe layer. Not surprisingly, depositing high quality CdTe, with long minority carrier lifetimes, is critical to obtaining high cell performance.

ACKNOWLEDGMENTS

The assistance of J. Ughrin and C. Narayanswami for some of the data acquisition and discussions with Y. Rajakarunanayake are acknowledged with pleasure. The support of NREL throughout this work is gratefully acknowledged (contract nos. ZN-1-19019-3 and ZAX-4-14013-01).

REFERENCES

[1] A.D. Compaan, et al, "CdS/CdTe Solar Cells by RF Sputtering and by Laser Physical Vapor Deposition," 23rd IEEE Photovoltaic Specialists Conference--1993 (Louisville, KY, May 10-14, 1993) pp. 394-9.

[2] K. J. Lesker Co, Clairton, PA

[3] AJA International, North Scituate, MA

[4] A.D. Compaan & R.G. Bohn, "Thin Film CdTe Photovoltaic Cells," Annual Subcontract Report, 1 Nov 1991--31 Oct 1993. [Available NTIS #NREL/TP-451-5813, DE94000201.]

[5] Z. Feng, C.N. Tabory, & A.D. Compaan, "Optical interference measurements of glass temperature," paper, this conference.

[6] R. Kubo, *Statistical Mechanics,* (North Holland, Amsterdam, 1965) p. 153.

[7] S.M. Rossnagel, "Magnetron Plasma Deposition Processes," in *Handbook of Plasma Processing Technology,* edited by Rossnagel, Cuomo, and Westwood. (Noyes Publications, Park Ridge, N.J., 1990) pp. 160-182.

HEALTH AND ENVIRONMENTAL HAZARDS OF CdTe PHOTOVOLTAIC MODULE PRODUCTION, USE AND DECOMMISSIONING

Paul D. Moskowitz, Brookhaven National Laboratory, Upton, NY 11973, USA
Hartmut Steinberger, Fraunhofer Institut fur Festkorpertechnologie, W-80686 Munchen, Germany
Werner Thumm, GSF-Institute of Ecological Chemistry, D-85758 Oberschleissheim, Germany

ABSTRACT

Health and environmental (H&E) risks presented by CdTe photovoltaic module production, use and decommissioning have been reviewed and discussed by several authors. Several H&E concerns exit. The estimated risks are based on extrapolations of toxicity, environmental mobility, and bioavailability data for other inorganic cadmium compounds. Little information, however, is available about CdTe itself. In response to the increased interest in CdTe, Brookhaven National Laboratory (BNL) has been engaged in a cooperative research program with the National Institute of Environmental Health Sciences (NIEHS), the Fraunhofer Institute for Solid State Technology (IFT), and the GSF Institute of Chemical Ecology to develop fundamental toxicological and environmental data for CdTe. This paper describes the results of these studies, and their potential implications with respect to the H&E hazards presented by CdTe module production, use and decommissioning.

Introduction

Health and environmental (H&E) risks presented by CdTe photovoltaic module production, use and decommissioning have been reviewed and discussed by several authors [1, 2]. Several H&E concerns exit. The estimated risks are based on extrapolations of toxicity, environmental mobility, and bioavailability data for other inorganic cadmium compounds. Little information, however, is available about CdTe itself. In response to the increased interest in CdTe, Brookhaven National Laboratory (BNL) has been engaged in a cooperative research program with the National Institute of Environmental Health Sciences (NIEHS), the Fraunhofer Institute for Solid State Technology (IFT), and the GSF Institute of Chemical Ecology to develop fundamental toxicological and environmental data for three compounds (CIS, CGS, and CdTe) of interest to the photovoltaics community. This paper describes the results of these studies for CdTe, and their potential implications with respect to the H&E hazards presented by CdTe module production, use and decommissioning.

Chemical and Physical Properties

Cadmium telluride (CdTe) is a black, dense cubic, crystalline compound that does not occur naturally. Cadmium telluride is insoluble in water; however, it decomposes on exposure to atmospheric moisture with the evolution of gaseous hydrogen telluride (H_2Te), a toxic, volatile acid that is very soluble in water. The presence of acids or acid fumes accelerates this decomposition. Under very dry conditions the compound will react with atmospheric oxygen to form oxides of cadmium and tellurium. It is unlikely, therefore, that the unencapsulated compound will persist for long periods of time as CdTe under natural soil conditions or within landfills. Reaction with H_2O and O_2 will lead to the decomposition of the compound and the release of new Cd and Te compounds and ions into the environment.

Toxicological Properties

No human clinical data are available on the potential health effects associated with exposure to CdTe. In the absence of human data, animal data are used by regulators to assess potential health hazards presented by chemicals used in commerce or found in the environment. At the request of BNL and others, the NIEHS National Toxicology Program studied the systemic and reproductive toxicity of CdTe in rats. In these studies rats were exposed to CdTe present in their food for a 28 day period. After the 28 days, the rats were sacrificed and clinical signs, including in-life body weights, terminal body weights and organ weights, histology, clinical chemistry, and hematology were evaluated. The studies, published elsewhere in greater detail [3], suggest that CdTe did not consistently cause adverse effects at the dosage levels evaluated; these dosage levels are orders of magnitude higher than the dosage that would be

 First WCPEC; Dec. 5–9, 1994; Hawaii

allowed under current U.S. Environmental Protection Agency (EPA) or Occupational Safely and Health Administration (OSHA) regulations. These studies also indicated that at the dosage levels tested, CdTe did not produce adverse reproductive or developmental toxicity.

Harris et al. [3] speculated that since no vascular, renal or teratologic damage was seen (indicative of cadmium poisoning), circulating blood levels of cadmium were low. The normal urinary Cd and protein values observed in the study support this conclusion. Additional studies are underway to measure levels of cadmium in blood and tissues in the animals used in the Harris et al. study and in other studies that used inhalation as the route of exposure. Cadmium is poorly absorbed by oral administration, and CdTe is virtually insoluble in aqueous solutions. Therefore, the CdTe may not be biologically available by this exposure route. Since occupational exposures are more likely to occur through inhalation or dermal absorption, not ingestion as was mimicked in these laboratory studies, caution must be exercised in extrapolating these findings to occupational settings. Further, other limitations exist including the short-term nature of these studies; plans are now being made to expand the scope of these studies to fill these gaps. Nevertheless, these initial studies suggest that CdTe may not be as toxic as other inorganic cadmium compounds.

Environmental Studies

At the GSF in Munich, Germany, studies are being conducted to develop physical-chemical data that could be used to evaluate the environmental mobility (i.e., transport through air, water, soil and biological organisms) of CdTe and by-products released during the combustion of CdTe modules in fires (e.g., building or incinerators) and during decay in landfills where manufacturing wastes and spent modules could be placed at the end of their useful life.

Module behavior in fires is mimicked by using modern thermo-analytical techniques. Results of a differential thermal analysis for pure CdTe are shown in Fig. 1. In air, the experiments showed that there was an increase in weight until about 750 - 800°C.; this is possibly due to oxidation. This oxidized product remains stable until about 1050°C. At higher temperatures, the compound begins to evaporate. The higher temperatures where material evaporation was observed are representative of the conditions present in a large municipal solid waste incinerator.

Module behavior, including metals leachability in landfills, is being characterized by two different elution tests: the EPA Toxicity Characterization Leachate Profile (TCLP), and the German DEV S4. In these tests, small ($<1cm^2$) pieces of a broken CdTe module are suspended and rotated in an elutent for 24 hours. The metals present in the elutent are then measured and compared with limits prescribed by each testing protocol. If the metal concentration in the elutent exceeds the limits, the modules are demonstrating metal leachability and may need to be recycled or disposed of in a hazardous waste landfill. if the metals are not leaching in excessive quantities the modules can be disposed of in a commercial landfill like many other consumer products. Tables 1 and 2 display the results of these tests for a CdTe photovoltaic module. To the extent that these data are representative of CdTe modules to be sold in the market place, spent and decommissioned modules might need to be treated as hazardous waste. More testing is required to confirm this conclusion. If this conclusion is correct, it will be important to develop cost-effective recycling strategies.

Table 1. EPA TCLP Leach Testing Results (n=4)

		TCLP	
Element	Results	Existing	Proposed
Cd	1.53±0.21	1.0	0.91
Te	0.47±.07		
Al	1.26±0.14		
Ni	0.39±0.05		5.0

Table 2. German DEV S4 Leach Testing Results (n=4)

		Class	
Element	Results	I	II
Cd	0.50±0.13	0.05	0.19
Te	0.50±0.13		
Al	0.87±0.20		
Ni	0.25±0.04	0.2	1

Industrial Hygiene Studies

In parallel with these experimental studies, BNL [4] has assembled industrial hygiene monitoring data from the photovoltaic industry. Because there is little published information on the monitoring of employee exposures in CdTe manufacturing of photovoltaic cells, we relied on personal communications [5, 6]. These personal communications from an industrial hygienist at the National Renewable Energy Laboratory reveal that no detectable levels of cadmium were reported from personal monitoring of maintenance operations (CdTe physical vapor deposition system and CdTe oven overhaul), daily operations in physical vapor deposition, or sandlbasting cadmium target operations [5]. However, area samples taken in line with exhaust flow during the maintenance of the physical vapor deposition system were measured at 30 ug/m^3 ; this is above the OSHA action level of 2.5 ug/m^3. Initial wipe samples taken in the deposition and sample preparatory laboratories ranged from below detection to 0.59 mg, but dropped below detectable levels after good housekeeping procedures were established. Wipe samples in the physical vapor deposition laboratory ranged from 1.7 - 92 ug.

At a pilot scale CdTe module production facility, personal samples were collected from the sandblasting/buffing operation, laser scribing room, and chemical treatment room [6]. There were no exposures to cadmium above the current OSHA action level of 2.5 ug/m^3. Area sampling found non-detectable levels of cadmium within and outside the cell production area (e.g., lunch room, module chemical treatment room center of work area, analytical laboratory, and laser room). Wipe samples ranged from a low of 0.0038 mg (secretary's desk) to a high of 2.4 mg within the work area. Unadjusted urinary cadmium levels from the workers at this same facility ranged from 0.2-0.9 ug Cd/L urine; this is well below the threshold limit of 5 ug Cd/L established by the American Conference of Governmental and Industrial Hygienists.

The literature contains limited industrial hygiene monitoring information on other semiconductor production processes; this literature is important because of some insights it provides. Ungers et al. [7] determined that arsenic can be released from freshly ion-implanted silicon wafers, manufactured under high vacuum conditions (10^{-6} torr). Although information is available about the behavior of dopant atoms on the crystal lattice of the substrate, there are little data on dopant atoms in the crystal lattice of the substrate, and nothing about the fate of atoms associated with this superficial deposition. The theory that these atoms diffuse out of the wafer and are released to the environment gained support from workers' complaints of odors in wafer-handling operations near the ion-implanter station. Odor problems have also been reported in CdTe and other photovoltaic thin-film production facilities [8-10]. The implication of these findings is that there is potential for metals accumulation in the air in any facility where modules are stored prior to encapsulation and the rooms are poorly ventilated.

Discussion/ and Conclusions

There is large interest and need to characterize, in a realistic fashion, the H&E risks presented by CdTe photovoltaic module production, use and decommissioning. This interest is fueled by the existing H&E regulations that focus on other inorganic cadmium compounds and the increasing investments being made to develop large-scale CdTe module manufacturing facilities. Arguments to date about the actual risks presented by CdTe have been based on the extrapolation of H&E data from other cadmium compounds and related industries. The experimental data developed for this project and the industrial hygiene observations collected from existing CdTe facilities suggest that the actual H&E risks from CdTe module production and use are smaller than a simple hazard analysis would suggest. Nevertheless, because of the inherent toxicity of the parent compounds, an ongoing effective industrial hygiene program is needed to ensure that workers' health is not compromised. Furthermore, recycling strategies for waste products from CdTe manufacturing facilities and spent modules may also be needed to protect public health and facilitate the marketing of these devices.

References

1. Moskowitz, P.D., 1992. Environmental, Health and Safety Issues Related to the Production and Use of CdTe Photovoltaic Modules. International Journal of Solar Energy, Vol. 12, pp. 259-281 (1992). BNL 47372, Upton, NY.

2. Patterson, M., A.K. Turner, M. Sadeghi, and R.J. Marshall, 1994. Health, Safety and Environmental Aspects of the Production and Use of CdTe Thin Film Photovoltaic Modules. Presented at the 12th EC PVSEC, The Netherlands.

3. Harris, M.W., et al., 1994. The general and reproductive toxicity of the photovoltaic material cadmium telluride (CdTe). Presented at the Society of Toxicology 1994 Annual Meeting, Dallas, Texas, March 13-17, 1994. The Toxicologist 14:267, 1994

4. Bernholc, N., P. Moskowitz, R. Chapin and M. Harris, 1994. "Cadmium telluride - summary report on industrial hygiene," in P.D. Moskowitz and K. Zweibel, Editors, "Understanding and Managing Health and Environmental Risks of CIS, CGS, CdTe Photovoltaic Module Production and Use: A Workshop Report. Brookhaven National Laboratory, Upton, NY.

5. Donley, J.,1994. Personal communication, National Renewable Energy Laboratory, Golden, Colorado.

6. Moyers, P., 1994. Personal communication, Solar Cells, Inc., Toledo, Ohio.

7. Ungers, L.J., J.H. Jones, A.J. McIntyre, and C.R. McHenry, 1985. Release of arsenic from semiconductor wafers. Am. Ind. Hyg. Assoc. J. 46(8):415-420.

8. Luke, D., 1993. Personal communication, National Renewable Energy Laboratory, Golden, Colorado.

9. Mosier, W., 1994. Personal communication, Martin Marietta, Denver, Colorado.

10. Gay, R., , 1994. Personal communication, Siemens Solar Industries, Camarillo, California.

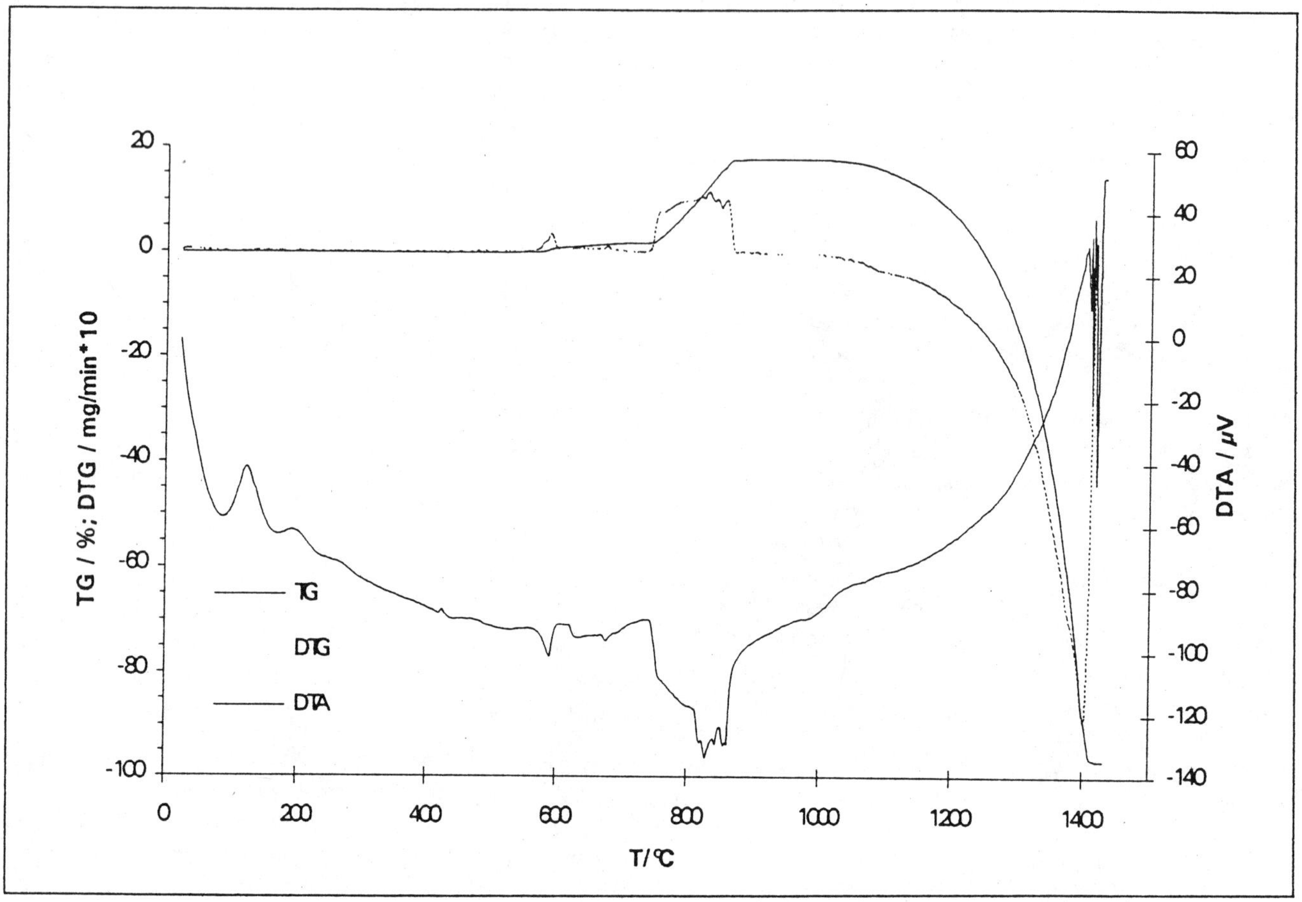

Fig. 1. Differential thermal analysis of pure CdTe in air.

SIX–YEAR EFFICIENCY GAINS FOR CdTe AND $CuIn_{1-x}Ga_xSe_2$ SOLAR CELLS: WHAT HAS CHANGED?

James R. Sites and Xiaoxiang Liu
Physics Department, Colorado State University, Fort Collins, CO 80523, U.S.A.

ABSTRACT

During a period of six years from roughly 1987 to 1993, a number of laboratories have contributed to truly remarkable efficiency gains for thin–film cells with polycrystalline CdTe and $CuIn_{1-x}Ga_xSe_2$ (CIGS) absorbers. Total–area efficiencies have increased from just above 10% to over 16%, and these cells can now be sensibly compared with single–crystal cells of comparable bandgap. A number of factors have contributed to these gains, but the most significant has been the reduction of recombination states in the depletion region and the resulting drop in forward–current density.

J–V COMPARISONS

Comparative current density vs. voltage (J–V) curves are shown in Fig. 1. In 1987, several CdTe cells in the 10–11% range had been reported. The best individual parameters, both then and more recently, were not always found in the same cell. The typical high–efficiency 1987 CdTe cell shown in Fig. 1 was made at Ametek by electrodeposition [cell 344AB–7]. The 1992 cell was fabricated by the University of South Florida using close–space sublimation [cell 5–19B–10A]. Also shown for comparison is a high–efficiency (25.7%) crystalline GaAs cell made at NREL [1] and the calculated room–temperature 100 mW/cm^2 maximum–efficiency curve for the CdTe or GaAs bandgap limit [2]. The CdTe curves show improvement in current, voltage, and fill factor, and clearly additional improvement with all three factors is necessary to approach the single–crystal curve.

The $CuIn_{0.8}Ga_{0.2}Se_2$ (CIGS) curves follow the same format. The 1987 cell [746] was made at Boeing by physical vapor deposition and the 1993 cell [C266–23–8] at NREL with the three–stage process. The value of $x = 0.2$ corresponds to a bandgap expansion from 1.0 eV for $x = 0$ to about 1.15 eV. This increase yields a significant potential efficiency gain for the terrestrial spectrum, whereas higher values of x have limited potential for further gain [3]. The curve labeled "Si" uses the data from New South Wales [4] for a 24% efficient silicon cell. A small modification was made to the data to account for the bandgap difference. In the CIGS comparison, the major efficiency improvement has clearly been due to the increase in voltage.

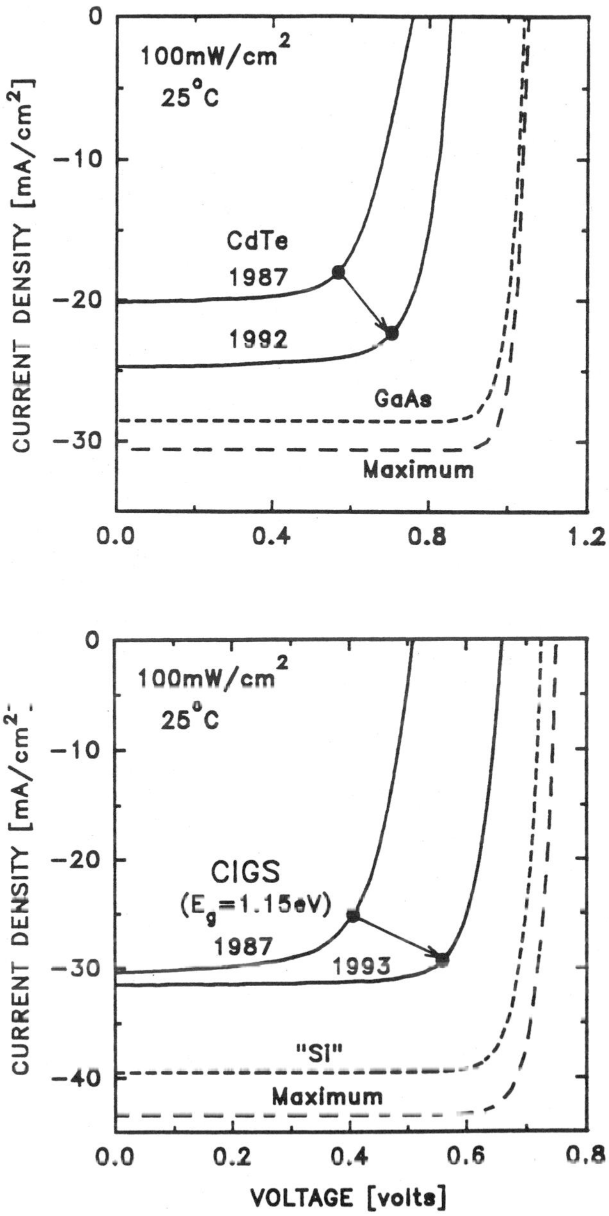

Fig. 1. Improvement of CdTe and CIGS total–area J–V curves with time. Also shown are single crystal curves and the calculated limit for comparable bandgap.

JUNCTION QUALITY

The details of the voltage and fill–factor increases for the CdTe and CIGS cells are shown in Fig. 2, which uses the same data as Fig. 1. Forward current in the light is defined as the difference between the measured current and the photocurrent at any voltage. The solid points highlight V_{OC} and V_{MP}. Again the corresponding curve for the calculated maximum efficiency is shown. The dashed–line fits eliminate series– and shunt–resistance effects.

Two features are clear in Fig. 2: The forward currents have decreased in magnitude and the slopes on the log plot are getting closer to that of an ideal diode. The latter corresponds to a decrease in diode quality factor under illumination from 2.4 to 2.0 for the CdTe cells and from 2.0 to 1.3 for the CIGS cells. These decreases translate into an increase in fill factor of approximately 10%.

The decreases in both forward–current magnitude and quality factor indicate a reduction in recombination states in the depletion layer of the cell. Physically, the more recent fabrication techniques are introducing fewer extraneous bandgap states as a result of tighter experimental control and growth processes that stay closer to thermodynamic equilibrium. Consequently grain boundaries are better passivated, and the polycrystalline structures are behaving more like single crystals. There is a significant difference seen in Fig. 2, however, between state–of–the–art CdTe and CIGS. The CIGS forward current has clearly gotten much closer to the target than has the CdTe curve.

There are other indicators that CIGS junctions have become very clean besides those apparent in Fig. 2.

1. Superposition. The superposition of light and dark CIGS J–V curves has narrowed to a voltage differential of less than 20 mV. For pure $CuInSe_2$ it has dropped to practically zero, and diode quality factors as low as 1.2 are observed.

2. Drift with time. The best CIGS cells show no measurable transient J–V effects following a change in illumination or bias. Earlier cells exhibited drifts on the order of 20 mV before stabilization.

3. Frequency dispersion in capacitance. The change in measured capacitance between 1 and 300 kHz is less than 4% for the newer cells. For older cells, it was often a factor of two and was a clear indicator of extraneous trapping states.

OTHER FACTORS

One other efficiency factor illustrated in Fig. 2 is the effect of series resistance. The horizontal difference between the dashed line and the V_{MP} point is the voltage loss at maximum power due to series resistance R_S. For

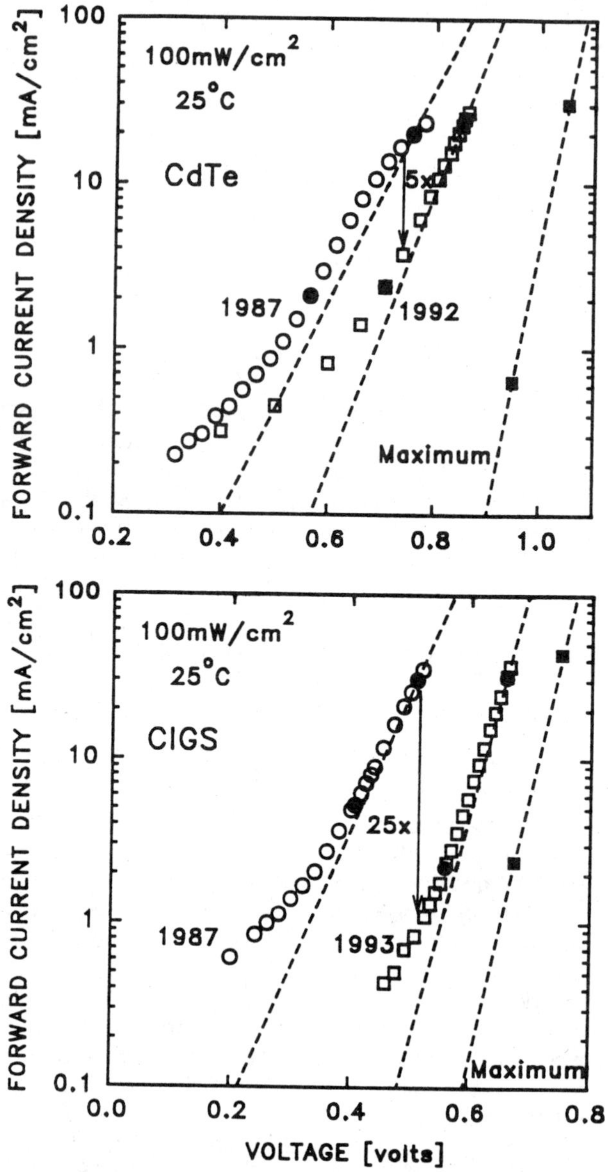

Fig. 2. Forward–current comparisons. Upper solid symbols denote V_{OC}, lower ones V_{MP}.

CdTe, R_S has dropped from near 2 to near 0.5 Ω–cm^2 due in large part to better contacting layers. For the earlier CIGS, the value was 0.25 Ω–cm^2, and since this value was already small, the subsequent improvement has been modest.

The major current improvement for CdTe cells has been the use of thinner CdS windows. Absorption by thick CdS removes as much as 6 mA/cm^2 from the potential photocurrent. That loss is now commonly reduced by a factor of two with thinner CdS. Other current improvements in CdTe have resulted from a less absorbant transparent conducting oxide (TCO) and from changing the glass superstrate, which can have significant absorption for wavelengths above 600 nm. Thinner CdS has also helped the CIGS efficiency, but proportionally less because of higher base current with the smaller bandgap and because the earlier Boeing cells used a $Cd_{1-x}Zn_xS$ window. CIGS current, however, has benefited a small amount from increases in bulk collection (presumably also related to reduction in recombination states), and from reductions in reflection.

SUMMARY

Fig. 3 summarizes the efficiency gains for CdTe. The format is a plot of maximum–power current vs. maximum-power voltage. The dashed rectangle is the calculated maximum for room temperature and 100 mW/cm^2. The smaller solid–line rectangle is for the earlier cell and the large one for the more recent cell. The expansion in area due to improved junction quality and thinner CdS are similar. Next most important increases come from the decrease in series resistance R_S and the superstrate absorption effects. Also shown in Fig. 3 is an approximate breakdown of the remaining differences between values achieved and the theoretical target. Again the largest difference is junction quality, and the second is window absorption. There is some coupling between maximum power current and voltage, so that an increase in one will automatically cause a small increase in the other. Graphically, this coupling is the area of small rectangles in the upper right.

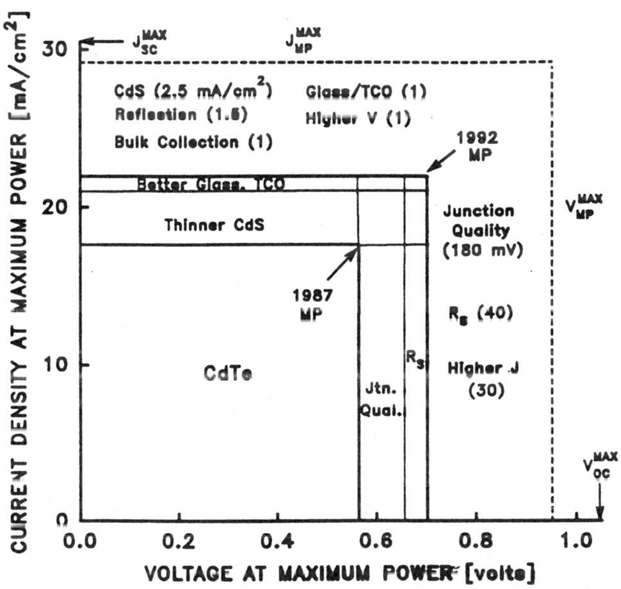

Fig. 3. Summary of CdTe efficiency increases and breakdown of remaining losses.

The summary for CIGS is given in Fig. 4. It shows the large gain due to junction quality and the smaller remaining voltage difference to the target. It also shows two remaining current losses that are larger for CIGS than CdTe: grid coverage, which was 4% for the CIGS cell, but 0 for typical CdTe, and CdTe, R_S has dropped from near 2 to near 0.5 Ω–cm^2 due in large part to better contacting layers. For the earlier bulk collection. The bulk collection loss is larger because the depletion, or high field, region for CIGS is typically 0.25 μm, compared to 2 μm for CdTe. A significant number of photons pass through the depletion region and rely on diffusion for collection. Typically, the quantum efficiency decreases continuously in the infrared, beginning some distance from the bandgap cutoff. The details of this decrease can be used to calculate the diffusion length and the collection loss [5}.

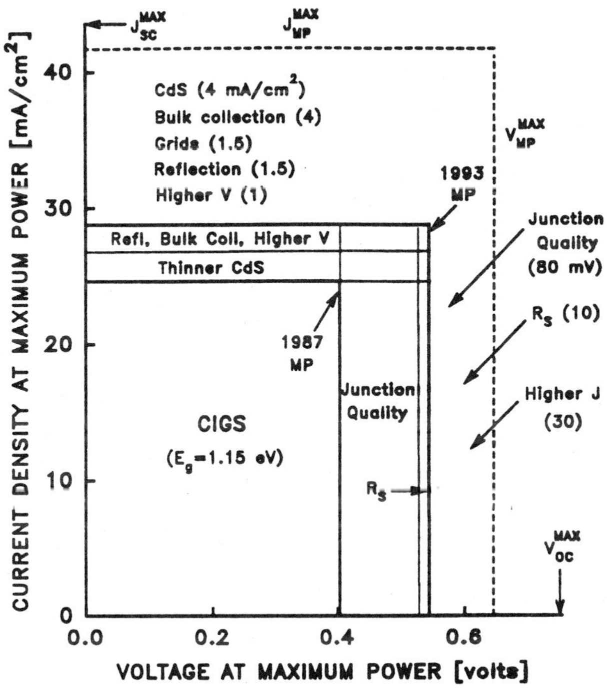

Fig. 4. Summary of CIGS efficiency increases and remaining losses.

ACKNOWLEDGEMENTS

The authors thank Ingrid Eisgruber, Gunther Stollwerck, and Jennifer Granata at Colorado State, Rick Sasala now at Solar Cells, Inc., and John Scofield at Oberlin College for various individual measurements. They are grateful to Ametek, Boeing, South Florida, and NREL for the cells used and to NREL for financial support.

REFERENCES

[1] S. R. Kurtz, J. M. Olson, and A. Kibbler, "High Efficiency GaAs Solar Cells Using $GaInP_2$ Window Layers," *Twenty-first IEEE PVSC*, 1990, pp. 128–140.

[2] J. R. Sites, "Calculation of Impact Ionization Enhanced Photovoltaic Efficiency," *Solar Cells* **25**, 1988, pp. 163-168.

[3] I. L. Eisgruber and J. R. Sites, "Studies of Polycrystalline Thin–Film Solar Cells," *AIP Conf. Proc.* **306**, 1993, pp. 411–413.

[4] A. Wang, J. Zhao, and M. A. Green, "24% Efficient Silicon Solar Cell," *Appl. Phys. Lett.* **57**, 1990, pp. 602–604.

[5] X. X. Liu and J. R. Sites, "Solar–cell Collection Efficiency and its Variation with Voltage," *J. Appl. Phys.* **75**, 1994, pp. 577-581.

NUMERICAL MODELING OF GRADED BAND GAP CIGS SOLAR CELLS

Jeffery L. Gray and Youn Jung Lee
School of Electrical Engineering
Purdue University
West Lafayette, Indiana 47907-1285

ABSTRACT

The high efficiency reported recently by NREL for CIGS solar cells demonstrates the potential of band gap grading in producing high efficiency thin film solar cells. In order to reap the full benefits of this design strategy, a clear understanding of the fundamental device physics of these structures is needed. The purpose of this paper is to examine the role grading of the band gap plays in achieving high conversion efficiencies. To aid in this examination, a detailed numerical device simulation program, ADEPT, is used.

INTRODUCTION

Graded band gap CIGS solar cells have proved to be very efficient. This increase in efficiency over that of CIS solar cells is attributed to an increase in open circuit voltage due to the larger band gad near the junction and an increase collection efficiency caused by the built-in electric field that results from the band gap grading. While this simple explanation is useful in understanding the basic device performance, a complete understanding of the physics of graded cells is needed in order to fully understand cell performance and to design and fabricate even higher efficiency cells.

Here, issues affecting cell performance are discussed and examined using ADEPT (A Device Emulation Program and Tool), a detailed numerical device simulation program developed at Purdue University.

ADEPT

ADEPT [1] and its developmental predecessors [2] have been used to model a wide variety of photovoltaic devices. It has been demonstrated to be a useful tool for interpreting the performance of existing solar cells. The accuracy of ADEPT in predicting cell performance depends on several factors. In well understood material systems, such as crystalline silicon and gallium arsenide, the predictive ability of detailed device simulation codes is quite good. In thin film devices such as CIS and CdTe, the predictive ability of these codes is not as good. Basic characterization of these materials is still an active area and is complicated by variations in the material parameters resulting from different fabrication methods and even different facilities. In a graded band gap structure these difficulties are even more severe.

In order understand these issues, it is necessary to examine the assumptions used in creating a detailed numerical model of a thin film solar cell.

Semiconductor Equations

ADEPT generates a numerical solution to a set of three coupled partial differential equations commonly referred to as the semiconductor equations. These are Poisson's equation and the hole and electron continuity equations and are shown below for steady state isothermal conditions.

$$\nabla \cdot D = q(p - n + N) \quad (1)$$

$$\nabla \cdot J_p = q(G - R) \quad (2)$$

$$\nabla \cdot J_n = -q(G - R) \quad (3)$$

Coupled with the following auxiliary equations, these are solved for the electrostatic potential, V, and the hole and electron concentrations, p and n, on a domain defined by the device structure subject to boundary conditions (at the contacts, for one-dimensional simulations) and the device operating conditions (the applied bias, operating temperature, and incident light flux).

$$D = -\varepsilon \nabla V \quad (4)$$

$$J_p = -qp\mu_p \nabla(V - V_p) - kT\mu_p \nabla p \quad (5)$$

$$J_n = -qn\mu_n \nabla(V + V_n) + kT\mu_n \nabla n \quad (6)$$

It is here, in the transport equations (5) and (6), that a significant assumption is made. Carrier transport within the device is assumed to be based on drift-diffusion mechanisms only. Tunneling and thermionic emission are neglected. This is not a severe restriction and further

CH3365-4/94/0000-0123 $4.[illegible] © 1994 IEEE

discussion of this issue is beyond the scope of this paper, but is addressed elsewhere in these proceedings [3]. Tunneling and thermionic emission at the contacts is modeled using effective carrier recombination velocities for the boundary conditions.

As can be seen by examining equations (1) through (6), there are only a modest number of parameters to be set in order model any device constructed from any semiconductor material, so long as the assumptions regarding transport are not violated. These parameters are: N, the trapped charge density; G, the optical generation rate; R, the net recombination rate; ε, the electric permittivity; μ_p and μ_n, the carrier mobilities; and V_p and V_n, the band parameters. While this is a convenient way to express the model equations, it is a somewhat misleadingly simple representation. All the information regarding device structure and material characterization is contained in these terms.

MODELING ISSUES

All the information relating to the device structure and material parameters of a graded band gap solar cell are contained in the eight parameters mentioned above. Accurate modeling of *any* device depends on detailed knowledge of these parameters. Even for a relatively simple CdS/CIS structure, there is debate about the exact structure of the device (i.e. presence of an OVC layer, interface states, contact characteristics, etc.) and the correct material characteristics (i.e. absorption coefficients, mobilities, trap/recombination centers, etc.). In a graded band gap structure, these difficulties are compounded. These issues are discussed below.

Absorption Coefficients

Even in CIS, there is presumably some variation in absorption coefficient with position within the CIS layer since the ratios of copper to indium to selenium tends vary (i.e. Cu-poor near the CdS interface). The addition of gallium would, of course, change the optical band gap as some function of gallium content. Any other material (such as sulfur) would have a similar effect. Characterization of the absorption coefficient as a function of gallium content is needed in order to produce accurate simulations.

Trapping/Recombination

The effective doping and carrier lifetimes in CIS cells are thought to be a result of native defects in the material. Detailed modeling of trapping and recombination can be quite complicated. However, reasonable modeling results have been obtained by assuming a fixed trapped charge density (effective doping) and by modeling recombination as a mid gap single level trap. Improved models are needed for a full understanding of trapping and recombination. ADEPT can be used to examine proposed models, but even a simple model presents a challenge in a graded band gap CIGS solar cell. How do the effective doping and lifetimes depend on band gap (gallium content)? How sensitive is device performance to these parameters?

Mobility

As with the other parameters, carrier mobility is sure to be a function of gallium content and represents another area of material characterization that must be addressed.

Band Parameters

The most explicit effect of band gap grading appears in the band parameters [4].

$$qV_p = -(\chi - \chi_{ref}) - (E_G - E_{Gref}) + kT\ln\frac{N_V}{N_{Vref}} + kT\ln\frac{F_{1/2}(\eta_V)}{e^{\eta_V}} \tag{7}$$

$$qV_n = (\chi - \chi_{ref}) + kT\ln\frac{N_C}{N_{Cref}} + kT\ln\frac{F_{1/2}(\eta_C)}{e^{\eta_C}} \tag{8}$$

Here, χ is the electron affinity, E_G is the band gap, and N_V and N_C are valence and conduction band densities of state. The last term in each band parameter expression accounts for degenerate (Fermi-Dirac) statistics. The 'ref' subscript refers to an arbitrary reference material. Note that we have assumed parabolic bands.

The band parameters are what allow the modeling of heterostructures and, specifically, graded band gap structures. The electron affinity, band gap, and densities of state can all vary with gallium content. The precise relationship must be determined experimentally.

Examination of equations (7) and (8) can give some insight into the effect the grading of gallium content may have on cell performance. Note that, neglecting degeneracy, the band parameters are constant with position in a uniform material and hence have no effect on carrier transport. In a graded structure, however, the effect is significant and can sometimes lead to nonintuitive results.

Consider the band structure of a hypothetical heterostructure as shown in Figure 1. If this were a uniform material, the local vacuum energy would be parallel to both the valence and conduction band edges. That is clearly not the case in a heterostructure. We see that the gradients of the band edges can be opposite in sign to the electric field. It is this gradient, or effective field, that represents the true 'quasi-drift' force acting on the carriers. This effective field is, in general, different for holes and electrons.

Even this can lead to an incorrect interpretation if the electron affinities and/or densities of state change significantly with band gap. It is the gradients of V_p and V_n that contribute to the effective fields acting on the carriers.

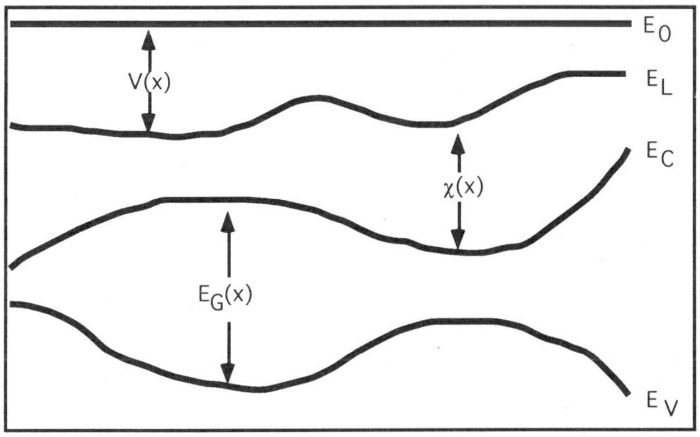

Figure 1. The band diagram of an arbitrary heterostructure.

We see in equations (7) and (8) that changes in the electron affinity and the densities of state with position can also contribute significantly to the effective fields. In fact, counter-intuitive predictions of performance can result. More useful in interpreting performance is the use of an effective band diagram, where the variations of electron affinity, band gap, and densities of state with position are all lumped into an effective band gap and effective electron affinity. Examination of this diagram correctly shows the influence of the effective fields on carrier transport. This is illustrated in Figures 2 and 3.

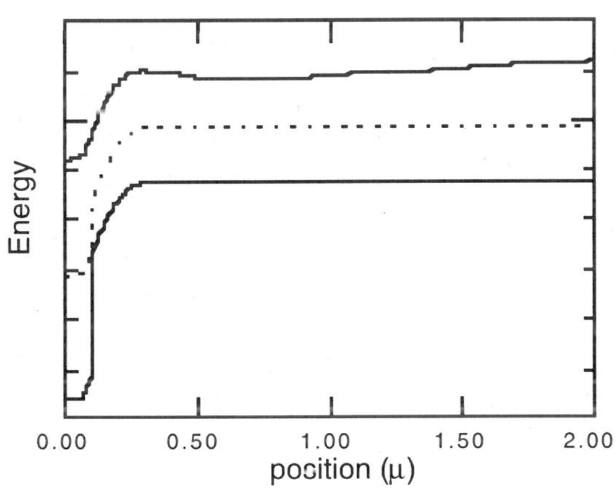

Figure 2. Energy band diagram of a simple CdS/CIGS graded band gap solar cell.

Figure 2 is the conventional energy band diagram of a simple CdS/CIGS solar cell. Examination of this plot might lead one to conclude that the band gap grading tends to keep minority carriers away from the CdS interface and the back contact. This would be true if only the band gap changed with position. In this ADEPT simulation, however, that is not the case. Here, the conduction band density of states was varied with position in such a way as to exactly cancel the effect of the graded band gap. This is shown in Figure 3. In fact, this device has a somewhat lower efficiency than a comparable CdS/CIS cell (14.1% versus 14.8% simulated efficiencies).

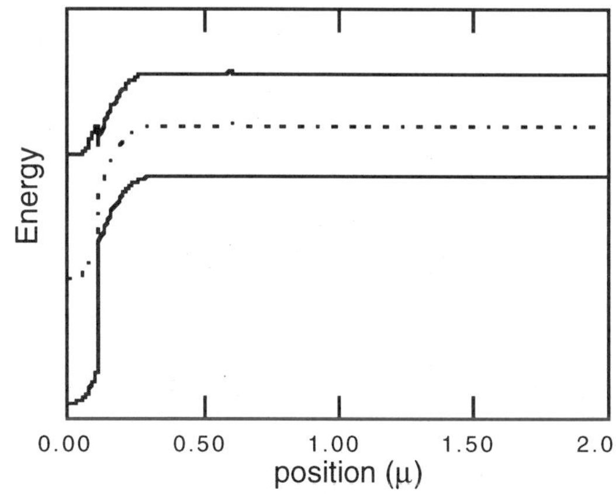

Figure 3. Effective energy bad diagram of same graded band gap cell as used in Figure 2.

SIMULATION RESULTS

It should be apparent that to gain a complete understanding of how gallium affects cell performance requires detailed characterization of CIGS material as a function of gallium content. In addition, the spatial variation of the gallium content must be known as well. A parameter study of how performance is affected over reasonable bounds on each parameter would require a very large number of simulations, and is beyond the scope of this paper. However, by limiting the parameters to be varied to a small set, some insight into graded band gap cell performance can be achieved.

We will limit ourselves to the simple cell structure for which the conventional energy band diagram is shown in Figure 2. It consists of a 0.05 μ n-type CdS layer on top of a 1.95 μ CIGS absorber layer. All parameters dependent on gallium content are assumed to vary linearly with position. The point of minimum gallium content is taken to be 0.5 μ into the CIGS layer. A simple CdS/CIS equivalent in structure is used for comparison purposes (Case 0). Reasonable materials parameters are used in the simulations and are assumed to be independent of gallium content. Only the band gap and absorption coefficients are allowed to vary with gallium content. It should be noted that simplified models were used and no attempt was made to match experimental data. The purpose here is to show trends in performance.

Several variations of the band gap grading were simulated. First, in Case 1 the CIGS layer is assumed to have a constant band gap for the first 0.5 microns and then a linearly graded band gap up to the back contact. In Case 2 only the first 0.5 microns of the CIGS layer is graded, and finally, in Case 3 the entire CIGS layer is graded, similar to the situation depicted in Figure 2. The maximum band gap in Cases 1 to 3 was 1.2 eV. The performance characteristics are summarized in Table 1.

Table 1. Computed performance characteristics for four examples of graded band gap CIGS cells.

	Case 0	*Case 1*	*Case 2*	*Case 3*
V_{OC} mV	515	519	590	595
J_{SC} mA/cm^2	41.1	41.7	39.0	39.6
FF	71.9	72.0	69.4	71.0
η	15.8	16.0	16.5	17.3

Case 1 basically corresponds to the addition of a back surface field. The (small) improvement in efficiency is due primarily to an increase in the collection efficiency. In Case 2, there is no BSF and the somewhat larger increase in performance is due to the bigger band gap in the junction region. However, the collection efficiency is reduced since the effective field tends to hinder collection. The collection efficiency can be recovered by now including a BSF, as in Case 3. There is an improvement in cell efficiency of 1.5 points.

It must be noted that the effect of band gap grading on cell performance is sensitive to many of the parameters that were not varied in this example. A comprehensive parameter study is needed if a good understanding of graded band gap solar cells is to be achieved.

SUMMARY

In this paper, the issues involved in modeling graded band gap CIGS solar cells were addressed. Experimental characterization of CIGS material is needed in order to improve our understanding of the performance of these devices. The value of detailed numerical device simulation programs as an analysis tool was demonstrated. Future work will involve a comprehensive examination of the sensitivity of the performance characteristics to parameters that might possibly vary with band gap.

ACKNOWLEDGMENT

This work was supported by the National Renewable Energy Laboratory under Subcontract Number XG-2-11036-2.

REFERENCES

[1] J. L. Gray, "ADEPT: A General Purpose Device Simulator for Modeling Solar Cells in One-, Two-, and Three-Dimensions", *Twenty-Second IEEE PVSC*, 1991, pp. 436-438.

[2] J. L. Gray, "Computer Simulation of Thin Film Silicon-Hydrogen Alloy Solar Cells", *IEEE Trans. Electron Devices*, **36**(4),1989, pp. 906-912.

[3] S. M. Durbin and J. L. Gray, "Considerations for Modeling Heterojunction Transport in Solar Cells", *First WCPEC*, 1994, (this conference proceedings).

[4] M. S. Lundstrom and R. J. Schuelke, "Numerical Analysis of Heterostructure Semiconductor Devices", *IEEE Trans. Electron Devices*, 1983, pp. 1151-1159.

MICRO THROUGH NANOSTRUCTURE INVESTIGATIONS OF POLYCRYSTALLINE CdTe: CORRELATIONS WITH PROCESSING AND ELECTRONIC STRUCTURES

D.H. Levi, H.R. Moutinho, F.A. Hasoon, B.M. Keyes, R.K. Ahrenkiel, M. Al-Jassim, and L.L. Kazmerski
National Renewable Energy Laboratory, 1617 Cole Blvd., Golden, Colorado 80401, U.S.A.

R.W. Birkmire, Institute of Energy Conversion, University of Delaware, Newark, Delaware 19716, U.S.A.

ABSTRACT

This paper provides first-time correlations of the nanoscale physical structure with the macroscale electronic and optical properties of CdTe/CdS thin films for several standard deposition techniques. Atomic force microscopy (AFM) was used to determine the micro and nanostructures of polycrystalline CdTe thin films used in photovoltaic (PV) cell fabrication. Photoluminescence (PL) was used to determine band gap, relative defect density, and photoexcited carrier lifetime. Nanostructural features (nanograins), beyond the spatial resolution of conventional scanning electron microscopy (SEM), were observed and characterized in as-deposited CdTe. The correlations of the proximal probe measurements of the physical structure with the optically determined electronic properties were used to show the effects of the chemical and heat processing, directly and conclusively. A particularly striking effect with important implications for PV applications is the diffusion of sulfur across the CdTe/CdS interface during heat treatment.

This work partially supported under DOE contract DE-AC36-83CH10093 with NREL.

Research CdTe thin-film solar cells have demonstrated efficiencies to 15.8% [1]. Improvements above the 10% efficiency regime have largely been accomplished through a combined understanding of materials properties and chemistry, related advancements in device/materials processing[2], and interface engineering. The active layer of this solar cell is subjected to various post-deposition processing, including chemical and thermal treatments, and these lead to improvement in the electro optical properties of the semiconductors. Morphology, inter grain properties, and grain configurations are very important, not only for optimal electro-optical characteristics of the films for PV applications, but also for junction formation, contacting, and photon engineering. Until now, the scanning electron microscope has served as the major analytical tool for investigating morphology and physical properties, but it has several limitations. First, it is limited in resolution to the 2-5 nm region, and finer features cannot be adequately resolved. Second, if the films are not conductive (typical for device-quality CdTe), charging effects associated with the high-energy electron beam impair or negate the imaging and/or resolution of this instrument.

Atomic force microscopy (AFM) overcomes these limitations, and can be utilized to provide previously unattainable information on the topography and electronic structure of these materials. Spatial resolutions in the nanometer and atomic ranges can be attained, as well as advantages of digital imaging (for the exact determinations of surface features, quantitative evaluations of roughness, and image analysis from different perspectives). In this paper, the morphology of polycrystalline CdTe films is evaluated using AFM, and correlated with conventional SEM images. The first direct observations and measurements of nanoscale structures in these PV semiconductors are reported as a function of the thermal and chemical processing. These are, in turn, correlated with the electronic structure as determined through photoluminescence and photoexcited carrier lifetime. Indications for improved cell behavior due to the thermal and chemical processing are reported for the first time based upon the correlations of nanoscale imaging with optical characterization.

1. MICRO AND NANOSTRUCTURAL PROPERTIES

A. Experimental

Polycrystalline CdTe thin films were deposited on CdS or conducting oxide substrates (SnO_2 or ITO) using physical vapor deposition (PVD), closed-space vapor sublimation (CSS), and radio-frequency sputtering (RFS). The CSS deposited samples were approximately 4.5μm thick while the PVD and RFS samples were in the 2-3μm range. Post-deposition heat treatments were carried out by depositing a thin layer of $CdCl_2$ from a $CdCl_2$ - saturated methanol solution. The samples were then heat treated for 30 minutes at 300°, 350°, or 400°C in a tube furnace.

AFM measurements were performed using an AFM Autoprobe LS from Park Scientific Instruments, with Si_3N_4 and Si cantilevers. The Autoprobe LS uses a cantilever with an attached tip as the analytical probe[3]. Voltages are transformed to distance through calibration of the piezoelectrics. The movements are calibrated using standards (e.g. graphite (HOPG) or gratings). Image dimensions may range from many microns down to several angstroms.

CH3365-4/94/0000-0127 $4.00 © 1994 IEEE

Fig. 1a. SEM image of a sputter grown CdTe/SnO_2 film

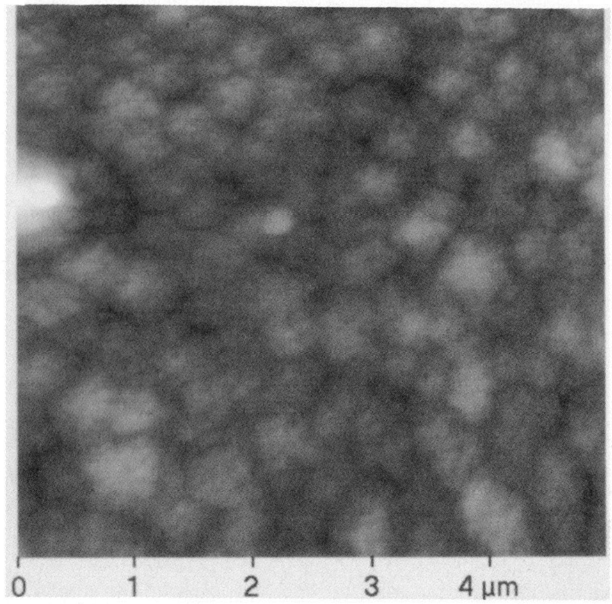

Fig. 1b. AFM image of the same CdTe/SnO_2 film as shown in Fig. 1a

Investigations of the surface morphology of CdTe thin films have typically reported 1000 Å to micron-sized grain dimensions evaluated by conventional SEM 4-6]. However, the enhanced resolution of the proximal probe techniques indicates that as-deposited films sometimes have two feature regimes: features with diameters in the SEM-determined ranges, and grains within these larger features, with diameters of hundreds of angstroms. Figures 1a and 1b show a comparison of the surface of an RFS CdTe on SnO_2 film, imaged by SEM and AFM, respectively. Both micrographs show identical surface structures in this resolution range, with grain sizes in the 0.1 - 0.3 µm range. Figure 2 shows an AFM image of the same film under higher magnification conditions—beyond those capable with the SEM. Much smaller grains, with 200-400 Å cross sections, are clearly resolved. The existence and nanostructure of this double grain structure depend upon the deposition method, the structure of the surface (substrate or sublayer) on which the film is grown, and on post-deposition processing. The thermal and chemical processing steps that are critical in producing higher efficiency CdTe thin-film devices have had mixed correlations with changes in the films' physical properties. We have conducted a comprehensive AFM study of the micro- and nanostructure of CdTe thin-films as a function of deposition technique, substrate structure, and processing conditions. An illustration of the changes in microstructure with processing is provided by Figs. 2 and 3. The as-deposited film is shown in Fig. 2, with 200-400 Å nanoscale grains. Fig. 3 shows the same film following heat-treatment with $CdCl_2$ at 400°C for 30 minutes. It is clear from the AFM image in Fig. 3 that the $CdCl_2$ heat treatment produces a completely recrystallized nanostructure, with elimination of the nanoscale grains .

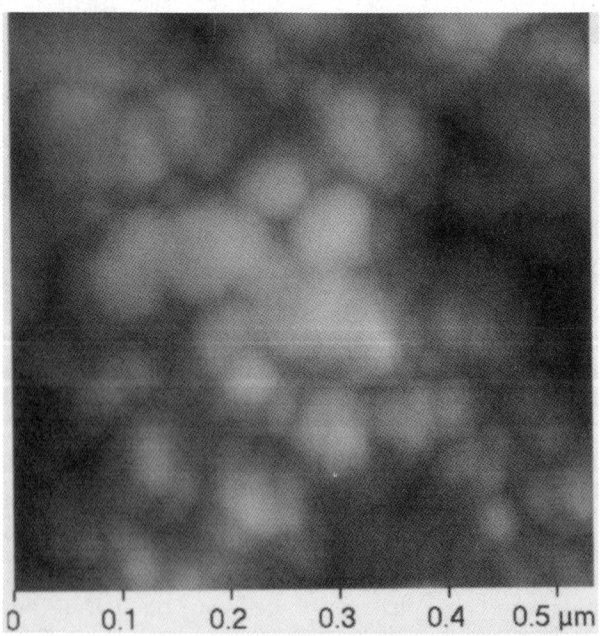

Fig. 2. High resolution AFM image of the same film as in Fig. 1 showing the 200 - 400 Å nanoscale grains within the larger grains detected at the coarser SEM resolution

The AFM images illustrate the dramatic evolution of the structure of sputtered CdTe films deposited on SnO_2 . We have studied the evolution of the structure of CdTe films under a total of five different deposition conditions and three post-growth processing conditions. These results are summarized in Table 1. The RFS-CdTe films deposited on SnO_2 are the only samples that produce the double grain structure as illustrated in Fig. 2. Films sputter-deposited on CdS/ITO substrates show only very small grains (700 Å) without the larger grain structure. The PVD-deposited films have an average initial grain size of 3000 Å, independent of substrate. CSS-deposited films show a relatively large initial grain size of 1.5 µm . We have studied the changes in microstructure using the standard $CdCl_2$ heat treatment at 300°, 350°, and 400° C. Grain sizes for PVD and RFS films

Table 1. Summary of evolution of CdTe grain sizes with processing for various growth techniques and substrates.

Growth Method	Film Composition	Grain Size* (μm)				Plot Symbol
		As deposited	$CdCl_2$ 300° C	$CdCl_2$ 350° C	$CdCl_2$ 400° C	
CSS	CdTe/CdS/ITO/Glass	1.5	1.5	1.5	1.5	■
PVD	CdTe/ITO/Glass	0.3	0.3	0.4	1.5	◆
PVD	CdTe/CdS/ITO/Glass	0.3	0.4	0.4	1.5	●
RFS	CdTe/SnO_2/Glass	0.4	0.4	0.5	0.5-0.8	▼
		0.04	0.04	0.04		
RFS	CdTe/CdS/ITO/Glass	0.07	0.04	0.07	0.2-0.7	▲

* Grain sizes are averages over a distribution of sizes.

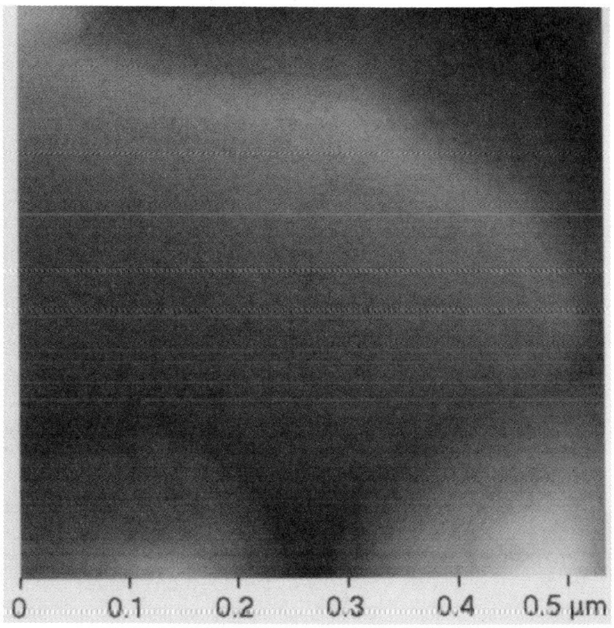

Figure 3. High resolution AFM image of RFS-CdTe/SnO_2 film following $CdCl_2$ heat treatment at 400° showing complete coalescence of the nanograins.

show little change at 300° or 350°, but heat treatment at 400° produces significant grain growth. The CSS films started with the largest grain sizes, and showed little change in their microstructure for any processing temperature.

2. OPTOELECTRONIC PROPERTIES

We have correlated our nanostructural investigations of there films with their electronic properties using optical methods. Photolunimescence (PL) spectroscopy and time-resolved photoluminescence (TRPL) have been used to evaluate the electronic structure and relative defect densities in these films as a function of deposition technique and processing. All of the films follow the general trend of improved PL efficiency and lifetime with processing. Differences in how the films respond to the standard chemical and heat treatments provide clues to the nature of the electronic and structural changes that are key to producing high efficiency CdTe/CdS PV cells.

A. Experimental

PL and TRPL measurements were conducted using the same experimental apparatus. Photoexcitation was provided by a mode-locked, cavity-dumped dye laser operating at repetition rates of either 1 Mhz or 38 Mhz at a wavelength of 600 nm. At room temperature, photoexcitation densities on the order of 10^{16} cm^{-3} were necessary to achieve adequate PL intensities for measurements. Photoluminescence was collected and then dispersed in a Spex model 1672S 0.25 m subtractive double spectrometer. Single photon detection was achieved using a Hamamatsu 2809U microchannel plate photomultiplier. Time resolved measurements were performed using standard time-correlated single photon counting techniques[7]. System resolution is approximately 20 ps using deconvolution techniques. Photoexcitation and photoluminescence collection were through the glass substrate. For photoexcitation at 600 nm, CdS is transparent whereas the absorption depth in CdTe is several thousand angstroms. Hence carriers are excited only in the CdTe absorber near the CdTe/CdS junction. Because of this the PL and TRPL measurements are sensitive to the material near the CdTe/CdS junction, in contrast to the AFM measurements, which probe the material at the top of the CdTe layer.

B. Results and Discussion

Photoluminescence lifetime and radiative efficiency are two optical properties that are quite sensitive to the concentration and type of defects present. Because defects are concentrated at the grain boundaries these parameters should also be quite sensitive to grain size. Figure 4 presents the relative radiative efficiency of the five types of samples as a function of processing conditions. It is clear that the radiative efficiency for as-grown and 300°C treated samples is virtually nil. This indicates a high concentration of defects which act as fast, non-radiative recombination centers. These defects are typically deep levels near mid gap which serve as centers for Shockley-Reed-Hall (SRH) recombination. With the exception of the RFS-CdTe/CdS, all samples treated at 350°C show a significant increase in radiative efficiency, with a much larger increase for the samples treated at 400°C. We believe the anomalous behavior of the RFS-CdTe/CdS samples is due to structural damage occurring during heat treatment, which is described in another publication.[8] The relation between radiative

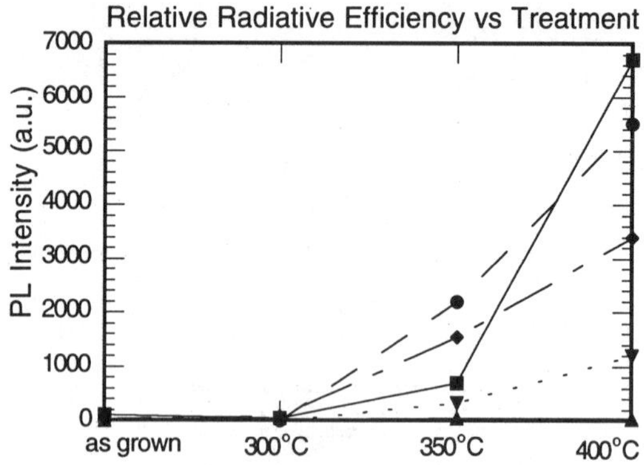

Figure 4. Room temperature photoluminescence intensities versus processing conditions for each of the five types of samples. ■ CSS-CdTe/CdS, ◆ PVD-CdTe/ITO, ● PVD-CdTe/CdS, ▲ RFS-CdTe/CdS, ▼ RFS-CdTe/SnO.

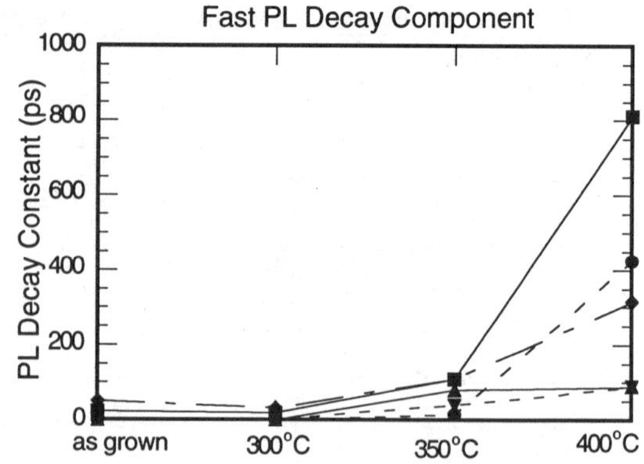

Figure 5. Photoluminescence decay time constant at room temperature versus processing conditions for each of the five types of samples. ■ CSS-CdTe/CdS, ◆ PVD-CdTe/ITO, ● PVD-CdTe/CdS, ▲ RFS-CdTe/CdS, ▼ RFS-CdTe/SnO.

efficiency and grain size for the PVD samples and the RFS samples is qualitatively as one would expect. It has been shown that if bulk and surface recombination rates remain unchanged with processing, there should exist a direct linear relation between PL efficiency and grain size[9]. Our results show that PL efficiency increases much more rapidly than grain size. This indicates that the $CdCl_2$ heat treatment not only promotes grain growth, but also passivates defects. This conclusion is reinforced by the results for the CSS samples where there is virtually no change in grain size, yet a large increase in radiative efficiency. The high optical quality of the 400° treated CSS film indicates that the $CdCl_2$ heat treatment is a very effective agent for passivation of the defects responsible for fast non-radiative recombination.

Photoluminescence decay in polycrystalline CdTe films typically shows a two-component decay. The early, fast component is produced by radiative and non-radiative recombination, and carrier trapping in shallow defect levels. The late, slow component is affected by all three of the former factors, yet is primarily governed by the rate of escape of carriers from the shallow traps. Because non-radiative recombination is the dominant decay mechanism in these films, the fast PL decay component is the best indicator of material quality. Fig. 5 illustrates the dependence of the fast decay component on processing for each of the five sample types. These results confirm our interpretation of the radiative efficiency data. PL lifetimes are quite short initially, but show dramatic improvement after the heat treatment. The percentage improvement in PL lifetimes corresponds well with improvements in PL efficiency. The final PL lifetimes and efficiencies correspond well with relative grain sizes among the various samples. As discussed in footnote [8], we believe the RFS samples suffer structural damage during heat treatment, leading to poor optical and electronic quality.

Table 2 lists the peak PL wavelength for each sample in the processing matrix. A trend worthy of note in this data is that the as-grown samples produce PL at 820nm, corresponding to 1.51 eV - the band gap of CdTe at room temperature. The intermediate processing temperatures produce intermediate wavelengths. The interesting result occurs for the 400° processed samples. All of the samples deposited on CdS substrates produce PL at 845 nm, or 1.47 eV. Samples deposited on conducting oxides still luminesce at or near the CdTe band gap - 820nm. To determine the origin of this shift we measured PL at 77K for the four CSS deposited samples. All of the samples produced two distinct emission bands, similar to spectra measured in MBE-grown single crystal CdTe at 77K. We have identified the higher energy emission at 1.55eV as band-to-band photoluminescence and the broad, lower energy emission at 1.4eV as photoluminescence associated with defects[10]. This assignment is also verified by the relative intensity depedences of the two PL emission bands. At high intensities the higher energy, band-to-band emission dominates, while at low intensities the defect emission dominates[11]. We conclude that samples deposited on CdS substrates have undergone a band-gap reduction of 45 meV during the 400° heat treatment. Data on band-gap versus composition or the ternary CdTeS indicatef that CdTeS with approximately 2% sulfur would have a band gap of 1.47 eV[12]. As discussed previously, our PL measurements are sensitive only to the material near the CdTe/CdS interface. The observed evolution of the PL

Table 2. Room temperature photoluminescence peak wavelength for each sample studied.

Growth Method	Film Composition	PL Peak Wavelength (nm)				Plot Symbol
		As deposited	$CdCl_2$ 300° C	$CdCl_2$ 350° C	$CdCl_2$ 400° C	
CSS	CdTe/CdS/ITO/Glass	820	820	828/845	845	■
PVD	CdTe/ITO/Glass	820	825/850	832/880	824	◆
PVD	CdTe/CdS/ITO/Glass	---	---	825	845	●
RFS	CdTe/CdS/ITO/Glass	---	---	840	845	▼
RFS	CdTe/SnO/Glass	---	---	820	820	▲

maximum for samples deposited on CdS indicates diffusion of sulfur into the CdTe near the interface. This result is consistent with a previous report of sulfur/tellurium interdiffusion occuring during $CdCl_2$ heat treatment. That study utilized x-ray diffraction, optical absorption, and quantum efficiency measurements to verify changes in band gap and lattice constant[13].

Although our measurement technique is not sensitive to the exact composition vs. distance from the junction, based on the physics of the diffusion process it is reasonable to assume that the highest percentage of sulfur occurs nearest the junction, and decreases away from the junction. We believe this mixing effect at the CdTe-CdS heterojunction could produce a graded bandgap in the CdTe absorber layer similar to that intentionally engineered into copper-indium-gallium-selenide thin films used for PV applications[14]. Such a graded bandgap could be a significant factor in the efficiency of carrier collection in CdTe/CdS PV cells.

Conclusions

Our proximal probe measurements of this sample matrix conclusively demonstrate that the nanostructure of polycrystalline CdTe films depends on both substrate and deposition method. We have observed a unique 'double-grain' structure in RFS CdTe deposited on SnO_2 substrates which eventually disappears with $CdCl_2$ heat treatment.

The correlations we have established between the nanostructure and optical properties of polycrystalline CdTe thin films prove that not only does the $CdCl_2$ heat treatment enhance grain size, but also passivates defects. In the case of the CSS deposited films, the $CdCl_2$ heat treatment passivates defects without producing any change in grain size. Finally, we have verified that sulfur diffuses across the CdTe/CdS interface during heat treatment. The resultant graded bandgap in the CdTe absorber layer may be an important element in the optoelectronic properties of the finished CdTe/CdS photovoltaic cell.

REFERENCES

[1] C. Ferekides, J. Britt, Y. Ma, and L. Killian, *Twenty-Third IEEE PVSC*, 1993, pp. 389.

[2] Art J. Nelson, F. Hasoon, and Dean Levi, *J. Vac. Sci. Technol. A*, **12**(5), 1994, pp. 2803.

[3] J. Frommer and E. Meyer, *J. Phys.: Condens. Matter*, **3**, S1 (1991).

[4] T.L. Chu, S.S. Chu, J. Britt, C. Ferekides, C. Wang, C.Q. Wu, and H.S. Ullal, *IEEE Electron Dev. Lett.* **13**, 1992, pp. 303.

[5] S. Albright, B. Ackerman, and J.F. Jordan, *IEEE Trans. Electron. Dev.* **37**, 1990, pp. 434.

[6] B.N. Baron, R.W. Birkmire, J.E. Phillips, W.N. Shafarman, S.S. Hegedus, and B.E. McCandless, *Tech. Rpt., Contract No. XL-9-19032-1*, SERI, Golden, Colo., 1990.

[7] R.K. Ahrenkiel, D.J. Dunlavy, T. Hanak, *Solar Cells* **24**, 1988, pp. 339.

[8] AFM microprobe results on the same set of samples which were reported earlier describe microscopic craters formed by apparent explosions of gas trapped below the surface in RFS materials. For more details see: H.R. Moutinho, F.S. Hasoon, F. Abulfotuh, and L.L. Kazmerski in the *Proceedings of the 1994 AVS Symposium*, in press.

[9] R.K. Ahrenkiel, B.M. Keyes, L. Wang, and S.P. Algright, *Twenty-Second IEEE PVSC*, 1991, pp.940.

[10] T.H. Myers, J.F. Schetzina, S.T. Edwards, and A.F. Schreiner, *J. Appl. Phys.*, **54**(7), July 1993, pp.4232.

[11] For a complete discussion of this effect see e.g. J.I. Pankove, *Optical Processes in Semiconductors*, Dover, New York, NY, 1971, Chapt. 6.

[12] R. Hill and D. Richardson, *Thin Solid Films*, **18**1973, pp.25.

[13] Brian E. McCandless and Robert W. Birkmire, *Solar Cells*, **31**, 1991, pp. 527.

[14] Miguel Contreras, J. Tuttle, D. Du, Y. Qu, A. Swartzlander, A. Tennant, R. Noufi, *Appl. Phys. Lett.* , **44**, 1993, pp. 1827.

OPTICAL CHARACTERIZATION OF $CuInSe_2$ GROWN BY MOLECULAR BEAM EPITAXY

S. Niki, Y. Makita, A. Yamada, H. Shibata, P. J. Fons and A. Obara,
Electrotechnical Laboratory, MITI, 1-1-4 Umezono, Tsukuba, Ibaraki 305, Japan
(phone) 81-298-58-5610, (FAX) 81-298-55-9664, (e-mail) sniki@qm.etl.go.jp

T. Kurafuji, S. Chichibu, and H. Nakanishi,
Science University of Tokyo, Noda, Chiba 278, Japan

ABSTRACT

$CuInSe_2$ epitaxial films with near stoichiometric compositions have been grown on GaAs (001) by molecular beam epitaxy, and radiative recombination processes in these films have been characterized by means of low temperature photoluminescence (PL) spectroscopy.

Temperature dependent PL measurements on such epitaxial films have made possible the identification of conduction band to acceptor and donor-acceptor pair transitions. Well-defined emission lines were present near the bandgap, and the emissions at 1.0386eV and at 1.0311eV were found to be due to the ground-state free exciton and the excitons bound to neutral acceptors, respectively. The bandgap of $CuInSe_2$ is also determined to be E_g=1.0462eV at 2K.

INTRODUCTION

Despite the lack of accurate material information, conversion efficiencies of the solar cells based upon $CuInSe_2$ (CIS) have significantly improved up to η=16.9% [1] predominantly by enhancing the bandgap with incorporating gallium and/or sulfur. One of the key issues to be addressed to further improve the CIS-based solar cells is control of defects in CIS. Defect-sensitive characterization techniques have been introduced to investigate the intrinsic defects in CIS, however considerable discrepancies still exist in their assignment and interpretation .

In our previous work, we reported the observation of sharp emission lines from epitaxial CIS and the separation of phonon replicas from substantive emissions [2]. In this work, radiative recombination processes in the epitaxial CIS films grown by molecular beam epitaxy (MBE) have been investigated further by low temperature photoluminescence (PL) spectroscopy.

EXPERIMENTAL

All the CIS films were grown on nominally undoped GaAs (001) by MBE at substrate temperatures of T_s=350-550°C. In-situ RHEED analysis indicated the growth of epitaxial films as well as the chalcopyrite structure [3]. Nominal thicknesses of the specimens were 0.7-0.8μm. Compositions of the CIS films were determined by means of electron micro probe analysis. Electrical properties of the films were characterized by the van der Pauw technique at room temperature. PL measurements were carried out with a one-meter single monochromator, Ar^+-laser excitation (λ=514.5nm), and a liquid nitrogen-cooled Ge detector. The wavelength resolution for the PL experiments was about 16Å. Liquid helium was used for fixed-temperature measurements at 2K by directly immersing the samples. Temperature dependent measurements were carried out using a flowing-type cryostat.

CH3365-4/94/0000-0132 $4.00 © 1994 IEEE

RESULTS AND DISCUSSIONS

General Features of PL spectra

PL spectra at 2K obtained from the CIS films with near stoichiometric compositions are shown in Fig. 1. Distinct emission lines including several band edge emissions can be observed, indicating high crystalline quality. The PL spectra were also found to be strongly dependent on growth temperature. All the films were strongly p-type, and the intensities of peaks c and d, and the hole concentrations of the films are plotted as a function of growth temperature in Fig. 2. Both the intensity of peak c and the hole concentration decrease with increasing growth temperature. Therefore, peak c is considered to be an acceptor-related emission. On the other hand, the intensity of peak d does not vary substantially with the change in growth temperature. Peak d is significantly weaker for the specimen at T_s=350°C, and strong a peak was correspondingly observed.

Temperature dependenct PL spectra obtained from a sample grown at 450°C are shown in Fig. 3(a), and the intensities of peaks a, c, and d are plotted in Fig. 3(b). Peak c has an activation energy of 80meV in good agreement with the bandgap energy assuming E_g=1.04eV. Peak d on the other hand has an activation energy of 55meV obviously different from the energy from the bandgap of 130meV. It indicates that defect-related donors with an activation energy of 55meV may be involved in this optical transition, thus peak d is considered to be a pair-type emission most likely a donor-acceptor pair emission. Peak a may be due to an exciton-related emission.

Excitonic Emissions

Bandedge emissions have been investigated further. Fig. 4 shows temperature dependent PL spectra of a sample grown at T_s=350°C. Two distinct emissions E_{X1} and I_{X1} appeared near the bandgap. The I_{X1} line was observed only for T_s=350°C, and it was found to be strongly correlated with that of peak d; the donor-type defects suppress the I_{X1} transition. I_{X1} is also very sharp at 7K, but disappears at T~35K. A peak at 1.0031eV, separated by 29meV from I_{X1}, shows a similar temperature dependence to I_{X1}. These results suggest the above emissions to be a bound exciton emission and its LO phonon replica [2], respectively. However, E_{X1} can be observed even at T=102K where the peaks c and d associated with donor- and acceptor-type defects disappear. The linewidth of E_{X1} becomes larger with increasing temperature from 11meV at 35K to 17meV at 102K. We attribute E_{X1} to the ground-state free exciton emission.

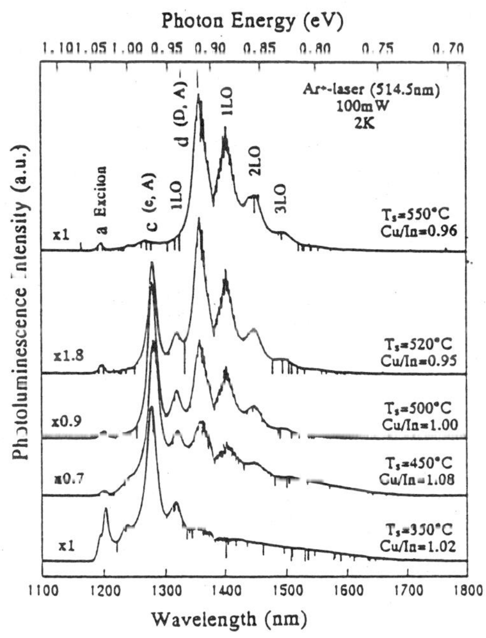

Fig. 1 PL spectra of $CuInSe_2$ epitaxial films grown at T_s=350-550°C.

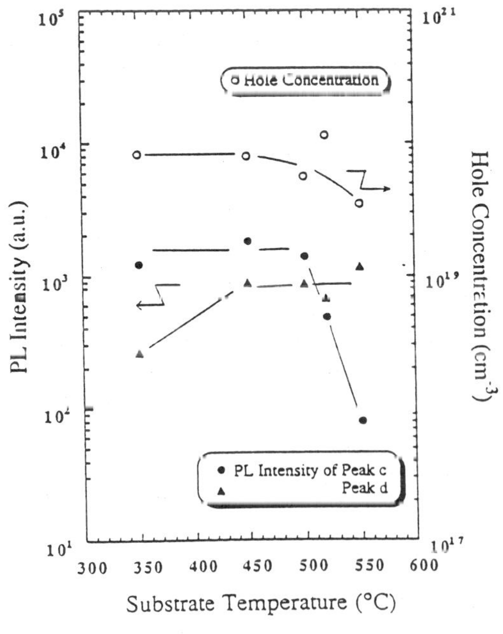

Fig. 2 Intensities of peaks c, d and hole concentrations as a function of the growth temperature.

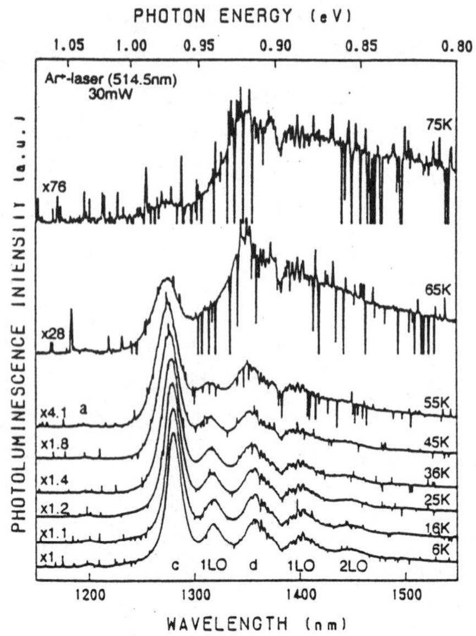

Fig. 3 (a) Temperature dependent PL spectra of a sample grown at Ts=450°C

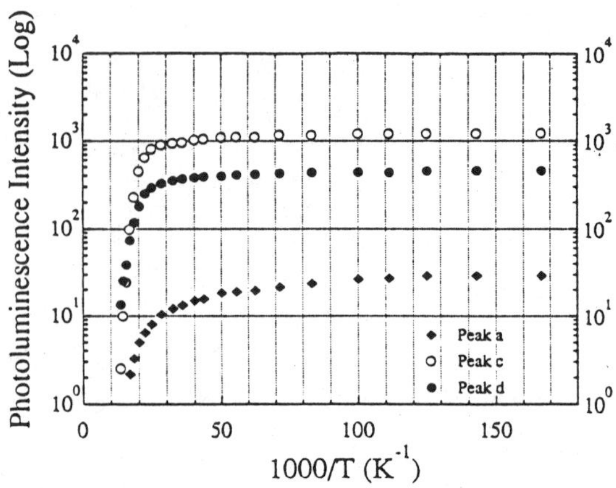

Fig. 3 (b) Intensities of peaks a, c, and d as a function of temperature

The bandgap of CIS can be calculated from the energy of the ground-state free exciton emission by $E_g=E_{X1}+E_{ex}$ where E_g is the bandgap energy of CIS and E_{ex} is the binding energy of the exciton.

E_{ex} can be calculated by the following expression [4]

$$E_{ex}=\frac{m_r^* q^4}{2h^2\varepsilon^2 n^2} \qquad (1)$$

m_r^* is the reduced mass, $m_r^*=m_e^*\cdot m_h^*/(m_e^*+m_h^*)$ with m_e^* and m_h^* being the electron and hole effective masses. ε is the dielectric constant, and h is Planck's constant. n is an integer, and is equal to 1 for the ground-state. The parameters used for the calculation were $m_e^*=0.09m_0$[9], $m_h^*=0.09m_0$ [10], and $\varepsilon=9.0\varepsilon_0$ [11]. E_{ex} is calculated to be 7.6meV, thus the bandgap energy of this sample is E_g=1.0462eV at 2K. E_{X1} blue-shifts with increasing temperature, and E_g correspondingly increases monotonically from E_g=1.0462eV at 2K to E_g=1.0488eV at 102K assuming that ε, m_e^*, and m_h^* in eqn. (1) are independent of temperature. Such anomalous behavior of the increase of the bandgap with respect to temperature up to about 100K has often been found in I-III-VI_2 semiconductors [5].

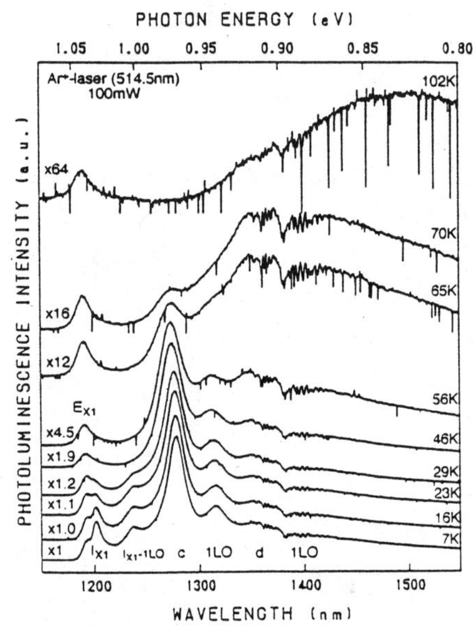

Fig. 4 Temperature dependent PL of a sample grown at Ts=350°C

In the case of CIS, the ratio between electron and hole effective mass, $\sigma=m_e^*/m_h^*$, is nearly equal to 1. Hopfield [6] and Sharma et al.[7] suggested that neither the exciton bound to ionized donors or ionized acceptors can be formed under this condition. On the other hand, the exciton-neutral acceptor complex and the exciton-donor complex are expected to be stable for all values of σ. The energy of the exciton-neutral acceptor complex, $[A^0, X]$, can be estimated adding another term of the required energy to dissociate the complex into free exciton, $C{\bullet}E_A$.

$$E[A^0, X] = E_g - E_{ex} - C{\bullet}E_A \qquad (2)$$

where C~0.1 using Haynes' rule [8]. The calculated energy of $[A^0, X]$ using E_g=1.0462eV, E_{ex}=7.6meV, and E_A=76.2meV is $E[A^0, X]$=1.0310eV in good agreement with the measured peak energy of I_{X1}=1.0311eV. Thus, I_{X1} is attributed to the exciton bound to neutral acceptor.

SUMMARY

The radiative recombination processes in CIS epitaxial films have been identified primarily by PL spectrocopy, and the results are listed in Table 1. Systematic control of MBE growth parameters and annealing conditions of the epitaxial films in conjunction with more detailed material characterization can lead to the reliable interpretation of defects in CIS, eventually making posssible the thorough u control of defects in CIS.

ACKNOWLEDGMENT

Authors (S. N, H. S. P. F, Y. M, A. Y, A. O) would like to thank Dr. H. Yajima for useful discussions and continuous support. We also would like to acknowledge Dr. S. Shirakata for useful discussions and comments.

REFERENCES

[1] J. Hedstrom et al., *Proc. 23rd. IEEE Photovoltaic Specialists Conf.*, Lousville, Kentucky, 1993.

[2] S. Niki et al., "Sharp Optical Emissions from $CuInSe_2$ Thin Films Grown by Molecular Beam Epitaxy", *Jpn. J. Appl. Phys.* **33**, 1994, p.L500.

[3] S. Niki et al., "Heteroepitaxy and Properties of $CuInSe_2$ on GaAs (001)", to be published in the *Proceedings of 8th International Conference on Molecular Beam Epitaxy*, Osaka Japan, August (1994).

[4] J. I. Pankove, *Optical Processes in Semiconductors*, Dover Publications Inc., Chap. 1, p12 (1971).

[5) P. W. Yu, "Radiative Recombination in Melt-Grown and Cd-implanted $CuInSe_2$", *J. Appl. Phys.* **42**, 1976, p.677.

[6] J. J. Hopfield, *Physics of Semiconductors* (Proc. 7th. Int. Conf.), Dunod, Paris and Academic Press, p.725 (1964).

[7] R. R. Sharma and S. Rodriguez, "Theory of Exciton Bound to Ionized Semiconductors", *Phys. Rev.* **153**, 1967, p.823.

[8] J. R. Haynes, "Experimental Proof of the Exisitence of a New Electronic Complex in Silicon"*Phys. Rev. Lett.* **4**, 1960, p.361.

[9] H. Weinert et al., *Phys. Status Solidi (b)* **81**, 1977, K59.

[10] G. P. Sanchez-Porras and S. M. Wasim, *Phys. Status Solidi (a)* **59**, 1980, K175

[11] V. Piede et al., *Solid State Commun.* **28**, 1978, 449.

Peak	E_{X1}	I_{X1}	c	d
Energy (eV)	1.0386	1.0311	0.9700	0.9126
Assignment	exciton bound to neutral acceptor	free exciton n−1	conduction band to acceptor	donor-acceptor pair

Table 1 Energies and assignment of emission lines, E_{X1}, I_{X1}, c, and d.

TAILORING THE ELECTRONIC PROPERTIES OF ONE STEP ELECTRODEPOSITED $CuInSe_2$ FILMS BY ANNEALING IN Se VAPOR UNDER CONTROLLED ACTIVITY CONDITIONS. ELABORATION OF EFFICIENT CIS / ZnO SOLAR CELLS

Daniel Lincot, Jean-François Guillemoles, Pierre Cowache, Sylvie Massaccesi,
Laurent Thouin, Kamel Fezzaa, Frederic Boisivon and Jacques Vedel

Laboratoire d'Electrochimie et de Chimie Analytique, CNRS, URA 216
Ecole Nationale Supérieure de Chimie de Paris
11 rue Pierre et Marie Curie - 75231 Paris Cedex 05, France.

ABSTRACT

Cu-In-Se alloys can be electrodeposited in a wide range of controlled composition. Annealing treatments under Se pressure transform these precursor films in large grained $CuInSe_2$ films with electronic properties suitable for solar cells, as shown by photoluminescence, resistivity, Hall effect, XRD and spectral responses measurements. Best CIS/CdS/ZnO cell yielded 6,5% efficiency.

INTRODUCTION

In the field of polycrystalline thin films solar cells, electrodeposition is among the most promising technique with respect to large area processing, low cost and high material yield. In the two emerging candidates, namely CdS/CdTe and $CuInSe_2$/ZnO cells, electrodeposition is already used for CdTe in CdS/CdTe cells [1]. However, for $CuInSe_2$/ZnO cells, until yet, electrodeposition has given poor results. We achieved a dramatic improvement by combining electrodeposition and annealing treatments in a reacting selenium atmosphere. In this paper we expose the properties of electrodeposited/selenized films.

RESULTS AND DISCUSSION

1 - One step electrodeposition of $CuInSe_2$

$CuInSe_2$ films were prepared by reducing copper (II) selenium (IV) and Indium (III) species in acidic solution at a fixed potential value, between -0,7 V and -1,1 V/MSE. The overall composition of the film is monitored by the deposition potential and the composition of the solution as shown in the zone diagram in figure 1.

As demonstrated by electrochemical studies [2] the composition is controlled by the ratio between Cu(II) an Se(IV) fluxes at the electrode surface ($\alpha = J_{Se}/J_{Cu}$, figure 1). For $\alpha < 2$ the film is copper rich and for $\alpha > 2$ the film becomes indium rich in a range of deposition potential. Changing the applied potential allows further possibilities to tune the morphology and the overall composition of the film to form a wide variety of precursor films. The precursor films studied in this work were prepared at -1 V (vs. MSE), with α = 2,1 for In-rich and 1,9 for Cu-rich films.

2 - Annealing treatments of the films

Annealing treatments dramatically improve the properties of the as-grown films. From our experience annealing treatments in vacuum or neutral (Ar, N_2) atmospheres result in poor electronic quality films despite correct structural and optical properties [3]. Our option has been to develop annealing treatments in an elemental selenium atmosphere. The originality of the process is to use a closed reactor allowing to impose well controlled Se pressures in a wide range from 10^{-5} to 10^{-1} atm. and to fix independently the substrate and the selenium source temperatures. This application-oriented approach has the great advantage to process the films in an uniform way, under near equilibrium conditions and with no loss of selenium from the reactor.

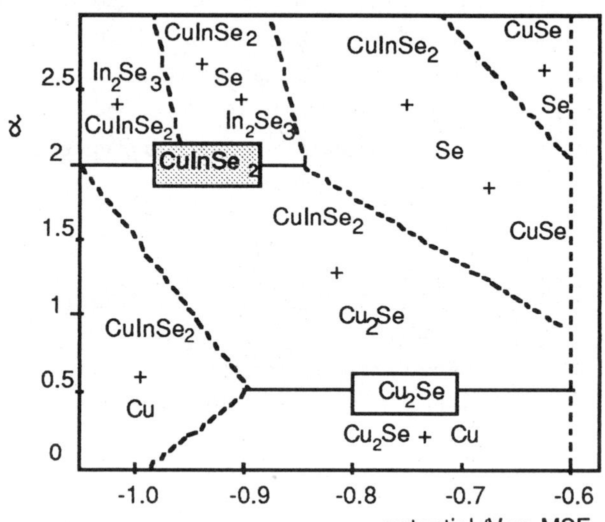

Fig. 1 : Qualitative phase diagram of electrochemically deposited compounds in the Cu-In-Se system.

3 - Structural studies

Annealing treatments dramatically improve the structural quality of the films. Before annealing, they are composed of very small grains (≈ 50 nm) and they are polyphasic with the presence of $CuInSe_2$ and excess In_2Se_3 or Cu_2Se. As compared to annealing treatments in inert atmospheres, annealing under elemental Se atmosphere enhances markedly the grain growth [4]. In the higher

CH3365-4/94/0000-0136 $4.00 © 1994 IEEE

First WCPEC: Dec. 5–9, 1994; Hawaii

pressure range, a spectacular recrystallization of the film can be achieved with grains extending all over the film with sizes up to 10 µm. It is important to note that this process takes place at rather low substrate temperatures (400-450°C). As a consequence the recrystallization is not due to Cu_2Se melting (T = 525°C) as believed in the case of evaporated films [5], but to a sintering effect induced by the selenium itself.

Fig. 2 : SEM micrographs of CIS films after selenization. Top, a Cu-rich film; bottom, an In-rich film.

In this process, the composition of the precursor film has a great influence, the onset of a strong recrystallization depending on the ratio Cu/In in the film. In figure 2 are shown micrographs of recrystallized films with different Cu/In ratios. Indium rich films remain very dense whereas going to Cu rich films leads to larger grain formation .

4 - Transport properties

The main problem in assessing the transport properties of electrodeposited films is the presence of the conducting substrate. Two methods were employed to circumvent this difficulty. Resistivity measurements of films annealed in an inert atmosphere for different Cu/In ratios are shown in figure 3. For the Cu rich film the decrease of the resistivity with the temperature indicates that transport is limited by phonon scattering. This is probably due to the presence of a quasi-metallic Cu_xSe phase. For indium rich or stoichiometric films, the behaviour corresponds to that of degenerate semiconductors ($N_A > 10^{19} cm^{-3}$) explaining the poor photovoltaic properties of these films.

Figure 4 shows the resistivity curve for an indium rich film before and after selenization. In that case the shape of the curve indicates a well defined semiconducting behaviour. With the help of Hall effect measurements, these films are of p type with mobilities comprised between 10 and 50 $cm^2/V/s$ in the 10^{16}-$10^{17} cm^{-3}$ range. The shape of the resistivity curve can be modellized by assuming a control of transport properties by barriers at grain boundaries, as it is expected for such doping levels, with a gaussian distribution of barriers heights [6.] centred at $E_b \approx 150$ meV, with a deviation of ± 40 meV, in good agreement with the model of Seto [7].

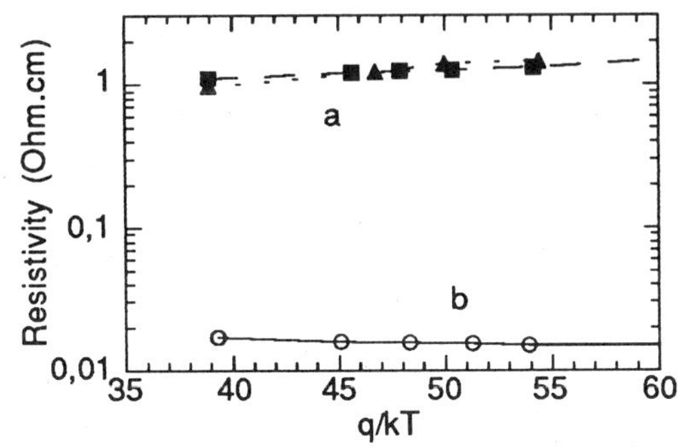

Fig.3: Resistivity vs. temperature plots for stoichiometric and In-rich films, squares and triangles resp., (a), and a Cu-rich film (b), annealed at 400°C in a nitrogen atmosphere.

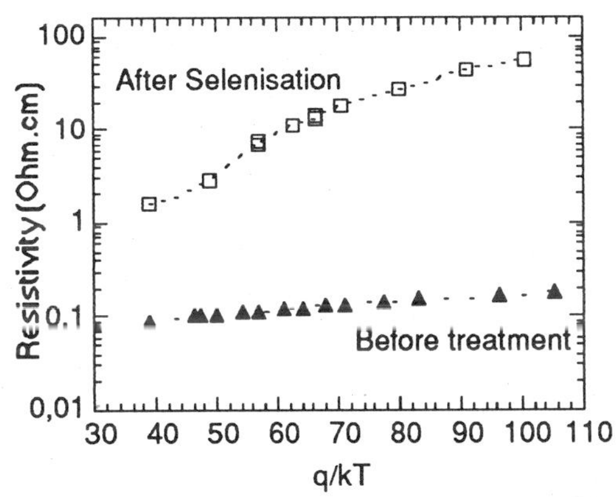

Fig.4: Resistivity vs. temperature plots for a stoichiometric film before and after selenization.

As a consequence, after heat treatment in selenium atmosphere the film becomes suitable for photovoltaic applications. For comparison the resistivity/temperature behaviour of the as-grown film is shown in figure 4. It appears also as a degenerate semiconductor. This explains why it can be deposited cathodically.

5 - Photoluminescence studies

Photoluminescence is a useful tool to assess the electronic quality of CIS or CIGS films [4,8,10,11]. Films annealed in neutral atmospheres have shown no luminescence signal but once annealing is carried out in a selenium atmosphere, a photoluminescence signal appears. It increases with the selenium pressure, and the position of the maximum is displaced towards higher energy as shown in figure 5. In that case the shape of the luminescence signal is similar to that of a coevaporated film.

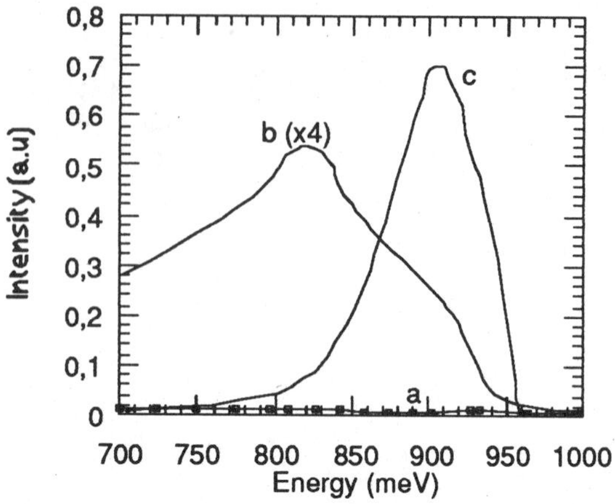

Fig.5: Photoluminescence of stoichiometric films annealed at 450°C under various selenium pressures. (a) P=0 , (b) P=10^{-4} atm. , (c) P=10^{-2} atm.

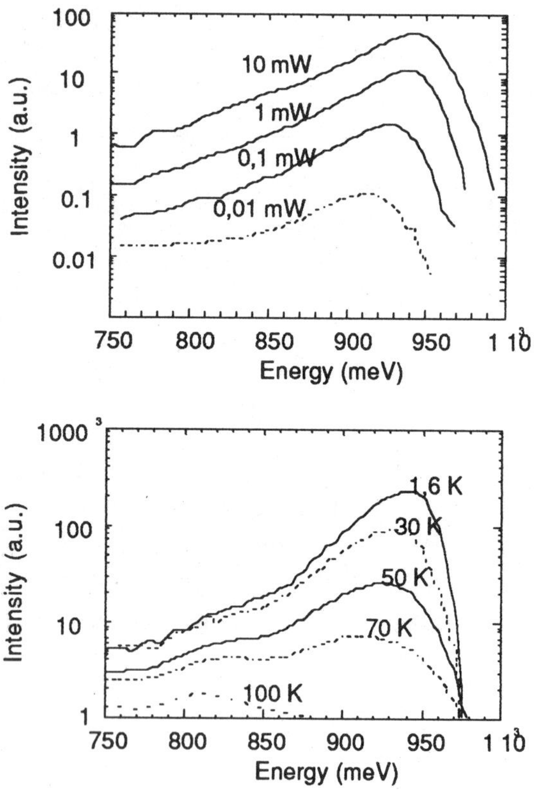

Fig.6 Photoluminescence of stoichiometric films . Top: under different power excitation density. Bottom : at different temperatures.

We have studied in detail the influence of temperature and light intensity on the luminescence spectra. In figure 6 we show for instance the influence of the laser intensity and temperature. It appears that the peak maximum is moving to higher energies with increasing laser intensity whereas the slope in the low energies remains almost intensity independent. This behaviour gives insights in the mechanism of emission in these films. We have interpreted it [4,11] as related to donor-acceptor pair recombination, probably through V_{Cu} and In_{Cu} entities. From the temperature dependence, we infer that the donor states are located at about 5-10 meV from the mobility edge.

A disorder parameter can be extracted from the slope of the low energy branch of the luminescence peak, corresponding to the Urbach slope in absorption studies. Values of 30 meV are obtained for our films, to be compared with ~ 20 meV for coevaporated films. These values, larger than for classical semiconductors (a few meV), indicate some alloying disorder effects in the films.

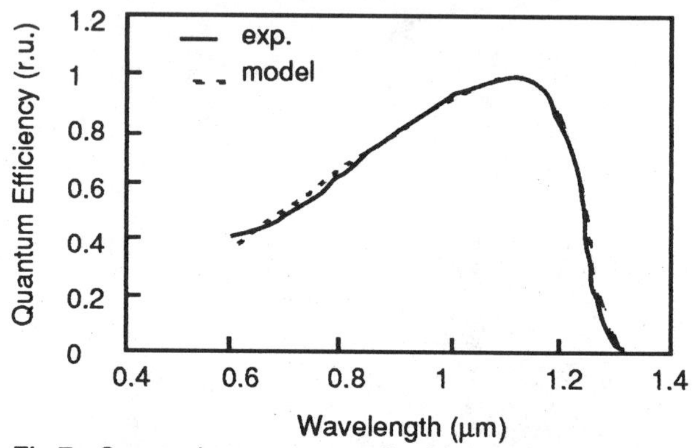

Fig.7: Spectral response (liquid junction) of a Se-annealed film and its fit with the model. Values found for L_n and dead layer thickness are about 2 μm and 150 nm resp. in this case.

6 - Photoelectrochemical characterisation

To assess the photovoltaic quality of the films, measurements have been carried out in with an electrolyte junction, using an acidic electrolyte. As compared to film annealed in an inert atmosphere, films annealed in selenium atmosphere give well defined p type behaviours. In figure 7 are shown typical spectral responses obtained with these films. The short wavelength decay of the spectral responses can be analysed by assuming the presence of a dead layer at the surface, as has been made for completed devices [9]. A fitting procedure on the relative shape of the curves allowed us to determine effective values of the dead layer thickness, space charge width and diffusion length. The absorption coefficients were determined previously [3]. A good agreement is observed between the estimated thickness of the dead layer and the measured value from etching experiments, indicating the consistency of the model. On the best measured devices, we found a dead layer of 0.035 μm with a diffusion length of 0.6 μm which is consistent with the present results [9,10.]

The most remarkable results is the large values of the diffusion length which are derived, up to 2 μm, indicating that the films really present a good quality. These films were generally associated with a thicker dead layer (above 100 nm). An extended model was used in these cases to consider the finite thickness of the layer, the possibility of reflection on the back surface and the recombination velocity on the back contact. It can be remarked that the back contact recombination velocity is best fitted by zero, contrariwise to the infinite value generally assumed for an ohmic contact. This suggest a back surface field effect in this structure that could possibly be associated with the presence of an $MoSe_2$ layer at the interface [11] since it is p type when produced under excess Se pressure.

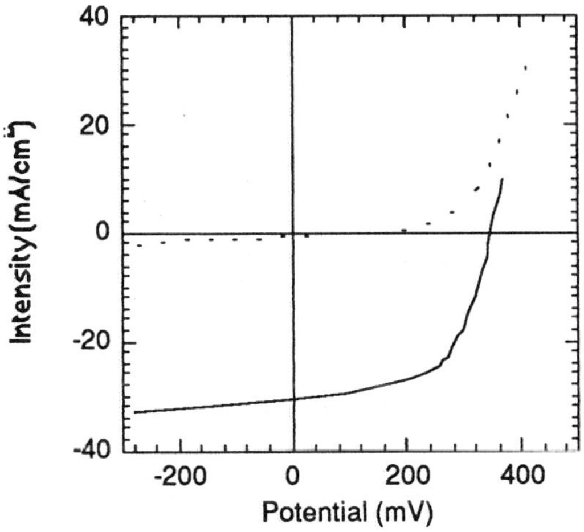

Fig. 8 I-V plot of the best cell elaborated under light (87,5 mW/cm²) and dark conditions. Total area : 0,1 cm². Efficiency is 6,5% under AM 1,5 without A.R. coating.

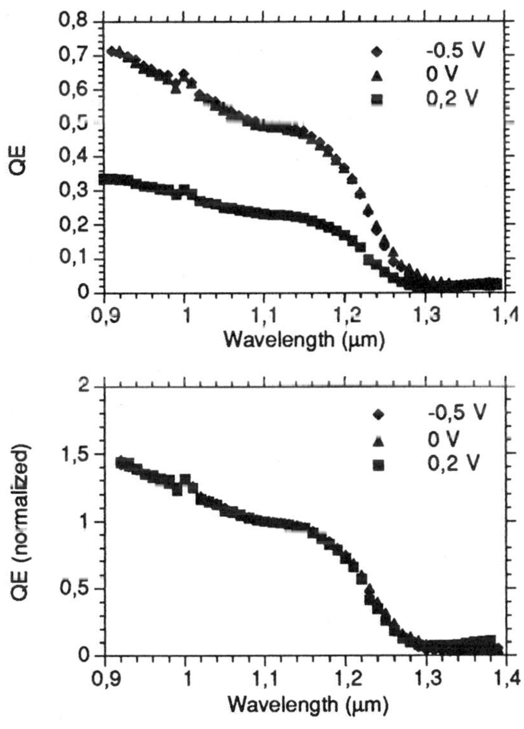

Fig. 9: Top, external Spectral responses of a cell at various bias. Bottom, idem but normalised.

7- Devices

Devices with the Mo/CIS/CdS/ZnO structure have been elaborated with these films [3,9,10,11]. Figure 8 shows the I-V curves of a 6.5 % cell (without antireflection coating). The low values of the open circuit voltage as compared to that of evaporated devices, are due to recombination losses at the interface [9,10,11], whereas the reasonable values of the short circuit current (30-35 mA/cm²) confirm the intrinsic quality of the semiconductor. The interface recombination effect is shown on figure 9 where spectral responses recorded at different biases are presented. They indicate clearly that the collection efficiency is a function of bias only, a characteristic of surface recombination.

CONCLUSION

Selenization of electrochemically deposited precursor films of CIS has been demonstrated as a efficient process to prepare high quality films. Controlling both the precursor composition and the selenium pressure allowed to tailor the characteristics of the film in a range suited for photovoltaic application : 10^{16}-$10^{17}cm^{-3}$, p type, large diffusion length.

Present limitation in cell efficiency appears to be related to losses at the interface (recombination, dead layer) and not to bulk properties of the CIS film. Further improvement needs to optimise surface treatments on selenized films before junction completion.

ACKNOWLEDGEMENTS

This work is supported by the Commission of the European Communities under a JOULE contract "EuroCis" (JOUR-0045-C) and the ECOTECH-ADEME (contract N°94N80/0020).

REFERENCES

1 J.M. Woodcock, & al., 12th Europ.Photov. Solar En. conf., Amsterdam, 1994, 948.

2 L.Thouin & al., *Electrochimica acta*, **38**(1993)2387
S. Massaccesi & al., *J.Electrochem. Soc.*,**140**(1993)2540
L.Thouin & al., *J. Electroanal. Chem.*, **374**(1994)81

3 L.Thouin & al. *Solid state Phenomena*, **37-38** (1994)527

4 J.F. Guillemoles & al., *Adv. Mat.* **6**(1994)376

5 R. Klenk & al. *Solid state Phenomena*, **37-38** (1994)509

6 J.H. Werner, *Solid state Phenomena*, **37-38** (1994)213

7 J.Y.W.Seto, *J. Appl. Phys.*, **49**(1978)5247

8 M. Tanda & al., *Jpn. J. Appl. Phys.*, **32**(1993)1913
R.E. Hollingsworth & al., 18th IEEE conf., 1985, 1409.

9 J.F. Guillemoles & al., *Adv. Mat.* **6**(1994)379

10 J.F. Guillemoles & al., 12th Europ.Photov. Solar En. conf., Amsterdam, 1994, 1550.

11 J.F. Guillemoles, PhD Thesis, Univ. Paris VI, 1994.

PROCESS DEVELOPMENT FOR $CuInSe_2$-BASED POWER MODULES

J. Britt, A. Delahoy, G. Butler, F. Faras, A. Muthaiah,
A. Sizemore, F. Ziobro and Z. Kiss
Energy Photovoltaics, Inc.
P.O. Box 7456, Princeton, NJ 08543, USA

ABSTRACT: We report further advances in $CuInSe_2$ photovoltaic technology aimed at a low cost manufacturing process. The incorporation of Ga into CIS has led to aperture area efficiencies over 8% for areas near one square foot. A modified selenization of Cu/In_xSe_y stacks appear to be insensitive to variations in the Cu/In ratio. In-line manufacturing equipment has been constructed to prepare 50 x 125 cm modules with a power output of 50 watts.

INTRODUCTION

Thin film photovoltaic devices based on a copper indium diselenide (CIS) absorber have been demonstrated capable of relatively high conversion efficiency. Laboratory test cells with efficiencies greater than 16% have been produced by several groups [1,2]. On another level, the production of an 0.4 m^2 CIS power module generating 40.8 watts is equally impressive [3]. These accomplishments strengthen CIS's position as leading candidate for commercial thin-film power modules.

EPV is committed to being among the first to introduce CIS photovoltaic modules to the commercial marketplace. Our approach is to identify and develop manufacturing techniques that are capable of high rate, uniform deposition over large areas, yet are relatively insensitive to variations in process parameters. All of this must be accomplished without sacrificing module power densities. EPV has assembled a square foot CIS module pilot production line to test concepts designed to overcome manufacturing hurdles. During the past year, EPV has designed, built, and installed in-line equipment capable of depositing all the films required to manufacture 50 x 125 cm 50 watt CIS modules using many of the concepts developed in the square foot pilot line.

Research is being conducted at EPV along various fronts to improve aspects of the CIS deposition process. One goal is to increase module aperture area efficiencies above 10% by alloying CIS with gallium. There are several motivations for incorporating gallium into CIS films. An obvious reason is to increase the energy bandgap and better utilize the solar spectrum. Lower current devices would allow wider and fewer interconnects, leading to higher aperture area efficiencies. It may even be possible to form graded CIGS junctions that have high-open circuit voltage but do not sacrifice short-circuit current density. There have also been reports that gallium in the precursor mixture can improve the adhesion of CIS films fabricated by a two-step technique [4].

Another goal is the replacement of metallic indium in the precursor with an indium selenide which may lead to improved uniformity, better adhesion, and higher throughput. The alloying of copper and indium, necessary in the two-step technique, is sensitive to the deposition process and the relative concentrations of the metals. The presence of undesirable liquid intermetallics in the precursor can result in segregation and non-uniform CIS. This effect will lower the overall efficiency and may lead to poor adhesion of CIS to the molybdenum-coated glass substrate. The formation rate of some CIS reactants is believed to be diffusion limited [5] and bypassing these particular reactions by starting with the "correct" compound may reduce the processing time.

STANDARD PROCESS

EPV has chosen to develop an all-vacuum process for CIS deposition because of the high degree of control possible. The standard CIS process begins by sputtering a molybdenum film onto a soda-lime glass substrate. Copper and indium are sputtered onto the molybdenum film and this multilayer structure is then exposed to an elemental selenium vapor while being ramped in temperature according to a precise schedule. A more complete description of this process is described in previous publications [6,7].

A number of different materials and deposition techniques have been investigated for window layers. The standard device structure is soda-lime glass/Mo/$CuInSe_2$/CdS/ZnO. CdS has been deposited by evaporation and chemical bath deposition (CBD). Modules have been fabricated using either CdS deposition technique and a lower short-circuit current density from the modules with evaporated CdS has been

CH3365-4/94/0000-0140 $4.00 © 1994 IEEE

observed as a result of increased absorption at energies greater than the bandgap of CdS. ZnO is deposited by RF-sputtering from a $ZnO{:}Al_2O_3$ target. A CIS device prepared by EPV's standard process has been measured by NREL to have a total area efficiency of 11.8% and we have previously reported a total area efficiency of 12.5% [8].

The fabrication of modules requires scribing operations be performed between specific material depositions. We have successfully developed a novel interconnect technology that functions very reliably. The technology has been validated by the fabrication of a 9.1% aperture area efficiency interconnected CIS module of area greater than 150 cm^2, verified by NREL. Several modules have been encapsulated and mounted in a outdoor test rack where they have been undergoing stability tests. The chronological record of efficiency for one such module, normalized to a CIS reference module not exposed to the elements, is shown in Fig. 1. The modules prepared by EPV's standard process appear to be durable and stable.

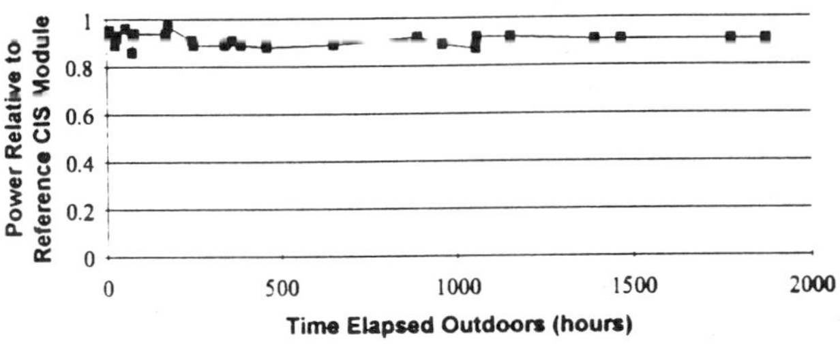

Fig. 1. Outdoor performance of an EPV CIS module after 1800 hours exposure

CIGS FILMS

For atmospheric selenization processes (in H_2Se) the most practical stage to introduce Ga is into the precursors. In an all-vacuum process, Ga can be introduced at virtually any stage of the process by evaporating Ga or an appropriate Ga-containing compound onto the substrate as it is being heated. In this work, we have investigated the effect of incorporating Ga into the CIS reactants at various stages of the selenization process. We have fabricated devices with the standard Mo/absorber/CdS/ZnO structure from absorbers with and without Ga. Devices have been characterized by current-voltage and spectral response measurements. Electron probe microanalysis (EPMA) was used to quantify the relative concentrations of Cu, In, Ga, and Se in the films and x-ray photoelectron spectroscopy (XPS) depth profile analysis was used to qualitatively estimate the distribution of these elements near the film surfaces.

The first attempts to introduce Ga into the absorber were made by evaporating Ga into various locations in the precursor stack (prior to selenization). All absorbers formed in this way showed poor adhesion. In most instances, a very adherent thin film remained on the Mo surface to which the CIS film adhered poorly.

Introducing Ga during the latter half of the selenization process, at temperatures greater than 350°C, proved more fruitful. In the majority of deposition experiments the amount of Ga reaching the substrate was sufficient to provide a maximum bulk Ga/(In+Ga) of approximately 5%. There was little or no difference in adhesion between Ga- and Ga-free absorbers as gauged from tape pull tests. There was, however, a noticeable improvement in the open-circuit voltage and fill factors of some of the devices fabricated from absorbers containing Ga. The illuminated current-voltage characteristics of devices formed from absorbers with and without Ga are shown in Table I. Spectral response measurements of Ga-containing devices show no shift in the cut-off wavelength from that of pure CIS devices. It may be that binary and ternary selenides are formed by this process, not CIGS. It is also possible that a thin layer of CIGS forms only at the surface and the underlying CIS functions as the primary absorber at longer wavelengths.

Table I. The effect of introducing Ga at various substrate temperatures on the illuminated I V characteristics.

Temperature	V_{oc} (Volts)	J_{sc} (mA/cm^2)	FF	Efficiency (%)
370°C	0.463	33.9	0.709	11.1
420°C	0.451	32.0	0.656	9.5
No Ga	0.442	34.7	0.697	10.7

Electron probe microanalysis (EPMA) of absorbers formed by introducing equal amounts of Ga onto identical precursors show a strong dependence of the ratio Ga/(In+Ga) on the substrate temperature at which Ga is introduced. For an amount of Ga reaching the substrate such that the ratio Ga/(Ga+In) is calculated to be 5.3%, the ratios Ga/(In+Ga) as measured by EPMA at 20 kV are 1.0% and 10.4% when the substrate temperatures are 370°C and 420°C, respectively. XPS depth profiles of the relative concentration of Cu, In, Ga, and Se near the surface for the latter two absorber films are shown in Fig. 2. There is significantly more Ga and less Cu near the surface when Ga is introduced at 420°C. A Cu-poor surface may indicate the presence of thin layer of $CuIn_3Se_5$ that has been found near the CIS surface in high-efficiency devices [2]. Introducing Ga at even higher temperatures leads to Ga profiles similar to those found when Ga is introduced at 420°C. There is evidence that CIS formation occurs at approximately 400°C during the two-step technique [5]. If this is the case, the relative diffusion of Ga may slow as other film components cease diffusing at about 400°C.

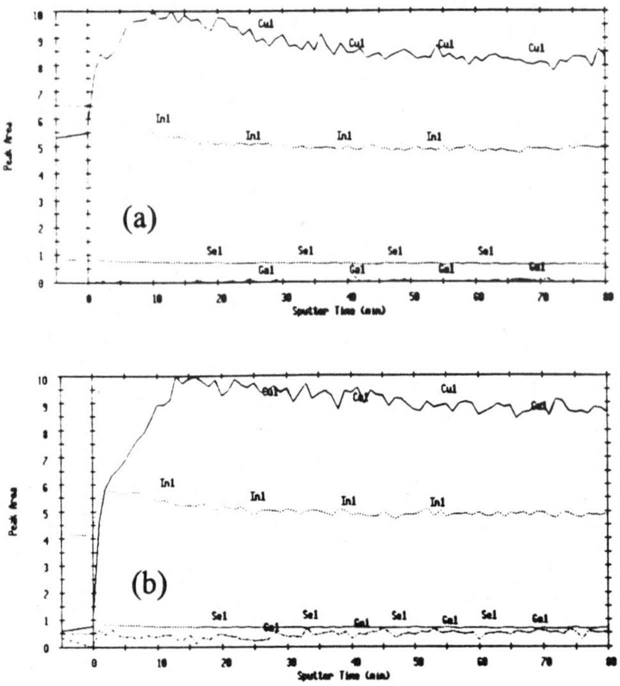

Fig. 2. XPS depth profiles near the surface of CIGS films with Ga introduced at (a) 370°C, and (b) 420°C

The improvement in device performance provided the initial verification of the value of incorporating Ga into the standard process. A marked improvement in uniformity as well as efficiency was noted after the modification of the square foot deposition line to include a step for introducing Ga. Early deposition experiments led to the fabrication of a CIGS-based submodule with a 9.0% aperture area efficiency and over 200 cm^2 area (Fig. 3). Further refinements have been made to the process, and we have recently succeeded in producing a near square foot unencapsulated CIGS-based module with an aperture area efficiency of 8.13%. The I-V curve for this device is shown in Fig. 4.

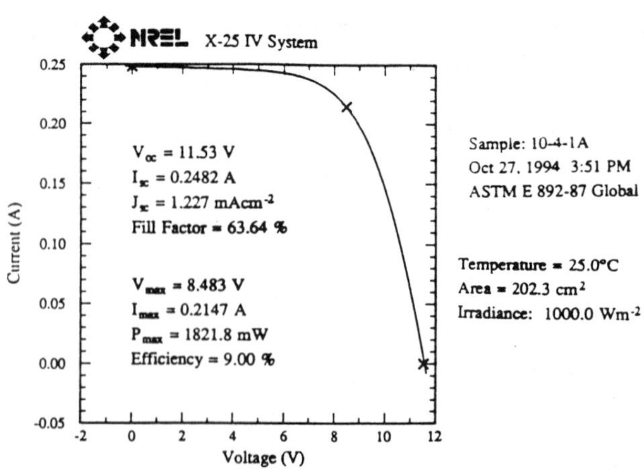

Fig. 3. I-V curve for a 9.0% CIGS-based submodule

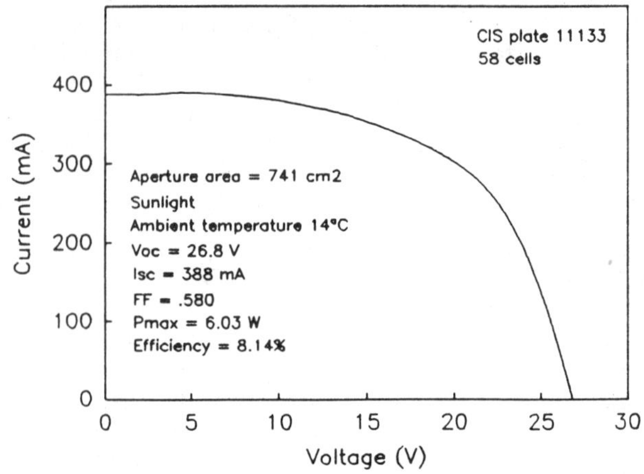

Fig. 4. I-V curve taken in sunlight for a 58 cell CIGS module (8.1% aperture area efficiency, 462 mV/cell V_{oc})

CIS FROM In_xSe_y/Cu STACKS

There are several issues to be considered when attempting to form CIS by selenizing an In_xSe_y/Cu stack. The compound form of In_xSe_y (and Cu_xSe_y), the deposition order, and the selenization thermal profile are all expected to play a significant role in absorber formation. A wide variety of reaction pathways to CIS exist and the reaction pathway taken will likely affect the material properties of the resulting CIS. In general, more thermally stable In_xSe_y compounds require greater external energy (heat) to react with Cu (or Cu_xSe_y) to form CIS [9]. For example, CIS formed by the two-step technique is believed to be the product of InSe and Cu_7Se_4 at about 400°C [5]. CIS formed by the reaction of In_2Se_3 with Cu and Se typically is done at >540°C [2]. Our next generation CIS deposition equipment has no difficulty transporting large glass sheets at temperatures up to 600°C with no deformation, so the final temperature is not an issue. It is necessary, however, that a chosen process have a sequential source approach to maintain manufacturing flexibility.

The use of a single source to deposit In_xSe_y is desirable because of its simplicity. It is imperative that the evaporation source must flux a constant ratio of its constituent components between material replenishment. In_2Se_3 has been previously identified as a "good" evaporation source capable of forming CIS with efficiencies of 13.7% [10]. We have performed many depositions with various source materials in our linear source deposition system to identify suitable indium compounds. One compound has demonstrated reproducible results and we are currently synthesizing it in our lab from the raw materials.

In a series of experiments, the In_xSe_y compound was evaporated onto a Mo/glass substrate heated to 300°C. The resulting films are small-grained and tightly adherent to the Mo substrate. The films have been analyzed by EPMA and XRD. EPMA at 20 kV shows the

In/Se ratio of the films to be 1.16. XRD indicates the presence of a number of In_xSe_y compounds but the predominant two are InSe and In_6Se_7.

CIS films have been formed on square foot Mo-coated soda-lime glass substrates by selenizing sputtered Cu/evaporated In_xSe_y stacked structures. The adhesion of the CIS deposited by this technique is excellent. Devices were completed by depositing thin CdS by CBD, sputtered ZnO, followed by metallization. The best device thus far had Voc, Jsc, and fill factor .385 volts, 35.7 mA/cm^2, 0.672, respectively, and an efficiency of 9.2%.

While trying to determine an optimum Cu/In ratio, a sample set with a large range of Cu/In mass ratios was generated. Surprisingly, the quantum efficiency at long wavelengths (>900nm) of devices fabricated from this set remained high even for the very In-rich samples (Fig. 5). Although the Cu/In ratio was varied by over 30%, the short-circuit current density varied by less than 2 mA/cm^2. The decrease in red response occurs over a much smaller range of Cu/In for CIS formed by the two-step technique [6]. It is possible that excess In is volatilized or remains near the back contact when CIS is formed in this way, far from the region of absorption. Further analysis is necessary to confirm this hypothesis. The insensitivity of this process is tantalizing from a manufacturing perspective. Although only a few deposition experiments have been conducted thus far, we are encouraged by these preliminary results.

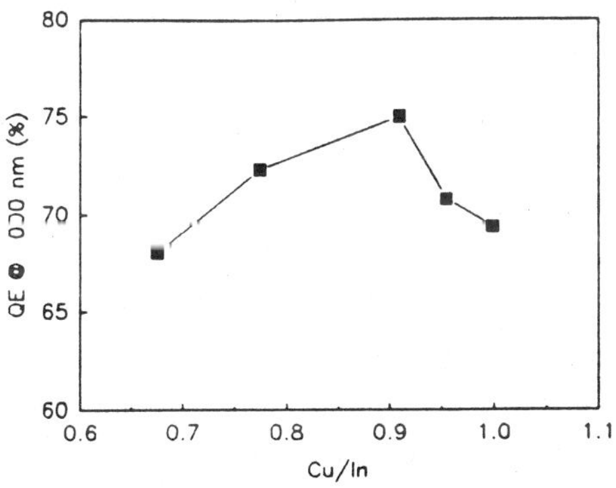

Fig. 5. The effect of the precursor Cu/In ratio on the quantum efficiency of CIS devices at 1000 nm

CONCLUSIONS

EPV has operated and gained experience from a CIS pilot line using square foot soda lime glass substrates. All of the techniques for module preparation including material deposition, scribing, and lamination have been developed. The CIS modules fabricated by our standard process have been demonstrated to have stable efficiencies to date during environmental testing. We have implemented a technique for incorporating Ga near the absorber surface in our square foot pilot line and have achieved module aperture area efficiencies above 8% for areas up to 740 cm^2. Preliminary results indicate that distinct advantages may be gained by replacing metallic indium in the precursor with In_xSe_y. We have begun applying many of these techniques to the manufacture of 50 x 125 cm 50W CIS modules in our new in-line system.

ACKNOWLEDGEMENTS

The authors are thankful to Alice Mason for EPMA measurements, Don Williamson for XRD measurements, Art Nelson for XPS measurements and useful discussion, and B. Tumler and C. Shurts for preparation of the manuscript. This work was partially funded by the U.S. Department of Energy through Subcontract No. XG-2-12051-1 with NREL.

REFERENCES

[1] J. Hedström, H. Ohlsén, M. Bodegård, A. Klyner, L.L.Stolt, D. Hariskos, M. Ruckh, and H W. Schock, *Proc. 23rd IEEE Photovoltaic Specialists Conf.* (IEEE, NY, 1993) pp.364-371.

[2] A.M. Gabor, J.R. Tuttle, M. Contreras, D.S. Albin, A. Franz, D.W. Niles, R. Noufi, *12th E.C. Photovoltaic Solar Energy Conference*(1994) pp. 939-943.

[3] C. Fredric, R. Gay, D. Tarrant, and D. Willett, *Proc. 23rd IEEE Photovoltaic Specialists Conf.*(IEEE, NY, 1993) pp. 437-440.

[4] G. Pollock, K. Mitchell, J. Ermer, U.S. Patent No. 4,915,745, (1990).

[5] S. Yamanaka, B. McCandless, and R.W. Birkmire, *Proc. 23rd IEEE Photovoltaic Specialists Conf.* (IEEE, NY,1993) pp. 607-612.

[6] A.E. Delahoy, J. Britt, F. Faras, F. Ziobro, A. Sizemore, G. Butler, and Z. Kiss, *AIP Conf. Proc.* Vol. 306 (AIP, NY, 1994) pp. 370-381.

[7] A.E. Delahoy, F. Faras, A. Sizomore, F. Ziobro, and Z. Kiss, *AIP Conf. Proc.*, **268** (AIP, NY, 1992; R. Noufi, ed.) pp. 170-176.

[8] A. Delahoy, J. Britt, G. Butler, F. Faras, A. Sizemore, F. Ziobro, and Z. Kiss, *12th E.C. Photovoltaic Solar Energy Conference* (1994) pp. 1612-1615.

[9] D. Albin, J. Cavapella, A. Gabor, A. Tennant, J. Tuttle, A. Duda, R. Matson, A. Mason, M. Contreras, and R. Noufi, *AIP Conf. Proc.*, **268** (AIP, NY, 1992) pp. 108-121.

[10] J. Kessler, D. Schmid, S. Zweigart, H. Dittrich, and H.W. Schock, *12th E.C. Photovoltaic Solar Energy Conference* (1994) pp. 648-652.

THE IMPACT OF CONTROLLED SODIUM INCORPORATION ON RAPID THERMAL PROCESSED $Cu(InGa)Se_2$-THIN FILMS AND DEVICES

V. Probst, J. Rimmasch, W. Riedl, W. Stetter, J. Holz, H. Harms, F. Karg
Siemens AG Corporate Research and Development
Domagkstr.11, D-80807 München, Germany

H. W. Schock
Universität Stuttgart, Institut für Physikalische Elektronik
Pfaffenwaldring 47, D-70569 Stuttgart, Germany

ABSTRACT

The alkali content of $Cu(InGa)Se_2$ thin films fabricated by RTP on a dense molybdenum backelectrode is below the detection limit of ESCA. Starting from this virtually sodium-free case, alkali impurities are successively added in order to study their impact on film morphology and device performance. Two novel techniques have been developed to add sodium to the chalcopyrite thin film: 1. By shifting the stress of the Mo-backelectrode from compressive to tensile its alkali permeability increases and in consequence so does the alkali content in the CIGS film. 2. Adding sodium compounds directly to the Cu-In-Ga-Se precursor film controls the final alkali content in CIGS independent from the substrate.

Along with the addition of sodium a significant increase in performance of CIGS/CdS/ZnO cells (1.85cm^2 active area) was found that peaked at 13.2%.

INTRODUCTION

Soda lime glass substrates have been noted as a key prerequisite for high efficiency chalcopyrite-based thin film solar cells. Conversion efficiencies up to 17% have been achieved [1,2]. Besides its low cost soda lime glass (SLG) shows a near optimum match in thermal expansion coefficient with $CuInSe_2$ (CIS) and $Cu(In_{0.84}Ga_{0.16})Se_2$ (CIGS). Another beneficial influence of alkali-containing substrates has been pointed out recently: During semiconductor formation sodium diffuses from the SLG into the CIS layer and improves grain growth and cell performance [1,3].

On the other hand ESCA measurements on our CIGS films formed by rapid thermal processing (RTP) [4] did not show any sodium signal if a dense molybdenum backelectrode on soda lime glass was deposited. We have demonstrated previously that CIS and CIGS films with a low alkali content show lower conductivity as compared to alkali enriched films [5]. By implanting sodium in alkali-free CIGS films on sapphire substrates we could increase the conductivity to values comparable to CIGS on soda lime glass. Thus sodium was identified as one possible impurity influencing the conductivity in chalcopyrite thin films.

In this work we examine the impact of alkali impurities on film morphology and the device properties. Our virtually alkali-free RTP processed CIGS thin films have been limited in solar cell conversion efficiencies to about 10%. The influence of impurities from the glass substrate can be made apparent by looking at pinholes in the back electrode. OBIC images of CIGS solar cells around defects in the molybdenum back electrode show areas of enhanced photocurrent (figure 1).

The photocurrent is increased inside a halo with a diameter of about 0.5mm around the pinhole in the Mo

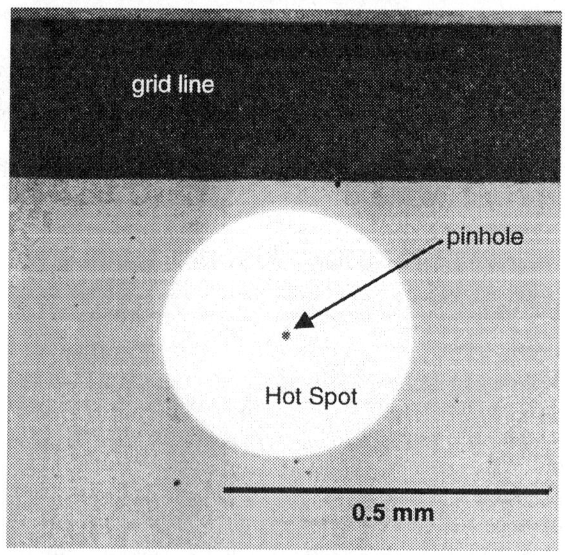

Fig. 1 OBIC-image of a CIGS solar cell showing a circular halo of enhanced photocurrent

CH3365-4/94/0000-0144 $4.00 © 1994 IEEE

First WCPEC; Dec. 5–9, 1994; Hawaii

back electrode. Alkali and earth alkali elements were detected only near the very center of such a halo by EDX measurements taking care to screen out alkalis still located in the glass substrate. As an obvious consequence of this observation we tried to incorporate alkali impurities in CIS in a more uniform fashion.

The method of sodium implantation after CIS-film formation has been used to determine the influence of sodium on electronic transport properties separate from its influence on the CIS morphology. Although very helpful for fundamental investigations this method is impractical for routine sodium incorporation on larger areas during cell fabrication. Hence two alternatives have been developed focusing on a controlled sodium content in CIGS films and solar cells.

EXPERIMENTAL

CIGS thin film formation

The Mo back electrode was DC-magnetron sputtered on 10x10cm^2 soda-lime glass or fused silica substrates. For CIGS film fabrication in a two-step process selenium was evaporated on Cu-Ga-In precursors. Formation of the $Cu(InGa)Se_2$ thin films was performed in a RTP furnace in an inert gas atmosphere as described elsewhere [4].

Solar cell devices were completed by a chemical bath deposition of a thin CdS buffer layer, subsequent CVD-deposition or RF-sputtering of the transparent ZnO-front-electrode and evaporation of a Cr/Al-Grid.

Controlled sodium incorporation

We examined different possible pathways to incorporate sodium into the CIGS films.

An indirect technique to increase the sodium concentration in the CIGS films on Mo/soda lime glass is the variation of the specific alkali permeability of the molybdenum layer. This was accomplished by sputtering Mo at different Ar pressures, which changed the state of stress in the thin film.

As a second even better controllable process option for sodium incorporation into the CIGS films we evaluated the evaporation of sodium selenide in between the molybdenum back electrode and the precursor. In this case both fused silica and sodalime substrates were used to evaluate the influence of sodium on the cell performance. A suitable range for the amount of sodium in CIS was determined empirically with respect to cell performance. We estimate the typical sodium dose to be in a range on the order of $10^{15}cm^{-2}$.

Sample analysis

The amount of sodium at the surface of CIS was determined by ESCA using a monochromatic Al-Kα radiation and the film morphology was examined by SEM. Stress in Mo and CIS was examined by determining the lattice constant with XRD as a function of the tilt angle Ψ. Stress is expressed as a relative change of the lattice constant $\Delta d/d$ as extracted from a plot showing d as a function of $\sin^2(\Psi)$.

Solar cell parameters were measured in the dark and under simulated AM1.5 global illumination (100mW/cm^2).

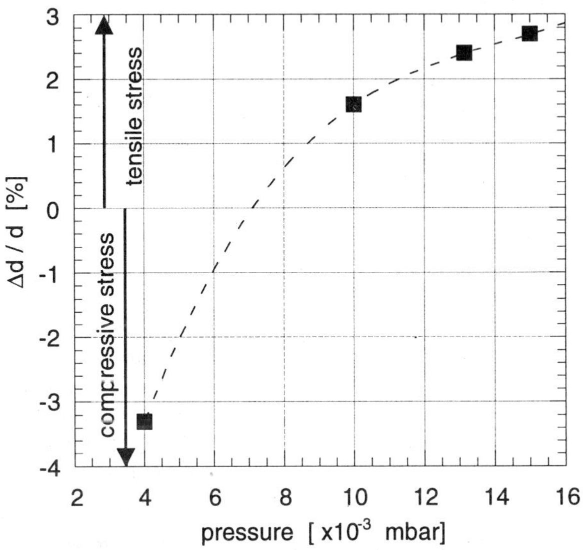

Fig. 2 Macroscopic stress of the molybdenum layer indicated by the change of the lattice constant as a function of the argon sputtering pressure

RESULTS

Sodium incorporation from soda lime glass

A series of CIGS films with various sodium contents was produced from identical Cu-In-Ga-Se stacks on molybdenum back electrodes sputtered at various Ar-pressures on soda lime glass substrates. Changes in the molybdenum sputtering conditions resulted in different stress conditions of the metal layer as determined by XRD measurements. Compressive and tensile stress is indicated by negative and positive values of $\Delta d/d$, respectively. Along with a transition of the molybdenum stress from the compressive to the tensile regime, the molybdenum shows a more open grain structure leading even to microcracks at the highest pressure values allowing more alkalis to penetrate during the high temperature film formation process. With increasing Mo-sputtering pressure and tensile stress, the amount of sodium as detected by ESCA on the CIS surface increases in parallel (figure 3). As one of our key observations we also noted increasing solar cell performance with increasing sodium concentration (refer to inset in figure 3). ESCA measurements were

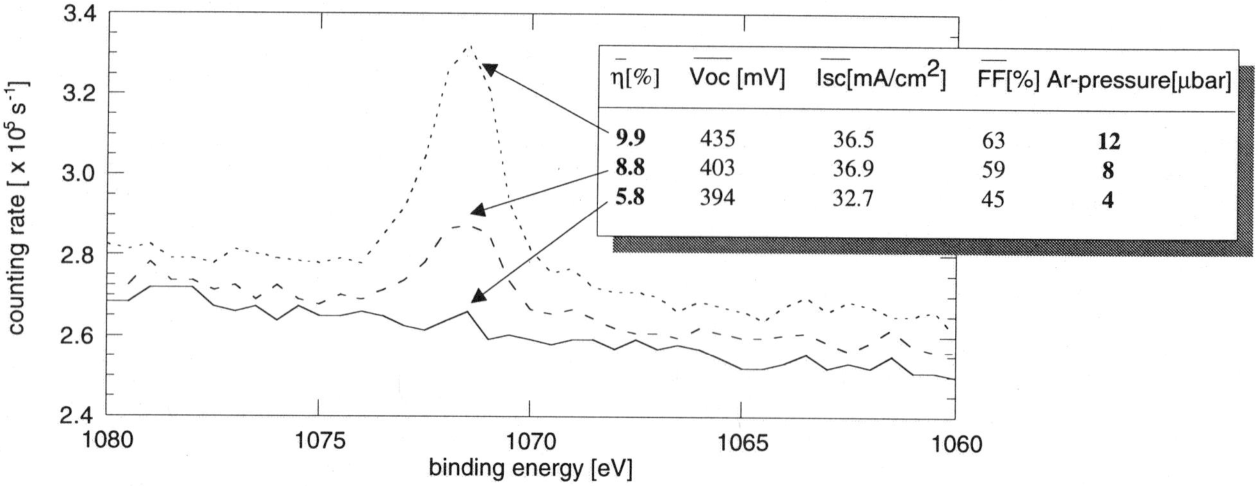

η[%]	Voc [mV]	Isc[mA/cm²]	FF[%]	Ar-pressure[μbar]
9.9	435	36.5	63	**12**
8.8	403	36.9	59	**8**
5.8	394	32.7	45	**4**

Fig. 3. ESCA measurements of the RTP-CIGS surface. The Na 1s emission varies in intensity as a function of the added amount of Na. Corresponding average cell parameters and Ar-sputtering pressure are displayed.

taken in the vicinity of the device.

In this case the soda lime substrate acts as a sodium source for the CIS thin film and the molybdenum exhibits a controlled sodium permeability. Although this method allows some improvements in process control, the amount of sodium ending up in the final CIS layer depends on the sodium concentration in the soda lime glass, the permeability of molybdenum and on the thermal budget used for film formation. Therefore we developed an alternative process to incorporate sodium independent from substrate and backelectrode properties.

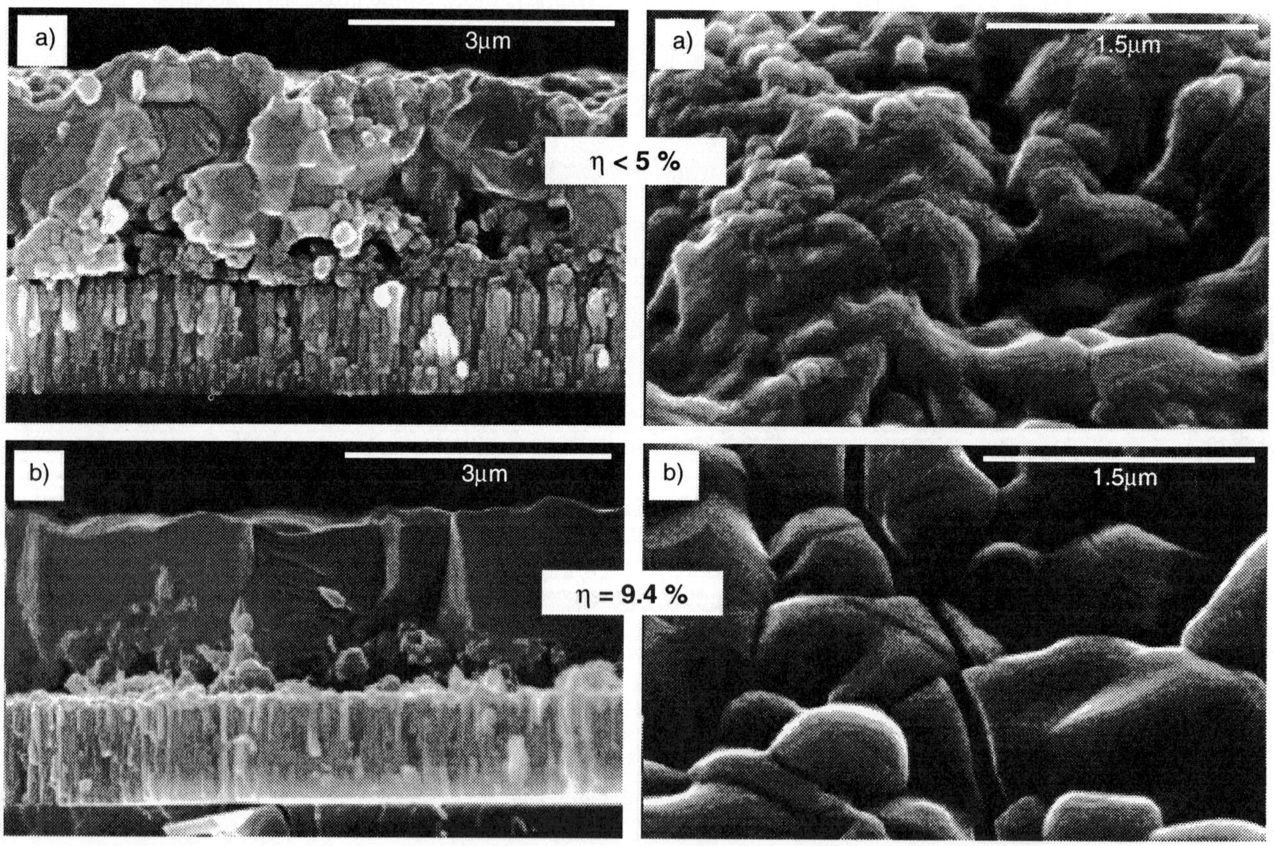

Fig. 4. SEM micrographs of CIGS on fused silica without (a) and with (b) additional sodium

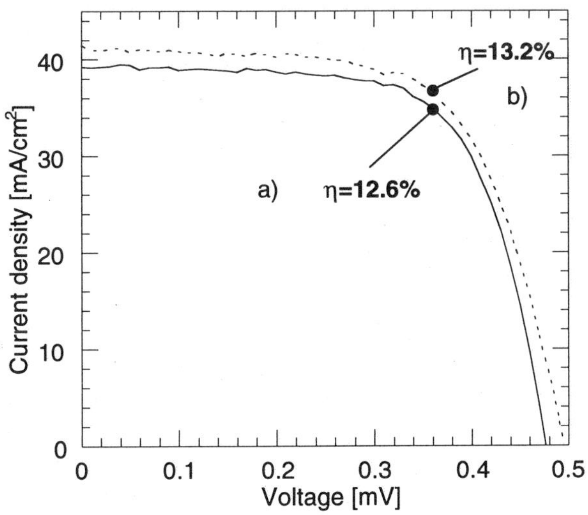

Fig. 5. Device characteristics of the best RTP-CIGS solar cells (a: CVD-ZnO; b: RF-sputtered ZnO), both with additional sodium in the precursor.

Sodium incorporation into the precursor

In order to decouple the alkali content in CIS from substrate and back electrode properties we used fused silica substrates. Sodium was added by inserting sodium selenide in between the molybdenum back electrode and the Cu-Ga-In precursor. By this process we were able to increase cell efficiency on silica substrates from less than 5% (sodium free) up to 9.4% (sodium containing). As demonstrated by the SEM micrographs in figure 4 sodium incorporation is clearly accompanied by a morphological change of the CIGS thin films i.e. significantly larger grains as demonstrated earlier with coevaporated films [3].

In the case of the sodium-containing film on fused silica substrates the observed cracks in the large CIGS grains are caused by the large thermal expansion mismatch of CIGS and fused silica. Those cracks are avoided by using soda lime glass as substrate. To separate out the influence of sodium added to the precursor from the one that may out-diffuse from the substrate we employed a dense molybdenum backelectrode sputtered at low pressure.

The optimization of the sodium amount in RTP-CIGS films on soda lime glass resulted in solar cells of improved performance. Conversion efficiencies partly averaging above 12% (11.1 cm^2 active area) and a peak efficiency of 13.2% (1.85 cm^2 active area) were achieved with sputtered ZnO window (figure 5). Employing a CVD-ZnO frontelectrode leads to a peak efficiency of 12.6%.

CONCLUSIONS

We developed various ways to incorporate sodium into CIS with the objective of quantifying its impact on grain morphology and solar cell performance. In a simple process version we varied the outdiffusion of alkali ions from the glass substrate by employing a partially permeable Mo-backelectrode combined with a low thermal budget process (RTP). In a second process alternative we decoupled the alkali content in the final CIS film from substrate and electrode properties by adding sodium selenide to the precursor .

A correlation of solar cell performance with the sodium content as determined by ESCA on the CIS surface was found. Controlling the sodium content in CIS finally leads to peak efficiencies beyond 13% .

Although there is now some good piece of evidence for a positive impact of sodium in chalcopyrite based solar cells, the detailed mechanism leading to improved morphology and increased conductivity still remains unclear. For example it would be of interest to know whether sodium acts as an extrinsic dopant (acceptor) or influences the intrinsic defect structure in the chalcopyrite lattice. In addition we consider further investigations concerning the role of sodium during growth of the compound as a very interesting field as it may open up new routes towards an understanding of the growth kinetics.

ACKNOWLEDGEMENTS

The authors gratefully acknowledge M. Ruckh and D. Hariskos (both from Stuttgart University) for ESCA analysis, respectively window preparation, B. Freienstein for electrical measurements and B. Jobst for XRD analysis.

This work has been supported by the Bundesministerium für Forschung und Technologie under BMFT contracts #0329218A and #0328059E.

REFERENCES

[1] J. Hedström, H. Ohlsen, M. Bodegård, A. Kylner, L. Stolt, D. Hariskos, M. Ruckh, H.W. Schock, "ZnO/CdS/Cu(In,Ga)Se_2 Thin Film Solar Cells with Improved Performance", *Proc. 23rd IEEE Photovoltaic Specialists Conf.*, 1993, p.p. 364-371

[2] A. M. Gabor, J. R. Tuttle, M. Contreras, D.S. Albin, A. Franz, D. W. Niles, R. Noufi "High-efficiency Thin Film Cu(In,Ga)Se_2 Solar Cells: Achievement of a Confirmed World-Record 16.4% Total Area Efficiency", *12th ECPV-SEC*,1994, p. 939

[3] M. Bodegård, L. Stolt, J. Hedström, "The Influence of Sodium on the Grain Structure of CuIn Se_2 Films for Photovoltaic Applications", *12th ECPSEC*, 1994, p. 1743

[4] F. Karg, V. Probst, H. Harms, J. Rimmasch, W. Riedl, J. Kotschy, J. Holz, R. Treichler, A. Milwalsky, A. Kiendl, "Novel Rapid-Thermal-Processing for CIS Thin-Film Solar Cells", *Proc. 23rd IEEE PVSC Conf.*,1993, p. 441

[5] J. Holz, F. Karg, H. von Philipsborn, "The Effect of Substrate Impurities on the Electronic Conductivity in CIS Thin Films", *Proc. 12th EC PVSEC*, 1994, p. 1592

STUDIES ON SUBSTRATES AND CONTACTS FOR CIS FILMS AND DEVICES

B.M. Başol, V.K. Kapur, C.R. Leidholm, A. Minnick and A. Halani
International Solar Electric Technology (ISET)
8635 Aviation Blvd., Inglewood, CA 90301

ABSTRACT

Nature of the glass/Mo substrates used in CIS solar cell structures influences the performance of the devices. Excessive Na diffusion from the commonly used soda-lime glass substrates through the Mo layer and its reaction with the processing atmosphere at the Mo/CIS interface may limit the efficiency of cells due to the formation of a high resistivity layer near this contact. A moderate level of Na diffusion, on the other hand, improves the efficiency of the cells, especially those fabricated on CIS films with low Cu/In stoichiometric ratios. The Na diffusion process depends on the type of the glass substrate used and the nature of the Mo layer.

INTRODUCTION

ISET has developed a selenization method for the growth of high quality CIS absorber layers and demonstrated 7% efficient submodules of 1 ft^2 area using this approach and the device configuration of glass/Mo/CIS/CdS/ZnO [1]. ISET's process involves deposition of a Mo layer on the soda-lime glass substrate, coating of the Mo surface with a thin layer of Te followed by In and Cu. The resulting Cu-In alloy precursor is then selenized at around 400 °C in a H_2Se atmosphere to form the CIS film.

Although, glass is commonly distinguished by its high chemical resistance at low temperatures, its reactivity may actually be extensive at temperatures that are employed in CIS film growth. The typical composition of a soda-lime glass sheet in weight percentages is 70% SiO_2, 15% Na_2O, and the balance, the oxides of Ca, Al, Mg, Ba and K [2]. The Na-O network in soda-lime glass is known to be rather reactive. Soda-lime glass, for example, leaches in acidic solutions through ion exchange which results in Na loss from the surface region. The glass surface also reacts with gases such as H_2, SO_2 and CO_2. These reactions may deplete the soda-lime glass surface of the alkali component provided that the reaction products are continuously removed from the surface.

Mo which is typically deposited on the glass substrates by the sputtering technique is the commonly used contact material in high efficiency CIS solar cells. Selection of Mo as the contact material to CIS is due to its relative stability under CIS film growth conditions. However, recent experience has shown that interaction of the glass/Mo substrate with the reactive CIS growth environment can be substantial especially in a selenization process involving H_2Se.

In this paper we will address the issue of the glass/Mo structure's reactivity with the selenizing atmosphere and its possible influence on the performance of solar cells.

EXPERIMENTAL

Mo films were grown on various types of glass substrates in a sputtering system with vertical geometry. The substrates were passed in front of a DC magnetron cathode at a controlled speed to deposit 1-3 μm thick layers. Soda-lime glass with 14%Na_2O, Corning 7740 containing 4% Na_2O, and Na-free Corning 7059 glass samples were used as the substrates. Deposition parameters such as the background pressure, sputtering gas pressure, background pressure, cathode to substrate distance and sputtering power level were changed and the resulting films were evaluated in terms of their stress, thickness and resistivity. Witness samples of 50 μm thick borosilicate glass were used to determine the type of the intrinsic stress (compressive vs. tensile) in the films by evaluating the curvature of these substrates after Mo depositions.

Plain glass samples as well as Mo/glass structures were selenized in H_2Se atmosphere at around 400 °C for about 30 minutes. Auger depth profiling was used to determine the degree of selenization near the surface regions of the Mo layers deposited on different types of glass substrates. XPS measurements were employed to investigate the reaction products on the Mo surfaces. Two-point resistance measurements were utilized to compare the resistive Mo surfaces formed after the selenization step. The measurement involved placement of two probes on the selenized Mo surfaces separated by a distance of about 1 cm. All of the experiments mentioned above were repeated on soda-lime glass samples coated with a 3000 Å thick evaporated SiO_2 film.

CIS films were grown on the various substrates using the selenization technique and solar cells were fabricated. Cell performance was correlated with the type of substrate used.

RESULTS AND DISCUSSION

After the selenization step a visual inspection of the various glass samples indicated that the soda-lime

CH3365-4/94/0000-0148 $4.00 © 1994 IEEE

glass surface was covered with a thin layer of reaction products. The surfaces of all the other glasses including the soda-lime sheet with the SiO_2 barrier layer were totally clear of any residues or reaction products. The reacted glass surface changed color to light brown when it was exposed to air and cooled to below 100 °C. Viewing under the microscope, one could observe that the surface layer absorbed moisture from the air forming small nodules.

The explanation for the observed phenomenon lies in the reaction of sodium on the surface of the soda-lime glass substrate with the H_2Se atmosphere at 400 °C in accordance with the reaction:

$Na_2O(s)$ + $H_2Se(g)$ ----- $Na_2Se(s)$ + $H_2O(g)$

which is favorable. Na_2Se is a deliquescent material and therefore it readily absorbs water vapor from air when taken out of the reaction chamber and cooled to below 100 °C. The resulting species are expected to be NaOH and Se precipitates.

Diffusion of Na from the glass substrate into the CIS layers grown by the co-evaporation technique has been previously reported and the influence of this diffusion process on the device efficiency has been discussed [3]. It should be noted that the results of the experiment described above is specific to the H_2Se selenization technique. In selenization approaches utilizing elemental Se as the source material, reaction of Na_2O with Se vapor is not favorable. In fact, we have not observed any reaction products on soda-lime glass substrates selenized with Se vapor in a vacuum chamber using Se evaporation. It is, therefore, important to distinguish between the selenization approaches while addressing the issue of glass reactivity.

After confirming the reactivity of the Na in soda lime glass substrates with the H_2Se atmosphere, we carried out experiments to study the possible influence of this behavior on the Mo/glass structures and solar cells.

The relationships between the micro-structure and stress of Mo films and the sputtering conditions employed for the deposition of such layers have been well established [4]. In general, the Mo layers obtained at low sputtering gas pressures tend to be dense and low-resistivity and they are under compressional stress due to the atomic peening process. As the Ar gas pressure is increased, film stress switches from compressional to tensile. Further increases in the working pressure may yield stress-free layers with porous micro-structures. The electrical resistivities of such layers may be orders of magnitude higher than the bulk resistivity (5.5×10^{-6} ohm-cm) of Mo depending upon the residual oxygen pressure in the sputtering chamber. A sketch qualitatively showing the relationships between the Mo film characteristics and the sputtering process parameters is shown in Fig. 1. This data was obtained from ISET's Mo sputtering process and it follows the trends observed by others. It should be noted that although the general behavior of the pressure-stress curves in the Mo sputtering process is well understood, the absolute values of the Ar gas pressures where the stress transition from compressional to tensile occurs may change from system to system because the film structure is also influenced by other factors such as the system geometry, ion bombardment of the substrate, deposition angles, substrate temperature and sputtering voltage/power etc.

The above mentioned structural and electrical properties of sputtered Mo layers play an important role in determining certain characteristics of the glass/Mo substrates and solar cells fabricated on them. Adhesion of the Mo film to the glass surface and the sheet resistance of the solar cell back contact are two factors that need to be taken into consideration. However, we should not forget that the sputter deposited Mo film is not utilized in the cell structure in its as-deposited form. Rather, it goes through the processing steps necessary for the fabrication of a complete device including a high temperature selenization step. Therefore, its properties measured right after deposition may not necessarily be the same as they are in the finished device structure.

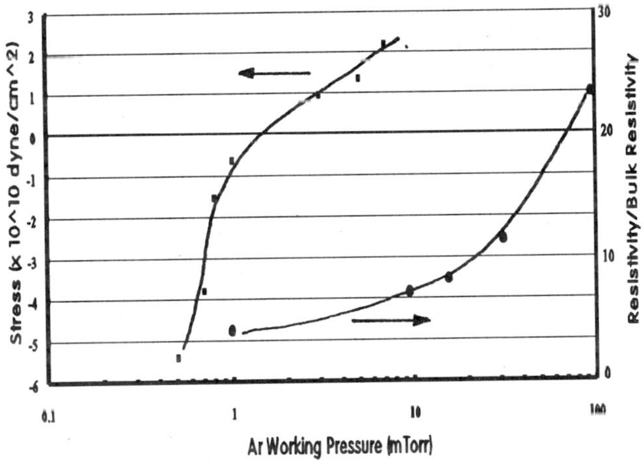

Fig. 1 ISET's sputtered Mo film characteristics as a function of Ar pressure. Background pressure was 1×10^{-6} Torr, power was 2.2 kW on a 5"x15" target.

The Auger surveys and depth profiles obtained from two different selenized Mo layers are given in Fig. 2. These films were both deposited on soda-lime glass substrates. Data taken from sample Mo-60 display about four times the Na signal observed at the surface of Mo-127. There is also a larger O signal. Auger depth profiles of Fig. 2 indicate that Mo-127 has a thin layer of selenide (about 1500 Å) on its surface, whereas, the selenized surface of Mo-60 is about five times thicker. It should be noted that Mo-60 film had a dark powdery layer on its surface which could be wiped off. After removing this layer both the Na and the O peaks observed in AES survey were reduced. However, they still were stronger than those for sample Mo-127. Two point resistance measurements made on the two selenized films showed resistance values of 2-10 Kohms for Mo-60 and 10-300 ohms for Mo-127.

The XPS survey spectra of the two Mo films are shown in Fig. 3a. This data indicates about 5 times more Na on the surface of Mo-60 compared to Mo-127. The Mo 3d spectra of the two samples (Fig. 3b) shows the double peaks associated with $MoSe_2$. But in the spectra of Mo-60 there is clearly a third peak shifted by 3.8 eV that can be attributed to Na_2MoO_4. The Se 3d spectra of Fig. 3c display the expected $MoSe_2$ peak for both samples. The data of Sample Mo-60, however, has additional peaks shifted by 4.6 eV (Na_2SeO_3 and/or SeO_2) and 8.9 eV (possibly higher order Se oxides).

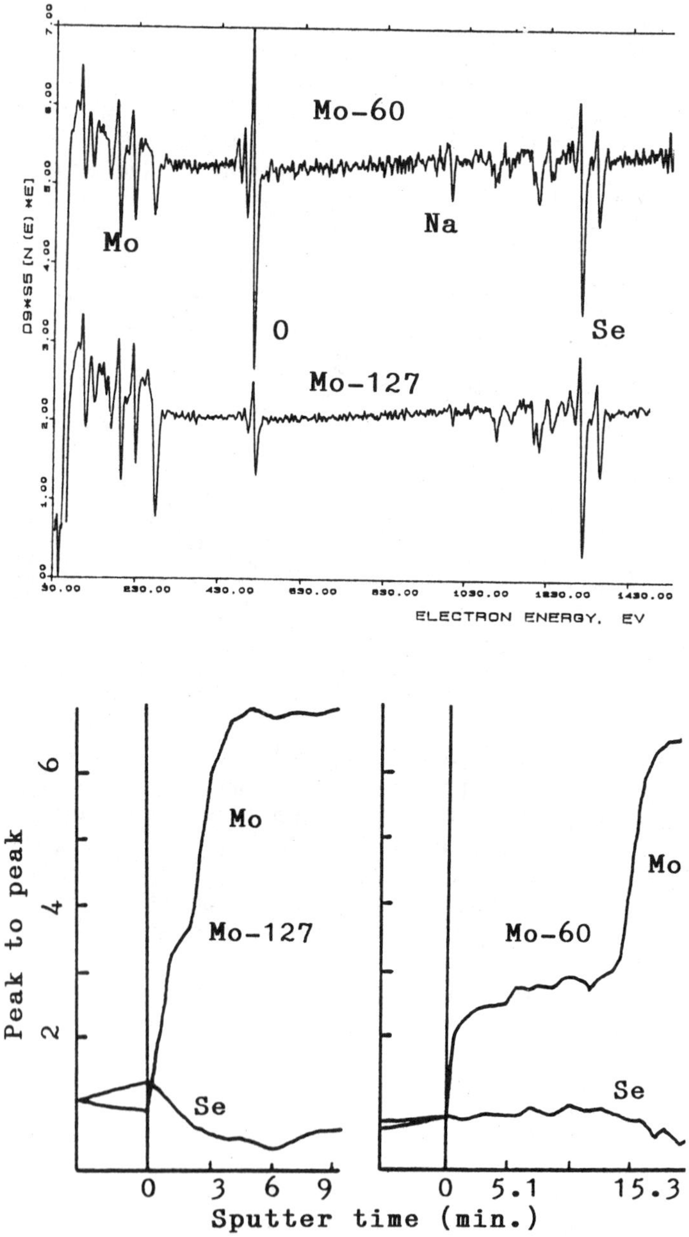

Fig. 2 Auger survey spectra and depth profiles of the Mo-60 and Mo-127 samples.

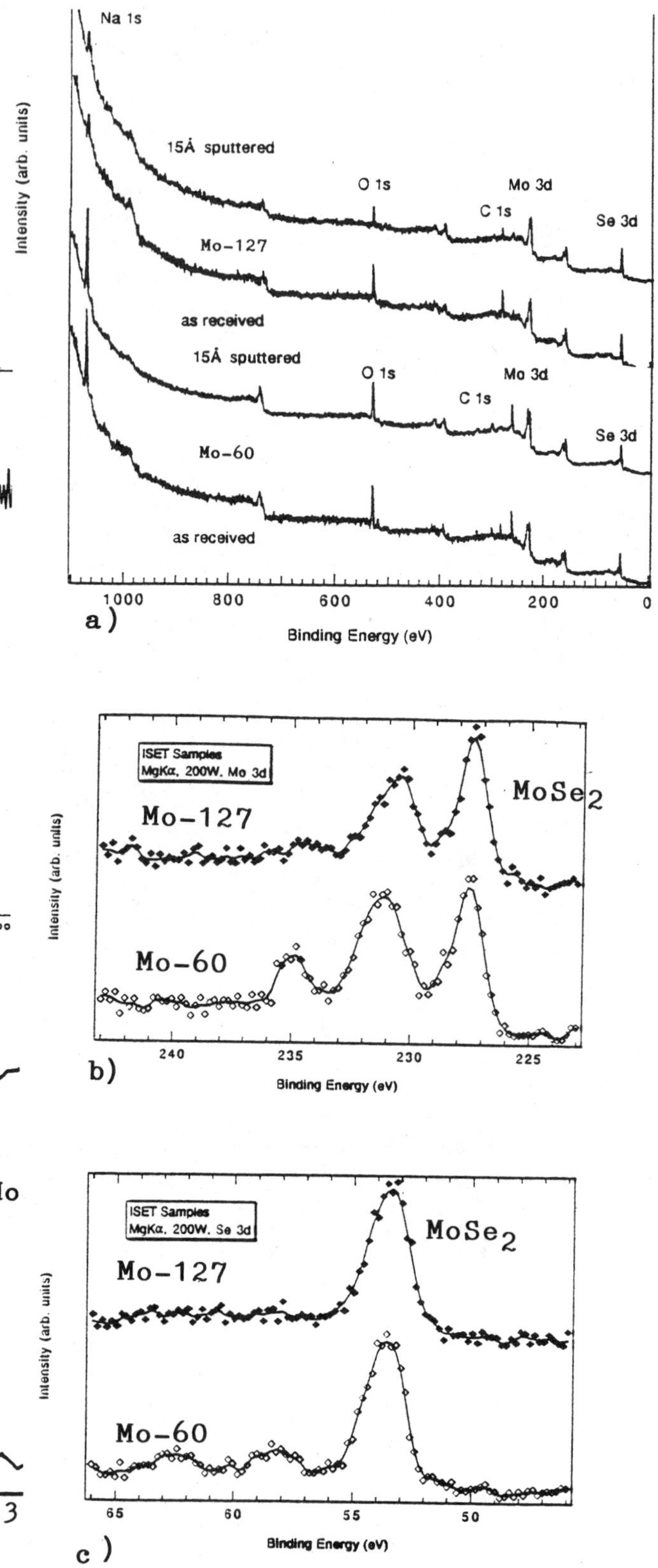

Fig. 3 a) XPS survey spectra, b) Mo 3d and c) Se 3d spectra taken from samples Mo-60 and Mo-127.

All the results discussed above point to the fact that two different Mo layers on the same soda-lime glass substrate may selenize differently. Excessive reaction yields a dull surface layer with high resistivity which is associated with the appearance of a strong Na signal as observed from the Auger as well as XPS data. Therefore, reactivity of the Mo surface is related to Na diffusion from the glass substrate to the Mo surface region. The degree of Na diffusion, on the other hand, depends on the nature of the Mo film itself.

To further correlate the surface reactivity of Mo film with Na diffusion, the selenization experiments were repeated on glass/Mo samples using glass sheets of various compositions. Mo films equivalent to sample Mo-60 did not show excessive reaction with the selenizing atmosphere when they were deposited on Corning 7044 and Corning 7059 glass substrates. Similarly, excessive reaction was not observed for Mo layers deposited on soda lime glass substrates with a surface barrier layer of 3000 Å thick SiO_2.

Influence of the excessive reaction of the Mo surface on CIS cell performance was studied in an experiment on two cells fabricated on soda-lime glass/Mo substrates. The Mo layer of the first device had developed a thick reacted surface layer when selenized separately. Two point resistance measurement gave a value of 8 K ohms for this sample. The Mo layer of the second device, on the other hand, was equivalent to the Mo-127 sample of Fig. 2. The Cu-In precursors and all the details of CIS film and device processing steps were identical for these two solar cells. The illuminated I-V characteristics of the devices showed that the one fabricated on the reactive Mo layer suffered from a poor fill factor value which can be attributed to the presence of a high resistivity region at the Mo/CIS interface.

The above example demonstrated the deleterious effect of excessive Na diffusion through the Mo layer on the cell characteristics. Na diffusion into the CIS film at moderate levels, however, can improve the device efficiency as shown in Fig. 4. This figure shows the I-V characteristics of a set of devices fabricated on soda-lime (A), Corning 7740 (B) and Corning 7059 (C) glass substrates. Device D was prepared on a soda-lime glass substrate with a SiO_2 surface barrier layer. This experiment was carried out for Cu/In stoichiometric ratios of 0.88, 0.92 and 0.98. The general observations from the data of Fig. 4 and the other experiments can be listed as follows:

i) The efficiencies of cells fabricated on the soda-lime glass substrate (without the SiO_2 barrier layer) are higher than those on other glass substrates.

ii) The difference between the efficiency of soda-lime glass based device and the others is the largest for the lowest CIS stoichiometries. For films with Cu/In ratios of 0.92 and 0.98 the differences between the fill factor values of various cells are not as dramatic as they are in Fig. 4.

iii) A thin layer of SiO_2 deposited on the soda-lime glass surface before Mo deposition lowers the fill factor of the CIS cells.

The above results suggest that Na diffusing through the Mo layer into the CIS film dopes the CIS layer p-type. Consequently, good devices can be obtained at low Cu/In ratios with Na doping. Presence of Na actually increases the processing window of solar cells in terms of usable CIS stoichiometries.

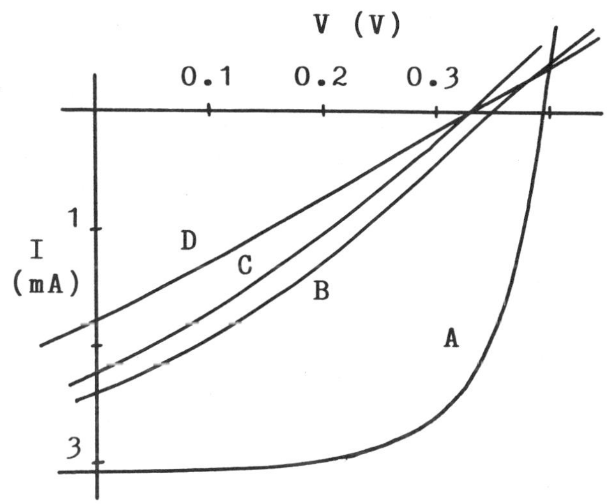

Fig. 4 Illuminated I-V characteristics of 0.09 cm^2 CIS cells fabricated on soda-lime glass (A), Corning 7740 glass (B), Corning 7059 glass (C), and soda-lime glass coated with SiO_2 (D). The Cu/In ratio for the CIS film was 0.88.

ACKNOWLEDGEMENTS

This work was funded by NREL contract #YI-2-12069-1. Contributions of A. Nelson, A. Swartzlander-Franz, D. Niles, and H. Ullal of NREL to the reported characterization work are gratefully acknowledged.

REFERENCES

[1] B.M.Başol, V.K.Kapur, C.R. Leidholm, A. Minnick and A. Halani, "Development of low cost CIS cells and modules", *12th NREL PV Prog. Rev., AIP Conf. Series 306*, 1994, p. 79.

[2] R. Brown, in "Handbook of thin film technology", *L. Maissel, R. Glang eds., Mc Graw Hill,* 1970, p. 6-1.

[3] M. Bodegard, L. Stolt and J. Hedstrom, *12th European PVSE Conf.*, 1994, p. 1743.

[4] T.J.Vink et al., "Stress, strain, and microstructure of sputter-deposited Mo thin films", *J. Appl. Phys.*, **70**, 1991, p. 4301.

SUBSTRATE INFLUENCES ON CIS DEVICE PERFORMANCE

David F. Dawson-Elli*, Chad B. Moore*, Robert R. Gay**, Cynthia L. Jensen**
*Corning, Inc., SP-FR-3-1 Corning, NY 14831,
**Siemens Solar Industries, P.O. Box 6032, Camarillo, CA 93011-6032

ABSTRACT

It has been reported that the substrate plays an active role in copper indium diselenide (CIS) devices. The physical properties of the substrate, i.e., thermal expansion coefficient, chemical composition, strain point, surface quality and cleanliness may all play a role in device efficiency and process reproducibility. For example, sodium is known to influence the conductivity of CIS, and that soda lime glass (SLG) introduces sodium into the CIS. In the experiments reported here, the sodium level in the CIS was varied by a) changing the substrate composition and b) the use of SiO_xN_y and SiN_x barrier layers. Cells were fabricated on the candidate substrate and barrier layer combinations. V_{oc} was observed to drop with the addition of barrier layers to sodium containing glasses. The films were also analyzed for sodium by SIMS. The cell performance and SIMS analysis are presented and their significance is discussed.

INTRODUCTION

In this paper, we specifically address the effects of sodium in the substrate, barrier layers to block the diffusion of sodium, and the physical properties of the glass.

The performance of CIS devices on soda-lime glass (SLG) substrates have been shown to be superior to that on Corning Code 7059 glass or Alumina substrates[1 ,2]. It has also been shown that a sodium dose of ~ 1×10^{16} cm^{-2} increases the conductivity of an absorber layer deposited on sapphire to the level found for an absorber deposited on SLG[3]. This increase in the conductivity may correspond to an increase in p-type doping of the absorber. This interpretation is consistent with several observations comparing devices grown on Code 7059 with those grown on SLG substrates. Devices grown on Code 7059 have lower open circuit voltages (lower built-in potential or reduced recombination) and have higher red response (greater depletion of the absorber). Devices grown on Code 7059 manifest a first-quadrant blocking behavior at relatively high temperatures (less ohmic CIS/moly contact). These observations are all consistent with an increase in p-type doping of the absorber.

But, this device performance sensitivity to the substrate could have its origin in other differences in the properties of these glasses. In addition to a difference in the amount of sodium, there are also differences in the thermal expansion coefficients and the strain points. The importance of the thermal expansion coefficient has been identified previously[4].

Glasses are characterized by rapid but smooth changes in viscosity as a function of temperature. Frequently used reference points (working, softening, annealing, and strain) on the viscosity-temperature curve have been defined by ASTM standards. At the anneal point, the viscosity is approximately 10^{12} Pa sec, and internal stress in the glass is reduced to an acceptable level in 15 minutes. The strain point is at the lower end of the annealing temperature range, and is generally considered an upper use temperature (viscosity of approximately $10^{13.5}$ Pa sec)[5].

EXPERIMENTAL DETAILS

To separate the contributions of the physical and chemical properties, an experiment was performed in which the composition of the substrates was selected to change these properties independently. The physical and chemical properties of the substrates that were chosen are shown in Tables 1 and 2, respectively. For comparison, CIS has a thermal expansion of about 90 x 10^{-7}, and melting point of 1250 C.

To further clarify the distinction between the contributions of the physical and chemical properties of the substrates to the characteristics of the CIS devices, barrier coatings of SiO_xN_y and SiN_x were applied by PECVD to the substrates to suppress the diffusion of impurities from the substrate to the cells. The barrier films were nominally 100 nm thick unless otherwise noted. Devices were then fabricated using the SSI graded-absorber process described previously[6].

RESULTS AND DISCUSSION

Device Performance

The cell results of this experiment are shown in Figures 1a through 1d. The mean value and standard deviation in the values are shown for the cell characteristics. The device efficiency, shown in Figure 1a, is highest for the

CH3365-4/94/0000-0152 $4.00 © 1994 IEEE

First WCPEC; Dec. 5-9, 1994; Hawaii

controls. Code 0317 also performed well. Code 7059 is clearly inferior, as is the thick SiN_x on SLG.

Table 1. Physical Properties of the substrates.

	SLG	0317	7059
Expansion (x 10^{-7})	90	87	46
Strain Point (C)	480	574	593
Anneal Point (C)	525	624	639

Table 2. Composition of the substrates.

	SLG	0317	7059
SiO_2	73.77	61	49
Na_2O	13.7	13	0.1
CaO	8.54	0.2	
MgO	3.78	3.7	
Al_2O_3	0.082	17	11
K_2O	0.03	3.6	
As_2O_3		0.8	0.8
BaO			24
B_2O_3			14

As seen in Figure 1b, within a lot, the short circuit currents for all of the devices are very similar (with the exception of thick SiN_x). However, Figure 1c shows that the open circuit voltage is very dependent upon the type of substrate and the presence of a barrier layer. Cells deposited on the control and uncoated Code 0317 are very similar in voltage, and much higher than the voltages obtained on barrier-coated Code 0317, SLG or any Code 7059-based substrate.

Figure 1d shows the fill factors for these devices. The fill factors obtained on the uncoated Code 0317 glass were highly variable, and are not highly correlated with the voltage data. The SiO_xN_y-coated Code 0317 glass has a high fill factor (with very little variance), yet it has a low open circuit voltage. The voltages and fill factors of these devices do not correlate, suggesting that independent mechanisms control these characteristics. The SiN_x coated SLG showed a marked performance loss with an increase in barrier thickness.

Figure 2 is a representative OBIC (optical beam induced current) image for the thick SiN_x on SLG. The characteristic features of the image are the bright spots and lines of enhanced current generation. The bright spots are believed to be the result of pinholes in the barrier, which create localized areas of high Na content. The bright lines are 'halos' around laser scribes in the Mo back contact. Not all cells have the laser scribe. The presence of the laser scribe did not affect average cell performance or standard deviation.

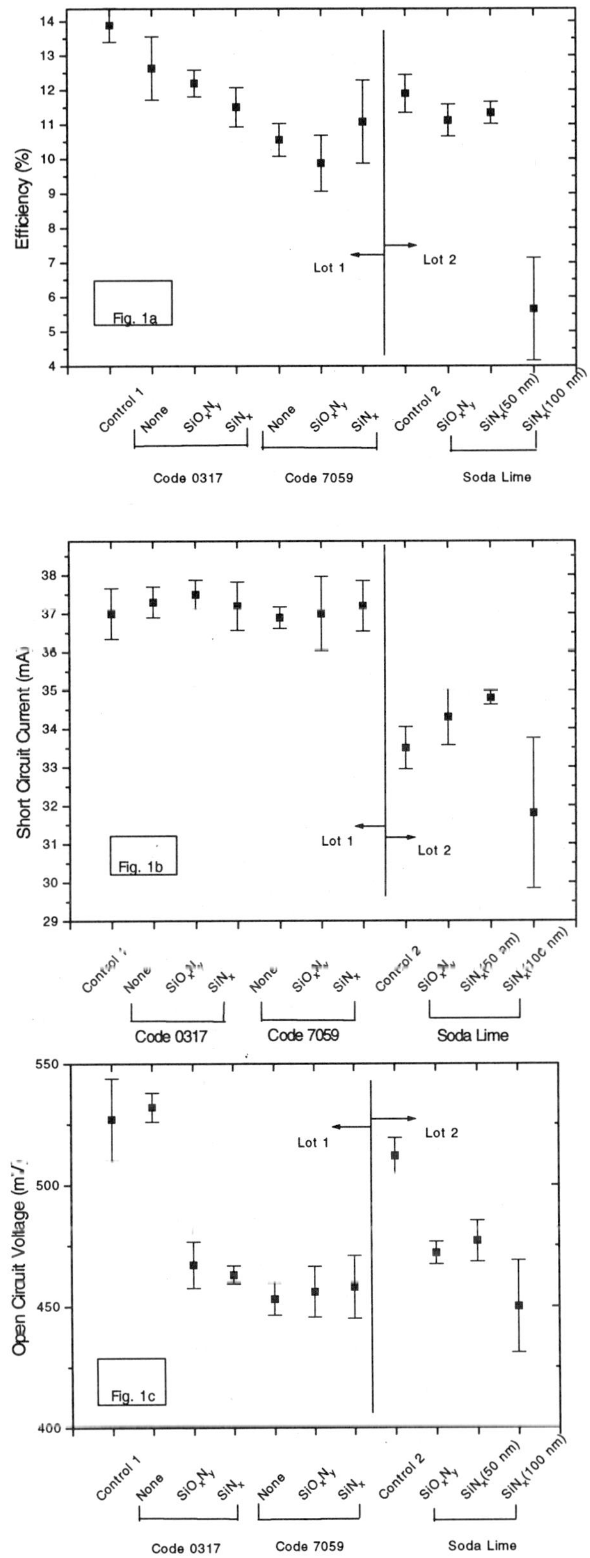

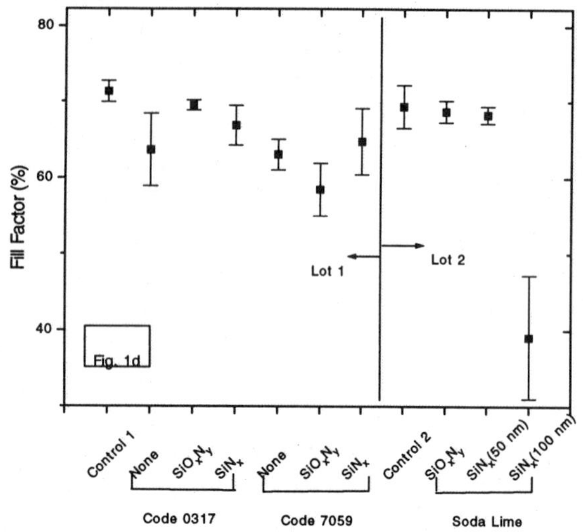

Figure 1: a) Efficiency, b) Short Circuit Current, c) Open Circuit Voltage and d) Fill Factor for devices fabricated on Corning Codes 0317, 7059, and soda lime glass, with controls. The marker is placed at the mean value. The error bars indicate one standard deviation. The presence of a barrier lowers V_{oc} to values found on low Na substrates.

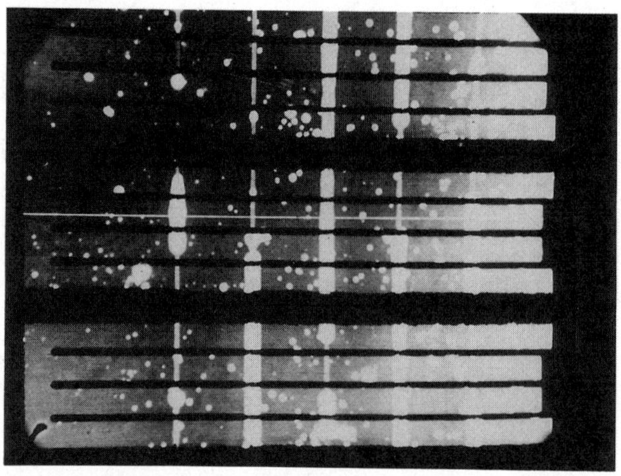

Figure 2: OBIC image of devices on SLG with 100 nm SiNx barrier. Note the bright spots of enhanced current generation, believed to be due to pinholes in the barrier. Laser-scribed cuts in the base electrode coincide with the bright lines across the cells.

SIMS Analysis

SIMS measurements were made to profile the sodium content of devices, using O^- ions in the positive mode. The shapes of the curves in Figure 3 are believed to be an artifact of the SIMS technique. The use of Cs^+ ions and detection of Cs^+M molecular ions (where M is the element of interest) have been shown to reduce matrix effects[7 ,8]. These data were taken previous to our adopting the Cs technique. The data presented have been normalized to the Cu signal by dividing the Na signal by the Cu signal, point by point, in the range which defines the CIS layer.

Since there is very little Na in Code 7059 glass, there was an expectation of low amounts of Na in absorbers formed on Code 7059. However, as seen in Figure 3a, there is more Na in absorbers on Code 7059 with barriers than the bare glass. For the Code 0317 glass, the presence of a barrier layer appears to lower the Na content of the films to (and in some cases, below) the level found in absorbers on Code 7059. The SIMS data in Figure 3b indicate that a SiO_xN_y barrier on Code 0317 glass is more effective in blocking Na diffusion than a SiN_x barrier, contrary to our expectation and experience. These results indicate that diffusion from the glass is not the sole source of Na found in the absorbers. The alternate source of Na has not been identified.

Soda Lime Physical Properties

The density of glass depends on its thermal history[5]. In general, one expects an as-formed glass to compact in an anneal approaching its strain point. The amount of compaction for SLG was measured by Dilatometer, in an thermal cycle to its anneal point. The results of this measurement are shown in Figure 4. The data indicates a 600 ppm compaction of the glass at its anneal point of 525 C. The degree of compaction will vary in SLG, as it is poorly annealed at manufacture. Code 7059 glass would compact approximately 250 ppm (unannealed) or 5 ppm (annealed), in a similar temperature profile. The effect of the compaction phenomenon on device performance is not known. It is speculated that compaction may induce a stress on the device films. However, the 'creep' behavior of hot glass may allow highly stressed films to relax.

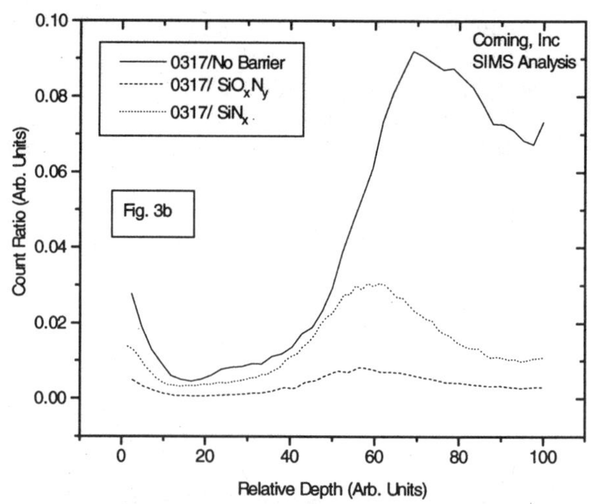

Figure 3a,b: Relative SIMS data for Codes 7059, 0317 samples, respectively. SIMS analysis was performed using O^- ions, sampling positive secondary ions. The data were generated by dividing the Na signal by the Cu signal in the range of 50% max. Cu on the leading edge of the profile to 50% max. Cu on the trailing edge, point by point. The shape of the curve is believed to be an artifact of the measurement.

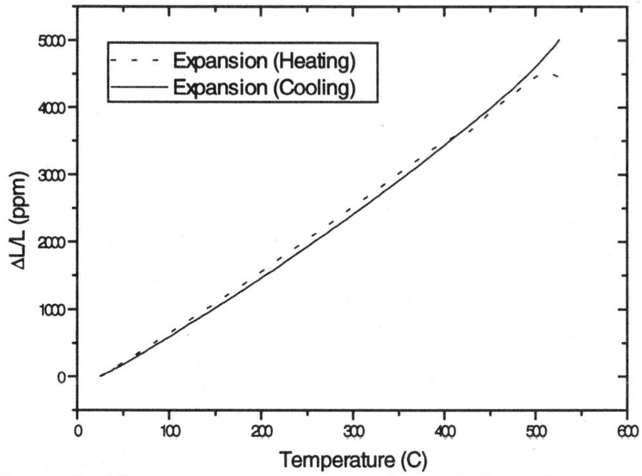

Figure 4: Dilatometer measurement for soda lime glass. A Dilatometer measures the overall length of the sample in comparison to a reference sample at the same temperature. In this case, the reference was Alumina. Note that the data has been offset such that the room temperature data points coincide. Thus, the 600 ppm compaction of the sample appears at the high temperature end of the curve.

Summary

The presence of Na in CIS appears to increase the effective hole concentration, causing an increase in open circuit voltage in cells with high Na content. It is unclear from these experiments whether Na affects the bulk carrier concentration or passivates grain boundaries. Fill Factor does not correlate with Na content or open circuit voltage. This leads to the conclusion that fill factor and V_{oc} are controlled by separate mechanisms. The use of a barrier layer on Na containing glass lowers the amount of Na measured by SIMS. The enhanced collection observed in the OBIC scan is believed to be caused by the absence of the barrier (pinholes or laser scribe lines). There appears to be a source of Na in the process aside from diffusion out ot the substrate. Identifying this source of Na is a current topic of investigation. Glass compaction is observed, and is ~600 ppm for SLG taken to its anneal point.

Acknowledgments

The efforts of Norbert Binkowski and Dr. Emmanuel Onyiriuka in obtaining the SIMS data, as well as Paul Sachenik and John Stempin for substrate preparation, are gratefully acknowledged.

References

[1] L. Stolt, M. Bodegård, J. Hedström, J. Kessler, M. Ruckh, K.-O. Velthaus, H.-W. Schock, 11th E.C. Photovoltaic Solar Energy Conf. (1992) 120.

[2] M. Bodegård, L. Stolt and J. Hedström, 12th E.C. Photovoltaic Solar Energy Conf., Amsterdam, 1743 (1994).

[3] J. Holz, F. Karg, H. von Philipsborn, 12th E.C. Photovoltaic Solar Energy Conf., Amsterdam, 1592 (1994).

[4] J. S. Chen, E. Kolawa, C. M. Garland, M. A. Nicolet, Thin Solid Films, 219 (1992) 183.

[5] D.C. Boyd and D.A. Thompson in "Kirk-Othmer: Encyclopedia of Chemical Technology", Volume 11, Third Edition, pp 807-880, John Wiley and Sons, 1980

[6] R. Gay, M. Dietrich, C. Fredric, C. Jensen, K. Knapp, D. Tarrant and D. Willett, 12th E.C. Photovoltaic Solar Energy Conf., Amsterdam, 935 (1994).

[7] H. Gnaser, J. Vac. Sci. Technol. A12, 452 (1994)

[8] S. Asher, NREL, Private Communication

INFLUENCE OF SUBSTRATES ON THE ELECTRICAL PROPERTIES OF $Cu(In,Ga)Se_2$ THIN FILMS

M. Ruckh, D. Schmid, M. Kaiser, R. Schäffler, T. Walter, and H. W. Schock
Institut für Physikalische Elektronik, Universität Stuttgart
Pfaffenwaldring 47, D-70569 Stuttgart, Germany

ABSTRACT

Thin films based on $Cu(In,Ga)Se_2$ prepared on alkali free substrates are compared to films prepared on soda lime glass. On the latter, the presence of sodium species as detected with X-ray photoelectron spectroscopy is correlated with an enhanced formation of Se-O, In-O and Ga-O bonds at the surface after several days air exposure. The electrical conductivity is also one order of magnitude higher on soda lime glass and solar cells prepared on molybdenum-coated substrates exhibit increased open circuit voltages. Capacitance-voltage characteristics on junctions prepared on alkali free substrates show an increased space charge width. The observations can be explained in terms of an increased net acceptor density in polycrystalline $Cu(In,Ga)Se_2$ when prepared on soda lime glass substrates. An alkali-metal-promoted oxidation of the surface is discussed.

INTRODUCTION

Polycrystalline thin films of ternary chalcopyrite semiconductors such as $Cu(In,Ga)Se_2$ are promising candidates as effective absorber materials for solar cells. Improved efficiencies over 15% have been achieved using soda lime glass as substrate material in combination with increased preparation temperatures (>550°C) during a coevaporation process [1]. The effects observation of sodium impurities on the structural properties as well as on the electrical conductivity are discussed in the literature [2-4]. Up to now no plausible explanation has been demonstrated for the influence of the substrate on the improved photovoltaic properties such as increased V_{OC}.

In this contribution we focus on the influence of different substrates on the electrical properties and composition of $Cu(In,Ga)Se_2$ thin films grown in a single layer coevaporation process. Special attention is paid to the question of how the observed differences can be related to the superior device characteristics of solar cells prepared on soda lime glass.
In order to influence the $Cu(In,Ga)Se_2$ absorbers more directly, we present a process with controlled addition of sodium species via a Na_xO precursor.

EXPERIMENTAL

All $Cu(In,Ga)Se_2$ thin films examined in this work were prepared in a single layer process at T_{sub}=600°C by coevaporation of the elements under constant preparation conditions. The Ga/(Ga+In) ratio of all films have been determined with energy dispersive X-ray spectroscopy (EDX) to about 25%.

Cu-poor films have been analyzed by X-ray diffraction (XRD) to investigate the crystal orientation and texture. X-ray photoelectron spectroscopy (XPS) was used to study the composition of the film surface as well as the chemical state of the detected elements. Bulk quantification was carried out by wavelength dispersive X-ray spectroscopy (WDX), in particular to investigate the incorporation of light mass impurities. The electrical conductivity was measured on films prepared on glass substrates without Mo-coating. C-V-characteristics of the completed $Cu(In,Ga)Se_2$ / CdS / ZnO heterojunctions were measured on substrates with Mo-coating.

RESULTS

Absorber materials with superior solar cell device properties show the existence of sodium species at the surface is examined with XPS. Sodium has only been detected at surfaces exposed to air; the signal increased with longer exposure time (see Fig.1). Samples kept 10 days under vacuum conditions (10^{-6} Pa) after the preparation do not exhibit a sodium signal. Analysis of the chemical state of the sodium species found on $Cu(In,Ga)Se_2$ surfaces after exposure to air suggests the formation of Na-C-O species at the surface of absorbers prepared on soda lime glass. The absence of a Na 1s signal in the XPS spectrum on absorbers prepared on sodium free substrates such as Corning 7059 glass excludes the atmosphere as the sodium source. A sodium diffusion from the substrate is assumed [3] which contains up to 15% Na_2O in the case of soda lime glass.

An enhanced oxidation of the surface has been observed associated with the higher sodium content. Fig. 1 shows the O 1s, In (MNN) and the Se 3d photoelectron signal of Cu- poor $Cu(In,Ga)Se_2$ films prepared on Corning 7059 (thin line) and soda lime glass (bold line). The formation of In-O and Se-O bonds is observed mainly on films prepared on soda lime glass. Ga-O has also been preferably detected at the surface of films

CH3365-4/94/0000-0156 $4.00 © 1994 IEEE

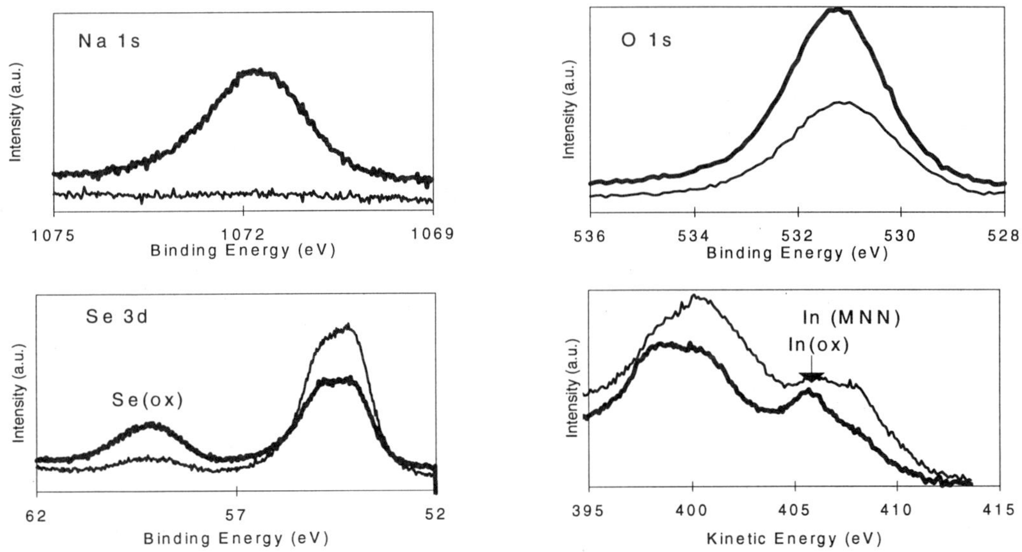

Fig. 1. XPS spectra of $Cu(In,Ga)Se_2$ surfaces prepared on Corning 7059 glass (thin line) and on soda lime glass (bold line) after several days air exposure. An enhanced formation of Se-O and In-O species is correlated with the sodium content at the surface.

prepared on soda lime glass as described in more detail in ref. [5].

The concentration of these impurities in the bulk was quantified with WDX. The sodium content as a function of the film composition is maximal for slightly Cu-poor films. The sodium signal decreases with increasing (In,Ga) content. No sodium can be detected with WDX in Cu-rich films (Fig. 2). A high oxygen signal is correlated with the sodium content while the Se signal is anticorrelated. One explanation could be a partial substitution of Se with O.

The electrical conductivity of the films strongly depends on the substrates. The sheet resistance of a series of identical $Cu(In,Ga)Se_2$ preparations on sodium free substrates such as Corning 7059, Pyrex, Vycor and

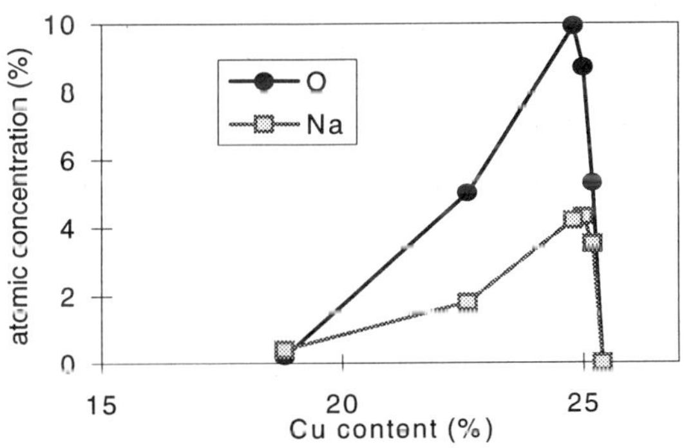

Fig. 2. WDX measurement of a $Cu(In,Ga)Se_2$ sample with a lateral Cu/In gradient. A significant amount of sodium and oxygen is found in the film. The measurement voltage of 5 kV leads to an information depth of about 100 nm. Most sodium can be found for stoichiometric $Cu(In,Ga)Se_2$ compositions.

Al_2O_3 is displayed together with the data obtained on the soda lime glass references in Fig. 3. Differences of about one order of magnitude between sodium free substrates and soda lime glass are measured.

The absorber layers are completed to solar cells with a $Cu(In,Ga)Se_2$ / CdS / ZnO structure. C-V characteristics were measured on these heterojunctions to determine the width of the space charge region. Junctions prepared on the alkali free substrates exhibit a three times larger space charge width as compared to soda lime glass (see Fig. 5). An increased V_{OC} is correlated with low values of the space charge width. The gain in efficiency of solar cells prepared on soda lime glass is mainly due to this improved V_{OC}.

At this point of the investigations there are several open questions for discussion. First of all, we need an explanation for the increased V_{OC} of solar cells prepared on soda lime glass. Afterwards the diffusion of sodium through the 1µm thick Mo back contact will be discussed. Finally we will describe an experiment with controlled addition of sodium species during the preparation process.

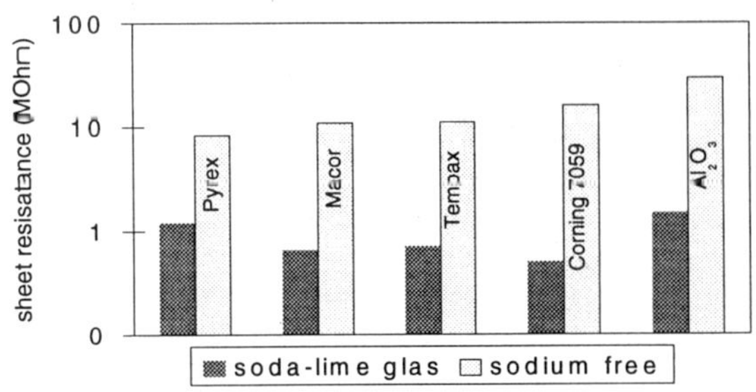

Fig. 3. Sheet resistance of $Cu(In,Ga)Se_2$ thin films on different substrates

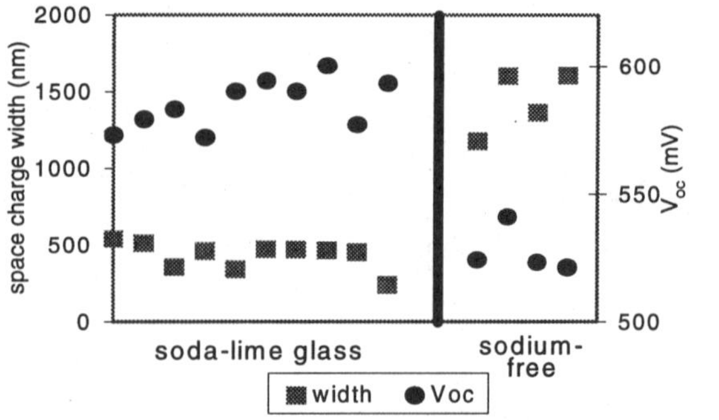

Fig. 5. Influence of substrates on space charge width and V_{oc} of $Cu(In,Ga)Se_2$ solar cells

DISCUSSION

The reason for the differences in V_{OC} of solar cells prepared on different substrates cannot be explained with morphological differences. XRD spectra do not indicate differences in the texture as the (112) / (204) intensity ratio for Cu-poor films prepared on Corning 7059 and soda lime glass is similar. The significant enhancement in grain size and orientation of films prepared on soda lime glass in a modified bilayer process [2] could be the result of a temporary Cu-rich phase during the film growth which does not exist in our single layer process.

Modelling of j-V-curves demonstrate that if the dominant recombination mechanism is in the space charge region [6] (neglecting the influence of tunneling mechanisms and surface recombination) an increased acceptor concentration N_A in the absorber material results in higher V_{OC}. According to the equation

$$\Delta V_{OC} = \frac{2kT}{q} \ln \sqrt{\frac{N_A}{N_{A0}}}$$

an increased concentration of one order of magnitude relative to a basic concentration N_{A0} leads to an increased V_{oc} of about 60 mV using a diode factor A=2. Assuming constant carrier mobilities, the observed differences in the electrical conductivitiy (Fig. 3) could represent higher effective doping concentrations and explain the high values of V_{oc} of films prepared on soda lime glass.
Assuming an abrupt n^+p junction, the width of the space charge region is inverse proportional to the square root of the carrier concentration in the absorber material. The three times larger width measured on alkali free substrates is compatible with the hypothesis that soda lime glass forces an increased net acceptor density, which could be responsible for the gain in V_{oc}.

The mechanism causing the increase in the effective doping is not yet clear. XPS measurements show a significant Na 1s signal at the surface after several days air exposure. No indications for an

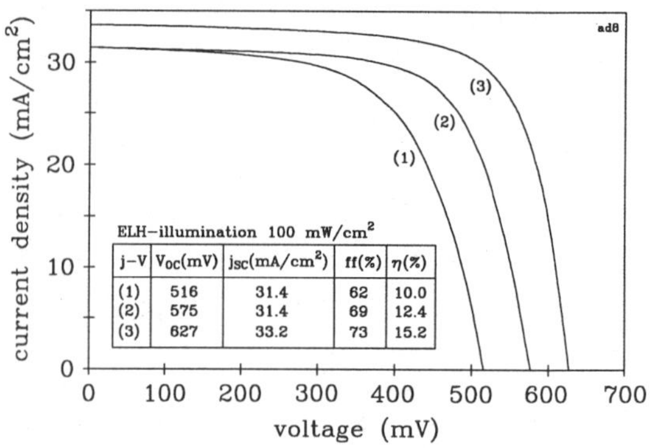

Fig. 4. j-V curves of $Cu(In,Ga)Se_2$ single layer based solar cells prepared on Corning 7059 (1), float (2) and float glass with an additional precursor (3). The efficiencies are active area; cell (3) has an AR coating.

incorporation of sodium in the $Cu(In,Ga)Se_2$ have been found with XPS measuring *in situ* prepared samples at different stages of film growth. This could mean that a sodium diffusion into the absorber material occurs mainly after the preparation process, possibly supported by the influence of air. The most likely path for such a diffusion is along defects at the grain boundaries. The XPS measurements further show the correlation of Na-species with an enhanced formation of oxides at the surface. It has been reported [7], that chemisorbed oxygen at Se vacancies can neutralize charged surface states. Because V_{Se} are donor-like defects, this could lead to an increase in the effective acceptor concentration.

It has been found that alkali-metals promote the oxidation of III-V semiconductors and silicon [8]. A number of mechanisms like an increased oxygen sticking coefficient or the dissociation of molecular oxygen are discussed. The alkali-metal promoted chemisorption of oxygen on the $Cu(In,Ga)Se_2$ surface and at grain boundaries could explain our observations.

How the sodium species diffuse from the substrate into the film through the molybdenum back contact is not yet clear. The substrate temperature (best values V_{oc} are achieved at higher temperatures) as well as the molybdenum preparation process seem to play an important role for the sodium transport. Recent XPS measurements [9] indicate that the existence of Mo-O species in the back contact can enhance the sodium transport via a solubility of sodium e.g. as Na_2MoO_4.

The undefined purity of the glass substrates as well as influences of the Mo preparation process are possible sources of nonreproducibility in the absorber preparation process. Therefore, we initiated experiments to control the sodium content by intentional addition of sodium species.

The presence of Na_2O in soda lime glass inspired the experiment of preparing a Na_xO precursor onto the unheated Mo-surface by thermal evaporation of Na_2O_2. The Na_xO film almost completely disappears while heating the substrate to 600°C. XPS measurements

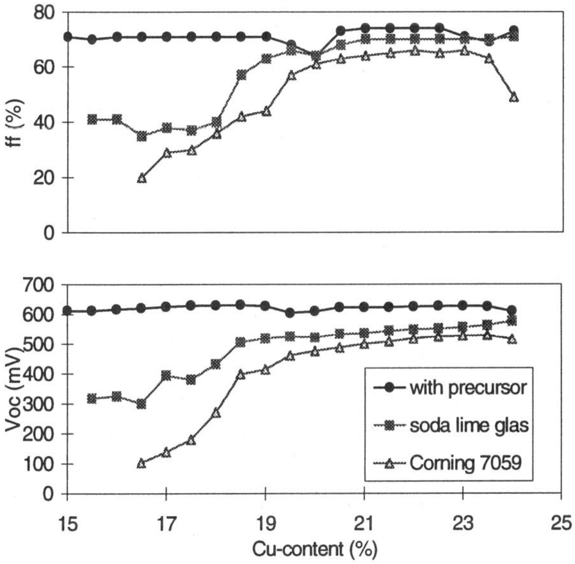

Fig. 6. Homogeneity in V_{oc} and FF of Cu(In,Ga)Se_2 solar cells prepared on different substrates

show only a small Na 1s signal left at the surface. A significant reevaporation at that low temperature can be excluded because a thin film coating the Mo reappears after several days air exposure. This indicates a notable solubility of Na-O species in Mo.

The conversion efficiency of a solar cell prepared on such a precursor improved by about 2% as compared to a cell without this precursor. Fig. 4 shows the j-V curves of Cu(In,Ga)Se_2 solar cells prepared on different substrates. The improvement cannot be explained by changes in texture or morphology, as shown by XRD measurements. The single layer process used here yielded an efficiency exceeding 15% without producing large crystals and compact films typical for other high efficiency processes [2,10].

The space charge width, as determined by C-V measurements, is about 300 nm in the junction with the precursor. This reflects the suggested influence of the precursor resulting in a higher effective doping concentration and increased V_{oc}. A further influence of the precursor has been observed in the homogeneity of cell properties over a large composition range (see Fig. 6). The fill factors have been higher than 70% for compositions down to 15% Cu content in the layer. An improved rectifying behavior indicates reduced shunts with increasing sodium content. It is assumed, that n-type shunt paths at grain boundaries limit the device performance. The use of the precursor seems to neutralize or type convert these shunt paths.

Stability tests carried out over several months show no changes in the composition range near stoichiometric Cu(In,Ga)Se_2, where the best device performances have been measured. We observe a degeneration of V_{oc} and ff of about 5% at the far Cu-poor side.

CONCLUSIONS

It has been demonstrated that the substrate material influences the electrical properties of Cu(In,Ga)Se_2 solar cells. We observed an increased electrical conductivity, reduced space charge width and higher V_{oc} for films prepared on soda lime glass as compared to those prepared on alkali free substrates. No influence on grain size, orientation or film density has been observed for films prepared by the single layer process used in this work. We favour the model that the increased V_{oc} is a consequence of a higher effective acceptor concentration in the absorber material. An explanation could be the neutralization of donor-like Se vacancies through a promoted chemisorption of oxygen in the presence of sodium species.

Experiments with a Na_xO precursor layer demonstrate the possibility of positively influencing the electrical properties of a solar cell. The increased V_{oc} and homogeneity versus the Cu-content signalize reduced losses through shunt paths, supposedly at grain boundaries. It opens a door for further studies of preparing solar cells with improved electrical properties, possibly at lower substrate temperatures, on various substrates.

ACKNOWLEDGMENTS

This work has been supported by the Bundesministerium für Forschung und Technologie (BMFT) contract 032 8059E and by the CEC contract No. JOU2-CT92-0141. We thank H. Dittrich for XRD measurements.

[1] L. Stolt, J. Hedström, J. Kessler, M. Ruckh, K.O. Velthaus, H.W. Schock, *Appl. Phys. Lett.* **62**, (1993) 597

[2] J. Hedström, H. Ohlsén, M. Bodegård, A. Kylner, L. Stolt, D. Hariskos, M. Ruckh, H.W. Schock, *Proc. 23rd IEEE Photov. Spec. Conf.* (1993) 364

[3] M. Bodegård, L. Stolt, J. Hedström, *Proc. 12th E.C. PVSEC* (1994) 1743

[4] J. Holz, F. Karg, H. von Philipsborn, *Proc. 12th E.C. PVSEC* (1994) 1593

[5] T. Walter, R. Menner, M. Ruckh, L. Käser, and H.W. Schock, *Proc. 22nd IEEE Photov. Spec. Conf.* (1991) 924

[6] T. Walter, R. Menner, Ch. Köbele, and H.W. Schock, *Proc. 12th E.C. PVSEC* (1994) 1755

[7] D. Cahen, and R. Noufi, *Solar Cells*, **30** (1991) 53

[8] H.I. Sternberg, P. Soukiassian, Z. Hurych, *Phys. Rev. B*, **39**, (1989) 12775

[9] D. Schmid, M. Ruckh, H.W. Schock, this conference

[10] A.Gabor, J. Tuttle, M. Contreras, D. Albin, A. Franz, D. Niles, and R. Noufi, 12th EC PVSEC (1994) 939

BINARY PHASE SEGREGATION IN $CuInS_2$ THIN FILMS FOR EFFICIENT SOLAR ENERGY CONVERSION

R. Scheer, I.Luck, S. Hessler, H. Sehnert, H.J. Lewerenz
Hahn-Meitner-Institut Berlin, Glienicker Str. 100, 14109 Berlin, Germany

ABSTRACT

We investigate Cu-In-S films with Cu/In > 1 grown by different preparation techniques such as coevaporation, sulphurization of stacked elemental layers and spray-pyrolysis. Using electron spectroscopy measurements, a preferential segregation of a CuS layer at the front surface of the films can be displayed by the structure of the Cu LVV Auger electron spectrum. This is observed for films grown by coevaporation and sulphurization. The CuS segregation is missing for as-deposited films which are grown by spraypyrolysis, however can be induced for these films by additional tempering. The results of the surface analytical study are discussed with respect to the junction performance of $CuInS_2$/CdS/ZnO solar cells.

INTRODUCTION

Recent progress in the formation of efficient hetero-junction solar cells on the basis of Cu-In-S thin films was achieved by using a new preparation approach [1]: Cu-In-S films were grown by the coevaporation technique with excess Cu throughout the whole deposition period. The Cu surplus is established in the film by a preferential segregation of the binary phase CuS which leads to an island-like capping of the growth surface with this quasi-metallic phase [2]. Before formation of the heterocontact with CBD:CdS/ZnO, the CuS capping is removed in a chemical etching step. Following this approach, for the first time an efficiency exceeding 10% under simulated AM 1.5 illumination could be obtained using selenium free chalcopyrite films, a value which has recently been improved to 12% by Walter et al. [3].

Since the chemical etching step can be considered as a post-deposition stoichiometry adjustment, this technique could be interesting for large area $CuInS_2$-based solar cell fabrication. Therefore the utilization of other deposition techniques for the growth of the absorber layers would be favourable. The CuS capping, however, which facilitates the post-deposition stoichiometry adjustment, seems to be a condition for the application of Cu-rich chalcopyrite layers in solar cell structures. For Cu-rich $CuInSe_2$ films, for instance, this approach had been shown to be less successful [4] since for these films, the binary Cu phases are assumed to be segregated <u>and</u> included in the films [5]. The reason for a preferential segregation of the CuS phase on coevaporated Cu-In-S is at present not clear. From the common observation that the morphology of Cu-rich chalcopyrite films is considerable improved compared to the In-rich, it is at least suggested that the segregation prevails during the film growth [6].

In this contribution we report on the investigation of the surface structure of differently prepared Cu-In-S layers by means of photoelectron spectroscopy. It is our intention to further substantiate the information about the binary phase formation in Cu-In-S films.

EXPERIMENTAL

Cu-In-S films with Cu/In > 1 were grown by coevaporation, sulphurization and spraypyrolysis techniques. As a substrate we used soda-lime glass which was covered by a 0.8 μm layer of sputtered Mo. The coevaporation of Cu, In, and S was performed in an high vacuum chamber at a substrate temperature of 550°C. The ratio of chalcogen to metal atoms was approximately 10. After a deposition period of 1 h using constant temperatures for the Cu and In sources, a film of 2-3 μm thickness is obtained. To avoid contamination of the samples, the transfer to the surface analysis system was performed under vacuum.

For the preparation of films by spraypyrolysis, we used a solution of 1×10^{-3} M CuCl, 0.9×10^{-3} M $InCl_3$ and 9×10^{-2} M thiorea which was bubbled with dry nitrogen before deposition. The spray-system consists of a glass containment which is evacuated by a water stream pump to reduce the oxygen partial pressure during the deposition. The solution was sprayed through an ultrasonic nozzle where a 200 sccm Ar flux served as the carrier gas. The substrate temperature was about 300°C. Typical deposition periods were 30 min which yielded a layer of 2μm thickness.

The sulphurization was done with stacked elemental layers of Mo, In, and Cu which were prepared by sputtering (Mo) and evaporation (Cu, In) onto unheated substrates. The precursor films were transfered through ambient air to an hot wall reactor for sulphurization. Before starting to heat the samples via radiation, a constant H_2S gas flux of 1 cm^3-s^{-1} was applied. For a reaction period of 3 h the samples were kept at 500°C. The sprayed and sulfurized films were transfered through air to the analysis chamber.

X-ray photoelectron spectra were recorded in a VG ESCALAB system using a MgK_α excitation source. The pass energy was 20 eV which produces a minimal line width at half maximum (FWHM) of 0.8 eV. Surface stoichiometries were determined by the intensity ratio of the In 3d and Cu LVV emission in comparison to a crystal standard. The preparation of the back surface of the films was performed in vacuo as described elsewhere [7]. Etching of

CH3365-4/94/0000-0160 $4.00 © 1994 IEEE

the front surface was done in a N_2 purged atmosphere. A 10% KCN solution served as the complexing agent which was followed by a rinse in deionized water and a drying step in nitrogen stream. The etched samples were transfered under N_2 atmosphere to the analysis system.

The stoichiometry of the samples was determined by energy dispersive x-ray spectroscopy (EDS). All samples were slightly Cu-rich before etching with a Cu to In ratio of 1.27, 1.25, and 1.37 for the evaporated (a), sulphurized (b), and sprayed (c) film, respectively. X-ray diffraction diagrams were recorded with CuK radiation using a $\theta/2\theta$ geometry.

RESULTS AND DISCUSSION

X-ray diffraction pattern of the films (a)-(c) are shown in Fig. 1. The structures at about 32° can be assigned to the (004), (200) reflections of $CuInS_2$ according to the JCPDS file No. 27-159. The tetragonal peak splitting is apparent for spectra (a)-(b) while for the sprayed film in (c) it is only indicated by a weak shoulder. In spectra (a) and (b) additional peaks at 2θ=29.3° and 31.8° can be discerned. They can be assigned to the CuS phase according to the JCPDS file No. 6-464 [8]. For spectrum (c) of the sprayed film, no additional diffraction signals are resolved. After etching in KCN, the films were again investigated by XRD (not shown). We found that the x-ray diffraction patterns of $CuInS_2$ are unaffected, however, the signals from CuS are distingueshed. After etching, film (a) was stoichiometric within the resolution limit of the EDS method of 1% while for the sulphurized and sprayed film we found [Cu]/[In] values of 1.08 and 1.32, respectively.

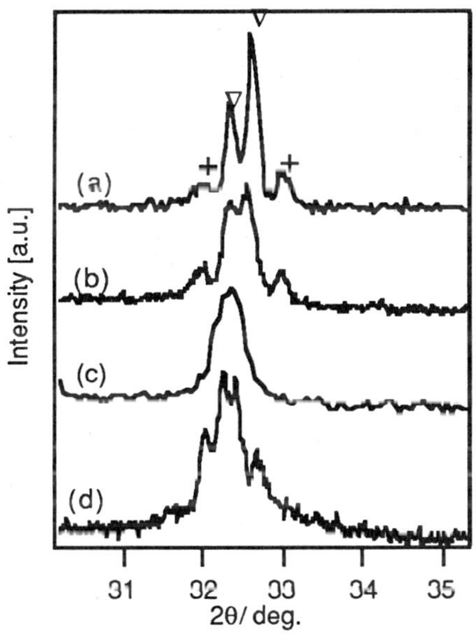

Fig. 1. X-ray diffraction diagrams of Cu-rich Cu-In-S films grown by (a) coevaporation, (b) sulphurization, (c) spraypyrolysis as-grown, and (d) spraypyrolysis with additional annealing at 400°C. The diagrams were recorded before etching of the samples. The signals can be assigned to the phases $CuInS_2$ (∇) or CuS (+).

In an additional experiment, an as-sprayed film was placed in the sulphurization reactor and exposed to the H_2S gas flux for 3h at 300°C or 400°C. The diagram (d) in Fig. 1 proves by the evolved CuS peaks, that after the annealing step, this phase has formed. Furthermore, the film has been recrystallized as depicted by the increased splitting of the (004) (200) reflections. After annealing, the KCN etch adjusts the stoichiometry as displayed by EDS of the annealed film to [Cu]/[In] = 1.0.

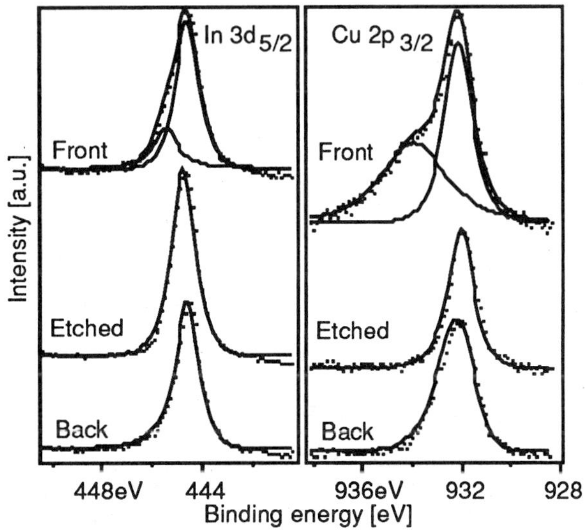

Fig. 2. Photoelectron spectra of the In $3d_{5/2}$ and Cu $2p_{3/2}$ emission (dots) from a sulphurized film before and after etching and a cleaved $CuInS_2$ crystal. The spectra are normalized and presented after removal of a Shirley type background. The solid lines represent results of a fit with a mixed Gaussian-Lorenzian curve.

In Fig. 2, photoelectron spectra of the In $3d_{5/2}$ and Cu $2p_{3/2}$ emission are shown for a sulphurized film which has been exposed to air for several days. The surface stoichiometry as determined by electron spectroscopy of [Cu]/[In] = 1.1±0.1 is smaller than the bulk value of 1.25. After etching this quantity is further reduced to 0.5±0.1 which equals to a cation ratio [In]/([In]+[Cu]) of 0.66±0.05. Approximately the same value has been found for the coevaporated film.

The binding energy of the Cu $2p_{3/2}$ electrons is 932.0±0.05 eV for the coevaporated and sulphurized films after etching. Since we were not able to acquire valuable ultraviolet photoelectron spectra of this sample in order to determine the energetic position of the valence band maximum E_{VBM}, we used a comparison with data from $CuInS_2$ crystals [2]. This procedure yielded E_F-E_{VBM} = 0.5±0.1 eV for the etched film, where E_F denotes the Fermi energy. The same value is found if In $3d_{5/2}$ lines are used for the binding energy evaluation. The peak widths of the In $3d_{5/2}$ and Cu $2p_{3/2}$ emissions are slightly reduced after etching. A broadened signal from the Cu core level is obtained from the back surface of the sample.

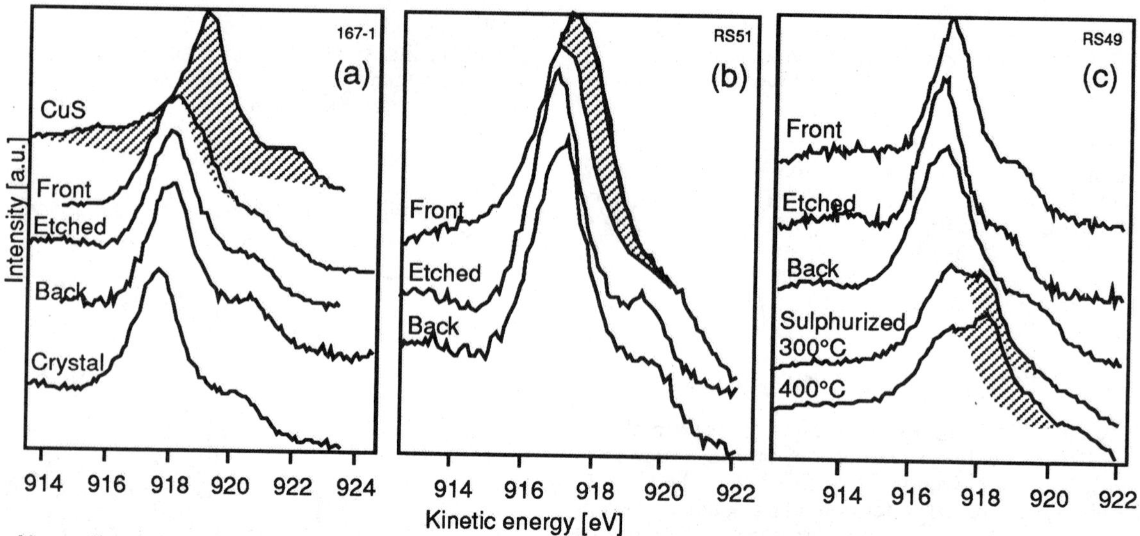

Fig. 3. Normalized Auger electron spectra of the Cu LVV transition for Cu-rich films grown by (a) evaporation, (b) sulphurization, and (c) spraypyrolysis. In (a) the spectra of a coevaporated CuS and a cleaved $CuInS_2$ crystal are included. The shaded regions denote additional emission components which are ascribed to the CuS phase.

The spectra of Cu $2p_{3/2}$ and In $3d_{5/2}$ emission in Fig. 2 from the as-grown front surface both exhibit second components at higher binding energy. These peaks can be ascribed to In_xO_y and Cu_2O phases. After etching, these components are distingueshed indicating the removal of oxides.

The most conclusive information about secondary phase formation in the Cu-rich films can be obtained from considering the Auger electron spectra from the Cu LVV transition. In Fig. 3a-c we plotted the normalized Auger electron count rates for an evaporated (a), a sulphurized (b), and a sprayed (c) film on the kinetic energy scale. Fig. 3a also includes the spectra of a cleaved $CuInS_2$ crystal and a CuS film which was grown by coevaporation.

A comparison of the spectrum of the front surface of the evaporated film with the spectrum of the crystalline sample (see Fig. 3a) points toward the occurrence of a second component (shaded). This Auger electron emission can be assigned to a CuS overlayer [2]. It was shown recently that the CuS layer forms an island like capping of the as-grown film. The removal of the second component was interpreted in terms of a decapping of the sample [2]. Since in the spectrum of the back surface this component is missing, it was concluded that the CuS formation mostly takes place at the front surface of the film [9]. If we consider the spectra of the sulphurized film, we find a largely similar sequence of spectra. The spectrum of the as-grown front surface is considerably broadened due to a second component which is removed after etching the film in KCN. The spectrum of the back surface, however is also slightly broadened.

The spectra of the sprayed film (see Fig. 3c) have a different appearance: Before etching, the emission of the front surface has approximately the same shape and FWHM as the spectrum from the crystal in Fig. 3a. Accordingly, after etching, the peak shape is maintained, however the peak maximum is shifted by 0.2±0.1 eV. We calculated the Cu Auger parameter α'_{Cu} which is 1849.0 eV for $CuInS_2$ [10]. In the case of the unetched film this value is 1849.5 eV and is somewhat reduced after etching to 1849.2 eV. The Auger electron spectra of films annealed at 300°C and 400°C in H_2S are also included in Fig. 3c. For these films, additional emissions can be observed which increases for higher annealing temperature. This can be considered as an indication, that the sulphurization step at elevated temperature gives rise to a segregation of CuS at the front surface of the sprayed film.

In order to understand the phase segregation in the Cu-In-S system, we consider the sprayed films as the most interesting example. For the as-sprayed film, no indication of a Cu binary phase has been obtained while the main diffraction peaks of the chalcopyrite phase can be detected. This indicates that additional Cu ions are included in the lattice, for instance, via the formation of Cu_{In} antisite defects. In terms of an extended homogenous region of $Cu_{1-x}In_{1-x/3}S_2$ with $x < 1$ this is a metastable state. If additional sulphur is provided at high temperature at the front surface of the films, the formation of CuS is energetically favoured and the binary phase starts to segregate, thereby approaching thermodynamic equilibrium. It should be mentioned that the conventional viewpoint which goes back to the work of Binsma et al. [11] assumes a one sided homogenous region with $x > 1$.

In the sulphurization experiment the samples are homogeneously irradiated. Therefore, a temperature gradient which was discussed in an earlier work [7] is not likely to explain the observed CuS segregation. During the coevaporation process, excess sulphur atoms are permanently adsorbing at and desorbing from the growth surface of the film. Since they exhibit a strong affinity to react with Cu, the formation of CuS could extract the excess Cu ions from the chalcopyrite via Cu-ion diffusion.

The same effect would be expected for the sulphurized elemental layers and is in fact indicated by the sequence of Auger spectra in Fig. 3b. During the spray process, however, the nucleation takes place at lower temperature, i.e. far from equilibrium.

In view of the CuS capping of evaporated and sulphurized films, it is interesting that after exposing the films to air, both Cu and In oxides have formed. This supports our earlier finding that the CuS phase forms islands on a surface where the uncapped regions are In-rich.

The Cu-In-S layer (a)-(c) after etching were used to fabricate $CuInS_2/CdS/ZnO$ cells as described elsewhere [1]. While with film (a) the best efficiency was 10.9 % (U_{ph}=705 mV, I_{sc}=22.7 mA, FF=68.5%, AM 1.5) without AR coating, no photoeffect was observed for film (b) and (c). This can tentatively be ascribed to the high density of pinholes in the sulphurized films which induce a shortance of the cell. Further work will concentrate on the reduction of morphological defects in the sulphurized films.

CONCLUSION

We showed that the formation of a CuS phase segregation at the front surface of Cu-rich Cu-In-S films can be investigated by Auger electron spectroscopy. For coevaporated and sulphurized layers, the CuS phase is found after growth. For the sprayed film, however, an additional annealing step is required. This indicates that the CuS segregation is induced by a supply of excess S at the front surface and is thermally activated.

ACKNOWLEDGEMENT

The preparation of the window layer by D. Hariskos from the IPE Stuttgart and R. Wendt is greatfully acknowledged. The authors wish to thank Dr. W. Jägermann and Dr. T. Walter for many fruitful discussions.

REFERENCES

[1] R. Scheer, T. Walter, H.W. Schock, M.L. Fearheiley, and H.J. Lewerenz, "CuInS2 based thin film solar cell with 10.2% efficiency", *Appl. Phys. Lett.*, **63**, 1993, p. 3294.

[2] R. Scheer, and H.J. Lewerenz, "Photoemission study of evaporated CuInS2 thin films. I. Surface stoichiometry and phase segregation", *J. Vac. Sci. Technol. A*, **12**, 1994, p. 51.

[3] T. Walter, M. Ruckh, K.O. Velthaus, and H.W. Schock, "Solar Cells based on CuIn(S,Se) - A Promising Alternative", *11th European Photov. Solar Energy Conf.*, 1992, p. 124

[4] J.R. Tuttle, M. Contreras, D.S. Albin, and R. Noufi, "Physical, chemical, and structural modifications to thin-film CuInSe2 based photovoltaic devices", *22nd IEEE Photovolt.Spec.Conf.*, 1991, p.

[5] J.R. Tuttle, D.S. Albin, and R. Noufi, "Thoughts on the microstructure of polycristalline thin film CuInSe2 and its impact on material and device performance", *Solar Cells*, **30**, 1991, p. 21.

[6] R. Klenk, T. Walter, H.W. Schock, and D. Cahen, "A model for the succesful growth of polycrystalline films of CuInSe2 by multisource physical vacuum evaporation", *Adv. Mater.*, **5**, 1993, p. 114.

[7] R. Scheer, I. Luck, and H.J. Lewerenz, "Photoemission study on CuInS2 thin films for efficient solar energy conversion", *12th European Photovoltaic Solar Energy Conference*, 1994

[8] T. Walter, A. Content, K.O. Velthaus, and H.W. Schock, "Solar cells based on CuIn(Se,S)2", *Sol. Energy Mat.*, **26**, 1992, p. 357.

[9] R. Scheer, and H.J. Lewerenz, "Formation of secondary phases in evaporated CuInS2 thin films: A surface analytical study", *J. Vac. Sci. Technol.*, submitted

[10] R. Scheer, and H.J. Lewerenz, "Photoemission study of evaporated CuInS2 thin films. II. Electronic surface structure", *J. Vac. Sci. Technol. A*, **12**, 1994, p. 56.

[11] J.J.M. Binsma, L.J. Giling, and J. Bloem, "Phase Relations in the system Cu_2S-In_2S_3", *J. Crystal Growth*, **50**, 1980, p. 429-436.

[12] A. Rockett et al., "Structure and chemistry of CuInSe2 for solar cell technology: current understanding and recommendations", *Thin Solid Films*, **237**, 1994, p. 1-11.

SODIUM DIFFUSION, SELENIZATION, AND MICROSTRUCTURAL EFFECTS ASSOCIATED WITH VARIOUS MOLYBDENUM BACK CONTACT LAYERS FOR CIS-BASED SOLAR CELLS

John H. Scofield
Physics Department, Oberlin College, Oberlin, OH 44074
and
S. Asher, D. Albin, J. Tuttle, M. Contreras, D. Niles, R. Reedy, A. Tennant, and R. Noufi
National Renewable Energy Laboratory, 1617 Cole Blvd., Golden, CO 80401

ABSTRACT

By varying the argon pressure during deposition, we have prepared a set of sputtered molybdenum films on soda-lime glass substrates with a range of mechanical and electrical properties. These films were subsequently exposed to several of the processing steps used in the fabrication of copper-indium-diselenide (CIS) solar cells. Processing steps of interest include heating in a vacuum, exposure to selenium vapor at elevated temperatures, and deposition of CIS and CIGS layers over the Mo. Resulting Mo films and structures were subsequently characterized using XPS, SEM, Auger, and SIMS. Here we describe the results of these experiments and their implications for CIS cell fabrication.

INTRODUCTION

Over the last several years the conversion efficiency achieved by the "best" polycrystalline thin film copper indium diselenide (CIGS) based small-area solar cells has risen to above 16% [1, 2, 3]. While the efficiencies of the "best" small-area cells have steadily improved, uncontrolled factors continue to thwart the reproducibility of these cells, and to a greater extent, degrade the quality and through-put of large-area CIS and CIGS module fabrication efforts. In connection with these problems, a variety of suspect areas have been suggested; these include 1) the nature and importance of surface defects on the glass substrates, 2) substrate cleaning processes, 3) the possible formation of $MoSe_2$ near the Mo/CIS interface, 4) the possible importance of sodium diffusion into the CIS, 5) the softening of the glass substrate at the high processing temperatures, and 6) adhesion failures at the Mo/glass and/or Mo/CIS interfaces [4].

A number of these issues appear to be related to the properties of the Mo back contact layer universally employed for CIS and CIGS devices. At most institutions, and particularly at NREL, considerable fabrication efforts have been devoted to improving the properties of the absorber layer. In contrast, significantly less effort has been devoted to understanding and shaping the properties of the Mo layer. In many cases, the Mo layer has been deposited by whatever means were available, with the only property of concern being the Mo sheet resistance. Only recently have we recognized importance of the Mo layer as the substrate upon which the CIS absorber layer is grown and that it may have significant impact on the absorber layer nucleation, growth, and morphology.

The electrical and mechanical properties of sputtered refractory metal films are known to vary with sputter pressure [5]. We have used this fact to fabricate a set of Mo films on soda-lime glass substrates with widely varying properties. The resulting films have been characterized using several analytical techniques including XPS, SEM, SIMS, XRD, and GIXRD (glancing-incidence XRD). We have also measured film resistivity and adhesion to the glass substrate. In order to shed light on some of the questions raised above we have subjected these Mo films to our usual solar cell processing stages, including the high-temperature deposition of CIS and CIGS and the completion of solar cells.

In this paper we describe the results of these experiments and discuss their implications for CIS and CIGS solar cell fabrication.

AS-DEPOSITED FILMS

Sample Fabrication

Films were sputter-deposited onto 2 in. x 2 in. x 0.095 in. soda-lime glass substrates held approximately 8 cm from a 12 cm x 20 cm, planar, dc magnetron, Mo target. Substrates were cleaned using our standard soap/DI water scrub and rinse process. After cleaning, substrates were blown dry with N_2 gas and immediately loaded into the vacuum chamber that was subsequently evacuated to a base pressure near 1×10^{-6} Torr. Two substrates were coated during one pump-down. The details of the deposition process may be found elsewhere [6]. For our first set of experiments a set of seven, roughly 500-nm-thick Mo films were sputtered at Ar pressures (P) of 0.2, 0.5, 1.0, 2.0, 5.0, 10, and 20 mTorr respectively. Deposition rates were approximately 1 nm/s.

Stress and Resistivity

Film resistivity (ρ) and in-plane stress (σ) are shown in Figs. 1 and 2, the latter determined from GIXRD measurements [6]. The figures show that films deposited at low Ar pressure have low resistivity and compressive stress. As the pressure is increased, the resistivity increases and stress becomes tensile. At the highest

CH3365-4/94/0000-0164 $4.00 © 1994 IEEE

pressures, the resistivity continues to increase while the magnitude of the stress falls back towards zero, again becoming compressive at the highest pressure. The

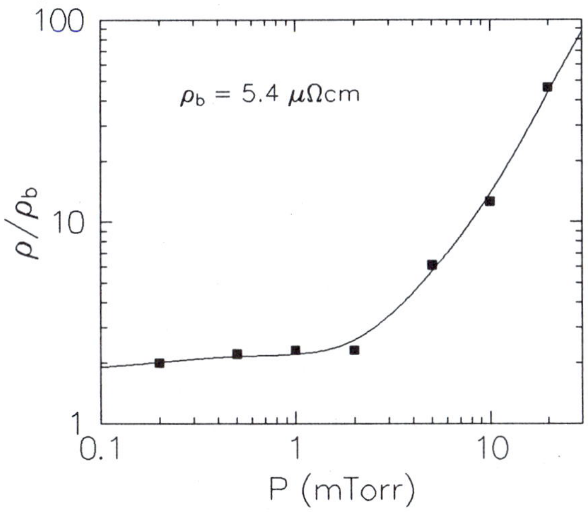

FIG. 1 Graph of the normalized, room-temperature film resistivity, ρ/ρ_b, versus Ar pressure during sputtering. The curve is a guide to the eye.

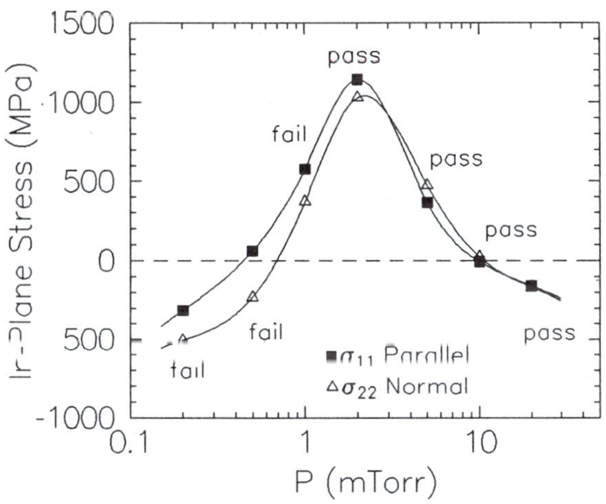

FIG. 2 In-plane stresses plotted against Ar pressure; negative values indicate compression while positive values indicate tension. Tape test results for adhesion are also indicated.

results are consistent with those reported for other sputtered refractory metal films [5]. It is also important to note that films sputtered at the three lowest pressures exhibited poor adhesion, failing the tape test (see Fig. 2).

As Figs. 1 and 2 show, films sputtered at low Ar pressure have the lowest resistivity while those sputtered at high pressures exhibit the best adhesion. We have developed a bilayer process that captures the best of both worlds. This process involves the deposition of a thin (~100 nm) layer at high Ar pressure immediately followed by a thicker (~900 nm) layer deposited at low Ar pressure. Such films achieve low resistivity and good adhesion and have been incorporated into NREL's high-efficiency cells [6].

SUBSEQUENT PROCESSING

Experimental Method

Since films sputtered at lower pressures did not adhere well to the glass, we decided not to submit the above films to subsequent processing steps, which would place even higher demands on film adhesion. To minimize variations associated with adhesion, we fabricated a second set of Mo films, each of these films having a bilayer structure, with the properties of the 900-nm-layer being varied. Also, so that differences in subsequent properties could be directly related to the original Mo films, we decided to restrict our attention to only four Mo films, and to process these four samples simultaneously in future experiments, thereby eliminating the effects of possible hidden, uncontrolled variables. This second set of samples consisted of 1-μm-thick Mo bilayers with the first, 100-nm-thick layer deposited at 10 mTorr and the second, 900-nm-thick layer sputtered at P = 0.5, 2, 5, and 10 mT respectively. Samples were subdivided into smaller pieces, and sets of four (one for each P) were processed together for subsequent experiments. Table I lists the various experimental conditions.

Results

SEM characterization of the as-deposited Mo films show that they all have columnar grain structure. Films deposited at low pressure are more dense with "closed" grain boundaries while those deposited at higher pressures are less dense with more "open" grain boundaries. These features qualitatively explain the variation of resistivity and stress with sputter gas pressure [5].

SIMS depth profiles were used to characterize the concentrations of Na, O, Se, and C or K impurities. Se levels in as-deposited films were negligible.

Expt.	Description	Subsequent Analysis
A1	as deposited	4-probe, SIMS, SEM, GIXRD, XRD
A2	vacuum anneal for 30 min. @ 450°C	4-probe, XPS, SIMS
B1	expose to 2 nm/s dep. rate of Se for 15 min @ 450°C	4-probe, XPS, SIMS
D1	deposit CIS using selenization	XPS, SIMS, SEM
E1	deposit CIGS using 3-stage process	XPS, SIMS, SEM

TABLE I List of processing steps to which the four Mo bilayer films were submitted along with the characterization methods to which they were subsequently submitted for analysis.

Depth profiles for Na and O impurities for as-deposited films are shown in Figure 3. (For comparison purposes, the depth scales for Figures 3 and 4 have been adjusted to account for variations in film thicknesses.) The O impurity level increases with sputtering pressure, being approximately 30 times higher in the 10 mT film than in the 0.5 mT film. Sodium levels varied with deposition pressure and depth, being lowest in the interior and highest towards the surfaces of the Mo films.

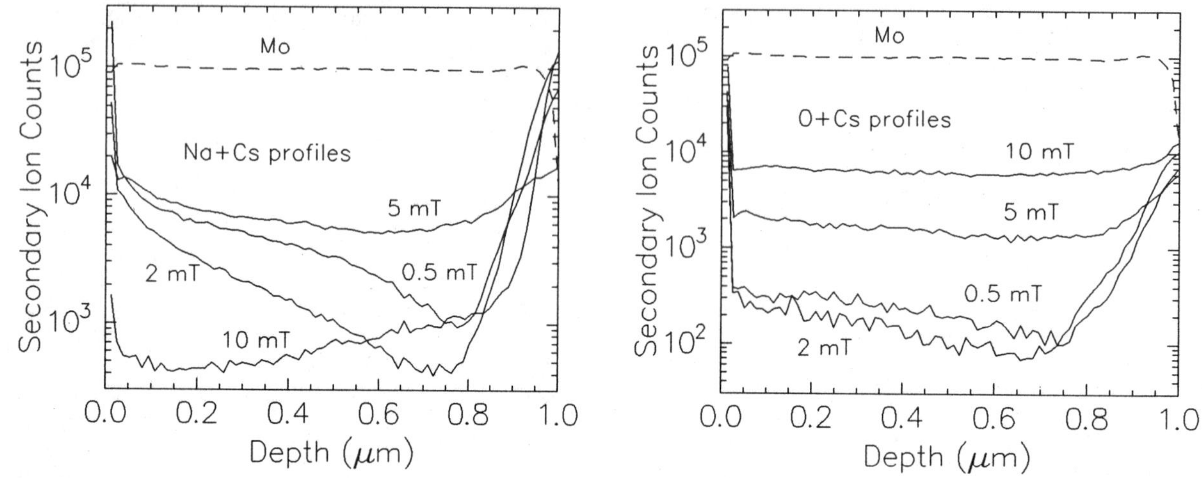

FIG. 3 SIMS data for as-deposited films showing depth profile for a) Na and b) O for the four bilayer Mo films. Also shown for reference is the SIMS signal for Mo.

With subsequent processing at elevated temperatures the Na and Se impurity concentrations changed significantly. Figure 4 shows (a) the Na- and (b) the Se-distributions for films exposed to Se flux at 450 °C. Na profiles changed significantly from those in the as-deposited films indicating significant diffusion, especially in the higher-pressure Mo films. For all films data show a significant build-up of Na within about 100 nm of the air/Mo interface. Figure 4(b) indicates substantially more selenization for the 10 mT than for the other Mo films.

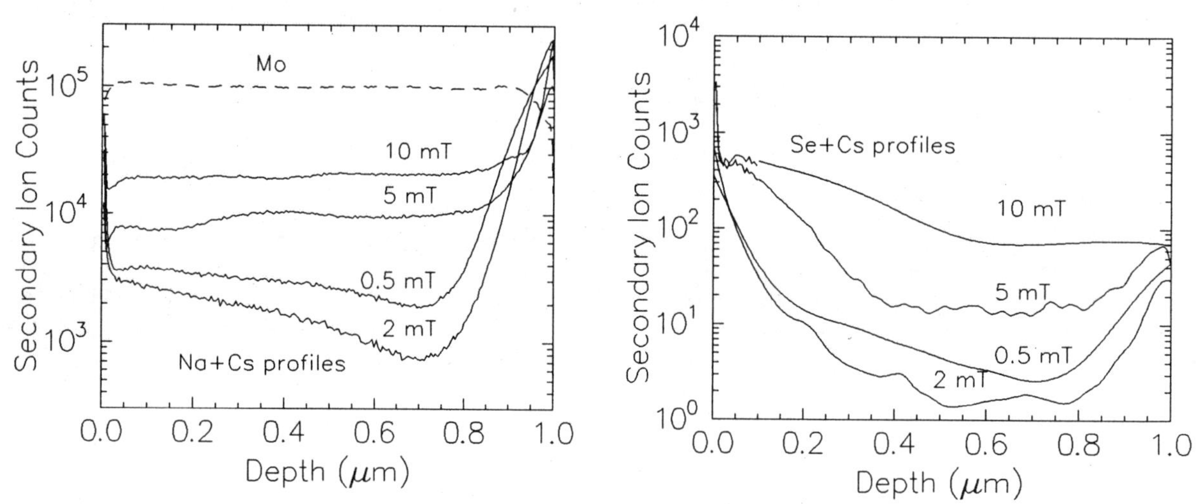

FIG. 4 SIMS data for films exposed to Se flux at 450°C showing depth profiles for a) Na and b) Se. Also shown, for reference, is the SIMS signal for Mo. 2 mTorr and 0.5 mTorr curves have been "smoothed" so that they are distinguishable.

Figure 5 shows the SIMS depth profiles for films following deposition of a CIGS absorber layer using NREL's 3-stage process (experiment E1). Similar data (not shown) have been obtained for films following deposition of a CIS absorber layer via selenization (experiment D1). Note that the depth scale has not been corrected for differences in sputter rates leading to distortion in the apparent absorber and Mo layer thicknesses.

Of particular interest is the Na depth profile. In the Mo back contact, Na impurity levels and distributions are similar to those of Figure 4, again showing that the low-pressure Mo contains significantly less Na than the high-pressure Mo. Despite these differences, the Na levels in the four absorber layers appear to be the same, independent of the underlying Mo layers. For all four samples there is significant build-up of Na at both CIGS surfaces. XPS analysis (not shown) indicates that CIS thin films grown with the selenization process have < 1% Na in the 20 nm region near the air/CIS surface while CIGS films grown using NREL's 3-stage process have more than 10% Na there. Both absorber layers show the build-up of Na at the absorber/Mo interface.

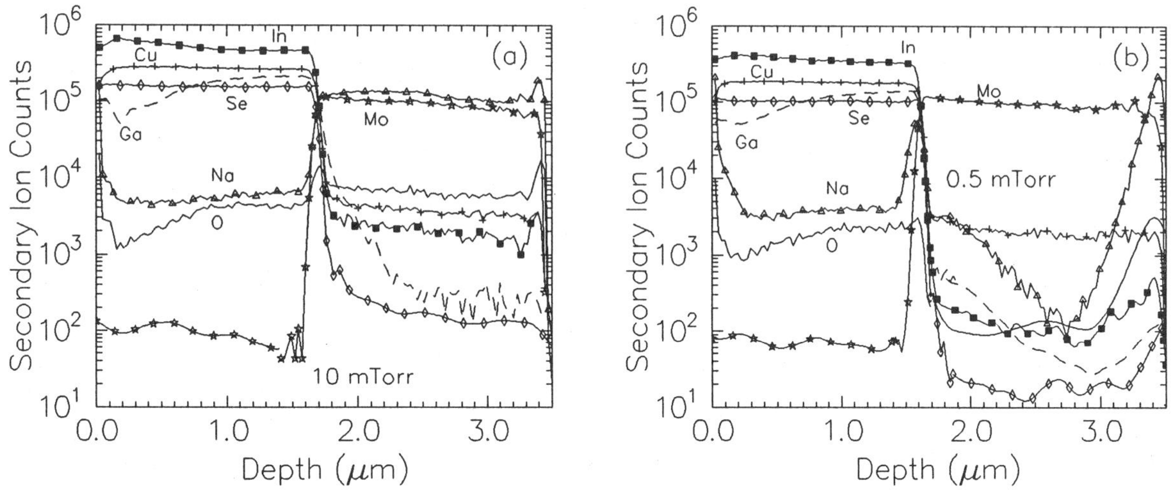

FIG. 5 SIMS data following CIGS deposition using NREL's 3-stage process on Mo sputtered at a) 10 mTorr, and b) 0.5 mTorr Ar pressure.

In addition to the data shown, SEM photos suggest that absorber grain size tends to increase with lower Mo deposition pressure, the effect being more pronounced for CIS deposition using selenization than for CIGS deposition using the NREL 3-stage process. In addition, selenized CIS layers appear to adhere better to the high-pressure Mo than to the low-pressure Mo.

DISCUSSION AND CONCLUSIONS

The O, Na, and Se depth profiles, electron micrographs, and Mo resistivity and strain data all suggest that the high-pressure Mo films have smaller grains, a more open grain structure (i.e., larger gaps between grains), and are more porous than the low-pressure Mo films. For the high-pressure Mo this leads to greater inclusion of oxygen (probably in the grain boundaries), higher resistivity, higher Na levels, and more rapid impurity diffusion. The morphology of the high-pressure Mo provides a greater number of nucleation sites for the absorber layer, leading to better absorber adhesion and smaller, less faceted absorber grains. These differences lead to greater selenization of the high pressure Mo during absorber formation (see Figs. 4 and 5).

The Na impurity levels in the absorber, however, appear to be independent of the underlying Mo properties, suggesting that they are determined by thermodynamic rather than kinetic factors. The relatively high Na concentrations at the two CIGS surfaces (air and Mo) are particularly interesting. There is some evidence that Na may play an important role in cell performance [7].

ACKNOWLEDGMENTS

The authors would like to thank J. Dolan for assistance configuring the Mo deposition system, P. Perdicki and B. Ballard for GIXRD measurements, A. Duda for bulk XRD measurements, and R. Matson for SEM characterization. One of the authors (JHS) was supported, in part, by an Associated Western Universities - Department of Energy Faculty Sabbatical Fellowship. This work was performed at NREL under contract No. DE-AC36-83CH10093 to the U. S. Department of Energy.

REFERENCES

[1] A. M. Gabor, J. R. Tuttle, D. S. Albin, A. L. Tennant, M. A. Contreras, and R. Noufi, in *Proc. 12th NREL P. V. Program Review Meeting,* Oct. 13-15, 1993, Denver, CO (AIP, 1994).

[2] M. A. Contreras, et al., to appear in *Progress in Photovoltaics.*

[3] J. R. Tuttle, M. A. Contreras, J. S. Ward, A. M. Gabor, K. R. Ramanathan, A. L. Tennant, L. Wang, J. Keane, and R. Noufi, "High efficiency Cu(In,Ga)Se_2-based thin-film solar cells: 16.8% total-area one-sun and 17.2% total-area 22-sun device performance," *this conference.*

[4] *2nd EPRI Sponsored Workshop on CuInSe$_2$ Materials,* Estes Park, CO, October 17-19, 1993, unpublished.

[5] T. J. Vink, et al., "Stress, strain, and microstructure of sputter-deposited Mo thin films," *J. Appl. Phys.* **70**, 4301-8 (Oct. 1991).

[6] J. H. Scofield, A. Duda, D. Albin, B. L. Ballard, and P. K. Predicki, "Sputtered molybdenum bilayer back contact forcopper indium diselenide-based polycrystalline thin-film solarcells," to appear in *Thin Solid Films.*

[7] J. Hedström, H. Ohlsén, M. Bodegård, A. Kylner, and L. Stolt, "ZnO/CdS/Cu(In,Ga)Se_2 thin film solar cells with improved performance," *Proceedings 23rd IEEE Photovoltaics Specialists Conference*, Louisville, KY (IEEE, New York, 1993) pp. 364-71.

PREPARATION OF $CuIn_xGa_{1-x}Se_2$ THIN FILMS ON Si SUBSTRATES

Yukio Yamamoto
Fukui College of Technology, Geshi, Sabae 916 JAPAN

Toshiyuki Yamaguchi and Masayoshi Suzuki
Wakayama College of Technology, Nada, Gobo 644 JAPAN

Yasutaka Demizu and Akira Yoshida
Toyohashi University of Technology, Tempaku, Toyohashi 441 JAPAN

ABSTRACT

For fabricating efficient tandem solar cells, $CuIn_xGa_{1-x}Se_2$ thin films have been prepared on Si(100), Si(110) and Si(111) substrates in the temperature range (R.T.~400°C) by rf sputtering. From EPMA analysis, these sputtered thin films are found to be nearly stoichiometric over the whole substrate temperature range, irrespective of the azimuth plane of the Si substrate. XPS studies showed that the compositional depth profile in these thin films is uniform. X-ray diffraction analysis indicated that all the thin films had a chalcopyrite structure. $CuIn_xGa_{1-x}Se_2$ thin films were strongly oriented along the (112) plane with increasing the substrate temperature, independent of the azimuth plane of the Si substrate, suggesting the larger grain growth.

INTRODUCTION

The ternary I-III-VI_2 chalcopyrite semiconductor $CuInSe_2$ has received considerable attention as an absorber in heterojunction photovoltaic devices because of its stability and high efficiency [1-6]. When replacing $CuInSe_2$ with its quaternary analog $CuIn_xGa_{1-x}Se_2$ or $CuIn(S_zSe_{1-z})_2$, the band gap increases with better match to the incident solar spectrum, resulting in higher efficiency close to 17% as reported by Hedström *et al* [7].

In order to absorb the incident solar beam more properly, a tandem cell is proposed. There are many combinations of cells which make a tandem cell. $CuIn_xGa_{1-x}Se_2$/Si tandem solar cell is one of the promising candidates. Also, CdS with hexagonal structure is usually used as *n*-type layer with *p*-$CuInSe_2$. However, it is a problem whether excellent $CuIn_xGa_{1-x}Se_2$ thin films with the (112) orientation can be fabricated onto Si substrates, which are commercially available.

In this study, we have prepared $CuIn_xGa_{1-x}Se_2$ thin films by rf sputtering onto Si(100), Si(110) and Si(111) substrates from a stoichiometric target and investigated the structural properties of these thin films.

EXPERIMENT

$CuIn_xGa_{1-x}Se_2$ thin films were deposited by sputtering in argon gas (13.56 MHz rf power). Powder targets were obtained by reacting high purity (5N) Cu, In, Ga and Se in a vacuum-sealed quartz tube. In this experiment, In content was fixed at 20 at%. The compound was then ground to a fine powder and cold-pressed at 4.9×10^7 Pa. Si(100), Si(110) and Si(111) wafers were used as substrates. They were ultrasonically cleaned in organic solvents and soaked in a hydrofluoric acid in order to remove the oxides. The deposition was carried out at substrate temperatures varying from room temperature (R.T.) to 400°C.

Quantitative and compositional analyses were performed using electron probe microanalysis (EPMA). Compositional depth profiles were obtained using X-ray photoelectron spectroscopy (XPS). X-ray diffraction (XRD) patterns using Cu Kα line (λ=1.5405 Å) and RHEED patterns were examined to investigate the crystalline structures and the orientation factor.

RESULTS AND DISCUSSION

Deposition rates were in the range of 48~58(Å/min) and found to decrease, as the substrate temperature was increased. No dependence of the deposition rate on the plane azimuth of Si substrate was found.

Figure 1 shows the variation of the composition of thin films deposited on Si(111) substrate as a function of substrate temperature. The atomic ratios, [I]/[III], [VI]/[I]+[III] and [In]/[In]+[Ga], are found decrease gradually, as the substrate temperature increases. This is due to the fact that Cu, In and Se atoms are re-evaporated at higher temperatures. Szot *et al.* [8] reported that In_2Se was re-evaporated when $CuInSe_2$ thin films were prepared by selenization onto glass. In the case of the films deposited on Si(100) and Si(110), the results were almost the same. This may be the cause of low deposition rates at higher temperatures. Since the ratios, [I]/[III] and [VI]/[I]+[III], lie between 0.92~1.03 and, thus, nearly stoichiometric $CuIn_xGa_{1-x}Se_2$ thin films can be grown on any Si substrate.

CH3365-4/94/0000-0168 $4.00 © 1994 IEEE

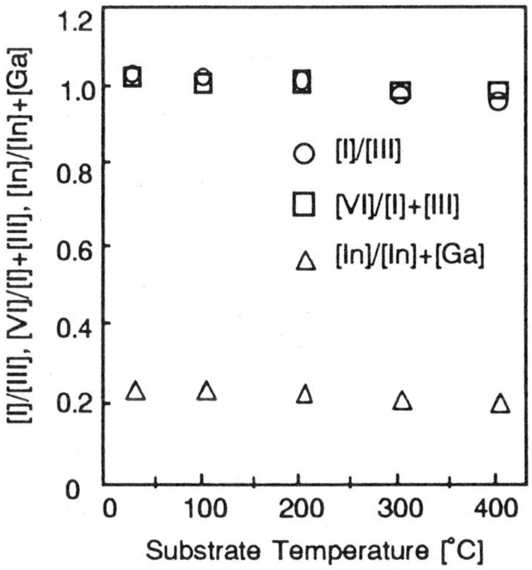

Fig. 1. Substrate temperature dependence of the composition of the thin films deposited on Si(111).

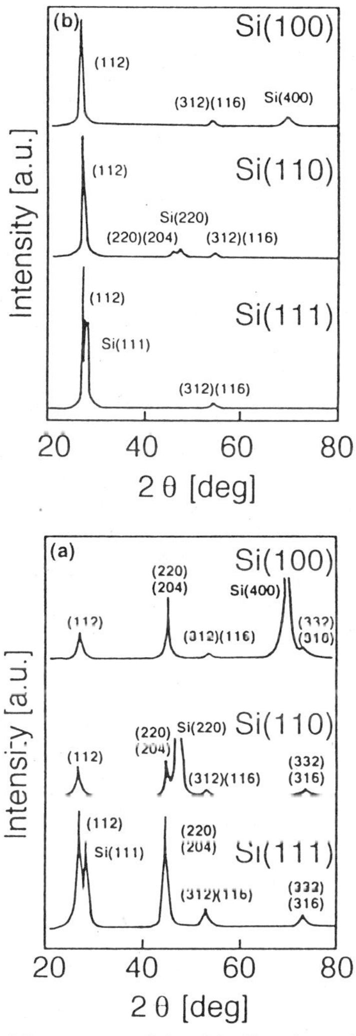

Fig. 2. XRD spectra of the thin film deposited (a) at room temperature and (b) at 400°C.

The compositional depth profile of Se, Cu, Ga and In in the films was found to be nearly constant from the XPS analyses. The XPS peak positions were found to be in good agreement with those in the case of the chalcopyrite $CuIn_xGa_{1-x}Se_2$ thin films as reported before [9]. The XPS signal from Si was not observed, even when the thin films prepared at 400°C were etchted for 100 minutes by Ar sputtering. Thus, it is considered that the Si atoms do not diffuse into the thin film from the Si substrate.

X-ray diffraction patterns of the thin films deposited at different substrate temperatures are shown in Fig. 2 (a) and (b). All peaks except those for single crystal Si correspond to the diffraction lines of $CuIn_xGa_{1-x}Se_2$. Thus, all these thin films are found to have the single phase chalcopyrite structure for any plane azimuth of Si substrates and for any substrate temperature. In comparison with the XRD pattern of $CuIn_xGa_{1-x}Se_2$ powder, thin films deposited at R.T. seem to be oriented to (220) and (204), because the diffraction peaks from those planes are observed. On the other hand, the XRD pattern of the film deposited at 400°C shows that the film is strongly oriented to (112) plane of chalcopyrite $CuIn_xGa_{1-x}Se_2$. The same behavior is observed in the case of films deposited on glass substrate. The (112) plane is the close packed one of the chalcopyrite structure. This shows that the Bravais's empirical law for crystal growth is confirmed even in $CuIn_xGa_{1-x}Se_2$ thin films deposited on Si substrates.

It is important to evaluate quantitatively the (112) orientation of the $CuIn_xGa_{1-x}Se_2$ thin film for lattice matching with CdS. The authors have employed the orientation factor p which is used for $CuIn_xGa_{1-x}Se_2$ and other materials. It can be calculated as the ratio of the (112) diffraction peak intensity, I_{112}, to the sum of all (hkl) peaks I_{hkl} including the (112) peak. Figure 3 shows the variation of p, of these films deposited by rf sputtering as a function of substrate temperature. The value of p reaches up to nearly 1.0 with increasing the substrate temperature. In particular, p shows the highest value of 0.983 in the thin films prepared on Si(111) at 400°C. Compared with a value (0.898) for the thin films sputtered on glass [9], the value p is considerably improved. In the case of fabricating $CuIn_xGa_{1-x}Se_2$ solar cell onto the single crystal Si solar cell, n-type semiconductor CdS with hexagonal wurtzite structure can be normally used because of the highly (112) orientation. Since $CuIn_xGa_{1-x}Se_2$ thin films deposited at 400°C were strongly oriented to (112) plane independent of the plane azimuth of the Si substrate, it is considered that polycrystalline Si solar cell can be used as the bottom part of tandem solar cell.

The relationship between the full width at half maximum (FWHM) of XRD peaks diffracted from (112) plane and the substrate temperature is shown in Fig. 4. The FWHM is found to decrease with increasing substrate temperature. This indicates that the grain size in the films becomes much larger. Also, it is observed that the plane azimuth of Si substrate seems to give no contribution to FWHM.

Figure 5 shows the RHEED pattern of $CuIn_xGa_{1-x}Se_2$ film deposited on Si(100) substrate at 400°C. The pattern was not observed for films deposited at the substrate temperature ≤200 °C, but was found for those deposited at ≥300 °C. As can be seen from the figure, diffracted spots from the (112) plane are observed. Thus, the crystallinity of $CuIn_xGa_{1-x}Se_2$ thin film is improved with the increase of substrate temperature. This is in agreement with the XRD studies.

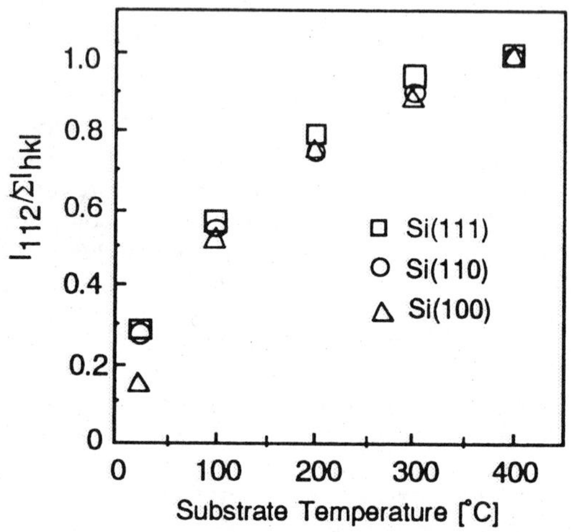

Fig. 3. The orientation factor as a function of substrate temperature.

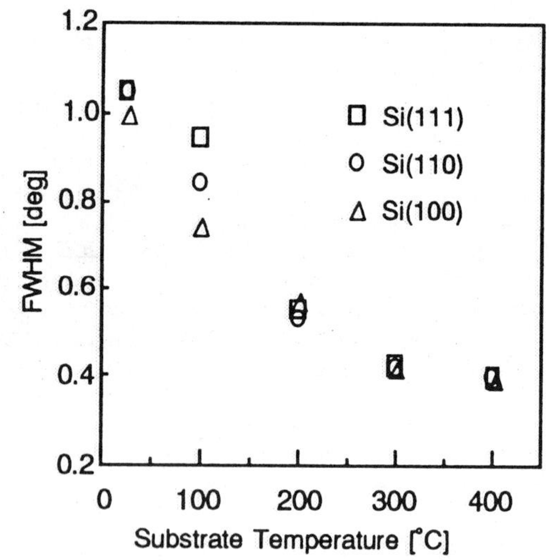

Fig. 4. FWHM from the $CuIn_xGa_{1-x}Se_2$ (112) diffraction peak as a function of substrate temperature.

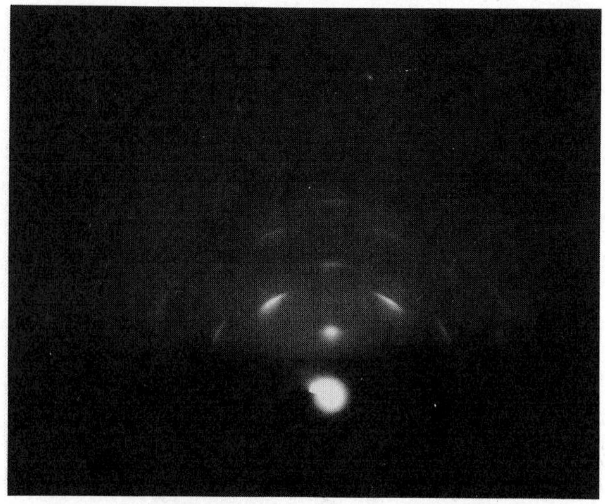

Fig. 5 RHEED pattern of the $CuIn_xGa_{1-x}Se_2$ thin film deposited on Si(100) at 400°C.

CONCLUSIONS

For fabricating high performance tandem photovoltaic devices, $CuIn_xGa_{1-x}Se_2$ thin films were prepared on Si(100), (110) and (111) substrates by rf sputtering in the temperature range (R.T.~400°C). From EPMA analyses, these thin films had nearly stoichiometric composition, but the slight deficiencies of Cu, In and Se were observed with increasing the substrate temperature. XPS studies showed that the compositional depth profile of these thin films was uniform. XRD analyses showed that all thin films had a chalcopyrite structure. The $CuIn_xGa_{1-x}Se_2$ thin films were strongly oriented to (112) plane and the crystallinity was improved with increasing the substrate temperature. The orientation was found to be independent of the plane azimuth of Si substrates. Therefore, hexagonal semiconductor CdS can be used to fabricate $CuIn_xGa_{1-x}Se_2$ solar cell onto the single crystal Si solar cell. Moreover, it is possible to use polycrystalline Si solar cell as the bottom part of tandem solar cell.

ACKNOWLEDGMENTS

The authors wish to thank the staff of the Instrument Center at the Institute for Molecular Science for the XPS measurements.

This work was partly supported by a Grant-in-Aid from Japanese Ministry of Education, Science and Culture.

REFERENCES

[1] S.Wagner, J.L.Shay, P.Migliorate and H.M.Kasper, "$CuInSe_2$/CdS heterojunction photovoltaic detectors" *Appl. Phys. Lett.*, **25** (8), 1974, pp. 434-435.

[2] L.L.Kazmerski, P.J.Ireland, F.F.White and R.B.Cooper, *Proc. 13th IEEE Photovoltaic Specialists Conf.*, 1978, p. 184.

[3] R.A.Mickelsen and W.S.Chen, *Proc. 13th IEEE Photovoltaic Specialists Conf.*, 1981, p. 800.

[4] K. Mitchell, C. Eberspacher, J. Ermer and D. Pier, "Single and Tandem Junction $CuInSe_2$ Cell and Technology", *Proc. 20th IEEE Photovoltaic Specialists Conf.*, 1988, pp. 1384-1389.

[5] T. Yamaguchi, J. Matsufusa and A. Yoshida, "Optical Transitions in RF Sputtered $CuIn_xGa_{1-x}Se_2$ Thin Films" *Jpn.J.Appl.Phys.*, **31**, 1992, pp. L703-705.

[6] T. Walter, M. Ruckh, K.O. Velthaus and H.W. Shock, "Solar Cell based on $CuIn(S,Se_2)$ - A Promising Alternative", *Proc. 11th EC Photovoltaic Solar Energy Conf.*, 1992.

[7] J. Hedström, H. Olsen, M. Bodegard, A. Kylner, L. Stolt, D.Hariskos, M. Ruckh and H.W. Schock, "ZnO/CdS/Cu(In,Ga)Se_2 Thin Film Solar Cells with Improved Performance", *Proc. 23rd IEEE Photovoltaic Specialists Conf.*, 1993, pp. 364-371.

[8] J. Szot and U. Prinz, "Selenization of Metallic Cu-In Thin Films for $CuInSe_2$ Solar Cells", *J.Appl.Phys.*, **66**, 1989, pp. 6077-6083.

[9] T. Yamaguchi, J. Matsufusa and A. Yoshida, "Structural Properties of $CuIn_xGa_{1-x}Se_2$ Thin Films Prepared by RF Sputtering", *J.Appl.Phys.*, **72**, 1992, pp. 5657-5662.

STRUCTURE AND OPTOELECTRONIC PROPERTIES OF SINGLE CRYSTAL EPITAXIAL $Cu(In_{1-x}Ga_x)Se_2$ AND ORDERED DEFECT COMPOUNDS

A. Rockett, G. Berry, D. Schroeder, H.-Z. Xiao, and L. Chung Yang
1-107, Engineering Sciences Building, University of Illinois
1101 W. Springfield Ave., Urbana, IL 61801

ABSTRACT

Epitaxial $CuIn_{1-x}Ga_xSe_2$ was grown on As-terminated (111) GaAs between 550 to 735°C with $0 \leq x \leq 1$. The Cu/[In+Ga] ratio, y, ranged from y=0.3 to 1.3. Analysis of the deposited films showed an ordered defect structure that was homogeneous throughout the epitaxial layers when group III rich for all Ga contents examined. Films grown with y=0.3 had energy gaps of ~1.2 eV and showed evidence by both cathodoluminescence and optical absorption of band tails. Stacking faults affect both the growth rate and the luminescence but can be converted to dislocations by rapid thermal annealing. The highest hole mobilities to date, >1500 cm^2 /V-sec, were measured at 50-75K by Hall-effect in near-stoichiometry samples. Room temperature hole mobilities were >200 cm^2/V-sec and increased at low temperatures. Hole concentrations showed evidence of a level 80 meV above the valence band edge at a concentration in excess of 10^{17} cm^{-3} in all p-type samples. A composition-dependent level at ~40 meV and type conversion at ~100K was also observed.

INTRODUCTION

One of the critical problems facing the $CuInSe_2$ solar cell community is a lack of fundamental understanding of $CuInSe_2$, related alloys (particularly Ga alloys), and ordered defect layers thought to result from segregation of excess group III constituents to the surfaces of $CuInSe_2$ polycrystalline layers. This work continues previous efforts to improve this understanding through characterization of single-crystal epitaxial layers of $Cu(In_{1-x}Ga_x)Se_2$ on GaAs substrates. A review article describing the properties and applications of $CuInSe_2$ in thin film photovoltaic devices may be found in Reference 1. Papers describing the critical issues in materials and devices based on $CuInSe_2$ and related compounds may be found in References 2 and 3. The principle goal of the work reported here is to study the effect of point, line, and planar defects in $CuIn_{1-x}Ga_xSe_2$ on the optical and electronic properties of the semiconductor.

EXPERIMENTAL

The single crystal $CuIn_{1-x}Ga_xSe_2$ layers investigated in this study were deposited directly on pieces of two-inch diameter epitaxy-ready (001) or Ga or As-terminated (111) GaAs wafers. The deposition process used magnetron sputtered Cu, Ga, and In fluxes and evaporated molecular Se. The depositions were carried out in a stainless steel vacuum system with a base pressure of $1x10^{-4}$ Pa ($1x10^{-6}$ Torr). Details of the vacuum system and growth procedures may be found in References 4-6. The GaAs substrates were cleaved into smaller pieces, clamped to silica or In-bonded to Mo plates, and inserted into the deposition system through a load lock. The backing plates and substrate holders were cleaned between each deposition. During sputter cleaning of the targets, the substrates were covered by a main shutter. An additional shutter isolated the Se source during warm-up. To initiate deposition, the target currents were raised to the appropriate values, the Se shutter was opened, and finally the main shutter was opened. Sputtering was carried out with 99.999% pure Ar at a pressure of 0.1-0.2 Pa (1-2 mTorr). Sputtering targets consisted of pure Cu and pure In targets as well as vacuum hot pressed Cu-Ga alloy targets mounted on Cu backing plates in each case. Two Cu-Ga targets were used with compositions of 20.0 and 42.5 atomic % Ga. The Ga content could not be controlled arbitrarily to select values of x. However, outdiffusion of Ga from the substrate did result in changes in Ga content depending upon the deposition temperature. The Cu or Cu-Ga target sputtering current ranged from 0.10 to 0.35 A, at voltages from 440 to 510 V, respectively. For the In target, the current and voltage were 0.10 to 0.40 A at 410 to 480 V, respectively, and were adjusted to obtain the desired stoichiometry. The Se effusion-cell temperature was 270 °C during deposition corresponding to a Se/metal atomic flux ratio [Se/(Cu+In+Ga)] > 5. The deposition rate was 10 to 50 nm/min (controlled by the Cu and In sputtering currents).

The cathodoluminescence system used in this study is based on a Zeiss scanning electron microscope (SEM) and employs a Ge detector and a liquid-He cooled stage for low temperature measurements (≥4.3 K). Optical absorption, transmission, and reflection measurements were carried out at room temperature using a Perkin Elmer Lambda-9 double beam spectrometer equipped with an integrating sphere. Surface morphologies and compositions of the epitaxial layers were examined using a Hitachi S-800 SEM equipped with an energy dispersive X-ray spectroscopy (EDS) system. Standard specimens with a known composition were used to quantify the EDS data when determining the film compositions when possible. When standards were not available the "ZAF" numerical corrections alone were used. Chemical composition depth profiles were obtained by secondary

CH3365-4/94/0000-0172 $4.00 © 1994 IEEE

ion mass spectrometry (SIMS) in a Cameca IMS 5f instrument. Microstructural studies (electron diffraction and bright- and dark-field imaging analyses) were carried out in Philips EM420 and CM12 transmission electron microscopes (TEMs). High resolution TEM studies were performed using a Hitachi 9000 TEM operated at 300 kV. Plan-view and cross-sectional TEM samples were prepared by mechanical thinning to 50-80 µm and then were ion milled to perforation on a (77 K) cold stage using 3 kV Ar^+ ions. Hall-effect measurements were carried out in an evacuated chamber that could be cooled by a closed-cycle refrigerator to ~15 K. The temperature was measured by a thermistor mounted beside the sample and was controlled to ±1°C by a small heater. The magnetic field was 1 Tesla. Contacts were In dots or sputtered Au-Pd. The electrometer and current source were computer controlled as was data acquisition and analysis.

RESULTS

Single crystal epitaxial layers of $Cu(In_{1-x}Ga_x)Se_2$ with x~0.1, ~0.3, ~0.6, and 1 with Cu/(In+Ga) ratios, y, of 0.7 to 1.0 and with x~0 and 0.3≤y≤1.2 were deposited on GaAs (001), $(111)_{Ga}$, and $(111)_{As}$ at temperatures ranging from 550 to 735°C and at a rate of 10 to 50 nm/min. The thickness of the grown layers was typically ~1 µm. All compositions lay close to a "tie-line" connecting Cu_2Se and $(Group\text{-}III)_2Se_3$ on a pseudo-ternary phase diagram. Cu-rich films showed evidence of coherent second phase precipitates in the chalcopyrite matrix. The basic growth mechanism and structural and surface defects were similar to those of near stoichiometry $CuInSe_2$ described previously.[4-6] The surfaces showed rippling with a tendency to facet to form the $(112)_{Se}$ surface although at high growth temperatures very smooth surfaces were obtained on GaAs (001). Surface pits were found on Cu-rich samples and samples for which the GaAs surface was damaged during sample preparation. Triangular islands were often found on the surfaces of group-III-rich films.

Films containing little or no intentionally added Ga all included significant Ga near the substrate interface due to Ga out diffusion from the substrate. Diffusion of this Ga into the remainder of the films increased with increasing growth temperature. Concurrent vacancy in-diffusion generally led to the formation of highly (112)-faceted Kirkendall voids at the interface extending into the GaAs. For very high deposition temperatures and with no Ga intentionally added to the films, outdiffusion was so strong that homogeneous high Ga contents were found in the samples. Delamination from the substrates was common in this case due to coalescence of the Kirkendal voids.

Group III rich single crystals showed evidence of ordered defects (as determined from electron diffraction patterns obtained in the TEM). Unlike the polycrystalline films, epitaxial layers showed no surface segregation of the ordered defect layer. Ordered defects varied in concentration from sample-to-sample but were homogeneous within a sample. A $CuIn_3Se_5$ sample had an increased energy gap (>1.8 eV) compared to $CuInSe_2$ with significant subgap densities of states giving rise to both absorption and luminescence over a range of energies. Figure 1a shows the cathodoluminescence spectrum for the $CuIn_3Se_5$ sample. The subgap absorption and luminescence decreased with a characteristic slope of 80meV (from absorption measurements at room temperature) to 70 meV (from cathodoluminescence at 8K).

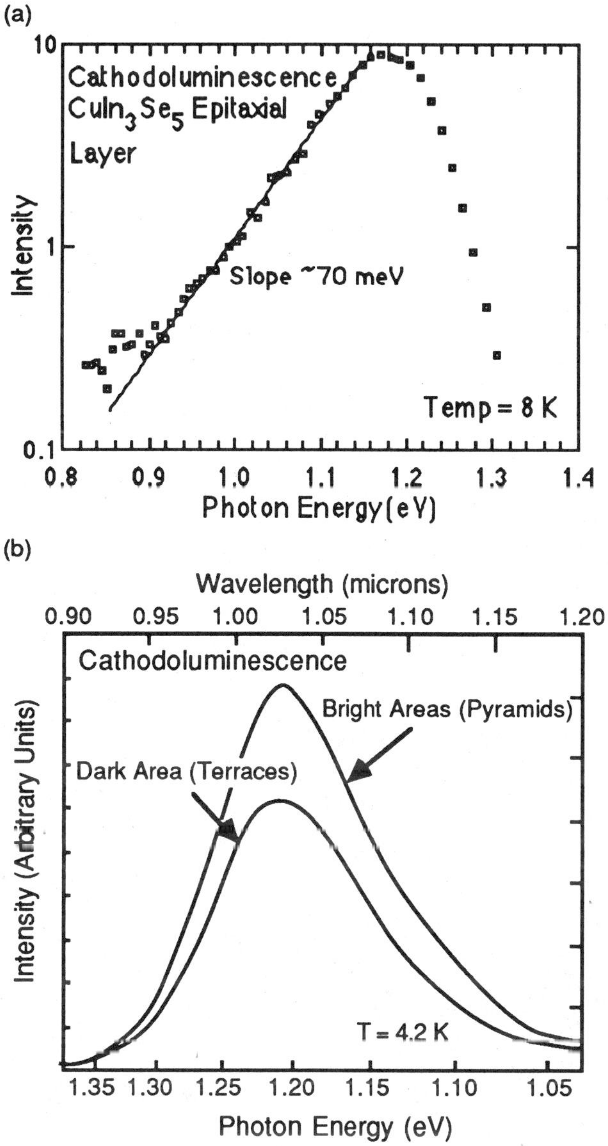

Figure 1. Cathodoluminescence spectra for (a) $CuIn_3Se_5$ and (b) from a flat surface region and from a pyramidal island on an epitaxial layer with x~0.3. Both samples were grown on GaAs $(111)_{As}$.

The diffraction results taken together with the observed composition and very high resistivity of the samples suggest the presence of ordered valence compensating defects on (001) planes, probably In on Cu site antisite defects and Cu vacancies. While the defects are ordered on (001) planes, they are disordered within these planes. The disorder within the planes should give rise to local variations in defect density. This should cause local variations in energy gap due to a range of subgap states. A more complete characterization of $CuIn_3Se_5$ may be found in Reference 7.

Cross-sectional TEM analyses showed the presence of stacking faults and microtwins in all films. The density of stacking faults was higher for high Ga content films but may have been due to an apparently greater sensitivity of high-Ga-content layers to the preparation of the GaAs surface prior to growth and growth conditions.

In dark field cross-sectional images, stacking faults were observable near the surface of the layers adjacent to but not underlying islands (see Figure 2). Apparently, when a stacking fault occurs during growth the surface polarity may be reversed (i.e., the surface changes from metal to Se terminated or vice-versa) and the growth rate of the crystal decreases abruptly. In the surrounding area where there is not fault the growth continues giving rise to a triangular island. Cathodoluminescence measurements showed increased luminescence from the islands (Fig. 1b) but little change in the spectrum shape. This implies that luminescence and minority carrier lifetime are quenched by the stacking faults just below the surface. Where there is an island, a larger volume exists near the surface that is fault free and thus provides stronger luminescence. Eventually another fault occurs and rapid growth resumes. The stacking faults were commonly a multiple of six (112) plane spacings wide, consistent with the repeat distance of the $CuInSe_2$ chalcopyrite unit cell.

Figure 2. A cross-sectional dark-field TEM image of a $CuIn_{0.7}Ga_{0.3}Se_2$ sample. The bright areas are stacking faults.

Hall effect measurements on stoichiometric and slightly group-III-rich epitaxial films deposited on semi-insulating GaAs substrates showed the films to be p-type with carrier concentrations of ~10^{17} cm^{-3} and mobilities of >200 cm^2/V-sec at 25°C. Cu-rich and stoichiometric (y≥1), films had resistivities of the order of 10^{-1} Ω-cm while group-III-rich resistivities were >10^1 Ω-cm. Strongly group-III-rich samples were n-type with resistivities of >10^5 Ω-cm. The results of temperature-dependent Hall effect measurements are shown in Figures 3 and 4. For clarity the number of data sets shown on each graph is limited. Different symbols correspond to different samples. Most samples could not be measured reliably below temperatures ranging from 90 to 140 K because of a large increase in resistivity accompanying a conversion to n-type conduction. The hole mobility increased and the carrier concentration decreased for all samples measured as the temperature was decreased. A remarkable similarity in the hole concentration (less than a factor of 4 variation) and mobility (less than a factor of two variation) was found over a wide range of samples with compositions ranging from stoichiometric to slightly group-III rich and for which x~0.3 (see Figs. 3 and 4).

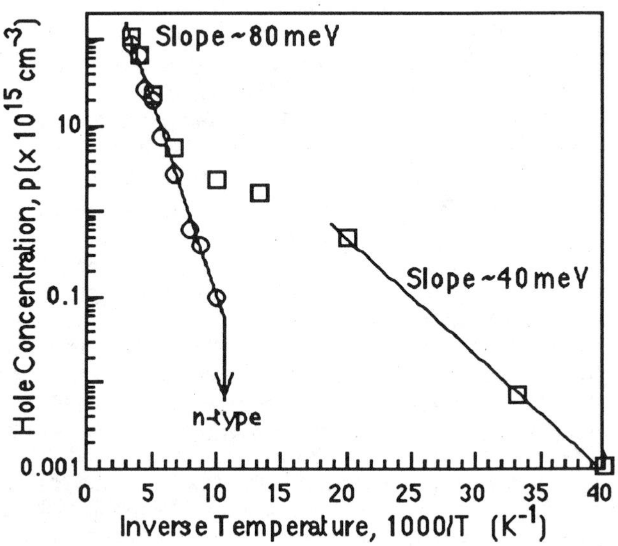

Figure 3. Temperature dependent hole concentrations for two epitaxial layers of $CuIn_{1-x}Ga_xSe_2$ for which x~0.3.

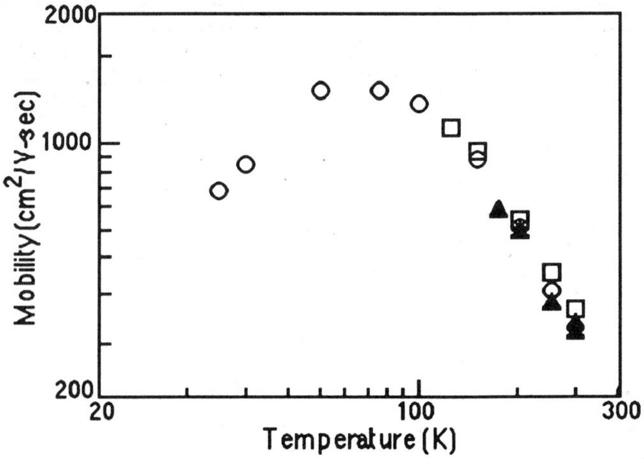

Figure 4. Mobility results for three p-type epitaxial layers for which x~0.3.

For higher temperatures (> ~150 K), all samples were dominated by a defect level at ~80 meV above the valence band edge at a concentration in excess of $2x10^{17}$ cm^{-3}. The sample with the highest value of y (very near stoichiometry) showed the presence of a second defect state at 40 meV with respect to the valence band edge and a concentration of ~$3x10^{15}$ cm^{-3} (see Fig. 3). No other samples showed this state but none was as close to stoichiometry. The absence of a type conversion and a shallow acceptor in this sample is consistent with a change from an n-type to a p-type minority defect associated with a change in stoichiometry. The hole concentration at which type conversion was observed suggests that the n-type minority defect concentration was ~10^{15} cm^{-3}. The high resistivity of strongly group-III-rich samples indicates that very far from stoichiometry the concentration of the donor defect rises and eventually fully compensates the 80 meV p-type defects. This explains the conversion from n to p-type on the group-III-rich side of stoichiometry generally found in polycrystals.

For the sample closest to stoichiometry the absence of type conversion allowed measurement of the mobility down to 20 K. The mobility curve follows a classic semiconductor behavior, increasing with decreasing temperature as ionized impurity scattering decreases; passing through a maximum; and then decreasing as neutral impurity scattering becomes more significant.

CONCLUSIONS

Epitaxial growth of $CuIn_{1-x}Ga_xSe_2$ across a wide range of compositions has been accomplished on GaAs. There is no evidence of significant general optical, electronic, or morphological differences other than changes in energy gap in $CuIn_{1-x}Ga_xSe_2$ as a function of x. The results indicate that point defect ordering occurs in group-III-rich samples on (001) planes. The defects give rise to increased energy gaps and band tail states below the band edge. Stacking faults can cause significant reductions in the growth rate along (112) directions and decreases in the cathodoluminescence intensity, presumably by radiative recombination of carriers. The very high hole mobilities observed indicate that the epitaxial layers are of high quality. Corresponding carrier concentrations indicate the presence of defect states at ~80 meV and ~40 meV with respect to the valence band edge. The former was present in all samples while the latter was only present in one sample with a composition very close to stoichiometry. Samples containing a larger group-III elemental composition showed type conversion to n-type upon cooling. The variation in the hole concentration at which this type conversion occurs suggests a composition-dependent n-type doping level.

ACKNOWLEDGEMENTS

The authors gratefully acknowledge the support of the Electric Power Research Institute (EPRI) (contract RP-2702-1) and the National Renewable Energy Laboratory (subcontract XAD-3-12114-1) and the Department of Energy under contract DEFG02-91-ER45439 which subsidizes the Center for Microanalysis of Materials at the University of Illinois which was critical to this work.

REFERENCES

[1] A. Rockett and R.W. Birkmire, Applied Physics Reviews, in J. Appl. Phys. **70(7)**, 1991, p. R81.

[2] A. Rockett, F.Abou-Elfotouh, D. Albin, M. Bode, J. Ermer, R. Klenk, T. Lommasson, T.W.F. Russell, R.D. Tomlinson, J. Tuttle, L. Stolt, T. Walter, and T.M. Peterson, Thin Solid Films **237**,1994, p. 1.

[3] J.R. Tuttle, J.R. Sites, A. Delahoy, W. Shafarman, B. Basol, S. Fonash, J. Gray, R. Menner, J. Phillips, A. Rockett, J. Scofield, F. Shapiro, P. Singh, V. Suntharalingam, D. Tarrant, T. Walter, S. Wiedeman, and T.M. Peterson, Progress in Photovoltaics, in press.

[4] A. Rockett, T.C. Lommasson, P. Campos, L.C. Yang, and H. Talieh, Thin Solid Films **171**,1989, p. 109.

[5] L. Chung Yang, L.J. Chou, A. Agarwal, and A. Rockett, Proc. of the 22nd IEEE Photovoltaic Specialists Conference, Las Vegas, October 7-11, 1991 (Institute of Electrical and Electronics Engineers, New York, 1992), p. 1185.

[6] L. Chung Yang, G. Berry, L.J. Chou, G. Kenshole, A.Rockett, C.A. Mullan, and C.J. Kiely, 23rd IEEE Photovoltaic Specialists Conference, Louisville, May 10-14, 1993 (Institute of Electrical and Electronics Engineers, New York, 1993), p. 505.

[7] H. Xiao, L.-Chung Yang, and A. Rockett, J. Appl. Phys. **76**, 1994, p.1503.

OSCILLATING TEMPERATURE SELENIZATION OF Cu, Ga, In ELEMENTAL LAYERS IN A Se VAPOUR

N. Romeo, A. Bosio, V. Canevari and R. Tedeschi
Department of Physics, University of Parma,
Viale delle Scienze, 43100 Parma, Italy

ABSTRACT

$CuGa_{0.3}In_{0.7}Se_2$ thin films have been grown by selenization of Cu, Ga, In elemental layers in a Se vapour by a novel method in which a first part of the film is prepared by making the substrate temperature to oscillate between 200-230°C and 480-550°C with a risetime of 1-2 minutes and cooling down time of 6-9 minutes. The film is completed by increasing the substrate temperature from 380°C to 480-500°C with a ramp of 1-2°C/min. X-ray spectra and scanning electron micrographs indicate that the films are monophasic and very well crystallized with a strong orientation along the (112) direction. $CuGa_{0.3}In_{0.7}Se_2$/CdS thin film solar cells made on both Corning 7059 or Soda Lime glass substrates exhibit an efficiency larger than 12%.

INTRODUCTION

Recently it has been reported that $CuGaInSe_2$/CdS thin film solar cells can exhibit a conversion efficiency close to 17% [1, 2]. Therefore $CuGaInSe_2$ materials can be considered as good candidates for large scale production of solar cells. However a single, reproducible and scalable process for the preparation of high quality $CuGaInSe_2$ films has not yet clearly identified. Here we propose an original process, entirely developed in our laboratory, which could meet the above requirements.

PREPARATION OF Cu, Ga, In ELEMENTAL LAYERS

As substrates both Corning 7059 and Soda Lime glass with a size of 1 inch-square have been used. These substrates have been covered by 1 μm thick Mo film deposited by sputtering at a substrate temperature of respectively 500°C and 100°C. In the case of Soda Lime glass the substrate temperature was kept at 100°C since at higher temperature the adhesion of Mo to the substrate becomes poor. The Cu, Ga, In elemental layers are deposited by Electron-Gun evaporation from a single rotatable crucible in order to have a good uniformity on the substrate. First of all, 3700 Å of In and 2400 Å of Cu are deposited in sequence on top of the Mo covered glass substrate which is kept at 120°C. Then the In-Cu bilayer is annealed at 290°C for 30-60min in order to form the $Cu_{11}In_9$ phase. After the annealing, 1200 Å of Ga are deposited at 240°C substrate temperature. At this temperature Ga reacts with the $Cu_{11}In_9$ phase forming the $CuGa_2$ phase and freeing a small amount of In. It has been chosen a substrate temperature of 240°C for the deposition of Ga since the $CuGa_2$ phase melts at 254°C [3]. It has been confirmed by X-ray diffraction that the Cu-Ga-In film is made up of the $Cu_{11}In_9$ phase, of the $CuGa_2$ phase and a small amount of In [4]. This method of preparing the Cu-Ga-In elemental layers gives the maximum possible intermixing between the layers without having melt droplets in the film.

SELENIZATION

Selenization is done in a vacuum chamber in which a vacuum of 10^{-6} mbar is maintained during the process. Se is evaporated from a graphite crucible at a typical rate of 100 Å/sec. The sample is mounted into a graphite holder and it is heated up indirectly by a Mo thin slab or Mo filaments through which a high current is allowed to pass. A peculiarity of this process is the oscillation of the substrate temperature at the first stage of $CuGaInSe_2$ growth. The temperature oscillation seems to be very effective in containing the surface tension of the liquid. At the beginning, the substrate temperature is raised up to 280°C in about 30 sec. and decreased to 200°C in about 3-4 min. During this first oscillation a small part of the film grows on the surface. This covering seems to be sufficient to contain the surface tension during the second oscillation up to 450°C, which is done with a risetime of about 1 min and a cooling down time of about 6 min. During the second oscillation part of the film grows covering further the film with a solid and allowing to do a

CH3365-4/94/0000-0176 $4.00 © 1994 IEEE

third oscillation and so on. It has been seen that the process depends on both the oscillation risetime which should be not larger than 2 min for reaching a maximum temperature of 550°C and the cooling down time which should be comprised between 6 and 9 min from 550°C to 200°C.

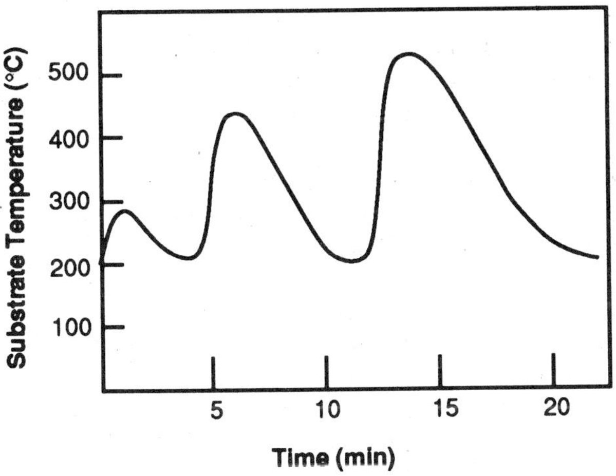

Fig. 1 Temperature oscillations during the first stage of the thin film growth. Here the first three oscillations are shown.

The risetime has to be contained below 2 min in order for the film not to lay at high temperature while the amount of the liquid is increasing.

Concerning the cooling down time, it should be larger than 6 min in order for the film to have enough time to grow, and less than 9 min in order for the surface tension of the liquid not to give rise to formation of microdroplets.

After about 10 oscillations, the temperature of the film is steadily increased from 380°C to 480-500°C with a ramp of 1-2°C/min (see Fig. 1).

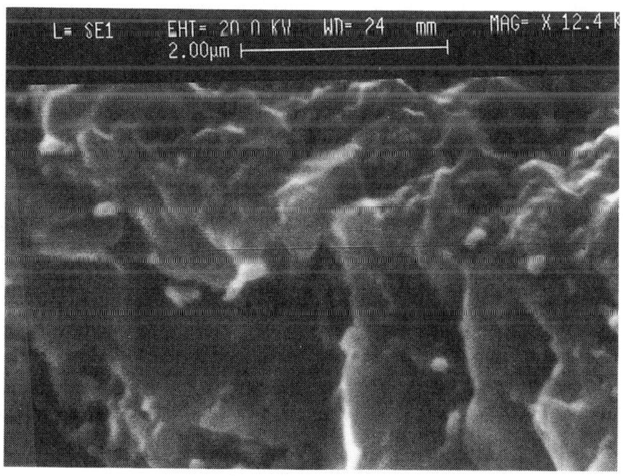

Fig. 2 Scanning Electron Micrograph of a $CuGa_{0.3}In_{0.7}Se_2$ film.

This last part of the process is necessary in order to complete the growth of the film. This process is very reproducible and can be easily controlled. The adhesion of the film to the substrate is very good when the parameters of the process are respected. Films prepared by this process are very well crystallized as it can be seen from the micrograph (see Fig.2) and very well oriented along the (112) direction (see Fig. 3).

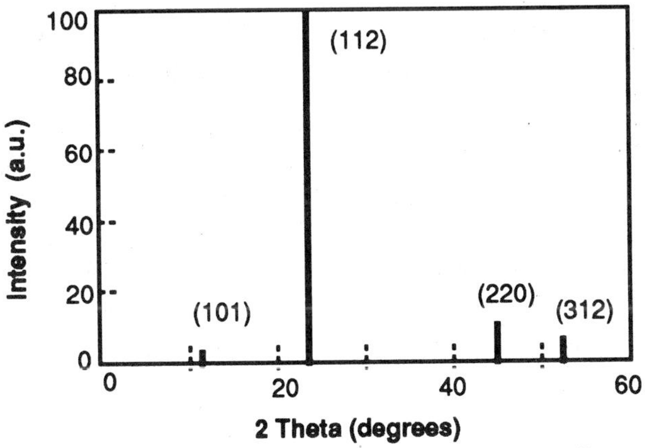

Fig. 3 X-ray Spectrum of a $CuGa_{0.3}In_{0.7}Se_2$ film.

X-ray spectra also indicate a monophasic film with a composition corresponding to $CuGa_{0.3}In_{0.7}Se_2$.

$CuGa_{0.3}In_{0.7}Se_2$/CdS SOLAR CELLS

Solar cells have been prepared by depositing on top of $CuGa_{0.3}In_{0.7}Se_2$ films a 300-500 Å thick layer of intrinsic CdS at 220°C substrate temperature both by sputtering or evaporation. The cell is completed by depositing a 2-3 µm thick film of In doped CdS or Al doped ZnO by sputtering. On 1 inch-square substrate cells with an area of 0.3-1cm^2 were prepared by using a suitable mask during the deposition of the low resistivity window or they were obtained by a mechanical scribing. Open-circuit voltages between 420 and 450 mV, short-circuit current close to 40mA/cm^2 and fill factors ranging from 0.65 to 0.7 are generally obtained in these cells. On average, better results are obtained when soda lime glass is used as a substrate, indicating that the diffusion of Na should play some role during the growth of the film. No much difference in the photovoltaic parameters has been observed when the junction is made by depositing the intrinsic CdS by sputtering or evaporation. The highest efficiency so far measured in these cells under 100 mW/cm^2 simulated solar light has been 12.4% with an open-circuit voltage of 450 mV, a short-circuit current of 40mA/cm^2 and a fill factor of 0.69. One should expect in these cells a larger open-circuit voltage since $CuGa_{0.3}In_{0.7}Se_2$ has a larger gap than that of $CuInSe_2$. This could depend more on the quality of CdS film rather than on the crystalline quality of $CuGa_{0.3}In_{0.7}Se_2$ film. A larger open-circuit voltage could be probably obtained if the junction is prepared by depositing CdS by chemical bath deposition. In conclusion we would like to stress that the process described in this paper to prepare the $CuGa_{0.3}In_{0.7}Se_2$ films does not use any poisoning gas such as H_2Se, is scalable since uniformity is assured by the use of elemental layers, is very reproducible and allows one to obtain very well crystallized $CuGa_{0.3}In_{0.7}Se_2$ films.

REFERENCES

[1] A. Gabor, J. Tuttle, M. Contreras, D. Albin, A. Tennant, Y. Qu, D. Du, J. Carapella and R. Noufi: "High Efficiency Thin Film Cu(In, Ga)Se_2 Solar Cells: Achievement of a confirmed World record 16.4% Total Area Efficiency",
Proc. of 12th European Photovoltaic Solar Energy Conf. and Exhibition,11-15 April 1994, Amsterdam, The Netherlands, pp. 939-943

[2] M. Bodegart, L. Stolt and J. Hedstrom:
"The influence of Sodium on the Grain Structure of $CuInSe_2$ Films for Photovoltaic Applications"
(same as ref. 1), pp. 1743-1746

[3] *Binary Alloy Phase Diagrams*:
Ed. by T.D. Massalsky, American Society for Metals, Metals Park, Ohio 44073, pp. 917

[4] N. Romeo, A. Bosio and V. Canevari:
"$CuGaInSe_2$ Thin Film Grown by Selenization of Metallic Elemental Layers in a Se Vapour"
(same as ref. 1), pp. 608-609

The Low Temperature Deposition of ICB-Grown $CuInSe_2$ Thin Films

Hiroyuki Sano, Susumu Nakamura, Ken'ichi Kondo and Katsuaki Sato*
Stanley Electric Co.,Ltd. Yokohama, Kanagawa 225
**Faculty of Technology, Tokyo University of Agriculture and Technology, Koganei, Tokyo 184*

Superstrate-type solar cells were successfully fabricated for the first time using the $CuInSe_2$ thin films prepared by the ICB(ionized cluster beam) technique. The substrate temperature was as low as 300 ℃. The films were characterized by EPMA, X-ray diffraction and PL spectrum. SEM micrographs and photovoltaic properties were also investigated on the solar cells. As the results, those films showed nearly stoichiometric composition with grain size about 1 μm. The superstrate-type solar cells showed the conversion efficiency of 2.06%.

KEYWORDS: $CuInSe_2$, chalcopyrite type compounds, superstrate-type solar sells, ionized cluster beam deposition

1. Introduction

The chalcopyrite ternary semiconductor $CuInSe_2$ has been attracting much attention as a polycrystalline thin film material for high efficiency solar cells application, since the material has a direct energy gap of around 1.0eV and the highest absorption coefficient reported to date for any semiconductors.[1)] Recently, a conversion efficiency exceeding 16% was reported in solar cells based on thin films of $CuInSe_2$.[2)]

However, most of the deposition techniques reported to date require high processing temperatures above 400 ℃ in order to obtain a good quality film. For realization of superstrate type solar cells, the substrate temperature lower than 300℃ is necessary for deposition of $CuInSe_2$. The ICB (ionized cluster beam) deposition has been known to be an excellent technique for preparation of high-quality thin films at relatively low substrate temperatures since it utilizes the effect of bombardment of ionized clusters and the effect of the charge of ionized particles.[3),4)] We have been working with ICB-deposition of $CuInSe_2$ films and have shown that this technique provides additional controllability of Cu/In ratios by changing the acceleration voltage for clusters and to be a promising technique for solar cells application.[5),6)]

In this paper, preparation of $CuInSe_2$ thin films by the ICB technique and characterization by electron probe micro analysis(EPMA), X-ray diffraction, scanning electron micro-scope(SEM) and photoluminescence(PL) spectra as well as I-V characteristics of fabricated solar cells will be described.

II. Experiment

$CuInSe_2$ films were prepared on CdS-coated glass substrates with simultaneous deposition of Cu-, In- and Se-clusters, of which Cu and In were ionized and accelerated, using an apparatus shown schematically in Fig.1. Source materials employed were Cu, In and Se of 99.9999% purity. Source temperatures of the three crucibles used to obtain films were 1320 ℃, 950 ℃, 300℃ for Cu, In and Se, respectively. Coating of CdS on the glass substrate was carried out by chemical bath deposition(CBD) technique.

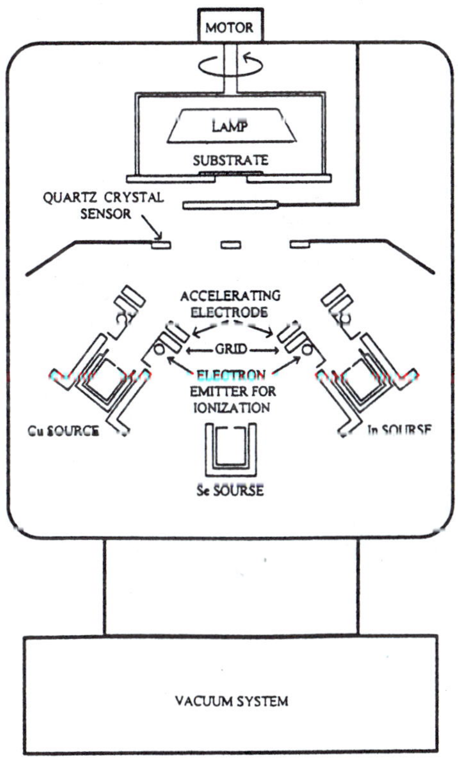

Fig.1 Schematic diagram of an ICB deposition system.

CH3365-4/94/0000-0179 $4.00 © 1994 IEEE First WCPEC; Dec. 5–9, 1994; Hawaii

During the film-deposition, the acceleration voltage of Cu- and In-cluster beams were maintained at Va=2kV and the electron current for ionization of the Cu- or In-cluster were varied between Ie=0 and 150mA. The substrate temperature was kept at 300℃ by an infrared lamp, and the pressure in the vacuum chamber was maintained at less than $1x10^{-5}$ Torr during deposition. The typical thickness of the films obtained was approximately 2μm for a deposition time of 120 minutes.

Characterization of $CuInSe_2$ films were performed by photoluminescence(PL) spectra which have been known to have a strong correlation with the photovoltaic effect of $CuInSe_2$ thin films. PL spectra were measured using a 514.5nm Ar-laser excitation at 20K. Detection of emitted light was performed by a liquid nitrogen-cooled Ge photodetector (Northcoast EO817). The penetration depth of the laser wavelength was compared using an absorption spectra of $CuInSe_2$ films: At 514.5nm (=2.41eV) the absorption coefficient is as large as $1.3x10^5 cm^{-1}$, which means the penetration of the light is approximately 80nm. The films were also characterized by X-ray diffraction for structural analyses and scanning electron microscope(SEM) for surface morphology observation. The chemical composition of the samples were determined by electron probe micro analysis (EPMA: JEOL,JXA-840).

Thin film solar cell devices based on these $CuInSe_2$ films were fabricated by using this technique. The superstrate-type solar cells were consisted of Glass/TCO/CdS/CIS/Au. The CdS films were grown from an aqueous solution containing CdI_2, NK_4OH, and thiourea at 80℃.(Chemical Bath Deposition)
I-V characteristics were measured by solar simulator.

Ⅲ.Results and discussion

A X-ray (Cu K α) diffraction pattern of $CuInSe_2$ film prepared on bare soda lime glass substrate is shown in Fig.2. The 112, 204 or 220, 116 or 312 and other diffraction lines characteristic of chalcopyrite structure were observed. And also, there was a small amount of In_6Se_7 line. The relative intensity suggests that the 〈112〉 axis is preferentially oriented perpendicular to the film surface.

Figure 3(a) shows the surface morphologies of the $CuInSe_2$ films on the CBD-CdS films and (b) shows the cross section of the films. The surface shows leaf-shape morphology which is often observed ICB deposition. The cross sectional micrograph shows that crystals are columnar with densely packed structure and tightly adhered to the substrate.

Figure 4 shows PL spectrum of $CuInSe_2$ films.(Cu/In ratio =0.88) PL peak was observed at 1360 nm (=0.91 eV) and 1490 nm (=0.83 eV). Such double-peaked PL

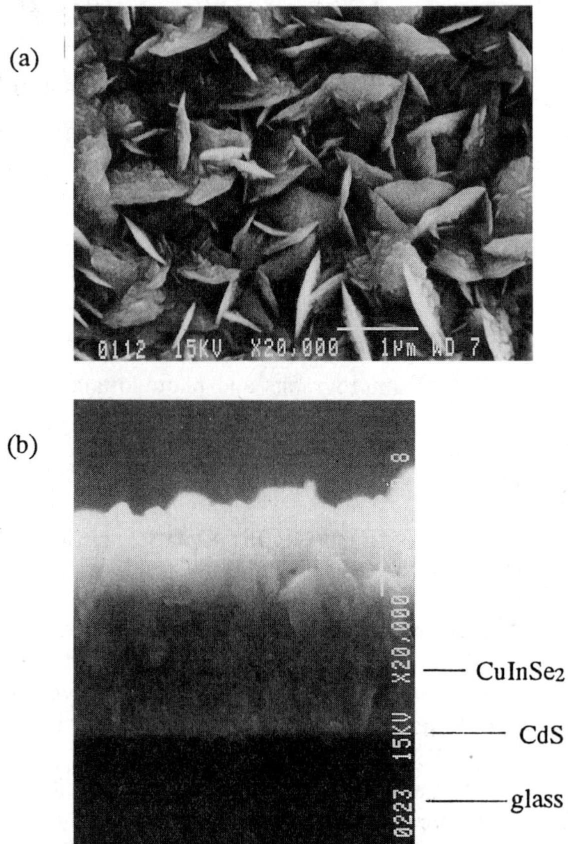

Fig.3 SEM micrographs of $CuInSe_2$ films on the CBD-CdS films: (a)surface morphology and (b) cross section

structure has been reported in $CuInSe_2$ films with a good solar cell performance.[7]

Figure 5 shows current-voltage characteristics of the superstrate-type $CuInSe_2$ solar cell with a structure of Glass/TCO/CBD-CdS/CIS/Au. The best device had a Voc of 0.22V, Jsc of 32.8 mA/cm^2 and fill factor(FF) of 0.29, corresponding to 2.06% efficiency. We have not

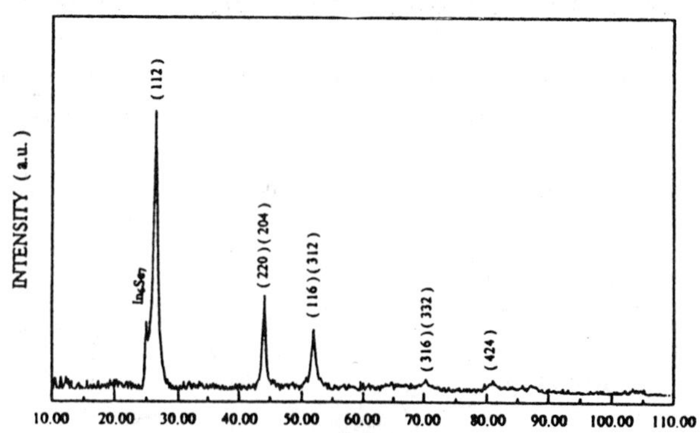

Fig.2 The X-ray diffraction pattern of $CuInSe_2$ film prepared on bare soda lime glass substrate.

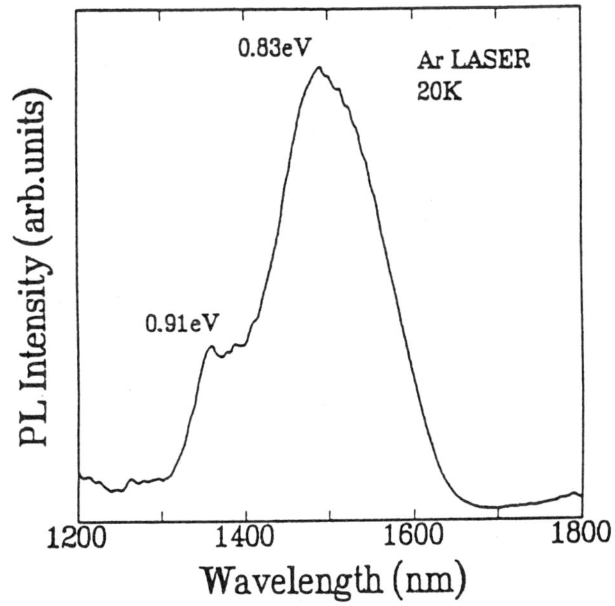

Fig.4 PL spectrum of $CuInSe_2$ films (Cu/In = 0.88) at Ar-laser excitation.

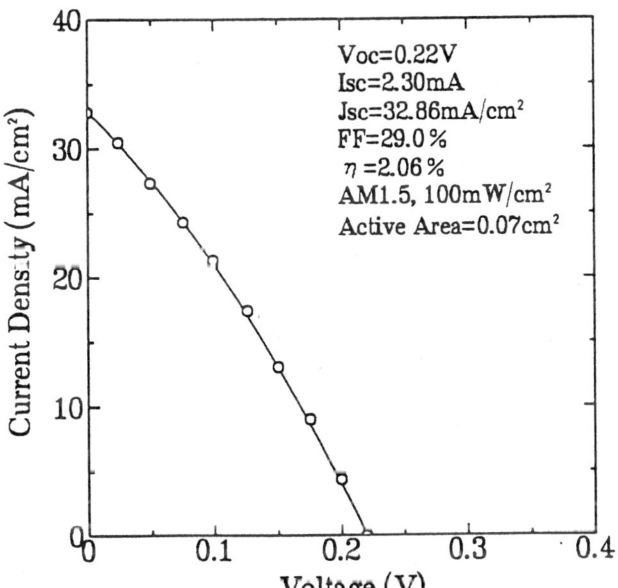

Fig.5 Current-voltage characteristics of the superstrate-type $CuInSe_2$ solar cell.

yet carried out the optimization of the cell structure, and we are going to investigate it in future.

It is often said that CdS and $CuInSe_2$ suffer interdiffusion at temperatures higher than 400℃. On the contrary, ICB technique provided $CuInSe_2$ with excellent crystalline properties, which we consider lead to better PV-performance, since the cross section of $CuInSe_2$ showed very dense structure. Crystallization of CBD-CdS's process seems to have a good influence on the growth of $CuInSe_2$ crystals.

IV. Conclusion

We succeeded in fabricating superstrate-type solar cells with the structure of Glass/TCO/CBD-CdS/CIS/Au at 300 ℃ and obtaining the photovoltaic characteristics with the efficiency of 2.06% by the ICB technique. This result suggests that the ICB technique is a promising technique for providing superstrate-type solar cells, because the ICB technique provides excellent $CuInSe_2$ films at a low temperature.

Acknowledgments

Authors are much indebted to Professor T. Saito of Tokyo University of Agriculture and Technology for the measurement of photovoltaic properties. This work was supported in part by the Agency of Industrial Science and Technology under the Sunshine Project.

References

1) *Copper Indium Diselenide for Photovoltaic Applications*, eds. T. J. Coutts, L.L. Kazmerski and S. Wagner (Elsevier, Amsterdam, 1986).

2) M. Bodegard, A. Kylner, L. Stolt, D. Hariskos, M. Ruckh and H.W. Schock: *presented at 23rd IEEE Photovoltaic Specialist Conference*, Louisville, 1993.

3) T. Takagi, K. Matsubara, H. Takaoka and I. Yamada: Thin Solid Films **63** (1979) 41.

4) T. Takagi: Thin Solid Films **92** (1982) 1.

5) S. Matsuda, Y. Kudo, T. Ushiki, H. Inoue and K. Sato: Jpn. J. Appl. Phys. **31** (1992) 999.

6) T. Ushiki, A. Ueno, T. Yano, H. Sano, H. Usui and K. Sato: *Proc. 9th Int. Conf. on Ternary and Multinary Compounds, Yokohama, 1993*, Jpn. J. Appl. Phys. **32** (1993) Suppl. 3. 103

7)T.Hama,Y.Ichikawa and H.Sakai: *Ext. Abstr. 40th Spring Meetng of Jpn. Soc. Appl. Phys., Tokyo,March,* No.0, 1383

DEVELOPMENT OF MANUFACTURABLE CIS PROCESSING

G. Attar, A. Muthaiah, H. Natarajan, D. Nierman, S. Karthikeyan,
S. Zafar, C. S. Ferekides, and D. L. Morel

Department of Electrical Engineering
University of South Florida
4202 E. Fowler Ave.
Tampa, FL 33620

ABSTRACT

We have developed a manufacturing friendly process for CIS solar cells that includes all-solid- state processing and an inert gas, atmospheric pressure selenization/anneal. Present efficiencies are in the 9 - 10% range with V_{oc}'s at 400+ mV. Correlations between processing and performance have been established. Recent V_{oc}'s have been advanced to 430 mV using an etching process. These results combined with insights provided by a recombination model indicate that V_{oc}'s in excess of 500 mV are within reach.

INTRODUCTION

Impressive progress continues to be demonstrated with CIS laboratory cell efficiencies. Groups at NREL, EuroCIS and Siemens Solar, Inc. have reported efficiencies in the 15-17% range[1]. While these high end devices typically are from processes involving optimized control, there is nevertheless advancement as well toward these performance levels with reduced control processing. Both the NREL and EuroCIS technologies are rumored to be undergoing transfer to commercial organizations. And, Siemens Solar, Inc. has continues to report significant advancements in laboratory cell yields and in the ability to transfer these advancements to module production[2]. These accomplishments continue to highlight the high potential for commercialization of CIS technology. While it is generally agreed that this will occur, there is also general agreement that further progress on process simplification and on fundamental understanding will hasten accomplishment of this goal. In this paper we report progress in our approaches to contribute to these two areas. Primary emphasis is on development of improved processing. However, we also include insights from our efforts to model device performance and to correlate performance with processing details.

PROCESSING APPROACH

Our approach to processing has been driven by two primary guidelines: 1. Non-H_2Se based processing, and 2. Non-critical process control. Our primary process(Process I) is thus one which we describe as "coat and cook"[3]. It involves sequential physical vapor deposition of Cu, In and Se under minimal control, and the anneal cycle runs at atmospheric pressure, features which make it very attractive from a manufacturing perspective. In recognition of the possibility that some amount of additional control might be appropriate as well as the need to expedite understanding of the growth mechanisms, we are also running additional fabrication processes. Thus for the results presented here we have also used a second process(Process II) which allows complete control of the Se flux throughout the anneal. We have found the orthogonal views which these provide to be essential to understanding film growth mechanisms.

RESULTS AND DISCUSSION

Process Sensitivities

To understand the sensitivity of the favored Process I to processing details we must first describe the baseline. By baseline in this case we mean device performance which is typical and reproducible. For Process I the key attributes of baseline devices are an efficiency of 8% with a V_{oc} of 350 mV. While J_{sc} is typically around 40 mA/cm^2, FF tends to be in the .58 - .6 range because the same mechanism that controls V_{oc} also controls FF. The attributes of the best devices from this process are shown below. J_{sc}'s are on the high side because the band

V_{oc}: 400 - 410 mV
J_{sc}: 35 - 40+ mA/cm^2
FF: .58 - .62
Eff: 9 - 10 %

CH3365-4/94/0000-0182 $4.00 © 1994 IEEE

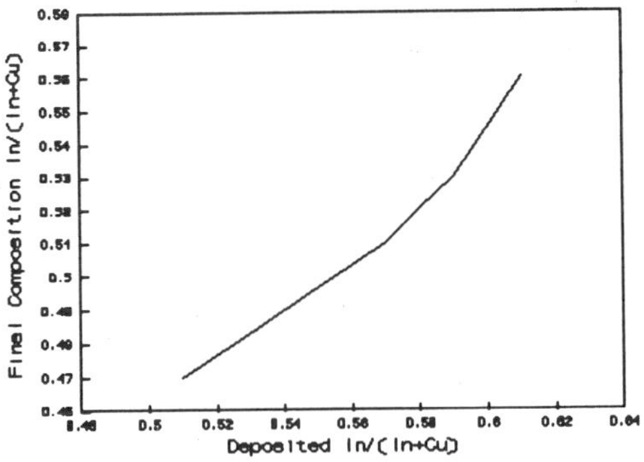

Figure 1. Final metal ratio vs. deposited metal ratio for Process II baseline.

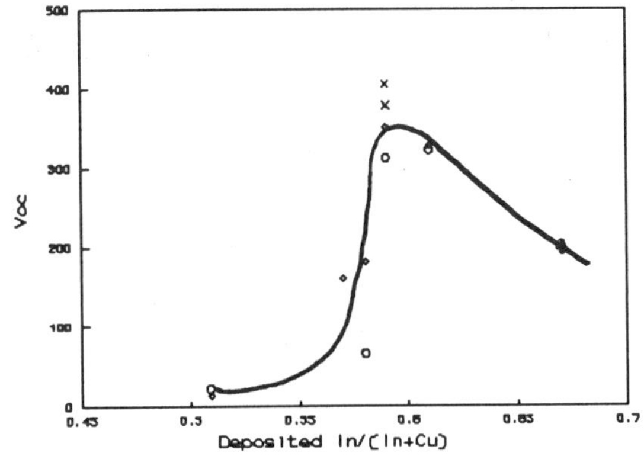

Figure 2. Voc vs. deposited metal ratio for Process I(o) and II(♦) baselines and for advanced selenization(x).

gap is about .95 eV. This in part also lowers the expected V_{oc}'s, although as we will discuss, they are further limited by other mechanisms. Also, we wish to note that these devices do not contain Ga. Though we expect that the addition of Ga will enhance performance, we have not yet included it to avoid complications that might hinder progress on understanding the fundamental mechanisms for CIS itself. While these efficiency levels can not be compared to those of state-of-the-art devices mentioned above, we feel that when placed within the context of the simplicity of the process used to achieve this performance, the results are significant. Furthermore, as we will show below, there are further developments under way which we believe will advance our performance levels toward the 15 - 17% range without compromising process simplicity.

An important issue regarding manufacturability is the sensitivity of performance to atomic composition. In all of our processes we deposit the metal precursor layer first and then add Se under various conditions. We are thus concerned with the sensitivity to the starting metal ratios and to the amount and method of delivery of Se. In figure 1 we show the relationship between the deposited and final(in the finished CIS film) metal ratio, and in figure 2 we show the dependence of V_{oc} on that ratio. To first order J_{sc} and FF track V_{oc}, and thus it is an indicator of overall performance. As seen, the two processes track very closely indicating that Process I, thus far, is able to match the performance of the controlled flux Process II. Sensitivity to metal ratio is high on the low In side with V_{oc} plummeting toward zero as film composition approaches In/(In+Cu)=.5. The In-rich side is less precipitous though still indicating the need for adequate control. The data points above the curve at In/(In+Cu)=.58 in figure 2 are for recent modifications to the baseline processes which have taken us through 400 mV. These advances are discussed in terms of improvements in recombination mechanisms below.

Another fertile area for process comparisons is understanding the relationship between selenization details and V_{oc} generation. In Process I we make a specific amount of Se available to the precursor, but have little understanding of the details of its incorporation. To address this issue, we have used Process II to mimic one aspect of the Process I reaction, that of the total amount of available Se. We can also deliver the Se during various stages of the anneal. Results from these experiments are shown in figures 3 and 4. One micron of Se is required for stoichiometry. We first provide one micron of Se to the surface prior to reaching the subject temperature range. Then in the indicated temperature range additional Se flux is provided as indicated. The data shown is a compendium of variations of these conditions. For the 300 - 400 C range the flux is terminated prior to proceeding to the ultimate anneal temperature above 430 C. As can be seen, there is a steady increase in V_{oc} with Se flux for both ranges. While this suggests that the stoichiometric quantity of Se which has previously been provided has not all been incorporated, we must also consider that this may be an entirely surface driven result having primarily to do with adjustment of the

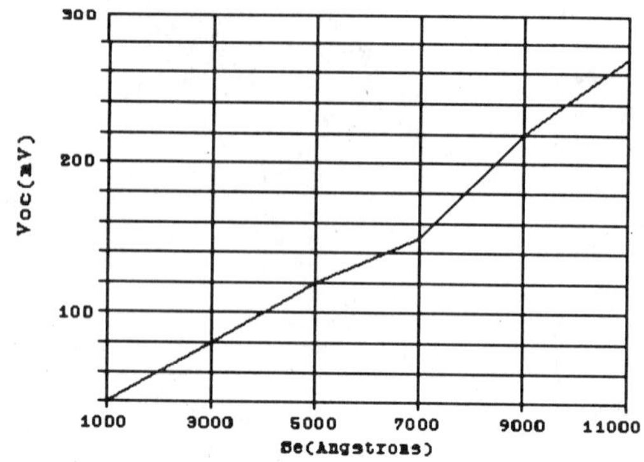

Figure 3. V_{oc} vs. fluxed Se thickness added to 1 μm base at 300<T<400.

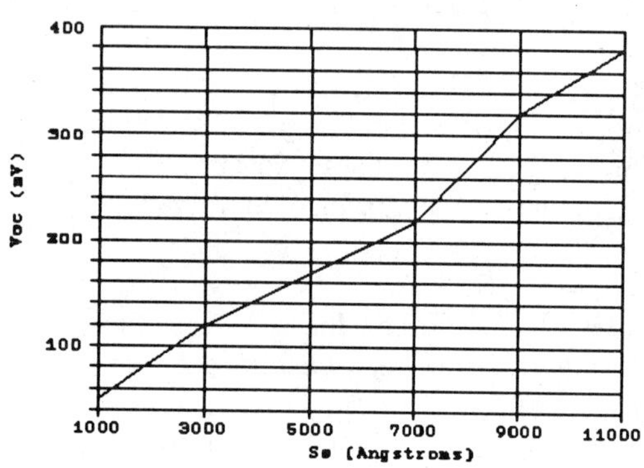

Figure 4. V_{oc} vs. Se flux thickness added to 1 μm base at T>430.

surface composition.

An important observation from these results in reference to Process I is the necessity of providing Se flux above 430 C to achieve 400 mV. Because of the intentional minimal control design of Process I we know little about the behavior of Se during the anneal. These results tell us that in Process I runs in which we are achieving 400 mV there is Se flux available even at the high temperature end of the anneal. This insight is providing critical guidance to Process I modifications which are expected to lead to further advances in performance.

Another issue of importance is to establish generic understanding of the relationships among processing, film properties and device performance. Table I below contains data that addresses this objective. Although the data is exclusive to Process I, we feel that the observations and conclusions will prove to be generic. The atomic compositions and grain size are determined from SEM/EDS analysis. While EDS is calibrated by a NREL supplied reference, we believe it to be accurate to only +/- 0.5%. As can be seen, the metal composition is essentially the same for all devices. We note however, that the initial Cu/In ratios were in the range .75 - .90. While this is certainly an important variable, the results presented here suggest an at least equally strong role for grain size. As can be seen, grain size varies directly with anneal temperature, and both V_{oc} and J_{sc} vary directly with grain size. Variations in the starting metal ratio certainly contribute to the range of V_{oc} and J_{sc} values observed for a given grain size. This result is what one might expect qualitatively. Higher growth temperatures normally result in larger grain size.

The dependence of V_{oc} and J_{sc} on grain size is also intuitively comfortable. J_{sc} depends on transport properties, and smaller grain size usually results in poorer transport. V_{oc} behavior suggests that grain boundaries may contribute to J_o(We note that even the lowest values of V_{oc} are not due to shunting). Further discussion of the implication of this data to the development of device models is provided below.

Model Implications

We have developed a recombination-based model for CIS[4] which is being used to help guide the development of our devices. We have found that the primary determiner of device performance is the density of recombination centers(N_t) in the space charge region. For baseline devices described above the V_{oc} value of 350 mV is due to a recombination center density of about 3 x $10^{17}/cm^3$. Since V_{oc} varies logarithmically with N_t as shown in figure 5, improvements in V_{oc} require rather large reductions in N_t. In our better devices with 400+ mV we have been able to lower N_t below the $10^{17}/cm^3$ level through improvements in the selenization/anneal profile. Also, we have recently applied an etching procedure to our baseline devices to determine their performance as we systematically lowered the recombination center density. In doing so we found that the V_{oc} increased to 430 mV which would correspond to a N_t of 5 x $10^{16}/cm^3$. However, as we varied the degree of etching, we also noted a decrease in J_{sc} proportional to the etching time. Through further analysis of device performance we determined that J_{sc} was decreasing due to a reduction in collection efficiency. A tentative explanation is that the etchant, while removing interface recombination centers, was also

TABLE I

Sample	Cu(%)	In(%)	Grain size(μ)	V_{oc}(mV)	J_{sc}(mA/cm^2)	Anneal max temp	Anneal ramp
11	24.3	25.7	.5	190	23	300	slow
21	23.3	25.0	.7	220	33	350	fast
40	24.5	26.9	.7	200	31	350	fast
45	24.1	25.3	.7	175	24	350	fast
148	24.8	24.8	>1	380	39	500	fast
175	24.9	24.8	>1	410	40	500	fast

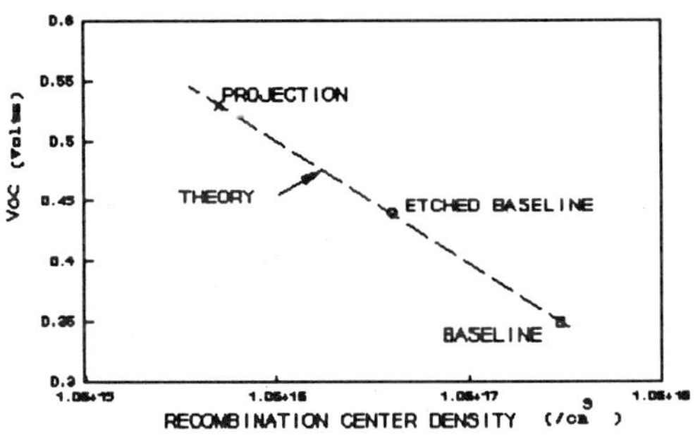

Figure 5. V_{oc} vs. recombination center density.

creating recombination centers deeper into the space charge region. While on balance this helped V_{oc}, it diminished J_{sc}. In recent experiments we have found that it is possible to decouple the effect of the etchant. We hope with further development to just remove the interface states without affecting current generation. In fact, in the example just cited if the original J_{sc} had been maintained, the V_{oc} would have risen to 530 mV as shown in figure 5(projection). Once this is accomplished with the etching procedure, we expect to apply the understanding which is developed toward modification of our deposition procedure so that these levels of performance can be accomplished without the extra etching step.

CONCLUSIONS

We have developed a manufacturing-friendly process for CIS which currently produces devices in the 9 -10% range. Key correlations have been established between processing and performance. In combining these with guidance from a recombination-based device model we are gaining insights to the performance limiting mechanisms and to means of further advancing performance. Through a combination of recent experimental results and simulation we project an increase in V_{oc}'s from 400+ mV for our current best devices to the 500+ mV level along with concomitant increases in efficiency.

Acknowledgements: This work was conducted under the sponsorship of NREL.

REFERENCES

1. This conference.

2. K. Knapp, C. Eberspacher, D. Tarrant and G. Pollock, Proceedings of the 23rd IEEE PV Specialists Conference, Louisville, KY, May, 1993, p. 1073.

3. G. Attar, L. Cai, J. Lai and D. L. Morel, Proceedings of the 22nd IEEE PV Specialists Conference, Las Vegas, October, 1991, p. 1190.

4. D. L. Morel and C. S. Ferekides, "Heterojunction Development and Optimization in Thin-Film Compound Semiconductor Solar Cells", Phase I Annual Technical Progress Report, NREL, May, 1994.

A NOVEL METHOD OF SYNTHESIZING P-$CuInSe_2$ THIN FILMS FROM THE STACKED ELEMENTAL LAYERS USING A CLOSED GRAPHITE BOX

F.O. Adurodija, M.J. Carter and R. Hill
Newcastle Photovoltaics Applications Centre, University of Northumbria, Newcastle upon Tyne, UK

ABSTRACT

The Stacked Elemental Layer (SEL) technique has been demonstrated as a method for producing films over a large area, but the films normally annealed in vacuum, mostly exhibited poor morphology with small grain sizes <1μm which result in poor devices. A novel method of fabricating $CuInSe_2$ films by annealing or selenization of the Cu, In and Se layers deposited onto Mo coated glass substrates in a closed graphite box was developed. SEM, EDX and XRD were used to characterize all films. Dense films with crystal sizes of about 4 μm were obtained over an area of 15 cm^2. The mechanical properties of $CuInSe_2$ formed from various SEL sequences of In, Cu and Se layers on different substrates was also investigated. A sequence starting with an In layer (In/Cu/In/Se...) was found to produce the best adhesion on Mo coated glass, however, the adhesion was found to depend on the thickness of the first In layer. A thin In layer produced the best adhesion, increasing the thickness of the In layer resulted in poor adhesion. Devices with efficiency of 6.5% have been achieved using the absorbers processed in the graphite box. The low efficiency was due to a high series resistance in the cells.

INTRODUCTION

Polycrystalline thin film $CuInSe_2$ is an important material which has been studied extensively for use in photovoltaic power generation. It is one of the leading thin film materials nearing commercial production.

Different techniques have been used successfully to fabricate $CuInSe_2$ thin film material. The best device efficiency of about 17% has been achieved over small areas using the coevaporation technique [1]. In order to make any breakthrough technologically, the fabrication of $CuInSe_2$ absorbers over large areas with good reproducible optoelectronic and mechanical properties is essential for commercial applications. The SEL technique is a very promising method for producing $CuInSe_2$ films over a large area. The major problem which has hindered development of this technique was the inability to produce films with quality crystalline structures suitable for efficient device fabrication using vacuum annealing of the elemental stack, even though strong adhesion could be obtained on Mo coated glass substrates. This paper presents a novel method for synthesizing device grade $CuInSe_2$ with improved crystal sizes from the stacked elemental layers of Cu, In, and Se using a closed graphite box. This processing method is simple, economical, non toxic and could easily be scaled up or integrated into an in-line deposition technique for large area production. Non vacuum deposition techniques, such as electrodeposition, may also be employed in producing the elemental layer precursors.

EXPERIMENTAL DETAILS

The elemental layers of Cu, In and Se were deposited on glass or Mo coated glass substrates by thermal evaporation. The sample was placed in a graphite box and covered with a graphite lid. Two cavities were provided at the end within the box to hold some solid Se shot, and the sample was arranged faced up, (see Fig. 1). The annealing or selenization was carried out at 500°C for 20 minutes within the graphite box loaded into an evacuated glass quartz tube in a furnace with a vacuum pressure of 10^{-2} Torr. Two types of annealing were investigated. The first one involved annealing of the elemental stack with and without extra Se shot using the graphite box. The second annealing was carried out in vacuum or selenium ambient at the same temperature and time. For a study of the mechanical/adhesion properties, different SEL sequences from Cu, In and Se layers were sequentially deposited on glass and Mo coated substrates and annealed at 500°C for 20 minutes. All the films were analysed using XRD, SEM and EDX to study the structural, morphological and the compositional properties respectively. The SEM and scotch tape pull test was used to study the mechanical properties of the films.

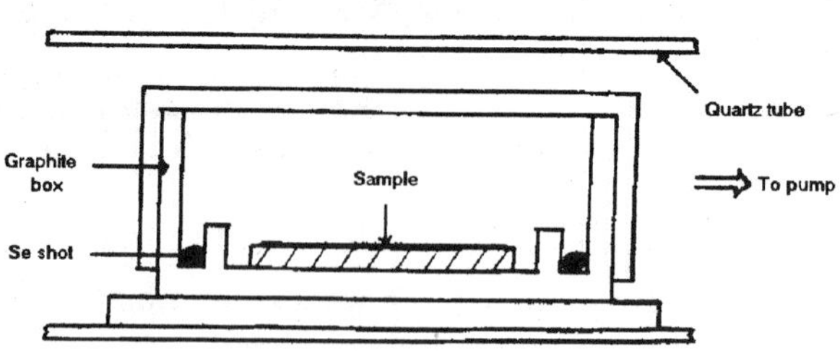

Fig. 1. The graphite annealing and selenization system.

RESULTS AND DISCUSSION

The SEL technique is the most direct and simple method of fabricating $CuInSe_2$ since good control of the composition can easily be obtained during the evaporation

CH3365-4/94/0000-0186 $4.00 © 1994 IEEE

First WCPEC; Dec. 5-9, 1994; Hawaii

(a) (b) (c)

Fig. 2. SEM of $CuInSe_2$ processed in (a) vacuum (b) the graphite box without extra Se (c) the graphite box with extra Se.

of the elements. It allows the freedom of choice of a suitable processing method for achieving improved film properties. Different elemental layer sequences and processing techniques have been investigated by various workers with some degrees of success [2,3,4]. Others have investigated metallic layers, binary selenides as precursors for $CuInSe_2$ fabrication [5,6,7]. However, the problems of non homogeneity, phase segregation, poor crystallisation etc, still remains in most of the processes. This paper focuses mainly on the use of a closed graphite box for the synthesis of device quality $CuInSe_2$ films from the elemental layers as an alternative potential method for achieving large area reproducible films compared to vacuum annealed or selenized films. Studies on the different sequential layers of Cu, In and Se layers for forming $CuInSe_2$ was also carried out to determine the sequence(s) which could enhance the mechanical properties of the films on the substrates.

SEM micrographs of all films synthesized using the graphite box revealed dense and large grain sizes of about 4μm measured over an area of 15 cm^2 (see Fig. 2). Improved adhesion on Mo coated substrates was also exhibited by the films.

The use of the graphite box in the synthesis of $CuInSe_2$ from the stacked elemental layers was found to play three major roles during the film formation. Firstly, it provided a controlled high vapour pressure (HVP) of Se due to a reduced loss of Se particularly at 210°C where $CuInSe_2$ formation starts with an increased heat of formation as a result of an exothermic reaction between ß-In_2Se_3 and liquid Se. The HVP of Se within the box was beneficial to $CuInSe_2$ formation by suppressing the exothermic reaction at this temperature which tended to cause poor crystallization and film delamination as reported earlier [8]. This may also prevent the poor adhesion experienced in the selenization of Cu/In with H_2Se gas due to a large increase in volume of the film during $CuInSe_2$ formation [9]. Secondly, it compensated for any excessive loss of Se which may leave the film n-type during the $CuInSe_2$ formation. Thirdly, it improved the surface uniformity of the films particularly when extra Se shot are included in the box during annealing.

The morphologies of the films annealed in vacuum or Se vapour were very poor with crystal sizes less than 1 μm. This poor morphology could be associated with an uncontrolled increase in the HVP of Se within the film at 210°C resulting in a great loss of Se during annealing thus leaving the films with voids and Se deficient. The SEM results of films produced using the graphite box has shown a noticeable improvement over films processed in vacuum even though films with similar composition of the elemental layers were used.

EDX analysis of the films synthesized in the graphite box showed good reproducibility of the compositional properties, (see Table 1). No significant difference was observed in the compositional properties of the films annealed with or without Se vapour using the graphite box. Films annealed in vacuum had lower Se content with an average [Se/(Cu+In)] ratio of 0.95 in contrast to a ratio of 1.03 for films synthesized in the graphite box. These results indicate that good control over composition could be achieved.

XRD studies of the films showed only single phase chalcopyrite $CuInSe_2$, with no additional phases, or oxides even though films were synthesized in a vacuum of 10^{-2} Torr. Splitting was observed from high resolution scans of the

Sample Number	Films annealed in Vacuum with or without Se ambient				Films synthesized in a closed Graphite Box without extra Se				Films synthesized in a closed Graphite Box with extra Se			
	Cu	In	Se	Se/M	Cu	In	Se	Se/M	Cu	In	Se	Se/M
F1	24.0	20.4	49.0	0.90	24.5	25.7	40.0	0.00	24.1	24.0	51.0	1.01
F2	23.8	27.5	48.7	0.95	24.3	24.6	51.1	1.04	23.2	24.6	52.6	1.11
F3	24.5	26.0	49.5	0.98	24.8	25.0	50.2	1.01	24.5	25.8	50.7	1.03
F4	24.8	26.3	48.9	0.97	24.4	25.3	50.3	1.01	23.6	24.9	51.5	1.06
F5	26.6	25.3	48.1	0.93	23.7	25.4	50.9	1.04	23.1	24.6	52.3	1.10
F6	26.0	25.4	47.0	0.92	24.1	25.1	50.5	1.03	23.6	26.5	49.9	1.00
F7	24.9	27.1	48.0	0.92	24.5	25.6	49.9	1.00	24.3	25.2	50.5	1.02
F8	24.7	25.9	49.4	0.98	23.6	26.7	49.7	0.99	24.1	25.9	50.0	1.00

Table 1. EDX analysis of $CuInSe_2$ films processed in vacuum and in the graphite box with and without Se ambient.

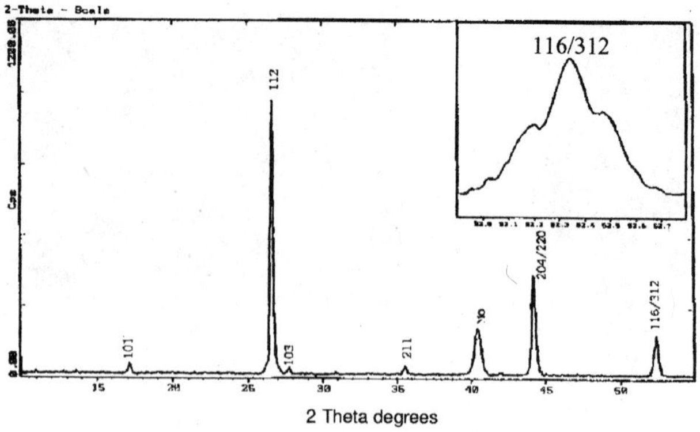

Fig. 3. XRD spectra of a chalcopyrite $CuInSe_2$ film synthesized in the graphite box.

204/220 and 116/312 peaks. This splitting is a characteristic of good quality material and is also observed by other workers [4], (see Fig. 3).

The problem of poor crystallization experienced in vacuum annealed films could be circumvented by annealing or selenizing the films in the graphite box. All films were strongly p-type.

Initial devices produced by films annealed in the graphite box yielded efficiencies of 6.5%, Voc = 400 mV, Jsc = 27 $mAcm^2$ and FF = 55% measured over an area of 0.25 cm^2. The low efficiency was due to a high series resistance

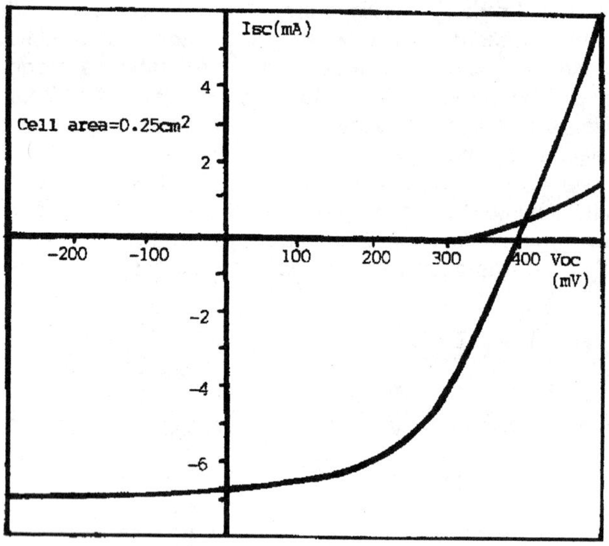

Fig. 4. I-V characteristics of a $ZnO/CdS/CuInSe_2$ cell with 6.5% based on $CuInSe_2$ synthesized using the graphite box.

in the cells. A typical I-V characteristic of the solar cells is shown, (see Fig. 4). It is believed that optimizing all the design and processing parameters required in the graphite system and using an appropriate SEL sequence, more efficient devices over large areas could be produced.

Adhesion of films to the substrates has been a major problems suffered by various $CuInSe_2$ deposition techniques.

Sequence No.	Sequence	Types of Substrate	Adhesion and Mechanical Properties
SEQ 1	In/Cu/In/Se... (thin In layer, 200nm, deposited first)	Mo coated Soda lime, 7059, Soda lime, SKAN glass	Films had very strong adhesion on all the substrates, grey colour with shiny surface and grain sizes larger than 2 microns.
SEQ 2	In/Cu/In/Se... (thick In layer, 690nm, deposited first)	Mo coated Soda lime, 7059, Soda lime, SKAN glass	Good adhesion of films on all the substrates, grey in colour with a shiny surface, poor morphology at the film substrate interface and small grain sizes of 1 micron.
SEQ 3	In/Cu/Se/In/Cu/Se (with a thin layer of In starting the sequence)	Mo coated Soda glass 7059, Soda lime, SKAN glass	Strong adhesion on the substrate, film had grey colour with a shiny surface. Good adhesion of films was obtained on all glass subtrates with morphology similar to that of Mo coated glass.
SEQ 4	In/Cu/Se/In/Cu/Se (thick In layer deposited first on the substrate)	Mo coated and Soda glass 7059, SKAN glasses	Films had voids at the substrate/film interface, good and uniform surface morphology with grain sizes of about 1 micron. Poor adhesion of film on the substrates with stress pattern.
SEQ 5 & SEQ 6	Cu/In/Se/Cu/In/Se (First Cu layer, 400nm - thick; First Cu layer, 100nm - thin)	Mo coated Soda lime, 7059, Soda lime, SKAN glass	Poor adhesion of film on all the substrates, patches of films with pinholes was left on the substrates after annealing.
SEQ 7	In/Se/Cu/In/Se/Cu (thick uniform layers in stoichiometric ratios)	Mo coated Soda lime, 7059, Soda lime, SKAN glass	Films had voids at the substrate/film interface, stress pattern, delaminations and poor surface morphology with grain sizes less than 1 micron.

Table 2. Adhesion and mechanical properties of $CuInSe_2$ films based on different elemental layer sequences on different substrates.

A critical study of the mechanical properties of different sequences of elemental layers was carried out and results are shown in Table 2. The various sequential layers were thermally evaporated on different types of glass and Mo coated glass substrates. All the films were annealed at 500°C for 20 minutes in a vacuum of 10^{-5} Torr. This study showed that the starting elemental material and the thickness composition in a sequence influence the properties of the film formed. Starting the sequence with Se or Se on Cu produced very poor adhesion and should be avoided. The layer sequence could either be started with an In layer followed by a Cu or Se layer or a Cu layer followed by an In layer. However, films whose sequences started with an In layer had better adhesion. The scotch-tape pull test and SEM have shown that the best adhesion on the substrate could be achieved by starting the sequence with a thin layer of In (200nm) followed by a Cu (450nm) layer, a thick layer of In (690nm) and a layer of Se (twice the thickness of (Cu+In)), (thin-In/thick-Cu/thick-In/(Se)...). The number of elemental layers in the sequence could be increased to enhance interdiffusion of the materials during film formation. The above sequence, excluding the last Se layer could also form a basic precursor for selenization process using elemental Se.

CONCLUSION

The graphite system has produced films with good morphological, structural and compositional properties suitable for solar cell fabrication. No segregation or anomalous growth occurred. This method of synthesizing

$CuInSe_2$ from the elemental layers of Cu, In and Se could also be applied to quaternary compounds such as $CuIn(SeS)_2$ or $Cu(InGa)Se_2$ and has the potential for in-line industrial-scale manufacture. Although this technique of producing $CuInSe_2$ is at its early stage of development, films produced have exhibited good uniformity and reproducibility over reasonably large areas. The adhesion of the films on Mo coated glass produced using the graphite box was found to be superior to those processed in vacuum and selenization of the metal precursors. The low vacuum (10^{-2} Torr) required and the absent of toxic gases during annealing makes this method economical and environmentally acceptable. Studies on the mechanical properties of $CuInSe_2$ produced by the different SEL sequences have shown that the arrangement of the elemental layers during evaporation determine the general properties of the films. The best adhesion was obtained from the sequence (thin-In/thick-Cu/thick-In/(Se)...). The stoichiometry of the film could easily be controlled within the desired range. Although, the initial device produced from these films had efficiencies less than 10%, it is believed that there is ample room for improvement and work is currently being carried out to increase the substrate area and the device efficiencies.

REFERENCES

[1] J. Hedestrom et al., "ZnO/CdS/Cu(In,Ga)Se_2 Thin Film Solar Cells with Improved Performance", 23rd IEEE PVSC, 1993, pp. 364-371.

[2] F. Karg et al., "Novel Rapid Thermal Processing for CIS Thin Film Solar Cells" 23rd IEEE PVSC, 1993, pp. 441-446.

[3] F.P. Baumgartner, M. Regula and E. Bucher, "Formation of CIS Layers in a Small Space Vacuum Chamber by RTP", 12th EC PVSEC, 1994, pp. 637-640.

[4] J. Kessler et al., "Low Pressure Vapour Phase Selenization of CuIn Films without H_2Se", 10th EC PVSEC, 1991, pp. 879-882.

[5] J. Kessler and H.W. Schock, "Selenization from Sequential Deposition of Cu:In(Se):Se", 11th EC PVSEC, 1992, pp. 838-841.

[6] T. Nakada et al., "Preparation and Characterization of $CuInSe_2$ Films for PV Applications by Low Pressure Vapour Phase Selenization", 11th EC PVSEC, 1992, pp. 794-797.

[7] J. Kessler et al., "$CuInSe_2$ Formation from Sequential Deposition of In(Se):Cu:Se", 12th EC PVSEC, 1994, pp. 648-652.

[8] F.O. Adurodija et al., "Reaction Mechanism and Structural Sturdies of $CuInSe_2$ Fabricated by the Stacked Elemental Layer (SEL) Technique", 12th EC PVSEC, 1994, pp. 621-624.

[9] D. Willet et al., "The Effect of Localised Defect Features on $CuInSe_2$ Device Performance", 10th EC PVSEC, 1991 pp. 939-942.

POLYCRYSTALLINE $CuIn_{1-x}Ga_xSe_2$ THIN FILM PV SOLAR CELLS PREPARED BY TWO-STAGE SELENIZATION PROCESS USING Se VAPOR

Neelkanth G. Dhere, Shanker Kuttath, and Kevin W. Lynn
Florida Solar Energy Center
300 State Road 401, Cape Canaveral, FL 32920-4099, USA

Robert W. Birkmire and William N. Shafarman
Institute of Energy Conversion, University of Delaware
Wyoming Road, Newark, DE 19716-3820, USA

ABSTRACT

Novel two-stage Se vapor selenization of magnetron sputtered metallic precursors and Ga incorporation using single Cu-Ga(22 at.%) alloy target have been developed for preparation of well-adherent, large, compact, well-faceted polyhedral grain $CuIn_{1-x}Ga_xSe_2$ thin films having optimum composition Cu:In:Ga:Se of 22.95:25.03:1.40:50.63. Cu-In-Ga precursor homogenization by *in situ* heat-treatment at ~90° C for 10 minutes prior to first selenization, selenization temperature of 550-560° C, and Se vapor incidence rate of 50 Å sec^{-1} resulted in improved morphology of completed $CuIn_{1-x}Ga_xSe_2$ thin films, and solar cells with open circuit voltage V_{oc} of 377 mV, short circuit current density J_{sc} of 34.8 mA, fill factor FF of 62.5%, active area efficiency of 8.2% (total area efficiency 5.8% limited by FF 49.1%) and fairly constant spectral response over the spectral range.

INTRODUCTION

PV modules based on $CuIn_{1-x}Ga_xSe_2$ and CdTe polycrystalline thin films have been targeted as candidates for low-cost energy production by the U.S. Department of Energy and a segment of PV industry. They offer low material cost, potential scalability and automation of the fabrication processes, and efficiencies competitive with the presently predominant crystalline-Si technology [1-5]. High efficiency $CuIn_{1-x}Ga_xSe_2$ solar cells are based on Cu, In, Ga metallic precursors coevaporated or sequentially evaporated in the presence of Se [4,5]. These processes are difficult to scale-up for large-scale, economic manufacture [1-5]. Sputtering of Cu-In-Ga metallic precursors, on the other hand, can be easily scaled-up and hence is the preferred industrial process for thin-film deposition [6-8]. $CuIn_{1-x}Ga_xSe_2$ is a pseudo-quaternary analog of the ternary compound $CuInSe_2$ in which gallium is substituted on some indium sites which serves to raise the bandgap towards the theoretical optimum value of 1.5 eV and also to benefit from bandgap tailoring such as built-in back-surface field. Gallium-indium form a low temperature eutectic with the eutectic point at ~16° C. Indium-gallium or gallium layers would invariably result in liquid or liquid-like behavior destroying the structural integrity. Residual stresses in refractory metal Mo films prepared by magnetron-sputter deposition are caused by peening effect of energetic backscattered neutrals and hence depend strongly on the working sputtering power and gas pressure [9]. A general feature is a transition pressure below which arriving atoms have higher kinetic energy and the resulting films are in compression, whereas above this pressure films have tensile stress. Higher working gas pressure and lower power are expected to moderate the flux and energy of these particles. This paper describes a novel two-stage selenization process using selenium vapor, and a new process for successful incorporation of gallium while avoiding deleterious indium-gallium interaction.

EXPERIMENTAL TECHNIQUE

Magnetron-sputter-depositions of Mo and Cu-In-Ga metallic precursor films were carried out in a 18" diameter and 24" height bell jar. Soda-lime float glass was used as substrate for fabrication of solar cells. Higher working gas pressure combined with lower sputtering power were employed to moderate the flux and energy of back-scattered neutrals and thus to obtain well-adherent Mo coatings reasonably free from residual stresses. Some Mo coated samples were obtained from Siemens Solar Ind.

Initially the depositions consisted of a Cu/In bilayer. Selenized films were found to be more homogeneous if there was a proper intermixing of Cu/In bilayer prior to placement for the first selenization. H_2Se gas is highly toxic with toxicity comparable to arsine. It is also highly flammable. Its use was avoided. Selenizations were carried out by evaporating Se from a Knudsen cell type molybdenum boat in high vacuum on samples mounted under a ceramic plate heater. Excessive selenization temperatures >565° C were found to be detrimental to uniform single phase formation. Hence selenization temperatures were varied in the range 450-560° C. Selenization temperatures were calibrated with a thermocouple attached to a backing copper plate. Recent

First WCPEC; Dec. 5-9, 1994; Hawaii

more accurate temperature measurements with a thermocouple attached to one of the substrates using a high-temperature glue showed the actual temperatures to be 20-30° C lower. Corrected values of temperature are reported in the following. A minimum rate of incidence of selenium vapor equivalent to 25 Å sec^{-1} was found to be essential for complete selenization. Selenization process was quite reproducible when the selenium vapor incidence rate were maintained between 50-300 Å sec^{-1}. However, the problems of peeling off of selenized films became more frequent at excessive (>350 Å sec^{-1}) rates of selenium vapor incidence. Recent high efficiency $CuIn_{1-x}Ga_xSe_2$ solar cells benefit from the fluxing action of $Cu_{2-x}Se$ having a melting point of ~523° C for the growth of compact, large-grain films from Cu-rich metallic species [4,5]. The novel two-selenization process using Se vapor developed at FSEC utilizes initially a Cu-rich precursor for achieving recrystallization during the first selenization followed by a Cu-poor precursor for conversion of $Cu_{2-x}Se$ phase at the surfaces and grain boundaries during the second selenization. A new process developed at FSEC solved the problem of incorporation of gallium by using a single Cu-Ga(22 at.%) alloy target and thus avoided the difficulties of low-melting point precursors and deleterious interaction between indium and gallium. The two-stage selenization process sequence for obtaining a 2.5 µm thick $CuIn_{1-x}Ga_xSe_2$ layer is shown in figure 1. In this process, firstly 2210 Å Cu-Ga layer was sputter-deposited on molybdenum followed by 981 Å of In and 736 Å Cu-Ga layers. The thin Cu-Ga over-layer is intended to avoid globule formation of In and In-Ga. The Cu-In-Ga films were homogenized by *in situ* heat treatment at 80-90° C for 10 min. First selenization was carried out using Se vapor. A second set of metallic precursors consisting of 2943 Å In and 736 Å Cu-Ga were sputtered on to the sample. Second selenization was then carried in Se vapor. Initial selenization temperatures were varied in the range 450-530° C. Later the temperatures were adjusted to the optimum range of 550-560° C.

The two-stage selenization process improved the adhesion to Mo back-contact layer considerably. Addition of gallium to prepare higher bandgap $CuIn_{1-x}Ga_xSe_2$ thin films improved adhesion still further. Combination of two-stage selenization process and Ga addition near Mo contact have eliminated the vexing problem of peeling off. Using two-stage selenization process, well adherent $CuIn_{1-x}Ga_xSe_2$ thin films free from pin-holes are being prepared routinely.

Structure and morphology of deposited films were studied by x-ray diffraction (XRD) and scanning electron microscopy (SEM). The composition of the films was analyzed by electron probe microanalysis (EPMA), x-ray photoelectron spectroscopy (XPS), and Rutherford backscattering (RBS). Rough estimates of the composition of $CuIn_{1-x}Ga_xSe_2$ thin films were made by measuring the contact resistance of the film by contacting molybdenum back layer and the front of $CuIn_{1-x}Ga_xSe_2$ films and by verifying its photoconductive behavior. These

Total thickness	
CuGa	3682 Å
In	3924 Å

Selenization II 450-560° C		
20%	CuGa	736 Å
75%	In	2943 Å
Selenization I 450-560° C		
20%	CuGa	736 Å
25%	In	981 Å
60%	Cu-Ga	2210 Å
Molybdenum		
Glass		

Fig. 1. 2.5 µm $CuIn_{1-x}Ga_xSe_2$ layer preparation process.

measurements are material and pressure sensitive. Comparison with similar measurements at NREL have shown that contact resistance values reported in the following are lower probably because of folded aluminium foil contact used to avoid scratching the film.

Solar cells were fabricated by deposition of 1-µm thick CdZnS:In\ITO\Al fingers or thin CdS\ZnO\Ni\Al fingers on glass\Mo\$CuIn_{1-x}Ga_xSe_2$ samples. Efficiency and properties of cells were studied by measurement of IxV characteristics and spectral response.

RESULTS AND DISCUSSION

Phase changes in the first Cu-In layers were studied by varying time-temperature profiles with the maximum temperatures in the range 80-160° C for *in situ* heat-treatments. Temperatures above 155° C were found to be detrimental to the formation of a uniform precursor phase and hence were avoided. A reaction zone was observed to pass through the film heated to temperatures approximately 10-30° C below the melting point of indium (T_m). Globule formation on the surface was observed in extreme cases. SEM photographs of Cu/In precursor bilayer pre-heat treated at >130° C showed surface protrusions and pitting. Such a transfer of mass below the melting point has been attributed to large surface velocities of a small (~10%) fraction of atoms at temperatures above 0.8 T_m. RBS spectra of Cu-In bilayers pre-heat-treated *in situ* at ~90° C

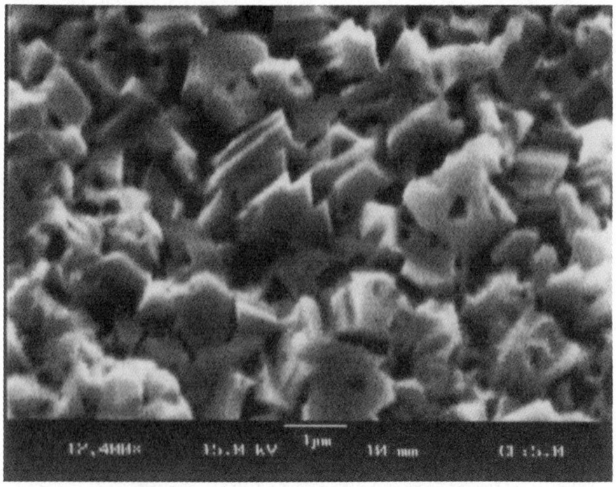

Fig. 2. X-ray diffraction pattern from a $CuIn_{1-x}Ga_xSe_2$ thin film

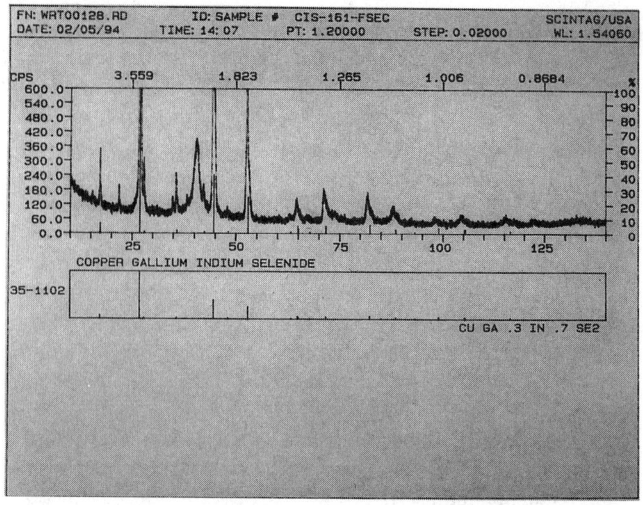

Fig. 3. Scanning electron micrograph of $CuIn_{1-x}Ga_xSe_2$ thin film

for 10 minutes showed completely intermixed precursor. Hence in most cases the Cu-In-Ga precursor films were subjected to a *in situ* heat treatment at ~90° C for 10 minutes prior to the first selenization.

As reported elsewhere, atomic force microscopy revealed a weak subgrain structure because of the tendency towards globule formation of In deposits and also due to secondary nucleation [10]. The subgrain structure disappeared after the first selenization which resulted in a weakly faceted growth of well-connected hillocks. The subgrain structure was seen to reappear when the second metallization layers were deposited. The subgrain structure disappeared completely after the second selenization [10].

Copper content in $CuIn_{1-x}Ga_xSe_2$ thin films was varied systematically over a wide range. Loss of indium was found to be negligible at selenization temperatures of 450-500° C. Contact resistance of $CuIn_{1-x}Ga_xSe_2$ thin films with Cu:In:Ga:Se atomic proportions of 18.9:28.1:2.5:50.5 approached 1 MΩ. XRD patterns of these films showed the formation of $CuIn_{1-x}Ga_xSe_2$ and $CuIn_2Se_{3.5}$ phases. Solar cells prepared using these films gave a open circuit voltage V_{oc} of ~300 mV and a very low short circuit current density J_{sc} of ~1 mA.

Contact resistance of $CuIn_{1-x}Ga_xSe_2$ thin film having Cu:In:Ga:Se atomic proportions of 19.3:27.6:1.3:51.8 was 100-200 kΩ. Solar cell fabricated using this film showed an open circuit voltage V_{oc} of 331 mV, a short circuit current density J_{sc} of 18 mA, with a very low fill factor FF of 28% due to the excessively large series resistance resulting from the copper deficiency. The copper content in $CuIn_{1-x}Ga_xSe_2$ thin films was adjusted to obtain near the optimum atomic proportion of Cu:In:Ga:Se of 23.5:25.5:1.5:49.5. These films have shown contact resistance in the range of 50-100 kΩ and were photoconductive. XRD pattern showed the formation of $CuIn_{1-x}Ga_xSe_2$ phase (Fig. 2). Study of the morphology by SEM of $CuIn_{1-x}Ga_xSe_2$ thin films with the near optimum atomic proportion showed the importance of proper intermixing of Cu:In:Ga precursors prior to selenization. SEM photographs of $CuIn_{1-x}Ga_xSe_2$ thin films in which the precursors were homogenized showed large, compact, well-faceted grains (Fig. 3). The most stable polyhedral growth form of chalcogenides is visible in many of the grains. Some pits were also visible. On the other hand the grain morphology was inferior in $CuIn_{1-x}Ga_xSe_2$ thin films prepared from unhomogenized precursor layers.

PV characteristics of cells completed with CdZnS:In\ITO\Al fingers reaffirmed the importance of proper intermixing of Cu:In:Ga precursors prior to selenization. Solar cells prepared from $CuIn_{1-x}Ga_xSe_2$ thin films in which the precursors were not homogenized showed V_{oc} of 378 mV, J_{sc} of 33.4 mA, a FF of 60.2%, and an active-area conversion efficiency of 7.6%. The best cell obtained with films prepared using precursors homogenized *in situ* prior to selenization gave V_{oc} of 377 mV, J_{sc} of 34.8 mA, a FF 62.5% and active-area PV conversion efficiency of 8.2%. Spectral response was found to be fairly constant over the entire spectral range. A few cells completed at NREL with thin CdS\ZnO\Ni\Al fingers showed V_{oc} of 345-404 mV, J_{sc} of 30.8-34.2 mA, FF of 41.6-49.1%, series resistance R_s of 11.2-29.7 Ω and parallel resistance R_p of 398-1200 Ω, and the best total area efficiency of 5.8% limited by series resistance R_s of 11.2 Ω and parallel resistance R_p of 398 Ω (Fig. 4).

An excessively high selenization temperature (>565° C) and a low or interrupted selenium vapor incidence rate resulted in a rough and uneven morphology which most probably arose from compositional non-uniformity and separation of Cu-rich phase with a rough morphology and In-rich phase with a smoother morphology [5]. Solar cells

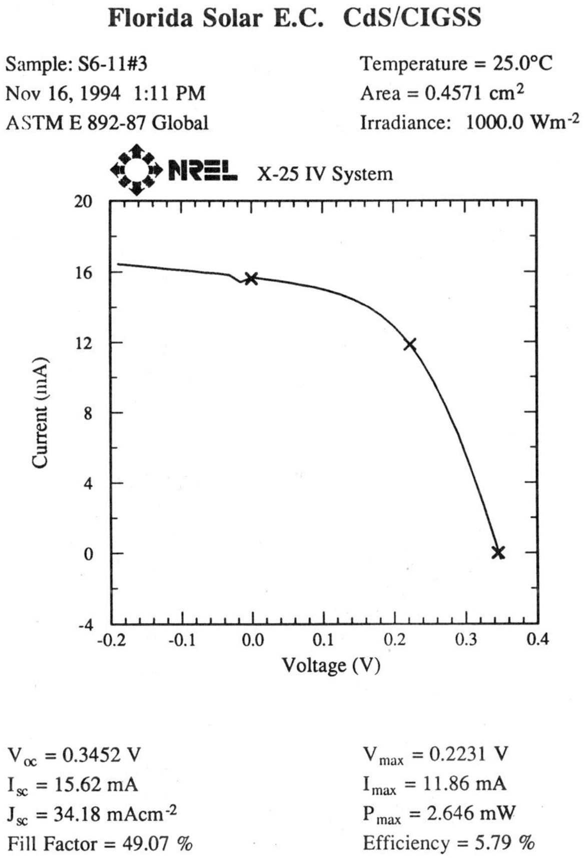

Fig. 4. ASTM E 892-87 Global IxV characteristic of a 0.4571 cm^2 $CuIn_{1-x}Ga_xSe_2$ thin-film solar cell

prepared from these films showed very low conversion efficiencies (<4%) because of the presence of rough, Cu-rich regions at the junction between $CuIn_{1-x}Ga_xSe_2$ and CdS.

CONCLUSIONS

A novel two-stage selenization process using Se vapor has been developed for achieving recrystallization of magnetron-sputtered Cu-rich precursor during the first selenization and conversion of $Cu_{2-x}Se$ phase at surfaces and grain boundaries during the second selenization of Cu-poor precursor. A new process developed at FSEC using a single Cu-Ga(22 at.%) alloy magnetron-sputtering target solves the problem of incorporation of low-melting point precursors without encountering the deleterious interaction between indium and gallium. Improvements in properties of $CuIn_{1-x}Ga_xSe_2$ thin films and solar cells were obtained by employing an optimum selenization temperature of 550-560° C. The novel two-stage selenization process improved the adhesion to Mo back-contact layer considerably. Gallium addition to form higher bandgap $CuIn_{1-x}Ga_xSe_2$ thin films improved film adhesion still further. The combination of two-stage selenization process and Ga addition near Mo contact have eliminated the vexing problem of peeling off.

The best cell obtained using a $CuIn_{1-x}Ga_xSe_2$ film having optimum composition Cu:In:Ga:Se of 22.95:25.03: 1.40:50.63 and large, compact, well-faceted grains prepared from *in situ* homogenized precursors gave V_{oc} of 377 mV, J_{sc} of 34.8 mA, FF 62.5% and active area conversion efficiency of 8.2%. Spectral response was found to be fairly constant over the entire spectral range. A few cells completed at NREL showed V_{oc} of 345-404 mV, J_{sc} of 31.5-34.2 mA, FF of 41-49.1%, and total area efficiency of 5.8%.

ACKNOWLEDGEMENTS

This work was supported by National Renewable Energy Laboratory Contract # XG-2-11036-5. Authors are thankful to Alice Mason, Dr. David Albin, Dr. Kannan Ramanathan, Dr. Rommel Noufi, Don Dunlavi and Keith Emery of NREL for EPMA analysis, and completion and measurements of solar cells. Dr. Orlando Melendez of Kennedy Space Center for XPS analysis, and Dr. Bijoy K. Patnaik and Dr. Nalin R. Parikh of University of North Carolina for RBS analysis.

REFERENCES

[1] H. Ullal, K. Zweibel, and B. G. von Roedern, "Thin-Film CdTe and $CuInSe_2$ Photovoltaic Technologies" Proc. ISES Solar World Congress, Budapest, Hungary, Aug. 23-27, 1993, 187-193.
[2] N. G. Dhere, "Recent Developments in Thin-Film Solar Cells", Thin Solid Films **194**, 1990), 757-768.
[3] N. G. Dhere, "Present Status of the Development of Thin Film Solar Cells", Vacuum, **39**, 1989, 743-747.
[4] M. A. Contreras, A. M. Gabor, A. L. Tennant, S. Asher, J. R. Tuttle, and R. Noufi, "16.4% Total-area Conversion Efficiency Thin-Film Polycrystalline MgF_2/ZnO/CdS/ Cu(In,Ga)Se_2/Mo Solar Cell" Prog. in Photovoltaics. Res. & Appl. **2**, 1994, 287-292.
[5] T. Walter and H. W. Schock, "Crystal Growth and Diffusion in Cu(In,Ga)Se_2 Chalcopyrite Thin Films", Thin Solid Films **224**, 1993, 74-81.
[6] R. Gay, M. Dietrich, C. Fredric, C. Jensen, K. Knapp, D. Tarrant, D. Willet, "Efficiency and Process Improvements in $CuInSe_2$-Based Modules", Proc. 12th European Photovoltaic Solar Energy Conference, Amsterdam, April 1994, 935-938.
[7] V. K. Kapur, B. M. Başol, A. Halanl, C. R. Leldhlm, and A. Minnick, "Technical and Business Factors Affecting Commercialization of Thin Film CIS Technology", Proc. 12th EPVSEC, Amsterdam, April 1994, 1608-1611.
[8] A. Delahoy, J. Britt, G. Butler, F. Faras, A. Sizemore, F. Ziobro, and Z. Kiss, "Recent Advances in Manufacturing Technology for $CuInSe_2$-Based Power Modules", 12th EPVSEC, Amsterdam, April 1994, 1612-1615.
[9] D. W. Hoffman, J. A. Thornton,"Effects of Substrate Orientation and Rotation on Internal Stresses in Sputtered Metal Films", Journal of Vacuum Science Technology, **16**, 1979, 134-137.
[10] N. G. Dhere, S. Kuttath, and H. R. Moutinho, to be published in J. Vac. Sci. & Technol. A.

NON-CADMIUM ZnO/CIS AND ZnSe/CIS SOLAR CELLS

Larry C. Olsen, F.William Addis, David Greer , Wenhua Lei
Washington State University /Tri Cities
And
Fuad Abulfotah
National Renewable Energy Laboratory

ABSTRACT

This paper concerns studies of CIS solar cells based on ZnSe and ZnO buffer layers. ZnSe and ZnO films are grown by MOCVD. These investigations have been primarily carried out with CIS substrates from Siemens Solar. CIS substrates have been characterized with photoluminescence and thermoelectric probe measurements. In particular, PL studies established that the Siemens standard CIS material has a segregated phase(s) on the surface, and thermoelectric probe measurements have established that this surface is n-type. Test cells formed on ZnSe/CIS film structures indicate that the ZnSe/CIS cell structure has a potential efficiency > than 14 %. Open circuit voltages are typically larger than 500 mV and the optimum range of ZnSe film thickness for maximum efficiency is between 100 Å and 250 Å. Deposition of ZnO onto a ZnSe/CIS structure typically leads to poor cell performance. Using a layer of MOCVD ZnO as a protective layer, a total area efficiency of 8.74 % was achieved for a ZnSe/CIS cell. Direct ZnO/CIS cells were fabricated with MOCVD ZnO buffer layers. A total area efficiency of 11.1 % has been achieved.

INTRODUCTION

This paper describes investigations of CIS solar cells based on MOCVD ZnSe and ZnO buffer layers. This research is important for several reasons: (1) Development of non-cadmium buffer layers may be essential for CIS cells to be accepted in the marketplace; (2) Development of direct ZnO/CIS cells could lead to a simplified cell and module fabrication; (3) Knowledge gained in investigations of ZnO and ZnSe buffer layers may help us understand the unique role CdS plays in CdS/CIS solar cells.

Other studies of ZnSe/CIS solar cells have involved ZnSe deposition by thermal evaporation and reactive magnetron sputtering [1,2]. The work reported in Reference 1 was based on sputtered ZnSe and resulted in cells with efficiencies of 7 to 8 %. Results for direct ZnO/CIS cells have been reported by the European CIS group [3]. Active area efficiencies of 10.6 % and V_{OC} values of 398 mV were achieved for ZnO/CIS structures for which the ZnO buffer layer was deposited with an O_2-free sputtering process. The work discussed in this paper has resulted in total area (active area) efficiencies of 8.7% (9.2 %) and 11.1 % (11.7 %) for ZnSe/CIS and ZnO/CIS cells, respectively. The remainder of this paper consists of discusions concerning MOCVD growth of ZnSe and ZnO, CIS substrates, ZnSe/CIS solar cells, ZnO/CIS solar cells and conclusions.

MOCVD GROWTH OF ZnSe and ZnO

Growth of ZnSe and ZnO are carried out in the WSU MOCVD system which consists of a SPIRE 500XT reactor with added gas handing capabilities. ZnSe is grown by reacting a zinc adduct with H_2Se. Conductive n-type ZnSe is grown by using iodine as a dopant. Ethyliodide was mixed with helium and installed on one of the gas lines to the system. The zinc adduct is formed by reacting triethylamine and DMZn. Growth rates of ZnSe are typically 1 Å/s and CIS substrate temperatures are between 100°C and 250°C. The WSU SPIRE reactor has been modified for ZnO growth. Tetrahydrofuran (THF) is utilized as a source of oxygen, while the zinc adduct used for the growth of ZnSe is also utilized for ZnO growth.

CIS SUBSTRATES

Siemens Substrates

This work has been conducted using two kinds of CIS substrates from Siemens Solar, namely, Siemens 'standard' CIS material and the so-called 'graded absorber' material [4]. CIS substrates consist of a CIS layer deposited onto Mo coated glass. As discussed in Reference 4, the standard material is essentially $CuInSe_2$ throughout the film except near the CIS-Mo interface where Ga is added. This material is denoted in this paper as simply CIS. The Graded absorber material has sulfur near the front face and so material is referred to as CIS(GA) in this paper. It is important to note that the photoresponse reported in Reference 4 for the CIS(GA) material is very similar to that for the standard CIS material. Thus, although sulfur and Ga are incorporated in the film, this material should not be regarded as CIGS with a higher bandgap.

Physical Characterization

Two approaches were used to characterize the CIS surface, photoluminescence and thermoelectric probe analyses. The PL measurements clearly establish that a segregated phase or phases exist at the CIS surface, and thermoelectric probe studies determined that the surface layer is n-type.

CH3365-4/94/0000-0194 $4.00 © 1994 IEEE

The PL measurements were conducted at NREL. Optical excitation was provided using 835 nm line of a diode laser . By varying the angle of incidence of the laser beam, it is possible to measure the PL emission from different depths from the sample surface. Fig. 1 shows a PL emission spectrum for a CIS sample that had been degreased and etched with KCN. The PL data were taken with an incident laser beam power of 20 mW, an angle of incidence of 25 ° and at a temperature of 15°K. Emissions due to the bulk CIS and due to a segregated phase(s) at the surface are evident. If the angle of incidence is decreased to 15°, the emission due to the surface phase(s) increases relative to the CIS emission. PL emission spectra measured for as-received CIS material (prior to any surface preparation) were very similar to that described by Fig.1.

Thermoelectric probe analysis involved the measurement of the potential between two points on the surface of a CIS sample while heat is appl;ied at one end of the sample. With the probe nearest the heated side grounded, the potential at the probe furthest from the heat became negative with respect to ground. This result is obtained consistently, and indicates that electrons are the dominant carrier at the surface. If the probes are allowed to penetrate into the material, p-type conductivity is indicated. Results of the PL and thermoelectric probe analyses are consistent with a buried heterojunction model.

ZnSe/CIS SOLAR CELLS

Al/ZnSe/CIS Test Cells

ZnSe/CIS structures have been studied by depositing an array of aluminum circular areas 2.8 mm in diameter onto the ZnSe film to serve as contacts. Al films are deposited with a thickness of 100 to 150 Å and at a rate ≥ 15 Å/sec so that light can pass through the film and to minimize the sheet resistance. Illuminated and dark characteristics for a test cell (defined as a single Al circular area) allow one to estimate the potential performance of CIS cells based on ZnSe windows. Illuminated characteristics of test cells are measured by illuminating the device until $J_{SC} \approx 40\ mA/cm^2$. The cell efficiency may be regarded as an estimated active area, AM1.5 efficiency, assuming that one can add a top contact layer and collector grid without degrading the junction properties. A plot of test cell efficiency versus ZnSe thickness is shown in Fig. 2. As indicated by Fig. 2, test cell efficiencies can exceed 14 % as long as the ZnSe thickness is in the range of 100 Å to 250 Å, where the thickness is a value determined by reflection analysis on a silicon witness (during MOCVD growth of ZnSe). Generally, it is found that for ZnSe thicknesses > 250 Å, inflected I-V curves result and values of J_{SC} approach zero.

The reduction of photocurrent can be interpreted to be a result of the formation of a significant conduction band spike at the ZnSe-CIS interface, as predicted (0.85 to 0.9 eV) in Reference 5 for an interface between relaxed, bulk layers of ZnSe and CIS. T-I-V analyses have been carried out for several test cells. Typically, I-V data can be interpreted in terms of two current mechanisms acting in parallel. At larger voltages (0.4 to 0.6 Volts), the dominant current mechanism appears to be due to depletion layer recombination with n = 1.5, and $J_o = 5 \times 10^{-8}\ A/cm^2$ at room temperature, and the lower voltage mechanism (0.2 to 0.4 Volts) appears to be due to multiple step tunneling.

Physical Characterization

PL studies have also been conducted on ZnSe/CIS structures. Fig. 3 shows a PL emission spectrum taken at a low angle of incidence with respect to the plane of the interface on a ZnSe/CIS structure. Test cells formed on the ZnSe/CIS substrate exhibited very good properties. The PL spectrum due to emission from the top surface layer is dominated by three emissions of 1.26 eV, 1.22 eV and 1.15 eV. Low energy emission from a deeper point (measured at a slightly larger angle) shows the typical emission from a bulk CIS film which is attributed mainly to donor-acceptor recombinations. Considering the PL results for CIS substrates and for ZnSe/CIS structures, it appears that the MOCVD growth of ZnSe on CIS does not alter the buried heterojunction.

Completed ZnSe/CIS Solar Cells

Solar cells based on the ZnSe buffer layers were fabricated by having highly conducting ZnO deposited onto ZnSe/CIS structures by other laboratories. Top contact layers of ZnO were deposited by sputtering and CVD processes. In every case, degradation of cell properties relative to the original ZnSe/CIS junction properties occurred, much as reported in Reference 3 for other non-cadmium window materials. We have fabricated ZnSe/CIS cells that incorporate a layer of MOCVD ZnO as a protective layer (for the ZnSe). A cell fabricated with this procedure was measured by NREL to have a total area efficiency of 8.74 %.

DIRECT ZnO/CIS SOLAR CELLS

Studies of direct ZnO/CIS cells were initiated once the approach to MOCVD growth of ZnO was developed. Although this effort is relatively new, significant progress has been made. To date the resistivity of ZnO films has been controlled by varying the substrate temperature as described by Figure 4. Additional control of the resistivity will soon be added by including a dopant in the growth process.

Solar cell studies based on MOCVD ZnO buffer layers have involved the use of Siemens CIS(GA) material. The approach to cell fabrication is described by Figure 5. After surface preparation, a thin layer of high resistivity ZnO is grown by MOCVD for the buffer layer. The structure is then sent to Siemens for deposition of a low resistivity ZnO top contact layer, and then returned to WSU for deposition of a Ni/Ag collector grid. After adding a capability for doping ZnO, the top contact layer will also be grown at WSU. The properties of the cell shown in Figure 5 are possibly the best yet reported for a direct ZnO/CIS cell. Note that the open circuit voltage is 492 mV which much higher than values reported by other workers.

CONCLUSIONS

These studies have resulted in the following key results: (1) Siemens standard CIS material has an n-type segregated phase at the surface; (2) ZnSe/CIS test cells exhibit high voltages, and efficiencies on the order of 14 %; (3) To date, deposition of a top contact layer of ZnO has consistently caused degradation of ZnSe/CIS cells; (4) Efficient direct ZnO/CIS(GA) cells have been fabricated using MOCVD grown ZnO buffer layers.

ACKNOWLEDGMENT

This work was supported by The National Renewable Energy Laboratory under Subcontract XG-2-11036-6. We are grateful to Siemens Solar for providing CIS material.

REFERENCES

1. A. Nouhi, R. J. Stirn and A. Hermann, *19th IEEE PVSC*, 1987, page 1461.

2. Ji-Beom Yoo, AlanL. Fahrenbruch and Richard H. Bube, *Solar Cells* **31**, 171 (1991).

3. J. Kessler, M. Ruckh, D. Hariskos, U. Ruhle, R. Menner and H. W. Schock, "Interface Engineering Between $CuInSe_2$ and ZnO," 23rd IEEE PVSC, 1993, pp 447 - 452.

4. D.Tarrant and J. Ermer, "I-III-VI_2 Multinary Solar Cells Based On $CuInSe_2$", 23rd IEEE PVSC. 1993, pp 372 - 378.

5. Larry C. Olsen, F. William Addis and Daniel A. Huber, "Investigation Of Polycrystalline Thin Film CuInSe2 Solar Cells Based On ZnSe Windows", 23rd IEEE PVSC. 1993, pp 603 - 606.

6. Art J. Nelson, et al., "Theoretical and Experimentral Studies of the ZnSe/CuInSe2 Heterojunction Band Offset",Appl. Phys. Lett. **62**, 2559 (1993).

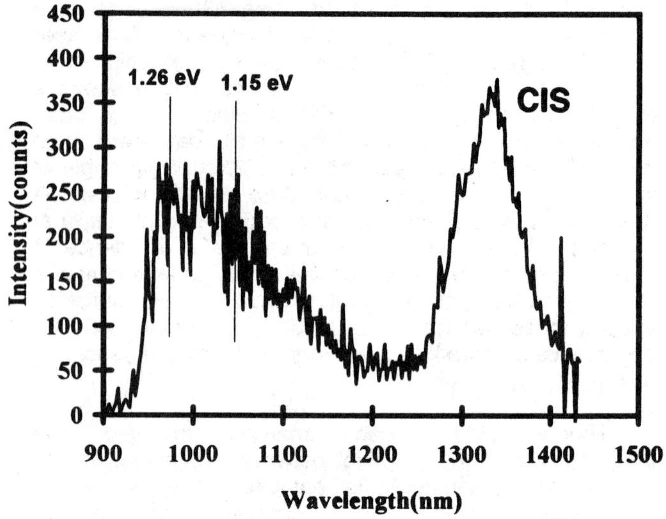

Fig.1 PL emission for Siemens CIS material taken at an incident angle of 25°.

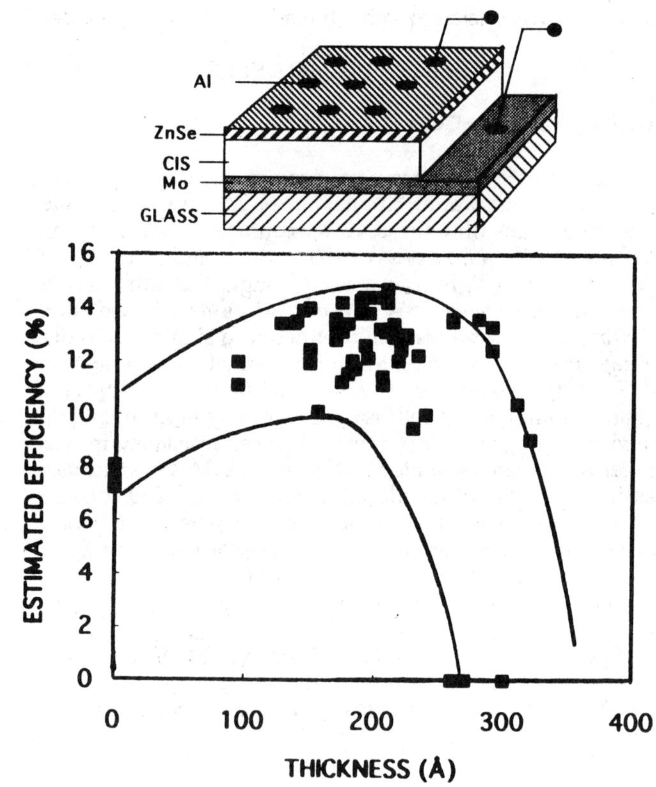

Fig. 2 Test cell efficiency vs ZnSe thickness (as measured on a Si witness).

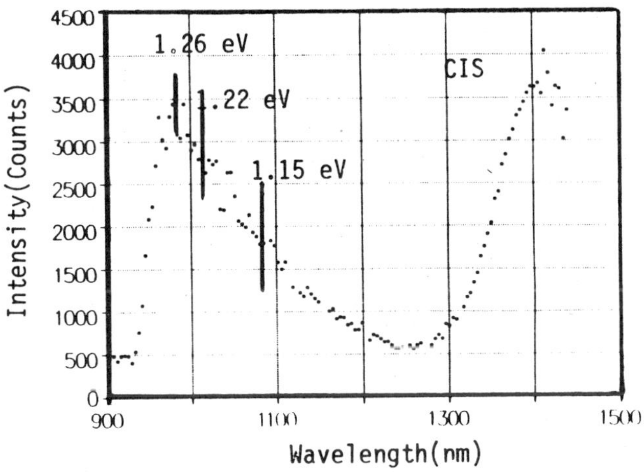

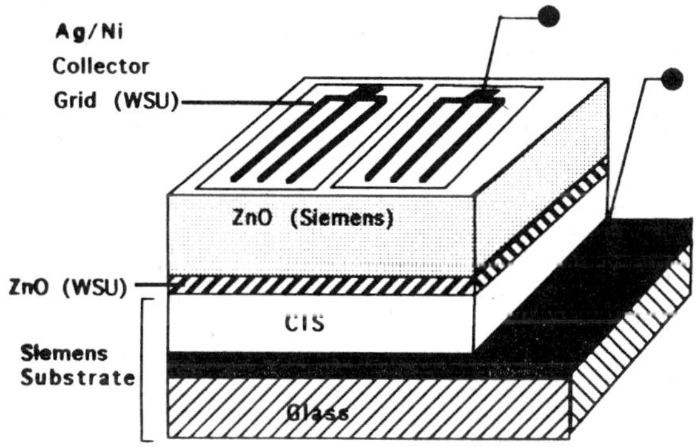

Fig. 3 PL emission for ZnSe/CIS structure taken at a low angle and at room temperature. Emission peaks indicate a segregated phase exists at the ZnSe/CIS surface.

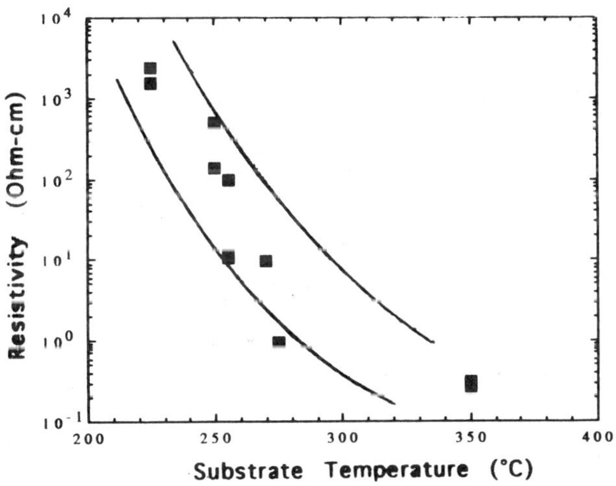

Fig. 4 Resistivity of ZnO films grown on quartz by MOCVD vs substrate temperture.

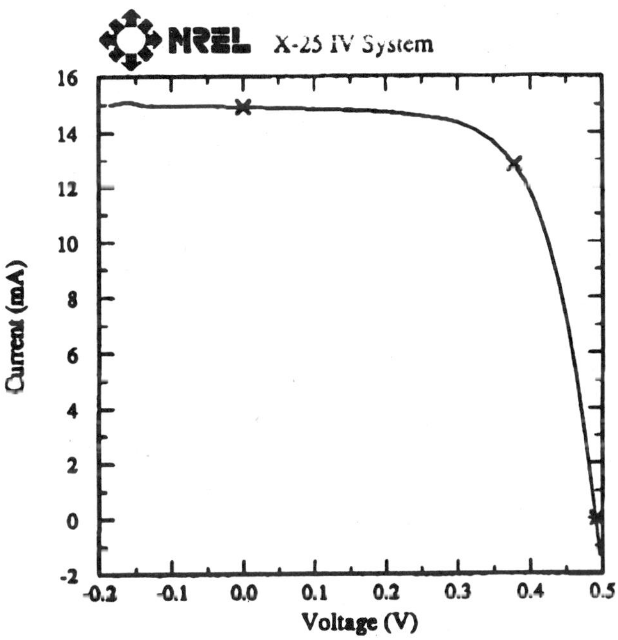

Sample 94XC016 Area = 0.435 cm^2
Isc = 14.91 mA Fill Factor = 66.0 %
Jsc = 34.26 mA/cm2 Voc = 0.492 Volts
Total Area Efficiency = 11.1 %

Fig. 5 Illuminated I-V characteristics measured by NREL.

A COMPREHENSIVE CHARACTERIZATION OF THE INTERFACES IN Mo/CIS/CdS/ZnO SOLAR CELL STRUCTURES

Dieter Schmid, Martin Ruckh, Hans Werner Schock
Institut für Physikalische Elektronik, Universität Stuttgart
Pfaffenwaldring 47, D-70569 Stuttgart, Germany

ABSTRACT

The alignment of energy bands in a Mo/CIS/CdS/ZnO solar cell structure is presented. Special attention is paid to the surface chemistry of the molybdenum coated substrate. In this study we have performed in situ analyses on polycrystalline thin films starting from the molybdenum coated soda lime glass. The different films have been deposited sequentially under ultra high vacuum conditions and analyzed with photoelectron spectroscopy techniques. We have found that Mo surface is significantly oxidized. Air annealing at 200°C leads to a diffusion of sodium through the Mo layer. Mo - O and Mo - Se compounds are formed and present during the evolution of the Mo-CIS interface. Band lineups have been determined for the CIS/CdS/ZnO interfaces of the solar cell structure. We have found no indication for band discontinuities deteriorating the solar cell device performance.

INTRODUCTION

Thin film solar cells on the basis of chalcopyrite thin films have gained increasing interest in the past few years due to the steady progress of conversion efficiencies. Some of the major improvements have been achieved by choosing the correct type of substrate as well as by depositing the thin films with apropriate techniques. In order to reach and surpass further limits it is neccessary to understand the mechanisms underlying these improvements.

One of the key parameters seems to be an effective transport of sodium through the Mo back contact material into the chalcopyrite thin film [1-3]. This can be studied by looking closely at the chemistry of the interface between Mo back contact and CIS during the growth with in situ photoemission spectroscopy.

Furthermore the electrical properties of the devices are dependent on the electronic matching of the interfaces i.e. band lineups. There exist already several data on these interface parameters. Some of them have been measured on single crystals, which have been pretreated by chemical etching and/or ion sputtering. We present data which have been collected using materials which are commonly used for state of the art solar cells, i.e. thin films. Therefore we do not deal with idealized systems but with sometimes rather complex ones. This is especially true for the Mo back contact, which turned out to be not only metallic Mo, but large amounts of oxygen are present on the surface and probably in the bulk of the Mo layers. Therefore experiments have been performed where the Mo layers have been modified concerning their oxygen content.

EXPERIMENT

Standard Mo coated float glass substrates were used for all experiments. The Mo was deposited by electron gun evaporation at substrate temperatures of about 400°C, the substrates have been air exposed for some time before further processing.

Deposition of the chalcopyrite CIS thin films and treatments of Mo coated substrates were performed in an ultra high vacuum system. The chemical state of the surfaces was investigated by X-ray photoelectron spectroscopy (XPS), valence bands and their electron densities of states were accessed by ultraviolet light photoelectron spectroscopy (UPS) without breaking the vacuum between deposition, treatment and analysis.

The ZnO base layer with a thickness of ~300 nm has been prepared by RF-sputtering from a ZnO target in a vacuum chamber not directly connected with the analysis system. After preparation, the layers have been quickly transfered under Ar atmosphere into the UHV chamber. The ZnO has been heated to 200°C to remove weakly bonded physisorbed species before depositing the CdS. This procedure has been chosen to avoid surface changes due to a sputter treatment.

Overlayers of CdS and Ge were deposited sequentially by thermal evaporation. After each deposition step the surface was analyzed with XPS and UPS. Core level binding energies and Auger structures have been examined carefully in order to detect any interface reactions between elements of the substrate and the overlayer. The positions of substrate and overlayer valence band maxima (VBMs) could be detected during the interface formation from the valence band spectra. The energetic positions of the VBMs relative to E_F are determined by the linear extrapolation of the leading edge of the valence band energy distribution curves (EDCs). The accuracy of the experimental determination of ΔE_v is in the range of ±0.1eV in our measurements.

CH3365-4/94/0000-0198 $4.00 © 1994 IEEE   First WCPEC; Dec. 5-9, 1994; Hawaii

RESULTS

Surface Analysis of the Mo-coated Substrate

The first step in the analysis of the different interfaces involved is the characterization of the surface of the molybdenum coated substrates. Two types of substrates have been used. Type A is a standard Mo coated soda lime glass which has been exposed to ambient air for several days. This type of substrate is usually used for solar cell fabrication without further treatment. Type B is a similar substrate with additional heat treatment in air at 200°C for one hour.

Spectra (1) and (3) in Fig. 1 show the Mo 3d corelines of films A and B, respectively, as measured with Mg Kα excitation. The oxidation states and chemical compound can be deduced from the spectra.

$3d_{5/2}$	Mo	MoO_2	MoO_3	$MoSe_2$
Binding Energy (eV)	227.85	229.6	232.8	228.3

Tab. 1. Core level binding energies of various Mo species listed in ref. [4]

According to Tab.1 type A surface shows metallic Mo together with a Mo - O compound. From quantification of Mo 3d and O 1s photoelectron signals we conclude that the Mo - O compound is MoO_2. The surface oxygen content is as high as 50 atomic % after removal of physisorbed contamination layers by heating to about 400°C for some minutes. Binding energy of the oxygen is close to 530.5 eV which is compatible with a Mo - O binding.

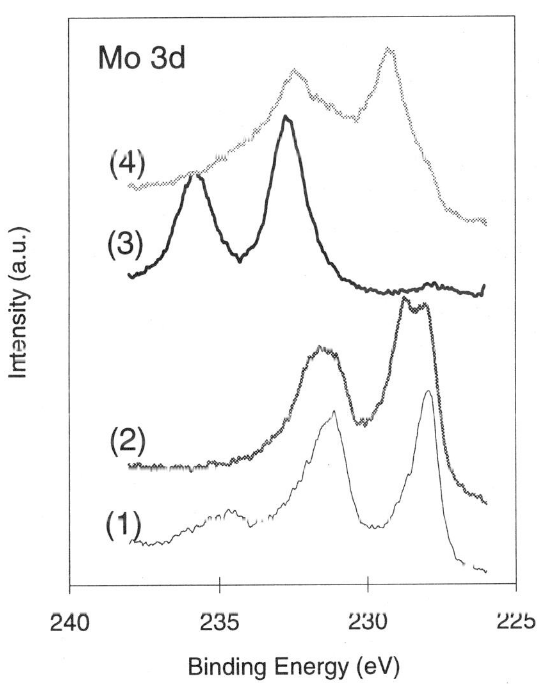

Fig. 1. XPS signals of the Mo 3d core levels of a standard air exposed Mo film (1) and after 1h air anneal at 200°C (3). Spectra (2) and (4), respectively, show the surfaces after selenization at 600°C.

Type B surface shows a totally different behavior. It shows exclusively the presence of a MoO_3 compound formed by the air anneal.

The most important finding for the different types of substrates is their sodium content. Fig. 2 shows the Na (KLL) Auger structure as measured on both types of surfaces. Only the air annealed surface shows significant amounts of Na. Bulk measurements of the Na content with wavelength dispersive X-ray spectroscopy (WDX) showed sodium only in the oxidized (Type B) substrates, in agreement with the surface results. This could be an indication that the oxidation state of the molybdenum is critical for the transport of the Na through the Mo layer. The solubility of sodium is reported for the Na_2O - MoO_3 - Na_2MoO_4 system [5].

Mo - CIS Interface

Influence of selenium vapor on the surfaces prior to the CIS deposition has been studied. Direct selenization for about five minutes at substrate temperatures of 600°C changes the chemical state of the surfaces. The metallic Mo is selenized to $MoSe_2$ at the surface of substrate A (see spectrum (2) in Fig. 1) The MoO_3 is converted to MoO_2 and $MoSe_2$ (Spectrum (4) in Fig.1). We did not observe the complete exchange of oxygen by selenium. Significant amounts of oxygen are present directly at the interface.

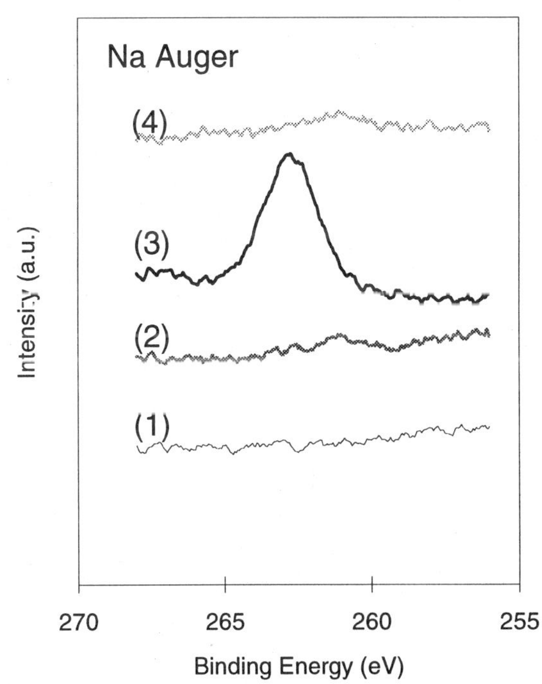

Fig. 2. The Na Auger structure observed at the surfaces illustrated in Fig 1.

From our data it is not clear, whether Mo, MoO_2, $MoSe_2$ form continuous layers or not. It is probable, that the CIS is in partial contact to all of the three different materials. Therefore, in principle, all three types of interfaces are possible.

Measurements of Valence Band Offsets

Valence and conduction band offsets are among the most important parameters which characterize a heterojunction. The valence band offsets (VBO) between two semiconductors A and B can be measured with UPS by successive deposition of semiconductor B on semiconductor A and step by step determination of the valence band maxima (VBM) of both. In many cases it is not possible to clearly distinguish between the VBM of the semiconductors A and B. In these cases it is necessary to use an indirect method to find the valence band energy maxima [6]. If the energy difference between a core level and the valence band maximum is constant during interface growth, i.e. no shifts appear due to chemical reactions, the VBM can be determined by measuring the core level energies of A and B during interface formation.

$$\Delta E_V(A-B) = (E_{cl}^B - E_V^B) - (E_{cl}^A - E_V^A) + \Delta E_{cl}(A-B)$$

The values of the conduction band offsets can be calculated if the bandgaps of the semiconductors are known.

CIS/CdS Interface

The CIS/CdS interface has been investigated in detail in ref. [7]. It has been found, that an ordered defect chalcopyrite (ODC) segregates at the surface of In-rich CIS. We determined the valence band offset at VBO=-0,85(15) eV (the negative sign indicates that the VBM of the CdS has a lower binding energy than the VBM of the ODC). With an assumed bandgap of the ODC (determined from optical absorption measurements) of E_g=1.3(1) eV the conduction band offset was calculated at +0.25(19) eV.

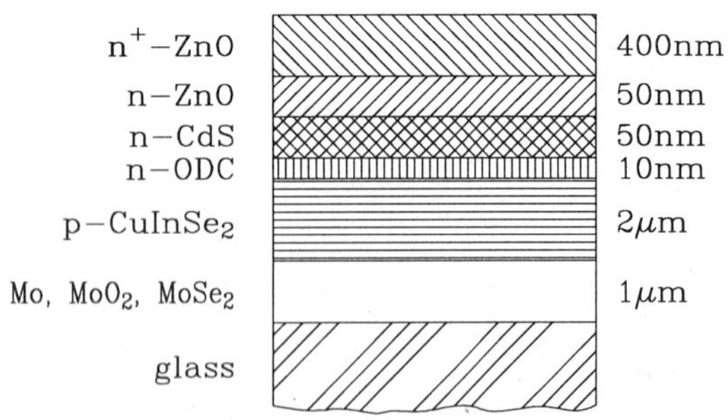

Fig. 3. Schematic representation of the layer sequence in a $CuInSe_2$ based solar cell

Neither the core level energies of Cu, In, Se, Cd or S nor the corresponding Auger parameters showed any changes during the interface growth. Therefore no chemical reactions at the interface between ODC and CdS were detected.

With the information of the chemical composition of the Mo backcontact and the ODC segregation found on top of the polycrystalline CIS we can draw the complete picture of the $CuInSe_2$based solar cell (see Fig. 3).

CdS/ZnO Interface

We evaporated CdS on sputtered ZnO substrates and investigated the evolution of the interface with XPS and UPS. Besides a small energy shift at low coverages, there is no significant change in the shapes of the photoemission lines when comparing the pure materials with the species located at the interface region. This indicates little or no chemical interaction when depositing CdS on ZnO at room temperature.

The fact that the leading edge of the energy distribution curve (EDC) of the overlayer (CdS) is closer to E_F than that of the ZnO substrate allows the direct determination of the valence band offset ΔE_v by photoemission measurements. For intermediate coverages (CdS thickness ≈10 nm) both valence band edges are visible (Fig. 5). We deduce a valence band offset ΔE_v=1.2(1) eV. Assuming values for the energy gaps E_g(CdS)= 2.4 eV and E_g(ZnO)= 3.3 eV the conduction band offset between polycrystalline ZnO and CdS is ΔE_c (ZnO / CdS) = 0.3 eV.

The determination with the indirect method leads to a valence band offset ΔE_v(ZnO / CdS) = 1.1eV. This correlation again indicates the absence of chemical reactions at the interface. The agreement in the data demonstrates the reliability of the indirect method for the system under investigation.

The results of the direct measurements are displayed in Fig. 4. For solar cell applications with an electron transport through the CdS / ZnO interface the carriers are not affected by a barrier in the conduction band.

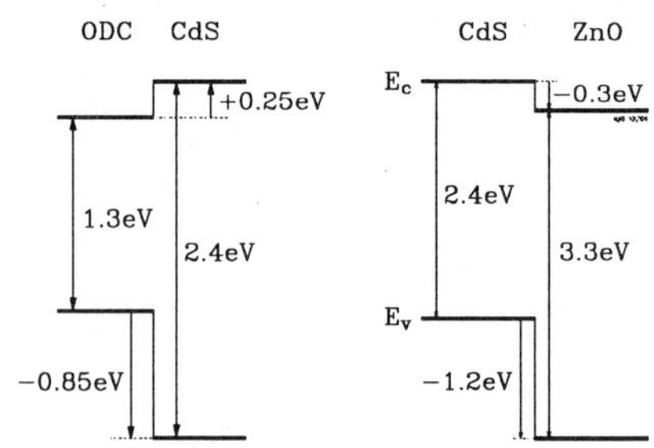

Fig. 4. Directly measured band alignment between the important interfaces of the $CuInSe_2$ based solar cell.

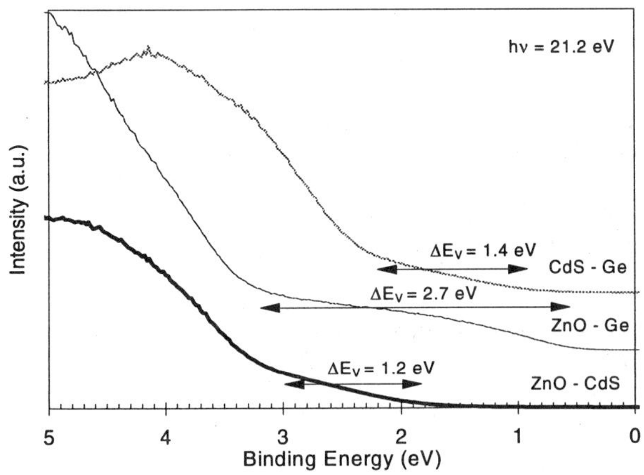

Fig. 5. Direct determination of valence band discontinuities from UPS spectra

Measurements Relative to a Common Ge Reference

In order to relate the data to a common reference level as suggested in Ref. [6] , Ge has been evaporated sequentially on top of polycrystalline ODC, ZnO and on CdS surfaces.

The EDCs around E_v are plotted in detail in Fig. 5 for intermediate coverages, where in all cases both valence band edges are visible. We deduced values ΔE_v(ZnO / Ge) = 2.7(1) eV and ΔE_v(CdS / Ge) = 1.4 (1)eV.

The data of the directly measured band discontinuities coincide with the values calculated using the Ge reference level. The ΔE_V measurements are in excellent agreement with the direct measurements (see Fig. 4) and differ by less than 150 meV which have to be considered.

With the additional information about the band discontinuity of the x-tal-$CuInSe_2$/Ge interface [8] it is possible to draw a picture of the band lineup from the $CuInSe_2$/ODC/CdS/ZnO solar cell (Fig. 6).

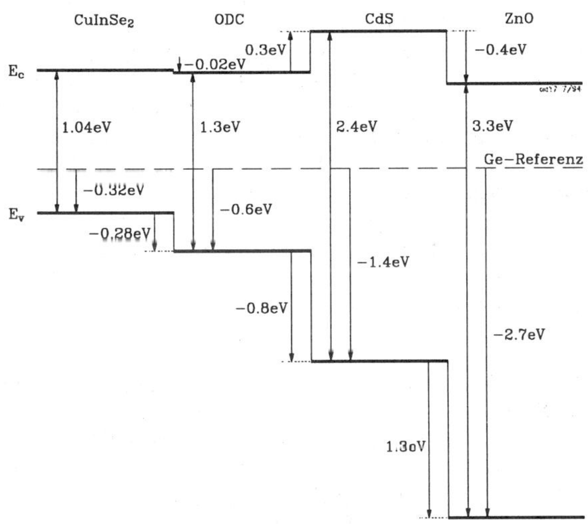

Fig. 6. Schematic representation of the band alignment in the $CuInSe_2$ based solar cell. The valence band discontinuities were measured relative to the valence band maximum of Ge.

As demonstrated by Turner et al. [9] there are some critical points if we want to reach good transport properties in the solar cells.

- a cliff in the conduction band (negative sign of ΔE_c) has to be avoided at interfaces with high interface recombination.
- in the case of a conduction band spike the barrier should be smaller than 400 meV to avoid losses in fill factor and short circuit current.

This is both true for all interfaces under investigation. Therefore we have no indication for band discontinuities affecting the solar cell performance.

CONCLUSIONS

We have investigated all interfaces of the Mo/CIS/CdS/ZnO solar cell in view of chemical reactions at the interfaces and the electrical band alignment. Mo-oxides and $MoSe_2$ may significantly contribute to the contact properties of the interface.

The absorber-window system CIS/CdS/ZnO seems to be sufficiently well aligned for optimal operation of solar cells.

ACKNOWLEDGMENTS

This work has been supported by the Bundesministerium für Forschung und Technologie (BMFT) contract 032 8059E and by the CEC contract No. JOU2-CT92-0141.

[1] J. Hedström, H. Ohlsén, M. Bodegård, A. Kylner, L. Stolt, D. Hariskos, M. Ruckh, H.W. Schock, *Proc. 23rd IEEE Photov. Spec. Conf.* (1993) 364

[2] M. Bodegård, L. Stolt, J. Hedström, *Proc. 12th E.C. PVSEC* (1994) 1743

[3] J. Holz, F. Karg, H. von Philipsborn, *Proc. 12th E.C. PVSEC* (1994) 1593

[4] D. Briggs und M.P. Seah (Hrsg.), *Practical Surface Analysis*, 2nd ed. Vol. 1; *Auger and X-ray Photoelectron spectroscopy* (Wiley, Chichester, 1990).

[5] D.H. Killeffer, A. Linz *in Molybdenum compounds, their chemistry and technology*, Interscience Publishers, New York (1952) pp. 79

[6] A.D. Katnani, G. Margaritondo, *Phys. Rev. B* **28**, 1944 (1983).

[7] D.Schmid, M.Ruckh, F. Grunwald, H.W. Schock, *J. Appl. Phys.* **73** (1993) 2902

[8] M. Turowski, G. Margaritondo, M.K. Kelly, R.D. Tomlinson, *Phys. Rev. B* **31**, (1985) 1022

[9] G.B. Turner, R.J. Schwartz und J.L.Gray, *Proc. 20th IEEE PVSC*, 1457 (IEEE, New York, 1988).

PREPARATION OF $CuInSe_2$ THIN FILMS USING ELECTRODEPOSITED In/Cu METALLIC LAYER.

Yi Hwan Kim, Yong Eui Lee, Hyeong Joon Kim and Jinsoo Song*

Department of Inorganic Materials Engineering, Seoul National University, Seoul 151-742, Korea
* Photovoltaic Research group, Korea Institute of Energy Research, P. O. Box 5,
Taedok Science Town, Taejon, 305-343, Korea

ABSTRACT

$CuInSe_2$ thin films were prepared by two-stage process: electrodeposition of In/Cu metallic layer and selenization. Effects of current density in In electrodeposition on properties of In/Cu metallic layer and those of $CuInSe_2$ thin films were investigated. For lower current densities, In/Cu metallic layer had rough surface with formation of droplets, but smooth surface appeared for higher current densities. And the compositional nonuniformity became severe for lower current densities, and the amount of loss of indium during the selenization decreased with increasing of current density. It was found that the compositional nonuniformity in the In/Cu metallic layer was caused by formation of droplets and was dependent on the composition nonuniformity in In/Cu metallic layers. Thus, in case of the highest current density of 80 mA/cm^2, homogeneous $CuInSe_2$ thin films without second phases was prepared. And (112) preferred orientation of $CuInSe_2$ thin films was influenced by its composition.

INTRODUCTION

$CuInSe_2$ semiconductor has attracted particular interest as an absorber for high efficiency photovoltaic devices because it possesses the following useful properties: (i) it has a direct bandgap of about 1.02 eV at room temperature, which is in the energy range for optimum solar conversion; (ii) its optical absorption coefficient, which is about 5 x 10^5 cm^{-1} near the bandgap for polycrystalline thin films, is the highest reported in the literature for any semiconductors to date; (iii) it has high quantum efficiency and good thermal stability.[1] While $CuInSe_2$ thin films have been commonly prepared by various deposition processes, the two-stage process is one of the most promising approaches to make these films with low cost and in large areas. It includes a deposition of the Cu-In metallic layer followed by a selenization.

However, a significant problem in such a selenization process was the difficulty in controlling of composition of $CuInSe_2$ thin films which was caused by the presence of a second phase, such as $Cu_{2-x}Se$, on the surface of thin films due to the reevaporation of volatile In_2Se.[2]

To overcome this problem, various approaches are being carried on. They can be classified into two categories. One is development of new selenization technique through optimizing processing parameters, such as heating history. The other is to optimize properties of Cu-In metallic layer by understanding the effects of its morphological and compositional properties on the formation of $CuInSe_2$ thin films. Nowadays, It has been known that the properties of Cu-In metallic layers are crucial for the $CuInSe_2$ compound and consequently for the final performance of the device.[3] But a clear correlation between the properties of the Cu-In metallic layers and those of the ternary compound has not well established yet.

In present study, $CuInSe_2$ thin films were prepared by two-stage process; the metallic layers were deposited by electrodeposition and then selenization of the layers was carried out in resistive heated two- zone tube furnace, using Se vapor as a source of Se. The aim of the present study was to investigate the properties of In/Cu metallic layers with various current densities and to establish correlation between properties of In/Cu metallic layers and those of $CuInSe_2$ thin films.

EXPERIMENTAL PROCEDURE

Preparation of In/Cu metallic layers.

Cu and In films were sequentially electrodeposited on the Mo-coated alumina substrates. Mo layer was deposited using electron beam evaporation. Cu films were electrodeposited using a bath of acidic copper sulfate (0.5 M $CuSO_4 \cdot 5H_2O$+2 M H_2SO_4) at constant current density of 40 mA/cm^2. And then In films were electrodeposited using a bath of acidic indium sulfamate ($In(SO_3NH_2)_3$) at various current densities from 1 mA/cm^2 to 80 mA/cm^2. In both cases, the baths were held at room temperature and the electrodepositions was carried out with stirring.

CH3365-4/94/0000-0202 $4.00 © 1994 IEEE

First WCPEC; Dec. 5-9, 1994; Hawaii

From Fig. 3, droplets were identified as an indium-rich alloy by EDAX measurements, as other researchers have reported.[3, 5] The composition of droplet was not changed with current density. However, the base region between droplets was Cu-rich and its composition was drastically changed with current density. But the composition in the overall area was near stoichiometric composition Cu/In ~ 1 regardless of varing current densities. Lateral compositional nonuniformity appeared in the In/Cu metallic layers electrodeposited in lower current densities, and it decreased with increasing of current density. Thus, In/Cu metallic layer with uniform composition was obtained in the highest current density of 80 mA/cm^2.

Annealing of In/Cu metallic layers.

Fig. 1(b) shows the XRD pattern of the In/Cu metallic layer annealed at 150℃ for 60 min in N_2 atmosphere. CuIn and $CuIn_2$ phases disappeared after annealing, but In, $Cu_{11}In_9$ and Cu_7In_4 phases appeared. However, the changes of surface morphology and lateral compositional nonuniformity, which were observed in as-deposited films, were not observed in the annealed films. Although annealing time decreased to 10 min and/or annealing temperature increased to 400℃, the

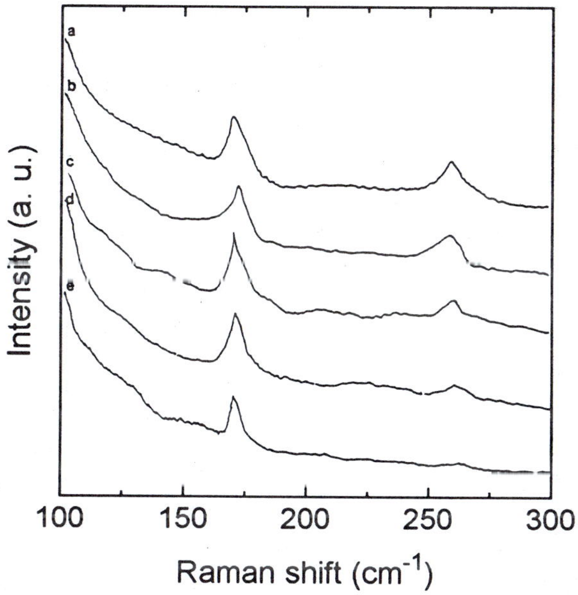

Fig. 4. Raman spectroscopy of $CuInSe_2$ thin films with various current densities in In electrodeposition: (a) 1 mA/cm^2, (b) 5 mA/cm^2, (c) 10 mA/cm^2, (d) 40 mA/cm^2 and (e) 80 mA/cm^2.

same behavior occurred. Therefore, it was concluded that lateral compositional nonuniformities of In/Cu metallic layer was maintained through the whole selenization process.

Properties of $CuInSe_2$ thin films.

XRD analysis showed that $CuInSe_2$ thin films had chalcopyrite structure regardless of varing current densities in In electrodeposition, but appearance of second phases was not clearly identified because interplanar spacings of second phase, $Cu_{2-x}Se$, was similar to those of $CuInSe_2$.

The appearance of second phases can be easily identified by Raman spectroscopy.[6] As shown in Fig. 4, intensity of second phase peak at 260 cm^{-1} decreased with increasing current density and only $CuInSe_2$ peak appeared at 173 cm^{-1} in the highest current density of 80 mA/cm^2. Fig. 5 shows Cu/In composition ratio in $CuInSe_2$ thin films with various current densities. As current density increased, Cu/In composition ratio decreased and deviation of composition also decreased in $CuInSe_2$ thin films. Such a result was consistent with that of Fig. 4. From the results of Fig. 4 and 5, it was concluded that Cu/In composition ratio was larger than 1, that is, appearance of second phases of Cu-Se, and compositional nonuniformity of $CuInSe_2$ thin films were caused by compositional nonuniformity of In/Cu metallic layers.

On the other hand, it is known that the presence of $Cu_{2-x}Se$ may have the influence on the preferred orientation and grain size of $CuInSe_2$ aggregates.[7] The variation of surface morphology with current density can be observed in Fig. 6. Large crystallites surrounded with nanocrystallites were observed on the surface. Large crystallites were identified as Cu-rich $CuInSe_2$ or Cu-Se second phases, while nanocrystallites were as In-rich $CuInSe_2$ by EDAX measurements. And the size difference between these two types of crystallites increased with compositional nonuniformity. Thus, in case of the highest current density of 80 mA/cm^2, $CuInSe_2$ thin film which was identified as slightly In-rich one showed uniform, smooth surface morpology. And degree of (112) preferred orientation was

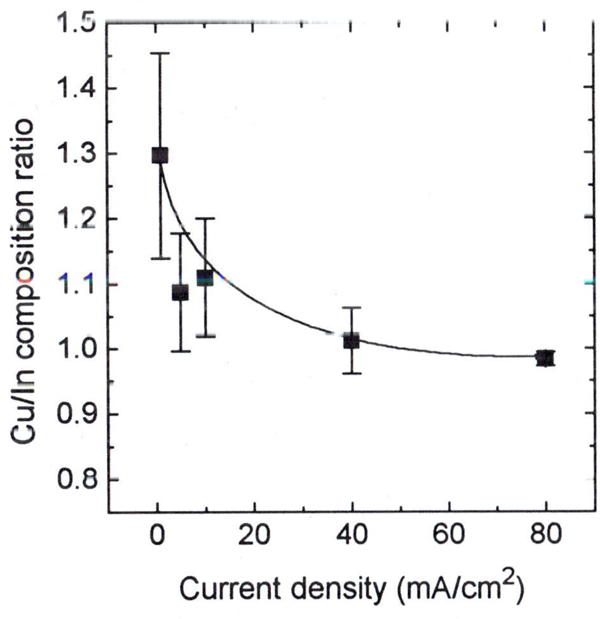

Fig. 5. Cu/In composition raio of $CuInSe_2$ thin films as a function of current density in In electrodeposition.

Selenization of Cu-In metallic layers.

The metallic layers were selenized by two-step chemical-vapor-transport process, a elemental Se source material (Se powder 99.99 %) was heated to 350℃. Ar was used as a carrier gas at flow rate of 1500 sccm. The sample was heated with two-step temperature profile; first heated at 150℃, just below the melting point of In, for 10 min. and then further heated at 400℃ for 60 min. Such a process was developed to restrict reevaporation of volatile In_2Se in the previous study.[4]

RESULTS AND DISCUSSION

Properties of electrodeposited In/Cu metallic layers

It is known that current density in galvanostatic electrodeposition determines surface morphology and structure of deposited films.

In order to determine the effects of current density on the phases of In/Cu metallic layers XRD measurements were performed. The results are shown in Fig. 1(a), where it can be seen not elemental In peak but peaks of CuIn and $CuIn_2$ phases. Such a behavior was observed regardless of varing current densities.

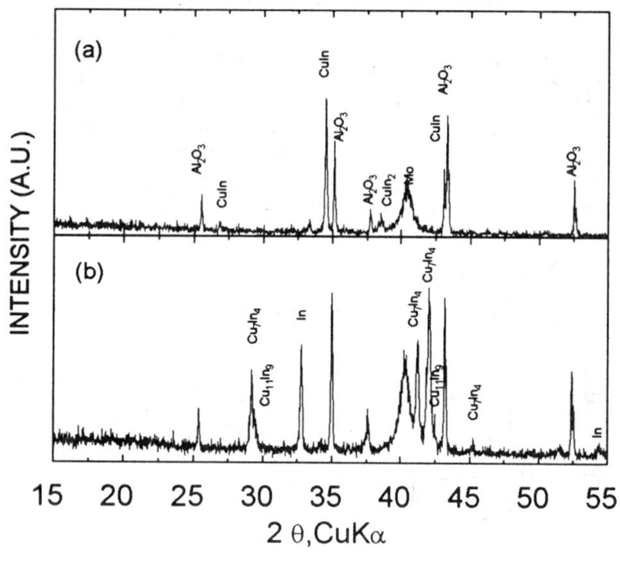

Fig. 1. XRD patterns of In/Cu metallic layer: (a) as-deposited and (b) annealed at 150℃ for 1 hr in N_2 atmosphere.

However, surface morphology was changed with current density. As shown in Fig. 2, In/Cu metallic layers deposited in lower current densities (1, 5, 10 mA/cm^2) contained clearly defined droplets. As the current density increased, the size of droplets became smaller but the number of droplets increased. Therefore, droplets could not be distinguished each other on the surface in the highest current density of 80 mA/cm^2. Such a behavior can be explained as follows; since growth rate is relatively larger than nucleation rate in lower

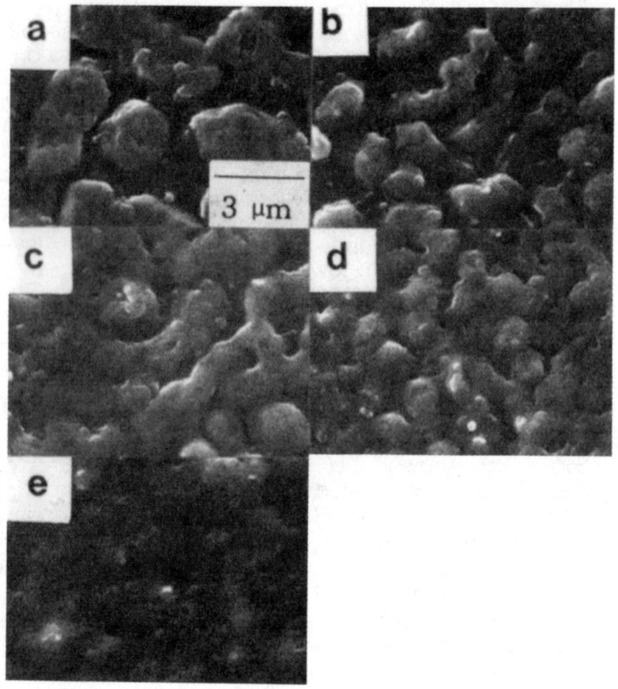

Fig. 2. SEM micrographs of In/Cu metallic layers with various current densities in In electrodeposition: (a) 1 mA/cm^2, (b) 5 mA/cm^2, (c) 10 mA/cm^2, (d) 40 mA/cm^2 and (e) 80 mA/cm^2.

current densities, the number of nucleation became less but vertical growth which resulted from lack of indium wetting, was dominated. As current density increased, the number of nucleation increased and droplets were easily coalesced to make smoother surface.

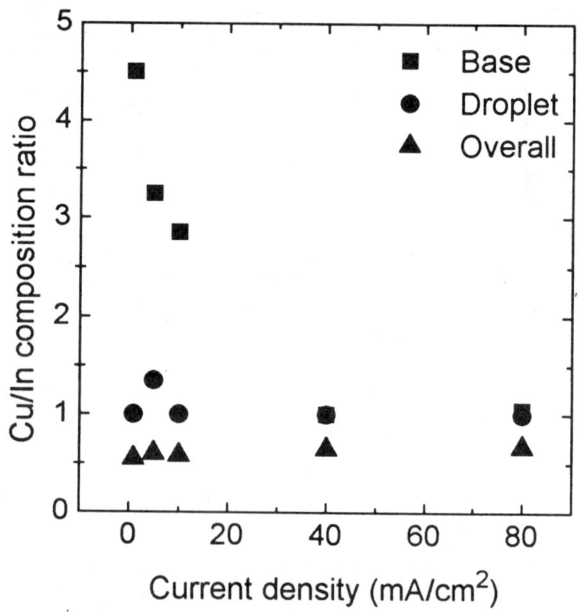

Fig. 3. Cu/In composition ratio of In/Cu metallic layers with various current densities in In electrodeposition.

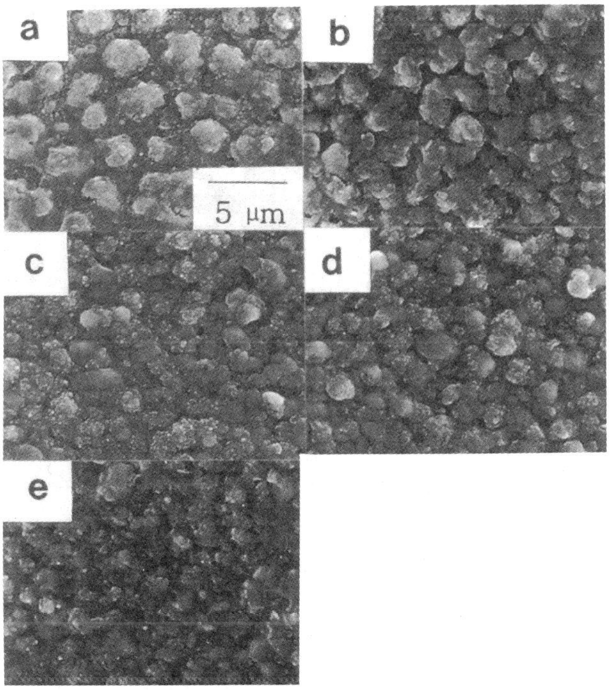

Fig. 6. SEM micrographs of $CuInSe_2$ thin films with various current densities in In electrodeposition: (a) 1 mA/cm^2, (b) 5 mA/cm^2, (c) 10 mA/cm^2, (d) 40 mA/cm^2 and (e) 80 mA/cm^2.

determined by XRD measurement. Fig. 7 shows that as current densityincreased, degree of (112) preferred orientation decreased. Such a behavior was due to variation of Cu/In compositional ratio with current densities. It was meant that Cu/In composition ratio was dependant on (112) preferred orientation of $CuInSe_2$ thin films.

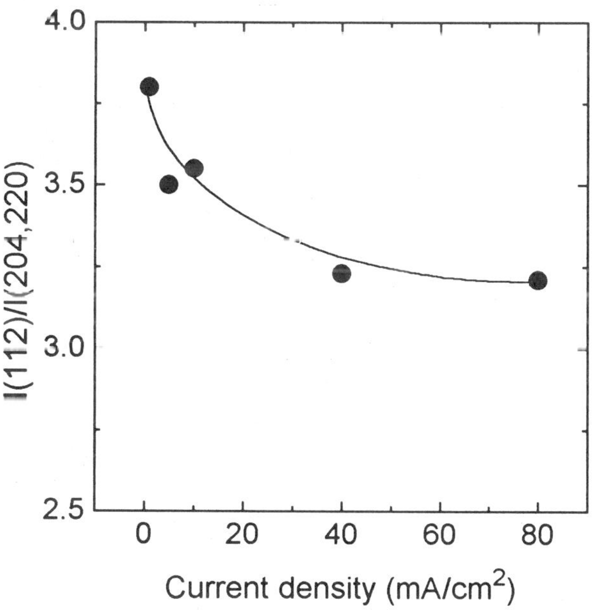

Fig. 7. The intensity ratio of (112)/(204) peak of $CuInSe_2$ thin films as a function of current density in In electrodeposition.

CONCLUSIONS

$CuInSe_2$ thin films were prepared by two-stage process. When In/Cu metallic layer was electrodeposited, morphology and surface composition were dependent on the current density in In electrodeposition. In case of lower current densities, In/Cu metallic layer was characterized by droplets formation, but smoother surface was observed for higher current densities. The compositional nonuniformity in the In/Cu metallic layer was caused by droplet formation. Therefore, as current density increased, the composition in the In/Cu metallic layer became more uniform. The amounts of loss of In during selenization process were determined by the compositional and morphological nonuniformity in the In/Cu metallic layer. Thus, the higher current density was used in In electrodeposition, the more homogeneous $CuInSe_2$ thin films can be prepared. In addition, dependence of Cu/In composition on the degree of (112) preferred orientation of $CuInSe_2$ thin films can be observed.

ACKNOWLEDGEMENTS

The authors acknowledge the financial support of this work by Photovoltaic Research Group, the Korea Institute of Energy Research,

REFERENCES

[1] K. L. Chopra and S. R. Das, "*Thin Film Solar Cells*", Plenum Press, New York, 1983, pp. 417-422.

[2] J. Kessler et al., "Low pressure vapor phase selenization of Cu-In films without H_2Se", *10th European photovoltaic solar energy conference*, 1991, pp. 879-882.

[3] A. Parretta, et. al., "Properties of sputtered metallic alloys for $CuInSe_2$ based solar cells", *11th European photovoltaic solar energy conference*, 1992, pp. 850-853.

[4] Y. E. Lee, et. al., "Preparation of homogenous polycrystalline $CuInSe_2$ thin films by a two-step chemical-vapor-transport process", *J. Electrochem. Soc.*, Vol. **141**, No. 2, 1994, pp. 558-561.

[5]. D. S. Albin, et. al., "The phase behavior of evaporated copper and indium precursors for selenization", *Solar Cells*, **30**, 1991, pp. 41-46.

[6] S. Yamanaka, et. al., "Study of $CuInSe_2$ formation kinetics in the selenization process by Raman spectroscopy", *Jpn. J. Appl. Phys.*, **30**, 3, 1991, pp. 442-446.

[7] J. R. Tuttle, D. S. Albin and R. Noufi, "Thoughts on the microstructure of polycrystalline thin film $CuInSe_2$ and its impact on material and device performance", *Solar Cells*, **30**, 1991, pp. 21-38.

PROGRESS IN $Cu(In,Ga)Se_2$ SOLAR CELLS AND SUBMODULES

J. Kessler, S. Wiedeman, L. Russell, T. Lommasson, S. Skibo, J. Fogleboch, T. Kloss, R. R. Arya
Solarex Corporation, Thin Film Division
826 Newtown-Yardley Road
Newtown, PA 18940 USA

ABSTRACT

$Cu(In,Ga)Se_2$ based solar cells with total area efficiencies of 14.4% have been achieved using processes that exhibit very promising uniformity over areas above 500 cm^2. In order to develop interconnect schemes and optimize various parameters for large area modules, 40 cm^2 submodules with 10 segments are used. Progress in the interconnect formation and the ZnO characteristics are discussed. Despite a persisting difference between the fill factors of typical grided devices and of typical interconnected devices, efficiencies of 9.55% and 11.6% have been achieved on 35.38 cm^2 (10 segment) and 14.46 cm^2 (6 segment) submodules respectively.

INTRODUCTION

Various $Cu(In,Ga)Se_2$ (CIGS) deposition processes are investigated, keeping in mind their potential for large area development. In this work the "proof of concept" for a CIGS deposition process is demonstrated not only by high efficiency device results but also by matrices of such results over reasonable areas. Once these device arrays show sufficient spatial uniformity of the CIGS quality, submodules are produced. Devices and modules have different demands and restrictions and it is necessary to evaluate the specifics of each deposition process for its use in module production. To date our best cell shows a total area efficiency of 14.4% and our best submodule shows an aperture area efficiency of 11.6%. The difference between these numbers is suspected to be mainly due to the quality of the interconnects.

Within the large spectrum of processes some clarity may be gained by contrasting them on the basis of: sequential versus simultaneous deposition; the use of evaporation versus sputtering; and the use of elemental versus multinary sources. Until recently the most successful devices were produced by simultaneous evaporation of elemental sources [1],[2]. When large area deposition is considered many efforts have concentrated on sequential elemental sputtering, at least for the metals. The selenium has often been sequentially delivered by elemental evaporation or in the form of a gaseous compound such as H_2Se [3],[4]. Sequential evaporation, either of elemental or binary sources has recently shown very promising small area results [5],[6] and has potential for larger area production.

In this paper we also address the question of module structures. Interconnection schemes as well as different layer optimizations are studied. The minimization of the differences between average cell results and interconnected cell results (i.e. the submodules) is the objective. In our first submodules the most penalized parameter was the fill factor. Therefore, a large part of this present work has been focused on obtaining fill factors (without significant V_{oc} losses) for submodules that are comparable to those obtained for devices.

ABSORBER PROCESSING

At Solarex we continue to pursue several film formation processes ranging from sequential sputtering to elemental co-evaporation, with and without the addition of Gallium [7]. Cross-sectional SEM micrographs of sequentially sputtered and of sequentially evaporated absorber layers are shown in Fig. 1. In both cases dense and relatively smooth films are found.

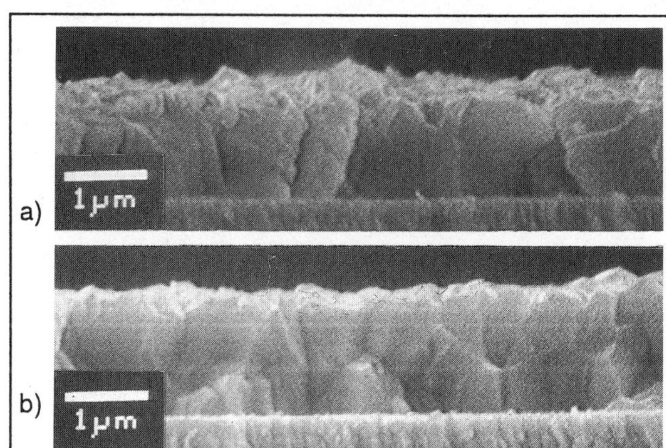

Fig. 1. Cross-Section SEM Micrographs of :
a) CIGS formed by sequential sputtering
b) CIGS formed by sequential evaporation

Our baseline remains the potential for "scale-up" as we only value processes that we will be able to perform at the sq. ft. (or larger) area level. Nevertheless, at this stage of our work, our preferred substrate size is 3"x3" (7.6cmx7.6cm) for reasons of greater experimental flexibility. In order to test larger area capabilities we simply use a number of 3"x3" substrates (covering an area over 500 cm²) in a single deposition. On each substrate we can use a matrix of 84 (6 rows of 14) small area (~0.4 cm²) cells in order to identify gradients in the photovoltaic parameters (V_{oc}, J_{sc}, and FF) as well as to localize "defects" that are not part of the continuous spatial trend.

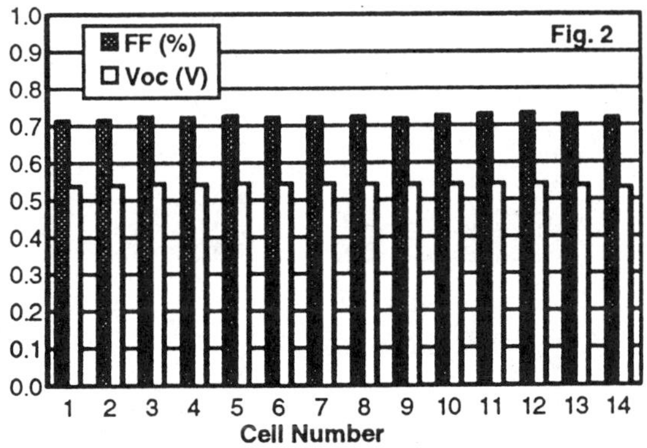

Fig. 2. V_{oc} and FF uniformity across the 14 cells (each 4.75 mm wide) of a 3" substrate.

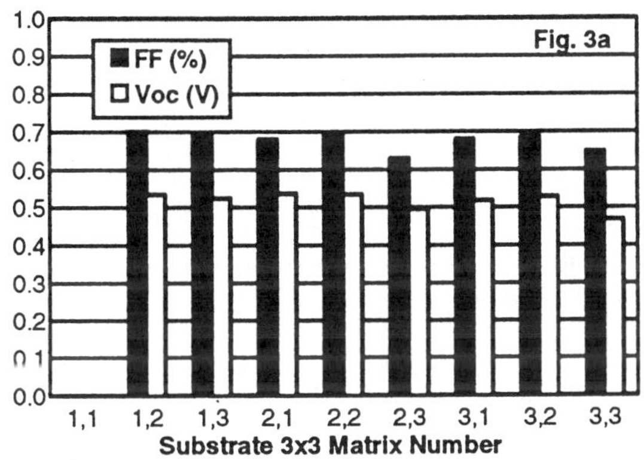

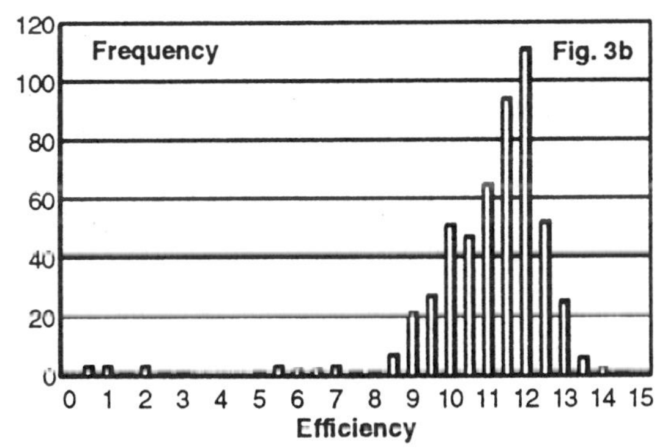

Fig. 3. Eight 3"x3" substrates from a single deposition
a) average V_{oc} and FF uniformity
b) total area efficiency distribution

We find that across most 3"x3" substrates, with the exception of a few random shunted cells, the photovoltaic parameters show only minor variations. An example is given in Fig. 2 where the V_{oc} and FF uniformity across a row of 14 cells of a 3" substrate is shown, the currents densities of these are all close to 31 mA/cm², giving total area efficiencies around 12%. In each CIGS deposition nine 3"x3" substrates are arranged in a 3x3 matrix and are numbered using the standard matrix notation. The uniformity across this array is shown in Fig. 3a (eight substrates only : one was lost in the processing) where the average V_{oc} and FF for each substrate is given as a function of its position. These averages are based on about 90% of the cell population, the 10% loss being due to imperfect handling or being shorted cells that exhibited V_{oc} values less than 100 mV. In Fig. 3b is given the total spread of the efficiencies for the population of cells used in Fig. 3a.

Shown in Fig. 4 is our best single cell result to date. It is for a Ga containing film followed by a thin CdS layer and a RF sputtered ZnO bilayer. The CdS is deposited by Chemical Bath Deposition (CBD) as has been reported previously [8], and the ZnO bilayer is the standard high resistive / low resistive structure sputtered from a pure ZnO and a ZnO-(2%)Al_2O_3 target. For these devices the window layer is usually comprised of 500 to 1000 Å of CdS, of 400 to 600 Å of undoped ZnO and of 3500 to 5500 Å of doped ZnO. Such ZnO films have sheet resistivity values in the order of 10 to 12 Ω/□.

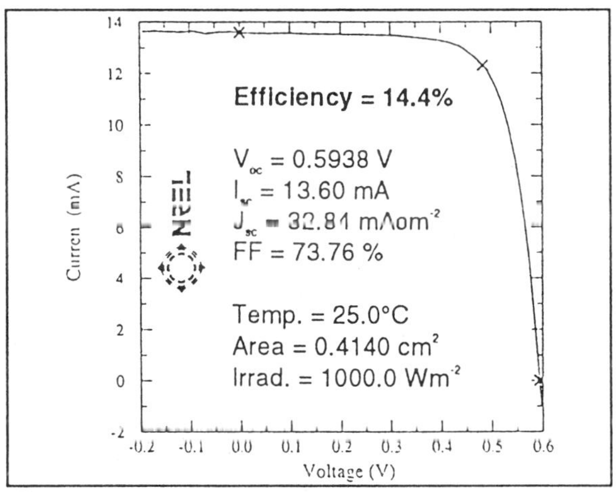

Fig. 4. Current-voltage characteristic of a 14.4% total area efficiency, Ga containing grided device.

MODULE PROCESSING

The most promising processes (efficiencies and uniformity) are used to produce submodules. As in the device work, nine 3"x3" substrates (each being a submodule) can be produced in a single deposition. The standard submodule substrate consists of 12 Mo pads 1/4" wide and 2.5" long. The two end pads are only the "top" and "bottom" contacts and thus this layout results in a ~40 cm² aperture area 10 segment submodule. The ratio of active area to aperture area depends on the interconnect scheme but is usually close to 95%. In Fig. 5 is shown the classic three scribe structure consisting of a Mo scribe, a CIGS contact scribe and a ZnO/CIGS isolation scribe. As can be seen our relative position control of each scribe is sufficient to assure that no "dead" CIGS areas are created.

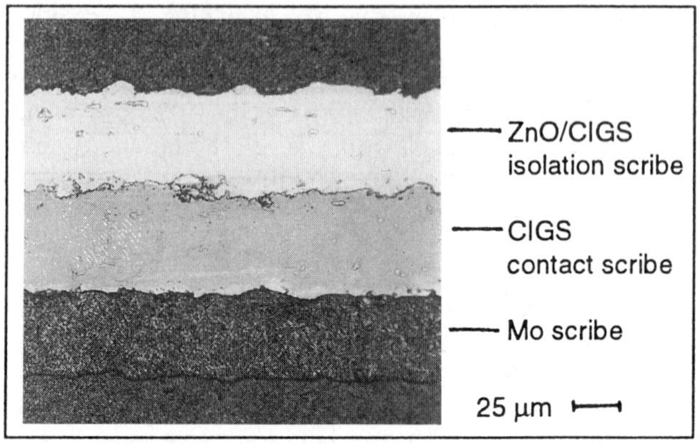

Fig. 5. Micrograph of the "classic" three scribe structure.

The question of the optimal width of each scribe is investigated. Such phenomena as the influence of the Mo scribe on the CIGS growth or the optimal contact area for the ZnO/Mo interconnect must be considered. It is also reasonable to question the quality of the ZnO/Mo contact as it is believed that scribing CIGS from the Mo (i.e. the contact scribe) does not reveal a "clean" Mo interface but more probably a $MoSe_2$ polluted surface [9]. Additionally, one must assure continuity of the ZnO across the "steps" formed by the Mo scribe and the CIGS contact scribe.

Considering the grided device measurements as reference values, our efforts in submodule production concentrate on achieving comparable parameters. Initially, the largest problems were found in the fill factors. We are now commonly measuring fill factors in the order of 65% for interconnected structures, though it is clear that the submodules still suffer from excessive series resistance. The areas of potential improvement are, the back Mo contact, the sheet resistance of the ZnO and the interconnection scheme. Using thicker ZnO layers (in the order of 1 µm) than for devices has helped with the "step coverage" situation, i.e. the continuity of the ZnO across the Mo scribe, and thus has helped with the fill factors. Maintaining ZnO sheet resistivity between 5 and 10 Ω/□ has also been beneficial, the difficulty being in the conservation of the optical transmission of these films. This has required a novel optimization of the process parameters in order to find a ratio of conductivity to transparency similar to that obtained with the thin ZnO films. The cross-sectional SEM micrographs and EDAX mappings in Fig. 6 show the behavior of the various layers on both sides of the Mo scribe. It can be seen that the ZnO layer appears to be continuous.

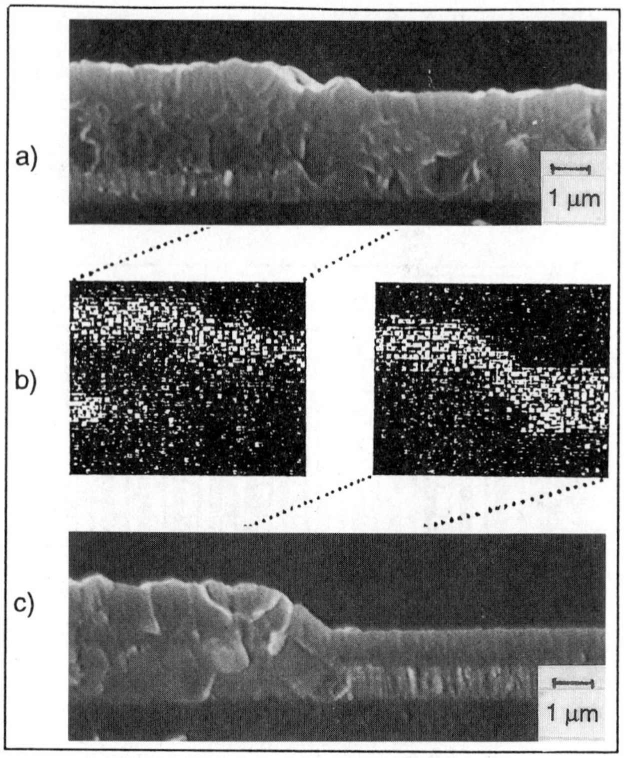

Fig. 6. The Mo scribe "step coverage" situation
a) and c) SEM micrographs
b) EDAX mapping : Zn and Mo signals

We are also actively pursuing alternative methods for the ZnO deposition since higher deposition rates and lower process costs are more important as the films become thicker. A "novel" process has been defined which can rapidly deposit 1 to 2 µm thick films of both high conductivity and high transparency at lower production costs. First devices and submodules using these ZnO layers appear to be of equivalent quality to those produced using the RF sputtered ceramic targets. Typical transmission spectra of ZnO films produced from the two processes are shown in Fig. 7.

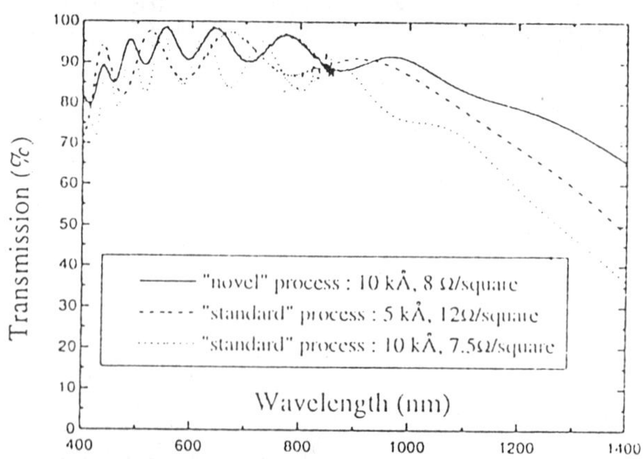

Fig. 7. Spectral transmission characteristics of different ZnO layers from different processes.

The best submodule I-V characteristics are shown in Fig. 8 for a 35.38 cm², 10 segment structure (NREL measurement : 9.55% aperture area efficiency) and in Fig. 9 for a 14.46 cm², 6 segment structure (Solarex measurement : 11.6% aperture area efficiency). These two submodules are from the same CIGS and window depositions and differ only in their interconnect quality. The current densities of both are similar and are comparable to the best device values (respectively 31.7 and 32.3 mA/cm² per segment versus 32.8 mA/cm²). On the other hand, both the V_{oc} per segment and the fill factors improve considerably as we progress in our interconnect quality. Nevertheless, relative to the average device values, there is need for further improvement.

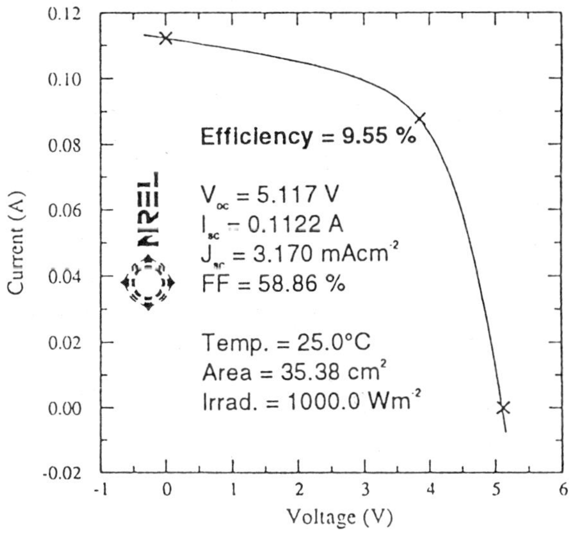

Fig. 8. I-V characteristics for a 35.38 cm² submodule.

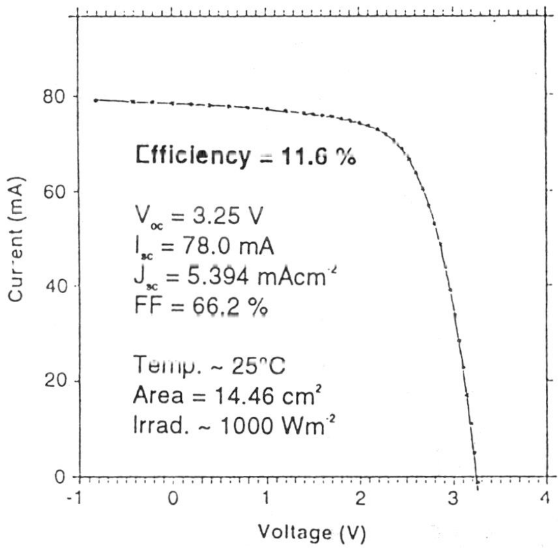

Fig. 9. I-V characteristics for a 14.46 cm² submodule with improved interconnect quality.

CONCLUSIONS

At Solarex the $Cu(In,Ga)Se_2$ deposition processes, as well as the other steps in device formation, have shown excellent results. Small devices exhibiting total area efficiencies in the range of 10 to 14.4% are found across large areas (500 cm²) from single CIGS depositions. Although our interconnected submodule results are still considerably behind their potential, we have processed small submodules with aperture area efficiencies up to 11.6%. Although our cell characteristics continue to improve, we expect that the efficiency difference between cells and submodules will decrease. Novel current collection schemes, more transparent window layers and smaller aperture to active area ratios are also under investigation.

In the development of CIGS based modules, cost and manufacturability are as important as conversion efficiency. The scale-up from 3"x3" to 12"x13" should be limited only by technological considerations.

ACKNOWLEDGMENTS

The authors would like to thank NREL for its help through the collaboration with A. Gabor, M. Contreras and R. Noufi. This work is partially funded by NREL under contract # ZN-1-19019-4.

REFERENCES

[1] W.S. Chen, J.M. Stewart, W.E. Devaney, R.A. Mickelsen and B.J. Stanbery Proc. of the *23rd IEEE PVSC*, 1993, p. 422

[2] M. Bodegård, L. Stolt and J. Hedstrom Statement at the *12th EC PVSEC*, 1994, Oral O10.2

[3] R.R. Arya, T. Lommasson, S. Wiedeman, L. Russell and A. Catalano Proc. of the *11th EC PVSEC*, 1992, p. 112

[4] V.K. Kapur, B.M. Basol, A. Halani, C.R. Leidholm and A. Minnick Proc. of the *12th EC PVSEC*, 1994, p. 1608

[5] J. Kessler, D. Schmid, S. Zweigart, H. Dittrich and H.W. Schock Proc. of the *12th EC PVSEC*, 1994, p. 648

[6] A.M. Gabor, J.R. Tuttle, D.S. Albin, M.A. Contreras, R. Noufi and A.M. Hermann *Appl. Phys. Lett.* **65**(2), 11 July 1994 p. 198

[7] T.C. Lommasson, S. Wiedeman, L. Russell, J. Kessler and R.R. Arya Proc. of the *12th EC PVSEC*, 1994, p. 1226

[8] L. Russell, B. Fieselmann and R.R. Arya Proc. of the *23rd IEEE PVSC*, 1993, p. 581

[9] S. Benhida, M. Morsli and J. C. Bernède Proc. of the *3 Jours de la Mat. Cond.*, 1992, p. 385

PHOTOVOLTAIC CELLS WITH EFFICIENCY EXCEEDING 10 % USING MONOCRYSTALLINE $CuInSe_2$ SUBSTRATES

L.S. Yip and I. Shih
Electrical Eng. Dept., McGill University
3480 University St., Montreal, PQ, Canada, H3A 2A7
Tel: (514)398-7121, Fax: (514)398-4470.

ABSTRACT

Photovoltaic cells with the structure of $ZnO/CdS/CuInSe_2$ have been fabricated on single crystal $CuInSe_2$ substrates obtained from p-type Bridgman-grown ingots. A chemical bath method was used to deposit the CdS film onto the abrasively-polished $CuInSe_2$ substrates while r.f. sputtering was employed for the ZnO deposition. It was found that the performance of these cells were dominated by the high density of surface defects, which could be reduced by a substrate annealing prior to the CdS deposition. A conversion efficiency of 10.3 % at about AM 1.5 under the sun has been achieved on cells with an effective area of 8 mm^2. Anti-reflection coating was not used in the present case.

INTRODUCTION

Since the pioneering work of Shay, Wagner and Kasper [1] in which a 12 % $CdS/CuInSe_2$ photovoltaic cell with an effective area of 0.8 mm^2 was fabricated using a single crystal $CuInSe_2$ substrate with a cleaved (112) surface, $CuInSe_2$-based photovoltaic cells have received considerable research interest. Although a substantial progress has been achieved on cells based on thin-film $CuInSe_2$, few work has been reported on cells fabricated on monocrystalline $CuInSe_2$ substrates and the 12 % cell has remained the best cell fabricated on single crystal $CuInSe_2$ for nearly two decades. Only a single crystal $CuInSe_2$ based solar cell was reported about a decade later by Arya et. al [2]. A solar conversion efficiency of 5.8% was obtained on this cell which had an active area of 4.5 mm^2. By eliminating regions with microcracks and low response, the conversion efficiency increased to 10.6% but the active area was reduced to 0.8 mm^2. At the meantime, the conversion efficiency of thin-film cells has increased from a mere few percents to as high as 15.4% for $CuInSe_2$ based cells and 16.9% for $Cu(In,Ga)Se_2$ based cells [3] for the same period of time. The lack of activities on bulk $CuInSe_2$ cells is mainly due to the difficulties in obtaining high-quality crystals and preparing a good $CuInSe_2$ surface with a reasonable size. On the other hand, monocrystalline $CuInSe_2$ cells are not considered to be practical for photovoltaic applications. Nevertheless, it can provide valuable insight for the improvements of thin-films $CuInSe_2$ cells.

Recently, we have fabricated photovoltaic cells having a $ZnO/CdS/CuInSe_2$ structure with conversion efficiency exceeding 10 % (under sunlight with an illumination near AM 1.5) on monocrystalline $CuInSe_2$ substrates without the use of an anti-reflection coating. The V_{OC}, J_{SC}, F.F and efficiency of our best cell are 0.47 V, 32.8 mA/cm^2, 0.67 and 10.3 % respectively. The effective area of this cell was about 8 mm^2, which is an order of magnitude larger than the cell prepared by Shay et. al [1].

EXPERIMENTAL METHODS

The cells fabricated in this work have a structure of $ZnO/CdS/CuInSe_2$. The surfaces of the monocrystalline $CuInSe_2$ substrates were prepared by abrasive polishing and etching, the CdS by chemical dipping and ZnO by r.f. sputtering. The cell structure is shown in Fig. 1.

Substrate Preparation

The substrates of the cells were obtained from p-type $CuInSe_2$ ingots grown by a horizontal Bridgman method using an one ampoule method. It should be noted that stoichiometric starting proportions of the elements Cu, In and Se were used for the growths, and yet almost all the ingots obtained were p-type. Using a technique [4] to avoid ingot adhesion to the quartz ampoule after the growth, high-quality void-free and microcrack-free ingots were obtained. $CuInSe_2$ wafers

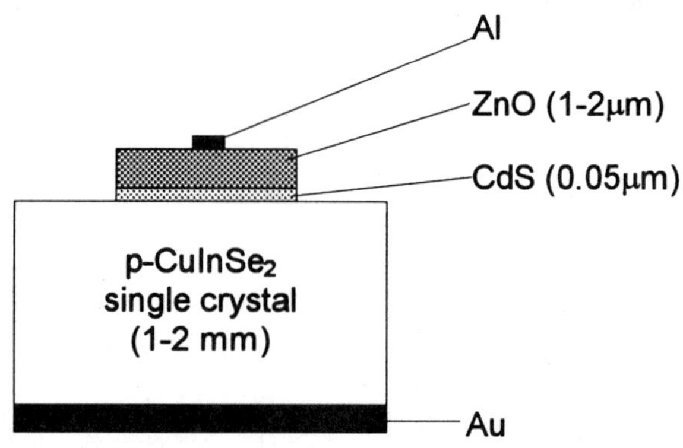

Figure 1 The structure of a CIS/CdS/ZnO cell using a monocrystalline substrate.

First WCPEC; Dec. 5–9, 1994; Hawaii

Table 1 Comparison of parameters between a cell (D38) without and a cell (D39) with substrate annealing in Ar at 350 °C

Sample	η (%)	V_{oc} (V)	J_{sc} (mA/cm^2)	F.F.	n	J_o (μA/cm^2)
D38	8.0	0.42	30.5	0.63	1.74	1.25
D39	10.3	0.47	32.8	0.67	1.87	3.75

with linear dimensions from several millimeters to up to one centimeter were cut from these ingots using a diamond saw. The wafers were generally randomly oriented although a few wafers were cut in the (112) orientation. After this, the wafers were lapped on both sides by using 600 grit silicon carbide paper and then polished on one surface with 1 μm diamond abrasive and 0.05 μm alumina powder, followed by an etching in a 0.5% bromine-methanol solution for one minute. The wafers were then heat treated in Ar at 350 °C for two hours in order to reduce the defects at the surfaces.

CdS Buffer layer

A high resistivity CdS thin film about 0.05 μm in thickness was deposited by a chemical bath method in a solution of $CdCl_2$ + NH_4Cl + $(NH_2)_2CS$ + NH_4OH for 10-20 minutes with the solution maintained at a temperature of 60 °C. A lot of effort has been put in to optimize this process. In the work reported in this paper, all the CdS films were deposited under the same conditions. The resistivities of these films were about 10^5 Ω cm.

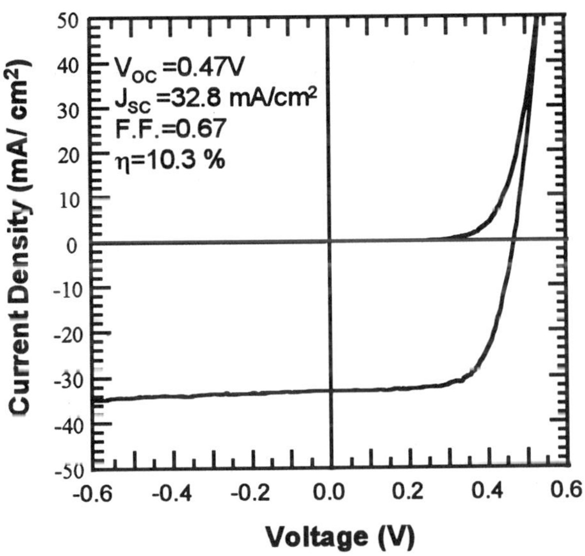

Figure 2 I-V characteristics of a ZnO/CdS/CIS(bulk) cell in dark and under sunlight with an illumination of about AM 1.5.

ZnO Window Layer and Metal Contacts

A layer of low-resistivity ZnO with a thickness of about 1-2 μm was then deposited on the CdS film by r.f. sputtering from an In_2O_3-doped ZnO target. Pure argon was used as the sputtering gas. The substrates were water-cooled during the sputtering process. The ZnO films deposited under such conditions had resistivities of about 10^{-3} to 10^{-4} Ω cm. Next, a smaller area of ZnO and CdS was defined on the wafer using photolithography with dilute HCl. An aluminum dot was then evaporated onto the ZnO and followed by the evaporation of a layer of gold to the back surface of the $CuInSe_2$ wafer. The effective areas of the cells were usually 8 mm^2.

EXPERIMENTAL RESULTS

Current-Voltage Measurements at Room Temperature

Fig. 2 shows the I-V characteristics of a cell illuminated with sunlight at about AM 1.5 and in the dark condition. The conversion efficiency of this cell is 10.3% without the use of a anti-reflection coating. The effective area of this cell is 8 mm^2, which is about 85% of the total area. Cells with comparable performance have been routinely obtained using the fabrication process described above. The diode ideality factors of these devices were in the range of 1.5 to 2 and the leakage currents were about a few microamps per square centimeter. The series resistances were usually smaller than 1 Ω cm^2.

Most of the substrates were randomly oriented but several samples were prepared in the (112) orientation. However, in this particular work, the crystal orientation has not been found to have a significant effect on the cell performance. Actually, provided that the $CuInSe_2$ surfaces were carefully polished, the most important factor affecting the cell performance in the current work has been the annealing of the substrates before the CdS deposition. A series of experiments have been carried out to determine the optimum conditions for the annealing. Samples were heat-treated at different temperatures from 200 to 400 °C in different

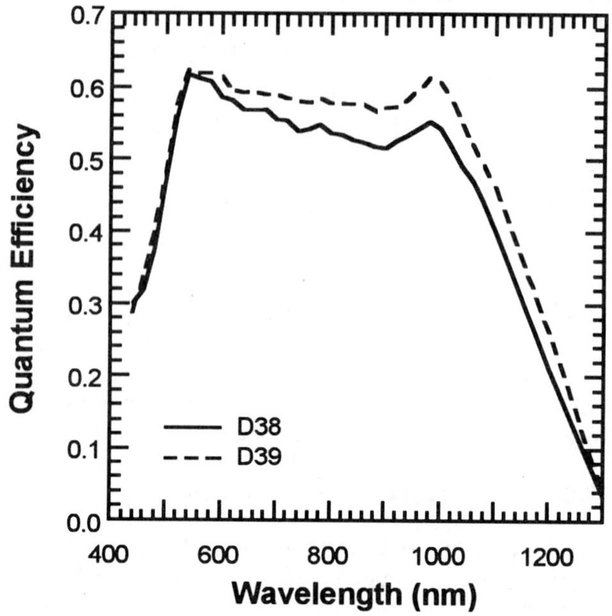

Figure 3 Spectral responses of two cells, one with substrate annealing (D39) and one without (D38).

environments such as air, vacuum, oxygen and argon. The optimum condition was 350 °C in argon. Cells with the optimum annealing generally had larger V_{oc}, J_{sc} and fill factor than cells without the annealing. The amount of improvements on each of the three parameters varies from a few percents to up to 20%. Table I shows the parameters of two cells, one with the annealing and one without. The two substrates were cut from the same ingot with the same orientation and properties. The improvements on cell performance can be clearly seen.

Spectral Response

The spectral responses of these two cells are shown in Fig. 3. The graph clearly shows a better spectral response of the annealed cell and this accounts for the higher short circuit current density of this cell. However, the increase in quantum efficiency is mainly in the infrared or long wavelength region in which the solar spectral irradiance is much less than in the visible region, the increase in short circuit current density of the annealed cell is, therefore, small, about 7.5 %. It is also noted that the quantum efficiency of these cells were around 60 % for most of the wavelengths it absorbed. These values are relatively low when compared to the more than 80 % of quantum efficiencies of other thin-film $CuInSe_2$ cells [3,5-7]. The small quantum efficiencies might reflect on the relatively small short circuit current densities of the cells, which can be as high as more than 40 mA/cm^2 for thin-film cells [3]. Because of the specular nature of the substrate surface after polishing and the fact that anti-reflection coating was not used in the present case, the reflection of the incident light can be very large. This may account for the low quantum efficiency values of these cells. With the use an effective anti-reflection coating, the quantum efficiency and, thus, the short-circuit current density can be expected to increase.

Capacitance-Voltage Measurements

The capacitance-voltage characteristics of the cell (D39) with substrate annealing at three different frequencies are shown in Fig. 4. The capacitance of the cells were measured using the parallel mode and the series resistances, usually under 1 Ω cm^2, of the cells should not cause any problems in the capacitance measurements. The frequency dispersion of the $1/C^2$-V curves observed in this cell are very common for $CuInSe_2$ based solar cell [7-8]. This indicates the existence of deep levels in the $CuInSe_2$. It can be seen that at 100 kHz, a straight line can be drawn on the $1/C^2$ vs V curve for reverse voltages larger than 0.4 V, giving an acceptor concentration of about 1.8×10^{17} cm^{-3}. At 1 kHz, the $1/C^2$ vs V becomes a little bit curved, which may result from deep levels or interface states in the $CuInSe_2$ starting to response at lower frequencies. For the cells without substrate annealing, all the $1/C^2$ vs V (not shown) were not straight and the curvature was large, indicating a much larger density of deep levels and/or interface states compared to the annealed samples.

DISCUSSION

It is believed that the improvement in the performance of cells with substrate annealing were mainly the result of the decrease in the density of the defects at the substrate surface. The long heat treatment seems to be able to reduce the number of defects which were created by the abrasive polishing of the substrate. The higher quantum efficiencies of the annealed cells can be explained as the result of less recombinations of light-generated carriers near the $CuInSe_2$ surface.

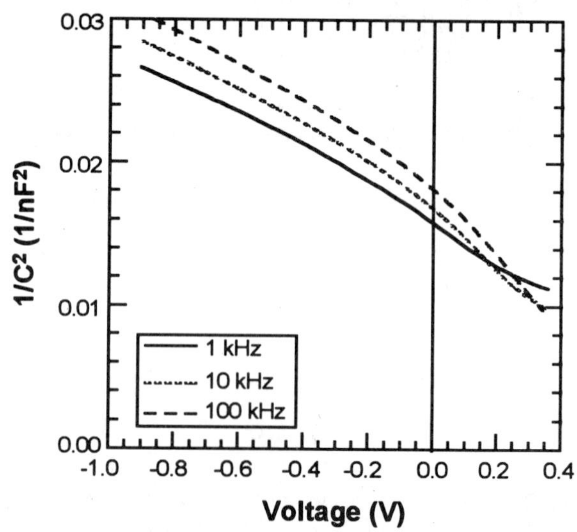

Figure 4 C-V Characteristics of cell D39.

Furthermore, capacitance measurements also seem to show a decrease in surface defects after the annealing. However, it is believed that the numbers of surface defects were still quite high even after the substrate annealing. This can explain the fact that the substrate orientation had no effect on the cell performance as contrary to thin-film cells, which showed a much better performance with (112) $CuInSe_2$ orientation. With a method to further improve the substrate surface, it is very probable that the substrate orientation will come into effect and a solar cell with even higher performance can be obtained.

CONCLUSIONS

Photovoltaic cells with a $ZnO/CdS/CuInSe_2$ structure have been fabricated on monocrystalline $CuInSe_2$ substrates. The active area of the cells was about 8 mm^2. Despite the low quantum efficiency, which was probably due to the large reflection of the incident light, conversion efficiencies exceeding 10 % have been obtained under sunlight with an illumination near AM 1.5. It should be noted that anti-reflection coating has not been used in this work. The $CuInSe_2$ substrates were prepared by abrasive polishing, the CdS by chemical dipping and ZnO by r.f. sputtering. It has been found that a substrate annealing at 350 °C in Ar for two hours prior to the CdS deposition could greatly improve the solar cell performance, probably resulted from a decrease of the density of surface defects. However, the reduced density of surface defects is still considered high. A further improvement of the solar performance may be realized by the use of an appropriate anti-reflection coating and a better surface preparation method.

REFERENCES

[1] J.L. Shay, S. Wagner and H.M. Kasper, "Efficient $CuInSe_2$/CdS Solar Cells", *Appl. Phys. Lett.*, **27**, 1975, pp. 89-90.

[2] R.R. Arya, T. Warminski, R. Beaulieu, M. Kwietniak, J.J. Loferski and W. Giriat, "Photovoltaic and Structural Properties of $CuInSe_2$/CdS Solar Cells", *Solar Energy Materials*, **8**, 1983, pp.471-481.

[3] J. Hedstrom, H. Ohlsen, M. Bodegard, A. Kylner, L. Stolt, D. Hariskos, M. Ruckh and H. Schock, "ZnO/CdS/Cu(In,Ga)Se_2 Thin Film Solar Cells with Improved Performance", *Proc. 23rd IEEE Photovoltaic Spec. Conf.*, 1993, pp. 364-371.

[4] L.S. Yip, I. Shih and C.H. Champness, "Method of Avoiding Ampoule Adhesion of Ingots in Bridgman Growth of $CuInSe_2$", *J. of Crystal Growth*, **129**, 1993, pp. 102-106.

[5] J.Kessler, M. Ruckh, D. Hariskos, U. Ruhle, R. Menner and H.W. Schock, "Interface Engineering Between $CuInSe_2$ and ZnO", *Proc. 23rd IEEE Photovoltaic Spec. Conf.*, 1993, pp. 447-452.

[6] T. Nakada, K. Migita and A. Kunioka, "$CuInSe_2$ Solar Cells by Three Source Magnetron Sputtering", *Proc. 23rd IEEE Photovoltaic Spec. Conf.*, 1993, pp. 560-565.

[7] J. Yoo, A.L. Fahrenbruch and R.H. Bube, "Transport Mechanisms in $ZnO/CdS/CuInSe_2$ Solar Cells", *J. Appl. Phys.*, **68** (9), 1990, pp.4694-4699.

[8] R.K. Ahrenkiel, "The Effect of Deep States on the Photovoltaic Performance of $CdZnS/CuInSe_2$ Thin Film Devices", *Solar Cells*, **16**, 1986, pp. 549-565.

RECENT DEVELOPMENTS IN ZNO/CDS/CUGASE$_2$ SINGLE CRYSTAL SOLAR CELLS

M. Saad *, H. Riazi-Nejad, E. Bucher and M. Ch. Lux-Steiner

Universität Konstanz, Fakultät für Physik, Postfach 5560, D-78434 Konstanz, Germany
* Permanent address: Damascus University, Energy Research Group, P.O. Box 9707, Damascus, Syria

ABSTRACT: An improvement of the power conversion efficiency of ZnO/CdS/$CuGaSe_2$ single crystal solar cells was achieved by repetitive low temperature annealing. This resulted in a new improved efficiency of 6.7 % (under 83 mW/cm^2 AM1.5 illumination), given by an open circuit voltage of 837 mV, a short circuit current density of 13.4 mA/cm^2 and a fill factor of 50 %. Diode characteristics with and without illumination were analysed. A reduced doping concentration within the $CuGaSe_2$ absorber close to the CdS interface as well as an improved dark saturation current are made responsible for the efficiency enhancement. Photoluminescence spectra showed recombination on both sides of this interface. The results indicate that tunneling assisted by the defect levels in the CdS layer and a high series resistance still restrict the optimum device perfomance.

1. Introduction

With its optical properties $CuGaSe_2$ belongs to the promising absorber materials for photovoltaic systems. Its lowest bandgap of 1.68 eV favors this material especially for high voltage single cell modules as well as for the top cell in tandem systems. As the optical absorption beyond the bandgap is high, thin film technology is applicable.
Up to now, Schottky diodes [1] and heterojunctions based on $CuGaSe_2$ single crystals [2-6] or polycrystalline thin films [7,8] have been prepared with power conversion efficiencies up to 6.2 % [8,9]. Preliminary annealing experiments on ZnO/CdS/$CuGaSe_2$ single crystals solar cells have shown that the power conversion efficiency could be increased from 3.5 to 6.0 % under 83 mW/cm^2 AM1.5 illumination [5].
Therefore, we paid special attention to the effect of device annealing. In this paper we will present the effects of repetitive low temperature annealing on the device parameters with and without illumination. Furthermore, as all the experimental efficiencies so far have been well below the theoretical limit of maximum efficiency and tunneling within the space charge region is made responsible for that, we focused our attention on the sources of current losses, in particular on radiant recombination processes in the device.

2. Experiments and results

Single crystal solar cells of ZnO/CdS/$CuGaSe_2$ were prepared by chemical bath deposition of CdS on $CuGaSe_2$ single crystals and subsequent magnetron sputtering of ZnO. Ohmic front and rear contacts were made by evaporation of indium and gold, respectively. Additionally, MgF_2 was evaporated as antireflection coating. Detailed information on the device preparation is described elswhere [5].
Device annealing at 200°C in vacuum results in an improvement of the solar cell performance. By single step annealing a solar cell efficiency of 6.0 % could be achieved under 83 mW/cm^2 AM1.5 illumination [5]. As shown in figure 1 the repetition of annealing led to a further improvement.
After three cycles of annealing the power conversion efficiency could be enhanced to 6.7 % under 83 mW/cm^2 AM1.5 illumination. A final short circuit current density of 13.4 mA/cm^2 and an open circuit voltage of 837 mV were obtained, whereas the fill factor of 50 % remained low. Table 1 summerizes the changes in the photovoltaic parameters as well as in the dark current voltage characteristics for this treatment.
The short circuit current density and the open circuit voltage increased dramatically by each annealing step, whereas the fill factor remained nearly unchanged.
In order to analyse the reduced photocurrent density compared to the theoretically achievable value of more than 20 % for $CuGaSe_2$ ($\approx$ 19 mA/cm^2 for E_g = 1.68 eV) spectral response measurements were performed. These measurements revealed (figure 2) that repeated annealing yields an increase of the spectral quantum efficiency, especially in the long wavelength region, so that a maximum external quantum efficiency of 83 % could be ob-

 First WCPEC; Dec. 5–9, 1994; Hawaii

tained at 700 nm. This improvement at long wavelengths indicates the formation of an increased depletion layer within the $CuGaSe_2$. This might be due to a reduced net acceptor concentration. Hall measurements, performed on annealed crystals, confirmed a reduced doping level.

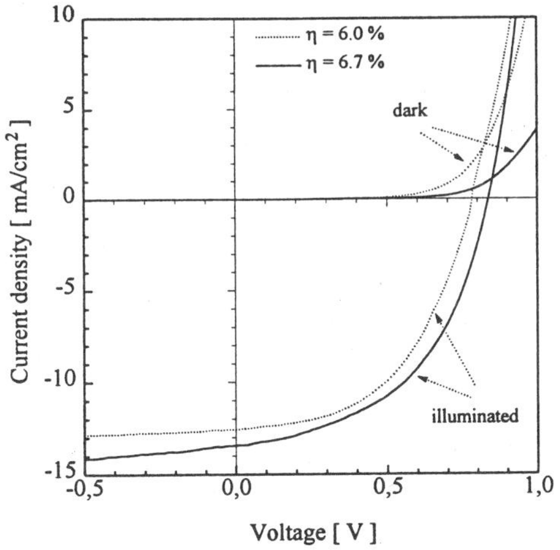

Fig. 1: Dark and illuminated I(V)-characteristics of the ZnO/CdS/$CuGaSe_2$ devices prior and after repetitive annealing.

Table 1: Photovoltaic parameters and dark I(V)-characteristics prior and after repetitive annealing.

Photovoltaic parameters	Ref [5]	this work
Open circuit voltage [mV]	782	837
Short circuit current density [mA/cm²]	12.6	13.4
Fill factor [%]	51	50
Solar conversion efficiency [%]	6.0	6.7
Dark characteristics		
Saturation current density [mA/cm²]	$6.6.10^{-4}$	$1.9.10^{-4}$
Diode quality factor	3.8	3.8
Series resistance [Ωcm²]	2.3	15.9
Shunt resistance [Ωcm²]	$3.8.10^{5}$	$2.5.10^{5}$

At 530 nm the quantum efficiency of 81 % of the repetitive annealed device coincides with the maximum value reached for the device exposed to a single step annealing, but exceeds the quantum efficiency of 57 % for an unannealed devices [5]. Optical losses by the absorption of the high bandgap materials, CdS and ZnO, and the reflection of the device still reduce the short circuit current density by approximately 4 mA/cm².

The higher value of the open circuit voltage of the device with η = 6.7 % compared to that of the device with η = 6.0 % [5] can be attributed to the lower saturation current density in the dark (table 1).

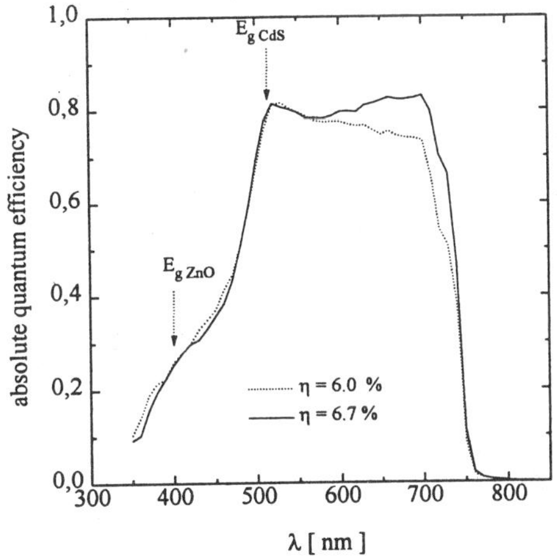

Fig. 2: Spectral quantum efficiency of the devices prior and after repetitive annealing.

Temperature dependent I(V)-measurements were performed in order to study the current transport mechanism in both devices, with 6.0 % and 6.7 % power conversion efficiency. For the latter device, illumination dependent characteristics were studied (using a halogen lamp at various illumination intensities). Figure 3 shows the temperature dependence of the open circuit voltage.

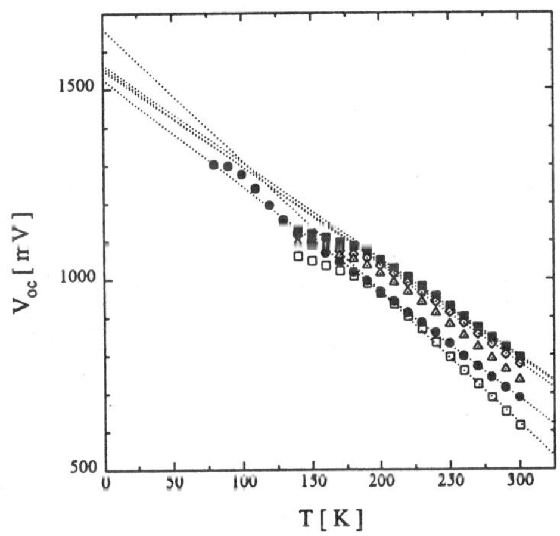

Fig. 3: Temperature dependence of the open circuit voltage of the device with η = 6.0 % (●) and the device with η = 6.7 % (at increasing intensities: □, Δ, ○, ■). Solid symbols correspond to the same illumination intensity.

In the higher temperature range the open circuit voltage increases linearly with decreasing temperature. The slopes vary from -2.6 to -3.6 mV/K with decreasing illumination intensity. The extrapolation of these curves to 0 K results in barrier heights from 1.56 to 1.69 V. The device with η = 6.0 % exhibits a similar behaviour, with a slope of -2.8 mV/K and a barrier height of 1.54 V at 0 K. All values of the barrier height are slightly smaller than the optical bandgap of 1.72 eV at 0 K [10]. The decrease of the slopes below 200 K for the device with η = 6.7 % and below 120 K for the device with η = 6.0 % is caused by the high series resistance at low temperatures.

Fitting of the temperature dependent dark current voltage curves by the single diode model using the equation

$$J(V) = J_0\left\{e^{\frac{q(V-JR_s)}{nkT}} - 1\right\} + \frac{V-JR_s}{R_{sh}}$$

revealed the series resistance R_s, the shunt resistance R_{sh}, the diode saturation current density J_0 and the diode quality factor n, shown in figure 4.

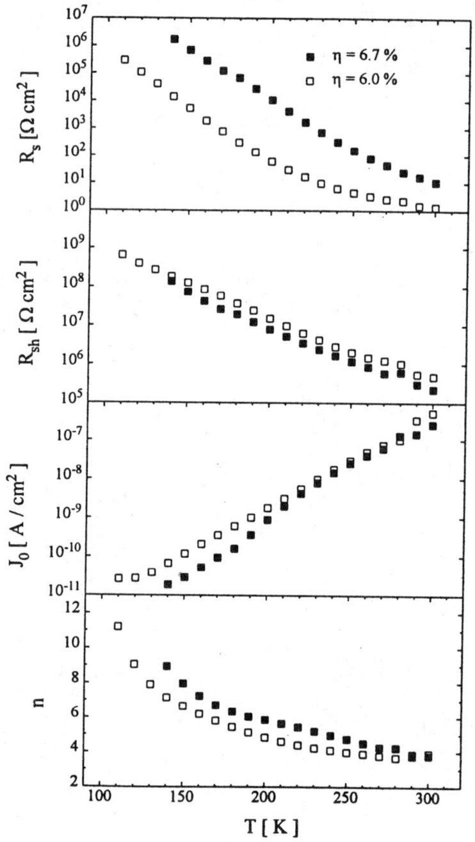

Fig. 4: Temperature dependence of the dark I(V)-parameters for the devices with η = 6.7 % (solid symbols) and η = 6.0 % (open symbols).

For both devices the analysis shows a drastic increase of the series resistance with decreasing temperature. Considering the electrical transport phenomena in $CuGaSe_2$ single crystals we can exclude that this behaviour is related to the bulk properties. Although, the repetition of annealing results in an improvement of the solar cell efficiency, the series resistance becomes worse. The series resistance seems to be the only dark I(V)-parameter which is negatively affected by the repeated heat treatment. At all temperatures the shunt resistance is high enough, that the second term in the diode equation can be neglected. The diode saturation current density decreases with decreasing temperature, in contrast to the shunt resistance. The large value of the diode quality factor as well as its temperature dependence indicate direct or trap assisted tunneling but we suspect that defect levels in the CdS layer acting as recombination centres are responsible for this behaviour.

Therefore, we paid special attention to recombination processes and studied the photoluminescence of the $CuGaSe_2$ devices and its components (figures 5a, b).

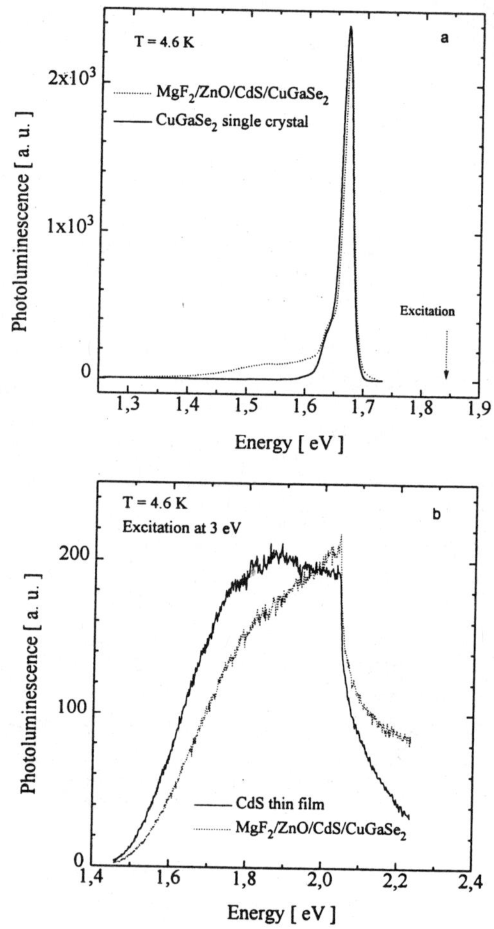

Fig. 5: (a) Photoluminescence spectra of the $CuGaSe_2$ crystal and device structure at 4.6 K. Excitation was performed at 671 nm.
(b) Spectra of the device structure and a CdS thin film deposited on glass. Excitation was performed at 404 nm and 4.6 K.

Figure 5a shows the photoluminescence spectra of the $CuGaSe_2$ crystal and the device structure at 4.6 K. The excitation was performed at an energy lying between the bandgaps of CdS and $CuGaSe_2$. Both spectra exhibit high emission at 1.67 eV [11], reflecting a transition in $CuGaSe_2$. For the device structure, additional emission is observed at lower energies between 1.45 and 1.63 eV. Figure 5b shows the photoluminescence spectra of the device structure and a CdS thin film deposited on glass, both excited at 406 nm. This higher energy excitation mainly took place within the CdS film. Both spectra show broad band emission beyond 1.4 eV, probably due to defects levels in the CdS layer. The comparison of spectra in figures 5a and 5b led to the conclusion that optical recombination of carriers, which are originally excited within the $CuGaSe_2$ base material, takes place within the CdS layer of the device. Thus, a model for radiant recombination losses is proposed as illustrated in figure 6.

As described earlier [5], low temperature annealing a $ZnO/CdS/CuGaSe_2$ device yields a structure consisting of $CuGaSe_2$ bulk material with a net carrier concentration of $p \approx 10^{17} cm^{-3}$, a $CuGaSe_2$ layer with a thickness of approximately 400 nm and a reduced p-type doping concentration of $N_A \approx 4 \cdot 10^{15} cm^{-3}$ (probably caused by interdiffusion; i.e., occupation of Cu vacancies by Cd atoms [12] or S interstitials), a 90 nm thin highly n-doped CdS layer with $N_D \approx 10^{18} cm^{-3}$ and a degenerated ZnO top layer. Due to the optical penetration depths in CdS and $CuGaSe_2$ an estimate of the localization of excitation and radiant recombination can be given.

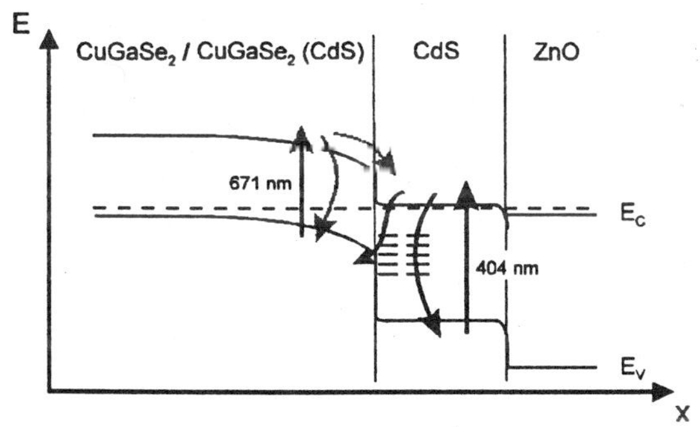

Fig. 6: Schematic representation (main paths) of optical excitation and recombination of carriers within the device structure (On x-axis dimensions are not properly scaled).

For blue light excitation (λ = 406 nm) 90 % of the incoming intensity is used to excite electrons within the CdS layer. They recombine there, with some of the recombination close to the $CdS/CuGaSe_2$ (CdS) interface. The remaining 10% of intensity will immediately be absorbed in the top interdiffused $CuGaSe_2$ layer and additional radiant recombination will occur in that region. Illumination with red light (λ = 671 nm) creates mainly excited carriers in $CuGaSe_2$, 80 % of them within the interdiffused layer. Due to the electrical field across the junction, the excited electrons will be driven towards the CdS and enhanced recombination will take place at the interface. The existence of defect levels in CdS will assist the recombination, and might be also responsible for the current loss mechanism of the solar cell devices.

3. Conclusion

Repetitive annealing of $ZnO/CdS/CuGaSe_2$ device structures at low temperatures led to $CuGaSe_2$ solar cells with the highest-to-date power conversion efficiencies, due to the increased photocurrent, particular in the infrared region, and the higher open circuit voltage. These improvements can be attributed to the reduction of the net doping concentration in a $CuGaSe_2$ interface layer and the lower saturation current density of the device. In these devices tunneling is found to be the dominant transport mechanism, preventing a high fill factor.

Based on the observed photoluminescence spectra we assume the defect levels in CdS, assisting the tunneling, to be the main reason limiting the fill factor.

Acknowledgments

We thank M. Keil and H.-P. Schweikardt for their help in this work. We are indebted to the Baden-Württemberg Stiftung Energieforschung for the financial support of the project (N° A00005193) and the Energy Research Group at Damascus University for the grant of the Ph.D. work of M. Saad.

References

[1] J. Stankiewicz, W. Giriat, *Appl. Phys. Lett.* **35** (1), 70 (1979)

[2] N. Romeo, G. Sberveglieri, L. Tarricone, C. Paorici, *Appl. Phys. Lett.* **30** (2), 108 (1977)

[3] W. Arndt, H. Dittrich, F. Pfisterer, H. W. Schock, *Proc. 6th EC PVSEC*, Kluwer Dordrecht, 260 (1985)

[4] B. Dimmler, H. Dittrich, R. Menner, H. W. Schock, *Proc. 19th IEEE PVSC*, IEEE New York, 1454 (1987)

[5] M. Saad, W. Simon, K. Friemelt, E. Bucher, M.Ch. Lux-Steiner, *Proc. 12th EC PVSEC*, Amsterdam, The Netherlands, 1546 (1994)

[6] W. Simon, *Konstanzer Dissertationen* **424**, Hartung-Gorrre Verlag, Konstanz, ISBN 3-89191-789-9, (1994)

[7] K. T. Ramakrishna Reddy, P. Jayarama Reddy, *Proc. 23rd IEEE PVSC*, IEEE New York, (1993)

[8] R. Klenk, *Dissertation der Universität Stuttgart*, (1993)

[9] R. Klenk, R. Mauch, D. Schmid, H. W. Schock, *Proc. 22nd IEEE PVSC*, IEEE New York, (1991)

[10] M. Quintero, C. Rincon, P. Grima, *Appl. Phys.* **65** (7), 2739 (1989)

[11] W. Simon, M.Ch. Lux-Steiner, S. Riazi Nejad, E. Bucher, *Proc. 11th EC. PVSEC*, Montreux, Switzerland (1992)

[12] A. Zunger, S. Wagner, P.M. Petroff, *J. Electr. Mat.* **22** (1), 3 (1993)

DEVELOPMENT OF $ZnO/ZnSe/CuIn_{1-x}Ga_xSe_2$ THIN-FILM SOLAR CELLS WITH BAND GAP OF 1.3 TO 1.5 eV

Yasutoshi OHTAKE, Katsumi KUSHIYA*, Akira YAMADA and Makoto KONAGAI

Department of Electrical and Electronic Engineering, Tokyo Institute of Technology
2-12-1 O-okayama, Meguro-ku, Tokyo 152, JAPAN
* Showa Shell Sekiyu K.K. Central R&D lab

ABSTRACT

From the standpoint of environmental safety, the establishment of the process to fabricate Cd-free buffer layers is significantly desirable. In this study, ZnSe has been employed as a buffer layer together with a new fabrication method which has been named as an atomic layer deposition (ALD) method in order to deposit ZnSe continuously on polycrystalline CIGS thin-film absorber in the form of an atomic layer epitaxy (ALE) method. The currently best efficiency of polycrystalline CIGS thin-film solar cells with ZnSe buffer layer by the ALD method is 11.6 %.

INTRODUCTION

As a buffer layer deposited on polycrystalline Cu-III-VI_2 thin-film absorbers, CdS is the most common II-VI compound in the polycrystalline Cu-III-VI_2 thin-film solar cells and a chemical bath deposition (CBD) method has been in general employed as its deposition process. However, from the standpoint of environmental safety, the establishment of the process to fabricate Cd-free buffer layers and the minimization and eventually elimination of the Cd-containing waste water discharged from the CBD process are significantly desirable. Therefore, ZnSe as an attractive alternative to CdS buffer layer is employed in this study. ZnSe can demonstrate several advantages as follows:

- The band gap of ZnSe (2.67 eV) is wider than CdS (2.42 eV), which can transmit higher-energy photons than the band gap of CdS,
- ZnSe has a better lattice-match to CIGS thin-film absorbers in the band gap range of 1.3 to 1.5 eV than CdS.

In this study, ZnSe buffer layers have been deposited on the CIGS thin-film absorber continuously in the same fabrication apparatus. Several deposition techniques to prepare a good-quality ZnSe for thin-film solar cells, such as CBD [1], metalorganic chemical vapor deposition (MOCVD) techniques [2], have been developed. The atomic layer deposition (ALD) method has been proposed, for the first time, in this study in order to prepare ZnSe buffer layers, which contributed to the improvement of the quality of the p-n heterojunction between the n-type ZnO and the p-type CIGS thin-film absorber. In this method, the remarkable low substrate temperature of 250 °C was employed. In addition, this method made it possible to attain both better coverage on the surface and better control of the thickness than the coevaporation method in this study. The thickness of ZnSe could be controlled until the range of less than 30 nm which was the lower limit by the coevaporation method in this study [3]. The quality of ZnSe buffer layers has been verified by fabricating the polycrystalline CIGS thin-film solar cells with ZnO/ZnSe/CIGS structure. Polycrystalline CIGS thin-film solar cells with ZnSe buffer layer have demonstrated, for the first time, high efficiency of over 11 %. Further improvement of efficiency will be strongly expected by the establishment of the continuous fabrication process for the ZnSe/CIGS interface, which requires the optimization of the transition period from CIGS to ZnSe as well as the fabrication process of ZnSe.

EXPERIMENTAL

The structure of ZnO/ZnSe/CIGS thin-film solar cells in this study is shown in Fig. 1. CIGS thin-film absorbers are deposited by a coevaporation method onto a heated Mo sputter-coated soda-lime glass. In order to

CH3365-4/94/0000-0218 $4.00 © 1994 IEEE

fabricate ZnSe buffer layers, the same fabrication apparatus is employed as shown in Fig. 2. In the ALD method, ZnSe buffer layers are deposited on CIGS thin-film absorbers at the substrate temperature of 250 °C and the Zn/Se beam-intensity ratio of 1/10 as shown in Fig. 3. The shutter of each Knudsen cell of Zn or Se is independently and sequentially controlled every 7 seconds with 1-second interval as shown in Fig. 4. The thickness of ZnSe is adjusted by the number of cycles.

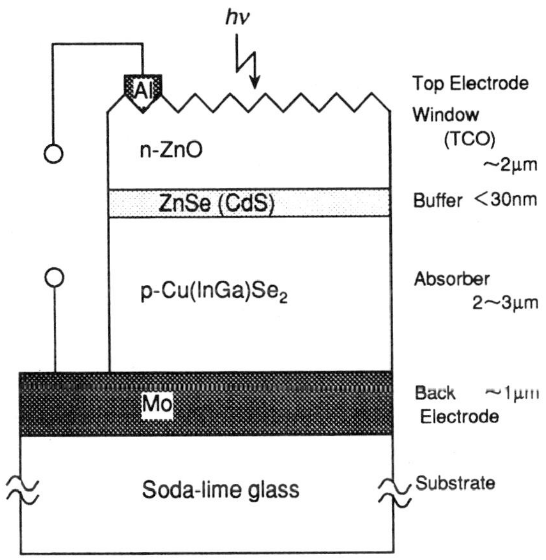

Fig. 1. Structure of polycrystalline CIGS thin-film solar cell with ZnSe buffer layer.

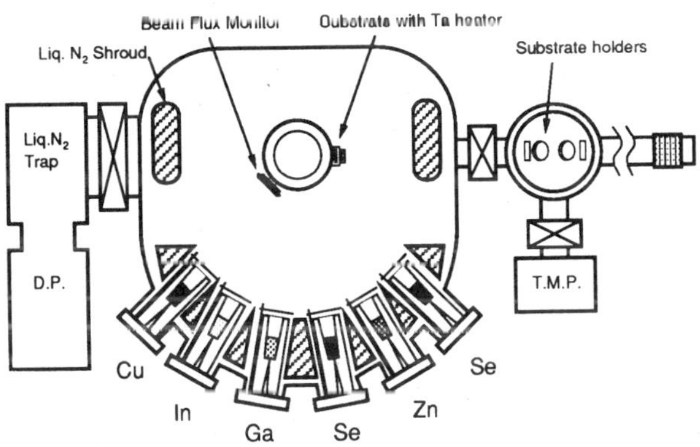

Fig. 2. Schematic diagram of fabrication apparatus.

On the other hand, about 10-nm-thick CdS buffer layers are deposited by a CBD method. After annealing of 200°C for 30 min in air, textured ZnO window layers are grown by an MOCVD technique described elsewhere [4]. Al top metal grids are applied by a vacuum evaporation method.

The X-ray diffraction (XRD) measurements are applied for thick ZnSe layers over 100 nm. Since most of ZnSe buffer layers deposited in this study are too thin to be detected by the XRD and the scanning electron microscope (SEM) measurements, the quality of ZnSe buffer layers is eventually evaluated electrically by fabricating polycrystalline CIGS thin-film solar cells with ZnO/ZnSe/CIGS structure.

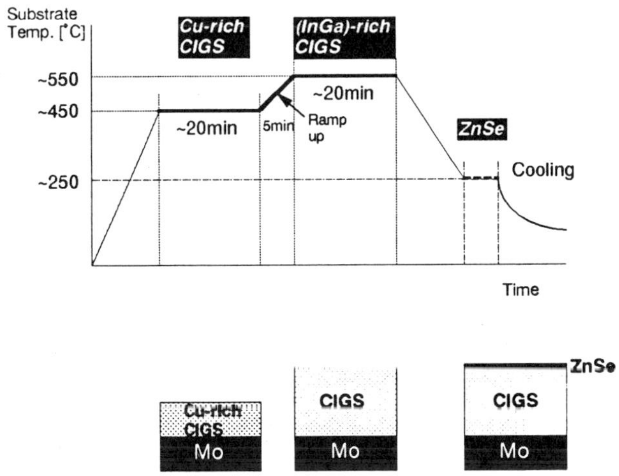

Fig. 3. Substrate-temperature program for ZnSe buffer layer deposited by the ALD method on the CIGS thin-film absorber continuously.

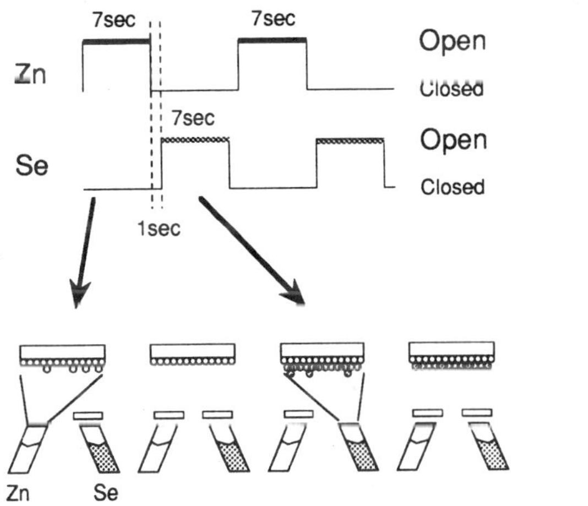

Fig. 4. Shutter sequence of the ALD method in this study.

RESULTS AND DISCUSSION

ZnSe layers, about 300 nm thick, were grown by

the ALD method for 1000 cycles on a GaAs substrate with (111) lattice plane and polycrystalline CIGS thin films in order to confirm the growth rate of one monolayer per cycle and the preferential growth direction of ZnSe with (111) lattice plane. The XRD pattern of ZnSe (111) lattice plane grown on the polycrystalline CIGS thin films were shown in Fig. 5. As seen in this figure, the XRD peak position of ZnSe (111) was confirmed at $2\theta = 27.22°$, which was consistent with the Joint Committee on Powder Diffraction Standards (JCPDS) card No. 37-1463.

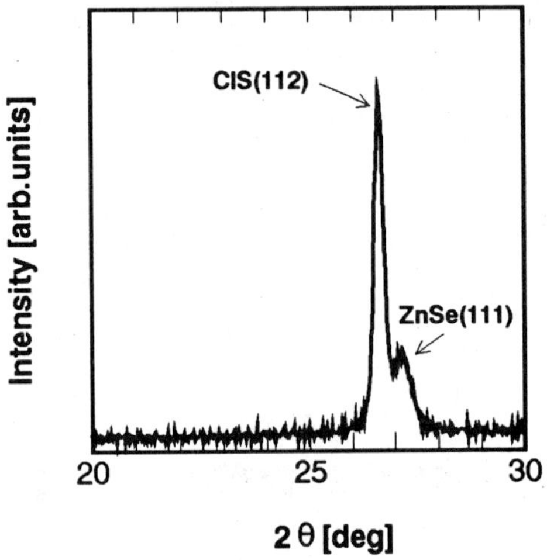

Fig. 5. XRD pattern of ZnSe grown for 1000 cycles by the ALD method on polycrystalline CIGS thin film.

FF in the case of ZnSe buffer layers was normalized to FF in the case of a CdS buffer layer as shown in Fig. 6. As seen in this figure, ZnSe buffer layers thinner than 30 nm can be prepared by this method, which is a good example that the ALD method proposed in this study has a rather high capability to control the thickness. At the 1-nm-thick ZnSe that corresponded to 3 monolayers of ZnSe, the best FF was attained, which was comparable enough to the FF with CdS buffer layers. In the case of direct deposition of ZnO window layer on the same lot of CIGS thin-film absorbers as the experiments for the buffer layers, FF was worse than that with a ZnSe buffer layer. This may be good evidence to prove that a current quality-level of polycrystalline CIGS thin-film absorbers requires a ZnSe buffer layer in order to improve the performance of polycrystalline CIGS thin-film solar cells.

The spectral response measurements of CIGS thin-film solar cells with about 10-nm-thick ZnSe buffer layer prepared by the ALD method are shown in Fig. 7. The gain at the shorter wavelength from 400 to 500 nm is clearly observed, which reflects wider band-gap of 2.67 eV for ZnSe than 2.42 eV for CdS.

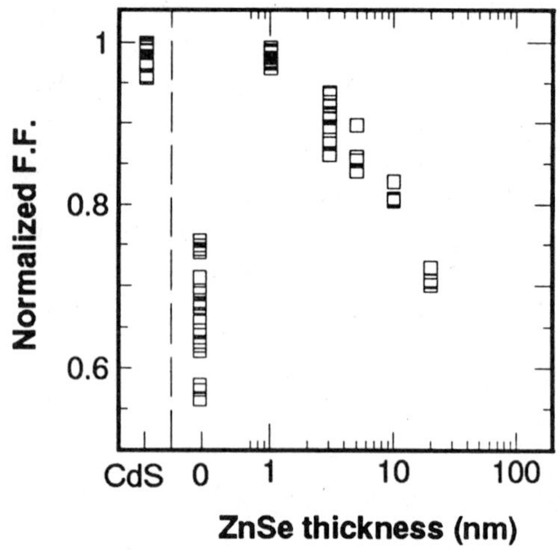

Fig. 6. Correlation between normalized FF and thickness of ZnSe buffer layer fabricated by the ALD method.

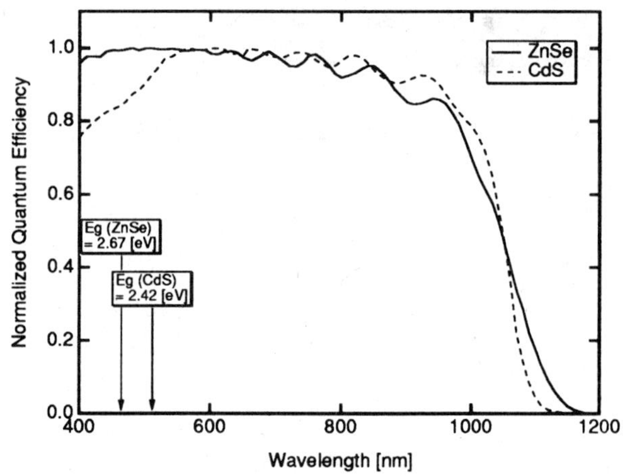

Fig. 7. Spectral responses of polycrystalline CIGS thin-film solar cell with ZnSe or CdS buffer layer.

The currently best efficiency of CIGS thin-film solar cells with about 10-nm-thick ZnSe buffer layer is 11.6 % as shown in Fig. 8. This considerable achievement is, for the first time, reported with the continuous growth of ZnSe on the p-CIGS thin-film absorber. Further improvement of the performance of polycrystalline CIGS thin-film solar cells with ZnSe buffer layer is most likely to be accomplished with the ALD method proposed in this study. It is necessary to increase in FF as well as V_{oc} in order to attain higher efficiency with ZnSe buffer layers. In order to establish the ALD method, the surface quality of CIGS thin-film absorber should be maintained sufficiently because it is understood as one of the key factors to attain a significantly high quality of the p-n heterojunction and enable better performance of polycrystalline CIGS thin-film solar cells with a ZnSe buffer layer rather than a CdS buffer layer. For that purpose, the control of the Se beam-flux intensity, for instance, should be performed carefully even in the transition period from the end of the fabrication for CIGS thin-film absorber to the beginning of the fabrication of ZnSe buffer layer when ZnSe is deposited continuously on the CIGS thin-film absorber.

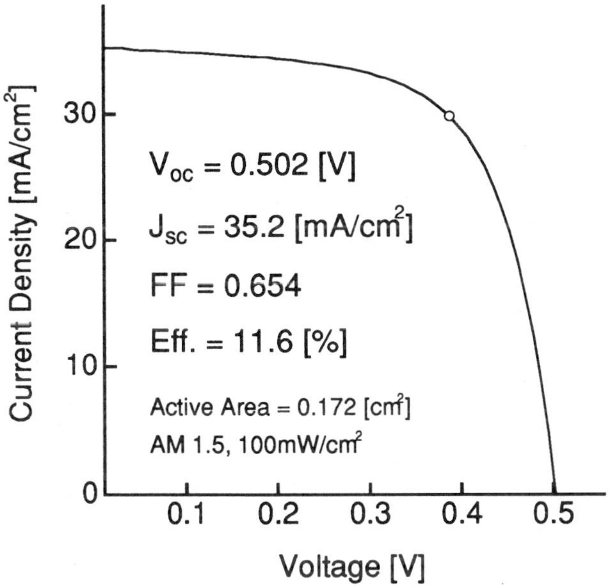

Fig. 8. I-V characteristic of the currently best polycrystalline CIGS thin-film solar cell with ZnSe buffer layer fabricated continuously by the ALD method.

CONCLUSIONS

ZnSe buffer layers, which have a wider band gap and better lattice match to CIGS thin films than CdS, have demonstrated considerable potential for forming good p-n heterojunction and achieving comparable performance to CdS buffer layers.

The ALD method has been proposed, for the first time, in this study to prepare ZnSe buffer layers and the remarkable advantages of this method have been verified such as considerably low substrate temperatures, sufficient thickness control and the continuous fabrication process for the ZnSe/CIGS interface in the same fabrication apparatus. In the case of the ALD method proposed in this study, the substrate temperature of 250 °C was applicable. Such a low substrate temperature should be favorable to maintain the surface quality of the CIGS thin-film absorbers during the ZnSe deposition.

The currently best efficiency of CIGS thin-film solar cells with ZnSe buffer layer is 11.6 %. The establishment of the fabrication process of ZnSe buffer layers is strongly linked to the understanding of the interface engineering of the n-ZnO/p-CIGS which is believed to be the most important key to improve the interface quality and, thus, the efficiency.

ACKNOWLEDGMENT

This study was supported in part by NEDO as a part of the New Sunshine Program under Ministry of International Trade and Industry.

REFERENCES

[1] J. Kessler, M. Ruckh, D. Hariskos, U. Rühle, R. Menner and H.W. Shock, "Interface Engineering Between $CuInSe_2$ and ZnO", *23rd IEEE PVSC*, 1993, p. 447.

[2] L.C. Olsen, F.W. Addis and D.A. Huber, "Investigation of Polycrystalline Thin Film $CuInSe_2$ Solar Cells Based on ZnSe Windows", *23rd IEEE PVSC*, 1993, p. 603.

[3] K. Kushiya, Y. Ohtake, A. Yamada and M. Konagai, "Photoluminescence Study of Polycrystalline $CuIn_{1-x}Ga_xSe_2$ Thin Films", *12th EC PVSEC*, 1994, p. 1580.

[4] W. W. Wenas, A. Yamada, M. Konagai and K. Takahashi, "Textured ZnO Thin Films for Solar Cells Grown by Metalorganic Chemical Vapor Deposition", *Jpn. J. Appl. Phys.*, **30**, 1991, L 441.

PHOTOVOLTAIC PROPERTIES OF $Cu(In,Ga)Se_2$ THIN FILM SOLAR CELL FABRICATED BY COEVAPORATION PROCESS

M.Nishitani, T.Negami, M.Ikeda, N.Kohara, M.Terauchi,T.Wada and T.Hirao
Central Research Laboratories, Matsushita Electric Ind. Co., Ltd.
3–4 Hikaridai, Seika–cho, Soraku–gun, Kyoto 619–02,JAPAN

ABSTRACT

Thin film solar cells based on $Cu(In,Ga)Se_2$ films were fabricated, and their junction and photovoltaic properties were investigated. Ga in $CuInSe_2$ thin films formed by so-called "bilayer process" was incorporated homogeneously. The fabricated cell structure was glass/Mo/$Cu(In,Ga)Se_2$/CdS/ZnO /ITO(/MgF_2). The incorporations of Ga into $CuInSe_2$ films up to about 20 mol% improved the photovoltaic performance. From the study of photoluminescence, capacitance–voltage and current–voltage characteristics, it was clarified that the incorporation of Ga not only widened the bandgap energy, but also played an important role in the effect which yield hole concentration. The best cell with an AR–coating(MgF_2) exhibited an efficiency of 15.2% ; Jsc=33.9mA/cm^2, Voc=0.616V, FF=0.730. The device performance can be improved by the development of the film with higher hole concentration, keeping the crystalline quality.

INTRODUCTION

$CuInSe_2$ is one of the most promising materials for photovoltaic devices with low cost and high efficiency. Thin film solar cells based on this material and the related compounds have been previously reported to show an efficiency of more than 15% [1–3].

Primarily, the incorporation of Ga into $CuInSe_2$ was tried in order to widen the bandgap energy more matched to the solar spectrum.[4,5] Other roles of Ga in $CuInSe_2$ were also reported ; Jensen et al.[6] reported about the improvement of the adhesion to the Mo back electrode, and Walter et al.[7] showed Ga was necessary to avoid a significant drop of the hole concentration in $CuIn(S,Se)_2$ alloys. For the development of higher performance cell, it is important to study such electrical and mechanical effect of Ga. Especially, it is very difficult, but significant works to research a correlation of optical and electrical film properties and device characteristics through the defect chemistry of $CuInSe_2$ and the related compounds which is mainly dominated by intrinsic defects.

So far, we have been continuing to investigate the feasibility of the efficiency on thin film solar cells based on $Cu(In,Ga)Se_2$ films, using a coevaporation process from its controllability. In this paper, we report the progress of our development for $Cu(In,Ga)Se_2$ thin film solar cells with the results of the analyses from photoluminescence and electrical measurements.

EXPERIMENTAL

The $Cu(In,Ga)Se_2$ films used in this experiment were prepared on a Mo–coated soda–lime glass by either multi–source coevaporation of Cu, In, Ga and Se. The films were prepared by the so–called "bilayer process" [8] based on our coevaporation process described elsewhere in detail [9]. The substrate temperature was about 500°C, a Ga flux was kept constant during the deposition. The total thickness of the film was in the range of 2.5–3.0μm. The heterojunctions were completed by the deposition of a thin CdS (CBD,40–60nm thickness), an undoped ZnO film (RF magnetron sputtering, 0.3μm thickness) and an ITO film (RF magnetron sputtering, 0.5μm thickness). Furthermore, a MgF_2 AR–coating (electron–beam evaporation) was deposited onto the best devices.

Photoluminescence (PL) measurements of the $Cu(In,Ga)Se_2$ films were carried out at 4.2K and 77K with Ar^+ laser on the various excitation intensity. For the estimation of hole concentration, capacitance–voltage (CV) measurements of the fabricated $Cu(In,Ga)Se_2$ cells were performed at a frequency of 1 MHz. For the observation of the junction properties, an electron beam induced current (EBIC) method was used at the accelerating voltage of 10keV under the low injection condition. Current–voltage measurements were also performed in the dark and under AM1.5, 100mW/cm^2 condition in the temperature range of 288–328K. Quantum efficiency of the fabricated cell was measured in the wavelength range of 300–1400nm.

RESULTS and DISCUSSION

Figure 1 shows PL spectra of $CuIn_{1-X}Ga_XSe_2$ (X=0.0,0.13,0.21,0.60) films measured at 4.2K. The PL peak was shifted to higher energy with increasing Ga content, mainly due to widening the bandgap energy. The dependence of PL spectra on excitation intensity (10–100mW), observed in $Cu(In,Ga)Se_2$ films, shows that every PL emission intensity have a tendency to saturate besides the emission peak energy doesn't a shift. However, in the PL spectra measured at 77K, we observed the blue shift of the PL emission peak with the increase of the excitation intensity even in the range of 10–100mW. It suggests that the observed emissions may be due to the donor–acceptor(DA) pair emissions. Each PL emission intensity at 77K was almost same level except the emission intensity from the $CuIn_{0.4}Ga_{0.6}Se_2$ film, which was approximately one order weaker. We suppose that the $CuIn_{0.4}Ga_{0.6}Se_2$

CH3365-4/94/0000-0222 $4.00 © 1994 IEEE First WCPEC; Dec. 5–9, 1994; Hawaii

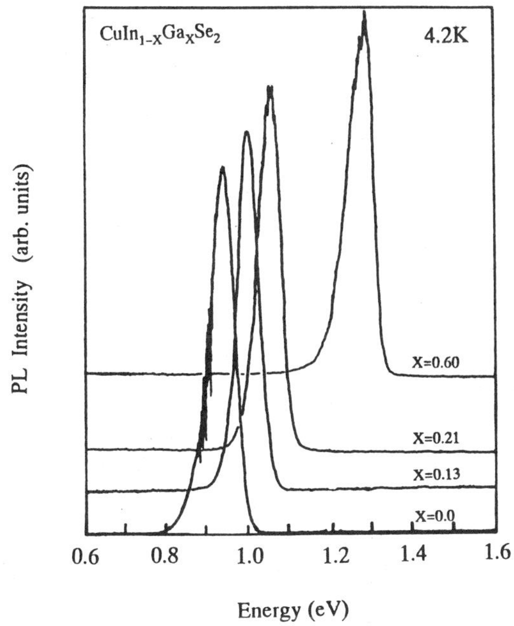

Fig.1 Photoluminescence spectra of $CuIn_{1-X}Ga_XSe_2$ films measured at 4.2K.

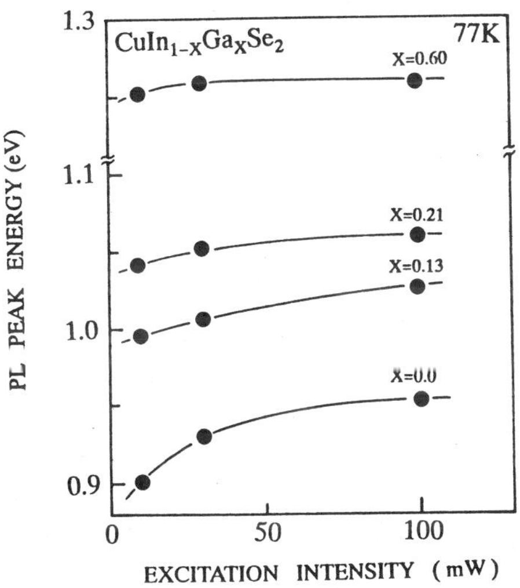

Fig.2 The photoluminescence peak energy of $CuIn_{1-X}Ga_XSe_2$ films measured at 77K as a function of the excitation intensity.

film was inferior to the other films in quality, and that the PL emission was weakened due to non-radiative recombination rather than having the lower concentration of the DA pair. Figure 2 shows the PL peak energy of $CuIn_{1-X}Ga_XSe_2$ (X=0.0,0.13,0.21,0.60) films measured at 77K as a function of the excitation intensity. We can find in Fig.2 that the peak energy was less dependent on the excited intensity with the increase of the Ga content. Those dependence suggests that Ga in $CuInSe_2$ not only widened the bandgap, but also played an important role in the native defect formation such as a decrease of donor concentration.

Figure 3 shows the values of hole concentration, or Na(acceptor density)–Nd(donor density), obtained from CV measurements of the CdS/Cu(In,Ga)Se_2 heterojunction as a function of Ga content. For the estimation of the hole concentration (P), an abrupt n^+p junction with constant doping levels was assumed. The values obtained should be regarded as relative ones to each other rather than absolute ones.[7] From the results, the $CuInSe_2$ films were considered to be significantly compensated and the addition of Ga seemed to yield the increasing of hole concentration. However, the larger amount of Ga such as $CuIn_{0.4}Ga_{0.6}Se_2$ film saturated the effect. These results coincided the result obtained from PL measurements. Therefore, as for our preparation process of Cu(In,Ga)Se_2 films, the film with high hole concentration and high quality can be produced in the region around Ga content of 20 mol%.

The characterized results obtained from PL and CV measurements were well reflected in the junction properties of Cu(In,Ga)Se_2 solar cell devices. Figure 4 shows junction EBIC profile observed in the $CuIn_{1-X}Ga_XSe_2$ (X=0.0,0.13,0.21,0.60) cells. According to Evehart et al.[10], the diameter of the generation volume (Rg) in $CuInSe_2$ at the accelerating voltage of 10kV is about 0.6μm. In addition, Shea et al.[11] suggest the necessary condition, Rg/4<L(the minority carrier diffusion length), for the validity of the point source approximation. In other words, the Rg at 10kV in $CuInSe_2$ is approximated by a point source as long as the L in $CuInSe_2$ is longer than about 0.15μm. However the profile of EBIC linescans are affected by surface recombination.[12] Ordinarily, the collective region of the generated carrier in the p–n junction consists of the space charged region and the diffusion region. The observed EBIC profile seemed to vary from the dominant profile of the space charged region to that of the diffusion region with the increase of Ga content. The variation of the profile corresponds to the variation of the hole concentration on Cu(In,Ga)Se_2 films shown in Fig.4, since the space charged region is proportional to the square root of 1/P on the assumption of the abrupt junction with the constant doping level. Furthermore, the minority carrier diffusion length of the $CuIn_{1-X}Ga_XSe_2$ (X=0.21,0.60) film was estimated from the EBIC profile by using the point source approximation, and the values of 1.5μm and 0.7μm were obtained in the films of

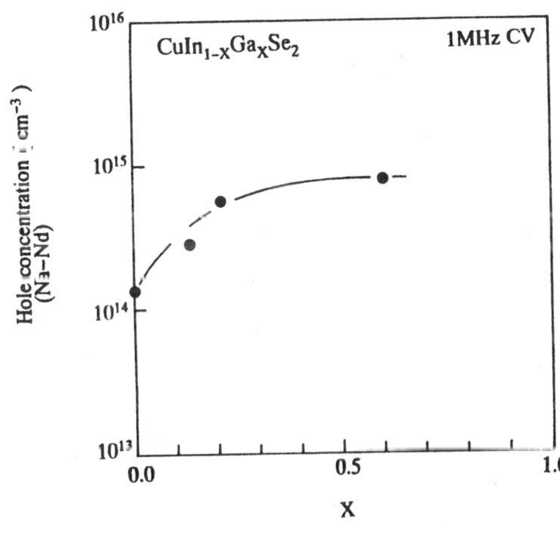

Fig.3 The hole concentration of $CuIn_{1-X}Ga_XSe_2$ films as a function of Ga content, obtained from capacitance–voltage measurements of the CdS/Cu(In,Ga)Se_2 heterojunction.

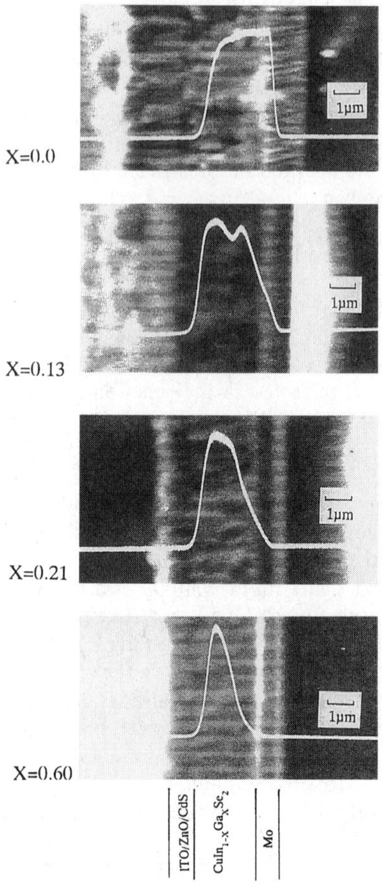

Fig.4 Electron beam induced current profiles of $CuIn_{1-X}Ga_XSe_2$ solar cell devices.

X=0.21 and X=0.6, respectively. Those values satisfied the criterion for the point source approximation (Rg/4<L), self-consistently. The difference of the diffusion length between the films of X=0.21 and X=0.60 would be reflected that of the quality of the two films, which is qualitatively understood from the PL measurements as already mentioned.

Current-voltage measurements of the $CuIn_{1-X}Ga_XSe_2$ (X=0.0,0.13,0.21,0.60) cells in the dark at the temperature range of 288–328K showed a thermally activated behavior. The diode quality factors (n) were approximately independent of temperature and came up to two with the increase of Ga content. The values of the activation energy of the saturated current (I_0) estimated by a Arrhenius plot of I_0 were almost independent of Ga content. Each value of the activation energy times the diode quality factor was close to the bandgap of $Cu(In,Ga)Se_2$. This implies that the fabricated junctions in this work was primarily determined by Shockley–Read–Hall (SRH) recombination in the space charged region.[7] Current-voltage measurements of the $CuIn_{1-X}Ga_XSe_2$ (X=0.0,0.13,0.21,0.60) cells under AM1.5, $100mW/cm^2$ at the temperature range of 288–328K showed the decrease of Voc due to thermally activated recombination process, however Jsc was independent of temperature. The dark and illuminated properties of the $Cu(In,Ga)Se_2$ cell were summarized in Table

Table 1 Dark and illuminated current–voltage characteristics of the $CuIn_{1-X}Ga_XSe_2$ cells.

X	Jsc (mA/cm^2)	Voc (V)	FF	Eff. (%)	J_0 (A/cm^2)	n	Ea (eV)
0.00	35.6	0.426	0.691	10.5	$7.96x10^{-7}$	1.63	0.602
0.13	34.2	0.494	0.705	11.9	$4.29x10^{-7}$	1.76	0.614
0.21	33.2	0.581	0.708	13.7	$3.82x10^{-8}$	1.84	0.643
0.60	23.1	0.691	0.676	10.8	$6.89x10^{-9}$	1.93	0.657

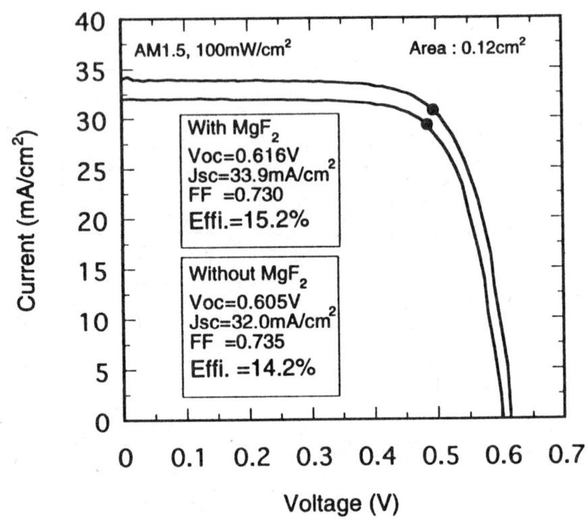

Fig.5 Current–voltage characteristics of the best $Cu(In,Ga)Se_2$ cell with and without an AR–coating under AM1.5, $100mW/cm^2$.

1. According to these results, the device performance can be improved by the development of the film with higher hole concentration, keeping the crystalline equality.[7]

As an anti–reflection (AR) coating, a MgF_2 film was deposited onto the device and Si substrate by electron beam evaporation. The value of the refractive index of the MgF_2 film was 1.39, obtained by ellipsometric measurement using He–Ne laser. From the calculation of the reflectivity on the structure of MgF_2/ITO, the optimized thickness of MgF_2 film was about 0.1μm. Figure 5 shows photovoltaic characteristics of the best device obtained so far, which was fabricated by using the $Cu(In,Ga)Se_2$ film with Ga content of about 20 mol%, with and without the AR coating. An active area efficiency of 15.2% (Jsc=$33.9mA/cm^2$, Voc=0.616V, FF=0.730, active area=$0.12cm^2$) was obtained in the device structure of glass/Mo/$Cu(In,Ga)Se_2$/CdS/ZnO/ITO/MgF_2. Figure 6 shows the external quantum efficiency of the cell measured in the wavelength range of 300–1400nm. A few percent recovering was observed in the range of 500–900nm by using the 0.1μm thickness of MgF_2 film as an AR–coating.

CONCLUSION

From the analyses of PL and CV measurement, the incorporation of Ga into $CuInSe_2$ not only widened the bandgap, but also played an important role in the defect chemistry such

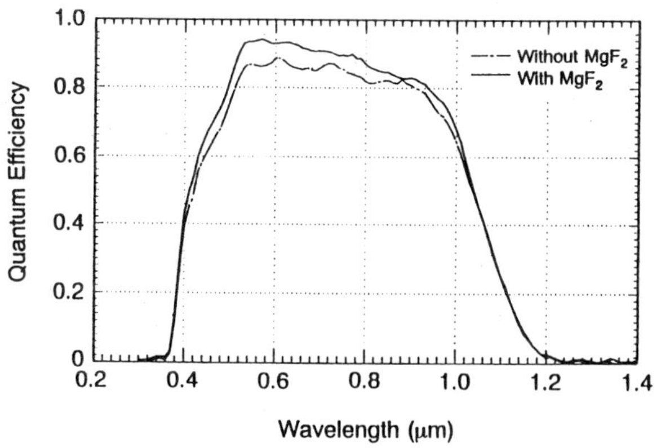

Fig.6 External quantum efficiency of the best $Cu(In,Ga)Se_2$ cell with and without an AR-coating

as a decrease of donor concentration which yield hole concentration as a result. These characteristics were well reflected in the EBIC profiles of $Cu(In,Ga)Se_2$ solar cell devices. As for our preparation process of $Cu(In,Ga)Se_2$ films, the film with high hole concentration and quality would be produced in the region around Ga content of 20 mol%.

Current-voltage characteristics show that the fabricated junctions in this work was primarily determined by Shockley-Read-Hall (SRH) recombination in the space charged region. Therefore, the device performance can be improved by the development of the film with higher hole concentration, keeping the crystalline equality.

The best cell fabricated so far with an AR-coating has an active area efficiency of 15.2% (Jsc=33.9mA/cm^2, Voc=0.616V, FF=0.730, active area=0.12cm^2).

ACKNOWLEDGEMENTS

The authors acknowledge Dr.T.Nitta for encouragement. This work was supported by the New Energy and Industrial Technology Development Organization as a part of the New Sunshine Program under the Ministry of International Trade and Industry.

REFERENCES

[1]J.Hedström et al., "ZnO/CdS/$Cu(In,Ga)Se_2$ thin film solar cells with improved performance", *Proceedings of the 23rd IEEE PV Specialists Conference*(1993).

[2]T.Walter et al., "Solar cells based on $CuIn(S,Se)_2$-A promising alternative", *Proceedings of 11th European PV Solar Energy Conference*(1992)124.

[3]D.Tarrant and J.Ermer, "I-III-VI_2 multinary solar cells based on $CuInSe_2$", *Proceedings of the 23rd IEEE PV Specialists Conference*(1993).

[4]W.S.Chen et al., "Development of thin film polycrystalline $CuIn_{1-x}Ga_xSe_2$ solar cells", *Proceedings of the 19th IEEE PV Specialists Conference*(1987)1445.

[5]R.Klenk et al., "Wide bandgap $Cu(In,Ga)Se_2$/(Zn,Cd)S heterojunctions", *Proceedings of the 20th IEEE PV Specialists Conference*(1988)1443.

[6]C.Y.Jensen et al., "The role of gallium in $CuInSe_2$ solar cells fabricated by a two-stage method", *Proceedings of the 23rd IEEE PV Specialists Conference*(1993).

[7]T.Walter et al., "Characterization and junction performance of highly efficient ZnO/CdS/$CuInS_2$ thin film solar cells", *Proceedings of 12th European PV Solar Energy Conference*(1994)1755.

[8]B.J.Stanbery et al.,IEEE Transactions on Electron Devices, **37**(1990)438.

[9]M.Nishitani et al., "Preparation and characterization of $CuInSe_2$ thin film by Molecular-Beam deposition method", Jpn.J.Appl. Phys. **31**(1992)192.

[10]T.E.Everhart et al., "Electron energy dissipation vs penetration in solids", J.Appl.Phys.**42**(13)(1971)5837.

[11]Shea et al.:Scan.Elect.Microsc. vol.I(1978)435f.

[12]D.Schmid et al., " Diffusion length measurement and modeling of $CuInSe_2$-(Zn,Cd)S solar cells", *Proceedings of 10th European PV Solar Energy Conference*(1991)935.

THE EFFECT OF SE CONTENT ON DEFECT LEVELS IN CUINSE$_2$ SINGLE CRYSTALS DETECTED BY PHOTOACOUSTIC SPECTROMETRY

A. Zegadi, M.V. Yakushev, E. Ahmed, M.A. Slifkin*, A.E. Hill and R.D. Tomlinson
Department of Electronic and Electrical Engineering,
Salford University, Salford M5 4WT, UK. Telefax +44 61 745 5999
*Department of Electrical Engineering, Jerusalem College of Technology, Israel.

ABSTRACT

The objective of this paper is to investigate the impact of changes in the selenium content of $CuInSe_2$ single crystals on their optical properties in the subgap region of the infrared spectrum. A high resolution near-infrared photoacoustic spectrometer of the gas-microphone type is used for room temperature analysis of non-radiative defect states in as-grown n and p-type CIS crystals. Samples with an excess and a deficiency of Se (5% off the stoichiometric composition), were grown from the melt by the vertical Bridgman technique. The absorption coefficient has been derived from photoacoustic spectra in order to establish activation energies for several defect-related energy levels.

INTRODUCTION

In recent years, $CuInSe_2$ (CIS) and related compounds have been the subject of a vigorous development and study programme stimulated by their potential for use in photovoltaic devices. Although CIS-based thin film solar cells have achieved efficiencies in excess of 16% [1], there is a lack of understanding of the complex intrinsic defect structure [2] in this compound. Numerous studies have shown that the properties of the compound are dominated by various types of electrically active intrinsic defects associated with deviations from stoichiometry [3]. It is known that CIS can be made either n or p-type by adjusting the elemental composition of the material, either during growth or by employing post-growth annealing procedures. Several levels have been attributed to Se-related defects, and different interpretations are reported in the literature.

Photoacoustic spectroscopy (PAS) has emerged over the last few years as a promising technique for analysing the optoelectronic properties of semiconductors. One of its major advantages, apart from being non-destructive and contactless, is that it is virtually unique in observing non-radiative defect populations in semiconductors. These are known to be associated with energy loss mechanisms in devices.

The subgap absorption spectrum obtained from as-grown CIS crystals using a high resolution near-infrared photoacoustic spectrometer has been studied. Samples were grown by the vertical Bridgman technique from melts with different Se contents, which varied from the stoichiometric value by approximately 5%. Absorption coefficients were derived from the photoacoustic spectra in order to determine the gap energy and to establish activation energies for several defect-related energy levels.

EXPERIMENTAL

The specimens used in this work were parallel sided, as-grown n and p-type crystals, cut from the middle regions of ingots grown by the vertical Bridgman technique [4]. Samples were mechanically polished, using a final 0.05 micron grade alumina, then etched in a 1%Br in methanol solution for 1 minute. The composition, conductivity type and the initial Se-content for the various samples used, are listed in Table 1.

Table 1 : Data related to the samples.

Sample	Cond. type	Se-content	[Cu] Atm %	[In] Atm %	[Se] Atm %
CIS-1	n	stoichio.	23.1	25.9	51.0
CIS-2	n	-5%	23.5	26.2	50.3
CIS-3	p	stoichio.	22.7	26.3	51.0
CIS-4	p	+2%	23.1	26.1	50.8
CIS-5	p	+5%	22.4	26.5	51.1

Details of the photoacoustic spectrometer are described elsewhere [5]. All spectra were recorded at room temperature using a light source modulation frequency of 112Hz. This relatively low frequency gives a thermal diffusion length μ_s=120 μm.

RESULTS AND DISCUSSION

Fig.1 shows PA spectra measured from two n-type CIS crystals. The spectra clearly show the fundamental band-to-band transition edge together with peaks at lower energies which are associated with either impurity-to-conduction

CH3365-4/94/0000-0226 $4 © 1994 IEEE   First WCPEC; Dec. 5-9, 1994; Hawaii band or valence band-to-impurity transitions. Five of those peaks (E_1 to E_5) are present in both samples. E_6 is observed in only CIS-2 and seems to be a composite of two different peaks. The fall-off of the absorption edge is seen to be much sharper for CIS-2 and is a reproducible characteristic of this type of sample. The difference between the two curves is probably due to the presence of a shallow defect level in CIS-1. Of all the samples analysed in this series, CIS-2 is closest to the stoichiometric composition.

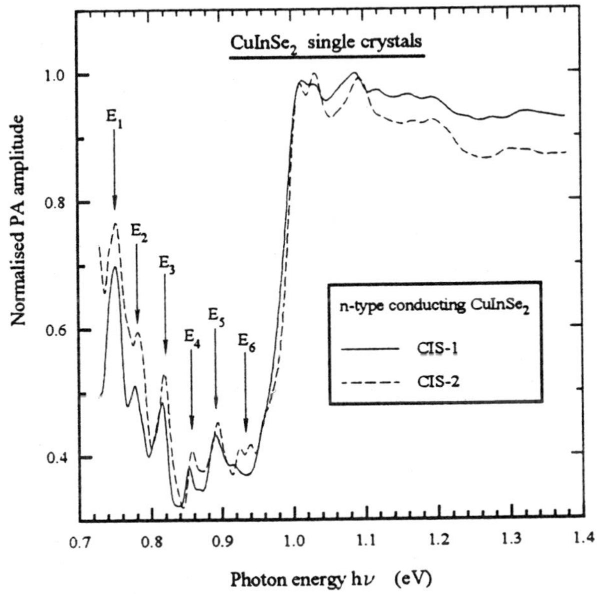

Fig. 1 : The normalised spectral dependence of the PA amplitude for as-grown n-type $CuInSe_2$ crystals.

Fig. 2 shows the PA spectra from three p-type crystals. The spectra again clearly show the band-to band transition, together with five of the main peaks already observed in the n-type samples. The variation in peak heights are associated with different levels of compensation. The shoulder detected close to the band edge maximum in these spectra is a reproducible characteristic of p-type samples. This feature is enhanced with increasing Se-content.

These PA spectra are more clearly resolved when the absorption coefficient is considered. Fig. 3(a) shows the absorption coefficient spectra which were derived directly from normalised PA amplitude data near the fundamental absorption edge for the two n-type samples. Three distinct regions can be observed. In the first region ($0.95 \leq h\nu \leq 0.962$ eV) both spectra exhibit near-identical behaviour. Both curves show an exponential tail due to a shallow level (D_1=49±1 meV) which could be either a shallow donor or a shallow acceptor. A study of the electrical and optical properties of as-grown n-type single crystals [3,6] indicate the simultaneous presence of both shallow donors and

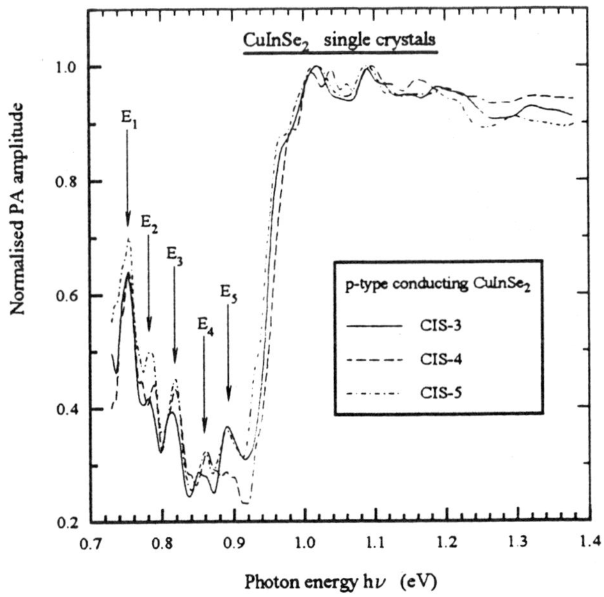

Fig. 2 : The normalised spectral dependence of the PA amplitude for as-grown p-type $CuInSe_2$ crystals.

acceptors in such crystals. The second region of the spectra ($0.962 \leq h\nu \leq 1.0$ eV) is characterised in CIS-1 by a broad peak, probably due to a second shallow defect level D_2. Such a level can be more accurately located by subtracting the absorption coefficient of CIS-2 from that of CIS-1. This is shown in Fig. 3(b). The analysis yields an activation energy of 24±1 meV for level D_2 and is most likely to be a shallow donor. For $h\nu \geq 1.0$eV, both spectra are very similar, and exhibit a steep increase towards 10^4 cm^{-1}. The gap energy of both samples is found from this data to be E_g=1.009±0.001 eV.

Fig. 4(a) illustrates the spectral dependence of the absorption coefficients of the p-type samples near the fundamental absorption edge. This part of the spectra can also be considered in terms of three distinct regions. For values of $h\nu \leq 0.96$, α is seen to exhibit an exponential dependence giving A_2=27±1 meV in the case of CIS-3, and A_3=19±1 meV for CIS-4 and CIS-5. A_2 and A_3 can also be associated with shallow levels in CIS [7]. Another feature which appears with increasing Se-content can be observed, - a small shoulder at $h\nu$=0.935 eV, masking A_2 (the shallow level observed in CIS-3). In the photon energy range $0.96 \leq h\nu \leq 1.0$ eV the p-type shoulder A_1 is clearly resolved. A more precise peak value is obtained when the absorption coefficient of CIS-2 is subtracted from those of the p-type samples. These, together with the small shoulder A_4 are shown in Fig. 4(b). The activation energies of these levels are A_1=35±3 meV (this level has been detected in many CIS samples and has been attributed to V_{Cu}) and A_4=72±2 meV, probably shallow acceptor states [8]. In a study of the electrical properties of CIS crystals [3] it has been argued that V_{Cu} give rise to an acceptor state with an ionisation energy in the range 78 to 90 meV. Above hν=1 eV the three curves exhibit steep increases with α values approaching 10^{+4} cm^{-1}. The gap energy for these samples is found to be E_g=1.006±0.002 eV.

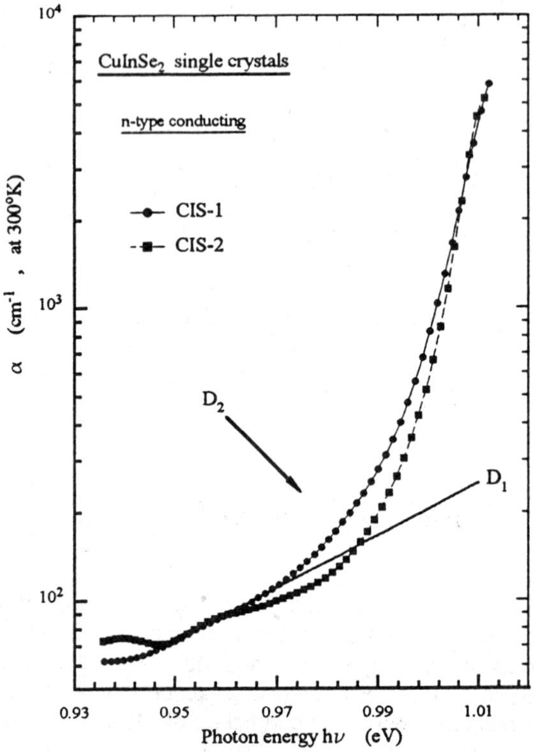

(a)

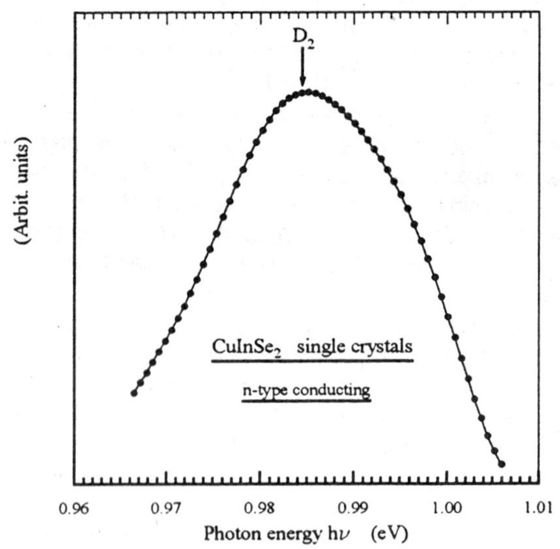

(b)

Fig. 3 : (a) The spectral dependence of the absorption coefficients near to the fundamental edge calculated from Fig. 1. (b) Spectral dependence of D_2.

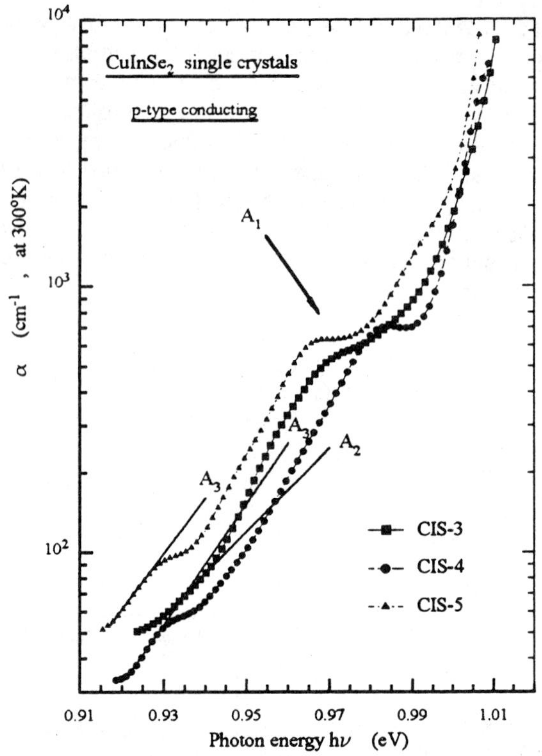

(a)

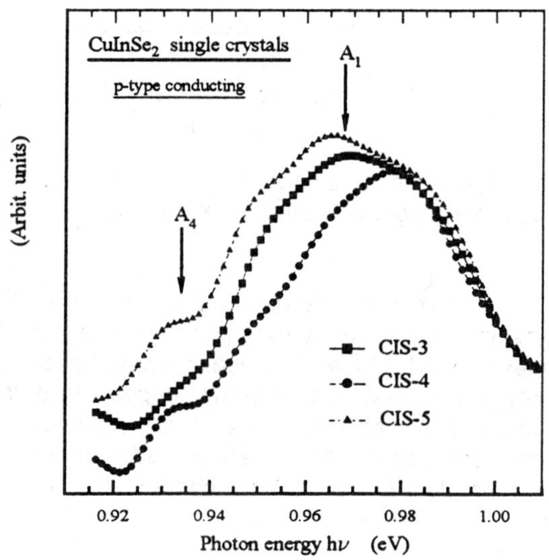

(b)

Fig. 4 : (a) The spectral dependence of the absorption coefficients near to the fundamental edge calculated from Fig. 2. (b) Spectral dependence of the shoulder (A_1).

Fig. 5 shows the absorption spectra of all samples below the fundamental edge. The spectra for CIS-1 and CIS-2 are shifted by 180 cm^{-1} for clarity. Five major peaks (E_1 to E_5) are observed in all spectra. E_6 observed only in CIS-2 has an activation energy of 80±10 meV (a shallow donor [8]) and could be due to a substantial population of V_{Se} defects [8]. The ionisation energies and possible activity of these peaks are listed in Table 2. The change in Se-content has to some extent affected the deeper levels, in particular E_1. The change in peak-heights which has been observed previously for E_4 and E_5 after annealing [7] is not apparent here, which could be explained by the fact that thermal treatment of CIS samples, if not carried out in the presence of CIS powder, will lead to major changes in the intrinsic defect equilibrium. The observed changes in this work can not be related to selenium defects only because the overall intrinsic defect equilibrium will be determined by the growth conditions.

Table 2 : Defect ionisation energies, and proposed electrical activity (**D** donor, **A** acceptor) of the corresponding defects.

Parameter	Energy (meV)	Proposed activity
D_1	49±1	D,A
D_2	24±1	D
A_1	35±3	A
A_2	27±1	D,A
A_3	19±1	D,A
A_4	72±2	A
E_1	258±6	A
E_2	232±8	D
E_3	190±6	D,A
E_4	154±7	A
E_5	120±4	D,A
E_6	80±10	D

CONCLUSION

In conclusion, we have reported on the effect of initial melt Se-content on the PA spectral distribution of bulk CIS single crystals. Shallow and deep non-radiative transitions have been detected and are found to be influenced by the selenium content of the samples. However, the interpretation of the data in term of specific defect states remains speculative and further systematic work is required with single crystal substrates in order to distinguish between surface and bulk states and their effect on opto-electronic characteristics.

ACKNOWLEDGEMENTS

Financial support of this work by the SERC, the Royal Society and the British-German ARC Programme is gratefully acknowledged.

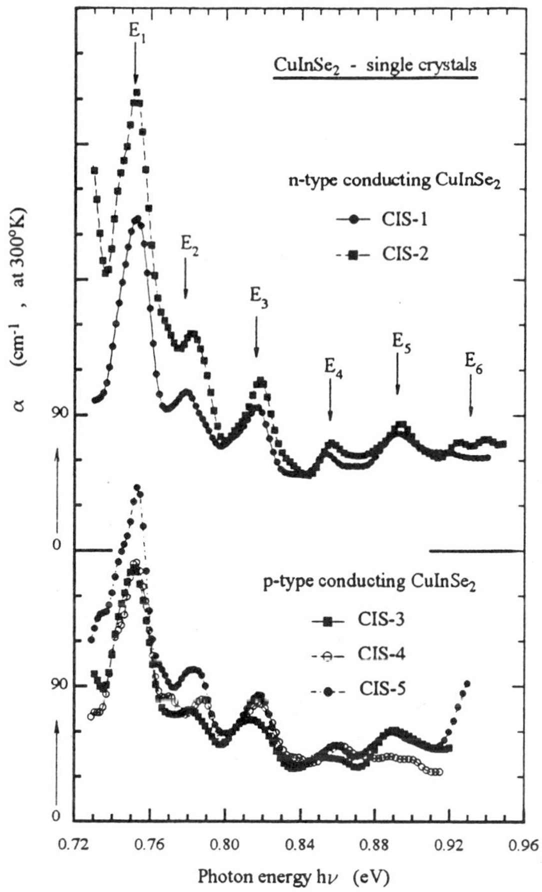

Fig. 5 : Comparative plot of absorption coefficients showing deeper state transitions in both n and p-type $CuInSe_2$.

REFERENCES

[1] H.W. Schock, "Solar Cells Based on $CuInSe_2$ and Related Compounds", Solar Energy Mats.& Solar Cells, **34**, 1994, pp.19-26.

[2] A. Rockett et al.,"Issues in the Structure and Chemistry of $CuInSe_2$ for Solar Cells",ThinSol.Films,**237**,1994,pp.1-11.

[3] H. Neumann and R.D Tomlinson, "Relationship between Electrical Properties and Composition in $CuInSe_2$ Single Crystals", Solar Cells, 28, 1990, pp.301-313.

[4] R.D. Tomlinson, "The Fabrication of $CuInSe_2$ Single Crystals using Melt-growth Techniques", Solar Cells, **16**, 1986, pp.17-26.

[5] A. Zegadi, M.A. Slifkin and R.D. Tomlinson, "A Photoacoustic Spectrometer for Measuring Subgap Absorption Spectra of Semiconductors", Rev. Sci. Instrum., **65**, 1994, pp.2238-2243.

[6] M.A. Slifkin et al., "Photoconductivity Spectra of n-type $CuInSe_2$ Crystals",Cryst. Res. Techn. **26**, 1991, pp.109-119.

[7] A. Zegadi, et al.,"Defect Levels in $CuInSe_2$ studied by Photoacoustic Spectroscopy", 12th E.C. Photovolt. Sol. Energy Conf., 1994, pp.1576-1579.

[8] F.A. Abou-Elfotouh et. al., "Characterisation of Defect Levels in $CuInSe_2$", Solar Cells **30**, 1991, pp.151-160.

GROWTH STUDIES OF $CuInSe_2$ USING Cu-Se FLUXES

Marc Casteleyn, Marc Burgelman, Ben Depuydt, Alex Niemegeers, Ivan Clemminck
Universiteit Gent - Vakgroep voor elektronica en informatiesystemen
St-Pietersnieuwstraat 41, B-9000 Gent, Belgium
tel.:+32-9-264.34.59 fax:+32-9-264.35.94

ABSTRACT

The latest developments in CIGS/CdS/ZnO cells show the importance of understanding the growth mechanisms of the absorber layer. For this purpose we studied the influence of Cu, CuSe, Cu_2Se and Se as additives to a $CuInSe_2$ base material. We found that in a selenium rich atmosphere the copper rich phase forms CuSe, which acts as a flux in a liquid phase sintering process (above its melting temperature of 523°C). Testing our findings to the selenization of glass/Mo/In(Se)/Cu structures lead to marked grain growth with grains up to 20μm.

INTRODUCTION

Recent improvements in $Cu(In,Ga)Se_2$ cell performances (partially) originate from a better morphology of the absorber layer [1,2]. Studies on how to improve this morphology, and to understand the mechanisms behind it are thus very useful. We have already shown in previous articles [3,4] that a Cu-Se phase promotes grain growth in copper rich $CuInSe_2$ layers, while for indium rich layers one can add $CuTlSe_2$ to improve the grain size. In this article we will try to explain the growth mechanisms in Cu-rich $CuInSe_2$ layers.

EXPERIMENTAL

To study these growth mechanisms we used the screen printing and sintering technique. Technological aspects of this technique are already described elsewhere [4,5]. We just mention here that as a base material for the paste we used indium rich $CuInSe_2$ grains (<2μm) to which CuSe, Cu_2Se and/or Se were added. These additives, called sintering fluxes, enhance grain growth of the base material at temperatures far below its melting point and in relative short times if the following conditions are fulfilled [4]:

- the flux material is liquid at the sintering temperature,
- the flux wets the CIS grains, so that a liquid shell is formed around the grains,
- and the CIS can be dissolved in the flux material.

This is known as liquid phase sintering.

Starting with very indium rich CIS (Cu/In=0.86) allows adding reasonable amounts (3 to 5 wt%) of CuSe or Cu_2Se, while still ending up with only slightly copper rich layers (Cu/In=1.05). Finally a composition adjustment should be done to make the surface indium rich. (We will use to notation Cu_xSe to refer to either Cu_2Se or CuSe).

RESULTS AND DISCUSSIONS

Heating of pure CuSe and Cu_2Se

To find out which Cu-Se phase is most likely to act as a growth promoter in Cu-rich CIS films, it is important to know which phase is preferentially formed at high temperatures. (The melting temperatures of the two important phases are 523°C for CuSe and 1130°C for Cu_2Se.)

It is in that perspective that we heated a CuSe powder in a sealed quartz ampoule to temperatures ranging from 450° to 650°C. XRD spectra of the treated powders (Fig. 1) showed that CuSe formed the main constituent, but also a small amount of Cu_3Se_2 appears at the higher temperatures. Heating the same powder, but now screen printed on an alumina substrate, in a box with 25 holes in the lid [5], gave a totally different result. All samples (temperature range: 450°-650°C) exhibit only $Cu_{2-x}Se$ and Cu_2Se peaks in the XRD spectra (Fig. 1). Also the weight lost during the heating matches a selenium loss as in

$$2CuSe \rightarrow Cu_2Se + \tfrac{1}{2}Se_2\uparrow \qquad (1)$$

perfectly. The same heat treatment done on a Cu_2Se powder gave no weight loss and only Cu_2Se peaks were seen in the XRD.

From this we can conclude that, at temperatures between 450° and 650°C, CuSe is a stable product, if a high Se-pressure is present. In a selenium poor atmosphere CuSe loses part of its Se and forms Cu_2Se. These results are in agreement with thermodynamical studies done by Moony and Lamoreaux [6] for the spray pyrolysis of CIS: in an H_2Se rich atmosphere CuSe, rather than Cu_2Se, is a stable product, but with less Se Cu_2Se is formed, and not CuSe.

It is also important to note the marked grain growth (Fig. 2) of the screen printed CuSe layer, when heated to 450°C (which is almost 75°C beneath its melting point!) in a box with 25 holes. This can be explained be a high selfdiffusion of the material near its melting point (solid state sintering), perhaps aided by the

CH3365-4/94/0000-0230 $4.00 © 1994 IEEE

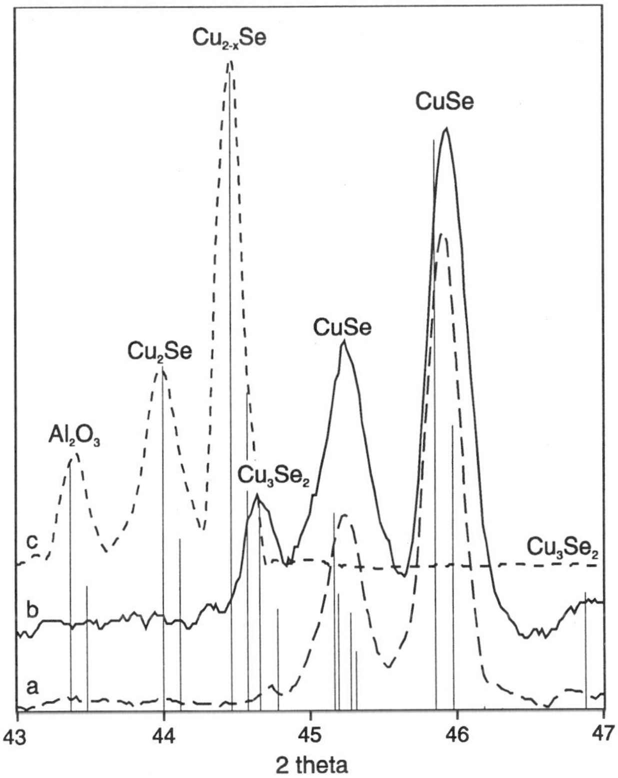

Fig. 1 : XRD spectra of a unsintered CuSe powder (a), a CuSe powder heated to 550°C in a closed ampoule (b), and a screen printed CuSe layer on a alumina substrate, sintered in a box with 25 holes at 600°C.

rearrangement effect of liquid selenium (=the first step in the liquid phase sintering [4]). This selenium is formed by that part of the CuSe that already formed the more stable Cu_2Se phase.

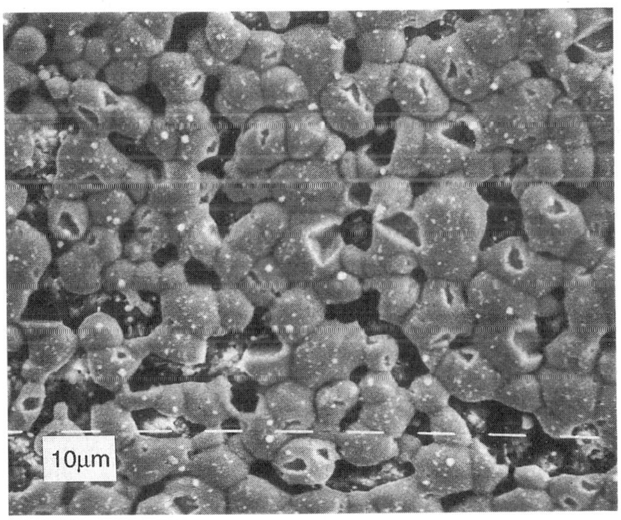

Fig. 2 : SEM picture of a CuSe powder screen printed on a alumina substrate and sintered at 450°C, which is almost 75°C beneath its melting point.

Sintering of CIS + Cu_xSe (x=1 or 2)

Adding only CuSe or Cu_2Se to an indium rich CIS base material does not improve the morphology when sintered at 625° or 725°C. Only if extreme large amounts of CuSe are added (final Cu/In=1.89) we see some coalescence of the grains. In this case the selenium loss is large enough (and therefore also the selenium partial pressure in the semi closed box) to temper the reaction (1). At least in the initial stages there will be some liquid CuSe left that can act as a flux. Due to the diffusion of Se out of the semi closed box this tempering is not lasting long, so that the growth only goes through its initial steps. Adding the same amount of Cu_2Se instead of CuSe gives no morphology improvement. XRD spectra of both these very copper rich films show Cu_2Se peaks, but no CuSe peaks, which is in agreement with the explanation given above (Fig. 3).

Sintering of CIS + Se

Sintering $CuInSe_2$ with only Se as an additive gives no improvement to the morphology, except if the CIS base powder is copper rich and its preparation enabled a Cu_2Se precipitation [4]. This case is described in the next paragraph.

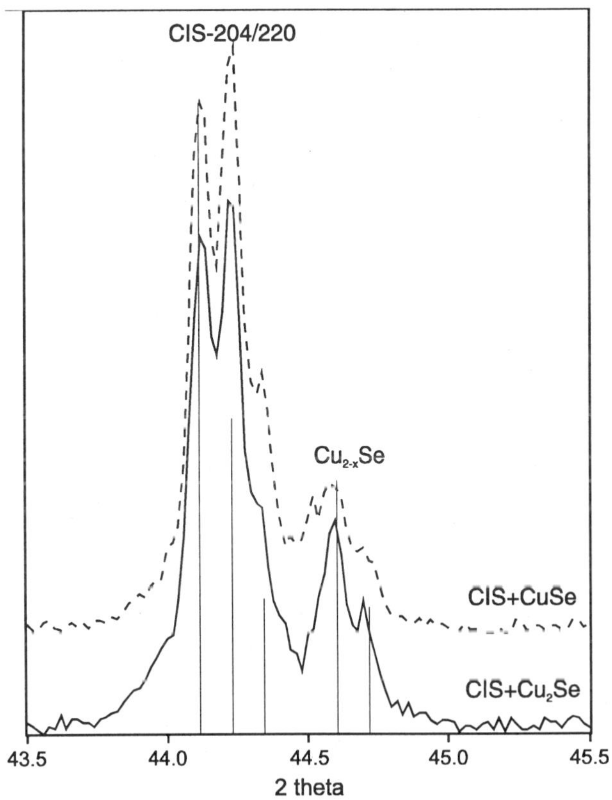

Fig. 3 : XRD spectra of a screen printed an sintered CIS layer to which large amount of Cu_2Se (a) or CuSe (b) where added. In both cases only CIS and $Cu_{2-x}Se$ peaks are visible.

Sintering of CIS + Cu_xSe + Se (x=1 or 2)

Both CuSe and Cu_2Se improve the morphology of indium rich CIS layers if the final composition is copper rich and a sufficient amount of Se is also added to the paste (10wt% or more) (Fig. 4). The added selenium makes CuSe the stable phase, so that Cu_2Se converts to this phase. Above 523°C CuSe becomes liquid, and acts as a flux in the liquid phase sintering process. As described earlier [4] this enhances grain growth in relative short times. Adding Se is thus necessary to stabilize the CuSe phase, so that liquid phase sintering can come at hand.

Because the starting grains are large in the screen printing and sintering technology, the grain growth is not so pronounced. Also, at temperatures between 525°C and 600°C the morphology improvement is only in islands. These islands are probably formed around the initial grains of Cu_xSe. The area of improved morphology can be enhanced if:

- the final Cu/In ratio is larger (with the same initial ratio), or the initial Cu/In ratio is smaller (with the same final ratio) (both indicate the necessity for larger amounts of added Cu_xSe)
- the sintering temperature is higher (the mobility and reactivity of liquid CuSe is larger, so that the liquid phase sintering is enhanced),
- the number of holes in the lid is smaller (only zero holes give reasonable sintering),
- the amount of Se added is larger,
- the heating up is slower,
- an extra naked substrate is placed at 0.7mm above the CIS layer, to ensure a higher selenium partial pressure close to the substrate.

The last four conditions indicate the necessity to keep the selenium present in the film for as long as possible.

In these conditions, we even noted rounding of the sharp edges of the grains and coalescence at temperatures down to 475°C, where CuSe is still solid (Fig. 5). This is probably related to the large grain growth seen when pure CuSe is heated to 450°C (see Fig. 2). The high diffusivity of CuSe and CIS in each other and the possible rearrangement effect of liquid selenium would then explain this possibility to improve the morphology even at temperatures beneath 525°C.

Fig. 4 : SEM picture of a CIS + CuSe + Se layer, unsintered (overview : a), and sintered at 625°C in a semi closed box (overview and cross section view : b).

Selenization of In(Se) + Cu layers

The selenization of bilayers with the ICSVT technique (Isothermal Close Spaced Vapour Transport) is a perfect way to test our findings. In this method a

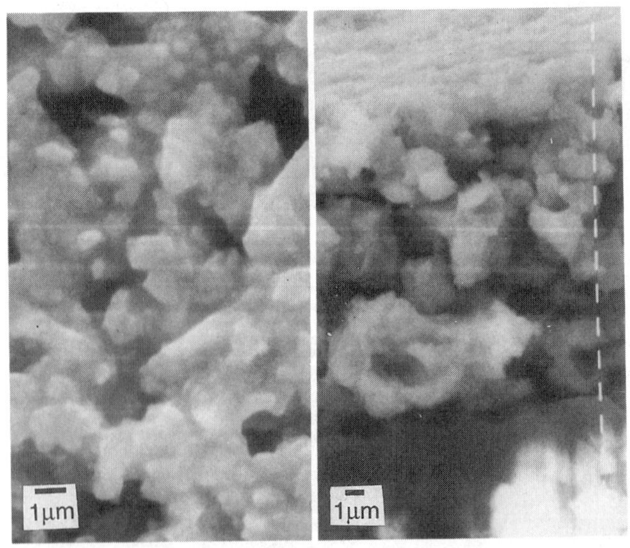

Fig. 5 : SEM picture of CIS + CuSe + Se layer sintered at 475°C in a semi closed box (overview and cross section view).

screen printed layer of selenium (≈ 18 μm thick) acts as a source to deposit selenium on an evaporated glass/Mo/In(Se)/Cu structure. Both substrates are put in an alumina box, with the Se facing the bilayer, and separated with spacers of 0.7mm. This box is then heated in an inert atmosphere.

As the initial grain size in the bilayer is much smaller than for screen printed layers, the densification and grain growth by liquid phase sintering should be more pronounced. Also the indium rich compound here is an In(Se) phase, so a maximum amount of Cu can be added as flux. We selenize at 525°C, which is above the melting point of CuSe. The selenium partial pressure will be high due to the high temperature of the source and the small distance between source and target.

Fig. 6 : SEM picture of an overview (a:x1250, b:x5000) and cross section view (c:x5000) of a evaporated glass/Mo/In(Se)/Cu structure, selenized at 525°C by ICSVT.

Fig. 6 shows a SEM picture of a slightly copper rich selenized sample. The layer is formed by closely packed big grains (up to 20μm) and is one grain thick (3μm). This marked grain growth proves that under the optimal conditions described above, the morphology improvement is maximum.

CONCLUSIONS

Copper rich $CuInSe_2$ layers can give good morphology films (e.g. monograin layers formed by large grains up to 20μm) if heated at temperatures above 525°C in a very selenium rich atmosphere. In this case the growth happens by a liquid phase sintering process with CuSe as liquid flux.

To obtain good photovoltaic layers a final composition adjustment will further be necessary to render the final composition indium rich.

ACKNOWLEDGEMENTS

We acknowledge the support of the Belgian NFWO (M.B., research associate), of the IWT (A.N., research assistant), and of the Commission of The European Union (JOULE Contract JOU2-CT92-141). We thank Sigmund Zweigart of the IPE at Stuttgart for the bilayer samples for our ICSVT experiments.

REFERENCES

[1] J. Hedström et al., "ZnO/CdS/Cu(In,Ga)Se_2 thin film solar cells with improved performance", *23th IEEE PVSC, Louisville,* 1993, pp. 364-371.

[2] A.M. Gabor et al., "High efficiency thin-film Cu(In,Ga)Se_2 solar cells: achievement of a confirmed world-record 16.4% total-area efficiency", *12th European PVSEC, Amsterdam,* 1994, pp. 939-943.

[3] M. Casteleyn et al., "Screen printed and sintered $CuInSe_2$ based chalcopyrite layers for solar cells", *12th European PVSEC, Amsterdam,* 1994, pp. 604-607.

[4] M. Casteleyn et al., "Sintering of screen printed $CuInSe_2$", *11th European PVSEC, Montreux,* 1992, pp. 826-829.

[5] A. Vervaet et al., "Screen printing of CIS films for CIS-CdS solar cells", *10th European PVSEC, Lisbon,* pp. 900-903.

[6] J.B. Mooney and R.H. Lamoreaux, "Spray Pyrolysis of $CuInSe_2$", *Solar Cells,* **16** (1986), pp. 211-220.

CHARACTERIZATION OF Cu-In-Ga PRECURSORS USED TO FORM $Cu(In,Ga)Se_2$ FILMS

M. Marudachalam, R. Birkmire, J. M. Schultz† and T. Yokimcus
Institute of Energy Conversion, University of Delaware
United States Department of Energy
University Center of Excellence for Photovoltaic Research and Education
(National Renewable Energy Laboratory)
Newark, Delaware 19716-3820 USA
†Department of Chemical Engineering, University of Delaware

ABSTRACT

The stacked precursors for the formation of $CuIn_{1-x}Ga_xSe_2$ were processed in sequences Cu/In/Ga and Cu/Ga/In for Ga/(Ga+In) ratio ranging from 0.2 to 1.0. The precursors, before and after a 1 hour heat treatment in an Ar atmosphere were found to contain only elemental and binary phases. The precursors were reacted with H_2Se or elemental Se at 400°C and 500°C. Reacted precursors contained mixtures of binary phases $Cu_{2-x}Se$, GaSe, Ga_2Se_3 and Cu-Ga, and, ternary phases $CuInSe_2$ and $CuGaSe_2$. The reacted precursors were heat treated in an Ar atmosphere and an H_2Se atmosphere. The heat treatment in an Ar atmosphere resulted in a single phase $CuIn_{1-x}Ga_xSe_2$ film, whereas, heat treatment in H_2Se atmosphere improved only the crystallinity of the phases present and does not effect homogenization.

INTRODUCTION

The $CuIn_{1-x}Ga_xSe_2$ films used for all high efficiency, >15%, solar cells where the Ga/(Ga+In) ratio is >0.2 have been deposited by physical vapor deposition [1,2]. Most approaches to fabricate modules based on $CuInSe_2$ and its alloys, however, react metal precursor layers in a Se atmosphere. This paper discusses the processing and characterization of the stacked Cu-In-Ga layers and the resulting films obtained by reacting the precursors with H_2Se or elemental Se at 400°C and 500°C.

EXPERIMENTAL

Cu/In/Ga and Cu/Ga/In stacked precursors with Ga/(Ga+In) ratios from 0.2 to 1.0 were deposited on molybdenum-coated soda lime glass substrates. Cu and In were sputtered from elemental targets in a DC magnetron sputtering unit and the Ga was deposited by physical vapor deposition to obtain Cu/(In+Ga) ≈1.0 The substrates were maintained at room temperature during the deposition of Cu, In and Ga. The Cu film thickness was chosen to yield 2 μm $CuIn_{1-x}Ga_xSe_2$ film. The Cu/Ga/In films were heat treated in an Ar atmosphere at 250°C for 1 hour to alloy the films. The films were characterized by X-ray diffraction (XRD) and the morphology was characterized by SEM before and after heat treatment. The composition of the heat treated films was evaluated by energy dispersive spectroscopy (EDS).

The as-deposited stacked precursors were reacted in a selenium atmosphere, by H_2Se[3] or elemental Se, at 400°C and 500°C. The reacted precursors were characterized by XRD, SEM and EDS. The films obtained by reacting Cu/In/Ga films in H_2Se were heat treated in an Ar atmosphere at 500°C to 600°C for 1 hour and in an H_2Se atmosphere at 500°C for 1 hour in separate experiments. The heat treated films were characterized by XRD and SIMS.

RESULTS AND DISCUSSION

Stacked Precursor Layers

Results of X-ray diffraction analysis of stacked elemental layers for deposition sequences Mo/Cu/In/Ga and Mo/Cu/Ga/In are summarized in Table.1. The phases formed in the as-deposited condition for both sequences are identical. The films contained only elemental phases of Cu and In and binary phases CuIn and $CuGa_2$. The morphology of the Mo/Cu/Ga/In films were more uniform than that of Mo/Cu/In/Ga (see Fig. 1) and were most likely controlled by the $CuGa_2$ phase. Since the morphology was uniform only for the sequence Mo/Cu/Ga/In, only these films were heat treated and their composition was evaluated after heat treatment. The morphology of the films did not change upon heat treatment. The heat treated films contained elemental phases Cu and In and binary phases $Cu_{11}In_9$, Cu_9Ga_4 and $CuGa_2$; no ternary phases were detected. The Cu_9Ga_4 phase was observed only in films with Cu/Ga > 2.0.

CH3365-4/94/0000-0234 $4.00 © 1994 IEEE

First WCPEC; Dec. 5-9, 1994; Hawaii

Reacted Precursors

The phases formed and the composition of the films reacted with H_2Se at 400°C and 500°C are shown in Table 2 for as-deposited precursors. The phases formed were found to be independent of the Ga and In deposition sequence and the morphology of the reacted films retained that of the precursors. Similar results were obtained for films reacted with elemental Se at the same temperatures. As shown in the table the films contained mixtures of $CuInSe_2$, $CuGaSe_2$ and secondary phases consisting of $Cu_{2-x}Se$, GaSe, Ga_2Se_3 and Cu-Ga depending on the starting Ga/(Ga+In) ratios. The secondary phases were observed only in films with Ga/(Ga+In) > 0.4. No binary phases containing In were observed. These two observations suggest that the reaction rates for the formation of the ternary $CuGaSe_2$ is lower than that of the $CuInSe_2$. The binary phases of Cu-In and Cu-Ga present in the precursors react with H_2Se or elemental Se to form the ternaries $CuInSe_2$ and $CuGaSe_2$. Further alloying between these ternaries was not favored for the reaction conditions investigated. The secondary phases were detected only for films with Ga/(Ga+In) > 0.4. Ga/(Ga+In) ratios measured by EDS, a technique that probes only 0.5 µm - 1.0 µm from the top surface, were less than expected in films containing In suggesting that in these films the Ga content was lower near the surface than at the Mo interface. This observation was supported by SIMS measurements.

The films reacted at 400°C were heat treated at temperatures from 500°C to 600°C in Ar and H_2Se atmosphere. Fig. 2 shows a detailed XRD scan of (220) and (204) peaks of the film with overall Ga/(Ga+In) ≈ 0.8 heat treated in an H_2Se atmosphere and of a film with the same composition heat treated in an Ar atmosphere. Heat treatment in a H_2Se atmosphere improved only the crystallinity of the $CuInSe_2$ and $CuGaSe_2$ phases. The Se content increased from 51.8 at. % to 54.2 at.% and the Ga/(Ga+In) ratio was effectively constant at 0.62. Single phase $CuIn_{0.22}Ga_{0.78}Se_2$ was obtained for the film heat treated in an Ar atmosphere. The Se content decreased from 51.8 at. % to 50.2 at. % and the Ga/(Ga+In) ratio increased to 0.78. The excess anion vacancies formed as a result of Se loss during the heat treatment in an Ar atmosphere could have promoted interdiffusion of In and Ga between the $CuInSe_2$ and $CuGaSe_2$ phases. Single phase $CuIn_{1-x}Ga_xSe_2$ films were also obtained for all Ga/(Ga+In) ratios by heat treating in an Ar atmosphere while no homogenization was obtained in H_2Se heat treatment. Some of the secondary phases were still present after heat treatment in films with Ga/(Ga+In) > 0.4 independent of the heat treatment.

Higher temperature heat treatments were required for producing single phase films with lower Ga content, i.e., Ga/(Ga+In) <0.4 . This can be explained by one or both of the following: 1) if the grain sizes in films with Ga/(Ga+In) < 0.4 are large, lattice diffusion is more dominant and higher temperatures would be required for complete interdiffusion, 2) if the diffusivities of In and Ga are low in $CuIn_{1-x}Ga_xSe_2$ films with lower Ga content then higher temperatures would be required for interdiffusion. Similar heat treatment results, in Ar and H_2Se, were obtained for films reacted with elemental Se at 500°C.

CONCLUSIONS

The stacked Cu-In-Ga precursor layers contained only elemental and binary phases before and after heat treatment. The morphology of the precursors with Mo/Cu/Ga/In sequence was found to be more uniform than that of Mo/Cu/In/Ga sequence. The reacted films contained mixtures of $CuInSe_2$, $CuGaSe_2$ and secondary phases. The presence of secondary phases in films with Ga/(Ga+In) > 0.4 suggests that reaction with Se was not complete in those films. Heat treatment in an Ar atmosphere produces single phase films whereas heat treatment in an H_2Se atmosphere does not yield single phase films. Excess Se vacancies which may have formed due to Se loss during heat treatment in Ar , which are unlikely to form during H_2Se heat treatment, probably aided the interdiffusion of In and Ga leading to single phase films.

ACKNOWLEDGEMENT

The authors thank Dr. Habib Hichri for his contributions and Dr. Angus Rockett for providing SIMS data. This work was supported in part by EPRI and NREL subcontract XAV-3-13170-01.

REFERENCES

[1] J. Hedstrom and H. Ohlsen , *Proc. of the 23rd IEEE PVSC*, **1993**, pp. 364.

[2] A. M. Gabor, J.R. Tuttle, M. Contrera, D.S. Albin, A. Franz, D.W. Niles and R. Noufi, *Proc. of the 12th E.C. PV Solar Energy Conf.*, **1994**, pp. 939.

[3] S. Verma, T.W.F. Russell and R.W. Birkmire , *Proc. of the 23rd IEEE PVSC*, **1993**, pp. 431.

Table 1. Composition and phases formed for different Ga/ (Ga+In) ratios.

Composition (at. %) (Determined for sequence Mo/Cu/Ga/In)	**Phases in Mo/Cu/In/Ga**	**Phases in Mo/Cu/Ga/In**	**Phases in Mo/Cu/Ga/In Annealed in Ar at 250° C for 1 hr.**
Cu= 0.52 Ga= 0.48	Cu, $CuGa_2$	Cu, $CuGa_2$	Cu, $CuGa_2$
Cu=0.45 In=0.12 Ga=0.43	Cu, In, $CuGa_2$	Cu, In, $CuGa_2$	Cu, In, $CuGa_2$
Cu=0.3 In=0.1 Ga=0.6	In , $CuGa_2$	In , $CuGa_2$	In , $CuGa_2$
Cu=0.43 In=0.38 Ga=0.19	Cu, In , CuIn , $CuGa_2$	Cu, In , CuIn , $CuGa_2$	In, Cu_9Ga_4
Cu=0.54 In=0.33 Ga=0.13	Cu, In , CuIn, Amorphous Ga	Cu, In , CuIn, $CuGa_2$	In , $Cu_{11}In_9$, Cu_9Ga_4

a) Cu/In/Ga

b) Cu/Ga/In

Fig. 1. Difference in morphology between Cu/In/Ga and Cu/Ga/In films with Ga/(Ga+In) ≈ 0.8. Spot EDS/ X-ray mapping revealed that the protruding regions on the surface of Cu/In/Ga films were In rich.

Table 2. Precursors reacted in H_2Se at 400°C.

Phases and Composition of Cu/In/Ga films reacted at 400°C for 90 mins.		**Phases and Composition of Cu/In/Ga films reacted at 500°C for 90 mins.**	
Compn. (at. %)	**Phases**	**Compn. (at. %)**	**Phases**
Cu=0.29 Ga=0.29 Se=0.42	$CuGaSe_2$, $Cu_{2-x}Se$, Cu-Ga	*	$CuGaSe_2$, $MoSe_2$
Cu=0.24 Ga=0.19 In=0.07 Se=0.5	Multiphase $CuIn_{1-x}Ga_xSe_2$, Cu-Ga	Cu=0.22 Ga=0.19 In=0.08 Se=0.51	$CuInSe_2$, $CuGaSe_2$ GaSe, Ga_2Se_3
Cu=0.20 Ga=0.18 In=0.10 Se=0.52	$CuInSe_2$, $CuGaSe_2$, GaSe, Cu-Ga	Cu=0.19 Ga=0.14 In=0.16 Se=0.51	$CuInSe_2$, $CuGaSe_2$, Ga_2Se_3,GaSe
Cu=0.23 Ga=0.04 In=0.19 Se=0.54	$CuInSe_2$, $CuGaSe_2$	*	$CuInSe_2$, $CuGaSe_2$, $MoSe_2$
Cu=0.24 Ga=0.01 In=0.18 Se=0.57	$CuInSe_2$, $CuGaSe_2$	*	$CuInSe_2$, $CuGaSe_2$, $MoSe_2$

Secondary phases $Cu_{2-x}Se$,Ga_2Se_3,GaSe and Cu-Ga were present in minor amounts.
* Film peeled

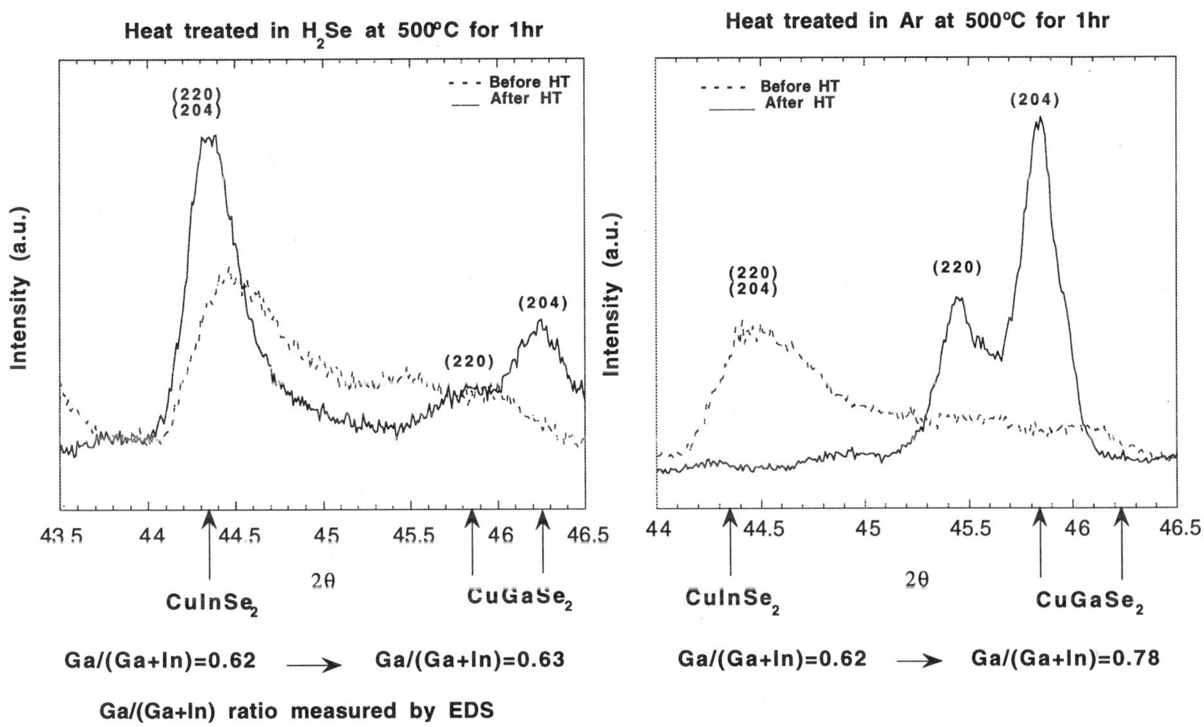

Fig.2. Detailed XRD scans of (220) and (204) peaks of the reacted films with Ga/(Ga+In)≈0.8 heat treated in an Ar atmosphere and in an H_2Se atmosphere.

A CHEMICAL KINETIC ANALYSIS OF THE FORMATION OF $CuInSe_2$ AT DIFFERENT TEMPERATURES

TW Fraser Russell, Nese Orbey and Robert W. Birkmire
Institute of Energy Conversion
University of Delaware

United States Department of Energy
University Center of Excellence
For Photovoltaics Research and Education
(National Renewable Energy Laboratory)
Newark, Delaware 19716

ABSTRACT

Understanding the mechanism of the growth of $CuInSe_2$ to obtain the pertinent kinetic data is essential for the effective design and real time control of processing equipment to produce $CuInSe_2$ modules. Species concentrations as a function of time and temperature are obtained from experiments in a tubular chemical vapor deposition selenization reactor and a physical vapor deposition selenization reactor.

In the present work, a chemical reaction analysis is carried out for both reactors.

Using this information, the numerical solution to the set of mass balance equations, estimates of the rate constants and the activation energies are presented. We believe that this reaction analysis can be used to interpret reactions which take place during formation of $CuInSe_2$ in a three source physical vapor deposition (PVD) reactor .

INTRODUCTION

Most high efficiency $CuInSe_2$ based solar cells have been made in multi-source PVD factories, but efforts to develop manufacturing facilities have emphasized selenization of metal precursors. A reaction analysis of film growth supplies the quantitative information needed to make $CuInSe_2$ by either reactor scheme.

A reaction analysis of the growth of $CuInSe_2$ films under isothermal conditions in a tubular CVD reactor at 400°C has been summarized in a previous paper by Russell et al. [1]. The experimental studies in the tubular CVD reactor [1,2] identified, by XRD analysis, the species which take part in the reaction to form $CuInSe_2$ film and produced chemical species concentrations as a function of time. A reaction analysis allowed the chemical equations to be postulated and the specific reaction rate constants for 400°C to be estimated from the experimental data. Previous studies in a PVD reactor [3] suggested that a slightly different set of chemical equations might be suitable.

In the present work we analyze experimental data of species concentration at different temperatures obtained in a PVD reactor to estimate energies of activation.

Rate constants are dependent upon temperature through the Arrhenius relation:

$$k_i = k_{oi} \exp(-E_i/RT). \qquad (1)$$

To obtain the energy of activation E_i, one would like to have chemical species concentrations as a function of time for a series of temperatures. This would allow the specific reaction rate constant k_i to be obtained for each temperature. A plot of ln k_i versus 1/T should yield a straight line with slope $-E_i/R$ and intercept k_{oi}.

Chemical species concentrations can not be obtained as a function of time in our PVD reactor so the k_i's are estimated using the species concentration measured in the film at the end of the experimental run. This, in our opinion, gives an estimate of the energy of activation which should be accurate within a factor of two.

EXPERIMENTAL

The time progressive experiments in the tubular CVD reactor have been described previously [1,2].

In the PVD reactor Cu-In layers were reacted by physical vapor deposition of Se from an open crucible [3]. The Cu and In were sequentially deposited at room temperature on 1μm-thick Mo-coated soda lime glass substrates by electron gun evaporation to form the starting Cu-In layers. The Cu thickness was fixed at 2500Å, while the In thickness was varied to control the initial Cu/In ratio from 0.8 to 0.9 and yield a final $CuInSe_2$ thickness of about 2μm.

Approximately 1.5μm of Se, enough to completely react the Cu-In layer, was evaporated onto the Cu-In substrates at room temperature. Following the Se deposition, the substrate temperature was ramped to the reaction temperature at a rate of ~35°C/minutes, then held for 30 minutes for reaction, while Se vapor is maintained at 10Å/sec.

CH3365-4/94/0000-0238 $4.00 © 1994 IEEE First WCPEC; Dec. 5–9, 1994; Hawaii

The selenized Cu-In layers were evaluated with XRD to determine species concentration as a function of temperature and initial Cu/In ratios. The data used in this analysis is shown as Table 1. All experiments were carried out with a reaction time of 30 minutes.

REACTION ANALYSIS

Examination of the experimental data from the PVD selenization reactor [3] and the CVD tubular reactor [1,2] (Table 1) suggests that the following set of chemical equations for $CuInSe_2$ film growth of Cu-In bilayers will allow film growth to be satisfactorily modeled.

$$2In + Se \Rightarrow In_2Se \qquad k_2$$
$$In_2Se + Se \Rightarrow 2InSe \qquad k_3$$
$$2Cu_{11}In_9 + 20Se \Rightarrow 11Cu_2Se + 9In_2Se \qquad k_5$$
$$Cu_2Se + 2InSe + Se \Rightarrow 2CuInSe_2 \qquad k_7$$

This set of chemical equations does not include a reaction, originally proposed between Cu_2Se and In_2Se to produce $CuInSe_2$. The simplified set of mass balance equations describing the film growth are shown in Table 2. The concentration versus time profiles predicated by solving this set of equations are shown in Figure 1 as solid lines. As can be seen from the figure, the data (solid dots) appears to fit well with the simpler set of chemical equations.

As discussed in our previous papers [1] the k_i's are related to the k'_i's as follows:

$$k'_i = k_i\,[Se] \qquad (2)$$

Equation 1 assumes that the Se concentration in the film is a constant.

The concentration-time profiles can't be obtained in our PVD reactor with its present configuration so a set of specific reaction rate constants (k_i's) must be estimated using only the initial conditions and the concentration measured at the end of the experimental run. The set of k_i's obtained for both the CVD and PVD reactor experiments are shown in Table 3.

The energy of activation, E_i and the frequency factor, k_{oi} are obtained by plotting ln k_i versus 1/T. A typical curve for k'_5 is shown as Figure 2. The values of k_{oi} obtained from this data are:

$$k'_{03} = 4.1 \times 10^3 \text{ min-1}$$
$$k'_{05} = 1.3 \times 10^3 \text{ min-1}$$
$$k'_{07} = 9.4 \times 10^3 \text{ cm}^3\text{/gmol-min}$$

CONCLUSIONS

1. Using the time progressive concentration profiles obtained in the tubular CVD reactor at 400°C:
   - a set of chemical equations simpler than the previously used is proposed;
   - the specific rate constants for these chemical equations at 400°C are obtained by solving the species conservation equations.

   The concentration profiles obtained with these rate constants gave good agreement with the experimental data.
2. Using the species concentrations measured at the end of the experimental runs in the PVD reactor, the specific rate constants at 250°C and 300°C are obtained by solving the species conservation equations. Good agreement is obtained for two different Cu/In ratios (0.8 and 0.9).

   Experiments are planned in the CVD reactor to obtain time progressive concentrations so more accurate estimates of the rate constants at 250°C and 300°C can be obtained.
3. The activation energies E_i and the frequency factors k_{oi} for the proposed reactions are obtained from the set of rate constants at 250°C, 300°C and 400°C using the Arrhenius relation.

ACKNOWLEDGEMENT

This work has been supported in part by the National Renewable Energy Laboratory under Subcontract number XAV-3-13170-01.

REFERENCES

[1] TWF Russell, S. Verma and R.W. Birkmire. *"Chemical Kinetic Studies of Copper Indium Selenization"*, 12th European PV Solar Energy Conf., Amsterdam , April 11-15, **1994**.

[2] S. Verma Ph.D. Thesis, University of Delaware **1993**.

[3] Satoshi Yamanaka, Brian E. McCandless and Robert W. Birkmire.*"Reaction Chemistry of $CuInSe_2$ Formation by Selenization Using Elemental Se"*, Proc. IEEE PVSC, **1993**.

Table 1. Experimental Results

T (°C)	$Cu_{11}In_9$	In_2Se	InSe	$CuInSe_2$
CuIn = 0.8				
250	0.57×10^{-6}	5.8×10^{-6}	0.7×10^{-6}	3×10^{-6}
300	0.2×10^{-6}	0	5.6×10^{-6}	9×10^{-6}
CuIn = 0.9				
250	0.63×10^{-6}	2.6×10^{-6}	2.5×10^{-6}	3×10^{-6}
300	0.11×10^{-6}	0	4.5×10^{-6}	9.5×10^{-6}

Table 2. Reaction Analysis Model Equations

$$\frac{d([In]V_f)}{dt} = -2k_2[Se][In]V_f = -2k_2'[In]V_f$$

$$\frac{d([In_2Se]V_f)}{dt} = (k_2[Se][In] - k_3[Se][In_2Se] + 9k_5[Se][Cu_{11}In_9])V_f = (k_2'[In] - k_3'[In_2Se] + 9k_5'[Cu_{11}In_9])V_f$$

$$\frac{d([InSe]V_f)}{dt} = (2k_3[Se][In_2Se] - 2k_7[Se][InSe][Cu_2Se])V_f = (2k_3'[In_2Se] - 2k_7'[InSe][Cu_2Se])V_f$$

$$\frac{d([Cu_{11}In_9]V_f)}{dt} = -2k_5[Se][Cu_{11}In_9]V_f = -2k_5'[Cu_{11}In_9]V_f$$

$$\frac{d([CIS]V_f)}{dt} = (2k_7[Se][InSe][Cu_2Se])V_f = (2k_7'[InSe][Cu_2Se])V_f$$

$$\frac{d([Cu_2Se]V_f)}{dt} = (11k_5[Se][Cu_{11}In_9] - k_7[Se][InSe][Cu_2Se])V_f = (11k_5'[Cu_{11}In_9] - k_7'[InSe][Cu_2Se])V_f$$

Table 3. Results

Reaction Rate Constants	CVD Reactor 400°C	PVD Reactor 300°C	PVD Reactor 250°C	Enegy of Activation, kJ/mol (factor of 2 estimate)
k2´ (1/min)	1.5	Not obtainable	Not obtainable	
k3´ (1/min)	0.3	0.04	0.02	$E_{a3} \approx 53$
k5´ (1/min)	0.2	0.05	0.02	$E_{a5} \approx 48$
k7 (cm^3/gmolmin)	26	8.0	5	$E_{a7} \approx 32$

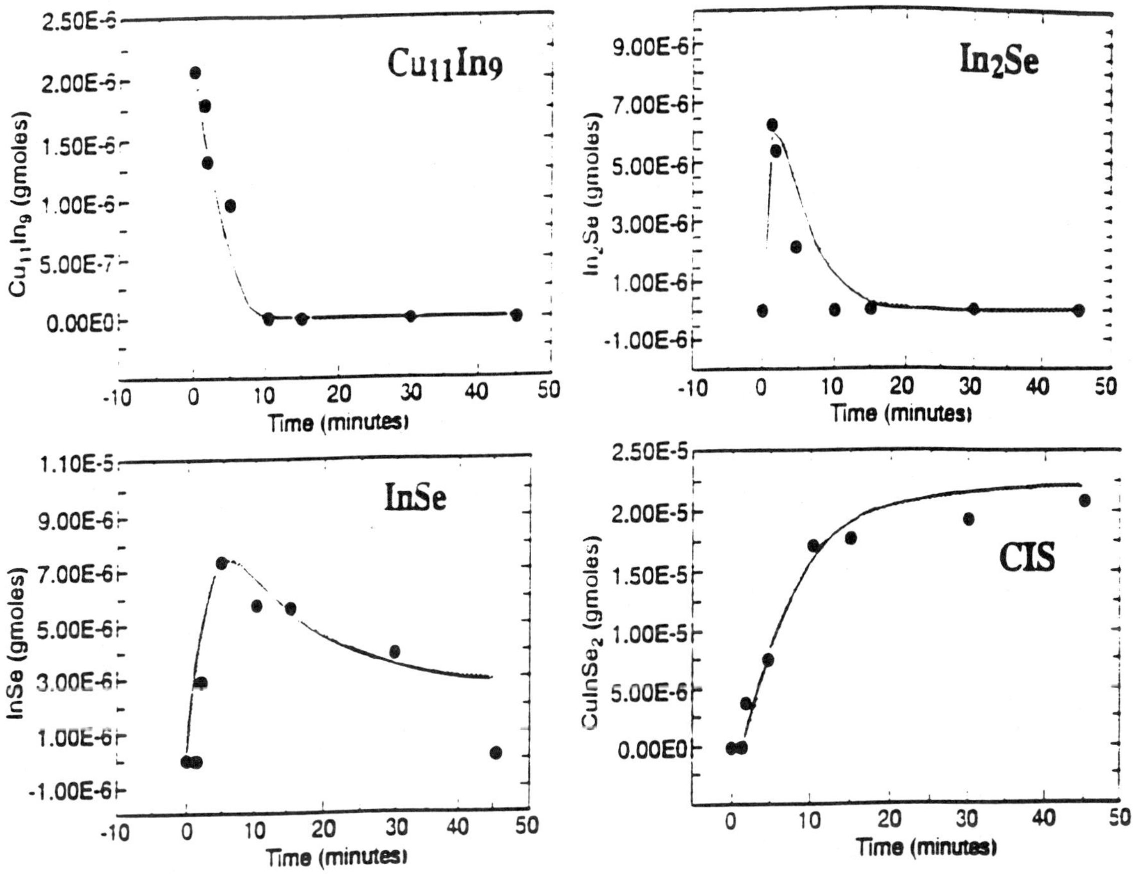

Figure 1. Time progressive selenization of Cu-In layers in the tubular CVD reactor (. experimental data; solid line: predictions).

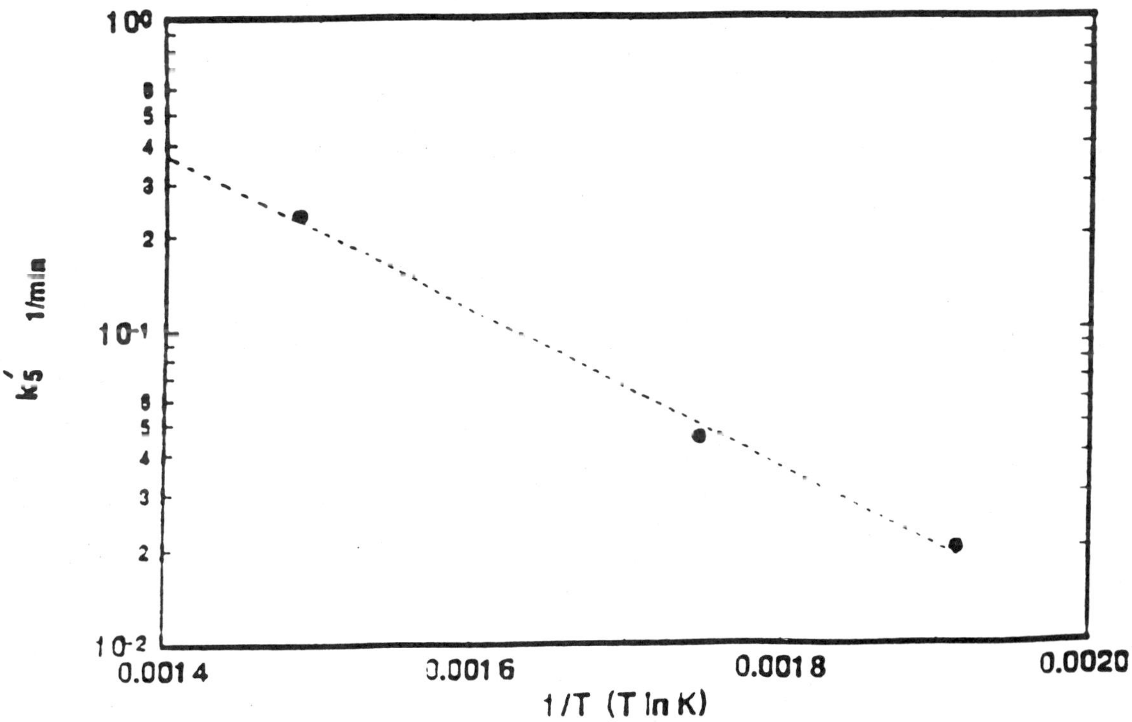

Figure 2. The Arrhenius plot to get the activation energy Ea_5 and frequency factor $k0_5$.

HETEROEPITAXY OF $CuIn_xSe_y$ ON Si AND $PbSe/CaF_2/Si$ SUBSTRATES

S. Blunier, A.N. Tiwari, K. Kessler, and H. Zogg
AFIF (Arbeitsgemeinschaft für Industrielle Forschung) at Swiss Federal Institute of Technology, ETH-Teil Technopark, Pfingstweidstr. 30, CH-8005 Zürich, Switzerland

M. Döbeli
Paul Scherrer Institute, c/o ETH-Hönggerberg, CH-8093 Zürich, Switzerland

M. Filzmoser
Laboratorium für Festkörperphysik, ETH-Hönggerberg, CH-8093 Zürich, Switzerland

ABSTRACT

Heteroepitaxial $CuInSe_2$, $CuIn_2Se_{3.5}$ and $CuIn_3Se_5$ ($CuIn_xSe_y$) layers have been directly grown on Si(111) substrates by molecular beam epitaxy. Reflection high energy electron diffraction was used to study the growth kinetics. (112)-oriented epitaxial layers free from impurity phases have been obtained. The positions of (448) x-ray diffraction peaks and vibrational mode frequencies of Raman spectra are used to identify different $CuIn_xSe_y$ phases. Rutherford backscattered ion channeling minimum yields of about 12% and 20% have been measured for about 0.5 μm thick $CuInSe_2$ and $CuIn_3Se_5$ layers, respectively.

INTRODUCTION

$CuInSe_2$ has emerged as an important material for high efficiency, stable thin film solar cells. A renewed interest has generated in the epitaxial growth of copper indium selenide (CIS) to understand the basic material properties. In high efficiency solar cells a part of the CIS layer near the CdS / CIS interface has a In-rich composition/phase [1]. This phase has been identified as $CuIn_3Se_5$. The pseudo binary phase diagram of Cu_2Se-In_2Se_3 suggests [2] the existence of $CuIn_3Se_5$ and $CuIn_2Se_{3.5}$ phases, and there are a few reports on the phase relationships and crystal structures of these phases [3,4].

Properties of $CuInSe_2$ are quite known but very little is known [5,6,7] about the physical properties of In-rich phases. Growth of $CuIn_xSe_y$ epitaxial layers with different composition will help in the measurement of different physical properties of these materials. Despite a large lattice mismatch of ~6% between tetragonal CIS and cubic Si, and ~4 times larger average thermal expansion coefficient of CIS compared to Si, we have grown [8,9] epitaxial CIS layers directly on Si substrates by molecular beam epitaxy (MBE). Here we describe the growth kinetics and characterization of heteroepitaxial $CuIn_xSe_y$ layers on Si(111) substrates.

EXPERIMENTAL TECHNIQUES

Heteroepitaxial CIS layers were grown on 3 inch Si(111) wafers by thermal evaporation of Cu, In, and Se from effusion cells at a substrate temperature of ~520°C with a growth rate of ~5 nm/min. The details of growth procedures have been described elsewhere [8,9]. Reflection high energy electron diffraction (RHEED) was used in-situ to monitor the clean Si surface and to study the CIS growth. The layers were characterized by θ-2θ x-ray diffraction (XRD), electron channeling (EC) in a scanning electron microscope (SEM), ion channeling in Rutherford Backscattering Spectroscopy (RBS), and Raman spectroscopy.

RESULTS AND DISCUSSION

RHEED has been extensively used to optimize the growth conditions and to study the growth kinetics. When CIS starts to grow on Si(111), the streaks in the RHEED pattern of Si transform to spots indicating growth of nucleated islands. Growth kinetics depends on the substrate temperature and Cu/In ratio. Fig. 1 shows the RHEED patterns of ~0.5 μm thick epitaxial $CuInSe_2$ and $CuIn_3Se_5$ layers. The spacing between diffraction spots and streaks for the two phases are not to be compared here, because the photographic camera was positioned at different distance for the growth runs. In the RHEED patterns of figs (a) and (b), in the [11-2] azimuth of Si, the characteristic reflections of the tetragonal chalcopyrite structure with lower intensity [10] are clearly observed for both phases. These patterns reappear after every 60° rotation around the substrate normal. The patterns of $CuInSe_2$ and $CuIn_3Se_5$ shown in figs. (c) and (d) are observed when the electron beam is along the [1-10] direction (30° rotation after the [11-2] azimuth). These patterns also reappear after every 60° rotation around the substrate normal. The pattern of $CuIn_3Se_5$ is different from that of $CuInSe_2$ in the [1-10] azimuth; in case of $CuIn_3Se_5$ additional spots are distinctly observed. We obtain streaky RHEED patterns with Kikuchi bands for thicker layers and CIS grown at higher temperature [8].

a: $CuInSe_2$, [11-2] azimuth

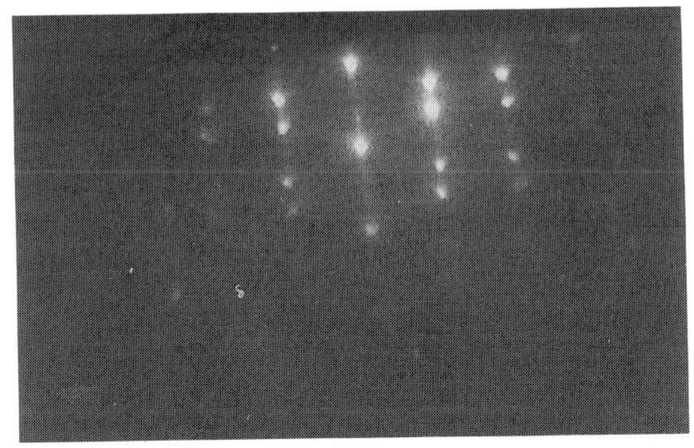

c: $CuInSe_2$, [1-10] azimuth

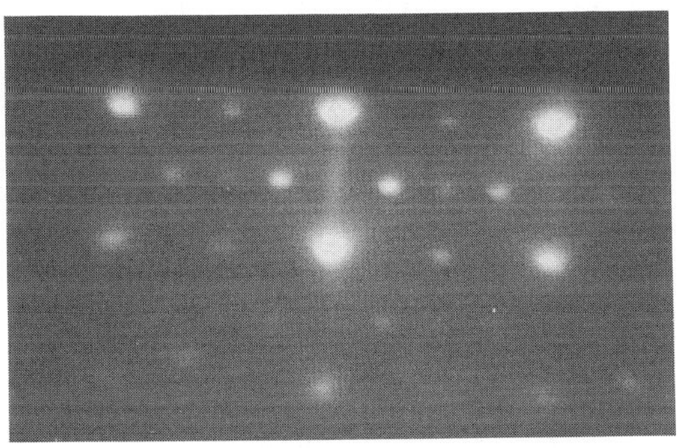

b: $CuIn_3Se_5$, [11-2] azimuth

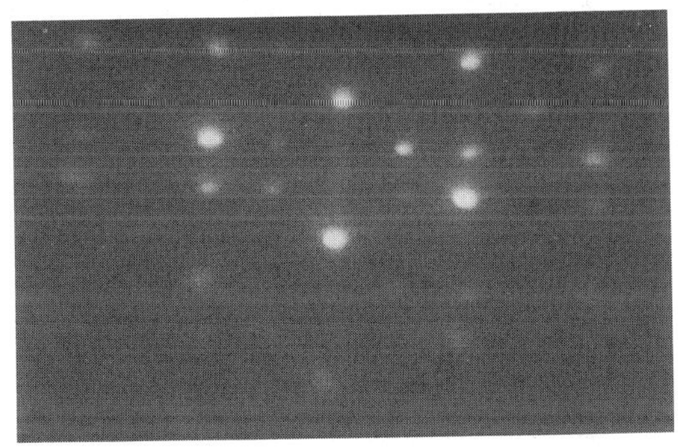

d: $CuIn_3Se_5$, [1-10] azimuth

Fig. 1: 15 kV RHEED patterns of ~0.5 μm thick layers in [11-2] azimuth of Si (left) and in [1-10] azimuth (right); (a) and (c) are for $CuInSe_2$/Si(111); (b) and (d) are for $CuIn_3Se_5$/Si(111).

EC patterns of $CuIn_3Se_5$ layers are sharp, well resolved and exhibit a three-fold symmetry. This is an indication for the epitaxial growth of (112)/(111)-oriented untwinned layers with a smooth morphology. The epitaxial relationship for the lattice arrangement is type-A compared to Si. Six-fold symmetry is observed for epitaxial $CuIn_2Se_{3.5}$ layers. For $CuInSe_2$ layers EC patterns exhibit a three-fold or six-fold symmetry depending on the growth conditions [8]. For Cu-rich $CuInSe_2$ epitaxial layers EC patterns are blurred because of rough morphology.

CIS layers are characterized by θ-2θ XRD scans using unfiltered Cu radiation (reflections from α1, α2 and β radiations are expected). Only CIS diffraction peaks corresponding to the (112) and its multiple reflections are observed along with the reflections of the substrate. It should be mentioned that the lattice constants of $CuInSe_2$ (a=5.78 Å, c=11.62 Å) and $CuIn_3Se_5$ (a=5.75, c=11.52) are almost same, and the XRD peaks for the epitaxial layers are expected at nearly the same 2θ value. However, a careful measurement and analysis of the (448) peak may be used to distinguish the two phases. The (448)α1 peak of $CuInSe_2$ is observed at θ=67.2°, while that of the $CuIn_3Se_5$ is shifted to 68.2°. The (448)/(444)α1 peak for an epitaxial (as evidenced by RHEED, ECP, and RBS) layer of intermediate composition (about $CuIn_2Se_{3.5}$ stoichiometry) is measured at 67.8°. The elemental composition of $CuIn_2Se_{3.5}$ is equivalent to the average composition of the $CuInSe_2$ and $CuIn_3Se_5$ mixture and the lattice constant is also intermediate [3,4]. The (448) XRD peak of $CuIn_2Se_{3.5}$ should be in between the (448) peaks of the $CuInSe_2$ and $CuIn_3Se_5$ phase. We analyzed the (448)α1,2 doublet and (448)β peaks. For the precise measurement a calibration was used by measuring the Si(444) diffraction peak. Fig. 2 shows a part of the θ-2θ XRD scans for the measurement of (448)β reflections of $CuInSe_2$, $CuIn_3Se_5$, and $CuIn_2Se_{3.5}$ layers. A comparison clearly establishes that the peaks for three phases are indeed at different positions. The peak widths, although slightly different (perhaps due to a difference in crystal quality) correspond to the singular nature of the material. This analysis (under certain assumptions for strain and difference in crystal quality) suggests the growth of the $CuIn_2Se_{3.5}$ phase.

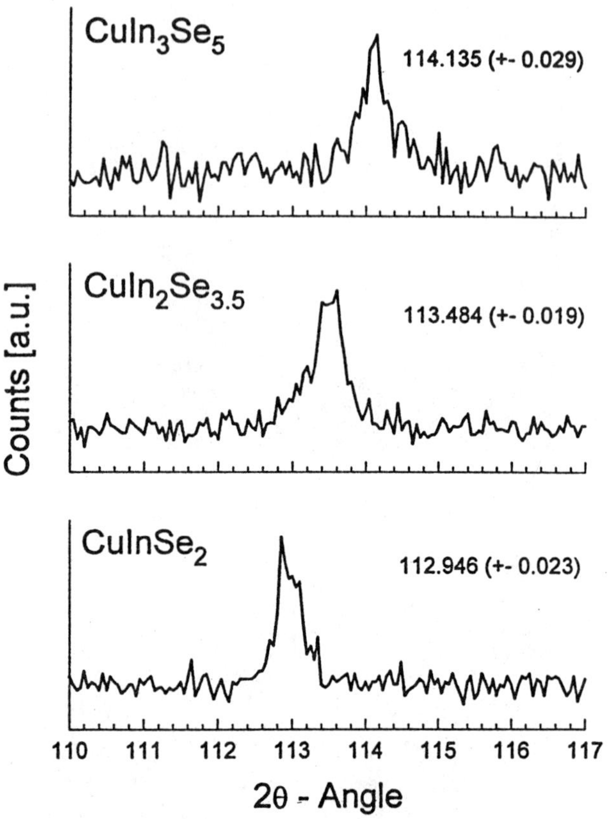

Fig. 2: θ-2θ XRD scans which show (448)β reflections for $CuInSe_2$/Si(111), $CuIn_3Se_5$/Si(111), and $CuIn_2Se_{3.5}$/Si(111) samples.

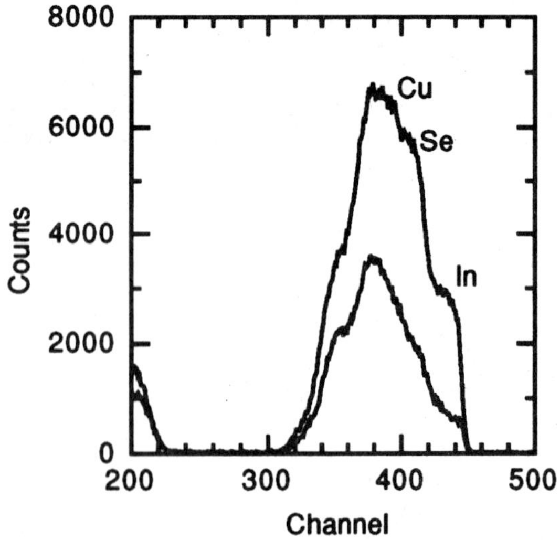

Fig. 3: RBS 2 MeV ion channeling spectra of a $CuIn_3Se_5$/Si(111) sample in random and aligned direction.

Ion channeling in RBS is used to evaluate the crystal quality. The RBS ion channeling minimum yield χ_{min} (defined as the ratio of the ion yield in the aligned to random direction) is a measure of the crystal quality. Fig. 3 shows the 2 MeV He ion RBS spectra of a 0.5 μm thick $CuIn_3Se_5$ layer on Si(111). χ_{min} is about 20% near the In related peak which suggests the epitaxial growth of reasonable crystal quality. For $CuInSe_2$ layers of similar thickness lower χ_{min} ~12% have been measured [8]. For $CuIn_2Se_{3.5}$ layers χ_{min} ~27% have been obtained. It is commonly observed that RBS χ_{min} values for epitaxial layers decrease with increasing layer thickness because of improvement in the crystal quality, lower values are expected for thicker layers.

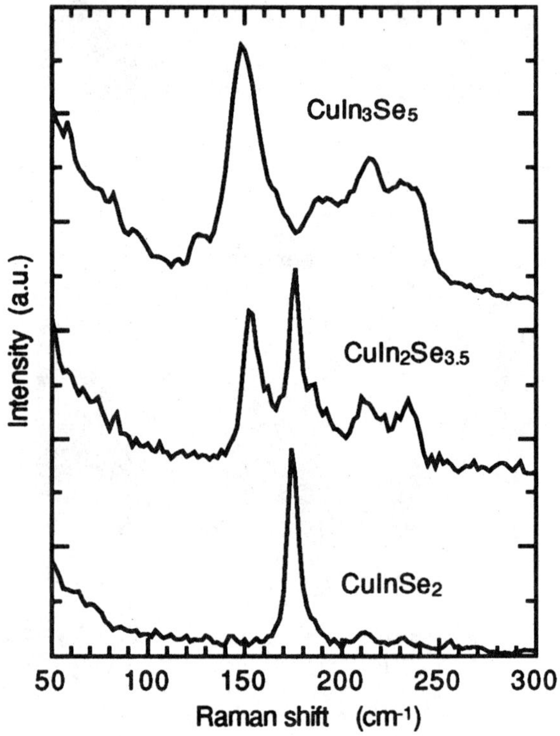

Fig. 4: Raman spectra of $CuInSe_2$, $CuIn_3Se_5$, and $CuIn_2Se_{3.5}$ layers on Si(111).

Fig. 4 shows the Raman spectra of epitaxial $CuInSe_2$, $CuIn_3Se_5$, and $CuIn_2Se_{3.5}$ layers. The most intense A1 vibrational mode is at 173 cm^{-1} for $CuInSe_2$ and at ≈150 cm^{-1} for the $CuIn_3Se_5$ phase. For $CuIn_2Se_{3.5}$ we observe two distinct, intense and sharp peaks at ≈150 and ≈170 cm^{-1}, and a broad band between 200 and 230 cm^{-1} superimposed with distinct peaks. In fact this spectrum seems to be the superposition of the $CuInSe_2$ and $CuIn_3Se_5$ spectra. The $CuIn_2Se_{3.5}$ spectrum is compared with the published Raman spectra of In-rich polycrystalline CIS layers where growth of mixed phase material is more probable, and also with the Raman spectra of different binary and ternary selenides of Cu and In. The spectrum of $CuIn_2Se_{3.5}$ is found to be quite different. The Raman spectra of chalcopyrite $CuInSe_2$ has been studied in detail but not much is known about the vibrational modes in CIS with defect chalcopyrite structure. The rearrangement of different atoms and vacancies in the crystal lattice should modify the vibrational modes, and for a disordered system the selection rules may be disobeyed giving rise to additional vibrational modes [11].

$CuInSe_2/PbSe/CaF_2/Si(111)$ for solar cells

A study of the CdS/CIS (single crystal) thin film solar cells will be interesting to understand the effect of grain boundaries on the junction transport mechanism. As a first step in this direction we have grown epitaxial $CuInSe_2$ on p- type epitaxial $PbSe/CaF_2/Si(111)$ substrates. The epitaxial PbSe layers (used for IR sensor fabrication in our lab) are doped with Se and have a carrier concentration of $\sim 6 \times 10^{17}$ cm^{-3}. They are expected to form an ohmic back-contact with CIS. We have grown CIS layers at ~500°C and performed in situ RHEED, surface morphology, XRD, and electron channeling measurements to characterize the layers. Fig. 5 shows the θ-2θ XRD scan for a CIS layer on $PbSe/CaF_2/Si(111)$ substrate.

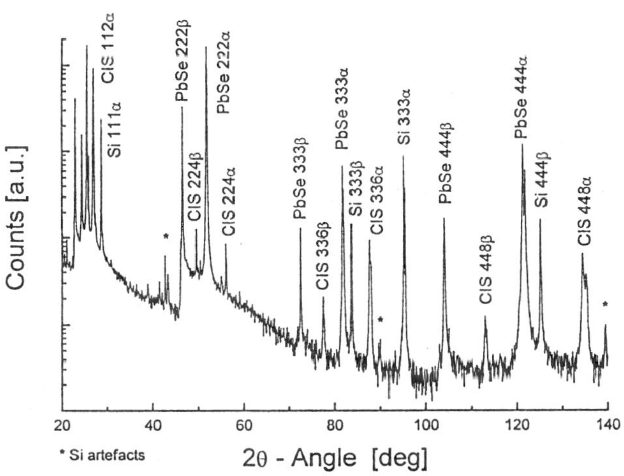

Fig. 5: θ-2θ XRD scan of $CuInSe_2/PbSe/CaF_2/Si(111)$

Diffraction peaks due to (112) oriented CIS are observed along with the peaks for the PbSe and Si substrate. CIS epitaxy is obtained but the PbSe layers become n-type due to Se out-diffusion. To preserve the p-type conductivity it is necessary to keep the substrate temperature below 400°C during the overgrowth of CIS layers. Presently conditions are being optimized for low temperature epitaxial growth.

Conclusions

In conclusion we have successfully grown epitaxial copper indium selenide layers of different compositions and phases on Si substrates. Properties of In-rich phases i.e. $CuIn_2Se_{3.5}$ and $CuIn_3Se_5$ have been studied and compared with $CuInSe_2$. Initial efforts have been made towards the development of solar cells based on epitaxial CIS layers.

ACKNOWLEDGEMENT

We thankfully acknowledge the generous supply of PbSe/Si substrates by A. Fach and J. John. A.N. Tiwari would like to thank Prof. H. Newmann for fruitful discussions on Raman measurements.

This work was performed within the EU project "EUROCIS", and supported by a grant from the Swiss Federal Office of Education and Science.

REFERENCES

[1]. D. Schmid et al., "Chalcopyrite/Defect Chalcopyrite Heterojunctions on the Basis of $CuInSe_2$", *J. Appl. Phys.*, **73**, 1993, pp. 2902-2909.

[2]. M.L. Fearheiley, "The Phase Relations in the Cu,In,Se System and the Growth of $CuInSe_2$ Single Crystals", *Solar Cells*, **16**, 1986, pp. 91-100.

[3]. U.C. Boehnke and G. Kühn, "Phase Relations in the Ternary System Cu-In-Se", *J. Mater. Science*, **22**, 1987, pp. 1635-1641.

[4]. W. Hönle et al., "Crystal Structures of Two Quenched Cu-In-Se Phases", *Cryst. Res. Technol.* **23**, 1988, pp. 1347-1354.

[5]. D. Schmid et al., "Investigation on the Defect Chalcopyrite Phase on the $CuInSe_2$ Surface", *12th E.C. Photovoltaic Solar Energy Conference and Exhibition*, 1994, pp. 050-057.

[6] A.J. Nelson, "Ordered vacancy compound $CuIn_3Se_5$ on GaAs(100): Epitaxial growth and characterization", *Appl. Phys. Lett.* **64**, 1994, pp. 3600-3601

[7] H.Z.Xiao et al., "Structural, optical and electrical properties of epitaxial $CuIn_3Se_5$-films", *J. Appl. Phys.*, 1994, pp. 1503-1510

[8] A.N. Tiwari et al., "Direct Growth of Heteroepitaxial $CuInSe_2$ Layers on Si Substrates", *Appl. Phys. Lett.*, **65**, 1994, pp. 2299-2301.

[9] A.N. Tiwari et al., "Characterization of Heteroepitaxial $CuIn_3Se_5$ and $CuInSe_2$ layers on Si Substrates", *Appl. Phys. Lett.*, **65**, 1994, to appear.

[10] O. Igarashi, "Epitaxial Growth of $CuInSe_2$ Single Crystal by Halogen Transport Method", *Journal of Crystal Growth*, **130**, 1993, pp. 343-356.

[11] H. Newmann, Personal communication

FABRICATION AND CHARACTERIZATION OF THIN FILMS OF $CuIn(S,Se)_2$ QUATERNARY COMPOUNDS OBTAINED BY CLOSE-SPACED VAPOR TRANSPORT.

F. Guastavino, A. Zouaoui and C. Llinarès
Centre d'Electronique de Montpellier, Université Montpellier II
F 34095 Montpellier Cédex 05 France

K. Djessas and G. Massé
C.E.F Semiconducteurs, Université de Perpignan
52, Avenue de Villeneuve
F 66860 Perpignan France

ABSTRACT

Close-spaced vapor transport in a vertical closed tube has been used and successfully developed for growing large area $CuInSe_2$ thin films. This method has been applied also to different $Cu(In, Ga)(Se,Te)_2$ ternary compounds (1)(2)(3). In this paper we present results concerning $CuInS_xSe_{2-x}$ with composition x varying from 0 to 2.00. The parameters of quasi-stoichiometric deposition are determined. We can obtain columnar grains of nearly 10µm diameter, well oriented in the (112) direction. Deviations from these parameters induce the presence of copper iodide in the samples (source temperature below the optimal one) or the formation of (In,Se) compounds (source temperature above the optimal one).

EXPERIMENTALS DETAILS

The reactor used is very simple and low cost. In Fig. 1 is shown this reactor: a quartz tube with an inner diameter of about 20 mm and a length of 10-20 cm. The source and substrate face each other and are separated by means of a quartz spacer. The distance between source and substrate can be varied from 0.3mm to 2mm. Iodine acting as a transport reagent is placed in the upper part of the tube. The heating system allows a temperature difference between source and substrate of the order of 25C and can be varied by modifying the area of the spacer in contact with the substrate. The rate of growth is 0.5 to 1µm/min.

The sources are plates of $CuInS_xSe_{2-x}$ powder pressed at 300 kg cm^{-2}. The powders are obtained from bulk syntheses prepared by chemical reaction at high temperature of the different elements: Cu, In, S and Se.

THEORITICAL CONSIDERATIONS

The transport reactions of the $CuInSe_2$ and $CuInS_2$ compounds by the iodine are:

$CuInSe_2$ (solid) + $2I_2$ (gas) = CuI (gas) +InI_3 (g) +Se_2 (g)

and

$CuInS_2$ (solid) + $2I_2$ (gas) = CuI (gas) +InI_3 (g) +S_2 (g)

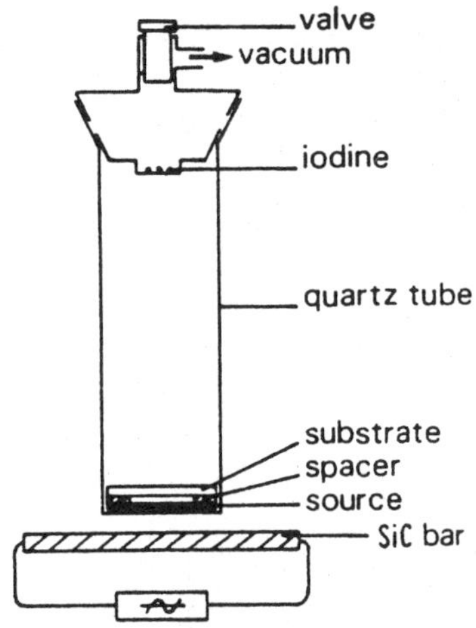

Fig.1 - Schematic diagram of the reactor.

First it is necessary to determine the temperature range at which films formed with only one phase can be deposited. According to the above chemical equations, the gaseous phase contains CuI, InI_3, S_2, Se_2 and I_2. The last three compounds are gaseous at T > 300°C under our pressure conditions (the tube is sealed under vacuum), but CuI might be solid at temperatures of the order of 500°C. The iodine pressure (in other words the iodine concentration) is also important for the stoichiometry and crystalline quality of the films.

From X-ray and electron microprobe analysis we have deduced the presence of CuI on films grown at source temperatures below 450°C and with iodine concentrations above 0.03 mg cm-3. The proportion of CuI increases as the temperature decreases or the iodine concentration increases.

For source temperatures above 450°C and iodine concentrations lower than 0.0.3 mg cm-3 there is no apparent CuI in the films, and the composition is nearly stoichiometric.

The formation of CuI is a way of modifying the film composition : copper involved in CuI is lacking in the material and copper vacancies are created. However, the

CH3365-4/94/0000-0246 $4.00 © 1994 IEEE
First WCPEC; Dec. 5-9, 1994; Hawaii

presence of CuI certainly modifies the electronic properties of the films.

EXPERIMENTAL RESULTS

X-ray studies

X-ray diffractographs indicate a strong orientation of the grains in the (112) direction (fig. 2). In fig. 3, is shown the position of this peak for films made with different sources, in good agreement with the position of the corresponding peaks of the sources used. The presence of a copper iodide (CuI) peak is detected for growth conditions below the temperature range for obtaining stoichiometric layers. When the temperature is too high peaks corresponding to InSe or In_2Se_3 appear.

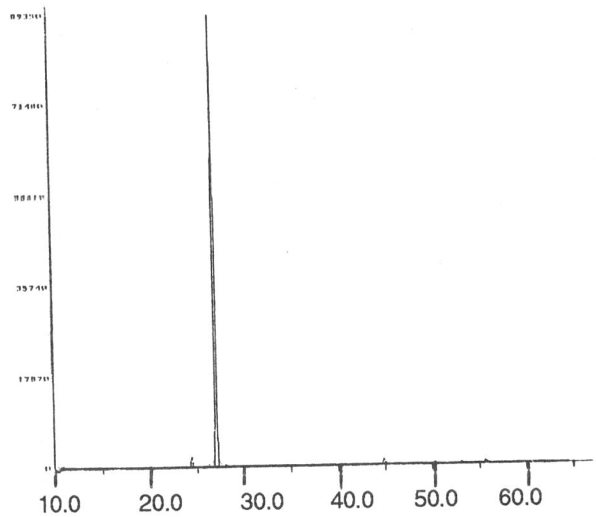

Fig. 2 - XRD of CuInSSe thin film showing a strong orientation in the (112) direction.

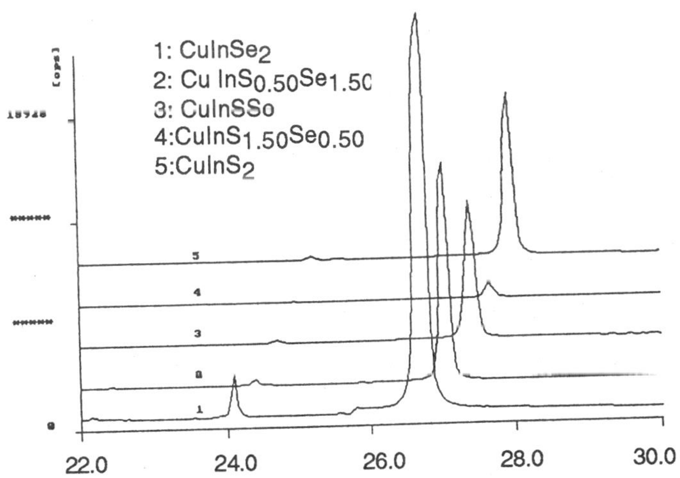

Fig. 3 - Shift of the (112) peak of the films with the composition of the starting material.

Film compositions

The composition and the morphology of the films depend essentially on three experimental parameters: temperature, iodine pressure and the composition of the source. The film compositions have been determined essentially by Electron Diffraction Spectroscopy (EDS). The morphology study has been done by Scanning Electron microscopy (SEM).

In tables I - a to d are indicated the compositions of films obtained at different temperatures from different source materials.

Table I - a: EDS composition of $CuIn(S_{0.25}Se_{0.75})_2$ films.

Sample	Temp.	Cu %	In %	S%	Se%
25-470	470 C	29.82	24.53	12.78	32.43
25-500	500 C	28.87	24.80	12.36	33.84
25-560	560 C	20.83	32.17	7.06	39.81
25-620	620 C	14.88	37.26	10.66	37.08

Table I - b: EDS composition of $CuIn(S_{0.50}Se_{0.50})_2$ films.

sample	Temp.	Cu %	In %	S%	Se%
50-470	470 C	26.96	24.93	26.46	21.09
50-500	500 C	27.96	25.15	25.54	21.07
50-560	560 C	27.87	25.48	25.04	21.24
50-620	620 C	25.15	28.72	19.21	26.91

Table I - c: EDS composition of $CuIn(S_{0.75}Se_{0.25})_2$ films.

Sample	Temp.	Cu %	In %	S%	Se%
75-470	470 C	24.48	24.64	40.54	9.79
75-500	500 C	27.74	24.41	37.82	9.66
75-560	560 C	25.43	25.11	39.12	9.96
75-620	620 C	22.66	24.93	37.58	14.46

Table I - d: EDS composition of $CuInS_2$ films.

Sample	Temp.	Cu %	In %	S%	Se%
S2-350	350 C	26.32	24.09	49.06	0
S2-500	500 C	26.93	23.98	48.32	0
S2-590	590 C	26.82	23.87	48.76	0
S2-620	620 C	26.21	24.93	48.83	0

At the lower temperature EDS detects the presence of iodine but at a non-measureable percentage. At the higher temperature a large amount of In and Se is present.

Film morphology

In fig.4a the SEM picture shows the cross-sectional morphology of a typical film obtained at 470C from a $CuInS_{0.5}Se_{1.5}$ source. We can distinguish easily the formation of two layers during the growth . The first layer with grains of <10µm diameter and a second one with very large grains.

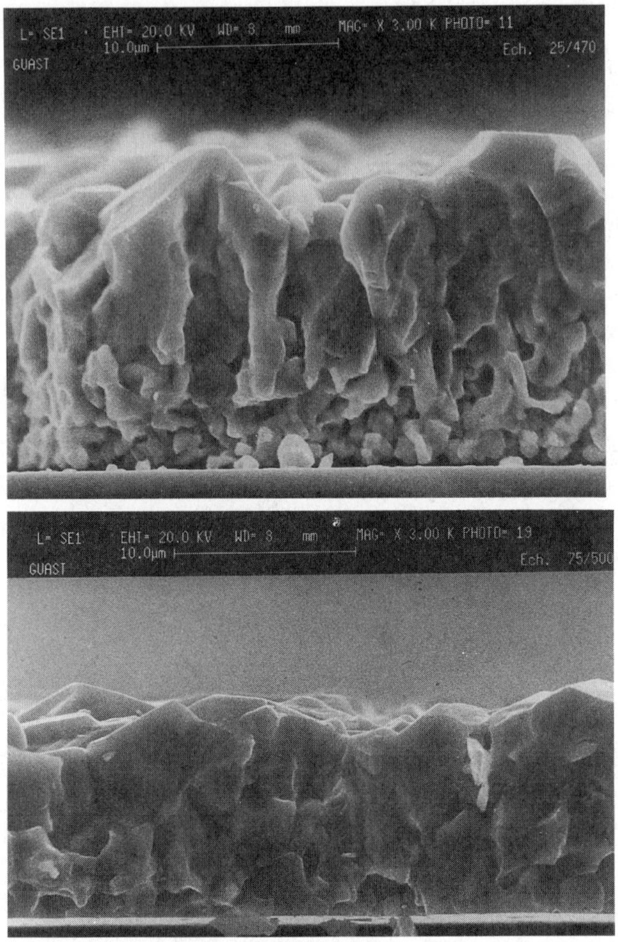

Figs. 4 a and b - SEM pictures for two typical films.

In fig.4 b the SEM picture shows the cross-sectional morphology of a typical film obtained at 500C from a $CuInS_{1.5}Se_{0.5}$ source.

Fig 5a shows the morphology of the surface of a film made at 500C. By observing the picture, we can see that the grains are polyhedral, with various shapes and well-crystallized faces.

An increase in the substrate (and source) temperature causes only a slight increase in the grain sizes, showing that the dependence of the grain sizes on the substrate temperature is weak in this temperature range.

When the iodine concentration increases, the grain sizes increase up to 50 µm or more for an iodine concentration greater than 1 mg cm-3. However, the samples are non-homogeneous, forming aggregates, since the high depositiion rates prevent the material from diffusing on the substrate surface .

For temperature higher than the range for obtaining stoichiometric films (T>560C), typical lamellar crystals appear in relation with the presence of a high content of In and Se in the film compositions (fig.5b).

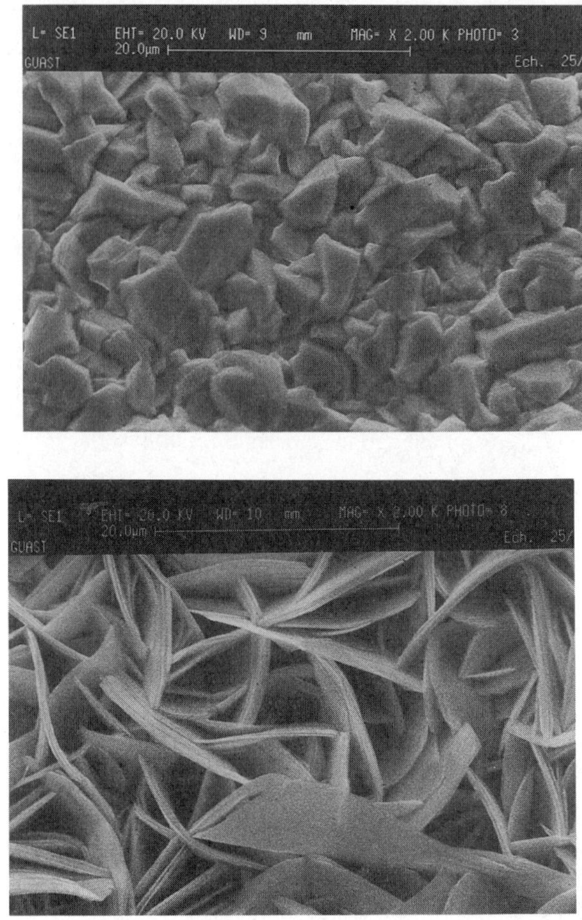

Figs. 5 a and b - Typical SEM pictures of two surfaces.

Analysises of the composition of the two layers in films made at 620C give the results presented in table II.

Table II: Topographical composition analysis of two films.

Sample	Layer	Cu %	In %	S%	Se%
00620	Top	10.56	41.29	0	47.86
00620	Bottom	11.49	35.25	0	52.64
25620	Top	16.61	37.12	9.83	36.42
25620	Bottom	8.02	33.02	8.02	43.12

Optical properties

In fig 6 and 7 are presented for two source compositions the transmission curves of films obtained at different temperatures. For a nominal composition of the starting material the positions of the transmission edges vs the temperature of growth are located at (or very near) the wavelength expected, i.e corresponding to the energy band gap of the source. But always at high temperature (T>560C) there is a noticeable shift toward higher wavelength corresponding to the presence of a high content of In and Se in the films. In fig. 8 is shown the shift of the transmission edge vs the composition of the starting material for films made at 500C.

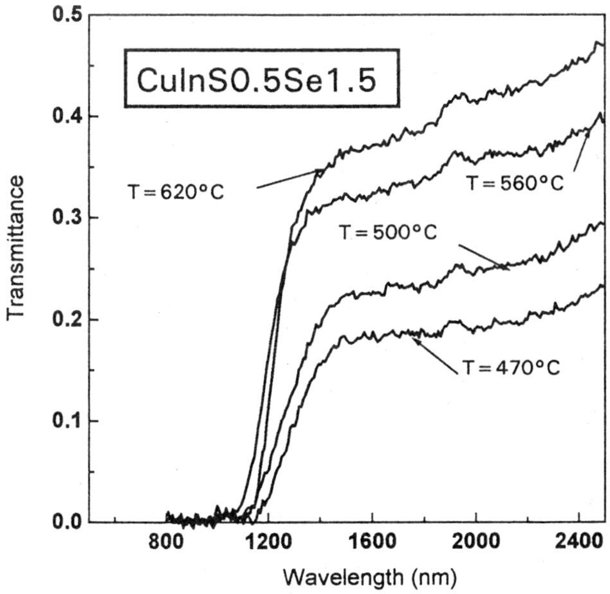

Fig. 6 - Optical transmission curves for a film grown from a starting material of $CuInS_{0.5}Se_{1.5}$.

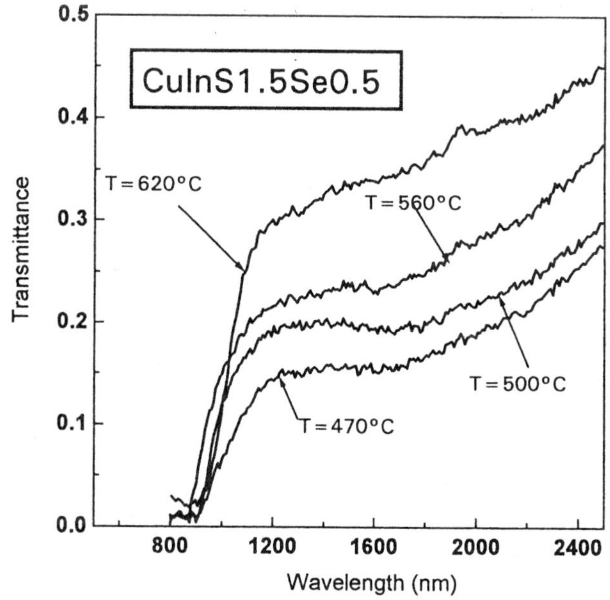

Fig.7 - Optical transmission curves for a film grown from a starting material of $CuInS_{1.5}Se_{0.5}$.

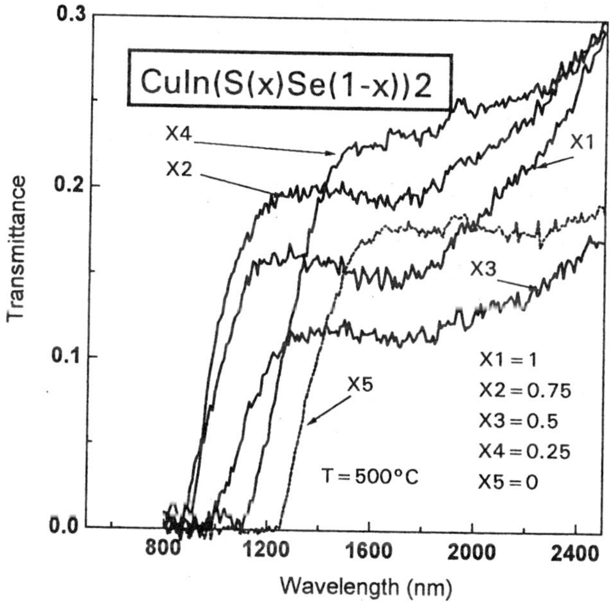

Fig.8 - Shift of the transmission edge with the composition of the starting material.

CONCLUSION

Close-spaced vapor transport (CSVT) is a simple and low cost method very promising for solar cellls fabrication. Our method is different from the one used in (4) because the reaction is done in a closed reactor. Taking into account the results concerning the quality of our films, we are now in the process of preparing cells using films of (Cd,Zn)S made by liquid bath chemical deposition and films of Cu In S_x Se_{2-x} made by CSVT.

REFERENCES

[1] G. Massé and K. Djessas. Thin Solids Films, 226 (1993) 254

[2] K. Djessas and G. Massé. Thin Solids Films, 232 (1993) 194

[3] G. Massé and K. Djessas. Thin Solids Films, 237 (1994) 129

[4] Chu T.L., Chu S.S., Chien C.P. and Lo D.H.J. Electrochem. Soc. 132 (1985) 2020.

CIS SOLAR CELLS WITH ZnO WINDOWS DEPOSITED BY ALE

Lars Stolt, Uppsala University, Teknikum, P.O. Box 534, S-75121 UPPSALA, Sweden
Jonas Hedström, Industrial Microelectronics Center, P.O. Box 1084, S-16421 KISTA, Sweden
Jarmo Skarp, Microchemistry Ltd., P.O. Box 45, FIN-02151 ESPOO, Finland

ABSTRACT

In an effort to find a process for deposition of the ZnO layer suitable to large scale fabrication of CIS cells the ALE method was investigated. In this study devices have been made with ZnO grown by ALE at temperatures which has been varied in the 125 - 200 °C interval. We find that high deposition temperatures have to be avoided in order not to degrade the open circuit voltage and fill factor of the devices. On the other hand the higher temperatures are needed to obtain high conductivity in the doped ZnO layer. In spite of that, devices with up to 14% efficiency were fabricated. Experiments where we omitted the CdS layer, normally used in CIS cells, were carried out. A promising result of 11% efficiency was obtained.

INTRODUCTION

$CuInSe_2$ (CIS) based thin film solar cells have demonstrated great promise in meeting the efficiency and the long term stability requirements for grid-connected electric power generation. No commercial manufacturing is going on today and among the major PV materials, CIS is the least developed in terms of production technology. Most of the development work has been devoted to upscaling the absorber deposition, expecting the other layers in the cell structure to cause less problems in this respect. The ZnO window layer is deposited by a CVD process in some of the reported work, especially that concerning module prototypes [1]. In the best reported $CuInSe_2$ and $Cu(In,Ga)Se_2$ cells the ZnO layer has been deposited by rf magnetron sputtering [2] or ion beam sputtering [3]. These processes cannot be scaled to large area production, where high throughput is essential. In this paper we report CIS cells with ZnO windows deposited by the ALE-process (Atomic Layer Epitaxy). Some of the features of the ALE technique are well known: good conformal coating/step coverage, uniform film thickness over large area substrates and high film quality (i.e. high density, low porosity). One important feature of ALE, which is not always realized, is the high productivity possible. In spite of the somewhat low growth rate, because the film is grown atomic layer by atomic layer, the technique allows compact parallell processing of ~50 large area substrates in one run [4].

ALE-ZnO

A brief summary of the ALE-ZnO process is given below. For details see Refs. [5] and [6].

Processes for ALE-ZnO has been developed using 5 × 5 cm^2 glass substrates in an ALE-reactor F-120 [7]. ZnO has been successfully grown on soda-lime glass substrates from organic zinc precursors, dimethyl zinc (DMZ) or diethyl zinc (DEZ), together with water or air. In order to enhance the conductivity, the material has been doped with boron or aluminium. The dopants have been introduced as e.g. trimethyl aluminium (TMAl) replacing DEZ in the DEZ + H_2O cycle. The process temperature has been varied from 120 °C to 350 °C.

A resistivity of $1 \cdot 10^{-3}$ Ωcm is obtained with two processes: either with DMZ/H_2O at 250 °C or with DEZ/H_2O at 200 °C, both using TMAl as the dopant agent. Hall mobilities as high as 30 cm^2/Vs has been measured.

Optical transmission measurements in the wavelength region 450 - 1100 nm have yielded average values of 84% to 90%, depending on film thickness and growth temperature. All the samples are stoichiometric ZnO, as confirmed by Rutherford backscattering spectrometry (RBS).

A large area ALE-reactor has been developed. The reactor is capable of handling plates of size 30.5 × 30.5 cm^2 (1 × 1 ft^2). Test runs have demonstrated uniform ZnO growth over the full substrate area.

EXPERIMENTAL

In order to study ALE-deposited ZnO window layers we fabricated a number of devices with the structure: soda lime glass/Mo/Cu(In,Ga)Se_2/CdS/ZnO. The absorber layers were deposited on the Mo-coated substrates by co-evaporation of the elements. The CdS layers were grown from solution (CBD) and about 50 nm thick. Except for the ZnO we used the same fabrication methods as reported earlier [2].

CH3365-4/94/0000-0250 $4.00 © 1994 IEEE

Table I. Sheet resistance of ZnO:Al films deposited with ALE at different substrate temperatures.

Growth temperature [°C]	125	150	175	200
Sheet resistivity [Ω/□]	400	120	40	35

Table II. Solar cell performance for different deposition temperatures [°C] of the ALE ZnO and the ALE ZnO:Al process. Open-circuit voltage and fill factor for reference devices with evaporated CdS:Ga windows are also included.

$T_{ZnO}/T_{ZnO:Al}$ [°C]	125/125	125/150	125/175	150/150	150/175	175/175	200/200
CIGS#	253B3	253B2	253B1	250B3	252A2	252A1	251B2
Eff. [%]	14.0	13.6	9.1	12.6	9.6	10.3	7.1
V_{oc} [mV]	584	586	498	595	497	497	348
J_{sc} [mA/cm^2]	35	34	31	32	33	31	33
FF [%]	68	70	59	67	59	66	62
Ref. device#				250B	252A	252A	251B
V_{oc}/FF [mV]/[%]				599/73	567/73	567/73	618/68

Device structures where the CdS-layer was omitted were also made by depositing the ZnO layers directly onto the chalcopyrite films. Having in mind that a sodium-containing layer resides on the chalcopyrite surface after the CIGS growth process [8], we made the CdS-free devices both on as grown CIGS films and on CIGS films immersed in 90 °C water just before the ZnO deposition. The water dip dissolves the sodium containing layer [8].

Thus, in each ALE-run three samples were processed in parallell; CIGS with CdS, as grown CIGS, and water-dipped CIGS. The number of ZnO cycles was the same in all runs. As in the state-of-the-art devices the ZnO windows were made in two steps, first a relatively thin (~50 nm) high resistivity layer and secondly, the conductive part of the window. We used 300 cycles for the undoped ZnO and 3000 cycles for the Al-doped ZnO. The doping was carried out by replacing each 30th cycle with TMAl + H_2O. The total thickness of the ZnO films is estimated to 400 - 500 nm.

We have investigated how the growth temperature in the ALE process influences the photovoltaic properties of the cells. A number of runs were made when the temperatures during the first and the second step of the ALE-growth were varied.

The heating-up times were 1.5 - 2 hours. A similar period of time was needed when the temperature was changed between the undoped and doped ZnO layers. The total deposition time was 66 minutes which corresponds to 1.2 seconds/cycle.

Reference devices were fabricated using an evaporated CdS/CdS:Ga window, in addition to the solution grown CdS. The total thickness of the evaporated CdS is 1.5 µm, of which the first 0.15 µm is undoped. Our experience is that this process give cells with similar V_{oc} and fill factor as devices with sputtered ZnO. Hence, they serve as a good indicator of the photovoltaic quality of the CIGS layer used.

Because of the low conductance of the ALE-ZnO:Al deposited at the lower temperatures, we also finished a few devices with rf-sputtered ITO (15 Ω/□). These devices have the window structure CdS/ALE-ZnO/ITO.

The cell performance was characterized with light J-V measurements. The cells were subjected to a one minute, 200 °C air anneal and 15 minutes of light-soaking before the characterization. The error in determining the current density is about ±2 mA/cm^2. This is due to inaccuracies measuring the area of the small test devices.

RESULTS

The growth temperature for the doped ALE-ZnO is governing the resulting resistivity. In Table I the measured sheet resistivities are presented for the growth temperatures used in this work. It is clear that only the two higher growth temperatures give reasonably good conductivities. The lower temperatures produce resistivities which are too high for module fabrication.

In Table II data from the cells with a CBD-CdS layer are shown. It is clear that the highest temperature allowed is 150°C in order to obtain good performance. The cell size is small and the contact arrangement is such that resistive losses even in the 125/125°C case are not excessive, but they will possibly affect the fill factor. The open circuit voltage obtained is at par with reference devices, or sputtered ZnO windows on comparable absorbers, but the fill factor is reduced.

In the low temperature cases, where the poor conductivity of the ZnO:Al layer may cause resistive losses, we can compare with devices where the ALE-ZnO:Al is replaced with ITO. The results from these cells are shown in Table III.

Table III. Solar cell performance for $Cu(In,Ga)Se_2$ cells with CdS/ALE-ZnO/ITO window structures.

T_{ZnO} [°C]	125	150
CIGS#	253C1	253C3
Eff. [%]	11.6	13.2
V_{oc} [mV]	521	563
J_{sc} [mA/cm^2]	33	33
FF [%]	68	71

The ALE experiments without using the CdS buffer layer were, with one exception, dissappointing. The exception is the 150/175 °C case on as-grown CIGS, see Table IV. An efficiency of 11% was noted on this device. The open-circuit voltage and short-circuit current were reasonably good, whereas the fill factor was on the low side. When the CIGS layers were water-dipped before the ALE-depositions the resulting conversion efficiencies were about one percent or lower, except for the lowest temperature sample (125/125 °C). It had an efficiency of 4.1% and good short-circuit current but very low V_{oc} and FF, 268 mV and 43%, respectively.

DISCUSSION

The open-circuit voltage results from the samples with CdS buffer layer give a quite clear picture. The 125 and 150 °C ALE growth temperatures give good values but they are lower by about 100 mV at 175 °C and at 200 °C the V_{oc} is very low.

Inspection of the fill factors give similar indications. A little bit surprising is the better performance of the 175/175 °C sample as compared to the 125/175 °C and the 150/175 °C ones. But one important factor appears to be the combination of time and temperature, and because of the temperature change during growth for the latter samples, they have been subjected to a longer heat treatment.

In order to check the influence of the temperature alone we heat treated a sample (125/150 °C) in air at 200 °C after characterization. After 1.5 hours the performance had deteriorated to a level comparable with the 200/200 °C ALE-ZnO devices.

It has already been said that the measurements are not strongly influenced by the high sheet resistivity in the low temperature cases. Looking at samples with 150 °C ALE-ZnO (undoped), and comparing ITO to ALE-ZnO:Al, it still seems as if excessive resistance lowers the fill factor of the all ALE device. Now, if that is the case the 125 °C ALE-ZnO:Al would be expected to yield even lower fill factors. This is apparently not what the results show, instead a slightly higher fill factor is noted for the 125/125 °C sample. Unfortunately, we have no reference devices for the CIGS material used for the 125 °C ALE samples, but it could be that this CIGS layer is of better photovoltaic quality.

Table IV. Solar cell performance for different deposition temperatures [°C] of the ALE ZnO and the ALE ZnO:Al process. The ALE ZnO is deposited on as-grown CIGS layers. Open-circuit voltage and fill factor for reference devices with evaporated CdS:Ga windows are also included.

$T_{ZnO}/T_{ZnO:Al}$ [°C]	125/125	125/150	125/175	150/150	150/175	175/175	200/200
CIGS#	230B2	238C1	250A2	251C1	251C2	252B2	258B1
Eff. [%]	0.2	0.2	0.0	2.3	11.0	4.1	0.6
V_{oc} [mV]	433	124	37	416	542	384	132
J_{sc} [mA/cm^2]	1.2	6.6	0.0	21.4	35.8	26.2	21.1
FF [%]	33	20	-	26	56.8	41	22
Ref. device#	230B	238A/B	250A	251C	251C	252B	
V_{oc}/FF [mV]/[%]	586/66	630/75	623/62	586/76	586/76	473/67	

In the search for a cadmium-sulphide or cadmium free CIS cell the most elegant solution would of course be to just omit this layer. When sputtering or CVD is used for depositing the ZnO one possible function of the CdS layer could be to protect the CIS surface. We had some hope before starting this work that the more gentle ALE process could work without the CdS buffer layer. One of the processes tried gives some support to this hope. Eleven percent efficiency was observed with ALE process temperatures of 150/175 °C. It is difficult to see why this combination gives this outstanding result when compared to the other. There is some indications that the 125 °C undoped ALE-ZnO is inferior to the 150 °C one, see Tables III and IV. The higher growth temperature is expected to yield a less porous film of higher crystalline quality. This may be important protecting the CIGS surface from a possibly deteriorating reaction with the metalorganic Al-precursor.

The last speculation is also applicable when explaining why the best of the 175 °C ALE-ZnO:Al devices with CdS has the highest (175°C) undoped ZnO growth temperature.

CONCLUSION

In summary, it is shown that the efficiencies obtained on ALE-ZnO devices are comparable to the performance obtained with techniques giving state-of-the-art cells. The best result we have achieved sofar is 14% (active area) on cells without an antireflective coating. These cells have a solution grown CdS layer.

However, the sheet resistivity of the ZnO:Al grown at tempertures producing good cell performance is too high for module fabrication.

Cadmium-free devices with 11% efficiency have been prepared by omitting the CdS layer and growing the ZnO directly on the $Cu(In,Ga)Se_2$ layer. The performance of these devices are limited by a low fill factor.

ACKNOWLEDGEMENT

This work was carried out as a collaborative effort within the EUROCIS, an E. C. JOULE project.

REFERENCES

[1] V.K. Kapur, B.M. Basol, A. Halani, C.R. Leidholm and A. Minnick, *Proc. 12th European Photovoltaic Solar Energy Conf.*, Amsterdam, The Netherlands, April, 1994, p. 1608.

[2]J. Hedström, H. Ohlsén, M. Bodegård, A. Kylner, L. Stolt, D. Hariskos, M. Ruckh, and H.W. Schock, *Proc. 23rd IEEE Photovoltaic Specialists Conference*, Louisville, Kentucky, May 1993, p. 364.

[3] M.A. Contreras, A.M. Gabor, A.L. Tennant, S. Asher, J. Tuttle and R. Noufi, *Progr. Photovolt.*, **2**, 1994, p. 287.

[4] T. Suntola, *MRS Fall meeting, Symposium E 1.1: Chemical Perspectives of Microelectronic Materials,* 1994.

[5]V. Lujala, J. Skarp, M. Tammenmaa, T Suntola and J. Wallinga, *Proc. 12th European Photovoltaic Solar Energy Conf.*, Amsterdam, The Netherlands, April, 1994, p. 1511.

[6] V. Lujala, J. Skarp, M. Tammenmaa and T. Suntola, *Appl. Surf. Sci.*, in press.

[7] ALE-reactor model F-120 is fabricated and marketed by Microchemistry Ltd, Espoo, Finland.

[8] M. Bodegård, L. Stolt and J. Hedström, *Proc. 12th Euoropean Photovoltaic Solar Energy Conf.*, Amsterdam, The Netherlands, April 1994, p. 1743.

EFFECTS OF Cd-FREE BUFFER LAYER FOR $CuInSe_2$ THIN-FILM SOLAR CELLS

T.Nii, I.Sugiyama, T.Kase, M.Sato, Y.Kaniyama, S.Kuriyagawa, K.Kushiya and H.Takeshita
Central R&D Laboratory, Showa Shell Sekiyu K.K.
123-1 Shimokawairi, Atsugi-shi, Kanagawa-ken, JAPAN

ABSTRACT

ZnO buffer layer by a chemical-bath deposition (CBD) method is developed in this study to improve the interface quality between n-ZnO window layer and p-$CuInSe_2$ (CIS) thin-film absorber in CIS thin-film solar cells as one of the approaches to the fabrication of Cd-free thin-film solar cells. The optimization of the fabrication conditions of CBD-ZnO leads to the efficiency of about 10 %. These results indicate the CBD-ZnO buffer layer has rather high capability to fabricate high-efficiency CIS thin-film solar cells.

INTRODUCTION

We have achieved considerably high efficiency of over 12 % in $CuInSe_2$ (CIS) thin-film solar cells with ZnO/CdS/CIS/Mo structure. However, some attempts to avoid the use of Cd-containing materials, from the stand-point of environmental safety, have been continued.

Application of ZnO as a buffer layer is performed to achieve this objective. Since ZnO is a wider band-gap material of about 3.3 eV than CdS with band-gap of 2.42 eV, the shorter-wavelength lights can transmit into the ZnO layer and are absorbed in the CIS thin-film absorber. The combination of ZnO window and ZnO buffer layers may be favorable to fabricate the CIS thin-film solar cells, because of the same material.

A chemical-bath deposition (CBD) method to prepare a high-quality ZnO buffer layer is developed and established in this study with rather high efficiency of 10 %. This fabrication method is considered to be applicable to the fabrication of large-area thin-film photovoltaic modules.

Since the CBD process proceeds in maintain the stoichiometry of cations and anions, the deposition rate is relatively slow due to the chemical reaction by way of ions. Better understanding of the CBD-ZnO fabrication process is closely linked to the improvement of device-quality ZnO buffer layers for CIS thin-film solar cells. The aim of this study is to identify and optimize the material parameters that are the most relevant to the performance of CIS thin-film solar cells with ZnO/ZnO/CIS structure.

EXPERIMENTAL

In this study, the CBD-ZnO buffer layers were grown on two types of thin-films, such as ZnO by a metalorganic chemical-vapor deposition (MOCVD) technique and CIS. The zinc hydroxide ($Zn(OH)_2$) deposition was performed by a CBD method, in which Zn salt and NH_4OH were employed as a reactant. The pH of the CBD solution was kept constant in the range of 7.0 to 12.0 during the deposition period at the solution temperatures of about 50 to 85 C. Annealing in air after $Zn(OH)_2$ deposition by the CBD method was carried out to form a ZnO layer for 5 to 30 minutes at 200 C. The film thickness of ZnO layer is in the range of 50 to 150 nm. The apparatus for CBD method in this study is shown in Fig. 1.

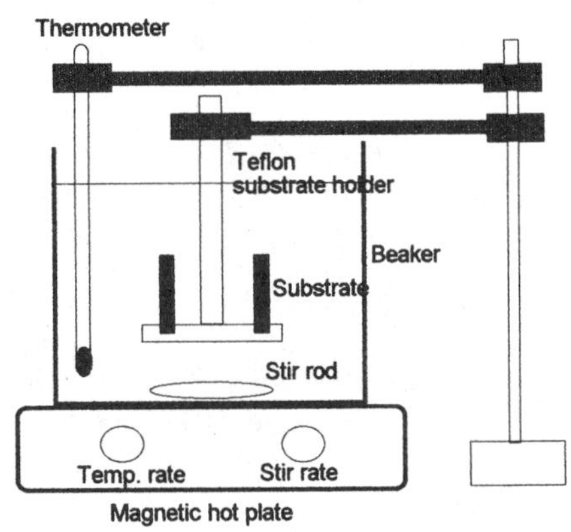

Fig. 1. Schematic diagram of the CBD apparatus for CBD-ZnO layers.

The characterization of CBD-ZnO layers was performed by X-ray diffraction (XRD), secondary Ion mass spectroscopy (SIMS), X-ray photospectroscopy (XPS) and scanning electron microscopy (SEM). In the case of the fabrication of CIS thin-film solar cells with ZnO/ZnO/CIS structure, the ZnO window layer was fabricated by a sputtering method or a MOCVD method. The effects of CBD-ZnO buffer layer were evaluated based upon the performance of CIS thin-film solar cells including the no-buffer case.

CH3365-4/94/0000-0254 $4.00 © 1994 IEEE
First WCPEC; Dec. 5-9, 1994; Hawaii

RESULTS AND DISCUSSION

Effects of substrate for the deposition of CBD-ZnO layers

Zn^{2+} ions from Zn salt and OH^- ions from NH_4OH can form a zinc hydroxide ($Zn(OH)_2$) in the CBD solution.

ZnO layer was not formed on a bare glass substrate, while it could deposit uniformly and steadily on MOCVD-ZnO and CIS layers. These results indicate that nucleation sites to grow $Zn(OH)_2$ as an intermediate zinc-compound are necessary on the substrate in the CBD method developed in this study.

Effects of pH in the CBD solution

The surface morphologies of CBD-ZnO layer grown on the CIS layer at the CBD-solution temperature of 55 C are shown in Fig. 2. As seen in this figure, the CIS surface is covered uniformly with granular particles of ZnO, which has been deposited in the pH 11 solution for 10 minutes.

The performance of CIS thin-film solar cells with ZnO/ZnO/CIS structure as a function of pH in the CBD solution is shown in Fig. 3. As seen in this figure, both the open circuit voltage (Voc) and fill factor (FF) are improved in the pH range of 10 to 12.

On the basis of these results, CBD-ZnO buffer layer is necessary to enhance the performance of the current level of CIS thin-film solar cells by increasing the Voc and FF.

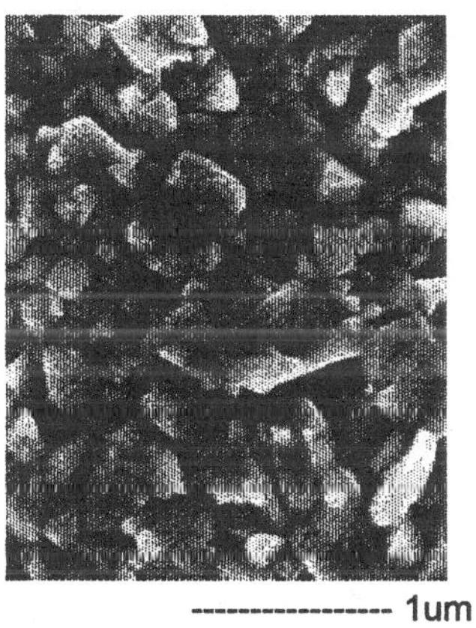

Fig. 2. Surface morphology of ZnO layer grown on CIS substrate by CBD method at pH 11.

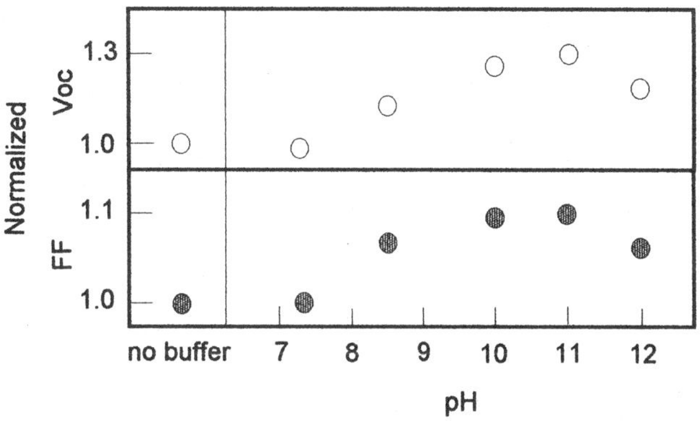

Fig. 3. Performance of CIS thin-film solar cells with ZnO/ZnO/CIS structure as a function of pH in the CBD solution at 55 C for 10 minutes. (Annealing condition : 180 C , 10 minutes)

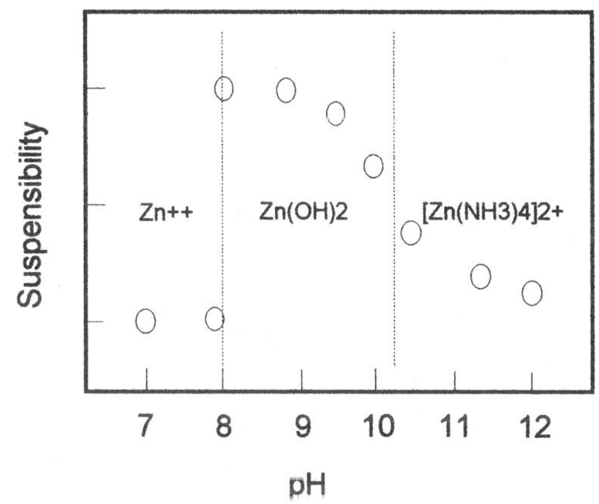

Fig. 4. Relationship between pH in the CBD solution and the formation of intermediate zinc-compounds.

In this study, the higher pH region of over pH 10 in the CBD solution is employed to form an intermediate zinc-compounds.

The formation reactions of CBD-ZnO are proposed in this study as shown schematically in Fig. 4, which is dependent on the pH in the CBD solution.

In this region of the CBD solution, the formation of $Zn(OH)_2$ would progress by way of the formation of zinc-ammonium complex ions as a precursor stage. This complex ion is formed through the reaction of zinc ions with ammonium radicals and considered to be necessary to form a better quality of $Zn(OH)_2$. Then, annealing of the zinc hydroxide in air leads to the formation of CBD-ZnO layers.

Effects of CBD-ZnO dipping-time on the performance of CIS thin-film solar cells.

In order to determine the optimized thickness of CBD-ZnO buffer layer for CIS thin-film solar cells, the dipping time was investigated in combination with the annealing temperature of 200 C. As shown in Fig. 5, the best performance of CIS thin-film solar cells was obtained at the dipping time of 10 minutes, although the performance becomes worse as the dipping time was longer or shorter. In this dipping time, the thickness of CBD-ZnO layer is estimated at about 10 nm.

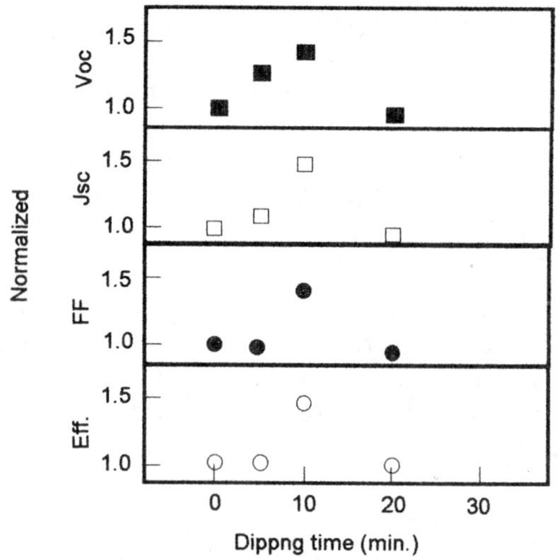

Fig. 5. Effects of the dipping time on the performance of CIS thin-film solar cells. (annealing temperature of about 200 C)

Effects of annealing temperature on the performance of CIS thin-film solar cells.

Table 1 shows inferable zinc-component substances on CIS surface by XPS measurements. As seen in this table, it is clearly indicated that CBD-ZnO layer consists of ZnO and $Zn(OH)_2$ in this study. $Zn(OH)_2$ was not completely converted to ZnO at lower annealing temperatures than 180 C, although $Zn(OH)_2$ is believed to be converted to stable ZnO at the annealing temperatures of over 125 C [1].

Table 1. Results of XPS measurement for CBD-ZnO thin films on CIS surface.

XPS analysis	ZnO (mol%)	$Zn(OH)_2$ (mol%)
sample A	72.1	27.9
sample B	56.1	43.9
sample C	74.5	25.5

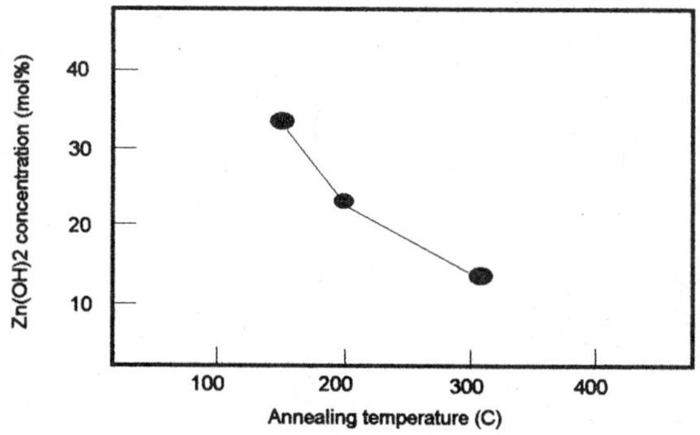

Fig. 6. $Zn(OH)_2$ concentration in the CBD-ZnO layer on the CIS substrate as a function of annealing temperature.

In order to reduce the amounts of $Zn(OH)_2$, the annealing temperature was increased. The concentration of $Zn(OH)_2$ in the CBD-ZnO layer are shown in Fig. 6 as a function of annealing temperature. As seen in Fig. 6, $Zn(OH)_2$ concentration decreased with the increase in the annealing temperature. On the basis of Figs. 6 and 7, the remaining $Zn(OH)_2$ may have no considerable effects on the performance of CIS thin-film solar cells with CBD-ZnO buffer layer. As seen in Fig. 7, the annealing temperature of 200 C is considered to be necessary to attain a better performance.

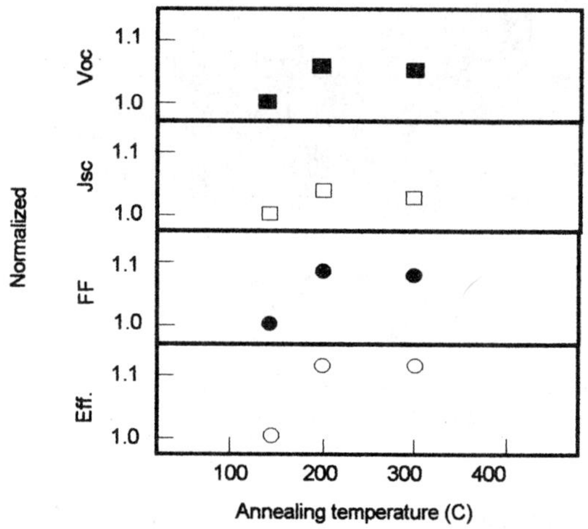

Fig. 7. Performance of CIS thin-film solar cells as a function of annealing temperature of CBD-ZnO buffer layer.

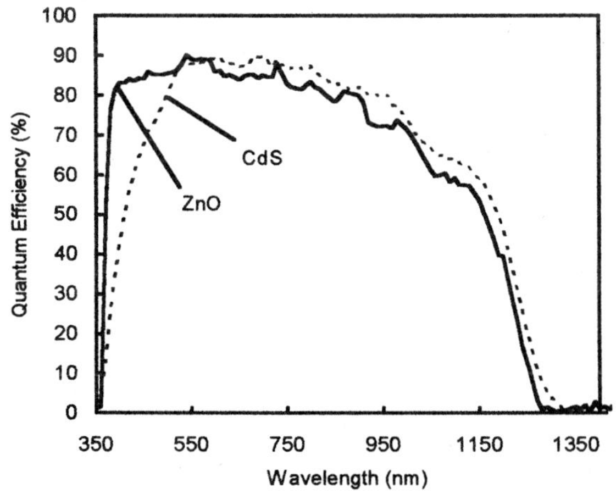

Fig. 8. Spectral response measurement of the CIS thin-film solar cells with ZnO/CBD-ZnO/CIS structure.

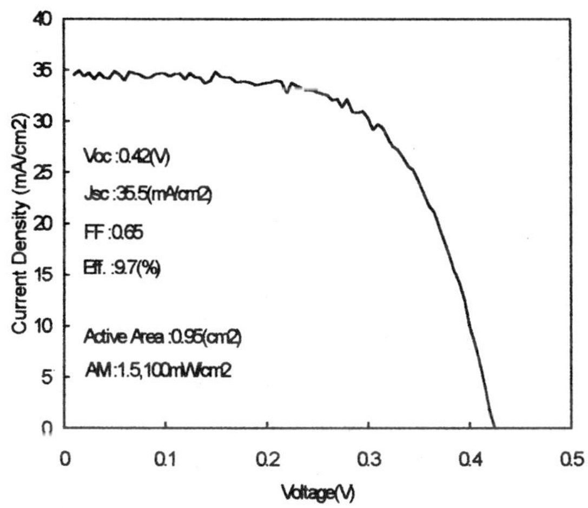

Fig. 9. I-V characteristic of the currently best CIS thin-film solar cell with ZnO/CBD-ZnO/CIS structure.

Performance of CIS thin-film solar cells with optimized CBD-ZnO buffer layer

On the basis of the above results, it is clearly understood that the performance of CIS thin-film solar cells with CBD-ZnO buffer layer is strongly dependent on the pH in the CBD solution and the annealing temperature.

As seen in Fig. 8, the gain at the shorter-wavelength region is clearly observed due to the wider band-gap of ZnO (about 3.3 eV) as a buffer layer than CdS (2.42 eV).

The currently best CIS thin-film solar cell with CBD-ZnO buffer layer is shown in Fig. 9. On the basis of rather high efficiency of 10 %, further improvement of the performance of CIS thin-film solar cells with CBD-ZnO buffer layer is strongly expected by the establishment of the CBD-ZnO process as well as the understanding of the effects of buffer layer on the interface between p-CIS thin-film absorber and n-ZnO window layer.

CONCLUSIONS

Some attempts to fabricate Cd-free CIS thin-film solar cells from the standpoint of environmental safety, led to the ZnO buffer layer prepared by the CBD method. About 10 % efficiency was achieved with CBD-ZnO buffer layer. Further improvement of the performance of CIS thin-film solar cells with ZnO buffer layer is strongly linked to the establishment of the CBD process and understanding of the effects of the buffer layer on the interface quality of p-CIS thin-film absorber / n-ZnO window layer.

ACKNOWLEDGMENT

This study was supported by the New Energy and Industrial Technology Development Organization as a part of the New Sunshine Project under the Ministry of International Trade and Industry.

REFERENCES

[1] M. Iguchi, " Zinc Hydroxide ", Chemical Dictionary, vol. 5 (KYORITSU-Shuppan, INC., Tokyo, 1963), p.16.

THERMAL OXIDATION OF $CuInSe_2$: EXPERIMENT AND PHYSICO-CHEMICAL MODEL

Michael E. Boiko and Gennadiy A. Medvedkin
A. F. Ioffe Physico-Technical Institute,
26 Polytekhnicheskaya, St. Petersburg 194021, Russia

ABSTRACT

The thermal oxidation of $CuInSe_2$ single crystals and bulk polycrystalline plates of n- and p-type conductivity with polished facets have been carried out at elevated temperatures 400 to 610°C in dry oxygen/air atmosphere. All the crystals demonstrated an enhancement in hole conductivity near the surface. Additionally, transparent layers of In_2O_3 chemical composition have been grown on the surface.

Electron-probe analysis, Rutherford backscattering, x-ray diffraction, optical and photoelectric experimental data have provided a basis to develop a physico-chemical model thermal oxidation of $CuInSe_2$ crystals. The model includes the In_2O_3 compound appearance, V_{Se}, V_{Cu} vacancy absorption, Se_i, Cu_i extra atom and Cu_xSe generation. The chemical reactions are accompanied by arising dislocations and mechanical stresses at the $In_2O_3/CuInSe_2$ interface.

INTRODUCTION

The thermal treatment in oxygen or in air is a very important post-deposition stage in thin-film $CuInSe_2$/CdS/ZnO-type solar cell technology to obtain more than 10% quantum efficiency [1]. The role of oxygen in increasing the photoelectric parameters is not clear till now. This may be associated with a crystallite structure improvement in the polycrystalline films, sensibilizing the heterolayers and perfecting the interface. The oxygen interaction with a multicomponent system of 6 to 7 chemical elements combining more than three compounds presents to be too complex for fast understanding. Therefore, the simpler experimental investigation was undertaken [2].

EXPERIMENT

Technology

$CuInSe_2$ single crystals and bulk polycrystalline plates of n- and p-type conductivity with carrier concentration of 10^{16}–10^{18} cm^{-3} and specific resistivity of 5 to 500Ω cm at 300 K were used. The thermal oxidation of flat plates with polished planes or with {112} natural facet were carried out at elevated temperatures 400 to 610 °C in an atmosphere of dry oxygen or air (standard composition 80% N_2 + 20% O_2) during 5 to 5×10^4 min. The conductivity type conversion (n→p, n→i) has been obtained as a result of as-grown n-type material re-compensation. P-type crystals demonstrated an enhancement in their hole conductivity near the surface. Surprisingly, some technological regimes allow to get surface layers of In_2O_3 chemical composition with various controllable thicknesses and excellent optical and electronic properties [3–6]. The selected treatment regimes for homogeneous samples provided uniform oxide layers with yellow, light blue or violet interference coloration and specific resistivity of 10^1–10^3Ω cm at 300 K.

Composition and structure

The composition of thermal oxide on $CuInSe_2$ samples were measured with "CAMECA" x-ray electron-probe microanalyzer. Cu, In, and Se element proportion were determined in reference to Cu, InP and ZnSe, oxygen content was estimated by difference of degree.

To study the process of selenium and copper evaporation outward the crystal surface, the specially abraded down to a half specimens were subjected to the electron-probing on preservation of a stoichiometric proportion and also to a control on absence of atomic Cu, In, Se and appropriate binary compounds in depth of the specimens. The x-ray spectral, diffraction and Rutherford backscattering experiments have confirmed the In/O = 40/60 proportion on the crystal surface and given the Cu, In, Se and O element distribution in depth.

Thermally oxidated at $T = 550$°C $CuInSe_2$ samples were investigated using alpha particle beam with initial energy of $E_0 = 4.6$ MeV. Fig. 1 presents spectra of backscattered He^{++} ions for the samples annealed during $t = 10$–70 min. The analysis of RBS curves shows the indium concentration [In] in an undersurface layer rises with the time increase, the selenium and copper concentrations decrease and become more spreaded of energies (curves 30 to 70). This may indicate not only the [Se] and [Cu] drop in the undersurface region, but the phase inhomogeneity rise in depth with the oxidation time. A computation for the component concentration profile in depth was carried out using an iteration program worked out for multicomponent systems [7]. The stoichiometric composition was found experimentally by the measurements on a crystal cleavage. After 10 min annealing the selenium concentration drops sharply reaching the value of ca 60%. The [Cu] decrease is observed at longer times, $t > 10$ min, yet but only the indium is identified on the surface at $t > 50$ min. An estimation for the oxide film thickness gets: $d = 90$ nm at 30 min and it increases up to 200 nm at 70 min oxidation.

Double-crystal fore circles spectrometry by MoK_α and CuK_α x-rays in combination with photomethod data were

CH3365-4/94/0000-0258 $4.00 © 1994 IEEE

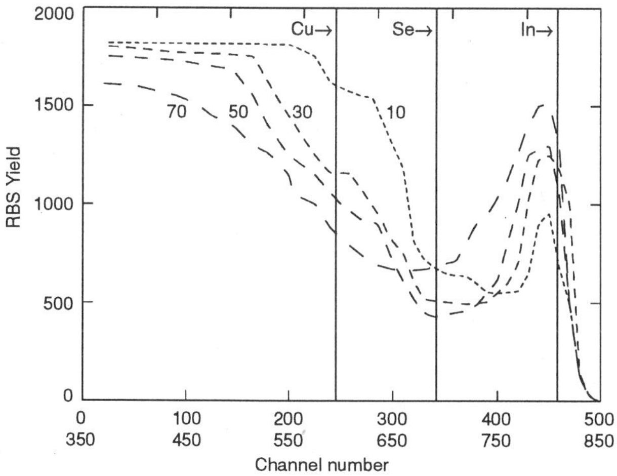

Fig. 1. Rutherford backskattered He^{++} distribution for thermally oxidated $CuInSe_2$ crystalls. $T = 550$ °C, t: 10, 30, 50, 70 min.

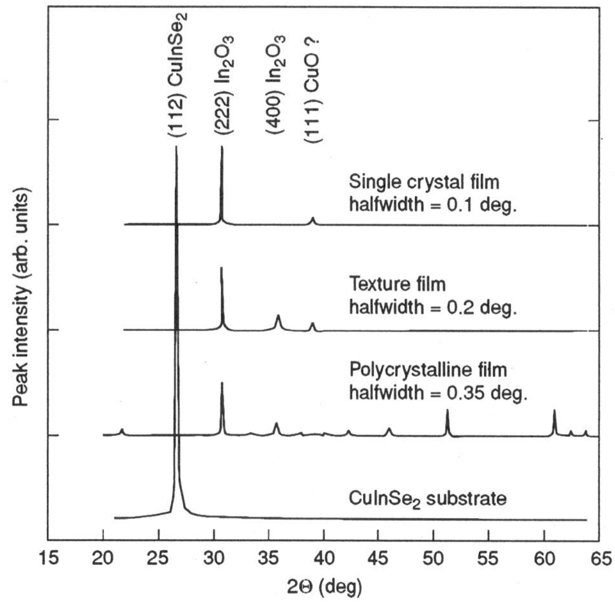

Fig. 2. X-ray diffraction spectra of thermal oxides on (112) $CuInSe_2$ substrate. The basal substrate spectrum is subtracted.

used to study a structure of prepared oxide layers. The substrate plates of various orientation treated at $T = 490$–550 °C were studied. There were additional peaks on x-ray diffraction spectra (see Fig. 2) and spots/rings on Debye x-rayogramms corresponding to In_2O_3 phase. X-ray diffraction patterns gave three diverse shapes of reflections: spots for monocrystalline films, prolate ellipse-like lentils for texture, and regular Debye rings for polycrystalline films in accordance with the x-ray spectra [2, 3]. The cubic lattice parameter of the prepared layers $a = 1.013 \pm 0.001$ nm agrees reference data for In_2O_3 [8]. The single crystal structure of films was observed after the first stage of oxidation ($d < 10$ nm), such thin films showed very weak diffraction peaks corresponded 0.291 and 0.252 nm interplanar distance. The second stage of oxidation gave rise to In_2O_3 texture growth and the more number of intensive peaks (see Fig. 2). The crystal orientation {111} was to be characteristic of In_2O_3 layers on {112} basal plane of $CuInSe_2$ substrate. At the third stage ($d > 10$ nm) of oxidation the polycrystalline growth starts both on oriented and unoriented crystal planes [3]. Note, the carried out x-ray phase analysis by JCPDS data did not exhibit the following components: Cu_2O, SeO_2, Se_2O_3, Cu_2Se (1 and 2), α-Cu_2Se, CuSe, CuSe*Cu_2Se, Cu_3Se_2, Cu_9In_4, Cu_4In, and also elements Cu, In, Se. A number of very feeble and spread bands indicated a possible existence of a little content of copper oxide CuO.

Optical and photovoltaic properties

The optical reflectance spectra measured in the range of 0.5–5 eV demonstrated a general decrease in R magnitude in passage over the clean $CuInSe_2$ ($R = 20$–25%) to oxidated surfaces (down to $R < 5$%). It was associated with anti-reflecting by an oxide layer [2, 4]. At the same time principal spectral maximums characteristic of the clean surface remained for the oxidated surfaces ($\hbar\omega = 2.6$, 4.3 eV).

The laser ellipsometric control showed constancy in refractive index value $n = 2.12$–2.14 and dynamic change in layer thickness depending on treatment time and temperature (490 to 560 °C) [4]. For example, d increases linearly from 30 to 70 nm at $T = 510$°C with $t = 10$ to 60 min but the curve $d(t)$ in the wider time range can not yet be described in terms of linearity. The estimation shows that the In_2O_3 layers of 50 nm thickness with an optical transmittance of $t = 0.77$ and reflectance of $R = 0.13$ possess a quite low extinction index $k = 10^{-3}$. The thermal oxides on $CuInSe_2$ do not give in optical quality to cathode flash deposited In_2O_3 films.

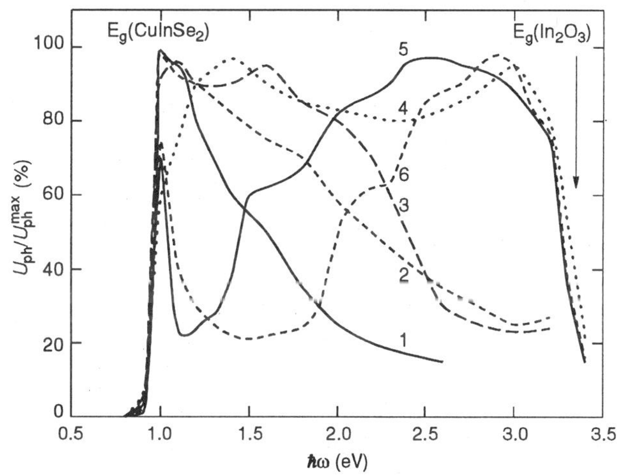

Fig. 3. Photoresponse spectra of $In_2O_3/CuInSe_2$ heterojuctions at room temperature (on illumination across the widegap oxide film). The heterojunctions were fabricated at temperature T (°C, first number) applied for duration t (min, second number): (1) 450, 240; (2) 450, 10; (3) 500, 40; (4) 564, 120; (5) 540, 30; (6) 540, 45.

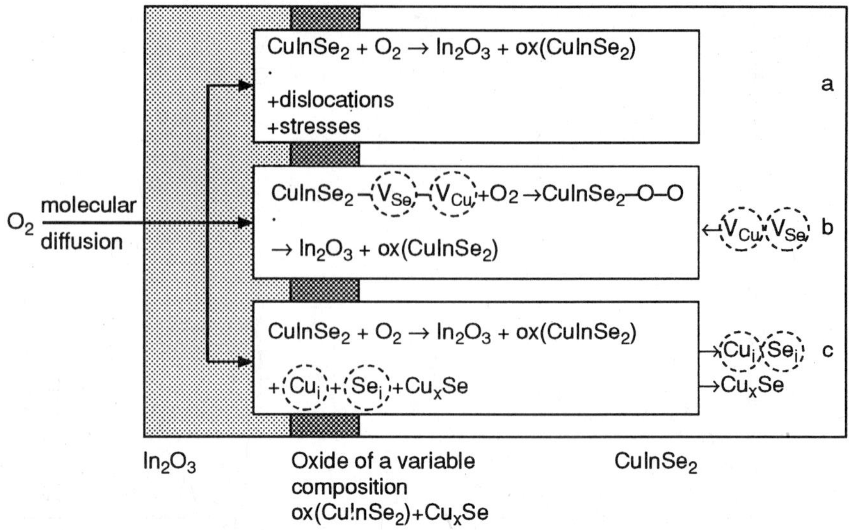

Fig. 4. Physico-chemical model of thermal oxidation of $CuInSe_2$ crystals.

The achieved anti-reflection at short wavelengths is responsible for extended photoresponse from 1.0 up to 3.2 eV [5, 6]. Photovoltaic spectra for a series of heterojunctions prepared by the thermal oxidation techniques are presented in Fig. 3. The $In_2O_3/CuInSe_2$ heterostructures obtained at low temperatures ($T = 450°C$) demonstrate a long wavelength maximum U_{ph} near the energy $\hbar\omega = E_g$ (1.04 eV). The oxide layer by the technological regimes for $t = 10$ to 240 min is quite thin ($d < 10$–20 nm) and comparable with ox($CuInSe_2$) interface and Cu_xSe layer (3–5 nm) containing in it. The T rise results in strengthening the U_{ph} at short wavelengths and attenuating it in the mediated range. It is explained by recombination of photogenerated carriers in the interface layer of variable composition and defect concentration, since the higher the annealing temperature the larger the thermodefect concentration. At higher temperatures ($T > 490°C$) the In_2O_3 layers of interference thickness are formating that appears distinctly in photoresponse spectra in deep absorption band of $CuInSe_2$ (see Fig. 3, curves 3 to 6). As the processing time rises the layer thickness grows linearly and the heterojunction layer structure can be described accurately in frameworks of unilayer model: $d_{In_2O_3} >> d_{Cu_xSe}$ [4]. The further increase up to $T = 540$ °C gives rise to more essential variations in photoresponse spectra (curves 5, 6). The absolute maximum $U_{ph}(\hbar\omega)$ realizes at 2.5–2.8 eV ($d_{ox} = 100$ to 200 nm) and the drop at $\hbar\omega > 3.2$ eV is responsible for beginning the intrinsic absorption in In_2O_3 material. The intrinsic absorption range of $CuInSe_2$ exhibits a narrow peak photoresponse near $\hbar\omega = 1.0$ eV. Such $U_{ph}(\hbar\omega)$ profile we associated with n-p junction imbedding [5]. Thus, $In_2O_3/CuInSe_2$ heterojunction spectral features can be given on the base of two space charge regions in layers connected in series: 1. heterojunction n-In_2O_3/p-$CuInSe_2$; 2. homojunction p-n(p-p$^+$) $CuInSe_2$.

PHYSICO-CHEMICAL MODEL

The experimental evidence of In_2O_3 compound existence on the oxidated $CuInSe_2$ surface and the hypothesis of the layer structure for heterojunctions allow to advance in understanding the oxygen interaction with the ternary compound.

The modeling of chemical reactions including phase boundary movement into a substrate and a new compound species formation on $CuInSe_2$ crystal have been developed. The oxidation results from oxygen transfer through the native oxide on the semiconductor surface on which the following transition takes place:

$$CuInSe_2 + O_2 \rightarrow In_2O_3 + Cu_xSe + ox(CuInSe_2),$$

here the ox($CuInSe_2$) is an oxide with a variable composition at interface. Owing to the movement of $In_2O_3/CuInSe_2$ heterophase boundary depthward the substrate, the new oxide phase formates, extra atoms of Se and Cu appear at every plane of the crystal lattice resulting in an essential inconsistence at the boundary that leads to mechanical stresses and dislocations (see Fig. 4,a). Alternative methods suggesting the removal of selenium and copper extra atoms consist in the use of V_{Se}, V_{Cu} vacancies (see Fig. 4,b) or in the Se_i, Cu_i atom generation into interstices and the Cu_xSe "molecule" production (see Fig. 4,c). Oxygen interaction with V_{Se} leads to a decrease of donor level concentration near the surface. Reacting Cu_i and Se_i atoms formate p-type precipitates of binary copper selenides primarily at the interface $In_2O_3/CuInSe_2$ with decreasing temperature of oxidation, $T < 350°C$.

Thus, physico-chemical interaction of oxygen with clean $CuInSe_2$ surface at high and lower temperatures consists of complex temperature-caused processes: decomposition of the ternary compound, growth of the In_2O_3 oxide layer with a controllable thickness on crystal surface and formation of a thin interface layer by copper selenides and oxides with a variable composition, and also movement of point defects (vacancies and interstitials) toward the surface and depthward in accordance with a diffusion front for the above defects interacting with ternary compound atoms. Apart from this the processes are accompanied by appearance of mechanical stresses and dislocations at the interphase boundary.

CONCLUSIONS

The presence of point defects, teared bonds and extended dislocations at the heterophase boundary is often undesirable because of recombination centers which can deteriorate electronic properties. However, meritorious SiO/SiO_2 oxides in silicon devices permits to be in hopes also of the promising native heterocouple $In_2O_3/CuInSe_2$ in device applications. For example, In-O precipitates at the interface can serve as gettering sites for unwanted fast-diffusing impurities and affect distortions during thermal treatment.

The above model approach can be applied to the more complex task solution under a study of thermal treatment for I–III–VI_2 based solar cells.

ACKNOWLEDGEMENT

We appreciate Prof. V. D. Prochukhan for the provided $CuInSe_2$ crystals and Dr. M. V. Yakushev for his RBS measurements.

REFERENCES

[1] T. J. Coutts, L. L. Kazmerski, and S. Wagner (Eds), *"Copper Indium Diselenide for Photovoltaic Applications"*, Amsterdam, Elsvier, 1986.

[2] G. A. Medvedkin, G. A.Ambrazevichius, and A. A. Yakovenko, "Investigation of Oxidation Processes on $CuInSe_2$ Crystals." *Phys. Chem. Mech. Surface* (GB) 1987, **2**, pp. 81–87.

[3] M. E. Boiko and A. A. Yakovenko, "Crystal Structure of $CuInSe_2$ and In_2O_3 Thin Layers". *5th USSR National Confer. "Ternary semiconductors and their applications".* Kishinev, Shtiintsa 1987, vol. I, p. 94.

[4] G. A. Medvedkin, R. N. Bekimbetov, T. L. Makarova, A. D. Smirnova, and V. I. Sokolova, "Optical Properties of a Thermal Oxide on $CuInSe_2$". *Sov. J. Techn. Phys.* **57**, 1987, pp. 960–964.

[5] G. A. Medvedkin, Yu. V. Rud, and M. A. Tairov, "The Photoeffect in $In_2O_3/CuInSe_2$ Heterostructures Prepared by the Method of Thermal Oxidation". *Sov. Phys. Semicond.* **23**, 1989, pp. 869–872.

[6] M. A. Magomedov, G. A. Medvedkin, Yu. V. Rud, and V. Yu. Rud, "The Preparation and Properties of Heterostructures Based on n-$CuInSe_2$". *Sov. Phys. Semicond.* **26**, 1992, pp. 556–558.

[7] L. Wen-Zhi, "Computer Program for the Analysis of RBS Spectra", *Nuclear Instr. Research* **B15**, 1986, pp. 241–245.

[8] A. V. Novoselova and V. B. Lasarev (Eds), *Handbook "Physico-Chemical Properties of Semiconductor Materials"*, Moscow, Nauka, 1979 (in Russian).

ENVIRONMENTAL AND HEALTH ASPECTS OF CIS-MODULE PRODUCTION, USE AND DISPOSAL

Werner Thumm, Andreas Finke, Birgit Neumeier, Barbara Beck, Antonius Kettrup,
Institute of Ecological Chemistry, GSF, Neuherberg, Germany
Hartmut Steinberger, Fraunhofer Institute for Solid State Technology, Munich, Germany
Paul D. Moskowitz, Brookhaven National Laboratory, Upton, New York, USA

ABSTRACT

Copper indium diselenide (CIS) is one of the most promising compounds in thin film technology. Since there is scant information available about environmental and health hazards, a study was initiated to characterize risks associated with the production, use and disposal of thin film photovoltaic modules. Data available from literature and developers of this technology contribute to an assessment of potential risks during production. In laboratory experiments the release of hazardous materials during operation caused by accidents or false handling and after disposal are simulated. In biological experiments the possible impact on living matter is established. These experiments comprise toxicity tests with aquatic organisms and rats representing mammals.

THEORETICAL STUDIES

The complete lifecycle of almost all products includes:

- the starting material
- the manufacturing
- the installation, operation and use
- the recycling and disposal.

Today, nearly any type of marketing concept is based on a linear product path starting with the raw material and ending after usage with the disposal of the product. However, most product life cycles can be closed by a recycling strategy based either on the product or material level.

In order to characterize all materials and material-paths into air, water and soil in the life cycle of our selected product for the thin-film modules we define the areas of investigations according to the following:

In the production of starting materials for the thin-film modules, solids, liquids, gaseous materials (elements and compounds) and prefabricated components must be considered. Solar module manufacturing uses these materials including the substrate, thin-films, laminates, cover glass, junction boxes, cables and frame materials. For the installation, operation or use we have restricted the photovoltaic (PV) modules to power application products like solar power plants or solar houses or to public accessible products like solar powered emergency call boxes on highways. Normal operation and accidents must be examined. Disposal including landfilling and incineration are issues to be investigated with their contribution of the material and material-flows of the products and byproducts into the environment e.g. drain-water, decomposition gas, slag or electro-filter dust and exhaust fume.

The principal material flow of input, output and auxiliary substances in a single processing step is shown in Fig. 1. Input materials are all materials partly or totally transformed into the product. Auxiliary materials are all materials taking part in the processing but not remaining in the product. Output materials are the rejected input materials or the rejected auxiliary materials [1].

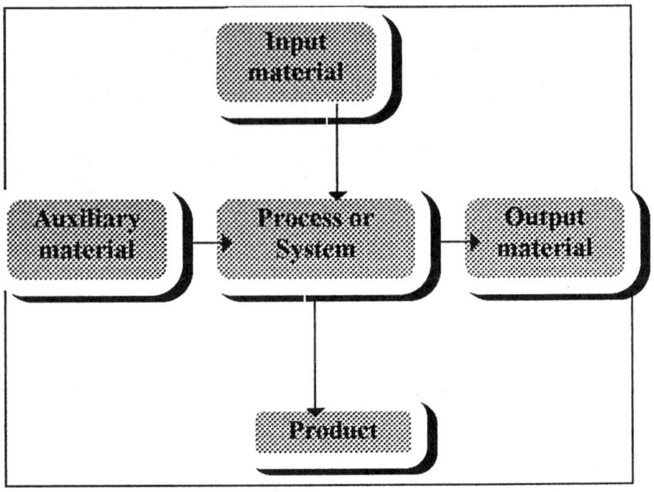

Fig. 1. Definition of input, output, auxiliary materials

Starting materials

The weight ratio of an element within the uppermost shell of the earth is called the total abundance. In order

CH3365-4/94/0000-0262 $4.00 © 1994 IEEE

to identify the highest need for the conservation of resources Table 1 compares total material abundance, annual production rate, and minimum material consumption of important materials needed to produce one GW_e of solar modules. This roughly correlates to a 10 km^2 module area. Based on this analysis, it seems that indium (In) and selenium (Se) will be the bottleneck materials in future.

Material	Abundance	Production t/a	Consumption t/GW
copper	$1 \cdot 10^{-4}$	9,300,000	1100
indium	$1 \cdot 10^{-7}$	134	30
selenium	$8 \cdot 10^{-7}$	1,500	41
cadmium	$3 \cdot 10^{-5}$	18,000	3
molybdenum	$1 \cdot 10^{-5}$	117,000	76
soda lime glass	$1 \cdot 10^{-1}$	21,000,000	125,000

Table 1. Abundance, production and material consumption of important elements for thin-film module fabrication

Manufacturing

For a CIS module the total material input of the critical element Cd is calculated for a 1 m² module area and a 1 µm layer thickness. CIS consumes roughly 1 g/m²/µm Cd for the cadmium-sulfide (CdS) layer formation as the total input material. In comparison CdTe uses 10 to 550 g/m²/µm Cd for the CdTe and CdS layer formation depending on the deposition technique.

Installation, operation, use

Several authors describe problems caused by accidents or incidents during installation [2-4]. Failures during operation due to weather and other reasons may lead to the destruction of modules which may in certain cases lead to a release of critical substances. Table 2 shows the different reasons for the failure of modules.as investigated by O´Riordan [5] on the Fota Islands in Ireland.

Reason for the problems	Numbers	% Ratio
Broken front glasses	22	28
Oxidation of front contacts	16	20
Color change of the substrate	15	19
Penetration of water	11	14
Broken substrate glasses	10	13
Broken ends	4	5
Laminate	1	1
In total for 2,772 modules	79	100

Table 2. Effects and problems of silicon solar modules on the Fota Islands in Ireland

Disposal

In Germany the "technical regulations for municipal waste" demand that any kind of waste should be pretreated before disposal. This usually implicates incineration in a municipal waste incinerator. Other industrial countries like USA also favour the treatment of waste in a comparable manner. For the final disposal of wastes (e.g. slag and filter dust) limit values are given for the content of certain elements and organic residues. The analytical tool for the determination of element content is the German leaching test DEV S4.

On the basis of a one year averaged distribution of input and output materials for a German municipal waste incineration facility [6], we have estimated the total released material in slag, filter dust and pure gas assuming a 1 kW module input of CIS-technology (see Fig. 2).

These estimations again take into consideration the materials both from the thin-film layers on the substrate (primary) and the frame materials (secondary). The main contribution in the case of CIS is Cu in the slag of around 200 g in total, of 43 g in filter dust and 4 g in the released gas dust. Cadmium is represented in slag as 1.5 g, as 0.5 g in the filter dust, and around 0.05 g in the released gas dust.

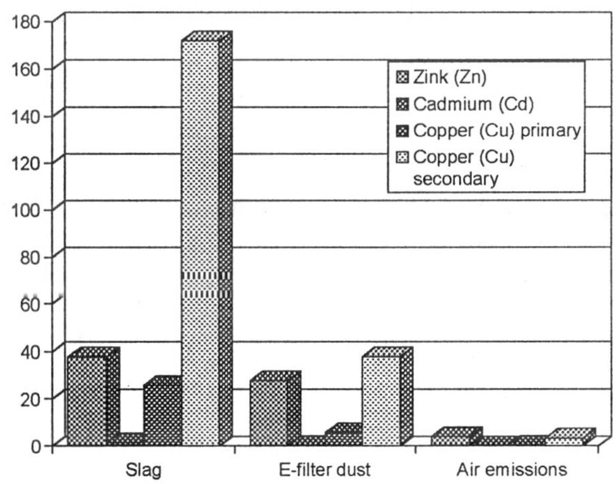

Fig. 2. Output material distribution of a 1 kW CIS-module input into a municipal waste incineration facility in Germany

Recycling

Preexperiments on commercially available CdTe-modules for recycling have been carried out using liquid nitrogen, sand blasting techniques and HCl etching. Liquid nitrogen has peeled off large areas of thin-film layers. The remaining residues on the surface of the glass substrate were then etched back by HCl, which left a clean transparent glass substrate showing small reflecting zones along the removed structuring lines. Sand blasting with different particle sizes seems also possible for recycling, if the working speed can be controlled to avoid surface cracks or holes of the glass substrate. Similar studies are being prepared for CIS modules.

EXPERIMENTAL WORK

Leaching tests

The leaching of substances from modules broken by accident or disposal in a landfill is simulated by na-tionally and internationally approved elution tests and in outdoor experiments. The outdoor tests started in 1992 and will be finished by the end of 1994 [6].

For the elution tests four different methods are used: The German DEV S4 (Deutsches Einheitsverfahren), the US TCLP (Toxicity Characteristics Leaching Procedure), the Swiss leaching test and a test developed by the University of Wisconsin for EPA. They have in common that fragmented samples are rotated end over end for a certain time period (usually 24 h; TCLP 18 h) with an eluent.

In Fig. 3 a part of the results with these tests is shown. Zn, Mo (molybdenum) and Se are the elements eluted in the highest amounts from CIS-modules. Mo and Se are supposed to elute as anionic species and this may explain their elution behavior at rather high pH. Zn concentrations increase with decreasing pH and increasing complexing agent content.

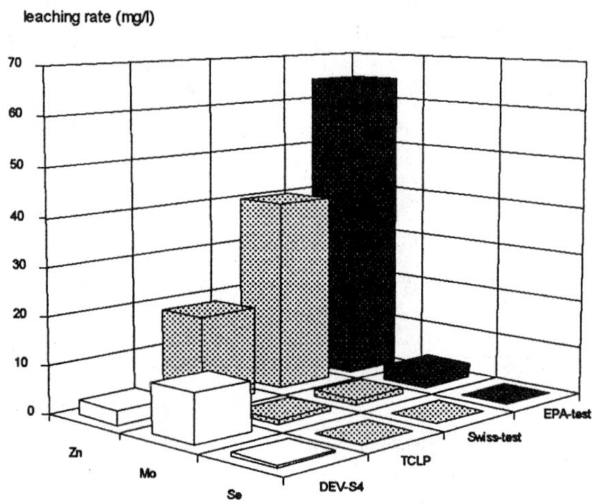

Fig. 3. Element concentration in the eluate of DEV S4, TCLP, EPA and Swiss test

With respect to element concentrations the limit values in Germany for a landfill class II are 5 mg/l Zn (found 2.8 mg/l); 1 mg/l Cu (found 15 µg/l) and 50 µg/l Cd (found 28 µg/l). In the US only Se and Cd are regulated to 1 mg/l (found 60 µg/l Se and 31 µg/l Cd) with respect to the TCLP. If only related to metal the tested modules would pass both tests

Thermal testing

Fig. 4 shows the results of a differential thermal analysis. Samples of pure CIS material were heated in air. Important information can be deduced from the thermogravimetry curve. For CIS, a weight loss of 27% is found starting at 300 ^{o}C. This corresponds to the replacement of the entire Se by oxygen (O) and the release of selenium dioxide (SeO_2).

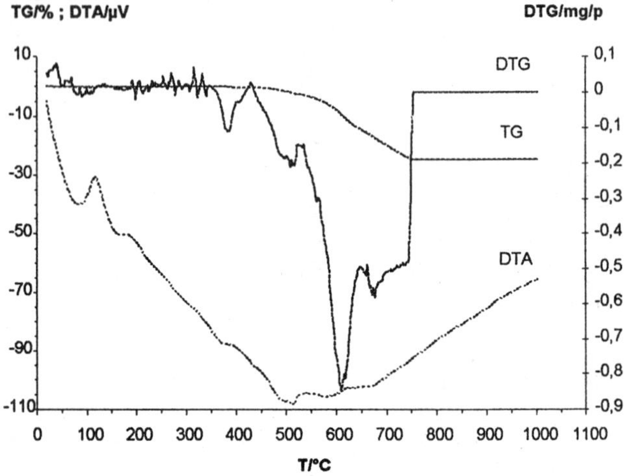

Fig. 4. Differential thermal gravimetry (DTG), thermal gravimetry (TG), differential thermal analysis (DTA) of pure CIS material in air

Sorption tests

In the sorption experiments with soils and clay minerals, usually a strong sorption is found for cationic species, whereas the sorption for anionic species like molybdate is rather faint (see Fig. 5). For Se in its two possible anionic species, the situation becomes more difficult, because selenite is a strong adsorbent in contrast to selenate which usually is a weak one.

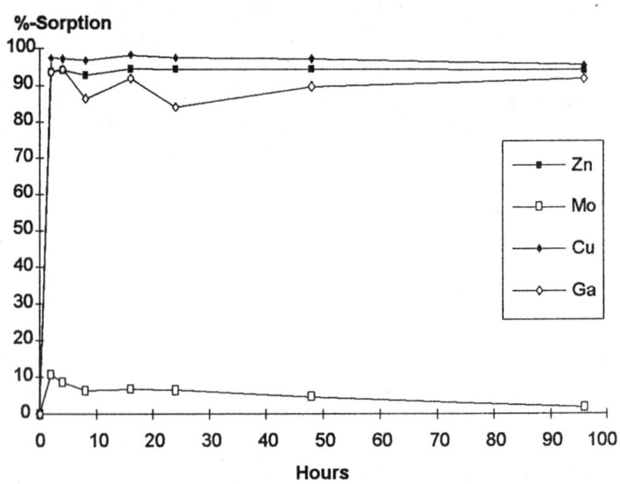

Fig. 5. Sorption tests with standard soil for CIS material

Toxicity tests with daphnia and algae

In toxicity tests, the effects found were dependent on the speciation of the single elements tested. That means, the oxidation state as well as the counterions influence toxicity. This effect is demonstrated in Table 3 for toxicity tests with daphnia magna (waterflea) for Se. The same effect can be observed in other toxicity tests.

Test substance	EC_{50} (mg/l of element)	element
$CuSO_4$	0.02	Cu
Na_2SeO_4	2.2	Se
Na_2SeO_3	3.7	Se
$ZnSO_4$	1.7	Zn

Table 3. Daphnia magna - acute toxicity (24h-Test)

Fig. 6 shows the result of an acute toxicity test with algae for sodium selenate. Selenium in its different speciations shows a non-classical behavior in some toxicity tests; at rather low concentrations, no toxic action can be detected, but a small stimulating effect can be observed.

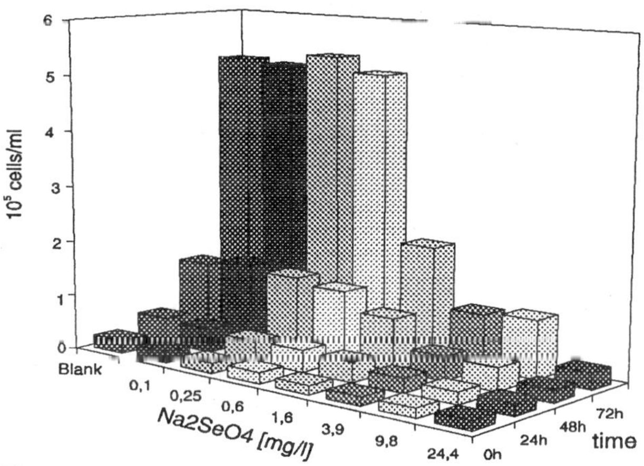

Fig. 6. Acute toxicity on algae

For future experimental work, it will be necessary to focus more on the speciation of the elements that are able to enter the environment, because the impact and the effect of these elements on compartments of the environment greatly depends on their speciation. Combinatory effects of compounds that may come into existence from solar cells also have to be taken into account.

Rat studies

To determine the toxicity of the thin-film materials of the PV modules, a short term toxicity study was undertaken using rats as the model species. First, a pilot study was performed to set doses for the main study. The main study evaluated a number of endpoints of systemic toxicity, as well as potential reproductive and developmental effects [8].

Groups of mature male and female Sprague-Dawley rats were administered an aqueous suspension of CIS in 0.5% methylcellulose by gavage at levels of 0, 50, 100, and 250 mg/kg of body weight.

The different groups were investigated with respect to histopathology, hematology, fertility and reproduction. Even at high doses no or only small effects were detected.

From the range finding studies with rats reference doses for man were derived from the lowest observed adverse effect level (LOAEL) using appropriate uncertainty factors. The LOAEL for CIS of 250 mg/kg rat/day results in a reference dose for man of 8.3 µg/kg/day. This means a daily uptake of a 70 kg man of 0.58 mg CIS is not expected to be harmful.

ACKNOWLEDGEMENT

This study is funded by the German Bundesministerium für Forschung und Technologie (BMFT). The rat studies were funded in parallel by the US Department of Energy (DOE) and the National Institute of Environmental Health Sciences (NIEHS).

LITERATURE

[1] Steinberger, Environmental and health aspects of Copper-indium-diselenide and Cadmium-telluride thin film Photovoltaic modules, *CIS, CGS, and CdTe Photovoltaic NREL-workshop report*, Denver 1994.

[2] van der Weiden, et al., PV/Diesel systems on house boats, *Tenth EC Photovoltaic Energy Conference*, Lisboa 1991

[3] Schaffrin, The solar boat "KORONA": Two years of experience , *Tenth EC Photovoltaic Energy Conference*, Lisboa 1991

[4] Chianese, et al., Tiso: 4-kW experimental amorphous silicon PV-plant, *Tenth EC Photovoltaic Energy Conference*, Lisboa 1991

[5] O'Riordan et al., Detailed analysis of a photovoltaic project after 7 years of operation, *Tenth EC Photovoltaic Energy Conference*, Lisboa 1991.

[6] Nottrodt, Die festen Verbrennungsrückstände, *Müll-Handbuch*, Loseblattsammlung, Band 4, 38 Lfg XI/75, Berlin, 1975.

[7] Thumm, et al. Experimental approach to asses environmental risks of thin-film solar cells, *CIS, CGS, and CdTe Photovoltaic NREL-workshop report*, Denver 1994.

[8] Chapin, et al. The general and reproductive toxicity of the Photovoltaic material Copper-indium-diselenide, *CIS, CGS, and CdTe Photovoltaic NREL-workshop report*, Denver 1994.

OVERVIEW OF THE U.S. DOE/NREL POLYCRYSTALLINE THIN-FILM PHOTOVOLTAIC TECHNOLOGIES

Harin S. Ullal, Kenneth Zweibel, Bolko G. von Roedern, Rommel Noufi, and Peter Sheldon
National Renewable Energy Laboratory
1617 Cole Boulevard
Golden, CO 80401 USA
Tel: (303) 384 6486; Fax: (303) 384 6481; Internet: ullal@nrel.gov

ABSTRACT

Thin-film solar cells based on CdTe, $CuInSe_2$ (CIS) and Si-Film have made rapid progress in the past 18 months. Total-area efficiencies of 16.8% for CIGS, 15.8% for CdTe, and 13.4% for Si-Film on graphite cloth, verified by the National Renewable Energy Laboratory, have been achieved thus far. The performance of thin-film modules for both CdTe and CIS have also improved markedly. Stability data for these emerging technologies is encouraging. Nominal 1-kW CIS and CdTe PV arrays have been installed and are undergoing testing. A 2-MW thin-film CdTe manufacturing plant has been installed, and another 10-MW thin-film CdTe plant is expected to be completed in 1996. A new Thin Film Partnership Program has been initiated to further advance the state-of-art of thin film technologies.

INTRODUCTION

Substantial technical progress has occurred in the past 18 months in thin-film copper indium diselenide ($CuInSe_2$, [CIS]) based solar cells. During this time, National Renewable Energy Laboratory (NREL) researchers have repeatedly improved the world record, total-area efficiency now 16.8%, for any thin film solar cells. New reaction pathways have been identified by NREL to improve the performance of these devices [1]. Improvements have also taken place at the module level in both CIS-based and CdTe modules in terms of efficiency and reliability. Critical research, development, and technical issues have been identified to improve the performance of solar cells and modules [1]. Nominal 1-kW thin-film CIS and thin-film CdTe photovoltaic (PV) arrays have been installed in the United States, and are undergoing reliabilty testing. A 2-MW thin-film CdTe manufacturing plant has been installed by Golden Photon, Inc. (GPI), Golden, Colorado. Another 10-MW thin-film CdTe manufacturing plant is expected to be completed in 1996 by Solar Cells, Inc. (SCI), Toledo, Ohio. Initial sales of 1-kW thin-film CdTe PV arrays have been made to Brazil and Tunisia.

THIN-FILM SOLAR CELLS

The most remarkable improvement in thin-film solar cell efficiency has been the total-area 16.8% efficient device achieved by NREL researchers for a thin-film $Cu(In,Ga)Se_2$ (CIGS) solar cell. A three-stage process for the growth of CIGS absorber layer was used to achieve this result. First, the precursor $(InGa)_2Se_3$ was deposited on a Mo-coated soda-lime glass, followed by the CuSe layer, and a third layer of $(InGa)_2Se_3$. Experimental details are given elsewhere [2,3]. Siemens Solar Industries (SSI), Camarillo, California has been successful in fabricating a reported active-area efficiency of 16.2% by grading the absorber layer with Ga and S [4]. Energy Photovoltaics (EPV), Princeton, New Jersey has fabricated a total-area efficiency of 12.5% for a 1-cm^2 thin-film CIS cell grown by selenization [5]. Se vapor has been used instead of the H_2Se gas in this device. Solarex, Newtown, Pennsylvania in collaboration with NREL, has fabricated a thin-film CIGS solar cell with a total-area efficiency of 14.4% [6]. In all cases, the device structure is ZnO/CdS/CIS/Mo/glass. MgF_2 is normally used as the anti-reflection coating. Washington State University (WSU), Richland, Washington, in collaboration with SSI has fabricated an 11.1%-efficient thin-film $Cu(In,Ga)(S,Se)_2$ (CIGSS) solar cell without the use of CdS as the heterojunction partner [7]. Using a non-vacuum proprietary process, International Solar Electric Technology (ISET), Inglewood, California, has fabricated a 10.4% thin-film CIS solar cell.

For thin-film CdTe solar cells, the best total-area efficiency achieved so far is 15.8% by researchers at the University of South Florida (USF), Tampa, Florida, by the close-spaced sublimation technique. The Institute of Energy Conversion (IEC) at the University of Delaware, Newark, Delaware has demonstrated 12%-efficient thin-film CdTe solar cells by the physical vapor deposition method. Innovative chemical and heat treatments using $CdCl_2$ vapor, which are critical for high performance thin-film CdTe solar cells, have been developed by IEC. A similar process has been developed by SCI

CH3365-4/94/0000-0266 $4.00 © 1994 IEEE

First WCPEC; Dec. 5-9, 1994; Hawaii

[8]. Colorado School of Mines (CSM), Golden, Colorado, has obtained an efficiency of 11% for a thin-film CdTe solar cell fabricated by electrodeposition [9]. In all cases, a superstrate configuration, that is, glass/SnO_2/CdS/CdTe/contact is used for high performance solar cells. Various contacting schemes have been used for these devices, such as graphite (Cu- or Hg- doped), Cu/Ni, Ni/Al, Ni, etc. Au contacts are highly unstable and are not recommended for practical devices.

A Si-Film device fabricated by AstroPower, Newark, Delaware on a graphite cloth substrate has resulted in a total-area efficiency of 13.4% for a 1-cm^2 solar cell, verified by NREL [10]. Al paste is used as the back contact and Ti/Pd/Ag as the front contact. A two layer anti-reflection coating of ZnS/MgF_2 is also used in this device structure. Progress of the polycrystalline thin film (PTF) solar cells is shown in Figure 1.

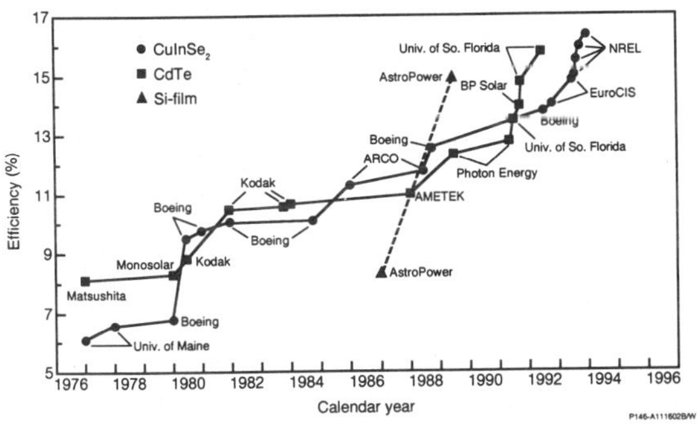

Fig.1 Progress of polycrystalline thin-film solar cells

Ten university groups are supported to advance the understanding of PTF materials and devices. Research efforts in modeling, analysis and characterization are underway. Purdue University, West Lafayette, Indiana, has developed a code "ADEPT" to model the various CIGS graded structures [11]. Colorado State University, (CSU) Fort Collins, Colorado, has developed a standardized procedure to quantify mechanisms that lead to less than ideal device performance. The analyses by CSU are currently scaled up to include submodules. Working with NREL, technique to identify loss mechanisms use is made of a laser spot scanning [12]. According to the CSU analysis, the performance of PTF devices is at two-thirds the value of single crystal devices [13]. Florida Solar Energy Center (FSEC), Cape Canaveral, Florida, has developed a two-stage selenization process to form high quality CIGS absorber films [14]. So far in collaboration with IEC, they have fabricated an 8.2% efficienct thin-film CIGS solar cell.

MODULE FABRICATION, PROCESSING AND TESTING

Progress at the module level has also been exceptional. Table I summarizes the performance of the various prototype PTF modules fabricated by seven groups worldwide. In the United States, EPV, ISET and SSI have fabricated thin-film CIS-based modules, whereas GPI and SCI have fabricated thin-film CdTe modules. Only two groups outside the United States, namely, BP Solar, England, and Matsushita Battery, Japan, fabricate prototype thin-film CdTe modules.

TABLE I Performance of Polycrystalline Thin Film Photovoltaic Modules

Group	Material	Area (cm^2)	Eff (%)	Power (W)
Solar Cells Inc.	CdTe	6862	8.1*	55.4*
Siemens Solar**	$CuInSe_2$	3830	11.2	43.1
Siemens Solar	$CuInSe_2$	3859	10.3*	39.7*
BP Solar***	CdTe	4540	8.4	38.2
Siemens Solar	$CuInSe_2$	3261	10.1*	35.7*
Golden Photon	CdTe	3528	7.7*	27.2*
Siemens Solar	$CuInSe_2$	938	11.1*	10.4*
Matsushita Battery	CdTe	1200	8.7	10.0
BP Solar	CdTe	706	10.1	7.1
Golden Photon	CdTe	832	8.1*	6.8*
ISET	$CuInSe_2$	845	6.9*	5.8*
Energy Photovoltaics	$CuInSe_2$	791	7.2	5.7
Energy Photovoltaics	$CuInSe_2$	722	6.0*	4.3*

*NREL measurements; All aperture-area efficiency; ** Unencapsulated; *** Non-monolithic integration

The highest efficiency obtained so far for a thin-film CdTe module was achieved by SCI, using a modified close-spaced sublimation method. The SCI module had an aperture-area efficiency of 8.9% and power output of 60.6 W. A similar number has been verified by NREL. A CdTe module fabricated at GPI by the spraying method (Figure 2), reached 7.7% and had a power output of 27.2 W, also verified by NREL. GPI currently has an installed manufacturing capacity of up to 2-MW for the fabrication of thin-film CdTe modules at its facilities in Golden, Colorado.

Fig.2 Thin-film semiconductor layers are deposited by spraying at GPI

SSI has fabricated a world record aperture-area efficiency of 10.3%, and a power output of about 40 W for a thin-film CIGSS module. EPV and ISET have also fabricated near-0.09-m^2 prototype thin-film CIS submodules with an aperture-area efficiency of about 7%. A minimodule of thin-film CIGSS with an aperture-area efficiency of 12.7% was also fabricated by SSI [4]. EPV has fabricated a thin-film CIS minimodule with an aperture-area efficiency of 9.1%. Se vapor was used for the CIS absorber layer formation. Recently, Solarex has fabricated a thin-film CIGS minimodule with an aperture-area efficiency of 10.4%.

CIS modules fabricated at SSI are undergoing reliability testing at NREL's Outdoor Test Facility. No change in module performance has been observed for the various modules being tested from 1988 to 1994. This is shown in Fig. 3. Typical module aperture-area efficiencies have improved from 7% to 9% during this time. Early SCI modules being tested outdoors at NREL also demonstrate fairly robust performance (Fig. 4). These are encouraging results for the emerging thin-film technologies. The research, development and technical issues have been discussed elsewhere [1].

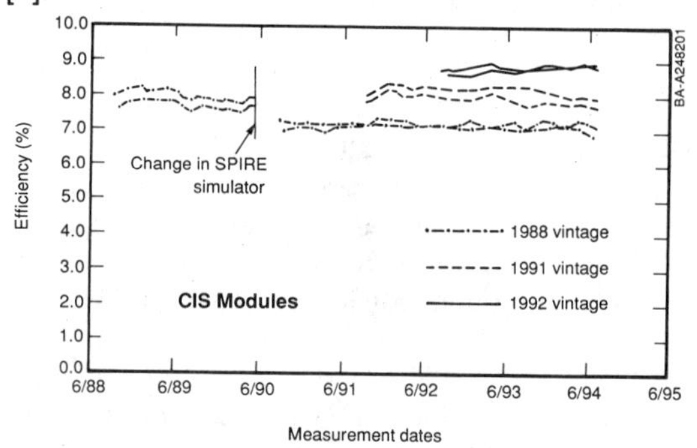

Fig.3 Reliability testing of thin-film CIS modules fabricated by SSI demonstrate no change in module performance.

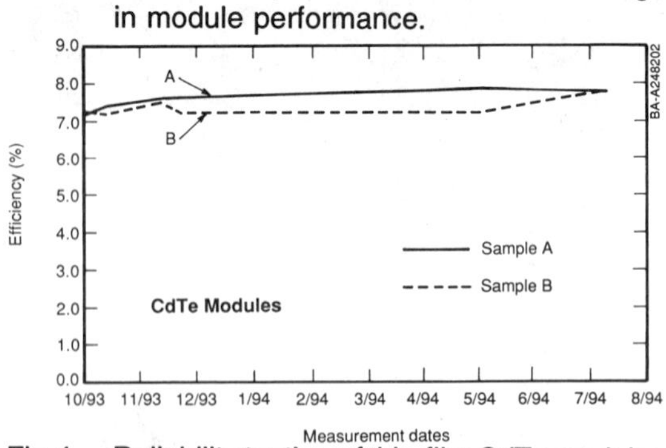

Fig.4 Reliability testing of thin-film CdTe modules fabricated by SCI demonstrate stable performance after several months.

THIN-FILM PV ARRAYS

A few thin-film PV arrays based on CIS and CdTe have been deployed so far. A nominal 1-kW thin-film CdTe array has been installed by SCI at its outdoor test facility for almost 2 years. A nominal 1-kW thin-film CIS array (Fig. 5) has been deployed by NREL at its outdoor test facility for several months. The thin-film CIS modules were supplied by SSI. Another 0.4-kW thin-film CdTe array has also been installed at NREL's outdoor test facility. The modules were supplied by SCI. Most recently, SCI has installed a 1.2-kW thin-film CdTe PV array at its outdoor test facility (Figure 6). Frameless modules have been used in this array configuration. The typical module output is about 50 W. GPI has also installed a 0.6-kW thin-film CdTe array at its outdoor test facility. All of these PV arrays are at various stages of data acquisition and testing. GPI has sold its first 1-kW thin-film CdTe PV array to a group in Brazil, while SCI has sold its first 1-kW thin-film CdTe PV array to a utility in Tunisia.

Fig.5 A nominal 1-kW thin-film CIS PV array being tested at NREL. Modules were supplied by SSI.

Fig.6 A 1.2-kW thin-film CdTe PV array being tested at SCI.

CRADA

Technology transfer is one of the key activities of the NREL mission. Towards this end, NREL has signed three cooperative research and development agreements (CRADAs) in the area of thin films with the various members of the U.S. PV industry. NREL is working with AstroPower in the area of Si-Film; with Martin Marietta, Denver, Colorado, in the area of CIGS absorber layer deposition; and with SCI for back contact-related research activities.

ES&H AND RECYCLING

One of the focuses of the PTF task is environmental safety and health (ES&H) issues associated with the commercialization of thin films. These span the spectrum from safety issues of gases (H_2Se) to the recycling of compound semiconductor-based modules such as CIGS and CdTe. An area of ongoing focus is recycling. We have begun the investigation of recycling Cd-bearing materials, both within the production plant (in terms of production wastes and rejects) and in terms of product recycling. It has become the norm for CdTe companies to have agreements with suppliers to take back in-plant wastes for recycling. This is accomplished by returning wastes to a supplier smelter. We are now investigating routes for product recycling. An important aspect of such plans is the fact that both In and Te are relatively rare. The potential thin-film manufacturers desire to have these materials returned to them for reuse in order to avoid availability and cost concerns. This should facilitate the process of assuring a "cradle-to-cradle" approach for these materials and modules. Some recent work in the toxicology of CIGS and CdTe is also of note. It is reported extensively by Brookhaven National Lab [15]. The new Thin Film Partnership has formed an ES&H "team" in order to address the key issues associated with the introduction of a large volume of thin film products.

FUTURE PLANS

The U.S. DOE/NREL thin film program has been reorganized in the following manner: all thin film R&D is consolidated under one program; and that effort is now based on teamed research within the the Thin Film Partnership Program. The program has been funded at the $18.3 million level in FY 1995, with a breakdown of those funds as follows: In-house NREL, $6.1 million; Technology Partners: $5.5 million; R&D Partners:, $6.7 million. Technology partners are companies that are leaders in near-term commercialization of thin-film technologies. They include: AstroPower, Energy Photovoltaics, Golden Photon, Siemens Solar Industries, and Solar Cells, Inc. R&D partner are those universities and companies which are adding significant technical contributions to thin films. They include International Solar Electric Technology, Solarex, Institute of Energy Conversion at the University of Delaware, Colorado School of Mines, Florida Solar Energy Center, Georgia Institute of Technology, Purdue University, University of Florida, University of South Florida, University of Toledo, Washington State University, and others. The operating principles of the partnership are (i) to enhance the support of leading companies in thin films in order to accelerate and improve the likelihood of success of near term commercialization and (ii) work, through a teamed approach, on critical technical issues identified above to assure mid- and longer-term progress. To that end, we have formed research teams in CIS, CdTe, and thin film ES&H.

CONCLUSIONS

Progress in PTF solar cells and prototype power modules during the past 18 months has been exceptional. A total-area conversion efficiency of 16.8% has been achieved for a thin-film CIGS solar cell fabricated by NREL scientists, while 15.8% has been achieved for a thin-film CdTe solar cell by USF. For Si-Film deposited on a graphite cloth substrate, a total-area efficiency of 13.4% was achieved. A world record aperture-area efficiency of 10.3% and a power output of about 40 W has been demonstrated for a thin-film CIGSS module fabricated by SSI. A thin-film CdTe power module of 8.1% efficiency and a power output of 55.4 W has been fabricated by SCI. Reliability testing of these emerging thin-film technologies has been encouraging. A few thin-film PV arrays based on these technologies have been deployed by the U.S. industry and NREL for data acquisition and reliability testing. A 2-MW thin-film CdTe manufacturing plant has been installed, and another 10-MW thin-film CdTe manufacturing plant is expected to be completed in 1996. Initial sales of 1-kW thin-film CdTe PV arrays have been made to Brazil and Tunisia. The polycrystalline thin-film technologies are on track to meet the U.S. DOE cost, performance, and reliability goals.

ACKNOWLEDGMENTS

This work was supported by the U.S. Department of Energy under Contract No. DE-AC36-83CH10093. We would like to thank Benjamin Kroposki, Laxmi Mrig, and Troy Strand for the NREL outdoor test data.

REFERENCES

[1] H.S. Ullal, K. Zweibel, and B.G. von Roedern, "Thin-Film CdTe and CuInSe$_2$ Photovoltaic Technologies," *Proceedings of the ISES Solar World Congress*, Budapest, Hungary, **3**, 1993, pp. 187-192.

[2] M.A. Contreras, J.R. Tuttle, D.A. Albin, A. Tennant, K. Ramanathan, A.M. Gabor, J. Scofield, and R. Noufi, "High Efficiency Cu(In,Ga)Se$_2$-Based Solar Cells: Processing of Novel Absorber Structures" *Proceedings of the First World Conference on Photovoltaic Energy Conversion*, Waikoloa, Hawaii, December 5-9, 1994, in press.

[3] A.M. Gabor, A.M. Hermann, J.R. Tuttle, M.A. Contreras, A. Franz, D.W. Niles, M. Bode, K. Ramanathan, R. Noufi, and D.G. Jensen, "Bandgap Engineering in Cu(In,Ga)Se$_2$ Thin Film Grown From (In,Ga)$_2$Se$_3$ Precursors," *Proceedings of the First World Conference on Photovoltaic Energy Conversion*, Waikoloa, Hawaii, December 5-9, 1994, in press.

[4] R. Gay, M. Dietrich, C. Fredric, C. Jensen, K. Knapp, D. Tarrant, and D. Willett, "Efficiency and Process Improvements in CuInSe$_2$-Based Modules," *Proceedings of the 12th European Photovoltaic Solar Energy Conference*, Amsterdam, The Netherlands, April 11-15, 1994, pp. 935-938.

[5] A.E Delahoy, J. Britt, F. Faras, A. Sizemore, G. Butler, and Z. Kiss, "Progress in CuIn(Se,S)$_2$ Module Development," *Eds. R. Noufi and H.S. Ullal, AIP Conference Proceeding 306*, Denver, Colorado, October 13-15, 1993, pp. 370-381.

[6] J. Kessler, S. Skibo, T. Lommasson, S. Wiedeman, J. Foglebock, L. Russell, and R.R. Arya, "Progress in CuInSe$_2$ Solar Cells ana Modules," *Proceedings of the First World Conference on Photovoltaic Energy Conversion*, Waikoloa, Hawaii, December 5-9, 1994, in press.

[7] L.C. Olson, F.W. Addis, D. Greer, H. Aguilar, W. Lei, and F. Abulfotuh, "Electrical and Physical Characterization of ZnSe/CIS Solar Cells," *Proceedings of the First World Conference on Photovoltaic Energy Conversion*, Waikoloa, Hawaii, December 5-9, 1994, in press.

[8] T.X. Zhou, R.C. Powell, R. Sasala, N. Reiter, P.V. Meyers, "In Situ CdCl$_2$ Vapor Treatment of Polycrystalline CdTe/CdS Films," *Proceeedings of the First World Conference on Photovoltaic Energy Conversion*, Waikoloa, Hawaii, December 5-9, 1994, in press.

[9] D. Kim, S. Pozder, Y. Zhu, and J.U Trefny, "Polycrystalline Thin Film CdTe Solar Cells Fabricated by Electrodeposition," *Proceedings of the First World Conference on Photovoltaic Energy Conversion*, Waikoloa, Hawaii, December 5-9, 1994, in press.

[10] J.A. Rand, Y.B. Bai, T.R. Ruffins, C.J. Thomas, A.M. Barnett, "Silicon on Graphite Cloth," *Proceedings of the First World Conference on Photovoltaic Energy Conversion*, Waikoloa, Hawaii, December 5-9, 1994, in press.

[11] J.L. Gray and Y.J. Lee, "Numerical Modeling of Graded Band Gap CIGS Solar Cells," *Proceedings of the First World Conference on Photovoltaic Energy Conversion*, Waikoloa, Hawaii, December 5-9, 1994, in press.

[12] I.L. Eisgruber and J.R. Sites, "Adaptation of Current-Voltage Analysis to Module-Scale Thin-Film Cell Geometries," *Proceedings of the First World Conference on Photovoltaic Energy Conversion*, Waikoloa, Hawaii, December 5-9, 1994, in press.

[13] J.R. Sites and X. Liu, "Six-Year Efficiency Gains For CdTe and CuIn$_{1-x}$Ga$_x$Se$_2$ Solar Cells; What has Changed?," *Proceedings of the First World Conference on Photovoltaic Energy Conversion*, Waikoloa, Hawaii, December 5-9, 1994, in press.

[14] N.G. Dhere and S. Kuttath, "Polycrystalline CuIn$_{1-x}$Ga$_x$Se$_2$ Thin-Film PV Solar Cells Prepared by Two-Stage Selenization Process Using Se Vapor," *Proceedings of the First World Conference on Photovoltaic Energy Conversion*, Waikoloa, Hawaii, December 5-9, 1994, in press.

[15] P.D. Moskowitz, R.E. Chapin, H. Steinberger, W. Thumm, "Characterizing Health and Environmental Hazards of CdTe Module Production, Use and Disposal," *Proceedings of the First World Conference on Photovoltaic Energy Conversion*, Waikoloa, Hawaii, December 5-9, 1994, in press.

EFFECT OF THIN FILM MODULE GEOMETRY ON SOLAR CELL CURRENT-VOLTAGE ANALYSIS

I.L. Eisgruber, J.R. Sites, Colorado State University

ABSTRACT

Extraction of diode quality factor and series resistance from current-voltage data may be complicated by voltage drops across the transparent conducting oxide. The effect of the voltage drops on IV analysis is negligible for small cells, but becomes important for larger cells such as those found in thin-film modules. Agreement between predicted and measured voltage drops is demonstrated. Numerical results are presented for CdTe and $CuInSe_2$ cells.

INTRODUCTION

Extraction of diode quality factor and series resistance is an important part of solar cell analysis, both for improvement of cell efficiency and for understanding of the physical current mechanisms. Traditional analysis of current-voltage characteristics to extract these parameters assumes that the voltage across the cell junction is constant with position. In reality, there is a voltage drop across both contact surfaces, particularly across the transparent conducting oxide, which typically has a higher sheet resistance than the metallised back contact. This voltage drop causes a subtle distortion of the IV curve which can skew the apparent values of diode quality factor and series resistance. The effect is usually negligible for research-sized cells, but becomes important for larger cells such as those typically used in thin-film modules. Methods to avoid this analysis problem have been proposed for crystalline cells [1], but such methods are problematic for polycrystalline cells, where cell parameters often vary with light intensity.

THE PROBLEM

Figure 1 compares the current-voltage (IV) curve of a research-sized cell with one from a module. The higher series resistance of the module-sized cell can easily be seen in the reduced slope of the IV curve near V_{oc}. Ignoring the difficulties of keeping cell quality high and uniform for a large-area cell, module-sized cells necessarily have higher series resistances than smaller cells. As cell dimensions increase, more lateral flow is required for current to be collected, resulting in more series resistance. Additionally, because more lateral flow is required, it makes sense to replace the standard solar cell model (shown in figure 2a) with a model that allows for voltage drop between

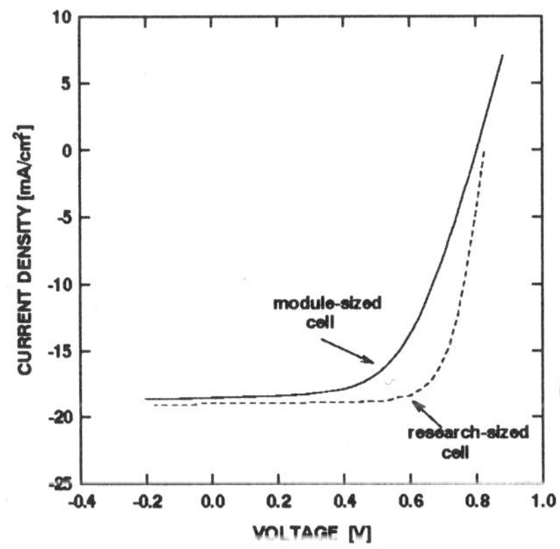

Figure 1: IV curves for CdTe research-sized and module-sized cells.

different parts of the solar cell due to lateral current flow, such as the model shown in figure 2b. We will call the standard model of figure 2a the "linear series resistance approximation" and will call the improved model of figure 2b a "distributed series resistance model".

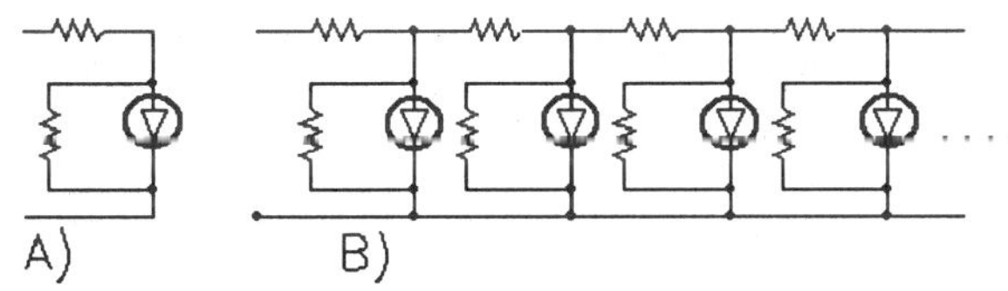

Figure 2: a) Standard model for solar cell with series and shunting resistances. b) Model for solar cell that incorporates multiple voltage drops due to lateral current flow.

It is fairly simple to find the effect of a voltage drop across the TCO (transparent conducting oxide) on the current-voltage characteristics of a solar cell if a few simplifying assumptions are made. We assume the following: (i) V is spatially constant on the back contact. Considering typical relative sheet resistances of the TCO and the back contact, this is usually a good assumption. (ii) The current-voltage characteristics of the junction are spatially uniform. (iii) All current flow in the semiconductor layers is perpendicular to the semiconductor layers, and all lateral flow occurs in the TCO and back contact. Since the semiconductor layers are thin, and the resistance of the semiconductor layers for lateral flow is relatively high, this assumption is generally good. (iv) Photocurrent is generated uniformly over the cell. (v) Series resistance due to the bulk of the semiconductor is negligible compared

CH3365-4/94/0000-0271 $4.00 © 1994 IEEE First WCPEC; Dec. 5–9, 1994; Hawaii

to that due to the TCO. Comparing typical semiconductor layer thicknesses and resistivities to typical TCO sheet resistances and cell dimensions, this assumption is also generally good. Figure 3 schematically shows the two dimensional current flow resulting from the preceding assumptions.

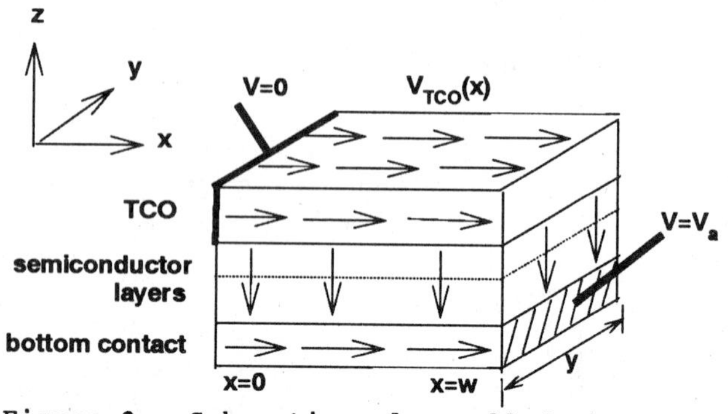

Figure 3: Schematic solar cell design. Arrows show assumed pattern of current flow.

It is then a simple physics problem to derive the relationship between the applied voltage and total current through the solar cell. One arrives at the following nonlinear differential equation:

$$-\frac{1}{\rho_s}\frac{d^2V_{TCO}(x)}{dx^2} = J_0 e^{\frac{V_a - V_{TCO}(x)}{V_0}} - J_L \quad (1)$$

with the boundary conditions

$$\left.\frac{dV_{TCO}}{dx}\right|_{x=w} = 0 \qquad \left.V_{TCO}\right|_{x=0} = 0 \quad . \quad (2)$$

Here, as indicated in figure 3, $V_{TCO}(x)$ is the voltage along the TCO as a function of position along the TCO, ρ_s is the TCO sheet resistance, J_0 is the diode saturation current density, V_a is the voltage applied to the cell, J_L is the photocurrent density, w is one half the distance between grid fingers (or the entire cell width of a gridless cell), and V_0=AkT/q. Nielsen [2] gives an in depth mathematical analysis of the behavior of this equation.

The differential equation does not have a closed form solution. Numerical solutions yield curves similar to those shown in figure 4. The curves labeled "distributed series resistance" are the numerically generated light and dark IV curves, taking into account voltage drop across the TCO. The curves labeled "linear series resistance approximation" are IV curves of an identical diode in series with a resistor, the model assumed for traditional extraction of diode quality factor and series resistance. The resistor's value is the first-order linear series resistance, ρ_sw/3y, which is the resistance as determined by power loss in the TCO assuming spatially uniform current density through the junction [3]. The difference between the distributed and linear series resistance curves is small at low voltages, but increases and becomes significant at higher values of forward current.

EFFECT ON CELL ANALYSIS

As can be seen in figure 4, the effect of the voltage drop across the TCO is small at operating voltages.

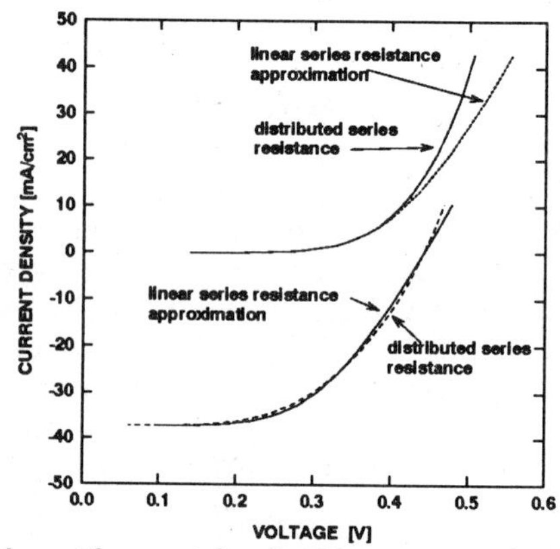

Figure 4: Theoretical IV curves for cell with J_L=37.3 mA/cm², V_{oc}=0.442 V, A=1.5, T=298°C, w=1 cm, and ρ_s=7.6 ohms/□.

Therefore, the linear series resistance can be used to find a good estimate of the power loss at the maximum power point (V_{mp}) due to TCO resistance. However, data used to extract diode quality factor A and series resistance R is at voltages higher than V_{mp}, where the effect is more pronounced than at V_{mp}. Furthermore, extraction of A and R is rather sensitive to small changes in the data. As a result, even though the effect of the voltage drop across the TCO is small at V_{mp}, the effect on the extraction of A and R can be substantial.

Figures 5 and 6 show the effect of voltage drop across the TCO on apparent A and R for typical $CuInSe_2$ and CdTe cells. The x-axis shows $w^2\rho_s$, the most important parameter in determining the difference between the apparent and true A and R. Typical research-sized cells have $w^2\rho_s$<1 ohm-cm², and therefore their analysis is not skewed by voltage drop across the TCO. Modules, however, may easily have values of $w^2\rho_s$ which imply significant error in the apparent A and R. The approximate value of $w^2\rho_s$ for modules from Solar Cells Inc. and Siemens Solar, as well as for research-sized cells from NREL and Solar Cells Inc. are marked on the x-axes in figures 5 and 6.

In figures 5 and 6, the difference between true and apparent A is more severe for for $CuInSe_2$ than for CdTe at the same value of $w^2\rho_s$. This occus chiefly because $CuInSe_2$ has a larger short-circuit current (smaller bandgap) than CdTe, and therefore more voltage drop across the TCO near V_{mp}. Thin film materials with a bandgap intermediate between $CuInSe_2$ and CdTe are likely to have

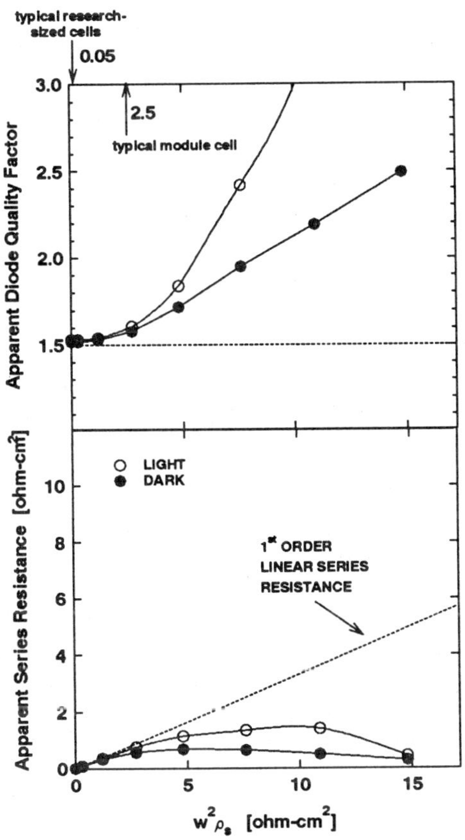

Figure 5: The effect of cell width and TCO sheet resistance on apparent A and R for a typical $CuInSe_2$ cell.

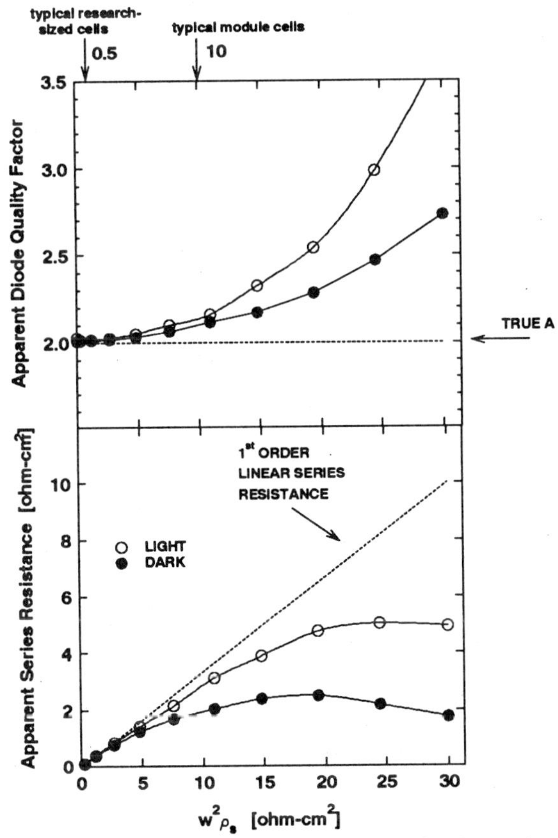

Figure 6: The effect of cell width and TCO sheet resistance on apparent A and R for a typical CdTe cell.

a difference between true and apparent A also intermediate between that of $CuInSe_2$ and CdTe. In addition, higher forward current tends to exacerbate the effect, e.g. higher values of A and lower values of shunting resistance result in a larger difference between the true and apparent A.

A and R in figures 5 and 6 were determined from the numerically generated IV curves in a standard method [4] - plotting dV/dJ versus $1/(J+J_l)$, which yields and intercept of R and a slope related to A. Examples of such fits, and the resulting A's and R's, are shown in figure 7. Note that since the data points for the "distributed series resistance" curve in figure 7 are somewhat nonlinear, the apparent A and R will depend slightly on the conditions of the fit, such as what range of data is used, how closely data points are spaced in voltage, and how data points are weighted in determining the fit. The conditions used in creating figures 5 and 6 are typical of those used in experiment. Voltage points in the IV curves were spaced .015 V apart, and the light data from V_{mp} to +10 mA/cm² were used to determine A and R. The range of dark data used to determine A and R was the same range of forward current as the light data. For figure 5, parameters typical for $CuInSe_2$ were used: J_L=37 ma/cm², V_{oc}=.442, A=1.5, and T=298 °C. For figure 6, parameters typical for CdTe were used: J_L=17 ma/cm², V_{oc}=.83, A=2, and T=298 °C. Deviations from these parameter values or from the conditions used to determine A and R may cause a slightly

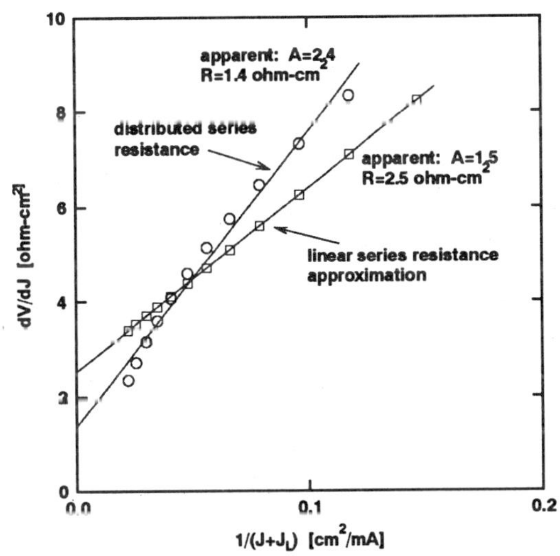

Figure 7: Fits used to determine A and R for IV curves shown in figure 4. In both cases, the true A=1.5 . Actual data from a large cell would resemble the distributed series resistance curve.

different apparent A or R than shown in figures 5 and 6. In both cases, cells were assumed to have an infinite shunting resistance, which is a good approximation for cells that do not shunt excessively (r>1kΩ-cm²). Leaky diodes, however, will show a larger difference between the apparent and true values of A.

Figure 8 shows agreement between the calculated and measured voltage drops across the TCO for a $CuInSe_2$ mini-module cell. Predicted voltage drop was calculated

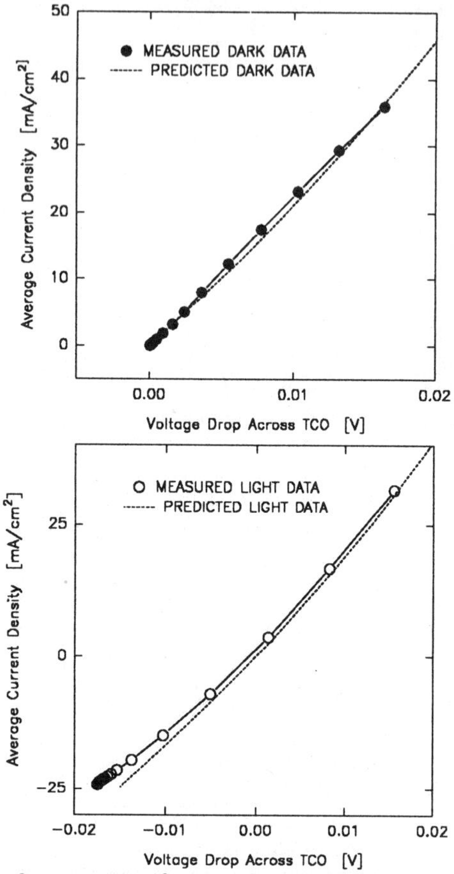

Figure 8: Comparison of calculated and measured voltage drop across TCO for CIS module cell. The top graph shows the dark case, and the bottom graph shows the illuminated case.

from IV characteristics of the cell and the manufacturer's estimate of TCO sheet resistance. Voltage drop was measured by placing voltage-sensing probes on either edge of the unencapsulated cell as current flowed through the entire module. A small difference between predicted and measured values is to be expected, as the voltage drop is sensitive to precise placement of the probes and nonuniformities in current flow due to localized defects.

PRACTICAL CONSEQUENCES

There are several practical consequences resulting from the distortion of the IV curve by voltage drop across the TCO. First, and most importantly, A is overestimated and the power loss due to series resistance is underestimated. If ρ_s is known, figures 5 and 6 can be used for correction. A further complication of the overestimation of A and underestimation of R occurs when making a group of similar cells, during which one switches from using a high ρ_s TCO to using a lower ρ_s TCO. The value of $\rho_s w^2$ decreases, so the apparent A becomes closer to the true A. Thus, switching from a high ρ_s TCO to a lower ρ_s TCO causes an apparent but unreal improvement in diode quality factor, and the apparent R does not decrease as much as it should to give a true picture of the decrease in power loss due to decreased resistance. Additionally, in analysing a single cell, there is an artificially large apparent difference between the light and dark apparent diode quality factors and series resistances. These artificial differences are superimposed upon any real differences between the light and dark diode quality factors and series resistances. In all cases, the analyst should be aware of the value of $\rho_s w^2$ for the cell being characterized and the distortion likely to occur.

CONCLUSIONS

Voltage drop across the TCO causes a subtle distortion of solar cell IV curves. This change is not likely to cause a large change V_{mp}, but can be important in determination of diode quality factor and series resistance. The effect is negligible in research-sized cells, but becomes noticeable in modules, where cells dimensions and distances between grid fingers are larger. For $CuInSe_2$ cells, the effect becomes important around $\rho_s w^2 \geq 3.5\ \Omega\text{-cm}^2$. For CdTe cells, the effect becomes important around $\rho_s w^2 \geq 10\ \Omega\text{-cm}^2$. The result is an overestimation of A and an underestimation of R.

ACKNOWLEDGEMENTS

The authors are grateful for samples and information provided by scientists at Solar Cells Incorporated, Siemens Solar, the National Renewable Energy Laboratory, and the University of South Florida. They particularly acknowledge the financial support of the National Renewable Energy Laboratory.

REFERENCES

[1] A.G. Aberle, S.R. Wenham and M.A. Green, "A New Method for Accurate Measurements of the Lumped Series Resistance of Solar Cells", *Proceedings of the 23rd IEEE Photovoltaic Specialists Conference*, 1993, pg. 133.

[2] L.D. Nielsen, "Distributed Series Resistance Effects in Solar Cells", *IEEE Transactions on Electron Devices*, 29, pg. 821, 1982.

[3] M.A. Green, *Solar Cells: Operating Principles, Technology, and Systems Applications*, Prentice-Hall, Englewood Cliffs, NJ, 1982, p. 145.

[4] J.R. Sites and P.H. Mauk, "Diode Quality Factor Determination for Thin Film Solar Cells", *Solar Cells*, 27, pg. 411, 1989.

ROOM-TEMPERATURE FT-LUMINESCENCE ANALYSIS OF $Cu(In,Ga)Se_2$ FILMS AND DEVICES

John D. Webb, Miguel Contreras, and Rommel Noufi
National Renewable Energy Laboratory (NREL)
Golden, Colorado (USA)

ABSTRACT

We report a rapid, convenient luminescence technique for quality control of polycrystalline $Cu(In,Ga)Se_2$ (CIGS) films and photovoltaic (PV) devices. The speed and convenience of the luminescence analyses were realized by using a Fourier transform (FT) Raman spectrophotometer which operates in the near-infrared (NIR) spectral region encompassing the band gap and defect levels of CIGS. With minor modifications to the FT-Raman spectrophotometer, we were able to detect both photoluminescence (PL) and electroluminescence (EL) from CIGS devices at room temperature. The FT-EL technique allows luminescence measurements to be made using this equipment at energies up to 1.3 eV, while the FT-PL technique is limited to energies below 1.15 eV. By increasing sensitivity and eliminating the need for sample cooling, this approach reduced the measurement time by an order of magnitude relative to comparable dispersive PL measurements. We used a fiberprobe accessory to the FT-Raman spectrophotometer to demonstrate that samples can be checked for uniformity at remote locations, e.g. online, using FT-PL spectroscopy. We also used a microscope accessory to obtain the PL spectra of visibly discolored regions some tens of microns in diameter on a CIGS device, and to show that these regions emit PL at significantly lower energy and intensity than nearby uniform regions of the device.

INTRODUCTION

Conventional dispersive spectroscopy has been generally unsuccessful in measuring PL or EL spectra of polycrystalline CIGS or $CuInSe_2$ (CIS) films, although dispersive PL spectra of CIS single crystals have been measured at room temperature. Photoluminescence (PL) and electroluminescence (EL) spectroscopic techniques have been used extensively to measure band gap, defect, and impurity levels in semiconductor materials for PV applications. Although PL measurements have been reported for some materials at room temperature,[1] detection of PL or EL using dispersive spectrophotometers has typically required cryogenic cooling of CIGS films to obtain adequate sensitivity. Cooling the samples is a slow process, and large or fragile samples are difficult to cool without breaking them to a smaller size.

In earlier work demonstrating FT-PL spectroscopy,[2,3] we presented what we believe to be the first room-temperature PL spectra of CIS films and devices. We also correlated higher CIS device efficiencies with the absence of low-energy PL from defect states in CIS devices.[3] Not counting sample cooling time, this approach reduced the measurement time by a factor of four relative to dispersive PL spectroscopy at similar resolution, while increasing sensitivity by a factor of two to three.[2] However, the upper energy limit for FT-PL measurements is about 1.15 eV, and is determined by the Rayleigh holographic notch filters for the fundamental Nd:YAG laser emission at 1.17 eV. The high-efficiency CIGS devices now being developed at the National Renewable Energy Laboratory[4,5] typically have band gaps near 1.15 eV, making it impossible to obtain complete FT-PL spectra from these devices. We solved this problem by turning off the laser and removing the Rayleigh line filters from the FT-Raman spectrophotometer. By applying forward bias to CIGS devices <u>in-situ</u>, we were able to measure complete FT-EL spectra from these devices at room temperature.

EXPERIMENTAL APPROACH

The CIGS devices analyzed consist of multi-layer films deposited in the sequence Al/Ni (grid)/MgF_2/ZnO/CdS/CIGS/Mo/glass substrate.[4] We used a Nicolet FT-Raman accessory coupled to a Nicolet System 800 FTIR spectrophotometer[2] to obtain the results shown in Figs. 1-3. The spectra shown in Figs. 4-6 were obtained using a Nicolet System 950 FT-Raman spectrophotometer[3] equipped with microscope and fiberprobe accessories. Spectral resolution was 4 cm^{-1}. The FT-Raman accessory imaged a 50-μm diameter spot for analysis, and laser power was kept at 10 mW. The 32X Caccograinian objective of the microscope accessory to the System 950 spectrophotometer collected PL emitted from a spot about 6 μm in diameter. Laser power of 24 mW was used with the microscope. The silica fiberprobe accessory to the System 950 spectrophotometer collected PL from an analysis area 0.5 mm in diameter. Laser power of 50 mW was used with the fiberprobe. Neutral-density (ND) filters were inserted between the samples and the cooled, Zn-doped Ge detector as needed to prevent saturation.

RESULTS

Figure 1 compares room-temperature FT-PL and FT-EL measurements on a 15.5%-efficient CIGS device (C-253). Note the discontinuity in the FT-PL spectrum between 9300 and 9600 cm^{-1} (1.15 - 1.19 eV) caused by the Rayleigh line filters. The FT-EL spectrum is continuous in this region. The PL observed in Figure 1 at energies above the 1.17-eV laser emission may be related to thermal broadening of the energy distribution of electron-hole pairs. Insertion of a 1.2-eV silicon low-pass filter into the laser beam to block any visible light, such as that emitted by the helium-neon laser in the interferometer of the FTIR spectrophotometer, did not reduce the high-energy PL, which was observed to be proportional to the intensity of the Nd:YAG laser emission. Also note the similarity

First WCPEC: Dec. 5–9, 1994; Hawaii

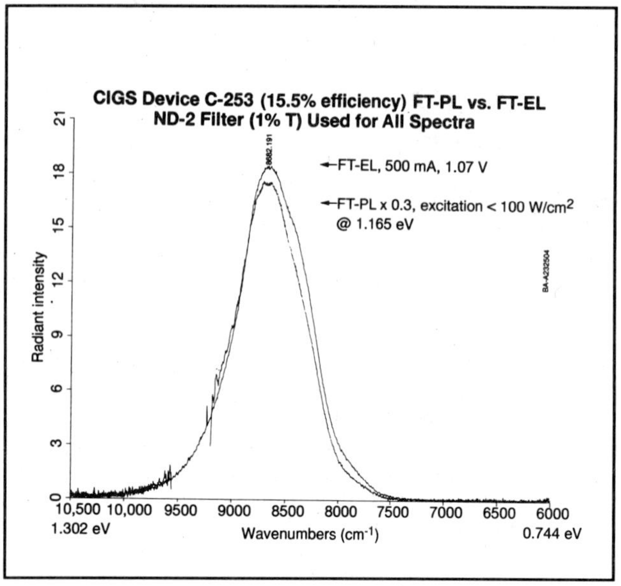

Figure 1

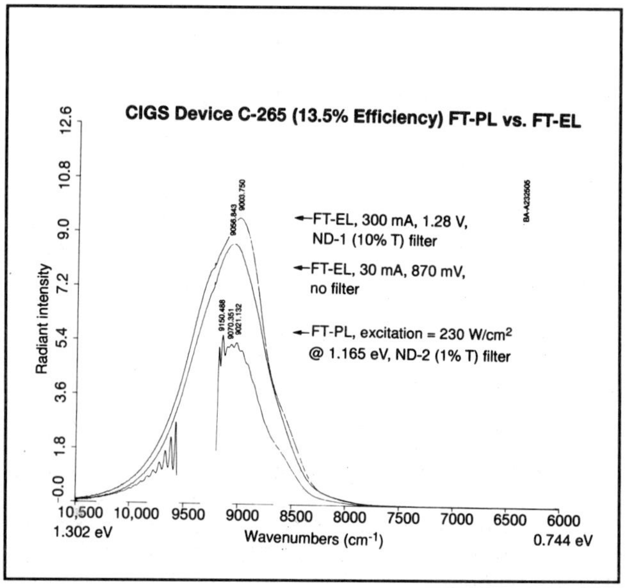

Figure 2

in the band shapes and maxima of the FT-EL spectra to those of the FT-PL spectra of this relatively low-band gap CIGS device, although a low-energy shoulder that may be related to the graded junction in this device is more prominent in the FT-EL spectrum. This device luminesced so strongly that it required an ND-2 (1% transmittance) filter to prevent detector overload in PL and EL. From the data in Fig. 1, the cell area (0.43 cm^2), and the FT-PL excitation power and area given previously, the power efficiency of FT-EL for this device would appear to be more than an order of magnitude greater than the power efficiency of FT-PL, but it is likely that the FT-EL measurement includes a larger excitation area than the FT-PL measurement.

The FT-PL and FT-EL spectra of a 13.5%-efficient device (C-265) are shown in Fig. 2. This device has a higher band gap, making the Rayleigh line filter artifact in the FT-PL spectrum more apparent. Note that device C-265 has a higher band gap but exhibits lower efficiency than device C-253, despite the expectation that higher band gap devices (up to 1.5 eV) allow a better match to the solar spectrum. Increasing the bias current to 300 mA from 30 mA resulted in a linear increase in EL intensity (note filter types given in Fig. 2), but some asymmetry of the FT-EL spectrum is evident at the higher current. This asymmetry, plus the prominent low-energy shoulder evident in the FT-EL and FT-PL spectra of this device may be related to its lower efficiency relative to device C-253. Similar spectral features were deconvoluted, have been related to lower efficiency in CIS devices and may represent defect states.[6]

Figure 3 shows the luminescence spectra of a 15.1%-efficient CIGS device (S-423). The FT-EL band is smooth and symmetrical, while the FT-PL band maximum is entirely missing. The small, sharp peak just below the FT-EL band maximum is a gain-ranging artifact of the A/D converter in the FT-Raman spectrophotometer. Raising the bias current much above 500 mA resulted in a decrease in FT-EL stability, and, ultimately, to a decrease in FT-EL intensity and appearance of thermal radiative features in the low-energy region of the FT-EL spectra of the CIGS devices. These problems may be related to contact resistance in the in-situ sample mount built for the FT-EL measurements, as well as to device instability.

The vertical scales on the first three figures are in radiance units (W/ster-cm^{-1}) and the luminescence peaks are about three orders of magnitude too intense given the excitation power used and a field of view of about 140 degrees. There is evidently an error in the instrument calibration procedure, which involves comparison of a white light reference spectrum to the black-body spectrum at the bulb's color temperature to obtain an instrument correction, and we are working with the manufacturer to resolve this problem.

Figure 4 is a comparison of FT-PL analyses of a visually uniform region of CIGS device C-253 which were performed using a microscope accessory capable of analyzing a 6-μm or larger diameter spot, and a 1.5-m (long) silica fiberprobe. The spectra overlay well when autoscaled to similar intensities, showing that the fiberprobe does not contribute spectral artifacts to the FT-PL spectrum of a similar region of the sample. Figure 5 shows that at the spatial resolution of 0.5 mm obtainable with the fiberprobe, variations in the FT-PL spectra of different regions of device C-253 are evident.

Figure 6 shows FT-PL microscope analyses of two visually discolored regions of CIGS device C-253 compared to that of a visually uniform region. Some variations from the FT-PL spectra of visually uniform regions of the device are observed two of a number of discolored regions visible through the microscope. The FT-PL spectra of the discolored regions contain a number of low-energy peaks, and resemble the FT-PL spectrum of low-efficiency CIS.[3] The discolored areas also emit much lower PL intensity than a visually uniform region of the device. These observations suggest that device efficiency,

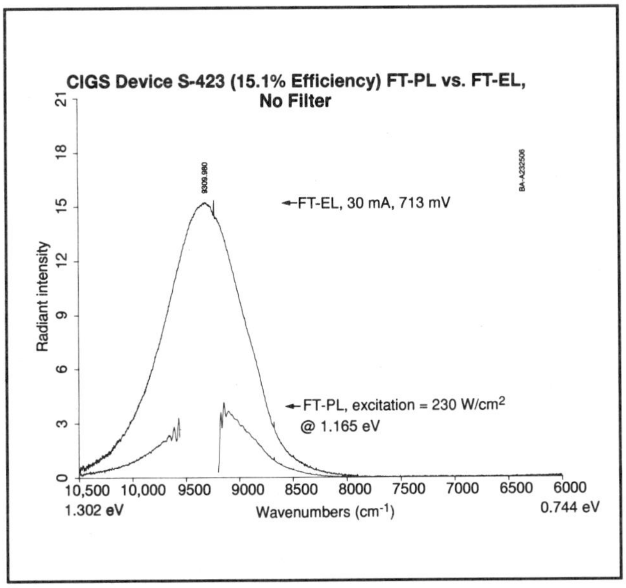

Figure 3

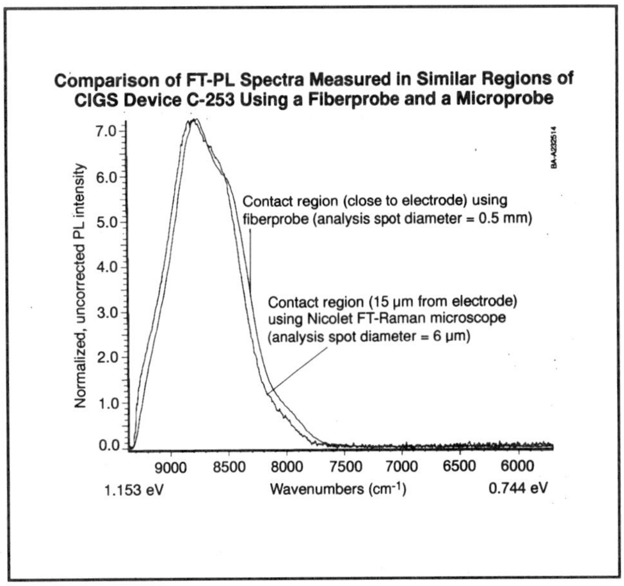

Figure 4

although high, could be further improved by increasing film uniformity. However, not all of the discolored spots on the device exhibited low-energy FT-PL spectral features.

The microscope and fiberprobe can also be used in the FT-EL mode, further increasing their utility as potential online diagnostic techniques for CIGS devices and modules. We are currently attempting to correlate the micro-FT-PL results with those obtained by using other microanalytical procedures on the anomalous regions of the CIGS films, and are examining other CIGS devices to assess the extent of non-uniformity. The FT-luminescence band structures could be correlated with defect levels in CIGS, as has been done for CIS.[6] We are also interested in using FT luminescence techniques to confirm the optical bowing data recently obtained for CIGS using total reflection and transmission measurements of band gaps.[7]

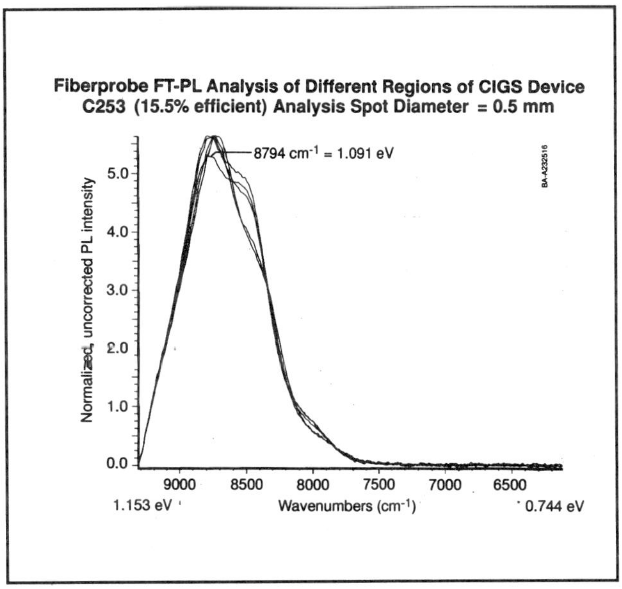

Figure 5

SUMMARY

The room-temperature FT-EL technique is novel and, at low bias currents, gives spectra similar to FT-PL spectra. It can be used to measure band gaps up to 1.3 eV in CIGS devices. We are continuing to develop room-temperature FT-PL and FT-EL spectroscopy as non-destructive techniques for quality control of CIGS devices and improved understanding of material properties. The detection of low-gap regions in the CIGS films promises that even higher efficiencies can be realized from CIGS devices if uniformity problems can be overcome.

ACKNOWLEDGMENTS

We gratefully acknowledge the assistance given by Gabor Kemeny, Chris Petty, and Bonnie Leimer in collecting some of the data presented and allowing the first author to use their Nicolet System 950 FT-Raman spectrophotometer at Nicolet Corporation (Madison, WI) to collect additional data. This work was supported by the U. S. Department of Energy under contract No. DE-AC36-83CH10093.

Micro-FT-PL Analysis of Different Regions of CIGS Device C-253, Cell #3 (15.5% efficient), Analysis Spot Diameter = 6 μm

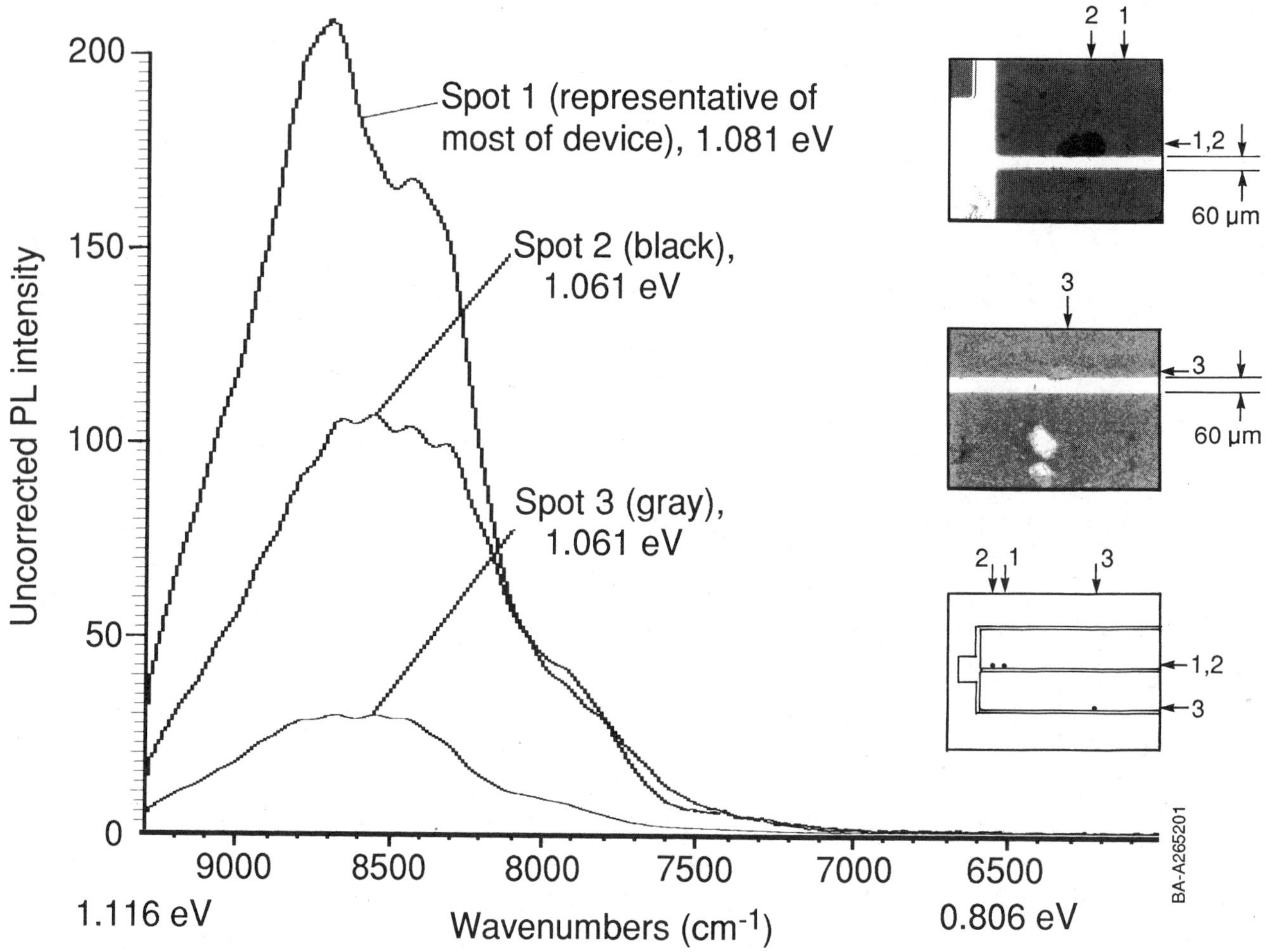

Figure 6

REFERENCES

[1] F. Abou-Elfotouh, D.J. Dunlavy, D. Cahen, R. Noufi, L.L. Kazmerski, and K.J. Bachmann, *Prog. Crystal Growth*, **10**, 1984, p. 365.

[2] J.D. Webb, D.J. Dunlavy, T. Ciszek, R.K. Ahrenkiel, M.W. Wanlass, R. Noufi, and S.M. Vernon, *Appl. Spectrosc.*, **47** (11), 1993, pp. 1814-1819.

[3] J.D. Webb, D.J. Dunlavy, T. Ciszek, R.K. Ahrenkiel, M.W. Wanlass, R. Noufi, and S.M. Vernon, *Mat. Res. Soc. Symp. Proc.*, **324**, 1994, pp. 233-239.

[4] A.M. Gabor, J.R. Tuttle, M. Contreras, D.S. Albin, A. Franz, D.W. Niles, R. Noufi, *Appl. Phys. Lett.*, **65** (2), 1994, pp. 198-200.

[5] J.R. Tuttle, M.A. Contreras, J.S. Ward, A.M. Gabor, K.R. Ramanathan, A.W. Tennant, L. Wang, J. Keane, and R. Noufi, *Proc. 1st World PVSEC Conf.*, 1994 (in press).

[6] F.A. Abou-Elfotouh, H. Moutinho, A. Bakty, T. J. Coutts, and L.L. Kazmerski, *Solar Cells*, **30**, 1991, p 151-160.

[7] D.S. Albin, J.J. Carapella, J.R. Tuttle, and R. Noufi, *Mat. Res. Soc. Symp. Proc.*, **228**, 1992, p. 267-272.

STUDY ON THE INTERFACIAL STRUCTURE AND JUNCTION DEPTH OF POLYCRYSTALLINE $Cu(In,Ga)Se_2$ DEVICES

Art J. Nelson, A.M. Gabor, M.A. Contreras, J.R. Tuttle, and R. Noufi
National Renewable Energy Laboratory, Golden, CO 80401

P. Asoka-Kumar and K.G. Lynn
Department of Physics, Brookhaven National Laboratory, Upton, NY 11973

ABSTRACT

X-ray photoemission spectroscopy (XPS) and positron annihilation spectroscopy (PAS) have been used to characterize the surface-versus-bulk composition, electronic, and physical structure of polycrystalline $Cu(In,Ga)Se_2$ thin-film interfaces. Angle-resolved high resolution photoemission measurements on the valence band electronic structure and Cu 2p, In 3d, Ga 2p, and Se 3d core lines were used to evaluate the surface and near-surface chemistry of $CuInSe_2$ and $Cu(In,Ga)Se_2$ device-grade thin films. XPS compositional depth profiles were also acquired from the near-surface region. PAS was used as a nondestructive, depth-sensitive probe for open-volume-type defects. Results of these measurements are related to device efficiencies to show the effects of compositional variations and defect concentrations in the near-surface region on device performance.

INTRODUCTION

The ternary $A^IB^{III}X_2^{VI}$ chalcopyrite semiconductor, $CuInSe_2$ (CIS)(E_g = 1.04 eV), has proved to be a superior absorber for heterojunction photovoltaic (PV) device applications [1]. Technical advances in polycrystalline thin-film technology have demonstrated PV devices with measured efficiencies of approximately 16% (total area) [2]. This has been realized by developments in physical vapor deposition and processing. Furthermore, these devices have exhibited non-heterojunction behavior, and thus it has been proposed that there exists a moderately doped n-type surface layer [3-5]. Proper engineering of this n type surface region could result in further improvements in device performance. Specifically, the presence of an optimized n-type surface region will effectively reduce recombination at the $CdS/Cu(In,Ga)Se_2$ (CIGS) heterojunction interface.

Deviations from the ideal CIS stoichiometry introduces defects in the chalcopyrite lattice, e.g., incorporation of excess In results in unoccupied Cu-sites [6]. These defect chalcopyrite phases have a tendency to segregate at the surface of polycrystalline thin-films [3,4]. In addition, the lattice defects (Cu vacancies) can have well-defined positions (ordered), which has lead to the term "ordered vacancy compound," or OVC. Two such phases have been identified in the literature, specifically, $CuIn_3Se_5$ [7,8] and $CuIn_2Se_{3.5}$ [9,10].

This paper presents and discusses the results of X-ray photoemission (XPS) and positron annihilation spectroscopic (PAS) characterization of the surface-versus-bulk composition, electronic and physical structure of polycrystalline CIGS thin-film interfaces. These results are related to device efficiencies to show the effects of compositional variations and defect concentrations in the near surface region on device performance.

EXPERIMENTAL

The enhanced-grain polycrystalline p-type $CuInSe_2$ and $Cu(In,Ga)Se_2$ thin-films were fabricated by a novel physical vapor deposition (PVD) process described in detail elsewhere [2]. These films were all grown on Mo-coated soda-lime glass substrates. Quantitative bulk compositional analysis was performed with an electron microprobe (EMPA) using X-ray wavelength dispersive spectroscopy (WDS). Results of these measurements provided the standard for quantification of the bulk XPS results.

Angle-resolved and high resolution XPS analyses were performed with monochromated Al Kα radiation and constant-pass energy of 11.75 eV, thus giving an overall energy resolution of 0.5 eV. Angle-resolved measurements were acquired at take-off angles of 15° (increased surface sensitivity) and 65° (bulk-like sensitivity) in order to nondestructively (i.e., no sputtering) probe the surface composition. To further study variations in the near-surface and bulk composition, ion beam sputter etching was performed with a differentially pumped ion gun operating with a 3 kV Ar^+ ion beam (10^{-2} Pa Ar pressure). Previous studies have shown that semiconductor films can be sputtered with up to 3 keV Ar ions without artificially changing the composition [11].

Positron annihilation spectroscopy is a nondestructive, depth-sensitive probe for open-volume-type defects with a sensitivity of 5 x 10^{15} cm^{-3}[12,13]. This technique examines changes in Doppler broadening of the annihilation gamma rays that originate from various depths of a test material. The Doppler broadening can be characterized by an S (for shape) parameter, defined as

CH3365-4/94/0000-0279 $4.00 © 1994 IEEE

First WCPEC; Dec. 5–9, 1994; Hawaii

the ratio of counts in a fixed central region (of width ≈1.59 keV and centered around 511 keV) of the annihilation photopeak to the total number of counts in the annihilation photopeak. When positrons annihilate after being localized near open-volume-type defect sites, annihilation occurs predominantly with low momentum electrons as opposed to the defect-free crystal where positrons are more likely to annihilate with core electrons. As a result, in the presence of open-volume defects, the annihilation photopeak is sharp and produces a higher S parameter.

RESULTS AND DISCUSSION

The CIS films chosen for this study were identified by M1201 and M1218. Devices made from different areas of these films had total area efficiencies of 13.2% and 11.3%, respectively. The CIGS films were identified by C266, S423, and S521, with corresponding total-area efficiencies of 16.4%, 15.1%, and 14.2%, respectively [2].

Tables I and II summarize the angle-resolved XPS compositional analysis of the near surface region of the CIS and CIGS polycrystalline thin-films, respectively. Remembering that the data acquired at a 15° take-off angle are extremely surface sensitive, the results clearly show that all of the films have a Cu-poor surface composition. Specifically, for the CIS samples, M1201 has a surface composition indicative of Cu:In:Se = 1:5:8 (or a two phase mixture of In_2Se_3 and $CuIn_3Se_5$), while M1218 has a surface composition more indicative of single-phase In_2Se_3. Since the Cu_2Se–In_2Se_3 phase diagram does not support the existence of a $CuIn_5Se_8$ phase, the mixed phase is more probable.

TABLE I. Angle-resolved XPS Compositional Analysis of $CuInSe_2$ Polycrystalline Thin-Films

	M12 01		M12 18	
Take-off Angle	65°	15°	65°	15°
Cu	7.3%	6.3%	3.7%	2.5%
In	31.6%	35.8%	36.3%	38.7%
Se	61.1%	57.9%	60.0%	58.8%
(Na)	(17.9%)	(19.5%)	(4.6%)	(6.7%)

In the case of the CIGS samples, the average surface composition is also representative of $Cu(In,Ga)_5Se_8$. However, the 16.4% device has a surface composition more indicative of $Cu(In,Ga)_3Se_5$ while the 15.1% and 14.2% devices have surface compositions more indicative of mixed In_2Se_3 and $Cu(In,Ga)_3Se_5$ phases. In addition, an accumulation of Na on all of the surfaces was observed with typical surface concentrations of ≥10 at. %. This is due to the fact that the films were grown on soda-lime glass at 550°C. However, Na concentrations in the bulk material were <1 at. %. This is in agreement with the observation by Bodegård and Stolt [14]. In any case, recent experimental investigations [15] have shown that the presence of Na on the surface does not significantly affect (<20 meV) CdS/CIGS heterojunction valence band offsets.

TABLE II. Angle-resolved XPS Compositional Analysis of $Cu(In,Ga)Se_2$ Polycrystalline Thin-Films

	C 266		S 423		S 521	
Take-off angle	65°	15°	65°	15°	65°	15°
Cu (at.%)	12.3	8.4	7.2	5.7	7.3	6.4
In (at.%)	20.5	22.5	26.2	28.5	26.5	35.4
Ga (at.%)	5.7	5.5	8.9	8.2	9.7	13.4
Se (at.%)	61.5	63.6	57.7	57.6	56.5	44.8
(Na) (at.%)	(2.1)	(1.8)	(11.6)	(10.6)	(18.5)	(15.1)

Figures 1 and 2 show the XPS compositional depth profiles for the CIS and CIGS devices with a sputter rate of 1 Å/s. These results show a remarkable trend. Specifically, a relation between the depth of the metallurical interface, defined as the region between the Cu-poor surface and bulk stoichiometric material, and the device efficiency was observed. This analysis revealed that the depth of the metallurgical interface was 320 Å for the 13.2% CIS device and >450 Å for the 11.3% CIS device. Similarly, the depth of the metallurgical interface was 115 Å for a 16.4% CIGS device, 240 Å for a 15.1% CIGS device, and 300 Å for a 14.2% CIGS device (Fig. 2). Photoemission results also show

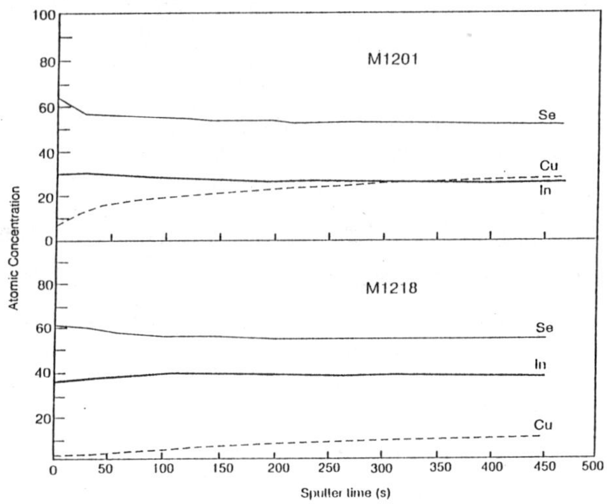

Figure 1. XPS compositional depth profiles for $CuInSe_2$ thin-films M1201 (η=13.2%) and M1218 (η=11.3%).

that the surface region is n-type and that the bulk material is p-type and that there is a 0.5-eV valence band offset between the n-type surface layer and the p-type bulk [4]. These results clearly show that the depth of the buried junction may be the determining factor in device

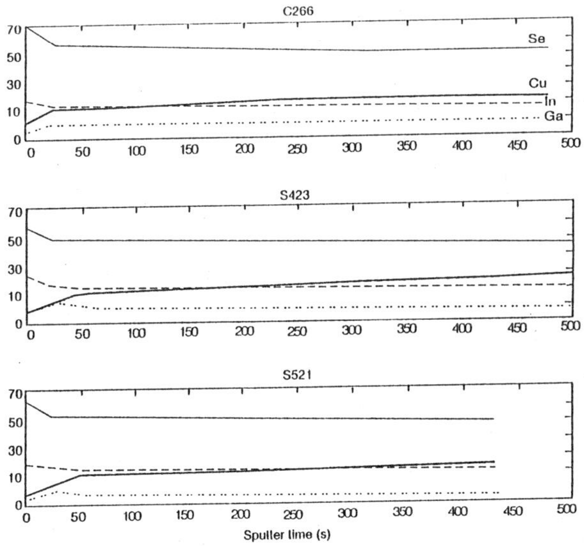

Figure 2. XPS compositional depth profiles for $Cu(In,Ga)Se_2$ thin-films C266 (η=16.4%), S423 (η=15%), and S521 (η=14%).

performance of these cells [2]. Because the bulk properties of the absorber differ, and can affect the overall device performance, we can only point to the consistency of the surface analysis results. Previous investigations indicate that the CdS/CIS heterojunction may have a minimal contribution to device performance due to the large conduction band (CB) offsets [4,17-19].

High resolution XPS core level spectra were used to determine chemical bonding differences between the surface and the bulk by evaluating changes in the photoelectron binding energies. However, analysis of these results reveals that differences in the In 4d/Ga 3d and Se 3d binding energies between the surface layers and the bulk material, i.e., the chemical shift, are ≤0.2 eV. These small differences are not sufficient for chemical discrimination.

Figures 3 and 4 show the positron results from the polycrystalline CIS and CIGS thin-film devices, respectively. Figure 3 is normalized to the bulk glass substrate. As previously mentioned, open-volume defects produce a higher S parameter. The measured value of the S parameter is directly related to the size of the open-volume defect and can be used to identify the defect.

As shown in Figs. 3 and 4, the S parameter is measured as a function of incident positron energy (E) to obtain depth resolved defect information. The mean positron depth (in nm) is given by

$$\bar{z} = \frac{AE^{1.6}}{\rho} \qquad (1)$$

where E is in keV, r is the density in gm/cm^3, and A is a material-dependent parameter. The S–E data can be analyzed by dividing the test system into a layered structure, each representing a region with a distinct S parameter and positron diffusion length, and solving a diffusion-annihilation equation to find the fraction of positrons annihilating in each of these regions. Such an analysis requires a good knowledge of the layer structures and their densities.

The interpretation of the S-parameter data is easier when it is normalized to a bulk material (in our case glass). The normalization of the S-parameter is performed only to compare results from different experimental geometries. In all of our experiments, the geometry was kept identical. In Fig. 3, because the substrate glass produced identical S-values, we normalized the data to show the changes as a percentage effect. However, this is not necessary to reach the conclusion that the CIS film used in the lower efficiency device (M1218) is defected.

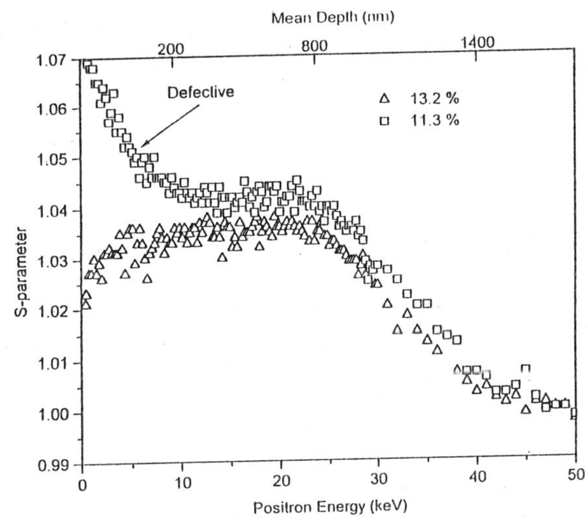

Figure 3. Normalized S-E data for polycrystalline $CuInSe_2$ thin-film devices M1201 and M1218.

Data for the CIGS films (Fig. 4) were not normalized because the glass substrates showed significant differences. This difference is not associated with changes in the experimental geometry, and is solely

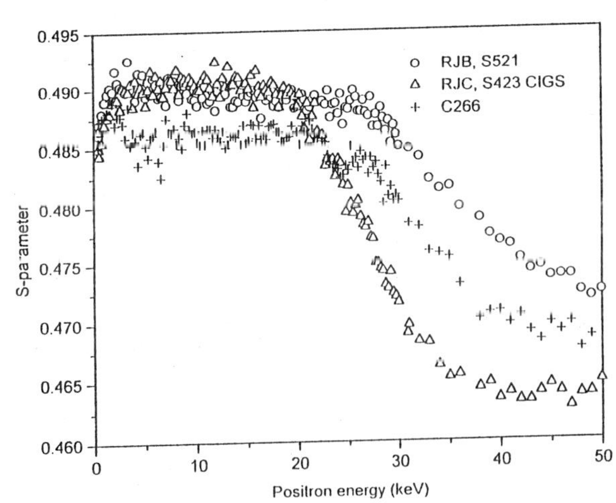

Figure 4. S-E data for polycrystalline $Cu(In,Ga)Se_2$ thin-film devices C266, S423, and S521.

due to the difference in the quality of the substrate material — substrate materials are not always identical! It is also important to notice that the bulk substrate value (for energies >30 keV) in no way influence the S-parameter measured at low energies. At low energies, the positrons are completely stopped in the film. Hence, the positron annihilation signal for energies less than 20 keV is not influenced by variations in the substrate material.

The surface S-parameter and bulk S-parameter are different for most materials. Positron surface dynamics and the annihilation process are different from bulk annihilations. In other words, the electron-momentum distribution sensed by the positron is different at the surface. A thin native oxide layer can contribute to a lowering of the S-parameter [16]. Positronium (a bound state of a positron and an electron) formation at the surface can also lower the S-parameter. It is worth noting that all samples (except the one showing defect in Fig. 3) show low surface values.

Comparing the XPS results with the PAS results for CIS device M1218-3 reveal that the In_2Se_3 surface layer is highly defective as evidenced by the large S value. The highest efficiency CIGS device (16.4% total area) had the lowest S value, which suggest more homogeneous composition and fewer defects near the surface.

Theoretical [18] and experimental [17,19] studies have shown that the traditional view [20] of $\Delta E_C < 0$ used for modeling the performance of heterojunction CdS/CIGS devices needs to be re-evaluated. Instead, the buried junction between the p-type bulk CIGS and the In-rich n-type defect chalcopyrite should be used for modeling the performance of these devices. These results clearly show that the consequences for successful solar cell performance need to be based on a revised model and reexamined.

CONCLUSIONS

The surface versus-bulk-composition, electronic and physical structure of polycrystalline $Cu(In,Ga)Se_2$ thin-film interfaces has been characterized using angle-resolved high resolution XPS and PAS measurements. Results of these measurements have been related to device efficiencies and clearly show that variations in composition in the near surface region affects device performance. In addition, fewer defects in the Cu-poor n-type surface layer correlates with better overall device performance.

ACKNOWLEDGEMENTS

This work was supported by the US Department of Energy under Contract No. DE-AC36-83CH10093.

REFERENCES

[1] A. Rockett and R.W. Birkmire, *J. Appl. Phys.* **70**, 1991, pp. R81.

[2] A.M. Gabor, J.R. Tuttle, M.A. Contreras, D.S. Albin, R. Noufi, and A.M. Hermann, *Appl. Phys. Lett.* **65**, 1994, pp. 198-200.

[3] J.R. Tuttle, M. Contreras, A. Tennant, A. Duda, J. Carapella, D. Albin, and R. Noufi, *Eleventh PV AR&D Review Meeting, AIP Conference Proceedings No.* **268**, 1992, pp. 186-193.

[4] A.J. Nelson, J.R. Tuttle, R.Noufi, D. Rioux, and Hartmut Hochst, *J. Appl. Phys.* **74**, 1993, pp. 5757-5760.

[5] D. Schmid, M. Ruckh, F. Grunwald and H.W. Schock, *J. Appl. Phys.* **73**, 1993, pp. 2902-2909.

[6] E. Parthe, *Seventh Int. Conf. Ternary and Multinary Compounds*, 1987, pp. 3-17

[7] L.S. Palatnik and E.I. Rogacheva, *Neorgan. Mat.* **2**, 1966, pp. 478-480.

[8] D.M. Ganbarov, G.G. Guseinov and Z.Sh. Karaev, *Neorgan. Mat.* **8**, 1972, pp. 2211-2213.

[9] B. Schumann, G. Kühn, U. Boehnke and H. Neels, *Sov. Phys. Crystallogr.* **26**, 1981, pp. 678-681.

[10] C. Djega-Mariadassou, A. Rimsky, R. Lesueur and J.H. Albany, *Jpn. J. Appl. Phys.* **19**, 1980, pp. 89-93.

[11] L.L. Kazmerski, A. Swartzlander, A.J. Nelson and S.E. Asher, *Nineteenth IEEE PVSC,* 1987, pp. 1315-1321.

[12] P.J. Schultz and K.G. Lynn, *Rev. Mod. Phys.* **60**, 1988, pp. 701.

[13] J.E. Kluin and H.J. Möller, *Eleventh European Photovoltaic Conference*, 1992, pp. 858-861.

[14] M. Bodegård, L. Stolt, and J. Hedström, *Twelveth European Photovoltaic Conference*, 1994. (in press)

[15] D. Niles, C.R. Schwerdtfeger, A.M. Gabor, R. Noufi, E. Colavita, and H. Höchst (to be published).

[16] P. Asoka-Kumar, H.-J. Gossmann, F.C. Unterwald, L.C. Feldman, T.C. Leung, H.L. Au, V. Talyanski, B. Nielsen, and K.G. Lynn, *Phys. Rev.* **B48**, 1993, pp. 5345-.

[17] A.J. Nelson, A. Rockett, E. Colavita, M. Engelhardt, and H. Hochst, *Phys. Rev.* **B42**, 1990, pp. 7518-7523.

[18] Su-Huai Wei and Alex Zunger, *Appl. Phys. Lett.* **63**, 1993, pp. 2549-2551.

[19] A.J. Nelson, C.R. Schwerdtfeger, Su-Huai Wei, A. Zunger, and Hartmut Hochst, *Fifth International Conference on Electron Spectroscopy, J. Electron Spectroscopy and Related Phenomena* **64/65**, 1994, pp. 185-193.

[20] G.B. Turner, R.J. Schwartz and J.L. Gray, Twentieth IEEE PVSC, 1988, pp. 1457-1460.

INTERPRETATION OF LASER SCANS FROM THIN-FILM POLYCRYSTALLINE PHOTOVOLTAIC MODULES

I.L. Eisgruber[1], R.J. Matson[2], J.R. Sites[1], K.A. Emery[2]
1: Colorado State University, 2: National Renewable Energy Laboratory

ABSTRACT

Laser scanning of photovoltaic devices and modules with submillimeter resolution can provide valuable information about localized defects and cell-to-cell variations in response. There are, however, sources of possible errors, or artifacts, in interpretation that must be addressed, especially when series-connected cells are scanned. These issues are addressed for thin film CdTe and $CuInSe_2$ (CIS) modules and cells using a recently developed computer-controlled large-scale laser scanner at the National Renewable Energy Laboratory (NREL). This instrument was designed to serve as a tool for research, device and module quality control, and module failure analysis. The thrust of the report is twofold: i) the development of reliable submillimeter module characterization techniques and analysis and ii) present the results of such techniques as applied to thin film CdTe and CIS modules.

INTRODUCTION

A large scale laser scanner has been developed at NREL to provide information about response as a function of position for areas ranging from 1 cm^2 to 4 m^2. This wide range of scan areas allows data to be taken on single cells or entire modules. A red laser beam (with 633 nm wavelength and a beam diameter variable from 100 μm to over 1 mm) is scanned over samples, and optical beam induced current (OBIC) at each point is recorded by computer. Data scans can be made under a variety of conditions, including voltage bias and white light bias. The system is described in more detail elsewhere [1]. Interpretation of the laser scanner data has several aspects. First, a recognition of artifacts and of all the variables affecting output is required for meaningful and reproducible results. Second, information regarding cell defects and characteristics may be evident in single cell measurements. Finally, one would like to apply interpretation of single cell measurements to cells in encapsulated modules, and to extract information regarding cell-to-cell comparisons in the module.

METHODOLOGY ISSUES

Several issues have been identified that are important to the accuracy and reproducibility of OBIC data. These include laser beam power, size, intensity, and chopping frequency, optical alignment, correct amplifier calibration, and accurate measurement of voltage bias.

Variations in beam size, intensity, and total power may alter data in several ways. Beam size is important for resolution. Features that are visible with a small beam may not be resolved and effectively disappear with a large beam. Beam intensity also plays an important role. For polycrystalline cells, the diode junction parameters may vary with intensity. In addition, even small resistances through the semiconductor layers cause non-negligible voltage drops when subjected to high current densities at the illuminated spot, thereby forward biasing the junction at the illuminated spot, even though the cell back and front surfaces are at zero bias. OBIC signal depends strongly on voltage bias, and the magnitude of the voltage drops depends on the current density at the laser spot, and hence the laser intensity. In module measurements, high total beam power introduces a further complication. The illuminated cell in a module measurement is forward biased. All the remaining cells are reverse biased. Forward biased cells produce a smaller OBIC signal than reverse biased cells. Therefore, at high beam powers, when the illuminated cell is significantly forward-biased, adjacent cells collect much more efficiently. As the beam nears the edge of the illuminated cell, any scattered light or light contained in the far edges of the gaussian beam is collected very efficiently in the adjacent cells. This produces an artificial increase in response near cell edges due to adjacent cell collection. The beam power at which this effect becomes significant depends on the characteristics of the module

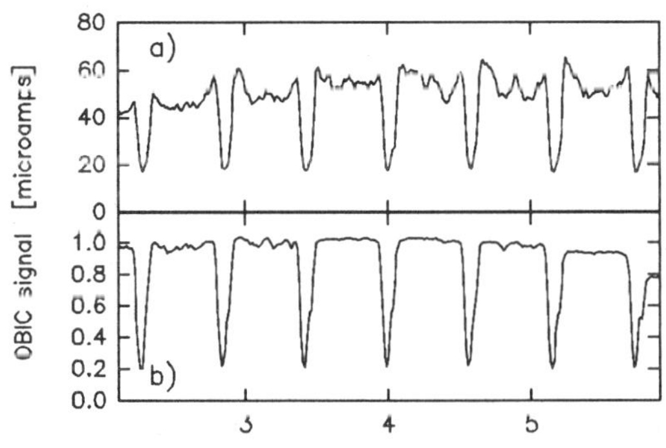

Figure 1: Line scan of module at a) high beam power, and b) lower beam power.

being measured. Figure 1 shows a section of a laser scan from a 52-cell CIS module taken at high and low beam powers. The high beam power scan shows adjacent cell collection.

CH3365-4/94/0000-0283 $4.00 © 1994 IEEE First WCPEC; Dec. 5-9, 1994; Hawaii

Beam power and size at the sample may vary if the sample is moved slightly, if focus is adjusted, or if optical components such as mirrors or choppers are moved. Because of possible scattering and absorption due to dust or dirt on components, special care must be taken when assuming a certain optical component changes only beam power or only beam size. Therefore, because of the importance of beam size and intensity, these quantities should be measured each time any part of the system is adjusted. A convenient manner of doing this is to step the laser beam over a sharp edge onto a single crystal Si cell of known quantum efficiency at the laser wavelength, and using this line scan to calculate beam profile and power.

Chopping frequency (if lock-in detection is used) is another variable that can have a drastic effect on OBIC data. In single cell measurements, OBIC signal will not be reduced by capacitive effects if

$$2\pi f C R << 1 \quad (1)$$

where f is chopping frequency, C is cell capacitance, and R is the resistance of leads and circuitry. For a typical cell and measurement setup, the requirement is that chopping frequency be less than ten kHz and is not an issue with most mechanical choppers. In module measurements, on the other hand, the requirements are more stringent. Equation (1) still holds, but now the relevant resistance in the RC time constant is the lead resistance **plus** the sum of the shunting resistances of all the cells in the module. This typically leads to the requirement that chopping frequency be less than around 100 Hz to operate in the DC limit, although the exact requirement must be determined by the characteristics of the module being tested. If the module is measured above this frequency, the signal from leaky cells is enhanced and the signal from good cells is diminished. In both single-cell and module measurements, particularly in the case of amorphous silicon, care should be taken that the cell's inherent photoresponse time is not decreasing signal magnitude.

Poorly aligned optical components can cause an artificial radial pattern to appear in OBIC data. If the laser is not in the center of the small scanner mirrors that position the beam on the sample, a portion of the beam may fall off the edge of the mirrors. How much of the beam falls of the edge depends on the orientation on the mirrors, i.e. the position of the laser on the sample. The same effect may occur if the beam is well-aligned, but the beam size at the scanner mirrors is comparable to the scanner mirror size.

Accurate amplifier calibration is important if conversion of OBIC signal to absolute collection efficiency is desired. For lock-in detection, the calibration factor between the lock-in reading and absolute quantum efficiency depends somewhat on the OBIC signal shape, since absolute quantum efficiency is calculated from peak-to-peak signal height, whereas the lock-in output is root mean square amplitude.

OBIC signal depends strongly on voltage bias,so it is important to know the exact voltage bias of a measurement for reproducible results. Therefore, using a four-point probe configuration (2 current-providing probes and 2 voltage-sensing probes) is extremely important. This is particularly true for gridless cells, or when large currents due to white light bias are present, since these factors often cause a significant voltage drop between the current probes and the cell.

SINGLE CELL MEASUREMENTS

Single cell measurements have been made on CdTe and CIS cells under short-circuit, forward voltage-biased, dark, and white light biased conditions.

Two important areas that can be investigated with short-circuit OBIC measurements are photocurrent defects and shunting defects. The signature of these defects can be calculated using the theory presented by Lehovec and Fedotowsky [2]. In general, photocurrent defects affect the OBIC signal only when the laser overlaps the defect, whereas shunting defects affect the OBIC signal from an area much larger than the actual defect size since current

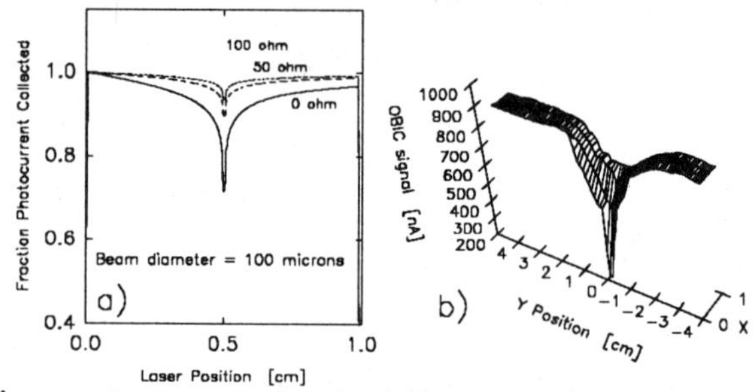

Figure 2: a) Calculated OBIC signal from cell containing 10Å shunt of various resistances. b) OBIC data from CdTe cell with 10Ω shunt at interconnect on left.

from the surrounding area is shunted through the defect.

Figure 2a shows the calculated defect signatures for a 10 Å radius shunt of various resistances. In each case, the OBIC signal rolls off into the shunt, although the roll-off may be hard to see for higher resistance shunts. Figure 2b shows OBIC data from a CdTe cell with an approximately 10 Ω shunt at the interconnect (on the left of the cell). The OBIC signal rolls off into the shunt over nearly the entire 1 cm by 8 cm cell.

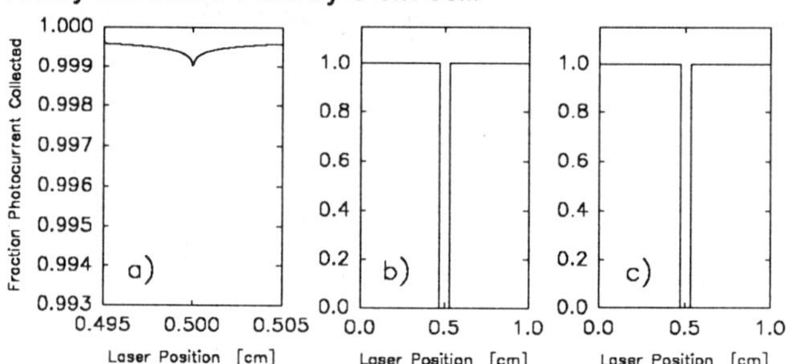

Figure 3: a) Calculated OBIC signal from 14 kΩ shunt. Note that axes have been expanded. b) Calculated signal from 300 μm diameter photocurrent defect. c) Calculated signal from combined 14 kΩ shunt and 300 μm photocurrent defect.

In many cases, both photocurrent reduction and

shunting may occur at a defect. Such a combined defect was intentionally introduced into a CIS cell by pricking the cell with a pin. Examining the impact of the defect on the I-V curve showed the defect to have a resistance of about 14 kΩ. The size of the defect measured under the microscope was 300 μm in diameter. Figure 3 shows the calcuted OBIC signature of a 14 kΩ shunt, a 300 μm photocurrent defect, and a combination of the shunt and photocurrent defects. The roll-off into the shunt is completely masked by the decrease in photocurrent. OBIC measurements on the CIS cell with the intentionally introduced defect support this conclusion. Thus, although the major impact of a defect on a cell's I-V curve may be shunting, the major impact on the OBIC signal may be the sharp drop-off associated with a photocurrent defect.

Single cell measurements have also been made under voltage and white light bias. Such measurements can in theory be used to measure TCO sheet resistance [2-3], and have been successfully used for this purpose with single-crystal Si [4]. Polycrystalline cells, however, present some special challenges. First, a lack of grid fingers on substrate configuration cells leads to a much more complicated analysis of the relation of the OBIC signal drop-off to the sheet resistance than addressed in [2-4]. Second, most polycrystalline cells tested show unexpected signal fluctuations when forward voltage biased. These fluctuations occur independently of white light bias. Figure 4 shows zero-biased and forward-biased line scans from a single crystal Si cell, and from a CdTe cell. The Si cell shows no unexpected signal fluctuations in forward bias. In both cases the quantity plotted is apparent quantum efficiency (AQE), the ratio of measured electrons to incident photons [5].

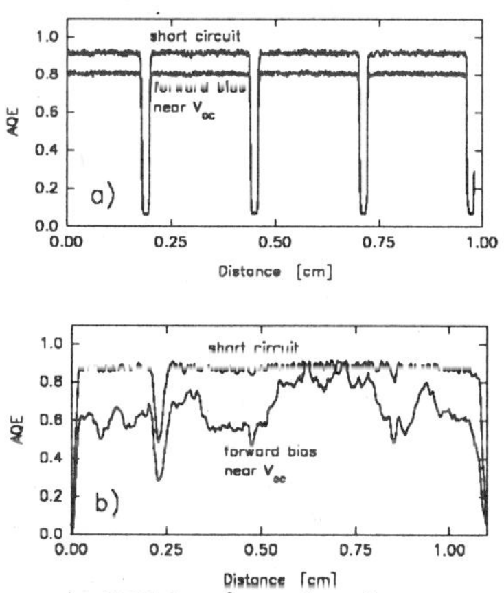

Figure 4: a) OBIC signals from a single crystal Si cell at short circuit and forward bias. The signal drops are due to grid fingers. b) OBIC signals from a CdTe cell at short circuit and forward bias. The forward biased signal is graphed as AQE*5.

Two possible explanations for these signal fluctuations are evident. Figure 5 shows apparent quantum efficiency as a function of beam intensity and resistance associated with the semiconductor layers. Examples of such resistances would include semiconductor bulk resistance and resistance between the semiconductor and back contact, but not TCO or interconnect resistances. Apparent quantum efficiency decreases as beam intensity or resistance increases. Thus, local fluctuations in either optical scattering or in the effective resistance of a back contact barrier could cause the observed signal fluctuations.

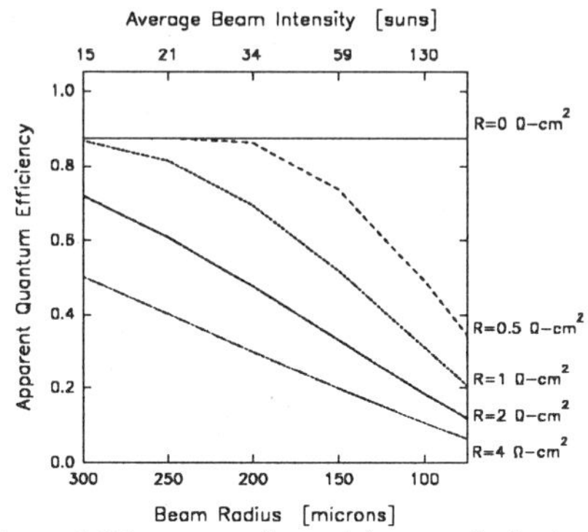

Figure 5: AQE as a function of intensity and resistance for a typical CdTe cell.

The effect of beam intensity and resistance on AQE are roughly equivalent: to first order a factor of two increase in beam intensity causes the same reduction in AQE as a factor of two increase in resistance. Additional experiments are planned to determine whether variations in optical scattering or resistance, or some combination of both, are the source of the forward-bias signal fluctuations. The dependence of AQE on voltage bias and beam intensity may eventually yield a quantitative measurement of resistances associated with the semiconductor layers.

MODULE AND MINI-MODULE MEASUREMENTS

OBIC measurements on encapsulated modules have two goals: (i) to learn about the characteristics of and variations over a single cell in the module, and (ii) to gain information about cell-to-cell comparisons in the module.

Near short-circuit, analysis of defects in one cell of a module is the same as for single cell measurements. When voltage bias is applied, however, analysis is complicated by the fact that the exact voltage across any particular cell is unknown. How the voltages distribute depend on the IV characteristics of each cell, as well as the illumination and voltage bias conditions.

OBIC measurements on modules at short circuit both in the dark and under white light bias have been analyzed. Scanning the laser over each cell in an otherwise dark module might appear to yield the short circuit current of each cell. Holding the module at short circuit, however, produces a different result than holding each cell at short circuit, since the dark cells are reverse biased and the laser-illuminated cell is forward biased. The photocurrent flowing from the illuminated cell causes a voltage drop ΔV across the dark cells. This voltage drop pushes the illuminated cell into forward bias, causing a reduction in the OBIC signal of (dI/dV)ΔV , where dI/dV is the slope of the IV curve of the illuminated cell at the voltage bias of the illuminated cell. A larger dI/dV yields a larger reduction in OBIC signal due to the voltage redistribution. The OBIC signal is roughly inversely proportional to dI/dV, i.e. proprtional to dV/dI. Near short circuit, dV/dI is just the cell shunt resistance, and the OBIC signal is the product of the photocurrent and shunting resistance of the illuminated cell, divided by the sum of the shunt resistances of all cells. Figure 6a shows the measured OBIC signal for a CdTe mini-module, as well as the OBIC signal predicted from independent I-V measurements.

White light bias adds an additional complication to short circuit mini-module measurements. Even small nonuniformities in white light bias change the distribution of voltage biases in the module enough to significantly affect the OBIC output. More brightly illuminated cells are more forward biased, and therefore have a smaller value of dV/dI and produce a smaller OBIC signal. In fact, relative signal heights of short-circuit OBIC measurements yield basically an intensity map of the light source. This is illustrated in figure 6b, where measured OBIC signal from a CdTe mini-module under white light bias with a small intensity

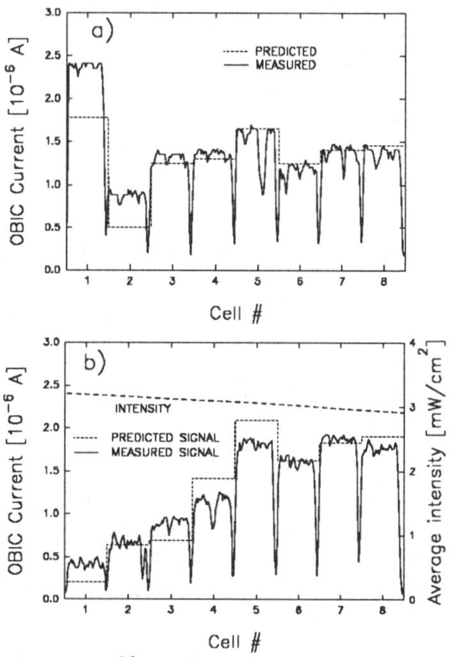

Figure 6: Predicted and measured OBIC signals from CdTe mini-module at short circuit a) in dark, and b) with white light bias.

nonuniformity is shown. Agreement with the OBIC signal predicted from independent intensity measurements and I-V characterizations is reasonably good. The predicted signal is extremely sensitive to intensity and cell parameters, so some disagreement is not surprising. The effect of intensity nonuniformities is more severe at higher intensities.

Numerous possibilities for future work in OBIC measurements on modules exist. These include whether or not capacitive effects can be used to decouple information about photocurrent and shunt resistance in short-circuit measurements, whether or not forward voltage biased measurements will yield useful cell-to-cell comparisons, and how the theory of single-cell voltage-biased measurements can be applied to cells in encapsulated modules.

CONCLUSIONS

This paper describes groundwork laid for the interpretation of laser scanning data from polycrystalline thin film modules. Artifacts and variables important in OBIC data have been identified. Issues include beam power and intensity, chopping frequency, alignment of optics, amplifier calibration, and bias measurement. Single cell short circuit measurements have shown shunt and photocurrent defects with the expected signatures. Forward biased measurements have shown unexpected behavior, possibly due to variations in optical scattering or resistances. Relative signal heights from individual cells in short circuit dark and light-biased module measurements appear well understood.

ACKNOWLEDGEMENTS

The authors would like to thank Tom McMahon of NREL, Rick Sasala of Solar Cells Inc., and Randy Rico of Laser Designs, for their assistance. This work was supported by the U.S. Department of Energy through contract DE-AC02-83CH10093 to NREL and contract XAX-4-14000-01 to Colorado State University.

REFERENCES

[1] R.J. Matson, K.A. Emery, I.L. Eisgruber, and L.L. Kazmerski, "The Large-Scale Laser Scanner: Milli-Characterization of Photovoltaic Devices and Modules", *Twelfth European Photovoltaic Solar Energy Conference and Exhibit*, 1994.

[2] K. Lehovec and A. Fedotowsky, "Scanning Light Spot Analysis of Faulty Solar Cells", *Solid State Electronics*, Vol. 23, pp. 565-576, 1980.

[3] D.E. Sawyer, and H.K. Kessler, "Laser Scanning of Solar Cells for the Display of Cell Operating Characteristics and Detection of Cell Defects", *IEEE Transactions on Electron Devices*, Vol. 27, no. 4, pp. 864-872, 1980.

[4] P. Kowalski, W.F. Lankford, and H.A. Schafft, "Nondestructive Measurement of Solar Cell Sheet Resistance Using a Laser Scanner", *IEEE Transactions on Electron Devices*, Vol. 31, no. 5, pp. 566-570, 1984.

[5] J.R. Sites, H. Tavakolian, and R.A. Sasala, "Analysis of Apparent Quantum Efficiency", *Solar Cells*, Vol. 29, pp. 39-48, 1990.

THE EFFECTS OF THE BAND BENDING CAUSED BY INTERFACE STATES IN CdTe/CIS SOLAR CELLS

Youn Jung Lee and Jeffery L. Gray
School of Electrical Engineering
Purdue University
West Lafayette, IN 47907

ABSTRACT

In this paper, the effects of interface states in the ZnO/CdS/$CuInSe_2$ and CdS/CdTe solar cells are presented. The effects are investigated through numerical modeling using ADEPT (A Device Emulation Program and Tool). The results show that donor-like interface states have very little effect, but acceptor-like interface states at the resistive ZnO/CdS can cause pinning of the bands at the interface, thus leading to non-exponential illuminated I-V curves when the interface state densities are high enough. High density of acceptor-like states between the CdS and In-rich CIS does not result in the two-diode like I-V curves. Instead they can significantly lower the fill factor.

In the CdS/CdTe solar cells, either donor- or acceptor-like interface states have little effect since almost all the depletion region lies in the CdTe. Thus, the metallurgical junction where the interface states are located is away from the electrical junction where the conductivity type changes.

INTRODUCTION

Polycrystalline thin-film $CuInSe_2$ (CIS) and CdTe based solar cells have a great potential for domestic use. They are economical, have ideal bandgaps and high absorption coefficients. Recently over 15% efficiencies have been reported for solar cells for both materials [1-2]. However, their transport mechanisms are not fully understood and existence of interface states at the heterojunction is suspected. In the CIS based solar cells, the law of superposition fails to hold especially at low temperatures. It has been suggested that a conduction offset and/or non-ohmic back contact may be the cause, resulting in two distinctively different non-exponential I-V curves. It has been previously demonstrated that the electron affinity differences can lead to a non-exponential, two diodes like illuminated I-V curves at low temperatures and if the difference is large enough, the same results can be obtained at room temperature [3]. Although the non-exponential illuminated I-V curves are experimentally measured usually at low temperatures, they are observed even at room temperature for some cells. It is unlikely that the magnitude of the electron affinity difference can change dramatically from device to device since it is one of the fundamental properties of the materials. However it is possible that interface states at the heterojunction whose density is heavily dependent on the fabrication conditions, may cause band bending or bend offset, thus having the similar effects.

In this paper, the effects of interface states, if any, are investigated in the CIS and CdTe solar cells. The investigation is carried out numerically using ADEPT (A Device Emulation Program and Tool). ADEPT which was developed at Purdue University, solves the fundamental semiconductor equations (Poison's and continuity equations for holes and electrons) simultaneously using finite box method [4]. The merits of the numerical modeling is that spatially varying parameters can be seen throughout the device and new theories and design concepts can be tested without actual fabrication.

APPROACH

A ZnO/CdS/CIS structure is used for the numerical modeling of CIS based solar cells. It is assumed that the ZnO consisted of 500 Å of highly conductive layer ($N_d = 10^{21}$ cm^{-3}) followed by a more resistive 500 Å thick layer ($N_d = 4\times10^{15}$ cm^{-3}). The CdS layer has the same thickness and doping level as the resistive ZnO layer. At the heterojunction, it is assumed that a very slightly n-type, almost intrinsic (presumably In-rich) CIS layer of 0.1 μm thick lies next to the stoichiometric CIS layer whose hole concentration is 10^{16} cm^{-3}. The whole structure is 2.2 μm thick. There are 4 places where interface states may exist, i.e., at the n^+-ZnO/n-ZnO, n-ZnO/CdS, CdS/In-rich CIS or at the In-rich CIS/stoichiometric CIS interface. If there were no interface states, a short circuit density of 42.86 mA/cm^2, open circuit voltage of 517.08 mV and conversion efficiency of 15.91% is predicted with this structure.

A very simple structure of CdS/CdTe is used since its predicted solar cell performance is similar to the best measured one to date. With 0.1 μm CdS ($N_d=10^{18}$ cm^{-3}) followed by 5 μm thick CdTe ($N_a=10^{15}$ cm^{-3}), a short circuit current of 25.79 mA/cm^2, open circuit voltage of 844.69 mV and efficiency of 16.98% are obtained without any series resistance or interface states.

During the simulations, interface states which are either donor- or acceptor-like, are introduced at one interface at a time. 10 interface states are spread throughout the bandgap equi-distantly and the total states

CH3365-4/94/0000-0287 $4.00 © 1994 IEEE

First WCPEC; Dec. 5-9, 1994; Hawaii

density is varied. It is assumed that the interface recombination velocity is 10^4 cm/s.

RESULTS

CIS

When interface states are introduced between the n^+-ZnO and n-ZnO, the I-V curves are unaffected whether they are acceptor-like or donor-like at all even the densities were varied from 10^8 to 10^{12} cm^{-2}. At the n-ZnO/CdS interface, donor-like states have no effect but acceptor-like states cause two-diode like I-V curves when their density is larger than 10^{11} cm^{-2} (see Fig. 1). With the interface state density higher than 10^{11} cm^{-3}, the position of the conduction band is fixed at the ZnO/CdS interface. It results in splitting of the built-in voltage into two potentials with different gradients, i.e., one across the n-ZnO with very steep gradient and the other across the CdS and CIS (see Fig. 2). If there were no splitting of the V_{bi}, the electrons in the emitter move to the base neutralizing N_d^+, and the holes to the emitter neutralizing N_a^- as the junction potential is lowered. However, with the high density of interface states, the steep potential across the n-ZnO acts as a potential barrier for the electrons making it harder for the N_d^+ to be neutralized. Thus, as bias is applied, V_1 is unaffected because $\nabla V = -\rho/\varepsilon$, while V_2 flattens. When more bias is applied, it is reflected in the lowered potential difference between the two contacts of the device. Since the very steep electron barrier, V_1 is still unaffected, the lowered potential difference appears in the opposite band bending of V_2 (see Fig. 3). Now the electrons in the CdS are depleted from the interface due to V_2 and some of the generated carriers are not collected as the field is in the opposite direction resulting in the increased recombination. With more bias, there is more drop in the carrier collection efficiency and the electron concentration in the base cannot increase exponentially with the bias due to the potential barrier which prevents supply of the electrons from the emitter. Therefore, the current increase rate is slowed with different slop, dI/dV, i.e., it is much more resistive. Once it passes the open circuit voltage, an opposite gradient appears in the F_n, and recombination current dominates over the generation current. Thus, more electrons exist in the CdS and (E_c-F_n) is reduced at the ZnO/CdS interface, lowering V_1. Therefore, the current resumes exponential increase. Note that there is little change in the short circuit current. This is because V_1 develops in the n-ZnO and V_2 is spread across the CdS and CIS. There is almost the same field in the CIS as in the case where there are no interface states and the collection efficiency is little affected at short circuit.

At the CdS/n-CIS interface, donor-like states does not have ant effects because there are a lot of electrons at the interface already and since the Fermi-level lies close to the conduction band making most of the interface states below the Fermi-level neutral, a few more donor-like states do not affect the electron concentration. However the acceptor-like interface states greatly reduces the short circuit current and fill factor and slightly increases open circuit voltage (see Fig. 4). When the density is high enough, the splitting of the built-in

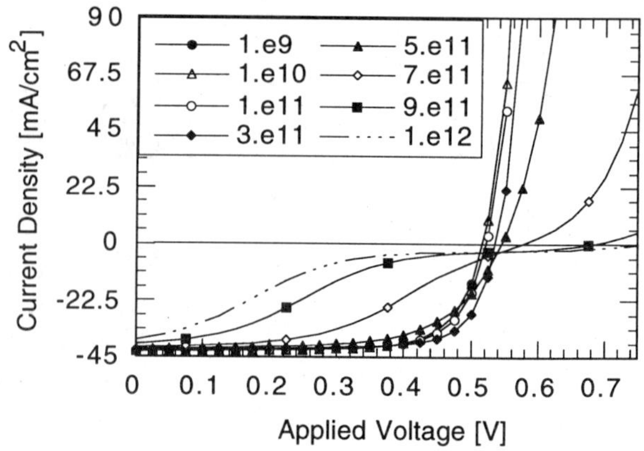

Fig. 1. Acceptor-like interface states are located between the resistive ZnO and CdS. The total interface state density is varied.

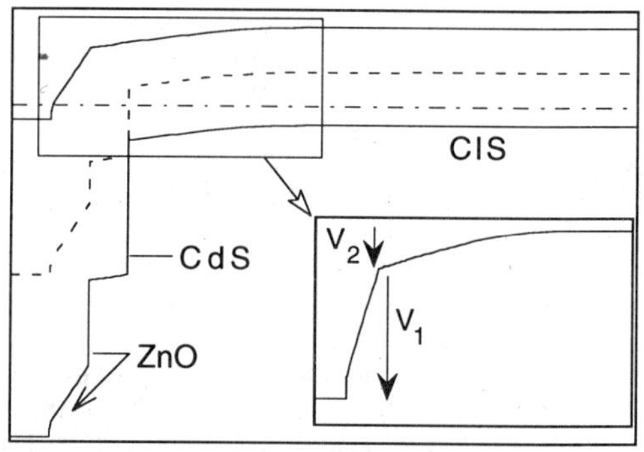

Acceptor-like interface States at n-ZnO/CdS

N_{ss} = 7.e11 cm^{-2}, at Eqilibrium

Fig. 2. The high density of the acceptor like states at the n-ZnO/CdS interface causes splitting of built-in voltage into two potentials with different gradient.

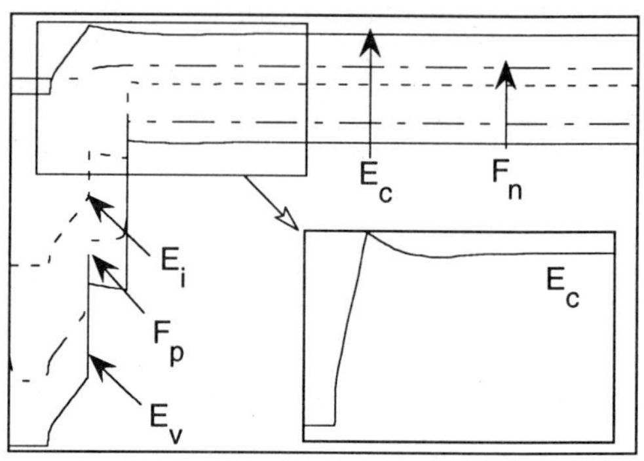

Acceptor-like Interface States at n-ZnO/CdS

N_{ss} = 7.e11 cm^{-2}, Applied Voltage = 0.4 V

Fig. 3. When bias is applied, it is reflected in the opposite band bending of V_2 instead of reducing V_1, which acts as a potential barrier for the electrons.

voltage occurs. However, V_1 is spread over the ZnO and CdS since the position of the conduction band is set at the CdS/CIS interface. When there are more acceptor-like interface densities, the position of the conduction band is further away from the Fermi level, and more of the built-in voltage appears in V_1 and less in V_2.. Therefore, more interface states result in less collecting field in the CIS and lower short circuit current. When bias is applied, a slight opposite band bending occurs in the CIS, increasing the recombination current. When more bias is applied, it lowers V_1 since the gradient is not as steep as in the case where the acceptor-like states are located at the n-ZnO/CdS interface, instead of increasing the opposite band bending of V_2. When the density of the interface states are low, the depletion region recombination dominates, but with a large number of interface states, electrons deplete from the interface due to the presence of V_2 which has opposite gradient than that of V_1, greatly reducing the space charge recombination. Since there is very little depletion region recombination occurring, and supply of electrons is somewhat disturbed by V_1, it becomes more resistive, reducing the fill factor. The open circuit voltage is also pushed further away because of the smaller dI/dV. It can be said that the increase of V_{oc} is the by-product of the large drop in the fill factor and does not result in the increased performance.

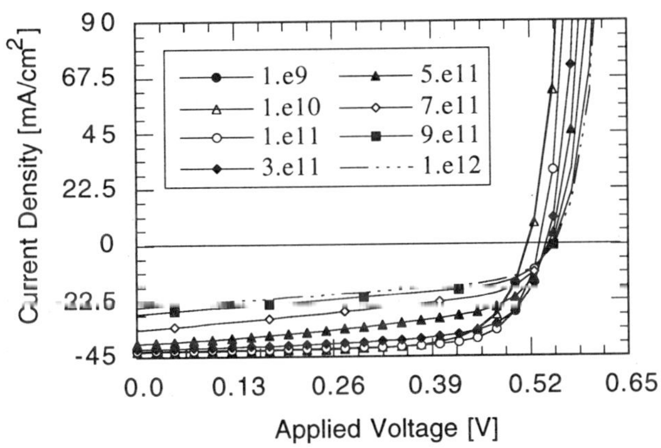

Fig. 4. With higher density of acceptor-like states at the CdS/In-rich CIS, the solar cell performance is greatly reduced. Since the interface states effectively make the layers less lightly doped, J_{sc} and FF are decreased.

Both the donor-like and acceptor-like interface states slightly affect the solar cell performance when located between the In-rich CIS and stoichiometric CIS. As it can be seen in Fig. 5, The open circuit voltage decreases until the donor-like interface state density reaches about 1.5×10^{11} cm^{-2} and starts to increase back. The trend of the short circuit current is the opposite but the range in which it changes is very minimal. When the interface state density is low, an electrical junction is formed in the CIS but as more donor-like interface states are introduced, the n-CIS becomes more n-type and the junction lies further into the CIS resulting in a wider depletion region across both the n-CIS and p-CIS, resulting in more depletion region recombination and smaller V_{oc} and higher J_{sc}. With even more donor-like states, there is hardly any depletion region in the n-CIS and a lot of generated carriers are not collected causing a drop in the J_{sc}. Most of the generated holes recombine at the donor-like states at the n-CIS/p-CIS interface and there is less recombination occurring due to the lack of the holes in the n-CIS and a very narrow recombination peak dominates where both the holes and electrons are substantially present in the p-CIS and resulting in increase of the V_{oc}. Note that with the interface states between the n-CIS and p-CIS, recombination through these states are significant. Holes from the CIS recombine at the interface and recombination through these donor-like interface states always dominates.

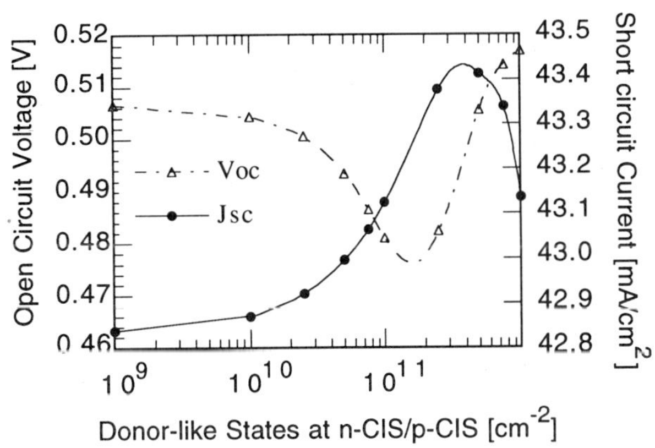

Fig. 5. The open circuit voltage is sensitive to the density of the donor-like states at the In-rich CIS and stoichiometric CIS interface. higher density of the interface states result in less field in the n-CIS and extended depletion region into the p-CIS.

Fig. 6 shows that there is a small drop in the J_{sc}, small increase in the V_{oc} with the acceptor-like interface states densities higher than 10^{11} cm^{-2}. The pn junction is formed at the CdS/CIS interface as the acceptor-like interface states effectively turned the n-CIS into p-type as well as making the p-CIS to be more doped at the interface. This results in a slightly narrower depletion region and thus a little smaller J_{sc} and a little higher V_{oc}. With applied bias, electrons from the emitter recombine with the holes at the interface and electrons are depleted from the interface in the p-CIS side. Most of the recombination occur in the n-CIS and at the interface. With interface state density as high as 10^{12} cm^{-2}, the recombination peak in the n-CIS becomes even more narrower due to the depletion of the electrons. Unlike normal solar cells, there is very little increase in the quasi-neutral region recombination with the applied bias since most electrons keep recombining at the interface, i.e., the depletion region recombination always dominates and it is occurring only in about a half of the n-CIS resulting in a slight increase in the V_{oc}.

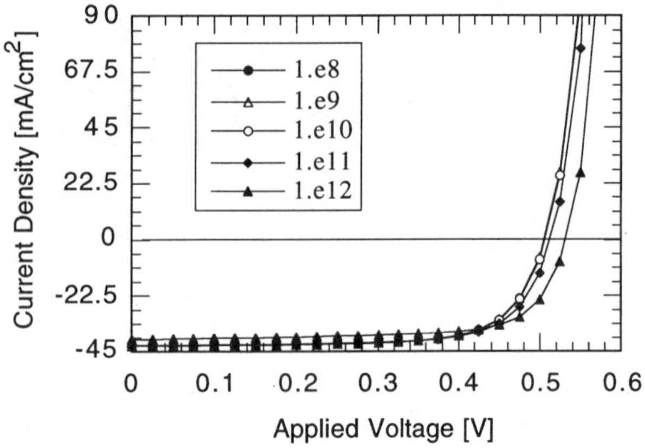

Fig. 6. As a result of the acceptor-like interface states between n-CIS and p-CIS, the open circuit voltage benefits a little, although there is a small decrease in the J_{sc} due to the reduced depletion region width.

CdTe

With the very simple structure, only one interface exists between the CdS and CdTe. Computer simulations show that whether they are donor-like or acceptor-like, the interface states have no influence on the I-V curves. Since it is assumed that the CdS is much more heavily doped than the CdTe, almost all the depletion region lies in the CdTe. Thus the metallurgical junction where all the interface states are located lies away from the true electrical junction where the conductivity changes from n- to p-type (see Fig. 7). Therefore the interfaces states have no effect in the transport mechanism which is depletion region recombination dominated [5]. However, some changes may occur if the CdS is not as heavily doped as assumed but in that case, the solar cell performance is also reduced.

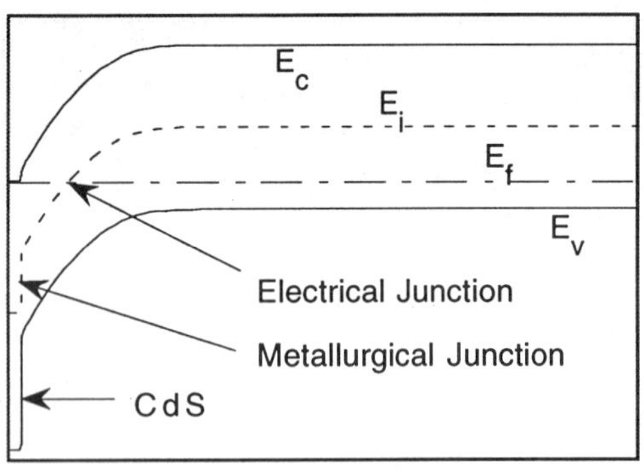

N_d in CdS = 1.e18 cm^{-3}

N_a in CdTe = 1.e15 cm^{-3}

Fig. 7. Since it is assumed that the emitter is much more heavily doped than the base, almost all of the depletion region lies in the base. Most of the recombination occur where the Fermi-level crosses the intrinsic level that are away from the metallurgical junction of the CdS and CdTe.

CONCLUSION

The numerical modeling results of existence of the interface states in the n^+-ZnO/n-ZnO/n-CdS/n-CIS/p-CIS solar cells shows that when acceptor-like states with densities higher than $10^{11}cm^{-2}$ are present at the n-ZnO/n-CdS interface, two diode-like illuminated I-V curves can be obtained at room temperature. The interface states cause splitting of the built-in voltage into two potentials with different gradients instead of a smooth gradual junction potential. The steep potential acts as a barrier for the electrons and non-exponential I-V relationship occurs. Although high density of acceptor-like interface states at the n-CdS/n-CIS (In-rich CIS) does not cause non-exponential current increase with the applied bias, it greatly reduces the fill factor and short circuit current resulting in poor cell performance.

In the CdS/CdTe solar cells, the interface states have very little effect since it is assumed that the CdS is much more heavily doped than the CdTe. Almost all the depletion region lies in the CdTe and the metallurgical junction where the interface states are located, is positioned away from the true electrical junction where change of the conductivity occurs.

ACKNOWLEDGMENT

This work has been supported by NREL under subcontract number XG-2-11036-2.

REFERENCES

[1] J. Hedstrom, H. Ohlsen, M. Bodegard, A. Klyner, L. Stolt, D. Hariskos, M. Ruckh and H. W. Schock, "*ZnO/CdS/Cu(In,Ga)Se$_2$ thin film solar cells with improved performance*", Proceedings of the 23rd IEEE Photovoltaic Specialists Conference, 1993, pp. 364-371.

[2] C. Ferekides, J. Britt, Y. Ma and L. Killian, "*High efficiency CdTe solar cells by close spaced sublimation*", Proceedings of the 23rd IEEE Photovoltaic Specialists Conference, 1993, pp. 389-393.

[3] Y. J. Lee and J. L. Gray, "*Numerical modeling of the temperature and illumination intensity dependent performance of CIS solar cells*", 12th European Photovoltaic Solar Energy Conference and Exhibition, Amsterdam, April 1994, in printing.

[4] J. L. Gray, "*Computer model for the simulation of thin-film silicon- hydrogen alloy solar cells*", IEEE Transaction of Electron Devices, **v 36**, n 5, 1989, pp. 906-912.

[5] R. J. Schwartz, J. L. Gray and Y. J. Lee, "*Design considerations for thin film CuInSe$_2$ and other polycrystalline heterojunction solar cells*", 22nd IEEE Photovoltaic Specialists Conference, 1991, pp. 920-923.

ADMITTANCE MEASUREMENTS ON $Cu(In,Ga)Se_2$ POLYCRYSTALLINE THIN-FILM SOLAR CELLS

John H. Scofield
Physics Department, Oberlin College, Oberlin, OH 44074

M. Contreras, A. M. Gabor, and R. Noufi
National Renewable Energy Laboratory, 1617 Cole Blvd., Golden, CO 80401, USA
and
J. R. Sites, Physics Department, Colorado State University, Ft. Collins, CO 80523, USA

ABSTRACT

We have measured the complex admittance for approximately 30 CIS and CIGS polycrystalline thin film solar cells having efficiencies of 12 % or better. Measurements were performed for frequencies ranging from 1 to 1000 kHz and bias voltages from -2.0 to 0.4 V and were usually at room temperature. Capacitance versus voltage data were used to extract effective charge densities. Frequency-dependent capacitance and conductance data revealed information about trapping. While results vary from cell to cell, several common features emerge from the data. These results are described in this paper.

INTRODUCTION

The complex admittance, $\mathbf{A}(\omega) = G(\omega)+j\omega C(\omega)$, of a reverse-biased diode contains considerable information about the junction [1]. Here $\omega = 2\pi f$ is the angular frequency and $G(\omega)$ and $C(\omega)$ are the diode conductance and capacitance respectively. The capacitance of a well-behaved reverse-biased diode, for instance, gives a direct measure of the depletion width. For an ideal, uniformly doped, one-sided junction, a plot of $1/C^2$ versus V (bias voltage) yields a straight line, the slope of which may be used to determine the lesser of N_a and N_d. For non-uniform doping, the doping density as a function of distance from the junction may be obtained from the derivative, $d(1/C^2)/dV$.

The ac capacitance measures the mobile charge density over a small distance δW (determined by the ac voltage δV) at the edge of the depletion width W (determined by the dc bias V). Trapped charge can respond to low frequencies but not to high frequencies, where high and low are determined by the trap's relaxation time. Thus, traps contribute to the low-frequency, but not the high-frequency, capacitance and, accordingly, trap distributions may be extracted from measurements of capacitance versus frequency [2]. Finally, zero-bias measurements of the conductance for various frequencies and temperatures may be used to further characterize trapping states [3, 4].

EXPERIMENTAL DETAILS

We have performed admittance measurements on twenty-nine, 9-16.4% efficient, 0.43 cm^2 area, polycrystalline, thin-film solar cells solar cells from 16 different $CuInSe_2$ (CIS) and $Cu(In,Ga)Se_2$ (CIGS) absorber layer depositions. Samples were fabricated at the National Renewable Energy Laboratory (NREL) on soda-lime glass substrates with the usual n^+-ZnO/i-ZnO/i-CdS/CIGS/Mo structure. Absorber layers were formed from precursor films evaporated from elemental sources (Cu, In, Ga, and Se), in most cases using a 3-stage coevaporation process [5]. We have also looked at cells with absorber layers fabricated using the recrystallization method [6], the in-line source process [7] and Se-vapor selenization [8]. The properties of several of these cells are summarized in Table I.

Admittance measurements were typically performed on two or more devices on the same substrate, in the dark at room temperature for frequencies ranging from 1 to 1000 kHz and bias voltages ranging from -2.0 to 0.4 V. In a few cases, measurements were performed at frequencies as low as 100 Hz and, in one case, for temperatures ranging from -20° to 75°C. Data were either logged at constant V while scanning log(f) or at constant f while scanning V. In both cases, devices were held at zero bias for a few minutes then the bias was switched to the desired value for the beginning of the scan. For bias scans, V was stepped from V_{min} to V_{max} in 25, 100 mV increments. Each scan took approximately 2 mintues.

RESULTS

The detailed results of measurements varied from substrate to substrate, but were similar for cells on the same substrate whenever the cells showed similar current-voltage (J-V) characteristics. Despite their many differences high-efficiency cells (η > 12%) shared several common features; these include: 1) a frequency independent capacitance for f > 1 kHz, 2) a voltage-dependent shunt resistance, 3) the existence of slow transient states, and 4) effective absorber carrier densities which increase away from the junction. Below we elaborate on these four topics.

CH3365-4/94/0000-0291 $4.00 © 1994 IEEE First WCPEC; Dec. 5-9, 1994; Hawaii

Cell ID	Absorber	J-V Measurements				Illuminated	
		η %	V_{oc} mV	J_{sc} mA/cm^2	FF %	R_s Ω cm^2	A
C217-12-5	CIS/selenize	11.0	473	36	65	0.8	1.8
M1187-23-1	CIGS/3-stage	10.0	529	28½	66	0.2	2.1
M1201-14-4	CIS/3-stage	13.2	485	36½	75	0.15	1.2
M1174-14-3	CIGS/3-stage	15.9	649	32	76½	0.25	1.55
C253-23-9	CIGS/3-stage	15.8	611	34	76	0.3	1.4
C266-23-8	CIGS/3-stage	16.4	660	31½	78½	0.1	1.3

TABLE I. Summary of cell properties including efficiency (η), open-circuit voltage (V_{oc}), short-circuit current density (J_{sc}), fill factor (FF), series resistance (R_s), and ideality factor (A).

Dependence of Capacitance on Frequency

The first common feature of high-efficiency cells was a capacitance that was relatively independent of frequency. Fig. 1 shows typical C(f) curves for several different bias voltages. Only the forward-bias scan shows a noticeable increase in C as f → 0. The high-frequency turn-up in the measured capacitance is an artifact associated with a resonance due to the self-inductance of the measurement circuit [9]. When compared with cells from several years ago these cells show significantly lower concentrations of traps that respond on the 1-1000 μs time scale [2]. Of course, significant numbers of traps might be present whose relaxation times are slower than 1 ms. In fact, transient behavior, discussed below, suggests the presence of traps with slower response times.

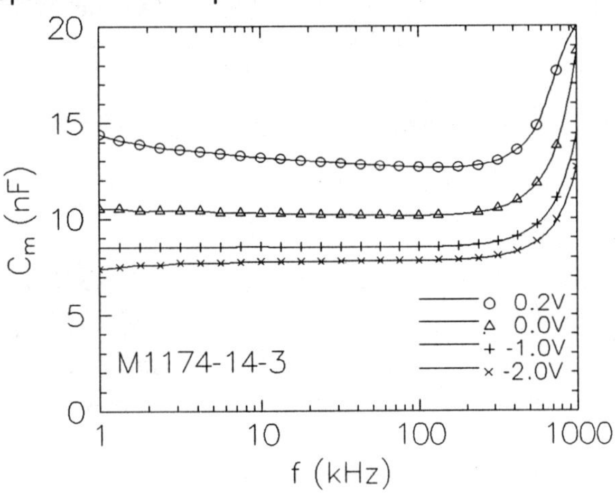

FIG. 1. Graphs of the measured capacitance versus log frequency for four different biases for a 15.9% CIGS cell. The high-frequency turn-up is associated with a resonance due to the self-inductance of the connecting cables.

Bias-Dependent Shunt Resistance

The second feature, common to most cells, was the observation of a voltage-dependent shunt resistance, $R_{sh}(V)$. To appreciate how this shows up in the admittance measurements it is useful to compare the data with the calculated admittance of a three-component circuit consisting of a small resistance (R_s) in series with a parallel combination of a capacitance (C) and a (shunt) resistance (R_{sh}). Even when R_s, R_{sh}, and C are ideal (i. e., voltage- and frequency-independent) the real part, $G_m(\omega)$, of the measured admittance is expected to vary with frequency. For the ideal case, $G_m(\omega)$ asymptotically approaches $1/R_{sh}$ as $\omega \rightarrow 0$ and $1/R_s$ as $\omega \rightarrow \infty$ [9]. Thus, we expect $G_m(\omega \rightarrow 0) = 1/R_{sh}$.

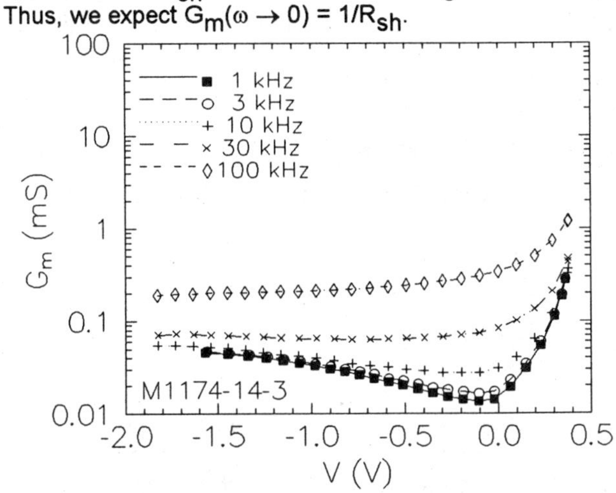

FIG. 2. Semilog graphs of the measured conductance versus bias voltage for various measurement frequencies for CIGS device M1174-14-3.

Typical voltage-scans of the measured conductance for different frequencies are graphed in Fig. 2. The higher-frequency data (e. g., f = 30 and 100 kHz) are consistent with expectations for an ideal circuit, allowing for the voltage-dependence of the depletion capacitance. At lower frequencies, however, measured conductances exhibit a voltage dependence consistent with calculations for the model circuit only if R_{sh} is assumed to change with V. For device M1174-14-3 shown, the shunt resistance has a maximum near zero bias and decreases by about a factor of 4 as V ranges from 0 to -2 V. For some cells, the shunt resistance varied by less than a factor of 2 over this range of bias. For an 11% efficient cell made by selenization, R_{sh} varied by about a factor of 30 over this range of bias (see Figure 3).

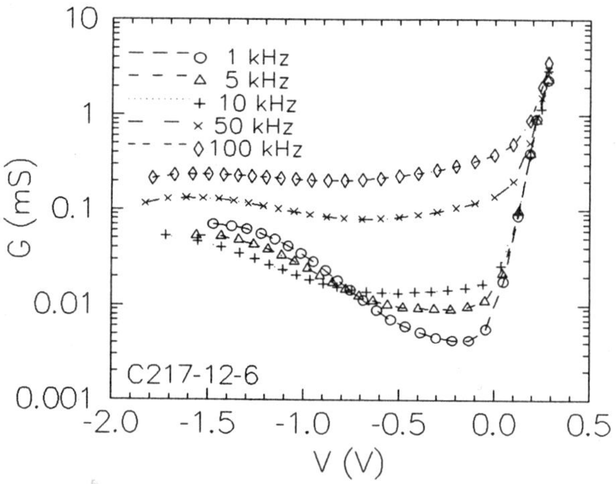

FIG. 3. Semilog graphs of the measured conductance versus bias voltage for various measurement frequencies for 11% efficient CIS device C217-12-6.

Slow Transient Response

The third common feature was the observation of transient current response, I(t), when cells were switched from zero to reverse bias, especially for V < −1 V. Prior to such a measurement, cells were typically held at zero bias for at least several minutes. When the applied voltage was rapidly switched to V_0, the magnitude of the resulting current was initially low and increased with time, asymptotically approaching some steady value. The importance of this effect became larger, the more negative V_0. For V_0 > −0.5 V, the effect was not noticeable. For V_0 < −2 V, transient effects were quite noticeable. This effect is related to the voltage-dependent shunt resistance, and may well be associated with other transient effects, such as those reported for V_{oc} [10].

Both the transient current and bias-dependent, low-frequency conductance ought to show up in J-V measurements. Most J-V measurements on these CIS and CIGS devices were been performed for -0.5 ≤ V ≤ 1.0, a range of bias in which both of these effects seem to have negligible consequences. We have performed additional J-V measurements on several devices to -2 V bias. For these measurements we have varied both the voltage sweep rate and direction. Figure 4 shows the results of such measurements for a device in which transient effects were particularly pronounced. Tabulated in the figure inset are the delay times between 10 mV changes in the bias voltage.[1] When the bias voltage was swept in the opposite direction (i.e., from 0.5 to -2.0 V) the reverse current did not show a minimum near -1.0 V, but instead, continued to increase the more negative the bias. We conclude that the zero J for -2V bias in Fig. 4 is due to the fact that the cell has not been allowed sufficient time to reach the steady-state current associated with this bias. It appears that all of these transient effects are associated with the capture and emission of carriers from very slow traps.

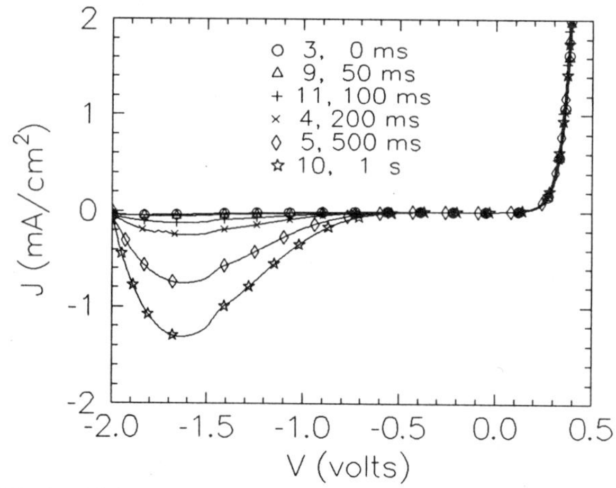

FIG. 4. J-V measurements for 11% efficient CIS device C217-12-6 for various sweep rates. Tabulated in the inset are delay times Δt, such that sweep rates were 10mV/Δt.

C-V Measurements

Room-temperature C-V measurements on all devices have been analyzed to determine the effective hole density, p(x), in the absorber as a function of distance x from the interface, assuming an abrupt, one-sided junction. The results for several cells with absorber layers fabricated using the 3-stage, indium-containing precursor method are shown in Fig. 5.

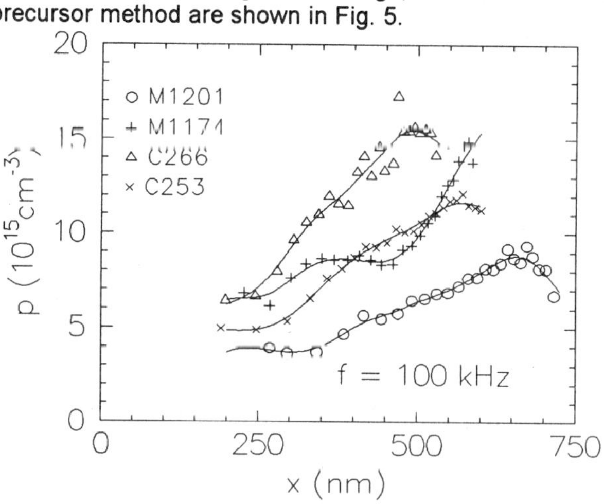

FIG. 5. Effective absorber layer carrier densities deduced from C-V measurements at 100 kHz.

Evidence for Localized Trapping State

Finally, admittance measurements from one CIGS cell show evidence of a narrow distribution of traps. Fig. 6 shows graphs of room-temperature conductance versus bias for various measurement frequencies. These curves exhibit peaks in the conductance superposed with G(V) behavior similar to that shown in Fig. 2. The peak voltage, V_p, was found to vary linearly with the logarithm

---

1 Prior to a measurement, the voltage was held at zero bias. A J-V measurement was subsequently performed by rapidly changing the bias voltage to -2.0 V, then increasing V in 10 mV steps, delaying a time Δt at each voltage step.

of the measurement frequency. Measurements, performed at temperatures between -20°C and 75°C, are consistent with thermally-activated behavior with an activation energy of 560 meV. The results suggest the presence of a trap near the interface at an energy close to midgap.

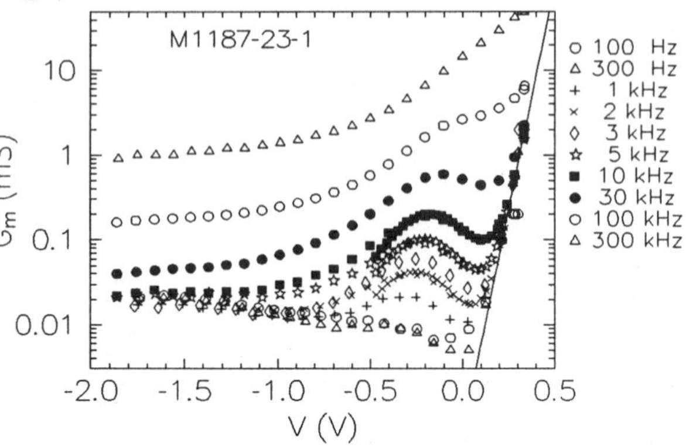

FIG. 6. Graphs of conductance versus bias V for various measurement frequencies for a CIGS device showing conductance peak indicative of a narrow distribution of traps.

DISCUSSION AND CONCLUSIONS

The first obvious conclusion is that the 14-16% efficient CIS and CIGS cells we have looked at generally have considerably fewer "fast" traps than did the 10-12% CIS cells of a few years ago, as evidenced from their frequency-dependent capacitance. This is consistent with the observation that these cells have higher conversion efficiencies and diode ideality factors closer to unity.

Second, the best CIS and CIGS cells all have effective charge densities, as determined from C-V measurements, which increase with distance from the junction. Also, more efficient CIGS cells appear to have higher carrier densities. This clearly has limits. however, as the highest carrier density was observed for a CGS cell which had very low conversion efficiency.

Third, very slow transient effects (10's of seconds to a few hours) continue to be observed, which appear to be associated with the capture and emission of carriers by traps. We suspect that these are related to time-dependent V_{oc} [10] but further work is required.

Finally, a voltage-dependent conductance has been observed for all devices in reverse bias. On the one hand this may simply be more evidence for the presence of a "2nd junction," previously reported for a variety of devices. On the other hand, it may be that careful comparison of conductance data with model calculations might shed light on critical device structures.

ACKNOWLEDGMENTS

The authors would like to thank I. R. Sasala, Eisgruber, and X. Liu for assistance with the admittance measurements and useful conversations, A. Tennant for assistance with J-V measurements, and J. Tuttle, B. Keys, and R. Ahrenkeil for helpful discussions. One of the authors (JHS) was supported, in part, by an Associated Western Universities - Department of Energy Faculty Sabbatical Fellowship. This work was performed at NREL and CSU under contract No. DE-AC36-83CH10093 to the U. S. Department of Energy.

REFERENCES

[1] See, for instance, S. M. Sze, *Physics of Semiconductor Devices, 2nd edition* (John Wiley & Sons, New York, 1981).

[2] Peter H. Mauk, Hossein Tavakolian, and James R. Sites, "Interpretation of thin-film polycrystalline solar cell capacitance," *IEEE Trans. Electron Devices* **ED-37** (2), 422-427 (1990).

[3] D. L. Losse, "Admittance spectroscopy of impurity levels in Schottky barriers," *J. Appl. Phys.* **46** (5) 2204-14 (1975).

[4] L. B. Fabick and K. L. Eskenas, "Admittance spectroscopy and application to $CuInSe_2$ photovoltaic devices," in *Proc. of the 18th IEEE PVSC,* Oct. 21-25, 1985, Las Vegas, pp. 754-757.

[5] A. M. Gabor, J. R. Tuttle, D. S. Albin, A. L. Tennant, M. A. Contreras, and R. Noufi, in *Proc. 12th NREL P. V. Program Review Meeting,* Oct. 13-15, 1993, Denver, CO (AIP, 1994).

[6] J. R. Tuttle, M. Contreras, A. Tennant, D. Albin, and R. Noufi, "High efficiency thin-film $Cu(In,Ga)Se_2$-based photovoltaic devices: progress towards a universal approach to absorber fabrication," in *Proc. 23rd IEEE PVSC,* May 10-14, 1993, Louisville, KY, pp. 415-421.

[7] M. A. Contreras, J. Tuttle, D. Albin, A. Tennant, and R. Noufi, "Validation of an in-line evaporation process for large-scale production of $CuInSe_2$-based solar cells," in *Proc. of the 23rd IEEE PVSC,* May 10-14, 1993, Louisville, KY, pp. 486-490.

[8] D. S. Albin, et al., "Progress towards higher efficiency $CuInSe_2$ and $Cu(In,Ga)Se_2$ solar cells by Se-vapor selenization," *in preparation.*

[9] J. H. Scofield, "Effects of series resistance and inductance on solar cell admittance measurements," *submitted for publication.*

[10] R. A. Sasala and J. R. Sites, "Time dependent voltage in $CuInSe_2$ and CdTe solar cells," in *Proc. of the 23rd IEEE PVSC,* May 10-14, 1993, Louisville, KY, pp. 543-548.

ELECTRONIC PROPERTIES OF INTIMATE Mo AND CdS JUNCTIONS IN-SITU FORMED ON $CuInSe_2$ (011) CLEAVAGE PLANES

T. Löher, C. Pettenkofer, and W. Jaegermann,
Abteilung Grenzflächen, Bereich Physikalische Chemie,
Hahn-Meitner-Insitut, Glienicker Str. 100, 14109 Berlin, Germany

ABSTRACT

Clean $CuInSe_2$ (011) surfaces are prepared in UHV by cleavage of single crystals. The structural and electronic properties of the cleavage planes and of the intimate heterocontacts to Mo and CdS are investigated by electron diffraction (LEED) and photoelectron spectroscopy (UPS, (S)XPS). The cleavage planes are stoichiometric and close to flatband conditions. Ion bombardement and annealing leads to structural disorder, changes in surface stoichiometry, and to band bending. The intimate p-$CuInSe_2$ (011)/Mo contacts is reactive and interfacial metallic In is formed. A Schottky type contact with a barrier height of about 0.8 eV is established. Deposited CdS forms a non-reactive overlayer at room temperature. The experimentally determined valence band and conduction band offsets are 0.8 and 0.7 eV, respectively, in deviation from the electron affinity rule. At sample temperatures above 490 K a Cu_xS interfacial reaction layer is formed and the valence band offset is reduced to 0.3 eV. The intimate heterocontacts investigated here do not correspond to the common knowledge of $CuInSe_2$ solar cells suggesting more complex interface processes there.

INTRODUCTION

The ternary chalcopyrite semiconductor $CuInSe_2$ has emerged to a very promising absorber material for solar cells [1]. Conversion efficiencies of 12% have been reported for single crystalline devices [2]. In thin film solar cells the efficiencies are already above 17% [3]. The energy converting interface is formed by deposition of an n-doped CdS window layer onto the p-doped substrate material at typical substrate temperatures above 150°C [1]. In a recent investigation the promising properties of thin film solar cells are attributed to the formation of an In rich defect chalcopyrite interfacial layer [4]. In spite of the evident importance of the $CuInSe_2$/CdS heterointerface its structural and electronic properties are still not understood in its microscopic details. For example, the valence band offset values are reported to be 1.4-1.6 [5,6] and 0.3 eV [7]. Buried $CuInSe_2$/CdS interfaces were studied by sputter-profiles using Auger and secondary ion mass spectroscopy [8] and an interdiffused interface layer was deduced. In addition, the $CuInSe_2$/ZnS heterointerface [5-7,9,10] and some metal contacts [11,12] have been investigated. No procedure is established to prepare clean and defined $CuInSe_2$ surfaces.

Therefore it seems worthwhile to reinvestigate the formation of metal contacts (Mo) and of the CdS/p-$CuInSe_2$ heterointerface with defined surfaces. The interface was formed stepwise in-situ by deposition of the contact phase from effusion cells and the interface was investigated by photoelectron spectroscopy and electron diffraction.

EXPERIMENTAL

The experiments were performed in an ultra high vacuum (UHV) system with a base pressure close to 5 x 10^{-11} - 1 x 10^{-10} torr. Photoemission measurements were performed with commercial spectrometers (VG ADES 500, Leybold). As excitation source either synchrotron light (SXPS: BESSY storage ring, TGM 7, hν = 10 - 120 eV, energy resolution 0.3 eV) or MgKα (XPS) and He I and He II radiation (UPS) was used. For experimental details refer to ref. [13]. The work function of the sample was determined from the cut-off of the secondary electrons. LEED experiments were performed with a back view LEED system (VG).

The $CuInSe_2$ single crystals were prepared by the gradient freeze technique. Their p-doping concentration was $N_A = 2 \times 10^{16}$ cm^{-3} as determined by Hall experiments. The crystals were oriented by Laue diffraction measurements along the (011) direction. Cleavage was performed in UHV by using prenotched crystals. CdS was deposited from home-made effusion cells and Mo by an electron beam evaporator and monitored by a quartz microbalance.

EXPERIMENTAL RESULTS

Surface Preparation

Clean $CuInSe_2$ are usually prepared by combining ion bombardment and annealing (IBA) as reported by Kahn et al. [14,15]. However, as discussed in the original work, but mostly neglected later on, IBA leads to nearly stoichiometric but structurally transformed surface layers. Therefore we performed experiments on surfaces cleaved in UHV. Fig. 1a shows the LEED diffraction pattern obtained for the (011) cleavage plane. It shows sharp spots in a centered rectangular lattice on

CH3365-4/94/0000-0295 $4.00 © 1994 IEEE

a rather low background. The a/b ratio of 2.25 agrees to the (011) plane [13]. Also Laue diffraction pattern confirm this assignment. After IBA the LEED pattern (Fig. 1b) shows an increased diffuse background and larger reciprocal lattice constants. In analogy to the LEED data of IBA cleaned $CuInSe_2$ (112) surfaces a randomization of the metal sublattice (Cu/In exchange) must be assumed [14,15].

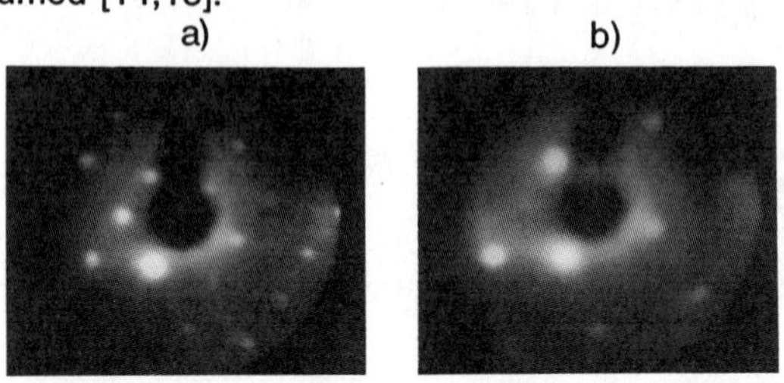

Fig. 1. LEED diffraction pattern of cleaved (011) faces (a) and after IBA treatment (b) (E=56 eV).

In Fig. 2. a comparison of He II valence band spectra after different surface treatment is shown. Ion bombardment leads to increased disorder (as is evident by the increased halfwidths also in the XPS core lines, not shown). Annealing at T=620 K leads to a considerable loss of Cu at the surface. The In/Cu XPS intensity ratio changes from 1 to 1.4. In addition, the Fermi level moves from a position 0.5 eV above the valence band edge for the cleaved surface by 0.3 eV towards the conduction band for the IBA treated surface.

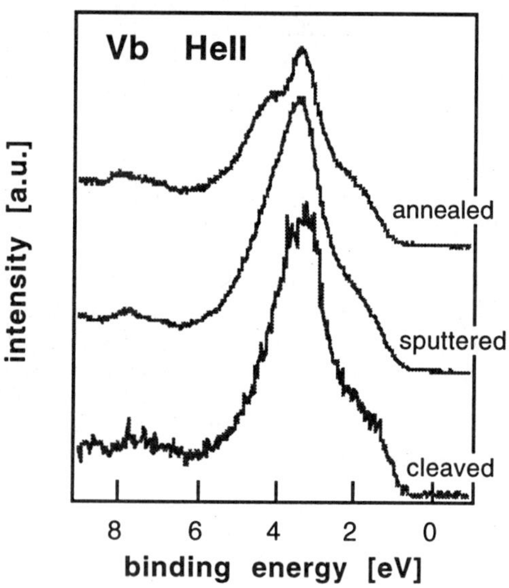

Fig. 2. He II valence band spectra of $CuInSe_2$ (011) after different surface treatments.

Mo deposition

Mo deposited onto UHV cleaved surfaces forms a continous metallic overlayer as derived from the XPS and UPS intensity attenuation data (Frank-van der Merwe growth mode). The Mo 3d core line does not show the formation of an interfacial reaction layer e. g. $MoSe_2$: there is only a slight asymmetric broadening at low coverages. The In 3d (Fig. 3) lines indicate a negatively shifted chemical component, which is assigned to metallic In situated at the phase boundary. The Se 3d and Cu 2p levels (Fig. 3) are broadened with increasing Mo coverages but do not indicate a reacted component. The BE values of the Se and In levels increase due to band bending by about 0.3 eV. The shift observed for Cu 2p is considerably larger (ΔBE=0.5 eV) and the Cu line is also stronger attenuated: evidently the surface stoichiometry is changed by Cu indiffusion as already observed after the IBA treatment. As a consequence a Schottky barrier is formed, which shows a bias light induced surface photovoltage of 0.3 eV. Evidently the intimate Mo contact formed for the above given conditions is no favourable ohmic back contact.

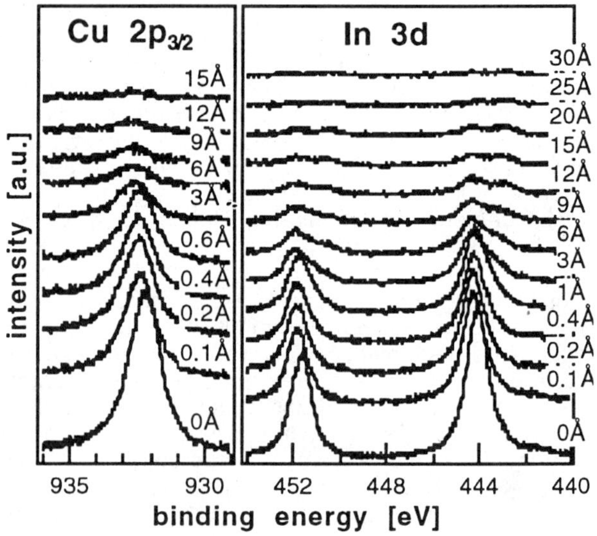

Fig. 3. Cu 2p and In 3d XP core level spectra of the $CuInSe_2$/Mo interface as function of increasing Mo coverage.

CdS deposition

CdS deposited at room temperature forms a non-reactive CdS overlayer (no chemically shifted emission lines in the SXPS data, Fig. 4) [16]. Also the In/Se ratio does not change with CdS coverage. The valence band spectra are superpositions of substrate and of weakly n-doped CdS overlayer emission lines. No epitaxial overlayer is formed as judged from LEED measurements. The attenuation dependence indicates initially a flat layer growth followed by the formation of three-dimensional clusters (Stranski-Krastanov growth). When the CdS overlayer is annealed at sample temperatures above 490K the deposited CdS is transformed to a Cu_xS surface layer and the Cd, Se as well as the In emission lines are lost from the surface. When the deposition is performed at elevated sample temperatures similar spectra are obtained for all deposition sequences. For the low fluxed used in our experiments there is no formation of a thick CdS overlayer. Only the Cu_xS layer builds up at the surface. The formed Cu_xS layer is a degenerate semiconductor as can be deduced from the position of the valence band maximum vs. the Fermi level which is close to BE=0 eV. The BE values of the substrate levels increase by about 0.6 eV with CdS deposition due to band bending. This band bending as measured for the unreacted interface is expected between the p-doped $CuInSe_2$ absorber material and the slightly n-doped CdS overlayer. However, after annealing and Cu_xS formation the band bending is drastically reduced again approaching flat band conditions for the substrate.

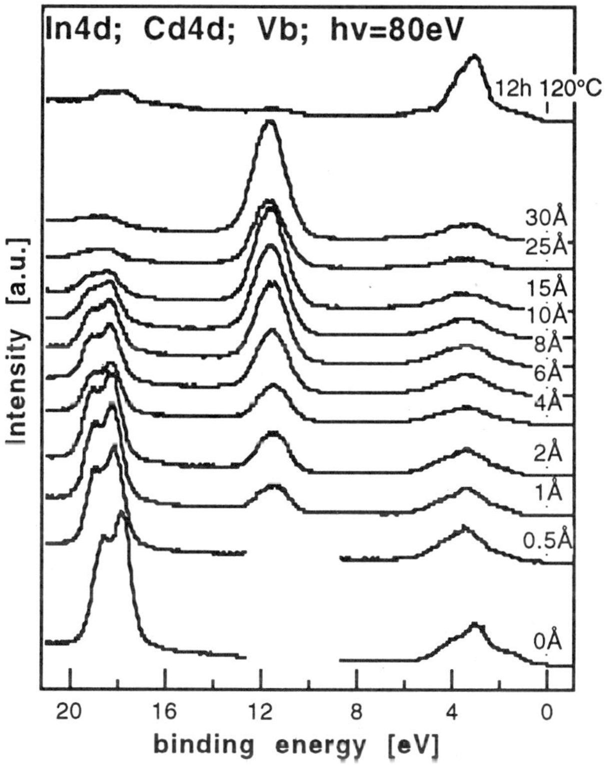

Fig. 4. SXP valence band, In and Cd 4d core level spectra of the $CuInSe_2$/CdS interface as function of CdS coverage and after annealing.

ENERGY DIAGRAMS

After cleavage in UHV the conditions of the interfaces as determined by our measurements are given in Fig. 5a. There is a slight band bending of 0.3 eV as evidenced by the difference of E_F-E_{VB} determined from the photoemission data compared to the results of the Hall experiments. The electron affinity is deduced from the measured work function Φ=5.2 eV and E_F-E_{VB}. The IBA treatment increases band bending to an overall value of 0.5 eV. n-doped samples do not show any band bending after cleavage [13].

The energy diagram of the intimate contact to Mo prepared on the cleavage plane is shown in Fig. 5b. As is expected based on the difference of metal work function and semiconductor electron affinity (Φ(Mo)=4.6 eV; χ($CuInSe_2$)=4.7 eV) a Schottky type junction is formed. This is supported by bias light induced surface photovoltages which are only obtained for the Mo covered surface, but not for the cleavage plane.

The valence band offset of the $CuInSe_2$/CdS heterointerface was determined from the valence band cutting edges as well as from the core level positions of the $CuInSe_2$ substrate and the CdS overlayer following a procedure by Grant [17], respectively. For the unreacted interface a value of 0.8 eV is deduced. With this value and the known bandgap and our measurement of work functions and electron affinities the band energy diagram as shown in Fig. 5c is constructed. It agrees to recent theoretical calculations of the band offset. However it is in disagreement to previously determined valence band discontinuities [5-7]. For the reacted interface, however, a valence band offset of 0.3 eV may be deduced which agrees to previous measurements, which also have been performed at elevated substrate temperatures [7].

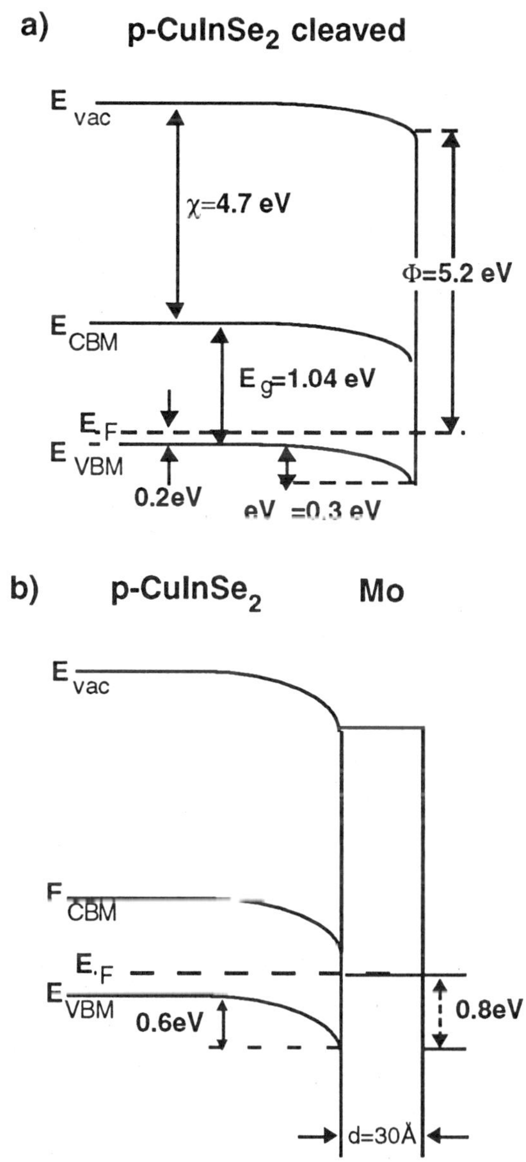

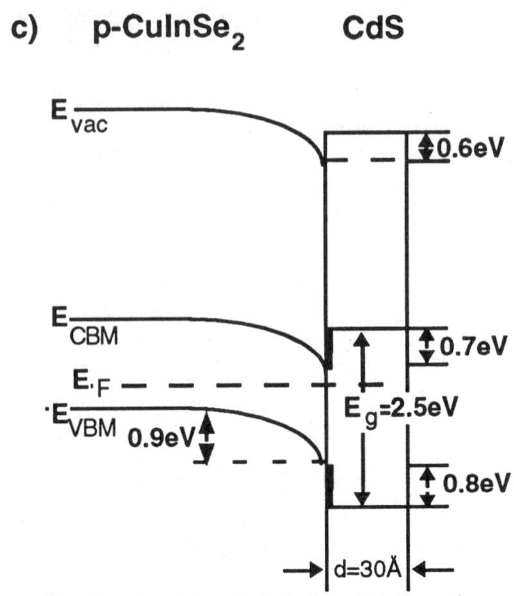

Fig. 6. Energy diagram of $CuInSe_2$ (011) surfaces and interfaces as determined from our experiments: a) after cleavage, b) Mo contact, c) CdS heterointerface.

In summary, our results obtained for the intimate heterocontacts of $CuInSe_2$ (011) cleavage planes to Mo and CdS disagree to the common assumptions about these contacts as derived from electrical measurements of solar cells. The discrepancy is evidently due to a modified and more complex interface morphology formed during preparation at elevated temperatures. Evidently the easy interdiffusion and reactivity at the interface, which are also suggested by our results, may induce favourable changes in barrier properties. Further and more complex surface science investigations are evidently needed to elaborate a detailed understanding of the barrier properties of $CuInSe_2$ based solar cells.

ACKNOWLEDGEMENT

This work was supported by the BMFT. The cooperation with Prof. Lewerenz is gratefully acknowledged.

REFERENCES

1. L. Kazmerski and S. Wagner in: *Current Topics in Photovoltaics*, Vol. 1; T.J. Coults and J.D. Meakin (eds.) Academic Press, London (1985)

2. J.L. Shay, S. Wagner, and H.M. Kaspar, "Efficient $CuInSe_2$/CdS solar cells", *Appl. Phys. Lett.* **27**, 89 (1975)

3. J. Hedström, H. Ohlsen, M. Bodegard, A. Kylner, L. Stolt, D. Hariskos, M. Ruckh, and H.W. Schock, *23rd IEEE Photov. Spec. Conf.*, Louisville, (1993)

4. D. Schmidt, M. Ruckh, F. Grunwald, and H.W. Schock, "Chalcopyrite/defect chalcopyrite heterojunctions on the basis of $CuInSe_2$", *J. Appl. Phys.* **73**, 2902 (1993)

5. M. Turowski, G. Margaritondo, M.K. Kelly, and R.D. Tomlinson, " Direct confirmation of the conduction-band lineup in the $CuInSe_2$-CdS heterojunction solar cell", *Appl. Phys. Lett.* **44**, 768 (1984)

6. M. Turowski, G. Margaritondo, M.K. Kelly, and R.D. Tomlinson, "Photoemission studies of $CuInSe_2$ and $CuInSe_2$ and of their interfaces with Si and Ge", *Phys. Rev.* **B 31**, 1022 (1985)

7. A.J. Nelson, et al., "Synchrotron-radiation photoemission study of CdS/$CuInSe_2$ heterojunction formation", *Phys. Rev.* **B42**, 7518 (1990)

8. L.L. Kazmerski, et al., "Formation, growth, and stability of the CdS/$CuInSe_2$ interface", *J. Vac. Sci. Technol.* **21**, 486 (1982)

9. A.J. Nelson, et al., "Theoretical and experimental studies of the ZnSe/$CuInSe_2$ heterojunction band offset", *Appl. Phys. Lett.* **62**, 2557 (1993)

10. S.-H. Wei and A. Zunger, " Band offsets at the CdS/$CuInSe_2$ heterojunction", *Appl. Phys. Lett.* **63**, 2549 (1993)

11. A.J. Nelson, et al., "Formation and schottky barrier height of Au contacts to $CuInSe_2$", *J. Vac. Sci. Technol. A*,978 (1991)

12. K.M. Jones, L.L. Kazmerski and B.G. Yacobi, *Thin Solid Films* **116**, L59 (1984)

13. P. Corvini, A. Kahn, and S. Wagner, "Surface order and stochiometry of sputter-cleaned and annealed $CuInSe_2$", *J. Appl. Phys.* **57**, 2967 (1985)

14. A. Kahn, et al., "First investigations of the atomic structure of (112) and (-1-1-2) $CuInSe_2$ surfaces", *Solar Cells* **16**, 123 (1986)

15. M. Sander, W. Jaegermann, and H.J. Lewerenz, "Site specific surface interaction of adsorbed $H_2O$and Halogenes on $CuInSe_2$ surfaces", *J. Phys. Chem.* **96**, 782 (1992)

16. T. Löher, C. Pettenkofer, W. Jaegermann, "Formation and Electronic Properties of the CdS/$CuInSe_2$ (011) heterointerface studied by synchrotron induced Photoemission", *J. Appl. Phys.*, in press

17. J.R. Waldrop and R.W. Grant, "Semiconductor heterojunction interfaces: nontransitivity of energy band discontinuities", *Phys. Rev. Lett.* **43**,1686 (1979)

SURFACE ENERGY BAND GAP OF POLYCRYSTALLINE $CuInSe_2$ THIN-FILMS IN TUNNELING SPECTROSCOPY

Z. Zhang, P. Heuell, M. Kulakov and B. Bullemer
Federal Armed Forces University Munich, Faculty of Electrical Engineering, Institute of Physics
D-85577 Neubiberg, Germany.

ABSTRACT

in this paper we demonstrate a tunneling spectroscopic technique to determine energy band gaps of the $CuInSe_2$ thin-film surfaces. The I-V characteristics are measured with STM in ambient air. Using the well-known theory of tunneling spectroscopy, the normalized differential conductivity corresponding to surface density of states is both evaluated from the measured I-V characteristics and simulated with help of two gap models with and without additional localised state tails in the gap. The poorly defined gap edges are explained by additional localised states near the top of the valence band and the bottom of the conduction band. These additional localised states could be true surface defect states or originate from deep atomic layers underneath the surfaces due to band bending.

1. INTRODUCTION

The polycrystalline CIS ($CuInSe_2$) thin-films for solar cells have a buried p-n heterojunction between the n-type surface layer called OVC (ordered vacancy compound) and the p-type CIS bulk [1]. The homogeneity of the n-type OVC surface layer appears to play an important role in device performance. It is therefore necessary to accurately monitor the electronic properties at the film surface. The only known experimental technique that is in principle capable of directly probing electronic properties with sufficiently high resolution is STS (scanning tunneling spectroscopy). A direct STS investigation in ambient air on polycrystalline CIS thin-films synthesized by RTP (rapid temperature processing) [2] showed different conductive types at the surface according to rectifying tunneling characteristics [3]. A problem worthy to be considered is how these characteristics are correlated with band structure at the surface layer. To answer this question, we apply tunneling spectroscopy in ambient air. From the measured I-V characteristics we derive the surface DOS(density of states) with the well-known formula $\rho(E) \propto (dI/dV)/(I/V)$. This derivation is verified by simulations of tunneling spectroscopy. The obtained result is that energy band gaps exist on the thin-film surface, but the gap edges are not sharply defined. Thus we believe that additional localised states exist at the top of valence band and the bottom of conduction band.

2. EXPERIMENT AND SIMULATION

The investigated samples are polycrystalline CIS thin-films synthesized by RTP. The associated solar cell efficiency is by 9%. I-V characteristics are measured with STM (scanning tunneling microscope) in ambient air. The tunneling current of 0.2 nA is used to control the tip-sample separation, and the corresponding tunnel voltages are properly selected so that a stable tunnel process proceeds. The typical tunnel voltages are smaller than 1 V. Our STM can continuously measure an I-V characteristic 256 times at any single point on the samples under a controlled constant tip-sample separation. The full set of curves is displayed on the screen and after inspection averaged to yield the I-V characteristic for the given point.

According to the well-known theory of tunneling spectroscopy , the tunneling current I is correlated with surface DOS $\rho(E)$ and tunnel voltage V as

$$I = \int_0^{eV} \rho(E)\exp(-2\kappa s\sqrt{1+(eV-2E)/2\phi})dE \qquad (1)$$

where e is the electron charge, V is the sample voltage, ϕ is the average tunnel barrier height, E is the electron energy relative to the Fermi-level, and κ is the decay

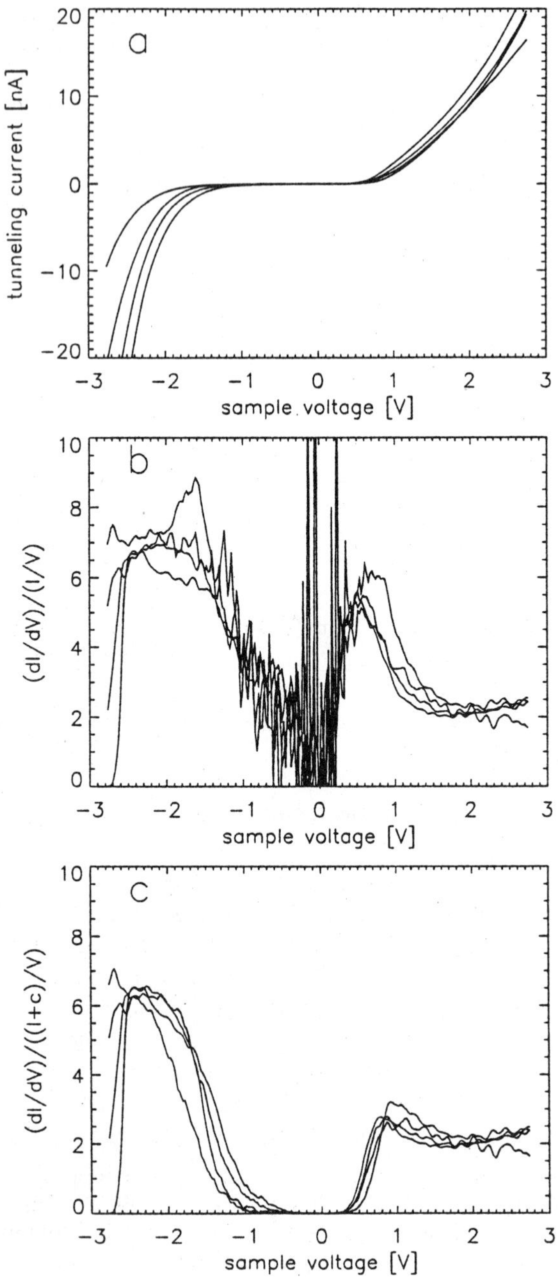

Fig.1 The measured I-V characteristics(a) with tunneling current 0.2 nA and tunnel voltages 0.6, 0.7, 0.8, 0.9V to control the tip-sample separation respectively, and the derived surface DOS(b,c) using Eq.(2) and Eq.(3) respectively.

constant of the tunneling current [4,5,6]. The surface DOS is given by the relation

$$\rho(E) \propto (dI/dV)/(I/V) \qquad (2)$$

which actually represents a normalized differential conductivity and is qualitatively correct [5,6]. We evaluate Eq.(2) and use the measured I-V characteristics shown in Fig.1a. The result is shown in Fig.1b. The serious noise at small sample voltages comes from the very small tunneling currents there. In order to restrain this noise, we modify Eq.(2) into

$$\rho(E) \propto (dI/dV)/((I+\varepsilon)/V) \qquad (3)$$

where ε is a relatively small constant which should be just big enough as to reduce the apparent noise effectivey. The addition of ε does not change the properties of Eq.(2) [7]. With ε =0.5nA, we obtain Fig.1c. Obviously, the curves in Fig.1c become smooth and a valley appears in the middle. This valley shows that the state densities in it are much smaller than those at both sides of it, it should therefore be the band gap at the surface. The poorly defined edges of this gap imply existence of additional localised states there.

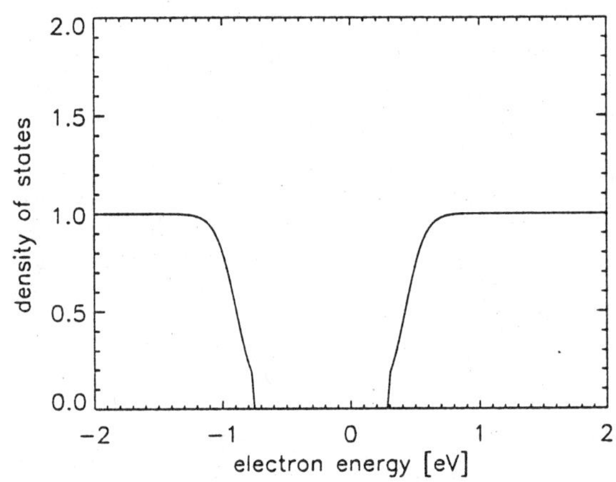

Fig.2 The assumed band gap without the state tails at the band edges.

In order to verify the above-mentioned deduction, we start from an assumed energy band gap without additional localised state tails in it, as shown in Fig.2, and simulate the measurement of the I-V characteristics with Eq.(1). The obtained I-V curves for different tip-sample separations are shown in Fig.3a, where each curve shown is the average of 256 curves with artificial statistical noise superimposed. The surface DOS is again calculated from each curve in Fig.3b using Eq.(2) and Eq.(3), respectively. The results are shown in Fig.2b and Fig.3c. A ε equal to 0.02 is enough to reduce noise so that the curves get smoothed. Incidentally the peaks become lower in particular for the larger tip-sample separations. Comparing Fig.2 and Fig.3, we recognize that although Eq.(2) and Eq.(3) cannot reproduce the

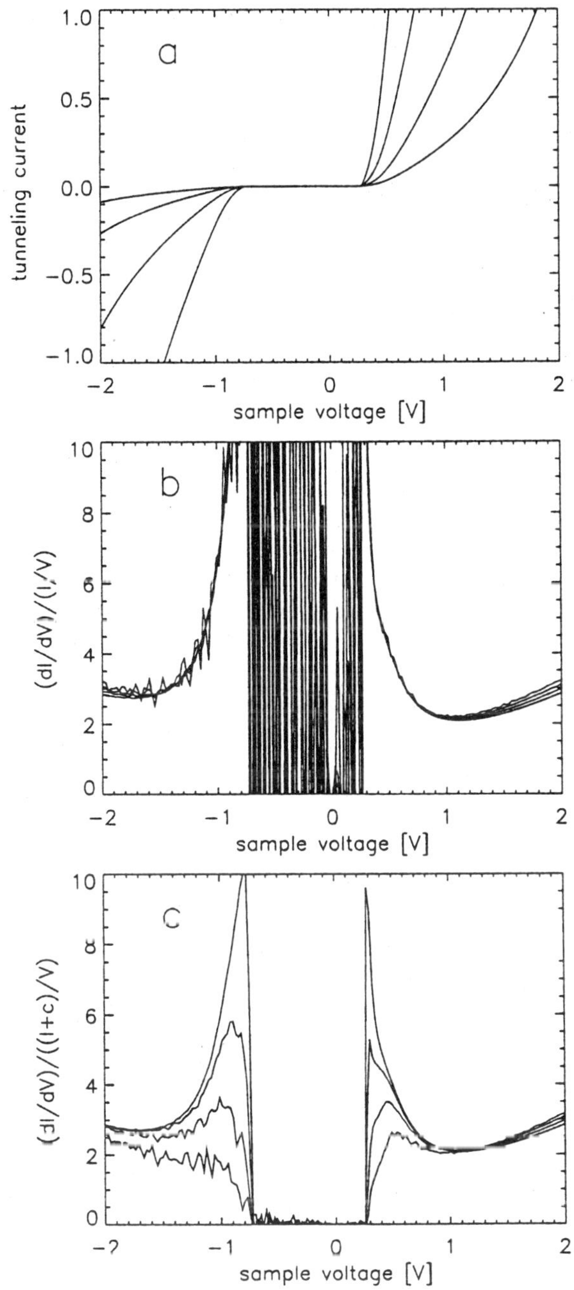

Fig.3 The simulated I-V curves (a) with tip-sample separations 10.5,11.0,11.5,12.0Å (K=1.1/Å, ϕ = 4.5eV) respectively and surface density of states(b,c) reproduced from the simulated I-V curves using Eq.(2) and Eq.(3) respectively.

assumed surface DOS quantitatively, they do indicate abrupt changes and relative distribution of it. For example, the peaks in Fig.3b and Fig.3c mark the gap edges and the gap in between. With these features, surface DOS can be qualitatively well described.

In order to clearly explain the effect of additional localised states in the gap, we start now with another assumed gap having additional localised state tails, as shown in Fig.4a, and repeat the simulation. The final result using Eq.(3) is shown in Fig.4b. In comparison to Fig.3c, Fig.4b shows poorly defined gap edges as if the gap itself had become larger with increasing tip-sample separation. We call this phenomenon edge state effect. Thus we come to believe that the valley in Fig.1c indicates the surface energy band gap and the poorly defined gap edges correspond to the edge state effect. The position of Fermi-level near the bottom of the conduction band indicates n-type conductivity at the point examined.

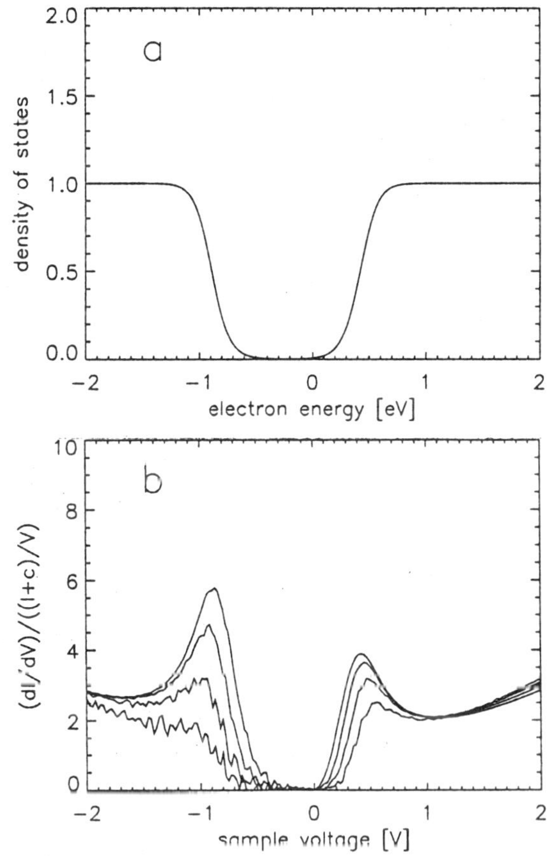

Fig.4 The assumed band gap with state tails at the gap edges(a), and the reproduced surface DOS(b) from it by the simulation.

Based on this method, we have done a lot of measurements at different points on the samples. In general, we obtain energy band gaps with poorly defined edges when the tip-sample separation is large. With small tip-sample separations we may obtain gaps with sharp edges, however, the gap widths are always smaller than 1 eV of p-type or 1.3 eV of n-type as reported in [1]. Fig.5a, for example, illustrates a sharp gap of 0.6 eV measured at a point, while this gap in Fig.5b and Fig.c appears larger with increasing tip-sample separation. We conclude therefore that additional localised states exist in the gap near the top of valence band and the bottom of conduction band.

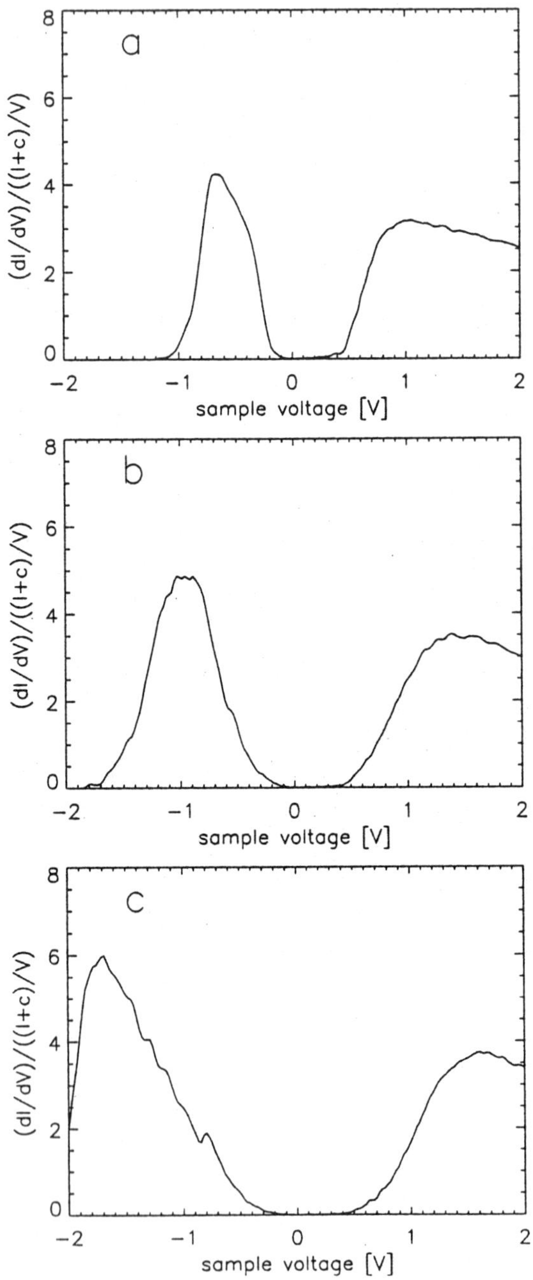

Fig.5 The measured gap larger with increasing the tip-sample separation

3. DISCUSSION

For ideal polycrystalline $CuInSe_2$ Thin-Films the surface conductivity should be homogeneously n-type. In reality, local stoichiometry variation may cause changes of surface conductivity type. In general, we obtain energy band gaps indicating n- or p-type conductivity on the thin-films. However, I-V characteristics without energy band gap are quite often measured. In addition to surface contamination, existence of metallic conductivity at the surface is possible too. We obtain poorly defined gap edges and define this phenomenon as existence of additional localised states at the gap edges. These states could stem from deep atomic layers underneath the surface activated by band bending or be true surface defect states. To decide this question, further investigation is needed. Summing up the comprehensive I-V characteristics measured, we find that the positive and negative polarities of the tunneling current in an I-V curve are representative for the n- and p-conductivities. Therefore, surface hetero-conductivity can be roughly mapped, according to different polarities of tunneling currents, by so called current imaging tunneling spectroscopy(CITS).

This work is supported by ICAMS e. V. Neubiberg.

References

[1] D. Schmid, M. Ruckh, F. Grunwald and H. W. Schock, "Chalcopyrite/defect chalcopyrite heterojunctions on the basis of $CuInSe_2$ ", *J. Appl. Phys.*, **73**, 1993, pp.2902-2909.

[2] F. Karg et al., "Novel Rapid-Thermal-Processing for CIS Thin-Film Solar Cells", *23rd IEEE PVSC*, 1993, pp. 441-446

[3] W. Riedl et al, "Surface Microstructure of CIS Thin Films Produced by Rapid Thermal Processing", International PVSEC-7, November 22-26 1993, Nagoya, Japan.

[4] N. D. Lang, "Spectroscopy of Single Atoms in the Scanning Tunneling Microscope", *Phys. Rev. B*, **34**, 1986, pp. 5947-5950.

[5] R. M. Feenstra, "Scanning Tunneling Microscopy: Semiconducter Surfaces, Adsorption, and Epitaxy", *Scanning Tunneling Microscopy and Related Methods,* ed. R. J. Behm et al., (Kluwer Academic Publishers, Netherland, 1990), pp. 211-240.

[6] R. M. Feenstra, J. A. Stroscio and A.P.Fein, "Tunneling Spectroscopy of the Si(111)2x1 Surface", *Surf. Sci.*, **181**, 1987, pp. 295-306.

[7] R. M. Feenstra and P.Martensson, "Fermi-Level Pinning at the Sb/GaAs(110) Surface Studied by Scanning Tunneling Spectroscopy", *Phys. Rev. Lett.*, **61**, 1988, pp. 447-450.

EVIDENCE FOR AMORPHOUS LIKE BEHAVIOR IN SMALL GRAIN THIN FILM POLYCRYSTALLINE SOLAR CELLS

J. E. Phillips, W. N. Shafarman and Enze Shan
Institute of Energy Conversion
University of Delaware

United States Department of Energy
University Center of Excellence
For Photovoltaics Research and Education
(National Renewable Energy Laboratory)
Newark, DE 19716 USA

ABSTRACT

Solar cells based on CdTe, $CuInSe_2$ or the alloys of $Cu(In,Ga)Se_2$ have grain sizes on the order of one micron. It has been suggested that materials with these size grains could have an exponential distribution of localized states within the bandgap [1,2] similar to amorphous materials such as a-Si:H. The current voltage characteristics of a number of CdTe, $CuInSe_2$ and $Cu(In,Ga)Se_2$ based solar cells made by various processes and manufacturers are measured as a function of light intensity and temperature. These measurements are then analyzed and compared for evidence of this amorphous like behavior.

INTRODUCTION

It appears that most high efficiency polycrystalline thin film solar cells such as CdTe, $CuInSe_2$ and its alloys operate with the diode current being controlled by recombination within the active light absorbing material. [3-6] This is due to the material being inverted, (i.e. changing from p to n-type) near the junction. This helps in eliminating interface recombination as a loss mechanism in these devices. The recombination is then controlled by states within the bandgap in this region. When these states are localized near the center of the bandgap, the diode or recombination current can be expressed as [7]:

$$J_d = J_{rec.} = J_0 \cdot \exp(qV/2kT) \qquad (1)$$

where:

$$J_0 = J_{00} \cdot \exp(-E_g/2kT) \qquad (2)$$

For additional levels the total recombination can be summed and expressed numerically. However, when the recombination is through an exponential distribution of recombination centers i.e. :

$$N_{rec}(E) \sim \exp[-(E_C-E)/kT^*] \qquad (3a)$$

or:

$$N_{rec}(E) \sim \exp[-(E-E_V)/kT^*] \qquad (3b)$$

Then, for $T >> T^*$, the diode or recombination current can be expressed in closed form as [1,8]:

$$J_d = J_{rec} = J_0 \cdot \exp(qV/AkT) \qquad (4)$$

where the diode quality factor (A) becomes slightly temperature dependent and not necessarily equal to one or two, i.e. :

$$A = 2/(1+T/T^*) \qquad 1 < A < 2 \qquad (5)$$

The reverse saturation current remains the same as before (equation 2). This means that the solar cell current-voltage characteristic with a series resistance term (R_S) becomes:

$$J = J_0 \cdot \exp[(V-J \cdot R_S)/AkT] - J_{SC} \qquad (6)$$

This is the standard exponential behavior with a diode quality factor that has a slight temperature dependence.

ANALYSIS

The current voltage characteristics of a number of CdTe, $CuInSe_2$ and $Cu(In,Ga)Se_2$ based solar cells made by various processes and manufacturers and measured as a function of light intensity and temperature have been analyzed and compared for evidence of this amorphous

First WCPEC; Dec. 5–9, 1994; Hawaii

like behavior. this section describes, with examples, the J-V analysis techniques used.

Rearranging and differentiating equation 6 gives:

$$dV/dJ = R_S + (AkT/q)\cdot(J+J_{SC})^{-1} \qquad (7)$$

Figures 1 and 2 show the slope of the J-V curve (dV/dJ) plotted as a function of the inverse total current for both a $CuInSe_2$ and a CdTe device. If the plot is linear, the diode quality factor (A) and a constant series resistance (R_S) can be extracted from the slope and intercept respectively.

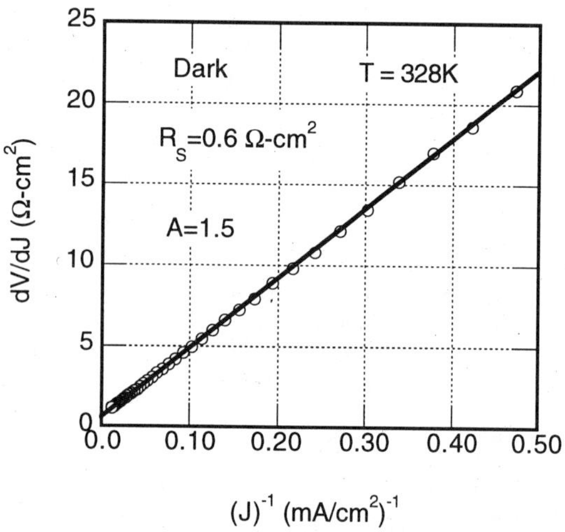

Figure 1. Graph of the slope of a J-V curve vs. 1/J for an IEC $CuInSe_2$ solar cell (IEC-32220-22-5).

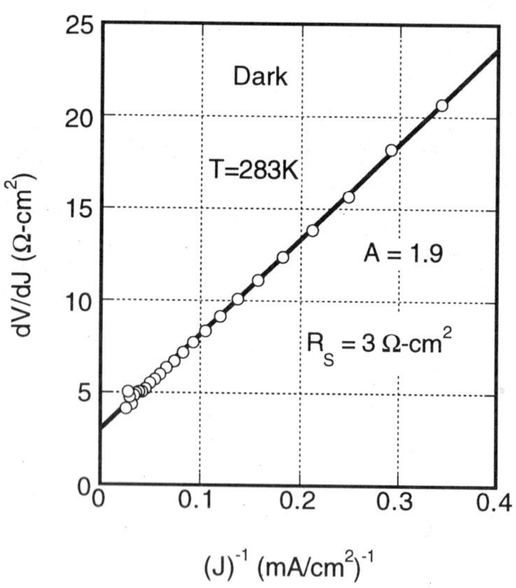

Figure 2. Graph of the slope of a J-V curve vs. 1/J for an IEC CdTe solar cell (IEC-40803-22-8)

After R_S has been determined, the J-V data can be plotted in another form, $\ln(J+J_{SC})$ vs. $(V-R_SJ)$ (see Figures 3 and 4). This graph is an additional check on the exponential diode behavior and the reverse saturation current (J_O) can be extracted since:

$$\ln(J + J_{SC}) = \ln(J_O) + (q/AkT)\cdot(V - R_S\cdot J) \qquad (8)$$

Finally, at lower temperatures, most of these devices have a current dependent series resistance $R_S(J)$ [3,6] due to non-ohmic contacts (as shown in figures 5 and 6).

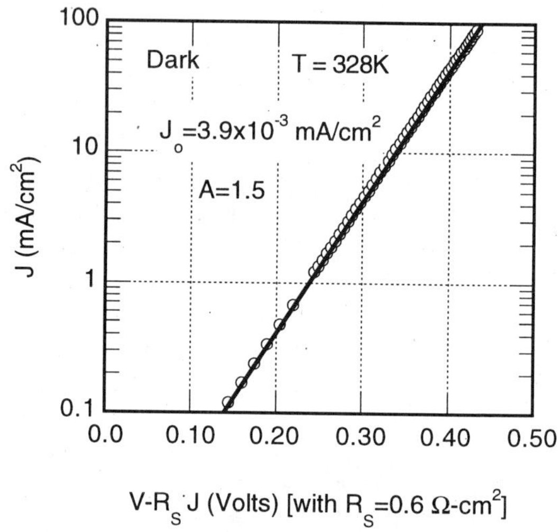

Figure 3. Plot of ln(J) vs. voltage across the diode for the previous $CuInSe_2$ solar cell (IEC-32220-22-5)

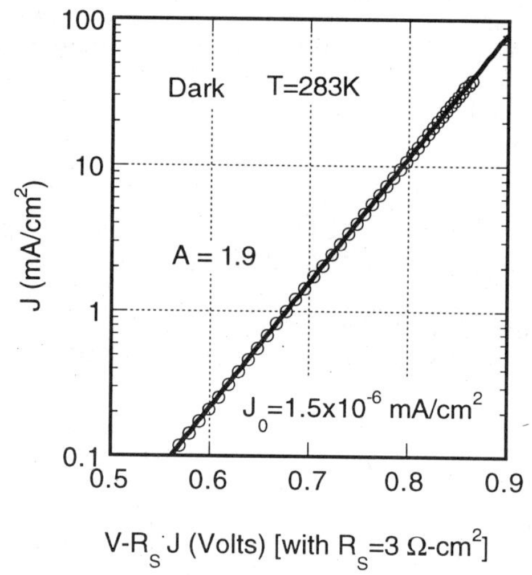

Figure 4. Plot of Log J vs. voltage across the diode for the previous CdTe solar cell (IEC-40803-22-8)

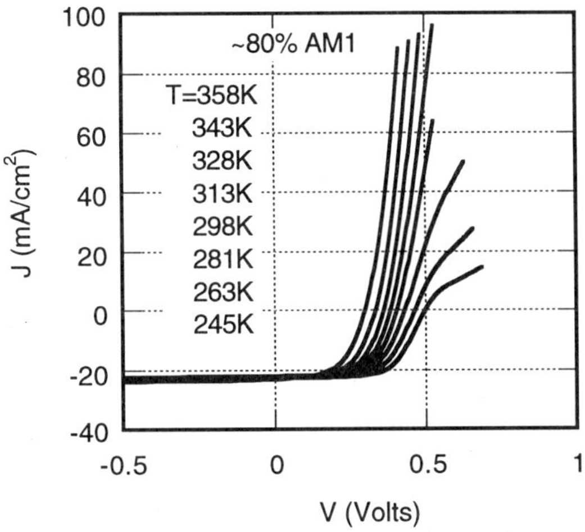

Figure 5. J-V data as a function of temperature for the $CuInSe_2$ device IEC-32220-22-5 (note the non-ohmic behavior at lower T in forward bias)

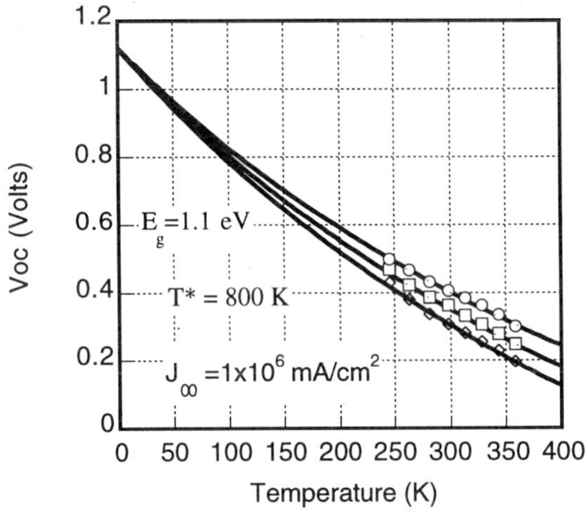

Figure 7 V_{oc} as a function of temperature and light intensity (~5% AM1, ~20% AM1 and ~80% AM1) for the $CuInSe_2$ device IEC-32220-22-5. The lines are calculated using the values for E_g, J_{00} and T* shown.

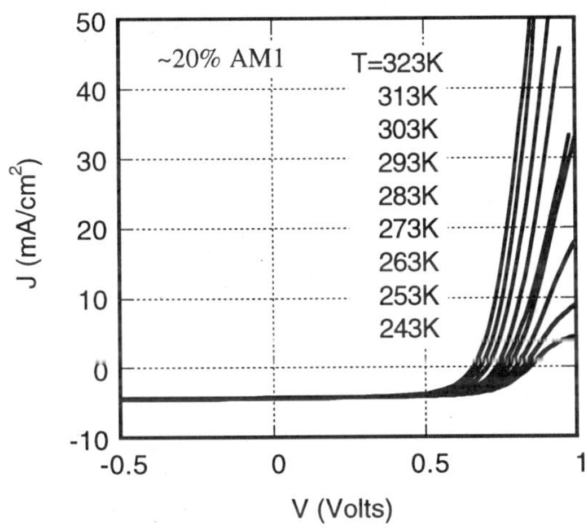

Figure 6. J-V data as a function of temperature for the CdTe device IEC-40803-22-8 (note the non-ohmic behavior at lower T in forward bias)

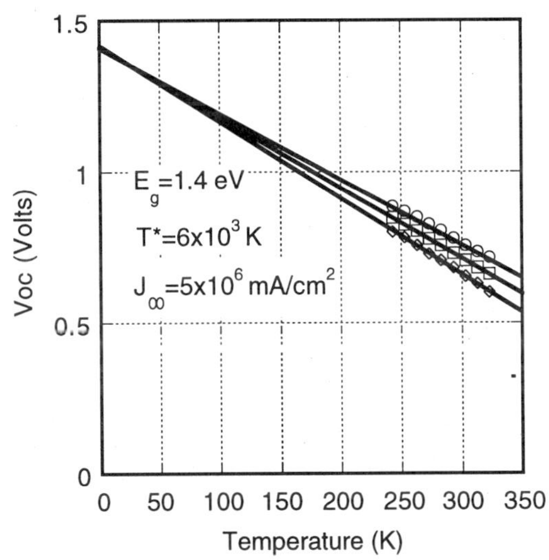

Figure 8 V_{oc} as a function of temperature and light intensity (~5% AM1, ~20% AM1 and ~80% AM1) for the CdTe device IEC-40803-22-8. The lines are calculated using the values for E_g, J_{00} and T* shown.

In this case, the analysis of V_{oc} and J_{sc} as a function of temperature and light intensity can be used to extract E_g, T* (and therefore A) and J_{00} since the series contribution is eliminated at J=0. From equations 2 and 6:

$$(1+T/T^*)\cdot V_{oc} = (E_g/q) + (2kT/q)\cdot \ln(J_{sc}/J_{00}) \quad (7)$$

Examples of this type of analysis is shown in figures 7 and 8.

RESULTS

In determining whether this type of model can account for the observed J-V behavior in polycrystalline thin film solar cells, 14 different devices made by various processes and manufacturers were examined. J-V data was measured over a range of temperatures (218K to 373K) and light intensities (dark, ~ 5% AM1, ~20% AM1 and ~80% AM1). This usually involved forty sets of J-V data for each device. The results of analyzing this data according to equation 8 (i.e. V_{oc} as a function of temperature and light intensity) are presented in table 1.

CONCLUSIONS

There are indications, in the case of the $CuInSe_2$ and CdTe based materials, that the diode characteristic can be described by recombination through a continuous exponential-like distribution of centers within the bandgap of the light absorbing material. Although not as pronounced, this behavior is quite similar to that of photovoltaic devices made from amorphous semiconductors. Possibly this is because the crystalline order in these polycrystalline thin films, with grain sizes on the order of about a micron in diameter, is part way between that of crystalline and amorphous material.

ACKNOWLEDGEMENTS

The authors would like to thank many people at Energy Photovoltaics, IPE of the University of Stuttgart, the National Renewable Energy Laboratory, Siemens Solar and Solar Cells Inc. for supplying their solar cells for analysis. This work was supported by the Department of Energy through the National Renewable Energy Laboratory under subcontract #X-AV-13170-1.

REFERENCES

[1] T. Walter, R. Menner, Ch. Köble and H. W. Schock, "Characterization and Junction Performance of Highly Efficient ZnO/CdS/$CuInS_2$ Solar Cells", *12th. Euro. PSEC*, 1994, pp 1755-1758.

[2] Yi Tang, R. Braunstein and Bolko von Roedern, "Determination of Drift Mobility and Lifetime for Dominant Charge Carriers in Polycrystalline $CuInSe_2$ by Photomixing", *Appl. Phys. Lett.*, **63** (17), 1993, pp 2393-2395.

[3] M. Roy, S. Damaskinos and J. E. Phillips, "The Diode Current Mechanism in $CuInSe_2$/(CdZn)S Heterojunctions", *Twentieth IEEE PVSC*, 1988, pp 1618-1623.

[4] G. B. Turner, R. J. Schwartz and J. L. Gray, "Band Discontinuity and Bulk Vs. Interface Recombination in CdS/$CuInSe_2$ Solar Cells", *Twentieth IEEE PVSC*, 1988, pp 1457-1460.

[5] D. A. Fardig and J. E. Phillips, "Characteristics of CdTe/CdS Solar Cells", *22nd. IEEE PVSC*, 1991, pp 1146-1150.

[6] W. N. Shafarman and J. E. Phillips, "Current-Voltage Characterization of $CuInSe_2$/CdS Solar Cells Deposited by Different Methods", *23rd. IEEE PVSC*, 1993, pp 453-458.

[7] A. L. Fahrenbruch and R. H. Bube, "Fundamentals of Solar Cells", Academic Press, New York, 1983

[8] A. Rose, "Concepts in Photoconductivity and Allied Problems", Wiley, New York, 1963.

DEVICE TYPE	MANUFACTURER	E_G (eV)	T* (K)	J_{00} (mA/cm^2)	A (@300 K)	DEVICE ID #
$CuInSe_2$						
elem. evap.	Inst. of Energy Conv.	1.1	800	$1x10^6$	1.5	IEC-32220-22-5
elem. evap.	Inst. of Energy Conv.	1.1	1300	$2x10^6$	1.6	IEC-32187-22-12
	Energy Photovoltaics	1.1	2000	$4x10^6$	1.7	EPV-11133-21-5
	IPE U. of Stuttgart	1.2	1000	$6x10^6$	1.5	STU-M243-4
selenized with Se	Inst. of Energy Conv.	1.2	800	$6x10^6$	1.5	IEC-61079-23-5
selenized with H2Se	Inst. of Energy Conv.	1.1	1700	$2x10^6$	1.7	IEC-89095-2-9
$Cu(In,Ga)Se_2$						
	Siemens Solar	1.2	1200	$7x10^6$	1.6	SSI-16543-4-3
elem. evap.	Inst. of Energy Conv.	1.3	2000	$4x10^6$	1.7	IEC-32356-33-5
elem. evap.	Nat. Ren. Energy Lab.	1.3	1000	$2x10^6$	1.5	NREL-S478-14-5
elem. evap.	Nat. Ren. Energy Lab.	1.4	1500	$2x10^6$	1.7	NREL-S416-23-5
CdTe						
phys. vapor dep.	Inst. of Energy Conv.	1.4	6000	$5x10^6$	1.9	IEC-40803-22-8
phys. vapor dep.	Inst. of Energy Conv.	1.4	1400	$1x10^5$	1.6	IEC-40821-21-7
	Solar Cells Inc.	1.4	4000	$1x10^6$	1.9	SCI-L9808-24-4
	Solar Cells Inc.	1.5	1500	$2x10^5$	1.7	SCI-L10304-26-6

Table 1 A summary table of the diode parameters extracted from measurements of V_{oc} as a function of temperature and light intensity.

PREPARATION AND CHARACTERIZATION OF CdTe THIN FILMS DEPOSITED BY CSS

G. Gordillo, J.M. Florez , L.C. Hernandez and P. Teherán
Departamento de Física, Universidad Nacional de Colombia, Bogotá, COLOMBIA

ABSTRACT

CdTe polycrystalline thin films were grown using a new procedure based on the C.S.S. Technique. With an adequate control of the deposition parameters, it was possible to grow CdTe films with grain sizes greater than 45 μm. The samples were characterized by XRD, SEM and EDS measurements. The studies revealed that the substrate and sublimation temperatures influence strongly the crystallographic orientation of the films, which grow with a mixed structure of cubic and hexagonal phases. EDS analysis carried out on different samples, showed a significant influence of some deposition parameters on the chemical composition of the CdTe films.

The effect of introducing O_2 during the deposition of the CdTe films, on the resistivity and crystallographic properties will also be discussed.

1. INTRODUCTION

CdTe has long been regarded as an attractive candidate for the preparation of thin film solar cells [1]. Conversion efficiencies greater than 13% have been reported for CdTe based solar cells with structure SnO_2/CdS/CdTe/Graphite [2].

Polycrystalline CdTe films have been deposited by several techniques including electrodeposition [3], close-spaced sublimation [3,4], direct combination of elemental cadmium and tellurium [5], and screen printing [6]; each technique has its own advantages. However, the best CdTe based thin film solar cells have been fabricated using the CSS method.

The purpose of this paper is to present results related to studies of structural and electrical properties of CdTe films deposited by the CSS technique, using a deposition system which have some modifications compared to the setups reported in the literature [2].

2. PREPARATION OF THE FILMS

The deposition of the CdTe films by CSS method is based on the reversible dissociation of CdTe at high temperature. The CdTe source and the substrate are separated by a small distance and the source is maintained at a higher temperature than the substrate. The CdTe dissociates into its elements ($2CdTe(s) \rightarrow 2Cd(g) + Te_2(g)$) which recombine on the substrate surface, to form the CdTe film The apparatus developed for the deposition of CdTe films by the CSS technique is shown in Fig. 1. The most significant characteristics of our deposition system are:

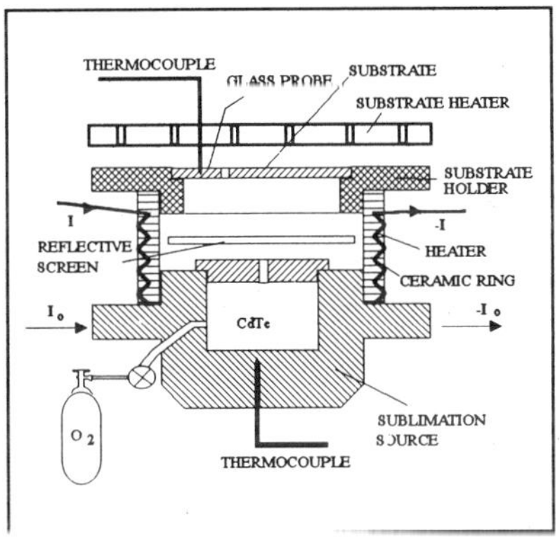

Fig. 1: System for deposition of CdTe thin films based on the CSS method

- The evaporation source and the substrate are communicated through a ceramic ring. In this way the CdTe vapor is confined to a close space. The ceramic ring was kept at a higher temperature than the substrate's to eliminate the condensation of material on its walls . With this, the evaporating materials will always have better access to the substrate.

- The gas is introduced directly into the reaction chamber, where it is mixed with the CdTe vapor. The chamber is evacuated to a pressure of $7x10^{-6}$ Torr and oxygen is introduced until the O_2--partial pressure is at the desired level ($9x10^{-6}$ -- $2x10^{-4}$ Torr). During the sample preparation the oxygen is continuously fed into the reaction chamber. The presence of small amounts of oxygen in the evaporation chamber enhances the p-type characteristics of CdTe and enables the obtainment of CdS/CdTe cells [7].

CH3365-4/94/0000-0307 $4.00 © 1994 IEEE

First WCPEC; Dec. 5–9, 1994; Hawaii

3. RESULTS AND DISCUSSION

3.1. Structural characteristics of CdTe films

X-ray diffraction measurements were performed in order to identify the crystalline structure and to study the influence of the deposition parameters on the crystallographic parameters of CdTe films. The measurements were made using a Philips diffractometer PW 1820 in the Cu k_{α} radiation region (1.54060 Å).

Correlation between the experimentally observed X-ray diffraction spectra and the theoretical simulation of the same spectra, indicates that the CSS deposited CdTe films, grow with a structure mix of cubic and hexagonal phases. Most of the reflections correspond simultaneously to the hexagonal and cubic structures; yet, two reflections at (331) and (400) were observed which only correspond to the cubic phase, and other three reflections came up at (106), (107) and (305) associated only to the hexagonal phase.

Fig. 2 displays typical X-ray diffraction spectra, showing the influence of evaporation and substrate temperature on the crystallographic orientation of CdTe films. These results indicate that, at low substrate and evaporation temperatures, the films grow disoriented; however, at evaporation temperatures greater than 650°C, the films become highly oriented in the direction H(002). CdTe layers deposited at low evaporation temperature grow also oriented if the substrate temperature is greater than 350°C.

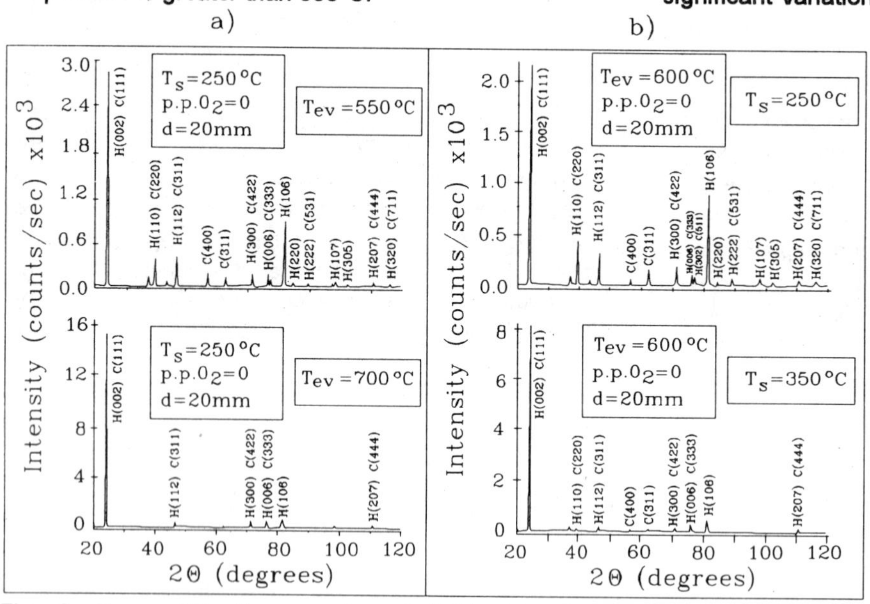

Fig. 2. X-ray diffraction spectra of CdTe thin films deposited by CSS method.

a) Source temperature,

b) Substrate temperature

Data corresponding to reflections associated to the hexagonal and cubic phases are displayed simultaneously in the diffraction spectra. XRD measurements, carried out on CdTe films deposited in an O_2 - atmosphere, indicate that the introduction of oxygen does not produce changes in the crystalline structure of the films. The formation of Te or Cd oxides, during the growth of the CdTe films, was not identified through the XRD measurements. However, the samples prepared in an O_2 - atmosphere, present a higher degree of preferential orientation along the H(002) direction than those prepared without O_2 indicating that the introduction of oxygen enhances the growth with preferred orientation along the c - axis, perpendicular to the substrate's surface. It was observed in the samples prepared in O_2 atmosphere, a small increase of the lattice constants, compared to the values of the lattice constants of CdTe films deposited without oxygen.

3.2 Electrical Characteristics of CdTe Films

Fig. 3 shows the influence of the oxygen partial pressure and substrate temperature, on the resistivity of the CdTe films. The results presented in the fig. 3 show that an increment in the substrate temperature leads to a decrease of about two orders of magnitude of the resistivity, which can be attributed to an enhancement of the preferential orientation along the direction H(002). From these results it can be concluded that the most significant variation on the resistivity of the CdTe film is caused by changes in the oxygen partial pressure . The presence of O_2 within the reaction chamber leads to an additional increment of the conductivity, probably because, under the conditions present in the chamber, the oxygen reacts with Te to produce TeO_2 which is evacuated from the chamber by the pumping system, favoring in this way the growth of Cd-rich CdTe layers.

3.3 Morphology Studies

The CdTe films were studied by SEM in order to determine the grain size and define the grain shapes and orientations. Fig. 4 shows the influence of the main deposition parameters on the grain size. The results indicate that the grains are large, with sizes greater than 1,5 μm and strongly dependent on the evaporation and substrate temperatures, whereas the O_2 partial pressure affects weakly the grain size. At substrate temperatures below 250°C, the samples present a large grain size, but are non-homogeneous in grain size (sizes range between 3 and 15 μm) and morphology. During this work, it was found that some samples deposited at source temperatures above 750°C (substrate temperature of 350°C) presented grain sizes of up to 50 μm. Figs. 5a and 5b show micrographs corresponding to CdTe films deposited at a substrate temperature of 200°C (T_{ev}= 600°C) and source temperature of 750 (T_S= 350°C) respectively. From the pictures we can see that the grains are polyhedral with well crystallized faces.

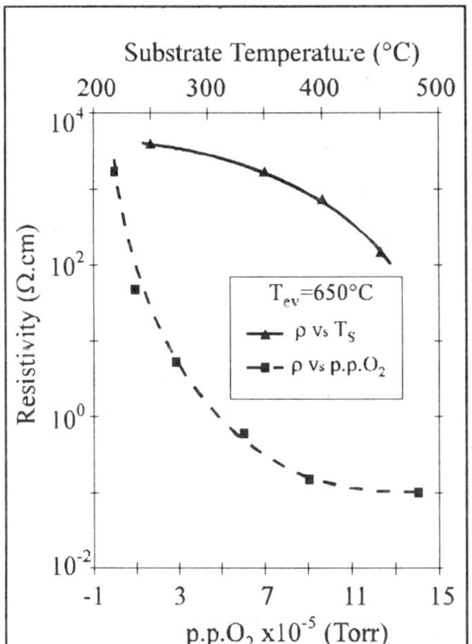

Fig. 3: Variation of the resistivity of CdTe films as a function of the substrate temperature and oxygen partial pressure.

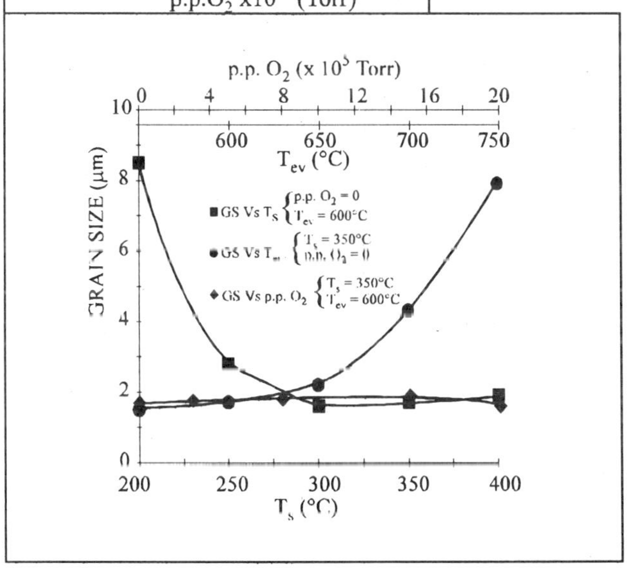

Fig. 4: Variation of the grain size of CdTe films deposited by CSS method, in dependence of the evaporation temperature, substrate temperature and O_2 - partial pressure.

a)

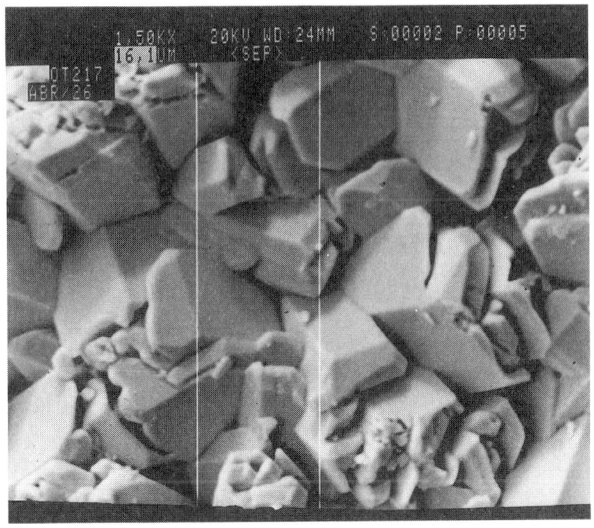

b)

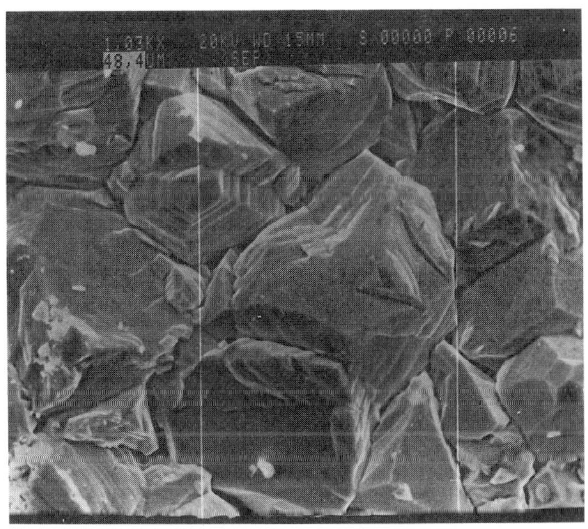

Fig. 5: SEM pictures of CdTe thin films grown by CSS method. a) T_{source}= 600°C, $T_{subst.}$=200°C b) T_{source}= 750 °C, $T_{subst.}$=350°C

3.4 Electron Microprobe Analysis

Quantitative analysis of the chemical composition of the CdTe films were carried out with an electron microprobe, using pure Cd and Te as standards. Scanning large areas on the films, it was observed that the composition is rather uniform and in some cases close to the stoichiometry. Variations in the concentration of Cd and Te of about ±1% were observed, which can be attributed to irregularities of the film surfaces.

Fig. 6 shows the influence of the substrate and evaporation temperatures and O_2 - partial pressure on the Te content in the CdTe films.

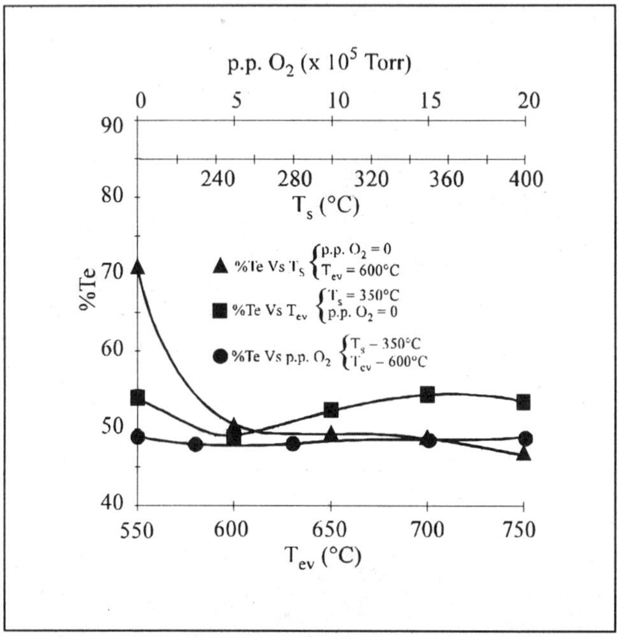

Fig. 6: Influence of evaporation temperature, substrate temperature and O_2 - partial pressure on the Te - concentration in the CdTe films deposited by the CSS method.

The results indicate that the chemical composition of the CdTe films is significantly influenced by the deposition parameters. At substrate temperatures lower than 250°C the CdTe thin films are Te-rich and the Te-concentration increases strongly by decreasing the substrate temperature .

The introduction of oxygen during the preparation of the samples, gives rise to a light reduction of the Te content in the CdTe thin films, which could be responsible for the reduction of the resistivity shown in fig. 3.

Changes in the stoichiometry of the samples were also observed varying with the evaporation temperature. Samples deposited at temperatures around 600°C present a composition close to the stoichiometry, whereas the CdTe deposited at temperatures different of 600°C are Te-rich. The changes in the chemical composition of the CdTe thin films, are caused by variations in the substrate and evaporation temperatures, and could be attributed to the different sublimation temperatures and condensation rates of the Cd and Te species.

4. CONCLUSIONS

CdTe polycrystalline thin films were grown by using a novel procedure based on the CSS technique, a sample method which can produce low cost, large area homogeneous samples. The films presented a crystalline structure composed of a mixture of the hexagonal and cubic phases. In general, at evaporation temperatures greater than 600°C and substrate temperatures over 350° C, the films grow with preferred orientation along the c-axis, perpendicular to the substrate surface.

The grain size and chemical composition of the CdTe thin films are strongly by the substrate (and source) temperature. Samples with grain size of about 50μm were obtained in this work.

The introduction of a small quantity of O_2 during the deposition of the CdTe thin films, improved the crystallographic properties and the electrical conductivity of the samples without deteriorating the optical properties of the films.

ACKNOWLEDGEMENTS:

This work was supported by COLCIENCIAS (Contract N° 075-93) and Universidad Nacional de Colombia

5. BIBLIOGRAPHY

[1] K. Zweibel, H. S. Ullal, R. L. Mitchell, *21st IEEE Photov. Spec. Conf.*, Kissimee, 1990
[2] T.L. Chu, et al., *J. Appl. Phys.* **70**(12), (1991)7608
[3] G. Fulop et al., *Appl. Phys. Lett.* **40**(1982)327
[4] T.L. Chu, et al., *Solar Cells*, **27**(1987)73
[5] T.L. Chu, et al., *J. Appl. Phys.* **58**(1985)1349
[6] H. Uda et al., Proc. *16th IEEE Photov. Spec. Conf.*, San Diego (1982)801
[7] Y.S. Tyan and E.A. Perez-Albuerne, *16th IEEE Photov. Spec. Conf.*, San Diego (1982) 794

ENVIRONMENTALLY RESPONSIBLE PRODUCTION, USE AND DISPOSITION OF Cd-BEARING PV MODULES

Richard A. Sasala and Theodore Zhou
Solar Cells, Inc., Toledo, OH 43607

Walter M. Kocher
Department of Civil Engineering, Cleveland State University
Cleveland, OH 44115

ABSTRACT

Cadmium telluride (CdTe) has many physical properties which contribute to its great promise as a solar cell material. In order to fully develop its market potential, however, it is critical that any concerns of an environmental nature be fully addressed. In this paper, a method is presented that simply and safely deals with off-specification modules and process by-products at an estimated cost of $0.05 / watt or less. The method includes an acid etch of the CdTe substrates and further recovery by means of an ion exchange process.

INTRODUCTION

Cadmium telluride (CdTe) exhibits many physical qualities which will establish its position as a leader among polycrystalline thin film PV materials.[1-6] In order to fully develop its market position, however, it will be necessary to satisfactorily answer any concerns that might exist of an environmental nature due to the utilization of cadmium in the manufacturing process and end product. In this paper, Solar Cells Inc. (SCI) presents evidence that CdTe modules can be safely produced and used throughout the entire life-cycle. It is also shown that process wastes, off-specification modules, as well as product at the end of its life cycle can be disposed in a simple, safe, and cost effective manner. To establish this position, SCI has conducted research of procedures for dealing with the technical aspects of environmental and regulatory concerns relating to the manufacture, deployment and ultimate disposition of Cd-bearing PV modules. A wide variety of treatment options exist including recycling, conversion of cadmium compounds to benign forms, and disposal as regulated wastes, each with their own environmental and economic ramifications. Regulations related to disposal of Cd-bearing wastes may apply to process by-products, work-in-progress modules, and substandard modules. In most cases the Cd content of these regulated wastes is only a few ppM - just enough to trigger the regulations but far too low for normal recycling operations. For example, a broken work-in-progress module may contain 4 to 8 grams Cd out of a 10 kg module. Thus, techniques are being developed to separate the Cd compounds from the non-regulated materials. This important step not only minimizes the cost of disposal, but also enables recycling of cadmium and of many other unregulated module component materials.

In summary, this paper will highlight our present ability to contain Cd within our facility and project this ability for containment in the field. The bulk of the paper will discuss completed exploratory experiments which show the possibility of potentially low cost and environmentally friendly processing of off specification modules.

PRESENT SAFETY AND HEALTH RECORD

From the beginning of its PV program, SCI has pursued a series of rigorous safety and environmental standards. This program includes monitoring Cd levels in the air, on work surfaces, in process exhausts and pump oils. Every attempt is made to maintain zero emission of regulated materials. Air exhaust ducts from process equipment and laboratory hoods are filtered and tested for Cd content. These measurements indicate that no measurable Cd powder has been released into the environment. In addition, all personnel are given a complete physical at the commencement of work at SCI and then periodic urine tests for Cd content. In the near future, the routine personnel medical tests will be amplified to include β_2-microglobulin in urine and blood tests in order to monitor possible long term accumulations in the body. To date, no personnel have been exposed to cadmium at or above the Personal Exposure Limits, PEL, established by OSHA. Furthermore, there have been no measurable discharges of regulated materials from SCI's facilities. This performance has been accomplished primarily through engineering controls but employee training and work practice procedures also contribute.[7]

PRODUCT RECYCLING EXPERIMENTS

Exploratory experiments of methods to effectively and cheaply recycle off-specification modules have been performed. The general strategy investigated is to first separate Cd-containing semiconductor films from the SnO_2- coated glass by means of an acid etch and to then treat the effluent. Nitric (HNO_3), hydrochloric (HCl), and ethylenediaminetetraacetic (EDTA) acids were tested as potential etchants. EDTA is a chelating agent which reacts with the metals to form soluble complexes. Samples were cut from 7200 cm^2 modules in six-inch by six inch squares - but these samples generally broke into three to six pieces in the process. Agitation was provided using a shaker table or a standard toxicity characteristic leaching procedure (TCLP) agitator unit.[8]

CH3365-4/94/0000-0311 $4.00 © 1994 IEEE

First WCPEC; Dec. 5–9, 1994; Hawaii

At the end of the agitation period, the glass was removed from the extraction vessel and the vessel contents were filtered.[9] This filtrate was collected and analyzed to determine the soluble Cd levels from the extraction. The filtered solids were then thoroughly washed with a measured amount of 15% acid solution to dissolve the metals. These solutions were analyzed to determine the amount of metals removed from the glass in solid form. Cd concentration in the solutions was measured by using a Perkin Elmer 3100 flame atomic absorption unit with a hollow cathode lamp as a light source. The ultimate resolution of the apparatus for Cd is 0.03 mg/l.[10] Further recovery of the Cd from the acid extraction fluid used is desirable. Two techniques were employed to accomplish this task, electrolytic deposition and ion exchange. The electrolytic deposition apparatus was a simple bath (~0.5l) with an infeed, an outfeed and carbon electrodes. The ion exchange experiment was conducted with a bench scale column (~ 1l) in an upflow mode and a flow rate per unit bed volume of 35 $l \cdot min^{-1} \cdot m^{-3}$.

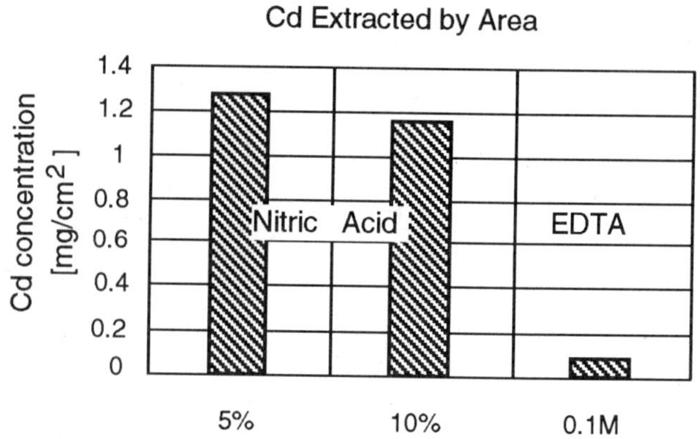

Fig. 1 Total Cd extracted from CdS/CdTe thin films on glass.

RESULTS

Stripping of Glass

The initial trials were performed with nitric acid and EDTA. The results are shown in Fig. 1. It is apparent that EDTA at 0.1 M concentration is a rather ineffective etchant. However, the nitric acid in both 5 and 10 % concentration levels is very effective and nearly all of the Cd is removed from the glass. For the 5 ± 1 μm films used in these experiments and assuming a CdTe density of 5.8 g/cm^3, 1.3 ± 0.3 mg/cm^2 of Cd is initially present on the glass. The fact that more Cd was removed with the 5% acid is due to nonuniformities of the film thickness among the samples tested and not due to any greater effectiveness of the 5% acid. An important point is that the graph displays the total Cd removed including both dissolved Cd and solid Cd. While Cd is more soluble in the more concentrated acid, the weaker acid is still effective in Cd removal due to CdTe flaking off the glass, presumably by dissolving the underlying CdS layer. This result is expected as HNO_3 is an oxidizing acid, when in concentrations greater than 2M, and therefore attacks the CdTe directly. Dilute HNO_3 and HCl, on the other hand, only dissolve the CdS. Fig. 2 shows the amount of soluble Cd removed with agitation for 30 and 60 minutes with 2, 5, and 10% concentration nitric acid on six separate samples. Note that the 10% solution dissolves the majority of the Cd within 30 minutes and that the amount dissolved is far greater than with the 2 and 5% solutions. Within the normal variations of the measurement, the samples at 60 minutes had no change in the amount of soluble Cd. Fig. 3 shows the kinetics of the nitric acid extractions for 1, 2, and 5% solutions showing the soluble Cd concentration. The 5% solution quickly saturates in 10 - 20 minutes, whereas the 1% solution dissolves only a small amount even for extended periods of time. The 2% solution, as expected is halfway between showing an increasing Cd concentration but at a far slower rate than the 5% solution. A similar experiment was performed using HCl with similar results except the rates were much closer for the three concentrations.

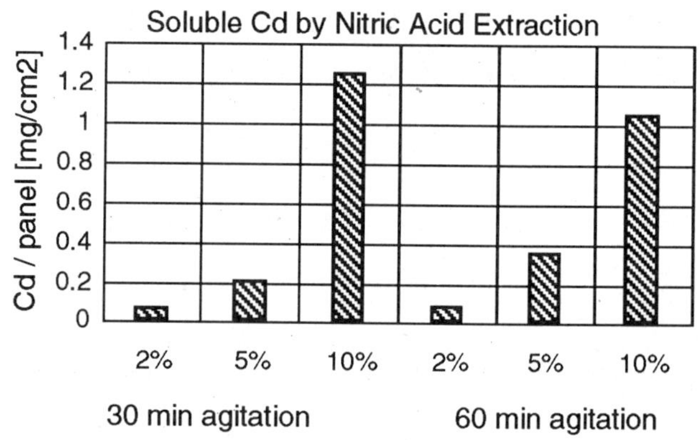

Fig. 2 Comparison of soluble Cd concentration for 2, 5 and 10% HNO_3 extraction with 30 and 60 minutes of agitation.

In addition to determining the rate that Cd dissolves into various concentrations of acid, it is also important to determine the total rate of removal of Cd. Fig. 4 shows the total and soluble Cd extracted over a 90 minute time period. The soluble concentration is similar to that shown in Fig. 3, but the total concentration is much greater due to solid CdTe flakes being removed from the glass. In contrast, Fig. 5 shows the same exercise but using 5% nitric acid and nearly all of the Cd extracted is soluble. The experiment was repeated for 2, 3, and 4% nitric acid as well and the results fall between the 1 and 5% solutions as expected. These results have important ramifications for subsequent processing steps. Specifically, removal of suspended solids can be accomplished by physical processes whereas extracting soluble Cd requires an electrolytic deposition, ion exchange, or some other similar process. Physical processes are , in principle, a more rudimentary technique and therefore more economical. However, removing the Cd as a solid is a slower process which itself has a cost which has to be considered.

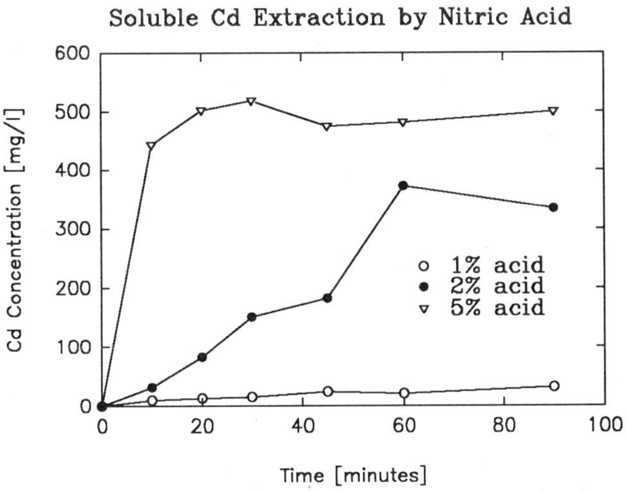

Fig. 3. Soluble Cd concentration vs. time for three CdTe/CdS samples using 1, 2, and 5% HNO_3 acid.

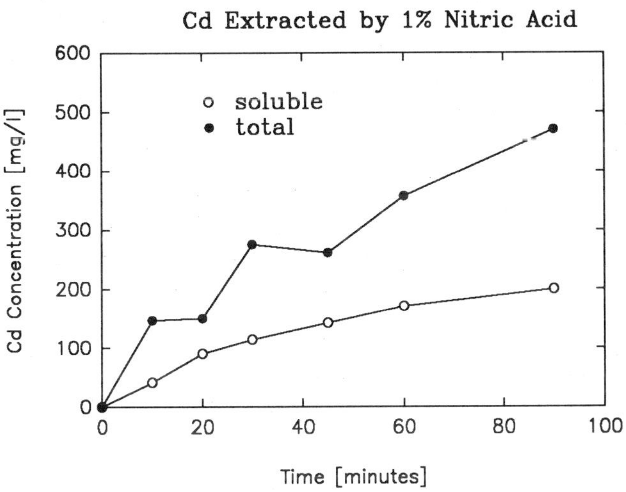

Fig. 4. Total and soluble Cd extracted from substrate using 1% HNO_3.

To determine if the acid etch removes enough Cd to allow the glass substrate to be classified as non hazardous, samples were subjected to TCLP tests using an acetic acid leachate, the more rigorous of the TCLP tests. Present regulations permit a maximum of 1.0 mg/l of Cd in the leachate.[11] From Fig. 6, it is clear that the treated panels pass the TCLP tests, all by a wide margin with the exception of the 0.1% nitric acid solution. It is further noted that the HCl treated glass has substantially lower TCLP readings for the same acid concentrations as the HNO_3 treated films. One should also note that the readings for the HCl treatment are well below the TCLP level proposed in Germany, which is 0.10 mg/l.[12]

Treatment of Extraction Fluids

Removal of the regulated materials from the glass substrate by acid etching leaves yet another challenge. The Cd concentration is still too low for economical recycling by conventional means; therefore, further processing is required. The two approaches taken were

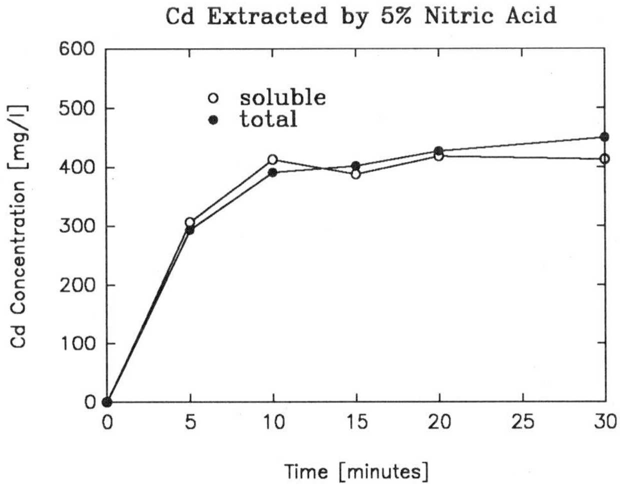

Fig. 5. Total and soluble Cd extracted from substrate using 5% HNO_3.

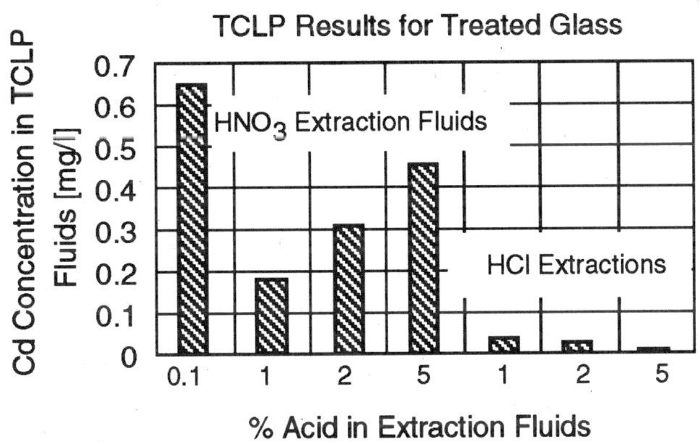

Fig. 6. Results of Toxicity Characteristic leaching procedure tests for glass treated with HNO_3 and HCl.

electrolytic deposition and ion exchange. The electrolytic deposition results were discouraging. Only a small fraction of the soluble Cd was recovered from the extraction fluids. A number of combinations of voltage to the electrodes and pH of the extraction fluids were tried with only marginal improvement in Cd removal. Since electroplating of Cd is commonly performed, further experiments are needed.

The results of the ion exchange are much more encouraging. Two commercially available cation exchange resins were tested with similar results. The first resin is Amberlite IRC-718 produced by Rohm and Haas Company and is a weakly acidic chelating exchange resin with a high affinity for heavy metal cations. The resin can only be used for aqueous solutions with no more than 0.5% nitric acid content. Fig. 7 shows a comparison of the Cd concentration of the influent and effluent for a period of 50 minutes. The resin removes over 99.9% of the Cd from the influent over the test period. The expected capacity of the resin is about 2 meq/ml resin. A second ion exchange resin produced by Sybron Chemicals Inc. was also tested. This resin has the advantage of tolerating higher influent acid concentrations. The performance was comparable to the Amberlite. The disadvantage of the ion exchange

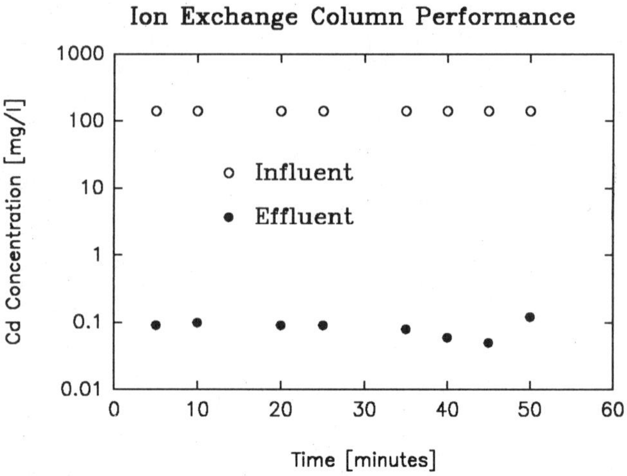

Fig. 7. Comparison of Cd concentration of influent and effluent through an ion exchange column with Amberlite resin.

process is the high cost of the resin. If, however, a weak acid is used to strip a high percentage of the CdTe off the glass in solid form that can be removed by physical processes, the amount of resin required can be minimized and therefore the cost minimized.

SUMMARY

Solar Cells Inc. intends to take a proactive stance to neutralize any market concern that may exist due to the use of Cd in the manufacturing of its product. SCI has managed and will continue to maintain a tight control on Cd in its facilities and among its personnel. In addition, the company is conducting research to explore the potential of recycling modules when that step becomes necessary due to off-specification or out-dated products. The procedure is to etch the CdTe and CdS with a combination of nitric and hydrochloric acids which removes the Cd in both solid and soluble forms. The solid Cd is then separated by a physical process such as filtering. The extraction fluid is then passed through an ion exchange column to remove the soluble Cd. The solid Cd and Cd concentrated in the ion exchange resin can then be recycled cost effectively in a conventional manner. SCI estimates that these procedures can be accomplished for $0.05 / watt or less which SCI believes to be the most cost effective technique to date.

ACKNOWLEDGEMENTS

The authors gratefully acknowledge Dr. Peter Meyers of IEC for his initiative on this work. The technical assistance provided by Al Paskonis at Cleveland Advanced Manufacturing Program (CAMP) Environmental Services Program is also sincerely appreciated. This work has been partially supported by the NREL subcontract ZR-1-11059-1.

REFERENCES

[1] Y. Tyan and E.A Perez-Albuerne, "Efficient Thin Film CdS/CdTe Solar Cells". *Sixteenth IEEE PVSC,* 1982, pp. 794-800.

[2] K.W. Mitchell, C. Eberspacher, F. Cohen, J. Avery, G. Duran, and W. Bottenberg, "Progress Towards High Efficiency Thin Film CdTe Solar Cells", *Eighteenth IEEE PVSC,* 1985, pp. 1360-1364.

[3] P.V. Meyers, Design of a Thin Film CdTe Solar Cell", *Solar Cells,* **23**, 1988, pp. 59-67.

[4] J.F. Jordan and S.P. Albright, "Large Area CdS/CdTe Photovoltaic Cells", *Solar Cells,* **23**, 1988, pp. 107-113.

[5] A.K. Turner, et. al., "Stable, High Efficiency Thin Film Solar Cells Produced by Electrodeposition of Cadmium Telluride", *Technical Digest of the International PVSEC-5,* Kyoto, Japan, 1990, pp. 761-764.

[6] S. Ikegami and A. Nakano, "Effect of Oxygen on Photovoltaic properties of Screen-Printed CdS/CdTe Solar Cell and Module", *Int. J. Solar Energy,* 1992, Vol. 12, pp. 53-65.

[7] P.V. Meyers and D.W. Sandwisch, "Polycrystalline Thin Film PV Panel Manufacturing Program", *AIP Conference Proceedings* **306** 12th NREL Photovoltaic Program Review 1993, pp. 265-71.

[8] 40 Code of Federal Regulations 261 "Toxicity Characteristic Leaching Procedure", FR Vol. 55, **61**, March 1990.

[9] A.E. Greenberg, L.S. Clesceri and A.D. Eaton, eds., "Standard Methods for the Examination of Water and Wastewater", 18th ed., American Public Health Association, New York, 1992.

[10] Perkin Elmer, "Analytical Methods for Atomic Absorption Spectrophotometry", January 1992.

[11] P.D. Moskowitz and K. Zweibel, "Recycling of Cadmium and Selenium from Photovoltaic Modules and Manufacturing Wastes: A Workshop Report" Brookhaven National Laboratory Informal Report 47787, p. 5.

[12] Hartmut Steinberger, private communication.

PHOTOVOLTAIC PROPERTIES OF SINTERED CdS/CdTe SOLAR CELLS WITH TCO ELECTRODES

Hye Dong Kim, Dong Seop Kim*, Jin Soo Song**, Byung Tae Ahn

Dept. of Electronic Materials Engineering, Korea Advanced Institute of Science and Technology, 373-1, Koosung-dong, Yusung-gu,Taejon, Korea

*Energy/Environment Lab. Materials and Devices Research Center, Samsung Advanced Institute of Technology, Suwon, Korea

**Korea Institute of Energy Research, Taejon, Korea

ABSTRACT

TCO electrodes such as ITO and tin oxide were applied to sintered all-polycrystalline CdS/CdTe solar cells for large-area cell fabrication. Adopting TCO to sintered CdS/CdTe cells caused the increase of leakage current. The leakage current was reduced by using (Cd+Te) mixed slurry instead of CdTe compound slurry to form CdTe layer because (Cd+Te) mixed slurry can make more improved CdS/CdTe junction during synthesis of CdTe compound. The CdS/CdTe cell with ITO electrode showed slightly larger leakage current and that with SnO_2 electrode showed much larger leakage current, compared to the cell without TCO electrode. However, the reduction of the series resistance of the cell with ITO improved the cell parameters such as J_{sc} and FF, resulting in maintaining high conversion efficiency of the cells with large area. As a result, the ITO/CdS/CdTe cell with 90 mm^2 area has an efficiency of 9.4% under irradiation with intensity of 50 mW/cm^2.

INTRODUCTION

The method of coating and sintering is reported [1, 2] to be one of the most efficient ways to produce low-cost CdS/CdTe heterojunction solar cells. The screen printed and sintered CdS/CdTe solar cells with an active area of 30 mm^2 had efficiencies over 12% [3, 4]. Kim and Im [5] reported that there hasbeen a limit to enlarge the size of active area of the CdS/CdTe solar cell because of the large sheet resistance of CdS layer. It is essential to enlarge the active area of solar cell for terrestrial application to increases total area efficiency. It is, therefore, interesting to use a transparent conducting oxide (TCO) as an electrode to reduce the series resistance. High efficiency thin film CdS/CdTe solar cells with TCO electrodes have been reported [6, 7, 8]. However, there was little investigation on the application of TCO for the CdS/CdTe solar cells fabricated by the coating and sintering method.

This paper reports the photovoltaic properties of the sintered CdS/CdTe solar cell with an ITO and SnO_2 electrodes and discusses factors that can affect cell parameters. The application of ITO has less degraded the cell efficiency with increasing cell length because of smaller series resistance, resulting in the improvement of the total area efficiency with the large-area cells.

EXPERIMENTAL

A slurry consisting of CdS and $CdCl_2$ powder mixed with propylene glycol (binder) was coated on to a borosilicate, ITO-coated and SnO_2-coated borosilicate glass substrates using a screen printer. The coated substrates were dried at 120 °C for 2 h in air to remove the binder and were sintered at 600 °C for 90 min in a tube furnace with a nitrogen flow rate of 250 $cm^3 \cdot min^{-1}$. The sintered CdS films were then annealed at 600 °C for 15 min in an open tray to remove the residual $CdCl_2$.

Either CdTe or (Cd+Te) slurry was coated on to the sintered CdS and TCO/CdS layers, using a screen printer. The samples were then pushed slowly into a tube furnace at the rate of 14 $°C \cdot min^{-1}$ up to 500 °C to form the CdTe compound and then were sintered at 625 °C for 60 min with a nitrogen flow rate of 250 $cm^3 \cdot min^{-1}$. The (Cd+Te) slurry is much cheaper than CdTe slurry and the resultant CdTe layer provides better electrical characteristics. The detailed preparation procedure had been reported in ref. [9, 10]. Scanning electron microscope was used to analyze microstructure. Transmission spectra were obtained with a Si photodiode, by placing it in

CH3365-4/94/0000-0315 $4.00 © 1994 IEEE

First WCPEC; Dec. 5–9, 1994; Hawaii

contact with the CdS and TCO/CdS films, and measuring the photodiode current with and without the sintered films in the optical path. Ohmic contacts were provided by applying In-Ag paint to n-type CdS layer and carbon paint to p-type CdTe films followed by annealing at 350 °C for 10 min in nitrogen.

Fig. 1(a) and 1(b) show the SEM cross sections of sintered glass/CdTe interface, prepared from CdTe slurry and (Cd+Te) mixed slurry, respectivley. More dense CdTe interface was formed when (Cd+Te) mixed slurry was used. We also expect that the denser CdS/CdTe interface can be formed from (Cd+Te) mixed slurry.

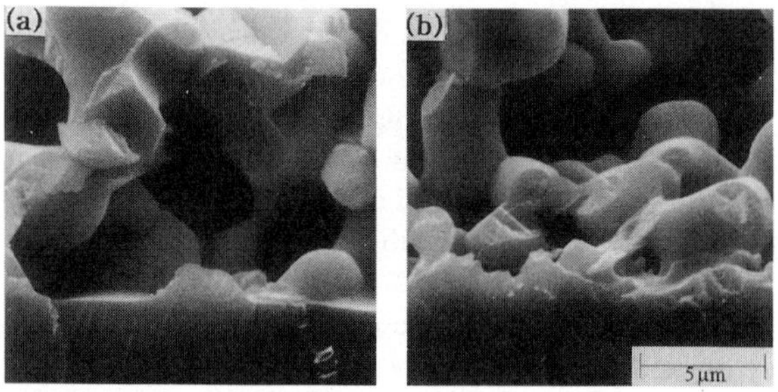

Fig 1. SEM cross sections of sintered glass/CdTe interface, prepared from CdTe slurry (a) and (Cd+Te) slurry (b).

Fig. 2 shows the schematic structure of cell fabricated in this study.

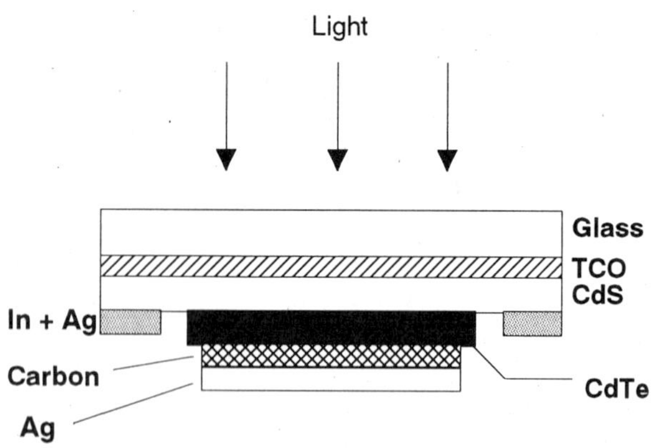

Fig. 2. The sructures of TCO/CdS/CdTe solar cell.

RESULTS AND DISCUSSION

Fig. 3 shows the dark J-V characteristics of CdS/CdTe, ITO/CdS/CdTe and F-doped SnO_2 /CdS/CdTe cells with the cell size of L X W = 10 X 3 mm^2. The deviations of the linear J-V curve at high bias voltage and at low bias voltage reflect the effects of series resistance and shunt resistance, respectively. The ITO/CdS/CdTe cell shows a smaller series resistance and also a smaller shunt resistance compared to the CdS/CdTe cell. The value of J_o and A of the cell with ITO electrode, obtained by the extrapolation from the linear part (about 0.5 V) of J-V curve, are similar to those of cell without ITO. However, the leakage current of the SnO_2/CdS/CdTe cell is very high and the values of J_o and A are somewhat different from those of CdS/CdTe and ITO/CdS/CdTe cells. The results suggest that the junction property itself is degraded slightly by applying the ITO, while it is degraded significantly by applying the SnO_2 electrode.

Fig. 4 shows the series resistances of the CdS/CdTe, SnO_2/CdS/CdTe and ITO/CdS/CdTe cells as a function of cell length. The cell width is fixed at 10 mm.

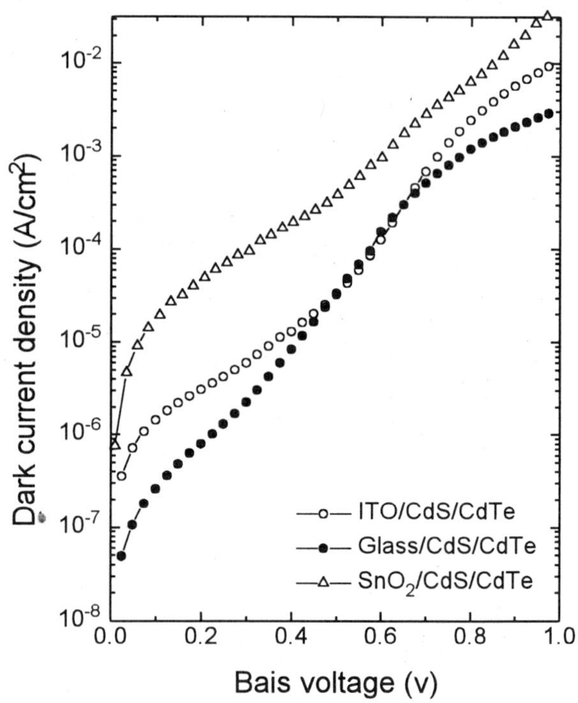

Fig. 3. Dark J-V characteristics of sintered CdS/CdTe solar cells with or without TCO layers with L x W = 10 x 3 mm^2.

The application of TCO electrodes significantly reduced the series resistance of CdS/CdTe cells. As the cell length increases from 3 to 9 mm, the resistance of the CdS/CdTe cells increases from 5 to 15 Ω·cm², whereas those of the ITO/CdS/CdTe and SnO_2/CdS/CdTe cells are nearly unchanged with the value of 4~6 Ω·cm² because of the low resistivity of the TCO films.

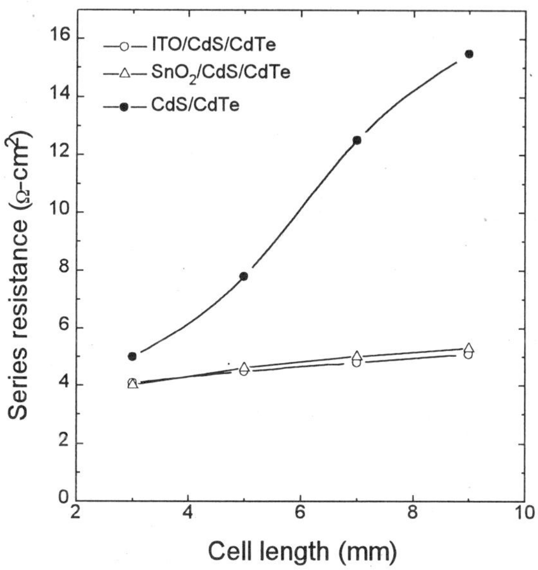

Fig. 4. Series resistances of sintered CdS /CdTe solar cells with or without TCO layers as a function of cell length. The cell width was 10 mm.

Fig. 5 shows the cell parameters and cell efficiencies of the sintered CdS/CdTe, SnO_2/CdS/CdTe and ITO/CdS/CdTe cells as a function of cell length measured under 50 mW/cm² tungsten light. The open circuit voltage (V_{oc}) and fill factor (*FF*) of SnO_2/CdS/CdTe cells are much lower than those of CdS/CdTe or ITO/CdS/CdTe cells. It was mainly due to the low shunt resistance of SnO_2/CdS/CdTe cells. According to the C-V analysis, it appears that the larger leakage current in SnO_2/CdS/CdTe cells is due to the higher concentration of interface states. The absolute values of the short-circuit current density and fill factor of the ITO/CdS/CdTe cells are higher than those of the CdS/CdTe cells. However, the open-circuit voltages of the ITO/CdS/CdTe cells are slightly lower than those of the CdS/CdTe cells. As the cell length increases, the short-circuit current and fill factor of the cells with ITO electrodes are changed little because of the low series resistances.

As the cell length increased from 3 to 9 mm, the efficiency of the CdS/CdTe solar cells with ITO electrode slightly decreased from 10.5 to 9.4%, while that of cells without ITO decreased from 10.4 to 8.4% under the illumination of 50 mW/cm². The efficiency of the cells with ITO was less degraded with increasing cell length because the short-circuit current and fill factor of the cell with ITO changed little.

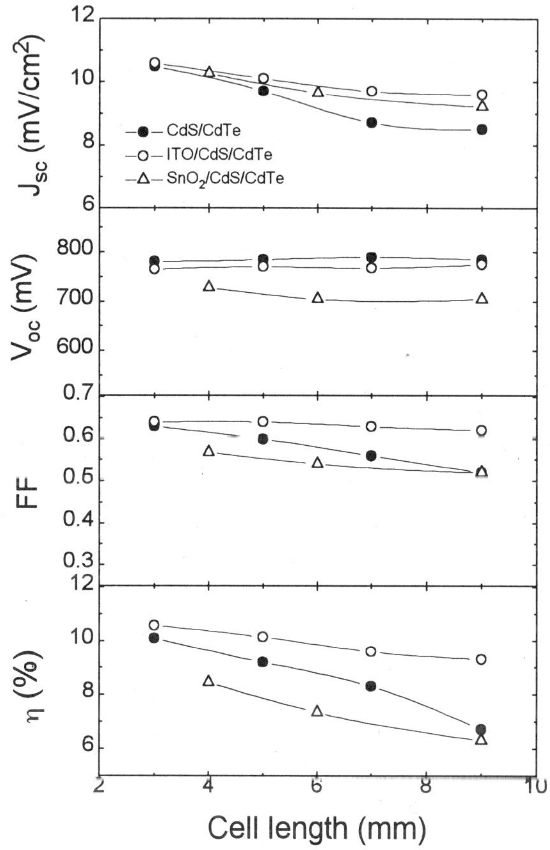

Fig. 5. Cell parameters of the sintered CdS/CdTe cells with or without TCO electrode as a function of cell length under 50 mW/cm² tungsten light. The cell width was 10 mm.

Fig. 6 shows the spectral response of the quantum efficiency of the size of L x W = 9 x 10 mm². The spectral response of the ITO/CdS/CdTe cells is significantly higher than that of the CdS/CdTe cells. It appears that the difference is due to the series resistance of the cell that increases with larger cell area.

Fig. 7 shows the total area (active + contact area) efficiencies of the sintered CdS/CdTe and ITO/CdS/CdTe cells as a function of cell length. The CdS/CdTe cells without ITO had a maximum value of

6.5% at only 4 mm length. However, that of the cells with ITO electrode increased to 7.6% at 9 mm length.

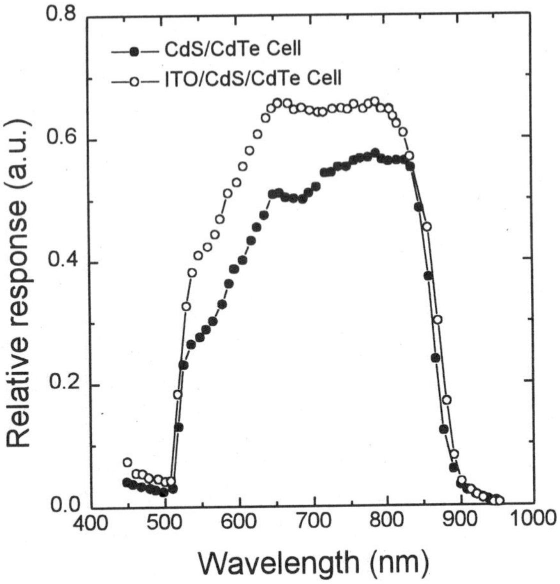

Fig. 6. Spectral responses of the CdS/CdTe and ITO/CdS/CdTe cells with WxL = 9x10 mm^2 as a function of wavelength.

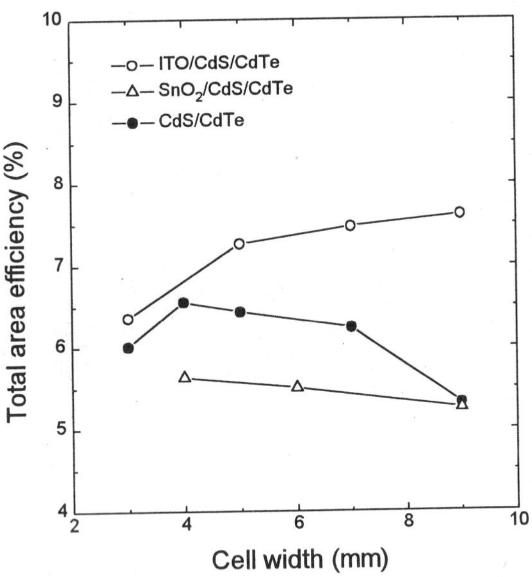

Fig. 7. Total area efficiency of the sintered CdS /CdTe cells as a function of cell length. The cell width was 10 mm.

Thus, adopting ITO electrode enabled us to improve the total area efficiency of the sintered CdS/CdTe solar cells with large active area.

CONCLUSION

To obtain high efficiency solar cells with large active area, we have adopted ITO and SnO_2 -coated glass as a TCO substrate for screen printed/sintered CdS/CdTe solar cells. The ITO electrode successfully reduced the series resistance with a slight increase in the leakage current, while SnO_2 electrode showed a significant increase in the leakage current. The reduction of the series resistance of the cell with ITO improved the cell parameters such as J_{sc} and *FF*, resulting in maintaining high conversion efficiency of the cells with large area. As a result, the total area efficiency of the ITO/CdS/CdTe cell with L X W = 9 X 10 mm^2 was improved to 7.6%.

ACKNOWLEDGMENT

Authors are greatly indebted to late professor Ho Bin Im at KAIST, who passed away on December 26, 1993. He has been work for sintered CdS/CdTe solar cell for more than 10 years.

REFFERENCE

[1] J. S. Lee and H. B. Im, *J. Mater. Sci.*, 21, 1986, p980

[2] S. Ikegami, *Solar Cells*, 23, 1988, p89

[3] Y. S. Seol and H.B. Im, *Proc. 9th European Photovoltaic Solar energy Conf.*, (Freiburg, Germany, November 25-29, 1989) p294

[4] J. T. Moon, K. C. Park and H. B. Im, *Solar Energy Materials*, 18, 1988, p53

[5] H. S. Kim and H. B. Im, *Thin Solid Films*, 214, 1992, p207

[6] S. P. Albright, B. Ackerman and J. F. Jordan, *IEEE Trans. on Electron Devices.,* 37, 1990, p434

[7] J. F. Jordan and S. P. Albright, *Solar cells*, 23, 1988, p107

[8] K. W. Mitchell, C. Eberspacher, F. Cohen, J. Avery, G. Duran and W. Bottenberg, *Solar Cells*, 23, 1988, p49

[9] D. S. Kim, S. Y. Kim, B. T. Ahn, H. B. Im, *J. Mater. Sci., Mater. in Electronics*, 5, 1994, p17

[10] D. S. Kim and H. B. Im, *Powder Metallurgy*, 35, 1992, p57

Surface Cleaning of the Screen Printed CdS Film

T. Aramoto, H. Higuchi, T. Arita, T. Nishio, H. Ohyama, A. Hanafusa, S. Kumazawa H. Takada
, K. Omura, N. Nakayama, H. Takakura* and M. Murozono
Matsushita Battery Industrial Co. Ltd., PV Research & Development Center
Matsushita-Cho-1, Moriguchi, Osaka 570, Japan
TEL 06-991-1141 FAX 06-994-7056
*Toyama Prefectural Univ.,Kosugi-cho,Toyama 939-03,Japan
TEL 0766-56-7500

ABSTRACT

Surface analysis has revealed the existence of many kinds of contaminations on the surface of screen printed and sintered CdS films. These contaminations are the main cause of restrictions on the performance of the solar cell. Rinsing and etching treatments of the sintered CdS film have been studied for the removal of these contaminations. 17mV increase on Voc and a 50% of reduction on Jo has been achieved from these treatments.

1 Introduction

Our solar cell producing process doesn't require any special facilities(i.e. clean manufacturing areas ,etc) whichmakes this process low cost. This method ,however ,causes the adsorption of carbon or the other contaminations from the air.

In this production process, the CdS film is sintered at a temperature of 700℃. This high temperature causes oxidation of the Cd on the surface of the CdS film, and this reaction makes the CdS film deviate from an ideal stoichiometry. We use the $CdCl_2$ for the melting point dropping agent of the CdS. The $CdCl_2$ separates and rises to the surface of the CdS film by the grain boundary diffusion during the sintering process. This unequal residue of the surface prevents the formation of an ideal CdS/CdTe junction , and acts as a recombination center of the hole electron pair. In order to solve these problems we analyzed the CdS film surface by the Electron Spectroscopy for Chemical Analysis (ESCA) method and the Thermal Desorption Spectroscopy (TDS) method.

2 Experiments

We analyzed the CdS film surface produced by the conventional screen printing and sintering process by the ESCA method. Table 1 shows the atomic concentration of the elements on the surface of the CdS film.

	(a)surface	(b)2nm below	(c)rinced surface
Cd	28.4	48.8	23.8
S	22.8	40.8	18.4
Cl	4.4	ND	3.0
C	25.1	4.9	37.9
O	11.3	4.5	13.2
Na	7.7	0.4	2.9
Ba	0.3	0.6	1.1

Table 1 (a) Atomic concentration of elements on the surface of non-treated CdS (b) 2nm below the surface
(c) deionized water rinced surface without N_2 blow .

First WCPEC; Dec. 5–9, 1994; Hawaii

Table 1 shows the atomic concentration of elements on non treated CdS surfaces.

The CdS films were produced on #7059 glass(Corning Co.Ltd). This table shows a high degree of contaminations on the surface, as was expected. It is interesting to note that we detected as much carbon as Cd and S. Oxygen was also detected. This may exist in the form of CdO. A high concentration of Ba, ingredient of the glass substrate (#7059 glass) and Na was also detected in the contaminations. However , if the thin surface layer of the CdS film was removed by a depth of a few nm by the ion sputtering, the impurity concentration would be reduced to 1/10. The distribution of the contaminations is caused by the sintering mechanism of the CdS film which deposits contaminations by the grain boundary diffusion. If we can remove the impurity by asimplified method like rinsing or etching, improved energy conversion efficiency can be expected at low cost.

We have already reported the production of the screen-printing and sintering method in detail[1].

CdS films are deposited on the glass substrates utilizing the screen printing method in the thickness of 40~ 60 μm. Ultrasonic treatment is then applied to the as-printed CdS film. [2] After the drying process the CdS films are sinterd at a temperature of 700 ℃.

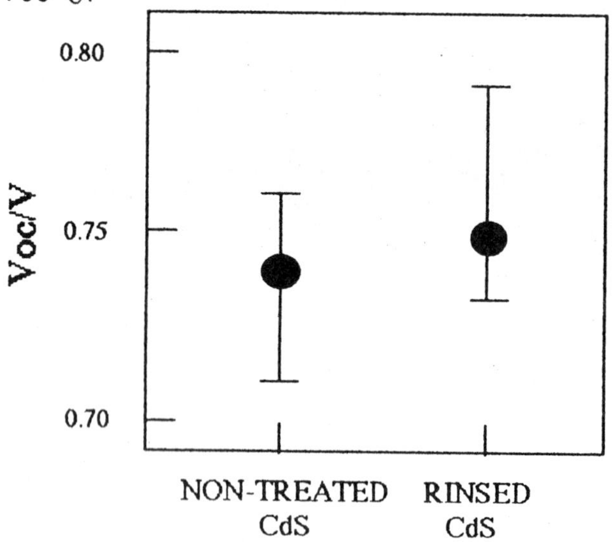

Figure 1 (a) open circuit voltage of CdS/CdTe solar cell by the conventional CdS and deionized water rinced CdS.

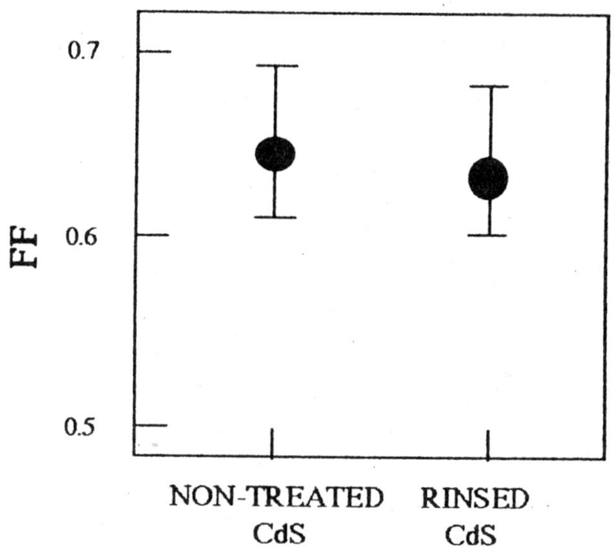

Figure 1 (b) fill factor of CdS/CdTe solar cell by the conventional CdS and deionized water rinced CdS.

In order to remove these contaminations, deionized water rinse is applied to as-sintered CdS films during the production of the CdS/CdTe solar cell,which has an area of 12cm^2. Fig. 1 shows Voc and fill factor of the CdS/CdTe solar cell by conventional screen-printed and sintered CdS films and deionized water rinsed CdS films. We can see an increase of Voc by 17mV at the water rinsed CdS films, and also a decrease of the fill factor,when compared to non-treated films under the same condition.

	non-treated CdS	rinced CdS
n	2.38	1.92
Jo	0.143e-7 A	0.764e-8 A
ΔV	32.6 mV	

Table 2 Jo,n and increase of the Voc on non- treated CdS and rinced CdS.

Table 2 shows the saturation curren t (Jo) ,diode quality factor (n) and increase of open circuit voltage(ΔV) of conventional CdS/CdTe solar cell and the one produced by deionized water rinsed CdS films. The increase of Voc is calculated from Jo and n. About 1/2 of Jo and 80% of n has been obtained by the deionized water rinse treatment. We attribute this to be an effect of impurity removal. The value of calculated increase of Voc and the observed difference of Voc in fig.1 are very close.

We think that the decrease of FF was caused by a carbon related species which was adsorped at the drying stage after rinsing. In order to eliminate the contaminations which adsorpe at this drying process ,we applied an N2 gas blow to remove the drops of water on the CdS film before the drying process. HCl and Br etching were also tried. The concentration of $HCl(H_2O)$ and $Br(CH_3OH)$ was 1% and 0.3% respectively.

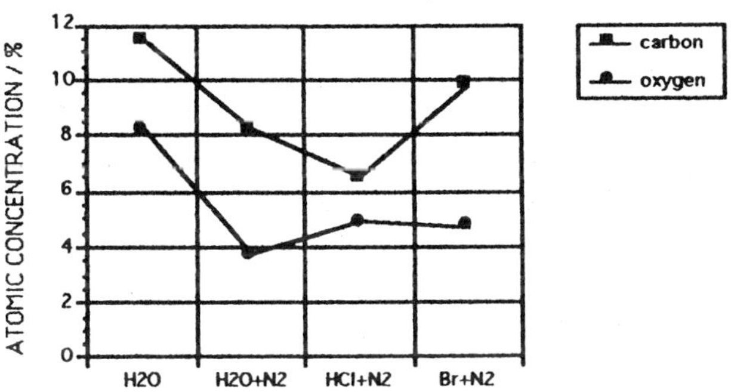

Figure 2 Atomic concentration of carbon and oxygen on the CdS surface by each treatment.

Figure 2 shows the atomic concentration of carbon and oxygen on the surface of CdS by four treatments. The ordinate shows the atomic concentration of carbon and oxygen,as an averaged value for the thickness of 100 nm from the surface. The N2 gas blow was used on the H2O rinse and two etching treatments. Each of the CdS surfaces which were treated with the N2 gas blow have fewer contaminations.

In order to inspect the impurity species, a TDS method was applied. Fig. 3 (a),(b),(c) shows the thermal profile of TDS method.

The ordinate shows the ion current that corresponds to the quantity of released elements from the surface of the CdS and the abscissa shows the temperature at which the elements desorped.

TDS is a useful method for determining what combination of contaminations exist on the surface at specific temperatures.

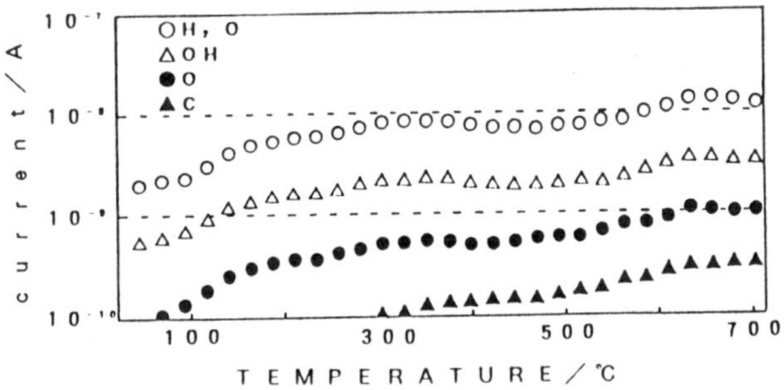

Figure 3 (a) non-treated CdS surface

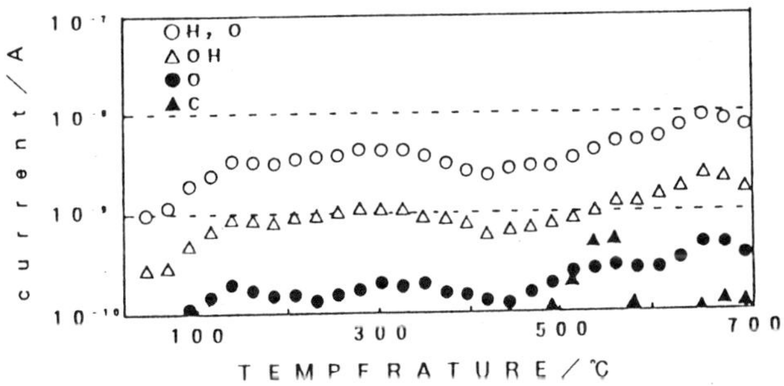

Figure 3 (b) CdS surface rinced by deionized water

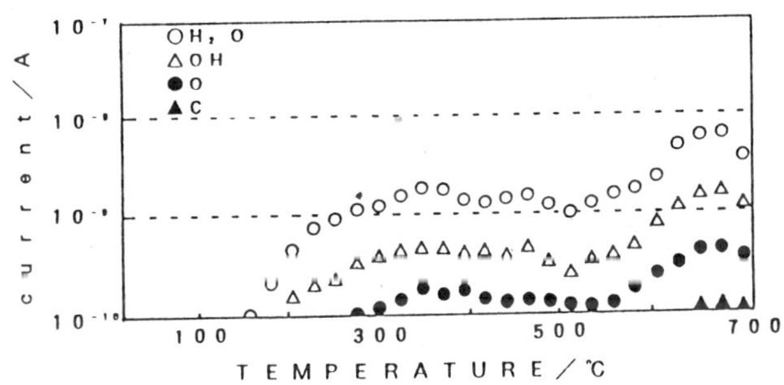

Figure 3 (c) CdS surface rinced by deionized water and N_2 blow

Fig. 3 (a) shows the TDS profile of a non treated CdS film, (b) shows a water rinsed CdS film, and (c) shows a CdS film that was water rinsed and then dried by N2 gas. We can observe a decrease in contaminations on the surface of the CdS film from (a)to (b)to (c). Fig. 3 (b) shows carbon being desorped for a wide range from 300 ℃ to 700 ℃. This profile reveals the existence of various types of carbon related species on the surface of the film. Meanwhile, Fig. 3 (b) shows the existence of specific carbon related species. Fig 3. (c) shows that there is little carbon related species. The quantity of contaminations on Fig. 3(c) are less than 1/10 that of Fig. 3 (a) throughout almost the whole temperature range . We are now trying to apply these treatments, deionized water rinse, N_2 gas blow ,HCl and Br etching, to the production of CdS/CdTe solar cells with an area of $\sim$1000cm^2. We expect to improve the solar cell caracteristics by these treatments.

3 Conclusions

The TDS analysis shows that a decrease in contaminations to a level of 1/10 can be achieved by applying the deionized water rinse, and afterwards thoroughly dried clean.

4 Acknowledgements

This work was supported by the New Energy and Industrial Technology Development Organization as a part of the New Sunshine Program under the Ministry of International Trade and Industry ,Japan.

5 References

[1] T. Arita , A. Hanafusa , N. Ueno , Y. Nishiyama , S. Kitamura , and M. Murozono Solar Energy Materials 23, p.371 (1991).
[2] T. Aramoto, H. Higuchi, T. Arita, T. Nishio, N. Nakayama, K. Hiramatu, H. Takakura, and M. Murozono , Conf. Record PVSEC-7, p513 (1993)

GROWTH OF EPITAXIAL CdTe/CdS HETEROSTRUCTURES FOR SINGLE CRYSTAL THIN FILM SOLAR CELL APPLICATIONS

K. Kessler, A.N. Tiwari, S. Blunier, and H. Zogg
AFIF (Arbeitsgemeinschaft für industrielle Forschung) at Swiss Federal Institute of Technology, ETH-Trakt Technopark, Pfingstweidstr. 30, CH-8005 Zürich, Switzerland.

ABSTRACT

Epitaxial CdTe/CdS heterostructures have been grown by molecular beam epitaxy onto BaF_2 covered Si (111) substrates. An epitaxial BaF_2 buffer is used for compatibility reasons, and because of easier dissolution during the lift-off processing. Epitaxy of cubic CdS (111) layers on BaF_2/Si (111) is achieved; electron channelling patterns exhibit a three-fold symmetry which is a characteristic for cubic crystal structures. The growth kinetics and structural properties of epitaxial CdS and CdTe/CdS have been studied with reflection high energy electron diffraction, Rutherford backscattering spectrometry and x-ray diffraction rocking curve measurements. The full width at half maximum of the (222) CdS and (333) CdTe x-ray peaks are ~1150 arcsec for 2.7 and 3.4 µm thick CdS and CdTe layers, respectively. To fabricate CdTe/CdS single crystal thin film solar cells, a lift-off process has been developed to remove the epitaxial layers from the Si substrates.

INTRODUCTION

CdTe/CdS solar cells with efficiencies ~10-16% have been fabricated with single crystal and polycrystalline CdTe layers grown by a variety of techniques [1,2]. Higher efficiencies (~20%) may be achieved with epitaxial CdTe/CdS solar cells [3]. Earlier work on single crystal cells was limited because CdTe and CdS bulk crystals of good crystalline quality and reasonable size were expensive and not easily available. Another problem is due to the different structures of bulk crystals; wurtzite for CdS and sphalerite for CdTe. Due to the substantial progress in thin film deposition technologies in the last years, high quality single crystal thin films (epitaxial layers) of CdTe, CdS and other II-VI compounds have been grown by a variety of techniques like metalorganic vapor phase epitaxy (MOVPE), molecular beam epitaxy (MBE), and atomic layer epitaxy (ALE). These growth processes yield single crystal layers with good control over stoichiometry variation, doping profiles, and large area as compared to bulk single crystals. Because of these improvements it is interesting to fabricate epitaxial CdTe/CdS solar cells for different reasons. Apart from the potential for high conversion efficiency, the study of material properties and junction characteristics of single crystal cells, free from grain boundaries and other structural defects, should lead to a better understanding of cheap polycrystalline solar cells. A variety of growth processes like ALE, close space sublimation, molecular beam deposition, metal-organic chemical vapor deposition, electrodeposition, spray pyrolysis, and screen printing have been used to fabricate high efficiency polycrystalline cells [2]. All these growth methods produce polycrystalline layers of different grain sizes, but yield solar cells with comparable efficiencies. Initial as-deposited efficiencies are rather low, but a large improvement is obtained when $CdCl_2$ and air annealing treatments are used [2,5]. These post-deposition treatments improve the grain size, and also affect the CdS/CdTe interface. A simultaneous study of polycrystalline and epitaxial CdTe/CdS heterostructures will be interesting to gain a better understanding of the role of structural defects; how they influence the material and junction properties when $CdCl_2$ and annealing treatments are used. We have grown heteroepitaxial CdTe [6,7] and CdS layers on Si substrates. Epitaxial IIa-fluorides (BaF_2, SrF_2, CaF_2) as intermediate buffer layers were used to overcome the lattice mismatch and because of other compatibility reasons [8]. Our objective is to fabricate single crystal CdTe/CdS thin film solar cells using a lift-off process for the removal of the epitaxial layers from the Si substrates, and compare their properties with polycristalline cells.

EXPERIMENTAL DETAILS

Heteroepitaxial CdTe/CdS structures were grown on BaF_2 and SrF_2 covered Si(111) substrates using a multiple chamber MBE system developed in-house. First, IIa-fluoride buffer layers are grown in a separate chamber. CdTe and CdS compound source materials are then sublimed from effusion cells in a second chamber. CdS is near lattice matched to SrF_2, while the lattice constant of BaF_2 is about 7% larger. BaF_2 buffers are preferred because of easier dissolution in water; this is advantageous for the lift-off process. Epitaxial CdTe layers were then grown onto the CdS layer in the same MBE system. The growth and structural properties of the epitaxial layers were studied with reflection high energy electron diffraction (RHEED) and electron channelling (EC) in a scanning electron microscope (SEM). The crystal qualities were characterized with Rutherford backscattering spectroscopy (RBS) and X-ray diffraction (XRD) rocking curve measurements. SEM and optical microscopy was used to study the morphology of the layers.

CH3365-4/94/0000-0323 $4.00 © 1994 IEEE
First WCPEC; Dec. 5-9, 1994; Hawaii

RESULTS AND DISCUSSION

We grow CdS at a substrate temperature of 380 °C which is subsequently raised to 420 °C after growth of a ~20 nm thick layer. The average growth rate is ~1 µm/h. As the growth starts, the streaky RHEED pattern of BaF_2 transforms to spots indicating that in the initial stage CdS nucleates as islands (3-d growth). The CdS surface is found to be sensitive to the e-beam exposure during RHEED measurements; due to the e-beam damage of the CdS surface, elongated spots appear on the RHEED streaks during the observation time. After ~100 nm growth, streaky RHEED patterns with weak Kikuchi bands are observed. This is shown in fig. 1 for a 0.4 µm thick CdS layer along [11-2] and [-110] azimuths of Si(111). It suggests epitaxial growth of smooth CdS with a layer by layer (2-d) like growth mode. Epitaxial CdS may also be grown at a lower temperature of ~170 °C, but the corresponding RHEED patterns are spotty and characteristic of heavy twinning. The RHEED patterns indicate growth of cubic CdS, but this may not be conclusive because with RHEED measurements it is difficult to distinguish between a cubic or hexagonal lattice [9].

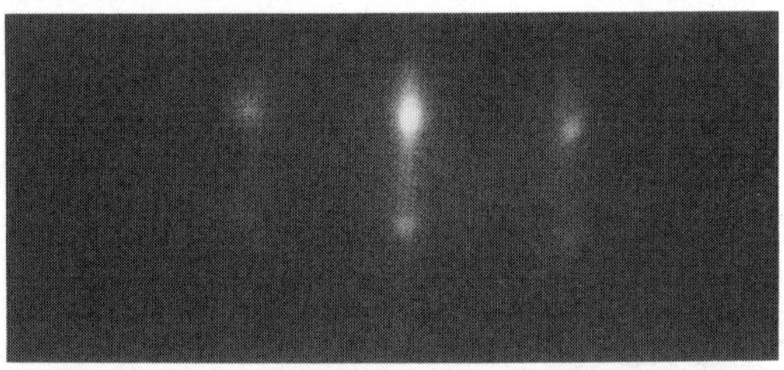
(a)

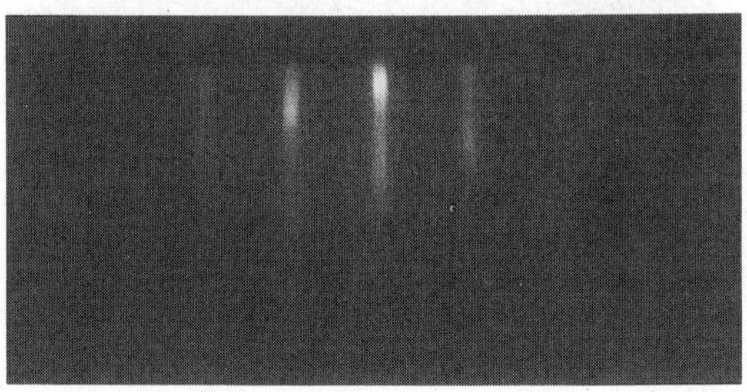
(b)

Fig. 1: 15 kV RHEED patterns observed during the growth of CdS layer on BaF2/Si(111); (a) in [11-2] and (b) in [-110] azimuth of Si(111)

To determine the crystal symmetry of the CdS layer, we performed EC measurements. A threefold rotational symmetry is expected for a (111)-oriented untwinned cubic lattice, while a hexagonal (wurtzite) lattice exhibits a 6-fold symmetry. As shown in fig. 2, sharp and well resolved electron channelling patterns with a three-fold symmetry are observed. This proves that untwinned cubic (111)-oriented epitaxial CdS layers have been obtained. Bulk single crystals of CdS normally crystallize in a stable wurtzite structure and some times in a metastable cubic zincblende structure.

Fig. 2: Electron channeling pattern of $CdS/BaF_2/Si(111)$

Recently it has been observed that heteroepitaxial CdS layers with hexagonal or cubic crystal structure can be grown on GaAs substrates by a variety of techniques [10-13]. These papers [10-12] describe the growth of cubic CdS(100) on GaAs(100) substrates. Growth of epitaxial CdS layers with cubic structure is important because the overgrown CdTe is zincblende. The CdS layers are crack-free, smooth, and do not show any pronounced morphology in the SEM and optical microscope.

RBS ion channelling is used to evaluate the crystal quality by measuring the channelling yield χ_{min} (defined as the ratio of ion yield in the aligned to random direction). Fig. 3 shows the random and <111> aligned spectra of a 0.5 µm thick CdS layer on $BaF_2/Si(111)$. The observed channelling yield of 25% near the Cd surface peak is a reasonable value of crystal quality considering the small thickness (~0.5 µm) of the CdS layer and the lattice mismatch between BaF_2 and CdS.

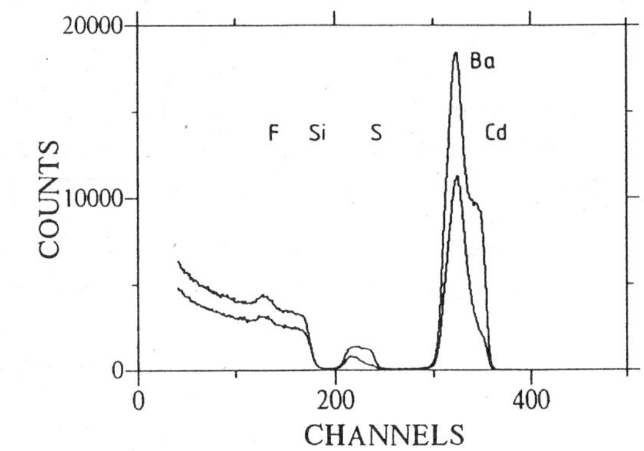

Fig. 3: 2 MeV random and <111> aligned spectra of CdS on BaF_2/Si

Lower χ_{min} values are expected for thicker layers. This has been already shown for a 5 µm thick epitaxial wurtzite CdS(0001) layer on CdTe(111), having a $\chi_{min} \approx 5\%$ [14]. As described below for our CdS layers of different thicknesses, XRD measurements support this observation. XRD rocking curve measurements (see fig. 4) are used to characterize the crystal quality, too.

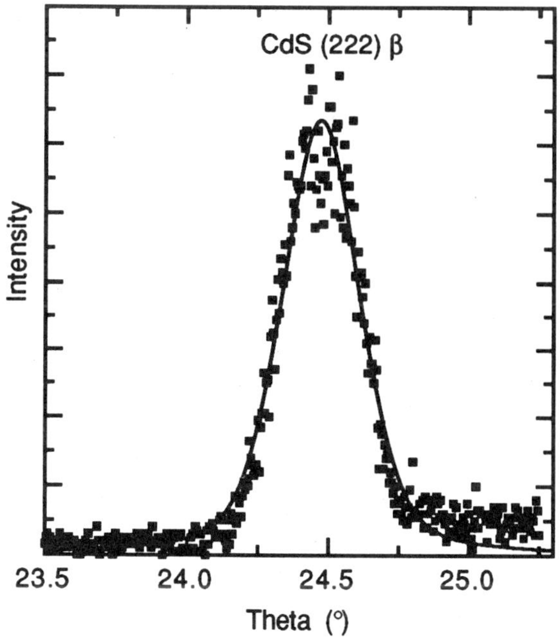

Fig. 4: XRD Rocking curve measurement of a 2.7 μm thick CdS layer on BaF_2/Si

The full width at half maximum of the CuK_β (222) CdS peak is ~1150 arcsec for a 2.7 μm thick CdS layer; a higher value of about 3000 arcsec has been measured for a 0.5 μm thick layer (for which χ_{min} ≈25% has been obtained from RBS-measurements). These XRD rocking curve values are comparable to the values (~2900 arcsec) reported for cubic CdS(100) layers grown on GaAs(100) substrates by MOVPE [11] and ALE/MBE [12]. Epitaxial growth of cubic CdS is a recent development, and only few measurements are available on the crystalline quality of cubic CdS layers [10-12].

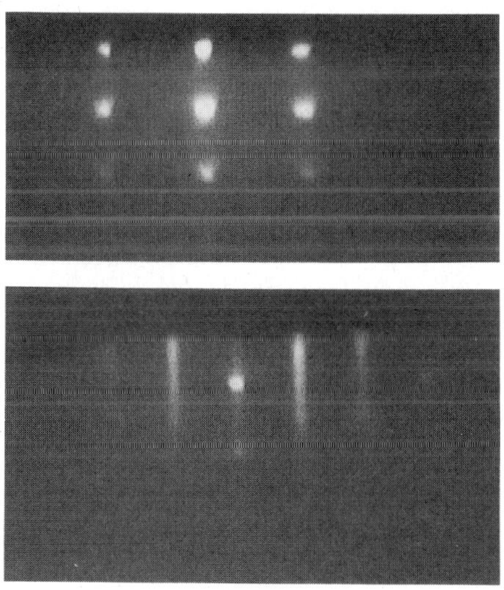

Fig. 5: 15 kV RHEED patterns observed during the growth of CdTe layer on CdS/BaF2/Si(111); (a) in [11-2] and (b) in [-110] azimuth of Si(111)

Epitaxial CdTe layers were grown on the CdS/BaF_2/Si(111) samples at a substrate temperature of 350 °C, and with a growth rate of about 1 μm/h. As shown in fig. 5, streaky RHEED patterns are observed in the [1-10] azimuth (of Si), while in the [11-2] azimuth the pattern is spotty. This suggests that the epitaxial CdTe layer has grown not in a 2-d layer by layer mode but in a mixed 2-d and 3-d (island) mode. Epitaxial CdTe layers of thickness ~4 μm are shiny, crack-free, and exhibit a smooth morphology in the SEM. The crystal quality of the epitaxial CdTe layers were also characterized by XRD rocking curve measurements. Fig. 6 shows the rocking curve for a 3.4 μm thick CdTe layer on CdS/BaF_2/Si(111). The full width at half maximum of the (333) peak is ~1150 arcsec. This suggest that the crystal quality of the overgrown CdTe layer is similar to the underlying CdS, and the lattice mismatch does not degrade the crystal quality of CdTe.

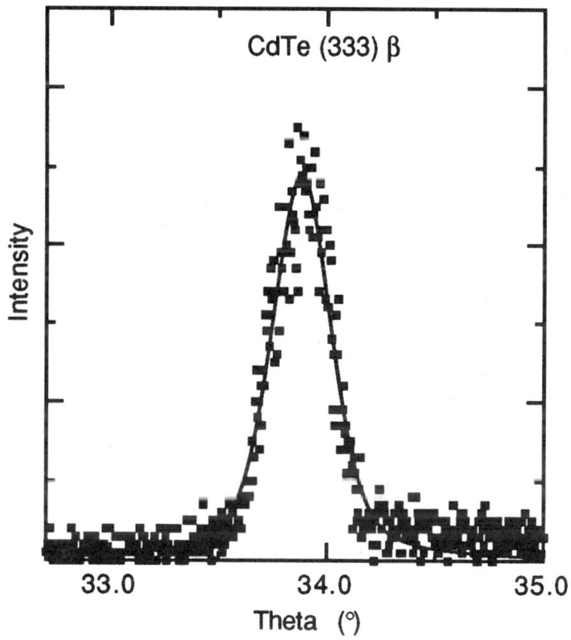

Fig. 6: XRD Rocking curve measurement of a 3.4 μm thick CdTe layer on CdS/BaF_2/Si

CdTe/CdS epitaxial lift-off process for solar cells

The optimized heteroepitaxial CdTe/CdS structures are suitable for solar cell fabrication. We already fabricated CdS(polycrystalline)/CdTe(single crystal) thin film solar cells with 7% efficiency with an earlier developed lift-off process [15]. Now we are involved in the development of single crystal CdS/CdTe thin film solar cells using an improved lift-off process for the removal of the CdS/CdTe stack from the Si substrate, and a method to subsequently transfer the CdS/CdTe structure to another support (glass, epoxy). For the fabrication of single crystal CdTe/CdS thin film solar cells we tried to develop a process based on the principle of Van der Waals bonding which has been used for the epitaxial lift-off of GaAs devices [16]. Here the device structure is separated from the substrate and transferred onto another substrate where it adheres due to the Van der Waals bonding between the atoms of the substrate (TCO/glass) and the device layer (CdS). The results of this approach were not satisfactory and reproducible for our CdTe/CdS layers. We have obtained good results with another epitaxial lift-off method which is also easier for sample handling. A cross-section of the solar cell structure is shown in fig. 7. Au is deposited on the bromine-methanol treated CdTe surface to form the ohmic contact. The structure is then covered with an epoxy (which acts as a support after the lift-off). The epoxy/Au/CdTe/CdS stack is removed from the Si substrate by dissolving the intermediate BaF_2 buffer layer. Finally, an Al grid is evaporated on the CdS to form the front contact. For the optical analyses and to measure the optical transmission of this single crystal CdTe/CdS structure a transparent epoxy is used as a lift-off support. Presently, we are studying the performance of such solar cells.

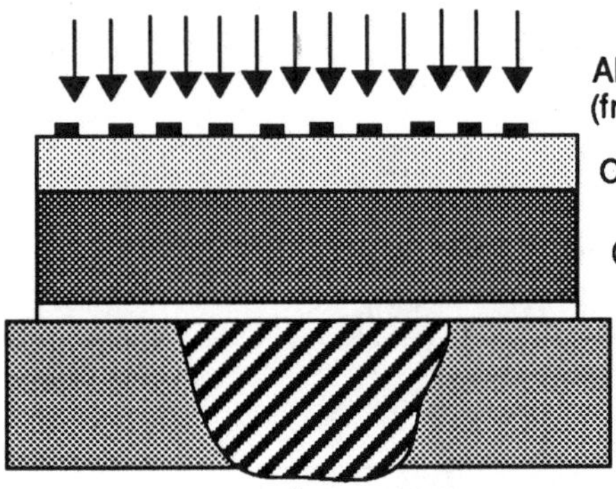

Fig. 7: Cross-section of the single crystal CdTe/CdS thin film solar cell on epoxy

In conclusion, we have grown epitaxial cubic CdS(111) layers and epitaxial CdTe/CdS heterostructures on BaF_2/Si(111) substrates. The growth kinetics and structural properties of CdS and CdTe layers have been characterized with a variety of techniques. A lift-off process has been developed to remove the epitaxial CdTe/CdS structures from the Si substrate to fabricate all thin film single crystal CdTe/CdS solar cells.

ACKNOWLEDGEMENT

This work was performed within the EU project "EUROCAD", and supported by a grant from the Swiss Federal Office for Education and Science.

REFERENCES

[1] J. Britt and C. Ferekides, "Thin-film CdS/CdTe solar cell with 15.8% efficiency", *Appl. Phys. Lett.* , **62** (22), 1993, pp. 2851-2852.

[2] For example see special issue of "The CdTe thin film solar cell", *International Journal of Solar Energy* , **12**, No. 1-4, 1992.

[3] K. Yamaguchi et al., "CdS-CdTe solar cell prepared by vapor phase epitaxy", *Jap. J. Appl. Phys.*, **16**, No. 7, 1977, pp. 1203-1211.

[4] A.W. Brinkman et al., "Epitaxial CdTe-based solar cells", *Int. J. Solar Energy,* **12**, 1992, pp. 233-245.

[5] S.A. Ringel et al., "The effects of $CdCl_2$ on the electronic properties of molecular-beam epitaxially grown CdTe/CdS heterojunction solar cells", *J. Appl. Phys.*, **70** (2), 1991, pp. 881-888.

[6] A.N. Tiwari et al., "Molecular beam epitaxial growth of (100) oriented CdTe on Si (100) using BaF_2-CaF_2 as a buffer", *Appl. Phys. Lett.*, **57** (11), 1990, pp. 1108-1110.

[7] A.N. Tiwari et al., "Study of As-doping and surface reconstructions in molecular beam epitaxy grown CdTe (100) on BaF_2-CaF_2/Si(100)", *Appl. Phys. Lett.* , **60** (5), 1992, pp. 621-623.

[8] S. Blunier et al., "Low temperature growth of group II-A fluoride layers on silicon as buffer for heteroepitaxial IV-VI and II-VI compound semiconductors", *Mat. Res. Soc. Symp. Proc.*, **220**, 1991, pp. 531-535.

[9] M.P. Halsall et al., "Growth and Assessment of CdS and CdSe layers produced on GaAs by metalorganic chemical vapour deposition", *J. Crystal Growth*, **91** (1988) pp. 135-140.

[10] T. Yodo and S. Tanaka, "Growth of CdS by atmospheric pressure metalorganic vapor-phase epitaxy at low temperature", *J. Appl. Phys.*, **72** (7), 1992, pp. 2781-2790.

[11] T. Tadokoro et al., "Growth and characterization of CdS epilayers on (100) GaAs by atomic layer epitaxy", *J. Crystal Growth*, **130**,1993, pp. 29-36.

[12] A.G. Cullis et al., "Structure of hexagonal and cubic CdS heteroepitaxial layers on GaAs studied by transmission electron microscopy", *Appl. Phys. Lett.*, **55** (20), 1989, pp. 2081-2083.

[13] S. Guha et al., "Microstructure and pseudomorphism in molecular beam epitaxially grown ZnCdS on GaAs(001)", *Appl. Phys. Lett.* **63** (15), 1993, pp. 2129-2131.

[14] A. Guerrieri et al., "Ion channeling Rutherford backscattering spectrometry structural characterization of CdS/CdTe heterostructure", *Materials Science and Engineering*, **B16** 1993, pp. 160-164.

[15] A.N. Tiwari et al., "P-type doping in heteroepitaxial CdTe and lift-off technique for making single crystal CdTe solar cell", *Proc. 10th European Photovoltaic Energy Conference*, Lisbonne (Portugal), 1991.

[16] E. Yablonovitch et al., "Van der Waals bonding of GaAs epitaxial lift-off films onto arbitrary substrates", *Appl.Phys. Lett.*, **56** (24), 11 June 1990.

OPTICAL AND ELECTRICAL CHARACTERIZATION OF CHEMICALLY DEPOSITED CADMIUM SULFIDE THIN FILMS

M.E. Özsan, D.R. Johnson, M. Sadeghi and D. Sivapathasundaram
B.P. Solar, 12 Brooklands close, Windmill Road, Sunbury on Thames, Middlesex, TW16 7 DX, UK
L.M. Peter, M.J. Furlong, G. Goodlet and A. Shingleton
School of Chemistry, University of Bath, Claverton Down, Bath BA2 7AY, UK
D. Lincot, B. Mokili and J. Vedel
Laboratoire d'Electrochimie et de Chimie Analytique, Ecole Nationale Supérieure de Chimie de Paris, 11 rue Pierre et Marie Curie, 75231 Paris Cedex 05, France

ABSTRACT

Optical and electrical properties of chemical bath deposited CdS films are characterized as a function of the bath chemistry and annealing treatments. Changes in band gap and broadening parameter are evident and discussed in terms of quantum size effects and structural defects. The films are n-type and photosensitive with carrier concentrations around 10^{17} cm^{-3}.

INTRODUCTION

Chemical bath deposition (CBD) is now widely used for the elaboration of low cost polycrystalline thin film solar cells, because it offers the advantages of economy, convenience and the capability of large area and multiplate deposition. CBD CdS layers have been used as the n-type window layer in high efficiency CdS/CdTe thin film solar cells [1,2]. In these cells CBD CdS layers were combined with CdTe layers deposited using either a "wet" electrodeposition [1] or a close space sublimation (CSS) technique [2]. The increased understanding of the properties of CdTe layers have led to efficiencies higher than 10%. Substantial potential exists, however, to improve cell performance, if the properties of CdS layer are better understood.

In this paper, the chemical formulation of the ammonia based CBD system will be presented together with some of the optical and electrical properties of the CdS films deposited on to SnO_2 coated glass substrates. The effect of heat treatments on the film properties is also presented.

Solution chemistry and growth mechanism

Chemical bath deposition is the analogue in solution of the well known chemical vapor deposition technique (CVD, MOCVD). It is based on the reaction at low temperature (<100°C) between precursors dissolved in solution. In the case of ammonia bath the deposition is based on the following overall reaction:

$$Cd(NH_3)_4^{2+} + SC(NH_2)_2 + 2\,OH^- \rightarrow CdS + CH_2N_2 + 4\,NH_3 + H_2O$$

Thiourea is the sulfur precursor. Film formation takes place in a wide range of deposition parameters. Solution chemitry and chemical kinetics analysis gives a good understanding of the mechanism and allows to model it [3]. The growth is mainly heterogeneous and controlled kinetically by successive surface reactions:

$$Cd(NH_3)_4^{2+} + 2\,OH^- + site \rightarrow Cd(OH)_{2,ads} + 4\,NH_3$$

$$SC(NH_2)_2 + Cd(OH)_{2,ads} \rightarrow C_{ads^*}$$

$$C_{ads^*} \rightarrow CdS + CN_2H_2 + 2H_2O + site$$

where C_{ads^*} represents an adsorbed reaction intermediate. When the precursor concentrations increase, the growth rate increases and a growth by colloid aggregation becomes possible. In the case of cadmium it is related to the formation of cadmium hydroxide clusters in the solution, which catalyse the formation of CdS particles in the bath. The boundary conditions are determined by calculating solubility diagrams of cadmium hydroxide. In Fig.1 is presented a calculated diagram which shows for example the influence of ammonia concentration.

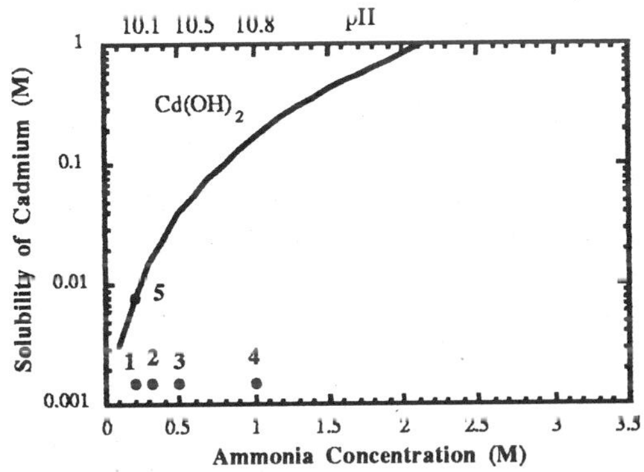

Figure 1: Calculated solubility diagram of cadmium hydroxide as a function of the ammonia concentration with $[NH_4Cl] = 0.025$ M. pH ≈ 9.2 + 0.5 log$[NH_3]/[NH_4^+]$. (1-4) : Experimental points with $[CdCl_2]=1.5\ 10^{-3}$M. Other parameters: [Thiourea] = 7.5 10^{-3}M, T = 80°C. (5), same as (1) with $[CdCl_2]=8\ 10^{-3}$M.

Experimental conditions used in this work are shown in the figure in order to investigate the influence of the distance to the precipitation line of cadmium hydroxide.

CH3365-4/94/0000-0327 $4.00 © 1994 IEEE

First WCPEC; Dec. 5-9, 1994; Hawaii

On the other hand the effect of increasing the thiourea concentration has been studied.

Experimental

The chemical bath solutions used here were based on ammonia concentrations in the range 0.2-4 M. The pH of the solution was adjusted between 10 and 12.5 by using ammonium chloride or sodium hydroxide. Cadmium and thiourea concentrations were respectively in the ranges 10^{-3}-2.10^{-2}M and 10^{-3}-2.10^{-1}M. The bath temperature was in the range 60-85°C. The substrates were commercially available SnO_2 coated glass. The thickness of the CdS layers was about 0.1 µm. The electrical and optical properties of the films were characterized by electroabsorption (EEA) and capacitance measurements in an electrolyte (NaOH (0.1M) / Na_2S(0.1M)) [4]. Schottky darriers with mercury contacts were also used. Thermal post-deposition treatments (400°C, 20mn) were carried out in air with or without the presence of $CdCl_2$ as a recrystallisation flux .

RESULTS AND DISCUSSION

Structure and composition

The structure of CdS films has been described previously [5,6]. High Resolution Transmission Electron Analysis of detached films revealed a close packed structure with grain size of 5 to 50 nm with predominantly hexagonal modification [5]. The presence of stacking faults is evident [5,6]. As measured by Rutherford Back Scattering the composition of the film is generally not the stoichiometric one. An excess of cadmium relative to sulfur is observed in most of the films. This excess, which can reach 10 to 20 %, is attributed to the presence of cadmium hydroxide. In some cases, in particular for low growth rates, the composition becomes closer to stoichiometry.

Optical properties

Electrolyte-electroabsorption (EEA) and photocurrent (PC) measurements

Electrolyte Electroabsorption (EEA) is a modulation spectroscopy where changes in the applied electric field perturb the refractive index and the absorption coefficient of a semiconductor. This gives rise to changes in the transmission (ΔT) which lead to an enhancement of features in the ordinary optical spectrum. For the CdS films studied here a spectrum of a third derivative type was obtained, see figure 2. It is analysed using the Aspnes three point fit method [7] assuming a "n" value of 2.5 (3D transition). This semi-empirical fitting procedure allows the determination of the band gap and the broadening parameter (Γ), where Γ can be related to the disorder or strain in the semiconductor structure.

Values obtained from EEA are shown in table I. These CBD-CdS films yield Γ values in the range 100-200 meV. In comparison Atomic Layer Epitaxial grown films (from Microchemistry Ltd) give values of 80-100meV. These differences in Γ can be attributed to differences in crystallinity or possibly to the presence of stacking faults. Increasing the thiourea concentration had a tendency to lower the Γ value in such a way that values as low as 100meV could be obtained. Γ was unaffected by variations in the pH or ammonia concentration (cf figure 1).

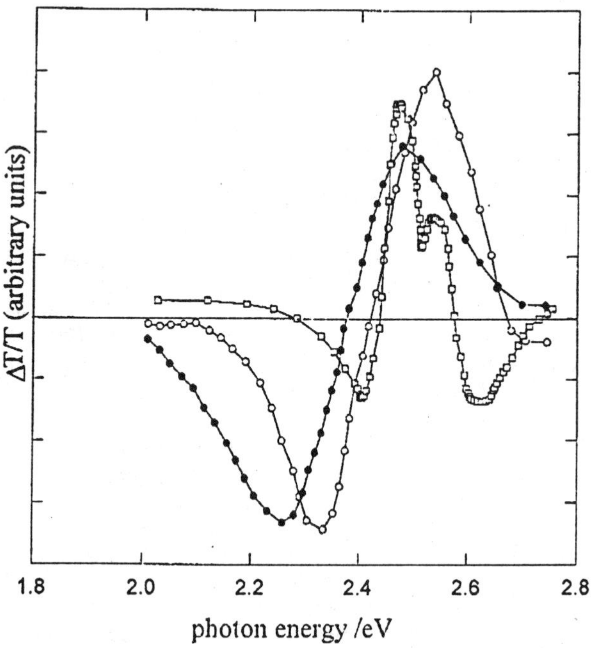

Figure 2: Electrolyte electroabsorption spectra at 250 mV/SCE, 100 mV pp, 270 Hz, of as deposited (○), air annealed (400°C, 20 mn) (●), and air annealed in presence of $CdCl_2$ (□)

Clear trends between the band gap and the deposition conditions have been found, see Table I. Figure 3 shows the variation produced by altering the ammonia concentration (also the pH in these conditions) in the bath. This was thought to be due to variations in the cadmium hydroxide content in the films.

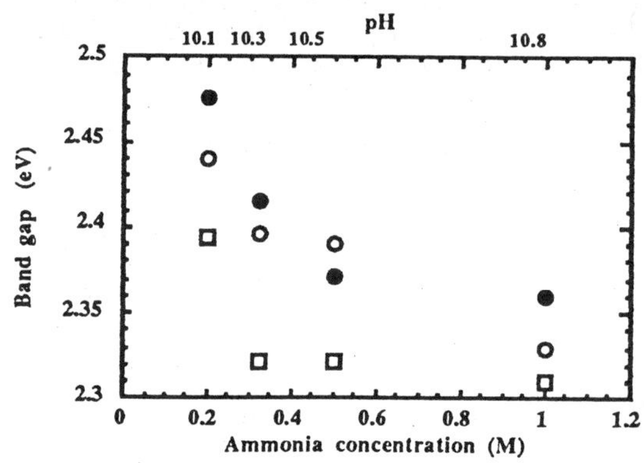

Figure 3: Band gap values for films deposited in conditions (1-4) determined from optical transmission (○), electrolyte electroabsorption (EEA ●) and electrolyte photocurrent spectroscopy (PCS □).

An increase in the band gap was also brought about by increasing the thiourea concentration, see figure 4. Despite cadmium excess is found also in these films, the presence of cadmium hydroxide does not seems to be the reason behind this phenomenon.
As in both cases the increase of the gap is obtained in conditions where the growth rate is also increasing, it is possible that a relation exists. As a general rule increasing the growth rate affects the crystallinity of the film, with a trend to reduce the grain sizes. As a consequence the increase of band gap may be related to quantum size effects as observed in colloids or particulate films. Calculations for CdS show that this effect is relevant for particle sizes < 10 nm [8]. An increase of BAND GAP of 0.1 eV is expected for particle sizes < 4 nm. These values are compatible with those measured by TEM in some conditions. In these conditions, A participation of growth from colloids is possible. In the case of point 5 (cf figure 1), where the deposition conditions are on the precipitation line of $Cd(OH)_2$ the band gap further increases to 2.55 eV.

Characterization of the films was also carried out using photocurrent spectroscopy (PC) in the electrolyte (between 2 to 2.8 eV). The band gap is then determined from the spectral response in the absorption edge. As shown in table I and fig. 3 and 4 the values were in good agreement with the values obtained from EEA and optical transmission spectroscopies.

N°	Eg EEA	Eg/eV PCS	Γ/meV n = 2.5	E_{fb}/V vs SCE	N_D/cm^{-3} X 10^{16}
1	2.47	2.39	178	-0.74	40
2	2.41	2.32	182	-0.73	5
3	2.37	2.32	161	-0.74	10
4	2.36	2.31	183	-0.93	2 00
5	2.43	2.39	166	-0.79	0.67
6	2.46	2.44	146	-0.9	0.57
7	2.5	2.47	143	-1.03	1.5

Table I : Band gap, broadening parameter, flat band potential and donor densities of CBD films as a function of ammonia concentration (1-4 in Fig. 1, Fig. 3) and thiourea concentration (5-7 Fig. 4).

Electrical properties

The electrical properties have been assessed by two techniques. One utilises an electrolytic junction, (as used for EEA and photocurrent spectroscopies), which is highly suitable for CBD CdS films [4,9] and the other employs a mercury probe, which mimics the solid state device. Good rectifying schottky barrier diodes are obtained in both cases. A typical I-V curve of a CBD-CdS/Hg Schottky diode is shown in figure 5.

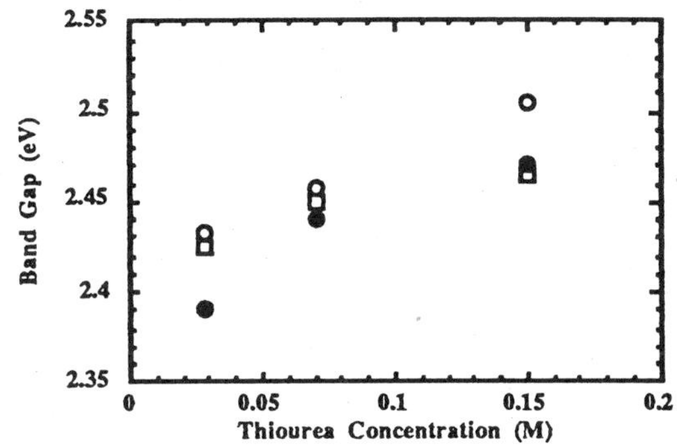

Figure 4 : Band gap values obtained from optical transmission (□), electrolyte electroabsorption (EEA ○) and electrolyte photocurrent (PCS ●) for films deposited with different thiourea concentrations at 60°C with $[NH_3]$ = 1.74 M, $[CdCl_2]$=1.4 10^{-2}M.

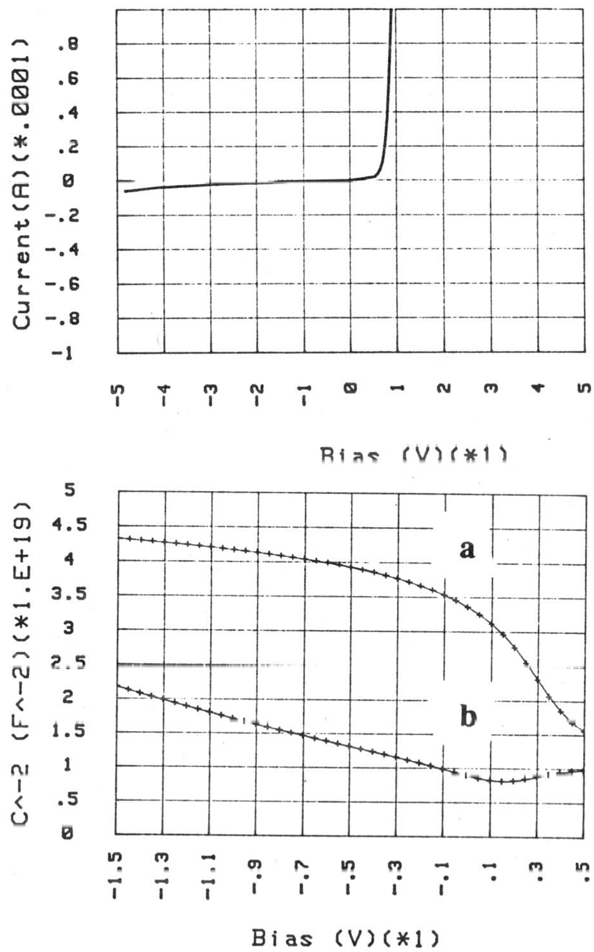

Figure 5 : Top : I-V curve of CdS-CBD/Hg Schottky diode in dark. Bottom : C^{-2}-V curve in dark (a) and under illumination (b). Measurement frequency: 1 MHz. Sample area ≈ 0.4 mm^2.

The reverse voltage biasing generally extends over several volts without passing appreciable current. The low thickness of the CdS layer (≈ 1000 Å) and the high roughness factor of the substrate (1.5 to 2) demonstrate the excellent covering properties associated to the CBD technique.

In figure 5 (bottom) are shown capacitance voltage curves, measured at 1 MHz, of the CBD CdS / Hg Schottky barrier in dark (a) and under illumination (b). In dark the C^{-2}/V curve is not linear, with a saturation of the value in the reverse bias direction. This behaviour, also observed with the electrolyte junction, is related to the limited thickness of the film. In this range the film is completely depleted and the capacitance is constant. From the slope of the curve towards direct bias, taking into account the roughness factor (1.5-2), it is possible to determine the doping level in the film. The values are generally in the range 10^{16}-$10^{17}cm^{-3}$ (see Table I) without clear correlations with the deposition conditions. Very high values obtained in some cases are probably not related to the true doping level but may indicate shunting effects with the substrate due to porosity of the film. An influence of the deposition conditions on the coverage properties has been previously reported [9]. Concerning the barrier height (from Table I) it seems that some variations appear when varying the thiourea concentration.

The films appear very sensitive to the illumination (Fig.5 curve b). From solid state measurements an increase of the doping level by a factor of 3 to 10 is generally observed under illumination. This bring the doping level to around 5.10^{17} cm^{-3}. As a consequence CBD films, prepared from ammonia bath, are fairly highly doped under illumination. It can be noted that the C^{-2}/V curve is almost linear under illumination (Fig.5 b). As compared to [2] no doping impurities were intentionnally introduced in the bath.

Influence of thermal annealing treatments

Thermal treatments in air (400°C, 20 min) in the presence or absence of CdCl2 flux have been studied. Figure 2 shows the EEA spectra obtained after these treatments. Table II summarizes typical values of their optical parameters. Air annealed films exhibit at shift of the band gap to lower energies by about 0.1 eV without a significant change of the broadening parameter. This indicates that the films are still disordered which is consistent with HRTEM experiments (not shown).

	Eg (TRS)	Eg (PCS)	Eg(EEA)	Γ / meV
As grown	2.41	2.44	2.43	164
air	2.31	2.27	2.22	177
$CdCl_2$	2.42	2.43	2.43	61

(eV)

Table II: Effect of annealing treatments on band gap and broadening parameters of CdS films.

CdCl2 annealed films show a much sharper band edge and Γ is reduced substantially. This points to a reduction in the defect density, in agreement with the strong recrystallization of the films, observed in the presence of a CdCl2 flux.

Concerning the electrical properties, the donor concentrations seem to be somewhat lower after annealing. The layers are still photosensitive and exhibit doping levels around 1-3 $10^{17}cm^{-3}$ under illumination .

CONCLUSIONS

The Optical properties of as-grown CBD CdS films , in particular the band gap, are strongly dependent on the deposition conditions. Increasing the deposition rate tends to increase the gap. This can be due to quantum size effects in the films related with its microcrystalline structure. On the other hand the broadening parameter is rather large and less affected by the deposition conditions. It decreases markedly after annealing treatments in presence of $CdCl_2$ flux.

The films are all n-type and exhibit doping levels around $10^{17}cm^{-3}$ under illumination without intentional doping in the bath.

Studies in progress show that the device characteristics are closely related to the properties of the as-grown CdS film.

Aknowledgements

This work has been supported by the European Joule II program, contract JOU2-CT92-0243.

References

[1] J.M. Woodcock, M.E. Özsan, A.K. Turner, D.W. Cunningham, D.R. Johnson, R.J. Marshall, N.B. Mason, S. Oktik, M.H. Patterson, J.M. Sherborn, D. Sivapathasundaram and I.A. Walls, Proc. 12th European PVSEC, Amsterdam (1994), 948-950.
[2] T.L. Chu, S.S. Chu, N. Schultz, C. Wang and C.Q Wu, J. Electrochemical Soc., 139(1992)2443.
[3] R. Ortega-Borges and D. Lincot, J. Electrochemical Soc., 140(1993)3464.
[4] M.E Özsan, D.R. Johnson, M. Sadeghi, D. Sivapathasundaram, L.M. Peter, G. Goodlet, M.J. Furlong and A.A. Singleton, to be published.
[5] M.E. Özsan, D.R. Johnson, M.H. Patterson, D. Sivapathasundaram and J. M. Woodcock, Proc. 12th European PVSEC, Amsterdam (1994),1604-1607.
[6] D. Lincot, R. Ortega-Borges and M. Froment, Phil. Mag. B, 68(1993)185. R. Ortega etal. Solid State Phenomena, 37-38(1994)497.
[7] D.E. Aspnes, Surface Science, 37(1973)418.
[8] Y. Wang and N. Herron, J. Phys. Chem., 95 (1991)525.
[9] A. Kampmann, P. Cowache, B. Mokili, R. Ortega-Borges, D. Lincot and J. Vedel, 12th European Photovoltaic Solar Energy Conference, 1994, 664.
[10] D. Lincot and R. Ortega-Borges, J. Electrochemical Soc., 139(1992)1880.

A COMPARATIVE STUDY OF COMPOSITIONAL AND STRUCTURAL PROPERTIES OF CdTe THIN FILMS PREPARED BY DIFFERENT ELECTROPLATING TECHNIQUES

P.D.Paulson, V.Dutta and *Chatar Singh

Photovoltaic Laboratory, Centre for Energy Studies
Indian Institute of Technology, New Delhi-110 016, INDIA.

* Industrial Tribology Machine Dynamics and Maintenance Engineering Centre.
Indian Institute of Technology, New Delhi-110 016, INDIA.

ABSTRACT.

The structural and compositional properties of CdTe films, deposited from an aqueous solution of CdSO and TeO by Pulsed Electrodeposition, Potentiostatic and Galvanostatic Electrodeposition, have been studied. Polycrystalline films with a preferred <111> orientation are obtained in all techniques. However films deposited by pulsed technique show the higher intensity for <111> peak. EDAX study shows that, for lower concentrations of $CdSO_4$ in the electrolyte, pulsed films are more stoichiometric compared to films deposited by other two techniques.

INTRODUCTION.

Thin film CdTe is one of the most promising materials for terrestrial photovoltaic[1]. The recent reports show that CdTe based solar cells deposited by different techniques like Electro- deposition, Screen printing, PVD, MOCVD [2] and Closed Space Vapor Transport (CSVT)[3] can yield efficiency more than 10% over small area. However the electrodeposition seems to be the ideal technique for the photovoltaic application, as it does not require any costly instrumentation for the vacuum and also can easily be scaled up for large area deposition[4].

CdTe films can be deposited by pulse, Potentiostatic and Galvanostatic Electrodepostion techniques. Since the solubility of TeO in HCl acid bath is less compared to sulphuric acid bath, we have used latter for the deposition of CdTe films. CdTe deposition occurs at a potential that is more positive than required for deposition of Cd as a separate phase. This is possible because of the negative free energy of formation of CdTe (-92 kJ/mole) from Cd and Te. There are two reactions (eq.(1) and (2)) occurring at this potential [5]

$$HTeO_2^+ + 3H^+ + 4e^- - Te + 2H_2O - (1)$$

and

$$Cd^{2+} + 2e^- \cdots Cd \qquad (2)$$

It has been reported that the CdTe films obtained with the pulse technique has reduced porosity and improved conductivity compared with that obtained by the Potentiostatic method[6]. In order to compare the compositional and structural properties of pulse plated films with Potentiostatically and Galvanostatically deposited films, we have conducted the EDAX and XRD studies, keeping the deposition conditions nearly identical.

EXPERIMENT

CdTe films were deposited on to the ITO/Glass substrates from a deposition bath contains aqueous solution of CdSO saturated with TeO ($\approx 10^{-3}$M) using a conventional three electrode system. The pH of the electrolyte is adjusted to 2.2 using HSO. The bath temperature is kept constant at 90°C. The deposition technique is described elsewhere[5,7]. CdTe films were deposited for different $CdSO_4$ concentrations. The $HTeO_2^+$ concentration is varied by speciating the TeO first with NaOH then brought in to 2.2 pH by adding H_2SO_4 to the solution[8]. Deposition is also carried out for different pulse frequencies with 33% duty cycle. The films were then characterised using XRD (Rigaku Rotaflex) and EDAX (EDAX PV9900)

RESULT & DISCUSSION

At high concentrations of Cd in the electrolyte, the CdTe deposition is independent of the CdSO concentration. It is only depend on the HTeO concentrations in the electrolyte. At lower concentrations of $CdSO_4$ (<0.2M), the deposition is found to be very slow. For pulse deposition, good quality CdTe films were obtained for a frequency range of 40 to 100 Hz. At lower frequencies, since the deposition rates are high, films were found to peel off from certain parts during deposition, after a certain thickness was obtained. At high frequencies (>150 Hz) however, no good film except a black deposit was obtained.

CH3365-4/94/0000-0331 $4.00 © 1994 IEEE

First WCPEC; Dec. 5-9, 1994; Hawaii

XRD Characterization:

Fig.1 shows the XRD pattern of three samples deposited by (a) Galvanostatic (b) Potentiostatic and (c) Pulse electrodepositions from a typical electrodeposition bath. All the three are films essentially cubic <111> preferential orientation. However the intensity of <111> peak is high for pulse deposited films suggesting a higher degree of preferential orientation. A low value FWHM of the <111>peak revealed that Potentiostatically deposited films have the higher grain size. This result is confirmed with the SEM and the grain size measured is $\approx 0.7 \mu m$. XRD pattern also shows the reflection from CdTe (220), (311) planes. It is also found that Cd (101) Te (112) and Te(101) phases are also present in the films.

EDAX Analysis

Fig.2 shows the variation in the stoichiometry with the concentration of $CdSO_4$. At lower concentration of $CdSO_4$, the pulsed films were found to be more stoichiometric. No such difference is observed at higher concentrations and films deposited by either technique is Te rich.

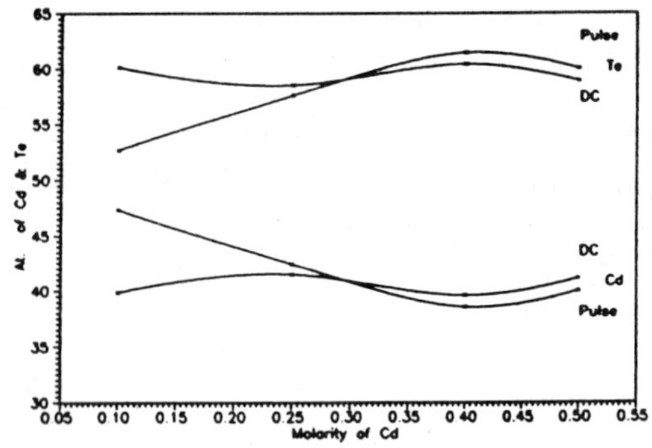

Fig.2 The variation of Cd and Te content in the film with the concentration of $CdSO_4$ in the electrolyte

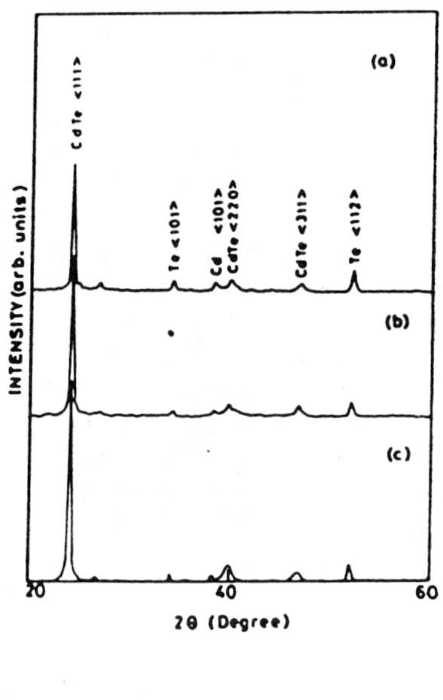

Fig.1.Diffraction patterns of different as-deposited CdTe films. (a) Galvanostatic (b) Potentiostatic and (c) Pulse.

This behavior can be understood using the two step process proposed by Vedel et.al [9]. A two step process, adsorption of $HTeO_2^+$ ions and the Cd ions followed by the reduction of $HTeO_2^+$ in to CdTe is involved in the CdTe electrodeposition. Adsorption is supposed to occur on special surface sites designated by s, which is supposed to be constant. The adsorption process of $HTeO_2^+$ and Cd^{2+} ions are governed by eq.(3) and (4) respectively.

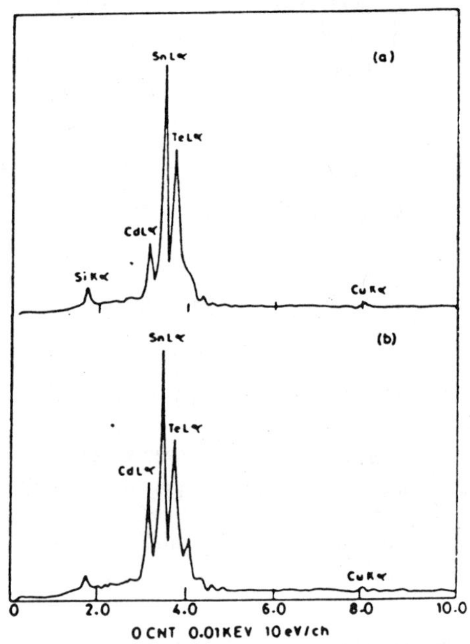

Fig.3.EDAX Spectra of a typical (a) dc electrodepoited and (b) Pulse electrodeposited films.

$$HTeO_2^+ (Surf.) + s \underset{k_1'}{\overset{k_1}{\rightleftharpoons}} HTeO_2^+, s \quad -(3)$$

$$Cd^{2+} (Surf.) + s \underset{k_2'}{\overset{k_2}{\rightleftharpoons}} Cd^{2+}, s \quad --(4)$$

Where k and k' are the rate constants respectively

for the adsorption and desorption reactions. The near-stoichiometric deposit obtained in the periodic pulse deposition might be due to the domination of either k' or k. However, a more detailed study will be required in this direction.

Table I EDAX data for CdTe films

Element	DC	
	With NaOH	Without NaOH
Cd	43.73	39.84
Te	56.27	60.16
	Pulse	
Cd	45.67	47.35
Te	54.33	52.65

Table.I summarises the EDAX results for effect of NaOH speciation. The introduction of more Te ions using NaOH seems to be advantageous for dc deposition, whereas it does not have any significant effect in Pulse deposited films. Thus a stoichiometric CdTe film can be obtained using DC technique by speciating Te with NaOH.

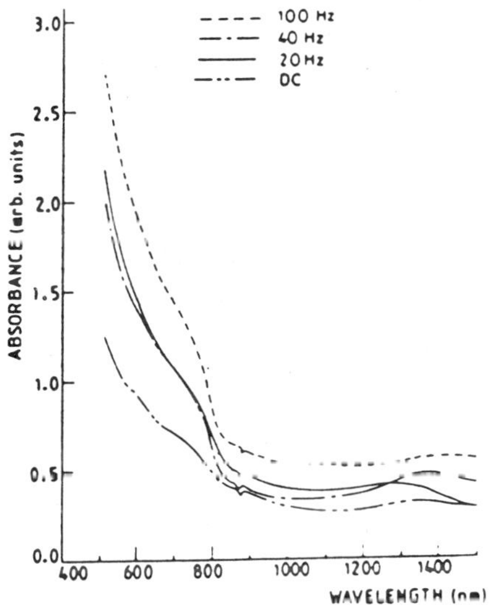

Fig.4. Absorption Spectra of as-deposited CdTe films

The optical absorption of these films is as shown in the Fig.4. The films deposited by pulsed technique have higher absorption than the films deposited by the direct current technique.

CONCLUSION

CdTe films were deposited by Pulse and Potentiostatic depositions. The structural, and compositional properties of these films have been studied. The Pulse deposition is found to be superior than the other electrodeposition techniques. Good near-stoichiometric films are obtained for low ($20 < f > 100$)pulse frequencies. The speaciation of TeO with NaOH was found to be advantageous for DC deposition in confirmation with the compositional data. annealing.

ACKNOWLEDGEMENTS

We are indebted to Prof.L.K.Malhotra, Thin Film Lab, IIT, New Delhi for providing the experimental facilities for these studies. The financial support from Council of Scientific and Industrial Research (CSIR), New Delhi is greatfully acknowledged.

REFERENCES.

[1] J.Skarp, Y.Koskinen, S.Lindfors, A.Rautianen and T.Suntola, "Development and Evaluation of CdS\CdTe Thin Film PV Cells", *Proceedings of 10 E.C.Photovoltaic Solar Energy Conference*, April 8-12, 1991, Lisbon, Portugal, pp567-569

[2] H.S.Ullal, J.L.Stone, K.Zweibel, T. Surek, R.L.Mitchel, " Polycrystalline Thin Film Solar Cells and Modules", *Proceedings of 6 PVSEC*, Feb.10-14, 1992, New Delhi, India, pp81-87

[3] J. Britt and C. Ferekides, "Thin Film CdS/ CdTe olar Cells With 15.8% Efficiency" *Appl.Phy Lett.*, **62**, 1993, pp2851-2852

[4] S.K.Das and G.C.Morris,"Preparation and Properties of CdS/CdTe Thin Film Solar Cells Produced by Periodic Pulse Electrodeposition Technique", *Solar Energy Materials and Solar Cells*, **30**, 1993, p107-118

[5] M.P.R.Panicker, M.Knaster and F.A.Kroger, " Cathodic Deposition of CdTe from Aqueous Electrolyte", *J.Electrochem.Soc.*, **52**, 1978, pp566-572

[6] S.Moorthy Babu, R.Bhaskaran and P. Ramaswami,"Electrodeposition of CdTe by Potentiostatic and Pulse Technique", *Thin Solid Films*, **202**, 1991, pp67-75

[7] M.W.Verbrugg and C.W.Tobbias, "The Periodic Electrochemical Codeposition of Cadmium Telluride", *AIChE Journal*, **33**, 1987, pp628-641

[8] R.N.Bhattacharya and K.Rajeshwar, " Electrodeposition of CdTe Thin Films", *J.Electrochem. Soc.*, **131**, 1984, pp2032-2037

[9] Catherine Sella, Pascale Boncorps, and Jacques Vedel, " The Electrodeposition Mechanism of CdTe from Acidic Aqueous Solutions", *J.Electrochem. Soc.*, **133**, 1986, pp2043-2047

POLYCRYSTALLINE THIN FILM CdTe SOLAR CELLS FABRICATED BY ELECTRODEPOSITION

Donghwan Kim, Scott Pozder, Yuming Zhu, and John U. Trefny
Department of Physics, Colorado School of Mines, Golden, Colorado 80401

ABSTRACT

Several issues regarding the fabrication of CdTe solar cells by electrodeposition have been investigated. The CdTe deposition parameters were optimized as follows: the deposition current density J_{dep} = 0.33—0.40 mA/cm², the anode current ratio $R_{Te/Cd}$ = 2.7, and the deposition potential V_{dep} = -600 mV against a Ag/AgCl reference electrode. $CdCl_2$ treatment and subsequent annealing made a significant change in the composition: the film became Cd-rich throughout its whole thickness, and the concentrations of both Cd and Cl increased near the surface. Electron beam-induced current (EBIC) data showed a peak inside the CdTe layer suggesting a buried homojunction. Use of CdS double layers composed of a $CdCl_2$-treated, annealed CdS underlayer covered with an as-deposited CdS overlayer resulted in a significant increase in open-circuit voltage V_{oc} up to 0.78 V. 11.0% CdS/CdTe/Au dot cells (0.033 cm²) have been produced in this work.

INTRODUCTION

Since Panicker et al. [1] laid the foundation for electroplating CdTe thin films, various groups have worked on the technique to produce CdS/CdTe solar cells. Of these, the British Petroleum (BP) [2], University of Queensland [3], Monosolar [4], and Ametek [5] groups are several that stand out. Using this technology, the BP group has demonstrated cells and panels with efficiencies in excess of 14% and 8%, respectively [6]. Although extensive studies have been carried out in this field, it is our impression that more systematic investigations of the effect of deposition conditions on cell performance are needed.

EXPERIMENTAL PROCEDURES

Commercially available SnO_2-coated sodalime glass substrates were used in this investigation. The SnO_2 layers were approximately 0.4 μm thick with a sheet resistance of 10Ω/sq. The substrate size was 5 cm x 7.5 cm (38 cm²).

CdS layers were deposited by chemical bath deposition using cadmium acetate, ammonium acetate, and ammonium hydroxide as reported by Chu et al. [7] The thickness of the CdS was 0.19 μm as deposited and changed to 0.12~0.13 μm after $CdCl_2$ treatment and annealing. The thickness uniformity was ±4% across the sample surface.

A $CdCl_2$ coating was applied to the surface of the CdS using a mist generator containing a 2.4 M $CdCl_2$ aqueous solution. A Devilbis™ ultrasonic nebulizer generated the mist which was carried though a tube to a preheated (~ 150°C) box where the sample was placed vertically to the mist flow. The sample was next annealed at 450°C for 50 min in an N_2 atmosphere and cooled in the furnace. Finally, the CdS substrates were rinsed in hot (~80°C) DI water and dried in N_2. Electrical leads were connected to the edges of the substrates using indium solder.

The CdTe deposition was done in an electroplating system composed of four electrodes: Cd and Te anodes, a Ag/AgCl reference electrode, and a cathode (sample substrate). The deposition current J_{dep} was controlled by adjusting the concentration of the tellurium ions [Te]. The experimental conditions used in this study were: pH, 2; temperature, 80°C; 1.2 M $CdCl_2$; J_{dep} between 0.26 and 0.40 mA/cm²; deposition potential -600 mV (measured against the Ag/AgCl electrode); [Te], 20~22 ppm; and anode current ratio, $R_{Te/Cd}$, between 2.2 and 3.5. J_{dep}, $R_{Te/Cd}$, and V_{dep} were varied in this study. A typical growth rate was 0.86 μm/hr and the Coulomb efficiency of the deposition process was on the order of 85~90%. After completion of the deposition, the CdTe samples were rinsed in hot DI water and dried in N_2.

A $CdCl_2$ coating was applied to the CdTe surfaces using the same procedure as used for the CdS layers. The samples were then annealed at 410°C for 45 min in air, removed from the furnace, cooled at room temperature, rinsed again, and dried. Etching was done immediately before the back contact metallization by dipping the samples in 0.1% Br-MeOH for 10 s. Back contacts were made by evaporating Au through masks to form dot cells (0.033 cm²). Front contacts were made by indium solder and silver paint.

CH3365-4/94/0000-0334 $4.00 © 1994 IEEE

Light current-voltage measurements were performed under an ELH lamp calibrated with a GaAs standard cell. For each sample, ten dot cells were measured and averaged. The best cells obtained as a result of this study were sent to the National Renewable Energy Laboratory (NREL) for standard measurements.

Auger electron spectroscopy (AES) measurements were performed at NREL (by Mr. Art Nelson) on a CdTe sample deposited with a varying potential from -400 mV to -650 mV over a period of two hours in order to investigate possible Cd/Te compositional variations caused by the change in V_{dep}. Electron beam-induced current (EBIC) measurements were carried out on Au/CdTe/CdS/SnO_2 film structures (by Dr. Rick Matson at NREL). Scanning electron microscopy (SEM) was performed at CSM using a JEOL Model 840 instrument. SEM images on both the CdTe surface and on cross-sections were taken.

X-ray diffraction (XRD) analysis was carried out on as-deposited and annealed sections of each sample using a Rigaku x-ray diffractometer with Cu radiation. The measurements were done between 20°~80° (2θ) with an increment of 0.02°.

RESULTS AND DISCUSSIONS

Table 1 shows the effect of changing J_{dep} upon the cell performance. It seems that there is an optimum cell efficiency at 0.33 mA/cm² but the difference of cell efficiencies between samples is small. This result suggests that the deposition rate may be increased beyond 0.40 mA/cm² without greatly sacrificing the cell performance.

Table 1. Change of solar cell performances according to the CdTe deposition current densities (or deposition rate). Other deposition paramters were: $R_{Te/Cd}$= 2.7, V_{dep} = -600 mV.

Sample#	J_{dep}, mA/cm²	V_{oc}, V	J_{sc},mA/cm²	FF, %	η, %
042694	0.26	.660	21.5	59.3	8.5
041493	0.33	.630	22.3	59.2	8.7
042194	0.40	.620	22.0	61.5	8.4

Changing $R_{Te/Cd}$ induced an appreciable change in cell characteristics. As shown in Table 2, there was an optimum current ratio of 2.7 between the Te and Cd anodes. We believe that the change in $R_{Te/Cd}$ should have caused a corresponding change in CdTe stoichiometry, but XRD measurements did not reveal any differences among the samples. We think that the change in Cd/Te composition due to the variation of $R_{Te/Cd}$ occurs only within the CdTe stoichiometry range (~ 10^{16} cm^{-3} either Cd or Te rich) [8] and is therefore not detectable by XRD. We speculate that even a sensitive analytical method such as AES cannot detect the difference because the changes will be shadowed by the Cd/Te matrix. This idea is supported by AES measurements on a CdTe sample deposited with a variation of V_{dep} (Fig 1).

Table 2. Change of solar cell performances with anode current ratio, $R_{Te/Cd}$. Other deposition parameters were: $J_{dep} \approx$ 0.33 mA/cm², V_{dep} = -600 mV.

Sample#	$R_{Te/Cd}$	V_{oc},V	J_{sc} mA/cm²	FF,%	η,%
042094	3.5	.650	20.4	60.0	8.0
041494	2.7	.630	22.3	59.0	8.7
050394	2.4	.630	22.2	58.7	8.2
041594	2.2	.630	21.5	59.0	8.0

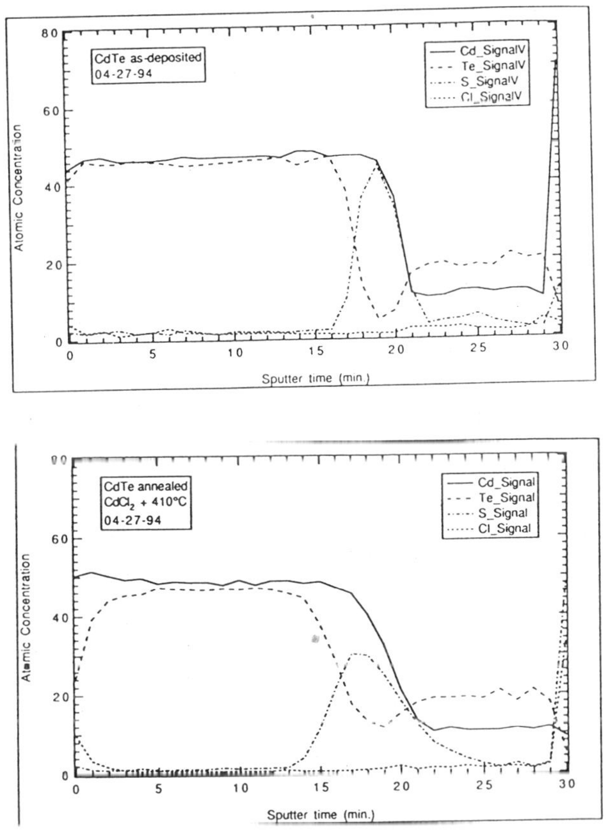

Fig. 1. Composition profile through the depth of a CdTe/CdS sample measured by Auger electron spectrosopy. The CdTe film was deposited with V_{dep} varied during the deposition: (a) as-deposited and (b) after $CdCl_2$ treatment and annealing.

According to Panicker et al. [1], the composition of CdTe should become Cd-rich as $|V_{dep}|$ increases. Sample 042794 was deposited with a changing Vdep from -400 mV to -650 mV over two hours. The AES data for the as-deposited CdTe in Fig.1 do not show the corresponding variations in composition, probably because of shadowing by the CdTe matrix. The $CdCl_2$ treatment and annealing induced significant changes, especially near the back surface. Our etching step for the metallization did not completely remove the Cd- and Cl-rich region at the surface. Chlorine and excess Cd tend to make CdTe n-type If there is an n-type region at the backside of the cell, this will cause a back-diffusion loss of photogenerated electrons. One must therefore optimize the amount of $CdCl_2$ coating or find an alternative fluxing agent.

SEM studies of the surfaces and cross-sections of CdTe samples did not reveal significant differences regardless of the growth conditions used here. As-deposited CdTe showed a faceted surface morphology and a quasi-columnar structure both of which were erased after annealing as shown in Fig. 2. Obtaining a columnar structure is important to minimize the recombination losses for the same material characteristics. Fig. 2 shows that one must carefully contrôl the annealing process to maintain columnar structure in the final cell.

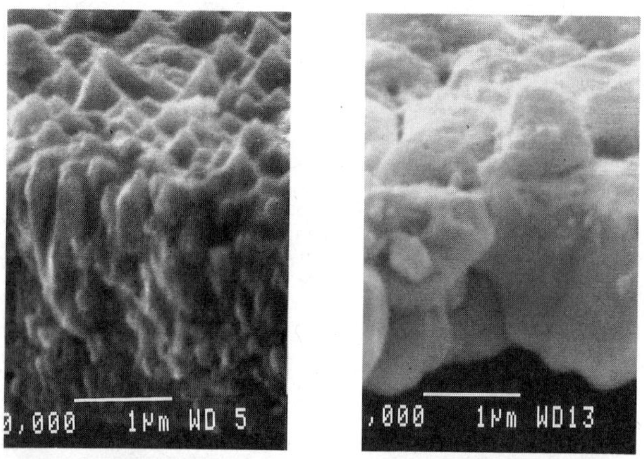

Fig. 2. Scanning elecron microscopy images of electrodeposited CdTe surface: (a) as-deposited, (b) as-annealed.

From the EBIC study (Fig.3), we found that the peak of the signal is inside the CdTe layer suggesting a buried homojunction. We suggest that the residual Cd and Cl on the CdS surface may have diffused into the CdTe making an n-type CdTe region at the CdS/CdTe interface. In order to have a stoichiometric CdS surface free of contamination, we put an overlayer of as-deposited CdS on the $CdCl_2$-treated, annealed CdS. The latter typically has large grains and a smooth surface. As a result of this approach, we observed a significant increase of V_{oc} up to 0.78 V as is presented in Table 3. We have demonstrated 11.0% CdS/CdTe/Au cells (0.033 cm²) from this study as shown in Figure 4.

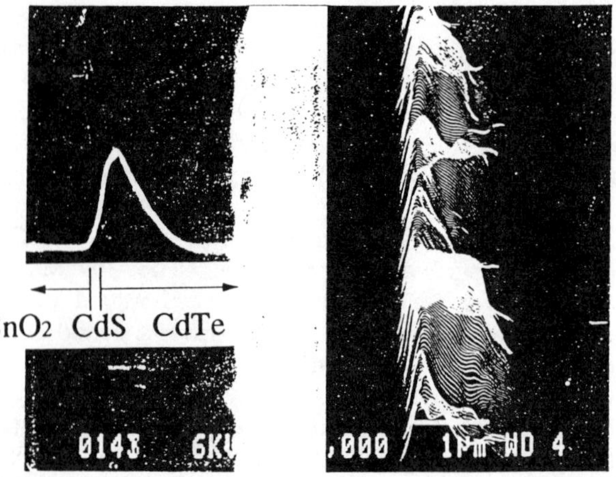

Fig. 3 Electron beam-induced current (EBIC) response for Au/CdTe/CdS/SnO_2 film. (Data were obtained at NREL by Dr. Rick Matson.)

Table 3. Effect of CdS double layers on V_{oc}. The best V_{oc} from CdS single layers is presented for comparsion.

Sample#	V_{oc}, V	J_{sc}, mA/cm²	FF, %	η, %
2-60	.780	18.9	58.4	8.61
1-60-2	.768	20.9	58.6	9.39
1-60-2	.774	20.7	57.9	9.26
120993 (single layer)	.772	21.3	67.7	10.4

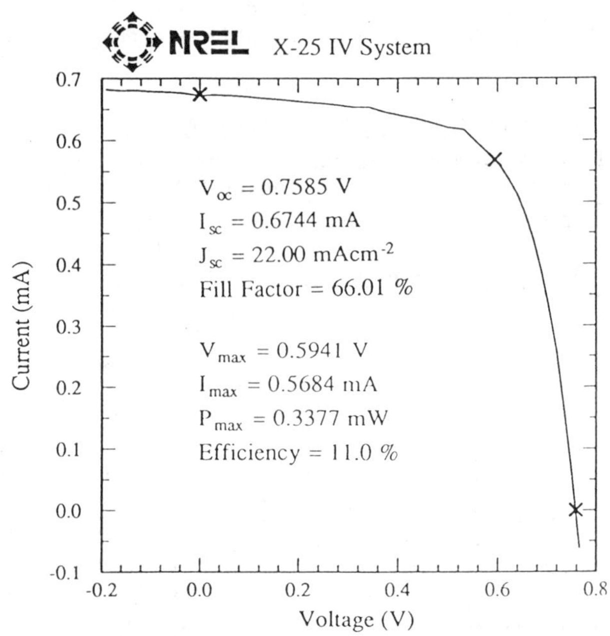

Fig. 4. Result of light current-voltage measurement on a CdS/CdTe/Au dot cell fabricated by electrodepostion (0.03 cm²). (Data furnished by NREL.)

SUMMARY

We have demonstrated 11.0% CdS/CdTe/Au dot cells (0.033 cm²) made by electrodeposition. The deposition parameters for CdTe layers were optimized: depostion current density J_{dep} = 0.33~0.40 mA/cm²; anode current ratio $R_{Te/Cd}$ = 2.7; and deposition potential V_{dep} = -600 mV against a Ag/AgCl reference electrode. Within the variations of the deposition parameters investigated in this study, XRD, surface SEM, and cross-sectional SEM did not reveal appreciable differences in the CdTe films. Annealing changed the CdTe surface morphology from faceted to rounded and increased the grain sizes from below 0.5 μm to 0.5~1 μm. Changes of V_{dep} did not show corresponding changes in AES profiling but $CdCl_2$ treatment and annealing resulted in a Cd- and Cl-rich surface. EBIC data showed a peak inside the CdTe layer suggesting a buried homojunction. CdS double layers composed of a $CdCl_2$-treated, annealed CdS underlayer covered with an as-deposited CdS overlayer were used and a significant increase in Voc up to 0.78 V was achieved.

ACKNOWLEDGMENTS

We thank the scientists at the National Renewable Energy laboratory (NREL) for various measurments, especially Dr. Rick Matson for SEM and EBIC, Mr. Halden Field for light J-V, and Mr. Art Nelson for AES. This work is supported by NREL under Subcontract No. XG-2-11036-4.

REFERENCES

[1] M.P.R. Panicker, M. Knaster, and F.A. Kroger, J. Electrochem. Soc. **125**, 566 (1978).

[2] J.M. Woodcock, et al., Proc. 12th. European PVSEC meeting, Amsterdam, 1994.

[3] S.K. Das and G.C. Morris, J. Appl. Phys. **72**, 4940 (1992).

[4] B.M. Basol, S.S. Ou, and O.M. Stafsudd, J. Appl. Phys. **58**, 3809 (1985).

[5] P.V. Meyers, Solar Cells **27**, 91 (1989).

[6] J.M. Woodcock, A.K. Turner, M.E. Oezan, and J.G. Summers, Proc. 22nd IEEE Photovoltaic Specialists Conference, (IEEE, New York, 1993) p. 842.

[7] T.L. Chu, Shirley S. Chu, N. Schultz, C. Wang, and C.Q. Wu, J. Electrochem. Soc. **139**, 2443 (1992).

[8] K. Zanio in *Semiconductors and Semimetals, v. 13 Cadmium Telluride* (Academic Press, 1978, R.K. Willardson and A.C. Beer, eds.), p. 5.

EFFECT OF ANNEALING ON MICROSTRUCTURE, RESIDUAL STRESS, AND PHOTOVOLTAIC CHARACTERISTICS OF ELECTRODEPOSITED CdTe FILMS

Donghwan Kim, Bin Qi, D. L. Williamson, J. U. Trefny
Department of Physics, Colorado School of Mines, Golden, Colorado 80401

ABSTRACT

The effect of annealing on the characteristics of electrodeposited CdTe was investigated. As-deposited CdTe films showed a preferred orientation of <111> normal to the surface of the films. During the annealing, recrystallization and grain growth occurred and the crystallographic orientation became more random. Examination of relative intensity ratios of x-ray diffraction (XRD) peaks for samples annealed at different temperatures for different times showed that the orientation changed toward random up to a certain point before it started to move back toward the textured state. An Arrhenius analysis of the XRD intensity changes yielded an activation energy of 2.5 eV which matches well with that of Cd diffusion in CdTe suggesting that the reaction should be Cd diffusion-limited. The residual stresses in CdTe layers measured by XRD were on the order of 50 MPa which is much higher than the yield stress of CdTe indicating that the stress should introduce crystalline defects.

INTRODUCTION

Fabrication of CdTe-based solar cells using electroplating is an attractive means to produce solar panels because: (i) cells with efficiencies higher than 14% have been demonstrated by this method [1], (ii) the technology is relatively well understood and established, and (iii) it is one of the most inexpensive methods for coating large areas.

Electroplated CdTe was pioneered and theoretically established by Panicker et al. [2] and since then the method has been applied to making CdTe/CdS solar cells by several groups [3, 4]. As-deposited CdTe films typically have poor microstructure with a high density of crystalline defects, so they are heat-treated at elevated temperatures in atmospheres containing oxygen. A $CdCl_2$ coating is usually applied to the surface of CdTe before the heat treatment [5]. During heat treatment, electrodeposited CdTe films undergo several reactions:(i) recrystallization, (ii) grain growth, (iii) changes in surface morphology and microstructure [6], and (iv) electrical type conversion from n-type to p-type [3]. The difference in thermal expansion coefficients of glass and CdTe is expected to induce stress in the CdTe layer when the samples are heated [7]. A high thermal stress may create crystalline defects and thus reduce the performance of CdTe as a solar cell material.

It is well accepted that the annealing step is critical for making high-efficiency solar cells based on electrodeposited CdTe films, but the mechanism of the reactions during annealing does not seem to be very well understood. Nor has the effect of annealing on the residual stress in CdTe been measured with the whole cell structure in place. In this study, we have investigated the effect of heat treatment on the recrystallization, grain growth, residual stress, and photovoltaic performance of CdTe layers.

EXPERIMENTAL PROCEDURES

CdTe films (2 μm thick) were grown on CdS/SnO_2/glass substrates by electrodeposition. The sample dimensions were 7.5 cm x 15 cm. The SnO_2 layer had a sheet resistance of 10 Ω/sq. and was manufactured by Nippon Sheet Glass. The CdS layer was approximately 0.1μm thick and deposited by spray pyrolysis. The CdTe was cut into 1 cm x 1 cm samples for annealing experiments. The annealing was done in a N_2 atmosphere after $CdCl_2$ treatment. A strip heater was used in our experiments and was composed of Mo sheet metal as the heating element and a graphite block as the sample holder. This system reaches temperatures above 400°C in less than 1.5 min and cooling from 400°C to 200°C takes approximately 2 min. The temperature was maintained within 2°C of the set temperature. After the annealing, the samples were rinsed in hot (about 80°C) water and dried in N_2 gas. For making solar cells, the samples were rinsed after x-ray diffraction (XRD) measurements, dried in N_2, and etched in 0.1% Br-MeOH solution for 10 s. Au contacts were evaporated through a mask to make 0.033 cm^2 dot cells. An ELH lamp was used as the light source for the light current-voltage (J-V) measurements. The lamp was calibrated with a GaAs standard cell for AM1.5 radiation.

The XRD measurements were performed in a Rigaku x-ray diffractometer using CuKα radiation. For this

CH3365-4/94/0000-0338 $4.00 © 1994 IEEE

study, we used a small scanning step ($\Delta 2\theta$= 0.02° or 0.01° per step) and a fixed time interval of either 2 s or 1 s for each step. A wide range of 2θ (from 20° to 80° or 20° to 100°) was scanned, so that all the major diffraction peaks could be detected.

RESULTS AND DISCUSSION

Heat treatments induce a recrystallization and grain growth in electrodeposited CdTe films and thus cause a large change in XRD patterns as presented in Fig.1. As-deposited films were textured in the <111> orientation normal to the film surface. The orientation changed to near random after annealing as ascertained by comparison with the pattern from a bulk powder. Changes in XRD patterns due to annealing have been reported by various other authors [8].

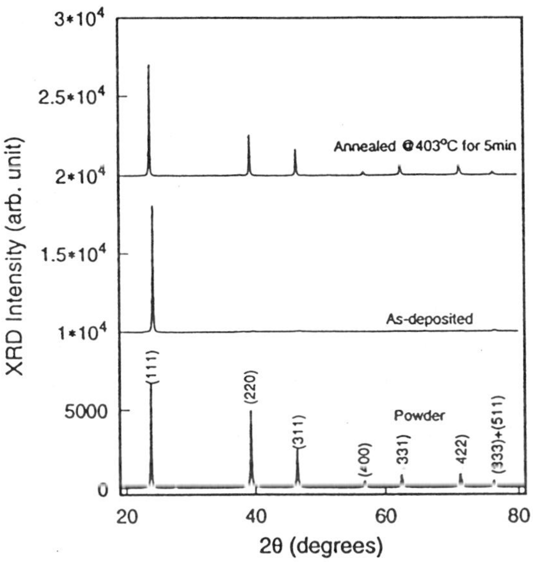

Fig.1 XRD patterns of different CdTe samples: annealed film, as-deposited film, and CdTe bulk powder.

In order to quantify the degree of the recrystallization/grain growth in CdTe, we plotted the ratio (R) between the intensity of the (220) peak and the (111) peak in the XRD patterns. This ratio was divided by the same ratio obtained from the bulk powder and called the "normalized intensity ratio" ($R_{normalized}$), i.e.,

$$R = I(220)/I(111) \quad (1)$$

$$R_{normalized} = R_{film} / R_{powder} \quad (2)$$

$R_{normalized}$ is a measure of the degree of texture in a film as referenced to that of bulk powder. It is assumed that the bulk powder is in a random orientation. For example, $R_{normalized}$ = 0 means that the film is completely textured along <111>, and $R_{normalized}$ = 1 indicates completely random orientation. Fig.2 shows the change of $R_{normalized}$ for CdTe samples annealed at various temperatures for different times.

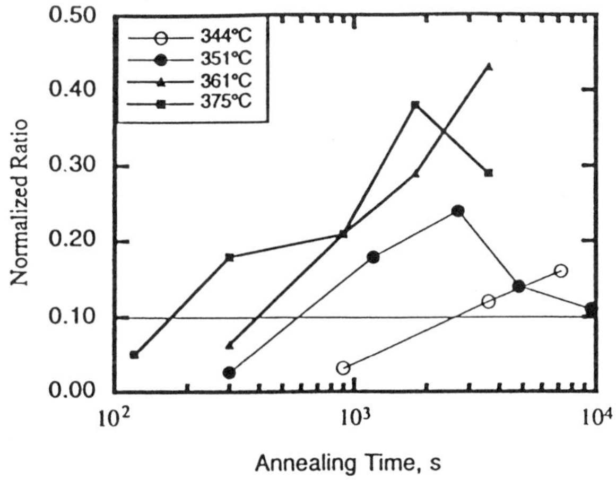

Fig. 2. Normalized intensity ratios of XRD peaks for CdTe films annealed at various temperatures for different times.

There is a trend that for shorter times the orientation of the films moves toward a random state to a certain degree probably because of recrystallization, and then back toward the textured state for even longer anneal times probably because of grain growth. Assuming that the recrystallization process is dominant at the beginning of the annealing, we chose $R_{normalized}$ = 0.1 and plotted the times at that point for different temperatures as is shown in Fig. 3. Using the equation:

$$R_{normalized} = A \exp(- E_a / kT) \quad (3)$$

we obtained an activation energy (E_a) of 2.5 ± 0.3 eV which matches very well with E_a for Cd diffusion in CdTe which is 2.67 eV at maximum Cd vapor pressure and 2.44 eV at minimum Cd vapor pressure [9]. This result indicates that the recrystallization of CdTe during annealing should be Cd diffusion limited.

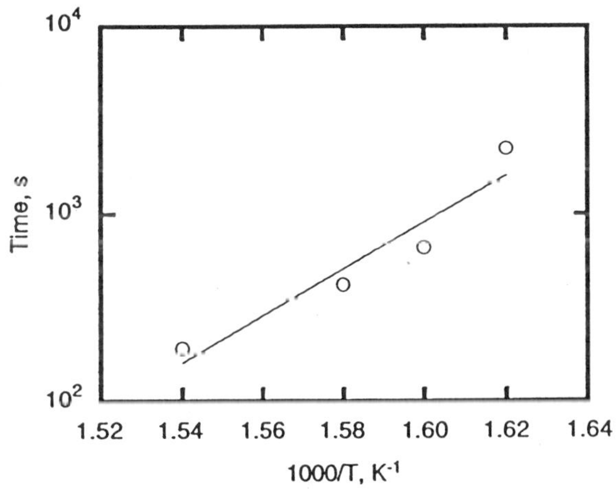

Fig. 3. Relationship between the annealing temperature and the time for 10% recrystallization.

Heat treatment should induce stress in CdTe films due to differences in the thermal expansion coefficients of CdTe and the glass substrates. In order to estimate the magnitude of the residual stress, we measured the lattice constants of the samples using standard XRD methods [10]. The results are summarized in Fig. 4. The lattice constants are measured along the direction normal to the sample surface when using a standard θ/2θ diffractometer as done here. A decreased lattice constant in the surface normal direction is a result of an increase of in-plane lattice spacing which in turn is a result of an in-plane tensile stress and vice versa for an in-plane compressive stress.

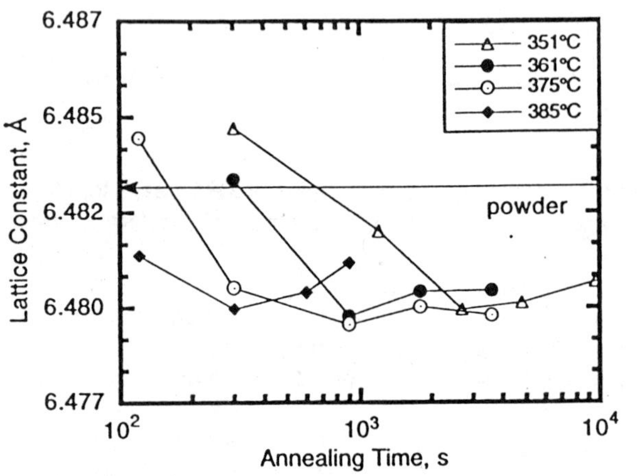

Fig. 4. Change of lattice constants of electrodeposited CdTe according to the annealing temperature and time.

Clemens and Bain [11] have developed a formula to calculate the in-plane stress σ for films textured along the <111> orientations. The strain at an angle ψ with respect to the surface normal (n) is,

$$\varepsilon_\psi^{n//<111>} = (d_\psi - d_o)/d_o$$
$$= \sigma[(2S_{11} + 4S_{12} - S_{44})/3 + (S_{44}/2)\sin^2\psi] \quad (4)$$

where d_ψ is the atomic plane spacing for a particular set of planes with their normal at an angle ψ with respect to the surface normal (ψ = 0 here), d_o is the unstrained plane spacing, and S_{ij} are the compliance elastic constants. The values of S_{11}, S_{12}, and S_{44} for CdTe are 4.27×10^{-12}, -1.73×10^{-12}, and 5×10^{12} cm²/dyn, respectively [12]. Positive and negative signs of stress correspond to tensile and compressive stress, respectively. The magnitude of the residual stress is given in Fig. 5. Both the sign and the magnitude of the stress are observed to change with the annealing time. The magnitude of the stress showed a maximum at approximately 50 MPa and the stress was relieved somewhat for longer anneal times. The residual stress was also measured by the curvature method using the Flexus™ system and there was a reasonable agreement between the two methods. The critical resolved shear stress of CdTe is less than 10 MPa [13]. This suggests that it is highly probable that the above stress may introduce crystalline defects such as stacking faults and dislocations in the crystal. Time-resolved photoluminescence (PL) data obtained at NREL and given in Fig. 6 support this view. As annealing time is increased, the decay time decreases indicating a defect density in CdTe films which grows with annealing time. However, the solar cell efficiency measured from CdS/CdTe/Au dot cells (0.033 cm²) made with these samples (Fig. 7) shows a monotonic increase with time indicating that the cell performance does not depend directly on the PL-determined quality of the CdTe in this case as has been observed by other authors [14].

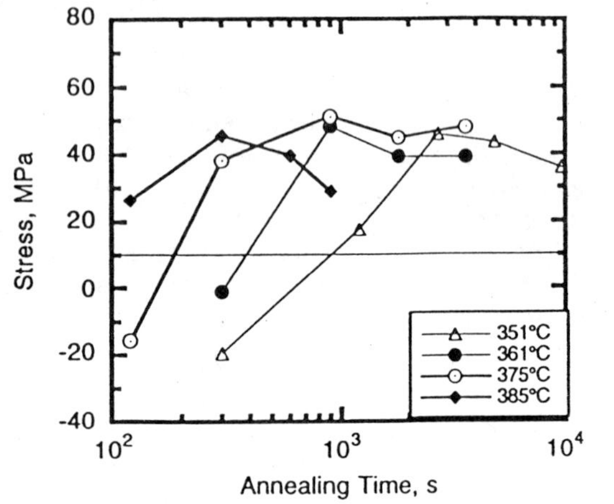

Fig.5. Residual stress in electrodeposited CdTe films as a function of annealing temperature and time.

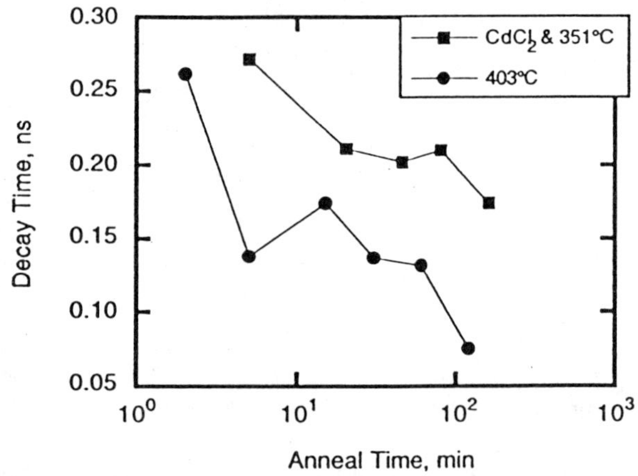

Fig. 6. Decay times obtained from electrodeposited CdTe films by the time-resolved photoluminescence method as a function of annealing temperature and time.

Our experimental evidence suggests that the magnitude of residual stress is related to the microstructure of the underlying CdS. The change of the band gap due to the stress was found to be minimal. If crystalline defects are introduced in polycrystalline films due to stress, it is expected that the distribution of the defects should vary from grain to grain according to the relative crystallographic orientation of the grains with respect to the stress direction. A non-uniform distribution of defects among grains has been observed by Al-Jassim et al. [15]. We speculate that the non-uniform distribution of defects should be responsible for the fluctuation of the electron beam-induced current (EBIC) response among CdTe grains observed by various authors [14,16]. We suggest that the residual stress must be controlled below the yield stress of CdTe in order to have a uniform photovoltaic performance among different grains in CdTe films and ultimately to get closer to the theoretical efficiency limit of CdTe solar cells.

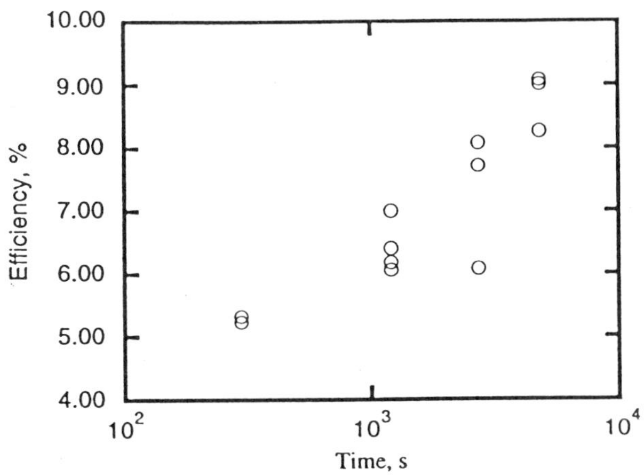

Fig. 7. Efficiency of CdS/CdTe/Au solar cells as a function of CdTe annealing time at 351°C.

SUMMARY

Effects of annealing on the microstructure and on the stress state of electrodeposited films have been investigated. As-deposited films showed a preferred orientation along <111> normal to the surface and changed to a more random orientation as a result of annealing indicating recrystallization. By plotting normalized XRD intensity ratios for different annealing times at different temperatures, it was shown that the crystallographic orientation changed only to a certain degree for intermediate annealing times and moved back toward the textured state for even longer times. Activation energy (E_a) for recrystallization was measured to be 2.5 ± 0.3 eV by plotting the time for "10%" recrystallization for different temperatures. The value of E_a for recrystallization matches well with that of Cd diffusion in CdTe, indicating that the recrystallization reaction is Cd diffusion limited. The residual stress in CdTe films was estimated from lattice constant changes during annealing as measured by XRD. Both the sign and the magnitude of the in-plane stress changed with annealing and the maximum stress was on the order of 50 MPa which is much larger than the critical resolved shear stress of CdTe (<10 MPa). Crystalline defects should be created in the film by the stress and this view is supported by the results of the time-resolved photoluminescence studies where the decay time decreased with annealing time. However, the cell efficiency increased with annealing time, suggesting that the efficiency in these devices is not limited by the defects that are affecting the PL decay time.

ACKNOWLEDGMENTS

We appreciate Mr. Brian Keyes of the National Renewable Energy Laboratory (NREL) for the time-resolved photoluminescence measurements. We thank Dr. Muhling Ger at the Department of Electrical Engineering, University of Michigan, for the Flexus measurements. This work is supported by NREL under Subcontract No. XG-2-11036-4.

REFERENCES

[1] J.M. Woodcock, A.K. Turner, M.E. Oezan, and J.G. Summers, Proc. 22nd IEEE Photovoltaic Specialists Conference, (IEEE, New York, 1993) p 842.

[2] M.P.R. Panicker, M. Knaster, and F.A. Kroger, J. Electrochem. Soc. **125**, 566 (1978).

[3] B.M. Basol, S.S. Ou, and O.M. Statsudd, J. Appl. Phys. **58**, 3809 (1985).

[4] S.K. Das, Thin Solid Films **226**, 259 (1993).

[5] A. Rohatgi, Int. J. Solar Energy **12**, 37(1992).

[6] B. Qi, *Effects of Post Deposition Heat-Treatment on Morphology and Microstructure of CdTe Grown by Electrodeposition*, M.S. Thesis, Department of Physics, Colorado School of Mines, Golden, Colorado 80401 (1994).

[7] S. Jimenes-Sandoval, M. Melendez-Lira, and I. Hernandez-Calderon, J. Appl. Phys. **72**, 4197 (1992).

[8] See, for example, M. Takahashi, K. Uosaki, H. Kita, and Y. Suzuki, J. Appl. Phys. **58**, 4292 (1985).

9] P.M. Borsenberger and D.A. Stevenson, J. Phys. Chem. Solids **29**, 1277 (1968), or *Atomic Diffusion in Semiconductors*, D. Shaw, ed., (Plenum Press, New York 1973), p.503.

[10] H. Klug and L. Alexander, *X-ray Diffraction Procedures for Polycrystalline and Amorphous Materials*, 2nd ed., (Wiley, New York 1974), p.594.

[11] B.M. Clemens and J.A. Bain, MRS Bulletin **17**, 47 (1992).

[12] A. Gavini and M. Cardona, Phys. Rev. **B1**, 672 (1970).

[13] R. Balasubramanian and W.R. Wilcox, Materials Science and Engineering **B16**, 1(1993).

[14] See, for example, A.D. Compaan, C.N. Tabory, Y. Li, Z. Feng, and A. Fischer, Proc. 23rd IEEE Photovoltaic Specialists Conference, (IEEE, New York. 1993) p.394.

[15] M.M. Al-Jassim, F.S. Hasoon, K.M. Jones, B.M. Keyes, R.J. Matson, and H.R. Moutinho, Proc. 23rd IEEE Photovoltaic Specialists Conference, (IEEE, New York. 1993), p.459.

[16] S. Pozder, *Effect of CdTe Electrodeposition Parameters on the Efficiency of CdS/CdTe/Au Solar Cells*, M.S. Thesis, Department of Physics, Colorado School of Mines, Golden, Colorado 80401 (1994).

HIGH CONDUCTIVITY CdS THIN FILMS FORMED BY GAS ASSISTED CHEMICAL VAPOR TRANSPORT (CVTG)

P.J. Sebastian and A. Sanchez
Photovoltaic Systems Group, Laboratorio de Energia Solar
IIM-UNAM, 62580 Temixco, Morelos, Mexico

ABSTRACT

Formation of high conductivity CdS thin films by a new chemical vapor transport (CVT) method, employing a carrier gas, is reported for the first time. The film deposition was carried out in a quartz tube kept inside a tubular furnace. The chemical vapor transport was done using a screen printed film of CdS with $CdCl_2$ as the flux incorporated in the film. The nature of the film depends on parameters like condensation temperature, cooling rate, substrate position etc. The extrinsic films showed an electrical conductivity of the order of 10^1 $(\Omega\ cm)^{-1}$ and the films which showed intrinsic nature exhibited a conductivity of the order of 10^{-8} $(\Omega\ cm)^{-1}$. The new CVTG system is designed and further studies are in progress to get thin film solar cell structures employing this technique.

INTRODUCTION

CdS films are studied with renewed interest due to its potential application as the window layer in heterojunction solar cell structures like CdS/CdTe, CdS/$CuInSe_2$ etc. Recently most of the high efficiency polycrystalline solar cell structures based on CdTe, $CuInSe_2$, CuIn(Ga)Se_2 etc. are formed with CdS film as the n layer. CdS films have been prepared by various techniques such as thermal evaporation in vacuum [1], sputtering, [2], CVD [3], chemical deposition [4] screen printing [5], electro- and electroless deposition [6] etc. The difficulty in obtaining high conductivity and stable films with good optical properties is a major concern of many of the above mentioned techniques. High conductivity CdS films are obtained by post-deposition treatments [7]. But post-deposition treatments in many occasions impart structural transformations [8] and impurity states in the films, which can affect the stability of the films and the devices.

In the present communication we report the initial results of the deposition and characterization of low resistivity and stable CdS films by a new CVT technique, called chemical vapor transport by gas (CVTG). In the present study Argon was used as carrier gas. The various parameters which control the film properties have been identified and the system to carry out CVTG is designed. This technique may successfully be used for chalcogenization of metals and alloys and deposition of other semiconductors like CdTe, CdSe, $CuInSe_2$ etc. By CVTG semiconductor films can be deposited by a one step process as in the present case (CdS) or by a two step process involving the metal or alloy deposition followed by chalcogenisation employing CVTG. This method may function as an alternative to chalcogenisation of alloys or metals by thermal evaporation in vacuum or chalcogenisation by toxic gases like H_2Se, H_2S etc.

EXPERIMENTAL

CdS films were screen printed using 99.995 % pure CdS powder as the pigment and $CdCl_2$ as the flux. Propylene glycol was used as the binder. The printing was done using a #120 screen. The screen printed film was kept at the middle of the tube and functions as the source in the present case. The screen printed film containing CdS and $CdCl_2$ was used as the source, instead of powder sample, for the ease with which the vapors of CdS can be transported due to the large surface area exhibited by the film. During the transport process the material was completely removed from the source. Argon was allowed to flow from one end of the tube to the other, controlled by a gas flow mechanism, and removed by a pumping system connected at the exit port of the tube. The quartz tube was heated by the tubular furnace around it and the temperature was monitored using chromel-alumel thermocouples. The glass

CH3365-4/94/0000-0342 $4.00 © 1994 IEEE First WCPEC; Dec. 5–9, 1994; Hawaii

substrates were kept at the exit port of the tube at a lower temperature than that of the source. The sublimation was carried out at a gas pressure of $3x10^{-1}$ mbar and the films were cooled down rapidly (cooling rate of the order of 20 ^{o}C/minute) using cool air.

The XRD patterns of the films were recorded by using a siemens D500 x-ray diffractometer. The optical transmittance of the film was measured with a shimadzu UV365 spectrophotometer. The photoresponse characteristics of the films were studied by using a computerized opto-electronic analysis system described elsewhere [4].

RESULTS AND DISCUSSION

The CVTG Reactor

Figure 1 shows the two dimensional sketch of the new CVTG reactor used for the experiment. It consists of a tubular furnace with 40 cm in length and facilities for temperature control along the length of the tube. The quartz tube with 60 cm in length and 5.4 cm in diameter is kept inside the furnace. The controlled gas flow mechanism is at the inlet port and the pumping system is connected at the exit port. There is a temperature gradient along the length of the tube with the maximum temperature at the center of the tube and minimum temperatures at the ends. The temperature at the exit port is influenced also by the air cooling mechanism at the exit port. The reactor is evacuated first to a pressure of the order of $1x10^{-3}$ mbar. Later on the reactor was fluxed with pure argon gas. During vapor transport the gas pressure was kept constant at 3×10^{-1} mbar at a constant flow rate of 100 SCCM (standard cubic centimeters per minute).

It was observed that film resistivity and film nature depend on the condensation temperature and rate of cooling. Generally the film is formed with three distinct zones along the length of the tube as follows;

1) Intrinsic CdS, close to the source but towards the gas flow direction and at a distance of 13-16 cm from the source (condensation temperature - 500 ^{o}C)

2) Extrinsic CdS (n-type), away from the source and towards the gas flow direction and at a distance of 20-21 cm from the source (condensation temperature - 150 ^{o}C)

3) $CdCl_2$,, in between the above two regions and at a distance of 16-20 cm from the source (condensation temperature - 230 ^{o}C)

The photoresponse of CdS films relating to cases 1 and 2 above is shown in figure 2, which is a plot of current versus time in dark and under illumination. The applied voltage in this case is 5V. The film was kept in dark till the dark current got stabilized. The measurement was done as follows; 30 seconds in dark, 30 seconds under illumination using a tungsten-halogen lamp of intensity 600 W/cm^2 and another 80 seconds of photocurrent decay in dark. This figure shows that CdS corresponding to case

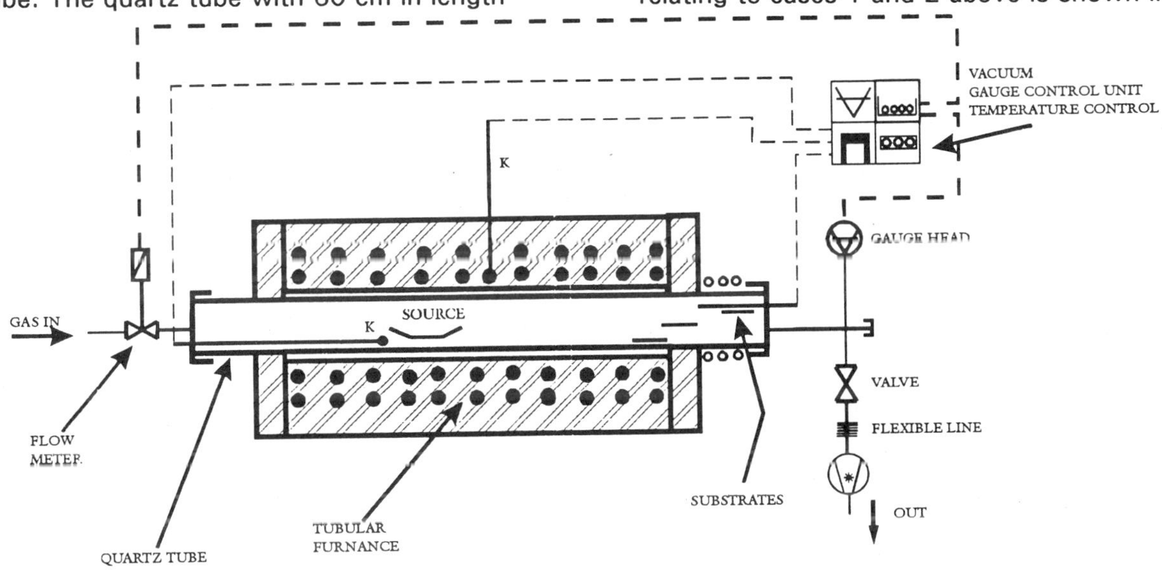

Fig. 1: The CVTG Reactor

1 exhibits very low dark current, the photo- to dark current ratio is a few orders of magnitude and the photocurrent decay is fast, which shows the intrinsic nature of the film. The CdS film corresponding to case 2 shows high dark current and does not show much photosensitivity due to the high conductivity, the type of conductivity was confirmed as n-type. The planar electrical conductivity was calculated in both the above cases. The ntrinsic film has a thickness of 1100 A^{o}, dark conductivity of 1.3×10^{-8} $(\Omega\ cm)^{-1}$ and photoconductivity of $2.5\times10^{-6}(\Omega\ cm)^{-1}$. The

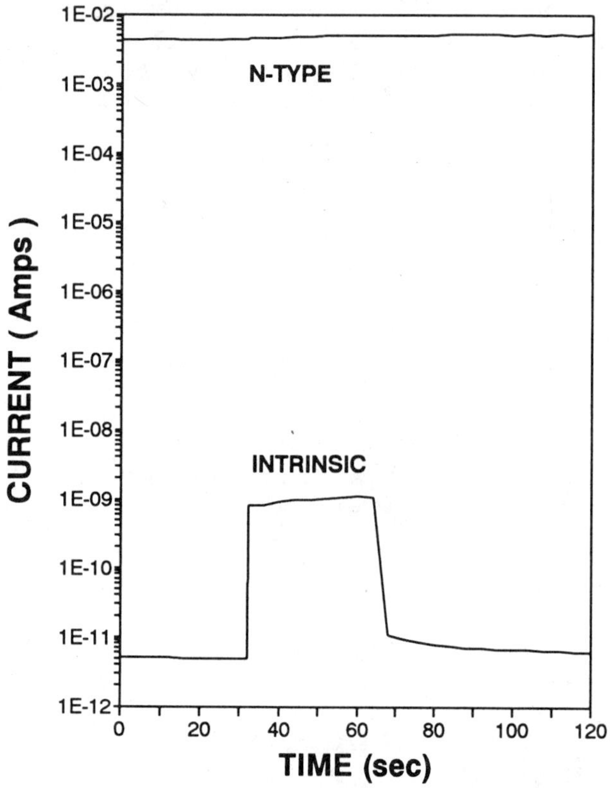

Fig.2: The dark and photocurrent characteristics of CVTG deposited CdS films

n-type film exhibited a dark conductivity of 1.2×10^{1} $(\Omega\ cm)^{-1}$ without appreciable variation upon illumination,

Figure 3 gives the XRD pattern for CdS films obtained by CVTG. This figure indicates that CVTG deposited both n-CdS and intrinsic CdS match very well with the standard hexagonal structure (Greenockite). The XRD peaks corresponding to CdS in the figure did not match with the standard cubic (Hawleyite) structure. The intrinsic sample is polycrystalline in nature and the extrinsic sample shows textured nature with preferred orientation along (002). It is certain that this film is c-axis oriented. The intrinsic film also show peaks corresponding to $CdCl_2$ due to the overlapping of the two films formed side by side. This can be avoided by properly placing the substrates at distances given in the previous section, where the nature of the film formation is described.

CdS films are known to form in the hexagonal or cubic modifications. Generally when it is formed in the hexagonal form, the

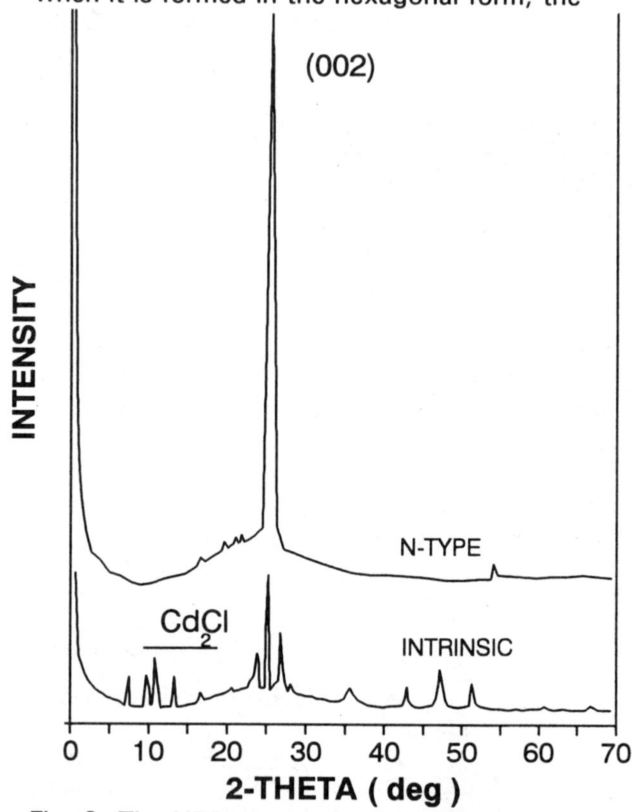

Fig. 3: The XRD patterns for CVTG deposited CdS thin films

film shows high conductivity due to close packing, but the cubic modification or a mixture of the two show poor conductivity [9]. In many occasions researchers have reported the high conductivity shown by CdS films due to the growth of the hexagonal phase at the expense of the cubic phase when there is chlorine doping in the film [10]. In the present case there is every chance of chlorine doping when the film is condensed, so also excess of Cd present in the film. Further studies are in progress to relate the high electrical conductivity with possible doping (Cl_2, Cd etc.) in the film. Another possibility is the effect of condensation temperature on film conductivity. The n-type film was formed when the film was formed at a low condensation temperature (case 2 above) than that of the intrinsic film (case 1 above).

Figure 4 displays the optical transmittance of the extrinsic film in the region 350 to 2500 nm. The spectra was recorded with air as reference. This figure shows that the film exhibits an optical transmittance of the order of 80 % above the band gap which is adequate for its application as the window layer in thin film solar cell. The optical spectra indicates considerable absorption or light scattering in the film. This

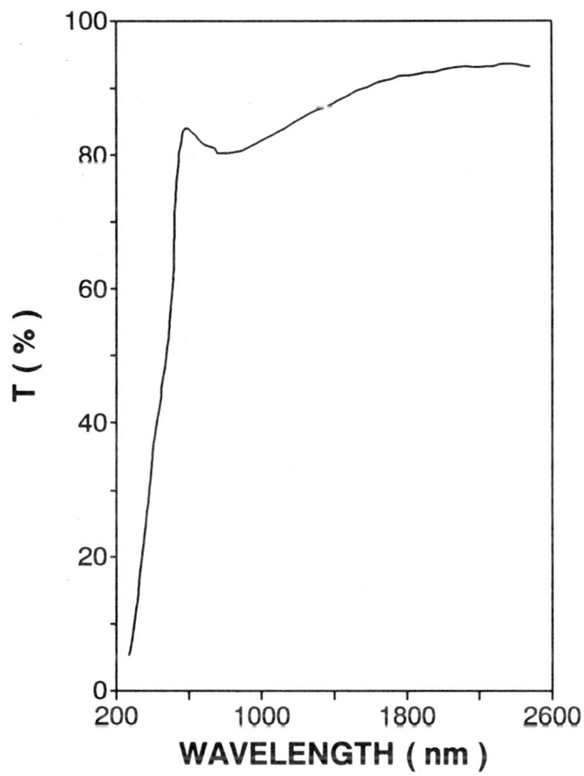

Fig.4: The optical transmittance spectra for CVTG formed CdS thin film

may be attributed to the extrinsic nature of the films and due to the absorption by the glass substrate.

CONCLUSION

n-CdS films were formed by a new CVT technique called chemical vapor transport by gas (CVTG). In the present case the chemical vapor transport was done employing argon as the carrier gas. The transport was performed by using a screen printed CdS film containing $CdCl_2$ as the flux. Three types of films are formed along the length of the tube. a) intrinsic CdS at a distance of 13-16 cm from the source and towards the gas flow direction, b) $CdCl_2$ at 16-20 cm from the source and towards the gas flow direction and c) n-CdS at 20-21 cm from the source and towards the gas flow direction. The intrinsic films exhibited high photosensitivity and extrinsic films showed a dark conductivity of the order of 10^1 $(\Omega\ cm)^{-1}$. Both intrinsic and n-CdS exhibited the hexagonal Greenockite structure, with a preferred orientation along (002) for n-CdS. This technique holds the prospects of developing low cost thin film solar cell and photoconductor structures based on different semiconductors. Further studies are in progress towards this direction.

ACKNOWLEDGMENT

The authors express their gratitude to Ms. Leticia Baños for assistance in XRD studies and O. Gomez-Daza in film deposition.

REFERENCES

[1] K.H. Norian and J.W. Edington, *Thin Solid Films,* **75**, 1981, p. 35

[2] W.H. Leighton, *J. Appl. Phys.*, **44**, 1973, p. 5011

[3] L.D. Partain, G.J. Sullivan and C.E. Birchenall , *J. Appl. Phys.,* **50**, 1979, p. 551

[4] P.J. Sebastian and P.K. Nair, *Advanced Mater. Opt. Electron.,* **1**, 1992, p. 211

[5] S. Ikegami , *Solar Cells,* **23**, 1988, p. 89

[6] B.M. Basol , *Solar Cells,* **23**, 1988, p. 69

[7] I. Martil, G. Gonzalez-Diaz and F. Sanchez-Quesada, *Solar Ener. Mater.,* **12**, 1985, p. 345

[8] S. Kolhe, S.K. Kulkarni and A.S. Nigavekar, *Solar Ener. Mater.*, **10**, 1984, p. 47

[9] R. Jayakrishnan, S.R. Kumar and R.K. Pandey, *Semicond. Sci. Technol.,* **9**, 1994, p. 97

[10] S. Bilodeau, D. Campbell, R.K. MacCrone and S.J. Nettel, *Phys.Rev. B,* **37**, 1988, p. 3275

OPTIMIZATION AND CHARACTERIZATION OF REACTIVELY SPUTTERED ZNO

S. Zafar, C. S. Ferekides, and D. L. Morel
Department of Electrical Engineering
University of South Florida
4202 E. Fowler Ave.
Tampa, FL 33620

ABSTRACT

Highly uniform films of reactively sputtered ZnO:Al with resistivity of 5×10^{-4} Ω-cm and good optical properties have been produced. The resistivity can be reduced to 1-2 x 10^{-4} Ω-cm with some loss in near-IR transmission. Al incorporation drops rapidly with increasing T_{ss}, while carrier concentration(n) increases resulting in increases in doping efficiency of up to x 30. Transport properties are controlled by bombardment induced damage and ionized impurity scattering. The later results in a correlation between mobility and n which combined with the effect of free carriers on optical properties leads to fundamentals-driven design tradeoffs for solar cell use.

INTRODUCTION

ZnO is becoming an increasingly important material for use in various solar cell technologies. Because of its attractive electro-optical properties it is commonly used as a window material and can also be applied as an optical reflector at the back of thin-film cells. It can be deposited by a number of deposition techniques, including CVD which has a high deposition rate desirable for manufacturing. However, there are concerns with uniformity and yields in scaling up these techniques. Sputtering is an attractive manufacturing process but typically suffers from low deposition rates. With highly doped ZnO targets very high rates can be achieved[1], however, uniformity and yields in a manufacturing environment have not yet been established. Reactive sputtering from a zinc target is an attractive alternative because of the high deposition rates that can be achieved. It is necessary, however, to demonstrate good electro-optical properties and the potential for these to be maintained in a manufacturing environment. In this paper we wish to report progress which we have made toward this objective.

PROCESS DESCRIPTION

We have made considerable progress in achieved performance with reactively sputtered ZnO and in understanding the correlations between processing and performance. The fundamentals of our deposition process have been described previously[2]. Basically we reactively sputter zinc in an argon/oxygen environment using a 3" diameter DC magnetron. We dope with aluminum by adding Al pieces to the target and with fluorine through gaseous CF_4. Our substrates are 4" x 4" soda-lime glass.

RESULTS/DISCUSSION

Since there is a tradeoff between electrical and optical properties due to free carrier effects and the need to optimize this tradeoff against a particular device, there is no single figure-of-merit to quote for performance.

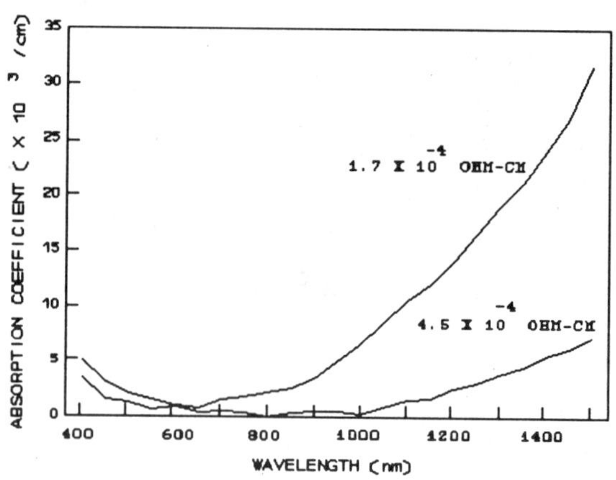

Figure 1. Absorption profile of ZnO.

We thus report that we can routinely achieve resistivities of 4.5×10^{-4} Ω-cm with good optical properties as shown in figure 1 and good uniformity as shown in figure 2b. As seen in figure 2a., conductivity as well as

CH3365-4/94/0000-0346 $4.00 © 1994 IEEE

First WCPEC; Dec. 5-9, 1994; Hawaii

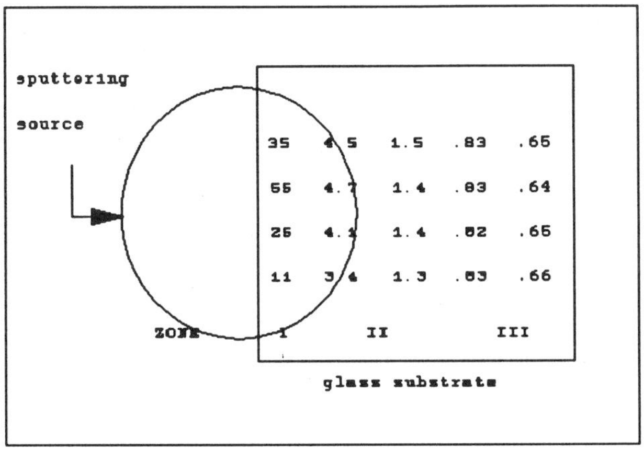

Figure 2a. Resistivity uniformity(value indicated x 10^{-3} Ω-cm) for reactively sputtered ZnO at T_{ss} = 150°C.

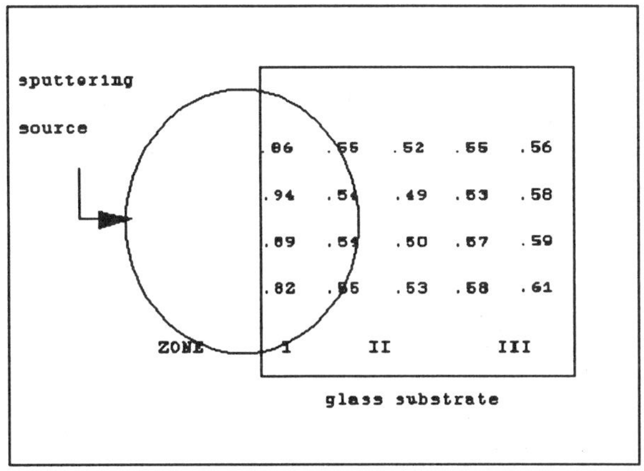

Figure 2b. Same as 2a. with T_{ss} = 350°C.

uniformity is lost at lower substrate temperatures(T_{ss}), particularly in the vicinity of the target racetrack. By increasing the doping concentration we can lower the resistivity to 1.7 x 10^{-4} Ω-cm but suffer increased absorption losses(figure 1) due to free carrier effects. These performance parameters are as good as the best reported in the literature for ZnO made by any process. We are now at the point where performance is limited by physical fundamentals rather than by limitations imposed by the deposition process. We thus now focus our efforts on furthering the understanding of these fundamentals to determine how additional improvements in performance can be achieved.

To consider the effect of deposition conditions on the underlying fundamental mechanisms we have divided our 4" x 4" substrates into three zones as shown in figure 2. The substrate is intentionally offset relative to the target because of the known effects of bombardment, especially over the racetrack region. Thus film properties are generally found to vary according to Zone location and sputtering conditions. Zone I generally exhibits poorer properties because it is closest to the source and most affected by ion bombardment. However, as will be discussed, the effects of bombardment are complex and can be manipulated to effect improved performance in Zone I.

Although growth rate is not significantly influenced by substrate temperature, we have previously reported the role of T_{ss} in achieving uniformity and good electro-optical properties[2]. We speculated that improvements with increasing T_{ss} were due to enhanced grain structure, but our most recent results indicate the influence and interplay of additional mechanisms. The dependence of conductivity on T_{ss} for each zone position is shown in figure 3 for ZnO:Al. As shown, the lower conductivity of Zone II relative to Zone III can be overcome by increasing T_{ss}. This is also the case for Zone I, but higher T_{ss} is required. Although the conductivity data is relatively well behaved the underlying details tell a more complex story.

In figure 4 is shown the dependence of Al incorporation as determined by EDS on T_{ss} along with the carrier concentration(n) from Hall effect measurements. As can be seen, Al incorporation decreases with increasing T_{ss}, while n generally increases. This indicates a rather strongly increasing Al incorporation efficiency with increasing T_{ss}. The most dynamic region is Zone I which shows an order of magnitude increase in n while Al incorporation is dropping by a factor of 3 which results in a x30 increase in Al incorporation efficiency. Zone II and Zone III indicate "relatively modest" changes in doping efficiency of about x6 in going from 150°C to 350°C, although Zone II exhibits strong deviations from the general trend between the limiting temperatures. The dependence of mobility(μ) on T_{ss} is shown in

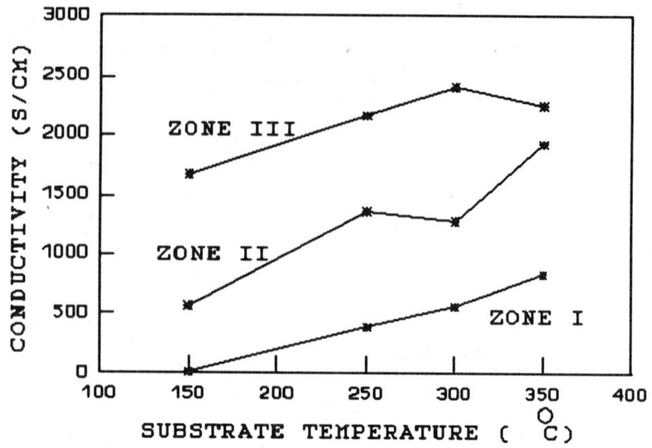

Figure 3. Conductivity versus substrate temperature for ZnO:Al.

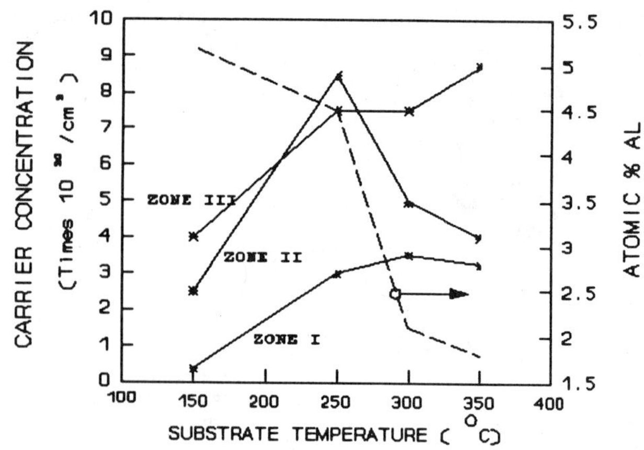

Figure 4. Carrier density and Al content versus substrate temperature for ZnO:Al.

figure 5. For Zone I the increase in μ with T_{ss} when combined with the increased carrier density results in the substantial increase in conductivity exhibited in figure 3. For Zone III, on the other hand, the decreasing μ with T_{ss} compensates the increase in carrier density resulting in the relatively flat behavior of figure 3. Although the conductivity dependence on T_{ss} for Zone II somewhat parallels that of Zone I, we see that the behavior of n and μ are quite different. Their non-monotonic behavior is due to the fact that Zone II is a transition Zone and hence more sensitive to small changes in other deposition parameters. It is noteworthy that their behavior is somewhat coupled resulting in the generally increasing conductivity dependence on T_{ss}.

In examining the data it is clear that there is a different response at a microscopic level to T_{ss} in the three zone regions. We attribute this to the fact that Zone I suffers intense ion bombardment damage which does not occur in Zone III. Consequently, in addition to the role T_{ss} plays in affecting Zone III properties, in Zone I it can also help remove bombardment induced damage. In fact, the significant rise in both n and μ with increasing T_{ss} in Zone I suggests that temperature induced defect removal is the primary driver of performance in Zone I. Once this damage is removed the mechanisms which control performance in Zone III then apply to Zone I. We have found that the transport mechanisms thus change from defect dominated scattering to ionized impurity scattering as the defects are annealed out[3] and that film performance is ultimately limited by these

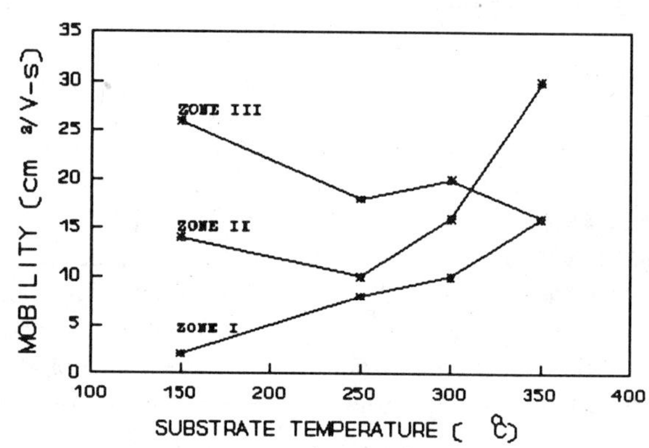

Figure 5. Mobility versus T_{ss} for ZnO:Al.

fundamental mechanisms.

The improved uniformity in going from T_{ss} = 150°C to 350°C in figure 2 can be explained in light of the above. While the resistivity is comparable in Zone III for the two, it increases significantly in moving toward Zone I at low T_{ss} due to dominance by bombardment induced defects. Thus films with good properties can be deposited at lower T_{ss}, but only with larger source-to-substrate distances and the concomitant loss in deposition rate.

The implication of these results to solar cell applications of this technology are important. In addition to the tradeoff between deposition rate and uniformity mentioned above, the interplay of two fundamental mechanisms must guide our design considerations.

* Optical properties are dominated by free carrier mechanisms in the near IR and perhaps into the visible.

* Carrier concentration and mobility are correlated.

Thus for a target spectral range as determined by the absorber and for a required sheet rho as determined by the grid spacing or cell width there is an optimum choice for the doping level. Since this choice is a tradeoff between optics and conductivity, neither will be optimized. It is thus necessary to explore ways of circumventing these fundamental limitations. What is desired is a reduction of the negative effect of dopants on optics and transport. A group at Kanazawa Institute of Technology[4] used Group IV dopants to attempt to produce two charge carriers for each dopant atom. Unfortunately the dopants bonded trigonally rather than tetrahedrally again producing one rather than two electrons each. Based upon these results and similar observations of our own we believe that it is necessary to better understand the effect of dopants on the microstructural properties of ZnO if we are to further improve its performance. Along these lines we have used XRD to analyze the series of films discussed above. Our analysis indicates that substrate temperature is playing a primary role in determining both grain size and orientation[3]. We have previously reported the effect of F on growth mechanisms, microstructure and electronic properties[2] and still feel that it may help decouple the mechanisms associated with conventional dopants. We have also tried Ga as a dopant in the hope that its closer size match to oxygen would result in reduced scattering. While our results thus far suggest a positive effect, it is rather small and will require further work to properly quantify. The insights which we are gaining are leading us to new dopant techniques that we feel will result in further improvements in performance.

Finally, we have initiated the use of reactively sputtered ZnO as the window layer in our CIS and CdTe solar cells. For CIS we have demonstrated equivalent device performance to our conventional ZnO layers which are sputtered from a ZnO target. This indicates that the excited oxygen species present during reactive sputtering do not damage the CdS/CIS surface. While the requirements for the window layer for these two technologies are quite different, we are finding that the properties of ZnO itself as well as the attributes of our deposition technology combine to provide a capability not available from other window layer options.

CONCLUSIONS

We have developed techniques for producing state-of-the art films of ZnO by reactive sputtering from a zinc target. The role of ion bombardment in film growth and performance and the effect of T_{ss} on the controlling mechanisms has been demonstrated. Increases in doping efficiency of up to x 30 have been observed. Transport mechanisms have been shown to be controlled by defect dominated scattering in the vicinity of the target and by ionized impurity scattering once this mechanism is defeated. The electro-optic properties of the films are thus ultimately dominated by the correlation between carrier concentration and transport and by the effects of carrier concentration on optical properties. The level of performance and the understanding of fundamental mechanisms that have been achieved combine to provide a promising technology for PV applications.

REFERENCES

1. A. Banerjee, D. Wolf, J. Yang and S. Guha, J. Appl. Phys., 70(3), 1692(1991).

2. A. Zafar, S. Karthikeyan and D. L. Morel, "Comparison of the Properties of F and Al Doped ZnO Deposited by Magnetron Sputtering from ZnO and Zn Targets", Proceedings of the XXIIIrd IEEE Photovoltaic Specialists Conference, May, 1993, Louisville, KY.

3. A. Zafar, C. S. Ferekides and D. L. Morel, to be published.

4. H. Sato, T. Minami, and S. Takata, J. Vac. Sci. Technol. A 11(6), 2975(1993).

IN SITU INTERFERENCE MEASUREMENTS OF GLASS SUBSTRATE TEMPERATURE

Z. Feng, C.N. Tabory, and A.D. Compaan
Dept. of Physics and Astronomy, Univ. of Toledo

ABSTRACT

We describe the use of optical interference with a small HeNe laser for the measurement of glass substrate temperatures. The measurement is insensitive to window transmission changes and requires only the monitoring of the beam reflected from the back of the substrate while the glass is heated or cooled from a known reference temperature. We find that the technique readily yields a temperature measurement sensitive to ± 10 C; however, good accuracy requires not only knowledge of the thermal expansion coefficient but also data on the temperature dependent changes in the index of refraction good to one part in 10^6. Since such information generally does not exist in the literature, we have obtained the data up to 400 C for four glasses: fused silica, Corning 7059, NEG 0A-2, and LOF soda-lime. Analysis of the data yields the direct temperature derivative as well as the indirect density derivative of the index of refraction.

INTRODUCTION

The reliable measurement of radiatively heated glass substrate temperatures during the deposition of thin films poses significant challenges. The low thermal conductivity of glass causes difficulties in establishing thermal equilibrium between the glass and any mechanically contacted sensor. Furthermore, front-side contact exposes the sensor to the deposition flux and backside contact may shadow the glass from the radiative heating flux. Earlier we have used fluorescence from a small ruby chip attached to the glass and monitored the wavelength shift of the R_1, R_2 red doublet [1]. However, this method also suffers from some of the same problems as, e.g., thermocouple sensors. A classic technique for glass temperature determination is optical pyrometry. However, for photovoltaics the deposition temperature is often below 400 C and the pyrometry must then be done in the infrared. This requires not only appropriate infrared-transmitting windows and knowledge of the emissivity of the glass and/or film surface, but also is subject to errors resulting from window transmission changes due to accumulating deposits on the window.

In this paper we describe the use of optical interference analysis using a small HeNe laser beam reflected from the back side of the glass substrate. This method measures a combination of the thermal expansion of the glass and thermal changes in the index of refraction of the glass. The measurement is insensitive to window transmission changes but does require constant monitoring of the reflected beam while the glass is heated or cooled from a known reference temperature.

EXPERIMENTAL

In order to facilitate *in situ* optical measurements, we included four small optical access ports into the design of our new double-gun, rf sputtering chamber [2]. The chamber design is shown in Fig. 1. The substrate is radiatively heated by a tantalum wire heater behind the glass substrate. In order to accomodate the laser beam, the radiation shield behind the wire heater was slotted to permit the entrance and exit of the laser beam. Note that this chamber design permits the substrate and heater assembly to be rotated by 180° to change from the first to the second deposition while maintaining substrate temperature. The four windows at 45° to the axis of the sputter guns permit backside interference measurements in either position. An additional benefit is that this geometry permits *in situ* transmission measurements to monitor the film thickness at the center *during* deposition. A small HeNe laser (λ=633 nm) worked well for the reflection measurements of glass temperature and also for the absorption measurement of CdTe film thickness. An argon laser line (λ=458 nm) was used for absorption measurements of CdS film thickness.

In order to calibrate the fringe shift vs. temperature, the various glasses were measured at near normal incidence in air in a specially constructed thermal enclosure with a window for the entering and exiting beams. In this case the glass was placed directly on a hot stage and a Type K thermocouple was bonded with silver paint to the top surface of the glass. Figure 2 shows the accumulated fringe count as a function of the substrate temperature for alkali-free glass, soda-lime glass, and for fused silica. The rate of fringe shift is, of course, linear in the glass thickness but varies with different glass compositions. The data are not explained simply by the different thermal expansion coefficients of the glasses. One must account also for the temperature dependence of the indices of refraction.

The precision needed for the temperature dependence of the index of refraction is one part in 10^6 and this is not available in the literature for these glasses over the

CH3365-4/94/0000-0350 $4.00 © 1994 IEEE

First WCPEC; Dec. 5-9, 1994; Hawaii

temperature range needed for PV film depositions. We have therefore measured this temperature dependence of the index of refraction using the hot stage and thermal enclosure described above. The thermocouple was placed within 1 cm of the point of incidence of the HeNe laser beam.

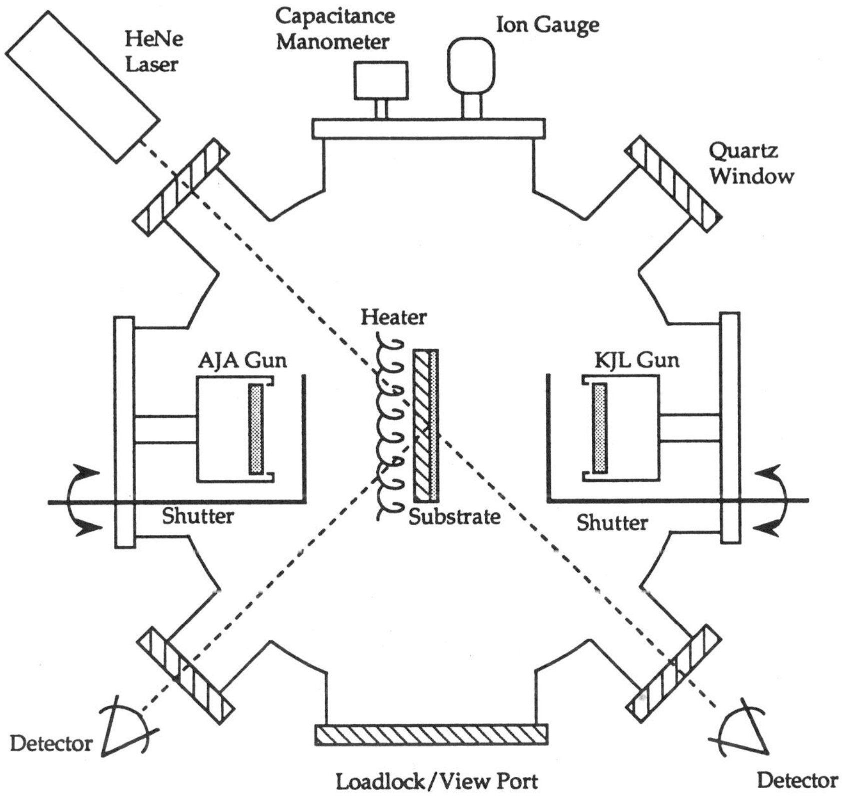

Fig. 1. Top view sketch of chamber used for sequential sputtering of CdS and CdTe with four optical access ports used for *in situ* interference measurements of glass temperature and film thickness.

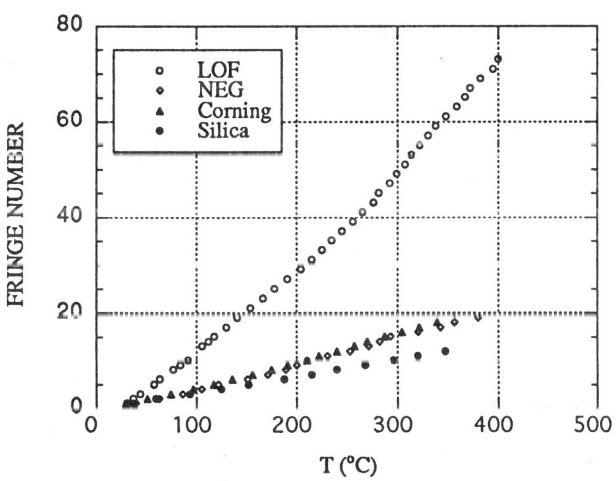

Fig. 2. Number of fringe shifts observed as a function of glass temperature for λ = 632.8 nm with $\theta = 5^{\circ}$ for four different glasses.

TEMPERATURE DEPENDENCE OF d AND n

It is readily shown that a shift of one fringe will occur when

$$2(n\Delta d + d\Delta n)\cos\theta = \lambda, \tag{1}$$

where $\Delta d = \alpha d \Delta T$, $\Delta n = \gamma \Delta T$, and θ is the angle of incidence. d is the glass thickness and α is the linear thermal expansion coefficient. The thermal coefficient of the index of refraction, γ, is composed of two terms, a direct temperature dependence of the index of refraction and an indirect term related to the density effect on the index of refraction. Thus

$$\gamma = dn/dT = \partial n/\partial T - \beta\rho\, \partial n/\partial \rho, \tag{2}$$

where ρ is the glass density and β is the volume coefficient of expansion [$\beta = 3\alpha = \partial V/(V\partial T)$]. The thermal expansion of the different glasses is shown in Fig. 3 where the data have been taken from the literature [3]. The work described in this paper provides the needed values for γ.

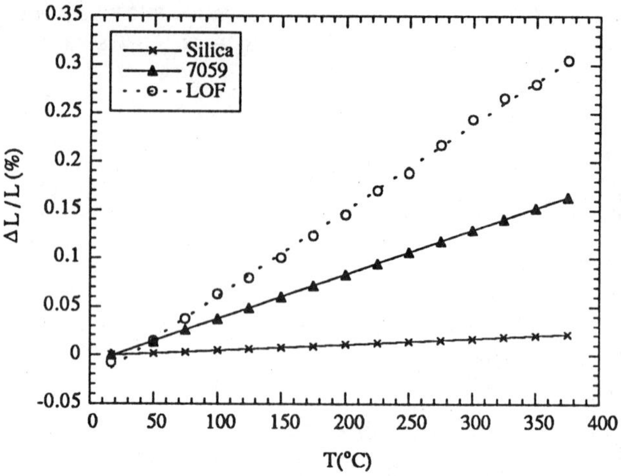

Fig. 3: Thermal expansion data from Ref. 3 used for the analysis of the thermal dependence of the index of refraction.

The measured changes in the index of refraction with temperature (dn/dT) are shown in Figs. 4, 5, 6, and 7 for Corning 7059, NEG 0A-2, LOF soda-lime, and fused silica. Note that over the temperature range from 0 to 400°C the value of γ = dn/dT is monotonically increasing with a linear dependence on temperature for all but the soda-lime glass. In this case, γ exhibits a significant quadratic component vs. temperature.

In general, γ may be fit by a quadratic function of the form

$$\gamma(^{o}C^{-1}) = A + B(T-T_0) + C(T-T_0)^2, \quad (3)$$

where, in our case T_0 = 273 K. The functions which fit the data of Figures 4-7 are summarized in Table I.

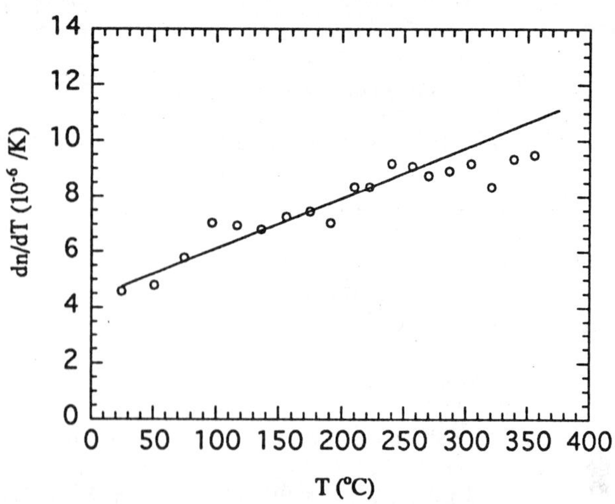

Fig. 4: dn/dT vs. T for Corning 7059 at 633 nm.

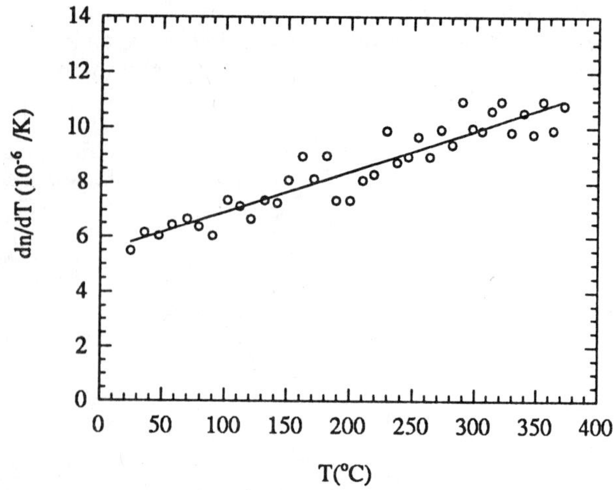

Fig. 5: dn/dT vs. T for NEG 0A-2 at 633 nm.

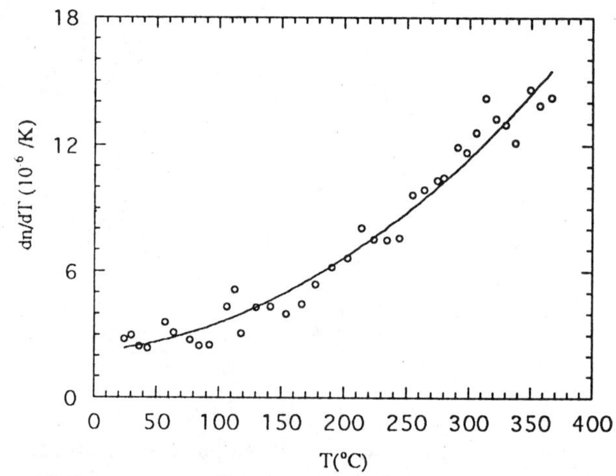

Fig. 6: dn/dT vs. T for LOF soda-lime glass at 633 nm.

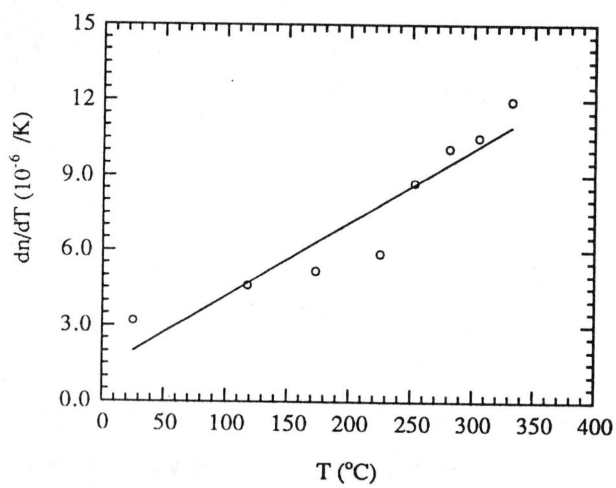

Fig. 7: dn/dT vs. T for fused silica at 633 nm.

Table I: Temperature Derivative of n, γ=dn/dT				
	silica	7059	0A-2	LOF
A($^oC^{-1}$)	0.7524	4.272	5.414	2.167
B($^oC^{-2}$)	0.0317	0.0182	0.0149	0.0054
C($^oC^{-3}$)	0	0	0	$8.4x10^{-5}$

DISCUSSION

Note from Eq. 2 that the density change with temperature will yield a negative contribution to γ for cases in which the volume coefficient of expansion $\beta = (1/V)\partial V/\partial T$ is positive. Since the measurements which were used to calculate the values of dn/dT involve simply monitoring the fringe changes, the measurement does not directly yield the sign of γ. However, we have taken advantage of the cos(θ) term in Eq. 1 to determine the relative sign. This involved comparison of the fringe change resulting from a small rotation of the glass with the fringe change caused by a small temperature change. dn/dT was found to be positive in all cases. Since dn/dT was found to be positive, the direct increase of index of refraction with temperature ($\partial n/\partial T$) at 633 nm is larger, apparently, than the density-dependent decrease ($-\beta\rho\ \partial n/\partial\rho$) for all of the glasses measured here.

The two parts to the index of refraction change with temperature (Eq. 2) arise in the following way: The indirect term arises from the fact that the index of refraction is directly proportional to the density of polarizable "molecular" units in the solid which decreases as the temperature rises (due to thermal expansion). Conversely the direct dependence of n on T arises from the facts that the band gap in most solids decreases with increasing temperature and the broadening of optical transitions increases with temperature. The real part of the index of refraction of these glasses has normal dispersion ($dn/d\nu>0$) which is tied essentially to the onset of absorption in the ultraviolet. Thus a band gap decrease with increasing temperature produces a positive dn/dT which is largest for the soda-lime glass due to its smallest photon energy for onset of absorption. Furthermore the broadening with temperature also produces a positive dn/dT which, again will be largest for the glass with the smallest optical gap.

There is existing index of refraction data in the literature for borosilicate glass which covers a similar temperature range as the work reported here. In Ref. 4, the index of refraction at λ = 587.7 nm was measured directly at four temperatures. The results at 20, 200, 400, and 600°C were, respectively n = 1.4760, 1.4772, 1.4790, 1.4824. The average values of dn/dT from these data agree very nicely with the data presented in Figs. 4 and 5 above.

As an illustration of the u[illegible]ity of this technique, we have measured the increase in substrate temperature after the sputter gun is ignited. We find the temperature increase to be 10 - 15 °C, for 60 W of r[illegible] power into a 50 mm diameter gun 70 mm from the substrate.

ACKNOWLEDGMENTS

We should like to thank R.G. Bohn and Y. Rajakarunanayake for helpful discussions. The support of NREL through contracts ZN-19019-3 and ZAX-4-14013-01 is gratefully acknowledged.

REFERENCES

[1] A.D. Compaan and R.G. Bohn, "Thin Film Cadmium Telluride Photovoltaic Cells," Annual Subcontract Report 11/91-10/92. Available NTIS: Order No. DE94000201.

[2] A.D. Compaan, M. Shao, C.N. Tabory, Z. Feng, A. Fischer, I. Matulionis, and R.G. Bohn, "RF Sputtered CdS/CdTe Solar Cells: Effects of Magnetic Field, RF Power, Target Morphology, and Substrate Temperature," this conference.

[3] *Thermophysical Properties of Matter*, vol. 13, "Thermal Expansion Nonmetallic Solids," edited by Y.S. Touloukian, R.K. Kirby, R.E. Taylor, and T.Y.R. Lee (IFI/Plenum, New York); R. McCurdy, private communication.

[4] A.J. Holland and W.E.S. Turner, J. Soc. Glass Technol. **25**, 164 (1941).

RF SPUTTERED FILMS OF Cu-DOPED AND N-DOPED ZnTe

R.G. Bohn, C.N. Tabory, Csaba Deak, M. Shao, and A.D. Compaan
Dept. of Physics and Astronomy, The University of Toledo
Toledo, OH 43606
and
Nicholas Reiter
Solar Cells, Inc, Toledo, OH 43607

ABSTRACT

Polycrystalline thin film solar cells using the CdS/CdTe structure have good efficiencies but the ideal low resistance ohmic contact to the p-type CdTe which is stable has yet to be developed. A good candidate for this contact would be p-type ZnTe produced by sputtering, which is a process that is scalable to large areas. We have successfully doped ZnTe with copper and nitrogen using rf planar magnetron sputtering. For the copper doping, resistivities as low as 0.01 ohm-cm have been achieved. Raman spectra have been used as an indicator of film quality. Nitrogen doping was achieved by introducing small amounts of molecular nitrogen into the argon sputter gas during ZnTe deposition. A minimum film resistivity of about 20 ohm-cm was obtained for films grown using approximately a 5% N_2/Ar ratio at a pressure of 18 mTorr and a substrate temperature in the vicinity of 400 C. Here, again, Raman spectra were used to check on film quality by comparing to that of pure ZnTe.

INTRODUCTION

Solar cells using the CdS/CdTe configuration have shown considerable promise exhibiting high efficiencies and offering the potential for thin film solar cell manufacturing[1]. However, the absorber material, CdTe, is notorious for its inability to form a stable, low resistance ohmic contact with metals. An intervening layer of ZnTe could serve as a solution to this problem since it has little valence-band discontinuity with p-CdTe and it can be heavily doped p-type with copper and possibly with nitrogen[2]. Copper doping has proven to be successful in our laboratory using cold-pressed targets and laser-deposited films[3]. Results using a single gun rf sputtering system with a two-inch diameter planar magnetron sputter gun point to the possibility of performing CdS/CdTe/ZnTe depositions in a single chamber with the potential of scale-up to larger areas.

Cu-DOPED ZnTe

Fig. 1 shows the rf sputtering system used for these films. The sputter gas pressure was 12 mTorr and the rf power was 60 W. In these preliminary studies the copper doping was controlled by placing small pieces of copper on the ZnTe target and measuring the surface area of the copper relative to the active target area. The amount of copper in the film was estimated using energy dispersive x-ray spectroscopy (EDS). These results were then calibrated using a pressed powder target with a known Cu content which was used for laser deposition of films. The calibration was also checked against the film grown by laser deposition from the same target[3].

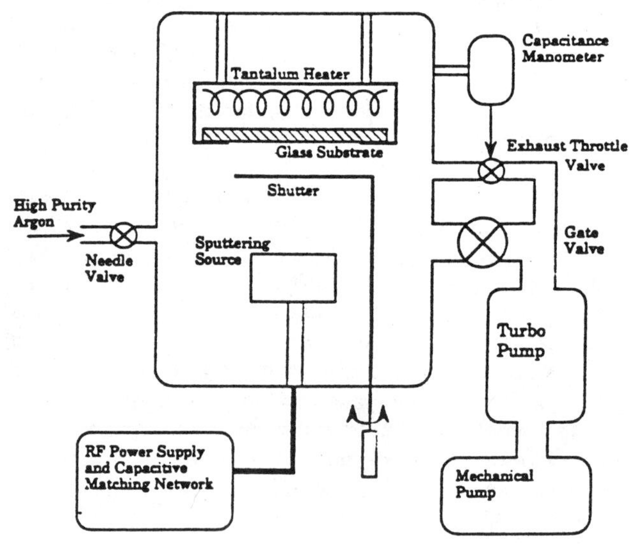

Fig. 1. RF sputter deposition system using two-inch diameter targets and 3-inch square substrates.

All films were grown on Corning 7059 substrates. Film thicknesses were measured using a Tencor Alpha-Step 200 profilometer and ranged from 0.14 to 0.72 micrometers. Scanning Electron Microscopy (SEM) was used to determine surface morphology. Generally, the grains were well defined with a typical grain size of about 0.2 to 0.3 micrometers. Electrical conductivity was measured for all the samples between 290 K and 350 K using either parallel contacts or the van der Pauw technique. (Additional data were taken for ZT4 down to 80 K.) The results are shown

CH3365-4/94/0000-0354 $4.00 © 1994 IEEE
First WCPEC; Dec. 5-9, 1994; Hawaii

in Table 1. There appears to be poor correlation between the area covered by the copper and the EDS measurements; however, copper is a very efficient sputterer and placement of the pieces relative to the magnetron sputter track is critical. There is also a dependence of copper incorporation on substrate temperature, with more copper in the film at higher temperature. Comparisons can be made to Gessert, et al[4,5], where our maximum conductivity is about 5 times larger than that of his 5% sample grown at 400 C. We believe that this could be in part due to a higher carrier density in our 6.4% Cu sample ($n \sim 10^{21}$ /cm^3). Hall measurement results have large uncertainties which are typically associated with low mobility films. However, they suggest that the mobilities are of the order of 1 cm^2/V-s or less. This can be compared to Gessert's mobility of about 0.5 cm^2/V-s[4]. The Hall results also suggest that the variation in conductivity is largely due to a variation in carrier density since the estimated carrier density for ZT7 is about 10^{17} /cm^3. (The values for n should be good to within a factor of 3.) It should also be noted that the conductivity for sample ZT4 changes very little over the temperature range from 80 K to 350 K, one indication of a degenerate semiconductor. (Fig. 2)

Table 1: Sputtered ZnTe:Cu

SAMPLE ID	Cu area (cm^2)	Cu (wt%)	T_s (°C)	σ (1/Ω cm)
ZT47	0	0	400	5.7E-8
ZT7	0*	< 0.2	420	7E-3
ZT5	0.085	0.67	420	0.059
ZT6	0.085	1.1	470	0.040
ZT11	0.16	0.80	420	18.6
ZT4	1	6.4	470	110

* ZT7 may contain some residual copper

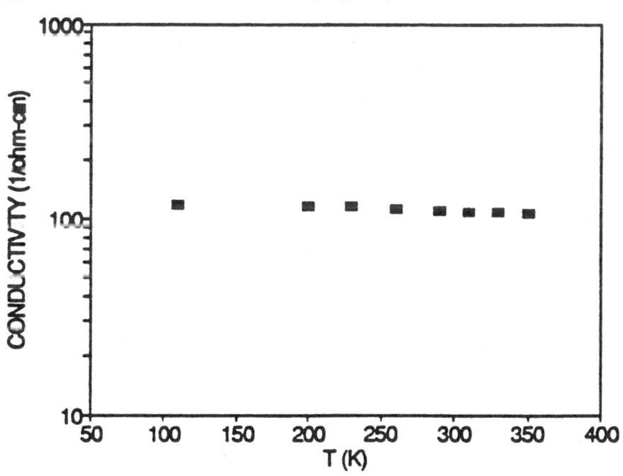

Fig. 2. Electrical conductivity versus temperature for ZnTe with 6.4% Cu grown at 470 C.

Raman scattering also provides some information on the film quality and bonding arrangements for these films. Fig. 3 shows Raman spectra for three sputtered films and, for comparison, a spectrum for a bulk crystal of copper-doped ZnTe. The single crystal sample was estimated to have approximately 0.8% Cu doping. Spectra were obtained at room temperature using a CCD detector on the ISA S3000 triple spectrometer with excitation at 514 nm (19,434 cm^{-1}). The strength of the first order longitudinal optic (LO) mode at 220 cm^{-1} decreases slightly with Cu content, but all samples show clearly the second order LO mode near 440 cm^{-1} indicating reasonably high quality films. All samples with Cu doping show an extra peak near 160 cm^{-1} which possibly may relate to a resonant mode associated with Cu, but this frequency has not been confirmed.

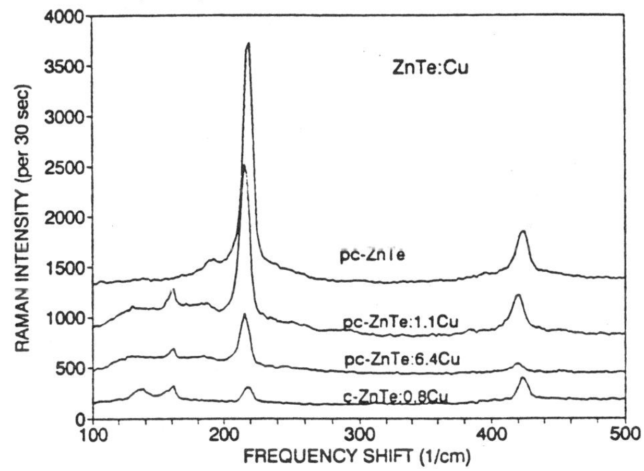

Fig. 3. Raman spectra of three sputtered films and one crystalline ZnTe:Cu sample.

N-DOPED ZnTe

At least one study has been done on doping ZnTe with atomic nitrogen using MBE[6]. Since rf sputtering produces atomic and ionic species in the plasma, we have introduced molecular nitrogen into the argon sputter gas in an attempt to dope the ZnTe films. Table 2 shows that the electrical properties of the reactively sputtered films are very sensitive to the nitrogen content of the sputter gas. The electrical conductivity was measured as a function of temperature between 290 K and 350 K. Over this limited temperature range the conductivity is described by an activation energy model, with the values for some of the samples listed in the table. With no nitrogen in the sputter gas the conductivity is low and the activation energy is of the order of 0.5 eV. This is true for films grown at both 300 C and 400 C. By introducing a small amount of molecular nitrogen (1%-5%), the conductivity increases by several orders of magnitude. The activation energy is reduced to about 0.2 eV. The highest temperature also seems to produce higher conductivity films. Note that with a growth temperature of 400 C and 5% nitrogen in argon not only do we get the highest conductivity, but also the lowest activation energy (0.09 eV).

Table 2: Reactive Sputtering of ZnTe with N_2					
SAMPLE ID	T_s (°C)	P_{rf} W	$\frac{N_2}{(Ar+N_2)}$	E_{act} (eV)	σ (1/Ωcm)
ZT39	300	60	0	0.50	1.0E-7
ZT47	400	60	0	0.47	5.7E-8
ZT46	300	60	0		1.4E-5
ZT48	400	75	0.05	0.09	5.9E-2
ZT49	400	75	0.01	0.20	1.3E-3
ZT50	300	75	0.03	0.21	3.7E-4
ZT51	300	75	0.05	0.26	4.0E-4
ZT40	300	75	1		4.8E-6

Raman scattering also shows that the nitrogen ratio has a strong effect on the vibrational properties of the film. Fig. 4 shows that in pure nitrogen there is no longer any ZnTe-like Raman peak at 220 cm^{-1} but there is a strong doublet at 140 and 160 cm^{-1}. These peaks are characteristic of excess Te which is known to be a very strong Raman scatterer. At 1% N_2 the spectrum is similar to that of pure ZnTe although weaker.

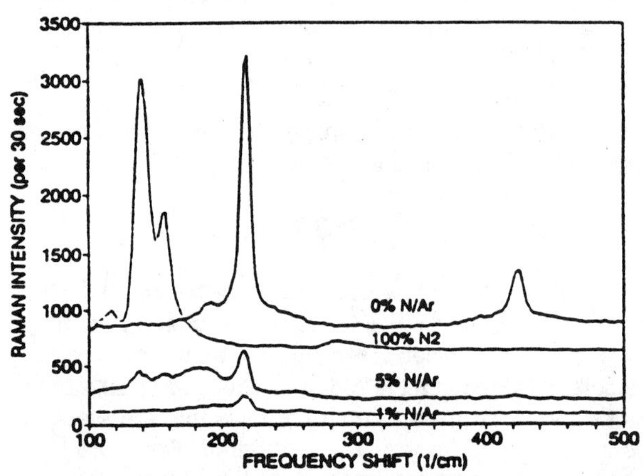

Fig. 4: Raman spectra of ZnTe:N films grown in N_2/Ar mixtures. T_s = 400 C except for pure N_2 where T_s = 300 C.

At 5% N_2 The spectrum still shows the intrinsic ZnTe LO mode but in addition there is some evidence for the elemental Te peaks at 140 and 160 cm^{-1} and a broad band centered near 180 cm^{-1}. This peak may arise from amorphous material in the film. Fig. 5 shows that the strength of this amorphous-like band is very sensitive to the deposition temperature. A deposition temperature of about 400 C appears to be optimum for achieving a strong ZnTe LO phonon mode. This result is consistent with the data in Table 2, especially samples ZT51 and ZT48, which show much higher electrical conductivity for samples grown at 400 C. Thus, there appear to be at least two important factors (N_2 partial pressure and substrate temperature) in achieving high electrical activity and good vibrational properties in reactively sputtered ZnTe.

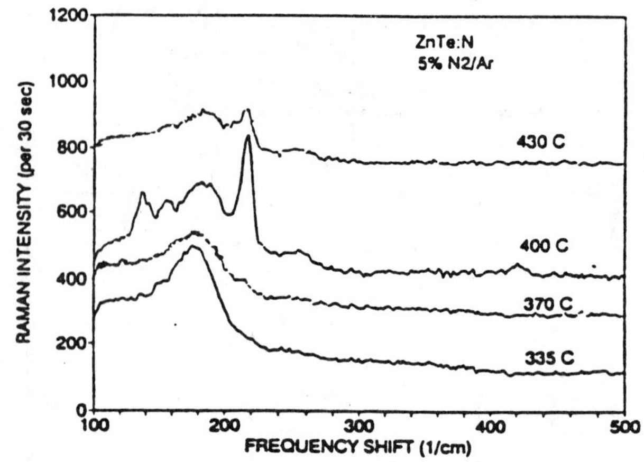

Fig. 5. Raman spectra of ZnTe:N films grown at different substrate temperatures. Gas mixture is 5% N_2 in Ar; P_{rf} = 75 W.

CONCLUSIONS

We have successfully grown ZnTe doped with copper and nitrogen in this preliminary study. Copper doping can be used to produce high conductivity films which become degenerate at high doping levels. We have yet to produce degenerate ZnTe:N by mixing molecular nitrogen with the sputtering gas; however, there is evidence that doping is taking place, but at a much lower level than in the ZnTe:Cu films.

ACKNOWLEDGMENTS

The support of NREL for this work is gratefully acknowledged (contract numbers ZN-1-19019-3 and ZAX-4-14013-01).

REFERENCES

[1] J.F. Nolan, *Proc 23rd IEEE Photovoltaics Specialists Conf.* (IEEE, Piscataway, NY, 1993) p. 34.

[2] D. Rioux, D.W. Niles, and H. Hochst, *J. Applied Physics*, **73**, 8381 (1993).

[3] A. Compaan and A. Bhat, *Int'l J. Solar Energy*, **20**, 155 (1992).

[4] T.A. Gessert and T.J. Coutts, *12th NREL Photovoltaics Program Review Meeting*, Denver, CO (Oct. 13-15, 1993).

[5] T.A. Gessert, X. Li, T.J. Coutts, A.R. Mason, and R.J. Matson, *J. Vac. Sci. Technol. A* (to be published)

[6] J. Han, T.S. Stavrinides, M Kobayashi, R.L. Gunshor, M.M. Hagerott, and A.V. Nurmikko, *Appl. Phys. Lett.* **62**, 840 (1993).

THIN FILM SOLAR CELLS BASED ON LAYERED CHALCOGENIDES: FUNDAMENTALS AND PERSPECTIVES OF VAN DER WAALS EPITAXY

W. Jaegermann, C. Pettenkofer, O. Lang, R. Schlaf, S. Tiefenbacher, Y. Tomm
Abteilung Grenzflächen, Bereich Physikalische Chemie,
Hahn-Meitner-Insitut, Glienicker Str. 100, 14109 Berlin, Germany

ABSTRACT

The preparation of thin films of layered chalcogenide semiconductors as MX and MX_2 (X=S, Se) based on the concept of van der Waals epitaxy (VDWE) is presented for different substrate/overlayer combinations as GaSe, InSe, $SnSe_2$, WS_2 on WSe_2, GaSe, $MoTe_2$, graphite and mica. In all cases stoichiometric films are formed either as epitaxial layers or strongly textured films with the c-axis aligned in spite of strong lattice mismatch. The interfaces are non-reactive and atomically abrupt. The electronic properties of the interfaces are mostly ideal showing band offsets according to the electron affinity rule and no operative interface states. However, doping of the films is still a problem which limits the band bending and the attainable surface photovoltage. The perspectives and preconditions for the further development of layered semiconductor VDWE films for solar cells will be critically discussed.

INTRODUCTION

Layered chalcogenide semiconductors as MX and MX_2 (X=S, Se) have been considered as interesting alternatives for solar energy conversion [1] because of their high absorption constants. The bandgaps are between 1.0 and 2.2 eV and the carrier mobilities reach typical values between 50 and 200 cm^2/Vs. The compounds are characterized by a two-dimensional structure with covalently bound X-M(-M)-X sandwich units bound to each other along the c-axis only by weak interactions (van der Waals like). The van der Waals surfaces are chemically saturated and free of dangling bonds. The physical properties of the different layered chalcogenides may vary considerably ranging from wide band gap semiconductors as SnS_2 or GaS, over semiconductors as MoX_2, WX_2 and InSe to metals as TaX_2 or WTe_2. This variety of naturally two-dimensional compounds allows many different combinations of materials for solar cells. Single crystal based solar cells with conversion efficiencies η well above 10% have already been reached: for n-WSe_2 η is above 14 % in photoelectrochemical cells [2]. PV-devices are usually based on p-WSe_2 with η of 8 % for p-WSe_2/ZnO and 6.2 % for p-WSe_2/CdS heterointerfaces [3,4]. The best solid state device based on layered materials is an Au/p-InSe/ITO heterodevice with η of 10.9 % [5].

For a practical use of layered compounds the preparation of thin films is unavoidable. All previous attempts to grow high quality films have not been successful because the films do not tend to grow with their basal plane oriented to the substrate [1], which leads to poor optoelectronic properties due to efficient charge carrier recombination at the non-van der Waals planes.

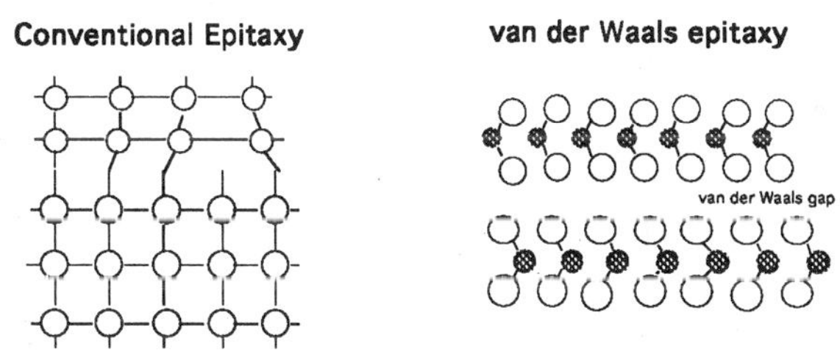

Fig. 1. Van der Waals epitaxy compared to conventional epitaxy: The lack of dangling bonds at the phase boundary allows to prepare epitaxial overlayers with ideal interface properties also for strongly lattice mismatched semiconductor combinations.

Recently the concept of van der Waals epitaxy (VDWE) has emerged for the deposition of layered chalcogenides [6] on van der Waals like substrates (Fig.1). The growth of different layered chalcogenides as InSe, GaSe, $SnSe_2$ and WS_2 on layered substrates as e. g. WSe_2, $MoTe_2$, graphite, and mica will be presented. The films and electronic interface properties are characterized in-situ by surface science techniques as photoelectron spectroscopy (XPS, UPS, SXPS) and electron diffraction (LEED) and ex-situ by X-ray diffraction (XRD) and scanning tunneling microscopy (STM).

EXPERIMENTAL

The growth experiments were performed in home-made ultra high vacuum (UHV) growth chambers which were coupled to different UHV-surface analysis systems (for details see [7]. LEED experiments were perfomed with a back view LEED system (VG). STM experiments were performed with a nanoscope II at air. For XRD a Siemens powder diffractometer was used.

The layered compound single crystals used as substrates were prepared by standard gas transport reactions. Graphite (HOPG) and mica was taken from commercial sources. The substrates were cleaved in UHV or cleaned at elevated temperatures prior to their use as substrates. For the molecular beam epitaxy (MBE)

CH3365-4/94/0000-0357 $4.00 © 1994 IEEE First WCPEC; Dec. 5-9, 1994; Hawaii

experiments the films were deposited from home-made effusion cells and monitored by a quartz microbalance. For the deposition of WS by metal organic (MO)-MBE $W(CO)_6$ as volatile precurser and a leakage valve was used. Some of the films were grown in conventional high vacuum evaporation systems.

EXPERIMENTAL RESULTS

VDWE by MBE experiments

Most of the epitaxial films of layered chalcogenides were grown in UHV MBE experiments as a good control and easy analysis of film properties is provided. Deposition techniques have been applied, which may be transferred to standard thin film deposition systems. In Table 1 a survey of MBE experiments and of typical results are given.

Table 1. Survey of typical substrate and overlayer combinations investigated for VDWE.

Substrate	Overlayer	Growth mode	Ref.
WSe_2	GaSe	FvdM, E	7
WSe_2	$SnSe_2$	FvdM, E	8
$MoTe_2$	InSe	FvdM, E	9
$MoTe_2$	WS_2	FvdM, E	*
$MoTe_2$, MoS_2	SnS_2, $SnSe_2$	FvdM, E	8
SnS_2 ($SnSe_2$)	$SnSe_2$ (SnS_2)	FvdM, E	8
GaSe	GaSe, InSe	FvdM, E	10,11
GaSe	$SnSe_2$	FvdM, E	8
InSe	InSe, GaSe	FvdM, E	11
InSe	SnS_2, $SnSe_2$	R	8
Graphite	GaSe, InSe	FvdM, E	10, *
Graphite	WS_2	FvdM, E	12
Mica	GaSe	VW, P, T	*
Mica	InSe	FvdM, T	*
Mica	SnS_2, $SnSe_2$	VW, P	8

FvdM: Frank-van der Merwe-Growth, E: Epitaxial Growth, R: Reaction, VW: Volmer-Weber-Growth, T: Textured Growth, P: Polycrystalline ; * This work

As typical result the deposition of GaSe on WSe_2 (0001) will be presented [7], which may be interesting as heterojunction solar cell with GaSe as window. Fig. 2 shows a sequence of LEED pattern obtained for different overlayer thicknesses. As the sticking coefficient of the deposited material is generally very low, the first deposition steps have to performed at rather low temperatures (for GaSe 570 K) and an annealing step (for GaSe 720 K) is necessary to improve the cristallinity. The LEED pattern of the clean substrate shows the hexagonal diffraction pattern of the WSe_2 (0001) surface. After 7 Å GaSe deposition (before the annealing step) a superimposed hexagonal diffraction pattern appears which is oriented to the substrate spots: the deviation of the lattice mesh of 14 % agrees to the lattice mismatch. For thicker layers (and after annealing) only the overlayer diffraction spots remain and the diffuse background is considerably reduced indicating a high quality epitaxial film.

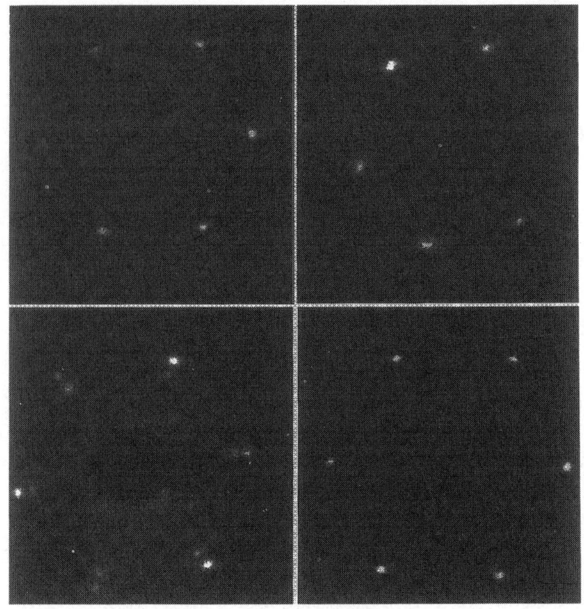

Fig. 2. LEED pattern of the WSe_2 (0001) substrate (a) and of GaSe overlayers of 7Å (b), 28Å (c) and 120 Å (d) thickness.

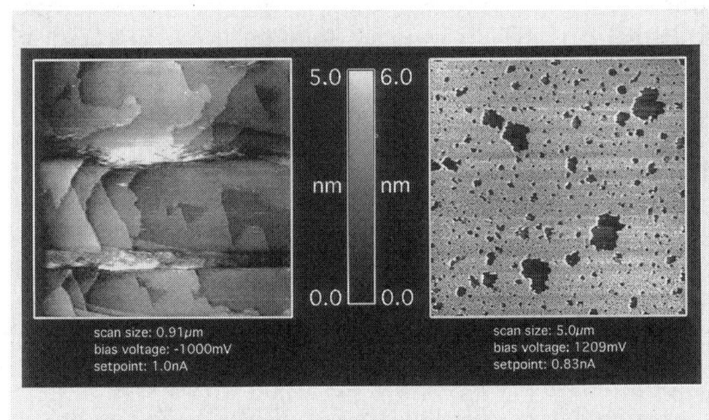

Fig. 3. STM pictures of GaSe films on WSe_2 (0001) substrate of 28 Å (right) and 140 Å (left) thickness

The ex-situ STM results are presented in Fig. 3. For the 28 Å film there are still some holes in a film of otherwise homogeneous thickness (4 GaSe layers). The 140 Å film shows triangular terraces of about 200 nm size which are separated by monolayer steps. On the terraces atomically resolved corrugations in agreement to the GaSe lattice constant are measured. The XPS and UPS analysis of the films [7] show nearly identical spectra as obtained for cleaved single crystals. However, the doping of the films cannot be controlled up to now. Without additional doping films are weakly p-doped (E_F-E_{VB}=0.5 ±0.1 eV). We tried to dope the films by adding up to 50 % of Se, P, As and CdSe during the deposition. However, only minor changes in doping were obtained. As a consequence the device is limited in the obtained conversion efficiency: the SPV with n-WSe_2 is in the range of 0.5 eV and the photocurrent is very low (< 2 mA/cm^2) due to the high series resistance.

$SnSe_2$ (E_G=1.0 eV) could also be deposited as high quality film on different layered chalcogenide substrates including mica [8]. Also these films show high quality LEED pattern and smooth film morphology in STM. The doping is strongly n-type (E_F-E_{VB}=0.9 eV) as deduced from the photoemission measurements. In contrast the morphology of the InSe (E_G=1.2 eV) films is not as homogeneous in spite of high quality LEED pattern [11]. The STM analysis indicates that the film is composed of large and oriented but separated two-dimensional clusters. Nevertheless first electrical experiments on the doping of these films and of SPV-spectra could be obtained which approach the values of undoped single crystals.

Deposition of WS_2 by MO-VDWE

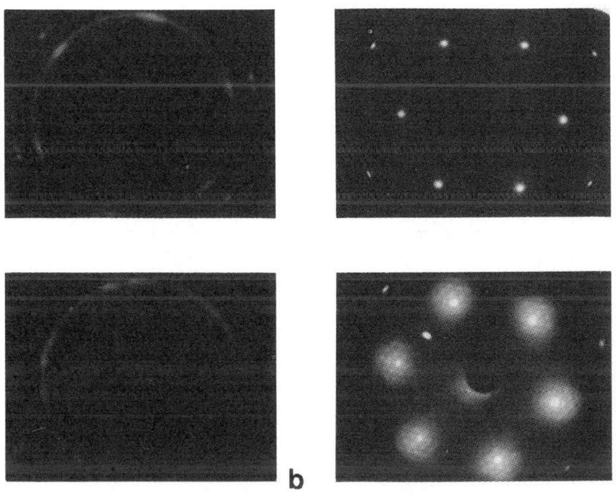

Fig. 4. LEED diffraction pattern of WS_2 films grown by MO-VDWE: a) on HOPG (10 and 50Å film), b) on $MoTe_2$ (0001) (clean substrate and 15 Å film).

As WX_2 (X=S,Se) are interesting absorber materials for solar cells we performed deposition experiments on $MoTe_2$ and HOPG as substrates using MO-VDWE [12]. With metal organic precursers ($W(CO)_6$) the obtained results may easily be adopted to standard MOCVD equipment. As the precurser turns out to be rather inert to chemical reactions only slow deposition rates were obtained leading to slow film growth. Stoichiometric films are obtained by a $S/W(CO)_6$ pressure ratio of about 1/100 at about 900 K. On HOPG [12] the LEED data (Fig. 4a) show only diffraction rings: as the substrate is polycristalline within the layers and preferentially oriented (textured) along the c-direction (outer ring), the deposited films are also statistically oriented in azimuthal direction and oriented along the c-axis (inner ring). The deviation of the ring radii (25 %) agrees to the lattice mismatch (28 %). On $MoTe_2$ (0001) substrates LEED diffraction spots are obtained (Fig. 4b). The high quality of the films is indicated by Moiré patterns, which agree to the lattice mismatch between the substrate and the overlayer of 11 %.

Deposition of InSe by PVD

In some preliminary experiments InSe was deposited on mica by physical vapour deposition (PVD). The XRD results (Fig. 5) show only the (00n) reflections as proof of preferential orientation of the deposited films. We also performed contactless conductivity experiments (TRMC) with these films. The results indicate that the films are photoactive. A comparison to single crystal data suggest that the difference in response is mostly due to the reduced absorption of light in the thin films.

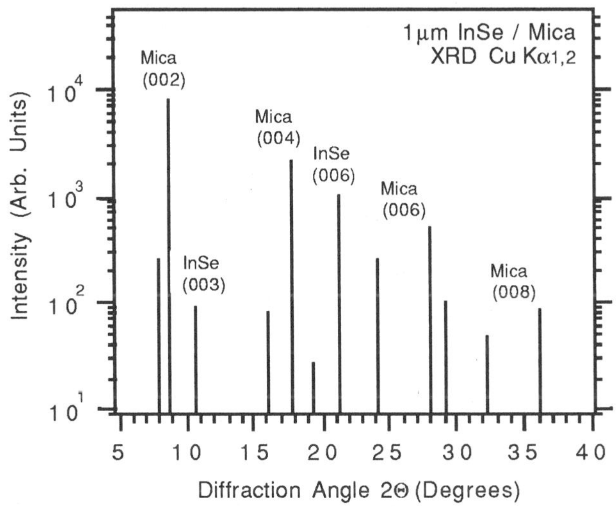

Fig. 5. XRD of InSe films deposited on mica by PVD.

PERSPECTIVES FOR SOLAR CELLS

The electronic properties of the different VDWE heterointerfaces (band line-up, band bending, surface photovoltage) have been completely characterized by photoelectron spectroscopy. The energy diagram of the GaSe/n-WSe_2 heterointerface is shown in Fig. 6a [7]. Also shown is the energy diagram of a GaSe/InSe/GaSe quantum well [11] (Fig. 6b). These diagrams are based on our experimental results, only the value of the bandgap has been taken from literature. The electronic interface properties of non-reactive heterointerfaces obtained by VDWE are mostly ideal. There are no dipole layers or interface states formed at the phase boundary. As a consequence the band line-up follows the electron affinity rule (Anderson model). This is different to most heterointerfaces with three-dimensional semiconductors and is related to the specific properties of the van der Waals surface with only saturated chemical bonds. In addition, the band bending and thus the diffusion voltage is just given by the bulk Fermi level position which is determined by the doping of the semiconductors.

It could already be shown that strongly textured photoactive thin films of InSe can be deposited by standard thin film deposition techniques as PVD on potentially cheap substrates as mica or graphite. Also WX_2 films may easily be achieved by MOCVD as suggested by our MO-VDWE results. However, further development of layered chalcogenide thin film solar cells is limited by problems with defined and durable doping of both the substrates and the deposited overlayers. Efficient doping during thin film growth has not been achieved. In addition, for promising device combinations based on the epitaxy experiments cheap substrates have to be further explored as a possible route to practical application. The most interesting device configurations for future experiments seem to be injection and p-i-n devices based on the layered compounds. Due to their mostly ideal interface properties it is expected that such devices may exhibit promising properties for novel photovoltaic device structures.

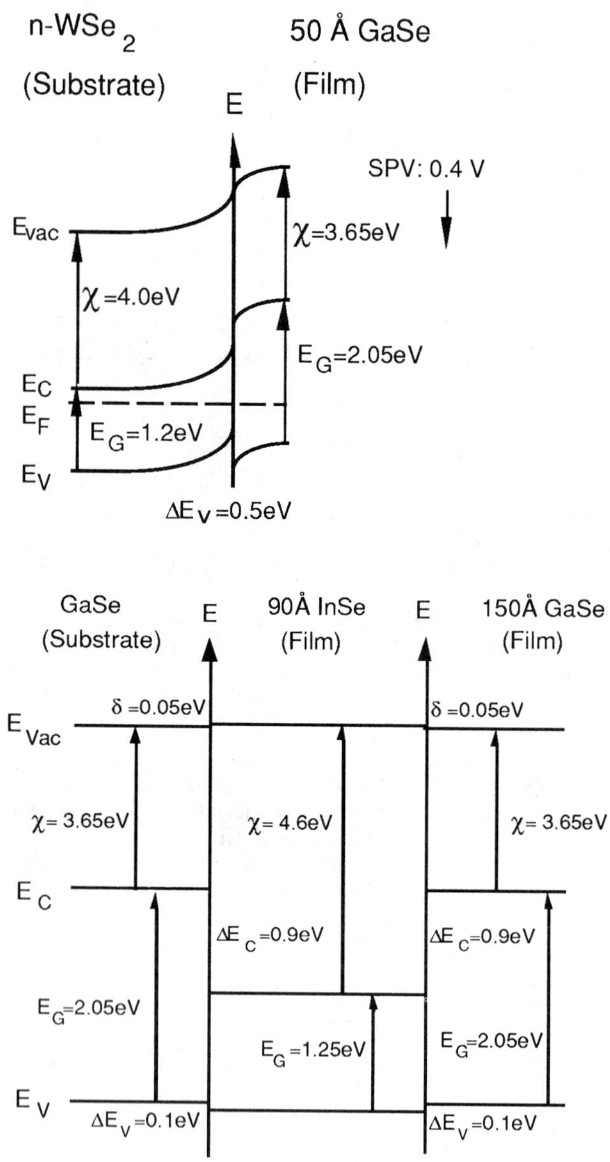

Fig. 6. Energy diagram of n-WSe_2/p-GaSe hetero-interfaces (top) and of GaSe/InSe/GaSe quantum wells (bottom).

ACKNOWLEDGEMENTS

The work presented here was supported by grants from the BMFT and the EC. We would like to thank our coworkers, who were involved in part of the work: they were mentioned in the refences.

REFERENCES

[1] A. Aruchamy (ed), "Photoelectrochemistry and Photovoltaics of Layered Semiconductors" (Kluwer, Dordrecht, 1992)

[2] R. Tenne, A. Wold, "Passivation of Recombination Centers in n-WSe_2 yield high efficiency (>14 %) photo-electrochemical cell" , *Appl. Phys. Lett.* **47**, 707 (1985)

[3] M. Vögt et al., "Stoichiometry and Doping effects in MoSe and WSe_2 Solar Cells", *Int. PVSEC-4, Sidney* , 493 (1989)

[4] M. Lux-Steiner et al., "n CdS/p-WSe_2 Heterojunction; a Candidate for Solar Cell Application?", *PVSEC-1, Kobe,* 259 (1984)

[5] J. Martinez-Pastor, et al, " Electrical and Photovoltaic Properties of Indium-Tin-Oxide/p-InSe/Au Solar Cells", *J. Appl. Phys.* **62**, 1477 (1987)

[6] A. Koma, "Van der Waals Epitaxy a New Epitaxial Growth method for a Highly Lattice-Mismatched System", *Thin Solid Films*, **216**, 72 (1992)

[7] O. Lang, et al, "Single Crystalline GaSe/WSe_2 Heterointerfaces Grown by Van der Waals Epitaxy: 1. Growth Conditions; 2. Junction Characterization" , *J. Appl. Phys.*, **75**, 7805; 7814 (1994)

[8] R. Schlaf, et al, "MBE Growth of Thin Films of SnS_2 and $SnSe_2$ on Cleaved Mica and the Basal Planes of Single Crystal Layered Semiconductors: RHEED, LEED, Photoemission and STM/AFM Characterisation" , *J. Vac. Sci. Technol.*, submitted

[9] R. Schlaf, et al., "Van der Waals Epitaxy of Thin InSe Films on $MoTe_2$", *Surf. Sci. Lett.*, **303**, L343 (1994)

[10] O. Lang, et al., "InSe/GaSe Heterointerfaces Prepared by van der Waals Epitaxy", *J. Crystal Growth*, in press

[11] O. Lang, et al, "Band Lineup of Lattice Mismatched InSe/GaSe Quantum Well Structures Prepared by van der Waals Epitaxy : Absence of Interfacial Dipoles", *Phys. Rev. B*, submitted

[12] S. Tiefenbacher et al. "Epitaxial Films of WS_2 by Metal Organic van der Waals Epitaxy (MO-VDWE)", *Surf. Sci. Lett.* **318**, L1161 (1994)

A THIN FILM SOLAR CELL BASED ON THE $CuIn_3Se_5$ ORDERED VACANCY COMPOUND

S. Menezes
Interphases Research, P.O. Box 1532, Thousand Oaks, CA 91358-1532
J. Kessler
Solarex, 826 Newtown-Yardley Rd, Newtown, PA 18940
D. Schmid, H.W. Schock
Institut für Physikalische Electronik, Universitat Stuttgart, Pfaffenwaldring 47, D-7000 Stuttgart 80, Germany
A. Nelson, R. J. Matson
National Renewable Energy Laboratories, 1617 Cole Blvd., Golden, CO 80401-3393

ABSTRACT

The special attributes of the $CuIn_3Se_5$ material are exploited for an absorber layer in a new thin-film photovoltaic (PV) cell. $Cu_XIn_YSe_Z$ thin films with x~11% are synthesized by co-evaporation or sequential evaporation. An electrochemical approach, first developed for $CuInSe_2$, is employed to construct an *n*-$CuIn_3Se_5$/*p*-$CuISe_3$ heterojunction. This approach leads to a remarkably uniform, clean *np*-heterointerface and a rough surface conducive to light trapping. EPMA , SIMS profiles and XRD analysis examine the $CuIn_3Se_5$ films and the $CuISe_3$ overlayer. EBIC and I-V analysis investigate the formation of an electrically active *n*-$CuIn_3Se_5$/*p*-$CuISe_3$ junction.

INTRODUCTION

The ordered vacancy compound (OVC), $CuIn_3Se_5$, was recently identified as the surface phase of In-rich *p*-$CuInSe_2$ films [1,2]. It is presumed to serve as the *n*-type heterojunction partner for the *p*-$CuInSe_2$ absorber, independent of the presence of the CdS window. The $CuIn_3Se_5$ material combines the PV properties of $CuInSe_2$ with a solar-matched bandgap (1.3 eV) and a self-stabilizing composition that leads to smooth, homogeneous, consistently *n*-type films [3]. This paper exploits these attributes in a new PV cell, *n*-$CuIn_3Se_5$/*p*-$CuISe_3$, where the $CuIn_3Se_5$ serves as the absorber layer.

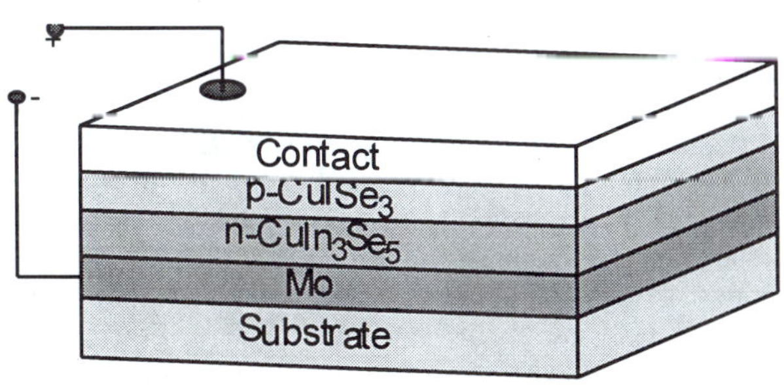

Fig. 1. Schematic of the n-$CuIn_3Se_5$ cell constituents.

Fig. 1 shows a schematic diagram of the cell components. OVC films with a range of stoichiometries and morphologies were synthesized and converted into heterojunctions. Electron probe microanalysis (EPMA) and Scanning electron microscopy (SEM) examine the composition and microstructure of the OVC films. Secondary ion mass spectroscopy (SIMS) and X-ray diffraction (XRD) characterize the heterostructures. Current(I)-voltage(V) and electron beam induced current (EBIC) analysis assess the junction quality and device parameters.

CELL CONSTRUCTION

$Cu_XIn_YSe_Z$ thin films with x~11% were synthesized by either co-evaporation of elements [3] or sequential evaporation [4] on Mo-coated sodalime glass. The latter involved sequential evaporation of Cu, In_2Se_3 and Se at substrate temperatures (T_S) between 400-500°C.

An electrochemical approach, previously developed for $CuInSe_2$ [5], was used to construct *n*-$CuIn_3Se_5$/*p*-$CuISe_3$ heterojunctions. Photoanodization of $CuIn_3Se_5$ surface layers in an acidic Cu-polyiodide electrolyte synthesizes the *p*-$CuISe_3$ window layer [6]. This medium also serves as a convenient photoelectrochemical (PEC) characterization tool to assess the OVC film properties and the surface conversion rates. The net photocurrent ($I_{h\nu}$) and the modulated photocurrent component ($\Delta I_{h\nu}$) may be monitored during anodization as a function of potential or time.

PEC measurements show considerable variation of PV output and interface stability as a function of the morphology and stoichiometry. The PV parameters ranged from relatively high photovoltage (V_{OC}) of 0.5V, stable $\Delta I_{h\nu}$ and low dark current (I_{dark}), as shown in Fig.2a, to low V_{OC}, rapidly decreasing or bi-directional $\Delta I_{h\nu}$, and increasing I_{dark}, Fig. 2b. Fortunately, the *n*-$CuIn_3Se_5$/*p*-$CuISe_3$ heterostructure formation is unaffected by variations of the PEC behavior, as evidenced by the surface analytical data.

First WCPEC; Dec. 5–9, 1994; Hawaii

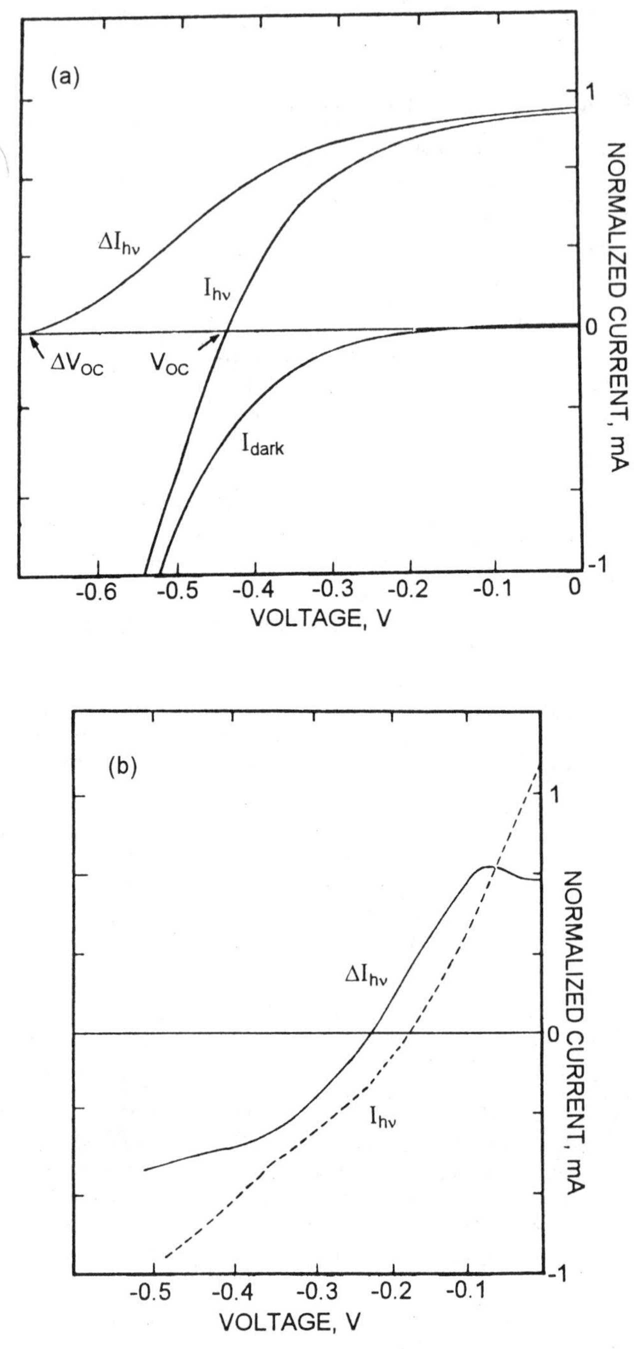

Fig.2. I-V, ΔI_{hv}-V and I_{dark}-V curves for OVC cells showing (a) stable, n-type photoresponse and (b) unstable, bi-directional photocurrent.

HETEROJUNCTION CHARACTERIZATION

The OVC film compositions, listed in Table I, were determined with EPMA. The measurements at two depths, obtained with 10 and 20 keV electron beam, depict the depth gradients for the films. The data suggests that the higher temperature co-evaporation produces relatively homogenous 135 compositions, series A and B films. The sequential deposition, on the other hand, produces a range of graded compositions (also referred to as OVC) with Cu-rich surfaces and T_S dependent stoichiometries, series C films. The bulk/surface compositions range from $CuIn_3Se_5/CuIn_2Se_{3.5}$ at high T_S, to $CuIn_2Se_{3.5}/CuInSe_2$ at intermediary T_S and Cu-doped In_2Se_3 at low T_S. The direction of the Cu gradient in the film is remarkable since it is in the reverse order of the deposition sequence (Cu-layer at the back).

Table I. Composition of Co-evaporated Films

Sp. No.	T_S °C	thick µm	depth µm	composition Cu	In	Se	Cu:Se:In	Conductivity photocurrent PEC	SS
A1	500	2	0.8	11	35	54	1:3:5	n	n
			2.5	11	34	55	1:3:5	unstable	stable
A2	500	1	0.8	11	34	55	1:3:5	bi*	
			2.5	11	32	57	1:3:5	unstable	
B1	500	2	0.8	12	34	54	1:3:5	bi*	
			2.5	12	33	55	1:3:5	unstable	
B2	500	2	0.8	11	34	56	1:3:5	bi*	
			2.5	11	33	56	1:3:5	unstable	
B3	500	2	0.8	12	32	57	1:3:5	bi*	
			2.5	12	32	56	1:3:5	unstable	
C1	500	4	0.8	18	30	52	1:2:3.5	n	n
			2.5	12	33	55	1:3:5	stable	stable
C2	400	6	10	10	35	56	1:4:6	n	n
			20	4	39	57	1:9:14	stable	stable
C3	450	6	0.8	13	33	54	1:3:4	n	n
			2.5	9	35	56	1:4:6	stable	stable
C4	450	6	0.8	21	28	51	1:1:2	n	n
			2.5	14	32	54	1:2:3.5	stable	stable
C5	450	6	0.8	22	28	51	1:1:2	n	n
			2.5	16	31	53	1:2:3.5	stable	stable

*bi-directional photocurrent

The surface and cross-sectional morphology of the OVC films and their heterostructures were examined with SEM. The micrographs in Fig. 3 show the cross-sectional views for three specimens from each of the series, A, B and C, deposited under different conditions. Film A2 in Fig.3a exhibits a flaky, porous structure. The porosity is consistent with the rapid change in the PEC characteristics, as shown in Fig. 2b. Film C3 is dense with large grains and a smooth surface. Film B1 is denser than A2 but has a more granular surface than film C3.

Upon anodization in the iodide electrolyte the topography for all the OVC-based films resembles that shown in Fig. 3d, comprising polycrystalline hexagonal platelets, as previously seen for the iodized $CuInSe_2$[6,7]. The density of the platelets increases while their size decreases with increasing anodization time. Energy dispersive X-ray analysis indicated decreasing In:I ratio with the anodization time [3].

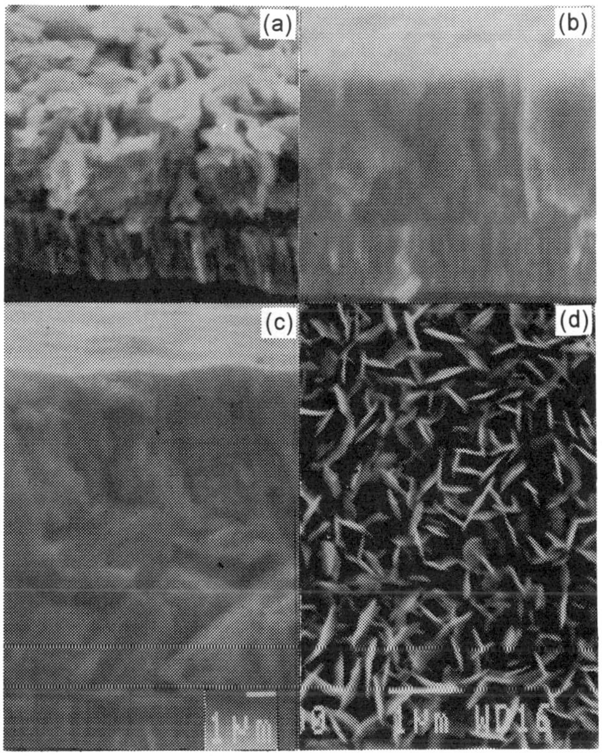

Fig. 3. *Cross sectional views of films (a) A2, (b) B1 and (c) C3; (d) topography of iodized OVC film*

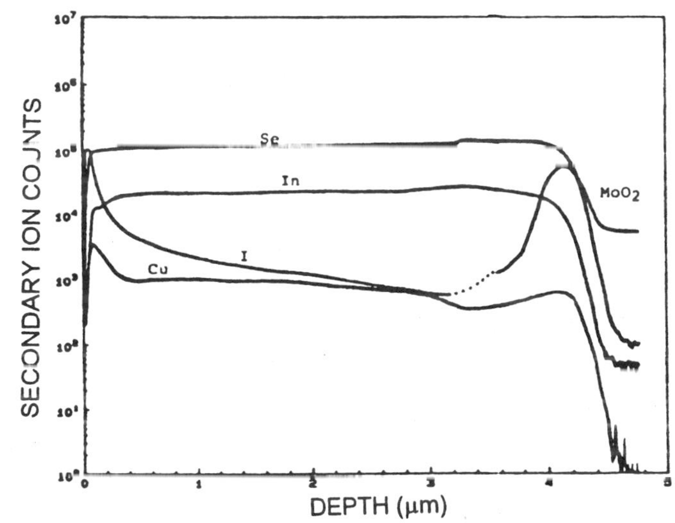

Fig.4. SIMS profile for iodized OVC film, C1₅

The SIMS profile in Fig .4 shows an altered surface composition for the iodized OVC films relative to the base. It also indicates a gradual decrease of Cu, corroborating the EPMA and thickness data in Table I regarding the composition gradient for film C1. The surface changes include an increase in the Cu concentration, depletion of In and incorporation of I. The peak at a depth of 4µm arises from a MoO_2 signal.

The surface composition changes are similar to those seen for $CuInSe_2$ films [5]. But the measured thickness for the $CuISe_3$ overlayer of 0.4 µm is less than that formed on similarly processed $CuInSe_2$ films. The overall substrate decomposition reactions leading to the formation of $CuISe_3$ are given by equations 1-2.

$$2CuInSe_2 + 12h^+ + 2I^- \rightarrow 2CuISe_3 + 3In^{3+} + Cu^+ \quad (1)$$
$$2CuIn_3Se_5 + 20h^+ + 2I^- \rightarrow 2CuISe_3 + 6In^{3+} + 4Se \quad (2)$$

Comparison of the equations 1 and 2 implies that nearly twice as many photons (photocurrent) are needed to grow a molecular layer of $CuISe_3$ from the $CuIn_3Se_5$ substrate relative to the $CuInSe_2$ substrate. Thus, the $CuISe_3$ growth rate is slower on the $CuIn_3Se_5$ material.

The XRD pattern for the iodized $CuIn_3Se_5$ resembles the spectrum for $CuISe_3$ rather than the $CuIn_3Se_5$ substrate, Fig.5, confirming the formation of $CuISe_3$ overlayer. The $CuISe_3$ overlayer is predominantly (116) oriented.

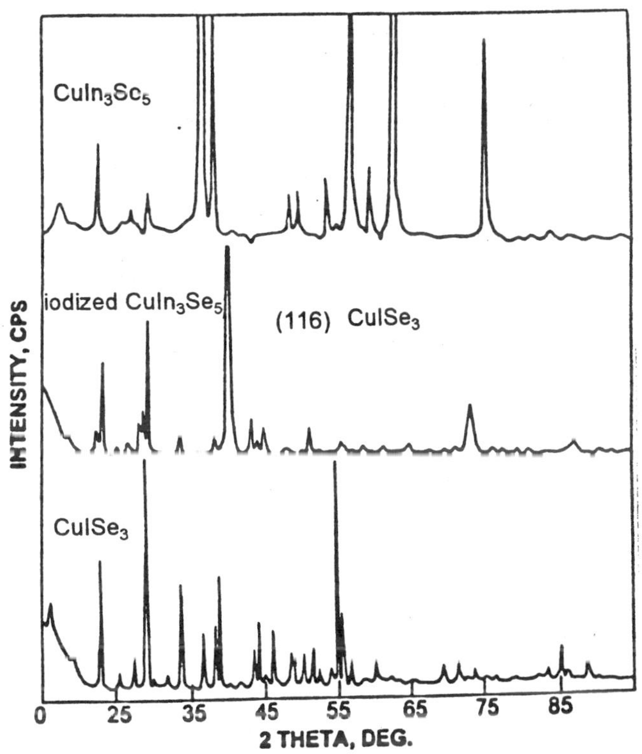

Fig.5. XRD patterns for $CuIn_3Se_5$, $CuISe_3$ and iodized $CuIn_3Se_5$

Thin-film Mo/$CuIn_3Se_5$/$CuISe_3$/Pt solid state(SS) devices, Fig.1, were fabricated by evaporating a Pt or Au film on the electrochemically formed *n*-$CuIn_3Se_5$/*p*-$CuISe_3$ heterostructure. The I-V curves obtained for the solid state cell were similar to those obtained with PEC cell, Fig 2a. However, the films exhibiting PEC instability or low photo-response were completely stable in the SS configuration. The V_{OC} for the SS cells was 0.31± 0.04V, irrespective of the PEC behavior. The ΔV_{OC} measured from the

ΔI_{hv}-V curve was 0.55V, despite a 500Å thick Pt or Au contact layer. The voltage loss (ΔV_{OC} - V_{OC}) is attributed to the microstructural defects for these first $CuIn_3Se_5$ films. It also indicates considerable room for improving the PV performance. The reflection and absorption losses in the metallization layer used for the test cells lower the I_{SC} and V_{OC}, precluding efficiency estimation.

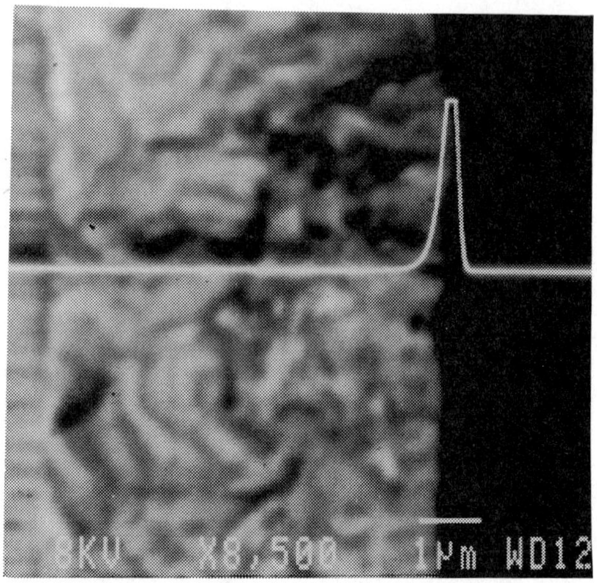

Fig. 6. EBIC profile of a Mo/$CuIn_3Se_5$/$CuISe_3$/Pt device

EBIC response for the Mo/$CuIn_3Se_5$/$CuISe_3$/Pt devices was detected, Fig. 6, confirming the existence of an electrical barrier at the *np* heterointerface. The EBIC linescan peaks at < 0.5μm from the outer surface beneath the hexagonal $CuISe_3$ crystallites in concordance with the SIMS data. The e-beam quantum efficiency for the $CuIn_3Se_5$-based devices was higher than that for *n*-$CuInSe_2$-based devices. The space charge region was fairly narrow and the diffusion length on both sides of the junction was equivalent. The EBIC response was remarkably uniform along the length of the device relative to that observed through the *p*-$CuInSe_2$/*n*-CdS cell development history; the latter uses a deposited window. The uniformity is attributed to the electrochemical interface processing. Material grading, inherent to this approach, removes interface defects from the active *np* junction region and effectively terminates the lattice to enhance its performance. Surface EBIC indicated lateral variation in junction depth and junction efficiency, which is attributed to the local defect chemistry.

CONCLUSIONS

Co-evaporation produces homogenous $CuIn_3Se_5$ films while sequential deposition produces graded compositions with Cu-rich surfaces. PEC surface reconstruction of the *n*-$CuIn_3Se_5$ films leads to the growth of a *p*-$CuISe_3$ overlayer similar to the one produced on the $CuInSe_2$ films [5]. The *n*-$CuIn_3Se_5$/*p*-$CuISe_3$ heterojunctions exhibit a range of I-V characteristics, with the microstructural quality rather than the stoichiometry of OVC determining the PV parameters. The porosity of some $CuIn_3Se_5$ films leads to the deterioration of photoresponse in the PEC cell but does not affect their performance in the SS configuration. The EBIC and I-V results unambiguously confirm that the *n*-$CuIn_3Se_5$/*p*-$CuISe_3$ heterostructure is an electrically active junction with better PV properties than its *n*-$CuInSe_2$ analogue. The device performance is remarkably insensitive to the window layer processing or morphology, which implies that the electrochemically formed *np* interface lies below these surface inhomogeneities. This conclusion is consistent with EBIC and SIMS results which indicate a junction depth of ~ 0.5 μm from the surface. The electrochemical processing leads to a remarkably uniform, clean *np* heterointerface and a rough surface conducive to light trapping.

AKNOWLEDGEMENT

The authors thank Alice Mason, for the EPMA analysis, Sally Asher, for the SIMS and Sharmila Menezes for the I-V measurements.

REFERENCES

1. D. Schmid, M. Rückh, F. Grunwald and H. W. Schock, *J. Appl. Phys.*, **73**, (1993), pp 2002.
2. D. Niles, A. Mason, A. B. Swartzlander, A. J. Nelson, J. Tuttle and R. Noufi, *MRS Spring Meeting* (1993).
3. J. Kessler, D. Schmid, R. Schaffler, H. W. Schock and S. Menezes, *23rd IEEE PVSC*, (1993), pp 549.
4. T.C. Lommasson, S. Wiedeman, L. Russell, J. Kessler and R.R. Arya, *11th Euro PVSEC Proc.* (1994) pp 1226.
5. S. Menezes, *Appl. Phys. Lett.*, **61** 1564 (1992)
6. S. Menezes, *Solar Cells*, **16**, 255 (1986).
7. S. Menezes, R. Noufi and R.J. Matson, *11th Euro PVSEC. Proc.* (1992) pp 779.

INVESTIGATION ON SnS FILM BY RF SPUTTERING FOR PHOTOVOLTAIC APPLICATION

Wei Guang–Pu, Zhang Zhi–Lin, Zhao Wei–Ming,
Gao Xiang–Hong, Chen Wei–Qun
Shanghai University of Science and Technology, Shanghai, 201800, China

Hiromasa Tanamura, Masaki Yamaguchi, Hidenori Noguchi,
Takao Nagatomo, Osamu Omoto
Department of Electronics, Shibaura Institute of Technology, Shibaura, Minato–ku, Tokyo 108, Japan

ABSTRACT

Tin sulfide (SnS) films were prepared by RF sputtering method. The component, crystal structure, optical properties and electrical properties of SnS films were measured. The relationship between the SnS films properties and the sputtering condition was discussed. Using Sb as a dopant and combining thermal annealing at elevated temperatures (about 400℃), the n type SnS films were obtained.

INTRODUCTION

Tin sulfide is a promising materials for low cost solar cell due to its high optical absorption and its suitable bandgap for solar cell [1–2]. According to Prince–Loferski diagrams [3–4], a high photovoltaic conversion efficiency of ≈ 25% can be obtained from it [5]. From the points of view of cost, availability, toxicity and stability, SnS is also a suitable material. Tin sulfide can also be used as a coating material for the glass tubular solar collector, and a solar absorptance of ≈ 0.93 is available [6].

We have reported some research results about SnS films prepared by the vacuum evaporation and made a n–CdS/p–SnS heterojunction solar cell with the structure of ITO/n–CdS/p–SnS/Ag [7].

In this paper we focus our attention on the relationship between the SnS film's properties and the sputtering conditions. We are also interested in the preparation method of n–type SnS films to obtain a full SnS p/n homojunction for photovoltaic devices.

SAMPLE PREPARATION

Tin sulfide films were deposited on slide glass substrates by RF (radio frequency 13.65MHz) sputtering method using SnS powder (with 99.99% purity) as a source material. The main control parameters are the RF power (include anode voltage and anode current), Arogon gas pressure of sputtering chamber, substrate temperature, target to substrate distance. In this work, the gas pressure was about 5×10^{-2} Torr, RF power, anode voltage and anode current were changed from 30W to 70W, 0.5kV to 1.4kV and 30mA to 85mA, respectively. The substrate temperature was changed from room temperature to 350 ℃, and the target to substrate distance was 80mm. The deposition rate was 3 ~ 5Å/sec.

CH3365-4/94/0000-0365 $4.00 © 1994 IEEE First WCPEC; Dec. 5–9, 1994; Hawaii

CHARACTERIZATION OF SnS FILMS

1. Component and Structure

Tin sulfide is a Ⅳ – Ⅵ compound semiconductor and belong to the orthorhombic system with a distorted NaCl type structure [2].

In every crystal unit cell of SnS crystal, there are four Sn atoms and four S atoms, they arranged with a Pnma (62) space group symmetry. If Sn/S atom ratio was changed from 1, the compounds Sn_2S_3, SnS_2 and their solid solution or amorphous material in component Sn_xS_y can be obtained.

We used electron probe microanalysis (EPMA) and Auger electron spectroscopy (AES) equipments to measure the conponents of SnS films. Figure 1 is a typical EPMA spectroscopy for a SnS film. From the measurement results, the component of SnS film change with the sputtering conditions in a large range. The Sn/S atom ratio can be changed from ≈0.1 to about 0.6.

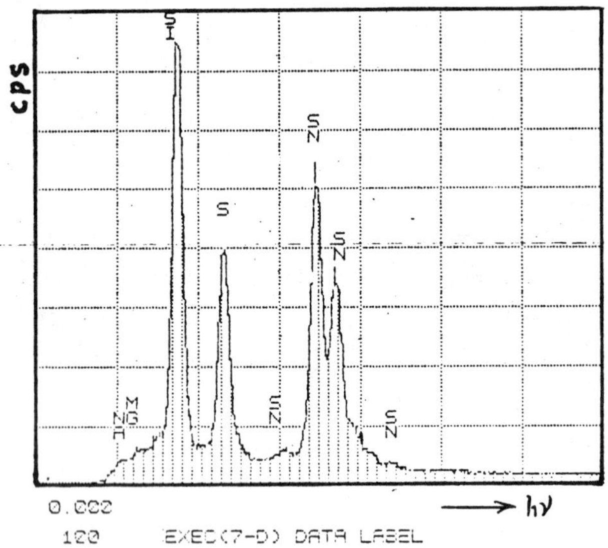

Fig.1 EPMA spectroscopy of SnS film on glass substrate

The crystal character of SnS film was examined by X–ray diffraction (XRD) technology. Figure 2 shows two typical patterns of two different samples with the Sn/S atom ratio about 1. The pattern of sample 1 shows the film has a amorphous structure, and the pattern of sample 5 shows the film has a crystalline structure mixed with amorphous structure. For the crystalline sample, the XRD pattern shows the (040) diffraction peak is very strong, it maybe mean that there is a preferential grows of (010) cryastal plane with its orientation parallel to the surface of glass.

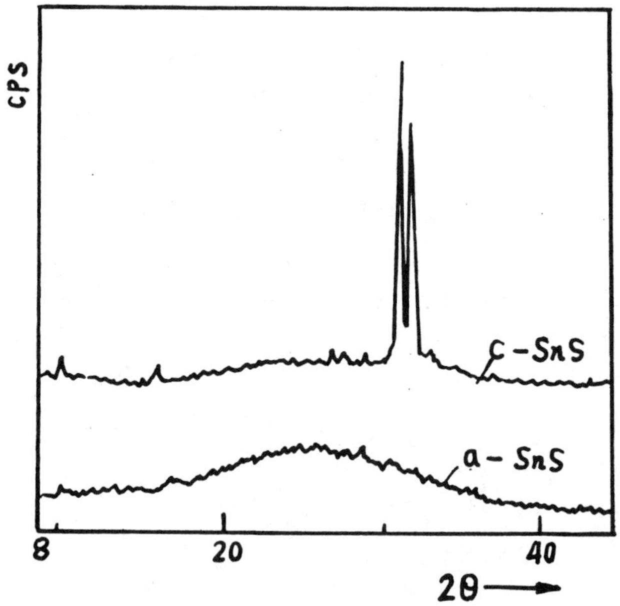

Fig.2 XRD patterns of a–SnS and c–SnS film

From the analysis of the sputtering condition, the component of SnS films mainly dependent on the anode voltage. The Sn/S atom ratio increase with the anode voltage. It maybe originate from that the Sn atom is more heavy than S atom, therefore the Sn atom is rather difficult to sputter on the substrate. If the anode voltage was increased, the Ar^+ ions can obtain more large sputtering energy, and can make more Sn atoms to sputter on the substrate.

2. Optical properties

The color of sample can be changed from light yellow to gray with the change of sputtering condition. The change of color can be related essentially to the Sn/S atom ratio. The light yellowish sample has a small Sn/S atom ratio and the heavy gray sample has a big Sn/S atom ratio.

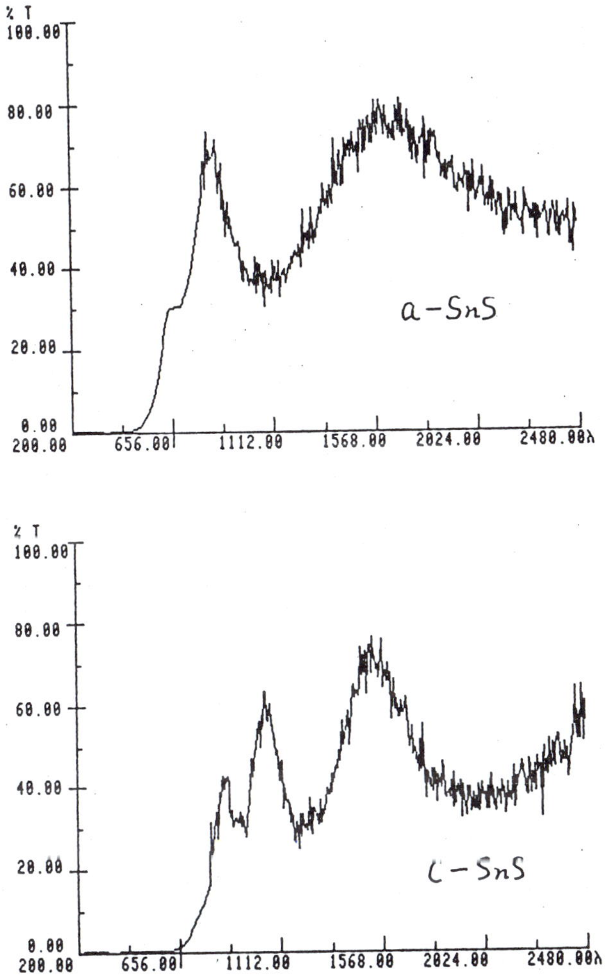

Fig.3 Transmission spectroscopy of a−SnS and c−SnS film

The transmissivity and absorptivity of SnS films were measured with a spectrophotometer. Figure3 shows a typical transmission spectroscopy of a−SnS and c−SnS film. The absorption coefficient α of films was calculated out from the data of transmissivity and the thickness of films. Plots of the square of absorption coefficient α^2 vs the photon energy hν were drawn out. Figure 4 is a typical plot. It can be seen, the square of absorption coefficient varies linearly with photon energy. This maybe mean that SnS is a direct transition semiconductor near the fundamental absorption edge. The energy bandgap Eg of SnS films was determined by extrapolating (α^2−hν) plots (as shown in Fig.4), typical data were 1.43 ≈ 1.46eV for different samples. It has been demonstrated that the energy bandgap change with the film component Sn/S atom ratio, therefore change with the sputtering conditions.

The absorption spectroscopy of amorphous SnS (a−SnS) films are different from that of crystal SnS (c−SnS) films. Figure 5 shows the difference of absorption spectroscopy of a−SnS and c−SnS film.

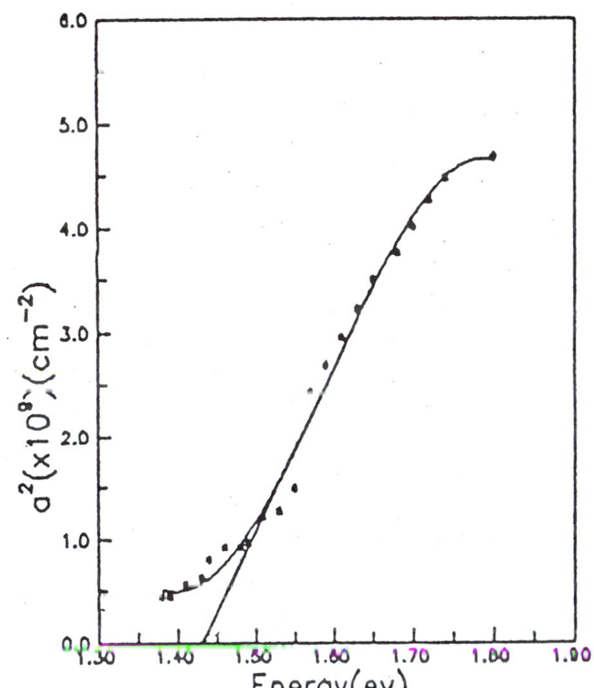

Fig.4 Plot of the square of absorption coefficient vs the photon energy

3. Electrical properties

The conduction type and resistivity of SnS films were measured. Non−dope as−grown SnS films exhibit always p−type conduction with a resistivity of about 20 Ω cm.

PREPARATION OF n−TYPE SnS FILM

The effect of annealing at elevated temperature (~500 ℃) on altering the SnS conductivity from p−type to n−type was discussed [8]. Using Sb as a dopant and combining thermal annealing at elevated temperature (350~450 ℃), we obtained the n−type SnS semiconductor film with the resistivity of about 2 Ω cm. Annealing in hydrogen atmosphere is suitable for the prevention of oxidation of SnS film. Because annealing can make the dopant atoms diffusing into the SnS crystal lattice, the conduction type of SnS film can be altered to n−type.

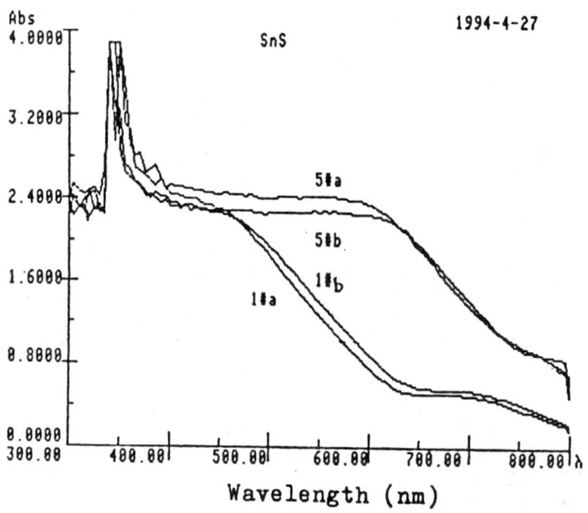

Fig.5 Absorption spectroscopy of a−SnS (sample 1#a, 1#b) and c−SnS (sample 5#a,5#b)

CONCLUSION

SnS films were prepared by RF sputtering method. The Sn/S atom ratio can be controlled by changing the sputtering anode voltage. Using Sb as a dopant and combining thermal annealing at elevated temperature (about 400 ℃), n−type SnS films with a resistivity of about 2 Ω cm were obtained.

REFERENCES

[1] M.Parenteau and C.Carlone : Phys.Rev. , B41(1990)P.5227

[2] Engelken R.D., McCloud H.E., Lee C., Slayton M. and Ghoreishi H. : J.Electrochem.Soc. 134(1987)P.2696

[3] Prince M.B. : J.Appl.Phys., 26(1955)P.534

[4] Loferski J.J. : J.Appl.Phys., 27(1956)P.777

[5] Nair M T S and Nair P K : Semicond.Sci. Technol., 6(1991)P.132

[6] Nair P K and Nair M T S : J.Phys.D.Appl. Phys., 24(1991)P.83

[7] H.Noguchi, A.Setiyadi, H.Tanamura, T.Nagatomo, O.Omoto : Solar Energy Materials and Solar cells., 35(1994)P.325

[8] Sharon M. and Basavaswaran K. : Solar cells, 25(1988)P.97

STUDY OF CdS_XTe_{1-X} SOLID SOLUTIONS OBTAINED BY SCREEN PRINTING

O. de Melo, M. Melendez-Lira, I. Hernandez-Calderon
CINVESTAV- IPN, Physics Dept., Apartado Postal 14-740, 07000, México D.F.
L. Baños
Instituto de Investigación en Materiales, UNAM, Apartado. Postal 70-360, 04510 México D. F.
Arturo Morales-Acevedo
CINVESTAV- IPN, Electrical Engineering Dept., Apartado. Postal 14-740, 07000, México D.F.

INTRODUCTION

Ternary solid solutions of II-VI compounds have been widely studied due to the possibility of their applications in semiconductor devices for specific utilization of tuning the band gap and lattice constant as a function of composition. In general, these solutions exist in the whole range of compositions when there is a structural coherence between the binary extremes. This is the case of CdTe-ZnTe, CdTe-HgTe and other II-VI ternary solutions. However a different situation is observed when the binary compounds present different stable crystallographic phases. In this case, one can expect solubility only in composition ranges near to the extremes. Around these regions the solid solution structure is the same as that of the closest binary compound. CdS-CdTe or CdS_xTe_{1-x} belong to this kind of solutions since CdTe and CdS exhibit the zinc blend and wurzite phases, respectively. Probably for this reason the study of this solution has been limited to a few works[1,2].

Recently, important progress has been reached in the elaboration of low cost solar cells based in CdS/CdTe heterostructures[3,4]. This was accomplished by using different techniques such as chemical bath deposition, close space sublimation, electrodeposition and screen printing and sintering[5-8]. In all these low cost techniques, at least a relatively high temperature step is necessary during the processing for the completion of the cell. Under these conditions, a CdS_xTe_{1-x} interfacial layer is formed which has great influence in the cell parameters and performance[9,10].

In this work we present a study of the properties of this solution obtained in thin film form by screen printing and sintering technique. Optimal annealing conditions were found where the films were single phase and homogeneous. The samples were characterized by optical microscopy, talystep profiles, x-ray diffraction patterns, optical transmission and photoluminescence (PL) spectra.

EXPERIMENTAL

CdS_xTe_{1-x} films were synthesized and sintered from CdTe and CdS powders using the screen printing and sintering technique[7,8]. 5N CdTe and CdS powders were milled and mixed at nine different CdS molar fractions, namely: $x = 0$, 0.03, 0.09, 0.15, 0.18, 0.88, 0.91, 0.97 and 1. In all cases 5 % weight of $CdCl_2$ was added as a flux. This amount is higher than that used in CdS/CdTe solar cells but it was used to help the synthesis of the solutions. Some drops of propylene glycol were added to form the paste that was screen printed on corning 7059 glass substrates. Polyester screens of 130 and 180 mesh were used. Prints were air dried in a belt furnace at 130 ^{o}C during one hour. The sintering was performed in forming gas. In order to obtain complete solution between CdS and CdTe it was necessary to find the optimum annealing conditions. Temperature, time and discovered area were varied. We used a graphite boat specially designed for this purpose. The main feature of this boat is that it has a double sliding cover with 24 holes of 1 mm of diameter in each cover. In this way one can change the total open area of the holes by slight modification of the position of one cover respect to the other. Full match of holes in both covers giving the higher exposed area, and complete mismatch null area.

Lattice parameters were measured in a D5000 Siemens powder diffractometer with zero point deviation of 0.0012 Å, using Cu-Kα_1 radiation. The samples were finestly ground and mounted in a special sample holder. The plane (400), with a d = 1.019 Å PDF reported value, was scanned at 0.5^{o}/min. Optical characterization of the films was carried out by room temperature transmittance and 14 ºK photoluminescence (PL) spectroscopies. The transmittance measurements were accomplished in a UV/VIS Unicam spectrometer. But because the films were very thick after annealing, the samples were mechanically lapped to a small thickness before the transmission measurements were made. The PL measurements were carried out employing a 85 cm focal length double monochromator with a GaAs photomultiplier as detector. Excitation of 100 mW/cm^2 of the 4579 Å line of an Ar^+ laser was used for the CdTe rich branch and 3 mW/cm^2 of the 3250 Å of a HeCd laser for the CdS rich branch.

RESULTS AND DISCUSSION

In optimal treatment conditions (annealing temperature and time of 670 ^oC and 3 hours respectively) the samples showed single phase x-ray diffraction pattern and homogeneous color. The annealing was also successful in enlarging the grain size and the thickness homogeneity as observed from talystep surface profiles and by optical microscopy of as-deposited and annealed samples. The films annealed at optimal conditions were scraped from the substrates and milled, and x ray diffraction patterns were taken to determine the lattice parameters of the solutions.

The dependence of the lattice parameters in composition for both zinc blend and wurzite structures is shown in figure 1. In the CdTe-rich branch a good agreement with reported data is found. The dependence of lattice parameter *a* on the CdS molar fraction x, was found to be: a = 6.488 - 0.669 x, according to the Vegard's law. The value at x = 1 for CdS in the zinc blend structure was taken from ref. [11]. The above dependence is comparable with that of ref. [1] for sintered pellets (a = 6.477 - 0.657 x). The solubility limit was found in the x range between 0.15 and 0.18 since sample with x = 0.18 was out of the Vegard's law behavior. This behavior was considered to be due to the higher CdS content that was not incorporated into the solution. The phase diagram of ref. 1, that starts at higher temperatures than those of our annealing, predicts a solubility limit of 0.2 at 700 ^oC. We consider that this value agrees with our results since at a lower temperature a lower solubility can be expected. In the CdS-rich branch the wurzite structure was found with solubility limit x ranging between 0.88 and 0.91 in relatively good agreement with 0.88 at 700 ^oC found in ref. [1]. However, the lattice parameter behavior was very different. We found: a = 4.28 - 0.15 x and c = 7.079 - 0.365 x, while in ref. [1] a = 4.566 - 0.431 x and c = 7.052 - 0.775 x. In this case, our linear fitting of the experimental data does not contain the literature parameter for wurzite CdTe. It can be supposed that, in spite of the closeness of the boat, excessive evaporation of CdS shifted the real composition to lower x values. However, if this was the case, the correction of the composition would lead to a higher disagreement. Then, it seems that Vergard's law is not obeyed in this range of composition and that, consequently, CdTe wurzite lattice parameter can not be obtained by extrapolating the linear fit of the experimental data for the wurzite CdS - rich solution.

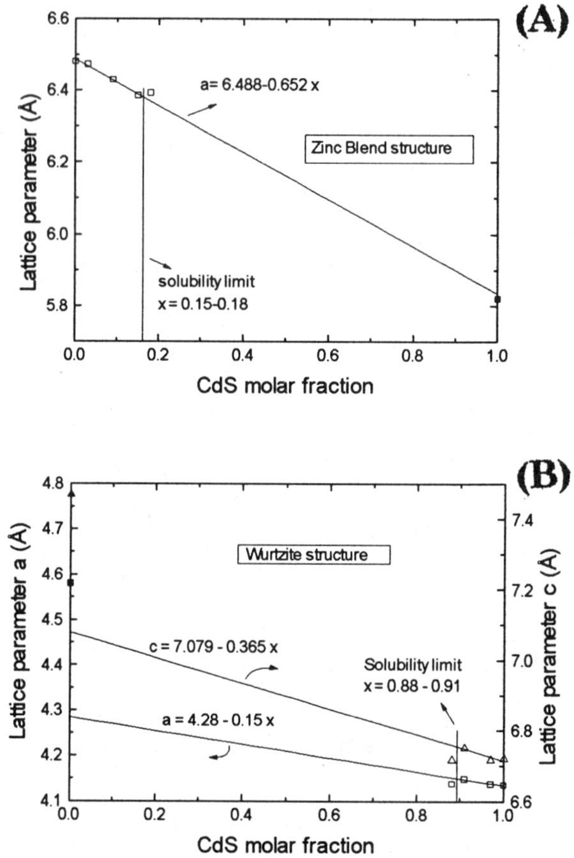

Fig. 1. Lattice parameter as a function of CdS molar fraction x for (A) Zinc Blend and (B) Wurzite structures respectively.

Due to lack of space we do not show here the room temperature transmittance spectra but we shall discuss the results. We could appreciate a shift toward lower energies for the absorption edge with the increase of CdS content up to x = 0.15. For the CdS rich region an increased shift toward high energies with the CdS content exists. It should be noted that we did not observe the presence of additional absorption edges in the transmittance spectra as we could expect if there was a mixture of different phases. In order to get the behavior of the band gap (Eg) as a function of the CdS content, the parabolic model for direct transitions was fitted to the absorbance obtained from the measured transmittance.

Figure 2 shows the E_g values along with those of a fitting proposed in ref. [2]. It is apparent the good agreement in values corresponding to the CdTe rich region. In the CdS rich region, however, the values obtained here are lower than those obtained in ref. [2]. A likely origin for these discrepancies is in the film preparation method used in that work. There, the right composition before the film deposition was well known, but the vacuum evaporation procedure employed might have shifted the composition of the obtained films. There is evidence that substrate or annealing temperatures around 250 °C give CdTe thin films with optical parameters near those for the bulk material, and higher temperatures produce a high rate of reevaporation of the deposited material[11]. So, because the substrate temperature was 200 °C during the flash evaporation, it could be possible to have good quality CdTe rich samples. Our samples suffered thinning because of the thermal annealing, but we may guarantee that the remaining material is rather homogeneous because of the 670 °C annealing temperature. For the CdS rich branch it could be argued that the relatively high temperature during the annealing should give us a material rich in CdTe, but we think that the relatively high thickness of our films and the use of a semi-closed graphite boat limits a strong evaporation of sulfur. Anyway, the bowing effect in the band gap behavior previously observed[2] is confirmed here.

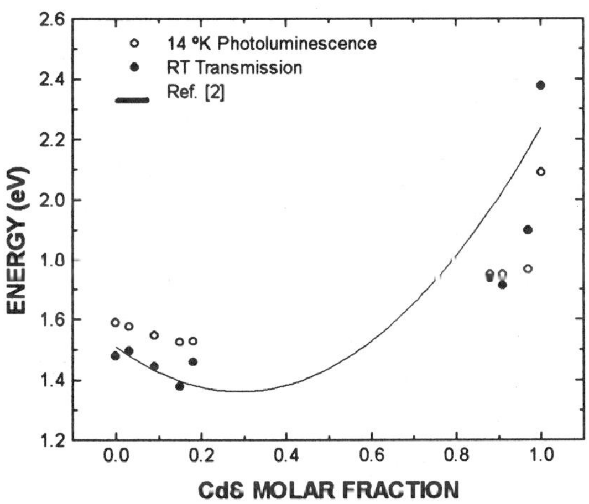

Fig. 2. Energy gap from transmittance measurements at RT and luminescence peaks (at 14 K) as a function of CdS molar fractions. Results of ref. [2] are also included as a full line.

Figures 3(A) and 3(B) show (up to our knowledge) the first PL spectra of the alloy CdS_xTe_{1-x}, corresponding to the CdTe and CdS rich branches, respectively. In figure 3 (A) it is clear the presence of well defined near band gap peaks giving us further support about the formation of big enough grains of the alloy. These peaks shift toward lower energies than that of CdTe with increasing x. In figure 2 the energies of the near band gap emission as obtained from the higher energy peak in the PL spectra for these samples are included. Because the high Cl content, these peaks could be attributed to Cl donor emission. For the CdTe thin films there is an additional structure. Other band is observed between 1.52 and 1.57 eV. The 1.47 eV band composed by the fundamental peak and LO phonon replicas is present. This band has been previously observed[12,13] and has been ascribed to excitons related to structural defects as twins, dislocations or stacking faults. The presence of this band in our samples deposited by screen printing on glass, confirms that molecular beam epitaxy growth method or Ga and As difunding from the substrates in heteroepitaxial CdTe/GaAs are not responsible of this emission. When the CdS content is increased many changes are observed in the PL spectra: it would seem that the structure in the near band edge peak is more resolved and the additional bands disappear. In order to identify these changes a detailed PL study is underway.

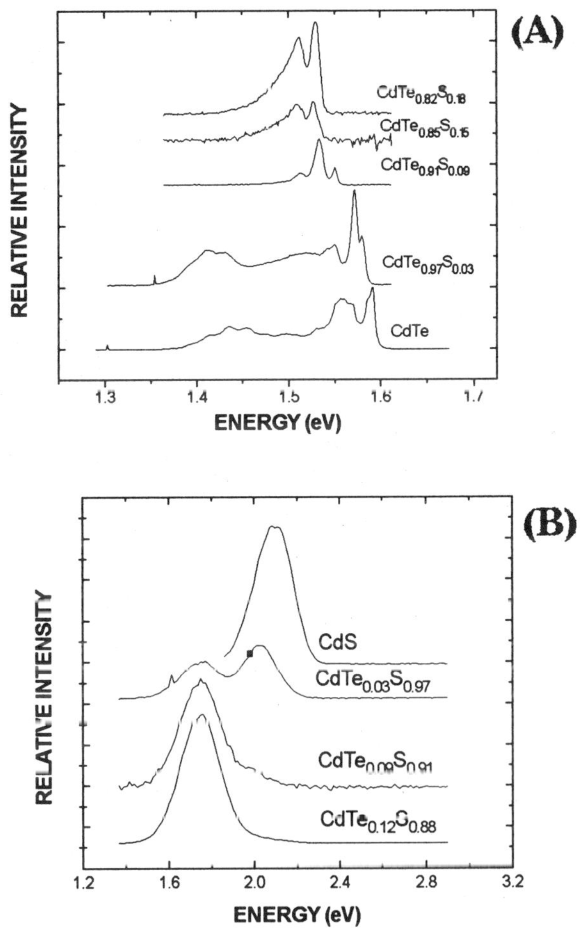

Fig. 3. Photoluminescence spectra of samples with different CdS composition. (A) CdTe rich branch and (B) CdS rich branch.

For the CdS rich branch near band gap emission was not observed in pure CdS sample. In figure 3 (B), the emission of the well-known red band of CdS is observed which has been associated to surface damage, sulfur vacancies or neutral complex $[V_{Cd}^{2-} + V_S^{2+}]$[14]. In the case of CdS there are room temperature PL studies about the behavior of melt growth Cl doped bulk material showing a broad band around 2.41 eV[15]; our CdS samples show a broad band at around 2.2 eV. Preliminary PL measurements as a function of temperature confirms the assignment as deep levels originated by defects, because there is not an appreciable change in its position. For x = 0.97 there are two broad peaks whose origin can be explained as follows. The band centered at 1.8 eV have the same nature as the red band in pure CdS but shifted in energy due to the composition change. Then, the other band center at around 2 eV would be a near band gap emission (probably due to Cl doping) taking into account the transmission gap measured above. For the same reason, the bands observed in the x = 0.91 and x = 0.88 would be also near band gap ones, while the corresponding "red band" was not observed or remained out of our range of measurements.

CONCLUSIONS

The Screen Printing Technique has been used successfully to obtain polycrystalline CdS_xTe_{1-x} solutions on glass substrates. Optimal annealing conditions were found where the samples were single phase and homogeneous. Several techniques were used to characterize the films. As a result, the solubility limits in both CdTe and CdS rich branch were obtained and the dependence of lattice parameter in x was determined and compared with previous results. Optical measurements allowed to study the behavior of the band gap and characteristic luminescence bands. The previous result of a bowing effect in the band gap dependence with composition was confirmed.

According to this results we can expect that the interdiffusion region in the CdTe/CdS interface would be composed by a double layer: one CdTe rich (near de CdTe) and another CdS rich (near de CdS). Both these layers would have a lower band gap than the closest pure constituent. These considerations may be important for understanding light absorption and recombination at the CdS and CdTe interface in CdS/CdTe solar cells.

ACKNOWLEDGEMENTS

This work was partially supported by CONACYT. The authors thank Miss. M. Guerrero-Cruz and Mr. M. Ortega-López for their technical assistance.

REFERENCES

[1] K. Ohata, J. Saraie and T. Tanaka, Japan. J. Appl. Physics 12, 1198 (1973).
[2] K. Ohata, J. Saraie and T. Tanaka, Japan. J. Appl. Physics 12, 1641 (1973).
[3] T. L. Chu, S. S. Chu, J. Britt, C. Ferekides, C. Wang, C. Q. Wu and H. S. Ullal, IEEE Electron Device Letters 43, 303 (1992).
[4] J. Britt and C. Ferekides, Appl. Phys. Lett. 62, 2851 (1993).
[5] O. de Melo, L. Hernández, O. Zelaya-Angel, R. Lozada-Morales, E. Vasco, Appl. Phys. Lett. (in press).
[6] C. Cohen Solal, M. Barbe, H. Afifi and G. Neu, J. of Crystal Growth 72, 512 (1985).
[7] I. Clemminck, M. Burgelman, A. Vervaet and J. De Poorter in: Proc. of the 10th European Photovoltaic Solar Energy Conference (Lisbon, 1991), p. 577.
[8] H. Uda, H. Matsumoto, K. Kuribayashi, Y. Komatsu, A. Nakano and S. Ikegami, Japan. J. Appl. Physics 22, 1832 (1983).
[9] A. Rohatgi, S. A. Ringel, R. Sudharsanan and H. C. Chou, Conference Records of the Twenty Second IEEE Photovoltaic Specialists Conference (Las Vegas, 1991), Vol. 2, p.962.
[10] B. E. McCandless and S. S. Hegedus, Conference Record of the Twenty Second IEEE Photovoltaic Specialists Conference (Las Vegas, 1991), Vol. 2, p.967.
[11] M. Meléndez-Lira, S. Jiménez-Sandoval, I. Hernández-Calderón, J. Vac. Sci. Technol. A7, 1428 (1989).
[12] S. Fujii, T. Terada, Y. Fujita, T. Iuchi, Japan. J. Appl. Phys. 28, L1712 (1989).
[13] J. S. Gold, T. H. Myers, N.C. Giles, K.A. Haris, L. M. Mohnkern, R. W. Yanka, J. Appl. Phys. 74, 6866 (1993).
[14] S. Achoar, G. H. Talat, Thin Solid Films 144, 1 (1986).
[15] U. Venkateswaran, M. Chandrasekhar and H.R. Chandrasekhar, Phys. Rev. B 30, 3316 (1984).

AMORPHOUS SILICON TECHNOLOGY: COMING OF AGE

C. R. Wronski
Electronic Materials and Processing Research Laboratory
The Pennsylvania State University, University Park, PA 16802, U.S.A.

ABSTRACT

This paper describes the current status of hydrogenated amorphous silicon (a-Si:H) based solar cell technology and the evolution to its "coming of age." Several features which are unique to the a-Si:H technology are reviewed as well as the advances in concurrently improving materials, cell design and manufacturing. Future developments in this technology, whose impact is not only felt in the photovoltaic field but also in the huge market of liquid crystal displays, are also discussed.

INTRODUCTION

In the last few years significant progress has been made in improving the stabalised efficiencies of hydrogenated amorphous silicon (a-Si:H) based solar cell panels. This has led to a steady increase in the production of these panels and the planned construction of manufacturing plants capable of producing up to ten megawatts per year. These advances together with the already established large scale mass production of liquid crystal displays based on a-Si:H thin film transistors, indicate a "coming of age" for amorphous silicon technology. It was twenty years ago that the first small area a-Si:H solar cells were fabricated with initial efficiencies of 1-2%.[1] Today one square foot panels, composed of a-Si:H based materials, have been fabricated with a stabilised efficiency greater than 10%.[2] This progress has been achieved in a way that is somewhat unique to the a-Si:H solar cell technology in that it has been based on concurrent developments in the areas of new materials, novel cell designs and large area deposition techniques suitable for mass production.

The advances made in materials and cell design could utilise the advantages offered by the outstanding optoelectronic properties of a-Si:H, its ability to form germanium and carbon alloys and the relatively low film deposition temperatures (250-300°C). However from the outset the efforts in these areas could not just focus on improving the intrinsic photovoltaic properties and solar cell efficiencies but also had to take into account the light induced degradation found in a-Si:H based materials.[3] As a consequence not only better but also more stable materials were sought and improvements in cell design had to take into account the material instabilities and to optimise not just the initial cell efficiencies but more importantly their stabilised performance. The development of the a-Si:H alloys and optical enhancement during this period have had a profound effect on the improvements achieved in the stabilised efficiencies.

This paper reviews some of the areas that have been important in the development of a-Si:H based solar cells, discuss their present status and future directions of research and development.

MATERIALS

Hydrogenated Amorphous Silicon

Although studies were carried out on a-Si:H, first deposited by Chittik[4] in 1969 by glow discharge decomposition of silane, world wide interest was not generated till 1976 after the publication of results on the n- and p-type doping of amorphous silicon[5] and the results on amorphous silicon solar cells.[1] This interest resulted in a wide range of studies which improved the understanding of the optoelectronic properties and the role of hydrogen. At first a large range of deposition conditions were investigated but once it was established that the hydrogen was needed to passivate the dangling bonds, attention was given to fabricate "hydrogen alloys" with the lowest density of midgap states.[6] By incorporating different amounts of hydrogen it is possible to change not only the gap of a-Si:H but also the density of states in the gap. Materials which are commonly called "device quality a-Si:H" are generally prepared by glow discharge decomposition of silane and have 5 to 15 at.% of hydrogen, Tauc optical gaps of around 1.7eV and densities of dangling bond midgap states on the order of 10^{15} to 10^{16} cm^{-3}. Such low densities of midgap states means that in 1μm p-i-n solar cells an electric field is present across its entire thickness.[7] Example of such materials is illustrated in Fig.1, where the a-Si:H materials having between 5 and 15 at.% hydrogen, Tauc optical gaps 1.7 to 1.78 eV, and dangling bond densities from 4 to 8×10^{15} cm^{-3}[8] In this case the films were prepared by rf discharge decomposition of silane with only the substrate temperature being changed.

The main thrust in the development of a-Si:H at first focused on optimising the performance of single junction cells. This meant improving and tailoring the optoelectronic properties of a-Si:H as an optimum single absorber layer for the generation and collection of carriers where an optimum gap is around 1.7eV. Highly

First WCPEC; Dec. 5-9, 1994; Hawaii

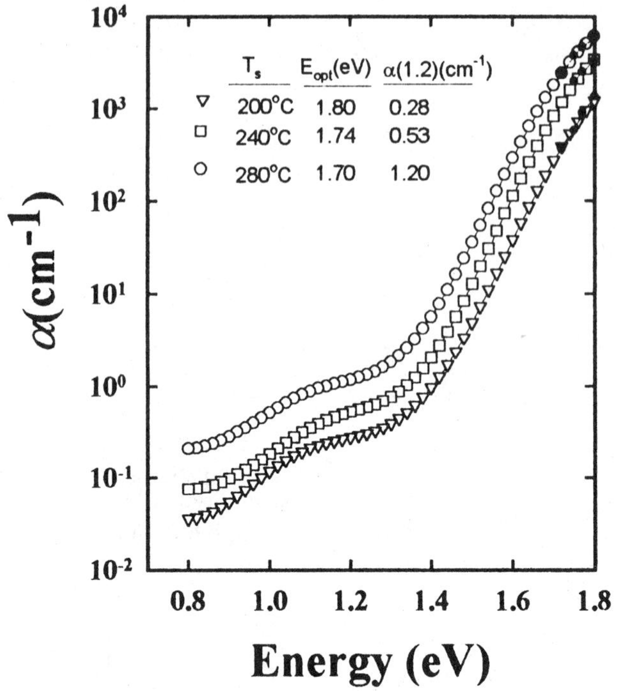

Fig. 1 Absorption coefficient vs. energy for a-Si:H films deposited at different substrate temperatures by RF decomposition of SiH_4

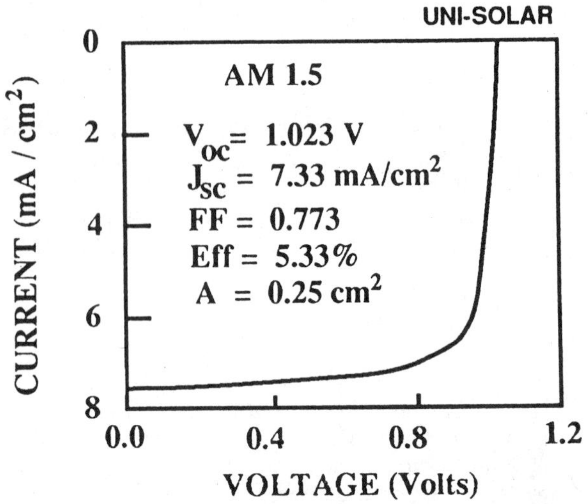

Fig. 2 Current-Voltage characteristic of a single junction a-Si:H/ p-microcrystalline solar cell on a stainless steel substrate (From Reference 10)

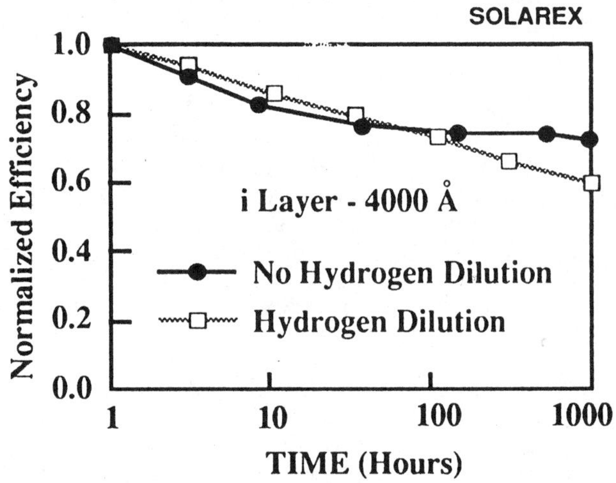

Fig. 3 Normalised efficiency as function of time of illumination (AM1.5) for a single junction a-Si:H solar cell whose absorber layers are fabricated with and without hydrogen dilution

electron injecting, n-type contacts can be made with phosphorus doped a-Si:H. However the p-a-Si:H achieved with heavy boron doping, exhibits gap shrinkage so that p-type hydrogenated amorphous carbon[9] and microcrystalline silicon[10] were developed. However with the introduction of multijunction cells this bandgap constraint was no longer valid and wider bandgap a-Si:H materials have been developed. This has been achieved in part by the use of hydrogen dilution, first studied by Matsuda et. al.,[11] which is extensively used in the fabrication of alloys. Using hydrogen dilution it is possible to deposit high quality a-Si:H with Tauc optical gaps in excess of 1.8eV. This has enabled efficient a-Si:H solar cells to be fabricated with open circuit voltages greater than 1 volt. The light I-V of such a a-Si:H n-i-microcrystalline p solar cell is shown in Fig.2 (The short circuit current of 7.3mA in this particular cell has been tailored for multijunction cell operation). Hydrogen dilution has also been used to obtain more stable materials and solar cells.

That sunlight creates light-induced metastable defects in a-Si:H was discovered very soon after the first cells were made[3]. The serious limitations that this Staebler-Wronski Effect (SWE) imposes on the performance of solar cells have led to many extensive studies on the exact origin of the effect and how it can be eliminated. Progress has been made in understanding the mechanism and quantifying the changes in materials in ways that are relevant to cell performance. An example of two cells where one is fabricated with hydrogen dilution and the other without hydrogen dilution is shown in Fig. 3.

It can be seen in the figure that not only does the hydrogen dilution result in a higher relative stabilised state but also one which is reached much more rapidly. Thus for the improvements in the stability of the materials have been empirical but this is becoming less so as the growth and microstructure of materials and cells is becoming better characterised.[13] In parallel with the improvements in materials, "engineering" approaches have been taken to minimise the effects of the SWE and these are discussed later.

Hydrogenated Amorphous Silicon Alloys

Alloys with bandgaps either lower or higher than a-Si:H have become part of the a-Si:H based technology. Hydrogenated amorphous silicon carbide (a-SiC:H) can be produced with optical gaps up to about 3eV and hydrogenated amorphous silicon germanium (a-SiGe:H) can have optical gaps as low as about 1.2eV. Both these alloys can be deposited at the same substrate temperatures as a-Si:H by plasma assisted CVD from silane mixed with carbon and germanium containing gases. Most often methane(CH_4) and germane(GeH_4) are used but other gases are also used to produce intrinsic and doped alloy materials. This simple extension of a-Si:H deposition technique allows the composition of these alloys to be readily tailored by changing the compositions of the gas mixtures.

Hydrogenated amorphous silicon carbide was initially introduced as a wide bandgap, p-type heterojunction contact for a-Si:H solar cells.[9] The wider bandgap contact offers the advantage of generating a higher open circuit voltage as well as higher optical transmission to the a-Si:H absorber layer. Although materials with optical gaps as high as 3eV can be obtained, the range of gaps that can be utilised in solar cells is limited by the increase in the gap states that accompanies the incorporation of carbon. Despite this, the widening of the gap achieved even after doping significantly increases the transmission of the p-type window layer. Recently attention has been given to using a-SiC:H as the absorber layer in the wide bandgap top cell of multijunction cells. Initially, as anticipated, serious problems were encountered with the effects of the higher densities of defects. However with improved deposition techniques, such as hydrogen dilution[11], 1.9eV a-SiC:H materials can be made with midgap densities as low as those in good a-Si:H and cell degradation has been drastically reduced.[14] An example of an a-SiC:H cell performance is given in Fig.4, where the light I-V of an a-SiC:H cell with a Tauc gap about 1.9eV is shown in the initial state and after 1700 hours of AM1.5 illumination.

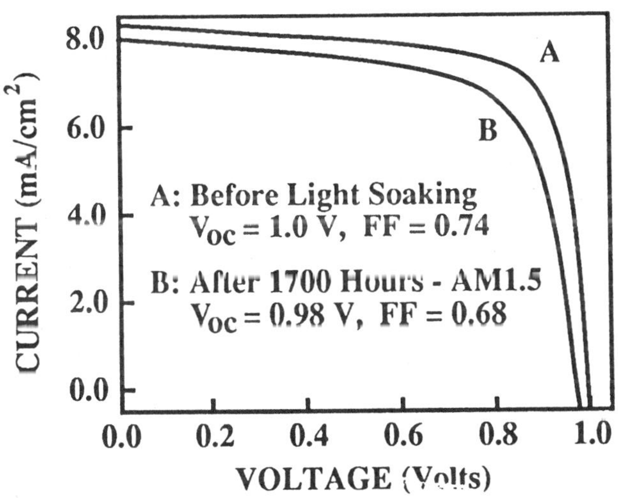

Fig. 4 The initial and degraded current-voltage characteristic of an a-SiC:H single junction cell. The a-SiC:H absorber layer has an optical gap of about 1.9eV (From reference 10)

The current in this cell is about 8 mA/cm^2 as it is designed as the top cell of the triple junction cells which are operating in such a current regime. As will be discussed later, this means the cell is thin (about 1000Å in this case) so that the degradation is reduced.

Interest in the low bandgap alloy, a-SiGe:H, also started quite early in the development of a-Si:H technology. It was seriously considered a key component of double and triple multijunction cells(discussed later) so that over the years significant progress has been made in improving its photovoltaic properties. As in the case of carbon alloys incorporation of germanium in the a-Si:H increases the density of intrinsic (dangling bond) defects. Because low bandgap a-SiGe:H alloys require large fractions of germanium the challenge has been to do this with a minimum increase in defects. The bandgap of useful a-SiGe:H has been consistently lowered so that today alloys with Tauc gap between 1.4 and 1.5eV have densities of midgap states less than $10^{16}cm^{-3}$. This enables efficient solar cells to be fabricated as illustrated in Fig.5, where the light I-V is shown for an a-SiGe:H/a-SiC:H single junction cell with the absorber layer having a Tauc optical gap of 1.45eV.[15]

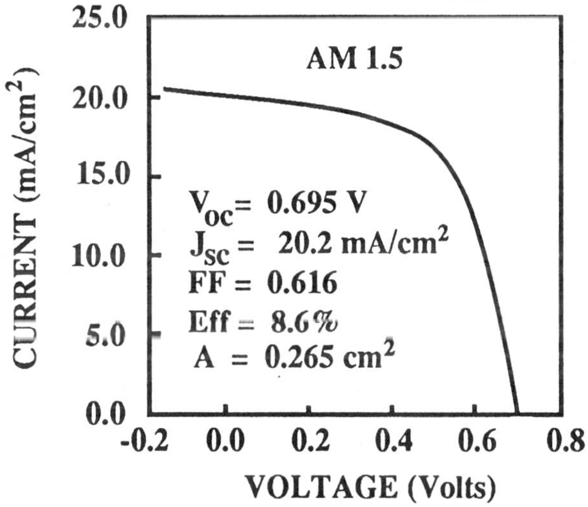

Fig. 5 Current Voltage characteristic of an a-SiGe:H single junction solar cell. The a-SiGe:H absorber layer has an optical gap of about 1.45eV (From reference 18)

As expected the open circuit voltage is lower than in a-Si:H/a-SiC:H cells but the short circuit current is higher. Although the cell characteristic shown here is for AM1.5, a-SiGe:H cells are designed for operating in multijunction cells where the currents are significantly lower so that there is an improvement in the fill factor. Similar results have been obtained by several groups[2,15,16] and there are efforts in further improving the properties of these alloy materials. This is being pursued along several fronts which appear more promising as the understanding of the role that deposition conditions and microstructure play in determining not only the intrinsic properties but also their stability to light-induced changes. The role of hydrogen is again proven to be important since, as in the case of a-Si:H, it plays a role in determining the optical gap of the alloys.[18] Also, the control of the bonding of hydrogen in a-SiGe:H alloys offers a path for not only tailoring the gap but also improving the microstructure and hence the stability further.

SINGLE AND MULTIJUNCTION CELLS

Single Junction a-Si:H Cells

The low processing temperature of a-Si:H solar cells allows a variety of substrates and cell structures to be used. Single(and multijunction) p-i-n and n-i-p cells are fabricated with either a-SiC:H or microcrystalline silicon p-layers on substrates such as glass/TCO, stainless steel and plastics . In all cases however, high efficiency cells utilise optical enhancement[17] obtained by texturing these substrates. This increases the already high optical absorption in the a-Si:H cells and efficient light trapping enables cells less than 1 μm thick to have AM1.5 short circuit currents around 17 mA/cm^2. Because the usual gap of high quality a-Si:H is around 1.7eV, open circuit voltages up 0.9 volts can be obtained as well as fill factors of around 0.75. The ability to generate such high currents with thinner absorber layers lowers the requirements on the carrier transport required for their efficient collection. The concomitant decrease in carrier recombination in these thinner structures reduces the rate at which SW defects are created and hence the degradation in the cell performance.

There have been continuous improvements in the materials and various components of a-Si:H junction cells. The densities of gap states have been lowered not only in the bulk but in the p-i interface regions. Improvements have also been made in the p-layers and the textured substrates as well as in the in-situ processing procedures developed to further enhance the performance of these cells. Single junction a-Si:H cells with efficiencies greater than 12% have been reported by a number of different groups and recently Sanyo has reported results on 100 cm^2 a-Si:H/a-SiC:H single junction module with an efficiency of 12%.[18] This module consists of 14 cells in series and its light I-V is shown in Fig.6. Several of the recent improvements in a-Si:H technology have been incorporated in the fabrication of this module. The processing of the large area deposition into the integrated a-Si:H solar cell module was carried out using laser scribing which has become part of the a-Si:H solar cell technology and allows aperture rations greater than 95% to be achieved. In parallel with these advances in single junction cells and modules there have also been continuous advances in the technology of double and triple junction cells and modules.

Multijunction a-Si:H, a-SiGe:H Cells

The development of the multijunction cells was initiated by the interest in improving high efficiencies not only of the initial state but particularly in the stabilised state. A two track approach was taken, with the emphasis

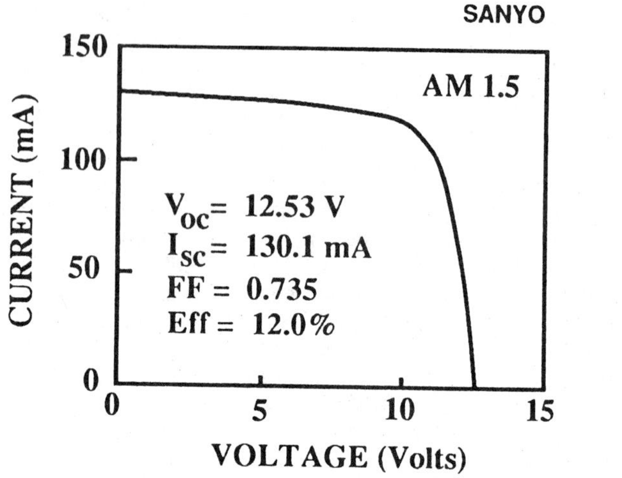

Fig. 6 Current-Voltage characteristic of an integrated type single junction a-Si:H module (100cm^2) with 14 cells in series (From reference 18)

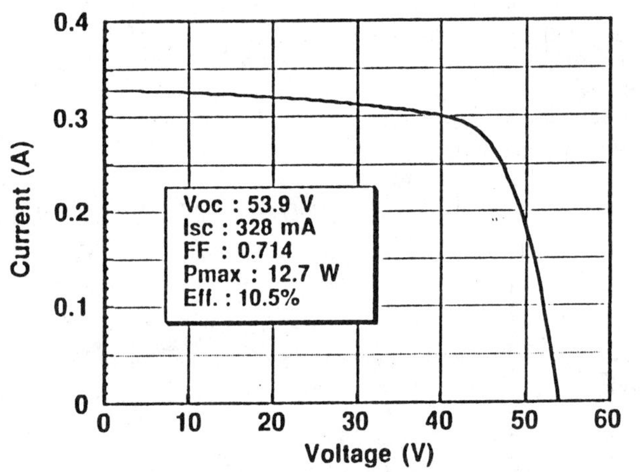

Fig. 7 Current-Voltage characteristic of a 10.5%, 30x40 cm^2 double junction a-Si:H/a-Si:H module (From reference 19)

of the first track being on improving the stability or the stabilised efficiency by basically dividing the a-Si:H cell into two thinner ones. In these double junction a-Si:H/a-Si:H stacked cells the carrier recombination is reduced because of the smaller thickness of each a-Si:H component cell. In order to match the currents the bottom cell is somewhat thicker but not by much since the optical enhancement is more efficient in increasing the absorption in the bottom component of the stacked cells. This approach has been successfully implemented into both cells and modules with initial efficiencies greater than 10% and stabalised efficiencies greater than 9% being obtained.[19] An example of the I-V characteristics of such an a-Si:H/a-Si:H tandem cell module is shown in Fig. 7. This 30x40 cm^2 Fuji module consisting of 30 a-Si:H/a-Si:H stacked cells in series whose initial efficiency is 10.5% and its degradation is less than 15%.

The second approach is the more traditional one where the stacked multijunction cells have different bandgaps in order to efficiently absorb different parts of the solar spectrum. As indicated earlier there are three materials in the a-Si:H based technology that can be usefully implemented as absorber layers in such cells; a-Si:H, a-SiC:H and a-SiGe:H. Modules composed of up to three stacked solar cells are considered to be practical. Thus far primary attention has been given to a-Si:H/a-SiGe:H tandem cells and a-Si:H/a-SiGe:H/a-SiGe:H triple junction cells. Consideration is also being given to the a-SiC:H/a-Si:H/a-SiGe:H triple junction cells. A key to having an efficient multijunction cell is the splitting of the sunlight absorption in such a way as to maximise the equal currents that are generated in each cell. The ability to vary the bandgaps of these materials gives flexibility in the combinations of the absorber layers that can be used to achieve the same overall efficiencies.

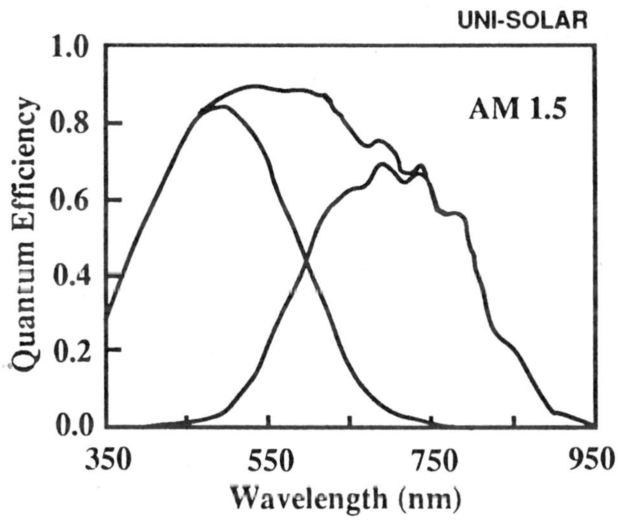

Fig. 8 Quantum efficiency as a function of wavelength in the degraded state, of a double junction a-Si:H/a-SiGe:H cell with a stabilised efficiency of 11.1%. See text for details (From reference 21)

Extensive work has been carried out on obtaining high-performance double junction cells where the component cells have high efficiency and stability. However, the quality of the narrow bandgap a-SiGe:H materials is still usually poorer than that of a-Si:H so it has been necessary to develop special cell designs to maximise performance and minimise degradation. This has included such things as bandgap profiling[20] and actually mismatching the initial performance of the two cells in order to maximise the stabalised overall performance. As a result tandem cells have been fabricated with absorber layers having different bandgaps, thicknesses and profiled bandgap structures.[21] The best reported result for the best stabalised performance of a double junction cell is an efficiency of 11.16%[21] obtained with an a-Si:H top cell having a bandgap of 1.75 eV and a-SiGe:H, whose bandgap in the narrowest region is 1.41eV. The initial efficiency of this cell was 12.61% and the parameter changes between this and the stabalised state are: J_{sc} from 10.67 to 10.61 mA/cm^2; V_{oc} from 1.65 to 1.61V; and FF from 0.716 to 0.655.

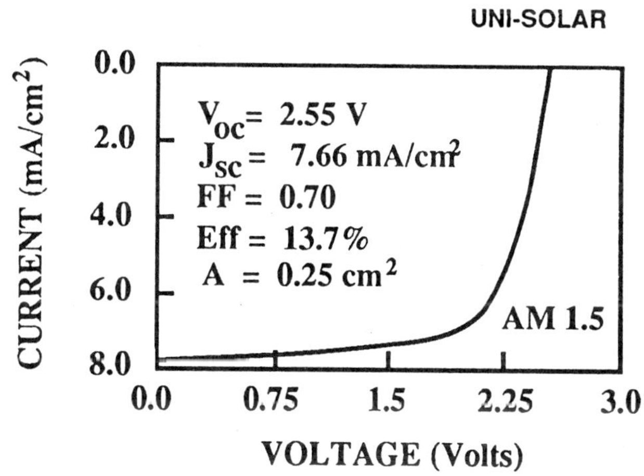

Fig. 9 Current-Voltage characteristic of a triple junction a-Si:H/a-SiGe:H/a-SiGe:H cell with an efficiency of 13.7% (From reference 20)

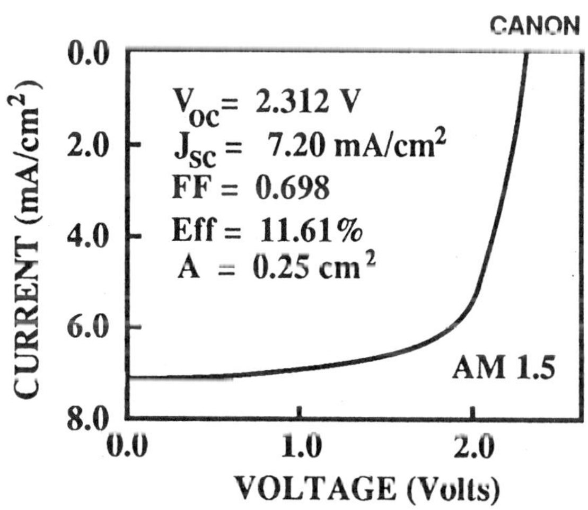

Fig. 10 Current-Voltage characteristic of a triple junction a-Si:H/a-SiGe:H/a-SiGe:H cell deposited at 100 Å/s (From reference 27)

It is interesting to note that the smallest change is in the short circuit current which is generated in the top and bottom cell by different parts of the solar spectrum.

The quantum efficiency versus wavelength is shown in Fig.8, where the left curve is for the a-Si:H and the right for a-SiGe:H cells, respectively. The result in Fig.8 also illustrates how the a-SiGe:H extends the sensitivity to much longer wavelengths which allows the high stabalised efficiencies to be attained. Such extensions in spectral response can also be obtained with triple junction cells, where the use of a middle cell with an intermediate bandgap raises the open circuit voltage of the multijunction cell. The current voltage characteristics of a 13.7% triple junction a-Si:H/a-SiGe:H/a-SiGe:H cell is shown in Fig.10.[20] Whereas the bottom a-SiGe:H cell has a bandgap similar to that of Fig.9, the middle a-SiGe:H cell is in between it and the top a-Si:H cell so that the open circuit voltage is 2.55 volts as compared to 1.65 volts in the tandem cell of Fig.9. The concomitant

reduction of Jsc from 10.67 to 7.66 mA/cm^2 is advantageous in reducing the degradation. One square foot a-Si:H/a-SiGe:H panels have been recently produced which have greater than 9% stabalised efficiencies.[17] One square foot a-Si:H/a-SiGe:H/a-SiGe:H panels have also been produced whose stabalised efficiencies are greater than 10%.[2] That the tandem a-Si:H alloy technology has come into its own is also demonstrated by the announcements by USSC and Solarex that they will both construct multi-megawatt a year manufacturing plants based on this technology.

MANUFACTURING

Single junction low efficiency a-Si:H cells have been mass-produced and commercialised primarily as power sources for calculators for over 10 years. During this time, the development of a-Si:H based solar cell panels has resulted in the manufacture of single and double junction a-Si:H panels. Currently single junction a-Si:H cell panels of 5 watts (1 ft2) and 50 watts (12.5 ft2) are produced by Solarex and APS, respectively. Double junction a-Si:H/a-Si:H tandem panels of 5 watts (1ft2) and 20 watts (4 ft2) are produced by UPG and USSC, respectively. Such all amorphous modules have been supplied to the demonstration at PV USA site by USSC, and APS; UPG and Solarex at PECO site. Amorphous panels produced in Japan and Europe have also been incorporated into similar demonstration sites there.

A big advantage of amorphous silicon technology is that it is possible to scale up the sizes of modules relatively quickly without significant losses in performance. For example, the stabilised (guaranteed) efficiencies of the single and double junction a-Si:H panels are essentially independent of the module size. Scale-up of a-Si:H/a-SiGe:H has also been successfully carried out , for example, from 11% 0.25 cm2 cells to 10.5% 1ft2 module to 9.5% 4 ft2 module in a relatively short time. As mentioned before, this technology has prompted two companies to construct very large manufacturing plants. The amorphous silicon technology having come of age still has to mature and there are a number of issues that need to be and are addressed.

FUTURE DIRECTIONS

The research and development programs are making and will continue to further advance the a-Si:H technology. Progress is being made in improving the photovoltaic properties of materials to be used in triple junction cells. Advances are also being made in making the cells more efficient, reliable and cost-effective. Improved understanding of the growth mechanisms and their relation to the microstructure of the materials is being obtained due to in situ monitoring such as spectroscopic ellipsometry [13,22] and IR spectroscopy.[23] This has led to the development of novel deposition methods based on the classical RF glow discharge deposition technique that have led to improved materials.[23,24] In parallel, other techniques are being applied to the deposition of a-Si:H and its alloys like hot wire decomposition of SiH_4[25], as well as very high (20-150MHz)[26] and microwave[27,28] frequency PACVD.

The latter two techniques have been able to deposit high quality materials at significantly higher deposition rates than the usual RF PACVD. In particular both Canon[27] and USSC[28] have been able to deposit highly efficient multijunction cells using microwave PACVD in which the a-SiGe:H is deposited at 100 Å/s. The current voltage characteristics of Canon's triple junction a-Si:H/a-SiGe:H/a-SiGe:H cell, where all the cells were deposited at 100 Å/s is shown in Fig. 10. Similar efficiency was achieved with a double junction a-Si:H/a-SiGe:H cell by USSC with the a-SiGe:H being deposited at 100 Å/s. Progress like this together with improvements in stability and tailoring of materials for optimum multijunction operation, have important technological and economical potential and will advance a-Si:H based solar cells to the forefront of terrestrial photovoltaics.

ACKNOWLEDGMENTS

The author would like to acknowledge the help given by Drs. Arya and Guha as well as other colleagues at Canon, Fuji, Sanyo, Solarex and USSC

REFERENCES

[1] D.E. Carlson and C.R. Wronski, *Appl. Phys. Lett.*, **28**, 1976, p. 671

[2] J. Yang, A. Banerjee, T. Glatfelter, K. Hoffman, X. Xu and S. Guha (This Conference)

[3] D.L. Staebler and C.R. Wronski, *Appl. Phys. Lett.*, **31**, 1977, p. 292

[4] R.C. Chittik, J.H. Alexander and H.F. Sterling, *J. Electrochem. Soc.*, **116**, 1, 1969, p. 77

[5] W. Spear and P. LeComber, *Solid State Comm.*, **17**, 1975, p. 1193

[6] C.R. Wronski, *Solar Energy Materials*, **1**, 1979, p. 287

[7] J.K. Arch and S. Fonash, *Appl. Phys. Lett.*, **60**, 1992, p. 757

[8] M. Gunes et. al. (This Conference)

[9] Y. Tawada, K. Tsuge, M. Kondo, H. Okamato and Y. Hamakawa, *Appl. Phys. Lett.*, **35**, 1979, p. 187

[10] S. Guha, J. Yang, P. Nath and M. Hack, *Appl. Phys. Lett.*, **49**, 1986, p. 218

[11] A. Matsuda et. al., *Japan J. Appl. Phys.*, **25**, 1986, L54

[12] J. Yang, X. Xu and S. Guha, *Proceedings of Materials Research Society 1994 Spring Mtg.*, San Francisco, April 1994 (In press)

[13] R.W. Collins, Ilsin An, Hien V. Nguyen, Youming Li and Yiwei Lu, in *Physics of Thin Films*, edited by M. Francombe and J. Vossen (Academic, NY, 1994), p. 49

[14] Y.M. Li, *Materials Research Symposium Proc.*, **297**, 1993, p. 803

[15] R. Arya et. al. (This Conference)

[16] A. Tarekawa et. al., *Proceedings of Materials Research Society, 1994 Spring Mtg.* San Francisco, April 1994 (In press)

[17] H. Deckman, C.R. Wronski and E. Yablanovitch, *Conf. Record of 17th IEEE Photovoltaic Specialists Conf.*, Orlando, FL., (IEEE, NY) 1984, p. 341

[18] S. Tsuda et. al., *J. Non-Cryst. Solids*, **164-166**, 1994, p. 679

[19] Y. Ichikawa et. al., *Conf. Record of the 23rd IEEE Photovoltaic Specialists Conf.*, Louisville, KY, (IEEE, NY), 1993, p. 27

[20] S. Guha et. al., *Appl. Phys. Lett.*, **54**, 1989, p. 2330

[21] J. Yang and S. Guha, *Appl. Phys. Lett.*, **61**, 1992, p. 2917

[22] K. Nakamura, K. Yoshino and I. Shimizu (This Conference)

[23] Y. Toyoshima et. al., *J. Non-Cryst. Solids*, **137-138**, 1991, p. 765

[24] G. Ganguly and A.A. Matsuda (This Conference)

[25] B. Nelson et. al., *Proceedings of 12th European PV Solar Energy Conf.*, Amsterdam, Netherlands (In press)

[26] A. A. Howling et. al., *J. Vac. Sci. Technol.*, **A**, 1991, p. 1080

[27] K. Sato, M. Sano, K. Ogawa and I. Kajita, *J. Non-Cryst. Solids*, **164-166**, 1993, p. 689

[28] S. Guha, X. Xu, J. Yang and A. Banarjee, *Appl. Phys. Lett.* (In press)

PROGRESS IN TRIPLE-JUNCTION AMORPHOUS SILICON-BASED ALLOY SOLAR CELLS AND MODULES USING HYDROGEN DILUTION

J. Yang, A. Banerjee, T. Glatfelter, K. Hoffman, X. Xu, and S. Guha
United Solar Systems Corp., 1100 West Maple Road, Troy, Michigan 48084 USA

ABSTRACT

We have achieved a stable module efficiency of 10.2% using amorphous silicon-based alloy in a triple-junction structure. This record performance has been accomplished by improving the component cells, the "tunnel" junction between the component cells, the back reflector, the uniformity over one-square-foot area, and the module design. Further advances have been made in the initial performance of the component cells by using high hydrogen dilution. The top cell, having Jsc > 8 mA/cm^2, exhibits Voc = 1.014 V and FF = 0.78. The middle cell, showing Voc = 0.757 V and FF = 0.719, exhibits Jsc = 9.5 mA/cm^2 under AM 1.5 illumination with a 530 nm cut-on filter. A record initial power output of 4.43 mW/cm^2 for the bottom cell has been obtained under AM 1.5 with a 630 nm cut-on filter. Light soaking experiments for the improved component cells also show higher stabilized values than previously reported ones.

INTRODUCTION

The low material cost and proven manufacturing technology of amorphous silicon (a-Si) alloy solar modules make them ideally suited for inexpensive generation of electricity [1]. A critical requirement for this technology to be used for large-scale terrestrial application is the achievement of 10% stable module efficiency [2]. It is now well recognized that multiple bandgap, multijunction a-Si alloy solar cells offer the attractive advantage of obtaining high stable efficiency [3]. In this paper we review the progress made in our laboratory in achieving a world record initial efficiency of 11.8% and a stable efficiency of 10.2% on a one-square-foot triple-junction module as confirmed by National Renewable Energy Laboratory (NREL). The stabilized efficiency was measured after 1000 hours of one-sun light soaking at 50 °C. Key factors leading to this achievement will be discussed. They include: (1) high quality component cells, (2) high conductivity, low optical loss "tunnel" junctions between component cells, (3) high quality back reflectors with enhanced light trapping, (4) good electrical and optical uniformity over a large area, and (5) optimum module design with low shadow and electrical loss.

As a continuous effort in further enhancing the module performance, we have been constantly reviewing and optimizing each critical area. In a recent study [4], we found that by using high hydrogen dilution during film growth, a-Si alloy cells showed improved performance both before and after light soaking. This approach has led to an improved top cell suitable for a triple structure. We have extended our hydrogen dilution study to a-SiGe alloy for the middle and bottom cells. Superior performance in both the initial and stabilized states was observed for these component cells as well. The results are reported in this paper. It is anticipated that when the improved component cells are incorporated in a triple-stack configuration, higher stabilized cell and module efficiencies can be achieved.

COMPONENT CELL OPTIMIZATION

In order to obtain high efficiency triple-junction modules, we have made component cells and studied their performance both in the initial and light-soaked states. A schematic diagram of a triple-junction structure is shown in Fig. 1. The top cell uses an a-Si alloy

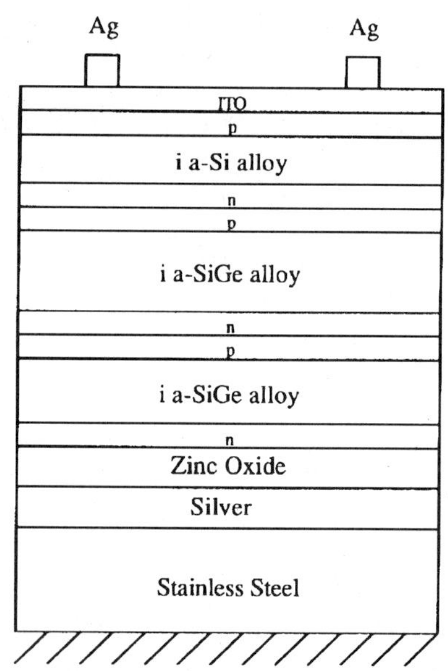

Fig. 1. Schematic diagram of a triple-junction cell structure.

First WCPEC; Dec. 5-9, 1994; Hawaii

intrinsic (*i*) layer, while the *i* layers of the middle and bottom cells use a-SiGe alloys. In order to obtain high open-circuit voltage (Voc) with good stability, we use hydrogen dilution of the gas mixture. Since the first report [5] of observation of improved stability with hydrogen dilution in a-Si alloy films, significant work has been carried out [6] to understand the role of excess hydrogen in the growth process. It is believed that hydrogen coverage of the growing surface gives the impinging species longer time to be incorporated and results in a denser structure. For a-SiGe alloy, we use hydrogen diluted gas mixture of Si_2H_6 and GeH_4 [7]. We have previously shown that since the dissociation rates of Si_2H_6 and GeH_4 are similar, this mixture produces higher quality material than that produced with SiH_4 and GeH_4. Although significant progress has been made to improve the quality of a-SiGe alloys, the best quality a-SiGe alloys still have poorer transport properties than a-Si alloys. To achieve better collection from a given material, profiling of the bandgap by varying the Ge-content in the cell has been used [8]. This approach not only provides for increased built-in field in the device but also produces a modified hole generation profile resulting in the holes being generated closer to the *p* contact for easier collection. This novel profiling approach has enabled us to fabricate a-SiGe alloy cells with improved performance.

IMPROVEMENT OF THE "TUNNEL" JUNCTION

The doped layers play an important role in terms of providing high built-in potential in the *i* layers and facilitating junctions between the adjacent cells without resistive loss. Since the doped layers are inactive as far as converting light to electricity is concerned, they must be optically transparent. Boron doping of a-Si alloys usually result in larger absorption in the material. The conductivity is also low, resulting in a large junction loss between the *p* and the *n* layers. We have developed [9] a microcrystalline *p* layer with low optical loss. The doped layers are also highly conducting so as to provide high built-in potential and low junction loss. Recently, a new technique [10] for evaluating the current-voltage characteristics of an *n i p n* structure, involving a single-junction *n i p* cell with an overlying doped *n* layer, was developed for the determination of losses at the "tunnel" junction. We found that the requirement on the properties of the *p* layer at the "tunnel" junction are more stringent than at the top of a cell. As a result of the study, "tunnel" junctions with improved properties were achieved.

HIGH QUALITY BACK REFLECTORS

A high quality back reflector should perform two important functions. It must be highly reflecting, and it also should scatter light at an angle higher than the critical angle for total internal reflection. Silver (Ag) is usually used to obtain high reflectivity. The interface between Ag and Si, however, is not highly reflecting; this is probably due to intermixing of the two elements. We use a transparent and conductive layer of zinc oxide (ZnO) as a buffer [11] to prevent intermixing. The required texturing is obtained by depositing Ag and ZnO at a high temperature in the range between 100 to 400 °C.

The use of Ag/ZnO back reflector has resulted in significant enhancement in short-circuit current density (Jsc) compared to a specular stainless steel substrate. For example, for a 4000 Å *i* layer of a-Si alloy, the gain is as much as 4 to 5 mA/cm^2. For a 3000 Å *i* layer of a-SiGe alloy with an optical band gap of $\sim$1.5 eV, the current corresponding to the long wavelength (> 630 nm) region is typically more than twice the value on bare stainless steel substrate. It is readily seen that high quality back reflector is a prerequisite for achieving high efficiency multijunction cell and module.

Theoretical calculations [12], however, show that there is scope for further enhancement if one can obtain Lambertian scattering without any loss at the reflecting surface. Experiments with different kinds of texture and different specular to diffusive reflection ratios show a remarkable insensitivity of Jsc to the degree of texture once certain level of texture is achieved [13]. It appears that there are parasitic losses at the Si/ZnO and ZnO/Ag interfaces that limit Jsc. Elimination or reduction of this loss will have a large impact on the cell efficiency, especially for the multijunction structure.

UNIFORM DEPOSITION OVER A ONE-SQUARE-FOOT AREA

In order to demonstrate high efficiency one-square-foot modules, we have taken special care to obtain good uniformity in terms of both the thickness and quality of the deposition. To evaluate the uniformity over the deposited area, we measured the performance of arrays of 132 small-area (0.82 cm^2) devices interspersed over the large area. The pertinent device parameters (Voc, Jsc, FF, and efficiency) are measured, and equal value contour characteristics of the above parameters are plotted. A systematic study of the correlation between the deposition parameters and corresponding contour plots has enabled us to achieve a good uniformity over the large area. The variation in multi-junction subcell efficiency over the one-square-foot area is within 2%.

MODULE DESIGN

A monolithic design on one single stainless steel substrate has been used for the fabrication of the modules. The unit cell is thus one large-area cell of one-square-foot area. The grid design and encapsulant are optimized to give an optical loss of ~ 2% and an electrical loss of ~ 2 to 3%. The total loss from average small-area efficiency to encapsulated module is thus ~ 4 to 5%.

MODULE PERFORMANCE

More than 200 multijunction modules have been made over the last three years incorporating the optimizations outlined above. This resulted in a significant improvement of module efficiency. The program on the fabrication of one-square-foot monolithic modules using Ag/ZnO back reflectors started in September 1991. In a period of about two and one-half years, the initial module efficiency has increased from 7.5% to 11.8%, a world record on large-area thin-film modules. The best results achieved to date are shown in Table I where the initial efficiencies for one-square-foot modules as measured at NREL under a Spire 240A simulator are shown. Also listed are the measurements at NREL on modules from the same batch which have been light soaked under one-sun condition for 1000 hours at 50 °C. The stable module efficiency of 10.2 % meets the important milestone of 10% as stipulated for thin-film photovoltaic modules to be acceptable for large-scale terrestrial applications. As we have demonstrated earlier [14], our multijunction modules show saturation in degradation after about 600 hours of one-sun light soaking at 50 °C. Because of thermal annealing effects, the degradation is less at higher light soaking temperature and more at lower temperature.

COMPONENT CELL OPTIMIZATION WITH HIGH HYDROGEN DILUTION

To further enhance the module performance, we have been constantly reviewing each critical area and conducting experiments for making improvements. As can be seen from the above analysis, one of the key areas is the quality of the component cells. We have recently reported [4] our success in achieving an improved a-Si alloy top cell using high hydrogen dilution during film growth. We have since extended the high hydrogen dilution work to a-SiGe alloys with the aim of improving the middle and the bottom cells. Hydrogen diluted gas mixtures of Si_2H_6 and GeH_4 were used for this study. The results are reported here.

Initial Status of Component Cells

We should first mention that for the design of the component cells, our goal is to obtain high efficiency triple-junction cells having a Jsc of about 8 mA/cm^2. This implies that each component cell should produce at least 8mA/cm^2 in current density. Since the top and middle cells in a triple-junction solar cell stack do not receive much reflected light, we deposited the component cells onto bare stainless steel substrates without the Ag/ZnO back reflectors for evaluation.

For the top cell, we used an a-Si alloy without incorporation of carbon. By using hydrogen dilution to improve the quality of the *i* layer and incorporating a microcrystalline *p* layer to enhance the built-in potential, we have obtained an initial Voc of 1.014 V, a FF of 0.78, and a Jsc of 8.65 mA/cm^2. The J-V characteristic is shown in Fig. 2. This is the best performance of a top

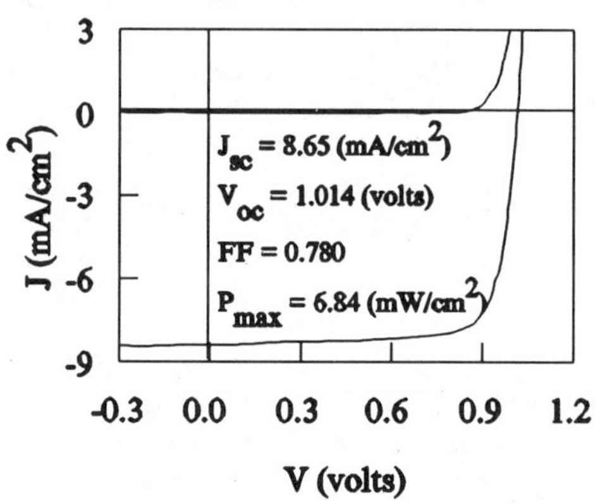

Fig. 2. Initial J-V characteristic of the top cell measured under AM1.5.

Table I Summary of module results as measured at NREL.

SAMPLE	STATE	Jsc(mA/cm^2)	Voc(V)	FF	η(%)
2452	Initial	7.25	2.400	0.675	11.75
2465	Initial	7.48	2.395	0.652	11.69
2437	Final	6.86	2.354	0.629	10.16
2445	Final	7.04	2.349	0.607	10.04
2447	Final	7.17	2.318	0.612	10.17

cell reported in the literature. We should mention that one could obtain a higher Voc if one uses a wider band gap *i* layer. In fact, we have made a wider band gap top cell with larger hydrogen incorporation in the *i* layer and achieved a Voc of 1.056 volts. However, the stability of the cell is poorer.

For the middle cell, we used a high hydrogen diluted gas mixture of Si_2H_6 and GeH_4 in the *i* layer. The J-V characteristic measured with AM1.5 through a 530 nm cut-on filter is shown in Fig. 3. One notes that Jsc is

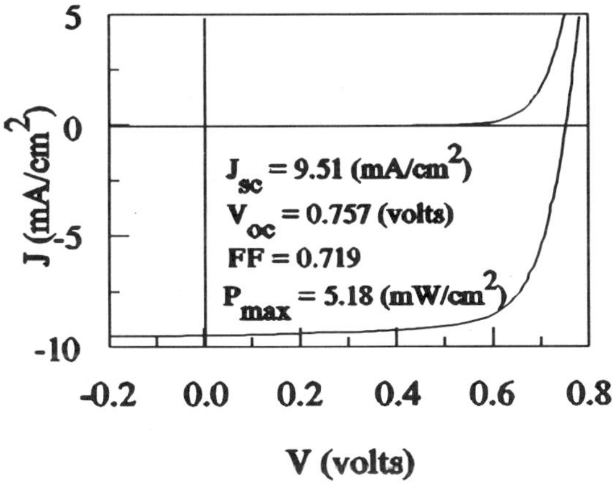

Fig. 3. Initial J-V characteristic of the middle cell measured under AM1.5 with a 530 nm cut-on filter.

greater than 9 mA/cm^2, and both Voc and FF show high values. The 530 nm filter was used to simulate the middle cell's insolation in a triple-junction structure. A 630 nm filter is used when evaluating a bottom component cell for like reasons. The bottom cell incorporated an a-SiGe alloy *i* layer with high hydrogen dilution and was deposited onto a Ag/ZnO back reflector. The J-V characteristic measured with AM1.5 through a 630 nm cut-on filer is shown in Fig. 4. The initial power output of 4.43 mW/cm^2 is the highest reported to date. These results also reflect the improvement that has been made on the quality of the a-SiGe alloy. Table II summarizes the initial photovoltaic characteristics of the component cells obtained with high hydrogen dilution. It is anticipated that when the improved component cells are incorporated in a triple-junction structure with proper

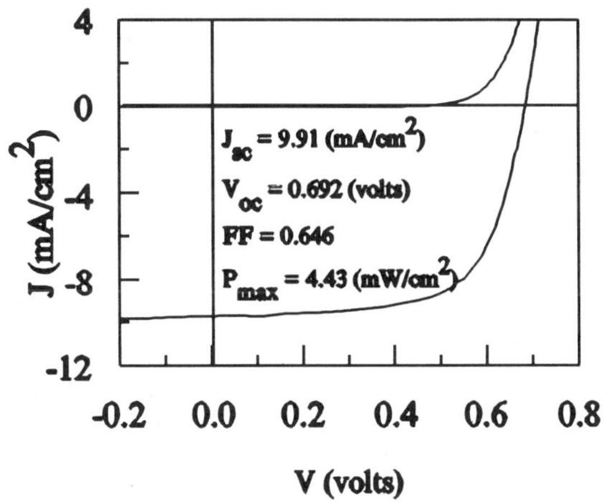

Fig. 4. Initial J-V characteristic of the bottom cell measured with a 630 nm cut-on filter.

current matching, an initial cell efficiency exceeding 14% can be obtained. This would certainly lead to higher module efficiencies.

Stability of Component Cells

To study the stability of the component cells, we used a metal-halide lamp with the intensity adjusted to one sun. Appropriate filters were used for the middle and bottom cells to simulate their operations in a triple-junction solar cell stack. We should also mention that the top and middle cells for the light soaking experiment were deposited on textured substrates overcoated with Cr, since they do not receive much reflected light in a triple structure. For J-V measurements, the top cell was measured with AM 1.5 light, while the middle and bottom cells were measured through 530 and 630 nm cut-on filters, respectively. All the component cells were light-soaked in the open-circuit mode and the temperature was maintained at 50 °C.

Table II Initial photovoltaic characteristics of the component cells with high hydrogen dilution.

COMPONENT CELL	Jsc(mA/cm^2)	Voc(V)	FF	Pmax(mW/cm^2)
Top*○	8.65	1.014	0.780	6.84
Middle*△	9.51	0.757	0.719	5.18
Bottom†□	9.91	0.692	0.646	4.43

* deposited on bare stainless steel
† deposited on textured Ag/ZnO
○ measured under AM1.5
△ measured under AM1.5 with a λ > 530nm filter
□ measured under AM1.5 with a λ > 630nm filter

Table III Previous and present stabilized component cell status at United Solar. Parameters for 16% goal are also listed.

COMPONENT CELL	STATUS	Jsc (mA/cm^2)	Voc(V)	FF	Pmax (mW/cm^2)
TOP *○	previous	7.2	0.98	0.71	5.0
	present	7.3	0.97	0.72	5.1
	goal	8.2	1.10	0.75	6.8
MIDDLE *△	previous	6.9	0.74	0.57	2.9
	present	7.6	0.76	0.60	3.5
	goal	8.4	0.89	0.70	5.2
BOTTOM †□	previous	7.7	0.65	0.56	2.8
	present	9.8	0.66	0.55	3.6
	goal	8.6	0.68	0.68	4.0

* deposited on a textured substrate coated with Cr
† deposited on textured Ag/ZnO
○ measured under AM1.5
△ measured under AM1.5 with a λ > 530 nm filter
□ measured under AM1.5 with a λ > 630 nm filter

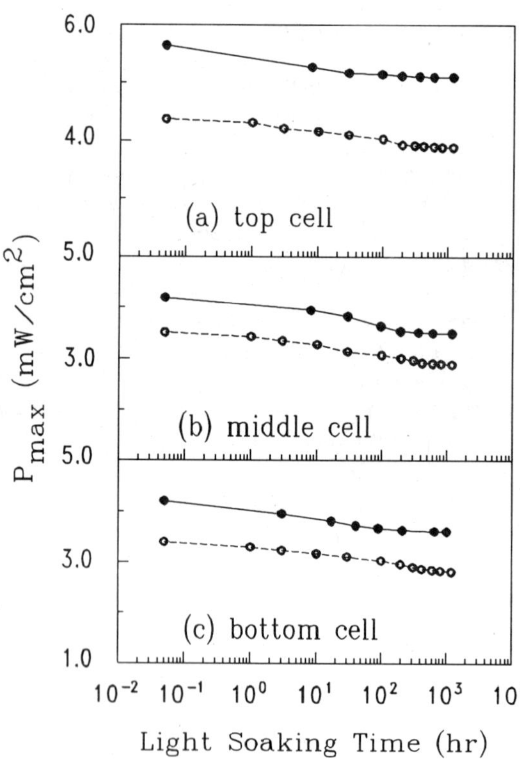

Fig. 5. Previously reported (O) and improved (●) stability data for the (a) top, (b) middle, and (c) bottom component cells.

Figure 5 plots the power output for the component cells versus light soaking time. We have also plotted the light soaking data for our previously reported component cells [15]. It is clear from the figure that the cells with high hydrogen dilution not only exhibit better initial performance but also retain higher stabilized power. All the component cells in this study show true saturation after prolonged light exposure. The normalized degradation for the improved top and middle cells are 8.9% and 16.7%, respectively, compared to 11% and 18% reported previously. Note that the improved bottom cell shows the tendency of saturation in a shorter time than the previous sample. There is very little degradation beyond 200 hours of light soaking for the improved bottom cell. The stabilized power of 3.6 mW/cm^2 measured through a 630 nm cut-on filter after over 1000 hours of light soaking is the highest reported to- date for a-SiGe alloy bottom cells.

DISCUSSION

Although we have achieved a 10.2% stabilized module efficiency, a major effort is still needed for thin-film photovoltaics to meet the 15% efficiency goal for large-scale utility-power applications. Guha [14] has estimated that a 16% stabilized small-area cell efficiency is necessary to obtain a 15% stable module efficiency. In Table III we list the previously established stabilized cell performances from our laboratory, the results from this study, and the values for reaching the 16% goal. One can see from Table III that the difference between the present status and the goal are mainly in the following two areas: (1) Voc in both the top and the middle cells, and (2) FF for the middle and the bottom cells. The two major challenges are therefore the enhancement in Voc of ~100 mV for the top and the middle cells and the further improvement in the quality of the a-SiGe alloy. These challenges can be met only by innovative material research and device design with focused effort. In addition, if one can further improve the quality of the back reflector by eliminating or reducing its parasitic loss, the requirement for the bottom cell can be relaxed. One may then either use a thinner *i* layer for higher FF or incorporate a smaller amount of Ge for higher Voc. The degree of back reflector texture also plays a role in the stability of the cell as discussed in detail in a separate paper [13].

CONCLUSION

The key factors leading to the achievement of 11.8% initial and 10.2% stabilized triple-junction one-square-foot modules using a-Si based alloys are reviewed. Significant improvement in both the initial and stabilized component cell performance has been obtained by using high hydrogen dilution in the gas mixture. When the improved component cells are incorporated in a triple-stack configuration, higher efficiency cells and modules can be expected. Major challenges for achieving 15% stable module efficiency are discussed.

ACKNOWLEDGEMENTS

The authors thank G. Hammond, M. Hopson, N. Jackett, and E. Chen for sample preparation and measurements, and V. Trudeau for help in manuscript preparation. The work was supported in part by NREL under Subcontract Nos. ZM-1-19033-2 and ZAN-4-13318 02.

REFERENCES

[1] S. Guha, "Amorphous Silicon Alloy Technology for Photovoltaics", *Proc. Sixth PVSEC*, 1992, p. 55.

[2] W. Luft, B. Stafford, and B. von Roedern, "The Amorphous Silicon Program Needs with Respect to Stabilized Module Performance", *Solar Cells*, **30**, 1991, p. 195.

[3] J. Yang and S. Guha, "Double-Junction Amorphous Silicon Based Solar Cells with 11% Stable Efficiency", *Appl. Phys. Lett.*, **61**, 1992, p. 2917.

[4] J. Yang, X. Xu, and S. Guha, "Stability Studies of Hydrogenated Amorphous Silicon Alloy Solar Cells Prepared with Hydrogen Dilution", *Proc. Mat. Res. Soc. Symp.*, 1994 (in press).

[5] S. Guha, K.L. Narasimhan, and S.M. Pietruszko, " On Light-Induced Effects in Amorphous Hydrogenated Silicon", *J. Appl. Phys.*, **52**, 1981, p. 859.

[6] See, for example, W. Luft and Y. S. Tsuo, *Hydrogenated Amorphous Silicon Alloy Deposition Processes*, Marcel Dekker, Inc., New York, 1993.

[7] S. Guha, J. S. Payson, S. C. Agarwal, and S. R. Ovshinsky, "Fluorinated Amorphous Silicon-Germanium Alloys Deposited from Disilane-Germane Mixture", *J. Non-Cryst. Solids*, **97-98**, 1987, p. 1455.

[8] S. Guha, J. Yang, A. Pawliekiewicz, T. Glatfelter, R. Ross, and S. R. Ovshinsky, "Band Gap Profiling for Improving the Efficiency of Amorphous Silicon Alloy Solar Cells", *Appl. Phys. Lett.*, **54**, 1989, p. 2330.

[9] S. Guha, J. Yang, P. Nath, and M. Hack, "Enhancement of Open Circuit Voltage in High Efficiency Amorphous Silicon Alloy Solar Cells", *Appl. Phys. Lett.*, **49**, 1986, p. 218.

[10] A. Banerjee, J. Yang, T. Glatfelter, K. Hoffman, and S. Guha, "Experimental Study of p Layers in "Tunnel" Junctions for High Efficiency Amorphous Silicon Alloy Multijunction Solar Cells and Modules", *Appl. Phys. Lett.*, **64**, 1994, p. 1517.

[11] R. Ross, R. Mohr, J. Fournier, and J. Yang, "Status of Fluorinated Amorphous Silicon-Germanium Alloys and Multijunction Devices", *Proc. Nineteenth IEEE PVSC*, 1987, p. 327.

[12] K. Hoffman and T. Glatfelter, "An Optical Study of Each Layer in a-Si:H Solar Cells", *Proc. Twenty-Third IEEE PVSC*, 1993, p. 986.

[13] A. Banerjee, K. Hoffman, X. Xu, J. Yang, and S. Guha, "Back Reflector Texture and Stability Issues in High Efficiency Multijunction Amorphous Silicon Alloy Solar Cells", *Proc. First World Conference on Photovoltaic Energy Conversion*, 1994.

[14] S. Guha, J. Yang, A. Banerjee, T. Glatfelter, K. Hoffman, and X. Xu, "Progress in Multijunction Amorphous Silicon Alloy-Based Solar Cells and Modules", *Solar Energy Materials and Solar Cells*, **34**, 1994, p. 329.

[15] X. Xu, J. Yang, and S. Guha, "Stability Studies of Component Cells for Multijunction Amorphous Silicon Solar Cells", *Proc. Twenty-Third IEEE PVSC*, 1993, p. 971.

APPROACHES FOR STABLE MULTI-JUNCTION a-Si SOLAR CELLS

Y. Hishikawa, K. Ninomiya, E. Maruyama, S. Kuroda, A. Terakawa, K. Sayama,
H. Tarui, M. Sasaki, S. Tsuda and S. Nakano
Sanyo Electric Co., Ltd, New Materials Research Center

ABSTRACT

Improvements in the stabilized efficiency of single-junction and multi-junction a-Si solar cells are studied. It has been found that the stabilized efficiency can be improved by suppressing the oxygen and nitrogen content in the i-layer. The defect density and conversion efficiency are strongly correlated with the content of hydrogen having an SiH_2 configuration rather than with the total hydrogen content in a-Si i-layers. A practical estimation method, which considers the variation of the I-V characteristics of each component cell, is also developed in order to optimize multi-junction cells. The world's highest stabilized efficiency of 8.8% for a single-junction a-Si solar cell and 10.6% for an a-Si/a-SiGe double-junction solar cell are achieved for 1cm^2 cells.

INTRODUCTION

Multi-junction amorphous silicon (a-Si) solar cells are superior to single-junction a-Si solar cells in terms of the stabilized conversion efficiency. However, it has been difficult to design the optimum structure such as the number of component cells and the optical gap of each layer, especially after light-induced degradation. On the other hand, efforts to improve the stability of single-junction solar cells is also important in view of their low cost and small material consumption, and because they are components of multi-junction solar cells.

In this study, improvement in the stabilized efficiency of multi-junction and single-junction a-Si solar cells is studied by extensively investigating the effects of impurities, hydrogen content, and the bonding configuration of hydrogen in the i-layer. A new procedure is introduced to simulate and optimize the I-V characteristics of multi-junction structures before and after light soaking, which is based on experimental illuminated and dark I-V characteristics of component single-junction cells.

EXPERIMENTAL

The a-Si and a-SiGe films and solar cells in this study were deposited with separated UHV capacity-coupled RF plasma-CVD reactors [1]. The deposition conditions for the i-layers were carefully chosen to avoid the undesirable incorporation of impurities, particle formation, and excess ion bombardment to the film-growing surface. Optical properties of the films were determined from transmittance (T) and reflectance (R) spectra, where optical interference effects were suppressed by using T/(1-R) in the analysis [4]. The optical gap (E_{opt}) was determined from $(\alpha h\nu)^{1/3}$ versus hν plots, in order to avoid experimental ambiguity in E_{opt} [2]. Other experimental details are described elsewhere [3,4].

a-Si SINGLE-JUNCTION CELLS

Recent efforts to improve the conversion efficiency of single-junction a-Si solar cells have resulted in an initial conversion efficiency of 12.7% for a 1cm^2 cell, and 12.0% for a 100cm^2 module [5-7]. Although the component materials such as TCO, p, i, n layers and the device structure appear to reach a certain level of optimization, many factors actually remain to be revealed. In particular, the fact that light-induced degradation tends to be more serious for cells with a relatively thick (~5000 Å) i-layer, which results in best initial efficiency, prevents a high stabilized efficiency for this structure. In most cases, a thinner (2000 ~ 3000 Å) i-layer is employed to obtain a better stabilized efficiency. Several special deposition techniques have been developed to give a-Si films better film properties [8-12]. Most of these techniques, however, have not been successfully applied

CH3365-4/94/0000-0386 $4.00 © 1994 IEEE

to solar cell devices, partly because they tend to increase damage to underlying layers, and partly because the relation among the deposition technique, initial material properties, and stabilized material properties is not known in a systematic way. It is therefore of crucial importance to understand the controllable range of the stability, or the defect density after light exposure, and to understand which material parameters are closely related to the stability of the material.

Hydrogen content and bonding configuration

There seems to be a general agreement that hydrogen atoms play an important role in the light-induced degradation of a-Si. However, the experimental information now available can not effectively answer the question: "*Which parameter is the effective measure to control and improve the stability of a-Si?* ". Figures 1(a) and (b) show the results of the present study on the defect density (N_d) of a-Si plotted versus the hydrogen content C_H and the hydrogen content with an SiH_2 bonding configuration C_{SiH2}. The figures clearly indicate that both the initial and stabilized N_d are more closely related to C_{SiH2} than to C_H, which agrees with previous reports on electric conductivities [13]. Although there are a few candidates for the specific microscopic structure represented by "SiH_2" and the actual microscopic structure is not well defined, the present results strongly suggest that some measures are still available to reduce the stabilized N_d of the a-Si i-layer while conserving the suitable optical properties, by modifying conventional deposition techniques. For example, as long as 100% SiH_4 is used as a material gas and deposition parameters are controlled in the conventional range to avoid particle formation in the plasma and excess ion bombardment, C_H (which has a strong correlation with the optical properties or E_{opt} of a-Si) and C_{SiH2} (which has a strong correlation with stabilized N_d) have nearly a one-to-one relationship [4]. Therefore, there seems to be little chance of reducing the stabilized N_d at a fixed C_H or E_{opt} if 100% SiH_4 and conventional deposition conditions are used. However, when SiH_4 is diluted with H_2 or He, or techniques such as H_2 plasma treatment is used, C_{SiH2} can independently be controlled at a fixed C_H in a certain range [4,14]. Therefore, low stabilized N_d can be achieved at a relatively high C_H.

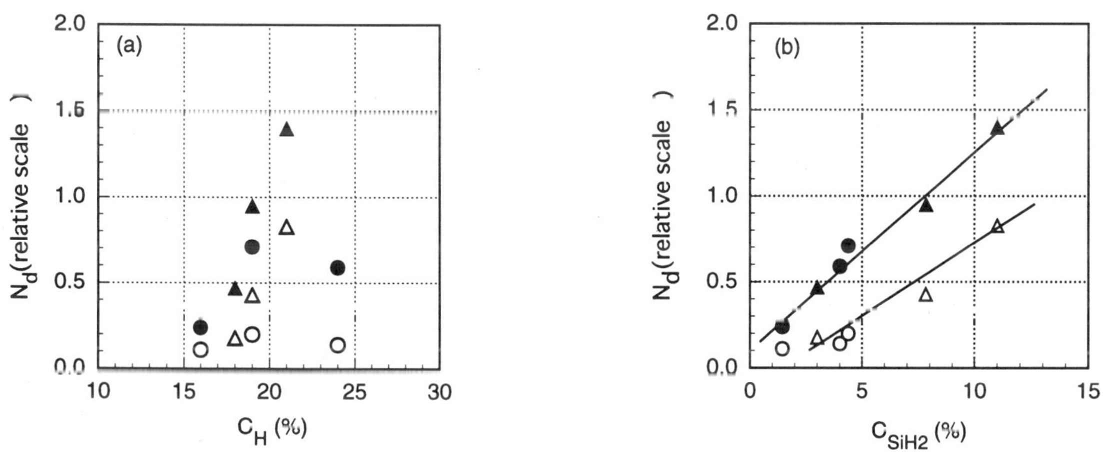

Fig.1 Defect density N_d of a-Si i-layer films as functions of (a) the hydrogen content C_H and (b) the hydrogen content with an SiH_2 bonding configuration C_{SiH2}. Here, C_{SiH2} is determined from the integrated infrared absorption peak at ~2100 cm^{-1}, and C_H is determined from the sum of the peaks at ~2000cm^{-1} and ~2100 cm^{-1}. N_d is calculated from the CPM (Constant Photocurrent Method). Open and solid symbols denote N_d at the initial state and after light soaking (5 suns, AM-1.5, 25°C, 160 min.). The triangles and circles indicate that the films are deposited from 100% SiH_4 and SiH_4 balanced with H_2 (2 < H_2/SiH_2 < 9), respectively.

Effects of impurities in the i-layer

It has been pointed out that even trace amounts ($\sim 1\times10^{19}$ cm^{-3}) of oxygen impurities in the i-layer affects the initial a-Si film properties and solar cell performance [5]. On the other hand, the effects of impurities on a stabilized solar cell performance are not well known. Figures 2(a) and (b) show the results of the present study on N_d of a-Si i-layer before and after light exposure, respectively, as functions of the oxygen and nitrogen concentrations. It is apparent that initial N_d increases as oxygen or nitrogen impurities are incorporated into the films. The stabilized N_d, on the contrary, is not appreciably affected by the impurities. It should be noted that the dark conductivity, photoconductivity, and the activation energy of the dark conductivity after light exposure are also nearly constant in the range of impurity concentrations in Fig. 2. This means that these film properties are dominated by the light-induced N_d, and that the effects of impurities are not significant.

Figures 3(a) and (b) show the initial and stabilized conversion efficiency of single-junction a-Si solar cells, respectively. The figures clearly show that the solar cell characteristics are significantly affected by the range of impurity concentrations where the film properties such as N_d are not affected (See Fig. 2). This is probably

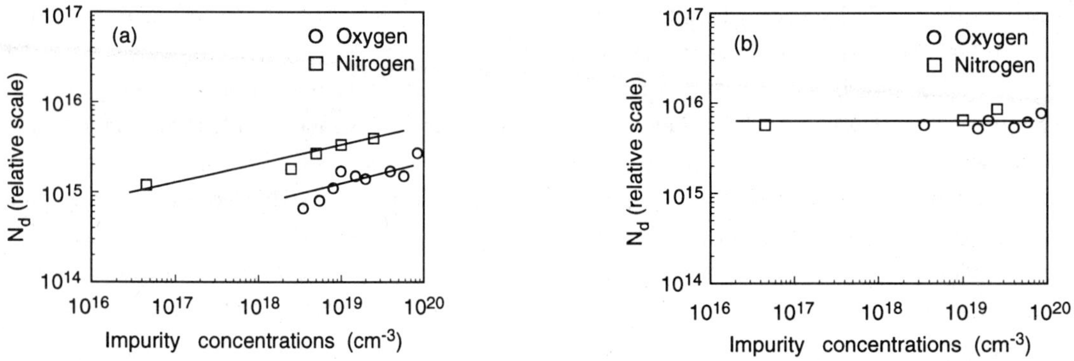

Fig.2 (a) Initial and (b) stabilized N_d of a-Si i-layer determined from constant photocurrent method (CPM). Oxygen and nitrogen atoms are incorporated to a-Si by mixing CO_2 and N_2 with SiH_4, respectively. The light soaking condition is 5 suns, AM-1.5, 25°C, 10 hours.

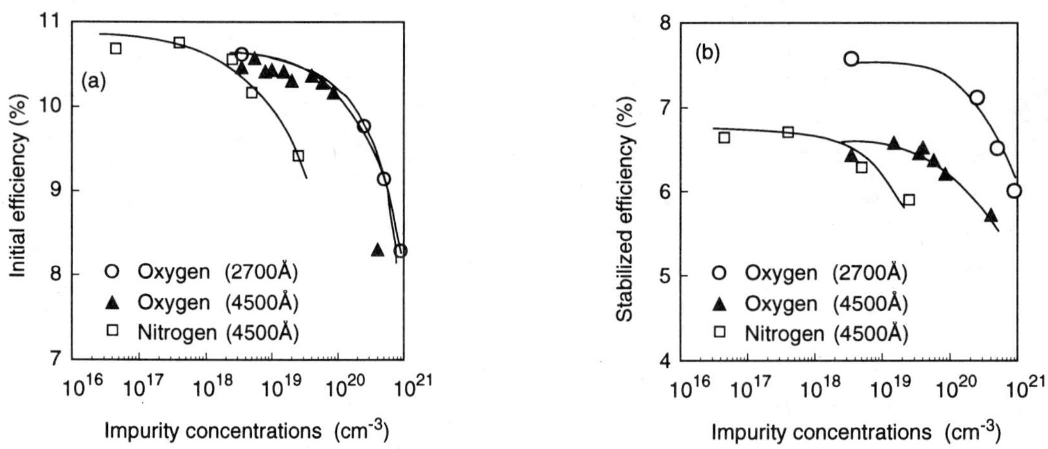

Fig.3 (a) Initial and (b) stabilized conversion efficiency of single-junction a-Si solar cells as functions of oxygen and nitrogen concentrations in the i-layer. Other parameters are identical for all the samples, and not optimized for high efficiency. The light soaking condition is 5 suns, AM-1.5, 25°C, 10 hours, and open circuit.

due to the fact that the Fermi level of the i-layer of an a-Si solar cell traverses most of the energy gap of the i-layer, and the built-in electric field in the i-layer is modified by gap states not influencing the films properties after light exposure.

The increase in N_d caused by light exposure modifies the distribution of the electric field in the i-layer, in addition to the reduction of the $\mu\tau$ product. The built-in electric field in the i-layer after light exposure is more concentrated near the p-i interface of the i-layer than at the initial state, which results in a lower electric field at the rest of the i-layer, and thus there is poor carrier collection for red light and a low fill factor. As nitrogen and oxygen impurities create positively charged defects, they further enhance the electric field near the p/i interface and reduce the electric field at the rest of the i-layer, which is not a favorable situation for solar cells. On the other hand, p-type impurities such as boron, which creates negatively charged defects, may be used to increase the electric field around the mid i-layer and reproduce the electric field similar to the initial state. In fact, improvements in the carrier collection near p/i interface is reported by graded boron doping to the bottom i-layer of stacked a-Si solar cells [15]. Figure 4 shows the results of the present study on the dependence of the initial and stabilized conversion efficiency of single-junction a-Si solar cells on the boron concentration of the i-layer. When boron is incorporated in the i-layer, the conversion efficiency firstly increases (5sun, AM-1.5, ~30min.), and then decreases. The stabilized efficiency becomes higher than the initial conversion efficiency when the boron concentration is $\geq 10^{17}cm^{-3}$. This result indicates that the distribution of the built-in electric field can be improved by light-soaking when boron is incorporated in the i-layer. Although the stabilized conversion efficiency of "boron-doped" cells does not exceed that of "undoped" cells, probably due to the presence of boron-related gap-states, this result suggests possibility to improve the stabilized efficiency by controlling the distribution of electric field.

Based on the present results, the stabilized efficiency of single junction a-Si solar cells is expected to be highest when C_{SiH2} is kept as low as possible while maintaining suitable E_{opt}, by controlling T_s and R_d, and without intentionally doping impurities to the i-layer. A stabilized efficiency of 8.8% has been achieved for a $1cm^2$ cell (Fig. 5), which was confirmed at JQA under a light-soaking condition nearly equivalent to 1-year outdoor exposure. This is the highest stabilized efficiency ever reported for a single-junction a-Si solar cell. Higher

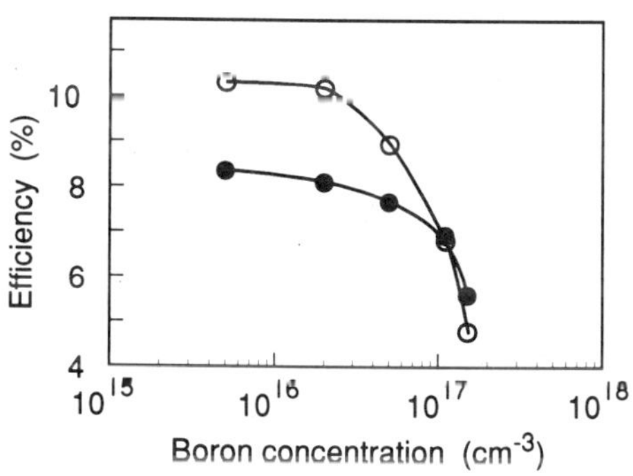

Fig.4 Initial (open circles) and stabilized (solid circles) conversion efficiency of single-junction a-Si solar cells as functions of the boron concentration in the i-layer. Boron is introduced by mixing B_2H_6 with SiH_4. Other parameters are identical for all the samples, and not optimized for high efficiency. The light soaking condition is 5 suns, AM-1.5, 25°C, 160 min., and open circuit.

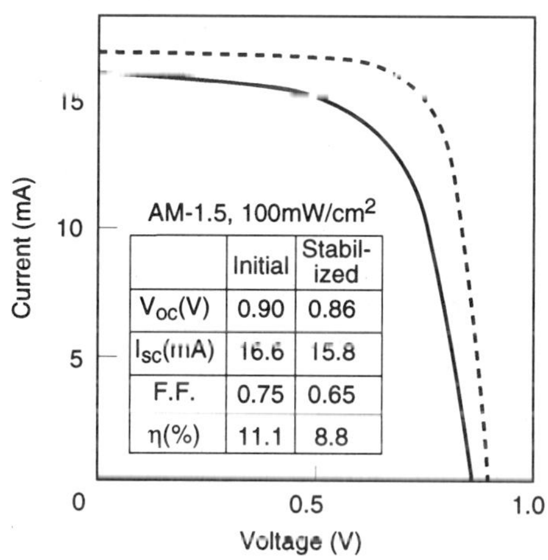

Fig.5 Initial (broken line) and stabilized (solid line) I-V characteristics of a single-junction a-Si solar cell, measured at JQA. The size of the cell is $1cm^2$. The light soaking condition is 1.25 suns, AM-1.5, 48°C, 310 hours, and open circuit.

stabilized efficiency is expected by optimizing other factors, such as TCO, doped layers and interface layers.

a-SiGe i-LAYER AND MULTI-JUNCTION CELLS

Multi-junction a-Si solar cells have advantages over single-junction cells concerning the stabilized efficiency. For example, each component cell can utilize a different spectral range of the sunlight by employing an a-SiGe bottom i-layer, the thickness of the i-layer of the top cell can be made thinner, and the top cell and the middle cell filter incident light to reduce the light-induced degradation of the thick bottom cell. However, these advantages also complicate the design of the material and the structure of the device. As a-SiGe is a ternary alloy of Si, Ge, and H, control of E_{opt} of a-SiGe is much more complicated than that of a-Si. Furthermore, the design of a multi-junction device is not straightforward because the I-V characteristics and the degradation ratio of the bottom cell strongly depend on E_{opt} and the thickness of the top and middle cells.

Controlling a-SiGe properties by C_{Ge} and C_H

Although variations in the a-SiGe properties were conventionally attributed to the germanium content (C_{Ge}), it has recently been pointed out that E_{opt} of a-SiGe is strongly dependent on both C_{Ge} and C_H, and a wide range of combinations of C_H and C_{Ge} can have the same E_{opt} [2]. This means that it is not enough to consider only E_{opt} when designing the a-SiGe material for solar cells. Figure 6 shows the C_H and C_{Ge} of various a-SiGe films having the same E_{opt} (=1.32 ± 0.005 eV) . N_d of the films and a-SiGe solar cell characteristics are shown in Fig. 7. It can be seen that the initial N_d has a minimum value at C_H ~ 9 atomic%, and the conversion efficiency is maximum at the same C_H. The difference in the conversion efficiency is mainly due to the difference in the fill factor. Although not shown, the slope of the Urbach tail as measured by CPM had minimum value at the same C_H. It can also be seen that there exists an optimum C_H after light soaking, and that this optimum value of C_H becomes smaller. In this study, a clear correlation of C_H, C_{Ge}, N_d, and solar cell characteristics

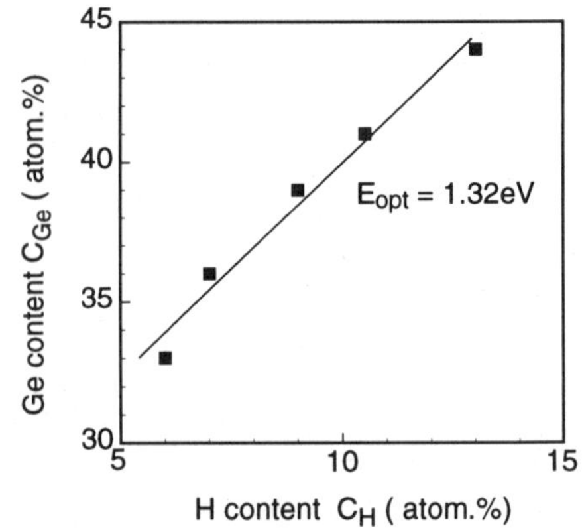

Fig.6 C_H and C_{Ge} of a-SiGe films having the same E_{opt} (=1.32 eV). C_H and C_{Ge} were determined by SIMS (Secondary Ion Mass Spectroscopy). Both the substrate temperature T_s and the flow rate of material gas such as SiH_4 and GeH_4 were varied to control C_H and C_{Ge} of the samples.

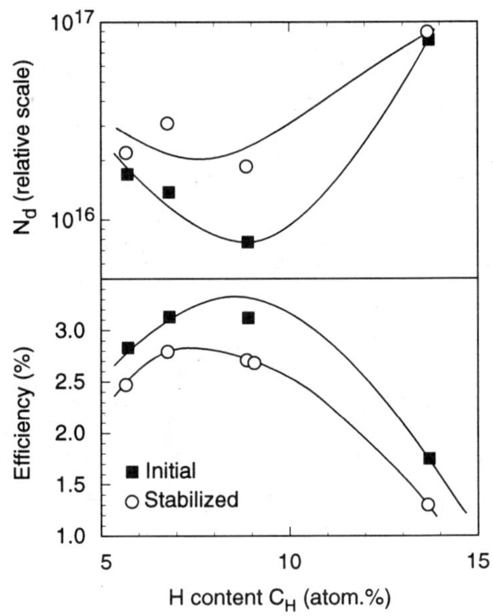

Fig.7 Defect density N_d of a-SiGe films and the conversion efficiency of a-SiGe single-junction solar cells under red light as functions of C_H of the i-layer. The thickness and E_{opt} of the i-layer are ~1500Å and 1.32 eV for all the samples. The light soaking condition is 5 suns, AM-1.5 filtered by R65, 48°C, 6 hours, and open circuit.

was observed in spite of a thin (~150 nm) i-layer thickness, because E_{opt} was kept constant. Results of the present study have also clarified that the light-induced instability of a-SiGe is not principally related to the energy gap or E_{opt}, but more probably to hydrogen-related properties. It should be noted that this result does not mean that the stabilized N_d and efficiency of a-SiGe films and solar cells well correlate with only C_H . Of course correlation with a microstructure like a-Si discussed above is expected, but the actual factor dominating the stabilized N_d is not known at the present stage.

Practical procedure to simulate and optimize the I-V characteristics of multi-junction solar cells

Because there are a lot more parameters to consider for multi-junction solar cells than for single-junction cells, simulation plays an important role in optimizing multi-junction devices. For this purpose, a numerical method based on a self-consistent solution of Poisson's equation and continuity equations is not practical, because a lot of physical parameters should be assumed.

In this study, a new practical method, which is based on the experimental I-V characteristics (I(V)) of each component cell, is proposed (Fig. 8). The I(V) of each component cell in a multi-junction cell before and after light exposure are precisely estimated while considering the contributions of variations in the intensity and spectrum of the incident light, which is dependent on the thickness and E_{opt} of the component cells. The illumination intensity dependence of I(V) is calculated from the dark I(V) and the I(V) at one illumination intensity by using the following relation,

$$I_{E2}(V) = I_d(V) + (E_2 / E_1)(I_{E1}(V) - I_d(V)). \qquad (1)$$

Here, $I_d(V)$, $I_{E1}(V)$, and $I_{E2}(V)$ denotes I(V) at dark, illumination intensities of E_1 and E_2, respectively. The spectral dependence of I(V) is estimated under proper color filters which reproduce multi-junction cells. Then, I(V) of the multi-junction cell are readily calculated as the sum of each component cell. When optimizing the i-layer thickness while E_{opt} is fixed or nearly fixed, experimental data from a set of component cells for the top, (middle), and bottom cells are enough for optimizing a multi-junction cell. Although component cells with different E_{opt} are also required if the thickness and E_{opt} have to be simultaneously optimized, the present procedure makes the optimization procedure more efficient than

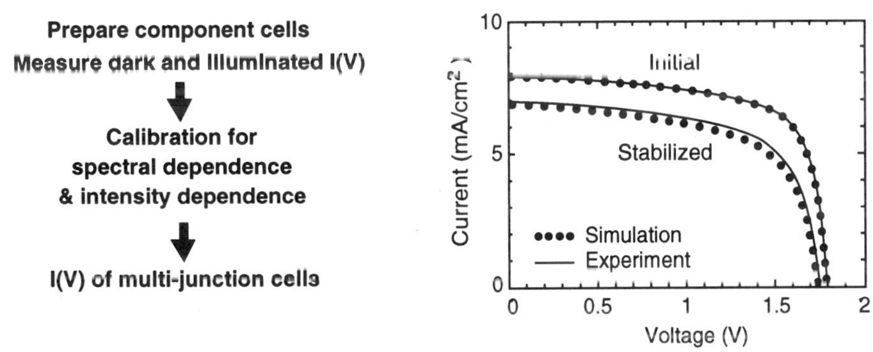

Fig. 8 Schematic procedure to simulate multi-junction a-Si solar cells. 1)Component single-junction cells are prepared. 2)Initial dark and illuminated I(V) of the component cells are measured by using appropriate color filters. 3)Light soaking is also performed by using the same color filters, and the stabilized I(V) are measured. 4)Dependence of the I-V characteristics of the component cells is calibrated for illumination intensity dependence and spectral dependence. 5)The I(V)'s of multi-junction cells having i-layers with variety of thicknesses and E_{opt} are calculated. Simulated and experimental I(V)'s of an a-Si / a-SiGe double-junction solar cell before and after light exposure are also shown. Simulation and experiments agree within an accuracy of about 5%.

conventional procedures. The estimated I(V) agrees well with the experimental result, as shown in Fig. 8. It is noted that if one tries to estimate the I(V) of each component cell from the I(V) around the short-circuit conditions of the component cells determined from the measured separated spectral response of a multi-junction cell, the error in the estimation is large. A separated spectral response of a multi-junction cell only offers information on the I(V) of each component cell near the short circuit condition [16]. There are no simple formulae to represent the entire I(V) from such limited data, especially in stabilized a-Si solar cells where the distribution of the electric field in the i-layer is not uniform.

Based on the above procedure, the stabilized efficiency of a-Si/a-SiGe double-junction solar cells could be optimized. Figure 9 shows a contour map of the simulated stabilized efficiency as functions of the thickness of the bottom i-layer and the I_{sc} of the top layer. Experimental I-V characteristics of component single-junction cells were used. E_{opt} of the component cells was fixed to 1.55eV and 1.32eV, respectively. A maximum stabilized efficiency close to 11% was expected from the simulation. Experimentally, a stabilized efficiency of 10.6% has been achieved by optimizing the double-junction cells around these values (Fig.10).

Simulation was also performed on triple-junction solar cells to explore the possibility of a higher stabilized efficiency. E_{opt} of the top and bottom i-layers was fixed to 1.75 eV and 1.28 eV, respectively. They are the widest and narrowest values where device-quality i-layers can be expected. The I-V characteristics of the corresponding component cells were estimated from conventional cells. Then, the thickness and E_{opt} of the middle cell were optimized based on experimental I-V characteristics. The result of the simulation is shown in Fig.11, which suggests that stabilized efficiency exceeding 12% is possible if the narrow-gap i-layer for the bottom cell and the wide-gap i-layer for the top cell are improved.

CONCLUSION

Effects of impurities, hydrogen content, and a hydrogen-related microstructure on the properties of a-Si and a-SiGe films and solar cells have been extensively investigated for improving the stabilized efficiency of single-junction and multi-junction a-Si solar cells. The defect density of a-Si before and after light exposure is more strongly correlated with hydrogen atoms with an SiH_2 configuration, implying the possibility of further improvement in the stabilized efficiency by modifying

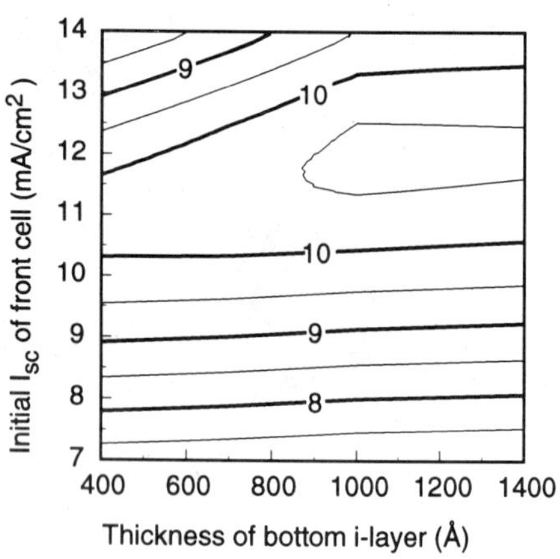

Fig.9 Contour map of stabilized efficiency of a-Si / a-SiGe double-junction solar cells as functions of the thickness of the bottom cell and the initial I_{sc} of the front cell. E_{opt} of the component cells is fixed to 1.55eV and 1.32 eV.

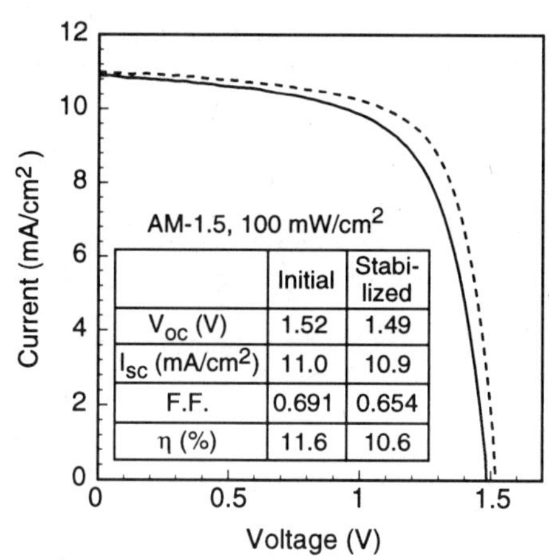

Fig.10 Initial (broken line) and stabilized (solid line) illuminated I-V characteristics of an a-Si / a-SiGe double-junction solar cell. The size of the cell is 1 cm^2. Light soaking condition is 5 suns, AM-1.5, 48°C, 6 hours, and open circuit.

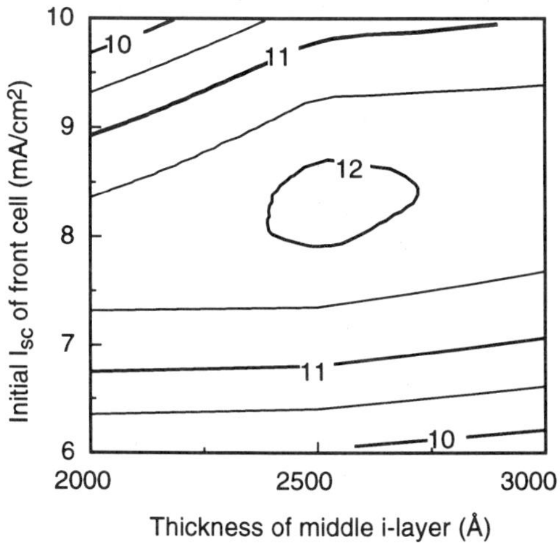

Fig.11 Contour map of the stabilized efficiency of a-Si / a-SiGe / a-SiGe triple-junction cells as functions of the thickness of the middle cell and the initial I_{sc} of the top cell. E_{opt} of each component cell is fixed to 1.75, 1.46, and 1.28 eV. The thickness of the bottom cell is fixed to 700Å.

conventional deposition techniques. A stabilized efficiency of 8.8% is achieved by employing an i-layer with a low impurity concentration, and reducing SiH_2 by controlling T_s and R_d while conserving the optimized E_{opt}.

When designing a-SiGe for multi-junction solar cells, considering E_{opt} is not simple, because a variety of combinations of C_H and C_{Ge} have the same E_{opt}. In fact, the stability of a-SiGe films and solar cells is improved in this study by optimizing C_H at a fixed E_{opt}. A practical procedure is proposed to simulate and optimize the structure and i-layer properties for solar cells, and a stabilized efficiency of 10.6% has been achieved. Further improvements are expected by improving the i-layer properties of the top and bottom cells.

ACKNOWLEDGEMENT

This work was supported by NEDO as a part of the New Sunshine Program under the Ministry of International Trade and Industry.

REFERENCES

[1] S. Tsuda, T. Takahama, M. Isomura, H. Tarui, Y. Nakashima, Y. Hishikawa, N. Nakamura, T. Matsuoka, H. Nishiwaki, S. Nakano, N. Ohnishi and Y. Kuwano, *Jpn. J. Appl. Phys.* **26**, 1987, pp. 33

[2] A. Terakawa, M. Shima, K. Sayama, H. Tarui and S. Tsuda, to be published in *Proceedings of the Materials Research Society 1994 Spring meeting*, San Francisco

[3] Y. Hishikawa, N. Nakamura, S. Tsuda, S. Nakano, Y. Kishi and Y. Kuwano., *Jpn. J. Appl. Phys.* **30**, 1991, pp. 1008

[4] Y. Hishikawa, S. Tsuda, K. Wakisaka and Y. Kuwano, *J. Appl. Phys.* **73**, 1993, pp. 4227

[5] Y. Hishikawa, M. Isomura, S. Okamoto, H. Hashimoto and S. Tsuda, *Sol. Energy Materials and Sol. Cells*, **34**, 1994, 303

[6] S. Okamoto, T. Takahama, Y. Hishikawa, S. Tsuge, M. Nishikuni, N. Nakamura, S. Tsuda, S. Nakano and Y. Kuwano, *Proceedings of the 11th E.C. Photovoltaic Solar Energy Conference*, 1992, Montreux, pp. 537

[7] M. Green and K. Emery, *Progress in Photovoltaics*, **2**, 1994, pp. 231

[8] G. Ganguly and A. Matsuda, *Mater. Res. Soc. Symp. Proc.* **258**, 1992, pp. 39

[9] D. Das , H. Shirai, J. Hanna and I. Shimizu, *Jpn. J. Appl. Phys.* **30**, 1991, pp. L239

[10] H. Matsumura, *J. Appl. Phys.* **65**, 1989, pp. 4396

[11] A. Mahan, J. Carapella, B.P. Nelson, H.S. Crandall and I. Belberg, *J. Appl. Phys.* **69**, 1991

[12] Y. Mishima et al., *Jpn. J. Appl. Phys.* **22**, 1983, pp. L41

[13] N. Nakamura et al., *Jpn. J. Appl. Phys.* **28**, 1989, pp. 1762

[14] S. Okamoto, Y. Hishikawa, S. Tsuge, M. Sasaki, K. Ninomiya, M. Nishikuni and S. Tsuda, *Jpn. J. Appl. Phys.* **33**, 1994, pp. 1773

[15] D. Fischer, N. Pellaton, A. Shah and C.M.Fortmann, *Proceedings of the 11th E.C. Photovoltaic Solar Energy Conference*, 1992, Montreux, pp. 560

[16] S. Moriuchi, Y. Inoue, H. Sannomiya, A. Yokota, M. Itoh, Y. Nakata and H. Itoh, *Proc. 22th IEEE Photovoltaic Specialists Conference*, 1990, pp. 1449

PROGRESS IN AMORPHOUS SILICON BASED MULTIJUNCTION MODULES

R. R. Arya, R. S. Oswald, Y. M. Li , N. Maley, K. Jansen, L. Yang, L. F. Chen,
F. Willing, M. S. Bennett , J. Morris and D. E. Carlson
Solarex Corporation, Thin-Film Division
826 Newtown - Yardley Road, Newtown, PA 18940, USA

ABSTRACT

Advances in amorphous silicon based alloys and device structure have led to the demonstration of 10.5% initial efficiency and 9.1% stabilized efficiency on 1 Ft^2 a-Si/a-SiGe tandem junction module. Scale-up to 4 Ft^2 in pilot production has resulted in the best initial efficiency of 9.75% as measured by NREL, which should result in a stabilized efficiency over 8%. Presently the 4Ft^2 modules have an average initial efficiency of 8.75% and a stabilized average efficiency of about 7.5%, as measured by Solarex.

INTRODUCTION

Large-scale application of amorphous silicon based thin-film modules requires stabilized conversion efficiencies in the range of 8-10% with variable manufacturing costs of <$1.00 / watt. Although triple junction device structures of the type a-Si/a-Si/a-SiGe or a-Si/a-SiGe/a-SiGe offer the potential for high stabilized conversion efficiency, tandem junction devices of the type a-Si/a-SiGe offer a more attractive cost advantage over triple junction device structures. This is largely due to its simplicity (six a-Si based layers versus nine layers) and shorter deposition time (almost half that of triple junctions). Large area (~ 8 Ft^2) tandem junction modules at 8% stabilized efficiency are deemed sufficient to compete in most of the present remote markets.

At Solarex we have been addressing various aspects of a-Si/a-SiGe multijunction technology. Two concurrent thrusts are in progress : (I) Advances in materials and device research are improving the device performance and (II) Research results from the laboratory are being scaled-up to cost-effective 4 Ft^2 modules in pilot production. In this paper we shall discuss the progress made in improving (a) the non-semiconductor layers, (b) the semiconductor layers and multijunction devices and (c) module development and multijunction scale-up.

NON-SEMICONDUCTOR MATERIALS

In the area of non-semiconductor materials, we have been developing textured ZnO by low-pressure chemical vapor deposition (LPCVD) to replace the textured tin oxide front contact because the former has superior optical and electrical properties. We have been addressing the ZnO/p interface problem and have succeeded in fabricating tandem junction devices with comparable V_{oc}'s and FF's and higher total J_{sc}'s as shown in Table I. Figure 1 shows a comparison of quantum efficiency versus wavelength measurements of the tandem devices on ZnO and on SnO_2 front contacts. Optimization of the front and the back cells to balance the current between the two junctions is underway .

The 'standard' rear contact for our multijunction devices consisted of sputtered ZnO and silver. While this rear contact was optimum for initial optical enhancement it was plagued with two problems : (a) the reflectance from this contact system degraded with time resulting in a 3-4% loss in J_{sc} after long-term light-soaking (> 1000 hours) [1] and (b) One Ft^2 modules were susceptible to initial shunts and light-induced shunts, resulting in an extra 3-5% module degradation compared to small area devices [2]. We have developed a new rear contact of LPCVD deposited ZnO and sputtered Al which is remarkably less susceptible to shunts. Figure 2 shows a comparison of shunted segments on a-Si/a-SiGe modules with ZnO/Ag and with ZnO/Al rear contacts.

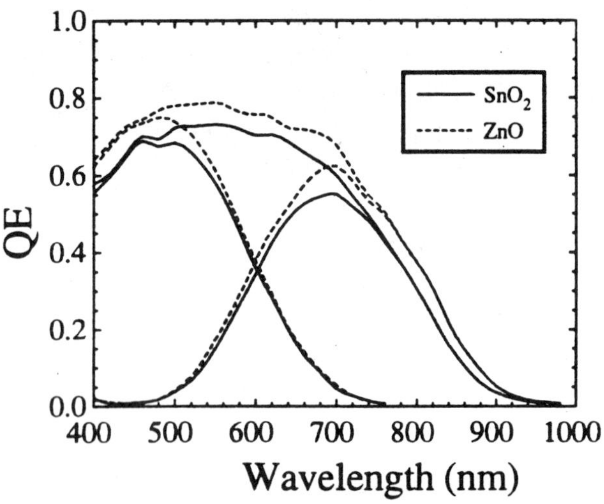

Figure 1: Quantum efficiency versus wavelength measurements on tandem devices with ZnO and with SnO_2 contacts.

First WCPEC; Dec. 5–9, 1994; Hawaii

Front Contact	V_{oc} (V)	Fill-Factor	J_{front} (mA/cm^2)	J_{back} (mA/cm^2)	J_{Total} (mA/cm^2)
SnO_2	1.50	0.714	9.20	10.99	20.18
ZnO	1.48	0.701	9.91	11.30	21.20

Table I :Comparison of photovoltaic parameters of a-Si/a-SiGe tandem junction devices with ZnO and with SnO_2 front contacts.

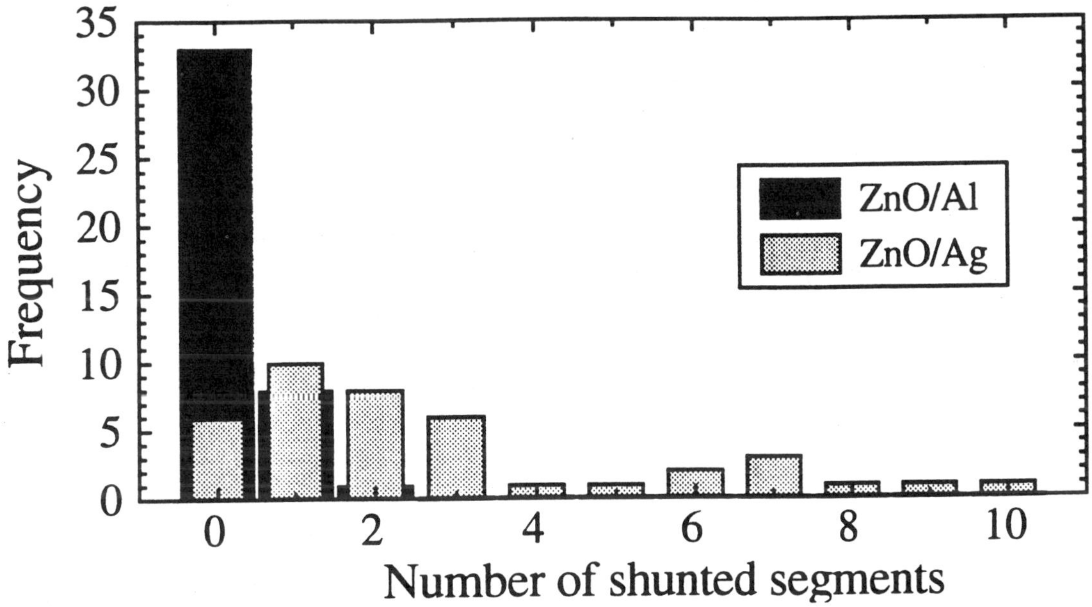

Figure 2: Shunt Distribution for a-Si/a-SiGe tandem modules with ZnO/Ag and ZnO/Al rear contacts.

SEMICONDUCTOR MATERIALS AND MULTIJUNCTION DEVICES

In semiconductor materials, recent advances have been made in the deposition of a-Si:H intrinsic layers from H_2 - diluted silane feedstock which has resulted in better initial performance and definite stabilization of single junction a-Si:H solar cells after a few hundred hours of light-soaking [3]. Deposition of high quality, undoped wide-bandgap a-SiC:H thin layers have been incorporated at the p/i interface which result in an increase in V_{oc} without a loss in stability [4]. This has led to the demonstration of a-Si:H i-layer single junction devices with V_{oc}'s higher than 1.0 V and fill-factors greater than 0.70. Improvements in the surface preparation technique used for the deposition of thin microcrystalline n-layers [5] has been extended to deposition on a-SiC:H n-layers resulting in high quality tunnel or recombination junctions. We have also made improvements in a-SiGe:H alloys, particularly in the bandgap tailoring of the i-layer which allows us to reduce the thickness of the narrowest bandgap portion in the composite a-SiGe:H alloy i-layer. This has resulted in single junction a-SiGe:H alloy i-layer cells with AM1.5 currents of over 20.5 mA/cm^2 with fill-factors greater than 0.64. When this device is employed as the bottom junction in an a-Si/a-SiGe tandem configuration it results in J_{sc2} ~ 10.25 mA/cm^2 [6].

Employing these advances we have demonstrated a-Si/a-SiGe tandem junction solar cells with initial efficiencies as high as 11.7%. Figure 3 shows the J-V characteristics of an a-Si/a-SiGe device before and after 1000 hours of light-soaking. The initial efficiency was 11.4% and the stabilized efficiency was 9.6% after 1000 hours of light-soaking. Figure 4 shows the corresponding quantum efficiency versus wavelength measurement of this device.

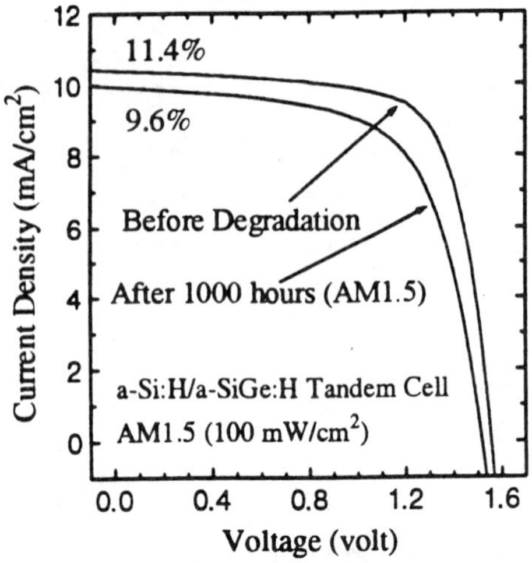

Figure 3:J-V Characteristics of an a-Si/SiGe device before and after 1000 hours of light-soaking.

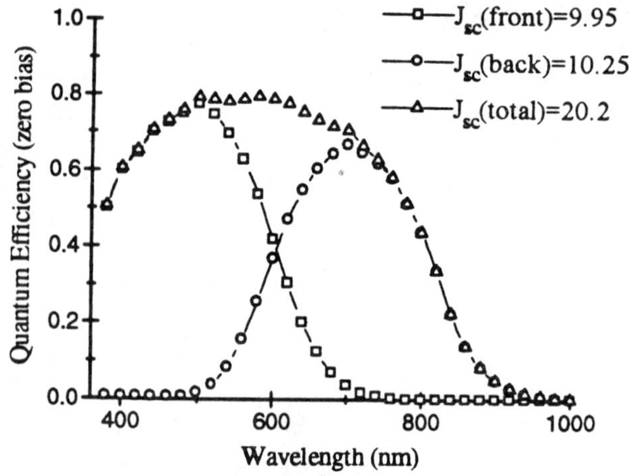

Figure 4: Quantum Efficiency versus wavelength of the device shown in Figure 3.

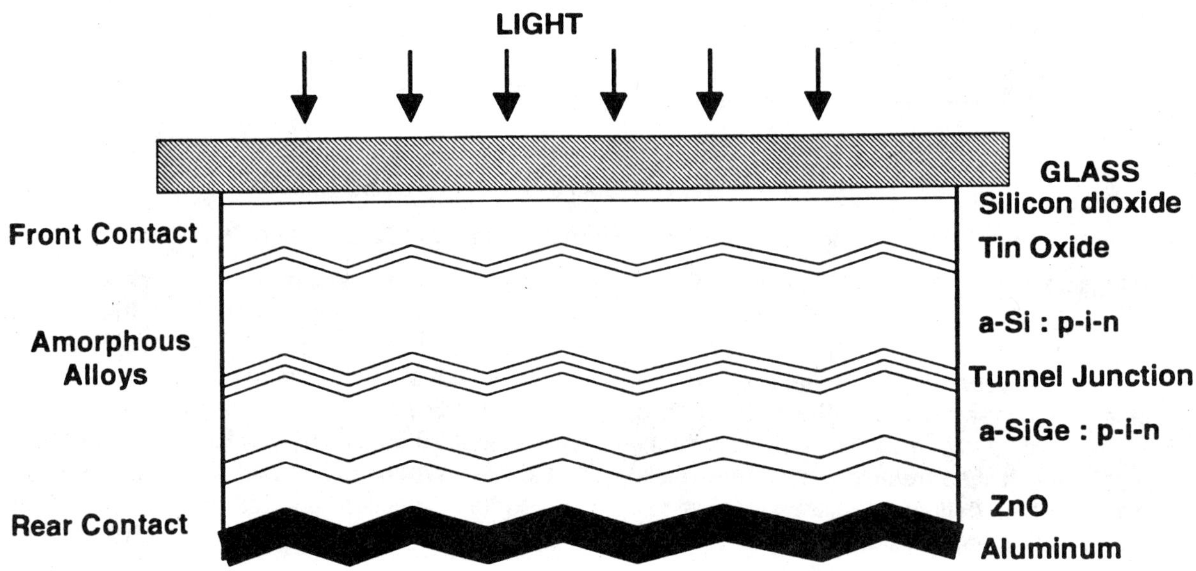

Figure 5: Device structure of a-Si/a-SiGe multijunction module.

MODULE DEVELOPMENT AND MULTIJUNCTION SCALE-UP

The device structure for a-Si/a-SiGe multijunction module is shown in Figure 5. This device structure resulted in a solar cell with initial conversion efficiency of 11.2 %, which became the starting point for module development. Module development involves scale-up of various thin-film deposition processes , such as, the textured tin oxide front contact, the amorphous silicon based alloys and the ZnO/ Al rear contact first to 1 Ft² and then to 4 Ft² along with optimization of front contact sheet resistance, transmission, haze, segment-width and development of closely-spaced laser scribing to minimize interconnect losses.

The best a-Si/a-SiGe module (~1 Ft^2) fabricated to date had an initial aperture area conversion efficiency of 10.5% and a stabilized aperture area conversion efficiency of 9.1% after 1200 hours of light-soaking. The J-V characteristics of of this module before and after light-soaking are shown in Figures 6(a) and 6(b) respectively. Figure 6 (c) shows the J-V characteristics of the stabilized module measured at NREL which is in excellent agreement with our measurement.

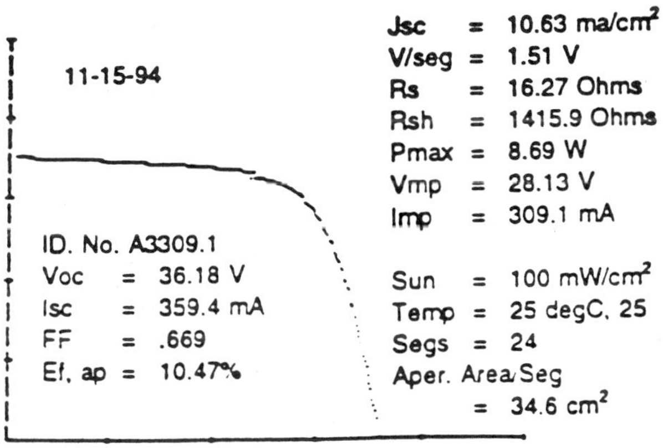

Figure 6 (a): Initial J-V characteristics of best a-Si/a-SiGe module.

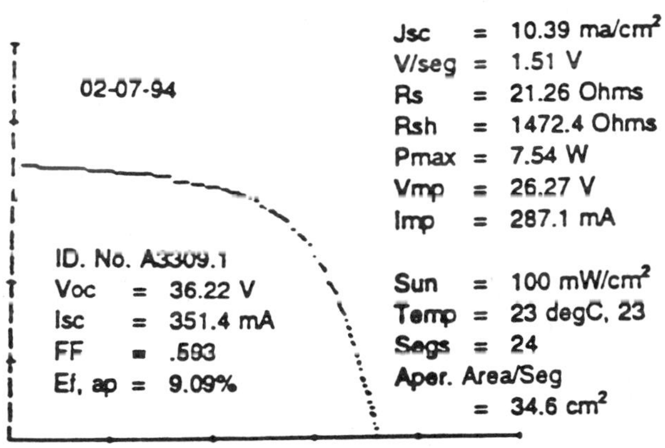

Figure 6(b): J-V characteristics of 1 Ft^2 shown in Figure 6 (a) after 1200 hours of light-soaking.

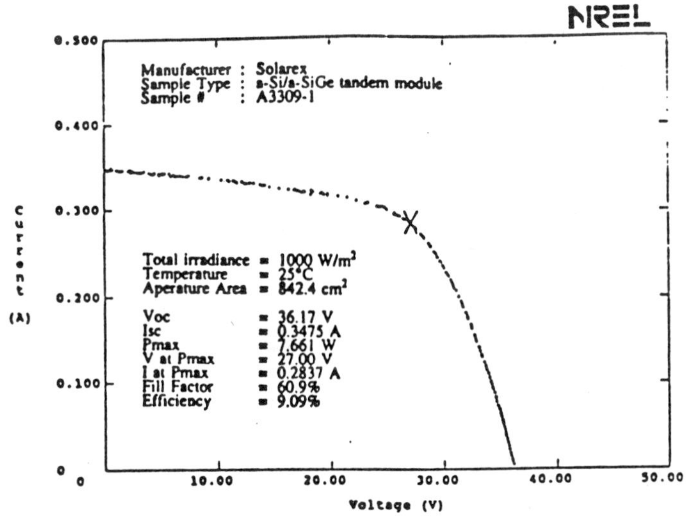

Figure 6(c) : J-V characteristics of the module in shown in Figure 6 (b) measured at NREL .

We have been concurrently addressing the scale-up and the manufacturability issues issues regarding a-Si/a-SiGe tandem junction modules on the 4 Ft^2 size.. This requires establishing process controls at each process step and fabricating a large number of modules in pilot production to determine the variation and yield in the manufacturing process. Moreover, the process steps have to meet the materials cost and module thruput criterion for large volume manufacturing. The major features that have to be discounted in manufacturing due to yield/thruput considerations are shown in Figure 7 which shows an expectation of over 10% initial efficiency with the present device structure.

11.2% Small area solar cell

-6% due to interconnect and Scale-up

10.5% One square foot Module

-1.5% due to AR coating

10.34% One square foot module

Scale-up to four square foot module

9.75% Present best four square foot module

Figure 7: a-Si/a-SiGe tandem junction initial efficiency changes from research to pilot production.

We have fabricated about 200 multijunction modules of 4 Ft^2 size. Figure 8 shows a run chart of over 180 modules which depicts the statistical distribution of the initial conversion efficiency. The average initial efficiency of all module is about 8.5% as measured at Solarex. Figure 9 shows a histogram plot of the same data with 85% yield. The average initial efficiency of 141 modules is about 8.75% with a standard deviation of 0.30.

Table II shows a comparison of I-V measurements made indoors at Solarex and at NREL on 4 Ft^2 a-Si/a-SiGe modules. While the measurements agree within the errors, NREL has consistently measured conversion efficiencies about 5% higher than those measured at Solarex. The differences are mainly in I_{sc} (~ 2%) and in fill-factor (~3%). The best 4 Ft^2 module measured by NREL had an initial aperture area efficiency of 9.75%. The I-V charecteristics of this module is shown in Figure 10.

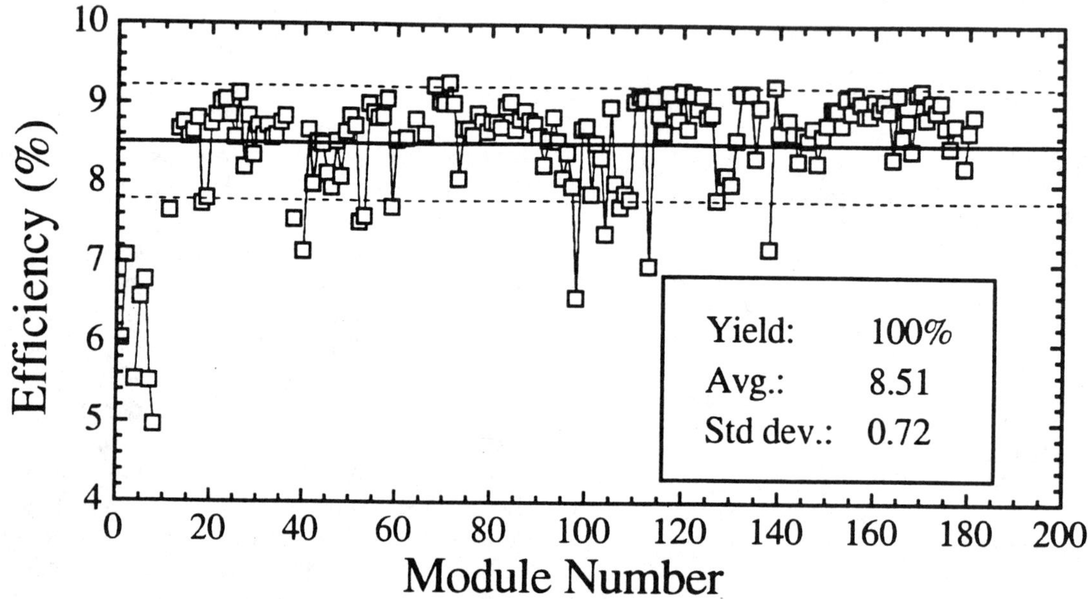

Figure 8: Run chart of initial conversion efficiency of 4 Ft^2 multijunction modules.

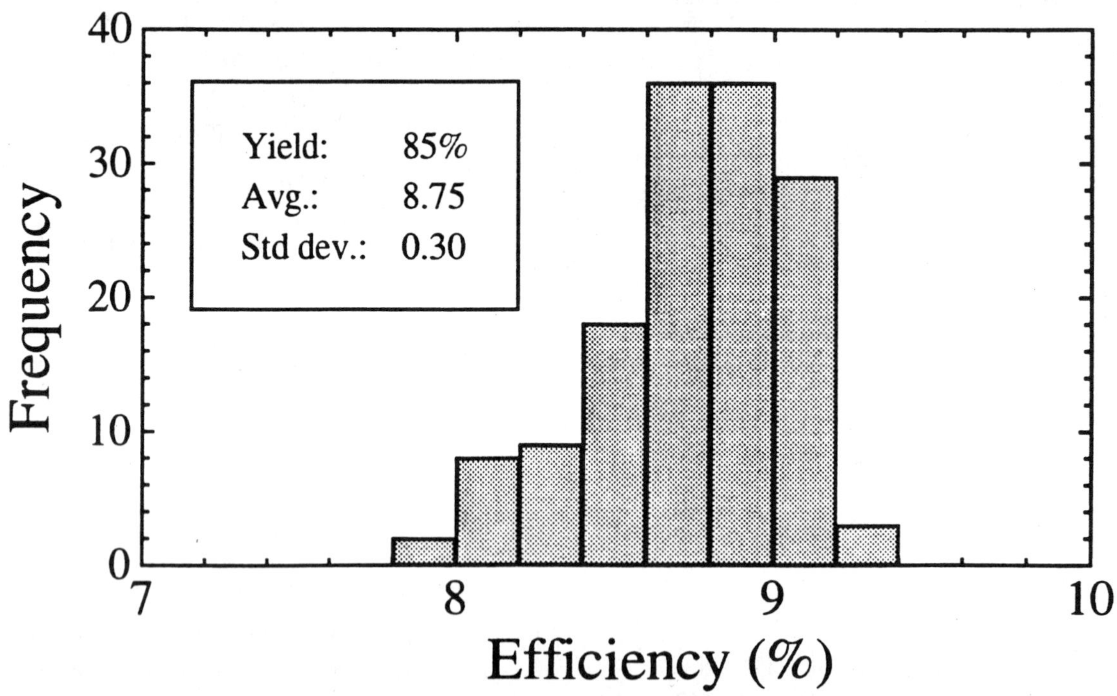

Figure 9: Histogram plot of 4 Ft^2 multijunction modules with 85% yield.

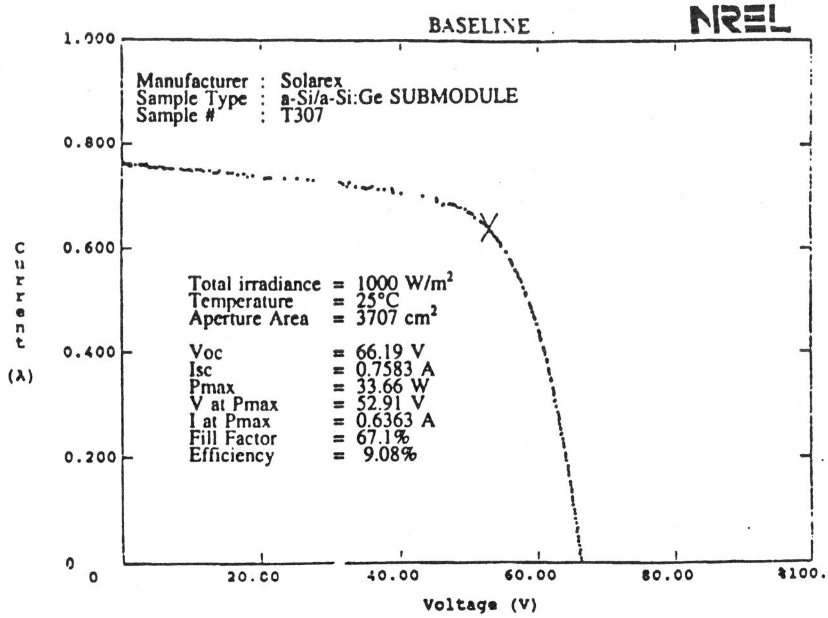

Figure 10 : J-V characteristics of best a-Si/a-SiGe multijunction module measured at NREL.

Module #	V_{oc} (V)	I_{sc} (mA)	F. F	Total Area Efficiency (%)	Aperture Area Efficiency (%)	Total Area (cm^2)	Aperture Area (cm^2)	Power (Watts)	Measured At
T 70	66.01	728.8	64.0		8.94		3442.46	30.64	Solarex
	65.94	743.5	66.2	8.76	9.43 *	3706.8		32.46	NREL
T 92	65.49	736.0	64.1		9.0		3432.0	30.90	Solarex
	65.98	753.0	65.9	8.83	9.50 *	3706.8		32.73	NREL
T322	64.85	734.5	62.8		8.66		3452.24	29.90	Solarex
	64.78	761.9	64.1	8.53	9.16 *	3707		31.63	NREL
T335	66.83	744.7	64.1		9.24		3452.24	31.90	Solarex
	66.54	751.3	66.2	8.92	9.57 *	3707		33.08	NREL
T336	66.98	741.0	63.8		9.17		3452.24	31.64	Solarex
	66.65	748.9	66.3	8.93	9.58 *	3707		33.10	NREL
T302	67.23	730.6	64.3		9.17		3452.24	31.58	Solarex
	67.43	742.5	66.6	8.99	9.65 *	3707		33.33	NREL
T307	66.63	737.9	64.9		9.24		3452.24	31.90	Solarex
	66.19	758.3	67.1	9.08	**9.75 ***	3707		33.66	NREL

* NREL used total area for measurements, aperture area efficiency calculated.

Table II : Initial I - V measurements made at Solarex and at NREL on 4 Ft^2 modules

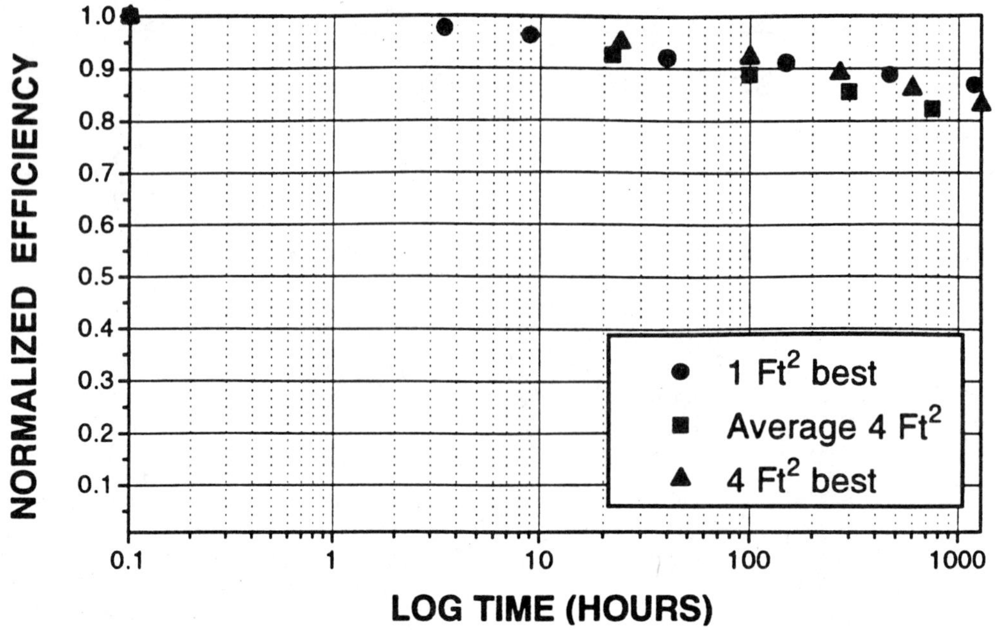

Figure 11: Normalized conversion efficiency as a function of time for a-Si/a-SiGe modules.

The stability of large-area a-Si/a-SiGe modules is reasonably close to the stability of small area cells. While the solar cell shown in Figure 3 degraded about 14% and the best 1 Ft^2 module shown in Figures 6 (a), (b)and(c) degraded about 15%, the average degradation of 4 Ft^2 modules is between 17% -18 %. This is shown in Figure 11 where the normalized efficiency is plotted as a function of time for the best 1 Ft^2 module, the best 4 Ft^2 module and the average of eleven 4 Ft^2 modules.

SUMMARY

Large strides have been made in the development of a-Si/a-SiGe multijunction technology. Stabilized efficiency of 9.1% on 1 Ft^2 has been scaled-up to 4 Ft^2 modules. The best 4 Ft^2 module had an initial conversion efficiency of 9.75% which should result in a stabilized efficiency of over 8%. At present, the average stabilized efficiency of the 4 Ft^2 modules in pilot production is about 7.5% with clear pathways to achieve 8% with this module design. Incorporation and scale-up of ZnO front contact and improvements in the a-SiGe back junction should lead to above 9% stabilized efficiency in large area modules (> 4 Ft^2) in the near future.

ACKNOWLEDGEMENTS

The authors would like to thank the entire technical staff at thin-film division for their efforts. This work has been partially funded by NREL under contracts ZM-2-11040-2 and RAN-4-13318-01 and by EPRI.

REFERENCES

[1] R. R. Arya, Proc. of the *7th Sunshine Workshop*, Tokyo, Japan (1993).
[2] M. Bennett, J. Newton, C. Poplawski, Proc. of the *23rd IEEE Photovoltaic Specialists Conf.*, p867 (1993)
[3] M. Bennett, K. Rajan and K. Kritikson, Proc. of the *23rd IEEE Photovoltaic Specialists Conf.* p. 845, (1993).
[4] R. R. Arya, L. Yang, M. Bennett, J. Newton, Y.M.Li, B. Fieselmann, L.F.Chen, K. Rajan, G. Wood, C. Poplawski and A. Wilczynski, Proc. of the *23rd IEEE Photovoltaic Specialists Conf.*, p (1993)
[5] L. Yang, L. F. Chen, S. Wiedeman, A. Catalano, *Proc. of the MRS Fall Meeting*, Boston, MA (1992).
[6] Y. M. Li, L. Yang, M. S. Bennett, L. F. Chen, F. Jackson, K Rajan and R. R. Arya, *Proc. of the MRS Spring Meeting*, San Franscisco, CA (1994).

Large Area Module Performance and Identification and Control of P-I Interface-Correlated Device Degradation and Further Improvement in Stabilized Efficiencies of Single-Junction a-Si Solar Cells.

Jianping Xi, Daniel Shugar, and Hermann Volltrauer
Advanced Photovoltaic Systems, Inc. (APS)

ABSTRACT

On a large scale, APS' PVUSA array has been in operation for two years now and continues to produce power in excess of its original specifications. Its output measured in October 1994 was 423 kW_{AC}. On a laboratory scale, in the last IEEE PVSC we reported our progress in improving the stability of a-Si based solar cells and in establishing 7.7% stabilized efficiency in single-junction devices. The best of these devices had a relatively low Voc of about 0.8 volts. By further optimization of the p layer, and p-i interface we have produced single junction devices with 0.86 volt Voc and 8.5% stabilized efficiency. Stable output in these devices was obtained after a very brief, 10 hour light soak.

INTRODUCTION

After 22 months of operation the APS nominal 400 kW_{AC} PVUSA field in Davis California is producing 423 kW_{AC} power and appears to have stabilized in the past year of operation. This array output corresponds to a 48 W_p per module output, in agreement with predictions made from small samples of modules that were individually measured. This 48 W_p corresponds to 4.5% active area efficiency, well below our recent laboratory achievement of 8.5% efficiency which is the main focus of this paper.

Since the last IEEE PVSC, more research efforts were focused on understanding the causes of light induced degradation and on the further development of p- and p-i buffer layers. We have found that incorporating the photo-conductivity-stable i-layer material in single junction devices with a thicker i-layer (5000-7000 Å) resulted in improved device stability. However, for devices with a thinner i-layer (3500 Å), we found no difference in stability between our base line i-layer material and the photo-conductivity-stable i-layer. This phenomenon indicates that for devices with i-layers up to 3500 Å thick using our baseline i-layer material, the remaining degradation is dominated by regions other than the bulk i-layer.

By replacing a-SiC p- and p-i buffer layers with a-SiO type materials, we obtained improvement of the spectral response in the short wavelength regime. Most interesting, we have found that these devices exhibit new characteristics in their light-induced degradation. The device degradation becomes a very fast process and saturates in about 10 hours with a loss of only about 10% in efficiency. Thus we have further improved the stabilized efficiency in single-junction devices to above 8.5% using only our base-line i-layer with a thickness of 3500 Å.

We have compared the light-induced changes in device performance and the photo-conductivity changes of the corresponding i-layer films and found, for the first time, that these changes saturate in the same time period. This indicates that device degradation actually involves two processes, a fast process, on the time scale of about 10 hours, and a slow progress, of about 1000 hours. Our experiment supports the hypotheses that the bulk i-layer is responsible for the fast process and the p-layer and p-i interface are responsible for the slow process.

In this paper, we will briefly review two years of real outdoor performance of modules made between two and three years ago, discuss in more detail the experimental results leading to 8.5% stabilized efficiency devices, draw a few conclusions, and discuss the technical approaches for future work.

LARGE AREA MODULE AND ARRAY PERFORMANCE

Module performance - The fact that a-Si PV modules undergo a power loss from the time of their initial exposure in the field has caused potential users to wonder if, in fact, a-Si modules do stabilize. This concern has been caused primarily by non Staebler-Wronski failures and by the lack of actual long term field data for second and third generation a-Si power modules used in terrestrial applications. Fortunately empirical long term field data on a statistically significant sample size is now available for APS modules from the PVUSA system. The output power of 240 modules (see Fig.1) that were installed at the APS PVUSA field in December of 1991 in a special test section was determined in 33 separate I-V measurements. These early modules were installed at the site seven months prior to the main build-up of the field of 9600 modules total, and were rated at 50 W_p.

These 240 modules were installed as three sub-arrays of 80 modules each - exactly as the remaining 117 sub-arrays are wired. I-V curves were taken approximately every two weeks for the first year and have been taken

CH3365-4/94/0000-0401 $4.00 © 1994 IEEE
First WCPEC; Dec. 5–9, 1994; Hawaii

roughly every month for the last ten months. The test data have been normalized to the standard PVUSA test conditions with interconnecting wire losses subtracted to yield individual module power levels.

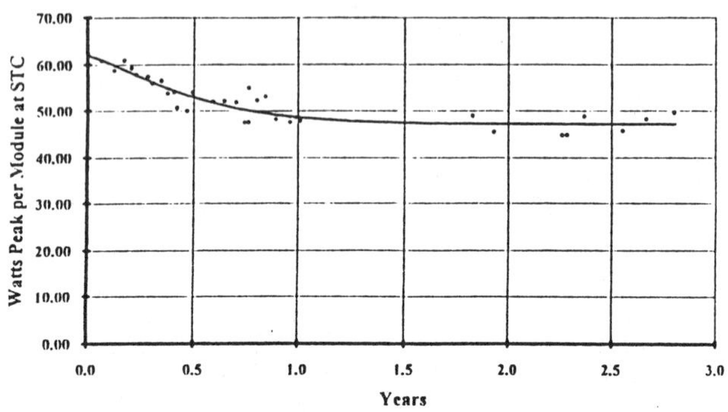

Fig. 1 Module Performance of initial PVUSA installation

As seen in the figure the modules reached their stabilized power level approximately one year after installation and are now fully stabilized with constant power output over the last year and a half.

PVUSA Array Performance - APS' contractual requirement for the PVUSA array was to deliver 400 kW_{AC} under PVUSA test conditions of 100 mW/cm2 plane-of-array insolation, 20 C ambient temperature, AM1.5 spectrum and one meter/second wind speed. Output power tests were conducted after the field was completed and solar soaked for a period of one month. IncOluding initial installation the system had experienced an average of 3 months light exposure when first measured.

The performance of the array during the acceptance test was computed using a regression analysis on the full months's data. The APS system was rated at 479 kW_{AC}, well in excess of the contractual requirement. The test was repeated in October 1993 after the array had operated for 15 months. The re-rating value of the stabilized system was 428 kW_{AC}, also well in excess of the contractual and design requirements, demonstrating the fact that a-Si systems can and do perform as well as expected. The maximum power produced during the first year of operation was 498 kW_{AC}.

In 1993 the net AC energy production was 530,048 kWh which yields an annual capacity factor for the year of 15.1% based on the designed and contracted array rating of 400 kW_{AC}. From January through August of 1994, the system produced 508,364 kWh which represents an annual capacity of 21.8%. This 1994 capacity factor is slightly skewed toward summer months since it does not include October through December; a final 1994 capacity factor is likely to be about 20.5%, with annual energy production of about 720,000 kWh. It should be noted that the APS array is not washed by PVUSA, unlike other systems at the same site, and that as much as 8% annual energy is lost due to soiling. It is estimated that the "energy production cost" of cleaning the arrays would be about 2.8c/kWh. This is estimate assumes four washings in the summer months [1].

TECHNICAL APPROACH TO SMALL AREA DEVICE DEVELOPMENT

At the 1993 MRS Symposium, we reported that a stabilized device efficiency of 7.5% was obtained in devices with a relatively low V_{oc} of about 0.80 Volts [2]. More recently 7.7% efficiency was reported [3]. One of the distinct features of these devices is that we used a thin p-layer of about 40 Å. By doubling the p-layer thickness, we can increase Voc of the device to nearly 0.9 Volts. However, these thicker p-layer devices will suffer from V_{oc} loss and higher FF loss upon light soaking and end up with a lower stabilized efficiency. We concluded that much of the remaining device degradation came from the light-induced changes in the p layer or p-i interface and decided to study this along three lines:

1. Optimizing carbon grading in the p-i buffer layer, with the goal of reducing the lattice mismatch between a-SiC p and a-Si, to increase both initial and stabilized V_{oc}.
2. Use of a-SiO:H to replace a-SiC:H, with the goal of obtaining a more transparent, better doped, and more stable window layer.
3. Use of a uc-SiC p layer to replace a-SiC:H p, with the same goal as technique 2.

The main effort was concentrated on characterizing the impact of these variations on device stability. By applying techniques 1 and 2 we succeeded in achieving a new record for stabilized efficiency of single-junction devices. Technique 3 helped in increasing the initial efficiency, but did not help in improving stabilized device efficiency and will not be further discussed.

SUMMARY OF EXPERIMENTAL RESULTS FOR SMALL AREA DEVICES

Carbon Graded Buffer Layer

We used a carbon graded buffer layer to replace the previously used constant carbon buffer layer. Table I compares initial and stabilized performance for these two types of devices.

By optimizing the carbon grading of the buffer layer, the initial V_{oc} was increased from 0.81 volts to 0.86 volts, and J_{sc} and FF were slightly improved. After light soaking improvement are seen in two major aspects:

1. The stabilized device efficiency has been increased to 8.19% from 7.39%, with higher stabilized values of both V_{OC} and J_{SC}.

2. The time of stabilization, under 1 sun illumination, has been shortened to about 150 hours from a previous value of about 600 hours.

Both devices seem truly stabilized in that we did not see measurable degradation after an additional few thousand hours of light soaking.

Table I. Initial and stabilized device performance with and without carbon grading in p-i buffer layer.

	V_{OC}	J_{SC}	FF	Eff.
Graded init. stab.	0.865 0.834	16.63 15.98	0.725 0.614	10.40 8.19
Constant init. stab.	0.814 0.806	16.56 14.98	0.710 0.616	9.58 7.39

Replacing SiC P- and Buffer Layers with SiO P- and Buffer Layers

It was reported that using CO_2 to deposit SiO type p layers results in better doping efficiencies and improves a-Si/a-Si same band gap device performance [4]. Since it was not clear whether this type of p-layer will affect the device stability, we undertook a study to determine the impact of such new p-layers on the device stability. We replaced CH_4 with CO_2, without change of flow rate ratios, in both p layer and p-i buffer layers. Our results have shown that the primary benefit of SiO is the further improvement of device stability, while the enhancement to the initial performance is secondary.

The evidence of improvement of device stability comes from three areas:

1. The time for stabilization of device performance has been reduced to about 10 hours. After that, the device efficiency become very stable (see Fig. 2).

2. The Fill factor (FF) stabilizes routinely at 0.63-0.65, for most devices fabricated. Devices with SiC p layers and buffer layers rarely have FF as high as 0.65, with a typical value being 0.60.

3. V_{OC} increase on light soaking are still observed, but this occurs at a higher initial V_{OC} of up to 0.84 volts, while such behavior in SiC p devices occurred at an initial V_{OC} of 0.78 volts.

With a combination of the above device features we recently improved the stabilized efficiency for single junction devices to 8.5% as confirmed by NREL measurements (see Fig. 3). The relative drop of device efficiencies is controlled within 10-13%. To our knowledge, this is the highest NREL confirmed stabilized device efficiency for single junction a-Si devices.

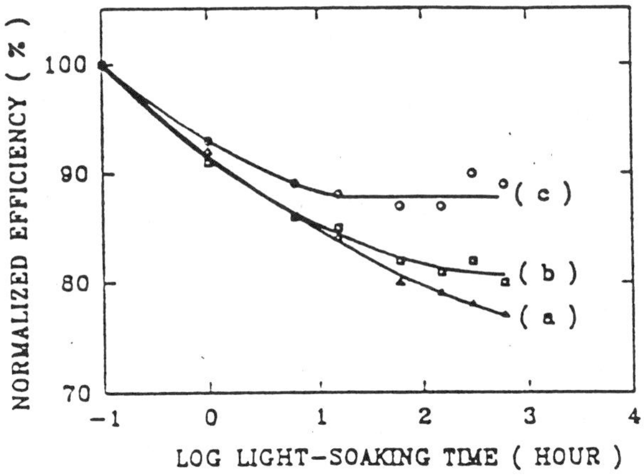

Figure 2. Efficiency at 1 sun and 50 C, for three type buffer layers: (a) constant CH_4 flow, (b) graded CH_4 flow, (c) graded CO_2 flow instead of using CH_4.

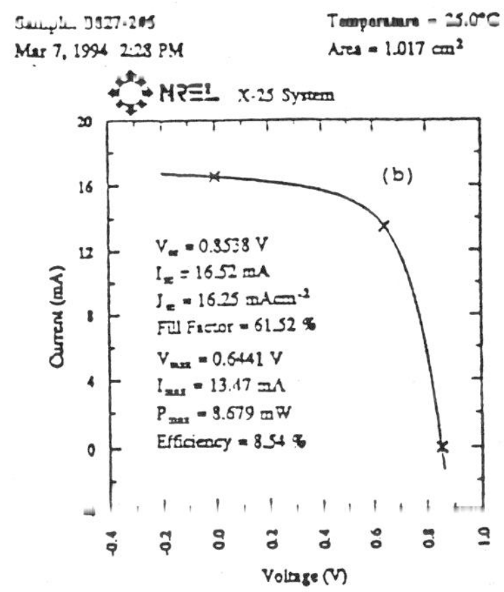

Fig. 3 JV curve for the device incorporating new techniques for stability enhancement

DISCUSSION

There is a widely recognized discrepancy that occurs between the saturation time of device and material stabilization [5]. By modifying the p and p-i interface structures, the saturation time for device stabilization has been shortened to 10 hours, under one sun illumination. For our devices, this coincides with the saturation time of photo-conductivity degradation of intrinsic a-Si material. Since all our recent improvements in device stability come from the changes made in the p layer and p-i interface, it is clear that in the conventional a-Si PV devices, the observed slow degradation is not due to the light-induced changes of i-layer, but caused by the light-induced changes of p layer and p-i interface. The negative impact of mc-p on the device stability further supports this concept.

The use of band gap-graded buffer layers will reduce the lattice mismatch. Replacing SiC by SiO in p and buffer layers further reduces the defects within these device regions. Perhaps the two bonds of oxygen offer more lattice flexibility than the four bonds of carbon, therefore resulting in more stable material. It is not appropriate to correlate the light induced changes of devices with light-induced changes of i-layers without considering the light-induced changes in p layers and p-i interfaces. The models used to interpret the high intensity light soaking results, including the model that "fast" and "slow" metastable defects co-exist in i-layers [6], should be more carefully re-examined.

PLANS FOR FUTURE WORK

An important question is whether the remaining 10% drop in device performance can be removed. The very short saturation time of device degradation strongly implies that this degradation is i-layer related. We currently do not use our photo-conductivity stable and H_2 diluted i-layers to make the 8.5% stabilized efficiency devices. We plan to incorporate these i-layers into new devices to see if we can further reduce their light-induced degradation.

On the larger scale we want to incorporate some of these changes into the manufacturing environment of our production facility. An important question will be to what extent the well controlled and lengthy laboratory process can be streamlined and adopted to the faster pace of the production floor.

SUMMARY

From this work, we are able to draw the following conclusions:

1. Large area a-Si can perform as expected and produce significant output yearround for many years.

2. In the laboratory, for high initial performance devices with high quality i-layers, the dominant factor in the remaining degradation is no longer the i-layer, but the p layer or p-i interface.

3. By appropriate carbon grading in the p-i buffer layer, we are able to raise the stable V_{oc} to 0.86 volts and establish 8.19% stabilized efficiency in single junction devices.

4. By use of SiO:H p layers and oxygen graded buffer layers, we are able to further improve the FF to the 0.63-0.65 level and establish an NREL confirmed 8.54% stabilized efficiency in single junction devices. The relative performance drop has been controlled at the 10-13% level. The saturation time has been reduced to about 10 hours.

5. It is clear that most light-induced degradation of a-Si devices is removable. The gain in stability of single junction devices will improve the technology of current a-Si PV production of which the single junction structure is dominant. Most of the improvements can be applied to multi-junction devices which can more fully utilize the solar energy spectrum.

ACKNOWLEDGEMENT

We thank the staff at the PVUSA site for making the measurements of our array and Dr. Bolko von Roedern at NREL for discussion and coordination of device evaluation, and acknowledge NREL for device performance calibration. We also thank Dr. Frank Kampas, Mr. Tongyu Liu, Dr. Jingya Hou, Dr. Satoshi Yamanaka, and Ms. Malathi Ghosh of APS for technical discussions.

REFERENCES

[1] D. Shugar and R. Matlin "Design and Two-Year Operational Performance of a 400 kWAC Amorphous Silicon Solar Photovoltaic Power Plant at PVUSA", Proc. IEA Task VI Workshop, Italy (1994).

[2] J. Xi, et al., "Major Improvement of Intrinisic a-Si:H Films for Solar Cell Applications", Mat. Res. Soc. Proc., 297, 785, (1993).

[3] J. Xi, et al., "Major Improvement in Material and Device Stability of a-Si:H and Establishment of High Stabilized Efficiency of Single-Junction a-Si Solar Cells", Proc. 23rd IEEE PVSC, p821,(1993).

[4] Y. Ichikawa, et at., "Large Area Amorphous Silicon Solar Cells with High Stabilized Efficiency and their Fabrication Technology", Proc. 23rd IEEE PVSC, p27, (1993).

[5] B. von Roedern, "Shortfall of Defect Models for Amorphous Silicon Solar Cell Performance", Appl. Phys. Lett., 62, 1368, (1993).

[6] L. Yang and L. Chen, Appl. Phys. Lett. 63, (1993)

APPLICATION OF P-TYPE MICROCRYSTALLINE SILICON TO a-Si ALLOY SOLAR CELLS

H. Sannomiya, K. Nomoto, A. Chida, Y. Nakata and Y. Yamamoto
Energy Conversion Laboratories, SHARP Corporation
282-1 Hajikami, Shinjoh-cho, Kitakatsuragi-gun, Nara 639-21, Japan

ABSTRACT

Detailed studies by the RHEED method have clearly indicated that microcrystalline silicon (μc-Si:H) films prepared with a conventional RF p-CVD system can be deposited from the thickness of 10nm even on a-SiC:H, a-SiGe:H alloy films and, furthermore, 25nm-thick films have almost the same crystallinity as the 200nm-thick μc-Si:H film.

Application of this μc-Si:H film to the p-layers of a-SiC:H and a-SiGe:H alloy solar cells have led to increases of 20~40 mV in Voc, compared to typical a-SiC:H p-layer cells. Analysis of the diode saturation currents have showed that these increases of Voc were due to the increases of the built-in potential of the cells.

INTRODUCTION

In order to improve the stabilized efficiency of hydrogenated amorphous silicon based solar cells, we have been studied and developed multi-junction, multi-bandgap a-Si alloy solar cells. As a result, by using the cell structure shown in Fig.1, 10.2% stabilized efficiency was achieved[1] up to now, which was confirmed by the JQA measurement.

To obtain solar cells with higher stabilized efficiency, the applicability of p-type μc-Si:H films to a-Si alloy solar cells has been investigated with expectation of built-in potential increase of the solar cells due to the nature of shallow Fermi level of the doped microcrystalline films. This application had been reported[2][3], but the question whether thin p-layer on a-Si:H alloy films becomes microcrystalline film is uncertain. This paper describes the results of the crystallinity study of thin p -type μc-Si:H on a-Si alloy films and the a-Si alloy cell characteristics to which the p-type μc-Si:H was applied.

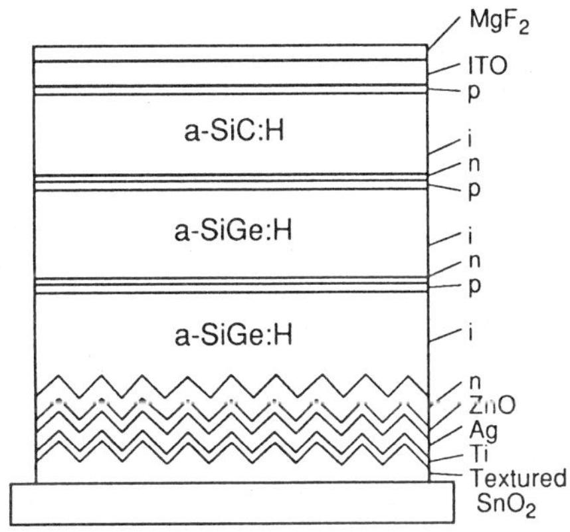

Fig.1. Structure of the multi-junction, multi-band gap a-Si alloy solar cell with 10.2% stabilized efficiency.

EXPERIMENTAL

P-type μc-Si:H films were prepared with a conventional capacitance-coupled RF p-CVD system using a gas mixture of SiH_4, H_2 and B_2H_6 under the condition of high hydrogen dilution (H_2/SiH_4=30~200) and high RF power (0.25~1.0W/cm^2). Samples were deposited on Corning 7059 glass for the optical and electrical measurement and on a-Si:H alloy films deposited on the stainless steel substrate for the RHEED analysis. The film thickness between 5 to 300nm. The typical bulk properties of the p-type μc-Si:H are shown in Table1 compared to the properties of typical p-type a-SiC:H film. Figure2 shows absorption coefficients of the μc-Si:H and a-SiC:H. The absorption coefficient of μc-Si:H is lower than that of a-SiC:H for short wavelength region in this comparison and is suitable for a window layer of a-Si solar cells.

CH3365-4/94/0000-0405 $4.00 © 1994 IEEE
First WCPEC; Dec. 5–9, 1994; Hawaii

Table 1. Film properties of a typical p-type μc-Si:H compared to the ones of a typical p-type a-SiC:H.

	σ_d (S/cm)	*Ea (eV)	**Egopt (eV)
p-type μc-Si:H	5×10^{-1}	~0.05	2.08
p-type a-SiC:H	2×10^{-6}	0.45	1.97

*Ea (eV):Activation energies of dark conductivity.
**Egopt (eV):Tauc's optical bandgap..

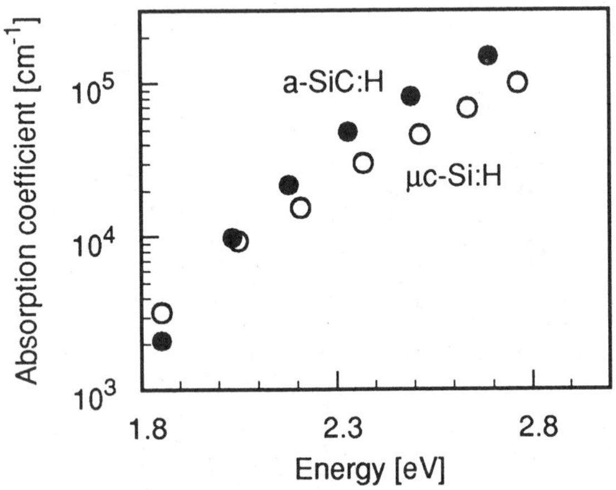

Fig. 2. Absorption coefficients in μc-Si:H and a-SiC:H
black symbol : a-SiC:H p-layer
white symbol : μc-Si:H p-layer

RESULTS AND DISCUSSION

Crystallinity of μc-Si:H

One of the most important point for the application of the doped μc-Si:H film to the amorphous solar cells is the crystallinity of the thin μc-Si:H film with the thickness of the order of 10nm on different kinds of underlying materials: in the cell structure shown in Fig.1, the crystallinity of μc-Si:H on the a-SiC:H and a-SiGe:H alloy films. Detailed studies by the RHEED method were carried out on the crystallinity of the μc-Si:H with film thickness of 5nm to 200nm on the a-SiC:H, a-Si:H and a-SiGe:H films deposited on the stainless steel substrate. The diffraction pattern examples are shown in Fig.3 and Fig.4.

Debye-Scherrer rings are observed from the thickness of 10nm and the rings of 25nm-thick films

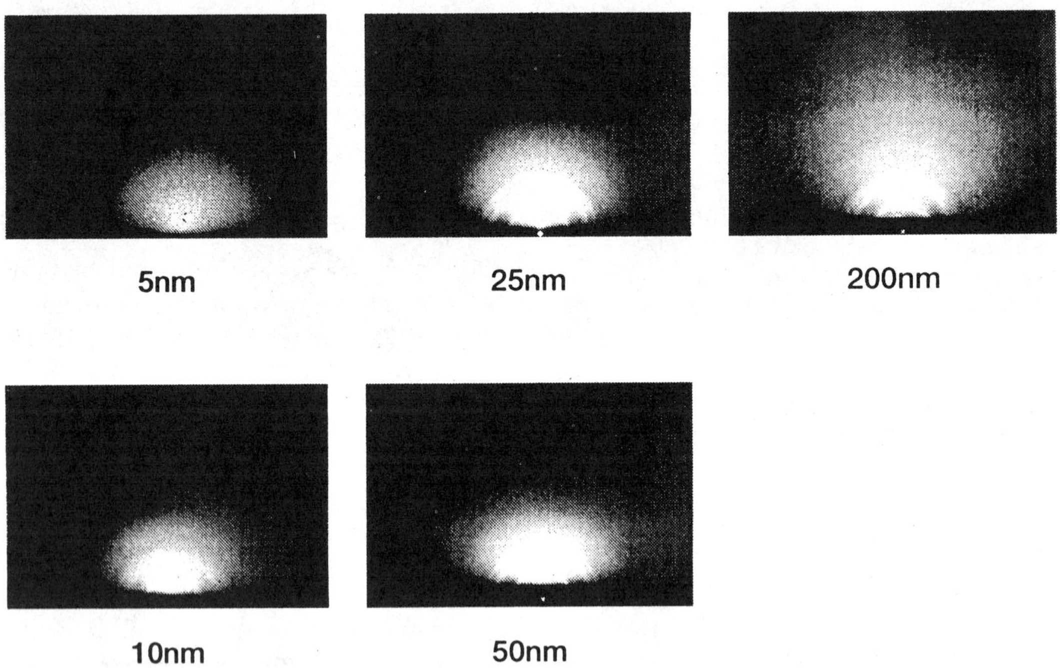

Fig.3. RHEED pattern examples of the μc-Si:H with film thickness of 5nm to 200nm on the a-Si:H alloy films deposited on the stainless steel substrate.

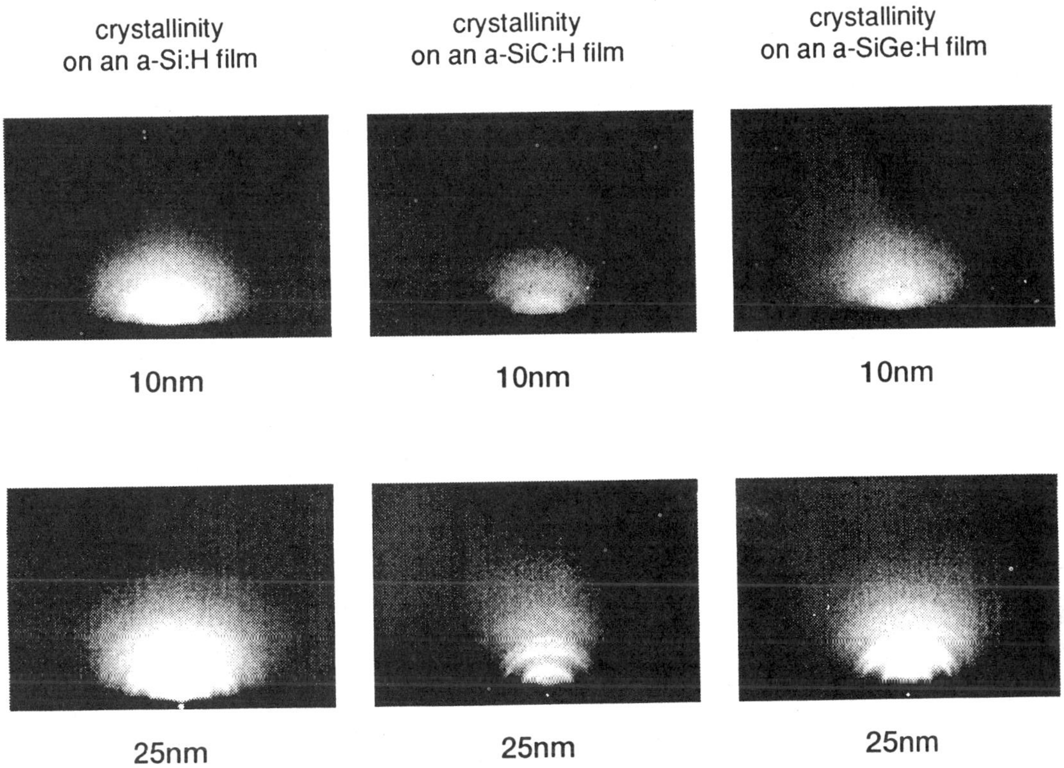

Fig.4. RHEED pattern examples of the μc-Si:H with film thickness of 10nm and 25nm on the a-SiC:H, a-Si:H and a-SiGe:H alloy films deposited on the stainless steel substrate.

become as clear as those of 200nm-thick μc-Si:H, regardless of the kind of underlying a-Si alloy films. These results indicate that microcrystalline films can be deposited with the thickness of 10nm even on a-SiC:H, a-SiGe:H alloy films.

Cell Characteristics

Table2 lists a-Si alloy single cell characteristics to which the p-type μc-Si:H film mentioned above were applied, together with the cell characteristics whose p-layer is a typical p-type a-SiC:H film. The bulk properties of the p-layer used in these experiments are shown in Table 1. The μc-Si:H p-layer solar cells exhibit increases of 20~40 mV in Voc compared to the a-SiC:H p-layer cells and absorption coefficients are shown in Fig.1.

Figure 5 is the Arrhenius' plot of the diode saturation current of the solar cells in Table 1 under light irradiation which were obtained by the measurement of Jsc-Voc characteristics under various light intensities at various measurement temperatures. Built-in potentials, obtained from the slope of this Arrhenius' plot[4], of the μc-Si:H p-layer cells are larger than those of the a-SiC:H p-layer cells. These increases of built-in potential are considered to be due to the film characteristics of shallow Fermi level and wide bandgap of p-type μc-Si:H.

CONCLUSIONS

From this study it was turned out that p type μc-Si:H film with the thickness of as thin as 10nm can be deposited even on a-SiC:H, a-SiGe:H alloy films and, furthermore, 25nm-thick films have almost the same crystallinity as the 200nm-thick μc-Si:H film in terms of the RHEED pattern.

Application of this p-type μc-Si:H film to a-SiC:H and a-SiGe:H alloy solar cells has successfully increased the built-in potentials of a-Si alloy solar cells. Further studies on the film quality and application to solar cells are now under the investigation to increase the stabilized efficiency, especially the open circuit voltage of multi-junction, multi-bandgap a-Si alloy solar cells.

Table 2. a-Si alloy single cell characteristics to which the p-type μc-Si:H were applied, together with the cell characteristics whose p-layer is a typical p-type a-SiC:H film.

i layer	p layer	Jsc (mA/cm^2)	Voc (volt)	F.F.	Pmax.(mW/cm^2)
a-SiC:H	μc-Si:H	9.65	0.986	0.691	6.57
	a-SiC:H	9.43	0.954	0.706	6.35
a-Si:H	μc-Si:H	15.16	0.927	0.633	8.90
	a-SiC:H	14.70	0.906	0.648	8.61
a-SiGe:H	μc-Si:H	6.64	0.600	0.592	2.36
	a-SiC:H	6.86	0.560	0.623	2.39

light source : for a-SiC:H and a-Si:H cells ; AM1.5,100mW/cm^2
for a-SiGe:H cells ; the light through band pass filter of λ>660nm of AM1.5,100mW/cm^2

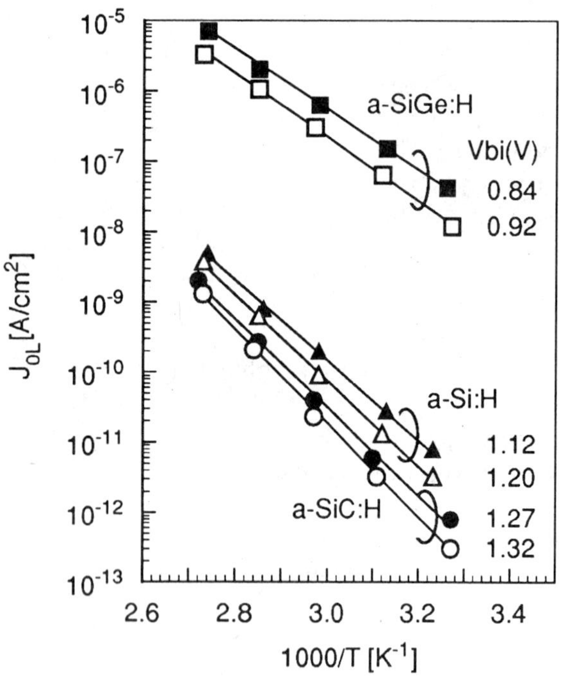

Fig.5. Arrhenius' plots of the diode saturation current of the solar cells in Table 1 under light irradiation, together with built-in potential values of each cell.
black symbol : a-Si alloy cells with the a-SiC:H p-layer
white symbol : a-Si alloy cells with the μc-Si:H p-layer

REFERENCES

(1) K.Nomoto, H. Saitoh, A. Chida, H. Sannomiya and Y. Yamamoto, Proc. 7th International Photovoltaic Science and Engineering Conference, Nagoya, 1993, p275

(2) Y.Uchida, T.Ichimura, M.Ueno and H.Haruki, Jpn,J. Appl. Phys. 21 (1982) l586

(3) S.Guha, J.Yang, P.Nath and H.Hack, Appl. Phys.Lett. 49 (1986) 218

(4) S.S. Hegedus, N.Salzman and E. Fagen, J. Appl. Phys. 63, 1988, 5126

INTRINSIC MICROCRYSTALLINE SILICON (μc-Si:H) - A PROMISING NEW THIN FILM SOLAR CELL MATERIAL

J. Meier, S. Dubail, R. Flückiger, D. Fischer, H. Keppner, A. Shah,
Université de Neuchâtel, Institut de Microtechnique (IMT), rue A.-L. Breguet 2,
CH-2000 Neuchâtel, Switzerland, phone: 038 20 54 59, fax: 038 25 42 76

ABSTRACT

"Compensated" microcrystalline silicon is obtained by adding 8 - 20 ppm diborane in the plasma gas phase. p-i-n cells with such i-layers have increased infrared sensitivity, when compared to a-Si:H p-i-n cells. The preparation of the world's first "mixed stacked" a-Si:H/μc-Si:H tandem cell with an initial efficiency of 9.1 % is reported. The μc-Si:H cells showed no degradation of the cell performance under intense light-soaking. Typical properties of the μc-Si:H cell indicate that the electronic transport is dominated by the crystalline phase of the material.

INTRODUCTION

Light-soaking experiments on amorphous (a-Si:H) and microcrystalline silicon (μc-Si:H) have generally shown μc-Si:H to be more stable or even completely stable under light-exposure [1-4]. This fact encouraged us to incorporate the intrinsic microcrystalline silicon in a p-i-n solar device [1, 5-7]. According to previous measurements, μc-Si:H turns out to be a defect-rich material with defect densities in the order of 10^{17} to 10^{18} per cm^{-3} [2, 8-10]. The defects are proposed to be located at the grain boundaries or in amorphous zones between the crystallites. The high defect densities were one of the reasons why μc-Si:H did not receive much scientific and technological interest sofar. In the present study we show that despite this intrinsic μc-Si:H has the potential to be a very promising thin film solar cell material.

Compared to the conventional 13.56 MHz plasma process, the Very High Frequency (VHF) technique has several advantages for the growth of μc-Si:H [11, 12], as e.g. higher deposition rate and larger grain sizes. Results from plasma diagnostics [13] and impendance analysis [14] suggest that lower ion energies and a higher atomic hydrogen flux to the growth surface lead to a favourable growth of μc-Si:H under VHF conditions.

As-grown undoped intrinsic (i) μc-Si:H is a n-type semiconductor. This is probably due to its high oxygen content and possibly also to internal structural effects. In the first part of the paper we will point out the effect on i μc-Si:H properties of adding small traces of diborane to the plasma phase. The absorption spectra of such "compensated" μc-Si:H will be compared with that of amorphous silicon. In the second part, the effect of "compensated" i-layers on the properties of fully μc-Si:H p-i-n solar cells as well as the world's first a-Si/μc-Si tandem cell will be presented. The light-induced degradation of μc-Si:H and a-Si:H p-i-n cells under 6 suns Illumination will be compared.

EXPERIMENTAL

All films and solar cells were deposited in a single VHF-GD chamber reactor at an excitation frequency of 70 MHz. The properties of p-doped μc-Si:H window layers and the less critical n μc-Si:H back contact for the μc-Si:H p-i-n cell were recently reported elsewhere [1, 15-16]. For the intrinsic μc-Si:H we used the following deposition parameters: T_{dep} = 220°C, p = 0.4 mbar, P = 7 W, SiH_4/total ~ 2-3 %, leading to a deposition rate of around 0.9 Å/s. To compare film properties directly with the cell devices, the film thickness in the compensation study was done on 1.5 and 1.8 μm films, i.e. relatively thick μc-Si:H films. In a previous temperature study [5], where the deposited μc-Si:H films were thinner (0.4 μm), the above parameters led to a material with a high cristalline fraction of over 90 % (Raman spectroscopy) and a crystallite size of around 220 Å (Scherrer's equation). To compensate the as-grown n-type character of μc-Si:H, diborane (B_2H_6) in the ppm range was added to the silane /hydrogen gas mixture. We call this the "microdoping technique". The films were deposited on Corning glass (7059), the p-i-n cells on SnO_2- and ZnO-coated glass. For dark conductivity measurements coplanar aluminum electrodes were used. Subgap absorption was measured by PDS (Photothermal Deflection Spectroscopy) and calibrated by transmission/reflection spectroscopy.

The microdoping technique was applied to the growth of the i -layer in the μc-Si:H cell. The back metallisation of the cells was made by an ITO/Ag contact. The tandem cell was realized by the open transfer of the a-Si:H top cell from the a-Si:H reactor to the μc-Si:H reactor. The cells were characterised under AM1.5 illumination at 100 mW/cm^2 in a two-source solar simulator (Wacom WXS-140S-10). The short circuit current densities (J_{SC}) were calibrated by integrating spectral response (SR) data in the range of 350 to 1000 nm times AM1.5 sun spectrum.

Previous degradation experiments with a long-term AM1.5 illumination source (outdoor conditions) sofar showed no degradation for the entirely μc-Si:H p-i-n device [7]. In order to apply a harder check for stability a 6 sun sodium lamp was used. Taking into account an absorption coefficient of $1.3x10^4$ cm^{-1} at the peak wavelength (590 nm) and the contribution of the ITO/Ag back reflector, almost uniform absorption can be

CH3365-4/94/0000-0409 $4.00 © 1994 IEEE

First WCPEC; Dec. 5–9, 1994; Hawaii

assumed. During the degradation experiment, the cells were kept at a constant temperature of 46 ± 2°C (controlled by an infrared thermometer).

Finally, films and cells were characterised by Scanning Electron Microscopy (SEM).

RESULTS AND DISCUSSION

1. Compensation of μc-Si:H

The microdoped films were analysed by dark conductivity measurements: in Fig. 1, dark conductivity σ_{dark} and dark conductivity activation energy E_σ are plotted in function of the B_2H_6/SiH_4 flux ratio.

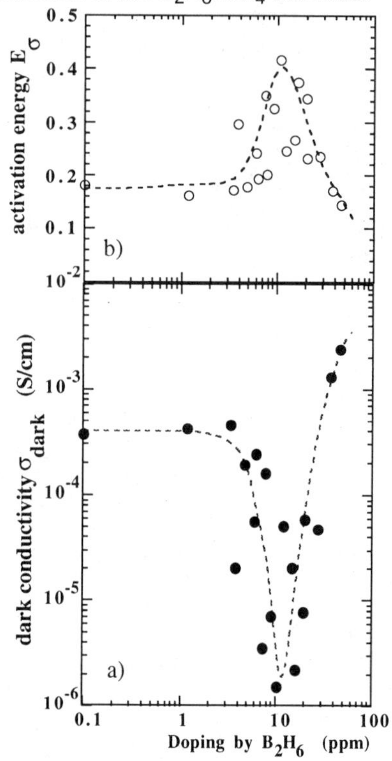

Fig. 1: Room temperature dark conductivity σ_{dark} and dark conductivity activation energy E_σ of μc-Si:H films (thickness: 1.5 - 1.8 μm), in function of the diborane microdoping of the gas phase. The values were obtained by coplanar evaporated Al electrodes. (Dashed lines are merely to guide the eye).

Fig. 1 shows that the electrical properties of the microcrystalline silicon react sensitively on the diborane partial pressure. The σ_{dark} value shows a sharp minimum in the regime of 8 to 20 ppm. The activation energy E_σ peaks in the same range of doping: Here, we find "compensated" μc-Si:H material. One has to notice that all obtained E_σ-values are smaller than half the gap of crystalline silicon. From our experience we observed that the compensation regime is also dependent on the outgasing rate, the dopant desorption of the chamber walls and the type of turbomolecular pump used. Furthermore, a previous study [17] showed that the oxgen content in the films is partially correlated with the deposition rate.

Investigations by SEM indicate a preferential columnar growth of the deposited films (Fig. 2). That means that the electrical properties obtained in Fig. 1 using a copla

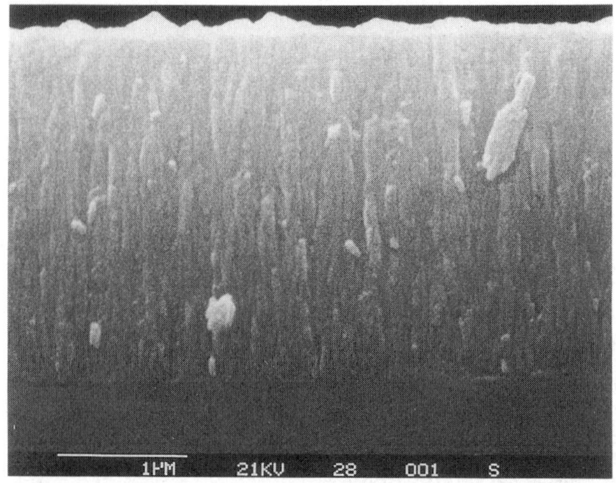

Fig. 2: Cross-section SEM picture of a 2.6 μm thick μc-Si:H film on glass.

planar electrode configuration may differ from those obtained for a sandwich configuration of the film. This fact has to be kept in mind, because the properties relevant to cells are those of vertical transport.

Microcrystalline silicon has a weaker absorption for the short wavelengths; however, in the near-infrared an enhanced absorption can be observed. Fig. 3 shows the absorption spectra of a "compensated" film measured by PDS. A similar absorption behaviour was found by different groups [18, 8]. However, our material seems to have a significant **lower subgap absorption** than that of sofar published μc-Si:H data: Our subgap absorption reaches the level of good quality a-Si:H material in the annealed state.

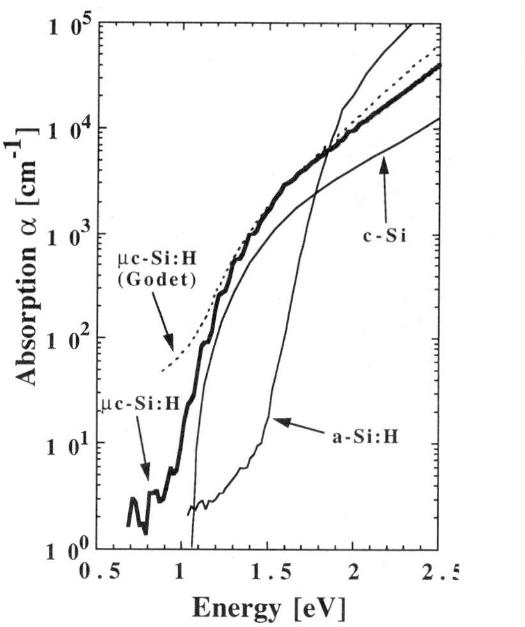

Fig. 3: Absorption spectra of a 1.8 μm thick compensated μc-Si:H film measured by PDS and transmission/ reflection spectroscopy. For comparison the spectra of another work [18], of amorphous and crystalline silicon are added.

Somehow, μc-Si:H combines both properties: in the high absorption range it seems to be dominated by the amorphous phase, and in the infrared range by the crystalline phase.

2. Fully μc-Si:H p-i-n solar cells

Microdoping of the i-layer has a very pronounced effect on the cells and on their spectral response (Fig. 4). All three samples presented have the same p-, n-layers, and the same i-layer except for microdoping. Fig. 4 confirms the enhanced infrared absorption of the μc-Si:H cell compared to a-Si:H: however the i layer thickness has to be increased to obtain sufficient J_{sc}.

The highest cell efficiency obtained sofar for the entirely μc-Si:H p-i-n solar cell is 4.6 % (V_{oc} = 375 mV, FF = 60 %, J_{sc} = 20.5 mA/cm^2, active area 5 mm^2, i-layer ~1.6 μm). By further increase of the i-layer thickness (1.8 - 2 μm) short circuit current densities of up to 24.4 mA/cm^2 were obtained. This is significantly above the highest values obtained for amorphous silicon. Further potential for increasing the efficiency lies in an optimization of the p-i interface for achieving higher V_{oc}-values. In a previous study [1] we have shown that V_{oc}-values of 600 mV can be obtained using very thin a-Si:H buffer layers.

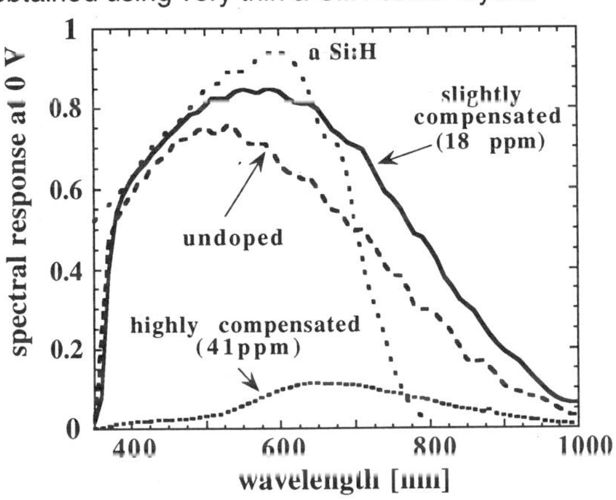

Fig. 4: Influence of microdoping on the spectral response of entirely μc-Si:H p-i-n cells (1.6 μm thickness). For comparison, an a-Si:H cell is also plotted.

3. "Mixed stacked" cells (a-Si:H/μc-Si:H)

The extended spectral response to wavelengths above 800 nm makes μc-Si:H an attractive material for a bottom cell in tandem structures. For the first time, we present a mixed stacked a-Si:H/μc-Si:H cell. At optimized current matching of the top a-Si:H cell (thickness 0.26 μm) to the μc-Si:H bottom cell (see Fig. 5), a current density of 11.5 mA/cm^2 for AM1.5 conditions was obtained (top 11.6 mA/cm^2). The I-V characteristics of this new tandem cell proves (Fig. 6) that the whole **growth process of the μc-Si:H** p-i-n cell is **fully compatible** with the amorphous p-i-n cell preparation. The results show that the substitution of conventional a-Si:H or costly a-Ge:H by the μc-Si:H bottom cell opens new ways to increase the efficiency of silicon thin film solar cell.

The morphological cross-section of our new a-Si:H/μc-Si:H tandem cell is documented by the SEM picture of Fig. 7. The different layers, as the first thinner amorphous top cell on the SnO_2 and the more "rocky" μc-Si:H bottom cell on the top, are clearly visible.

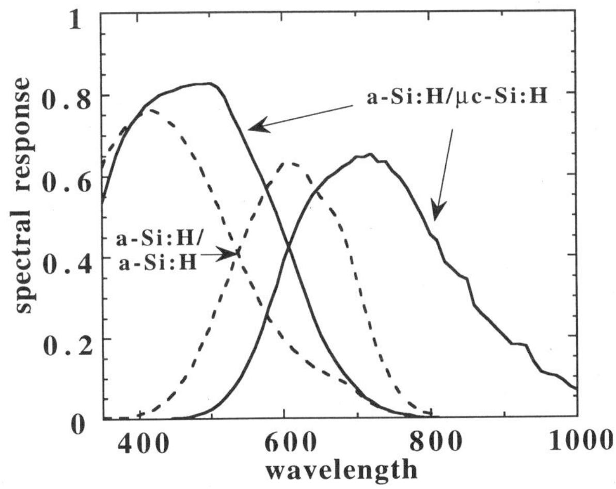

Fig. 5: Spectral response of the a-Si:H/μc-Si:H tandem solar cell in comparison with a standard a-Si:H/a-Si:H tandem cell. The J_{sc} values for AM1.5 conditions are 11.5 mA/cm^2 for the μc-Si:H bottom and 11.6 mA/cm^2 for the a-Si:H top cell (active area 8.7 mm^2).

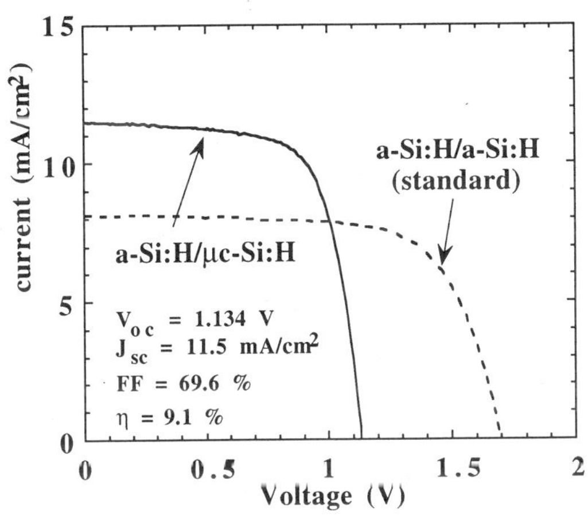

Fig. 6: I-V characteristics under AM1.5 conditions of the a-Si:H/μc-Si:H tandem cell, in comparison with a conventional a-Si:H/a-Si:H tandem cell.

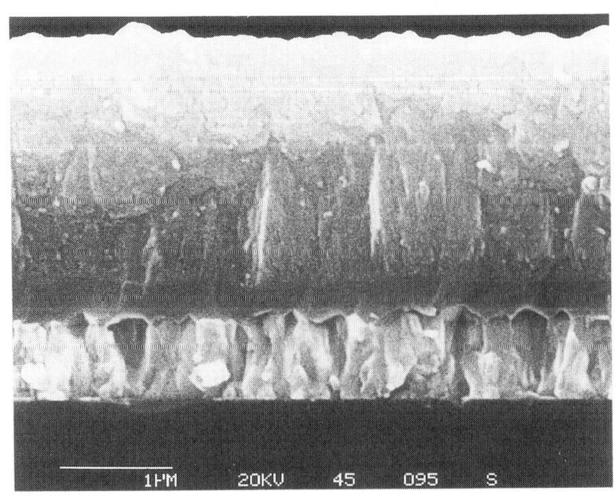

Fig. 7: SEM cross-section micrograph of the a-Si:H/μc-Si:H tandem cell deposited on SnO_2-coated glass.

3. Degradation experiments

As already presented [7], the induvidual μc-Si:H p-i-n cells showed no degradation effect after more than 640 hours of light exposure at AM1.5. In the experiment the 4.6% efficient cell was exposed to the high-intensity sodium lamp. In parallel, an amorphous p-i-n cell of 10 % initial efficiency was also exposed. The amorphous cell had a thickness of 0.6 μm, whereas the μc-Si:H 1.6 μm. The result is summarized in Fig. 8, where the kinetics of the degradations are shown: The intense illumination has a strong effect on the amorphous cell which decays already after one day of light exposure to less than 40 % of its initial efficiency. Efficiency values given in Fig. 8 are based on an evaluation of standard I-V (simulator) measurements (AM1.5 conditions) before and after the 280 h of intense light exposure.

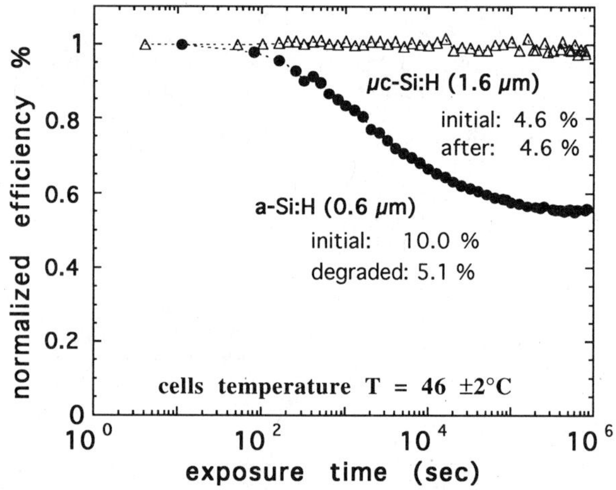

Fig. 8: Degradation kinetics of the normalized efficiencies of a μc-Si:H and an a-Si:H p-i-n solar cell under intensive light illumination (sodium lamp of ~6 suns). The cells were precisely characterized before and after 280 h of intense light soaking under AM1.5 conditions (numerical values).

The experiment presented in Fig. 8 confirms previous results from film studies: microcrystalline silicon is more stable than amorphous silicon. The degradation might be masked by the fact that up to now only an efficiency of 4.6 % efficiency could be obtained, on the other hand the μc-Si:H cell is much thicker than the amorphous one and should be more affected by light soaking.

Voltage-dependent SR measurements have recently [7] indicated that the carrier transport in complete μc-Si:H p-i-n cells is different from that of a-Si:H cells and might be partly supported by diffusion, in addition to drift transport; systematic studies have yet to be conducted.

CONCLUSIONS

Compensated microcrystalline silicon has the potential to become a promising new material for thin film solar cells. The high infrared absorption and the surprising stability favours the material as a bottom cell in a-Si:H/μc-Si:H tandems or even in triple cell structures. The first a-Si:H/μc-Si:H mixed stacked cell with an initial efficiency of 9.1 % and an enhanced short circuit current density of 11.5 mA/cm^2 was realized by our group. The results have shown that the growth of entirely microcrystalline cells is fully compatible with the growth of amorphous silicon, when using the VHF-GD technique at 70 MHz.

The electronic transport in μc-Si:H films turns out to be describable by the model used for polycrystalline silicon: This means that the crystallites determine the transport properties. Because of their higher infrared sensitivity, their improved stability against light-soaking, becaues of the fact that activation energies are smaller than the half gap of crystalline silicon, and because of the partly diffusion-like characteristics, we suggest that electronic transport of our **μc-Si:H cells is dominated by the crystalline phase** of the material. The role of the grain boundaries where the high defect densities are supposed to be located is still unclear and not understood. However, it seems that the overall cell behaviour is not affected by these high defect densities.

The limits of this new solar cell material, w.r.t. the V_{oc}, J_{sc} and FF are open and requires for further investigation. For future applications two problems have to be solved: a further increase in efficiency and a higher deposition rate.

ACKNOWLEDGEMENTS

This work was supported by Swiss Federal Research Grant EF-REN (93)032. The authors would like to thank M. Götz, C. Hof and U. Kroll of IMT Neuchâtel for helpful discussions and K. Ufert of Siemens Solar München for ZnO supply.

REFERENCES

[1] R. Flückiger, J. Meier, H. Keppner, M. Götz, and A. Shah, *Proc. 23rd IEEE PVSC*, 1993, p. 839.

[2] H. Liu et al., *Solid State Comm.*, **58**(9), 1986, p. 601.

[3] M.J. Williams et al., *AIP Conf. Proc.* , **234**, 1991, 211.

[4] R. Flückiger, J. Meier, M. Götz and A. Shah, to be published in *J. Appl. Phys.*, Jan. 1995.

[5] R. Flückiger et al., *Proc. 11th EC PVSEC*, 1992, 617.

[6] J. Meier, R. Flückiger, H. Keppner, M. Götz, A. Shah, *Proc. 12th EC PVSEC*, p. 1237.

[7] J. Meier, R. Flückiger, H. Keppner, A. Shah, *Appl. Phys. Lett.*, **65**, 1994, p. 860.

[8] H.N. Liu, Y.L. He, F. Wang, and S. Grebner, *J. Non-Cryst. Solids,* **164-166**, 1993, p. 1005.

[9] P.G. LeComber, G. Willeke, and W.E. Spear, *J. Non-Cryst. Solids,* **59/60**, 1983, p. 795.

[10] F. Finger et al., *Phil. Mag. Lett.*, **70**, 1994, p. 247.

[11] R. Flückiger et al., *Proc. MRS,* Boston 1994, in press.

[12] F. Finger, P. Hapke, M. Luysberg, R. Carius, H. Wagner, *Appl. Phys. Lett.,* **65**(20), 1994, p. 2588.

[13] A.A. Howling, J.-L. Dorier, Ch. Hollenstein, U. Kroll, F. Finger, *J. Vac. Sci. Technol.*, **A10** (1992), p. 1080.

[14] U. Kroll et al.,*Proc. MRS*, **336**, 1994, in press.

[15] R. Flückiger et al., *Proc. MRS,* **336**, 1994, in press.

[16] D. Fischer, H. Keppner, F. Finger, K. Prasad, A. Shah, *Proc. 10th EC PVSEC,* 1991, p. 201.

[17] U. Kroll, J. Meier, H. Keppner, A. Shah, to be published.

[18] C. Godet, Ph.D. thesis, Univ. Paris-Sud, Centre D'Orsay (1987).

HIGH EFFICIENCY a-Si SOLAR CELLS WITH ZnO FILMS

Wilson W. Wenas, K.Dairiki, A.Yamada, M.Konagai and K.Takahashi
Department of Electrical & Electronic Engineering, Tokyo Institute of Technology,
2-12-1 Ohokayama, Meguro-ku, Tokyo 152 Tel:+81-3-5734-2554, Fax:+81-3-5734-2897

J.H. Jang and K.S. Lim
Department of Electrical Engineering, Korea Advanced Institute of Science and Technology,
373-1 Kusong-Dong, Yusong-Gu, Taejon, 305-701 Korea

ABSTRACT

In this study special importances are attached to the ZnO/p and n/ZnO interfaces. A-Si solar cells with a structure of glass/ZnO/delta doped p/buffer/i/n/ZnO/Ag/Al have been fabricated. By optimizing the grain size of the films along with their electrical and optical properties, a high conversion efficiency of 12.5% was obtained under AM-1.5 illumination. The collection efficiency of the cell at 700nm reached 60%. Furthermore, effects of interface states represented by a barrier E_0 at the n/metal contact were investigated. It was found that this barrier affected the performance of the cells, particularly when n-layer became thin. Finally, to demonstrate the electrical role of the ZnO at back contact, p-i-n a-Si solar cells having thin n-layer combined with ZnO back contact were fabricated. It was surprisingly found that the performance of the cells with ZnO at back contact did not deteriorate though the thickness of n-layer was reduced to 3nm.

1.INTRODUCTION

By using the MOCVD technique so far we have established the growth of textured and low resistivity ZnO films with various type of surface morphologies. The films have been applied to the front and rear contacts in single junction p-i-n a-Si solar cells and a conversion efficiency of 12.1 % has been achieved[1].

In order to improve the performance of a-Si solar cells further, in this study special importances are attached to the ZnO/p and n/ZnO interfaces. For the ZnO/p contact, we examined the optimization of the light trapping effect by using the tetrapod-like surface morphology of ZnO films with various grain sizes.

In order to increase the conversion efficiency of a-Si based solar cells, many efforts have been done in improving both the material quality and the structure of the device. At present, the p-i-n a-Si solar cell in which the incident light enters through the p-layer side has become a standard device structure. To increase the conversion efficiency of p-i-n a-Si solar cells, significant improvements in p-layer quality, p/i interface as well as the quality of i-layer have been achieved.

Furthermore, in order to reduce the photo-degradation of a-Si solar cells, the recent trend now is to reduce the thickness of the absorber i-layer. However, since the use of thin i-layer may resulting in the decreasing of the light absorption in i-layer, TCO materials should be inserted between the n-layer and metal (Ag/Al) in order to increase the reflection from the rear contact[1]. Based on the above trends, the interface between the n- layer and the rear contact (n/ZnO or n/metal) which never received much interest for p-i-n devices, may become an important factor in determining the device performance, since the portion of light which passes the vicinity of the back contact increases as the i-layer becomes thin.

In this study, to obtain a clear information about that interface, the effects of interface states by means of a barrier height E_0 at the n/metal contact were investigated. Furthermore, by employing the ZnO layer at the rear contact, the fabrication of a-Si solar cells with a very thin n-layer was demonstrated.

2. EXPERIMENTAL

The growth of ZnO films was reported previously[1]. The a-Si solar cells fabricated in this work consist of TCO, delta-doped p, buffer, i and n a-Si layers, ZnO and a metal (Ag/Al) electrode. The delta-doped p-layer was prepared by alternating deposition of a thin boron layer of 0.1-0.5 atomic layers and a 2-3-nm-thick undoped a-SiC:H layer. A graded-bandgap a-SiC:H layer was employed as the buffer layer. The delta-doped p- and buffer layers were prepared by direct photo-CVD, and i- and n-layers were prepared by mercury-sensitized photo-CVD. The thicknesses of delta-doped p and buffer layers were fixed at 6nm, 10nm, respectively. The thickness of i-layer was varied from 200m to 500nm, while the thickness of n-layer was reduced from 20nm to 3nm. The size of the cells was $3x3mm^2$. The cell performances were measured under AM-1.5 ($100mW/cm^2$) illumination.

3. RESULTS AND DISCUSSION

3.1 ZnO at the front contact

In this work, the MOCVD technique was employed to grow ZnO films using diethylzinc (DEZ) and H_2O as source gases. The grain size of the film was successfully controlled by adjusting the growth rate of the films through the control of the flow rate of DEZ. Figure 1 shows the surface morphology of ZnO films grown at 130°C as a function of the films' growth rate. It is clear that the grain size becomes larger as the growth rate is increased, while the shape of the tetrapod-like morphology is unchanged. The obtained films have been used for the front contact of a-Si solar cells with a structure of

CH3365-4/94/0000-0413 $4.00 © 1994 IEEE First WCPEC; Dec. 5-9, 1994; Hawaii

Total thickness : 2μm

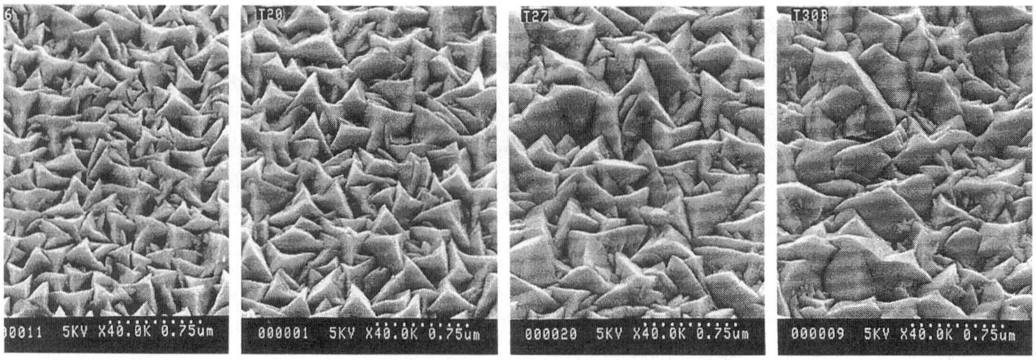

Fig.1 Surface morphology of ZnO films as a function of growth rate of the films (substrate temperature: 130°C).

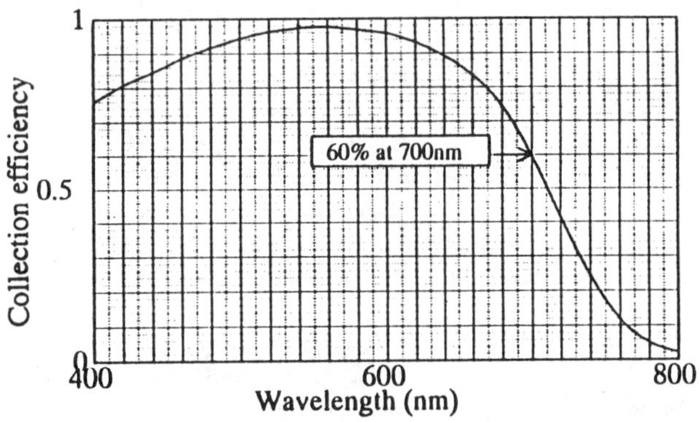

Fig.2 Collection efficiency of the fabricated cell with a structure of glass/ZnO/p/buffer/i/n/ZnO/Ag/Al.

glass/ZnO/delta doped p/buffer/i/n/ZnO/Ag/Al. By optimizing the grain size of the films along with their electrical and optical properties, a high conversion efficiency of 12.5% (V_{oc}:0.889V, I_{sc}: 19.46 mA/cm^2, FF: 0.721) was obtained under AM-1.5 illumination. The collection efficiency of the cell was measured and is plotted in Fig.2. The collection efficiency at 700nm reaches 60%. This result reveals that by combining the textured ZnO at the front contact and the highly reflective ZnO/Ag at the rear contact, the effective path length of the scattered light could be increased, resulting in a drastically increase of the absorbed light in a-Si layer, especially at long wavelength region.

3.2 n/metal interface

In this work, the effects of interface states at n/metal contact by using a barrier E_n were studied. The model used for the simulation was described previously[2], with an additional improvement to include the Gaussian-distributed defect states. The schematic diagram of the barrier E_n at the n/metal contact is shown in Fig.3. Figure 4 shows the cell characteristics as a function of the barrier height E_n for various thicknesses of n-layer. The thickness of i-layer was 500nm. For the values of

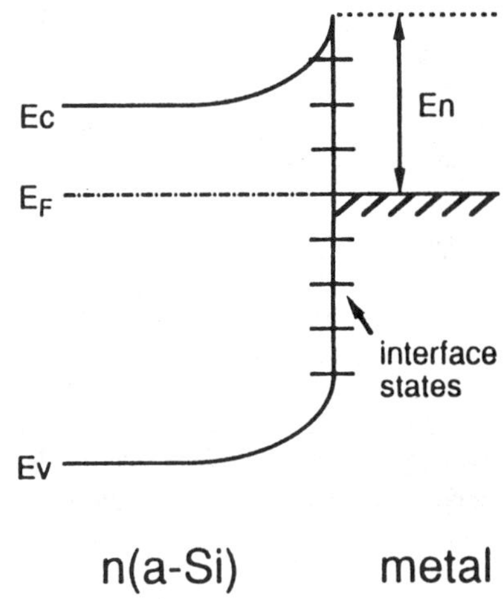

Fig.3 Schematic structure of the barrier E_n at the n/metal interface.

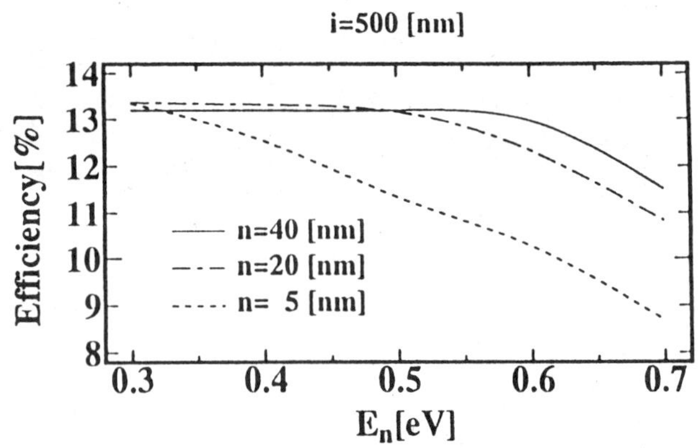

Fig.4 Dependence of the calculated conversion efficiency of the cells on the height of barrier E_n at the n/metal interface.

lower than 0.55 eV, the barrier does not affect the conversion efficiency of the cell when the thickness the n-layer is 40nm and 20nm. However, the conversion efficiency drops when the thickness of n-layer is reduced to 5nm even for a low barrier height. As the n layer becomes thinner, a part of n-layer near its interface with i-layer may become a depletion layer (Fig.5).

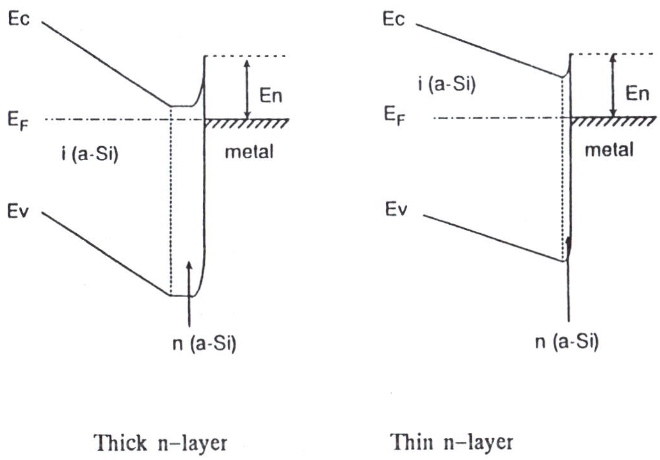

Fig.5 Band diagram of p-i-n a-Si solar cells with thick and thin n-layers.

As a consequence, the built-in potential which is basically determined by the difference of the quasi-Fermi levels between the p- and n-layers, can not be maintained high. Moreover, as a part of n-layer becomes a depletion layer, the internal electric field in i layer weakens, resulting in the degradation of the cell performance.

The same calculation was also carried out for the cell with a thinner i-layer of 100nm. The result is shown in Fig.6. As can be seen from the figure, the efficiency of the cells decreases as the barrier height increases. However, at a low barrier height range, the efficiency of the cell with an n-layer thickness of 5nm is higher compared with the cells with n-layer thicknesses of 20nm and 40nm. This is because the absorption

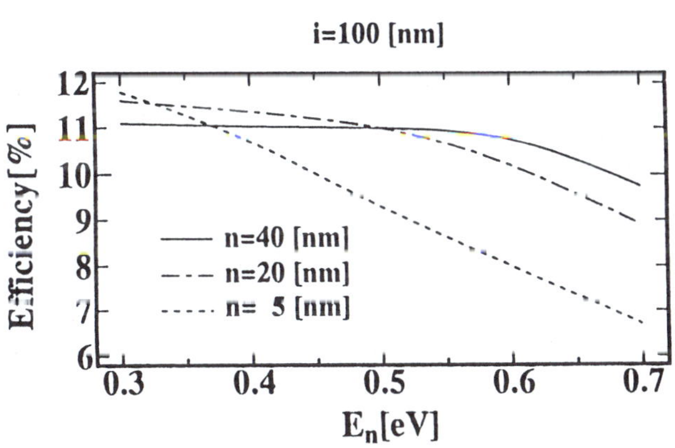

Fig.6 Dependence of the calculated conversion efficiency of the cells on the height of barrier E_n at the n/metal interface.

in n-layer decreases as it becomes thin. Therefore, it can be expected that the high efficiency should be achieved for the cells with a thin n-layer if the barrier height at the i/n interface can be maintained low.

3.3 The role of ZnO at the rear contact

Basically, the placement of ZnO between n-layer and metal is only for enhancing the optical reflection from the rear contact. However, since the ZnO is a degenerate n-type semiconductor with a high carrier concentration, ZnO may influence the pinning of the Fermi level and should modify the barrier height E_n. Based on this assumption, a-Si solar cells having a thin n-layer combined with ZnO at rear contact were fabricated. The cells have a structure of glass/SnO_2/delta-doped-p/buffer/i/n/ZnO/Ag/Al. Figure 7 shows the characteristic of the cells for various thicknesses of the n-layer with and without ZnO at the rear contact. The thickness of ZnO layer was 0.1μm. Performances of the cells without ZnO drop when the

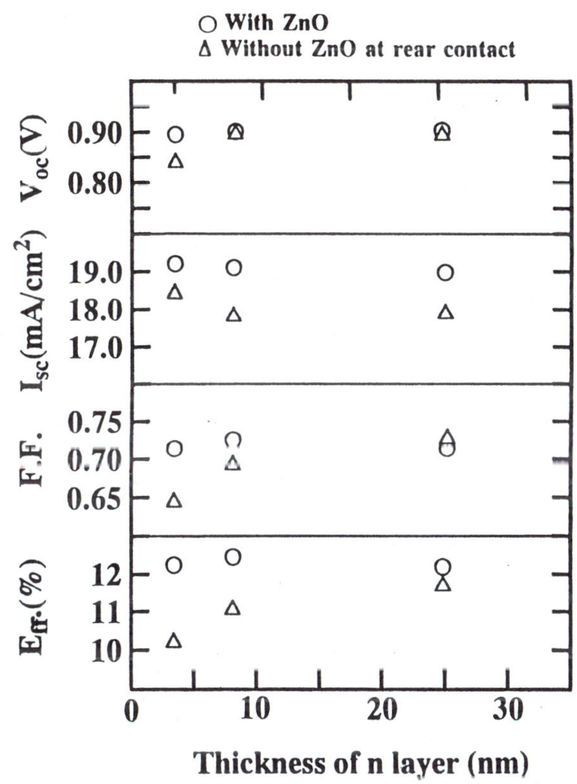

Fig.7 Performances of the fabricated cells as a function of the thickness of n-layer, with and without ZnO at rear contact.

thickness of n layer is reduced to less than 15nm. However, the cells with ZnO layer still show a good performance though the n layer is reduced to less than 5nm. This result implies an electrical-role of ZnO, in addition to the optical one. A similar result was also obtained for the cells with an i-layer thickness of 200nm. The effect of the thickness of ZnO layer at the back

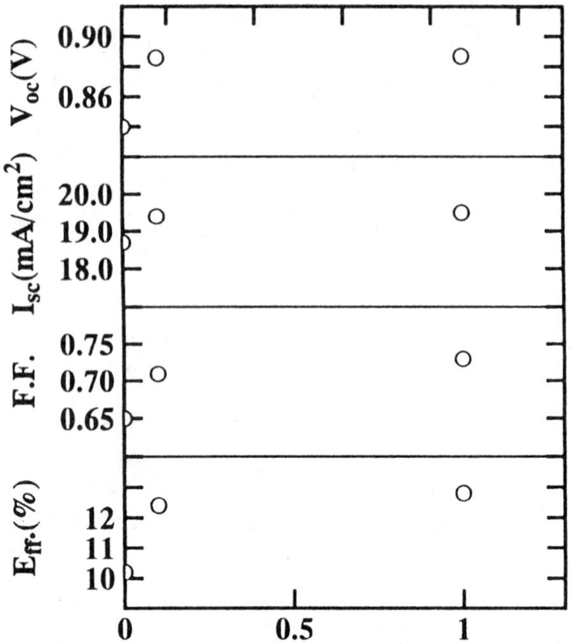

Fig.8 Performances of the fabricated cells for various thicknesses of ZnO layer at the rear contact.

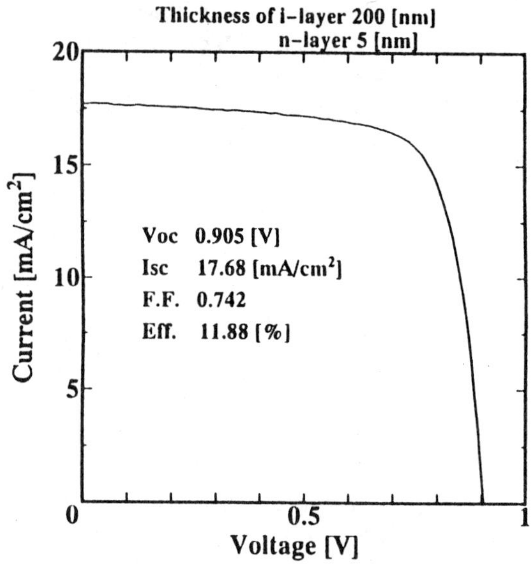

Fig.9 I-V characteristic of the cell with a thin microcrystalline n-layer.

contact was also examined. Figure 8 shows the performances of the cells with various thicknesses of ZnO layer. The thickness of n-layer was kept at 3nm. As shown in the figure, the efficiency is low for the cell without ZnO layer. However, the efficiency of the cells increases drastically when ZnO layer is inserted between n- and Ag layers. Almost no significant change in the cell performance is obtained when the thickness of ZnO layer is increased from 0.1μm to 1μm.

In a-Si solar cells with a thin i-layer, the improvement of the quality of n-layer is neccessary. In this work, to improve the quality of n-layer, the growth of micro-crystalline n-layer was developed by employing the hydrogen dilution technique. The obtained film was incorporated to a-Si solar cells. Figure 9 shows an I-V characteristic of the fabricated solar cells with a structure of glass/SnO_2/p/buffer /thin-i/micro-crystalline n/ZnO/Ag/Al. The thicknesses of i- and n- layers were 200nm and 5nm, respectively. A high efficiency of 11.9% is obtained. The improvements in the quality of both n-layer(high conductivity and wide-band gap) and the ZnO layer are required to improve the cell performance further.

4. CONCLUSIONS

A-Si solar cells with a structure of glass/ZnO/ p/buffer/i/n/ZnO/Ag/Al have been fabricated. By optimizing the grain size of the films along with their electrical and optical properties, a high conversion efficiency of 12.5% was obtained under AM-1.5 illumination. The collection efficiency of the cell at 700nm reached 60%.

Effects of the interface states at the n/metal contact was simulated by using a barrier height. It was found that the barrier affected the cell performance as the n-layer became thin. To demonstrate the electrical role of the high conductivity n-type ZnO at back contact, the unconventional structure of p-i-n a-Si solar cells having a thin n-layer (3nm) combined with ZnO back contact was fabricated. It was found that the performance of the cells with ZnO at back contact did not deteriorate though the thickness of n-layer was reduced to 3nm. This result indicates that the placement of ZnO layer at the back contact may modify the barrier at the n/metal interface through the pinning of the Fermi level.

ACKNOWLEDGMENT

This work was supported in part by NEDO as a part of the New Sunshine Program under the Ministry of International Trade and Industry.

REFERENCES

[1] W.W.Wenas, A.De,A.Yamada, M.Konagai and K.Takahashi, *Solar Energy Materials and Solar Cells*, 34(1994)313.

[2] H.Tasaki, W.Y.Kim, M.Hallerdt, M.Konagai and K.Takahashi, *J. Appl. Phys.*, 63(1988)550.

THE UTILIZATION OF MICROCRYSTALLINE Si AND SiC FOR THE EFFICIENCY IMPROVEMENT IN a-Si SOLAR CELLS

W. Ma, T. Saida, C.C. Lim, S. Aoyama, H. Okamoto and Y. Hamakawa
Faculty of Engineering Science, Osaka University
Toyonaka, Osaka 560, Japan

ABSTRACT

A systematic investigations on the mocification of interface potential by introducing doped microcrystalline Si (μc-Si) and it's carbon alloys (μc-SiC) at the TCO/p a-SiC interface has been made. It has been shown that a remarkable increase of the built-in potential with insertion of these materials. A series of technical data on the improvement of photovoltaic characteristics achieved by employing the microcrystalline materials are presented. The mechanism of improvements is discussed on the basis of systematic diagnostics on the interface properties.

INTRODUCTION

With tremendous recent R & D efforts, conversion efficiency of a-Si basis solar cells have been improved day by day. Among various factors which are closely tied with the photovoltaic performances, the interface properties both in the front and rear sides have been proved to play a predominant role in the determination of photovoltaic characteristics. For solar cells provided with microcrystalline (μc-) Si alloy electrode layers, however, little have been attempted to elucidate their functions in association with interface properties although the cell performances are found largely improved by their incorporation.

Our previous works on a-Si heterojunctions have shown that replacing p-type a-SiC with wide-window and highly-conductive μc-SiC results in a remarkable increase in the V_{OC}, which is ascribed primarily to an enhancement of the built-in potential [1]. Also found out recently is that a μc-SiC intermediate layer of a thickness as thin as 2 nm between TCO and a-SiC/a-Si heterojunction have significant effects on the Voc and FF [2]. This finding implies alternative mechanisms to be operated, that is, modification of TCO interface.

We have carried out a systematic experimental investigation on the electronic properties of μc-Si(C) related interfaces to shed light on their role played in the improvement of the photovoltaic performances of a-Si solar cell. The aim of this paper is to describe the results of experiments as well as to discuss the usefulness of μc-Si(C) alloy materials for a window-side contact layer.

MICROCRYSTALLLINE SiC ALLOYS AS A WINDOW CONTACT LAYER

It is often experienced in a-Si basis solar cells that the open circuit voltage increases upon increasing the p-layer thickness. The behavior is likely linked with a variation of the built-in potential V_b. An example of such dependence is shown in Figure 1 for a-SiC/a-Si heterojunctions formed on SnO_2/ZnO glass substrates. The electroabsroption technique has been employed for the measurement of V_b [3]. As found in this figure, V_b sharply increases with p a-SiC layer thickness and then saturates at about 0.89 V for the thickness above 25 nm. A change of the observed size can never be explained without taking into account the existence of the Fermi level clamping effect at the TCO/p a-SiC layer interface, which tends to induce a downward band bending in the p-layer adjacent to the TCO/p-layer interface, or in other word, an energy barrier $\Delta\phi_p$ indicated in inset of Fig.1.

The presence of interface energy barrier inferred above give some adverse affects on the photovoltaic performance. A lowered built-in potential naturally limits the open circuit voltage, while the transmission of photogenerated holes are impeded and the recombination of photogenerated electrons diffusing back toward the TCO interface are enhanced, leading to a deterioration of the curve fill factor. These problems are expected to be removed if a thin layer of highly conductive is inserted between TCO and p a-SiC layer, since the width of the energy barrier, even if it exists, shrinks to have no appreciable influence on hole transmission. A lifting of the band arising from the Fermi level located closer to the valence band edge would prevent photogenerated electrons from flowing toward the TCO interface. Both the mechanisms contribute an improvement of the built-in potential.

Microcrystalline (μc-) SiC alloys are raised as the most possible candidate for the window-side contact layer described above. Figure 2 summarizes the hole concen-

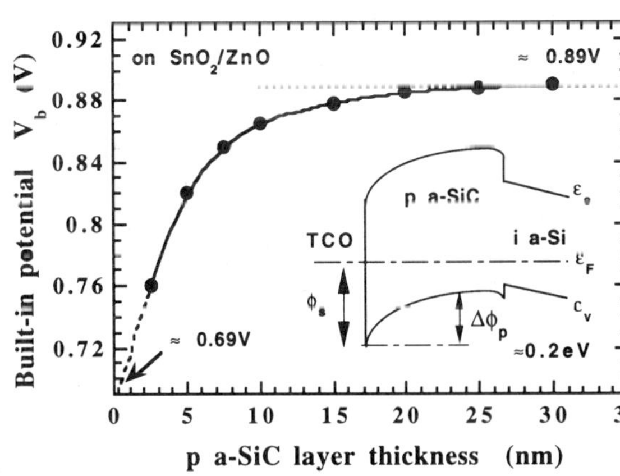

Fig. 1. The built-in potential V_b as a function of p a-SiC window layer thickness.

CH3365-4/94/0000-0417 $4.00 © 1994 IEEE

First WCPEC; Dec. 5–9, 1994; Hawaii

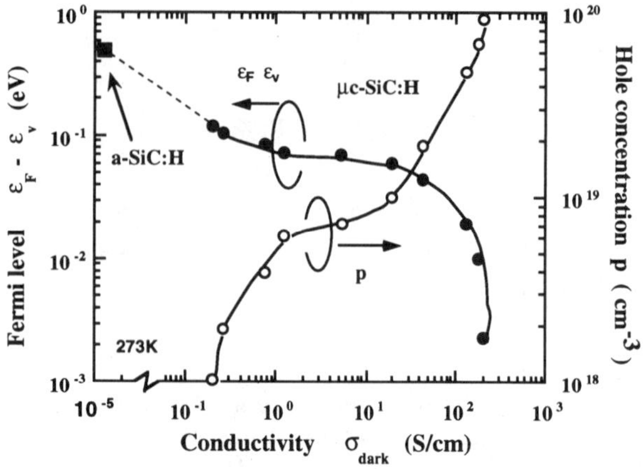

Fig. 2. The hole concentration p and Fermi level position ε_F - ε_V against the dark conductivity of p μc-SiC material.

tration and Fermi level position of Boron doped p-type μc-SiC alloys fabricated by ECR plasma CVD. The data were taken by combined measurements of conductivity and Hall coefficient. The hole concentration is translated to the Fermi level position upon assuming the effective density-of-states at the valence band at 10^{20} cm^{-3}. The conductivity ranges from 10^{-1} and 10^2 S/cm depending on apparent Tauc optical gap (2.2 to 2.6 eV) and doping level [4], in which the hole concentration varies from 10^{18} to almost 10^{20} cm^{-3}. The number is by more than 5 orders of magnitude greater than that in typical p-type a-SiC of the Tauc optical band gap 2 eV. The separation between the Fermi level and the valance band edge ε_F-ε_V appears to be less than 0.1 eV, being below one fifth of that in a-SiC estimated from the conductivity activation energy. It should be remarked here that p-type μc-Si possessing identical electrical properties with ECR CVD μc-SiC can be fabricated by conventional RF plasma CVD if the conductivity range from 10^{-1} to 10 S/cm is concerned.

Another requirement for the window-side contact layer is the optical transparency for solar radiation spectrum covered by a-Si photovoltaic active layer. A simple measure of the optical transparency may be the magnitude of Tauc optical band gap, however, this is hardly applied to microcrystalline materials although the apparent number can be attached for them. Instead of comparing the optical gap, in Figure 3, the absorption loss spectra are displayed for various p-type materials including μc-Si and a-SiC. The layer thickness is chosen at 10 nm. Clearly found is almost identical or rather lower absorption loss especially in the shorter wavelength region for microcrystalline alloys compared with a-SiC. The experimental results demonstrated here fully convince us of the usefulness of microcrystalline materials as a window-side contact layer for the improvement of photovoltaic properties.

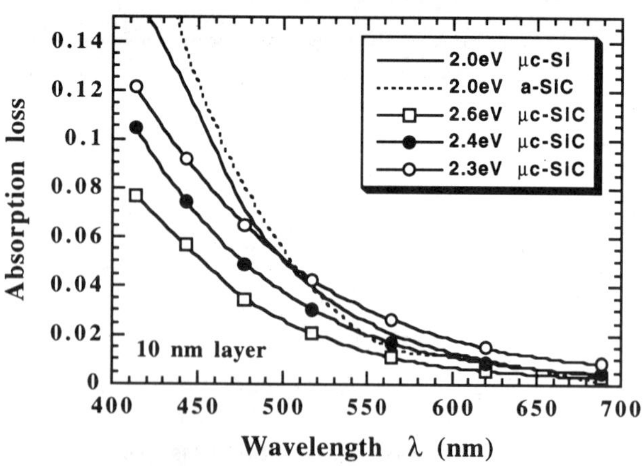

Fig. 3. The absorption loss spectra for various p-type μc-Si(C) and a-SiC materials.

PHOTOVOLTAIC CHARACTERISTICS OF SOLAR CELLS WITH μc-Si(C) INTERFACE LAYER

The device employed in the present work consists of TCO/p a-SiC/i a-Si/n μc-Si/ITO/Ag heterojunction structure, in which p type μc-Si(C) layers of the thickness form 1 nm to 15 nm prepared by the RF PCVD (μc-Si) and ECR PCVD (μc-SiC) methods are inserted at the TCO/p a-SiC interface. The photovoltaic characteristics are summarized in Fig.4, comparing with those on the cell without μc-Si(C) interface layer. Improvements are found in V_{OC} by nearly 0.1V and FF of 15% with an incorporation of 5 nm thick p μc-Si(C) layers. The simplest explanation of an increased V_{OC} would be to attribute it to an enhancement of the built-in potential. By the optimization procedure on each constituent layer, an efficiency of 12.3% has been achieved with the Voc greater than 0.9V for both .he cases using p μc-Si and μc-SiC materials.

Figure 5 displays typical normalized collection efficiency spectra $J_{ph}(V_a)/J_{ph}(0)$ under two specific bias voltages (V_a = +0.4 and -1.0 V) for the cells with and without μc-Si(C) interface layer. At a first glance, the $J_{ph}(V_a)/J_{ph}(0)$ values appear to be closer to unity for the cell pro-

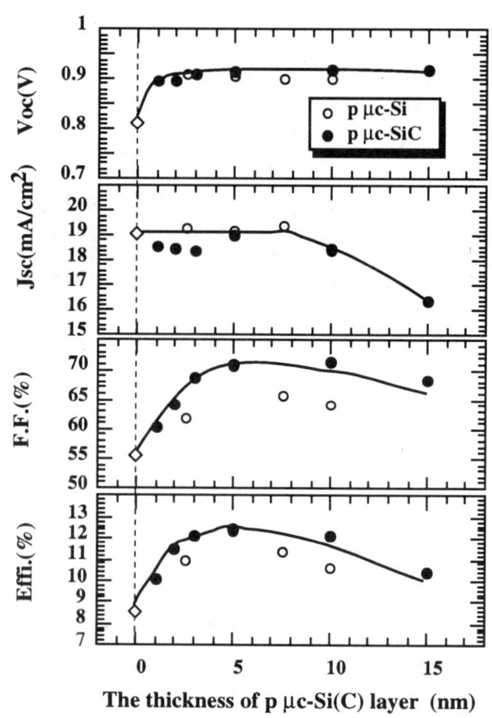

Fig. 4. The dependence of the device photovoltaic characteristics on the p μc-Si(C) layer thickness.

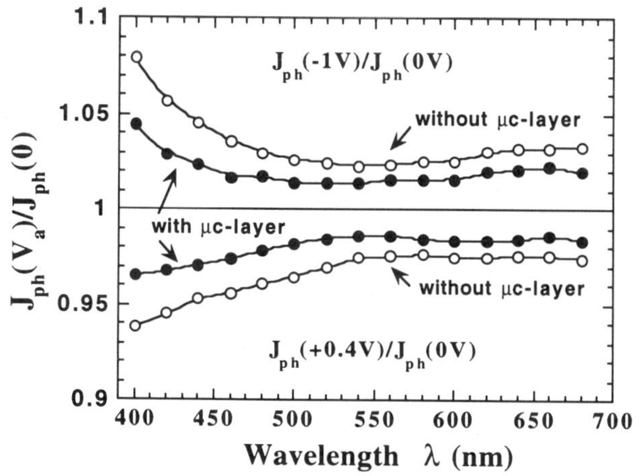

Fig. 5. Normalized collection efficiency spectra under two specific bias voltages (Va =+0.4V, -1.0V) for the cells with and without μc-Si(C) interface layers.

vided with μc-Si(C) interface layer over the entire spectral range of practical importance, which is reflected in the better curve fill factor in this type of cell than normal cell without interface layer. The trend observed here is expected for the case in which a larger built-in field is provided in the photovoltaic a-Si layer [5], being consistent with a conclusion described above. Another important feature of the $J_{ph}(V_a)/J_{ph}(0)$ spectra to be addressed here concerns with those in the short wavelength region, that is, the deviation of $J_{ph}(V_a)/J_{ph}(0)$ values from unity gets smaller in this spectral range for the cell with the interface layer, if compared with those of the normal cell. It is now well accepted that $J_{ph}(V_a)/J_{ph}(0)$ spectra in the short wavelength region is determined predominantly by carrier recombination effects at the front side contact and junction region. Then, the experimental result implies reduced carrier recombination in such regions for the cell incorporating μc-Si(C) interface layer.

CHARACTERIZATION OF BUILT-IN POTENTIAL AND INTERFACE ENERGY BARRIER

To identify the role of μc-Si(C) interface layers played in the improvement of photovoltaic characteristics in more quantitative manner, it is required to get direct information on the built-in potential as well as interface properties. This chapter is devoted to the description of systematic characterization on related issues. Before demonstrating the results, it would be best to illustrate the schematic band diagram of TCO/p/i interface region in order to clarify physical meanings of what to be characterized. That is Fig.6, in which ϕ_S denotes the surface potential barrier between TCO and adjacent p-layer being a constant specific to the material combination and chemical nature of the interface. When ϕ_S is larger than the bulk Fermi level energy of the p-layer, $\Delta\varepsilon_p$ (measured from the valence band edge; ε_F-ε_V), then a potential barrier $\Delta\phi_p$ (seen by holes) occurs at the interface to impede the transmission of photogenerated holes. In an ideal case, the built-in potential V_b is determined by $\Delta\varepsilon_p$- Δ_V (valence band offset at the p/i interface) if only the p/i junction side is concerned. However, in the practical case, the existence of positive ϕ_S - $\Delta\varepsilon_p$ (referred as $\Delta\phi_S$) prevents the Fermi level of the p-layer from reaching the bulk Fermi position, that inevitably causes the lowering of V_b from that determined by $\Delta\varepsilon_p$- Δ_V [3]. It might be straightforward to find that $\Delta\phi_p$ reaches its maximum, $\Delta\phi_S$, for the p-layer sufficiently thicker than its specific Debye length, while the maximum lowering of V_b is also just $\Delta\phi_S$.

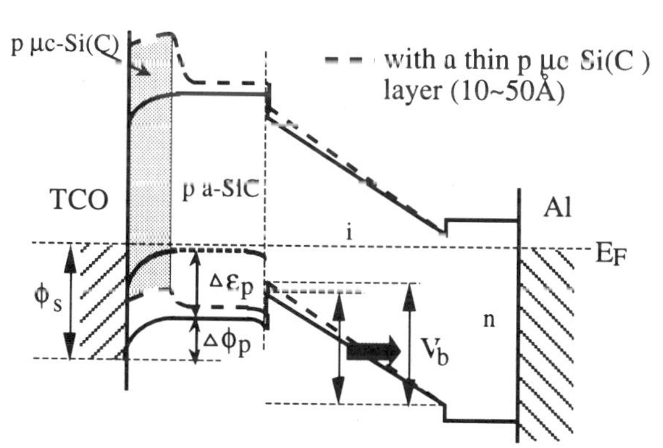

Fig. 6. The schematic band diagram of a-Si solar cell with and without p μc-Si(C) interface layer.

Figure 7 displays what happens on the $\Delta\phi_p$, V_b and V_{OC}, when the thickness of p μc-Si(C) layers inserted between the TCO/p a-SiC (7.5 nm) is varied from 0 to 15 nm, where $\Delta\phi_p$ and V_b were measured by the transient photopotential method [6] and small signal illuminated J-V characteristics, respectively. The magnitudes of V_b and $\Delta\phi_p$ for TCO/p a-SiC contact agree well with those reported so far [3, 6]. The V_b is found to sharply grow from 0.85 V to about 1.05 V, while $\Delta\phi_p$ decreases from 0.16 eV down to 0.06 eV with incorporating thicker μc-Si(C) layers. Since $\Delta\phi_p$ saturates for the thickness larger than about 5 nm, the ϕ_S for TCO/p μc-Si(C) is estimated to be 0.06 eV, while that for TCO/p a-SiC is likely greater than 0.16 eV, or presumably about 0.2 eV according to the discussion described above and also the result in Fig.1.

If only the interface potential determines the behav-

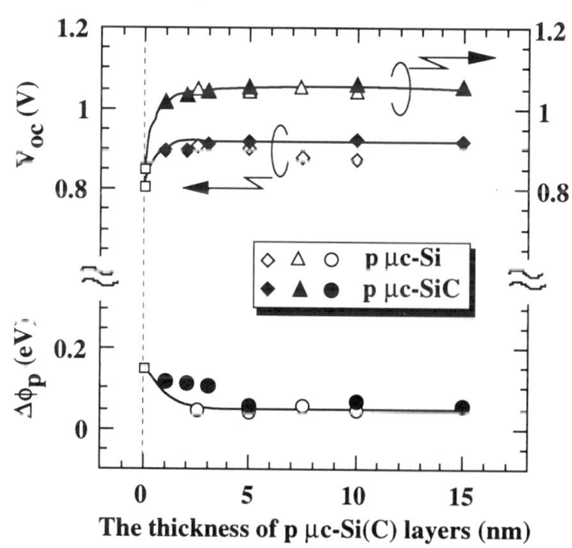

Fig. 7. The surface potentail barrier $\Delta\phi_p$, built-in potential V_b and open-circuit voltage V_{oc} vervus the thickness of μc-Si(C) interface layers.

ior of V_b, then the difference between TCO/p a-SiC and TCO/p μc-Si(C)/a-SiC structures would be at most 0.1 V, that is the difference in ϕ_S, whereas the experimental data in Fig.7 demonstrates an increase by 0.2 V. Then, it may be permitted to attribute the remainder 0.1 V to the increase in the $\Delta\varepsilon_p$- Δ_V factor. Such an effect would be possible if the entire region of p a-SiC layer is positively charged due to the existence of adjacent p μc-Si(C) layer of the smaller $\Delta\varepsilon_p$. Both the mechanism would operate to increase V_b by about 0.2 V with the incorporation of p μc-Si(C) layer. Consequently, the results of characterization lead us to confirm some ideas for the mechanism of performance improvements due to p μc-Si(C) layers described in previous chapters.

UTILIZATION OF n μc-Si CONTACT LAYER

As described above, by introducing the p μc-Si(C) layers into the TCO/p a-SiC interface, the cells performances have been large improved due to the decreasing of surface potential barrier. However, there still remains the surface barrier of a magnitude around 0.05 eV, which may have non trivial adverse affect on the hole transmission. To solve this problem, n type μc-Si layers with a high dark conductivity of 20~30S/cm and wide band gap of 2.0eV are inserted at the TCO/p μc-Si(C) interface. This is expected to promote the hole current transmission through

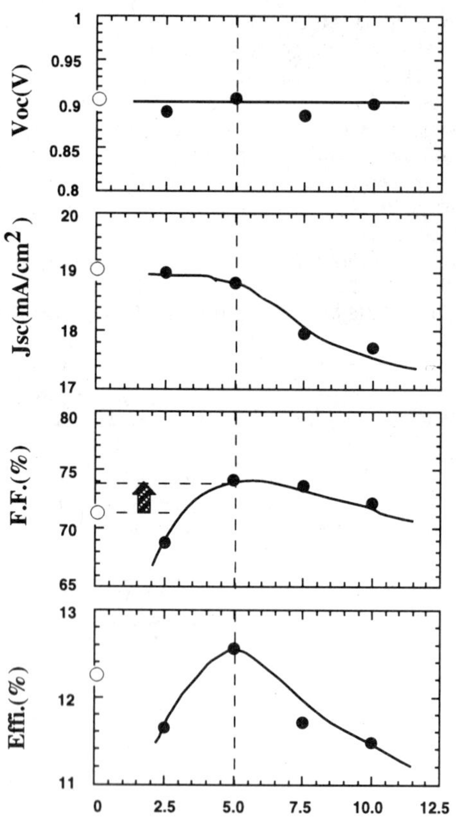

Fig. 8. Photovoltaic performances of a-Si solar cell with n μc-Si layer of various thicknesses at the TCO/μc-Si(C) interface region.

the tunnel effect between the n μc-Si/p μc-Si(C) interface. Figure 8 shows the photovoltaic performances of a-Si cells with the structure of TCO/n μc-Si/p μc-Si/p a-SiC/i a-SiC/i a-Si/n μc-Si/ITO/Ag with various thickness of n μc-Si layer. As shown in this figure, at the optimum thickness of 5.0 nm, the FF is increased to about 3% without any deterioration on V_{oc} and J_{sc}. As the result, conversion efficiency is increased to 12.5% with V_{oc}=0.905V, J_{sc}=18.8mA/cm^2 and FF=73.6%.

CONCLUSION

High conductivity and transparent materials of microcrystalline Si(C) were utilized for a-Si solar cell process with an aim of modifying the TCO/p a-SiC interface. As the experimental results, remarkable improvements on V_{oc} and FF have been achieved with the incorporation of p μc-Si(C) layers. It was proved by a systematic characterization that the surface potential barrier at the TCO/p-layer is reduced by more than 0.1 eV and built-in potential is increased by almost 0.2 V, and thereby leading V_{oc} greater by 0.1 V. A large improvement of FF have been obtained because of the combined effect of larger V_b, reduced interface recombination as well as better Ohmic contact between TCO and p interface. By inserting an n type μc-Si layer between the TCO/p μc-Si interface, the FF is further improved and a conversion efficiency of 12.5% has been achieved.

REFERENCES

[1] Y.Hattori, D. Kruangam, K.Katoh, Y. Nitta, H. Okamoto and Y. Hamakawa: "High Efficiency amorphous heterojunction solar cell employing ECR-CVD produced p-type microcrystalline SiC film", *Technical Digest of Int. PVSEC-3*, Tokyo (1987) 171-174.

[2] W. Ma, T. Horiuchi, C.C. Lim, K. Goda, H. Okamoto and Y. Hamakawa: "An optimum design of a-Si // poly-Si tandem solar cell", *Proc. 23rd IEEE PVSC*, Louisville, Kentucky (1993) 833-838.

[3] S. Nonomura, H. Okamoto and Y. Hamakawa: "A study of built-in potential in a-Si solar cells by means of back-surface reflected electroabsorption ", *Appl. Phys. A32* (1983) 31-38.

[4] W. Ma, T. Horiuchi, H. Okamoto and Y. Hamakawa: "a-Si // poly-Si four terminal stacked solar cell having an efficiency more than 19%", *Proc. PVSEC-6*, New Delhi (1992) 463-468.

[5] H. Okamoto, S. Nonomura and Y. Hamakawa: "Mobility-lifetime product and interface property in amorphous silicon solar cells", *J. Appl. Phys. 54* (1983) 3236-3245.

[6] K. Itoh, H. Matsumoto, T. Kobata, A. Fujishima, O. Nakamura and K. Fukui: "A leaky schottky barrier in the SnO_2/a-SiC(p^+) junction monitored with the pulsed laser-induced transient photopotential technique", *Technical Digest of Int. PVSEC-3*, Tokyo (1987) 167-170.

REAL TIME CHARACTERIZATION OF THE PREPARATION OF AMORPHOUS SILICON-BASED SOLAR CELLS

Yiwei Lu, Sangbo Kim, Ing-Shin Chen, Yeeheng Lee, C.M. Fortmann, C.R. Wronski, and R.W. Collins
Materials Research Laboratory, The Pennsylvania State University, University Park, PA 16802, USA.

ABSTRACT

Real time spectroscopic ellipsometry (RTSE) has been applied to characterize the p and i layers that compose amorphous silicon-based solar cells prepared in the superstrate configuration using a single-chamber reactor system. In this study, 106-point spectra (1.5-4.5 eV) in the ellipsometry angles (Ψ, Δ) are obtained during solar cell preparation with acquisition and repetition times as short as 160 ms and 1 s, respectively. With the spectroscopic capability, the evolution of the microstructure can be determined, including void densities, and surface roughness and bulk layer thicknesses, the latter with submonolayer sensitivity. In addition, the complex dielectric function and optical gap of the individual layers can be determined. An ultrathin (mass thickness ~4 Å) narrow gap interface layer is observed to form during an Ar flushing period between the growth of the p and i layers, which may reduce the open-circuit voltage and short-circuit current of the solar cell.

INTRODUCTION

It is well known that the characteristics of very thin films (for example, the ~200 Å p-layers used in solar cells) may differ substantially from those of thick films (for example, the corresponding 0.5-1 µm layers used for electrical characterization) [1]. In addition, interactions between transparent conducting oxide (TCO) substrate layers and the hydrides in the plasma used to deposit the overlying p-layer can modify the properties of the TCO, for example, by increasing the near-surface doping level in the case of the ZnO/a-$Si_{1-x}C_x$:H [2]. Finally, p-type dopant-related contaminants are often incorporated at the p-i interface due to their persistence in the reactor [3]. Effects such as these yield solar cell performances that depend sensitively on the details of the processing procedure. Thus, the performance of cells cannot be predicted solely on the basis of information obtained on the thick counterparts of each of the layers that make up the devices. In order to extend our knowledge of the relationships between the preparation and properties of very thin layers and interfaces, as well as their effects on devices, monolayer-sensitive real time probes such as RTSE are needed for monitoring solar cell fabrication.

EXPERIMENTAL DETAILS

The p and i layers of the solar cells were deposited in a high vacuum (1.0×10^{-7} Torr base pressure) single-chamber system and monitored by a rotating polarizer-type multichannel ellipsometer mounted on the deposition chamber as shown in Figure 1. The ellipsometer is equipped with an optical multichannel analyzer system that has data acquisition and repetition times as short as 16 and 32 ms, respectively. This capability makes real time monitoring of the growth of p and i layers at submonolayer (~2 Å) resolution readily achievable.

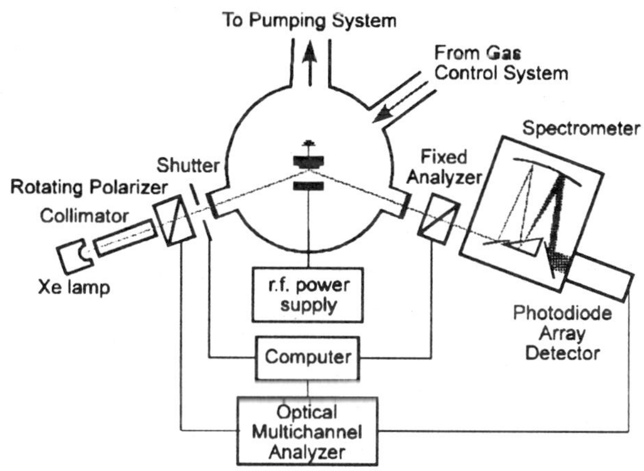

Fig. 1. Schematic of the plasma enhanced chemical vapor deposition system and the real time spectroscopic ellipsometer.

The p/i structures were co-deposited on two types of substrates, SnO_2-coated glass and SnO_2-coated Si wafers, for which the TCO is deposited under similar conditions. The latter structure utilizes the reflectance of Si to enhance sensitivity to p-layer properties in the RTSE studies. The former structure is used for device measurements. Because the specular SnO_2 onto which the solar cell is deposited is quite rough on the scale of the doped layer thicknesses, rather sophisticated optical models are required for data analysis. Figure 2 shows the optical model used to interpret the evolution of surface and interfacial roughness layers in this case. The effective complex dielectric functions of these layers are calculated using the Bruggeman effective medium approximation [4].

CH3365-4/94/0000-0421 $4.00 © 1994 IEEE First WCPEC; Dec. 5-9, 1994; Hawaii

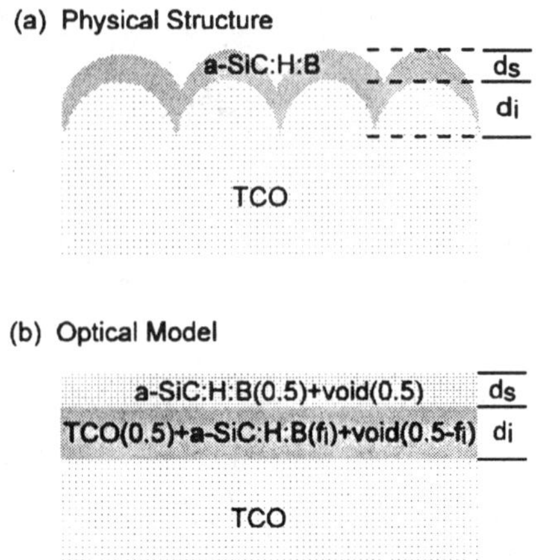

Fig. 2. The physical structure and optical model of film deposited on a rough TCO substrate before the formation of a bulk layer. The thicknesses of the a-$Si_{1-x}C_x$:H surface roughness layer and the TCO/a-$Si_{1-x}C_x$:H interface roughness layer are given by d_s and d_i, respectively. The volume fraction of a-$Si_{1-x}C_x$:H filling the TCO roughness is given by f_i. The a-$Si_{1-x}C_x$:H surface roughness is modeled as a 0.5/0.5 mixture of a-$Si_{1-x}C_x$:H and void. When f_i reaches 0.5, then the TCO roughness is completely filled and the bulk layer begins to form.

The deposition temperature (T_s) and r.f. power density ($P_{r.f.}$) were fixed at 240 °C and 70 mW/cm^2 for both p and i layers. In our earlier RTSE studies of the SnO_2/p-layer interface, we determined quantitatively the degree of SnO_2 reduction to Sn metal (and the associated interface optical absorption) that occurs when boron-doped a-$Si_{1-x}C_x$:H ($x \approx 0.09$) is deposited on SnO_2 [2]. In the present study, we have designed our p-layer growth process such that the mass thickness of Sn generated at the SnO_2/p-layer interface is below detectivity limits (<1 Å). Gas flows were set at [SiH_4]:[CH_4]:[B_2H_6]:[H_2]=6:4:0.01:0.99 (in sccm) for the p-layer and 20:0:0:0 for the i-layer. A 55 minute Ar flush was performed after the deposition of the p-layer in an attempt to reduce the B-contamination in the i-layer.

In order to complete the device, the TCO/p/i structure was exposed to air and transported to another deposition system for n-layer growth. A 200 Å thick n-type μc-Si:H was deposited on the top of i-layer under the following conditions: T_s=240 °C, $P_{r.f.}$=500 mW/cm^2, and [SiH_4]:[PH_3]:[H_2]=0.99:0.01:100 (in sccm). No treatment was performed on the i layer surface, based on the belief that any oxide that may form during sample transport will be removed by the strong H_2 plasma in the μc-Si:H n-layer deposition. In separate controlled studies, we found that the device performance is not adversely affected by the sample transport. Finally, a 1000 Å thick Cr top contact was deposited by thermal evaporation. The initial performance of the solar cells are measured under AM 1 illumination.

RESULTS AND DISCUSSION

Figure 3a shows part of a single photon energy trajectory (at 2.45 eV), plotted in terms of the ellipsometric angles (Ψ, Δ), as an illustration of the p/i deposition process on the SnO_2 surface at 240 °C. Data are also available for ~100 other photon energies. Because we established conditions of p-layer deposition that avoid reduction of the SnO_2, the TCO/p-layer interface is not of particular concern in this study.

p-Layer Growth

Figure 3b shows the optical properties of p-type a-$Si_{1-x}C_x$:H, including the complex dielectric function and optical gap, as extracted from the RTSE data. The gap is obtained by the extrapolation of $(\varepsilon_2)^{1/2}$ to zero ordinate [5] and is characteristic of the substrate temperature of 240 °C. (Note that the $(\varepsilon_2)^{1/2}$ gap at 240 °C is ~0.20 eV less than the Tauc gap at 25 °C. Thus the 1.65 eV gap for the p-layer shown in Fig. 3b leads to a conventional room temperature Tauc gap of ~1.85 eV.) Figure 3e shows the microstructural evolution of the p-layer as it grows on the rough SnO_2 substrate, information also extracted from the same RTSE data set. In the first 55 s of film growth, the p-layer is filling in the modulations of the rough TCO surface (98 Å thick, as obtained from the t=0 extrapolation below). After 55 s, the substrate roughness is filled in by depositing material, and a well-defined bulk layer grows at a linear rate of 1.5 Å/s. In the time it takes to grow the 111 Å bulk p-layer, the substrate-induced roughness on the p-layer decreases in thickness from 82 Å to 67 Å. Extrapolating this trend back to t=0 yields an estimated 98 Å roughness layer on the SnO_2 surface, a result in good agreement with atomic force microscopy measurements [6]. Including the interface and surface roughness layers, the final effective (or mass) thickness of the p-type a-$Si_{1-x}C_x$:H is given by $d_{mass}=f_i d_i+d_b+0.5d_s=0.5(98)+111+0.5(67)$ Å=194 Å.

B-Rich Contamination Layer Growth

For reference, Figure 3c shows the complex dielectric function and optical gap of a B-rich contamination layer obtained in a continuous in situ exposure of the p-type a-$Si_{1-x}C_x$:H surface to the reactive gas mixture used to deposit the p-layer. The predicted room temperature Tauc optical gap for this layer is ~1.5 eV, 0.35 eV narrower than the p-layer gap. Figure 3f shows the structural development of a layer of similar composition, but much thinner, that forms during the 55 minute Ar flushing between the p-layer and i-layer steps

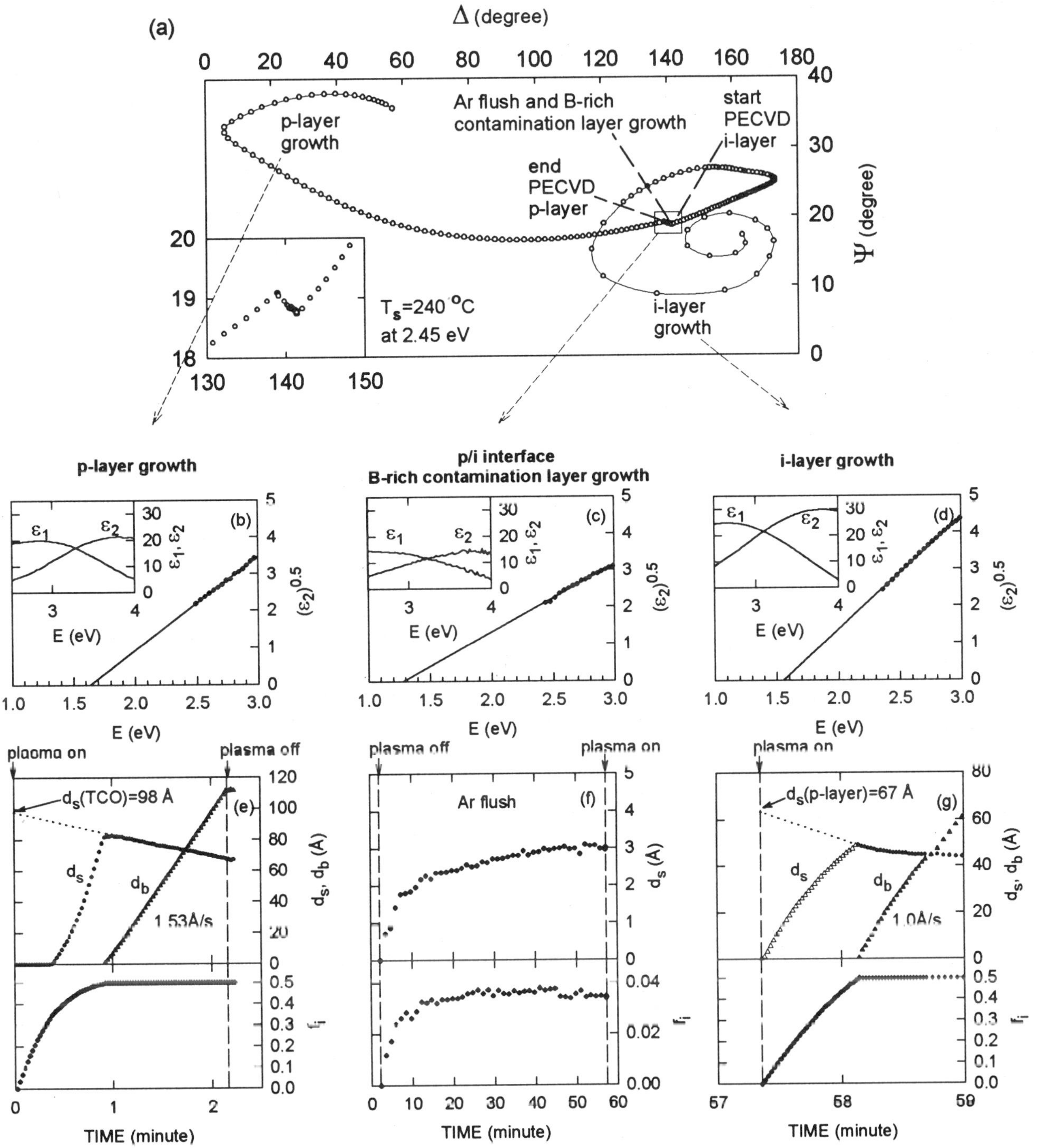

Fig. 3. (a) Single photon energy ellipsometry trajectory (2.45 eV) for the growth of the p-layer, p/i interface layer (expanded in the inset), and i-layer; (b), (c) and (d) Complex dielectric functions (insets) and optical gap determinations for the individual layers: (e), (f) and (g) Evolution of microstructure of the individual layers as measured by RTSE. ε_1 and ε_2 are the real and imaginary parts of the complex dielectric function. The optical gaps are determined by extrapolating the linear trend in $(\varepsilon_2)^{1/2}$ to zero ordinate. d_b and d_s are the bulk layer and surface roughness layer thicknesses, respectively. f_i is the volume fraction of the film that deposits within the roughness layer of the substrate.

of the solar cell fabrication. The results of Fig. 3f show an average 0.5 Å/min growth rate in the first 5 minutes of Ar flushing and an exponential reduction in rate thereafter. Because the thickness of this layer is less than that of the roughness layer on the PECVD p-type a-$Si_{1-x}C_x$:H onto which it deposits, a bulk layer does not form. The mass thickness of the layer after 55 min is given by $d_{mass}=f_i d_i+0.5d_s=0.035(67)+0.5(3)$ Å=3.8 Å. This layer most likely forms from the desorption of B-containing species from the reactor walls, followed by redeposition on the p-layer surface. A depletion of the species from the chamber accounts for the exponential reduction in the growth rate. This narrow band gap material cannot be avoided during device preparation even in a multiple-chamber system, owing to its high growth rate in the initial stage. Thus, it may be necessary to apply (i) an interface treatment such as exposing the p-layer surface to filament-generated atomic H to passivate the defects and widen the band gap, or (ii) a process procedure change such as terminating the B-containing gas a minute or more before extinguishing the plasma, thus forming a buffer layer.

i-Layer Growth

Figure 3d shows the dielectric function and optical gap of the i-layer as extracted from the RTSE data. These results show that the p-layer gap is 0.1 eV wider than the i-layer gap. Figure 3g shows the first 100 s of i-layer growth. In the first 45 s, the i-layer fills the roughness in the p-layer surface. Once the p-layer is filled ($f_i \rightarrow 0.5$) then the bulk layer forms, and increases in thickness at a rate of 1.0 Å/s. In the first ~60 Å of bulk i-layer growth, the substrate-induced roughness on the i-layer relaxes from 49 Å to 44 Å. By the end of the deposition of a 0.5 μm bulk i-layer, the surface roughness layer has decreased to 21 Å.

Solar Cell Performance

The device of Fig. 3 yielded the following initial cell characteristics: V_{oc}=0.74 V, J_{sc}=13.7 mA/cm^2, and FF=0.65 for an efficiency of 6.6%. As a reference point, devices fabricated in a multiple-chamber system under the same conditions, exhibit better initial V_{oc} (0.85 V) and similar J_{sc} (12.3 mA/cm^2) and FF (0.67) for an efficiency of 7.0%. The decrease of V_{oc} is probably caused by B-contamination expected in the initial stage of i-layer growth in the single-chamber system. Since the narrow band gap interface layer will also affect the V_{oc} and J_{sc} for cells made in both single and multiple-chamber systems, interface treatments and process procedure changes will be explored to passivate this layer, remove it, or avoid its formation. RTSE will be used to assess the efficacy of the procedures.

SUMMARY

Real time spectroscopic ellipsometry has been applied to characterize the individual layers during solar cell fabrication. The information extracted includes the surface roughness and bulk layer thickness evolution, and thus the instantaneous growth rates. In addition, the complex dielectric functions from 2.5 to 4.0 eV and the optical gaps of each of the individual layers can also be determined. A ~4 Å thick interface layer is also identified due to the high sensitivity of the optical technique. This information provides a better understanding of the relationships between the processing procedure and device performance, and thus may also lead to guidelines for optimizing device.

ACKNOWLEDGMENTS

This research has been supported by NREL under subcontract No. XAN-4-13318-03, NSF under grant No. DMR-9217169, and NEDO (Japan).

REFERENCES

[1] R.W. Collins, "Ellipsometric Study of a-Si:H Nucleation, Growth, and Interfaces", *Amorphous Silicon and Related Materials*, edited by H. Fritzsche, World Scientific Publishing, 1988, pp. 1003-1044.

[2] Y. Lu et al., "Effect of Gas Phase Hydrogen-Dilution on the Nucleation, Growth, and Interfaces of a-$Si_{1-x}C_x$:H", *Mater. Res. Soc. Symp. Proc.*, **297**, 1993, pp. 31-36.

[3] R.W. Collins, "In Situ Study of p-Type Amorphous Silicon Growth from B_2H_6:SiH_4 Mixtures: Surface Reactivity and interface Effect", *Appl. Phys. Lett.*, **53**, 1988, pp. 1086-1088.

[4] D.E. Aspnes et al., "Investigation of Effective-Medium Models of Microscopic Surface Roughness by Spectroscopic Ellipsometry", *Phys. Rev. B*, **20(8)**, 1979, pp. 3292-3302.

[5] G. Cody, "The Optical Absorption Edge of a-Si:H", *Semiconductors and Semimetals*, **21(B)**, 1984, pp. 11-81.

[6] Y. Lu et al., "Spectroscopic Ellipsometry and Atomic Force Microscopy Characterization of Surface Roughness on Amorphous Semiconductor Thin films", to be published.

[7] F.A. Rubinelli et al., "Performance and Degradation of The (p)a-SiC:H/(i)a-Si:H/(n)a-Si:H Solar Cell; A Computer Modeling Study", 22nd IEEE Photovoltaic Spec. Conf., 1991, pp. 1405-1408.

ATOMIC FORCE MICROSCOPY OF ULTRA THIN AMORPHOUS SILICON FILMS DEPOSITED ON VARIOUS SUBSTRATE SURFACES

Mitsutaka MATSUSE, Masashi KAWASAKI
and Hideomi KOINUMA

Research Laboratory of Engineering Materials, Tokyo Institute of Technology
4259 Nagatsuta, Midori-ku, Yokohama, 226, Japan

ABSTRACT

The surface morphology and interface structure of ultra-thin amorphous silicon (a-Si:H) films deposited on glass, Si, SnO_2 and graphite substrates were investigated by *in-situ* X-ray photoelectron spectroscopy (XPS) and *ex-situ* atomic force microscopy (AFM). The growth mode varied between two dimensional homogeneous coverage and three dimensional island formation depending on the substrate material, deposition temperature and reactive plasma treatment of substrate. The phenomena are interpreted by taking into account the surface diffusion of adsorbed precursors and the wettability of a-Si:H with the substrate surface. Chemical modification of substrate surface is shown to exhibit a remarkable effect on the growth mode of a-Si:H

INTRODUCTION

Hydrogenated amorphous silicon (a-Si:H) is a key material for future solar energy conversion system. The improvement in the conversion efficiency and stability are the key issues in current research. From material point of view, the microscopic structure of the amorphous network should be closely related with these problems. Since the network structure is primarily determined at the growing surface of a-Si:H films during plasma chemical vapor deposition (CVD), it is important to understand and control the surface reaction for improving the quality and stability of a-Si:H. One of the factors to control the surface reaction is the substrate temperature. The improvement of photoconductivity and the decrease of defect density in a-Si:H film are achieved by optimizing the substrate temperature. So far, the deposition of a-Si:H on such Si based amorphous underlayers as a-Si:H, a-SiN and a-SiO_2 has been extensively studied and various microscopic mechanisms of the surface reaction have been proposed. The surface bonding states of both silicon and hydrogen were analyzed by using *in-situ* infrared reflection absorption spectroscopy (IR RAS) [1] and high resolution electron energy loss spectroscopy (HREELS) [2]. The initial stage of film growth was examined by spectroscopic ellipsometory and Raman-scattering [3,4]. Matsuda et. al. proposed that the defect density in a-Si:H could be reduced by the enhancements of precursor supply and surface diffusion at high deposition temperature [5]. From these studies, the surface migration of adsorbed precursors (mainly SiH_3 radical) is considered to play a key role for controlling the film quality. However, there is no direct evidence in real space for the enhanced surface migration of adsorbed precursors at high substrate temperature. In practical devices, not only the film quality itself, but also the interface structure should dominate the performance. However, little is known about the atomic scale mechanism of interface formation relating to a-Si:H.

We have investigated the initial growth mode and surface morphology of a-Si:H on Corning #7059 glass, Si wafer, SnO_2 (transparent conductive electrode) coated glass, and highly oriented pyrolytic graphite (HOPG) by means of *in-situ* X-ray photoelectron spectroscopy (XPS) and atomic force microscopy (AFM). HOPG is especially useful for the propose, since high resolution surface structure is easily observed even in air by AFM. Microscopic pictures of the interface formation process are presented here in relation to our interest in the role of substrate in a-Si:H plasma CVD.

EXPERIMENTAL

A capacitivly-coupled rf (13.56MHz) plasma apparatus was used to deposit a-Si:H thin films from pure SiH_4 gas. The growth conditions were rf power density

First WCPEC; Dec. 5–9, 1994; Hawaii

40mW/cm^2, SiH_4 gas flow rate 3sccm, reaction pressure 100mTorr, and substrate surface temperature ranging from 30 to 230°C. We used Corning #7059 glass, crystal silicon, SnO_2 transparent conductive electrode, and HOPG as substrates. Prior to the film deposition, HOPG substrate was cleaved in high vacuum to prepare fresh surface. Other substrates were treated with Ar ions (500eV) so as to remove carbon contaminating adsorbates.

The reaction chamber was directly connected to an XPS analyzer through a gate valve as shown in Fig.1. *In-situ* XPS analysis and AFM observation in air were performed for ultra thin (0.5-20nm thick) a-Si:H films deposited on the above substrates. The thickness of a-Si:H film was determined as the product of deposition rate and time.

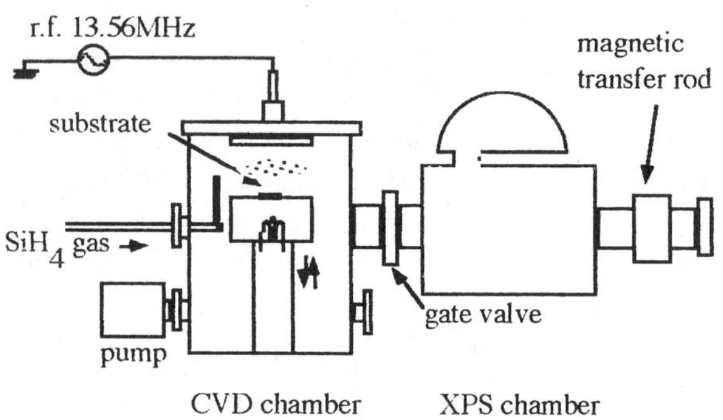

Fig.1. Schematic representation of a-Si:H film deposition and *in-situ XPS* analysis system.

RESULTS and DISCUSSION

1. a-Si:H growth on Corning #7059 and Si substrates

First, we shows the initial growth mode at 100°C on Corning #7059 glass, which is the substrate most popularly used for the film deposition. The signal intensity (I(d)) of photoelectrons generated from oxygen in the substrate and detected through the ultra thin a-Si:H film is plotted in Fig. 2 as a function of film thickness (d). The straight line expressed as

$$I(d)=I(0)\bullet\exp(-d/\lambda) \qquad (1)$$

where λ is mean free pass of photoelectron, indicates uniform coverage of a-Si:H film and atomically abrupt interface [6]. The AFM image of a film surface (d=4nm) is shown in Fig. 3. The surface is as smooth as the substrate. Thus, uniform growth of a-Si:H films occurred on glass substrate which is covalently bonded amorphous material. Such growth mode was commonly observed regardless the substrate temperature ranging from 30 to 230°C. Essentially the same surface morphology was observed on a-Si:H films deposited on Si substrates.

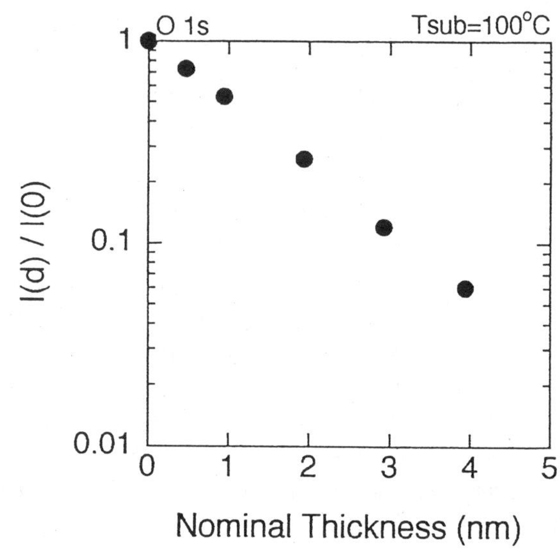

Fig.2. Photoelectron intensity for O 1s core level in the glass substrate covered with a-Si:H layers of various thicknesses.

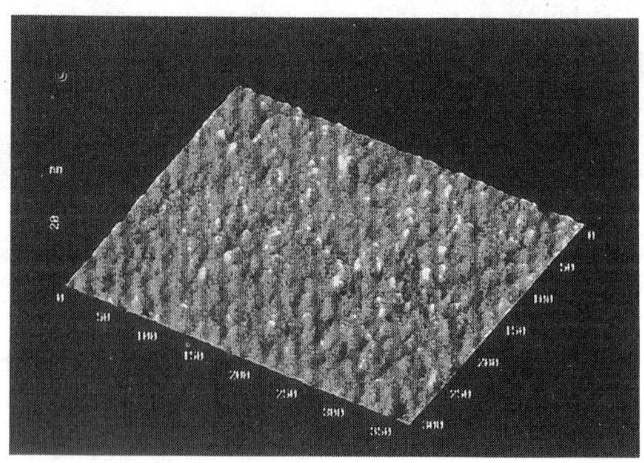

Fig.3. The AFM image (350x300nm^2) of a-Si:H on Corning #7059 glass substrate at a substrate temperature of 100 °C.

2. a-Si:H growth on SnO_2 coated glass substrate

Figure 4 shows the AFM images of 4nm thick a-Si:H films deposited on SnO_2 coated glass substrates. At a low substrate temperature (110°C), the surface is flat and smooth (Fig. 4(a)). At a high substrate temperature (230°C), three dimensional growth takes place to form islands with their diameter and height of about 50nm and 5nm, respectively (Fig. 4(b)). The above experimental results shown in Fig. 4 can be understood as follows. When a precursor adsorbed weakly on the surface can migrate long enough to reach a-Si:H nucleus achieved to a defect, it will be incorporated in the nucleus. If the nucleus is not wettable with the substrate, island-like nucleation and growth should take place to reduce the interface energy. This is presumably the case because SnO_2 is an ionic crystal and bonding nature is quite different from that of a-Si:H. If the mobility of the adsorbed species is restricted due, for instance, to the decrease in temperature, as in the case of 110°C deposition on SnO_2 substrate, the impinging precursors are frozen to form apparently homogeneous coverage of the surface.

(a)

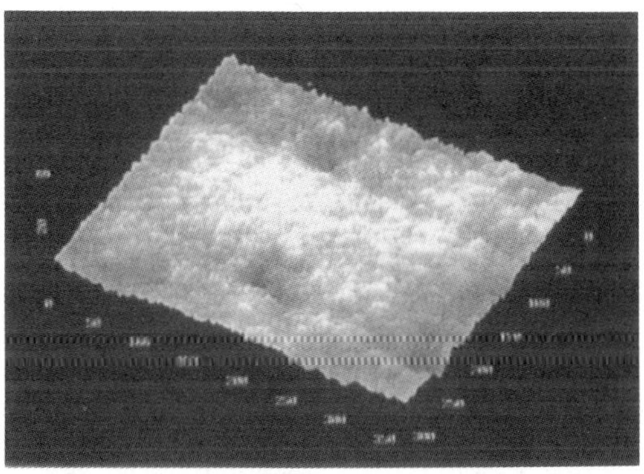

(b)

Fig.4. The AFM images (350x300nm^2) of a-Si:H on SnO_2 coated glass substrates at substrate temperatures of (a) 100°C and (b) 230°C.

3. a-Si:H growth on HOPG substrate

Now, we show the variation of the initial growth mode of a-Si:H on HOPG depending on the deposition temperature. The AFM images for the films deposited at 110 and 230°C are depicted in Fig. 5(a) and (b), respectively. The morphology of these films changed similarly to the deposition on SnO_2. The clean surface of HOPG should be covered by π orbitals of carbon atoms and quite different from the surface of other materials studied in this paper. The initial growth of a-Si:H is apparently dominated by the homogeneous nucleation of islands at 110°C. At 230°C, a-Si:H deposits with their heights higher than 1 nm are arrayed along the steps. The surface coverage by a-Si:H deposits decreased and

(a)

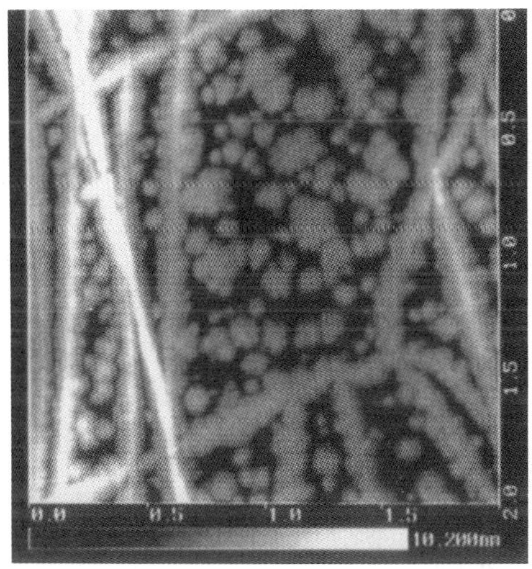

(b)

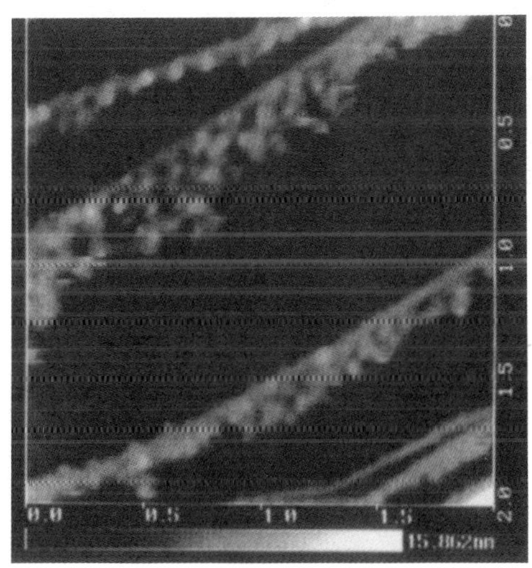

Fig.5. The AFM images (2x2μm^2) of a-Si:H on HOPG at substrate temperatures of (a) 110°C and (b) 230°C.

the ratio of island height to island width increased with increasing the substrate temperature. These observations clearly indicate that the surface migration length of adsorbed precursors increased with the increase of substrate temperature.

The initial growth mode of a-Si:H on HOPG substrate was found to change drastically when the substrate surface was exposed to H_2 plasma prior to the a-Si:H deposition. Figure 6 shows the AFM image of a-Si:H deposited at 110°C on HOPG treated with H_2 plasma. By comparing with Fig. 5. (a), we can see smoother and more homogeneous coverage of HOPG by a-Si:H overlayer. Thus, the initial growth mode is extremely sensitive to the chemical state of substrate surface.

(a)

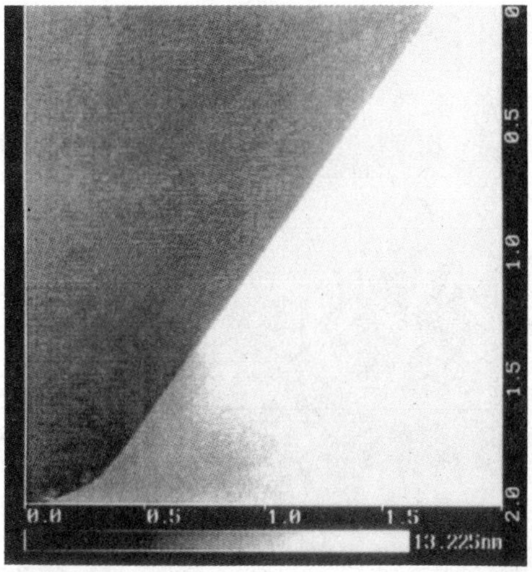

(b)

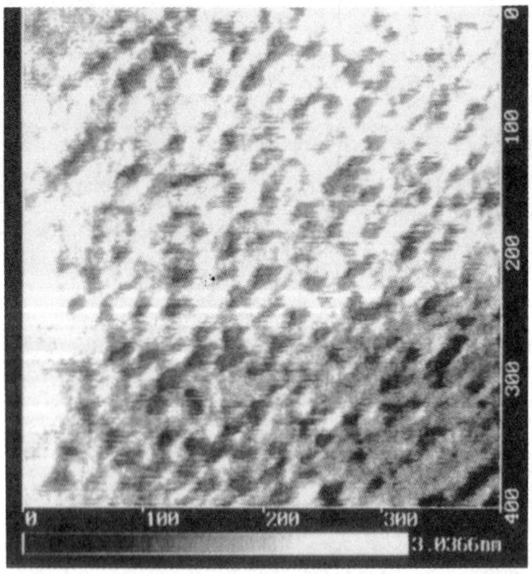

Fig.6. The AFM images of a-Si:H on HOPG substrate treated with H_2 plasma. (a) Scan area: 2x2μm^2 (b) Scan area: 400x400nm^2. The deposition temperature was 110°C.

CONCLUSION

We have used AFM to study the initial growth mode and the surface morphology of a-Si:H overlayers on various substrate surfaces. The surface morphology drastically changed depending on the material and chemical state of substrates as well as on the deposition temperature. The observed behaviors are considered to reflect the mobility of precursors adsorbed on the surface. A remarkable change was observed for the growth of a-Si:H on HOPG when HOPG was treated with H_2 plasma. The substrate surface plays a significant role at least in the initial growth stage of a-Si:H.

REFERENCES

[1] Y. Toyoshima, K. Arai, A. Matsuda, and K. Tanaka, *Appl. Phys. Lett*, **57**, 1990, pp. 1028

[2] H. Sasaki, M. Deguchi, K. Sate, and M. Aiga, *J. Non. Cryst. Solids*, **114**, 1989, pp. 672

[3] B. Drevillon, *J. Non. Cryst. Solids*, **114**, 1989, pp. 139

[4] H. Tanino, G. Ganguly, and A. Matsuda, *Phys. Rev. B*, **46**, 1992, pp. 15277

[5] G. Ganguly, and A. Matsuda., *Phys. Rev. B*, **47**, 1993, pp. 3661

[6] M. Kawasaki, Y. Matsuzaki, and H. Koinuma, *Phys. Rev. B*, **39**, 1889, pp. 13316

CORRELATION BETWEEN IMPROVED STABILITY AND MICROSTRUCTURAL PROPERTIES OF a-Si:H AND a-Ge:H

T. Haage, S. Bauer, B. Schröder, H. Oechsner

Fachbereich Physik und Forschungsschwerpunkt Materialwissenschaften, Universität Kaiserslautern, Postfach 3049, D-67653 Kaiserslautern, F. R. Germany Tel.: (49)-631-205-4050, Fax: (49)-631-205-2996

ABSTRACT

In this paper we report on comparative investigations concerning the stability of glow discharge and hot wire deposited a-Si:H films. Using spectroscopic ellipsometry measurements and a data interpretation basing on a tetrahedron model we investigated the microstructural properties of the films. We found that the stability of a-Si:H films is significantly increased by a deposition process guided in a way that high quality material can be grown with high density and low hydrogen content. The a-Si:H results are confirmed by similar results obtained on sp-a-Ge:H. For the first time we have shown that metastability is not connected with the existence of surface like (SiH_2, $(SiH)_x$) hydrogen bonding configurations.

INTRODUCTION

In the last few years many efforts have been made to improve the stability of hydrogenated amorphous silicon (a-Si:H) and the quality of germanium (a-Ge:H) films. The hot-wire-technique enables the deposition of a-Si:H with device quality and low hydrogen content [1,2]. Recently we reported that these films show an improved stability against keV-electron irradiation [2] which can now also be confirmed in this paper by light soaking experiments. Since it was demonstrated that a-Si:H material deposited by the hot wire technique can be used for solar cell application [2,3] there opens up a possibility to prepare more stable solar cell collector material. Changes in microstructure were suggested to influence the stability of the hot wire material in a series of papers [4,5]. In this study we present a detailed comparison of the structural properties and the degradation behaviour of glow discharge (gd)- and hot wire (hw)-a-Si:H and show the first results on a-Ge:H, which seem to support the a-Si:H results as well as stability calculations presented. Spectroscopic ellipsometry (SE) has been used as a nondestructive method to determine the density and the void volume fraction of the thin films [6]. The last mentioned quantities are used in connection with the knowledge about hydrogen incorporation to characterize the microstructure. The resulting study should provide a deeper understanding of the degradation process and therefore show how to improve stability of amorphous semiconductors.

SAMPLE PREPARATION AND CHARACTERISATION

The a-Si:H samples were grown on Corning glass or c-Si substrates using two different deposition techniques. Glow discharge (gd) films were deposited in a capacitively coupled conventional rf glow discharge reactor operated at 13.56 MHz. Additionally, high quality a-Si:H films have been deposited using thermal decomposition of silane at a heated tungsten filament. This hot wire (hw) method is described in detail elsewhere [2]. For both techniques, mainly the variation of the substrate temperature results in a change of hydrogen incorporation as well as the film's microstructure. High quality a-Ge:H samples were deposited by dc magnetron sputtering in an argon-hydrogen atmosphere. The defect properties of our samples have been investigated by photothermal deflection spectroscopy (PDS) and photoconductivity spectroscopy in the constant photocurrent mode (CPM). The hydrogen content c_H of the samples has been determined by Fourier transform infrared (FTIR) spectroscopy. Metastability of the a-Si:H and a-Ge:H films was investigated by keV-electron irradiation which has been demonstrated to be a powerful tool for fast degradation tests [7]. The results of these investigations were confirmed by light soaking experiments.

ANALYSIS OF ELLIPSOMETRIC DATA

Ellipsometry measures the ratio of the complex Fresnel reflection coefficients for light polarized parallel and perpendicular to the plane of incidence. From this quantity the real and imaginary part of the dielectric function $\varepsilon_1(E)$ and $\varepsilon_2(E)$ are calculated as a function of photon energy E [6]. Fits to the SE data were performed using a three layer structure consisting of substrate/bulk/surface overlayer. The SE data interpretation was based on the tetrahedron model in order to take into account the effect of hydrogen incorporation. For details we refer to Mui and Smith[8] as well as to our previous work [6]. The surface roughness has been modelled as a density deficient overlayer consisting of bulk material with an increased void volume fraction. Since the measurements on gd-a-Si:H were performed immediately after deposition in the ultra high vacuum chamber, the presence of an oxide overlayer can be excluded for the gd-samples. On the other hand we assumed the existence of an oxide overlayer in the case of the hw-samples which have

CH3365-4/94/0000-0429 $4.00 © 1994 IEEE

First WCPEC; Dec. 5–9, 1994; Hawaii

been investigated by *ex situ* measurements in contrast to the analysis of our gd-films.

We have calculated the film densities using the results of the SE data analysis. The derivation was based on the volumes and atomic masses associated with each Si-centered tetrahedron along with the volume fractions of the tetrahedra and voids. The obtained film densities have been shown to be in very good agreement with the densities determined by broad band phonon spectroscopy [6], a method which enables the most accurate measurement of thin film density known to us.

RESULTS AND DISCUSSION

In Fig. 1 stability results are presented for differently prepared and differently degraded a-Si:H films. It is clearly demonstrated that the metastable defect density ΔN induced either by keV-electron irradiation (saturation defect density ΔN_{sat} [7]) or by light soaking (defect density ΔN_{1000} obtained after 1000h exposure to AM1 illumination) strongly depends on the hydrogen content c_H. However, since the different c_H mainly results from different preparation temperatures, the changes of c_H are also correlated with different film growth which should lead to different microstructures. But there are also significant differences in the $\Delta N_{sat} = f(c_H)$ dependence for gd- and hw-a-Si:H films. For hw-films ΔN_{sat} obviously decreases much stronger with c_H than for gd-films. The values of ΔN_{sat} and ΔN_{1000} reported here are measured with the photoconductivity spectroscopy (CPM). CPM measurement was used since some hw-films are quite inhomogeneous, but the general results are confirmed by PDS (Photothermal Deflection Spectroscopy) measurements, too.

The experimental results have been supported by model calculations carried out in our group [7] to explain the saturation value ΔN_{sat} of metastable defect creation after keV-electron irradiation in terms of film properties. It has been shown that ΔN_{sat}- values calculated by a model basing on an irradiation disturbed thermodynamic equilibrium agree very good with the measured values of ΔN_{sat} [7]. The results of the theoretical calculations can be summarized as follows: For device grade a-Si:H (Urbach energy $E_0 \approx 50$ meV) the stability mainly depends on c_H, while ΔN_{sat} for a-Ge:H depends likewise on c_H as well as on E_0. Both predictions of the model calculations are in very good agreement with the measured ΔN_{sat} values. Since the c_H and E_0 dependence of the stability is understandable in terms of defect creation models, the differences observed for gd- and hw-a-Si:H films have to be explained. Therefore the correlation between stability and microstructural properties of the differently prepared films provided by the SE data analysis has been investigated.

As shown in Fig. 2, the film density varies significantly with the temperature controlled hydrogen content c_H. Increasing hydrogen content leads to a reduced film density for all samples. With respect to the photoelectri-

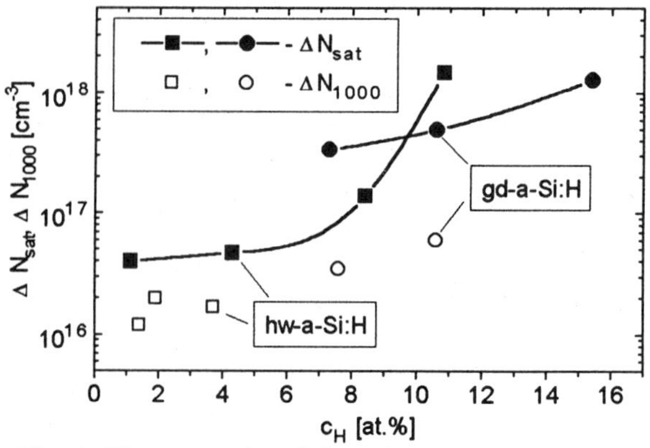

Fig. 1: The saturation defect density ΔN_{sat} after keV-electron irradiation (full symbols) and the defect density ΔN_{1000} after 1000h exposure with AM1 illumination (open symbols) of gd-a-Si:H (circles) and hw-a-Si:H (squares) films versus the hydrogen content c_H.

cal properties (defect density) glow discharge films have not been deposited at substrate temperatures above 370°C, whereas the hot wire technique yields a-Si:H material with low defect densities even at $T_S \approx 500$°C. We assume the surface mobility of the adsorbed precursors of a-Si:H deposition to influence film growth decisively, as suggested by Tanaka and Matsuda [9]. The substrate temperature along with the hydrogen coverage of the growing surface which decreases with increasing T_S control the surface mobility. Therefore, we expect a densification due to an enhanced mobility with increasing substrate temperature. But this densification will be partially compensated at high substrate temperatures, since the hydrogen coverage decreases with increasing T_S. This effect is confirmed by our experimental results which yield a saturation of the film density at high substrate temperature and correspondingly low hydrogen content, as shown in Fig. 2.

The different structural properties of the a-Si:H films can be characterized as follows: At a certain substrate temperature a-Si:H films prepared by the hw technique exhibit a smaller density as well as a smaller hydrogen content compared to gd-a-Si:H (see Fig. 2 and Table 1). The growth conditions resulting from the individual preparation techniques are quite different. In contrast to the plasma deposition the hot wire deposition is a process in which only particles with thermal energy are involved. Moreover, as can be seen from table 1, for the deposition of hw-a-Si:H the growth rate r has been larger (in the range of 4.0 to 5.7 Å/s) compared to the glow discharge deposition (r varies between 1.7 and 3.2 Å/s). These differences in the growth rate may also be a possible source for the different densities of hw- and gd-a-Si:H described above, since we expect a densification of the a-Si:H network with a decreasing growth rate.

To fit the ellipsometrical results of less dense material the introduction of a void volume fraction f_V was necessary for data evaluation. However, as we know

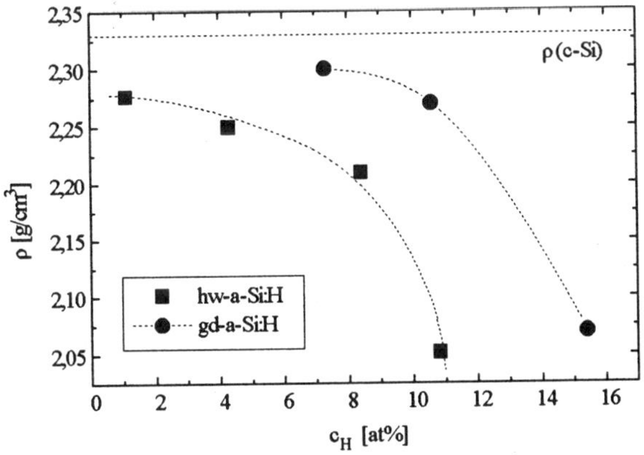

Fig. 2: Dependence of the film density ρ determined by spectroscopic ellipsometry on hydrogen content c_H for a-Si:H films prepared by the gd and hw technique.

from FTIR-spectra of low dense hw-a-Si:H (see also Fig. 4) no IR-absorption line at 2090 cm^{-1} must be present for such less dense material. From this fact it can be concluded, that this less dense material not really contains larger voids, since such larger voids would be connected either with a high dangling bond density or with the presence of surface like ($(SiH)_x$ or SiH_2) Si-H bonding configuration leading to an absorption around 2090 cm^{-1}. Therefore we have to conclude that less dense material with low defect density has a more loose network or microvoids. We observed a considerable increase of the void volume fraction f_V with increasing hydrogen content. Therefore hydrogen incorporation along with a change of the temperature-controlled growth process leads to a less dense network.

In the following we describe the correlation between the structural properties and the degradation behaviour of a-Si:H films. We take the saturation value of the created metastable defects $\Delta N_{sat}=N_{sat}-N_0$ after keV-electron irradiation as a measure for the stability of our films (N_0 is the initial (stable) defect density). The saturation defect densities are summarized in Table 1, too. As shown in Fig. 3, stability (indicated by reduced values of ΔN_{sat}) in general increases with decreasing void fraction (and correspondingly increasing density ρ) as well as with decreasing hydrogen content. However, the hydrogen content does not exclusively determine the degradation behaviour, as indicated by the different saturation defect densities we observed e.g. for gd- and hw-a-Si:H with c_H= 10.8 at.% (see Fig. 3).

Additionally, we examined the hydrogen bonding configuration by FTIR-spectroscopy. The IR-absorption spectra in the range of 2000 cm^{-1} to 2090 cm^{-1} show that the films contain predominantly isolated Si-H bonds except for the gd-sample with the highest hydrogen content (c_H= 15.4 at.%) (see Fig. 4).

In addition to the clear differences in the dependence $\Delta N_{sat} = f\,(f_V)$ between gd- and hw-a-Si:H films, which will

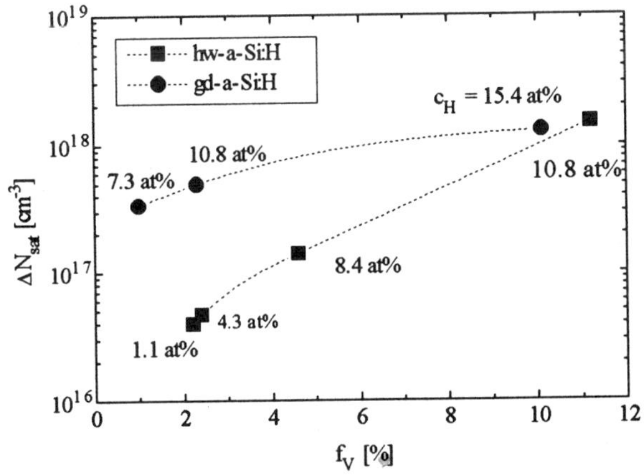

Fig. 3: Saturation defect density ΔN_{sat} (determined by CPM) as a function of the void volume fraction f_V for a-Si:H films prepared by gd- and hw-technique

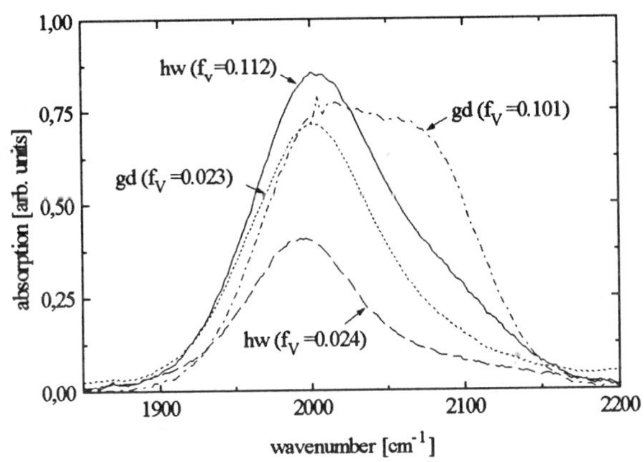

Fig. 4: Infrared spectra of hw- and gd-samples.

be discussed later, it is seen in Fig. 3 and Table 1 that a gd-sample with large hydrogen content (c_H = 15,4 at%) has about the same stability as a hw-sample containing much less hydrogen (c_H = 10,8 at%) while both samples show about the same f_V. As indicated from the IR-spectra in Fig. 4 the gd-sample obviously contains a large fraction of hydrogen bonded in SiH_2- or $(SiH)_x$-configurations (absorption near 2090 cm^{-1}). If metastability would be connected with the surface like bonding configurations an inherent instability of this sample is expected. Since this is not the case and instability is rather connected with the amount of disolved/isolated monohydrid for films with the same density or void fraction f_V, we suppose that metastability is not directly connected with the existence of SiH_2 or $(SiH)_x$-bonding configurations.

Coming back now to the differences between hw- and gd-a-Si:H films being obvious from Fig. 3 and the fact that instability is connected with film density. Due to the weak bond - dangling bond conversion model stressed (weak) Si-Si and nearby Si-H bonds are involved in metastability. In a less dense film the bond reconstruction necessary to obtain a defect poor network obviously leads to more weak (stressed) bonds giving rise to metastability. This interpretation which can explain the $\Delta N_{sat} = f(f_V)$ dependence in Fig. 3 is also correlated with the observation we made, that the more dense a-Si:H and a-Ge:H film show a smaller Urbach energy E_0, which characterizes the topological disorder. If we add more hydrogen to films with a given density (f_V) it seems very clear that we increase the stress in the film. Since gd-a-Si:H films contain much more hydrogen at a given density or void fraction compared to hw-films (see Fig.3 and table1) it is not surprising that they show a higher instability.

To deposit more stable amorphous hydrogenated semiconductor films it seems necessary to guide the deposition process in such a way that a low defect density and a dense network is obtained together with a low hydrogen content. As mentioned already earlier in this chapter the growth and the densification process must be controlled by the deposition parameters T_S, r_{dep}, bombardement, surface mobility of adsorbed precursors etc. in this direction.

The results obtained on a-Si:H films can be confirmed by the results measured on sp-a-Ge:H films. It should be remembered that a-Ge:H films need a powerful (sputtered or cathodic CVD) deposition to get good photoelectrical quality. We have prepared a-Ge:H films in a magnetron sputtering process under improved bombardment conditions (low $p_{dep} \cdot d_{TS}$-product, p_{dep} = deposition pressure, d_{TS} = target-substrate-distance) [10]. We obtained a-Ge:H films with a higher density than cathodic CVD films, a lower hydrogen content, smaller Urbach energies, smaller SAXS void volume and in consequent correlation to the results reported for a-Si:H films before, a much higher stability. The minimum values ΔN_{sat} observed for the most dense sp-a-Ge:H film, was 3 x 10^{17} cm^{-3} [10], which was a factor of four lower than the value of ΔN_{sat} we measured for the most stable gd-a-Ge:H film, a sample we obtained from W. Paul's group from the Harvard University.

SUMMARY

The ellipsometric studies reveal a different microstructure of a-Si:H deposited by the hot wire technique compared to the glow discharge samples. The film density varies significantly with substrate temperature and hydrogen content depending on the individual preparation technique. Microstructural properties are shown to determine the degradation behaviour of a-Si:H films. Our comparing studies reveal that only films which exhibit a high density as well as a low hydrogen content show an improved stability. For the first time we have shown that metastability is not connected with the existence of $Si\text{-}H_2$ bonding configuration. The a-Si:H results can also be confirmed by first measurements on high quality a-Ge:H films. Sputtered a-Ge:H films with high density and low hydrogen content show a strongly improved stability.

Tab. 1: Survey of substrate temperature T_S, hydrogen content c_H, deposition rate r_{dep}, densities ρ void, volume fractions f_V and the saturated defect density ΔN_{sat} for the different a-Si:H samples under investigation.

	hw-a-Si:H			
T_S (°C)	525	500	275	200
c_H (at.%)	1.1	4.3	8.4	10.8
r_{dep} (Å/s)	5.7	4.4	5.4	4.0
ρ (g/cm³)	2.28	2.25	2.21	2.05
f_V	0.022	0.024	0.046	0.112
ΔN_{sat} (cm^{-3})	$3 \cdot 10^{16}$	$4 \cdot 10^{16}$	$1.4 \cdot 10^{17}$	$1.5 \cdot 10^{18}$

	gd-a-Si:H		
T_S (°C)	370	250	150
c_H (at.%)	7.3	10.8	15.5
r_{dep} (Å/s)	1.7	2.6	3.2
ρ (g/cm³)	2.30	2.27	2.07
f_V	0.010	0.023	0.101
ΔN_{sat} (cm^{-3})	$3 \cdot 10^{17}$	$5 \cdot 10^{17}$	$1.3 \cdot 10^{18}$

Therefore, to deposit more stable a-Si:H or a-Ge:H films it seems to be necessary to guide the growth process in a way that low defect density and a dense network is obtained together with a low hydrogen content.

ACKNOWLEDGEMENTS

The authors are grateful to P. Papadopulos and B. Müller for providing them with samples. This work has been financially supported by the German Ministry of Research and Technology.

REFERENCES

[1] A.H. Mahan, J. Carapella, B.P. Nelson, R.S. Crandall, and I. Balberg, *J. Appl. Phys.* ***69***, 1991, p. 6728

[2] P. Papadopulos, A. Scholz, S. Bauer, B. Schröder, H. Oechsner, *J. Non-Cryst. Solids* ***164-166***, 1993, p. 87

[3] B.P. Nelson, E. Iwaniczko, R.E.I. Schropp, A.H. Mahan, E.C. Molenbrock, S. Salomon, and R.S. Crandall, *Proc. 12th ECPVSEC*, Amsterdam 1994, p. 679

[4] M. Vanecek, A. H. Mahan, B. P. Nelson, and R. S. Crandall, *Proc. 11th ECPVSEC*, Montreux 1992, p. 96

[5] M. Vanecek, Z. Remes, J. Fric, R. S. Crandall, A. H. Mahan, *Proc. 12th ECPVSEC*, Amsterdam 1994, p. 354

[6] T. Haage, U.I. Schmidt, H. Fath, P. Hess, B. Schröder, H. Oechsner, *J. Appl Phys.* ***76***, 1994, p. 4894

[7] A. Scholz and B. Schröder, *J. Non-Cryst. Solids* ***137-138***, 1991, p. 259

[8] K. Mui and F. W. Smith, *Phys. Rev. B* ***38***, 1988, p. 10623

[9] K. Tanaka and A. Matsuda, *Mater. Sci. Rep.* ***2***, 1987, p. 139

[10] B. Müller, *Diploma thesis*, Universität Kaiserslautern, 1993, to be published

H_2 DILUTED LOW TEMPERATURE DEPOSITED a-Si:H, INITIAL DEGRADATION AND MODELING

A. Rothwarf, Electrical and Computer Engineering Department, Drexel University, Philadelphia, PA 19104

Y. Lubianiker and I. Balberg, The Racah Institute of Physics, the Hebrew University, Jerusalem, 91904, Israel

R.R. Arya and J. Keane*, Solarex, Thin Film Division, Newtown, PA 18940

ABSTRACT

In a previous study of the hydrogen diluted low temperature deposited (HDLTD) a-Si:H materials no difference was found between the phototransport properties of these materials and those of the standard "undiluted" materials deposited at a substrate temperature of 260°C. However, solar cells made of HDLTD materials exhibited higher open circuit voltage, initially faster light-induced degradation, followed by saturation of this degradation. Here we report a study of the initial degradation of the standard and the HDLTD materials using the steady state photocarrier grating (PCG) technique. The results reveal differences in the degradation of the minority carrier, mobility-lifetime product, $(\mu\tau)_h$, that are consistent with the degradation of the cells. However, the degradation of the majority carrier product, $(\mu\tau)_e$, of the two materials is the same. Our model suggests that the probable cause for the higher V_{oc} and saturated degradation in HDLTD material is a lower density of valence band tail states.

INTRODUCTION

Bennett et al.[1] reported recently, that a-Si:H solar cells produced by using hydrogen diluted silane deposited at low temperatures, showed a significant increase in the open-circuit voltage, and improved long-term degradation properties, compared to their usual cells. The latter cells are made with undiluted silane and are deposited with a substrate temperature, T_S, of about 260°C. The advantage of the higher V_{oc}, and saturated degradation of the HDLTD cells, is that these cells have a higher efficiency than the standard cells, which continue to degrade at long exposure times. While it had been demonstrated previously that low temperature deposition of "undiluted" films and cells produced both a higher band gap and a higher V_{oc}, prior experience also showed that such low temperature deposited "undiluted" films and cells degraded more rapidly than the standard material. Attempts[1] to identify how the HDLTD material differed in its properties from standard material were not successful.

In this paper we report a comparison between the standard and HDLTD i-layer materials using the PCG technique. Both the minority and majority mobility-lifetime, $\mu\tau$, products were determined as a function of degradation time, with either red light irradiation in a wavelength range between 6300 and 7000A°, or with 5145 A°, Ar^+ laser irradiation. Both were with nominal intensities of 100 mW/cm^2. Previous studies on cells showed that with degradation under Na vapor lamps, the HDLTD cells degraded more rapidly than the standard material during the first 20 hours, and then appeared to saturate.[1]

It was previously inferred from the IR spectra of the HDLTD films that they have less hydrogen in them than the standard films. Bennett et al.[1] proposed without providing supporting evidence that the higher V_{oc} was associated with an improved p-i interface. Below, we will present arguments that show how such an improved p-i interface is related to the hydrogen content, and to both the faster initial degradation and long term saturation of the degradation.

PREPARATION OF FILMS AND CHARACTERIZATION

The undoped a-Si:H film used in the present study were deposited under exactly the same conditions as the cells described in Ref. 1. Our films were deposited on 7059 Corning glass and are 1 μm thick. As in Ref. 1, the i-layers were deposited without dilution or with H_2 diluted SiH_4, using substrate temperatures of 260°C and 210°C respectively. The degradation of the samples was achieved as described above and was interrupted only for the PCG measurements

The characterization of the phototransport properties of the films was carried out using the PCG technique that was explained in detail previously.[2,3] Applying this technique, the values of the majority carrier i.e. the electron $(\mu\tau)_e$ products were derived simply from the photoconductivity, while the corresponding results for the minority carrier, i.e. the hole $(\mu\tau)_h$ products were derived simply from the application of the PCG technique[2,3]. The illumination used in the PCG measurements was from a He-Ne laser with an illumination intensity of 8 mw/cm^2.

EXPERIMENTAL RESULTS

In Fig. 1. we show the $(\mu\tau)_e$ and $(\mu\tau)_h$ dependences on the degradation time under red light exposure for the standard material. The $(\mu\tau)_e$ of this material degrades at a rate of $t^{-0.16}$, while there is no degradation within our experimental accuracy for $(\mu\tau)_h$. In

First WCPEC; Dec. 5–9, 1994; Hawaii

Figure 2 we show the same quantities for the HDLTD material. For this material $(\mu\tau)_e$ degrades at a rate of $t^{-0.15}$, while $(\mu\tau)_h$ degrades finitially (up to 10^4 sec) at a rate of $t^{-0.04}$, and then at a faster rate of $t^{-0.12}$. The striking feature of these results is that for the $(\mu\tau)_e$ products, exactly the same exponent of the degradation is found for the two materials. This is not necessarily the most common degradation rate reported in the literature[4] but is

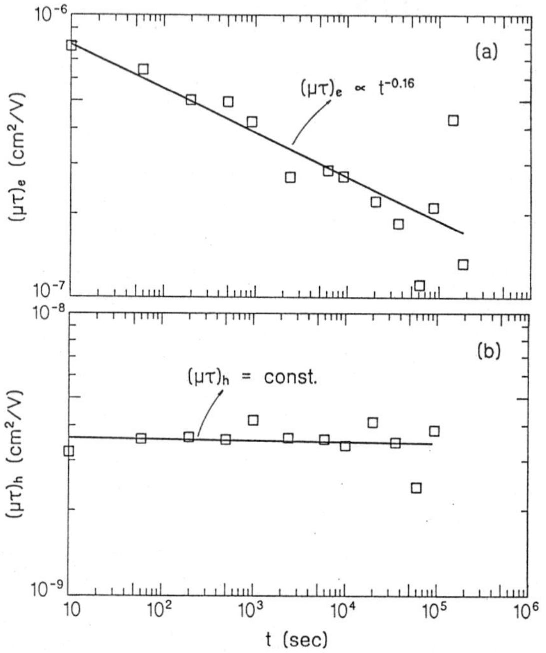

Fig. 1. $\mu\tau$ products as a function of degradation time under red light (100 mw/cm^2) for standard i-layer a-Si:H material.

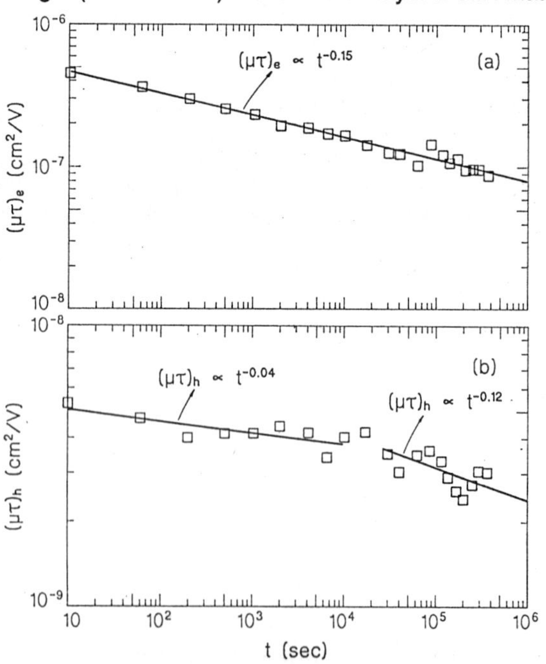

Fig. 2. $\mu\tau$ products as a function of degradation time under red light (100 mw/cm^2) for HDLTD i-layer a-Si:H material.

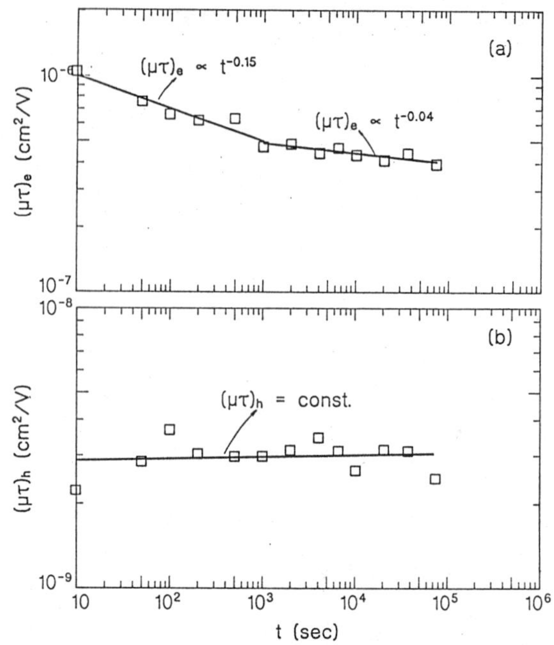

Fig. 3. $\mu\tau$ products as a function of degradation time under green (Ar^+) laser light (5145A°, 100 mw/cm^2) for standard i-layer a-Si:H material.

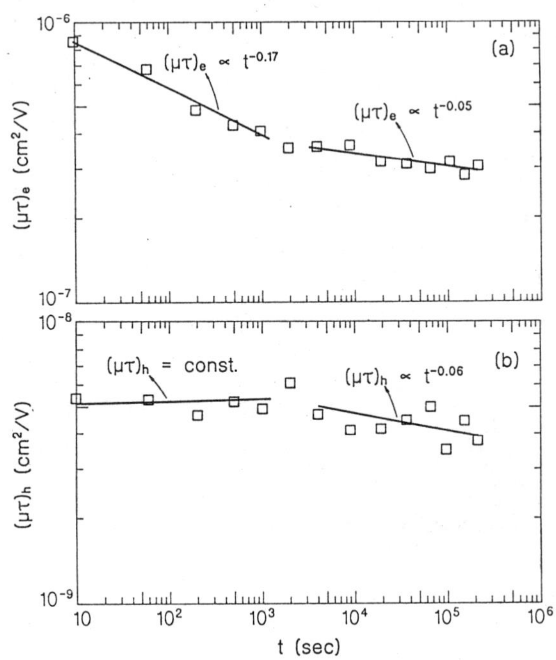

Fig. 4 $\mu\tau$ products as a function of degradation time under green (Ar^+) laser light (5145A°, 100 mw/cm^2) for HDLTD i-layer a-Si:H material.

definitely within the accepted range.

In Fig. 3 and Fig. 4 we show the same type of results, but these were obtained from measurements of the degradation under Ar^+ laser irradiation of 100 mW/cm^2. The results for the material deposited with "no dilution" are shown in Fig. 3. These results demonstrate that the $(\mu\tau)_e$ product degrades with a time exponent of -0.15 up to 1000 sec and then with a time exponent of 0.04. The $(\mu\tau)_h$ product is, within experimental error, independent of irradiation time within the time interval studied. In contrast, the results for the HDLTD material shown in Fig. 4 indicate, again, that while $(\mu\tau)_e$ degrades similarly to the standard materials (with a time exponent 0.17, in the first 1000 sec. and then as $t^{-0.05}$), $(\mu\tau)_h$ does not degrade in the first 1000 sec. and then degrades as $t^{-0.06}$ roughly parallel to $(\mu\tau)_e$.

ANALYSIS OF EXPERIMENTAL RESULTS

Both our results and those in the literature show differences between the initial rates of degradation of the $\mu\tau$ products for electrons and holes[5]. In addition, we find differences between the corresponding degradation of the standard material and the HDLTD material for $(\mu\tau)_h$. We present below a self consistent picture that may be useful for interpreting both our data and that of others.

In Fig. 1a for standard material, the degradation of $(\mu\tau)_e$ with exposure time to red light is shown on a log-log scale up to 10^5 sec. It follows a $t^{-0.16}$ dependence, while in Fig. 1b, $(\mu\tau)_h$ shows, in agreement with other reports[5], no degradation on this time scale. Several phenomena are occurring that can account for this behavior. First, dangling bonds are being created from valence band tail states. The creation of extra states below the Fermi level lowers the Fermi level.[6] The lower Fermi level decreases the effective mobility of the electrons, and hence $(\mu\tau)_e$. For the holes, as long as the density of the valence band tail states N_{ts} times their capture cross-section, σ_{ts}, remains much larger than the density of the dangling bonds N_{db} times their capture cross-section, σ_{db}, $(\mu\tau)_h$ will remain constant. At longer degradation times $(\mu\tau)_h$ will show a time dependence[5] equal to the creation rate of dangling bonds , which will dominate both $(\mu\tau)_e$ and $(\mu\tau)_h$ at long degradation times.

In Fig. 2 we see the dependence on red light degradation time of $(\mu\tau)_e$ and $(\mu\tau)_h$ for the HDLTD material. Here, $(\mu\tau)_e$ shows the same behavior as the standard material, while $(\mu\tau)_h$ shows an immediate slow decrease, and an early onset of the steep degradation. A consistent picture for this comes by assuming that the HDLTD material has a lower total density of valence band tail states (VBT), so that the density of dangling bonds has an immediately observable effect on $(\mu\tau)_h$ for the intensity of the testing light. With a low density of VBT states at a given intensity, trapped holes quickly saturate these states, and recombination through the dangling bonds dominates. For a large density of VBT states saturation doesn't occur and hence the dangling bond recombination path remains low until the density of dangling bonds has reached the value for which $\sigma_{dB}N_{dB} = \sigma_{ts}N_{ts}$.

In Figs. 3 and 4 the dependence on degradation time of the $(\mu\tau)$ products under green light degradation is shown. Here, we have apparently a difference in the degradation of the $(\mu\tau)_e$ products in comparison to the red light case. This apparent difference arises from the fact that the green light is absorbed more strongly than the red light, and hence the degradation occurs more intensively near the top surface. As the resistance of the top layer increases it may be considered to drop out of the photoconductivity path since it is "connected in parallel with the deeper, less degraded, material. A further complication is that with different degrees of degradation as a function of depth, a built-in voltage develops from the Fermi level shift with degradation, that tends to push electrons toward the bottom of the film and holes towards the illuminated top layer of the film. In solar cells such an effect might actually retard the degradation of a cell with the light entering through the p-side and speed the degradation of one illuminated through the n-side.

The fact that the $(\mu\tau)_e$ products for both standard and HDLTD materials behave in the same way for the green light implies that the processes are essentially the same in the two materials. To account for the reduction in the $(\mu\tau)_e$ degradation under green light after $\approx 10^3$ sec., it is required that the rate of degradation under the penetrating residual low light intensity, is smaller. This residual light is expected to dominate the long term decay. This is consistent with the known behavior[4]. The $(\mu\tau)_h$ result under green light degradation shows no difference from the red light degradation for standard material (both show no degradation), while for the HDLTD material a somewhat slower degradation under green light is observed. This is consistent with the fact that the characterization is based on the PCG technique, which applies a red light that senses at a depth that has been degraded by the residual lower intensity light. While the $(\mu\tau)_h$ under green light for the standard material shows no change, it is expected that if the degradation is extended to longer times $(\mu\tau)_h$ will remain constant for a longer time than under red light degradation

CELL RELATED MODELING

The results derived by the PCG technique show that the minority carrier $(\mu\tau)_h$, product initially degrades faster in the HDLTD material, and this is consistent with the faster initial degradation of cells made of this material, which is manifested by the degradation of the solar cell fill-factors and efficiency. On the other hand there is no correlation with the behavior of the $(\mu\tau)_e$ product. These two results constitute a strong confirmation of the limiting carrier model of cell efficiency[7], that attributes the cell performance to the phototransport properties of the minority carrier. These results also establish a physical difference between the HDLTD and standard materials that was not previously detected.

Turning to the reasons for the observed behavior of the cells, we note that Rothwarf[8] has pointed out that the number of weak tail state bonds near the p-i interface sets the number of dangling bonds that will be present there, in equilibrium. In the HDLTD material the density of such weak interface bonds is expected to be reduced for two reasons. The first is that this material has less hydrogen, and hence potentially less stress with fewer stressed bonds that constitute the valence band tail states.

It is these stressed bonds that are the sites of potential dangling bonds. Second, by keeping the material at a lower temperature there is less likelihood that an existing weak bond with a hole in it will break, and produce dangling bonds. This second aspect is consistent with previous studies of the dependence of V_{oc} on deposition temperature by Takahama et al.[9] They found that the thermal history of the p-i interface was important, and that with higher deposition temperatures the density of defects at the p-i interface increased.

The relation of V_{oc} to the density of dangling bonds, at the p-i interface, has previously been treated by Rothwarf.[8] One finds that V_{oc} is given by:

$$qV_{oc} = qV_D - kT \ln(J_{Lo}/qN_cS) \quad (1)$$

with V_D the built-in (or diffusion) voltage in the i-layer and J_{Lo} is the light-generated current density, which is equal to the short circuit current density in a good cell, N_c is the density of states in the conduction band, and

$$S = v_{th}\, \sigma_{db} N_{dbl}, \quad (2)$$

with v_{th} the thermal velocity of the carriers, and N_{dbl} the density per unit area of dangling bonds at the interface. From Eqs. (1) and (2) we expect V_{oc} to increase by about 60 meV for every factor of ten reduction in N_{dbl}. The excess V_{oc}, after taking into account the increase in V_D due to the larger bandgap in the HDLTD material, is of this order of magnitude. Hence one might conclude that N_{dbl} is on the order of a factor of ten below that in the standard material. The p-i interface has a high density of dangling bonds intrinsically, since the Fermi level there falls in the distribution of the valence band tail states, where equilibrium holes in the weak bonds are converted to dangling bonds by thermal vibrations[8].

If the lower N_{dbl} stems from a lower density of weak bonds which is present throughout the material, then the saturation of the degradation may also be explained on this basis, since it is these bonds which can be converted to dangling bonds under illumination. The physical picture for the conversion to dangling bonds, of a trapped hole in a weak bond of the valence bandtail states, has been proposed previously[8]. A hole in such a state tends to remain there much longer than a hole in a higher density of states region, since it has few states to which it can move.

A further cause of the saturation of the degradation may be another aspect of the lower hydrogen content of the films. There is strong evidence[10] that hydrogen motion is involved in forming the Staebler-Wronski defect. Hence materials with reduced hydrogen, or reduced hydrogen of the specific type revealed by the IR studies, may not be able to create as many defects.

Bennett et al.[11] have shown that at least two types of degradation sites exist in a-Si:H, those that degrade and recover rapidly, and those that degrade and recover more slowly. If the weak bonds in the HDLTD material have a narrower distribution in energy, due to less stress in the material, then there may only be fast sites which quickly saturate. Hence, the more rapid initial degradation may indicate that the HDLTD material is more homogenous than the standard material.

CONCLUSIONS

Using the PCG technique we have compared initial degradation of both standard and HDLTD materials under red and green light. We find that while there is no difference in the $(\mu\tau)_e$ degradation of the two materials, $(\mu\tau)_h$ degrades only in the HDLTD material. This explains the difference between cells made of these two materials, and is consistent with the limiting carrier model.

Based on these PCG results, and the previously reported lower hydrogen content of the HDLTD material, we have proposed a plausible picture for the improved open-circuit voltage, and the saturation of degradation in the HDLTD material. This model uses two aspects of the low hydrogen content of the HDLTD material. The first is that with the material under less stress there are fewer dangling bonds at the p-i interface leading to a higher V_{oc}, as well as fewer weak valence band tail states throughout the material. The second is that since hydrogen appears to have an active role in the formation of the Staebler-Wronski defect, the lower hydrogen content of the HDLTD material limits the total number of defects that can be created during degradation.

The initial higher degradation rate in the HDLTD material indicates a preponderance of fast degradation sites or simply the more rapid effect of the degradation due to the low density of the valence band tail states. However, once those sites degrade, no further sites seem to be present for additional degradation. This suggests that the HDLTD material is in a sense more homogeneous than the standard material, and this accounts for its superior performance in solar cells.

REFERENCES

[1] M. Bennett, K. Rajan and K. Kritikson, *Proc. of the 23rd IEEE Photovoltaic Specialist Conf.* p. 845, (IEEE, New York, 1993).

[2] I. Balberg, *Mater. Res. Soc. Symp. Proc.* **258** **693** (1992).

[3] I. Balberg and S.Z. Wiesz, *J. Appl. Phys.* **70** **2204** (1991).

[4] Amorphous Silicon Materials and Solar Cells, edited by B.L. Stafford, *AIP Conf. Proc.* **234** (1991).

[5] E. Souvain, J. Hubin, A. Shah, and P. Pipoz, *Phil. Mag. Lett.* **63** **327** (1991).

[6] J-H. Yoon, Phys. Rev. B 49 8000 (1994).

[7] K. Misiakos and F. A. Lindholm, J. *Appl. Phys.* **64**, **383** (1988).

[8] A. Rothwarf, *Proc. of the 20th IEEE Specialist Conf.*, p. 199 IEEE, New York, (1988).

[9] T. Takahama, S Okamoto, K. Ninomiya, M. Nishikuni, N. Nakamura, S. Tsuda, M. Ohnishi, A. Nakano, Y. Kishi, and Y. Kuwano, *Tech. Digest of International PVSEC-5* (Kyoto, 1990) p. 375.

[10] D. E. Carlson, *Appl. Phys.* **A 41**, **305** (1986).

[11] M. S. Bennett, J. L. Newton, K. Rajan, and A. Rothwarf, *J. Appl. Phys.* **62** **3968** (1987).

*Present address: NREL Golden, CO.

PRECISE INVESTIGATION OF THE LIGHT-SATURATED DEFECT DENSITY IN AMORPHOUS SILICON VERY THIN FILMS AND SOLAR CELLS

J. Kočka, M. Vaněček, P. Macháček, A. Fejfar, E. Šípek,
Ho-The-Ha, I. Pelant, J. Fric, J. Rosa, Z. Remeš, A. Poruba
Institute of Physics, Czech Academy of Sciences, Cukrovarnická 10, 162 00 Prague 6, Czech Republic

M. Konagai
Tokyo Institute of Technology, 2-12-1 O-okayama, Meguro-ku, Tokyo 152, Japan

W. Kusian
Siemens Central Research Laboratories, Otto-Hahn Ring 6, 81730 Munich, Germany

ABSTRACT

We present progress in the three regions important for optimization of the highest long term a-Si:H solar cells efficiency:

1) In order to achieve light saturated state rapidly and controllably a new method of the accelerated degradation by combined AM1 and pulsed ruby laser illumination is proposed.

2) Diagnostic of the degraded state is then made possible by a new CPM setup combining standard and transmission modes which gives absorption coefficient α in the absolute units and undisturbed by interferences.

3) The possibility to measure deep defect density of states directly on a-Si:H solar cells by space-charge-limited-current time-of-flight is demonstrated.

INTRODUCTION

For further progress in still continuing effort to improve the stability of amorphous hydrogenated silicon (a-Si:H) based solar cells there are from our point of view **three important tasks.**

First of all, we need to improve the reliability of the accelerated degradation [1], i.e. to exclude the influence of temperature and light intensity on the saturated a-Si:H properties at the high speed degradation.

For responsible selection of more stable a-Si:H material we need to improve the absolute precision of the widely used CPM (Constant Photocurrent Method) [2,3] which is the **second task**, together with the aim to minimize the influence of the interference fringes [4,5] namely for thin layers of a-Si:H, used in solar cells.

The **third** important **task** is to introduce new technique(s) suitable for finding the deep defect density of states (DOS) directly on a-Si:H solar cells, such as Space-Charge-Limited-Current Time-of-Flight (SCLC-TOF) method.

ACCELERATED DEGRADATION

We suggest to use combined 100 mW/cm^2 AM1 illumination (which defines realistic conditions, namely the temperature of the actual operation of the cell) with the limited application of the pulsed light which speeds up the initial degradation. We have applied combined use of the AM1 with the pulsed ruby laser (which ensures homogeneous degradation for a-Si:H up to 2 μm thickness; CW power $\leq$ 10 mW/cm^2) to undoped a-Si:H sample (see Fig. 1). During the whole degradation we monitor the AM1 induced photoconductivity. If after a given number of the ruby laser pulses the DOS is still lower than saturated one (and so the photoconductivity $\sigma_{ph} \approx 1/\mathrm{DOS}$ higher than the saturated one) the continuing AM1 illumination degrades the sample still further. If after the laser pulses the DOS is higher (and so σ_{ph} lower) than the saturated value, the AM1 light induced annealing takes place and σ_{ph} increases to the saturated value (as shown in Fig. 1). Relatively quick saturation can be achieved if the DOS (and therefore σ_{ph}) after combined AM1 and the laser pulses are near to the saturated values (upper curve in Fig. 1).

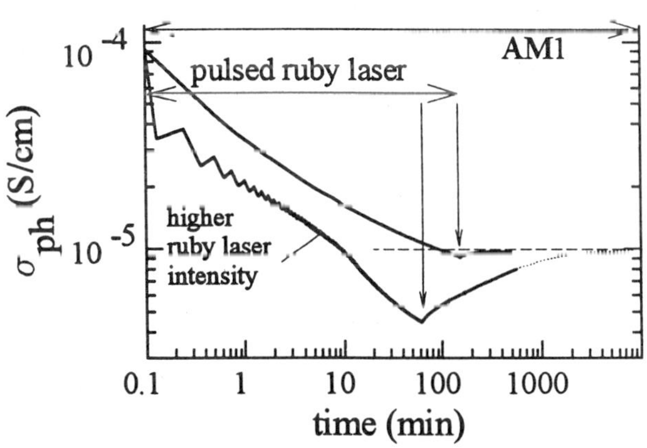

Fig. 1: The decay of the photocurrent due to AM1 illumination during accelerated degradation. The arrows indicate the switch off of the pulsed ruby laser.

ABSOLUTE CPM

The great advantage of the CPM, widely used for the measurement of small values of subgap optical absorption and the determination of the DOS in a-Si:H, is that it measures bulk properties of material and is (contrary to PDS) almost insensitive to the surface states. This is especially important for films thinner than 1 micron which are used in solar cells. On the other hand the drawback of standard (S) CPM is that it gives just the relative values of the optical absorption coefficient α and separate transmittance/reflectance measurement has to be used to set α to the absolute [cm^{-1}] scale. Because of strong interference fringes (see curve I in Fig. 2b) in thin a-Si:H layer the deconvolution of the resulting α into the deep defect DOS is seriously hampered.

The first proposal to reduce the influence of the interference fringes by deducing α from the ratio A/T, where A is absorptance and T transmittance of the film/substrate system, was made by Ritter and Weiser [4]. Recently Sasaki et al [5] achieved substantial reduction of the interference fringes with the help of the CPM in so called "transmission (T) mode" (see Fig. 2a). In this case the CPM measures directly the spectrum of the ratio A/T instead of A in the case of (S)-CPM. However, we have found that while the measured (S)-CPM on high energy side "underestimates" α (because of surface recombination and non-homogeneous absorption) the (T)-CPM apparently "overestimates" it (see curve II, Fig. 2b) due to T approaching zero.

Here we suggest the additional logical step: to **measure simultaneously** CPM in both "standard" and "transmission" modes, it means with two identical detectors, as shown in Fig.2a. (T)-CPM gives us the A/T relative values, with greatly reduced interference fringes and the ratio of signals of both detectors gives us the spectral dependence of the transmittance T (see curve III in Fig 2b). Important point is that the transmittance maxima T_{max} in the low absorption region (below approx. 1 eV) are given just by the optical properties of substrate (Corning glass no. 7059) and so we can use it for setting the transmittance to the absolute scale. T_{max} below 1 eV equals 0.92. Then we set A/T into the absolute scale in one reference point (T=0.05), in which the transmittance T is free from the interference fringes. Now, using the Ritter-Weiser formula [4]

$$\alpha d = \ln\left(\frac{1}{2}\left[(1-R_2)\left(1+\frac{A}{T}\right)+\sqrt{(1-R_2)^2\left(1+\frac{A}{T}\right)^2+4R_2}\right]\right)$$

where R_2 is the reflectivity of the back interface of a-Si:H film, we calculate from A/T (in the absolute scale) the optical absorption coefficient $\alpha(E)$ in the absolute scale [cm^{-1}] (see curve IV in Fig. 2b) under the assumption of approximately constant film reflectance R, the same as used in PDS. We use $R=0.26$ and $R_2=0.33$ for sample illuminated from the substrate side, as was the case for the measurement presented in the Fig.2b. The

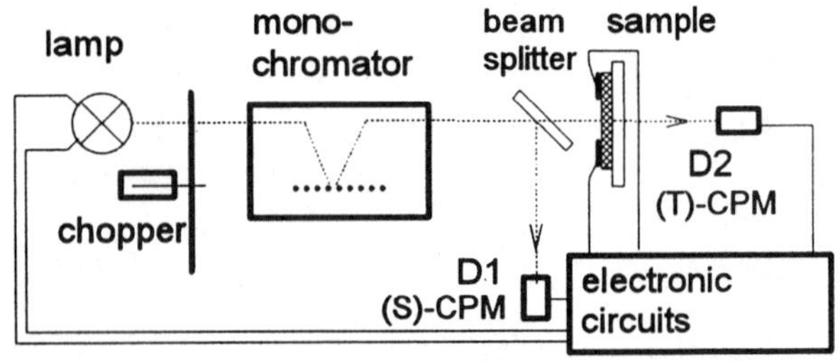

Fig.2a: The experimental set-up for (a.c.) measurement of the "absolute CPM". The standard mode (S)-CPM is measured using detector D1 and the transmission mode (T)-CPM is measured using detector D2. For "absolute CPM" both detectors are used simultaneously.

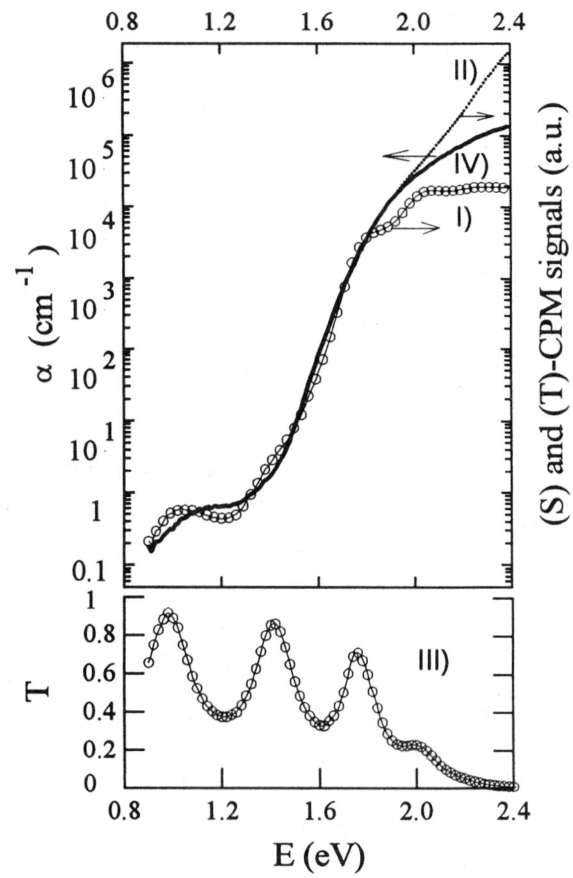

Fig 2b: The curves denoted by I-IV correspond to:

I) The standard (S)-CPM spectrum (in arbitrary units) proportional to the absorptance A measured with detector D1 on 0.32 μm thin a-Si:H layer.

II) The "transmission" mode (T)-CPM measured using the detector D2. The signal is in this case proportional to the ratio A/T.

III) Transmittance T of this layer deduced from the signals of the both detectors recorded simultaneously.

IV) Resulting spectral dependence of the absorption coefficient α[cm^{-1}] in **absolute** units under the assumption of constant R $\doteq$ 0.26 in a strong absorption region (similar assumption as usually used in PDS).

corresponding values for illumination from the side of the a-Si:H film are R=0.4 and R_2=0.18. Hence, in the reference point A=1-T-R=0.69 in the first case and 0.55 in the second case.

By this way we easily get the **"absolute CPM"**, i.e. α(E) in absolute scale (full curve IV in Fig. 2b) in one CPM measurement using the experimental setup with two detectors (Fig. 2a), without any need for calibration by additional measurement of transmittance by a spectrophotometer or by PDS.

SCLC-TOF

For the TOF technique, described in detail elsewhere [6], we have used top quality glass/TCO/p-i-n junction (0.45 μm thick solar cell with the initial 11.3 % efficiency) prepared by photo CVD and 0.43 μm thick glass/TCO/n-i-p junction prepared by plasma CVD [7].

It is rather difficult to deduce the defect density of a real a-Si:H **solar cell by conventional small signal** TOF method because the transit time is extremely short and so the interaction with traps is very limited (see Fig. 3a).

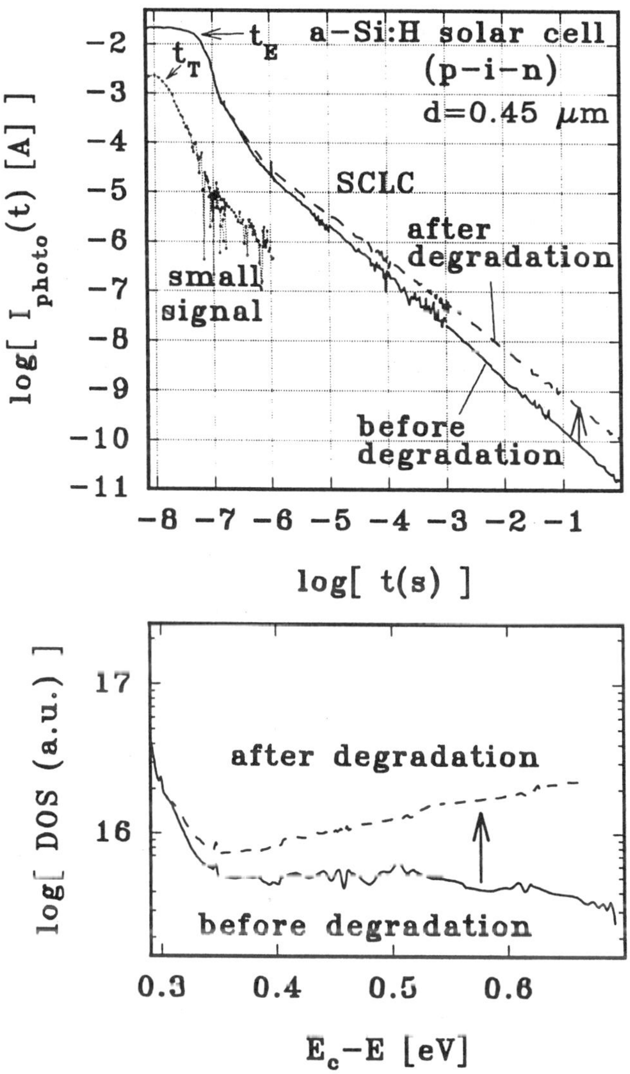

Fig.3 a) The current transients I(t) measured by TOF technique on a real a-Si:H solar cell with 11.3% efficiency (0.45 μm thin p-i-n junction) at room temperature and 1 V reverse bias in annealed and partially degraded state. The transit (t_T) and extraction times (t_E) are marked.
b) Density of states calculated from the current transients as DOS(E) ≈ I(t)·t and E = kT·ln(νt).

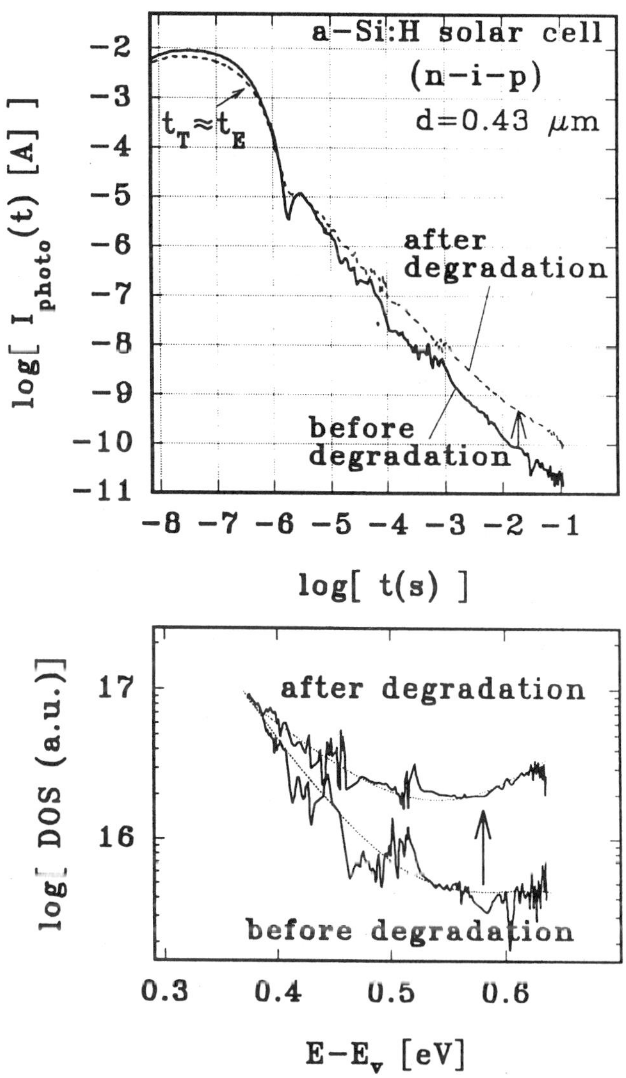

Fig. 4 a) The current transients I(t), measured by TOF technique on a real a-Si:H solar cell (0.43 μm thin n-i-p junction) at room temperature and 0.5 V reverse bias in annealed and partially degraded state. The extraction time $t_E \approx t_T$.
b) Density of states calculated from the current transient I(t) in the same way as in Fig 3b.

By the use of the Space-Charge-Limited Current (SCLC) TOF we create much larger charge, the extraction of which takes much longer time (t_E instead of t_T). Due to the increase of the time "window" for interaction with the traps we can relatively easily deduce the total defect density as we have shown recently on 1 μm thin p-i-n junction [8]. When the SCLC-TOF is used in a post-transit mode we can get also the energy profile DOS(E) of the defect density of states [9,10].

In Fig. 3a the SCLC-TOF pre- and post-transit spectra are presented for the first time on a real solar cell - 0.45 μm thin p-i-n junction. The DOS evaluated on the basis of simple expression $DOS(E) \approx I(t) \cdot t$, where $E = kT \ln(\nu \cdot t)$ [11], is shown in Fig. 3b.

The great advantage of this new technique is, that it can be used for both electrons and holes.

In Fig. 4a there are the SCLC-TOF transit and post-transit spectra measured on n-i-p structure [12] illuminated through the n-layer, so that the transient should be dominated by the extraction of holes.

In addition to the atypical "dip" at about 1 μs [12] there are two clear differences in comparison with the (probably [13]) electron dominated transients in Fig. 3. First of all there is no difference between SCLC-extraction time (t_E) and the small signal transit time (t_T) and, moreover, in this region ($t \leq t_T \approx t_E$) the hole transients are insensitive to degradation (while in the post-transit region $t > t_E$ the degradation is clearly visible).

Within the weak bond (WB) → dangling bond (DB) model [14] the increase of the DB should be accompanied by the decrease of the WB, it means the DOS in annealed and degraded state should cross in the deep valence band tail. The DOS, evaluated on the basis of Fig. 4a is in Fig. 4b and surprisingly no crossing is observed. Further study is necessary to clarify this point.

The SCLC-TOF post-transit technique is very perspective as it can be used directly on the solar cell structures and it allows to follow either electrons or holes and therefore DOS above and below the Fermi level. Naturally some new questions have to be answered, for example what is the role of the "minority" carriers (holes in the transients "dominated" by electrons [13]) or how to explain the conflict with the WB-DB model.

CONCLUSIONS

Very simple but substantial improvements of the accelerated degradation by combined AM1 with the pulsed ruby laser and the absolute measurement of the light induced defect density by "absolute" CPM were presented. Moreover, it was demonstrated that the newly introduced SCLC-TOF in the post-transit time region is very perspective for direct application to solar cells.

Acknowledgements

The authors want to acknowledge the partial support by the EC-PECO CIPD3510PL92 3029 and by the associated contract CT930312 to the EC grant ERBJOU2CT920027.

References

[1] J. Kočka et al, EC-PECO 3029 Progress Report No. 3, *Appl. Phys. Lett.* **62**, 1993, p. 1082, and to be published.

[2] J. Kočka, M. Vaněček, A. Tříska, in *Amorphous Silicon and Related Materials*, (ed. H. Fritzsche), World Scientific, Singapore, 1989, p. 297.

[3] N. Wyrsch, F. Finger, T.J. McMahon, M. Vaněček, *J. Non-Crystal. Sol.* **137&138**, 1991, p. 347.

[4] D.Ritter and K. Weiser, *Optics Communications* **57**, 1986, p. 336

[5] M. Sasaki, S.Okamoto, Y. Hishikawa, S. Tsuda and S. Nakano, Tech. Digest of the Int. PVSEC-7, Nagoya , 1993, *Solar Energy Materials and Solar Cells* **34**, 1994, p. 541.

[6] J. Kočka, *Proc. of the ISCMP,* Varna 1992, World Scientific Publ. Co., Singapore, 1992, p. 129.

[7] W. Kusian, R. D. Plättner, J. Furlan, G. Conte, *Proc. of the 23rd IEEE PVSC*, 1993

[8] J. Kočka, G. Juška, O. Klíma, E. Šípek, G. Nobile, E. Terzini and G. Conte, *J. of Non-Cryst. Sol.* **164-166**, 1993, p. 489.

[9] J.Zapletal, A. Fejfar, S.Fouad Abdel Hamied, E.Šípek, I. Pelant, J. Kočka, G. Juška, T. Yokoi, M. Azuma, I. Shimizu, H. Mahan and R.S. Crandall, *Proc. of the 12th E-PVSEC*, Amsterdam, 1994, p. 104.

[10] F. Schauer, A. Eliat, M. Nesládek and G.J. Adriaenssens, *Appl. Phys. Lett.* **64**, 1994, p. 3009.

[11] G.F. Seynhave, R.P. Barclay, G.J. Adriaenssens, J.M. Marshall, *Phys. Rev. B* **39**, 14, 1989, p. 10196.

[12] G. Juška, M. Viliunas, S. Vengris, A. Fejfar and J. Kočka, *Proc. of the ISCMP*, Varna 1994, to be publ. by Taylor & Francis.

[13] R. Brüggeman, ibid. ref. [12].

[14] M. Stutzmann, *Phys. Rev. B* **35**, 1987, p. 9735.

A NEW STRUCTURE A-SI SOLAR CELL WITH PLASTIC FILM SUBSTRATE

Y.Ichikawa, K.Tabuchi, S.Kato, A.Takano, T.Sasaki, M.Tanda, S.Saito, H.Sato, S.Fujikake, T.Yoshida and H.Sakai
Fuji Electric Corporate Research and Development, Ltd.
2-2-1 Nagasaka, Yokosuka, Kanagawa 240-01Japan

ABSTRACT

An amorphous silicon solar cell deposited on plastic film substrate has been studied. The solar cell has a new monolithic series-connected structure called *SCAF*, and will be fabricated by a newly developed fabrication process based on "Stepping Roll" film deposition system.

The process was compared with the conventional glass solar cell process. Preliminary Results small and large area *SCAF* solar cells were presented and discussed.

INTRODUCTION

An amorphous silicon (a-Si) solar cell is suitable for low-cost production because it has the following advantages: a) it demands less semiconductor material than crystalline silicon (c-Si) solar cells; b) it has a shorter energy payback time c) it is harmless for environment; d) monolithic series-connection are formed on a large glass substrate: e) various materials are applicable. From these advantages, we believed that the production cost of the a-Si solar cells will decrease in the mass production.

Typical a-Si solar cells are formed on glass substrate or stainless steel substrate nowadays. Glass substrates are bulky and heavy. Thus the substrates themselves can be an obstacle against cost reduction of the a-Si solar cells. For example, conveyance, storage, and packing of the glass substrates are all costly. On the other hand, stainless steel substrate is easier to handle, but the monolithic series-connected structure is not able to be formed on the conductive film.

We have, therefore, been developing a-Si solar cells on flexible plastic substrates, instead of glass or stainless substrates, with new device structure named "SCAF"(Series-Connection through Apertures on Film).

The market of PV (photovoltaic) houses, meanwhile, starts growing as an attractive market for PV industries in Japan. To apply the a-Si solar cells to the PV houses, both shape and weight of the a-Si modules are very important because they should be easy to install and nice looks. From this viewpoint, our flexible a-Si modules fit for the PV house application. The details of this application are reported elsewhere [1].

In this paper, we introduce the SCAF structure and its production process; we compared it with the conventional glass solar cell process. In addition, we present results of preliminary results for small and large area solar cells with the SCAF structure.

DESIGN OF SUBMODULE STRUCTURE AND PROCESS FOR SCAF

The greatest advantage of using flexible substrate for production process is that a roll-to-roll process is applicable; this process is considered to have the best throughput among the a-Si solar cell processes. In this process, however, the conventional series-connected structure used for glass substrate solar cells is not able to apply because it requires a laser-scribing process after respective layer formations of transparent electrode, a-Si and metal electrode formations.

To overcome this problem, we invented a new structure and a process for making series connection by one-step laser-scribing after completing all the deposition processes. The structure and process of the new solar cell named SCAF is explained below.

SCAF Structure

The schematic diagram of "SCAF" solar cell is shown in Fig. 1. A metal electrode/a-Si:H/transparent electrode layer is formed on one side of a film substrate to make a solar cell, and a metal layer named back-side electrode is formed on the other side of the film. The current collected by the transparent electrode flows into the back-side electrode through a plurality of apertures made in the region of transparent electrode. Both the solar cell side and the back-side electrode are divided into a number of unit cells. The apertures made on both sides out of the transparent electrode are used to contact the back-side electrode of a unit cell with the metal electrode of the next unit cells to make a series connection between these unit cells.

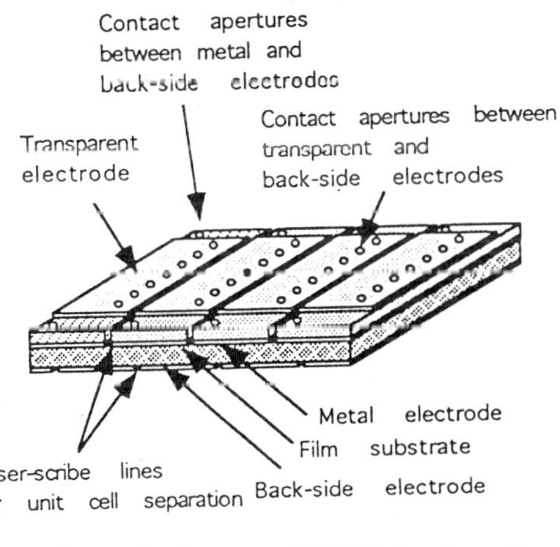

Fig. 1 The schematic diagram of "SCAF" solar cell.

Process Design for SCAF

CH3365-4/94/0000-0441 $4.00 © 1994 IEEE

First WCPEC; Dec. 5–9, 1994; Hawaii

Figure 2 shows the process flow for fabricating the SCAF cell in comparison with the conventional glass substrate solar cells. An advantage of this structure is to simplify the fabrication process. As seen from the figure, the conventional process consists of eight steps to fabricate a large area submodule. On the other hand, the process for the SCAF solar cell requires seven steps. Some processes are, however, integrated in a roll-to-roll apparatus and the process practically reduces to three steps in the production line.

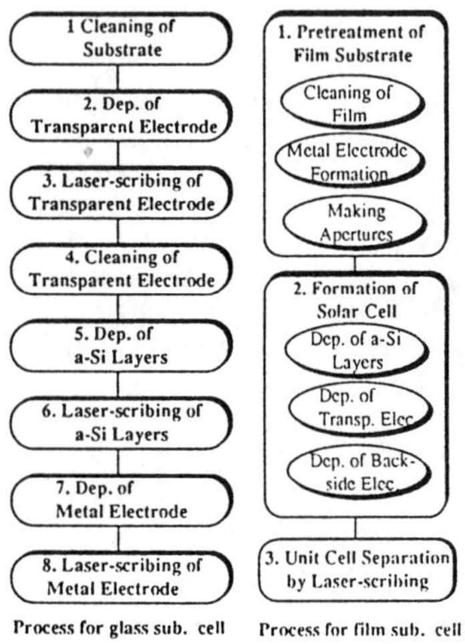

Fig. 2 Comparison of fabrication process flows between conventional glass substrate solar cell and SCAF structure solar cell.

The reason it is realized is due to one step laser-scribing process besides the roll-to-roll process. This laser-scribing process is made after all the film deposition processes are completed. Hence, the a-Si:H and transparent layers can be formed in a successive process to avoid generation of short-circuit points. Moreover, the function required for the laser-scribing is to separate the unit cells electrically, not to make series-connection. This means that precise control of position is not required. This is a great advantage for film substrate solar cells because the flexibility of the film makes fine position-control very difficult.

Film deposition process is the most important part for improvement of the performance in the SCAF structure a-Si solar cell. The difficulties in the conventional roll-to-roll deposition system are as follows:

i)Interdiffusion of raw material gases among CVD chambers cannot be prevented.

ii)Gas pressure in the CVD chambers cannot be controlled independently.

iii)Precise control of the composition in the depth direction cannot be made.

We designed a new roll-to-roll deposition apparatus to overcome the above difficulties. Figure 3 shows the schematic diagram of the apparatus named "Stepping Roll Film Deposition (SR) Apparatus." The basic configuration of the apparatus consists of a common chamber and several film deposition reactors in it. During deposition each reactor is completely isolated from the common chamber to prevent interdiffusion of gases, and then deposits each layer simultaneously. After film deposition required is completed, the substrate is released from the reactor and is wound up by one frame like film in a camera. Adopting this stepping roll system, we can apply almost all the techniques developed for high performance, large-area, glass substrate solar cells such as bandgap grading to the SCAF solar cell.

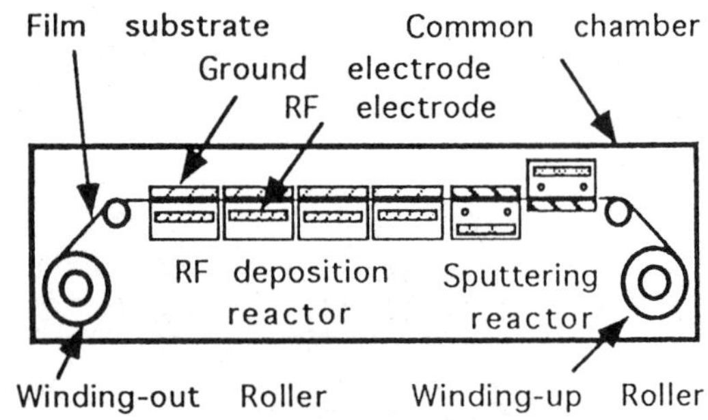

Fig. 3 Schematic diagram of Stepping Roll Apparatus.

EXPERIMENTAL AND RESULTS

Apparatus for Cell Fabrication

To verify the new process for the SCAF solar cell, we developed a prototype process line, which was composed of a prototype SR deposition apparatus, a roll-to-roll laser-scriber, and a substrate cleaner with mechanical punch units for making holes on the film. As shown in Fig. 4, the SR apparatus has two deposition chambers: one is for plasma-CVD and the other is for sputtering. Both chambers have an effective deposition area of 40cm x 40cm, where a-Si:H or electrode is deposited on plastic film substrates with a width of 50cm. Figure 4 also shows the distribution of thicknesses for a-Si:H and ITO deposited on film substrates: the variation of a-Si:H film ranged within ±5% and that of ITO within ±7%. The photo- & dark- conductivities of a-Si:H films are also quite uniform, as shown in Fig. 5.

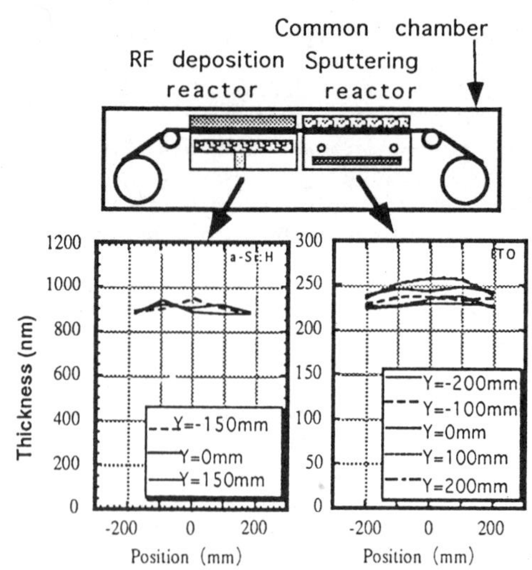

Fig. 4 Prototype SR apparatus and its distributions of thicknesses for a-Si:H and ITO deposited on film substrates: variation of a-Si:H film is ranged within ±5% and that of ITO within ±7%

This suggests that key parameters such as temperature of the film are uniform. Moreover, the photo-sensitivities of a-Si:H exceed 6 figures, that is sufficient quality for solar cells. The resistivity of ITO varied between $2x10^{-4}$ and $3x10^{-4}$ (ohm•cm).

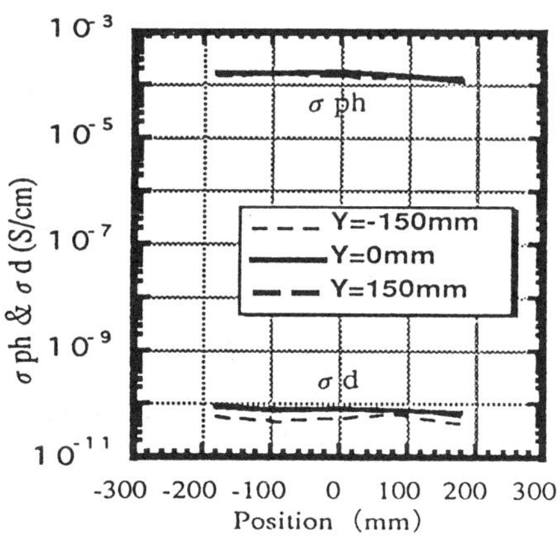

Fig. 5 Photo- & dark- conductivities of a-Si:H films deposited by SR apparatus.

Small Area Cells

As the prototype SR apparatus is a single deposition chamber system, it is not suitable to study the potential performance of film substrate based solar cells. Thus we used our IVE apparatus[2] for a-Si deposition of small-area devices to study the performance of the substrate type a-Si solar cells formed on the film. The best efficiency obtained to date is 11% (Voc=1.81V, Jsc=8.45mA/cm^2 , FF=0.72) in a film substrate/metal/a-Si(n-i-p/n-i-p)/ITO double-junction cell (1cm^2).

It was reported that, in the n-i-p devices, phosphorous contaminant was incorporated in the i-layer during deposition from the n-layer [3]. We tried compensation doping with a trace amount of boron in the i-layer to increased the fill factor, and found that the counter doping increases fill factor of the top cell of the double-junction device. Further investigation was necessary to improve the efficiency.

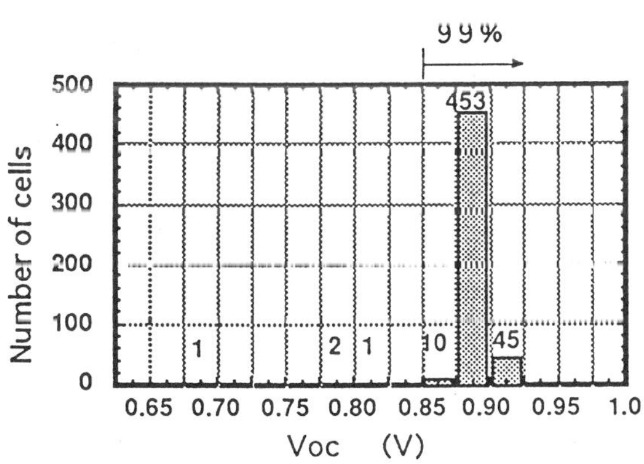

Fig. 6 Open-circuit voltages (Voc) of 512 cells deposited on a 40 cmx 40 cm film substrate.

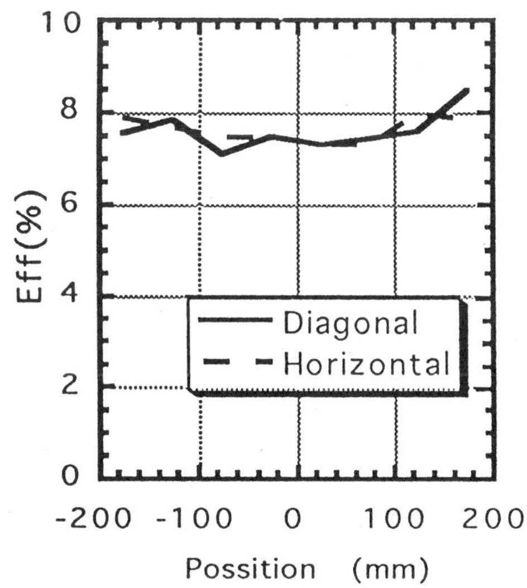

Fig.7 Horizontal and diagonal distributions of efficiency on a 40cm x 40cm substrate.

Large-area devices were studied with the prototype SR apparatus. As already mentioned, this apparatus has only one reactor for a-Si film deposition. Thus, with this apparatus, we do not expect high efficiencies, but can study the process technology for the SCAF solar cell and fundamental structure of the stepping roll deposition apparatus for designing a full specification apparatus.

Generation of the short-circuit point due to pin holes of the a-Si layer was one of the typical defects for glass substrate cells. At the beginning, we were anxious about the shunt in the SCAF cells formed through roll-to-roll conveyance. Hence we fabricated 64 pieces of 1 cm^2 single-junction solar cells on 40 cm x 40 cm area by the prototype SR apparatus. Figure 6 shows the histogram of open-circuit voltages, Voc, of 512 cells; we fabricated 8 samples and the summation of them was shown. As seen from the figure, 99% of the cells exceed 0.85V. This means that the numbers of short-circuit points are negligibly small.

Figure 7 shows the horizontal and diagonal distributions of efficiency on a 40cm x 40cm substrate; they are uniform enough to fabricate large-area devices.

From the above results, we concluded that large-area a-Si solar cells can be formed on plastic film by the stepping roll apparatus in principle, though the efficiencies of the small area cells are lower than those deposited by the multi-chamber system. The full specification SR apparatus under construction will have a multi-reactor system. Thus we will improve the efficiency in the near future.

Fabrication of SCAF Solar Cells

In this experiment, all the thin film layers such as metals, a-Si, and ITO were deposited by using the prototype SR-apparatus. The layers deposited were laser-scribed at the final process by the roll-to-roll laser-scriber. Figure 8 shows the picture of the film substrate solar cells deposited on the plastic films. We fabricated a series of 40cm x 40cm solar cells with the SCAF structure on a 50cm-wide rolled film. On the cell four rectangular cell-regions were fabricated, each of which was composed of 70 series-connected unit cells in 7 cm x 35 cm area. The diameter of the holes was 1 mm and 13 holes were punched in a unit cell.

Fig. 8 Photograph of 40cm x 40cm solar cells deposited on plastic film.

Figure 9 shows the simulated efficiencies of the SCAF cell and the typical I-V characteristics of the fabricated cells. In the simulation, there are three important keys; which are the area efficiency (the ratio of active area to the total area), the sheet resistance of transparent electrode and the contact resistance at the contact apertures. The model developed took all the keys into account. The efficiencies were plotted as a function of ITO sheet resistance. As seen from the figure, it was found that the efficiency was not affected by the sheet resistance. The design of module enables us to use ITO with a sheet resistance of up to 40ohms/square. The aperture efficiency obtained is smaller than calculated value but is within a reasonable range. In the present design, the discrepancy of efficiency between the small area cell and the SCAF cell are mainly due to the area efficiency; in the prototype cell, we designed to be an area efficiency of 89%. This will be improved to 96% in the next design.

From these results, we confirm the SCAF structure works very well on the newly developed prototype process.

For further improvement in conversion efficiency, the small area efficiency is the most important key, which will be improved by using the next SR apparatus as multi-chamber system. As seen from Figure 9, the conversion efficiency of 11% for small area cell, which was already obtained for the multi-chamber system named IVE, will increase the module efficiency over 10%.

SUMMARY

We proposed a new structure for film substrate a-Si solar cells to simplify the fabrication process. To fabricate the cells, we developed a prototype process line, including Stepping Roll Film Deposition (SR-) Apparatus, and confirmed that structure worked very well on the newly build process line.

Area(cm2)	Effi.(%)	Voc(V)	Isc(mA)	FF	notes
1.0	11	1.81	8.45	0.72	IVE (tandem)
980.0	5.75	60.72	182.0	0.51	pro.-SR (single)

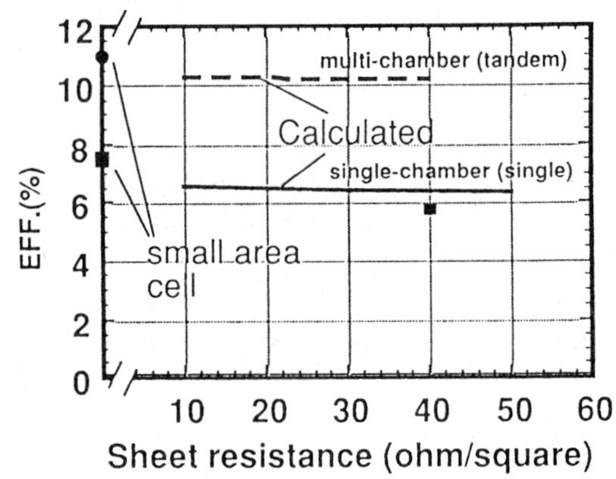

Fig. 9 Simulated submodule efficiencies and the typical I-V characteristics of the module fabricated.

ACKNOWLEDGMENT

This work was supported by the New Energy and Industrial Technology Development Organization under the New Sunshine Program of the Ministry of International Trade and Industry.

REFERENCES

[1]Y. Ichikawa et al., "A New Flexible a-Si PV Modules and Its Application to Rooftop PV Systems", 1st WCPEC, 1994, to be published.

[2]K. Maruyama et al.1, "Fabrication and characteristics of large-area a-Si solar cells", 8th IEEE PVSC, 1985, p 883.

[3]H. Haruki et al., "Effect of boron doping and its profile on characteristics of p-i-n a-Si:H solar cells", Solar Energy Materials, 8, 1983, p 441.

DIAGNOSING TANDEM CELLS

Jingya Hou and Stephen J. Fonash,
Electronic Materials and Processing Research Laboratory, The Pennsylvania State University
Malathi Ghosh[1], Jianping Xi[2], Tongyu Liu and Frank Kampas[3],
Advanced Photovoltaic Systems Inc.

ABSTRACT

In this paper we report three new diagnostic techniques for identifying problem areas in tandem cells. These techniques can be easily extended to diagnosing triple junction cells. We applied these techniques to diagnose an a-Si:H based tandem cell having an extraordinarily low Voc. As we demonstrated, the assessments of the problem in this cell produced by each of our diagnostic approaches agree well with each other.

INTRODUCTION

Multijunction a-Si:H based solar cells provide better energy conversion efficiency and stability than single junction a-Si cells; hence, they are currently the subject of intense manufacturing development. In manufacturing multijunction cells, it is important to be able to identify which component cell is the problem area when yield losses occur. We report on three new techniques we developed for this purpose. Intuitively, one would expect that quantum efficiency(QE) measured under different color bias light can identify which component cell is causing a problem in a bad multijunction cell. We show with an example that this is not true. We designed three new techniques which we call QE switching, balanced QE, and mixed color light current-voltage measurement (JV), We apply these techniques to diagnose a tandem cell with extraordinary low Voc to test these techniques. The results are positive.

COLOR LIGHT BIAS QE

Color light bias QE (CLBQE)[1] is widely adopted by multijunction solar cell manufacturers and research groups. This technique can measure different QEs corresponding to different component cells of a multijunction cell. For example, for a tandem cell using a red bias light one can measure a QE curve peaked in the blue spectral region, which one interprets as being indicative of the quality of the top cell. Using a blue bias light one can measure a QE peaked in the red spectral region, which one interprets as being indicative of the quality of the bottom cell. The correspondence between the QE curves and the light absorption of component cells seems to indicate that the red light bias QE and the blue light bias QE are determined exclusively by the top cell quality and the bottom cell quality, respectively. But this is not the case. Fig. 1 presents two CLBQE measurements for two nominally identical same band gap a-Si/a-Si tandem cells. One of them has an open circuit voltage (Voc) of 1.6 V, which is normal, and the other, cell D341, has a Voc of only 0.98V. Fig. 1(a) and 1(b) are the CLBQEs of the two cell measured under 0.8 V forward bias, respectively. Cell D341 obviously has a severe problem. However, comparing the CLBQE in the two figures, the blue light biased QE and the red light biased QE of cell D341 are similarly balanced as those of the normal cell. The only noticeable difference is that the heights of the two QE peaks in the Fig. 1(b) are lower than the corresponding peaks in Fig. 1(a). This does not give us a clue as to which component cell has a severe problem. The CLBQE of the two cell measured at zero bias voltage are less distinguishable, because both QEs are saturated.

Later we will show that there is a problem in the top cell of cell D341. Why does this problem not show up in the CLBQE curve comparison between the two cells? The answer is a phenomena we called the electric field shift[2] in multijunction cells. In a tandem cell under color light bias the number of photo-carriers generated in its two component cells is very much out of balance. The component cell which absorbs less bias light photons and generates less photo-carriers limits the output photo-current. This component cell is a limiting cell, and the other component cell is a non-limiting cell. To balance the photo-currents generated in the two component cells, the electric field of the non-limiting cell shifts to the limiting cell. In this way the non-limiting cell weakens its electric field and eliminates the excessive photo-current by recombination, and the limiting cell increases its electric field to maximize the output photo-current. The electric field in the non-limiting cell is very weak, and the QE corresponding to the light absorbed by the non-limiting cell is almost zero. The electric field in the limiting cell is very strong, and the QE corresponding to the light absorbed by limiting cell is very high. Here, one should notice that the electric field which

---

[1]Current address: Epitaxx Inc. 7 Graphic drive West Trenton, NJ 08628
[2]Current address: Intersolar Group Ltd. United Kindom
[3]Current address: SunGard Capital Market Inc. 510 Walnut St -Suit 900 Philadelphia, PA 19206

CH3365-4/94/0000-0445 $4.00 © 1994 IEEE

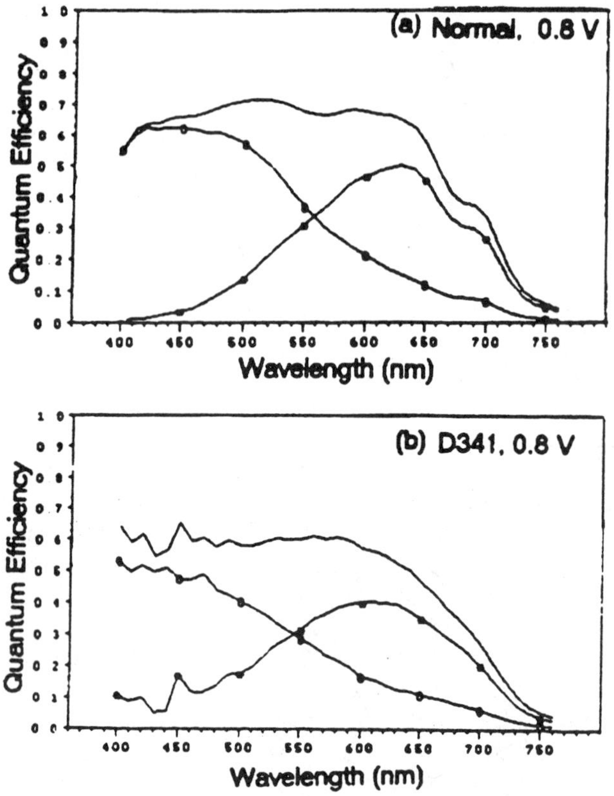

Fig. 1 are the color light biased QEs of two tandem cells under 0.8 V forward bias voltages Figure (a) are the QE curves of a normal tandem cell with a Voc of 1.6 V, and figure (b) is the same measurements for cell D341 which nominally has same structure as the normal cell but with a Voc of only 0.98 V. The difference of the CLBQE of the two cells does not indicate the problem area of cell D341.

actually determines the CLBQE is not the original electric field in the limiting cell under thermal equilibrium condition, but the electric field enhanced by the electric field shift. If a bad cell is a limiting cell, its quality directly affects the QE; if it is a non-limiting cell its bad quality reduces the electric field shifted to the limiting cell, it indirctly affects a CLBQE. Therefore, a bad component cell can equally affect both CLBQEs no matter if it is a limiting cell or a non limiting cell. This is seen in Fig. 1(b), where both QE peaks are lowered compared to Fig. 1(a) of the normal cell.

From the above analysis, we see that because of the electric field shift, the CLBQE is actually determined by the electric fields of all the component cells. Hence, it is not an exclusive indicator of the quality of a particular component cell. To diagnose the quality of each component cell we have to find new techniques.

QE SWITCHING

The first new technique we introduce is QE switching. To understand this technique, let us think what will happen if we apply blue and red bias lights, simultaneously to a tandem cell while measuring a QE? There are three possibilities: 1) the red light is much stronger than the blue light, then the field shifts from the bottom cell to the top cell, and we will see the blue QE peak; 2) if the blue light is much stronger than the red light, then the field shifts from the top cell to the bottom cell, and we will see the red QE peak; 3) the two light intensities are balanced, namely the photo-currents generated in the two component cells are equal without the field shift. Then no electric field will shift and the QE curves are flat. If we fix one bias light intensity and vary another light intensity from very weak to very strong, we will see the three states consecutively, the first QE peak, flat, and then the second QE peak. The balance state is the QE peak switching state. Why does the QE peak switch? Because the photo-current that would be present in each of the two component cells if there were no electric field shift, which we name "if" current, changes their relative amplitude. The electric field always shifts from a high "if" current component to a low "if" current component. When the "if" current of the component cell which strongly absorbs the variable light exceeds the "if" current of the component cell which strongly absorbs the fixed light, the electric field shift changes its destination. So does the QE peak. The "if" current depends on the intensity of the light absorbed and the quality of the component cell. If the component cell which absorbs the variable light is deteriorated then the QE switch will happen at higher variable light intensity, and if the component cell absorbing the fixed light is deteriorated, then the QE switch will happen at lower variable light intensity. Therefore one can identify a deteriorated cell by comparing the switching light intensity of the tandem cell to a normal tandem cell. To identify the QE switching light intensity, there is no need to see the whole QE curve. One needs measure only the QE value at one of the two QE peak wavelengths.

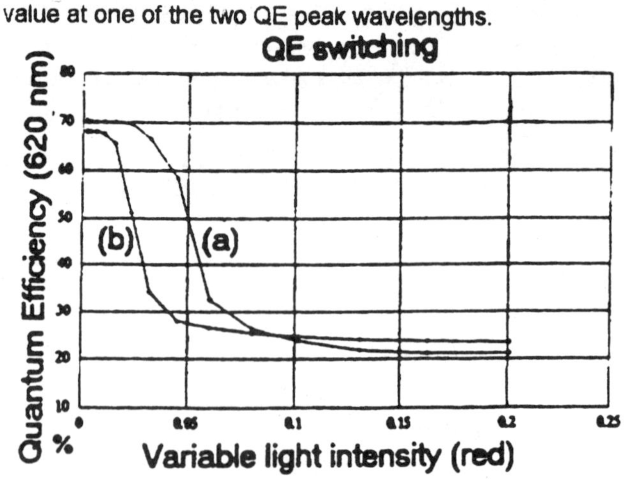

Fig. 2 is a plot of QE switching. The curves are the 620 nm QE vs. the variable light intensity. The fixed light is a blue light and the variable light is a red light. The curve (a) is the QE of the normal cell, the curve (b) is the QE of cell D341.

Fig. 2 is measured QE switching data for cell D341 and the normal cell. The fixed light is a blue light and the variable light is a red light. The QE is measured at wavelength of 620 nm(red peak QE) vs. the variable light intensity. Fig. 2 shows that cell D341 switches at lower variable red bias light intensity than the normal cell does. This indicates that the top cell is a problematic cell. Because the top cell of cell D341 generates less "if" current than the top cell of the normal cell does. the bottom cell of cell D341 needs less red ligth to reach the "if" current balance than that of normal cell does. We also measured QE switching using fixed red light and varing blue light. The result shows that cell D341 switches at higher blue bias light intensity than the normal cell does. This also indicates the top cell being a problematic cell, because the top cell needs more light to balance the bottom cell photo-current. We also used the blue QE peak probe light beam instead of 620 nm probe beam to monitor the QE switching, and had same assessment. We did all four possible QE switching measurements to verify the consistency of this technique.

In general, if in comparing to an otherwise identical normal cell, a bad tandem cell needs more variable bias light to switch QE, the bad component is the component cell strongly absorbing the variable bias light; and if less variable bias light can switch the QE then the bad component is the component cell strongly absorbing the fixed bias light.

BALANCED QE

The second new diagnostic technique we introduce is what we term balanced QE. The principle of this technique is closely related in the principle to the technique of QE switching, but it is an easier measurement to implement. In this technique only one white bias light is needed. A special filter is made to adjust the white bias light spectrum so that the QE for a normal cell becomes quite flat, as shown in Fig. 3. When this is done, it creates a situation where for a normal tandem cell the "if" currents of the two component cells are balanced. If one of the two component cells deteriated, then the "if" current balance is broken. The electric field has to be shifted from the good component cell to the bad component cell to balance the output photo-current. The QE in the spectrum region corresponding to the absorption of the deteriated component cell peaks up. Therefore, the bad component cell can be spotted by looking for a peak in the QE. In Fig. 3 QE curves of two normal cells and cell D341 measured under balanced bias light are presented. The two normal cell QEs have little difference and both are flat. The QE of cell D341 increases dramatically in the blue spectrum region, indicating that the top component cell is deteriated.

The special filter used in this experiment is made simply by using a piece of glass with blue ink on it.

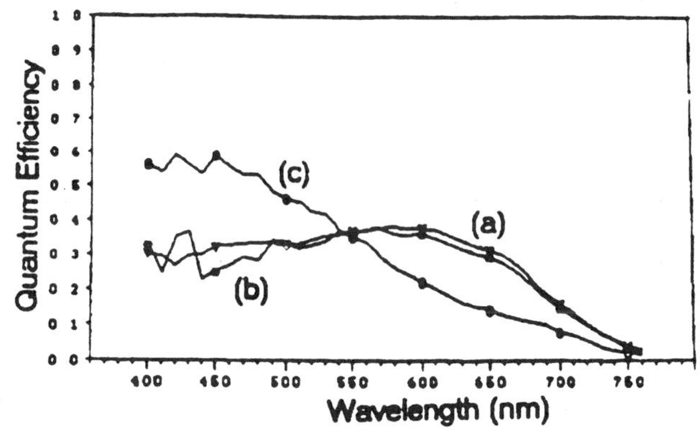

Fig. 3 is the plots of balanced QE. The two flat curves (a) and (b) are the measurements on two normal tandem cells, and the curve (c) is a measurement on cell D341. Its peak in the blue region indicates its top cell has severe problem,

Because the conventional tungsten bias light source is red light rich, we use blue ink to correct the spectrum. To fine tune the spectrum to balance normal cell QE, we use a fine needle to scratch the ink on the glass. The white light through scratches adds red light into the spectrum. The filter is positioned to make the bias light homogenous at the location of the sample cell.

MIXED COLOR LIGHT JV

The most clear way to show which component cell is a bad component, is to show the light JV curve of each component cell without breaking the multijunction cell. This is possible and, in fact, quite simple. This is the third technique -- mixed color light JV. In this technique, we use mixed red-rich or mixed blue-rich light to illuminate a tandem cell. The light is a mix of a red light and a blue light, in red-rich light the red light intensity is at least three times more than the blue light intensity, and in the blue-rich light the ratio is opposite. The very unbalanced ratio puts one component cell as a limiting cell and the other as a non-limiting cell. In this case the measured JV curves essentially is the limiting cell JV. The measured current certainly is the current through both component cells. The measured voltage is the total output voltage of the two component cells(assuming the contact junction between two component cells is a good junction and gives no effect to the output voltage). But the output voltage of the non-limiting cell is approximately a constant; approximately is its Voc under this particular illumination condition. This is because for common a-Si based pins, when the output current is about 1/3 of its Jsc or less, the output voltage is fairly close to its Voc. Therefore, we see the difference of the limiting component cell JV to the JV measured under unbalanced mixed color light is only a constant voltage. Even though we may not know this constant voltage, the curvature and the slope of the measured JV curve still provides the insight into the limiting cell.

Fig. 4(a) and 4(b) show the mixed color light JV of the normal cell and of cell D341, respectively. Each figure contains two JV curves under blue-rich light and red-rich light, respectively. They corresponding to the bottom cell JV and top cell JV respectively. The two JV curves of the normal cell in Fig. 4(a) are in similar shape indicating the two component cells have comparable quality. The difference of the Jsc is not a essential character of the cell. It is due to different light intensity of the weak component light in the two light mixtures. The two JV curves of the cell D341 in Fig. 4(b) have an essential difference. The large slope of the JV curve measured under red-rich light indicates the output current drops quickly as the voltage increases. This means the top cell has a severe problem in its structure. This assessment agree with the assessments provided by the other two techniques.

SUMMARY AND DISCUSSION

We have shown that the conventional CLBQE is not a proper diagnostic tool for identifying the problem area in multijunction cells, and introduced three new diagnostic techniques for this purpose. We succesfully applied these three techniques to cell D341 to identify the problem area. This indicates these techniques, at least, provide reliable diagnostic approaches for problems similar to the problem in cell D341. The general applicability of these techniques to a wider variety of problems needs further study and verifacation.

In these three techniques the balanced QE is the one easiest to implement and to operate. It is also the one easiest to extend to triple junction cells. It only needs a filter to balance all three component cell QEs. For other two techniques the third light corresponding to the third component cell is needed, and the intensity of it should be the strongest so that the third component cell is saturated and out of the picture. We then treat the remaining two component cells as a tandem cell. By changing the color of the strongest light we can diagnose any two component cells.

The first two techniques are sensitive to the changes of thickness and the absorption coefficients of each component cell. Because they affect the balance of the "if" current among the component cells. This may lead to mis-interpretation of the measured results. In this case the mixed color light JV is a good alternative which is not sensitive to the change of the component cell thickness and absorption coefficients. The sensitivity of the first two techniques to the component cell thickness and the light absorption can be used in advantageously to monitor the component cell thickness configurations. The mixed color light JV can be used to understand the relative quality among the component cells in a normal multijunction cell to guide further improvement.

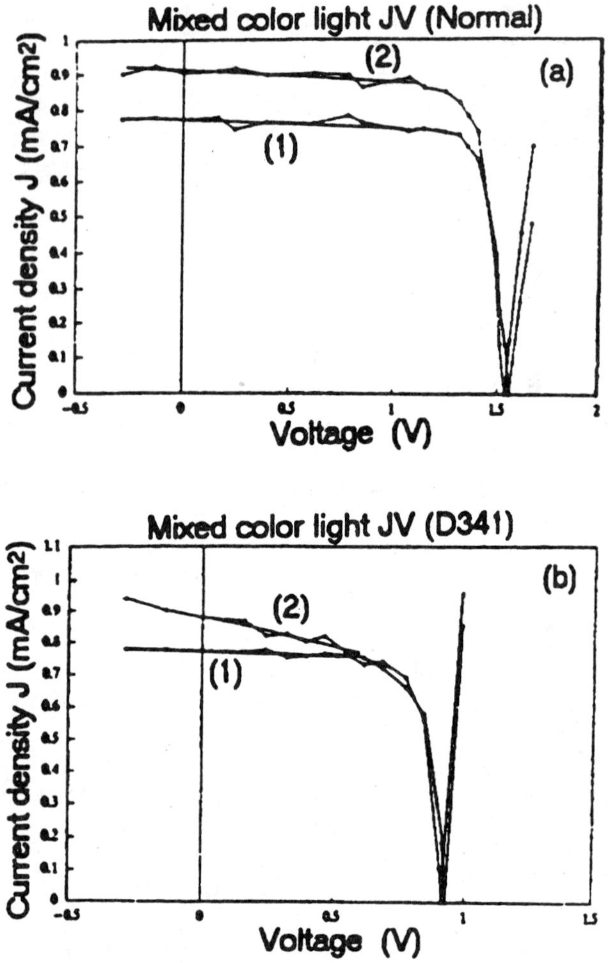

Fig. 4. is the plots of mixed color light JV. Fig. (a) is the measurements on the normal cell, and Fig. (b) is the measurements on cell D341. The two curve (1)s in the two figures are measured under same blue-rich light, and the two curve (2)s measured under same red-rich light. Curve (1)s and curve (2)s are essentially the JV curves of the bottom cell and the JV curves of the top cell, respectively.

REFERENCES

1. Josph Burdick and Troy Glatfelter, *Proceedings of Photovoltaics and Insulation measurements Workshop* pp151-- pp167. 1985 Vail, Colorado.

2. Jingya Hou, Ph.D. thesis, 1993, The Pennsylvania State University.

METASTABLE SHUNT PATHS IN a-Si SOLAR CELLS

T.J. McMahon* and M.S. Bennett**
*NREL, Golden CO 80401 **Thin Film Division, Solarex Corp., Newtown, PA 18940.

ABSTRACT

The transitory and erratic nature of shunt currents, whether caused by light-soaking or electrical biasing, in amorphous Si (a-Si) single- and triple-junction solar cells has been a real puzzle and made the study of these cells difficult. We present a careful study of the time/voltage dependence of these current transients in several different a-Si solar cell structures and find they reveal more about the basic shunt mechanism. In single-junction cells, we see stepwise current changes that increase in size and number with reverse bias and can be removed with forward bias. This stepwise, on and off switching suggests a discrete shunt path conduction mechanism. The kinetics of these metastable shunt paths show that both the "on-state" and "off-state" possess memory. Cells without (Al)ZnO show no metastable switching. We associate the stepwise features with the textured substrate and the switching metastability with contact to (Al)ZnO. Switching in triple-junction cells occurs with hundreds of oscillations at each step.

BACKGROUND

The engineering solution to reduce light-induced degradation of amorphous silicon (a-Si) solar cells is to use thin, tandem p-i-n cells with light trapping provided by a textured transparent conductive oxide (TCO) coated substrate. The motivation for thin cell designs dates back to 1983 when Tsuda et al. studied the effect of reverse bias and thinner intrinsic layers on cell stability [1]. However, when a textured TCO substrate is used for either thin single p-i-n or thin multilayer p-i-n structures, a new mode of cell degradation occurs which has been associated with the increased shunting [2]. Two degradation mechanisms now occur: (i) the classical bulk Staebler-Wronski defect generation mechanism causes an increase in diode quality factor. (ii) In cells exhibiting more as-grown shunting behavior, light-soaking greatly enhances shunting. Enhanced shunting becomes the primary mode of degradation in thinner cells [3]. At that time, we also noticed reversible incremental current steps occurring as the I-V curve was being measured. We concluded that the increase in current is due to an increase in the number of shunt pathways and not the size of existing shunts [3]. Laser scribing of small-area cells (0.25 cm^2) into smaller areas reveals that shunt currents are due to small defect groups localized within a portion of the cell area and are not necessarily associated with visual defects or flaws [4]. This bias-dependent switching and a voltage-squared dependence of the reverse current after the onset of switching is being studied by others [5].

The abrupt steps and the ability to switch between a conducting "on-state" and an "off-state" by changing the sign of the bias is very similar to the switching mechanism found by LeComber et al. [6]. In their switching devices, very small metal shunt paths are formed that act as switchable current paths. Recently, bias-dependent switching on a-Si p$^+$/metal devices "formed" with various metal contacts was attributed to charged metal inclusions. The authors guess that V, W, Ni,or Co contacts form different geometrical chains yielding "quantum" or "analog" switching [7].

SINGLE-JUNCTION CELLS

In this study of the time/voltage dependence of these current transients, we are able to find a pattern in the erratic nature of these shunt currents. Figure 1 shows the range of excess currents found in these single junction cells. Cells were deposited on textured InSnO (ITO) coated glass superstrates to form an ITO/p$^+$/i/n/(Al)ZnO/Ag structure. Cell A behavior is representative of most as-grown cells with nanoamp leakage and no switching or hysteresis. With use, switching develops and cell B behavior sets in with noisy erratic currents and a gradual increase in the reverse current at each voltage. The "additional" shunt paths give the greater-than-linear ohmic voltage dependence for current. These currents decrease with forward bias at greater speed; hence, the smaller hysteresis loop. Cell C behavior is the next stage in the

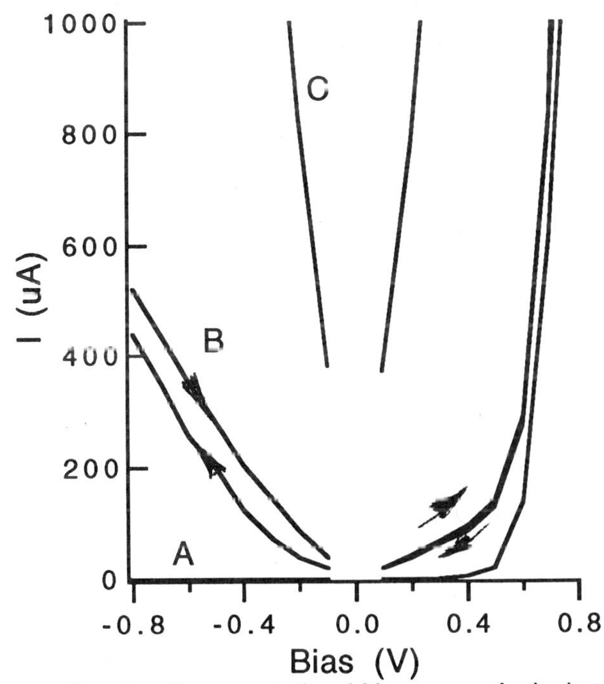

Fig. 1 Representative I-V curves of single junction cells: A - as-grown, B - after voltage stressing, and C - after a lot of stressing.

CH3365-4/94/0000-0449 $4.00 © 1994 IEEE First WCPEC; Dec. 5–9, 1994; Hawaii

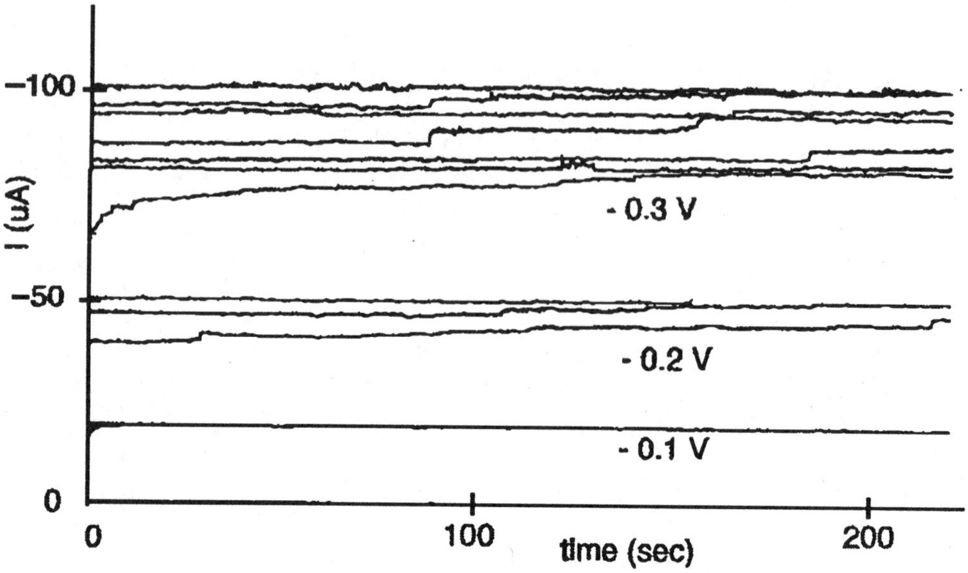

Fig. 2 Chart recorder traces of single-junction reverse-current transients starting at -0.1 (1.5 traces), -0.2 (2.7 traces), then -0.3 V (7 traces).

progression towards failure resulting in reduced efficiency at full sunlight. An ohmic I-V appears with shunt paths stabilized and very little, if any, hysteresis.

Figure 2 shows how reverse bias applied to a cell like A produces stepwise increases in the reverse current. Higher reverse biases produce more steps. The fact that these steps are abrupt can be explained by invoking the concept of localized "shunt paths," which after a period of applied reverse bias, switch to a conducting "on-state." Some shunt paths in the "on-state" can be switched to the "off-state" by decreasing the amount of reverse bias, and all of these shunt paths can be turned "off" by resting the cell at 0-bias. After taking data for Fig. 2, 0.1 V forward bias was applied to turn off these shunts, and the biasing conditions of Fig. 2 were repeated with identical step-for-step results. The current steps that are observed to switch on and off are 3 μA or less at 0.3V. The size of a step can depend on the size of the shunt path, *as well as how close on the cell it is to another path already in the "on-state"*. For this reason, the last step out or the first step in should be among the largest, as is usually observed!

Figure 3 shows how the "off-state" is achieved by applying a forward bias, allowing us to observe stepwise decreases in current. This stepwise activation and deactivation of shunt paths is reversible through at least 10 cycles. By this time, the "on-state" had been on for a while and the stepwise increases occur more rapidly than in Fig. 2. Results in Fig. 4 were taken after the results of Fig. 3 and show prolonged "on-time" results in a longer "removal time" and, subsequently, an even shorter "turn-on" time. In addition, the longer a shunt is off, the longer it takes to turn on; i.e. the on-state and off-state possess memory. Light exposure at 100 mW/cm^2 does not deactivate these shunts.

We also find it informative to carefully monitor the "turn-offs" that occur on dropping the reverse bias from 0.3 V to -0.1 V. Initially there is a large undershoot in the current followed by a slow(10s of sec) recovery to some steady state current that is below the current preceding the "turn-off." Always, if you are willing to wait say 10 h, there is an equilibrium value for the current at each voltage whether that voltage is approach from the high or low side. Only once on this cell did the current not fully recover with forward bias.

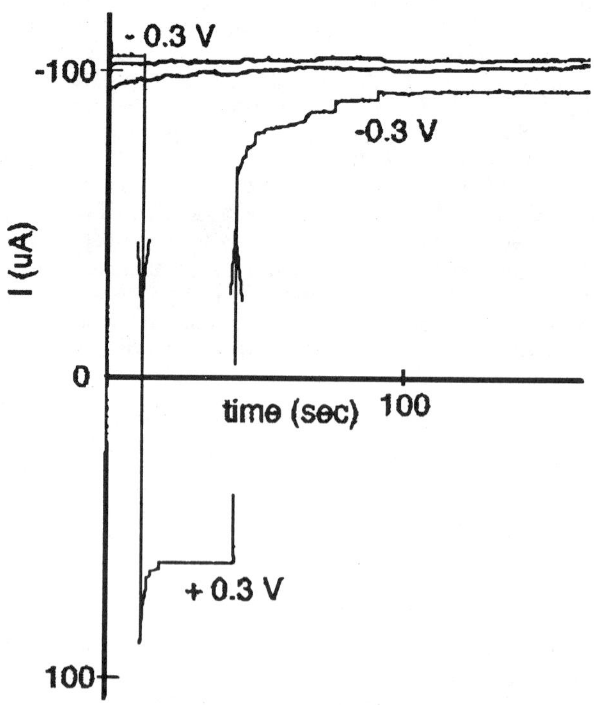

Fig. 3 Stepwise turn-offs at +0.3 V followed by a series of stepwise turn-ons at -0.3 V.

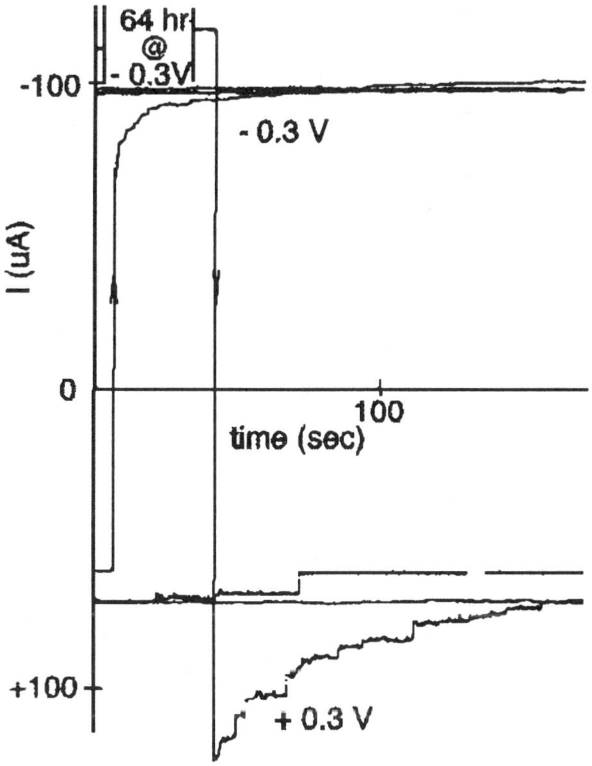

Fig. 4 Stepwise turn-off at +0.3 V after 64 h on time, followed by a series of stepwise turn-ons at -0.3 V.

We studied p-i-n-cells deposited on smooth (Al)ZnO-coated glass, where the p^+ layers were 8-nm and 48-nm thick. Here the p^+ is in contact with the (Al)ZnO. We found the reversible increases and decreases in current with bias as before, but without the steps, as shown in Fig. 5; this behavior started within minutes of applying reverse bias. In Fig. 5 the + 0.1V forward bias was applied as the increasing reverse current approached 30 nA. With the thin 8-nm p^+ layer, current increases into the µA range occurred even faster and were still found to be fully reversible.

Smooth and textured InSnO substrates were also used with the surprising result that the shunting behavior was not reversible. Cell structures not including (Al)ZnO showed no metastable switching properties!

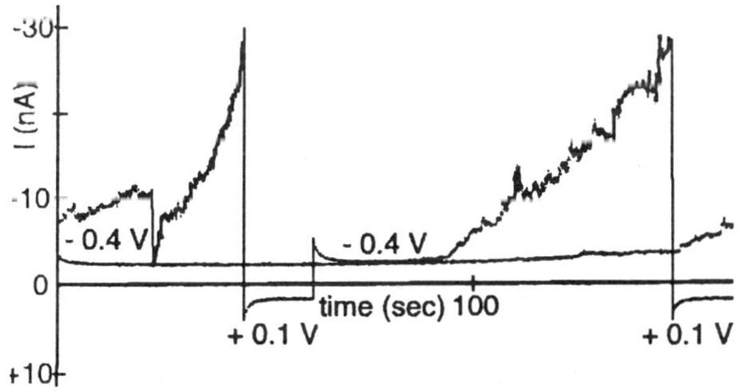

Fig. 5 Turn-on at -0.4 V followed turn-offs at +0.1 V for the (Al)ZnO/48 nm p^+-i-n cell structure.

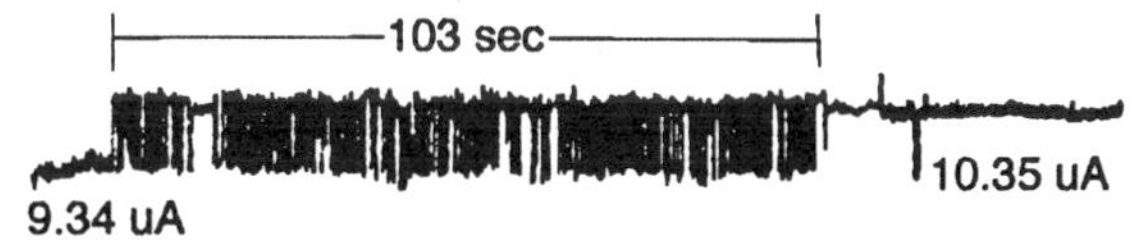

Fig. 6 Typical triple-junction "turn-on" at -0.7 V.

TRIPLE-JUNCTION CELLS

Observations of shunts made on *triple-junction cells* show a similar switching on and off with applied reverse and forward biases; but during the switching process, very erratic bistabilities with hundreds of switching events occur for a minute or two before the shunt converts to the on-state, as shown in Fig. 6. Another important distinguishing feature in the triple cell is that the forward bias switching to the off-state is much slower. Both of these differences can be explained by noting that the sheet resistance of the intermediate p/n tunnel junctions in a triple cell are much higher than either of the contacting layers. When the shunt first turns on, the voltage at the plane of the internal p/n tunnel junction is reduced across the i-layer of that cell containing the shunt path in the area near the shunt path and the voltage across the shunted layer drops allowing the shunt to switch off until the voltage is re-established. These bistable shunt paths oscillate until the shunt is established in a metastable configuration. The oscillating behavior means it does not penetrate the entire triple cell. The impeded forward bias deactivation is also explained, since the applied forward bias potential is reduced in the region of the shunt path when it is in the on-state.

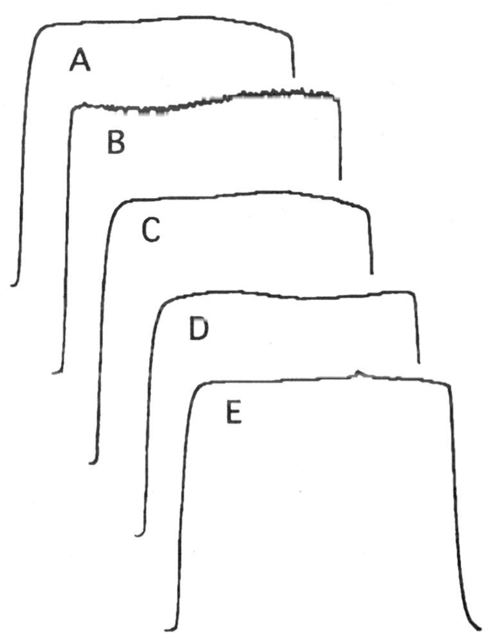

Fig. 7 Photoresponse line scans across the same line of the 0.25-cm^2 triple-junction cell; A - no shunts, B - 34-µS bias induced shunt, C - shunt off, D - 150-µS light-induced shunt, and E - shunts off.

Spot scanning shows a 1-mm radius area of reduced photoresponse when metastable shunt paths are "on," whether it is caused by light-soaking or electrical biasing. The 5% depression in the red photoresponse for the 10^{-5} ohm^{-1} – 10^{-4} ohm^{-1} sized shunt paths is localized in the triple junction again because of the high sheet resistance of the internal tunnel junctions. As before, locally near a shunt path, the plate voltage is reduced on the shunted cell of the triple and collection of carriers generated by red laser light is not as good as in surrounding areas: metastable shunt paths in the triple-junction cells are in one of the back cells. Spot scans of shunted single cells never showed localized depressions, presumably because of the low sheet resistances of the conductive oxides.

The oscillating steps and spot scan observations can be explained only if the sheet resistance of an intermediate (depleted) p/n tunnel (recombination) junction is as high as 10^5 ohms/sq. This also explains why the front cell on a triple-junction solar cell can still operate effectively at thicknesses of 60 nm to 80 nm when we know that a single-junction cell of this thickness is likely to be shorted. Even though the front cell of a triple is loaded with shunt paths, our spot scans and sound reasoning show that for a metastable path to form, one of the back cells must contain the switching defect. Very likely, the metastable shunt paths occur in the back cell nearest the (Al)ZnO contacting layer.

MODELS

To explain this switch-like behavior, we have a choice of the metal inclusion chain (MIC) model [7] and the metal/semi-insulator/semiconductor switch (MISS) model [8]. The switching mechanism in MISS devices is well-understood, but the physics of an MIC device is only conjecture at this time.

Applying the MISS model to the discrete shunt currents in our cells would involve locating the "shunt paths" on the tops of the ITO crystallites. Electric fields are enhanced at these sharp points. Additionally, the cell's film layers are thinnest at these positions, allowing for the condition known as "punch-through." In reverse bias, the conducting "on-state" occurs when the first p-layer depletes to the extent that electrons are injected and swept through the p-layer into the i-layer. The "on-state" is achieved only after detrapping has removed all the blocking effects of the reverse-biased p/i interface. The conductance of the current path is limited by the i-layer. Each localized area on top of a ITO pinnacle along with its adjacent p/i interface composes a unique switching device with its own threshold voltage and conductance properties.

Using Pennsylvania State University's AMPS program, representative parameters are applied to the layers of the ITO/p/i/n cell. The result of the calculation is that the p-layer is completely depleted for an acceptor density of 10^{18} cm^{-3} and a thickness of 5 nm. Each ITO pinnacle along with its adjacent p/i interface could, indeed, comprise a unique localized switching device with its own threshold voltage and conductance properties.

Alternatively, the MIC model could be applied with the understanding that something diffuses out of the (Al)ZnO layer forming these switchable conducting paths [7]. Because Al diffuses more easily than the metals used in Ref. 7, it is possible that Al inclusions in the sputtered ZnO form conducting paths in the a-Si as a result of diffusion. Reversible shunts resulting from diffusion are consistent with all our observations, including the relative amount of shunting in cells with thick and thin p^+ layers.

The p-i-n device deposited on the smooth (Al)ZnO layer on glass produces hundreds of switching defects that turn on in a relatively gradual "analog" fashion with time as shown in Fig. 5. The devices deposited on textured ITO with the (Al)ZnO layer on the back form hundreds of switching devices as before. In this case the tips of the higher ITO crystallites provide the individual switching defects.

CONCLUSIONS

The switching metastability in a-Si cells is due to something (diffusing) from the (Al)ZnO and the discrete, stepwise quality is due to shunt paths formed on peaks of the textured TCO. Shunt path on- and off-states are observed to have memory: Some sort of defect relaxation process must stabilize the conduction paths with biasing, e.g. a trapped charge becomes stabilized on a defect through a relaxation in the lattice surrounding that defect.

ACKNOWLEDGEMENTS

We wish to thank S. Hegedus and V. Dalal for some of the cell structures used in this study. This work was conducted under U.S. Dept. of Energy contract no. DE-AC36-83CH10093

REFERENCES

[1] S. Tsuda, N. Nakamura, K. Watanabe, T. Takahama, H. Nishiwaki, M. Ohnishi, and Y. Kuwano, "Light-induced instability of amorphous silicon photovoltaic cells", Solar Cells **9**, 1983, pp. 25-36.

[2] R.D. Wieting, M. Bolingen, J. Anderson and D. Willett, "Performance of large area thin film Si:H modules: a link between shunting and stability", Proc. of the 21st IEEE Photovoltaic Specialists Conference, 1990, p. 1569.

[3] T.J. McMahon and M.S. Bennett, "Film morphology, excess shunt current and stability in triple-junction cells", Proc. of the MRS **258**, 1992, p. 941.

[4] M. Bennett, J. Newton, C. Poplawski, and T. McMahon, "An investigation of the effect and nature of shunts in a-Si solar cells", Proc. of the 23rd IEEE Photovoltaic Specialists Conference, 1994, p. 867.

[5] K. R. Lord, M.R. Walters, and J.R. Woodyard, "Investigation of shunt resistances in single-junction a-Si:H alloy solar cells", Proc. of the MRS Spring '94 Mtg., 1994.

[6] P.G. LeComber, A.E. Owen, W.E. Spear, J. Hajto, A.J. Snell, W.K. Choi, M.J. Rose and S. Reynolds, "The switching mechanism in amorphous silicon junctions", J. Non-Cryst Sol. **77** & **78**, 1985, p. 1373.

[7] M. Jafar and D. Haneman, "Switching in a-Si devices", Phys. Rev. **B49**, 1994, pp.13611.

[8] J.G. Simmons and A. El-Badry, "Theory of switching phenoma in metal/semi-insulator/n-p^+ silicon devices", Solid-St. Electron. **20**, 1977, p. 955.

PREPARATION OF HIGH QUALITY WIDEGAP a-Si:H FILMS AND ITS APPLICATION TO a-Si SOLAR CELLS

Y.Ashida, N.Ishiguro, K.Saitoh, T.Miyashita,
N.Yanagawa, H.Tanaka, M.Sadamoto and N.Fukuda
Central Research Institute, Mitsui Toatsu Chemicals, Inc.
1190 Kasama-cho, Sakae-ku, Yokohama 247 Japan

Absrtact

The novel film formation method of alternately repeating a-Si deposition and hydrogen plasma treatment (ADHT method) was studied. The optical bandgap could be changed from 1.67 to 1.98 eV, the defect densities under $1x10^{16}$ cm^{-3} were obtained at wide range of optical bandgap, and the saturated defect densities after light soaking were about $3x10^{16}$ cm^{-3}. Single junction pin a-Si solar cells with i-layer formed by ADHT method were stable to light soaking. Applying the these widegap a-Si films to the i-layer of top cell of 2-stacked tandem solar cells, high open circuit voltage of 1.85 V and high efficiency of 10.5 % was attained.

Introduction

Main items concerning amorphous silicon films and solar cells are the bandgap engineering and the improvement of stability against light soaking. There are many reports on the improvement of stability by the use of new deposition method [1-4]. We have developed the novel deposition method in order to control the optical bandgap and to raise the light stability of a-Si films [1,5]. By the application of these widegap a-Si films to p/i buffer layer, a-Si single-junction solar cells with high Voc and FF were obtained [6-8]. Also, the degradation of solar cells were suppressed by utilizing 2- or 3-tandem type solar cells, and the stabilized efficiency was attained more than 10% [9-11]. Moreover, to obtain the higher stabilized efficiency, high quality and stable wide gap and narrow gap a-Si alloy films are needed because these materials are applied to tandem type solar cells.

In this paper, high quality a-Si films were prepared by using new deposition method and the light stability and the application of these film to intrinsic layer were examined.

Experimental

The deposition system in this paper was a conventional plasma CVD system with capacitively coupled electrodes and glow discharge was produced by a 13.56 MHz rf power source through a matching network. The novel deposition method of alternately repeating a-Si deposition (thin film deposition process) and hydrogen plasma treatment(modification process), abbreviated by ADHT method, was explained in other reports [1,5]. Absorption coefficients, reflectance and transmittance of films were measured by spectrophotometer. Hydrogen content and its bonding configuration were estimated by the IR spectroscopy. Cell structures used in this study were glass/SnO_2/p/p-i buffer/i/n/Al or glass/SnO_2/p/p-i buffer/i/n/p/p-i buffer/i/n/Al, whose area was 1 cm^2. And, we have also uesd ITO/Ag as a back reflector, to obtain the high efficiency [12]. Preparation of p, p/i buffer, i-layers and n-layers were reported in other literatures [9,11,12]. The i-layer was deposited from pure SiH_4 for conventional method and SiH_4/H_2 mixture gas for ADHT method.

Results and discussion

<u>Hydrogen plasma treatment effect</u>

In figure 1, hydrogen content is plotted against substrate temperature for the cases of a-Si films prepared by conventional plasma CVD (conventional films) and ADHT method (ADHT films). ADHT films were deposited under the conditions of the unit thickness of 4 nm and hydrogen plasma treatment time of 20 seconds in this case. Hydrogen content increased with lowering the substrate temperature in each method as same as general experimental results. Difference (ΔH) of hydrogen content of conventional films and ADHT films at each substrate temperature indicates the change of bonded hydrogen content by hydrogen plasma treatment process (modification process). Differential hydrogen content (ΔH) was increased with substrate temperature and had a maximum at near 200°C and then decreased to zero at about 380°C. This trend is explained as follows. It is previously reported that hydrogen atoms bonded in a-Si network during a-Si deposition process was exchanged by the hydrogen atoms of hydrogen plasma during plasma treatment process [8]. At higher temperature, hydrogen evolution from a-Si network and intrusion of hydrogen atoms from hydrogen plasma was balanced, as a result, the additional hydrogen content became zero. So, at higher temperature more than 400°C, hydrogen evolution from a-Si network was domain and the

CH3365-4/94/0000-0453 $4.00 © 1994 IEEE First WCPEC; Dec. 5–9, 1994; Hawaii

additional hydrogen content got minus, in consequence, the accumulated hydrogen content decreased compared with that of conventional films [1]. At medium temperature, according to larger intrusion rate of hydrogen, the accumulated hydrogen content was increased. At lower temperature, diffusion rate of intrusion and effusion were low and the change of hydrogen content by hydrogen plasma treatment process was not large. It is appropriate that these behaviors can be explained by the balance of desorpion of hydrogen atoms from a-Si network and the intrusion of hydrogen atoms from plasma into a-Si network. Because the activation energy of diffusion of hydrogen atoms in the solid phase of a-Si network is reported as 1.5 eV [13] and activation energy of hydrogen diffusion in the case of thin a-Si films is 0.31 eV [14]. The change of accumulated hydrogen content is explained by the difference in the activation energy of two diffusion processes.

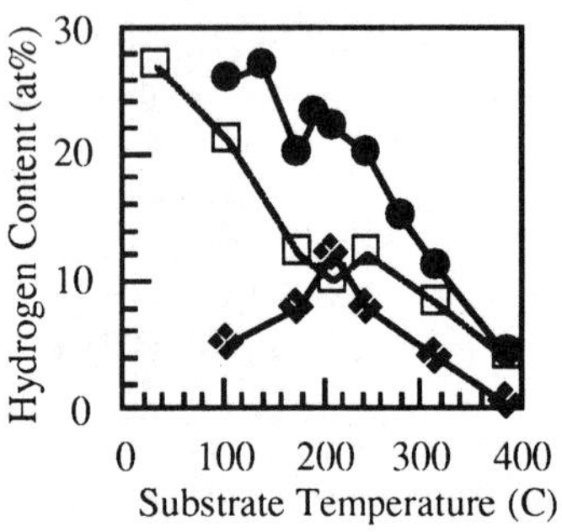

Fig.1 Hydrogen content vs substrate temperature. □; conventional films, ●; ADHT films, ◆; ΔH.

Figure 2 shows the relationships between the bonded hydrogen content and the optical bandgap. The relation had a good linearity regardless of deposition method, and the relation was expressed as

$$Egopt = 0.012[H] + 1.60$$

In the case of ADHT films, bonded-hydrogen content could be increased to more than 30 % and wide range of the optical bandgap (1.64~1.96 eV) was obtained. On the other hands, in conventional CVD, lower substrate temperature less than 100°C must be used to obtain the hydrogen content of 30 %, as a result, quality of conventional films at lower temperature was deteriorated further. So, the advantage of ADHT method is to get wide bandgap and large hydrogen content at usual substrate temperature such as 180°C.

Concerning the bonding configuration, it is reported in other literature [9]. Monohydride (SiH) and dihydride (SiH_2) mode were increased by hydrogen plasma treatment and this increasing tendency of these mode was similar to that of hydrogen content. However, SiH_2/SiH ratio of ADHT films was smaller than that of conventional films at same hydrogen content [9].

Figure 3 shows the relationships between the defect density, measured by constant photocurrent method, and the optical bandgap for ADHT films and conventional films. As it is shown, ADHT films had lower defect density than conventional films for all range of optical bandgap, and low defect densities under 10^{16}cm^{-3} was obtained from 1.65 to 1.9 eV. Especially, defect density was lower at a region of widegap. And also, though the defect density of conventional films had a minimum at optical bandgap of about 1.75 eV, in the case of ADHT films, the optical bandgap at lowest defect density was shifted to widegap and was about 1.84 eV. Hydrogen plasma treatment process is effective to lower the defect density of films and shift the optimum optical bandgap to widegap.

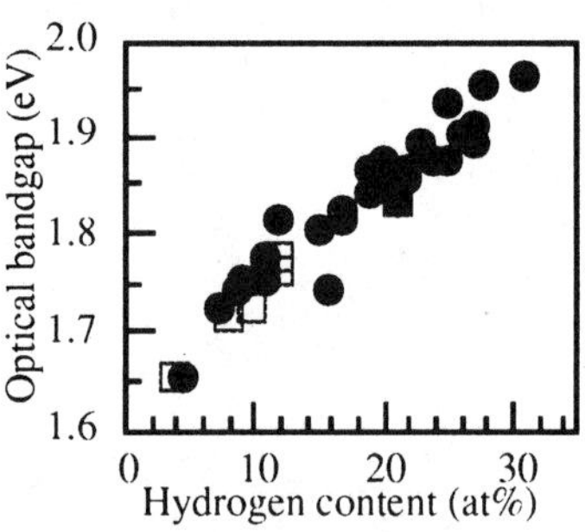

Fig.2 Relationship between hydrogen content and optical bandgap. ●; ADHT films, □; conventional films.

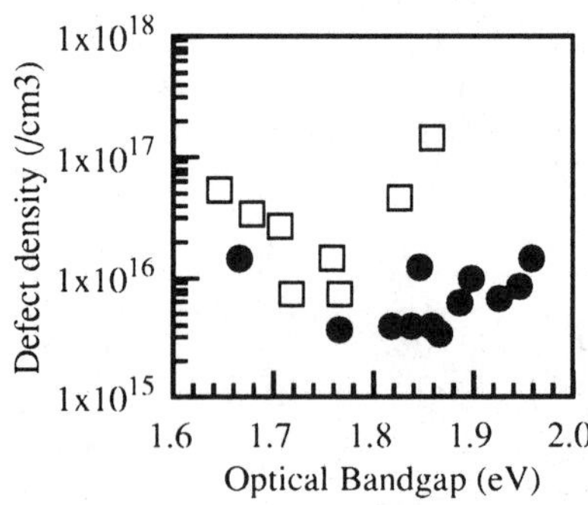

Fig.3 Relationships between defect density and optical bandgap. ●; ADHT films, □; conventional films.

Next, the relationship between the change of hydrogen content and defect density was examined. In figure 4, the change in defect density of a-Si films by hydrogen plasma treatment is plotted against the hydrogen content at various substrate temperature. In this case, the unit thickness of 4nm and hydrogen plasma treatment time of 20 sec were adopted for the preparation of ADHT films. Change in bonded hydrogen content of a-Si films was large for the moderate temperature (~ 210°C) and was small for the higher and lower temperature. These results were already explained in figure 1. The behavior of defect density against the accumulated hydrogen content is discussed. Defect density of ADHT films was always lower than that of conventional films at same substrate temperature. These show that hydrogen plasma treatment process is the reduction process of defect density. Theses results indicate that the hydrogen atoms introduced to a-Si network from hydrogen plasma bond to Si atoms (resulting the bonding configuration of SiH2 and SiH), compensate the dangling bonds existed in a-Si network and act as a terminator. The relation between hydrogen content and defect density (Nd) of the conventional films is not same as that of ADHT films.

$$Nd = 5.7\times10^{18}(Egopt-1.75)^2+6\times10^{15}$$
(conventional films)

$$Nd = 7.0\times10^{17}(Egopt-1.86)^2+3.5\times10^{15}$$
(ADHT films)

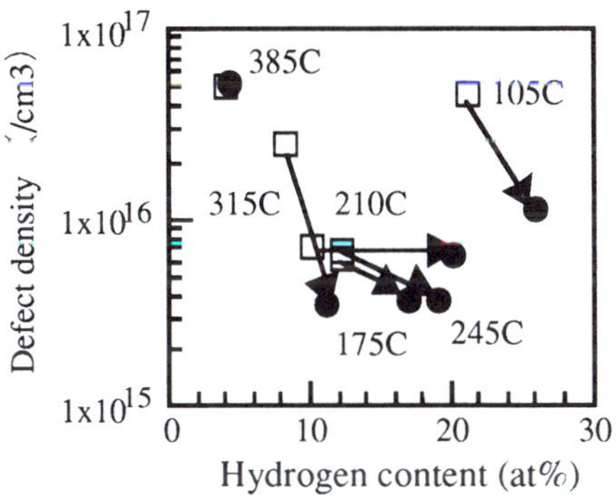

Fig.4 Change in defect density plotted against hydrogen content. ●; ADHT films, □; conventional films.

That is thought that defect of the conventional films is mainly determined by the a-Si surface reaction process, however, defect of ADHT films is controlled by another process such as post-hydrogenation. Therefore, hydrogen plasma treatment process is the compensation process of dangling bonds by post-hydrogenation in solid phase. In addition, it is indicated that the defect density is not always determined by surface reaction process during deposition process, and post-annealing in solid phase (e.g. additive hydrogen atoms injection) has the possibility of lowering the defect density in a-Si network.

The light stability of these ADHT films was tested and the results were shown in figure 5. High stability of photoconductivity was reported about low hydrogen content films by ADHT method [1]. In figure 5, initial and saturated defect density after light soaking were plotted against the optical bandgap for the ADHT films and conventional films. This figure shows the relation between light stability and the optical bandgap, namely hydrogen content. Concerning the ADHT films, initial defect density were lower and saturated defect density after light soaking of 900 hr were very low and had about $3\times10^{16}cm^{-3}$, not depending on the optical bandgap (hydrogen content).

According to high stability of these ADHT films, light soak degradation is not always dependent on the hydrogen content and bonding configuration represented by SiH_2/SiH ratio. High stability might be owing to the elimination of weak-bonds in a-Si network ,or the change in defect recovery rate by thermal and light-induced annealing process by hydrogen atoms modification of a-Si network. However, the origin of high stability of these films is not clear at present.

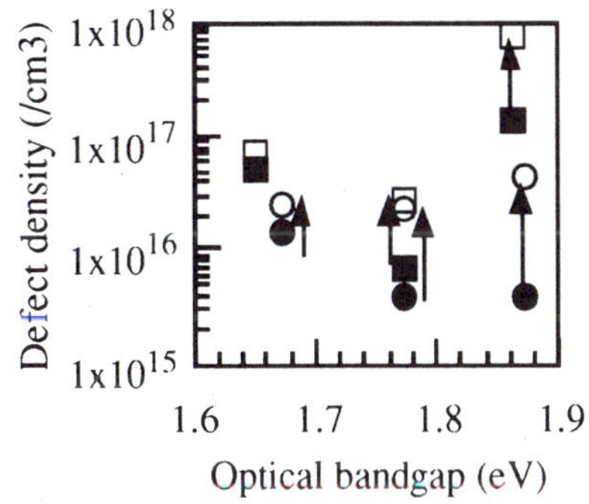

Fig.5 Initial and saturated defect density vs optical bandgap. ●O; ADHT films, □■; conventional films. O□; saturated, ■●; initial.

Application to single and 2-tandem cells

Next, we report about the results of application of these intrinsic films to pin a-Si single and 2-tandem solar cells. In figure 6, the results of application to pin single junction cell are shown for the case of i-layer of ADHT films and conventional films after and before light soaking. In these single junction cells, thickness of i-layer were 100 nm in order to test the possibility of application to a-Si/a-Si 2-tandem cells. Fill factor and efficiency were represented for after and before light soaking and the saturated values were plotted for open circuit voltage. Voc were increased with the optical bandgap and the high Voc of 0.94 V was obtained for the i-layer of the optical bandgap of 1.90 eV. About light soaking degradation, FF of cells with i-layer of ADHT films changed from 0.73 to 0.68, and the degradation rates were about 5 %. And also, the degradation rate of efficiency of about 7% were also stable, comparing with the case of conventional films. These degradation rates are not dependent on the optical bandgap shows that the light instability is not always related the hydrogen content and the bonding configuration. These results are consistent to the one of light stability of a-Si films, above mentioned. It is clear that ADHT films are comparatively stable and applicable to a-Si/a-Si 2-tandem cells.

Figure 7 indicates the Voc and FF of a-Si/a-Si 2-tandem cell with top i-layer of ADHT films and conventional films against the optical bandgap of top i-layer. Voc was increased with the optical bandgap as well as the results of a-Si single junction cells, and extra high Voc of 1.85 eV was obtained for the structure of a-Si/a-Si 2-tandem type solar cells. Moreover, fill factor of 2-tandem cell with i-layer of ADHT films were over 0.70 and higher compared with cells widegap with i-layer of conventional films. High Voc and FF are according to the high quality of ADHT films. These results indicate that the ADHT films could be applicable to light stable structure of a-Si/a-Si 2-tandem solar cells.

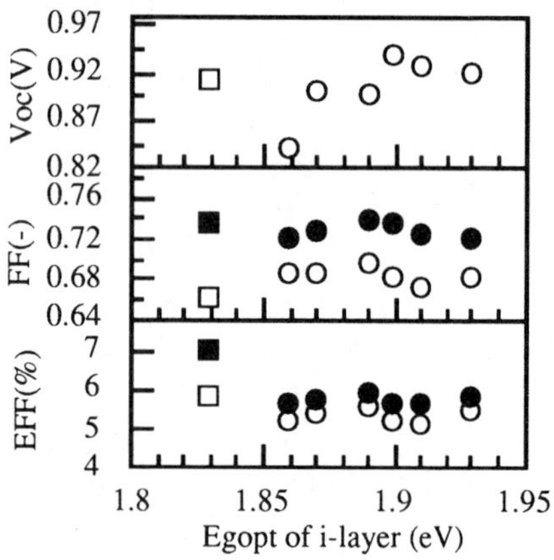

Fig.6 Results of light soaking of pin single junction cell. ●○; ADHT films, □■; conventional films. ○□; saturated, ■●; initial

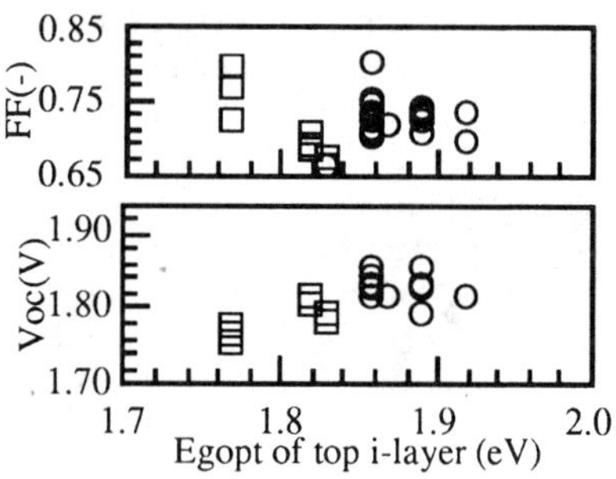

Fig.7 Voc and FF vs Egoptof top i-layer.(a-Si/a-Si 2-tandem cell) ○; ADHT films, □; conventional films.

Using the high reflection back electrode such as ITO/Ag double layer and highly textured SnO_2 to the above mentioned materials and technology[12], high open circuit voltage of 1.85 V and high conversion efficiency of 10.5 % was obtained for the solar cell structure of glass/texture SnO_2/p-layer/p-i a-SiC layer/i-layer(ADHT a-Si 1.86eV, 80nm)/n-layer/p-layer/p-i a-SiC layer/i-layer(conventional a-Si, 1.75eV, 400nm)/n-layer/ITO/Ag (Fig. 8).

Conclusion

Alternately repeating a-Si deposition (thin film deposition process) and hydrogen plasma treatment (modification process) method, ADHT method,was studied to obtain high quality and stable a-Si films and the solar cells. The initial defect densities under 1×10^{16} cm^{-3} and the saturated defect densities of about 3×10^{16} cm^{-3} were obtained at wide range of optical bandgap from 1.67 to 1.96 eV. Hydrogen plasma treatment process is the compensation process of dangling bonds by post-hydrogenation in solid phase. These ADHT films and solar cells were stable to light soaking and their stability were not dependent on the optical bandgap and the hydrogen content. Applying the these widegap a-Si films to the i-layer of top cell of a-Si/a-Si 2-stacked tandem solar cells, high open circuit voltage of 1.85 V and high fill factor of 0.75 was obtained, as a result, the high efficiency of 10.5 % was attained.

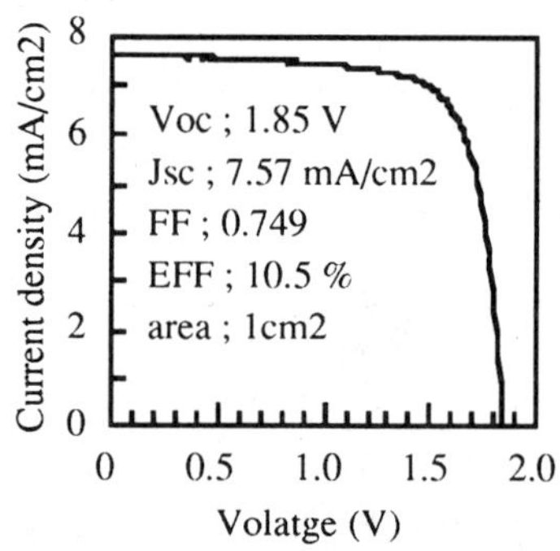

Fig.8 I-V characteristics of a-Si/a-Si 2-tandem cell .

Acknowledgment

This work was supported by the New Energy and Industrial Technology Development Organization as a part of the New Sunshine Program under the Ministry of International Trade and Industry .

References

[1] K. Miyachi et al., Tech. Dig. Int'l. PVSEC-5, (1990) 63
[2] A.H.Mahan et al., J. Appl. Phys., 69,(1991) 6728.
[3] H. Shirai et al., Appl. Phys, Lett., 59 (1991) 1443.
[4] N. M.Johnson et al., Appl. Phys, Lett., 59 (1991) 1443.
[5] Y. Ashida et al., Proc of 22nd IEEE PVSC, (1991) 1352.
[6] Y. Ashida, Solar Energy Mat. Solar Cells,34 (1994) 291.
[7] H. Tanaka et al., Proc of 23nd IEEE PVSC, (1993) 811.
[8] H. Tanaka et al., Solar Energy Mat. Solar Cells,34 (1994) 493.
[9] Y. Ichikawa et al., Proc of 23rd IEEE PVSC, (1993) 27.
[10] K.Nomoto et al., Tech. Dig. Int'l. PVSEC-7,(1993) 275.
[11] S.Guha, Tech. Dig. Int'l. PVSEC-7,(1993) 43.
[12] K.Miyachi et al., Proc. of 11th ECPVSEC, (1992) 88.
[13] D.E. Carson and C.W.Magee, Appl. Phys. Lett., 33 (1978) 81.
[14] M.Nakamura et al.,J.Appl. Phys.,65 (1989) 5238

CONSIDERATIONS FOR LIGHT SOAKING OF AMORPHOUS SILICON PHOTOVOLTAIC MODULES

Werner Luft and Bolko von Roedern
National Renewable Energy Laboratory
Golden, CO 80401

ABSTRACT

To estimate the stabilized performance *a priori*, laboratory tests under controlled, simulated 1-sun, conditions have been performed on hydrogenated amorphous silicon (a-Si:H) multijunction modules by various organizations. Such tests generally use an arc light source and maintain the module temperature at some fixed value, e.g., 50°C. Because the spectra of various light sources used for light soaking do not match that of the sun and because the operating temperature for real systems is not constant and differs from the fixed temperature in laboratory tests, the validity of the laboratory light soaking tests has been questioned. The paper examines the most significant parameters in light-soaking tests to determine their effect and to indicate test strategies to minimize any errors caused by these parameters that may tend to invalidate the results. There is a significant difference (up to a factor of 3) in the relative (to AM1.5) intensity to which various layers of solar cells are subjected depending on the light source spectra. The effect of temperature during light exposure on the stabilized performance varies between a-Si devices of different structures. The aim of this paper is to determine whether a specific standardized, continuous light-soaking test in principle can predict the stabilized performance that will be observed when modules with different a-Si:H-based layers are deployed outdoors.

INTRODUCTION

It is now recognized that the significant parameter for amorphous silicon photovoltaic module performance is the *stabilized* efficiency [Luft et al. 1992]. The stabilized performance is realized after one or more years of actual system operation in a given location. To predict the stabilized performance *a priori*, laboratory tests under controlled conditions have been performed by various organizations. Such tests generally use an arc light source and maintain the module temperature at some fixed value, e.g., 50°C. To shorten the testing time, light intensities as high as 70 suns (70,000 W/m^2) have been used, but is was soon realized that higher light intensities than 1 sun lead to excessive degradation and that there is no general adjustment factor to predict 1-sun performance from higher intensity results [Guha, 1990]. In fact, it appears that different defects are created at the higher intensities [Guha, 1993] and that the fill factor self-anneals at room temperature after high-intensity illumination [Hegedus and Phillips, 1993].

Because the spectra from various light sources used for light soaking do not match that of the sun, and because the operating temperature for real systems is not constant and differs from the fixed temperature in laboratory tests, the validity of the laboratory light soaking tests has been questioned. It is the purpose of this paper to examine the most significant parameters in light-soaking tests to determine their effect on stabilized performance and to indicate test strategies to minimize any errors caused by these parameters that may tend to invalidate the results.

The following items have been identified as potentially affecting the validity of light soaking tests:

A. Illumination — Spectra and Intensity

- o Spectrum and intensity of the light source
- o The spectral response on the subcells in a multijunction structure
- o The mismatch between subcells in a multijunction structure and its change during light exposure
- o The fluctuating component of the light
- o The uniformity of the illumination.

B. Temperature

- o The temperature of the semiconductor layer of the modules
- o The temperature uniformity over any module.

C. Other

- o The loading point of the modules
- o History of temperature and light exposure

ILLUMINATION EFFECTS

Light Sources Used

The light sources used by various organizations for light soaking vary greatly. To obtain the proper magnitude of the

CH3365-4/94/0000-0457 $4.00 © 1994 IEEE
First WCPEC; Dec. 5–9, 1994; Hawaii

light intensity, the light sources are calibrated. The National Renewable Energy Laboratory (NREL) has used either an unfiltered or a filtered silicon reference cell, United Solar Systems Corporation (USSC) and Advanced Photovoltaic Systems, Inc. (APS) use KG-5 (Schott Glass)-filtered silicon reference cells, and Solarex Corporation uses a radiometer.

The specific light sources used by these organizations are:

A. Argon plasma (Vortek) lamps operated at different currents, calibrated with an unfiltered Si reference cell of 350-1100-nm range (I_{sc} = 127.05 mA for 4 cm^2) or calibrated with a KG-5-filtered Si reference cell (12.50 mA/cm^2) used by NREL;

B. Metal (Hg and Xe) arc lamps calibrated with a KG-5-filtered Si reference cell of 330-800 nm range (I_{sc} = 35.7 mA) used by USSC;

C. Sodium vapor lamps calibrated with a spectroradiometer having a 350-1100-nm range used by Solarex;

D. Metal halide lamps (Phillips model #1000/U, 1000 W) calibrated with a KG-5-filtered Si reference cell with a 330-800-nm range (J_{sc} = 14.22 mA/cm^2) used by APS.

The spectral radiances for these calibrated (using a filtered reference cell) light sources and that of AM1.5 sunlight are shown in **Figure 1**. Examining these spectra and comparing them to AM1.5 sunlight, we see that they all have strong peaks in the range of interest (320-1000 nm) that do not occur in the solar spectrum. To determine the effect of these peaks on the performance of amorphous silicon solar cells, we use the methodology described below.

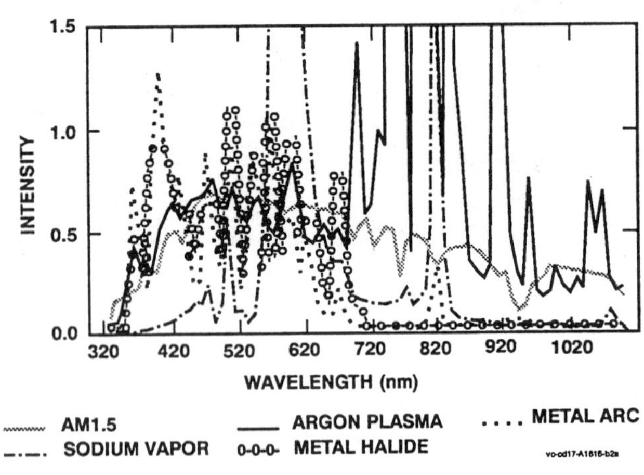

Figure 1. Normalized spectral radiance for four light sources and the global solar AM1.5 spectrum. [Guha, 1993; ASTM E892; Newton, 1993; Myers, 1993]. Normalized with a filtered APS reference cell.

Methodology

We compare the light intensity absorbed by each layer in a multijunction structure from the various light sources and compare them with the corresponding absorbed intensity from sunlight. We obtain the absorbed light intensity at each layer by integrating the product of the normalized source spectrum and the spectral quantum efficiency of the test cell under consideration, $\Sigma E_\lambda(QE)_\lambda$.

The spectral quantum efficiencies for various amorphous silicon cell structures are shown in **Figure 2 (a-c)**. The integral of the product of source spectra and spectral quantum efficiency, $\Sigma E_\lambda(QE)_\lambda$, was calculated for the device structures of interest, namely, same-bandgap dual-junction, different-bandgap dual-junction, and two different-bandgap, triple-junction modules. The ratios of $\Sigma E_\lambda(QE)_\lambda$ for each light source, when calibrated with two different KG-5-filtered reference cells to that for AM1.5 sun, is shown in **Table 1**. The quantum efficiencies for three reference cells, a filtered and an unfiltered silicon cell used by NREL, and a filtered silicon cell used by APS, are shown in **Figure 3.**

In the past, it has been stipulated that the degradation at stabilization for a given cell under illumination (or the stabilized efficiency reached) is dependent on the *intensity* of the illumination, spectrum, loading point, and temperature [Guha, 1990]. The degradation at stability at 50°C as a function of light intensity in the range of 0.5 to 2 suns for a-SiGe:H as derived from NREL experiments [von Roedern, 1994] and for a-Si:H [Guha, 1990] is shown in **Figure 4.** The light intensity in that figure is that at the surface of a triple-junction device. Yang et al. (1991) have proposed a proportionality constant between light intensity (I) and light-induced degradation. However, we claim that the proportionality of the degradation and the light intensity [$D \propto I^{1.8}$] applies only to *equivalent times* for a given degradation *prior to reaching stability* but that the function is not appropriate to determine the ultimately stabilized performance.

Discussion

From **Table 1** we see two things: 1) that none of the light sources match sunlight for all component cell layers, and 2) filtered reference cells having different spectral responses will result in differences in the relative intensity to which any one layer is subjected. The first factor results in up to a factor of 3 in the relative (to AM1.5) intensity to which various layers are subjected. The difference in relative intensity on a given layer may vary by up to 30% depending on the filtered reference cell used to set the light intensity.

The NREL unfiltered argon plasma arc (Vortek) light source is rich in the red part of the spectrum, especially in the 740 -860-nm range, as compared to AM1.5 sunlight. There is also a peak in the 560-600-nm range. The light source gives a good match for the same-bandgap dual-junction

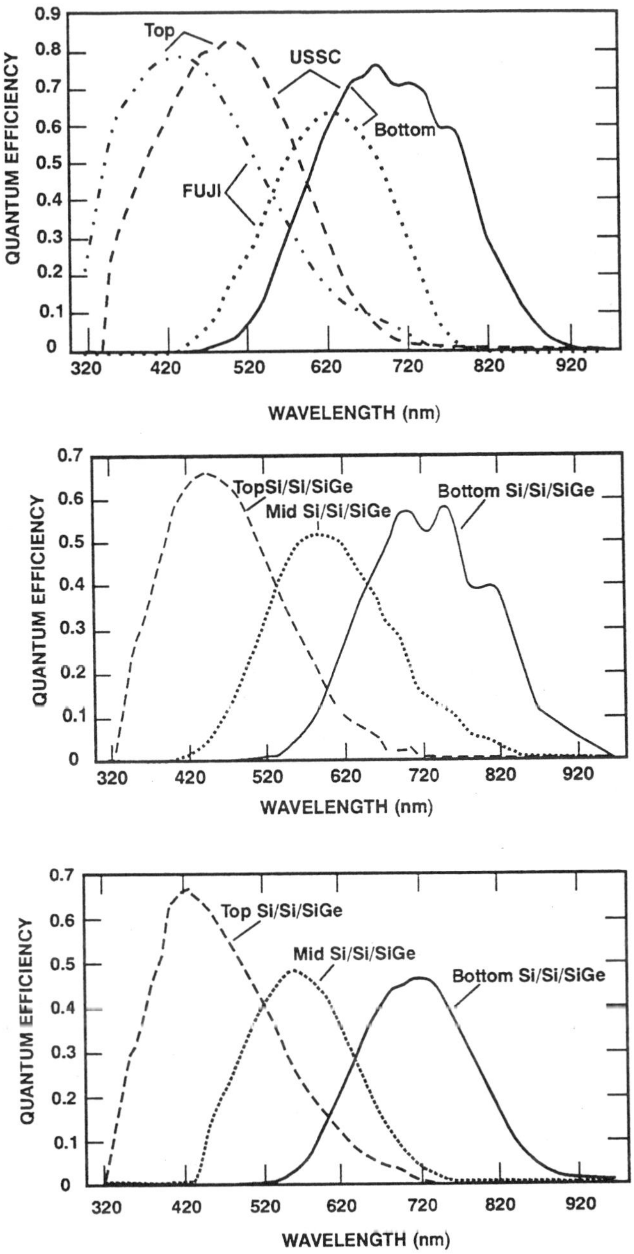

Figure 2. Spectral quantum efficiencies for (a) dual-junction, same bandgap [Ichikawa, 1993] and dual-junction, different bandgap [Guha, 1993]; (b) triple-junction USSC [Guha, 1993], and (c) triple-junction Solarex [Newton, 1993] amorphous silicon modules.

cell, but not for the bottom cell of a different-bandgap dual-junction cell. It is quite good for the top and middle cells in triple-junction structures, but gives too much intensity for the bottom cells in such structures.

The metal arc lamp used by USSC does not give a very good match for the bottom cells in dual-junction and triple-junction structures; it provides insufficient intensity.

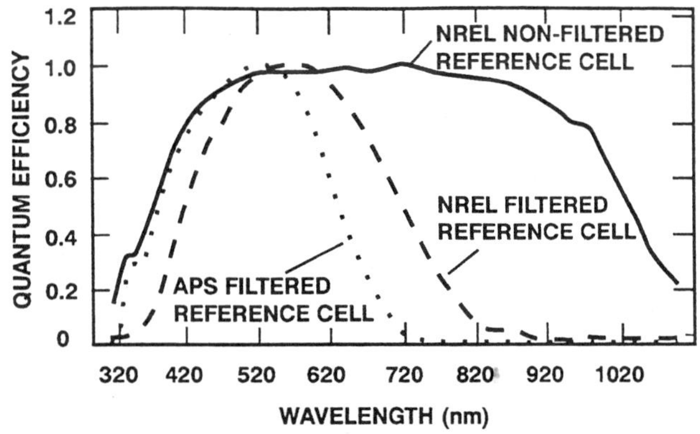

Figure 3. Spectral quantum efficiencies for reference cells used in the calculations [Burdick, 1993].

The sodium vapor lamp used by Solarex has a similar peak as the Vortek lamp in the 580-630-nm range and a small peak in the 810-830-nm range, but much less than the argon plasma. The light source provides insufficient intensity for the top cells for all structures investigated and an intensity about 40% too high for the mid cell of triple-junction structures.

Finally, the metal halide source used by APS gives insufficient intensity for the bottom cells of all structures.

If we match the light source and cells by manufactures, we find that USSC using the metal arc source has the proper illumination light level for the top cell in their different-bandgap, dual-junction case, but only 40% of the required intensity for the bottom cell. For their triple-junction cell, the ratios are 1.07, 0.78, and 0.25 for the top, middle, and bottom layers, respectively. Nevertheless, USSC has

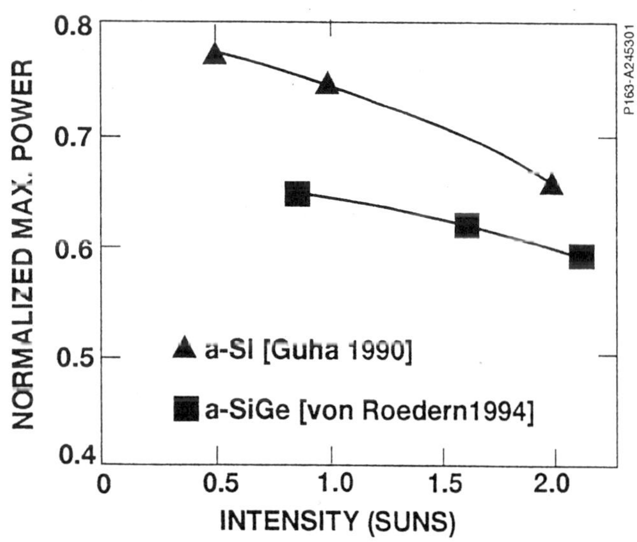

Figure 4. Degradation of relative maximum power at stability vs. light intensity [Guha, 1990; von Roedern, 1994].

Table 1. Relative intensity to AM1.5 on the subcells for various modules for four light sources when calibrated with two types of filtered reference cells

Module Type	Cell in Structure	Light Source							
		Argon Plasma		Metal Arc		Sodium Vapor		Metal Halide	
		1	2	1	2	1	2	1	2
Same-Bandgap, Dual-Junction (Fuji)	Top	1.04	0.81	1.06	1.34	0.61	0.56	1.01	1.19
	Bottom	1.15	0.89	0.69	0.87	1.42	1.32	0.81	0.97
Different-Bandgap, Dual-Junction, (USSC)	Top	0.99	0.77	1.04	1.31	0.84	0.78	1.01	1.20
	Bottom	2.49	1.94	0.40	0.50	1.15	1.07	0.49	0.58
Triple-Junction (USSC), [Guha, 1993]	Top	1.02	0.79	1.07	1.35	0.61	0.57	1.02	1.21
	Mid	1.24	0.97	0.78	0.97	1.41	1.32	0.86	1.02
	Bottom	3.28	2.57	0.25	0.32	0.99	0.92	0.30	0.36
Triple-Junction (Solarex), [Catalano et al., 1993]	Top	1.03	0.80	1.08	1.36	0.61	0.57	1.02	1.22
	Mid	0.96	0.75	0.91	1.13	1.40	1.32	0.96	1.14
	Bottom	2.88	2.24	0.25	0.32	0.91	0.84	0.33	0.39

1. APS filtered reference cell; 2. NREL filtered reference cell

reported good correspondence of degradation under the metal arc light source and under a AM1.5 solar simulator for different-bandgap dual- and triple-junction modules [Guha, 1994a].

The ratios of intensities for Solarex triple-junction cells with their sodium vapor lamp are 0.61, 1.40, and 0.91. However, a comparative study of degradation of Solarex triple-junction modules outdoors and under sodium vapor lamps found that the degradation was quite similar under both conditions [D'Aiello and Newton, 1992].

During earlier light soaking tests by NREL, an unfiltered Si reference cell was used to set the argon plasma light source intensity. To estimate the effect of that procedure as compared to using a filtered reference cell, we compare the intensity seen by the various layers of test cells relative to that under one-sun illumination. The ratio of integral product of source spectral radiance (when calibrated with an unfiltered and a KG-5-filtered reference cell) and spectral quantum efficiency for various modules for the argon (Vortek) light source (operated at 62 amperes) relative to that for AM1.5 sun is shown in **Table 2**. Depending on the structure, for the unfiltered reference cell case, the top cell receives only about 50% of the light as compared to AM1.5, while the bottom cell receives 120%-157% of that for AM1.5 for different-bandgap, dual-junction and triple-junction cells. For the filtered reference cell case, the intensity seen by each layer is essentially doubled relative to the unfiltered case.

As the intensity on the various cell layers was still unsatisfactory, even when using a filtered reference cell, NREL filtered the light from the argon light with an HAG-003, tempered, heat absorbing, 6.3-mm-thick filter by Precision Glass & Optics (PGO) and increased the lamp current to 120 A [Mrig et al., 1994]. This filtering gives a nearly perfect match to AM1.5 for all structures and layers, as seen in **Table 3.**

Figure 4 shows the relative performance at stability for a-Si:H [Guha, 1990] and a-SiGe:H [von Roedern, 1994]. The a-Si:H shows a decline in the stabilized power of 5% from 0.5 to 1 sun, and 12% from 1 to 2 suns. The a-SiGe:H decreases 9% from 0.56 to 2.1 suns. Usually, the top layer in multijunction structures is the current-limiting layer. In that case, the relative intensity endured by that layer should determine the resulting overall device degradation. That means when the top layer sees a relative intensity of unity, the device degradation should essentially match that experienced in sunlight. From Table 1, we see that the relative intensity experienced by the top layer for the four light sources investigated ranges from 0.56 to 1.36. Based on Figure 4, the corresponding

Table 2. Ratio of integral product of source spectral irradiance and spectral quantum efficiency for various modules for argon (Vortek) light source (operated at 62 A) relative to that for AM1.5 sun when calibrated with an unfiltered and a filtered NREL reference cell

Type structure	Cell in structure	Unfiltered NREL reference cell	Filtered NREL reference cell
Same-bandgap, dual-junction (Fuji)	Top	0.50	1.04
	Bottom	0.55	1.15
Different-bandgap, dual-junction (USSC)	Top	0.48	0.99
	Bottom	1.20	2.49
Triple-junction (USSC)	Top	0.49	1.02
	Mid	0.60	1.24
	Bottom	1.57	3.28
Triple-junction, (Solarex)	Top	0.49	1.03
	Mid	0.46	0.96
	Bottom	1.38	2.88

Table 3. Ratio of integral product of source spectral radiance and spectral quantum efficiency for various modules for the filtered argon (Vortek) light source (operated at 120 A) relative to that for AM1.5 sun when calibrated with a filtered NREL reference cell

Type structure	Cell in structure	Filtered NREL reference cell
Same-bandgap, dual-junction (Fuji)	Top	0.99
	Bottom	0.89
Different bandgap, dual-junction (USSC)	Top	0.98
	Bottom	0.97
Triple-junction (USSC)	Top	1.00
	Mid	0.91
	Bottom	1.06
Triple-junction, (Solarex)	Top	1.00
	Mid	0.92
	Bottom	1.02

variations in stabilized power output are less than ±5%. Consequently, the ranges of relative intensities for the different component cells in dual- and triple-junction structures do not matter greatly. This explains why the degradation for USSC and Solarex cells under their light sources and under sunlight give essentially the same degradation.

TEMPERATURE EFFECTS

The stabilized performance is significantly affected by the temperature *during light exposure.* There are two sets of data generated at NREL to support this point.

1. Four commercial 0.09-m^2 dual-junction a-Si/a-Si UPM-880 modules (SEB1-1002, -1029, -1028, and 1032) from the United Solar Systems Corp. were exposed to outdoor condition in Colorado for 11 months from September 17, 1991 to August 26, 1992. They dropped to as low as 73% of initial efficiency by mid-January 1992, and then recovered to about 80% by mid-June through mid-August. Thus, the performance actually improved during summer months from that obtained during winter months. The module temperature in January was 24°C and by June-August was 34°-44°C. This means that the degradation increases by 2.5 to 4 percentage points for each 10°C lower temperature than the 50°C generally used in the NREL stability testing.

2. Lightsoak test No. 1 was a controlled-exposure test for a-Si modules from three manufacturers performed at a 50°C module temperature [Luft et al., 1991]. After

the controlled exposure to 1000 W/m² (unfiltered Si reference cell) for 2000 h and subsequent annealing to 60°C and 70°C, the modules were deployed outdoors at the NREL test facility in Colorado starting in April 1992. Measurements of the performance were done at various intervals of outdoor exposure.

Although the performance for same-bandgap, dual-junction modules from two manufacturers, Chronar Corporation [Group 1] and USSC [Group 3], had stabilized at the end of the controlled 50°C exposure, all three sets of modules exhibited further degradation as a result of the outdoor exposure as illustrated in **Figure 5** in the extension of the linear time scale of the insert. We have equated 1 month outdoors to 175 h on that time scale. A remarkably fast adjustment (less than 300 h of equivalent 1000 W/m² illumination) of the previously stabilized performance was observed to a new, lower, stabilized level. This adjustment occurred for all three types of modules (same-bandgap, dual-junction on glass; same-bandgap, dual-junction on stainless steel; and triple-junction on glass). The additional degradation from the 70°C annealed state after 2000 h light exposure ranged from 8%-9% for APS dual-junction, same-bandgap modules; to 11%-12% for 3 Solarex triple-junction modules; and 13% for one USSC dual-junction, same-bandgap module.

It might be speculated that some of this further degradation is due to other causes (e.g., corrosion) rather than due to Staebler-Wronski-type degradation. However, we found by annealing at >140°C that the degradation encountered was reversible. This finding suggests that rapid adjustment of the stabilized state does indeed occur, and that the stabilized level essentially establishes itself within a relatively short time (~100 h) according to the actual temperature and illumination under operating conditions. This suggests that the constant 50°C condition of our test may noticeably underestimate the degradation observed under real outdoor conditions in Colorado even during summer months.

USSC has made controlled temperature experiments on its a-Si/a-Si modules at 30°, 50°, and 70°C and found 5%-8% less degradation at stabilization per step as the temperature is increased in 20°C steps. In another USSC test for dual-junction and triple-junction cells lightsoaked for 1360 h at 30°C and 50°C, the lower temperature resulted in 4% lower stabilized power output [Guha, 1994b].

OTHER EFFECTS

It has been observed that the degradation co-varies with the voltage imposed across the modules [Catalano et al., 1992, p. 14]. Thus, the degradation is least under short-circuit

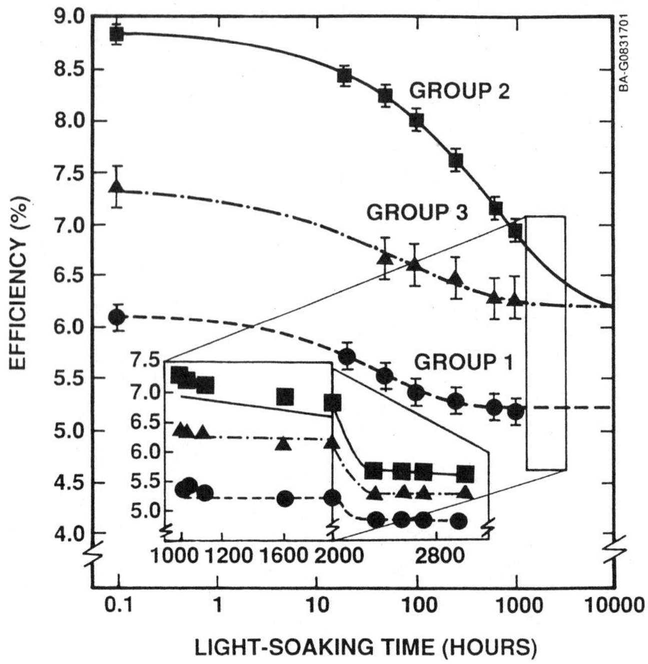

Figure 5. Additional degradation for three sets of amorphous silicon modules after controlled light soaking at 50°C when exposed to outdoor conditions [based on Luft et al., 1991]

conditions and highest under open-circuit voltage conditions. Tests are generally conducted with fixed resistors across the terminals to approximate operation near the maximum power point. In multijunction structures, where the source spectrum degrades the various layers at different rates from that in actual sun operation, additional, secondary errors may be introduced [Catalano et al., 1992, p. 33-38].

CONCLUSIONS

We have shown that all light sources used by the U.S. PV industry to evaluate the stabilized efficiencies of a-Si modules lead to erroneous light intensities on some layers in multijunction devices as compared to those experienced in AM1.5 global sunlight. Overall, the relative light intensities range from 0.25 to 3.3. The errors are strongly affected by the spectral response of the reference cell used for light source calibration. Even crystalline Si reference cells filtered to match the spectrum of a-Si cells will result in differences in the light intensity ratios from one reference cell to the next depending on their spectral response. One way to avoid the erroneous light intensities is to filter the light source to match its spectrum in the region of a-Si response to that of sunlight. NREL has done this.

Usually, the top layer in multijunction structures is the current limiting layer. In that case, the relative intensity endured by that layer should determine the resulting degradation. That means when that layer sees a relative intensity of unity, the device degradation should essentially match that in sunlight. (However, this depends how strongly the top layer is current limiting and how serious the mismatch in relative intensities is for the various layers.) For top-cell current-limiting structures, the range of relative intensities is 0.56 to 1.36. Since the stabilized performance is essentially flat (within ±5%) over the range of 0.56 to 1.36 suns intensity, the observed errors in intensity for the top layer are of little importance to the stabilized performance. *The error in stabilized performance because of that range is only ±5%.*

The additional degradation experienced under outdoor conditions after controlled indoor exposure can be explained by the generally lower temperatures to which the devices deployed outdoors were subjected. What is important to the stabilized power output appears to be the device temperature *during illumination,* not the temperature when the device is not illuminated. The drop in output at the lower temperature during sunlight exposure as shown in Figure 5 is much greater than the effect of any annealing experienced in the dark at 70°C. Once a device has stabilized after illumination at a low temperature, the performance does not increase again unless the device is subjected to annealing at a high temperature of 150°C or greater.

Thus, temperature during illumination rather than light source is the biggest factor in differences in stabilized performance between indoor light soaking and outdoor exposure.

ACKNOWLEDGEMENTS

We want to thank D. Myers for measuring the spectra for the NREL light source, J. Newton for the Solarex Na source spectrum, J. Burdick for the quantum efficiencies for NREL reference cells, Y. Ichikawa for the Fuji test cell quantum efficiency, and F. Kampas for the quantum efficiency for the APS reference cell. This work was supported by the U.S. Department of Energy under contract No. DE-AC36-83CH10093.

REFERENCES

Burdick, J., Personal communication 1993.

Catalano, A., et al., "Research on Stable, High-Efficiency Amorphous Silicon Multijunction Modules", Annual subcontract report, 1 May 1991 - 30 April 1992, by Solarex Thin Film Division, NREL/TP-411-4995, 1992, NTIS Accession No. DE92010599.

D'Aiello, R.V., and J. Newton, "Outdoor Performance of Triple-Junction Amorphous Silicon Modules," *Proceedings Photovoltaic Performance and Reliability Workshop,* Golden, CO, 1992, SERI/CP-411-5184, p. 149 ff and Figure on p. 155.

Guha, S., "Research on High-Efficiency, Multiple-Gap, Multijunction Amorphous Silicon-Based Alloy Thin-Film Solar Cells," Final Subcontract Report SERI/TP-211-3918, 3/1/1987-2/28/1990, under contract with SERI, 1990, p. 108 ff. NTIS Accession No. DE90000364.

Guha, S., "Research on Stable, High-Efficiency Amorphous Silicon Multijunction Modules," Annual Subcontract Report NREL/TP-411-5654, 1/1/1992-2/28/1993, under contract with NREL, 1993, p. 17. NTIS Accession No. DE93010046.

Guha, S., "Research on Stable, High-Efficiency Amorphous Silicon Multijunction Modules," Semi-Annual Subcontract Report NREL/TP-411-6396, 3/1/1993-11/30/1993, under contract with NREL, 1994a, pp. 28-29. NTIS Accession No. DE94000298.

Guha, S., Personal communication 1994b.

Hegedus, S.S., and J.E. Phillips, "Stable, High Efficiency Amorphous Silicon Solar Cells with Low Hydrogen content", Final Subcontract Report NREL/TP-451-5756, 3/1/1992-3/31/1993, under contract with NREL, 1993, NTIS Accession No. DE93018214.

Ichikawa, Y., Private communication 10/25/1993.

Luft, W., B. von Roedern, B. Stafford, D. Waddington, and L. Mrig, "Controlled Light-Soaking Experiment for Amorphous Silicon Modules," *Conference Record of the 22nd IEEE PV Specialists Conference,* Vol. II, 1991, pp. 1393-1398.

Luft, W., B. Stafford, and B. von Roedern, "Perspective on Photovoltaic Amorphous Silicon," *AIP Conference Proceedings* No. 268, 1992, pp. 347-356.

Kampas, F., Personal communication, 1993.

Mrig, L., J. Burdick, W. Luft, and B. Kroposki, "Outdoor performance stability and controlled light-soaking testing of multijunction amorphous silicon PV modules at NREL," *This conference.*

Myers, D., Personal communication, 1993.

Newton, J. Personal communication, 11/17/1993.

von Roedern, B., unpublished data, 1994.

Yang L., L. Chen, and A. Catalano, "Intensity and temperature dependence of photodegradation of amorphous silicon solar cells under intense illumination," *Appl. Phys. Lett.* **59**, No. 7, 1991, pp. 840-842.

A CRITICAL REVIEW OF THE GROWTH AND PROPERTIES OF A- (SI,GE):H

Vikram L. Dalal, Sanjeev Kaushal,Jun Xu and Kay Han
Dept. of Elec. Engr. and Microelectronics Research Center
Iowa State University, Ames Iowa 50011

ABSTRACT

We review the status of a-(Si,Ge):H technology for tandem cell applications. In particular, we show that the chemistry of growth and deposition conditions play a major role in determining the properties of a-(Si,Ge):H. We explain why dilution is necessary to make good films, and show that ion bombardment plays a major role in improving the properties of the material.Recent work on films deposited using ion bombardment in a controlled ECR reactor environment is discussed, and it is shown that fundamental material properties such as Urbach energies and mid-gap defects can be improved using controlled bombardment. It is also shown that high ion bombardment in glow-discharge deposited devices improves the fundamental material properties. We also discuss the unknown parameters in the a-(Si,Ge):H system, and suggest some research avenues that should be pursued.

INTRODUCTION

Amorphous Silicon Germanium [a-(Si,Ge):H] is an important photovoltaic material for tandem junction solar cells. In the past, this material has been shown to have poorer electronic properties, such as higher defect densities and higher Urbach energies compared to a-Si:H[1]. It also has a higher void density than a-Si:H.Recent results[2,3] indicate that by using significant H dilution, the material can be improved somewhat. In this paper, we discuss alternative growth techniques which may lead to better material, and also discuss appropriate device designs that may improve the devices.

GROWTH CHEMISTRY

Growth chemistry plays a major role in the deposition of high quality material using plasma deposition. In a plasma, several chemical reactions can take place, and many species can be generated. As an example, Silane can be fractionated into SiH_3, SiH_2 and SiH radicals, and it is also possible for these radicals to interact with each other, and with the parent molecule. Moreover, if the concentrations of the different species of radicals are comparable, then the growth of the film can become complicated. As an example, if both SiH_3 and SiH_2 are present, then the insertion reaction leading to growth can be as shown in Fig. 1. As is obvious from Fig. 1, after an insertion reaction, the surface of the material becomes non-homogeneous, with some sites are empty, ready to accept another radical, but other sites are passivated by H, and this H has to be removed before another radical can insert itself. Thus, multi-species reactions can lead to non-homogeneous growth, which will lead to excessive hills, valleys and voids. At the low growth temperatures, the structure cannot homogenize by diffusion of radicals, and these defects will be frozen in. Therefore, it is necessary to achieve a growth chemistry where one can lead to a predictable growth from one species.

For the growth of a-(Si,Ge):H, these conditions become even worse. The two primary precursors, silane and germane, fractionate at very different rates. Therefore, it is essential that the chemical reaction proceed in a predictable way, or that extra energy be imparted to the surface to improve the diffusion of radicals, and hence the properties of the material. One predictable way of forcing the reaction is to use an excess of hydrogen. With high concentrations of H present in the plasma, the predictable chemical reactions would be :

$H + SiH_4 = H_2 + SiH_3$

$H + GeH_4 = H_2 + GeH_3$

Thus, excess of H will force the production of silyl and germyl radicals, and we will get a somewhat homogeneous insertion reaction.

We still have a problem with the different mobilities of germyl and silyl radicals on the surface. To overcome this problem, it is useful to use ion bombardment, either with H or with an inert gas such as He.
Therefore, it is not surprising that the use of highly diluted H plasmas, with high ion bombardment

CH3365-4/94/0000-0464 $4.00 © 1994 IEEE

energies, has led to the growth of better a-(Si,Ge):H films. Another advantage of H dilution is that the presence of excess H prevents the reactions between radicals, thereby preventing formation of higher polymers, which will produce highly nonuniform films.

Another aspect of the growth chemistry deals with the relationship between bandgap and Ge content. The bandgap of a-(Si,Ge):H alloy depends not only on the Ge content, but also on what happens during growth. Excess H etching during growth can lead to a widening of the gap, thereby requiring more Ge to achieve a given bandgap. In general, more Ge incorporation will lead to greater non-homogeneities in the film, and poorer film properties. To overcome this, it may be useful to limit the amount of H-induced etching during growth. A technique such as neutral ion bombardment may allow both increased surface mobility and reduced etching, thereby leading to a lower gap and yet a better material for a given Ge concentration.

PROPERTIES OF FILMS GROWN IN ECR REACTOR

To achieve growth of a-(Si,Ge):H films using controlled ion bombardment, we used an electron-cyclotron-resonance (ECR) reactor to achieve growth. The reactor geometry has been described previously[4]. The films were grown using a remote ECR low pressure plasma beam of Helium. The He ions and radicals stream towards the substrate, and react with silane and germane to produce the a-(Si,Ge):H film. Typical dilution ratio of He:(Silane + Germane) was 15:1. Typical growth rates were between 1-1.5 A/sec. The films were grown at temperatures of 350 C, and the pressure was 10 mT.

In Fig. 2, we show the photo-conductivities of the ECR generated films,and compare them with the properties of glow-discharge generated films. It is apparent from this figure that the ECR generated films have higher photo conductivities for almost all the Tauc bandgaps of interest ($>$1.4 eV).In Fig. 3, we plot the valence band Urbach energy of the films, measured using sub-gap photo-conductivity techniques[5] as a function of Tauc gap. Once again, the ECR-generated films appear to have lower Urbach energies than the glow-discharge films. In Fig. 4, we plot the subgap absorption coefficient(α) for one of these films, as a function of photon energy. Finally, in Fig.5, we plot the subgap α as function of Tauc gap. It is obvious from Fig. 3,4 and 5 that the ECR-generated films have better electronic properties than comparable glow

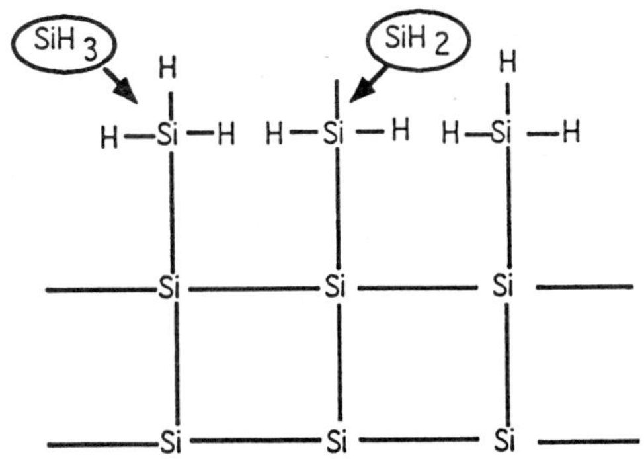

Fig. 1
Schematic diagram showing how insertions by different radicals can lead to non-homogeneous film growth

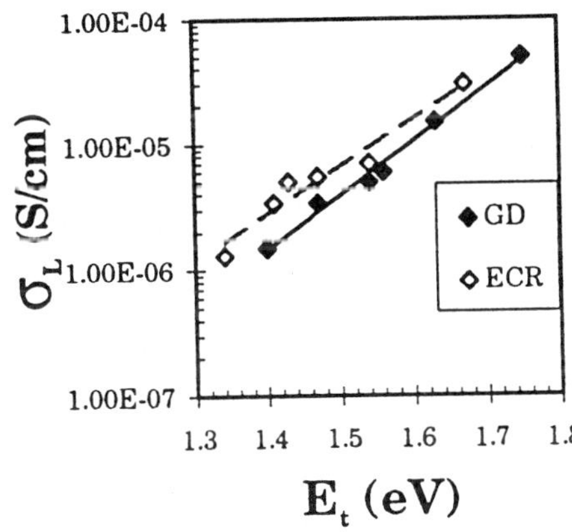

Fig. 2
Photo-conductivity of ECR and GD deposited a-(Si,Ge):H films as a function of Tauc gap

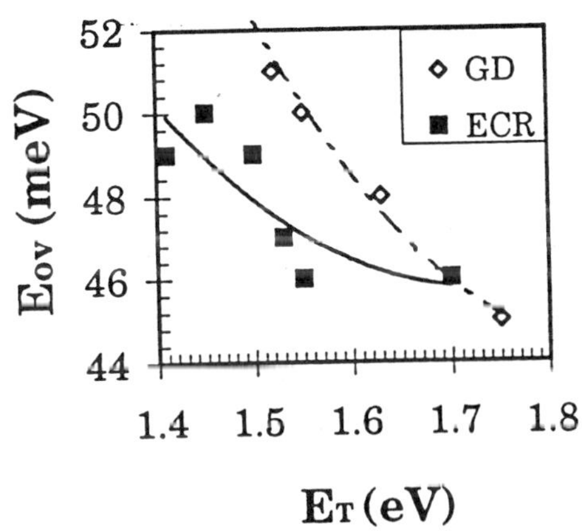

Fig. 3
Urbach energy of valence band as function of Tauc gap for ECR and GD deposited a-(Si,Ge):H films

discharge films. We speculate that this improvement is a consequence of ion bombardment during growth. To show that a low pressure ECR discharge does produce significant ion bombardment, we show, in Fig.6, the plasma potential as a function of pressure in the discharge. At the pressures we are using, the He ions are bombarding the film with 10-20 V energies.

A particularly noteworthy feature is the lower bandgap(1.67 eV) for a-Si:H that we can achieve using He-ECR technique. This is a manifestation of the absence of H-induced broadening of the gap when the material is grown using a He-ECR plasma. The low gap in a-Si:H is similar to the results achieved by et al[6] using highly diluted(100:1) He remote plasma discharge, except that we have achieved this at much higher growth rates than the previous work.

DEVICE PARAMETERS IN A-(SI,GE):H RESEARCH

To design the best device, it is necessary not only to know the defect densities, but also to know the necessary parameters for designing devices. Among the parameters we need are : electron and hole mobility-lifetime products, and the positions of the conduction and valence bands for different alloys. Just knowing these quantities as functions of Tauc gap or Ge content will not do, since we have shown above that materials such as H and He in the discharge have a profound influence on the material properties, including Tauc gaps and subgap α. Unfortunately, no comprehensive set of measurements of electronic properties , or electron affinities exist for these materials. Worse, some of the measurements that exist (such as measurements of electron affinity mismatches [7]) contradict the usual device design and data. As an example, the device design proposed by both Guha et al[8] and Dalal et al[9] proposes that the bandgap be graded from a low gap near the p-i interface to a larger gap near the i-n interface. This design is based on the assumption that changing the gap introduces at least significant changes in the valence band position, and that this movement of valence band will help the holes by introducing additional electric field. But the data of et al[7] show, instead, that most of the movement as we change the gap is in the conduction band,and not in the valence band. If that is the case, it is not clear why the proposed device design works. Actually, the best device would be a reverse graded device, with the bandgap decreasing away from the p interface. But, when one makes such a device, its performance is

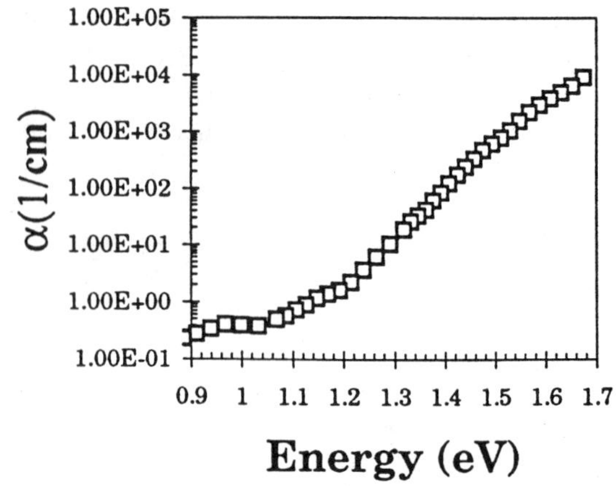

Fig. 4
Subgap absorption coefficient α of an a-(Si,Ge):H film deposited using ECR deposition

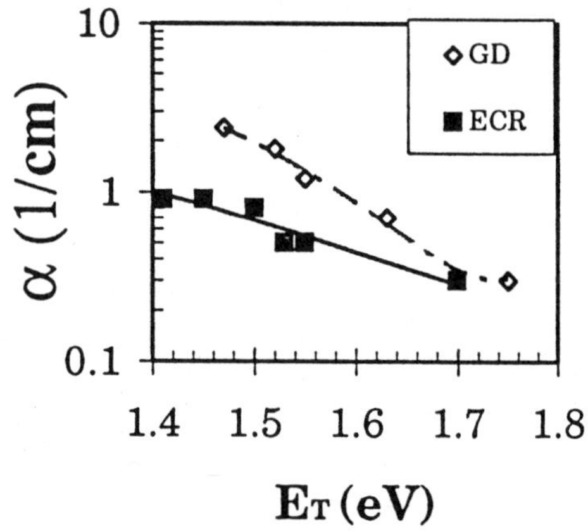

Fig. 5
Subgap a at shoulder vs. Tauc gap for a-(Si,Ge):H films deposited using ECR and GD techniques.

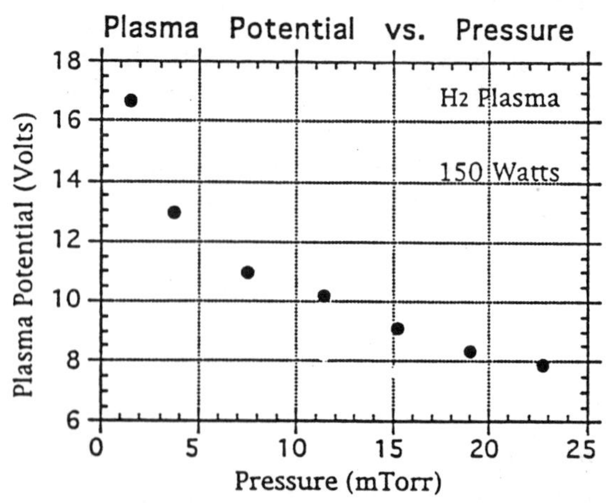

Fig.6
Plasma potential in ECR reactor as function of pressure

inferior to that of the standard or normal graded gap cell. This discrepancy between two sets of experimental results has to be taken into account somehow. Clearly, much more work on determining the positions of Fermi energy,and the valence and conduction band edges as functions of deposition conditions for the a-(Si,Ge):H alloy has to be done. This is a major deficiency in the data base we have.

Another major deficiency in the data base is the lack of reliable data on hole properties in the material, when measured in a device. We have shown before that careful measurements of quantum efficiency of a p-i-n cell can yield useful information about Urbach energies[10] and hole mobility-lifetime products[11]. There is very little data of this kind on the different types of a-(Si,Ge):H materials which are being made into devices. Just to illustrate the importance of making these measurements on devices, we show, in Fig. 7, the data for Urbach energy measured in a p-i-n device as a function of reactor pressure. The devices were all made at about 275 C, using a triode glow-discharge reactor, using a high H dilution plasma process. The substrate was biased to - 100 V. Quite clearly, there is a systematic change in Urbach energy as pressure increases, for reasons that we do not understand at present. Similarly, in Fig.8, we show the influence of changing the temperature while keeping the pressure constant at 50 mT. Since the device, being on a conducting substrate, does not charge up under ion bombardment whereas a film does, it is essential that such measurements be made on devices, and not only on films. And yet, it is rare to see such measurements. In the absence of such measurements, it is not a surprise that researchers find significant differences between the stability of films and devices. It is possible that the material in the device is not the same material found in the film, especially for a complicated material system such as a-(Si,Ge):H.

CONCLUSIONS

In conclusion, we have shown that the deposition chemistry of a-(Si,Ge):H plays a major role in its properties. We have shown that H_2 dilution can be expected to significantly improve the material properties by controlling the reaction chemistry in a predictable way. We have also shown that controlled ion bombardment, as can be achieved using an ECR source, may help in improving the material properties. We have stressed the need for making many more measurements in devices.

The tin oxide substrates were provided by Solarex. This work was partially supported by NREL.

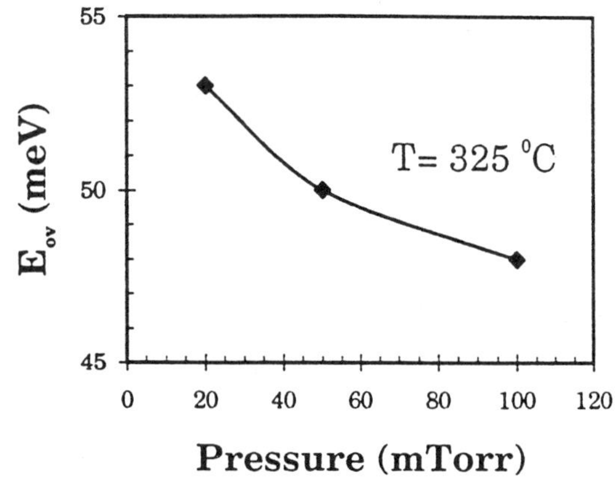

Fig. 7
Variation of Urbach energy of devices with pressure

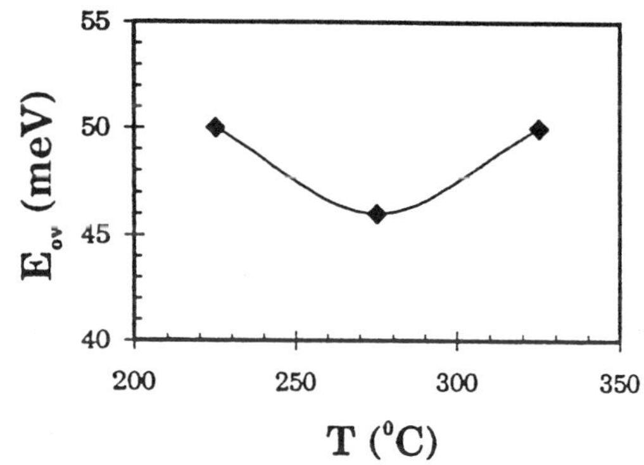

Fig. 8
Influence of temperature on Urbach energy of devices

REFERENCES

1. V. L. Dalal and G. Baldwin, Proc. of 23rd. IEEE Photovolt.Conf.,p.1037(1993)
2. S. Jones et al, Proc. Matl. Res. Soc.,297,815(1993)
3. W. Paul, J. Non-cryst. Sol.164-166,1(1993)
4. V. L. Dalal, M. Leonard and G. Baldwin, J.Non-Cryst.Sol.,164-166,71(1993)
5. S. Lee, S. Kumar and C. Wronski, J.Non-Cryst.Sol.,114,316(1989)
6. K. Park et al., Proc. Mat'l. Res. Soc.,297, 767(1993)
7. J.P.Conde, V. Chu,D. Shen and S. Wagner,J.Appl.Phys. 75,1638(1994)
8. S. Guha, J. Yang, A. Pawliekwicz, T. Glatfelter, R. Ross and S. Ovshinsky, Appl.Phys.Lett. 54,2330(1989)
9. V. Dalal and G. Baldwin, Proc. Mat'l. Res. Soc. 297,833(1993)
10. V. L. Dalal, R. Knox and B. Moradi, Solar Energy Mat'l. and Solar Cells, 31,349(1993)

MOBILITY AND OPTICAL GAPS IN DIFFERENT a-Si:H BASED MATERIALS AND THEIR IMPACT ON CELL PERFORMANCE

Ing-Shin Chen, Taraneh Jamali-beh, Yeeheng Lee, Ching-Yi Li, and C. R. Wronski
Electronic Materials and Processing Research Laboratory
Pennsylvania State University, University Park, PA 16802

ABSTRACT

This work investigates the correlation between the "mobility gaps" (E_μ) and the "optical gaps" (E_{OPT}) of a-Si:H materials and the open circuit voltages (V_{OC}) of their standard solar cell structures. The study was carried on both RF and DC glow discharge materials using *p-i*-metal Schottky barrier and *p-i-n* cell structures having abrupt junctions formed with *p*-a-Si:C:H layers. Because the correlation between E_{OPT} and E_μ in amorphous materials has not been established, a self-consistent approach is presented here which can be applied to both a-Si:H and a-Si:C:H materials. The self-consistency of the E_μ results obtained using internal photoemission with the activation energies of the dark reverse currents in the *p-i*-metal structures is also presented. It is found that hydrogen dilution has a large effect on "displacing" E_μ further away from E_{OPT}, which is consistent with higher V_{OC} observed with such materials.

INTRODUCTION

Both the mobility gap, E_μ, and the optical gap, E_{OPT}, are key parameters of a-Si:H based materials that determine the performance of their solar cells. Unlike crystalline materials, where a single gap determines the optical and carrier transport properties, E_μ in amorphous materials can be significantly different from E_{OPT} and in principle this provides more freedom in tailoring the appropriate material for given solar cell performance. An absorber layer having a lower E_{OPT} which can absorb photons over a wider spectrum range results in higher short circuit currents. On the other hand, a higher E_μ is desirable since it has lower bulk generation/ recombination currents which thus can lead to higher V_{OC}'s. Consequently it is important to understand the correlation, if any, between E_{OPT} and E_μ in intrinsic a-Si:H materials, and to determine how the band gaps of the absorber layer affect V_{OC} in solar cells.

Recently several groups have reported on a-Si:H based solar cells which have open circuit voltage of ~ 1V [1,2]. It was found in these studies that the improvements in the open circuit voltages were substantially greater than the corresponding increase in the absorber layer's optical gap. This difference could be in part due to the greater mobility gaps in these materials. It has been found [1] that the open circuit voltages of solar cells, keeping the *p* and *n* contacts the same, appear to increase faster than the corresponding optical gaps of the absorber layers. We found that with our materials and solar cell structures the open circuit voltage can be directly correlated with the mobility gaps of absorber layers and the generation/recombination currents which are related to these gaps.

EXPERIMENTAL DETAILS

Sample Preparation

Studies ere carried out on intrinsic films and both *p-i*-metal Schottky barrier structures and *p-i-n* solar cells fabricated with the same *i* layer materials. The films of around 1 micron were deposited onto Corning 7059 glass substrates with and without hydrogen dilution at temperatures from 170 to 240°C where H_2 to SiH_4 flow ratio for hydrogen diluted films was 10:1. Optical transmission and reflection measurements were made on the films for wavelengths from 500 to 1000nm using a Perkin Elmer Lambda 9 UV/VIS/NIR Spectrophotometer.

The *p-i*-metal Schottky barrier structures, 0.5 to 0.8 micron thick, were deposited on textured tin oxide coated glass. Thermal evaporation of nickel provided the good Schottky barrier contacts to the intrinsic layers to form the glass/tin oxide/*p*-a-Si:H(250Å)/*i*-a-Si:H or *i*-a-Si:C:H/ Ni(250Å) structures. Textured tin oxide substrates were used in order to eliminate interference fringes in the measurements of internal photoemission (IPE). The results of this were used to measure the barrier heights for holes, ϕ_{Bp}, and electrons, ϕ_{Bn}, and thus determine the mobility gap of the i-layers, $E_\mu = \phi_{Bp} + \phi_{Bn}$.

Single junction *p-i-n* solar cells were fabricated on specular tin oxide coated glass having the structures glass/tin oxide/*p*-a-Si:C:H(250Å)/*i*-a-Si:H/*n*-a-Si:H (250Å)/ Cr(1000Å). The thick *p* layers and specular tin oxide structures were used to have reproducible *p* contacts. The intrinsic layers were from 0.5 to 0.8 micron thick for both *p-i*-metal and *p-i-n* cell structures which were fabricated by a multiple-chamber system. The doped and

CH3365-4/94/0000-0468 $4.00 © 1994 British Crown Copyright First WCPEC; Dec. 5–9, 1994; Hawaii intrinsic layers were deposited in different chambers to prevent cross contamination and thus reduce the effects of *p-i* interface. Current-Voltage characteristics were measured between 30 to 170°C on both types of structures and the cell characteristics were measured using an Oriel solar simulator.

Measurement Of Optical Gap

Since the different analysis techniques yield different E_{OPT}'s of given absorption spectra, three forms of analyses were examined. The Tauc [3], Cody [4], and Klazes [5,6] plots were compared for a wide range of materials, an example of which is given in Fig.1. Tauc's method is most widely used but the generally seen non-linearity not only causes ambiguity in the extrapolation of the data but also indicates its inadequacy in determining the optical gap in a-Si:H [6]. Klazes' method, while giving good fits to the data, assumes a constant *momentum* matrix element in optical transitions for which there is no experimental evidence. We decided the best criterion of E_{OPT} is that proposed by Cody [4] which defines the optical gap E_{Cody} from the extrapolation from $(\alpha/E)^{1/2}$ vs. E plot. This model assumes a constant *dipole* matrix element which has been experimentally justified by Jackson *et al.* [7] and Collins *et al.* [8]. It also gives a linear plot over a photon energy range from ~1.8 to 3.0eV [8] which is larger than any of the other plots.

Measurement Of Mobility Gap

Wronski *et al.* [9] were the first to use internal photoemission measurements on a-Si:H *p-i*-metal Schottky barrier structures to measure ϕ_{Bp} and ϕ_{Bn}, of the metal contacts onto a-Si:H, and thus E_μ of a-Si:H as $\phi_{Bp} + \phi_{Bn}$. In determining the barrier height ϕ_B from the IPE results they used the simple Fowler analysis. Recently a revised analysis was proposed for the interpretation of IPE spectra from metal contacts to *amorphous* materials [10]. The relaxation of ***k*** selection rule in amorphous materials such as a-Si:H was properly taken into account. It is found that for a barrier height ϕ_B the IPE yield Y at a photon energy $h\nu$ is given by:

$$Y(h\nu) \propto \left(h\nu - \phi_B\right)^{2+c} \tag{1}$$

This is for a density of extended states in the amorphous material, $N(E)$, which has the following energy dependence,

$$N(E) \propto \left(E - E_{edge}\right)^{c} \tag{2}$$

where E_{edge} is the "band edge" for which Eq. (2) is valid. It is only required that the mobility edge is above E_{edge} for Eq. (1) to be valid [10].

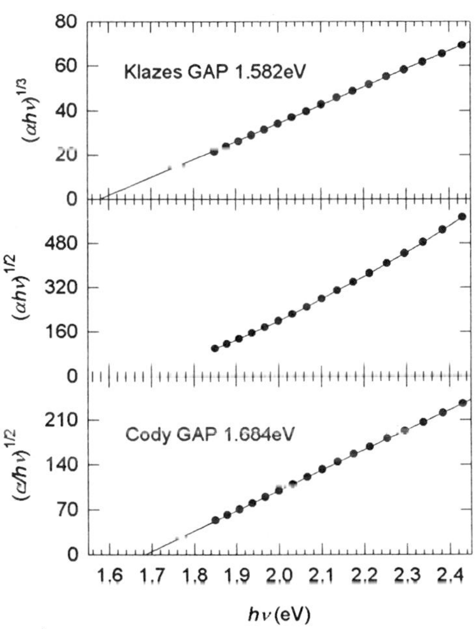

Fig. 1. The extrapolation plots of Tauc, Cody, and Klazes methods for the optical absorption coefficients of a typical a-Si:H film.

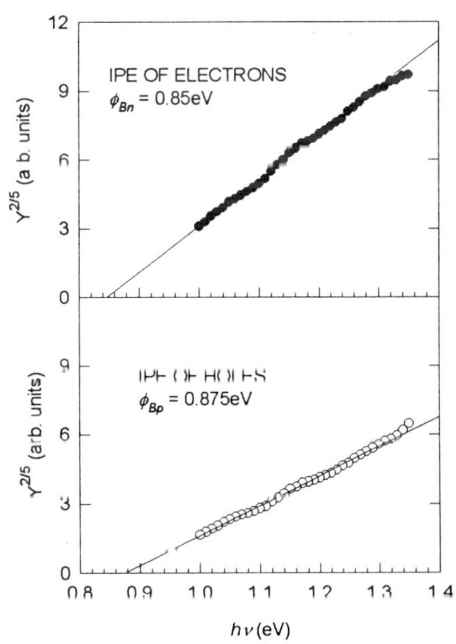

Fig. 2 Plot of $(Yield)^{2/5}$ versus photon energy for photoemitted holes and electrons from a *p-i*-Ni structure with the corresponding barrier heights being indicated.

Since the extended states in a-Si:H have a quadratic dependence on energy [4]; then $c = 1/2$ and $Y^{2/5}$ versus $h\nu$ should be linear with an extrapolation that gives ϕ_B at zero ordinate. This analysis is applied to the internal photoemission yields of a-Si:H Schottky barrier structures and typical results for a p-i-Ni structures are shown in Fig. 2. In this case the hole and electron barrier heights are 0.875eV and 0.85eV, respectively, thus giving the mobility gap E_μ = 1.72eV.

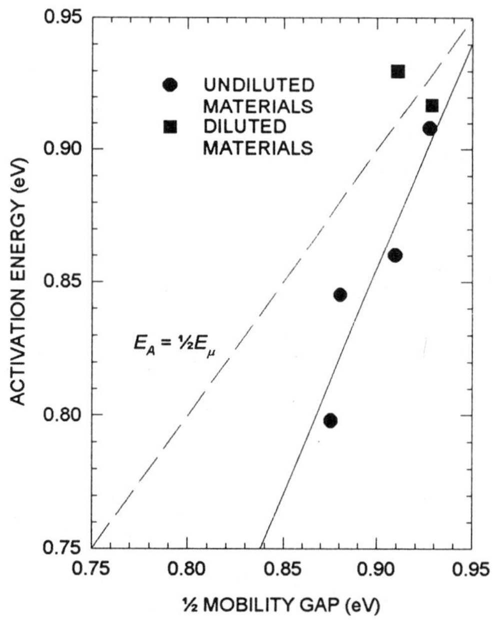

Fig. 3 The activation energies of the reverse bias generation current versus half of the mobility gaps as measured by IPE for a number of *p-i*-metal structures.

Another parameter that is directly related to the mobility gap are the generation recombination currents in *p-i*-metal and *p-i-n* strictures. Generation currents and their activation energy, E_A, were obtained by reverse biasing the *p-i* junction in these structures and the activation energies extracted from Arrhenius plots. When the reverse current is determined primarily by the generation kinetics in the bulk *i* layer, their activation energies should be close to half of the mobility gap. In Fig. 3 the results obtained for E_A on a number of *p-i*-metal structures are plotted versus half of E_μ as determined from IPE on the same structures. There is a clear correlation between E_A and E_μ. As can be seen from Fig. 3, E_μ is larger than $2E_A$ for undiluted materials but ~ $2E_A$ for hydrogen diluted materials. This offset may be attributed to the different distributions of recombination centers (defects) in these different types of materials. It was also found that the change in E_a is far larger than that in E_{OPT} and in many cases E_{OPT} is substantially smaller than $2E_a$, signaling that E_{OPT} cannot even properly account for the observed E_a.

DISCUSSION

Measurements were carried out on a large number of intrinsic films and their corresponding *p-i*-metal Schottky and *p-i-n* solar cell structures. The correlation found between the optical and mobility gaps depends on preparation conditions and, in particular, whether the material is prepared with or without hydrogen dilution. The results are shown in Fig. 4 where the measured mobility gaps for a large number of materials are plotted versus their optical gaps. It can be seen in Fig. 4 that the materials prepared with hydrogen dilution have a region where the displacement of mobility gaps above optical gaps is significantly larger than that for the materials prepared from pure silane. The two correlations can be expressed as

deposited with hydrogen dilution

$$E\mu = 1.42\ E_{Cody} - 0.64\text{eV} \quad (3)$$

deposited without hydrogen dilution

$$E\mu = 0.82\ E_{Cody} + 0.35\text{eV} \quad (4)$$

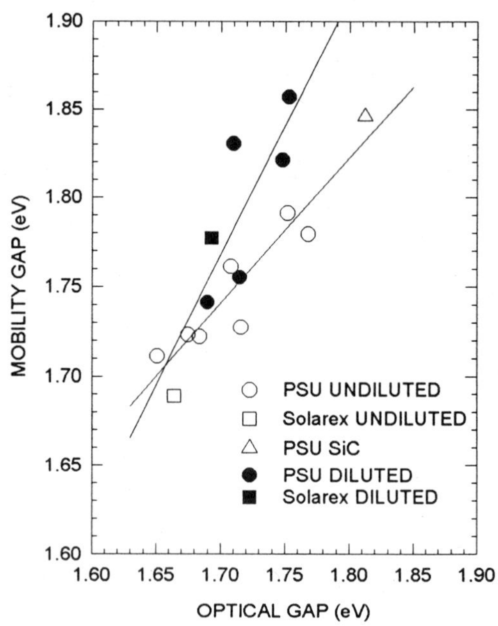

Fig. 4 The mobility gaps plotted versus their (Cody) optical gaps for differently prepared a-Si:H and a-Si:C:H materials prepared by Penn State Univ. (PSU) and Solarex.

The results obtained on the *p-i-n* solar cells also showed a linear dependence between V_{OC} and $E\mu$ which indicates that the open circuit voltage is controlled by the mobility gap of the absorber layer for the materials and cells studied. This together with the results in Fig. 4, which shows that for the hydrogen diluted materials $E\mu$ - E_{OPT} increases as E_{OPT} gets larger, is consistent with the results of Bennett *et al.* [1] who reported that the increase in V_{OC} were significantly higher than those in *i* layer E_{OPT}. Consquently, when high open circuit voltages are desired (*e.g.*, the high band gap cell component of the triple junction modules), wide band gap absorber layers should be used. While using wider band gap absorber layers would cause lower short circuit currents, it is not as adverse in tandem solar cells where other (low band gap) component cells can absorb the photons that leak through the top wide band gap component cell and generate photocurrents, thus providing the better overall performance.

CONCLUSION

The correlation between $E\mu$ and E_{OPT} of a-Si:H based materials was investigated using thin film and *p-i*-metal structures. Solar cells with either hydrogen diluted or undiluted materials as absorber layers were also fabricated and the dependence of V_{OC} on $E\mu$ and E_{OPT} investigated. For a-Si:H with E_{OPT} greater than about 1.7eV the displacements of $E\mu$ above E_{OPT} are greater for materials prepared with hydrogen dilution than those without dilution. Those results on $E\mu$ and E_{OPT} explain the published results by Bennett et al. [1] who reported that for the diluted films V_{OC} increases more rapidly than E_{OPT}. It is demonstrated that wide mobility gap absorber layers should be employed if a high open circuit voltages are desired. Further advantages can be taken of this enhancement in the difference between E_{OPT} and E_μ in designing cells for maximum performance, particularly as the high band gap cell component of the triple junction modules.

ACKNOWLEDGMENTS

This work is supported by Electrical Power Research Institute (EPRI) Thin Film Solar Cell program, National Renewable Energy Laboratory (NREL) under subcontract No. XAN-4-13318-03 and NEDO International Joint Research Grant.

REFERENCES

[1] M. Bennett, K. Rajan, and K. Kritikson, "Amorphous Silicon Based Solar Cells Deposited from H_2-Diluted SiH_4 at Low Temperatures", *Proc. 23rd IEEE PVSC,* 1993, pp. 845-849.

[2] Y. -M. Li, "Amorphous silicon-carbon alloys for solar cells", *Mat. Res. Soc. Symp. Proc.*, **297**, 1993, pp. 803-814.

[3] J. Tauc, R. Grigorovici, and A. Vancu, "Optical Properties and Electronic Structure of Amorphous Germanium", *Phys. Stat. Sol.*, **15**, 1966, pp. 627-637.

[4] G. D. Cody, B. G. Brooks, and B. Abeles, "Optical absorption above the optical gap of amorphous silicon hydride", *Sol. Energy Mater.*, **8**, 1982, pp. 231-240.

[5] R. H. Klazes, M. H. L. M. van den Broek, J. Bezemer, and S. Radelaar, "Determination of the optical bandgap of amorphous silicon", *Phil. Mag.*, **45**, 1982, pp. 377-383.

[6] Y. Hishikawa, N. Nakamura, S. Tsuda, S. Nakano, Y. Kishi, and Y. Kuwano, "Interference-Free Determination of the Optical Absorption Coefficient and the Optical Gap of Amorphous Silicon Thin Films", *Jpn. J. Appl. Phys.*, **30**, 1991, pp. 1008-1014.

[7] W. B. Jackson, S. K. Kelso, C. C. Tsai, J. W. Allen, an S. -J. Oh, "Energy dependence of the optical matrix element in hydrogenated amorphous and crystalline silicon", *Phys. Rev. B*, **31**, 1985, pp. 5187-5198.

[8] R. W. Collins, I. An, H. V. Nguyen, and Y. Lu, "Real time spectroscopic ellipsometry for characterization of nucleation, growth, and optical functions of thin films", *Thin Solid Films*, **233**, 1993, pp. 244-252.

[9] C. R. Wronski, S. Lee, M. Hicks, and S. Kumar, "Internal Photoemission of Holes and the Mobility Gap of Hydrogenated Amorphous Silicon", *Phys. Rev. Lett.*, **63**, 1989, pp. 1420-1423.

[10] Ing-Shin Chen and C. R. Wronski, "Internal Photoemission on a-Si:H Schottky barrier structures revisited", *J. Non-Crys. Solid*, 1995, to be published.

IMPROVEMENT IN STABILIZED EFFICIENCY OF a-Si:H SOLAR CELLS THROUGH OPTIMIZED p/i-INTERFACE LAYERS

B. Rech, C. Beneking, H. Wagner
Forschungszentrum Jülich GmbH, ISI-PV, D-52425 Jülich, Germany

ABSTRACT

This paper focuses on the relation between p/i-interface layer properties and the light stability of the corresponding solar cells. In a series of cells containing different p/i-interface structures large differences in relative degradation were found. These differences are explained by a redistribution of the electric field due to the insertion of different p/i-interface layers leading to different collection from the i-layer volume in the course of i-layer degradation. Based on this hypothesis, we developed an optimized design of the p/i-interface region, which increases the initial efficiency without introducing additional degradation. a-Si:H/a-Si:H stacked cells including this new design exhibit only 12 % degradation after 300 hours of one sun light-soaking. A stabilized efficiency of 9 % was achieved.

INTRODUCTION

The insertion of a wide bandgap p/i-interface (buffer) layer between the a-SiC:H p-layer and the a-Si:H i-layer increases the initial conversion efficiency of a-Si:H based solar cells [1,2,3]. Many groups, however, find that these cells degrade strongly during light exposure [3,4,5]. In contrast to this, a-Si:H/a-Si:H stacked cells with the highest reported stabilized efficiency of 10.0 % do contain buffer-layers [6].

This paper addresses primarily the additional FF degradation introduced through the buffer. The V_{oc} degradation, as reported and explained by Sakai et al. [3], also occurs for our cells [5], but is not treated here. Our experimental efforts served to study the physical relations between cell design and light stability with the goal to avoid the additional degradation.

The degradation behaviour of different cell types was investigated in a comprehensive light soaking study. The results of this study were the starting point for a new approach to increase the stability. Additional layers, inserted between the p- and the buffer-layer, served to modify the p/i-interface region. Further degradation studies demonstrate the influence of these layers on the stability of the corresponding solar cells. Finally, we transferred the improvements in single-junction cell design to our stacked cell development.

EXPERIMENTAL

Our deposition system is a commercial 3-chamber PECVD-System (manufacturer Glasstech Solar Inc., supplied by Materials Research Group). The TCO of the single-junction solar cells is textured tin oxide with ≈ 6 % haze, the back contact is thermally evaporated silver. The TCO of the stacked cells is Asahi Type "U". For these cells the back contact is ZnO/Ag [7]. The p-layer is boron doped (doping gas: B_2H_6 or TMB) a-SiC:H with 10 nm thickness (Tauc-Gap: 2.0 eV), the n-layer is phosphorous doped a-Si:H with 30 nm thickness.

In the following we distinguish between two groups of a-SiC:H p/i-interface layers: i) buffer (b)-layers consisting of intrinsic material with high photoconductivity, prepared with hydrogen dilution, and ii) p/buffer-interface layers (IL), prepared without hydrogen dilution of the source gases during the process. These interface layers are either undoped (intrinsic IL) or boron doped (doped IL). The gas flows used during the deposition of the interface layers and the corresponding Tauc-gaps (E_G) are listed in table 1. The cell types (single-junctions) prepared for this study were p-b-i-n, p-intrinsic IL-b-i-n, p-doped IL-b-i-n, and simple p-i-n for comparison.

The I-V measurements were performed with a double-source WACOM solar simulator (AM1.5, 100 mW/cm^2, 25 °C). The cells were exposed to continuous light soaking (spectrum close to AM1.5, 100 mW/cm^2) at 50 °C and under open-circuit conditions.

	SiH_4 (sccm)	CH_4 (sccm)	H_2 (sccm)	TMB (sccm)	E_G (eV)
buffer	2-6	2-4	170	-	1.85-2
intrinsic IL	50	30	-	-	1.9
doped IL	50	30	-	0.05	≈1.9

Table 1. Deposition parameters of various layers and corresponding optical gaps.

CH3365-4/94/0000-0472 $4.00 © 1994 IEEE

RESULTS

Light Soaking Studies

In previous experiments we found a strong degradation (not only in V_{OC}, but particulary in FF) of cells containing buffer-layers (p-b-i-n). Our first goal was to determine if this strong degradation is due to an enhanced degradation of the a-SiC:H material used for the buffer-layers or if it is related to the device structure. Therefore we performed light soaking tests on solar cells with 100 nm thick i-layers. The i-layer of the first cell (p-i-n) is a-Si:H (E_G=1.75 eV), the i-layer of the second cell (SiC-p-i-n) is a-SiC:H (E_G=1.95 eV). The i-layer of the third cell (p-b-i-n) consists of a 15 nm thick front part a-SiC:H (E_G=1.95 eV) and a 85 nm thick back part a-Si:H (E_G=1.75 eV). This cell is equivalent to a p-b-i-n cell with a thick, high bandgap buffer-layer. For these cells the relative degradation of FF is plotted as a function of light exposure time in figure 1. The cells with a-Si:H (p-i-n) or a-SiC:H (SiC-p-i-n) i-layers both show a similar, low relative degradation. However, the p-b-i-n cell degrades strongly. This indicates: The enhanced degradation of p-b-i-n cells can not be explained by an enhanced degradation of the a-SiC:H material.

In a second step, we increased the i-layer thickness of p-b-i-n cells and compared the results with an i-layer thickness variation of p-i-n cells. The FF of p-i-n cells is high (>70 %), even if the i-layer exceeds 500 nm, whereas the FF of the p-b-i-n cells strongly decreases with increasing i-layer thickness (see figure 2). From these results we conclude that buffer-layers reduce the electric field in the bulk of the i-layer (accompanied by a field enhancement in the p/i-interface region). This leads to low FF for thick i-layers of the corresponding cells, and to strong degradation even for thin i-layers.

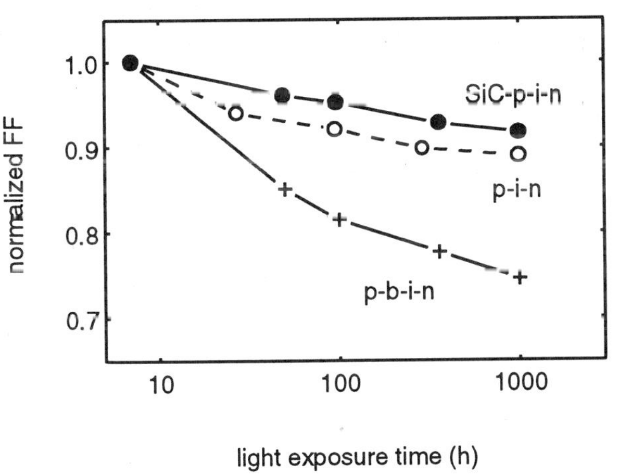

Fig. 1. Normalized fillfactor FF as a function of the light soaking time for cells with 100 nm thick i-layers.

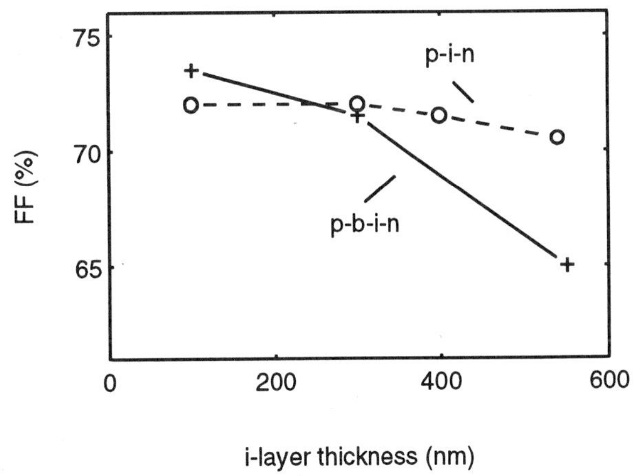

Fig. 2. Initial fill factor FF as a function of the a-Si:H i-layer thickness for p-i-n cells and p-b-i-n cells with a 15 nm thick a-SiC:H buffer-layer.

Approach for Higher Stabilized Efficiency

According to the hypothesis of field redistribution through buffer-layers mentioned above, we searched for a p/i-design which avoids the field loss in the i-layer. As a monitor we took the FF of solar cells with thick (550 nm) i-layers.

Our strategy for the p/i-interface optimization was to introduce thin (2 nm) a-SiC:H layers (prepared without hydrogen dilution of the process gases) between the p- and the buffer-layer. These p/buffer-interface layers (IL) should not only modify the p/i-interface region, but also protect the p-layer against the hydrogen-rich plasma of the following buffer layer deposition. The buffer-layer itself was kept constant[1] (thickness=10 nm, E_G=1.85 eV).

Figure 3 demonstrates the dependence of the initial FF of cells with 550 nm i-layer thickness on the design of the p/i-interface region. The insertion of the buffer-layer decreases the FF compared to the p-i-n cell. An undoped IL further decreases the FF. However, doping of the IL significantly increases the FF, which is even higher than the FF of the p-i-n cell.

The efficiency of these four cells as a function of light soaking time is shown in figure 4. All cells with buffer-layer increase the initial efficiency, but only the p-doped IL-i-n cell significantly increases the stabilized efficiency.

An i-layer thickness variation shows, that the improvement in stabilized efficiency is obtained for a wide range of i-layer thicknesses (see Figure 5).

---

[1]Remark: The 1.85 eV buffer led to higher stability of p-b-i-n cells than the 1.95 eV buffer, but still enhances degradation compared to p-i-n cells (see figure 4).

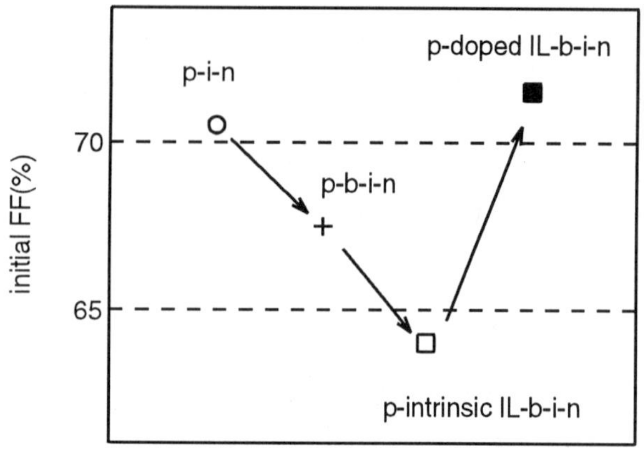

Fig. 3. Fillfactor FF of cells (thickness: 550 nm) with differently designed p/i-interface regions.

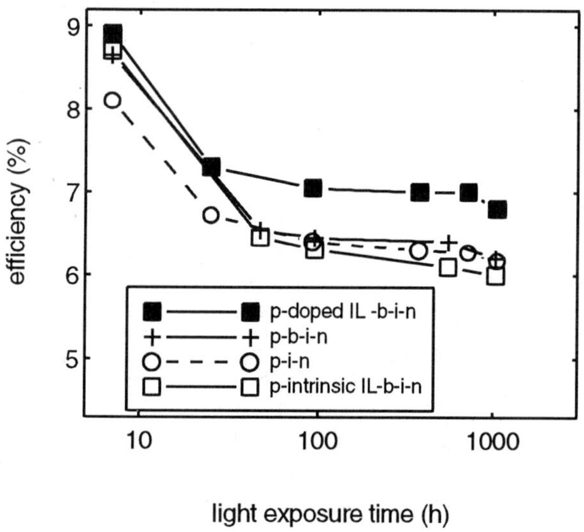

Fig. 4. Efficiency of cells (i-layer thickness: 550 nm) with differently designed p/i-interface region plotted versus the light-exposure time.

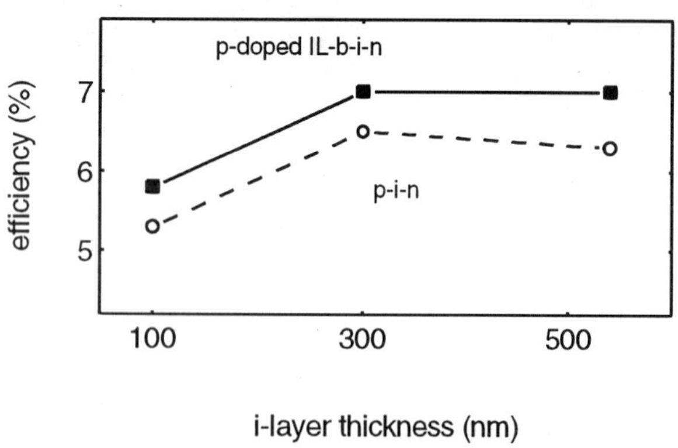

Fig. 5. Stabilized efficiency (after 300 h light soaking) of cells with doped p/buffer-interface layer (p-doped IL-b-i-n) and p-i-n cells as a function of the i-layer thickness.

Application in Stacked Cells

We prepared stacked cells, including these optimized design in both, top and bottom cell. Figure 6 shows the efficiency as a function of light-exposure time for such a cell ("new design"), compared to a stacked cell with our former device design ("former design") [7]. The stacked cell including the new p/i-design degrades only 12 % and stabilizes at a efficiency close to 9.0 % after 300 h light soaking. The further degradation of our cells after 700 h light soaking is only in short-circuit current and probably related to corrosion of the unprotected back contacts.

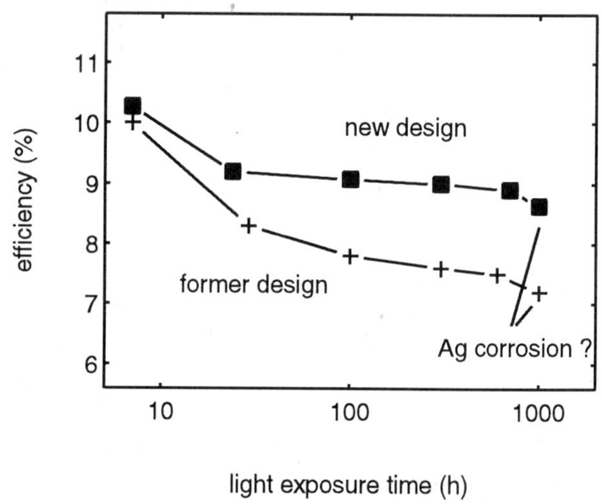

Fig. 6. Efficiency of stacked cells (total i-layer thickness: 400 nm) as a function of light-soaking time.

DISCUSSION

We explain the additional degradation due to the insertion of p/i-interface layers as follows: High bandgap a-SiC:H material has a high density of midgap states. By inserting this material between the p- and the i-layer, these defects become positively charged (D^+-states) because the Fermi-level is close to the valence-band edge in this cell region. Hence, it follows that the electric field is enhanced in the p/i region, but is lowered in the i-layer. After degradation, the defect density in the i-layer is increased. Now the electric field is too low to effectively separate the carriers. This reduces the fillfactor and by that the efficiency of the corresponding cell.

Figure 7 illustrates our idea to maintain a higher electric field in the volume of the i-layer by reducing the electric field in the p/i-interface region. Especially at the beginning of the buffer-layer a high density of D^+-states is expected, responsible for a high electric field peak. The negatively charged acceptor states (A^-) in the doped p/buffer-interface layer partly compensate these D^+-states and thereby reduce the field peak in the p/buffer-interface region. This enhances the electric field in the i-layer. Necessary conditions for this concept are: the doping concentration in the p/buffer-interface layer must be high enough to significantly influence the electric field, but the p/buffer-interface layer must not act as an extensed p-layer.

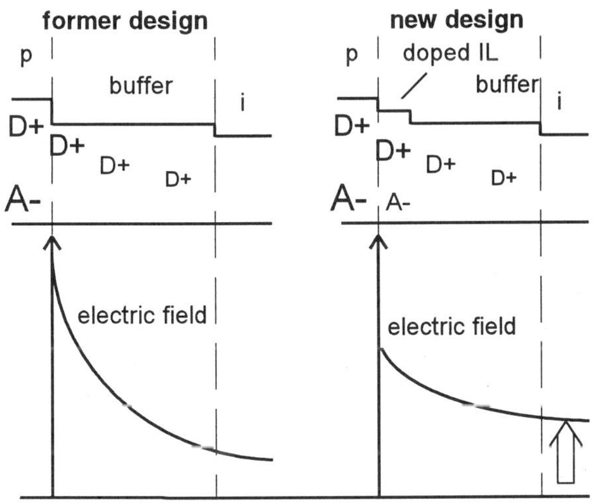

Fig. 7. Schematic electric field distribution in the p/i-interface region for a p-b-i-n cell and a p-doped IL-b-i-n cell. The arrow indicates the field enhancement in the i-layer due to the new p/i-interface structure.

CONCLUSIONS

i) The enhanced degradation of cells including buffer-layers is due to a reduction of the electric field in the volume of the i-layer. We found no evidence for an enhanced degradation of the a-SiC:H material or of the interfaces.

ii) Thin p/buffer-interface layers significantly influence the performance of the corresponding solar cells. The combination of a doped p/buffer-interface layer and a 10 nm relatively low bandgap (1.85 eV) a-SiC:H buffer-layer increased the stabilized efficiencies of our single- and stacked junction cells. We explain this improvement by compensation of positively charged dangling bonds by negatively charged acceptor states in the p/buffer-interface region.

iii) We have realized this idea of space charge compensation by an addional layer. However, optimized boron profiles introduced through doping or boron cross contamination, could also serve for the space charge compensation. This could be one reason for the differences in light stability of cells containing buffer-layers reported from different groups.

ACKNOWLEDGEMENTS

The authors are grateful to W. Appenzeller and F. Birmans for extensive experimental assistance and M. Kolter, M. Kubon and N. Schultz for many helpful discussions.

This work was supported by the BMFT (German Federal Ministry of Research and Technology). B. Rech acknowledges support from the Bundesland Nordrhein-Westfalen for a doctoral fellowship.

REFERENCES

[1] R.R. Arya, A. Catalano, and R.S. Oswald, "Amorphous Silicon p-i-n Solar Cells with Graded Interface", *Appl. Phys. Lett.,* **49**, pp. 1986-1089.

[2] H. Tasaki et al.," Computer Simulation Model of the Effects of Interface States on High Performance Amorphous Silicon Solar Cells", *J. Appl. Phys.,* **63**,1988, pp. 550-560.

[3] H.Sakai et al.,"Effect of p/i Interface Layer on Dark J- V Characteristic and V_{oc} in p-i-n Solar Cells", *J. Appl. Phys.* **67**, 1990, pp. 3494-3499.

[4] P. Lechner et al., "Comparison of Prolonged Light Soaking of Single and Stacked Junction a-Si:H Solar Cells", *Mat. Res. Soc. Proc. Vol.,* **149,** 1989, pp. 583-588

[5] B. Rech et al., "Preparation and Characterization of Intrinsic a-SiC Alloys and a-SiC Solar Cells", *Twelfth EC-PVSEC, Amsterdam*, 1994, pp. 1245-1248.

[6] H. Sakai and Y. Ichikawa, "Process Technology for a-Si/a-Si Double Stacked Tandem Solar Cells with Stabilized 10 % Efficiency", *J. Non-Cryst. Solids,* **137&138,** 1991, pp 1155-1160.

[7] C. Beneking et al.,"Preparation and Light Stability of a-Si/a-Si Stacked Solar Cells", *Twelfth EC-PVSEC, Amsterdam*, 1994, pp. 683-686

DETERMINATION OF THE QUALITY OF a-Si:H FILMS: "TRUE" TRANSPORT PARAMETERS

N. Beck, A. Shah, N. Wyrsch,
Institute of Microtechnology, University of Neuchâtel, Rue Breguet 2, 2000 Neuchâtel, Switzerland,
tel. +41-38-20 54 52, fax. +41-38-25 42 76, e-mail: Beck@imt.unine.ch

ABSTRACT

For the characterisation of a-Si:H layers, steady-state photoconductivity (SSPC) and steady-state photocarrier grating (SSPG) are currently used. But the $\mu\tau$-products deduced from these measurements are a function of the prevailing dangling bond occupation in the film and, thus, are not a measure of material quality. In the present paper the authors introduce the product $\mu^o\tau^o$, which is independent of dangling bond charge and which monitors material quality in terms of band mobility μ^o and capture time τ^o,the latter being, in its turn, inversely proportional to dangling bond density. They present a simple evaluation procedure to deduce $\mu^o\tau^o$ from simultaneously performed SSPC and SSPG measurements. In undoped, but slightly n-type a-Si:H, the quality of the material will typically be related to the ambipolar diffusion length rather than to photoconductivity. The application of this method to a series of undoped a-Si:H films and solar cells incorporating these materials exhibited a good correlation between film properties and cell performances.

INTRODUCTION

Extensive research on the characterisation of a-Si:H films was carried out in the past with the aim to increase the basic understanding of a-Si:H material and, thus, improve the performance of a-Si:H based solar cells.

Nevertheless, a lack of quantitative correlation between the measured characteristics of a-Si:H films and cell performance persists [1], [2], [3]. In the present paper we contend that this lack of correlation is partly related to the type of methods normally used to characterise the transport properties of a-Si:H films. In fact, transport properties like photoconductivity and ambipolar diffusion length are measured, and these are strongly dependent on the occupation of dangling bond states. The dangling bond occupation in uniform a-Si:H films has little bearing to the (strongly non-uniform) distribution of dangling bond charge throughout the i-layer of a solar cell. Thus, simply transfering the results obtained from such transport measurements from films to solar cells must fail. Nevertheless, we will show in the present paper that the simultaneous measurement of both photoconductivity **and** ambipolar diffusion length, can indeed be a solution to this issue.

In fact, both electron and hole tranport, and, thus, both photoconductivity (σ_{ph}) and the ambipolar diffusion length (L_{amb}) are related to the same recombination centers (dangling bonds D^o, D^- and D^+)[4]. This correlated behaviour of σ_{ph} and $L_{amb}{}^2$ enables one to determine the influence of the dangling bond charge on transport.

"TRUE" TRANSPORT PARAMETERS: DEFINITIONS AND BASE FOR THEIR EVALUATION

It is well known and widely accepted that the $\mu\tau$ products obtained by photoconductivity (σ_{ph}) and by the ambipolar diffusion length (L_{amb}) measurements are band mobility x recombination time products ($\mu^o\tau^R$) of the form:

$$\sigma_{ph} = e \cdot G \cdot \left(\mu_n^o \tau_n^R + \mu_p^o \tau_p^R\right) \qquad (1)$$

$$L_{amb}^2 = \frac{kT}{e} \cdot \frac{\mu_n^o \tau_n^R \cdot \mu_p^o \tau_p^R}{\mu_n^o \tau_n^R + \mu_p^o \tau_p^R} \cdot C \qquad (2)$$

where kT is the thermal energy, e the elementary charge, G the generation rate and C a correction factor between 1 and 2 [5]. Combining both σ_{ph} and L_{amb} measurements, allows one to determine the parameter $b = \mu_n^o \tau_n^R / \mu_p^o \tau_p^R = \mu_n^o n_f / \mu_p^o p_f$; b reflects the position of the two quasi Fermi-levels with respect to midgap [6]. b can be evaluated by an expression deduced from equations (1) and (2):

$$\frac{b}{(b+1)^2} = \frac{L_{amb}^2 e^2 G}{kT\sigma_{ph}C} \qquad (3)$$

The experimentally measured recombination times, as used in (1) and (2), are determined not only by the **density of deep defects**, but also by the **dangling bond charge state**. As dangling bond occupation is not a material characteristic but merely a function of experimental conditions, device structure and doping, it is not correct to directly use these $\mu^o\tau^R$-products as a quality parameter for the material. However by measuring **both,** σ_{ph} and L_{amb} we can define a $\mu\tau$-product which is a "true" quality parameter.

Thereby, we will take advantage of the fact that the transport properties of majority carriers (monitored by σ_{ph}) and minority carriers (monitored by $L_{amb}{}^2$) are coupled via the dangling bonds, the latter being the common recombination centers for both of them [4]. Thus, electron and hole $\mu^o\tau^R$-products are not independent from each other, but display an anticorrelated behaviour (see Fig.1). This behaviour makes it possible to evaluate and eliminate the contribution of the dangling bond occupation on $\mu^o\tau^R$ measurements and, finally, to find an

CH3365-4/94/0000-0476 $4.00 © 1994 IEEE

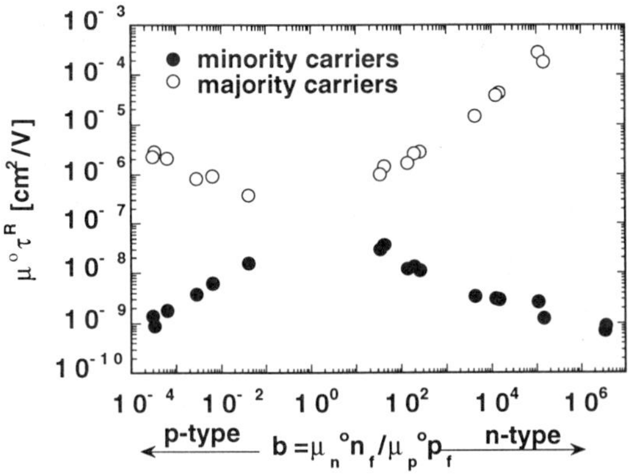

Fig.1. Anticorrelated change of majority and minority carrier $\mu^o\tau^R$-products with lowlevel doping, as measured by SSPC and SSPG.

$\mu^o\tau^o$-product (band mobility × capture time) that is independent of dangling bond charge, and has the form:

$$\mu_n^o\tau_n^o = \mu_n^o \frac{1}{v_{th}N_{db}\sigma_n^o} \quad \text{and} \quad \mu_p^o\tau_p^o = \mu_p^o \frac{1}{v_{th}N_{db}\sigma_p^o} \tag{4}$$

Thus, the product $\mu^o\tau^o$ is a monitor for the deep defect density as well as for the band mobility, but it is independent of the dangling bond charge.

Consider now a dangling bond recombination model (DBR), which takes into account the three possible charge states of the dangling bonds: D^o, D^-, D^+ [7,8]. Assume further a simplified description where recombination of free carriers **only** occurs through dangling bond states and where thermal emission from deep defects as well as recombination through bandtail states are neglected. The recombination times in (1, 2) are then:

$$\tau_n^R = \frac{n_f}{r_{cn}^o + r_{cn}^+} = \left\{ v_{th}N_{db}\sigma_n^o\left(f^o + \frac{\sigma_n^+}{\sigma_n^o}f^+\right)\right\}^{-1} \tag{5}$$

$$\tau_p^R = \frac{p_f}{r_{cp}^o + r_{cp}^-} = \left\{ v_{th}N_{db}\sigma_p^o\left(f^o + \frac{\sigma_p^-}{\sigma_p^o}f^-\right)\right\}^{-1}, \tag{6}$$

r_{cn}^o, r_{cn}^+, r_{cp}^o, r_{cp}^- are capture rates of free carriers on the different dangling bond states, n_f, p_f densities of free electrons and holes, σ_n^o, σ_n^+, σ_p^o, σ_p^- capture cross-sections of dangling bonds, N_{db} the defect density, v_{th} the thermal velocity and f^o, f^-, f^+ the dangling bond occupation functions.

f^o, f^- and f^+ in (5) and (6) are governed by the recombination traffic of free carriers n_f and p_f only and, thus, can be expressed as a function of them [8]:

$$f^o = 1\Bigg/\left(\frac{p_f}{n_f}\frac{\sigma_p^o}{\sigma_n^+} + 1 + \frac{n_f}{p_f}\frac{\sigma_n^o}{\sigma_p^-}\right) \tag{7}$$

$$f^- = \frac{n_f}{p_f}\frac{\sigma_n^o}{\sigma_p^-}\Bigg/\left(\frac{p_f}{n_f}\frac{\sigma_p^o}{\sigma_n^+} + 1 + \frac{n_f}{p_f}\frac{\sigma_n^o}{\sigma_p^-}\right) \tag{8}$$

$$f^+ = \frac{p_f}{n_f}\frac{\sigma_p^o}{\sigma_n^+}\Bigg/\left(\frac{p_f}{n_f}\frac{\sigma_p^o}{\sigma_n^+} + 1 + \frac{n_f}{p_f}\frac{\sigma_n^o}{\sigma_p^-}\right) \tag{9}$$

Let us now introduce an additional hypothesis (10), which is not at all mandatory, but which can help the reader gain a better physical insight:

$$\mu_n^o\tau_n^o \approx \mu_p^o\tau_p^o \tag{10}$$

Hypothesis (10) can be justified on the basis of Time-of-Flight measurements performed by our group earlier [9,10] on compensated and strongly degraded a-Si:H films, where almost all dangling bonds are neutral. The measurements (Fig. 2) show $\mu_n^o\tau_n^o$ and $\mu_p^o\tau_p^o$ to be almost equal ($\mu_n^o\tau_n^o/\mu_p^o\tau_p \approx 1.2$). Note that the ratio given in the text of our previous papers [9, 10] was $\mu_n^o\tau_n^o/\mu_p^o\tau_p^o$ = 3...5: however, this value had not been correctly derived from the experimental data.

Thanks to hypothesis (10) we may now define a single quality parameter for a-Si:H material; we will call it $\mu^o\tau^o$:

$$\mu^o\tau^o = \mu_n^o\tau_n^o = \mu_p^o\tau_p^o \tag{11}$$

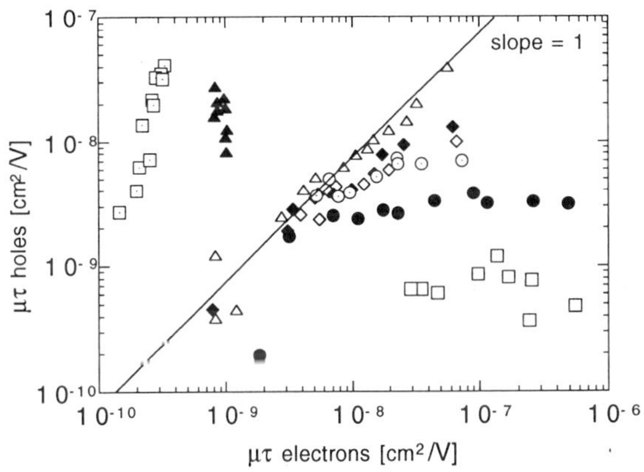

Fig. 2. Mutual change of electron and hole $\mu^o\tau$-products (see text) obtained by TOF measurements during degradation [9, 10]. The measurements show the tendency of $\mu_n^o\tau_n$ and $\mu_p^o\tau_p$ to approach each other under light-soaking and then, when degrading further, to degrade with equal rates (line with slope 1).

EVALUATION OF $\mu^o\tau^o$ FROM σ_{ph} AND L_{amb}

Starting with the expressions for the experimentally acessible recombination times (5), (6), and using the dangling bond occupation functions (7), (8), (9) and introducing the second hypothesis (10) in the calculation, $\mu^o\tau^o$ can now be deduced. The following results are obtained:

$$\mu_n^o\tau_n^R = \mu^o\tau^o \frac{b}{b+1}\left(\frac{\sigma_n^o}{\sigma_n^+}\frac{1}{b} + 1 + \frac{\sigma_p^o}{\sigma_p^-}b\right) \tag{12}$$

$$\mu_p^o\tau_p^R = \mu^o\tau^o \frac{1}{b+1}\left(\frac{\sigma_n^o}{\sigma_n^+}\frac{1}{b}+1+\frac{\sigma_p^o}{\sigma_p^-}b\right) \qquad (13)$$

The expressions (12) and (13) show that the measured $\mu^o\tau^R$-products generally need to be corrected by a certain factor, which is a function of the parameter b and, thus, is influenced by doping (also by low-level, non-intentional doping due e.g. to [O] contamination). Nevertheless, there is a problem with these expressions: they cannot be applied to "truly" intrinsic (i.e. "compensated" or "midgap") a-Si:H, where both electrons and holes contribute to the photoconductivity. Thus, it is more convenient rewrite the expression for photoconductivity (1) and obtain:

$$\frac{\sigma_{ph}}{eG} = \mu^o\tau^o \cdot z \quad \text{with} \quad z=\left(\frac{\sigma_n^o}{\sigma_n^+}\frac{1}{b}+1+\frac{\sigma_p^o}{\sigma_p^-}b\right) \qquad (14)$$

This formula is now valid for all types of a-Si:H; the character of the material is monitored by the correction factor z. Furthermore, it is not necessary to know the different contributions of electrons and holes to σ_{ph}. The information gained by L_{amb} measurements appears here only through the parameter b.

Let us now evaluate $\mu^o\tau^o$ for some particular situations:

When the material is "**truly intrinsic**" and the Fermi energy level is near midgap (i.e. in compensated a-Si:H), b becomes close to unity and the correction factor z approaches 1. Thus, $\mu^o\tau^R$ (deduced from σ_{ph}) and $\mu^o\tau^o$ become identical. In this case, σ_{ph} is the proper parameter to describe material quality.

On the other hand side, in **n-type** material (even in undoped slightly n-type films) b»1 and the correction factor takes on the value $z \approx b(\sigma_p^o/\sigma_p^-)$. In this situation, according to equation (13), $\mu^o\tau^o$ becomes proportional to $\mu_p^o\tau_p^R$. This means, that for n-type a-Si:H with $b > \sigma_p^-/\sigma_p^o$, material quality is monitored by the ambipolar diffusion length and not by σ_{ph}. The same also applies to p-type a-Si:H. Thereby the capture cross section ratios σ_n^+/σ_n^o and σ_p^-/σ_p^o, when compared to the experimental parameter b, will mark the limits between intrinsic a-Si:H and n-type or p-type material. Unfortunately, the values of σ_n^+/σ_n^o and σ_p^-/σ_p^o are still a controversial topic; values given in literature vary between 1.5 [11] and ≥100 [12]. Here, we will use for the evaluation of $\mu^o\tau^o$-products, $\sigma_n^+/\sigma_n^o=\sigma_p^-/\sigma_p^o=50$. This estimate is based on the behaviour of $\mu^o\tau^o$-products with degradation and doping (to be presented elsewhere).

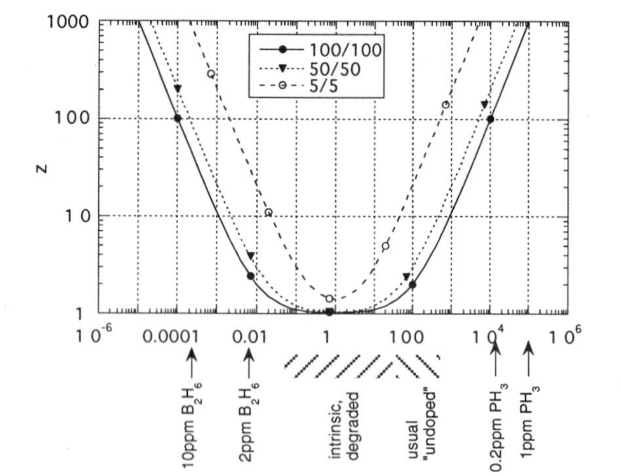

Fig. 3. Calculated magnitude of the correction factor z for experimental $\mu^o\tau^R$-products deduced from σ_{ph} for different values of b. The insert indicates the three different ratios of capture cross sections σ_n^+/σ_n^o and σ_p^-/σ_p^o used. On the x axis some typical values of b are indicated.

Fig.3. shows the calculated magnitude of the correction factor z for the parameter $\mu^o\tau^R$ deduced from σ_{ph} , plotted as a function of b. "Typical" values for b, as obtained by low-level PH_3 and B_2H_6 doping are also indicated.

CORRELATION BETWEEN $\mu^o\tau^o$-PRODUCTS OF a-Si:H FILMS AND SOLAR CELL PERFORMANCES

The quality-related parameter $\mu^o\tau^o$ was mainly introduced by us with the aim to correlate film properties and cell performances. For that purpose, a series of films and cells was produced by the VHF-GD deposition technique at 70 MHz [13], while varying thereby the deposition temperature for the undoped layer. For each **film - cell** couple, first a ~2.5 μm thick, undoped a-Si:H film was deposited on a glass substrate; subsequently, using the same set of deposition parameters, the "same" a-Si:H material was incorporated as an i-layer in a standard, ~0.6 μm thick p-i-n solar cell. Before measuring the characteristics, both cells and films were annealed for 2 hours at 180°C. The a-Si:H films were then characterised by SSPG, SSPC, Photothermal Deflection Spectroscopy (PDS), Constant Photocurrent Method (CPM) [14] and dark conductivity measurements. For the transport measurements, a Krypton laser (λ=647 nm at 3mW/cm^2 to 90 mW/cm^2) was used, in order to obtain, in the film, a generation rate of electron hole pairs of $2\text{-}60 \cdot 10^{19}$ 1/cm^{3}s. For the cells, the efficiency (η) were detemined, under standard AM1.5 illumination at 100 mW/cm^2, using a two-source solar simulator (Wacom WXS-140S-10).

The Annealed State

The $\mu^o\tau^o$-products evaluated for this series of films as well as the corresponding solar cell efficiencies are plotted in Fig. 4. The quality of the film material, monitored by $\mu^o\tau^o$, is almost constant for temperatures above 220°C. Lowering the deposition temperature leads to a deterioration of material quality. Most films exhibited a clear extrinsic behaviour in the annealed state.

Cell efficiencies slightly peak at 220°C and exhibit the same strong decrease of quality with lower deposition temperatures. As the reduction of optical absorption in the device due to the increased bandgap ($E_g \approx 1.76$eV from Tauc's plot for T_{set}=150°C compared to $E_g \approx 1.73$eV for T_{set}=270°C) can be estimated to be less than 5%., the observed deterioration of the cells for low deposition temperatures can mainly be attributed to the bad quality of the material. On the other hand, the slight reduction of cell efficiencies for higher deposition temperatures is not consistent with the observed good film quality. As for these cells the value of V_{oc} is also lower, chemical and

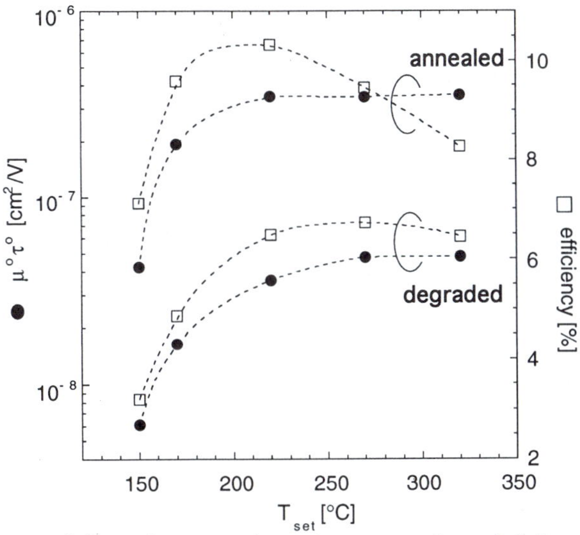

Fig.4.$\mu^o\tau^o$-products evaluated on a series of 2.5 μm thick, undoped films and efficiencies of solar cells incorporating the "same" films as an i-layer, in the annealed state and the degraded state. T_{set} is the setpoint of the deposition temperature. (For evaluation we assumed $\sigma_n^+/\sigma_n^o=\sigma_p^-/\sigma_p^o=50$).

structural problems (e.g. interdiffusion or interface deterioration) may be responsible for cell deterioration.

The degraded state

The cells were degraded during 3 weeks by an AM1.5 light source of 100mW/cm^2 (array of Philips PL-L 24W/95/4P lamps) at 47°C. The films were light-soaked during 4 weeks by a 6-sun high pressure sodium lamp (spectral maximum at 590nm with a FWHM of around 40nm) at 46°C. Though films and cells were light-soaked by different light sources, they show a very good correlation after light-soaking (see fig. 4): both exhibit the same strong deterioration of quality for lower deposition temperatures. Thereby, the quality of a-Si:H films was reduced by a factor of ≈ 10 and cell efficiencies by 20% - 50%. After degradation, all films showed to be almost intrinsic.

CONCLUSIONS

We have introduced and defined a new $\mu^o\tau^o$-product, which can be used as a quality parameter for a-Si:H material optimisation. $\mu^o\tau^o$ is determined by band mobility and deep defect density. This quality-related $\mu^o\tau^o$ product does not depend - in contrast with direct experimental data - on the dangling bond occupation prevailing in the material. Furthermore, we have shown that $\mu^o\tau^o$ can easily be determined from SSPC and SSPG measurements according to the following procedure: first, and most important in order to get a correct information about the material quality, one has to identify if the material is intrinsic (midgap) or if it has an extrinsic character (n-type or p-type). Thereby, the following criterion for the experimentally determined parameter b applies:

1) **intrinsic** (midgap): $$\frac{\sigma_n^o}{\sigma_n^+} < b < \frac{\sigma_p^-}{\sigma_p^o} \quad (15)$$

2) **extrinsic:** $$b > \frac{\sigma_p^-}{\sigma_p^o} \quad \text{or} \quad b < \frac{\sigma_n^o}{\sigma_n^+} \quad (16)$$

If the material fits into category 1), the quality is simply given by the $\mu\tau$ product deduced from photoconductivity: $\mu^o\tau^o = \sigma_{ph}/(eG)$. On the other hand, if condition 2) is fullfilled, the material quality is monitored by the ambipolar diffusion length L_{amb}. In this case, L_{amb} can be related to $\mu^o\tau^o$ by the following expression ($\sigma_p^\pm/\sigma_p^o \approx 50$):

$$\mu^o\tau^o = L_{amb}^2 \frac{e}{kT}\frac{1}{C}\frac{\sigma^\pm}{\sigma^o} \quad \text{with } C \approx 1.5 \text{ and } \sigma_p^\pm/\sigma_p^o \approx 50 \quad (17)$$

In limit cases, where the sample can neither clearly be attributed to 1) nor 2), the general expression (14) must be used in order to obtain $\mu^o\tau^o$.

When applying $\mu^o\tau^o$ to correlate film quality and cell performances one finds: the proposed quality parameter $\mu^o\tau^o$ correlates in general well with the measured cell performances, above all in the degraded state. However, this correlation may be "hidden" in certain cases by more severe problems in the solar cell structure, which are not related to the quality of the intrinsic layer.

ACKNOWLEDGMENTS

The authors acknowledge the help of S. Dubail for sample preparation, as well as financial support from the Swiss National Science Foundation under grant (FN-39377) and from the Swiss Federal Renewable Energy Program under mandat (EF-REN(93)032).

REFERENCES

[1] C. Wronski et al., *MRS Proc.* **297**, 1993, p. 443.

[2] R. Schropp et al., *J. Non-Cryst. Sol.* **164-166**, 1993, p. 709.

[3] J. Yang et al., *MRS Proc.* ,1994, to be published.

[4] A. Shah et al., *J. Non-Cryst. Sol.* **164-166**, 1993, p. 485.

[5] E. Sauvain, *Ph. D. Thesis*, University of Neuchâtel, 1992.

[6] P. Pipoz et al., *MRS Proc.* **258**, 1992, p. 777.

[7] F. Vaillant et al., *Phys. Rev. B* **34**, 1986, p. 4088.

[8] J. Hubin et al., *Phil. Mag. Lett.* **66**, 1992, p. 115.

[9] N. Beck et al., *Proc. 11th ECPVSEC*, 1992, p. 625.

[10] N. Beck et al., *MRS Proc.* **297**, 1993, p. 479.

[11] K. Hattori et al., *J. Non-Cryst. Sol.* **137&138**, 1991, p. 363.

[12] N. Wyrsch et al., *J. Non-Cryst. Sol.* **137&138**, 1991, p. 431.

[13] H. Curtins et al., *Plasma Chem. & Plasma Proc.* **7**, 1987, p. 267.

[14] M. Vanecek et al., *Solid State Com.* **39**, 1981, p. 1199.

THE ROLES OF ATOMIC HYDROGEN PLAYED IN "THE CHEMICAL ANNEALING" FOR FABRICATION OF a-Si:H

Kenjiro Nakamura, Kunihiko Yoshino and Isamu Shimizu
The Graduate at Nagatsuta, Tokyo Institute of Thechnology
4259 Nagatsuta, Midori-ku, Yokohama 226, Japan

ABSTRACT

A systematic study has been performed to reveal the role of atomic hydrogens in "Chemical Annealing", where the deposition of thin layer and treatment with atomic hydrogen are repeated alternately for fabrication of stable structure. Relaxation resulting from impingement of atomic hydrogen on the growing surface are distinguished into two pieces, viz., relaxation on the surface and the sub-surface depending on the conditions for deposition of thin layer and the flux of atomic hydrogen. The structural changes within sub-surface resulted in either formation of wider gap films or crystallization at rather low substrate temperature (Ts:100--150 C). At high Ts, on the other hand, the structural relaxations are mainly promoted on the growing surface, resulting in formation of the narrower optical gap a-Si:H and stimulating the grain growth.

INTRODUCTION

Hydrogen treatments during the growth of a-Si:H or μc-Si:H have attracted great attentions since hydrogens were considered to play some important roles in making Si-network at rather low substrate temperature (Ts C). We proposed first time "CA" where the deposition of very thin film and treatment with atomic hydrogens were repeated alternately for the aim of making a-Si:H exhibiting more stable characteristics for light soaking.[1] Prior to our work, Asano made μc-Si:H films by means of the similar technique.[2] In recent years, Sanyo and Mitsui-Toatsu group succeeded independently to make high quality a-Si:H with wider gaps by hydrogen treatment, which is inconsistent with our results where the films with narrower gaps were mainly achieved.[3],[4]

In this study, a systematic investigation was performed to reveal these puzzles concerned to the roles of atomic hydrogens played in construction of S-network. The real time observation with an ellipsometer was adopted to monitor the growing surface. In addition, we fabricated films by varying widely the conditions; i.e., Ts, flux of atomic hydrogens, duration for deposition (T_1 s) and hydrogen treatment(T_2 s).

EXPERIMENTAL

Fig.1 shows the preparation apparatus used in this study. SiH_4 was decomposed in the reactor by RF glow (13.56 MHz) and atomic hydrogen generated by the microwave plasma (2.45 GHz) was supplied into the reactor.

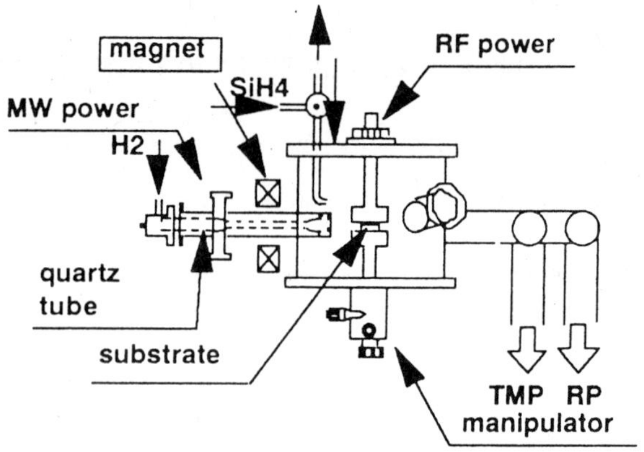

Fig.1 Sketch of the apparatus for fabrication of silicon films in order to supply enough atomic hydrogens with microwave generation system.

The durations (T_1,T_2 s) for supplying source and hydrogen treatment were controlled by a set of the

First WCPEC: Dec. 5-9, 1994; Hawaii

air-valves. During preparation of films, the growing surface was monitored by an ellipsometer (Mijoziri DVA-36L) as the trajectories on a plane of ellipsometric angles (Δ,Ψ). Typical conditions are summarized in Table 1.

Table 1. preparation conditions.

SiH4 (sccm)	20
GeH4 (20%)+H2 (sccm)	1-10
Prf (W/cm2)	0.3
Ptotal (mTorr)	72-80
T1 (s)	2-20
T2 (s)	0-40
Substrate Temperature (C)	room temperature ----- 400

The electrical conductivity in the dark and under light illumination was measured using the cells with the gap-type Al electrodes having a space of 0.2 mm. The surface of the cells were illuminated by monochromatic light, 600 nm in wavelength and the intensity of 0.2 mW/cm^2. The defect density was measured by CPM. Measurements of the degradation for light soaking were made by illuminating the cells with light emitted from a Xe lamp fitted with the filters (500 mW/cm^2) or a pulsed light of a strobe flash light (duration: 150μs, intensity: 140 W/cm^2 repetition 1 Hz). The temperature of sample was kept constant (40 C) by circulating water.

RESULTS AND DISCUSSION

In our previous paper, we classified the structural relaxations of Si-network by the aid of atomic hydrogens into two processes; i.e., the one caused on growing surface and the other promoted in sub-surface.[5] In both cases, microcrystalline silicon was finally made resulting from these structural relaxations. However, some conspicuous differences in the structure of a-Si:H were found between the films made by stimulating the relaxation on the surface and in the sub-surface.

In Fig.2, the optical gap energies (Eg eV) are plotted as a function of the hydrogen contents for films fabricated by "CA". The solid line shows a linear relation obtained in the films deposited continuously at various Ts. The closed circles and squares indicate the films made by the relaxations on the surface at Ts of 300 C and in the sub-surface at low Ts of 100 C, respectively. The optical gaps are shifted to narrower in the former along to the solid line, whereas they are moved towards wider in the latter apart from the line.

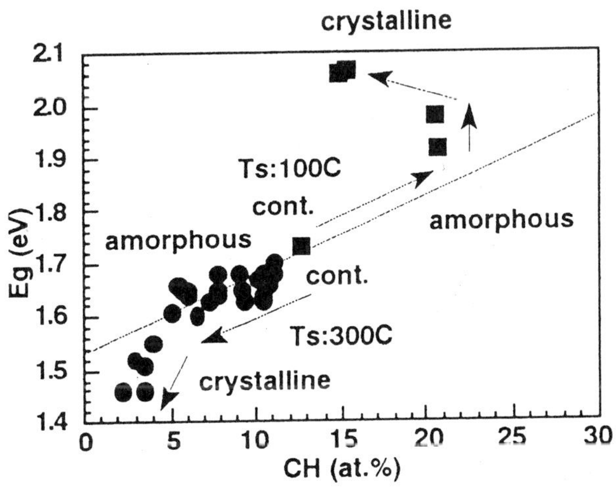

Fig.2 The correlation between hydrogen concentration (CH at.%) and optical bandgap (Eg eV) as a function of T2 in Ts=100, 300C, respectively.

These marked different behaviors suggested us that the structural relaxation induced by the hydrogen treatment on the surface is entirely different from that in the sub-surface. The parameters in making thin layer; i.e., thickness of the layer deposited during T_1 c, deposition rate and Ts C, respectively are considered to determine the thickness of sub-surface. We kept the deposition rate at constant in this study. Consequently, the thickness of sub-surface is mainly ruled by Ts. In practice, the optical gaps became narrower accompanying with monotonous reduction in C_H at Ts=300 C, while the gap energies were expanded due to an increase in C_H at Ts=100 C. In addition, bonding configurations of SiHn (n=1,2) measured from the the decomvolution of IR bands at around 2000 cm^{-1} were markedly in the films made at low Ts.

In Fig.3, the changes in the ellipsometric angles (Δ) monitored during the deposition and the hydrogen treatment are shown. The conditions are kept constant except the Ts (top: Ts=300 C, bottom Ts=100 C). Marked differences are seen among them, namely, large reduction in the (Δ) values is observed during hydrogen treatment at low Ts but not high Ts. These changes depending strongly on the surface and within the high Ts but were kept constant at low Ts give a support to this explanation.

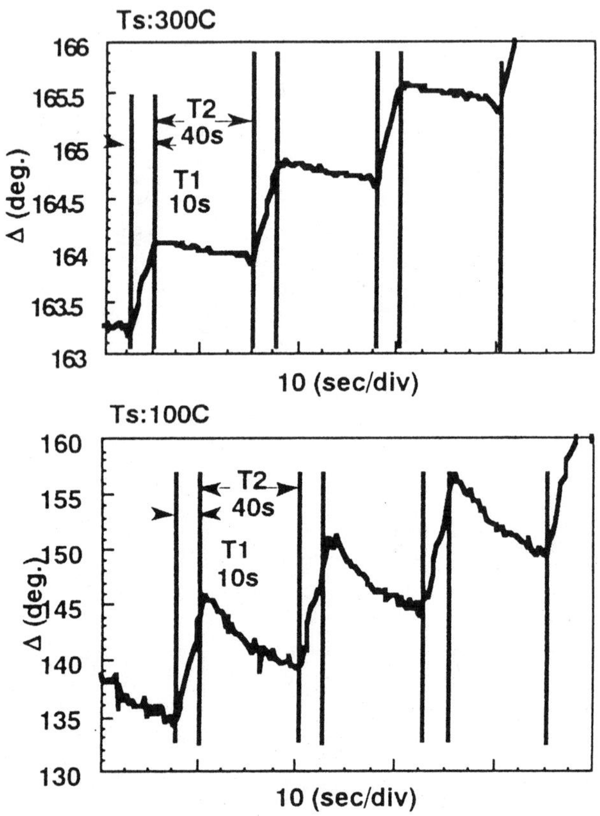

Fig.3 Ellipsometric trajectories for deposition of films for films at high Ts (300C: on the top) and low Ts (100C:on the bottom).

Fig.4 shows the ratio of SiH_2 to the total of SiHn (n=1 or 2) plotted as a function of T_2 for the films deposited at Ts=100 C by varying the thickness of thin layers; 10 A thick and 30 A thick, respectively.

No marked changes in the ratio are observed in the films fabricated by "CA" at high Ts (300 C),while the ratio. fall obviously with the T_2 for the films made at low Ts (100 C). Besides, the slopes of the curves corresponding to the velocities of changes in the local structures for the films made at low Ts are proportional to the reciprocal of the thickness of later deposited during T_1. These two curves are, however, described in the one by normalizing the thickness of the sub-surface. In the inset of the figure, the ratio are plotted as a function of T_1/T_2. Consequently, the structural relaxation is considered to be promoted by accumulation of a certain amount of hydrogens within the sub-surface.

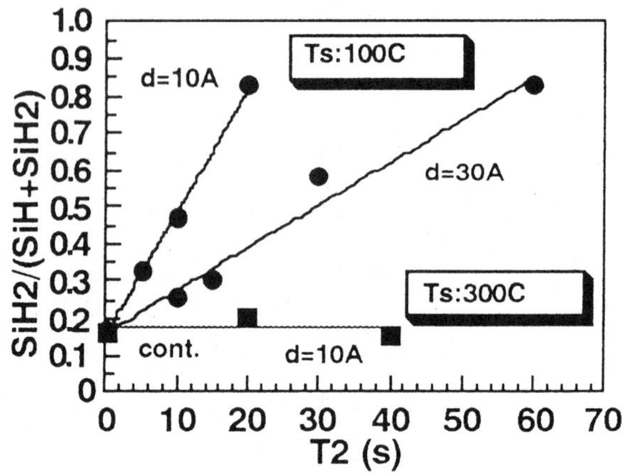

Fig.4 The ratio of SiH2 to the total of SiHn (n=1,2) plotted as a function of T2.

In Fig.5 the electrical conductivities at room temperature in the dark σd (S/cm) and under light illumination with σd (S/cm) are plotted as a function of optical gap (Eg eV).

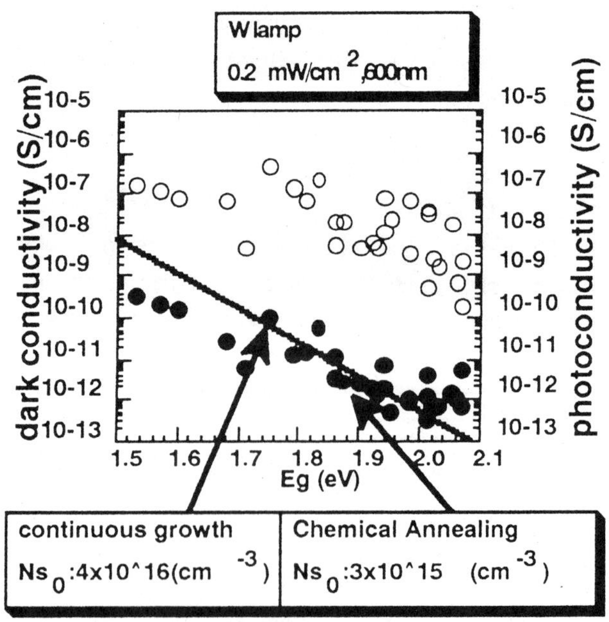

Fig.5 Dark, photoconductivity plotted as a function of optical band gap (Eg eV). The light is monochromatic in 0.2 mW/cm2 at 600 nm.

Though there are scattering, the σd values are represented by a simple equation as follows:

$$\sigma d=\sigma d_0 \exp(-Eg/2kT) \quad (1)$$

where σd_0, k and T indicate the pre-factor, Boltzmann constant and absolute temperature. The value of 10^4 (S/cm) applied to this calculation is generally accepted as the pre-factor. From these facts, the Fermi-levels are located at the mid-gaps for whole films by "CA". The activation energies obtained from Arrehnius plots of the electrical conductivities give a strong support to this explanation. σp values are obtained under illumination of light at a constant wavelength (600 nm) and intensity (0.2 mW/cm^2) as well. The solid line illustrates the σp values normalized by the absorption for the light of 600 nm in wavelength under the assumption of a constant $\eta\mu\tau =10^{-6}$ (cm^2/V). There are fairly good agreements between the observed values and the estimated ones, although there are large scattering in these experimental data. Finally, we address that some obvious improvement of the stability for light soaking was observed for a-Si:H films promoted the relaxation in the sub-surface by atomic hydrogen.

CONCLUSION

1. Structural relaxation caused by CA is distinguished obviously into two processes: the one is promoted on the surface and the other is enhanced in the sub-surface. The films with narrower gaps are obtained in the former, while the wider gaps films are made in the latter.
2. The thickness of sub-surface is determined by either T_1 or the substrate temperature Ts.
3. Very low defect densities as low as $(2\text{-}5)\times10^{15}$ cm^{-3} for the as-grown films made by CA with the gaps in the wide range from 1.8 to 2.1 eV independent of the optical gaps due to well passivation with hydrogens.
4. The defect densities are also independent of the SiH_2 ratio in Si notworks.
5. Obvious improvement in the stability for light soaking was observed in the films with wider gaps prepared by CA.

ACKNOWLEDGEMENT

This work was supported by NEDO International Research Grant in part and by a Grant-in-Aid for Scientific Research on Priority Areas of "Free Radical Science" in part.

REFERENCES

[1] H.Shirai, D.Das, J.Hanna and I. Shimizu, *Jpn. J. Appl. Phys.*, **30** (1991) L239

[2] Asano, *Appl.Phys. Lett.*, **56** (1990) 533

[3] S.Okamoto, et. al. *Jpn. J. Appl.Phys.*, **33** (1994) 4373

[4] Ashida and Fukuda, *Solar Energy and Solar Cells*, **34** (1994) 291

[5] K. Nakamura, K.Yoshino and I.Shimizu, *Jpn. J. Appl. Phys.*, (1994) in press

EXAMINATION OF OPTIMIZED STRUCTURES FOR a-Si:H BASED TRIPLE JUNCTION SOLAR CELLS

Sanghoon Bae and Stephen J. Fonash
Electronic Materials and Processing Research Laboratory
The Pennsylvania State University, University Park, PA 16802

ABSTRACT

We have examined a-Si:H based triple junction solar cells by using the AMPS computer program. We have found that, although the a-SiC:H material has wider band-gap, it does not yield better performance than a-Si:H absorber as the first absorber in the triple junctions because of the severely degraded material quality of today's a-SiC:H in the stabilized state. Moreover, for the stabilized state, we find that the optimized thickness distributions of a-Si:H based triple junctions are thinner than those determined in the annealed state. Our results for optimized stable efficiencies show that (1) with today's a-SiC:H materials, stabilized cell efficiencies can be better with a-Si:H top cells, (2) sub-cell layers need to be thinner than generation current matching would suggest and (3) because of the need for thinner i layers, the window losses become critical (to keep Jsc up).

INTRODUCTION

We have examined the design of a-Si:H based triple junction solar cells by using the AMPS computer program. We use the light stabilized density of states for the triple junction structure because our goal is to design for optimized structures in the stabilized state. In this process of the multi-junction modeling, we first used AMPS simulations to fit various pieces of experimental data such as QEs, light I-Vs, and dark I-Vs of component sub-cells. Using these simulations, we extracted material parameters and then used these material parameters for the sub-cells in the multi-junction simulations. [1], [2]

A critical problem in triple junction design studies is to determine the generation rate for the various parts of these multi-junction structures. Thus, one must account for absorption in the doped and inter-cell contact layers and for light trapping. [3] Moreover, one must account for front (light entry) structure reflection and absorption. In this report we account for losses in doped and inter-connect layers and make an attempt at accounting for light trapping. Our focus here, however, is in accounting for reflection and absorption losses in the front entry (window) structure of the triple junction cells. In focusing on this, an ideal front window is first assumed for 100 % transmittance of AM 1.5 incident light at all wavelengths; in this mode we do not have any optical loss at the front window. In our second approach, a very thin ~ 700 Å) indium tin oxide film is chosen as a front window material with negligible absorption loss. [4] In our third approach, we assume a ~ 2 mm glass/~ 7000 Å textured tin oxide for the front window. In this third case, the optical loss data was taken from measurements using a refractive index matching liquid to prevent the effect of scattering and diffuse transmission. [5] Fig. 1 shows the transmittance data for these three assumed front windows.

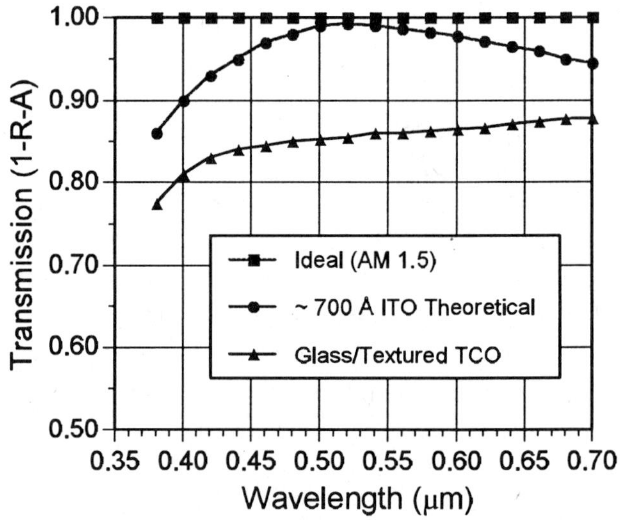

Fig. 1. Transmittance data (T = 1-R-A) from three different transmission modes. Square data (■), circle data (●), and triangle data (▶) correspond to ideal situation (100 % transmission of AM 1.5 incident light), ~ 700 Å indium tin oxide, and ~ 2 mm glass/~ 7000 Å textured tin oxide, respectively. [4], [5]

RESULTS AND DISCUSSION

With this approach we have examined both a-SiC:H/a-Si:H/a-SiGe:H and a-Si:H/a-Si:H/a-SiGe:H triple junctions using our assessment of current material parameters (See Table 1). [1], [2], [6] Fig. 2 gives an example of the results of our analysis. It shows light J-V curves for two different triple junctions structures: (a) a-SiC:H/a-Si:H/a-SiGe:H and (b) a-Si:H/a-Si:H/a-SiGe:H structures for the case of 100 % transmittance through the front window. The light J-V characteristics of the 1st, 2nd, 3rd sub-cells are displayed with the solid, dotted, and dot-dashed lines, respectively. The overall triple junction cell performance is displayed with dashed lines. The thickness of each absorber for both structures has been optimized for the highest efficiency with the given material parameters (See Table 1). The most prominent difference between Fig. 2 (a) and (b) is the shape of light J-V curves from the 1st sub-cells (solid lines); Fig. 2 indicates that light J-V curve fill factor of a-SiC:H sub-cell is even lower than that of a-Si:H sub-cell. Simulations using the current material parameters find that the stable energy conversion efficiency of the a-Si:H/a-Si:H/a-SiGe:H triple junction structure is ~ 1.5 % greater than that of the a-SiC:H/a-Si:H/a-SiGe:H triple junction structure for 100 % transmittance for incident light. Also, for the other two transmission modes (i.e., ~700 Å ITO and glass/~7000 Å textured tin oxide) the a-Si:H/a-Si:H/a-SiGe:H triple junction always has 1.4 ~ 1.6 % higher stabilized efficiency than the a-SiC:H/a-Si:H/a-SiGe:H triple junction.

Table 1. Stabilized material parameters of a-SiC:H and a-Si:H for the top absorbers

Parameters \ Absorbers	Optical gap Eop (eV)	Mid-gap defects (cm^{-3})	Cha. Energy of tail states Ed (eV)	Cha. Energy of tail states Ea (eV)	Thickness (Å)
a-SiC:H top absorber	1.90	4.0E17	0.060	0.045	~ 900
a-Si:H top absorber	1.72	9.0E16	0.050	0.030	~ 450

Although the a-SiC:H absorber material has a wider band-gap, it does not yield better performance in the stabilized state than an a-Si:H absorber in a triple junction because of its stabilized material properties. To be specific, its lack of performance is due to the severely degraded material quality of today's a-SiC:H absorber (i.e., it is due to the high stabilized density of states values). This underscores the importance of material quality of the absorber in the consideration of wide-gap materials.

We performed thickness optimization for stabilized a-Si:H/a-Si:H/a-SiGe:H triple junction structures with three different transmission modes. Employing the ideal, ~ 700 Å ITO, and glass/textured TCO as a front window, we compared stabilized energy conversion efficiencies of a-Si:H/a-Si:H/a-SiGe:H triple junctions. The cell performance of the a-Si:H/a-Si:H/a-SiGe:H triple junctions with these three different transmission modes, represented in Table 2, indicates that open circuit voltage(Voc) and fill factor (FF) are not affected by the

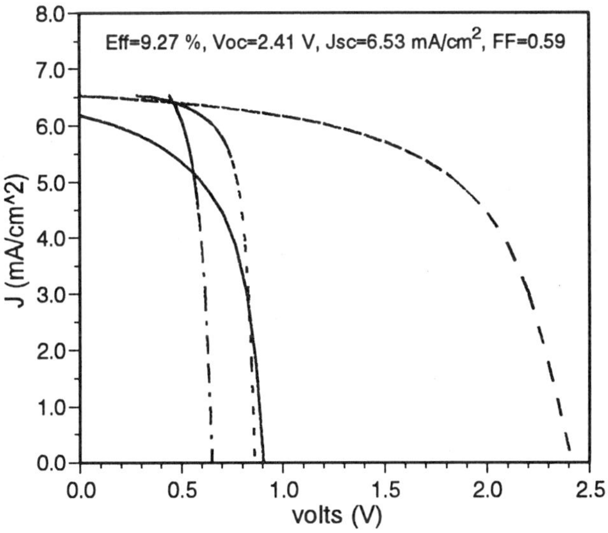

(a)

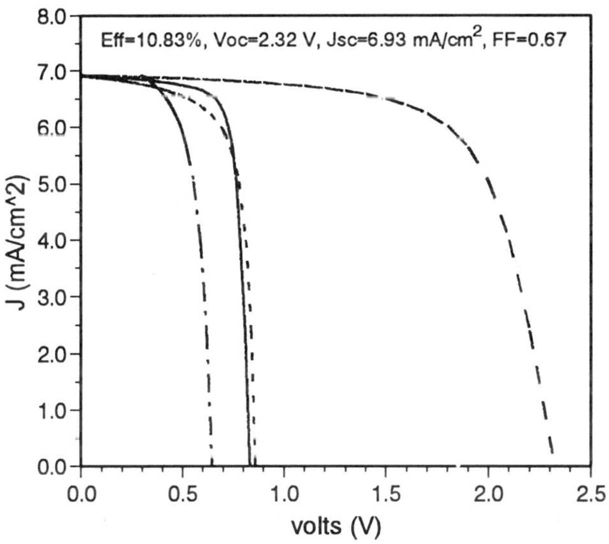

(b)

Fig. 2. Light J-V characteristics of (a) a-SiC:H/a-Si:H/a-SiGe:H and (b) a-Si:H/a-Si:H/a-SiGe:H triple junction structures. Solid, dotted, and dot-dashed correspond to the 1st, the 2nd, the 3rd sub-cells, respectively. Dashed lines correspond to overall characteristics of the triple junctions.

transmittance of front window while short circuit current (Jsc) is decreased with lower transmittance of incident photon fluxes. According to our simulation results, ~ 3 mA/cm^2 current is lost at glass/textured TCO front window whereas ~ 0.7 mA/cm^2 current is wasted at ~ 700 Å ITO front window. Since these optical losses at the front window structure obviously decrease short circuit currents of a-Si:H based solar cells, this current loss at the glass/textured TCO front window should be reduced to produce high current performance of a-Si:H solar cells.

Table 2. Stabilized cell performance of a-Si:H/a-Si:H/ a-SiGe:H triple junction with three different transmission modes.

Performance / Front window	Eff (%)	Jsc (mA/cm^2)	Voc (V)	FF
Ideal (100 %)	10.8	6.93	2.32	0.67
~ 700 Å ITO	10.5	6.71	2.31	0.68
Glass/textured TCO	9.38	5.91	2.30	0.69

As we stated earlier, we carried out thickness optimization of absorber layers for stabilized state a-Si:H based triple junctions. Our simulation results indicate that for the stabilized triple junction cells the thickness of each individual absorber should be thinner than the thickness values used for the initial state triple junction cells. In the stabilized state, thinner absorbers provide less opportunity for the recombination of photo-generated carriers because the number of mid-gap states is proportional to the thickness of absorbers. This fact is well reflected in the fill factors in Table 3; the thickness optimized triple junction has a fill factor that is 9 % higher than the non-optimized case. Therefore, we emphasize thickness optimization of multi-junction configurations is something that must be undertaken using stabilized state parameters.

According to our simulations, the optimized thicknesses of a-Si:H/a-Si:H/ a-SiGe:H triple junction are ~450 Å/~2900 Å/~2500 Å with the glass/textured TCO transmission mode data (See Table 3). For instance, the ~ 2900 Å optimized thickness of the 2nd absorber layer is much thinner than the thickness (e.g., ~ 5000 Å) based upon generation current matching; the optimized triple junction case with this thin 2nd absorber layer produces ~ 5.9 mA/cm^2 short circuit current. This short circuit current is even lower value than one would expect from the non-optimized a-Si:H triple junctions based on generation current matching.

Table 3. Comparison of stabilized a-Si:H/a-Si:H/a-SiGe:H triple junction performance between thickness optimized and non-optimized cases. These two case are examined with glass/textured transmission mode.

Performance / Cases	Eff (%)	Jsc (mA/cm^2)	Voc (V)	FF
Non-optimized (600/5000/3000 Å)	8.69	6.34	2.29	0.60
Thickness optimized (450/2900/2500 Å)	9.38	5.91	2.30	0.69

Fig. 3 displays the stabilized short circuit current of the a-Si:H 2nd sub-cells with three different thickness of the a-Si:H top absorber layer. The 2nd sub-cells with thin top absorber accomplish higher short circuit current than the 2nd sub-cells with thick top absorber. For example, to generate ~ 6.0 mA/cm^2 short circuit current from the 2nd sub-cell, we need to have only ~ 3000 Å thick absorber layer with 450 Å top absorber while ~ 8000 Å is required with ~ 900 Å top absorber layer (See Fig. 3). This arises because the thinner top cell structures pass more photons into the 2nd sub-cell. Such thin 2nd sub-cell structures utilize more photons and generate higher current. Moreover, this thin 2nd sub-cell will lead to a high stabilized fill factor since thin structure provides less opportunity for carrier recombination.

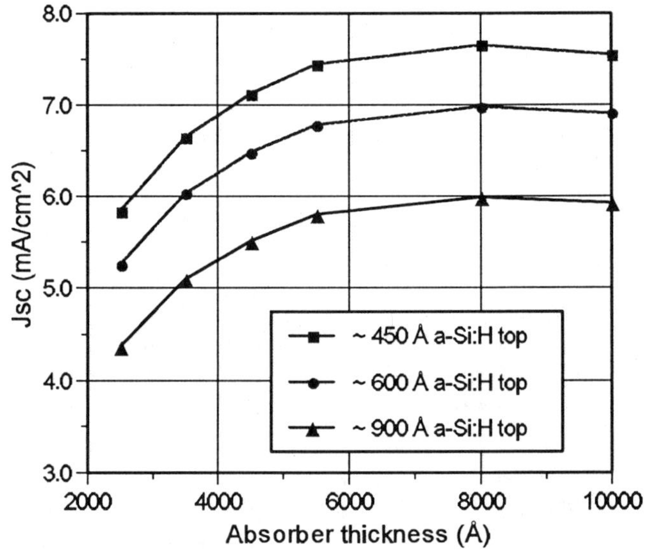

Fig. 3. Stabilized short circuit currents of the a-Si:H 2nd sub-cell with three different thicknesses of the a-Si:H top absorber layer. Glass/textured TCO front window was used for these data.

CONCLUSIONS

In general, our results for optimized stable efficiencies show that (1) with today's a-SiC:H materials, stabilized cell efficiencies can be better with a-Si:H top cells, (2) sub-cell structures need to be thinner than generation current matching would suggest and (3) because of the need for thinner i layers and of the need to utilize more photons, the window losses become critical to keep short circuit current up.

ACKNOWLEDGMENTS

This research was supported by amorphous silicon thin film program of The Electric Power Research Institute.

REFERENCES

[1] S. Bae, J. Hou, V. Suntharalingam, and S. J. Fonash, Proc. 23rd IEEE PVSC, 1011 (1993).
[2] J. Hou, V. Suntharalingam, S. Bae, and S. J. Fonash, Proc. 23rd IEEE PVSC, 1021 (1993).
[3] S. Wiedeman, J. Morris, and L. Yang, Proc. 21st IEEE PVSC, 1529 (1990).
[4] Kevin Hoffman, private communication (the data provided) (1994).
[5] Steve Hegedus, private communication (the data provided) (1994).
[6] M. Gunes and C.R. Wronski, private communication (1994).

THE ROLE OF INTERFACES IN THE END-OF-LIFE PERFORMANCE OF a-Si:H SOLAR CELLS

K. Vasanth and S. Wagner
Department of Electrical Engineering, Princeton University, Princeton NJ 08544

D. Caputo
Dipartimento di Elettronica, Universita "La Sapienza", 00184 Rome, Italy

M. Bennett
Solarex Thin Film Division, Newtown PA 18940

ABSTRACT

We address the question of interfacial contribution to the end-of-life performance of a-Si:H solar cells by focusing on the p/i interface with a comprehensive set of experimental and modeling tools. To separate interfacial from bulk defects we light-soaked a thickness series of cells with i-layers from 0.2 μm to 1.0 μm. The degradation of the thickness series shows that there is no extra interfacial contribution to efficiency loss. We also soaked cells with 1.0 μm i-layers on specular substrates with blue and red light. The change in their quantum efficiency spectra also shows that the p/i interface does not degrade more than the bulk. Using the numerical model AMPS we show that light-soaking can be modeled without invoking extra interface losses.

INTRODUCTION

The physics of hydrogenated amorphous silicon (a-Si:H) solar cells is becoming sufficiently well understood to encourage the computer-aided design of cells. It is generally agreed that a-Si:H cells should be designed with high end-of-life performance. Doing so requires modeling parameters that include the cell geometry and the end-of-life material parameters. This approach has been questioned because of reports on a lack of correlation between materials parameters and cell performance [1]. Such a lack of correlation suggests that additional parameters, beyond geometry and bulk material properties, affect the cell performance. One possibility which has been invoked is that interfaces change during cell operation in ways that cannot be predicted from bulk properties [2].

We approach this question of interfacial contribution to cell performance loss using both experimental and modeling tools. If interfaces do indeed contribute to cell loss, their effects must be seen clearly in measurements that measure cell performance, namely I-V characteristics and quantum efficiency (QE) spectra. It is believed that the p/i interface controls cell performance to a greater degree than the n/i interface. Therefore, for this paper we have designed the experiments to bring out the effects of the p/i interface, if any. To add to the experimental verification of the effects of interfaces, we have to be able to extend the model too to match these results. Therefore, we discuss model parameters controlling the light-soaking, along with how they can be used with our experimental results to answer the question about interfacial effects.

EXPERIMENT AND MODEL

Blue and Red Light-soaking

Solar cells with the structure Glass/Specular SnO_2/p a-Si,C:H/i a-Si:H/n a-Si:H/ZnO/Ag were used for the blue and red light-soaking. The i-layer was 1 μm thick. Specular cells were used in this experiment to increase the reliability of computer simulations used to interpret the results. The cells were light-soaked with blue (486 nm, 0.8 Wcm^{-2}) and red (648 nm, 1.2 Wcm^{-2}) light from a Kr^+ ion laser in open circuit and at room temperature. The I-V characteristics and quantum efficiency (QE) spectra were measured after each soak. The QE was measured without bias light to make the cell response sensitive to the defects induced in the soaking step. We used the numerical model AMPS to calculate the expected generation profiles, free carrier densities and other device parameters established in the cells during light soaking.

Thickness Series Light-soaking

Cells with the structure Glass/Textured SnO_2/p a-Si,C:H/i a-Si:H/n a-Si:H/ZnO/Ag were used for the thickness series light-soaking experiment. The i-layer thicknesses were 0.20, 0.24, 0.36, 0.48, 0.59 and 0.79 μm, respectively, and the initial efficiencies lay above 8.5%. These cells were light-soaked in open circuit using the Kr^+ laser (1.5 Wcm^{-2}) at room temperature for 30 minutes. I-V performance was measured before and after light-soaking, using a solar simulator. The specular cells used in the blue and red light degradation experiment were made using the same deposition conditions as the textured cells, and hence are made of high quality a-Si:H.

Model

Several numerical models are in use for simulating cell performance [3,4]. We use the numerical model AMPS

CH3365-4/94/0000-0488 $4.00 © 1994 IEEE

(Analysis of Microelectronic and Photonic Structures) developed at Penn State University [5]. AMPS solves the continuity equations for electrons and holes along with the Poisson equation using a number of material parameters as inputs. Using both measured and calculated material properties, we have been working on developing a parameter set which can be used to model both a-Si:H and a-Si,Ge:H alloy cell performance [6].

RESULTS

Blue and Red Light

a-Si:H has different absorption coefficients for blue (486 nm) and red (640 nm) light. Blue light with an absorption coefficient α of ~ 1×10^5 cm^{-1} is largely absorbed within the first 0.20 µm of the i-layer. Red light, on the other hand, has an absorption coefficient α of ~ 4×10^4 cm^{-1} and therefore is absorbed more uniformly throughout the i-layer [6]. The modeled electron-hole pair generation profiles and carrier densities are very different for the two colors and hence the degradation, too, can be expected to be different. Separately studying the interfacial loss mechanism for blue and red light, the boundary colors of the useful spectrum, simplifies the extension to white light degradation. The cells were soaked with 0.8 Wcm^{-2} blue light for 5 minutes, and 1.2 Wcm^{-2} red light for 20 seconds. Further soaking, till saturation does not change the results discussed below.

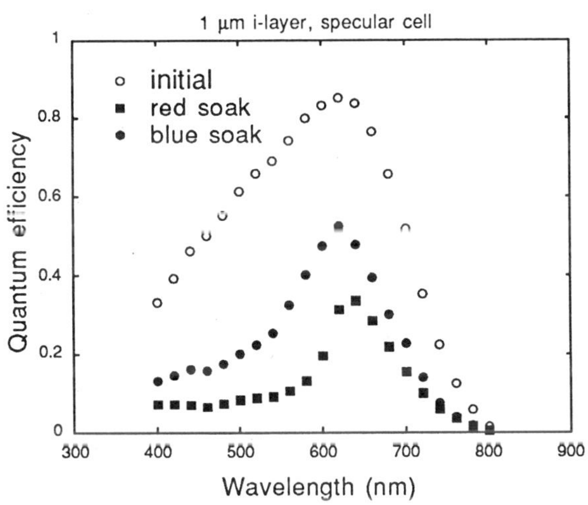

Fig. 1. Quantum efficiency spectra of cells before and after blue and red light-soaking.

Shown in Fig. 1 is the QE plot for the specular 1 µm cells irradiated with blue or red light. We define the relative change at every wavelength in the QE spectra as:

$$\text{Relative change} = (QE_{initial} - QE_{final}) / QE_{initial} \qquad (1)$$

It is important to note that a large loss of QE at a given wavelength corresponds to a large increase in this relative change at that wavelength. Forming this relative change ratio removes the effects of spectral reflectance and absorption in the SnO_2 and p-layers, thereby allowing us to focus on the effects of the i-layer on the QE [7].

Fig. 2 shows the relative change in QE vs. penetration depth $1/\alpha$ for cells soaked with blue and red light. The zero of the x-axis corresponds to the p/i interface. To obtain Fig. 2, we have converted the wavelength scale of Fig. 1 to a penetration depth scale using the corresponding absorption coefficients [6] as conversion factor. For example, the penetration depth of 330 Å corresponds to a wavelength of 400 nm of Fig. 1, and so on. Fig. 2 shows how a given wavelength of light absorbed mostly within its penetration depth from the p/i interface is affected by the blue or red light-soak. From Fig. 2 it is clear that the relative response in regions within 500 Å of the p/i interface is less affected by light-soaking than that in the layers below. This means that the collection of carriers generated close to the p/i interface is not affected as much by soaking as the carriers generated deeper in the bulk. If p/i interface defects were dominant this would not be the case, because newly generated electron-hole pairs would recombine before becoming separated by the electric field. The explanation for the shape of the relative change profile is given elsewhere [7].

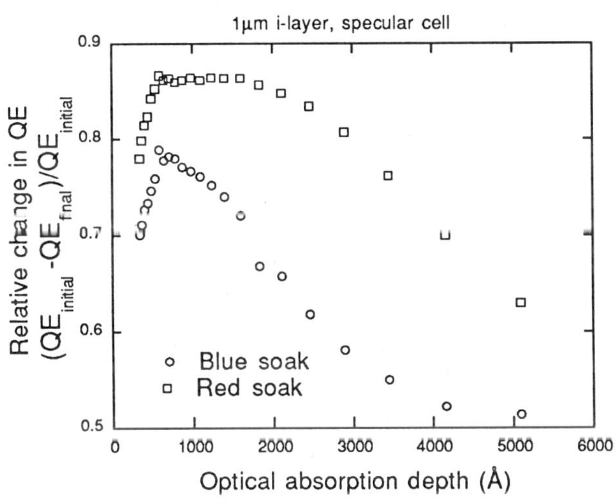

Fig. 2. Relative change in QE vs. penetration depth, for cells soaked with blue or red light.

Thickness Series

The motivation behind the thickness series degradation was to separate out the interface and bulk effects. As the i-layer thickness shrinks interface effects may come to dominate the degradation. Therefore, in a plot of degradation vs. i-layer thickness, the y-intercept (i-layer thickness -> 0 μm) will give the interfacial contribution to cell performance loss. Cells with very thin i-layers (< 0.15 μm) are not suited for intense light-soaking because they tend to become shorted during intense light-soaking steps. We therefore restricted the lower bound of the thickness series to 0.20 μm. The 30 minute high intensity soak is enough to bring the cells close to saturation.

Shown in Fig. 3 is the percentage degradation vs. the i-layer thickness for the thickness series after the red light-soaking step. The percentage degradation is calculated from the measured initial and final efficiencies (η_{init} and η_{final}) using:

$$\% \text{ degradation} = 100 * (\eta_{init} - \eta_{final}) / \eta_{init} \qquad (2)$$

While the efficiencies of the cells with i-layer thicknesses over 0.40 μm drop about 30%, that of the cell with i-layer thickness 0.20 μm drops only 14%. Reasonable extrapolation of the degradation vs. thickness curve to zero i-layer thickness gives a zero or even negative interface contribution to degradation. This again is experimental proof that no extra degradation mechanism is at work at the interfaces.

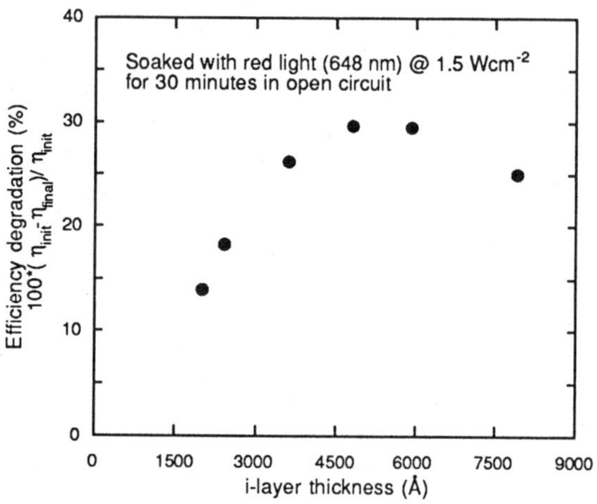

Fig. 3. Efficiency degradation vs. i-layer thickness.

Modeling of light-soaking

One of the main aims of modeling is to design cell structures that show little light-induced degradation and hence high end-of-life efficiency. To achieve this goal it is important that the model be able to predict the light-soaked performance of cells in addition to their as-deposited performance. One of the main parameters that is changed in models to simulate light-soaking is the i-layer midgap defect density. We have [8] been using the measured light-soaked midgap defect density of separately deposited i-layer films as an input parameter to model light-soaking of cells. But in several cases the modeled cell performance obtained using this method has been found to lie above the measured cell performance [2,8]. This discrepancy suggested that other loss mechanisms, for example interfacial loss, may be at work, besides losses due to an increase in the i-layer midgap defects. The interface loss has been modeled by placing thin, highly defective, layers between the doped and i-layers [2]. Doing so adds to the number of variables without direct experimental measure, and makes the model less reliable. In the light of the experimental results discussed above, we have been attempting to reproduce light-soaking results without invoking any extra interfacial loss mechanism [9].

We model light-soaking by changing only the midgap defect density in the i-layer of the solar cell. The cell performance depends on the number and spatial location N(x) of defects in the i-layer. N(x) controls the charge in the i-layer which affects the electric field, which in turn controls carrier collection. Furthermore, the cell performance depends on carrier recombination through the Nσ product (σ is the capture cross section of midgap defects). The Nσ product controls the recombination kinetics and hence carrier populations. The value of σ depends on the charge state of the defect.

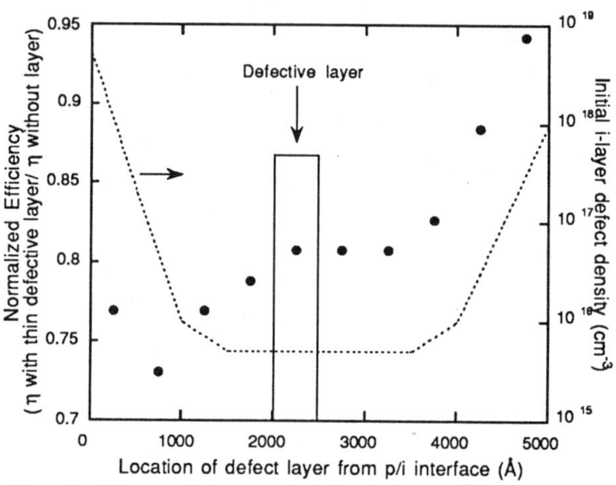

Fig. 4. Effect of the position of a 500 Å thick defect layer (shown here at 2250 Å) on cell efficiency. Grown in defect profile from Block.

The sensitivity analysis of Fig. 4 shows how cell efficiency depends on the spatial location of defects in the i-layer. In this simulation we used the initial defect profile as calculated [10] which is also shown in Fig. 4. Table I shows some of the simulation parameters used in this simulation, a detailed list is given in [6]. We then varied the position of a defective layer of thickness 500 Å and defect density $5x10^{17}$ cm^{-3} through the i-layer, all other parameters of the defective layer being the same as the i-layer. The points of Fig. 4 correspond to the center position of the defective layer as it is moved through the i-layer. As can be seen from Fig. 4 the maximum damage is caused by the thin layer when its center is 750 Å from the p-layer, but not at the p/i interface (see Fig. 2). This again shows that defective interfaces do not affect the cell as much as defects located elsewhere in the i-layer.

Shown in Fig. 5 is the simulated efficiency vs. N in the i-layer for three σ values. This figure will help understand the effect defects have on performance because of their number N (affects field) and their Nσ product (affects recombination). From Fig. 5 it is clear that for a small defect density ($5x10^{15}$ cm^{-3}) the effect of σ on the performance is small. At high defect density the effect of σ is much more pronounced. It is clear that the σ value is a very important parameter in the modeling of light-soaking. Fig. 5 demonstrates how using too low a value of σ can give modeled results above measured parameters, results which may be interpreted as evidence for an extra loss mechanism.

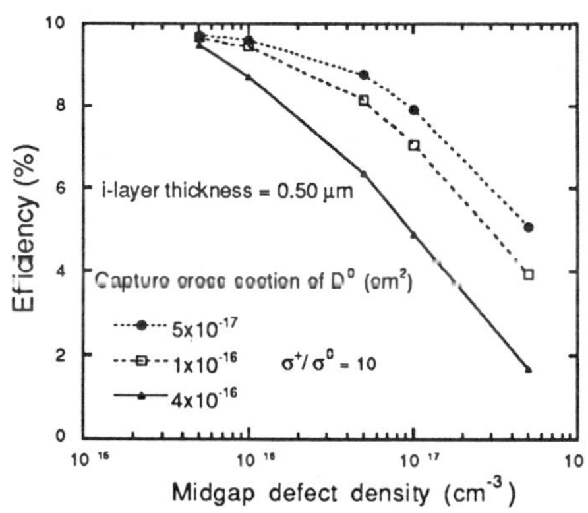

Fig. 5. Efficiency vs i-layer midgap defect density for three capture cross sections.

It is clear that the cell performance depends on the position, number and capture cross section of the defects. But, we do not find a need for assuming extra defective interface layers. We have also been able to simulate the light-soaking of a series of a-Si,Ge:H cells with varying band-gap [9] without invoking the existence of a defective layer.

ACKNOWLEDGMENTS

This work is supported by the Amorphous Thin-Film Solar Cell Program of the Electric Power Research Institute and the Princeton Plasma Physics Laboratory through a Graduate Tuition Fellowship. We would also like to thank Dr. John N. Bullock for providing us with the solar cells used in the thickness series measurements and Dr. R.S. Crandall for helpful discussions.

Table I: Simulation parameters

Parameters	i-layer
Thickness (nm)	500
Optical (Tauc) gap (eV)	1.72
Mobility gap (eV)	1.80
Electron affinity (eV)	4.00
Valence band tail slope (eV)	0.045
Conduction band tail slope (eV)	0.027
Midgap defect density (cm^{-3})	10^{16}-10^{17}
σ for neutral midgap defects (cm^2)	$4x10^{-16}$
σ for charged midgap defects (cm^2)	$4x10^{-15}$

REFERENCES

[1] X. Xu, J. Yang and S. Guha, *Appl.Phys.Lett.*, **62**, (1993), 1399.
[2] X.R. Li, S. Wagner, M. Bennett, J.Y. Hou, F. Rubinelli and S.J. Fonash, *Proc. 11th European Photovoltaic Solar Energy Conference*, (1992).
[3] A.H. Pawlikiewicz and S. Guha, *IEEE Transactions on Electron Devices*, **37**, (1990), 403.
[4] J.W. Park, R.J. Schwartz, J.L. Gray and G.B. Turner, *Proc. 20th IEEE Photovoltaic Specialists Conference*, (1988), 55.
[5] J.K. Arch, F.A. Rubinelli, J.Y. Hou and S.J. Fonash, *J.Appl.Phys.*, **69**, (1991), 7057.
[6] S. Wagner, K. Vasanth, M. Nakata, J. Yang and S. Guha, *Solar Energy Materials and Solar Cells*, **34**, (1994), 373.
[7] K. Vasanth, D. Caputo, S. Wagner, M. Bennett, S. Bae and S. Fonash, *Mat.Res.Soc.Symp.Proc.*, **336**, (1994), 693.
[8] K. Vasanth, M. Nakata, S. Wagner and M. Bennett, *Mat.Res.Soc.Symp.Proc.*, **297**, (1993), 827.
[9] S. Wagner, K. Vasanth, X. Xu, J. Yang and S. Guha, *Proc. 12th E.C. Photovolatic Solar Energy Conference*, (1994).
[10] M. Block, *Ph.D. Thesis*

IMPROVEMENT OF a–Si SOLAR CELLS FABRICATED BY MERCURY–SENSITIZED PHOTO–CVD USING H_2 DILUTION METHOD

Pavan Siamchai, Akira Yamada and Makoto Konagai
Tokyo Institute of Technology, Department of Electrical and Electronic Engineering
2–12–1 Ohokayama, Meguro–ku, Tokyo 152 Tel: +81–3–5734–2554 Fax:+81–3–5734–0963

J. H. Jang and K. S. Lim
Korea Advanced Institute of Science and Technology, 373–1 Kusong–dong, Yosong–gu, Taejoen 305–701 Korea. Tel: +82–42–869–3427 Fax: +82–42–869–8530

ABSTRACT

The high quality–wide bandgap i–layers prepared by hydrogen dilution method were applied to a–Si solar cells fabricated by mercury–sensitized photo–CVD. It was found that the initial efficiency and stability of the cells were significantly improved compared to the undiluted one. This improvement was due to the increase in the quality and stability of an i–layer as confirmed by the dark I–V characteristics measurement of the solar cell. Two techniques for increasing the stabilized efficiency of the H_2–diluted i–layer–based solar cell were employed. First, the use of appropriate bandgap profile leads to the increase in J_{sc} while maintaining high FF, resulting in the increase in the stabilized efficiency. Another one is the improvement of V_{oc} of the solar cell by the optimization of the p– and buffer layer. By lowering the temperature of p– and buffer layers, the solar cell with V_{oc} as high as 0.990 V has been obtained.

INTRODUCTION

In order to increase the initial efficiency and stability of a–Si solar cells, improvement in i–layer quality is one of the most essential issue[1]. One possible solution is the hydrogen dilution method, which is known to have beneficial effects for a–Si and a–Si alloy films prepared by plasma–enhanced chemical vapor deposition[2].

The mercury–sensitized photo–chemical vapor deposition is known to be able to produce a–Si films and solar cells with the quality comparable to those fabricated by plasma CVD[3]. Moreover, in the solar cell fabrication process, it is believed that by employing this technique, the cross–contamination during deposition can be reduced due to the absence of ion bombardment to the underlying layers and the chamber wall.

In this study we showed that the application of H_2 dilution method in the mercury–sensitized photo–CVD can significantly improve the quality of the a–Si film. The solar cell with a H_2–diluted i–layer, also showed an increase in initial effiiency and stability. To confirm the results, we examined directly the quality of the i–layer by measuring dark I–V characteristics of the cell. For further improvement of the stabilized efficiency of the H_2–diluted solar cell, we introduced two techniques which are the bandgap profiling and the p– and buffer layers optimization.

EXPERIMENTAL

A–Si:H films were deposited using the separated–chamber photo–CVD system[3]. SiH_4 and H_2 reactant gases were fed through the Hg bubbler. The low–pressure mercury lamp with the 253.7 nm emission line was used as a light source. The substrate temperatures during deposition were 150°C and 180°C which correspond to the heater temperatures of 200 and 250°C, respectively. Before deposition, the background pressure was kept at 2.7×10^{-5} Pa by the turbo molecular pump.

Amorphous silicon solar cells were fabricated using an AR–coated glass/TCO(textured SnO_2)/delta– doped p–layer/buffer layer (a–SiC)/i (a–Si)/n (a–Si)/ZnO back reflector/Ag/Al structure. The light degradation experiment was done at AM 1.5, 1 sun insolation for 24 h with the cell temperature of 50°C under the open–circuit condition.

RESULTS AND DISCUSSION

Properties of hydrogen–diluted a–Si films

The effect of H_2 dilution on the undoped a–Si film is summarized in Fig. 1. The hydrogen content is found to increase with the H_2 flow rate. The reduction in the SiH_2/SiH bond ratio for the substrate temperature of 150°C is observed. The optical band gap, in consistent with the hydrogen content, increases along with the H_2 flow rate. The defect density of the diluted film was found to be less than the undiluted one, and the photo– and dark conductivity were not significantly changed. These results indicate that we can prepared a wider bandgap film without any trade off in the film quality for the normal i–layer deposition temperature (150°C).

a–Si solar cells with diluted i–layer

We expect improvement in the stability of the cells with a hydrogen–diluted i–layer since lower defect (dihydride bonding) in the films usually relates to improvement in cell degradation. It is observed that hydrogen–diluted cells are higher in both the initial and light–soaked efficiencies. Moreover, cells with higher hydrogen dilution ratio (H_2/SiH_4 flow rate ratio r is from 0 to 1.5) show less percentage of degradation. The improvement of the stability of the solar cell is mainly due to the lowering of the

CH3365-4/94/0000-0492 $4.00 © 1994 IEEE First WCPEC; Dec. 5–9, 1994; Hawaii

short circuit current and fill–factor degradation while retaining a higher open–circuit voltage as shown in table I.

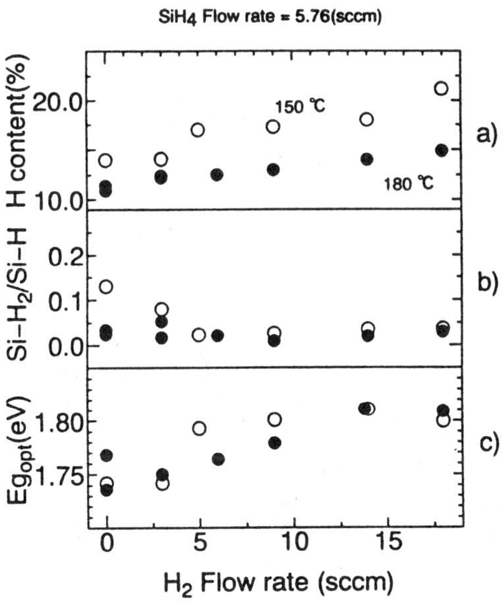

Fig. 1 (a) Total bonded hydrogen content as calculated from 2000 cm^{-1} and 2090 cm^{-1} peak, (b) the ratio of Si H_2 bond to Si–H bond and (c) the optical bandgap as a function of the hydrogen flow rate.

Table I Photovoltaic performance and degradation of the solar cells with various dilution ratios. (Measured at T.I.T.)

Sample	V_{oc}(V)	J_{sc}(mA/cm^2)	F.F.	Eff.(%)
SiH_4 100%				
Initial	0.905	18.74	0.721	12.23
Degraded	0.886	16.62	0.571	8.41
Degradation(%)	2.1	11.3	20.8	31.2
H_2 3 sccm				
Initial	0.930	18.18	0.740	12.52
Degraded	0.907	16.72	0.612	9.27
Degradation(%)	2.5	8.0	17.3	26.0
H_2 9 sccm				
Initial	0.944	17.80	0.740	12.45
Degraded	0.930	16.94	0.628	9.89
Degradation(%)	1.5	4.8	15.1	20.6

* SiH_4 flow rate 5.76 sccm

In order to measure stabilized efficiency, the accelerated degradation test of the cells was carried out at Sanyo Electric Corp. with AM 1.5, 5 sun, and 25 °C for 160 min in the open-circuit condition. The results are shown in Fig. 2. As the i-layer becomes thinner, the initial efficiency decreases but the open-circuit voltage and FF increase. The highest stabilized efficiency is obtained in the cell (1.0 cm^2 area) with a 190–nm–thick i-layer. This cell has an efficiency of 8.5 % with V_{oc} of 0.94 V, I_{sc} of 13.6 mA/cm^2, and FF of 0.67. The percentage of degradation is only 14 %.

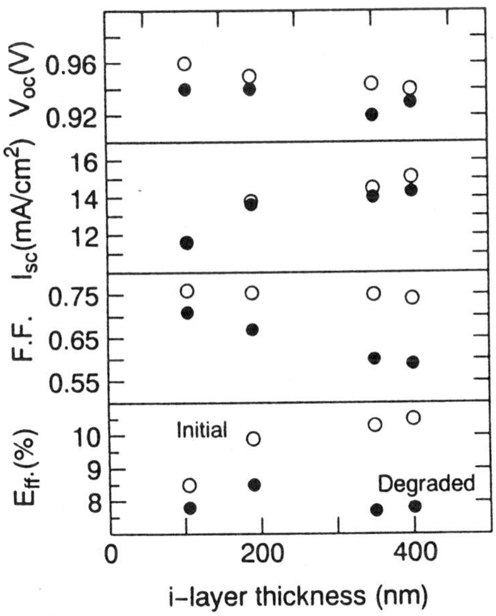

Fig. 2 The relationship between the hydrogen–diluted i–layer thickness and initial (O) and stabilized (●) performance of the solar cell (1.0 cm^2). The dilution ratio (r) is 1.5. (Measured at Sanyo Corp.)

We also examined the dark I–V characteristics of the solar cells as another direct tool to evaluate the properties of the i–layer of the cell. Figure 3 shows the simulation results of the reverse dark I–V characteristics of the cells. These calculations are based on our solar cell simulation model[4], with further

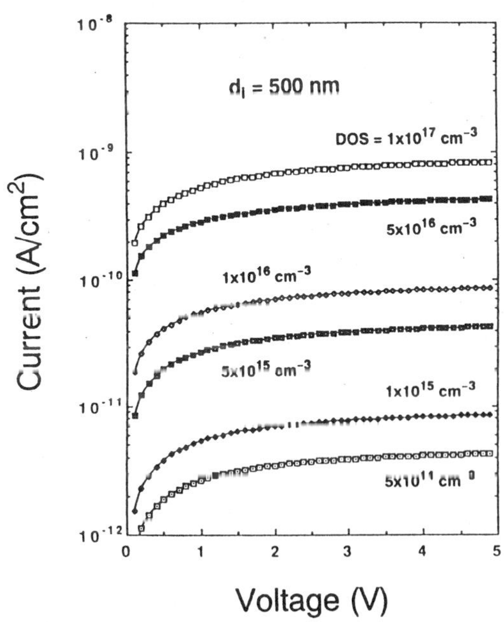

Fig. 3 Simulation results of the reverse–bias dark I–V characteristics of p–i–n solar cells with the i–layer defect density as the parameter. The i–layer thickness is 500 nm.

improvement to include the Gaussian-distributed defect states. It is observed that the reverse saturation current has a linear relationship with the defect density. Thus by measuring the reverse saturation current we can get the information about the relative value of the defect density. The experimental results shown in Fig. 4 are the reverse-bias dark I-V curves of two cells before and after light soaking. As can be seen from the figure, the level of initial state saturation current of the diluted cell is slightly lower than the undiluted one, and a significant difference in the saturation current can be observed after 24 hours of light soaking. The undiluted solar cell shows one order of magnitude increase in saturation current compared with the initial states, while the diluted cell exhibits only 1–2-fold increase. This result indicates that the increase in defect density in the i–layer is less for the hydrogen–diluted solar cell.

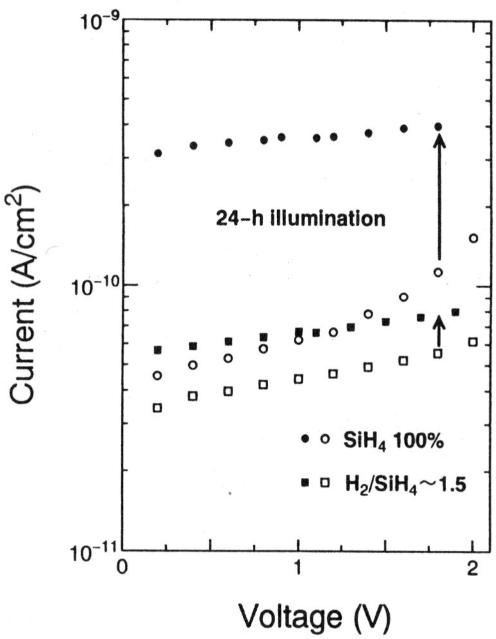

Fig. 4 The reverse-bias dark I-V curves of the diluted and undiluted solar cells before and after light soaking for 24 h at AM 1.5, 1 sun. The i-layer thickness is 430 nm.

Stability improvement techniques

Since the use of the wide bandgap H_2–diluted i–layer may result in the reduction of the short-circuit current of the cell, in order to increase the I_{sc} without deteriorating other properties of the cell, the bandgap profiling technique which is accomplished by varying the hydrogen flow rate during i–layer deposition was introduced. We found at first that the normal profile (high–to–low bandgap from p– to n–layer) gives high V_{oc} and I_{sc} but the worst FF, so the initial efficiency is lowered. This result indicates the weakening of the internal field for holes. Thus, we concentrated our study on the reverse profile (low–to–high bandgap) and the double profile (high–to–low–to–high bandgap) i–layers which can give higher initial efficiency. We investigated three types of bandgap profile as shown in Fig. 5.

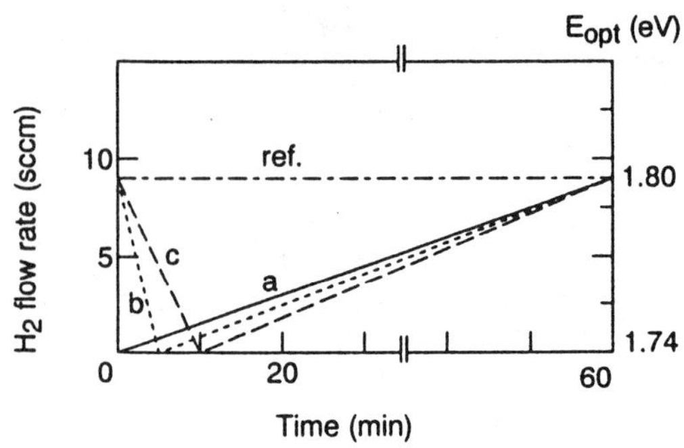

Fig. 5 Three profiles of H_2–diluted i–layers fabricated in this work, a) is the reverse profile, b) and c) are the double profiles and the reference cell is the one with a uniformly diluted i–layer. The i–layer thickness is about 300 nm.

From the results shown in Fig. 6 we observed that the reverse profile gives the best stabilized efficiency (measured at Sanyo) of 8.1% (for 300–nm i–layer) which is better than the stabilized efficiency of uniformly diluted cell (7.7%). We can understand from this result that although the undiluted part in the i–layer has lower quality, the improvement in the internal field by the bandgap profiling contributes more effect on the cell stability.

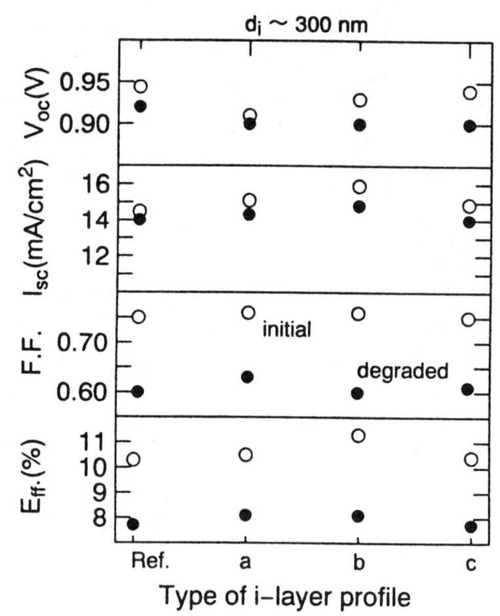

Fig. 6 The relationship between the profile type of the i–layer according to Fig. 5, and initial and stabilized performance of the solar cell. (Measured at Sanyo Corp.)

One interesting characteristic of H_2–diluted solar cell is a high open–circuit voltage. Since the degradation of the V_{oc} is very small, it can be thought that the stabilized efficiency should be improved if we can obtain higher V_{oc} from the cell with same i-layer quality. In this work we have tried to increase the bandgap of the p-layer by lowering the deposition temperature. By lowering only the deposition temperature of SiC undoped layers in the delta-doped p-layer, no improvement in V_{oc} was observed. However, when the deposition temperatures of both the SiC undoped layers and the graded bandgap buffer layer were lowered, the considerable improvement in V_{oc} and initial efficiency was obtained as can be seen in Fig. 7. This results implies that the bandgap of the film at the p-i interface plays a significant role in determining the V_{oc} of the solar cell. Anyway, as the carrier generation and recombination in a buffer layer is high when the cell is illuminated, the quality and stability of a buffer layer can affect cell stability considerably, particularly in the cell with a thin i-layer. From the light soaking experiment results, it was observed that the cell with the buffer layer prepared at a low temperature shows slightly higher degradation of FF, resulting in the same stabilized efficiency as the cell with the buffer layer prepared at a higher temperature. More studies are underway in order to improve the quality of a buffer layer.

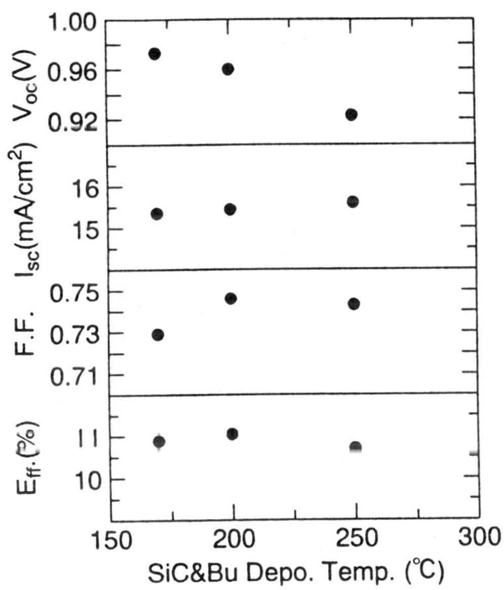

Fig. 7 The relationship between the deposition (heater) temperature of the p- and buffer layers and the solar cell performance (measured at T.I.T). The i-layer thickness is 180 nm, and the dilution ratio (r) is 1.5.

By using the low-temperature p- and buffer layer combined with higher dilution ratio (r=2.5) which gives the wider bandgap i-layer, the solar cell with V_{oc} as high as 0.990 V and FF of 0.721 has been obtained (Fig. 8).

CONCLUSION

High-quality a-Si film with a wide band gap has been successfully fabricated by the mercury-sensitized photo-CVD combined with the hydrogen dilution technique. The application of i-layer with the hydrogen dilution ratio of 0 to 1.5 resulted in the improvement of initial efficiency and stability of the cell. For the 1 cm^2 solar cell, the highest stabilized efficiency of 8.5 % has been obtained with an i-layer thickness of 190 nm. From the forward- and reverse-bias dark I-V characteristic measurements, it was found that the increase in the i-layer defect density after light soaking is considerably lowered for the hydrogen-diluted cell compare to the undiluted one.

By using the bandgap profiling technique, the stabilized efficiency of the H_2-diluted cell could be further increased due to the improvement of short-circuit current and fill factor. Moreover, a considerable higher open-circuit voltage (0.990 V) has been achieved by the better matching between the bandgap of p-, buffer and i-layer.

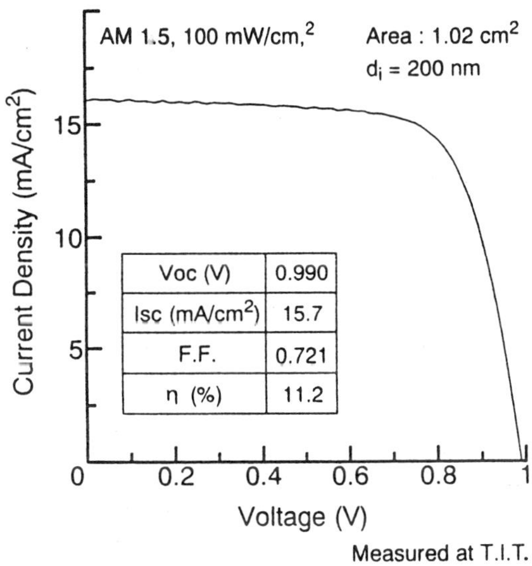

Fig. 8 I-V characteristic of the H_2-diluted solar cell with the high open-circuit voltage.

ACKNOWLEDGEMENT

The authors would like to thank Sanyo Electric Co., Ltd for the light soaking experiment. This work was supported in part by NEDO as a part of the New Sunshine Program under the Ministry of International Trade and Industry.

REFERENCES

[1] S. Yamanaka, M. Konagai and K. Takahashi: Jpn. J. Appl. Phys. **28**(1989)1178.

[2] S. Okamoto, Y. Hishikawa, S. Tsuge, M. Sasaki, K. Ninomiya, M. Nishikuni and S. Tsuda: Jpn. J. Appl. Phys. **33**(1994)1773.

[3] W. W. Wenas, M. Yoshino, K. Tabuchi, A. Yamada, M. Konagai and K. Takahashi: *Proc. 23rd IEEE Photovoltaic Specialist Conf.*, Louisville, 1993, p. 935.

[4] H. Tasaki, W. Y. Kim, M. Konagai and K. Takahashi: J. Appl. Phys. **63**(1988)550.

CORRELATION BETWEEN TCO/p AND p/i HETEROJUNCTION AND EFFECT OF n/TCO HETEROJUNCTION ON a-Si:H SOLAR CELL PERFORMANCE

Franc Smole, Marko Topič, and Jože Furlan
Faculty of Electrical and Computer Engineering, University of Ljubljana
Tržaška 25, 61000 Ljubljana, Slovenia

ABSTRACT

The influence of front TCO/p(a-Si:C:H) heterojunction in correlation with p(a-Si:C:H)/i(a-Si:H) heterojunction on TCO/p-i-n/TCO/metal a-Si:H solar cell performance is examined using the ASPIN computer simulator. The results indicate that defect states at both heterojunctions affect the potential barrier profile at the TCO/p heterojunction, and that their increase lowers the free hole concentration in the p-layer. However, defect states at the p/i interface change the potential distribution in the i-layer more strongly, and hence also photoelectric properties.

At the improved TCO/metal back contact, the effect of n(a-Si:H)/TCO heterojunction on the electrical properties is also examined. The analysis of n(a-Si:H)/TCO structures shows that the best electrical properties can be achieved with ZnO, which enables the reduction of the n-layer thickness, resulting in further enhanced quantum efficiency and short-circuit current.

INTRODUCTION

Despite remarkable progress in the conversion efficiency of single p-i-n a-Si:H solar cells, there are still many factors that limit cell performance. Apart from a-Si:H film quality, interfaces play the most significant role. The interface between a transparent conducting oxide (TCO), and the front a-Si:H layer plays an important role in affecting photovoltaic properties of a-Si:H solar cells. Although TCO layers (SnO_2, ZnO, ITO) exhibit good visible light transmittance and low ohmic resistivity, the work function difference between TCO and a-Si:H builds a potential barrier [1], which results in an imperfect ohmic contact. The potential barrier impedes the extraction of photogenerated carriers, and thus affects the photoelectric properties of the complete glass/TCO/p-i-n/TCO/metal solar cell.

The film quality and the p/i interface have recently been improved using hydrogen plasma treatment [2]. The detrimental effects of H_2-dilution on TCO/p(a-Si:C:H) heterojunction, such as chemical oxidation of a-Si:H (reduction of TCO) by plasma treatment and cross-contamination by interdiffusion, have been extensively examined [3]. Scientific interest was mainly focused on optical properties; much less effort was directed toward the determination of how a-Si:C:H deposition conditions affect the electrical properties of the interface.

In the first part of our paper, our computer model ASPIN was used for an analysis of TCO(SnO_2 or ZnO)/p-i-n structures; results for different TCO/p and p/i interface states distributions are presented. The correlation between TCO/p and p/i heterojunction interfaces is examined. The implication of these effects for the photoelectric properties of a p-i-n a-Si:H heterojunction solar cell is demonstrated.

To achieve high efficiency, a back contact of the a-Si:H cell must provide high optical reflectance. Either a thin-layer of indium tin oxide (ITO) [4] or ZnO [5] - inserted between a-Si:H and a metal electrode - is effective in the enhancement of the back reflectance. In the second part of our paper, the electrical properties of the inserted TCO layer at the back contact are examined using the ASPIN computer model. The results for different n-layer/TCO heterojunction structures are presented. The possibility of a thinner n-layer is discussed, and its beneficial effect on the photoelectric properties of the p-i-n a-Si:H heterojunction solar cell is presented.

NUMERICAL MODELLING INCLUDING THE DEFECT POOL MODEL

In numerical models for a-Si:H solar cells, thermal equilibrium boundary conditions, as well as surface recombination velocities at the a-Si:H front surface attached to a TCO layer, are questionable due to relatively low p-layer quality and strong influence of the TCO/a-Si:C:H heterojunction. In order to define more accurate boundary conditions, and to bring effects at the TCO/a-Si:H heterojunctions into consideration, ASPIN has been extended to incorporate a TCO layer with a TCO/a-Si:H interface, enabling the analysis of TCO/p-i-n/TCO heterostructures [6].

The ASPIN model is based on an iterative method of solving a set of coupled linearized differential equations: Poisson's, continuity and transport equations. It calculates the internal electric properties and external characteristics of a multi-layer heterostructure. The density of states (DOS) model in a-Si:H consists of a tail of donor-like states above the valence band edge, a tail of acceptor-like states below the conduction band edge, and three Gaussian-shaped defect states densities D^-, D^0, D^+, whose density depends on the Fermi level according to the defect pool model [7].

A typical 0.5 μm thick p(a-Si:C:H)-i(a-Si:H)-n(a-Si:H) solar cell structure with a p(a-Si:C:H) window layer and a graded a-Si:C:H/a-Si:H heterojunction was chosen as the initial structure (parameters in [8]).

ANALYSIS OF TCO/p-i-n STRUCTURE: CORRELATION BETWEEN TCO/p and p/i HETEROJUNCTION

The calculated energy band diagram of an SnO_2/p-i-n structure in thermal equilibrium is shown in Fig. 1.

CH3365-4/94/0000-0496 $4.00 © 1994 IEEE

First WCPEC; Dec. 5-9, 1994; Hawaii

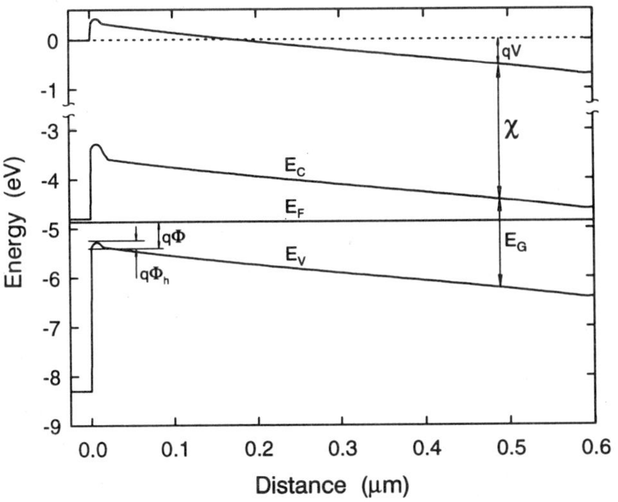

Fig. 1. Calculated profile of energy bands in SnO_2/p-i-n structure in thermal equilibrium. qΦ denotes a surface energy barrier and Φ_h denotes a potential barrier.

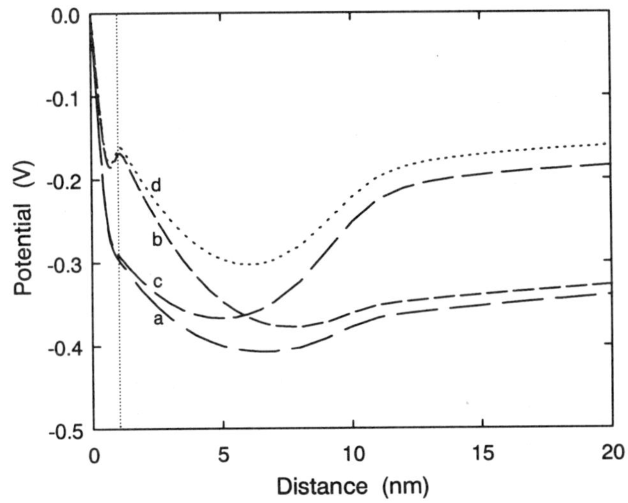

Fig. 2. Calculated potential profiles in SnO_2/p-i-n structure for different surface state densities in a-Si:C:H and different p/i interface defect concentrations (see text).

The contact potential between SnO_2 and the p(a-Si:C:H) layer, which is defined as the work function difference, is much higher (0.44 eV) than the measured potential barrier Φ_h of 140 mV in the p-layer [9].

The work function of an n-type ZnO (4.2 eV) is lower than that of an n-type SnO_2 (4.8 eV); thus the contact potential between ZnO and p(a-Si:C:H) is higher (1.1 eV). Due to a lack of experimental results for the potential barrier at the ZnO/a-Si:C:H heterojunction, we can only assume its value from a comparison of the V_{OC} results. These indicate that with a ZnO front contact, the V_{OC} is 20 - 50 mV lower than with an SnO_2 front contact. Therefore, a higher potential barrier is expected (around 0.2 V).

In addition to the contact potential, interface defect states may play an important role in determining band bending at the TCO/a-Si:C:H heterojunction [8].

In this analysis, a p(a-Si:C:H) layer with activation energy of 0.4 eV and with 10^{19} cm^{-3} defect states concentration was chosen.

In order to achieve 140 mV potential barrier at the initial SnO_2/p-layer heterojunction, density of interface defect states was set to $2 \cdot 10^{13}$ cm^{-2} (Fig. 2, plot a). The rest of the contact potential (300 mV) builds a steep potential drop in the interface region, which does not impede the transport of free carriers. To approach realistic values of Φ_h in the p-layer at the ZnO/p(a-Si:C:H) heterojunction, the density of interface defect states was set to $6 \cdot 10^{13}$ cm^{-2}, resulting in a 200 mV potential barrier.

In the analysis of the relations between SnO_2/p and p/i heterojunction, the effect of chemical oxidation (reduction) of a-Si:C:H (SnO_2) at the early stage of a-Si:C:H growth on SnO_2 [10], and the influence of p/i heterojunction quality were examined.

The density of surface states of the nonoxidized a-Si:C:H (Fig. 2, plot a) was 10^{12} cm^{-2}. The detrimental effects of oxidation were modelled by an increased density of surface states (4 nm in depth) of $8 \cdot 10^{12}$ cm^{-2}. The Φ_h in the p-layer increased by 120 mV (Fig. 2, plot b). This increased density of surface states also affected the potential distribution in the i-layer.

An equal increase in density of defect states at the p/i interface affects the potential distribution far more strongly (Fig. 2, plot c). The built-in voltage and the resulting electric field in the i-layer are lowered. Although the Φ_h was reduced, a depletion of hole concentration in the p-layer occurred up to the SnO_2/p interface, where the potential is nearly pinned due to the high interface state density. Therefore, the defect states at the p/i heterojunction have relatively small influence on the situation at the (SnO_2/p) surface. Confirmation can be seen in Fig. 2, plot d, where both increased p-layer surface state density (as in plot b) and equivalently increased density of defect states at the p/i interface (as in plot c) were taken into account.

Analysis under AM1.5 illumination showed that an undamaged SnO_2/p-i-n/ZnO/Ag structure would achieve efficiency of 12.9 %, with 17.1 mA/cm^2 short-circuit current density J_{SC}, 987 mV open-circuit voltage V_{OC} and fill factor FF of 0.764.

A higher potential barrier in the oxidized p-layer (case b) reduced mainly the V_{OC} (by 9.7%). The FF dropped by 3.5%, while the J_{SC} stayed almost unchanged. However, collection efficiency at positive bias tended to fall in the short-wavelength region up to 500 nm (Fig. 3, plot b).

The increased defect states at the p/i interface (case c) affected the solar cell more strongly. V_{OC} and the FF dropped by 10.9% and 11.8%, respectively. J_{SC} fell by 3.3%. The collection efficiency decreased up to the long-wavelength region (Fig. 3, plot c). With both oxidized p-layer surface and increased p/i interface defect states (case d), V_{OC} and FF dropped by 17.5% and 23.2%, respectively. The collection efficiency is even worse, than predicted (Fig. 3, plot d).

Analysis of the relation between ZnO/p and p/i heterojunction in ZnO/p-i-n structure revealed very similar behaviour. However, cell performance (especially FF and V_{OC}) was more sensitive to the quality of the p/i heterojunction. The results indicate that with a ZnO/p-i-n structure different optimization of p-layer doping (higher) and p/i buffer layer is necessary [11].

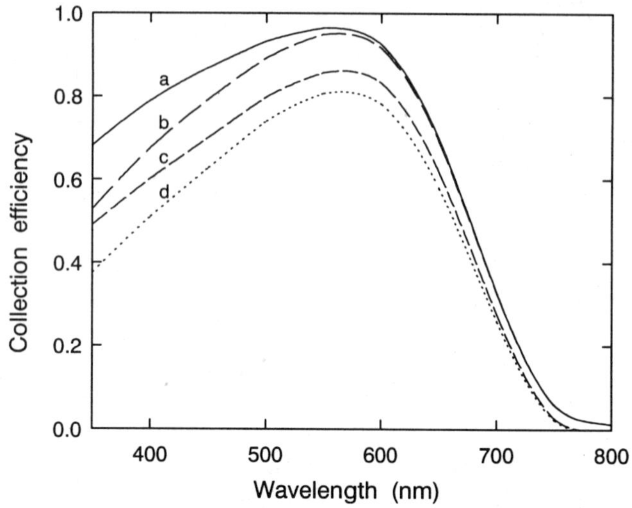

Fig. 3. Calculated collection efficiency at 0.5 V forward bias for structures: a - SnO_2/p-i-n (as in Fig. 2, case a); b - SnO_2/p-i-n with oxidized a-Si:C:H surface (case b); c - SnO_2/p-i-n with increased p/i interface defect states (case c); d - SnO_2/p-i-n with oxidized a-Si:C:H surface and increased p/i interface defect states (case d).

ANALYSIS OF pin/TCO STRUCTURE

Insertion of a thin TCO layer at the back contact enhances reflectance in the long wavelength region (λ>600 nm) [4,5], resulting in improved quantum efficiency in this wavelength region (Fig. 4, plot b). Therefore, in the calculation, it increased the J_{SC} from 16.04 mA/cm^2 to 17.07 mA/cm^2. Much less scientific attention was paid to how the inserted TCO layer affects the electrical properties.

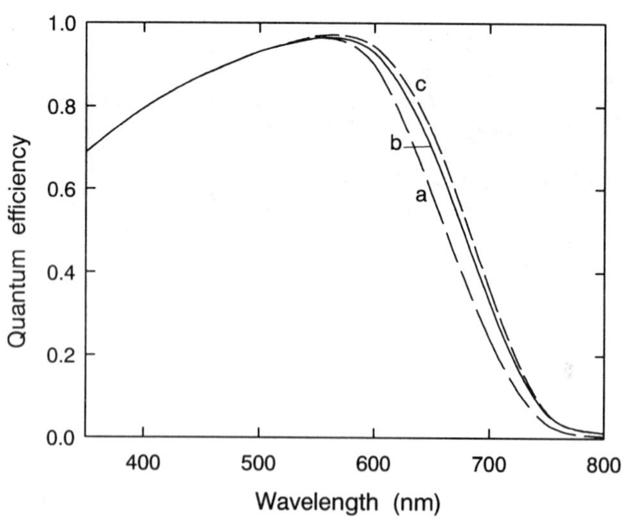

Fig. 4. Calculated quantum efficiency for structures:
a - SnO_2/p-i-n/Ag; b - SnO_2/p-i-n/ZnO/Ag;
c - SnO_2/p-i-n/ZnO/Ag with 3 nm thick n-layer.

The work function of n(a-Si:H) is much lower (typically around 4.2 eV) than the work function of p(a-Si:C:H). Of SnO_2, ITO and ZnO, ZnO has the lowest work function difference with n(a-Si:H). With respect to the potential barrier in the n-layer, the most appropriate TCO for an n(a-Si:H)/TCO heterojunction is therefore ZnO. Because there is a strong pinning of the Fermi level at the n(a-Si:H)/ZnO heterojunction, and almost no contact potential with a negligible potential barrier in the n-layer, the thickness of the n-layer can be reduced. With a thinner n-layer, the losses present in the defect-rich n-layer are reduced, and the cell performance can thus be improved. An ideal heterojunction between a-Si:H and ZnO would even not require the n-layer, since there would be the same pinning of the Fermi level at the i(a-Si:H)/ZnO heterojunction. In an ideal heterojunction, the same built-in potential would therefore be created throughout the p-i/ZnO structure as throughout the p-i-n/ZnO structure.

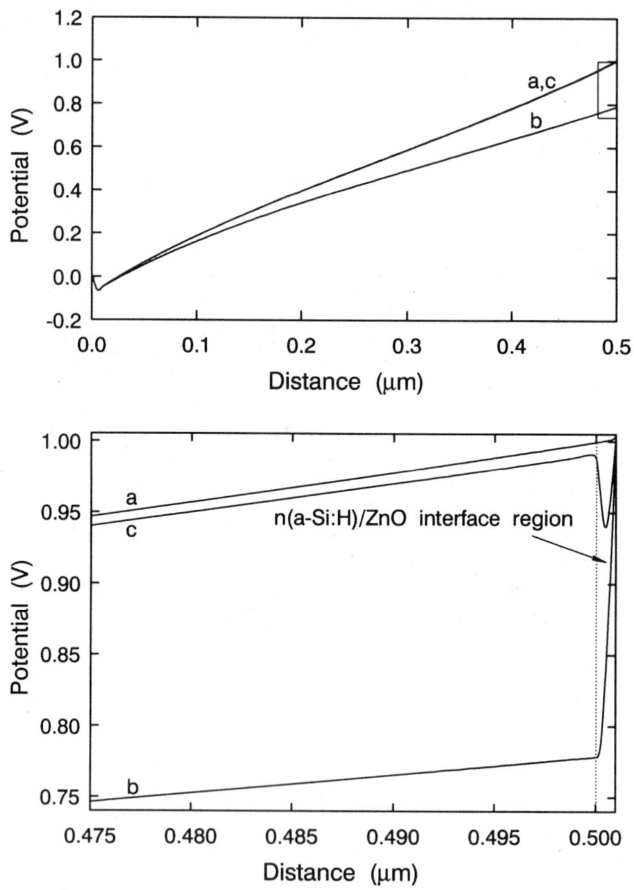

Fig. 5. Calculated built-in potential throughout the solar cell structures with ZnO/Ag improved back contact: a - SnO_2/p-i with ideal i(a-Si:H)/ZnO heterojunction; b - SnO_2/p-i with interface defect states that cause a strong negative space charge; c - SnO_2/p-i-n structure with an n layer (1 nm) that compensates for the interface negative space charge.

To examine whether the n-layer is necessary or not, n(a-Si:H)/ZnO and i(a-Si:H)/ZnO structures were analysed using the ASPIN simulator. Analysis showed that the increased defect states at the a-Si:H/ZnO heterojunction due to the large difference in lattice constants between the adjacent materials strongly affect the electrical properties of the junction. Trapped electrons on the interface defect states and on D^- defect states at the a-Si:H surface build up a strong negative space charge. This charge results in a steep potential drop in the vicinity of the heterojunction, thus reducing the built-in potential in the i-layer (Fig. 5, plot b).

A thin n-layer is therefore necessary between the i-layer and ZnO layer so that the ionised phosphorous atoms with positive charge compensate for the negative charge of trapped electrons, thus reducing the unwanted steep potential drop at the heterojunction (Fig. 5, plot c). Analysis showed that the minimum thickness of the n-layer would be a few nanometers. Beyond this thickness, the open-circuit voltage and especially, the fill factor starts to decrease. In the p-i structure (without the n-layer), calculated V_{OC} drooped by 5% and FF by 15%.

A thinner n-layer yielded an improvement in quantum efficiency in the wavelength region 550 nm $< \lambda <$ 750 nm (Fig. 4, plot c), and thus in an improved J_{SC} of p-i-n a-Si:H solar cell by additional 0.3 mA/cm^2 (17.4 mA/cm^2).

CONCLUSIONS

In the analysis of relations between TCO/p and p/i heterojunctions, the results indicate that worsening of TCO/p interface, e.g. due to the oxidized p-layer surface at SnO_2/p heterojunction, mainly reduces V_{OC} and the collection efficiency in the short wavelength region. Equally increased defect states at the p/i interface affect cell performance more strongly, because it depletes the p-layer and lessens the built-in electric field in the i-layer. Both V_{OC} and FF are lowered and collection efficiency under forward bias tends to decrease more strongly. To conclude, the quality of TCO/p heterojunction dominantly determines the potential barrier at the p-layer surface, the quality of the p/i heterojunction reflects in good FF, and together they set the limits of V_{OC}.

Improved back contact from electrical point of view is done by insertion of ZnO, which creates almost no contact potential at n(a-Si:H)/ZnO heterojunction. Despite the E_F pinning at the a-Si:H/ZnO heterojunction, the n-layer is necessary to compensate for the negative space charge at the a-Si:H/ZnO interface. However, analysis showed that the thickness of the n-layer can be reduced to few nanometers. A thinner n-layer enhances quantum efficiency in the long-wavelength region and consequently results in an increased short-circuit current.

ACKNOWLEDGEMENT

This work was supported by the Commission of the European Communities, Program Joule II.

REFERENCES

[1] F.S.Sinencio and R.Williams, "Barrier at the interface between amorphous silicon and transparent conducting oxides and its influence on solar cell performance", *J. Appl. Phys.*, **54**, 1984, pp. 2757-2760.

[2] K.Miyachi, M.Koyama, H.Tanaka, Y.Ashida, N.Fukuda and A.Nitta, "Preparation of a-Si:H films by alternately repeating deposition and hydrogen plasma treatment", *Technical Digest of 5th International Photovoltaic Science and Engineering Conference*, Kyoto, 1990, pp. 63-66.

[3] Y.Lu, I.An, M.Gunes, M.Wakagi, C.R.Wronski and R.W.Collins, "Effect of gas phase hydrogen-dilution on the nucleation, growth, and interfaces of a-$Si_{1-x}C_x$:H", *MRS Sym. Proc.,* **297**, Pittsburg, 1993, pp. 31-36.

[4] G.Tao, B.S.Girwar, G.E.N.Landweer, M. Zeman, and J.W.Metselaar, "Enhanced light absorption in a-Si:H layers of solar cells by applying TCO/metal back contacts", *Material Research Society Symposium Proceedings,* **297**, 1993, pp. 845-849.

[5] K.Tabuchi, W.W.Wenas, M.Yoshino, A.Yamada, M.Konagai and K.Takabashi, "Optimization of ZnO films for amorphous silicon solar cells", *Proc. 11th E. C. PVSEC*, Montreaux, 1992, pp. 529-532.

[6] F.Smole, M.Topic, and J.Furlan, "Amorphous silicon solar cell computer model incorporating the effects of TCO/a-Si:C:H junction", *Solar Energy Materials and Solar Cells,* **34**, 1994, pp. 385-392.

[7] G.Schumm and G.H.Bauer, "Equilibrium and non-equilibrium gap state distribution in a-Si:H", *J. Non-Crystalline Solids* , **137&138**, 1991, pp 315-318.

[8] F.Smole, M.Topic, and J.Furlan, "The influence of defect states in TCO/a-Si:C:H interface on p-i-n a-Si:H solar cell performance", *Proc. of 12th European PVSEC,* Amsterdam, 1994, pp. 77-80.

[9] K.Itoh, H.Matsumoto, T.Kobata and A.Fujishima, "Determining height of a leaky Schottky barrier existing in the junction between SnO_2 and a highly doped *p*-type amorphous SiC by using the pulsed laser-induced transient photopotential technique", *Appl. Phys. Lett.*, **51**, 1987, pp. 1685-1686.

[10] H.Koinuma, M.Nakano and S.Gonda, "Structure of a-Si:H/SnO_2 interface characterized by XPS", *Material Research Society Symposium Proceedings,* **192**, 1990, pp. 201-206.

[11] M.Kubon, N.Schultz, M.Kolter, C.Beneking and H.Wagner, "Modification of TCO-p interface in a-Si:H solar cells by intermediate layers and plasma treatments", *Proc. of 12th European Photovoltaic Solar Energy Conference,* Amsterdam, 1994, pp. 1268-1271.

SOLUTION OF THE ZnO/p CONTACT PROBLEM IN a-Si:H SOLAR CELLS

M. Kubon. E. Boehmer, M. Gastel*, F. Siebke, W. Beyer, C. Beneking, and H. Wagner
Institute of Thin Film and Ion Technology (ISI-PV),*Central Department for Chemical Analysis (ZCH)
Research Centre Juelich, D-52425 Juelich, Germany

ABSTRACT

This paper addresses the problem of preparing a good p-layer contact with zinc oxide as TCO. Our approach was to deposit pin cells with different p-layer recipes on ZnO coated SnO_2:F and on uncoated SnO_2:F in one run. The pin cells prepared on the ZnO surface exhibit a lower fill factor (FF). Our experiments demonstrate that the hydrogen interaction with the ZnO surface plays the most decisive role for the ZnO/p contact. We explain the observed effects using a band diagram of the ZnO/p interface and show that the accumulation layer at the ZnO surface -caused by atomic hydrogen in the plasma- is responsible for the low FF in pin cells. Based on this model the contact problem is solved by introducing a μc-n-Si intralayer between ZnO and p layer resulting in an identical high FF on both ZnO and SnO_2 substrates.

1. INTRODUCTION

In order to obtain further improvement in stabilized efficiency of a-Si:H solar cells, growing attention is being paid to the transparent conductive oxide (TCO) material that forms the front electrode. Especially the TCO/p contact properties and their influence on solar cell parameters have been the object of a number of investigations.

Fluorine doped SnO_2 is a well known material for TCO films, exhibiting favourable surface textures that enhance the conversion efficiency. However, SnO_2 is reduced during the early stages of solar cell deposition, leading to additional absorption [1]. In contrast, aluminum or boron doped ZnO shows higher transparency in the visible range [2] and higher plasma stability, but ZnO/p contact problems have been reported [3]. Especially free hydrogen in the plasma has an influence on the ZnO surface [4].

In this study we coated one half of a SnO_2 substrate with thin ZnO:Al and ZnO:B overlayers, thus being able to compare directly the properties of the different TCO materials and their influence on solar cell performance [5]. The dependence on p layer deposition parameters was studied. Furthermore we investigated the stoichiometry of thin SiC layers on flat, half coated substrates by XPS to study the TCO reduction. Based on these results we develop a model of the ZnO/p interface that explains all our results and leads the way to the solution of the ZnO/p contact problem.

2. EXPERIMENTAL

We prepared 15 nm ZnO coating layers by RF magnetron sputtering from ZnO:Al_2O_3 (2wt%) and ZnO:B_2O_3 (4wt%) targets on standard SnO_2:F substrates (≈ 6% Haze) [5]. The thin ZnO layers grow conformally on top of the SnO_2 thus conserving the surface roughness [6]. For the XPS investigation we used flat SnO_2 substrates and measured in a Perkin Elmer ESCA 3057 system. Using a transfer system it was possible to transfer the samples between XPS chamber and plasma CVD system without breaking the vacuum. In order to remove a possible surface adsorption layer we sputtered the sample surface with an Ar^+ ion gun prior to the XPS measurement.

The pin-cells (structure: 7 nm p-SiC:H/ 400 nm i-Si:H / 25 nm n-Si:H) and the amorphous p-SiC:H layers (for XPS) were deposited in a commercial multichamber plasma CVD system, which is routinely used for the preparation of single-junction and stacked solar cells with >10% initial efficiency [7]. For I-V-measurements we used a calibrated solar simulator (AM 1.5, 100 mW/cm^2). The results given are representative values averaged over a number of cells prepared on each sample.

3. RESULTS AND DISCUSSION

In order to compare the properties of pin cells directly on the different substrates, we used the cells on uncoated SnO_2 as a reference and calculated ΔV_{oc}, ΔJ_{sc} and ΔFF between the cell properties on ZnO and SnO_2.

The ZnO coating causes mainly differences in J_{sc} and FF, but also a slightly decreased V_{oc} ($\Delta V_{oc} \approx -15$ mV).

CH3365-4/94/0000-0500 $4.00 © 1994 IEEE First WCPEC; Dec. 5-9, 1994; Hawaii

The dependence of ΔJ_{SC} and ΔFF on ZnO coating thickness is shown in figure 1.

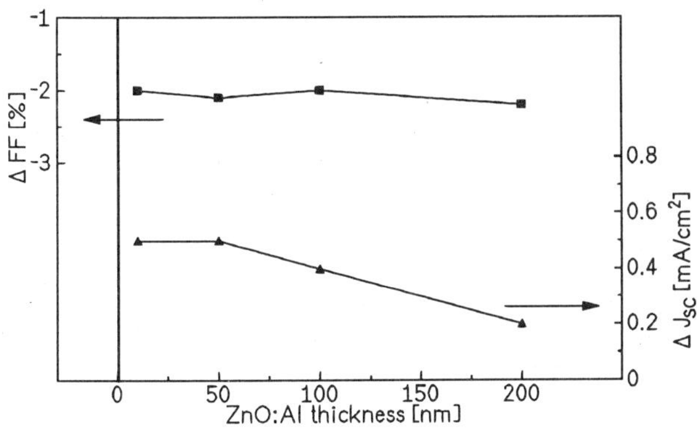

Fig. 1: Differences in fill factor and short-circuit current density between cells on ZnO coated substrate and cells on uncoated SnO_2 in dependence on ZnO coating thickness.

The increase in J_{SC} ($\Delta J_{SC}>0$) is interpreted as a consequence of a protection of SnO_2 by the ZnO against roduction in tho p layor plasma [6]. However, the FF of the cells on ZnO is lower ($\Delta FF<0$) corresponding to an increased series resistance of the solar cell. But the FF drop is almost independent of the ZnO thickness and therefore points towards a poor ZnO/p interface [5].

3.1 Influence of p-layer preparation parameters on pin cell properties on SnO_2 and ZnO

Atomic hydrogen is known to have a strong influence on tho ZnO ourfaoo [4]. In order to oxamino wother the FF drop in a pin-cell is due to this hydrogen impact, we

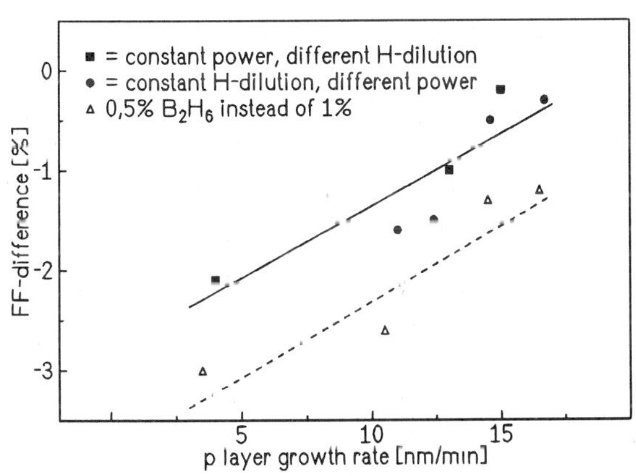

Fig. 2: Dependance of FF drop (ΔFF) on p-layer growth rate (on ZnO coated substrate relative to SnO_2 substrate). The growth rate was varied by changing the H dilution and plasma power. Additionally two different doping levels were used.

changed the p-layer deposition parameters by using different plasma powers and hydrogen dilutions. Figure 2 shows that the FF decrease is strongly sensitive to the growth rate of the p-layer which can be interpreted by different durations of hydrogen impact because of different times to reach a full coverage with the first monolayer of p-SiC:H.

This result supports the hypothesis that the interaction of hydrogen with the ZnO is responsible for the low FF in pin cells. Figure 2 also shows that the FF drop on ZnO is increased when a lower p doping is used so that even at highest growth rates the FF on ZnO does not recover.

3.2 Hydrogen interaction with TCO surfaces

The most prominent influence of a hydrogen plasma on TCO surfaces consists in the reduction of the material, as it is well studied for SnO_2:F surfaces [1]. Our approach to compare the hydrogen-reactions on the different TCO materials was to perform XPS measurements. Instead of hydrogen plasma exposed surfaces we investigated very thin p-SiC layers on the TCO surfaces, thus obtaining conditions that are identical with a real pin cell deposition. The results on flat SnO_2:F, that was again half-coated with ZnO:Al, are shown in figure 3.

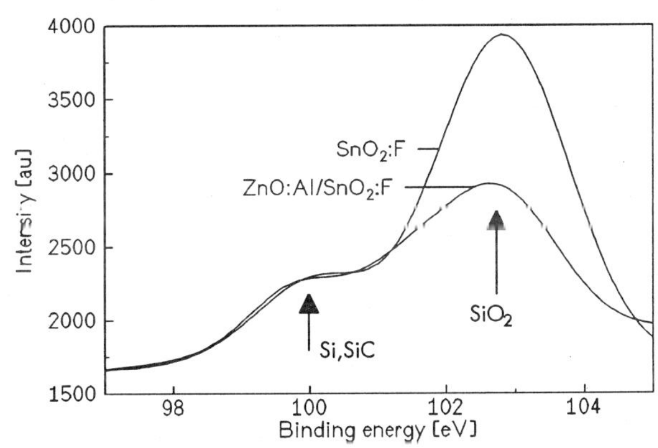

Fig. 3: XPS-Si signals of 3 nm p-SiC:H on flat, ZnO:Al coated and uncoated SnO_2:F substrates. Prior to the measurement the sample surface has been cleaned by sputtering with an Ar^+ ion gun for 40 sec, corresponding to a sputter depth of ≈ 1 nm.

Even on the ZnO coated substrate a remarkable amount of SiO_2 is formed, but much less compared to the uncoated substrate. This is the first direct evidence for the existance of SiO_2 species at the ZnO/p interface, probably due to ZnO reduction, as proposed by Bilger, Eicke and Roca i Cabarrocas et al [8,9]. The total content of Si and C (not shown here) is identical on both substrates. This indicates that the film stoichiometry is the same except for differences in the formation of SiO_2 due to TCO reduction. However, such an insulating SiO_2 layer can not cause the FF differences, because in this case the FF on SnO_2 should be decreased.

Apart from possible reduction there are other influences of hydrogen on TCO surfaces. In the case of ZnO atomic hydrogen impact on the surface leads to a higher conductivity due to the creation of an accumulation layer [4,10]. By SIMS measurements we have proven that there is no observable diffusion of hydrogen into the ZnO even at highest temperatures as proposed by An et al [4]. The hydrogen influence is in fact limited to the very surface of the ZnO.

3.3 Model of the ZnO/p interface including the hydrogen impact

Taking into account the results so far the ZnO/p contact properties can be consistently explained using the band diagram shown in Figure 4.

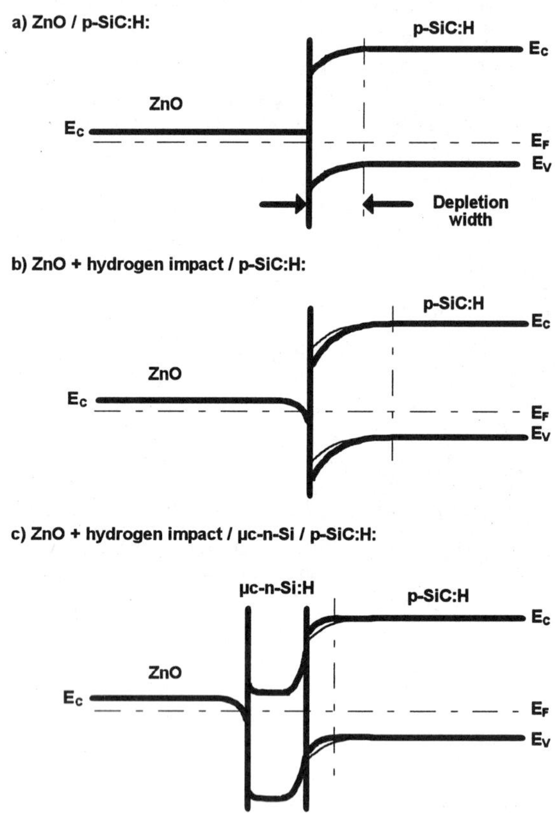

Fig. 4: Schematic band diagram of the ZnO/p-SiC:H interface (with thick p-layer): a) without change at ZnO surface but with resulting band bending and depletion zone at the p-layer surface; b) with hydrogen influence on ZnO creating a surface accumulation and an enhanced depletion in the p-layer (thin line = case a) for comparison); c) case b) plus µc-n-Si intralayer causing a thinner depletion layer (thin line = case a) for comparison).

ZnO is a degenerate n-type semiconductor with a bandgap of 3.3 eV (left side, valence band edge not shown), and the contacting p-layer exhibits a depletion layer similar to the well known SnO_2:F/p-SiC:H interface [11] (Figure 4 a). Depositing p-SiC:H on top of a ZnO substrate with accumulation layer results in a modified Schottky contact between both materials as depicted in Figure 4 b). The accumulation layer in the ZnO causes a stronger downbending of the bands in the p-layer resulting in a thicker depletion zone and a lower FF of the solar cells.

This model explains the results presented in section 3.1: If the p-layer growth rate is too low, the ZnO is covered too slowly by the SiC and a hydrogen exposure of the ZnO surface takes place, resulting in a poor ZnO/p contact. At higher growth rates the coverage is fast enough to protect the ZnO surface and the FF recovers. Accordingly, a higher p doping of the SiC enables a stronger band bending and the depletion layer becomes thinner, leading to an improved ZnO/p contact (top curve in fig. 2). Using a lower p doping, however, results in a thicker depletion layer causing a poor ZnO/p contact (bottom curve in fig 2).

3.4 Solution of the ZnO/p contact problem

The band diagram in fig. 4 shows that this contact problem only arises when a p-layer is brought into contact with the ZnO. Thus, preparing an n-doped layer on the ZnO, the depletion width in this layer should be smaller and even more decreased in the presence of an hydrogen induced accumulation layer in the ZnO (figure 4 c). A possible solution for the drop in FF - apart from a faster deposition or higher doping of the p-layer - is hence the deposition of an intermediate, n-doped layer. Using a microcrystalline n-Si layer, the n-p-contact is of high quality as known for example from our contact studies [12]. The FF differences of the pin-cells on coated and uncoated SnO_2:F prepared with and without a µc-n-Si layer between TCO and p-layer are shown in table 1.

These results demonstrate that the contact to the p-layer is remarkably improved by the µc-n-Si layer even on differently (Al- or B-) doped ZnO substrates which results in almost identical fill factors (between 70% and 72%) compared to cells on SnO_2:F substrates. However, a disadvantage of the intralayer is an additional absorption causing a loss in J_{SC}.

TCO material:	ΔFF of pin-structure [%]	ΔFF with μc-n-Si / pin [%]
SnO_2 + 15nm ZnO:Al	−3.3	0 ±0.5%
SnO_2 + 15nm ZnO:B	−16.2	0 ±0.5%
ZnO:B (Siemens)	−3.1	0 ±0.5%

Table 1: Fill factor difference between pin cells on ZnO (coated substrate or CVD-ZnO) and pin cells on uncoated SnO_2:F substrate (reference) before and after insertion of a 10 nm μc-n-Si layer between TCO and p-layer. The FF of the pin reference cells on the uncoated substrates was always between 70 and 72%.

4. CONCLUSION

Using thin ZnO:Al and ZnO:B overlayers on SnO_2:F substrates we were able to compare the properties of these materials as TCO substrates and their influence on solar cell performance. It was demonstrated that the interaction of the ZnO surface with plasma-generated atomic hydrogen is responsible for the ZnO/p contact problem. This hydrogen treatment and the mechanism that leads to a decreased FF in pin cells are desribed in a model of the ZnO/p interface, taking into account the influence of the change in ZnO work function due to a hydrogen induced accumulation layer on the ZnO/p Schottky contact. This model gives no comprehensive description of the TCO/p interface (as it is tried by other groups [13]), but explains the observed dependence of the ZnO/p contact properties (as monitored by the FF of pin cells) on variations of the p-layer deposition parameters. The model also leads the way to the solution of the ZnO/p contact problem by insertion of a μc-n-Si layer. Further work will be addressed to an improvement of this structure in order to avoid the encountered loss in short-circuit current density.

ACKNOWLEDGEMENT

The authors wish to thank B. Rech and M. Kolter for many helpful discussions, W. Appenzeller for his extensive technical assistance, F. Birmans for performing the I-V-measurements and Siemens for supplying their CVD ZnO:B substrates. This work was supported by the BMFT (German Federal Ministry of Research and Technology).

REFERENCES

[1] H. Schade, Z. E. Smith, J. H. Thomas III, and A. Catalano, "Hydrogen Plasma Interactions With Tin Oxide Surfaces", *Thin Solid Films*, **117**, 1984, p. 149

[2] R. Schäffler, and H. W. Schock, "High Mobility ZnO:Al Thin Films Grown by Reactive DC Magnetron Sputtering", *Proc. 23rd IEEE PVSC, Louisville KY*, 1993, p.1026

[3] H. Sakai, and Y. Ichikawa, "Process Technology for a-Si/a-Si Double Stacked Tandem Solar Cells With Stabilized 10% Efficiency", *J. Non-Cryst. Solids*, **137&138**, 1991, p. 1155

[4] I. An, Y. Lu, C. R. Wronski, and R. W. Collins, "Real Time Spectroellipsometry Study of the Interaction of Hydrogen with ZnO During ZnO/a-SiCH Interface Formation", *Appl. Phys. Lett.*, **64**, 1994, p. 3317

[5] M. Kubon, N. Schultz, M. Kolter, C. Beneking, and H. Wagner, "Modification of TCO-p Interface in a-Si:H Solar Cells by Intermediate Layers and Plasma Treatments", *Proc. 12th European PVSEC, Amsterdam NL*, 1994, p. 1268

[6] K. Sato, Y. Matsui, K. Adachi, Y. Gotoh, Y. Hayashi, and H. Nishimura, "Hydrogen Plasma Treatment of ZnO coated SnO_2 Films", *Proc. 23rd IEEE PVSC, Louisville KY*, 1993, p. 855

[7] C. Beneking, B. Rech, T. Eickhoff, Y.G. Michael, N. Schultz, and H. Wagnor, "Preparation and Light Stability of a-Si/a-Si Stacked Solar Cells", *Proc. 12th European PVSEC, Amsterdam NL*, 1994, p. 683

[8] A. Eicke, and G. Bilger, "XPS and SIMS Characterization of Metal Oxide/Amorphous SiC Interfaces", *Surface and Interface Analysis*, **12**, 1988, p. 344

[9] P. Roca i Cabarrocas, N. Layadi, A. Hadjadj, M. Favre, and J. Dutta, "Characterization and Modification of SnO_2 – Effects of SnO_2 Treatments on Solar Cell Parameters", *Proc. 12th European PVSEC, Amsterdam NL*, 1994, p. 1272

[10] G. Heiland, and D. Kohl, "Interpretation of Surface Phenomena on ZnO by the Compensation Model", Phys. Stat. Sol. (a) **49**, 27, 1978, p. 27

[11] K. Itoh, H. Matsumoto, T. Kobata, A. Fujishima, O. Nakamura, and K. Fukui, "Determining Height of a Leaky Schottky Barrier Existing in the SnO_2/p-SiC Junction", *Appl. Phys. Lett.*, **51**, 1987, p. 1685

[12] M. Kolter, C. Beneking, D. Pavlov, T. Eickhoff, P. Hapke, S. Frohnhoff, H. Münder, and H. Wagner, "Highly Conductive Microcrystalline n-Layers for Amorphous Silicon Stacked Solar Cells: Preparation, Properties, and Device Application", *Proc. 23rd IEEE PVSC, Louisville KY*, 1993, p. 1031

[13] F.Smole, M. Topic, and J. Furlan, "The influence of defect states in TCO/a-Si:C:H interface on p-i-n a-Si:H solar cell performance", *Proc. 12th European PVSEC, Amsterdam NL*, 1994, p. 77

LESSONS LEARNED FROM A HYDROGEN EXPLOSION AT A PHOTOVOLTAIC RESEARCH FACILITY

Paul Moskowitz, Biomedical & Environmental Assessment Division, Brookhaven National Laboratory, Upton, NY, 11973
Wayne Buchanan and William Shafarman, Institute of Energy Conversion, University of Delaware, Newark, DE, 19716

Abstract

At the Institute of Energy Conversion (IEC), University of Delaware, a hydrogen explosion occurred on September 1, 1992. A formal investigation was conducted. Suggestions provided by the Accident Investigation Team formed the basis for a complete review and upgrade of the compressed gas handling systems and procedures. In this paper, the causes of the explosion are discussed along with resulting modifications made to the gas handling system, facility, and procedures to prevent a recurrence of this type of incident and to improve the safety of this facility. Lessons learned from this incident and the resulting investigation and response are highlighted, including the need to regularly review both hardware and procedures to ensure that all aspects of the research program and facilities are up-to-date.

Introduction

Hazardous compressed gases, including some that are toxic, explosive, pyrophoric or flammable, are used in photovoltaic module research and manufacturing facilities. The hazards to life and property presented by these gases demand that safety always be given a high priority. At the Institute of Energy Conversion, University of Delaware, hydrogen gas is used in research reactors to deposit amorphous silicon thin-films. On September 1, 1992, an explosion in this facility occurred due to an accidental release of hydrogen. A formal investigation was conducted. Suggestions provided by the Accident Investigation Team formed the basis for a complete review and upgrade of the compressed gas handling systems and procedures. In this paper, the causes of the explosion are discussed along with resulting modifications made to the gas handling system, facility, and procedures to prevent a recurrence of this type of incident and to improve the safety of this facility. Lessons learned from this incident and the resulting investigation and response are highlighted. In particular, there is a need to regularly review both hardware and procedures to ensure that all aspects of the research program and facilities are up-to-date.

Description of the Incident

An explosion occurred during the replacement of a hydrogen cylinder which was stored in a ventilated and sprinklered gas cabinet. This and other hazardous gases flowed to an enclosed reactor in an adjacent room. The gas handling system in use at the time of the accident is shown in schematic format in Fig. 1. During a cylinder change-out, a quantity of high pressure hydrogen gas was accidentally released through a line used for vacuum purging. The explosion occurred when a vacuum bellows burst releasing the gas into the reactor enclosure. The Plexiglas walls of the reactor enclosure blew outward causing extensive damage to the laboratory. Fortunately, no one was in the room at the time of the incident and no one was injured.

The Accident Investigation

A formal accident investigation was conducted, at the request of the Director of the IEC, by an Accident Investigation Team comprised of P. Moskowitz, H. Schulman, and C. Scott from Brookhaven National laboratory and W. Griffing from the National Renewable Energy Laboratory. The investigation was conducted following Department of Energy accident investigation protocols. In this investigation, the Team's findings centered around facts developed from the following sources:

- Physical inspection of the site
- Physical disassembly and inspection of selected piping/gas distribution components (i.e., excess flow valve, pneumatic valve, and single stage regulator)
- Review of incident reports
- Review of documents, photographs, manuals, etc.
- Interviews with personnel

 First WCPEC; Dec. 5–9, 1994; Hawaii

Presented below are summaries of the Team's findings divided into three separate sections: Root cause, contributing factors and direct causes.

Root cause

The basic underlying reason for a transient to occur is called the root cause. Root causes may relate to failures in administrative procedures, engineering devices, or in personal protective equipment. Root cause is defined as, "the most basic cause(s) of an event that, when eliminated or compensated for, will prevent accident recurrence." The intent of a root cause analysis is not to place blame, but to determine how to classify responsibilities and reduce errors. In this accident, the root cause was related to the absence of a formal policy for the systematic oversight of gas handling and safety systems.

Contributing factors

Contributing factors are defined to include those factors that did not necessarily directly result in the incident, but presented an environment for an incident to occur. In this incident, several contributing factors were identified including the presence of at least one poorly operating component; the excess flow valve in the gas handling train; the absence of written procedures or checklists for purging, cylinder changing, startup or shutdown of gas system or reactor; and, the lack of formal training and testing programs for gas system operators.

Direct cause

The direct cause arises from the failure of the final protective barrier which isolates the worker from the biologically or physically harmful agent. In this accident the direct cause resulted from an operator error in opening a valve in the wrong sequence resulting in the discharge of hydrogen at pressure >1100 psi through an open valve into a vacuum system not rated for this amount of pressure. Overpressure resulted in the bursting of a vacuum bellows releasing ~0.5 m^3 of hydrogen gas into the reactor enclosure. The resulting overpressure of the enclosure, either due to mechanical pressure or chemically induced explosion, caused a blow-out of the enclosure Plexiglas panels, exhaust ductwork and room ceiling tiles.

Safety Modifications

The Accident Investigation Team developed a list of suggestions to prevent a similar event from occurring in the future and to improve the safety of the hazardous gas operations. A representative sample of these suggestions is shown in Table 1.

Table 1. Suggested Modifications to the IEC Compressed Gas Safety Program.
Engineering
Move gas cabinet panel vacuum pump out from the high pressure to the low pressure side of the regulator.
Install isolation valve on the high pressure side of the regulator in the gas purge panel.
Provide monitoring points for flammable gases in cabinets, cabinet rooms, and reactors. Interface these detectors to the alarm system.
Administrative
Develop written check-lists for all compressed gas systems change-outs, purging, running and shut-down.
Establish training and authorization requirements for all gas system operators.
Personnel
Develop/update training proficiency testing for all gas system operators.
Develop/update training requirements for all "first responders" to emergency incidents.

In the aftermath of this incident, a number of modifications were made to the gas handling system as a follow-up to the accident. The modifications were based on suggestions of the Accident Investigation Team, supplemented by information compiled by the IEC staff from discussions with researchers at other laboratories, and review of assorted documents, especially the "Safety Analysis Report For the Use of Hazardous Production Materials in Photovoltaic Applications at the National Renewable Energy Laboratory." Fig. 2 displays a schematic of the modified gas handling system. Some key engineering aspects of the upgraded gas handling system include:

- Flow restricting orifices are required for all gas cylinders. The sizes and pressures of the cylinders are limited to ensure that the low pressure line and all components are able to withstand the pressure in the case of a catastrophic regulator failure, provided the new excess flow valve functions properly.

- New tied-seat regulators were installed. These regulators allow the cylinders to be purged through the regulator so that vacuum purge lines could be moved to the low pressure side of the regulator.

- Isolation valves were installed on the high pressure side of the regulator.
- Each hazard class of gas is stored in separate gas cabinets with dedicated purge cylinders.

Many facility improvements were made. Two critical features, largely in place before the accident, were upgraded:

- Interlock of the building fire alarm system and automatic gas flow shut off circuits with on-line monitoring for hydride gases and hydrogen in the laboratories, gas cabinets, and reactor enclosures.
- Interlock circuits to ensure that all hazardous gas flows shut off automatically in the event of a loss of ventilation, gas detection, power loss, emergency shut-off buttons, or other alarms in the building.

Finally, written safe operating procedures were developed for cylinder change-out, system start-up, shut-down, operation, maintenance, and emergency response. Each of the procedures was reviewed and approved by the IEC Safety and Chemical Hygiene Committee.

Discussion and Conclusions

Many of the factors which contributed to this accident could potentially have been averted with a formal and regular review of the administrative, engineering and personnel programs associated with the compressed gas systems in place at the IEC. Perhaps most important is the establishment of policies to ensure the continuing surveillance of maintenance operations and review of the condition of all components in the gas handling train. In addition, routine reevaluation of the equipment and procedures should be conducted in light of the latest advances in administrative, engineering and personnel programs. For example, at the time the original system was constructed, the standard and recommended placement of the vacuum purge line was on the high pressure side of the regulator to enable complete purging of the line up to the cylinder valve. Since that time, new tied-seat regulators which enable complete purging through the regulator have become available, so the purge line should now be moved to the low pressure side of the regulator.

Finally, It is useful to visit other facilities and talk with other researchers using similar materials and equipment. Both formal and informal programs to share critical safety information must be encouraged. The costs associated with these efforts and the necessary maintenance and upkeep can be significant, particularly for small laboratories with limited resources such as university-based laboratories. These must be factored into all research and manufacturing programs. Clearly, a failure to accommodate these priorities can be costly.

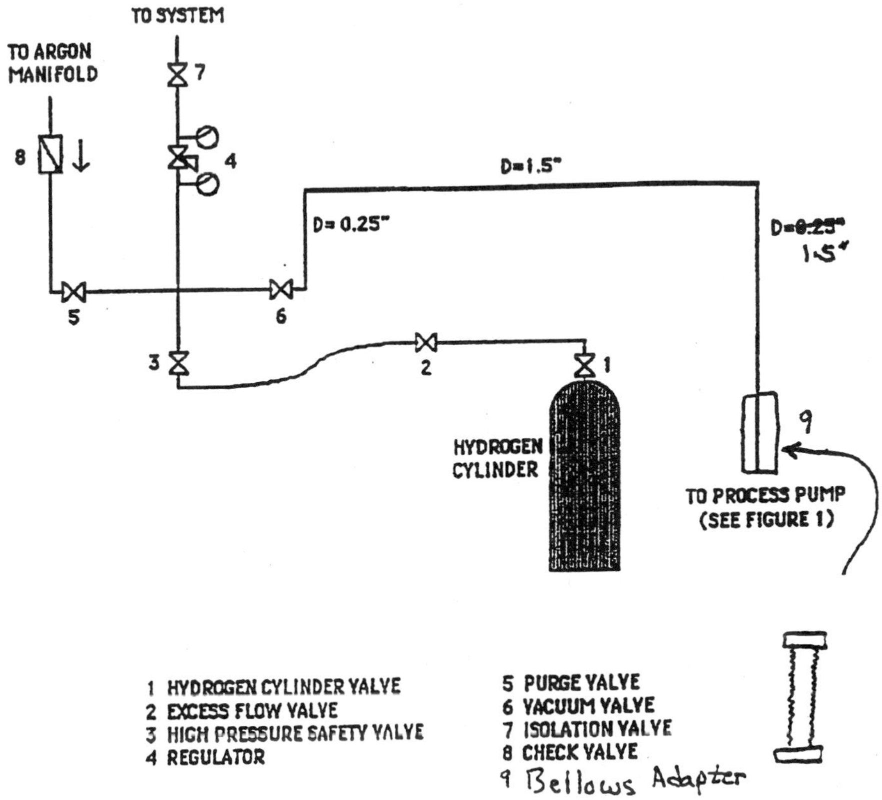

Fig. 1. Schematic of the gas handling system in use at the time of the explosion.

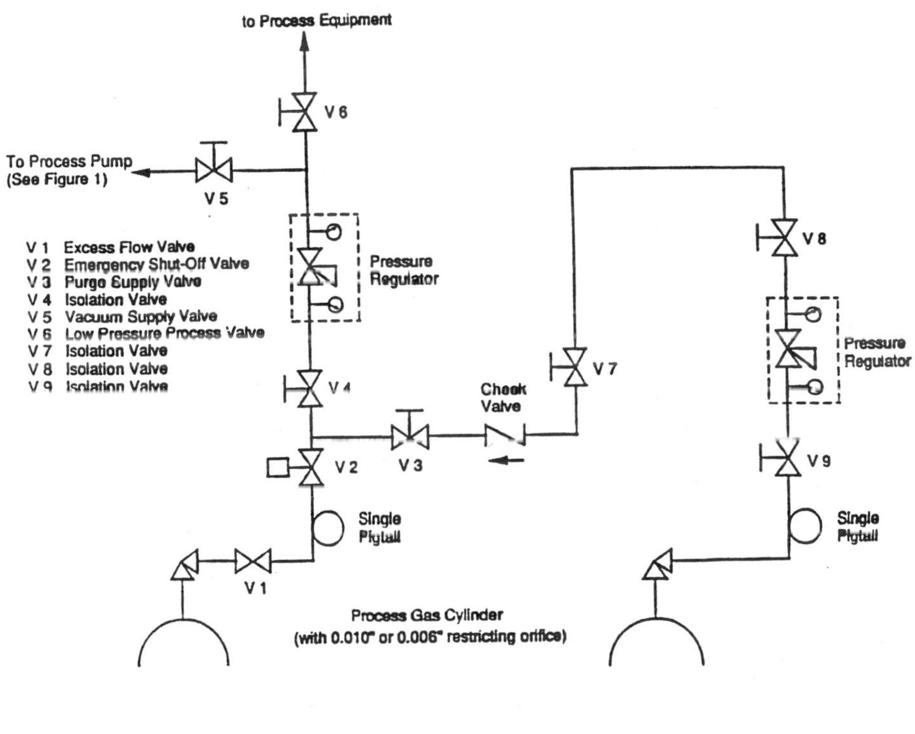

Fig. 2. Schematic of the gas handling system in use after the explosion.

TRANSPORT PROPERTIES OF DOPED SILICON OXYCARBIDE MICROCRYSTALLINE FILMS PRODUCED BY SPATIAL SEPARATION TECHNIQUES

Rodrigo Martins, Manuela Vieira*, Isabel Ferreira, Elvira Fortunato and L. Guimarães

Materials Science Department, Faculty of Science and Technology of New University of Lisbon,
*CEMOP/UNINOVA, Quinta da Torre, P-2825 Monte de Caparica, Portugal.

ABSTRACT

This paper presents results of the role of the oxygen partial pressure (p_{O2}) used on the properties exhibited by doped μc silicon oxycarbide films produced by a Two Consecutive Decomposition and Deposition Chamber (TCDDC) system [1], where a spatial separation between the plasma and the growth regions is achieved. The films produced are highly conductive and transparent with suitable properties for optoelectronic applications.

INTRODUCTION

Wide band gap μc silicon carbide based alloys have aroused considerable interest since they combine some advantages of amorphous and crystalline materials. However, the conventional r.f. glow discharge process does not satisfy these requirements, since it does not allow the production of μc silicon films with carbon (C) content above 10 at%. In previous work [2, 3], high quality silicon carbide materials, using a strong beam of excited hydrogen [H] were used to decompose the silicon-and-other mixed-containing gases as it is the case of the production of doped silicon oxycarbide films (Si:C:O:H) presenting a high chemical stability and optical gaps (E_{op}), with the required conductivity (σ_d).

EXPERIMENTAL PROCEDURE

The films were prepared in a TCDDC system whose main characteristics are described elsewhere [1]. Table I summarises the deposition conditions used, where p_{O_2} varied from 10^{-3}-10^{-1} mbar.

TABLE I - Deposition conditions used.

	μc n-type		μc p-type	
Substrate temp., T_s (°C)	220		220	
Gas pressure, p (mbar)	0.5		0.5	
Gas doping level (at.%)	0.5		2.0	
Power, dp (mW/cm^3)	100	100	120	135
H_2 dilution ratio (at.%)	95	90	90	85
CH_4/SiH_4, X (at.%)	25	40	25	40

The film's structure, composition, morphology and optoelectronic properties were analysed by means of grazing X-ray diffraction, Electron Secondary Chemical Analysis (ESCA), Secondary Ion Mass Spectroscopy (SIMS) σ_d as a function of temperature (T) and optical absorption ($\alpha(T)$) (ranging from 300 to 80 K) measurements.

ANALYSIS OF THE RESULTS

Structure and morphology

The X-ray diffraction analysis were performed on films deposited on alkali free substrates, showing the presence of Si microcrystals with average sizes in the range of 100-50 Å as p_{O_2} varies from 10^{-3} to 10^{-1} mbar. The crystal's volume fraction (f_c) was inferred through the ratio between the experimental peak intensity recorded on the μc-films and that one obtained in a c-Si sample, taken as reference [4]. Figs. 1 a. and b. show the dependence of (crystallite dimensions) S_c and f_c on p_{O_2}, for p-and n-type films produced. The results indicate that f_c and S_c decrease as p_{O_2} increases.

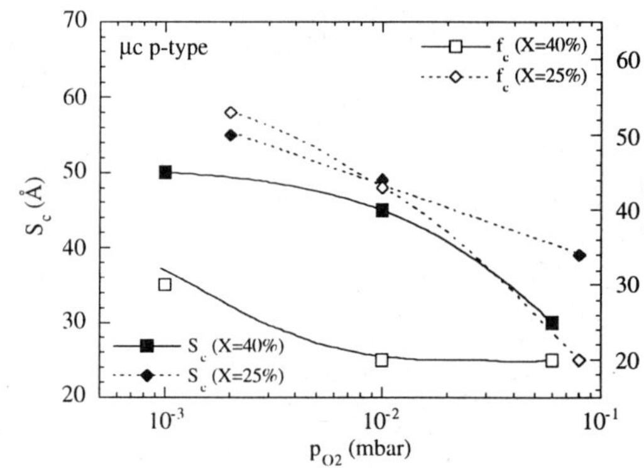

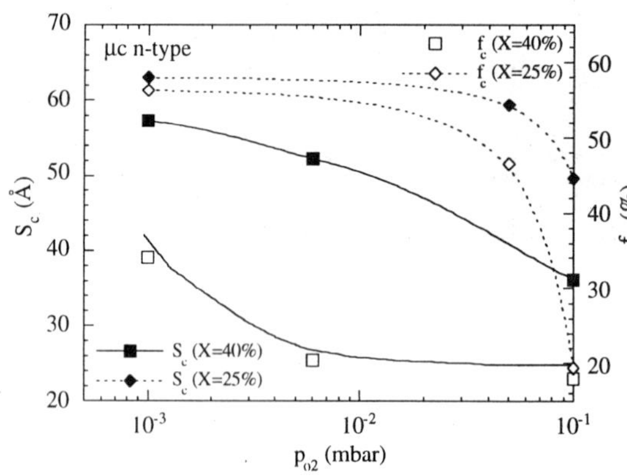

Fig 1. Dependence of S_c and f_c on P_{O2}. for a. μc p- and b. p-type films.

Dependence of film composition on P_{O2}

The film's chemical composition were analysed by ESCA (Fig. 2) and SIMS (Figs. 3.a. and b.). Changing p_{O_2} from 10^{-3} to 10^{-1} mbar the film's composition (without taking into account the amount of H

CH3365-4/94/0000-0508 $4.00 © 1994 IEEE

incorporated) changes from $Si_{89}C_7O_4$ to $Si_{56}C_{19}O_{25}$, for n-type films produced with an H_2 dilution ratio of 95% and X=25%, whilst it changes from $Si_{80}C_{16}O_4$ to $Si_{45}C_{35}O_{20}$, for n-type films produced with an H_2 dilution ratio of 90% and X=40%. The p-type films show the same behaviour, but for different power density and H_2, dilution as indicated in table-I.

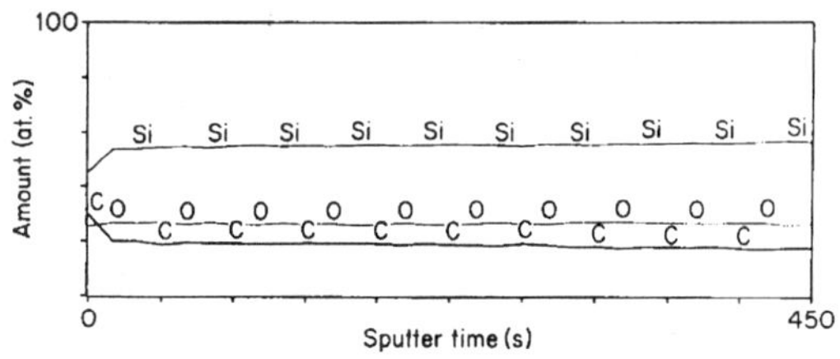

Fig. 2. Example of an ESCA depth Si 2p, O 1s and C 1s spectra for the μc n-type sample.

The SIMS measurements were performed to determine the total amount of phosphorous (P), boron (B) and H. Fig.s 3.a. and b. represent the SIMS depth profiles corresponding to P, B and H distributions, for two typical μc p- and n- type films ($Si_{89}C_7O_4$ and $Si_{80}C_{16}O_4$, respectively).

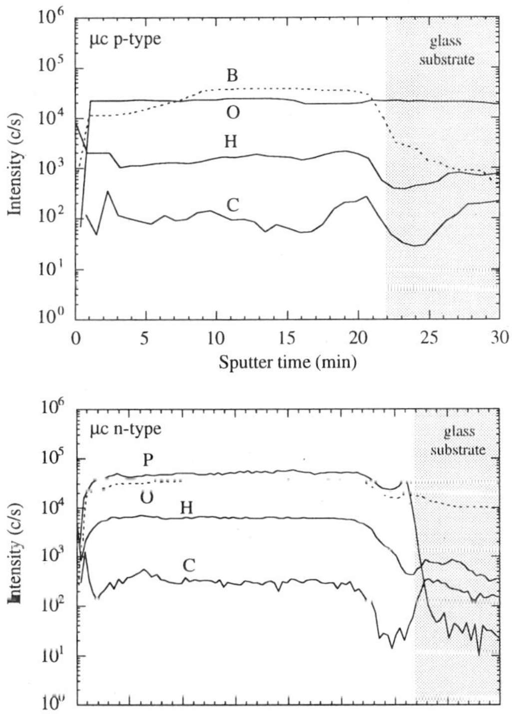

Fig. 3. Depth profile distribution of P, B, C and H for a. p-type μc and b. n-type μc samples.

P and B concentrations were respectively equal to 1×10^{21} (≈ 2%) and 3.4×10^{20} at.cm^{-3} (≈ 0.68%). These data indicate that for the p- and n-type doped films, respectively 80% and 40% of the gas phase species containing P and B are effectively incorporated, about twice as much as it was observed in a-Si:H films [5].

Role of pO_2 on the material's optoelectronic properties

The optical transmittance and reflectance measurements allow the determination of α(T) at room temperature (RT) and for 80 K ≤ T ≤ 250 K, as well as the film's refractive index (n). σ_d measurements, were performed in the range of 300 to 80 K allowing the determination of σ_d at RT and the activation energy. Fig. 4 shows the dependence of α(T) on photon energy, for n-type and p-type μc films produced with X=25% and pO_2≈ 10^{-3} mbar.

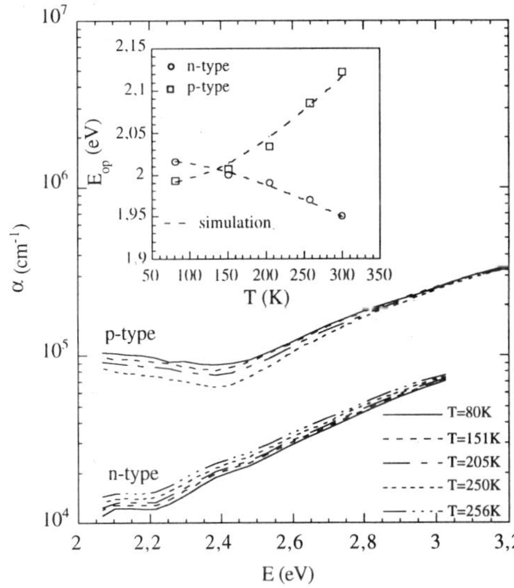

Fig. 4. α(T) as a function of energy for n- and p-type μc films, with X=25% and 40%. The insert shows the E_{op}(T) dependence for both films.

The p-type films reveal an enhancement of α(T) as T decreases, which is opposite to that one of n- type films.

The film's E_{op} were determined by the classical Tauc's plot, which is valid for non-direct transitions, as it is the case of amorphous semiconductors [6]. Though the values obtained by this method can be highly questionable since we are in presence of a quite inhomogenous material, they allows to compare the obtained results with those ones published and where E_{op} was determined by the same method. Nevertheless, independent of the rules that govern the optical transitions in these structures, the obtained E_{op} indicates at least the degree of the film's transparency, which are ruled by the amorphous tissue [7].

The role of T on E_{op} was determined by following a similar procedure as that one proposed by Cody [8] (E_{op} is the energy value for which $(\alpha/E)^{0.5}$, either by isoabsorption or linear extrapolation, intercepts $\alpha \geq 10^4$cm^{-1}) to determine the effect of thermal and structural disorder on the absorption edges of amorphous semiconductors. The insert in Fig. 4 shows the experimental recorded E_{op}, as proposed by Cody [8], showing a T dependence that fits an Einstein oscillator [9], given by:

$$E_{op}(T) = E_{op}(0) \pm \frac{A}{e^{\frac{\theta}{T}} - 1} \qquad (1)$$

where to the p-type films, the signal in the second term of the above equation is positive. A and θ is for n- and p-type samples are about 0.2 eV and 400 K and 0.42 eV and 445 K, respectively.

The α(T) measurements have been performed below RT and so the role of thermal disorder on the values recorded can be neglected, as observed by Weiser et al. [10] in a-Si:H and c-Si films. The data recorded lead to $E_{op}(0)$ values of the order of 2 eV, showing that the optical transitions are ruled by the amorphous tissue [7].

By knowing the value of n for the films under analysis, the volume fraction of voids (f_V) can be estimated, using the expression [11]:

$$f_v = \frac{(N^2 - n^2)(1 + 2n^2)}{3n^2(N^2 - 1)} \qquad (2)$$

with N(c-Si)=3.458.

In Figs. 5 a. and b. we plot the dependence of f_V on pO_2, for both types of films analysed, that show similar behaviour, for X=25% and X=40%.

Fig. 6 shows σ_d versus T^{-1}, for films produced at X≈40 %. There, we observe that σ_d is almost non-activated and that the Arrhenius plot does not fit the obtained experimental data. However, σ_d versus $T^{-1/4}$ plot fits well the experimental data, for T < 250 K. This indicates that the conduction is dependent on the percolation channels established between crystallites (hopping jump varying as $T^{-1/4}$ not implies necessarily variable ranging hopping around E_F [12]). These data indicate that the number of phonons available to assist the conduction process is very low. Under these conditions, the carriers are "trapped" in localised states from which they can travel, by hopping, to the closest available site, function of the capability of the state in retaining (or not) the carriers. This corroborates the conductivity time dependence data obtained at T=80 K, where for the n- and p-type samples studied σ_d follow a time (t) dependence behaviour given by:

$$\sigma_d(T,\tau) = \sigma_d(T)\exp\left(-\frac{t}{\tau_r}\right) \qquad (3)$$

where, $\sigma_d(T)$ is the conductivity at T, for t = 0 and, τ_r is the relaxation time whose value is dependent on film's composition (for an n-type film with $E_{op} \approx 2.5$ eV, $\tau_r \approx$ 476 min. and for a p-type film with $E_{op} \approx 2.3$ eV, $\tau_r \approx$ 625 min.).

Such high values for τ_r clearly indicate that the amorphous tissue controls the transient behaviour observed (carrier's transition between crystallites governed by the amorphous tissue), and so, that percolation is the most suitable conduction mechanism for this type of μc films. Lucovsky et al. [13] also found a similar behaviour in highly doped μc-Si films, where they found that the transport at high temperatures is determined by the amorphous encapsulate tissue that determines the position of the Fermi level, deep in the tails of the conduction bands.

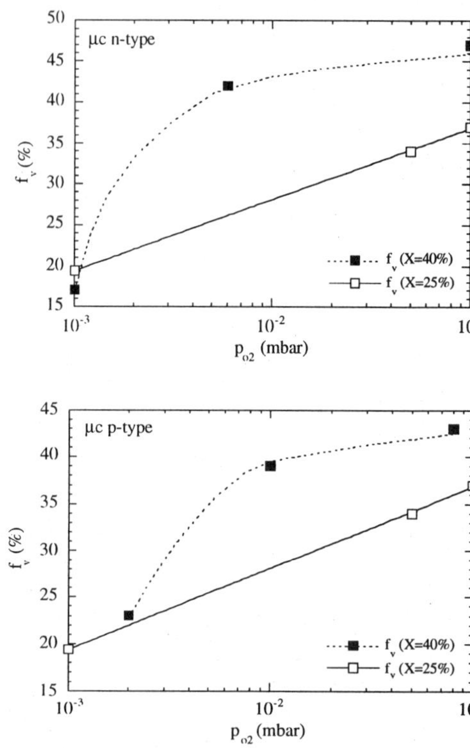

Fig. 5. Dependence of f_V on P_{O2} for a. μc n- and b. μc p-type samples.

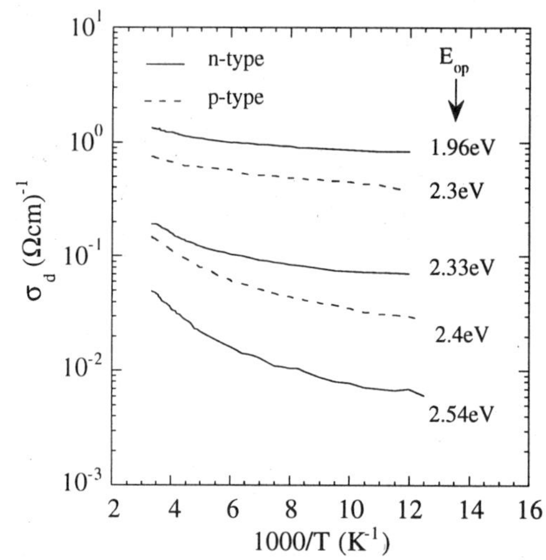

Fig. 6. Dependence of σ_d vs T^{-1} for μc p- and n-type films.

Figs. 7a. and b. show the role of pO_2 on the optoelectronic properties of the doped μc films analysed. There, it is plot the dependence of the E_{op} and σ_d at RT on pO_2, for the set of n- and p-type films investigated. Comparing these results with those ones obtained for silicon carbide n- and p-type μc films produced by ECR [3] (where the n- and p-type doped films have two orders of magnitude difference in σ_d), the films produced by TCDDC technique show similar behaviour for both n- and p-type films. The experimental data also show that the variation of E_{op} depends on X used and it is directly proportional to pO_2, whilst σ_d shows a reverse dependence behaviour on pO_2.

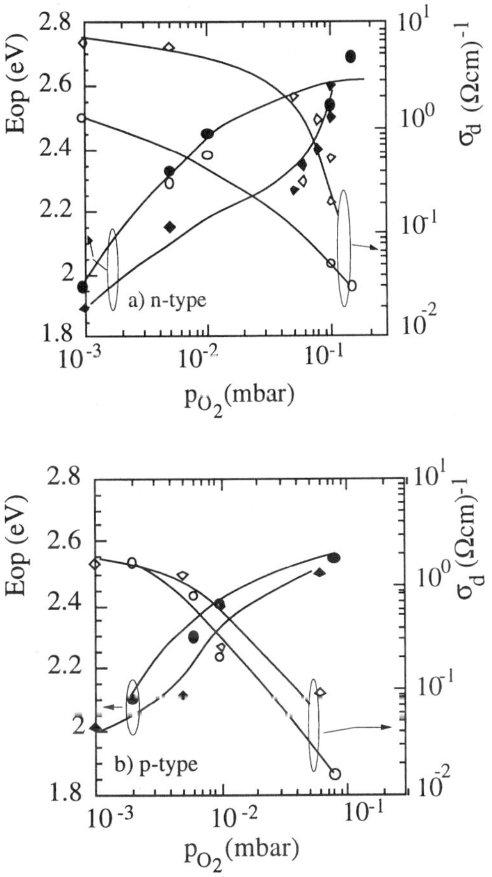

Fig. 7 E_{op} and σ_d dependence on pO_2 for p- and n-type a. and b.] films produced at X = 25 % (◆ for E_{op} and ◇ for σ_d) and X = 40 % (● for E_{op} and o for σ_d).

CONCLUSIONS

The deposition technique and the pO_2 used lead to the preparation of μc n- and p-type weakly absorbing and highly conductive Si:O:C:H thin films. The data show that the preparation technique, a spatial decoupling of the decomposition and deposition areas, leads to the formation of μc silicon phase even at high O and C incorporation ratios. The films under analysis do not show the presence of Si:C microcrystals, in contrast to those ones produced by ECR [3], with similar optoelectronic properties.

The presence of oxygen during the deposition process controls S_c, f_c, and f_v, enhancing also the chemical reactivity on the growth surface, and so, the C incorporation.

These films can be used either in the production of p.i.n solar cell devices (enhancement of the blue response of the device and the light optical path [2]) or in TFT's [14], taking advantage of the optoelectronic characteristics presented by this material.

ACKNOWLEDGEMENTS

The authors are grateful to A. Maçarico and M. Santos in the help given during film's preparation as well as to G. Willeke for the inestimable contribution during the first phase of this study. We also would like to thank F. Lopez and J. Puigdollers of Univ. of Barcelona for performing the SIMS analysis. This work was supported by NATO SfS programme (PO-THINFILM project), by JNICT, project N° STDR /661/CTM/92/A and by PEDIP, project N° 0250. The authors also thanks the co-funding of "Fundação Luso Americana para o Desenvolvimento" for the grant given.

REFERENCES

[1] R. Martins I. Ferreira, N. Carvalho and L. Guimarães, J. of Non.Cryst. Solids, **137-138**, 757 (1991).

[2] L. Guimarães, R. Martins, E. Fortunato, I. Ferreira, M. Santos and N. Carvalho, Proc. Mat. Res. Soc., **118**, 617 (1988).

[3] Y. I. Hatori, D.Kruangan, T. Toyama, H. Okamoto and Y. Hamakawa, J. of Non.Cryst. Solids, **87-98**, 1079, (1987).

[4] M. Vieira, "Optoelectrónica no Silício Amorfo Higdrogenado-Técnicas Estáticas e Transientes", New University of Lisbon, (1993).

[5] W. E. Spear and P. G. LeComber, Sold State State Commun. **17**, 1193 (1975).

[6] J. Tauc in "Photo and Thermoelectric Effects in Semiconductors", Pergamon, (1962).

[7] M. Taniguchi, M. Hirose, Y. Osaka, S. Hasegawa and T. Shimidzu, J. Appl. Phys. **19**, 655 (1980).

[8] G. D. Cody, in Semiconductors and Semimetal Vol. 21 Part B, edited by J. I. Pankove (Academic Press, Inc., New York, 1984).

[9] W. Bludeau, A. Onton and H. Heinke, J. Appl. Phys., **45**, 1846 (1974).

[10] G. Weiser and H. Mell, J. of Non-Cryst. Solids, **114**, 298, (1989).

[11] L. Landau and E. Lifshitz, in "Electrodynamique des Millieux Continus", MIR, (1969).

[12] D. Adler, L. P. Flora and S. D. Senturia, Solid. State Comm., **12**, 9, (1973).

[13] G. Lucovsky and H. Overhof, J. Non-Cryst. Solids **164-166**, 973, (1993).

[14] R.L. Weisfield and C.C. Tsai, Proc. MRS, **192**, 423, (1990).

DIRECT CORRELATIONS OF BULK CHARGED AND NEUTRAL DEFECT DENSITIES OF STATES IN a-Si:H FILMS WITH CHARACTERISTICS OF SCHOTTKY BARRIER SOLAR CELL STRUCTURES

Mehmet Güneş, Hongyue Liu, C. M. Fortmann, and C. R. Wronski
Electronic Materials and Processing Research Laboratory
The Pennsylvania State University, University Park, PA 16802.

ABSTRACT

We report a systematic study on the a-Si:H thin film materials with different densities of charged defects which are incorporated in Schottky barrier solar cell structures. The densities of charged and neutral defects were obtained from photoconductivity and sub-bandgap absorption measurements and using the sub-bandgap absorption model(SAM). Current-voltage characteristics at different temperatures and internal quantum efficiencies were measured on the corresponding SnO/a-Si:H(n^+) / a-Si:H(i)/Ni Schottky barrier structures and the results were modeled using AMPS. It was found that for structures thicker than ~1.5 μm, Schottky barrier characteristics can be fitted using the same densities and parameters for the charged and neutral defects as obtained from SAM without having to invoke effects from interface layer.

INTRODUCTION

There is a great interest in improving our understanding of the nature and origin of defects in undoped hydrogenated amorphous silicon(a-Si:H) in order to develop more stable solar cells. In spite of the extensive research, there is still no general agreement about the exact nature, densities and energy distributions of the gap states in different a-Si:H materials. This has seriously limited the ability to control improvements in the materials and the characterization of solar cell in terms of their bulk material properties.

Analysis of thin film and solar cell results has in the past been based on two Gaussian distributions of gap states associated with the neutral, D^0, and negatively charged dangling bonds, D^-. However, this distribution cannot be used to fit results of photoconductivity and sub-bandgap absorption when the Fermi level is above the midgap[1] nor light induced changes in films[2] or solar cells[3]. Recently self-consistent modeling of these results was obtained for a-Si:H films both in the annealed and light soaked states using gap states consisting of three Gaussian distributions of defect states[4,5] similar to those proposed in the charged defect models[6,7]. In the thermal equilibrium there are neutral, the donor-like D^0 states located below E_F, the donor-like, D^+, states located above E_F which are positively charged, and negatively charged acceptor-like, D^-, states located below midgap.

In this work the results on a variety of thin film materials and their corresponding Schottky barrier solar cells have been analyzed using the charged defect distribution of states[4] with same the densities of states, energy distributions and carrier capture cross-sections.The steady-state photoconductivities and sub-bandgap absorptions, obtained using dual beam photoconductivity (DBP), were measured over a wide range of generation rates and self-consistent fits to the data obtained using the detailed numerical sub-bandgap absorption model(SAM)[8]. Detailed characterization of the results for annealed and light soaked states of different films was obtained using the same kind of distributions of states. In the case of the Schottky barrier cells, dark J-V and quantum efficiency(QE) characteristics were analyzed using the AMPS(Analysis of Microelectronic and Photonic structures) computer model[9] with the results obtained in the a-Si:H films as inputs.

EXPERIMENTAL DETAILS

The undoped a-Si:H thin films and the corresponding TCO/a-Si:H(n^+)/a-Si:H(i)/Nickel Schottky barrier solar cell structures were prepared under the same deposition conditions using a multi-chamber PECVD system. In Schottky barrier solar cells, the thickness of n^+ layer was 250Å and that of Ni metal contact 200Å. Both the films and Schottky barrier solar cells were annealed in a vacuum chamber($<10^{-4}$ torr) at 170°C for 2 hours before carrying out the measurements. The 1μm thick a-Si:H films were characterized in the annealed state with bulk photoconductivities measured with the generation rates from 10^{15} to 10^{20} $cm^{-3}s^{-1}$ and the sub-bandgap absorption using the dual beam photoconductivity(DBP) technique[8] with generation rates from 10^{15} to 10^{18} $cm^{-3}s^{-1}$. Both these measurements were carried out in the ohmic regimes of currents[10]. The dark current-voltage characteristics of the Schottky barrier solar cells were measured between room temperature and 130°C and the internal quantum efficiency measurements were performed in the dark(without bias light).

CH3365-4/94/0000-0512 $4.00 © 1994 IEEE

RESULTS AND ANALYSIS

Undoped a-Si:H thin films

An example of the results on a-Si:H films,the steady-state photoconductivity(inset) and sub-bandgap absorption, $\alpha(h\nu)$, are shown in Figure 1 for an a-Si:H thin film deposited at substrate temperature 240°C. This film has an optical gap of 1.72 eV and Fermi level is located 0.83 eV from the conduction band edge. The sub-bandgap absorption spectra are normalized to the optical absorption coefficient obtained from transmission and reflection data. In this material the characteristic energy of the valence band tail states, E_{0VB}, is 48 meV and the sub-bandgap absorption coefficient at 1.2 eV, α(1.2eV), measured at the low generation rate of $2x10^{16}$ $cm^{-3}s^{-1}$ is 0.33 cm^{-1}. When the generation rate is increased to $2x10^{18}cm^{-3}s^{-1}$, the sub-bandgap absorption at lower energies in Fig. 1 increases because of the greater increases in the occupation of the states by electrons above the Fermi level. The photoconductivity and the dependence of $\alpha(h\nu)$ on generation rate are significantly higher than those in intrinsic a-Si:H film[5]. These results cannot be explained in terms of the D^-, D^0 states with the positive correlation energy[11].

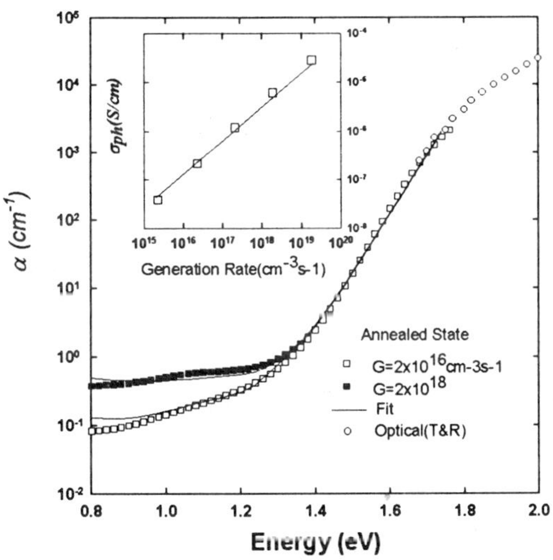

Fig. 1. The sub-bandgap absorption measured by DBP at two generation rates for a film deposited at 240°C in the annealed state. In the inset, photoconductivity versus generation rate are shown. Symbols are the experimental results and the solid lines are the best fits.

However, It was recently reported that the results of electron spin resonance(ESR) and light induced ESR(LESR)[12,13] indicate the presence of charged as well as the neutral defect states in device quality a-Si:H films. Their energy, densities and capture cross-sections of these states were determined from the self-consistent analysis of photoconductivity and sub-bandgap absorption spectra in intrinsic a-Si:H films[4,5]. In this study, the same distribution of defect states was applied to the analysis of results such as shown in Figure 1. Also the same values, as those obtained for the intrinsic films, were used for the extended states, the density of states at optical absorption edge, capture cross-sections of the tail states and neutral, D^0, and charged defect states, $D^+$and D^-. For this film the mobility gap is 1.90 eV[14] and the free carrier mobilities are 30 cm^2/Vsec and 1 cm^2/Vsec for electrons and holes respectively used in the analysis. The best fits to the data of Fig. 1 are shown as the solid lines.

The densities of the D^+, D^0, and D^- sates derived from these best fits are $8x10^{15}$, $3x10^{15}$, and $8x10^{15}$ and their location is at energy 1.4 eV, 0.80 eV, and 0.6 eV from the valence band edge respectively. The location of the neutral and charged defect states derived here are very close to those of the improved defect pool model[7]. It is interesting to note here that the ratio of the charged to neutral defect densities, R, is 5.3, which is significantly higher than those obtained in intrinsic a-Si:H films[5]. Such higher values of R were proposed in the LESR studies[11] and the defect pool models[6,7]. Examples of results obtained on different films are shown in Table 1.

It is also important to note here that, in the case of higher R, the sub-bandgap absorption measured by the constant photocurrent method(CPM) or using DBP at low generation rate is not only determined by the densities of the D^0 states but also depends on the densities of the charged defect states[5]. The densities of states derived here from the relation, α(1.2eV)x$3x10^{16}$[4], is $1x10^{16}$ cm^{-3} about three times higher than that derived from the detailed characterization. This indicates that in general the densities of the neutral defect states cannot be obtained directly from a single sub-bandgap absorption spectrum where R greater than about 2.0.

Schottky barrier solar cells

The QE and dark J-V characteristics at different temperatures were investigated on corresponding n/i Schottky barrier cells. On thick structures(> 1.5 µm), it is possible to fit the experimental results , using AMPS and the distribution and densities of states obtained from the results on the corresponding thin films. Thick i-layer devices are necessary to obtain sensitivity to bulk material parameters. This is illustrated in Figures 2 and 3 for 3 µm thick Schottky barrier cell corresponding to the thin film whose results are shown in Fig. 1. The J-V characteristics at different temperatures are shown in Figure 2 where the symbols are the experimental results and the solid lines are the best fits obtained using the AMPS.

In AMPS, the electric field profile and the positions of the quasi-Fermi levels depend on the densities and energy location of the neutral and charged defect states in the thick solar cell structures. The dark J-V characteristics over the wide range of temperatures are fitted with the energy locations for the neutral and charged defect states

Samples	E_{opt}(eV)	E_{0VB}(meV)	N_{D0}(cm^{-3})	N_{D+}(cm^{-3})	N_{D-}(cm^{-3})	$R=(N_{D+}+N_{D-})/N_{D0}$
MB2	1.72	49	$4.5x10^{15}$	$3x10^{15}$	$3x10^{15}$	1.4
YM36	1.74	48	$4.5x10^{15}$	$3x10^{15}$	$3x10^{15}$	1.4
YM33	1.76	49	$7.5x10^{15}$	$6x10^{15}$	$6x10^{15}$	1.6
200A	1.80	49	$8x10^{15}$	$9x10^{15}$	$9x10^{15}$	2.25
240A	1.74	48	$8x10^{15}$	$1.6x10^{16}$	$1.6x10^{16}$	4.0
240B	1.72	49	$3x10^{15}$	$8x10^{15}$	$8x10^{15}$	5.3
280A	1.70	48	$8x10^{15}$	$6x10^{16}$	$6x10^{16}$	15

Table 1. The densities of the neutral and charged defect states obtained from the analysis of photoconductivity and sub-bandgap absorption for different a-Si:H thin films in the annealed state.

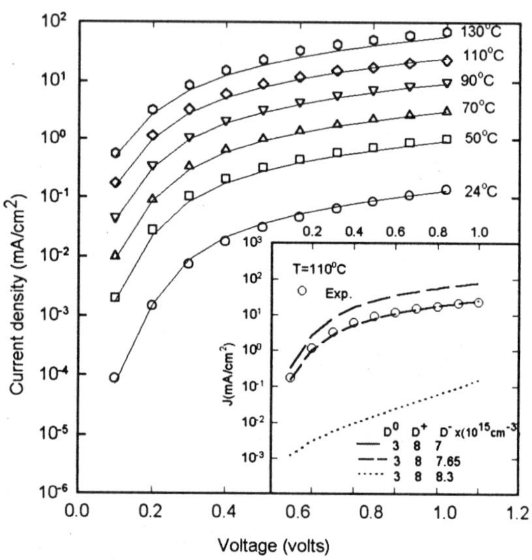

Fig. 2. Dark J-V characteristics at different temperatures for a Schottky barrier solar cell structure with 3 μm i-layer. Symbols are experimental data and solid lines are the fits with parameters of Fig. 1. In the inset, the sensitivity of J-V at 110°C to the ratio of D^+/D^-.

,the same as those obtained from the film results and the densities for the D^0, D^+, and D^- of $3x10^{15}$, $8x10^{15}$, and $7.65x10^{15}$ cm^{-3} respectively. Using these parameters, excellent fits are obtained from room temperature up to 130°C by just changing the temperature and the corresponding barrier height. Although a small difference in the densities of the D^+ and the D^- states such as obtained here does not affect the results of photoconductivity and sub-bandgap absorption, it is very important in the forward J-V characteristics since the charged defects directly influence the fermi level position in order to maintain the charge neutrality. The effect of the ratio of the D^+ to the D^- states on the J-V characteristics at 110°C is shown in the inset of Fig. 2. As the density of the D^- states decreases relative to D^+ states, to maintain charge neutrality, the Fermi level moves closer the conduction band edge and neutralizes some of the D^+states which results in an increase in the forward J-V. If the density of the D^- states is higher than the D^+ states, the Fermi level moves closer to the valence band edge and this results in a decreases in the forward J-V. This decrease is drastic because the densities of D^0 states is much lower than the D^+ and D^- states so that the Fermi level has to move far in order to maintain the charge neutrality. This illustrates the importance of having reliable gap state parameters for the bulk in understanding the causes that limit solar cell performance. To test the assumptions that the characteristics of this thick Schottky barrier solar cell are determined by bulk properties, a 200Å n/i interface regionwas inserted in the model with 10^{18} cm^{-3} defects(as has been found necessary in the past[15]) with the results of only about a 10% change in the far forward currents.

It has been proposed that internal quantum efficiencies(QE) of thick Schottky barrier solar cells with thick i-layer are determined by the bulk defect states[16]. The results of the QE measurements without bias light measured on the 3μm thick solar cell in the annealed state, are shown in Figure 3 as symbols. The solid lines represents the fit using the same parameters for the neutral and charged defect states, the same as those used to fit the data in Fig. 2. The QE's are also modeled without introducing any interface layers. It is also found that the QE, especially in the shorter wavelength, is sensitive to the densities of bulk charged defect states. This is illustrated in Figure 3 as the dashed lines where the same parameters are used as in the characteristics in the inset of Figure 2. The excellent fit to the QE shown as the solid line in Figure 3 confirms a self-consistency between the nature, densities and distribution of defect states derived for a-Si:H thin films and those in the bulk of these cell structures.

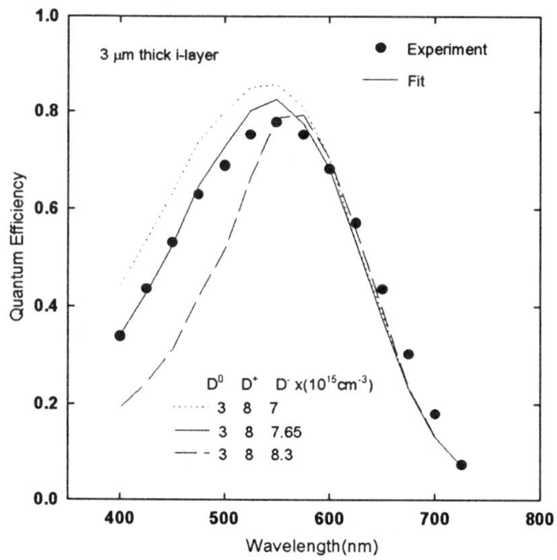

Fig. 3. Internal quantum efficiency versus wavelength measured without bias light for a Schottky barrier as in Figure 2. Symbols are experimental data and lines are the fit and sensitivity of blue response to the ratio of $D^{+}$to D^{-} states using with the same parameters as used in Figure 2.

CONCLUSIONS

It is found that the neutral and charged states that are required to explain photoconductivity and sub-bandgap absorption in undoped a-Si:H films have an important effect on the characteristics of solar cell structures. The reliable gap states parameters derived from the analysis of thin film results have been directly correlated with the characteristics of thick Schottky barrier solar cell structures having the same undoped i-layer. Using the same distributions and parameters for the gap states, the far forward J-V characteristics over a wide range of temperatures and internal quantum efficiencies(QE) were self-consistently modeled. Excellent agreement is found between the densities of states obtained from the results on the films and those on thick Schottky barrier solar cells. It is also found that the ratio of the D^{+} to the D^{-} states has a very large effect on certain Schottky barrier characteristics and these can then be used to fine tune the densities of states. However, for devices with i-layer thinner than about 1.5 μm, it is necessary to introduce an interface layer either at the n/i or at metal-i layer interface to account for the observed changes. Using the correct bulk gap states parameters, it is possible to probe the effects of the contacts and interfaces by fitting the thin cell data with AMPS modeling.

ACKNOWLEDGEMENTS

The authors would like to thank to Drs. S. J. Fonash, R. W. Collins, and Vyshnavi Suntharalingam for helpful discussions. This research was supported by EPRI Thin Film Solar Cell Program, NEDO International Joint Research Grant and NREL under subcontract No. XAN-4-13318-03.

REFERENCES

[1] M. Gunes, Y. M. Li, R. M. Dawson, C. M. Fortmann, and C. R. Wronski, in *Proc. of the XXIII IEEE PVSC*, p. 885 (New York, 1993).
[2] M. Gunes and C. R. Wronski, *Appl. Phys. Lett.* **61**, 678(1992).
[3] C. T. Malone et al. in *Proc. of the XXIII IEEE PVSC*, p. 1219 (New York, 1993).
[4] M. Gunes, C. R. Wronski, and T. J. McMahon, *J. Appl. Phys.***76**, 2260(1994).
[5] M. Gunes, R. W.Collins, and C. R. Wronski, in *Mat. Res. Soc. Symp. Proc.* (1994)(in press).
[6] G. Schumm, *J. Non-Crst. Solids* **164-166**, 317(1993).
[7] M. J. Powell and S. C. Dean, *Phys. Rev.* **B 48**, 10815(1993).
[8] S. Lee, M. Gunes, C. R. Wronski, N. Maley, and M. Bennett, *Appl. Phys. Lett.* **59**, 1578(1991).
[9] P. J. McElheney, J. K. Arch, H. S. Lin, and S. J. Fonash, *J. Appl. Phys.***64**, 1254(1988).
[10] R.M. Dawson, et al. *Mat. Res. Soc. Symp. Proc.* **Vol 258**, 747(1992).
[11] See Review paper by P. Nagels, in *Amorphous Semiconductors*, edited by M. H. Broadsky (Springer, New York, 1979), **Vol. 36**, p. 117
[12] T. Shimizu, et al. in *Mat. Res. Soc. Symp. Proc.* **vol.118**, 665(1988).
[13] G. Schumm, W. B. Jackson, and R. A. Street, *Phys. Rev.* **B 48**, 14198(1993).
[14] C. R. Wronski et al. *Phys. Rev. Lett.* **63**, 1420(1989).
[15] D. Heller et al. Mat. Res. Soc. Symp. Proc. **vol. 258**, 1031(1992).
[16] P. J. McElheney, S. S. Nag, S. J. Fonash, and C. R. Wronski, in *Proc. of the XXI IEEE PVSC*, p. 1459 (New York, 1990).

INVESTIGATION OF TCO/p-INTERFACE AND SURFACE MORPHOLOGY OF a-Si:H SOLAR CELLS BY IN-SITU ELLIPSOMETRY

H.N. Wanka, M.B. Schubert and E. Lotter, Universität Stuttgart,
Institut für Physikalische Elektronik, Pfaffenwaldring 47, D-70569 Stuttgart, Germany.

ABSTRACT

Different concepts for optimizing the TCO/p-interface in a-Si:H based solar cells have been studied in order to avoid metal layer segregation, and hence considerable reductions in short circuit current. Therefore we analysed structural and chemical changes which occur at the surface of transparent conductive oxides - TCOs (SnO_2, ZnO, Indium Tin Oxide - ITO) in silane, hydrogen and CO_2 plasmas. We also used a-SiO:H instead of a-SiC:H in the p-doped layer. In-situ ellipsometry and SIMS/XPS depth profiling show that room temperature as well as fast deposition easily overcome all detrimental effects. TCO deterioration by ion and radical bombardement at high rates is more than compensated if the surface is protected by a rapidly growing a-Si:H film. Using ZnO as a TCO, or as a 20 nm buffer layer only, completely suppresses metal formation. In-situ ellipsometry in conjunction with atomic force microscopy reveals significant changes in surface morphology, namely filling of TCO-texture during deposition, which is of crucial importance for light trapping in solar cells.

INTRODUCTION

Although subject of numerous investigations [1-4] the optimum choice of materials and cell design for the crucial TCO/p region in a-Si:H based solar cells is still a matter of debate. Successful concepts include the use of a higher bandgap intrinsic buffer layer for improved electrical field profiling, lowering of the substrate temperature during p-layer deposition, replacement of the standard a-SiC:H by a-SiO:H [5], CO_2 plasma treatment of tin oxide front contacts [6], or the use of ZnO for enhanced chemical stability of the TCO (transparent conductive oxide) [7].

Especially in-situ ellipsometry in conjunction with SIMS and XPS depth profiling revealed that optical losses at the TCO/p interface are mainly due to the formation of metallic phases originating from a chemical reduction of the TCOs upon exposure to silane or hydrogen plasmas. Whereas the standard layer sequence glass/tin oxide/p-a-SiC:H is well characterized, no systematic study of the above listed options for improvement is available as yet. Benefits for solar cell performance due to the implementation of these options can only be assessed from the data of different laboratories and from rather different overall cell structures.

Therefore we present a study of in-situ ellipsometry and SIMS/XPS/UPS analyses for addressing the problem of optical loss by the investigation of chemical and structural changes at the interface. We consider this to be a prerequisite for studying interface states, electrical field profiling, and finally optimizing solar cell efficiency on safe grounds.

EXPERIMENTAL

We have characterized different types of commonly used substrates: i) ZnO, sputtered from an Al-doped ZnO target at room temperature (rf-power 1.1 W/cm^2, thickness (10-230) nm, room temperature conductivity 2000 S/cm); ii) commercially available tin oxide (delivered by Nippon Sheet as well as Asahi Glass, thickness 450 nm), referred to as SnO_2. Amorphous silicon films (a-Si:H) were deposited from a dc glow discharge (substrate temperature 300 or 520 K, resp., SiH_4 flow 6 sccm at 20 Pa); growth rate has been controlled by adjusting the plasma power (3 Å/s at 30 mW/cm^3 - referred to as *fast* deposition, 0.16 Å/s at 2 mW/cm^3 - *slow* deposition). Plasma treatment of the TCOs has been performed by using H_2 or CO_2 (at 300 as well as 520 K, 10 sccm, 20 Pa and power levels of 27 and 14 mW/cm^3, resp.). a-SiO:H has been deposited from a mixture of SiH_4 (6 sccm) and CO_2 (9 sccm); in order to qualify TCO interaction, the deposition rate has been changed as mentioned above (2.4 Å/s at 20 mW/cm^3 and 0.12 Å/s at 1.25 mW/cm^3).

In-situ ellipsometry has been performed by using a phase-modulated instrument (Jobin-Yvon, UVISEL [8]). Kinetic measurements (10 data points per sec) were usually carried out at a photon energy of 4 eV, for optimum surface sensitivity. A proximity shutter system has been employed in front of the substrate for providing stable plasma conditions at the very beginning of the depositions. XPS (MgK_α x-ray source, Ar^+ sputtering, E_{kin} = 5 keV) and SIMS (Ar^+ sputtering, E_{kin} = 2 keV) depth profiles have been recorded from a multi-purpose surface analysis system (Leybold-Heraeus, LHS 10). For atomic force microscopy (AFM) we used the contact mode of a TopoMetrix instrument.

CH3365-4/94/0000-0516 $4.00 © 1994 IEEE

First WCPEC; Dec. 5-9, 1994; Hawaii

RESULTS AND DISCUSSION

TCO/p-Interface

Fig. 1 comprises real-time trajectories of the ellipsometric angles recorded during a-Si:H deposition on SnO_2 at different growth rates and temperatures, plus one obtained from a H_2 plasma treatment of the TCO. 'Slow' deposition data exhibit a sharp kink after 60 sec deposition time which is characteristic of the formation of metallic interlayers [2]. Initially this trajectory is governed by a strong increase in extinction coefficient and by a lowering of the real part of the refractive index. This results from a chemical reduction of SnO_2 and the corresponding formation of metallic tin close to the TCO surface. The same effect clearly occurs during H_2 plasma treatment at deposition temperature. The solid line in Fig. 1 represents the

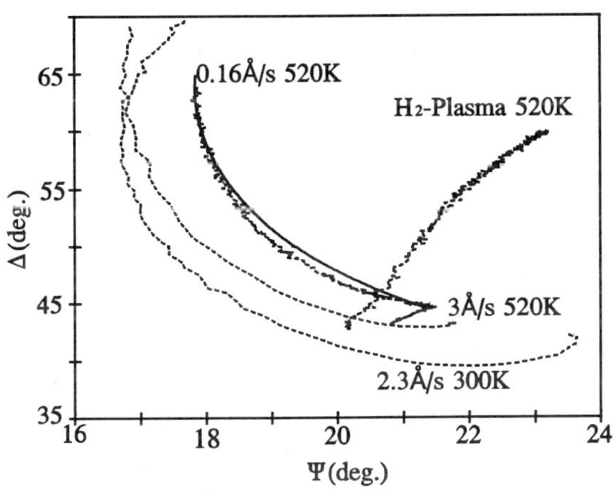

Fig. 1. Real time ellipsometric trajectories at an incident energy of 4 eV, taken during deposition of a-Si:H on top of SnO_2, and during H_2-plasma treatment of SnO_2, resp.

calculated time evolution of the effective optical parameters of a-Si:H growing on top of a 10 Å thick film containing 50% metallic Sn and 50% SnO_2. The amorphous silicon deposition after 60 sec can consistently be modelled by homogeneous growth in conjunction with a decrease in surface roughness (details see below). The 'fast' deposition data in Fig. 1 do not give direct hints at a reduction of the SnO_2 surface. The 'fast' trajectory is well reproduced if the change in surface roughness is properly accounted for.

Photovoltaic applications suffer from remarkable transmission losses if there is metallic tin present at the interface, which reduces light incidence into the active layer of pin solar cells. Most interesting with Fig. 1, however, fast deposition at room temperature gives clear evidence of nucleation behavior instead of homogenous growth. During slow deposition, and even H_2 plasma treatment, chemical reduction could completely be avoided at 300 K. Although this kind of remedy is not feasible for application in real production, it points to a threshold activation of SnO_2 reduction, independent of the energy impact from the plasma. In order to trace the formation of metal layers by chemical analyses, XPS and SIMS depth profiling has been applied. An SiO_x barrier forming at the interface, which results from the reduction of SnO_2, is evident from Fig. 2. For better comparison the maxima positions of the SiO_x spectra have been aligned, clearly showing a larger thickness of the oxide barrier in slowly deposited films. It should be mentioned that the depth profiles are broadened since the ion beam used for sputter etching hit the film at an angle of 30 degrees, and the surface roughness of our samples therefore leads to shadowing of ions and results in non-uniform etching. Nevertheless all spectra in Fig. 2 should be affected in a similar way due to comparable roughness of the samples.

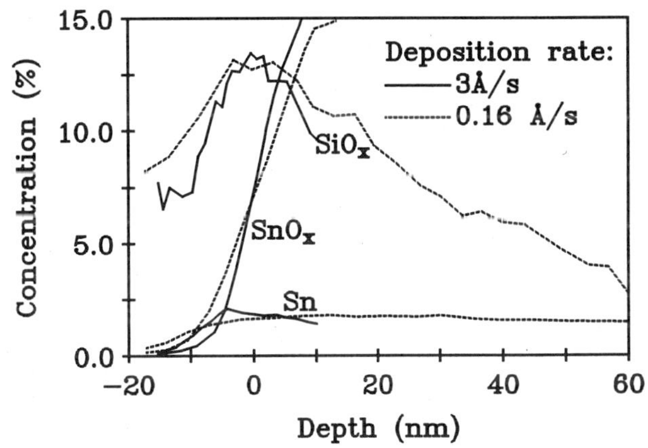

Fig. 2. XPS depth profiling of SnO_2/a-Si:H interfaces; depth zero has been assigned to the peak position of the SiO_x concentration

The XPS data reveal that even in the fastly deposited samples SnO_2 has been reduced to metal close to the interface, although no signature of such a layer is detected from Fig. 1. Based on the knowledge of room temperature deposition this contradictory evidence can easily be explained: Nucleation processes (initial upward bending of the 300 K data in Fig. 1) and the formation of metallic tin (downward kink) compensate each other in the optical parameters and the trajectory starts smooth. Nevertheless the fastly grown interface is significantly sharper, and there is less Sn present owing to the quick coverage of the SnO_2 surface, which is thereby being protected against impact of energetic particles from the discharge.

It is well known, that ITO and SnO_2 are substantially being reduced in H_2 discharges, and metallic layers are formed at the TCO surfaces. In order to find out, whether TCO reduction in SiH_4 plasmas is mainly due to hydrogen present in the discharge, ITO as well as SnO_2 have been exposed to H_2 plasmas (at 520 K) for 15 min, which deteriorated the optical transmission of both TCOs by about 50 %. On the other hand no reduction effects could be observed upon exposure to an H_2 discharge at 300 K. ZnO on the contrary has not been reduced at all in an H_2 plasma, its optical transmission as well as spectroscopic ellipsometry data remaining unchanged.

Fig. 3 nicely demonstrates from ellipsometric data that ZnO can serve as a protection layer for the more common TCOs like SnO_2 or ITO [7]. Layers as thin as 20 nm of room temperature rf sputtered ZnO efficiently suppress metal formation at the TCO/p interface.

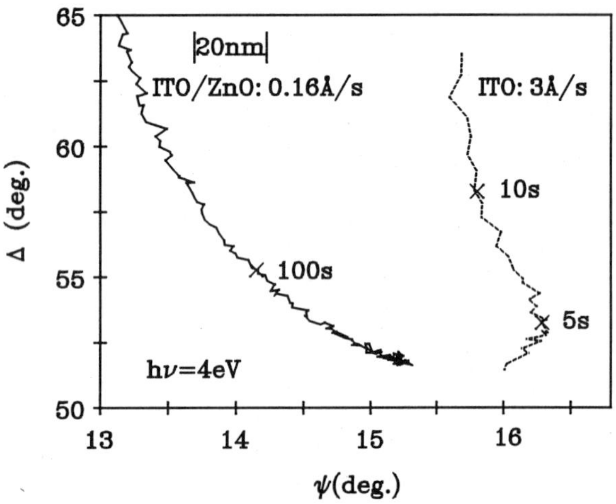

Fig. 3. Effect of 20 nm rf-ZnO on top of ITO

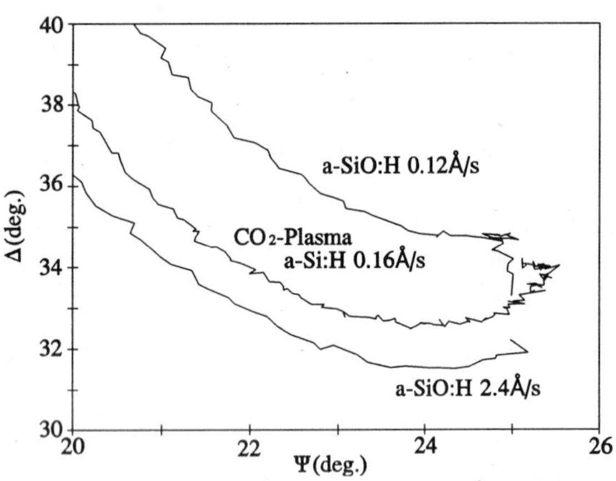

Fig. 4. Recent approaches: i) a-SiO:H deposition, and ii) a-Si:H growth after CO_2-plasma treatment of SnO_2, monitored by in-situ ellipsometry

More recent approaches for improved TCO/p performance have been investigated, too: real-time ellipsometry data, taken during a-SiO:H deposition at high and low rates, are plotted in Fig. 4. Slow deposition trajectories are well reproduced by adding an interlayer of 6 Å with equal contents of Sn and SnO_2 to the numerical modelling. Compared with a-Si:H, or even a-SiC:H, the use of a-SiO:H reduces metal formation at the interface, but TCO reduction cannot completely be avoided. Most promising, after CO_2 plasma treatment of SnO_2 the trajectory does not indicate any reduction at all, even for slow a-Si:H deposition; on the contrary the data exhibit a strong nucleation comparable to room temperature growth.

Surface Morphology

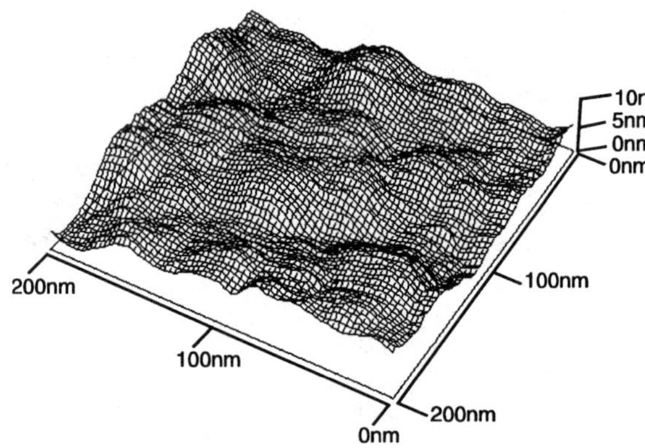

Fig. 5. Surface morphology of 1 μm thick a-Si:H on smooth ZnO from AFM

The surface morphology of a-Si:H films, growing on top of smooth ZnO (Average ZnO surface structures are about 5 nm high and 60 nm wide), has been examined with atomic force microscopy (AFM). It does not change with increasing film thickness (Fig. 5), and corresponding ellipsometry spectra also remain unchanged once the layers have become opaque. Combining both analyses we are able to re-calculate the spectroscopic ellipsometry bulk data, and hence to determine the surface roughness of similarly deposited films merely from spectroscopic ellipsometry.

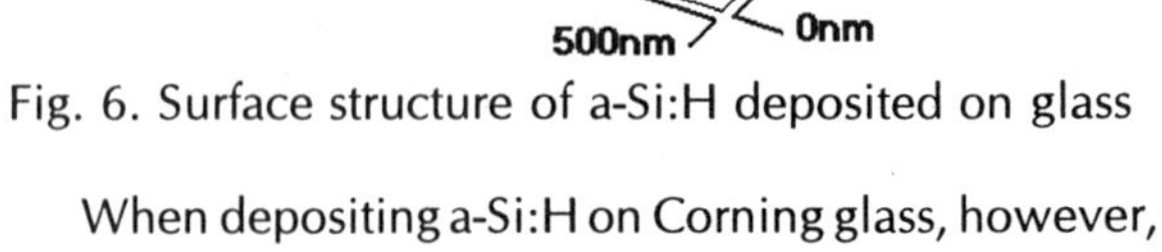

Fig. 6. Surface structure of a-Si:H deposited on glass

When depositing a-Si:H on Corning glass, however, the surface roughness is enhanced; the thickness of the surface roughness layer amounts to approx. 8 Å. In Fig. 6 the final morphology of a-Si:H on Corning 7059 is shown, which is close to that of Fig. 5. The lateral dimensions of the structures (20-30 nm) can be correlated to the surface diffusion length.

Unlike a-Si:H on top of smooth ZnO or glass, the surface of amorphous silicon growing onto texturized SnO_2 substrates is partly planarized with increasing thickness. The imaginary part of the dielectric function is enhanced during growth as a result of decreasing surface roughness (Fig. 7) [2,3]. The spectroscopic data from Fig. 7 yield an exponential drop in surface roughness like $R(d) = (R_s-R_f)\exp(-d/c) + R_f$ (film thickness d, SnO_2 roughness R_s = 45 nm) with c = 330 nm and a final roughness of R_f = 5 nm. This smoothing of initial TCO texture during a-Si:H deposition deteriorates light trapping in solar cells, and thereby the cell efficiency. Further work has been started in order to quantify the loss of TCO texture under production-like growth conditions in real solar cells.

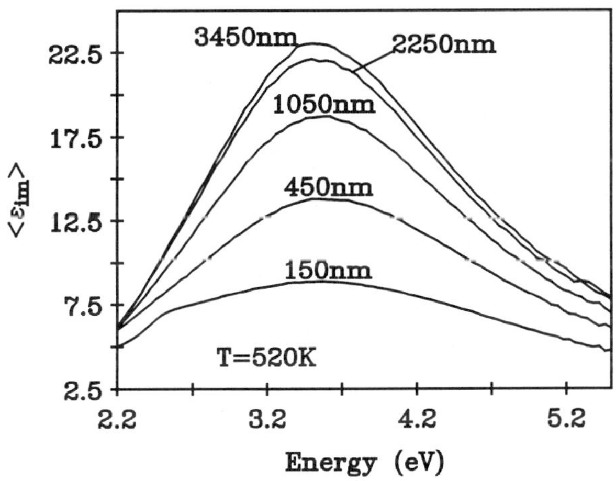

Fig. 7. Imaginary part of the dielectric function of a-Si:H growing on SnO_2

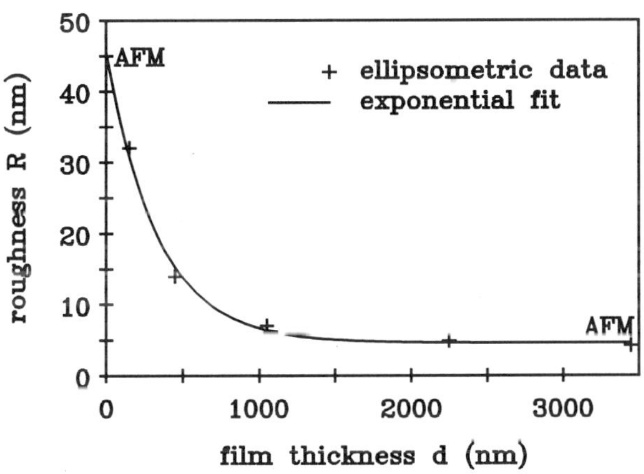

Fig. 8. Surface roughness as a function of film thickness

ACKNOWLEDGEMENTS

The authors wish to thank Dr. G. Bilger for XPS analyses and helpful discussions, and M. Wieber for valuable help in preparing the figures and the poster presentation. Atomic force microscopy would not have been possible without the cooperation and assistance of U. Griesinger.

Financial support by the German 'Bundesminsterium für Forschung und Technologie' under contract No. 0329521A is also gratefully acknowledged.

REFERENCES

[1] G. Bilger, H.C. Weller, A. Eicke and G.H. Bauer, "Study of the a-SiC:H/TCO Interface in Amorphous Silicon Solar Cells by Combination of SIMS and XPS", in *International Symposium on Trends and New Applications in Thin Films*, Strasbourg, France [*Le Vide Les Couches Minces*] **235**, 1987, pp. 91-95.

[2] S. Kumar and B. Drevillon, "A Real Time Ellipsometry Study of the Growth of Amorphous Silicon on Transparent Conducting Oxides", *J. Appl. Phys.* **65**, 1989, pp. 3023-3033.

[3] B. Drevillon, "In Situ Studies of the Growth of Hydrogenated Amorphous Silicon", *J. Non-Cryst. Solids* **114**, 1989, pp. 139-144.

[4] R.W. Collins, "Dielectric Functions of Thin Interface Layers in a-Si:H-Based Device Structures by Spectroscopic Ellipsometry", *J. Non-Cryst. Solids* **114**, 1989, pp. 160-162.

[5] Y.-M. Li et al., "An Exploratory Survey of p-Layers for a-Si:H Solar Cells", *Mat. Res. Soc. Symp. Proceed.* **3 36**, 1994, in print.

[6] J. Bauer et al., "Manufacturing of Large Area Single Junction a-Si:H Solar Modules with 10,7% Efficiency", *J. Non Cryst. Solids* **164-166**, 1993, pp. 685-688.

[7] G. Bilger, A. Eicke and G.H. Bauer, "BSI and ZnO for Blocking of Impurity Migration in Amorphous Silicon Solar Cells", ***19**th IEEE PVSC*, 1987, pp. 615-620.

[8] S.N. Jasperson, D.K. Burge and R.C. O'Handley, "A Modulated Ellipsometer for Studying Thin Film Optical Properties and Surface Dynamics", *Surface Science* **3 7**, 1973, pp. 548-558.

CHARACTERIZATION OF AMORPHOUS SILICON PIN SOLAR CELLS BY TRANSIENT MEASUREMENT TECHNIQUES

G. Schmid, M.B. Schubert, H. Brummack and N. Bernhard; Universität Stuttgart,
Institut für Physikalische Elektronik, Pfaffenwaldring 47, D-70569 Stuttgart, Germany.

ABSTRACT

In order to provide more specific access to solar cell limitations (and to faults during line production as well) we report on space-charge limited time-of-flight (SCL-TOF) studies in the voltage mode, and on forward bias transient switching experiments in 500 nm thick pin structures. Complementing I/V characteristics and quantum yield, both techniques are easy to use and give a clear and reproducible measure of the degradation state of a-Si:H based solar cells. An effective carrier lifetime can be deduced from the extraction time of SCL-TOF. Time and temperature dependent forward current transients show a distinct voltage threshold in the onset of double-injection recombination currents which is related to the built-in voltage of the devices. Enhanced bulk recombination due to light or current induced degradation reduces saturated forward currents and increases the time initially needed for building up space charge in the pin-diodes. Differences upon light-soaking by front or back side illumination have also been investigated.

INTRODUCTION

Amorphous silicon (a-Si:H) and its alloys (a-SiGe:H, a-SiC:H) have extensively been characterized from a materials' point of view, many investigations being concerned with photovoltaic applications. Keeping track of single effects which during line production may influence the solar cell performance is a major challenge which has to be met for realizing continously high production yield. Current-voltage (I/V) characteristics as well as quantum yield provide valuable information, but for some purposes transient measurements offer a more specific access to the device properties. Characterization of real solar cells, however, often results in somewhat general conclusions, since steady-state techniques like charge collection [1] tend to yield information in a complex form, often requiring a complete understanding or numerical modelling of the pin-diodes for proper interpretation [2] [3]. Many of the techniques used are limited to experiments on rather thick ($> 2\,\mu$m) pin-structures, significantly different from standard solar cells and not easily available from the line production of those.

In order to complement the I/V- and quantum efficiency measurements usually being performed, we have tested the potential of several methods for providing detailed information about standard a-Si:H solar cells of (250-500) nm thickness. Results on Modulated (MPC) as well as Constant Photocurrents (CPM) will be published elsewhere while a critical assessment of transient measurements, namely space-charge limited time-of-flight and forward bias transient switching (in the dark), is presented here.

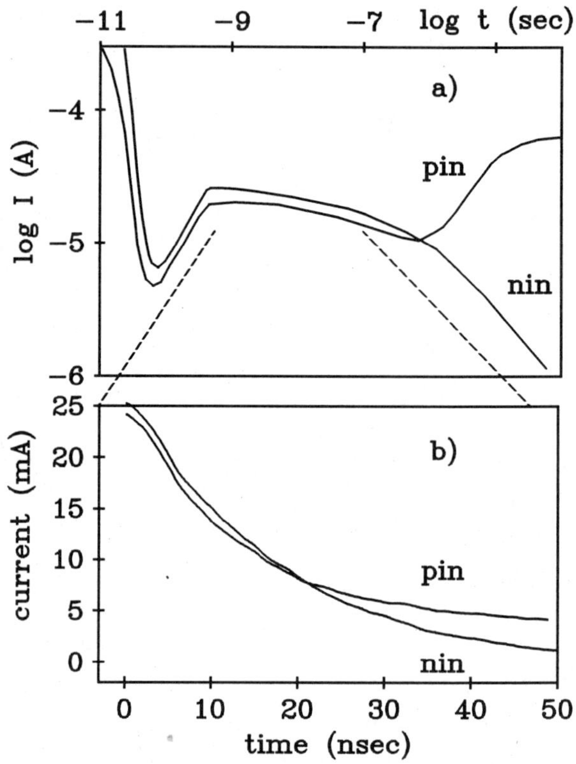

Fig. 1. Typical forward current switching transients; schematically after [4] (a), and comparing nin vs. pin structures (b)

General features of forward current transients in pin structures have recently been studied (see Fig. 1, [4], [5]) emphasizing the major influence of space charge initially piled up close to the p/i- and i/n-interfaces [6], and proving the final current rise (on a msec scale) to be recombination limited [7]. By examining different time and temperature regimes, and various degradation states as well, we now try to separate interface effects from recombination limited bulk carrier transport in the i-layer.

Although the TOF technique [8] has extensively been used for transport studies in disordered semiconductors [9], a successful application to real a-Si:H solar cells has only recently been demonstrated [10], [11].

CH3365-4/94/0000-0520 $4.00 © 1994 IEEE First WCPEC: Dec. 5–9, 1994; Hawaii

EXPERIMENTAL

Amorphous silicon pin solar cells have been deposited from a dc glow discharge (single chamber with load lock) at optimized conditions, yielding initial efficiencies of (7-8) % from 500 nm thick devices. Boron doped a-SiC:H was used as a window layer, intrinsic a-SiC:H buffers have not been employed, bandgap profiling was also not applied here. Back contacts were Al or Cr without special optimization, but some of them semi-transparent for front and back side illumination. For some of these experiments smooth Al-doped ZnO, rf-sputtered at room temperature, has been used as front and back contacts of the same cells. SCL-TOF was measured from a standard set-up under vaccum, with a N_2-laser (3 nsec pulses) pumping a dye for adjusting absorption depth [11]. Forward bias current transients have been recorded from a 330 kHz A/D converter in a fully computer-controlled setup, and with a digitizing storage oscilloscope (300 MHz bandwidth) which has also been used for SCL-TOF. Light soaking has been achieved either by applying white light similar to AM1-illumination, or by strongly absorbed laser light from an Ar^+-laser. For comparison also double-injection current degradation has been performed at (50-100) mA/cm^2.

RESULTS AND DISCUSSION

Forward current transients

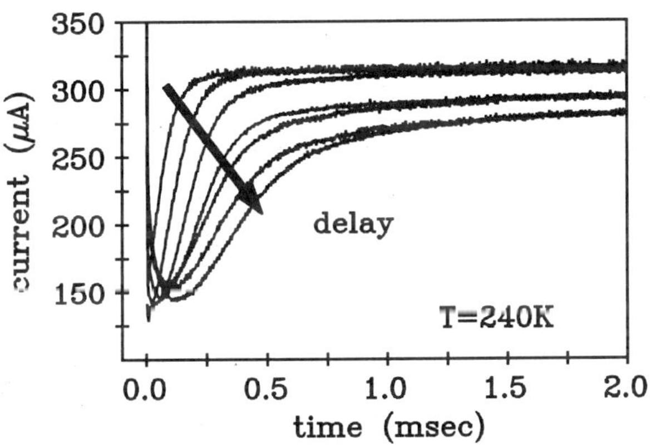

Fig. 2. Forward current transients in a 500 nm thick pin-diode (after 50 msec pulses at delay times from 0.1 msec to 1000 sec)

Typical current transients from double-injection experiments are plotted in Figs. 1 and 2. After initial charging of the condensor plates the samples represent, the current in 500 nm thick solar cells drops to a space charge limited injection current due to accumulation of space charge barriers close to the p/i and n/i interfaces. The similarity of transient signals from nin and pin structures in the initial phase, the single layers of which have been grown under identical conditions, clearly demonstrates this current to be mainly carried by electrons (Fig. 1). Thus the leading edge of the electrons has to cross the i-layer from the n- to the p-side before bulk recombination of the injected electrons and holes starts near the p-interface, resulting in a reduction of the space charge barrier that prevents hole injection from the p-layer. Changing the repetition rate of 50 msec pulses from 10^4 Hz to 10^{-3} Hz makes

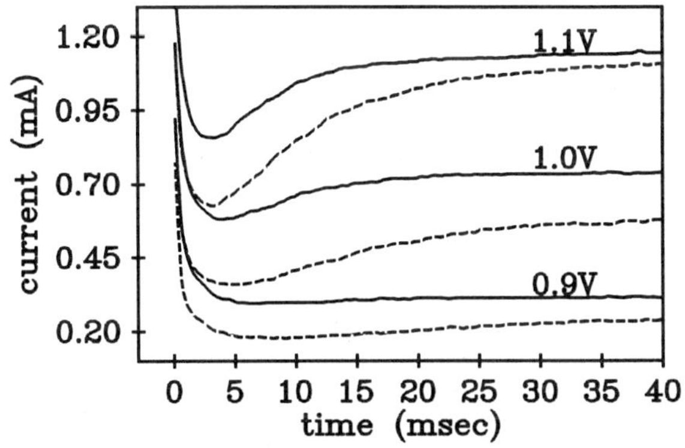

Fig. 3. Onset of the recombination current as a function of the pulse height applied, in the annealed state (solid), and after 15 min of AM1 degradation from the n-side (dashed lines)

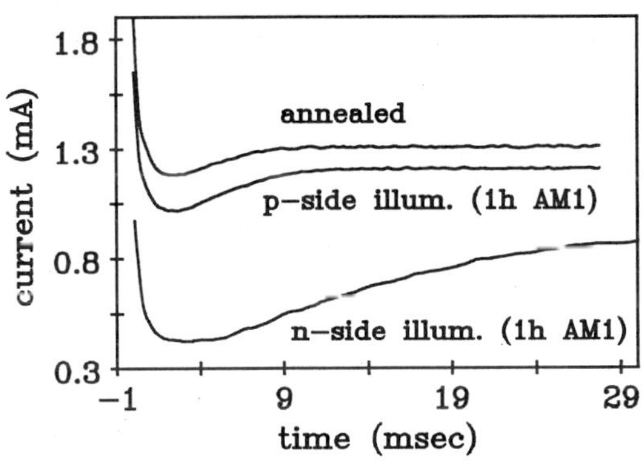

Fig. 4. Front or back side illumination with white light (AM1) results in significantly different forward current transients

the transients in Fig. 2 slow down, since less charge will remain stored in trap states when the interval between individual pulses is increased. The initial decay of the current to the space charge limited value is significantly slower in 'thin' pin-diodes because the internal field, which is much stronger here, has to be compensated first. Varying the pulse height clearly demonstrates this effect, too: Fig. 3 shows a distinct onset of recombination currents at voltages above 0.9 V and therefore yields information about the built-in potential, and about barrier heights at the interfaces as well. Because the current increase is due to recombination, the transients are strongly dependent on the degradation state of the solar cells (dashed lines in Fig. 3, and Fig. 5). Whereas the general shape of the time evolution seems to be unaffected, the initial space-charge limited current as well as the saturated recombination current decrease upon degradation, both in agreement with an increase in defect density due to light soaking.

The rise time of recombination currents is limited by the speed of hole transport across the i-layer. Therefore the rise time is determined by multiple trapping of holes in shallow traps rather than subject to light-induced generation of deep defects. This is in agreement with numerical calculations [6]. Furthermore the onset time of the current increase should be temperature activated via the hole drift mobility. An evaluation gives an activation energy of 0.41 eV, in good agreement with hole drift mobility measurements by the TOF method [12].

Monitoring the delay time dependence for different degradation states yields information about the energetic position of the respective defect states involved, which is 0.77/0.86/ 0.90 eV for the delay times of 1/30/150 sec used in Fig. 5. Finally Fig. 4 gives an idea, to which extent degradation effects at or close to the interfaces can be distinguished.

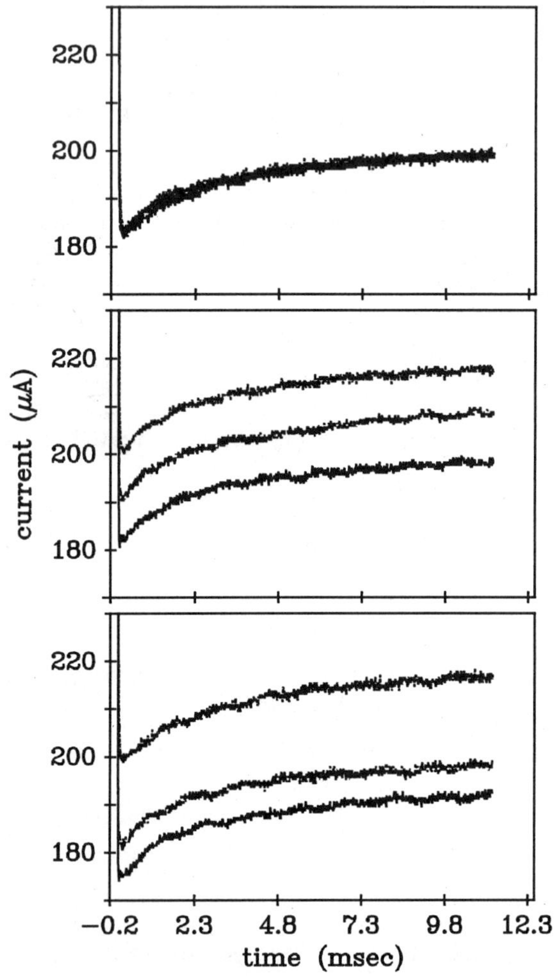

Fig. 5. Change of forward current transients at 1, 30, and 150 sec delay after n-side AM1 light-soaking at different illumination times (from top to bottom: 0/5/30 min)

Space charge limited time-of-flight (SCL-TOF)

Due to high RC time constants of real solar cells in comparison to the transit time space-charge limited time-of-flight measurements are restricted to an integral 'voltage' mode, i.e. integrating rather than directly monitoring current transients [10], some of which are plotted in Fig. 6. Choosing SCL instead of small signal mode provides a remarkable increase in signal strength and, even more important, slows carrier transit and extraction down to a time scale which is accessible with standard nsec time-of-flight equipment. The photo-generated charge in this case amounts to some multiples of the geometrically stored charge CU_o on the capacitor formed by the pin with its electrodes, and therefore immediately screens

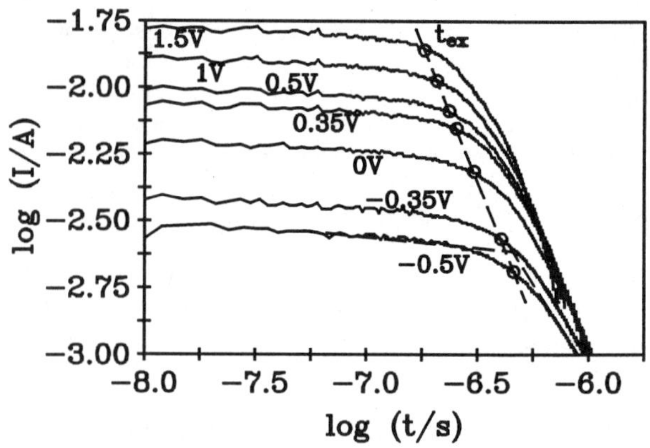

Fig. 6. Field dependent transients of the space-charge limited photo-current in a solar cell (500 nm thick)

the electric field in the i-layer. Up to a certain extraction time t_{ex} only CU_o portions of the charge can be extracted at a time. A relation between t_{ex} and the extraction current i_{ex} had been given by [10] and was modified to

$$t_{ex} = \tau \ln\left(\frac{Q_{Ph} - Q_{res} - CU_o}{Q_{res} + \tau i_{ex}} + 1\right) \qquad (1)$$

in order to account for residual charge Q_{res} still stored in the sample at $t > t_{ex}$. Q_{res} cannot be evaluated analytically, therefore we integrate in first approximation the measured current for times larger than t_{ex}:

$$Q_{res} = \int_{t_{ex}}^{\infty} i_{ex}(t)\, dt - CU_o \qquad (2)$$

Unless correcting for Q_{res} the photo-injected charge is strongly underestimated, which can be seen from the intensity dependence in Tab. I. At moderate intensities ($3 \leq Q_{Ph}/CU_o \leq 30$, $CU_o \approx 0.5$ nC) an effective carrier lifetime has been deduced, the validity of which can easily be cross-checked by comparing the calculated amount of collected charge $Q_{Ph}(fit)$ to the one, which has in fact been photo-injected (first column in Tab. I). The resulting lifetime is clearly degradation dependent (see Fig. 7), and due to field screening mainly reflects bulk i-layer properties, quite similar to a deep trapping lifetime in standard time-of-flight. In our samples this eff. lifetime drops from 440 ns to 270 ns after 75 min of double-injection current degradation at 100 mA/cm^2, and it is fully restored upon annealing.

Unfortunately a straightforward interpretation of the physical meaning of the measured lifetimes τ can hardly be given, since the complex phenomena taking place during multiple trapping carrier transport in amorphous silicon do not only involve deep trapping of photo-generated charge, but also true recombination due to non-negligible hole contributions.

Therefore SCL-TOF seems useful for qualitatively characterizing the degradation state of solar cells, a detailed understanding, however, requires full numerical modelling.

	Q_{res} neglected [10]		accounting for Q_{res}	
Q_{Ph} (real)	Q_{Ph} (fit)	τ	Q_{Ph} (fit)	τ
1,4 nC	828 pC	18 ns	1,4 nC	53 ns
2,8 nC	1,15 nC	25 ns	2,3 nC	57 ns
7 nC	1,8 nC	29 ns	3,9 nC	58 ns
14 nC	3,6 nC	27 ns	6,2 nC	56 ns
28 nC	2,6 nC	39 ns	5,4 nC	78 ns
70 nC	2,4 nC	50 ns	5,9 nC	88 ns
140 nC	2,7 nC	54 ns	6,4 nC	93 ns

Tab. I. Intensity dependence of effective lifetimes τ from SCL-TOF extraction times, evaluated after [10] and after (1).

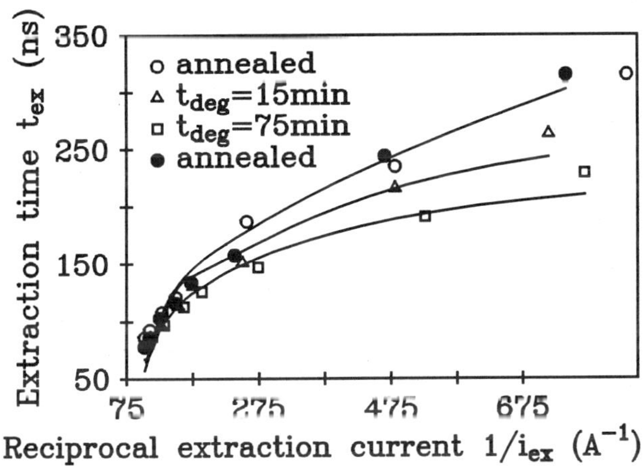

Fig. 7. Evaluation of the extraction time t_{ex} from Fig. 6 after (1), yielding degradation dependent lifetimes τ.

CONCLUSION

In conclusion both transient methods discussed here are easily applicable to real solar cell investigations, and they show a clear and reproducible dependence on the degradation state of the devices. Effective lifetimes have been deduced from SCL-TOF extraction times t_{ex} by accounting for residual charge, remaining in the sample after t_{ex}. This lifetime is mainly affected by bulk i-layer defect densities and their degradation induced changes. Forward bias switching experiments support the current interpretation in terms of double-injection recombination currents, which are initially hampered by space-charge barriers close to the p/i and n/i interfaces. These transient currents are furthermore subject to metastable changes upon light-soaking, with differing signals for front or back side degradation. The temperature and field dependence finally offer a promising potential for separating bulk and interface effects in real solar cells.

ACKNOWLEDGEMENT

The authors would like to thank M. Wieber for special efforts and skillful help with the presentation of the results. We also gratefully acknowledge financial support by the German 'Bundesministerium für Forschung und Technologie' under contract No. 0329521A.

REFERENCES

[1] C.-D. Abel, G.H. Bauer, Evaluation of the steady-state photocarrier grating technique with respect to a-Si:H and its application to a-$Si_{1-x}Ge_x$:H alloys, *Progress in Photovoltaics* **1**, 1993, p. 269

[2] M. Block, Defect distribution in a-Si:H pin solar cells before and after degradation, *J.Non-Cryst.Sol.* **164-166**, 1993, p. 701

[3] R. Brüggemann, C.-D. Abel, G.H. Bauer, Implications of the Defect-Pool-Model for the Simulation of a-Si:H Solar Cells, *Proc. 11th EC PVSEC* (Kluwer Academic Publications, Dordrecht, NL), 1992, p. 676

[4] F. Shapiro et al, Interpretation of Transient Currents in Amorphous Silicon Hydride p-i-n and n-i-n Devices, *IEEE Transactions on Electron Devices* **36**, 1989, p. 2785

[5] D. Han, K. Wang and M. Silver, Transient forward bias in a-Si:H p-i-n-devices, *J.Non-Cryst.Sol.* **164-166**, 1993, p. 339

[6] M. Hack and R.A. Street, Analysis of double injection transients in amorphous silicon p-i-n diodes, *J. Appl. Phys.* **72**, 1992, p. 2331

[7] R. Carius et al., Transport and recombination in a-Si:H p-i-n diodes under forward bias conditions, *J. Non-Cryst. Solids* **164-166**, 1993, p. 551

[8] W. E. Spear, Transport and tail state interactions in amorphous silicon, in *Amorphous Silicon and Related Materials*, edited by H. Fritzsche (World Scientific Publ. Co., Singapore), 1988, pp. 721-765

[9] C.E. Nebel, G.H. Bauer, Tail-state distribution and trapping probability in a-Si:H investigated by time-of-flight experiments and computer simulations, *Phil. Mag. B* **59**, 1989, p. 463

[10] J. Kočka et al., Demonstration of space-charge limited time-of-flight as a new tool for characterization of real a-Si:H solar cells, *J. Non-Cryst. Sol.* **164-166**, 1993, p. 489

[11] H. Brummack, N. Bernhard and M.B. Schubert, The time-of-flight technique applied to amorphous silicon p-i-n solar cells, *Mat. Res. Soc. Symp. Proceed.* **336**, 1994, in print

[12] C.E. Nebel et al., Valence band tail and defect distribution in a-Si:H, *Proc. 9th EC PVSEC* (Kluwer Academic Publications, Dordrecht, NL), 1989, p. 267

INDUCED DEFECTS IN a-Si:H AND a-Ge:H CALCULATED BY MOLECULAR ORBITAL THEORY

Akira Yoshida and Yuji Matsuda
Toyohashi University of Technology
Tempaku, Toyohashi, Aichi 441 JAPAN

Koichi Nakanose and Hiroshi Ogawa
Saga University
Honjyo, Saga 840 JAPAN

ABSTRACT

Gap states induced by weak bond and hydrogen movement in a-Si:H have been investigated on the basis of DV-Xα molecular orbital theory. Two energy levels appear near both sides of band edge, and shift to the gap center, while the weak bond is stretched and the bonded H atom is moved to the neighboring Si or Ge atoms. These additional states deep in the energy gap are due to the modification of bonding states, decreasing the electrical conductivity of a-Si:H. These facts give a possible explanation of light-induced degradation, when the weak bond and the movement of bonded H atoms are induced by prolonged exposure to intense light.

INTRODUCTION

Light-induced changes in the optical properties of hydrogenated amorphous silicon (a-Si:H) thin films, known as the Staebler-Wronski (SW) effect, have received much attention, because they are closely related to the long-time stability of solar cells[1].

It is well known that the SW effect in a-Si:H is due to the additional gap states deep in the energy gap, created by the prolonged exposure to light. In order to explain the increased gap states, several kinds of microscopic models based on photo-induced changes of amorphous structures have been reported. One of the simple changes of amorphous structure is to stretch the bond length (weak bond), yielding the gap states. Also, most of them are associated with hydrogen movement, because photo-induced changes decrease as the hydrogen content decreases in a-Si:H thin films. Hydrogen atom migrates easily through the amorphous network at low temperatures, modifying the hydrogen bonding configuration. Light-enhanced motion of H atoms creates a metastable state, which is responsible for the degradation of a-Si:H films. However, the theoretical or experimental support for these models is not always sufficient.

Recently, the investigation of hydrogen movement is calculated by ab-initio molecular dynamics and other methods[2-5], and it has been tried to apply this theory to the microscopic mechamism of SW effect. However, the results that have been reported have not estimated the changes in electronic states.

In this paper, we have estimated the changes in electronic states of a-Si:H and a-Ge:H induced by weak bond and hydrogen motion by DV-Xα molecular orbital method, based on the Hartree-Fock-Slater (HFS) approximation.

CLUSTER MODELS

In the present work, we have accepted a cluster, as shown in Fig.1, consisting of eight Si atoms. Each outermost Si atom is terminated with three hydrogen-like atoms. The bond length (d) between Si(1) and Si(2) was changed to obtain the weak bond. The Si-Si bond length (d_0:4.44[a.u.]) and bond angles (109.47 °) were consistent with those in ctystalline solids, and the Si-H bond lengths (2.80[a.u.]) and bond angles (109.47 °) were chosen from the reported values in monosilane molecules (SiH_4). When the Si atoms and the related parameters are replaced by the Ge atoms and the counterparts, a cluster model of a-Ge:H is reproduced. We calculated the energy levels for each arrangement in bond lengths.

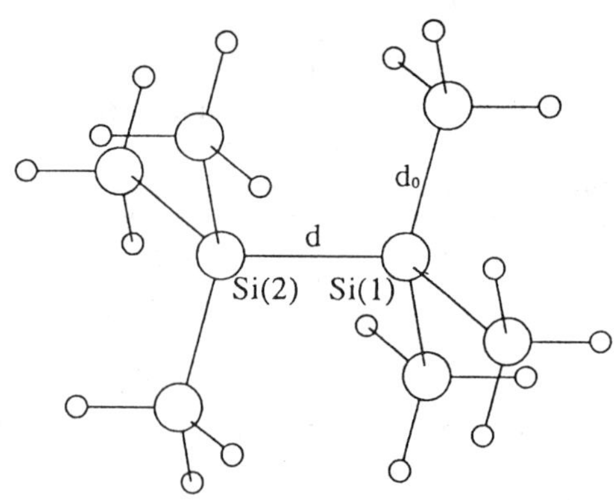

Fig.1. The 26-atom cluster Si_8H_{18} for calculating the gap states induced by the weak bond. (Large circles:Si, small circles:H)

CH3365-4/94/0000-0524 $4.00 © 1994 IEEE

First WCPEC; Dec. 5-9, 1994; Hawaii

In order to investigate the gap states induced by the movement of bonded hydrogen atom from the normal position, one hydrogen atom bonded to Si atom is fixed at the crossing point in the mesh shown in Fig.2. The motion of the H atom is expressed on each mesh point denoted as c0, b0, ..., -b2, -c2. A series of mesh points from c0 to -c0 are called LINE0, as shown in Fig.2. We calculated the electronic states for each position of the H atom.

These calculations were carried out both at the Computer Center of the Institute for Molecular Science and of Toyohashi University of Technology.

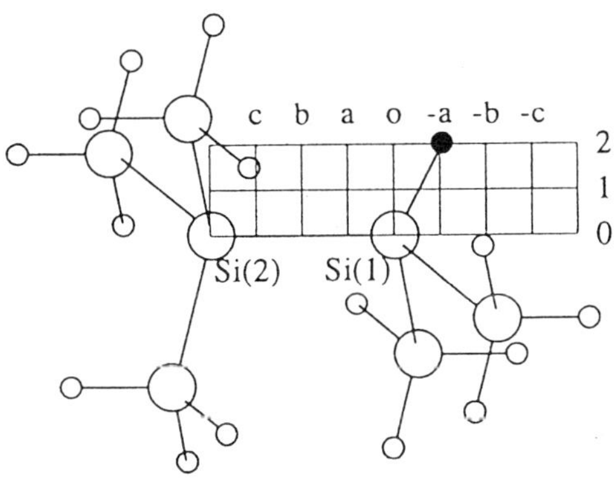

Fig.2. The 23-atom cluster $Si_7H_1H_{15}$ for calculating the gap states induced by the hydrogen movement. (Large circles:Si, small circles:H. Black circle:moved H)

RESULTS AND DISCUSSION

Fig.3 shows the calculated results of weak bond states for bond length ($\Delta d=d-d_0$) equal to 0, 0.2, 0.5, 1, 2, 4, 6, 8 and 10[a.u.], respectively. The two states appear just under and over the band edge and are contracted to shift to the gap center, as the weak bond is stretched.

Fig.4 shows the wavefunction contour plot of weak bond states at Δd=0.5[a.u.] and Δd=10[a.u.]. Fig.4(a) shows that the weak bond states consist of Ge(1)4p-Ge(2)4p bonding and anti-bonding orbitals, respectively. Fig.4(b) shows that the two states are related to the isolated dangling bond states.

Similary, we have calculated the weak bond states in a-Si:H (see Fig.5). The calculated results of a-Si:H are found to be essentially similar to those of a-Ge:H. These states are distributed near both sides of band gap and are contracted to shift to the gap center, while the weak bond is stretched. Two states are related to the isolated dangling bond states at Δd=10[a.u.].

However, the weak bond is not always so suitable for the explanation of SW effect, because only the weak bond states are hard to exist and the two kinds of weak bond must be induced. It is well known that the bonding states with H atom are crucial in the hydrogenated amorphous solids, and that hydrogen diffusion modifies the hydrogen bonding configuration, yielding a metastable state responsible for the degradation.

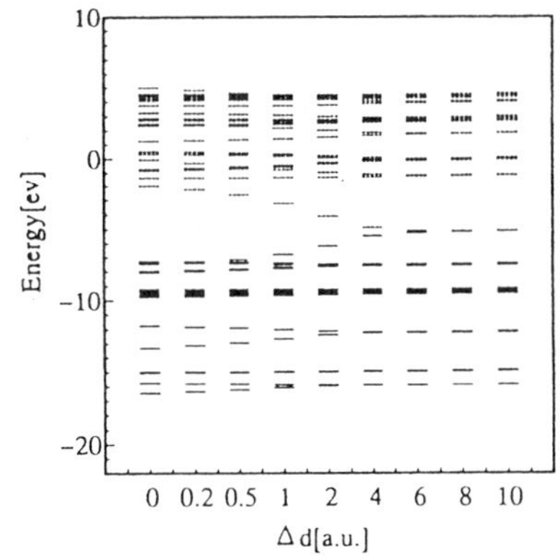

Fig.3. Energy level diagrams of weak bond in the cluster Ge_8H_{18}.

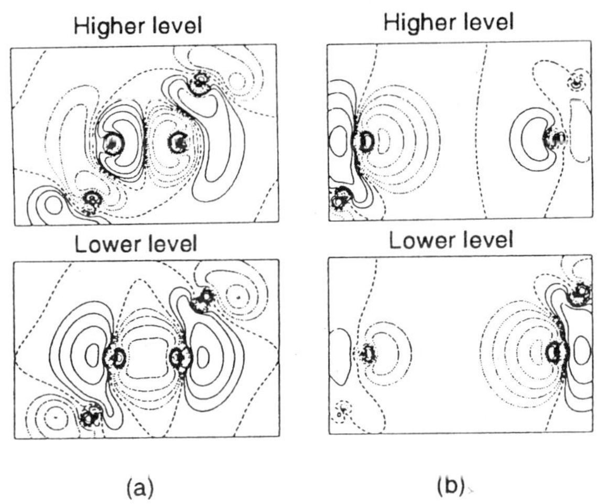

Fig.4. The wavefunction contour plot of weak bond states in the cluster Ge_8H_{18} at (a) Δd=0.5[a.u.] and (b) Δd=10[a.u.].

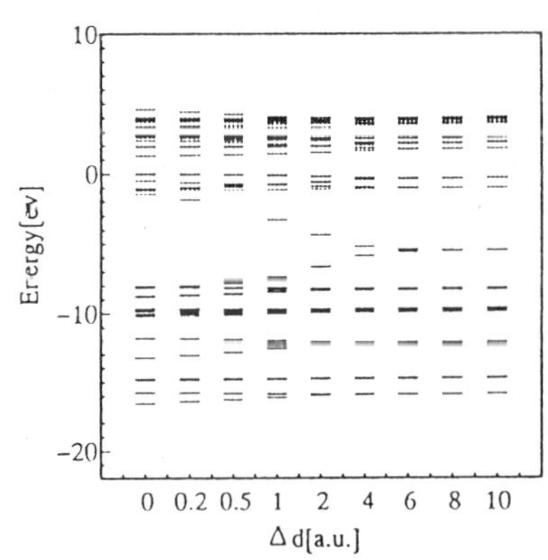

Fig.5. Energy level diagrams of weak bond in the cluster Si_8H_{18}.

Fig.6 shows the gap states induced by the movement of the position of H along LINE2. If the position of H is o2 or -a2, the gap states disappear because H1s orbital are terminated by the dangling bond of Ge(1). However, two gap states shift to the center of band gap by bringing the H atom far from the Ge(1). The variation of shift of gap states from o2 to c2 is larger than from o2 to -c2.

Fig.7 shows the wavefunction contour plot of gap states when the position of H is c2 and -c2. The lower gap state and higher gap state consist of Ge(1)4p-H1s bonding and anti-bonding states, respectively.

Fig.8 shows the overlap population between the Ge(1), Ge(2) and H atoms, based on the population analysis. If the position of H is close to Ge(1), Ge(1)-H bond is a bonding state, and Ge(2)-H bond is a nonbonding state or weak anti-bonding state. However, in the position of H is close to Ge(2), Ge(1)-H bond changes to the weak bond, and Ge(2)-H bond changes to the tight bond. If the position of H is c2, Ge(2)-H bond is stronger than Ge(1)-H bond. These behaviors of Ge (1)-H and Ge(2)-H bond correspond to the variation of the shift of gap states. Therefore, the bond transfer of H from Ge(1) to Ge(2) results in a non-termination of the dangling bond of Ge(1).

The calculated results due to the movement of the position of H atom along LINE1 are similar to the results of LINE2. However, the variation of the shift of the gap states along LINE1 is smaller than that along LINE2. As the H atom is near to Ge atoms, the interaction between them is large, and it is necessary to relax the structure with the movement of H atom. Because of the strong interaction on the LINE0, it is more necessary to relax the structure.

Similarly, electronic states of a-Si:H were calculated for a series of positions of H in the $Si_7H_1H_{15}$ cluster. The calculated results of a-Si:H are found to be essentially similar to those of a-Ge:H.

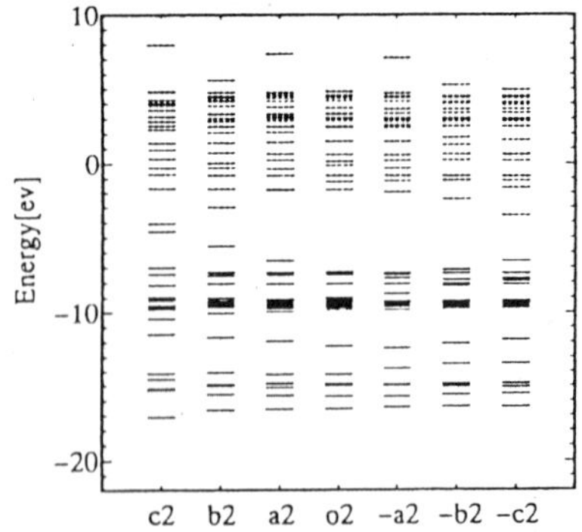

Fig.6. Energy level diagrams in the cluster $Ge_7H_1H_{15}$, when the position of H is along LINE2.

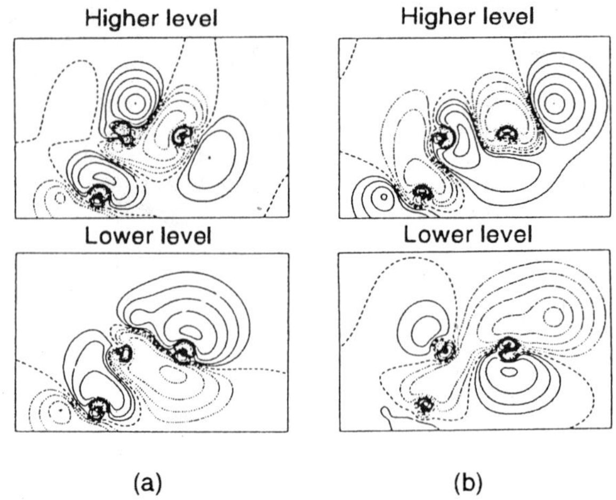

Fig.7 The wavefunction contour plot of gap states in the cluster $Ge_7H_1H_{15}$ when the position of H is (a)c2 and (b)-c2.

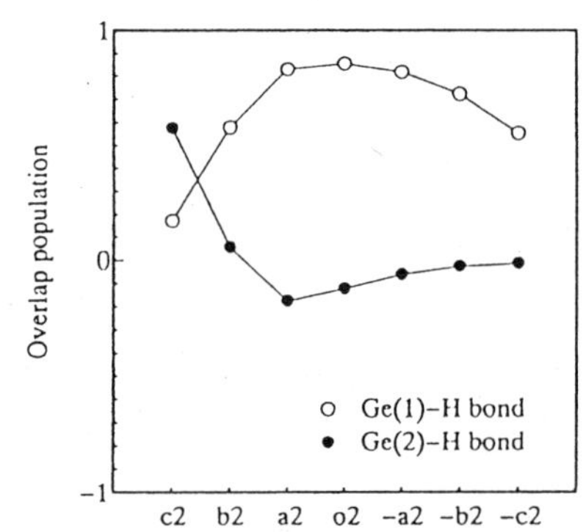

Fig.8 The overlap population between the Ge(1), Ge(2) and H atoms in the cluster $Ge_7H_1H_{15}$, based on the population analysis.

In the estimation of the gap states due to the hydrogen movement, the gap states are found to be localized around the three-coordinated Ge or Si. The termination of the dangling bond of Ge or Si becomes poor by bringing the H atom close to the neighboring Ge or Si. It should be noted that it is not important for these neighboring atoms to have dangling bonds or weak bonds. This result shows that the deep gap states are caused by the comparatively short movement. Conversely, those H atoms close to the dangling bond or weak bond are able to terminate these gap states.

We have tried to explain the decrease in the dark conductivity and the photoconductivity under the prolonged exposure to light. The fermi level is changed by localized states in the gap. As the gap state density increases, the fermi level shifts to the gap center, and the carrier density decreases, giving the decrease of the dark conductivity. The conductivity is also changed because of the decrease of mobility and lifetime of electrons. In the microscopic mechanism, these H motions easily create the gap states, and the lifetime is decreased.

In order to discuss the more detailed model, we have to calculate the exact total energy and excited states. However, from these results, we conclude that the additional gap states are introduced through the H atom motion, and that the energy is dependent on the H atom position. If the bonded H atom migration is light-enhanced, deep gap states are created, suggesting a possible explanation of light-induced degradation.

SUMMARY

The SW effect is believed to the generated defects associated with the weak bond and the H atom movement. We have found the gap states induced by these defects in a-Si:H and a-Ge:H, and investigated the electronic structure of the states on the basis of the DV-Xα molecular orbital method. These results suggest a possible explanation for the SW effect in hydrogenated amorphous solids.

REFERENCES

[1] D.L.Staebler and C.R.Wronski, "Reversible conductivity changes in discharge-produced amorphous Si", *Appl. Phys. Lett.*, **31**, 1977, pp. 292-294.

[2] R.Zhang, "Origin and nature of gap states in a-Si:H alloys", *Solid State Commun.*, **69**, 1989, pp. 681-684.

[3] R.Biswas, I.kwon, and C.M.Soukoulis, "Mechanism for the Staebler-Wronski effect in a-Si:H", *Phys. Rev.*, **B44**, 1991, pp. 3403-3406.

[4] R.Jones and G.M.S.Lister, "Ab initio calculations on metastable defects in a-Si:H : the Staebler-Wronski effect", *Philos. Mag.*, **B61**, 1990, pp. 881-894.

[5] P.V.Santos, N.M.Johnson, and R.A.Street, "Light-enhanced hydrogen motion in a-Si:H", *Phys. Rev. Lett.*, **67**, 1991, pp. 2686-2689.

OUTDOOR PERFORMANCE STABILITY AND CONTROLLED LIGHT-SOAK TESTING OF AMORPHOUS SILICON MULTIJUNCTION MODULES AT NREL

Laxmi Mrig, Joseph Burdick, Werner Luft, and Benjamin Kroposki
National Renewable Energy Laboratory
Golden, Colorado 80401 U.S.A.

ABSTRACT

The National Renewable Energy Laboratory (NREL) has been testing amorphous silicon (a-Si) Photovoltaic (PV) modules for more than a decade. NREL has been conducting controlled light-soak testing of multijunction a-Si modules to characterize their performance for stability evaluation as well as to benchmark the technology status. Some of the test modules, after controlled light-soak testing, have been installed outdoors. We have observed that under outdoor exposure, the modules further degrade in performance, possibly due to lower outdoor temperatures and varying spectra. The paper presents data on the light-induced degradation for the third controlled light-soak test on multijunction a-Si modules as well as outdoor performance data on single- and multijunction modules under prevailing conditions.

INTRODUCTION

The National Renewable Energy Laboratory (NREL) has been testing amorphous silicon (a-Si) photovoltaic (PV) modules since 1984, when the modules first became commercially available. Over the last few years, the technology focus, to a large extent, has been shifting from single-junction cell structure to multijunction configuration for reasons of higher efficiencies and improved stability. NREL has been conducting both outdoor and indoor controlled light-soak testing of multijunction a-Si modules to characterize their performance for stability evaluation and to benchmark the status of the technology. The controlled testing has been conducted in an environmental chamber with a fixed module temperature of 50°C and an irradiance of 1 sun, for periods ranging from 600 h to 2000 h under fixed maximum-power resistive loads. Two series of controlled light soak tests have been conducted on vintage 1990 and 1991 multijunction a-Si modules.[1],[2] After controlled light-soak testing, some of the test modules were installed outdoors at NREL's PV Outdoor Test Site to study their outdoor performance. We have observed that under outdoor exposure, the modules degrade further in performance, possibly due to lower outdoor temperatures and varying spectra.[3] A third series of controlled light-soak testing at NREL was initiated in 1994 and encompasses test specimens from major multijunction a-Si module manufacturers. In parallel with this controlled indoor light-soak testing, some of the modules were deployed outdoors. The controlled light-soak test was performed, and 250 h data, in addition to one year of outdoor performance data, have been collected.

This paper presents the above-described long-term data on the outdoor performance of commercially procured single- and multijunction modules under prevailing conditions, as well as data on the performance of multijunction a-Si modules under outdoor exposure after the modules had been light-soaked under controlled conditions for a period of 2000 h.

COMMERCIALLY PROCURED SINGLE- AND DUAL-JUNCTION a-Si MODULES

Single-Junction a-Si Modules:

Figure 1 illustrates the performance of two single-junction, large-area, power modules by a U.S. manufacturer. It shows the normalized (to 1000 W/m^2) outdoor performance of the modules after an outdoor exposure of approximately 2 years. The measurements were taken under prevailing outdoor conditions. The modules experienced approximately 25%-30% degradation in normalized power during the first 5 months of exposure. The data show that the modules have "stabilized" around 48-W normalized power. No temperature corrections have been applied to the data shown in the figure. Approximately 10% seasonal performance variation has been observed. Whether this seasonal variation is due to temperature or spectral effects is still being investigated.

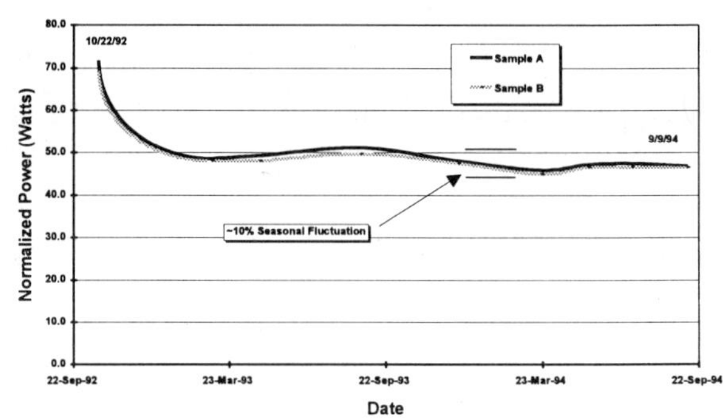

Figure 1. Single-junction a-Si performance

Dual-Junction a-Si Modules:

Figure 2 illustrates the performance of three commercially procured dual-junction a-Si modules under long-term outdoor exposure, since September 1991, at the NREL PV Outdoor Test Site. Current-Voltage (I-V) curve data were taken once every day around noon, when the irradiance was above 750 W/m^2. The power output is normalized to 1000 W/m^2; however, no temperature corrections are applied. The modules experienced an initial light-induced degradation of approximately 20%-25% in normalized power during the first few months of exposure. Since then, the modules appear to have stabilized in output, with seasonal performance variations of approximately 10%.

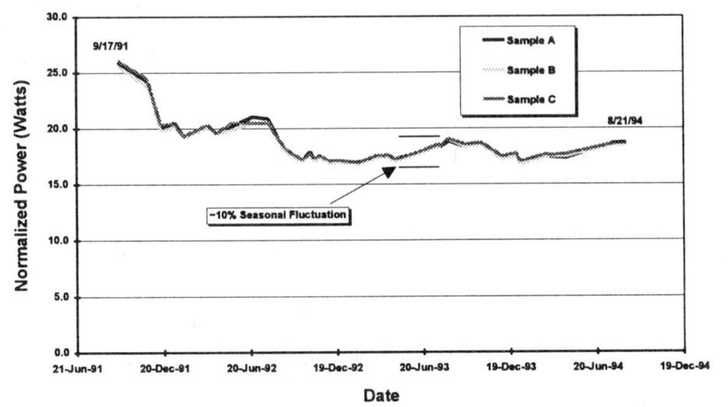

Figure 2. Dual-junction a-Si performance

LIGHT-SOAK TESTING OF PROTOTYPE a-Si MODULES:

The results of controlled light-soak tests #1 and #2 have previously been reported.[1,2] A few of the test modules from these experiments were then deployed outdoors for exposure testing at the NREL Outdoor Test Site. These modules are monitored periodically for stability performance characterization. The following paragraphs describe their outdoor exposure performance.

Outdoor Exposure of a-Si Modules after Light-soak Test #1:

Figure 3 illustrates the performance of one triple-junction and two dual-junction a-Si modules selected from the batch of modules in controlled light-soak test #1, in which the modules were light-soaked for a period of 2000 h. In NREL's light-soaking equipment, an environmental chamber with a filtered argon-arc light source simulating the solar AM1.5 spectrum (see Fig. 4) is utilized.[1],[2] The temperature uniformity of the test modules has been measured to be ±4°C, and the irradiance uniformity is within ±5%.

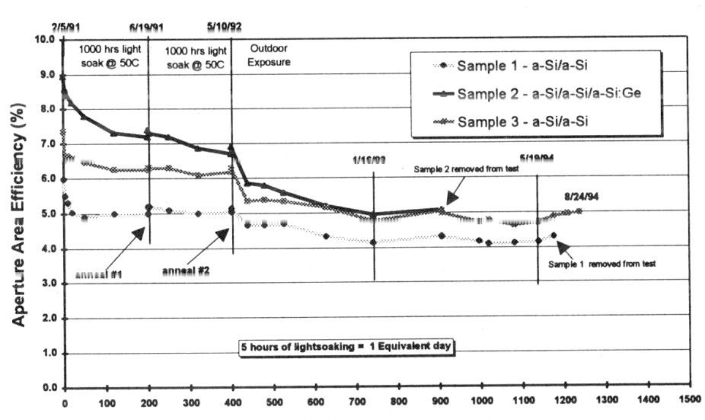

Figure 3. Performance of Light Soak #1 a-Si modules

These three modules were then deployed outdoors under maximum-power load resistors. The graph indicates a decrease in performance of approximately 10%-15% of initial efficiency during an outdoor exposure period of approximately 2 yr after the controlled light-soak exposure of 2000 h. The data highlight the differences in stabilized performance between the indoor and outdoor light-soak testing of multijunction a-Si modules. It indicates a potential need to revise the present indoor light-soak criteria of 1000 h of constant 1 sun illumination at the module operating temperature of 50°C for determining a-Si module stability.

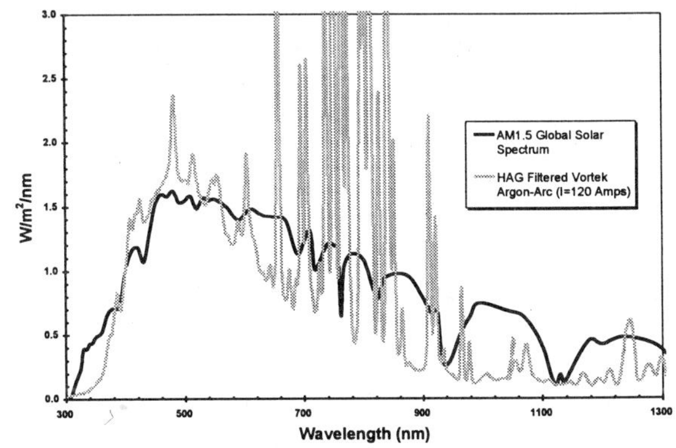

Figure 4. Comparison of Vortek argon-arc light source to ASTM standard solar Spectrum at AM1.5 global

Outdoor Exposure of a-Si Modules after Light-soak Test #2:

Figure 5 illustrates a similar difference in the indoor vs. outdoor stability performance for a dual-junction a-Si module selected from the batch of modules that completed the controlled light-soak test #2. This module, after completing 2000 h of controlled light-soak testing, was deployed outdoors at the NREL test site to determine the outdoor performance. Although the module had stabilized in performance under controlled light-soak test indoors, it experienced approximately 10% additional degradation outdoors. Though the module, whose performance is illustrated in Fig. 5, is of newer vintage than the modules used in light-soak test #1, the indoor vs. outdoor degradation differences are similar to the ones observed on the modules from light-soak test #1.

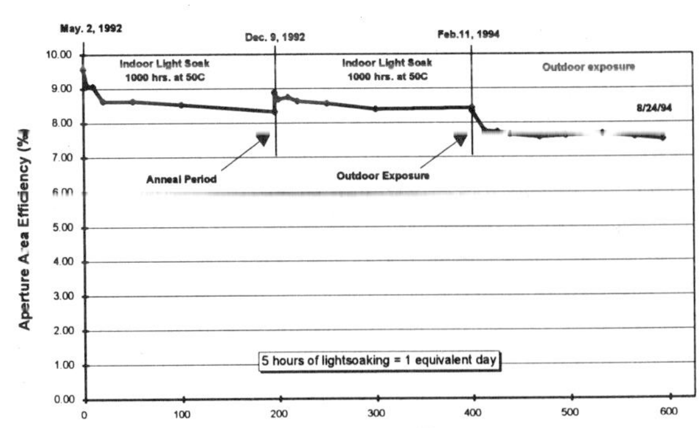

Figure 5. Performance of a Light Soak #2 a-Si module

Light-Soak Test #3 Results:

The controlled light-soak test #3 has completed 250 h of indoor light soaking. Several of the module types undergoing light-soak test #3 have also been deployed outdoors and have completed 1 year of outdoor testing. The multijunction a-Si modules that are part of this test are from several U.S. and foreign a-Si manufacturers. The comparison of indoor vs. outdoor stability performance of five selected modules from one manufacturer is presented in Fig. 6. These data also indicate higher outdoor light-induced degradation as compared to the controlled indoor light-soak test results similar to that observed from modules of light-soak tests #1 and #2.

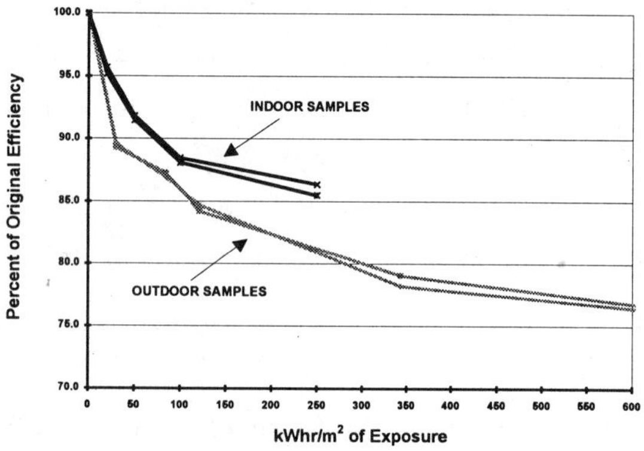

Figure 6. Indoor and outdoor performance comparison of triple-junction a-Si modules from Light Soak #3

CONCLUSIONS

The a-Si modules that were commercially procured a few years ago and have been under long-term outdoor testing at NREL have degraded approximately 25%-30% prior to "stabilization" outdoors. Most of the degradation observed is due to the initial light-induced degradation that occurs during the first few hundred kWh/m^2 of exposure. The modules show an approximately 10% seasonal power variation in Golden, Colorado. This may be due to temperature variations, spectral influences, or a combination of both.

Another issue that needs further investigation and resolution is that additional degradation is observed in the stabilized performance of a-Si modules when they are deployed outdoors (in Golden, Colorado) after having undergone controlled light-soak testing indoors at 50°C for 1000 h at 1-sun constant illumination. This could be due to light exposure at lower temperatures experienced outdoors or due to the cyclic nature of the outdoor exposure (in terms of both illumination and temperature). Several options of controlled light-soak testing are planned to reduce the differences in degradation between indoor and outdoor stability characterization of a-Si PV modules.

REFERENCES

1. W. Luft, B. von Roedern, B. Stafford, D. Waddington, and L. Mrig, "Controlled Light-Soaking Experiment for Amorphous Silicon Modules," 22nd IEEE Photovoltaic Specialists Conference, 1991, p. 1393-1398.

2. W. Luft, B. von Roedern, B. Stafford, and L. Mrig, "Second Controlled Light-Soaking Experiment for Amorphous Silicon Modules," 23rd IEEE Photovoltaic Specialists Conference, p. 860-866.

3. W. Luft and B. von Roedern, "Considerations for Light-Soaking of Amorphous Silicon Photovoltaic Modules," 1st World Conference on Photovoltaic Energy Conversion, (in press).

ACKNOWLEDGEMENTS

The authors gratefully acknowledge the contributions of S. Rummel, Y. Caiyem, L. Ottoson, and J. Pruett in conducting the tests and performing the measurements.

This work was supported by the U.S. Department of Energy under contract No. DE-AC36-83CH10093.

MONITORING OUTDOOR PERFORMANCE AND PHOTODEGRADATION OF a-Si:H MODULES BY EVALUATION OF CONTINUOUSLY MEASURED I-V CURVES

G.Kleiss*, K.Bücher*, P.Ragot**, M.Chantant**
*Fraunhofer Institute for Solar Energy Systems (FhG-ISE) Oltmannsstr. 5, D-79100 Freiburg, Germany
**GENEC, C.E. Cadarache, Bat.351, F-13108 St.-Paul-lez-Durance, France

ABSTRACT

The performance of a-Si:H modules is frequently and continuously monitored during outdoor operation. The simple constant field approximation (3 parameters) is applied to fit the I-V curves. Indoor measurements demonstrate the sensitivity of the proposed method to monitor photo degradation. The evaluation of outdoor measurements proves, that with the proposed method the effect of degradation can be isolated from the effects of temperature and irradiance dependent outdoor behavior of the a-Si:H module. The proposed method relates the overall defect density in the module with its outdoor performance and degradation state.

INTRODUCTION

When regarding light-induced Staebler-Wronski degradation of a-Si:H devices, usually two approaches can be found in literature: first, laboratory studies have been frequently performed for a-Si:H cells and films, showing the performance of e.g. the defect density or the cell's fill factor after 1000 hours illumination [1],[2],[3],[4]. Despite the fact that these steady-state investigations (constant temperature and irradiation during the experiment) are necessary for the aim of understanding the physical processes involved, one has to keep in mind that outdoor degradation of a-Si:H modules has to be treated dynamically. Degradation state and degradation speed depend on several quantities that are continuously changing under outdoor operation such as irradiation, temperature, number of already existing defects, etc. In this respect it is worth to note, that the commonly used stability criterion, namely degradation state after 1000 hours illumination (as a typical lab experiment), is not necessarily a good figure of merit for the outdoor stability of a-Si:H modules, since the capability of annealing is not taken into account. Equivalently, it has been found [5] that a total saturation behavior on a logarithmic time scale of the saturated defect density [6] (as observed in laboratory) will not occur during outdoor operation of a-Si:H modules. Another major problem to apply the results of this quoted type of investigations is, that the degradation experiments are often performed under concentrated light, to accelerate the degradation speed. Outdoors, however, PV modules operate always at concentration ratios smaller than unity and the extrapolation from data acquired at concentration ratios between 10 and 100 delivers not necessarily the best data. Second, there have been many long-term field experiments performed, giving examples of the changes in the photovoltaic performance (such as fill factor, open circuit voltage) after a certain time interval (typically in the range of years) of outdoor light exposure [7],[8]. These investigations have the clear advantage of exhibiting typical realistic and product specific results. However, the information is site-dependent as well as subject to the period of investigation and manner of installation. Thus, these data will be hard to be transformed to different situations with respect to installation [18] and meteorology in order to predict the time evolution. The aim of this study is to fill the gap between the two existing approaches. We perform continuous long-term outdoor measurements (thus acquiring data under realistic operation conditions) of a-Si:H modules. By studying the frequently acquired data we try to separate the module behavior under realistic reporting conditions [5],[9] (I_{SC}, V_{OC} FF and P_{MAX} of any semiconductor solar cell are dependent on temperature, irradiation etc.) from degradation phenomena. Thus, on the basis of the applied straightforward model, a tool is developed for investigating the dynamics of degradation of realistically operating modules whereas many models describe the dynamics of the defect density in a-Si:H films [10], [11])

METHOD

Theoretical Approach

The exact theoretical modeling of an a-Si:H cell's I-V curve requires the full simultaneous solution of the fundamental semiconductor equations (continuity equation, current density equation and Poisson equation) and a distinct knowledge of defect density profiles and cell structure. However, since there are plenty of free parameters and the required computational effort is immense [12], this method is not applicable to fit the I-V curves continuously acquired outdoors. Further, the Shockley equation has often been used to fit the I-V characteristics of a-Si:H modules. However one has to be careful when applying

CH3365-4/94/0000-0531 $4.00 © 1994 IEEE

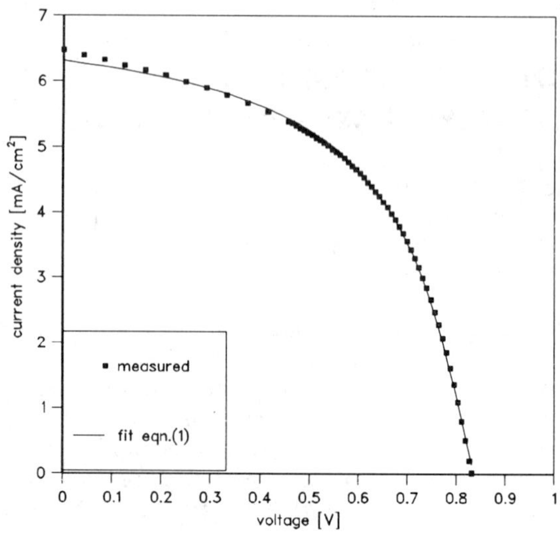

Fig. 1: Example of a fit of eqn. (1) to an I-V curve of the a-Si:H module studied in the indoor solar test.

this equation derived for crystalline devices to amorphous silicon solar cells, dominated by a drift type photo effect. Reports on the applicability of the Shockley equation exhibit contradictory results [13], [14]. In order to both reduce the number of fit parameters (thus benefiting with respect to the computational effort) and to have a straightforward physical interpretation, we applied a 3 parameter equation proposed by Crandall as uniform field approximation [15]. For our purposes, the I-V curve may be expressed as:

$$I(V) = I_0 D (V_0 - V) [1 - \exp(-1/D(V_0 - V))] \qquad (1)$$

We are aware of the fact, that eqn. (1) yields only a rough fit to the I-V curve as can be seen in Fig. 1. However, a major advantage of (1) lies in the straightforward interpretation of the fitted parameters with respect to degradation phenomena: I_0 being a saturation photocurrent is proportional to the module's short circuit current, V_0 is the built-in voltage and is proportional in this picture to the open circuit voltage of the cell. D is a parameter corresponding to the overall defect density N_R according to [15]:

$$D \propto 1/N_R \qquad (2)$$

Thus, in this simple picture photo degradation is expected to affect mainly D, reducing the cell's fill factor whereas V_{OC} and I_{SC} should only be slightly changed. This behavior has been repeatedly reported for cells fabricated without p/i interface layer [16],[17]. However, if V_{OC} is enhanced by the p/i interface layer, one encounters degradation effects in this cell region and so finally a considerable reduction of V_{OC}. However the modules we studied did not show this behavior.

Experimental and Numerical Approach

Indoor results to be presented here have been acquired for an amorphous silicon sub-module (15 cells in series, total device area 200*200 mm), which has been at a pre-degraded state. Outdoor results have been obtained for commercial a-Si:H modules of the same type (*pin* technology) exposed during a period of 3 months at two different geographical locations: Freiburg (Germany, lat. 48N) and Cadarache (France, lat. 44N). The modules have a peak power of about 9Wp and had not been exposed to light before the start of the experiments The frequency of module measurement in the outdoor experiment is one measurement per 12 minutes (Cadarache) and 4 minutes (Freiburg), resp. During two I-V acquisitions the modules have been operated at open circuit. This is expected to lead to a degradation performance similar to MPP operation [3]. The three parameters (I_0,D,V_0) of eqn. (1) have been extracted by a least square fit of (1) to the measured I-V data using a simplex algorithm. The data have been normalized (voltages divided by the number of cells in series, currents divided by the cell area) to obtain the corresponding data for a single cell with area 1cm². Since according to the theory (1) will give only a poor description of the cell's I-V curve in the neighborhood of the V_{OC} point, we checked the stability of the fit performing several fits omitting different numbers of points in this region. All I-V curves have been treated equally (i.e. fitted with the same starting values). The large number of I-V-curves to evaluate the outdoor experiments (more than 10000 per module) explains the fact, that the fitted parameters of the outdoor experiment appear somewhat more noisy than they would be, if every single fit could be attended with maximum effort.

RESULTS

Indoor Results

We tested the degradation behavior on a short time regime (5 hours) under STC irradiation (1000 W/m²) and STC spectrum (AM1.5), while the temperature was stabilized at 40°C. This investigation serves as a first test for the proposed theoretical concept, as described in the preceding section. Fig. 2 shows the extracted results obtained by the indoor photodegradation experiment with the pre-degraded a-Si:H module. As can be clearly seen in this indoor experiment, the change in parameters corresponds to the expected behavior. Neither I_0 nor V_0 are vastly affected, but a significant decrease in D is detected, forced by an increase in the defect density N_R (see eqn. (2)). However, since the effects under investigation are small, temperature instabilities might play a major role in this analysis. Thus, one has to take care that the system is in thermal equilibrium before starting the experiment.

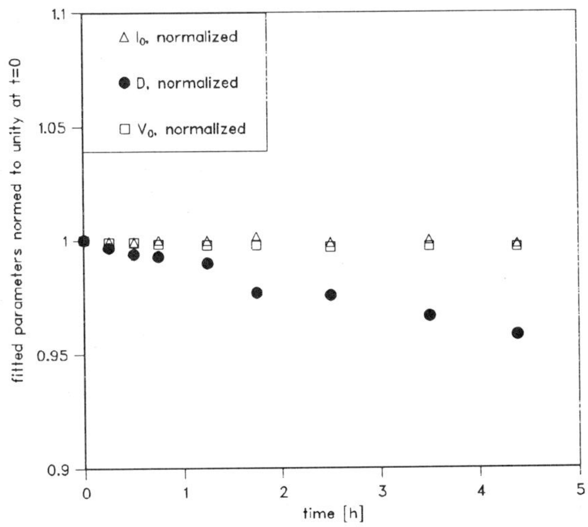

Fig. 2: normalized time evolution of the fitted parameters I_0, V_0 and D due to photo degradation for the a-Si:H sub-module tested in the laboratory.

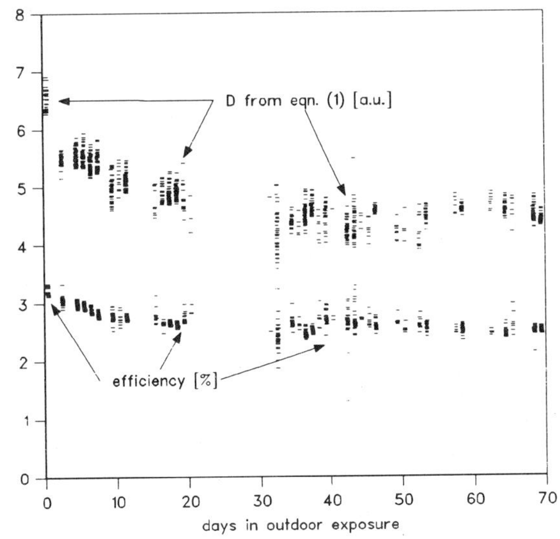

Fig. 3: extracted parameter D (upper part of the graph) and module efficiency η (lower part) during the first 3 months of light exposure at Freiburg test site; see text.

Outdoor Results

Outdoor measurements are subject to continuously changing operating conditions as irradiation, temperature and spectrum: Thus, in this section it will be tested, whether the model provides a possibility to separate the effects of degradation from temperature-, irradiance- and spectrum-dependent performance of the solar cell (the conventional behavior under Realistic Reporting Conditions" [5]). Fig. 3 demonstrates extraction results for the degradation parameter D as well as the efficiency η. Note, that these points are shown as measured and have not been corrected to standard testing conditions. Points shown in this graph have been selected from irradiations larger than G>750W/m² in the plane of array and in a two hour interval round solar noon. Thus we can minimize the known large impact of spectral effects on a-Si:H devices. The noisy structure of the signal is due to the influence of module temperature, spectrum and level of the irradiation. A clear trend of the degradation phenomenon can be seen analogous to the indoor results presented in the preceding subsection. The change observed in parameter D is in the order of magnitude of 25% and D proves once again to be a sensitive quantity for measuring degradation. The relative changes are even higher than those observed for the loss in efficiency (order of magnitude for change in η: 10%). For all 8 modules investigated the relative loss in D was between 25% and 35%. Further, it has to be verified, whether or not the remaining two parameters (I_0, V_0) change strongly due to degradation. Fig. 4 presents the change in the ratio of the extracted I_0 and the plane of array irradiation as measured by a filtered mono-crystalline silicon cell. Since the spectral response of the filtered irradiance sensor is similar to the spectral response of an a-Si:H module one can minimize spectral effects. However, one should be aware of the fact that the spectral response of an amorphous cell may not be assumed to be independent of the irradiating light as it could be for a crystalline cell. To avoid these problems, we plotted only points for irradiations larger 900 W/m² ensuring spectra not too different for the points measured. A relative change can hardly be detected, we estimate an upper limit for a decrease in I_0 (in other words for I_{SC}) as low as 5% for the modules investigated. The changes of the extracted built-in voltage (that means V_{OC}) are harder to visualize since V_0 depends strongly on the continuously changing module temperature. Fig. 5 depicts the time evolution of V_0 as a function of (if necessary interpolated) module temperature. For a fixed temperature V_0 is decreased only very little during the degradation experiment. The maximum observed relative degradation loss of V_0 is estimated as less than 3%

DISCUSSION AND CONCLUSIONS

Summarizing all single effects, it is clear, that D changes most due to light-induced degradation. Whereas I_0 depends principally on the level of irradiation and V_0 is principally subject to temperature, it has been found, that D is a good monitor for the degradation state and depends only slightly on irradiation and temperature, at least in relatively large intervals (e.g. [750W/m²;1000W/m²],[35°C;65°C]). The observed decrease in D (25...35%) can be interpreted as increase of the overall defect density N_R

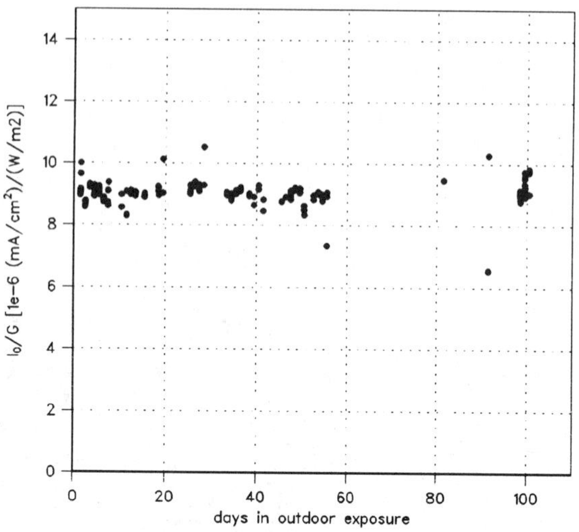

Fig. 4: ratio of extracted photocurrent saturation density I_0 and the filtered plane of array insolation for an a-Si:H module measured at Cadarache (see text).

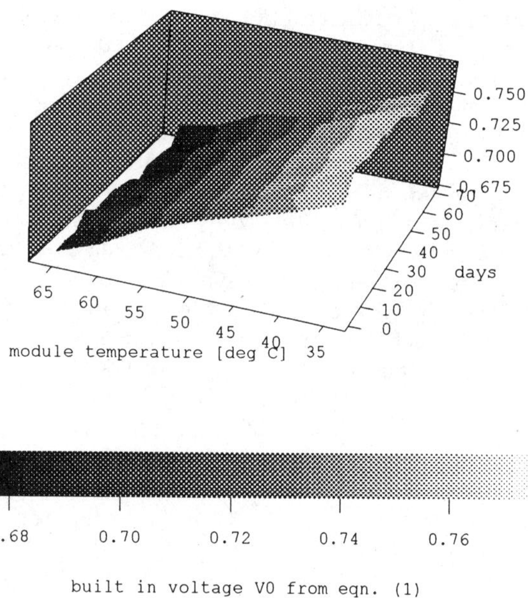

Fig. 5: observed trend in the temperature dependence of the built in voltage V_0 (vertical axis) during 2 months of outdoor degradation at the test site Freiburg. Time intervals are averaged to 5 days, temperature intervals are averaged to 5 K intervals.

We tried to apply known dynamics on the defect density N_R (namely stretched exponentials[4],[11]) to confirm the physical interpretation presented, but since there are several free parameters, one has to wait for a stabilization of the module´s degradation state, to extract the limit for large times. In conclusion, the application of the constant field approximation provides a sensitive tool for monitoring degradation effects in the investigated single junction amorphous silicon PV modules.

ACKNOWLEDGMENTS

We are indebted to H.G.Wagemann from the Technical University Berlin, Germany, for valuable discussion and to K.Reiche for technical assistance. Work reported here is being funded by the German Ministry for Research and Technology BMFT under contract number 0328650E/F. G.Kleiss gratefully mentions financial support from the German Academic Exchange Service DAAD for enabling research at GENEC, C.E.Cadarache, France.

REFERENCES

[1] H.S.Ullal, et al., *17th IEEE PVSC, Kissimmee,* p.359 (1984)

[2] M.S.Bennett, et al., *7th EC-PVSEC, Sevilla*, p.544 (1986)

[3] A.Yokota, et al., *Techn. Digest Intn'l. PVSEC-5,Kyoto*, p.637 (1986)

[4] L.Chen, L.Yang, *Journ.NonCryst.Sol.* **137&138**, p.1185 (1991)

[5] G.Kleiss, K.Bücher, A.Raicu, K.Heidler *11th EC-PVSEC, Montreux*, p. 578 (1992)

[6] H.R.Park, J.Z.Liu, S.Wagner, *Appl.Phys.Lett.*, **55** (25), p.2658 (1989)

[7] J.C.Joshi, P.K.Konar, *10th EC-PVSEC, Lisbon*, p.399 (1991)

[8] P.Ragot, A.Chenevas-Paule, H.S.Costa, et al., *10th EC-PVSEC, Lisbon,* p.403 (1991)

[9] A.Raicu, K.Heidler, G.Kleiss, K.Bücher *22nd IEEE PVSC,* Las Vegas, p.744 (1991)

[10] D.Redfield, *Appl.Phys.Lett.*, **52** (2) p.492 (1988)

[11] M.Stutzmann, W.B.Jackson, C.C.Tsai, *Phys.Rev.B*, **32** p.23 (1985)

[12] M.Hack,M.Shur, *J.Appl.Phys.* **58** (2) p.997 (1985)

[13] A.Mittiga, P.Fiorini, M.Falconieri, E.Evangelisti, *9th EC-PVSEC, Freiburg,* p.971 (1989)

[14] M.C.Alonso, F.Chenlo, *12th EC-PVSEC,* Amsterdam, April 11-15 1994, in print

[15] R.S.Crandall, *J.Appl.Phys.* **54** (12) p.7176 (1983)

[16] H.Sakai, T.Yoshida, S.Fujikake, T.Hama, Y.Ichikawa, *J.Appl.Phys*; **67** (7) p.3494 (1990)

[17] P.Chaudhuri, et. al. , *Techn. Digest Intn'l PVSEC-5, Kyoto*, p. 641, (1990)

[18] G.Kleiss, A.Raicu, K.Bücher,*23th IEEE PVSC, Liousville* p.896 (1993)

NEW TYPE OF LARGE-AREA a-Si MODULE PRODUCED USING A POLYMER ENCAPSULATION METHOD

Katsuhiko Hayashi, Atsuo Ishikawa, Toshihito Endho, and Hideo Yamagishi
Electronic Materials Research Laboratories,Kaneka Corporation
1-1,Hieitsuji 2 Chome,Otsu,Shiga.520-01,Japan

ABSTRACT

A new type of amorphous silicon (a-Si) solar module has been developed, the back surface of which is encapsulated with a thermosetting polymer by using a spreading method. For this purpose, poly-olefin origomer was chosen as an encapsulation layer and the process have been investigated. The modules show very little change in electrical performance under several types of acceleration test.

INTRODUCTION

Several types of module structure have been proposed for single-crystal and polycrystalline silicon solar modules[1]. Among these structures, the super-straight-type is one of the most popular method, and has been applied to other types of solar module, such as those incorporating a-Si. However, this is not always the optimum structure for a-Si solar modules.

In comparison to crystalline silicon (c-Si), a-Si has the advantage that it can be fabricated very easily as a thin, large-area film on various types of substrate. In the case of a-Si solar modules, a glass sheet can be employed not only as the substrate of the solar cell, but also as the cover glass of the solar module. If a large area a-Si solar cell was to be directly fabricated on a tempered cover glass, the back surface of the module would be much smoother than that of the standard c-Si solar module, and it would be possible to encapsulate the module with a thermosetting polymer using a spreading method.

As a liquid encapsulant polymer for our solar module coating process, we have developed a poly-olefin origomer which has PIB as main chain. In this study, we will describe some features and the reliability of modules produced using this polymer as an encapsulating layer.

SOLAR MODULE STRUCTUERES

Fig.1 shows three types of module structure. Recently, super-straight-type module A, which has been developed as a substitute for the substrate-type module, has become the most popular. The back surface of the module is covered with EVA and fluorocarbons polymer films. Vacuum thermal laminators are generally used to adhere of these films in production lines.

In order to reduce the cost of the solar modules, the structure B is now being widely investigated. Since a large area a-Si solar cell can be fabricated on various types of substrate such as glass, stainless steel and

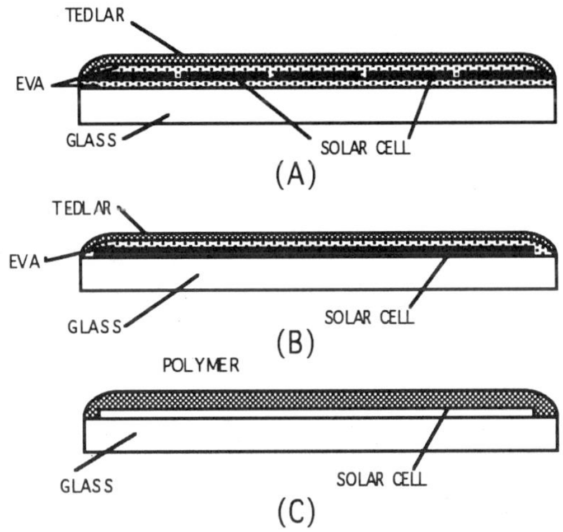

Fig. I Schematic diagrams of three types of module structure.

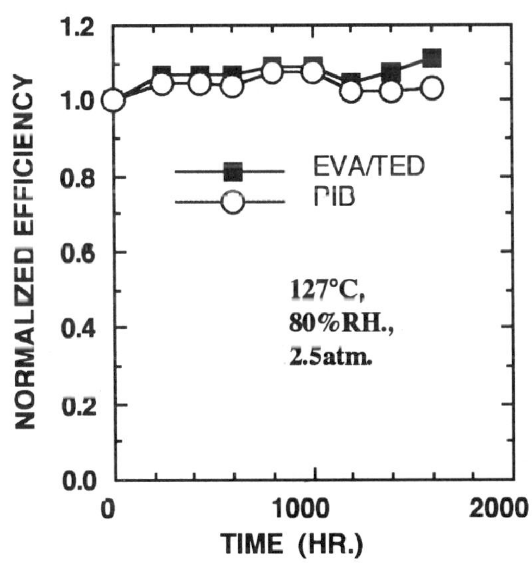

Fig.2 Changes in output power of new and conventional type solar modules by pressure cooker testing.

CH3365-4/94/0000-0535 $4.00 © 1994 IEEE First WCPEC; Dec. 5–9, 1994; Hawaii

several types of heat proof polymer film, it is expected to be one of the most promising low cost solar cells. In structure B, TCO, a-Si and metal layers are directly deposited onto the cover glass. Following this, using a vacuum thermal laminator, the back surface is covered with fluorocarbon polymer film by an EVA film adhesive layer.

If the back surface can be encapsulated by a thermosetting polymer that has high stability, it will be possible to considerably simplify the construction of the a-Si solar module. Type C is a module, where EVA layer is replaced with a thermosetting polymer layer, fabricated by using a spreading method.

The main aim of this work is to achieve a complete cost-reduction, by developing low cost materials, structure and technology which are suitable for large-area amorphous solar modules.

Nine types of poly-olefin origomers synthesized by various methods (Mn: 2500-6500) were chosen and the viscosity, thermosetting time, thermal resistivity, UV resistivity, impurity, mechanical properties, and permeability were measured. Taking the result of pressure cooker test three types of origomers were selected as a candidate of substitute article.

The amounts of plasticizer, adhesive, antioxidant, curing agent, catalyst, and UV stabilizer were optimized and the viscosity was controlled to be convenient to plaster the compound by curtain-coating method (around 1000 PS with small dependence on the shear rate).

ACCELERATION TESTING

Several tens of modules having a size of 127 mm x 152 mm were prepared for the acceleration testing. On the SnO_2 coated flat glass, p-a-SiC:H/i-a-Si:H/n-a-Si:H heterojunction solar cells were fabricated by rf glow discharge decomposition of silane gas or its mixture using a separated chamber system. Typical deposition conditions are described in our previous paper[2]. Separation of SnO_2 and a-Si:H layers was patterned by a laser-scribing method. Aluminum was deposited under vacuum as a back metal contact layer.

After completion of the wiring, the back surfaces of the modules were coated with polyisobutylene that is one of the candidate of encapsulating polymer chosen in this work. The viscosity-controlled polyisobutylene polymer was extruded as a highly viscous fluid sheet on to the solar cells. One advantage of the extrusion method is that it can be applied to the encapsulation of uneven

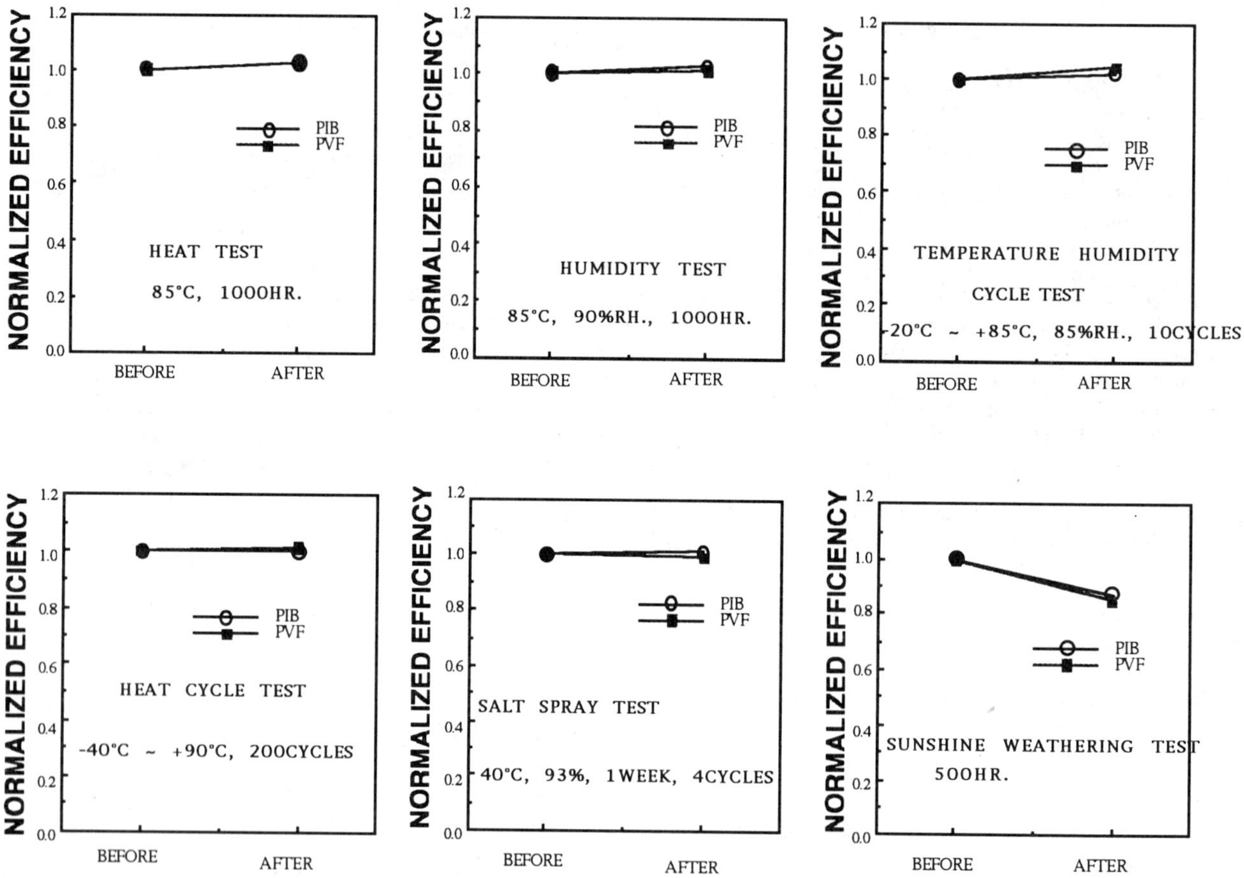

Fig.3 Output power of new and conventional type solar modules measured before and afrer six types of endurance test legislated by the Japanese Industrial Standards Commitee.

surface such as the cell surface on tempered distorted glass. Details of the coating process in this work will be discussed in next section. The thickness of the encapsulation layer was between 0.4 and 1 mm. This value can be easily controlled by changing the extrusion conditions.

The modules were then kept at 130 °C for 1 hour to cure the encapsulation polymer. The temperature and/or time required to cure the polymer can be controlled by modification of the polymer composition. Following encapsulation of the module edges by thermal plastic butyl rubber, aluminum frames were furnished. Finally, terminal boxes were installed using silicone resin.

Pressure cooker testing was applied for the initial acceleration test. The pressure, humidity and temperature during testing were controlled at 2.0 kg/cm^2, 100%, and 120.6 °C, respectively, which is the most common condition used. These conditions are severe for the durability testing of electronic devices, but the influence of oxygen in the air cannot be determined. In addition to the general pressure cooker test, conditions under which the influences of oxygen and water vapor could be checked were also chosen. In this case, the humidity, temperature, and partial pressures of water vapor and air were controlled to be 80 %, 126.8 °C, 2.0 kg/cm^2, and 0.5 kg/cm^2, respectively.

Fig. 2 shows the changes in output power during testing of modules prepared using the processes described above. This figure also includes the results of the reference samples shown in Fig. 1 A. The changes in output power of both types of sample were very small. The samples encapsulated with polyisobutylene showed very small changes not only in output power but also in an outside appearance.

Among many endurance tests legislated by the Japanese Industrial Standards Committee[3], six tests were chosen, and applied to the two types of solar module shown in Fig. 1 A and C. The tests chosen in this work were the heat cycle test, the temperature humidity cycle test, the heat test, the humidity test, the sunshine weathering test, and the salt water spray test. The observed output power before and after testing is compared in Figs. 3 A-F. Similarly to the results obtained by pressure cooker testing, the changes in output power throughout these tests were very small except in the case where the samples were exposed to a carbon arc lamp through the front glass during a sunshine weathering test. In this case, the changes in cell performance are thought to be determined by Staebler-Wronski effects in the a-Si layers.

DEVELOPMENT OF LOW COST ENCAPSULATION PROCESS FOR SOLAR MODULES

We have determined the fundamental module production process consists of a soldering by ceramic solder or metallic paste, wiring by copper ribbon and filling polymer there, encapsulation, lamination of cover film, framing. Fig. 4 shows the outline of this process.

Copper ribbons on the metallic paste are dotted with solders to avoid the pealing by thermal stress. Some types of silver paste showed good bond strength in the results of tensile and shearing strength tests.

As the thickness of the copper ribbon is several hundred micrometers, it becomes an origin of air bubbles when the thermosetting polymer is directly coated on them. In order to avoid the occurrence of bubbles, encapsulating polymer is filled by a dispenser previously. It was confirmed that the polymer can be filled there without air bubbles with a scanning rate of 8 cm/sec.

Although cured poly-olefin compound is rubbery and suitable for encapsulation, it is weak in mechanical stress such as scratch. To protect modules from mechanical stress, glass clothes were laminated as cover films tentatively after encapsulation of poly-olefin origomer.

Encapsulation process applied in the work, consists of premixing of resin, dispersing, degas, coating by curtain coat method, cover film lamination and thermosetting processes. Resin and other additives are preliminary mixed by spiral blade mixer, then the compound is mixed again to get fully dispersed compound material by three roll mill. After dispersion process, air included during mixing process and low boiling point materials such as solvents are excluded after 30 minutes' treatment by vacuum degas machine with spiral blade.

The tank of vacuum degassing machine filled with processed compound instantly attached to the coating

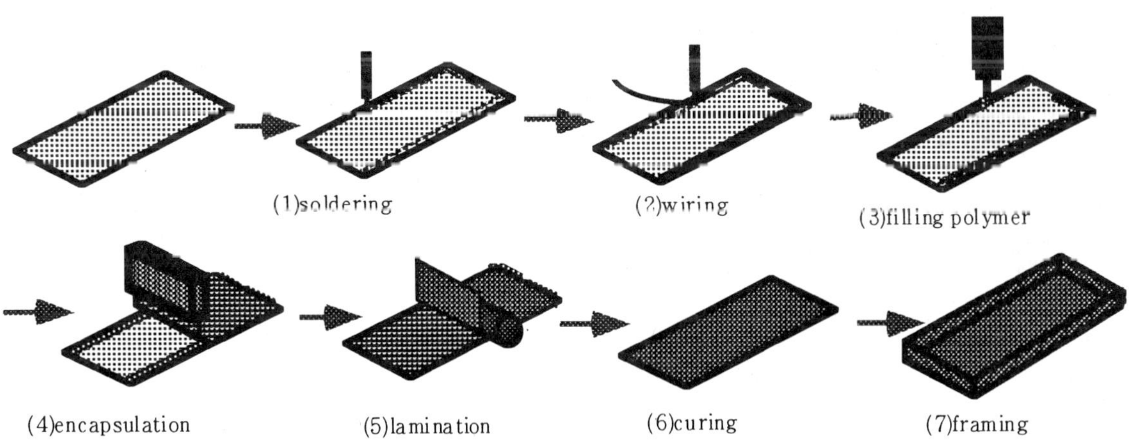

Fig.4 Outline of new type solar module production process.

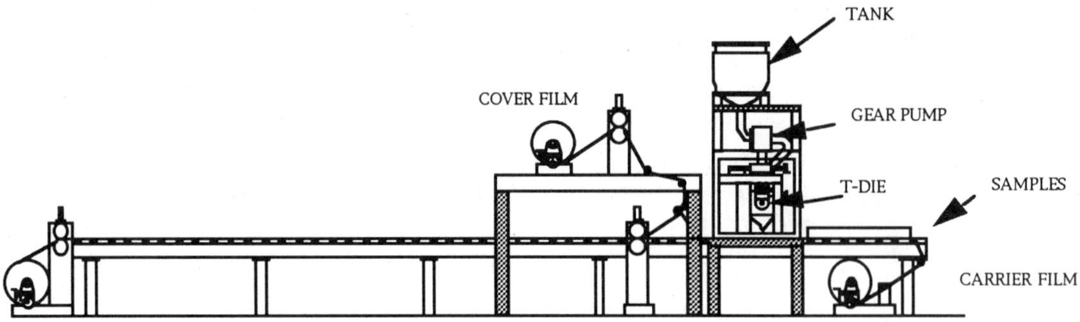

Fig. 5 Schematic diagram of the coating machine.

machine after degas process and used as a server tank of coating machine. Fig. 5 is a schematic diagram of the coating machine. The resin in the server tank transferred to T-die by gear pump. A carrier film transfers solar cells at a constant speed, and they are laminated with cover film after resin coating by T-die.

The design of T-die was optimized to compound resin viscosity, i.e., 500-1000 PS. The coating process parameters such as rotation speed of gear pump, transfer speed of sample, lip clearance of T-die, distance between T-die and sample were controlled to adjust coating speed and coating thickness. The process stability was investigated at each condition. At high speed condition, bubbles tend to be included between solar-cell surface and coating resin. Up to now, the maximum coating width and speed within 5% thickness variance are 30 cm/min and 50 cm, respectively.

OUTDOOR TESTING

From FY1991 we began outdoor testing of these new modules at four different locations. Fig. 6 shows a photograph of the 500 mm x 600 mm modules kept in Choshi city, Chiba Prefecture, Japan. To date, no large changes in output power have been observed during a one-year outdoor exposure except for the changes caused by light-induced degradation of the a-Si solar cells. Concerning the reliability, these new modules passed all the examinations carried out. We will continue these tests in order to confirm the long-term reliability of this new type of module.

Fig.6 Photograph of the new type solar modules in Chiba.

SUMMARY

As a liquid encapsulant polymer for our solar module coating process, we have developed a poly olefin origomer which has PIB as main chain. Viscosity of the resin compound was controlled to as low as 1000 PS. About the same performance as conventional one' s was confirmed with our new compound after the first screening tests on impurity density, hardening behaviors, weathering stability and encapsulation performance (electrical change of encapsulated materials after pressure cooker test). We have also finished the first step of developing new coating machine, by deciding coating process of new curtain coat method and getting a condition which achieves 30 cm/min speed with 50 cm coating width.

For the time being, we will investigate the topics shown below.

1) To decide compound recipe optimized for higher process speed and product reliability reflecting our experience of module fabrication

2) To select low cost cover film witch guarantees long range stability. To investigate the encapsulation method which suppresses back side metal peeling, migration, Si/backside metal inter-layer corrosion regarding utilization of high reflective electrode such as ZnO/Ag.

3) To confirm stability of the module made by mass productive process by accelerated weathering tests and out door field test.

By investing these topics, we will promote the evaluation of new module in respect of long range reliability, productivity and cost effectiveness.

ACKNOWLEDGMENTS

This work was partially supported by the New Energy and Industrial Technology Development Organization as part of the New Sunshine Program. The authors would like to thank members concerned.

REFERENCES

[1] D.C.Carmichael et.al., *Proc. 12nd IEEE Photovoltaic Specialist Conf.*, (1976),317.

[2] H. Yamagishi et.al., *Proc. 4th International Photovoltaic Sci.&Eng. Conf.*, (1989),415.

[3] "*JIS Handbook for Electronic Devices*", (1992), published by the Japanese Industrial Standard Committee.

BACK REFLECTOR TEXTURE AND STABILITY ISSUES IN HIGH EFFICIENCY MULTIJUNCTION AMORPHOUS SILICON ALLOY SOLAR CELLS

A. Banerjee, K. Hoffman, X. Xu, J. Yang, and S. Guha
United Solar Systems Corp., 1100 W. Maple Road, Troy, MI 48084, U.S.A.

ABSTRACT

The role of the back reflector texture on the initial and stabilized efficiency of high efficiency triple-junction triple-bandgap amorphous silicon alloy based cells has been investigated. The devices have been deposited on Ag/ZnO back reflector possessing three different textures. The performance of the bottom single-junction a-SiGe alloy *n i p* cell on the three textures has also been analyzed. The value of the short-circuit current density of the *n i p* cell initially increases and then decreases with increasing texture. Light soaking results show that the higher textures exhibit superior device stability. Degradation as low as 8% has been obtained on the triple-junction cells.

INTRODUCTION

The highest values of initial cell and module efficiencies using amorphous silicon alloy technology are 13.7% [1] and 11.8% [2], respectively. One critical component that has led to the success of these triple junction multi-bandgap structures is the high quality textured Ag/ZnO back reflector [3] deposited on stainless steel (ss) substrate. The back reflector helps in reducing the required thickness of the intrinsic layers especially for the narrow bandgap a-SiGe bottom cell since it is located adjacent to the back reflector. The use of thin intrinsic layers leads to superior light stability of the multijunction device.

The effect of the back reflector is manifested in the quantum efficiency, Q, versus wavelength curve of the cell. Specifically, the value of the integrated Q in the red part of the spectrum (such as Q ≥ 600 nm and Q ≥ 630 nm) are a direct measure of the quality of the back reflector. The Q and reflection, R, curves of the completed cell show interference effects in the higher wavelength region. This effect is due to interference between the light reflected from the top and bottom surfaces of the cell. The nature and extent of the modulations between the maxima and the minima contain information about the specularity and parasitic losses in the back reflector. Higher textures are generally characterized by smooth Q and R curves.

It has been reported [4] that the extent of light-induced degradation [5] also depends on back reflector texture. In the referenced work, it was shown that a back reflector exhibiting Lambertian scattering characteristics resulted in reduced degradation. Though the initial characteristics of the device on the Lambertian back reflector were lower, the stabilized values were higher. Practical high efficiency cells are not usually made on the highly textured Lambertian back reflectors. Based on the above, the purpose of this work is twofold: (1) to study the effect of the texture of back reflectors on the short-circuit current density, J_{sc}, and (2) to determine if the light-induced degradation is dependent on the substrate texture.

EXPERIMENTAL

Ag/ZnO back reflectors of three different textures, A, B, and C, were fabricated on ss substrates by sputter deposition process. Texture A had the least texture, texture C had the highest, and B had an intermediate one. The total reflectance as a function of wavelength of the back reflectors was measured by using an integrating sphere. The texture (or specularity) of the substrate was calculated by measuring the fraction of light reflected at the angle of incidence compared to the total reflected light integrated over all angles. Small values imply high textures. The surface morphology or feature size was determined by using scanning electron microscopy (SEM).

The triple-junction triple-bandgap cells and the corresponding component *n i p* a-SiGe bottom cells were fabricated on the three back reflectors A, B, and C using conventional glow discharge technique. The glow discharge machine and the deposition parameters of all the devices were similar to the corresponding ones used to obtain the world-record 10.2% stabilized module efficiency [2]. The current-voltage (I-V) characteristics, Q and R data were obtained for all the cells. The devices were light-soaked under one-sun, open-circuit, and 50 °C conditions for more than 1000 hours. The bottom cell was exposed through an appropriate cut-on filter in order to simulate both the value of J_{sc} ~ 8.0 mA/cm² and the red light it would actually see at the bottom of the triple-junction structure. For completeness, the corresponding top a-Si and middle a-SiGe component *n i p* cells of the triple-junction structure, made directly on specular ss

First WCPEC; Dec. 5-9, 1994; Hawaii

substrates (no back reflector), were also degraded under similar conditions. The middle cell was exposed through an appropriate cut-on filter to simulate its application in the multijunction structure.

RESULTS AND DISCUSSION

The feature size, as determined by SEM, increases from 200 nm to 500 nm as the texture is increased from A to C. The reflectance and texture of the back reflectors are summarized in Table I. The results of both the Ag and Ag/ZnO surfaces are shown. The total reflectance of the Ag film decreases from 98.5 to 97.0% as the texture increases from A to C. This decrease is attributed to the occurrence of multiple reflection in the case of a textured surface. Since there is a small loss (1-2%) at each reflection, the loss gets amplified as the light bounces off the textured surface several times before it can escape and be measured. Presence of the ZnO film on top of the Ag film reduces the reflection even further. Table I gives the values of the reflection minimum and maximum at 850 nm extrapolated off the envelope of the reflection minima and maxima versus wavelength curve. The reflection values for textures A and B lie in the range 81-88%. The corresponding values for the best texture are 87-88%. The lowering of the reflectance with the addition of the ZnO film is attributed to the superior light trapping inside the ZnO film and the associated losses. The exact cause for the losses is not well understood. The specular component of reflectance from the Ag surface decreases from 82% to 13% as a function of increasing texture. On the completed Ag/ZnO back reflector, the corresponding decrease is from 50% to 8%. The loss in specular component after the addition of the ZnO film for each Ag texture can be explained on the basis of (a) an additional fine texture introduced by the ZnO film as seen in the SEM micrographs and (b) internal scattering taking place in the bulk of the ZnO film [3]. It may be noted that all the Ag/ZnO films exhibit similar reflectance versus angle curves except for the suppression of the superimposed specular component with increasing texture.

Table I. Reflection and texture results of the back reflectors.

Texture	Reflectance at 850 nm (%) Ag	Reflectance at 850 nm (%) Ag/ZnO	Specular Component (%) Ag	Specular Component (%) Ag/ZnO
A	98.5	81-87	82	50
B	97.8	81-88	29	18
C	97.0	87-88	13	8

Initial Results

The Q and R data of the a-SiGe cells made on the three back reflectors are shown in Fig. 1. For clarity, only results in the wavelength range 600-950 nm are shown. The light interference effect is clearly visible in both the Q and R curves. The effect decreases with increasing texture. Further, the maxima of the reflection curve correspond to the minima of the Q curve and vice versa for all the cases. The effect of the increasing texture is the suppression of the reflection maxima due to the enhanced diffused scattering from the back reflector. This, in turn, leads to enhanced absorption in the i-layer and thereby enhanced Q at the wavelengths corresponding to the reflection maxima.

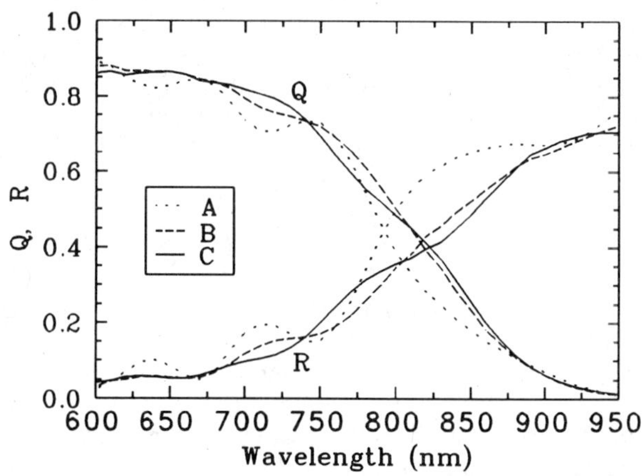

Fig. 1. Q and R data of a-SiGe cells on textures A, B, and C.

The values of J_{sc} for the cells in the wavelength range 600-950 nm has been obtained by the integration of the Q curve, and the results are summarized in Table II. The values of J_{sc} for wavelength $\geq$ 600 nm and $\geq$ 630 nm for the least texture A are 11.68 mA/cm^2 and 9.80 mA/cm^2, respectively. For texture B, the corresponding values increase to 12.44 mA/cm^2 and 10.54 mA/cm^2, respectively. The highest texture C does not give any further gain.

Table II. J_{sc} measurements from Q curves.

Texture	Bottom Cell Jsc (mA/cm^2) Q$\geq$600 nm	Bottom Cell Jsc (mA/cm^2) Q$\geq$630 nm	Triple Cell Jsc (mA/cm^2) Top	Mdl.	Btm.	Mdl.+Btm.	Total
A	11.68	9.80	7.03	7.41	8.04	15.45	22.48
B	12.44	10.54	7.00	7.67	8.45	16.12	23.12
C	12.43	10.55	6.94	7.67	8.49	16.16	23.10

A similar behavior is observed in the case of the triple-junction cells. The values of the maxima of the reflection curve reduce with increasing texture. Consequently, the Q curve improves at the corresponding wavelengths. The individual values of J_{sc} of the component cells and their sum obtained from the Q curves are summarized in Table II. The top cell does not see much reflected light and the value of J_{sc} is expectedly similar ~ 7.0 mA/cm^2 for all the cases. The sums of the values of J_{sc} of the middle and bottom cells, a good measure of the back reflector quality, are 15.45 mA/cm^2, 16.12 mA/cm^2, and 16.16 mA/cm^2 for the textures A, B and C, respectively. The higher sums for the two higher textures is primarily attributable to the enhanced red response. However, there is no difference in the value of J_{sc} for textures B and C.

The results in Table II show that a certain degree of texture as in B is desirable to obtain an optimized value of J_{sc}. Further increase does not lead to any additional gain. In fact, in the extreme case of a Lambertian scatterer, we have observed [4] a decline in the value of J_{sc}. We have estimated that the experimental value of the enhancement factor, EF, for the a-SiGe cells is only 15-20 in the red/infrared regime. The EF is defined as the increase in the effective absorption due to light trapping. It has been shown [6] that when the light is completely randomized within the material, the thermodynamic limit for the absorption increase is given by EF (max) = $4n^2$, where n is the refractive index. For a-SiGe, EF (max) is approximately 60. The difference between the experimental and the theoretical values of EF is attributed to (a) parasitic optical losses and (b) nonrandomized photons. Increase in substrate texture helps in the process of randomization of the photons. The accompanying increase in the value of EF and the concomitant increase in J_{sc} are composed of two components. The more important one is the enhancement of the red/infrared response of the cell due to the enhanced optical path. The second, an artifact of the first, is the gain associated with suppressed reflection losses due to the reduction of interference effects of the light reflected from the top and bottom surfaces of the cell. The first component is observed when the substrate surface is changed from a specular to the textured case: the Q curve of the device exhibits an extended red response. Our experimental results show that for the level of texture investigated in this work, only the second component is effective. The nonexistence of the first component may be attributed to the competing parasitic optical losses associated with increasing texture. The trade-off might explain the occurrence of the optimum texture B. In texture C, the expected gain is offset by the enhanced optical losses. The explanation is consistent with the observed decrease in J_{sc} for very high textures [4]. More work needs to be done to understand the exact cause of the parasitic optical losses which would elucidate the discrepancy between the experimental and theoretical values of EF.

Stability Results

The normalized efficiency of the bottom and triple-junction devices measured under AM1.5 conditions as a function of exposure time are shown in Figs. 2 and

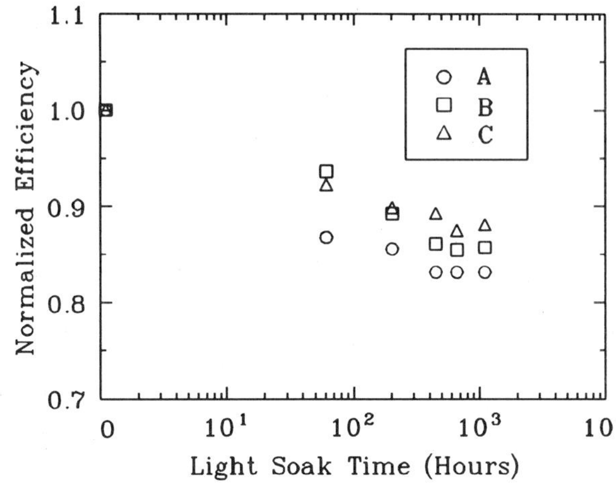

Fig. 2. Normalized efficiency versus light soak time of a-SIGe cells on textures A, B, and C.

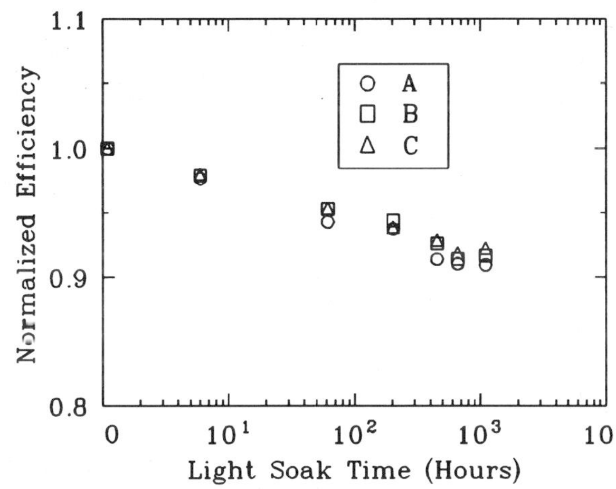

Fig. 3. Normalized efficiency versus light soak time of triple-junction cells on textures A, B, and C.

3, respectively. All the devices exhibit saturation after an exposure of 400-600 hours. The stability improves with increasing texture for both sets of samples. Texture C shows the least degradation. The triple-junction devices exhibit better stability.

The percentage degradation for saturation is summarized in Table III. Referring first to the cells on back reflectors, the bottom cell degradation decreases from 16.9% to 11.9% as the texture increases from A to C. The corresponding numbers for the triple-junction devices are 9.1% and 7.7%. The degradation of the V_{oc}, J_{sc}, and fill factor, FF, of the bottom cell is also tabulated. The values of degradation for both types of cells on textures B and C are the best reported for high efficiency devices. For the cells on ss substrates, the degradation

Table III. Percentage degradation for saturation of top, middle, bottom, and triple-junction cells.

Device	Texture	Exposure (hours)	Degradation (%) Eff.	Voc	Jsc	FF
Top	Spec.(ss)	1060	10.7			
Middle	Spec.(ss)	1530	19.0			
Bottom	A	1100	16.9	4.3	3.6	9.9
	B	1100	14.3	4.3	3.6	8.0
	C	1100	11.9	4.3	3.2	6.0
Triple	A	1100	9.1			
	B	1100	8.3			
	C	1100	7.7			

for saturation for the top and the middle cells are 11% and 19%, respectively. The relatively larger degradation for the ss cases is attributed to the lack of texture of the substrates.

There are two major highlights of the stability work. First, the remarkable stability of the triple-junction devices constitutes a technological milestone in combating Staebler-Wronski degradation [5]. This result is attributed to both the substrate texture and the high initial quality of the cells. The second is the explicit dependence of stability on the substrate texture. Earlier work [4] has shown that Lambertian back reflectors exhibit superior values of stabilized FF than normally textured substrates. This was explained on the basis of nonconformal growth of the a-Si films which in turn leads to superior current collection. The present work shows that even for textures away from the Lambertian extremum, the stability is dependent on the extent of the texture. The attribution of the superior stability on higher textures to the nonconformal growth process is still valid. Thus, as shown in Table III, the reduction in degradation of efficiency of the a-SiGe bottom cell with increasing texture is primarily due to the corresponding reduction in the degradation of the value of FF from 9.9% to 6.0%. The degradation of the V_{oc} and J_{sc} is reasonably independent of texture. The degradation of the triple-junction case is also dependent on other factors.

One-square-foot triple-junction modules, fabricated using a similar recipe, have exhibited stabilized efficiency of 10.2% and light-induced degradation of approximately 12%. In contrast, the unencapsulated cells degrade only 8% (see Table III). This difference may be attributed to module-related degradation losses stemming from reliability issues. An improved encapsulation technique is expected to reduce the module degradation to 8% which in turn would lead to greater than 10.2% stable module efficiency.

CONCLUSIONS

There is an optimum texture for best back reflector performance as regards initial results. Use of an appropriate texture has been shown to offer a partial engineering solution to the Staebler-Wronski effect. Degradation as low as approximately 8% has been observed on triple-junction cells.

ACKNOWLEDGEMENTS

We thank J. Noch, N. Jackett, and E. Chen for experimental help and V. Trudeau for the preparation of the manuscript. Partial funding for the project by NREL under Subcontract Numbers ZM-1-19033-2 and ZAN-4-13318-02 is acknowledged.

REFERENCES

[1] J. Yang, R. Ross, T. GLatfelter, R. Mohr, G. Hammond, C. Bernotaitis, E. Chen J. Burdick, M. Hopson, and S. Guha, "High Efficiency Multi-Junction Solar Cells Using Amorphous Silicon and Amorphous Silicon-Germanium Alloys", *Proc. Twentieth IEEE PVSC*, 1988, pp. 241-246.

[2] S. Guha, J. Yang, A. Banerjee, T. Glatfelter, K. Hoffman, S.R. Ovshinsky, M. Izu, H.C. Ovshinsky, and X. Deng, "Amorphous Silicon Alloy Photovoltaic Technology--From R&D to Production", *Mat. Res. Soc. Symp.*, **336**, 1994, pp. 645-655.

[3] A. Banerjee and S. Guha, "Study of Back Reflectors for amorphous Silicon Alloy Solar Cell Application", *J. Appl. Phys.* **69**, 1991, pp. 1030-1035.

[4] A. Banerjee, J. Yang, K. Hoffman, and S. Guha, "Characteristics of Hydrogenated Amorphous Silicon Alloy Solar Cells on a Lambertian Back Reflector", *Appl. Phys. Lett.*, **65**, 1994, pp. 472-474.

[5] D.L. Staebler, and C.R. Wronski, "Reversible Conductivity Changes in Discharge-Produced Amorphous Si, *Appl. Phys. Lett.*, **31**, 1977, pp. 292-294.

[6] E. Yablonovitch and G. Cody, "Intensity Enhancement in Textured Optical Sheets for Solar Cells", *IEEE Trans. Electron Dev.*, **ED-29**, 1982, pp. 300-305.

CHARGED AND NEUTRAL DEFECT STATES IN A-SI:H DETERMINED FROM IMPROVED ANALYSIS OF THE CONSTANT PHOTOCURRENT METHOD

F. Siebke, H. Stiebig, A. Abo-Arais, and H. Wagner
Forschungszentrum Juelich GmbH, Institute of Thin Film and Ion Technology (ISI-PV), D-52425 Juelich, Germany

ABSTRACT

A numerical model has been developed to simulate constant photocurrent method spectra. It takes into account the full set of optical transitions between localized and extended states under sub-bandgap illumination, capture, emission and recombination processes as well as the energetic position of the Fermi level. In order to obtain information about the density of localized gap states in a-Si:H, i.e. the valence band tail, the integrated defect density, the energetic defect distribution, and the charge state of the defect states we compare simulated and measured spectra. In the annealed state the defect absorption of undoped, n- and p-type a-Si:H is dominated by a charged defect states. In undoped material the defect density increases upon light soaking but the charged-to-neutral defect ratio does not change.

INTRODUCTION

The interpretation of solar cell characteristics and design optimization demand information about the defect distribution in the gap and the charge state of the defect states. The defect-pool model suggests that in the annealed state there are more charged than neutral defect states even in undoped a-Si:H while the charged-to-neutral defect ratio decreases to 1 in the degraded state [1,2]. In order to examine these predictions of the defect-pool model we extend a photoconductivity modelling approach to simulate constant photo current mothod (CPM) spectra taking into account the full set of optical transitions between localised and extended states, capture, emission and recombination processes. Comparing simulated and measured CPM spectra we derive the defect distribution of doped and undoped a-Si:H.

MODEL

Our simulation of CPM spectra is based on the density of localized states g(E) illustrated schematically in Fig. 1. E_V and E_C denote the valence band edge and

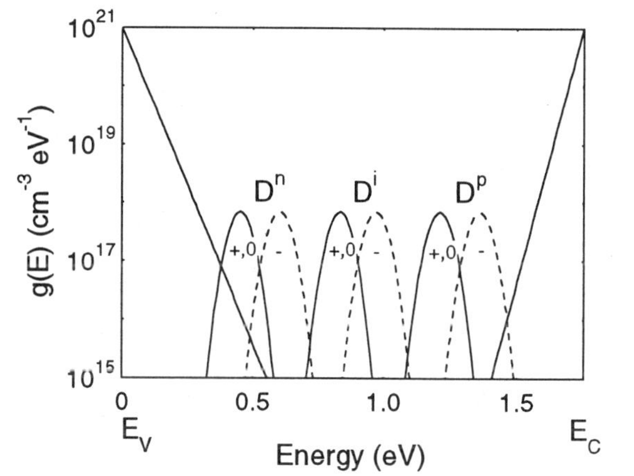

Fig. 1. Density of localized states in a-Si:H according to the defect pool model.

the conduction band edge, respectively. The distribution of dangling bond states is chosen according to the defect pool model. In order to describe a non-equilibrium situation, e.g. the occupation of defect states under sub-bandgap illumination it is extremely useful to divide the density of dangling bond states into three Gaussian defect peaks D^n, D^i, and D^p. Calculations of Deane and Powell [1] predict that in the case of thermal equilibrium D^p, D^i, and D^n correspond to unoccupied, singly, and doubly occupied dangling bonds. Taking into account a correlation energy U = 0.2 eV the midgap states D^p, D^i, and D^n are divided into three further peaks (+,0,-) since each state can be occupied with zero, one or two electrons. According to the defect-pool model the energetic position of the D^n is chosen to be in the lower half of the bandgap, D^i lies near midgap and D^p is positioned in the upper half of the bandgap. The density of states is completed by two exponentially increasing band tails. In the calculation each midgap peak is divided into nineteen discrete levels whereas the tail states are described by continuous distributions.

In general the absorption coefficient α can be expressed by

CH3365-4/94/0000-0543 $4.00 © 1994 IEEE First WCPEC; Dec. 5–9, 1994; Hawaii

$$\alpha(h\nu) \propto h\nu\, R^2(h\nu) \int g_{occ}(E-h\nu)\, g_{unocc}(E)\, dE \qquad (1)$$

where $R^2(h\nu)$ denotes an average dipole transition matrix element. Following Pierz [3] we assume $h\nu \cdot R^2(h\nu)$ to be energy independent between 0.8 eV and 2.0 eV. The density of occupied states g_{occ} and the density of unoccupied states g_{unocc} are modelled by the density of states shown in Fig. 1 taking into account the occupancy functions given below. Our model includes all possible optical transitions between localized states and extended states but neglects optical transitions between localized states because of the smaller transition probabilities and smaller densities of states. For a single defect state at the energetic position E_{db} the resulting transition processes under sub-bandgap illumination are shown in Fig. 2, where c_n and c_p are the electron and hole capture coefficients, e_n and e_p describe the thermal emission of electrons and holes, and Φ_i denote the optical transitions. The capture coefficients are determined by an extensive calculation of the temperature and light intensity dependent photoconductivity. The optical transitions Φ_i are given by

$$\Phi_1 \propto \phi \cdot N_V^* \cdot h \cdot \nu \cdot \frac{R^2}{2 \cdot \Delta} \cdot \int_{h\cdot\nu-\Delta}^{h\cdot\nu+\Delta} \Theta(E_{db} - E_V - E)\, dE \qquad (2)$$

$$\Phi_2 \propto \phi \cdot N_C^* \cdot h \cdot \nu \cdot \frac{R^2}{2 \cdot \Delta} \cdot \int_{h\cdot\nu-\Delta}^{h\cdot\nu+\Delta} \Theta(E_C - E_{db} - E)\, dE \qquad (3)$$

$$\Phi_3 \propto \phi \cdot N_V^* \cdot h \cdot \nu \cdot \frac{R^2}{2 \cdot \Delta} \cdot \int_{h\cdot\nu-\Delta}^{h\cdot\nu+\Delta} \Theta(E_{db} - E_V - E)\, dE \qquad (4)$$

$$\Phi_4 \propto \phi \cdot N_C^* \cdot h \cdot \nu \cdot \frac{R^2}{2 \cdot \Delta} \cdot \int_{h\cdot\nu-\Delta}^{h\cdot\nu+\Delta} \Theta(E_C - E_{db} - U - E)\, dE \qquad (5)$$

where ϕ describes the photon flux and N_C^* and N_V^* the energy independent density of states in the conduction and valence band, respectively. Θ is a step function defined as $\Theta(x) = 0$ for $x > 0$ and $\Theta(x) = 1$ for $x \leq 0$. Taking into account the finite energy resolution of the monochromator used in the experiment we calculate the optical transitions for photon energies in an energy interval $h\nu \pm \Delta$ instead of using a discrete photon energy $h\nu$.

The occupancy functions f^+, f^0, and f^- of amphoteric dangling bond states in each charge state, i.e. occupied by zero, one, or two electrons, can be obtained from the solution of the rate equations for the transitions between the defect state and extended states taking into account that $f^+ + f^0 + f^- = 1$. The resulting steady state occupation of a singly occupied defect state is given by

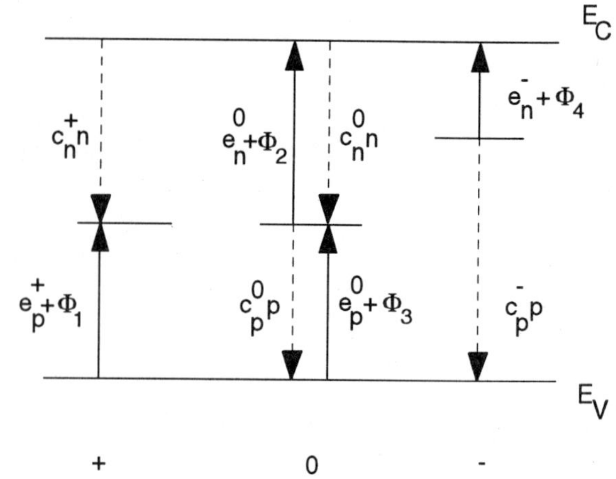

Fig. 2. Electron and hole transitions for a discrete defect state under sub-bandgap illumination.

$$f^0 = \frac{CC1 \cdot CC3}{CC1 \cdot CC3 + CC1 \cdot CC4 + CC2 \cdot CC3} \quad \text{with}$$

$$CC1 = \left(c_n^+ \cdot n + e_p^+ + \Phi_1\right) \qquad CC2 = \left(c_p^0 \cdot p + e_n^0 + \Phi_2\right) \qquad (6)$$

$$CC3 = \left(c_p^- \cdot p + e_n^- + \Phi_4\right) \qquad CC4 = \left(c_n^0 \cdot n + e_p^0 + \Phi_3\right).$$

The tail state occupation is described following Hattori [4]. Using this occupation statistics the density of free electrons (n) and holes (p) is calculated and the photoconductivity $\sigma_{phot} = e\, \mu_n\, n + e\, \mu_p\, p - \sigma_d$ is determined, where $\mu_n = 10\ cm^2\ V^{-1}s^{-1}$ and $\mu_p = 1\ cm^2\ V^{-1}s^{-1}$ denote the electron and hole mobility and σ_d the dark conductivity. For photon energies between 0.8 eV and 1.75 eV we calculate the photon flux ϕ_{CPM} required to keep the photocurrent on a constant level and determine the absorption coefficient $\alpha(h\nu)$ by $\alpha(h\nu) = C / \phi_{CPM}(h\nu)$. This is exactly the quantity measured in the experiment.

RESULTS AND DISCUSSION

In the following simulated CPM spectra will be shown and compared with experimental results. The experimental details are described elsewhere [5]. In the simulation only the defect density, inverse logarithmic slopes of the valence band tail E_0, and the Fermi level position taken from dark conductivity measurements are varied. Input parameters as capture cross sections, etc. have been kept constant in all simulations. According to total yield photoelectron spectroscopy results the density of states at the band edges was chosen to be $8 \cdot 10^{20}\ cm^{-3}eV^{-1}$ [5]. The slope of the conduction band tail was assumed to be 25 meV. Corresponding to CPM measurements in the ac mode the photocurrent in the simulation was chosen to be 1% of the dark current.

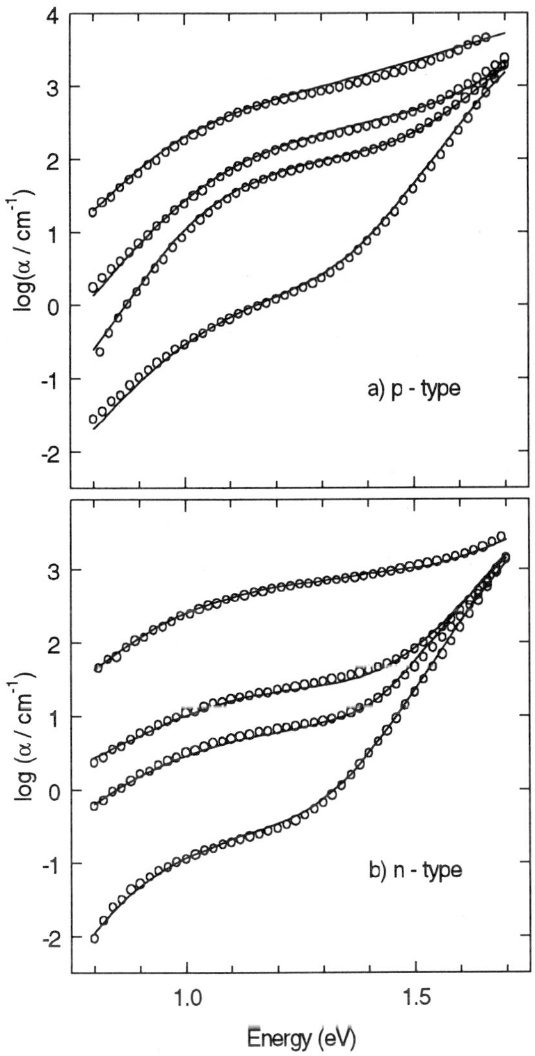

Fig. 3. Measured (-) and simulated (o) CPM spectra of a-Si:H with various doping levels; a) between 1ppm and 0.2% diborane in the gas phase, b) between undoped material and 1% phosphine in the gas phase.

Fig. 3 shows simulated and measured CPM spectra of p- and n-type samples with various doping levels up to 0.2% diborane and 1% phosphine in the gas phase, respectively. To simulate CPM spectra of p-type a-Si:H we introduced a demarcation level into the simulation above which the valence band tail truncates as proposed by the weak-bond dangling-bond conversion model [6]. Under the assumption that in p-type a-Si:H the demarcation level has an energetic position close to the Fermi level as predicted by the defect-pool model [1] we achieve a good fit between simulation and experimental data. The simulation reveals that the defect absorption is dominated by the D^p peak. For n-type samples the simulation is also in good agreement with the experiment. The defect absorption is dominated by the D^n peak. No contribution of D^i to α can be observed.

We now consider the defect distribution of undoped amorphous silicon. In spite of the success of the defect pool model describing the defect density in doped amorphous silicon the predictions concerning the defect structure of undoped material are still debated. The standard model implies that the majority of dangling bonds is occupied by one electron, i.e. nearly all defect states are neutral. On the other hand the defect pool model suggests a simultaneous formation of defects in all three charge states yielding more charged than neutral defect states even in undoped material.

Fig. 4a shows measured and simulated CPM spectra of an undoped sample in the annealed state. The spectra are calculated for various defect distributions, i.e. various heights of the D^i, D^n, and D^p peak, keeping the integrated defect density constant. The energetic position of the D^n and D^p peak was taken from the simulations for doped a-Si:H whereas the position of the D^i peak is derived from the condition of charge neutrality: For a certain Fermi level position the energetic position of the D^i peak is determined by the assumptions about the height and standard deviation of the distribution of D^i states if no unintentional dopants are taken into account. If we assume a defect distribution which is dominated by the D^i peak, as predicted by the standard model, the majority of defect states is occupied by one electron. But in the simulation an enhanced defect absorption at low photon energies followed by a flat region is found which does not fit the experimental data. On the other hand we achieve a good agreement if we adjust the defect distribution to have more charged than neutral defect states. The best fit to the experimental data is obtained for a charged to neutral ratio $(D^+ + D^-) / D^0 = 4$. This result confirms the predictions of various calculations based on the defect-pool model which propose a charged to neutral ratio between 3 and 7 for equilibrium defect densities [1,2].

Beside the density of states in thermal equilibrium the knowledge of the defect structure after light degradation is essential to understand the properties of light soaked devices and to design more stable solar cells. It is widely assumed that light soaking creates defect states which are occupied by one electron. Schumm [2] developed a generalized concept of chemical equilibration including electronic occupation statistics. This description should apply to both equilibrium and metastable defect densities. It predicts an increase of charged and neutral defect states and a decrease of the charged to neutral ratio to ≈ 1. In contrast to Schumm we do not ob-

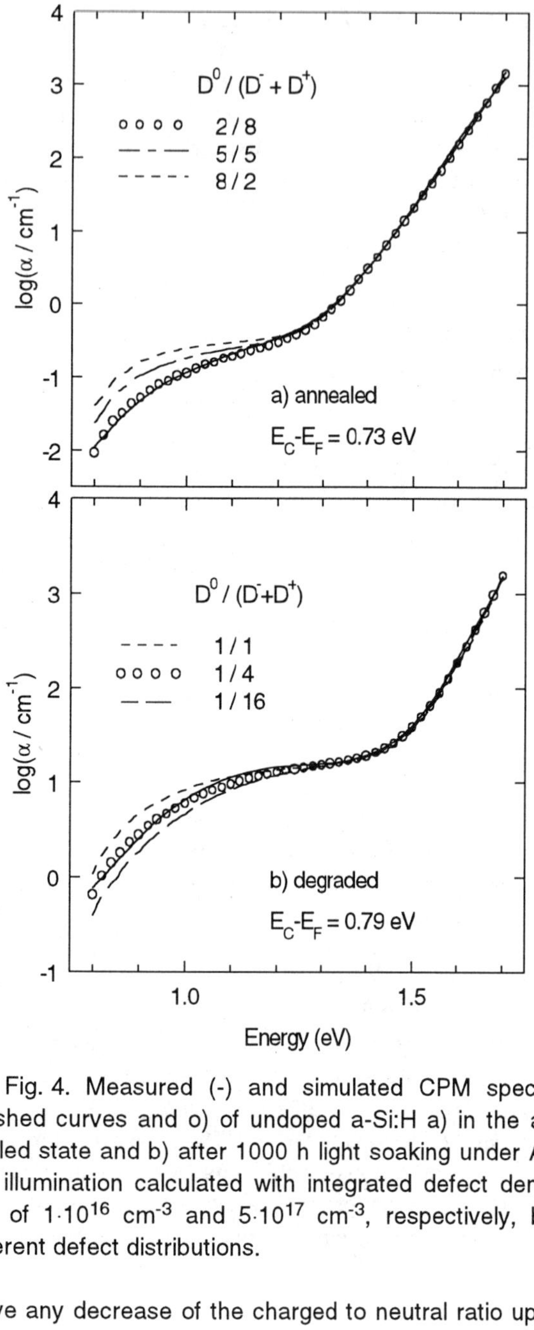

Fig. 4. Measured (-) and simulated CPM spectra (dashed curves and o) of undoped a-Si:H a) in the annealed state and b) after 1000 h light soaking under AM 1.5 illumination calculated with integrated defect densities of $1{\cdot}10^{16}$ cm^{-3} and $5{\cdot}10^{17}$ cm^{-3}, respectively, but different defect distributions.

serve any decrease of the charged to neutral ratio upon degradation (see Fig. 4b). The defect density increases and the charge to neutral ratio remains constant. This behaviour is in very good agreement with the results of electron spin resonance and photothermal deflection spectroscopy measurements on annealed and degraded samples where a linear correlation between the spin density and the integrated defect density determined from sub bandgap absorption was found [7]. Our observation reveals that further work is needed to achieve a description of metastable defects within the framework of the defect-pool model.

CONCLUSIONS

We have developed a numerical model to simulate CPM spectra taking into account all emission, capture processes, and optical transitions between extended and localized states. For n- and p-type samples we achieved a good agreement between experimental data and simulated spectra. We used our model to examine the predictions of the defect-pool model with regard to the defect distribution and the charged to neutral defect ratio in a-Si:H. In the annealed state we find a good agreement between the experimental data and the density of states proposed by the defect-pool model for n- and p-type a-Si:H. The defect absorption is dominated by charged defect states. Even in undoped a-Si:H we observe more charged than neutral defects. Upon light soaking the defect density in undoped a-Si:H increases but we do not find a decrease of the charged-to-neutral ratio. Further work on the description of the metastable defect density by an equilibration model is necessary.

REFERENCES

[1] M. J. Powell, and S. C. Deane, "Improved defect-pool model for charged defects in amorphous silicon", *Phys. Rev. B*, **48**, 1993, 10815-10827.

[2] G. Schumm, "The defect-pool model and charged defects in amorphous silicon", *J. Non-Cryst. Solids*, **164-166**, 1993, pp. 317-322.

[3] K. Pierz, B. Hilgenberg, H. Mell, and G. Weiser, "Gap-state distribution in n-type and p-type a-Si:H from optical absorption", *J. Non-Cryst. Solids*, **97&98**, 1987, pp. 63-66.

[4] K. Hattori, S. Fukuda, K. Nishimura, H. Okamoto, and Y. Hamakawa, "Interpretation of CPM measurements in amorphous semiconductors", *J. Non-Cryst. Solids*, **164-166**, 1993, pp. 351-354.

[5] F. Siebke, W. Beyer, J. Herion, and H. Wagner, "In-situ determination of the surface and bulk density of states of a-Si:H", *11th E.C. Photovoltaic Solar Energy Conference*, 1993, pp. 100-103.

[6] M. Stutzmann, "The defect density in amorphous silicon", *Phil. Mag. B*, **60**, 1989, pp. 531-546.

[7] M. S. Brandt, A. Asano, and M. Stutzmann, "Are there charged dangling bonds in device quality amorphous silicon?", *Mat. Res. Soc. Symp. Proc.*, **297**, 1993, pp. 201-206.

IMPROVEMENT OF FILM STABILITY OF A-Si:H DEPOSITED BY RPCVD USING SiH_2Cl_2

Mi Kyung Chu, Jae Seong Byun, Hong Bin Jeon, Kyung Ha Lee, Min Park, Kyu Chang Park and Jin Jang
Department of Physics, Kyung Hee University, Dongdaemoon-ku, Seoul 130-701, Korea

Hyuk Ryeol Park
Department of Physics, Mokpo University, Muan-kun, Chullanam-do 534-729, Korea

Dae Won Kim and Kyung Shik Yoon
Yukong R&D Center, Daeduck, Yoosung-ku, Taejon 305-606, Korea

ABSTRACT

Hydrogenated amorphous silicon (a-Si:H) films were prepared by remote plasma chemical vapor deposition (RPCVD) using $SiH_4/SiH_2Cl_2/He/H_2$ mixtures. The defect density and Urbach energy of a-Si:H(:Cl) were found to be little affected by a small addition of SiH_2Cl_2 in SiH_4. And a small addition of H_2 in SiH_2Cl_2/SiH_4 mixture improves the film quality. We deposited a-Si:H(:Cl) films with 6 at.% of hydrogen content and with 3×10^{15} cm^{-3} defect density which exhibited very small photoconductivity degradation during AM-1 illumination.

INTRODUCTION

In order to realize high efficiency, stable a-Si:H solar cells, it is the most important to make low defect density a-Si:H which exhibits no defect generation during light illumination. It seems to be almost impossible to make such a film in near future in view of the present thin film deposition technology. Therefore, it is important to suppress the Staebler-Wronski effect in a-Si:H film in order to produce stable, high efficiency a-Si:H solar cells. Recently, a-Si:H(:Cl) films have been prepared by various deposition methods using SiH_2Cl_2 mixture to improve film quality and stability [1-3]. The addition of chlorine(Cl) in a-Si:H is a new technique to prepare a-Si:H solar cell with improved stability. In this study we concentrated our work to deposit low defect density a-Si:H exhibiting small Staebler-Wronski effect by remote plasma chemical vapor deposition (RPCVD) using SiH_4/SiH_2Cl_2 /He/H_2 mixtures. And we investigated the effects of the Cl addition on optical and structure properties of a-Si:H by using FT-IR (Fourier-Transformed InfraRed) spectrophotometer and SIMS (Secondary Ion Mass Spectroscopy).

DEPOSITION OF a-Si:H(:Cl) BY RPCVD

The a-Si:H(:Cl) films were deposited in a RPCVD reactor, where plasma generating region was remote from a deposition chamber. We used He as a nondepositing inert gas. Helium was passed through the plasma generating region, inside of the cylindrical quartz tube of a diameter of 3.8 cm, and some of them were ionized and/or excited into metastable state by RF power at 13.6 MHz applied to the induction coil, wrapped around the outside of quartz tube. A mixture of SiH_4, H_2, and SiH_2Cl_2 was introduced downstream into the deposition reactor. The detailed deposition parameters are shown in Table Ⅰ.

Table Ⅰ. Deposition conditions of a-Si:H films using $SiH_4/SiH_2Cl_2/H_2/He$ mixtures.

H_2 dilution effect	
RF power	30W
Gas flow rates (sccm)	
He	100
SiH_2Cl_2	0.2
SiH_4	1.0
H_2	0 - 10
Substrate temperature	220℃
SiH_2Cl_2 flow rate dependence	
RF power	20W
Gas flow rates (sccm)	
He	100
SiH_2Cl_2	0 - 1.0
SiH_4	1.0
H_2	2.0
Substrate temperature	220℃

At first, we varied H_2 flow rate with fixing the other deposition parameters as shown in Table Ⅰ, and obtained the optimum flow rate of 2 sccm. Then we varied SiH_2Cl_2 flow rate. The substrate temperature was fixed at an optimum value of 220℃. The Corning 7059 glass plates and Si wafers were used as substrates for depositing a-Si:H(:Cl).

RESULTS AND DISCUSSIONS

Figure 1 shows the sub-band gap absorption

CH3365-4/94/0000-0547 $4.00 © 1994 IEEE First WCPEC; Dec. 5-9, 1994; Hawaii

coefficients for a-Si:H(:Cl) films prepared with a variation of H_2 flow rate. The sub-band gap absorption coefficients were obtained by measuring both CPM (constant photocurrent method) and optical transmittance by Cary 17D spectrophotometer. The interference peaks in the optical absorption coefficients in the sub-band gap region, caused by multiple reflections, are shown because the film thickness is 400 nm ~ 600 nm.

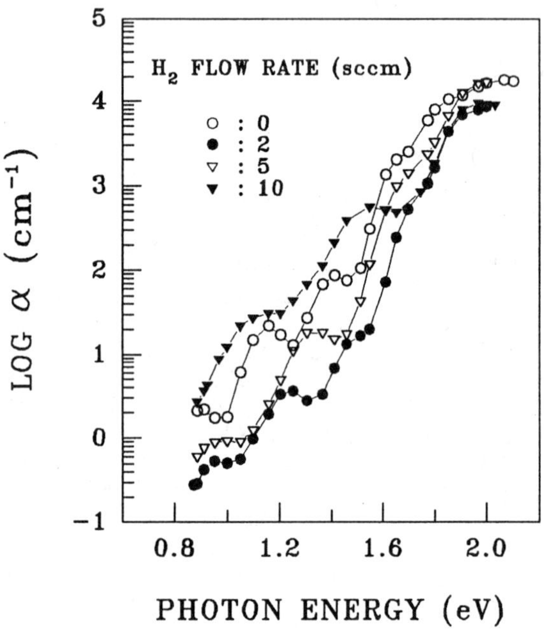

Fig. 1. Optical absorption coefficients for a-Si:H(:Cl) films deposited with various H_2 flow rates.

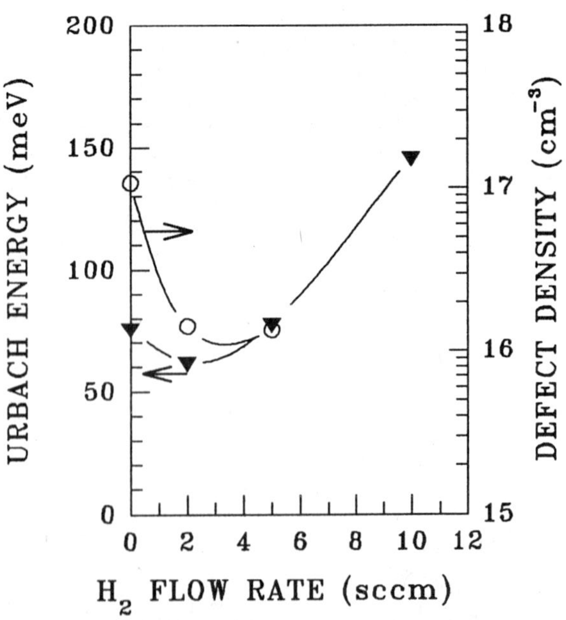

Fig. 2. Urbach energy and defect density of a-Si:H(:Cl) films versus H_2 flow rate.

From Fig. 1, the Urbach slope and the defect density can be obtained from the slope of the interferenceless, exponentially increasing absorption coefficients and from the integration of the absorptions due to defect transitions, respectively[4]. The results are shown in Fig. 2. Both the Urbach slope and defect density are decreased by adding 2 sccm of H_2 in $SiH_4/SiH_2Cl_2/He$ mixture. And they start to increase by adding more than 2 sccm of H_2. The addition of 10 sccm of H_2 to $SiH_4/SiH_2Cl_2/He$ mixture results in the growth of microcrystalline silicon, which was confirmed from the measurements of Raman scattering and X-ray diffraction spectra. The Urbach slope and the defect density are 60 meV and 10^{16} cm^{-3}, respectively, when H_2 flow rate is 2 sccm.

Figure 3 shows the degradation of AM-1 (100 mW/cm^2) photoconductivity under light illumination. The degradation is slow for the a-Si:H films deposited with H_2 flow rates of 0, 2, and 5 sccm, however, it is relatively high for the a-Si:H(:Cl) film deposited with 10 sccm of H_2. The degradation factor, γ, obtained from the time (t) dependence of photoconductivity (σ_{ph}) : σ_{ph} = const. $t^{-\gamma}$, is 0.045 for the film deposited with H_2 flow rate of 2 sccm and is 0.053 for the film deposited with 0 sccm of H_2, but it increases to 0.1 when H_2 flow rate is 10 sccm. The stability of a-Si:H is directly or indirectly related with hydrogen content in the film.

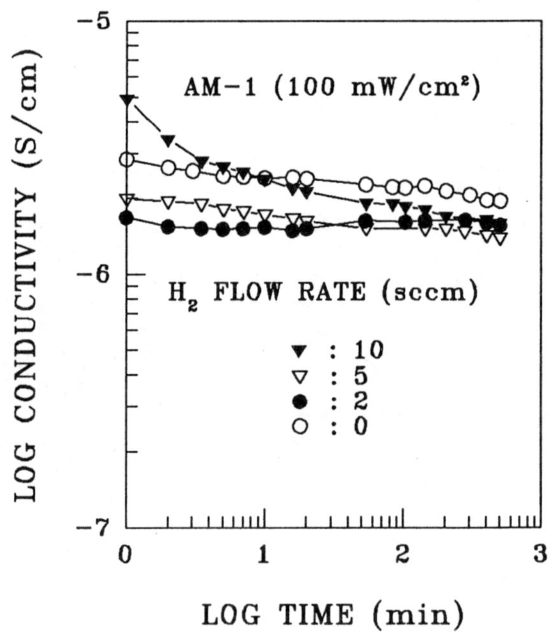

Fig. 3. Photoconductivity degradation under AM-1 of a-Si:H(:Cl) films deposited with various H_2 flow rates.

The hydrogen content in a-Si:H(:Cl) films is shown in Fig. 4. The hydrogen content in the film was obtained from the integration of stretching mode absorption coefficients of Si-H_n modes[5]. As shown in the figure, the hydrogen content is around 6 at.% when H_2 flow rate is less than 6 sccm, but it decreases to 4.3 at.% when H_2 flow rate is increased to 10 sccm. The decrease of hydrogen content is related with the change in the film structure from amorphous to microcrystal. It is noted that the hydrogen content in μc-Si is lower than that of a-Si:H when the substrate temperature is the same[6].

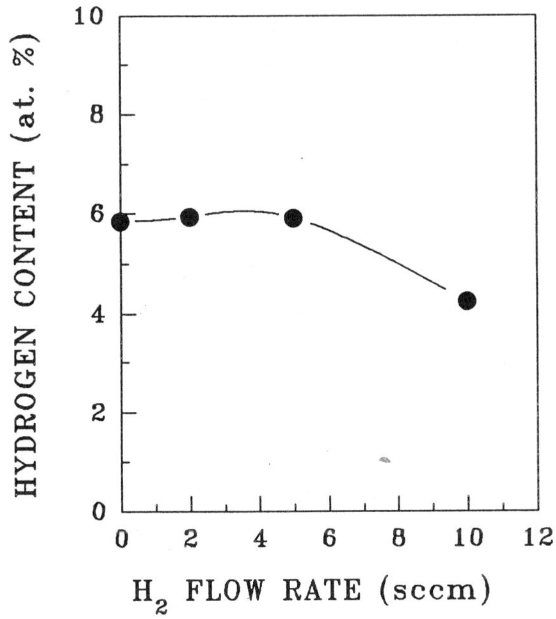

Fig. 4. Hydrogen content in a-Si:H(:Cl) films versus H_2 flow rate.

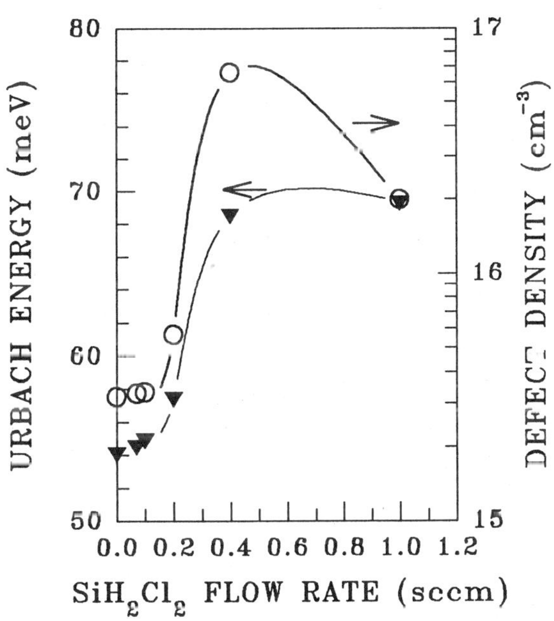

Fig. 5. Urbach energy and defect density in a-Si:H(:Cl) films plotted against SiH_2Cl_2 flow rate.

Figure 5 shows the Urbach energy and defect density in a-Si:H(:Cl) films deposited with various SiH_2Cl_2 flow rates. The Urbach energy and defect density are 55 meV and 3 x 10^{15} cm^{-3}, respectively, when the flow rate of SiH_2Cl_2 was less than 0.1 sccm, and increase to 58 meV and 6 x 10^{15} cm^{-3}, respectively, when SiH_2Cl_2 flow rate is 0.2 sccm. The defect density and Urbach energy, increase little with Cl content when a small amount of SiH_2Cl_2, up to 10 %, is mixed in SiH_4.

Figure 6 shows the absorption coefficients for Si-H(2000cm^{-1}) and Si-H_2(2100cm^{-1}) stretching modes in the a-Si:H(:Cl) films deposited at various flow rates of SiH_2Cl_2. The hydrogen content in the a-Si:H(:Cl) decreases with increasing SiH_2Cl_2 flow rate. And the structure of the film was changed from amorphous to microcrystalline by increasing SiH_2HCl_2 above 0.5sccm. This points out the possibility of eliminating weak Si-Si bonds and Si-H bonds by exposing Cl radicals on the growing surface.

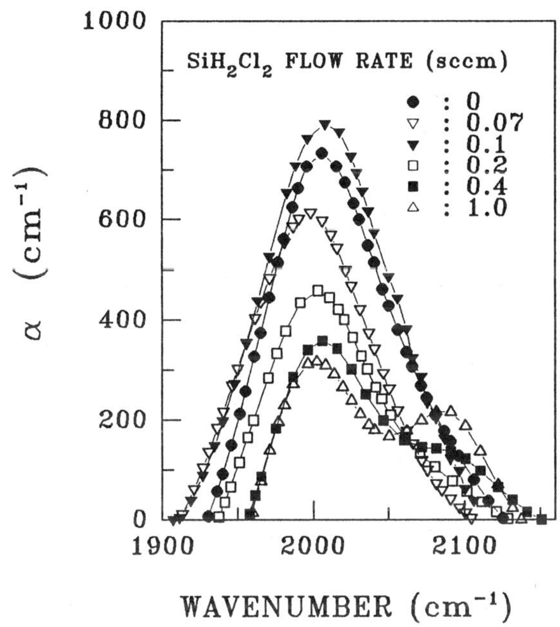

Fig. 6. The absorption coefficients of Si-H and Si-H_2 stretching modes for a-Si:H(:Cl) films deposited with various SiH_2Cl_2 flow rates.

Figure 7 depicts the degradation of photoconductivity for a-Si:H(:Cl) films under AM-1 illumination. The slow decay of photoconductivity can be seen for a-Si:H(:Cl) films prepared with the addition of SiH_2Cl_2 between 0.07 and 0.1 sccm in 1 sccm SiH_4. The degradation factor γ is less than 0.06 for these films, which is exhibited in Fig. 8. However, the γ is above 0.1 when SiH_2Cl_2 flow rate is less than 0.1 sccm or above 0.4 sccm. The role of SiH_2Cl_2 on the growth of a-Si:H can be considered as follows. Because the bond strength of H-Cl (4.5eV) is higher than that (3.1eV) of Si-H[7], the hydrogen in the film can be removed from the growing surface during the deposition. And the weak Si-Si bond can be broken since bond strength (4.7eV) of Si-Cl is higher than that of Si-Si (3.4eV)[7] and thus Si can be etched by Cl radicals. This is deduced from the observation that the deposition rate decreases with increasing SiH_2Cl_2 flow rate, as observed in the experiments. The content of Cl atoms in a-Si:H(:Cl) increases with SiH_2Cl_2 flow rate. The Cl content in a-Si:H(:Cl) was observed to be $\sim 1 \times 10^{19}$ cm^{-3} in a-Si:H(:Cl) deposited with 0.1 sccm of SiH_2Cl_2.

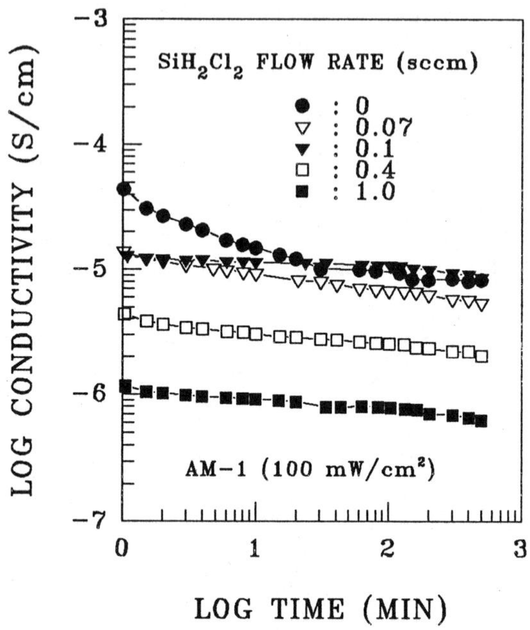

Fig. 7. Degradation of photoconductivity for a-Si:H(:Cl) films under AM-1 deposited with various SiH_2Cl_2 flow rates.

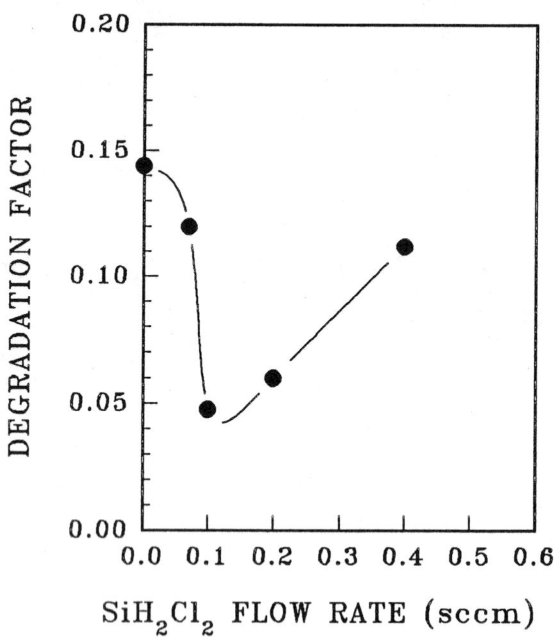

Fig. 8. Degradation factor, obtained from the photoconductivty decay, of a-Si:H(Cl) films deposited with various SiH_2Cl_2 flow rates.

The good stability of photoconductivity in the deposited a-Si:H(:Cl) may be arised from both small amount of hydrogen content and small amount of incorporated Cl in the film. However, the film quality rapidly degrades when SiH_2Cl_2 flow rate is higher than 0.1 sccm. This is related with the increase of both Si-H_2 bond density and Cl concentration when SiH_2Cl_2 flow rate is above 0.1 sccm. The Cl concentration was observed to be 5 x 10^{19} cm^{-3} when SiH_2Cl_2 was 0.2 sccm. It is noted that the stretching mode absorption of Si-H_2 is negligible when SiH_2Cl_2 flow rate is less than 0.1 sccm in the experiments.

CONCLUSIONS

We have deposited the a-Si:H(:Cl) film with defect density of 3 x 10^{15} cm^{-3}, Urbach slope of 55 meV and hydrogen content of 6 at. % at the substrate temperature of 220℃. The optimum flow rate of SiH_2Cl_2 is around 10% of SiH_4 flow rate. The a-Si:H(:Cl) exhibited a very slow degradation in photoconductivity under AM-1. The degradation factor γ of the film was found to be less than 0.05 at room temperature.

ACKNOWLEDGEMENT

This work was supported by the G7 Project of Korea and by the Korea Science and Engineering Foundation through the Semiconductor Physics Research Center.

REFERENCES

[1] J.N. Bullok and S. Wagner, "Amorphous Silicon Films from Dichlorosilane and Hydrogen", *Mat. Res. Soc. Symp. Proc.*, 1994, to be published.

[2] T. Oshima, K. Yamaguchi, A. Yamada, M. Koganai and K. Takahashi, "Improvement of Film Quality of a-Si:H Deposited by Photo-CVD Using SiH_2Cl_2", *Mat. Res. Soc. Symp. Proc.* 1994, to be published.

[3] T. Yokoi, M. Azuma, and I. Shimizu, "Fabrication of Stable Hydrogenated Amorphous Silicon from SiH_2Cl_2", *Tech. Dig. Int'l PVSEC-7,* 1993, pp. 295-296.

[4] W.B. Jackson and N.M. Amer, "Direct Measurement of Gap-State Absorption in Hydrogenated a-Si by Photothermal Deflection Spectroscopy", *Phys. Rev.* ***B25***, 1982, pp. 5559-5564.

[5] A.A. Langford, M.L. Fleet, B.P. Nelson, W.A. Lanford, and N. Maley, "Infrared Absorption Strength and Hydrogen Content of Hydrogenated Amorphous Silicon", *Phys. Rev.* ***B45***, 1992, pp. 367-377.

[6] I. Shimizu, J. Hanna and H. Shirai, "Control of Chemical Reactions for Growth of Crystalline Si at Low Substrate Temperature", *Mat. Res. Soc. Symp. Proc.* **164**, 1990, pp.195-204.

[7] *CRC Handbook of Chemistry and Physics* (59th edition, CRC Press, Florida, 1978), pp. F224-F230.

LOW BANDGAP a-Si:H FILM WITH BETTER STABILITY PREPARED BY RF PECVD METHOD USING HELIUM DILUTION

Swati Ray, Sukti Hazra, A.R.Middya and A.K.Barua
Energy Research Unit, IACS, Jadavpur, Calcutta 700 032, India

ABSTRACT

Low bandgap a-Si:H films have been prepared by RF PECVD method (13.56 MHz) using helium as diluent to silane gas. Helium dilution and chamber pressure play an important role to reduce optical gap at substrate temperature ~210°C. This highly photosensitive low bandgap material shows less light induced degradation compared to that for normal bandgap a-Si:H material. Single juntion solar cell having efficiency 7.2% (1cm^2 area) and spectral response upto 850 nm have been fabricated using 1.6 eV a-Si:H film.

INTRODUCTION

In the present stage of amorphous silicon solar cell technology, development of double and triple junction solar cell is of prime interest[1,2]. To attain its full potential it is very important to develop good quality material having band gap lower than that of usual a-Si:H, to be used as bottom cell of multijunction solar cell. Intensive research during the last decade on the development of low bandgap a-SiGe:H has resulted in considerable improvement in its properties[3-5] leading to fabrication of multijunction cells having efficiency ~12-13%[6]. However there are some basic problems with this material and it may not be possible to further improve its properties significantly. The development of low bandgap a-Si:H film under conditions suitable for solar cell fabrication is an attractive alternative in this regard. Since most of the light induced degradation in a multijunction solar cell comes from the bottom cell , it will be very useful to have low bandgap a-Si:H film with lower light induced degradation. However low bandgap a-Si:H, which have been prepared still now, are deposited at substrate temperature higher than 300°C[7]. In this paper we have reported the development of low bandgap a-Si:H films by the usual RF PECVD method (13.56 MHz) and preliminary results of fabrication of solar cell with this material.

EXPERIMENTAL DETAILS

The films have been prepared from a mixture of silane and helium in a Plasma-CVD system (Anelva Corp. Japan) with the substrate temperature at 210°C. The chamber pressure has been varied between o.5 to 2.0 torr. and r.f. power was 15 mW/cm^2. The optical bandgap has been calculated from Tauc's Formula. Mobility lifetime product ($\eta\mu\tau$) measurement has been done under same electron hole generation rate 3 X 10^{18} cm^{-3}sec^{-1}. The bonded hydrogen content has been estimated from infrared vibrational spectra. The midgap defect density and Urbach energy have been determined by Electron Spin Resonance Technique and Constant Photocurrent Method respectively. The solar cell has been

CH3365-4/94/0000-0551 $4.00 © 1994 IEEE

fabricated in multichamber deposition system with ultrahigh vacuum facility (Glass Tech. Solar Inc.) on textured TCO coated glass substrate.

RESULTS AND DISCUSSIONS

In the optimization of deposition conditions, the chamber pressure (P_r) has played a dominant role in controlling the optical gap (E_{opt}) of a-Si:H films. The variations of E_{opt}, growth rate (R_D) and bonded hydrogen content (C_H) with P_r have been shown in figure 1. As P_r increases above 1.0 torr, R_D values increase sharply. E_{opt} decreases monotonically from 1.68 eV to 1.51 eV

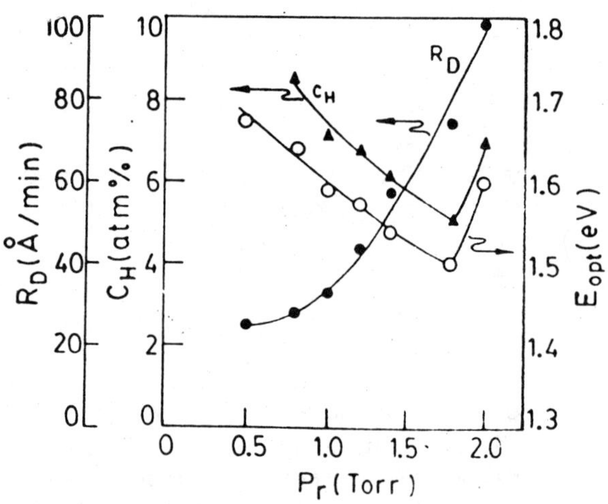

Fig.1. Variations of R_D, E_{opt} and C_H with P_r.

as P_r increases from 0.5 torr to 1.8 torr. However for P_r > 1.8 torr both E_{opt} and C_H increases. Variations of $\eta\mu\tau$, σ_D and σ_{Ph}/σ_D of the films with the chamber pressure have been shown in fig 2. Defect density (N_S), as measured by ESR technique and Urbach energy, calculated from the slope of constant photocurrent measurement are given in table I. Values of N_S at 1.51 eV is at least one order less compared to that of a-SiGe:H alloy of same bandgap[8].

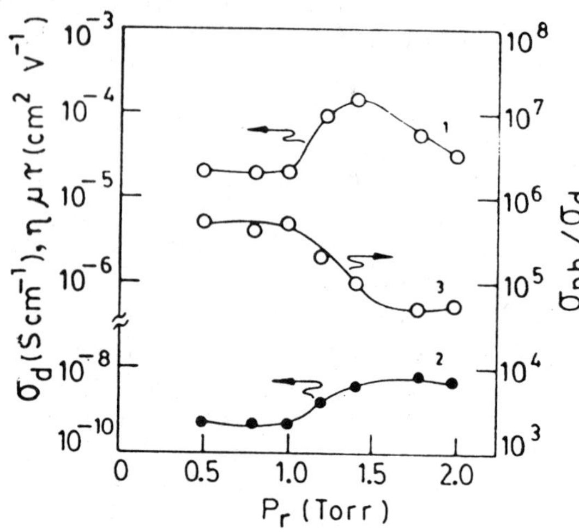

Fig.2. Variations of $\eta\mu\tau$, σ_D and σ_{Ph}/σ_D with P_r.

Table 1. Defect densities and urbach energies of a-Si:H films having different optical gaps.

optical gap (eV)	Defect density ($cm^{-3}eV^{-1}$)	Urbach energy (meV)
1.68	$9.0X10^{15}$	51
1.57	$7.5X10^{15}$	52
1.51	$7.8X10^{15}$	54
1.60	$9.0X10^{15}$	59

The phenomenon of lowering of optical gap of a-Si:H films prepared with helium dilution may be explained as follows. With the increase in chamber pressure the energy gaining mechanism of the electrons in plasma moves from α-regime to γ-regime. These highly energetic electrons dissipate their energy by exciting helium atoms to their metastable state (He^*,20eV) and ionization level (He^+,24eV) instead of producing ionic precursors. On deexcitation of these He^*, He^+ species energy is transfered to the reactive species as well as to the growing surface which enhances surface mobility to the precursors. This leads to better structural relaxation and more compact network of a-Si:H. As a consequence bonded hydrogen concentration is lowered and optical gap becomes low. It may be seen from the optoelectronic properties and defect density data that the quality of the material is good even at 1.51 eV optical gap.

In fig.3 the change of σ_{Ph} of low bandgap a-Si:H films under light soaking (100 mW/cm^2) is shown as function of light soaking time. It may be seen that σ_{Ph} values degrade initially quite fast but saturate very quickly (within 10 hrs). It has been observed that with increase of P_r as E_{opt} of the material decraeses to 1.54eV saturated values of σ_{Ph} remains higher than $10^{-5} Scm^{-1}$. After 300 hrs of light soaking which at least one order higher than the

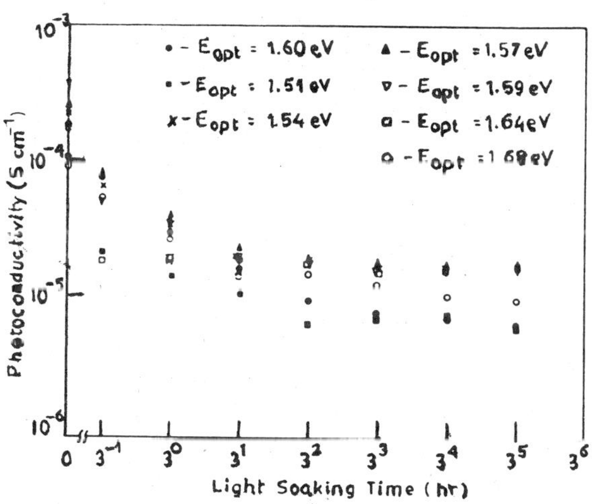

Fig.3. Light induced degradation of σ_{Ph} at different optical gap.

values attained in case of a-SiGe:H. The degradation is less also compared to normal bandgap a-Si:H[7]. This shows that the network of a-Si:H prepared by helium dilution contains less weak bonds and microstructural defects. For films prepared at $P_r \geq$ 1.8 torr the degradation is slightly more. Low bonded hydrogen content and dominant monohydride peak at 2000 cm^{-1} of IR vibrational spectra of low bandgap a-Si:H material also support the suggestion of less macroscopic defects.

Preliminary work on the fabrication of single junction a-Si

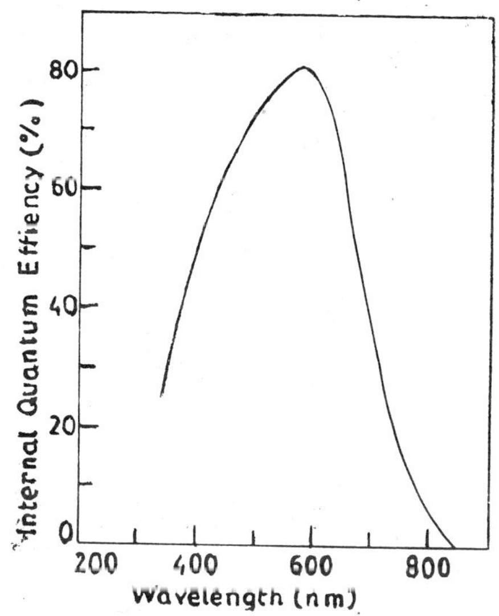

Fig.4. Spectral response of solar cell with intrinsic layer having optical gap 1.6eV.

solar cell having the structure glass /TCO / p-a-SiC:H / a-SiC:H / a-Si:H / n-a-Si:H /Al has yielded efficiency 7.2% ($1 cm^2$ area) under AM1.5 ($100 mW/cm^2$) radiations with V_{oc}=.850V, J_{sc}=14.4mA/cm^2 and F.F=.59. The 1.6eV bandgap intrinsic silicon layer has been used and the internal quantum efficiency plot is shown in fig.4. The spectral response is extended upto 850 nm whereas in case of usual a-Si:H solar cell it extends upto 760 nm.

CONCLUSIONS

Low bandgap (≥ 1.5eV) amorphous silicon thin films with high values of $\eta\mu\tau$ and photosensitivity have been developed at comparatively low temperature i.e, 210°C for the first time. These a-Si:H films prepared from SiH_4+ He mixture have low defect density and low hydrogen content. Degradation is less compared to that for a-SiGe:H of same bandgap. The quantum efficiency of the solar cell fabricated with low bandgap a-Si:H shows higher response at longer wavelength compared to that with usual a-Si:H layer. As the preparation temperature of the film (210°C) is compatible with the fabrication of solar cell, this low bandgap a-Si:H film can be used successfully as the bottom cell of double or triple junction solar cell.

ACKNOWLEDGEMENT

This work was supported by Ministry of Nonconventional Energy Sources and Bharat Heavy Electrical Limited.

REFERENCES

[1] S.Guha et.al., "Progress in Based Solar Cells and Modules" Technical Digest of the International PVSEC-7, Nagoya, Japan, 1993, pp. 43-46.

[2] K.Nomoto et.al., "a-SiAlloy Three Stacked Solar Cells with High Stabilized Efficiency" Technical Digest of the International PVSEC-7, Nagoya, Japan, 1993,pp. 275-276.

[3] A.Matsuda et.al., "Guiding Principle for preparing Highly Photosensitive Si based Alloys" J.Non-Cryst. Solids, **97&98**, 1987, pp. 1367-1374.

Properties and the Gap State Distribution in a-Si,Ge Alloys",

Amorphous Silicon and Related Materials, edited by H.Fritzsche, 1988, pp. 887-938.

[5] A.R.Middya and S.Ray, "influence of Deposition Rate and Hydrogen/ Helium dilution on the Structural relaxation of a-SiGe:H Network Prepared at Low Substrate Temperature", J. Appl. Phys., **75**, 1994,pp.7340-7348.

[6] S.Tsuda et.al., "a-Si Technologies for High Efficiency Soalr Cells" J. Non-Cryst. Solids **164-166**, 1993, pp. 679-684.

[7] H.Shirai et. al.,"A Novel Preparation Technique for Preparing Hydrogenated Amorphous Silicon With a more Rigid and Stable Si Network", Appl. Phys. Lett., **59**, 1991, pp. 1096-1098.

[8] M.Stutzman et. al., "Structural, optical and Spin Properties of Hydrogenated Amorphos Silicon-germanium alloys" J. Appl. Phys., **66**, 1989, pp. 569-592.

NEW EVALUATION TECHNIQUE FOR THIN-FILM SOLAR CELL BACK-REFLECTOR USING PHOTOTHERMAL DEFLECTION SPECTROSCOPY

Xunming Deng and K. L. Narasimhan
Energy Conversion Devices, Inc., 1675 West Maple Road, Troy, Michigan 48084

ABSTRACT

We report a new technique to determine the optical loss of a thin-film solar cell back-reflector using Photothermal Deflection Spectroscopy (PDS). We modified a conventional PDS technique to measure the reflection loss of highly reflective and highly textured back reflector surfaces. PDS has demonstrated its advantages in such measurements because, unlike conventional reflectance measurements, it measures the fraction of the light that is converted to heat.

We then used this technique for developing new back-reflectors. We developed a Hot-Ag/Cold-Ag two-layer system in which the texture is provided by the bottom Ag layer deposited at high temperature and low deposition rate, and the reflectance is provided by the top Ag layer deposited at low temperature and high deposition rate. Such a highly reflective and highly textured back-reflector system enhanced the quantum efficiency (QE) of a-Si solar cells both in the red and in the blue regions.

INTRODUCTION

A high performance back-reflector plays an important role in achieving high solar cell efficiency. In particular, for a-Si solar cells deposited on a stainless steel substrate with a structure such as a SS/back-reflector/triple-n-i-p/TCO, a high performance back-reflector must provide both reflectance and surface texture [1]. This type of back-reflector enhances the solar cell efficiency in two ways: 1) the highly reflective surface reflects unabsorbed light back into the solar cell; and 2) the highly textured surface provides efficient light trapping.

In order to further improve back-reflector performance, we need methods to reliably measure its reflection and light trapping capability. The optical loss of a back-reflector is conventionally evaluated by measuring the reflectance using a spectrophotometer equipped with an integrating sphere, and then subtracting the reflectance from the incident light intensity to get the optical loss. However, for a good back-reflector, which is highly reflective and textured, this technique becomes less reliable.

Back-reflector performance is also often evaluated with a complete solar cell device, such as an SS/back-reflector/single n-i-p solar cell/ITO. The long wavelength response in the solar cell quantum efficiency (QE) curve depends sensitively on the back-reflector. A solar cell device is indeed one of the most important tools to evaluate the gain of a back-reflector, however, variation in many parameters, such as the semiconductor layer thickness also strongly affects the solar cell quantum efficiency. Furthermore, it cannot be determined where the optical loss occurs in a complete device from the QE curve measurement.

In this work, we report a new technique to determine the back-reflector optical loss using PDS. Using this technique, we developed a new back-reflector structure with improved texture and reflectance, as will be described.

PDS TECHNIQUE FOR BACK-REFLECTOR EVALUATION

PDS is used extensively in the field of amorphous silicon (a-Si) research for measuring the density of defect states of a-Si thin films [2]. Recently, we modified the PDS technique to evaluate the performance of a solar cell back-reflector deposited on a stainless steel substrate. PDS has proven advantages in the measurement of optical loss, especially for highly reflective and textured surfaces, because (unlike conventional reflectance measurements) PDS measures the fraction of the light that is absorbed and converted to heat.

In a conventional PDS system, monochromatic light is focused on the sample which is immersed in CCl_4 in a cuvette cell. The portion of light that is absorbed by the sample and substrate is converted to heat while the remainder of the light, either transmitted through or reflected back, does not contribute to heat generation. The absorbed light heats up the local region of CCl_4 around the illuminated spot and forms a semisphere of CCl_4 with an increased local temperature and hence a different refractive index. This semisphere acts as a lens, and a HeNe probe beam that passes through the CCl_4 in front of the sample surface is deviated by the CCl_4 lens. A position sensor is then used to detect the shift of the HeNe light spot and the deviation of the light spot on the position sensor is proportional to the amount of heat absorbed by the sample surface. We measure the light intensity of different wavelengths using a thermopile at the position of the sample. Absorption as low as 0.01% can be easily measured by this method.

In current back-reflector research, a back-reflector, such as a textured Ag reflector, is mounted in the cuvette cell. The monochromator light beam is focused onto the back-reflector. With a small portion of the monochromatic light being absorbed by the sample, the majority of the incident light is specularly reflected, or in the case of a textured surface, scattered. The reflected or scattered light does not contribute to the signal that PDS measures since it is not converted into heat inside CCl_4. Therefore, with appropriate calibration, PDS is a powerful tool in measuring

CH3365-4/94/0000-0555 $4.00 © 1994 IEEE First WCPEC; Dec. 5–9, 1994; Hawaii

the light absorption loss in highly reflective and highly textured surfaces. Since the back-reflector is reflective over a wide spectrum, we cannot calibrate the PDS absorption measurement using the band-to-band absorption as in the case of a-Si sub-bandgap absorption measurement. Special procedures are therefore followed to achieve reproducibility of PDS signal for different samples.

The output signal we measure includes magnitude and phase information. With every new sample, we adjust the distance between the HeNe laser beam and the sample front surface to maximize the magnitude of the signal. When the PDS signal is maximized, the phase of the signal is always the same for the same type of substrate. The PDS phase is used as an independent check for the signal maximization.

In this new technique, the PDS signal for different wavelengths is calibrated with a Perkin-Elmer 330 Spectrophotometer using plain specular stainless steel (SS). Table 1 lists the PDS signal (V_{pds}) read from the PDS lock-in amplifier and absorption (A) measured with a Perkin-Elmer Spectrophotometer, for a plain stainless steel substrate, as well as the method we used to obtain the calibration factor (C), which is the ratio of absorption (A) to PDS signal (V_{pds}). The calibration factor C in Table 1 was obtained after averaging the measurements for the samples. A plain stainless steel sample is used for the calibration because it is not textured and the absorption is around 30%, which is within the range that the spectrophotometer measures most accurately. When we measure other samples with PDS, we obtain the absorption loss by multiplying V_{pds} with the calibration factor C ($=A/V_{pds}$), which should be the same for all back-reflector samples on stainless steel substrate.

In Table 2, we list the absorption measurement of specular Ag (sample BR13) and the reference sample (BR00) using PDS and using the calibration factor we obtained above. We have identified that the optical loss at 827nm in the reference back-reflector (BR00) is 7.0%. The PDS signal also depends on the substrate. The V_{pds} of the sample on glass is approximately a factor of two higher than the V_{pds} for the same back-reflector sample on stainless steel due to the different thermal constants of the substrate materials.

We routinely check the system reproducibility of the PDS setup by measuring the reference sample BR00. The measurement of this sample has always been reproducible, within a few percent. The absolute PDS signal calibration depends, to some extent, on the surface conditions of the sample to be measured. We are continuing to further our understanding of these conditions in order to calibrate more reliably and accurately.

PDS can also be used to measure the light loss of different solar cell layers without a complete cell structure. In Table 2, we also include the PDS measurements of SS, SS/specular-Ag, SS/textured-Ag/ZnO, SS/specular-Ag/ZnO, SS/specular-Ag/ZnO/n^+ layers.

Besides the optical loss, another important characteristic of a back-reflector is its texture. We complement PDS measurement with 1) the measurement of angular dependence of scattering and 2) a scanning electron microscope. In the angular dependence of scattering measurement, we direct a HeNe laser beam to the back-reflector surface at approximately normal incidence, and measure the light intensity of the specularly reflected light and the scattered light. When the surface is more textured, the percentage of the specularly reflected light will decrease, and the percentage of scattered light at high scattering angle will increase. The texture of the surface is also studied with SEM photographs, taken at 60° electron beam incidence. Such pictures reveal the size and the depth of the texture. We then use these back-reflector evaluation techniques for developing new improved back-reflectors.

DEVELOPMENT OF HOT-Ag/COLD-Ag/ZnO BACK REFLECTOR

The Ag/ZnO back-reflector system that we currently use is highly reflective and textured. When Ag is deposited at high temperature and low deposition rate, it is

Table 1. Calibration of PDS signal with Perkin-Elmer 330 Spectrophotometer using a plain stainless steel sample.

Photon Energy hv(eV)	2.1	1.9	1.7	1.5	1.3	1.1
Wavelength(nm)	590	653	729	827	954	1127
Lock-in output Vpds(mV)	1.32	2.8	4.04	4.69	8.3	9.3
Absorption measured with Perkin-Elmer A	35.5%	34.5%	33.%	31.%	30.%	24.%
Calibration Factor C C=A/Vpds	25.9	12.1	8.4	6.9	3.7	2.6

Table 2. Reflection loss of various layers measured by PDS.

Sample	PDS data	Wavelength(nm)					
		590	653	729	827	954	1127
Stainless Steel (SS)	Vpds(mV)	1.32	2.80	4.04	4.67	8.30	9.30
	Ref'n Loss	34.1%	33.9%	33.9%	32.3%	30.3%	24.4%
SS/Specular-Ag	Vpds(mV)	0.09	0.18	0.23	0.23	0.39	0.41
BR13	Ref'n Loss	2.3%	2.2%	1.9%	1.6%	1.4%	1.1%
SS/Textured-Ag/ZnO	Vpds(mV)	0.57	1.00	1.03	1.00	1.57	1.50
BR00, reference	Vpds(mV)	14.8%	12.1%	8.7%	7.0%	5.7%	3.9%
SS/Specular-Ag/ZnO	Vpds(mV)	0.17	0.40	0.52	0.37	0.67	0.91
	Ref'n Loss	8.5%	4.8%	4.3%	2.6%	2.4%	2.4%
SS/Specular-Ag/ZnO/	Vpds(mV)	1.25	1.82	1.84	1.92	3.22	3.05
/thin n	Ref'n Loss	32.0%	22.0%	15.4%	13.2%	11.8%	8.0%

textured due to the Ag segregation. In order to further enhance the texture and reflectance, we analyze the optical loss using PDS, and analyze the texture using SEM, angular dependence of scattering and solar cell quantum efficiency. An evaporation system is used for exploring the different deposition conditions for the Ag layer. It was found that Ag that is evaporated at high substrate temperature and at low rate is more textured but less reflective, while Ag evaporated at low substrate temperature and at high rate is more reflective but less textured. Table 3 lists the PDS measurement for both Ag evaporated at high temperature (BR12) and at low temperature (BR13).

We have developed a Ag/Ag two-layer system in which the texture and reflectance are achieved by different Ag layers. We utilize the Ag, evaporated at high temperature and low deposition rate (Hot-Ag) to obtain ideal light trapping since the grain size of such Ag is around 0.3-0.4 μm, which is about the same as the wavelength of red light in ZnO, of which the refractive index is 2.0. Because the Ag evaporated at high temperature has relatively lower reflection compared to Ag evaporated at room temperature at high deposition rate (Cold-Ag), we cover the Hot-Ag with a layer of Cold-Ag to form a Ag/Ag double layer structure. In this way, the combined structure (Hot-Ag/Cold-Ag) can provide both efficient reflection and light trapping in solar cells. Table 3 includes the optical loss measured by PDS for a Ag/Ag two-layer system. Although the texture is much improved compared to BR12, as we found from SEM and measurements of the angular dependence of scattering, the reflection loss is smaller than BR12. Fig. 1 shows the SEM picture of a Ag layer deposited at 350°C and a Hot-Ag/Cold-Ag two-layer system.

On the other hand, the reflectance at 827nm for Hot-Ag/Cold-Ag (BR66) is 94.3% while that for specular Ag (BR13) is 98.4%, according to Table 3. It is likely that this slightly reduced reflectance is intrinsic due to the texture. The reflectance of highly specular Ag can be as low as 80% when the incident light is away from normal. For a textured surface, the incident light is microscopically at an angle, resulting in an increased reflection loss.

From Fig. 1 we also find that the size of the texture from Hot-Ag is around 0.4 μm. Therefore, in an SS/Ag/Ag/ZnO/n-i-p/ITO device, the red light is most efficiently scattered and trapped in the device due to the comparable texture size with the wavelength.

To find out whether the reflectance of the Cold-Ag layer degrades when the two-layer back-reflector system is subsequently subjected to high temperature, which it will be during the subsequent ZnO and n-i-p solar cell deposition processes, we annealed the Hot-Ag/Cold-Ag two-layer system at 400°C for 30 min. We observed that the reflectance is not reduced, in fact, it is even slightly improved.

We then deposited an n-i-p solar cell (with i layer thickness of approximately 4000 Å) on the Hot-Ag/Cold-Ag/ZnO back-reflector structure, where approximately 4000 Å of ZnO was deposited using RF sputtering from a ZnO target in a RD Mathis sputtering system. The QE curve of this solar cell is shown in Fig. 2. As a comparison, the QE curve of the same sample on the reference back-reflector, BR00, is also shown in Fig. 2. The J_{ph} from QE for the device on Hot-Ag/Cold-Ag is 17.2 mA/cm^2, while that for the same device on the reference back-reflector is 16.8 mA/cm^2. There are improvements in QE response in both the red and blue regions. The increased red response is due to the reduced reflectance loss and improved light trapping. The increase in the blue response is probably due to the decrease in the reflection loss of the ITO coating when the surface is textured, since 1) normal incident light microscopically enters the ITO at a non-normal incident angle; and 2) the ITO is effectively thinner, which shifts the reflection minimum toward the blue. For this reason, we obtained QE=73% at 400nm for a solar cell deposited on the Hot-Ag/Cold-Ag/ZnO back-reflector, as shown in Fig. 3.

SUMMARY

We have developed a new back-reflector evaluation technique using PDS. This technique is very effective in measuring the optical loss of highly reflective and highly textured surfaces. We used this technique to further improve the back-reflector system. A Hot-Ag/Cold-Ag/ZnO back-reflector system was developed which uses two Ag layers to achieve high texture and high reflectance separately. The improved back-reflector performance is demonstrated in solar cells.

ACKNOWLEDGMENTS

We thank S. R. Ovshinsky, M. Izu, H. Fritzsche, S. Guha, and J. Yang for stimulating and important discussions, and A. Myatt and J. Evans for experimental help. This work was partially supported by DOE/NREL PVMaT 2A Program under subcontract No. ZM-2-11040-7.

REFERENCES

[1] S. Guha, "Research on High-Efficiency, Multiple-Gap, Multi-Junction Amorphous Silicon-Based Alloy Thin-Film Solar Cells", ECD Final Subcontract Report to SERI, under Subcontract ZB-7-06003-4, (1990).

[2] W.B. Jackson, N.M. Amer, A.C. Boccara, and D. Fournier, Appl. Opt. 20, 1333 (1981).

Table 3. PDS measured optical loss at 827 nm for different Ag layers.

Sample	Type		Ts (C)	d (A)	Dep. Rate (A/sec)	PDS data	827nm 1.5eV
BR12	Ag		350	1500	2.0	Vpds(mV)	1.28
						Loss(%)	8.8
BR13	Ag		25	1500	2.0	Vpds(mV)	0.23
						Loss(%)	1.6
BR66	Ag/Ag	top Ag	25	1500	15	Vpds(mV)	0.83
		bottom Ag	400	1000	1.3	Loss(%)	5.7

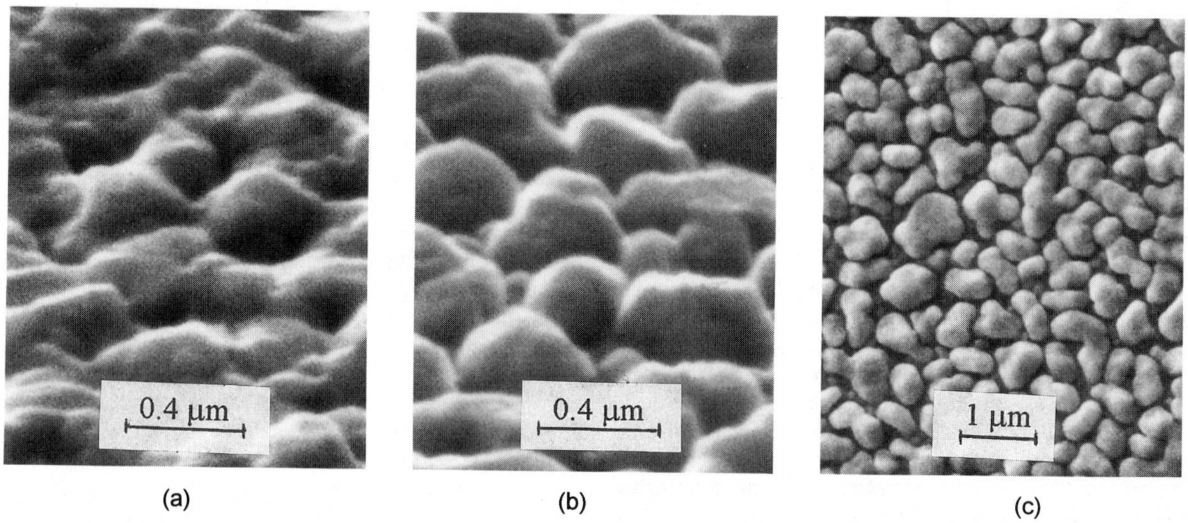

Fig. 1. SEM photographs of different Ag layers. (a) Ag evaporated at 350 C, taken at 60° angle with 40,000x magnification; (b) Hot-Ag/Cold-Ag two layer back-reflector system, taken at 60° angle with 40,000x magnification; and (c) Hot-Ag/Cold-Ag two layer system, taken at normal incidence with 10,000x magnification.

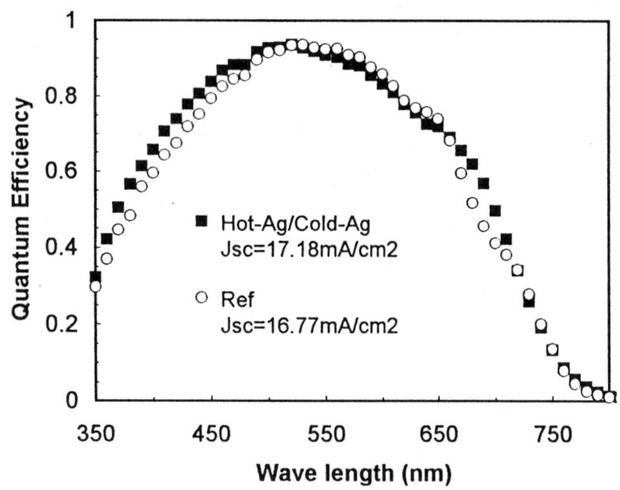

Fig. 2. Quantum efficiency curve of a-Si n-i-p solar cells on the Hot-Ag/Cold-Ag/ZnO back-reflector and on the Ag/ZnO reference back-reflector.

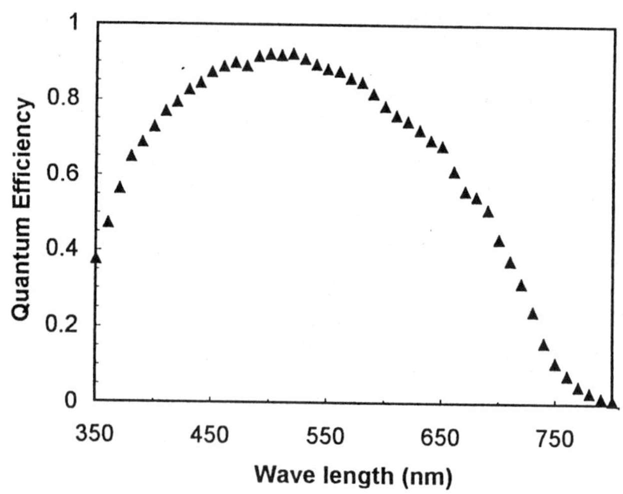

Fig. 3 Quantum efficiency curve of an a-Si n-i-p solar cell with QE=73% at 400 nm.

High Electron and Hole Drift Mobility in a-Si:H

Gautam Ganguly and Akihisa Matsuda
Electrotechnical Laboratory, Tsukuba City, Ibaraki, 305, Japan

ABSTRACT

The mobility of electrons and holes are crucial to the performance of a-Si:H based solar photovoltaic devices. We have found that the mobilities of both electrons and holes can be varied over 2 orders of magnitude by control of the ion bombardment energies using the triode type PECVD. The carrier mobilities in the best samples so obtained are an order of magnitude higher than those available hereto, which should considerably enhance the performance of a-Si:H solar cells.

INTRODUCTION

The hole mobility and the width of the valence band tail states distribution are important parameters for determining the performance of a-Si:H photovoltaic solar cells [1]. There have been reports of a wide variation in the mobility and its activation energy [2]. However, the connection between the growth conditions and the mobility have not been identified. Recently, we have systematically scanned the effect of most common deposition parameters but have found little variation in the carrier mobilities [3]. In this paper we report the effect of controlling the bias voltage applied to a mesh type third (control) electrode in a triode type parallel plate radio frequency plasma enhanced chemical vapour deposition apparatus [4]. The electron and hole drift mobilities measured using the time-of-flight (TOF) photo-current technique [4] have been found to vary by over two orders of magnitude when the bias voltage applied to the control electrode is varied. We interpret this as being a result of that of variation of the energy of ions bombarding the growth surface which modifies the material structure and hence the density of carrier trapping states.

EXPERIMENTAL

The schematic configuration of the reactor used for film fabrication is shown in Fig.1. The electrodes are aligned vertically to avoid dropping of flakes during the extended deposition period required to fabricate the 3-5mm thick samples required for the TOF measurements. The samples were deposited on n+ and p+ crystalline silicon wafers which had been precoated with gold and annealed to obtain ohmic back contacts. Semi-transparent Schottky diodes were formed by evaporating 0.5mm^2 gold electrodes on the a-Si:H films. The TOF measurement apparatus was conventional [4] and the transit time was identified from the break point in a double logarithmic plot of photocurrent versus time. The measurements were carried out over the temperature range -30°C to 80°C for hole transport and between -190°C and -50°C for electrons using fields of 5 - 100 kV/cm.

CH3365-4/94/0000-0559 $4.00 © 1994 IEEE

First WCPEC: Dec. 5–9, 1994; Hawaii

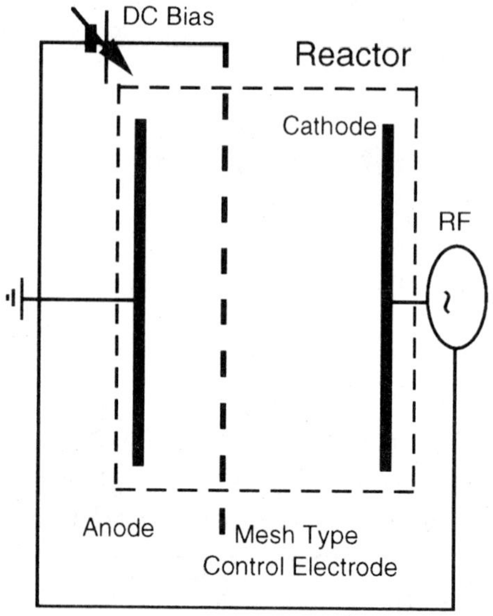

Fig. 1.Schematic representation of the reactor configuration for sample preparation.

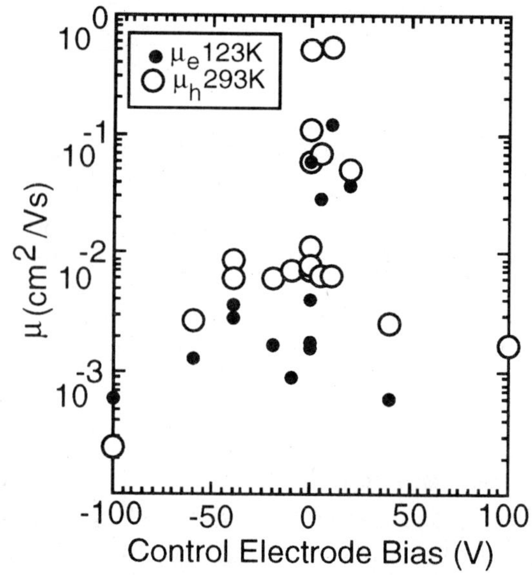

Fig. 2. The electron (•) and hole (O) drift mobilities at fields of 50kV/cm, plotted as a function of the bias applied to the control electrode.

RESULTS AND DISCUSSION

In Fig. 2 we have plotted the drift mobility of holes measured at room temperature(293K) and of electrons at 123K using fields of 50kV/cm, as a function of the bias applied to the control electrode. We find that as the bias voltage is changed there is a sharp variation of the drift mobility of both types of carriers over more than two orders of magnitude. The ion densities in the kind of conventional plasma conditions used here are two to three orders of magnitude smaller ($\approx 10^9$ cm^{-3}) [5] than the neutral radical densities (10^{12} cm^{-3}) [6], and therefore make only a negligible contribution to the film growth in terms of numbers, even when a factor of 10 difference in the sticking probability is taken into consideration [8]. However, the energy of the ions are much larger than the thermal energies of the neutral radicals. Therefore, it can reasonably be expected that ion energies would alter the structure of the material.

We know that the dangling bond densities in the material is determined byprecursor reaction kinetics on the growth surface in the absence of heavy ion bombardment [7]. However, Fig. 2 suggests that the mobility, which measures the effect of medium range order on the carrier transport, can be significantly altered by the energy of ion bombardment. In fact, it can be seen that at optimum ion energies, the structure can be made significantly better than in conventional material. The ion energies in a conventional plasma is determined by the 'plasma potential'. The plasma is at a positive potential with respect to the surrounding space to prevent the out diffusion of electrons which escape faster than the ions that are heavier. The anode in a parallel plate reactor carries the substrates and is usually grounded and thus electrons can escape. Hence the plasma potential is positive with respect to the anode which causes positive ions to bombard the film growth surface. When an insulating substrate is used the electrons

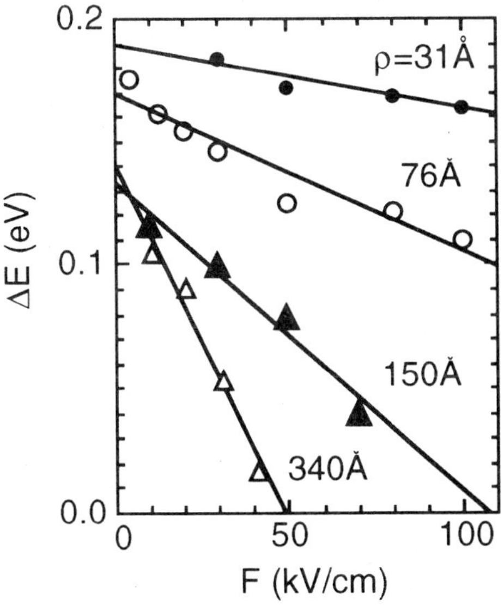

Fig. 3(a) The activation energy of the electron mobility versus the applied field for samples with 123K electron mobilities of (●) 4×10^{-4}, (○) 2×10^{-3}, (▲) 6×10^{-2} and (△) 1.2×10^{-1} cm^2/Vs.

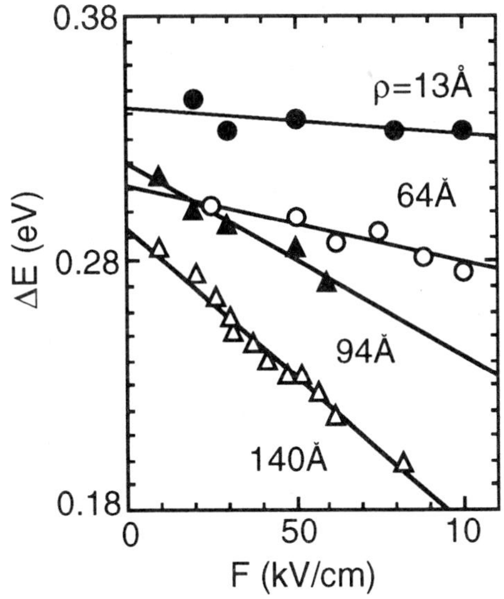

Fig. 3(b) The activation energy of the hole mobility plotted versus the applied electric field for samples with 293K hole mobilities of (●) 2.5×10^{-4}, (○) 2×10^{-2}, (▲) 8×10^{-2} and (△) 5×10^{-1} cm^2/Vs.

cannot escape so that the ion energies become smaller. Thus, in order to control the ion energy it is important to use conducting substrates. The control electrode is able to modify the plasma potential profile to alter the acceleration of the ions [8]. Therefore, in Fig. 2 the large bias voltage region gives rise to low mobilities while biases less than 40V of either polarities give rise to conventional material with high mobility material being formed close to zero bias. This optimum bias is a consequence of the plasma potential in the reactor we have used and would be different for another reactor. Thus for most reactors with plasma potential of a few tens of volts, the ion bombardment energies would give rise to conventional mobilities. However, if the plasma potential exceeds about 40V, the material produced using a diode system would be low mobility material irrespective of other deposition conditions. This explains the observation of low mobility in otherwise device quality material. [2]

In order to obtain information on the changes in medium range order associated with the changes in mobility we have analyzed the temperature dependence of the mobility at different applied fields. Theactivation energy of the electron drift mobility decreases as the applied field increases as shown in Fig. 3(a) for several samples. Similar data for holes are shown in Fig. 3(b). According to the model of dispersive carrier transport by variable energy/distance hopping amongst traps, the average distance between hopping sites (ρ) is related to the rate of decrease of the average hopping barrier height (reflected in the activation energy, ΔE) with the applied electric field (F)

$$\Delta E = \Delta E_o - (e\rho F/2\alpha_i) \quad (1)$$

where α_i is the dispersion parameter [9]. The values of ρ calculated using equation 1 from the data shown in Fig. 3 are indicated in the figure. We find that the increase of ρ correlates with an increase of the mobility. This implies that the samples with higher mobilities have fewer traps so that the average distance between hopping sites, ρ becomes longer. Thus, as ρ for electrons increases from 31Å to 340Å, the corresponding density of traps decrease from $\approx 10^{18}$ cm^{-3} to $\approx 10^{16}$ cm^{-3}. A similar variation is observed for the hole trap density.

This decrease of the carrier trap density is a feature that is observed even when the trap limited band transport model [10] is used to analyze the mobility data. In that case the reduction of the trap density is reflected as a sharpening of the band tail density of states distribution.

CONCLUSIONS

It has become clear from the above results that the ion bombardment energy is capable of altering the shallow trap density which determines carrier mobility in a-Si:H. The material having high carrier mobility and reduced trap density should yield significant improvement of the carrier collection in a-Si:H solar cells. The same properties also imply reduced carrier trapping and recombination in the solar cell which should reduce the light induced degradation of these devices.

The origin of the above mentioned shallow traps could be 'potential fluctuations' or 'weak bonds' which are believed to be the latent sources for as deposited and light induced defects. It would therefore be interesting to verify whether the stable and metastable dangling bond density are reduced in the material having a reduced trap density. These aspects of this new material are presently under investigation.

REFERENCES

[1] M. Hack and H. Shur, 'Analysis of light-induced degradation in amorphous silicon solar cells', *J. Appl. Phys.* **58**, 1985, pp.1656-1661.

[2] J. M. Marshall et. al., 'Band Tail States in a-Si:H - Influence of preparation conditions and techniques for characterization', *J. Non-Cryst. Solids*, **114**, 1989, pp. 339-341.

[3] G. Ganguly and A. Matsuda, 'Control of electron and hole drift mobilities in a-Si:H', *MRS Proc.* **336**, 1994, to be published.

[4] W.E. Spear, 'Drift mobility techniques for the study of electrical transport properties in insulating solids', *J. Non-Cryst. Solids*, **1**, 1968, pp. 197 - 214

[5] A. Matsuda and K. Tanaka, 'Plasma Spectroscopy - Glow Discharge Deposition of Hydrogenated Amorphous Silicon', *Thin Solid Films* **92**, 1982, pp. 171 -187.

[6] N. Itabashi et. al., 'Spatial Distribution of SiH_3 Radicals in RF Silane Plasma', *Jpn. J. Appl. Phys.*, **29**, 1990, pp. L505 - L507.

[7] G. Ganguly and A. Matsuda, 'Defect formation during growth of a-Si:H', *Phys. Rev.* **B 47**, 1993, pp. 3661 - 3670; 'Importance of surface processes in defect formation in a-Si:H' *J. Non-Cryst. Solids*, 164-166, 1993, pp. 31-36.

[8] N. Sano et. al., 'A new technique for diagnostics of a rf parallel plate remote plasma', *Appl. Phys. Lett.* **65**, 1994, pp. 162 -164.

[9] K. Murayama, et. al. 'Electron time of flight experiment under high electric field in a-Si:H', *Solid State Commn.* **81**, 1992, pp. 887 - 890.

[10] J.M. Marshall et. al., 'Electron Drift Mobility in a-Si:H', *Philos. Mag.* **B54**, 1986, pp. 51 - 60.

In situ CHARACTERIZATION OF GROWING HYDROGENATED AMORPHOUS SILICON THIN FILMS BY P-POLARIZED LASER LIGHT REFLECTION MEASUREMENT

N. Naito, M. Sumiya, M. Kawasaki, and H. Koinuma

Research Laboratory of Engineering Materials, Tokyo Institute of Technology
4259 Nagatsuta-cho, Midori-ku, Yokohama-shi, 226, Japan

ABSTRACT

In situ p-polarized laser light reflection measurement was applied to the determination of optical constants of growing a-Si:H thin films as well as to the detection surface transient process in plasma CVD. Turning off the plasma induced, a slight but noticeable change in the reflectance. This reflectance change can be well explained by taking into account the surface relaxation process. When 2MHz supersonic vibration was applied to the substrate during the plasma CVD of a-Si:H, it was revealed that the refractive index and absorption coefficient of a-Si:H increased as compared with those deposited without the supersonic vibration. The result indicates that the supersonic vibration works to form densified a-Si:H network, (presumably by improving the surface reaction.)

INTRODUCTION

For achieving highly efficient and stable solar cells, much effort has been focused on the fabrication of high quality and stable a-Si:H (hydrogenated amorphous silicon) thin films by silane plasma CVD. The understanding and controlling the surface reaction in plasma CVD should be primarily important for this purpose. Previously, we have reported that *in situ* optical reflectance measurement of p-polarized light obliquely impinged on growing a-Si:H film surface not only determined the optical constants of the growing film [1] but also detected the transient process in which the surface growth zone [2] was relaxed into bulk amorphous network by halting the plasma decomposition of SiH_4 [3]. For controlling the surface reaction, as a decisive step for determining the quality of a-Si:H films [4], we recently proposed and demonstrated a novel surface excitation method, i.e., supersonic vibration of the substrate during plasma CVD. The method was verified to be useful for the formation of a-Si:H films with low defect density [5] and high stability against photo-degradation [6].

Here we report the validity of *in situ* light reflection measurement for investigating the surface reaction with focuses on the surface reaction model and the supersonic vibration effect.

EXPERIMENTAL

a-Si:H films were deposited from pure monosilane by a triode plasma CVD, which was employed to avoid the particle formation in the light path [1]. The flow rate and pressure of SiH_4 were 10sccm and 50mTorr, respectively. The supplied radio frequency (13.56MHz) power density was 26mW/cm^2. During the deposition, linearly p-polarized laser light (He-Ne laser 633nm, AEROTECH OEM05SF) was impinged on a film surface at an incident angle of 72°~80° and the intensity of reflected laser light was detected by a silicon photo diode. The system and substrate setups are depicted in Fig.1 (a) and (b).

(a)

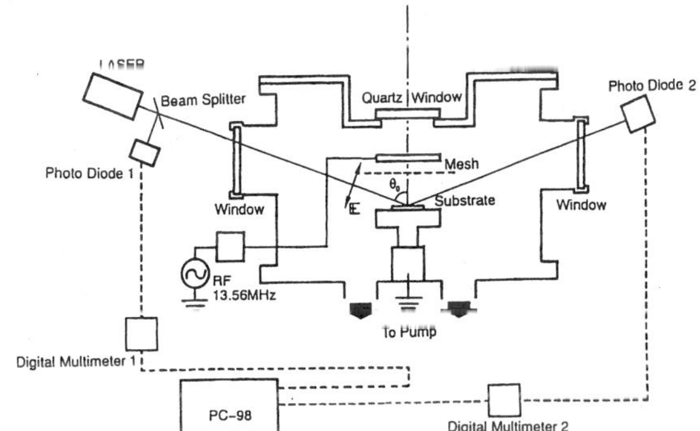

(b)

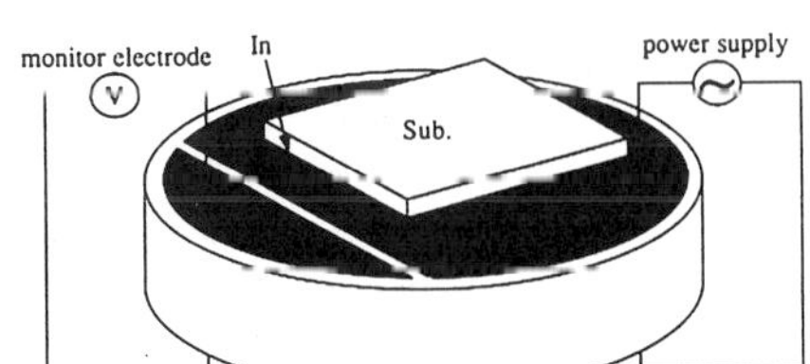

Fig.1. (a) Film preparation apparatus with *in situ* monitoring system. (b) The schematic of a substrate and PZT supersonic generator. This was placed on a substrate heater.

CH3365-4/94/0000-0563 $4.00 © 1994 IEEE

First WCPEC; Dec. 5–9, 1994; Hawaii

The stability of reflectance was tuned up to ±0.01%, by optimizing the experimental conditions, being high enough to determine not only the deposition rate, refractive index (n) and absorption coefficient (α) of the growing film surface but also the variation of these parameters when the film growth was interrupted. The details of optical measurement were described in our previous paper [1]. Corning 7059 glass and crystal silicon were glued with indium on a lead zirconium titanate oxide (PZT) disk (5cm in diameter and 2mm thick) placed on a substrate heater. By applying sinusoidal voltage of 30V peak to peak to the PZT disk, the substrate was vertically vibrated at a resonant frequency of 2MHz with an amplitude of 30nm peak to peak. The substrate surface was kept constant at a temperature between 90°C and 225°C. More details on the supersonic vibration experiments were described in a previous paper [5].

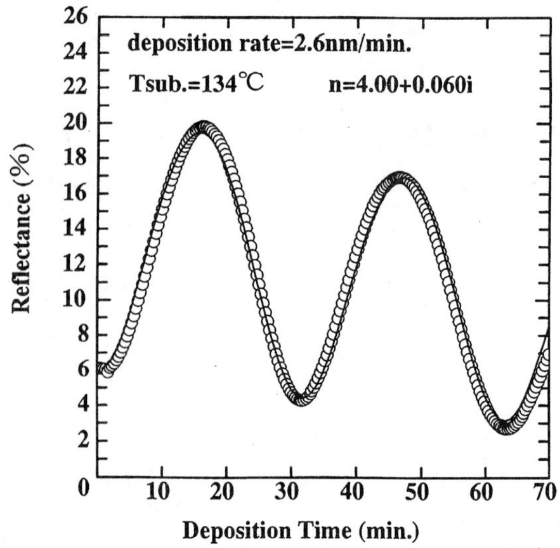

Fig.2. The p-polarized light reflectance observed during the deposition of a-Si:H film (open circle). Solid line shows the result calculated by using the optical constants as variable parameters.

RESULTS AND DISCUSSION

1. Reflectance measurement

First, we show in Fig.2 a typical reflectance variation during a-Si:H film deposition. The reflectance of the laser light oscillated due to the interference between the lights reflected at the top surface of the film and at the film/substrate interface. The reflectance R of p-polarized light is represented by the following equation considering multiple reflections.

$$R=\frac{l^2e^{2v\eta}+m^2e^{-2v\eta}+2lm\cos(\phi_{12}-\phi_{01}+2u\eta)}{e^{2v\eta}+l^2m^2e^{-2v\eta}+2lm\cos(\phi_{12}+\phi_{01}+2u\eta)} \quad (1)$$

where l, m, u, ν, ϕ_{01}, ϕ_{12} and η are functions of incident angle, film thickness (d) and optical constants (n, α) of vacuum, film and substrate. R is a function of n, α and d of the depositing film and the other parameters are constant. The oscillating reflection pattern was least-square fitted with the pattern (solid line in Fig.2) calculated by considering multiple reflections and using optical constants of the growing film and the deposition rate as variable parameters [1].

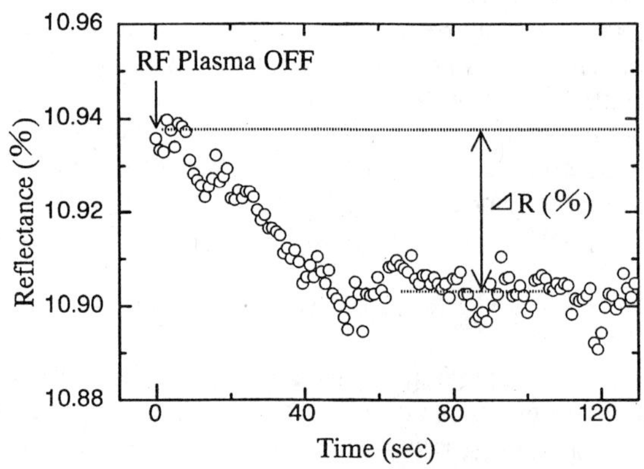

Fig.3. The time dependent variation of reflected laser light intensity after turning off the rf power. (0(sec) corresponds to the time of plasma off.)

2. Detection of growth zone relaxation

At first we detected a slight and gradual change in reflectance when the plasma was turned off. The incident angle of laser light was set to be 75°, close to the Brewster angle of a-Si:H (76°, 633nm). Figure 3 shows the time dependent variation of reflected laser light intensity when the film deposition at 200°C was halted by turning off the plasma. We constantly detected such slight and gradual changes in the reflected light intensity after terminating the film growth at various stages. For a reproducible observation of this slight change, tuning the reflectance stability up to ±0.01% was crucially important. The reflected intensity converged to a constant value in about 40 seconds at 200°C. The rate of the reflectance variation increased with the increase of substrate temperature to give an activation energy of 0.05eV [7]. Although the difference (ΔR in Fig. 3) between the reflectances during the deposition ($R_{growing}$) and after the relaxation ($R_{relaxed}$) was small, it was obviously dependent on the a-Si:H film thickness, namely the total period of film growth, as shown in Fig. 4.

The observed ΔR can hardly be explained without taking into account the presence of a transient layer at the growing top surface and its relaxation to bulk a-Si:H network when the film deposition was halted [3]. The variation of optical constants in this region is assumed to be represented by the following equations (2) and (3).

$$n_{obs}=n_{bulk}+(n_{growth\ zone}-n_{bulk})\exp(-d/d_{growth\ zone}) \quad (2)$$

$$\alpha_{obs}=\alpha_{bulk}+(\alpha_{growth\ zone}-\alpha_{bulk})\exp(-d/d_{growth\ zone}) \quad (3)$$

Here, n_{obs}, α_{obs}, n_{bulk}, α_{bulk}, $n_{growth\ zone}$, $\alpha_{growth\ zone}$, and $d_{growth\ zone}$ are the refractive indices and absorption coefficients of growing films, bulk films and the growth zone, and the specific thickness of the growth zone, respectively. The growth zone relaxes into the bulk network when the plasma is turned off. Under these assumptions, the ΔR value is calculated by the following equation.

$$\Delta R=R_{relaxed}(n_{bulk},\alpha_{bulk})-R_{growing}(n_{obs},\alpha_{obs}) \quad (4)$$

The solid line in Fig. 4 shows the reflectance difference ΔR plotted as a function of d. The calculation was made by assuming $n_{bulk}=4.0$, $\alpha_{bulk}=1.8\times10^4 cm^{-1}$, $n_{growth\ zone}=3.85$, $\alpha_{growth\ zone}=2.0\times10^2 cm^{-1}$, and $d_{growth\ zone}=$ 1.0nm. The calculated ΔR value agrees well with the observed data. The data obtained by changing the incident angle θ_0 to 80° were again fitted well with the calculation using the same parameters (n, α and d) and the same growth model [7]. These results strongly support the existence of the growth zone at the growing top surface. The growth zone has a lower density and a higher hydrogen content (lower absorption coefficient) than those of relaxed bulk amorphous network.

Figure 5 shows a model of a-Si:H film growth and surface relaxation derived from the results described above. Within a few atomic layers from the topmost growing surface, there should be a loosely bound growth zone originating from the radiation damage and high hydrogen content (see Fig. 5 (a)). Halting the film deposition should liberate the hydrogens and restructure the growth zone into a rigid (densified) bulk amorphous network (see Fig. 5 (b)).

3. Supersonic effect

By applying a supersonic vibration during the deposition under the conditions mentioned in this paper, the photoconductivity of a-Si:H was increased appreciably. In contrast, there were little effect on the hydrogen content and optical band gap. IR and UV spectra gave no detectable difference between a-Si:H films deposited with and without the vibration. For further elucidation of possible origin of the photoconductivity improvement in a-Si:H deposited with the supersonic vibration, we evaluated the optical constants by using *in situ* reflectance measurement. The incident angle was changed from 75° to 72°, a little more deviated from the Brewster angle (76°, 633nm), to make more sensitive to the bulk property rather than the surface. The refractive index n and the absorption coefficient α of growing films were evaluated from the simulation of observed oscillation pattern in the reflectances. The results

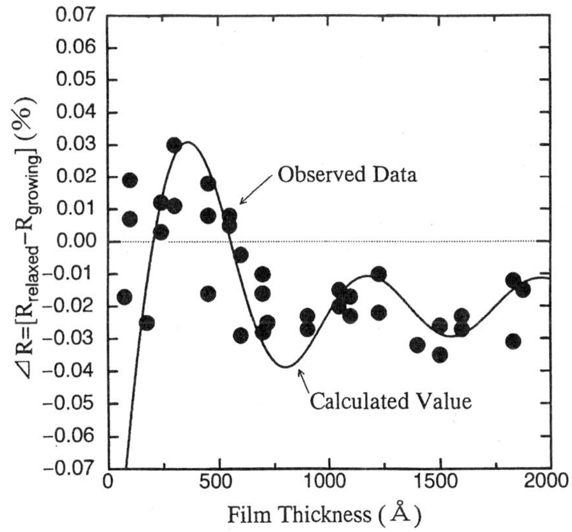

Fig.4. The reflectance variation change ΔR when the film growth was terminated. (closed circle: observed data, solid line: calculated data assuming the relaxation of the growth zone into the bulk network after halting the deposition)

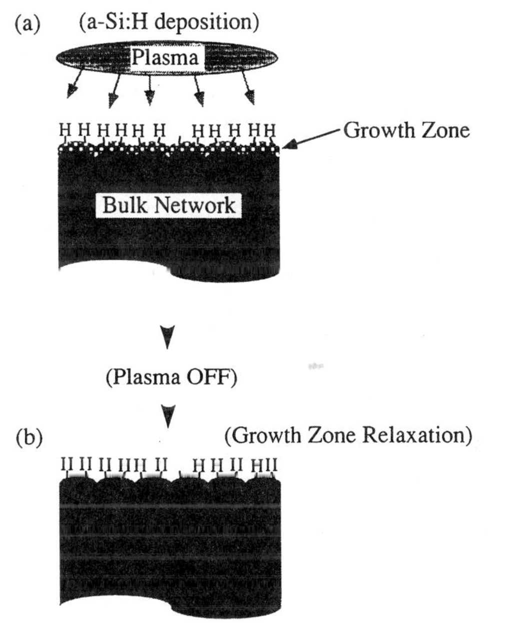

Fig.5. Schematic representation of a-Si:H growth during the deposition (a) and the growth zone relaxation after halting the plasma (b).

are plotted as a function of deposition temperature in Fig.6 (a) and (b), respectively. The values of n and α of a-Si:H increased with the increase of the deposition temperature (90~225°C). This is consistent with the fact that hydrogen content was decreased from 30% to 15% by increasing the substrate temperature in this range. Substantial differences exist both in n and α values between the films prepared with and without the supersonic vibration. The refractive index of film increased by the application of supersonic vibration, especially when the film was deposited at a temperature below 200°C. The absorption

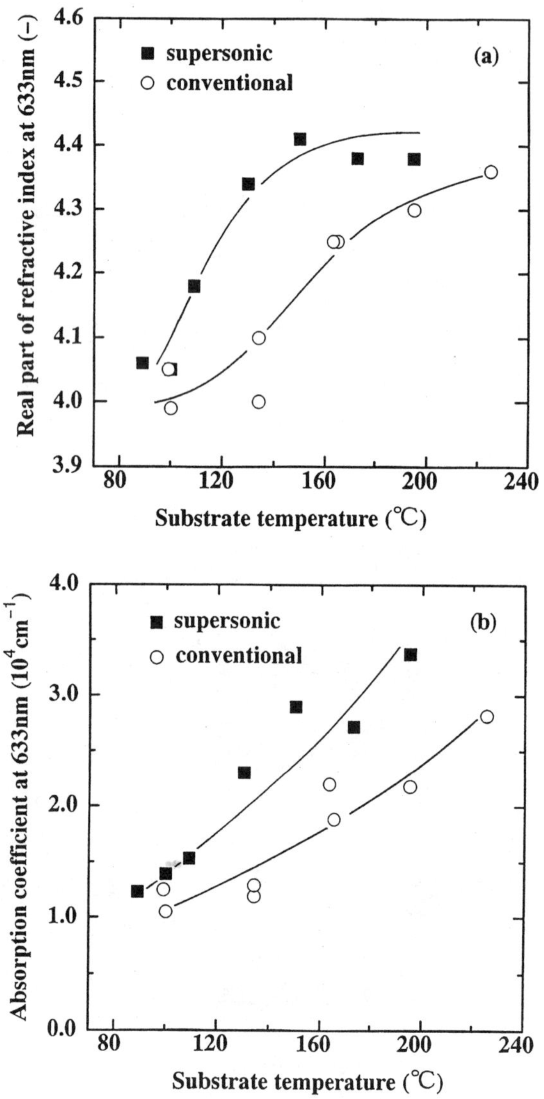

Fig.6. (a) The substrate temperature dependence of the real part of refractive index at the wavelength of 633nm. (b) The substrate temperature dependence of absorption coefficient at the wavelength of 633nm. (○:a-Si:H deposited without the supersonic vibration, ■: a-Si:H deposited with the supersonic vibration)

coefficient of film deposited with the vibration was improved as compared with that without the vibration in the entire temperature range examined in this study. This distinct difference in the refractive index strongly suggests that the supersonic vibration modulated the surface reaction to densify the amorphous network structure, even if the hydrogen content was unchanged. The increase in the absorption coefficient is presumed to come from an increase of density of extended states in the valence and conduction bands, since the photon energy of laser light is about 2eV, larger than the band gap of a-Si:H.

CONCLUSION

In situ p-polarized laser light reflection measurement was applied to plasma CVD of a-Si:H for examining surface relaxation process. The growth zone of several atomic layers thick should exist during the deposition of a-Si:H. By applying the supersonic vibration to the substrate during the plasma CVD, the refractive index and absorption coefficient in the resulting films increased presumably due to an improved in the surface relaxation process.

REFERENCES

[1] A. Takano, M. Kawasaki and H. Koinuma, "*In situ* Determination of Optical Constants of Growing Hydrogenated Amorphous Silicon Film by p-Polarized Light Reflectance Measurement on the Surface", *J. Appl. Phys.*, **73**, 1993, pp. 7987-7989.

[2] M. Kawasaki, Y. Matsuzaki and H. Koinuma, "Interface-Structure Analysis of Amorphous Semiconductor Heterojunctions by *in situ* X-ray Photoelectron Spectroscopy", *Phys. Rev. B.*, **39**, 1989, pp. 13316-13322.

[3] A. Takano, N. Naito, M. Kawasaki, and H. Koinuma, "Optical Detection and Dynamics of Surface Transient Processes in Plasma CVD of Hydrogenated Amorphous Silicon", *J. Non-Cryst. Solids*, **164-166**, 1993, pp. 115-118.

[4] G. Ganguly and A. Matsuda, "Defect Formation during Growth of Hydrogenated Amorphous Silicon" *Phys. Rev. B.*, **47**, 1993, pp. 3661-3670.

[5] M. Kawasaki, M. Sumiya, and H. Koinuma, "Piezoelectric Effect on Plasma Chemical Vapor Deposition of Hydrogenated Amorphous Silicon Films", *Mat. Res. Soc. Symp. Proc.* **297**, 1993, pp. 139-144.

[6] M. Sumiya, M. Kawasaki, and H. Koinuma, *Jpn. J. Appl. Phys.* in press.

[7] A. Takano, N. Naito, M. Kawasaki, and H. Koinuma, "The Observation of Growth Zone in Plasma CVD by *in situ* Optical Reflection Measurement", *Technical Digest of the Int'l PVSEC-7*, 1993, pp. 291-292.

OPTIMIZATION OF MULTIJUNCTION a-Si:H SOLAR CELLS USING AN INTEGRATED OPTICAL/ELECTRICAL MODEL

Richard E. Rocheleau and Matthias Vierthaler
Hawaii Natural Energy Institute, School of Ocean & Earth Science and Technology
University of Hawaii at Manoa, Honolulu Hawaii 96822

ABSTRACT

An phenomenological model for multijunction amorphous silicon solar cells which integrates a detailed optical model and an equivalent circuit model with a voltage-dependent photocurrent was developed. The model equations accurately describe the light J-V curves for a-Si:H and a-SiGe:H single junction cells using carrier transport properties comparable to values reported in the literature and which agree closely with values derived from quantum efficiency measurements. Closed form expressions for collection efficiency are derived by approximating the carrier generation profile by an exponential distribution. This new model can be used to predict cell behavior with changes in i-layer thickness and bandgap, carrier mobility-lifetime, and illumination. The model has been combined with a powerful multi-variable optimization routine to analyze alternative cell designs. The model is described and representative results are presented.

INTRODUCTION

The development of monolithic multijunction a-Si solar cells requires proper matching of the subcells. Several approaches of varying degrees of complexity have been used. Detailed optical models (e.g. 1-3) in conjunction with QE measurements allow i-layer thicknesses to be experimentally optimized. Optical models have been integrated with simple electrical models to characterize behaviors such as spectral dependence (e.g. 4) but these simple electrical models are generally not suitable for predicting cell behavior with changes in parameters such as i-layer quality. Hegedus has incorporated a voltage dependent light generated current into his equivalent circuit model (5), but does not include the effect of the carrier generation profile. Sophisticated microscopic models (e.g. 6-8), allow detailed analysis, but their complexity renders them inconvenient for many engineering and optimization studies.

We have developed a phenomenological model which integrates a detailed optical model and an equivalent circuit model with a voltage-dependent photocurrent (9,10). In this paper, we summarize the key elements of this new model and describe the validity of several simplifying assumptions. Examples of its use for cell optimization, and to evaluate the effects of spectral changes and i-layer quality on cell performance are presented.

OPTICAL MODEL

Our optical model, described in detail by Vierthaler (9), is based on models developed by Wiedeman et. al. (1) and Ellis et. al. (3) and accounts for all of the adsorption, reflection, and scattering losses included in the prior works. We obtained excellent agreement between simulated and experimental quantum efficiencies using optical parameters close to those suggested by Wiedeman (1) for the two cells used for calibration, Solarex L1324(4000Å, 1.72eV) and L1331(1000Å, 1.45eV). Of particular importance for our application is carrier generation rate as a function of position within each i-layer, G(x), which is also an output of the optical model.

COLLECTION EFFICIENCY

Equivalent circuit models based on superposition of a constant current source do not adequately describe the J-V behavior of amorphous silicon solar cells due to their poor transport properties. The electron (ξ_n) and hole(ξ_p) collection efficiencies, the ratio of the number of carriers that reach the interface to the number generated, depends on the carrier generation profile, carrier lifetime and mobility, and electric field. The photocurrent can be described as the product of the total generated current (J_g) times the collection efficiency.

$$J_{ph}(V) = J_g\xi(V) = \xi(V)q\int_0^W G(x)dx \qquad (1)$$

Previous models for collection efficiency have usually assumed either strongly or uniformly absorbed light (11-14). Unfortunately, neither assumption provides an accurate representation of the behavior of amorphous silicon solar cells operating under realistic spectra such as AM1.5 sunlight. Realistic generation profiles must be included in any analysis being used to evaluate the effects of i-layer quality (9).

The continuity and transport equations can be solved for an arbitrary distribution of the electric field (E) and carrier mobility lifetime product ($\mu\tau_n$ and $\mu\tau_p$) yielding

$$\xi_n = \left(1-\frac{\upsilon V_t}{E_p}\right)\int_0^W H(x)\exp\left(\int_x^W \frac{-d\kappa}{\mu\tau_n(\kappa)E(\kappa)}\right)dx \qquad (2)$$

CH3365-4/94/0000-0567 $4.00 © 1994 IEEE

First WCPEC; Dec. 5–9, 1994; Hawaii

$$\xi_p = \int_0^W H(x)\exp\left(\int_0^x \frac{-d\kappa}{\mu\tau_p(\kappa)E(\kappa)}\right)dx \quad (3)$$

where W is the i-layer thickness, x and κ represent the position within the i-layer, and H(x) is an arbitrary carrier generation profile. A correction term to account for electron back diffusion into the p-layer has been included in Eq. 2 where υ is the absorption constant and E_p the electric field at the p-i interface (11).

Collection efficiency can be calculated by numerical integration of Eqs. 2 and 3 with appropriate values of $\mu\tau$ and functions for H(x) and E(κ). In the next section, we develop an analytic expression for H(x) that accurately represents the real generation profile but which still allows symbolic integration of Eqs. 2 and 3 under the assumption of a uniform field. We discuss the validity of these simplifications by comparing results to numerical solutions. The analytic expressions for the voltage-dependent light generated current are incorporated into the equivalent circuit model which can be solved for single or multijunction devices using conventional techniques .

MODEL SOLUTION AND BEHAVIOR

It is reasonable to expect that the absorption of mixed wavelength light, the sum of many discrete exponential functions, can be approximated by a similar dependence with a suitably chosen characteristic absorption coefficient.

$$G(x)\alpha\exp(-\gamma x) \quad (4)$$

Normalizing Eq. 4 for an i-layer of thickness, W, yields:

$$H(x) = \frac{\gamma}{1-\exp(-\gamma W)}\exp(-\gamma x) \quad (5)$$

where γ, the generation coefficient, is a system value which depends on the spectra of the incident light, i-layer thickness and bandgap, and back contact and internal reflections. In this work γ was determined by minimizing the deviation between generation rates derived using Eq. 5 and the generation profile from the optical model. For AM1.5 illumination, generation coefficients for typical a-Si solar cells range from 3×10^4cm^{-1} for thin i-layers with wide bandgaps to 3×10^5cm^{-1} for thicker cells with narrower bandgaps. A value of 0 represents uniformly absorbed light while systems with values greater than 5×10^5cm^{-1} can be considered to absorb light strongly .

Fig. 1 shows the generation profiles for the low bandgap test cell (Solarex L1331) under AM1.5 illumination calculated using the optical model and using Eq. 5 (γ=1.38×10^5cm^{-1}). The exponential approximation clearly yields a better representation of the real generation profile than does either the uniformly or strongly absorbed light assumptions.

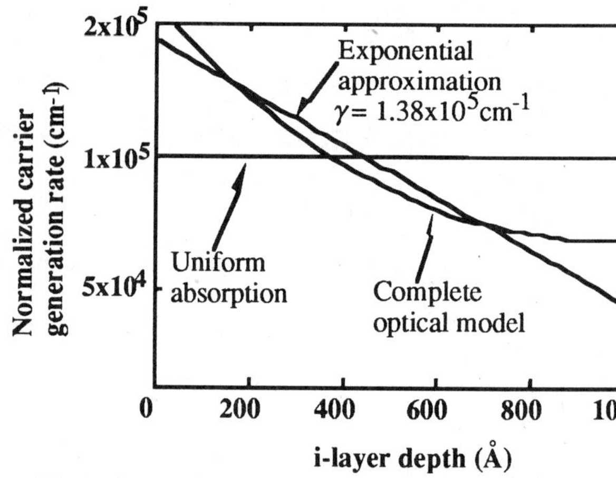

Fig.1 Comparison of the exponential approximation and results of the complete optical model for a 1000Å thick1.45 eV i-layer (solar cell L1331) under AM1.5 illumination.

Integration of equations 2 and 3 with H(x) given by Eq. 5 assuming constant electric field (E = (V_{bi}-V)/W)) and $\mu\tau$ yields:

$$\xi_p = \frac{\gamma\mu\tau_p E}{(1-\exp(-\gamma W))(1+\mu\tau_p E\gamma)}\left(1-\exp\left(\frac{-W}{\mu\tau_p E}-\gamma W\right)\right) \quad (6)$$

$$\xi_n = \left(1-\frac{\upsilon V_t}{E_p}\right)\frac{\gamma\mu\tau_n E}{(1-\exp(-\gamma W))(1-\gamma\mu\tau_n E)}\left(\exp(-\gamma W)-\exp\left(\frac{-W}{\mu\tau_p E}\right)\right) \quad (7)$$

The validity of using the exponential generation profile and uniform field approximations to represent cell behavior were evaluated by comparing collection efficiencies calculated using Eqs. 6 and 7 to those determined by numerical integration of equations 2 and 3 with various field and absorption behaviors assumed. These are discussed in some detail by Vierthaler(9). The following are some of the key conclusions from these comparisons:

1) For a-Si:H films, hole collection efficiencies increase as γ increases and are substantially higher for AM1.5 light than for uniformly absorbed light (γ = 0).
2) For a-Si:H films, the electron collection efficiency decreases only slightly as γ increases.
3) When $\mu\tau$ values representative of those reported for a-Si:H are used (9), equations 6 and 7 show holes to be the limiting carrier under AM1.5 illumination .
4) Hole collection efficiencies in a-Si:H calculated assuming a constant field with the exponential approximation showed excellent agreement with those calculated using the exact generation profile.
5) Collection efficiencies calculated using the uniform field approximation were essentially identical to those determined by numerical integration of equations 2 and 3 with exponential field profiles (6,7) as long as realistic generation profiles were used. The magnitudes of the fields were adjusted to balance electron and hole current collection efficiencies.

6) The equivalent circuit model with collection efficiencies calculated using Eq. 6 agreed exceptionally well with the light JV curves of both the high and low bandgap test cells. The $\mu\tau_p$ values yielding the best agreement were consistent with diffusion lengths reported in the literature (15,16) and agreed to within 10% of the values derived from quantum efficiency measurements. A good fit could not be obtained without a $J_{ph}(V)$ term unless an unrealistic diode factor of 3.6 to 4 was used (9).

MODEL APPLICATION

One obvious application of the model is its use to determine optimal cell designs for a given set of performance related properties. The model equations were combined with a powerful optimization routine (17) to find the best combinations of i-layer bandgap and thickness. Figure 2 shows the calculated maximum efficiencies of triple junction cells for different combinations of the bottom cell thickness and bandgap.

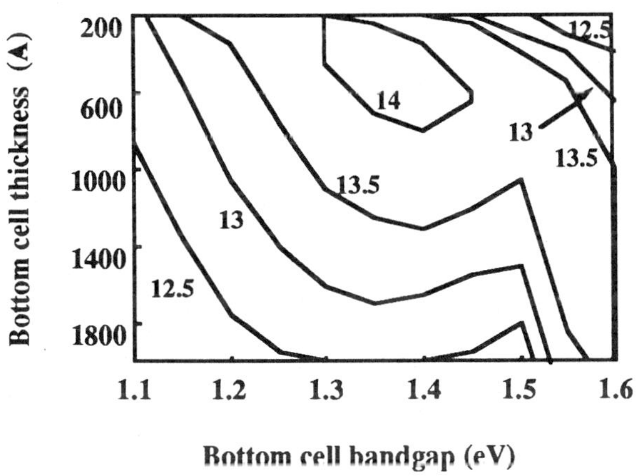

Fig. 2 Maximum cell efficiency for different bottom cell thicknesses and bandgaps ($\mu\tau$ from Fig 3).

For each specification of the bottom cell, the bandgap and thickness of the other two i-layers are independently optimized including adjustments for changes in the absorption profiles of each layer. Values of $\mu\tau$ shown in Fig. 3, representative of today's high quality materials (15,16), were used for this base case simulation. Other materials and cell parameters used in these simulations were derived from analysis of the test cells (9,10). A single junction cell with a 1.75eV bandgap with the assumed properties would be 11.5% efficient under AM1.5 illumination at 1000 W/m^2.

Efficiencies greater than 14% were found with bottom cell bandgaps between 1.3 and 1.45eV. In this range, the optimal thickness of the bottom cell increased with bandgap. A maximum efficiency of 14.2 % was calculated for a bottom cell with a 1.35eV, 450Å thick i-layer while efficiencies in excess of 13.5% were found for a relatively broad range of bottom cell thicknesses and bandgaps. The apparently unusual behavior near 1.5eV (Fig 2) results from the sharper improvement in transport properties with bandgap which occurs around this point. A preliminary study of the effect of a four-fold improvement in the $\mu\tau$ of the low bandgap materials (1.1 to 1.5eV), assuming no change in built-in potential, resulted in less than a 5%

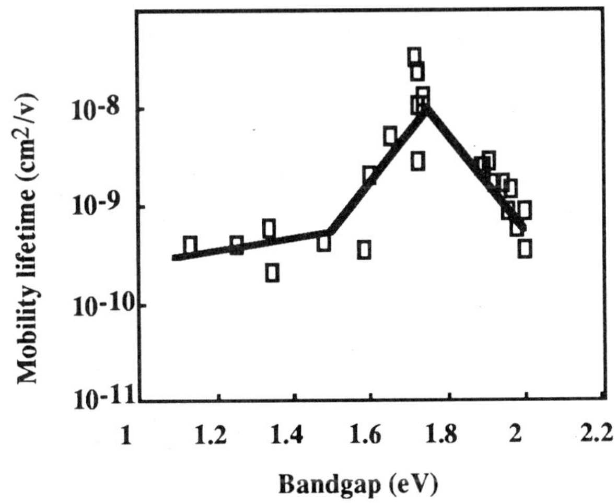

Fig. 3 Mobility-lifetime vs. bandgap for the base case simulation. Data derived from diffusion length ref. 15, 16.

improvement in the maximum efficiency but did result in a substantial increase in the range of bandgap-thickness combinations exhibiting efficiencies greater than 14%. With the assumed improvement in i-layer quality, there was a shift toward higher efficiency with thicker i-layers. Even though these results indicate that maximum achievable efficiencies may not increase substantially with simple improvements in the bulk transport properties of low bandgap materials, higher efficiency with thicker i-layers could have important impacts on the manufacture of cells due to reduced shorting and shunt problems.

With the fairly broad efficiency maxima (>13.5%) exhibited in Figure 2, it is interesting to pose the question, "is any one design more robust than another to unexpected changes in i-layer quality or thickness that may occur due to disturbances in the deposition process?" This model is particularly well suited to this type of assessment. Figures 4 and 5 show the calculated efficiencies of two different triple junction cells as a function of independent changes in the quality of the top, middle or bottom i-layer. The two cells exhibited virtually the same (13.75%) efficiencies under the base case conditions. Interestingly, neither design shows very significant increases in efficiency with a 2 to 3-fold improvement in any single i-layer $\mu\tau$ value but significant decreases in efficiency are found for similar reductions in i-layer quality especially with degradation of the middle or bottom cell. Comparison of Figs 4 and 5 also shows that the cell optimized for the wider bandgap bottom cell was less sensitive to an unanticipated reduction in transport properties in any of the i-layers. For example, the cell with the 1.35eV bandgap bottom cell lost 32% of its output with a sixfold reduction in the quality of the middle cell i-layer while that with the 1.55eV bottom cell lost only 12%.

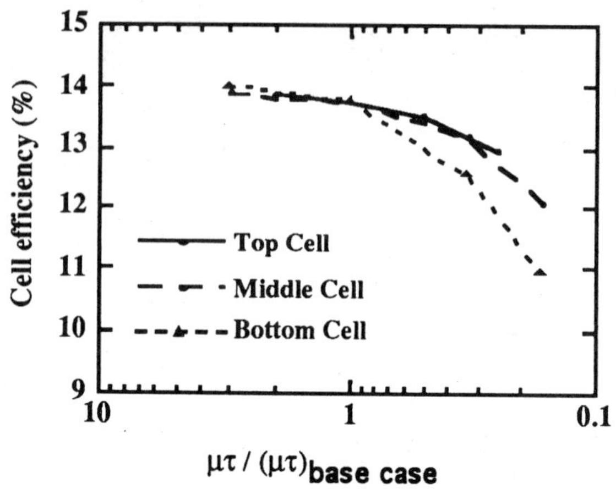

Fig. 4 Effect of changes in i-layer quality on cell efficiency (top cell - 1.92eV, 1000Å; middle cell - 1.70eV, 1900Å; bottom cell - 1.55eV, 1000Å).

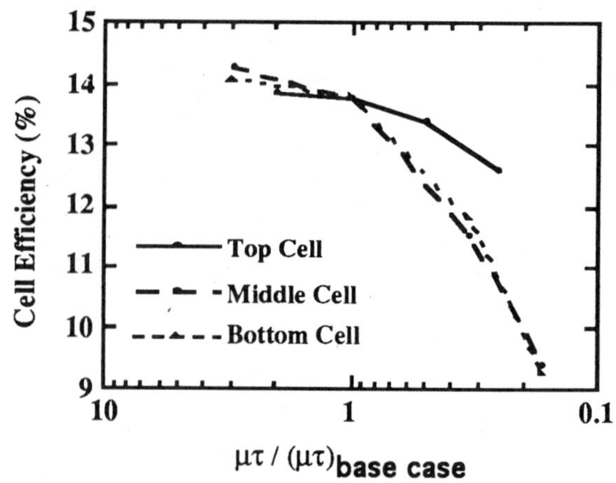

Fig. 5 Effect of changes in i-layer quality on cell efficiency (top cell - 1.89eV, 1290Å; middle cell - 1.62eV, 2535Å; bottom cell - 1.35eV, 1000Å).

Amorphous silicon solar cell development has undergone several changes in emphasis from maximum efficiency to stabilized efficiency to manufacturability. We believe that an important issue in manufacturing will be the robustness of the product to process changes. We also believe that models such as that presented here can contribute significantly to a better understanding of the manufacturing process. Based on the model results to date, it appears that improvements in the quality of the low bandgap materials may be less important than for its contribution to higher efficiency than its contributions to a more robust cell design and improved control of the manufacturing process. We believe that further analysis of this type, including graded bandgap devices, will provide additional insights into better product designs.

ACKNOWLEDGMENTS

This work was supported in part by the National Renewable Energy Laboratory under Subcontract No. XAR-3-13514-01. We also thank Liu Yang of Solarex Thin Film Corporation for the J-V and quantum efficiency measurements of the solar cells and Steve Hegedus of the Institute of Energy Conversion, University of Delaware for many helpful discussions.

REFERENCES

[1] S. Wiedeman et al., "Optical Losses in Multi-junction a-Si:H based Solar Cells and Modules", IEEE PVSC, 1990.

[2] J. Morris et. al. , Absorption Enhancement in Hydrogenated Amorphous Silicon based Solar Cells. J. Appl. Phys. 67, 1989, pp 1079-1087.

[3] F. B. Ellis , A. E. Delahoy, "Optical Properties of Hydrogenated Amorphous Silicon based Solar Cells", Solar Energy Materials", 1986, pp. 109-132.

[4] Z. E. Smith and S. Wagner, "Spectral Dependence of Performance in Multijunction Amorphous Silicon Solar Cells" IEEE PVSC, 1987.

[5] S. S. Hegedus , "Design and Analysis of a-Si Triple Junction Solar Cells using SPICE" , this conference.

[6] M. Hack, and M. Shur, "Theoretical Modeling of Amorphous Silicon-based Alloy p-i-n Solar Cells", J. of Appl. Phys., 54 (10), 1983, pp. 5858-5863.,

[7] M. Hack, and M.Shur, "Physics of Amorphous Silicon Alloy p-i-n Solar Cells", J. Appl. Phys. 58, 1985, pp. 997-1020.

[8] S. Bae, J. Hou, V. Suntharalingam, and S. J. Fonash, "A Payback Assessment for Materials Development for the Low-Gap Unit of a Multi-Junction Cell: A Numerical Modeling Study. IEEE PVSC, 1993.

[9] M. Vierthaler, "Equivalent Circuit Modeling of Amorphous Silicon Solar Cells Incorporating Realistic Carrier Generation Profiles and Transport" MS Thesis, Univ. of HI May 1994.

[10] M. Vierthaler and R. E. Rocheleau, in preparation.

[11] K. Misiakos, F. A. Lindholm, "Analytical and Numerical Modeling of Amorphous Silicon p-i-n Solar Cells", J. Appl. Phys., 64 (1), 1988, pp. 383-393,.

[12] H.P. Maruska, T.D. Moustakas "Influence of the Wavelength of incident light on Shunt Conductance and Fill Factor in Amorphous Silicon Solar Cells", IEEE Transaction on Electron Devices, Vol. Ed-31, No. 5, 1984, pp. 551-558.

[13] Richard S. Crandall, "Modeling of Thin Film Solar Cells: Uniform Field Approximation", J. Appl. Phys., 54 (12), 1983, pp. 7176-7186.

[14] Richard S. Crandall, "Transport in Hydrogenated Amorphous Silicon p-i-n Solar Cells", J. Appl. Phys., 53(4), 1982, pp. 3350-3352

[15] J. Z. Liu et.al. "Ambipolar Diffusion in a-Si,Ge:H:F Measured with the Steady State Photocarrier Grating Technique", IEEE PVSC, 1990, pp. 1606 - 1609.

[16] Y.M. Li et. al., " A Study of Amorphous Si-C Alloy Solar Cells", IEEE PVSC, 1993, pp. 850-854.

[17] J. A. Nelden, "A Simplex Method for Function Minimization" The Comp. Jour., 1964, pp. 308-313.

a-Si:H OPTICAL SPEED DETECTOR BASED ON THE FLYING SPOT TECHNIQUE

M. Vieira, A. Fantoni, A. Maçarico, F. Soares and R. Martins
FCT–UNL/UNINOVA, Quinta da Torre, 2825 Monte da Caparica, PORTUGAL

ABSTRACT

We have developed in the past a transient technique called Flying Spot Technique (FST)[1], based on the lateral photoeffect. It allows to determine the ambipolar diffusion length and the effective lifetime of the photogenerated carriers, once the light spot velocity and geometry of the structure are known. We propose to apply this technique backwards in order to detect the path and velocity of an object that is moving toward a light source direction. The light back reflected is analyzed by a p.i.n structure measuring the transient transverse photovoltage which is dependent on the object movement (position and velocity). Details concerning material characterization and device geometry will be presented.

1. INTRODUCTION

The p.i.n devices based on hydrogenated amorphous silicon (a-Si:H) are fundamental elements of many different types of radiation sensors. Most of this devices are based on either the transverse or the lateral photovoltaic effect [2]. It is known that when two dissimilar materials are locally illuminated by a light spot a lateral photovoltage is set up in addition to the transverse one. This lateral photovoltage vary in magnitude and polarity as the light spot is moving along the structure. The division current in the lateral direction (dependent on the light spot velocity and position) influence directly the potential profile along the junction and leads to a transient photovoltage during the light spot movement. In the past we have developed a transient technique, the FST, based on this lateral phototoeffect. This technique is a powerful tool to characterize solar cells, optical sensors and other optical devices. It allows to infer the ambipolar diffusion length and the effective lifetime of the photogenerated carriers once the light spot velocity and geometry of the device were known. In this paper we propose to apply this technique backwards in order to detect the path and velocity of an object that is moving in a light source direction. The light reflected back from the object is analyzed through a p.i.n structure being the transient transverse photovoltage dependent on the object movement (position and velocity). Assuming known the transport properties of the material and the geometry of the device and using a triangulation method we show that it is possible to map the object movement.

2. PHYSICAL MODEL

2.1 Main principles of the FST

Illuminating a semiconductor junction leads to the appearance of a photovoltage effect between the two sides. If the illumination is nonuniform, an additional effect arises resulting in a lateral photovoltage parallel to the junction, $V_{ph,l.}$ in addition to the usual transverse one, $V_{ph,t}$.

Fig.1 shows the charge movement, potential distribution and geometrical configuration of a p.i.n. structure illuminated perpendicularly with localized monochromatic light.

Under open circuit condition minority carriers builds up at the surface, while majority carriers

CH3365-4/94/0000-0571 $4.00 © 1994 IEEE

First WCPEC; Dec. 5–9, 1994; Hawaii

are accumulated in the material beyond the depletion layer edge, causing a reduction in the band bending across this region. If the light spot is incident on the center of the cell two lateral photovoltages will set up, of opposite polarity but equal magnitude, so that the lateral output voltage, V_{outl} is null.

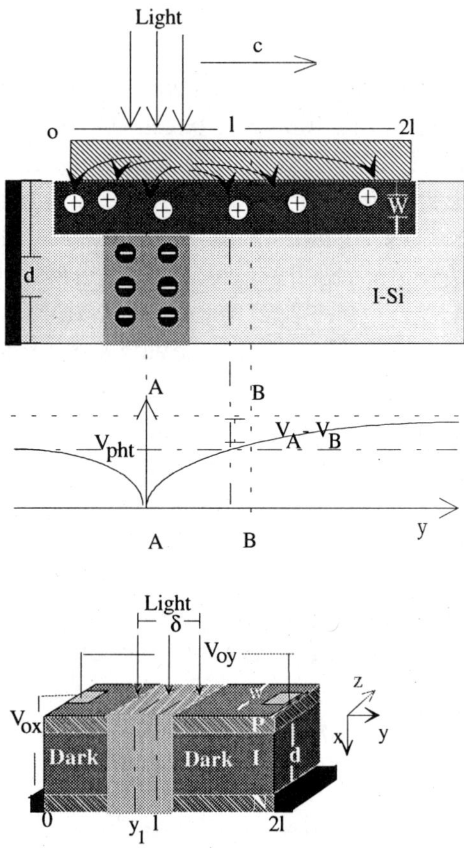

Fig.1- Charge movement, potential distribution and device configuration of a p.i.n. structure illuminated perpendicularly with localized monochromatic light.

When the light spot moves with velocity c, towards one side of the cell, the net transient output voltage equals the difference between the two lateral photovoltages (between the two pairs of opposing electrodes) [3]:

$$V_{oy}(t) \approx \frac{2\rho}{w.d} I\left(1 - y_p(t)\right) \tag{1}$$

$$I = w \int_0^{2l} J_x dx \tag{2}$$

Where ρ is the resistivity, $y_p(t)$ is the light spot position and I is the total photocurrent collected in the transverse direction. The transverse photocurrent density, J_x is evaluated from the conventional junction theory for pin devices [4].

We assume that the carriers generation function G is split into the product of a time dependent function f(t), describing the light spot movement and the usual space dependent function, $g(x)=G\alpha e^{-\alpha x}$. It is possible, by solving the transport and the Poisson's equation to determine the spatial distribution of the generated carriers during the light spot movement [5].

$$\frac{\partial p}{\partial t} = -\frac{1}{q} \operatorname{div} J_p + G - R$$

$$\frac{\partial n}{\partial t} = \frac{1}{q} \operatorname{div} J_n + G - R \tag{3}$$

$$\operatorname{grad} E = \frac{\rho}{\varepsilon_0 \varepsilon_r}$$

where n (p) is the free electron (hole) density, Jn (Jp) the sum of the drift and diffusion current density for electrons (holes), E the electrostatic field, R (G) the recombination (generation) rate, ρ the electric charge density, namely:

$$\rho = q\left(-N_A^- + N_D^+ + p - n + p_T - n_T\right) \tag{4}$$

where μ_n (μ_p) is the electron (hole) band mobility, D_n (D_p) the electron (hole) diffusion, N_A^- (N_D^+) is the concentration of ionised acceptors (donors), n_T (p_T) is the density of electrons (holes) trapped in tails and dangling bond states function of the density of states used. The above equations represent a system of partial differential equations completely defined by a set of suitable boundary conditions: the carrier and the current density distribution must be continuos along the entire device and carriers transport at the contacts is modelled by effective surface-recombination velocities. The complete solution of this system (eq.3) is not trivial due to the non linearity which is implicit in the terms $R(n,p)$, $p_T(n,p)$, $n_T(n,p)$ and can only be reached using numerical techniques [6]. However, assuming that the movement of the spot is similar to an applied electric field ($E_t=c/\mu$) the equations are simplified considering the carrier distribution during the spot movement, in the illuminated and dark regions, respectively, approximated by:

$$J_x \propto p(t,x) \approx g(x)(1 - f(\Delta_1)) \cdot \left[\frac{L_1}{L_1+L_2} e^{y(t)/L_1} + \frac{L_2}{L_1+L_2} e^{-y(t)/L_2}\right]$$

$$J_t \propto p(x,t) \approx g(x)\left[f(\Delta_1) - f(\Delta_2)\right]\frac{L_{1,2}}{L_1+L_2} e^{-y(t)/L_{1,2}}$$

$$1/L_{1,2} = \left| \frac{c}{2D^*} \pm \sqrt{\left(\frac{c}{2D^*}\right)^2 + \frac{1}{L^{*2}}} \right| \qquad (4)$$

where $f(\Delta_{1,2})=e^{-\delta/L_{1,2}}$ is dependent on the spot width, δ, D^* is the ambipolar diffusion coefficient and L^* the ambipolar diffusion length.

Carrier distribution during the spot movement is split into two distinct exponential regions dependent on L^*, D^* and on the light spot velocity, c.

Taking into account equations 1 and 2, the induced transverse photovoltage is asymmetric and depends on the spot velocity through $1/L_{1,2}$ (the slopes of the transient signal before and after the peak is reached) [7].

Fig.2 shows the experimental and the simulated (eq.4) transverse photovoltage signal for a pin structure having a L^* of 235 nm and $\tau^*=4.4\ 10^{-9}$ s and different light spot velocities. The V_t (t) asymmetrical signal have a peak dependence on the transit time of the light spot. From the slopes it is possible to infer $L^*(L^* = \sqrt{L_1 L_2}$ and $\tau^*=(L_2-L_1)/c$

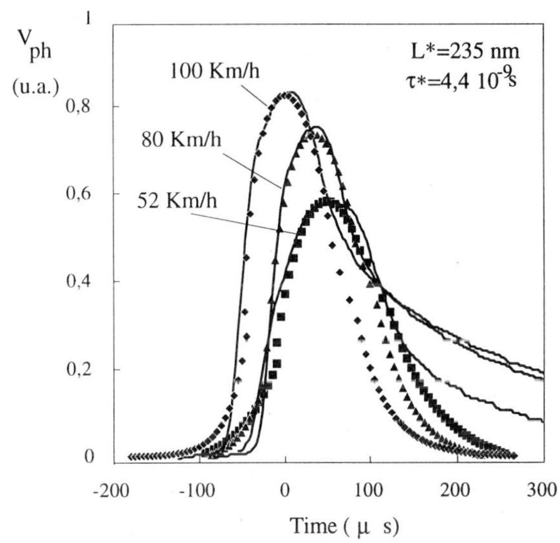

Fig.2- FST signal obtained on a a-Si:H structure, under different light spot velocities.

2.2 Detection method

Optical velocity sensors detect the distance and the speed vector of an object while it is approaching (without touching it). Fig.3 illustrates the method used as well as the formulas that allows to determine the position and the velocity of the moving object. The object is illuminated with a spot of visible light and the light reflected back is analyzed through a pin sensor that detects simultaneously a transverse and a photolateral voltage. From the lateral output signal and using eq.1 the spot position can be inferred and from the slopes of the transverse photovoltage (eq.4) the light spot velocity is determined (ω>2l/c).

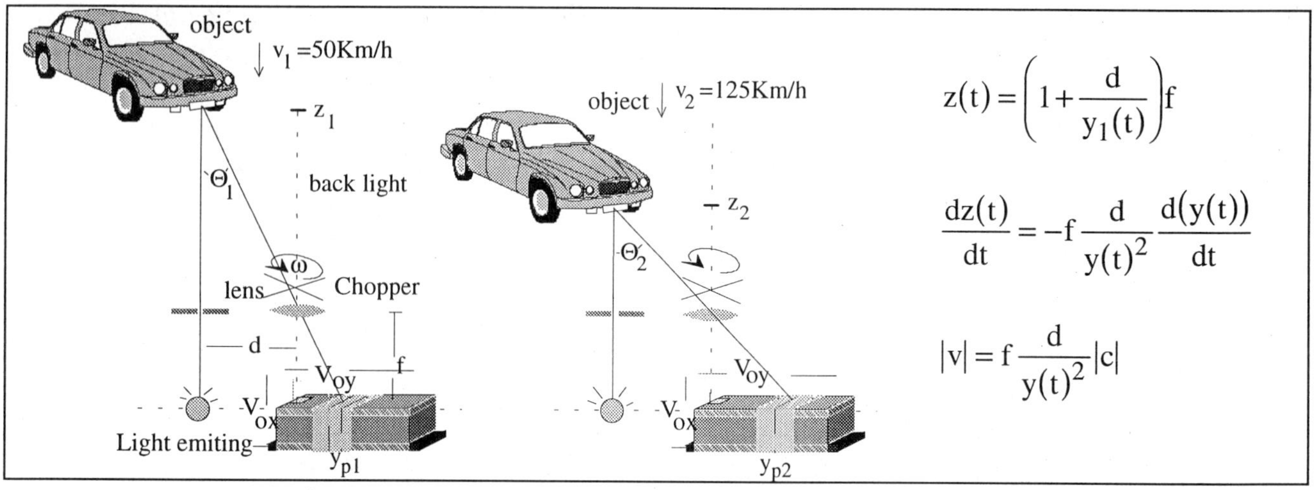

Fig.3- Principle of the optical speed and proximity a-Si:H sensor.

3. RESULTS, FUTURE APPLICATIONS AND DISCUSSION

The a-Si:H sensors were deposited on ITO covered glass substrates, followed by p^+a-Si:H/i-a-Si:H/n^+a-SiH. The geometrical configuration is summarized on Fig.1. The a-Si:H devices were prepared by glow discharge method, in a PECVD system [8].

Preliminary results show that the main sources of error are ascribed to non-uniformities in the material layers, edges effects and high dark currents. The lower limit for the detection are set by the thermal noise, proportional to the resistance of the intrinsic layer and dependent on the light spot position ($S/N \propto (1-y_p/l)$), temperature and background light level. The advantages of these sensors are as follows:

The sensor element is simple, lightweight, small sized and the signal processing continuous.They have a large field of applications that go from automobile industry to robotics. They can be used regardless of what material the object is made of, either in the ground or attached to a moving object. They can be employed, in automatic controller tolls and vehicle identification (if more than one is used at different levels); as vector speed sensor (attaching sensors in the direction of progress and to the sides of the vehicle and detecting the light reflected back from the road surface) or in object handling which other sensors as well as in guidance systems. A lot of work could be done in order to develop this new kind of sensors.

ACKNOWLEDGEMENTS

The co-funding by JNICT and "Fundação Calouste Gulbenkian" are gratefully acknowledged.

REFERENCES

[1] R. Martins, M. Vieira, F. Soares and L. Guimarães: in Proc. *PVSEC–5* (Kyoto, 1990) p.975.
[2] J. Torkel, Wallmark, *Proc. IRE*, **43**, p.474 (1956)
[3] D. J. W. Noorlag,, *Ph D thesis*, Delft , University of Technology (1982).
[4] J. Reichman, *Appl. Phys. Lett,* **38** p.251(1981)
[5] M. Vieira, *Ph D thesis,* FCT/UNL (1993)
[6] A. Fantoni, to be publisched
[7] M. Vieira, R. Martins, E. Fortunato and L. Guimarães, *J. of Non Cryst. Sol.* **137&138**, p.479 (1991).
[8] R. Martins, L. Guimarães, E. Fortunato, M. Santos and I. Ferreira, *Proc. 8 th EPVSEC,* p. 653 (1988) .

COMPARISON OF DIFFUSION LENGTH MEASUREAMENTS FROM THE FLYING SPOT TECHNIQUE AND THE PHOTOCARRIER GRATING METHOD IN AMORPHOUS THIN FILMS

M. Vieira, A. Fantoni, R. Martins, FCT–UNL/UNINOVA, Monte da Caparica, Portugal
V. Chu, INESC, Lisbon, Portugal
S. Koynov, CL-SENES, Bulgarian Academy of Sciences, Sofia, Bulgaria
F. Wang, S. Grebner, R. Schwarz, Technical University of Munich, Garching, Germany

ABSTRACT

Using the Flying Spot Technique (FST) we have studied minority carrier transport parallel and perpendicular to the surface of amorphous silicon films (a-Si:H). To reduce slow transients due to charge redistribution in low resistivity regions during the measurement we have applied a strong homogeneously absorbed bias light.

The defect density was estimated from CPM measurements. The steady-state photocarrier grating technique (SSPG) is a 1-dimensional approach. However, the modulation depth of the carrier profile is also dependent on film surface properties, like surface recombination velocity. Both methods yield comparable diffusion lengths when applied to a-Si:H.

1. OBJECTIVES

The ambipolar diffusion length, which is a measure of minority carrier transport properties, is a crucial parameter for the performance of thin film solar cells. Together with a knowledge of the internal field profile it provides the basis for the measurement of the collection efficiency.

The aim of this contribution is the comparison of the ambipolar diffusion length measurements using two different methods, the Flying Spot Technique, where a transient photovoltage signal is monitored, and the Steady-State Photocarrier Grating Method, where a lateral photocurrent under coherent laser illumination is detected. The application to amorphous and microcrystalline silicon thin films is discussed.

2. FLYING SPOT TECHNIQUE (FST)

The effect of illuminating a semiconductor junction is the appearance of a photovoltage effect between the two sides. If the illumination is nonuniform, an additional effect arises resulting in a lateral photovoltage parallel to the junction.

Fig. 1 shows the charge movement, potential distribution and geometrical configuration of a p-i-n structure illuminated perpendicularly with localized monochromatic light. When the spot is moved (with velocity c) towards one side of the cell, the net transient output voltage $V_{ph,t}$, measured under open circuit condition, is given by:

$$V_{ph,t}(t) - \frac{2\rho}{wd} I\left(1 - y_p(t)\right)$$

$$\text{with} \quad I = w \int J_t dy \qquad (1)$$

where ρ is the resistivity, $y_p(t)$ is the light spot position, 2l is the junction length and I is the total photocurrent collected in the transverse direction (see Fig.1). The transverse photocurrent density, J_t, is evaluated from the conventional junction theory for p-i-n devices [1].

CH3365-4/94/0000-0575 $4.00 © 1994 IEEE First WCPEC; Dec. 5–9, 1994; Hawaii

Assuming that the generation function is split into a product of a space and a time dependent function, $G(t,x)=g(x)f(t)$, it is possible, by solving the two transport equations, for electrons and holes, and the Poisson equation to determine the spatial distribution of the generated carriers during the light spot movement [2].

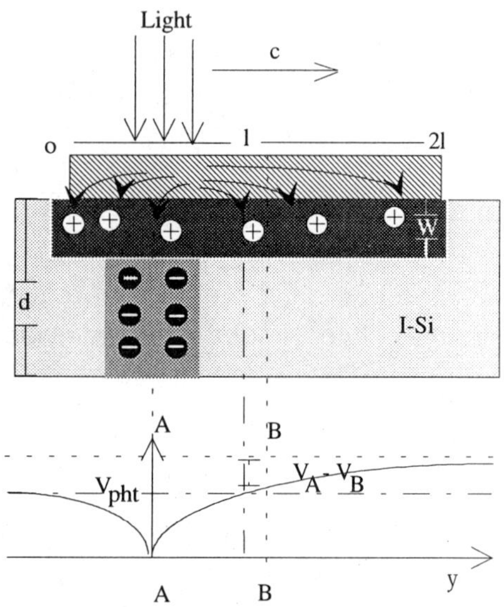

Fig.1. Charge movement, potential distribution and device configuration of a p-i-n structure illuminated perpendicularly with localized monochromatic light.

Due to the inherent non-linearity of the system of these equations, an exact solution can only be reached by numerical techniques. However, assuming that the movement of the spot is similar to an applied electric field, $E_l=c/\mu$, one can show that the current defined in eq. (1) depends exponentially on time with typical time constants defined by:

$$\frac{1}{L_{1,2}} = \left| -\frac{c}{2D^*} \pm \sqrt{\left(\frac{c}{2D^*}\right)^2 + \frac{1}{L_{amb}^2}} \right| \qquad (2)$$

where D^* and $L_{amb} = \sqrt{D^* \tau^*}$ are the ambipolar diffusion coefficient and diffusion length, respectively. Taking equation (1) into account, the induced transverse photovoltage is asymmetric and depends on the spot velocity c through $1/L_{1,2}$ which are the slopes on the transient signal before and after the peak is reached [3]. Fig. 2 shows typical results for a amorphous and microcrystalline p-i-n structure with L_{amb} = 120 nm and 105 nm, respectively.

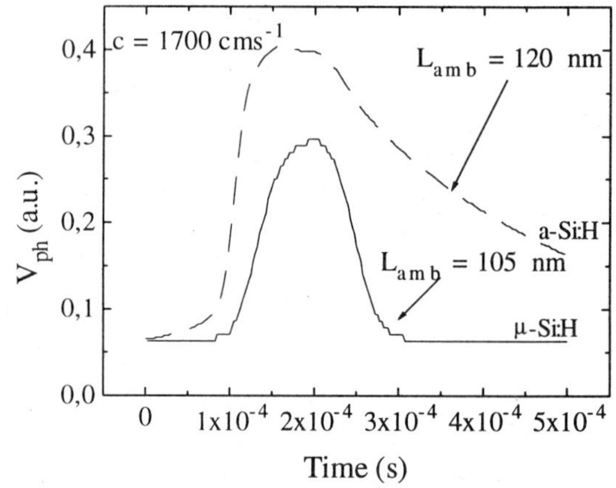

Fig.2. FST photovoltage transient from two p-i-n structure

3. PHOTOCARRIER GRATING TECHNIQUE (SSPG)

In the steady-state photocarrier grating technique, as initially proposed by Ritter et al. [4], two coherent HeNe laser beams (1.96 eV), I_1 and I_2, interfere on the surface of the film (see Fig. 3(a)). It results in a periodic distribution of carriers in space (the photocarrier grating). The final distribution of the carrier density is determined by (a) the ambipolar diffusion length L_{amb}, (b) the grating period Λ , (c) the surface quality. Usually one chooses $I_1 >> I_2$ in order to render both the carrier lifetime τ and L_{amb} independent of position.

What one measures is a value β, which is defined as the change of the photoconductivity under coherent illumination, normalized to the photoconductivity increase under homogeneous illumination with the same average intensity. β will then vary between -1 and +1. It can be shown that β contains the desired ambipolar diffusion length as expressed by the following equation:

$$\beta \approx 1-2\gamma_{pc}\gamma_d\gamma_0^2\gamma_s^2 / \left[1+(2\pi L_{amb}/\Lambda)^2\right] \qquad (3)$$

where γ_{pc}, γ_d, and γ_o are constants related to the intensity dependence of the photoconductivity, to the dark conductivity, and to the quality of the grating, respectively, γ_s is a factor related to surface and interface recombination.

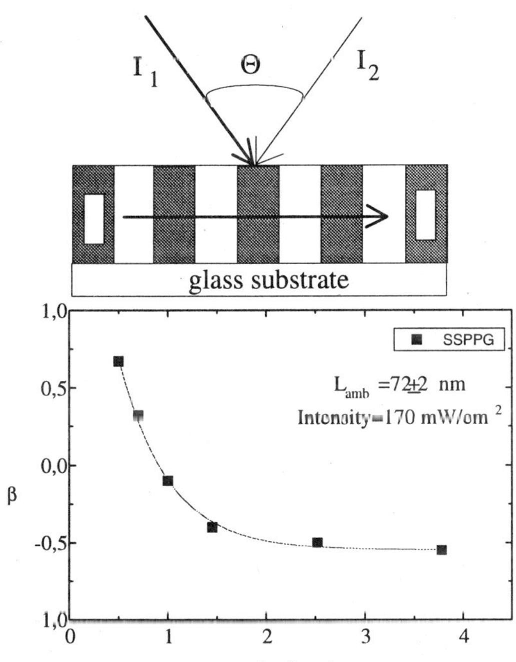

Fig.3. (a) Layout of SSPG measurement. (b) Typical β–Λ curve of an a-Si:H sample.

Fig. 3(b) shows a typical Λ–β curve of an a-Si:H sample. The ratio I_1/I_2 is 15, and the intensity of I_1 is 85 mW/cm^2. A fit of eq. (3) to the data yields a diffusion length of 72 nm.

4. COMPARISON OF DIFFUSION LENGTH MEASUREMENTS

The film quality strongly depends on the density N_d of recombination centers. This was measured by the Constant Photocurrent Method. Usually the (majority carrier dominated) photocurrent is inversely proportional to N_d, whereas L_{amb} varies only slowly with N_d [5.] For example, L_{amb} was 64 and 78 nm for films with N_d of 5.1 and 1.5 x 10^{16} cm^{-3}, respectively. This behavior is found with both the FST and the SSPG method.

Fig. 4 shows L_{amb} as a function of light intensity. Again, both methods show the same trend. As the light intensity increases, L_{amb} decreases slightly. This observation is well understood and typical for a continuous density of states distribution in the band gap.

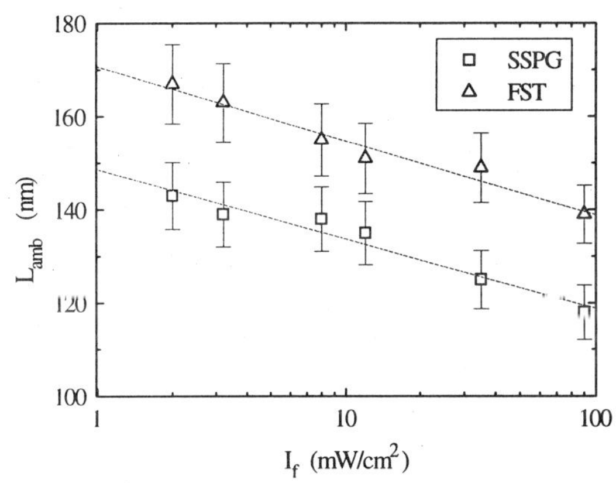

Fig.4. Intensity dependence of the ambipolar diffusion length measured by FST and SSPG.

The resulting diffusion length values from the FST measurement tend to be larger in some cases than the SSPG results. This is to be expected if the assumption of a field-free region below the junction in the Reichman analysis is not fully met due to the contributions to the total photocurrent from perpendicular drift currents.

We have also added transient measurements of majority carrier transport. The response time τ_R

of the photocurrent, measured after switch-off of an additional light beam, is directly related to the lifetime of the photogenerated carrier and the trap carrier density. For the a-Si:H a typical value of τ_R is about 1 μs under a strong light bias while in microcrystalline silicon films the transient photocurrent shows an initial fast decay followed by a slow decay with a time constant of up to about 1ms [6]. We attribute the fast decay component to recombination in the amorphous tissue and the high defect regions of the grain boundaries. The long tail is suggested to come from the long lifetime inside the crystallites.

5. NUMERICAL SIMULATION

We have performed a numerical analysis using the simulation program ASCA (Amorphous Silicon Solar Cell Analysis) [7]

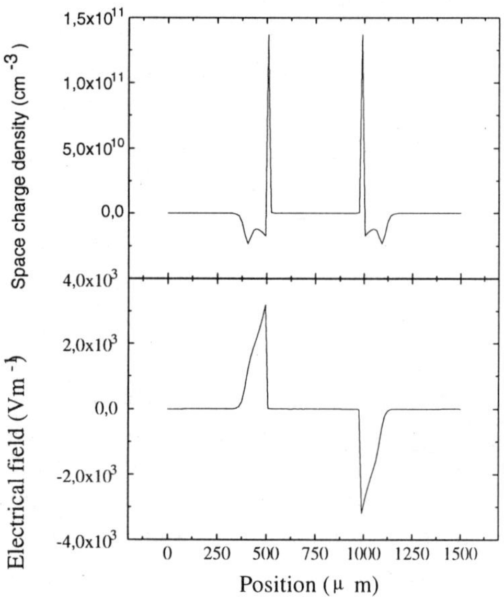

Fig.5. Electrical field and space charge density lateral profiles for an intrinsic a-Si:H film.

Fig. 5 shows the electrical field and the space charge density lateral profiles of an intrinsic a-Si:H film (1.5 mm length) illuminated with a spot of 0.5 mm diameter. The absorption coefficient is assumed to be 5×10^{3} cm^{-1}. At the borders of the illuminated region two symmetric electric field peaks, of opposite sign, appear due to the start of the photogenerated carrier diffusion. When the carriers reach the neutral region, charge neutrality is lost and an electric field distribution is built up to equilibrate the charged carriers diffusion. The irregularity in the charge density profile is due to the trapped charge into the midgap localised states. Data show that in the amorphous silicon, as in the crystalline one, a local illumination leads to the appearence of a lateral electric field. If a p-i-n structure is used, such a field is responsible for the asymmetric transverse photovoltage observed in the FST (transient condition).

6. CONCLUSION

Both the FST and the SSPG method lead to comparable values of the ambipolar diffusion length in the amorphous and microcrystalline silicon films. Care has to be taken to fully suppress contributions due to drift currents in the FST technique. A important advantage of this method over the SSPG technique is its applicability to full p-i-n solar cell structures.

ACKNOWLEDGEMENTS

The co-funding by JNICT and "Fundação Calouste Gulbenkian" and DFG through Bulgarian-German research is gratefully acknowledged.

[1] J. Reichman, *Appl. Phys. Lett*, **38** (1981) 251

[2] M. Vieira, *Ph D thesis*, FCT/UNL (1993).

[3] M. Vieira, R. Martins, E. Fortunato and L. Guimarães, *J. of Non-Cryst. Sol.* **137&138**, (1991) 479.

[4] D. Ritter, E. Zeldov and K. Weiser, *Phys. Rev.* **B38** (1988) 8296 .

[5] F. Wang and R. Schwarz, *J. Appl. Phys.* **71** (1992) 791

[6] S. Koynov, R. Schwarz, T. Fischer, S. Grebner and H. Munder, *Jap. J. Appl. Phys* **33** (1994) 4534.

[7] A. Fantoni, M. Vieira, and R. Martins, Proc. *MRS Spring Meeting* (1994) in print.

EFFECT OF "SOFT START" i-LAYER DEPOSITION ON THE PERFORMANCE AND STABILITY OF a-Si:H p-i-n SOLAR CELLS

W.Herbst, U.I.Schmidt*, H.Schade**, B.Schröder, H.Oechsner

Fachbereich Physik und Forschungsschwerpunkt Materialwissenschaften, Universität Kaiserslautern, Postfach 3049, D-67653 Kaiserslautern, Germany, Tel.: (49)-631-205-4050, Fax: (49)-631-205-2996

* Present address: University of California at Berkeley, Dept. of Chemical Engineering, Berkeley, CA 94720, USA

** Phototronics Solartechnik GmbH (PST), Hermann-Oberth-Str. 11, 85640 Putzbrunn, Germany

ABSTRACT

In this study, we investigate the effect of an ignition procedure for the i-layer deposition on cell performance and stability. Starting with low rf power ("soft start"), the hydrogen concentration and the defect density at the p-i interface is reduced leading to an increase of the conversion efficiency. This is due to the reduction of particle formation in the discharge and of its detrimental effects on initial film growth under "soft start" conditions. Light soaking experiments performed with solar cells and intrinsic films simultaneously deposited indicate that the relative degradation of the efficiency is mainly determined by the bulk i-layer rather than by the p-i interface region properties. However, due to the increase of efficiency in the initial state an improvement of the efficiency after degradation could be achieved by this p-i interface optimization.

INTRODUCTION

Many experiments and computer simulations show that the performance of a-Si:H solar cells is not only determined by the i-layer bulk properties, but is also strongly influenced by the interfaces. In particular, the p-i interface region is of crucial importance for the properties of a-Si:H solar cells. This can be attributed to the asymmetry in the density of states and the different mobilities of electrons and holes in a-Si:H [1,2]. As indicated by various experiments (e.g. [3]), an interface region within the order of 15nm near to the p-i heterojunction greatly influences the device performance. In the standard device structure (glass/TCO/p/i/n/contact) this region is deposited directly after the ignition of the silane discharge. It is therefore strongly affected by disturbances, due to the particle-induced initial transient stage (ITS), which become more pronounced with increasing power [4].

In the present study, we investigate the potential of an ignition procedure for the i-layer deposition starting with low rf power ("soft start") for p-i interface improvement in a-Si:H p-i-n solar cells. Furthermore, we examine, whether there is also an effect of the ignition procedure on the bulk i-layer material. In the case of a negligible effect on the bulk, light soaking experiments on solar cells containing different p-i interface regions and similar bulk properties can be performed. The results are discussed with regard to the controversy of interface and bulk degradation.

EXPERIMENTAL

The influence of the "soft start" deposition is compared with the conventional procedure ("constant start") by the incorporation of different intrinsic glow-discharge (gd) a-Si:H layers in an otherwise identical p-i-n structure: glass/TCO/p-doped a-SiC:H/ intrinsic layer ("soft start" or "constant start")/n-doped a-Si:H/Al-contact. For the "soft" deposition, the first 15nm of the i-layer were deposited at a low rf power P_{soft} under particle free conditions. The bulk part of this i-layer was then grown at a constant rf power P_{const}. In the "constant start" cell the whole i-layer was deposited at the constant rf power P_{const}. These experiments were performed in two different glow-discharge (gd) reactors. In the following they are termed gd-reactor I (University of Kaiserslautern) and gd-reactor II (PST company), the corresponding solar cells type I and type II respectively. Since gd-reactor I is only used for deposition of intrinsic layers, the doped layers are prepared separately. For the cells a glass/TCO/a-SiC:H:B-substrate supplied from

Table 1: i-layer deposition conditions and thicknesses (P_{const}: constant nominal rf power during bulk i-layer deposition; P_{soft}: nominal rf power in the case of "soft start").

	gd-reactor I	gd-reactor II
P_{const} rf power/ r_{dep}	10W/ 3Å/s	50W/ 1Å/s
P_{soft} rf power/ r_{dep}	1W/ 0.3Å/s	20W/ 0.4Å/s
SiH_4 pressure	0.133 mbar	0.25 mbar
T_{substr}	150°C	200°C
electrode area	165 cm^2	6000 cm^2
thickness	500 nm	300 nm

CH3365-4/94/0000-0579 $4.00 © 1994 IEEE

First WCPEC; Dec. 5-9, 1994; Hawaii

PST company was used. The n-layers (P-doped) were deposited by dc magnetron sputtering. The active area of the corresponding cells (type I) is 0.082 cm². The cells of type II were deposited in a single chamber system gd-reactor II at PST company. This reactor is of the same size as the deposition chambers used in the PST pilot production line. The solar cells type II were testing structures cut out of the 0.6 m² substrate with an active area defined by the aluminium rear contact (0.073 cm²). The deposition conditions for the i-layers are listed in table 1.

The separate deposition of the doped and the intrinsic layers in gd-reactor I offers the possibility for a more extensive investigation of the i-layer properties. Samples for CPM (Constant Photocurrent Method), PDS (Photothermal Deflection Spectroscopy), photo-conductivity (632.8nm, 9mW/cm²) and SIMS (Secondary ion mass spectrometry) measurements and FTIR spectra were grown simultaneously on glass substrates and crystalline silicon (c-Si), respectively. Furthermore, the a-Si:H growth behaviour during i-layer deposition in gd-reactor I was investigated by kinetic in situ ellipsometric measurements using a spectroscopic phase-modulated ellipsometer coupled to the deposition chamber via silica windows [4]. The solar cells were characterized by I-V characteristics. Light soaking of the cells and films was performed using fluorescent lamps (Phillips PLL 24-95, colour temperature ≈ 5000K). By means of fan cooling the sample temperature was kept near room temperature during light exposure.

RESULTS AND DISCUSSION

i-layer properties

Figure 1 shows the ellipsometric trajectories measured during a-Si:H film growth in gd-reactor I. The kinetic data were taken at a photon energy near maximum absorption. The trajectories describe the transition of the measured optical properties from that of the substrate towards those of the growing film. As indicated by the larger values of ε_2 the substrate-film-interface exhibits a significantly larger film density in the case of the "soft start" deposition. Since this region corresponds to the p-i interface region in the solar cells, a strong influence on the solar cell properties is expected. A weak effect of the starting procedure on the bulk i-layer material is also indicated. However, the ellipsometric trajectories converge nearly into the same region in the ε_2=f(ε_1)-plot indicating similar optical properties of the bulk material.

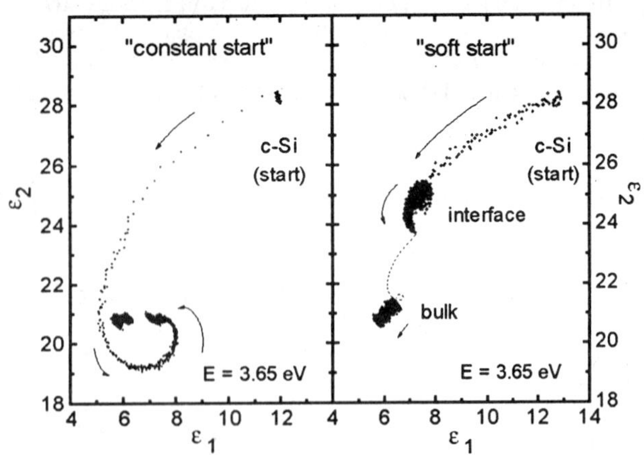

Figure 1: Ellipsometric trajectories for the growth of a-Si:H layers on c-Si substrates under "soft start" and "constant start" conditions. (Dashed line: data derived from measurements at a different photon energy).

The different properties of the interface region seen in the ellipsometric measurements are also noticed in the SIMS depth profiles, displayed in figure 2. According to Gnaser et al. [5], the hydrogen concentration can be calculated from the HCs⁺ secondary ion intensity. A significantly lower hydrogen concentration at the interface region is found in the case of the "soft start" ignition procedure. The comparison to the "constant start" procedure shows a slightly reduced hydrogen content even in the bulk i-layer material. An evaluation

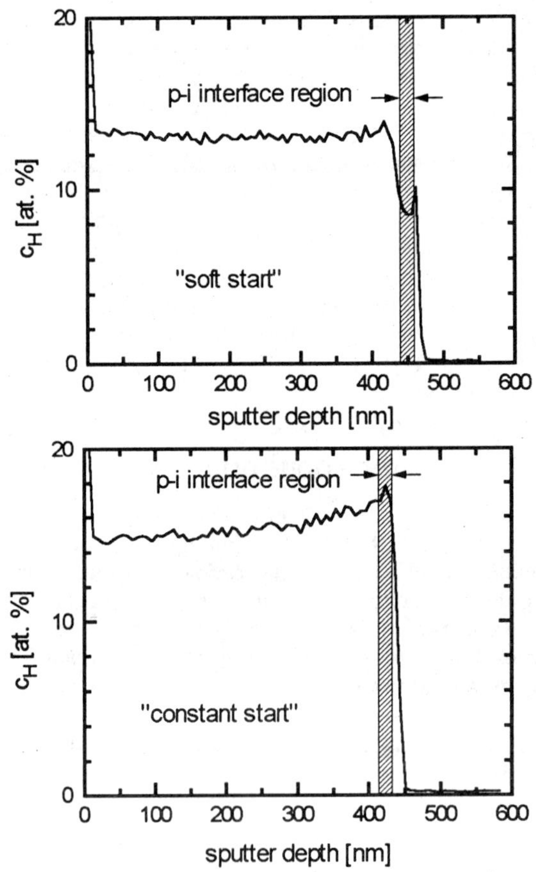

Figure 2: Hydrogen concentration deduced from SIMS depth profiles of a-Si:H layers (deposited in gd-reactor I) grown on crystalline silicon substrates using HCs^+ and $SiCs^+$ secondary ions. The samples were simultaneously deposited during solar cell preparation applying the two different ignition procedures. The interface region marked corresponds to the p-i interface region within the solar cells.

Table 2: Defect densities (determined by PDS and CPM) and photoconductivities of the corresponding a-Si:H layers obtained for the two different ignition procedures in the initial state (A) and after light soaking (B). The defect densities in the degraded state were measured after 1000h light exposure, photoconductivity after 306h.

A	"soft start"	"constant start"
N_d (PDS) [cm^{-3}]	$8.1 \cdot 10^{16}$	$1.0 \cdot 10^{17}$
N_d (CPM) [cm^{-3}]	$1.4 \cdot 10^{16}$	$1.8 \cdot 10^{16}$
σ_{ph} [$(\Omega cm)^{-1}$]	$1.2 \cdot 10^{-5}$	$1.2 \cdot 10^{-5}$

B	"soft start"	"constant start"
N_d (PDS) [cm^{-3}]	$1.4 \cdot 10^{17}$	$2.4 \cdot 10^{17}$
N_d (CPM) [cm^{-3}]	$5.7 \cdot 10^{16}$	$8.0 \cdot 10^{16}$
σ_{ph} [$(\Omega cm)^{-1}$]	$2.3 \cdot 10^{-6}$	$2.3 \cdot 10^{-6}$

Table 3: Solar cell performance parameters (initial state) grown under the two different i-layer ignition procedures.

	solar cell type I		solar cell type II	
	"soft start"	"const. start"	"soft start"	"const. start"
j_{sc} [mA/cm^2]	15.8	16.2	13.6	12.2
V_{oc} [V]	0.827	0.803	0.785	0.804
FF	0.6	0.53	0.72	0.67
η [%]	7.8	6.9	7.6	6.6

of the average hydrogen content deduced from the SIMS measurements agrees with the results obtained from FTIR spectra. Hydrogen contents of about 14 - 16 at.% for the a-Si:H films deposited in gd-reactor I and about 9 at.% for films deposited in gd-reactor II have been determined. SIMS - measurements performed on solar cells structures deposited in gd-reactor II, however, show no influence on the hydrogen content in the bulk material. This result is not surprising since the deposition rate and the hydrogen content are generally smaller in gd-reactor II.

Further information about the bulk properties can be obtained from the results of the photoconductivity and defect density measurements listed in table 2. No differences are found for the photoconductivity. Slightly lower defect densities are measured by PDS and CPM in the "soft start" case. This indicates a possible influence of the interface on the measurement or an effect of the ignition procedure on the bulk material. However, the small differences in the defect densities cannot explain the observed large differences in the solar cell parameters discussed in the following.

Solar cell performance

The results of the I-V characteristics obtained for the different experimental procedures are displayed in table 3. In general, the short circuit currents of the cells of type I are larger, since a textured TCO with a larger haze ratio (supplied from Asahi Glass Company) was used in that case. In both gd-reactors, a significantly larger efficiency is achieved if the "soft start" procedure is applied. Since only weak differences in the bulk properties have been found, the effect has to be attributed to a reduced defect density in the p-i interface region. In general, the reduced p-i interface defect density should lead to an increased open circuit voltage (The connection between V_{oc}, j_{sc} and the p-i interface defect density has been extensively discussed in a recent publication [2]). However, V_{oc} is determined by two competitive factors in this experiment. In addition to the reduction of the defect density, the hydrogen content and therefore the optical bandgap near the p-i interface are decreased as well in the case of the "soft start" ignition procedure. The latter causes a decrease of V_{oc} in contrast to the reduction of the p-i interface defect density. Hence, no clear dependence of the open circuit voltage on the ignition procedure is found here.

Stability

In order to investigate the degradation behaviour, the solar cells were exposed to "AM1"-like illumination for more than 1000 hours. The corresponding degradation kinetics of the cell efficiencies are displayed in two different ways. In figure 3a, the conversion efficiency is normalized to its initial value. In figure 3b, the absolute values during long term light exposure are given. As expected from their larger hydrogen content and their larger i-layer thickness, the cells type I show a much larger degradation compared to the cells type II. The plot of the normalized efficiencies indicates that the relative degradation η / η_0 is mainly determined by the bulk i-layer properties. For the solar cells described here, η / η_0 is correlated with the i-layer hydrogen content. There is no difference in the i-layer hydrogen concentration and therefore a very similar kinetics of η / η_0 is observed for the cells type II. The slightly smaller bulk hydrogen concentration of the "soft start"

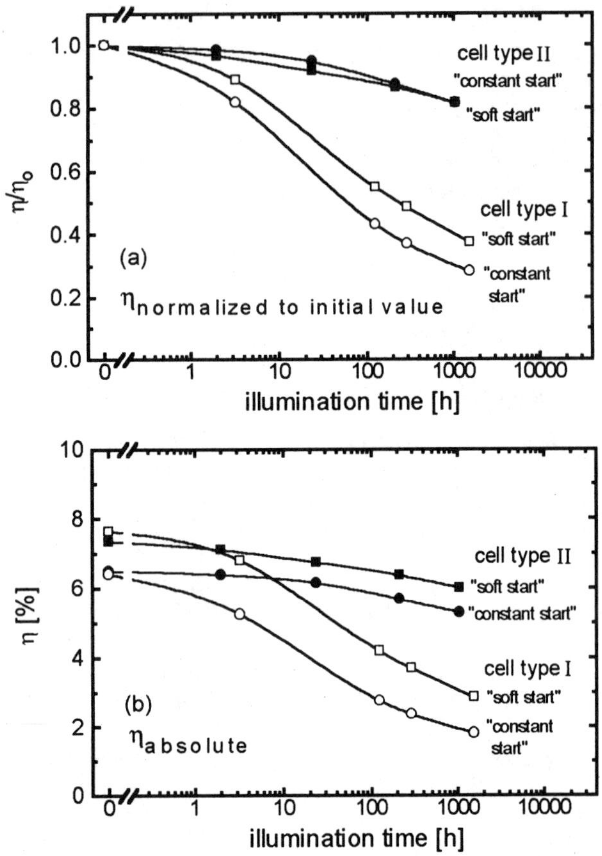

Figure 3: Degradation kinetics of the cell efficiencies during "AM1"-like light exposure. The cells of type I differ from the cells of type II by a larger hydrogen content and a larger i-layer thickness.

cell compared to the "constant start" in the case of cells type I could explain the small difference in η / η_0 observed.

However, in both cases the "soft start" solar cells reach a larger efficiency after degradation than the corresponding "constant start" solar cells (figure 3b). This shows that the recombination at the p-i interface is a very significant loss mechanism in the initial as well as in the degraded state. We point out here, that this is not in contradiction to our statement above that the degradation mainly occurs in the i-layer bulk material. The terms "interface degradation" and "interface recombination" are not to be confused. An increase of the interface recombination after light soaking is imaginable, but cannot be examined by this experiment. The degradation of the p-i interface region as the dominant degradation mechanism, however, can be excluded because the degradation kinetics of η / η_0 is determined by the bulk i-layer material even though the properties of the interface regions differ significantly. These different interface properties manifest themselves in the difference of the absolute values of the efficiency.

Another interesting detail of the degradation experiment gives further support to the conclusions above and our earlier investigations [2]: The degradation of the open circuit voltage V_{oc} is reduced significantly in the case of the "soft start" solar cells. This can be explained by a reduced degradation of the p-i interface region, e.g. due to the lower hydrogen concentration or different microstructure of the low deposition rate material [6]. However, the reduced interface degradation only changes the portions of the relative degradation of open circuit voltage and of the short circuit current, but does not significantly influence the degradation of the efficiency.

Hence, the improvement of stabilized efficiencies requires both the optimization of the interface quality and the stability of the bulk i-layer material.

SUMMARY

The experimental results emphasize the significant influence of the p-i interface properties on the solar cell performance. Applying the "soft start" ignition procedure, the defect density and the hydrogen concentration near the p-i interface is reduced. This leads to an improvement of the conversion efficiency in the initial as well as in the degraded state. The experiment demonstrates the pronounced influence of particle formation in the discharge on initial film growth. The results of the light soaking experiment can be explained by the degradation of the bulk i-layer material rather than by a predominant degradation at the p-i interface region. An influence of the ignition procedure on the bulk i-layer properties and therefore on stability is possible under specific conditions.

ACKNOWLEDGEMENTS

The authors are grateful to H.Gnaser (IFOS, Kaiserslautern) carrying out the SIMS-measurements. Financial support by the German Ministry of Research and Technology is gratefully acknowledged.

REFERENCES

[1] Z.E.Smith and S.Wagner, *MRS. Symp. Proc.* Vol.**70**, 1986, 525

[2] W.Herbst, J.Dudel, A.Scholz, B.Schröder, and H.Oechsner, *Solar Energy Materials and Solar Cells,* in press

[3] H.Sakai, T. Yoshida, S. Fujikake, T.Hama, and Y.Ichikawa, *J.Appl.Phys.* **67** (7), 1990, 3494

[4] U.I.Schmidt, W.Herbst, B.Schröder, and H.Oechsner, *MRS. Symp. Proc. Vol.* **336**, 1994, 85

[5] H.Gnaser and H.Oechsner, *Nucl. Instr. Meth. Phys. Res.* **B64**, 1992, 646

[6] T.Haage, S.Bauer, B.Schröder, and H.Oechsner, this proceedings

DEGRADATION OF A-SI:H SOLAR CELLS: NEW EVIDENCE FOR A BULK EFFECT

N. Wyrsch and A. Shah
Institut de Microtechnique, Université de Neuchâtel, Rue Breguet 2, CH-2000 Neuchâtel, Switzerland,

ABSTRACT

Well defined sequences of non-uniform and uniform degradation steps are applied on several high quality bifacial p-i-n solar cells with different thicknesses of the intrinsic layer. This procedure allows one to separate interface and bulk effects in the degradation of a-Si:H solar cells, or more precisely to separate the influence of the modification of the internal electric field profile (due mainly to effect near the interface) from that of an increase in the deep defect density in the bulk; according to present data the latter is clearly responsible for the major part of a-Si:H cell degradation.

INTRODUCTION

One of the most important topics in the present research on a-Si:H concerns the problem of light-induced degradation of layers and of solar cells. In spite of many studies, it is still not clear which part of the cell (doped layers, interfaces, bulk of intrinsic layer) limit the stable efficiency. The problem is complicated by the fact that both internal field distribution as well as recombination centre profile control cell performance. Although a majority of the members of the a-Si:H community probably agrees with the opinion that light-induced degradation of solar cells is a bulk effect, and is due to an increase of the defect density within the whole of the intrinsic layer [1-3], but there are also some arguments in favour of an interface effect (at the p/i and n/i interfaces) [4].

In a recent study [5], we tried to separate interface and bulk effects of the degradation on the solar cell performance (open-voltage, short-circuit, fill-factor and efficiency) and, in particular, to check the significance of the internal field distribution for the saturated efficiency of the cell. In this context the following procedure was applied on cells of different thicknesses (0.5 to 2 μm): We degraded both sides of bifacial cells one after the other, using strongly absorbed light (at 450 nm, to deform the field) and then degraded them further using a uniformly absorbed light beam (at 650 nm). Measurements of the I-V curve, as well as of spectral response (SR) and of the internal field profile were performed before and after every degradation step. The deep defect density in the central part of the i-layer was monitored using a wavelength-dependent measurement of the $\mu\tau$ product by time-of-flight (TOF).

Although, in thick cells, saturated efficiencies were clearly controlled by the defect density created within the i-layer (see Fig. 1a), the saturated efficiency of the 0.5 μm cell seems to be reached already after the first degradation step, namely after a strong blue light soaking through the p-layer (see Fig. 1b). However, since the generation profile under blue light extends well outside the interface region for this 0.5 μm thick cell, a bulk effect for the degradation could not be ruled out by this experiment alone. One should also note that each degradation step was performed until a (near-) saturation of the cell performance could be observed.

To refine our analysis (for the 0.5 μm thick cells), we devised a **new strategy** for the present study. The three different degradation steps, already mentioned above (with a well defined dose of light exposition), were,

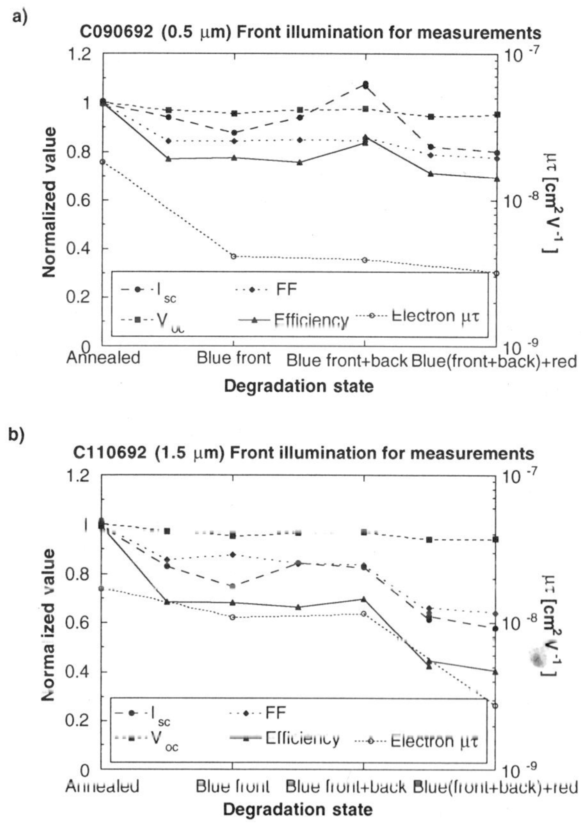

Fig. 1: Normalised I-V characteristics (short-circuit current I_{SC}, open-circuit voltage V_{OC}, fill-factor FF and efficiency) and electron $(\mu\tau)_e$ for (a) a 0.5 μm thick cell, (b) a 1.5 μm thick cell; measured under front illumination with an ELH lamp, in the annealed state and after each degradation step (as well as intermediate state).

CH3365-4/94/0000-0583 $4.00 © 1994 IEEE

First WCPEC; Dec. 5-9, 1994; Hawaii

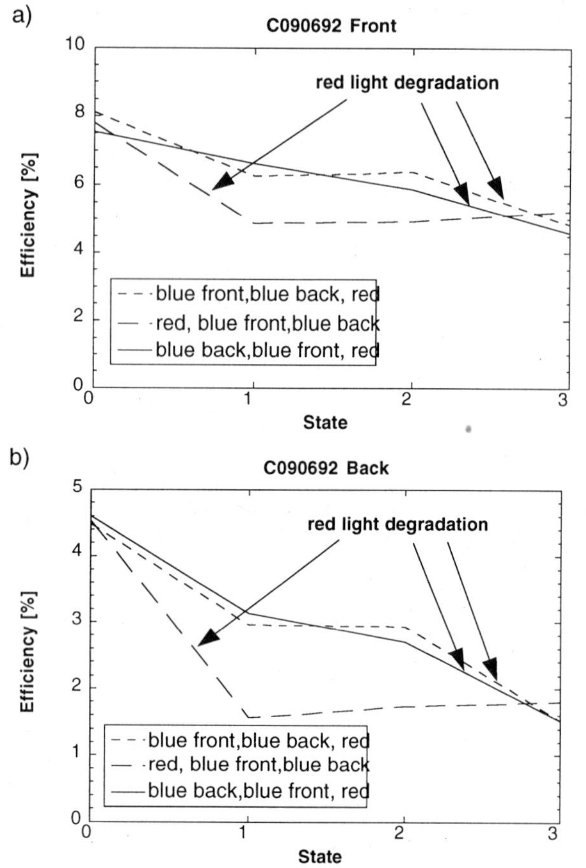

Fig. 2: Efficiency of a 0.5 μm thick bifacial cell (sample C090692) characterized in a solar simulator, with a front (a) and back (b) light exposure, after each step in three different sequences, composed each of three degradation steps. The same light-soaking time (100 hours) as well as the same light intensity were maintained for each individual step.

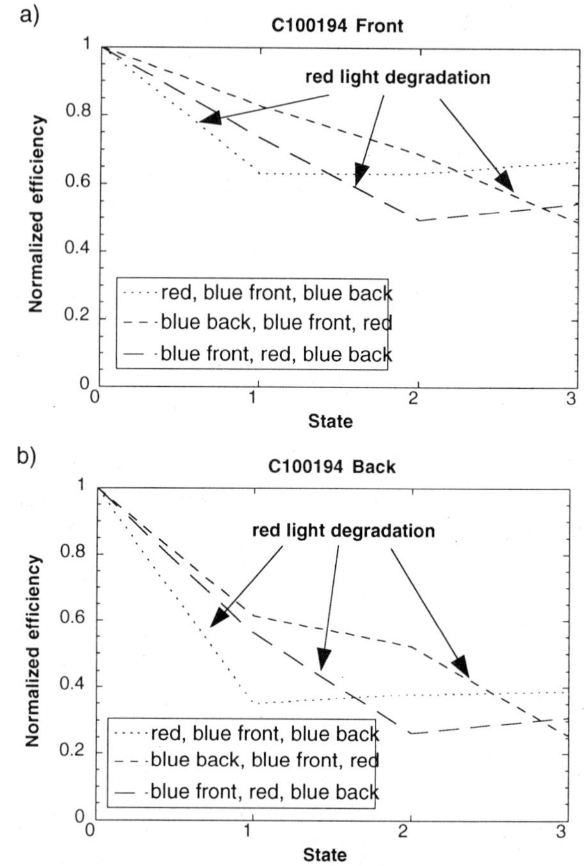

Fig. 3: Normalized efficiency of a 0.5 μm thick bifacial cell (sample C100194) characterized in a solar simulator, with a front (a) and back (b) light exposure, after each step in three different sequences, composed each of three degradation steps. The same light-soaking time (100 hours) as well as the same light intensity were maintained for each individual step.

this time, applied in different sequences, for example:

(1) blue light soaking from the front (through the p-layer), blue light soaking from the back (through the n-layer) and then red light soaking (through the p-layer)
(2) blue light soaking from the back, blue light soaking from the front and then red light soaking
(3) red light soaking, blue light soaking from the frontand then blue light soaking from the back

To achieve a better discrimination between the effect of the red and blue light degradation, the degradation doses (for each step) were kept much lower than in the preceding experiment [5]. Cell performance was recorded in-situ during each degradation step, as well as between the degradation steps in a solar cell simulator. Spectral response measurements were also performed to further analyse the evolution of the collection after each steps. Sequence (1), (2) and (3) were then compared to each other in order to determine the most harmful step for the degradation of the solar cell efficiency.

EXPERIMENTAL

All cells used in this study were deposited with the VHF-GD technique described elsewhere [6]. The structure used consists of: a glass substrate coated with TCO, a 120 Å thick p-window layer, a 60 Å thick buffer layer, an approximately 0.5 μm thick i-layer and a 300 Å thick n-layers. A thick layer of ITO was employed to provide transparent back contact. The doped and intrinsic layers were deposited at a plasma excitation frequency of 70 MHz; the deposition rate for the i-layer was 5 $Ås^{-1}$.

The cells were annealed at 180 °C for 5 hours before every degradation sequence and then characterized on each side with I-V measurements, on a double source Wacon solar simulator (under 100 $mWcm^{-2}$), and by determining spectral response at differe nt bias voltages, with and without bias light (50 $mWcm^{-2}$ white light). The initial efficiencies of the cells were between 8 and 8.5 %; these values were found to be lower than those of corresponding conventional cells because of the bifacial design (poor ITO reflectance properties).

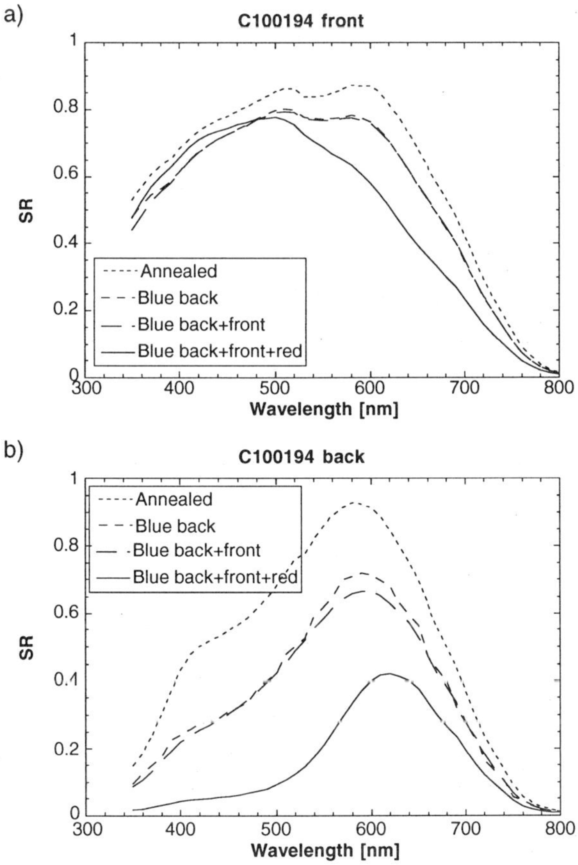

Fig. 4: Spectral response at 0 V under front (a) and back (b) illumination (with bias light) of a 0.5 μm thick bifacial cell after each step of a degradation sequence (second sequence of Fig. 3).

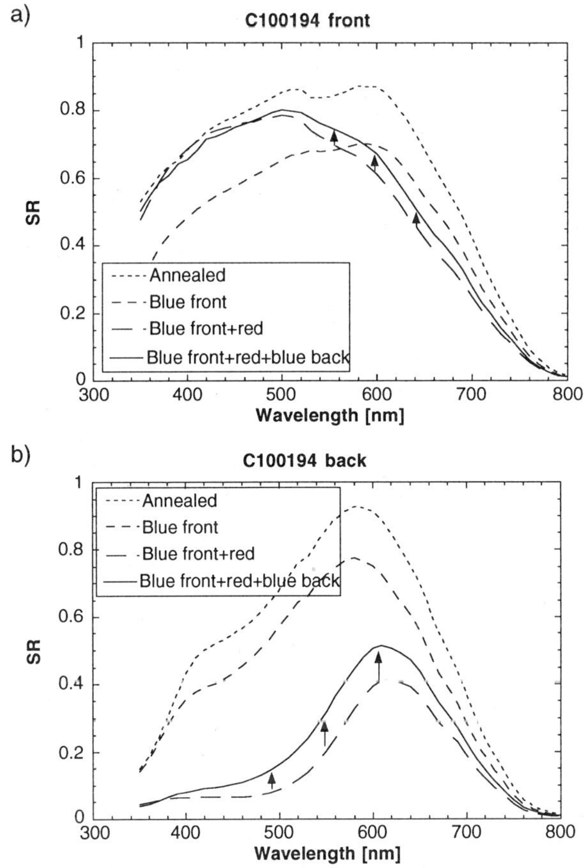

Fig. 5: Spectral response at 0 V under front (a) and back (b) illumination (with bias light) of a 0.5 μm thick bifacial cell after each step of a degradation sequence (third sequence of Fig. 3). Enhancement of the response by back-side blue light is indicated by arrows.

For these light-soaking experiments, a 2000 W halogen lamp was used, filtered with a water filter and with interference filters with 50 nm bandwidth; wavelengths were 450 nm (75 $mWcm^{-2}$) for the blue light and 650 nm (170 $mWcm^{-2}$) for the red light degradation. Each individual degradation step was carried out for 100 hours.

RESULTS AND DISCUSSION

Three different sequences of degradation steps were applied on two different 0.5 μm thick diodes. The evolution of the normalized (to the initial annealed) efficiency is given on figures 2a and 3. As we can see, red light soaking of the cell (indicated with an arrow) is much more damaging than the other steps. Taking into account the fact that generation rate for red light was much smaller in the interface regions than for blue light, and that most of the recombination (under red light) occurs in the central part of the i-layer, we can easily conclude that creation of deep defects within the i-layer (far from the interfaces) critically control the cell efficiency. The same picture is observed when characterizing the cell under back-side illumination (see Fig. 2b and 3b). Characterization of the cell performance under front or back illumination is (qualitatively) equivalent. Any degradation step affects, at the same time, the cell performances for front and back illumination, even though the latter is quantitatively more affected.

Thus, the final state of the cell is approximately given by the one reached after the red light degradation step. However, as we can observe on figures 2a and 3a, further back side degradation with blue light, slightly improves the cell performance (this fact is also observed on an in-situ recording of the degradation). This effect is related to an enhancement of the field close to the n/i interface, which improves hole collection, as is confirmed by spectral response measurement: Blue light soaking of the back-side of a red light soaked cell enhanced the red response for front illumination as well as the blue response for back illumination (see Fig. 4). Note, that blue light soaking of the back-side is only beneficial when done after the red light soaking.

Decrease of the performance during any degradation step is related to a decrease of both short-circuit current I_{SC} and fill-factor FF. However, in-situ recording of the I-V characteristics shows that FF drops faster than I_{SC} at the beginning of every step. The origin of this difference remains so far unknown.

Another interesting result of this study is the fact that the final behaviour of the cell does (almost) not depend on the degradation history, but rather on the accumulated time of light-soaking at the most damaging wavelength.

Spectral response measurements after each degradation step (see Figs. 4 and 5) are consistent with the I-V characteristics. The final state of the cell is obtained after the red light soaking step, independently of the previous degradation steps; responses of the cell after the first degradation step differ considerably from each other depending on the wavelength and the direction (front or back) of the light soaking used for that step. Fig. 4 offers a good illustration of the effect of "interface degradation" (by blue light soaking from both sides) as compared with "bulk degradation" (by red light soaking). The drop of the spectral response seen after the blue light soaking from both sides occurs for the full range of wavelengths and can be attributed to the reduction of the internal field (in the central part of the i-layer), whose profile becomes close to its final form and value after the red light soaking [5] (see [7] for details about the internal field measurement technique).

One should note that spectral response measurement must be performed with bias light in order to be compared with cell I-V characteristics. Spectral response curves measured without bias light (see Fig. 6) are determined by the internal field profile of the cell in the dark (see also Ref. [8]). The response without bias light always shows a strong decrease in the blue, even if the actual deterioration in the operating solar cell is in the red light region (compare Figs. 4a and 6). This can lead to a misinterpretation of the spectral response data, and is one of the reasons which lead some researchers to locate the part of the cell controlling the stable efficiency in the interfaces regions. On the other hand, as the internal field profile of the cell directly depends on the defect density in the i-layers, the spectral response in the dark is a good indicator of the degradation stage and can e.g. be used to verify the fully annealed state of the cell.

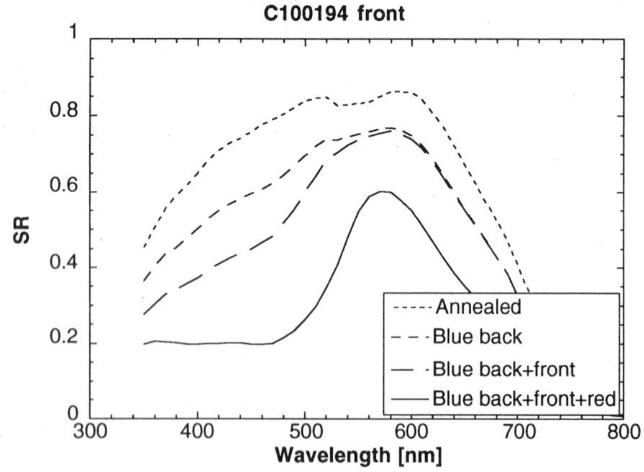

Fig. 6: Spectral response at 0 V bias voltage, under front illumination without bias light for a 0.5 μm thick bifacial cell after each step of degradation (corresponds to Fig. 4a).

CONCLUSIONS

From extensive characterisation of bifacial a-Si:H solar cells after well defined sequences of degradation steps using alternatively both strongly absorbed and uniformly absorbed light, the authors were able to demonstrate the following points:

1) The stable efficiency of a-Si:H solar cells is controlled mainly by an increase of the deep defect density in the bulk of the i-layer, far from the interface region. This statement remains valid for cell thicknesses down to 0.5 μm. The interfaces regions may thereby yet play a marginal role.

2) No history effect was observed. The final state of the cell is given by the accumulated time of degradation at the most damaging wavelength (i.e. the wavelength at which the generation of defects far from the interfaces is the most efficient).

3) Combining spectral response measurements (with bias light) with illumination from both front and back side helps one to interpret the cell behaviour. Spectral response without bias light does not reflect the situation of an operating cell.

ACKNOWLEDGEMENTS

We would like to thank S. Dubail for sample preparation and D. Fischer for valuable discussions. This work was supported by the Swiss Federal Office of Energy (OFEN) under contract EF-REN 93(032).

REFERENCES

[1] C.M. Fortmann, S. Lange, M. Hicks, C.R. Wronski, *J. Appl. Phys.* **64**, 1988, pp. 4219-4222.

[2] W. Herbst, A. Scholz, B. Schröder, H. Oechsner, *Proc. of the 23rd IEEE Photovoltaic Specialists Conf.*, Louisville, 1993, pp. 913-918.

[3] F. Smole, J. Furlan, M. Topic, *MRS Symp. Proc.* **297**, 1993, pp. 631-636.

[4] W. Kusian, H. Pfeiderer, *AIP Conf. Proc.* **234**, 1991, pp. 291-297.

[5] N. Wyrsch, H. Keppner, N. Pellaton, Y. Ziegler, A.Shah, *Proc. of the 11th EC Photovoltaic Solar Energy Conf.*, Montreux, 1992, pp. 742-745.

[6] H. Curtins, N. Wyrsch, M. Favre, A.V. Shah, *Plasma Chem. and Plasma Proc.* **7**, 1987, pp. 267-273.

[7] N. Wyrsch, D. Fischer, A. Shah, *Proc. of the 12th EC Photovoltaic Solar Energy Conf.*, Amsterdam, 1994, pp. 73-76.

[8] D. Fischer, N. Wyrsch, C.M. Fortmann, A.V. Shah, *Proc. of the 23rd IEEE Photovoltaic Specialists Conf.*, Louisville, 1993, pp. 878-884.

ROLE OF THE LATERAL LEAKAGE CURRENT ON AMORPHOUS SILICON SOLAR CELLS

Rodrigo Martins, Elvira Fortunato, Ana Bicho and Guilherme Lavareda

Materials Science Dep, Faculty of Science and Technology of New University of Lisbon - DCM/FCT-UNL and Centre of Excellence for Microelectronics and Optoelectronic Processes - CEMOP
Quinta da Torre, P-2825 Monte de Caparica, Portugal

ABSTRACT

The aim of this work is to interpret the role of the lateral leakage current on the a-Si:H solar cell performances (J-V characteristics, responsivity and the apparent device degradation behaviour), under low illumination conditions.

INTRODUCTION

Amorphous Silicon solar cells are used in several low and medium power generation systems [1], due to the possibility to deposit them in a large variety of substrates, up to sizes of the order of $1m^2$, at low production costs [2]. Nevertheless, one of the most constrains in spreading their massive application deals with the light degradation effect observed [3] and the difficulty in scaling up their characteristics when large area devices are required. Up to now, most of the experimental data is interpreted based on the peculiarities presented by this disordered device (e.g., density of states (DOS) of the i-layer; unstable bulk defects under light exposure; electric field distribution [4], etc.), and little attention is given to the role of the device's design and engineering on their performances. In this work, we propose a model able to explain the observed experimental behaviour of the a-Si:H solar cells under low illumination, where the lateral leakage current plays an important role. The model also explains the behaviour observed in devices where film's crystallisation on the device edges occurs.

IDENTIFICATION OF THE PROBLEM

The a-Si:H solar cell's analysed are shown in Fig. 1, where the doped and i- layers present thicknesses ranging from 10 nm to 100 nm and 200 nm to 700 nm, respectively. The doped front layer has the thickness just required to establish the diffusion potential without enhancing the optical losses and the back doped layer presents the required conductivity to improve the potential contact, enhancing so the carrier's collection.

On the other hand, the i-layer might have a low DOS and Urbach energy (related to the tail states) and, to be as thin as possible to decrease the carrier's recombination losses and the degradation phenomena.

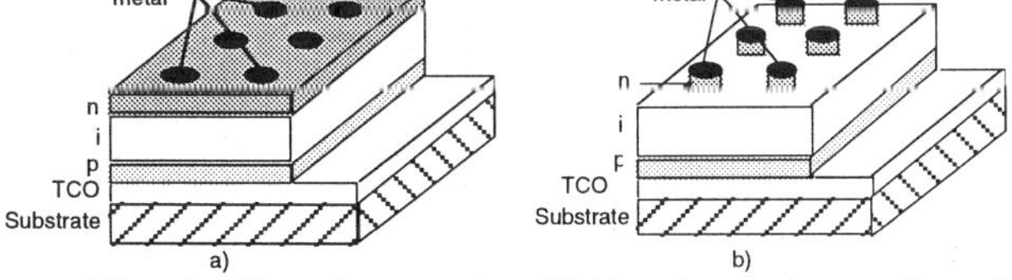

Fig. 1. Structures of a-Si:H p.i.n solar cells. a) unetched and b) etched device, respectively.

The above conditions are mainly followed as standard recipes during the production of a-Si:H solar cells, either at a laboratory or at an industrial level. At a laboratory level, to characterise the devices, small metal contacts are deposited on the top of the doped layer. Under these conditions, the experimental recorded data exhibit a high dark saturation current density J_S, the short current density (J_{SC}) exhibits a degradation behaviour with time and the fill factor (FF) can be as high as 0.7, depending on the device's size [5]. On the other hand, large area integrated device's produced using the laser scribing technique present lower performances than the small area ones though they are produced following the same recipes. This is usually attributed to pinholes and to the loss of film's uniformity as the device area increases. However, these devices present, generally, a ratio between the covered and uncovered metal areas higher than one or the conductivity of the uncovered area is higher than the covered one (recrystallization). Thus, we can consider the devices as having asymmetric structures where a lateral leakage current density (J_L) plays an important role in the device performances [6].

Under illumination, J_L depends on the photon flux F_{ph} impinging, the devices and on the lateral potential developed, since the regions underneath the contact and outside it do not deplete equally. To proof this, small metal dotes were deposited on the top layer and the J-V curves; the photocurrent dependence on the light intensity, $I(\lambda)$ and on time; and, the device responsivity, $R(\lambda)$, measured, before and after etching the top doped layer surrounding the metal contact (see Fig. 1).

FORMULATION OF THE MODEL

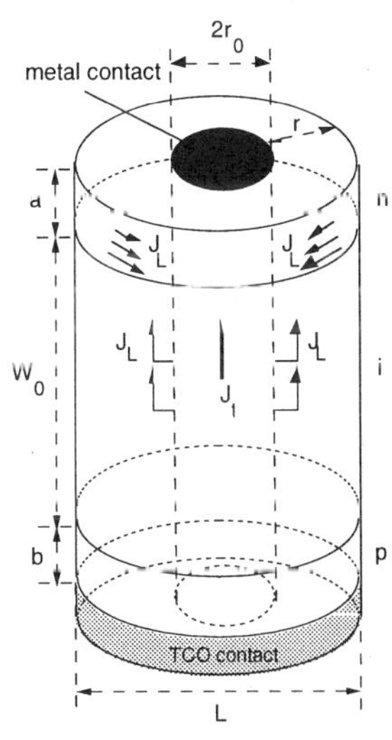

Fig. 2. Spatial carriers flow distribution within the unetched device.

Under steady state, the total carriers flow density ($J_T[V(r)]$) in a-Si:H asymmetric diodes has two main components, as shown in Fig. 2. One, is the transverse current density (J_t), due to carriers flowing within the area defined by the metal contact of radius r_0. The other is ($J_L[V(r)]$), related to the total current density ($\Sigma dJ_L[V(r)]$) collected by the top layer (outside the contact), ascribed to an annulus of height a (the thickness of the top doped layer), internal radius r_0 and external radius L/2 (the length of the deposited layer on the substrate):

CH3365-4/94/0000-0587 $4.00 © 1994 IEEE

First WCPEC; Dec. 5–9, 1994; Hawaii

$$J_T = J_t + \left[\left(\frac{L}{2r_0}\right)^2 - 1\right] J_L \tag{1a}$$

In Eq. (1a), J_t is given by the classical diode equation and the contribution of the photocurrent density, J_{ph}, dependent on $I(\lambda)$. Here, $J_{ph} \approx AI(\lambda)^{\gamma}$ where A is a constant dependent on the generation process and $0.5 < \gamma < 1.0$ [7]:

$$J_t = J_s\left[\exp\left(\frac{V + J_t R_s}{2\phi}\right) - 1\right] - \frac{V}{R_p} - AI(\lambda)^{\gamma} \tag{1b}$$

In Eq. (1b) R_s and R_p are the series and the shunt resistances, respectively, in Ωcm^2. Under low $I(\lambda)$ and close to the short circuit condition, $\gamma \approx 1$ and the exponential term of the above equation can be neglected, leading to:

$$J_t = -\left(J_s + AI(\lambda) + \frac{V}{R_p}\right) \tag{1c}$$

Evaluation of J_L

J_L depends on the resistivity, ρ, of the top layer and on the spatial distribution of the lateral voltage [6], being related to the transverse voltage, V(y) by $dV(y)/dy = -\rho J_L[V(y)]$ and to J_t by the charge carrier's flow conservation condition $\nabla \bullet \left[\frac{1}{\rho}\nabla V(y)\right] = -\frac{J_t}{a}$. Once known V(y), J_L is determined taking into account that:

$$\frac{\delta J_L}{\delta y} = \frac{dJ_L}{dV}\frac{dV}{dy} \quad \Rightarrow \quad \rho J_L \frac{dJ_L}{dV} = -\frac{J_t}{a} \tag{2}$$

Solving Eq. (2) (if $I(\lambda)$ is constant - use of monochromatic light. Otherwise, $I(\lambda)$ corresponds to the integral of $F_{ph}(\lambda)$ over the entire spectrum range) we get:

$$J_L \approx \sqrt{\frac{2}{a\rho} V J_t} \tag{3}$$

Dependence of the dynamic components on J_L

As the doped layers can not be considered as true equipotentials, under uniform illumination and close to the short circuit condition, the i-layer underneath the metal contacts is completely depleted whilst the outside region is not. This leads to a leakage current collected through a channel established in the collecting doped layer. Under these conditions, J_L is related to a "dynamic shunt" (R_d) resistance in parallel with a "static shunt" (R_p) resistance, as shown in Fig. 3, leading to:

$$\frac{1}{R_d} \approx \frac{1}{2}\sqrt{\frac{2}{a\rho}}\left[\sqrt{\frac{J_t}{V}} + \frac{1}{R_p}\sqrt{\frac{V}{J_t}}\right] \tag{4}$$

Fig. 3 also shows the capacitance's ascribed to the covered (fixed bulk capacitance, C_B) and uncovered (dynamic capacitance, C_d) metal regions, respectively.

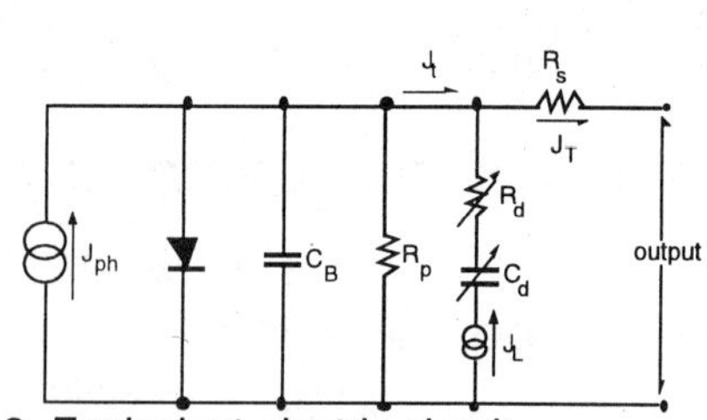

Fig. 3. Equivalent electric circuit.

From Eq. (4) we see that $R_d^{-1} \propto I(\lambda)^{0.5}$ when J_L has considered. Otherwise, J_L is neglected and from Eq. (1c) we get $R_p^{-1} \propto I(\lambda)$.

C_d, depends on the carriers stored outside the metal region (function of J_L and of the potential difference, ΔV between the covered and uncovered areas), given by:

$$\frac{C_d}{R_d} \approx \frac{\varepsilon\varepsilon_0}{W - \Delta W} \approx \frac{J_L(\tau')}{2\Delta V} \tag{5}$$

In Eq. (5) $J_L(t)$ is the lateral leakage current built up when the device is illuminated; τ' is the relaxation time of the uncovered region; ΔW is the thickness of the undepleted region outside the covered region and A is the area of the uncovered region. Thus, as a first approximation, it is expected that $J_L(t)$ varies as:

$$J_L(t) = J_L(0)\left[\exp\left(-\frac{t}{\tau_d}\right)\right] \tag{6}$$

where $\tau_d \approx R_d C_d$, and depends on $I(\lambda)$. Thus, as $I(\lambda)$ varies, the unbalanced charge flow between the static ($\tau_B \approx R_p C_B$) and dynamic (τ_d) components try to re-establish the previous steady state condition and so, the system tends to recharge up and discharge again, until reaching an almost stationary regime. This behaviour explains the higher decrease observed in J_{sc} for the unetched devices than for etched ones.

Role of J_L on the device responsivity

$R(\lambda)$ is also ruled by J_L since $R(\lambda) = J_T / h\nu F_{ph}$ where $h\nu$ is the energy of F_{ph} impinging on the surface of the device. Therefore, when J_L is present, $R(\lambda)$ depends on the contribution of two components, differently related to $I(\lambda)$. Thus, as the reverse bias increases, for $h\nu < E_{op}$ there are a double contribution from the covered and uncovered metal regions, leading to an overall enhancement of $R(\lambda)$. For $h\nu > E_{op}$ as the depleted region is mainly increased in the region sustained by the metal contact, it is expected an enlargement of $R(\lambda)$, controlled mainly by J_t.

EXPERIMENT

The a-Si:H p.i.n solar cells were deposited onto SnO_2/glass substrates with 25 mm $\times$ 25 mm size, by means of the PECVD technique described elsewhere [8]. The p-and n-layers were produced in the same reactor by mixing diborane (0.6%) or phosphine (1.5%), with pure silane, using growth rates of the order of 2 Å/s; substrate temperatures of about 210°C and, an r.f. power of 10 W. The final layer's thicknesses for a non optimised device was of the order of 20 nm, 1000 nm and 60 nm, respectively for the p, i and n layers. Once deposited the p.i.n structure on the SnO_2/glass substrate, Al 1% Si dots (r_0 = 1 mm) 1 µm thick were evaporated by the electron gun technique onto the top doped layer.

The top layer was etched using a Reactive Ion Etching system (Alcatel GIR 300) with end point detection, employing as reactive gas the SF_6. Laser scribing (Permanova, YAG Nd dual beam laser) was also applied to isolate the devices and to observe the role of the edge layer's crystallisation on the final device performances. Here, the conductivity was enhanced by more than two orders of magnitude, over a width of about 0.2 mm, using laser spots of 60 µm in diameter.

The J-V curves were determined by a programmable power supply/electrometer (Keithley 236) at room temperature, under dark and uniform illumination conditions, using a monochromatic light with λ = 632.8 nm. The device was reverse bias using a controllable dc power source, allowing so the determination of R_p^{-1} and R_d^{-1} as a function of I(λ). To perform the J_T dependence on I(λ), with and without reverse biasing the device, neutral density filters (Melles Griot 03 FSG 017) were used. The J_T time dependence was computer monitored.

ANALYSIS AND DISCUSSION

The experimental illuminated J-V curves are shown in Fig. 4, for unetched and etched p.i.n (dashed and solid lines) solar cells. The J-V curves depicted in Fig. 4 for the unetched solar cells (dashed line) show high J_{SC}, low shunt resistance's and FF, when compared with the etched devices (solid lines).

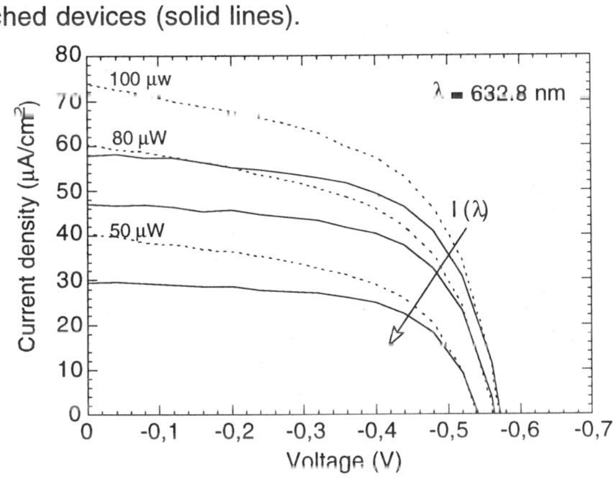

Fig. 4. J-V curves for p.i.n solar cells unetched (dashed line) and etched (solid line).

Fig. 5 shows the dependence of the relative variations of J_{SC} (ΔJ_{SC}) and FF (ΔFF) between the same unetched and etched device, on I(λ). The data show that ΔJ_{SC} is enhanced as I(λ) increases, saturating for I(λ) ≥ 80 µW, whilst the reverse behaviour is observed for ΔFF. The relative variation of J_{SC} ($\Delta J_{SC}/J_{SC}$) above 80 µW saturates around 20% whilst ΔFF/FF varies with I(λ) decreasing as I(λ) increases. For unetched devices, FF is strongly dependent on I(λ), saturating around 0.6.

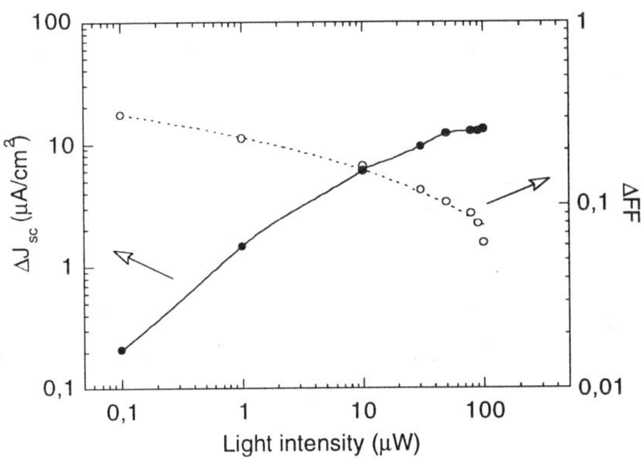

Fig. 5. Variation of ΔJ_{SC} and ΔFF on I(λ).

On the other hand, the etched devices present a FF almost independent on I(λ), reaching a value around 0.7. These results agree with the prediction of the model discussed above.

The role of J_L on the device performances was inferred plotting the slope of the J-V curves at V < 0 V, as a function of I(λ), for both types of devices under analysis (see Fig. 6).

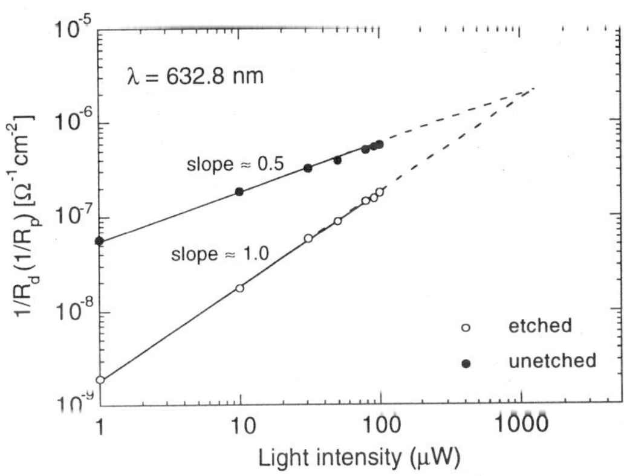

Fig. 6. Dependence of R_d^{-1} and R_p^{-1} on I(λ).

The data fit a power law dependence equal to 0.5 and 1, respectively for the unetched and etched devices, as predicted by Eqs. (4) and (1c), respectively. The above fitting intercept at I(λ) ~ 10^3 µW, above which the main role of J_L is in ruling the total device's shunt resistance.

After light exposure and reverse biased (V=-0.5V), J_{SC} decreases around 20% and 7%, for the unetched and etched devices, respectively. To explain this behaviour, let we refer to Figs. 7 and 3. There, J_T is ruled by J_t and J_L, respectively related to τ_B and τ_d. Thus, when the device is initially illuminated, J_L also contributes to J_T up to the establishment of a quasi stationary regime. After that, part of J_{ph} is lost through the lateral leakage path, (dependent on the ratio between the covered and uncovered metal areas, conductivity of the doped collecting layer, as well as on R_d and C_d), as suggested by Eq. (6). The experimental points presented in Fig. 7 follow an exponential fitting, where $\tau_d \approx$ 1537 s for the unetched device, agreeing well with the expected value given by the proposed model.

Therefore, besides the conventional degradation phenomena attributed to the Staebler-Wronski effect [3], it must be included the role of the lateral leakage current, related to the inappropriate design and engineering of the device.

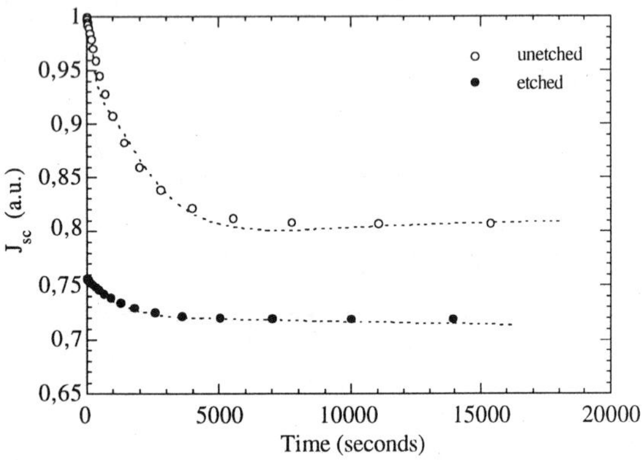

Fig. 7. Dependence of J_{SC} on time, for the p.i.n solar cells reverse biased (V=-0.5V), etched and unetched, respectively. The dashed line corresponds to an exponential fitting, as predict by Eq. (6).

R(λ) for the etched and unetched devices are shown in Fig. 8. There, when $\lambda < \lambda_m$ or $\lambda > \lambda_m$ (where λ_m is the wavelength for which R(λ) is maximum), R(λ) is enhanced for unetched devices whilst for the etched devices R(λ) is mainly enhanced when $\lambda < \lambda_m$. This behaviour is attributed to the role of J_L on the device performances, agreeing well with the model proposed.

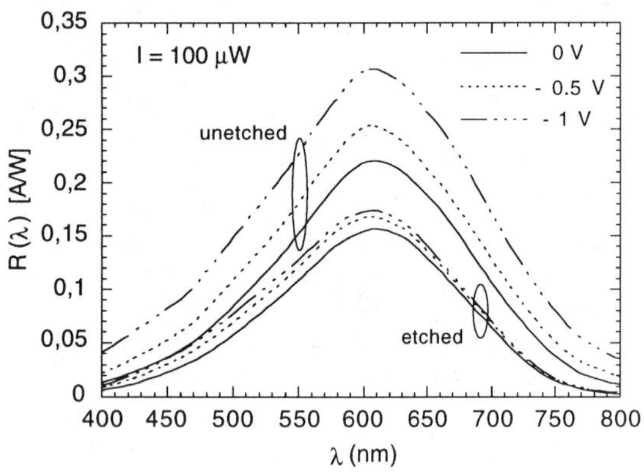

Fig. 8. Dependence of R(λ) on λ for the p.i.n solar cells, etched and unetched, respectively.

When the uncover metal region was laser scribed in conditions that lead to the crystallisation of the material onto the cut edges, J_S and J_L are enhanced and so, a behaviour similar to that one of the unetched devices is observed, as predicted by the model proposed.

SUMMARY

In summary, we investigated the role of the lateral effects on a-Si:H p.i.n solar cell performances. The experimental data fits a model, where J_L plays an important role in the low light flux regime in determining the shunt resistance (and so, the FF), the J_{SC} and R(λ). The model also explains the "apparent" light instability of the unetched devices, indicating that the device's lateral effect has to be taken into account during the determination of the device degradation behaviour. The obtained results also indicate that J-V characteristics reflecting only the behaviour of the device are obtained in samples where the metal contact covers more than 95% of the area of the doped collecting layer or when the uncovered top doped layer is etched away by means that do not lead to structural changes of the material that surrounds the metal contact.

ACKNOWLEDGEMENTS

The authors are grateful to C.N. Carvalho for its contribution to sample elaboration, T. Chee who perform the sample etching and to J. Carvalho for the sample laser scribing. This work is supported by the Ministry of Energy and Industry (PEDIP nº 00250), by NATO-POTHINFILM and BRE2-CT94-0944 projects.

REFERENCES

[1] See for instance the proceedings of *12th EPVSEC*, 1994.

[2] Y. Hamakawa, "Recent Advance in Solar Cell Technology", *PVSEC-3*, 1987, pp. 147-152.

[3] M. Vieira, E. Fortunato, G. Lavareda, C.N. Carvalho and R. Martins, "Role of Photodegradation on the μτ Product and Microstructure of the a-Si:H PIN Devices", *Mat. Res. Soc. Proc.*, **297**, 1993, pp. 637-642.

[4] F. Smole, J. Furlan and M. Topic, "Internal Electric Field in Light-Degradation p.i.n a-Si:H Solar Cells", *Mat. Res. Soc. Proc.*, **297**, 1993, pp. 631-636.

[5] A. Catalano, in "Amorphous & Microcrystalline Semiconductor Devices", J. Kanicki, ed. (Artech House, London), 1991.

[6] R. Martins and E. Fortunato "The Lateral Photoeffect in Large Area 1-D Thin Film Position Sensitive Detectors Based in a-Si:H P.I.N Devices", submitted for publication at *Journal of Scientific Review*, 1994.

[7] R.S. Crandall, in "Semiconductors and Semimetals", 21, part B, J. Pankove, ed. (Academic Press, New York), 1984.

[8] R. Martins, I. Ferreira, N. Carvalho and L. Guimarães, "Engineering of Plasma Deposition Systems Used for Producing Large Area a-Si:H Devices", *J. of Non-Cryst. Solids*, **137-138**,1991, pp. 757-760.

RECOMBINATION IN a-Si:H : RESULTS FROM THE STANDARD AND DEFECT POOL MODELS

Eduardo Morgado

Technical University of Lisbon, Instituto Superior Técnico, Centro de Física Molecular, Av. Rovisco Pais, Complexo I - IST, 1000 Lisboa, Portugal

ABSTRACT

Results from numerical calculations of the recombination kinetics in a-Si:H involving either the standard or the defect pool models, including band tail states, are reported. Fermi level, light intensity and temperature dependences of $\mu\tau$ products are investigated. Major features of experimentally observed majority carrier behavior can be described on the basis of the standard model. Photo-quenching of both $(\mu\tau)_e$ and $(\mu\tau)_h$ for Fermi level positions around mid-gap can be reproduced on the basis of the defect pool model. Photo-enhancement and thermal quenching of $(\mu\tau)_e$ are associated with a capture cross section of the band tails smaller than the capture cross sections of the dangling-bonds.

INTRODUCTION

In previous publications we have reported a number of results from numerical calculations with a recombination model for a-Si:H involving one class of positively correlated dangling-bonds states [1, 2], the so-called standard model, as well as exponential band tails of localized states [3, 4]. This approach has been successful in reproducing, in a qualitative way, major features of photoconductivity [1, 4] and $\mu\tau$ products [2, 3, 4] in a-Si:H, such as the anticorrelated behavior of the majority and minority carriers with respect to the movement of the Fermi level [2, 3, 4] and the corresponding sensitization of the photoconductivity [1], the power law light intensity dependence of the photoconductivity with exponents in the range $0.5 \leq \gamma \leq 1$ [1, 3] and the photo-enhancement of $(\mu\tau)_e$ and associated thermal quenching (or, equivalently, the supralinearity and thermal quenching of photoconductivity) [3, 4].

Recently, the so-called defect pool model [5] has received increasing attention. Calculations based on the defect pool model have been reported concerning the temperature dependence of photoconductivity in intrinsic material [6, 7] and the Fermi level dependence of $\mu\tau$ products at room temperature [7].

In the present paper, new results from calculations based on the defect pool model, including band tails effects, are reported and compared with experimental data and with results from the standard model.

MODEL EQUATIONS

The basic equations are:

$$G = \sum_i R_i \tag{1}$$

$$Q(n, p) = Q(n_o, p_o) \tag{2}$$

Equation (1) is the equilibrium between photo-generation rate G and the total recombination rate. R_i are net recombination rates, i.e., they include capture and emission terms, defined according to Shockley-Read theory. Only electronic transitions between extended states bands and localized gap states are considered.

Equation (2) is the conservation of charge Q from dark to illumination, where n(p) is the free electron (hole) density under illumination and $n_o(p_o)$ is the respective density in the dark.

The model equations have been solved numerically taking into account the occupation statistics of correlated defect levels in the dark [8] and under illumination [9] and the electronic occupation functions for an arbitrary distribution of traps deduced by Simmons and Taylor [10].

RESULTS

In the following, we report results of the calculations for different models of increasing complexity. We focus our attention on the Fermi level, light intensity and temperature dependences of the electron and hole lifetimes.

The power law light intensity dependences of the lifetimes are defined as:

$$\tau_e \propto G^{-\phi_e} \tag{3}$$

$$\tau_h \propto G^{-\phi_h} \tag{4}$$

The light intensity dependence of the majority carrier is usually deduced from the measurement of the photoconductivity σ_{ph}. The minority carrier mobility lifetime product is experimentally determined from the measurement of the ambipolar diffusion lenght L_d [11]. The light intensity dependences defined in equations (3) and (4) can, therefore, be represented in an alternative way:

$$\sigma_{ph} \propto G^{\gamma_{majority}} \tag{5}$$

$$L_d \propto G^{-S_{minority}} \tag{6}$$

Taking into account the well known expressions defining the photoconductivity and the ambipolar diffusion lenght [11], when electrons are majority carriers one has

First WCPEC; Dec. 5–9, 1994; Hawaii

$(1 - \gamma_e) = \phi_e$ and $2S_h = \phi_h$. Conversely, when holes are the majority carriers one has $(1 - \gamma_h) = \phi_h$ and $2S_e = \phi_e$.

In all the Figures throughout this paper, the points are results of the calculations and the lines are smooth fits which serve as guidelines for the eyes.

Standard Model

In the so-called standard model, it is assumed that recombination involves one class of positively correlated dangling bond levels in the gap. We have already reported some results from this approach [1, 2, 3, 4]. For a comparison with the results presented in the next sections obtained from more complex models, calculations have been made for the parameter values: energy distance of the $D^{o,+}$ level from the conduction band $E_C - E_{Do,+} = 0.95$ eV, correlation energy $E_U = 0.2$ eV, mobility gap $Eg = 1.7$ eV, defect density $N_D = 10^{16}$ cm^{-3}, capture cross sections for electrons or holes of the state of charge indicated in superscript $\sigma_n{}^o = \sigma_p{}^o = 10^{-15}$ cm^2, $\sigma_n{}^+ = \sigma_p{}^- = 10^{-14}$ cm^2 and T = 300 K. The calculated light intensity dependences of the lifetimes can be approximated by the power law relations (3) and (4) and the dependence of the exponents ϕ_e and ϕ_h on the Fermi level position is shown in Fig. 1 for the photogeneration range 10^{19} cm^{-3} s^{-1} $\leq G \leq 10^{20}$ cm^{-3} s^{-1}.

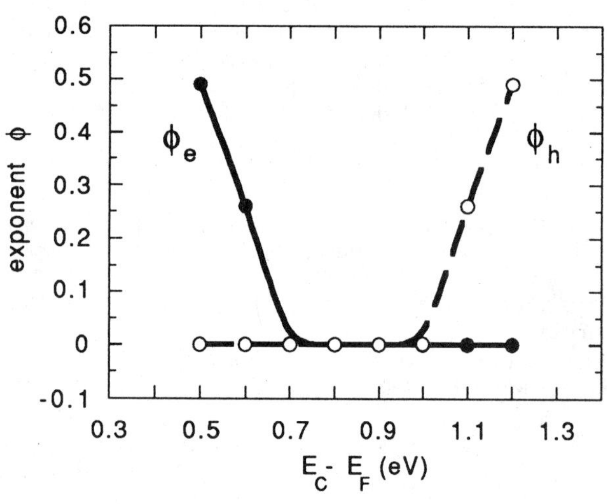

Fig.1 - The exponents of the light intensity dependence of the lifetimes as functions of the Fermi level position (Standard Model)

Standard Model plus Band Tails

Exponential band tails of localized states with characteristic energies $E_{oc} = 0.025$ eV and $E_{ov} = 0.045$ eV have been incorporated in the preceding model. Fermi level, light intensity and temperature dependences calculated from this approach are reported in references [3, 4] and we reproduce here the main results.

Negative values of ϕ_e appear for E_F in the middle of the gap leading to a superlinear photoconductivity ($\gamma > 1$) and associated thermal quenching [3, 4]; these effects are promoted by the asymmetry of the band tails and occur for a capture cross section of the band tails smaller than the capture cross section of the dangling bonds [4]. The maximum of the electron lifetime versus reciprocal temperature plot shifts to higher temperatures when E_F approaches the conduction band [3] in agreement with photoconductivity data [12] .

In spite of all these achievements, which are essentially associated with the behavior of the majority carrier lifetime, the model could not describe the simultaneous S>0 and γ<1 values found in some experiments with undoped a-Si:H [11, 13].

Defect Pool Model

The defect pool model arises from a thermodynamic analysis of defect formation and involves three classes of positively correlated dangling bond defect levels [5]. For the calculations we use the parameter values [14]: $E_{U1,2,3} = 0.2$ eV, $E_C - E_{Do,+1/2/3} = 0.63/0.95/1.27$ eV, $N_{D1} = N_{D3} = 10^{16}$ cm^{-3}, $N_{D2} = 5 \times 10^{15}$ cm^{-3} and the capture cross sections are the same as in the standard model calculations above.

As can be seen in Fig. 2 , values of $\phi > 0$ appear for E_F around the middle of the gap, contrarily to the standard model result in Fig.1 where $\phi = 0$ for that E_F range.

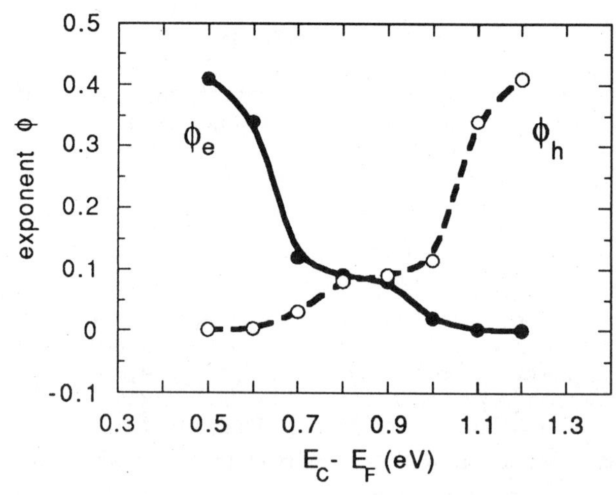

Fig.2 - The exponents of the light intensity dependence of the lifetimes as functions of the Fermi level position (Defect Pool Model)

In Fig. 3 one can see the corresponding Fermi level dependence of the majority carrier dominated photoconductivity exponent γ and of the minority carrier dominated diffusion lenght exponent S defined in equations (5) and (6). The results of S>0 and its decrease when E_F shifts away from the middle of the gap are consistent with experimental data [11, 15]. The need of a recombination model with, at least, two defect levels in order to explain the S≠0 result was pointed out by Balberg [16] on the basis of a semi-quantitative analysis.

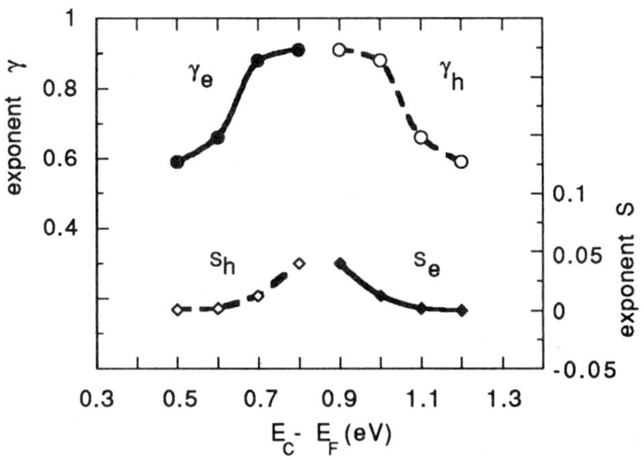

Fig.3 - The exponents of the light intensity dependence of the photoconductivity, γ, and of the diffusion lenght, S, as functions of the Fermi level position (Defect Pool Model)

Defect Pool Model plus Band Tails

By adding to the preceding picture two exponential band tails with characteristic energies E_{oc} = 0.025 eV and E_{ov} = 0.045 eV, and with capture cross section $\sigma_t = 10^{-16}$ cm^2, negative ϕ values appear for E_F in the valence band side of the gap, as can be seen in Fig.4.

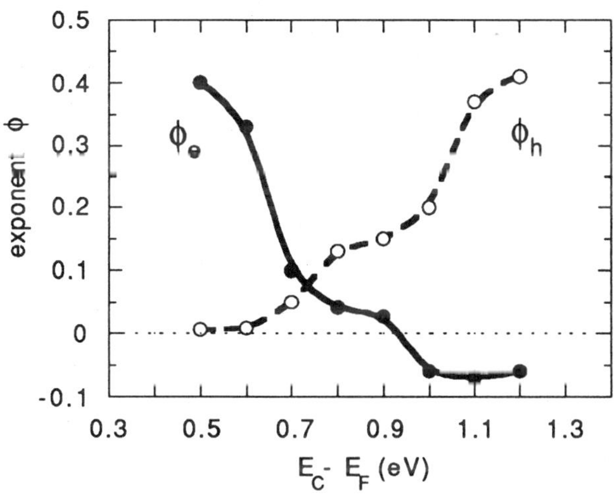

Fig.4 - The exponents of the light intensity dependence of the lifetimes as functions of the Fermi level position (Defect Pool Model + Band Tails).

In Fig. 4, the feature of $\phi>0$ (S>0) for E_F around mid-gap, characteristic of the "more than one level model" [16], and the feature of $\phi<0$ (S<0) in the lower half of the gap characteristic of the existence of band tails with different reciprocal slopes [3, 4], occur simultaneously. Considering that $\mu_e \approx 10\ \mu_h$, the $\phi<0$ result is consistent with superlinear photoconductivity and associated thermal quenching for undoped or slightly doped material. The entire pattern in Fig. 4 is also consistent with the light intensity dependences of the photoconductivity and of the ambipolar diffusion lenght found in boron doped samples [11] and field effect (MOS) experiments [15].

Like in the simulations with the standard model plus band tails [3, 4], $\phi_e < 0$, superlinear photoconductivity and thermal quenching appear associated with a capture cross section of the asymmetric band tails smaller than the capture cross section of the dangling bonds, as is shown in Fig.5.

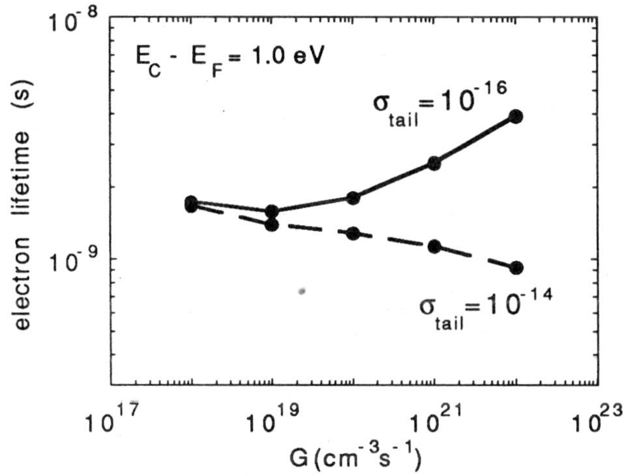

Fig.5 - Light intensity dependence of the electron lifetime for different capture cross sections σ_t of the band tails (Defect Pool Model + Band Tails).

The Fermi level dependence of the lifetimes is shown in Fig.6. The electron and hole lifetimes are anticorrelated and the hole lifetime for p-type like material is less sensitized by the Fermi level shift than the electron lifetime for n-type like material as a consequence of $E_{ov} > E_{oc}$.

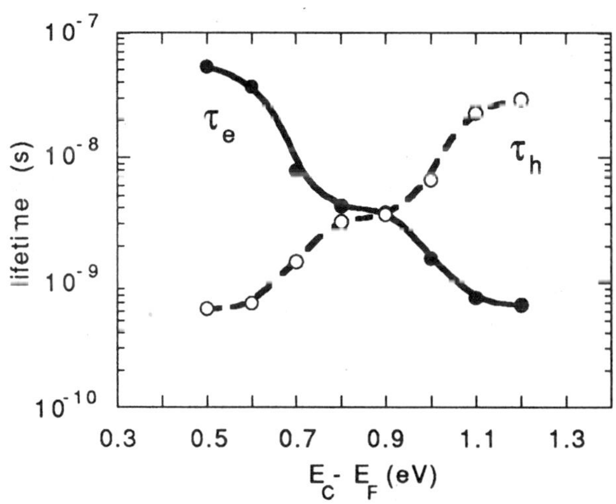

Fig. 6 - Fermi level dependence of the lifetimes (Defect Pool Model + Band Tails)

CONCLUSIONS

In the present paper, new results from calculations based on the defect pool model, including band tails effects, are reported and compared with experimental data and with results from the standard model. The main results are:

- From photocarrier grating technique experiments the exponent 2S of the light intensity dependence of the ambipolar diffusion lenght $L_d \propto G^{-2S}$ has been found to be greater than zero for undoped and slightly doped samples while, for the same Fermi level range, the exponent $\phi=(1-\gamma)$ of the majority carrier $\mu\tau$ product, $\mu\tau \propto G^{-(1-\gamma)}$, remains also positive [11, 13, 15] . This is an experimental result which has not been reproduced by previous simulations with the standard model [2, 3]. *The present calculations show that $S>0$ and $(1-\gamma)>0$ can be found if the defect pool approach is used.* This result is consistent with a recent theoretical qualitative analysis of Balberg involving two occupation-independent recombination levels [16]

- By adding, to the defect pool distribution, band tails of localized states with *different reciprocal slopes and capture cross sections smaller than the capture cross sections of the dangling-bonds,* the entire pattern of the experimentally observed light intensity dependences of the majority and minority carrier $\mu\tau$ products as functions of the Fermi level position [11, 15] can be qualitatively reproduced (see Fig.4).

The above results were calculated for a constant density of the three classes of dangling bond states in the defect pool model and should therefore be compared with data from MOS experiments or from undoped or slightly doped material.

In conclusion, although the standard model, in conjunction with band tails states, can explain major features and trends of the Fermi level, light intensity and temperature dependences of photoconductivity and majority carrier $\mu\tau$ products in a-Si:H, the minority carrier light intensity dependence seems to require a more complex picture and is not inconsistent with the defect pool approach.

AKNOWLEDGEMENTS

The present work is supported by JNICT under the contract PBICT/C/CTM/1385/92.

REFERENCES

[1] E. Morgado, "Recombination at correlated dangling bonds and the effect of Fermi level position on steady-state photoconductivity in amorphous silicon", *Phil. Mag.* , **B 63**, 1991, pp. 529-542.

[2] E. Morgado, "Mobility-lifetime products in a-Si:H and the dangling bond recombination model", *Amorphous Silicon Technology - 1992,* Mat. Res. Soc. Symp. Proc. Vol. **258**, ed. M.J. Thompson, Y. Hamakawa, P.G. Le Comber and A. Madan, (MRS, Pittsburgh, 1992), pp. 765-760.

[3] E. Morgado, "Electron and hole $\mu\tau$ products in a-Si:H and the standard dangling bond model", *J. Non-Cryst. Solids,* **164-166**, 1993, 627-630.

[4] E. Morgado, "Thermal quenching and photo-enhancement of $\mu\tau$ products in a-Si:H : the role of dangling bonds and band tails", *Amorphous Silicon Technology - 1994,* Mat. Res. Soc. Symp. Proc. Vol. **336**, ed. E.A. Schiff, M. Hack, A. Madan, M. Powell and A. Matsuda (MRS, Pittsburgh, 1994), pp. 419-424.

[5] K. Winer, "Defect formation in a-Si:H", *Phys. Rev.* **B 41**, 1990, pp.12150-12161.

[6] C.- D. Abel, G.H. Bauer, "Modelling of transport and recombination of photoexcited carriers in a-Si:H and a-SiGe:H", *Proc. 21st IEEE P.V. Spec. Conf.* (IEEE, New York, 1990), pp.1550-1554.

[7] G. Schumm, C.- D. Abel and G.H. Bauer, "Photoconductivity and $\mu\tau$-products in a-Si:H - compatibility with various defect models", *J. Non-Cryst. Solids,* **137&138**, 1991, pp. 351-354.

[8] H. Okamoto and Y. Hamakawa,, *Solid State Commun.,* **24**, 1977, pp. 23

[9] F. Vaillant and D. Jousse, "Recombination at dangling bonds and steady-state photoconductivity in a-Si:H", *Phys. Rev.* **B 34**, 1986, pp. 4088-4098.

[10] J.G. Simmons and G.W. Taylor, "Non-equilibrium steady-state statistics and associated effects for insulators and semiconductors containing an arbitrary distribution of traps" , *Phys. Rev.* **B4**, 1971, pp. 502-511.

[11] L. Yang, A. Catalano, R.R. Arya and I. Balberg, "Effect of low-level boron doping and its implication to the nature of gap states in hydrogenated amorphous silicon", *Appl. Phys. Lett.,* **57**, 1990, pp. 908-910.

[12] P.E. Vanier, A.E. Delahoy and R.W. Griffith, "New features of the temperature dependence of photoconductivity in plasma-deposited hydrogenated amorphous silicon alloys", *J. Appl. Phys.,* **52**, 1981, pp. 5235-5242.

[13] E. Sauvain, J. Hubin, A. Shah and P. Pipoz, "Effect of the dangling bond charge on the ambipolar diffusion lenght in a-Si:H", *Phil. Mag. Letters,* **63**, 1991, pp. 327-333.

[14] M.J. Powell and S.C. Deane, "Improved defect-pool model for charged defects in amorphous silicon", *Phys. Rev.* **B 48**, 1993, pp. 10815-10827.

[15] I. Balberg and Y. Lubianiker, "Evidence for the defect-pool model from induced recombination level shifts in undoped a-Si:H", *Phys. Rev.* **B 48**, 1993, pp. 8709-8714.

[16] I. Balberg, "The two carriers' mobility-lifetime products and their light intensity dependencies in hydrogenated amorphous silicon", *J. Appl. Phys.,* **75**, 1994, pp. 914-923.

EFFECT OF A WIDE BAND GAP EMITTER LAYER AND ITS PATTERNING ON THE PERFORMANCE OF a-Si:H BASED HETEROJUNCTION SOLAR CELLS

Parsathi Chatterjee
Energy Research Unit,
Indian Association for the Cultivation of Science,
Jadavpur, Calcutta - 700 032, India

ABSTRACT

Simulation of carrier transport in heterojunction p-i-n structure solar cells has been used to study the effect of a wide band gap p-layer and its contact barrier height (ϕ_{bo}) on cell performance. We find that the recombination loss at the front contact is small when the emitter layer has a wider band gap. Thus the short circuit current in these structures is high and practically independent of ϕ_{bo}. However, if ϕ_{bo} is held constant, the effect of increasing the p-layer band gap is actually to depress the open circuit voltage (V_{oc}). The performance of heterojunction cells deposited on ITO can then only be explained by assuming a larger ϕ_{bo}. A novel patterning of the p-a-Si-C:H layer has been explored and found to increase V_{oc} by 0.1 volts and the efficiency by over 10%.

INTRODUCTION

The heterojunction p-i-n structure solar cell having a wide band gap emitter layer, has emerged as the structure of great promise for solar cell applications. In such devices the p/i interface has been widely studied. However the TCO/p contact has not received as much attention. It is the aim of this report to use computer modelling to analyse the effect on solar cell performance of increasing the band gap of the emitter layer and the role the TCO/p barrier height (ϕ_{bo}) plays under this condition. The paper goes on to suggest a novel patterning of the p-a-Si-C:H layer that is capable of increasing the open circuit voltage (V_{oc}) by 0.1 volts and the cell conversion efficiency (η) by over 10%.

SIMULATION MODEL

Our computer model [1,2] solves the Poisson's equation and the two continuity equations under non-equilibrium steady state conditions. Very general boundary conditions are used [1]. The solution technique is the one developed by Schrafetter and Gummel [3]. In modelling the recombination term and the trapped carrier densities, we have made use of the Schockley-Read-Hall recombination model and taken proper account of the ambient temperature.

The gap state density of states (DOS) model consists of two Urbach tails and two Gaussian distribution functions to simulate the donor-like and the acceptor-like dangling bond (DB) states.

In a heterojunction structure the gradient of the electron affinity (χ) must also be taken into account for calculating the effective force field on the carriers. It can then be shown that the "effective field" causing drift of the electrons is given by the gradient of the conduction band edge and that on the holes by the gradient of the valence band edge.

First WCPEC; Dec. 5–9, 1994; Hawaii

RESULTS AND DISCUSSION

The input parameters (Table 1) used in our calculations have been standardised with respect to the wide-band-gap-emitter-layer p-i-n cells of Tasaki et al [4]. Our computer results for the illuminated current density-voltage characteristics for the heterojunction cells are shown in Table 2 for ϕ_{bo}= 1.15 eV and 1.30 eV. The calculations assume a 150 Å graded band gap buffer layer at the p/i interface and the band gap discontinuity equally divided between the two band edges. Also shown are the results for the homojunction p-i-n cell, where the front barrier height is 1.15 eV, the p-layer band gap 1.9 eV and the band tails considerably sharper. Optical enhancement has not been considered. The recombination losses at the front contact are also shown.

Our results reveal that the increase in the short circuit current (J_{sc}) in the p-i-n heterojunction structure relative to the homojunction device is due to a decrease of the considerable front surface loss observed in the latter. Thus J_{sc} here is high and practically independent of ϕ_{bo}.

However with ϕ_{bo} held constant at 1.15 eV (the TCO/p-a-Si:H value [1]), the widening of the emitter layer band gap is seen to have a detrimental influence on V_{oc}. To understand the situation we refer to Fig.1, which plots the effective field on the electrons as a function of position in the device. We see that the strong negative field corresponding to the gradient in the electron affinity at the graded buffer interfaces, leads to a collapse of the field in the bulk (dot dashed curve). As a result [1], V_{oc} deteriorates. Fig.1. also indicates that when ϕ_{bo} increases to 1.3 eV (solid line), the bulk electric field improves and surpasses the field in the homojunction (dashed line). We thus obtain the familiar improvement in cell performance.

PATTERNING OF THE p-LAYER

One important factor that prevents holes from being easily drawn off is the positive field caused by the TCO/p Schottky barrier. This causes an illumination-driven hole build-up in the p-region that reduces the net negative trapped charge in the p-region and leads to a decrease of the electric field across the device. Hence [1] V_{oc} decreases. If a band gap discontinuity is introduced in such a way that the positive field near the TCO contact is reduced, the efficiency of hole collection and hence ultimately V_{oc} and fill factor (FF) are expected to improve. Computer modelling reveals that this can be achieved if a 20Å p-layer of band gap lower than the main p-layer is introduced adjacent to the TCO. The results with ϕ_{bo} held constant are shown in Table 2. V_{oc} is seen to improve by about 0.1 volts and the cell efficiency (η) by more than 10%. However reduction in the band gap is expected to reduce ϕ_{bo} and hence also V_{oc}. Therefore this design will only be successful if we are working in a region of ϕ_{bo} (ϕ^c_{bo}), where cell performance is fairly insensitive to ϕ_{bo}. Modelling shows that ϕ^c_{bo} is 1.45 volts.

CONCLUSIONS

1. The introduction of a wide band gap emitter layer results in a strong improvement of J_{sc}, as a result of the decrease in front surface loss. J_{sc} is almost independent of ϕ_{bo} in these structures.
2. The effect of increasing the band gap of the emitter layer with ϕ_{bo} held constant is to reduce V_{oc} on account of the gradient in χ.

Table 1. Principle input parameters for the "standard" p-i-n heterojunction structure solar cell

Parameters	p-a-SiC:H	buffer	i-aSi:H	n-aSi:H
Length (Å)	80	150	4875	250
Band gap (eV)	2.02	2.02-1.9	1.9	1.9
Mobility (cm^2/V/S)				
electrons	25	25	25	25
holes	6	6	6	6
Effective DoS in bands (cm^{-3})	$2X10^{20}$	$2X10^{20}$	$2X10^{20}$	$2X10^{20}$
Tail exponential prefactor (cm^{-3}/eV)	$4X10^{21}$	$4X10^{21}$	$4X10^{21}$	$4X10^{21}$
Charac. energy (eV)				
Valence band tail	0.090	0.070	0.046	0.050
Conduction band tail	0.060	0.040	0.027	0.030
DB DoS (cm^{-3})	$2X10^{18}$	$1X10^{17}$	$2X10^{15}$	$4X10^{18}$
Doping density (cm^{-3})				
Acceptor	$9X10^{18}$	0	0	0
Donor	0	0	0	$6X10^{18}$
Capture cross-sec. (cm^2)				
Charged (tails)	10^{-15}	10^{-15}	10^{-15}	10^{-15}
neutral (tails)	10^{-17}	10^{-17}	10^{-17}	10^{-17}
charged (DB)	10^{-14}	10^{-14}	10^{-14}	10^{-14}
neutral (DB)	10^{-15}	10^{-15}	10^{-15}	10^{-15}

Table 2. J_{sc}, V_{oc}, FF, η and the recombination loss (R_{front}) at the TCO/p contact for p-i-n a-Si:H based solar cells of different structures.

Solar cell structure	ϕ_{bo} (eV)	J_{sc} (mA/cm^2)	V_{oc} (volts)	FF	η (%)	R_{front} (% of light absorbed)
p-i-n homojunction	1.15	11.363	0.811	0.763	7,28	23.54
p-i-n hetrojunction with buffer	1.15	14.184	0.709	0.725	7.29	3.03
p-i-n heterojunction with buffer.	1.30	14.224	0.856	0.754	9.17	2.83
patterned p-i-n with buffer.	1,30	13,816	0.960	0.767	10.17	4.78

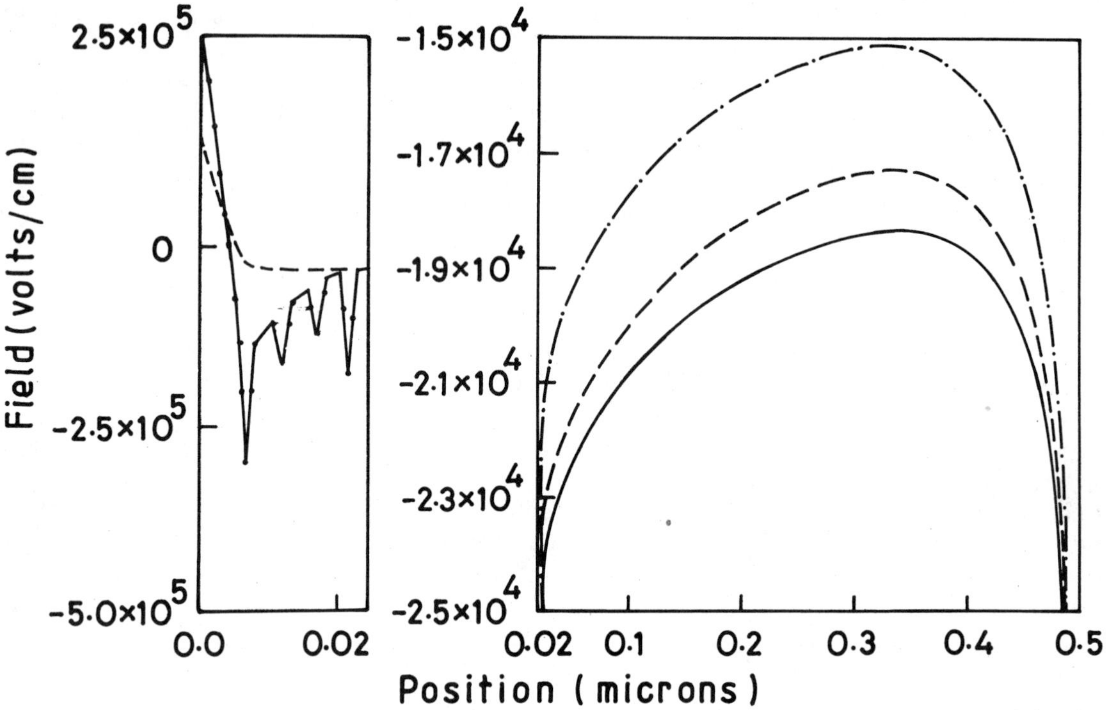

Fig. 1. The effective electric field on the electrons as a function of position in the device in the presence of AM 1.5 light for the p-i-n homojunction cell (- - - -), & the wide-band-gap-p-layer heterojunctions with ϕ_{bo}= 1.15 eV (—·—·—) and ϕ_{bo}=1.30 eV (——). The field in the first 0.02μm of the device is plotted on a different scale.

3. The observed improvement in cell performance can only be explained by the fact that ϕ_{bo} for a wide band gap emitter layer is larger than for p-aSi:H.
4. Introduction of a 20Å p-layer of band gap lower than the main p-layer adjacent to the TCO, leads to an improvement in V_{oc} and η.

ACKNOWLEDGEMENTS

The work was carried out under a project supported by the Bharat Heavy Electricals and the Ministry of Non-conventional Energy Sources, Govt. of India. The author is indebted to Prof. A.K. Barua, Head, Energy Research Unit, IACS, for his constant encouragement during the course of this work.

REFERENCES

[1] Parsathi Chatterjee, "Photovoltaic Performance of a-Si:H Homojunction p-i-n Solar Cells: A Computer Simulation Study", J.Appl.Phys. 76, 1994, pp.1301-1313.

[2] Parsathi Chatterjee, "Origin of the Modulated Photocurrent in the Presence of Bias Light in Quantum Efficiency Measurements", J. Appl.Phys. 75, 1994, 1093-1097.

[3] D.L. Schrafetter and H.K. Gummal, "Large Signal Analysis of a Silicon Read Oscillator", IEEE Trans. of Electron Devices ED-16, 1969, 64-77.

[4] H. Tasaki, W.Y. Kim, M. Hallerdt, M. Konagai and K. Takahashi, "Computer Simulation Model of the Effects of Interface states on High Performance Amorphous Silicon Solar Cells", J.Appl.Phys. 63, 1988, 550-560.

COMPUTER MODELING OF AMORPHOUS SILICON TANDEM CELLS

J.A. Willemen, M. Zeman, J.W. Metselaar,
Laboratory for Electronic Components, Technology and Materials,
Delft University of Technology, The Netherlands

ABSTRACT

We have used a new modeling approach for simulation of a-Si:H tandem cells. The Trap Assisted Tunneling model and an enhanced effective extended state mobility were used to model the recombination and transport in the high field region of the tunnel recombination junction. With this approach realistic tandem cell characteristics could be obtained. We illustrate how modeling can be used for the analysis of multi-junction cells.

INTRODUCTION

Hydrogenated amorphous silicon (a-Si:H) based solar cells are considered potential candidates for cost-effective, high-efficiency, large area photo-voltaic modules. In particular, the multi-junction a-Si:H solar cells offer good perspectives for stable, high-efficiency cells. As the structure of these cells becomes increasingly complex, computer modeling will be an important tool for analysis and optimization of multi-junction solar cells. In this paper we will present a new approach for computer simulation of stacked solar cells. We aim at a comprehensive model by which a complete stacked solar cell can be simulated in a physically justifiable way.

The main difficulty of simulation of multi-junction solar cells is the junction between the different *pin*-cells of the stacked structure, which are generally called the tunnel recombination junctions (TRJ). In literature several approaches have been published for the modeling of the TRJ in a stacked cell. However, these models show some shortcomings and have only restricted applicability.

The most widespread and simplest modeling approach is the separate simulation of the component cells of a multi-junction solar-cell [1, 2]. To find the multi-junction cell characteristics the I-V characteristics of the sub-cells are added by series connection of the component cells. With this modeling approach the band bending in the TRJ caused by the space charge in the doped layers is completely ignored. Furthermore, all contacts of the component cells are modeled as ideal ohmic contacts, thus ignoring any possible losses in the TRJ. This approach does not take into account the physical carrier transport and recombination processes in the TRJ, and therefore does not permit an in-depth study of multi-junction solar cells.

In the second approach reported in literature, multi-junction cells are simulated as single, stacked structures [3]. The recombination region in the TRJ is modeled by a heavily defective layer with a reduced bandgap. The potential barriers for mobile carriers moving towards this region are lowered by band grading; these graded layers ease the carrier transport towards recombination centers in the defective layer. However, neither physical background for the strongly reduced bandgap nor a quantitative model for the amount of barrier lowering is provided.

In this paper we present a new approach for the modeling of TRJs in multi-junction cells. We show that it is in principle possible to do realistic simulations of tandem cells by using the Trap Assisted Tunneling model and enhanced carrier transport in the high field region of the TRJ. Although we have not yet developed a detailed model for the transport in the the TRJ, we think this approach provides a more physically based description for the processes in the TRJ. Unlike the modeling strategies described above, it can be used for simulation of a wide variety of a-Si:H based structures, like single *pn*-junctions, *pinp*-structures, and multi-junction solar cells under dark and illuminated conditions. To our opinion a good model is characterized by the capability to simulate all these structures realistically.

MODEL DESCRIPTION

Solar cell device modeling

We have used our comprehensive simulation program Amorphous Semiconductor Analysis (ASA) for the modeling of tandem cells. ASA is a 1-D steady state device simulator, based on the solution of Poisson's equation and the electron and hole continuity equations. It contains models for the amorphous material features like the tail states and the defect states. As independent variables the vacuum energy level and the free electron and hole concentrations are used. The continuity equations are discretized according to the integral expressions of Scharfetter and Gummel. The resulting set of equations is solved by a fully coupled Newton Raphson iteration method.

In ASA the conduction and valence band tails are modeled by exponentially decreasing acceptor-type resp. donor-type states; the amphoteric dangling bond (defect) states are modeled by two Gaussian distributed density of states. The charge occupation and the recombination in the gap states are computed by Shockley-Read-Hall statistics (tail states) or by Sah-Shockley statistics (multiple state defects).

Trap assisted tunneling

To account for the increased recombination in regions with strong electric fields, we have implemented the Trap Assisted Tunneling model of Hurkx *et al.* [4] in ASA. This

CH3365-4/94/0000-0599 $4.00 © 1994 IEEE First WCPEC; Dec. 5–9, 1994; Hawaii

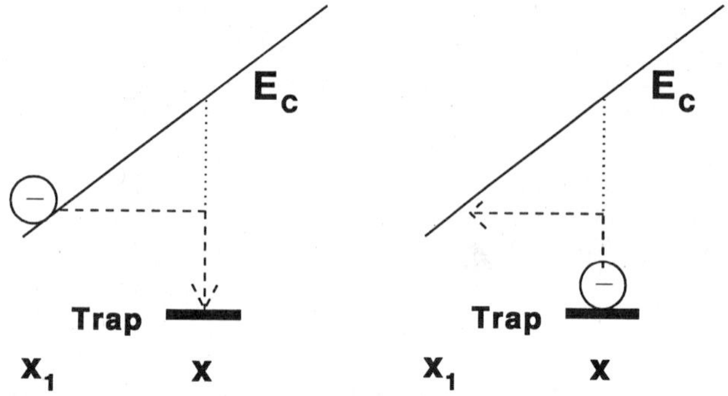

Fig. 1: Trap Assisted Tunneling: a). Electron capture. b). Electron emission

model was successfully developed for the modeling of the excess current in heavily doped crystalline *pn*-junctions at low forward biases. The model yields a much higher recombination in space charge regions by taking into account an increased carrier concentration at a certain recombination center, resulting from tunneling from nearby locations. At the same time the emission probabilities are enhanced by phonon assisted tunneling. This model gives a modification of the well known Shockley-Read-Hall formula for recombination:

$$R_{srh} = N_t \frac{c_n c_p n p - e_n e_p}{c_n n + e_n + c_p p + e_p} \quad (1)$$

In this formula R_{srh} is the recombination rate, N_t the trap concentration, n and p are the electron and hole concentration respectively, and c_n (c_p) and e_n (e_p) are the capture and emission rates for electrons (holes).

In our theoretical treatment of the trap assisted tunneling model we limit ourselves to the electrons; however, the same principle applies to the holes. In Fig. 1 we show schematically the tunneling electrons contributing to the recombination process in regions with strong electric fields. Fig. 1.a shows an electron at position x_1 which tunnels towards the position of the trap x. Similarly, the trapped electrons can be emitted from the trap by phonon assisted tunneling as depicted in Fig. 1.b. By integration over all positions x_1 from which an electron can tunnel to the trap at x, taking the tunneling probability into account, we can determine the concentration of electrons n_t that can effectively contribute to recombination through the trap at position x.

We can express the electron concentration available for recombination at position x as function of the conduction band electrons at this position n_0 and a field dependent factor Γ_n.

$$\frac{n_t}{n_0} = 1 + \Gamma_n \quad (2)$$

The emission coefficients $e_{n,t}$ are changed by the same factor

$$\frac{e_{n,t}}{e_{n,0}} = 1 + \Gamma_n \quad (3)$$

Under the assumption of a triangular potential barrier (uniform field), analytic approximations for $\Gamma_{n,p}$ are derived; these can be found in Ref. [4].

Introducing Eq. 2, 3 and their equivalents for holes into Eq. 1 yields

$$R_{tat} = \frac{N_t(1+\Gamma_n)(1+\Gamma_p)(c_n c_p n_0 p_0 - e_{n,0} e_{p,0})}{(1+\Gamma_n)(c_n n_0 + e_{n,0}) + (1+\Gamma_p)(c_p p_0 + e_{p,0})} \quad (4)$$

where R_{tat} is the Trap Assisted Tunneling recombination rate.

We have applied the same modeling principle to the correlated dangling bond state model. For the amphoteric dangling bond model we have four different capture and four different emission processes, which are modified in a similar way as described above. In this case we get four field dependent Γ factors: two for each transition energy level.

The model does not only result in increased recombination rates in high field regions, but it affects the charge occupation in the localized states too.

Enhanced carrier transport in high field regions

Preliminary simulations of tandem cells showed that no realistic I-V characteristics could be obtained with only the application of the Trap Assisted Tunneling model. This is due to the fact that this model accounts for higher recombination rates in high field regions but not for tunneling transport towards the recombination sites.

We have looked for an approach to take the tunneling transport into account effectively, without having to change the structure of the device simulation model drastically. We found that this could be accomplished by increasing the extended state mobility μ_{ext} in high field regions.

Usually, device simulators only account for the transport of carriers, which can move freely in the extended states. The electron current density J_n is driven by the gradient in the electron quasi-Fermi level $E_{f,n}$

$$J_n = \mu_{n,ext} \cdot n_{ext} \cdot \frac{dE_{f,n}}{dx} \quad (5)$$

where $\mu_{n,ext}$ is the extended state electron mobility and n_{ext} the extended state electron concentration. Here we assume that the trapped electrons in the tail states have zero mobility. However, in strong electric fields the trapped electrons can move between localized states by tunneling transport. In this case the electron current can be described by

$$J_n = \int_{-\infty}^{+\infty} \mu_n(E)\, n(E)\, dE \cdot \frac{dE_{f,n}}{dx} \quad (6)$$

where we take into account non-zero mobilities of electrons below the mobility edge. In this equation we assume that the behavior of electrons below the mobility edge can still be described by the free electron quasi-Fermi level. By integration over all electron states both below and above the mobility edge, we arrive at a much larger $\mu \cdot n$ product, than given in Eq. 5. For use in a device simulator this can be translated into an increased *effective* extended state mobility $\mu_{n,eff}$, modifying Eq. 5 into

$$J_n = \mu_{n,eff} \cdot n_{ext} \cdot \frac{dE_{f,n}}{dx} \quad (7)$$

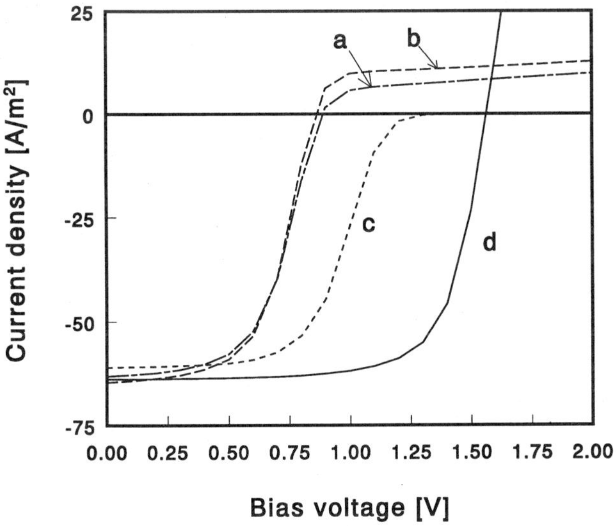

Fig. 2: Simulated tandem cell I-V characteristics with different models for the TRJ: a) neither Trap Assisted Tunneling nor enhanced carrier mobilities in high field regions; b) enhanced carrier mobilities; c) Trap Assisted Tunneling; d) both Trap Assisted Tunneling and enhanced carrier mobilities in high field regions.

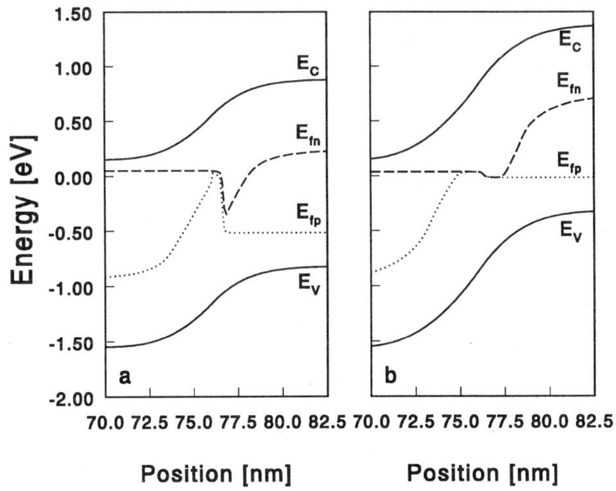

Fig. 3: Band diagram in the tunnel recombination junction, showing the influence of the present modeling strategies. a). Only Trap Assisted Tunneling. b). With both Trap Assisted Tunneling and enhanced carrier mobilities.

We have not yet developed a detailed model for the effective extended state mobility, but used a *constant*, increased mobility in high field regions. Based on experimental evidence that the room-temperature drift mobility increases exponentially in fields above 10^5 V/cm [5, 6], we applied the following formula to the usual extended state mobilities μ_{ext}

$$\mu_{eff} = \mu_{ext} \cdot \exp\left(\frac{|\vec{E}|}{E_0}\right) \tag{8}$$

where we took for the electric field $|\vec{E}|$ the thermal equilibrium values. With the parameter value $E_0 \approx 2 \cdot 10^5$ V/cm we obtained good results.

RESULTS AND DISCUSSION

The material parameters used for the simulation of tandem cells are based on the parameter set that we have determined by inverse modeling of single junction cells [7]. A tandem cell structure is used in which all doped layers are 10 nm thick; the first *i*-layer is 56 nm and the second *i*-layer is 424 nm. We found that the doping concentration of the *n*-layers in our basic set was too low to create sufficient band bending in the *np*-junction and a strong field; therefore, we had to increase this concentration.

In Fig. 2 we show current-voltage (I-V) characteristics of this tandem cell structure. In this figure the impact of the different models on the I-V characteristics is clearly visible. Curve a) is without extra models to account for the physics in the TRJ. In curve b) we only applied the model with the enhanced effective mobility in the high field regions, which has very little influence on the tandem cell characteristics. Curve c) is the result with the Trap Assisted Tunneling model, which shows some beneficial influence on the open circuit voltage; however, V_{oc} is too low in comparison with experimental values. For curve d) both models are applied and gives the most realistic tandem cell characteristic. In this case the open circuit voltage is almost twice the open circuit voltage of corresponding single junction cells. Furthermore, this curve does not show the current saturation effect at higher forward bias voltages, which is present in the three other curves.

Fig. 3 shows the band diagram around the TRJ in the short circuit case. These figures clearly show the impact of the application of the Trap Assisted Tunneling model and the enhanced mobility model. In Fig. 3 a) only Trap Assisted Tunneling is used; in Fig. 3 b) both models are used.

To obtain the largest open circuit voltage the majority carrier quasi Fermi levels in the *n* and *p* region of the TRJ should be as close as possible to each other. In this case the field over both *pin*-junctions is the largest and the tandem cell characteristics approach the sum of two single junction cells best. The difference between the majority carrier quasi Fermi levels can be considered as the loss in the TRJ. As shown in Fig. 3 this loss is much higher if *only* the Trap Assisted Tunneling model is applied. As illustrated in Fig. 2 in that case the open circuit voltage is much lower (curve c.).

With this modeling work we can study the influence of thickness variations of component cells on the efficiency and on internal quantities, like the band diagram, recombination rates, electric field distribution, etcetera. As an illustration some preliminary results are shown in Fig. 4 and 5. Fig. 4 shows the normalized efficiency as function of thickness ratio of top and bottom cell. Only the i-layer thicknesses are varied; the doped layer thicknesses and total cell thickness are kept constant. The influence on the band diagram is depicted in Fig. 5. Fig. 5 a). corresponds to the optimum ratio of Fig. 4. In this case the internal potential differences over both component cells are approximately the

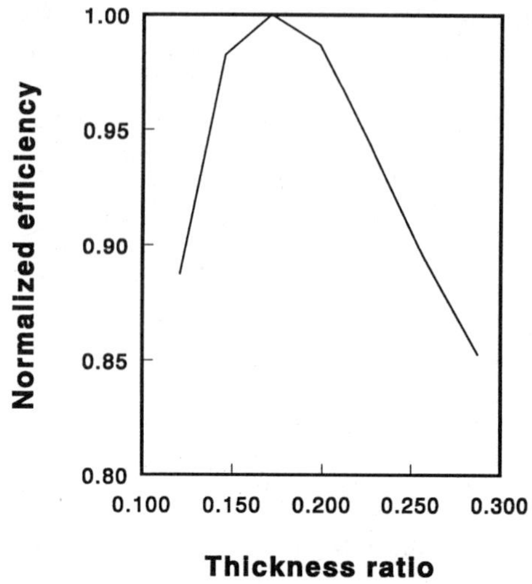

Fig. 4: Influence of thickness ratio variations of the top and bottom cell on the efficiency of a tandem cell

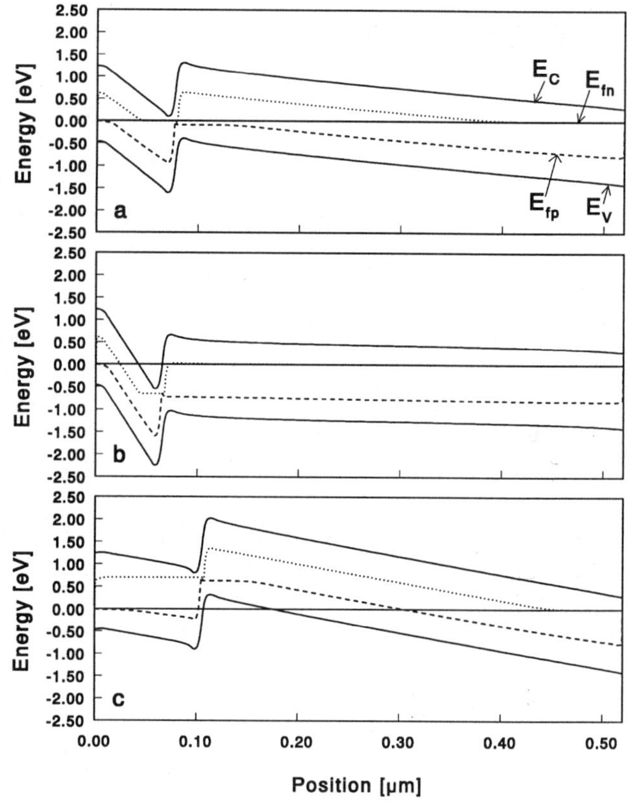

Fig. 5: Influence of thickness variations on the band diagram of a tandem cell. a). Matched cell. b). Top cell too thin. c). Bottom cell too thin.

same. Fig. 5 b). and 5 c). show the band diagram with a too thin top and bottom cell, respectively. here we clearly see the redistribution of the internal potential: the internal potential difference over the limiting component cell, which generates too little photocurrent, is increased, whereas the potential difference over the other part is decreased. In terms of the electric field in the component cells, this means an increase in the limiting part and a decrease in the other part, to ensure a constant total current density in the tandem cell.

CONCLUSIONS

In this paper a new approach for simulation of a-Si:H based tandem cells was presented. For simulation of complete multi-junction solar cells the recombination processes and charge carrier transport in the high field region of the tunnel recombination junction (TRJ) between the component cells were modeled:

• the strong recombination in the TRJ is modeled by the Trap Assisted Tunneling model.

• the tunneling transport in the TRJ is effectively accounted for by increased extended state mobilities.

We found that both models should be applied to get realistic tandem cell characteristics. We have illustrated the possibility of this approach to investigate thickness variations on tandem cell characteristics.

ACKNOWLEDGMENTS

This research was funded by the Netherlands Agency for Energy and the Environment.

REFERENCES

[1] Pawlikiewicz, A.H. and S. Guha, "Performance Comparison of Triple and Tandem Multi-Junction a-Si:H Solar Cells: a Numerical Study", *IEEE Trans. on ED*, **37**, 1990, pp. 1758-1762.

[2] Block, M., D. Bonnet and F. Zetzsche, "Modeling of Multilayer Amorphous Thin Film Silicon-Germanium Single and Tandem Solar Cells", *22th IEEE PVSC*, 1991, pp. 1275-1280.

[3] Hou, J.Y., J.K. Arch, S.J. Fonash, S. Wiedeman, and M. Bennet, "An Examination of the "Tunnel Junctions" in Triple Junction a-Si:H Based Solar Cells: Modeling and Effects on Performance", *22th IEEE PVSC*, 1991, pp. 1260-1264.

[4] Hurkx, G.A.M., D.B.M. Klaassen and M.P.G. Knuvers, "A New Recombination Model for Device Simulation Including Tunneling", *IEEE Trans. on ED*, **39** (2), 1992, pp. 331-338.

[5] Antoniadis, H., R.I. Devlen, S. Esipov, S. Guha, E.A. Schiff, and J. Tauc, "High-Field Electron Drift in a-Si:H", *ICAS-14*, 1991.

[6] Juška, G., J. Kočka, K. Arlauskas, and G. Jukonis, "Electron Drift Mobility in a-Si:H under Extremely High Electric Field", *Solid State Comm.*, **75**, 1990, pp. 531-533.

[7] M.Zeman, J.A. Willemen, S. Solntsev, J.W. Metselaar, "Extraction of Amorphous Silicon Solar Cell Parameters by Inverse Modelling", *Solar Energy Materials and Solar Cells*, **34**, 1994, pp. 557-563.

SPECTRAL RESPONSE MODELLING OF A-SI:H SOLAR CELLS USING ACCURATE LIGHT ABSORPTION PROFILES

H. Stiebig, A. Kreisel, K. Winz, N. Schultz, C. Beneking, Th. Eickhoff and H. Wagner
Forschungszentrum Juelich GmbH, (ISI-PV), D-52425 Juelich, Germany

M. Meer
Universität - GH - Siegen, Institut für Halbleiterelektronik, D-57068 Siegen, Germany

ABSTRACT

An optical model calculating the generated carrier profile in a-Si:H solar cells deposited on hazy and non-hazy transparent conducting oxide (TCO) has been integrated into a numerical transport and recombination program. A comparison of simulated and measured reflectances of pin structures opens up a possibility to check the prepared layer thicknesses and the optical device properties. Furthermore, the examination of the current gain employing a ZnO/Metal back contact reveals the importance using accurate light absorption profiles for spectral response modelling and device characterisation of pin solar cells.

INTRODUCTION

Numerical modelling is widely used for design optimization of a-Si:H solar cells and color sensors. In spite of an improved treatment of the defect states by the defect pool model, the still inadequate description of the generation profile makes the interpretation of device characteristic and especially spectral response data more difficult. Comparing simulated and experimental reflectance and spectral response data of pin structures with different i-layer thicknesses of 340, 455, 585, and 875 nm results in a better description of the optical multilayer and therefore, to a more realistic generation profile in the device which is necessary to calculate the spectral response curves. The investigated samples are described elsewhere [1].

TRANSPORT AND RECOMBINATION MODEL

The numerical transport and recombination model describes the dangling bond distribution according to the defect-pool model. The energetic position of the dangling bonds D^n, D^i and D^p are kept constant at 0.65, 0.90, and

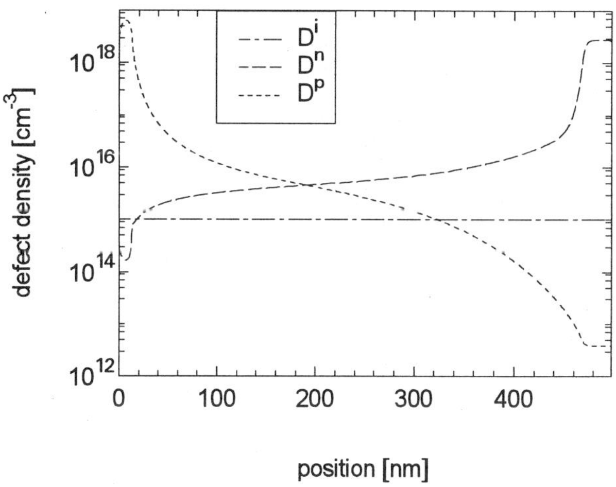

Fig. 1. Spatial defect state distribution (D^p, D^i, and D^n) of a pin structure with an i-layer thickness of 455 nm.

1.15 eV above the valence band edge. The Fermi level dependent defect state distribution is adapted to the results of CPM data [2,3]. Figure 1 shows the spatial dependence of the dangling bond density for a pin structure with an i-layer thickness of 455 nm. It is calculated self consistently for the thermal equilibrium. More insights into the numerical model are given elsewhere [4].

OPTICAL MODELS

The optical properties of nearly all layers in a solar cell have an influence on the device characteristics. An optical layer stack simulation program has been developed to calculate the generated carrier profile for solar cells deposited on non-hazy (coherent wave propagation) and hazy (incoherent wave propagation) TCO. The optical model requires the wavelength dependent complex refractive indices for the glass substrate, TCO front contact, p-, i-, n-layer, and the back contact which

CH3365-4/94/0000-0603 $4.00 © 1994 IEEE

have been determined from single film reflectance - transmittance and ellipsometric measurements.

Coherent wave propagation

In order to calculate the coherent wave propagation in the solar cell electromagnetic wave theory and boundary conditions which conserve the tangential components of the electric and magnetic field vector at every interface has been used. The electric field vector E is calculated using a matrix formalism to solve the equation system. The power flux of the electromagnetic wave through the solar cell is calculated from the Poynting vector:

$$I = \frac{1}{2} \cdot \mathrm{Re}\left[E \times H^{*}\right] \qquad (1)$$

Re refers to the real part of the complex vector, and * to the conjugation operator.

Including the thick glass substrate (d_{glass} = 1.1*10^6 nm) into the optical simulation formalism yields narrow Fabry-Perot oscillations in the reflectance which are not observed in the experiment. These Fabry-Perot oscillations are generally found in the simulation as long as one sufficiently transparent layer of the multilayer structure is thicker than several wavelengths. The result demonstrates that the assumptions of plane and parallel surfaces and a monochromatic light source are not realistic. Averaging reflectances calculated for various glass substrate thicknesses (d_{glass}, $d_{glass} \pm \lambda/4$, and $d_{glass} + \lambda/2$) results in a total reflectance which is inde-

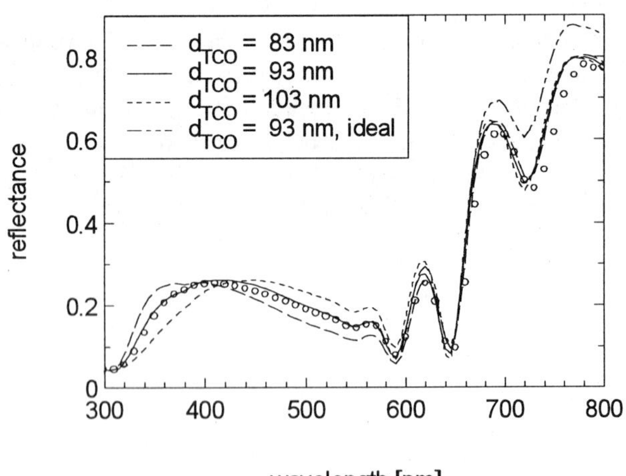

Fig. 2: Simulated (lines) and measured (o) reflectance of a pin structure (d_i = 445 nm) for different TCO-layer thicknesses taking into account a real and an ideal (mixed dashed line) description of the rear contact.

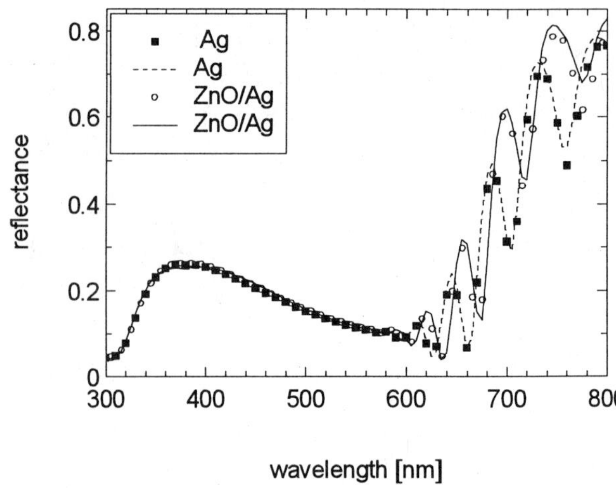

Fig. 3. Calculated (lines) and measured (symbols) reflectance of an a-Si:H solar cell (d_i = 875 nm) with and without employing a highly reflective ZnO/Ag back contact.

pendent of the phase propagation at the glass/TCO -interface and yields a good agreement with the experimental data. The averaging procedure used is based on the Nyquist theorem.

In contrast to the expected good reflectance of the ideal silver back contact (d_{Ag} = 800 nm) the simulation results indicate that the optical properties of the thermal evaporated Ag back contact (deposition rate 5 - 10 nm/s) differs strongly from the values determined from ellipsometric measurements of a single deposited Ag-layer on corning glass. Evaporated thin silver films are in general aggregated in structure. The aggregation depends on the rate of deposition, temperature and type of substrates [5,6]. In order to solve this problem the real Ag rear contact is composed in all simulations of an additional thin layer with higher absorption in the near infrared and a subsequent ideal Ag-layer. The introduction of a 1.5 nm Al-layer fits very well to the back contact properties observed. A good agreement between simulated and measured reflectance and spectral response data of pin structures with different i-layer thicknesses of 340, 455, 585, and 875 nm has been achieved. However, this result implies that the complex refractive indices determined from single layers can be different from the ones within a multilayer stack and an improved preparation of the Ag back contact is necessary.

Simulated and measured reflectance spectra (see Fig. 2) of pin structures show significant interference effects depending strongly on the TCO, p- and i-layer thicknesses. For d_{TCO} = 83, 93, and 103 nm 20 - 30 % of the photons at λ = 400 nm are reflected. The interference is caused by the different wavelength dependent complex refractive indices of the TCO, p- and i-layer. The maximum value depends on the p-layer thickness and the peak position is dominated by the TCO-layer thickness. Further, in the near infrared, i-layer interference fringes can be observed. The layer dimensions have been determined comparing simulated and measured spectra to fit the interference fringes.

A highly reflective ZnO/Ag rear contact is widely used to enhance the efficiency of solar cells [7]. For an optimized structure a ZnO-layer thickness of 90 nm has been determined. For this case the interference fringes in the near infrared are shifted to greater wavelengths (see Fig. 3).

Incoherent wave propagation

Up to now there is no theoretical model that incorporates both, the description of coherent and incoherent wave propagation. For this reason a linear combination of the two propagation models is used as an approach, knowing that the photon scattering decreases for higher wavelengths. The incoherent wave propagation is calculated assuming Lambertian scattering (scattering intensity $\cos(\vartheta)/\pi$) at the TCO/p-interface, Fresnel reflection and transmission at the interfaces except for the p/TCO-, glass/air-, and TCO/glass-interface. The reflection at the p/TCO-interface describes the fraction of light reflected into the p-layer. This factor has been pragmatically chosen, according to Bennett and Mattsson [8]:

$$\frac{R_t - R_s}{R_t} = \left(1 - \exp\left(-(4 \cdot \pi \cdot \delta / \lambda)^2\right)\right) \qquad (2)$$

taking qualitatively into account the roughness of the TCO. R_s, R_t and δ denote the specular reflectance (determined by Fresnel formalism), the total reflectance, and the rms surface roughness. For the description of the TCO/glass and glass/air-interface, the reflection factors have been modified, too. Above a critical angle light is totally reflected at the glass/air, TCO/glass and, p/TCO-interfaces, respectively.

The simulation (see Fig. 4) shows, in comparison to samples with a flat TCO front contact, a reduced reflectance for blue light, caused by the modified reflection factors and a thicker TCO-layer (d_{TOO} = 500 nm) for the rough substrates. Furthermore, in the near infrared enhanced absorption in the solar cell can be observed, as a result of a greater effective absorption length and higher reflection factors, too. So, the light trapping effect for red illumination condition in solar cells with rough TCO front contact can be attributed to photon multiple scattering. The measured reflectance exhibits weak interference fringes over the whole spectral range.

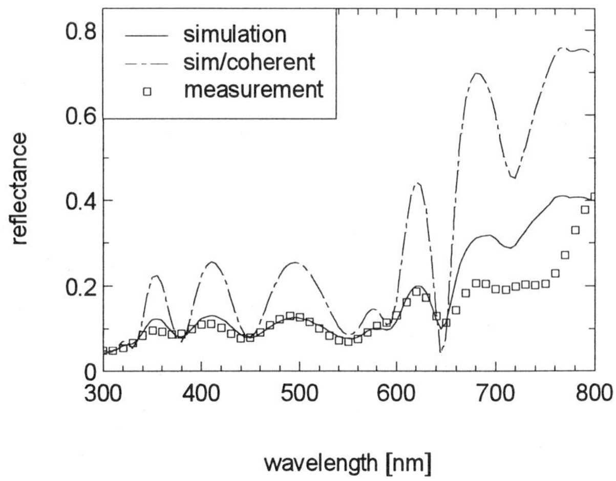

Fig. 4. Measured and simulated total reflectance of a pin structure (d_i = 445 nm) deposited on hazy TCO, in comparison with reflectance calculated using coherent wave propagation only.

The reflectance is simulated taking into account coherent and incoherent wave propagation. A constant ratio of the coherent light intensity to the total intensity has been estimated by means of the amplitude of the interference fringes to 25%.

SOLAR CELL MODELLING

Figure 5 shows the simulated and measured spectral response data of a sample (d_i = 875 nm) with and without a ZnO/Ag rear contact deposited on flat TCO. For non-hazy substrates only a small improvement of the

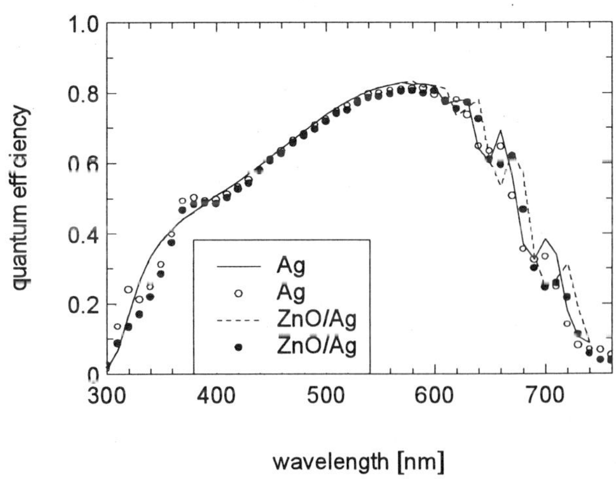

Fig. 5. Calculated (lines) and measured (symbols) quantum efficiency with and without ZnO/Ag rear contact (d_i = 90 nm) of a pin structure (d_i = 875 nm) deposited on flat TCO.

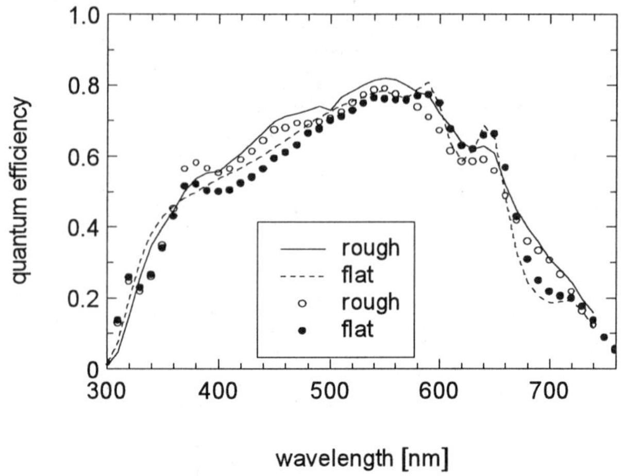

Fig. 6. Calculated (lines) and measured (symbols) quantum efficiency of a solar cell (d_i = 455 nm) with Ag rear contact deposited on flat and rough TCO.

short circuit current (0.8 - 2.6 %) has been observed in the experiment as predicted by our model. The introduction of ZnO shifts the interference fringes, yields a better optical "adjustment", and leads to a highly reflecting back contact resulting in an enhanced carrier generation rate in the cell. The results reveal that a ZnO-layer thickness around 90 nm yields a maximum of the short circuit current. The relative net gain becomes more pronounced with decreasing i-layer thickness and depends on the higher reflectance of the ZnO/Ag contact. The worse the real metal back contact reflectance the better the relative improvement by inserting ZnO. The current gain decreases from 2.3 % to 1 % for the simulation of a solar cell (d_i = 455 nm) using the ideal optical values for the Ag rear contact. Comparing the quantum efficiency of the samples with d_i = 455 nm and d_i = 875 nm deposited on flat TCO an enhanced response around 550 nm can be observed. This can be attributed to the reduced reflectance in that wavelength range.

For substrates coated with a hazy front contact a higher short circuit current has been found in the experiment and in the simulation for a pin structure of an 455 nm i-layer. The current gain is caused by an enhanced blue and red response (see Fig. 6) as an result of the light trapping effect.

CONCLUSIONS

In order to investigate the electronic solar cell properties it is necessary to examine the optical behavior, too. The results indicate that one problem of numerical modelling is the treatment of the optical carrier generation profile, which is normally derived from the Lambert-Beer formula. Further, the short circuit current gain using the ZnO/Ag rear contact depends on the preparation conditions of the Ag back contact which is aggregated in structure. For samples deposited on rough TCO the enhanced short circuit current can be attributed to the reduced reflectance caused by light trapping.

ACKNOWLEDGEMENTS

The authors would like to thank F. Birmans and W. Reetz for solar cell characterisation. This work was supported by the Bundesministerium für Forschung und Technologie.

REFERENCES

[1] Th. Eickhoff, C. Ulrichs, H. Stiebig, W. Grünen, W. Reetz, H. Wagner, "Quantum efficiency of a-Si:H p-i-n solar cells in the 250 to 375 K range - insights from the defect-pool modelling", *this volume.*

[2] F. Siebke, H. Stiebig, A. Abo-Arais, H. Wagner, "Charged and neutral defect states in a-Si:H determined from improved analysis of the constant photocurrent method", *this volume.*

[3] K. Pierz, B. Hilgenberg, H. Mell, G. Weiser, "Gap-state distribution in n-type and p-type a-Si:H from optical absorption", *J. Non-Cryst. Sol.*, **97&98**, 1987, pp. 63 -67.

[4] H. Stiebig, A. Kreisel, J.-L. Nicque, Th. Eickhoff, C. Beneking, H. Wagner, "Experimental and numerical study of a-Si:H solar cells", *12th E.C. Photovoltaic Solar Energy Conference*, 1994, pp. 164 - 167.

[5] R. S. Sennett, G. D. Scott, "The structure of evaporated metal films and their optical properties", *J. Opt. Soc. Am.*, **40,** 1950, pp. 203 - 211.

[6] H.-U. Finzel, P. Wissmann, E. Wittmann, "Characterisation of the structure of porous silver films by optical measurements in the near infrared", *Surface Science,* **166**, 1986, pp. 1226 - 1230.

[7] S. Yoshida, M. Yoshino, S. Kitahara, K. Seki, S. Katayama, K. Nabeshima, K. Nozue, A. Yamada, M. Konagai, "Improvement of performance of a-Si/a-Si double junction solar cells by utilizing the ZnO/Ag rear contact", *11th E.C. Photovoltaic Solar Energy Conference*, 1992, pp. 590 - 593.

[8] J. M. Bennett, L. Mattsson, "*Introduction to surface roughness and scattering*", Optical Society of America, Washington, D.C., 1989.

QUANTUM EFFICIENCIES EXCEEDING UNITY IN AMORPHOUS SILICON SOLAR CELLS

D. Vanmaekelbergh, and J. van de Lagemaat,
Utrecht University, Debye Institute, Dept. of Condensed Matter Physics, P.O. Box 80,000,
3508 TA Utrecht, The Netherlands

R.E.I. Schropp
Utrecht University, Debye Institute, Dept. of Atomic and Interface Physics, P.O. Box 80,000,
3508 TA Utrecht, The Netherlands

ABSTRACT

The experimental observation of internal quantum efficiencies above unity in crystalline silicon solar cells has brought up the question whether the generation of multiple electron/hole pairs has to be taken into consideration also in solar cells based on direct gap amorphous semiconductors. To study photogenerated carrier dynamics, we have applied Intensity Modulated Photocurrent Spectroscopy (IMPS) to hydrogenated amorphous silicon p-i-n solar cells. In the reverse voltage bias region at low illumination intensities it has been observed that the low frequency limit of the AC quantum yield Y increases significantly above unity with decreasing light intensity, indicating that more than one electron per photon is detected in the external circuit. This phenomenon can be explained by considering trapping and thermal emission of photogenerated carriers at intragap amphoteric dangling bond defect centers.

INTRODUCTION

Recently, the fundamental aspects of the generation of carriers by the absorption of photons with an energy larger than the band gap have attracted renewed attention. The generally accepted assumption that only one electron/hole pair is produced for each photon and that the excess energy is dissipated as lattice vibrations, was called in question by the experimental observation of internal quantum efficiencies (IQE) above unity at high photon energies in float zone (100)-oriented crystalline silicon solar cells [1,2]. The IQE values above unity were attributed to carrier multiplication by impact-ionization, which is an Auger-type process. In contrast to the classical impact-ionization process, where the excess energy is supplied by an electric field, the process of generation of multiple electron/hole pairs is brought about by the excess energy due to *direct* optical excitation of photons with a large energy in the band structure E(k) of an *indirect* semiconductor. The effective creation of a second electron/hole pair by impact ionization requires conservation of momentum and thus dictates a specific band structure of the direct band gap semiconductor. In crystalline Si, the minimum energy for primary photon absorption leading to multiple carrier generation is 3.4 eV. Carrier relaxation by phonon emission is still a competing process, so that no considerable contribution to the photocurrent in solar cell applications can be expected.

The question arises whether the generation of multiple electron/hole pairs also has to be taken into consideration in the theoretical description of solar cells based on amorphous semiconductors, such as amorphous silicon. The mechanism for internal quantum efficiencies exceeding unity will be quite different from Auger generation described above. First, amorphous silicon does not have a clearly defined band structure and behaves much like an direct semiconductor since the conservation of momentum is not required. This leads to a large absorption coefficient for photons with an energy larger than the bandgap energy. Therefore, additional generation by the collective thermalization of a large concentration of carriers is more likely than by the relaxation of a hot carrier in the energy-momentum space. Second, a continuous distribution of energy levels is present within the band gap. This opens up the possibility of creating electron/hole pairs by subbandgap photon absorption or other processes which supply an energy less than the bandgap energy.

EXPERIMENTAL DETAILS

We have developed and applied Intensity Modulated Photocurrent Spectroscopy (IMPS) to hydrogenated amorphous silicon (a-Si:H) p-i-n solar cells fabricated in our group at Utrecht University. These cells were deposited by Plasma Enhanced Chemical Vapor Deposition (PECVD) in our UHV multichamber system called PASTA, on SnO_2:F coated glass substrates, supplied by Asahi Glass Company. The p^+-type layer was compositionnally graded in order to maximize the transmission and the built-in internal electric field [3]. Nevertheless, the radiation admitted to the absorber layer is restricted to the wavelength region above 380 nm due to absorb-

First WCPEC; Dec. 5–9, 1994; Hawaii

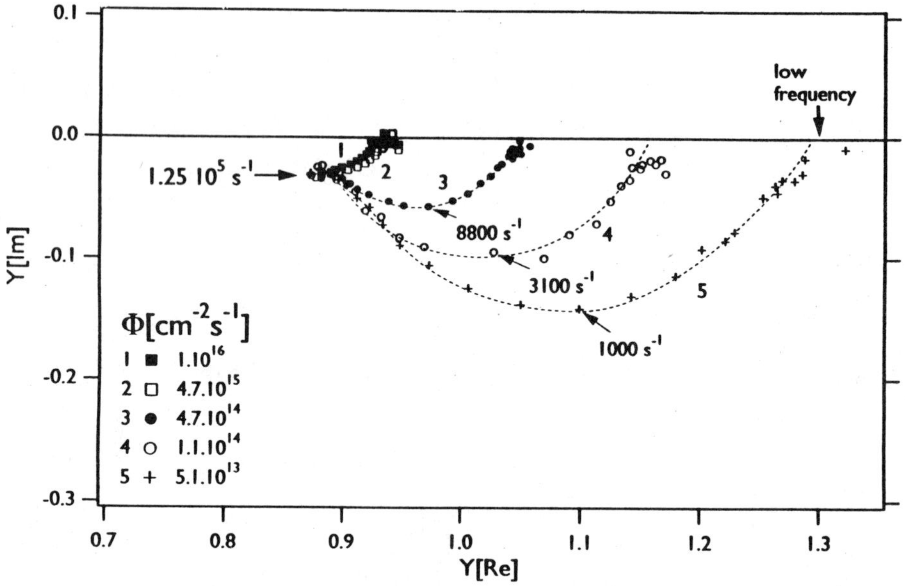

Figure 1: Impedance at reverse voltage bias for different light intensities with the frequency as the parameter.

tion in the substrate. The intrinsic absorber layer is a device-quality material, 0.5 μm thick, with a low impurity content and a low density of states (5×10^{15} cm^{-3}eV^{-1} at mid gap). The back contact consists of silver deposited directly on the n$^+$- type layer by thermal evaporation. The 100 mW/cm^2 AM1.5 conversion efficiency of these cells is 10 %. The fill factor is 0.70 under AM1.5 conditions but increases up to 0.76 under low intensity monochromatic light conditions [4], in the red region (λ = 600 nm) as well in the blue region (λ = 400 nm).

In the IMPS measurements, the solar cell is illuminated with light from a HeNe laser with a constant intensity on which a harmonically varying light intensity (with frequency ω) is superimposed. This is achieved by an acousto-optical modulator, such that the superimposed intensity is given by

$$\tilde{\Phi} = \Phi_{\mathrm{m}}\exp(\mathbf{i}\omega t). \tag{1}$$

The photon energy of the monochromatic incident beam (1.9 eV) is only slightly higher than the band gap energy of the active layer (1.78 eV). The textured TCO front electrode in combination with the highly reflective back contacts constitutes an effective light trapping structure, so that the generation rate profile is homogeneous [5]. The photocurrent response $\tilde{\imath}$ in the external circuit also varies harmonically,

$$\tilde{\imath} = i_{\mathrm{m}}\exp[\mathbf{i}(\omega t - \Theta)] \tag{2}$$

and is characterized by a frequency dependent amplitude i_{m} and a phase shift Θ. The phase shift is determined by trapping, emission and recombination of the photogenerated carriers during their transport in the intrinsic layer. The opto-electrical impedance $Y(\omega)$ is defined as

$$Y(\omega) = \tilde{\imath}/e\tilde{\Phi} = (i_{\mathrm{m}}/e\Phi_{\mathrm{m}})\exp(-\mathbf{i}\Theta) \tag{3}$$

and can be plotted in the complex plane with the frequency as a parameter. $Y(\omega)$ can be considered as the AC-equivalent of the quantum efficiency and reflects the interaction of free carriers with the density of states in the gap.

RESULTS

In order to study carrier *generation*, we performed IMPS measurements using a wide range of modulation frequencies (1 Hz $< \omega/2\pi <$ 10^4 Hz), bias illumination intensities (5×10^{13} cm^{-2}s^{-1} $< \Phi <$ 1×10^{16} cm^{-2}s^{-1}, as measured by a crystalline silicon reference photodiode), and external bias voltages (-1.5 V $< V_{\mathrm{bias}} <$ 0.7 V).

For the measurements taken at a voltage bias of -1.5 V it can be assumed that all carriers are collected and we can thus exclude recombination. In figure 1, $Y(\omega)$ is shown for different incident illumination intensities. At high light intensities a small semicircle is found in the (+,–) quadrant with Re (Y) = 0.95 in the low frequency limit and Re (Y) = 0.85 in the high frequency limit. This can be interpreted as being due to trapping of holes in valence band tail states and emission back to the valence band. The inverse of the frequency where Im (Y) shows a minimum is indicative of trapping and emission from states at the hole quasi-Fermi level. The characteristic time is equal to the number of hole trapping events in the defect states at the quasi-Fermi level times the life time of a localized hole at this level before re-emission.

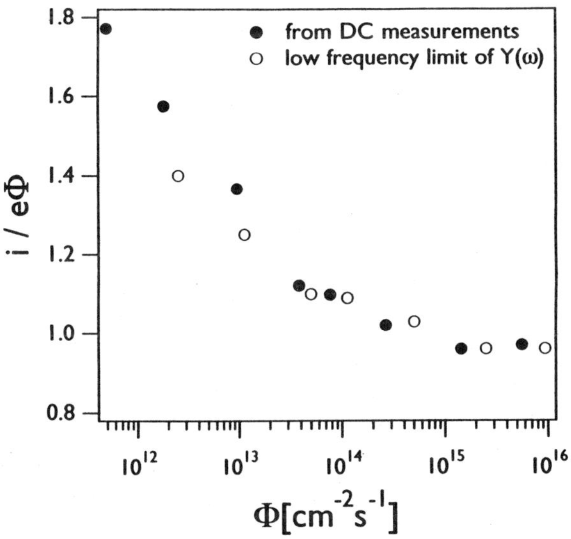

Figure 2: External quantum efficiency determined by IMPS and by DC measurements as a function of the photocurrent of a reference detector, which is proportional to the incident light intensity.

An interesting effect is clearly seen at low illumination intensities. For intensities lower than 1/100 of the maximum intensity, the low frequency limit of Y increases significantly above unity, indicating that more than one electron per photon is detected in the external circuit, whereas the high frequency limit remains the same. In order to confirm this, also DC measurements were performed, which showed that $i/e\Phi$ increases above 1 for decreasing light intensity. Figure 2 shows that the external quantum efficiency increases up to values of 1.8 for the lowest incident intensity. Photocurrent quantum efficiencies considerably higher than unity have previously been observed at illuminated n-type semiconductor/electrolyte junctions under conditions of photoanodic corrosion [6,7]. This could be explained by a model which assumes trapping of a photogenerated hole in a surface bond, leading to an electron deficient surface bond, and subsequent thermal emission of the remaining electron to the conduction band. This then results in a totally broken surface bond.

In order to explain the enhanced quantum efficiency in amorphous silicon we consider the amphoteric nature of the dangling bond defect center. Dangling bonds can be neutral, or can be positively or negatively charged. A negatively charged dangling bond state can capture a hole and thus become neutral. From a neutral dangling bond state D^0, an electron can be thermally emitted to the conduction band, thus creating a D^+ state. An extra electron/hole pair which can be measured in the external ciruit is created if the D^+ state is converted to a D^- state and thermally emits two holes into the valence band.

The fact that this only happens at low light intensity can also be explained with this model. The conversion of the D^0 state to the D^+ state occurs via two competing processes; namely trapping of a hole from the valence band or thermal emission of an electron from the D^0 state to the conduction band. At sufficiently low light intensities the thermal emission of electrons to the conduction band is the dominant process resulting in a quantum efficiency higher than one. At increasing light intensities, the concentration of free holes is increased and the quasi-Fermi levels will split up further. Trapping and emission of holes takes place near the quasi-Fermi level for holes. The capture of holes in the D^0 states at the hole quasi-Fermi level will be much faster than the emission of electrons to the conduction band. As a result of this competing process the excess photocurrent will be absent and the quantum efficiency returns to unity at an increased light intensity.

CONCLUSION

These measurements show that quantum efficiencies in excess of unity can be obtained in amorphous silicon solar cells, due to the presence of amphoteric midgap defect levels associated with dangling bonds. Although the additional generation of electron/hole pairs does not take place under normal operating conditions (i.e. high light intensity and forward voltage bias), the presence of the effect shows that capture and emission of carriers from intragap electronic defect levels can, under certain conditions, effectively enhance the quantum efficiency to values above unity and lead to extra photocurrent. Therefore, if the dynamics of electronic defect levels are accurately included in solar cell models, defect engineering can in principle lead to better photovoltaic energy conversion in amorphous semiconductor solar cells.

ACKNOWLEDGEMENTS

This work was financially supported by the Netherlands Agency for Energy and Environment (NOVEM).

REFERENCES

[1] S. Kolodinski, J.H. Werner, and H.J. Queisser, *Techn. Digest of the International PVSEC-7*, Nov. 1993, p629-630.

[2] S. Kolodinski, J.H. Werner, T. Wittchen, and H.J. Queisser, *Appl. Phys. Lett.* **63** (1994) pp2405-2407.

[3] R.E.I. Schropp, J. Daey Ouwens, M.B. von der Linden, C.H.M. van der Werf, W.F. van der Weg and P.F.A. Alkemade, *Materials Research Society Symp. Proc.* **297**, 1993, pp.797-802.

[4] A. Madan, P. Rava, R.E.I. Schropp, and B. von Roederen, *Applied Surface Science* **70/71**, 1993, pp.716-721

[5] F. Leblanc, J. Perrin, and J. Schmitt, *J. Appl. Phys.* **75**, 1994, pp.1074-1087

[6] H.J. Lewerenz, J. Stumper, and L.M. Peter, *Phys. Rev. Lett.* **61**, 1988, p1989

[7] B.H. Erné, D. Vanmaekelbergh, and I.E. Vermeir, *Electrochim. Acta* **38**, 1993, p2559.

MONITORING A-SI:H SOLAR CELLS DEGRADATION BY TWO SIDES BLUE LIGHT COLLECTION

G. Masini*, D. Fischer, J. Hubin, F. Palma* and A. Shah
Institute of Microtechnology, University of Neuchatel
Rue Breguet 2, CH-2000 Neuchatel, Switzerland (fax ++41-38-254276)
*Dipartimento di Ingegneria Elettronica Università 'La Sapienza'
Via Eudossiana, 18 00184 ROMA, Italy (fax ++39 -6-4742647)

ABSTRACT

In view of the persistent lack of correlation between single layer material properties and solar cell performance, one is at present interested in experimental techniques which allow the characterisation of material properties within a full p-i-n cell. In this paper we introduce a novel method to monitor i-layer properties during degradation of a-Si:H p-i-n solar cell. Two blue, strongly absorbed light beams shine from both p- and n- side on the cell, giving rise to an approximately constant density of charge trapped in the deep states over the i-layer. The sign and magnitude of the constant space charge can be systematically varied by the ratio of the two light beams. The deformation in the shape of the built-in electric field within the cell, as induced by this charge, dramatically affects the collection of photogenerated carriers in a very distinctive way. A simple analytical model, based on the 'standard' DOS model for a-Si:H and involving a reduced set of parameters is presented; it allows the determination of the density of the deep defects by fitting experimental collection data. A more detailed modelling of the experiment, using a finite element a-Si:H device simulator, is also presented.

EXPERIMENTAL

The experimental set-up is sketched in Fig.1. An interferential filter (50 nm bandwidth centred on 400 nm wavelength) is used to obtain blue light from a halogen lamp.

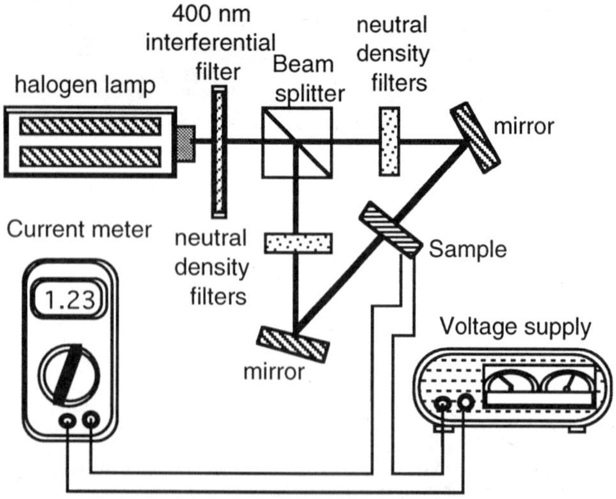

Fig.1: Experimental set-up.

Different light intensities from the two sides are adjusted by interposing neutral density filters.
Experimental data consists of internal collection efficiency values represented as a function of the ratio between the two photocurrents, as generated by each one of the two lights individually (I_{bn}/I_{bp}); I_{bn} and I_{bp} are measured under high reverse bias in order to ensure unity collection. The collection is evaluated as the ratio between the total photocurrent measured at short circuit (I_{sc}) and at high reverse bias (I_{rb}). Each data point consists, thus, of four independent measurements.

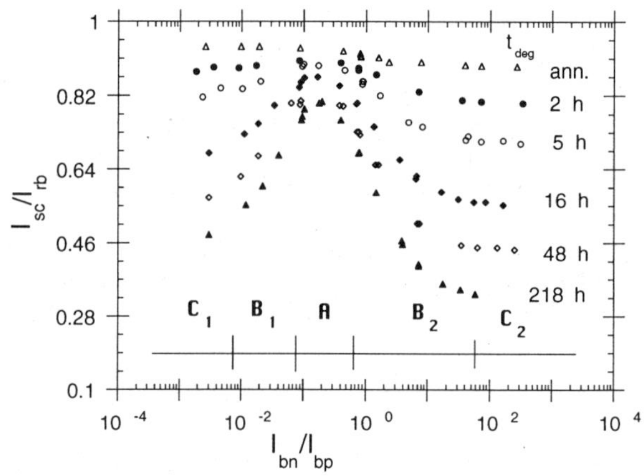

Fig.2: Internal collection at short circuit measured as a function of the relative intensity of the photocurrents generated by blue light shining from n (I_{bn}) and p (I_{bp}) side of the sample.

Samples used are 550 nm thick p-i-n solar cells with a transparent back contact realized by sputtered ITO. A set of measurements is taken in the annealed (2h @ 180 °C) and light-soaked (Sodium Lamp adjusted to give approx. 100 mW/cm^2) states. Experimental results for sample c221193 are reported in Fig.2: different symbols refer to different light soaking times. Clearly, the collection data is a strong function of the degradation time. The main features shown by the curves are: the asymmetric 'bell' shape with the right branch always extended towards lower values of collection, the position of the maximum shifted to the left half plane where the photocurrent generated by light shining from the p-side is stronger and the slopes of the two sides of the 'bell' which increase with increasing degradation. Also the maximum of the curve decreases with increasing degradation.
From a qualitative point of view, strongly absorbed light shining from the p-side is expected to fill the intrinsic layer

CH3365-4/94/0000-0610 $4.00 © 1994 IEEE First WCPEC; Dec. 5-9, 1994; Hawaii

with electrons, while holes are injected by the light which enters the n-side. The two populations tend to charge the deep defects and to give rise to a deformation of the built-in electric field. The collection of photogenerated carriers, which in amorphous cells is strongly dependent on the electric field, is expected to have a maximum when the two light beams are adjusted to produce a net zero charge within the intrinsic layer; this corresponds to minimum electric field deformation (region A in Fig.2). On the other side, a strong predominance of electrons produces a net negative charge in the intrinsic layer which, in turn, lowers the electric field in proximity of the p-i interface, thus reducing the collection probability of the photocarriers generated there (region B1). A similar reasoning can be made for excess hole population and electric field at the i-n interface (region B2). The higher the density of deep states, the stronger the electric field quenching for a given ratio of electrons to holes ratio, the smaller the collection probability. Moreover, as long as the tail states are not involved in the space charge, once all deep states are charged no further electric field depression takes place for an increasing unbalance between electrons and holes and collection eventually saturates (regions C1 and C2).

NUMERICAL SIMULATION

With the scope of obtaining a quantitative idea of charge distribution, field deformation and recombination profiles, numerical simulation of the experiment has been performed using the finite elements amorphous silicon device simulator developed at University of Rome. This programs allows the computation of current flowing through an arbitrary monodimensional semiconductor structure in presence of light shining from the two sides. A good insight of the mechanism regulating the p-i-n structure behavior under the different illumination conditions of the experiment, can be gained from the calculated charge and recombination profiles. In Fig.3 the charge distribution within the i-layer for different illumination conditions is reported assuming a density of deep states of 5e15 $[cm^{-3}]$. Three zones are easily distinguished: two at the extrema of the structure, of variable thickness, a third one in the middle in which the charge is approximately constant at a value fixed by the relative importance of the two light intensities. The thickness of this zone remains fixed even though its position changes. Comparing this results with recombination profiles reported in Fig.4, one can still individuate the same three zones; moreover the central one is characterized by a level of recombination which is always negligible compared to the other two. Even though the relative extension of the zones is found to change for different choices of capture cross sections of dangling bonds, the structure itself, i.e. the presence of three well defined zones, remains the same.

The change in the position of the charged, central, zone with different illumination conditions, gives a possible explanation for the difference, experimentally found, between the saturated asymptotic values of collection (zones C1 and C2 of Fig.2). In presence of a non-homogeneous space distribution of dangling bonds with the maximum near the p-i interface, the charge stored when I_{bn} is predominant over I_{bp} is higher than the one accumulated when I_{bp} prevails on I_{bn}: this results in a larger electric field quenching and a lower measured collection.

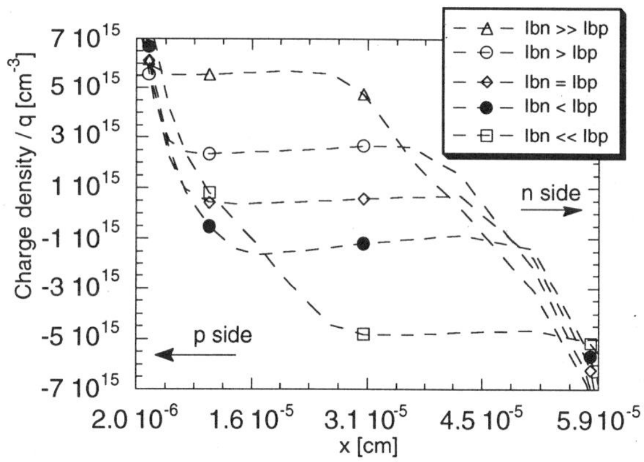

Fig.3: Charge distribution within the intrinsic layer of a p-i-n device under different blue light intensities, I_{bp} and I_{bn}, shining from p and n sides, respectively.

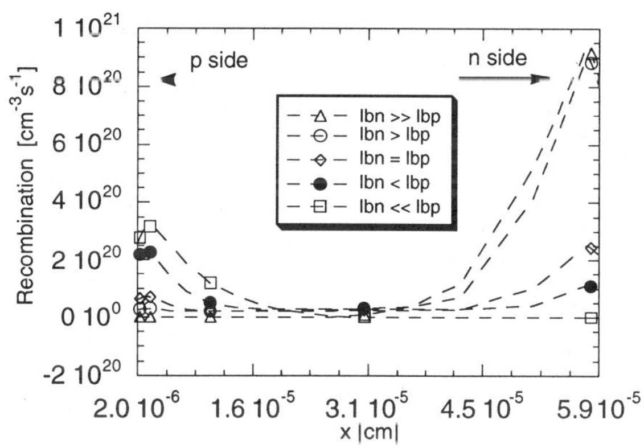

Fig.4: Recombination profiles within the intrinsic layer of a p-i-n device under different blue light intensities, I_{bp} and I_{bn}, shining from p and n sides, respectively

ANALYTICAL MODEL

The development of an analytical model can help one to explain experimental data on the basis of a small amount of parameters underlining, thus, in a non-ambiguous way the role played by each of the parameters in determining the device beahvior. On the other hand, it requires a simplified picture of complex mechanisms such as the ones which, in general, regulate photocurrent collection in an a-Si:H p-i-n device: the particular structure of the two blue beam experiment helps in drawing this picture.

Following the general behavior of charge distribution and recombination profiles, as can be seen from the numerical simulation, it seems justified to divide the intrinsic layer of the p-i-n cell into three zones. The thickness of the central zone is W and it can move within the intrinsic layer for different experimental conditions enlarging or collapsing the other two zones Continuity equations can be written for each zone, generation (G_1, G_2) and recombination (R_1 , R_2) taking place only in the outer zones, Fig.5.

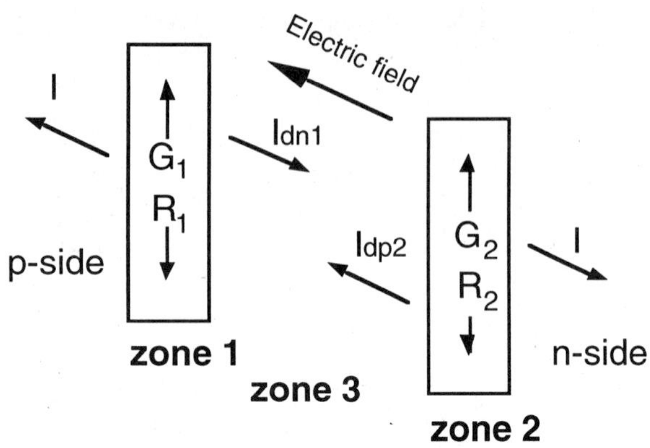

Fig.5: Three zones model for the intrinsic layer of the p-i-n cell.

At high reverse voltage, with unity collection, the photocurrent is given by

$$I_{rb} = (G_1 + G_2) * q \qquad (1),$$

where G_1 and G_2 are the generation currents, integrated all over zone 1 and 2 respectively, due to the light entering the p- and n- sides respectively.
At short circuit, neglecting diffusion currents, one obtains

$$I_{sc} = (G_1 + G_2 - R_1 - R_2) * q = I_{dn1} + I_{dp2} \qquad (2),$$

in which R_1 and R_2 are the total, integrated, recombination in zones 1 and 2 respectively. Eq.2 is the key point of the model: the assumption of a negligible level of recombination in zone three results in the possibility to solve the equations in term of the drift currents which flows there, thus avoiding the need for a description of the recombination terms in zone 1 and 2. Drift currents can be expressed as:

$$I_{dn1} = q\, \mu_n\, n_1\, E_0 \quad , \quad I_{dp2} = q\, \mu_p\, p_2\, E_W \qquad (3)$$

In which E_0 and n_1 (E_W, p_2) are the electric field intensities and the density of free electrons (holes) at the border between zone 1(2) and 3; μ_n and μ_p are band mobilities.
Electric field evolution in zone 3 is a linear function of the distance as the charge trapped there is assumed constant (Fig.3); the average value of the electric field multiplied by the zone thickness (W) gives the built-in voltage (V_{bi}).

$$E(x) = V_{bi} / W + S\, (W/2 - x) / \varepsilon \qquad (4)$$

Charge S is the charge trapped in the dangling bonds; non-homogeneity in the spatial distribution of dangling bonds, which, as noted in the discussion of the numerical simulation, results in an asymmetry of the maximum charge which can be stored under opposite condition of illumination, are taken into account by the offset term Q_o

$$S = Q_{db} + Q_o \qquad (5)$$

Charge trapped in dangling bonds can be related to the amount of photogenerated free carriers by [1]

$$Q_{db} = qN_{db}\left(\frac{p\,\sigma_p^0}{n\,\sigma_n^+} - \frac{n\,\sigma_n^0}{p\,\sigma_p^-}\right) \Bigg/ \left(\frac{p\,\sigma_p^0}{n\,\sigma_n^+} + 1 + \frac{n\,\sigma_n^0}{p\,\sigma_p^-}\right) \qquad (6)$$

in which N_{db} is the average density of dangling bonds and $\sigma_{n,p}^{-,0,+}$ are their capture cross-sections for electrons and holes in different states of charge. Using eqs. 1-6, the internal collection, given by the ratio of the photocurrent at short circuit and at high reverse bias, is

$$\frac{I_{sc}}{I_{rb}} = \frac{\frac{V_{bi}}{W}\left(n_1\mu_n + p_2\mu_p\right) + \frac{SW}{2\varepsilon}\left(n_1\mu_n - p_2\mu_p\right)}{G_1 + G_2} \qquad (7).$$

Assuming a proportional relationship between generation and free carrier concentration at the borders of zone 3

$$G_1 = K_1\, n_1\, ; \; G_2 = K_2\, p_2 \qquad (8)$$

and defining

$$X = G_2 / G_1 = K_2\, p_2 / K_1\, n_1 = I_{bn} / I_{bp} \qquad (9)$$

one obtains

$$\frac{I_{sc}}{I_{rb}} = \frac{\mu_n}{K_1}\, \frac{\frac{V_{bi}}{W}(1 + AX) + \frac{SW}{2\varepsilon}(1 - AX)}{1 + X} \qquad (10)$$

with

$$A = K_1\, \mu_n / K_2\, \mu_p \qquad (11)$$

Expressions for K1 (K2) can be found by evaluating the limit, while letting the p/n ratio go towards infinity (zero), for the continuity equations in zone 1 (2); in such a condition the electric field is strongly distorted, the thickness of the zone dramatically shrinks and the only contribution to the recombination current remains the interface recombination. Interface recombination is not dependent on the bulk density of states and can be assumed to be proportional to the minority carrier concentration via the 'surface' velocity S_v [2]

$$\lim_{p/n \to \infty} R_1 = \lim_{p/n \to \infty} G_1 - \frac{I_{dn1}}{q} = S_{r1} = n_1 S_{v1} \qquad (12)$$

from which

$$K_1 = \mu_n\left(\frac{V_{bi}}{W} + \frac{W}{2\varepsilon}\left(qN_{db} + Q_o\right)\right) + S_{v1} \qquad (13).$$

With a similar reasoning:

$$K_2 = \mu_p\left(\frac{V_{bi}}{W} + \frac{W}{2\varepsilon}\left(qN_{db} - Q_o\right)\right) + S_{v2} \qquad (14)$$

FIT OF EXPERIMENTAL DATA USING THE ANALYTICAL MODEL

The Levenberg-Marquardt algorithm [3] has been used to fit the analytical model to experimental data. The result is reported in Fig.6: the analytical model follows fairly well experimental data over five orders of magnitude of the abscissa. The parameters yielded by the fit procedure are reported in Tab.1 and Fig.7. All the experimental curves, as referring to different degradation stages, are fitted together: the number of dangling bonds (N_{db}) is the only quantity which is allowed to vary as a function of degradation time.

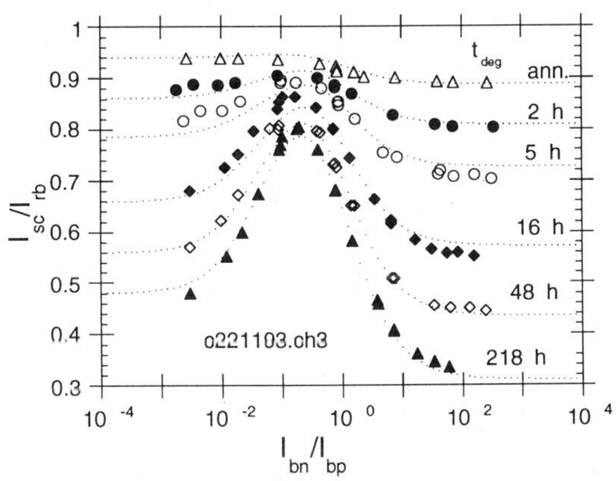

Fig.6. The analytical model fitted over experimental data: the fit of different degradation stages is obtained changing only the number of dangling bonds.

The determination of the absolute value of the deep density of defects (N_{db}) relies on the knowledge of the thickness of the zone 3 of the model (W). In the numerical simulation this thickness has been found to vary from 1/3 to 1/2 of the whole intrinsic layer thickness when using different capture cross section ratios for the dangling bonds. Moreover, W does not vary when one changes the number of dangling bonds, thus it can be assumed constant through the whole degradation process allowing the independent determination of the relative increase of N_{db}. In Fig.7 the evolution of N_{db} is reported as a function of degradation time; W is thereby assumed to be 1/3 of the intrinsic layer thickness. Assuming $\mu_n = 10$, $\mu_p = 1$ [$cm^2/(Vs)$] and $V_{bi} = 1.1$ [V], surface velocities of 22200 and 5520 [cm/s] are obtained for the p-i and i-n interfaces, respectively. from parameters 2 and 3. Parameter 4 indicates a non-homogeneity in the spatial dangling bonds distribution of around 50% of the average value (N_{db}). This can be originated by the non-uniform absorption profile of the light used for the degradation.
Finally, a value of around 3 for the capture cross section ratios of the dangling bonds (charged over neutral) can be deduced from parameters 5 and 6. This is surprisingly low [4] although it is still physically reasonable. As a first presentation of the method we stress the fact that still a large amount of work must be done in order to fully understand the effect of the assumptions, which are used in the analytical model, on the specific values of parameters obtained by curve fitting. This is true, in particular, for the capture cross section ratios.

Parameter	Value
1) $qN_{db}W^2 / 2\varepsilon V_{bi}$	(see Fig.7)
2) $S_{v1}W / \mu_n V_{bi}$	0.037
3) $S_{v2}W / \mu_p V_{bi}$	0.092
4) Q_o / qN_{db}	0.46
5) $\mu_n\sigma^0_p / \mu_p\sigma^+_n$	3.3
6) $\sigma^0_n\sigma^0_p / \sigma^-_p\sigma^+_n$	0.09

Tab.1. Parameters of the analytical model and their values obtained by the fit of experimental data

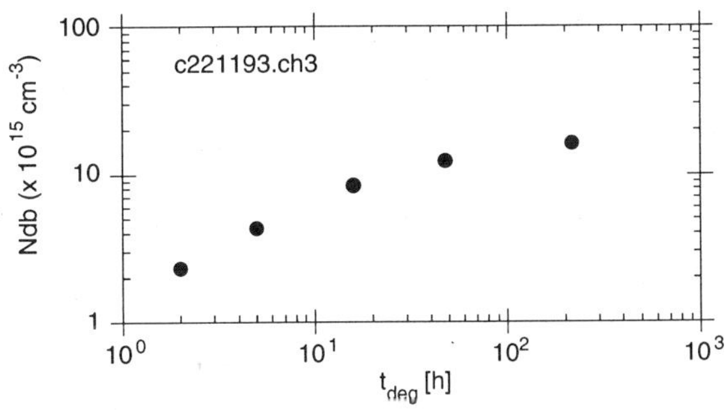

Fig.7. Density of dangling bonds as a function of the degradation time.

CONCLUSIONS

The collection of two sides blue light is shown to be an interesting tool for monitoring p-i-n solar cell degradation. The simple set-up used allows to probe the collection over a wide range of photocarrier populations, and thus yields a very detailed response of the solar cell operation as a function of degradation.
Experimental data is analized based on an analytical model, and the number of danglind bonds during degradation is estimated. The good agreement obtained between theory and experiment confirms the important role played by the dangling-bond charge state and its effect on the electric field shape in the amorphous silicon p-i-n solar cell. The detailed effect of model assumptions on the specific values of the fitted parameters requires, however, further investigation.

ACKNOWLEDGEMENTS

The great assistance of Sebastian Dubail is gratefully acknowledged for preparation of the samples. This work has been supported by Swiss research grant OFEN-EF-REN93/032.

REFERENCES

[1]. J. Hubin, A.V.Shah and E. Sauvain *Phil. Mag. Lett.* **66**,p. 115,(1992)
[2]. S.M. Sze 'Physics of Semiconductor Devices' Wiley p.56
[3] W.H.Press et al.; 'Numerical Recipes', p.521, Cambridge, (1988)
[4] N. Wyrsch, A. Shah *J. Non-Cryst. Sol.* **137&138,** p. 431, (1991)

INVESTIGATION OF CURRENT MECHANISMS IN a-Si:H ALLOY SOLAR CELLS [1]

Kenneth R. Lord II, Fazal UrRahman Syed, Michael R. Walters and James R. Woodyard
Institute for Manufacturing Research and Department of Electrical and Computer Engineering
Wayne State University, Detroit, MI 48202

ABSTRACT

The characteristics of a-Si:H solar cells depend on their past histories, namely, voltage and light soaking times and thermal annealing. The effect of light soaking is known to reduce cell efficiencies at the ten percent level. The details of the role of cell history on the basic current mechanisms are not well understood. The objective of this work is to identify the current mechanisms in I-V characteristics and elucidate their roles in cell stability. Commercial-grade single-junction cells with 0.35 cm^2 active areas and 200, 500 and 800 nm thick intrinsic layers were the focus of the investigations. The cells were characterized with light and dark I-V measurements. The effects of temperature and annealing cycles were investigated. A parametric fitting model was used to quantify four current mechanisms under forward-bias conditions, namely, shunt and series resistances, current injection and depletion. The current mechanisms under reverse-bias conditions are not understood but have been characterized. The investigations show the greatest instability under forward-bias conditions occurs in the shunt resistance mechanism.

INTRODUCTION

The first report of the role of shunt current in hydrogenated a-Si:H single-junction cells was reported by R. Wieting, M. Bolinger, J. Anderson and D. Willett [1]. They used a current measurement at a single reverse-bias voltage to infer the shunt resistance of modules. The shunt resistance was correlated with stabilized module efficiency. Shunt resistivities below 5.0 kΩ cm^2 degraded module efficiencies. Wieting et al. pointed out that prevention of shunting during module manufacture has always been recognized to be important. However, their study points to the role of thickness of the p-i-n layers on shunting. The authors concluded that while thinner layers clearly reduce Staebler-Wronski losses, they increase the likelihood of shunting.

M. Bennett, A. Catalano, J. Newton, C. Poplawski, R. Arya, K. Rajan, G. Wood and S. Wiedeman reported a mode of shunts which are thermally reversible [2]. Accelerated testing on small area triple-junction a-Si:H cells exhibited degradation dependent on initial shunting. They found as long as the shunt value was not small enough to lower the initial performance, the efficiency loss resulting from light soaking scaled with shunt resistance. The efficiency loss in most cases was thermally reversible to within 1% of the initial efficiency and fill factor. There was a shunting effect which could be cured by passing a current through the cell thereby "burning out" the shunt. However, they reported that these shunts can reform, thereby apparently enhancing degradation. Bennett et al. speculated that it is unlikely that the thermally reversible shunts are the direct cause of the increased, but reversible, photodegradation. More likely, they suggested, the shunts were facilitating a mechanism which in their absence simply would not occur. Small area devices studied by Bennett et al. lost 15% or less of their initial efficiency during 1000 hours of light soaking. However, modules generally lost 17 to 23%. They suggested that as the causes of shunts are identified and eliminated, photodegradation of modules will decrease to that of small area cells.

M. Bennett, J. Newton, C. Poplawski and T. McMahon reported on studies with triple-junction a-Si:H modules and cells [3]. A curve tracer scanning at about one kHz was used to observe dark I-V characteristics of cells measuring about 0.25 cm^2 and 1.0 cm^2; the cells were cut from 30 x 33 cm^2 modules. No obvious correlation was found between the amount and position of debris, and the number, position and magnitude of the shunts. Occasional cases of large individual shunts due to tin oxide debris were observed. It was found that laser scribing methods are rarely, if ever, a problem. However, their effects on stability were harder to predict. The most stable modules experienced about a 15% efficiency loss. Typical modules experienced a loss of 18-25% with light soaking. They indicated it should be possible to reduce the losses to those resulting solely from the Staebler-Wronski effect. Initial efficiencies of modules which are in excess of 11% should stabilize around 9.5%.

T. McMahon and M. S. Bennett reported they were able to establish that cells with enhanced initial shunting degraded more than cells without initial shunting when the cells were exposed to light [4]. They observed shunting increased incremently and associated it with an increase in the number of shunt pathways. The shunting was found to decrease incrementally with reverse bias; the effect was interpreted as being caused by localized defects in the junction of cells.

K. Lord, M. Walters and J. Woodyard reported on investigations of current mechanisms in a-Si:H 0.35 cm^2 single-junction cells [5]. They showed agreement between a parametric model and measured dark I-V characteristics was <4% for the shunt and injection regions, and <8% in the depletion region. Current switching was observed in the dark I-V characteristics which was associated with changing shunt resistances. The investigations showed voltage history was important; the current switching behavior in some cells reversed during measurements, i.e., the shunt resistance decreased under reverse bias and increased under forward bias. The changing behavior of the I-V hysteresis characteristic with voltage history was not the same for all the

---

[1] Portions of this work were supported by the TRW Spacecraft Technology Division and NASA.

CH3365-4/94/0000-0614 $4.00 © 1994 IEEE

First WCPEC; Dec. 5-9, 1994; Hawaii

cells. Lord et al. noted current in the reverse-bias region varied as the square of the voltage following the onset of switching; they did not identify the mechanism. Switching from a voltage region where shunt current was the dominant mechanism, to a region where the current varied approximately as the square of the voltage, depended on the rate at which voltage was scanned. They concluded that characterizing cells requires understanding the role of voltage history on the measurements.

The work reported in the literature shows the current mechanisms in hydrogenated amorphous silicon modules and cells are not well understood. The measurement techniques and histories of the devices play an important role in the results which are reported. There is evidence that shunt resistances decrease under light soaking; reduction in the shunt resistance may contribute to some of the degradation in large modules which, in the past, has been associated with the Staebler-Wronski effect. A detailed understanding of the current mechanisms in thin-film photovoltaic devices has the potential for reducing manufacturing costs and enhancing module efficiencies. The objective of this work is to identify the current mechanisms in I-V characteristics and elucidate their roles in cell stability.

EXPERIMENTAL

The current mechanisms were investigated using a parametric fitting model [5] to analyze I-V characteristics of cells measured under various conditions. I-V characteristics were measured in the light and dark with probe and contact resistances of less than 0.1 Ω. The measurements were made with a computer-interfaced Keithley Instruments model 236 source measurement unit (SMU); the unit was programmed to source bias voltage while measuring current. The instrument and software have been configured to carry out 320 current measurements at each source voltage which takes 2.5 seconds without a delay loop in the program; the voltage step was 0.025 V [5].

Light I-V measurements were carried out using an ORC model 331000 solar simulator with a current stabilizer. Annealing of cells was carried out in vacuum at a temperature of 200 °C. The cells were obtained from two fabricators; single and triple-junction cells were investigated with both superstrate and substrate device structures. The structures of the cells used in these studies have been reported previously [5,7]. Single-junction cells with intrinsic layer thicknesses of 200, 500 and 800 nm were investigated. Cell areas ranged from 0.35 to 1.0 cm^2; cells were cut from larger area cells produced for commercial applications. The cell structure does not appear to be important in the current mechanisms we have characterized; the thicknesses of the thin films in the cells does appear to be important.

Our experience shows smaller cells cut from a larger area cell may have very different I-V characteristics. There does not appear to be a spatial correlation of the differences with deposition conditions [3]. The smaller the area of the cell, the more striking the differences in the I-V characteristics; the reason, in part, is due to the magnitude of the shunt current of a large area cell. The larger the area of a cell, the higher the shunt current both because of shunt resistivity and areal non-uniformities. A large shunt current due to a non-uniformity in one area of a large area cell can mask current density instabilities in other areas of a cell; these current density instabilities play an important role in the net efficiency of large area cells. Cutting a large area cell into smaller area cells results in cells with small shunt currents thereby increasing the sensitivity for observing lower level currents produced by the various current mechanisms.

RESULTS AND DISCUSSION

Triple-junction cells cut from larger area cells used in these investigations had fill-factors ranging from 0.37 to 0.68 under as-received or virgin conditions. Fig. 1 shows light I-V characteristics for a cell under two conditions; the filled circles are for virgin conditions and result in a fill-factor of 0.37. The

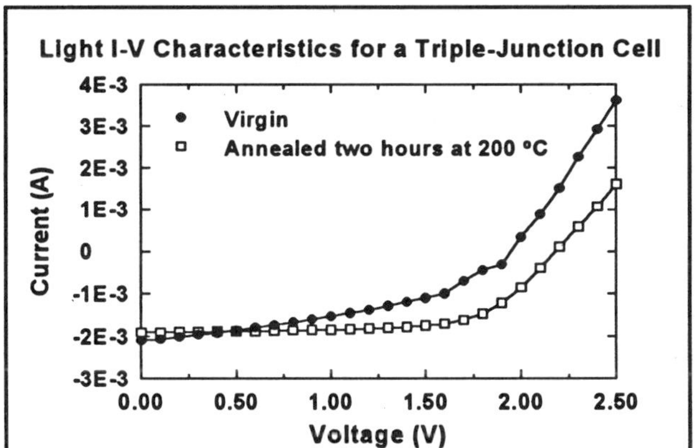

Figure 1: Light I-V characteristic for a triple-junction cell measured under virgin and annealed conditions.

virgin cell is characteristic of one of the poorer as-received cells. Note the current has an anomalous structure between 1.50 and 2.00 V; the structure is concave down instead of concave up as would be expected. The structure in the light I-V is a signature of current instability. The signal to noise ratio for current measurements of stable cells is of the order of 1000; it is limited by the stability of the solar simulator. We find the current signal to noise ratio decreases to about 100 during measurement of light I-V characteristics when current instabilities are observed. Cells with instabilities in light I-V characteristics may also have unstable dark I-V characteristics.

Following a two hour anneal at 200 °C, a light I-V measurement produced the open squares in Fig. 1; the corresponding fill-factor is 0.66. The anneal changed the light I-V characteristic to one which is like that of the better as-received cells.

It is difficult to investigate current mechanisms in triple-junction cells because of their complex structure. For this reason, single-junction cells were investigated in an effort to determine if the behavior observed in Fig. 1 is also characteristic of single-junction cells. Fig. 2 shows light I-V measurements for a single-junction cell under light-soaked and annealed conditions. The filled squares correspond to light I-V measurements for the cell annealed at 200 °C for two hours. I-V measurements following thirty two hours of room temperature light soaking are represented by open circles. An anneal of the cell following the light soak resulted in I-V values which were the same as before the light soak. A comparison of Fig. 1 and Fig. 2 shows similar structures in the light I-V characteristics; this suggests the current mechanisms resulting in the initial poor performance of the triple-junction cell may be elucidated with investigations of single-junction cells.

The effect of light soaking on the forward-bias dark I-V characteristic for a single-junction cell is shown in Fig. 3 by

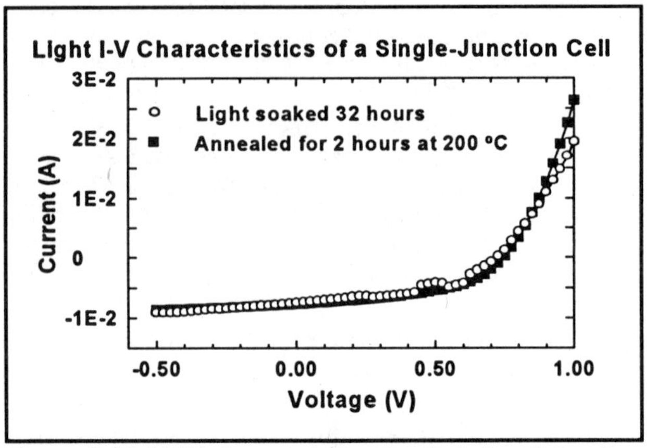

Figure 2: Light I-V characteristics of a single-junction cell under light soaked and annealed conditions.

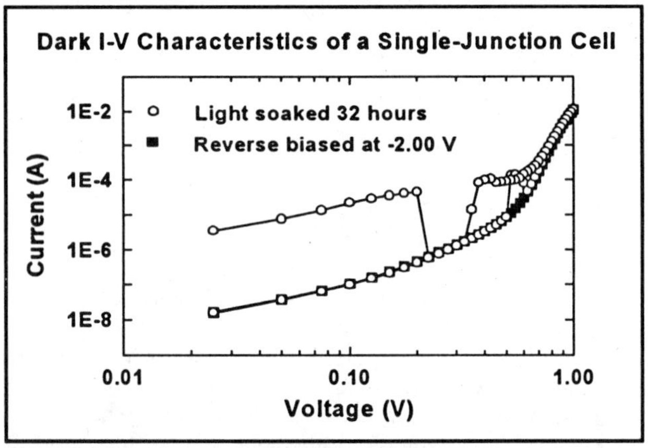

Figure 3: Dark I-V characteristics of a single-junction cell following light soaking and reverse-bias voltage soaking.

the open circles. The open circles were produced by stepping the voltage from 0.025 to 1.00 V and then back to 0.025 V. The cell exhibits the first sign of current switching at about 0.5 V while stepping the voltage from 0.025 to 1.00 V; the current increased from about 1E-5 A to 1E-4 A at 0.5 V. The current decreased at 0.6 V back to a level characteristic of forward-bias current injection. While stepping the voltage from 1.00 to 0.025 V, the current becomes dominated by a 5.0 kΩ shunt resistance down to about 0.38 V; the cell switches to a lower level current which corresponds to 1.0 MΩ shunt resistance. The 1.0 MΩ shunt resistance prevails as the voltage is decreased to 0.225 V when the cell current switches back to a level characteristic of a 5.0 kΩ shunt resistance. The switching effect has been observed in several cells; it occurs under both forward and reverse-bias conditions. An analysis of several cells shows the current switching effect increases in frequency with light and voltage soaking. The switching effect tends to decrease with annealing at 200 °C.

The filled squares in Fig. 3 represent a dark I-V characteristic measured following the application of a -2.00 V reverse-bias voltage for five minutes. It was also produced by stepping the voltage from 0.025 to 1.00 V and then back to 0.025 V. The currents measured while increasing the voltage from 0.025 to 1.00 V are the same as the currents measured stepping the voltage from 1.00 to 0.025 V; this results in a single forward-bias I-V characteristic represented by the filled squares. The cell current is stable following the application of a -2.00 V reverse-bias voltage for five minutes. The I-V characteristic is dominated by a 1.0 MΩ shunt resistance at lower voltages where the slope of the curve is one, and by current injection above 0.30 V. This characteristic is the same as one which was measured before light soaking and after annealing the cell at 200 °C for two hours. Dark I-V measurements on this cell, as well as several single-junction cells with 200, 500 and 800 nm thick intrinsic layers, show instabilities can be explained by changes in shunt resistance. Currents are stable at voltages where series resistance, current injection or depletion are the dominate current mechanisms. This is confirmed by current signal to noise measurements in the voltage regions where the various current mechanisms dominate. The current signal to noise measurements are typically greater than 1000 in regions where the current is dominated by series resistance, current injection or depletion. The current signal to noise has been observed to degrade to less than 10 in cells for voltages where unstable shunt resistances dominate the current.

Three time-dependent dark I-V hysteresis characteristics were measured on a single-junction cell with an intrinsic layer thickness of 800 nm. The results of the measurements are shown in Fig. 4. The measurements were

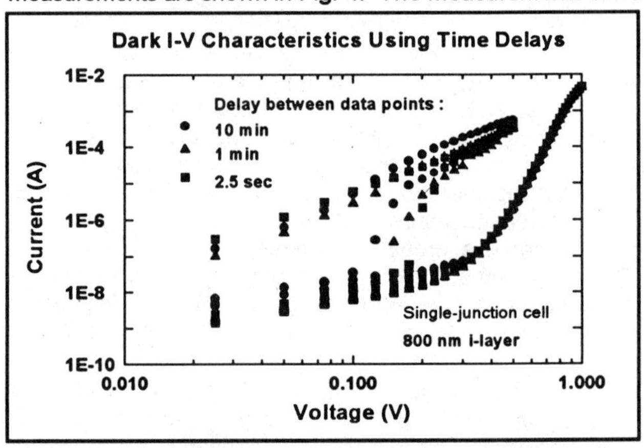

Figure 4: Dark I-V characteristics of a single-junction cell with an 800 nm intrinsic layer thickness; the measurement time delays were 2.5 seconds, and one and ten minutes.

made by starting at zero volts and stepping the voltage from zero, in 0.025 V steps, to 1.00 V and then stepping the voltage in -.025 V steps back through zero volts to -0.50 V; the voltage was then stepped to zero volts in 0.025 V steps. The delay time after stepping the voltage and measureing the current was varied for the three dark I-V characteristics shown in Fig. 4. The I-V characteristic represented by the filled squares was measured with a 2.5 second delay time. A one minute delay time was used for the curve represented with the filled triangles; filled circles represent the measurement with a ten minute delay time. The absolute values of the voltages and currents are plotted in Fig. 4.

Fig. 4 shows the current in the forward-bias injection region of the dark I-V characteristic is not dependent on the measurement delay time [5]. The current is dependent on the measurement delay time in the reverse-bias region of the I-V characteristic; the shunt current is higher and reverse-bias switching occurs at lower reverse biases with longer delay times. The currents under reverse bias are in the microampere and milliampere ranges; the levels of these currents preclude a time dependence of the current which is due to either the capacitance of the cell or trapped charge in the intrinsic layer. Even if the density of sub-band-gap states were as high as 1E17 cm^{-3} and all the states were emptied in one second, the current transient would only be of the order of one microampere. The measurements suggest the reverse-bias current is due to a stress-bias mechanism. The measurements show the importance of bias time and voltage history in measurements of dark I-V characteristics. In a search for current mechanisms which influence dark I-V hysteresis characteristics, it is important to distinguish between transient and steady-state mechanisms.

Single-junction cells with 500 nm intrinsic layers were voltage soaked under a 1.00 V forward bias in an effort to determine the role of defect states on I-V characteristics, current mechanisms and cell stability. Voltage soaking is expected to introduce sub-band-gap electronic states in the intrinsic layer of a cell. Injection currents were in excess of 1.0 milliamperes with a 1.00 V forward bias. Parametric fitting of dark I-V characteristics was carried out using the methods reported earlier [5]. The results of the measurements are shown in Fig. 5. The open squares show the dependence

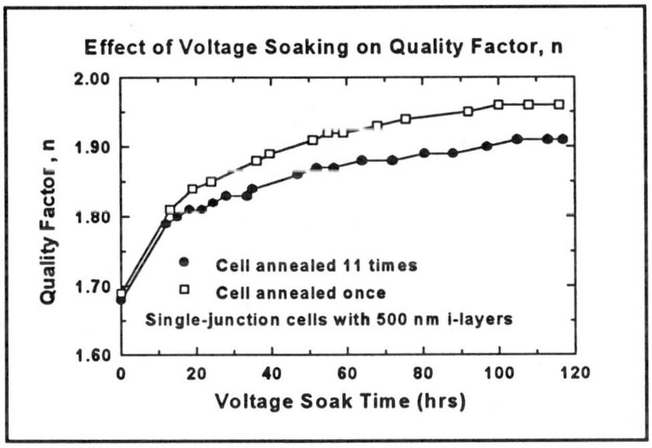

Figure 5: Quality factor dependence on 1.00 V forward-bias soak for two hour anneals at 200 °C.

of the quality factor, n, on voltage soaking time for a cell annealed once in a 1E-6 Torr vacuum for two hours at 200 °C. The initial quality factor was about 1.69; it increased and stabilized at about 1.96 during 120 hours of room temperature voltage soaking. Another cell was cycled eleven times between room temperature and a two hour anneal at 200 °C. The filled circles show n increased from 1.68 to 1.91 during the 120 hour voltage soak. The quality factor for the cells is restored with one anneal at 200 °C for two hours. Fig. 5 shows prolonged annealing at temperatures of about 200 °C reduces the steady-state defect density produced by voltage soaking. We have also observed that the two hour 200 °C annealing cycles improve the shunt resistance stability.

We are in the process of carrying out a systematic study with a large number of single-junction cells of various intrinsic layer thicknesses in order to determine the role intrinsic layer defect states on cell stability and the effect of prolonged annealing at 200 °C on cell parameters and stability.

CONCLUSIONS

The signal to noise ratio for light I-V current measurements of stable cells is of the order of 1000; it is limited by the stability of the solar simulator. The current signal to noise ratio decreased to about 100 during the measurement of light I-V characteristics when current instabilities are observed. Cells exhibiting instabilities in light I-V characteristics may also have unstable dark I-V characteristics. Annealing of some virgin cells with poor efficiencies resulted in major improvements in the cells' efficiencies. A current switching effect has been observed in several cells; it occurs under both forward and reverse-bias conditions. An analysis of several cells shows the current switching effect increases in frequency with light and voltage soaking. The switching effect tends to decrease with annealing at 200 °C. Dark I-V signal to noise measurements of cell current are typically greater than 1000 in regions where the current is dominated by series resistance, current injection or depletion. The current signal to noise has been observed to degrade to less than 10 in cells for voltages where unstable shunt resistances dominate the current. The investigations show the importance of bias time and voltage history in measurements of dark I-V characteristics, and suggest reverse-bias currents in some cells may be due to a stress-bias mechanism. In a search for current mechanisms which influence dark I-V hysteresis characteristics, it is important to distinguish between transient and steady-state mechanisms. Annealing cells at temperatures of about 200 °C reduces the steady-state defect density created by voltage soaking. 200 °C annealing cycles appear to improve the shunt resistance stability. We are in the process of carrying out a systematic study with a large number of single-junction cells of various intrinsic layer thicknesses. The objectives of the studies are to determine if prolonged annealing at 200 °C improves cell stability and to further elucidate current mechanisms.

REFERENCES

1. R. D. Wieting, M. Bolingen, J. Anderson, and D. R. Willet, *IEEE Proceedings of the 21st Photovoltaic Specialists Conference*, (Kissimee, 1990) p. 1569.
2. M. S. Bennett, A. A. Catalano, J. L. Newton, C. Poplawski, R. R. Arya, and K. Rajan and S. Wiedeman, *Amorphous Silicon Technology - 1991*, edited by A. Madan, Y. Harnakawa, M. J. Thompson, P.C. Taylor and P. G. Lecomber, Mat. Res. Soc. Proc. **219**,1992, p. 445.
3. M. S. Bennett, J. Newton, C. Poplawski and T. McMahon, *Proceedings of the IEEE 23rd Photovoltaic Specialists Conference*, (Louisville, 1993) p. 867.
4. T. J. McMahon and M. S. Bennett, *Amorphous Silicon Technology - 1992*, edited by M. J. Thompson, Y. Harnakawa, P.G. Lecomber, A. Madan, and E. Schiff, Mat. Res. Soc. Proc. **258**, 1992, p. 941.
5. K. R. Lord II, M. R. Walters and J. R. Woodyard, *Amorphous Silicon Technology 1994*, edited by E. A. Schiff, M. Hack, A. Madan, M. Powell and A. Matsuda, Mat. Res. Soc. Proc. **336**, 1994, p. 729.
6. Kenneth R. Lord II, Michael R. Walters and James R. Woodyard, *Proceedings of the IEEE 23rd Photovoltaic Specialists Conference*, (Louisville, 1993) p. 1448.
7. James R. Woodyard, *Amorphous Silicon Technology - 1992*, edited by M. J. Thompson, Y. Harnakawa, P.G. Lecomber, A. Madan, and E. Schiff, Mat. Res. Soc. Proc. **258**, 1992, p.1151.

THE P/I INTERFACE LAYER IN AMORPHOUS SILICON SOLAR CELLS; A NUMERICAL MODELING STUDY

V. Suntharalingam, C. M. Fortmann, S. J. Fonash
Electronic Materials and Processing Research Laboratory
The Pennsylvania State University, University Park, PA 16802, USA.

F. A. Rubinelli,
Instituto Tecnologico para el Desarrollo de la Industria Quimica, INTEC, Conicet, Universidad Nacional del Litoral, Guemes 3450, 3000, Santa Fe, Argentina.

ABSTRACT

The **A**nalysis of **M**icroelectronic and **P**hotonic Structures (AMPS) computer program was used to study the influence of the p/i interface layer on stabilized (p)a-SiC:H/(i) a-Si:H/(n) μc-Si heterojunction solar cell performance. The band gap, defect density, and doping variations in the p/i interface and their effects on the light J-V characteristic were examined. The greatest performance benefit could result from control of the recombination (mobility) gap of the interface region. Cells with interface layers with a recombination gap that is close to that of the p-doped layer have the best Fill Factors and Voc.

We suggest that in designing high quality buffer layers the emphasis be placed on increasing the recombination band gap of the interface at the expense of concerns about interface defect density. Hydrogenated interface layers or microcrystalline SiC:H layers can provide increased band gap at the p-i heterojunction. Further, it may be possible to produce high quality cells in single chamber systems. Preliminary studies to determine the influence of interface layers on stability were initiated.

INTRODUCTION

The boundary region between the p-doped (p) and intrinsic (i) layers in amorphous silicon based heterojunction solar cells is highly dependent on deposition conditions. The term "interface layer" is used for boundary regions between the p and i-layers. In some cases, layer properties completely different than either the p or the i-layer could be created. These interface layers may result from interruptions in the fabrication sequence, cross contamination in single chamber systems, outdiffusion, morphology created defects such as bonding mismatch, or frozen in defects [1]. These layers may be different from the i-layer in mobility gap, doping (from heavy to intrinsic), defect density, and lifetime.

In other cases, an interface layer is intentionally deposited (buffer layer) between the doped and intrinsic material. These buffer layers have been shown to improve the cell's performance. The buffer layer between the p-doped and intrinsic layers is thought to mitigate the detrimental effects of an interface layer by reducing carrier back diffusion, recombination, and trapping. The focus of this work is to further elucidate the role of the p/i interface layer on device performance to aid solar cell optimization.

APPROACH

In particular, computer program AMPS [2] was used to examine the illuminated forward voltage dependence of current in (p)a-SiC:H/(i) a-Si:H/(n) μc-Si heterojunction solar cells. The electric field profiles, recombination rate, space charge densities, and potential energy distributions were analyzed as a function of voltage bias and illumination condition for various interface structures.

THE DEVICE STRUCTURE

To examine the influence of the p/i interface on solar cell characteristics, AMPS was used to construct a reference cell to which interface layers were subsequently added. The general structure of this "No-Interface cell" is listed in Table I; it consists of a 150Å p-doped Silicon Carbide layer, a 4000Å intrinsic amorphous silicon layer, and a 250Å n-doped microcystalline silicon layer; the thicknesses of these doped layers are sufficient to provide electrostatic shielding of the intrinsic layer from the contacts. Neutral (flat energy band) boundary conditions for electron potential were imposed at the front and back metal contacts of the device; therefore, there was no band bending and no electric field at the contacts. In these simulations the entire p-i-n structure, rather than just the intrinsic layer, was modeled. These simulations used a density of states (DOS) which included exponential Urbach tail states and gaussian distributions of midgap states.The parameters used for this reference cell as well

Table I: Structure of the No-Interface Reference Cell

	p SiC:H	i a-Si:H	n μc-Si
Thickness (Å)	150	4000	250
Activation Energy (eV)	0.50	--	0.10
Mobility Gap (eV)	1.90	1.80	1.3
Optical Gap (eV)	1.88	1.72	2.0
Doping {Na | Nd} (cm^{-3})	$2.2x10^{19}$	--	$7x10^{17}$
Mid-Gap Defect Density (cm^{-3})	$1.5x10^{18}$	$9x10^{16}$	$3x10^{16}$
Charac. Energy of Ev Tail (meV)	200	50	10
Charac. Energy of Ec Tail (meV)	70	27	10

CH3365-4/94/0000-0618 $4.00 © 1994 IEEE

as for all others were chosen to be indicative of the stabilized (light soaked) state.

Figure 1 shows the band diagram of this No-Interface cell in thermodynamic equilibrium. The bandgap discontinuity between the p-doped SiC:H layer and the intrinsic layer was placed entirely in the conduction band since experimental measurements at Penn State show that there is no variation in Fill Factor between cells with SiC:H or a-Si:H p-doped layers [3]. The bandgap discontinuity between the intrinsic and the n-doped layers was placed entirely in the valence band since experimental measurements at Penn State have also shown that there is no difference between cells with microcrystalline n-layers or with amorphous silicon n-layers [3].

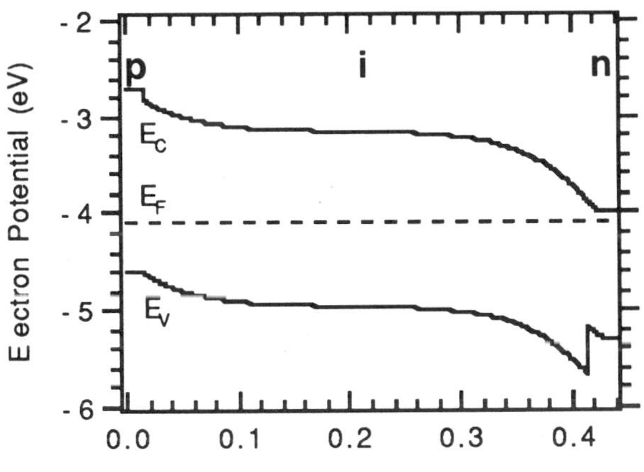

Fig.1. Thermodynamic Equilibium Band Diagram of the No-Interface Cell.

Next a 50Å interface layer was inserted between the p-doped and intrinsic layers. The band gap, the doping, and the defect density in this p/i interface layer were varied and the resulting changes in the illuminated J-V characteristic were examined. The Urbach tails of the interface layer were taken to be the same as those of the p-layer. When changing the band gap of this interface layer, the gaussians were shifted equally with respect to both the conduction and valence band edges since experimental measurements do not indicate a constant energy position relative to the band edges [4]. Note here that each of the cells having a p/i interface layer is 50Å longer than the No-Interface cell.

As described in Table II, three classes of interface layers were studied. Each class is defined by the recombination band gap of the p/i interface layer: Buffer Gap, Graded Gap, and Notched Gap. The p/i interface dopant and defect densities were also varied within each of these classes.

Table II: The Three Cell classes Examined

50Å p/i Interface	Buffer Gap	Graded Gap	Notched Gap
Recombination Gap (eV)	1.90	1.85	1.65
N_A (cm^{-3})	$2.2x10^{17}$ to $2.2x10^{19}$	$2.2x10^{17}$ to $2.2x10^{19}$	$2.2x10^{17}$ to $2.2x10^{19}$
$N_{defects}$ (cm^{-3})	$1.5x10^{16}$ to $1.5x10^{19}$	$1.5x10^{16}$ to $1.5x10^{19}$	$1.5x10^{16}$ to $1.5x10^{19}$

The Buffer Gap cell class is defined by having a p/i interface with the same band gap as the p layer of the No-Interface cell.

The Graded Gap cell class is defined by having a p/i interface with a bandgap which is smaller than that of the p layer, but larger than that of the intrinsic layer.

Finally, the Notched cell class is defined by having a p/i interface layer with a smaller bandgap than either the p or the intrinsic layer. This case is likely to occur when no special precautions are used during cell preparation as this bandgap shrinkage may be due to deposition of residual boron from the chamber walls prior to intrinsic layer deposition or to dopant outdiffusion into the i-layer material [5].

RESULTS AND DISCUSSION

The p/i interface was examined by changing three critical parameters: the recombination (or mobility) gap, the doping, and the density of dangling bond states. The absorption coefficients of the p/i interface layer were shifted appropriately to model an optical gap that was 3kT less than the recombination gap. It was assumed that photocarriers created between the optical gap and mobility gap quickly thermalize from the optical band edges to the recombination gap.

Recombination Gap

The No-Interface cell was compared to Graded Gap and Notched Gap cells. In both of the latter cases a doping density of $2.2x10^{19}$ cm^{-3} and a mid-gap density of states value of $1.5x10^{18}$ cm^{-3} was used. In Figure 2 we illustrate the effect of changing the recombination gap of the 50 Å p/i interface on the illuminated J-V characteristic. The No-Interface cell, by definition, has no p/i interface, the Graded Gap and Notched Gap cells have an interface with recombination gaps of 1.85eV and 1.65eV, respectively.

The Graded Gap cell's short circuit current (Jsc) is slightly less than that of the No-Interface cell due to the loss of absorption from the addition of a 50Å interface layer. The graded band gap of this cell acts to prevent a significant decrease in open circuit voltage (Voc). The effectiveness of interface treatments found in experimental work may be due to a better interface layer than that modeled here or due to treatment that actually removes the deleterious band gap notch.

As noted earlier, the Notched cell is defined by having an interface layer with narrower bandgap than either the p-SiC:H or i-a-Si:H regions. The Notched cell and its variations have been observed experimentally [5] and the trend in the terminal characteristics predicted by AMPS is consistent with experimental measurements. This type of p/i interface layer dramatically reduces the cell's Jsc, Voc, and efficiency (η) as seen in Fig. 2. The Notched cell has a lower Jsc than either the No-Interface or the Graded cells because electron back diffusion into and subsequent recombination in this narrowed bandgap interface limits the collected photocurrent.

Further, AMPS simulations predict that the electron concentration throughout much of the the intrinsic layer may be up to two orders of magnitude lower at Voc in a

Notched Gap cell than in a No-Interface cell for the range of p/i interface doping and defects listed in Table II. Thus, the Notched Gap cell may have a stability advantage in that the bulk of the i-layer would have fewer carriers under illumination if theories concerning the relationship between carriers (or quasi Fermi levels) and stability are correct [6].

Dopant and Defect Density - Buffer Gap Cell

We now discuss the influence of the p/i interface's doping and mid-gap defect density on the illuminated J-V characteristic. The range of values examined is listed in Table II. The effect of changing dopant and defect densities is highly dependent on their relative values since they compete in the interface; i.e, acceptor-like doping (N_A) introduces negative charge into the interface, while dangling bond defects above the Fermi Energy introduce positive charge. This can be seen the active acceptor (or donor) density saturates with increased doping [7]. Therefore, to simulate the bandgap shrinkage (due to boron accumulation at the interface), both the excess ionized boron and the additional defects created were used in this model.

In Figure 3 a Buffer Gap cell (i.e, the cell in which the p/i interface has the same recombination gap as the p-doped layer) was used to demonstrate the effect of varying the p/i interface's doping and defect density on the illuminated J-V characteristic.

First, the effect of p/i interface doping on the cell's illuminated J-V characteristic was examined. Fig 3 shows that Cases C1 and C2 (which have p/i interface doping of 2.2×10^{17} cm^{-3}) suffer from the greatest loss of photocurrent and Fill Factor relative to the reference No-Interface cell. Cases C1 and C2 have a p/i interface defect density of 1.5×10^{16} and 1.5×10^{18} cm^{-3}, respectively. Although these cell have interface doping at the 2.2×10^{19} cm^{-3} level, they still have some positive charge developed in their p/i interfaces due to the high level of defect states. This positive charge weakens the built-in electric field across the intrinsic layer. Photogenerated carriers are thus less effectively swept out of the i-layer by the built-in electric field -- those carriers recombine or become trapped. Therefore, these cells with trapping in the defect states of the p/i interface have a diminished Jsc relative to the No-Interface cell. The positive charge trapped in the p/i interface of these cases also reduces the i-layer electric field under bias and consequently limits collection of photogenerated carriers -- thus, the Fill Factor of these cases deteriorates. AMPS plots which confirm the loss of electric field and recombination under illumination are not shown due to space limitations.

Next, we examine the effect of the p/i interface defect density on the cell's performance. In Case C4 of Fig. 3, we see that increasing the p/i interface defect density to 1.5×10^{19} cm^{-3} places enough positive charge in the p/i interface to decrease the cell's Fill Factor. The trapping and electric field distortion caused by the interface are also sufficient to cause a decrease in Case C4's Voc.

In Case C1, although the interface defect density has been reduced to 1.5×10^{16} cm^{-3}, the decrease in Jsc relative to the No-Interface cell is <u>not</u> due to the interface defect density, but rather is due to reduced absorption from the addition of a 50Å thick interface layer. The Voc of Cell C3 is unaffected by the addition of the 50Å interface.

Dopant and Defect Density - Notched Gap Cell

Figure 4 displays the illuminated J-V characteristic of the Notched Gap class of cell. This Notched Gap cell is defined by having a recombination gap which is smaller than that of either the p-doped or the intrinsic layers. The dopant and defect densities modeled are within the ranges described in Table II.

Fig. 4 demonstrates that decreasing the doping density in the notched p/i interface from 2.2×10^{19} cm^{-3} (Case N1) to 2.2×10^{17} cm^{-3} (Case N2) weakens the built-in electric field and results in a loss of Fill Factor and Voc; this is similar to the Buffer Gap comparison.

Now investigating the p/i interface defect density, we find that decreasing it from 1.5×10^{18} cm^{-3} (Case N1) to 1.5×10^{16} cm^{-3} (Case N3) also improves the cell's performance . The decreased p/i interface defect density traps less positive charge and strengthens the built-in field across the intrinsic layer. Therefore, although there is still a loss of Jsc due to back diffusion of electrons into the notched gap interface, the Fill Factor improves.

In Case N4, we simultaneously decrease the p/i interface doping to 2.2×10^{17} cm^{-3} and the defect density to 1.5×10^{16} cm^{-3} and see in Fig. 4a that the cell's Jsc is dramatically improved relative to Case N1; this increase in photocurrent is due to reduced recombination in the p/i interface.

Although Case N4 demonstrates that if it was possible to have a low band gap notch region with a low defect density, the effect on Voc would be minimal. Notched region with low defect densiGap cell is physically created by high doping and defect densities in the p/i interface, then these improvements from decreasing doping and defect density are not achievable.

CONCLUSION

A variety of (p)a-SiC:H/(i) a-Si:H/(n) μc-Si heterojunction solar cells which have different p/i interface parameters have been simulated and their terminal characteristics determined. To improve the design of the p/i buffer layer the greatest benefit comes from controlling the recombination gap of the interface. Increased doping alone in the p/i interface layer will improve the cell's Fill Factor; however, doping related defects will limit it. Cells having interface layers with a recombination gap that is closest to that of the p-doped layer have the best Fill Factors and Voc since the larger bandgap is most effective at limiting electron back diffusion.

Since the No-Interface cell is simply a analytic construct that is not physically viable, the results of this investigation suggest that in order to design high quality buffer layers the emphasis should be placed on increasing the recombination band gap of the interface rather than the interface defect density. Hydrogenated interface layers or microcrystalline SiC:H could provide increased bandgap at the p-i heterojunction. Further, with this emphasis, high quality cells could be produced in single chamber systems. Since the efficacy of improving the cell performance by changing the dopant and defect densities is highly dependent on the recombination gap of the interface, it is suggested that these parameters only be used to "fine tune" the optimized cell performance. It is possible that if bulk electron densities influence the stability, some interface measures that increase the Voc will speed the degradation rate. Conversely, it is possible that solar cells that have low blue response due to strong interface recombination may have enhanced stability as the lower bulk carrier densities may reduce the driving forces for degradation.

ACKNOWLEDGMENTS

We thank Sanghoon Bae and Yi-wei Lu for helpful discussions. AMPS and this work have been supported by the amorphous silicon thin film program of The Electric Power Research Institute. V.S. was supported by a NSF Graduate Engineering Education Fellowship.

REFERENCES

[1] Z.E. Smith, S.Wagner, Phys. Rev. B **32,** 5510 (1985).
[2] F.A. Rubinelli, S.J. Fonash, J.K. Arch, 6th International PVSEC Proc., 851 (1992).
[3] Y. Lee, private communication.
[4] M. Gunes, private communication.
[5] Y. Li, S. Kim, I.Chen, Y.Lee, C.M. Fortmann, C.R. Wronski, R.W. Collins, 1st WCPVEC 1994.
[6] For example, M.J. Powell, S.C. Deane, I.D. French, J.R. Hughes, W.I. Milne, Phil. Mag. B **63,**325 (1991).
[7] N. F. Mott and E.A. Davis. 2d ed. Oxford, Clarendon Press; New York, Oxford University Press,1979.

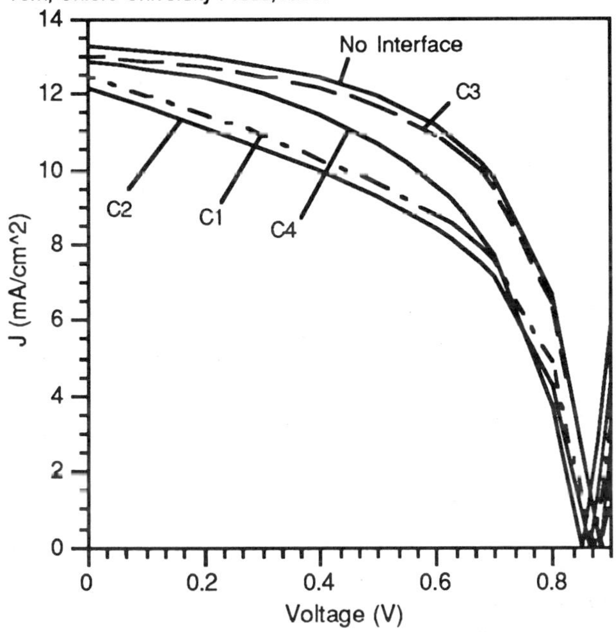

Fig. 3. Illuminated JV for the Buffer Gap cell as a function of p/i interface defect density and doping. The terminal characteristics and interface parameters of these cells are listed in Table III.

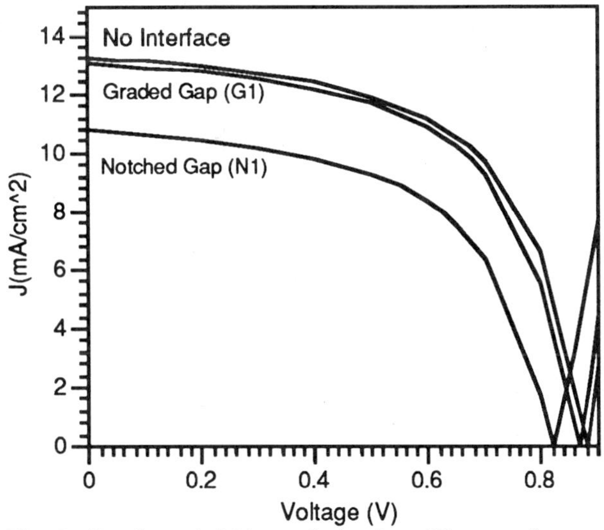

Fig. 2. Illuminated JV for cells having different p/i interface recombination gaps. The terminal characteristics and interface parameters of these cells are listed in Table III.

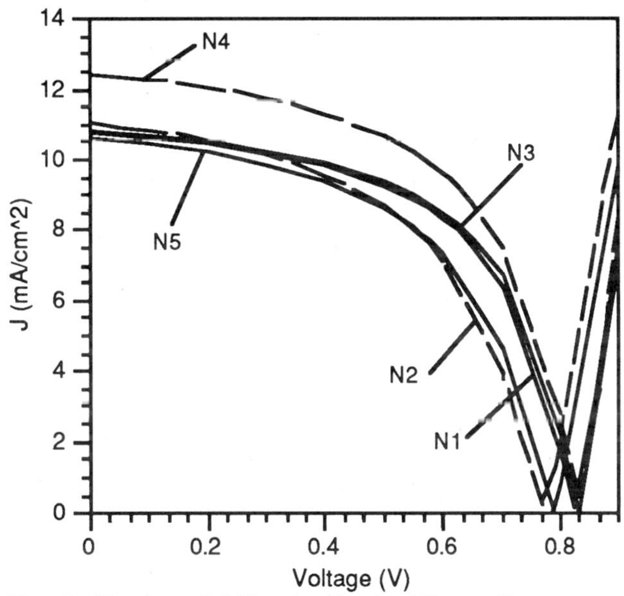

Fig. 4. Illuminated JV for the Notched Gap cell as a function of p/i interface defect density and doping. The terminal characteristics and interface parameters of these cells are listed in Table III.

Table III: Terminal characteristics and p/i interface parameters for the cells in Figs. 2-4.

Case	Eg(p/i) (eV)	Na(p/i) (cm^{-3})	Ndefects(p/i)(cm^{-3})	Jsc(mA/cm^2)	Voc (V)	FF	Eff.
No Interface	---	---	---	13.26	0.881	0.589	6.88
G1	1.85	2.2E19	1.5E18	13.06	0.867	0.586	6.64
C1	1.90	2.2E17	1.5E16	12.40	0.872	0.497	5.38
C2	1.90	2.2E17	1.5E18	12.14	0.863	0.488	5.11
C3	1.90	2.2E19	1.5E16	13.00	0.884	0.584	6.71
C4	1.90	2.2E19	1.5E19	12.88	0.849	0.526	5.76
N1	1.65	2.2E19	1.5E18	10.79	0.822	0.564	5.01
N2	1.65	2.2E17	1.5E18	11.04	0.771	0.519	4.42
N3	1.65	2.2E19	1.5E16	10.85	0.830	0.569	5.13
N4	1.65	2.2E17	1.5E16	12.42	0.833	0.556	5.76
N5	1.65	1.5E19	2.2E19	10.62	0.788	0.534	4.47

QUANTUM EFFICIENCY OF A-SI:H P-I-N SOLAR CELLS IN THE 250 TO 375 K RANGE - INSIGHTS FROM DEFECT POOL MODELLING

Thomas Eickhoff, Carsten Ulrichs, Helmut Stiebig, Werner Grünen, Wilfried Reetz, and Heribert Wagner
Forschungszentrum Jülich GmbH, ISI-PV, D-52425 Jülich, Germany

ABSTRACT

A variation of the temperature T of a p-i-n solar cell shifts the Fermi level and thereby changes the occupation of the defect densities within the device according to the defect model used. Thus the T dependence of the quantum efficiency QE promises to give a clearer insight into the defect distribution and recombination within the device. With decreasing T both the red and the blue response of an a-Si:H p-i-n solar cell decrease, the loss in the red being due to the i layer gap shift. The drop on the short wavelength side with decreasing T is attributed to a change of charge state of the D^+ in the non-equilibrium case within the first 100 nm of the i-layer. This negative space charge diminishes the field spike near the p/i interface and reduces the hole concentration there by recombination.

INTRODUCTION

Typically the reverse-bias photocurrent of a p-i-n solar cell is expected to be of primary nature and thus independent of temperature T. However, as was reported by Hoheisel et al. [1], the temperature coefficient TC of the reverse-bias photocurrent of p-i-n solar cells turns out to be positive as well for the red as for the blue part of the spectrum. We show that this behavior can be qualitatively described by the change of charged midgap states predominantly at the p/i interface if the energy position and occupation of defect states are treated according to the defect-pool (DP) model. Variation of T allows the shift of the quasi-Fermi levels within the cell thereby "internally" changing the charge state of defects independent of "external" parameters like voltage bias or irradiance level. Fixed (not Fermi-level dependent) defect peaks near midgap as considered in the standard (ST) model do not result in any TC of the simulated photocurrent. To achieve a very high reliability of the interpretation the quantum efficiency (QE) over the whole spectrum is compared with modelling results.

EXPERIMENTAL

Quantum efficiency QE(λ = 400-800 nm) measurements from 250 to 375 K have been performed on 10x10 mm^2 single cells without buffer (1.1 mm glass /~90 nm flat SnO_2:F/~13 nm p/i/~30 nm n/Ag, deposited in a MRG 3 chamber PECVD system) with different i-layer thicknesses. Substrates with flat TCO were used to assure correct optical multilayer calculation of the generation rates within the cells - avoiding light scattering [2]. The nominal i-layer thicknesses of 300, 400, 500 and 800 nm estimated from the deposition rates were checked by comparison of measured and simulated reflectance spectra and turned out to be 340 (sample A), 455 (B), 585 (C), and 875 nm (D), respectively. The samples were annealed at 180°C for 30 min prior to the measurements. The initial efficiency was about η = 6 %, the low fill factor at AM1.5 of only 55 % being due to the TCO series resistance. Additional shunt resistance contributions from the fringe region and the surrounding of the back contact were suppressed by etching the p-i-n area not coated with Ag. QE was measured without optical bias with chopped probe light, illuminating the cell through the p-layer. So the DC dark current contribution which is strongly T dependent, is suppressed using the lock-in technique.

MEASUREMENT RESULTS

Figs. 1 and 2 show the QE(T) spectra for samples B and D, respectively. Generally, with decreasing T a monotonous reduction of the response all over the spectrum can be observed. In the modelling section it will be shown that the loss in the red part is caused mainly by the shift of the optical gap of the absorbing layer. The blue light response has a similar tendency with the exception of the QE(375 K) of the thick cell D (see Fig. 2). The relative change of QE(λ = 400nm) with T (see Fig. 3) is different for the thin and the thick cell. Below room temperature the decrease of the response with decreasing T is more prominent for the thick cell. At elevated temperatures the QE of this sample shows a

CH3365-4/94/0000-0622 $4.00 © 1994 IEEE

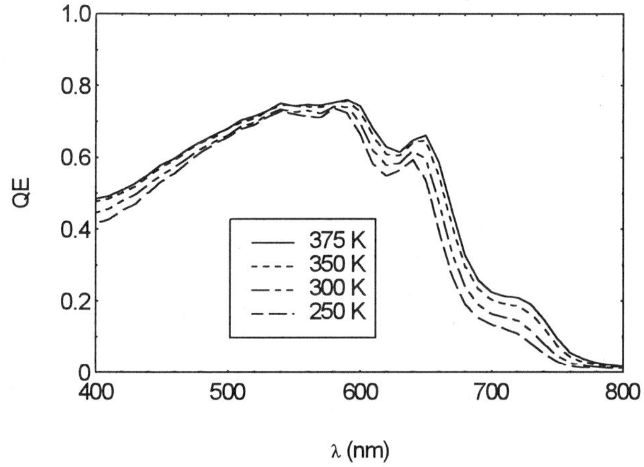

Fig. 1. Temperature dependent short circuit quantum efficiency of sample B (d_i = 455 nm).

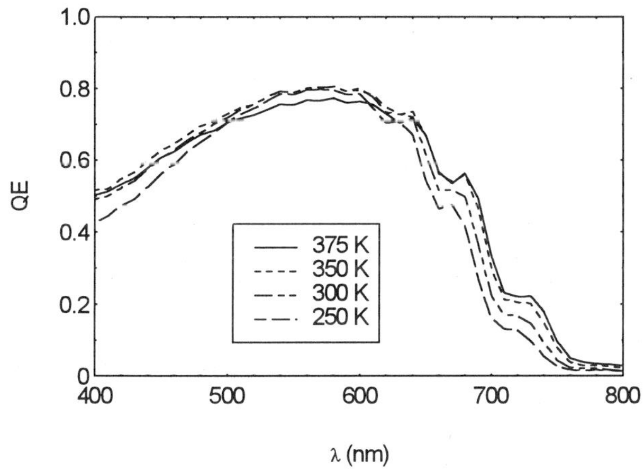

Fig. 2. Temperature dependent short circuit quantum efficiency of sample D (d_i = 875 nm).

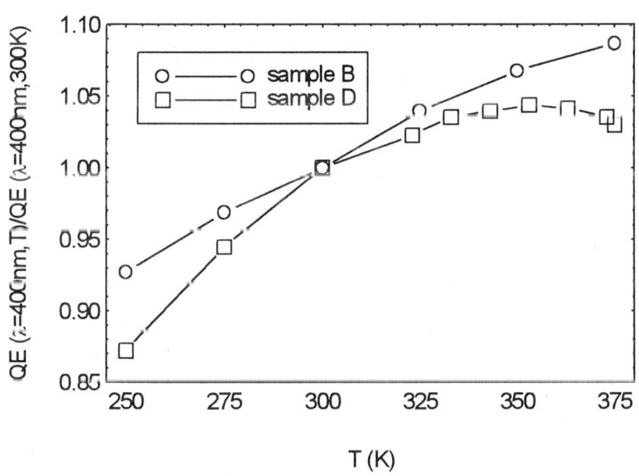

Fig. 3. Temperature dependent blue response relative to QE(300 K) of samples B (d_i = 455 nm) and D (d_i = 875 nm).

maximum at T ~ 350 K, and beyond that the blue response drops again.

From Fig. 3 one can extract that the TC of the QE(λ = 400 nm, T) is about 10^{-3}/K for sample B and $3*10^{-3}$/K for sample D, which is in good agreement with the values published for the TC of the reverse-bias photocurrent in [1]. It seems not very useful to define a TC for the red response at λ = 600 nm as the QE spectra are dominated by interference fringes in this region. A meaningful quantity for comparison is probably the photocurrent decrease with T (integration over QE(λ,T)*AM1.5 in the range λ = 400 to 800 nm) which is considerably higher in the red than in the blue part of the spectrum.

QE(λ,T) measurements have also been performed with moderate reverse-bias voltage (-0.5 V) applied to the samples. The results show that there is no 100 % carrier extraction at short-circuit condition and that a small part of the blue response decrease with lower T can be restored under reverse-bias.

MODELLING RESULTS

For the device simulation a modelling program [3] using accurate light absorption profiles and defect-pool model statistics (tested with defect absorption/CPM measurements [4] on a-Si:H films) was used. All material properties (absorption coefficients α(E), refractive indices n(E), mobility gaps E_g, defect concentrations N_d etc.) were determined from measurements on single layers. The temperature dependence of the i-layer gap was chosen according to [5]:

$$E_g(T)=E_g(300K)+0.135eV-4.5*10^{-4}eV/K*T. \quad (1)$$

The defect energy positions above VB were linearly scaled with the relative gap shift with T. A possible small variation of the band tail slopes with T [6] was neglected. The temperature dependent i-layer optical absorption was taken from [7]. As the mobility gap and the optical gap can be varied independently, the program allows to study the effect of the accurate T dependent generation rate on the QE separately. For the mobilities standard values (μ_e = 10 cm^2/Vs, μ_h = 1 cm^2/Vs) were used. The simulation program describes multiple trapping transport paths within the bands and at the band edges only and discards hopping and tunneling transport. This seems to be a valid approximation as long as T > 200 K.

Fig. 4 shows the calculated spatial variation of the quasi-Fermi levels for free electrons and holes within cell B for illuminated with different wavelengths of λ = 400 nm and λ = 600 nm. The levels are plotted in that way that the abscissa represents the valence band edge, the conduction band (CB) edge in the i-layer is approximately 1.74 eV above and parallel to VB. In this notation (reference level is VB) the equilibrium E_F is not at a constant position. In the QE measurement the incoming photon flux is 10^{13} to 10^{14} $cm^{-2}s^{-1}$ which is orders of magnitude below AM1.5 excitation. Comparing the two graphs of Fig. 4, one observes that the split of the quasi-Fermi levels is larger with lower T. For almost homogeneous generation profile in the sample (λ = 600 nm) the quasi-Fermi levels are nearly parallel to the band edges. With inhomogeneous generation (λ = 400 nm) the quasi-Fermi levels are asymmetric as the back part of the i-layer is still in the dark. The main reason for the blue response loss can now be explained according to the DP model as follows: the lower T, the larger the split of the quasi-Fermi levels which causes the large number of D^+ states (see Fig. 1 of [2]) at 1.15 eV above VB (marked at the energy axes in Fig. 4) to be charged with electrons. As in the standard (ST) defect model the defect peaks are close to midgap for an i-layer, no further charging of the defects with splitting of the quasi-Fermi levels can occur in the temperature range considered here. Indeed, the ST model does not lead to a T dependent blue response as long as the defect peaks are not shifted to non-realistic energy positions.

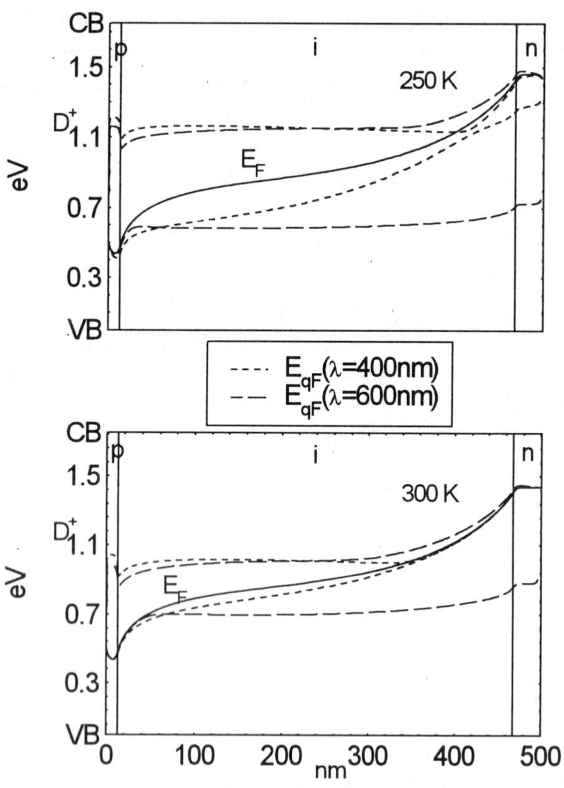

Fig. 4 Spatial variation of E_F and quasi-Fermi levels of electrons and holes in sample B. Top: 250 K, bottom: 300 K.

But in the DP model after altering their charge state, the D^+ act as more effective recombination centers for holes. The concentration of free holes within the first 30 nm of the i-layer is nearly one order of magnitude lower at 250 K than at 300 K. There is a field redistribution within the cell: the electric field becomes more asymmetric - it is lower within the first hundreds of nanometers with a reduced electric field spike at the p/i interface and higher (about 20 %) in the "dark" back part of the i-layer where it has no improving effect on the extraction of holes. Within the thicker cell the quasi-Fermi levels split more with decreasing T resulting in a larger relative drop in the blue response. In Fig. 5 the change in the electric field at λ = 400 nm illumination for both samples with T is plotted. The thick cell shows a more pronounced field spike reduction.

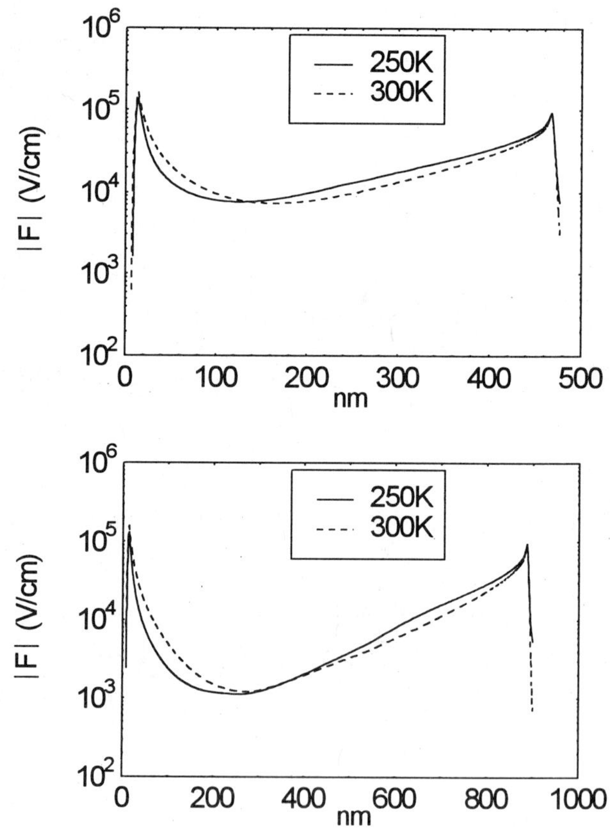

Fig. 5. Calculated electric field distribution for sample B (top) and sample D (bottom) for 250 K and 300 K (realize different abscissa).

QE(T) modelling calculations reveal qualitatively the behavior measured. As the exact TC of QE(T) is critically depending on the defect densities particularly at the interfaces no absolute comparisons of measurents and fits are presented here. But the general trend is observed in the simulations as well: with lower T the blue response loss is larger for thicker cells, at T > 325 K there is a "saturation" of QE(λ = 400 nm). At higher T the downshift of the gap of the p-layer which is not taken into account here, will probably reduce the QE again. A precise determination of the T dependence of the optical constants of all layers and their inclusion into the multilayer optics calculation of the solar cell is a prerequisite to draw final conclusions.

At λ = 600 nm the electric field profile (not shown here) is nearly symmetrical. The field redistribution with decreasing T favors the extraction of carriers generated in the main part of the i-layer and worsens the extraction of carriers at the p/i and i/n interfaces. The net effect on the "red" response is nearly zero - with the exception that less carriers are overall generated within an i-layer with wider (T dependent) optical gap.

Fig. 6 compares QE(T) simulations for sample B considering the mobility gap shift but with and without taking the optical gap shift into account. The plot clearly demonstrates that the main loss in the red response with decreasing T is due to lower absorption and not caused by a reduction of thermal emission of trapped carriers into the bands.

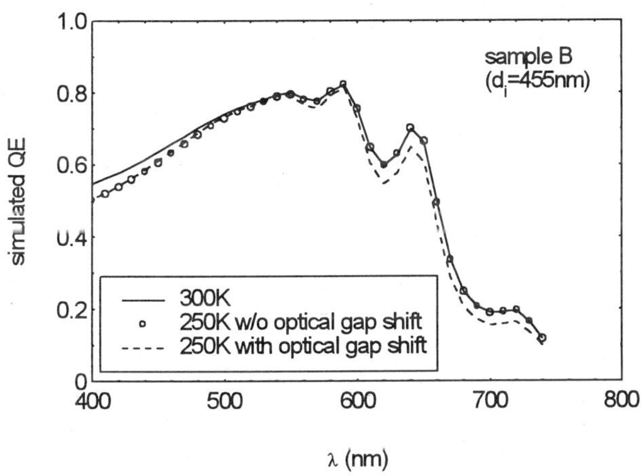

Fig. 6. Comparison of simulated QE(T) for sample B. The decrease in red response is mainly due to optical gap shift.

CONCLUSIONS

The measured and modelled quantum efficiency of a p-i-n solar cell has been found to be temperature dependent with a positive temperature coefficient. The physical reason for the change in the blue response is a change of the charge of the defect states (D^+) in the first 100 nm of the i-layer. The T dependent variation of the red response can be nearly attributed to a reduced absorption coefficient caused by the optical gap shift. In particular the reduction of the blue response with decreasing T cannot be described by the standard defect model.

ACKNOWLEDGEMENTS

Financial support by the Bundesministerium für Forschung und Technologie is gratefully acknowledged.

REFERENCES

[1] M. Hoheisel, E. Reichle, H. Harms, J. Kotschy, "Temperature Dependence of the Photocurrent in pin and nip Solar Cell Structures made from a-Si:H", *J. Non-Cryst. Solids* **137&138**, 1991, pp. 1181-1184.

[2] H. Stiebig, A. Kreisel, M. Meer, K. Winz, N. Schultz, C. Beneking, Th. Eickhoff, H. Wagner, "Spectral Response Modelling of a-Si:H Solar Cells Using Accurate Light Absorption Profiles", *this volume.*

[3] H. Stiebig, A. Kreisel, J.-L. Nicque, Th. Eickhoff, C. Beneking, H. Wagner, "Experimental and Numerical Study of a-Si:H Solar Cells", *Proc. 12th European Photovoltaic Solar Energy Conf.*, 1994, pp. 164-167.

[4] F. Siebke, H. Stiebig, A. Abo-Arais, H. Wagner, "Charged and Neutral Defect States in a-Si:H determined from improved Analysis of the Constant Photocurrent Method", *this volume.*

[5] H. Overhof, P. Thomas, *"Electronic Transport in Hydrogenated Amorphous Semiconductors"*, Berlin 1989, p. 133.

[6] M. Stutzmann, "A comment on thermal defect creation in hydrogenated amorphous silicon", *Phil. Mag. Lett.* **66**, 1992, pp. 147-150.

[7] G. Weiser, H. Mell, "Temperature Dependence of the Optical Absorption Edge in a-Si:H", *J. Non Cryst. Solids* **114**, 1989, pp. 298-300.

PROGRESS IN INEXPENSIVE a-Si:H/a-Si:H TANDEM MODULE TECHNOLOGY

R.E.I. Schropp, M.B. von der Linden, J. Wallinga, D. Knoesen*
Utrecht University, Debye Institute, Dept. of Atomic and Interface Physics,
P.O. Box 80,000, 3508 TA Utrecht, The Netherlands
*) on leave from University Western Cape, South Africa

J. Hyvärinen, J. Skarp, and T. Suntola,
Microchemistry Ltd., P.O. Box 45, Espoo 02151, Finland

J.A. Willemen, M. Zeman, and J.W. Metselaar,
Delft University of Technology, Dept. of Electrical Engineering, Mekelweg 4, 2628 CD Delft, The Netherlands

W. Loyer, D. Guillardeau, and E. Fabre
NAPS France, Z.I.Nord - Route de la Bassée, B.P. 66, F-62302 LENS CEDEX, France

ABSTRACT

This paper describes a research and technology development project within the framework of the E.U. JOULE II program in order to implement a-Si:H/a-Si:H tandem structures developed in the laboratory in the manufacturing procedure for large area integrated modules at NAPS France. In less than 20 test runs utilizing a stacked a-Si:H/a-Si:H tandem structure we have obtained current-matched PV panels (30 cm × 90 cm) with an initial power output equal to conventional single junction panels. Stability tests performed outdoor showed that tandem modules are indeed more stable than their single junction counterparts. This is accomplished without adding production steps with respect to single junction module production. Therefore the tandem fabrication concept will lead to a drastic reduction in the cost per peak Watt, even at small production levels.

INTRODUCTION

Amorphous silicon thin film technology can take a large market share if modules can be produced with a conversion efficiency of 8 % at a variable cost (direct manpower plus materials) of 1 Ecu/Wp [1]. Presently, the variable cost at NAPS France is close to 2 Ecu/Wp. This cost figure is based on a guaranteed stabilized aperture area conversion efficiency of 4.5 % for the modules as an average over the total production.

Laboratories around the world have accomplished higher stabilized efficiencies by developing multijunction cells based on amorphous silicon. With partial financial support from the E.U. JOULE II program, four groups have been put in a position to continue the effort to implement the 2-terminal tandem device concept in the production environment. These groups and their main tasks are:

- **Utrecht University, Debye Institute**: development of thin film tandem solar cells; optical, electronical, structural and chemical analysis of materials and devices, comparative studies of materials and diagnostic devices from the production line and from the laboratory; overall coordination of the project.
- **Delft University of Technology**: determination of the structural and electronic parameters and identification of the loss mechanisms in single and tandem solar cells by numerical simulation.
- **Microchemistry Ltd.**: development of the plasma equipment and the process for the production of 30 cm × 90 cm panels and development of the other process steps needed for demonstration of integrated 30 cm × 30 cm tandem plates.
- **NAPS France:** implementation of technical modifications in production to enhance the performance of new tandem structures and provide statistical data on the effectiveness of each innovation.

The main objective of this project is to demonstrate *in the production equipment* 8 % efficiency on a 30 cm × 30 cm panel using a 2-terminal tandem device structure at a variable cost level of 1 Ecu/Wp.

RESEARCH APPROACH

In a previous paper [2] we have evaluated the advantages and disadvantages of all possible configurations for thin film tandem solar cells. We decided to pursue the 2-terminal approach, because it is the lowest cost solution and because it has great potential if spectrum splitting

First WCPEC; Dec. 5–9, 1994; Hawaii

components are used in the near future. As a consequence we took on the challenge of solving the technological problems related to uniformity and n/p middle junction optimization. Specifically in a-Si:H/a-Si:H tandem junctions, an additional research issue is in dealing with effects of the rough TCO surface morphology on the performance of the very thin top junction.

RESEARCH ISSUES

In order to address the real bottlenecks in large area a-Si:H/a-Si:H tandem production, we have identified and initiated four main research activities.

Front Electrode Technology

In order for two-terminal structures to be stable it was found that the second junction should have an absorber thickness of less than 400 nm. Under the prevailing spectral light scattering conditions, this leads to the requirement of a very thin ($\approx$ 60 nm) first junction in a-Si:H/a-Si:H tandem cells. Without optimization of the morphology of the TCO surface, it appeared that partial shunting of the first junction is likely if the intrinsic layer thickness is less than 65 nm. These shunting defects can be very clearly observed by cross sectional Transmission Electron Microscopy (TEM) in the narrow valleys between surface facets of neighbouring grains. Two different approaches are used to make thinner cells possible: (i) Modify the TCO morphology to meet specific requirements for stacked tandem cells. NAPS France is the only PV module manufacturer in Europe who has the capability of producing low-cost TCO by the Atmospheric Pressure Chemical Vapor Deposition (APCVD) method. (ii) Adapt the multilayer structure of the first junction in order to prevent the shunting pathways.

n/p Tunnel/Recombination Junction (TRJ)

The TRJ should form a low ohmic contact between the two components of the tandem structure. This implies that the TRJ should have a sufficiently high density of recombination centers in order to let charge carriers recombine effectively via a tunneling process. In a heavily defective region at the n/p interface the band bending will be strong enough so as to allow tunneling assisted recombination. On the other hand, an insufficient density of recombination centers might lead to a reverse parasitic voltage due to accumulation of charge from excess photogenerated carriers. At the same time optical losses should be limited by using doped layers which are very thin. However, the doped layers should not be made too thin since this could lead to depletion and a loss in the built-in voltage in the component cells.

So far we have restricted ourselves to amorphous n^+- and p^+-type layers and we minimized the thicknesses in order to limit the optical losses. It appeared that 25 nm is needed to avoid n-layer depletion and that an n/p interface treatment can improve the ohmic behaviour of the TRJ.

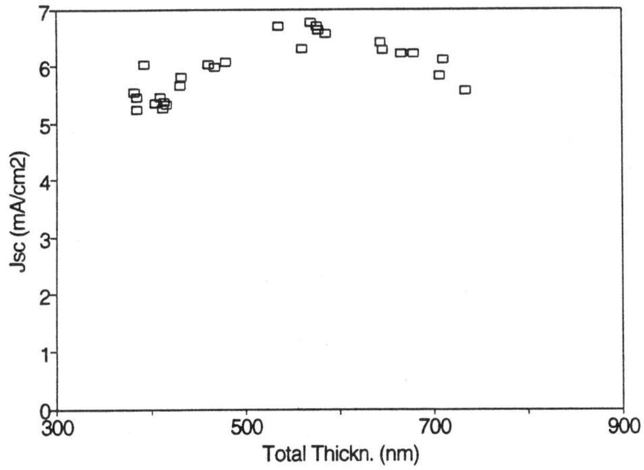

Figure 1: Dependence of the short circuit current density of a-Si:H/a-Si:H tandem cells on the total thickness of the structure, at a constant component thickness ratio.

Uniformity and Formation of Graded Layers

In a-Si:H/a-Si:H tandem modules uniformity of the layers is crucial because (i) the current generated in each tandem subcell must be equal, and (ii) the current generated in each of the junctions of the tandem must be equal everywhere on the panel. Although a overall thickness profile would proportionally affect the thicknesses of both junctions and would thus leave the thickness ratio unaltered, the importance of the latter point should not be underestimated. In Figure 1, a series of solar cell performance data is shown for tandem cells with varying total thickness, but with a constant ratio of the absorber layer thicknesses. It can be seen that the short circuit current density shows an optimum at one particular total thickness. This implies that the optimum thickness ratio is dependent on the total thickness of the structure. In order to improve the thickness uniformity of the layers on 30 cm $\times$ 90 cm substrates, an experimental reactor is available at Microchemistry in which various designs for the substrate holders are tested.

In this project, the carefully designed graded layers developed in the laboratory in a UHV system on 10 cm $\times$ 10 cm substrates are transferred to the large area (30 cm $\times$ 90 cm) production reactor. The deposition of very thin

Table 1: Effect of modifications to the p^+/i region in a single junction cell on 30 cm $\times$ 90 cm.

Modification	V_{oc} (V)	J_{sc} (mA/cm^2)	ff -	η (%)
a) basic structure	0.702	14.1	0.63	6.2
b) wider gap p-layer	0.719	14.9	0.63	6.8
c) p/i buffer + b)	0.779	15.0	0.66	7.7
d) further optimization	0.854	15.6	0.63	8.4

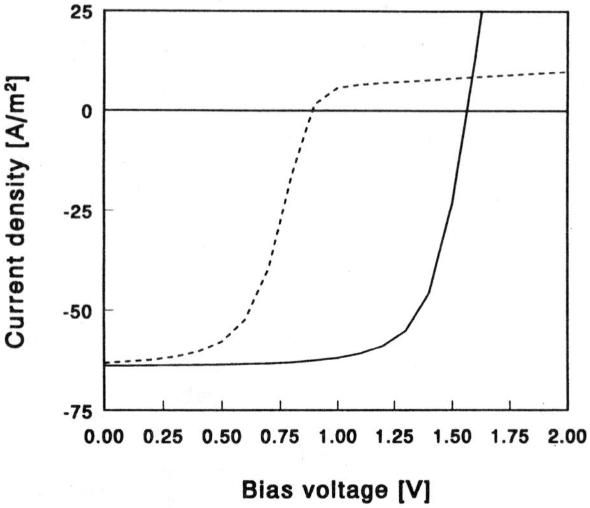

Figure 2: Simulated I-V characteristics for a $p^{\pm}$i-n^+/$p^{\pm}$i-n^+ a-Si:H/a-Si:H tandem cell. The dashed curve is without any recombination enhancing features, the continuous curve includes a model to account for tunneling effects and a carrier supply mechanism to recombination centers at the n/p junction.

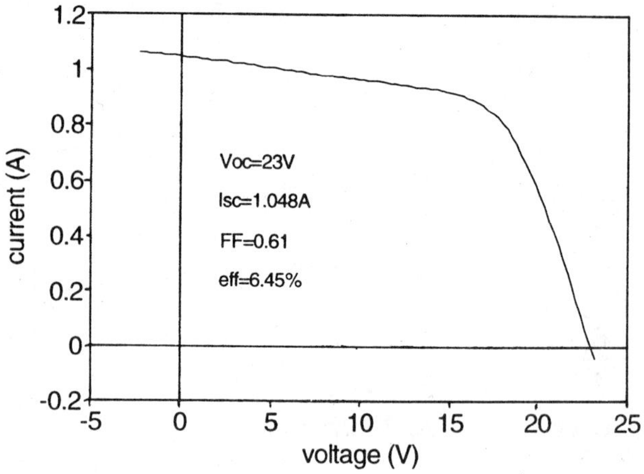

Figure 3: I-V curve of an integrated 30 cm × 90 cm a-Si:H/a-Si:H tandem panel. The panel contains 15 subcells.

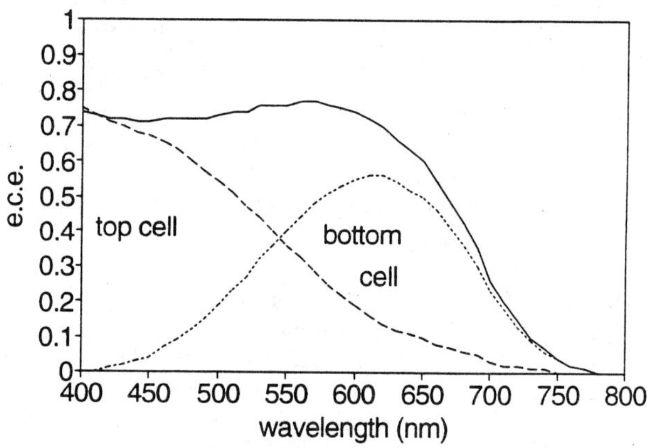

Figure 4: External collection efficiency of the two components of the tandem structure and the sum of both contributions. The top cell is measured by using red bias illumination and the bottom cell by using blue bias illumination. The forward voltage bias for both measurements was 0.6 V.

composition-graded layers uniformly over a large area is not trivial, as pumping speed, background doping and flow geometry are entirely different in research and production systems. Nevertheless, as seen in Table 1, the effect of a functional p^+/i buffer layer is clearly observable on a 30 cm × 90 cm substrate. The effect on individual solar cell parameters is similar to what is observed for cells deposited in a small area research reactor.

Computer Modeling and Identification of the Losses

The emphasis in computer modeling is currently in three areas: (i) Matching of the photogenerated currents from the two components. This is a delicate matter, since it depends on the properties of incorporated materials, the total thickness of the structure, the specific spectral scattering properties of the TCO substrate, and on the content on the incident light spectra. The complete modeling of a-Si:H/a-Si:H tandem cells on textured TCO now includes both the simulation of electrical performance of the structure (ASA software package) and the calculation of the generation profile taking into account the light trapping effects. (ii) Using the recent theory of defect formation in amorphous silicon, known as the defect pool concept, the dependence of the total concentration of defects as a function of Fermi level position is implemented in the solar cell simulator. (iii) Modeling of a tandem cell requires a good description of the physical processes in the n/p tunnel/recombination junction (TRJ). The currently adopted approach includes two aspects: 1) the high recombination rate in the TRJ is modeled using a trap assisted tunneling model. In this model the concentrations of charge carriers involved in the recombination process in the high field region are enlarged, to include the contribution of tunneling carriers; 2) the actual tunneling transport towards the recombination centers is modeled by an increased *effective* extended state mobility [3]. Fig. 2 shows simulated I-V curves for a $p^{\pm}$i-n^+/$p^{\pm}$i-n^+ tandem cell under study in this project.

4. RESULTS

Using a stacked a-Si:H/a-Si:H tandem structure, we have obtained current-matched PV panels (30 cm × 90 cm) with an initial active area conversion efficiency of 6.2 - 6.5 %. The I-V curve is shown in Fig. 3. The matching of the currents has been experimentally verified by calculating the individual current densities from each component from spectral response measurements (see Fig. 4).

The initial power output of the tandem PV panel is

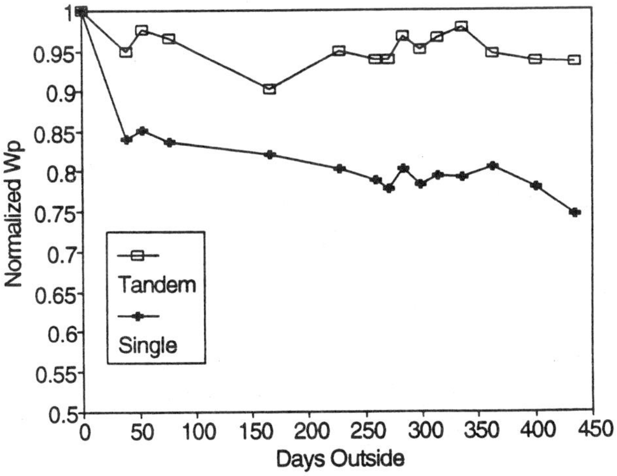

Figure 5: Outdoor stability test showing the changes in average performance of four tandem modules and of two reference single junction modules.

comparable to the initial power obtained in conventional single junction panels made with the same equipment. This is achieved in spite of additional optical absorption losses that exist in same bandgap tandem panels with respect to single junction panels due to the presence of the TRJ. This is possible because the interconnect pattern was adapted such that the width of the strip tandem subcells was increased. In this way the area loss was reduced by 50 %. The present tandem module contains 15 subcells.

The first tandem panels have been subjected to outdoor tests simultaneously with conventional single junction modules in order to evaluate light-induced and environmental degradation. A prolonged outdoor test (more than 400 days) revealed that these modules degrade by less than 7 % whereas the reference single junction modules degrade by 25 % within the same time frame.

COST-EFFECTIVENESS

An important aspect of this preliminary development is that there are essentially no additional production steps with respect to single junction modules to fabricate these tandem modules. The same glass substrates, the same front and back electrodes, the same encapsulation and the same frames were used for the tandem modules. Just as in single junction structures, the a-Si:H multilayer structure requires only one load/unload step in one reaction chamber and it is deposited merely by varying the gas composition versus time. The extra cost due to longer run times is marginal since this only involves machine time, not manpower. The extra run time in addition to the formation of typical single junction structures is typically 12 minutes, which constitutes a process time increase of approximately 10 % compared to the single junction process. Taking into account the expected stabilized efficiency improvement, the anticipated cost figures are listed in Table 2. Note that these

Table 2: Current and projected cost per peak Watt.

Production level 1.2 MW/yr	1993 Single junction 4.5 %	1996 Dual junction 8 %
Variable cost (matls + manpower)	2.0 Ecu/Wp	1.0 Ecu/Wp
Equipment depreciation	0.25	0.15
Indirect cost (utility)	0.25	0.15
Total	2.5 Ecu/Wp	1.30 Ecu/Wp

figures are obtained for a low production volume and that they will decrease significantly for larger sized plants.

6. CONCLUSION

In conclusion, we have fabricated large area (30 cm × 90 cm) a-Si:H/a-Si:H tandem modules using a processing sequence which is — apart from the gas composition sequence during a-Si:H deposition — essentially identical to that for single junction a-Si:H modules. Only amorphous semiconductor layers were used. The optical loss in the n/p tunnel/recombination junction could be compensated by changing the interconnect pattern, so that the initial output power is comparable to that of single junction panels.

A prolonged outdoor test (more than 400 days) revealed that these modules degrade by less than 7 % whereas the reference single junction modules degrade by 25 % within the same time frame. The low cost fabrication method in combination with enhanced stabilized efficiencies will lead to a sharp reduction in the cost per peak Watt, even at small production levels.

ACKNOWLEDGEMENTS

The preliminary studies until January 1st, 1994 were supported by the Netherlands Agency for Energy and Environment (NOVEM). From January 1st, 1994 the effort is supported by the E.U. programme JOULE II, contract number JOU2-CT93 0419.

REFERENCES

[1] Exchange rate 25 Nov. 1994: 1 Ecu = 1.22 US$.

[2] R.E.I. Schropp, D. Leone, J Hyvärinen, J. Skarp, T. Suntola, J.W. Metselaar, M. Zeman, D. Guillardeau, W. Loyer, and E. Fabre, "Cost-effective Large area Integrated Multijunction Amorphous silicon solar Cell Structures (CLIMAX)", *12th International E.C. Photovoltaic Solar Energy Conference 1994* (H.S. Stephens and Associates, 1994) pp.1253-1256.

[3] J.A. Willemen, M. Zeman, J.W. Metselaar, "Computer Modeling of Amorphous Silicon Tandem Cells", *This conference.*

PROPERTIES OF ZnO:Al THIN FILMS FOR SOLAR CELLS BY PYROSOL METHOD

Jinsoo Song, Woo-Yeong Cho*, Koeng-Su Lim*, Kyung-Hoon Yoon and Gwon-Jong Yu

Korea Institute of Energy Research,71-2, Jang-dong, Yusong-gu, Taejon 305-343, Korea
*Department of Electrical Engineering, Korea Advanced Institute of Science and Technology,
Kusong-dong, Yusong-gu, Taejon 305-701, Korea

Abstract

Undoped and aluminum-doped zinc oxide(ZnO:Al) thin films have been prepared on heated Corning 7059 glass substrates by pyrosol method under various deposition conditions. The effects of the different deposition variables and annealing in vacuum of ZnO thin films have been investigated by analyzing the electrical, optical, and compositional changes of the films. The results indicate that it is possible to obtain ZnO:Al thin films with a transmittance at about 80% and a resistivity as low as 3.5×10^{-3} Ωcm. Furthermore, it was also found that the films are stable in hydrogen plasma (within the exposure time of 10 minutes).

1. Introduction

TCO(Transparent Conducting Oxide) must have not only high transparency and conductivity but also durability of high process temperature and hydrogen plasma atmosphere. ITO(In_2O_3:Sn) and FTO(SnO_2:F) thin films are most frequently used as transparent oxide electrode of thin film solar cells due to their excellent electro-optical properties. However reduction of In, Sn elements in hydrogen plasma atmosphere results to deteriorate their electrical, optical properties and it is diffused into the a-Si, then the performance of a-Si solar cells are affected[1][2]. Recently ZnO thin films which have c-axis oriented hexagonal wurzite structure[3], have been attracted as a new transparent and conductive material due to its excellent stability in hydrogen plasma atmosphere as well as its good electro-optical properties prepared by sputtering[4] and MOCVD[5]. In this paper, we select pyrosol method which is a simple, easy to applicate large area solar cells and lower the production cost than the other methods for ZnO thin films deposition Optimum deposition conditions of ZnO thin films are obtained by analyzing electro-optical properties, and crystallite structure of the obtained films.

2. Experimentals

The pyrosol process is a kind of spray pyrolysis that the aerosol is produced by ultrasonic vibration instead of pneumatic means[6]. In the pyrosol method, carrier gas, substrate temperature, flow rate of the carrier gas, starting solution concentration and solvent can be experimental variables. And the air or nitrogen was used as the carrier gases with the flow rate of 10 l/min and the starting solution mixture was consisted of 0.4M solution of zinc acetate in water or the mixture of methanol and water (3:1). The starting solution was sprayed onto the Corning 7059 (50×50×1.1 mm) glass substrate with the substrate temperature of 375°C ~ 475° C. Four-point probe measurement and Hall measurement by using Van der Pauw method were employed to measure the electrical resistivity, mobility and carrier concentration. Specular transmittance in the wavelength range of 300 ~ 800 nm was measured by using UV-VIS spectrophotometer (Varian Superscan IBE) and the specular transmittance at 550 nm was used as the optical transmittance in this works.The structural properties were studied with XRD(Philips PW 1340) spectra.

3. Results and Discussion

Fig. 1 shows the effect of the solvent on the resistivity and transmittance of undoped ZnO films with variation of the substrate temperature from 350°C to 475 °C. In using the water as the solvent, there is not any significant changes of the resistivity with increasing substrate temperature up to 425°C and then falls rapidly to the value of 50 Ωcm at above 450°C, whereas optical transmittance increases continuously from 78% to 85% with increasing substrate temperature.

The lower resistivity is due to more complete decomposition of zinc acetate on the substrate in the higher substrate temperature and causes enhancement of grain size and uniformity of deposited films with substrate temperature. On the contrary, for the case of using the methanol and water mixture solvent, its

CH3365-4/94/0000-0630 $4.00 © 1994 IEEE

resistivity was 10 times less than that of the water solvent at 425°C substrate temperature. The methanol and water mixture solvent is seem to have less excess oxygen chemisoprtion in the ZnO thin films than the water only solvent and this result looks like to decrease the resistivity of ZnO thin films in the case of using the water only solvent with increasing substrate temperature. But the difference of optical transmittance between water solvent and mixture solvent is due to difference of film thickness (i.e. different deposition rate). Consequently, the methanol and water mixture is suitable for zinc acetate solvent because of its high solubility and controllability of resistivity.

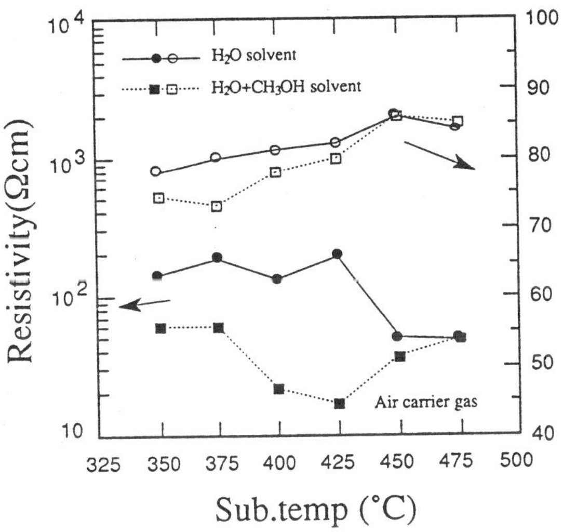

Fig.1 Variation of resistivity and transmittance of ZnO thin films as a function of solvent and substrate temperature.

Fig. 2 represents electro-optical properties dependence on carrier gases with substrate temperature.

In the case of using air carrier gas, resistivity was independent of substrate temperature in the range of 10 Ωcm ~ 80Ωcm. Optical transmittance was increased with increasing substrate temperature from 75% to 85%, but in using nitrogen carrier gas, resistivity was strongly influenced by substrate temperature. Resistivity was about 10^{-1} Ωcm up to 425°C, and a little improvement of the optical transmittance was observed. These results come from using air carrier gas which has much oxygen contents and causes oxygen chemisorption on the surface or grain boundary than using nitrogen carrier gas and this oxygen chemisorption leads to high resistivity[7][8]. And optical transmittance was increased due to its reduced thickness with decreased deposition rate. On the other hand, by using nitrogen carrier gas, oxygen contents of deposited ZnO thin films are depend on only oxygen in the solvent and excess oxygen chemisorption is occurred less in ZnO thin films with increasing substrate temperature and then resistivity was decreased with increasing substrate temperature.

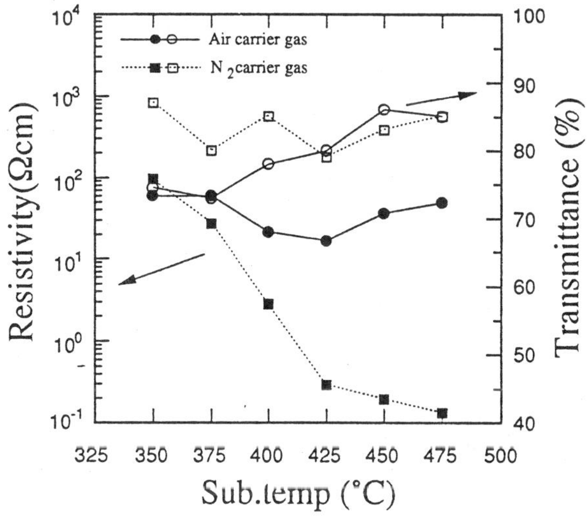

Fig.2 Variation of resistivity and transmittance of ZnO thin films as a function of substrate temperature.

No evident variation of optical transmittance is considered at nearly constant deposition rate. Specially in the case of using nitrogen carrier gas, low resistivity and high transmittance was obtained above substrate temperature of 425°C. From these results, nitrogen is considered more effective for carrier gas than air. Aluminum doping which was introduced by adding $AlCl_3$ into the starting solution with Al/Zn mole ratios of 1 ~ 4 at %. Fig. 3 shows that the resistivity of ZnO:Al thin films decreased by one order of magnitude with improvement of Hall mobility and carrier concentration and the optical transmittance reached from 89% to 93% compare with transmittance about 80% of undoped ZnO thin films. This result can be explained by low deposition rate by adding Al which affect on reaction between Zn and O.

In the XRD spectra of ZnO:Al thin films as shown in Fig. 4, (002) peak was more dominant than the other peaks and c-axis orientation was enhanced with the Al doping concentration increasing. Above 2 at % of Al/Zn mole ratio, no significant changes were observed in resistivities. But high doping ratio such as 4 at % of Al/Zn, Hall mobility decreasement and disordering of crystallite structure were occurred by impurity scattering. Thus, appropriate selection of doping concentration may be demanded. In this works, with considering of proper deposition rate and sufficient doping effect, 2 at % of Al/Zn doping condition with deposition rate of 120 /min, resistivity of 10^{-2} Ωcm, optical transmittance of 80% and less impurity scattering were selected as a compromised doping condition.

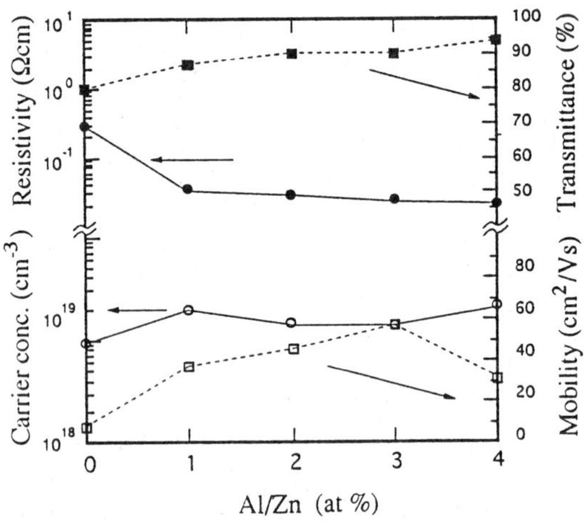

Fig.3 Electro-optical properties of ZnO:Al thin films as a functions of Al/Zn mole ratios

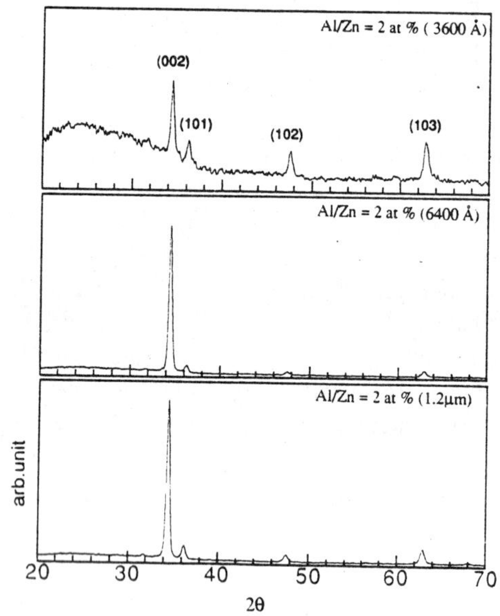

Fig.4 Variation of XRD spectra as a function of thickness of ZnO:Al thin films.

The dependence of resistivity on Al/Zn mole ratios after annealing in 10^{-4} torr vacuum at 200°C for 1 hour, are shown in fig. 5. This annealing procedure is known to reduce the resistivity of ZnO thin films[9][10]. In the case of undoped ZnO, resistivities were changed from 3×10^{-1} Ωcm to 9×10^{-2} Ωcm after annealing and also the ZnO:Al thin films had nearly the same order of magnitude of resistivities decreasement with changes of Al/Zn mole ratio. Fig.5 shows the dependence of resistivity on film thickness with doping of 2 at % Al/Zn. For the case of film thickness of 3600, resistivity changes from 3×10^{-2} Ωcm to 9×10^{-3} Ωcm after annealing. But resistivity decreasement was not observed after annealing above 6000 film thickness. For the case of undoped ZnO and ZnO:Al films with thickness less than 4000 revealed nearly the same order of magnitude of resistivity decreasement whereas ZnO:Al films with thickness more than 6000, no significant change of resistivities. These results may be considered as the electrical resistivity decreasement is not expected by crystallite structure enhancement or reactivation of dopant by low annealing temperature of 200°C ,which is much lower than that of deposition, for a rather thick films. But for the case of a rather thin films, resistivity dependence on the surface state and the grain boundary near the surface may be more remarkable than that of thick films.

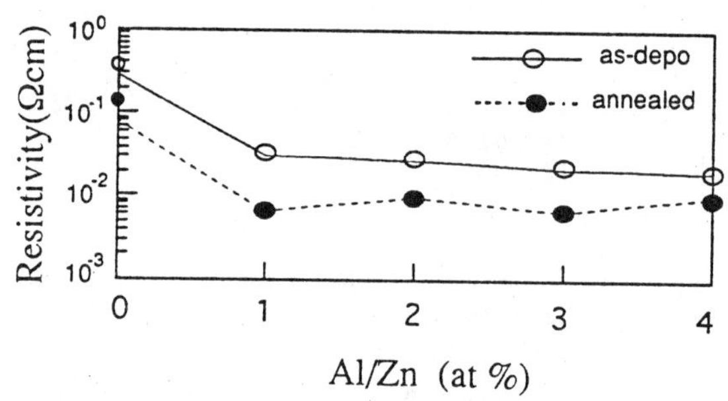

Fig.5 Dependence of resistivity on Al/Zn mole ratios.

Therefore this thickness dependent resistivity reductions are inferred as desoprtion of excess oxygens at the surface and/or the grain boundary beneath the surface, not at grain boundary in the deep from the surface. And these desorptions of excess oxygens which trap free electrons result to increasing the free electrons and reduce electrical resistivity.

Hydrogen plasma durability was tested by using PECVD with hydrogen plasma exposure time(see Fig. 6). Electrical resistivity was a little bit decreased after 3 minutes and then it was returned to the initial value of 3×10^{-3} Ωcm after 10 minutes. Optical transmittance was linearly decreased from 81% to 79% after 10 minutes.

Resistivities which had initially a little different

resistivity values, did not reveal the difference between before and after annealing after 3 minutes exposure. These results may be explained as the desorption of oxygens at the surface and/or the grain boundary beneath the surface by hydrogen plasma and this has nearly the same effect of annealing in a vacuum. But no significant changes were investigated in electrical and optical properties.

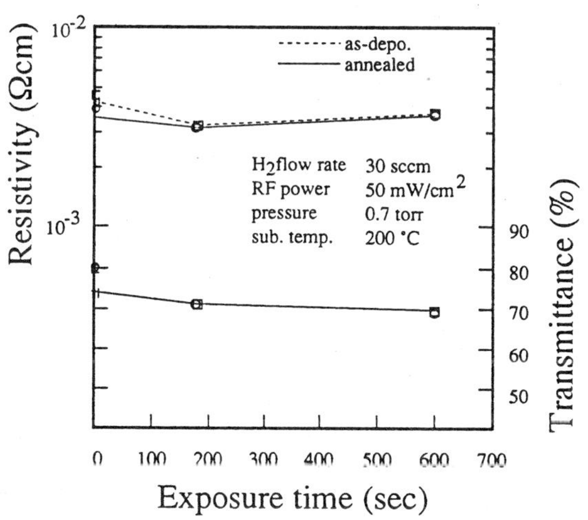

Fig. 6 Variation of resistivity and transmittance of ZnO:Al thin films as a function of hydrogen plasma exposure times.

Therefore ZnO:Al thin film was proved as having a good hydrogen plasma durability than ITO and FTO thin films.

4. Conclusion

It has been revealed that the electrical and optical properties of ZnO thin films prepared by pyrosol method are affected largely by the different deposition variables. The electrical resistivity of the films obtained with methanol+water mixture solvent and nitrogen as a carrier gas always leads to much lower value whereas the optical transmittance exhibits slight changes. In contrast, the poor electrical conductivity of the films obtained with the water solvent and air carrier gas apparently caused by the oxygen chemisorption on the surface or grain boundary in the excess oxygen atmosphere. Higher substrate temperature and post-deposition annealing in a vacuum also leads to substantial reduction in electrical resistivity and cause enhancement of grain size and uniformity of deposited films with the strong (002) preferred orientation of crystallite. The optimum deposition conditions, which yield the lowest electrical resistivity of 4×10^{-3} Ωcm and the optical transmittance of about 80% for 0.6mm thick ZnO:Al films with methanol+water(3:1) mixture solvent and nitrogen carrier gas, are substrate temperature of 425°C, Al/Zn doping ratio of 2 at % and post-deposition annealing at 200°C for 1 hour in a vacuum.

Furthermore, it has been found that there are not any significant changes of electrical and optical properties in hydrogen plasma. Al-doped ZnO thin films by pyrosol method will be very promising materials for photovoltaic application as low-cost TCO particularly in thin film solar cells.

Reference

[1] N. Fukuda, T. Imura, A. Hiraki, Y. Tawada, K. Tsuge, H. Okamoto and Y. Hamakawa, *Jpn. J. Appl. Phys.*, **21** (1982)271

[2] M. Kitagawa, K. Mori, S. Ishimura, M. Ohno, T. Hirado, Y. Yoshioka and S, Kohiki, *J. Appl. Phys.*, **54** (1983)3269

[3] Sorab K. Ghandhi and Rpbert J. Field, *Appl. Phys. Lett.*, **37**(5)449

[4] H. Nanto, T. Minami, S. Shooji and S. Taketa, *J. Appl. Phys.*, **54**(4)(1984)1029

[5] Wilson W. Wenas, A. Yamada and K. Takahashi, *J. Appl. Phys.*, **70**(11)(1991)7119

[6] J. -F. Guillemoles, D. Lincot, P. Cowache and J. Vedel, *10th European Photovoltaic Solar Energy Conference*, (1991)609

[7] A. P. Roth and D. F. Williams, *J. Appl. Phys.*, **52**(11)(1981)6685

[8] M. N. Islam, M.O. Hakim and H. Rahman, *Journal of Material Science* **22**(1987)1379

[9] J. Aranorich, A. Oritz and R. H.Bude, *J. Vac. Sci. Technol.*, **16**(1979)994

[10] R. P. Eischens, W. A. Pliskin and J. D. Low, *J. Catal.*, **1**(1962)180

CHARGED AND NEUTRAL DEFECTS IN LIGHT-SOAKED a-Si:H

Minoru Kumeda, Jiang-Huai Zhou and Tatsuo Shimizu
Faculty of Technology, Kanazawa University, Kanazawa 920, Japan.

ABSTRACT

The defect density measured by electron spin resonance (ESR) is increased by 50 times by light soaking, while that measured by the constant photocurrent method (CPM) is increased by only 4 times in N-doped a-Si:H. This large discrepancy between the CPM and ESR results is attributed to the presence of charged defects which are detected by the CPM but not by ESR. Light-induced ESR is thought to be able to detect the charged defects, but quantitative analysis is difficult because whether an increase or decrease of the neutral defect density is observed under illumination depends on the ratio of charged and neutral defect densities, light intensity and the cross-section for charge trapping.

INTRODUCTION

High-quality a-Si:H films with low defect density and long-time stability are required for application in solar cells. The main hindrance for the long-time stability is the light-induced creation of neutral dangling bonds (Staebler-Wronski effect[1]). Although ESR is the most effective technique for characterizing neutral dangling bonds, the constant photocurrent method (CPM) is also used as a simple and convenient means for estimating the density and energy distribution of defects. In some cases CPM results exhibit a good agreement with the ESR results[2], and the CPM is widely used for investigating the light-induced defect creation process. However, the CPM is considered to detect not only neutral dangling bonds D^0, which are observed by dark ESR, but also negatively charged dangling bonds D^-, which can be estimated by light-induced ESR (LESR). Therefore, CPM results can be largely different from ESR results[3] if charged and neutral dangling bonds react to light soaking in different manners. LESR is considered to detect not only D^- but also positively charged dangling bonds D^+. So far D^0, D^+ and D^- have been estimated by combining the results of ESR, CPM and LESR for annealed a-Si:H and its alloys with C, N and O[4]. But whether a similar estimation is also valid for light-soaked a-Si:H is not known.

In the present work, we compare the results of CPM with those of ESR for undoped and N-doped a-Si:H in both the annealed and light-soaked states. We also carried out LESR measurements for undoped and N-doped a-Si:H over a wide range of light intensities and in both the annealed and light-soaked states.

EXPERIMENTAL

Undoped and slightly N-doped (~2 at.%) a-Si:H films were prepared in a hot-wall glow-discharge (HWGD) system which can largely reduce the densities of unintentional impurities [5] and in a convensional glow-discharge (GD) system, at a substrate temperature of 300 ^{o}C. The film thicknesses were about 3 μm and 4 μm for HWGD and GD samples, respectively.

The annealed state was attained by annealing the sample at 200 ^{o}C in flowing N_2 gas for 5 h, and the light-soaked (LS) state was attained by light-soaking the sample at room temperature using a Xe lamp through an IR-cut filter (3.8 W/cm^2).

The CPM measurements were performed using a halogen-tungsten lamp coupled with a monochromator and interference filters. The relative optical absorption coefficient α obtained from the CPM measurements is made coincide with α obtained from optical transmission to get the absolute values of α. A Gaussian energy distribution for defects and an exponential function for valence band tail were used to fit α [6].

The ESR measurements were done both at room temperature and at liquid N_2 temperature. The LESR measurements were done at liquid N_2 temperature using a Xe lamp with a band-pass filter centered at 1.75 eV with a width of 0.2 eV.

RESULTS AND DISCUSSION

Figure 1 shows the illumination-time dependence of defect densities deduced by the CPM and ESR measurements. Here the defect densities are normalized to the values in the annealed state which are 2.5×10^{15} cm^{-3} (ESR) and 9.0×10^{15} cm^{-3} (CPM) for the undoped sample and 2.5×10^{15} cm^{-3} (ESR) and 1.3×10^{17} cm^{-3} (CPM) for the N-doped sample. Since the CPM is known to be insensitive to surface defects[6, 7], the ESR defect density in the bulk is compared with the CPM defect density in Fig. 1. The ESR spin density in the bulk was deduced by subtracting the contribution of the surface spin density ($\sim 2 \times 10^{12}$ cm^{-2}) which was estimated from the thickness dependence of the ESR spin density per area for undoped a-Si:H.

CH3365-4/94/0000-0634 $4.00 © 1994 IEEE

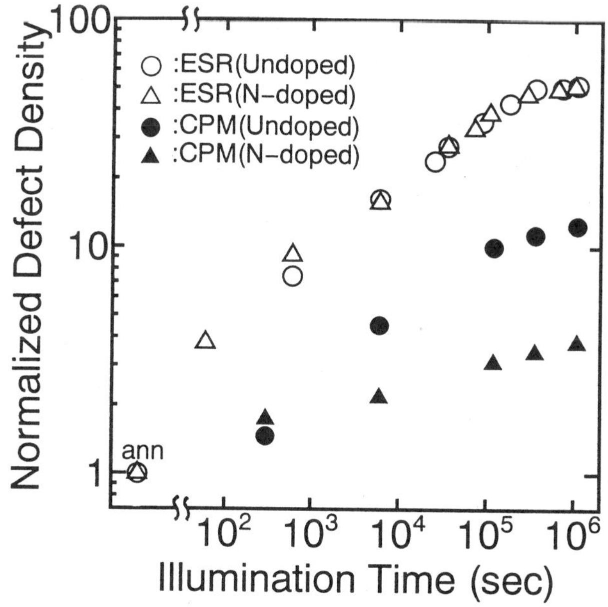

Fig. 1. Normalized defect densities measured by ESR and CPM as a function of illumination time for undoped and N-doped HWGD a-Si:H. The light source was a Xe lamp with an IR-cut filter (3.8 W/cm^2).

The same surface spin density was assumed also in the N-doped sample. For both samples, the further assumption was made that the surface spin density is not changed by light-soaking, because the increase in the surface spin density by light-soaking is known to be small compared with the bulk spin density[8]. It should be noted in Fig. 1 that the defect density in the light-soaked state is fifty times larger than that in the annealed state according to the ESR measurement for both undoped and N-doped a-Si:H, while only about ten- and five-fold increases are observed according to the CPM measurement for undoped and N-doped a-Si:H, respectively. This large discrepancy between ESR and CPM can be explained qualitatively as follows: In the annealed state, the density of charged defects is larger than that of neutral defects, making the CPM result larger than the ESR result since the CPM detects both D^0 and D^- while ESR detects only D^0. The density of charged defects does not change largely compared with that of neutral defects which increases significantly by light-soaking. Therefore, the ratio of the defect density in the light-soaked state to that in the annealed state is smaller when it is deduced using the CPM than when it is deduced using ESR. The D^- density can be obtained as a difference between the CPM and ESR defect densities, but it is not so reliable because the CPM is apt to underestimate the charged defect density if the ratio of the capture cross-sections of charged and neutral defects is large[9]. Recent theoretical analyses of the CPM predict that electron transition from the valence band to an empty gap state (in other words, hole transition from the gap state to the valence band) followed by thermal excitation of that electron to the conduction band is also included in the CPM result[10, 11]. This contribution of the hole transition is larger for samples with deeper Fermi levels. In our N-doped sample, the Fermi level moves toward midgap on light-soaking, making the effect of the hole transition larger. This should make the CPM defect density increase by a larger factor than the ESR defect density. The large discrepancy between CPM and ESR in Fig. 1 is not due to this effect, because the direction of the discrepancy observed is the opposite. Santos et al. also compared the CPM data with the ESR data in the light-induced defect creation process in a 15 μm thick a-Si:H sample and reported that the former were scaled to match the latter for long illumination times [2]. The difference between their results and the present results may be due to a difference in the samples; the ratio of the charged to neutral dangling bonds is probably small in their sample compared with our samples. The important point is that careless use of the CPM can lead to an erroneous estimation of the density of neutral dangling bonds, depending on the sample.

Charged defects can be observed by LESR in principle. But some difficulties can occur in estimating the charged defect density in the light-soaked state. Figure 2 shows an LESR signal (ESR signal under illumination) together with a dark ESR signal for undoped GD a-Si:H in both the annealed and light-soaked states. It can be noticed that the height of the main peak of the ESR signal is increased under illumination in the annealed state but is decreased under illumination in the light-soaked state. The LESR signal can be decomposed into three components; the dark ESR (D^0) component with g = 2.0055, a broad component with g = 2.01 and a narrow component with g = 2.004. The densities of these three components are shown as a function of the intensity of probe light in Figs. 3 and 4 for annealed and light-soaked states, respectively, for both N-doped and undoped GD samples. Here the light-soaked state was obtained by light soaking the sample from both sides, 1 h for each side, and the spin density is of the total spins including both surface and bulk spins. Although qualitatively similar results for the light-intensity dependence of the LESR signal were obtained for HWGD samples [12], we will present here only the results for the GD samples to avoid complication.

One can see from Fig. 3(a) that the D^0 density is increased under illumination over the entire range of light intensities for the annealed samples. Comparing the N-doped with undoped sample, we find from the figure that at low light intensities for annealed samples

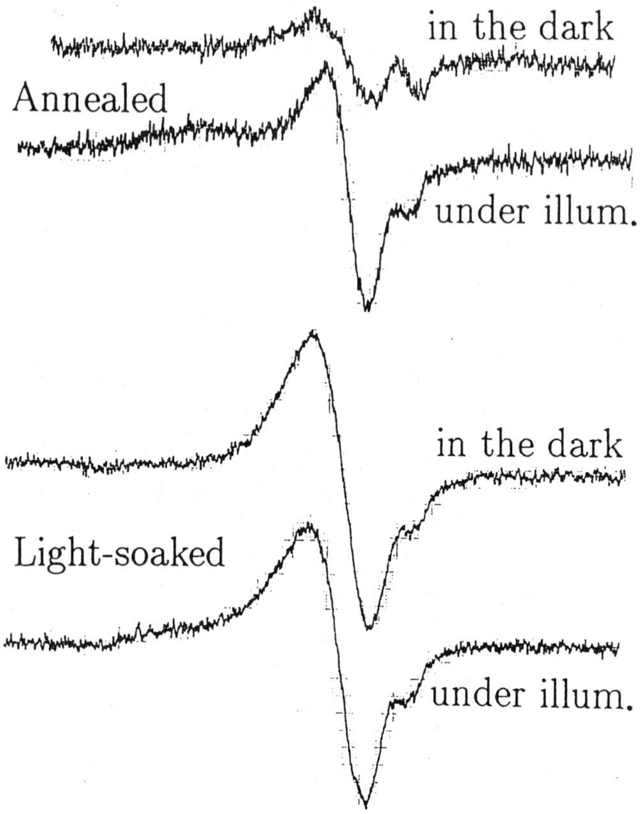

Fig. 2. ESR signals in the dark and under illumination taken at 77 K with the same gain for annealed (two upper spectra) and light-soaked (two lower spectra) undoped GD sample. The light-soaked state was obtained by illuminating the sample with 3.8 W/cm^2 at room temperature for 4 h from each side.

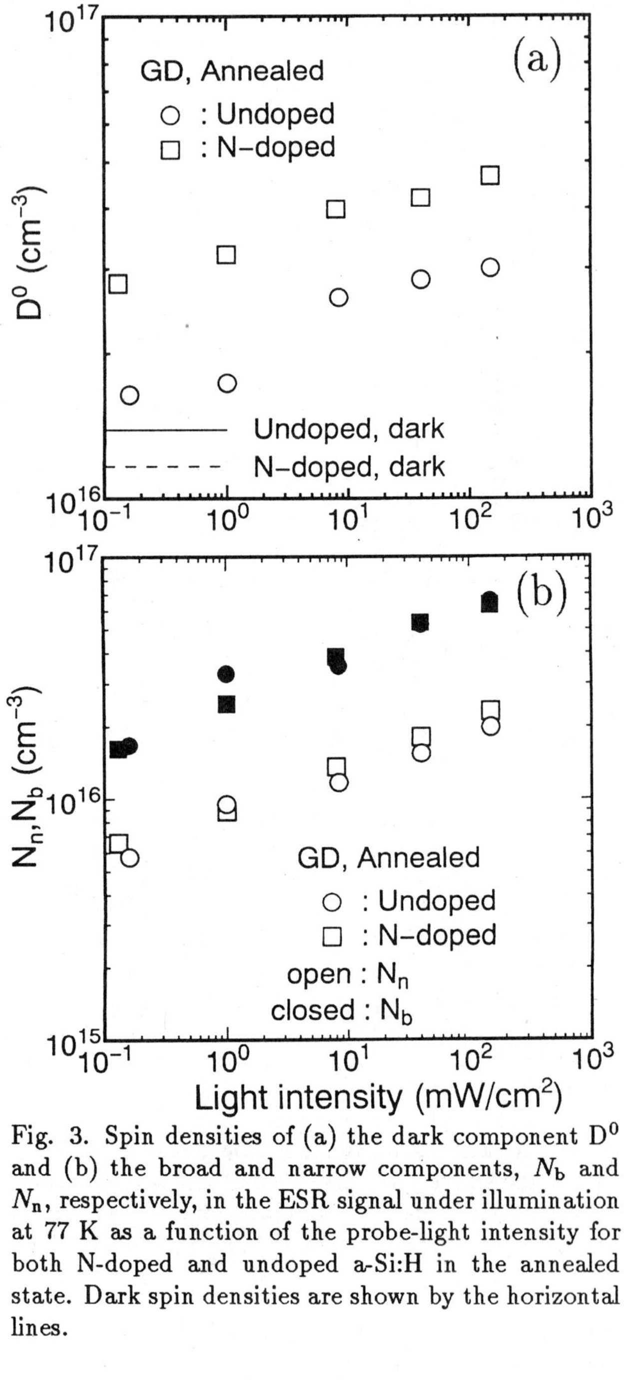

Fig. 3. Spin densities of (a) the dark component D^0 and (b) the broad and narrow components, N_b and N_n, respectively, in the ESR signal under illumination at 77 K as a function of the probe-light intensity for both N-doped and undoped a-Si:H in the annealed state. Dark spin densities are shown by the horizontal lines.

the D^0 density is increased much more in the former. This is consistent with the CPM results mentioned above, considering that N-doped sample has a larger density of charged defects than undoped sample, and charged defects are dominant over neutral ones so that the D^0 density under weak illumination is a measure of the charged defect density. However, the difference between the N-doped and undoped samples decreases with increasing light intensity. For the broad and narrow components in the LESR signal, one finds from Fig. 3(b) that the spin density of the broad component is significantly larger than that of the narrow component, consistent with previous work[4, 13, 14].

In the light-soaked state, one can see from Fig. 4(a) that the D^0 density is slightly decreased in the undoped sample at intensities lower than a few tens of mW/cm^2. A similar decrease appears to have been observed in the early studies[15, 16] although no decomposition of the LESR signal was done. The experimetal result can be understood by considering that the density of neutral defects overwhelms that of charged defects in the light-soaked state so that a net conversion of D^0 to D^+ or D^- is observed as a decrease in the ESR signal. Whether a net increase or decrease in D^0 occurs depends on the cross-section for carrier trapping and the carrier densities. In fact, an increase in the ESR signal was observed even in the light-soaked state when the light intensity was increased.

In conclusion, the increase in the density of de-

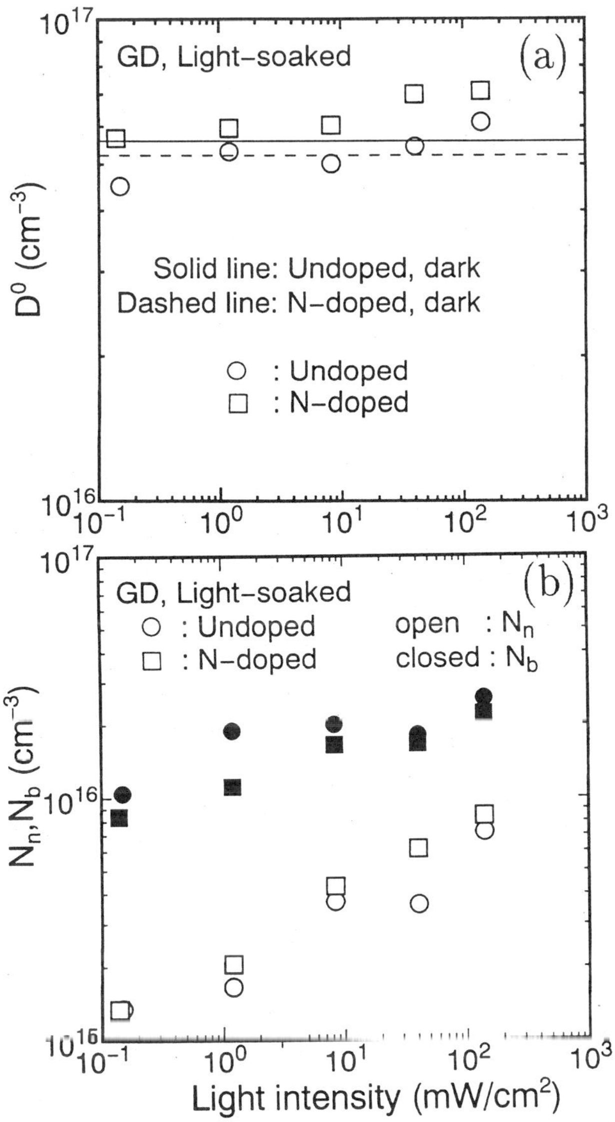

Fig. 4. Spin densities of (a) the dark component D^0 and (b) the broad and narrow components, N_b and N_n, respectively, in the ESR signal under illumination at 77 K as a function of the probe-light intensity for both N-doped and undoped a-Si:H in the light-soaked state. Dark spin densities are shown by the horizontal lines.

fects by light-soaking deduced by the CPM can be largely different from that deduced from ESR. The difference is especially large for N-doped a-Si:H in which the charged-defect density is much larger than the neutral-defect density. The charged defects can be estimated from LESR in principle, but in the light-soaked state where neutral defects overwhelm charged defects a quantitative estimation of the density of charged defects becomes difficult.

ACKNOLEDGEMENTS

The authors wish to thank Dr. A. Morimoto for helpful discussions and T. Okagawa, M. Takeda and S. Yoshimoto for experimental help. This work was partly supported by the New Sunshine Project of the Ministry of International Trade and Industry of Japan.

REFERENCES

[1] D. L. Staebler and C. R. Wronski, *Appl. Phys. Lett.*, **31** (1977) 292.

[2] P. V. Santos, W. B. Jackson and R. A. Street, *Phys. Rev. B* **44** (1991) 12800.

[3] M. Günes, C. R. Wronski and T. J. McMahon, *J. Appl. Phys.*, **76** (1994) 2260.

[4] T. Shimizu, H. Kidoh, A. Morimoto and M. Kumeda, *Jpn. J. Appl. Phys.*, **28** (1989) 586.

[5] A. Morimoto, M. Matsumoto, M. Kumeda and T. Shimizu, *Jpn. J. Appl. Phys.*, **29** (1990) L1747.

[6] T. Shimizu, X. Xu, H. Kidoh, A. Morimoto and M. Kumeda, *J. Appl. Phys.*, **64** (1988) 5045.

[7] L. Chahed, M. L. Theye, D. Fournier, J. P. Roger, A. C. Boccara, Y. M. Li, W. A. Turner and W. Paul, *Phys. Rev. B* **43** (1991) 14488.

[8] T. Shimizu, H. Kidoh, M. Matsumoto, A. Morimoto and M. Kumeda, *J. Non-Cryst. Solids*, **114** (1989) 630.

[9] P. Fiorini, M. Petracca, S. Korepanov, L. Fornarini, A. Mittiga and G. Grillo, *J. Non-Cryst. Solids*, **164-166** (1993) 379.

[10] K. Hattori, S. Fukuda, K. Nishimura, H. Okamoto and Y. Hamakawa, *J. Non-Cryst. Solids*, **164-166** (1993) 351.

[11] R. Platz, R. Brüggemann and G. H. Bauer, *J. Non-Cryst. Solids*, **164-166** (1993) 355.

[12] J-H. Zhou, T. Okagawa, M. Kumeda and T. Shimizu, *Jpn. J. Appl. Phys.*, **33** (1994) L1135.

[13] Z. M. Saleh, H. Tarui, N. Nakamura, M. Nishikuni, S. Tsuda, S. Nakano, Y. Kishi and Y. Kuwano, *Jpn. J. Appl. Phys.*, **31** (1992) 3801.

[14] G. Schumm, W. B. Jackson and R. A. Street, *Phys. Rev. B*, **48** (1993) 14198.

[15] A. Friederich and D. Kaplan, *J. Non-Cryst. Solids*, **35/36** (1980) 657.

[16] I. Hirabayashi, K. Morigaki and M. Yoshida, *Solar Energy Mat.*, **8** (1982) 153.

OPTIMUM TECHNIQUE OF DEPOSITING HIGH-RATE FILMS BY NORMAL PCVD METHOD

Xuanying Lin, Kuixun Lin, Yunpeng Yu, *Y W Lam, *Y C Chan, *Shunhui Lin, *Y. M. Chan
Department of Physics, Shantou University, Shantou, Guangdong, China 515063
***Department of Electronic Engineering, City Polytechnic of Hong Kong**

ABSTRACT

High quality hydrogenated amorphous silicon films with high resistivity (~10^{11} $\Omega\cdot$cm), high photosensitivity (>10^5) have been prepared at high deposition rate (>1.0 nm/s) by means of a conventional radio frequency glow discharge apparatus under optimum operation conditions. The results measured on the film properties exhibit that the stability of the high-rate deposited films is more than that of the normal low-rate deposited films.

I. INTRODUCTION

The investigation on hydrogenated amorphous silicon (a-Si:H) films deposited at high-rate has been paid considerable attention for realizing the low cost, large scale production, wide popularization and application. It is known that the deposition rate is increased by means of increasing the radio frequency (RF) power density and the concentration of a pure silane (SiH_4) feed gas. However, in a conventional plasma chemical vapor deposition (PCVD) system, the increase in the deposition rate leads to rapid gas phase reactions resulting in flake formation on the film and deterioration of the film properties. It is generally agreed that the optimum growth rate of high quality a-Si:H films prepared by normal glow discharge technique in a PCVD system is limited in the range of 0.1—0.3 nm/s.

A number of groups in the world have made effort to fabricate high quality a-Si:H films at high deposition rate using high silane or changing deposition technique.[1,2,3,4,5,6,7] Howerer, it is true that these techniques above mentioned could not be applied in production because of expensive disilane or complex apparatus and technology.

In this paper, we advance a new technique by increasing flow rate of silane and matching corresponding RF power density and increasing substrate temperature to obtain good quality a-Si:H films at high-rate of more than 1.0 nm/s without accompanying any appreciable deterioration in the electronic and structural properties of material.

The deposition process has been studied using a mass spectrometer, a floating double probe and a tuned single probe. The light-induced Staebler-Wronski (S-W) effect was also performed and discussed in references. [8,9,10,11,12]

II. EXPERIMENT

The deposition apparatus employed in this study is a PCVD system with a couple of parallel electrodes. The electrode area is $100\times100(\text{mm})^2$. The distance of electrodes can be moved from 1.0 cm to 5.0 cm for changing the plasma states. The RF power of 13.56 MHz is designed as capacitive-coupled. The substrate temperature, the RF power, the pressure and the flow rate of feed gas can be controlled and measured automatically by the temperature controller, automatic pressure controller and multi-channel gas controller. For understanding the space reaction processes and searching the optimum deposition conditions, a quadripole mass spectrometer, a floating double probe and a tuned Langmuir probe have been set on the line for detecting the discharge process and diagnosing the plasma parameters. The photoconductivity, darkconductivity and IR absorb spectra and ESR spectra of deposited a-Si:H films have been measured for understanding the surface reaction processes.

CH3365-4/94/0000-0638 $4.00 © 1994 IEEE

III. RESULTS AND DISCUSSIONS

A. Effect of the matching relation between flow rate and RF power on the deposition rate.

The deposition rate (DR) measured as a function of feed gas flow rate (FR) is shown in Fig. 1 for the samples prepared under conditions of pressure 60 Pa, substrate temperature 330 ° C. The curves A, B and C are corresponding to the RF power density of 0.6 W/cm^2, 0.3 W/cm^2 and 0.2 W/cm^2, respectively. In curves A, B and C, DR increases rapidly with increasing FR at low flow rate then increases slowly and finally reaches a saturation value DRs. It is found that DRs and corresponding flow rate FRs increase with the increase in RF power density. The saturation deposition rate in high flow rate region is sensitive to the change of the power density. There is an appropriate matching relation between FRs and corresponding power density PDs, which are necessary for achieving a high photosensitivity (PS) of a-Si:H films deposited at high-rate of more than 1.0 nm/s. Fig. 2 shows the dependence of photoconductivity σ_{ph}, darkconductivity σ_d and deposition rate on RF power density. Under identical flow rate of 20 sccm conditions, the deposition rate is increased linearly with the increase in RF power density resulting in the remarkable decrease in PS (the ratio of σ_{ph}/σ_d) from 3×10^5 down to 2×10^2. However, an excellent PS of more than 10^5 exists near FRs of 20 sccm, 30 sccm and 50 sccm for corresponding RF power density of 0.2 W/cm^2, 0.3 W/cm^2 and 0.6 W/cm^2 as showing in Fig. 3. The experimental results are ascribed that a large flow rate plays both roles of increasing the concentration of silane gas in a twinkling and preventing from the formation of high silane $(SiH_4)_n$

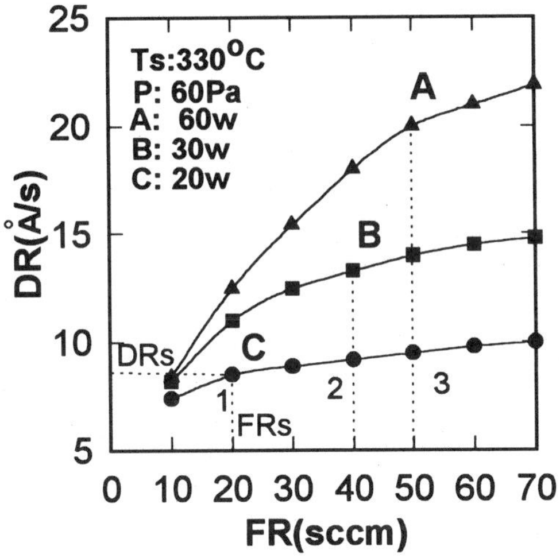

Fig.1 Deposition rate versus flow rate for different RF power(electrode area of 100x100 mm^2

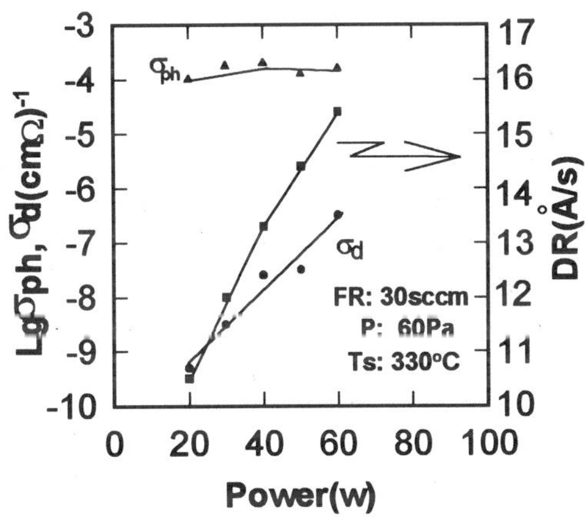

Fig.2 Darkconductivity,Photoconductivity and deposition rate as a function of RF Power

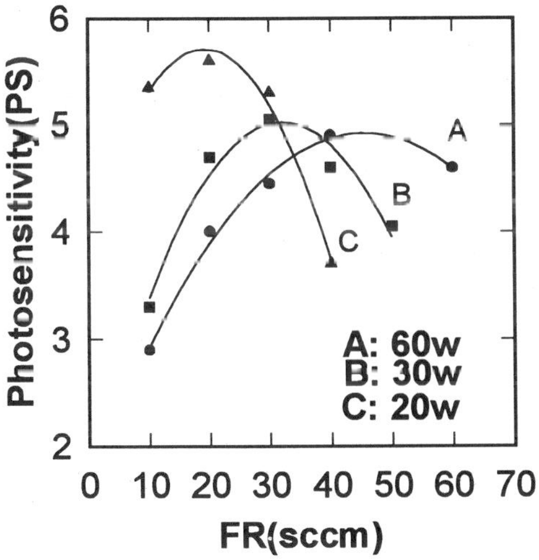

Fig.3 The dependence of photosensitivity on the relation between flow rate and RF power

powder, because the large flow rate means that the reaction gas can be renewable quickly. The growth rate for different flow rate is basically limited by supported RF power, under which silane gas is fully decomposed resulting in a optimizing discharge state. The matching relation between FRs and PDs for different DRs is necessary.

The deposition process and the discharge parameters strongly effect on the deposition rate. A quadripole mass spectrometer, a symmetrical floating double probe and a tuned single probe have been used for diagnosing the monosilicon radicals, the electron temperature, the electron concentration and the electron energy distribution. The experiments give the dependence of silane depletion fraction (the decomposed silane fraction discussed elsewhere [8]) on the RF power density for constant flow rate. It is indicated that the depletion fraction increases with increasing RF power density in the low power range and tends to saturation when the RF power is in high range, which explains why existing a DRs. The strong dependence of the neutral radical SiH_n (n = 0, 1, 2) density on RF power density is agreed with above conclusions.

On the other hand, the measured results using electron probe technique demonstrate that the electron concentration and the mean energy increase with increasing RF power density which is reported in details. [9,10,11] The means of adjusting the matching relation between parameters of flow rate and RF power is a very useful method for deposition of a-Si:H film with a good quality at high deposition rate.

B. Effect of substrate Temperature on Film Growth

The deposition rate and the properties of the a-Si:H film are dependent on the substrate temperature T_S closely. Fig.4 shows the dependence of the DR on the T_S. The DR increases and up to a maximum value then decreases down rapidly with increasing T_S continuously. This is suggested that the adsorptive power of the surface on the radicals and the bonding power of the radicals increase with increasing the substrate temperature and this accelerates the film growth. With increasing the substrate temperature further, the adsorbed moleculars (including hydrogen molecular) relieve

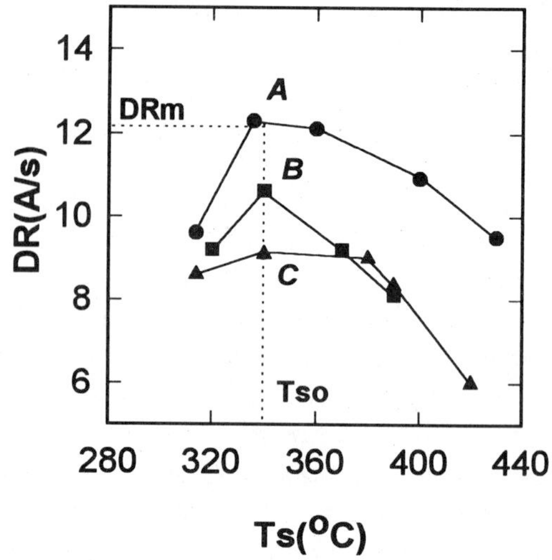

Fig.4 The deposition rate as a function of the substrate temperature

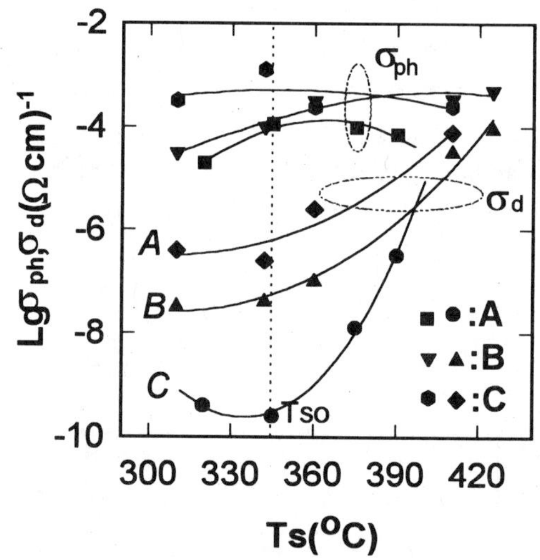

Fig.5. Darkconductivity, Photoconductivity dependence of substrate temperature for DR=1.3nm/s

rapidly in the surface procedures resulting in the decrease of deposition rate. In Fig. 4 the curves of A, C and B, C are corresponding to the different power density (0.6 W/cm², 0.2 W/cm²) at a same flow rate (20 sccm) and to the different flow rate (30 sccm, 20 sccm) at a same power density (0.2 W/cm²). It is interesting that all the maximum deposition rate DRm are corresponding to the substrate temperature of about 350 °C no matter how to change the flow rate and the discharge power. This demonstrates that the discharge power and the gas flow rate are key parameters on determining the space procedures while the substrate temperature is a key parameter on effecting the surface procedures in the PCVD process. The curves of photoconductivity σ_{ph} and darkconductivity σ_d versus the substrate temperature are displayed in Fig. 5. The curves of A, B and C were measured on the samples prepared at the same technology parameters as those shown in Fig. 1. It is found that both of the optimum photosensitivity and the highest deposition rate are corresponding to a same substrate temperature. The experimental results demonstrate that there is an optimum substrate temperature T_{sop}, which strongly effect on the deposition rate and the properties of the film prepared by means of PCVD.

Experiments of long time light exposing and high temperature annealing on the films have been made.[12] The measured density of the dangling bond is accordant with that calculated by the theory of steady-state photoconductivity. The densities of dangling bond in annealed and light soaking states are 5×10^{15} cm^{-3} and 1.0×10^{16} cm^{-3} ,respectively. The dangling bond caused by light soaking increases by a factor of 2 for high rate film and a factor of 3 for low rate film, respectively. The density of dangling bond of $(5—10)\times10^{15}$ cm^{-3} is agreed with that calculated by theory. The experimental results demonstrate that the photo-thermal properties of high rate film are more stable than those of the low rate film.[12]

IV. CONCLUSION

Experimental results and theoretical analysis demonstrate that it is possible to prepare an a-Si:H film with a high quality at a high deposition rate by means of adjusting matching relation between flow rate and discharge power, and optimizing the substrate temperature using a conventional PCVD system and method for preparing a low deposition rate film. This provides a real capability for producing a-Si:H solar cells and various devices successively at a large scale and low cost.

REFERENCES

1. B.A. Scott, M.H. Brodsky, D.C. Green, P.B. Kirby, R.M. Plecenik and D. F. Simony, Appl. Phys. Lett. 37,725(1980)
2. K. Ogawa, I. Shimgu and E. Lnoue, Jpn. J. Appl. Phys. 20, L639 (1981)
3. T. Shimada, N. Nakamura, S. Matsubara, H. Itoh, S. Muramatsu and M. Migitaka, The Technical Digest of the 1st International Photovoltaic Science and Engineer Conference, Kobe, Japan, 445 (1984)
4. T. Hamasaki, M. Ueda, A. Chayahara, M. Hirose and Y. Osaka, Appl. Phys. Lett. 44, 600 (1984)
5. T. Hamasaki, M. Uedak, A. Chayahara, M. Hirose and Y. Osaka, Appl. Phys. Lett. 44, 1049 (1984)
6. M. Zhang, Y. Nakayama, S. Nonoyama and K.Wakita, J. Non-Cryst. 164 - 166, 63 (1993)
7. P. Papadopulos, A. Scholz, S. Bauer, B. Schroder and H. Oechsner, J. Non Cryst. Solids, 164 -166, 87 (1993)
8. Quixun Lin, Xuanying Lin, Yunpeng Yu, Xu Zeng, Ping Wu, Chinese Physics Letters Vol. 10,338 (1993)
9. Quixun Lin Xuanying Lin, Yunpeng Yu, Hong Wang,, Chinese Physics Letters Vol. 10, 338 (1993)
10. Xuanying Lin, Quixun Lin, Yunpen Yu, Jiayi Chen, Hong Wang, Ping Wu, Chinese Physics Letters Vol. 10, 334 (1993)
11. Kuixun Lin, Xuanying Lin, Yunpeng Yu, Hong Wang, J. Appl. Phys. Vol. 74 (8) 4899 (1993)
12. Xuanying Lin, Hong Wang, Kuixun Lin, Yunpeng Yu and Shiliu Fu, Thin Solid Films, 237, 310 (1994)

DEGRADATION AND FAILURE MECHANISMS OF a-Si:H SOLAR CELLS WITH ALUMINUM CONTACTS

M. S. Haque, H. A. Naseem, and W. D. Brown
Department of Electrical Engineering/3217 BELL
University of Arkansas, Fayetteville, Arkansas 72701

ABSTRACT

Low temperature (100°C-300°C) annealing effects on a-Si:H solar cells with Al contacts have been investigated. A reduction in shunt resistance upon annealing in the temperature range of 150-180°C for thirty minutes was observed indicating degradation of the cells. Visible morphological changes of the aluminum surface and the a-Si:H film underneath initiate at approximately 170°C. Thirty minute annealing at 180°C decreased the shunt resistance by a factor of 50 and the solar cells showed a considerable decrease in efficiency and fill factor. Annealing at 200°C and higher degraded the solar cells to the point that no conversion occurred. The degradation and failure mechanisms of the solar cells can be explained using a simple model involving aluminum diffusion into the silicon followed by crystallization and type conversion of the n^+ a-Si:H due to aluminum counterdoping.

INTRODUCTION

The instability problem associated with thermal and light induced changes in a-Si:H material properties is a major impediment to its utilization in large scale power generation. Instability causes degradation in solar cell efficiency which decreases substantially within a few years. Light-induced degradation, the Staebler-Wronski effect, is reversible [1]. The a-Si:H solar cell efficiency can be restored by annealing the cell at temperatures above 150°C for some length of time [2]. However, there is evidence that annealing cannot restore the original efficiency for aged solar cells indicating the presence of an irreversible degradation mechanism along with the reversible light-induced degradation [3,4].

Since a-Si:H consists of a random network of atoms, distorted considerably from the minimum energy sites of its crystalline counterpart, it can easily interact with other materials such as aluminum which has been used extensively as the contact metal in a-Si:H solar cells. Aluminum forms a good ohmic contact with n^+ a-Si:H and has good photoreflecting characteristics. It has been reported, however, that a-Si:H interacts with aluminum at low temperatures [5]. In fact, metal-induced low temperature crystallization of a-Si:H is a well-known phenomenon [6,7]. In this work, a-Si:H solar cell degradation during low temperature annealing has been investigated and possible degradation mechanisms have been identified.

EXPERIMENTAL

P-i-n solar cells were fabricated using Ultra High Vacuum (UHV) Plasma-Enhanced Chemical Vapor Deposition (PECVD) with SnO_2 coated Corning 7059 glass as the substrate. Substrate temperature, pressure, and power density were kept at 250°C, 500 mTorr, and 50 mW/cm^2, respectively. Other parameters used to deposit the various layers of the solar cell, which were obtained after extensive optimization, are given in Table 1. Following fabrication of the p-i-n structure, 8000 Å thick aluminum contacts were thermally evaporated. The cells were then annealed at temperatures between 100°C and 300°C for 30 minutes in a nitrogen ambient. Dark and light I-V plots were used to electrically characterize the solar cells while x-ray diffraction and micro-Raman spectroscopy were used to detect the crystallinity of the a-Si:H under the aluminum contacts.

Table 1 - Optimized deposition parameters for fabricated p-i-n solar cells.

Layer	Thickness	Gases and Flow Rates (sccm)		
		SiH_4	B_2H_6	PH_3
p	150 Å	29.82	0.18	0.00
i	5000 Å	30.00	0.00	0.00
n	200 Å	29.70	0.00	0.30

RESULTS AND DISCUSSION

No significant change in efficiency (η), fill factor (FF), open circuit voltage (V_{oc}), and short circuit current density (J_{sc}) was observed for annealing up to 170°C as shown in Table 2. However, a considerable decrease in shunt resistance (R_{sh}) for 150 and 170°C annealing is evident from the table. This decrease, caused by low temperature annealing for 30 minutes, is clearly an indication of degradation in the device. From the log I-V

CH3365-4/94/0000-0642 $4.00 © 1994 IEEE

plot of Fig. 1, a gradual decrease in shunt resistance with increasing anneal temperatures is clearly observable. This degradation is believed to be due to enhanced diffusion of Al into the a-Si:H through pathways similar to grain boundary triple points. An activation energy of 1.2 eV was determined from the shunt resistance change which is similar to results published by other groups [3,8]. For annealing at 180°C, the shunt resistance decreases by a factor of fifty. Also, a considerable decrease in η, FF, and J_{sc} is observed at this temperature. For annealing at and above 200°C, the solar cells degraded to a point where meaningful output parameters and shunt resistance could not be extracted since the devices were essentially a short circuit.

Table 2 - Annealing effects on a-Si:H solar cell output parameters and shunt resistance.

Anneal Temp	status	Eff. %	FF	V_{oc} (Volts)	J_{sc} (mA/cm^2)	n	R_{sh} (Megohm)
100°C	before	6.73	0.634	0.74	11.777	1.78	33.3
	after	6.88	0.634	0.74	12.032	1.78	33.3
150°C	before	7.30	0.679	0.74	12.095	1.74	32.5
	after	7.42	0.684	0.74	12.191	1.74	14.0
170°C	before	7.17	0.683	0.74	11.777	1.76	33.3
	after	6.96	0.677	0.74	12.191	1.81	10.0
180°C	before	7.28	0.640	0.74	12.764	1.78	50.0
	after	5.74	0.578	0.74	11.141	2.46	1.0

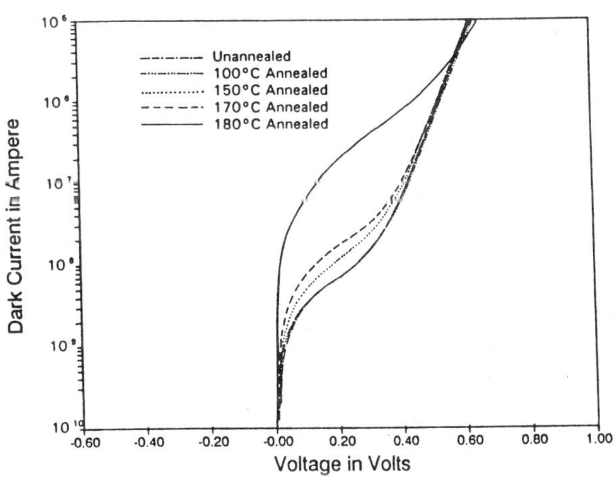

Fig. 1 - Annealing effects on a-Si:H solar cell shunt resistance

Figures 2a, 2b, 2c, and 2d show dark and light I-V characteristics for devices annealed at 170°C, 180°C, 200°C and 250°C, respectively. Upon annealing at 200°C, the solar cells show almost resistive behavior with a very low resistance. The characteristics in this figure indicate that even a brief annealing at temperatures above 200°C can degrade a-Si:H solar cells to the point that no energy conversion is observed. In degraded solar cells, the leakage current becomes so high that I-V curves under light and dark conditions are essentially the same as seen in Figs. 2c and 2d.

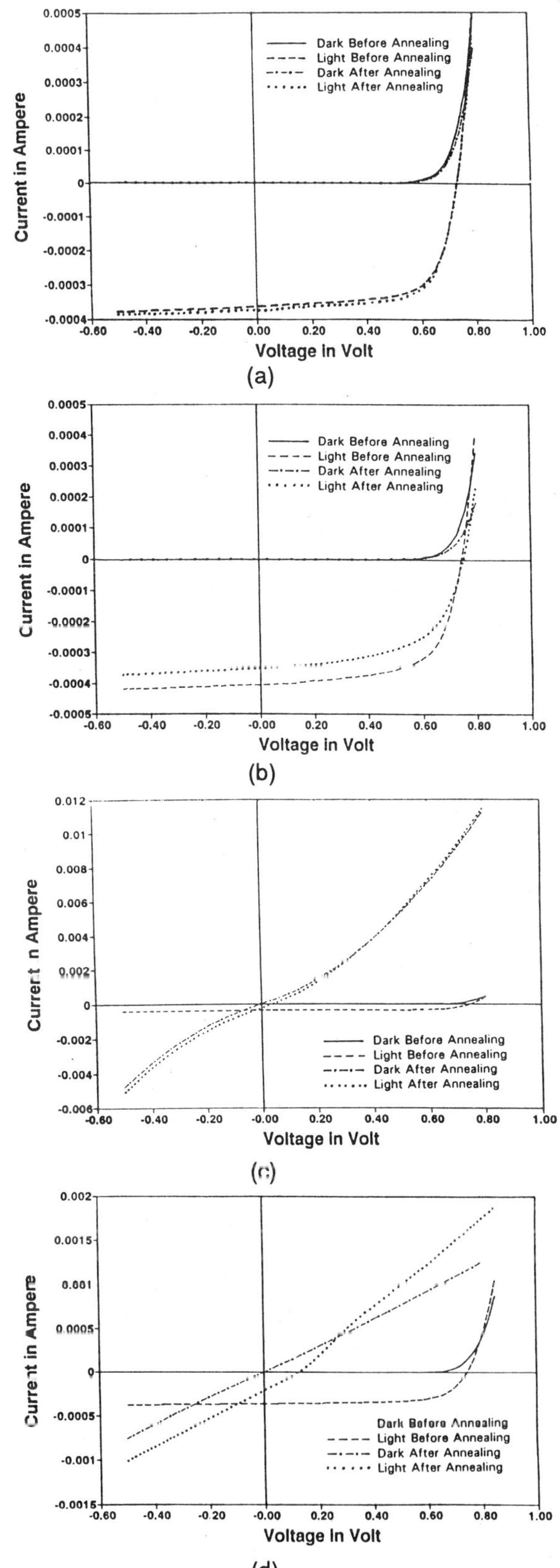

Fig. 2 - Dark and light I-V plots of solar cells annealed at (a) 170°C, (b) 180°C, (c) 200°C, and (d) 250°C.

In order to further characterize the failure mechanism, effects on photoresponse of the solar cells were observed. Figure 3 shows the effects of annealing on the light-generated current. These plots were obtained by subtracting the dark current from the light current. It is seen that, for annealing at 170°C, the light generated current is negligibly affected. For the 180°C annealed solar cell, V_T (the voltage at which light generated current apparently crosses the zero current point) remains unchanged, but a 13% decrease in light-generated current is observed. For annealing at 200°C and 250°C, the light generated current decreases by 38% and 41%, respectively. Also, V_T changes from 0.8 volts to 0.35 volt and 0.27 volt, respectively. Since a reduction in the depletion region width causes the light generated current to decrease, a sudden reduction in V_T indicates a catastrophic failure of one of the space charge regions. These results suggest that the n^+/i interface has been damaged as a result of excessive Al interdiffusion. This also causes the leakage current to increase by several orders of magnitudes as seen in the I-V plots (Figs. 2c and 2d).

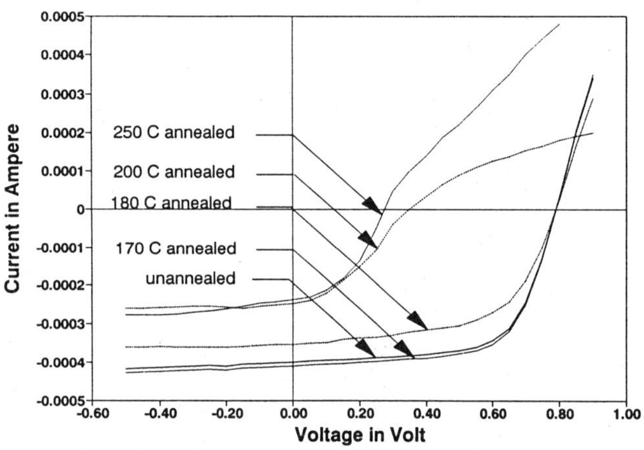

Fig. 3 - Light generated current in solar cells annealed at different temperatures.

The aluminum was removed from the annealed solar cells in order to analyze the interacted a-Si:H region. Figure 4 shows stacked Raman spectra for interacted a-Si:H annealed at different temperatures. For the 170°C annealed sample, a broad a-Si peak with a maximum around 470 cm^{-1} is seen. This spectrum is identical to that obtained for a sample before annealing. In both cases, no crystalline silicon peak (520 cm^{-1}) was observed. However, for the 180°C annealed sample, a crystalline silicon peak starts to appear along with the broad a-Si peak. Table 3 shows wave number shift (from the single crystal silicon peak position), FWHM, and corresponding grain size for crystallized samples of this study. Grain sizes were estimated using data reported by Campbell et al. and Richter et al. [9,10]. The crystalline peak position of the interacted a-Si:H material moves toward the wave number for single crystal silicon and the FWHM decreases with increasing anneal temperature. At the same time, the broad amorphous peak decreases with annealing temperature and completely disappears following annealing at 250°C.

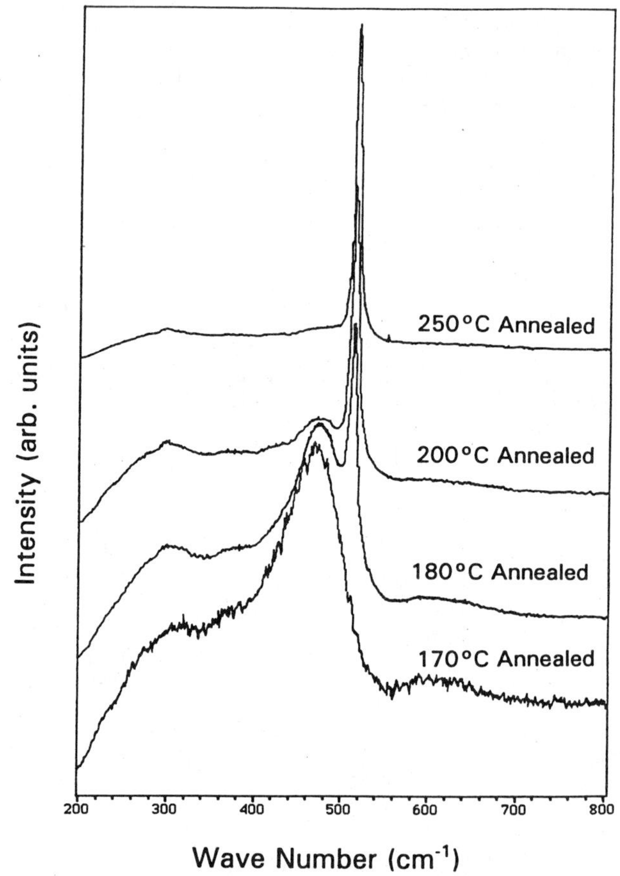

Fig. 4 - Raman spectra of interacted a-Si:H surface of solar cells annealed at different temperatures.

Table 3 - FWHM, peak shifting, and corresponding grain size of interacted, crystallized a-Si:H films annealed at different temperatures.

Annealing Temperature (°C)	Wave Number Shift (cm^{-1})	FWHM (cm^{-1})	Grain Size (Å) (ref 9 & 10)
180	4.163	10.30	10 - 30
200	3.832	9.535	20 - 40
250	2.056	7.055	30 - 80
300	0.525	6.586	70 - 100

Raman analysis was also performed on uninteracted a-Si:H samples annealed at different temperatures (i.e., in a location not covered with Al during annealing). Figure 5 shows Raman spectra for as-deposited and 300°C annealed, but uninteracted, a-Si:H material. It is obvious that crystallization of the a-Si:H does not occur, even at this high anneal temperature, unless the a-Si:H is in contact with Al.

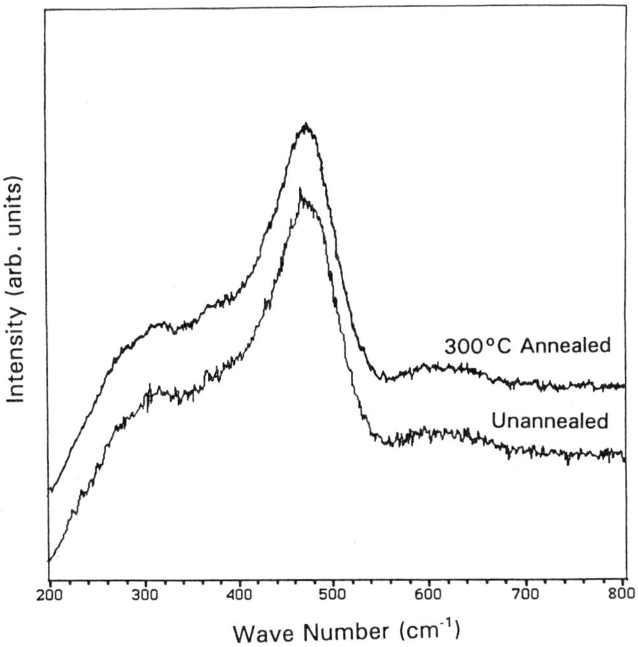

Fig. 5 - Raman spectra of uninteracted a-Si:H surface before and after annealing.

Interaction kinetics causing the degradation and failure of these solar cells can be explained using a simple model involving aluminum silicon interdiffusion, a-Si:H crystallization, and type conversion. As noted previously, metal-induced low temperature crystallization of a-Si-H is a well-known phenomena [6]. Crystallization at an a-Si:H/Al interface can be described briefly as follows. A metal-rich Al-Si mixed phase (Al_xSi) is formed at the interface at low annealing temperatures (150°C-200°C). The thickness of this layer increases with increasing anneal temperature and time. This mixed phase is a metastable, defect-rich material that undergoes a transformation upon annealing above a certain temperature (crystallization onset temperature). Silicon atoms from the metastable metal-rich layer coalesce at the mixed phase/a-Si:H interface to form silicon crystallites. Aluminum incorporation in the crystallized silicon gives rise to a counterdoped film.

SUMMARY AND CONCLUSIONS

Degradation of Al metal/a-Si:H p-i-n solar cells as a result of annealing for 30 minutes in nitrogen at low temperatures (100°C-300°C) was investigated. No degradation of cell output parameters (η, FF, J_{sc}, V_{oc}) was observed for annealing below 170°C. For 150°C and higher temperatures, however, a significant reduction in shunt resistance was observed indicating degradation of the cell. Annealing at 180°C and higher produced considerable degradation in output parameters. Catastrophic failure of the cells occurred for annealing at temperatures of 200°C and higher. The degradation and failure of these solar cells are explained by Al/a-Si:H interaction kinetics involving interdiffusion, a-Si:H crystallization, and type conversion.

REFERENCES

[1] D. E. Carlson and C. R. Wronski, "Amorphous silicon solar cells", *Appl. Phys. Lett.*, **28**(11), p. 671, 1976.

[2] D. L. Staebler and C. R. Wronski, "Optically induced conductivity changes in discharge-produced hydrogenated amorphous silicon", *J. Appl. Phys.*, **51**, p. 3262, 1980.

[3] F. Willing, N. Bennett, and J. Newton, "Thermal Degradation of a-Si:H Solar Cells", *Proc. 19th IEEE Photovolt. Specialists Conf.*, p. 1086, 1987.

[4] J. W. Larthop and P. A. Williamson, "Failure Mechanisms in Amorphous Silicon Solar Cells", *Proc. 19th IEEE Photovolt. Specialists Conf.*, p. 262, 1987.

[5] S. Ishihara, T. Hirao, K. Mori, M. Kitagawa, M. Ohnos, and S. Kohiki, "Interaction between n-type amorphous hydrogenated silicon films and metal electrodes", *J. Appl. Phys.*, **53**(5), p. 3090, 1982.

[6] C. C. Tsai, R. J. Nemanich, M. J. Thompson, and B. L. Stafford, "Metal-Induced Crystallization of Hydrogenated Amorphous Si Films", *Physica* **117B & 118B**, p. 953, 1983.

[7] M. S. Haque, H. A. Naseem, and W. D. Brown, "Interaction of aluminum with hydrogenated amorphous silicon at low temeratures", *J. Appl. Phys.*, **8**, p. 3928, 1994.

[8] A. F. Delhoy, "Task #2: Non-semiconductor Materials - Amorphous Silicon Aluminum Interface",SERI/STR-211-2814/DE 58016886, Nov. 1985.

[9] I. H. Campbell and P. M. Fauchet, "The Effects of Microcrystal Size and Shape On the One Phonon Raman Spectra of Crystalline Semiconductors", *Solid State Communications*, **58**, p. 739, 1986.

[10] H. Richter, Z. P. Wang, and L. Ley, "The One Phonon Raman Spectrum in Microcrystalline Silicon", *Solid State Communications,* **39**, p. 625, 1981.

ACKNOWLEDGMENT

This work was partially supported by a grant from the NASA/Lewis Research Center.

EFFECT OF DIFFERENT TCO INTERFACES ON THE PERFORMANCES PRESENTED BY HYDROGENATED AMORPHOUS SILICON P-I-N SOLAR CELLS

Elvira Fortunato, Carlos N. Carvalho, Ana Bicho and Rodrigo Martins

Materials Science Dep, Faculty of Science and Technology of New University of Lisbon - DCM/FCT-UNL and Centre of Excellence for Microelectronics and Optoelectronic Processes - CEMOP
Quinta da Torre, P-2825 Monte de Caparica, Portugal

ABSTRACT

In this paper we report results concerning the effect of the TCO interface on hydrogenated amorphous silicon (a-Si:H) p-i-n homojunction solar cells. Its correlation with dark current density-voltage (J-V) characteristics and spectral response, before and after white light-soaking degradation, is analysed. From this study, we conclude that the properties and stability of these devices are not only influenced by the a-Si:H film properties, but also by the properties of the transparent conductive electrode and its interface with the a-Si:H layer.

INTRODUCTION

Beyond the increasing progress observed in the last years, related to the improvement of the conversion efficiency and stability of a-Si:H p-i-n solar cells, there are still many factors which limit their performances. It is now well established that the most important parameter that contributes to the performance of the solar cells is the quality of the intrinsic layer [1]. Nevertheless, an a-Si:H solar cell has many interfaces, as the TCO-p, p-i, i-n and n-metal. To improve the devices performances it is necessary to consider their role on the device transport mechanism [2]. In this paper we mainly refer the influence of the TCO-p interface on the device characteristics.

Although the TCO layers exhibit good visible light transmittance, associated with low sheet resistance, there is always a work function difference between the TCO and the p-layer. This difference, which originates an imperfect contact [3], is responsible for the enhancement of the recombination losses and the appearance of a potential barrier which blocks the extraction of photogenerated carriers. So, the optoelectronic properties of the device are affected. Ideally, the contacts should not have Schottky barriers. However, it has been reported [4] that high densities of interface states, at the contacts, cause the pinning of the Fermi energy.

EXPERIMENTAL PROCEDURE

To investigate the role of the TCO on the a-Si:H solar cells, we chose two types of TCOs: ITO and SnO_2:F. From each one, we selected two different surface morphologies: transparent and milky-like.

Their characteristics are summarised in Table I. The haze ratio (H_r) is defined as the ratio between diffused/integral transmittance light.

The a-Si:H devices were deposited by a conventional rf glow discharge single chamber system. The main characteristics are described elsewhere [5]. Pure SiH_4 gas was used to produce the intrinsic layer and a gas mixture of SiH_4 with B_2H_6 (0,7%) and PH_3 (1,5%), where used to deposit the p and n layers, respectively. Once p.i.n (n.i.p) structures were deposited, Al dot contacts were evaporated, by electron gun technique, onto the top doped layer. To eliminate the lateral leakage currents [6], the doped layer which surrounds the Al contacts was etched using a Reactive Ion Etching system (Alcatel GIR 300, with end point detection). The reactive gas used was SF_6. The cells (deposited in the same run) were not optimised for efficiency, but for evaluating the effects of the different TCOs.

The transmittance (T) and reflectance (R) measurements, were performed using a Shimadzu spectrophotometer equipped with an integrated sphere. The p-i-n and n-i-p solar cells have been characterised through dark J-V measurements (under vacuum in a temperature controlled probed station) and by spectral response, SR (400 to 800 nm). Both measurements were performed before and after degradation (under AM1.5 white light exposure for a time period of 80 h).

The p.i.n and n.i.p structures (with thick doped layers), have been also analysed through Secondary Ion Mass Spectrometry (SIMS), in order to evaluate the distribution of impurities (and i-layer contaminants) in the solar cells.

RESULTS AND DISCUSSION

Transmittance and reflectance measurements

The T and R spectra of glass/TCO, as a function of the wavelength (λ), are shown in Figs. 1a. and b., from which it was also possible to infer H_r. Such data reveal that the TCO layers that present the best optical properties exhibit the lowest H_r.

TABLE I - Main characteristics of the TCOs coated glass substrates

TCO substrate	**A** (ITO)	**B** (ITO)	**C** (SnO_2:F)	**D** (SnO_2:F)
Glass thickness	1.8 mm	1.1 mm	1.1 mm	2.0 mm
TCO thickness	1200 Å	1200 Å	10000 Å	12000 Å
TCO sheet resistance	20 Ω/sqr	9 Ω/sqr	10 Ω/sqr	12 Ω/sqr
TCO average resistivity	2.4×10^{-4} Ωcm	1.1×10^{-4} Ωcm	1.0×10^{-3} Ωcm	1.4×10^{-3} Ωcm
Haze ratio, H_r at 550 nm	less than 1%	30%	6%	20%
Transmittance, T at 550 nm	80%	55%	80%	75%
Visual appearance	Transparent	Milky-like	Transparent	Milky-like

CH3365-4/94/0000-0646 $4.00 © 1994 IEEE

First WCPEC; Dec. 5–9, 1994; Hawaii

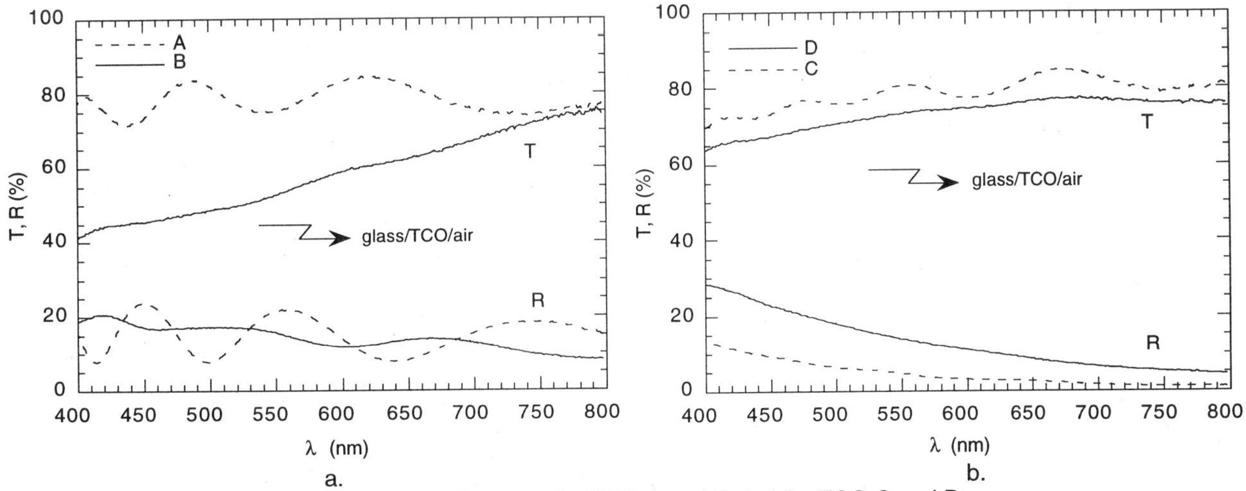

Fig. 1. (a.) Integral transmittance and reflectance for TCO A and B; (b.) for TCO C and D.

Dark current density-voltage characteristics

The J-V characteristics of p-i-n and n-i-p structures are related to the junction formation [7], and can be expressed by the following approximate expression [8]:

$$J = J_0 \left[\exp(qV/\eta kT) - 1\right] \quad (1)$$

J_0 is the saturation current (to whom contribute the carriers recombination, tunnelling and generation in the depletion regions), q the electron charge, k the Boltzamn constant, T the temperature and η the diode quality factor (dependent on the recombination and generation mechanisms). Previous work [9] related to the excessed J_0 observed in p-i-n diodes, ascribed it to: (1) inherent gap state recombination and/or extra defect states due to residual boron in the i-layer; (2) tunnelling at the interfaces or (3) macrostructural shunt paths.

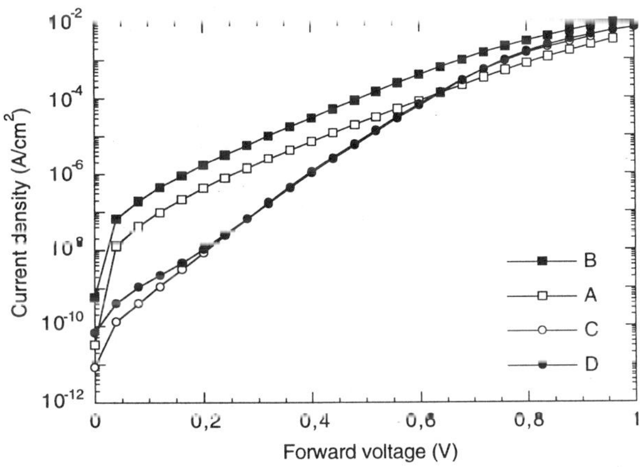

Fig. 2a. Forward J-V characteristics for the same p.i.n solar cell deposited on the different TCOs.

Figs. 2a. and b. show the forward (J_F) and reverse (J_R) J-V curves of p-i-n diodes, at room temperature, for the different TCOs used, before light-soaking degradation. The corresponding values of η, J_0 (at V = -1V) and the rectification ratio (R) for |V| = 1V, before and

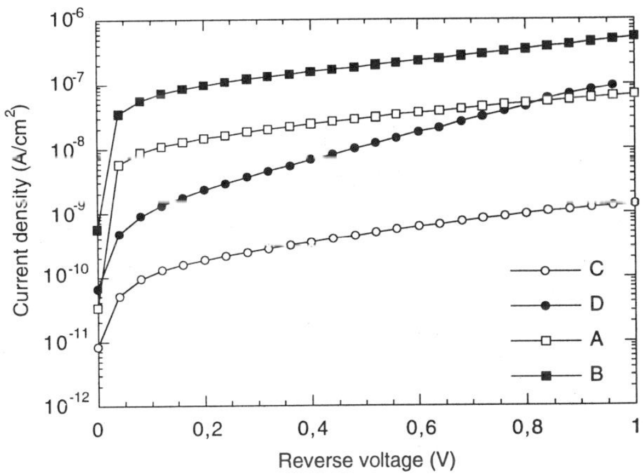

Fig. 2b. Reverse J-V characteristics for the same p.i.n solar cell deposited on different TCOs.

after degradation, are shown in Table II. Here we observe that the forward current density increases exponentially with the voltage V, for all the samples. Nevertheless, the dependence of J on V clearly indicates two main groups: one with $\eta > 2$ (for ITO) and another with $\eta < 2$ (for SnO_2:F). The recorded J_0 values, for devices with ITO, reflect a high density of states at the p/i interface before and after degradation. The data also reveal that the lowest J_0 is obtained for the device deposited on TCO C while the highest J_0 was obtained on TCO D, before and after degradation. These results indicate that devices deposited on SnO_2:F present better characteristics (η and J_0) than the ones deposited on ITO. This behaviour can be attributed to In diffusion formed at the interface [10].

Once the TCO that leads to the best device performances is selected (TCO C), we produced p.i.n and n.i.p diodes with thick doped layers, to observe the effect of the i-layer contamination on the device performances.

Fig. 3 shows the measured J-V characteristic of these devices deposited on TCO C. The data show that η and J_0 are larger in p.i.n than in the n.i.p ones (see Table III). This is attributed to recombination boron centers in the interface that diffuses towards the i layer (see Fig. 3.).

TABLE II - Main electrical characteristics of the p.i.n devices

Sample	Before degradation				After degradation			
	J_F (1V) (A/cm^2)	J_R (-1V) (A/cm^2)	η	J_F/J_R (|V|=1V)	J_F (1V) (A/cm^2)	J_R (-1V) (A/cm^2)	η	J_F/J_R (|V|=1V)
A (ITO)	3.3×10^{-3}	6.8×10^{-8}	2.73	4.9×10^{4}	2.3×10^{-3}	4.7×10^{-7}	2.96	4.9×10^{3}
B (ITO)	9.0×10^{-3}	5.3×10^{-7}	2.69	1.7×10^{4}	8.2×10^{-3}	8.6×10^{-7}	2.91	9.5×10^{3}
C (SnO_2:F)	7.0×10^{-3}	1.4×10^{-9}	1.57	3.6×10^{6}	6.7×10^{-3}	8.2×10^{-9}	1.82	8.2×10^{5}
D (SnO_2:F)	7.0×10^{-3}	9.2×10^{-8}	1.66	7.6×10^{4}	6.0×10^{-3}	3.4×10^{-7}	1.91	1.7×10^{4}

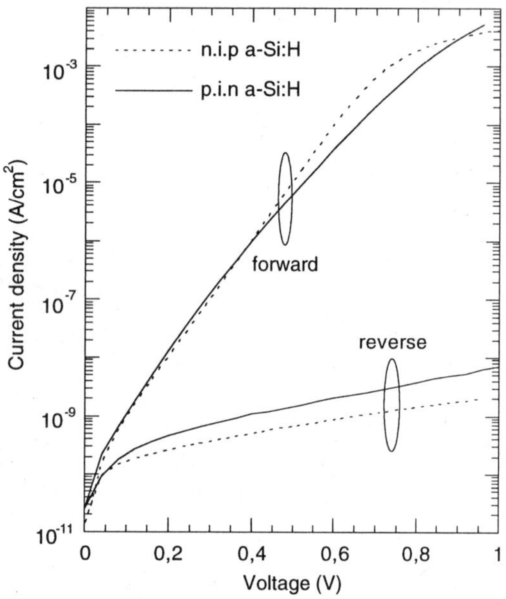

Fig. 3. Comparison of the forward and reverse J-V characteristic for p.i.n and n.i.p diodes deposited on SnO_2:F with the same i-layer thickness.

TABLE III - Electrical properties of p.i.n and n.i.p diodes

	p.i.n	n.i.p
J_0 (|V|=1V)	7.3×10^{-9}	2.0×10^{-9}
Rectification ratio |V|=1V	6.8×10^{5}	2.0×10^{6}
Diode quality factor, η	1.8	1.6

SIMS analysis

The effect of i layer contamination was analysed by SIMS on p.i.n and n.i.p devices whose electrical characteristics are shown in Table III. The SIMS results are depicted in Figs. 4.a and b. and the concentration of boron (B) and phosphorous (P), detected in the three layers, shown in Table IV. These data reveal a less contamination of the i-layer when the first deposited layer is n-type. The SIMS depth profiles also show an abrupt profile of P compared to B, which lead to an enhancement of the interface states towards the i-layer in p.i.n devices.

TABLE IV - Impurities in p.i.n and n.i.p solar cells.

Concentration (at.cm^{-3})	p.i.n	n.i.p
p - layer	7.5×10^{19}	1.0×10^{20}
n - layer	2.0×10^{21}	1.7×10^{21}
i - layer		
boron	1.0×10^{18}	$<1.0 \times 10^{17}$
phosphorous	1.0×10^{19}	2.3×10^{19}

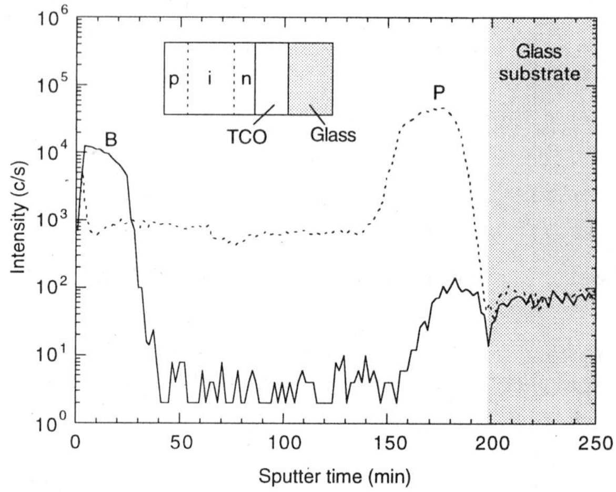

a.

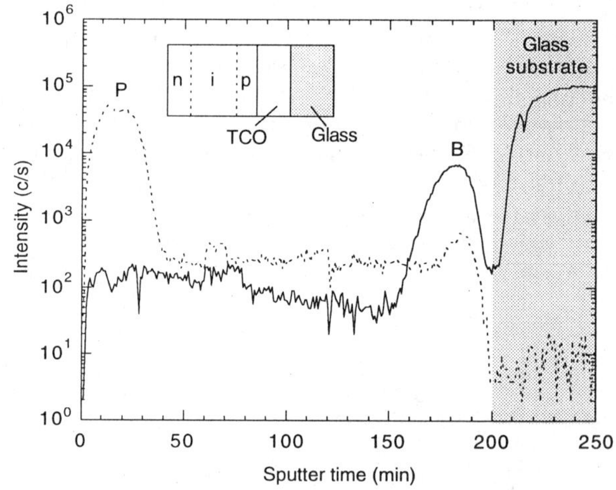

b.

Fig. 4. a. Distribution of impurities in a n.i.p and b. p.i.n solar cell, by SIMS analysis.

Spectral response

Fig. 5 a. and b. show the spectral response of the p-i-n solar cells studied, before and after degradation (80h). The highest peak efficiency, before degradation, was recorded on the devices deposited on high transparent TCOs, with small H_r, independent of their composition (TCO C and A). This clearly indicates that the high carriers collection is not always related to an enhancement of H_r. After light degradation, the best performances are obtained in the devices deposited on SnO_2:F substrates, proving so a better chemical stability of SnO_2:F when compared to the ITO ones. This confirms our previous data concerning the electrical performances of the devices deposited on the TCOs under analysis. The data also reveal that the device deposited on TCO B (ITO) show the highest change of the SR (peak shift towards high λ, with a sharp decrease on the SR towards the blue range). This behaviour is an add in proof of the deleterious effect of high H_r and In presence, on device performances.

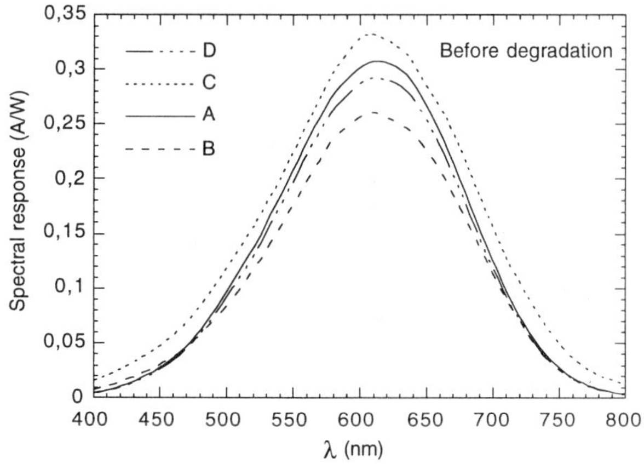

Fig. 4a. SR as a function of wavelength for p.i.n solar cells (with different TCOs) before degradation.

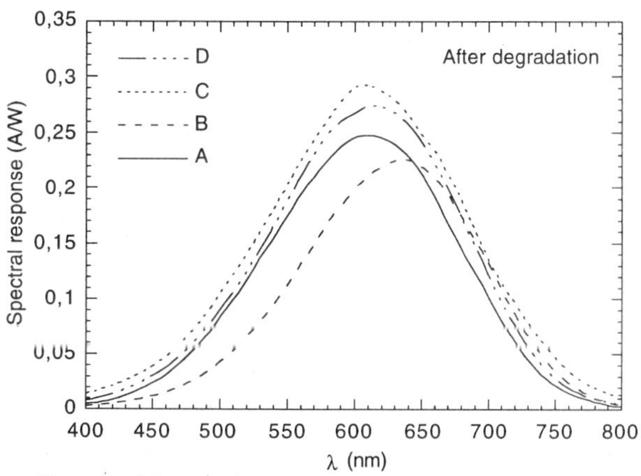

Fig. 4b. SR as a function of wavelength for p.i.n solar cells (with different TCOs) after degradation.

The results show a better stability for the devices deposited on milky-like TCOs. Besides this, a high degradation effect is observed for devices deposited on ITO (19,5% and 14,3%), than those ones deposited on SnO_2:F (10,2% and 5%), for the transparent and milky-like substrates, respectively.

CONCLUSIONS

The J-V characteristics, under dark conditions, are strongly influenced by H_r. The data also show that device electrical characteristics are less affected when SnO_2:F films are used instead of ITO. By degradating the devices on TCO C and D, η still remains <2 (changes about 15%) whilst for the TCO A and B, η>2 indicating an abnormal J-V behaviour. This can be explained by the enhancement of the recombination mechanism at the interface and by the presence of an intermetallic phase in the interface [10]. The J_F/J_R decreases about one order of magnitude on devices produced on TCOs A and B, whilst for those ones deposited on TCOs C and D, only changes about 22%. This indicates a better chemical stability of SnO_2:F, when compared to ITO. This type of behaviour is also confirmed by the spectral response data, that also reveal good optical stability on the devices deposited on SnO_2:F, with milky-like structure with a relative H_r (≤20%).

ACKNOWLEDGEMENTS

The authors are grateful to F. Lopez for the SIMS analysis, J. Puigdollers for the helpful discussions as well as T. Chee for the samples etching. The authors also thanks "Fundação Luso Americana para o Desenvolvimento" for the grant given. This work is supported in part by the Ministry of Industry and Energy (PEDIP nº 0250) and by NATO POTHINFILM project.

REFERENCES

[1] D. Staebler & C. Wronski, "Reversible Conductivity Changes in Discharge-Produced Amorphous Si", *Appl. Phys. Lett.*, **31**, 1977, pp. 292-293.

[2] H. Tasaki, W.Y. Kim, M. Hallerdt, M. Konagai & K. Takahashi, "Computer Simulation Model of the Effects of Interface States on High-Performance Amorphous Silicon Solar Cells", *J. of Appl. Phys.*, **63**,1977, pp. 550-560.

[3] F. Smole & M. Topic, "Effects of TCO/a-Si:C:H Interface Defect States on p.i.n a-Si:H Solar Cell Performance", *Mat. Res. Soc. Symp.*, 1994, in print.

[4] F.S. Sinencio & R. Williams, "Barrier at the Interface between a-Si:H and Transparent Conducting Oxides and its Influence on Solar Cell Performances", *J. of Appl. Phys.*, **54**,1983, pp. 2757-2760.

[5] R. Martins, I. Ferreira, N. Carvalho & L. Guimarães, "Engineering of Plasma Deposition Systems Used for Producing Large Area a-Si:H Devices", *J. of Non-Cryst. Solids*, **137-138**,1991, pp. 757-760.

[6] R. Martins, E. Fortunato, A. Bicho & G. Lavareda, "Role of Lateral Leakage Current on Amorphous Silicon Solar Cells", this conference.

[7] T. J. McMahon. B. G. Yacobi, K. Sadlon, J. Dick & A. Madan, "Excess Dark Currents in a-Si:H p.i.n Devices", *J. of Non-Cryst. Solids*, **66**, 1984, pp. 375-380.

[8] P. Fiorini and A. Mittiga, in "Amorphous & Microcrystalline Semiconductor Devices, Vol. II - Materials and Device Physics", J. Kanicki, ed. (Artech House, London), 1992.

[9] T. J. McMahon and A. Madan, "Photoconductivity and Electronic Doping Effects in a-Si:H", *J. Appl. Phys.*, **57**, 1985, pp. 5302-5305.

[10] J. M. M. de Nijs, C. Carvalho, M. Santos and R. Martins, "A Thin SiO Layer as a Remedy for the Indium Reduction at the $In_2O_3/\mu c$–Si:C:H Interface", *Ap. Surf. Science*, **52**, 1991, pp. 339–342.

IMPROVING THE STABILITY OF AMORPHOUS SILICON TANDEM CELLS

J. Daey Ouwens and R.E.I. Schropp
Utrecht University, Debye Institute, Dept. of Atomic and Interface Physics
P.O. Box 80.000, 3508 TA Utrecht, The Netherlands

ABSTRACT

We present an optimization procedure for *a*-SiC:H/*a*-Si:H tandem cells to limit light induced degradation effects. Optimum deposition conditions are combined with an optimum light trapping configuration, that has been determined by optical modelling. Upon light soaking, a remarkable light-induced behaviour is observed. The open circuit voltage (V_{oc}) in the degraded state is higher than in the annealed state.

INTRODUCTION

It is generally known that the efficiency of *a*-Si:H solar cells decreases in time due to the so-called Staebler-Wronski effect. Several attempts to improve the material quality using other deposition techniques like hot-wire or heated-mesh so far have not led to an improved stability of the cells incorporating these materials [1,2].

A better alternative is to reduce the cell thickness, leading to higher electric fields thereby limiting the recombination rates. For this reason, *a*-Si:H/*a*-Si:H tandem cells are more stable. However, we have recently shown that the performance of a (current) matched tandem with intrinsic layers of 65 and 500 nm respectively, is still quite unstable [3]. By separating the degradation of the top and bottom cell using spectral response measurements, the degradation behaviour appeared to be determined by the bottom cell. Consequently, the thickness of the bottom cell must be reduced. Since current matching is required and the top cell cannot be made thinner due to increased risk of shunts, this can only be done by increasing the band gap of the top cell. This is accomplished by adding methane and hydrogen to the silane gas.

OPTIMIZATION

First of all, we optimized our materials with respect to the band gap. The band gap of *a*-Si:H can be increased by increasing the hydrogen content. This leads however to an increased number of $Si-H_2$ bonds [4]. It is widely believed that electronic properties are deteriorated by these bonding configurations. To avoid these configurations, it is proposed to use carbon. However, it has been reported that the properties of *a*-SiC:H are determined by the hydrogen content rather than by the carbon content [5]. Moreover, the presence of carbon also leads to an increased density of $Si-H_2$ bonds [6].
It is concluded from infra-red measurements [7] that *a*-SiC:H films deposited without H_2 dilution consist of an *a*-Si:H phase with a low density of Si-C bonds and a high density of $Si-H_n$ bonds and an *a*-C:H phase where C-C bonds dominate. The properties of these films are determined by the *a*-Si:H like phase. This is why the optical and structural properties of *a*-Si:H and *a*-SiC:H are similar.

The bonding configurations in *a*-SiC:H, deposited using H_2 dilution are different. For these films, the amorphous network consists of a dense network of Si-C bonds.
Figure 1 shows that the band gap is also determined by the carbon content and that, at certain deposition conditions, *a*-SiC:H films can have significant higher gaps at the same hydrogen content as *a*-Si:H films. This is not valid for hydrogen diluted *a*-Si:H films.

Next, we have modelled the spectral response of the tandem cells using a statistical approach, taking light scattering due to the rough TCO into account [8]. In brief, the light is scattered at the TCO/p^+ interface under certain angles. At the n^+/silver interface at the back, the scattering is assumed to be random. To take the effect of index matching into account, two 1 nm thick buffer layers are added to the glass/TCO/p^+–i–n^+/p^+–i–n^+/metal structure, one between the TCO and p^+ layer and another between the p^+ and i layer. Interference effects are not taken into account, which is justified by the absence of an interference pattern in the spectral response of our cells. It is assumed that only light absorbed in the intrinsic layers contributes to the spectral response. The calculated and measured (by applying bias-light) spectral responses are in agreement for several test structures without changing experimental scattering input parameters.

As an example, the modelled spectral response of a 95/500 nm equal band gap tandem cell is shown in Fig. 2. The most interesting aim of the modelling procedure is to estimate the losses due to parasitic absorbances in the whole 340-800 nm range. For the structure in Fig. 2, the current losses in the

CH3365-4/94/0000-0650 $4.00 © 1994 IEEE

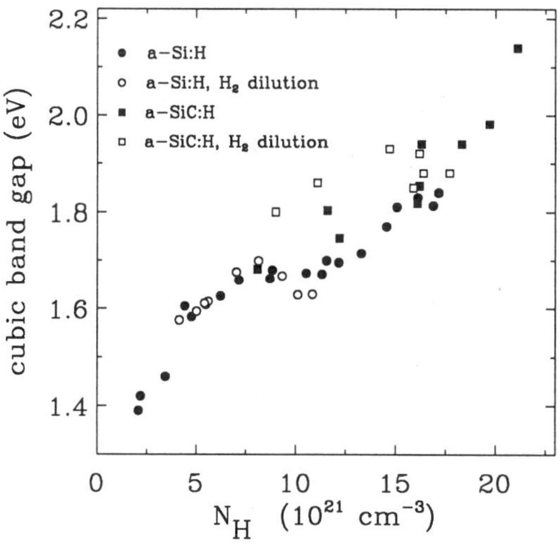

Figure 1: Band gap versus the hydrogen concentration for diluted/undiluted a-Si:H and a-SiC:H films.

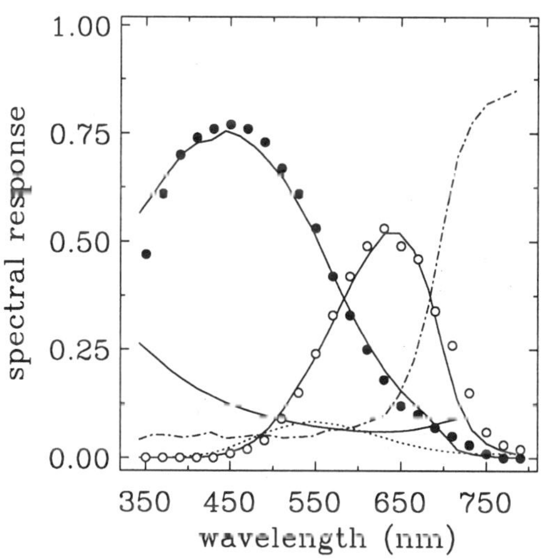

Figure 2: Modelled spectral response of bottom and top cell for a 95 nm/500 nm a-Si:H/a-Si:H tandem cell versus wavelength. The markers are the measured values. Also absorption losses in the inner n^+/p^+ junction (dotted curve), due to reflection (dashed curve) and in the TCO (continuous curve) are shown.

glass/TCO structure

equal 1.16 (glass) and 2.34 mA/cm^2 (Asahi 600 nm $SnO_2:F$). These values are comparable to the results of Wiedeman *et al.* [9]. They found 1.31 and 2.72 mA/cm^2 respectively. The low value for the loss in the TCO reported by Leblanc and Perrin [10], 1.16 mA/cm^2, could not be confirmed.

The losses in the doped layers are rather small, i.e. 0.50 mA/cm^2 for the first p^+ layer (including the buffer layers), 0.81 mA/cm^2 for the inner junction and 0.15 mA/cm^2 for the n^+ layer at the back. Wiedeman proposed 0.61, 0.56 and 0.17 mA/cm^2 respectively. Although these results are similar, we regard this fortuitous, since their structure is totally different from ours, e.g. their bottom cell is a SiGe:H alloy. We find that the losses in the doped layers strongly depend on the thicknesses of the intrinsic films. Nevertheless, these results show that the major optical absorption losses are in the glass and TCO. Efforts to increase the current in the cell should therefore be directed to the glass/TCO structure rather than minimizing absorption in the doped layers.

In addition, we have estimated the effect of applying a thin (10 nm) indium tin oxide (ITO) film between the n^+ layer and the silver back contact. The current densities in the two intrinsic layers were reduced by $\sim$ 0.05 mA/cm^2 due to extra absorption in the ITO film. This is in agreement with results of Miyachi *et al.* [11], namely that the reflection of a ITO/Ag contact is lower than of a single Ag contact. The suggestion of Beneking *et al.* [12] that the current density could be enhanced by $\sim$1 mA/cm^2 by applying a highly reflecting backcontact could not be confirmed. On the other hand, this discrepancy may be related to the assumption that the light at the back is randomly scattered.

The optimum dilution ratio, defined as $[H_2]/[CH_4]+[SiH_4]$, depends on the silane/methane ratio in the gas phase. If the methane flow is zero, the optimum ratio is close to 0.5 [13] and if the methane/silane ratio is 4, the optimal ratio is $\approx$20 [7]. We have deposited several films varying the methane/silane ratio between 0 and 0.7. We tried an 'average' optimum dilution ratio of 10. The optical parameters of these film were measured using transmission/reflection measurements. Then, with the aid of the model, we match our cells. Figure 2 shows the matched cases versus the methane flow. The short circuit current density was fixed at 7 mA/cm^2. This figure shows the enormous potential for tandem cells using unequal band gap components, i.e. the thickness of the bottom cell can be reduced by more than 200 nm on the cost of only a small increase of the thickness of the top cell. Consequently, the stability is dramatically improved. Another advantage is that these structures are less sensitive to thickness as well as spectral mismatches.

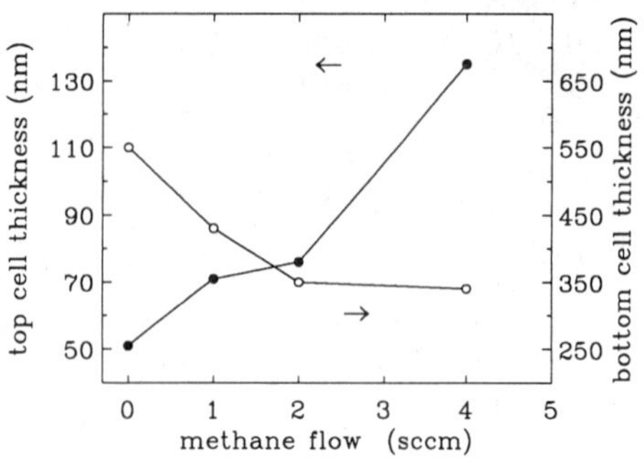

Figure 3: Calculated thicknesses of top and bottom cell for matched cases versus methane flow. The hydrogen flow is fixed at 100 sccm and the sum of the silane and the methane flow is 10 sccm. No hydrogen dilution is applied for the bottom cell.

This is caused by a more equal distribution of the spectral absorbance over the two cells. For very high methane flows ($[CH_4] > 2$ sccm), the thickness of the top cell gets too thick to be applied in the tandem; the fill factor will dropp.

In summary, the optimised material is deposited using silane (8 sccm), methane (2 sccm) and hydrogen (100 sccm) as source gases. These films combine a band gap of 1.9 eV with a low density of $Si - H_n$ bonds. The best films were applied in *a*-SiC:H cells. In single junctions, we have reached a V_{oc} of 0.92 V for intrinsic layers having a (cubic) band gap of 1.9 eV. The V_{oc} is ~0.1 V higher than for a cell with an intrinsic layer having a gap of 1.7 eV. The short circuit current of the *a*-SiC:H cells with an intrinsic layer thickness of 100 nm exceeded 7 mA/cm^2. The fill factor *ff* is equivalent to the *ff* of an *a*-Si:H cell (*ff*~0.66). This cell is used as the top cell in a *a*-SiC:H/*a*-Si:H tandem.

RESULTS

Figure 4 shows the degradation behaviour of a 100/300 nm *a*-SiC:H/*a*-Si:H tandem (cell A) under open circuit conditions (AM1.5G). The inner n^+/p^+ junction consist of a microcrystalline n^+ layer and an amorphous p^+ layer. The efficiency increases upon light soaking, caused by an increase of the V_{oc}.
The efficiency of this tandem after 350 hours of light soaking was 6.51 % (V_{oc}=1.66 V), which is still higher than the initial efficiency, η=6.21 % (V_{oc}=1.57).

First, we have verified that the effect is reversible. After 5 hours annealing (T = 150°), the V_{oc} was again lower than in the degraded state, but the efficiency was not completely restored. Upon light soaking, the degradation behaviour is reproduced (cell B). Second, we checked the reproducibility. The deposition was repeated (cell C) and a similar light soaking effect is observed.

Third, we note that the increase of the V_{oc} upon light soaking is also observed for thin single junctions, irrespective of the alloy composition of the i layer. Sometimes, the fill factor increases too. Therfore, the observed trends in Fig. 4 are not unique for tandem cells. This suggests that the increase of the efficiency is not related to the inner junction. To confirm this, the degradation of a $a-n^+/p^+$ incorporated tandem was monitored (cell D). Again, at the beginning, the efficiency increases.

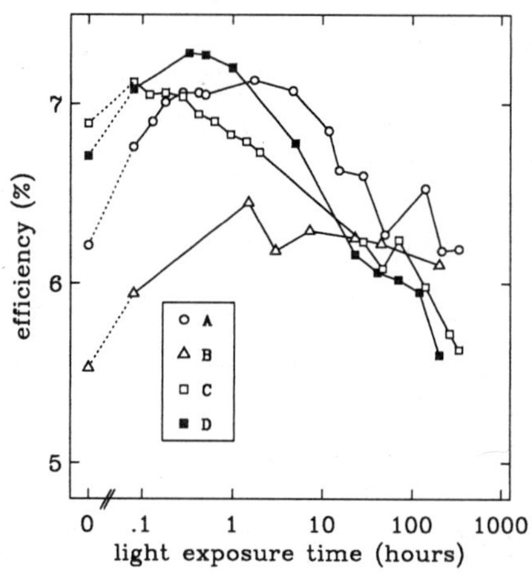

Figure 4: Efficiencies at T = 40° versus light soaking time for tandem cells. (note that the initial values were measured at T = 25°). **A**: 100 nm (SiC)/300 nm tandem, $\mu c-n^+/p^+$, **B**: cell A after annealing. **C**: repeated cell A and **D**: as A but *a*-n^+/p^+ inner junction. Note the inverse trend; the higher the efficiency in the annealed state, the lower the degraded efficiency.

These observations demonstrate that the annealed state is not always the optimal state.

Then, we measured the spectral response of cell C in the annealed state and in the degraded state (after 50 hours soaking). Figure 5 shows the ratio of the two spectral responses. For the top cell, the collection of the blue part of the spectrum is improved in the degraded state while the collection of the red has decreased. The opposite is observed for the bottom cell. Modelling [14] has shown that the electric field close to the p^+/i and n^+/i interfaces increases due to light induced charged defects. A high density of D^+ (D^-) centers at the front (back) of the cell enhances the collection of the holes (electrons). As a result, the field in the bulk is reduced. In a tandem cell, red light

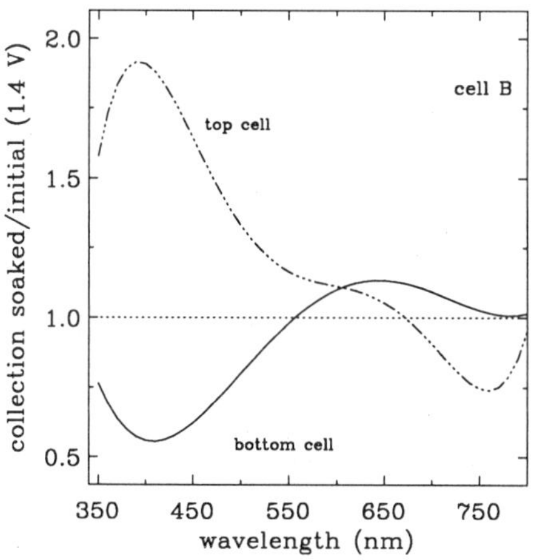

Figure 5: Ratio of the spectral response of cell B after 50 hours light soaking and initial (annealed state), respectively. The bias voltage is 1.4 V.

is collected by the top cell at the back and blue light is colllected by the bottom cell at the front. Fig. 5 thus suggests that the field in the regions on both sides of the central junction is reduced. The conclusion of this surprising result is that the field in the regions near the junction is reduced as if this was the bulk of a single junction.

This mechanism can also explain why the increase of the efficiency is observed only for thin cells (d <500 nm), namely because the relative contribution of the bulk of the i layer is small compared to thick cells.

In summary, using unequal band gap tandem cells, the stability is improved. While for a-SiH/a-Si:H tandem cells, the degradation yielded more than ≈30% [3], the degradation of all the a-SiC:H/a-Si:H tandem cells is less than ≈15%. A part of this improvement is caused by an increase of the V_{oc} upon light soaking.

ACKNOWLEDGEMENTS

We gratefully acknowledge C.H.M. van der Werf and F. Achten for technical assistance and J. Wallinga for the TCO characterisation. We thank the Asahi Glass Company for supplying the TCO substrates. This work has been supported by The Netherlands Agency for Energy and Environment (NOVEM).

REFERENCES

[1] B.P Nelson, E. Iwaniczko, R.E.I. Schropp, H. Mahan, E. Molenbroek, S. Salamon and R. Crandall, *Proc. of the 12th E.C. Photovoltaic Solar Energy Conference, 11-15 April 1994, Amsterdam, The Netherlands*, p683-686.

[2] P. Stradins, H. Fritzsche and M.Q. Tran, *to be published in: Proc. of the MRS spring meeting, a-Si:H Technology, San Fransisco* (1994)

[3] J. Daey Ouwens, H. de Gooijer and R.E.I. Schropp, *Proc. of the 12th E.C. Photovoltaic Solar Energy Conference, 11-15 April 1994, Amsterdam, The Netherlands*, p140-143.

[4] J. Daey Ouwens, R.E.I. Schropp and W.F. van der Weg, *Appl. Phys. Lett.* **92**, 204 (1994)

[5] W. Beyer and H. Mell, in *Disordered materials, edited by M.A. Kastner, G.A. Thomas and S.R. Ovshinsky (Plenum, New York, 1987)*, p641-657.

[6] R.A.C.M.M. van Swaaij, A.J.M. Berntsen, W.G.J.H.M. van Sark, H. Herremans, J. Bezemer and W.F. van der Weg, *J. of Appl. Phys.* **2**, 251 (1994).

[7] J. Daey Ouwens, R.E.I. Schropp and W.F. van der Weg, *Sol. State. Comm.* **92**, 853 (1994).

[8] J. Daey Ouwens, R.E.I. Schropp, J. Wallinga, W.F. van der Weg, M. Ritala, M. Leskelä and J. Hyvärinen, *Proc. of the 12th E.C. Photovoltaic Solar Energy Conference, 11-15 April 1994, Amsterdam, The Netherlands*, p1296-1299.

[9] S. Wiedeman, J. Morris and L. Yang, *Proceedings of the 20th IEEE PV conference (Kissimmee, Florida, 1990)*, p1529-1534.

[10] F. Leblanc and J. Perrin, *J. Appl. Phys.* **75**, 1074 (1994)

[11] K. Miyachi, N. Ishiguro, T. Miyashita, N. Yanagawa, H. Tanaka, M. Koyama, Y. Ashida, and N. Fukuda, *Proc. of the 11th E.C. Photovoltaic Solar Energy Conference, 12-16 October 1992, Montreux, Switzerland*, p88-91.

[12] C. Beneking, B. Rech, T. Eickhof, Y.G. Michael, N. Schultz, and H. Wagner, *Proc. of the 12th E.C. Photovoltaic Solar Energy Conference, 11-15 April 1994, Amsterdam, The Netherlands*, p683-686.

[13] H. Meiling, W. Lenting J. Bezemer and W.F.van der Weg, *Phil. Mag. B* **62**, 19 (1990)

[14] D. Fischer, N. Pellaton, A. Shah and C.M. Fortmann, *Proc. of the 11th E.C. Photovoltaic Solar Energy Conference, Montreux, Switserland*, p560-563.

PARAMETRIC ANALYSIS OF a-SI SOLAR CELLS FROM CURRENT VOLTAGE MEASUREMENTS

Stêven S. Hegedus and J.E. Phillips
Institute of Energy Conversion, University of Delaware

United States Department of Energy
University Center of Excellence
for Photovoltaic Research and Education
(National Renewable Energy Laboratory)
Newark, Delaware 19716 USA

ABSTRACT

A procedure is presented for accurately representing the current voltage (JV) characteristic of a-Si solar cells. The measured JV data is separated into the forward diode current and photocurrent. The forward diode current is characterized to obtain standard diode parameters A, J_o, and R. The AM1.5 photocurrent $J_L(V)$ is characterized using the uniform field model to determine the ratio of collection length to thickness L_c/D, flat band potential VB, and total photocurrent J_{LO}. Very good agreement is found between measured and calculated currents from reverse bias to V_{oc} in a variety of cells. We find that degradation of FF under light soaking from 69% to 49% can be explained solely by a decrease in a single parameter L_c/D, from 14 to 3. By examining cells with D from 0.2 to 0.6 μm we find that $\mu\tau$ ranges from $2\text{-}4 \times 10^{-8} cm^2/V$.

INTRODUCTION

Analysis of illuminated current voltage (JV) and quantum efficiency (QE) measurements of a-Si solar cells is complicated by the voltage dependent photocarrier collection. Assuming that the field and generation profiles are spatially uniform leads to a single analytical expression for the voltage dependence of the photocurrent dominated by a single parameter, the ratio of the collection length to thickness L_c/D [1,2,3]. Previous applications of the collection length model have been based on analyzing QE measurements taken at a single wavelength with a chopped detection system thereby neglecting the influence of the forward diode current. In this paper, we analyze the standard AM1.5 JV data, including the forward diode current. Our goal is to obtain a simple but accurate parametric description of single junction a-Si solar cell JV behavior which can be extended to modeling multijunction cells.

EXPERIMENT

Single junction p-i-n cells were deposited by plasma CVD at IEC and Advanced Photovoltaic Systems (APS). The IEC devices were deposited in a small area research system and the APS devices were from modules produced in a manufacturing system. The device structure was glass/textured SnO_2/p (a-SiC)/i(a-Si)/n(a-Si)/metal. The i-layer thickness D was 0.4 μm for the cells made at IEC and varied from 0.2 to 0.6 μm for the cells made at APS. The IEC devices were exposed to 90 hours of AM1.5 light soaking. Illuminated JV characteristics were measured with a AM1.5 global simulator at 30°C. The only data required for the analysis are measurements of the JV characteristic from reverse bias to beyond V_{oc} in the light and in the dark (or at very low intensity) at a fixed temperature.

ANALYSIS

The analysis is based on applying established models of the a-Si material and device behavior. Applicability of these models is tested by the agreement between calculated and experimental data. The forward diode current is assumed to be exponential with voltage and determined by recombination with an exponential density of gap states. It is also assumed that the forward bias diode current is independent of light intensity. The AM1.5 light-generated photocurrent is analyzed by applying the model of Crandall [2] which assumes a uniform field F and uniform generation, leading to a constant collection length $L_c = \mu\tau F$.

The net current through the device is the difference between the dark and light currents,

$$J(V) = J_D(V) - J_L(V). \quad (1)$$

$J_D(V)$ is the forward bias diode current and $J_L(V)$ is the photocurrent. The diode current is given by the standard expression

$$J_D(VD) = J_o \{\exp(qVJ/AkT) - 1\} \quad (2)$$

where the voltage V_J across the junction (i.e. across the i-layer) is reduced from the terminal voltage V by the series resistance R

$$V_D = V - \{R * J(V)\} \quad (3)$$

CH3365-4/94/0000-0654 $4.00 © 1994 IEEE First WCPEC; Dec. 5–9, 1994; Hawaii

The diode factor A should be between 1 and 2 for an exponential distribution of gap states. The derivative dV/dJ from Eqs. 1-3 yields (for $J_L(V) << J_D(V)$)

$$dV/dJ = R + (AkT/q) * \{1/J(V)\}. \quad (4)$$

Equations 1-4 were applied to analyze dark JV data. When dV/dJ is plotted against 1/J(V), a straight line is found whose intercept is R and slope is AkT/q, as seen in Figure 1. Using the value of R from Figure 1, a graph of log J_D vs V_D yields a straight line whose slope is 1/AkT and whose intercept is J_o, as seen in Figure 2. Values are given in Table 1. Figures 1 and 2 verify that the diode behavior is exponential with the junction voltage V_D and limited by a fixed series resistance R. Note that A found by the two methods is equivalent.

The photocurrent $J_L(V)$ in a-Si solar cells decreases with increasing bias because of the reduction in the internal electric field. The field assumed to vary as

$$F = (V_B - V_D)/D. \quad (5)$$

The field goes to 0 when the junction voltage equals the flat band voltage V_B. Experimentally, V_B is the voltage at which $J_L(V) = J_D(V) - J(V) = 0$. In comparison, V_{oc} is the voltage at which J(V)=0.

Crandall [2] derived an expression for $J_L(V)$ as

$$J_L(V) = J_{LO} * X * \{1 - \exp(-1/X)\} \quad (6)$$

where J_{LO} is the maximum (optically limited) photocurrent at far reverse bias, X is

$$X = (L_o/D) * \{1 - (V/V_B)\} \quad (7)$$

and

$$L_c/D = (\mu\tau F)/D = \mu\tau V_B/D^2. \quad (8)$$

The photocurrent $J_L(V)$ was obtained by subtracting the diode current $J_D(V)$ from the measured current under illumination and was fit using equation 6. The parameters L_c/D and V_B were varied until an optimum fit was obtained. (Occasionally, J_L was adjusted slightly to get better matching with the measured current at reverse bias.) The V_B value which resulted was always within 10 mV of the value determined experimentally. Thus, L_c/D is the primary fitting parameter, in agreement with others [1-3]. Figure 3 shows the measured and calculated J(V) data under AM1.5 illumination for the same cell (IEC#4326-22) whose dark $J_D(V)$ is analyzed in Figs. 1 and 2. $J_L(V)$ was calculated with $L_c/D=17$ as shown in Table 1. Agreement is within 0.3 mA/cm^2 from -1V to V_{oc}. This demonstrates that the simplified analysis is valid for devices with high FF.

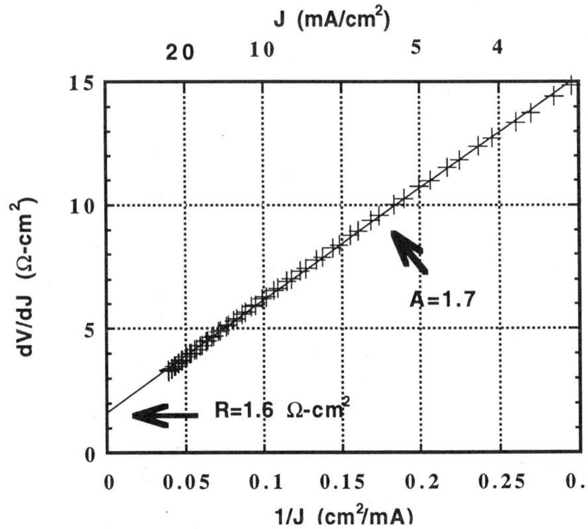

Fig. 1. dV/dJ vs. 1/J for dark diode.

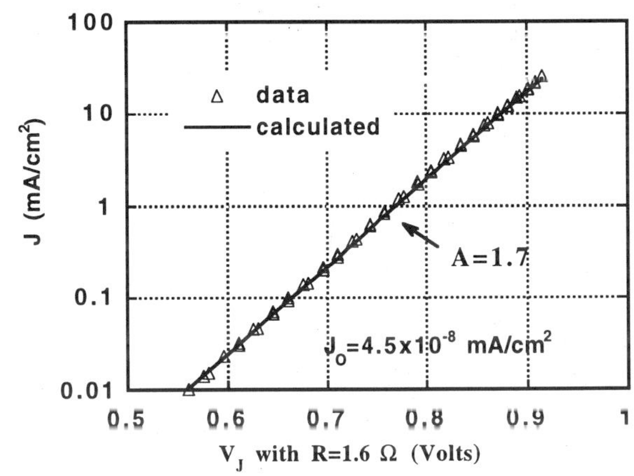

Fig.2. Log J vs. V_J for dark diode, with V_J calculated using R=1.6 Ohms.

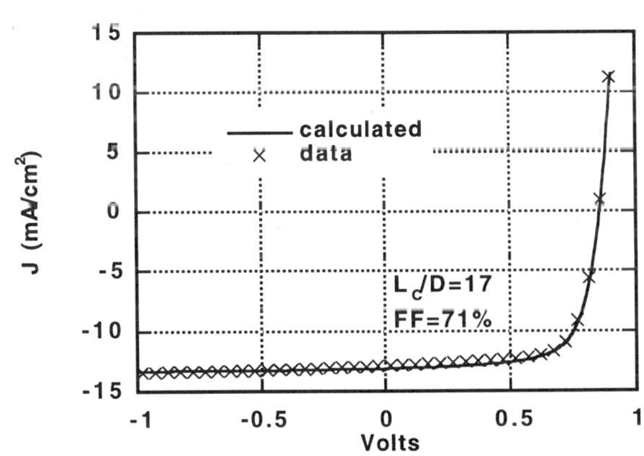

Fig. 3. Measured and calculated light JV values using parameters from Figs. 1,2 for J_D and fitting L_c/D and V_B.

EFFECT OF I-LAYER THICKNESS

The analysis was applied to devices deposited at APS having D from 0.2 to 0.6 μm. Results are shown in Table 2. Note that as D increased, FF decreased from 74% to 63% and L_C/D decreased from 27 to 8.5. Figure 4 shows the $\mu\tau$ value determined from L_C/D using Eq.8, suggesting an improvement in i-layer quality with increasing D. This may be due to greater defect density in material grown near the substrate, or due to the relatively greater influence of interface recombination in thin i-layers. The technique presented here yields a single value of $\mu\tau$ for each device representing the average throughout the i-layer. Values of $\mu\tau$ for holes of 0.5-3x10^{-8} cm^2/V are commonly reported [3-6] while values of $\mu\tau$ for electrons are in the range from 10^{-7} to 10^{-6} cm^2/V. Thus, L_C as determined here is the hole collection length. These measurements support the "limiting carrier" argument [7], i.e. the carrier with the shorter collection length or $\mu\tau$ (ordinarily holes) limits the collection.

EFFECT OF LIGHT SOAKING

It is well known that light exposure increases the defect density in a-Si material and reduces the $\mu\tau$ product of electrons and holes. Therefore, it is expected that L_C/D should change with light soaking.

A device deposited at IEC was characterized in the annealed state and after 90 hours light exposure as shown in Table 3. The FF decreased from 69 to 49% corresponding to L_C/D decreasing from 14 to 3. The dark diode and photocurrent parameters changed with light exposure as shown in Table 3. Figure 5 shows the excellent agreement between measurement and calculation using the appropriate parameters from Table 3. To determine if effect of light soaking could be described solely by changes in L_C/D, the AM1.5 illuminated JV data after light soaking were also fit with the same five parameters (A, J_O, R, J_{LO}, and V_B) as determined for the annealed case but with L_C/D=3 as found from light soaking. Figure 5 shows a very good fit to the light soaked data changing only L_C/D from 14 to 3. Clearly, the effect of light soaking can be represented primarily by changes in L_C/D.

CONCLUSIONS

The method presented here allows routinely measured JV data to be analyzed to yield parameters containing fundamental information about the i-layer material quality and its relation to the device structure and performance. The uniform field approach [1,2] contains broad approximations and assumptions. Yet, surprisingly good agreement is found between AM1.5 JV measurements and calculation on cells with very good and very poor FF. The method is closely linked to experiment. Five parameters are obtained directly from the measured data, requiring only one parameter L_C/D to be fitted. The method has been successfully applied to cells with FF from 74 to 49% and with thicknesses from 0.2 to 0.6 μm The $\mu\tau$ values (2-4 x10^{-8} cm^2/V) are consistent with hole transport. Effects of light induced degradation can be modeled with only one parameter, L_C/D.

ACKNOWLEDGEMENTS

This work was supported by NREL under subcontract #X-AV-13170-01. We thank Herman Volltrauer of APS and Frank Kampas for providing the production cells with a range of thickness, and Jeff Janukowicz and Sally Buchanan for measurements.

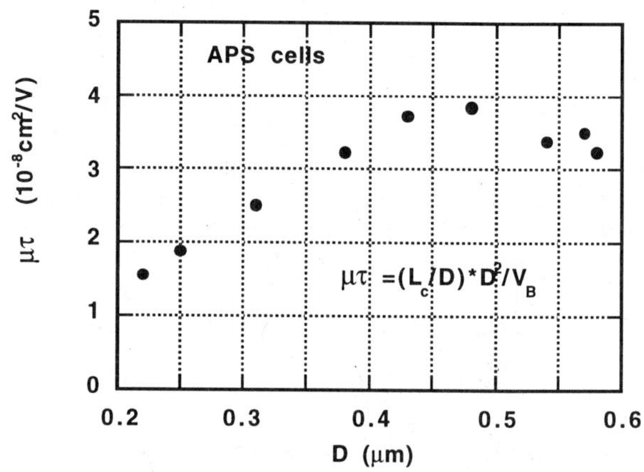

Fig. 4. $\mu\tau$ from L_C/D vs thickness for APS cells

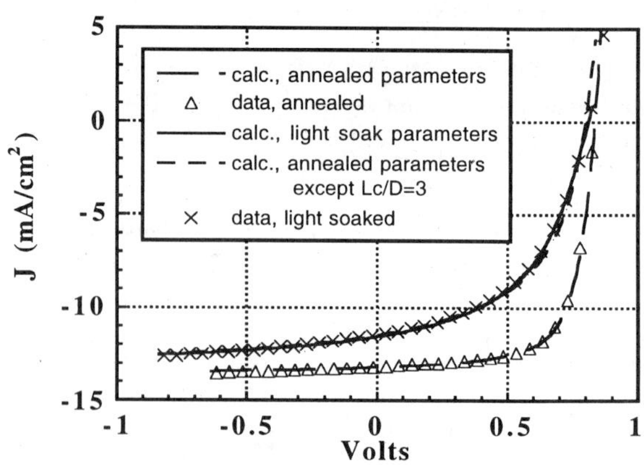

Fig. 5. Calculated and measured JV for annealed and light soaked states. Solid light soaked curve calculated with appropriate parameters from Table 3, dashed curve using same parameters as annealed except L_C/D=3.

REFERENCES

1. B.Faughnan, R. Crandall *APPL. PHYS. LETT.* **44**, 1984, p. 537.
2. R. Crandall, J. APPL. PHYS. **54**, 1983, 7176.
3. R. Crandall, K. Sadlon, J. Kalina, A. Delahoy, *Proc. Mat. Res. Soc. Symp.* Vol. **149,** 1989, p. 423.
4. R. Arya, A. Catalano, R. Oswald, *APPL. PHYS. LETT.* **49**, 1986, p. 1089.
5. I. Balberg, A. Delahoy, H. Weakliem, *APPL. PHYS. LETT.* **53**, 1988, p.992.
6. J. Liu et al., *Proc. 21st IEEE PVSC,* 1990, p. 1606.
7. K. Misiakos, F. Lindholm, *J. APPL. PHYS.* **64**, 1988, p. 383.

Table 1. Extracted device parameters and measured and calculated solar cell performance for IEC#4326-22-1 (Figs. 1-3).

A	J_O (mA/cm^2)	R_s (Ω-cm^2)	J_{LO} (mA/cm^2)	V_B (V)	L_c/D	V_{oc} (V)	J_{sc} (mA/cm^2)	FF (%)	Eff. (%)	
1.7	4.5x10-8	1.6	13.5	0.89	17	0.853	13.0	70.8	7.9	calc
						0.854	12.9	71.7	7.9	data

Table 2. Extracted device parameters and measured and calculated solar cell performance for APS devices having 0.2<D<0.6 µm.

D (µm)	R_s (Ω-cm^2)	A	J_O (mA/cm^2)	J_{LO} (mA/cm^2)	V_B (V)	L_c/D		V_{oc} (V)	J_{sc} (mA/cm^2)	FF (%)	Eff. (%)
0.22	2.1	1.54	$7x10^{-9}$	10.4	0.837	26.8	calc	0.815	10.2	73.8	6.1
							data	0.816	10.1	74.2	6.1
0.25	2.4	1.48	$5x10^{-9}$	10.6	0.827	24.9	calc	0.803	10.3	74.1	6.1
							data	0.803	10.3	73.5	6.1
0.31	2.8	1.54	$8x10^{-9}$	12.6	0.863	22.5	calc	0.827	12.3	71.7	7.3
							data	0.829	12.2	72.2	7.3
0.38	3.5	1.54	$7x10^{-9}$	13.6	0.850	19	calc	0.823	13.2	69.5	7.6
							data	0.826	13.2	69.5	7.6
0.43	2.5	1.54	$9x10^{-9}$	11.1	0.853	17.2	calc	0.814	10.8	70.5	6.2
							data	0.812	10.7	71.3	6.2
0.48	2.6	1.67	$1x10^{-8}$	12.4	0.867	14.5	calc	0.851	12.1	69.5	7.1
							data	0.853	12.0	69.5	7.1
0.54	2.2	1.67	$1x10^{-8}$	12.1	0.880	9.8	calc	0.852	11.5	65.0	6.4
							data	0.854	11.6	64.2	6.4
0.57	3.5	1.67	$1x10^{-8}$	12.1	0.861	10	calc	0.840	11.5	63.8	6.2
							data	0.844	11.6	63.2	6.2
0.58	2.7	1.67	$1x10^{-8}$	12.1	0.883	8.5	calc	0.853	11.4	63.2	6.1
							data	0.853	11.5	62.8	6.1

Table 3. Extracted device parameters and measured and calculated solar cell performance for IEC #4304-22-2 in annealed and light soaked states.

State	A	J_O (mA/cm^2)	R_S (Ω-cm^2)	J_{LO} (mA/cm^2)	V_B (V)	L_c/D		V_{OC} (V)	J_{SC} (mA/cm^2)	FF (%)	Eff. (%)
Ann.	1.7	$7x10^{-8}$	1.7	13.7	0.87	14	calc.	0.834	13.2	68.9	7.6
							data	0.835	13.2	68.7	7.6
L.S.	2.3	$2x10^{-6}$	2.6	13.6	0.83	3.1	calc.	0.809	11.5	49.4	4.6
							data	0.809	11.5	49.0	4.6

INTERNAL OPTO-ELECTRIC PROPERTIES OF p-i-n a-Si:H SOLAR CELL ON GROOVED TCO TEXTURE

J.Furlan, S.Amon, P.Popović, F.Smole
Faculty of Electrical Engineering, University of Ljubljana
Tržaška 25, Ljubljana, Slovenia

ABSTRACT

Effects of textured TCO layer on a-Si:H solar cell electric properties are investigated by computer modelling of thermal equilibrium profiles of charge carriers, electric field and potential distribution. 2D analysis of p-i-n a-Si:H cell on TCO layer, simulated by V-grooved morphology, was performed using the commercial program MEDICI, for which input data were modified in order to account for continuous distribution of states in the gap of a-Si:H.

Results of calculations show strongly increased trapped hole concentration at the p-i interface around the peaks of the illuminated side of V-shaped grooves. At these locations, the built-in electric field is lowered so that the collection of light-generated carriers in these regions is reduced.

The paper also demonstrates the dependence of the light generation profile on the wavelength of incident light and on V-groove tilt angle.

INTRODUCTION

Low resistivity texturized transparent conducting TCO films deposited on glass substrates are used as front-side electric contacts to a-Si:H solar cells. The results of the first experimental studies of light trapping effects in a-Si:H cells connected to the grain size of SnO_2 TCO film were reported a decade ago [1,2]. Several studies which followed have revealed many details pertaining to electrical and optical properties of TCO layer contacting the semiconductor a-Si cell structure. The dependence of the output characteristics of a-Si cells is a complex function of various parameters, which depend on the choice of the TCO material (ITO, SnO_2, ZnO) and the morphology of the texturized TCO film on glass substrate [3,4,5].

In order to relate output a-Si cell characteristics to internal optoelectric properties, computer modelling and simulation exploring various effects of the TCO/a-Si interface has been introduced. One part of these simulations pertains to electrical properties of the interface between a-Si and TCO of different compositions, assuming a smooth TCO surface [6]. Optical properties of texturized cells were also analyzed using different approaches. One way of modelling the light trapping effect is based on Lambertian angular distribution of the scattered light at "randomly rough" interfaces [7,8]. Given that the thickness of p-i-n device is nearly the same as the grain size of the TCO film, a three-dimensional (or at least two-dimensional approximation) optical analysis of a regularly texturized surface would better simulate light trapping in thin film a-Si cells. This approach has been followed by Savada et al. [9], developing 2D numerical simulation of a-Si cell on texturized TCO substrate following a regular and continuous cosine function.

In contrast with the cited reference [9], this paper presents a two-dimensional analysis of a-Si cell on a regular V-grooved texture of TCO substrate, where the electric properties were calculated using the commercial program MEDICI [10], for which the input data were modified in order to account for the continuous distribution of states in the gap of a-Si:H.

ANALYSIS OF OPTO-ELECTRIC PROPERTIES

The morphology of texturized TCO layers depends strongly on the deposited transparent conductive material (ITO, SnO_2, ZnO and combinations) and the conditions of deposition technology. All possible surface shapes, ranging from spherical to pyramidal, including highly irregular forms can be seen in SEM photographs of various deposited TCO films. However, it seems that a pyramid-like TCO texture is the most promising one. When applied to a-Si:H solar cells, it yields the most favourable solar cell characteristics [3,5].

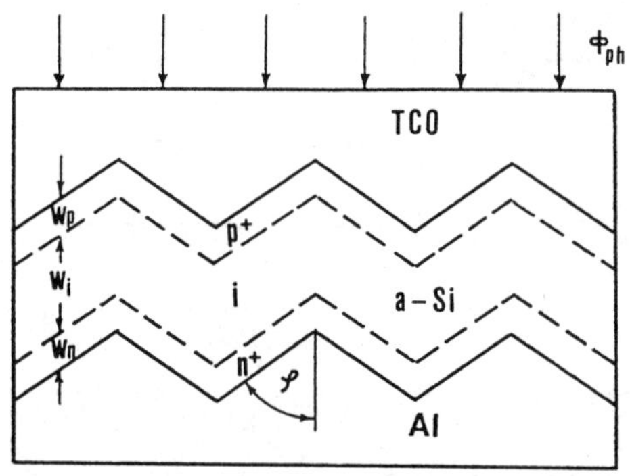

Fig.1. Cross-section of p-i-n a-Si:H cell on texturized V-grooved TCO substrate for 2D analysis.

In order to examine first order effects which can be expected in a-Si:H cells with a pyramidal TCO morphology, optoelectric properties were analyzed in a two-dimensional structure, shown in Fig.1, in which the texturized SnO_2 TCO layer was simulated by a regular V-grooved morphology. The structural parameters of this cell were as follows:

- impurity concentration in p^+ and n^+ layer, $N_A = N_D = 10^{18}\ cm^{-3}$,
- vertical thickness of p^+ and n^+ layer, $W_P = W_N = 0.02\ \mu m$,
- vertical thickness of the cell, $W_P + W_I + W_N = 0.7 \mu m$,
- tilt angle of grooved texture, $\varphi = 30°$ to $45°$,
- V-groove elevation $0.7 \mu m$

CH3365-4/94/0000-0658 $4.00 © 1994 IEEE First WCPEC; Dec. 5-9, 1994; Hawaii

Modelling of thermal equilibrium properties

Poisson's equation

$$\varepsilon \nabla^2 E = -q(p + p_t - n - n_t + N_D^+ - N_A^-) \qquad (1)$$

was solved using 2D program Medici by replacing free electron and hole concentrations with total concentrations of electrons, $n + n_t$, and holes, $p + p_t$, respectively.

For density of states in the gap of a-Si:H a symmetrical double exponential distribution of acceptorlike and donorlike tail states [11] was assumed

$$g(E) = g_A(E) + g_D(E) = A_{ct} e^{\frac{E - E_C}{W_C}} + A_{vt} e^{\frac{E_V - E}{W_V}} \qquad (2)$$

with selected parameter values $W_C = W_V = 0.052$eV, $A_{ct} = A_{vt} = A_t = 10^{21}$ eV^{-1} cm^{-3}, and mobility gap E_G=1.72eV.

Total concentrations of electrons and holes for such DOS distribution can be easily obtained from the formula

$$n + n_t = n^* = \int_0^\infty g_A(E) f_{FD}(E) dE = = A_{ct} W_C \frac{\frac{kT\pi}{W_C}}{\sin\frac{kT\pi}{W_C}} e^{\frac{E_F - E_C}{W_C}} = N_C^* e^{\frac{E_F - E_C}{kT_C^*}} \qquad (3)$$

$$p + p_t = p^* = \int_0^\infty g_D(E)(1 - f_{FD}(E)) dE = = A_{vt} W_V \frac{\frac{kT\pi}{W_V}}{\sin\frac{kT\pi}{W_V}} e^{\frac{E_V - E_F}{W_V}} = N_V^* e^{\frac{E_V - E_F}{kT_V^*}} \qquad (4)$$

where n* and p* denote the equivalent free carrier concentrations in program Medici, expressed by the equivalent effective densities of states, N_C* and N_V*, and by equivalent temperatures, T_C* and T_V*. Using the selected parameter values for tail state distributions, the calculated input data for program Medici were obtained:

$N_C^* = N_V^* = N^* = 8.2 * 10^{19}$ cm^{-3}

$T_C^* = T_C^* = T^* = 600$K.

Using these data, the 2D-solution of Poisson's equation generated the profiles of free and trapped carrier concentrations, electric field distribution, conduction and valence band energy profiles, and potential distribution inside the examined p-i-n cell on V-grooved TCO texture.

The most significant conclusions can be drawn from graphs in Figs.2.,3. and 4., showing the electric field in the p-i-n cell represented in three different ways. Fig.2 displays lines of equal absolute values of electric field strength. This graph is further illuminated by vector presentation of the electric field in Fig.3, pointing out the direction of electric force on charge carriers in the cell. The third mode of presentation, shown in Fig.4, depicts the electric field strength along preselected vertical cross-sections through a p-i-n cell. In accordance with these graphs, the following conclusions can be made :

Electric field reduction in the vicinity of p-i and n-i junction due to space charge of trapped carriers is clearly visible. Strongly pronounced charge carrier crowding exists around the peaks and troughs of the grooved texture (trapped holes at p-i junction on one side, trapped electrons at n-i junction on the other side), resulting in a correspondingly higher drop of electric field in these regions. This detrimental effect increases when the tilt angle of grooved texture decreases.

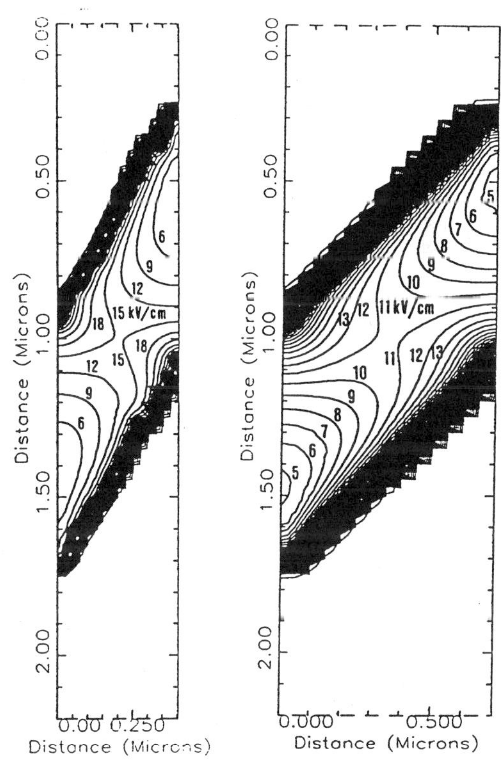

(a) (b)

Fig.2. Lines of equal values of electron field strength for V-groove tilt angle 30° (a) and 45° (b).

At the peaks and troughs of grooves, the electric field strength is directed normally to the flat glass substrate. Along the slopes of grooves however, the electric field is perpendicular to the tilted TCO surface, making the carrier paths shorter than the vertical thickness of the deposited i-layer.

Decreasing the tilt angle of grooved texture from 45° to 30° and keeping constant vertical thickness of p-i-n cell, the maximum electric field at p-i and n-i junction increases by about 20%, but it decreases more forcefully in the regions of increased trapped carrier concentrations around the peaks of the grooved texture.

Light generations

Since light generations equal the product of absorption coefficient α_{ph} and photon flux Φ

$$G_L = \alpha_{ph}\Phi \tag{5}$$

two-dimensional profiles of light flux were calculated considering light propagation across the interface between two media. Refraction angles and transmittances for monochromatic light at various wavelengths and angles of incidence were calculated first. Next, the transmitted fluxes in a-Si:H cell were obtained by a superposition of all contributions in x/y plane, not taking into account the interference effects of light.

Photon flux is greatest around the troughs of the V-shaped TCO on the illuminated side of the cell. In these regions, the light flux is composed mainly of rays refracted at both slopes of the grooves, so that the light flux in x/y plane, shown in Fig.5, can be expressed as

$$\Phi(x,y) = \Phi_0 2Tch(\alpha_{ph}\frac{\cos\varphi}{\cos\beta}x)e^{-\alpha_{ph}\frac{\sin\varphi}{\cos\beta}y} \tag{6}$$

where Φ_0 is the incident flux on TCO/a-Si interface, angle φ is tilt angle of grooves, α is incident angle, β is refraction angle given by

$$\sin\beta = \frac{n_1}{n_2}\sin\alpha = \frac{n_1}{n_2}\cos\varphi \tag{7}$$

and T is the transmittance

$$T = 2n_1 n_2 \cos^2\alpha\left[\frac{1}{(n_1\cos\beta + n_2\cos\alpha)^2} + \frac{1}{(n_1\cos\alpha + n_2\cos\beta)^2}\right] \tag{8}$$

with refractive indices of SnO_2 and a-Si, n_1 and n_2, respectively.

As an illustration, the calculated normalized flux $\Phi(x/y)/\Phi_0$ at wavelength λ=600nm for tilt angle 45° is shown in Fig.6. In graph, the calculated reflected light from Al/a-Si:H interface is also taken into consideration.

On the basis of performed calculations, the effects of V-grooved texture on light flux and light generations in p-i-n cell can be summarized as follows:

-Light generations at various locations in p-i-n cell strongly depend upon the incident angle of light.

-In case of normal incidence of light on glass substrate resulting in an incident angle to the TCO/a-Si:H interface, the greatest light generations occur at the bottom of grooves, where refracted fluxes from both slopes add up. The effect of this superposition grows with a decreasing tilt angle φ.

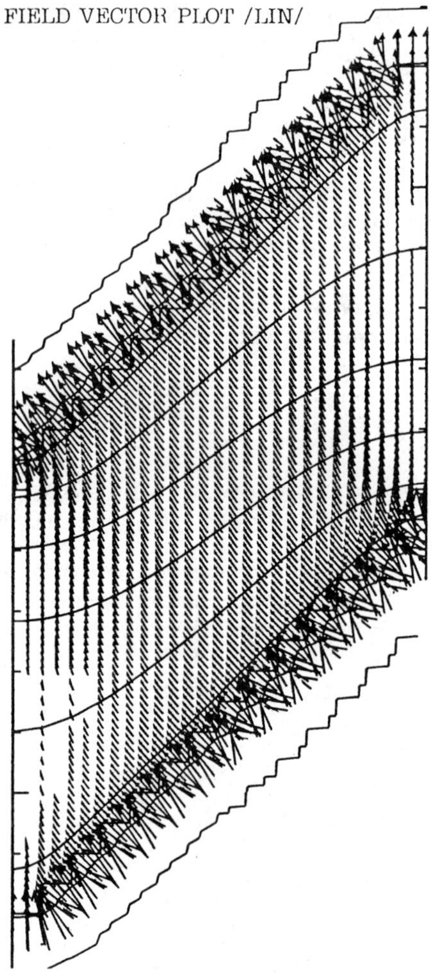

Fig.3. Electric field vector plot for structure with tilt angle 45°.

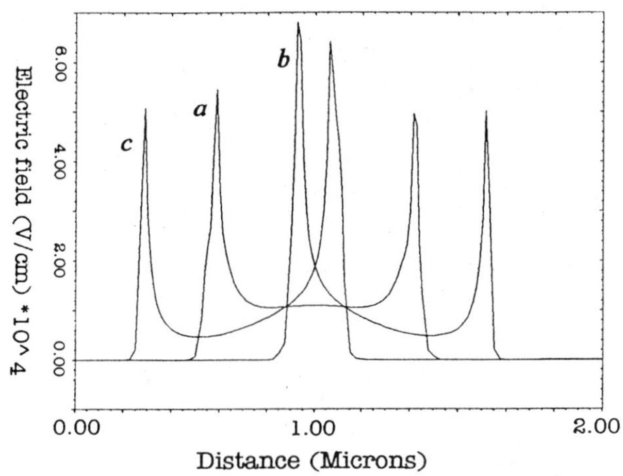

Fig.4. Electric field along three vertical cross-sections through p-i-n cell: center of 45° tilted slope (a), 0.35μm left from slope center (b), 0.35μm right from slope center (c).

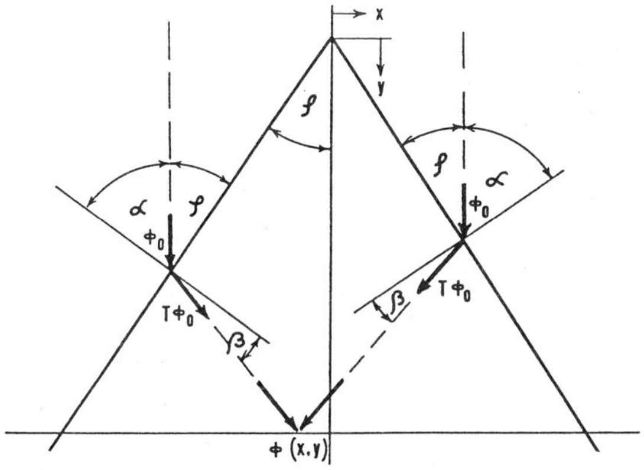

Fig.5. Light flux on illuminated side of p-i-n cell in grooves of TCO texture is the sum of fluxes refracted from both slopes of TCO/a-Si:H interfaces.

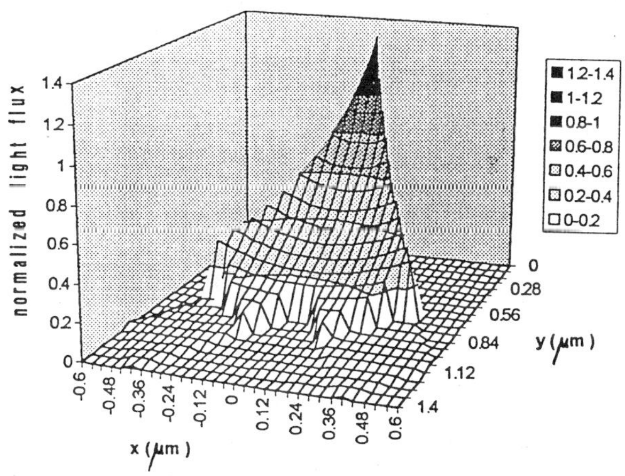

Fig.6. Plot of normalized flux Φ/Φ_0 in x/y plane at wavelength 600 nm and tilt angle 45°.

The location of greatest light generations almost coincides with the region of strongly reduced electric field, so that the collection of light generated carriers at these locations is attenuated.

The consequence of light refraction are nearly dead regions (visible also in Fig.6), where only back-reflected fluxes contribute to light generations.

Light generations depend on the wavelength of light. It affects the absorption of light in the p-i-n cell, and to a smaller extent the refraction angle and the transmittance of light. Calculations showed that changing the wavelength from 400 to 800nm increases the transmittance T by less than 7%, and the refraction angle β by about 25%.

Since the incident angle of light on TCO/a-Si:H interface changes with V groove tilt angle, this angle affects both, the refraction angle and the transmittance. Changing the tilt angle φ from 45° to 30° results in about 30% increase of transmittance T and about 20% increase of refraction angle β.

CONCLUSIONS

A two-dimensional numerical simulation of p-i-n a-Si:H solar cell on V-grooved texture showed that space-charge distribution and electric field in the cell are strongly influenced by texturized TCO surface. It was noted that the locations of greatest light generations coincide with regions where the accelerating electric field is largely reduced. This detrimental effect can be reduced by increasing the V-groove tilt angle. Along the slopes of the grooves, the electric field is aligned almost perpendicular to the tilted TCO surface, making the charge carrier collection paths shorter than the vertical thickness of the p-i-n cell.

ACKNOWLEDGEMENTS

The authors thank the TMA, Palo Alto, for providing the simulator MEDICI. This work was supported by the Commission of the European Communities, Programme Joule II.

REFERENCES

[1] H.Iida et al., "Efficiency of the a-Si:H Solar Cell and Grain Size of SnO_2 Transparent Conductive Film" IEEE Electron Device Letters, EDL-4, 1983,pp. 157-159.

[2] H.Iida, T.Mishuku, I. Sakata, Y. Hayashi, "A Milky TinOxide on Glass (MTG) Substrate Thin Undoped Layer p-i-n Amorphous Silicon Solar Cell with Improved Stability and Relatively High Efficiency", IEEE Electron Device Letters EDL-5, 1984, PP.65-67.

[3] S.Tsuge et. al., "Effect of the Texture of Transparent Electrodes on the p-layer and V_{OC} of Amorphous Silicon Solar Cells", 5th Int. PVSEC, Kyoto, 1990, pp. 261-264.

[4] F.S.Sinencio, R.Williams, "Barrier at the Interface between Amorphous Silicon and Transparent Conducting Oxides and its Influence on Solar Cell Performance", J. Appl. Phys., 54, 1983, pp.2757-2760.

[5] K.Tabuchi et al., "Optimization of ZnO Films for Amorphous Silicon Solar Cells", 11th E.C.PSEC, Montreux, 1992, pp.529-532.

[6] F.Smole, M.Topic, J.Furlan, "The Influence of Defect States in TCO/a-Si:C:H Interface on p-i-n a-Si:H Solar Cell Performance", 12th E.C.PSEC, Amsterdam, 1994, pp.77-80.

[7] H.Shade, Z.E.Smith, "Optical Properties and Quantum Efficiency of a-$Si_{1-x}C_x$:H/a-Si:H Solar Cells", J.Appl.Phys., 57, 1985, pp.568-547.

[8] G.Tao, M.Zeman, J.W.Metselaar, "Accurate Generation Rate Profiles in a-Si:H Solar Cells with Texturized TCO Substrates", 7th Int. PVSEC, Nagoya, 1993, pp.13-14.

[9] T.Savada et. al., "Numerical Approach for High Efficiency a-Si Solar Cells", 7th Int. PVSEC, Nagoya, 1993, pp. 15-16.

[10] TMA Medici: Two-Dimensional Device Simulation Program, Technology Modelling Associates, Inc., 1992.

[11] M.Hack, M.Shur, "Theoretical Modeling of Amorphous Silicon-based Alloy p-i-n Solar Cells", J.Appl.Phys., 54, 1983, pp. 5858-5863.

GENERATION LIFETIME IN AMORPHOUS SILICON AS DETERMINED BY MOS ADMITTANCE MEASUREMENTS

W. R. Fahrner, St. Löffler, FernUni Hagen (Germany)
H. C. Neitzert, CSELT Torino (Italy)
E. Klausmann, IAF Freiburg (Germany)

ABSTRACT

Capacitance-voltage measurements are performed in order to investigate dynamical properties of the undoped layer of an amorphous silicon pin solar cell. A special MOS diode is prepared which allows the determination of the resistivity, the net carrier concentration, and the doping density in the i-layer. The CV measurements indicate two clearly distinguishable dispersion behaviors in accumulation and inversion of the (slightly n-type doped) i-layer of the diode and inversion built-up. A numerical simulation computes the potential variation inside the sample, taking into account the trap density in the a-Si:H. A complete set of CV curves under quasistatic conditions is simulated and fitting to the experimental data yields the actual trap density. This information granted, we can calculate the generation lifetime of carriers in the depleted region of this diode biased to inversion.

INTRODUCTION

In the last years amorphous hydrogenated silicon has gained great interest not only for solar cell applications but in the fields of sensors, detectors and thin film transistors. The measurement procedures developed for crystalline material in roughly 5 decades are not necessarily useful for the characterization of a-Si:H due to its completely different structure. E.g. traps and electronic states in the forbidden band, playing a minor role in high quality crystalline semiconductors, control the properties in a-Si:H. Even for fundamental parameters like doping density and carrier mobility there is a big scatter in the published data which and it is not clear whether this is solely a result of different deposition and preparation techniques. The work presented here is intended to fit techniques, existing for years in the field of crystalline semiconductor, to a-Si:H. The parameters we are interested in, are basic for the operation of solar cells.

SAMPLE PREPARATION AND CHARACTERIZATION

An MOS sample has been prepared following a procedure described in [1]. Based on a heavily doped and thermally oxidized crystalline wafer two layers of amorphous silicon have been deposited using a standard parallel plate PECVD reactor. The thermal oxide of a heavily doped silicon wafer has been used as a substrate because of its outstanding electrical properties concerning traps and trapped charges. The p^+-layer just acts as an electrical contact. The 650nm thick i-layer was deposited first on top of the oxide and a subsequent n^+-layer of a few nm was deposited to assure a good ohmic contact to the Al metallization evaporated on this layer. The Al dots and the a-Si:H mesas have been defined using photolithography. The crossection of the

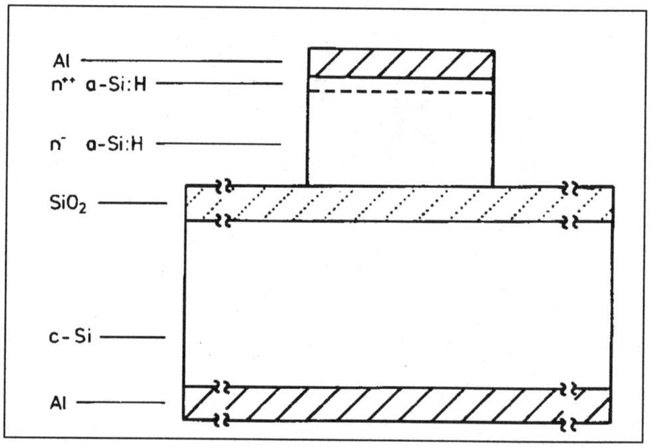

Figure 1: Cross section of the samples used for the dispersion and quasistatic CV measurements. According to the inverted structure (compared to conventional MOS diodes) all potentials and voltages have a reversed sign.

sample resulting from these procedures can be seen in fig. 1. CV curves of this structure are shown in fig. 2. Due to the high resistivity of the undoped layer the capacitance does not reach the value of the oxide capacitance in both voltage regimes but is lowered by resistors in series even for the lowest frequency of the admittance bridge. The value for the oxide capacitance C_{ox} has been obtained from quasistatic and ellipsometric measurements. In inversion and at high frequencies, the capacitance tends to the geometrical capacitance of two dielectric layers, namely the oxide and the two a-Si:H layers. Equivalent circuits, best suited to model the behavior in accumulation and inversion are shown in the next figure (3). The dispersion in accumulation has been used to determine the resistivity and the net carrier density.

CH3365-4/94/0000-0662 $4.00 © 1994 IEEE

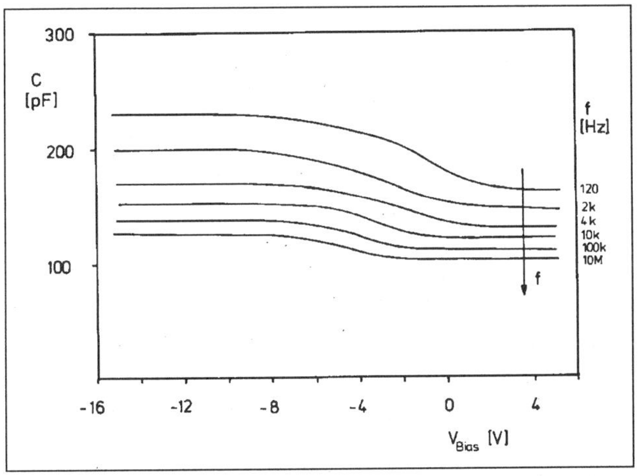

Figure 2: Dispersion of the capacitance of a sample according to fig 1. Note the different time constants of the dispersion in acculumation (left) and inversion (right) (taken from [1]).

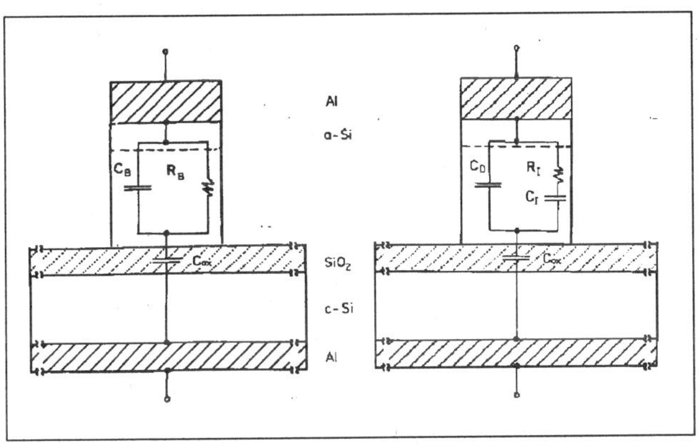

Figure 3: Equivalent circuits of the diode specialized to accumulation (left) and inversion (right).

CV-CURVE COMPUTING

The base for all subsequent calculations is the determination of the potential distribution in the a-Si:H layers. This has been done by a computer code using a finite element approach. The applied voltage V_a splits over the oxide and the semiconducting layer:

$$V_a = V_{ox} + \varphi_s \tag{1}$$

(The oxide voltage in turn is determined, once the charge distribution in the a-Si:H layer has been calculated). We compute all charges in the semiconductor and solve Poisson's equation

$$\frac{d^2\varphi}{dx^2} = -\frac{\rho}{\varepsilon}. \tag{2}$$

Due to the large number of charged traps in the mobility gap and the mutual dependency of charge concentration and potential, an integration of Poisson's equation is not feasible but a numerical iterative algorithm must be used. The second order Taylor approximation of the potential in the vincinity of any point x_0 delivers

$$\varphi = \frac{1}{2} \cdot \left[\varphi(x_0 + dx) + \varphi(x_0 - dx) - (dx)^2 \frac{d^2\varphi}{dx^2}\right] \tag{3}$$

so the left hand side potential of a subsequent iteration at a given point can be deduced from the values of its neighbors and its second derivation. The last term in the brackets of eqn. (3) can be substituted using eqn. (2). The interior potential distribution of the first iteration can be chosen arbitrarily. Now the charge density can be calculated and each iteration delivers a next better solution to the potential and charge distribution. We include a trap density with a $\cosh$-like potential dependency with its minimum centered in the mobility gap (see [2]). The complete expression for the charge density in the undoped layer to be solved is

$$\begin{aligned} \rho \;=\; & \left(\frac{n_i}{N_{D-}}\right)^2 \left(\exp\left[-u(\xi) - \log\left(\frac{N_{D+}}{N_{D-}}\right)\right] - 1\right) \\ & - \left(\exp\left[u(\chi) - \log\left(\frac{N_{D+}}{N_{D-}}\right)\right] - 1\right) \\ & - \left(\frac{N_{b0}\psi_c}{N_D}\right) sinh\left[\left(u(\xi) + log\left(\frac{N_{D+}}{n_i}\right)\right)\frac{V_{th}}{\psi_c}\right] \\ & + \left(\frac{N_{b0}\psi_c}{N_{D-}}\right) sinh\left[log\left(\frac{N_{D-}}{n_i}\right)\frac{V_{th}}{\psi_c}\right] \end{aligned} \tag{4}$$

(n_i: intrinsic carrier concentration; $N_{D-/+}$: doping density in the undoped and doped a-Si:H layers; u:potential in reduced units; ξ: length in reduced units). A low-frequency capacitance simulation of the structure at a given surface potential can now be performed (cf. fig. 4): (1) calculation of the distribution at a given surface potential $\varphi_s = \varphi_0$; (2) calculation of the distribution at a given surface potential $\phi_s = \phi_0 + \delta\phi$; (3) Integration of all charges in the semiconductor to calculate the applied voltage over the complete structure (incl. oxide); (4) integration of the charge difference of (1) and (2); (5) calculation of the capacitance $C = dQ/dU$.
An example of a simulation for the charge distribution of the two a-Si:H layers biased to inversion is seen in fig. (5). The change of the sign of the charge stems from the built-in band bending caused by the different doping concentrations. Fig. (6) shows the result of the simulation,

DYNAMICAL PROPERTIES

Using the above simulated data for the potential distribution for an applied bias in inversion, we can calculate the generation lifetime in the semiconductor. Starting from eqn. (1) and the expression for the charges

$$Q_a = Q_I + Q_{sc} \tag{5}$$

(Q_I: charge in the inversion layer; Q_{sc}: sum of all other charges in the semiconductor;) we can write the expressions for a small test signal superimposed to the applied voltage V_a:

$$\begin{aligned} Q_a + dQ_a &= Q_I + dQ_I + Q_{sc} + dQ_{sc} && (6) \\ V_a + dV_a &= (Q_i + dQ_I + Q_{sc} + dQ_{sc})/C_{ox} \\ & \quad +\varphi_s + d\varphi_s && (7) \end{aligned}$$

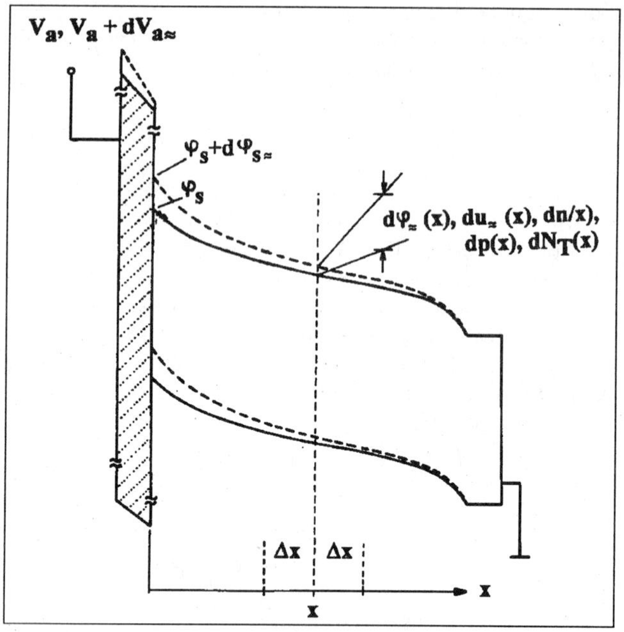

Figure 4: The calculation of the potential distribution at two surface voltages φ_0 and φ_1 simulates the low frequency test signal with an amplitude $V_{\sim} \propto (\varphi_1 - \varphi_0)$.

Assuming a sinoidal test signal we get for the time derivatives

$$
\begin{aligned}
dV_a/dt &= j\omega V_{a\sim} e^{j\omega t} && (8)\\
dQ_a/dt &= j\omega Q_{a\sim} e^{j\omega t} && (9)\\
dQ_{sc}/dt &= j\omega Q_{sc\sim} e^{j\omega t} \quad \text{and} && (10)\\
d\varphi_s/dt &= j\omega \varphi_{s\sim} e^{j\omega t}. && (11)
\end{aligned}
$$

The determination of the generation lifetime requires the calculation of the time derivative of the inversion charge Q_I. This can be done by establishing the rate equations for majority and minority carriers and expanding them at a fixed inversion bias to get the admittance. We assume the validity of the theory of Shockley, Read and Hall for generation and recombination of excess carriers and write

$$\frac{dn}{dt} = -c_n N_T (1-f) n + e_n N_T f \qquad (12)$$

$$\frac{dp}{dt} = -c_p N_T f p + e_p N_T (1-f) \qquad (13)$$

From equilibrium and neutrality considerations we get with $dn/dt = dp/dt$ (first order approximation):

$$
\begin{aligned}
\frac{dp}{dt} \cdot &\left[\tau_n (p_0 + p_1) + \tau_p (n_0 + n_1)\right] \\
&= -p_0 n_0 du - \frac{p_1 (n_0 + n_1)}{p_0 + p_1} \delta p \qquad (14)
\end{aligned}
$$

(p_0, n_0: values in equilibrium; $\tau_{n,p} = 1/(c_{n,p} N_T)$; N_T trap density; $c_{n,p}$: capture cross sections of electrons/holes). We assume that the majority carrier concentration will follow any variation of the applied potential and so does the occupation

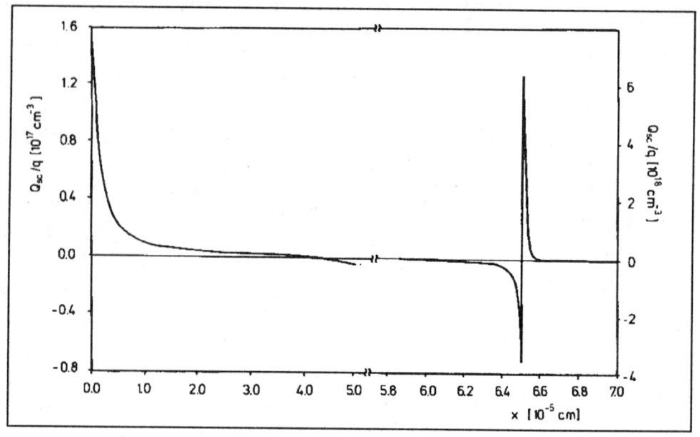

Figure 5: Charge in the a-Si:H layers for a surface potential of $-50k\theta/q$ without the charge of the minority carriers (holes). The interface between the high und undoped layer is at a depth of $650nm$ (taken from [3]).

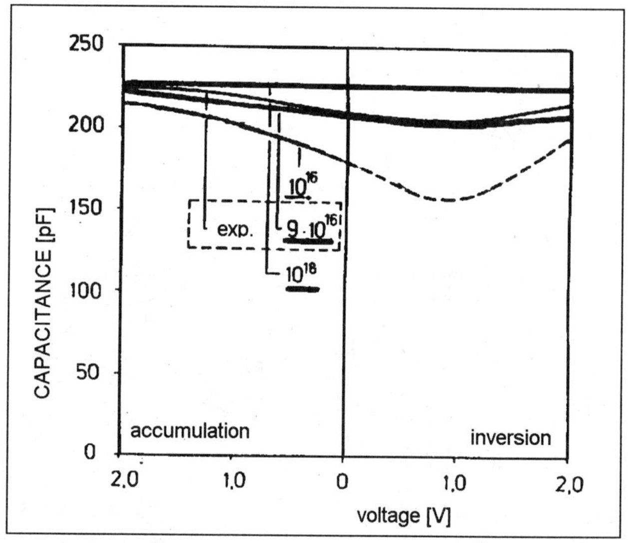

Figure 6: Simulated quasistatic CV curves with the trap density N_{b0} as a parameter. The best fit to the experimental one is for $N_{b0} = 9 \cdot 10^{16} cm^{-3}$ (taken from [4]).

of trap levels within the time scale of our experiment (typically in the range of 100Hz to 1MHz) whereas the contribution of the the minority carriers and the term proportional to the variation δp are small and can be neglected. Like for interface traps most efficient for generation and recombination processes are traps located at or near midgap, so we find $n_1 = p_1 = n_i$ which leads to the following expression

$$\frac{dQ_I}{dt} = q \int_0^{x_p} \frac{n_i^2 du(x)}{\tau_p (n_0 + n_i) + \tau_n (p_0 + n_i)} dx \qquad (15)$$

Eqn. (15) depends on both lifetimes, τ_n and τ_p. A comparison of closely related parameters, the diffusion lengths, leads to a ratio τ_n/τ_p of about 0.1 so we set $\tau = \tau_p = 10\tau_n$

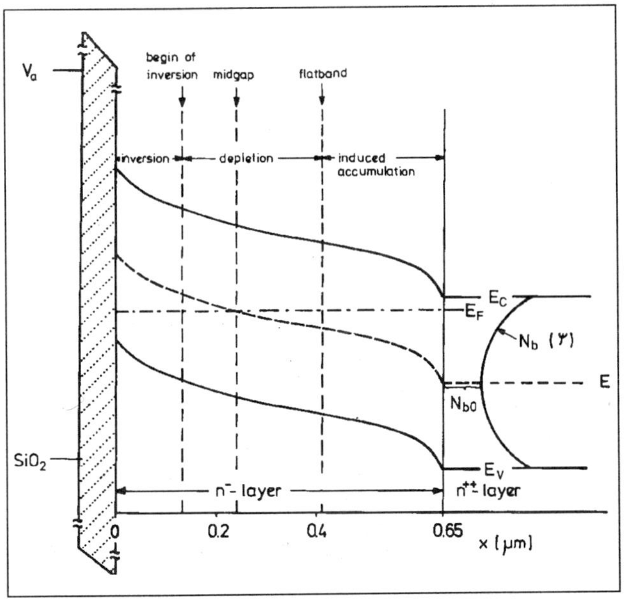

Figure 7: Band bending of the complete MOS diode for a surface potential of $-50k\theta/q$ (at the interface between oxide in the undoped a-Si:H layer). On the right hand side, a sketch of the assumed trap concentration is depicted (taken from [3]).

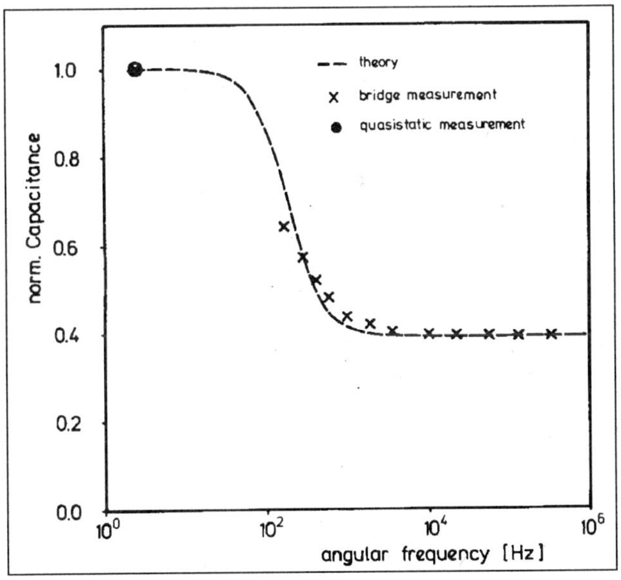

Figure 8: Dispersion of the MOS capacitance in inversion ($V_a = -10V$) (taken from [3]).

and rewrite eqn. (15) with $du(x) = u_\sim e^{j\omega t}$:

$$\frac{dQ_I}{dt} = \frac{qn_i}{\tau} e^{j\omega t} \int_0^{x_p} \frac{n_i^2 du_\sim(x)}{(\frac{n_0}{n_i}+1) + \frac{\tau_n}{\tau_p}(\frac{p_0}{n_i}+1)} dx$$

$$= \frac{qn_i}{\tau} e^{j\omega t} B \tag{16}$$

B is a material and geometry dependent factor in the order of one

$$B = \int_0^{x_p} \frac{n_i^2 du_\sim(x)}{(\frac{n_0}{n_i}+1) + \frac{\tau_n}{\tau_p}(\frac{p_0}{n_i}+1)} dx \tag{17}$$

and can be determined numerically, once the potential distributions for two different surface potentials have been computed as described in section *CV-curve computing*. The ratio of the small signal components of eqns. 6 and 7 delivers the small signal admittance and its time constant T:

$$\frac{Y}{j\omega C_{ox}} = \frac{\frac{qn_i}{\tau} B + j\omega Q_{sc\sim}}{\frac{qn_i}{\tau} B + j\omega(Q_{sc\sim} + \varphi_{s\sim} C_{ox})} \tag{18}$$

$$T = \tau \cdot \frac{\frac{Q_{sc\sim}}{q} + \frac{\varphi_{s\sim} C_{ox}}{q}}{n_i B} \tag{19}$$

$$T_1 = \tau \cdot \frac{Q_{sc\sim}}{q} \frac{1}{n_i B} \tag{20}$$

According to eqn. (18) the frequency dependent capacitance is given by

$$C = C_{ox} \cdot \frac{1 + \omega^2 T T_1}{1 + \omega^2 T^2} \tag{21}$$

and the lifetime can be determined by measuring the dispersion of the capacitance and the frequency ω^* half way between C_{ox} and $C(\omega \to \infty)$.

The band bending and assumed trap concentration can be seen in fig. (7). The built-up of the inversion layer (left hand side) leads to an increasing capacitance with decreasing frequency as seen in fig. (8). The final result is a lifetime of $\tau = 3.5 \cdot 10^{-13} s$.

CONCLUSION

In conclusion, we have presented two methods for the determination of material properties of undoped a-Si:H, namely $\rho = 9.1 \cdot 10^9 \Omega cm$, $N_D = 7 \cdot 10^9 cm^{-3}$, $\tau = 3.5 \cdot 10^{-13} s$, $N_{b0} = 9 \cdot 10^{16} cm^{-3} eV^{-1}$(data for sample of fig. 8).

ACKNOWLEDGEMENT

One of the authors (W.R.F) dedicates this article on the occasion of the 60. birthday of his former teacher, Prof. Dr. H. G. Wagemann.
This work has been made feasible by financial support of the AG Solar, NRW, Germany.

References

[1] H. C. Neitzert et al., "Metal-Oxide-Semiconductor Capacitance Measurements an Amorphous Silicon", *J. Electrochem. Soc.*, **141**, No. 9, p. 2474, (1994)

[2] M. Böhm et al.: "The Influence of Bulk and Interface Gap States on the Performance ..." *Mat. Res. Soc. Symp. Proc.* **95**, p. 481 (1987)

[3] W. R: Fahrner et al.: "Determination of Generation Lifetimes in Trap-Rich and Layered Semiconductors ...", *J. Electrochem. Soc.* **141**, No. 8, p. 2151 (1994)

[4] St. Löffler et al.: unpublished

OPTICAL MODELING OF A-SI:H BASED SOLAR CELLS ON TEXTURED SUBSTRATES

G. Tao, M. Zeman and J.W. Metselaar,
Delft University of Technology, P.O.Box 5031, 2600 GA Delft, The Netherlands

ABSTRACT

We have developed a new GENPRO2 computer program package to meet the needs of optical modeling of a-Si:H based solar cells on textured substrates. The new program takes into account the light scattering effects at the rough interfaces. We are now able to do optical modeling of a-Si:H based solar cells with textured substrates. We present the state of the art of optical modeling and our recent results on textured solar cells with different structures, such as: glass/TCO/p-i-n(a-Si:H)/Ag, ITO/p-i-n(a-Si:H)/Ag/SS, tandem cells and cells with TCO/Ag back contacts.

INTRODUCTION

The structure of a-Si:H based solar cells is getting more complex. Computer modeling is becoming more important in the optimization of solar cells, both for materials and device design. In the past years, much effort has gone into the electrical modeling of a-Si:H solar cells. Various computer program packages such as ASA [1], AMPS [2] and ASPIN [3] have been developed. However, the carrier generation profile in the solar cells under illumination, as an input parameter for the electrical modeling, is usually not accurately calculated. (ASA requires an external generation profile, while AMPS uses a simple Lambert-Beer formula to calculate the carrier generation profile). The accurate carrier generation profile is very important in simulating opto-electronic devices like solar cells. The generation profile can be calculated by optical modeling.

We have paid continuous effort to the optical part of the modeling of a-Si:H based solar cells in our group. A PC version of GENPRO (a program which simulates both the reflection and transmission and the absorption profiles of coherent light in a multi-layer system) was developed in 1989 [4] for solar cell structures with flat substrates. Similar programs have been reported recently from other groups [5-7].

To meet the needs of optical modeling of the solar cells with textured substrates, we developed a new GENPRO2 computer program package. It simulates the optical behavior (reflection, transmission and absorption profile) of a multi-layer system with rough interfaces. Such a program is badly needed not only for optical modeling of solar cells, but also for analyzing measured spectral response data (for example: DICE calculations). The GENPRO2 is the first program which takes into account the light scattering effects at the rough interfaces [8-10]. We are now able to do optical modeling of a-Si:H based solar cells with different layer structures on textured substrates.

In this paper, we present the state of the art of optical modeling of a-Si:H based solar cells on textured substrates.

THE OPTICAL MODEL USED IN GENPRO2

When light strikes a rough interface, scattering occurs. The light scattering at a rough interface (see Fig. 1) is determined by the light wavelength (λ), the interface roughness (σ_r) and morphology (m), the refractive indices ($\tilde{n_1}$, $\tilde{n_2}$) of the media and the light incident angle (Ω_{in}). In general, we can describe the reflection and transmission (both specular and diffused) by:

$$R(\lambda, \tilde{n_1}, \tilde{n_2}, \Omega_{in}, \Omega_{out}, \sigma_r, m)$$

$$T(\lambda, \tilde{n_1}, \tilde{n_2}, \Omega_{in}, \Omega_{out}, \sigma_r, m)$$

Here $\Omega = \Omega(\theta, \varphi)$, and ranges from 0 to 2π in the solid angle. If we take F_1^- and F_1^+ to be the up- and down-going light intensity in the first medium, and F_2^- and F_2^+ in the second medium, then they are related as:

$$F_1^- = R_{1,1}F_1^+ + T_{2,1}F_2^- \quad (1)$$

$$F_2^+ = T_{1,2}F_1^+ + R_{2,2}F_2^- \quad (2)$$

Here the Fs have a solid angle distribution. The Rs and Ts are functions of the wavelength (λ), the refractive indices ($\tilde{n_1}$, $\tilde{n_2}$), the incoming and out-going direction (Ω_{in}, Ω_{out}), the rms roughness σ_r and the rms slope m. The first subscript of the Rs and Ts represents the medium of incident light, and the second subscript represents the medium of the outgoing light.

CH3365-4/94/0000-0666 $4.00 © 1994 IEEE

First WCPEC; Dec. 5-9, 1994; Hawaii

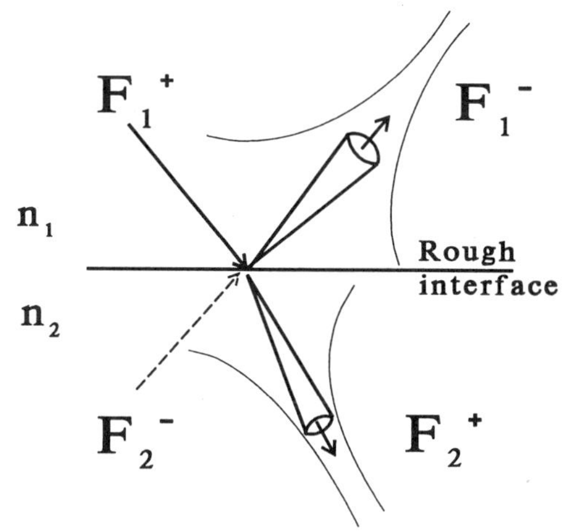

Figure 1: Light scattering at a rough interface

a-Si:H based solar cells with textured substrates contain several rough interfaces: TCO/p, p/i, i/n, and n/metal (or n(a-Si:H)/TCO/metal). Optically, the a-Si:H based textured cell is a multi-rough-interface system. In such a system, every rough interface can be represented by its reflection and transmission (with solid angle distributions). And the light absorption in every layer should be taken into account:

$$F^-|_{x=0} = MF^-|_{x=d} \quad (3)$$

$$F^+|_{x=d} = MF^+|_{x=0} \quad (4)$$

Here M is a diagonal matrix with elements of $\exp(-\alpha\, d/\cos(\theta))$, α is the absorption coefficient and d the thickness of the medium.

We can calculate the solid-angle resolved up- and down-going light intensities (F^- and F^+) in every medium by solving the matrix equations. From the solid-angle resolved up-going light intensity in the first medium (air) and the down-going light intensity in the last medium, we can calculate the angle-resolved reflection and transmission. The light absorption profile in the j-th medium for wavelength λ can be calculated by:

$$A_j(\lambda, x_j) = \int \frac{\alpha_j}{\cos(\theta)} F^-_{j,x_j}(\Omega)\, d\Omega + \int \frac{\alpha_j}{\cos(\theta)} F^+_{j,x_j}(\Omega)\, d\Omega \quad (5)$$

Here, F^+_{j,x_j} and F^-_{j,x_j} are the down- and up-going monochromatic light intensity at depth x_j of the j-th medium. For non-monochromatic light, equation 5 should be summed over all wavelengths.

- Data Input (optical system, mesh points, light spectra)
- Absorption profile initialization
- For every λ do
  - For every interface do
    - Calculate R and T
  - For every layer do
    - Calculate M
  - Calculate F^+ and F^-
  - Calculate absorption profile for this λ
- Summing the absorption profiles for all λ
- Data (absorption profile) output

Figure 2: The structure of GENPRO2.

GENPRO2 simulates the optical behavior of a multi-rough-interfaces system provided that the optical properties of every solar cell layer ($n(\lambda), k(\lambda)$) and of every interface (roughness and scattering) are known. Figure 2 shows the structure of GENPRO2.

APPLYING GENPRO2

Single junction cells

Figure 3 shows the calculated light absorption profiles in a single junction textured cell and in a flat cell with the same layer thicknesses. The structure of such solar cells is glass/TCO/p-i-n(a-Si:H)/Ag. The TCO layer is 900 nm thick. The p-i-n layers are 10 nm, 500 nm, 30 nm thick respectively. The p-layer is a-SiC:H which has a wider bandgap. The calculated results show that the light absorption has been significantly improved by texturing. This improvement is more pronounced in the long-wavelength region.

For the textured cell, we calculated the angle-resolved reflectance and compared with the measured one (see Fig. 4. They show a good agreement in the trends.

For two textured cells with different p-i-n layer thicknesses (cell A: 10 nm, 500 nm and 30 nm; cell B: 20 nm, 500 nm and 20 nm) on textured TCO glass substrate, we calculated the light absorption profiles under AM1.5 global spectrum. By assuming the collection efficiency to be unity in the i-layers, and zero in other layers, the integrated value of $q \times \int_i Abs.Pro\, dx$ should reflect the short-circuit current density. Table 1 shows the measured short-circuit current density values and the $q \times \int_i Abs.Pro\, dx$ values. These values agree with each other very well.

Tandem cells

One problem in developing the tandem solar cells is: the thickness of the intrinsic layers in the top cell and bottom

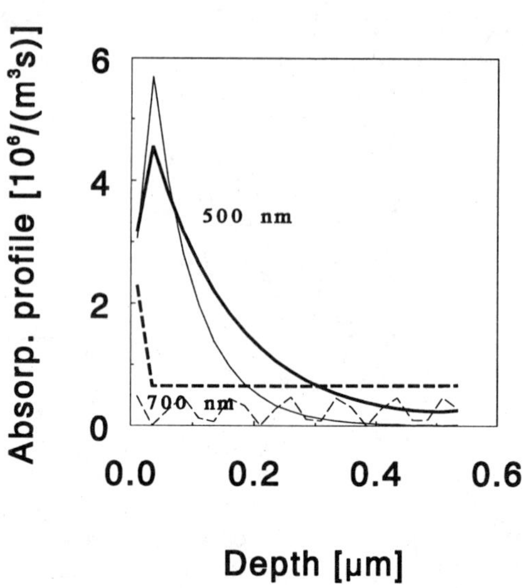

Figure 3: Light absorption improvements caused by interface texture. Thick lines are for a textured cell, and thin lines are for a flat cell with the same layer thicknesses.

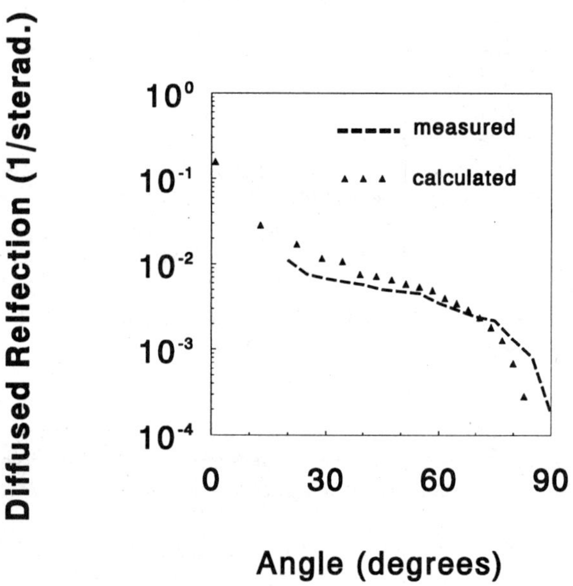

Figure 4: The calculated and measured angle-resolved reflection of a textured cell.

Table 1: The measured Jsc and the calculated integrated absorption.

Sample	measured J_{SC} (A/m^2)	calculated $q \times \int_i Abs.Pro\, dx$ (A/m^2)
A	151	152.0
B	136.5	137.1

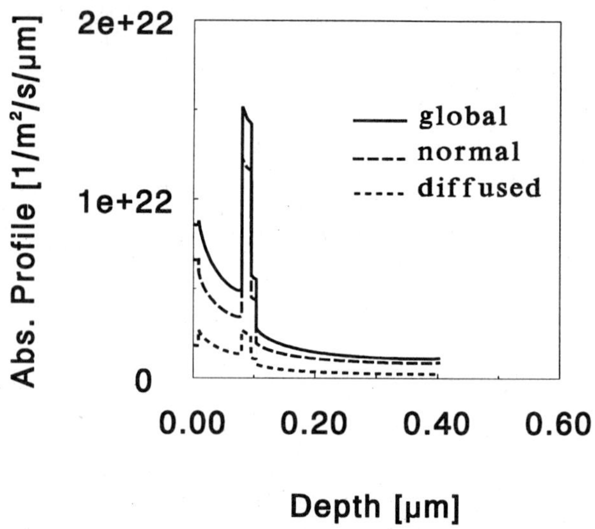

Figure 5: Light absorption profiles in a tandem a-Si:H solar cell under different AM1.5 illumination. The integrated absorption in the top i-layer and bottom i-layer are matched for global spectra.

cell should be optimized to achieve current matching under a certain illumination. We matched the values of integrated light absorption in the top i-layer and bottom i-layer by varying the one of the i-layer thickness while keeping the other i-layer thickness constant. One of the numerically matched pair of i-layer thicknesses of a glass/TCO/p-i-n/p-i-n/Ag structure tandem cell under AM1.5 global illumination are: 71.3 nm (top) and 300 nm (bottom). This result is in agreement with the experimentally optimized ones reported in literature [11–13]. The light absorption profiles in such a tandem cell is shown in figure 5.

Solar cells with inverted structure

Solar cells fabricated in n-i-p sequence on stainless steel substrates are often reported in literature. We calculated the light absorption profiles in two of such solar cells. These cells have an ITO (300 nm) front contact and an Ag back contact. We assume that these two cell have the same p-i-n layers as in cell A and B in the section of a single junction cell. We also assume that the p/i, i/n, n/back contact are all rough interfaces. The light absorption profile in such cells under AM1.5 global illumination at normal incidence are shown in figure 6. The integrated value of $q \times \int_i Abs.Pro\; dx$ for cell A and B are $155.7A/m^2$ and $138.3A/m^2$ respectively.

Highly reflective TCO/metal back contacts

The absorption of light in a thin film solar cell can be improved by applying a highly reflective TCO/metal back con-

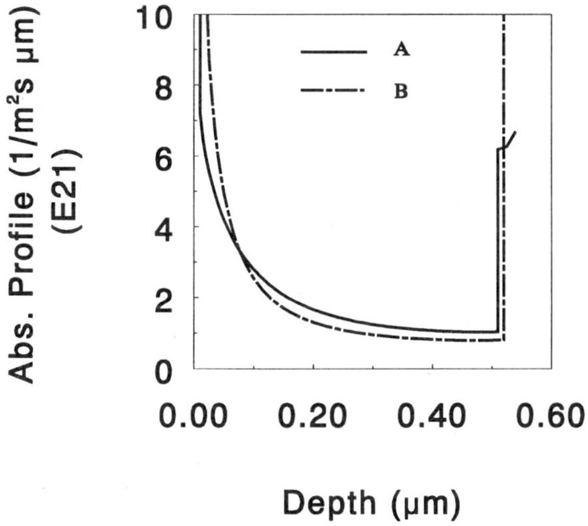

Figure 6: Light absorption profiles in inverted solar cells under AM1.5 global illumination. The p-i-n layer thicknesses are: 10 nm, 500 nm, 30 nm for cell A; 20 nm, 500 nm, 20 nm for cell B.

Table 2: The calculated integrated absorption.

Sample	$q \times \int_i Abs.Pro\ dx$ Ag back (A/m^2)	$q \times \int_i Abs.Pro\ dx$ TCO/Ag back (A/m^2)
Inv-A	155.7	162.5
Inv-B	138.3	146.0
Norm-A	152.0	166.1
Norm-B	137.1	143.9

tact. In the case of a flat cell, such improvement can be easily calculated and the highly reflective back contact can be optimized [14]. We calculated the light absorption in the p-i-n layers of solar cells with rough interfaces and highly reflective back contacts. Table 2 gives the integrated value of $q \times \int_i Abs.Pro\ dx$ under AM1.5 global illumination for inverted cells and normal cells with silver and TCO/silver back contacts. The A and B are for different p-i-n layer thicknesses.

CONCLUSIONS

With the present state of the art of optical modeling, it is possible to simulate the optical behavior of a-Si:H based solar cells on textured substrates. The possibility of such optical modeling opens, together with the electrical modeling part, the era of CAD (computer aided design) for a-Si:H solar cells!

Further study should be devoted to solar cell behavior with graded bandgap and interface buffer-layers in order to achieve stable cells with high efficiency.

Acknowledgments This work is funded by the Netherlands Agency for Energy and the Environment NOVEM.

REFERENCES

[1] H.M. Wentinck, B.F.P. Jansen, F. van Rijs, C.L.M. van der Klauw and W. Crans, *Digest of the Nascode VI Software Forum*, Editor W. Crans, Boole Press, Dublin (1989), 114.

[2] J.K. Arch, F.A. Rubinelli, J.Y. Hou, and S.J. Fonash, *J. Appl. Phys.* **69** (1991) 7057.

[3] F. Smole, M. Topič, and J. Fulan, *Technical Digest of the INT'L PVSEC-7*, Nagoya, Japan, (1993) pp 205–206.

[4] J. van den Heuvel, PhD thesis, Delft University of Technology, (1989).

[5] P. Popović, J. Fulan, and W. Kusian, *Technical Digest of the INT'L PVSEC-7*, Nagoya, Japan, (1993) pp 207–208.

[6] H. Stiebig, A. Kreisel, J.-L. Nicque, Th. Eickhoff, C. Beneking and H. Wagner, *The proceedings of the 12th ECPVSEC*, R. Hill *et al.* eds., Amsterdam, the Netherlands, (1994), pp 164–167.

[7] F. Zhu and J. Singh, *Solar Energy Materials and Solar Cells*, **31** (1993) pp 119–131.

[8] G. Tao, M. Zeman and J.W. Metselaar, *The Technical Digest of PVSEC-7*, Nagoya, Japan, (1993), pp 13–14.

[9] G. Tao, M. Zeman and J.W. Metselaar, *Solar Energy Materials and Solar Cells*, **34** (1994), pp 359–366.

[10] G. Tao, Zs. Makaro, M. Zeman and J.W. Metselaar, to be published in *Amorphous Silicon Technology – 1994*, MRS. Symp. Proc. Vol. **336**, E.A. Schiff, *et al.* eds., (1994).

[11] Liyou Yang, L. Chen, S. Wiedenman and A. Catalano, *Microcrystalline Semiconductors: Material Science& Devices*, Mater. Res. Soc. Proc. Vol. **283**, (1992), pp. 463–470.

[12] P. Lechner, H. Schade, B. Schoppat, A. Bubonzer, *The proceedings of The 11th EPVSEC*, Montreux, Switzerland, (1992), pp. 568–571.

[13] Y. Ichikawa, T. Ihara, S. Saito, H. Ota, S. Fujikake and H. Sakai, *The proceedings of the 11th EPVSEC*, Montreux, Switzerland, (1992), pp. 203–206.

[14] G. Tao, PhD thesis, Delft University Press, (1994), ISBN 90-407-1063-5

APPLICATIONS OF AMORPHOUS SILICON PHOTOVOLTAICS: MYTHS AND FACTS

Daniel S. Shugar, P.E., Member IEEE
Advanced Photovoltaic Systems, Inc.
150 Tehama Court
San Bruno, CA 94066 USA
(415) 871-2275, fax -2279

ABSTRACT

This paper presents assessment results of the technical and economic suitability of amorphous silicon (a-Si) solar photovoltaics (PV) to terrestrial applications of power generation and commercial remote power. The assessment covered a spectrum of a-Si production, PV system design and economics, operational performance, and market acceptance. Primary data were gathered from the operational PV production facilities, historical experience in the construction of utility-scale turnkey PV power systems and smaller industrial remote power systems, and large commercial users of PV systems. Performance reliability and conversion efficiency were given particular attention in the assessment of a-Si.

Findings were 1) a-Si has attained stability and environmental reliability comparable to conventional PV technologies; 2) PV conversion efficiency has a significant impact on system applications and economics only when there is limited area or very high transportation costs; 3) the low area-related costs and suitability of glass-on-glass a-Si as a building material enable a variety of building-integrated PV applications that provide high value to end-use grid-connected customers; 4) the completion of large scale, automated production facilities has allowed a-Si to attain a significant manufacturing cost reduction.

INTRODUCTION

The low material requirements of amorphous silicon (a-Si) photovoltaic (PV) panels created expectations over a decade ago that very low costs would soon be attainable for PV systems. By the late 1980's, however, such expectations remained unfulfilled due to poor technical performance of fielded a-Si PV systems, and negligible cost advantages in a-Si modules resulting from limited production and industry inability to produce large area modules. These failures, combined with an overemphasis on product efficiency in an industry still focused on research and technical milestones rather than economics, applications, and markets, left negative impressions of the capability of a-Si with many PV customers.

Fortunately, improved stability, encapsulation, and efficiency of a-Si PV modules, qualified by the National Renewable Energy Laboratory's (NREL) *Interim Thin-Film Qualification Test* and the European Solar Test Installation's *Ispra Test Specification 503*, became available from Advanced Photovoltaic Systems (APS) and other manufacturers. This paper evaluates a-Si suitability for terrestrial power applications, using technical performance and cost accounting data of several prominent installed systems. In doing so, we will address many of the lingering *myths* of a-Si applications and the role of PV conversion efficiency with *facts* from manufacturing, PV systems analysis, and long-term performance data.

RELIABILITY OF A-SI PV POWER MODULES

Myth: *a-Si is not Stable*

Fact: Amorphous silicon PV modules do undergo an initial power loss from the time of their fabrication and factory test to their stabilization in the field under conditions of sustained solar irradiation. This phenomenon is called the Staebler-Wronski effect and is well documented in scientific literature [1]. However, first generation a-Si had encapsulation, back contact, isolation and other problems which caused excessive instability and failures at elevated array voltages. This was evidenced by first generation thin-film utility systems installed in the 1980's as well as various consumer modules. These non Staebler-Wronski failures combined with a lack of actual long term field data to cause potential users of a-Si concern as to whether in fact a-Si modules do stabilize.

Fortunately empirical long term field data on a statistically significant sample size of second generation a-Si technology are now available from a 400 kW_{AC} APS system that is part of the *Photovoltaics for Utility Scale Applications (PVUSA) Project* [2]. APS's system is the largest a-Si PV system in the world (or of any thin-film) by over an order of magnitude, utilizing 9600 modules. Figure 1 plots the output power of 240 modules that were installed at APS PVUSA field in December of 1991 in a special test section. These output power readings resulted from 33 separate I-V tests performed on the test section over a three-year period. Current-voltage (I-V) curves were taken approximately every two weeks for the first year and have been taken roughly every month for the last year. The test data have been normalized to the Standard Test Conditions of 100 mW/cm^2 irradiance and 25°C cell temperature, with interconnecting wire loses subtracted to yield individual module power levels.

CH3365-4/94/0000-0670 $4.00 © 1994 IEEE
First WCPEC; Dec. 5-9, 1994; Hawaii

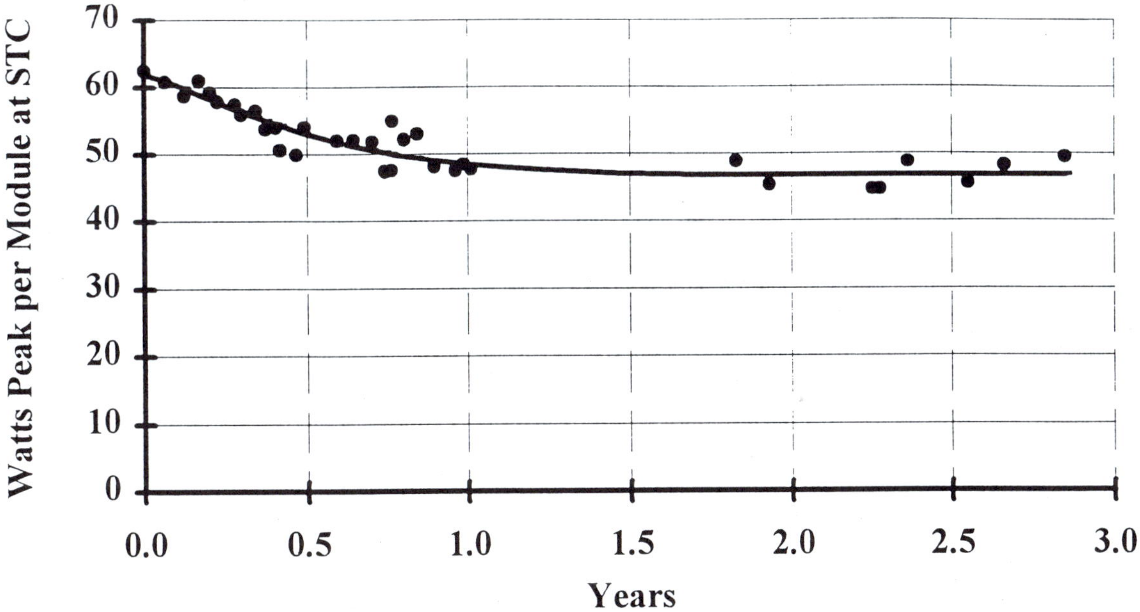

Figure 1. Stability of 45 Wp APS Modules at PVUSA: Power of First 240 Modules, Installed 12/91

Fig. 1 indicates modules reached their stabilized power 10 months after installation and are fully stabilized with constant power over the last two years. These early modules were installed seven months prior to the main field construction and carry a 45 Wp rating by APS's current rating method. All of the modules installed during the following summer carried a 50 Wp rating. APS rates it's modules for design, sales, pricing, and 12 year warranty at the *stabilized,* not initial value. For future consideration, APS has made considerable progress toward reducing light induced degradation through optimizing the p layer and p-i interface (with 8.54% NREL confirmed stabilized single-junction cells) which will have long-term benefits toward increasingly stable commercial a-Si modules [3].

Myth: *a-Si is not Appropriate for Utility PV Power Plants*

Fact: This is a corollary of the "stability myth". In fact there are several a-Si characteristics besides low cost which are quite beneficial for utilities. In particular, a-Si has a minimal power temperature coefficient of (-0.27%/°C) which enables the entire system to suffer minimal performance degradation with temperature. For example, increasing module temperature from 25°C to 60°C results in a performance decrease of only 9.5%, whereas decreases of 18% to 20% would occur for crystalline PV. The low temperature coefficient enables the system to have strong summer peak performance, which can yield high-value for summer peaking utilities [4,5] and their customers [6].

The best indication of appropriateness of a-Si for utility systems is again the performance of the APS PVUSA system. Monthly energy generation of the system is plotted in Fig. 2. Overall 1994 energy production exceeds 1993 values due to higher inverter availability, but in months with high availability in both years similar energy was produced, giving further indication of system stability. Low production in Nov. 1994 is due to an exceptionally rainy month. The APS system was the first utility-scale system at PVUSA to generate one million kWh, and with the exception of an inverter outage during the summer/early fall of 1993, has operated consistently. The system has generated a total of 1440 MWh through Nov. 1994 and should complete 1994 with a capacity factor over 20%. The APS system had the lowest operations and maintenance (O&M) costs of any utility-scale PVUSA system in 1993 [7], and is expected to have even lower O&M costs in 1994 due to virtually no system failures and high system energy production.

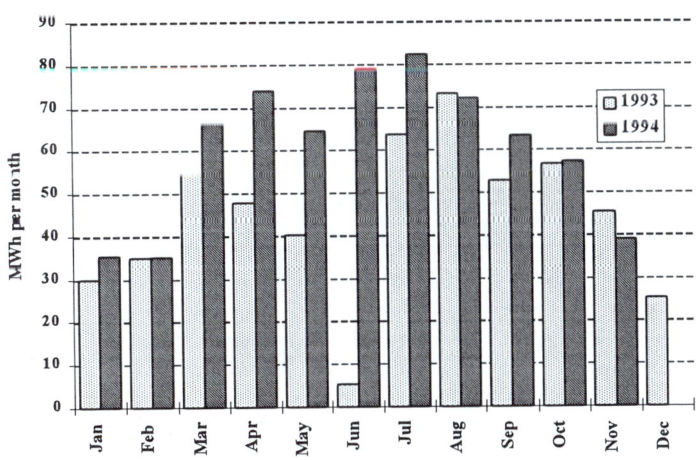

Fig 2. Energy Generation of APS 400 kW_{AC} PVUSA Sys.

THE ROLE OF EFFICIENCY IN a-Si ECONOMICS

***Myth**: Low Efficiency Incurs High Land Costs for Ground-Mounted Systems*

Fact: With crystalline silicon technologies costing \$50/ft^2 to \$60/ft^2 and a-Si technology costing \$12/ft^2 to \$14/ft^2 (in volume), land costs have to be extremely high to become a factor in PV system economics. For example, a fixed flat plate system with a 5% modules will require approximately 1.1 acres/100 kW, while a system with 13% efficient modules will require about 0.5 acres/100 kW. The effect on total PV system price of different efficiencies is plotted in Fig. 3. Actual prices paid for a commercially procured PV parcel (the PVUSA Kerman Project) of \$9000/acre are also noted on the graph, which indicates there is less than a two percent difference in total system price attributable to increased land cost between a "high" and "low" efficiency PV system even with land cost as high as \$20,000/acre.

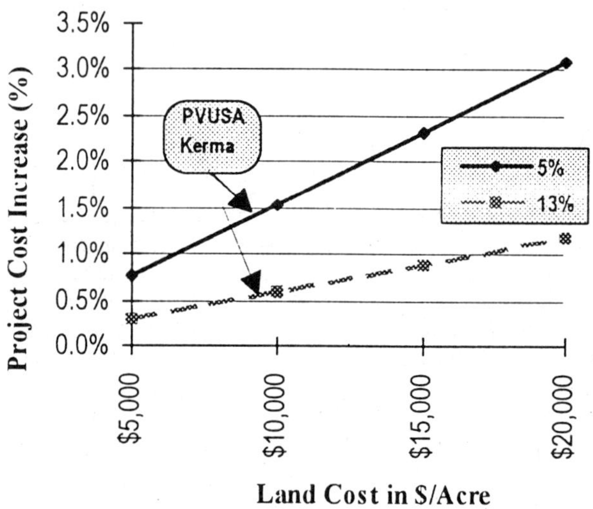

Fig 3. PV System Project Cost Increase as a Function of Land Cost for Different PV Conversion Efficiencies

***Myth:** Area-Related Balance-of-Systems (BOS) Costs Make Low Efficiency PV Economically Impractical in Ground-Mounted Systems*

Fact: This myth is true only if very expensive structures (on an area-related basis) are used to support the PV arrays. For this reason, APS completed a two year development effort for minimal-cost PV array support structures, culminating with the construction of the PVUSA system in 1992. This system had by far the lowest area-related costs (\$51/m^2 in 1994\$) validated to date in the project [8], and received the award of two U.S. patents for its unique design [9]. Using this minimal cost structure, the total area-related costs on the APS PVUSA system were only 23% of the total project costs, as depicted in Fig. 4. The costs reported are based on actual competitively bid subcontract awards.

Fig. 4 indicates if PV module efficiency had *doubled*, system costs would have been reduced by only about 11%. Now that APS can produce 60 Wp PV modules at the new, operational Fairfield manufacturing facility, area-related costs have decreased about 20% since PVUSA, which used 50 Wp modules manufactured three years ago. Thus ground mounted a-Si are not only feasible, but are among the most competitive systems available. This was evident during "bid openings" in May 1994 by the Sacramento Municipal Utility District, where APS's system was the low-price bidder for a 108 kW$_{AC}$ turn-key PV system for SMUD's Hedge Substation. APS/Bell Products were awarded the project in Sept. 1994, which is under construction.

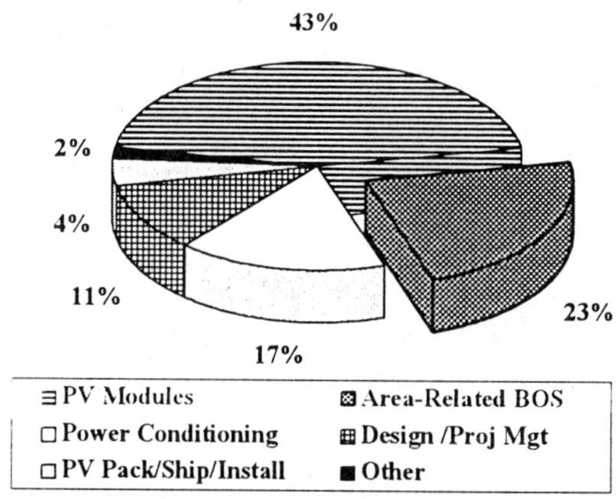

Fig. 4 Cost-Elements of APS's PVUSA System

***Myth:** Area-Related BOS Costs Make Low Efficiency PV Impractical for Building-Integrated Systems*

Fact: Actually, because PV is typically priced by power (\$/Wp) and not area (\$/ft^2), higher efficiency can be *disadvantageous* for BIPV. This somewhat counter-intuitive statement is best explained by the following example. A building owner is considering two 10 kW$_{AC}$ PV systems;

- a high efficiency PV system which may cost \$8000/kW$_{AC}$ and require 1000 ft^2 of roof area, or
- an a-Si PV system which may cost \$7000/kW$_{AC}$ and also save the owner 2200 ft^2 of building material that would otherwise have to be purchased and installed.

Clearly a-Si PV offers the owner a high-value option.

Fortunately, large-area glass-on-glass a-Si PV has many characteristics that permit suitability as a building material [10]. These characteristics include size, weight/strength, visual aesthetics, compatibility with conventional construction assemblies, adequate safety, partial visible light transmissivity, and excellent heat shading properties. Applications in buildings include skylights, curtain walls, awnings, and roofing [11]. Values of replaced conventional building materials are \$2/ft^2 to \$5/ft^2 for roofing systems, and \$7/ft^2 to \$10/ft^2 for glazing systems. When compared to turnkey a-Si systems costing \$20/ft^2 to \$30/ft^2 (including electrical interconnection), these avoided materials savings can have a meaningful impact on BIPV economic feasibility.

Myth: *Tracking is Impractical with Low Efficiency PV*

Fact: Tracking a-Si is clearly cost-effective in climates with reasonable insolation, providing minimal cost trackers such as APS's *Delta Tracker™* [12], a novel tilted one-axis (polar) tracker are used. APS constructed and operated prototypes of this system in late 1993 and early 1994, and has contracts to deliver Delta Trackers™ in early 1995. Firm subcontractor quotes indicate the tracker system will increase overall project costs by only 2.2%. Simulation of tracker performance with PVGRID [13] gives annual energy increases of 29% beyond fixed orientations, using hourly Fresno, CA weather data (5.4 $kWh/m^2/day$ global horizontal). Greater gains are expected for areas of higher insolation. The net impact of 2.2% higher costs to track and 29% increased energy production are a reduction in the levelized energy production costs (¢/kWh) of almost 27%.

PRODUCTION COSTS OF A-SI PV MODULES

As mentioned in the *Introduction*, one of the problems a-Si faced in the 1980's was that no cost advantages were apparent to customers, due to limited manufacturing capability in the industry. Fortunately this problem is behind us. In the case of APS, a 1991 analysis (based upon measured material/labor/energy costs and measured yield/efficiencies and from APS's Trenton Pilot Line) indicated direct manufacturing costs of $1.00/$W_p$ for the 1991 stabilized efficiencies of 4.5%, and costs of $0.71/$W_p$ for stabilized efficiencies of 6.0% ($1991) [14]. These results were based upon manufacturing large area modules at high volume production with multiple shift operations.

Today APS has completed and qualified such capability in its new Fairfield, CA plant [15]. This new 69,000 ft^2 facility was designed using operating experience from Trenton, and features world-class automation, in-line statistical process control, and higher power modules (60 Wp stabilized). Annual production capacity is 200,000 modules or 12 MWp at current efficiency. The plant was qualified in early 1994 and produced modules for the 108 kW_{AC} Bell/APS SMUD project. While Fairfield's specific production costs are proprietary, APS believes this plant has achieved a major reduction in a-Si manufacturing costs, enabling new markets to be reached for the industry and a-Si in particular.

CONCLUSIONS

1) a-Si has attained stability/environmental reliability comparable to conventional PV technologies;
2) PV conversion efficiency has a significant impact on system applications and economics only when there is limited area or very high transportation costs;
3) the low area-related costs and suitability of glass-on-glass a-Si as a building material enable a variety of building-integrated PV applications that provide high value to end-use grid-connected customers;
4) the completion of large scale, automated production facilities has allowed a-Si to attain significant manufacturing cost reductions.

REFERENCES

[1] Staebler, C.R. Wronski: *Appl. Phys, Lett.*, vol. 31, pg. 292, 1977

[2] *Design and Two-Year Performance of the World's Largest (400 kW_{AC}) Amorphous Silicon Photovoltaic Power Plant at PVUSA*, Shugar, D.S., Matlin, R., IEA Task VI, Paestum, Italy, July 1994

[3] *Identification and Reduction of P-I Interface Instability and Establishment of 8.5% Stabilized Efficiency for Single Junction a-Si Solar Cells*, APS, Inc., Xi, J., et al. MRS Proceedings, April 1994

[4] *Photovoltaics in the Utility Distribution Systems: The Evaluation of System and Distributed Benefits*, Shugar, D.S., IEEE PVSC, p. 836 - 844, May 1990

[5] *PVUSA 500 kW Grid-Support Photovoltaic Project: Interim Results on Value*, Wenger, H., Hoff, T., Shugar, D.S., Farmer, B., 12th European PVSEC and Exhibition, April 1994, Amsterdam

[6] *Photovoltaics as a Demand-Side Management Option: Benefits of a Utility-Customer Partnership*, Wenger, H.J., et al, World Energy Engineering Conference, Atlanta, GA., Oct. 1992

[7] *1993 PVUSA Progress Report,* PVUSA Project Staff; PG&E

[8] *PVUSA - Performance, Experience and Cost,* T.R. Candelario et al; 1991 IEEE PVSC, Las Vegas, NV

[9] *Support for Photovoltaic Arrays,* U.S. Patents 4,966,631 and 5,143,556, Matlin, R.W.

[10] *Photovoltaic Building Materials*, Paradis, A., Shugar, D.S., Solar Today, May 1994, p 34-38.

[11] *Optimizing Roof-Integrated Photovoltaics: A Case Study of the PowerGuard™ Roofing Tile*, Dinwoodie, T., Shugar, D.S., Proceedings of the 24 IEEE PVSC, Kona, Hawaii, Dec. 1994

[12] APS Technical Report 9401, *Design and Prototype of a Modular, Minimal Cost Tracking PV Array Structure*, Matlin, R., Feb. 1994.

[13] *PVGRID: A Micro-Computer Based Software Package for Central Station Photovoltaic System Analysis,* Wenger, H., Univ.of Colorado, June 1987

[14] *Recent Improvements in Very Large Area a-Si PV Module Manufacturing*, APS, J. Macneil, et al., 10th European PVSEC, Lisbon, Portugal, April 1991, pp 1188 - 1192

[15] *Power Plant*, Architecture, June 1993, pp 74 - 81

ZnO/Ag SPUTTERING DEPOSITION ON a-Si SOLAR CELLS

Katsuhiko Hayashi, Masataka Kondo, Atsuo Ishikawa, and Hideo Yamagishi
Electronic Materials Research Laboratories, Kaneka Corporation
1-1,Hieitsuji 2 Chome,Otsu,Shiga, 520-01,Japan

ABSTRACT

In order to produce large area amorphous silicon solar cell modules and to simplify the module production process, a continuous ZnO/Ag sputtering deposition process has been applied. We found that by means of a continuous ZnO/Ag sputtering deposition method an adhesive a-Si/electrode contact can be realized. We compared short circuit currents of Al ,ZnO/Al and ZnO/Ag back side contact cells and confirmed short circuit current increase by application of ZnO/Ag back side contact. We found that the series resistance is severely depend on the conditions during the first stage of ZnO deposition. We confirmed the reliability of ZnO/Ag structure as the back side contact through high temperature high humidity test. After 310 hours accelerated light induced degradation test which corresponds to one year light exposure our 100cm2 integrated a-Si tandem solar cell kept the efficiency higher than 8.5%.

INTRODUCTION

The most effective method to reduce the light induced degradation of a-Si solar cell is to make thinner its i-layer. But to make thinner the i-layer leads to decrease short circuit current and efficiency. Therefore several methods have been proposed to increase short circuit current even in thin i-layer. One of the methods is application of a high reflective back side contact structure of ZnO/Ag. And it is now widely investigated.In order to produce large area amorphous silicon solar cell modules and to simplify the module production process, a continuous large area ZnO/Ag sputtering deposition process has been applied. Back side contact fabrication by sputtering is suitable for producing practical solar cell modules, because the reliability of solar cell modules strongly depends on the adhesion of the a-Si/back side contact. In comparison with a-Si layer, the back side contact layer is intimately connected with the module fabrication process, because it contacts directly with the encapsulation layer. Several problems have become apparent and were overcome in these application trials.

SAMPLE PREPARATION

We prepared 1cm2 a-Si solar cells and 100cm2 integrated a-Si solar cells for this study. Electrical solar cell performance and back side metal adhesion was measured on 1cm2 a-Si solar cells, and reliability was tested on 100cm2 integrated a-Si solar cells. On the textured SnO_2 coated flat glass substrates, p-a-SiC:H/i-a-Si:H/n-a-Si:H heterojunction solar cells were fabricated by rf glow discharge decomposition of silane gas or its mixture using a separated chamber system. Typical deposition conditions are described in our previous paper[1]. In the case of 100cm2 integrated solar cell separation of SnO_2 and a-Si:H layers was patterned by using a laser-scribing method. ZnO layer was deposited as a transparent back side contact between n-layer and metal layer. Al and Ag were sputtering deposited or vacuum evaporated as a back metal contact layer. The sputtering deposition system we used is magnetron sputtering. The vacuum evaporation was carried out by electron beam or resistive heater method. Fig.1 shows the typical cell configuration in this work. For reliability test 100cm2 intgrated solar cells were laminated with EVA and TEDLAR films after completion of the wiring. For accelerated light induced degradation test executed by JMI Institute edges of the samples were encapsulated by thermal plastic butyl rubber and aluminum frames were furnished.

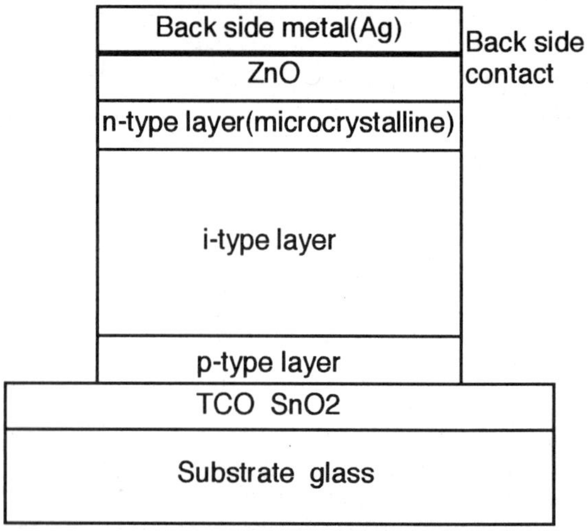

Fig.1 Typical cell configuration

First WCPEC; Dec. 5-9, 1994; Hawaii

ADHESION OF ZnO/Ag CONTACT

It is well known that the adhesion between a-Si and Ag is weak compared with that of Al. In order to fabricate reliable a-Si/back metal contact we have applied ZnO middle layer between a-Si and metal layer. And we also applied sputtering deposition for ZnO/Ag back side contact fabrication in this study. We think the sputtering deposition method is simlper than vacuum evaporation methods. In particular, as a large area deposition method, sputtering deposition is superior to vacuum evaporation because of its uniformity. We also found that back side contact fabrication by sputtering is suitable for producing practical solar cell modules, because the reliability of solar cell modules strongly depends on the adhesion of the a-Si/back side contact. We found that by means of a continuous ZnO/Ag sputtering deposition method an adhesive a-Si/back side contact can be realized. We have concluded that this strong adhesion is derived from the sputtering deposition method, because the deposited particle energy is large in the sputtering deposition case. Table.1 shows the results of tape stripping tests on several kinds of back side contacts.The adhesion strength of Al and Ag layers prepared by electron beam evaporation, resistance heater evaporation and sputtering deposition on a-Si/ZnO samples were compared. The highest value was observed in the sputtering deposition case. It was decided to use a Ag/ZnO back side contact structure because from the view point of increasing short circuit current, the Ag/ZnO back side contact is more favorable. Moreover a ZnO/Ag continuous deposition method simplifies the back side contact fabrication process. In the case of the ZnO/Ag back side contact system, continuous ZnO/Ag deposition after a-Si deposition and a-Si laser scribe patterning is possible because a good ZnO/TCO contact is easily formed.

Table.1 Results of tape stripping tests

Adhesion(g/10cm)	32	160	520	800
EB Al	O	O	O	O
EB Ag	X	—	—	—
Rh Al	O	O	O	O
Rh Ag	X	—	—	—
Sputter Al	O	O	O	O
Sputter Ag	O	O	O	O

O :not stripped
X :stripped
— :not tested

EB:Electron beam evaporation
Rh:Resistive heater evaporation

SHORT CIRCUIT CURRENT OF ZnO/Ag BACK SIDE CONTACT CELL

As a back side contact ZnO/Al is better than Al, because ZnO suppress light absorption in Al layer. We calculated and experimentally obtained optimum ZnO thickness 80nm for that purpose. The spectral responses of tandem cell with ZnO/Al and Al back side contacts are compared in Fig.2. There are two curves for each tandem cell and the influence of back side contact appers mainly in bottom cell responses. In the long wavelength range the spectral response of ZnO/Al back side contact cell is higher than that of Al back side contact cell. The short circuit current for typical single type a-Si solar cell with ZnO/Al back side contact is 1mA/cm2 larger than that with Al back side ocntact. As a back side contact Ag is better than Al, because its light absorption is less. The spectral response of tandem cell with ZnO/Ag and ZnO/Al back side contacts are compared in Fig.3. In these samples the ZnO thickness is 80nm. In the long wavelength range the spectral response of ZnO/Ag back side contact cell is higher than that of ZnO/Al back side contact cell. The short circuit current for typical single type a-Si solar cell with ZnO/Ag back side contact is 0.4mA/cm2 larger than that with ZnO/Ag back side contact.

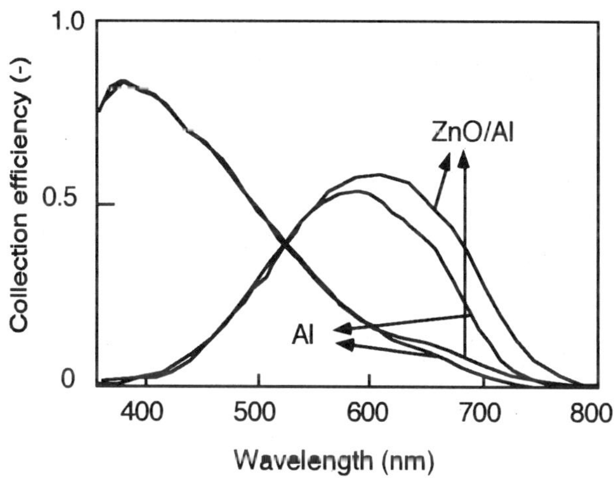

Fig.2 Spectral responses of ZnO/Al and Al back side contact cells

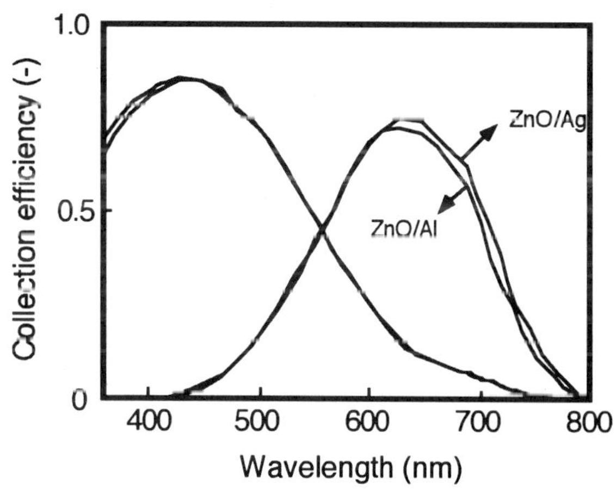

Fig.3 Spectral respomses of ZnO/Ag and ZnO/Al back side contact cells

Table.2 summarizes the short circuit currents of cells with several kind of back side contacts. Table.2 also shows the difference of short circuit current between vacuum evaporation deposited(Rh) back metal and sputtering deposited metal on a-Si/ZnO. The difference is smaller in the Ag back metal case. This result is thought to be derived from the difference of the tendency how easily Al and Ag is oxidized. With respect to the short circuit current Ag is less influenced by deposition method than Al. This is one of the reason why we selected ZnO/Ag as the practical large area cell back side contact material candidate.

Table.2 Single cell short circuit current comparison of several kinds of back side contacy on a-Si pin

Back side metal	ZnO/Ag		ZnO/Al		Al
Deposition method	Rh	Sp	Rh	Sp	Rh
Normalized Jsc	1.0	0.99	0.97	0.90	0.91

Rh:Resistive heater evaporation

Sp:sputtering deposition

ELECTRICAL CONTACT PERFORMANCE

At first we have examined the influence of ZnO sputtering conditions on the cell performance. Despite good ZnO electrical conductivity, the cell performance,especially the fill factor was strongly dependent on the sputtering conditions. Fig.4 shows the fill factor of a-Si single cell with an area of 1cm2 as a function of ZnO sputtering power. Fill factor increased with sputtering RF power. Therfore we concluded that the series resistance is severely altered by the nature of the n/ZnO interface, in other words the conditions during the first stage of ZnO deposition.

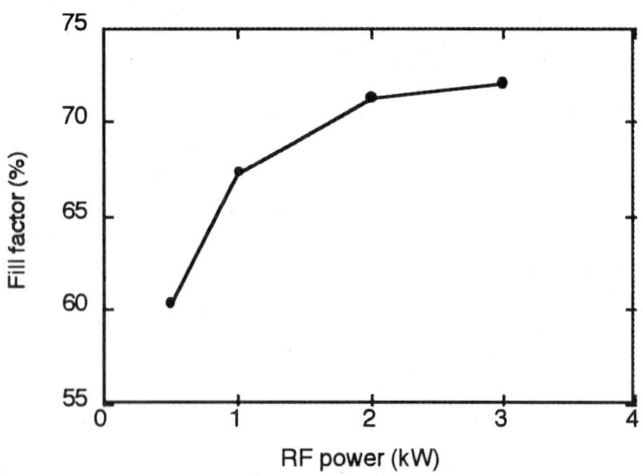

Fig.4 Fill factor of 1cm2 a-Si single solar cell as a function of ZnO sputtering power

We also investigated the electrical contact performance between ZnO and n-type amorphous or crystalline silicon. There is no problem in the electrical contact performance between metal and ZnO. Table.3 shows the cell performance comparison between samples which contain n-type microcrystalline-silicon-/ZnO contacts and samples which contain n-type amorphous-silicon/ZnO contacts. The ZnO deposition condition in Table.3 samples are same.Table.4 shows the deposition conditions of n-type microcrystalline-silicon and n-type amorphous-silicon. To deposit n-type microcrystalline-silicon relatively high RF power and high H2 dilution condition is applied. In the case of n-type amorphous-silicon/ZnO contacts, the solar cell performance is lower with respect to the fill factor. It is well known that the electrical conductivity of microcrystalline-silicon layer is higher than that of amorphous-silicon layer. But in this case n-layer/ZnO series resistance is the main reason of this difference.

Table.3 Fill factor comparison

	Back electrode	Fill factor (%)
microcrystalline Si n-type	ZnO/Ag	72
	Al	73
amorphous Si n-type	ZnO/Ag	60
	Al	70

Table.4 Deposition conditions of n-type layer

	microcrystalline	amorphous
Substrate temp.	200C	200C
SiH4 flow rate	1	1
H2 flow rate	50	10
Pressure	1torr	1torr
RF power	10	1

Flow rate and RF power is normalized by amorphous value.

RELIABILITY TEST

Fig.5 shows the results of high temperature 85C, high humidity 90% test of 100cm2 ZnO/Ag back side contact cells laminated with fluorcarbon polymer film by an EVA adhesive layer. During first 1000 hours no change in output power is observed. Therefore we concluded that no back side metal peeling ,migration and Si/back side contact inter-layer corrosion is observed. It is inferred that Ag migration is suppressed since Ag layer is sandwiched between ZnO layer and EVA lamination film and Si/back side contact inter-layer corrosion is suppressed by ZnO layer. In the case of amorphous silicon solar cell, stabilized efficiency is decreased by light induced degaradation so-called Staebler-Wronski effect. Therefore those samples mentioned below were so made to increase stabilezed efficeincy. The a-Si deposition condition of these cells are described elsewhere. Fig.6 shows the result of accelerated light induced degradation test of these 100cm2 integrated tandem solar cell measured by JMI Institute. The condition of this acceleration test is 48C,1.25kW/m2. After 310 hours which corresponds to one year light exposure the cell kept the efficiency higher than 8.5%. And this value is almost stabilized efficiency.

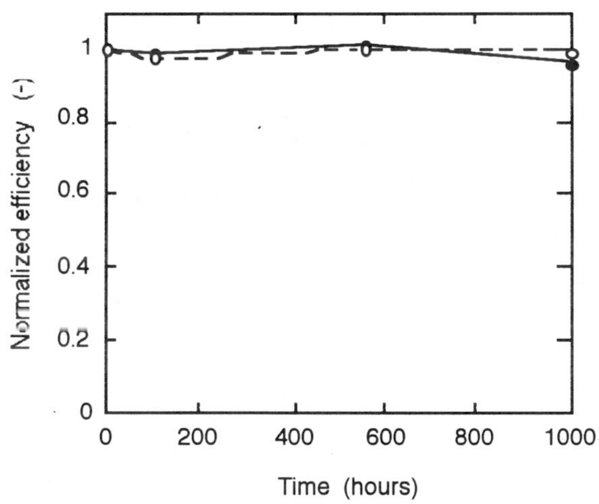

Fig.5 High temperature high humidity testing

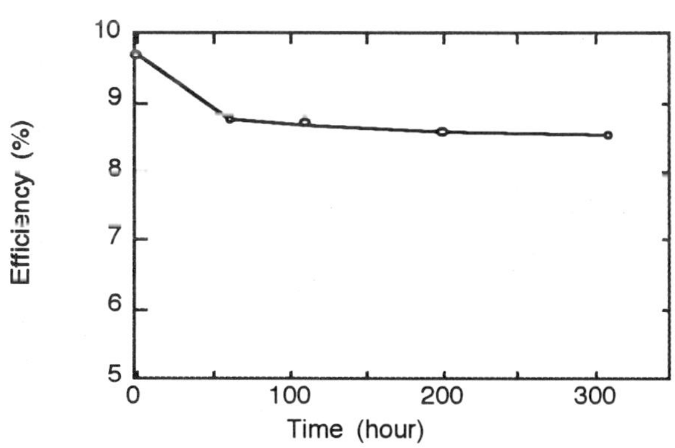

Fig.6 Result of accelerated light induced degradation test

SUMMARY

In order to produce large area amorphous silicon solar cell modules and to simplify the module production process, a continuous large area ZnO/Ag sputtering deposition process has been applied, and we obtained the following results.

1) By means of a continuous ZnO/Ag sputtering deposition method an adhesive a-Si/back side contact can be realized.

2) We compared short circuit currents of Al ,ZnO/Al and ZnO/Ag back side contact cells.

3) With respect to the short circuit current Ag is less influenced by deposition method than Al.

4) The series resistance is severely altered by the nature of the n/ZnO interface, in other words the conditions during the first stage of ZnO deposition.

5) We confirmed the reliability of ZnO/Ag structure as a back side contact through high temperature high humidity test.

6) After 310 hours accelerated light induced degradation test which corresponds to one year light exposure our 100cm2 integrated a-Si tandem solar cell kept the efficiency higher than 8.5%.

For the time being, we will investigate the following topics

1)To make clear the relationship between ZnO/Ag contact and its adhesion.

2)To make clear the relationship between n/ZnO contact and its series resistance.

By investigating these topics, we will promote production of large area amorphous silicon solar cell modules.

ACKNOWLEDGMENTS

This work was partially supported by the New Energy and Industrial Technology Development Organization as a part of the New Sunshine Program under the Ministry of International Trade and Indusry. The authors would like to thank members concerned.

REFERENCES

[1] H.Yamagishi et.al., *Proc. 4th International Photovoltaic Sci.&Eng. Conf.*,(1989),415

DEPENDENCE OF a-Si SOLAR CELL V_{oc} ON DEPOSITION TEMPERATURE

X. Deng, K. L. Narasimhan, J. Evans, M. Izu and S. R. Ovshinsky
Energy Conversion Devices, Inc., 1675 West Maple Road, Troy, Michigan 48084

ABSTRACT

We have investigated variations in the a-Si deposition process to achieve a high V_{oc} a-Si solar cell to be used as the top cell of a triple-junction stacked solar cell. In this investigation, a-Si n-i-p solar cells were deposited on stainless steel substrates in a multichamber RF PECVD deposition system. Different deposition conditions were explored for the intrinsic a-Si layer with the goal of obtaining high V_{oc}. After process optimization, we have achieved a V_{oc} of 1.04 V for a-Si n-i-p solar cells with the intrinsic layer deposited at 150°C at high hydrogen dilution. This V_{oc} is higher than any V_{oc} achieved for a single-junction solar cell using a-SiC as the intrinsic layer. This high V_{oc} a-Si material is a better choice for the top cell i layer of a triple-junction a-Si alloy solar cell than high bandgap a-SiC alloys.

INTRODUCTION

Energy Conversion Devices, Inc. (ECD) has pioneered the development of high efficiency a-Si alloy solar cells and significantly advanced the development of materials, device designs, and manufacturing processes [1-5]. In ECD's photovoltaic technology, a-Si alloy solar cells are deposited on stainless steel substrate using a triple-junction device structure: SS/metal/metal-oxide back-reflector/a-SiGe bottom cell/a-Si (or a-SiGe) middle cell/a-Si top cell/TCO. Using this device structure, ECD has fabricated a small-area solar cell with 13.7% initial efficiency [3] and 4 ft^2 production modules with 8% stable efficiency [4]. Recently, ECD's joint venture with Canon, United Solar, has achieved a small-area solar cell with 11.2% stable efficiency [6] and a 1 ft^2 PV panel with 10.2% stable efficiency [7]. To further improve the solar cell efficiency, we have studied the top cell deposition process to improve its V_{oc} so that the blue portion of the solar spectrum is better utilized.

Wide bandgap alloys such as a-SiC material have been extensively studied for use as the intrinsic layer of the top cell of an a-Si alloy triple-junction solar cell in many laboratories [8]. When a-Si is alloyed with carbon, oxygen or nitrogen, its bandgap widens. However, when the bandgap is widened, the absorption of the material decreases. Thicker intrinsic layers (approximately 2000 Å, depending on the alloying ratio) are therefore needed when compared with a-Si top cell intrinsic layer, so that current matching can be maintained in a triple cell. Furthermore, as a-Si is alloyed with carbon, the density of defect states increases rapidly. As a consequence, a-SiC top cells cannot replace a-Si top cells because the a-SiC top cells are thicker and have higher defect densities.

The bandgap of a-Si can also be widened by increasing the hydrogen content. When the a-Si deposition temperature is lower, the a-Si contains more hydrogen and its bandgap increases. However, when the deposition temperature is reduced to below 200°C, the material quality begins to deteriorate due to the increased void fraction and defect density. Strong hydrogen dilution of the gas mixture during a-Si deposition has been found to reduce the density of defect states and improve the material light stability [9,10]. When a low deposition temperature is combined with high hydrogen dilution during deposition, a-Si materials with wider bandgaps and low electronic defect densities can be achieved [11,12]. We have systematically investigated the growth of a-Si at low temperature with strong hydrogen dilution, and have incorporated this material into a n-i-p top cell to improve the solar cell open circuit voltage. In this paper, we report our studies on the improvement of high V_{oc} a-Si alloy solar cells.

EXPERIMENTAL DETAILS

We have primarily used an a-Si n-i-p solar cell devices for the optimization of the a-Si intrinsic layer deposition. The device structure principally used for this study is SS/a-Si n^+/a-Si intrinsic/μc-Si p^+/ITO. A bare stainless steel substrate without back-reflector is primarily used in this work, since the top a-Si cell is grown on the middle cell in a triple-cell structure and does not benefit from optical enhancement from the back-reflector. a-Si n, i and p layers are deposited in the dedicated chambers of a multi-chamber load-lock deposition system. The a-Si intrinsic layers were deposited at different substrate temperatures ranging from 150 to 250°C with a wide range of hydrogen dilution, deposition pressure of 1 Torr, and a power density of 20 mW/cm^2. The p layer is a heavily boron-doped microcrystalline silicon [13]. When such a μc-Si p^+ is used for the p layer, the V_{oc} of the solar cell is limited mainly by the properties of the intrinsic layer. The ITO layer was deposited using reactive evaporation of an indium-tin alloy in oxygen.

CH3365-4/94/0000-0678 $4.00 © 1994 IEEE

Table 1. V_{oc} and other solar cell performance data for a-Si n-i-p solar cells deposited on bare stainless steel with the intrinsic layer deposited at different temperatures

Sample No.	Dep. Temp (C)	Voc (V)	Jsc (mA/cm2)	FF	Thickness of i layer (nm)
712	250	0.951	12.1	0.585	320
723	230	0.959	11.3	0.608	320
722	210	0.980	11.7	0.615	310
731	210	0.986	12.0	0.626	330
732	175	1.026	10.0	0.670	285
746	150	1.040	9.4	0.652	264

RESULTS AND DISCUSSIONS

In Table 1, we list the solar cell performance we obtained with the intrinsic layer deposited at different temperatures, at an optimized hydrogen dilution for each deposition temperature. When the deposition occurs at low temperature and higher hydrogen dilution, the V_{oc} of a solar cell progressively increases. At 150°C, a high V_{oc} of 1.04 V is achieved for an a-Si nip solar cell. This is the highest V_{oc} reported for any single-junction a-Si solar cell. Figure 1 shows the J-V curve of a 1.04 V V_{oc} a-Si solar cell device deposited on a bare stainless steel substrate. From Table 1, we also see that the device with 1.04 V V_{oc} also has a higher FF than a device made in the same system but with an i layer deposited at 250°C. This indicates the high quality of this a-Si material which was deposited at a low temperature using high hydrogen dilution.

We also deposited a single layer of this optimized intrinsic a-Si material on a Dow Corning 7059 glass substrate for optical measurement. Its transmission spectrum was measured using a Perkin-Elmer spectrophotometer, and from the Tauc plot, $(\alpha h\upsilon)^{1/2}$ vs $h\upsilon$, we obtained a bandgap of 1.8 eV. We, therefore, conclude that the intrinsic a-Si material deposited at 150°C with high H dilution does not have a wide bandgap, as one would have expected for an amorphous material that leads to 1.04 V V_{oc}. We are unable to measure the hydrogen content of this material at this time since the film which was deposited onto a

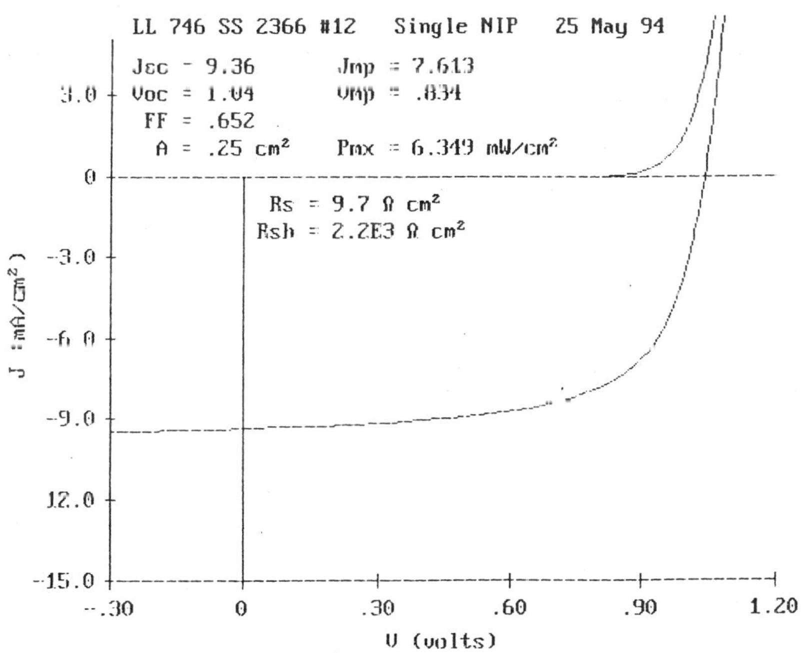

Fig. 1. J-V curve of an a-Si solar cell with the intrinsic layer deposited at 150 C with strong hydrogen dilution, showing 1.04 V open circuit voltage.

crystalline silicon substrate has a tendency to peel, probably due to high compressive stress. Fortunately, this material does not peel when it is deposited on stainless steel substrates.

When the a-Si deposition temperature is reduced, more hydrogen dilution is needed to achieve high quality material. However, when the hydrogen dilution during the a-Si layer deposition increases beyond a certain limit, V_{oc} as well as other parameters of the solar cell drastically decrease, probably due to the microcrystalline formation. Figure 2 shows the J-V curve of a device with the i layer deposited at 230°C using the high hydrogen dilution optimized for 150°C deposition. One can see that V_{oc} has dropped to 0.86 V. The FF is slightly higher than for device 723 in Table 1 because it is thinner.

We have also deposited a high V_{oc} solar cell on stainless steel coated with a textured Ag/ZnO back-reflector. To maintain high V_{oc}, the p layer was made thicker due to the texture in the under layer. A device with the same i layer as in device 746 demonstrates a V_{oc} of 0.997 V on this back-reflector. However this drop in V_{oc} is less critical since in a triple-junction device, the top cell is made on middle and bottom cells.

To demonstrate the improvement in V_{oc} in a multiple-junction device, we deposited this high V_{oc} device on a standard a-Si cell to form an a-Si/a-Si tandem device. The complete device structure is: SS/Ag/ZnO/a-Si n-i-p/high V_{oc} a-Si nip/TCO. Figure 3 is the J-V curve of such a tandem device. As a comparison, we have also deposited a tandem with both intrinsic layers deposited using regular conditions, i.e., 250°C and regular hydrogen dilutions. The J-V curve of the reference sample is shown in Figure 4. By comparing Figure 3 and 4, we find that the V_{oc} is increased from 1.768 V for the reference tandem, to 1.836 V for the tandem cell with the top cell intrinsic layer made at 150°C with high dilution. The gain in V_{oc} is around 0.07 V, similar to the gain in V_{oc} from device 712 to 746 as shown in Table 1. The slightly lower FF in the first tandem cell is due to different mismatch.

A similar approach to achieving a high V_{oc} a-Si alloy solar cell using low temperature and high hydrogen dilution has also been used in other laboratories [11,12]. J. Yang et al. of United Solar reported the achievement of single junction solar cell with a V_{oc} of 1.023 V and a FF of 0.773 with the intrinsic a-Si layer deposited at 175°C and with high hydrogen dilution [11]. They have also found that the solar cell fabricated with this intrinsic layer is relatively more stable after prolonged light soaking.

SUMMARY

A V_{oc} of 1.04 V has been achieved in an a-Si solar cell using an a-Si intrinsic layer deposited at relatively low temperature with high hydrogen dilution. Compared to the alternative a-SiC high bandgap intrinsic layer, this a-Si material absorbs light more efficiently and the material quality is higher. We thus conclude that the high V_{oc} a-Si material is a better choice as the top cell intrinsic layer of a triple-junction a-Si alloy solar cell than high band gap a-SiC alloys.

ACKNOWLEDGMENTS

We thank Drs. H. Fritzsche, S. Guha, J. Yang, and S. J. Jones for important discussions. This work was supported DOE/NREL PVMaT 2A program under subcontract No. ZM-2-11040-7 and NREL Amorphous Silicon Research Project under subcontract No. ZAN-4-13318-11.

REFERENCES

[1] S. R. Ovshinsky, New Scientist, November 30, 1978.

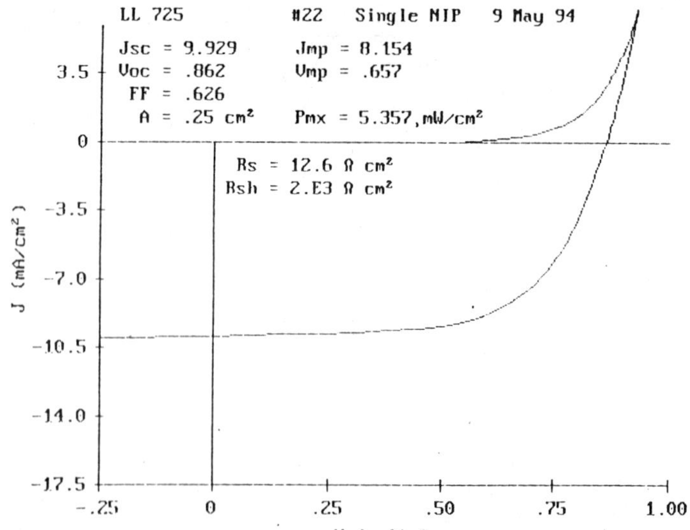

Fig. 2. J-V curve of an a-Si solar cell with the intrinsic layer deposited with the hydrogen dilution exceeding the optimum for the deposition temperature.

[2] M. Izu and S.R. Ovshinsky, SPIE Proc.**407**,42 (1983).

[3 J. Yang, R. Ross, T. Glatfelter, R. Mohr, G. Hammond, C. Bernotaitis, E. Chen, J. Burdick, M. Hopson and S. Guha, *Proc. 20th IEEE PVSC*, 241 (1988).

[4] M. Izu, X. Deng, A. Krisko, K. Whelan, R. Young, H. C. Ovshinsky, K. L. Narasimhan and S. R. Ovshinsky, *Proc. 23rd IEEE PVSC*, 919 (1993).

[5] X Deng, M. Izu, K. L. Narasimhan and S. R. Ovshinsky, MRS Proc. **336**, 699 (1994).

[6] J. Yang and S. Guha, *Appl. Phys. Lett.* **61**, 2917 (1992).

[7] S. Guha, J. Yang, A. Banerjee, T. Glatfelter, K. Hoffman, S. R. Ovshinsky, M. Izu, H. C. Ovshinsky, and X. Deng, MRS Proc. **336**, 645 (1994).

[8] Y. M. Li, MRS Proc., **297**, 803 (1993).

[9 S. Guha, K. L. Narasimhan, and S. M. Pietruzko, *J. Appl. Phys.* **52**, 859 (1981).

[10] K. Tanaka and A. Matsuda, Mat. Sci. Report, **2**, 139 (1987).

[11] J. Yang, X. Xu, and S. Guha, MRS Proc. **336**, 687 (1994).

[12] L. Yang and L. Chen, MRS Proc. **336**, 669 (1994).

[13] S. Guha, J. Yang, P. Nath, and M. Hack, *Appl. Phys. Lett.* **49**, 218 (1986).

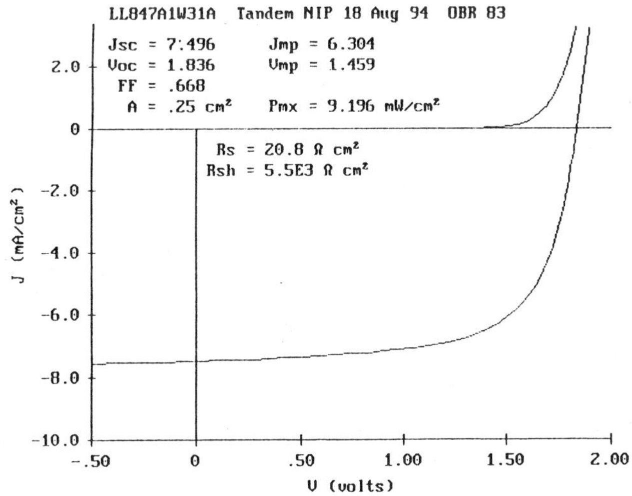

Fig. 3. J-V curve of an tandem device with the high V_{oc} a-Si top cell.

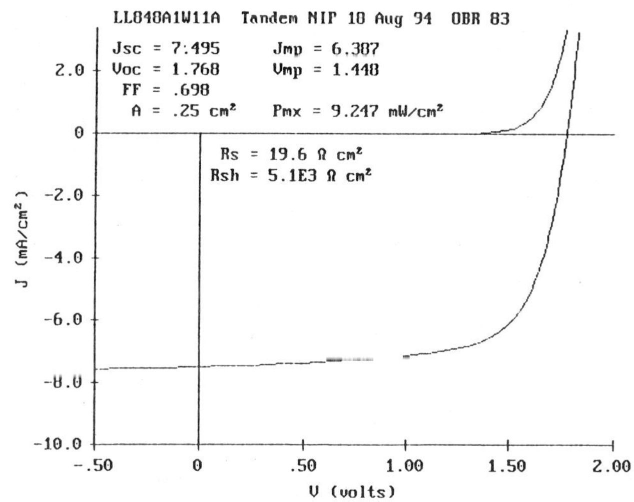

Fig. 4. J-V curve of an tandem device with regular a-Si top cell.

Fabrication and Properties of the Flexible a-Si:H Solar Cells on Plastic Films Using Roll-to-Roll Method

Hisato Shinohara and Masayoshi Abe
TDK Corporation R&D Center
2-15-7 Higashi-Ohwada, Ichikawa, Chiba 272, Japan
Kazuo Nishi and Yasuyuki Arai
Semiconductor Energy Lab. Co., Ltd.
398 Hase, Atsugi, Kanagawa 243, Japan

Abstract

The flexible a-Si:H solar cells were fabricated on plastic films using roll-to-roll method. The films used in this study were 0.075 mm thick and 230 mm wide. These solar cells were fabricated in a sequential process. We studied distribution of characteristics in the longitudinal direction of the film. A conversion efficiency of 8.02% was obtained under AM-1.5 100 mW/cm^2. Stability of submodule was estimated, and we developed a new organic resin material. The submodule was obtained high reliability at high temperature and humidity.

Introduction

Production of a-Si:H solar cells started in the beginning of 1980. The solar cells have been widely used for consumer goods, for example, calculators, watches, and radios. Thereafter production of the solar cells has increased year by year [1]. The needs have multiplied as their market became bigger. In the field of portable electric machines, solar cells as a battery source have been required to be lighter and thinner. On the other hand, in the field of the electric power source, reduction of cost for power generation has been required. Thus, further improvement in conversion efficiency and reduction in production cost has been required.

The a-Si:H solar cells have been prominent as a low-cost material even in the early stage of its development. An a-Si:H solar cell was prepared on a glass substrate. In addition, flexible a-Si:H solar cells were fabricated on a stainless steel or plastic film substrate [2,3]. The flexible a-Si:H solar cell is thinner, easier to make in a wader area and isn't easily broken. Thus it is expected that such solar cell is applicable to wide usage. Furthermore, it is considered that the roll-to-roll method is appropriate to mass production of solar cell because sequential production and automation is easily performed by this method.

Based on the above background, we fabricated flexible a-Si:H solar cells using a plastic film as a substrate. The fabrication process of those solar cells used roll-to-roll method. This paper reports characteristics of flexible a-Si:H solar cells.

Fabrication Processes for Flexible Solar Cells

A polyethylene naphthalete(PEN) film was used as a substrate of the flexible solar cells. Films used in this study were 0.075 mm thick and 230 mm wide. The roll-to-roll method was used for every fabrication process for deposition and patterning. The integrated flexible solar cell was fabricated by this method.

In this process, a metal electrode layer was prepared by a sputtering method. The electrode has a bi-layered structure of Al/ZnO. The semiconductor layers were deposited on the electrode by plasma CVD system with three reaction chambers. Figure 1 shows schematic diagram of the plasma CVD system. This layer has n-i-p junction with an n-type μc-Si:H, an i-type a-Si:H and a p-type μc-Si:H layer.

A patterning process for the integrated structure was employed. Figure 2 shows the cross section structure of series connection for integrated solar cell. The laser scribe method is usually utilized for this process [4]. However, it is difficult to make a flexible solar cell with plastic film. The substrate of plastic film was easily damaged by laser scribe process. Generally, plastic film have lower heat resistance than glass substrate, so that the condition of laser scribe is limited. This limits acceptable process conditions and worsens the process

CH3365-4/94/0000-0682 $4.00 © 1994 IEEE First WCPEC; Dec. 5–9, 1994; Hawaii

yield. To prevent thiseffect, we added a screen printing process in a laser scribe method. In this process, the integrated structure was prepared by the manufacturing processes as show in the following:

1) The first laser scribe removes the semiconductor and the metal electrode layers from substrate. Each layer of these divided a large area solar cell into some small area unit cells. The laser scribe system used Nd:YAG laser with at wavelength of 1.06 um.

2) The second laser scribe makes a groove of contact area for the metal and transparent electrode.

3) A patterned insulating layer on the first laser scribe lines and on the semiconductor layer. This layer was prepared by the screen printing method using organic resin ink.

4) An ITO layer was deposited on the semiconductor layer and the insulating layer by a sputtering method.

5) A grid electrode was prepared by the screen printing method using Ag paste.

6) The third laser scribe divide the ITO layer at on the insulating layer.

The submodules were continuously fabricated on the film by the process as above mentioned. Finally, each submodules were cut out from film. The characteristics of these solar cells were estimated from illuminated I-V characteristics. Stability of submodule was estimated from heat resistance and wet resistance test.

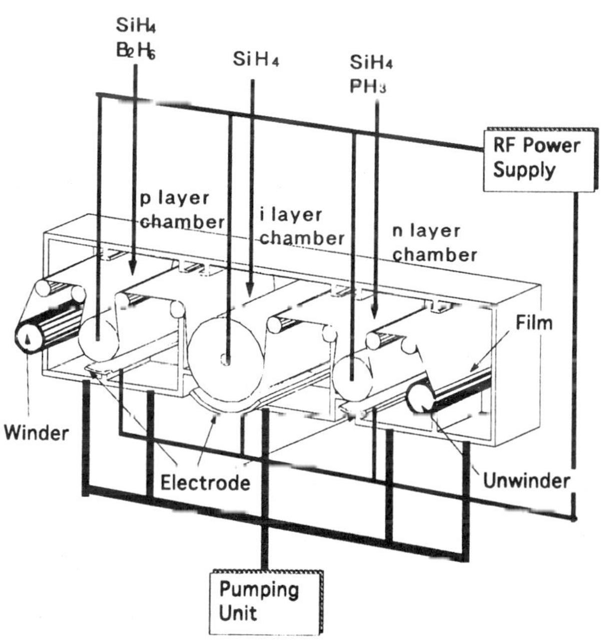

Fig. 1 Schematic diagram of the roll-to-roll PCVD system.

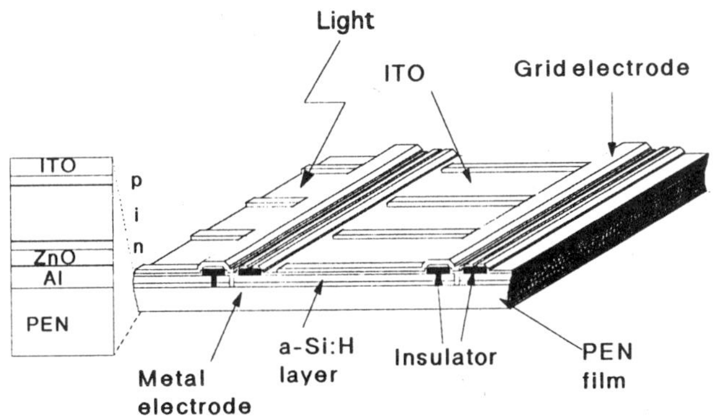

Fig. 2 Cross section of series connection for integrated solar cell

Results

I-V Characteristics

Single junction flexible solar cells having a metal/n-i-p/ITO structure were fabricated on the plastic film. In this structure, properties of metal/semiconductor interface layer were very important. The metal electrode was made of Al with 150 nm thick. A thin ZnO layer was deposited on the Al electrode to reduce reaction of Al and a-Si:H. The metal electrode needs high reflectance to make the most of long wavelength. Then the thickness of ZnO layer was studied to optimum conditions for stability and optics. Figure 3 shows the best I-V characteristics of a small area flexible solar cell in this work. The conversion efficiency was obtained 8.02 % for a 1 cm^2 solar cell under AM-1.5 100 mW/cm^2. Optimum thickness of ZnO layer was 80 nm.

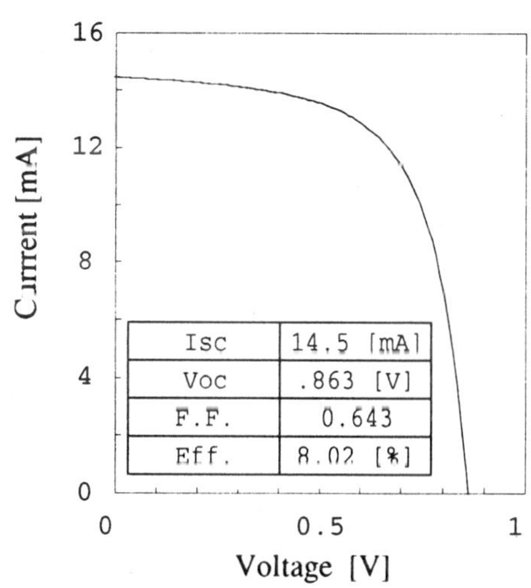

Fig.3 Illuminated IV characteristics of a flexible a-Si:H solar cell.

Stability of Submodules

The laser scribe and the screen printing processes was used in this roll-to-roll method. The screen printing processes printed a organic resin insulator on an a-Si:H layer. Conditions necessary for the insulator were as follows.

1) Stability of resin.
2) Baking temperature and time.
3) Contact on a-Si:H.
4) Heat resistance and wet resistance.

These conditions are important in this process. Conditions 1) and 2) affect the yield of production,and conditions 3) and 4) affect the stability of submodule. We used some organic resin materials and examined the stability of submodules.

Figure 4 shows changes of characteristics for the submodules under the condition of high temperature and humidity, i.e., 80 ℃, 90%. The organic insulator was well known as a polyester group material. Samples using this insulator do not have any problem in their initial characteristics, but degradation of these samples was observed in this test. In addition, we observed the semiconductor layers peel off the substrate. This results brought us assuming that the degradation of samples was due to the change of quality for the organic insulator. This organic insulator containing slightly uncured ingredient and hydrolyzed ingredient. These ingredient diffused into the semiconductor layer and degraded the film quality.

We developed a new organic insulator and applied submodules to it in consideration of the above mentioned issues. This organic insulator consisted of polyurethane resin. Figure 3 also shows result of these samples. The degradation was not observed even after the test of 2000 hours. The submodules were obtained high reliability at high temperature and humidity.

Property of Dispersion for submodules

The submodules were fabricated over 50 m in the lengthwise direction of a film. In deposition process of semiconductor layers, the deposition conditions (distance of electrodes, reaction pressure and rf power) were optimized to reduce generation of particles and flakes which cause a short of cells. In patterning process, on the other hand, a screen printing machine having working area of 250 mm × 200 mm at the maximum was used for printing of an insulator and a grid electrode. To improve the technique continuously and to prevent damages during printing, viscosity of organic resin ink was controlled by selecting solutions. In this work, we prepared the integrated type submodule (6 cells in series) having an area of 60 cm^2.

Figure 5 shows typical I-V characteristics of the submodule. The conversion efficiency was obtained 7.48 % under AM-1.5 100 mW/cm^2.

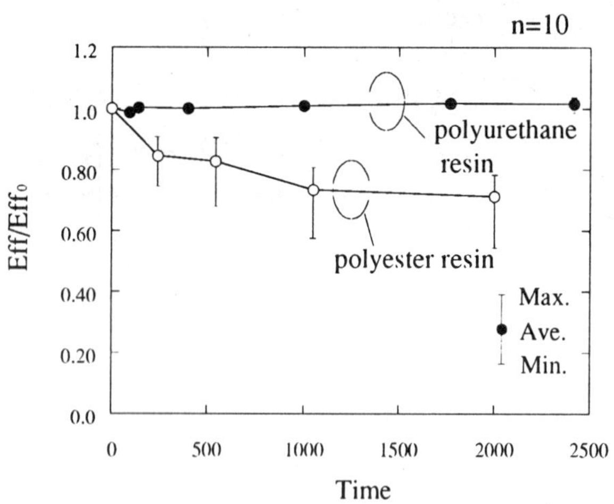

Fig.4 Characteristics of reliability for submodules.(80℃,90%RH)

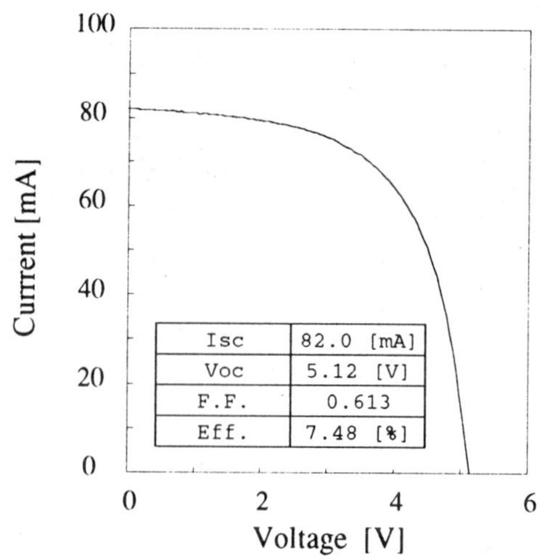

Fig. 5 Illuminated IV characteristics of a flexible a-Si:H solar cell submodule.

Figure 6 shows property of dispersion for I-V characteristics in the lengthwise direction of a film. The dispersion of Isc is considered to be caused by change in quality of semiconductor or ITO layers, and Voc is considered to be caused by pinholes of a semiconductor layer and/or degradation of an insulator. However, property of dispersion was few and high yield rate was obtained. Figure 7 is a photograph of a flexible solar cell submodule. This flexible submodule is thinner and lighter.

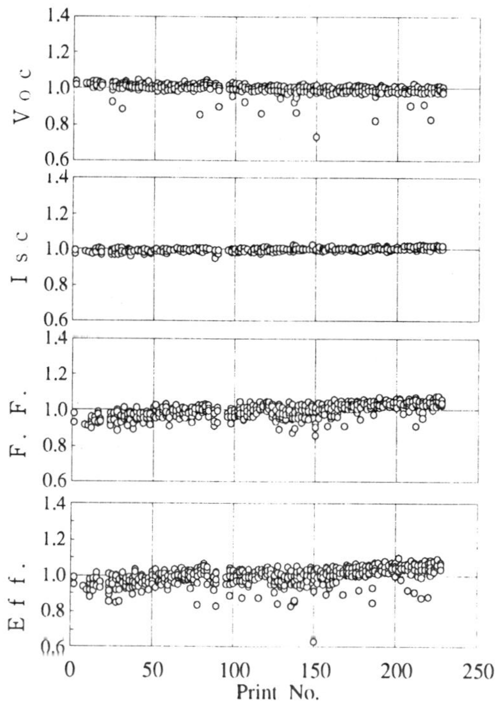

Fig. 6 Characteristics of dispersion for submodules in the film.

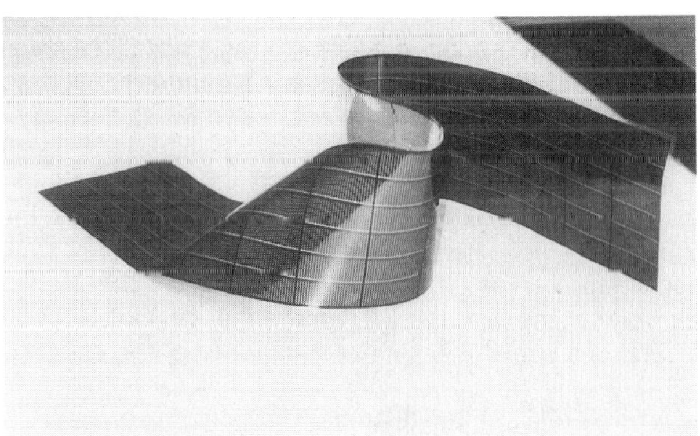

Fig. 7 Photograph of flexible solar cell submodule.

Conclusion

A flexible solar cell was fabricated on a plastic film using a roll-to-roll method. Process technology and component materials were optimized. The conversion efficiency of 8.02 % for a 1 cm^2 solar cell and 7.48 % for a 60 cm^2 submodule under AM-1.5 100 mW/cm^2 light were obtained. We developed a new organic insulator material and applied submodules. That submodule was obtained high reliability at high temperature and humidity condition. Furthermore, the roll-to-roll method confirmed the excellent productivity and capability of the mass production process.

Acknowledgment

We would like to thank Teijin Ltd. for their helpful discussions.

References

[1] Y. Hamakawa, "Recent Progress of Amorphous Silicon Solar Cell Technology in Japan", 22nd IEEE PVSC, 1991, pp. 1199.

[2] K. Hoffman, P. Nath, J. Call, G. DiDo, C. Vogeli and S. R. Ovshinsky, "Yield and Performance of Amorphous Silicon Based Solar Cells Using Roll-to-Roll Deposition", 20th IEEE PVSC, 1988, pp. 293.

[3] K. Nakatani, K. Suzuki and H. Okaniwa, "A Monolithic Series Connected a-Si:H Solar Cell on an Organic Polymer Film", PVSEC-3, Tokyo, Japan, 1987, pp. 391.

[4] S. Nakajima, M. Abe, H. Shinohara, Y. Arai, N. Ishida, A. Satake, K. Nishi, S. Kugawa, Y. Uehara, M. Ishii and S. Yamazaki, "High Effective Area Amorphous Silicon Solar Cell Using Excimer Laser Precess", 21st IEEE PVSC, 1990, pp. 1400.

LIGHT-INDUCED CHANGES IN HYDROGENATED AMORPHOUS SILICON: ANOMALOUS RECOMBINATION BEHAVIOUR AT LOW TEMPERATURES

J C L Cornish, Subaer, P Jennings, and G T Hefter
School of Mathematical and Physical Sciences,
Murdoch University, Murdoch, WA 6150, Australia.

ABSTRACT

The open circuit voltage decay (OCVD) method has been used to characterise a-Si:H, p-i-n photovoltaic cells over a range of temperatures from 228 K to 291 K.

We have found that the form of the OCVD voltage-time curves is sensitive to light soaking: the curves at different temperatures for as-deposited samples are nested inside each other with similar overall decay times, whereas, for the light soaked samples, as the temperature is decreased the decay time increases markedly and successive curves overlap.

Our preliminary results indicate that the phenomena observed may be related to the overshoot in the transient photocurrent observed by other researchers under similar conditions.

INTRODUCTION

In a semiconductor, excess carriers produced by transient illumination or forward bias, decay as a result of recombination with a characteristic time; the recombination lifetime. This is the average time of the existence of an excess electron-hole pair. It is normally referred to as the minority carrier lifetime when the minority carriers dominate the recombination process [1]. When electron-hole pairs are created in the intrinsic region of a p-i-n cell, the carriers diffuse to the p and n regions respectively, where they contribute to the photocurrent if the cell is connected to an external circuit, or produce the photovoltage if the cell is open-circuit. Some of the carriers are lost before this happens by the processes of bulk or surface recombination.

Recombination takes place preferentially at trap sites such as dangling bonds. The number of dangling bonds is a measure of the quality of the photovoltaic material. The greater the concentration of dangling bonds, the faster the rate of recombination and the shorter the minority carrier lifetime. Thus the minority carrier lifetime is an important parameter used in characterising photovoltaic materials and predicting photovoltaic performance. There are a number of methods used for the determination of the minority carrier lifetime, but the most popular appears to be the open circuit voltage decay (OCVD) technique [2,3,4].

This present study is an extension of a project carried out for an honours thesis [5] and not previously published and is, to our knowledge the first report of temperature dependence of minority carrier lifetime in a-Si:H.

Berry and Longrigg [3] have described the form of the OCVD response of an a-Si:H cell as shown in Fig. 1.

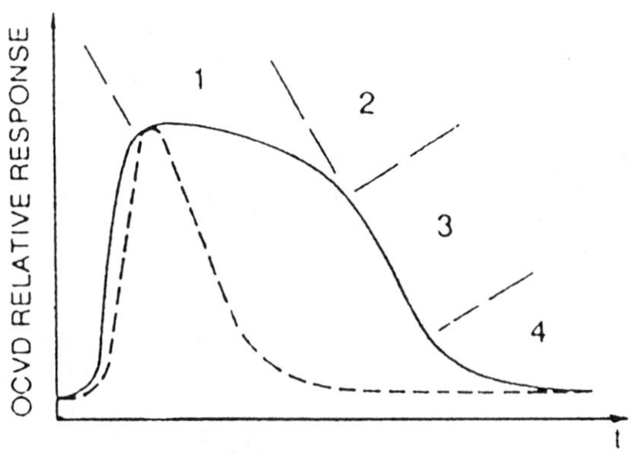

Fig. 1. A comparison of a typical OCVD response for an amorphous silicon module to the excitation pulse showing the four decay modes: 1, intrinsic; 2, trap activation; 3, trap-assisted intrinsic decay; 4, final non-linear exponential decay.

There is some dispute over the magnitudes of the minority carrier lifetime values obtained from such responses curves [6], but the form of the curve, particularly the voltage at which the break-point (2) occurs, is a useful measure of the quality of the cell [3].

EXPERIMENTAL

The samples measured consisted of single homojunction p-i-n a-Si:H cells produced by plasma enhanced CVD on ITO coated glass substrates, with substrate temperatures

CH3365-4/94/0000-0686 $4.00 © 1994 IEEE First WCPEC; Dec. 5–9, 1994; Hawaii

of 225°C. The i-layer was approximately 400 nm thick. Back contacts were aluminium dots 0.049 cm² in area. The cells were produced without any efficiency enhancement modifications and gave efficiencies up to 7%.

Open circuit voltage decay measurements, were made using the experimental arrangement shown in Fig. 2.

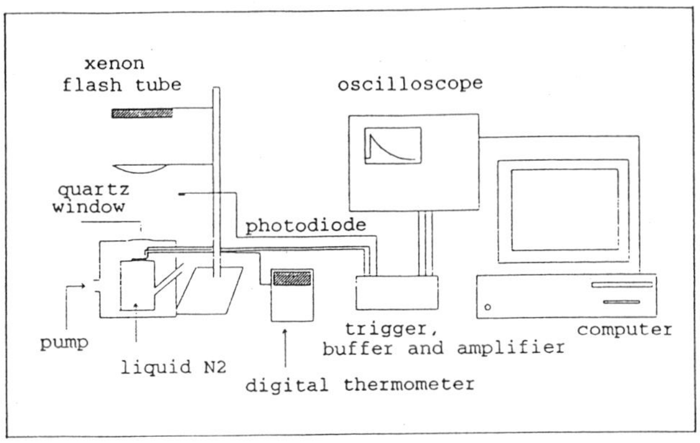

Fig. 2. Experimental arrangement for the measurement of the open-circuit-voltage-decay

A Xenon flash lamp with a repetition rate of 12 Hz and a total pulse length of 80 μs was used as the excitation source. A fast photodiode sensed the stray light from the lamp and was used to trigger the Phillips PM 3323 500MS/s digital storage oscilloscope. The open circuit voltage was fed via a high impedance buffer/amplifier to the oscilloscope. The oscilloscope was interfaced with a PC.

The sample was cooled using either a thermoelectric device or by the use of liquid nitrogen. Electrical contact to the sample was made using gold wire probes. The sample temperature was measured with a thermocouple. Heating and temperature control could be performed using either the thermoelectric device or a resistive heater.

RESULTS

The typical form of the measured OCVD curves is shown in Fig 3. The curves at low temperature display all four regions, band-to-band decay (1), trap activation (2), linear decay (3) and exponential decay (4), described by Berry and Longrigg [3]. At the highest temperatures the intrinsic region (1) is absent or very much attenuated, due to the comparatively low intensity of the flash lamp; the V_{oc} signal at room temperature is about 0.4 volt which is smaller than the V_{oc} obtained under AM1 illumination and decreases with decreasing temperature as shown in Fig. 4. At lower temperatures, the intrinsic region is clearly evident. At temperatures below room temperature, the decay curves neatly nest inside the room temperature curve. Region 1, attributed to band-to-band decay is enhanced at low temperatures, while the overall decay time is slightly reduced.

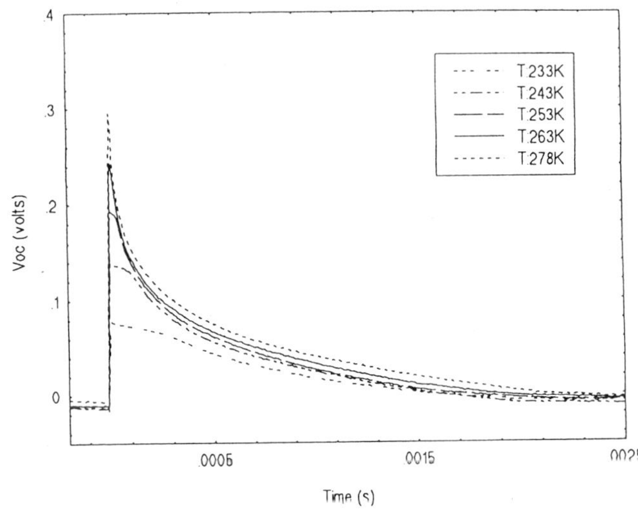

Fig. 3. V_{oc} decay curves for an a-Si:H p-i-n cell at various temperatures.

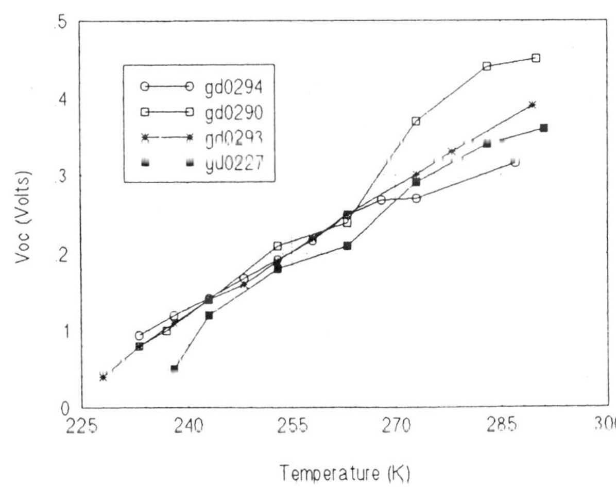

Fig. 4. V_{oc} vs. temperature for a-Si:H, p-i-n cells.

A set of curves for a sample following 24 hours of light soaking is shown in Fig. 5. In this series of curves the low temperature band to band region is enhanced and occurs over sufficiently long periods of time that the decay curves at successively lower temperatures cross over the higher temperature curves and the overall decay times increase with decreasing temperature. This is in complete contrast to the form of the as-deposited curves.

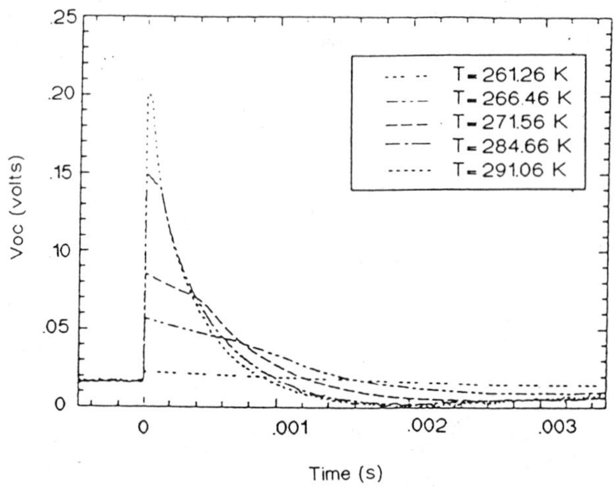

Fig. 5. V_{oc} decay curves for an a-Si:H p-i-n cell following light soaking.

It appears in this case that similar numbers of charge carriers are excited at each temperature but that these take much longer to recombine as the temperature is reduced.

OCVD behaviour intermediate between the two forms shown in figures 3 and 5 occurred for a sample following exposure to light during a number of IV measurements but with no explicit light soaking, this is shown in Fig. 6.

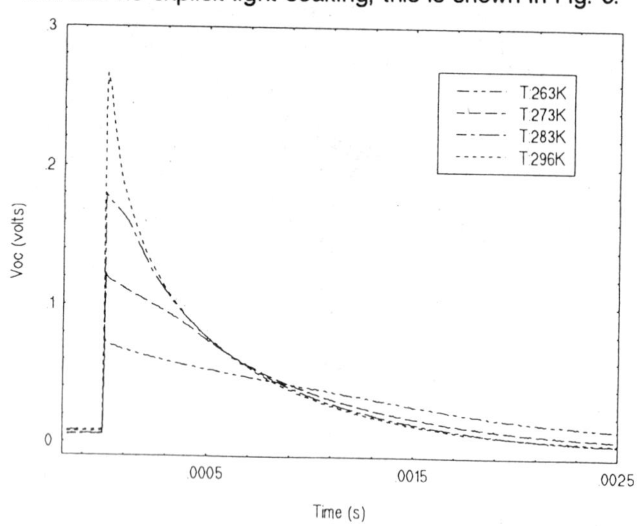

Fig. 6. V_{oc} decay curves for an a-Si:H, p-i-n cell after exposure to AM1 illumination.

For this sample the number of dangling bonds is expected to be somewhere between those for the previous two samples.

DISCUSSION

The different regions in the decay curves shown above can be related to different decay mechanisms. Berry and Longrigg [3] describe the response in terms of decay in the intrinsic layer of the cell, and in the p-i and i-n barrier layers. Intrinsic band-to-band decay occurs when all the traps within the mobility gap are saturated. Illumination of the a-Si:H creates excess carriers (electrons and holes) which populate the extended and localised states at the band edges. At room temperature the light intensity used in this investigation was not adequate to fulfil this condition and the band-to-band region is not seen on the curve. However, at lower temperatures, region 1 does appear and becomes more significant as the temperature is decreased. The onset of decay via trap states is increasingly delayed as lower temperatures are reached and intrinsic band-to-band like decay is observed.

It is expected that both the number of traps and the number of carriers generated should be independent of temperature. Thus saturation of the traps and intrinsic band-to-band transitions should not occur at any temperature.

If the trap filling process is thermally activated, then at low temperatures, the production rate of free carriers may exceed their loss rate to traps resulting in a quasi-saturation and band-to-band decay takes place which persists until the traps are activated. Evidence for this process is seen in transient photocurrent data, where at low temperatures an initial overshoot in the current has been observed [7]. This is explained as due to the rate of deep trapping of the current carriers being less than their generation rate. Ulrich, Eickhoff and Wagner [7] have observed such overshoots persisting up to several hundreds of milliseconds after the onset of illumination, adequately covering the band-to band decay times observed in the OCVD results for as-deposited samples.

In the OCVD results for light soaked material, the band-to-band transitions are significant for even a small reduction in temperature. This effect increases as the temperature is further reduced. Since light soaking produces an increased number of dangling bonds, these should give rise to more trap states, and hence saturation of such material should be less likely and band-to-band transitions are unlikely to be observed. This is clearly not so.

Ulrich, Eickhoff and Wagner have investigated the effect of light soaking on the transient photocurrent, and have shown that the overshoot is considerably reduced [8]. Since their carrier generation rate is much less than that used in the present OCVD measurements, an effect may be present but onobserved, it would have to tke place in the first 100 ms or so. Fig. 5. shows the region 1 decay extending up to about 100 ms. Wieczorek and Fuhs [9] have demonstrated overshoot in the hole current generated in a n-i-p diode even at room temperature. In this case the time for the overshoot to decay was over 100 ms. It is expected that this effect would be enhanced at lower temperature and could adequately describe the curves of Fig. 5.

CONCLUSIONS

Our preliminary investigations have revealed a distinctive effect which may have considerable value in determining the quality of a-Si:H material and in monitoring the degradation of solar cells and modules. Systematic investigations have commenced to obtain quantitative data suitable for a basis on which to model the processes responsible for the observations. In addition simultaneous transient photocurrent measurements will be made, using a pulsed laser source with a rapid rise time to provide further information on the initial processes involved.

REFERENCES

[1] G.E. McGuire, *Characterisation of Semiconductor Materials: Principles and Methods*, Noyes Publications, 1989.

[2] J. Mahan, T.W. Ekstedt, R.I. Franck and R. Kaplow, "Measurement of minority lifetime in solar cells from photo-induced open-circuit voltage decay", *IEEE Trans. on Electron Devices*, **26**, 1979, pp.733-739.

[3] W.B. Berry and P. Longrigg, "Open-Circuit Voltage Decay measurements of amorphous silicon material stability and module degradation", *Solar Cells*, **24**, 1988, pp. 321-328.

[4] I. Sakata and Y. Hayashi, "Open-circuit Voltage Decay (OCDV) measurement applied to hydrogenated amophous silicon solar cells", *Japan. J. Appl. Phys.* **29**, pp. l27-l29.

[5] Carr, A.J., *A Study of Minority Carrier lifetime of silicon solar cells*, B.Sc.(Honours) Degree Thesis, Murdoch University, 1992.

[6] M.A. Green, "Solar cell minority carrier lifetime using open-circuit voltage decay", *Solar Cells*, **11**, 1984, pp. 147-161.

[7] C. Ulrichs, Th. Eickhoff and H. Wagner, "Transient Photocurrent Spectroscopy on Amorphous Silicon Solar Cells", *Twentythird IEEE PVSC*, 1993, pp. 981-985.

[8] C. Ulrichs, Th. Eichhoff and H. Wagner, "Transient response of the photocurrent in a-Si:H layers and solar cells", *J. Non-Cryst. Solids*, **164-166**, 1993, pp. 705-708.

[9] H. Wieczorek and W. Fuhs, "Deep Trapping of Carriers in a-Si:H Solar Cells Studied by Transient Photocurrents", *phys. stat. sol. (a)*, **114**, 1989, pp. 413-418.

METALIZED POLYMER SUBSTRATE FOR FLEXIBLE AMORPHOUS SILICON SOLAR CELL

Lee-Ching Kuo,Yu-Tai Tsai, Wen-Yueh Hsu and Neng-Hsing Lu
Industrial Technology Research Institute, Material Research Laboratories

ABSTRACT

The properties of a deposited metal layer is quite important for the performance of a flexible amorphous silicon solar cell. The properties of deposited Al, Cr and Mo metal films on polyimid substrate are studied. Among the measured surface roughness, reflectance and stress of these films, the Al/Mo double metal layer has a smooth surface, higher reflectance, and better device performance. The Al/Mo metal film has higher Jsc and Efficiency than that made by the Al/Cr metal film.

INTRODUCTION

A flexible amorphous silicon solar cell is thin, lighweight, flexible and unbreakable . Therefore, it is a favorate product. An a-Si solar cell on a polymer substrate is especially thin and light. Since polyimide has high insolation resistance, mechanical strength and good dimemsion stability. It has been used in the packaging of microeletronics, and is an idea substrate for flexible solar cell. The procedure for making solar cell is 1.deposite metal film on a substrate, 2 GD-CVD three a-Si layers, and 3. deposite an ITO layer. The temperatare for deposition a-Si layers is at about 250 ℃ . The properties of metal/PI interface are importart, and have been reported by many authors(1-4). The high temperature may cause crack on metal surface. The right choice of metal film for electorde is quite important.

EXPERIMETAL

Biphenyl tetracarbonic acid (BPDA), benzophenone tetracarbonic acid (BTDA), p-phenylene diamine (PPDA), 4,4'-diaminophenyl ether (ODA) in NMP/Xylene solvent formed the polyamic acid varnish. The varnish was than coated onto a glass substrate by using a Doctor Blade method . The coated substrate was cured at 350℃ for the imidization process. The surface of the polyimide film was sputter cleaned by a RF plasma. The metals were deposited onto the cleaned surface. Two methods have been used, ie, E-gun evaporation and sputtering . The adhesion strengths of the metal film were measured by the conventional 90 ° peeling test. The surface morpholog were measured by a Atomic Force Metrology (AFM). The reflectances were meassured by using a UV-VIS spectrometer.

The structure of the flexible a-Si solar cell is shown in Figure 1. After deposite the metal layer on PI substrate, the p-i-n amorphous silicon layers were deposited by a RF glow-discharge CVD process. Finally, a transparent conductive ITO film as an electrode was deposited by a sputtering process. During a-Si deposition process, cracks were created on the Cr/PI or Mo/PI substrate due to high thermal stress. Therefore, before the Cr or Mo deposition, a layer of Al was coated on PI to decrease the crack. The thicknesses of deposited Al, Cr, and Mo were the same and was 2000Å .The three layers of p-i-n a-Si were 180Å ,4500Å and 300Å respectivily. The deposition conditions were pressure, 0.4 torr; temperature 250 ℃ ; power density, 0.01w/cm^2.

ITO
P, a-SiC:H
I, a-Si:H
N, a-Si:H
Metal
PI

Fig.1. The structure of the flexible a-Si solar cell

RESULT AND DISCUSSION

The roughness of the deposited metal are shown in Figure 2. The surface roughness of Al/Cr on PI, is measured by using a AFM mechine. Figure 2 shows that the E-Gun deposited metal film is more rough than the sputtered surface. The E-Gun created metal film has 1074Å from peak to

CH3365-4/94/0000-0690 $4.00 © 1994 IEEE

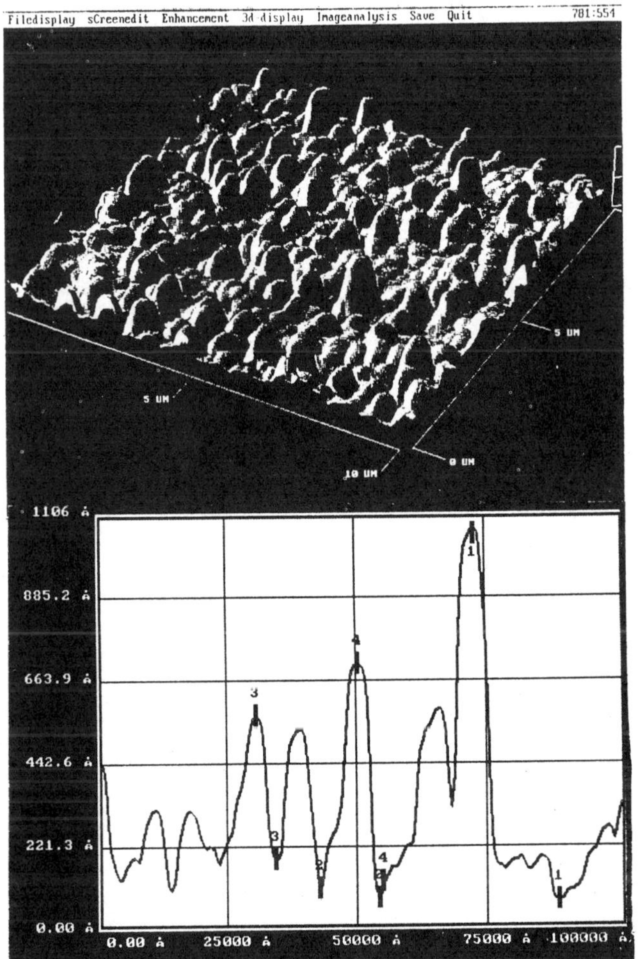

Pts.	Distances (Å)	Heights (Å)
1 - 1	13972.219	1017.103
2 - 2	11342.594	32.228
3 - 3	3703.705	402.847
4 - 4	4629.633	639.184

Fig.2(a) Non-contact AFM image of E-Gun deposited PI/Al/Cr

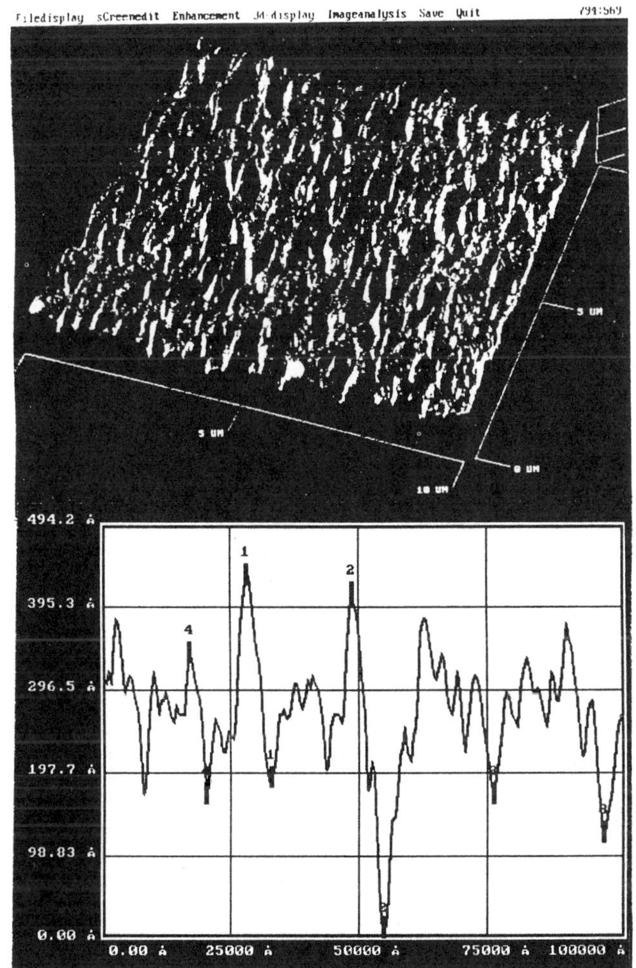

Pts.	Distances (Å)	Heights (Å)
1 - 1	4861.111	252.451
2 - 2	6250.000	424.332
3 - 3	21296.297	53.713
4 - 4	3240.741	150.396

Fig.2(b) Non-contact AFM image of sputtered PI/Al/Cr

valley, but the sputtered surface only 295Å. The sputtered metal film has a better adhesion with PI. Therefore, the sputtering was used in the rest work. Figure 3a is for Al film; 3b, Cr; 3c, Mo; 3d, Al/Cr; 3e, Al/Mo. Where Zavg is the average height of peaks, Zmax is the distance of maximum height to lowest valley, Ra is the average roughness. The deposited surfaces of Al and Cr are more rough than Mo, and double metal the Al/Mo surface is flater than Al/Cr , which indicates the deposited Mo can smoothen a rough surface. Figure 4 shows the reflectances of PI/Al/Mo and PI/Al/Cr, where PI/Al/Mo has a higher refelectance. Table 1 gives the obtained Voc, Jsc, F.F. and efficiency data for the flexible solar cells. For the a-Si solar cells of same area, materials processes and structure, and the only difference is the metal electrode, one is Al/Cr the other is Al/Mo. The a-Si solar cell of Al/Mo electrode has a better Jsc and conversion efficiency than that of Al/Cr. This is due to the flatness and higher reflection of Al/Mo electrode.

In general, the effeciency of a polymer flexible a-Si solar cell is lower than that of a glass substrate(5-6). It is due to the roughness, nonuniformity, out gasing and the higher mismatch of thermal expansion coefficiency between the

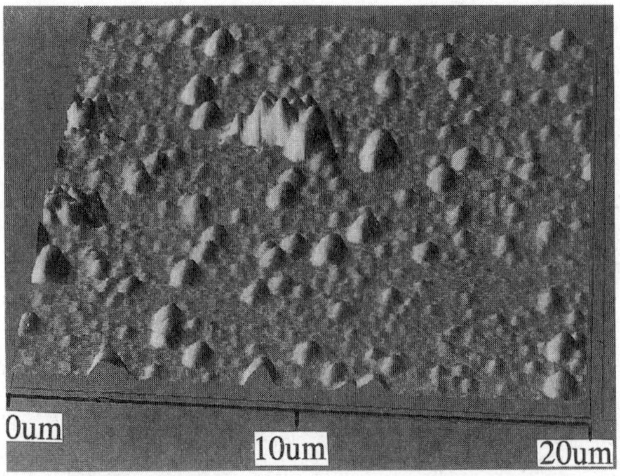

3a. PI/Cr ,
Ra:42.05Å, Zavg:207.5Å, Zman:536.05Å

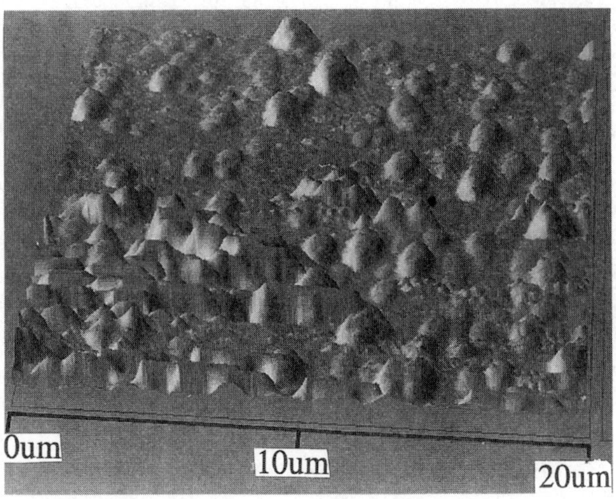

3b. PI/Al ,
Ra:16.4Å , Zavg:99.01Å , Zman:411.98Å

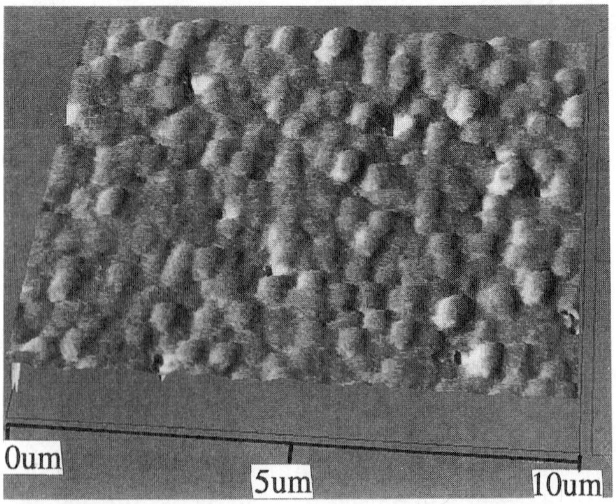

3c. PI/Mo ,
Ra:3.97Å , Zavg:132.8Å , Zman:157.38Å

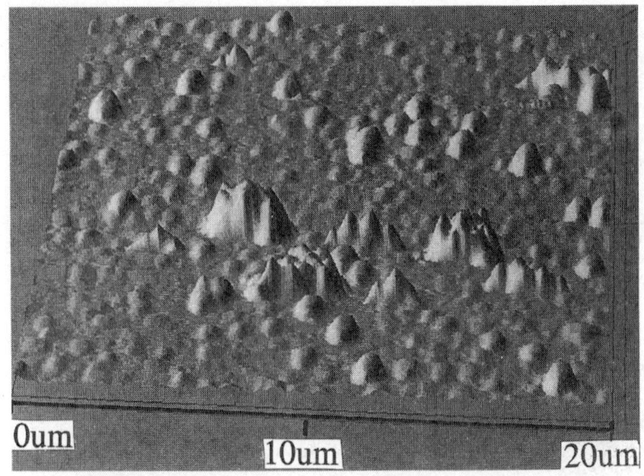

3d. PI/Al/Cr ,
Ra:12.79Å , Zavg:65.17Å , Zman:297.57Å

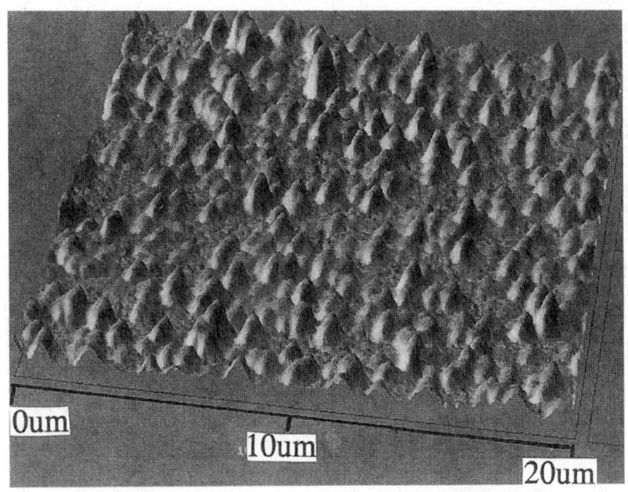

3e. PI/Al/Mo ,
Ra:12Å , Zavg:50.38Å , Zman:132.67Å

Fig.3.Non-contact AFM image of sputtered metal/PI surface (a)PI/Al,(b)PI/Cr,(c)PI/Mo, (d)PI/Al/Cr,(e)PI/Al/Mo

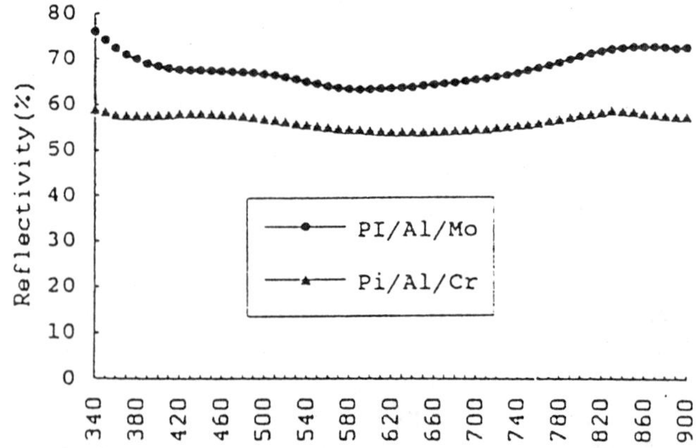

Fig.4.The reflectances of PI/Al/Mo and PI/Al/Cr

polymer to metal. The mismatch cause high stress porblem. The thermal expension coefficieney of the polyimide ,in this work, was 25×10^{-6}°C^{-1} ,which is much less then 45×10^{-6}°C^{-1} of the Kapton film. Table 2 shows the results of film stress measurements for PI/Al, PI/Al/Mo and PI/Al/Cr. For PI coated Si wafer measurable stress can be obtained. The tensile stress is greatly enchanced after the metal film being deposited . The measured PI/Al/Mo stress is still too high. Our future work will emphasis on to improve the process condition to decrease the stress.

Table1. The Voc,Jsc,F.F and Eff. of flexible a-Si solar cells with Al/Cr and Al/Mo metal electrode.

cell performace metal	Voc (V)	Jsc (mA/cm^2)	F.F. (%)	Eff. (%)
Al/Cr	0.702	14.39	49.9	5.1
Al/Mo	0.72	22.62	52.2	8.5

Table2. The film stress measurements for PI/Al, PI/AlCr,and PI/Al/Mo(Rd:radius of the bent curve, T:tensile stress, C:compressive stress)

	stress(dyne/cm^2)	T/C	Curve Rd(m)
PI	2.29E+08	T	9.64
PI/Al	2.44E+08	T	8.99
PI	1.71E+08	T	10.19
PI/Al/Cr	2.83E+08	T	6.12
PI	3.52E+08	T	6.26
PI/Al/Mo	4.13E+08	T	5.29

CONCLUSION

To produce a-Si solar cell on polymer film is more difficult than the conventional method, that is on a glass substrate. The key technology is how to metalize the polymer film . This work study the Al/Cr and Al/Mo films as electrodes on polyimide substrate. The morphology of Al/Mo is smooth, and has high reflectance than Al/Cr. Therefore, the Al/Mo produce a better a-Si solar cell than Al/Cr. To increase the reliability and reduce the metal film crack by decrease the stress between PI and metal film will be emphasised in our future work.

ACKNOWLEDGEMENTS

The authors would like to thank Mr.Jim-Shing King and Miss Ching-Juag Chiang for polyimide falirication. This work was supported by the Energy Commission Ministry of the Economic Affairs

REFERENCES

[1] American "Aluminum-Polyimides Interface" A.Selmani., *Chemical Society,* **1990,**P.345-351.

[2] "XPS Study of the Interface of Polyimide on Cr And Ni," Marco GARRILLI, Giovanni MARLETTA, Orazio PUGLSI,etc., *Mat. Res. Soc. Symp. Proc.* **Vol.154**,1989,P,317-334.

[3] P.S.HO, "Chemistry And Adhesion of Metal-Polymer Interfaces", *Applied Surface Science* **41/42** (1989) 559-566.

[4] S.E.MOLIS,T.S.OH,S.P.KOWALCZYK and J.KIM., " J *Adhesion Sci, Technol.* " **Vol. 5.** No.7, PP,509-521 (1991).

[5] Technical Digest of the International PVSEC-5, Kyoto Japan, 1990, P.1645.

[6] A.Ishikawa, D.Kruangam, H. Okamoto and Y. Hamakawa, *Proceedings of IEEE Photovoltaic Specialists Conference*, 1985, P.487-492.

INTERNATIONAL PHOTOVOLTAIC MARKETS, DEVELOPMENTS AND TRENDS FORECAST TO 2010

Paul D. Maycock, President
PV Energy Systems, Inc.

ABSTRACT

The 1993 world PV module shipments increased 6.7 percent from 57.9 MW in 1992 to 60.09 MW in 1993. European shipments increased one percent from 16.4 MW in 1992 to 16.55 MW in 1993. U.S. shipments increased 24.0 percent to 22.44 MW and Japanese shipments decreased 11 percent to 16.7 MW. The status of all major PV module producters will be summarized. New entrants in the thin film area include: Golden Photon (CdTe), Solar Cells Inc. (CdTe), United Solar Systems (triple stack amorphous silicon), B.P. Solar (CdTe), and Astropower (silicon film). In the silicon crystal area, plant expansions have been made by Siemens Solar Industries, B.P. Solar. and Solec International. New crystal silicon entrants include: Astropower, Texas Instruments, Nukem, and Gallivare. Potential capacity increases by 1996 of 75-100 MW are described. Module performance (efficiency and warranty), and manufacturing costs will be summarized and forecast to 2010. In the limit, the sliced silicon product tends to approach profitable prices of $2.00 per watt, while several thin film options and concentrators can be manufactured with profitable prices less than $1.50 per watt. The world market for PV modules is summarized by market sector and forecast to 2010 under two scenarios -- "Business as Usual" and "Acelerated". Under the Business as Usual assumptions, world PV world PV module shipments are forecast to be 195 MW in 1995 and 800 MW in 2010. The accelerated scenario forecasts world module shipments in 2000 to be 440 MW and 4000 MW in 2010. Module shipments for the last four years have been growing at a rate well below the BAU scenario.

Preliminary 1994 data indicate module shipments of 70-74 MW, a 20 percent increase over 1993.

WORLD PV MODULE SHIPMENTS -- 1993

1993 was not the excellent year that most had forecast. The goal of 65MW was not reached. Figure 1. shows the PV cell and module shipments from 1987 to 1993. 1993 cell and module shipments of 60.09 MW increased 6.7% over the 57.9 MW in 1992. U.S. shipments increased 24% to 22.44 MW. 1993 Japanese shipments decreased 11% to 16.7 MW from.18.8 MW while European module shipments languished at 16.55 MW. The Rest of the World shipments decreased 4% to 4.4 MW.

Fig. 1 World PV Module Shipments

WORLD PV MODULE SHIPMENTS*
(CONSUMER & COMMERCIAL) (MW)

COUNTRY	1987	1988	1989	1990	1991	1992	1993
UNITED STATES	8.7	11.1	14.1	14.8	17.1	18.1	22.44
JAPAN	13.2	12.8	14.2	16.8	19.9	18.8	16.70
EUROPE	4.5	6.7	7.9	10.2	13.4	16.4	16.55
REST-OF-WORLD	2.8	3.0	4.0	4.7	5.0	4.6	4.40
TOTAL	29.2	33.6	40.2	46.5	55.4	57.9	60.09

*Excludes modules made from purchased cells

Shipments by Cell Technology

Figure 2. shows the 1993 world module shipments by cell technology. The single crystal option continued its dominant position in the market with rapid gains made by single crystal silicon. Virtually all of the increased capacity in 1993 came from the Siemens Solar expansion of their U.S. capacity from 7MW to well over 15MW. The amorphous silicon option lost ground owing to the delay in opening the ten megawatt plant at Advanced Photovoltaic Systems and decreased sales in Japan. When the sliced single crystal and the cast-ingot polycrystal silicon products are combined, they account for 77% of the world PV product. Despite the intense R&D effort on thin films of all kinds, there is little evidence that the dominance of sliced silicon crystal product will be challenged unless the new plants being built have significantly reduced costs. Over 50 MW of new thin film plants are being constructed for delivery of product in 1995 to

CH3365-4/94/0000-0694 $4.00 © 1994 IEEE

New Capacity Additions

Table 1 shows new capacity or plant expansions which will be completed in the 1995-1996 time frame. Virtually all new capacity being installed for 1995-1997 production, nearly 60 MW (75%), is based on thin films of amorphous silicon, cadmium telluride or copper indium diselenide.. The past five years have seen single crystal and polycrystal silicon capacity double while the thin film options remained flat in shipments.

All of the plant construction decisions for the thin film options, the Texas Instruments spheres, the Astropower silicon film on low cost substrate are based on analysis that predicts module costs of $1.00 to $1.50 per watt, leading to profitable prices at the $2.00 per watt level. (Note: this price compares to the 1994 prices of $3.50-$4.50 per watt.)

Fig. 2 Shipments by Cell Technology

WORLD MODULE SHIPMENTS(1993) BY CELL TECHNOLOGY (MW)

TECHNOLOGY	U.S.	JAPAN	EUROPE	R.O.W.	TOTAL	PERCENT
SINGLE CRYSTAL SILICON	14.7	1.4	8.95	3.6	28.65	47.68
POLYCRYSTAL SILICON	6.0	5.4	6.0	0.2	17.6	29.30
AMORPHOUS SILICON	1.5	8.9	1.6	0.6	12.6	20.97
CRYSTAL SILICON CONCENTRATORS	0.02	--	--	--	0.02	0.03
RIBBON (SILICON)	0.2	--	--	--	0.2	0.33
SILICON FILM	0.02	--	--	--	0.02	0.03
CADMIUM TELLURIDE	--	1.0	--	--	1.0	1.66
TOTAL	22.44	16.7	16.55	4.4	60.09	100.00

TABLE 1. POTENTIAL MODULE PRODUCTION CAPACITY ADDITIONS (1995-1997)

COMPANY	1993 CAPACITY	NEW CAPACITY	CELL TECHNOLOGY	COMMENT
SIEMENS(US)	15MW	5-10	SINGLE CRYSTAL	VANCOUVER WASH
SIEMENS(US)	0	??	COPPER INDIUM DISELENIDE	IN GERMANY?
SOLAREX(US)	6	6-12	POLYCRYSTAL	NEW PLANT 1996
SOLAREX(US)	1	5-10	AMORPHOUS-SI	DOUBLE STACK A-SI
EUROSOLARE(IT)	3	2-3	POLYCRYSTAL	CRYSTALLOX CAST INGOT
BP SOLAR (UK)	5	2	SINGLE CRYSTAL	SLICE CAPACITY
	.1	?	CADMIUM TELLURIDE	PRODUCTION '95?
KYOCERA	6	2-4	POLYCRYSTAL	WACKER INGOT TECH
USSC	2	10	AMORPHOUS-SI	TRIPLE STACK PLANT'95,'96
ECD/RUSS KVANT	O	3	TRIPLE A-SI	SHIPPED TO USSR,'92
SIEMENS (GER)	0	10+	"THIN FILM"	ANNOUNCED-NO DETAILS-CIS?
ASTROPOWER	.5	3-5	POLYCRYSTAL SI FILM	IN PILOT,'94-95
ADV.PV SYSTEMS	2	10	AMORPHOUS SILICON	COMPLETED'93,SHIP '94
TEXAS INST'S	.1(PILOT)	10+	SILICON SPHERES	PILOT 94,PRODUCT '95-96
GOLDEN PHOTON	.1	3+	CADMIUM TELLURIDE	PRODUCTION 95
SOLARCELLS,INC	.2	10+	CADMIUM TELLURIDE	PRODUCTION '96,'97
EBARA	0	2-3	DENDRITIC WEB	PRODUCT '96,'97
EPV	0	10	COPPER INDIUM DISELENIDE	'96,'97
TOTAL	41	+90-115 MW		

Profile of the U.S. PV Industry

Figure 3. shows the US module shipments by company for the years 1988-1993.. It is clear that Siemens Solar Industries and Solarex are the dominant forces in the US PV industry with 84% of the U.S. shipments and 31% of the world shipments. Siemens shipped 12.5MW of single crystal silicon modules. A new crystal pulling and slicing operation in Vancouver Washington is reported to have capacity for 18-20MW of single crystal slices. Virtually all of the new capacity is being processed into cells, shipped throughout the world for module production. Solarex shipped 6.0MW of polycrystal cast-ingot silicon. Solarex has a one megawatt pilot plant for amorphous silicon and shipped about .5 MW in 1993. There is evidence that Solarex is close to a decision to build a 5-10 MW amorphous silicon, amorphous silicon/germanium stacked cell plant. Solarex is also performing extensive research on copper indium diselenide. Advanced Photovoltaic Systems (APS), Princeton, New Jersey has completed the new ten MW amorphous silicon module plant in California. Over 600 kW of modules where shipped in 1992 from the APS pilot plant in Trenton, New Jersey (400 kW was installed at the PVUSA site in California). The success of this system has opened the door for volume shipment of amorphous silicon modules. However. 1993 shipments were reduced by the delay in production from the new plant. The 1994 shipments were delayed by a temporary restraining order awarded to Solarex for possible infringement of the amorphous silicon patents purchased from RCA. Solec has installed a new crystal pulling and slicing operation that is capable of over 2 MW of 125 by 125 cm single crystal silicon wafers. Solec 1993 shipments were 1.3MW. In November, 1994 all Solec assets were purchased by Sumitomo-Sanyo. Considerable expansion is planned. United Solar Systems has reworked the Sovonics 1 MW amorphous silicon plant and expanded capacity to 2-3 MW. In 1993, USSC announced a new 10 MW triple-stack amorphous silicon plant to be built in NewPort News, Virginia. Present plans call for completion in 1994. Mobil Solar shipped about 200 kW of modules in 1993. Late in 1993 Mobil announced the termination of all PV operations. The octagon sheet technology has been offered for sale and all Mobil operations will ceased by mid 1994. In July, 1994 ASE Gmbh, Germany (a combination of Deutsche Aerospace and Nukem) purchased all Mobil Solar assets. Astropower is in pilot production of their new polycrystalline silicon film on low cost substrate. 1993 shipments of 900 kW were from reclaimed single crystal silicon wafers and purchased polycrystal wafers. 20 kilowatts of Astropower's new silicon film on low cost substrate product. Possible new entrants in the US module market include: Texas Instruments with spherical cells of crystal silicon, Golden Photon (Photon Energy) with cadmium telluride (2-3MW-1995). Solar Cells Inc. with cadmium telluride (production 1996/7), EPV with copper indium diselenide and International Solar Energy Technology (ISET) with copper indium diselenide and cadmium telluride.

Fig.3. U.S. PV Module Shipments by Company

U.S. MODULE SHIPMENTS
BY COMPANY (MW)

COMPANY	1988	1989	1990	1991	1992	1993
SIEMENS SOLAR	5.5	6.5	7.0	9.0	9.0	12.5
SOLAREX	3.2	5.0	5.4	5.6	5.7	6.5
SOLEC INTL	0.6	0.9	0.9	1.2	1.3	1.3
ADVANCED PV SYSTEMS		-	-	0.2	0.8	0.5
ASTROPOWER	0.1	0.2	0.4	0.45	0.6	0.92
USSC	0.4	0.5	0.6	0.2	0.3	0.5
MOBIL SOLAR	0.1	0.05	0.05	0.2	0.3	0.2
ENTECH	-	0.3	0.03	0.03	0.05	0.01
OTHER	1.2	0.65	0.42	0.2	0.1	0.01
TOTAL	11.1	14.1	14.8	17.1	18.15	22.44

INCLUDES CELLS AND MODULES

Profile of the European PV Industry

European module shipments increased less than 1% from 16.4MW in 1992 to 16.55MW in 1993. (Figure 1). 15.3MW of the European modules were sliced polycrystalline or single crystal silicon. 1.2MW was amorphous silicon modules based on Chronar (US) technology. Over 70% of the European production was sold for use in Europe. Germany, Switzerland, Italy, and Spain account for most of the consumption. Sales to former colonial interests in Africa, Indonesia, South Africa, and Polynesia accounted for most of the exports. Figure 4. shows the 1993 PV module shipments by European companies. BP SOLAR has grown to be number one in Europe with 4.5MW of PV cells being made in the plants in Spain and Australia. BP purchases single crystal slices from other suppliers. BP Solar opened a new plant in Spain to manufacture ultra high efficiency silicon cells using the licensed technology from Prof Martin Green, University of New South Wales, Australia. BP Solar has completed extensive research on Cadmium Telluride and is near to introducing CdTe as a product. Eurosolaire (Italy's combination of Italsolar and Ansaldo) took over second place in Europe with 3.2MW. of module shipments in 1993. Photowatt decreased shipments from 2.0MW to 1.7MW of polycrystalline modules. Photowatt has ingot casting and is using wire saws from Shaping Systems. Photowatt is slicing polycrystalline material very thin, 250 microns with 200 micron kerf. This reduction in the silicon usage and improved efficiencies have reduced costs so that Photowatt can have the lowest price of the European suppliers. Siemens A.G. shipped only .5MW of locally produced product in 1993 of single crystal material. Siemens Germany sold several MW of U.S.-made modules and cells throughout Europe. Isophoton shipped .5MW of single crystal silicon modules in 1993.

Fig. 4. European PV Module Shipments by Company

EUROPEAN MODULE SHIPMENTS
BY COMPANY (MW)

COMPANY	1988	1989	1990	1991	1992	1993
DEUTSCHE AEROSPACE	1.3	1.2	1.7	2.1	2.6	2.6
BP SOLAR INTERN'L	1.3	1.4	1.4	2.2	3.5	4.5
NAPS FRANCE	1.0	0.7	0.6	1.0	0.6	0.5
INTERSOLAR (UK)**	0.9	0.7	0.6	0.2	0.0	0.1
PHOTOWATT (FRANCE)	0.8	0.8	1.5	1.8	2.0	1.7
EUROSOLAIRE(ITALY)	0.4	0.8	1.0	1.5	2.6	3.2
HELIOS (ITALY)	0.3	0.8	1.2	1.5	2.0	1.0
ISOPHOTON (SPAIN)	0.2	0.3	0.5	0.5	0.6	0.5
SIEMENS (GERMANY)	0.2	0.4	0.6	0.8	0.6	0.5
R&S (NETHERLANDS)	--	0.4	0.0	0.0	0.0	0.0
OTHER	0.4	0.4	0.6	1.3	1.1	1.45*
TOTAL	6.8	7.9	10.2	13.4	16.4	16.55

* Nukem(Germany),Conkar(Croatia), Phototronics(Germany)
Western Silicon(Russia) ** Chronar Wales Plant

New activity in Europe includes a one MW plant by Nukem to produce metal/ insulator/ semiconductor (MIS) single crystal silicon cells; a new 3MW module assembly plant by Gallivare. Switzerland, a cast ingot/slicing plant by Bayer, and the operation of the amorphous silicon plant by Phototronics. Late in 1994, new companies were started as the result of mergers. Deutsche Aerospace and Nukem combined to form ASE GmbH and three companies have merged to create the largest silicon wafer company in Europe. The merger of Freiberger Elektronikwerkstoffe, Wacker Chemitronic,GmbH and the Solar division of Bayer is called BAYER SOLAR, GmbH. These new companies will provide sufficient depth and economy of scale to permit competitive status .

The Photovoltaic Industry in Japan.

Figure 5. shows the PV module shipments of the Japanese PV industry by company. In 1993, Sanyo shipped 5.5MW of amorphous silicon product and .7MW of cast-ingot polycrystal silicon modules. 3.8MW went to calculators, watches, trickle chargers, and other electronic products. 1.7MW was used in the outdoor "power" market. Most of the "power" modules were used in government projects. These include ventilators for vacant houses, PV assisted air-conditioners and other products to be subsidized by the Japanese government in the new ten year plan. Present "stable" efficiencies for Sanyo modules is about 6%. Sanyo goals for amorphous silicon are to have 10%, stable modules with costs that will permit the sale of modules for $1.00 per Watt. Sanyo amorphous silicon module shipments have been flat for five years. No expansion plans have been announced. Kyocera is the second largest producer of PV modules in Japan. Kyocera shipped 4.8 MW of modules in 1993. 4.6 MW were cast ingot, polycrystalline silicon and .2 MW were amorphous silicon. They have expanded their production capacity from 2 MW to 4 MW to 6 MW in successive years. Kyocera exports over 80% of their product and has made major inroads in all markets including the United States and Germany.

Fig. 5. Japanese PV Module Shipments by Company

JAPANESE MODULE SHIPMENTS
BY COMPANY (MW)

COMPANY	1988	1989	1990	1991	1992	1993
SANYO	4.8	4.8	4.9	6.0	6.5	6.2
KANEKA	2.2	2.4	2.5	3.1	3.0	1.7
KYOCERA	1.7	2.5	4.5	5.8	5.1	4.8
TAIYO YUDEN	1.3	1.5	1.6	1.6	1.6	1.6
SHARP	0.0	1.0	1.0	1.0	1.0	1.0
HOXAN	0.8	1.0	0.8	0.8	0.6	0.4
FUJI	0.5	0.1	0.1	0.1	--	--
MATSUSHITA*	--	--	0.6	0.8	1.0	1.0
OTHER	0.7	0.9	0.8	0.6	0.0	0.0
TOTAL	12.8	14.2	16.8	19.8	18.8	16.7

* MATSUSHITA - CdTe

Kaneka is the third largest producer of modules in Japan. Kaneka shipped 2.2 MW of amorphous silicon product in 1993. 1.9 MW went to the "electronic" products: calculators, watches, trickle chargers, etc.. Like Sanyo, Kaneka is developing roofing products for the Japanese residential program. Several small consumer products- ventilators, small lights, signals (using light emitting diodes) are under test. So far none have been sold to the international market. Kaneka is a division of Kanegafuchi Chemical Industries and has not yet developed an export or product marketing strategy. Kaneka uses the amorphous silicon/ amorphous silicon carbide technology and holds the basic patent on the process.

Fuji has done extensive government funded research on large amorphous silicon power modules. There is renewed evidence to indicate that Fuji will re-enter the market in the next two years. SHARP is the leading supplier of solar cells for space, including crystalline silicon and gallium arsenide. Sharp shipped a megawatt of single crystal terrestrial cells and modules in 1993. Most of Sharp product goes to communications systems throughout Asia. There is no evidence of Sharp modules in the U.S. or European markets. Matsushita produces about 1MW of cadmium telluride cells for calculators. They do not serve the power market.

Rest of World Module Shipments

Figure 6. shows the module shipments for the rest of world producers. CEL in India is the largest producer in the developing world. The Indian market, led by government-owned producers -- Central Electronics and Bharat, is poised for rapid expansion with reduced import duties, the possibility of foreign controlled PV manufacturing plants and a Prime Minister committed to photovoltaics. Russia is selling ingot silicon, slices, cells and modules to hard-currency countries. Over 300kW of Russian PV product was shipped in 1993. China has several megawatts of potential PV production.. However, we could not identify over one megawatt of product shipped in 1993.

Figure 6. Rest of World Module Shipments

REST-OF-WORLD MODULE SHIPMENTS BY COMPANY (MW)

COMPANY	1988	1989	1990	1991	1992	1993
CEL (INDIA)	1.3	1.3	1.4	1.4	1.5	1.8
BHARAT (INDIA)	0.4	0.4	0.4	0.4	0.8	1.0
HELIODINAMICA (BRAZIL)	0.5	0.6	0.6	1.0	0.5	0.5
CHINA	0.2	0.4	0.5	0.8	1.0	1.0
OTHER	0.6	1.3*	1.8*	1.4*	0.8*	0.1
TOTAL	3.0	4.0	4.7	5.0	4.6	4.4

* MAY REPRESENT SOME DOUBLE COUNTING DUE TO MODULES MADE FROM PURCHASED CELLS IN SEVERAL ARAB COUNTRIES

Heliodinamica in Brazil produces its own ingot using the Crystal Systems HEM process and also ships over a megawatt of slices to countries like India. There are several module assembly plants with 100-500 kW annual production. Since they buy cells, we do not include them in order to avoid double count. We expect the ROW sector to become the fastest growing area as more countries build and operate their own plants.

FORECAST OF PV MODULE MARKETS

Figures 7 and 8 show our latest forecast of PV module shipments (Ref. 2). Space does not permit detail. The PV market forecast is developed under two scenarios -- Business as Usual and Accelerated. The BAU scenario involves continued low prices for gas and oil (less than 2 percent real growth), low priority for renewables in government funding, no internalization of the social costs, and a general malaise concerning energy. The Accelerated scenario assumes 3 percent real escalation of gas and oil prices, rapid expansion of government funding, internalization of social costs by paying more for renewable energy than hydrocarbon and nuclear-fueled generation, real concern for energy and the environment and emphasis on renewable energy by the international funding agencies for the developing world. The United States has been the BAU mode since 1980, and Germany is in the Accelerated mode. The process in preparing the forecast involved several steps. First, a PV module performance (module efficiency and stability) forecast was developed. Figure 9 summarizes this forecast. The 2010 forecast is considered to be the practical limit of the technology. Second, a manufacturing cost versus time and plant size for each cell technology was proposed. Figure 10 summarizes these costs for most of the cell options. The market was analyzed and segmented into seven segments. The price/volume relationship for each sector was sought as a function of module and installed system prices.

Given this analysis, the two forecasts were compiled under their sets of assumptions. In the BAU scenario, the PV module market grows at about 12-15 percent per year. In the Accelerated scenario, the market grows at about 25 percent per year. The key issue confronting the PV industry is to grow with profit to recoup the enormous R&D investment to date (over $2 billion) and to stimulate capital to build large,efficient manufacturing plants with reduced costs. As we have seen, the last four years have seen module shipments increase less than ten percent per year. Since 1990, the environment for the world PV market has been below our conservative assumptions for the business as usual scenario. The cost of oil and gas has not increased for over ten years. (Prices actually decreased in real dollars). The world economy, especially Europe and, for the first time, Japan are in a major multi-year recession. Europe has become very serious in funding renewable energy development, while the United States continues to be luke-warm with respect to renewables.

Fig. 7. World Module Shipments to 2010-- BAU

PV WORLD MARKET FORECAST
BUSINESS AS USUAL (MWp)

MARKET SECTOR	1990	1993	1995	2000	2010
CONSUMER PRODUCTS	16	18	25	50	150
US OFF GRID RES	3	5	8	20	50
WORLD OFF-GRID,RURAL	6	8	10	30	200
COMMUNICAT'N,SIGN'L	14	16	21	30	150
PV/DIESEL,COMMCL	7	10	16	40	120
GRID-CONN RES,COMMCL	1	2	5	10	80
CENTRAL STATION	1	2	5	10	50
TOTAL(MW/YEAR)	48	61	90	190	800
AVERAGE PRICE($/W)	4.50	4.00	3.50	3.00	2.00

2010 FORECAST: 15%/YEAR 2000-2010

Fig. 8 World Module Shipments to 2010-- ACCEL

PV WORLD MARKET FORECAST
ACCELERATED SCENARIO

MARKET SECTOR	1990	1995	2000	2010
CONSUMER PRODUCTS	16	70	80	500
US OFF GRID RES	3	20	40	100
WORLD OFF GRID	6	15	40	600
COMMUNICATION,SIGNAL	14	40	60	200
PV/DIESEL,COMMCL	7	50	100	1000
GRID CONN RES,COMMCL	1	5	20	600
CENTRAL STATION	1	10	100	1000
TOTAL (MW)	48	210	440	4000
AVERAGE PRICE ($/W)	4.50	3.00	2.00	1.50

2010 FORECAST: 25%/YEAR 2000 TO 2010

REFERENCES

Reference 1. PV NEWS, VOL 12, NO 2, pp 1-6, Feb 1993, published by PV ENERGY SYSTEMS, USA.

Reference 2 . "PHOTOVOLTAIC TECHNOLOGY, PERFORMANCE, COST AND MARKETS TO 2010. April, 1993, published by PV ENERGY SYSTEMS,INC., PO BOX 290, CASANOVA VA, U.S.A., 22017. $750.00

Fig. 9 PV Module Performance Forecast: 1995-2010

PV DC MODULE EFFICIENCIES (%)
PRODUCTION MODULES IN MARKET

CELL TECHNOLOGY	1995	2000	2010
SINGLE CRYSTAL	15	18	22
POLYCRYSTAL INGOT	14	16	20
RIBBON (SHEET)	14	17	21
CONCENTRATORS (SILICON CELL)	22	25	30
AMORPHOUS SI (INCLUDES STACKED CELL)	7-9	10	14
COPPER INDIUM DISELENIDE	7-9	12	14
CADMIUM TELLURIDE	7-9	12	14
SI FILM	8-10	12	15
TI "SPHERAL"(TM) CELLS	10	12	14

2010 FORECAST,PRACTICAL LIMIT OF TECHNOLOGY

Fig. 10 PV Module Manufacturing Cost: 1995-2010

MANUFACTURING COST OF
TERRESTRIAL MODULES

OPTION	1990 COST/PRICE	1995 COST/PRICE	2000 COST/PRICE	2010 COST/PRICE
SINGLE XTAL	3.25/5.40	2.40/4.00	2.00/3.33	1.50/2.50
POLYCRYSTAL	3.00/5.00	2.25/3.75	1.95/3.25	1.45/2.40
RIBBON	3.60/6.00	3.00/5.00	2.40/4.00	1.50/2.50
CONCENTRATOR	3.00/5.00	2.00/3.33	1.20/2.00	1.00/1.67
AMORPHOUS SI STACKED CELL	3.00/5.00	2.00/3.33	1.20/2.00	0.75/1.25
LOW PRICE (BAU)	5.00	3.75	2.50	2.00
LOW PRICE(ACCEL)	5.00	3.00	2.00	1.25

NOTE: TOO EARLY TO FORECAST CIS, CdTe, TI "SPHERAL",etc.

(1995$) - 2010 FORECAST IS LIMIT OF TECHNOLOGY

NON-TECHNICAL BARRIERS TO THE COMMERCIALISATION OF PV IN DEVELOPING COUNTRIES

J A Gregory & B McNelis

IT Power Ltd, The Warren, Bramshill Road, Eversley, RG27 0PR, UK

Tel. +44 1734 730073, Fax. +44 1734 730820

ABSTRACT

PV is quoted as being a 'mature technology, one which is proven technically and economically for stand-alone applications in areas remote from the grid'. There are many thousands of working examples of PV systems in developing countries. However, this is minute compared to the potential market size.

Research has highlighted a number of non-technical barriers which have adversely affected the market development for PV, particularly for stand-alone systems in developing countries. This paper discusses these institutional barriers in developing countries, and ways in which they could be removed to allow for the large-scale commercial deployment of PV systems.

INTRODUCTION

Energy has long been regarded as the key to economic development. The assumption is that expanding the energy supply leads to increased industrialisation, expanded job creation and higher standards for living. Following the energy model of the developed countries, developing countries have quadrupled their energy consumption since 1960 [1]. However, basing their industrial development on fossil fuel technologies, has left them vulnerable to oil price rises and with increasing debt burdens (from importation of fuel and generation equipment).

The World Energy Council (WEC) predicts that global energy demand could double by 2020, most of the growth being in developing countries, which will account for 85% of the increase, particularly those with large populations, such as India and China [2]. The WEC also reports that solar energy could provide 10% of the demand in 2020 (all renewable energy technologies accounting for 30%) [3].

Renewable Energy: Sources for Fuel and Electricity was prepared for the 1992 UN Conference on Environment & Development (UNCED) [4]. This presents a scenario in which more than half the world's primary energy would be provided by renewable sources during the second quarter of the next century. The use of PV would be significant. An accelerated PV development strategy has been further developed by Williams and Terzian [5], which would lead to 400 GWp of PV being installed by 2020, half in the USA and half in developing countries. This would require large-scale market development, including subsidies, and accelerated R&D to bring PV costs down to a level for bulk electricity to be competitive in the USA.

Between 1 - 2 billion people in the world are without electricity. These people are mainly in rural areas, areas which present the highest costs for grid extension and which have the lowest economic development. Additional to the need for electricity in the home, are requirements for water pumping, light industry, community infrastructures (such as schools, halls, health centres and hospitals), outdoor lighting, communications etc.

MARKET DEVELOPMENT S

In 1993, about 90% of the PV market was for stand-alone systems installed in rural areas [6]. However, compared to the total *potential* market in developing countries, it is insignificant. This is despite over 20 years' of commercial activity in developing countries, millions of dollars being spent in international donor programmes, and millions of dollars allocated to R&D programmes.

Until around 1993, there was little hope that investments and financing large enough to even make a start to bring PV power to people in developing countries would ever be possible. Since then, a number of initiatives and programmes have started which have the potential to bring about the commercialisation of PV systems, particularly stand-alone system in developing countries.

In *Power for the World*, a 1993 initiative of the European Commission to introduce PV systems into unelectrified villages in developing countries, per capita needs are identified separately from

CH3365-4/94/0000-0700 $4.00 © 1994 IEEE

survival and basic needs. What this means for PV is identification of a huge range of applications which are technically feasible and economically-attractive now. *Power for the World* has identified the poorest of the poor in rural areas as the target for this programme. The aims of the programme include decreasing infant mortality rates and illiteracy and increasing life expectancy and quality of life through the deployment of PV systems [7].

The World Bank's *Solar Initiative,* which commenced in 1994, is looking to facilitate some 50 - 70 MWp of PV for small-scale applications over the next three years. The type of projects identified as a priority are rural electrification, water pumping, rural health clinics, street lighting and school buildings. It also includes other solar technologies and wind power. It identifies the need for increased funding for R&D to improve PV technologies suitable for use in both developed and developing countries [8].

Williams [9]., however, says that financing made available through the *Solar Initiative* is not nearly enough to accelerate market growth consistent with the Williams and Terzian target of 400 GWp by 2020. The rationale behind this scenario this is that business as usual, i.e. global market growth averaging 16 - 17% per annum (based on the yearly growth of the world market for the last 10 years), would only lead an installation rate of 200 MWp/year by 2000 and 900 MWp/year by 2010, representing an installed capacity of 1 GWp and 6 GWp respectively. Williams and Terzian further argue that this level of growth would only see PV accounting for about 2% of total global electricity generation by 2025, way below what is possible with and desirable for PV. The *Solar Initiative*, they conclude, is only enough to fund the market development of one company, rather than being enough to stimulate the large-scale manufacturing and product development which is required if the PV market is to become fully commercialised.

In the past, the applications identified as being primary targets were in developing countries and included: vaccine refrigerators, lighting systems for homes and community lighting, grid-support systems at end of line, water pumping village water supply, irrigation and stock watering. *Power for the World* and the *Solar Initiative* also concentrate on these markets for PV.

The *Global Environment Facility* (GEF) finances projects which contribute to improvements in the global environment, with one project area being the reduction of carbon dioxide and other greenhouse gas emissions. The current, second fund, stands at $2 billion, of which about half is for global warming projects [10]. PV fits into this category.

There are two multi-million dollar programmes for facilitating the installation of PV home systems currently underway with funding from the GEF (India and Zimbabwe). The government of Indonesia is currently negotiating with the GEF for funding for a 1 million PV home systems programme. These programmes have been slow to develop and it has been generally assumed that there were insurmountable institutional difficulties in applying GEF funds for PV projects. *The Solar Initiative* has indicated that this is no longer the case.

These programmes and initiatives, plus others which are coming onstream through national initiatives, have the collective ability to change perceptions about PV at a political as well as a local level. If they do not work, then there is the possibility that PV will never reach its market potential.

CURRENT STATUS

Technical readiness

If PV is to provide power to the billions of people without electricity, and to provide the power source for millions of stand-alone applications (in the short term) and grid-connected applications in the long-term, then it has to be technically ready and capable.

There are millions of dollars being spent each year for R&D on PV cell structures and improved efficiencies. However, much less is being spent on improving system performances. It has long been recognised that the batteries and the inverters are two of the weakest technical links in PV systems for stand-alone applications, particularly in developing countries. Preliminary results from an international survey being undertaken as part of the International Energy Agency (IEA) Photovoltaic Power Systems programme on Remote and Island Systems (Task III) show that batteries in particular, are still the weak link for stand-alone systems [11].

Yet, cell and module research still gets the majority of money, even though PV modules are technically capable and reliable today. This focus on cell and module R&D rather than systems R&D and the perceptions that cell technologies must be further developed before PV can find a mass market are holding the industry back.

Economic readiness

PV generated electricity is very expensive at present, compared to traditionally-provided grid electricity. A lot of work is underway to bring down the cost of PV modules. However, if cost were the main barrier to PV sales, there would be no sales in developing countries, outside of aid programmes.

There have been many studies which show that PV electricity is less expensive than grid-extension or diesel generators. For instance, a study has shown that, for small loads, PV electricity around $1.00 per kWh, but this is less than grid-extension, and is similar in cost to running a diesel generator for a small load - without the adverse environmental effects. A 1 kWp PV system in a place with good daily insolation costs about the same as extending the grid about 1 km [12].

However, even when environmental and societal costs are taken into account, the economics for PV become meaningless if the infrastructure is not there to train local installers and maintenance technicians; and it is meaningless if there is no infrastructure to supply spare parts and replacement batteries. In Mongolia, for example, there is a very large potential market amongst the nomadic herders. However, this market has never materialised, one of the reasons being that there is no infrastructure in use outside Ulaanbator to supply equipment [13].

Financial viability

Even if it can be shown that PV is economically viable and technically ready, it does not mean that people will buy PV. The slow market development to date clearly illustrates this! Householders in developing countries do not want to learn about technical ability and economics, they want TV, radio, lighting, fresh water etc. If they do not perceive that PV can supply these services easily and effectively, then they are not likely to purchase PV systems.

Most people in the rural areas do not have the capital to purchase a PV system (usually costing several month's income or more). However, this does not mean that they cannot afford to purchase PV if it is financed in a way which they can afford.

INSTITUTIONAL BARRIERS

There are many barriers which are impeding the development of the PV market which do not directly relate to technical problems or economics. The following issues are those which are currently having the most impact.

Ignorance of the technology

Lack of knowledge, mis-understanding and bias have acted as deterrents to the wider utilisation of PV. Thoughts such as 'it doesn't work'; 'the technology is too complicated and too high risk for developing countries'; 'it's not appropriate' are just some of the arguments which have been used against PV in the past.

Because energy is a cross-cutting issue, relevant to many ministries and disciplines (such as agriculture, resource planning, health, environment) professionals in sectors outside energy often do not know of PV or know enough about it to specify project applications. Even within energy ministries in most countries, there is a low level of knowledge on the technical advances and different applications for PV.

International loan, development and aid institutions contribute significantly to sectorial programmes within developing countries. Programme decision-makers within these organisations are often unaware of the environmental and social benefits of PV systems, such as possible employment generation in remote areas. A programme which ignores PV, or has the equipment wrongly specified, can have a significant negative effect.

Lack of utility involvement

The traditional development model for electricity utilities is to electrify the areas with the highest number of customers first, thereby receiving the highest return on capital. This means that rural areas do not usually have access to a reliable source of energy if the country is poor and/or has a sparse rural population (14). If utilities in developing countries do not have access to electricity distribution models which incorporate PV scenarios, and access to related PV information, how can they ascertain if/where PV would be relevant and cost-effective for them?

The attitude and role of utilities can significantly influence the market for PV. The IEA Photovoltaic Power Systems (PVPS) programme has identified the potential of PV for utilities, and now has a number of PV initiatives with them.

However, there is still a long way to go to convince utilities that PV can be a significant contributor to their energy mix. For instance, in South Africa, the national utility, Eskom, has responsibility to electrify the country. There are many areas which have a high population, but no grid extension. There are other areas which are very sparsely populated and are far away from the grid. PV has been installed in some homes for lighting and TV, and has the potential to be very effective in bringing electricity to these remote areas. However, Eskom is not convinced of this, and has not to date published their plans for grid extension. The result is that villagers are not keen to purchase PV systems in case the grid arrives in the next few years. For Eskom, on an economic and infrastructure basis, it would be much more expedient to implement a PV programme for these rural communities than extend the grid. However, their corporate culture at the moment is not as open as it could be to the benefits of PV.

Policy and price distortions

For governments of developing countries, the challenge is to provide a reliable supply of electricity to the rural population without comprising the industrial development of the urban areas. In most countries, governments have not made PV an integral part of their energy strategies.

The use of subsidies is an issue which the PV industry needs to address urgently. There are many studies which have been carried out on the effect of subsidies in the agriculture sector, but very few relating to PV electricity. This policy area is having a large impact upon the dissemination of PV systems in many countries.

Prior to the mid-1980s, subsidised credit loans were a normal part of many aid and development programmes. In the 1980s the use of subsidies was challenged and many studies were undertaken to ascertain the real effects of subsidies on development programmes. Today, the prominent view is that development subsidies do not help those people targeted within a project, but rather it is better policy to combine normal financing mechanisms with grants rather than heavily subsidising interest rates [15]. The reasons for these conclusions are:

i. Subsidies can lead to over-optimistic expectations, and are very hard to phase out.

ii. Low interest rates give banks and other financial organisations low margins, and reduces the services which they are able to offer.

iii Low-priced credit can generate excess demand, necessitating rationing. This can lead to corruption and political intervention, and reduce the borrowers repayment level.

iv Subsidised credit can lead to avoided repayments, either through default or cancellation of the debt (e.g. in the run-up to an election).

v. Not all subsidised credit goes to those who need it most. For instance, in Bolivia a review of agricultural credit over the past decade showed that only 9% of available credit reached the peasants who made up 95% of the producers, 91% of the credit going to agri-businesses [16].

Similarly, those charging market-based interest rates have historically been accused of denying especially the poorer section of communities the opportunity to take out loans. However, Aburge implies that this is not borne out by empirical evidence, i.e., there is no evidence to say that high interest rates per se discourage borrowing or affect repayments [17]. What discourages borrowing is the thought of being unable to repay the loan due to the variability of uncontrollable social and climatic factors (drought, lowered produce prices, civil conflict). Appropriate loan sizes combined with achievable repayment schedules and realistic interest rates encourage good repayment records and do not discourage borrowing in general.

On the other side of the debate about subsidies are the arguments by people in rural areas that they should get their electricity at the same price as those in urban areas (i.e. at grid prices).

Diesel is subsidised in many developing countries which also distorts the economics away from non-subsidised PV. PV electricity can bring increased economic development to rural areas and, as Williams argues, subsidised PV would mean that more people would be able to afford it. As the market grows, then the cost reduces until PV is affordable for all.

Fossil and nuclear fuels and large hydro projects enjoy subsidies from national governments, soft loans from the World Bank to build power stations, and large R&D funds, particularly from national governments and the European Union. PV should receive the same encouragement, particularly as lower PV costs will mean affordability for more people and lower greenhouse gas and pollutant emissions.

Inadequate programme structures

In the past, many PV programmes have not included provision for non-technical issues. These include setting up local infrastructures to supply spare parts; training programmes for local installers and maintenance technicians; information dissemination programme to inform the users and potential customer about the characteristics of using a PV system.

Information dissemination is another important aspects which is forgotten. Appropriate information needs to be given to potential buyers before they buy PV systems, so that they do not have false expectations of the systems and know how to operate them satisfactorily.

Lack of feedback

There are many instances where institutions, such as the World Bank, have commissioned studies and research, but have not made that information readily available to other institutions or organisations. The time lag between the commissioning of studies and publications and their distribution can be prohibitively long, so that when the publication is available, the information contained in it is out of date.

These are not isolated incidents or particular to just the World Bank. Many of the papers at international conferences do not convey the true facts, or omit to tell of the problems with both technology and systems. This lack of accurate and timely information dissemination within the industry is having a number of adverse effects:

i. Valuable knowledge which could assist in future projects is not known or unobtainable.

ii. The PV 'wheel' is recreated many times over.

FINANCIAL CONSTRAINTS

There are a number of financing schemes now in operation to facilitate the purchase of PV systems. However, these are still very few in number, and represent a very small fraction of PV programmes in terms of monetary outlay.

Availability & appropriateness

Studies have shown that even very poor people can often afford purchases, if financed in a way which allows them to make repayments without compromising their standard of living (18). Often it is not the price of a system that is the major restraint, but the access to appropriate financing which provides the critical barrier.

Within PV programmes in the past, no/little thought was given as to how the systems were to be purchased. Rural people, who typically rely on seasonal crops and/or livestock sales, do not usually have the cash to finance purchases. So other mechanisms need to be found to finance PV systems.

During the 1960s permanent revolving funds were tried by national and international agencies and governments for use within the agriculture sector. These were primarily loans for agricultural production.

A review was carried out by US Agency for International Development in 1972 [19]. Of the conclusions, the following are relevant for possible future PV funding schemes:

i. No particular type of institutional system is applicable for all situations in all countries.

ii. Successful institutions were those most able to reach the maximum of people with minimum costs.

iii. Group activities with an element of compulsory participation had a greater degree of success.

iv. Shortage of local leadership and fear of local/central government manipulation acted against many local co-operatives.

v. All rural credit agencies within a country, regardless of type of organisational structure, should be co-ordinated as a mutually supportive system, retaining their individual attributes but able to share lessons to be learnt from problems and successes.

vi. These financial operations must be viable.

vii. Trained professionals are essential for lending, understanding local issues (such as seasonal ability to repay loans) and understanding technology requirements.

viii. There is a need for self-evaluating mechanisms to monitor progress and evaluate achievements.

Types of appropriate credit facilities for PV

There are a variety of financing structures which operate for PV. Include the following schemes.

Revolving funds are typically started with seed money from an outside agency and organised by an existing or new local organisation. This seed money is used both to set up the revolving fund and to loan out as funds. These funds, once repaid by the original borrower, are then relent to other borrower(s).

A number of small revolving funds have been established to finance the purchase of small PV systems, set up by organisations such the Solar Electric Light Fund (SELF). Experiences with interest rates vary. In Sri Lanka, for instance, the PV experience has shown that people are willing to pay in excess of their current energy expenditures for kerosene and batteries because they value the quality of light, the cleanliness of the system and the ability to power small electronic systems. Over 2,500 of the 3,000 systems installed in Sri Lanka have been provided on commercial terms, at interest rates of up to 34% per year and requiring substantial down payments. The repayments for these credit schemes have been tailored to the disposable income of the target customer [20]. In other resolving fund schemes, the interest rate is low, often only 3 - 4% per annum.

Credit co-operatives are indigenous to many rural areas, and can be used for PV. For instance in Bolivia and the Dominican Republic, revolving funds set up through local credit co-operatives have facilitated the installation of many thousands of PV home systems.

Renting a PV system can be an attractive option in that it off-sets all capital costs and avoids financial contractual obligations. There is such a system operating for battery charging in Senegal [21].

TOP TEN FACTORS FOR SUCCESSFUL PV PROGRAMME IN A DEVELOPING COUNTRY

The authors suggest that, to set the right conditions for achieving the objectives of a programme, the following conditions need to betaken into account:

1. Market research is undertaken prior to programme and/or project formulation, to assess:
   - local resources (human resources, local industry, wealth of the communities);
   - needs of the population (e.g. water pumping, health care, home lighting, TV, radio).
2. Priorities need to be established at the outset of the project. These should be geographical and by application.
3. PV applications must be economic compared to the alternatives.
4. Project must be replicable in other areas of the country.
5. Co-ordination must be continuous, all people affected by the programme/project should be kept informed and consulted throughout. If this is done, the possibilities for technical and social failure are lessened.
6. Demonstration of the application need to have occurred before large programmes/projects begin. Therefore, any problems should already have been identified and fixed
7. Appropriate local financing needs to be available for a large number of small PV borrowers.
8. Local training and infrastructure has to be developed, as part of the project, not as an afterthought. This includes:
   - maintenance and repair
   - spare parts
   - repayment collections
9. The programme/projects need adequate funding to cover all aspects (not just equipment), and risk needs to be allocated between all project participants.
10. Information dissemination activities need to be undertaken before the programme/project begins, and during it. This need for information will not cease with the project, so infrastructures need to be set up to ensure its supply (e.g. literature distributed through community centres). This information needs to address:
    - users
    - potential users
    - installers and repairers
    - financiers

CONCLUSIONS

PV systems can contribute significantly to increased standards of living of people in particularly rural areas of developing countries. There are technical areas where improvements are needed, but these do not prevent the satisfactory operation of the systems.

Economically, when compared to grid extension and diesel generators, PV is cost effective for many applications.

However, there exist a number of institutional barriers which are affecting the wider dissemination and commercialisation of the PV industry. These barriers include lack of appropriate financing mechanisms, poor perception by the utilities and lack of knowledge about PV by decision-makers as well as potential users. Unclear energy policy and pricing strategies by governments, agencies and the PV industry do not help in gaining market share away from traditional fossil fuels.

A concentrated efforts needs to be made to remove these barriers, using new PV programmes, such as *Power for the World*, the *Solar Initiative* and the PV programmes in the Global Environment Facility portfolio, as successful demonstrations and instruments to facilitate a large-scale commercial PV market.

REFERENCES

[1] Nicholas Lenssen, *Empowering Development: The New Equation*, Worldwatch Paper 111, November 1992.

[2] World Energy Council, Energy for Tomorrow's World, London, 1994.

[3] World Energy Council, New Renewable Energy Resources: A Guide to the Future, London, 1994.

[4] T B Johansson, M Kelly, A Reddy, R H Williams, Renewable Energy: Sources for Fuel and Electricity, Island Press/Earthscan, London, 1993.

[5] R H Williams and G Terzian, *A Benefit/Cost Analysis of Accelerated Development for PV Technology*, Princeton, 1993.

[6] R Barlow, A Derrick & J A Gregory, *The World Market: 1993 Status Report & Future Prospects,* 12th EC PV Solar Energy Conference, Amsterdam, 1994.

[7] W Palz, *Power for the World*, Harmony with Nature, International Solar Energy Congress, Budapest, 1993.

[8] Denis Anderson and Kulsum Ahmed, *The Case for a Solar Initiative*, Draft, The World Bank, September 1994.

[9] R H Williams, *Proposal for a Global Accelerated PV Development Program*, Princeton University, September 1994.

[10] *The World Bank Operational Report on the Global Environment Facility*, Washington, 1994.

[11] International Energy Agency Agreement on Photovoltaic Power Systems, Task III, Stand-alone and Island Systems, *Questionnaire*, Cadarache, 1994.

[12] Jenniy Gregory & Bernard McNelis, *The growing role of solar electric power production in developing countries*, Power Generation Technology, 1993.

[13] *Power for the World - Common Concept*, Study 1, IT Power, Draft, November 1994.

[14] R Turvey and D Anderson, Electricity Economics, World Bank Research Publications, John Hopkins University, Baltimore, 1977.

[15] J D Von Pischke, *The Exit problem in credit projects*, Small Enterprise Development, Vol. 3 No. 4, 1992.

[16] K P Padmanadhan, Rural Credit. Lessons for Rural Bankers and Policy Makers, IT Publications, London, 1988.

[17] Charles Aburge, *When credit is not due - financial services by NGOs in Africa*, Small Enterprise Development, Vol 4, No.4, 1993.

[18] Marilyn Carr (ed), The Barefoot Book. Economically appropriate services for the rural poor, IT publications, London, 1989.

[19] Sharon Holt & Helena Ribe, Developing Financial Institutions for the Poor and reducing Barriers to Access for Women, World Bank Discussion Paper 117, USA, 1990.

[20] Philippe Callier (ed), Financial Systems and Development in Africa, Economic Development Institute of the World Bank, Washington, 1990.

[21] J A Gregory, *Solar Photovoltaic Power: Institutional Factors*, IT Power Ltd for the Asian Development Bank, 1994.

POSITIVE ATTRIBUTES FOR SUCCESSFUL PV PROJECTS	CONTRIBUTING ISSUES
Political will	• high level of support • promotion assistance
PV knowledge by decision-makers	• government • utilities • bankers
Good co-ordination between participants	• from users to funders
Clear objectives & work plan	• support of all parties • clear direction and progress
Genuine user interface at all stages	• with all partners involved in the ultimate use of the application.
Appropriate financing systems	• appropriate for the local conditions • adequate fee collection structure
Needs-orientated approach	• credible product • real need for the application • local maintenance and service capacities, adequate spare parts
Dialogue and consensus	• at all levels at all times. NO dumping
Use of existing resources	• local manufacture where possible for components
Flexibility	• to change the project when the original specification do not meet real requirements and local conditions
Training	• training of local technicians for service and repair • locally-based application engineering and development • support warranties
Organisation-building	• does the technology mesh well with local resources • adequate post purchase support
Pricing	• is the technology/application economically deliverable/ • can it be replicated in other areas • is it affordable for the local population without high subsidies
No legal disincentives	• electricity pricing • taxes on solar equipment lowering of grid connection costs
Technology transfer programmes	• university-level research • field technicians • project managers • information disseminators
Appropriate market information	• research into needs undertaken before project formulation • adequate marketing activities undertaken to support the programme (e.g. literature, discussion groups)
Allocation of risk	• by international agency which has experience in this • shared between all the partners
Previous good demonstration projects	• show that the technology works • find out any problems which could occur and rectify them

A PHOTOVOLTAIC-DIESEL HYBRID GENERATION SYSTEM FOR SMALL ISLANDS

Toshikazu Tadokoro (*1), Kouichi Taira (*2), and Masahisa Asaoka (*3)
*1 Solar Energy Department, New Energy and Industrial Technology Development Organization
*2 Research & Development Dept, The Okinawa Electric Power Co. Inc.
*3 Power Electronics Engineering Section, Development Department, Itami Works, Mitsubishi Electric Corporation

ABSTRACT

As a part of the New Sunshine Project carried out by the Agency of Industrial Science and Technology of MITI, the New Energy and Industrial Technology Development Organization has commissioned Okinawa Electric Power Co. Inc. and Mitsubishi Electric Corporation to jointly develop the practical technology for a photovoltaic-diesel hybrid generation system for small islands.

The project involves constructing, on a small island, a stand-alone system which uses photovoltaic power (solar arrays of 750 kW) as the main source and diesel power (300 kW) as the auxiliary source.

It is designed to supply electric power to private houses which are isolated from existing distribution lines.

RESEARCH AND DEVELOPMENT SCHEDULE

The research and development schedule for this system is shown in Table 1.

We have now completed the construction and adjustment of a pilot plant. In October of this year, we started the study of its operation as a stand alone system.

Table 1. Schedule

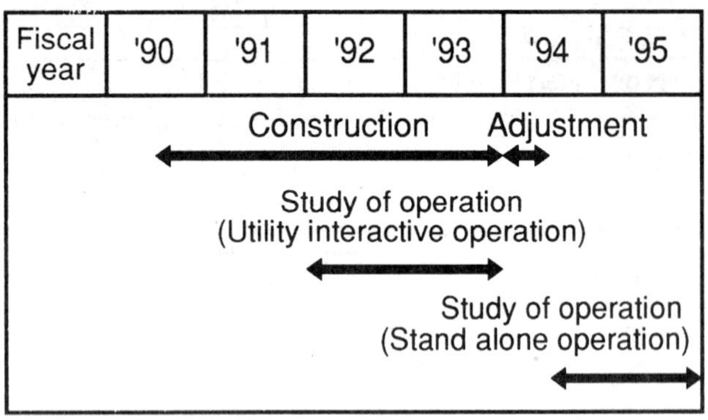

OUTLINE OF THE SYSTEM

System Configuration

Figure 1 shows a block diagram of system configuration. This system is a hybrid generation system using PV power as the main source and diesel power as the auxiliary source, with the two power sources connected in parallel to the AC line.

The hybrid system achieves a higher PV power utilization factor than a basic PV system using PV generation as the sole power source. By the supplementary use of diesel generation, it can reduce the capacities of the solar arrays and storage batteries.

In fine weather, the system supplies sufficient power by PV generation. If cloudy or rainy weather continues, the batteries lack the energy to fulfill the load demands, and the system then supplies power using the diesel generator.

The system features a unique operation method of charging electricity from the diesel generator to storage batteries at a high load rate in order to reduce fuel consumption rate.

The charging of the batteries by the diesel generator is implemented through inverters in charging mode .

Table 2 shows an example of the fuel consumption characteristics of the diesel generator.

The fuel consumption rate increases as the load declines.

When power of 90 kW, the same amount as the average power consumption of the houses, is supplied by the diesel generator, the average load rate is 30 percent.

However, this system can reduce the fuel consumption rate by increasing the load rate close to the rated value. This can be accomplished by charging the batteries with the diesel generator.

Table 2. Fuel consumption characteristics of the diesel generator

Rated capacity: 300kW

Load rate [%]	25	30	50	75	100	
Fuel consumption rate [ℓ/kWh]	0.456 (1.57)	0.421 (1.45)	0.346 (1.19)	0.307 (1.06)	0.290 (1.00)	← Ratio
		↑ Average load power			↑ During charging operation	

Test Site and House Loads

The system has been installed on Miyako Island, Okinawa Prefecture (lat. 24° 43' N and long. 125° 27' E). The average energy consumed by the houses was determined to be 2,170 kWh per day (including a plant utility

CH3365-4/94/0000-0708 $4.00 © 1994 IEEE
First WCPEC; Dec. 5-9, 1994; Hawaii

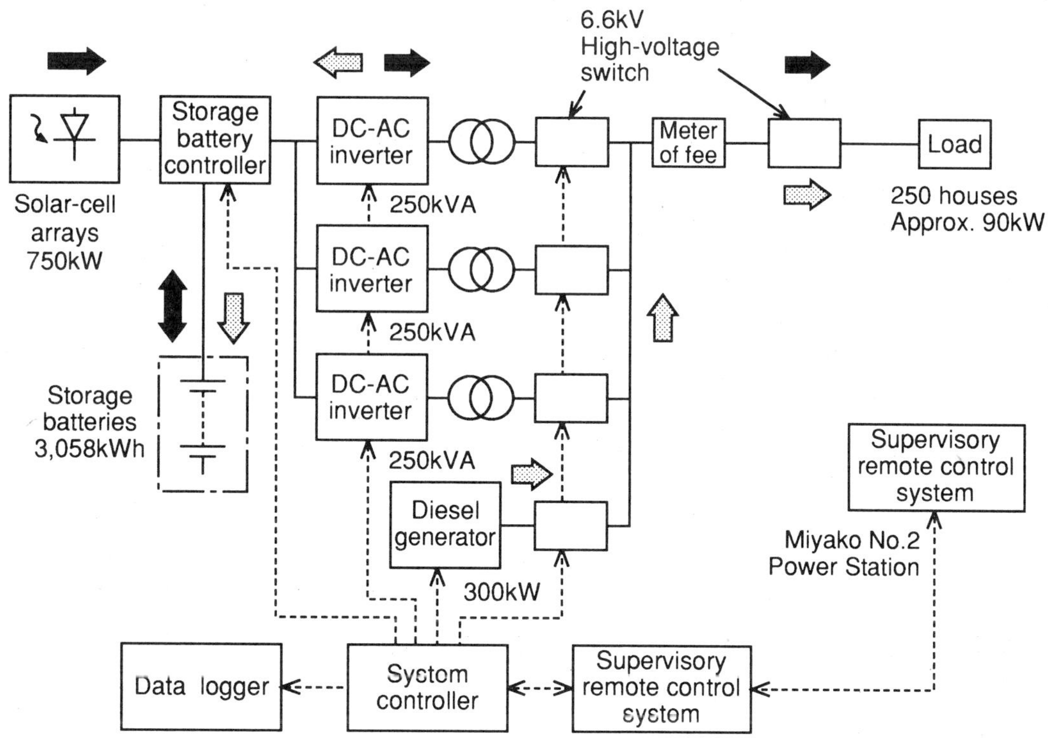

Fig. 1. Block diagram of system configuration

energy of 34 kWh per day) (see Table 3). Figure 2 shows the daily load pattern averaged over the year.

Table 3. House loads (tentative)

Item	Average power [kW]
House loads (250 houses)	80.9
Distribution loss (Note 1)	8.1
Plant utility power (Note 2)	1.4
Total	90.4

Note 1. Loss in the distributing transformer, for example.
2. Includes power of 0.5 kW for data logger.
3. The control system consumes an additional 1.75kw (not indicated in table above).

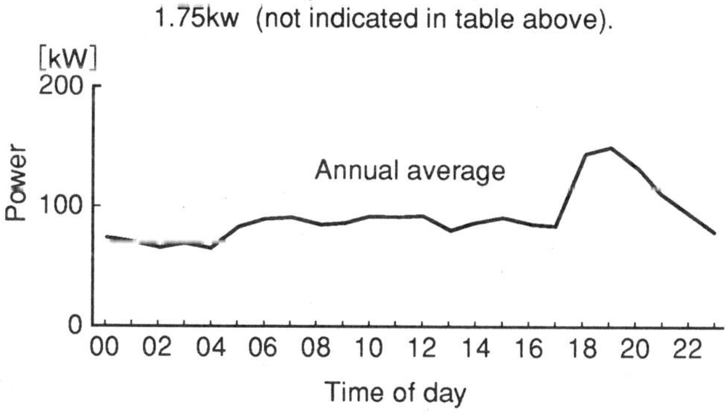

Fig. 2. Annually averaged load power (average power consumption for each hour)

System Capacity

The capacities of the solar arrays and storage battery for a house load of 90 kW were selected through a simulation of the system's operation and an economic evaluation.

The inverters are designed to provide a total capacity of 750 kVA to allow for utility interactive operation. Of the three units installed, one is used as a standby during stand-alone operation.

The capacity of the diesel generator is 300 kW, assuming that the maximum load is about three times the average load.

ANALYSIS BY SIMULATION

Energy Flow

Figure 3 shows a comparison of the annual average energy flows in two cases.

The upper diagram indicates an energy flow when the diesel generator charges the batteries. The lower diagram shows the flow without this charge.

The loss increases in the inverters and storage batteries due to charging, and thus the supplementary energy from the diesel generator increases by about 19 percent.

On the contrary, however, the fuel consumption is reduced by about 20 percent thanks to the increase in fuel efficiency by improving the load rate.

This system can thus save fuel effectively.

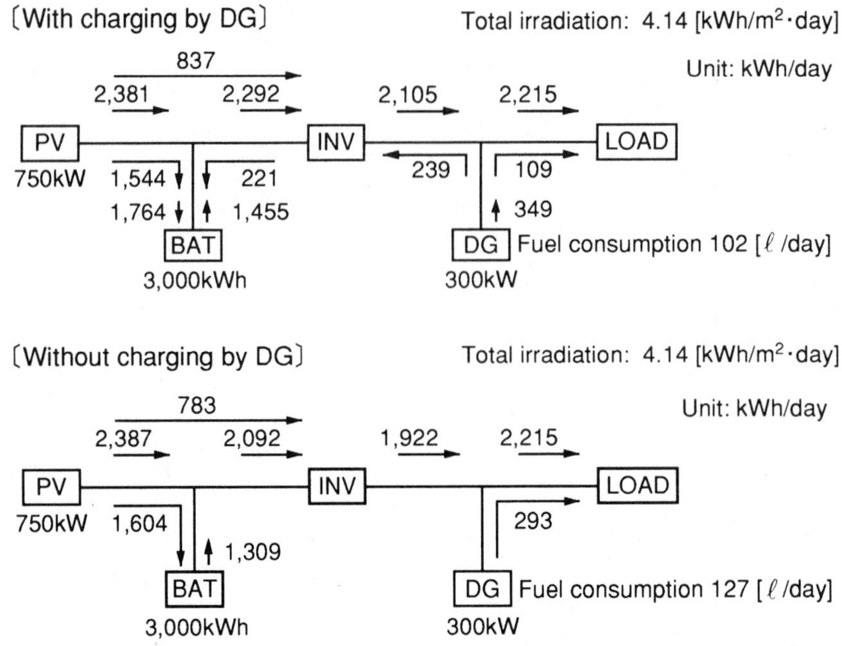

Fig. 3. Annual average energy flow (simulation)

Daily Operation pattern

Figure 4 shows the system's daily operation pattern obtained through simulation.

When the battery ampere-hour energy drops below 20%, the diesel generator is started to supply enegy to the houses, while at the same time chargeing the batteries with the surplus which is equal to the difference between the load power and the rated power.

Monthly Output Energy and Load Energy

Figure 5 shows the monthly output energy obtained through simulation throughout the year and estimated load energy.

A large amount of excess array energy is produced in summer, when the ideal transmission energy exceeds the load energy. On the contrary, a great deal of supplementary energy is needed for the diesel generator in winter, when the ideal transmission energy is smaller than the load energy.

System Capacity and Excess/supplementary Energy Rates

Figure 6 shows the simulated output characteristics for varying the capacities of the solar arrays and batteries when the load energy is kept constant.

The percentage of the supplementary energy from the diesel generator decreases with an increase in the solar arrays' capacity.

But, since the batteries don't have enough capacity to store all the energy generated by the solar arrays, the percentage of the excess energy from the solar arrays goes up.

This leads to a decrease in the output energy utilization factor. Thus, the percentage of supplementary energy does not decrease in proportion to the increase in the capacity of the solar arrays, resulting in high costs.

If the battery capacity is reduced, both the excess and supplementary energy rates will increase.

The simulation result shows that the rate of excess array energy is about 6 percent and that the rate of supplementary energy from the diesel generator is about 17 percent for the complete system.

CONSTRUCTION OF THE PILOT PLANT

The construction of the stand-alone system with 750 kW solar arrays was completed by the end of fiscal year 1993. Table 4 shows the major specifications of the whole system, Table 5 the major specifications of the solar arrays, Table 6 the major specifications of the storage batteries, Figure 7 an aerial view of the pilot plant, and Figure 8 shows the installation of the solar arrays.

The smaller the inclination of the solar arrays, the less vulnerable they are to wind load and the less the loss in the shade of the arrays. For this reason, an inclination of 15 degrees was selected, which is smaller than the latitude of the site, 25 degrees.

RESULTS OF TEST GENERATION

Figure 9 shows the results of test generation in an interactive utility operation system with 519 kW solar arrays and two 250 kVA DC-AC inverters. The entire system was shut down from January 11, 1994 for expansion work.

A full-scale study of operation as a stand alone system started in October 1994.

Figure 10 shows the actual daily operation pattern in a stand alone operation.

Operational instructions for the diesel generator	Starts up when the Ah is 20 % or less. Shuts down when the Ah is 30 % or more.

Ah: Battery ampere-hour energy

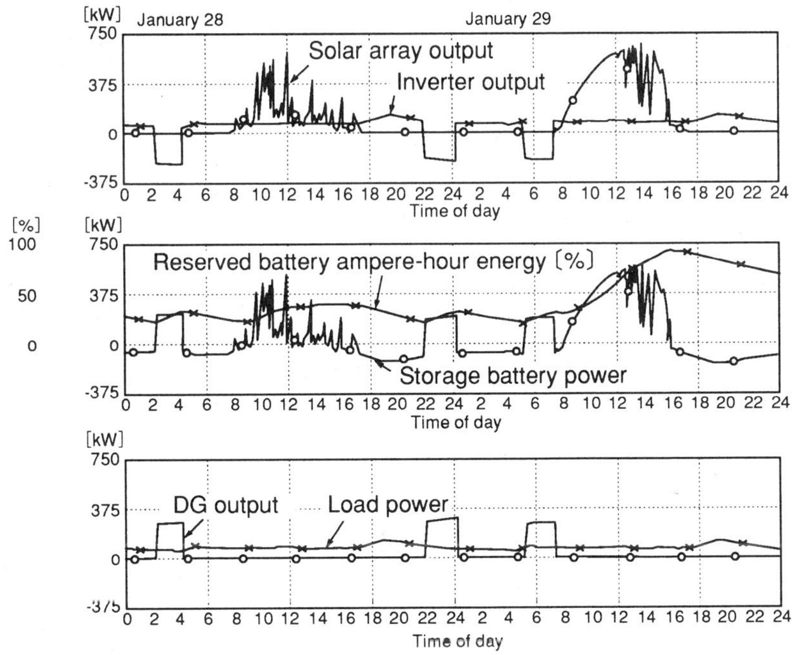

Fig. 4. Simulated patterns of operation

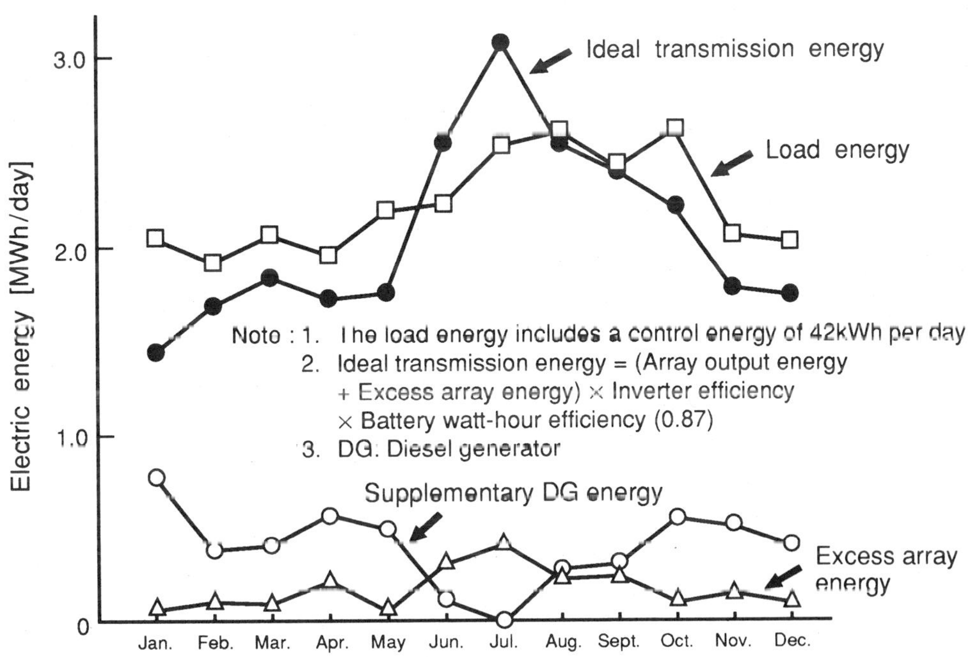

Fig. 5. Monthly output energy and load energy

Table 4. Major specifications of the whole system

Total capacity of solar arrays	750kW
Site	Site area: about 16,000 m^2 Array field area: about 10,300 m^2
Annual output energy (Note 1)	PV : 671,200kWh (76.6kW) DG : 136,400kWh (15.6kW)
Annual load energy	807,600kWh (92.2kW) (Note 2)
AC output voltage (nominal)	3 ϕ 6.6kV 60Hz
DC input voltage (nominal)	430V
Power conditioners	250kVA×3 PWM control by transistor Rated energy efficiency: 94% Effective energy efficiency: 92% (at 40% load)
Control method	Constant-voltage, constant-frequency control
Storage battery capacity	3,058kWh (10HR)
Capacity of auxiliary diesel generator	300kW
Overall configuration	Solar arrays → Storage batteries → Power conditioners → Load The diesel generator is connected in parallel to the AC line.

Note 1. PV: Photovoltaic power generation DG: Auxiliary diesel generation
2. Includes a control energy of 15,330 kWh (1.75 kW).

Table 5. Major specifications of solar arrays

Pm 〔kW〕			Module efficiency [%]			Vpm 〔V〕	
Single crystal	Polycrystal	Total	Single crystal	Polycrystal	Total	STC	SOC
548.5	201.8	750.3	12.2	11.2	11.9	488	423

Pm : Maximum power Vpm: Maximum power voltage
STC : Standard test conditions (cell temp. 25℃)
SOC: Standard operating conditions (1 kW/m2, temp. 25℃, wind 1 m/s)

Table 6. Major specifications of storage batteries

Type	Tubular-type lead-acid battery for photovoltaic application SLB-3000
Capacity	3,058 kWh (1,950 Ah x 4 P 10HR)
Number of storage batteries	196 in series and 4 in parallel
Voltage	392 V nominal (single cell: 2 V)
Depth of discharge	80 % or less (expected life: 1,600 cycles or more)
Auxiliary devices	Specific gravity detector, battery temperature detector, electrolyte level detector, and electrolyte agitator

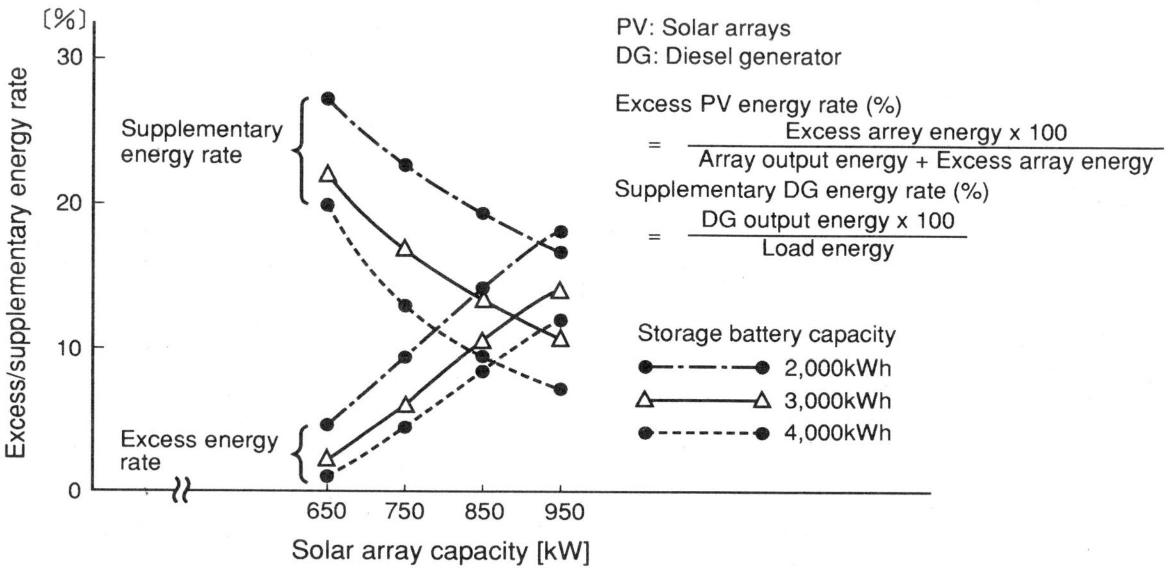

Fig. 6. Solar capacity vs. excess and supplementary energy rates

Fig. 7. Aerial view of the pilot plant

Fig. 8. Installation of the solar arrays

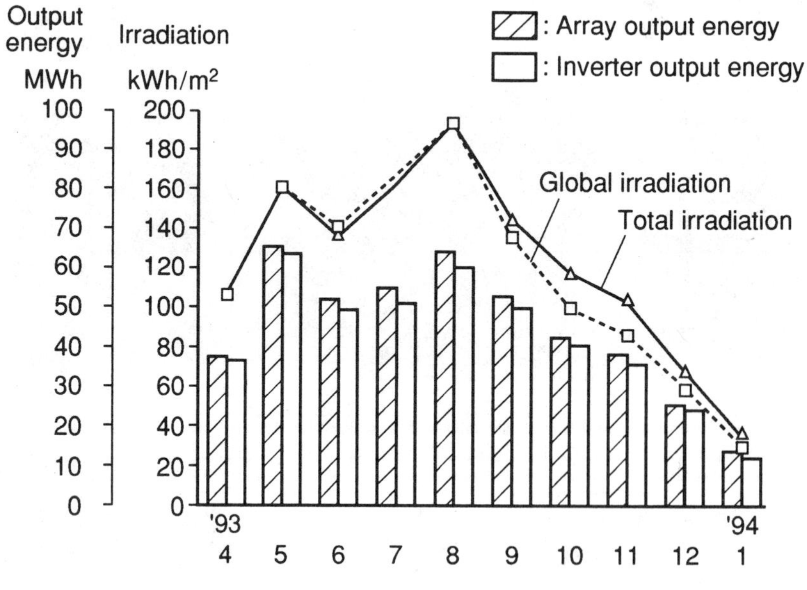

Fig. 9. Results of test generation in fiscal year 1993

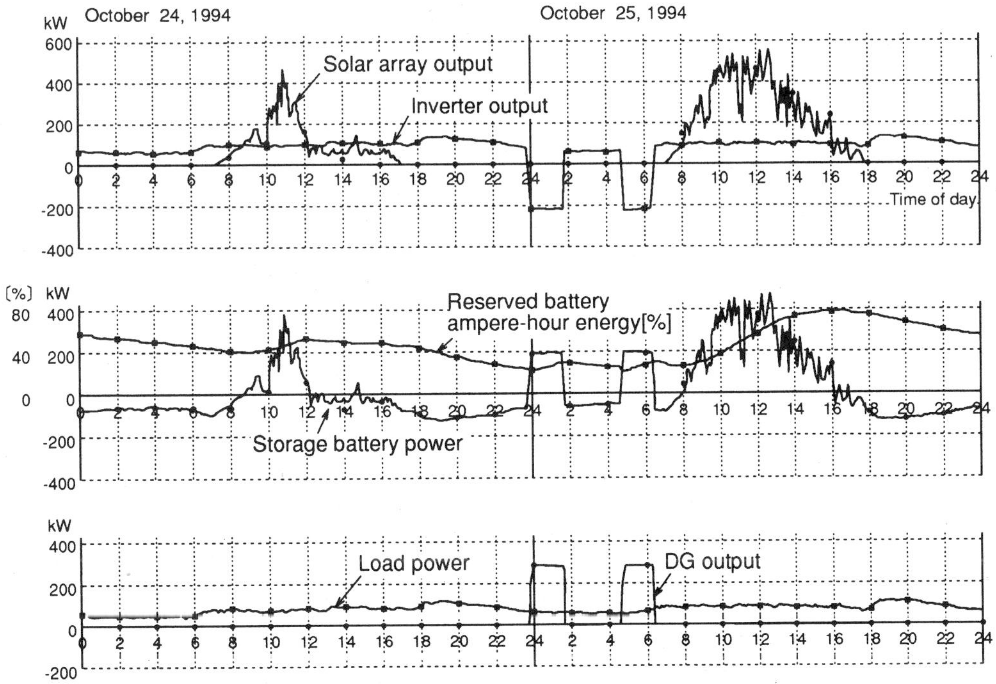

Fig. 10. Actual daily operation pattern

Sunplicity: from Complex System Design to Standard Product Level

Markus G. Real, Alpha Real AG, Zürich
Urs Bühler, Alpha Real AG, Zürich
Peter Toggweiler, PMS Energy AG, Mönchaltdorf
Gabriel Roth, Glas Trösch Solar AG, Ursenbach
Ernst Bolliger, Timonta AG, Mendrisio

ABSTRACT

In 1989 Alpha Real launched the project Megawatt to promote and demonstrate that standardised PV-systems in the kW-range are cost effective in comparison to large scale installations. Project Megawatt was a huge success and the project concept was adopted by many others around the world. Within this system standard new components had been developed and tested, such as an arc detector to improve safety against fire, roof tile for integration in dwellings and flat roof support for integration in commercial buildings.

Exactly five years later, the project Gigawatt was launched by Alpha Real to promote and demonstrate that small independent 200 W AC modules will be cost effective compared with DC wired PV-systems. Several developments were made to achieve a unique product. Some of the inventions were a new hot spot detector and control device, and the low cost DC to AC converter "edisun®". Three different types of PV-products were developed and called accordingly Gigalino Handy, Gigalino Top and Gigalino Max.

1. Introduction

Allmost all electricity today is produced by means of rotating machines. In general, they are powered by steam turbines, to a lesser extent by water turbines, and to a still negligible percentage by wind turbines. These conventional generating sets have two things in common, namely that they are driven by a volumetric flow of a medium such as steam, air, or water and that the power of the installation is more or less proportional to the flow rate, which is about proportional to the cube of the installed turbine volume.

Therefore, the cost of generated electricity is very much dependent on the size of the installation. In the past, this has led to larger and larger power plants, following the law of "economy of scale". The bigger the installation, the more cost-effective the generated power. Power plants of up to 1000 MW have been the consequence of this straight'forward physical correlation between power and volume.

Rotating turbo-sets generating electricity requires maintenance, as all rotating machinery does. Furthermore, the complexity of the installation requires a highly educated maintenance team. Since the size of the maintenance team is more dependent on the technology and the number of sub-systems within a given power plant than on its actual size, maintenance cost is another driving factor for building large power plants. The "cost per maintenance crew" divided by the actual power suggests that larger installations produce less expensive energy.

PV systems, however, are very much different in their nature. Since PV fields harvest a uniformly distributed energy source - solar irradiance- and since PV systems are almost 100 % linear in their correlation of the conversion of input - to output power, the actual required surface area to produce a certain amount of energy doesn't become smaller with larger systems. For example, given a system efficiency of 10%, a 1 kW PV system will require, under Standard Test Conditions (STC) a PV array area of 10 m2. If 10 kW of power are required, 100 m2 are necessary. There is no gain to reduce the amount of necessary PV modules by making the system larger.

The other basic difference is that PV systems require - at least in theory - almost no maintenance. This of course is not really true for the very first generations of PV systems, but it is foreseeable that, by progressing along the learning curve, PV systems will become more and more reliable. The ultimate reason why PV systems will be almost maintenance free is that there are no rotating machines involved in the process of generating electric power. In fact, most people haven't realized yet that PV is a revolution in terms of producing

CH3365-4/94/0000-0716 $4.00 © 1994 IEEE

First WCPEC; Dec. 5–9, 1994; Hawaii

electricity, since for the first time in history it is possible to generate large amount of electric power with absolutely no moving parts involved in the process, except for the electric meter, which is used to measure the energy delivered to the load.

Both, project Megawatt and project Gigawatt are based on this principle of "economy of volume". Developments on both components for PV-systems and on the PV product level have been successfully developed and tested.

2. Project Megawatt, standardised system level

Project Megawatt was launched in 1989 to promote and demonstrate that standardised PV-systems are cost competitive compared with large scale PV installations. Project Megawatt was extremely successful, and the concept of standardised PV-systems on dwellings had been adopted and also very much improved by other agencies and companies around the world. Especially SMUD has lately proved that a cost reduction potential is feasible when production and installation are carried out on an industrial level.

Under the project management of Alpha Real, several new components had been developed and successfully tested in the field, which may make system design even safer and simpler.

2.1. Arc detector

- Arcs in DC systems are extremely dangerous for various reasons: DC generates arcs which are more intense than arcs from AC sources.

- PV panels are current sources and arcs in PV arrays cannot induce high currents which may in normal systems activate protective devices, such as fuses.

- Current leakages due to failures in isolation may develop over the years. Such failures may lead to arcs, and since PV-systems are normally working fully automatic and unattended, development of such arcs may not be noticed until they have developed into a fire. Alpha Real and others have in the last years developed control routines, which should detect such a situation by analysing the current voltage situation of the array. Unfortunately, this attempt has been very unsatisfying.

The arc detector developed by Alpha Real with supports by Prof. H. Häberlin from the engineering school in Burgdorf is based on the idea that such an arch must develop electromagnetic interference which may be detected by any field detector. First prototypes have shown that the principle is working extremely well. As suggested by Prof. Häberlin, the arc as a negative impedance may exitate the resonance of the detector and thus generate a very clear signal far above noise.

Fig. 1 shows the arc detector. Most of the associated Hardware within the enclosure is related to the disconnected switch which, in case of an arc alarm, automatically disables the PV array.

So far, the arc detector has successfully been tested in systems with inverters which fulfil the IEC-standard on electromagnetic interference on the DC side. Tests have shown that any arc simulated at any place in the array will activate the detector and lead to an alarm signal or to a signal to disconnect the array from the inverter. Tests have been carried out on 3 kW and 60 kW installations. Arcs from other devices such as electric or conversion motors or switches could well be distinguished from the signal from the arc within the PV-array. A patent has been filed for .

2.2 Plug system

DC wiring is very costly because time consuming process, since every single module has to be connected. Normal standard procedure using junction boxes and connecting devices have proved to be reliable and are used in most systems. However, when the concept of roof tile was developed years ago, it became clear that on roofs only plugs for the interconnecting cabling system is a feasible solution.

A simple plug interconnecting system has been developed and has proved to be a very reliable solution. Around 3000 roof tiles have been successfully connected throughout Europe in countries as Holland, Germany, Liechtenstein, Austria, Italy and Switzerland. No failures have been reported. Extensive testing had been performed to simulate accelerated aging. Since roof tiles are relatively small units, the cost of the interconnecting devices become a major issue and after careful evaluation of various plug systems, it became clear that only a plug being used in cars would meet our cost goal.

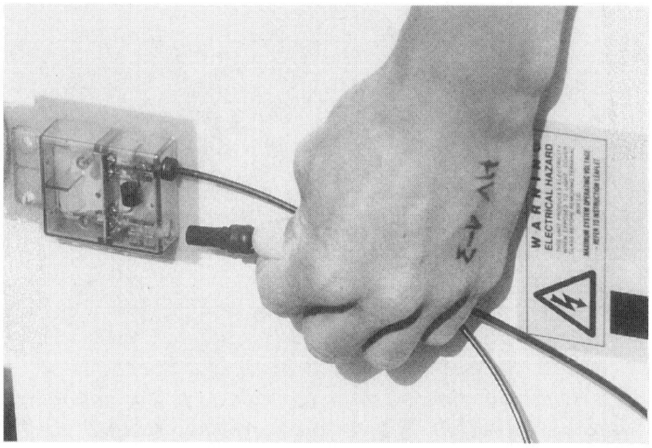

Fig.2 shows the low cost plug to connector, closed on the back of a module. The plug system has been developed for the PV roof tile for easy installation.

2.3 Roof tile for tilted roofs

Throughout the world small tiles are used to cover tilted roofs. The reason why roofer prefer small elements over large size devices to cover the roofs are manifold. One reason is certainly the better resistance against windloads. But then, smaller elements are easier to handle on a tilted roof. It was therefore our opinion to follow the standard practice when developing a PV roof tile. Figure 3 shows a roof tile on a 10 kW array.

The development of the PV roof tile has been reported earlier on EC PV Conference de Montreux. On some installations, the various parameters have been checked over the years to evaluate if there are no hidden errors in the design. For example water tightness against rain had been monitored over 15 months. Special water indicators had been installed in roughly 20 places and being checked every month. Comparison with water tightness of the normal roof has shown that the PV roof tile is even more water tight than the normal roof tile.

Fig. 3 shows an installation using PV roof tiles. The roof tile has been developed to achieve optimal integration and to reduce the required time for mounting the system on a roof.

2.4 Support design for flat roof systems

So far, most installations had been done with steel supports which have been screwed down on heavy supports to carry the windloads. These steel supports have proved to be time consuming on installation and quite expensive on manufacturing. A new design has been developed using only concrete and no metal frame structure. The result has been extremely positive. Installation time could be reduced, the wiring is completely covered, and the esthetic of the whole installation has been improved compared with other designs.

Fig. 4 shows the layout of the new design for flat roofs, using all concrete and no metal frame structure. Installation time could be reduced and costs for the structure were smaller compared with more traditional solution with steel structure.

3. Project Gigawatt, product level

In the last years, components which are used to combine PV-systems have remarkably been improved. This has led to simpler standardised PV design and therefore requires less know-how and skill, both from the designer and the installer. Still, it remains a system with two independent electric circuits, one DC and one AC. Another disadvantage is that system levels have to be decided in the very beginning of the planning process. Changing system size is, once a system is installed, very expensive and not very practical.

The new approach now presented does not replace the system level, but presents another option for the customer. And a very interesting option indeed. It allows him to start small, even at a 200 Watt level. And it allows him to enlarge his PV power potential at every moment he wishes. And he can do that by himself. The product developed for this purpose is easy to handle and uses just plugs for the final connection to the grid. No engineering skills are required any more.

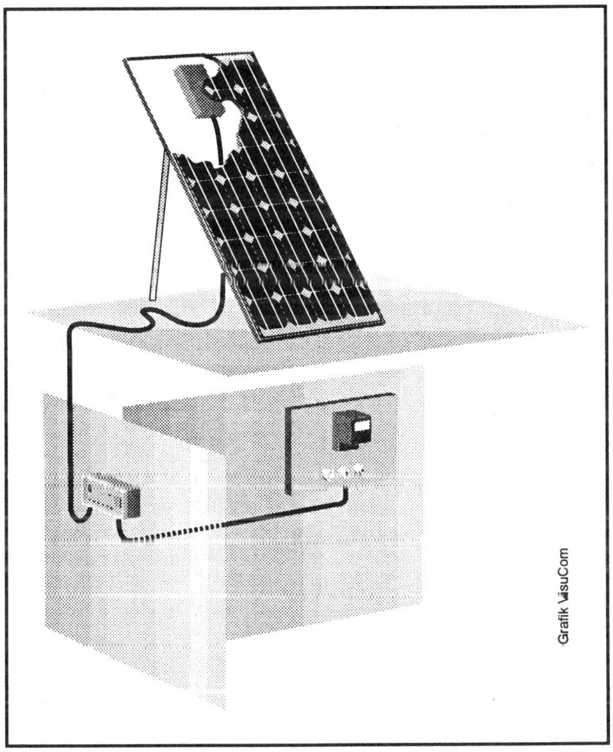

Fig. 5 shows the schematic installation of the Gigalino handy with the AC cable being plugged into a specially designed outdoor socket. Gigalino top is designed for tilted roofs, Gigalino max for commercial roofs.

A larger volume can be induced which will lead faster to large power scale. A 200 W element, of course, is a small construction unit when aming at Gigawatt. That's why they are called Gigalinos®. But mind you that on the other side we are used that a lot of small consumers require large power stations to meet their energy needs. It is said for example that just the stand-by losses of the TVs in the United States combine to the accumulated power of one Gigawatt. With the same approach that standardised 3 KW units led to MW-range, it is not unlikely that large volume of smaller products amount even faster to GW range.

Of course, Gigalinos can be used from the beginning to combine in larger installations. For example, 1000 Gigalinos at the same spot will lead to a 200 kW installation. And there are even some advantages of the concept: high redundancy and thus high availability over the entire life time of the system, no mismatch losses, reduced power cabling losses and almost no electrical engineering is required. This advantages have to be balanced against a maintenance concept which still have to be developed and a slightly reduced efficiency in the power conversion unit in comparing a 200 W inverter with a 200 kW inverter.

Project Gigawatt is based on three innovations:

3.1 Solar Inverter "edisun®"

In order to make PV power from project Gigawatt comparable with PV power from project Megawatt, the conversion unit on the small scale must be comparable with larger units used in normal systems. Of course, this is a very tough goal to achieve. Inverters so far have been one of the only items in PV-systems where economy of scale leads to cost advantages when building large inverters. There are, however, a few points which are interesting and which are also in favour of the small unit:

- In project Gigawatt no extra room or housing is required for the power conditioner system. This is a remarkable cost item.
- No DC switch gears are required. This is a again a remarkable cost item.
- AC wiring is more cost effective against DC wiring.
- At the outlet of the PV module some connecting box is required anyhow. The saving of this

conventional box can be directly balanced with the extra cost for the box with for the inverter.

- By trading each module individually, zero mismatch may be balanced to satisfy higher conversion efficiency (The edisun® has a max conversion efficiency of 93%).

Fig. 6 shows the module integrated inverter. Its dimensions are 20 x 10 x 5 cm. This is slightly larger than conventional connector boxes for the DC wiring.

3.2 Hot spot detector and hot spot controller

Hot spots are generated in crystalline silicon modules when one or more cells are partly or entirely shaded. Under these operational conditions, the shaded cells will operate as a load and dissipate heat which may lead to damage of the cells or the module. The inverse voltage which will develop when the cell is shaded is normally limited by using by-pass diodes. This method has the disadvantage that it is quite expensive to wire the module using by-pass diodes across every 15 to 20 cells. Especially in larger modules up to 4 and more diodes will be used, and the connection between the diodes and the cells requires outlets at the back of the module which are not only expensive but are also always a source of problems where current leakage may develop or where undesirable particles such as water or oxygen may get into the inner side of the module.

The new concept to avoid damage from hot spot situation is to limit the current instead of limiting the voltage across the shaded cell. This can easily be done when controlling the current controller (MPT-Tracker) of the module integrated inverter.

Of course, it remains the problem to know when actual hot spot situation occurs. Various possible solutions have been investigated and have been patented. The easiest one, however, is to control the voltage over two parts of the module. Under hot spot situation, voltage comparison will become unbalanced by up to 10 V and more. This is quite an easy signal to detect and active the control mechanism to limit the current under hot spot situation. The material required for the detector and controller are in the order of US $ 2 to 5. Compared with the cost of the by-pass diodes, another US $ 5 to 10 can be subjected from the cost of the module integrated concept.

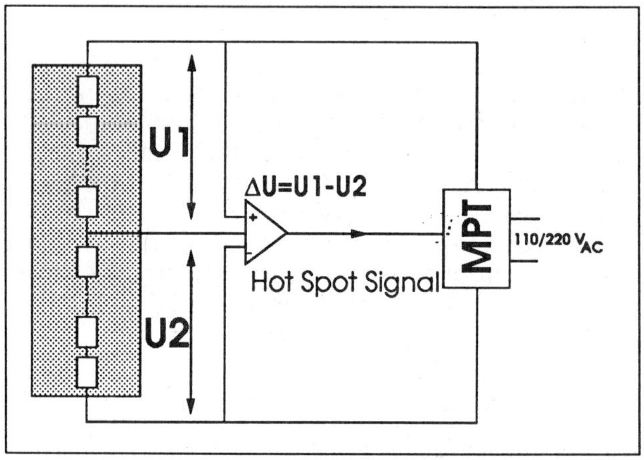

Fig. 7 shows the principle of the hot spot detector developed by Alpha Real (Pat. ppending. A very simple comparator is used to detect any unbalanced voltage over a module which will indicate hot spot situation.

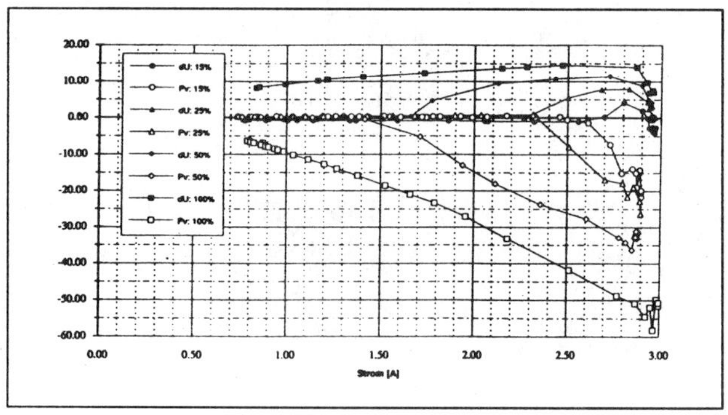

Fig. 8 shows for various shading conditions the difference in voltages and the dissipated power in correlation to the actual current flowing through the module.

3.3 Combiner box

In order to induce that customers will not make their own wiring from the electrical installation inside the house to the place where they put their Gigalinos and want to plug them in, a special combiner box has been developed. It is essentially a box with six sockets where up to 6 Gigalinos can be plugged in. The combiner box has, however, some additional features which will allow the customer to check whether his Gigalino is actually performing as expected and will provide some additional disconnect safety features to prevent Gigalinos to produce electric power under islanding conditions. This double feature is especially important when considering that there are not only utility people concerned about islanding condition but that home owners will, for whatever reasons, take out their fuses when working on their installation (for example connecting a lamp). Safety features must therefore insure that under any circumstances islanding will be prevented.

The project Gigawatt has been launched and various installations have been installed to check the technical feasibility under different climatic conditions in Switzerland. It is envisaged that more Gigalinos will be sold to more clients at the end of March 1995. Full field test may be reported at the next PV conference. Official introduction into the market place is planned for July 1995.

Acknowledgement

The work for the development of the arc detector was supported by the National Energy Foundation NEFF. Development of the inverter "edisun®" and the hot spot control device was supported by the Federal Ministry for Energy and the state of Zurich (ATAL) and Bern (Ministry for Energy and Water).

RESEARCH AND DEVELOPMENT OF EVALUATION TECHNOLOGY OF PHOTOVOLTAIC POWER SYSTEMS

Hisaichi MINE, Hiroyuki YOKOUCHI, Toshimitsu OSHIRO, Motonobu YUKAWA

Solar Energy Laboratory, Japan Quality Assurance Organization / JQA
1084–2 Hatsuoi–cho, Hamamatsu–shi, Shizuoka, JAPAN *433*

Kosuke KUROKAWA

Electrotechnical Laboratory / ETL
1–1–4 Umezono, Tukuba–shi, Ibaraki, JAPAN *305*

ABSTRACT

Although parameter analysis methods [1] [2] have been tendered as sizing designs for photovoltaic power systems, research and development is currently underway to establish specific figures for each parameter. These figures were quantified by analyzing measurement data obtained from experimental equipments [2] and through simulations, and primary estimates for design parameters were calculated. A summary of design parameter values is shown in Table 1.

1. Annual irradiation deviation factor, K_{HD}

The annual irradiation deviation factor compemsates for power deficiencies caused by weather–related fluctuations in the estimated amount of insolation.
Assuming that the probability of insolation deficiencies occurring is once in six years,

- Yearly fluctuation coefficients at 155 locations within Japan are

  $K_{HD} = 0.930$ to 0.976.

- Monthly fluctuation coefficients at 155 locations within Japan are

  $K_{HD} = 0.806$ to 0.957.

2. Shadowing factor, K_{HS}

The shadowing factor compensates for shade on array surfaces caused by shadows from mountains, hills, adjoining buildings, trees, the stands of front arrays or other objects. To generalize shadows from mountains and hills, the compensation coefficient was determined by assuming hills and mountains of the same height surrounded the array position for 360°, as shown in Fig. 1.

3. Photovoltaic conversion efficiency modification factor, K_P

The Photovoltaic conversion efficiency modification factor compensates for various factors produced by setting conditions which reduce (or on occasionincrease) the out–put of solar modules measured under standard conditions.

(1) Soil factor, K_{PDS}

The soil factor compensates for reduction in incident radiation caused by dirtiness on the module surface, and is obtained by comparing the short–circuit current of a module which is not washed with that of a module which is washed every day. The Soil factor K_{PDS} for the one year period from April, 1993 to March, 1994 was *0.99*.

CH3365-4/94/0000-0722 $4.00 © 1994 IEEE

First WCPEC; Dec. 5–9, 1994; Hawaii

(2) Degradation factor, K_{PDD}

Since degradation was not reported for crystal–type cells, this coefficient was set at *1.0*.

Since the degradation mechanism is not clear for amorphous–type cells and analysis will require some time, this will be left as a future issue.

(3) Cell temperature factor, K_{PT} [3]

The Cell temperature factor compensates for decreases in power generation output accompanying rises in the array temperature.

The cell temperature factor is defined by the following equation.

$$K_{PT} = \int P_{AS}\{1+\alpha P_{max}(T_c-T_s)\}G_A d\tau \Big/ \int P_{AS} G_A d\tau = 1+\alpha P_{max}(T_{CR}-T_S) \quad (1)$$

$$T_{CR} = \int T_C G_A d\tau \Big/ \int G_A d\tau \quad (2)$$

where P_{AS} is Array capacitance (kW) understandard conditions, G_A is Irradiance (kW/ m²), αP_{max} is Maximum power temperature coefficient (℃$^{-1}$).

in case of crystal–type $\alpha P_{max} = -0.0041$

in case of amorphous–type $\alpha P_{max} = -0.002$

T_C is Typical array temperature (℃), T_S is Standard condition temperature (25 ℃), T_{CR} is Weighted average temperature (℃)

The array temperature is affected by not only the irradiance but also by wind speed and direction, making the temperature distribution over the array field.
Cell temperatures over the array field at the experimental equipments are measured by attaching thermocouples to the backs of cells in modules located at the both ends and in the center of the sub–arrays.
The average sub–array cell temperature was calculated from these measured temperature values, and was then used in combination with irradiance data to calculate the weighted average temperature and temperature compensation coefficient.
Monthly and yearly average figures were then calculated.

Weighted average temperatures and compensation coefficients for systems by cell type over the one–year period from September, 1992 to August, 1993 were as follows. Fig. 2 shows weighted average temperatures and average ambient temperatures.

① Crystal–type Systems and amorphous–type systems exhibited similar monthly weighted average temperatures T_{CR}. Temperatures rose from *12 ℃* to *19 ℃* above average ambient temperatures, yielding an average yearly rise of *15 ℃*.

② When cell temperature factor K_{PT} were assessed in terms of energy using weighted average temperatures,

- Crystal–type systems showed a monthly average of *0.93* to *1.03* and yearly averages of *0.98*.
- Amorphous–type systems showed a monthly average of *0.97* to *1.01* and yearly averages of *0.99*.

(4) Array circuit unbalance factor, K_{PAU}

Photovoltaic arrays consist of solar modules with different electrical characteristics connected in series–parallel. The overall array capacity caused by unbalances in electrical characteristics to become smaller than the sum of the individual modules. This percentage is defined as the array circuit unbalance compensation coefficient, and was calculated through simulations. [4][5]

The deviation in the *Pmax* of solar modules which can currently be purchased from manufacturers is calculated as the standard deviation σ divided by the average value, which equals approximately 0.020. In case of a typical average output of 48.5 W, maximum and minimum values ranged from 51.5 W to 46.1 W respectively. Approximating the relationship between the array circuit unbalance factor and this deviation with a primary straight line produces the following equation.

Array circuit unbalance factor K_{PAU}

$$= 1 - 0.127 \cdot (\text{std. deviation } \sigma / \text{avg.value}) \quad (3)$$

(5) DC circuit loss factor, K_{PAL}

The loss of array circuits is the sum of the loss produced by resistance in the array's DC circuit wiring and the loss produced by voltage drop in the reverse current prevention diode.
The ratio of this loss to the array output is the DC circuit loss factor.

In many cases, designs for DC circuits determine the wiring size so that the DC circuit loss is less than a certain percentage of the rated output. (3% in most cases)

Therefore, methods of expressing the DC circuit loss during operation as a percentage of the loss during rated output were investigated. Fig. 3 shows measured average loss normalized with rated output loss. The equations used to calculate compensation coefficients can beexpressed approximately as follows.

- DC circuit loss ratio
  = *0.6 · (DC circuit loss during rated output / rated output)* (4)
- Reverse current prevention diode loss ratio
  = *1.0 / (DC rated voltage)* (5)
- DC circuit loss factor K_{PAL}
  = *1 − DC circuit loss ratio − reverse current prevention diode loss ratio* (6)

(6) Load matching factor, K_{PM}

The load matching factor indicates what extent the array output operating point determined by its correspondence with the load is offset from the optimal operating point for the solar−array.

The load matching factor for the one−year period from April, 1993 to March, 1994 is shown in Table 1.

4. Battery circuit factor, K_B

(1) Battery contribution ratio, γ_{BA}

The battery contribution ratio is defined as the ratio of the battery input energy to the over all array output energy. (Energy supplied directly to the load without passing through the battery is also taken into consideration.) This ratio was calculated by analyzing data over the one−year period from April, 1993 to March, 1994 excluding overcharged conditions.

(2) Energy storage efficiency, η_{BA} at the battery terminal

This coefficient refers to the output electrical energy as a percentage of the input electrical energy while the system is operating.

Overcharged conditions are not included in calculations of the input electrical energy.

5. Power conditioner circuit factor, K_C

(1) Inverter energy efficiency, η_{INO}

The inverter energy efficiency is obtained by calculating the ratio of the electrical energy output from the inverter to the input electrical energy based on actual values measured each minute. Inverter energy efficiencies by type of system are shown in Table 1.

ACKNOWLEDGMENTS

This work is supported by NEDO (New Energy and Industrial Technology Development Organization) under the New Sunshine Project by the Japanese Ministry of International Trade and Industry.

REFERENCE

[1] K.Kurokawa, Hierarchical Integration of Design Factors of Photovoltaic Systems, Reserches of the Electrotechnical Laboratory No.947, March, 1993

[2] M.Yukawa, K.Kurokawa, Research and Development of Evaluation Technology of Photovoltaic Power Systems, Solar Energy Materials and Solar Cells, Vol.35(1994), pp.453−459, (Proceedings of the PVSEC−7, Nagoya, Japan, 1993)

[3] H.Yokouchi, M.Yukawa, and K.Kurokawa, Temperature Distribution of Photovoltaic Array, Proceedings of the Fourth Annual Conference of Power & Energy Society, IEE of Japan,

July, 1993
[4] H.Mine, M.Yukawa, and K.Kurokawa, Simulation of Power Degradation on Series and Parallel Modules Connection, National Convention Record, IEE of Japan, 1993
[5] H.Mine, M.Yukawa, and K.Kurokawa, Simulation of Power Degradation on Series and Parallel Modules Connection, Proceedings of the Fourth Annual Conference of Power & Energy Society, IEE of Japan, July, 1993

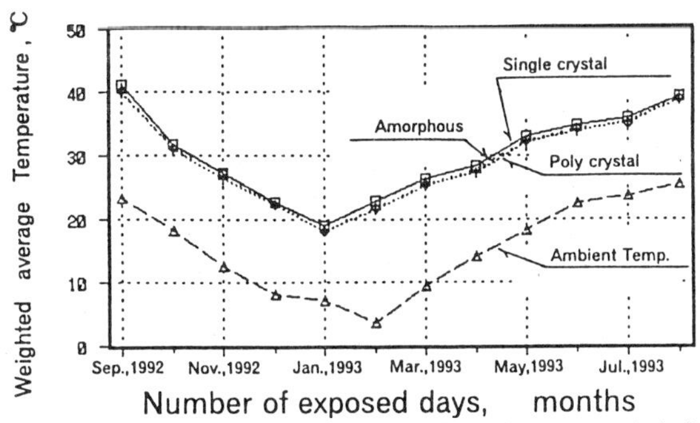

(an average weighted by irradiance profile at Hamamatsu test site)

Fig. 2 Weighted average Temperature

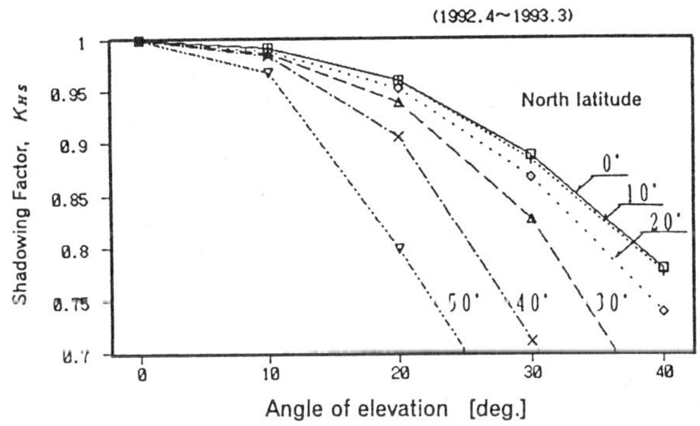

Fig. 1 Shadowing Factor, K_{HS}

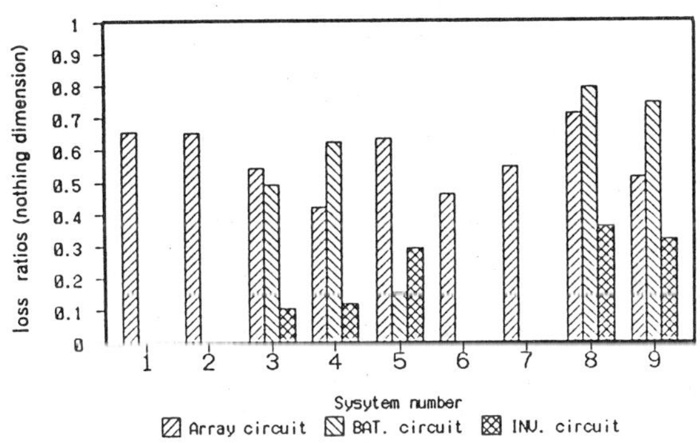

Fig. 3 Mesured wiring loss ratios normalized compared with loss ratios of rated output for DC circuit

Table 1. Primary Values of Design Parameters (estimated Provisionaly)

Design Parameters		Values of Design Parameters	
K_H	(Incident irradiation modification factor)	$K_H = K_{HD} \cdot K_{HS} \cdot K_{HC}$	
K_{HD}	(annual deviation)	0.94 ~ 0.97(yearly), 0.81 ~ 0.96(monthly)	155 sites in Japan
K_{HS}	(shadowing)		read from graph
K_{HC}	(contribution for tracking)	$K_{HC} = K_{HCD} \cdot K_{HCT}$	
K_{HCD}	(normal direct irr.)	0.66	focus,tracking
K_{HCT}	(flat plate tracking)	1.22	flat plate tracking
K_P	(PV conversion effeciency modification factor)	$K_P = K_{PD} \cdot K_{PT} \cdot K_{PA} \cdot K_{PM}$	
K_{PD}	(efficiency deviation factor)	$K_{PD} = 0.99$	(soil)
K_{PT}	(cell temperature)	$1 + \alpha P_{max} \cdot (T_{CR} - T_S)$ at Hamamatsu test site, T_{CR} = AVG.atmospheric + AVG.15 ℃,	Crystalline type 0.98 Amorphous type 0.99
K_{PA}	(array circuit)	$K_{PA} = K_{PAU} \cdot K_{PAL}$	
K_{PAU}	(unbalance)	$1 - 0.127 \times$ (STD.deviation σ /average)	0.997
K_{PAL}	(DC circuit loss)	$1 - (0.6 \times$ rated wiring loss rate $+ 1.0 \times$ rated diode loss rate)	
K_{PM}	(load matching)	0.94	utility-connected
		0.91	stand-alone,stable power supply (except overcharging)
		0.89	stand-alone,controlled load by irr.(except overcharging)
K_B	(Battery circuit factor)	$K_B = (1 - \gamma_{BA}) \cdot + \gamma_{BA} \cdot \eta_{BA}$	
γ_{BA}	(battery contribution)	0.8	stable power supply
		0.37	controlled load by irrdiance
η_{BA}	(energy storage eff.)	$\eta_{BA} = 0.83 \sim 0.86$	
K_C	(Power conditioner circuit factor)	$K_C = K_{ACC} \cdot \eta_{INO}$	
K_{ACC}	(INV. AC circuit)	$K_{ACC} = 0.99$	at Hamamatsu test site
η_{INO}	(INV. energy eff.)	0.92, 0.93	10kVA utility-connected
		0.75, 0.79	8kVA stand-alone,stable
		0.83	8kVA stand-alone,controlled load
		0.91	8kVA water pump
		0.64, 0.47	3kVA stand-alone,DC58V
		0.93	2kVA utility-connected
		0.93	3kVA utility-connected

DATA SAMPLING SPEED VERSUS ENERGETIC MEASUREMENT ERRORS IN PHOTOVOLTAIC SYSTEM MONITORING

Kosuke Kurokawa
Electrotechnical Laboratory

ABSTRACT

To measure solar irradiation and photovoltaic array output energy a measuring accuracy cannot be guaranteed unless data sampling interval is appropriately selected. From this viewpoint, actual irradiance had been measured by comparatively high speed sampling of 1 to 4 seconds for 44 months and the daily errors of numerical integral has been estimated for various step sizes. Approximation formulae of the error versus the step size have been statistically obtained as well as their probability density function covering $\pm 4\sigma$. Finally a nomograph is presented to decide an appropriate sampling interval. A concluding example shows that the deviating component of the error exceeding ±1% can happen once for every 1 month or 6 months if the step size is selected as 105 seconds or 65.5 seconds, in each of which the total error becomes −0.0485±1% or −0.0336±1% including an average error component.

INTRODUCTION

In a photovoltaic system the input solar energy has a basic day-and-night cycle and an uncertain, time varying factor caused by meteorological conditions. When a measurement is performed to obtain long term performance parameters such as irradiation, the amount of generated energy, etc., the accuracy of the measurement may not be maintained if the sampling speed of data acquisition is decided in no relation to the fluctuating rate of solar radiation.

Since simplified measurements may have to be adopted in the future along with the spread of photovoltaic systems, some instructions must be clearly prepared as a system monitoring standard. The authors present the analysis of this kind.

For this purpose, actual irradiance profile had been measured by comparatively high speed sampling for 44 months. By using these data, the daily errors of numerical integration have been evaluated for various time step sizes. Then, the errors for each step size have been analyzed for a certain term to obtain statistical parameters such as standard deviation σ and mean value. Approximation formulae have been also given to express relationships between these values and the step sizes. In addition, the probability density function profile of the errors has been also estimated. Finally by applying the approximation formulae and the probability density function, nomograph is presented for determining a maximum allowable step size to assure monitoring precision.

DATA ACQUISITION

In order to know true values, original irradiance data had been taken at comparatively a high speed sampling speed for 44 months from September, 1986 to April, 1990 at Tsukuba region. Actually used sampling period was 1 seconds but it was made longer up to 4 seconds in case of slower fluctuating speed of the irradiance to reduce data volume. Daily data were stored in a 2HD disk every day. Noises which may be generated in signal

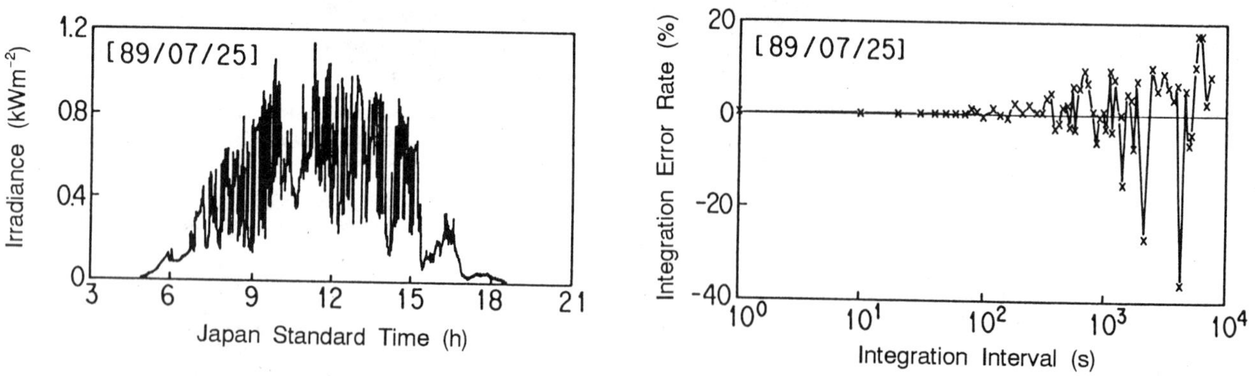

(a) Irradiance profile

(b) Integration error rate

Fig.1 An example of integration interval versus integration error rate for typical rapidly changing weather

First WCPEC; Dec. 5–9, 1994; Hawaii

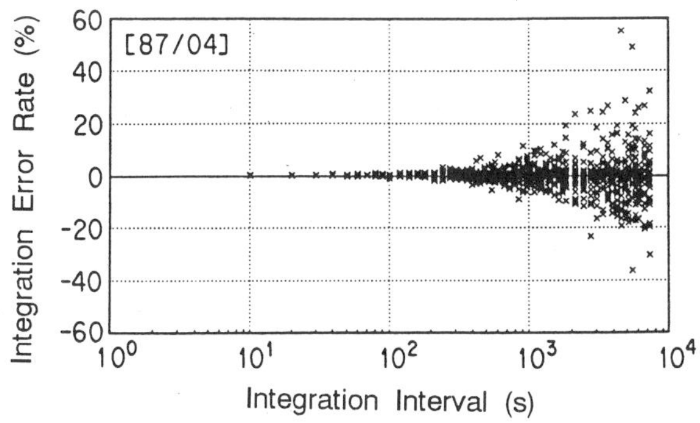

Fig.2 An example of scatter diagram

conditioners were also suppressed very carefully.

DEFINITIONS AND STATISTICS

By applying trapezoidal rule to the measured irradiance data as shown in Fig.1(a), the integrated value of irradiance, *i.e.* irradiation, s_i is given by

$$s_i = s_{i-1} + \frac{1}{2} (X_i + X_{i-1}) (t_i - t_{i-1}) \qquad (1)$$

where t_{i-1}: previous sampling time, X_{i-1}: measured irradiance at t_{i-1}, t_i: *i*-th sampling time, X_i: measured irradinace at t_i, h : integrating step slze ($= t_i - t_{i-1}$). s_{i-1}: integral till t_{i-1}.

Iteration by Eq.(1) gives S as daily integral of s_i until t_i= sunset.

The step sizes are denoted as h_j for j=1 to 60. The sampling interval of original data is h_0. The range of h_j corresponds to 10, 20, 30, ... seconds to 3 hours logarithmically. S_j is given as S for $h = h_j$. Then, per cent integration error, e_j is defined by

$$e_j = (S_j - S_0) / S_0 \times 100 \qquad (2).$$

An example of e_j is shown in Fig. 1(b). A set of e_j's are plotted on the scatter diagram for a reference period as shown in Fig. 2. Then, the mean value m, standard deviation σ and max./min. values for each interval are calculated. The overall result for 44 months is presented by Fig. 3. As shown in Figs. 4 and 5, approximation formulae showing m and σ versus h_j are determined by

$$\left. \begin{array}{ll} h \leq 652 & m = -1.30 \times 10^{-3} \cdot h^{0.778} \\ h > 652 & m = -4.07 \times 10^{-5} \cdot h^{1.31} \end{array} \right\} \qquad (3)$$

$$\left. \begin{array}{ll} h \leq 419 & \sigma = 3.28 \times 10^{-3} \cdot h^{1.03} \\ h > 419 & \sigma = 1.15 \times 10^{-2} \cdot h^{0.819} \end{array} \right\} \qquad (4)$$

PROBABILITY AND NOMOGRAPH

The frequency of the error classes covering $\pm 4\sigma$ at

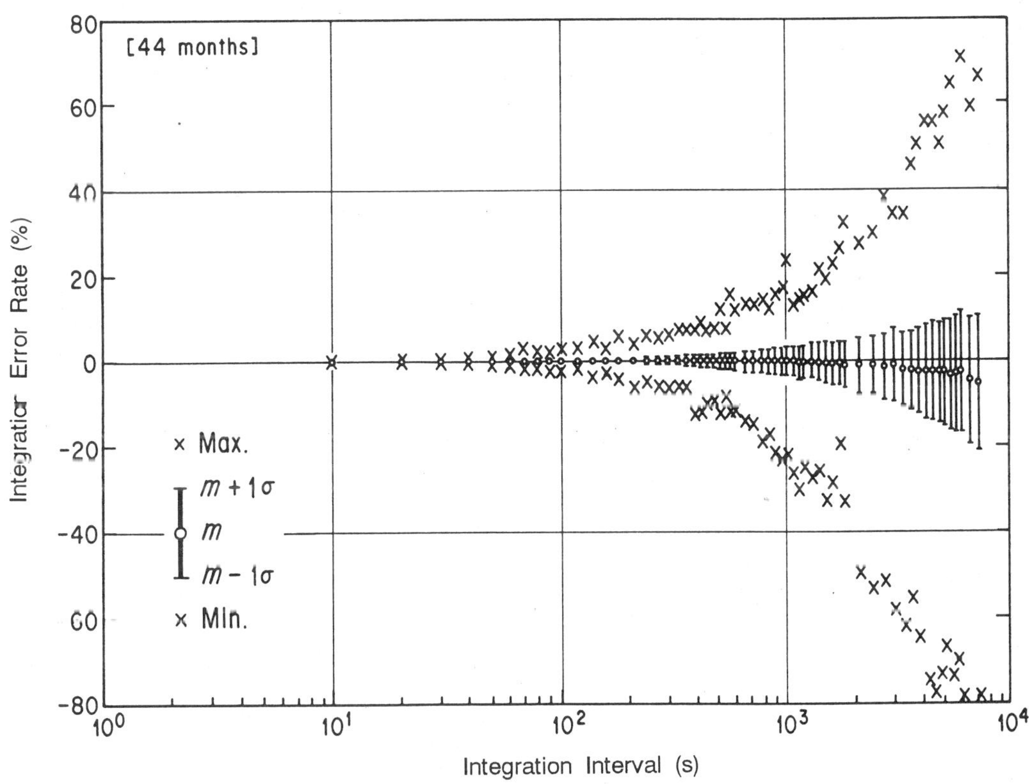

Fig.3 Integration intervals versus statistical values of integration error rate
m : mean value σ: standard deviation

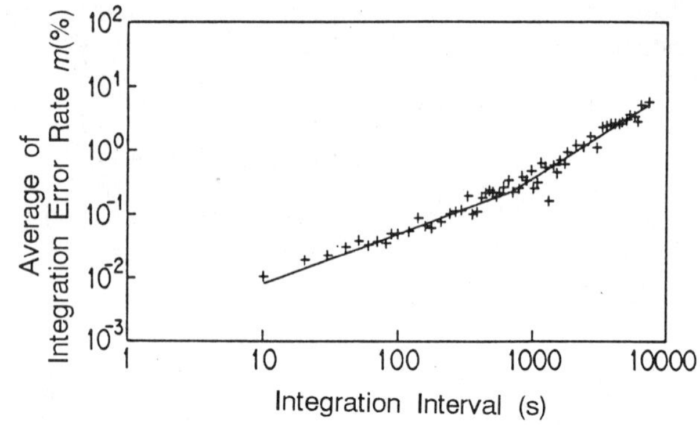

Fig. 4 Approximation of relationship between integration intervals and average of integration error rate

respective interval has been plotted by a histogram as shown in Fig. 6. It was observed that its shape is considerably similar among all intervals. Therefore, each histogram is superimposed by normalizing the integration error with the standard deviation σ. This result is given in Fig. 6. A solid line in the figure and Table 1 are an approximated probability density function p(x) expressed by

$$p(x) = \sum_{i=1}^{n} \frac{a_i}{\sqrt{2\pi}\ \sigma_i} \exp(-\frac{x^2}{2\ \sigma_i^2}) \qquad (5)$$

where,

$$\sum a_i = 1\ , \quad \sum a_i\ \sigma_i^2 = 1$$

x : integration error normalized by overall standard deviation,
σ_i : standard deviation of i-th function,
a_i : coefficient of i-th function.

Parameters a_i and σ_i were decided by least square fitting as follows: a_1 = 0.317, a_2 = 0.512, a_3 = 0.171 and σ_1 = 0.250, σ_2 = 0.826, σ_3 = 1.921. The precision of the approximation is indicated in Fig. 7.

Finally, a nomograph has been composed in order to select an step size appropriate for a target precision. If one can specify an allowable error rate on horizontal axis and select its guarantee probability on curve parameter

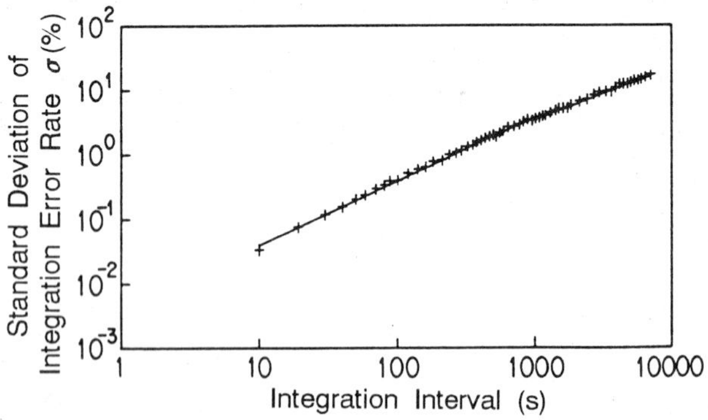

Fig. 5 Approximation of relationship between integration intervals and standard deviation of integration error rate

respectively in Fig. 8 or 9, a recommended step size is easily found on vertical axis. The guarantee probability is described as an outside probability of p(x), which means a probability exceeding specified error rate. For instance, "1/30" presents possibility that error larger than specified one may happen in one day per month.

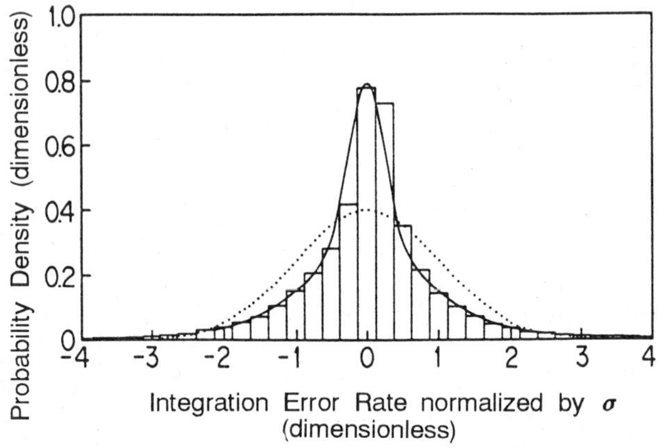

Fig.6 Appearing probability of integration errors for the whole period and for all integration intervals
bar graph: measured distribution
solid line: estimated probability density function
dotted line: Gaussian distribution

Table 1 Outside probability of p(x) versus x

outside probability	$\pm x$	outside probability	$\pm x$
1/2	±0.399	1/30	±2.528
1/3	±0.686	1/60	±3.187
1/4	±0.903	1/90	±3.547
1/5	±1.065	1/120	±3.787
1/7	±1.308	1/180	±4.107
1/14	±1.837	1/365	±4.626

According to this nomograph, as a concluding example it can be shown that the deviating component

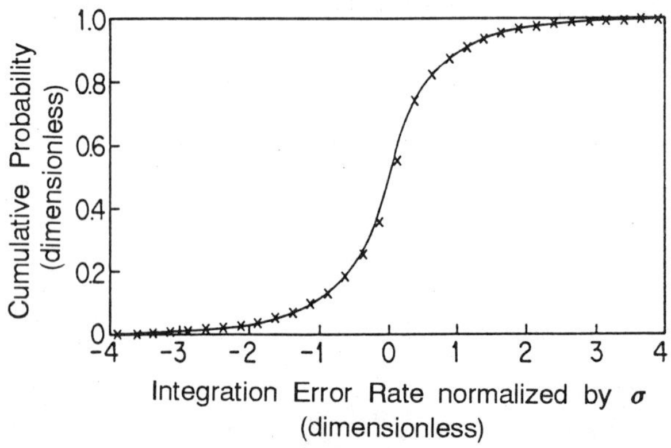

Fig. 7 Approximation of probability function

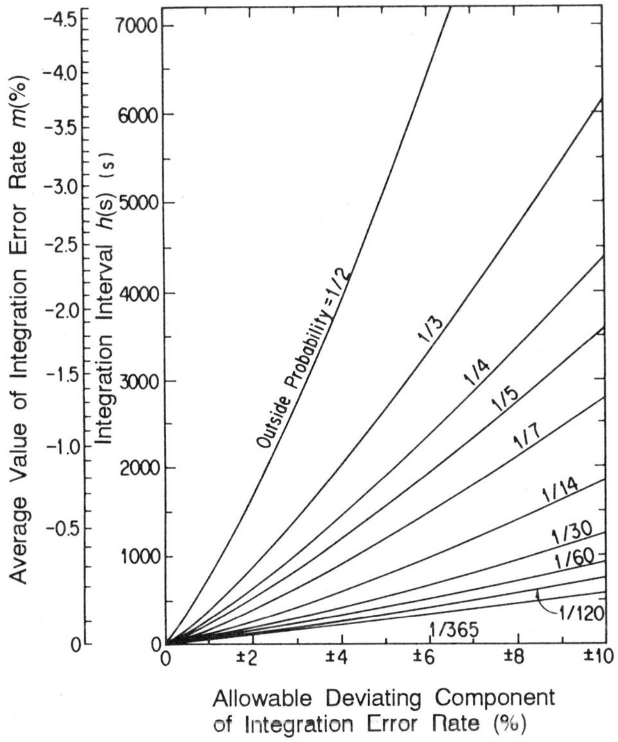

Fig.8 Nomograph for deciding appropriate integration interval

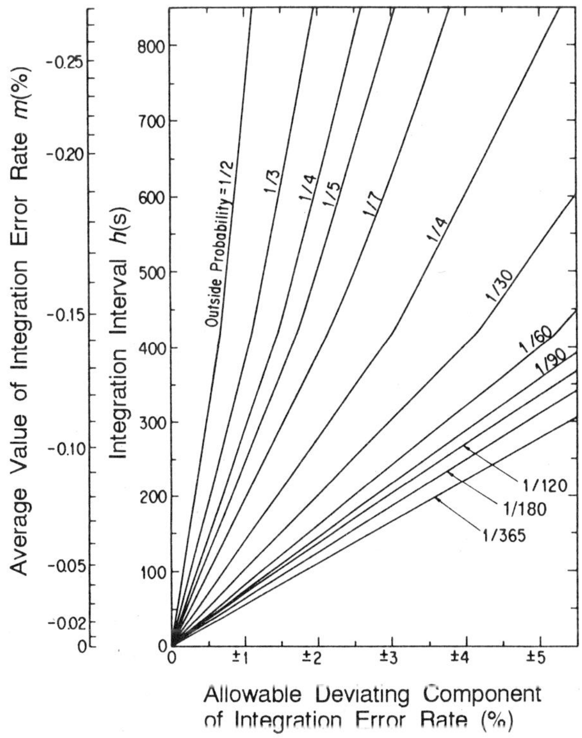

Fig.9 Nomograph for deciding appropriate interval integration interval (Enlarged)

of an integration error exceeding ±1% can happen once for every 1 month or 6 months if the integrating step size is selected as 105 seconds or 65.5 seconds. In each case, the total estimated integral error becomes -0.0485±1% or -0.0336±1% including an average error component.

CONCLUSIONS

To give guidelines for assuring measurement precision of PV system monitoring, the author examined the relationship of data sampling speed versus energetic measurement errors considering its probabilistic nature. From this view point he made the following study.

- Original irradiance data had been measured with comparatively high speed sampling of 1 to 4 seconds for 44 months at the Tsukuba region in Japan.
- By using the data, the daily errors of numerical integral has been estimated for 60 kinds of step sizes.
- Approximation formulae of the error versus the step size have been statistically obtained as well as their probability density function covering $\pm 4\sigma$.
- A nomograph is presented to decide an appropriate sampling interval.

A concluding example shows that the deviating component of the error exceeding ±1% can happen once for every 1 month or 6 months if the step size is selected as 105 seconds or 65.5 seconds, In each of which the total error becomes -0.0485±1% or -0.0336±1% including an average error component.

Japan Industrial Standard (JIS) referred this typical example. It is recommended in the standard that the sampling interval shall be 1 min. or below[1]. The same value is also described in the working draft of IEC TC82.

This work has been performed as a part of the New Sunshine Program.

REFERENCES

[1] JIS C8906, "Measuring Procedure of Photovoltaic System Performance", July, 1993, p.1 (in Japanese).

EFFECT OF INSTALLATION CONDITIONS ON THE OUTPUT CHARACTERISTICS OF PHOTOVOLTAIC MODULES

Kholid Akhmad, Frank Belley, *Akio Kitamura, *Fumio Yamamoto, *Shigeo Akita, **Hideyuki Takakura, Hiroaki Okamoto and Yoshihiro Hamakawa

Faculty of Engineering Science, Osaka University,Toyonaka, Osaka 560 Japan
*Technical Research Center, Kansai Electric Power Inc., 3-11-20 Nakoji, Amagasaki, Hyogo 661 Japan
**Faculty of Engineering, Toyama Prefectural University, Kosugi, Toyama 93903 Japan

ABSTRACT

A systematic investigation has been carried out on the tilt and azimuth angle dependencies of output characteristics of photovoltaic modules. The measurements have been made on some combinations of two prototype modules of poly-Si solar cells, installed at different tilt angles and azimuth angle. Output parameters were measured in the form of I-V characteristics.

The analysis of the results shows that (i) provided output voltage is appropriate to the system operation, no serious problems are likely to arise out of the parallel connection of two modules at different tilt and/or azimuth angles, if protected by blocking diodes (ii) in the case of series connection equipped with by-pass diodes, however, the max-power voltage (Vmp) becomes instable if the azimuth angle is 15 degrees or more different from each other. That is because the Vmp drops down to the Vmp level of the higher output-current module.

INTRODUCTION

In order to be able to obtain maximum output from a photovoltaic (PV) system, it is necessary to understand the nature of dependence of solar cell output on installation conditions. Theoretically, from the view of maximum annual output, the ideal position of PV arrays is one which faces due south with a tilt angle equal to the latitude of the installation site [1]. However, due to deviations in local weather conditions, specific application purposes or practical conditions e.g. installation on the rooftop, the theoretical method is not always viable. Moreover, a certain inclined surface that meets the requirements in some periods of the year is inadequate in other periods [2].

PV-arrays divided lengthwise into two or more parts might be effective and give a solution to the problem. One part is installed in the optimum inclination for a specific season, and the other parts are optimized for other seasons.

Photovoltaic arrays are made up of a suitable combination of PV modules to maintain specific voltage, current and power levels. Series connection of PV modules results in increased system voltage and the number of modules connected is determined by the rating of the PV system. However, the current output of an array string, connected in series, is controlled by the least current output in the string. If an array is made up of modules installed at different inclinations, as suggested above, mismatching of photocurrent causes dramatical loss in the overall output.

To find solutions to the issues mentioned above, and in search of other possible installation conditions for specific application purposes, the output performances of PV modules installed at different tilt and azimuth angles, connected in series or parallel, were measured and studied with special interest devoted to reveal as much as possible about maximum output conditions.

EXPERIMENTAL SETUP

Outdoor measurements have been performed with the experimental setup given in fig. 1. The array was constructed from two modules (A and B) of poly-Si solar cells. Module A was (defined as standard module) tilted at 35° (latitude of the experiment site) and positioned facing due south, while module B was installed at different tilt and azimuth angles. Specifications of the modules are shown in table 1.

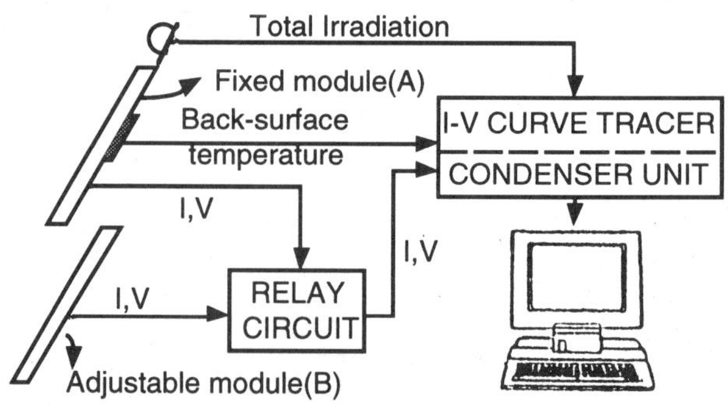

Fig.1. Experimental setup

CH3365-4/94/0000-0730 $4.00 © 1994 IEEE

First WCPEC; Dec. 5-9, 1994; Hawaii

Cell temperatures were measured using constantan foil resistors which were laminated on the surface of the back cover of the modules. Total irradiation in the plane of the standard module was measured with MS 801, a high precision pyranometer (EKO Instruments), which is sensitive in the wavelength range from 300 nm to 3,000 nm. Relay Circuit was used to connect the two modules in series, parallel or separate them automatically. I-V characteristics of the modules were measured by the I-V Curve Tracer unit (EKO Instruments). Nearly the whole I-V curve in one quadrant can be obtained with 40 data points. In every 2 minutes, automatically, I-V curves were drawn, with temperature and total irradiation measured. All data was processed by an NEC PC-9801 ns/L computer equipped with 20 MB HD memory.

The experiment was conducted at the Rokko Advanced Energy Experiment Center, in Kobe, from August to November 1993.

Table 1. Specifications of the poly-Si solar cell modules (AM 1.5, 1 kW/m^2, 25°C)

Number of cells	Voc (v)	Isc (A)	Vmp (V)	Imp (A)	Pm (W)	Eff. (%)	Area (m*2)
36	21.2	3.25	16.9	3.02	51	11.6	0.4383

OUTPUT CHARACTERISTICS OF PV MODULES

Typical daily variations of output parameters of the standard module and the adjustable module on a clear day, Aug. 25,1993, are shown in fig.2(a)~(c).

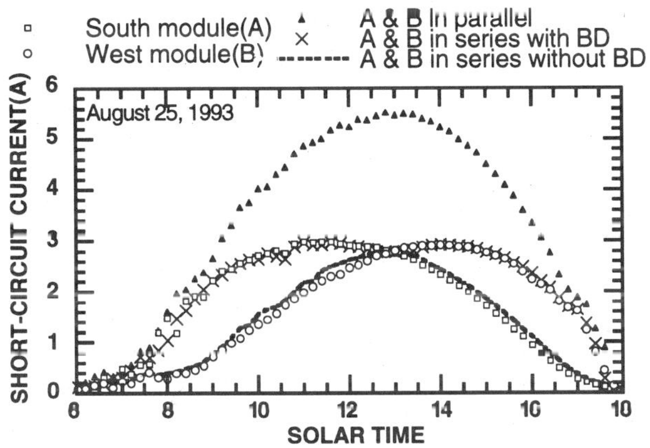

Fig.2(a). Characteristics of short-circuit current(Isc)

Fig. 2(a) shows transfigurations of the short-circuit current (Isc) curve, the result of connecting the modules in series, parallel or separate. Isc of the parallel string is the sum of the Isc of individual modules,as shown by closed triangles. In the series string, there are two patterns of Isc depending on whether by-pass diodes (BDs) are inserted or not. When BDs are bridging over individual modules, Isc of the string is almost equal to the highest Isc value among the individual modules in the string. On the contrary, without BDs Isc is restrained by the Isc of the weakest module.

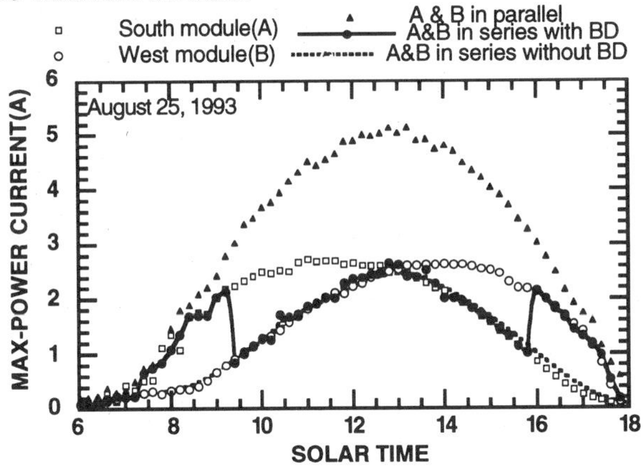

Fig.2(b). Characteristics of max-power current(Imp)

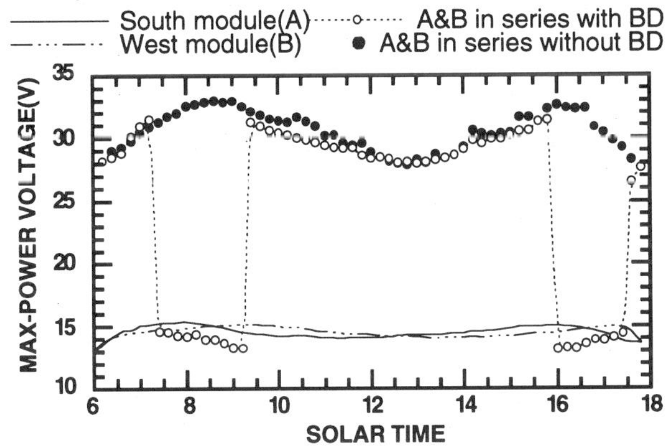

Fig.2(c). Characteristics of max-power voltage(Vmp)

Fig.2(b) shows the daily variation of the current at maximum power operation, Imp. In the parallel string, characteristics of Imp are very similar to those of Isc. Correspondingly in the series string without BDs, the Imp of the string is controlled by the weakest module in the string, but the Vmp stays stable at a high value, as shown in fig.2(c). When BDs are bridging over individual modules, Imp may "jump", which is observed four times in the figure (closed circles). The max-power voltage, Vmp in fig.2(c), also shows similar characteristics. BDs allow the string current to by-pass a module, when the reverse voltage in the relevant module (the west module in the morning, or the south module in the early evening) exceeds a certain voltage,usually 0.6 volt. The "switching" effects can be easily understood from the I-V and P-V characteristics, as shown in Fig.3.

Generally, if Isc discrepancy exists among the solar cells, the I-V characteristics curve, of serially connected solar cell strings with BDs, is shaped like two steps of a stairs and the P-V curve has two peaks. Fig.3(a) shows I-V and P-V curves measured in the afternoon (14:40). At that time, both modules received direct radiation from the sun and the ratio of Isc, of each module, was less than two. In this case, the output power is obviously larger at higher operating voltage, as shown by a cross sign. Whereas at 16:40 (fig.3(b)), direct radiation was only received by the module facing due west, and the south module was lit only by diffused sunlight. In such a case the ratio of Isc can be greater than 2 and output power is larger at lower operating voltage. The "jumping", in this experiment, occurs when the ratio of Isc of the modules is larger than 2 and the Vmp suddenly reduces to its half, as the string consists of two identical modules. This period with low Vmp is called "by-pass hole" in this report.

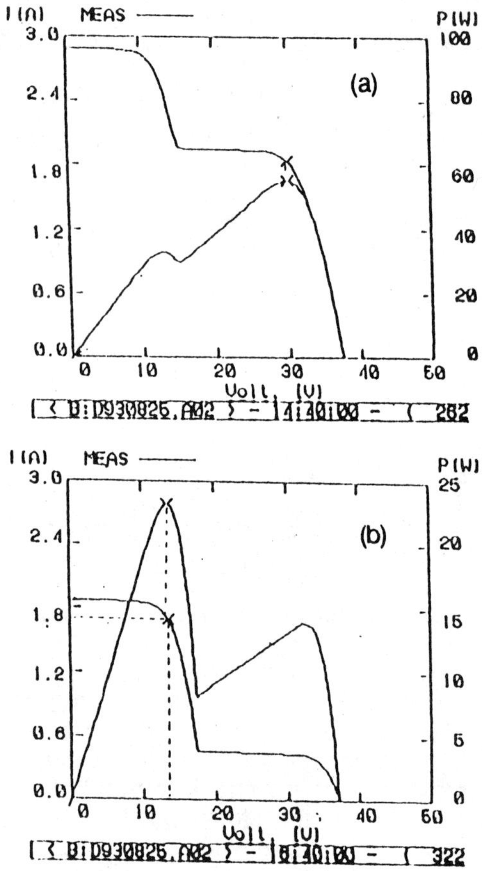

Fig.3. I-V and P-V characteristics of two modules in series connection equipped with by-pass diodes

Fig.4(a) and 4(b) show Vmp characteristics of series strings (equipped with BDs) with various tilt angles and azimuth angles, respectively. In fig.4(a), (35;0) indicates that the module A is tilted at 35° and the module B at 0° (horizontal plane). In this figure by-pass hole, clearly, appears when the module B is tilted at 90°, the reason for that is the lack of direct irradiation, on the module B in the morning, and the module A in the evening.

In fig.4(b) both modules are tilted at 35°. Module A is fixed due south (azimuth 0°), while module B is varied in positions from due south to due west (azimuth angles 0°~90°). By-pass hole, already, appears when the azimuth angle pair is (0:15). As the azimuth angle approaches to 90° the by-pass hole becomes wider and wider.

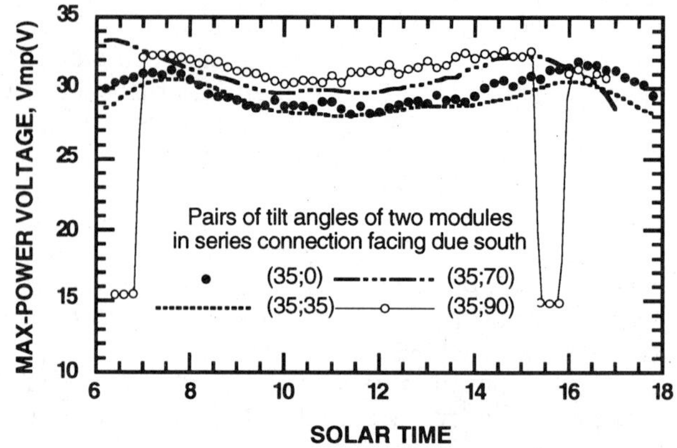

Fig.4(a). Vmp of the series string with the second module at various tilt angles

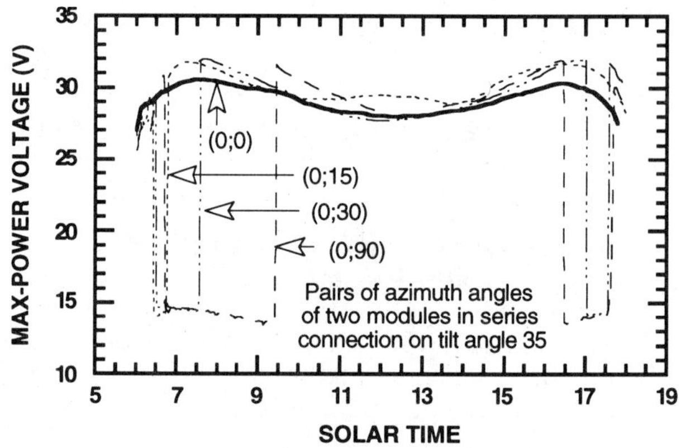

Fig.4(b). Vmp of the series string with the second module at various azimuth angles

Fig. 5 shows the ratio of daily integrated max-output power (Pmax) normalized to, the Pmax of, the standard module. Pmax outputs of series, parallel and individual modules are expressed in WH/m^2.

The first (south) module (A) as the unit

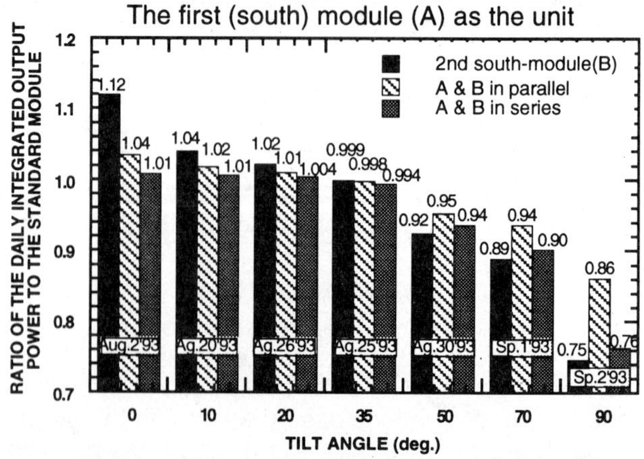

Fig.5(a). Max-output power with the second module at various tilt angles

As in fig.5(a), the Pmax of module B at the tilt angles from 0° to 20° is higher than the Pmax of module B at the same position as module A. This results in higher output power of modules in series and parallel connections compared to that of the standard position. This results indicates that the output power of the divided PV-array at different tilt angles in certain periods is higher than the PV-array located at the same tilt angle. In this figure we can also see that the output power of modules in series or parallel connections is always less than the sum of the output power of the two individual modules. That is because of the mismatching of the I-V characteristics, resulting in power loss. It can, also, clearly be seen that the power loss in the parallel string is lower than in the series string.

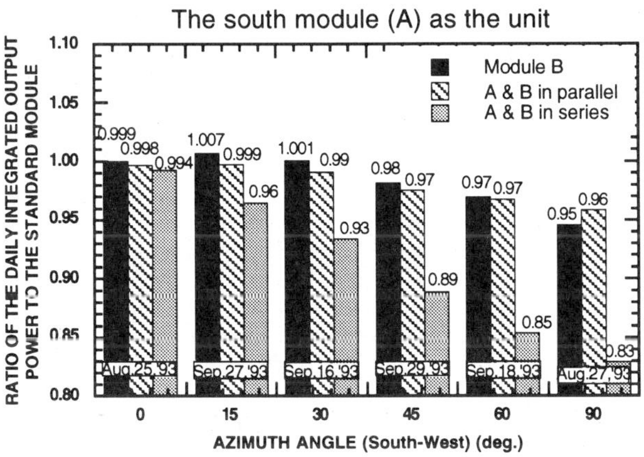

Fig.5(b). Max-output power with the second module at various azimuth angles

In fig .5(b), when the module B is positioned at the azimuth angles from 15° to 30°, the output power is higher than others. In this position the Pmax of parallel string are very closed to that of the standard position. Contrary to the parallel string, the Pmax of the series string sharply decreases when increasing the azimuth angles. As a result of increasing the azimuth angle, shadowing effects the two modules in different times, the module B in the morning and module A in the evening. This mismatch drastically reduces the output current of the series string, thus resulting in lower output power.

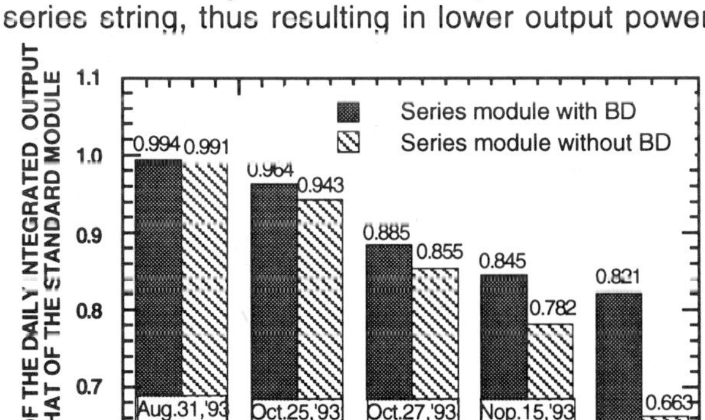

Fig.5(c). Max-output power of series string and effect of by-pass diodes

From a computer simulation we also got the result, that the output power of two modules in parallel connection installed at the same tilt angle (35°) with one module facing due east and an other module facing due west is 75% compared to the output power of both modules facing due south [3].

Fig.5(c) shows the Pmax of modules in series connection placed on 35° tilt angles, and the effects of BDs. The first module is fixed facing due south, while the azimuth angle of the second module is varied (from south to west). It can clearly be seen that the Pmax of the series string equipped with BDs is unexceptionally higher than the Pmax of a similar series string without BDs.

SUMMARY

The output characteristics of PV arrays constructed from two modules, which were installed at various tilt and azimuth angles, have been systematically studied. Special attention was paid to the dependence of maximum-output parameters on the modules' connections (parallel or series), and the effects of by-pass diodes.

In the case when modules are in series connection equipped with by-pass diodes, with the same tilt angle, but different azimuth angles, depending on the short-circuit current of the individual modules, the maximum-output voltage may "jump" several times in a day. This instability occurs when the difference of azimuth angles is 15° or more and the Vmp drops down to the Vmp level of the higher output-current module. Such instabilities may impede the operation of both maximum power tracking and fixed voltage inverters. If BDs are not installed in the string the above mentioned instabilities do not occur and Vmp stays at a higher value, but the output current gets repressed by the weakest module resulting in lower output power.

In case of parallel connections, connected with blocking diodes, no serious problems are likely to arise. If required voltage levels can be maintained by all the PV arrays in different inclinations, output of these arrays should be connected in parallel.

REFERENCES

[1] K.Takahashi, Y.Hamakawa and A.Ushirokawa, *"Taiyoukou Hatsuden"*, (Morikita Book Co., Tokyo, 1980), p.337 (In Japanese).
[2] Y.Kita, I I.Takakura, M.Murozono, M.Kubota, S.Yamaguchi, Y.Hamakawa: *Tech. Digest 1st Int. PVSEC*, Kobe, Japan, 1984, pp.319~322
[3] H.Takakura, Kholid A., F.Belley, A.Kitamura, Y.Hamakawa: *Proc. First World Conf. on PV Energy Conversion*, Waikoloa, Hawai, 1994. Elsewhere in this volume.

ACTUAL OPTICAL AND THERMAL PERFORMANCE OF PV-MODULES

S. Krauter, R. Hanitsch
Technical University Berlin
Electrical Machines Institute
Einsteinufer 11, Sec. EM 4
D-10587 Berlin, Germany

ABSTRACT

Actual efficiency of photovoltaic generators is often lower than predicted by standard test conditions (STC [1]) or standard operating conditions (SOC). This is caused mainly by an underestimation of reflection losses and solar cell temperature in the module. To get more accurate results in predicting the performance of PV-modules, the parameters influencing incoming (optical parameters [3]) and outgoing power flow (electrical and thermal parameters [4]) were investigated by simulation and some verifying experiments at the University of New South Wales and the Australian desert [5].

OPTICAL PARAMETERS

Losses by reflection at glass surface laminated PV-modules are 4-5 % only at perpendicular incidence of sun-radiation. For most applications incidence angles can vary significantly from normal, which results in increased reflection losses according to FRESNEL's laws [2] at each slab (see Fig.1).

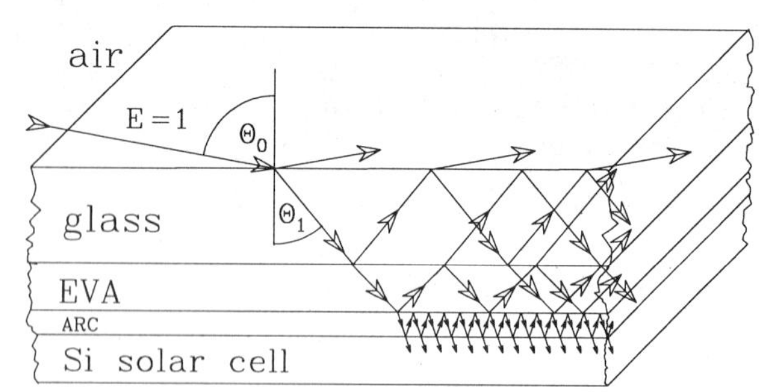

Fig. 1. Ray tracing through the module layers.

Also non-homogenous diffuse components (Fig.3) [7] and polarization of skylight (Fig.4) [6] have been accounted for. The difference to previous models is quite significant. As shown in Fig.2, the usually used perpendicular incidence gives only a peak value of the real transmittance of the encapsulation system during a day, especially if only the air/glass transition is regarded.

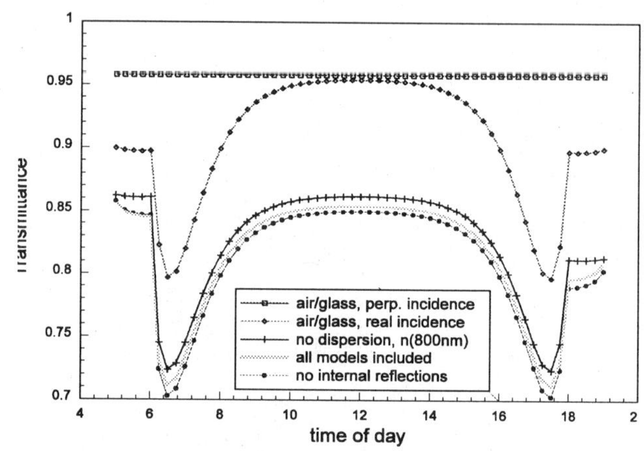

Fig. 2. Transmittance of an encapsulation of a standard PV module (PQ 40/50) over a day (21/3 at 34°S) calculated with different models.

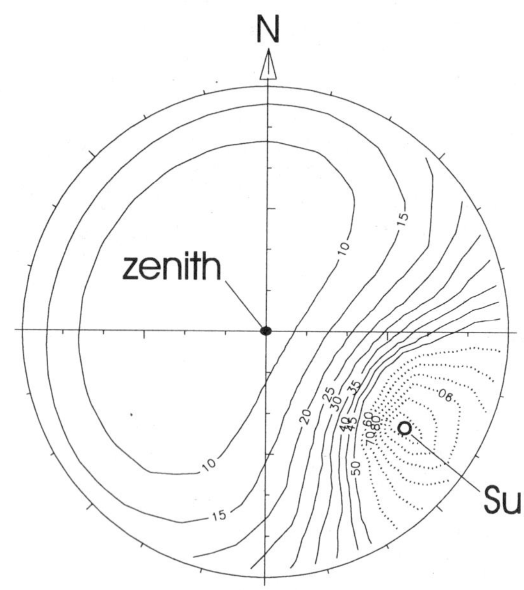

Fig. 3. Non-isotropic distribution of diffuse irradiance (according to DIN 5034) for a sun elevation of 30° and clear sky conditions in W m^{-2} sr^{-1} (projection from sky sphere).

CH3365-4/94/0000-0734 $4.00 © 1994 IEEE

First WCPEC; Dec. 5–9, 1994; Hawaii

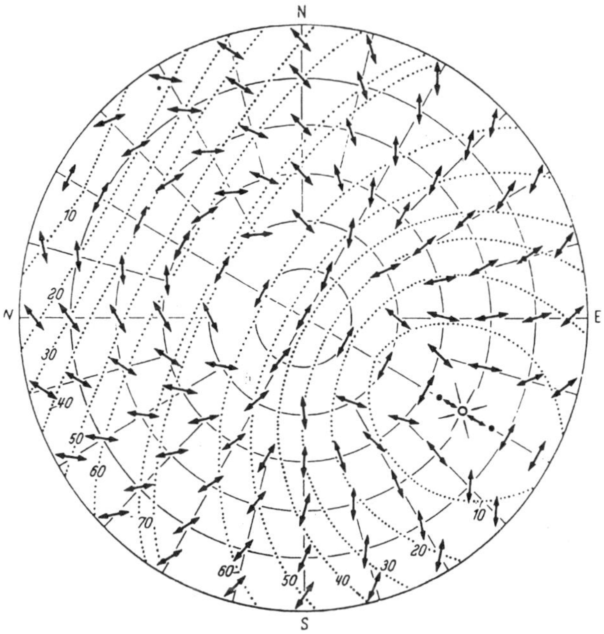

Fig. 4. Degree of polarization (in %) and directions of planes of polarization (arrows) of diffuse irradiance at a sun elevation of 30° (projection from the sky sphere).

Total reflection losses of the optical system consisting of module encapsulation and solar cell are in the vicinity of 20 % (Fig. 5 and Fig. 6).

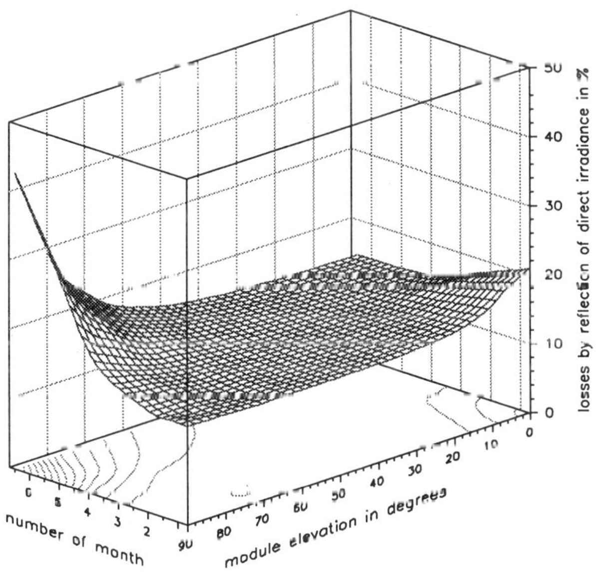

Fig. 5. Daily reflection losses of a non-tracking module for direct irradiance (Site: 34.5°N, Module: PQ 40/50, tilted southward).

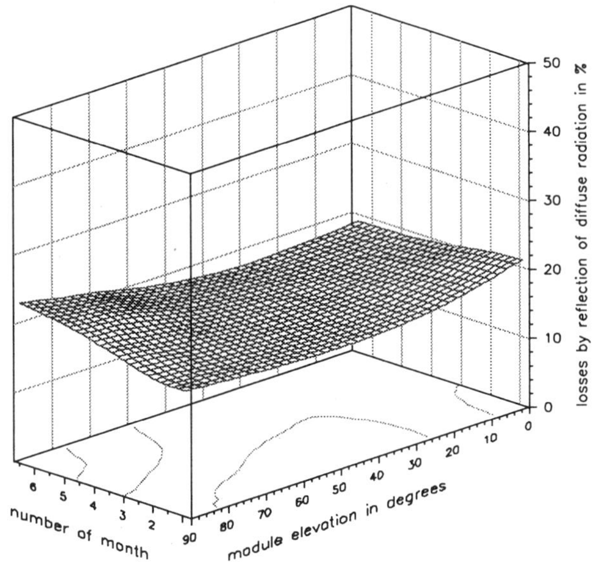

Fig. 6. Daily reflection losses of a non-tracking module for diffuse irradiance (Site: 34.5°N, Module: PQ 40/50, tilted southward).

An optimal matching of the refractive indices of the front layers allows an improvement of 4 % of PV output generated over a day [3]. To decrease further reflection losses, a V-structure on the surface is suggested [9].

THERMAL PARAMETERS

PV modules loose voltage and efficiency at elevated cell temperatures by 0.4-0.5 % K^{-1} for crystalline silicon solar cells. The cell temperature at a certain heat flow (absorbed irradiance minus generated PV power) is determined by the heat radiation exchange (among module and environment) and the convective heat transfer coefficient (see Fig. 7.).

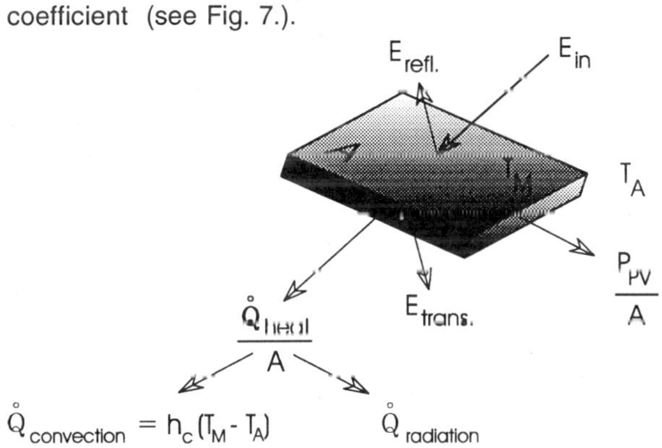

Fig. 7. Balance of energy flows at a PV module

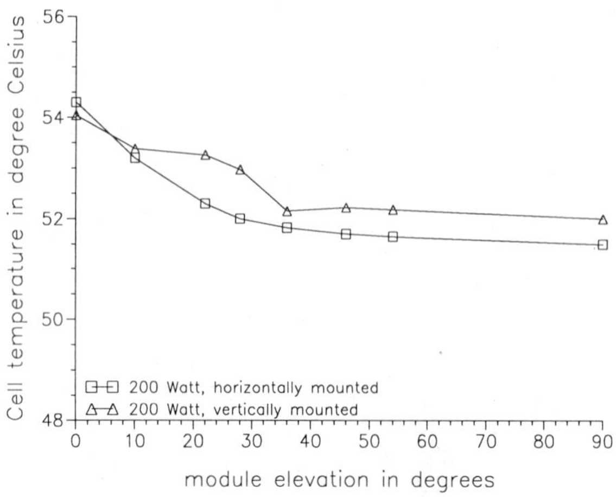

Fig. 8. Cell temperature at a heat flow of 200 W in a M 55 module (400 W m^{-2}) for $T_{ambient}$ = 24.9 °C.

The elevation angle of the module has little influence on the natural convective heat transfer coefficient in the 20°-80° elevation angle range (< 0.2 %). However, at almost horizontal position (0°-20° angle of elevation) the energy output of the module decreases by approximately 0.7 % due to elevated cell temperature. Horizontal, instead of vertical mounting of a standard module increases efficiency by 0.3 % (see Fig. 8). Heating was done by forcing an electrical power dissipation in the cells (using cells as diodes in forward direction).

SIMULATION

The process of insolation, absorption, power dissipation by electrical load, heat transfer to the front and back surface of the module, heat dissipation by radiation exchange with the ground and the sky, natural and forced convection was simulated by a computer program [5]. The results for a simulation of the module efficiency during a day are shown in Fig. 9.

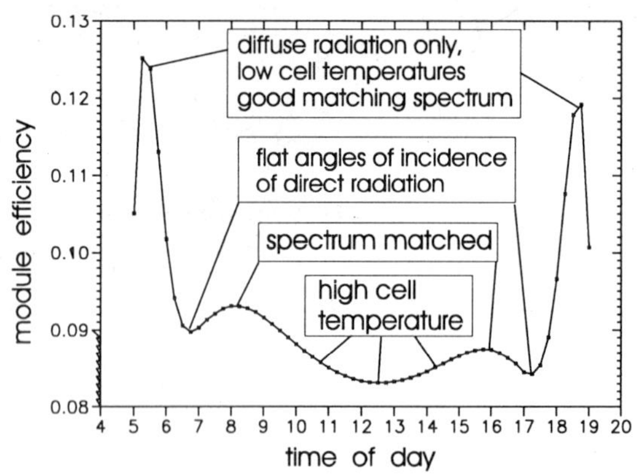

Fig. 9. Module efficiency during a day (21/6 at 34.5°N, PQ 40/50).

For every solar position during a day, each point on the sky sphere was described by its specific spectrum, polarization and incidence angle towards the module surface. All components were traced separately until their absorption in the solar cell. After adding up of all absorbed components the balance of energy flows was carried out.

VERIFICATION

The model used for the determination of the cell temperature was verified by measurements in the Australian desert. For the simulation a constant wind speed of 2 m/s was used as input and the ambient temperature as shown in Fig. 9. It can be seen that the predicted cell temperature follows accurately the actual measured one with a peak deviation of 3 K, which occurred only when the actual wind speed was quite different from 2 m/s.

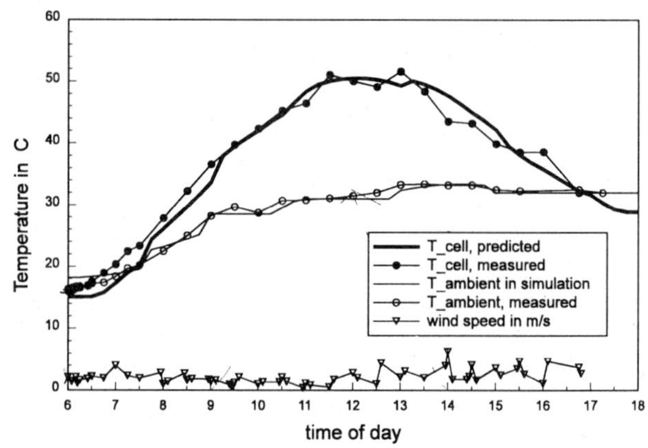

Fig. 10. Measured cell temperatures versus simulation at 4/3/94 in Fowlers Gap (31°5'S, 141°40'E, module elevation: 30°).

IMPROVEMENTS

A tilt of the module azimuth towards east allows a better performance in the morning time while lower ambient temperatures occur (maximum at + 0.4 % for 35°).

A gain of 2.6 % was achieved by attaching a small water tank (12.3 l) to the backside of a module to increase the total heat capacity. During a day the minimum of efficiency is then delayed two hours from the maximum of insolation at lower peak temperatures. A latent heat storage foil is under development and should allow an improvement of 8-10 %.

Continuous studies led us to the TOEPVIS-module (Thermal and Optical Enhanced PV module with Integrated Stand). Here the thermal capacity, a water tank made out of recycled PE, was enlarged in such a way, that the tank also serves as the module stand. The cell temperatures are lowered decisive, which results in a gain of energy output of 11-14 % (see Fig. 11). A photo of the prototype module is shown in Fig. 12.

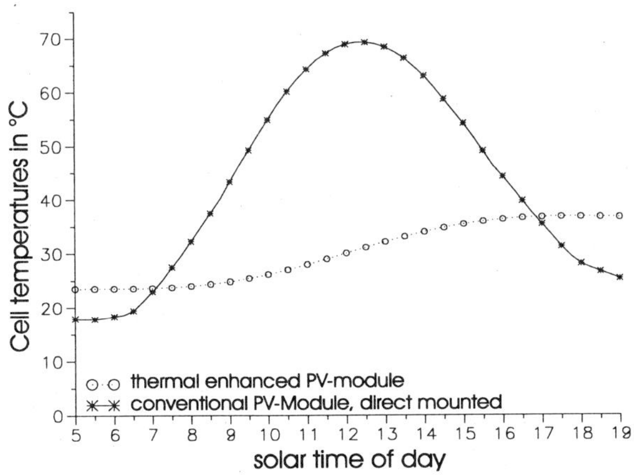

Fig. 1. Cell temperatures of a conventional PV module compared to TOEPVIS module during a day.

Fig. 12. TOEPVIS prototype

Together with an encapsulation with better matched refractive indices and a partial surface structure [9] an increased energy output of 18 % under arid conditions is possible. Comparing the costs of a conventional PV installation with framing, mounting, installation and fundament, the TOEPVIS installation will cost only 60-80 % of it. In total the TOEPVIS module will allow a cost reduction of 10-15 %, plus an increased output, so a total gain in the 30 % region could be achieved.

CONCLUSIONS

The model and the simulation program developed allow us to predict optical and thermal performance under realistic operating conditions, and they are promising tools for evaluating PV power plants with the aim of increasing efficiency.

LITERATURE

[1] Terrestrial Photovoltaic Measurement Procedures, *Report ERDA/NASA/1022-77/16*, June 1977.

[2] M. Born and E. Wolf: *Principles of Optics* (5th ed.) Oxford: Pergamon (1976).

[3] S. Krauter, R. Hanitsch, P. Campbell and S. R. Wenham: Optical Modelling, Simulation and Improvement of PV Module Encapsulation, *Twelfth European Photovoltaic Solar Energy Conference,* 1994, pp.1194-1197.

[4] S. Krauter, *Betriebsmodell der optischen, thermischen und elektrischen Parameter von PV-Modulen*, Berlin, Köster Press, 1993.

[5] Ph. Strauß, K. Onneken, S. Krauter and R. Hanitsch: "Simulation Tool for Prediction and Optimization of Output Power Considering Thermal and Optical Parameters of PV Module Encapsulation", *Twelfth European Photovoltaic Solar Energy Conference*, 1994, pp. 1198-1202.

[6] S. Krauter and R. Hanitsch, "Calculating the Influence of Skylight-Polarization on the Transmission of Encapsulations of PV-Modules" *Proceedings of Cairo Int. Conference of Renewable Energy Sources,* 1993.

[7] *DIN 5034* part 2, German Institute for Standardization, Berlin: Beuth, 1985.

[8] Commission Internationale d'Éclairage, *Publication No. 85,* 1990.

[9] S. Krauter, and R. Hanitsch, "Performance of a Partly Structured Surface at a PV-Module", *Proceedings of the Eleventh European Photovoltaic Solar Energy Conference,* 1992, pp. 1351-1354.

DC FLUORESCENT LAMP FOR OV. HOME POWER SYSTEM

By Prof. Wang, Anhua President of Gansu PV. Co. Ltd.
Mr. Zhang, Danping Engineer of Gansu GNERI PV. Co.
Mr. Wang, Yu Engineer of Gansu PV. Co. Ltd.

* Following the popularization of PV. Home Power Systems in developing contries,the application of DC fluorescent lamps grows rapidly. Estimated consumption in 93' is over 1 million.
* Lighting is the major load in PV. Home Power Systems (Fig. 1) and functuion of lighting is also the main evaluation of PV. Home Power System from the point of view of most rural users.
* reduction of load from the lamps could result either in system economy or prolonged daily electricity provision and battery lifetime.(Table 1.)
* Responce from rural users on PV. lights according to survey during the popularition of PV. Home Power System in Wetern China in the recent 3 yeas could conclude as:

1. Reliable & Duable:

Light tube lifetime	Users Response
less than 4 months	Much complain
near 1 year	no complain
near 2 year	very satisfaction

2. Low cost :
   * Price of light set should not exceed US$5;
   * Price of the tube should not exceed US$0.25;
3. Eay to buy the tube and self handling in replace it.

* Through 3 years efforts of R&D approach, Gansu PV.Co.Ltd. is now producing AHL-1206/8(12V 6/8W) DC fluorescent lamp (Chinese parten No:)and PV. Home power set (Fig. 2, 3) which can meet the users requirment. On-Off turns test (2 min on; 1 min off cycling) could reach 43,000 turns in laboratory products (the tube could last 2 years in protical use) and 21,000 turns for batch production (the tube last 1 year in proctical use). The consumption of electricity are 0.3 amps for AHL-1206 and 0.4 amps for AHL-1208 with 12 V rated voltage while ordinary products are around 0.5-0.7 amps under same rated voltage. Straight tube type is popular in Western China could be found in any electric device shop and easy to replace.

Table 1: PV. Home Power System Load Balance

Load Device	Daily Using Time	Ampere Rate Ordinary/AHL-1206	Consumption Amp-Hours Ordinary/AHL-1206
1.Light Liv. R.	4 hr.	0.5A/0.3A	2/1.2
2.Light Kit.	1.5 hr.	0.5A/0.3A	0.75/0.45
3.Light Bed R.	2.5 hr.	0.5A/0.3A	1.2/0.75
4.14" TV. (BW)	4 hr.	1A	4
Daily Total			8/6.4
4 days Total (3 Overcast days considered)			32/25.6
Battery Size (DOS < 80%)			40/32

* 36AH battery used instead of 32 AH with DOS 71.1% and prolonged battery life

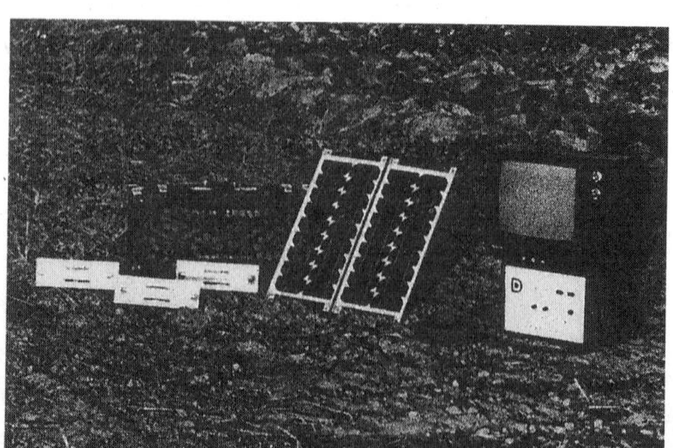

Fog.1 AHS-1202B PV. Home Power For Herdsman

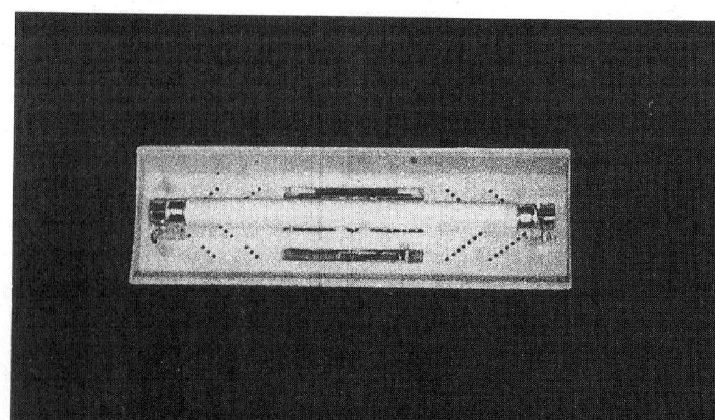

Fig. 2. Picture of AHL-1206/8 DC Fluorescent Lamp

First WCPEC; Dec. 5-9, 1994; Hawaii

Fig. 3 The Principle of AHL-1206/8
1.DC Switch; 2.Oscillator; 3. Transformer; 4. Controller; 5. Tube.

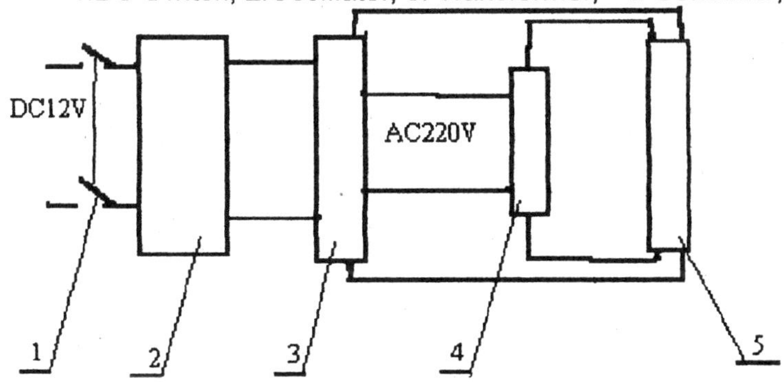

ONE YEAR EVALUATION OF THE PV-WIND-DIESEL HYBRID PROJECT IN X-CALAK, MEXICO

Jorge M. Huacuz, Jaime Agredano, Gonzálo Munguia, Marco A. Borja
Instituto de Investigaciones Eléctricas (IIE), Mexico
Steven Durand, Robert Foster
Southwest Technology Development Institute, USA
Ron Pate
Sandia National Laboratories, USA
Arturo Romero, Carlos Flores, Enrique Hill
CONDUMEX, S.A. de C.V., Mexico

ABSTRACT

A photovoltaic (PV)-Wind-Diesel hybrid system has been operating for 27 months, in Southeast Mexico. Performance of the system, level of users satisfaction and project sustainability are being studied since early 1993. Results from one year of project monitoring are reported, including electricity generated, load pattern and overall system performance. It is concluded that the hybrid is performing well, but a set of non-technical factors threaten the long-term sustainability of the project.

INTRODUCTION

X-Calak is a remote mexican fishing village on the shore of the Caribbean sea, next to the Mexico-Belize border (18°16' North latitude, 87°50' West longitude). The site is characterized by daily average global insolation of 6 kWh/m^2-day and wind resource with average speed of 4.7 m/sec. Availability of solar and wind resources complement each other, both in the daily and seasonal cycles, making the place suitable for operation of PV-wind hybrid systems.

Total population is about 200 people. Due to the natural beauty of the place, tourism is an important economic activity which, together with fishing, make the only sources of income for the local population.

Electricity was brough to X-Calak for the first time after the 1955 hurricane Janet, which partially destroyed the town. A mini-grid was constructed to deliver electricity produced by means of a diesel gen set to the population, which at that time was about 1400 people. The local fishermen cooperative was in charge of the electrical system, having an expedient scheme of money contributions, that helped defray operation and maintenance costs. In the years after, the cooperative broke up and most people fled town. Those who remained were unable to cover the expenses associated with the relatively large electrical system. Although smaller gen sets were subsequently installed, in the end electricity was generated for only 2 or 3 hours a day, due to economic constraints.

THE HYBRID SOLUTION

In 1992, a PV-wind hybrid systems was installed to reduce the running costs of the diesel gen set, and to improve the electrical service. The system was designed and built by the mexican company CONDUMEX, and totally financed by the Quintana Roo State Government. Main component characteristics are shown in table 1; the schematic diagram is shown in figure 1. Design capacity was set at 150 kWh/day. The already available diesel gen set was programmed to be used as back up capacity only. Figure 2 shows a photograph of PV and wind generators.

TABLE 1

X-CALAK HYBRID SYSTEM CHARACTERISTICS

SUBSYSTEM	# UNITS		BRAND	MODEL	TOTAL NOM. CAPACITY
PV GENERATOR	234	MODULES	SIEMENS	M-75	11.2 kW
WIND GENERATOR	6	MACHINES	BERGEY	EXCEL-S	60 kW
BATTERY BANK	216	CELLS	GNB	6-7C23	1738 kW
CURRENT INVERTER		ONE	AES	SINEMAX	40 kW
DIESEL GEN SET		ONE	SELMEC	OC-16663	125 kW

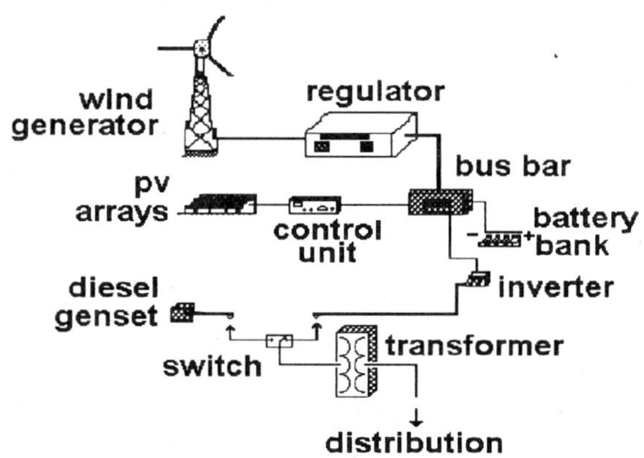

Figure 1. Simplified diagram of hybrid system

CH3365-4/94/0000-0740 $4.00 © 1994 IEEE

Figure 2. Photograph of the PV arrays and wind generators

PROJECT MONITORING

The project is being monitored since early 1993, with respect to: a) Physical and operational status of the hybrid system; b) level of users' satisfaction with the service; and c) organizational, institutional and other factors affecting project sustainability. Through a cooperation agreement between CONDUMEX, the US Sandia National Laboratories, and Mexico's Electrical Research Institute (IIE), a data acquisition and teletransmission system was installed. The hybrid system was thoroughly instrumented, as described elsewhere [1]. A data bank on the performance of the installation is being integrated since March, 1993.

RESULTS

A load inventory was carried out. Table 2 shows the loads in the commercial, domestic and services sectors, as of August, 1993. Figure 3 shows a typical daily load curve. The imbalance on one of the phases is due to the uneven distribution of loads on the electrical network.

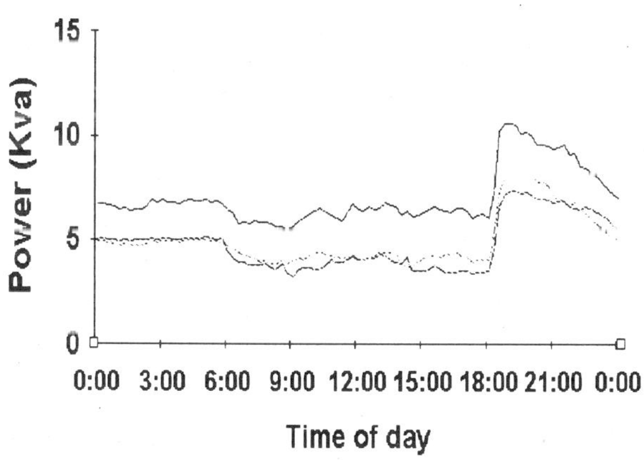

Figure 3. Typical daily load curve for the three circuit phases of the X-Calak system

TABLE 2
LOAD INVENTORY (August, 1993)

LOAD TYPE	TYPICAL POWER (W)	TOTAL POWER (W)
a) Domestic		
Lamps		
Flourescent	20	404
Incandescent	100	15,815
Radio 10	125	
Tape-recorder	15	642
BW TV	30	306
Color TV	77	1,180
Blender	350	8,675
Fan	75	2,160
VCR	18	341
Refrigerator & freezer	190	2,269
Water pump	190	1,140
Iron	800	20,682
Microwave oven	500	500
Washig mashine	635	3,525
Video tape rewinder	5	60
Power tools	---	1,435
Domestic appliances	---	2,633
Other	--	210
b) Commercial		
Lamps		
Flourescent	20	820
Incandescent	60	2,955
Radio, Tape-recorder	---	463
Color & B/W TV	55	290
Blender	---	1,375
Fan	75	645
Refrigerator & freezer	---	1,305
Other		2,105
c) Services		
Public light (sodium vapor)	70	3,995
Fluorescent lamps	78	156
Other	---	670

Figure 4 shows the monthly electricity generation from 03/21/93 through 03/31/94. Daily average generation has been about 227 kWh/day; that is, 51% more than the design capacity. Also shown is the monthly electricity consumption, which almost equals generation. This phenomenon has its origin on the nature and dynamics of the load, as explained below.

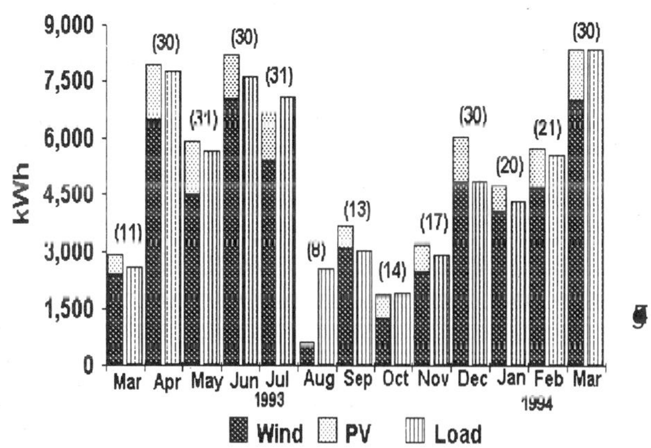

figure 4. Monthly energy produced and consumed (numbers in parenthesis on top bars represent days with data in the month) in X-Calak

Total measured generation was 65,905.53 kWh during the 6,944 hours of recorded data. The PV array had a contribution of 12,235.75 kWh, while the rest was supplied by the wind genreators. Total wind generation was only 4.4 times larger than total PV generation, even tough nominal installed wind capacity is almost 5.4 times larger than nominal installed PV capacity . This may be attributed to a number of factors, including higher forced outage rates for some wind generators.

System availablility to the load was 69%, which is not small number considering the diesel gen set operated for only 8 hours during the period reported here. Low battery votlage was the most frequently (0ver 80%) cause for system outage. This follows as a consequence of electricity consumption patterns in town and the fact that no provision was made in the construction of the system to recharge the battery bank with the diesel gen set. Diesel generation has been kept artificially nil by a decision of the community to save on fuel expenses.

Table 3 presents a summary of the results for the period reported here. It is clear from the number of hours of operation that wind machine #3 had a poor performance. Description of failures is outside the scope of this paper.

TABLE 3

X-CALAK HYBRID SYSTEM. SUMMARY OF OPERATION
FROM 03/21/93 TO 03/31/94

	Time of operation (Hours)	Energy Produced (Consumed) [kWh]
Diesel gen set	8.00	
Supply to load	4234.75	(63,921.59)
Wind generators		
#1	4,881.50	9,377.52
#2	5,134.00	9,045.20
#3	2,476.25	5,503.70
#4	5,376.75	10,135.60
#5	5,431.25	9,532.11
#6	5,317.50	10,075.65
PV generators		
#1	3,541.00	6,584.31
#2	3,541.00	5,741.44
Production		
Total wind		**53,669.78 kWh**
Total PV		**12,235.75 kWh**
Total system		**65,905.53 kWh**

PROJECT SUSTAINABILITY

Aside from a number of short-term power outages, clearly derived from technical failures, the hybrid system has performed beyond reasonable expectations. Nevertheless, a sense of dissatisfaction prevails among the local population, which complain about the quality of the electrical service. This situation, however, derives from a combination of non-technical factors, some of which originate within the population itself. For instance, trends in electricity consumption (including wasteful practices) reflect previous habits of the population. Even though electricity consumption is being metered in most cases, the population is notbeing charged for the electricity used, which in turn reveals a lack of organizational infrastructure to manage the electrical system. Hence, practically all the electricity produced gets immediately consumed (or wasted), as shown in figure 4. The net result has been that in around 26% of the time, the battery bank was not allowed to recover the charge previously delivered, thus limiting the system service due to low battery voltage, which the population interprets as a technical failure. There are also other pre-existing conditions, such as the lack of maintenance of the mini-grid, which contribute to the perceived low quality of the electrical service, and whose solution is strongly conditioned to the availability of organizational structures within the community.

CONCLUSION

The main lesson learned after one year monitoring of the PV-wind hybrid project in X-Calak is that, technically, the system has performed reasonably well; however, non-technical factors such as lack of community organization, threaten the long term sustainability of the project. However, such factors need to be endogenously resolved, as they are mostly of a political nature.

REFERENCE

[1]R. Foster, S. Durand, "Instrumentation for the Hybrid Generation System, X-Calak, Quintana Roo, Mexico". SWTDI report, October 1992.

The Solectrogen House with a 3kW PV array, wind generators, & thermal collectors

LIVING ON PV POWER IN COMFORT AND STYLE

Anoosh Mizany, President
Solar Depot Inc.

ABSTRACT

The Solectrogen House near San Francisco was conceived, designed and built by Solar Depot as an off-grid renewable energy house. It is powered by a Solectrogen Hybrid Power System consisting of a 3kW PV array, two 1kW wind powered generators, a 50kWH storage battery bank and a 6.5kW LP-fueled engine generator for back-up. Two 4kW sinewave DC to AC inverters are utilized, with 24VDC input and 120/240 VAC 60Hz output. The inverters are equiped with built-in battery chargers and microprocessors for controlling the engine generator operation. The water supply of the house is a well pumped by a PV powered submersible DC pump. The swimming pool filtration system utilizes a 3/4HP DC pump powered by a separate dedicated PV array of 640 Watts at 90V. Energy efficient lighting and refrigeration used in the house minimize the power requirements of the house. The house also utilizes a solar thermal system consisting of 640 sq. ft. of glazed solar collectors and a 500 gallon storage tank. The thermal system is utilized for heating water year round, for space heating in the winter and for swimming pool heating in the summer.

GENERAL DESCRIPTION OF THE HYBRID POWER SYSTEM

The "Solectrogen House" in Nicasio, California, about 40 miles to the northwest of San Francisco, California was built in 1992 as a showcase for energy independence and efficiency. This 4500 sq. ft. residence utilizes a hybrid power system consisting of a 3KW PV array and two 1KW wind-driven generators for electric generation and a 50kWH storage battery bank. A 6.5kW engine generator serves as a backup source for electricity. It also utilizes an active solar thermal system for heating of domestic water, space heating and swimming pool heating. In addition, the house was designed as a passive solar house. This house utilizes compact fluorescent and fluorescent lighting throughout, and utilizes highly efficient refrigerators. In order to demonstrate the compatibility of PV power systems with the standard design and construction of homes, this house was wired as a standard AC house.

THE PV ARRAY

The 3KW PV array is comprised of fifty MSX-60 polycrystalline Solarex PV modules. The modules are

CH3365-4/94/0000-0743 $4.00 © 1994 IEEE   First WCPEC; Dec. 5–9, 1994; Hawaii

arranged in panels of 5 modules each, wired in parallel. Each pair of adjacent panels are wired in series to produce a sub-array of ten modules with 600W output at 24V nominal. The system consists of five such sub-arrays, each wired directly to a separate breaker in the DC power center. The ten panel assemblies were installed on the garage roof using pipe/plate brackets that were fastened to blocking between rafters and weather-proofed with roof flashings. The 45° roof slope was chosen as a compromise between the optimum angle for winter and architectural considerations.

THE WIND GENERATORS

Two Whisper 1000 wind-powered generators by World Power Technologies, each rated for 1kW output at 25mph windspeed, each installed on a 20ft guyed pipe tower on the top of the hill about 300ft away from the house. The 30VAC output of the generators are converted to 30 VDC, combined and wired to a breaker of the DC power center. In this location the winds frequently reach speeds of 15 to 20mph, producing 200-400 Watts of power by each generator.

BATTERY BANK

The storage system consists of two Pacific Chloride 24V, 1055 AmpHour batteries(rated at 20 Hr discharge rate), connected in parallel. The batteries are lead-acid, flooded electrolyte, electro-motive type, each weighing about 1600 pounds. A special bracket was built into the garage roof rafters for use with a hoist, to load and unload batteries to and from a truck.

THE INVERTERS

Two 4KW sinewave inverters by Trace Engineering, convert the 24VDC to 120VAC 60Hz. The use of an interface cable allows the two inverters to operate in parallel/series configuration to produce 120/240VAC output which powers the house. Each inverter has a built-in battery charger that can provide up to 100A at 24V, for charging the batteries. The sinewave output of the inverters is stable and noise-free, allowing smooth operation of pumps, controls, electronic gear and audio and video equipment. The inverter controller is a sophisticated menu-driven unit with LCD display that can start and stop the generator based on the battery voltage set points. The controller can limit the power drawn by the charger, from the generator to field programmable levels. It can also allow the inverter to help the generator, if the power demand exceeds the output of the generator. "Quiet Hours" can be programmed to postpone the generator starting. "Must Start" battery voltage can also be programmed, at which the generator starts even during the programmed quiet hours. the generator can also be started and stopped by a timer built into the inverter controller. The digital voltmeter and ampmeter feature can display the DC and AC voltages of the system as well as the AC amps of the inverter and the generator. A digital clock can display time of day. The display can also display the cause of the inverter shutdown. Low voltage disconnect and reconnect for the inverter are also programmable.

DC SWITCHGEAR

An Ananda DC Powercenter ties the various components of the system together and provides safety fusing and NEC compliance for the system. A Heliotrope General Taper Charge Regulator of 90 Amp capacity, two 400 Amp fuses, eight breakers, several current shunts, two surge arrestors and many terminal blocks are incorporated into the DC Powercenter.

SYSTEM MONITORING

A DC/AC system monitor provides digital display of DC currents for the PV array, windpower, battery charger and inverter draw as well as AC currents for the inverter output and generator output. It also provides DC voltages for the PV array, system battery and genstart battery, as well as AC voltages and frequency indication for the inverter and generator outputs.

A six-channel digital thermometer allows monitoring of the thermal system by providing temperature readouts of the solar collectors, solar storage tank top and bottom, radiant floor system feed and return and the hot water tank temperature and gas backup boiler tank.

A digital weather monitor provides readouts of the indoor/outdoor temperature and humidity, barometric pressure, wind speed and direction, wind chill and rainfall.

BACK-UP GENERATOR

A 6.5kW LPG fueled Kohler Engine generator serves as a back-up source for the system. Whenever the energy consumption of the house exceeds the energy stored in the batteries, the generator controller built into the inverter automatically turns on the engine generator which charges the battery bank through the inverter's built-in charger. Simultaneously, an automatic transfer switch throw the AC house load onto the generator, allowing the batteries to charge without a load. Once the battery bank is charged up, the generator controller turns off the engine generator, allowing the transfer switch to shift the house load back onto the inverter, being supplied by the battery bank.

THE ELECTRICAL LOADS AND ENERGY EFFICIENCY

In order to reduce the size of the alternative energy system required, one has to consider and apply all available and practical energy conservation measures.

We have used fluorescent and compact fluorescent lighting throughout the interior and the exterior of the house, including the swimming pool lighting. We have also used two highly efficient refrigerators. Two disposal pumps, one 1.5HP and the other 1HP, operate at 240VAC. they represent a significant power demand but since the run time is limited to an average of less than half an hour per day, the overall energy consumption is reasonable. The central vacuum cleaner also has a 240V motor. Other loads include microwave oven, garbage disposal, dishwasher, clothes washer and dryer (LPG fueled), audio visual center, electronic intruder and fire alarm system, telephone answering machine, three garage door openers, circulator pumps for solar thermal collection and radiant floor heating system and the associated microprocessor-based controllers.

ENERGY EFFICIENT LIGHTING

There are fifty 32W fluorescent T-8 lamps with electronic ballasts by Osram throughout the house. Forty-one of them are placed inside soffits in various rooms, built-in for the purpose of direct lighting. The combination of soffits and vaulted ceilings produce a wonderful quality of light with the T-8 lamps of 3000K. There are 50 downlight cans with specular reflectors recessed into the ceilings and soffits in various rooms, each with two PL7 compact fluorescent lamps. There are a number of wall sconces with Osram EL15 electronic compact fluorescent lamps. Four-lamp and six-lamp strip vanity fixtures by Del Rey with 5W compact fluorescent globes are used in all the bathrooms.

Outdoor lighting is accomplished by sconces, flood lamps or pagoda lights, all containing compact fluorescent lamps. Swimming pool and spa lights are also 30W compact fluorescent by Spectrum. Since fluorescent lights can not be dimmed economically, we used multiple switching in most rooms to vary the level or mood of lighting.

ENERGY EFFICIENT REFRIGERATION

Refrigerators represent one of the biggest energy consumers in the average house because they are on line 24 hours a day. Most refrigerators with automatic defrosting are not very efficient and consume a great deal of energy.

We used two 16 cu. ft. Cold Mizer refrigerators in a side by side configuration. These refrigerators use about a quarter of the energy required by most refrigerators because they are surrounded by 4" of insulation on all sides, the compressors are on the top to prevent their heat from circulating around the refrigerator and finally they do not have automatic defrosting, where heating and cooling fight each other and waste energy. Manually defrosting the refrigerator twice a year by place a pan of boiling water in the turned off freezer is a small price to pay for all the energy saved and for extending the compressor life. These refrigerators are available with DC or AC compressors. We chose the AC model because we wanted the house to be wired for AC only.

PV-POWERED WELL PUMP

The water supply consists of a 150 ft. deep well with a 5" diameter casing, water level between 45 and 100 feet and can supply up to 4 gpm. A submersible DC diaphragm pump by Shurflo pumps water at a rate of about 2 gallons per minute into two 1700 gallon underground tanks made of corrugated polyethylene. Since the tank is between 60 to 70 feet above the house, the gravity feed from the tank produces from 30 to 35 psi pressure at the house, which is ideal. There is no need for a booster pressurizing pump. A 120W, 24V PV array directly powers the well pump from sun up to sun down. A float switch in the tank interrupts power to the pump when the tanks are full.

PV-POWERED POOL FILTRATION

The house has a 600 sq. ft. swimming pool and a spa. In order to separate the power requirements of the pool and spa filtration system from those of the house, we built a 640W PV array that powers a 3/4HP DC pool filtration pump directly, during the daylight hours. The use of a DC filtration pump, directly driven by its dedicated PV panels requires no electronics to control the operation of the pump. The pump starts at low speeds in the morning, builds up to full power around noon, slows down in the afternoon and shuts off as the sunshine leaves the PV array. This is a natural fit for the pool filtration requirements as the daily operation hours of the pumps varies with the seasons in the same way that the algae growth varies with the seasons. In the summer the pump operates for about 8 hours per day. This allows the use of a smaller filtration pump which reduces the PV array size requirement. For example in an application that requires a 2HP AC pump operating for 4 hours per day, one can use a 1HP DC pump because it will operate for twice as long yielding the same volume of filtration. In the winter with shorter days and cooler water temperatures, the algae growth slows down, reducing the need for filtration.

During the middle part of the day, the pump develops enough power to operate a Jandy pool vac. We recommend the use of a somewhat oversized corthridge type filter or sand filter. The DE filters, however, produce too much back pressure for effective operation with down-sized DC pumps.

One disadvantage of a DC filtration pump directly coupled to a dedicated PV array is that it can not operate at night. In order to be able to use the spa and power its jets at night, we installed a 1.5HP AC pump in parallel with our DC pump. With the used of two diverting valves, either the DC or AC pump can be placed in circulation and utilized. In this way we were able to use the spa at night, by powering the AC pump through the inverter.

THE THERMAL SYSTEM

Ten 4' x 8' glazed solar collectors, with selective black chrome coating, heat 500 gallons of water in a well insulated storage tank, which acts as the storage and transfer medium for the thermal energy. A highly efficient LPG fueled boiler provides the additional thermal demand of the house. The space heating system consists of a radiant floor system. Motorized valves controlled by microprocessors allow the system to switch between the solar tank and the LPG fueled boiler, depending on the availability of solar energy. The house is divided into 9 heating zones, each with its own thermostat. Naturally all the heat collection and transfer pumps and the associated electronic controls are powered by the PV via the inverter. In the summer, when space heating is not required, the thermal energy collected is used to heat the swimming pool and spa through another heat exchanger. Three energy efficient, air-tight, wood burning fireplaces serve as a supplemental source for space heating.

THE EXPERIENCE

The "Solectrogen House" utilizes a well as the source of water and has its own sewage system. The house is therefore a true experiment in self-sufficiency in a setting where aesthetics, comfort and traditional features are not sacrificed. With the exception of the LPG delivered to the fuel tank of the house, a phone line is the only connection of the house to the outside world. All systems have functioned well, over the two year period that the house has been occupied by the author and his family. The LPG fueled generator operated about 400 hours during the first year, which had an unusually wet winter. During the second year, when the winter was milder, it operated about 250 hours. The quality of AC power has been quite good. The audio visual equipment, intruder and fire alarm system, phone answering machine, and microprocessor based controls for the thermal system of the house all function without any problems. In addition, the house enjoys the luxury of two side-by-side refrigerator/freezers with a total capacity of 32 cu. ft., a microwave oven, a central vacuum cleaner system and electric garage door openers.

REMOTE POWER - INTEGRATED, CUSTOMER ORIENTED SOLUTIONS

D.Jordan, M.Twidell, J.Patel, F.Teofilo, R.Soin
BP Solar Australia
Unit 1, 100 Old Pittwater Road
Brookvale NSW 2100 Australia

ABSTRACT

Throughout its history, BP Solar has focussed strongly on the production of integrated power solutions to it's customers, using photovoltaics as the central, but not unique, technology.

This paper discusses the customer driven development of two 'new generation' hybrid power systems, one AC the other DC. Both involve the combination of renewable energy with a conventional power source in a fully integrated system, as well as providing remote electronic system access, advanced data and event logging, and comprehensive computer graphic analysis software.

INTRODUCTION

Over the past decade BP Solar Australia has earned a solid reputation in the Asia / Pacific region for the supply of reliable and technically advanced remote area power supply systems.

This reputation has been gained principally by maintaining a close relationship with current and prospective customers, by concentrating on the balance between technology and market requirement, and by striving to provide products with a high performance to cost ratio.

This paper examines two recent product development programs within BPSA as examples of the match of the company's skills and resources to its customer's requirements in a changing market.

THE RECENT PAST

Historically, BPSA's *customer oriented* approach has led to the development of a range of systems, utilising a mix of energy sources and a portfolio of support products.

Products incorporated into this system range include both conventional and 'Saturn' laser buried grid (very high efficiency) modules up to 85Wp power, 'Apollo' thin film modules, the highly successful PVSTOR deep cycle battery and a family of integrated electronic management systems.

The recent system provision of 1,000 or so solar electric systems into rural homes, hospitals, schools and community centres under the Sri Lankan rural development program is a good example of the type of work undertaken by BP Solar in this area.

Traditionally, however, BP Solar has designed and supplied AC and DC power supplies for professional users as a set of discrete components, integrated in design but separately housed.

This paper focuses on the development of two types of fully packaged systems; the first a DC system providing power to telecommunication authorities in the Asian-Pacific region, the second generating abundant high quality AC power for Australian domestic users.

DC HYBRID INTEGRATED POWER SYSTEMS

The move by the Post and Telecommunications Corporation in Papua-New Guinea (PTC PNG) from analogue to high capacity digital networks, and plans for significant expansion of their terrestrial network, left existing stand-alone solar systems struggling to handle the increased power demands, a problem exacerbated by limited real estate for array and battery bank expansion.

An optimised mix of solar and diesel was clearly required, with a large capacity battery bank and highly reliable control electronics.

Such a ***hybrid*** power system involves the integration of solar energy with a second power source, typically a diesel genset, supporting a common load.

This system allows the solar content (or fraction) to be a design variable, providing the option of deferring capital (solar) cost for running (fuel) cost. In addition,

CH3365-4/94/0000-0747 $4.00 © 1994 IEEE

system design uncertainty, due to unpredictable weather or future load variation, can also be handled, with the genset offering power as required.

Locating any such system, comprised of discrete components, onto high mountain tops and ridges, accessible only by helicopter and often in adverse weather conditions, presented PTC with formidable logistical and deployment difficulties, as did the problem of a proven multi power source control strategy.

BP Solar Australia's expertise in integrated solar / diesel hybrids in railway signalling installations in Western Australia led to an exchange of ideas with PTC in the early 1990's, culminating in BP Solar's success with the proof of concept PTC tender for the first helicopter delivered (heli-transportable) DC Hybrid Integrated Power System (the 'DCHIPS').

Under this contract, the first system was installed on the 3,000 metre Mt Mission microwave repeater station, on the Pt Moresby to Lae trunk, accommodating the 1.5kW continuous load resulting from the analogue to digital upgrade .

Since this installation, systems have been provided at a further four mountain sites. A typical heli-transportable hybrid power system configuration comprises a 10kW array (BP275 modules), a 7,000AHr / 24V PVSTOR battery bank, a 5kVA genset driven alternator (with significant derating for altitude) and, of course, BP Solar's proprietary electronic control systems, all except the array contained and integrated within a fully self contained transportable enclosure.

Recently, PTC's plans for a major digital trunk link between Alotau in the island's south and Wewak in the north, coupled with their experience and confidence in the existing DCHIPS technology, have allowed them to work with BP Solar, cooperatively formulating future power supply specifications.

The resultant product, while maintaining the overall configuration of a fully integrated heli-transportable solar / diesel hybrid DC power supply, has as the heart of the hub and spoke system a new range of ***electronic management systems.***

These powerful microprocessor based systems provide the user not only with sophisticated electronic charge control, but also with comprehensive remote system management. This allows complete operational interrogation, and multiple levels of short and long term data and event logging. Interrogation is via local or remote modem RS232C link.

Field proven control techniques have been incorporated into the system software. These algorithms include recently developed battery charging algorithms suited to light and heavy current charge regimes, which help to optimise system operation by minimising diesel run time, maximising component life and obviating expensive site visits.

Physically, the enclosure design allows not only for full system installation and factory test prior to helicopter shipment, but also for 'entombment' within the base of the high security tower, with the array securely atop the tower, avoiding the threat of theft or vandalism.

The technology provides the telecommunication's control centre with a comprehensive network management facility, enabling remote monitoring and maintenance of the network's power system, allowing it to operate at peak efficiency and reliability.

The first volume production run of this new generation equipment is now complete, with systems due for delivery before year end.

Salient features of the electronic management system include -

Level 1 Solar Controller

- autonomous IP65 or 19" rack mounted µP based 120A / 6 stage / C-B protected regulator, positive or negative earth, 12 - 48V with electronic isolation via switch mode power supply.
- generator and rectifier control.
- Extensive LED status indication, and RS232C serial port.
- Potential free alarm outputs and multiple analogue input channels .
- 7 day history data logging via battery backed on board RAM.

Level 2I Solar Interrogation Unit

- Extended monitoring of up to 6 Level 1 units with on-site 4 line / 40 character LCD display with 4 x 4 matrix keyboard.
- Extended 30 day data / event logging via high capacity on board RAM.
- Multiple access level password protection, allowing on-site system installation, control or monitoring.

Controller Interface Software

- Proprietary menu driven interface software package allowing multiple system central monitoring via modem and IBM compatible PC interface.
- Password protected interrogation, adjustment and control of all system operating parameters.
- Automatic polling of multiple network sites via modem link, with data and event retrieval and

storage in ASCII text format compatible with commercial data analysis software.

Customised hybrid power systems have indeed found their place in the telecommunications world over the last decade - by end 1994, BP Solar Australia has over 100 hybrid power systems supporting critical telecommunication sites across the Australasian region.
The 'new generation' system concept, developed in collaboration with one end-user, continues to find ready acceptance with others. Already in production are additional systems for Malaysia Telekom and Queensland DoT - others are currently being offered into tenders within the region. The future for this leading edge technology appears bright.

AC HYBRID POWER PACKAGES

Like users of packaged DC hybrid power systems, customers for remote AC power are increasingly requiring a similar range of features.

While the range of customers is varied, the requirement is very similar - the provision of an integrated and portable hybrid AC power system (of variable solar fraction), fully factory tested and offering comprehensive remote communications capability.

In products developed and supplied to date, transportable packaging has varied - from a converted shipping container for the South East Queensland Electricity Board, complete with sofa and coffee table, to a skid mounted system with canopy, access doors and panelling for the Urlampe Aboriginal Community in the Northern Territory, a project supported by Federal (ATSIC) funding.

Systems usually require an integral solar array, however the physical limit to solar fraction imposed by such an enclosure mounted array can be readily overcome by the addition of ground mounted arrays as and when required.

End users are typically rural properties or communities (increasingly less remote from the established grid) or electricity authorities reluctant to expand their existing grid networks.

System sizing is more subtle in domestic applications, given uneven, unpredictable and often increasing demand, however system intelligence and genset availability allow for substantial operating flexibility.

In packaged AC hybrid systems supplied by BP Solar to date incorporation of PV, wind and diesel is available as standard, with a battery bank comprising flooded lead acid deep cycle PVSTOR batteries and µP based load and battery electronic system management. Remote modem or local RS232 interrogation, data logging and control are standard.

In skid mounted systems, a base fuel tank is standard, with the option of a long range visual low fuel alarm and low fuel cut-out.

Options include the use of high efficiency laser buried grid modules, gelled electrolyte batteries and LPG generators.

Operational ability of the bi-directional inverter to rapidly synchronise with and supplement the diesel genset during short term peak loads provides genuine no-break load sharing and reduces the individual component size, helping to reduce system cost.

CONCLUSION

Because of difficulty in site access and installation in an increasing number of critical load applications, and the associated high cost of field engineering, customers are increasingly opting for transportable, fully assembled and factory tested power systems.

Lack of real estate, unpredictable weather, variable load and critical load application all combine to make a dual energy source system appealing, especially when balanced against the real cost of system unreliability.

Remote system management increasingly provides reassurance to customers that field troubleshooting of ever more sophisticated equipment need no longer present major concerns - barring the unusual event of component failure, system problems can be usually be remedied by the customer, or BP Solar if necessary, via the integral modem link

Additionally, the provision of comprehensive data and event logging allows the luxury of post installation operational optimisation, providing a sound basis for improved future design.

Initial and life cycle costs can potentially be further driven down by the use of high power density laser buried grid solar modules, which optimise real estate and balance of system costs, and by the reduction in battery bank size by the provision of a demand genset coupled with improved battery management.

The use of gelled (immobilised electrolyte) batteries is also available to provide improvement to system transportability, and to alleviate customer concerns with health and safety issues associated with sulphuric acid, albeit at a significant cost penalty.

ACKNOWLEDGMENTS

BP Solar Australia acknowledges the cooperation and assistance of the Post and Telecommunications Corporation, Papua-New Guinea (D. Collins & S. Reckman) in the preparation of this paper.

THE 3.3 MW-PEAK PHOTOVOLTAIC POWER STATION AT SERRE

A. Previi, A. Iliceto, and G. Belli
Electrical Research Centre
ENEL S.p.A.
Cologno Monzese
Via A. Volta, 1-20093
Italy

V. Arcidiacono, S. Corsi, and L. Lambri
Automatica Research Centre
ENEL S.p.A.
Cologno Monzese
Via A. Volta, 1-20093
Italy

ABSTRACT

The photovoltaic power station at Serre (Salerno - Italy) is the biggest plant connected to the grid which exists in Europe today.
The preliminary planning was carried out by ENEL in 1990.
Procurement of materials for the plant started in 1992 and the building will be completed in December 1994.
The plant consists of ten sub-fields, 330 kWp each, working in parallel; they are completely automated and able to operate unmanned.
This paper illustrates constructional aspects of the plant, provides details on real costs and on tests carried out both at the manufacturers and on site before accepting components.
The yearly performance of the plant is also provided.

INTRODUCTION

ENEL decided to build the photovoltaic power station at Serre (see Fig. 1) in 1990 with the main aim of checking the technical and economical feasibility of high power plants to be connected in parallel with the medium voltage electric grid.
ENEL has had considerable experience in constructing and running small size photovoltaic plants (a few kW) over the past ten years [1], but the knowledge acquired cannot be extrapolated to plants of the size of a few megawatts like the Serre power station, especially concerning aspects of capital cost.
After a few years of plant operation it will also be possible to quantify the maintenance costs. This aspect is extremely important since it should allow us to confirm our expectations of extremely moderate costs, and almost total operating reliability and safety.

Fig. 1. View of the power station. In the foreground the pv car parking.

TECHNICAL CHARACTERISTICS OF THE POWER STATION

The main characteristic of the power station architecture is its and subdivision into ten completely independent sub-fields (each one providing 330 kWp, totalling an overall 3.3 MWp) connected to a medium voltage open ring joined to a main control switchboard.
Nine sub-fields are mounted on supporting structures with a fixed tilt angle of 20 degrees.
The tenth tracks the sun by rotating from east to west on a N-S horizontal axis.
The latter sub-field will be constructed by the end of 1995 in the framework of the Thermie Program.

PLANT CHARACTERISTICS

Location	Serre (Salerno), Italy
Latitude	40° 35' N
Altitude	sea level
Annual Solar Radiation (Horizontal)	1620 kWh/m^2
Array tilt angle	20°
Total module surface	32000 m^2
Total Land area	70000 m^2
Array power, rated	3315 kWp
Total inverter power	n.10x550 kVA
Operation mode	Grid connected
Annual Energy Production	4800 MWh
Reactive power compensation	2x800 kVAr

TECHNICAL CHARACTERISTICS OF THE SUB-FIELDS

Photovoltaic field

The photovoltaic modules used have 36 or 72 cells, measuring 10 x 10 cm or 12 x 12 cm, made of multi-crystalline silicon (except for one sub-field made of mono-crystalline silicon).
The plant's total number of modules is about 46.000 - equivalent to more than 2,600,000 cells [2].

DC circuits

Source circuits are realized by connecting in series 10 or 20 modules.
No. 8 source circuits are connected in parallel inside each parallel box (see Fig. 2).
From these boxes, the connections are carried in cable ducts to the sub-field DC board, which feeds the DC/AC converter, and is situated in center of each sub-field (see Fig. 3).
The total number of source circuits for each sub-field is either 180 or 360 according to the specific lay-out of the modules.
The DC electrical system is insulated to earth.

CH3365-4/94/0000-0750 $4.00 © 1994 IEEE

First WCPEC; Dec. 5-9, 1994; Hawaii

DC/AC converters

Each sub-field is provided with an independent DC/AC converter and a step-up transformer.
The converter and transformer are installed in a prefabricated cabinet in the centre of the sub-field.
The ten DC/AC converters are of the line-commutated type, with a nominal unitary power of 550 kVA.
Each converter comprises two three-phase thyristor bridges, galvanically insulated and connected on the low voltage side to the two secondaries of the step-up transformer.
The converter efficiency (including the transformer) reaches 97% of the nominal power, and is close to 93% at 1/4 of the nominal power.
The converter's innovative control system, based on microprocessor technology and touch-screen monitor, provides the MPPT function of each section of the photovoltaic sub-field when the operating conditions vary (irradiance, temperature, cleaning of the panels, etc.).
The control was entirely designed by ENEL [3].

AC electrical circuits

In each sub-field, the DC/AC converter is connected, on the AC side, to the double secondary winding, star-triangle, of a 550 kVA transformer. The primary winding (triangle) of the transformer is connected, by means of a medium-voltage switch, to a buried cable, which forms a 20 kV ring terminating at the main MV board, where interconnection with the power grid is achieved.
The main MV cabin also houses the reactive power compensation system and filtering of the harmonics, and the auxiliary service in AC and DC.

CENTRALIZED CONTROL, SUPERVISION AND DATA ACQUISITION SYSTEM

Supervision

As it is an experimental and demonstration plant, the power station has a particularly innovative centralized

Fig. 2. Detail of a parallel box for 8 source circuits.

Fig. 3. DC switchboard of one sub-field.

management system, with a wealth of functions for the detailed visualization of the plant's operational state, and a wide range of features (see Fig. 5).
More precisely, the system makes it possible to monitor the plant in real time; to implement sophisticated diagnostics on the state of the plant and to archive significant data and events through the acquisition and processing of the data coming from the converters' control systems, the MV cabin, the auxiliary services and the meteorological station. Furthermore, it makes possible the centralized control of the plant, both manually and automatically, for the day-time start-up and night-time shut-down of the converters, the control of the reactive power compensation filters, the sweeping of the I-V sub-fields' electrical characteristics, the comparison of the performance of the various sub-fields, etc.
By means of the centralized command and supervision system, it is also possible to control the plant remotely, using concise commands and signals, and the full remote monitoring of the plant via modem.

Remote control

The photovoltaic power station will be manned for an initial period of two years, with the aim of carrying out activities concerning research and demonstration. On termination of the research phase, the plant will be managed completely automatically, unmanned and under remote control.

Remote control of the plant is achieved with a standard system for primary ENEL stations, suitably adapted, and consists basically of just the start-up and shut-down commands, and the on and off commands to the switch for connecting to the MV grid.
The signals transmitted from the power station to the ENEL control center of Naples are basically the MV switches triggerings, intervention of the main protections, control system failures, and fire alarms.

CONSTRUCTION TIMES AND COMPLETION OF THE PLANT

Work at the plant, which was planned in 1991, began in September 1992.
The preparation of the site required limited earth movements in order to create a flat area with a total surface of 70,000 m^2. The coefficient of photovoltaic-field land coverage is about 0.45. The supply of the modules took place progressively, according to the construction capacities of the manufacturers, and was completed in the late autumn of 1994.
Supply of the electro-mechanical equipment was completed at the end of 1993, and has made it possible for the power station to start running with the connection of the first sub-field to the utility grid on 30.12.93 (see Fig. 4).
The power station started its operation in June 1994, with 2 MW in operation, while the third MW will enter into service by December 1994. The full capability will be reached by the end of 1995 with the construction of the tracked sub-field.
The typical annual production diagram of the plant (sub-fields with fixed structures and sun-tracked sub-field) is shown in Fig. 6.

ANCILLARY ARRAYS

Two "service" photovoltaic fields have also been constructed at the plant.
The first, with a total power of about 7 kWp, is installed on the roof of the main building and will produce energy for the power station auxiliary services. The tilt angle is 10°.

Fig. 4. Partial view of one sub-field. On the right modules preassembled ready for mounting.

Fig. 5. Control and supervision room.

The second, of about 7.5 kW, is located on the roofing of the car park and will be used for recharging some electric vehicles used for moving around inside the plant. The tilt angle is 5°C.
It will therefore be possible to compare in identical environmental conditions the electricity production of fields with different tilt angles (tracking, 20°, 10° and 5° from the horizontal).

COST ANALYSIS

The final cost of the power station is about 8000$ per kWp (i.e., 6500 ECU/kWp).
Table 1 shows the sub-division of the total capital cost for the various classes of expense; as can be seen, the modules account for about 60% of the total.
As the plant has been designed also with research and demonstration purposes in mind, it includes conversion equipment, measuring instruments, a data acquisition and supervision system, remote monitoring, and a conference room and other facilities (which can be omitted from any replica power stations that might be built).
A reduction of the total BOS cost can be expected in the range of 40% for industrial replicas, at identical photovoltaic module efficiency, sub-divided as shown in Table 1, which highlights both the reduced cost of the BOS, which is unlikely to be significantly further reduced, and the future dependence of these plants on photovoltaic modules.

Tab. 2, 3 give the main characteristics of a optimised power station that could be replicated in the future.

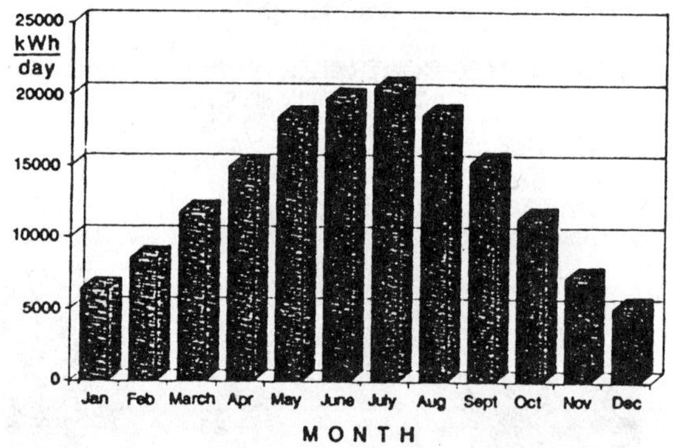

Fig. 6. Yearly energy production of the Serre power station.

Table 1: Subdivision of the total capital cost for multi-MW PV plants

	SERRE PLANT		POSSIBLE REPLICAS	
Capital cost items	Cost [$/kW]	%	Cost [$/kW]	%
PV MODULES	4700	58.6	4700	69.7
DC electrical system	800	10.0	600	8.9
Supporting structures and foundations	580	7.2	480	7.1
Ground preparation and fencing	440	5.5	280	4.2
Power conditioning	400	5.0	240	3.6
Engineering, tests and commissioning	300	3.7	100	1.5
AC electrical system	300	3.7	200	3.0
Buildings and cabins	260	3.2	100	1.5
Supervision, automation and DAS	240	3.0	40	0.6
BOS TOTAL	3320	41.4	2040	30.3
TOTAL CAPITAL COST	8020	100.0	6740	100.0

Table 2 : Characteristics of a optimized sub-field

	Serre	Replica (optimized)
Array power (kWp)	330 (2x165)	600 (2x300)
Max Array bus voltage (V)	440	600
Module power (Wp)	50/100	120
Module open circuit (V)	20/40	20
Array electrical configuration	bipolar, floating	bipolar, floating
Conversion bridge	six pulses	Two six pulses bridges in series
No of bridges per inverter	2	4
Type of inverter	line-commutated	line-commutated (*)
Inverter output (Vac/kV)	20	20
MPPT	Yes	Yes
Interconnection of modules	Manual wiring	Plug-in connectors
Modules assembling	pre-assembling at the site	idem
Supporting structures	fixed	fixed
Inverter and dc/ac interfaces housing	Barycentric prefabricated cabinet	idem

(*) In alternative, if profitable, PWM force commutated

Table 3 : Characteristics of a optimized power station

	Serre	Replica
No of subfields	10	up to 20
No of inverters	10	up to 20
Interconnections of subfields	MV cablo, (underground)	MV cable, (aerial or underground)
Filtering of harmonics	centralized	centralized
Main building	Yes	No
Automation and remote control	Yes	Yes
Supervision and data monitoring	Yes	No

CONCLUSIONS

The architecture of the Serre power station could be used again to create in the short and medium term, with the above mentioned changes, other PV stations sized between 0.5 and 10 MWp.
The smallest size could in fact be used as support to a MV feeder which is close to overload, whereas the 10 MWp size could be used as a booster in a primary substation where the HV/MV transformers are close to overloading.
During the construction of the Serre plant, some criteria for optimizing the original project have already been singled out which would allow the cost of the BOS to be reduced from the present 3.3$/Wp to about 2 $/Wp (including inverters, excluding modules).

FINAL REMARK

The Serre plant has been conceived and realized by the ENEL's R&D Department with the assistance of the ENEL's Construction Department and Research Institutes Conphoebus, CISE, ISMES and CESI.

Manufacturers involved were:

modules: ANIT and Helios (Italy)
Kyocera (Japan)
Photowatt (France)
Solarex (USA)

power conditioning: SEIRA (Italy)

BOS: Carlo Gavazzi (Italy)

REFERENCES

[1] G.Belli, A.Iliceto, and A.Previ - "ENEL's Ten Years Experience in the Use of PV Generation Units to Supply Isolated Dewellings with Electricity" - *11th EC PV Solar Energy Conf.* - Montreux, Oct. 1992.

[2] S. Flores, M. Scuto, A. Previ, and A. Iliceto - "Testing experience of photovoltaic modules for a multimegawatt power plant" *To be presented at WCPEC* - Hawaii - Dec., 1994.

[3] V. Arcidiacono, S. Corsi, and L. Lambri - "The control system of ENEL's 3,3 MWp PV Plant" - *To be presented at WCPEC* - Hawaii - Dec., 1994.

PVUSA UTILITY-SCALE SYSTEM CAPITAL AND MAINTENANCE COSTS

Christina Jennings
Pacific Gas and Electric Company
San Ramon, California USA

Antonio B. Reyes and Kerry P. O'Brien
Bechtel Corporation
San Francisco, California USA

ABSTRACT

PVUSA utility-scale systems are based on innovative, supplier-optimized turnkey photovoltaic designs, selected for their potential to produce low-cost energy for utility applications. Capital and maintenance costs of PVUSA utility-scale systems at Davis and Kerman, California, were analyzed as part of PVUSA's ongoing assessment of PV costs.

Available data from the system suppliers indicate that actual capital costs were as low as $10/W. Preventive maintenance costs range from 0.1¢/kWh to 0.5¢/kWh. Failure-related maintenance costs have been about 1¢/kWh for the fixed and tracking systems at Davis and 0.6¢/kWh for the tracking system at Kerman. Power conditioning units are the key driver for failure-related maintenance costs.

INTRODUCTION

Photovoltaic (PV) system capital and maintenance cost data are of great interest to utilities evaluating the proper timing of PV implementation. While PV module prices are reported regularly by PV industry researchers, minimal data are available to determine utility-scale (US) PV system cost trends.

Photovoltaics for Utility Scale Applications (PVUSA), a national cooperative research and development project, provides an opportunity to acquire US PV system cost data. This paper reports the capital and maintenance costs of PVUSA US systems at Davis and Kerman, California [1-4].

PVUSA US systems are based on innovative, supplier-optimized turnkey PV designs, selected for their potential to produce low-cost energy for utility applications. PVUSA specifications required that each PV system be designed for a 20-year life and that the design and installation comply with the National Electric Code (NEC), state of California regulations, and local codes. Because the power conditioning units (PCUs) were not certified by Underwriters Laboratories (UL), they underwent a safety design review to meet NEC requirements.

The Davis and Kerman US system contracts were completed by Advanced Photovoltaic Systems (APS), Integrated Power Corporation (IPC), and Siemens Solar Industries (SSI). APS assumed the contract initiated by Chronar, which declared bankruptcy during system construction. ARCO Solar was purchased by Siemens GmbH and renamed SSI during system construction at Davis. Ranging from 67 to 498 kW, the systems (Table 1) were accepted by PVUSA after a 30-day performance

Table 1. PVUSA utility-scale systems at Davis and Kerman

PV System	Technology	Reason for selection	System acceptance	Rating (kW_{ac})[a]	Efficiency ($\%_{ac}$)[a]
Chronar/APS Davis	9,600 1.23-m^2 APS frameless amorphous silicon modules, fixed 30° tilt, 4 APS PCUs	Lowest price per watt; innovative, low-cost support structure	October 1992	479[b]	4.2[b]
IPC Davis	1,100 2.21-m^2 Mobil Solar EFG ribbon silicon modules, one-axis active tracking, 1 Omnion PCU	Improved modules, innovative tracker design, IGBT PCU	July 1993	196	8.0
ARCO Solar/SSI Davis	3,968 0.426-m^2 SSI single-crystal silicon modules, one-axis passive tracking, 2 BluePoint PCUs	Innovative, promising PCU technology	June 1994	67[c]	7.9[c]
SSI Kerman	12,240 0.426-m^2 SSI single-crystal silicon modules, one-axis passive tracking, 2 Omnion PCUs	Low price, peak load correlation, proven experience	August 1993	498	9.8

[a] Based on PVUSA Test Conditions (PTC) and total array area, where PTC are defined as 1,000 W/m^2 plane-of-array irradiance, 20°C ambient temperature, and 1 m/s wind speed.
[b] Performance when system was accepted; the rating and efficiency would be about 400 kW and 3.5% if based on 1994 data.
[c] Based on 50% of the installed array; one of the two PCUs is not operational.

CH3365-4/94/0000-0754 $4.00 © 1994 IEEE
First WCPEC; D... 5-9, 1994; H...

period, during which data were collected to determine system rating and final payment.

The three Davis systems experienced multi-year schedule delays due to technical problems or supplier transitions. The Kerman system has fewer unproven components and was completed on schedule.

PV SYSTEM PRICES AND SUPPLIER COSTS

The PV system price paid by PVUSA does not define an upper bound on the actual PV system cost. For highly visible and research-oriented projects like PVUSA (i.e., inherently encouraging innovation at attractive prices), contract prices may be less than actual costs if suppliers underestimate or intentionally underbid the cost. Therefore, PVUSA contract prices may not be the most appropriate data on which to base cost trends, but availability of PV system cost data is at the discretion of the suppliers.

Cost data are available for three systems (Table 2): IPC Davis costs (except for PCU), Omnion-provided IPC Davis PCU cost, and published SSI Davis and Kerman costs [5]. Given the unanticipated problems and PVUSA's high visibility as a research project, PVUSA believes that the four contract prices were less than the PV system suppliers' actual costs. For the purposes of this reporting, all $/W values are based on ac power output ratings at PTC, which better reflect outdoor performance and are lower than ac ratings at Standard Test Conditions (STC).

PVUSA believes that the Chronar/APS Davis costs exceeded the $4.18/W price, attributable primarily to Chronar's low-bidding strategy. Chronar's bankruptcy caused a one-year break midway through construction. No installed US system except for APS Davis has been priced below $6/W at PTC, including subsequent APS bids. The Sacramento Municipal Utility District (SMUD) procured a 200-kW nominal Utility Power Group system in 1993 for $7.70/W and in mid-1994 awarded APS a $6.68/W contract for a 108-kW nominal system.

The IPC Davis contract price was 10% less than the actual cost. The $10.40/W actual cost includes $5.19/W for modules (50%), $1.12/W for the support structure (11%), $1.10/W for system installation (11%), $1.00/W for the PCU (10%), and $1.00/W for engineering and project management (10%).

Contract prices for SSI Davis and Kerman were, respectively, 20% and 6% less than actual costs. Unresolved PCU and ground-fault protection problems significantly delayed the SSI Davis schedule. SSI's progress in narrowing the price-cost difference can be attributed to lessons learned at Davis and the use of a more developed PCU at Kerman.

The $2,252,800 SSI Davis cost includes $1,027,400 for modules (46%), $358,600 for engineering and project management (16%), $299,200 for the support structure (13%), $250,800 for panelization (11%), and $149,600 for the PCUs (7%). It does not include 1993-1994 system start-up costs. The $10.19/W SSI Kerman actual cost includes $6.34/W for PV panels (62%), $1.12/W for the support structure (11%), $1.05 for the PCUs (10%), $0.59/W for sales tax (6%), and $0.16/W for engineering and project management (2%). Budgeted and actual cost breakdowns indicate that the SSI Kerman cost overrun is primarily attributable to increased overhead rates during the contract term that raised the SSI panel cost; this was partially offset by engineering and project management underbudget costs [5].

MAINTENANCE COSTS

Preventive maintenance for PV systems varies depending on system design, supplier recommendations, and owner preferences. To maximize energy production and minimize failure-related maintenance events and

Table 2. PVUSA utility-scale system capital cost and construction data

PV System ac rating at PTC	Chronar/APS Davis 479 kW[a]	IPC Davis 196 kW	ARCO/SSI Davis 67 kW[b]	SSI Kerman 498 kW
Contract price[c]	$2,000,081 ($4.18/W)	$1,836,364 ($9.37/W)	$1,805,136	$4,772,489 ($9.58/W)
Cost to PV system supplier[c]	Not available	$2,043,000 ($10.40/W)	$2,252,800	$5,072,293 ($10.19/W)
Duration from original contract award to system acceptance	40 months	49 months	60 months	15 months
Primary cause of schedule delays	Chronar bankruptcy, module development	Module termination disk problems	PCU problems	None (on schedule)

[a] Performance when system was accepted; the rating would be about 400 kW if based on 1994 data.
[b] Based on 50% of the installed array; one of the two PCUs is not operational.
[c] Includes sales tax and a total of 15 PV modules shipped to National Renewable Energy Laboratory, Sandia, and PG&E.

associated costs, PVUSA complies with system supplier recommendations for preventive maintenance, which may be considered excessive for commercial PV plants.

Preventive maintenance at PVUSA includes monthly resetting of the IPC tracker clock; monthly inspection of APS PCU container window filters, IPC metal oxide varistor fuses, IPC PCU air filters, and IPC and SSI tracker alignments; quarterly adjustment of APS PCU operating limits; quarterly inspection of SSI Kerman PCU air filters; semi-annual inspection of PV arrays, trackers, and PCUs; semi-annual lubrication of SSI bearings; and two-year replacement of IPC tracker oil [6,7].

Preventive maintenance costs incurred from system acceptance through October 1994 are 0.1¢/kWh for APS Davis, 0.5¢/kWh for IPC Davis, 0.2¢/kWh for SSI Davis, and 0.2¢/kWh for SSI Kerman.

Failure-related maintenance costs reflect maintenance activities required to address PV system component failures since system acceptance (Table 3). These costs exclude preventive maintenance, research-related maintenance (e.g., power quality testing), and travel time and expenses. Failure-related maintenance costs primarily reflect PCU problems, distributed throughout the months since system acceptance (Figures 1 and 2). Each maintenance event took 15 minutes to several hours; non-productive time is not included.

PCU problems had the greatest impact on maintenance costs for IPC Davis, which relies on a single PCU. The IPC PCU experienced insulated gate bipolar transistor (IGBT) and fuse failures on May 28, 1994, the third occurrence since system acceptance. The system was placed back in service on August 30, 1994. Failure-related maintenance took less than one week during the three months of downtime. The energy that might have been produced during this period would have significantly reduced the IPC failure-related maintenance cost per kilowatt-hour.

Failure-related maintenance costs incurred from system acceptance through October 1994 are 1.0¢/kWh for APS Davis, 5.0¢/kWh for IPC Davis, 1.0¢/kWh for SSI Davis, and 0.6¢/kWh for SSI Kerman. Labor accounts for 59-77% for the three systems with more than one year of operation (SSI Davis was accepted in June 1994). If there were no PCU problems, failure-related maintenance costs would have been 0.3¢/kWh for APS Davis, 0.05¢/kWh for IPC Davis, 0¢/kWh for SSI Davis, and 0.4¢/kWh for SSI Kerman.

Passive tracker reliability is another unresolved issue, based on SSI Kerman failures since system acceptance and SSI Davis failures during the extended system start-up. Additional passive tracker failures, requiring refrigerant recharges, are expected.

Module failures are rare, and minimal long-term, failure-related PV module maintenance costs are anticipated. APS Davis PV modules cracked due to unknown causes (26 replaced in 1994), and SSI Kerman PV modules had burned junction boxes (5 in 1993, 1 in 1994), as indicated in Figures 1 and 2.

CONCLUSIONS

The available data indicate that actual capital costs for PVUSA Davis and Kerman systems were as low as $10/W. About half of the actual capital cost is attributable to the PV modules. PVUSA encouraged and procured innovative and unproven PV systems at Davis to demonstrate utility-grade designs with low-cost potential; unanticipated problems caused supplier schedule delays and contributed to actual costs exceeding contract prices. PVUSA's requirement for proven components at Kerman resulted in a lower cost overrun. PV system bids submitted to SMUD indicate that PV system capital costs continue to decrease.

PCUs are the key driver for PVUSA failure-related maintenance costs. Failure-related maintenance costs

Table 3. PVUSA utility-scale system failure-related maintenance data

PV System ac rating at PTC	APS Davis 479 kW	IPC Davis 196 kW	SSI Davis 67 kW	SSI Kerman 498 kW
Failure-related maintenance costs (¢/kWh)[a]	0.4 (10/92-12/92) 1.2 (1993) 1.0 (1/94-10/94)	6.3 (7/93-12/93) 4.3 (1/94-10/94)	1.0 (7/94-10/94)	1.3 (8/93-12/93) 0.4 (1/94-10/94)
Failure-related maintenance costs if no PCU problems (¢/kWh)[a]	0 (10/92-12/92) 0 (1993) 0.7 (1/94-10/94)	0 (7/93-12/93) 0.1 (1/94-10/94)	0 (7/94-10/94)	1.1 (8/93-12/93) 0.2 (1/94-10/94)
Failure-related maintenance events (number)	12 (10/92-12/92) 24 (1993) 23 (1/94-10/94)	42 (7/93-12/93) 89 (1/94-10/94)	34 (7/94-10/94)	12 (8/93-12/93) 44 (1/94-10/94)
Energy production (MWh)	148 (10/92-12/92) 530 (1993) 594 (1/94-10/94)	75 (7/93-12/93) 142 (1/94-10/94)	41 (7/94-10/94)	369 (8/93-12/93) 940 (1/94-10/94)

[a] Assuming fully loaded labor rate of $50/hour.

Figure 1. Davis failure-related monthly maintenance costs since system acceptance.

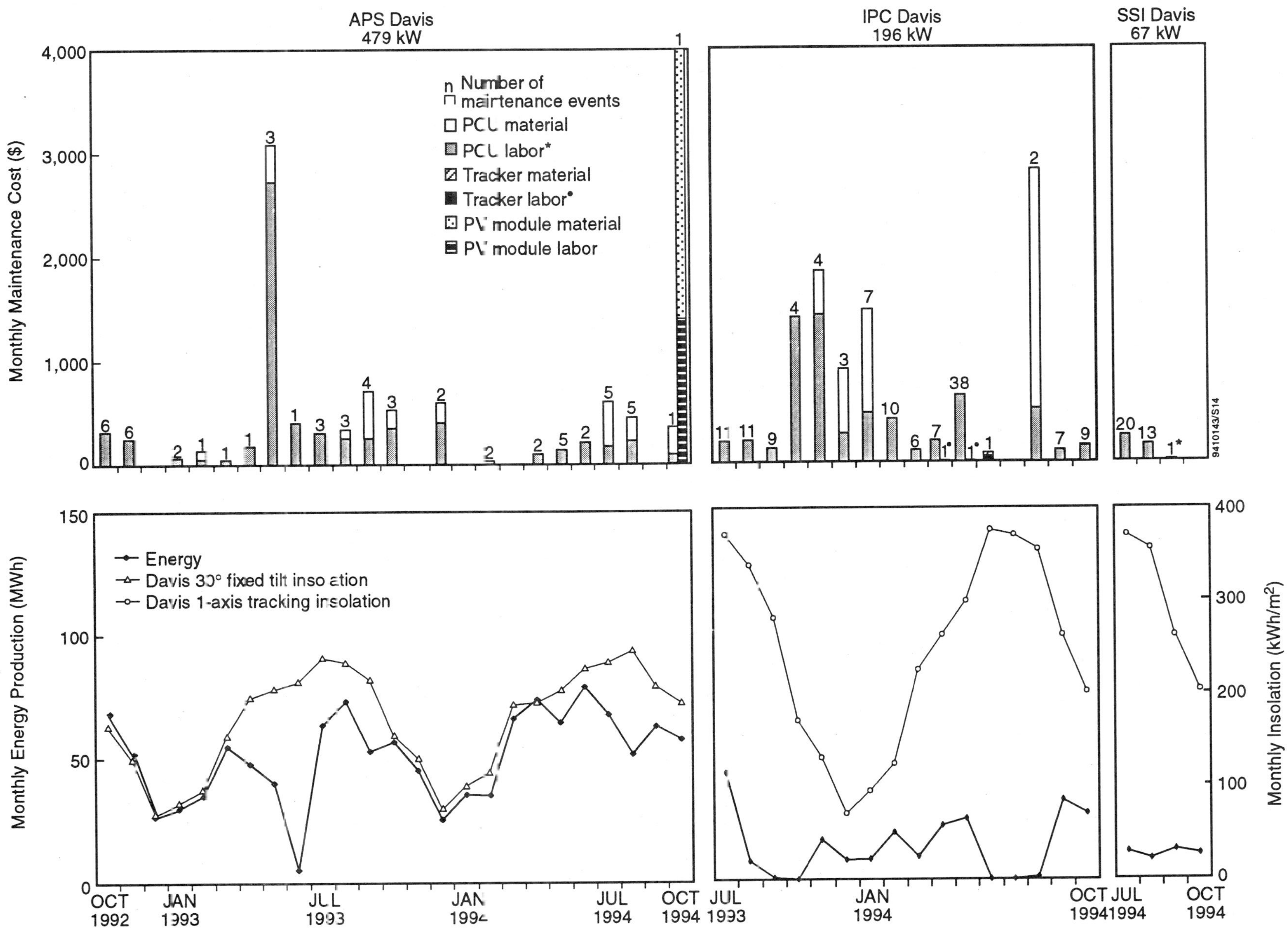

Figure 2. Kerman failure-related monthly maintenance costs since system acceptance.

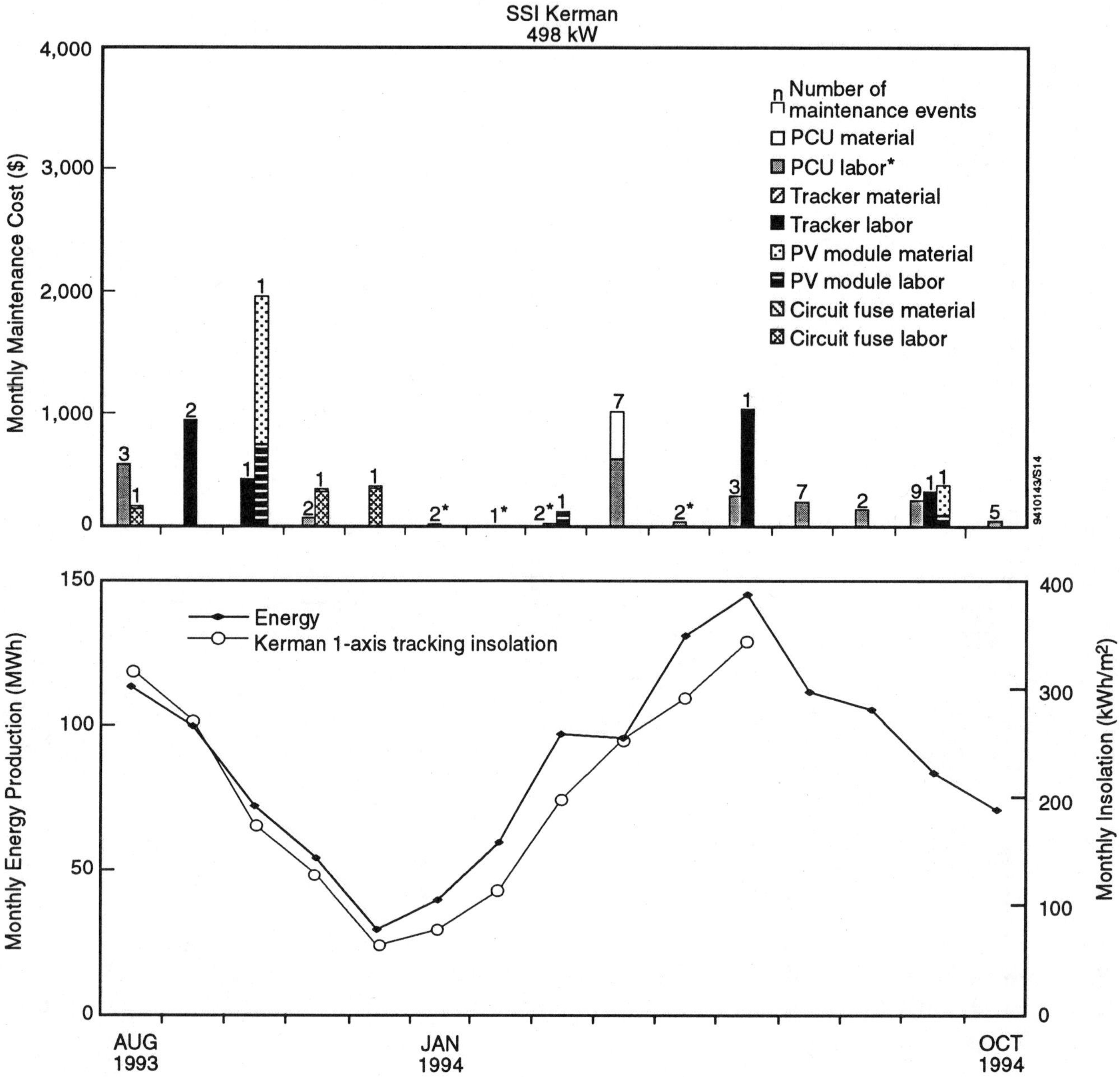

have been about 1¢/kWh for the innovative fixed and tracking systems at Davis and 0.6¢/kWh for the second-generation system at Kerman. Complying with system suppliers' maintenance recommendations results in preventive maintenance costs ranging from 0.1¢/kWh to 0.5¢/kWh. Total maintenance costs incurred from system acceptance through October 1994 are 0.8¢/kWh for SSI Kerman, 1.1¢/kWh for APS Davis, 1.2¢/kWh for SSI Davis, and 5.5¢/kWh for IPC Davis.

REFERENCES

[1] PVUSA Project Team, *1993 PVUSA Progress Report*, PG&E R&D Report 007.5-93.8, 1994.

[2] PVUSA Project Team, *1992 PVUSA Progress Report*, PG&E R&D Report 007.5-93.4, 1993.

[3] PVUSA Project Team, *1991 PVUSA Progress Report*, PG&E R&D Report 007.5-92.6, 1992.

[4] PVUSA Project Team, *1989-1990 PVUSA Progress Report*, 1991.

[5] E. Cunow, "Large PV Power Plants—Past Experience and Future Direction," IEA Workshop, Paestum, Italy, July 1994.

[6] PVUSA Project Team, *Davis PV Plant Operation and Maintenance Manual*, PG&E R&D Report 007.5-94.2, 1994.

[7] PVUSA Project Team, *Kerman PV Plant Operation and Maintenance Manual*, PG&E R&D Report 007.5-93.5, 1994.

ISLANDING PHENOMENON ELIMINATION STUDY AT ROKKO TEST CENTER

(CHARACTERISTICS COMPARISON OF THE REACTOR METHOD AND CAPACITOR METHOD)

A.Kitamura, M.Okamoto, F.Yamamoto, K.Nakaji, H.Matsuda, K.Hotta

The Kansai Electric Power Company Inc., (KEPCO), Japan

ABSTRACT

When a certain part of the utility grid, with substantial number of interconnected PVs, gets cut off from the grid islanding phenomenon may occur. There are two ways to suppress the islanding phenomenon, one of them is performed on the utility side, the other one on the PV side. On the utility side the reactor insertion method and the capacitor insertion method are advocated. The islanding phenomenon suppression characteristics of these two methods are compared in tests. Further, it is proved that the process, leading to the suppression of islanding phenomena, can be described in general power supply theory.

INTRODUCTION

When due to a breakdown or maintenance a certain part of the utility grid, with a substantial number of interconnected PVs, gets cut off from the grid the PVs of that section may continue to operate. This phenomenon is called "islanding phenomenon" and, generally it has to be suppressed as quick as only possible. That is because if the islanding phenomenon continuous for a considerable time the following reactions may occur [1]:

(1). If single phase PV systems maintain three phase system, phase-failure may occur.
(2). Frequency instability.
(3). The proportion of harmonics content may be high.

Further, even in the case of a short islanding, problems can occur [2].
If high-speed reclosure is performed on an islanding part of the utility grid, then a forced connection results. That forced connection not only affects the PVs, but also loads and the grid itself. It has been reported earlier, that the closer the ratio, of overall load to that of the overall output, is to 1:1 the longer the islanding lasts. That means it is possible to suppress islanding phenomena by changing this ratio.

ISLANDING PREVENTION METHODS

There are two main types of islanding prevention methods: one of them detects characteristics features of the phenomenon and stop the operation of the PVs, i.e. passive method.
The other one tips the balance between load and output by an impact on the circuit, i.e. active method. The active method can further be divided into two subtype:
One is performed on the power conditioner side,

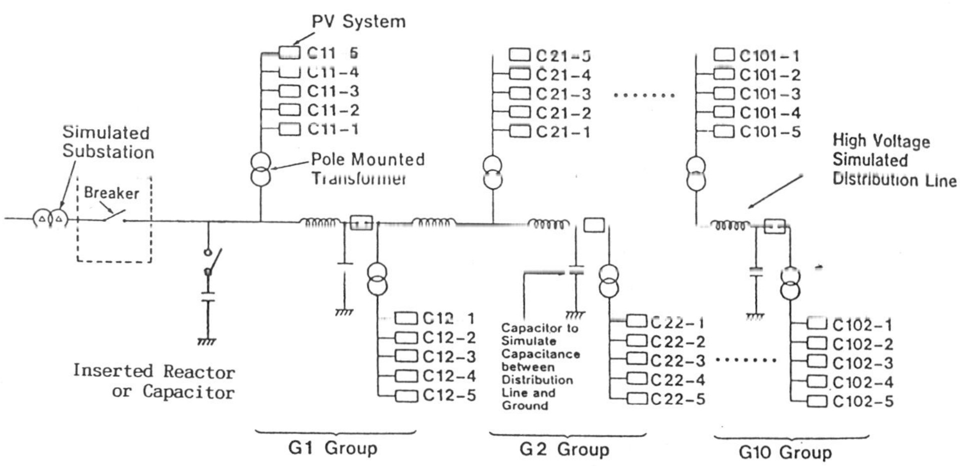

Fig.1. Experimental circuit of 200 kW PV test facility

CH3365-4/94/0000-0759 $4.00 © 1994 IEEE

First WCPEC; Dec. 5–9, 1994; Hawaii

the other one the utility side. We are going to see particulars of the later method, when a certain impedance is inserted into the relevant part of the grid during islanding. This method tries to tip the load to output balance by the used of impedance. in the tests we have used reactors and capacitors as impedance.

TEST METHOD

A modeled 10 km long 6 kV supply line is used for the test (see Fig.1). 100 PV systems, with outputs of 2 kW, are connected to the modeled grid. Islanding phenomenon is created by opening the circuit breaker. During islanding reactors and capacitors are inserted and the phenomenon is interrupted. To change the balance between the overall load and the overall output of PVs, the following method is used. Load consists of rotating machines and general load, the overall load is fixed during the tests. The overall output of the PVs is entrusted to Mother Nature. That is, PV output increases in the morning after sunrise, reaches its climax at noon, and decreases in the afternoon. Consequently, the load-output balance changes.

TEST RESULT

(1). Changes in the islanding characteristics when inserting reactors, capacitors.

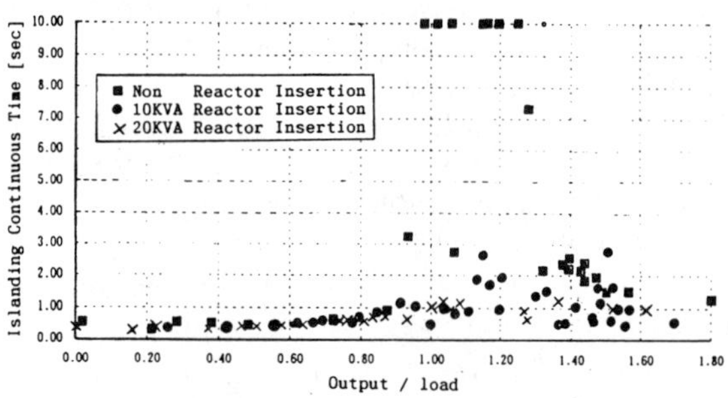

Fig.2. Islanding Characteristics (Reactor Insertion Method)

Fig.2. shows the islanding characteristics of a 65kW grid (overall load, including 25 kW rotating motor) when maintained by 100 sets of 2 kW PVs. It can be seen, that in the region where the load-output proportion is in the range of 0.95~1.30, islanding can last for more than 10 seconds.
Under similar circumstances 0.3 seconds were enough to detect that islanding had occurred and insert reactors.
With 10 kVA reactor capacity we could suppress islanding in 2.8 second as a maximum. That was 1.2 seconds with 20 kVA.
This characteristics showed the same pattern when capacitors were used.
Fig.3 shows the results of capacitor insertion, when the ratio of load (including 50 kW rotating motor) to output is 1, i.e. the condition of lasting islandings.

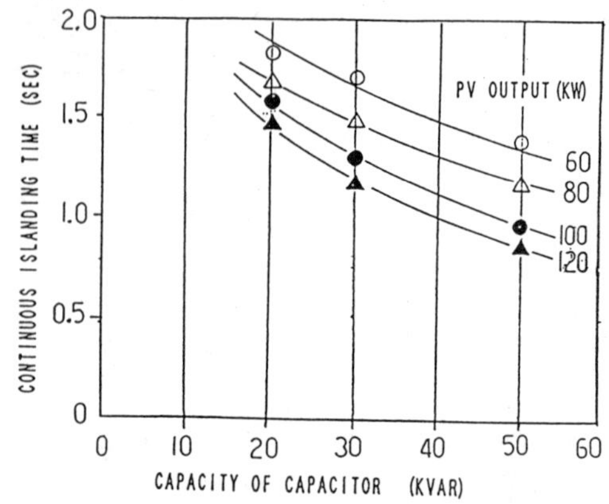

Fig.3. Islanding Elimination Effect
(Capacitor Insertion Method)
Note; PV : 100 units
Load : Including Motor Load

The more powerful the capacitor is the shorter the islanding becomes. If the load and output are in balance the greater the PV output is the shorter the islanding becomes. This characteristics showed the same pattern when the reactors were used.

(2). Behavior of the low voltage side when inserting reactors, capacitors.

Islanding was detected in 0.3 seconds after the grid had been isolated, and reactor was inserted.
In fig.4, it can be seen, where reactor is inserted, that voltage drops and frequency rises.

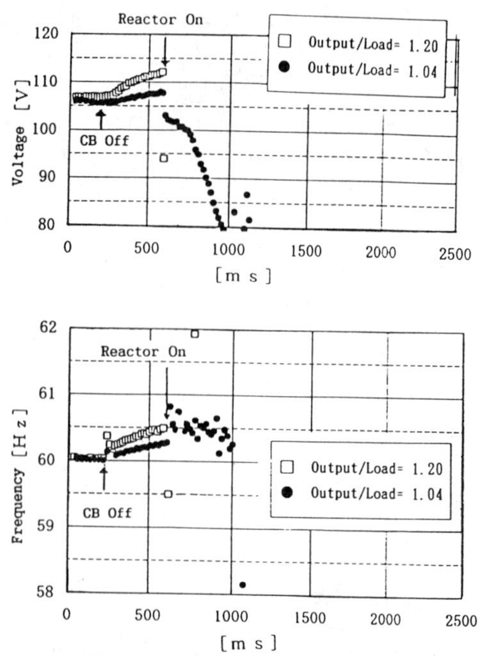

Fig.4. Characteristics of Reactor Method

In fig.5, however, where capacitor is inserted, voltage increases and frequency falls.

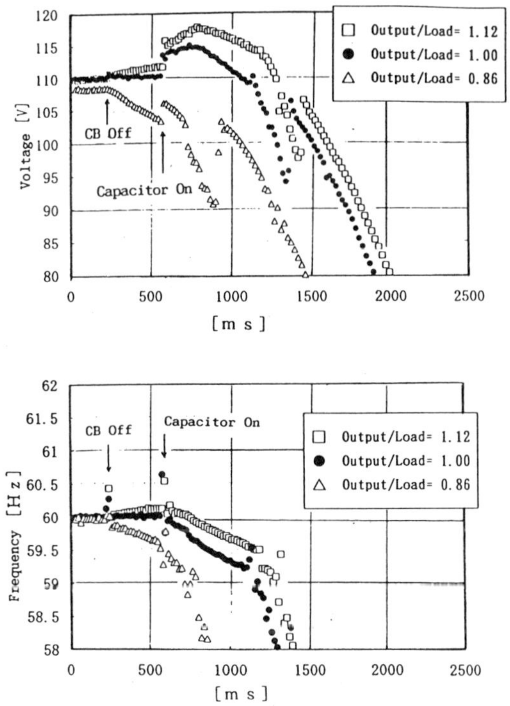

Fig.5. Characteristics of Capacitor Method

DISCUSSION OF THE TEST RESULTS

Following is the equivalent circuit of a grid during islanding (see Fig.6).

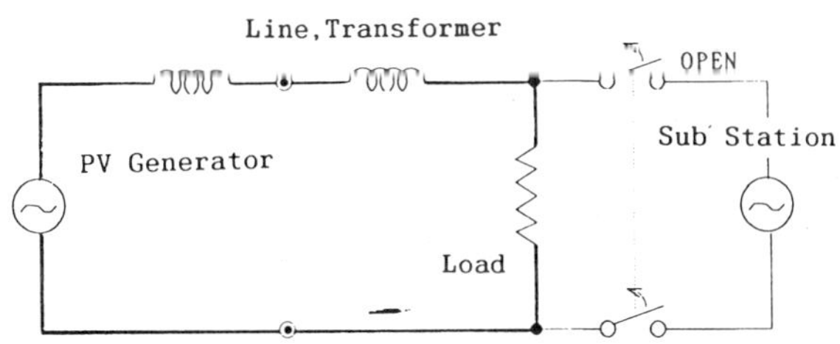

Fig.6. Equivalent circuit of a grid during islanding

If reactor is inserted into this circuit, voltage drops and, frequency rises as a result of decreased power consumption of resistance load (see Fig.7).

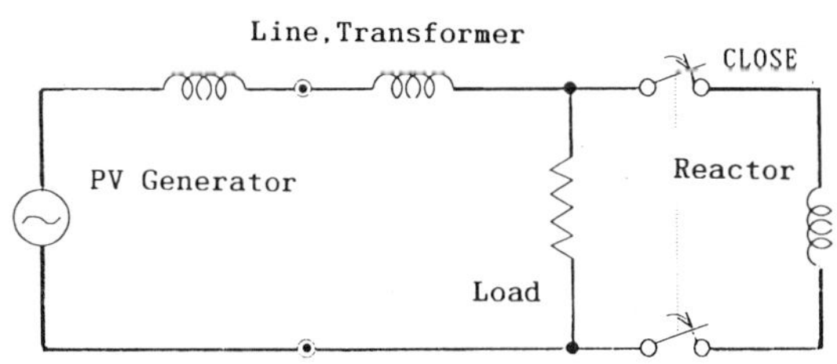

Fig.7. Equivalent circuit of Reactor insertion

If capacitor is used the reaction is the opposite (see Fig.8)

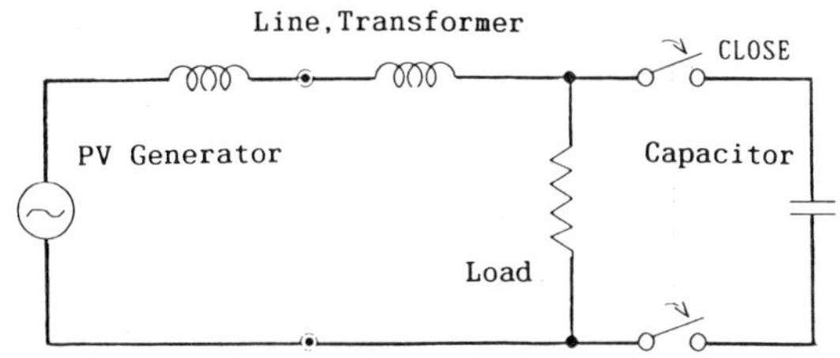

Fig.8. Equivalent circuit of Capacitor insertion

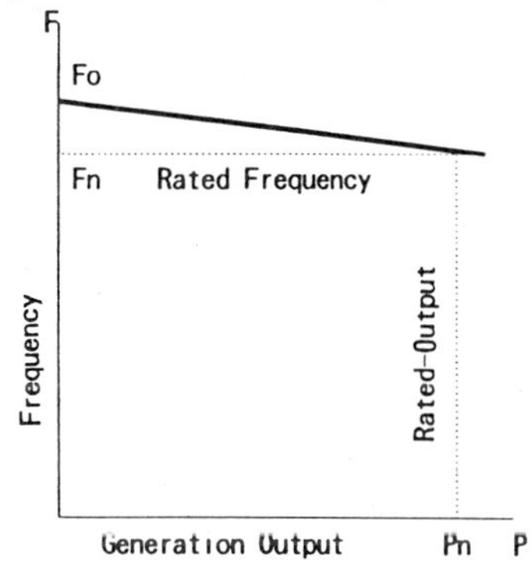

Fig.9. Relationship between Output and Frequency

Following is the mathematical expression related to the frequency changes. Assuming that the power characteristics produced by PV systems closely resemble those of rotating motors (fig.9 relationship between generating output and frequency) the following expression can be established:

$$\delta = \frac{F_o - F_N}{F_N} \times 100(\%) \qquad (1)$$

where F_o : Frequency of no-load operation
F_N : Frequency of rated-output

$$-\frac{\Delta P}{\Delta F} = \frac{P_N}{F_o - F_N} = \frac{P_N}{F_N \delta} \times 100(\%) \qquad (2)$$

where ΔP : Output variation of generator
ΔF : Frequency variation
P_N : Rated output of generator

$$-\frac{\Delta P}{\Delta F} = \sum_{i=1}^{n} \frac{P_{Ni}}{F_{Ni}\delta_i} \times 100 = K_G \qquad (3)$$

where P_{Ni} : Rated-output of i generator
δ_i : Speed regulation of i generator
F_{Ni} : Rated-output frequency of i generator
K_G : Frequency characteristics of generator in the grid

Load in the experiment includes rotating load (see Fig.10)

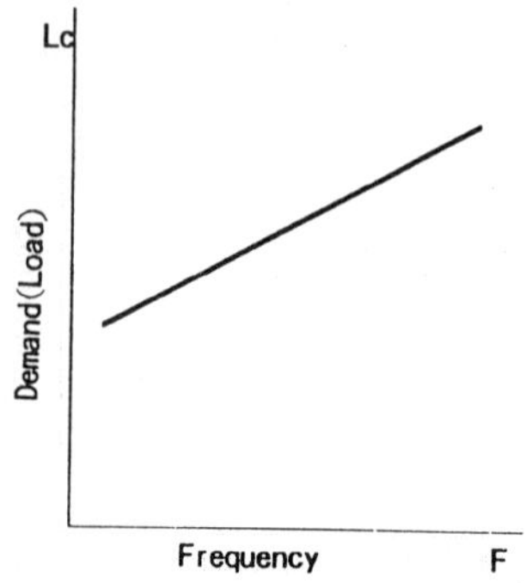

Fig.10. Relation between Load and Frequency

Resultingly, based on the electrical characteristics the following expression can be established.

$$\frac{\Delta L_c}{\Delta F} = K_L \qquad (4)$$

where ΔL_c : Variation of load consumption
K_L : Characteristic of Load and Frequency

In a grid during islanding, the total consumption and the output power of generator increase by ΔL and ΔP_G, respectively, while the consumption load decreases by ΔL_C. In this condition the ΔL can be expressed by equation (5).

$$\Delta L = \Delta P_G - \Delta L_c \qquad (5)$$

At the same time frequency of the grid also changes by ΔF. The following is the expression describing these frequency changes when the relative power of the generator decreases and its relation to equation (3)~(5).

$$-\frac{\Delta L}{\Delta F} = K_G + K_L = K \qquad (6)$$

where K shows the system constant of power-frequency characteristic.
The mathematical expression as mentioned above suggest that frequency falls if the power consumption of resistance load increase, and frequency rises if consumption decreases.

CONCLUSION

To suppress islanding, capacitor and reactor insertion was suggested as method performed on the utility side. These two types of suppression methods were tested, using the modeled grid and 100 sets of 2 kW PVs, on Rokko Test Center for Advanced Generation System.
The tests proved the following:
If capacitor is inserted into the substation side of islanding part of the utility grid, voltage increases and frequency falls.
If reactor is inserted, however, frequency rises and voltage drops. In the case of the 100 sets of PVs both methods, mostly on the basis of detecting irregular frequency, facilitated the suppression of islandings.
The process leading to suppression can be described in general power supply theory.

ACKNOWLEDGMENT

This study was carried out by the Kansai Electric Power Co.,(KEPCO) and Nissin Electric Co., under funding by the New Energy and Industrial Technology Development Organization (NEDO).
We thank all persons involved with this work.

REFERENCES

[1]. A.Kitamura, et al., "Islanding Prevention Measures: Demonstration Testing at Rokko Test Center for Advanced Energy System", *23 rd IEEE PVSC* ,1993.

[2]. A.Kitamura, et al., "Demonstration Test on PV-Grid Interconnection at Rokko Test Center (Reclosing of Distribution Circuit During Islanding)", *PVSC-7,* 1993.

DEVELOPING, MANAGING AND ASSESSING A UTILITY SPONSORED, INDUSTRY SUPPORTED PV OFF-GRID PROGRAM

Wayne R. Gould, Manager of Photovoltaic Applications
Southern California Edison Company

ABSTRACT

Working in conjunction with the Cal PV4U Working Group, the Southern California Edison Company (Edison) developed a unique off-grid pilot program in which Edison offers individuals 15-year photovoltaic (PV) service contracts.

This presentation will (1) describe Edison's program, (2) trace the steps which were taken to develop the program in a collaborative process comprised of a diverse set of stakeholders, (3) provide an evaluation of the program's experience since receiving regulatory approval, and (4) describe how an off-grid pilot program can accelerate the commercialization of PV technologies and provide the foundation for the future integration of PV technologies on the utility system.

PV UTILITY MARKETS

Photovoltaic technology is not new--It has been utilized in both space and in remote terrestrial applications for years. It is a dependable technology which has provided a primary power source for spacecraft and remote telecommunication applications for decades. It is a safe and environmentally benign technology whose active ingredient, silicon, is as common as sand and computer chips and is in great abundance in the world.

PV sales volume is increasing at 20-25% a year, and prices of PV systems are decreasing at an annual rate of 9%. PVs are commercially competitive with other energy sources in remote locations today, and a continuation of these trends may make residential rooftop systems commercially competitive in urban applications within the next five years. What is needed to accelerate the commercial deployment of this technology is (1) greater market development, (2) system standardization/ certification, and (3) increased financing.

The utility industry has been slow to recognize that there is a significant amount of customer interest to purchase or contribute to the installation of PV systems. Nonetheless, many utilities such as the Sacramento Municipal Utility District, Idaho Power and Light, Sierra Pacific, City of Austin, Texas, and Edison itself have been active in PV demonstration programs and are entering into introductory pilot programs. Currently, some 86 municipal, private, and cooperative utilities have formed the Utility Photovoltaic Group (UPVG). The UPVG was founded in 1992 with the support of the U.S. Department of Energy. Its main purpose is To build Technology Experience to Accelerate Markets in Utility Photovoltaics (TEAM-UP). In addition to the UPVG, there is the national organization of PV For Utilities State Working Groups which, like the UPVG, has as its goal the acceleration of PV commercialization through the utility sector.

Edison's involvement with PVs has been extensive and began 16 years ago with the installation of the solar-powered fountain in its general office courtyard. Since that time, Edison has demonstrated many PV applications. Among these are the 10 kw Desert Studies Center near Baker, California; the 24 kw SCAQMD solar chargeport; and recently, Edison was awarded a $2.5 million DOE contract to test the applicability of using PVs to relieve existing circuits which have reached their loading limits and thus defer costly circuit upgrades. Most recently, Edison has received CPUC approval to conduct a three-year, 1 MW pilot program to test the feasibility of a utility-financed off-grid PV service.

Edison's new off-grid PV pilot program offers a customer a cost-competitive option to installing an uneconomic extension of an electric distribution line. The program's PV systems can be designed to meet the customer's needs and are portable for easy installation. Edison's motivation in pursuing this activity is to (1) be more responsive to its customers' interests, (2) attract new customers, and (3) retain existing customers.

THE DEVELOPMENT OF A PV PROGRAM THROUGH A COLLABORATIVE PROCESS

In September of 1992, representatives from the California Public Utilities Commission (CPUC), the California Energy Commission, consumer groups, the solar industry, environmental groups, and California utilities formed the California State Working Group of the Photovoltaics for Utilities Project (Working Group). The Working Group is chaired by Mike DeAngelis of the California Energy

CH3365-4/94/0000-0763 $4.00 © 1994 IEEE
First WCPEC; Dec. 5-9, 1994; Hawaii

Commission, and its purpose is to accelerate the commercialization of PV in the utility sector.

As part of the Working Group activities, Edison volunteered (January 1993) to sponsor an off-grid pilot program. Throughout 1993, workshops were held in which Edison's current pilot program was proposed, debated, and developed. It was not an easy process. At each of the workshops, members of the Working Group freely voiced their support for, or opposition to, aspects of the program. It came as something of a surprise that there was a minority but vocal segment of the PV industry that opposed utility participation in the PV market. The chief and ongoing concern of this group is that utilities will dominate the marketplace and drive out the smaller service providers.

California Law requires that before a regulated utility or an affiliate can engage in commercial PV activities, it must obtain CPUC approval. The approval criteria include that (1) the PV market remain freely competitive and (2) the utilities' participation must accelerate the commercialization of PV technologies. Edison submitted its pilot program for regulatory approval in November of 1993; and although it enjoyed the majority support of the Working Group, there were protestors who feared utility domination of new markets. The CPUC approved the program, with some qualifying stipulations, on April 6, 1994.

The collaborative process for the development of Edison's pilot program was not easy. And not everyone is pleased with the outcome. However, the resulting program and the lessons learned are very much worth the effort. Edison recommends other utilities to undertake a similar approach should they wish to enter into the PV marketplace among and with the support of independent service providers. Further, Edison continues to look towards the Working Group for guidance in modifying its existing off-grid programs and in the development of its future on-grid programs as well.

EDISON'S PV OFF-GRID PILOT PROGRAM

Edison's program is a pilot program. In approving the program, the CPUC included significant limitations: (1) the program was limited to three years' duration or until 1 MW of PV system is sold, (2) each system sold must exceed 1 kW, (3) each system must be individually and competitively bid, (4) before beginning the marketing of the pilot projects, Edison was required to perform a market survey to provide the baseline for determining whether or not Edison's pilot program is successful and complies with the approval criteria associated with California law.

The program was designed to make PV technologies available to the customers in a manner that benefits everyone--the customer, independent contractors, and Edison. To accomplish this, our role is one of facilitation, integration, and financing. In this program Edison does not compete with independent contractors. Instead, Edison manages the procurement, installation, and maintenance activities of PV systems. As such, It's a win-win arrangement. Homeowners and businesses in remote locations receive clean, quiet electricity based on a low bidder's cost. Independent contractors acquire new construction projects, and Edison is able to serve new customers who otherwise would not have dependable power. Here's how the process works:

1. A remote customer calls Edison and requests electric service. If the customer is located beyond the economic reach of existing utility wires, Edison will suggest PV as an alternate to conventional service.

2. If the customer is interested, Edison will conduct an energy survey and receive permission to conduct a standard credit check.

3. Upon determination of credit worthiness and with a $500 deposit, Edison will send a bid package, based on the customer's energy survey, to a group of licensed, experienced contractors that have been pre-qualified by Edison to perform the work under the program.

4. Following bid evaluation, Edison will review with the customer the three lowest qualifying bids. At this point the customer will decide whether to proceed with the procurement and installation of the PV system by selecting one of the three bids, or whether to discontinue the process. If the customer determines to go ahead with the project, the $500 deposit will be credited toward the initial monthly payments. Should this process end without a system being installed, half of the $500 deposit will be refunded.

5. From this point, Edison will manage the installation and maintenance of the system for the customer for a 15-year period. At the end of 15 years, the customer will have the option of owning the system at no additional cost, or he may choose to have the system removed. Additionally, the customer may choose to purchase the system at any time during the term of the agreement for its depreciated value. Also, the customer may choose to terminate the Agreement upon 30 days' notice and the payment of termination costs.

Monthly energy and maintenance payments are 1.6% of the installed system cost. A 1 kw PV energy system may range from $150 to $300 per month, depending on the customer's backup power options.

SUCCESS TO DATE

The baseline market survey was completed in mid-September 1994; therefore, our active marketing has been under way for only two months. We have had approximately 250 inquiries in our off-grid program. To date we have made 12 site visits and have 5 projects in the bid phase.

Although Edison has yet to install its first PV system, the program has had many significant successes and has advanced the commercialization of PV technologies significantly. Some of these successes include:

1. The regulatory approval of Edison's three-year off-grid PV pilot program itself--demonstrating the CPUC's support for the utilization of PV technologies and the involvement of utilities in this new and emerging market.

2. The development of a collaborative working group, comprised of all the PV stakeholders, whose object is to review, monitor, and assess the progress of the pilot program. This group meets semi-annually and held its first meeting on November 17, 1994 with 57 people in attendance.

3. The collaborative working group has initiated the development of technical system standards, designed to allow for residential ownership through compliance with local building codes.

4. The increased interest in obtaining component and system certification by national certification labs such as UL (e.g., Underwriters Laboratories).

5. The increased national attention which Edison's PV programs have created in the PV industry, the utility industry, the construction/developer community, and in the regulatory arena.

UTILITY-SPONSORED OFF-GRID PROGRAMS CAN ACCELERATE PV DEPLOYMENT

It was earlier noted that the successful commercial deployment of PV technologies requires (1) accelerated market development, (2) system standardization, and (3) a source of ready financing. Though modest in current impact, Edison's off-grid program joins in the contributions of other utilities in the satisfaction of these three requirements.

Edison fully anticipates the successful sales of 1 MW of off-grid systems through its pilot program. Although this amount of sales may not be considered as significant on a global basis, it is fully expected that Edison's off-grid programs will be emulated by other utilities. This cumulative increase in demand is expected to create greater sales volumes which in turn will increase the manufacturing capacities necessary to lead to greater manufacturing efficiencies and lower costs.

Edison's efforts, in conjunction with those of others, to standardize and certify system components is leading to the development of PV technologies which will meet the strict building code requirements in off-grid and on-grid applications.

Edison's current pilot program is largely based upon Edison's willingness to finance the individual ownership of PV systems. To our knowledge Edison is the first major corporate entity to offer comprehensive financing for the deployment of PV technologies. Since the implementation of our program, banks and other lending institutions have shown greater interest in this market, and some have announced active lending programs for the financing of PV systems.

SUMMARY

Edison is proud of its PV Off-Grid pilot program and continues to pursue the accelerated commercial development of PV technologies in its off-grid, on-grid and utility system markets. Our pilot program is allowing us and the PV industry to develop the technology which will ultimately be deployed in on-grid applications. Some may wonder why large utilities should want, or in the case of California utilities, be allowed to participate in these new and highly competitive markets. The answer is easy and straight-forward. We wish to continue our proud heritage of providing energy services to our existing and potential customers. By assisting in the development of this emerging technology, we intend to earn the support of our customers, the PV industry, and thereby establish our right to be active participants in this marketplace. We encourage all utilities and PV industry participants to join us in the successful deployment of this wonderful technology.

MOTORWAY SOUND BARRIERS: RECENT RESULTS AND NEW CONCEPTS FOR ADVANCEMENT OF TECHNOLOGY

Thomas Nordmann TNC Consulting AG, Asylstrasse 84/P.O. Box 152, CH-8708 Männedorf / Switzerland and
TNC Energie Consulting GmbH,Oltmannsstrasse 5, D-79100 Freiburg / Germany
Prof. Dr. Adolf Goetzberger, TNC Energie Consulting GmbH,Oltmannsstrasse 5, D-79100 Freiburg / Germany and
Fraunhofer-Institut für Solare Energiesysteme, Oltmannsstrasse 5, D-79100 Freiburg / Germany

ABSTRACT

Since the first grid-connected 100kW photovoltaic plant on sound barriers along the N13 went into operation in autumn 1989 (Mark I, see Fig. 1), comprehensive monitoring data hase been collected and analysed. In over fiveyears of operation, the plant fed an annual average of 110'00kWh into the grid. The average performance ratio was 80% [Ref. 1]. Findings concerning this plant and a subsequent project located along the railway line in Canton Ticino (Mark II, 100kWp, in operation since autumn 1992) are being systematically analysed [Ref. 2].

A further step towards the improvement of the economic viability of photovoltaic installations associated with sound barriers can be seen in the integrated approach to building these. In the framework of a long-term research and development project in Switzerland and Germany, the possibility of integrating photovoltaic units into sound barrier components will be explored.

1. THE AIMS OF THE PROJECT ARE:

1. Evaluation of two new concepts
   a) Bifacial modules on north-south oriented sound barriers.
   b) Improvement of module-surface concerning reflections and dirt.
2. Measurement of the PV yield in relation to location, positioning of the modules and level of dirt (test-installation in Switzerland).
3. Development of a map of potential sites in the Federal Republic of Germany.
4. Launching of an international „ideas competition" in co-operation with the sound barrier industries and photovoltaic industries.
5. Building of several prototype 10kW plants.

Figure 1: The first grid-connected 100kWp-photovoltaic plant on sound barriers along the N13 in Switzerland.

CH3365-4/94/0000-0766 $4.00 © 1994 IEEE

Analysis of the findings of this project should make it possible at a later date to plan and consruct several 100kW PV/Motorway-installations dealing with integration of sound barriers and photovoltaic.

2. NEW CONCEPTS WILL BE INTRODUCED IN FUTURE PROJECTS

The first stage deals with the problem of orientation of the motor ways. So far only those routes oriented east-west were deemed suitable as PV-substrates. A much larger potential, however, could be utilised if north-south and intermediate directions could also be used. For this purpose a new concept has been developed by TNC. It is based on modules mounted vertically on top of a sound barrier structure. The modules best suited for this purpose contain bifacial solar cells. These cells have been proposed and manufactured some years ago and now manufacture is resumed by the ASE-Company based on MIS-cells.

For a north-south orientation the yearly energy yield can easily be estimated from solar irradiance data developed for building facades. Assuming a bifacial module with equal efficiency on both sides one can simply add the energy yield of the east and west face and thus finds that the annual yield is 106% compared to a regular module mounted in a south orientation with optimal inclination. This means that a bifacial array mounted vertically in the east-west orientation can even out perform a conventional array in an optimised setting. The extension of the technical potential is quite remarkable: An estimate for Switzerland shows that the potential along presently realised east-west routes is 45 MW. If all the other parts of the national motor way system that are technically suitable (50% of total length) are included, then the potential increases to 300 MW.

Figure 2: Application of Bifacial solar-modules on north-south oriented sound-barriers

The second improvement now in planning refers to reduced reflection losses of the glass surfaces of modules. Presently 4% of the perpendicular and 17% of the diffuse light is lost by reflection at the air-glass interface of a module. In a detailed investigation it was demonstrated that by a special structure involving saw-tooth ridges in the north-south direction the integrated reflection loss is reduced by 10% [Ref. 3]. The ridges were designed not only to minimise reflection-losses but also to avoid accumulation of dirt. A short-term experiment with a small sample has shown that no soiling takes place. For a first study in sound barrier systems a selected number of modules will be equipped with glued-on plastic saw-tooth film and compared with an adjacent reference area.

3. MONITORING AT THE MOTORWAY

Monitoring of the Mark I,100kW power plant along the motorway has shown that the level of dirt on the PV modules has a slight influence on the yield. Dust from the street and other forms of dirt are regularly and throughly washed away by rain. The extent to which this also happens if the modules are mounted differently (height and angle) will be examined in the comming months.

Figure 3: Mounting of the modules for the PV/sound barrier monitoring project: Four groups vertical, at different levels and three groups slanted at an angle.

In a project undertaken by TNC Consulting AG, with funding from the Swiss National Energy Research Fund NEFF, monitoring equipment has been installed at the above-mentioned plant. The influence of dirt along the motorway on the yield of the PV modules is now being measured. In order to do so, groups of vertically-mounted modules at different heights are compared to groups of modules which are positioned at the original angle of Mark I (see Fig. 3). The findings from this monitoring process will supplement the technical requirements of the submission which will serve as a basis for the international idea-competition, to be launched subsequently (spring 1995).

At the west end of the PV plant Mark I, a group of modules has been replaced by a mounting frame, which comprises four groups, each with four modules. These groups of modules are vertically mounted and serially wired.

The original neighbouring group, slanted at an angle, has been rewired, in the same way as the others, so that four modules are connected in series. In the long-term monitoring, the short-circuit current of the seven groups will be compared to three ESTI reference cells which are mounted in the corresponding plane.

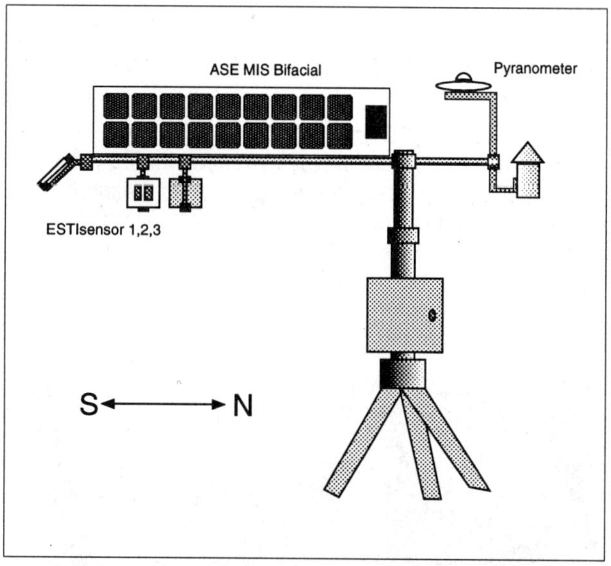

Figur 4: Measurement installation with bifacial module

4. MEASUREMENT WITH BIFACIAL PROTOTYPE

A bifacial prototype was produced by ASE in Germany. This module was integrated into a measurement installation that is used to investigate on an experimental basis the potential of bifacial applications (see Fig. 4). Three ESTI sensors are mounted on a tripod. They measure the irradiance on the surface with the following orientations:

- East, vertical
- West, vertical
- South, angled at 45°

The bifacial module is operated at MPP by a controllable load. The global irradiance and the ambient temperature are also recorded.

5. FIRST RESULTS OF THE EXPERIMENTAL INVESTIGATION

The initial measurement results for the bifacial measurement installation (see Fig. 5) points out the principle of an application with dual-sided activ PV-modules. The daily profile shows the east and west components of irradiation on a bifacial module and compares their sum with the south-facing plane. The daily total on a good november day (low sun) is highest for the south-facing plane. However, it is

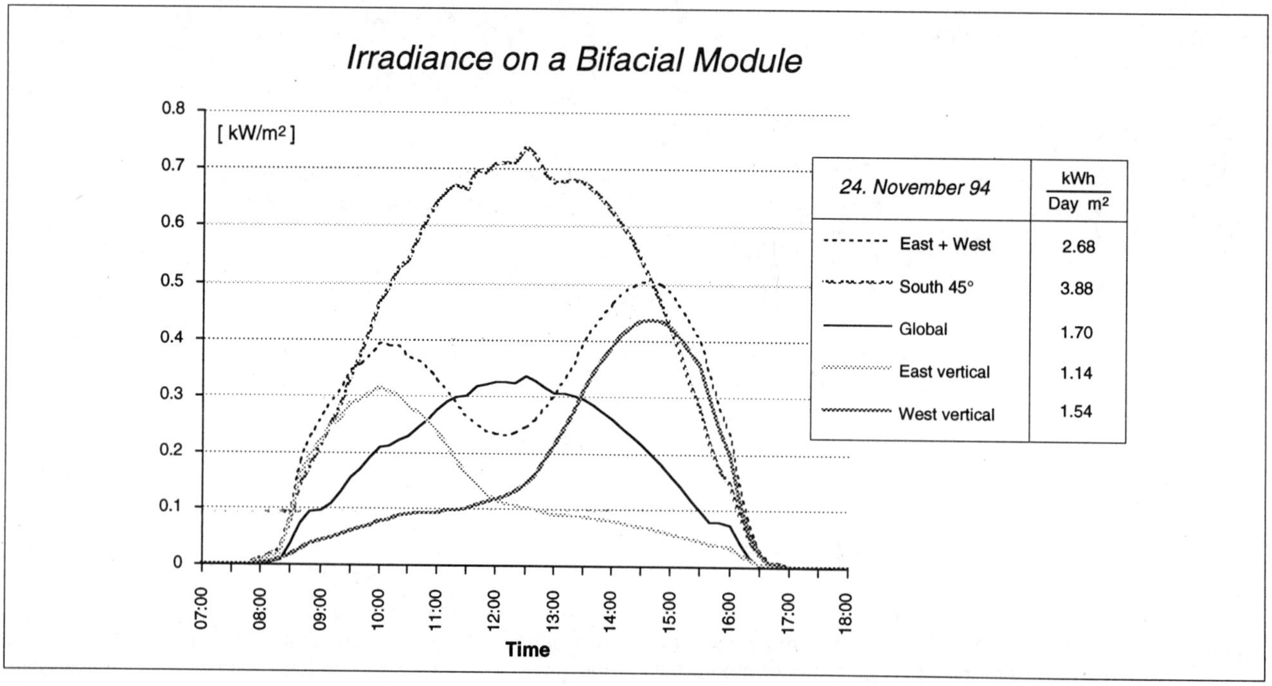

Figure 5: A day-profile of the irradiance on a bifacial module compared with the south-facing plane

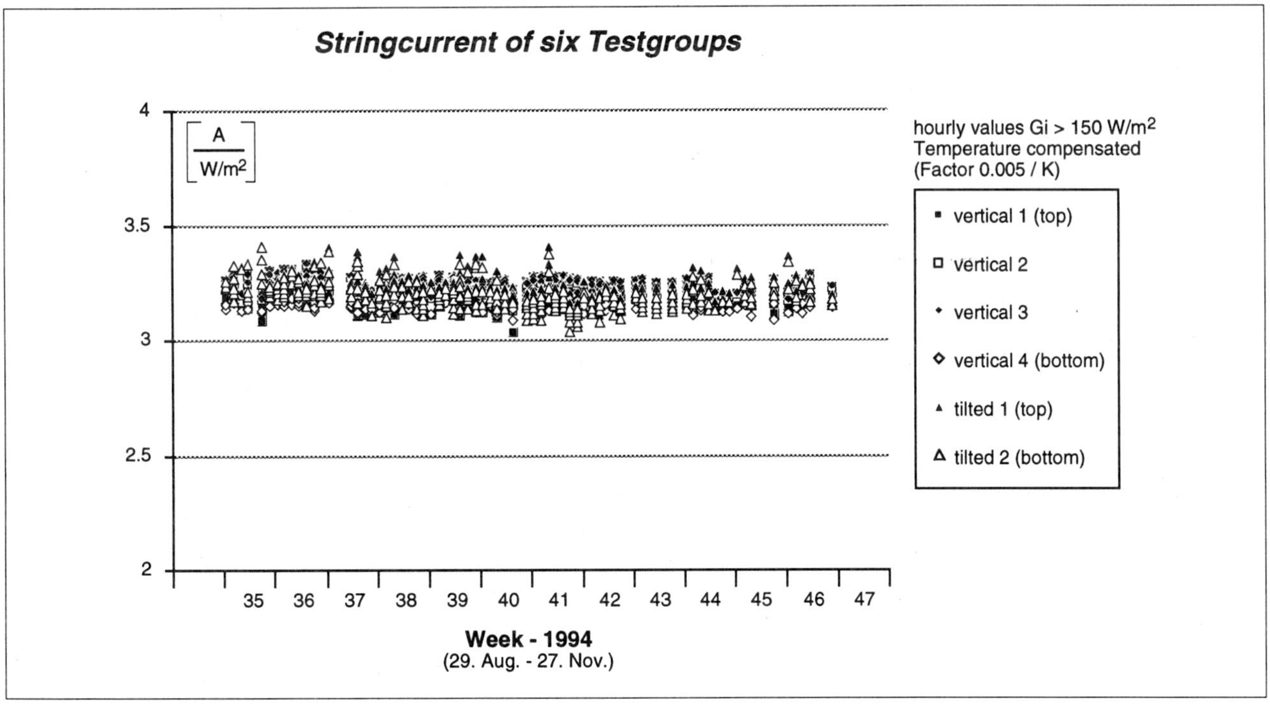

Figure 6: Hourly values of the stringcurrent of the test-module-groups along the motorway

anticipated that during the summer months, the values for the east/west sum will be higher.

With foggy or cloudy weather, the yields for the south-facing plane, the east/west sum and the horizontal plane are all roughly the same.

The following conclusion can be drawn from measurements of module groups along the motorway:

1. After four month of measurements, dirt levels on the module surfaces are not significant (see Fig. 6).
2. The difference in irradiation between the highest and the lowest of the vertically mounted groups is only 1-3%.
3. There are no signs of damage to the modules.

However, the interesting period for these measurements is just beginning. Immission levels on the motorway are at their highest in winter and spring. Dirt on the surface of the modules caused by traffic on wet and snow-covered roads and by pollen in the spring will be shown in measured results. Winter traffic conditions will also reveal the mechanical resistance of the vertically mounted modules to snow clearing operations.

REFERENCES

[1] «Prediction and Effective Yield of a 100 kW Grid-Connected PV-Installation» L. Clavadetscher and Th. Nordmann, Solar Energy Vol. 51, No. 2, pp. 101-107, 1993.

[2] «Evaluation of Grid-Connected PV-Installations» L. Clavadetscher and Th. Nordmann, 12th European Photovoltaic Energy Conference, Amsterdam, 1994.

[3] A. Scheydecker, A. Goetzberger, V. Wittwer: Proc. 10th EC PV-Conference 1991, p.39.

BUILDING INTEGRATION OF AN AMORPHOUS SILICON PHOTOVOLTAIC FACADE

Heinz A. Ossenbrink, Luigi Rigolini
European Commission DG XII - Joint Research Centre Ispra Establishment, Italy
European Solar Test Installation (ESTI)

Oussama Chehab, Olaf van der Venne
FlagSol Flachglas Solartechnik GmbH, Köln, Germany

ABSTRACT

A large photovoltaic facade was retrofitted to an industrial building complex of the Joint Research Centre. Amorphous Silicon technology was chosen to cover the maximum possible area at a minimum of costs. The a-Si Substrates are laminated within a double-glass facade element. Total area of the facade is 770 m², probably the largest PV facade presently in operation. Total power output fed into the grid is approximately 25 kWp after stabilisation. Emphasis was put on the architectural design and the visualisation of the function of the system to a wider public by a interactive user-display.

Fig.1 View of the a-Si Photovoltaic Facade at JRC Ispra, with 505 m² active area probably world-wide the largest PV-Facade

First WCPEC; Dec. 5-9, 1994; Hawaii

INTRODUCTION

The Joint Research Centre is currently investing in a pilot project, named ECOCENTRE Ispra. It aims to the application of modern technologies for the modernisation, under ecological and economical aspects of ageing premises of research centres, where results are intended to be transferable to other sites. Among other projects, the demonstration of photovoltaic technology in its application to a building facade was chosen as an example of future renewable energy technologies.

Larger Photovoltaic systems as currently installed in grid-connected applications suffer from relatively high costs for substructure, mounting and land use. These costs may contribute to the total systems costs with more than 30% and are mainly related to expenses which will probably not decrease by technology or mass production. An opportunity to solve this issue which is critical to the success of photovoltaic solar energy on a broader scale is to hide these costs in applications where they would occur in any case. Apart from utilising for instance highway sound-barriers or roof-tops, integration into building facades is a very promising solution. When compared to a traditional glazing, the additional photovoltaic use of the faced may cost only some 20% more for modules, cabling and inverters. Moreover, when designed well in terms of aesthetics, photovoltaic facade elements provide an opportunity for architects, to apply building components which are well in line with environmental concerns and pleasing for the eye.

OBJECTIVES

When initiating the photovoltaic facade project for the ECOCENTRE, the objectives were set out as follows:

- Apply photovoltaic solar energy on facades to demonstrate architects how to use a renewable energy source as an additional design element
- Keep area related costs are lower in comparison to a conventional photovoltaic array installed on plain ground
- At a later stage, make use of thermal heat recovery
- Improve the appearance of the building.

REALISATION

Within the frame of the ECOCENTRE project emphasis was put on visibility and design features, rather than maximum power output.

As the amount of financial resources was limited, the main design constraint was the lowest possible cost per area. Only amorphous silicon technology could be considered feasible under this condition without sacrificing the goal to cover the whole available area of the southern building facade.

Dimensions

The dimensions of area foreseen for the applications were 65 m wide • 25 m height; only the top half of the building was available for photovoltaic use, due to an office building located in front. Total area available resulted to be 61•12.6 = 770 m². Taking into account the uncovered window areas, framing losses, non-optimal distribution of modules, initial degradation and an minimum efficiency of 3.5 %, the design peak power was 21 kWp. Accordingly, the grid-connected inverter were designed for this load, in four blocks of 5 kVA each.

Array

The facade modules (Optisol®) were made from a-Si substrates delivered by Advanced Photovoltaic Systems (APS), Princeton (NJ), which were incorporated by a proprietary process in a double-glass structure by Flachglas Solartechnik, Köln (FlagSol). The dimension of one module is 0.82 • 1.58 m². A total of 420 modules were fitted to the facade substructure, yielding a module area of 544 m². As for wiring optimisation 4 of the modules were unconnected, the active area is 505 m², wired as to 8 modules in series and 52 in parallel for the DC-busbar of 468 V open-circuit voltage and 107 A short circuit current.

Subconstruction

Aluminium profiles as used in building industry were attached to internal H-bars of the building and provided the mounting and cabling structure for the PV modules. A clearance of 14 cm was kept between the modules, and the existing insolation cover material of the building to allow later thermal use of the facade. As the total weight of the retrofit is approximately 20 tons, additional reinforcement of the building was necessary. Fig.[1] shows the building with all modules installed.

Inverter

The four inverters are identical and operating in a master-slave configuration. The master controls the operation of the remaining units for optimisation of the overall inverter efficiency at lower irradiance conditions. In addition to the redundancy in case of failure, the maximum-power tracking of each inverter can be synchronised for the operating point of 330 V and 91 A at 1000 W/m² and 25°C.

Monitoring System

A dedicated data monitoring and visualisation system was installed to measure the relevant parameters of the photovoltaic facade. Apart from the capability to remote control the system, emphasis was put on a interactive visualisation of the operation of the plant as the target group of the demonstration plant probably are not experts in the field. This visualisation is realised by a touch-panel computer system inside the building, where visitors interactively can discover the photovoltaic facade and recall a number of real-time and stored performance data. At the building front, a huge display shows the instantaneous power fed into the grid and the total kWh produced since the start-up of the plant.

The data monitoring system records irradiance in the horizontal, 45° and vertical plane, as well as the module temperature at the bottom, in the centre and the top of the facade. Horizontal irradiance is measured by a K&Z CM11 pyranometer, the other two planes of irradiance by ESTI-Sensors. An ultrasonic device records wind speed and direction.

MEASUREMENT RESULTS

Acceptance Test

An acceptance test was performed to verify the installed power. The current-voltage characteristics 6 main strings, were measured by the capacitor-type IV-tracer developed at ESTI and corrected to Standard Test Conditions of Air Mass 1.5 (IEC 904), 1000 W/m² and 25°C. In correcting the measured data, reference made to laboratory measurements between 25°C and 65°C, and 200 W/m² and 1000 W/m², as the method of IEC 891 seemed to be not reliable enough for a-Si technology over this wide range of conditions.

As demonstrated in Fig.[2], the initial performance of 30 kW exceeded the design values, as expected. This corresponds to an DC efficiency of 6 % and includes cabling and wiring losses.

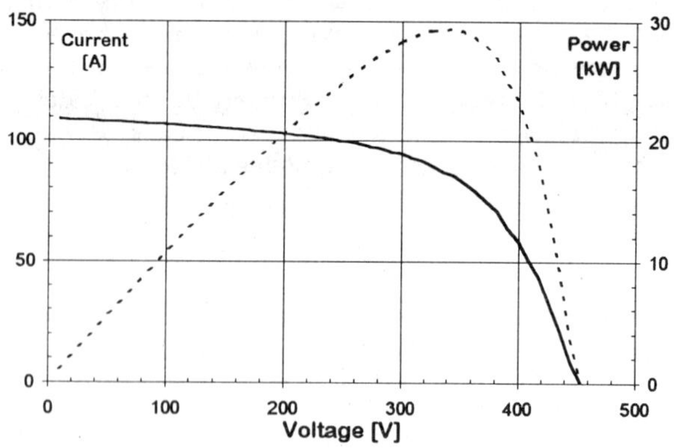

Fig.2. Acceptance Test: Current-Voltage and Power-Voltage characteristics of the 416 modules

Performance

The facade is in operation since July 27th, and has since then produced up to Nov. 15th 6.8 MWh of electric energy. The raw data of the monitoring system measured in 5-min intervals are transferred by remote calls through modem lines and off-line averaged to 1 hour values. For the period in operation, the performance ratio (Energy delivered relative to STC power, PR) was calculated in weekly intervals, as shown in Fig.[3]. It can be seen that after a very high DC-PR of 80% at the beginning of operation, the values drop to 65..70% later in the year, clearly verifying the light-induced degradation found in a-Si technology. We hope to verify the annealing effect, which should recover 5%..10% of the degradation with higher module temperatures (about 65°C) during the next summer. This should result in an average PR of 70..75% throughout the year. It should be noted that an AC performance ratio of 70% corresponds to the contracted power.

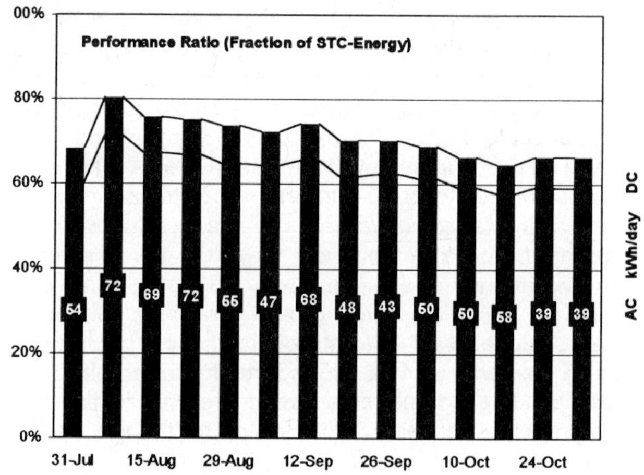

Fig.3 Performance Ratio of AC and DC Energy, Weekly Data.of the facade-array.

Inverter Efficiency

The figure gives also an indication of the inverter losses, which are, based on hourly values depicted in Fig.[4].

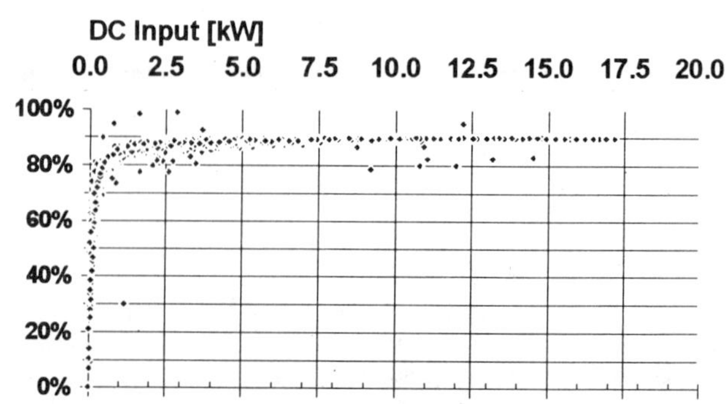

Fig.4 Inverter Efficiency, hourly values

The inverter efficiency at higher input power is about 90% and not outstanding high, but efficiency at low input power is very remarkable. Already at 100 W DC an efficiency of 60% is reached. Analysing the data in detail as in Fig.[5], it was found that the inverter starts up already at input power below 50 W, corresponding to 0.25% or 2.5 W/m² of in-plane irradiance. In fact, the monitoring data show energy production at moonlight!

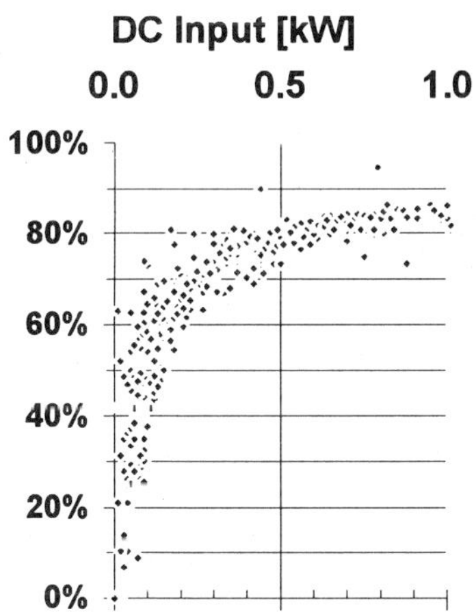

Fig.5 Inverter Efficiency, enlarged DC Input scale

Reliability

The system is since the start-up in continous operation with no failures. In September and October a total of 7 thunderstorms with lightning strokes in the region caused the stand-by mode of the inverters, when the utility grid failed. After return of the grid the inverter took over to normal operation. No shut-down by lightning strokes on the DC (Module) side of the inverter was observed.

NEXT STEPS

The remaining window areas are foreseen to be covered by semi-transparent a-Si modules during the next year, to gain experience with these devices and to improve the appearance. To assess the lifetime of the substrate-resin-glass compound, two sample facade-modules are presently exposed at the ESTI-facilities a long-time damp-heat (85°C and 85% relative humidity) tests which will last up to visible damage of the samples. This should allow ESTI to establish the acceleration constants for testing of such modules, if similar failure mechanisms could be found on the facade.

The facade modules have been mounted on its substructure such as to allow a chimney effect to occur between the modules and the isolation of the building wall. It is planned to make best use of this effect by heat exchangers on the top edge of the facade. Also the possibility of venting the building interior by placing air inlets in the wall is considered. For this goal theoretical modelling is presently undertaken, and a model of the facade will be tested in a wind-tunnel.

CONCLUSIONS

The goal to demonstrate a-Si technology as a design element for industrial building complexes has been achieved. The problem of retrofit and the related necessity of enforcing existing structures was underestimated at the beginning of the project. The experience with thin-film technology is so far positive, both from the aesthetics and the performance point of view. Since start-up, about 25 groups have visited the installation.

HIGH EFFICIENCY PHOTOVOLTAIC ROOF TILES WITH STATIC CONCENTRATORS

S. Bowden, S.R. Wenham, M.R. Dickinson and M.A. Green
Centre for Photovoltaic Devices and Systems
University of New South Wales
Sydney NSW 2052 Australia.

ABSTRACT

The use of static concentrators provides a level of concentration sufficient for significant cost reductions, without any of the problems of high concentration systems. There is no need for expensive tracking, most of the diffuse light is collected and passive cooling can be used for the cells. Following the construction of prototypes that confirm computer ray tracing predictions a structure has been designed for the manufacture of a 1 kW roof top array in association with the state power authority. Two manufacturing techniques for the array and potential commercial production are described. The pilot line at the University of New South Wales has commenced production of bifacial buried contact solar cells for the project. These cells provide the necessary high efficiency and low cost. Measurements have been made on the amount of dust likely to be present on a typical module, the effect on total internal reflection and thus the concentrator performance.

INTRODUCTION

The cost of current commercial modules is dominated by the cost of the solar cells. Indeed, in a typical module the cost of the wafers themselves is about 50% the total module cost. [1]. Thus a real reduction in wafer costs is necessary if the cost of photovoltaic power is to come down. While various technologies exist for reducing the cost of the silicon substrate these suffer from either low efficiencies [2] or are still in the early stages of development [3]. This has lead to the development of tracking concentrating systems which attempt to obtain a cost reduction by concentrating light onto a small area of high efficiency cells. The apparent advantages of high concentration tracking systems are offset by the cost of the tracking system, lower reliability, poor response to diffuse light and difficulty in installing the systems in building structures.

The use of static (non-tracking) concentrators with non-imaging optics provides a level of concentration sufficient for significant cost reductions, without any of the problems of high concentration systems. With low levels of concentration, there is no need for expensive tracking, most of the diffuse light is collected and passive cooling can be used for the cells. By using a three dimensional design for the concentrator with total internal reflection (TIR) and bifacial solar cells, the module can be made thin enough to fit within the dimensions of building structures such as roof tiles. Incorporating the cells into a roof tile further reduces the total system cost as the existing roof provides the support structure for the modules.

The concentrating scheme developed at the University of New South Wales (UNSW) using cheap lenses on either side of a strip of solar cells is shown in Figure 1. A glass coversheet forms the front surface of the structure as in a conventional module. The rear surface of the module is reflective and grooved, so light is directed towards the front surface of the module at an angle to the normal greater than the critical angle for total internal reflection. The light is now trapped inside the module structure and is steered towards the solar cells.

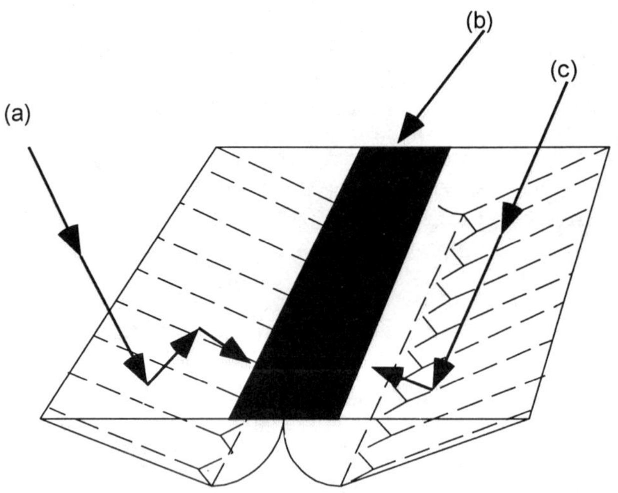

Figure 1. View of static concentrator showing the path of a number of light rays. (a) The light is reflected from the rear surface of the module undergoes TIR at the top surface and then strikes the cell. (b) The light ray enters the module and strikes the top of the cell directly. (c) The light is reflected from the module rear surface and then strikes the underside of the solar cell.

The grooves on the rear surface enable the module to be much thinner whilst maintaining the same

CH3365-4/94/0000-0774 $4.00 © 1994 IEEE

performance [4]. The photovoltaic cells cover only 25% of the module area, but the module still has an efficiency greater than that of a standard commercial module. The high module performance is obtained by using a very efficient collector design combined with high efficiency bifacial solar cells.

The module concentrates 2 suns of irradiance onto the rear surface of the cells and 1.5 suns of irradiance onto the front surface. The similar levels of illumination on both cell surfaces requires bifacial cells with similar efficiencies on the front and rear surfaces. The bifacial cells developed at the UNSW have demonstrated the necessary high efficiencies from both cell surfaces [5]. The pilot line production facility at UNSW has recently been converted from the production of monofacial cells to bifacial cells for static concentrators. The bifacial cells being manufactured are an advanced form of the laser grooved buried contact solar cell developed at UNSW, with potentially lower costs and higher efficiencies than the monofacial cells currently being manufactured by industry. The monofacial buried contact cells have been costed on commercial production to be substantially cheaper than screen-printed cells in \$/Wp[6].

1 KW ROOF TOP INSTALLATION

The ray tracing program used to design and analyse the module has been confirmed by initial prototypes[7]. Given the enormous promise of the technology indicated by these initial studies a 1kW roof top array is being installed with completion expected at the end of 1995. The array is being installed in association with the state power authority, Pacific Power. Each of the roof tile modules have an active area for photovoltaic collection of 750 cm^2. The primary purpose of the 1 kW array is to evaluate manufacturing techniques for full scale commercial production. For this project two manufacturing techniques are being implemented the first using glass lenses and the second a lamination technique with a pottant filler.

Glass Lenses

The first structure uses solid glass lenses on either side of the cell strip which are laminated to the glass cover sheet. Glass has several excellent properties for the production of static concentrators. The glass provides the necessary optical transmission for the module bulk and can be mirrored very easily and cheaply. The glass and the mirroring are very stable ensuring long operational life with no UV degradation. Glass also has a very high thermal conductivity so cell temperature increases are kept to a minimum (see below). Unfortunately shaped glass is not commonly manufactured in Australia although it is possible overseas. For this reason a second manufacturing technique is being developed.

Laminated mould

An improved method of construction [8] for the mass production of the static concentrator roof tile is shown in Figure 2. This consists of a plastic backing plate that has been silvered on the top surface. The plastic is moulded to form the necessary optical shape for the concentrator plus the tile surrounds needed for attaching the tile to the roof.

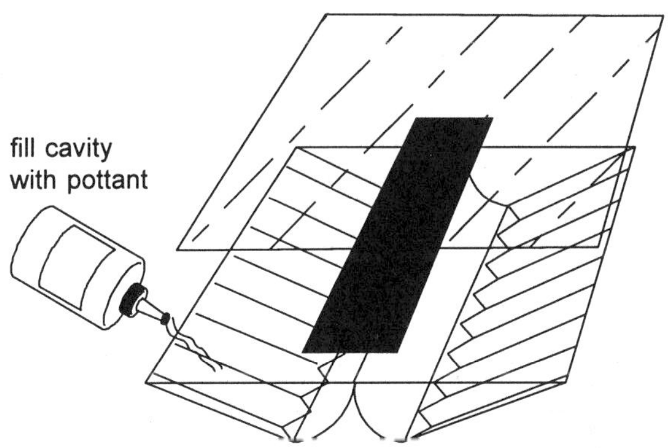

Figure 2. Manufacturing technique for the roof tile using a modified form of current commercial laminations.

A low iron glass cover plate forms the front surface of the module and a pottant material is injected into the cavity. The pottant material encapsulates the cell as in a standard module and also provides the optical transmission medium This method of construction is very similar to standard commercial laminating techniques, the only difference being the textured backing plate and extra pottant material required. Forming the structure in this manner allows the complete roof tile to be fabricated in one operation.

PERFORMANCE

Initial prototypes have indicated that the roof tile modules for the 1kW array will be 16% efficient. This allows for absorption and reflection losses, and imperfections in the moulds.

There are two critical areas for light loss in the static concentrator modules. The first is the reflection loss from the module rear surface. Each light ray strikes the rear surface an average of 1.4 time. However various silvering techniques can produce reflectivities of 90-95% keeping reflection losses to a minimum. The second area for light loss is absorption in the optical medium of the module bulk with each ray passing through an average of 6.8 cm of material. However, many materials exist with low absorptions (less than 0.05 cm^{-1}) so these losses can also be kept to a manageable level.

In most concentrator modules such as tracking systems the concentration of sunlight is increased but the area for heat dissipation is dramatically reduced. This causes heating of the cells and a corresponding drop in efficiency. For the static concentrator design there is no actual concentration of light onto the module surface. The incident irradiance and module area for heat dissipation are the same as a standard flat plate module. While the cells are subjected to 3.5 suns of irradiance the module bulk is able to conduct away the heat to other parts of the module if the thermal conductivity of the optical medium is high. For a material such as glass with a thermal conductivity of 1.2 $Wm^{-1}K^{-1}$ the thermal resistance between the cells and the module surface is dominated by the thermal resistance of the air/glass interface. Under these conditions the cells are not expected to run significantly hotter than in a standard module

EFFECT OF DUST ON TOTAL INTERNAL REFLECTION

The static concentrator design is dependent upon TIR for its operation. It is assumed in the ray tracing analysis that when the light strikes the top surface of the module at an angle greater than the critical angle then perfect reflection occurs. That is, the reflectivity is 100% and no light is transmitted out of the module. In practice there is a penetration of energy into the second medium, in the form of an evanescent wave. The evanescent wave falls exponentially in intensity as it moves away from the module surface. The penetration depth of the evanescent wave is also greatly reduced for incident angles of light away from the critical angle. If there is a disturbance in the second medium, such as a dust particle, within the penetration depth then TIR will no longer be total giving frustrated TIR. For the solar spectrum the penetration depth of the evanescent wave is of the oredr of 0.3 μm. However the size of dust particles in the atmosphere ranges from 0.01 to 10 μm in size [9], thus we can expect that some of them will interfere with the total internal reflection.

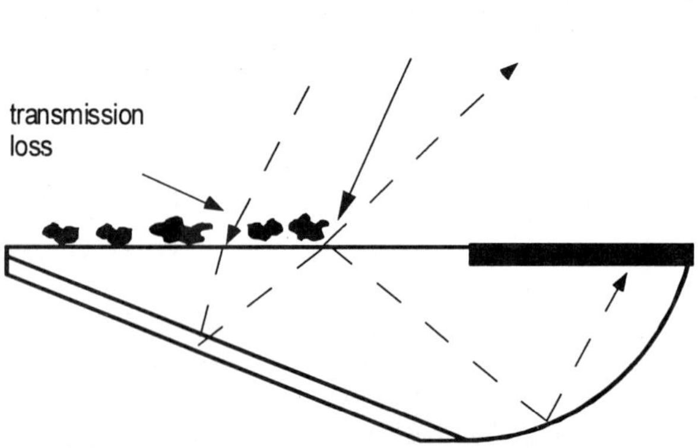

Figure 3.Effect of dust on static concentrator module

A number of samples were exposed for 6 months the the atmosphere at different locations in the Sydney region. The amount of TIR was measured before and after exposure at various incident angles (Figure 4). As expected the dust has the greatest effect on TIR at angles close to the critical angle, for large angles the effect of dust is minimal

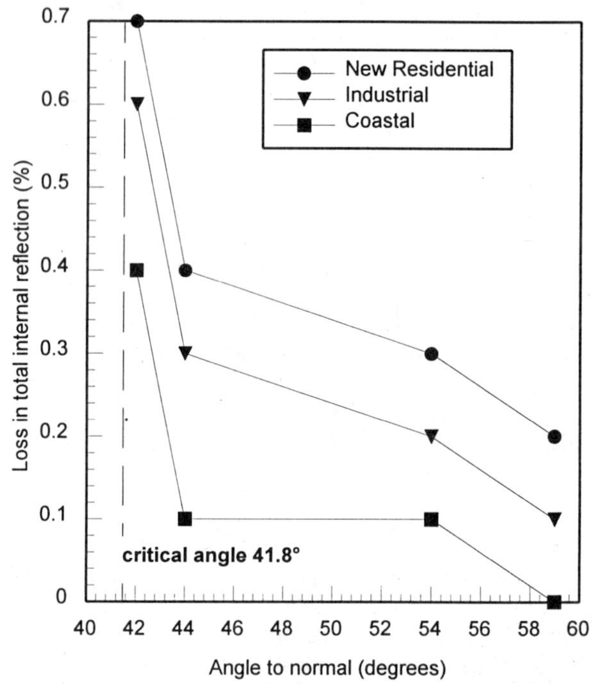

Figure 4. Loss in reflectivity of TIR as a function of angle from the normal. The critical angle is at 41.8 for glass with a refractive index of 1.5.

The loss of transmission of light into the module was also measured as this affects all modules, with or without static concentrators. The overall effect of the frustrated TIR loss on module performance was calculated (Table 1). The table shows that the effect of dust on module performance is minimal and that the efficiency drop caused by frustrated TIR is less than half of the loss experienced by light as it enters the module.

Location	Total Loss	Transmission loss	TIR loss
New Residential	1.3%	0.9%	0.4%
Coastal	0.9%	0.6%	0.3%
Industrial	0.7%	0.5%	0.2%

Table 1: The effect of dust on module performance.

For other locations and longer periods of exposure it is expected that greater quantities of dust would accumulate on the top surface of the module. Even in extreme environments the loss in perfmance of a standard module is less tha 10%[10]. Thus it is expected that the loss in performance of the static concentator module caused by frustrated TIR would be less than 5%.

CONCULSIONS

The static concentrator roof tile can produce significant cost savings by reducing the number of silicon wafers in a module by a factor of four. By using high efficiency bifacial buried contact solar cells module performance will not decline despite the massive reduction in cell area. Two manufacturing techniques are being pursued for a 1kW roof top array and the main performance loss mechanisms have been identified

While dust does effect total internal reflection, the corresponding loss in performance is minimal and much less than the loss experinced by light as it enters the module.

ACKNOWLEDGEMENTS

We would like to thank other members of the Centre for Photovoltaic Devices and Systems who have contributed to this work particularly Matt Boesen, Rob Largent and the laboratory manager Mark Silver. Thank you to the undergraduates that worked on the project this session Phillip Garland, Ling Ling Hui and Rashika Dias. We acknowledge the direct support for this work from Pacific Power, the Energy Research and Development Council and the Australian Research Council.

REFERENCES

[1] K.W. Mitchell, "Renaissance of Czochralski Silicon Photovoltaics", *Progress in Photovoltaics,* Vol. 2, No. 2, 1994, pp. 115-120

.[2] A.M. Barnett et al., "Polycrystalline Silicon-Film™ Solar Cells: Present and Future", *Progress in Photovoltaics,* Vol. 2, No. 2, 1994, pp. 163-170

[3] S.R. Wenham et al. "Innovative Strtuctures for Thin Film Crystalline Silicon Solar Cells to Give High Efficiencies from Low Quality Silicon", *Conf. Record, 24th IEEE PVSC*, December, 1994.

[4] S. Bowden, S.R. Wenham and M.A. Green, "High Efficiency Photovoltaic Roof Tiles with Static Concentrators", *12th European Photovoltaic Solar Energy Conference,* April, 1994.

[5] C.B. Honsberg, F. Yun, A. Ebong, M. Taouk, S.R. Wenham and M.A. Green, "685 mV Open Circuit Voltage Laser Grooved Silicon Solar Cell", *Technical Digest, 7th PVSEC*, Tokyo, Japan, November, 1993, pp. 89-90

[6] T.M. Bruton "Fabrication of Laser Grooved Buried Contact Silicon Solar Cells", *Proceedings of Workshop on Crystalline Silicon Solar Cells*, Spain 1994.

[7] S. Bowden, S.R. Wenham, P. Coffey, and M.A. Green, "High Efficiency Photovoltaic Roof Tile with Static Concentrator", *23rd IEEE PV Specialists Conference, Louisville*, May, 1993, pp. 1068-1072.

[8] S.R. Wenham,. and M.A. Green, "Improved Optical Design for Photovoltaic Cell Encapsulation", *Australian Patent Application PK8787*, 8th October, 1991.

[9] C.E. Lapple (1961), quoted in "*CRC Handbook of Chemistry and Physics*", 71st Edition, pp15-35.

[10] R. G. Ross and M.I. Smokler *"Flat-Plate Solar Array Project - Final Report, Vol VI: Engineering Sciences and Realiability"*, JPL Publication 86-31, 1986.

R&D of PHOTOVOLTAIC MODULES INTEGRATED with CONSTRUCTION MATERIALS

Fumihiro SHINJO

Solar Energy Department
New Energy and Industrial Technology Development Organization(NEDO)
Sunshine 60, 29F, 1-1, 3-chome Higashi-Ikebukuro, Toshima-ku, Tokyo, 170 JAPAN

ABSTRACT

In an attempt to promote the widespread introduction of photovoltaic power generation systems, it is necessary to achieve utility sales price-equivalent power generation costs through implemented system cost reduction. For this purpose, photovoltaic modules integrated with construction materials will be developed so that photovoltaic power generation module costs and installation work expenses can be substantially reduced by including them in the materials and general construction work costs.

Introduction

Research and development will be undertaken concerning three types of modules integrated with construction materials(six different methods) which have widespread applications, including private house and high-rise building uses. Specific development is being carried out during the period of four years from FY 1993 to 1996.

The three types of modules integrated with construction materials are:

(1) Photovoltaic modules integrated with roofing materials
(2) Photovoltaic modules integrated with wall materials
(3) Photovoltaic flexible modules applicable for construction materials

With maintenance, which involves repair, replacement, etc, and inter-module connection procedures in mind, installation and anchoring methods will be studied for successful development of highly reliable, low-cost modules.

Modules integrated with roofing or wall materials will be incorporated into a private house or building on an experimental basis and a demonstration test will be conducted.

Work program

(1) PV modules integrated with roofing materials

There are three methods for integrating PV modules with roofing materials: the first is the exchangeable PV shingle method; the second is the prefabricated PV roof panel method and the third is the heat insulated PV roof panel method. Table 1 indicates the targets for PV modules integrated with roofing materials.

Table.1 Targets for modules integrated with roofing material

Cost (Yen/W)	About 170 as a PV increment
Life (years)	About 20 years
Markets	Newly built, extended or improved, and existing detached private houses

Exchangeable PV Shingle Method

Feasibility studies were implemented as a preliminary step in developing modules integrated with roofing materials (exchangeable PV shingle method), and basic specifications were set based on cost, design, construction and maintenance characteristics.

The exchangeable PV shingle method includes such characteristics as BM (back metal) structure, frameless structure, EWC (exchangeable watertight connection) structure, and JFC (just fit connection) structure. These were adopted for the basic specifications.

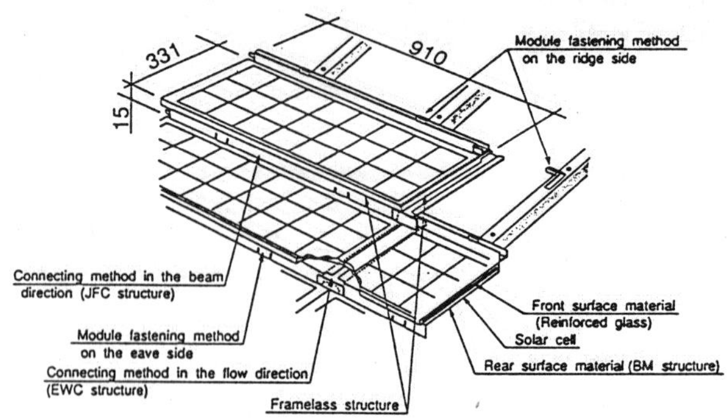

Fig 1. Basic specification for modules integrated with roofing materials

CH3365-4/94/0000-0778 $4.00 © 1994 IEEE First WCPEC; Dec. 5-9, 1994; Hawaii

Prefabricated PV Roof Panel Method

The prefabricated PV roof panel method shapes a modular unit by arranging usual slates and photovoltaic tiles side by side.

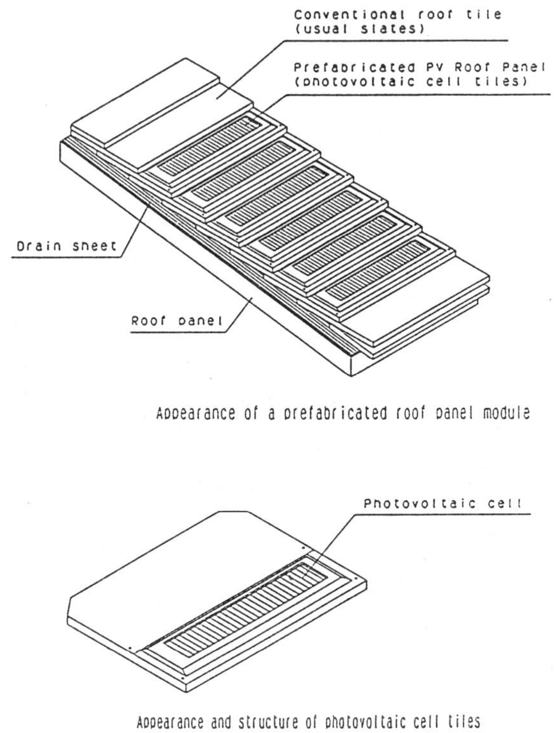

Fig 2. Appearance of a prefabricated roof panel module

Heat Insulated PV Roof Panel Method

The heat insulated PV roof panel method consists of surface glass, solar cells, water-proof sheets, heat insulating materials and base frames.

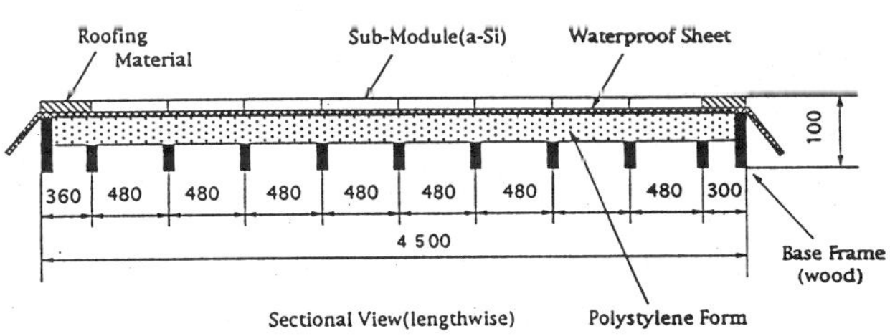

Fig 3. Heat insulated PV roof panel method

(2) Photovoltaic Modules Integrated with Wall Materials

There are two methods for integrating photovoltaic modules with wall materials: one is the PV glass curtain wall method, and the other is the PV metal curtain wall method.Table 2 indicates the targets for PV modules integrated with wall material.

Table.2 Targets for modules integrated with wall material

Cost (Yen/W)	About 85 as a PV increment
Life (years)	About 20 years
Markets	Office buildings and large detached private house's outer and curtain walls

PV Glass Curtain Wall Method

One of the objects of this study is to stress the design of building-material-integrated modules, which will help secure market share and, as a result, lower the cost. Module samples with different surfaces and colors of were produced and an opinion survey of designers and other construction-related people was conducted. As a result, it was found that non-AR-coated colors (the original color of crystal silicon) and relatively dark red and yellow made good impressions, besides the conventional shades of blue.

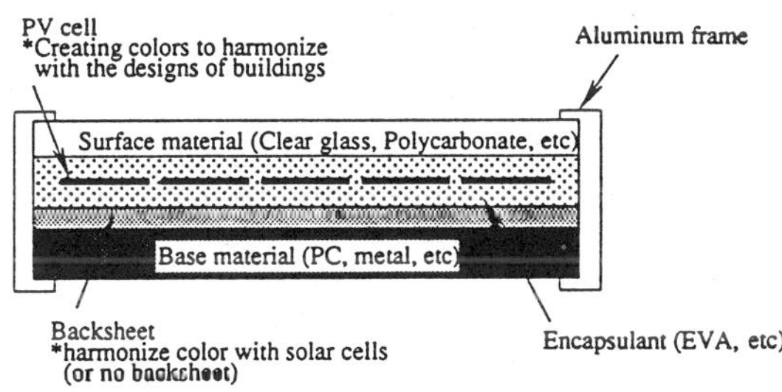

Fig 4. Structure of building-material-integrated module

PV Metal Curtain Wall Method

Successful R&D of photovoltaic modules integrated with a metal curtain wall to be applied to a building wall will promote the widespread introduction of photovoltaic power generation systems.

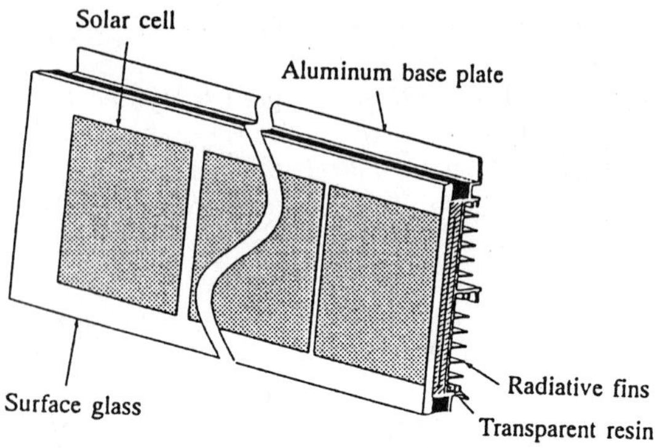

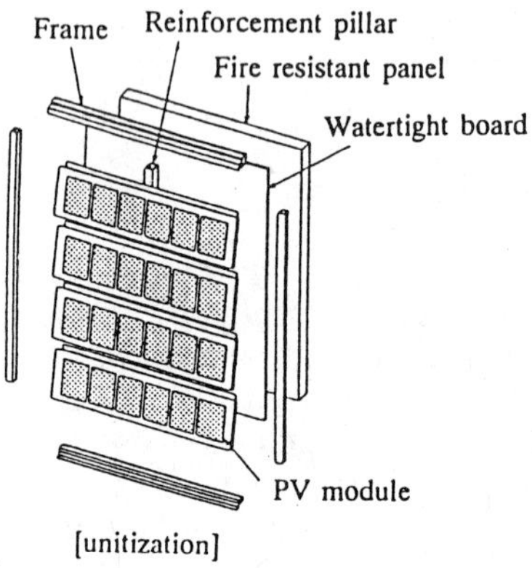

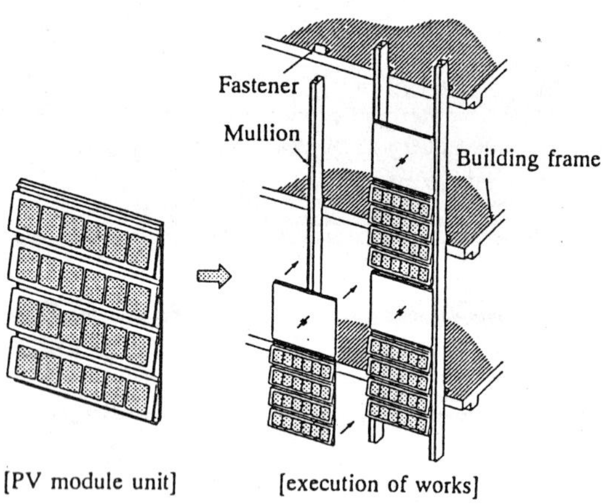

Fig 5. Initial schematic diagram of PV unit

(3) Photovoltaic Flexible Modules Integrated with Unspecified Construction Materials

Table 3 shows the targets for Photovoltaic Flexible Modules Integrated with Unspecified Construction Materials.

Table.3 Targets for flexible modules integrated with unspecified construction materials

Cost (Yen/W)	About 190 as a PV increment
Life (years)	About 10 years
Markets	Existing private houses or simple structures' roof and wall cladding materials, mobile structures, and leisure purposes

Flexible PV Module for Roofing Method

Development of photovoltaic modules for roofs with flexible amorphous silicon solar cells and technology of roof construction with these modules.

Characteristics of an ideal flexible PV roofing would be the ability to commonly apply the PV roofing to any kind of house, because the installation method would be the same as conventional roofing sheets, and that the structure of the PV roofing allows it to be cut, similar to conventional roofing sheets, so it can be applied to any size roof.

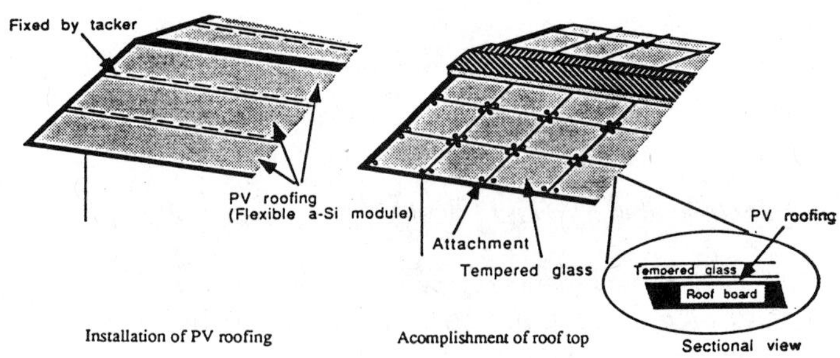

Fig.6 Roof structure with flexible PV roofing

Themes in the Future

In the preceding pages I described three types of modules integrated with construction materials (six different methods). Many problems remain to be solved in relation to modules integrated with construction materials, including the development of construction methods, water-proof structures and means of maintenance. However, successful results are expected to contribute to the introduction and spread of photovoltaic power generation.

BUILDING OPPORTUNITIES IN THE U.S. FOR PV (PV:BONUS) A PROGRESS REPORT

Roger W. Taylor, National Renewable Energy Laboratory
Golden, CO 80401 USA

ABSTRACT

Five contract teams are developing photovoltaic (PV) products that will have a significant impact on building-integrated PV systems. The product lines that these teams are pursuing include roofing materials, building facade materials, PV integrated into modular homes, ac-PV modules, and utility- dispatchable PV systems. The objective of these efforts is to develop product and market opportunities that can provide for the introduction of PV into the buildings market sector at higher allowable installed systems costs than conventional ground- or roof-mounted systems. Each of the teams has a unique approach, and synergistic opportunities among teams are beginning to emerge. This paper reviews the product and market development efforts of these teams and describes the links between the product efforts and parallel analytical work to develop PV as a demand-side management option.

INTRODUCTION

Major market opportunities exist for photovoltaic (PV) systems that can be effectively integrated into residential, commercial, institutional, and industrial buildings. The objective of the Building Opportunities in the U.S. for PV (PV:BONUS) program is to develop PV products that can facilitate the commercial introduction of PV into buildings.

Through a competitive procurement, the U.S. Department of Energy (DOE) and the National Renewable Energy Laboratory (NREL) selected five teams to develop PV-buildings products. The five teams that have begun work each include a lead contractor and up to 10 other organizations representing building materials manufacturers, building contractors, PV suppliers, utilities, colleges, system designers, architecture and engineering firms, and building owners. Additional efforts focus on PV/demand-side management (PV-DSM) economic analysis, solar resource/utility load correlation, and education of architectural students.

PROGRAM SCOPE

Flexible, lightweight roofing materials are being developed by Energy Conversion Devices, Inc. (ECD), and United Solar Systems Corporation (USSC). PV systems that double as electrical generators and roofing material have extra value to the building owner. ECD and USSC, along with the National Association of Homebuilders, Solar Design Associates, Arizona Public Service Company, Bechtel Corp., Detroit Edison, and Minoru Yamasaki Associates, are developing a flexible lightweight PV roofing product to be integrated into rooftops. Two parallel paths are being pursued one directed at a PV shingle that can be offered as an alternative to conventional asphalt shingles, the other at adapting PV to the existing batten-and-seam metal roofing industry. The metal roofing modules are being designed as approximately 0.5 m X 6 m (20-in by 20-ft) metal roofing pans with all electrical connections to be made along the roof ridge, under the roofing cap. A prototype system is being designed for incorporation into one of the National Association of Home Builders Research Center Townhouses (see Fig. 1). The team will also verify technical and economic performance of the products, develop a marketing plan, and lay the foundation for high-volume commercialization.

Dispatchable peak-shaving systems are being developed by Delmarva Power and Light (DP&L), an electric and gas utility located in Delaware. Although PV output has been shown to correlate well with the peak demand periods of many utilities, others have peaks that extend into the late afternoon and early evening. To address power needs at these times, DP&L and its team members are designing a PV system with batteries that can be called upon (dispatched) by the utility, much as power from conventional power plants is managed. Studies have shown good correlation between daily solar energy and peak loads, even though the time-matching may be off by several hours. The goals of the DP&L program are to develop a dispatchable PV system for commercial and utility applications, develop modular construction techniques to reduce field labor, develop a simple roof-mounting system, and establish that the system is in compliance with building and safety codes. The DP&L team will also perform a detailed market assessment and determine how they should design and sell PV systems within its service territory.

A preliminary test system was installed by DP&L in 1993 in time for summer operation. Those test results have lead to further system refinements, such as PV array size versus storage size, addition of a dc-dc converter for maximum power tracking, simplified user interface controls and interfaces with energy management systems (EMS) and utility SCADA (system control and data acquisition) systems, and improved battery management techniques. Work is also underway on array mounting options and identification of market niches, such as dual-use (emergency lighting, uninterruptible power supplies) and off-grid systems. During Phase 2, four prototypes will be installed and operated.

Homes manufactured with PV, passive solar design features, solar hot water, and energy-efficient appliances built in before they ever leave the factory is the vision of Fully

Figure 1 Artist rendering of PV townhome (on right)

Independent Residential Solar Technology, Inc. (FIRST, Inc.), a non-profit organization teamed with Bradley Builders and Avis America, a builder of manufactured homes in central Pennsylvania. Modular, manufactured houses are increasingly popular as a way to reduce home construction costs and improve quality. Modular houses, large boxes constructed in a factory and assembled at the building site, will incorporate PV designed for both grid-connected and stand-alone markets. Two prototype homes were constructed during 1994, with additional preproduction models anticipated in 1995 (see Fig. 2 - 4).

PV modules that produce ac electricity are being developed by Solar Design Associates, Inc., an architectural and systems engineering company, along with Solarex Corporation (and, previously, Mobil Solar Energy Corp.). Today's PV modules produce dc (direct current) electricity, which must be converted to ac (alternating current) for grid-connected applications. A module producing ac electricity frees architects and designers from dc wiring constraints and inverter sizing issues. PV ac modules would also eliminate constraints on orientation, shading, and the number of PV modules that can or need to be used. In this system, each module generates ac power on its own. Solar Design Associates is leading the effort to apply advances in power electronics and integrated circuits to deliver utility grade ac power from the back of individual PV modules. Economies in the mass production of inverters and modules in standard sizes should reduce component and installation costs. The technology is applicable to large-area roofing retrofits as well as emerging window-wall opportunities.

These new ac modules will also include an integrated communications capability allowing individual unit identification and operating status checks that will vastly simplify problem detection and repair. Other advantages include individual module maximum power point tracking (no siting constraints due to complex building geometries, shading, and string mismatch), no high-voltage dc wiring (safety issues) on roofs and vertical building facades, and a higher degree of system reliability through large-scale parallel system redundancy.

At present, five "inverters-in-a-junction-box" are under test. Twenty five preproduction units will be manufactured in early 1995. The units, which measure about 20 cm x 15 cm x 4 cm (8 in x 6 in x 1.5 in) are expected to demonstrate an overall

Figure 2 Modular homes are easily and rapidly assembled on site using large boxes constructed in a factory

efficiency of greater than 95% with good low-power efficiency, silent operation, very low harmonic distortion, and full radio-frequency interference (RFI) filtering (see Fig. 5).

Architectural PV glazing systems are the goal of Advanced Photovoltaic Systems, Inc. Glass is increasingly used as building facade material as well as for light and control of heating and cooling. A system of matching glazing products that include PV panels will be designed with a combination of opaque, semitransparent, or clear units. Such products will be incorporated into windows, skylights, and panels for vertical and sloped enclosures that provide supplemental power when fans and air conditioners are needed.

ADDITIONAL SUPPORTING ACTIVITIES

Educating the next generation of architects about the emerging application of PV in buildings is the goal of a project that NREL is conducting with the American Institute of Architects/Associated Collegiate Schools of Architecture Research Corporation (AIA/ACSA RC). The course was developed and pilot tested in the Fall of 1993. During 1994, the course materials are being distributed to North American schools of architecture. In 1995 the materials will be reworked into a form useful to practicing architects. As part of the work funded under this effort, the AIA/ACSA RC coordinated the United States' participation in a European-sponsored International Architectural Ideas Competition, to generate building designs that integrate PV.

Assessing the benefits of PV as a demand-side management (DSM) option has been the focus of researchers at the University of Delaware. They started by looking at the incentives utilities offer for other DSM options, such as demand control devices or fluorescent lights. Working closely with DP&L, they outlined the incentive options and proposed strategies for considering PV as a demand-side option. The University of Delaware has recently developed a computer spreadsheet program to assist in the evaluation of PV performance as a DSM option, again using DP&L for the case study. (This effort is closely integrated with the DP&L dispatchable PV system design effort funded under PV:BONUS.) The analytic effort is now being expanded to test the spreadsheet program with six additional utilities to better understand the concepts involved in considering PV in DSM configurations.

Figure 3 Completed PV-integrated modular home

Figure 4 PV-integrated modular home

Solar resource/utility load matching is another area where the DOE/NREL PV Program activities are contributing to the understanding of the PV-buildings opportunity. Working with 20 utilities, contractors at the State University of New York at Albany compared utility load profiles over the year with the output of PV power plants. They calculated output using solar resource data derived from satellite measurements. This *match* between PV output and the individual utility's load requirements is crucial to determining the effective capacity (hence the value) of PV. The load-matching study concluded that, for many of the utilities in the study, PV generation, either on the demand or supply side, could help meet capacity requirements. Load-matching studies have recently been completed with a second group of utilities.

PV power generation and its coincidence with individual building loads is being investigated at NREL using data being collected from the Environmental Protection Agency's (EPAs) recent cofunded program of utility-interconnected PV systems on residential and commercial buildings. The wide variety of building types and solar climates represented by the EPA data base should lend additional insights into the load correlations and energy storage requirements necessary to minimize consumer bills and utility system costs.

SUMMARY

Today, PV is cost effective for many applications. The concept of PV systems mounted on buildings or as a part of the building envelope has been demonstrated. However, for sustainable use of PV in the building sector, more work is needed. Architects, builders, utilities, and building owners need to be participants. DOE and NREL have begun a program that incorporates the key stakeholders working together to make energy-producing buildings a reality. The development of these products for both domestic and international markets can help define higher value, intermediate market opportunities that may eclipse large-scale grid-connected markets for many years.

This work was supported directly by the U.S. Department of Energy and by NREL under DOE Contract No. DE-AC36-83CH10093.

Figure 5 The ac modules will vastly simplify the installation and wiring of PV window walls. Solar Design Associates (President Steven Strong pictured) is working with Solarex Corporation to develop this building-integrated technology.

MULTIFUNCTIONAL SOLAR FACADES - A NEW CHALLENGE FOR PHOTOVOLTAIC

Joachim Benemann
Managing Director
FLAGSOL, FLACHGLAS SOLARTECHNIK GmbH
- A member of the PILKINGTON Group -
Muehlengasse 7, D-50667 Cologne (Germany)

During the last decades glass has become a very attractive material for sophisticated design on large buildings such as office or public buildings, hotels, shopping malls or manufacturing buildings. To protect employees in those buildings from light, heat or noise the glass industry has developed a number of multifunctional glass systems for windows, facades and roofs. As a logical step, the glass industry got the idea of including photovoltaics into facade elements to generate electricity in addition to the above mentioned functions. FLACHGLAS SOLARTECHNIK GmbH, Germany, developed this new type of photovoltaic facade elements into a building owned by a city utility. In May 1991 the facade was inaugurated and this system received the tradename OPTISOL®.

It created a widespread interest by architects, investors and public funding institutions. The façade has been designed together with a metal frame manufacturer. The technical perfection and reliability as well as the aesthetical design has positively influenced several decisions on new projects. At least fifteen buildings have been realized since then and demonstrate the various applications of photovoltaic systems.

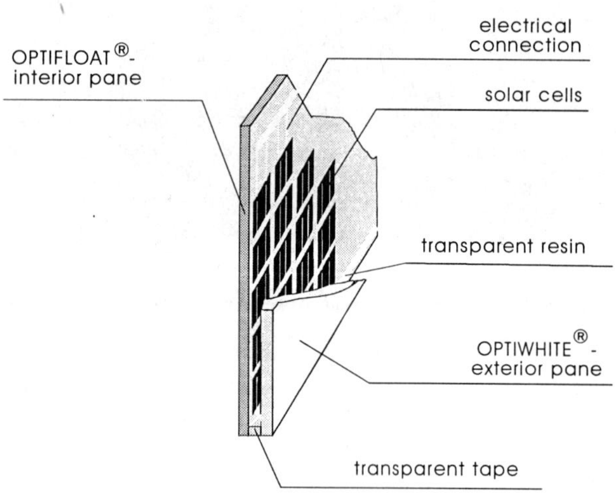

Fig. 1. The basic OPTISOL® element is using standard float glass for the backside and low iron, heat tempered float glass on the front side.

In the specially designed production line for OPTISOL® facade elements the solar cells are embedded with a special clear cast, UV-resistant resin between two glass panes. OPTISOL® elements can be produced in sizes up to 2.1 m x 3.2 m. The light permeability can also be adjusted to the respective requests by arranging the solar cells in different ways.

Depending on the application, the photovoltaic facade elements can be built as insulation glass design with high heat insulation, with noise protection or with security (break-up) protection.

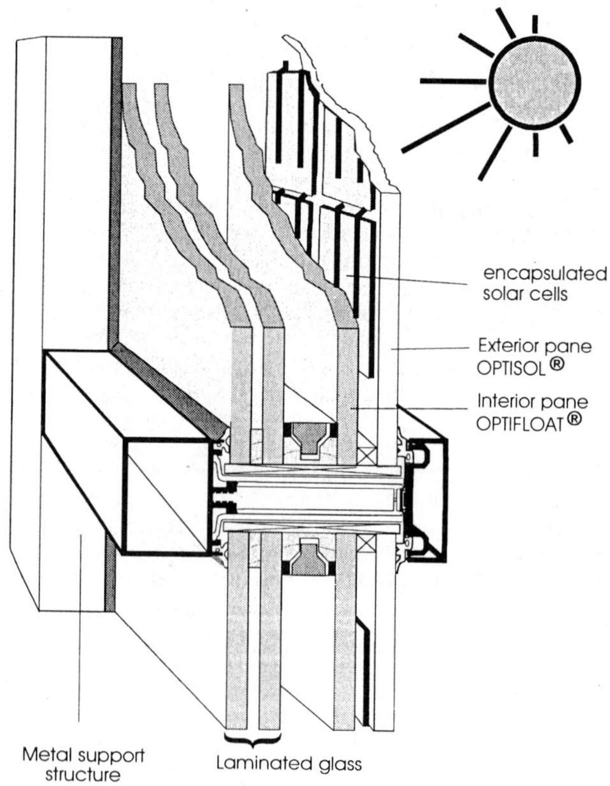

Fig. 2. To combine the basic OPTISOL® element with additional features, it is realized in double composite panes - similar to standard insulating windows - and fixed into metal frames which also host the cabling of the PV-elements.

CH3365-4/94/0000-0784 $4.00 © 1994 IEEE

Permanent improvement of technical details in respect to simple installation, safety and electrical cabling instructions have been made and have already established this technique as a semi-commercial product. Besides, several other items have been developed such as quality standards, safety authority standards, power purchase agreements with utilities, insurance policies for installation, start up and warranty insurances as well as performance warranties. Some additional activities are proceeding to grant tax incentives for investors using this innovative and environmentally friendly technique.

Fig. 3. Very interesting light effects can be produced inside a building by variation of the distance between the cells or by use of amorphous thin-film photovoltaic elements.

For the application of photovoltaic for a facade of a production facility, FLAGSOL received the European Solar Prize 1994 for the best integration of photovoltaic into buildings. The OPTISOL® elements are using polycrystalline cells. The north-west facing elements of the building were silk screen printed instead of using photovoltaic active cells. This ensures an architectural uniform appearance.

Fig. 4. The vertical screen facade of the FLACHGLAS AG production facility in Wernberg, Germany, with polycrystalline solar cells.

Fig. 5. OPTISOL® elements in vertical facade integration enables new architectural design possibilities.

Fig. 7. Frameless facade elements enable the mounting with almost invisible mounting points.

Fig. 6. A laboratory building at the ISPRA Research Center received a 720 m² photovoltaic facade. After completion, a large display is showing the actual and cumulated power of the OPTISOL® facade.

Fig. 8 Opaque amorphous photovoltaic solar cells made by APS, USA, were used to manufacture OPTISOL® elements for the European Research Center ISPRA, Italy. The backside vented post-peg construction is currently the world wide largest photovoltaic facade in thin-film technology.

Fig. 9. Solar insulation on south facing facade walls is often a problem and can be reduced by photovoltaic shading systems which can probably improve the design of the building.

Fig. 10. The light permeability of the modules depends on the area coverage with opaque solar cells.

Fig. 11. PV curtain wall facade of the Central library of the Research Centre Jülich

PHOTOVOLTAICS IN THE LIGHT OF SUSTAINABLE DEVELOPMENT AND QUALITATIVE GROWTH

Jochen Moesslein
Center for Solar Energy and Hydrogen Research Baden-Wuerttemberg (ZSW)
Hessbruehlstrasse 21 c; 70565 Stuttgart / Germany
Tel.: + 49 - 711 - 7870 - 261; Fax.: + 49 - 711 - 7870 - 100
new address: Regiolux / Solar Technology Division
Philipp - Reis - Strasse 15; 97437 Haßfurt / Germany
Tel.: + 49 - 9521 - 5410; Fax.: + 49 - 9521 - 64481

ABSTRACT

Although many people say that there is already a market for photovoltaics (PV), from the economical point of view there is at best, apart from the remote power "market", some demand by a pilot market, that consists of idealistic, in the economic sense non-rational consumers. In essence, an ideal market with full cost accounting does not exist in the energy sector. Concerning the future of PV, the question is, how the decision-making on electricity generating technologies will change in the future. The most probable path should be anticipated by the people of the PV community and R&D directions should be adopted in such a way to be able to realize the full potential of PV.

In this paper it is first shown that the decision-making in the future will be primarily governed by the adopted energy policy then by ideal markets. The catalogue of aims of this energy policy will shift towards new paradigms of economic development, because this is the most practical way to avoid a further increase of external effects. The new paradigms, for which in this paper I quote "sustainable development" [1] and "qualitative growth" [2] impose certain criteria for energy systems that have to be anticipated as early as possible.

INTRODUCTION

If we want to look closer at the competition of PV with conventional power sources, which are mostly distributed by a grid, we first have to define the technical and economical situtation in which the two energy options face each other and where the decision-making between those two options takes place. This is the objective of the first part of this paper.

The second part of this paper then shows that the the decision-making in the energy sector will most likely be governed by new paradigms that underlie the executed energy policy. Some example paradigms are then applied to the technology of PV.

FIRST THESIS: THE ELECTRICITY GRID WILL EXTEND FURTHER

This means that island PV systems are only temporary solutions. The ratio of grid-connected to remote land will become larger.

In the niche of "remote power" PV systems are already competitive today on the basis of internal costs only.

While the size of this niche is very small in industrial countries it is still huge in developing countries. The question is whether the electricity grid will further expand in these countries as it has in the industrial countries. According to the opinion of the author the answer is "yes". The reasons are as follows.

First of all, we have to be aware that any kind of industrialization so far, and especially in the first periods, calls for more electricity. This finding is empirically reflected by the coupling of the gross domestic product and the electricity consumption of all industrialized countries. Only very recently, due to the relocation of energy intensive productions and increased overall energy efficiencies, we can observe a decoupling.

Second, it is rather important that for efficiency and productivity reasons industrial processes require a certain availability, quality and quantity from the supplied electricity. All of these three criteria are harder to be met by island PV systems. If PV systems wanted to meet the demand characteristics of industrial production, they would have to be backed by a storage system, such as a hydrogen grid.

A third aspect is that as the grid spreads out in the industrial parts of the countries, it will reach closer to the parts that are not yet connected to the grid. Hence grid extensions into these parts will become cheaper.

As a fourth point the fact that also in developing countries the level of comfort rises should be quoted. As the level will grow it will be more difficult for PV systems to meet the demand, unless the consumers adopt to the electricity supply of the PV system.

In conclusion, supported by several reasons we should expect that at the cost of a growing remote power market, the industrial development and economic growth in developing countries will lead primarily to an expansion of the electricity grid. This tendency is empirically underlined by the share of financial funding that goes into grid extension projects.

OBSERVATION: THE ONLY INDEPENDENT, MARKET-LIKE DECISION THAT NON-REMOTE CONSUMERS ARE LEFT WITH IS THE DECISION BETWEEN THE GRID AND THE GRID PLUS A PV SYSTEM

CH3365-4/94/0000-0788 $4.00 © 1994 IEEE

The expansion of the grid means that the remote power market becomes smaller and once it is satisfied PV will be left to compete with grid-connected electricity. Then two issues are important:
First, being grid-connected the PV system does not constitute the only power source any more. The PV system can rather be considered as a fuel saver.
Second, the choice that the consumer has is between the grid only and the grid plus a PV system. There is no choice for the consumer between different energy systems that generate the grid electricity. Only if the electricity consumer had an independent choice over all available options, it could be expected that the outcome of all market transactions reflected the best compromise of individual preferences.

SECOND THESIS: THE MAJORITY OF CONSUMERS BEHAVE RATIONALLY

Economic theories generally assume that in most market decisions consumers behave rationally. This means that consumers most likely choose the option with the lowest cost-benefit ratio for them. This implies two important aspects:

First, the ratio is lowest when all the benefits for which the consumer has paid are felt by him exclusively.

Second, benefits and goods that do not cost are to be maximized (free rider system).

Concretely, this behaviour means that most consumers would choose the cheapest energy option and hence choose the grid only. The benefits that PV would have, namely a better environmental quality cannot be felt by the consumer exclusively.

THE REMAINING OPTIONS

The technical and economical situation following from the two thesis and the observation above are as follows. The remote power market has been successfully satisfied by PV systems. Due to the industrialization in developing countries, the total electricity grid constantly expands. The remaining market for PV is thus seen as an optional electricity generating system that can be connected to the grid and serve as a fuel saver.

At this point three major developments in terms of the future for PV can occur..

First, PV producers can still try to sell to as many non-rational, idealistic consumers as possible (the pilot market). However this number is relatively small. In Germany, sociological studies have shown, that the consumers might impede a further development of a market for ecologically benign products themselves, because the speed of the growth of the pilot market is smaller than the potential speed of the growth of the production capacities.

Both PV producers and enthusiasts wait until the production costs have travelled down the learning curve. From an engineering point of view this thinking is very honorable, from an economic point of view it means nothing else than accepting internal cost accounting and behaving according to the neoclassical paradigm of quantitative growth. In this case we also have to consider the power of conventional electricity producers and their potential for cost reductions in an economic competition. It is then questionable whether PV can become competitive on the basis of internal prices at all. Furthermore, most of the published cost projections for the production of solar modules only give proportional production costs and capital costs for the production equipment. They do not consider the overhead costs, which account for rather large shares of today∩s production costs.

Internalise External Costs or Adopt another Development Paradigm as the Source of a Criteria Catalogue

Third, we can try to influence the price or criteria system on which investment decisions in the energy sector are based. The justification for this is the increasing significance of external effects, on which I do not elaborate here. I assume that measures have to be taken. I also assume that the state plays the key role, which is a result of the following considerations.

So far, the current price system does not credit for features like sustainabilty. Not only that it does not honor most of the positive characteristics of photovoltaics (PV), it further neglects the damage (external costs) caused by other energy systems.

Once external costs [3] are included in the energy prices the decision-making at the utility or consumer level remains practically unchanged. Both utility and consumer will still perform a rational cost-benefit-analysis. However, the problem with external costs is the quantification of the costs. Essentially, quantifying external costs means to value goods in monetary units that have not been marketed so far (i.e. clean air, healthy forest). The fact that these goods have not been marketed or traded makes it extremely difficult to assess their price. In addition, goods that are marketed usually have owners, who are motivated by their own rationality to ensure that their goods yield a high price and preserve their potential value. Hence, if it was desired that the values of externally effected goods are negotiated in the marketplace, an owner of these goods is needed. From reasons of equity, this could most likely only be the society as a hole, represented by the state. However, since the state has to advocate many other interests, it will be difficult for him to play the ownership-role efficiently.

The basic idea of external costs is to charge for the damages induced by a certain technology and likewise prevent further damages or overuses of certain goods. To adopt a growth or development paradigm is a more strategic and less market-oriented approach. However, it is the approach that is essentially done when changes in the energy policy of a country are made nowadays. The reason for this proceedure is that it is more practical than the quantification of external costs. In a first step a catalogue of desired criteria reflecting essentially a paradigm is developed. In a second step one has to assess which technology fullfills these criteria at best. For this step a quantification of external effects is necesarry, which is one step less then the quantification of external costs and therefore substantially less difficult. In the following, two paradigms are presented.

PRINCIPLES OF SUSTAINABLE DEVELOPMENT FOR PHOTOVOLTAICS

There is not a generally accepted set of principles. However, most of them cover comparable ideas. In the following, a choice of principles that cover the main ideas is presented.

The first principle is obtained when requesting intertemporal justice and the balance of the ecosystem: the resource extraction rate should not be larger than the regeneration rate.

Among current PV materials, tellurium and indium are the most scarce [4]. Based on the resource estimates of the US Bureau of Mines [5] we obtain the following situation: if all indium and tellurium ressources are used for the fabrication of CIS and CdTe modules, 2% and 12%, respectively, of the current world electricity consumption could be met. Hence, these devices alone cannot form a sustainable energy supply system, especially since the electricity demands will most likely increase mainly due to the industrialization of third world countries. To deploy the full potential of generating 2 and 12% of the current electricity demands by CIS and CdTe solar cells, researchers have to aim at highest material utilization efficiencies (> 50%). These aims are indeed successfully pursued i.e. at the Center for Solar Energy and Hydrogen Research. Further, high recycling efficiencies are necessary to open perspectives beyond the lifetimes of the modules. Especially in the case of CIS the aspect of building up a knowledge capital stock by investigating this material seems important, because many other familiar chalcopyrite materials as potential PV materials are available.

A second effect of the scarcity of raw materials should also be anticipated: the rise in price as the demand exceeds the supply or as the reserves deplete. However, the price increase might also make it economic to mine for further reserves that were not considered as resources before. In order to maintain intertemporal equity, the theory of resource economics requires that an additional charge has to be imposed on the price of depletable resources to feed an investment fund for a back-stop technology. This could be i.e. a silicon solar cell technology.

Raw materials for other solar cells, mainly silicon, gallium and arsine are sufficiently available. Nevertheless, these modules should also be recycled [6] to protect resources and maintain a nearly constant resource capital stock.

The second principle: renewable energies should substitute depletable energy sources.

Shifting to renewable energies with harvesting factors larger than one this is true for the resource "energy". However, we also have to keep in mind that the production processes we develop should eventually be powered by renewable energies.

The third principle: emissions and waste depositions should be smaller than the assimilative capacity of the ecosystems.

To admit, damage functions and subsequent environmental limits are generally difficult to establish. Some studies on environmental effects [7, 8, 9, 10, 11] and life cycle analysis [12, 13, 14], primarily on CdTe and CIS solar cells, have been prepared. In essence, the studies indicate that a systematic or potential threat to the ecosystem does not exist. The given reason is mainly that the risks of CdTe and CIS solar cells can be effectively controlled.

The fourth principle: the land use should not expand.

Hence, PV systems should be installed on roofs only. Such a decentralized structure would also minimize the land use for power lines. Another issue is the mining for resources. At best, materials from already existing mines should be used.

PRINCIPLES OF QUALITATIVE GROWTH FOR PHOTOVOLTAICS

While quantitative growth focuses on the quantitative growth of the gross product, qualitative growth requires a change from the structure of the gross product. Whereas the paradigm of sustainable development focuses on the environmental compatibility, the paradigm of qualitative growth includes more goals: The four central aims are economical efficiency (EE), environmental compatibility (EC), social compatibility (SC) and international compatibility (IC). Values (V) and prices (P) for a product or service can be expressed to a first approximation as the sum of four factors which themselves break down into more factors according to the different criteria of the aims:

V, P = a1 EE + a2 EC + a3 SC + a4 IC.

The relative sizes of the coefficients are set by society.

A catalogue of criteria that is based on these aims and designed for PV systems virtually includes any possible aspect that should be addressed for a comprehensive technology assessment of PV and other technologies. Some of the criteria are given in the following.

1. Economical Efficiency:

Micro-economical efficiency:

What are the dependences between production volume and production cost for different solar cell technologies? What are the harvesting factors of the systems?

Macro-economical efficiency:

What are the CO2 reduction costs? Compared to other energy technologies does the manufacture of PV systems give rise to positive employment effects?

2. Ecological Compatibility:

How do the cumulated material in- and outputs for the production, use and recycle sequence look like for the different solar cell types? Is an energetically reasonable system linking production, use and recycling to a closed lifecycle possible? Most favorable is if the devices have high harvesting factors while the energy requirements for the recycling steps remain small.

3. Social Compatibility:

Fairness of distribution:

What benefits do the different social groups and actors have?

4. International Compatibility:

How do the developing countries benefit from PV systems in comparison to other electricity generating technologies? Is the technology of PV systems linking or further separating industrial and developing countries?

PROBABLE TENDENCY OF THE PRICING OF THE PRODUCTION INPUT FACTORS

Let us finally look at two of today's general problems caused by the prevailing economic structures: the degradation of environment due to resource and energy-intensive, emissive and open product life cycles and a non-vanishing base of unemployment due to the predominantly "Harrod"-neutral technological progress of the last decades ("Harrod"-neutral technological progress has no influence on the production factor capital, but leads to the liberation of working power. "Solow"-neutrality is the contrary.). In micro-economics, given a desired output volume, the allocation of the input factors - resources, land, energy and labour - is determined by their relative costs. Hence, in order to support efficient use of resources and energy leading to environmental benefits and on the same occasion enhance the demand for labour the former will have to become more expensive whereas the latter will have to become lower priced. For the product developer of today this means that future technological progress should be rather "Solow"-neutral and aim at high resource and energy-efficiency.

CONCLUSIONS

Most likely PV will further have to compete on non-ideal markets, without the external costs being internalised. Since the grid will further expand, grid-connected PV will have to compete with grid electricity. If energy policies adopt a paradigm comparable to the ones of sustainable development and qualitative growth, the focus of R&D has to shift from primarily aiming for increasing the efficieny towards developping devices and processes that

- realize a closed production-use-recycle loop
- are highly material efficient (especially CdTe and CIS technologies)
- preferentially substitute energy and material Input by labour.

In order to expand long-term perspectives of CIS and CdTe solar cells investigations should be started whether more indium and tellurium reserves are available and economic to win. Silicon solar cells are a possible back-stop technology for CIS and CdTe solar cells, which most likely have the highest cost reduction potentials in the close future.

REFERENCES

[1] Stephan Schmidheiny & Business Council for Sustainable Development, Changing Course - A Global Perspective on Development and Environment, Massachusetts Institute of Technology, 1992

[2] Helge Majer, Wirtschaftswachstum Paradigmen-wechsel vom quantitativen zum qualitativen Wachstum, Oldenbourg, Muenchen, 1992

[3] O. Hohmeyer, Social Costs of Energy Consumption - External Effects of Electricity Generation in the Federal Republic of Germany,Springer Verlag, Heidelberg, 1988

[4] J. Moesslein, Physical, Technical, Ecological and Economical Aspects of Thin Film Solar Cells and their Photovoltaic Systems as an Energy Option, ZSW/University of Stuttgart - IPE, 1994

[5] US Bureau of Mines, Mineral Commodities Summary 1992, Washington, 1992

[6] T.M. Brutton et al., Recycling of High Value, High Energy Content Components of Silicon PV Modules, 12th PVSEC, 1994

[7] E.A. Alsema and B.C.W. van Engelenburg, Environmental Risks of CdTe and CIS Solar Cell Modules, Proc. 11th PVSEC 1992, p. 995

[8] M. Karus et al., 8. Int. Sonnenforum, 1992, Berlin, p. 591 - 600

[9] P.D. Moskowitz, Environmental, Health and Safety Issues Associated with the Production and Use of CIS and CdTe-Based Photovoltaic Modules, Symphosium KFA Juelich, 11/12/91

[10] E.A. Alsema and B.C.W. van Engelenburg, Environmental Aspects and Risks of Amourphous Silicon Solar Cells, Dep. of Science, Technology and Society of the University of Utrecht

[11] M.H. Patterson et al, Health, Safety and Environmental Aspects of the Production and Use of CdTe Thin Film Photovoltaic Modules, 12th PVSEC, 1994

[12] G. Hagedorn and E. Hellriegel, Umweltrelevante Masseneintraege bei der Herstellung verschiedener Solarzellentypen, German Minister for Research and Technology: 426 - 3590 - PLI 14120, 1992

[13] W. Hirtz et al, KFA Juelich - STE, Umweltvorsorgepruefung bei Forschungsvorhaben - Am Beispiel von Photovoltaik, 1993

[14] K.M. Hynes, A.E. Baumann and R. Hill, Life cycle analysis of PV modules based on cadmium telluride, 12th PVSEC, 1994

MEASURING THE VALUE OF DISTRIBUTED PHOTOVOLTAIC GENERATION: FINAL RESULTS OF THE KERMAN GRID-SUPPORT PROJECT

Howard J. Wenger & Thomas E. Hoff
Pacific Energy Group
Walnut Creek, California, USA

Brian K. Farmer
Pacific Gas & Electric Company
San Ramon, California, USA

ABSTRACT

This paper presents results from the world's first power plant designed and built to measure the benefits of grid-support photovoltaic generation. The research results provide concrete evidence that non-traditional benefits are measurable and significant, doubling the overall value of the plant relative to a traditional resource planning evaluation.

A primary function of electric resource planning is to select new supply- and demand-side resources to satisfy growing demand. Traditionally, resource planning has been limited to the evaluation of avoided energy and capacity costs, and does not capture the unique benefits of distributed resources such as grid-support PV. The Kerman grid-support project provides further evidence that current resource planning practices need to be updated if distributed resources are to receive the credit they deserve.

INTRODUCTION

The primary objective of this research is to determine the magnitude of benefits, or value, provided by the Kerman grid-support photovoltaic (PV) power plant to Pacific Gas and Electric Company (PG&E). In essence, the objective is to answer the questions: Do distributed generation benefits exist and, if so, are they significant? Evaluation methods, benefits analyses, plant performance, and test results have been previously documented [1]. This paper focuses on the final results.

BACKGROUND

The Kerman PV power plant began commercial operation in June, 1993, and is reported to be the first plant designed and built to measure the benefits of grid-support PV. The plant was purchased by competitive bid, with part of the selection criteria allocated to the projected economic value the system would provide to the utility, including site-specific non-traditional benefits. Siemens Solar Industries was selected with a single-axis tracker design to enhance the capture of the afternoon solar resource for peaking power.

The Kerman plant is located several miles outside the city of Kerman, which is about 15 miles west of Fresno in California's Central Valley. The plant is rated at a nominal 500 kWac[i] and is connected to a semi-rural 12 kV distribution feeder about 8 circuit miles downstream of the Kerman substation. A 10.5 MVA transformer bank located in the Kerman substation maintains feeder voltage and supplies current to customers.

The Kerman feeder was selected after screening a total of 600 distribution feeders and 175 substation transformers in the San Joaquin Valley area [2]. The screening process was conducted primarily on the basis of the match between the solar resource and transformer and feeder loads during peak load hours. Secondary criteria were that transformer loading was nearing its rating and that load growth was sufficiently small to significantly defer the transformer replacement with a moderate PV investment.

A series of non-traditional benefits were identified and estimated prior to installing the Kerman plant [2]. Subsequently, this work was documented in a 1992 PG&E report referred to as the Kerman Case Study [3]. Prior to this work, utility planners evaluated generation resources solely on the basis of the traditional benefits of avoided energy and capacity costs. The Kerman Case Study broke new ground by developing methods to evaluate non-traditional benefits that a distributed generation resource can provide if appropriately sited. The study hypothesized that these non-traditional benefits would triple the Kerman plant's value.

The Kerman plant was built, in large part, to validate this hypothesis. A data acquisition system archives over 130 different parameters on a real-time basis, covering the Kerman solar resource, PV plant performance, and utility distribution system operation. The benefits validation is based on these data recorded over the one-year period July 1, 1993 to June 30, 1994, and on data collected during a series of special tests.

BENEFITS VALIDATION

The benefits validation is essentially a two-step process. Step 1 is technical. Measured data are combined with existing utility engineering models and improved evaluation techniques to determine the operational effect of the grid-support PV plant on the utility system. Step 2 is economic. The technical results are combined with

CH3365-4/94/0000-0792 $4.00 © 1994 IEEE
First WCPEC; Dec. 5-9, 1994; Hawaii

economic models to estimate the plant's value to the utility.

Table 1 presents an overview of the benefits evaluated and results. These results are specific to the Kerman plant and its location within the distribution system, as well as PG&E's present economic and regulatory operating environment.

Technical Results

The Kerman PV plant provides the PG&E system every benefit listed in Table 1. Data analysis and testing confirm that both traditional and non-traditional benefits are measurable and predictable for grid-support PV. By locating generation near customer loads, the plant is able to deliver benefits spanning the entire utility system; from the distribution feeder to the substation, and from the high voltage transmission system to the generation fleet.

The technical (and economic) benefits are driven in two predominate ways. First, the PV plant's annual energy production, aggregated as a function of time of day and season, drives externalities, loss savings, and energy generation benefits. Second, the extent to which PV plant capacity coincides with peak loads drives the remaining benefits of substation, transmission, minimum load, and system capacity. These two drivers are discussed below.

Table 1. Kerman PV Plant Benefits Evaluated and Validation Results ($1995).

Non-Traditional Benefits	Definition & Economics Driver	Technical Validation Results	Economic Results: Nominal ($/kW-yr)	Economic Results: High ($/kW-yr)
EXTERNALITIES	*Fossil fuel emissions reduction.* Driver: Generation fleet fuel mix and externality valuation method.	Pollution is reduced by 155 tons of CO_2 and a half a ton of NO_X each year.	31	34
RELIABILITY	*Local reliability enhancement.* Driver: Postpone planned expenditures to improve reliability.	Voltage support is predictable and almost 3 volts provided (on a 120 V base). Testing proves customer outage time can be reduced.	4	4
LOSS SAVINGS	*Real and reactive loss savings.* Driver: PV plant capacity factor and interconnection location.	Real energy losses reduced by 58,500 kWh/yr (or 5 percent of plant output). Reactive power losses reduced by 350 kVAR.	14	15
SUBSTATION	*Transformer replacement and load-tap-changer maintenance deferral.* Driver: Magnitude of planned upgrade expenditures and load growth rate.	Transformer cooled by more than 4 °C and its capacity increased by 410 kW on peak day. Load-tap-changer maintenance interval extended by more than 10 years.	10	88
TRANSMISSION	*Transmission capacity deferral.* Driver: Marginal cost of transmission capacity.	Transmission system capacity increased by 450 kW on peak.	45	45
MINIMUM LOAD	*Power plant dispatch savings.* Driver: Marginal cost of keeping peak load-following units on-line.	Minimum load savings confirmed. PV plant delivers 90 percent PV capacity coincident with peak load-following unit dispatch.	28	28
Traditional Benefits	**Definition & Economics Driver**	**Technical Validation Results**	**Nominal ($/kW-yr)**	**High ($/kW-yr)**
CAPACITY	*System reliability enhancement.* Driver: Utility need for capacity to improve system reliability.	Generation system capacity increased by 385 kW (ELCC about 77 percent.)	12	53
ENERGY	*Energy generation displacement.* Driver: Fuel price of avoided energy generation resource.	Plant achieved about 25 percent capacity factor, over 1080 MWh/yr, highly correlated to PG&E loads.	143	157
TOTAL VALUE			**293**	**424**

Technical Benefits Driver #1: PV Energy Production. Fig. 1 presents monthly energy production and Performance Index for the one year validation period. The Performance Index (PI) is a dimensionless ratio defined as actual plant energy production divided by expected plant production. The PI is an indication of how well the plant is operating. A PI of less than 90 percent, for example, generally indicates under-performance due to problems such as excessive plant downtime and malfunctioning tracking devices.

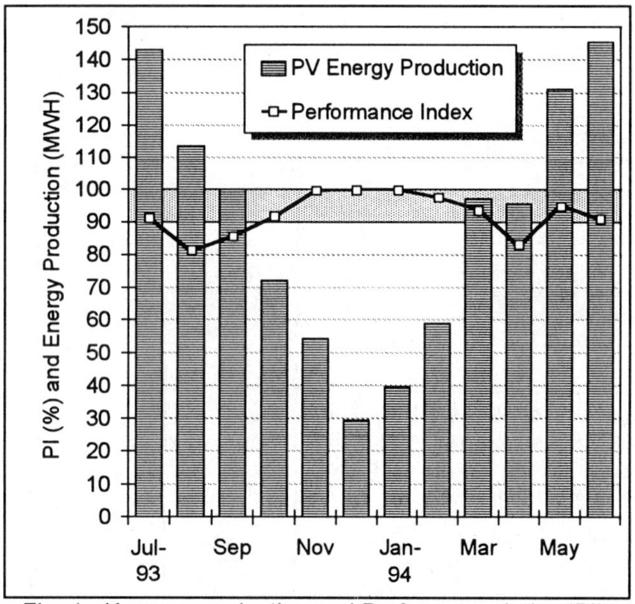

Fig. 1. Kerman production and Performance Index (PI).

The plant achieved a 91 percent Performance Index during the validation period. This indicates very good performance, especially for a PV plant's first year of operation. The monthly PI shown in Fig. 1 indicates, however, at least 3 months of substandard performance. This was due, primarily, to inverter downtime.

The plant achieved an annual capacity factor of about 25 percent, delivering over 1080 MWh of electricity. If the plant had performed as expected and achieved a PI of 100 percent, then it would have produced about 1190 MWh (a 27 percent capacity factor). Loss savings, externalities, and energy generation benefits would also be proportionately higher.

Technical Benefits Driver #2: PV-Load Match. Fig. 2 presents an example of the degree to which the Kerman PV plant's output matches peak utility loads. The Kerman Substation transformer load is shown with and without PV plant output on the summer peak day. The load peaks around 4:00 pm, at which time about 430 kW of PV plant load reduction capability, including loss savings, is available. Of equal importance is the fact that the plant provides load relief before the absolute peak occurs. This pre-cools the transformer bank. The two effects of pre-cooling and delivering power during the absolute peak combine to lower the maximum transformer temperature by 4 °C and boost transformer capacity by about 410 kW.

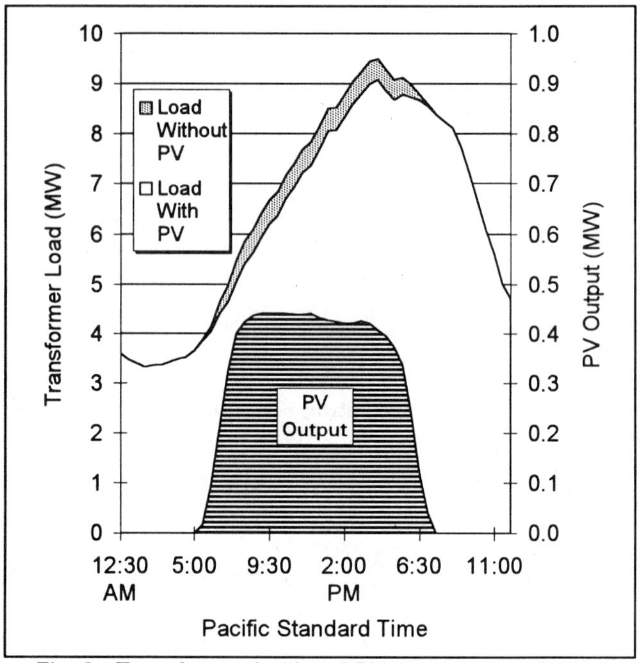

Fig. 2. Transformer load and PV output on peak day.

The favorable PV-load match extends to the transmission and generation systems. For example, the Kerman plant has a 77 percent effective load carrying capability (ELCC), or peak capacity availability, with respect to PG&E's generation system. A simplified definition of the ELCC is the fraction of the plant's rated capacity that is credited as being fully dispatchable. The ELCC allows PV to be directly compared with fully dispatchable systems. The plant increases transmission capacity by about 450 kW, reflecting an ELCC of 90 percent. The higher the ELCC, the higher the utility value.

In summary, the Kerman plant received high performance marks during the validation period from both energy and PV-load match perspectives. The details of translating the plant's real-time supply of energy and capacity into technical benefits to the PG&E system have been documented [1]. Table 1 summarizes the technical validation results.

Economic Results

The second essential step of the validation process is to translate each of the technical benefits into economic values. These values are then summed to derive the total value of the Kerman PV plant. The economic analysis is based on revenue requirements and net present value approaches commonly used by utilities to value supply- and demand-side resources. The results are presented in year 1995 dollars based on the following economic parameters: 9 percent cost of capital, 3.5 percent general inflation rate, and 30-year project life.

Economic Benefits. Fig. 3 presents the final economic value results for the Kerman PV plant. The results are presented in the form of levelized annual revenue requirements (LARR). The LARR, expressed in \$/kW-year, is an average of the actual year-by-year revenue required to recoup all plant-related investments and expenses, including a reasonable return.

Two sets of results are presented in Fig. 3. The "nominal" value of the plant, at \$293/kW-yr, represents the baseline evaluation. The "high" value of \$424/kW-yr represents a sensitivity to the Nominal Case. Because of the high degree of variation in some of the basic economics over time, it is desirable to present a realistic scenario that can help bound plant value.

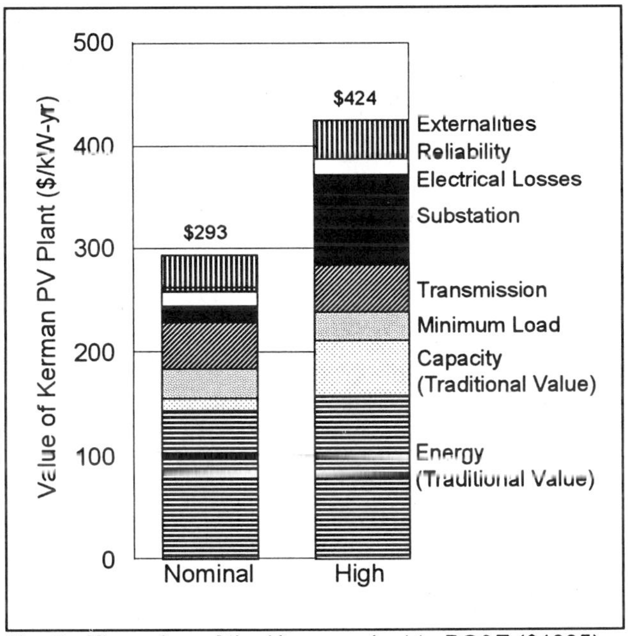

Fig. 3. The value of the Kerman plant to PG&E (\$1995).

The Nominal results are lower than anticipated due to three principal reasons:

(1) Traditional energy and capacity value components are lower because of low natural gas prices and a 1992 regulatory assumption that PG&E does not need capacity for the foreseeable future, respectively;
(2) Local reliability enhancement value is small because a capacitor bank could be added to the Kerman circuit and provide the same operational benefits at much lower cost than the value-of-service estimate used previously in the Kerman Case Study [3]; and
(3) The substation transformer value is lower because it is relatively easy to switch load in the Kerman area.

The High Case considers three factors that are different from the Nominal Case:

(1) PG&E will need bulk system generation capacity for reliability purposes sooner than forecasted in the Nominal Case;
(2) The benefits from deferring the transformer bank upgrade are derived from a "substation perspective" as delineated in the Kerman Case Study [3]; and
(3) Kerman PV plant production is increased to 1190 MWh/year reflecting that the plant will perform as expected (i.e., the plant Performance Index is 100 percent).

Regardless of which set of parameters are used, one predominant conclusion is clear: The value of the Kerman plant is *doubled* by capturing non-traditional benefits. In other words, the distributed PV plant is worth twice what it would be if evaluated as a traditional central station resource. The equivalent levelized value of the PV generation is about 14 to 20 cents/kWh as a distributed resource versus 7 to 10 cents/kWh evaluated as a central station resource.

Economic PV Price. The economic PV system price can be calculated once the value of the plant is established. The economic PV price, often referred to as the "break-even" price, is the PV system price required to just balance the value and cost of ownership, including cost of capital, O&M, rate of return, depreciation, property tax, and insurance.

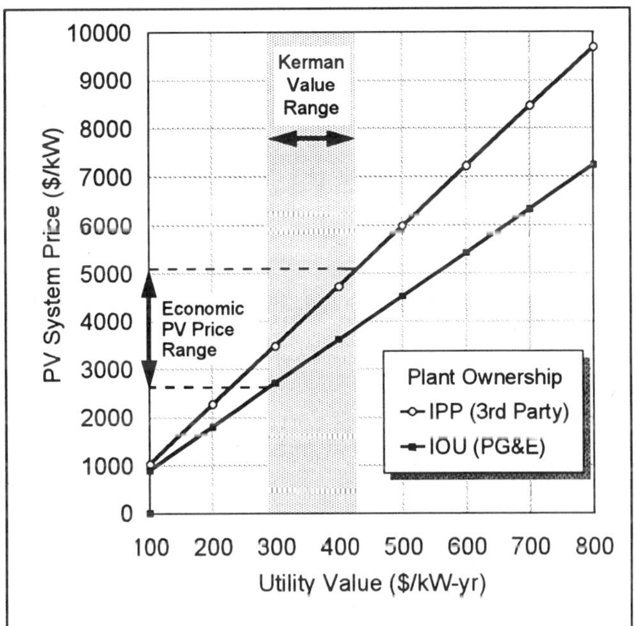

Fig. 4. Economic PV price depends on plant ownership.

Fig. 4 presents the economic PV price as a function of plant value for two ownership scenarios: Investor-Owned Utility (IOU), represented by PG&E, and Independent Power Producer (IPP).[ii] The economic PV price for the Kerman PV plant ranges between \$2,700/kW and \$3,800/kW for PG&E ownership, and from \$3,400/kW to over \$5,000/kW for IPP ownership.

The IPP scenario is considered to be as likely as utility ownership in the near-term. At least one California IOU will no longer build and own new regulated power plants in its service area in response to electric utility restructuring. A very plausible scenario is that third party IPPs, perhaps even unregulated subsidiaries of IOUs, will finance and construct distributed generation resources as this market develops.

IPPs presently have several unique advantages over IOUs to finance and build power plants. Advantages include access to tax credits, accelerated depreciation, flexibility with debt to equity ratios and financing, and, in many cases, access to sources of lower cost capital. These advantages enable the IPP to afford a 30 percent more expensive plant than an IOU, while maintaining profitability. Electric co-operatives and municipal utilities enjoy similar ownership advantages and would, in general, have an economic PV-cost-line comparable to the IPP line in Fig. 4.

CONCLUSIONS AND LESSONS LEARNED

The implications of the technical and economic results presented in this paper may be profound. Although the market price of grid-support PV systems is presently higher than the economic, or break-even, price of PV at Kerman, the validation process provides concrete evidence that non-traditional utility benefits are measurable and significant.

The following conclusions can be drawn from the Kerman Validation Project:

- Data analysis and testing confirm that both traditional and non-traditional benefits are measurable, predictable, and significant for grid-support PV;

- Non-traditional benefits double the overall value of the Kerman plant relative to a traditional central station resource planning perspective; and

- Methods to evaluate the Kerman grid-support plant are repeatable and generally applicable to other forms of distributed resources and applications.

Lessons Learned

While the technical evaluation methods are transferable to other utilities and applications, the technical results are largely dependent on the Kerman site and application. Moreover, the economic results should not be used as an absolute benchmark. These results are highly dependent on rapidly changing parameters reflected by California electric utility restructuring.

The Kerman validation results are promising from a renewable and distributed resource perspective. The research should help non-traditional benefit evaluation gain acceptance by utility and regulatory personnel for resource planning and regulatory oversight.

The Kerman site was ideal from the perspective of validating the technical benefits of distributed PV grid-support. The site was not ideal, however, from an economic benefits perspective. This is because of the relatively small economic benefit to the substation transformer due to the interconnectedness of the Kerman area allowing load transfers. Screening criteria to better optimize the economic value of future systems have been identified [1].

Finally, the Kerman Validation Project enabled the development of improved and simplified methods to evaluate any distributed generation technology. Further evolution of these methods is needed to provide credible and efficient economic justification for distributed resources.

ACKNOWLEDGMENTS

The construction of the Kerman PV plant, and this work, were funded by the PVUSA project (PV for Utility Scale Applications.) The U.S. Department of Energy and Pacific Gas and Electric Company are the primary sponsors of PVUSA, with support from other electric utilities, research organizations, and agencies. We thank the members of the PVUSA Technical Review and Steering Committees for their insight and support.

REFERENCES

[1] T. E. Hoff and H. J. Wenger, "The Value of Photovoltaics in the Distribution System: The PVUSA Kerman Grid-Support Project", PG&E and U.S. DOE, Draft Final Report 007.5-94.15, October, 1994.

[2] D. S. Shugar, "Photovoltaics in the Distribution System: The Evaluation of System and Distributed Benefits", *Proceedings of the 21st IEEE PV Specialists Conference*, Kissimmee, FL, May, 1990.

[3] D. Shugar, R. Orans, A. Jones, M. El-Gassier, and A. Suchard, "Benefits of Distributed Generation in PG&E's Transmission and Distribution System: A Case Study of Photovoltaics Serving Kerman Substation", PG&E R&D, Report 007.5-92.9, November, 1992.

---

[i] PVUSA has assigned a plant rating of 498 kWac at PVUSA Test Conditions of 1000 W/m^2 irradiance, 20 °C ambient temperature, and 1 m/sec windspeed.

[ii] IOU economic PV prices are based on a cost-of-ownership fixed charge rate of 11.04 percent. IPP economic PV prices are based on achieving a 20 percent return on equity and 10 percent return on debt, 60:40 debt to equity financing, 10 percent cost of capital, 10 percent federal solar energy tax credit, and 15 year loan life (source of assumptions and chart concept: Mike Lotker, formerly of Siemens Solar Industries.)

POLLUTION MITIGATION AND PHOTOVOLTAIC DEMAND-SIDE RESULTS FROM THE U.S. ENVIRONMENTAL PROTECTION AGENCY PHOTOVOLTAIC DEMONSTRATIONS

Edward C. Kern, Jr. and Daniel S. Greenberg
Ascension Technology, Inc.
P.O. Box 314, Lincoln Center, Massachusetts 01773 USA

ABSTRACT

The U.S. Environmental Protection Agency and 21 electric power companies in the United States are conducting two nationwide *Photovoltaic (PV) Demand-Side Management* (DSM) projects. Ascension Technology (AT) is responsible for system design, balance-of-systems equipment design and fabrication, installation supervision, field instrumentation, and data acquisition, monitoring and evaluation. The principal goal of the two projects is to investigate the environmental and demand-side-management (DSM) benefits of distributed, grid-tied PV systems as pollution mitigating energy replacements for fossil fuels. In 1993-4, 11 electric utilities installed PV systems on homes, schools, hospitals, office and light commercial buildings. In 1994-5, 12 electric utilities are installing PV systems on larger commercial, educational and industrial building roof-tops.

INTRODUCTION

Perhaps the most promising and potentially ubiquitous solar energy energy application is based upon photovoltaic (PV) conversion. Tremendous opportunities exist for PV technologies to help meet the electric energy needs in the 21st century. Costs have come down dramatically (about 70% per decade) since solar cells were first used in spacecraft applications. The current cost of energy from PV systems is 30 to 40 cents per kWh, but technology advances are expected to drop costs to under 10 cents per kWh by the year 2010. Competition with fossil-fuel-fired power generation is difficult in current bulk power markets, but PV will almost certainly be cost-effective for interconnection to electric grids within 15 years.

For grid-connected photovoltaic systems, the most immediate opportunity is on the customer-side, or demand-side, of the meter, where customer's peak demands can be partially met using PV systems. For example, in some utility service areas effective summer peaking rates range to over 40 cents per kWh. This implies that demand-side PV used to shave peak building loads might be cost-effective at today's PV prices. Additional PV value can often be found at locations where PV offsets requirements for new investment in electric transmission and distribution equipment.

U.S. ENVIRONMENTAL PROTECTION AGENCY PROJECTS

Because of solar energy's huge potential and clear environmental benefits, the U. S. Environmental Protection initiated its PV demonstration program by a competitive solicitation in the fall of 1991. Their approach has been to install and monitor solar photovoltaic energy systems at different geographic locations in the United States.

EPA Project One

In August 1992 a contract was awarded to Ascension Technology to conduct a PV DSM project. The goal of the EPA PV DSM project is being accomplished through case studies of PV systems installed by the project in 17 cities and by comparing PV DSM system performance with the actual operating characteristics of 11 project co-sponsor electric power companies: Arizona Public Service, Atlantic City Electric, City of Austin Electric Department, New England Electric, New York Power Authority, New York State Electric and Gas, Northeast Utilities, Northern States Power, Pacific Gas and Electric, Southern California Edison, and Wisconsin Public Service. Each utility has installed a 3.5-, 7.0-, or 10.5-kW rated PV system and the project PV capacity totals 101 kW_{ac}.

EPA Project Two

In September 1993 a second contract was awarded to Ascension Technology to conduct a DSM assessment of PV systems on commercial, institutional and light industrial buildings. This contract expands and builds upon the prior project by broadening the geographical coverage and by refining PV system designs for ease of installation, system safety, and electrical code compliance. Arizona Electric Power Cooperative, Atlantic City Electric, Boston Edison, Duke Power Company (North Carolina), Consolidated Edison (New York), Florida Power Corporation, Idaho Power Company, Los Angeles Department of Water and Power, Nevada Power Company, New York State Electric and Gas, Public Service Company of Colorado, and Public Service Company of Oklahoma are each installing an 18-kW PV system, a combined ac nameplate capacity of 216 kW.

CH3365-4/94/0000-0797 $4.00 © 1994 IEEE

Together with EPA project one, these systems will cover all but one of the North American Electric Reliability Council districts (the East Central Reliability Council being the exception). The twelve electric utility co-partners are being coordinated by the Electric Power Research Institute.

FIELD TESTING DISTRIBUTED PV GENERATION

The EPA approach has been to install and monitor DSM solar photovoltaic systems at locations from northern New England to Southern California. These systems supply power directly to the user, where they compete with the retail price of electric service rather than the bulk power cost to utilities. These systems are monitored for performance and reliability, with data collected to support studies of air pollution mitigation, utility generation and DSM impacts.

The EPA PV DSM monitoring began in July 1993. Electric demand and PV electric supply data are transferred daily by modem from sites across the country to Ascension Technology's offices in Waltham, Massachusetts. Example data for a 3.5-kW system on an office building in Minnetonka, MN is shown in Figure 1. Since the building load is typically greater than the PV generation, each has been individually normalized for comparison purposes. Rooftop PV's contributions range from 1 to 3 percent of daily building electric energy use on multi-story commercial buildings, to 10 to 15 percent for single story commercial buildings, to over 50 percent on single family dwellings.

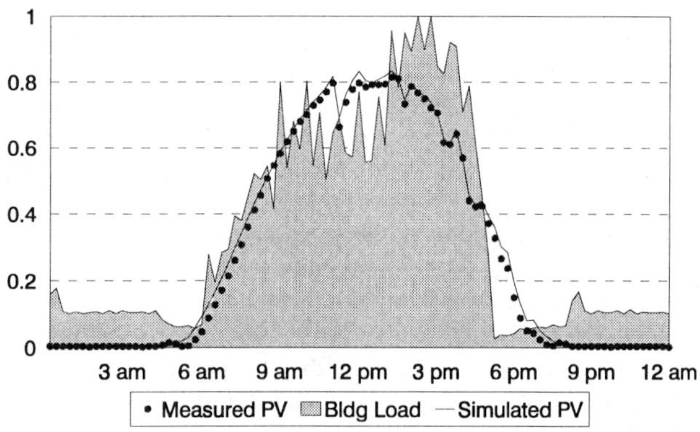

Figure 1. Normalized summer PV generation and building demand profiles for an office building in Minnetonka, MN.

POTENTIAL ENERGY AND ENVIRONMENTAL BENEFITS

Residential, commercial and industrial buildings account for 96% of the electric energy consumed in the Unites States. Depletable coal, oil and gas fossil energy reserves are used to generate 70% of this electricity and represent a significant contribution to U.S. global warming and environment-degrading gas emissions. Fossil fuels used for electric power generation account for 34% of the CO_2, 67% of the SO_x, and 37% of the NO_x emissions into the atmosphere from controllable sources. Photovoltaic generation installed where posssible in the U.S. inventory of residential, commercial and industrial buildings could produce 20% of the nation's electricity in an environmentally benign, distributed, and secure way. Table I provides a quantitative basis for considering the environmental implications of a *photovoltaic buildings* energy future.[1]

The environmental benefits of DSM using PV power supplies are significant. Determining environmental benefits requires information about the energy produced by the PV system, avoided power plant fuel consumption, and stack emissions from a mix of gas-, oil-, coal-fired generating stations. EPA intends to help quantify these benefits using the measurements from these projects. Assuming a 20% PV capacity factor and offsetting emissions from electric power generation using today's mix of power plants, we estimate that each megawatt of PV generating capacity will mitigate approximately 1,100 metric tons of carbon dioxide emissions to the atmosphere each year. If all U.S. residences were fully equipped with rooftop PV generators over 170 million metric tons less CO_2 would be emitted annually (representing 3.1% of controllable CO_2 emissions). Though PV market projections are rough, the potential environmental benefits of PV are clearly large.

Table I. Potential emissions mitigation from U.S. PV installations on buildings (percent of controllable emissions)

Emission	RES	COM	IND	Total
CO_2	3.1%	1.3%	1.5%	5.9%
SO_x	6.2%	2.7%	2.9%	11.9%
NO_x	3.7%	1.6%	1.7%	7.1%

ENERGY AND ENVIRONMENTAL BENEFITS

EPA PV DSM monitoring began in July 1993. Two primary data sets are being recorded for each PV system: energy produced by the PV system and energy demanded by the host building. In order to verify the performance of the PV array and the metering data acquisition system, both a plane-of-array irradiance pyranometer and a rotating shadowband pyranometer to measure global, direct and diffuse irradiance have been installed. This allows for cross-checking system performance through the use of array and power conditioner simulation algorithms. It also provides verification of the translation algorithm from direct normal and diffuse irradiance to plane-of-array irradiance, an essential tool in the translation of the results of the project to other potential PV system designs and to sites with similar solar resource data.[2]

Participating utilities provide hourly records indicating which load-following power plants (marginal or "swing" plants) are operating as well as total utility system load. They also supply emissions data for their load-following generating stations. Currently the project tracks carbon dioxide, nitrous oxides, sulfur dioxide and particulate emissions.

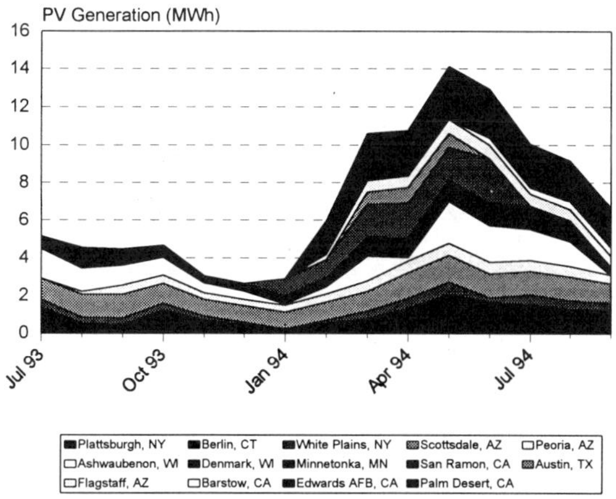

Figure 2. Photovoltaic power generation from the EPA Project 1 sites from July 1993 through September 1994.

Figure 2 illustrates the energy that has been produced by the EPA Project One systems beginning in July 1993. Variations in energy production result from several factors: the rating of the systems which range from 3.5 to 10.5 kW-rated, the date of system installation and start-up, the solar resource available for conversion to electricity, and the system reliability. The largest systems in Pleasantville (NJ), Scottsdale (AZ), and Ashwaubenon (WI) have demonstrated the greatest energy production. These have shown comparable generation, but it is interesting to note that the Arizona system is rated at 7 kW while the other two are 10.5 kW. This results from the significantly larger solar resource in the Southwest. The Plattsburgh (NY) system is also 10.5 kW, but has suffered from several power conditioner failures and heavy snow coverage.

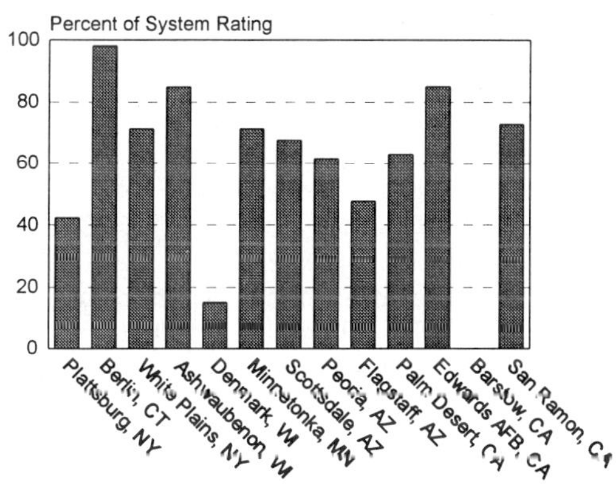

Figure 3. PV generation as a percent of rated capacity available at the time of building summer peak demand.

Figure 3 presents the building peak demand reductions resulting from the connection of PV on the customer's side of the demand meter. Note that these values are expressed as a percent of the rated PV generation capacity rather than as a percent of the building peak demand. In this way the value of the photovoltaic system as a demand reduction technology can be made by multiplying the rating by the percentage. The reader should use care interpreting this figure. While the figure does represent the actual behavior of the individual PV systems, it is based on a preliminary analysis of 12 or fewer months of data. The percent of PV system indicated is that observed at the time of the highest summer peak load recorded for each individual building.

In areas of high electric power generation cost, a 1 kW reduction in demand for a month can be worth up to \$10. Since the PV generation is, to a first approximation, symmetrical about solar noon (1 PM for daylight savings time), the best correlation of PV generation to building loads occurs when the building load peaks in the early afternoon. This happens routinely in commercial buildings with significant air conditioning demands that peak in the mid-afternoon. Many studies of photovoltaic system worth show this capacity value is comparable to the energy value. With the addition of energy storage, the capacity worth can be increased to 100% or even more than the PV array rating, depending on the sizing of PV arrays, batteries, and the dc to ac converter.

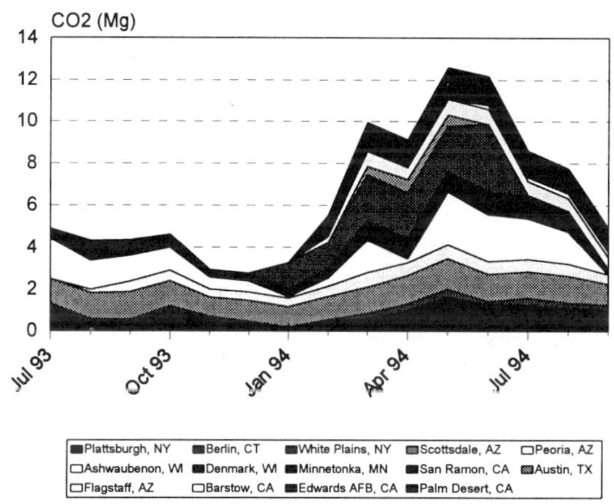

Figure 4. Estimated reduction in CO_2 emissions by the EPA Project 1 installations.

Figure 4 illustrates the CO_2 emission reduction benefit of photovoltaic power generation. These are preliminary results based on performance data from the various photovoltaic systems and CO_2 emissions data provided by the co-sponsor electric utilities. Monthly reductions in CO_2 emissions range between 70 and 180 kg per kW of PV system rating during the summer months. The variance among the different sites depends on the same factors that cause variations in total electricity production and the two-to-one variation in CO_2 emissions between coal-fired and gas-fired generation.

Similar results for SO_2 emissions are shown in Figure 5. Again, the results vary widely (from 15 to 800 grams per kW of system rating) because regional fuel-use differences. Where PV displaces coal-fired generation the environmental benefits are the greatest.

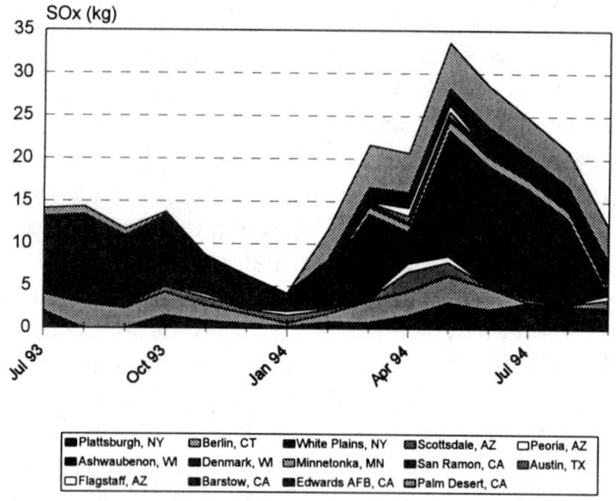

Figure 5. SO_x emissions reductions for the EPA Project 1 systems installations.

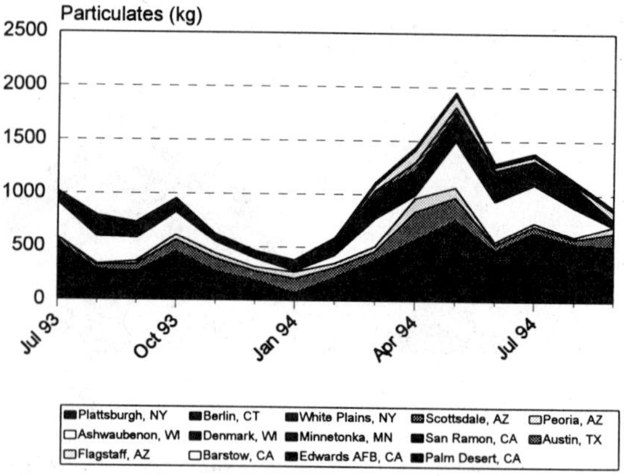

Figure 7. Reduction in particulate emissions from the the EPA Project 1 installations.

Finally, Figures 6 and 7 present the results for NO_x and particulate emissions respectively. Preliminary indicatations are that avoided NO_x emissions range from 150 to almost 1,000 grams per month per kW of PV capacity, and that particulate emissions are reduced by between 6 and 90 grams per month per kW of capacity.

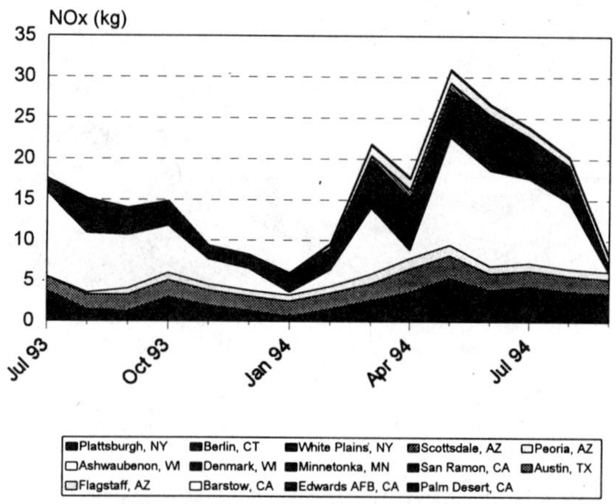

Figure 6. NO_x emission reductions for the EPA Project 1 installations.

Data reduction techniques for assessing these avoided emissions are still under development and testing, so it is important to recognise these as *preliminary* results. Further, there is considerable room for debate as to the correct algorithm to use for determining environmental benefits. For example, many of the emission rates used in this study are for coal plants that are currently being equipped with "end of the stack" emissions control technology. Future emissions savings from PV will not materialize if the current high emissions rates are substantially reduced by these and other measures to improve existing fossil electric power generation stations.

Another consideration is the dispatch order assumed for this study. It has been assumed that PV represents a small fraction of the electric generation capacity so that plant unit commitment and dispatch are not effected. Accordingly, we have assumed that those power plants that are operating under automatic generation control are the ones whose emissions will be reduced. When and if photovoltaic power achieves a substantial fraction of its potential, electric power companies should be able to make accurate short run (e.g. 24-hour) projections of PV generation. At such time, PV will likely displace intermediate plants with slower response (ramping) characteristics and potentially greater air emission offsets since these plants tend to be oil- or coal-fired, whereas load following plants tend include cleaner gas and hydro-electric stations. This suggests that the current estimates might underpredict the emission savings.

Finally, this study is based on the presumption of economic, rather than environmental, dispatch. If power systems are controlled so as to minimize harmful emissions from fossil-fired plants, the emissions reductions from the PV plants' operations would be reduced.

Acknowledgements

This project has been funded in part by the Environmental Protection Agency under contracts 68D20148 and 68D30144. It *is preliminary and* has *not* been subject to the Agency's review, and it has *not* been approved for publication as an EPA document. Mention of trade names or commercial products does *not* constitute endorsement or recommendation for use. Readers are advised that the results presented are preliminary and subject to revision.

Notes

1. The figures presented in Table 1. are Ascension Technology estimates based on data presented in the *Statistical Abstract of the United States*, 1992

2. For a discussion of performance modelling and verification algorithms please see a companion paper, Russell, M.C., "Grid-Tied PV System Modelling: How and Why", *First World Conference on Photovoltaic Energy Conversion*, Waikoloa, Hawaii, December 5-9, 1995

DISTRIBUTED PHOTOVOLTAIC DEMAND-SIDE GENERATION: AN ECONOMIC EVALUATION FOR ELECTRIC UTILITIES

Gregory J. Leng
Energy Diversification Research Laboratory, CANMET, Natural Resources Canada
1615 Montee Ste-Julie, P.O. Box 4800, Varennes, Quebec J3X-1S6 CANADA

José Martín
University of Massachusetts Lowell, Department of Chemical and Nuclear Engineering
1 University Avenue, Lowell, MA 01854 USA

ABSTRACT

From an electric utility's viewpoint, distributed photovoltaic demand-side generation (PV-DSG) systems can be evaluated in a similar manner to other demand-side management (DSM) technologies. This study evaluates the economic benefits of PV-DSG systems, using hourly utility cost and performance data, as a function of the utility's load duration curve (LDC). The analysis focuses on one utility, New England Electric System (NEES). Actual utility hourly system load data for 1991 and corresponding PV output data from a 2.2 kW (DC) grid-connected residential PV system, installed as part of the NEES Gardner project, are used for this study. At NEES's weighted average cost of capital of 8.78% the energy and capacity benefit values calculated equate to an allowable installed PV system cost of $2.41/watt. A social discount rate of 3% allows for an even higher installed cost-per-watt figure of $4.72. The higher the allowable installed cost, the easier it will be for the PV industry to enter the utility market. These results are represented graphically as a target to emphasize the importance of the utility target market to the PV Industry. [1,2]

INTRODUCTION

Conventional methods for measuring the financial value of utility connected photovoltaic power generators fail to capture the true economic benefits of these systems. Typically, only the annual energy production and generation capacity value for central power stations are considered by the utility when evaluating new generation technologies. However, with the implementation of Integrated Resource Planning (IRP) in several U.S. states and with the success of DSM, this situation is changing. IRP methods enable utilities to compare the relative costs (including environmental costs) of traditional supply resources with reductions in customer demand through efficiency and customer load management improvements . [3]

In this study, the economic benefits of photovoltaic power generators that are distributed throughout the utility network and located at or near the site of the utility customer are evaluated. PV-DSG will have more value to the electric utility when the distributed, demand-side management and environmental benefits of these systems are considered. By evaluating the benefits of PV-DSG, a PV firm can identify higher value applications for PV technology within the utility sector. [4,5] Also, by considering lower discount rates in the evaluation of PV-DSG systems, the kWh cost of PV produced energy is reduced considerably. Another way to look at this is that the allowable installed cost in $ per Watt peak capacity for these systems can be increased considerably resulting in a more obtainable $ per Watt installed cost target for the PV Industry.

METHODOLOGY

The analysis focuses on one utility, New England Electric System. Actual utility hourly system load data for 1991 and corresponding PV output data from a residential grid-connected PV-DSG system (2.2 kW DC) installed as part of the NEES Gardner project are used for this study. This study incorporates methods used by NEES to evaluate other DSM technologies. [6]

The benefits evaluated are calculated as a function of the utility's annual load duration curve. By looking at the photovoltaic generator output at major blocks (i.e. base load, intermediate and peak power load) along the LDC, one can estimate more accurately the true economic benefits of distributed PV-DSG, by capturing the time dependency of overall energy and capacity costs. [7]

Energy Benefit Calculation

The energy benefit is calculated for different load blocks along NEES's load duration curve as a function of the kWh saved by the PV-DSG system, the avoided cost of energy to the utility, including an environmental externalities adder and line loss multiplier for the distributed generator. The energy benefit of a DSM measure to NEES is defined as follows: [6]

$$EB = dkWh \times (ACE + EEA) \times LLM \qquad (1)$$

where,
- EB = energy benefit
- dkWh = energy saved by DSM measure
- ACE = avoided cost of energy
- EEA = environmental externalities adder
- LLM = line loss multiplier

CH3365-4/94/0000-0801 $4.00 © 1994 IEEE

First WCPEC; Dec. 5–9, 1994; Hawaii

Capacity Benefit Calculation

The capacity benefit is calculated as a function of the kW reduction capability of the PV system during the utility peak period, the utility's avoided generation capacity costs, the marginal T&D costs, including a reserve margin and a T&D loss multiplier for the distributed generator. The capacity benefit of a DSM measure to NEES is defined as follows: [6]

$$CB = dkW \times (T\&DLM) \times ((RMM \times ACC) + (MT\&DC)) \quad (2)$$

where,
- CB = capacity benefit
- dkW = kW reduction (capacity saved)
- T&DLM = T&D loss multiplier
- RMM = reserve margin multiplier
- ACC = avoided capacity costs
- MT&DC = marginal T&D costs

An estimate of the average generation capacity "available" from the PV system during NEES's peak demand period is calculated by taking the average PV system output during the peak 60 utility load hours. [5]

Dollar Per Watt Calculation

Given the annual stream of energy and capacity benefits calculated for NEES, the maximum capital cost ($ per Watt installed peak capacity) that the utility should be willing to pay for such an investment can be calculated. The present worth of a series of annual benefits are calculated for three different discount rates. These are a social discount rate of 3%, the 8.78% weighted average cost of capital used by NEES, and a higher risk related discount rate of 12%. The inflation rate for energy is assumed to be 8% based on NEES's IRP projections for fuel costs. A 3% inflation rate is assumed for the capacity benefit stream, the same rate NEES used to calculate the avoided cost of capacity in their IRP. In addition, operation and maintenance costs for the PV-DSG system are assumed to be zero for this analysis given some uncertainty in the benefit values. The formula for this calculation is as follows: [8]

$$\text{Allowable cost} = (1/d\text{-}i)x[1-\{(1+i)/(1+d)\}\hat{\ }N]x(\text{Benefit\$})\}/(DC) \quad (3)$$

where,
- d = discount rate
- i = inflation rate
- N = number of years benefit provided
- Benefit$ = benefit value calculated above
- DC = PV power rating (DC, Watts)

RESULTS

Energy Benefit

The energy benefit of the PV-DSG system is calculated for the different load blocks along NEES's annual load duration curve, considering the corresponding PV output along this curve (see Fig. 1). [1]

A summary of the energy benefit calculation for the different load blocks is presented in Table 1. [1,2] Utilities typically operate different generators during different load blocks. This is reflected in the different values for avoided energy costs and estimates for environmental externalities. [9] In addition, line energy losses are greater during peak demand periods.

Fig. 1. NEES's load duration curve & corresponding PV output.

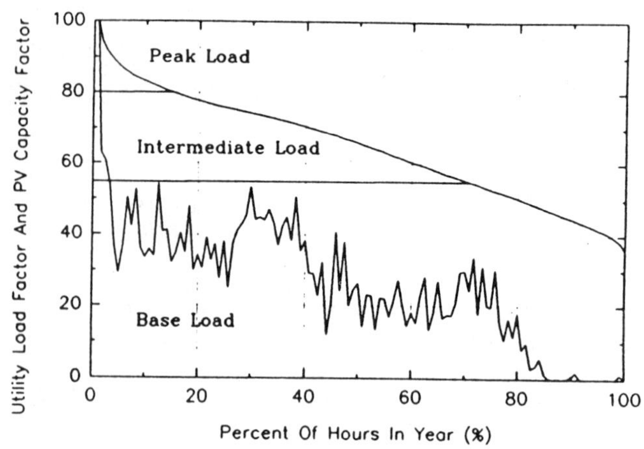

Table 1. NEES energy benefit calculation .

UNIT DESCRIPTION	PEAK	INTERMEDIATE	BASE LOAD	TOTAL
PV kWh SAVED	272	1,638	356	2,265
AVOIDED COST OF ENERGY (CENTS/kWh)	4	2	1.5	1.9
ENVIRONMENTAL EXTERNALITIES ADDER (CENTS/kWh)	2	2	2.5	2.2
LINE LOSS MULTIPLIER	18%	9%	5%	8%
ANNUAL ENERGY BENEFIT VALUE ($/YEAR)	$ 19.26	$ 71.42	$ 14.95	$ 105.63

Capacity Benefit

The utility is interested in the kW reduction (capacity) capability of the DSM measure during the utility peak demand period. NEES is a summertime peaking utility. Fig. 2. presents the normalized peak 60 annual load hours for NEES and the corresponding PV output for these hours during 1991. The average capacity available during NEES's peak load hours is used to determine the capacity credit of the PV-DSG system. [1]

NEES only anticipates a need for peak generation capacity addition during their current IRP. Therefore, the capacity value of the PV-DSG system was calculated for peak periods only. The distributed generator (like DSM) allows the utility to avoid or postpone additions to both generation capacity (including reserve margins) and the transmission and distribution (T&D) system. In addition, by generating power at or near the customer site capacity losses during peak periods are also reduced. The results of the capacity benefit calculation are presented in table 2. [1,2]

Fig. 2. Peak 60 NEES load hours & corresponding PV output.

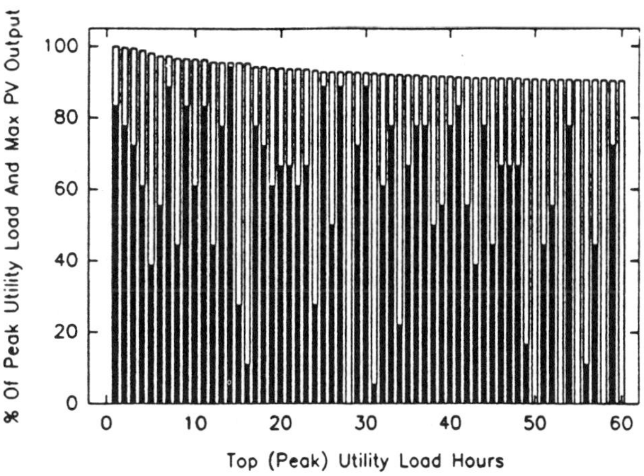

Table 2. NEES capacity benefit value calculation.

UNIT DESCRIPTION	PEAK
PV DESIGN CAPACITY	1.8 kW
PV CAPACITY FACTOR	55%
CAPACITY SAVED	0.99 kW
AVOIDED COST OF CAPACITY	$ 77.12/kW-yr
MARGINAL T&D COSTS	$ 118.84/kW-yr
T&D LOSS MULTIPLIER	15%
RESERVE MARGIN MULTIPLIER	15%
ANNUAL CAPACITY BENEFIT	$ 236.27/year

The Utility Target Market Model

The allowable installed cost of the PV-DSG system based on an annual stream of benefits to the utility (energy benefit of $105.63/year and capacity benefit of $236.27/year) for different discount rates are presented in table 3. [1,2]

Table 3. PV Allowable Installed Cost Per Watt

DISCOUNT RATE	ALLOWABLE SYSTEM COST (Energy + Capacity = Total $/Watt)
3%	2.15 + 2.57 = $ 4.72/Watt
8.78%	1.02 + 1.39 = $ 2.41/Watt
12%	0.72 + 1.05 = $ 1.77/Watt

These results can be represented graphically in the form of a target which emphasizes the focus of the PV industry on the utility target market. The values ($ per Watt) calculated for both the energy and capacity benefits at the 3% and 8.78% discount rates can represent radial distances from the center of a target. For example, the energy benefit at the 8.78% discount rate will have a radial unit of 1.02 ($ per Watt) while the capacity benefit at this discount rate ($1.39 per Watt) added to the energy benefit radius gives the total allowable installed cost of $2.41 per Watt for PV-DSG systems. In other words, the PV industry is "shooting the installed system cost arrows" (currently in the $7 to $8 per Watt range) at this target. When PV manufacturers reduce their costs down to the $2.41 per Watt range, the utilities should be willing to consider investing in PV-DSG.

Reducing the discount rate to 3%, the utility target or allowable installed system costs increases considerably. At this discount rate the energy benefit in radial units is $2.15 per Watt and the capacity benefit allows for a installed cost of an additional $2.57 per Watt. This brings the outer rim of the target to $4.72 per Watt!

Figure 3 represents this utility target for the energy and capacity benefits converted into allowable installed PV-DSG system costs. The outer radii of the two white areas shown represent the energy and capacity benefits at the 8.78% discount rate. The axis at the bottom is in $ per Watt radial units. Superimposed on this (white circles) figure is the energy and capacity benefits represented by the radii of the two outer black areas. This is done to emphasize how the utility target increases in size with the use of the lower discount rate. In other words, the PV industry has an easier target to "hit" (i.e. larger allowable installed PV-DSG system costs) based on their current price of systems to utilities. [1,2]

Fig. 3. Utility Target Market Model.

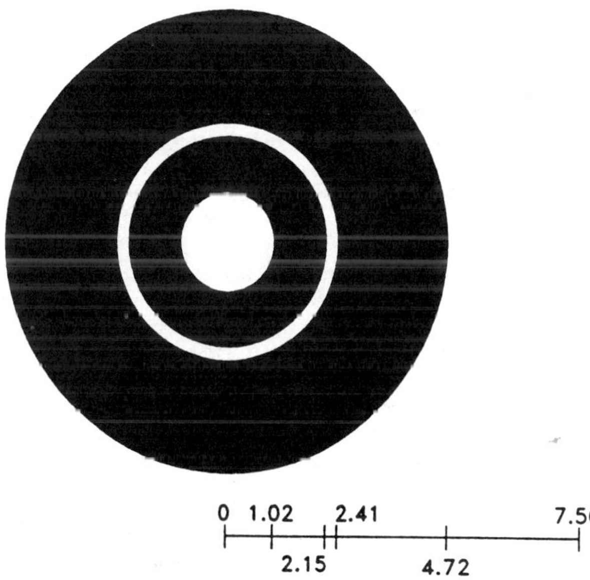

DISCUSSION

With a 3% "social" discount rate and including both the energy and capacity benefits of PV-DSG, the allowable installed cost of PV-DSG systems are roughly $4.70 per Watt to NEES! Given that the price of recent installations by the Sacramento Municipal Utility District (SMUD) were in the $7 to $8 per Watt range, the PV industry is not that far away in terms of cost-effective (higher value PV-DSG) utility applications of PV. However, with NEES's 8.78% discount rate, the allowable cost is in the range of $ 2.40 per Watt, which is more difficult to obtain. Of course, from a business stand-point it is reasonable for NEES to use a higher discount rate than a social discount rate of 3%; however, based on the argument for risk reduction because of more stable fuel costs, etc. [10,11] from PV, a discount rate somewhere between the utility's weighted average cost of capital and the social discount rate may be more appropriate for evaluating PV systems. Certainly, the discount rate issue warrants further study. [1,2]

In this model the energy and capacity benefits are represented with different circles. This is done to emphasize the importance of capturing the capacity value for PV-DSG. If only the energy benefits of PV systems are considered, the energy benefit at the 3% discount rate provides a smaller target with a radius of $ 2.41 per Watt. At the 8.78% NEES discount rate the target only has a $ 1.02 per Watt radius. That is a hard target for PV manufactures to hit! [1,2] Determining the capacity value of intermittent generators such as PV warrants further study.

A majority of the utility cost information used here is available in NEES's Integrated Resource Plan and supporting documents. This is public information. The reader can usually access this sort of information through their local utility regulatory body (if applicable). This model provides a means for electric utilities, utility regulators, and the PV industry to visualize the cost-effectiveness of PV-DSG systems in the utility market. In addition, the model can be used to evaluate other renewable energy and conservation technologies. [1,2]

ACKNOWLEDGMENTS

This study is based on the Masters Degree Thesis prepared by Gregory J. Leng at the University of Massachusetts Lowell as part of the requirements for the degree of Master of Science in Energy Engineering. Professor José Martín, Chairperson for the Department of Chemical and Nuclear Engineering, was Greg's thesis advisor.

There are several people who we would like to thank for their help and support of this project. First, Professor Supriya Lahiri of the Economics department and Professor John Duffy of the Mechanical Engineering department at UMass Lowell for providing guidance as members of the thesis committee. Their comments and suggestions have been very helpful. Also, Dr. William Berg of the UMass Lowell Center for Sustainable Energy and Dr. Douglas Smith and Mr. Michael Lesser of deLucia and Associates in Cambridge, MA. They were very helpful in developing the initial methodology for this study. Dr. John Bzura of New England Electric System was gracious to provide the utility data and Mr. Dan Shugar of APS (formerly with Pacific Gas and Electric) offered inspiration for this study topic.

REFERENCES

[1] G. Leng, Photovoltaic Demand-Side Generation: An Economic Evaluation For Electric Utilities, Masters Degree Thesis, University of Massachusetts Lowell, November (1993).

[2] G. Leng and J. Martín, The Utility Target Market Model, Proc. of the Solar Energy Society of Canada Annual Meeting, October (1994).

[3] United States Energy Association, Getting Down To Business: A Strategy For Energy Efficiency In The United States, Strategic Planning For Energy And The Environment, **Vol.12, No.2**, Fall (1992).

[4] D.S. Shugar, Photovoltaics In The Utility Distribution System: The Evaluation Of System And Distributed Benefits, Proc. of the 21st IEEE PV Specialists Conference, Kissimimee, FL (1990).

[5] D.S. Shugar, Personal Communication, employee PG&E, San Ramon, CA, August (1993).

[6] Integrated Least Cost Resource Plan For The Fifteen Year Period 1992-2006, Composite Version, Volume 1: Integrated Resource Plan, submitted to New Hampshire Public Utilities Commission by GSECO a NEES Company, April (1992).

[7] T.E. Hoff, The Value of Photovoltaics: A Utility Perspective, Proc. 19th IEEE PV Specialists Conference, May (1987).

[8] J. A. Duffie and W.A. Beckman, **Solar Engineering Of Thermal Processes**, John Wiley and Sons, Inc., New York, (1991).

[9] **Environmental Costs of Electricity**, PACE University Center For Environmental Legal Studies, Oceana Publications, Inc., New York, NY, (1991).

[10] J. Martín, H. de la Fuente, and P. Mulas, Nuclear Cost Uncertainties and Energy Choices, Trans. American Nuclear Society - Vol.21 (1975).

[11] S. Awerbuch, The Role of Risk and Discount Rates in Utility Integrated-Resource Planning, **The Electricity Journal, Vol.6**, No.2, March (1993).

SOLAR ENERGY MAPPING BY USING CLOUD IMAGES RECEIVED FROM GMS

Kenji Otani*, Tadashi Saitoh*, Izumi Tsuda**, Kosuke Kurokawa**
* Tokyo A & T University, ** Electrotechnical Laboratory

ABSTRACT

Solar energy maps that indicate the wide-ranging spatial distribution of solar irradiation are required by the researchers of the solar power systems. However, the irradiation measurement networks at ground level are not enough to obtain reliable information of the solar energy distribution in the world. On the other hands, Geostationary Meteorological Satellites (GMS) have provided the images of cloud fields over the whole surface of the earth. The main cause of an irregular change in the irradiation at the ground level is effect of the clouds; therefore, the methods for estimating the irradiation by using the GMS images may be very useful. The solar irradiation for 14 sites in Japan was estimated by the semi-physical model we have developed. The rms errors, averaged over 14 stations, were 0.0804kWh/m^2 for hourly estimates, 0.534kWh/m^2 for daily estimates and 7.14kWh/m^2 for monthly estimates. There were little regional differences of estimate accuracy. An example of solar energy maps for Eastern Asia in 1992 is presented.

INTRODUCTION

The knowledge of the available solar irradiation at the earth's surface is essential to many solar power systems in terms of their design, site selection, and performance efficiency. Solar energy maps that indicate the wide-ranging spatial distribution of the irradiation are required by the researchers not only of the solar power systems but also of the meteorology and the agronomy. However, the irradiation measurement networks at ground level are not enough to get reliable information on the solar energy distribution. On the other hands, Geostationary Meteorological Satellites (GMS) have provided the images of cloud fields over the whole surface of the earth. The main cause of an irregular change in the irradiation at the ground level is effect of the clouds; therefore, the methods for estimating the irradiation by using the GMS images may be very useful.

The models of estimating the irradiation by the cloud images have been already proposed by several organizations[1][2][3]. Authors also have developed the following two different approaches of estimating the irradiation: (i) simple regression approach with the knowledge of actually measured irradiation, (ii) two steps approach by estimating ground albedo in parallel[4][5]. In terms of treatment of the interaction between the irradiation and atmosphere or between the irradiation and the ground albedo, these approaches may be classified into statistical and semi-physical models respectively[6]. The latter model has the advantage of generality because of the removal of regional differences in model parameters. It may be applied anywhere without knowing actually measured irradiation data on the ground level. This approach seems to be preferable for making the solar energy map.

This time, the solar irradiation cumulated hourly, daily and monthly was estimated by the above semi-physical model for 14 sites in Japan. Estimated values were calculated with the same regression coefficients for any locations; nevertheless, there was little regional difference of estimate accuracy. The correlation coefficients, averaged over 14 stations, were 0.942 for hourly estimates, 0.959 for daily estimates and 0.985 for monthly estimates; the rms errors were 0.0804kWh/m^2, 0.534kWh/m^2 and 7.14kWh/m^2, respectively.

An example of solar energy maps for Eastern Asia in 1992 is presented. These maps may compensate for the lack of solar irradiation measurements, where the solar power systems will develop; such as on the sea and on desert areas.

METHODOLOGY

Estimating hourly irradiation

The GMS images in the visible and in the infrared are available at hourly intervals in the daytime. Hourly irradiation was estimated by using a visible image observed instantaneously by GMS VISSR (Visible Infrared Spin Scan Radiometer), whereas the ground truth data of the hourly irradiation measured at pyranometer stations were obtained by accumulating instant measurements

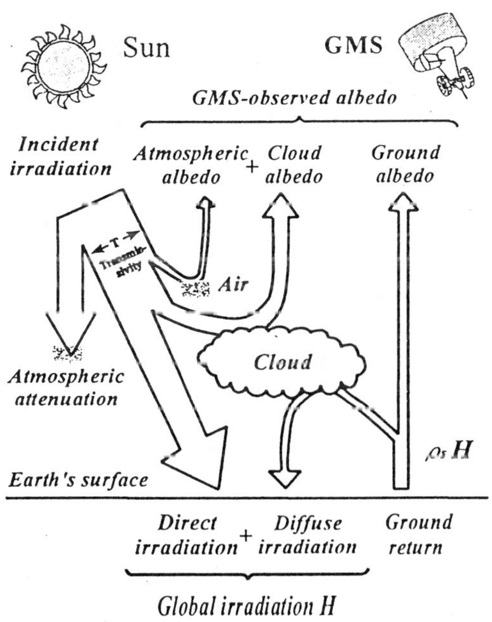

Fig. 1. Semi-physical model

CH3365-4/94/0000-0805 $4.00 © 1994 IEEE
First WCPEC; Dec. 5-9, 1994; Hawaii

during an hour. Fig. 1. shows the solar irradiation balance in the visible spectral band, and the semi-physical model was induced by this model.

In the absence of cloud, global irradiation H, received at the ground, is supposed to consist of only direct irradiation that is a residual after extraterrestrial irradiation has been attenuated by atmospheric absorption and scattering. The hourly amount of the direct irradiation transmitted after absorption and scattering under the clear sky is

$$H_{clear} = \int_0^1 T(t)\, I_0 \cos z(t)\, dt \approx T(0.5)\, I_0 \cos z(0.5) \times 1^h \quad [kWh/m^2] \qquad (1)$$

where, I_0: solar constant ($1.367 kW/m^2$), z : solar zenith angle and T: atmospheric transmittance under clear sky. The expression, t=0.5, represents half an hour. This time, we gave the assumption that the atmospheric transmittance T is obtained with following two-parameters empirical relationship:

$$T(t) = (a_1 + a_2 \cdot \cos z_{noon})^{1/\cos z(t)} \qquad (2)$$

where, z_{noon}: solar zenith angle at noon, a_1 and a_2: model parameters. a_2 is the corrective factor of the seasonal variation of the transmittance.

Under cloudy sky, two effects of clouds are considered: attenuation of incident irradiation by cloud reflection, and downward reflection referred to as the multiple scattering. The global irradiation is the sum of the direct irradiation and the diffuse irradiation. In our semi-physical model, hourly global irradiation is given by the equation:

$$H = \frac{T(0.5) - \rho_p}{1 - \rho_s} I_0 \cos z(0.5) + a_3 \quad [kWh/m^2] \qquad (3)$$

where, ρ_p: GMS-observed albedo, ρ_s: ground albedo and a_3: model parameter. ρ_p is given by converting the GMS brightness data with the correlation of the cosine law of illumination, and by averaging over 3 × 3 pixels surrounding their corresponding areas. Since the ground albedo ρ_s varies according to the location and the season, reference maps of the ground albedo must be produced and be daily updated in order to acquire its exact information. a_3 compensates for the loss of estimated values in terms of sky diffuse irradiation when the solar zenith angle is close to a right angle.

A total of 3 model parameters, a_1, a_2 and a_3, to be determined by regression, may be independent of the site where the estimation of the irradiation is operated, because of the removal of the regional singularity by considering the ground albedo.

Estimating daily and monthly irradiation

Since GMS provides the visible images ten times per day in summer and eight times per day in winter, more images are usable to estimate daily and monthly irradiation than to estimate the hourly irradiation. However, the images were not available every hour because of the matter of a GMS system, the failure in receiving and so on. In order to fill gaps between the images, an interpolation technique (the trapezoidal integration method) was used for estimating the hourly irradiation. The exact time of sunrise and of sunset for each day were calculated to decide limits of the integration. The monthly cumulative irradiation was calculated as the sum of daily irradiation within a certain month.

Estimating ground albedo

On the clear sky, the dominant element of GMS-observed albedo is ground albedo, which depends on the ground conditions, such as moisture content and vegetation. In order to estimate the irradiation on various ground conditions, the ground albedo must be estimated exactly. If the ground albedo can be detected by GMS images, the semi-physical model may be available for anywhere, with deciding model parameters at a certain site at the least.

Since the ground albedo is much lower than the cloud one (except on a snow-covered surface), and the life-time of cloud is generally shorter than that of snow, it is proposed that the minimum value of the albedo derived at a certain site within a certain period is taken as the index of the ground albedo. Fig. 2. shows, as an example, the plot of daily GMS-observed albedo over *Asahikawa*, Japan. Solid lines indicate snow depth and estimated ground

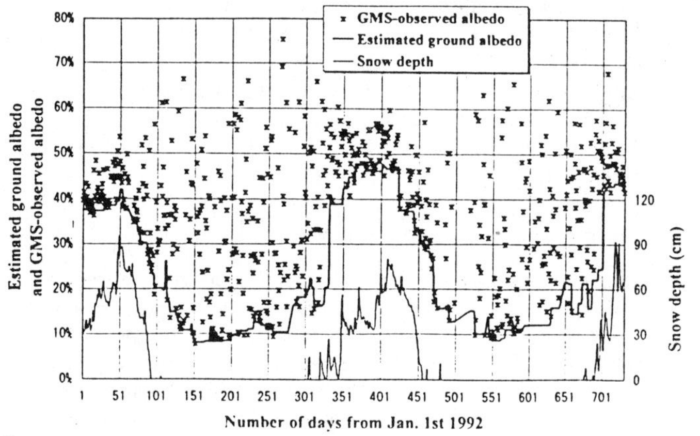

Fig. 2. Variation of estimated ground albedo and of snow depth from Jan. 1992 to Dec. 1993

albedo respectively. The ground albedo was obtained by connecting the minimum value of the GMS-observed albedo for a term of previous 14 days. Agreement with the variation of the ground albedo and of the snow depth indicates the validity of this simple method for estimating the ground albedo. Therefore, estimated ground albedo ρ_s was assigned the minimum value of ρ_p observed within a certain period. In terms of the simple technique of remote sensing, not many ground-based measurements must be used. This approach needs only a time series of GMS-observed albedo, and was effective to estimate the ground albedo in spite of its simplicity.

RESULTS

Our semi-physical model was tested at 14 pyranometer stations, which are listed in Table 1. In order to verify that little difference in the value of the model parameters

Table 1. Geographical coordinates of 14 stations

Station	Latitude(° N)	Longitude(° E)
Asahikawa	43° 25'	142° 22'
Obihiro	42° 55'	143° 13'
Morioka	39° 42'	141° 10'
Yamagata	38° 15'	140° 21'
Fukushima	37° 45'	140° 29'
Maebashi	36° 24'	139° 4'
Chichijima	27° 5'	142° 11'
Matsumoto	36° 15'	137° 58'
Kohfu	35° 40'	138° 33'
Nagoya	35° 10'	136° 58'
Hikone	35° 16'	136° 15'
Osaka	34° 41'	135° 31'
Nara	34° 42'	135° 50'
Saga	33° 15'	130° 18'

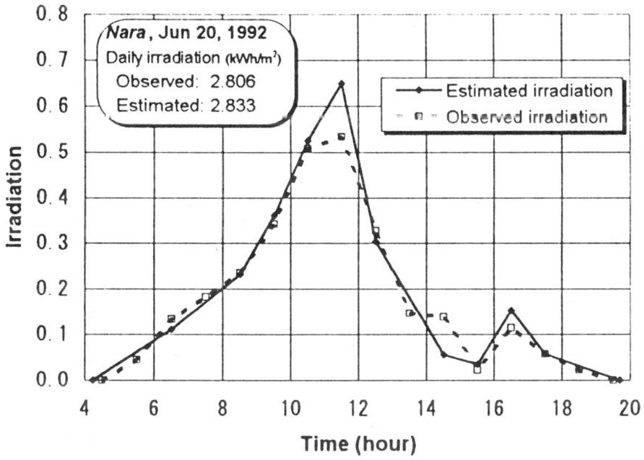

Fig. 3. Variation of estimated and measured hourly irradiation in the daytime

between 14 stations is negligible, each station is assigned common values as the model parameters to calculate the hourly irradiation by equation (2) and (3); which values are determined by regression analysis with the data of all stations. Determined parameters were a_1 = 0.9079, a_2 = -0.2694 and a_3 = 0.0531 in 1992.

As shown in Table 2., the resulting correlation coefficient for estimating the hourly irradiation was 0.942, and the rms error was 0.0804kWh/m^2 (25.1%), on average. Since there was little difference in estimation accuracy between 14 stations, the model parameters may be applicable to anywhere without knowing actually measured irradiation. Accumulation of the hourly irradiation by the trapezoidal integration method for daily estimates increased the correlation coefficient from 0.942 to 0.959, and decreased the rms error from 25.1% to 15.5%, on average. An example of the behavior of the trapezoidal integration is presented in Fig. 3. Similarly, accumulation of the daily irradiation for monthly estimates, increased the correlation coefficient from 0.959 to 0.985, and decreased the rms error from 15.5% to 6.84%, on average. Fig. 4. shows the correlation between estimated and measured irradiation for (a)hourly, (b)daily and (c)monthly cumulative periods at *Chichijima*, 1992.

Solar energy distribution in Eastern Asia was calculated by extrapolating the common model parameters mentioned above. An example of solar energy maps is illustrated in Fig. 5.

CONCLUSION

In this paper, hourly visible images broadcasted by GMS have been used for composing solar energy maps. The method for estimating the irradiation by using GMS images and for composing the solar energy map are described. The methodology, which may have the advantage of generality, was examined for 14 sites in Japan. The resulting correlation coefficients were 0.942 for hourly estimates, 0.959 for daily estimates and 0.985 for monthly estimates, on average; the rms errors were 0.0804kWh/m^2, 0.534kWh/m^2 and 7.14kWh/m^2, respectively. For designing solar power systems, these solar energy maps may be useful to where the solar irradiation have not been measured .

Table 2. Comparison statistics of observed and estimated irradiation for each station

	Hourly cumulative irradiation				Daily cumulative irradiation				Monthly cumulative irradiation			
Station	Number of cases	correlation coefficient	rms error (kWh/m^2)	rms error (%)	Number of cases	correlation coefficient	rms error (kWh/m^2)	rms error (%)	Number of cases	correlation coefficient	rms error (kWh/m^2)	rms error (%)
Asahikawa	2919	0.936	0.0820	27.04	363	0.963	0.574	17.64	12	0.990	8.53	8.67
Obihiro	2921	0.943	0.0728	24.96	363	0.955	0.512	16.36	12	0.983	5.68	6.01
Morioka	2937	0.915	0.0925	31.68	363	0.942	0.631	20.08	12	0.988	5.24	5.52
Yamagata	2941	0.939	0.0791	26.03	363	0.963	0.491	14.99	12	0.993	4.99	5.03
Fukushima	2942	0.932	0.0848	28.10	363	0.951	0.558	17.11	12	0.991	4.27	4.33
Maebashi	2944	0.954	0.0731	24.28	363	0.960	0.555	17.08	12	0.962	11.02	11.21
Chichijima	2983	0.946	0.0858	22.21	363	0.961	0.512	12.05	12	0.994	4.10	3.19
Matsumoto	2923	0.939	0.0944	26.03	361	0.949	0.701	17.86	12	0.993	12.60	10.67
Kohfu	2945	0.953	0.0740	21.44	363	0.959	0.507	13.54	12	0.988	7.62	6.73
Nagoya	2935	0.957	0.0692	20.58	363	0.970	0.442	12.15	12	0.986	5.68	5.17
Hikone	2922	0.957	0.0705	21.95	362	0.972	0.453	13.03	12	0.987	6.08	5.79
Osaka	2925	0.933	0.0863	27.92	363	0.959	0.559	16.74	12	0.976	10.29	10.18
Nara	2928	0.941	0.0799	25.38	363	0.964	0.494	14.55	12	0.972	8.37	8.15
Saga	2879	0.940	0.0806	24.08	362	0.965	0.488	13.62	12	0.985	5.49	5.08
MAX	2983	0.957	0.0944	31.68	363	0.972	0.701	20.08	12	0.994	12.60	11.21
MIN	2879	0.915	0.0692	20.58	361	0.942	0.442	12.05	12	0.962	4.10	3.19
MEAN	2932	0.942	0.0804	25.12	363	0.959	0.534	15.49	12	0.985	7.14	6.84

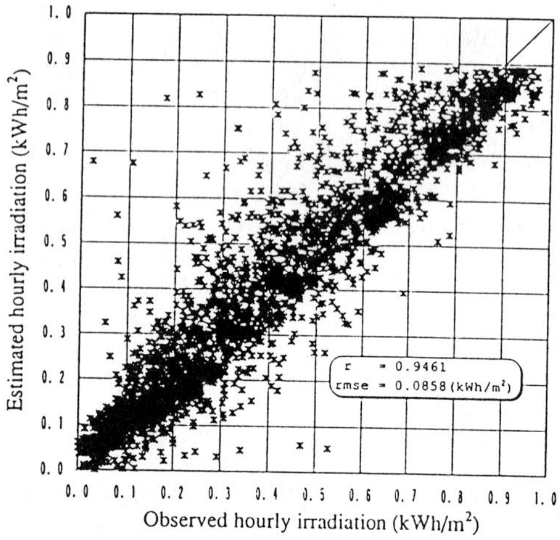

(a) Hourly irradiation

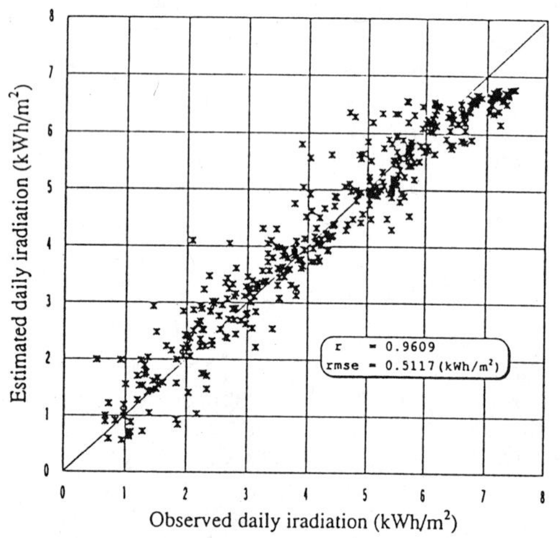

(b) Daily irradiation

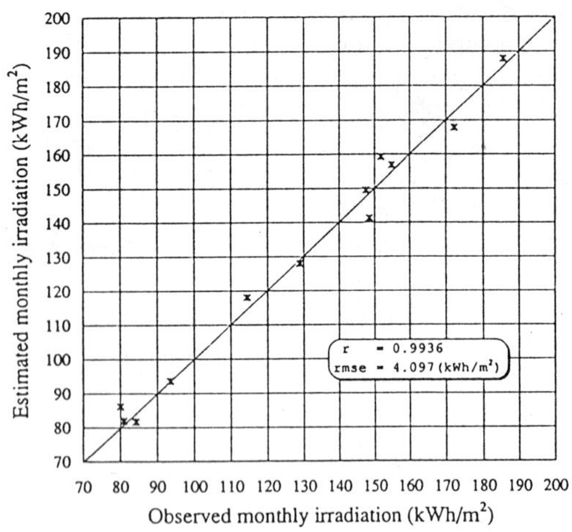

(c) Monthly irradiation

Fig. 4. Correlation between estimated and measured irradiation at *chichijima*, Japan, 1992

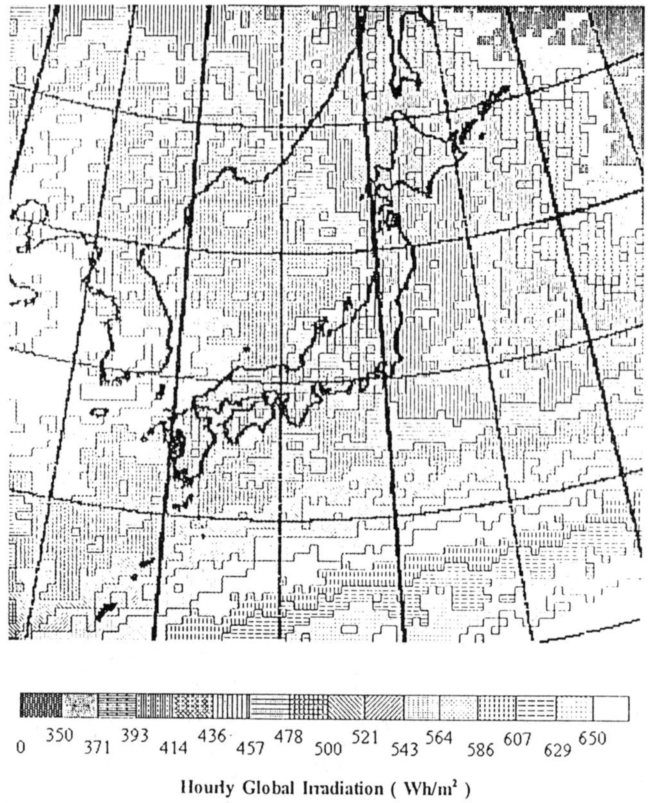

Fig. 5. Example of solar energy map (in part) at noon, 1992

REFERENCES

[1] D. Cano et al., "A method for the determination of the global solar radiation from meteorological satellite data", *Solar Energy*, **37**, 1986, pp. 31-39.

[2] L. Diabate et al., "Description of an operational tool for determining global solar radiation at ground using geostationary satellite images", *Solar Energy*, **42**, 1989, pp. 201-207.

[3] A. Ben Djemaa et al., "A comparison between one year of daily global irradiation from ground-based measurements versus METEOSAT images from seven locations in Tunisia", *Solar Energy*, **48**, 1992, pp. 325-333.

[4] K. Kurokawa et al., "Statistical formation of solarenergymap by cloud images received from GMS", *Pacific Ocean Remote Sensing Conf., PORSEC'94*, 1994, pp. 435-442

[5] K. Otani et al., "Estimation of ground albedo by GMS images for solar irradiation monitoring", *Solar Energy Materials and Solar Cells*, **35**, 1994, pp. 395-400

[6] M. Noia et al., "Solar irradiance estimation from geostationary satellite data", *Solar Energy*, **51**, 1993, pp. 449-465

ANALYSIS OF GROUNDED AND UNGROUNDED PHOTOVOLTAIC SYSTEMS

Ward I. Bower, Sandia National Laboratories, Photovoltaic Applications Department, Albuquerque, NM 87185-0753
John C. Wiles, Southwest Technology Development Institute, New Mexico State University, Las Cruces, NM 88003

ABSTRACT

Grounding has always been a subject of controversy during discussions of electrical systems. Grounding techniques and requirements, like language, vary from region to region and country to country. Optimized grounding for personnel protection does not optimize the fire safety of a system and grounding for fire safety does not optimize personnel safety. Grounding to provide protection for equipment requires a third set of requirements. Photovoltaic (PV) systems are current sources and require different grounding techniques than conventional voltage sources. Distributed leakage paths, multiple fault paths, and new roles for fuses and circuit breakers are among a few of the new issues that need careful consideration. This paper presents and analyzes the grounding issues associated with PV energy sources. Grounding configurations, faults, personnel safety, fire safety, and surge protection are addressed.

INTRODUCTION AND BACKGROUND

When early power generation and electrical distribution systems were being developed in the late 1890's to the early 1900's, grounding issues were the subject of many debates. The requirements were not agreed upon at that time and the result has been different requirements and codes in different countries throughout the world [1]. In the United States, the *National Electrical Code* (NEC) was first published in 1897. Most countries independently developed their own version of an electrical code that addressed the installation and grounding issues for electrical power systems.

The first grounding requirement in the NEC was a mandatory earth-grounding requirement to minimize lightning damage. The grounding debate began a few years later when it was suggested that the secondary distribution circuit of ac utility grids be grounded. It was immediately argued that grounding secondary circuits carrying hazardous voltages increased the probability of shocks. Counter-arguments cited the added safety features that limited the maximum voltage imparted on the secondary distribution circuits should a primary circuit be accidentally connected to a secondary circuit. The discussions and arguments have continued, and international standards have never been written. The PV industry is now poised to further complicate the issue of grounding electrical systems throughout the world.

PV codes and PV system grounding requirements have followed the codes for electrical power systems. In the USA, Article 690 on PV systems was added to the NEC in 1984, but NEC grounding practices were used before 1984. European codes have also followed the national practice of the individual countries for ac power, but the PV codes are being developed as stand-alone documents rather than being integrated in existing codes.

PV systems range in size from single-module, 12-volt systems through hundreds-of-watt systems with storage batteries, to multi-megawatt systems operating at thousands of volts connected to the electrical utility grid. System and equipment grounding practice and legal requirements vary widely between countries and sometimes within a particular country. Codes in the USA require equipment grounding of all systems, and system grounding for systems with voltages over 50 volts (open circuit module voltage) [2]. European codes require equipment grounding, but do not require systems grounding, and most European PV systems do not have a grounded current-carrying conductor [3].

GROUNDING ISSUES

General: Grounding or "earthing" various metal parts and conductors of an electrical system is intended to reduce the effect of faults by minimizing electric shock and fire hazards, to reduce damage to equipment from induced surges, and to reduce the incidence of electromagnetic interference [4]. System grounds provide grounding paths for the current-carrying conductors, whereas equipment grounds provide grounding paths for the metallic surfaces that may be unintentionally energized. Equipment grounds ensure that those surfaces remain at or near ground potential.

Systems are solidly grounded to limit the magnitude of the voltage to ground during normal operations and to reduce induced voltages due to lightning, switching transients, or line surges. Grounding of electrical systems stabilizes the system voltage during surges and also provides a path to trip protection devices if there is unintentional contact with higher voltage lines.

PV Systems: PV system electrical schematics rarely show all of the electrical parameters. Stray inductance, capacitance, and resistance abound and are distributed throughout the system. Leakage currents associated with the array, wiring, surge protection, and conduits often make fault detection difficult. The leakage currents associated with all of the system elements pose unseen and unfamiliar hazards to personnel, and may contribute to fault currents and fires. Improperly selected or improperly installed cables, diodes, PV modules, and other components have failed in PV systems.

PV systems are frequently connected to other sources of

power or energy storage devices such as battery banks, standby generators, hydo generators, wind turbines, and the utility grid. The grounding of the PV system must be consistent with the grounding used on the connected power system.

The interface between connected power systems may allow unanticipated currents to flow in the PV system. These fault conditions must be accounted for in the design of the PV grounding system [4]. Utility interconnected PV systems are often installed in close proximity to utility power lines, and accidental cross connection is a possibility that must be addressed.

There has been only one well documented electrical shock case relating to PV systems in the USA [5], and there have been several fires resulting in substantial damage [6]. General grounding practices were followed in all cases, but the uniqueness of the PV energy source contributed to conditions that allowed fire dangers to exist.

PV SYSTEM GROUNDING CONFIGURATIONS

General: In this paper, grounding is defined as connecting a conductor or piece of equipment to earth or ground potential with a low-resistance and low-impedance conductor. The grounding conductor must be sized to carry the same fault current that any other conductor in the system is sized to carry. System grounding with single-point grounding, and equipment grounding with solid, low impedance bonding of multiple grounds, provides an equipotential ground for surge protection and minimizes circulating ground currents that can lead to corrosion and ground-fault detection problems. The grounding configurations for PV systems may differ greatly [7]. Figure 1 shows a PV system with single-point system and equipment grounding.

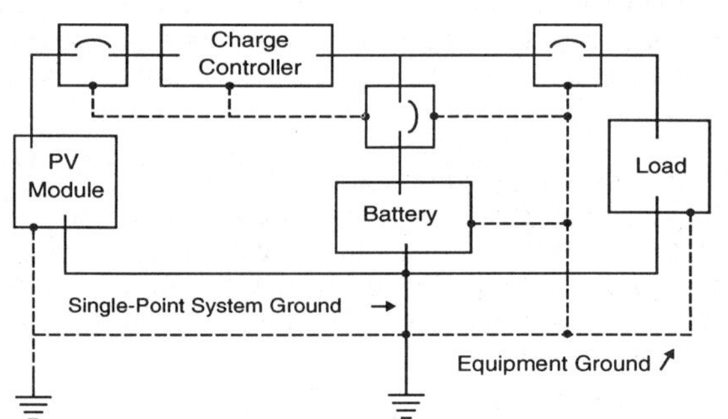

Figure 1. Equipment and System Grounding

IEEE JW 12/94

Ungrounded: A totally floating or double insulated system (Type II insulation) has none of the current-carrying conductors grounded, and any exposed metal conducting surfaces are effectively double insulated from the current-carrying conductors. There is no requirement for the exposed conducting surfaces to be grounded.

An ungrounded PV system with only equipment grounding has all of the exposed metal surfaces grounded that could possibly come into contact with energized sources. These surfaces typically include the module frames, metal enclosures, distribution panels, inverter housings, and the chassis of end-use appliances. Most codes require equipment grounding in all PV systems [3].

Grounded: A PV installation with system grounding has one of the current-carrying conductors grounded. System grounding almost always is used with equipment grounding since the two systems complement each other in interrupting fault currents.

FAULTS IN PHOTOVOLTAIC SYSTEMS

General: Faults are defined as currents flowing in other than the normally designated current-carrying conductors. The magnitude of fault currents depends on the impedance of the fault path. Faults are generally localized and not distributed. Fault currents may pose fire and electrical shock hazards. Leakage currents are small in magnitude, are distributed throughout the PV array and system components, and are associated with high voltages (>100V). The leakage currents in an ungrounded PV system create a distributed high-resistance ground connection, and can allow sufficient ground fault current to injure personnel.

PV modules can develop leakage currents as they age, especially in wet conditions. The leakage comes from module currents flowing to the outside frame or surface of the module through the edge seal or through other deteriorated insulating membranes. Faults can evolve from continuous leakage when carbon or metal is released or deposited by the minute leakage current. Module junction boxes can accumulate water, either from condensation or rain, and this water, when coupled with airborne debris or corrosion products, creates leakage paths within or external to the junction box. Leakage currents also occur as the insulation on conductors ages or is damaged due to abrasion [6].

Insulation failures between current-carrying conductors of opposite polarity cause bolted, line-to-line faults. The current flow into bolted faults can be from PV modules in the faulted circuit, from modules connected in parallel with the faulted circuit where blocking diodes are absent or shorted, or from external sources such as batteries or inverters. Blocking diodes have failed frequently in PV installations and have allowed multiple PV strings to contribute to the fault current [6]. Inverters even under normal operating conditions can feed ac utility currents into faults in the dc array wiring.

Insulation failures between current-carrying conductors and the ground are known as ground faults. Ground faults can develop within the PV array, in circuits that have electrically combined the array, or in switches and inverters. Ground fault detectors must be used to sense ground faults in both grounded and ungrounded PV systems. The sensitivity of ground-fault detectors for ungrounded systems should be higher than for grounded systems. The practical limit for ground-fault sensitivity is limited by wet-weather leakage currents. More than one ampere has been measured in a 300-kW system in the USA. Single-point grounds are necessary to minimize circulating ground currents that may interfere with ground fault detectors and pose hazards.

Ungrounded: On ungrounded systems, a single ground fault will not cause fault currents, but a second ground fault in conductors of a different voltage will allow fault currents that circulate through circuits associated with the two faults [6]. PV array modules and conductor capacitance to ground can contribute significant shock hazards in the form of a capacitive discharge on ungrounded systems. Resistance grounding can eliminate the charge on these distributed capacitances.

Grounded: On grounded systems, the first ground fault will cause currents to flow. The location of the fault and the source and magnitude of these currents may cause overcurrent devices to function and interrupt the fault currents. Distributed capacitances are a lesser problem because they are already ground referenced and act in the same manner as the other grounded current sources.

PERSONNEL SAFETY

General: The requirement for the safety of users and maintainers of electrical power systems has been the basic impetus for the study and application of grounding to electrical power systems. Much of the research on electrical shock has been derived from studies on human volunteers and animals [8].

Research has defined the term "let-go" as the current level, either ac or dc, where the subject in contact with the electrical circuit experiences involuntary muscle contractions to the point where he or she is unable to disengage from the circuit. The effect of currents through the body are a function of the applied voltage, the resistance of the body, and the path through the body. Although varying greatly, the average body resistance is assumed to be about 1000 ohms. Biological and situation variations can cause this value to vary over a ten-to-one range [9]. While the "let-go" level of current may not be fatal, higher levels of current can cause the heart to go into a rapid useless beating called ventricular fibrillation.

The values of "let-go" and ventricular fibrillation currents are higher for dc circuits than for ac circuits. Men are less susceptible to these currents than are women. DC "let-go" currents are in the range of 90 milliamps (dc) for men and 60 ma. for women. Ventricular fibrillation currents are between 500 and 1300 ma. for men and women [9].

Ungrounded: It is thought that a floated, ungrounded, electrical system is safer to work on and service than is a system in which one of the current-carrying conductors is grounded [1]. With an ideal ungrounded system, contact between one of the conductors and the earth, or a grounded conductor, cannot result in a current path. This is true only of theoretically ideal systems with no leakage and no capacitance to ground. However, the average PV installation has many leakage resistance paths to ground and distributed capacitance between module wiring and ground.

With resistive leakage paths to ground, accidental contact with the conductors of the ungrounded system results in a closed circuit for shock currents. Even with no leakage to ground, the ungrounded system, with its distributed capacitance to ground, can impose a shock experience during the capacitive discharge which can cause involuntary muscle contractions that result in other injuries.

Leakage paths and capacitance generally result in ungrounded systems that have the voltages referenced to ground and leakage currents to ground The Class II systems proposed in parts of Europe will minimize these factors by ensuring a more ideal ungrounded system. If Class II guidelines and hardware are not used in ungrounded systems, both conductors can have voltages that are theoretically higher than the normal operating system voltage. For example, a 300-V system would typically have balanced line voltages that are approximately 150 volts to ground. If one conductor develops a leakage to ground, the second conductor will have 300 volts to ground. Additionally, static charge may build up in an ungrounded system and the resulting conductor voltage to ground may be greater than the operating voltage of the system.

Grounded: In a grounded system, in which one current-carrying conductor is grounded, or the system is center-tap grounded, the system voltage is stable, and the highest voltage appearing on the ungrounded conductors, except when large surge currents are induced into circuits, is the normal system operating voltage. During nearby lightning strikes, the conductor voltages may rise to a clamped voltage determined by the surge protection devices for a short time during the surge pulse.

With grounded systems using grounded equipment enclosures, there is a greater possibility of a service person completing a fault path when working within the enclosure. The grounded system presents energized conductors in close proximity with substantial amounts of grounded exposed metal in the enclosures. Non-conductive enclosures can minimize this problem.

FIRE SAFETY

General: The fault currents supplied by the PV array are determined and limited by the array size and the incident irradiance. A properly designed system uses cables and overcurrent devices sized to carry the highest short-circuit in each protected circuit. The current limited PV source normally will not supply enough current to open protection devices on faults. In cases where more current was supplied from a second source or PV string, the blocking diodes had failed. Other sources of current that may cause the fuses or circuit breakers in PV circuits to blow or trip include inverters connected to the utility grid or batteries [10]. Even bolted faults (line-to-line) in PV array conductors may create arcs that will burn insulation at current levels that will not trip an overcurrent device. Multiple faults in grounded or ungrounded systems result in circulating currents that may not be detected, but can result in loss of power and create fire hazards.

Ungrounded: Ungrounded systems offer clear advantages for avoiding fire hazards. The ungrounded system must develop two ground faults before a ground-fault fire hazard exists. Ground-fault detectors, which are easier to install on ungrounded systems than on grounded systems, can detect the first ground fault and disable the system before dangerous multiple ground faults develop. Disabling an ungrounded system can be achieved by short-circuiting the array. Although the fault is not eliminated, the voltage on the conductors is substantially reduced. Excessive leakage to ground may, however, cause ground-fault detection sensitivity problems.

Grounded: A single fault or leakage to ground results in circulating ground current whenever the PV array is illuminated in a grounded system. This is a dc current and can result in corrosion of dissimilar metal connections and, if sufficiently high, can pose fire hazards. Grounded systems are very difficult to disable because of the division of currents. Short circuiting a faulted array will not eliminate the fault current. The disabling circuit may include many meters of wire, several disconnect devices, fuses or circuit breakers, and blocking diodes between the fault and the disabling device. The current division between the fault resistance and the array disable circuit can approach a value that allows enough current to flow through the fault with enough power dissipated to cause a fire [6]. The fault current, voltage, and power can actually increase unless the fault current is interrupted by ungrounding the system. If all of the conductors are open circuited, the fault path is opened. The ungrounded array then can be disabled by short circuiting.

SURGE SUPPRESSION

Surge suppression devices may be added to either of the grounding configurations and are independent from the grounding system used. Suppression devices are used to control surges induced by nearby lightning strikes and any other inductively or capacitively coupled surges that might be introduced to the system. Surge suppression devices are connected between each of the current-carrying conductors for line-to-line (differential mode) surge reduction. They are connected between each current-carrying conductor and earth for line-to-ground (common mode) protection.

SUMMARY

The grounding of systems is complicated by the introduction of current-limited PV systems interconnected with batteries and conventional voltage-source electro-mechanical generators. There is still no international agreement for grounding PV systems, and there is little likelihood of agreement. The USA requires system grounding of all electrical systems with voltages greater than 50 volts. Many other countries require ungrounded PV systems. Two universal rules for grounding are a) most codes and standards require equipment grounds for all metal surfaces that might become energized, and b) when system grounds are used, single-point grounds are required.

The ungrounded system provides the best fire hazard reduction because multiple faults are needed to create a fire hazard. The ungrounded system might also provide the best protection for preventing shock if the system were ideal, with reliable, sensitive ground-fault detection and no leakage. The proposed Class II European systems operating at voltages below 120 volts will increase fire and personnel safety in PV systems.

The grounded PV system provides the best personnel protection from electrical shock because voltages to ground are well defined. Since PV systems inherently have leakage to ground, they are not solidly grounded or ungrounded. The system ground ensures a solid or known PV array ground through properly sized conductors. With proper design, both grounded and ungrounded PV systems can achieve personnel, fire, and equipment safety.

ACKNOWLEDGMENTS

This work was sponsored by the U. S. Department of Energy under Contract DE-AC04-94AL85000.

REFERENCES

[1] Simmons, P.J. *International Association of Electrical Inspectors Soares Book on Grounding,* 5th Edition, Richardson, TX, 1993.

[2] *National Electrical Code-1993*, ANSI/NFPA-70, National Fire Protection Association, Quincy, MA, 1992.

[3] Herig, C., Real, M. "Grounding and Lightning Protection of PV Systems," Soltech '92 BOS Workshop, Sandia National Laboratories, Albuquerque, NM, Feb. 1992.

[4] *IEEE Recommended Practice for Grounding of Industrial and Commercial Power Systems*, ANSI/IEEE Std 142-1982, IEEE, NY, NY, 1982.

[5] Jacobson, L., Moskowitz, P.D., Garrett, J.O., and Tyler, R., *Accident Investigation of The Electrical Shock Incident at the PG&E PVUSA Site at Davis, CA*, Analytical Sciences Division, Brookhaven National Laboratory, Upton, NY Feb. 3, 1992.

[6] Steitz, P., *Photovoltaic Power Conditioning: Status and Needs*, Sandia National Labs and the Electric Power Research Institute GS-7230, Palo Alto, CA, 1991.

[7] Stolte, W. J. *Photovoltaic System Grounding and Fault Protection Guidelines*, Bechtel Group, Inc. Sandia Report, SAND83-7025, Albuquerque, NM, Reprinted Jan. 1988.

[8] Dalziel, C. F., "Effects of Electric Shock on Man," IRE Transactions, Medical Electronics, Vol. **PGME-5**, pp. 44-62, July 1956, Reprinted as USAEC Safety Bulletin 7.

[9] Cool, J.K., "Electrocution and the Human Body," Power Quality Assurance, July/Aug. 1994, pp. 84-86.

[10] Wiles, J.C., *Photovoltaic Power Systems and the National Electrical Code: Suggested Practices*, Southwest Technology Development Institute for Sandia National Laboratories, August 1994.

SINGLE CELL CONVERTER SYSTEM (SCCS)

Markus Wuest & Peter Toggweiler
PMS Energy Ltd., Lindhofstrasse 13, CH-8617 Moenchaltorf
Jon Riatsch
ETHZ, Power Electronics Laboratory, CH 8092 Zurich

ABSTRACT

Among the different concepts for module constructions, there is a new way of cell and module production proposed. An entire module-sized thin-film layer is used as one cell. This large cell delivers a voltage range of about 0.5-1.5V, depending on the type of cell. To get an appropriate voltage level of 12V DC or 230V AC, it must be transformed by specially designed converters. Investigations on a modular power converter have already been started at the power electronics laboratory of the ETH Zurich. These investigations proved the feasibility of such a converter. Because of the large area of this single cell, it produces a high current of about 100A, depending on its efficiency and the module area. Alucobond®, a compound of Aluminium - Polyethylene - Aluminium, fullfils the needs for high current contacts perfectly, because the two Aluminium plates can be used for this purpose.

INTRODUCTION

The bright future predicted for building integrated photovoltaics, results in a growing number of development activities. In Switzerland, Alusuisse-Lonza Services recently started the Alucosol® project, which aims at façade / roof integrated pv-products based on their Alucobond® material.

This compound material (Aluminium-Polyethylen-Aluminium) has used so far in building elements, vehicle coachwork and other related purposes. The product offers a wide range of design capabilites at low prices to architects and engineers. Alucobond® reached a good market position with annual sales of 2 Mio. m^2 worldwide.

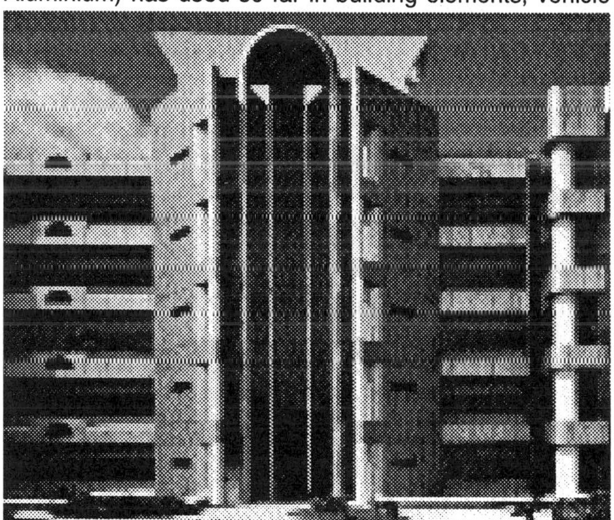

Fig. 1 Example of an Alucobond®-façade

Five catch words shall give an idea of the Alucosol® project and its goals:

- façade and roof integrated solar cells
- full integration of cells and building elements
- large area, cheap and ecological solar cells
- easy mountable building system with inexpensive integration of electrical components and cables
- automated module production process

From the different aspects of such a product development, this paper wants to introduce the new approaches in the electrical system and the possible simplification in the manufacturing process of the specific modules.

THE ALUCOSOL® MODULE

Among the different cell technologies, the amorphous silicium (a-si) seems to be the most promising technology for ecological and cheap large area cells and modules. The multilayer thin film technology of Uni-Solar allows large areas of metal or plastic layers to be covered with amorphous silicium. To get the tightest possible integration onto the Alucobond® sheet, it would be very attractive to use a a-si aluminium-plate as the upper part of the Alucobond® compound. Due to unsolved problems of direct a-si and aluminium contacts, ALS did reintroduce its earlier Steelobond® know-how. Steelobond® is the same compound material with steel in place of aluminium. The proposed module composition contains the Aluco- or Steelobond® material on which the upper part is covered with amorphous silicon layers.

The SCCS-principle allows the upper metal plate to be used as a contact to the electrically negativ part of the a-si covering. The contact layer on the front-side is composed of a large amount of thin lateral contacts,

THE ALUCOSOL® PROJECT TEAM

- Alusuisse-Lonza Services Ltd., Dr. M. Weng
- Electrowatt Engeneering Ltd., T. Kaelin
- PMS Energy Ltd., P. Toggweiler
- Paul Scherrer Institute, Dr. H. Kiess

CH3365-4/94/0000-0813 $4.00 © 1994 IEEE

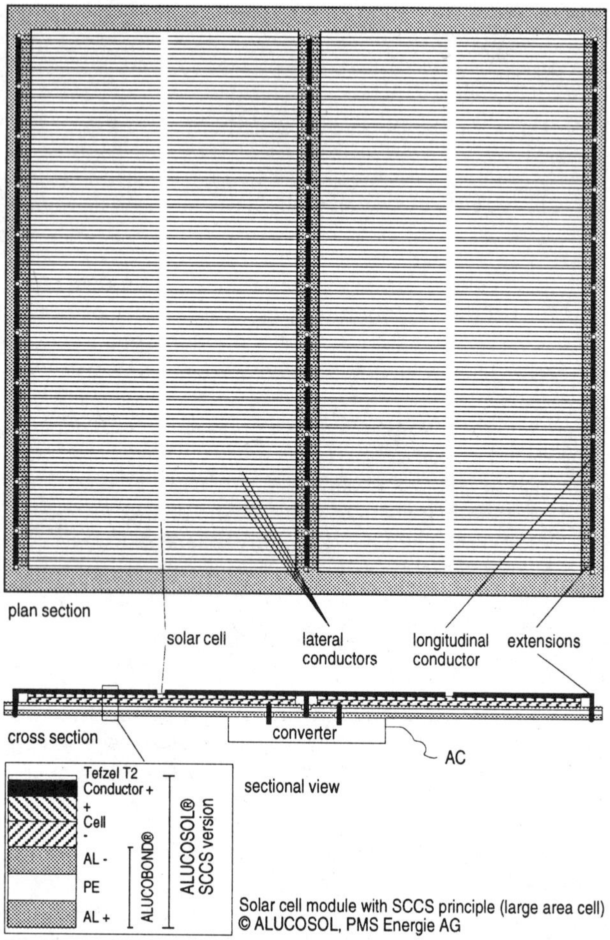

Fig. 2 Detailed drawing of an Alucosol® module

which lead to the longitudinal contacts on the side of the cell. One way to realize these lateral contacts is to use printed structures of a conducting material. Another way is to use a transparent layer of transparent metal oxides or a combination of both variants. The longitudinal contacts lead in regular distances trough the whole compound structure below to the backside metal-plate. The extension of the contacts themselves must of course be isolated and the contacts themselves must be designed to work with the high amount of resulting current. To protect the cells surface a transparent plastic protection or a thin glass layer completes the module compound.

SIMPLIFIED MANUFACTURING PROCESS

All the steps to produce such a module compound are suitable for continous flow processing. This is one well known advantage of amorphous over crystalline cell technology. In addition, the complexity of the conventional a-si manufacturing process is further reduced by the single cell converter approach. Caused by the unnecessary laser cutting and metal conductor depositioning, the single cell approach promises to be easier, faster and with significant cost savings than today's a-si cell and module manufacturing. The mentioned extensions can be done mechanically and must absolutely assure the isolation of the two electrical potentials, which should be easily achievable with the lowvoltage level of a single cell.

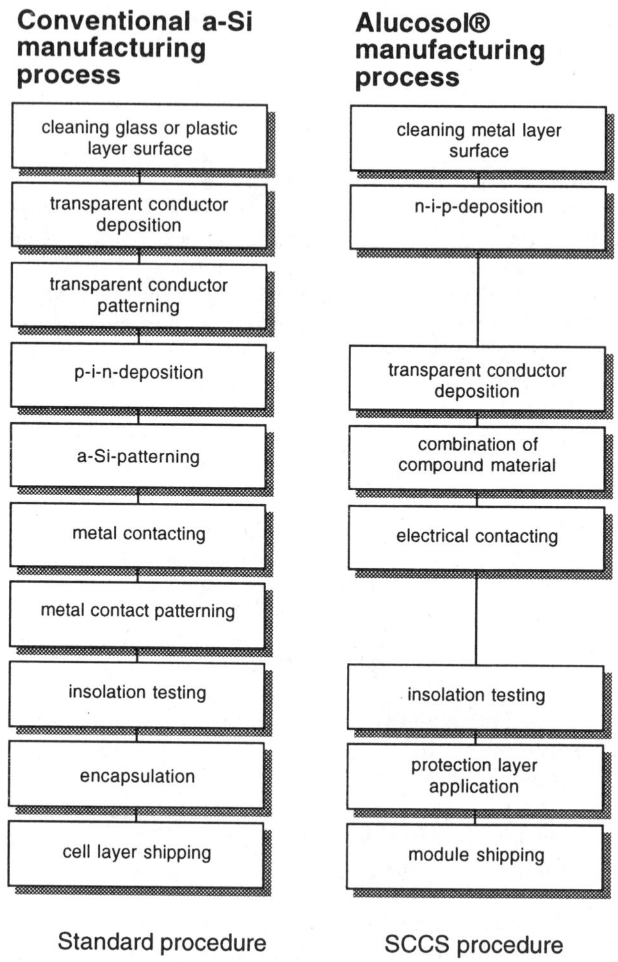

Fig. 3 Comparison of process steps

ELECTRICAL CONFIGURATION

The use of an entire large area thin film layer as one cell will achieve the above mentioned advantages, but it needs new approaches in the electrical configuration to get a complete photovoltaic power unit. The produced power will be in a voltage range of 0.5 to 1.5V and deliver easily an amount of 100A and more. The small voltage level helps to simplify isolation and reduces the danger for the mounting personnel. The large amount of current is no problem in the cell area due to the combination with the metall-plastic-metall compound, but it is impossible to transport this current to a centralized power converter outside the cell area.

CONVENTIONAL AND MODULAR POWER CONVERSION

Important drawbacks of conventional solar power generation are the DC-wiring and fusing needed for the serial and parallel connection of solar panels. These problems can be reduced to AC-wiring and fusing by using a modular set-up of panels with integrated single phase DC-AC converters on a power level of approximately 200W. In addition to AC-wiring and fusing, another important advantage of a modular converter is the possibility to build any size of power plant with identical modules. On the AC-side any number of such converters can be connected in parallel for the supply of single and three phase mains.

SINGLE CELL POWER CONVERTER

Investigations on a modular power converter have already been started at the power electronics laboratory of the Swiss Federal Institute of Technology in Zurich (ETHZ). The single cell converter system (SCCS) is a special case of a modular power converter, where each solar cell has its own power converter. The power level of one of these cells corresponds to that of today's conventional solar modules. The goal of further investigations is to find appropriate converter topologies for the given special requirements of the single cell converters application, which are high current and low voltage on the converters input side.

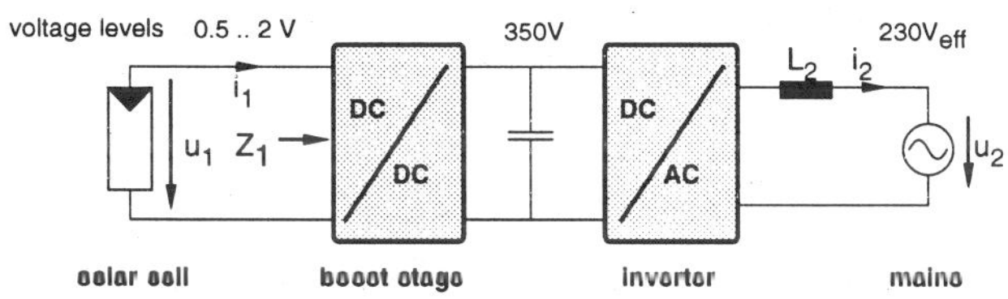

Fig. 4 Electrical Scheme of a Single Cell Converter

Extended investigations have proved that single phase converter topologies are the best suited for the given power and voltage levels. They consist of a boost stage, a DC-link, and a single phase inverter and can go over 85% efficiency with more than 1V DC input.

The main difficulty of the converter topology is to develop a boost-stage of high efficiency for the unfavourable input impedance of the SCCS. With reference to the invertor it should be noted that good and efficient solutions are state of the art today. For the given power level and the desired switching frequency, MOSFET's are preferred as switching elements.

Some goals of the power converter to be developed are:

- high efficiency
- low volume, high power density
- high reliability
- low costs for components and manufacturing
- low 100 Hz power pulsation on the solar cells
- electrical isolation between grid and cell
- easy integration procedure on the module

In addition to the hardware intelligent control strategies for the Maximum Power Tracking (MPT) and the mains current have to be found to increase the efficiency of the whole system as much as possible. The whole investigation on the power converter are part of a doctoral thesis started in June 1994 at the power electronics laboratory of the ETH in Zurich.

CONCLUSIONS

The Single Cell Converter System uses a new combination of known PV-technologies. A patent is pending. Primarily SCCS allows to achieve advantages in the manufacturing process of large cells. It will allow a faster and cheaper production of a module which is based on the same compound materials as is Alucobond®. The know-how in designing building elements with Alucobond® can easily be transfered to the new Alucosol® material. Electrical configuration is simplified by the integrated DC-AC power converter. The mounting procedure will be the same for Alucobond® and Alucosol® elements. Due to the low voltage level, no dangers are raised for mounting personnel in any situation. The photovoltaic behaviour will be characterized by the amorphous thin film technology, but with less inactive area, no cell miss-match and therefore a better efficiency. As there is no partial-shading problem without serial connection of cells, this technology is perfectly suitable for the promising market of building integrated photovoltaic power units.

FUTURE EFFORTS

As there are several steps to be optimized, two proposal for European development projects are in preparation. There, the Frauhofer Institute in Freiburg (G) and the ETHZ are involved. High development potential is predicted in the switching technology and semiconductor optimisation.

Because the SCCS principle allows a wide margin of possible products, other foundation materials are also under investigation. Efforts are made in the fields of concrete integration and steel-only versions of the SCCS principle.

ACKNOWLEDGMENTS

Thanks to the NEFF (National Energy Research Foundation), the ETHZ - out of it specially Dr. F. Jenny, Prof. H. Stemmler and J. Riatsch - and to the ALUCOSOL® Project Team for their creative and supporting collaboration during the ongoing project.

REFERENCES

The Project Proposals for the NEFF and the BEW (Federal Office of Energy).

J. Riatsch: Single Cell Converter System; Feasibility study; 1993, ETH Zürich

M.Weng, H. Kiess, T. Kälin, P. Toggweiler: Development of a facade system with integrated PV-modules, 12th European PV Conf., Amsterdam,1994

FOURTH-GENERATION CONCENTRATOR SYSTEM: FROM THE LAB TO THE FACTORY TO THE FIELD

M.J. O'Neill and A.J. McDanal
ENTECH, Inc.
P.O. Box 612246 (1015 Royal Lane)
Dallas-Ft. Worth Airport, TX 75261 USA

ABSTRACT

In our May 1993 IEEE PVSC paper (1), we described improvements in manufacturing technology which our team was developing to mass-produce our new fourth-generation concentrator module. In the past 18 months, we have completed the development of this new module. We are currently producing over 600 modules of the new type, equivalent to more than 250 kW of rated power output. Most of these modules are presently being deployed in two utility-scale power plants owned by the two largest utility companies in Texas: (i) a 100 kW system near Ft. Davis, Texas, at the Solar Park being developed by Central and South West (C&SW); and (ii) a 100 kW system near Dallas, Texas, at the Energy Park being developed by TU Electric (TUE). Thus, the module technology we described at the last PVSC has progressed from the laboratory to the factory floor to the field.

OBJECTIVE & ACKNOWLEDGEMENTS

The objective of this paper is to describe the key developments, related to our fourth-generation concentrator technology, which our team has achieved since the last PVSC. These recent developments include manufacturing technology improvements; independent testing of modules by Sandia National Labs; improved sun-tracking approaches; and implementation and verification of all required production processes for the new module. The impact of these developments on module production cost and system energy economics will also be presented.

The authors gratefully acknowledge the contributions of our international team, especially the U.S. Department of Energy, National Renewable Energy Lab, and Sandia National Lab.

LATEST MODULE DESCRIPTION

Figure 1 shows a cross-sectional view of our new fourth-generation module. The acrylic lens (90% measured optical efficiency) is made via a continuous plastic process by 3M. The lens aperture is 85 cm wide by 366 cm long. The aluminum heat sink is made via a continuous extrusion process by Columbia Aluminum. The lens and heat sink are integrated together to form a module with a marine-grade aluminum housing structure comprising two sidewalls and two endplates. The silicon concentrator cells (presently 17-19% electrical efficiency) are made in two-per-wafer form from conventional 125 mm diameter wafers by ASE (previously Deutsche Aerospace), BP Solar, and other one-sun cell manufacturers. As shown in Figure 2, each cell (9.7 cm length by 4.1 cm active width) is integrated into a package containing two integral bypass diodes, four solder-plated copper ribbon interconnectors, and a prismatic cell cover to eliminate gridline shadowing loss. Thirty-seven cell packages are connected in series and attached to the heat sink via a proprietary dry film laydown process. The entire cell string is then encapsulated using another proprietary dry film process. The final module thus comprises one lens, two sidewalls, two endplates, one heat sink, and thirty seven cell packages, integrated together to form the world's most powerful photovoltaic module, with a rating of 430 W_{peak} (using 17% efficient cells) to 480 W_{peak} (using 19% efficient cells).

MODULE FABRICATION AND TESTING

In January of this year, we built, tested, and delivered to Sandia National Labs our first two modules of the final fourth-generation configuration. These two modules were identical, except for the Fresnel lens thickness. One module used a conventional 3.5 mm thick lens, while the other module used a new, thinner, impact-resistant 2.0 mm thick lens. The new 2.0 mm lens is made in final form by 3M using a continuous roll-to-roll lamination process. Both lenses used the same prismatic pattern (SCL-1000), which provides both a mild irradiance profile over the solar cell and a ± 0.8 degree sun-pointing error tolerance (1). Both modules used typical production-version cells from ASE, with about 17% conversion efficiency, after soldering and prism covering. (ASE and other suppliers have delivered 19% efficient cells but production cells are not yet averaging this performance level.) Both modules used our proprietary new dry film laydown and encapsulation process. Both modules used our new heat sink, which is the largest extruded heat sink yet produced in terms of fin area per unit length. Prior to delivering the modules to Sandia, we measured IV curves for both modules, with results shown in Figure 3. Note that both modules provide essentially identical performance levels. In the Spring of this year, Sandia thoroughly tested both modules for both electrical and thermal performance. Sandia has confirmed a power level of about **430 W_{peak} at standard test conditions** and about **390 W_{peak} at standard operating conditions** for each of these modules. We believe that both of these power outputs are the highest yet achieved for any type of photovoltaic module produced to date.

CH3365-4/94/0000-0816 $4.00 © 1994 IEEE

Sandia has also performed side-by-side tests of our latest module and conventional, frame-mounted (open-back) flat plate modules to determine the relative operating temperature levels under various conditions. Sandia's data show that the measured module temperatures (at the heat sink base for ours, on the back of module for flat plates) are essentially identical, even under relatively low wind conditions. These results are not surprising, since our new heat sink provides fin area equivalent to 4 times the lens aperture area. For a frame-mounted flat-plate module, the heat transfer area is about 2 times the module aperture. For a roof-mounted flat-plate module, the heat transfer area is about 1 times the module aperture. Thus, our new module has a substantial heat transfer advantage over flat-plate modules.

MANUFACTURING IMPROVEMENTS

In the past year, our team has streamlined our module production in many ways. One of the key improvements is the production of die-cut prism cover tape in label form, as shown schematically in Figure 4. This remarkable new 3M process provides continuous rolls of individual prism covers (3,500 covers per roll), precisely cut to size and coated with pressure sensitive adhesive (PSA). To use the new cover, it is merely removed from a continuous paper release liner and stuck down to the cell. A semi-automated prism cover application station has also been developed to enable rapid microscope-aided alignment of the cover to the cell, and roll-on application of the cover to the cell. We have also developed a proprietary new dry film receiver laydown and encapsulation approach (patent pending). We can now pour water over our cell string in Niagara Falls fashion while maintaining a 2,200 VDC potential between the cell string and heat sink, with less than 1 μA of leakage current. In addition, compared to our previous approach, our new cell string laydown and encapsulation approach eliminates more than 300 parts per module and is implemented with a very simple process.

IMPROVED SUN-TRACKING APPROACHES

In the past year, our team has developed a new microprocessor-based controller, new DC-powered motors and drives for both sun-tracking axes, and new galvanized steel frames. We have implemented these new approaches for both large-scale sun-trackers which include 72 modules per tracker (**SolarRow**, shown in Figure 5), and for small-scale sun-trackers which include 2 modules per tracker (**SunLine**, shown in Figure 6). We expect these new sun-tracking improvements to eliminate many of the problems which have plagued concentrator systems in the past, including relatively high maintenance costs associated with controls, drives, and paint systems.

PRODUCTION IMPLEMENTATION AND VERIFICATION

To produce the 600 modules needed for the two utility jobs mentioned above, our team has set up production lines to implement the newly developed production processes for our fourth-generation module. ASE has produced over 21,000 cells. BP Solar has produced over 42,000 diodes. 3M has produced over 600 module-ready lenses and over 21,000 prism covers in label form. DuPont has produced the required quantity of dry film material for laydown and encapsulation of 600 cell strings. E. Jordan Brooks Co. has produced all of the necessary solder-plated copper ribbon interconnectors for 21,000 cell assemblies. Columbia Aluminum has extruded more than 600 full-size heat sinks. Consumers, Inc., has produced more than 600 sets of module housing parts (endplates and sidewalls). ENTECH is presently assembling these key components into production modules. Thus, we have implemented and verified all of the new module production processes on the factory floor.

In addition, Consumers, Inc., has produced 8 sets of SolarRow galvanized steel tracking structures. Prime Manufacturing has produced 8 sets of microprocessor-based sun-tracking controllers. ENTECH has installed 8 SolarRow tracking structures of the latest design. Omnion has produced 2 power conditioning units, each rated at 100 kW, which are each compatible with the power output of four SolarRows. Thus, our team has demonstrated the producibility and practicality of our fourth-generation **system** technology.

IMPACT OF RECENT DEVELOPMENTS

The developments described above have favorably impacted both our production rates and our manufacturing costs. Our team can now produce lenses, cells, diodes, copper ribbon interconnectors, aluminum heat sinks, dielectric films, sun-tracking controllers, etc., at rates well over 10 MW/year. Indeed, scale-up to production rates of 100 MW/year could be achieved very quickly.

Figure 7 shows the costs and economics of our fourth-generation concentrator system technology, as functions of annual production rate. One key data point is a levelized electricity price range of **7-15 cents/kWh** for systems produced at a rate of 30 MW/year. We know of no other photovoltaic technology which can realistically reach this level of cost effectiveness without major breakthroughs in efficiency, stability, or production processes.

The developments described above are evolutionary in nature (2-3). For example, Figure 8 shows the long-term performance of our **third-generation** concentrator technology, compared to other leading photovoltaic technologies, as measured by the PVUSA project in side-by-side field tests at Davis, California. Note that our concentrator array has outperformed the other technologies over a 4-year period. Thus, our fourth-generation technology is founded on proven previous versions of our line-focus concentrator systems.

REFERENCES

1. M.J. O'Neill and A.J. McDanal, "Manufacturing Technology Improvements for a Line-Focus Concentrator Module," 23rd IEEE-PVSC, Louisville, May 1993.
2. T.M. Bruton, M.J. O'Neill et al., "Recent Developments in Concentrator Cells and Modules Using Silicon Laser Grooved Buried Grid Cells," 11th European PVSEC, Montreaux, Switzerland, October 1992.
3. M.J. O'Neill, "Fourth-Generation, Line-Focus, Fresnel Lens Photovoltaic Concentrator," 4th Sunshine Workshop on Crystalline Silicon Solar Cells, Tokyo, Japan, November 1992.

Fig. 1

Fourth Generation Concentrator Module

ENTECH, Inc.

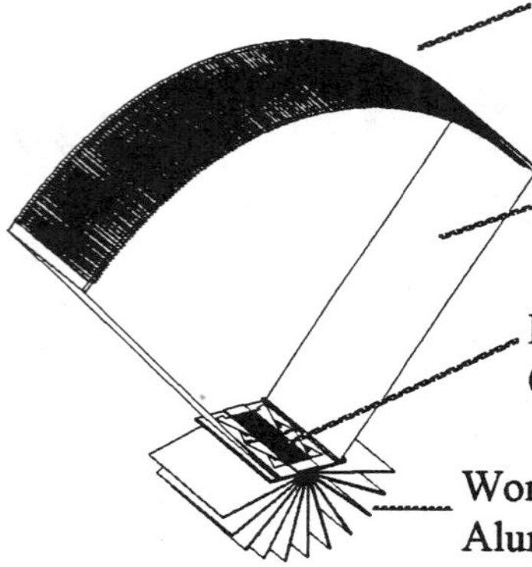

Fig. 2

Cell Package: Simple and Inexpensive

ENTECH, Inc.

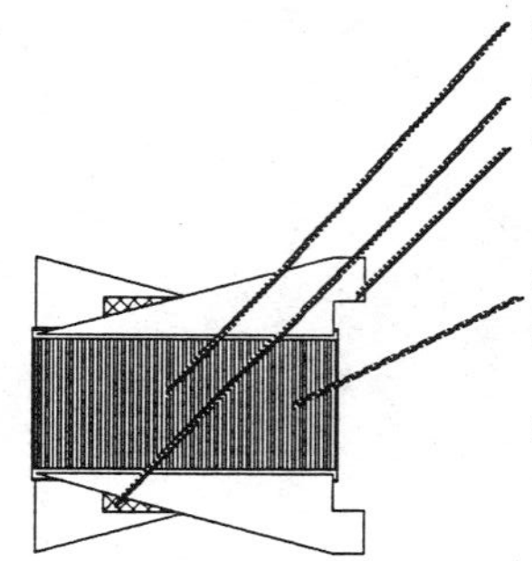

Fig. 3

World's Most Powerful PV Modules

ENTECH, Inc.

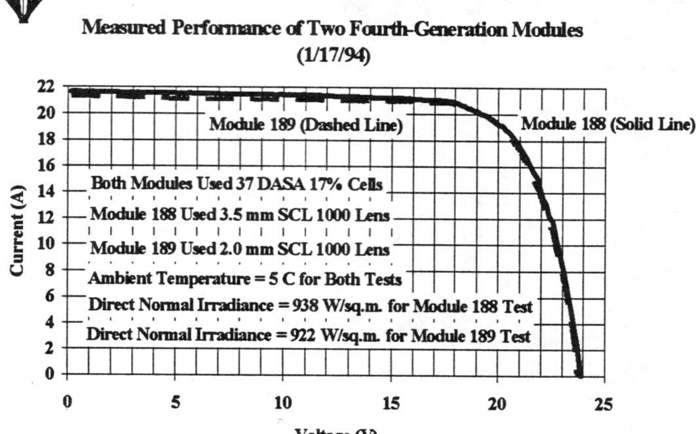

Fig. 4

Cross Section of New Prism Cover Tape

ENTECH, Inc.

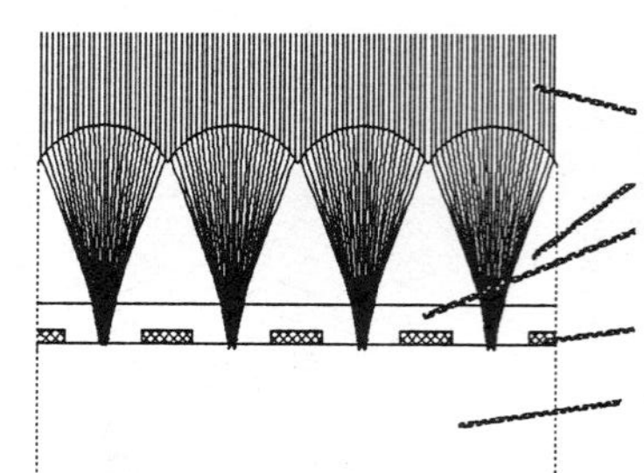

Fig. 5

SolarRow Array

ENTECH, Inc.

- 72 Modules
- > 220 m² Aperture
- Full Sun Tracking
- Galvanized Structure
- > 25 kW Output
- Easy to Install

Fig. 6

SunLine Array

ENTECH, Inc.

- 2 Modules
- Full Sun Tracking
- Galvanized Structure
- > 1 Horsepower
- Easy to Install

Fig. 7

Module and System Costs and Economics

ENTECH, Inc.

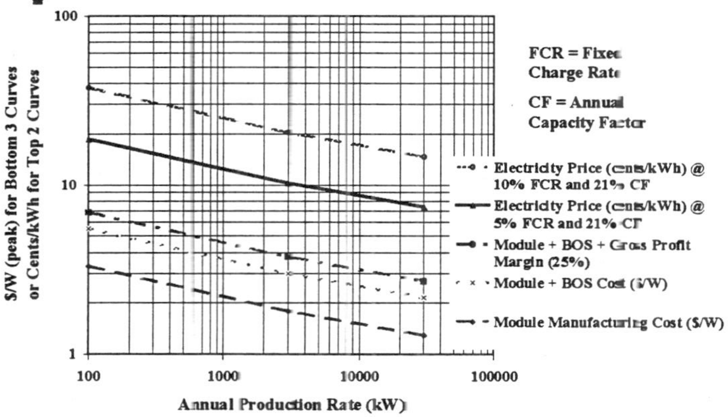

Fig. 8

Long-Term Independent Test Results for Leading Photovoltaic Technologies at PVUSA

ENTECH, Inc.

Elec. Power Out / Solar Power In

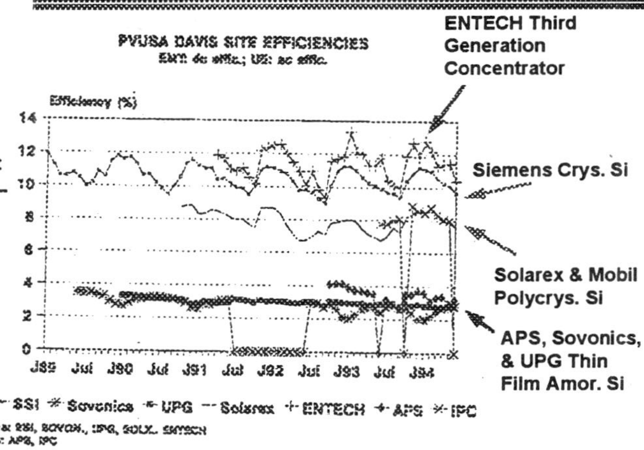

CONTINUOUS ROLL-TO-ROLL SERPENTINE DEPOSITION FOR HIGH THROUGHPUT a-Si PV MANUFACTURING

M. Izu, H. C. Ovshinsky, X. Deng, A. J. Krisko, K. L. Narasimhan,
R. Crucet, T. Laarman, A. Myatt and S. R. Ovshinsky
Energy Conversion Devices, Inc., 1675 West Maple Road, Troy, Michigan 48084

ABSTRACT

In order to further improve the economies of scale which are inherent in ECD's continuous roll-to-roll amorphous silicon alloy solar cell manufacturing process, we have developed a concept for a serpentine web plasma CVD deposition process to maximize throughput while keeping the size of the deposition chambers small. When this technique is incorporated into a continuous roll-to-roll PV manufacturing process, it will maximize the throughput for a high volume production plant, reduce the machine cost, improve gas utilization, reduce power consumption, and improve the solar cell stability. To demonstrate the serpentine web deposition concept, we have constructed a single loop serpentine deposition chamber to deposit a-Si for n-i-p structure solar cells. During the initial process of optimization, we have produced single-junction a-Si solar cells with 8.6% efficiency, and triple-junction a-Si solar cells with a 9.5% initial efficiency, where the top cell intrinsic layer was deposited in the serpentine deposition chamber.

INTRODUCTION

Continuous roll-to-roll manufacturing technology, utilizing a thin flexible stainless steel substrate, offers a number of advantages in a fully automated high throughput PV module production plant [1-18]. The stainless steel substrate is rugged and this improves production yield by eliminating substrate breakage--a significant problem in many glass substrate amorphous silicon alloy PV module manufacturing plants. ECD's transport mechanism on a thin flexible stainless steel substrate is also mechanically simple and significantly less expensive than the transport mechanism for conventional glass substrates. Heating and cooling of the substrate during solar cell deposition can also be accomplished quickly compared to glass substrate manufacturing. As a result, capital equipment costs for a large volume plant utilizing ECD's roll-to-roll technology will be significantly less than for other PV manufacturing processes.

SERPENTINE DEPOSITION DESIGN CONCEPT

In order to further improve the economies of scale of the continuous roll-to-roll amorphous silicon alloy solar cell deposition process, we have developed a concept for a serpentine web deposition process to maximize throughput while keeping the size of the deposition chambers small. Such a deposition process , when incorporated into continuous roll-to-roll PV manufacturing process, will maximize the throughput for a high volume production plant, reduce the machine cost, improve gas utilization, reduce power consumption, and improve the solar cell stability. A schematic diagram for a RF plasma processor that deposits amorphous silicon alloy solar cells utilizing a serpentine web roll-to-roll process in shown in Figure 1. As can be seen in the diagram, deposition takes place on the substrate as it travels vertically through the deposition chamber in contrast to the current process in which the substrate travels horizontally. Also, in the serpentine process, the thin stainless steel substrate travels up and down through the use of rollers in the deposition chamber to create, in effect, a large deposition area. In the deposition chamber, a single RF cathode will generate a plasma which will produce film deposition on regions of substrate facing both sides of the cathode. Substantial equipment cost savings will be achieved from the compactness of this design which has half the volume of a non-serpentine design. The surface area of this machine is 3 to 4 times lower than that of a non-serpentine design and thus will reduce heat loss and consequently electric power consumption. Furthermore, the use of a mostly open perforated cathode configuration will improve the gas utilization efficiency since deposition of a-Si alloy material on the cathode is reduced. The hardware design for the internal components of the chamber is simpler for the serpentine design because the cathode heating and dark-space shields can be eliminated. In addition, for constant throughout, the number of rf power supplies, pumps, feedthroughs and cathodes is reduced by about a factor of two.

We will minimize the Staebler-Wronski efficiency degradation by depositing amorphous silicon alloy intrinsic materials with a minimum density of microvoids at a low deposition rate [19] in the serpentine roll-to-roll process. We anticipate the current 15-18% degradation in the current commercial product can be reduced to less than 8%.

RESULTS

To demonstrate serpentine web deposition concept described above, we have constructed a single loop serpentine deposition chamber to deposit a-Si for the

CH3365-4/94/0000-0820 $4.00 © 1994 IEEE

First WCPEC; Dec. 5-9, 1994; Hawaii

n-i-p solar cells. This serpentine chamber has been incorporated into ECD's continuous roll-to-roll multipurpose deposition machine, as shown in Figure 2. The tall vertically standing box in the picture is the serpentine chamber.

One critical area which requires engineering attention in the serpentine web continuous roll-to-roll PV manufacturing process, is the top roller. When the web passes over this roller, the deposition side of the web faces the roller. We used a dumb-bell shape design for the top serpentine roller, so that the deposition side of the web, except for the edges which are outside the active area, does not touch any hard surface. No deformation of the web is observed after the high tension test. Furthermore, no film peeling or other mechanical damage is observed when stainless steel web, coated with back-reflector, triple solar cell and top conductor, passes through the machine.

We have carried out an initial solar cell deposition using the serpentine chamber. During the initial process optimization, we achieved 8.6% initial efficiency for a single-junction serpentine solar cell. Figure 3 is the J-V curve of a single-junction n-i-p solar cell with the intrinsic layer deposited in the serpentine chamber. The J-V curves under the weak blue and red illumination are shown in Figure 4

We then incorporated this single-junction n-i-p solar cell into a-SiGe/a-SiGe/a-Si triple-junction solar cells. The top cell intrinsic layer of the triple cell was deposited in the serpentine chamber. We optimized the cell current matching between each of the component cells and achieved initial efficiency of 9.5%, as shown in Figure 5.

SUMMARY

A serpentine web deposition concept has been described. Such a deposition configuration, when incorporated into a continuous roll-to-roll PV manufacturing process, will maximize throughput for a high volume production plant, reduce the machine cost, improve the gas utilization, reduce power consumption, and improve the material stability. A single loop serpentine deposition chamber has been constructed and incorporated into ECD's roll-to-roll deposition machine. During the initial optimization, we have demonstrated a single-junction serpentine solar cell with 8.6% initial efficiency and a triple-junction solar cell with 9.5% initial efficiency.

In summary, the serpentine technology will provice the following benefits:

- For a given throughput, serpentine web technology will decrease the machine footprint significantly and reduce the number of rf cathodes and the required pumping speed by a factor of two. We expect that serpentine web technology will provide at least a factor of two reduction in capital and maintenance costs for the deposition machines in high volume plants.
- Serpentine web technology should increase the gas utilization efficiency by 85%, and correspondingly reduce the gas cost.

ACKNOWLEDGMENTS

The authors would like to thank Drs. S. Guha, J. Yang, and H. Fritzsche for important discussions and suggestions. Also, the authors would like to thank other project members who contributed to this work. This work supported by DOE/NREL PVMaT 2A Program under subcontract No. ZM-2-11040-7.

REFERENCES

[1] "Production of Tandem Amorphous Silicon Alloy Cells in a Continuous Roll-To-Roll Process", M. Izu and S. R. Ovshinsky, SPIE Proc. 407, 42 (1983).

[2] "Roll-To-Roll Plasma Deposition Machine for the Production of Tandem Amorphous Silicon Alloy Solar Cells", M. Izu and S. R. Ovshinsky, *Thin Solid Films* 119 55 (1984).

[3] "Amorphous Silicon Solar Cells Production in a Roll-To-Roll Plasma CVD Process", H. Morimoto and M. Izu, JARECT 16 (1984); *Amorphous Semiconductor Technology & Devices*, North Holland Publishing Company, Edited by Y. Hamakawa, 212 (1984).

[4] "Roll-To-Roll Mass Production Process for Amorphous Silicon Solar Cell Fabrication", S. R Ovshinsky, *Proc. International PVSEC-1*, 577 (1988).

[5] "High efficiency Multiple-Junction Solar Cells using Amorphous Silicon and Amorphous Silicon Germanium Alloys", J. Yang, R. Ross, T. Glatfelter, R. Mohr, G. Hammond, C. Bernotaitis, E. Chen, J. Burdick, M. Hopson and S. Guha, *Proc. 20th IEEE PV Spec.* Conf. 241 (1988).

[6] "Physics of High Efficiency Multiple-Junction Solar Cells", J. Yang, R. Ross. R. Mohr, and J. P. Fournier, *Proc. MRS Symp.* **95**, 517 (1987).

[7] "Performance of Large Area Amorphous Silicon Based Single and Multiple-Junction Solar Cells", P. Nath and M. Izu, *Proc. of the 18th IEEE Photovoltaic Specialists Conf.*, Las Vegas, Nevada, 939 (1985).

[8] "1MW Amorphous Silicon Thin-Film PV Manufacturing Plant", P. Nath, K. Hoffman, J. Call, C. Vogeli, M. Izu and S. R. Ovshinsky, *Proc. of the 3rd International Photovoltaic Science and Engineering Conf.*, Tokyo, Japan, 395 (1987).

[9] "Conversion Process for Passivation Current Shunting Paths in Amorphous Silicon Alloy Solar Cells", P. Nath, K. Hoffman, C. Vogeli and S. R. Ovshinsky, *Appl. Phy's. Lett.* **53** (11), 986 (1988).

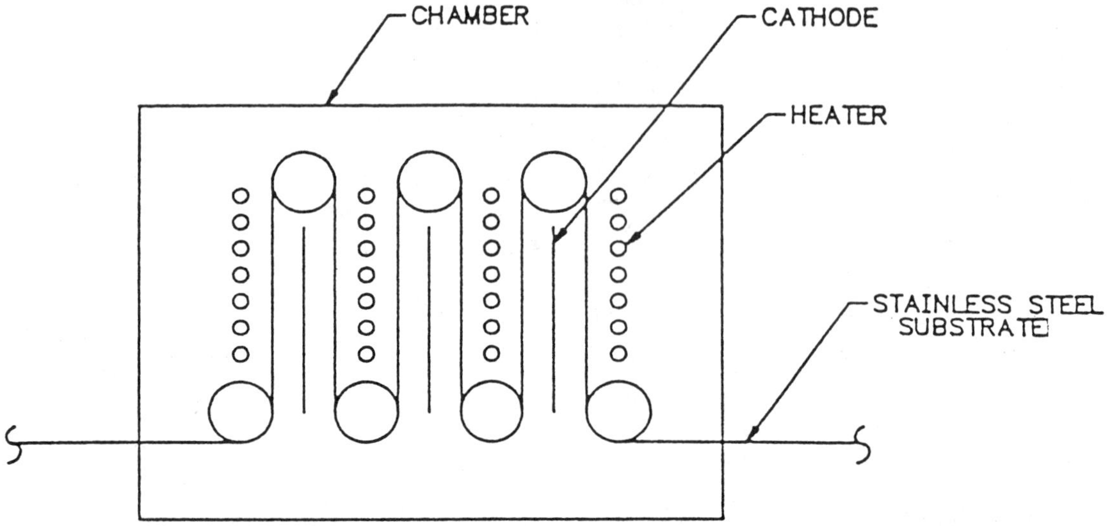

Figure 1. A Schematic Diagram of Serpentine Web, Roll-to-Roll RF PECVD Deposition Chamber.

Figure 2. ECD's Multipurpose Continuous Roll-To-Roll A-Si Solar Cell Deposition Machine. The Tallest Chamber Is The Serpentine Deposition Chamber.

[10] "Yield and Performance of Amorphous Silicon Based Solar Cells Using Roll-To-Roll Deposition", P. Nath, K. Hoffman, J. Call, G. DiDio, C. Vogeli and S. R. Ovshinsky, *Proc. 20th IEEE PV Spec. Conf.,* 293 (1988).

[11] "A New Inexpensive Thin Film Power Module", P. Nath, K. Hoffman, C. Vogeli, K. Whelan and S. R. Ovshinsky, *Proc. 20th IEEE PV Spec. Conf.,* 1315 (1988).

[12] "Fabrication and Performance of Amorphous Silicon Based Tandem Photovoltaic Devices and Modules", P. Nath, K. Hoffman, and S. R. Ovshinsky, 4th International PV Science and Engineering Conf., Sydney, Australia, (1989).

[13] "Advances in High-Efficiency, Multiple-Bandgap, Multiple-Junction Amorphous Silicon Based Alloy Thin-Film Solar Cells", S. Guha, A Paper Presented at MRS Spring Meeting, San Diego, April (1989).

[14] "Amorphous Silicon-Germanium Alloy Solar Cells with Profiled Bandgaps", J. Yang, R. Ross, T. Glatfelter, R. Mohr, and S, Guha, MRS Symposium Proc. Vol. 149, 435 (1989).

[15] "Manufacturing of Triple-Junction 4 ft^2 a-Si Alloy PV Modules", M. Izu, X. Deng, A. Krisko, K. Whelan, R. Young, H. C. Ovshinsky, K. L. Narasimhan, and S. R. Ovshinsky, *Proc. of 23rd IEEE PVSC,* 919, (1993).

[16] "Continuous Roll-to-Roll Amorphous Silicon Photovoltaic Manufacturing Technology", M. Izu, S. R. Ovshinsky, X. Deng, A. Krisko, H. C. Ovshinsky, K. L. Narasimhan, and R. Young, *AIP Conf. Proc.* **306,** 12th NREL PV Program Rev., 198, (1993).

[17] "Amorphous Silicon Alloy Photovoltaic Technology - From R&D to Production", S. Guha, J. Yang, A. Banerjee, T. Glatfelter, K. Hoffman, S. R. Ovshinsky, M. Izu, H. C. Ovshinsky, and X. Deng, Proc. of MRS Spring meeting (1994).

[18] "Stability Test of 4 ft^2 Triple-Junction a-Si Alloy PV Production Modules", X. Deng, M. Izu, K. L. Narasimhan and S. R. Ovshinsky, Proc. of MRS Spring meeting (1994).

[19] "Effects of Microvoids on Initial and Light-Degraded Efficiencies of Hydrogenated Amorphous Silicon Alloy Solar Cells", S. Guha, J. Yang, S. Jones, Y. Chen, and D. Williamson, *Appl. Phys. Lett.* **61,** 1444 (1992).

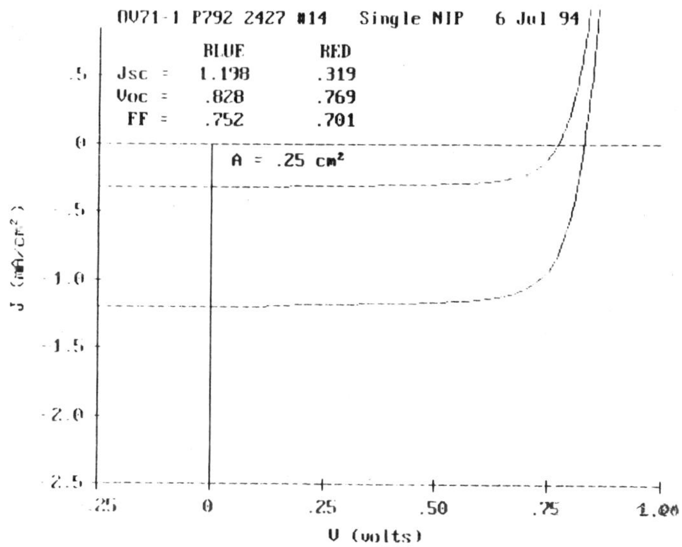

Figure 4. J-V Curve of the Single-junction Serpentine Solar Cell Under Weak Blue and Red Light.

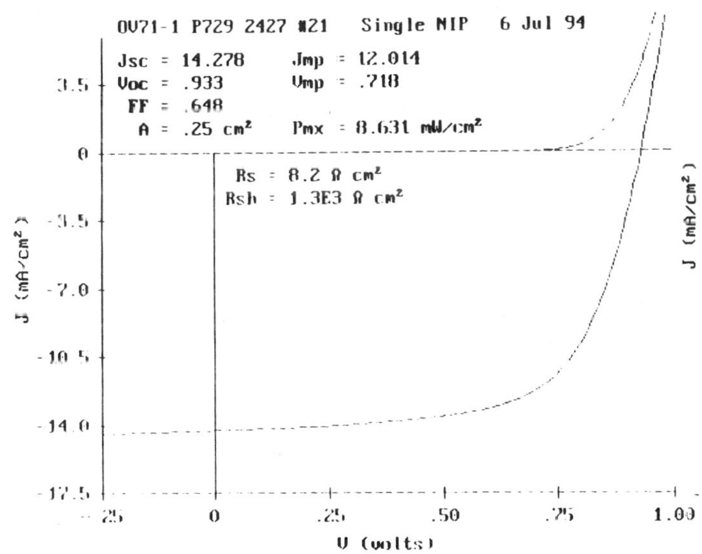

Figure 3. J-V Curve of a Single-junction Serpentine a-Si Solar Cell.

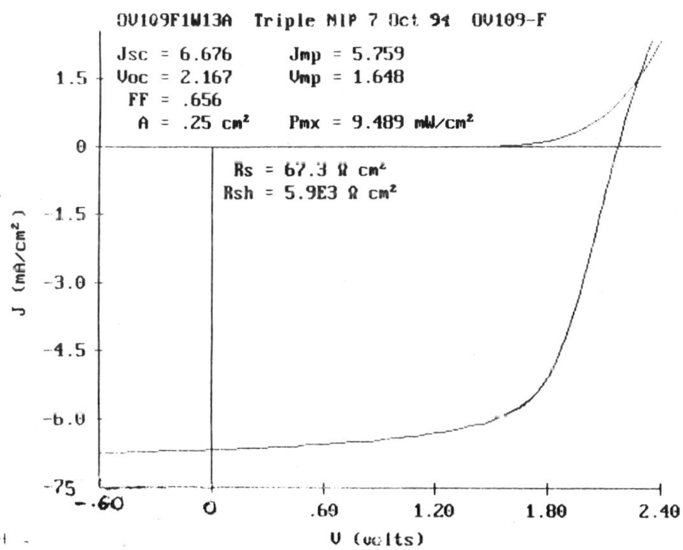

Figure 5. J-V Curve of a Triple-junction a-Si Alloy Solar Cell with the Top Intrinsic Layer Deposited in the Serpentine Chamber.

PHOTOVOLTAIC Cz SILICON MANUFACTURING TECHNOLOGY IMPROVEMENTS

Theresa L. Jester, Kim W. Mitchell, Rob Probst, Dave Tanner and Sergio Vasquez
Siemens Solar Industries, Camarillo, CA 93011
This work was supported in part by the National Renewable Energy Laboratory

ABSTRACT

Siemens Solar Industries (SSI) began a three-year, three-phase cost shared contract in March 1992 to demonstrate significant cost reductions and improvements in manufacturing technology. The work has focused on near-term projects for implementation in the SSI Czochralski (Cz) manufacturing facility in Camarillo, California.

The work has been undertaken to increase the commercial viability and volume of photovoltaic manufacturing by evaluating the most significant cost categories and then lowering the cost of each item through experimentation, materials refinement, and better industrial engineering.

The program has concentrated in the areas of crystal growth, wafer technology, module development and environmental, safety and health issues.

INTRODUCTION

The objective of the program is to reduce costs in photovoltaic manufacturing by approximately 10% per year. The program consists of three focused tasks relating to cost reduction. The silicon wafer itself contributes about half of the total module cost and has the most potential for cost reduction. The cell processing costs are about a quarter of the module costs, with cell efficiency results being most important. Module assembly and packaging costs are the balance, with the module design, both materials and labor, contributing significantly.

Crystal growing costs are driven by material growing yields and indirect manufacturing costs such as electricity and machine parts used each time an ingot is fabricated. Wafering costs are driven by labor and the number of good slices yielded per length of crystal processed. This first task of reducing the wafer costs has focused on the polysilicon material used for ingot growth, and the evaluation of wire saw machines to improve the yielded wafers per inch of ingot.

Cell processing costs are driven by the electrical contacts used, and the labor required for the process steps to clean the wafer, form the semiconductor junction, and the contacts. The cell tasks have been focused on better junction formation and reduced contact resistance. Process automation for lower labor costs have begun.

Module costs are highly sensitive to labor and materials. The module design tasks are driven by high reliability in the field and lower costs.

Environmental, safety and health issues govern the ability to manufacture products commercially. This program has focused on chlorflorocarbon (CFC) elimination and caustic chemical waste reduction.

CRYSTAL GROWTH AND THIN WAFERS

The Crystal Growing operation increased production significantly during the last two years. The increase is attributable to several factors including improved polysilicon quality and cleaning, upgrades to the growing equipment, specifically diameter controls, and upgrades to the hotzone graphite parts which have increased reliability and reduced cycle time.

A wide range of poly silicon starting materials have been evaluated. Crystal Growing yields, cell performance and impurity analysis have been conducted. Overall crystal growing yields improved by 4% as a result of the study.

One thousand kilograms of Semi-Prime virgin poly material were processed with the standard crystal growing parameters. The yields for thirty experimental crystal growth runs were compared to the standard baseline process (see Fig.1). The slight yield improvement observed with Semi-Prime material could not justify the added cost.

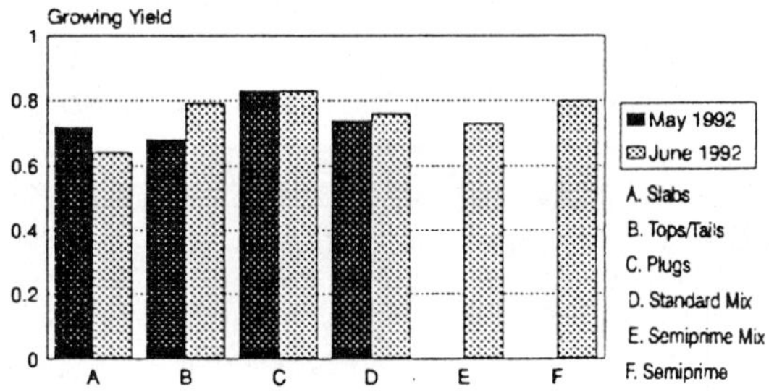

Fig. 1. Semi-Prime Poly Silicon Study

Remelt silicon from outside sources was investigated. The resistivity of the starting material plays a role in the quality of the grown ingot and has a direct affect on the crystal growing yield (see Fig.2). Through this study it was determined that 1-3 Ohm-cm, n-type material consistently produced the lowest crystal growing yields. This material must be heavily counter doped in order to grow 1-2 Ohm p-type ingot. In comparison, remelt material which is >10 Ohm-cm, n or p-type, repeatedly produces the highest yields for remelt material purchased from outside vendors.

CH3365-4/94/0000-0824 $4.00 © 1994 IEEE

SSI uses in-house remelt material as feed stock for ingot growth. In-house remelt used includes tops, tails, slabs and selected plugs, those which are not suspected of containing quartz chips or other foreign material. Historically, the yields obtained using the slab material were low, believed caused by the heavy oxide buildup on the sides of the ingot. Use of slab material was stopped until a cost effective method of cleaning could be found.

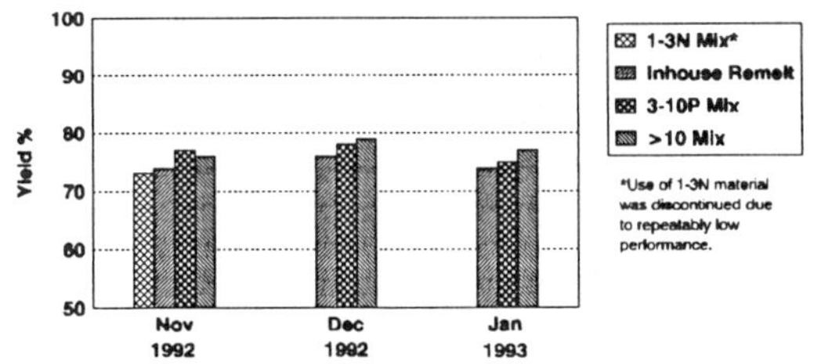

Fig. 2. Resistivity Effects of Poly Silicon on Growth Yields

Using an abrasive tumbler gave the ability to abrade the outside layer of the slab material, increasing the crystal growing yields. Further studies were conducted which showed that an overall yield improvement was obtained by placing all remelt material, inhouse and vendor material, through the tumbling process.

Crystal growth improvements focused on better techniques for growing ingots from remelt material. Two types of necking were tested: "step" necking, which produces the fewest dislocations but is slower, and "paintbrush" necking, which is faster and easier from an operations standpoint. The concern during faster pull speed necking is the creation of "slip" dislocations in the crystal structure. The electrical effect of slip dislocation has shown to reduce short circuit current in processed cells[1]. Cells made from these two different types of ingots did not show a significant difference in electrical performance, thus the faster technique is desirable for production throughput considerations.

A study of oxygen concentration was conducted. The study included different types of crucibles and different growth parameters such as pull speed and rotation speed to improve the electrical properties of the as grown ingot.

The crucible is the major source of the oxygen which is incorporated into a silicon ingot during growth. Crucibles with a denser, bubble free layer on the inside wall have been shown to devitrify less and to contribute less oxygen to the silicon melt. A set of "low bubble" crucibles was compared to standard crucibles and samples of the silicon ingot taken from the top and the bottom of the ingots. Fourier Transform Infrared, FTIR, analytical technique was used to measure the oxygen concentration on each sample. The results show the oxygen concentration for the low bubble crucible to be lower at the bottom of the ingot than for the standard crucible.

Larger diameter ingots have been implemented successfully and larger crucibles have also proven effective in reducing crystal growth costs.

A prototype, commercially available wiresaw was tested and a process developed for manufacturing silicon wafers for photovoltaic use. The average number of wafers per inch obtained to date on the wiresaw is 44, which is a 50% yield improvement over the results obtained for ID sawn wafers (see Fig. 4).

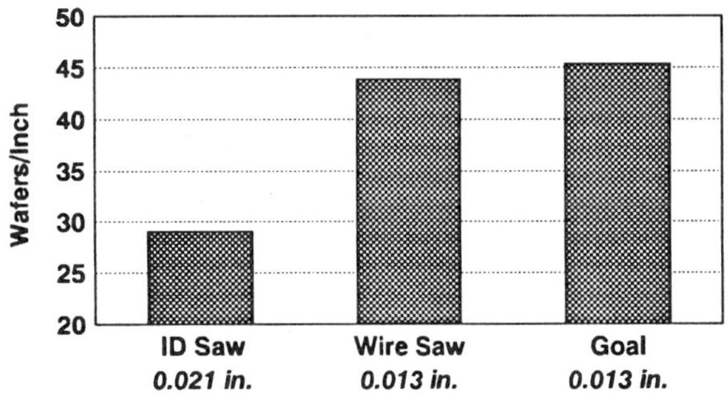

Fig. 4. Wire Saw Evaluation

Several critical issues have been identified for high yield of the wiresaw process. These include:

- Wire material specification , such as strength and temper
- Slurry and solvent usage
- Wafer cleaning
- Ingot mounting

The most significant of these has been the combination of wire usage and slurry usage. The yield on the wire saw process has improved from 70% to over 90% in a one year time frame through better understanding of these interactions (see Fig. 5).

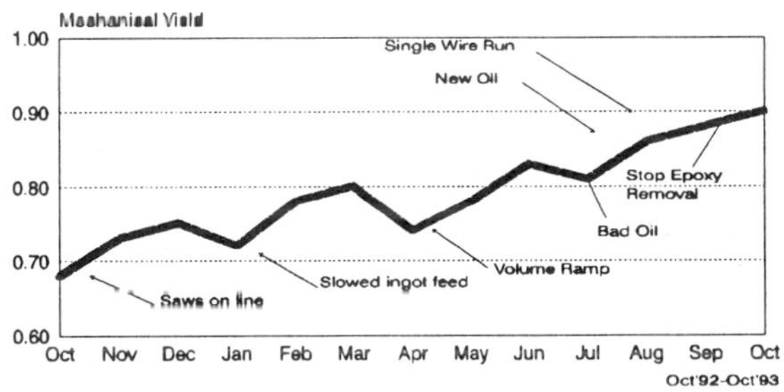

Fig 5. Wire Saw Yields

The quality of wafers sliced varies significantly in thickness control and taper (non-parallel sides) if these parameters are not tightly controlled. The goal is to maximize the number of runs obtained with each wire change and slurry batch used. The quantity of these materials consumed per run or per wafer is a very important parameter for cost effectiveness in manufacturing high volumes of wafers.

CELL PROCESSING

Cell processing costs are driven by the amount of silicon used per cell (wafer thickness), the electrical performance of the cells, and the labor required for the processing of wafers into active cell devices. Work in this area has been focused on the reduction of labor through automation of various material transfer steps and the improvement of the electrical output of the cells processed by improving the contact resistance and diffusion processes.

A study to examine the effect of wafer thickness, mechanical yield, and electrical output was performed. The experiment tested four groups of wafers, two cut by both ID and wire saws. The experimental groups, and the test results are shown below:

Table 1. Cell Thickness v. Performance

Saw Type	ID	WIRE	ID	WIRE
Thickness (.001")	14	14	17	17
% loss (Breakage)	6.7	8.3	3.0	2.9

The results of the tests can be summarized as follows:

- Breakage was highest with thin wafers, especially thin wire saw wafers. Most of the breakage occurred at material handling processes.
- Saw type used for slicing appeared insignificant.
- The Electrical $mean_{(amps)}$ did not show any significant differences.

The increased yield of wafers per inch with wire saws is a significant improvement in cost. Decreasing wafer thickness in high volume is a more critical process change affecting yield. Optimizing the mechanical yield with thin wafers requires automated wafer and cell transfer mechanisms. Automated upgrades to cell processing equipment have been designed and installed in the manufacturing line.

The cell electrical distribution has improved significantly as a result of improved contact and diffusion processes. Through experimentation with process changes, an interesting and significant cell performance sensitivity was discovered. Certain etch processes showed sensitivity to environmental exposure, particularly humidity exposure. Results of humidity exposure v. process type showed a significant change in the amount of Phosphorus present before and after exposure (see Fig. 6). This larger amount of Phosphorus coincided with visible delaminations in the modules made with these cells. The process which was insensitive to humidity exposure was chosen for production implementation. Through these studies, coupled with contact firing process changes, contact coverage re-design and diffusion changes, a significant shift in the electrical performance was effected (see Fig. 7).

Larger area cells have been developed and implemented which have an area improvement of almost 40%.

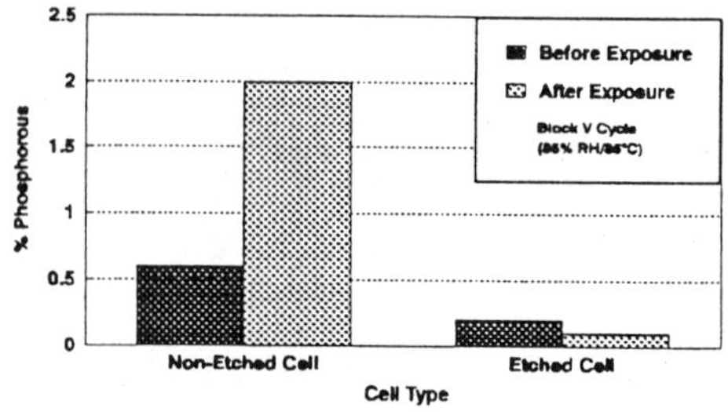

Fig. 6. Humidity Sensitivity of Cells

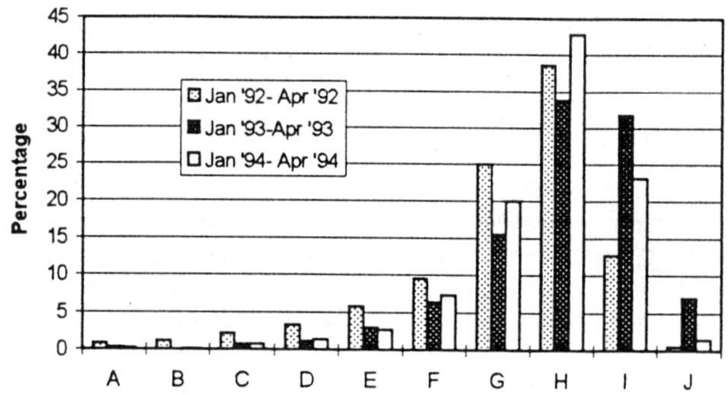

Fig. 7. Cell Electrical Improvement

MODULE DEVELOPMENT

Significant progress in cost reduction in module design has been made. Several items have been identified and implemented in manufacturing to reduce the final cost per watt of modules manufactured. The most leveraging of these has been the introduction of a larger module with larger cells. The IV characteristics of this new module shows 75 watts produced at Standard Test Conditions (see Fig. 8).

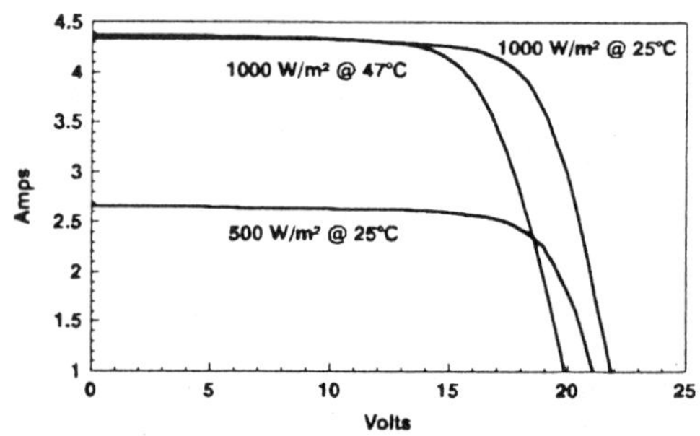

Fig. 8. I-V Curve for Standard Model 75W Module

The second of these improvements showed the use of anti-reflective glass etching to decrease the amount of reflected light on a module surface. Greater than a 1 Watt improvement in electrical power was seen on M55 (53 Watt)

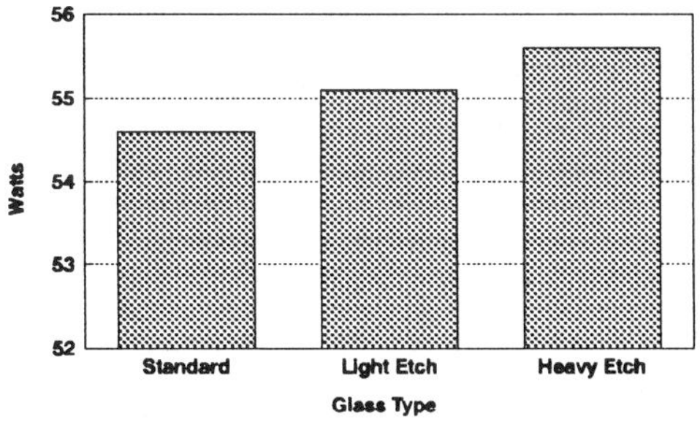

Fig. 9. Anti-Reflective Glass Improvement

modules. This gain varies with the degree of etch (see Fig. 9). Although this proved to be a significant electrical performance boost at low cost, the commercial availability of this glass is quite limited, and use of this material in high volume manufacturing is not feasible.

The final module improvement has been in the development of a new junction box for modules (see Fig. 10). This improvement in manufacturing has been approved by Underwriters Laboratory.

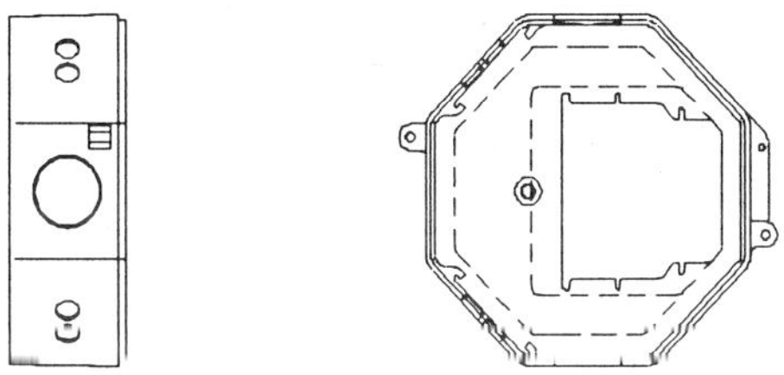
Fig. 10. New J-Box

Safety, health and environment issues have been addressed with the complete elimination of CFC's during May 1993. Significant reduction in costs and volume of caustic chemicals have also been achieved. Costs per cell have been reduced 20% and volume by over 10% (see Fig. 11).

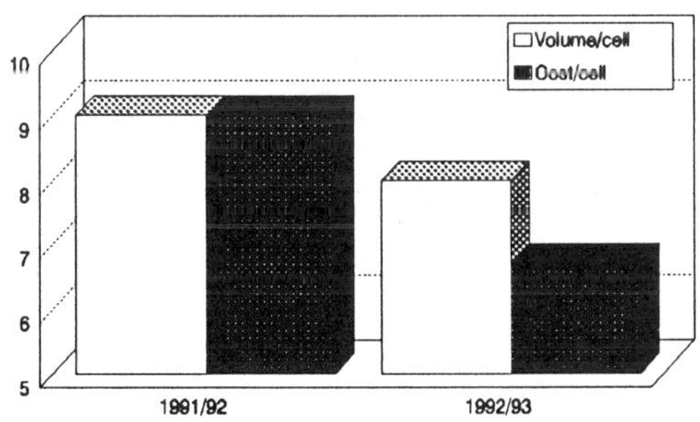

Fig. 11. Caustic Waste Cost and Volume Per Cell

SUMMARY

Reduction in Cz photovoltaic manufacturing costs are driven by several parameters. Low wafer cost, high electrical and mechanical yields in fabricating cells, low labor in cell fabrication and module fabrication, and materials used for making modules are all significant factors. Siemens Solar has reduced costs by over 25% in two years by addressing each of these items. Table 2. shows the categories and total savings for manufacturing cost improvements effected in this program. Further improvements in each area are planned.

Table 2. Cost Reduction Summary

Category	*% Reduction in Cost*
1. Crystal Growth/Thin Wafers - Large crucible - 43 vs. 29 wafers/inch - Large ingot	10%
2. Cell Improvements - 4% electrical improvement - Large Cell - Automated System	4%
3. Module Improvments - Automated Assembly - Large Module - New J-Box	14%
Total	***28%***

Goal: 30% 3rd Year

REFERENCES

[1] K.L. Pauls et al., "The Effect of Dislocations on the Performance of Silicon Solar Cells", *23rd IEEE Specialist Conference, 1993*

ADVANCED AUTOMATION TECHNIQUES FOR INTERCONNECTING THIN SILICON SOLAR CELLS

M.J. Nowlan, S.J. Hogan, G. Darkazalli, S.F. Sutherland, W.F. Breen, J.M. Murach, and J.S. Patterson
Spire Corporation, One Patriots Park, Bedford, MA 01730, USA

ABSTRACT

The objective of this work is to reduce the cost and improve the quality of terrestrial photovoltaic modules by developing automated high-throughput (5 MW/yr) processes for interconnecting thin silicon solar cells. New low-stress, high-throughput processes have been developed for cell loading, alignment, and inspection, interconnect ribbon handling, flux application, ribbon-to-cell soldering, cell string handling, and I-V testing of assembled cell strings. Both standard thickness (350 μm) and thin (200 μm) cells have been used to evaluate and refine these processes.

INTRODUCTION

Improvements in ingot slicing techniques (wire sawing) allow significantly more thin (*e.g.,* 200 μm) wafers to be produced from a silicon ingot than standard thickness (350 to 450 μm) wafers, thus reducing wafer cost. However, unacceptably high breakage in subsequent handling and processing normally makes production with such thin cells uneconomical. In this effort, low mechanical and thermal stress processes are being developed to achieve high production yields with thin cells. Other areas being addressed include flexible automation, higher process throughput, and improved quality control through in-line inspections and tests.

METHOD OF APPROACH

A fully automated solar cell assembly system is being developed for interconnecting thin and standard solar cells (Fig. 1). The materials loaded into the system are stacks of cells, reels of solder coated copper interconnect ribbon, and liquid soldering flux. The system assembles multiple rows of cell strings, where each string comprises a number of cells soldered together in series.

Flexible automation techniques have been incorporated into the system to enable the production of a variety of module designs without tooling changes. Flexibility is provided through software control of ribbon length and bend location, cell spacing, the number of cells per string (up to 2m long), string spacing, the number of strings per module (up to 1.4m wide), and the arrangement of strings in a series or parallel configuration.

In-line automated inspections and tests are provided for improved yield and quality control. Prior to assembly, cells are inspected with a machine vision system for chips, upside-down placement, and cell type. After string completion, current-voltage (I-V) curves are measured with a string I-V tester to verify that the strings function properly prior to final placement.

Fig. 1. Automated solar cell assembler.

CH3365-4/94/0000-0828 $4.00 © 1994 IEEE

RESULTS

A prototype cell assembly system was designed, fabricated, and is now undergoing tests with a variety of cell types. The processing system has the capacity to handle round, square, or pseudo-square (round with four flats) cells up to 15 cm across. Cell strings up to 2.0m long can be assembled and arranged side-by-side to create two-dimensional module-size arrays up to 1.4m wide. A layout drawing that identifies key elements of the integrated system is provided in Fig. 2. The subsystems are described in the sections that follow.

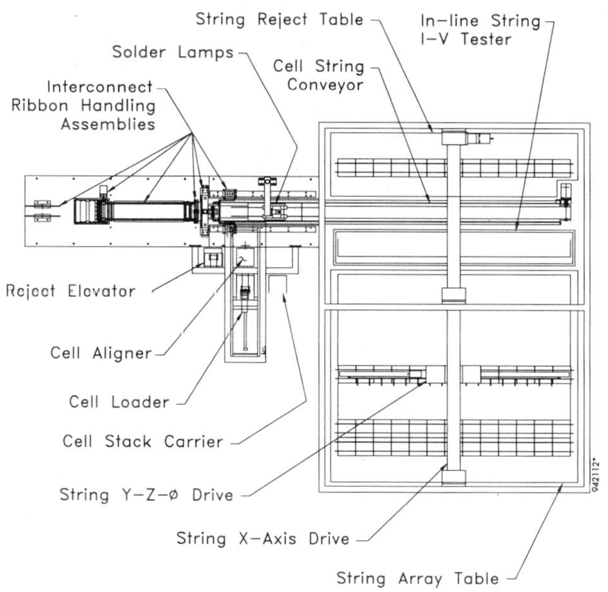

Fig. 2. Lay-out drawing of cell assembly system (plan view).

Cell Loading

An extensive survey of commercial wafer handling equipment was conducted, since this equipment is commonly used in the semiconductor industry. No equipment was found that satisfies the requirements for both speed (4 s/cell) and cell size capacity (15 cm), so a custom system was designed and built.

The loading system consists of carriers for holding stacks of cells, a shuttle that transports two carriers, an elevator with an optical sensor for maintaining a constant stack height during loading, air knives to separate cells on pick-up, and a manipulator to transport cells from the elevated stack to the cell aligner. Carriers for cell stack loading were selected over slotted cassettes because of the greatly reduced volume and frequency of carrier loading. For example, a slotted cassette that holds 25 cells has the same volume as a carrier that holds a stack of 650 cells (with a cell thickness of 200 μm), equivalent to 26 cassettes.

A two-stack shuttle system was developed that allows continuous cell feeding without interruptions for stack loading by the operator. The loader manipulator is a two-axis (x-z) linear actuator with a vacuum pick-up end effector. The vacuum pick-up and its floating mount were specially designed for handling thin cells.

Cell Alignment and Inspection

A cell alignment and inspection system was developed, consisting of a machine vision system, an alignment stage, and a four-axis (x-y-z-θ) robot with a vacuum pick-up end effector. Vision system alignment is gentler on the cells than mechanical alignment, since no contact is made to the cell edges. The vision system also allows alignment to be done to the cell's contact pattern rather than the wafer edges, a more accurate approach when contact patterns are not precisely aligned to the silicon wafer.

After a cell is placed on the stage by the loader, a video camera transmits an image of the cell to a vision processor board in a personal computer (PC). The image is analyzed to inspect the cell and determine its location in x, y, and θ. The PC then directs the robot to pick up the cell, transport it, and place it in the properly aligned position for assembly. Cells that fail the visual inspection due to chips, upside-down placement, or incorrect cell type are placed in a reject carrier.

Initial work with the vision system identified problems with lighting and background which prevented the cell edges from being detected reliably. Since the cell edges are dark, a white shutter was added behind the cell to provide high contrast. Testing with the shutter showed that shadows were produced at the cell edges which the vision system sometimes selected instead of the true cell edges. Experiments with a wide range of lighting configurations failed to eliminate the edge shadows. Finally, the shutter was replaced with a white translucent material which was illuminated from behind. The transmitted light successfully eliminated the edge shadows.

The addition of the shutter allows the aligner stage to be inspected with the vision system for cell fragments or other debris that could be left on the aligner. Such debris could damage cells or cause errors in subsequent cell alignments or inspections. A debris check routine has been implemented that inspects the stage after each cell is picked up. If debris is found, the cell is rejected (since the debris could be a piece of that cell), the debris image is displayed on a monitor, and the operator is notified to clear the stage.

Ribbon Handling and Flux Application

A number of process subassemblies for ribbon feeding, fluxing, bending, cutting, and placing were developed. A ribbon shuttle assembly with pneumatic grippers pulls the ribbons through a bending die and a cutter to form stress-relief bends and cut the ribbons to length. Linear stepper motors drive the shuttle at high speed and provide programmability for cutting and forming ribbons to user-specified dimensions.

Various automated solder flux application processes were considered, including coating ribbons in a flux bath and dispensing flux onto the ribbons and cells with syringes. The ribbon coating process was selected because it requires less complex hardware, it adds no process time to the ribbon-cell assembly sequence, and it coats both sides of the ribbon with flux for soldering to the cells' front and back contacts. Copper ribbon, obtained with a 12 μm thick solder coating, provides sufficient solder for the joint, eliminating the need to dispense solder paste.

A flux station was developed for coating the ribbons with flux as the ribbons are dispensed from the reels. Several commercially available fluxes were evaluated. Most had problems with either poor solderability to cell contacts, excess post-soldering residues, or tackiness, which caused the ribbons to stick to the tooling. One flux was found that has good solderability, a low solids content for minimal residues, and low tackiness for reliable ribbon feeding.

Ribbon-to-Cell Assembly and Soldering

A unique conveyor belt system with vacuum capability was invented for holding ribbons and cells securely in position and transporting them as the string is assembled. An assembly process was devised involving ribbon placement, cell placement, and belt indexing in the proper sequence so that ribbons connect the back contact of one cell to the front contact of the next, thus creating a series string. The belt is driven by a stepper motor to produce the desired spacing between cells and number of cells per string.

A heated platen supports the first section of belt to preheat the cells and ribbons. Preheating reduces the thermal stress on the cells during soldering, allowing shorter cycle times and higher yields. A solder head with high intensity lamps is mounted over the belt at the end of the heated section. The lamps provide a non-contact means for rapidly heating the cell, avoiding the thermal and mechanical shocks generated by conduction heating. Both front and back cell contacts are soldered simultaneously, eliminating one of the two heating cycles used in previous methods, in which front and back contacts were soldered separately.

Cell String Handling

A string transfer mechanism has been developed for picking up completed cell strings from the conveyor belt and transporting them to an I-V tester and to an array table. The mechanism comprises a linear array of vacuum pick-ups suspended from a programmable four-axis (x-y-z-θ) manipulator. The manipulator can rotate the pick-up array 180° to place strings with their ends alternately positive and negative for series connections. Fig. 3 shows the mechanism with three 12-cell strings produced by the system.

941032

Fig. 3 String transfer mechanism and three completed strings of 10 cm square cells.

String I-V Testing

Cell string performance is measured after assembly with the string I-V tester. Designed for high throughput, the tester illuminates a cell string with a single pulse of simulated sunlight during which a complete current-voltage curve is measured. Strings up to 15 cm wide by 200 cm long can be tested. The string transfer mechanism carries strings that fail the test to the reject table; strings that pass are placed on the array table in the series or parallel arrangement selected for the module.

The light source is a pulsed xenon lamp filtered to provide an Air Mass 1.5 spectrum at one sun (100 mW/ cm^2). A new pulse forming network was designed that generates a stable light pulse with a flat peak of 7 to 8 ms. Electronics and software were developed for sweeping the string's I-V curve in 3 ms, acquiring 100 data points from I_{sc} to V_{oc}, and acquiring lamp intensity data from a calibrated reference cell during the measurement. String data for V_{oc},

I_{sc}, and P_{max} are compared to minimum values selected for the particular string type to determine acceptance.

System Performance

A range of different cell sizes (up to 125 mm square) and thicknesses (down to 200 μm) have been processed in the assembly system. Two trial runs with several thousand 300 μm cells have been run. Yields ranging from 97.5% to 98.5% have been recorded. Cell-to-cell cycle times, initially clocked at 5.0s, were reduced to 3.9s through software enhancements. String-to-string cycle times were measured at 68s per 12 cell string, or 5.7 s/cell. This cycle time is higher than the cell-to-cell cycle time due to the string fabrication sequence, in which each string is soldered and transported by the conveyor for pick-up before the next string is begun. Sequencing improvements are expected to reduce the string-to-string cycle time.

Solder joint quality was monitored with pull strength measurements made with a digital force gauge. Good pull strengths are routinely achieved, as shown by the typical pull test distribution plotted in Fig. 4.

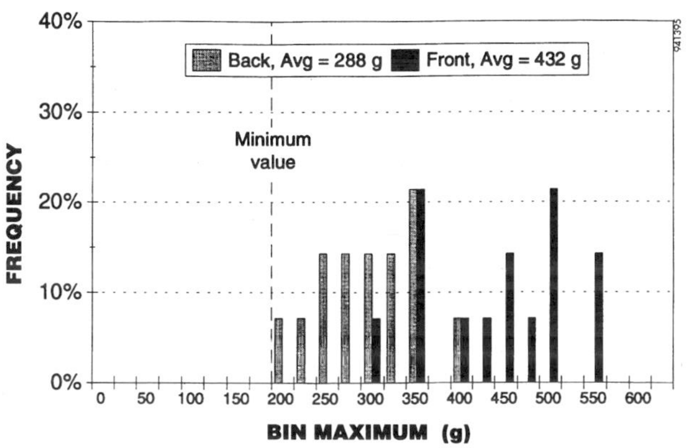

Fig. 4. Pull test results for ribbons soldered to 200 μm thick x 102 mm square Solec cells. Pull angle = 90°.

Environmental Testing

Accelerated environmental testing was done to assess the long-term reliability of strings produced by the automated assembly system. Ten 36-cell modules were fabricated at Spire from strings made with 10 cm square Siemens Solar cells. The small amount of flux residue left on the strings after soldering was not cleaned prior to encapsulation. The modules were laminated between a glass superstrate and a back cover film with standard cure ethylene vinyl acetate (EVA) encapsulant in a Spi-Laminator™ 350. The modules were sent to Siemens Solar for environmental testing.

Three modules were subjected to 200 thermal cycles (T-200) from +90°C to -40°C; three received 50 thermal cycles (T-50) followed by 20 humidity-freeze cycles (HF-20) from +85°C/85% RH to -40°C; three were subjected to 1000 hours of damp heat soaking (DH-1000) at +85°C/85% RH; one was a control module. Test results show that the maximum module degradation was 3.8%, well within the 10% acceptance limit.

CONCLUSIONS

New low stress techniques have been developed for automated solar cell assembly, including a machine vision cell alignment process that eliminates mechanical handling of cell edges, and a soldering process that uses preheating and high intensity light to solder interconnect ribbons to cell front and back contacts in one step. The soldering process eliminates one of the two heating steps normally used for interconnecting cells, thus improving process yield.

Flexible automation techniques have been incorporated to enable the production of a variety of module designs with minimal mechanical adjustments or tooling changes. This flexibility is provided through software control of the ribbon length, the ribbon stress-relief bend location, the cell spacing, the number of cells per string, the string spacing, the number of strings per module, and the arrangement of strings in a series, parallel, or combined configuration.

Trial runs with 300 μm thick cells demonstrated process yields of 98% and cell-to-cell cycle times of 3.9s. Additional work is planned for 200 μm cells. Improved string fabrication sequencing is expected to reduce the string cycle time from the current value of 5.7 s/cell.

Modules fabricated from strings made by the assembly system passed accelerated environmental testing, including thermal cycling, humidity-freeze cycling, and damp heat soaking. This testing provides initial verification of the reliability of the soldering process and the ability to use a no-clean flux. Since no flux residue removal step is required, the expense, the potential for cell breakage, and the effluents associated with such a step are eliminated.

ACKNOWLEDGEMENTS

This work was supported by the U.S. Department of Energy's National Renewable Energy Laboratory (NREL) through the Photovoltaic Manufacturing Technology (PVMaT) program. Additional funding was provided by Spire Corporation. Environmental test data were provided by J. Hummel and T. Jester, Siemens Solar Industries.

SOLAREX CRYSTALLINE PVMaT PROGRAM

J. H. Wohlgemuth, D. Whitehouse, T. Koval, J. Creager, F. Artigliere and M. Conway
Solarex Corporation, Frederick, MD 21701

ABSTRACT

Solarex has begun a process and product development program under the NREL sponsored Photovoltaic Manufacturing Technology (PVMaT) Program. The two major objectives of this program are to reduce the manufacturing cost for polycrystalline silicon PV modules to less than $1.20/watt and to increase the manufacturing capacity by a factor of three. To meet these objectives Solarex has proposed: 1) to increase the polycrystalline wafer size to 15 cm by 15 cm, 2) to use wire saws to cut wafers on 400 μm centers, 3) to increase cell efficiencies to 15%, 4) to improve the automated assembly of modules, 5) to develop a frameless module with a low cost backsheet and electrical termination system and 6) to develop the equipment necessary for automated handling of thin 15 cm by 15 cm wafers and cells. This paper will discuss the first year program efforts in the first five technical areas.

CASTING

The goal of the casting task is to develop the ability to cast ingots that yield four 15 cm by 15 cm bricks with at least equivalent material quality as now achieved for 11.4 cm by 11.4 cm bricks. The new ingot will represent a 67% increase in the useable silicon obtained from each cast.

The goal of the effort was to increase the capacity of the casting stations with minimum changes to the present production units. A 20% increase in charge weight was achieved by redesigning the ceramic crucible to have an elliptical rather than a cylindrical cross-section to fit around the heating elements. This change has been successfully demonstrated and already transferred to manufacturing. To further increase the size of silicon charge that the station can melt, an extension collar was design and fabricated. The crucible was then enlarged by adding an insert to the top. This has two advantages over a single piece crucible. Fabrication of 2 smaller ceramic pieces is easier and has higher yield than one large piece. The second advantage is in assisting the operator to load the silicon charge. The bottom piece can be loaded first and then the extender attached and loaded.

New insulation packages and heaters have been designed and fabricated and are ready for installation in a station. Experiments to cast intermediate sized ingots have begun. Material quality as measured by cell performance is equivalent to the smaller ingots, but the yield of usable material from the experimental ingots is a few percent lower. Modifications to the insulation package and the casting program are under evaluation.

WIRE SAWS

The goal of this task is to develop the wire saw technology for cutting 15 cm by 15 cm polycrystalline wafers on 400 μm centers at lower cost per cut than achieved today on the ID saws. This represents a 50% increase in the useable silicon obtained from each cast and a 50% increase in the yield of wafers per purchased kilogram of Si feedstock.

Solarex identified four vendors of wire saws, visited each vendor and prepared a detailed comparison of the characteristics. A summary of this data appears in Table 1. HCT Shaping Systems has the most experience building wire saws. The test run on Solarex material provided the best uniformity of surface texture and wafer thickness. Probably the most important factor, however, was the potential of the HCT machine to cut 4-15 cm by 15 cm bricks at the same time. Solarex selected and purchased an HCT wire saw.

The HCT saw has been operational since July, 1994. It has now been used to cut more than 200 bricks at an average yield of 96%. Developments with the slurry and grit have resulted in demonstrating the ability to cut 20 bricks with a single batch of slurry. We have achieved a 30% increase in the feed rate which allow us to achieve a per wafer cost that is 5% below the initial wire saw goal and 13% lower than the present Internal Diameter (ID) saw cost per wafer. Preliminary results indicate that the wire saw wafer quality and cell efficiencies are at least equivalent to wafers cut on the ID saw.

CH3365-4/94/0000-0832 $4.00 © 1994 IEEE

Table 1
Comparison of Wire Saws

Company	Type of Business	Wire saw Experience	Solarex Material	Capacity 15x15	Cost
Meyer-Berger	Semiconductor Equipment	> 10	good	2	mod
HCT	Wire saws	> 60	excellent	4	mod
M. Setek	Crystal growing & wafering	10 to 20	good	1	mod
Nippei Toyama	Machine tool manufacturer	~ 10	NA	2 or 3	high

CELL PROCESS

The goal to this task is to increase cell efficiencies to 15%, while decreasing the cost per watt at the module level. While many approaches to high efficiency have been reported, many of these results are on small area high quality float zone silicon and utilize processes and material that are not likely to be cost effective when applied to cast polycrystalline silicon in a manufacturing environment. Solarex has selected the following 4 areas to work on in the program.

1) Optical Coupling

The highest silicon cell efficiencies have been achieved using textured surfaces to increase the amount of incident sunlight that is coupled into the solar cell.

Chemical texturing is used by most manufacturers of single crystal silicon solar cells. In this process an alkaline etch is used to etch the 100 silicon surface leaving the exposed 111 planes that etch at a slower rate, resulting in a surface covered with small pyramids. This technology has not worked well on polycrystalline silicon because this material contains crystal grains of different orientations. A large fraction of the grains are typically oriented close enough to the 111 plane that they etch slowly and do not produce pyramids. However, recent work has shown that electrochemical etching of polycrystalline silicon can produce a highly textured surface with low reflectivity [1]. In our program, we were able to texture the polycrystalline silicon using an electrochemical etch, reducing the bare silicon reflection to approximately 28%. However, the process produces a reactant oxide on the surface. If this oxide is left on the surface when a TiO_2 anti-reflective coating is applied, the resultant system produces approximately 6% less short circuit current than the same single layer AR coating on a planar surface. If the oxide is stripped before the TiO_2 is applied, the reflection and the resultant short circuit current are similar to that observed for a cell with a single layer AR. However, the open circuit voltage is lower because of the junction damage caused by the etch. This process has not been successful at producing improved cell efficiencies.

Both lasers and/or beveled blade dicing saws have been used to mechanically texture silicon, resulting in increased absorption and improved cell efficiency. Solarex has demonstrated the use of a beveled blade to increase the efficiency of CZ buried contact solar cells from 15.2% to 17.9% as shown in Table 2. In this experiment mechanical texturing resulted in an 18% increase in short circuit current, more than the 16% achieved using the more traditional chemical texturing. These increases are much larger than ordinarily achieved with texturing because the SiO_2 AR coating used in the buried contact cell process is far from optically ideal. This experiment also showed that mechanical texturing could be utilized without loss of open circuit voltage or fill factor.

Table 2
Comparison of Chemically and Mechanically Textured CZ Cells

Cell Type	Efficiency (%)	Voc (mV)	Jsc (mA/cm^2)	FF (%)
Planar	15.2	619	31.5	78
Chemical Texture	17.2	618	36.6	77
Mechanical Texture	17.9	618	37.4	78

The challenge of this task is to apply mechanical texturing to the polycrystalline thick film paste cell process and to do it cost effectively. By leaving planar islands for the screen printed metal, we have fabricated mechanically textured cells using both beveled blades and a texturing tool. Use of the beveled blades results in current gains of 6 to 7 % over the planar controls (both cells have a single layer TiO_2 AR coating), but both the fill factor and voltage were lower for the textured cells, resulting in an efficiency gain of 3 to 4%. Utilizing a texturing tool to groove large areas at one time resulted in a current gain of 3 to 4%, but with no loss of voltage and fill factor, resulting in a similar 3 to 4% efficiency gain. When encapsulated the surface textured with the tool had exactly the same integrated absorption and predicted short circuit current as the best chemically textured and the best beveled blade textured cells [2]. Use of a tool to mechanically texture an entire cell in one pass holds promise as a low cost route to texturing of polycrystalline cells.

2) Back Surface Field

Aluminum pastes have been used successfully to improve the efficiency of solar cells for years. Table 3 compares the results for non-BSF cells with those made using Al paste from two different vendors. Back surface fields usually increase cell efficiency by 5 to 6%. The issue once again is that of cost. Program efforts have involved development of a paste and a process that does not require post fire clean-up. In addition, we have worked with the paste vendors in an effort to reduce the cost of the material and the amount required per cell. Both of these efforts have been successful. This new process must now be environmentally qualified before being transferred to manufacturing.

Table 3
Al Paste Results

Paste	Efficiency (%)	Voc (mV)	Jsc (mA/cm^2)	FF (%)
None	13.13	580.9	29.4	76.8
Vendor 1	13.92	592.2	31.2	74.8
Vendor 2	13.99	589.6	31.6	75.1

3) Hydrogen Passivation

The literature contains a number of reports on the use of hydrogen passivation to increase the efficiency of polycrystalline silicon cells [3-6]. Various approaches including implantation [5], annealing in forming gas [4] and plasma enhanced CVD of silicon nitride [3] have all been found to improve cell efficiencies. The effectiveness of a particular method of hydrogen passivation depends upon the substrate material and the balance of the cell process sequence.

The first area of experimentation in this program was an attempt to duplicate the PECVD SiN process. The result was a 3 to 4% increase in short circuit current and a similar increase in cell efficiency over cells with the standard Solarex single layer TiO_2 AR coating. However, half of this increase is due to the difference in index of refraction between SiN and TiO_2 which will disappear upon encapsulation. The remaining 1.5 to 2% increase in current came from improved blue response not red response, so there was no measurable bulk passivation using the PECVD SiN process.

The second area of experimentation was an attempt to duplicate the improvements reported using a forming gas anneal [4]. In all cases when the anneal was performed after the front Ag metallization had been fired, the cell efficiency decreased due to an increase in series resistance of the cell. The time at temperature required to anneal the cells in forming gas damages the front contacts, while the process produced no measurable improvement in current collection. Initial results indicate that annealing in forming gas before front firing may lead to improved short circuit current. Investigation of longer times and higher temperatures is now underway.

4) Phosphorus Gettering

Phosphorus gettering has been utilized to improve the minority carrier diffusion length and increase cell efficiency in some polycrystalline material [7-9]. We have successfully utilized phosphorus gettering as part of a buried contact cell process leading to improved cell efficiencies, although the results depended upon the initial quality of the silicon [7]. Material from the bottom of the ingot, that produces 12 to 13% efficient cells without gettering, showed 3.5 to 4.5% improvement in cell efficiency from gettering. The better material from the middle and top of the ingot, that typically produce cells with greater than 13% efficiency without gettering, showed only marginal (0 to 2%) improvement in efficiency. So far attempts to take advantage of phosphorus gettering in the thick film paste process have not given consistent efficiency gains.

MODULE ASSEMBLY

The goal of this task is to modify Solarex's present module assembly system to increase throughput by 100% and decrease the labor requirement by 50%.

Solarex is presently using automated module assembly equipment that was designed in-house. While this equipment has been extremely successful at reducing labor and producing reliable product, there is always room for further improvement. Solarex has subcontracted to the Automation and Robotics Research Institute (ARRI) at the University of Texas at Arlington to review the current automated module assembly system and to make recommendations for improving the equipment and/or process flow. As a first step a detailed flow chart of the module assembly system was prepared. A simulation model of the Solarex module manufacturing process was developed using the MPX queuing analysis tool. The model was used to identify resource limits that prevent additional output capacity. The model was able to identify those process stations that limit the factories ability to produce a larger quantity of modules. The model was also useful in developing "what-if" scenarios, to determine what additional resources were necessary to increase the manufacturing capacity by 50%.

FactoryFlow simulation was used to determine materials handling costs and to identify potential bottlenecks in the factory. Combined with a resource capacity study, these models lead to proposed changes in a number of processes and lay-outs. Several of these changes have been implemented. One of these changes resulted in an improvement in the cycle time of the matrix robot, so that it is no longer a limit on factory output.

FRAMELESS MODULE DEVELOPMENT

The goal of this task is to develop and qualify a frameless module design incorporating a lower cost back sheet material (less than $0.05/square foot) and user friendly, low cost electrical termination (less than $1.00/module). Since PVMaT is designed for large systems, the modules were designed to mount directly onto the support structure without integral frames. The first step has been the design of a compatible support structure and the identification of 3M Very High Bond Tape for mounting the modules to the structure. A sample array structure has been constructed and modules mounted using the tape. So far the tape has proven to be an excellent adhesive candidate, having withstood winds in excess of 60 miles per hour from the north or back of the array, the worst case condition.

A module backsheet specification has been prepared and utilized in discussion with potential vendors. Three experimental back sheet materials have been selected and all of these materials have successfully passed environmental qualification to IEC-1215, "Crystalline Silicon Terrestrial Photovoltaic Modules - Design Qualification and Type Approval". All three materials also successfully passed an in-house simulated UL fire test and wet high-pot tests. Mini-modules with these backsheets are now undergoing accelerated outdoor exposure tests with the backsheet exposed directly to concentrated sunlight.

AUTOMATED PRODUCTION

In the final year of the contract all of the improvement as well as development of automated handling equipment for 15 cm by 15 cm wafers will be combined into the design of the next generation automated module factory.

ACKNOWLEDGMENTS

This work was supported in part by NREL contract ZAI-4-11294-01. The authors wish to thank the technical review committee of R. Mitchell, B. Sopori and J. Gee for their support and assistance. The authors would also like to acknowledge the support and assistance of the other members of the technical staff at Solarex, as the program described in this paper required the efforts of many of the members of the technical staff.

REFERENCES

[1] Y. S. Tsuo, Y. Xiao, M. J. Heben, X. Wu, F. J. Pern and S. K. Deb, "Potential Applications of Porous Silicon in Photovoltaics", *Twenty Third IEEE PVSC*, 1993, p. 287.
[2] J. H. Wohlgemuth, T. Koval, D. Whitehouse and J. Creager, "Evaluating the Effectiveness of Optical Coupling Techniques", *First World Conference on PV Energy Conversion*, 1994.
[3] K. Shirasawa, K. Okada, K. Fukui, M. Hirose, H. Yamashita and H. Watanabe, "Large Area High Efficiency Multicrystal Silicon Solar Cells", *Third PVSEC*, 1987, p. 97.
[4] P. Sana, A. Rohatgi, J. P. Kalejs and R. O. Bell, "The Effect of Aluminum Treatment and Forming Gas Anneal on EFG Silicon Solar Cells", *Twenty Third IEEE PVSC*, 1993, p. 111.
[5] B. L. Sopori, X. Deng, S. Narayanan, and Roncin, "Hydrogenation for Impurity/Defect Passivation of Solar Cells", *Eleventh E.C. PVSEC*, 1992, p. 246.
[6] J. Coppye, J. Szlufcik, H. E. Elgamel, M. Ghannam, P. De Schepper, J. Nijs and R. Mertens, "Effect of Hydrogen Plasma Passivation on the Efficiency of Polycrystalline Silicon Solar Cells", *Twenty Second IEEE PVSC*, 1991, p. 873.
[7] J. H. Wohlgemuth and S. Narayanan, "Buried Contact Solar Cells", Technical Digest of the Sixth Sunshine Workshop on Crystalline Silicon Solar Cells, 1993, p. 145.
[8] I. Perichaud, F. Floret and S. Martinuzzi, "Limiting Factors of Phosphorous External Gettering Efficiency in Multicrystalline Silicon", *Twenty Third IEEE PVSC*, 1993, p. 243.
[9] S. Narayanan, S. R. Wenham and M. A. Green, "High Efficiency Polycrystalline Solar Cells", *IEEE Trans. Electron Devices*, Vol. **ED-37**, 1990, p. 382.

DEVELOPMENT OF CdTe MODULE MANUFACTURING

Dan W. Sandwisch
Solar Cells, Inc., Toledo, Ohio

ABSTRACT

Solar Cells, Inc. (SCI) has produced a 60.3 watt module with total area efficiency of 8.4% (NREL confirmed). The average total area efficiency of 60cm x 120cm modules manufactured on a developmental pilot line has increased to greater than 6.5% while the relative standard deviation has decreased to less than 10%. Modules routinely pass extensive stress testing. This paper summarizes the development of CdS/CdTe module manufacturing including product definition, process definition, and support programs. In addition the paper discusses our main conclusions to date and our future projections.

INTRODUCTION

Cadmium telluride (CdTe) has been recognized as one of the leading contenders for low-cost thin-film photovoltaic (PV) modules. One of the most attractive features of the technology is that it is relatively insensitive to the method of deposition - devices with efficiency greater than 10% have been made using at least seven different techniques.

In late 1990, a systematic evaluation of these techniques by SCI resulted in the selection of close-spaced sublimation. This technique demonstrated very high deposition rates (>5 µm/min). The high deposition rate results in total deposition cycles five to ten times faster than those demonstrated by typical thin film a-Si by plasma enhanced vapor deposition. In 1991, SCI began a PV module manufacturing development program to take advantage of this deposition process and other significant advantages such as superior stability and process flexibility. This paper describes the results of this program including product definition, process definition, quality assurance programs and other support programs. Conclusions and projected next steps are also provided.

PRODUCT DEFINITION

Solar Cells, Inc. has developed products to address remote and grid-connected applications. Table 1 describes the attributes of these two products. Both of these products incorporate glass/EVA/glass module construction. However, alternative potting and mounting designs are offered to address market installation requirements. The grid connected product utilizes a molded pigtail and support pad design to reduce module and panelization costs. The remote product provides more installation flexibility by offering a junction box and framing. The remote product will also be available in a half-size (30 watts) if market conditions warrant.

Table 1. Product characteristics for grid-connected and remote applications.

	Grid-Connected	Remote
Power (8.0%)	58W	58W
Voltage (Vmax)	65V	17V
Connection	Pigtail	Pigtail/J-Box
Panelization	Molded Mounts	Mounts/Frame
Size	60cm x 120cm	60cm x 120cm 60cm x 60cm

The long-term baseline SCI product is the 60 cm x 120 cm high-voltage module targeted at grid-connected applications. Development efforts on this product have resulted in greater than a 40% reduction in product finalization costs including encapsulation, potting, connections and panelization (see Fig. 1). SCI projects that this baseline product along with a patented support structure will reduce the cost of photovoltaic installations to below $3.00 per watt by the year 2000.

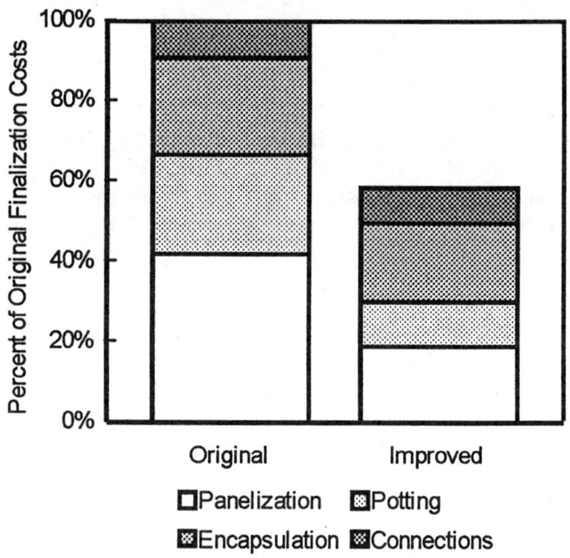

Fig. 1. Product finalization cost reductions.

This product has been installed in a 1.2kW array near SCI's plant (see Fig. 2). A similar array will be

CH3365-4/94/0000-0836 $4.00 © 1994 IEEE First WCPEC; Dec. 5-9, 1994; Hawaii

shipped to Tunisia in December 1994 as part of a PV evaluation project sponsored by their national utility.

Fig. 2. Photograph of two thin film CdTe test arrays installed at SCI in Toledo, Ohio. 1.2kW (foreground) and 1.0kW (background).

PROCESS DEFINITION

Table 2 outlines the major processes used to manufacture the thin film cadmium telluride modules. These processes are demonstrated daily on a 100kW pilot production line. This line has advanced module aperture area efficiency to 8.8% (see Fig. 3).

Table 2. Major manufacturing processes.

TCO Inspection	Scribe 3
TCO Preparation	Bus Bar Application
Scribe 1	Encapsulation
Semiconductor Deposition	Potting
Film Thickness Check	Safety Checks
Post Deposition Treatment	Pmax Test
Scribe 2	Final Inspection
Metal Deposition	

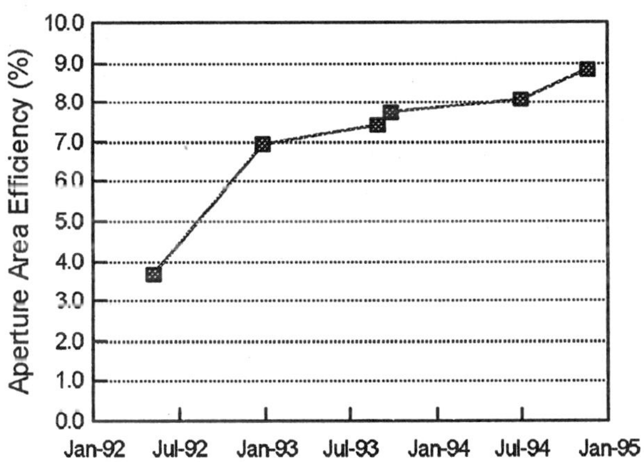

Fig. 3. History of aperture area efficiency on champion 60cm x 120cm modules (NREL confirmed).

Even though these processes have proven to be adequate for manufacturing introduction, SCI has concentrated on further development of the semiconductor deposition process and post deposition treatments.

In 1994, SCI built and began to debug a high-throughput deposition system which incorporated several advantages over the pilot deposition system including increased throughput and steady-state, continuous operation (see Table 3). The manufacturing system is designed for throughput of one module/minute. This advanced deposition system will be integrated with approximately 40 other pieces of equipment to produce modules at an annual capacity of 10-20MW.

Table 3. Comparison of pilot deposition system to high-throughput deposition system.

System:	Prototype	Production
Year Built:	1991	1994
Purpose:	Demonstrate Process Optimize Process	High Volume Manufacturing Limited Production
Features:	100kW 10 Modules/Day Manual Batch	10-20MW 480 Modules/Shift Automatic Continuous, Steady-State

SCI has developed a post deposition vapor treatment which eliminates the traditional wet dip $CdCl_2$ treatment, thereby, mitigating the environmental impacts of the traditional process and providing a more efficient process for manufacturing implementation [1]. Table 4 compares the vapor process to the wet dip process.

Table 4. Comparison of the wet dip $CdCl_2$ treatment to an alternative vapor process.

	Wet Dip	Vapor
Device Efficiency	11.9%	10.7%
Module Efficiency	8.1%	7.0%
High Voc (mV)	820	800
Process Time (min)	>60	<30
Process Steps	6	2
Materials	Solvents	No Solvents
Material Utilization	Low	High
Post Processing	Rinse	No Rinse

QUALITY CONTROL AND ASSURANCE

SCI has initiated a quality assurance program specifically focused on the 20MW manufacturing line. This program is under evaluation and refinement on a pilot production line. Most of the current efforts are focused on process control and module testing. Process control efforts have paid dividends in three basic areas;

1) module performance, 2) module performance variation and 3) process improvements. Module performance has steadily improved over a two and one-half year period with significant increases in the past twelve months covering production batches 30-50 (see Fig. 4). Concurrently, the relative standard deviation of the performance decreased substantially to below 10% (see Fig. 5). The target for the next twelve months is to demonstrate sustained variation of less than 5%.

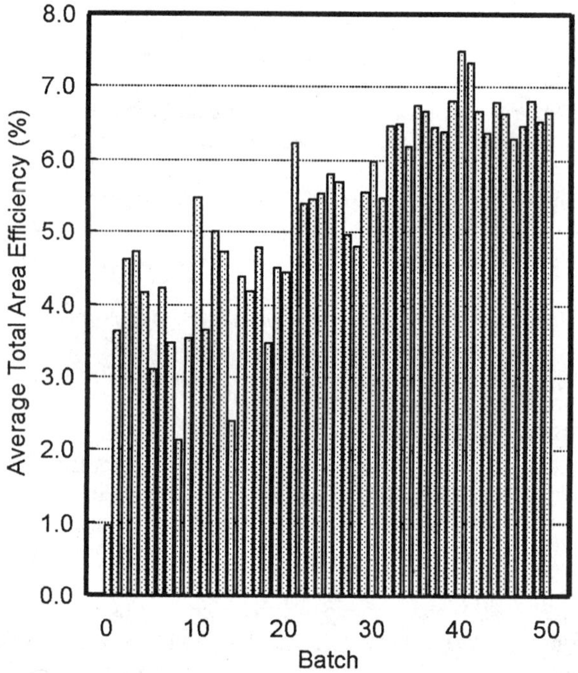

Fig. 4. Average total area efficiency vs. production batch. A batch contains an average of 15 modules.

These improvements are the result of several quality programs including standard operating procedures, process charting, designed experiments, and process changes. In addition, pilot line operators are involved in capability studies of several processes. One such study on the laser scribing process identified the opportunity to reduce the distance between adjacent scribes. More than 100 modules were utilized in identifying the laser process capability and tracking the results of a reduction in separation distance (see Fig. 6). This change resulted in a 0.5% relative increase in module output due to increased active area. These incremental improvements are important to meeting long term performance objectives as outlined in Table 5.

Table 5. Projected manufacturing total area efficiency targets.

1997	7.5%
1998	8.0%
2000	9.0%
2005	12.0%

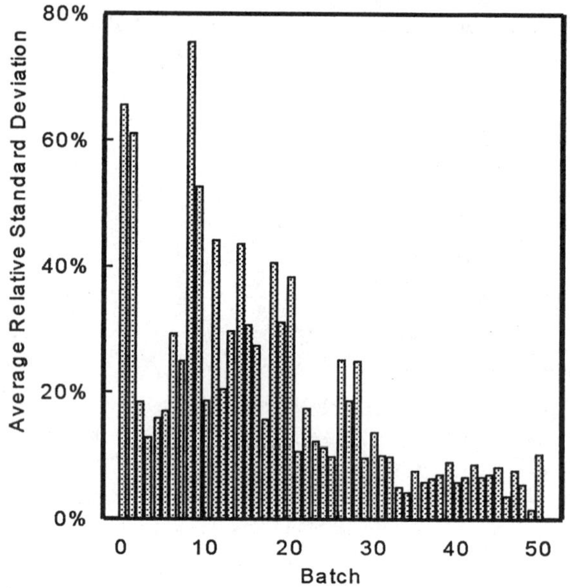

Fig. 5. Average relative standard deviation of total area efficiency vs. production batch. A batch contains an average of 15 modules.

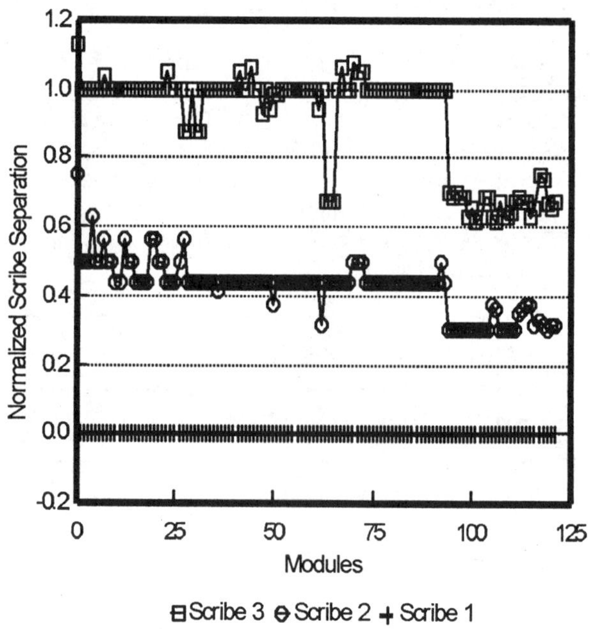

Fig. 6. Laser scribing process capability study resulting in a reduction of scribe separation distance. By definition Scribe 1 is at the 0.0 normalized position.

Many of the tests described above focus on process improvements which impact the initial performance of the module. Additional module stress tests are incorporated to understand the effects of these process improvements on long-term durability. The majority of these stress tests are based on the protocols established in the "Interim Qualification Tests and Procedures for Terrestrial Photovoltaic Thin-Film Flat-Plate Modules" (IQT) [2]. SCI conducts one IQT series per month. This testing has demonstrated significant improvement in overall pass rates from below 10% in early 1993 to well over 80% currently. SCI also extends certain elements of the IQT to evaluate module lifetime. For instance, three modules have undergone over 650 thermal cycles and demonstrated less than a 10% decrease in performance. This result is one indication that the current module design has a 30-year lifetime [3].

ENVIRONMENTAL, SAFETY, AND HEALTH PROGRAMS

Production of PV modules by the process outlined above involves regulated materials including cadmium. An important part of the development effort is to establish programs which effectively handle safety, health and environmental issues that accompany the production, deployment and disposal of these modules.

SCI has engaged outside agencies and consultants to conduct safety and health audits of the manufacturing facilities and to formulate appropriate programs and corrective actions. These programs include basic training programs as well as specific operational plans such as industrial hygiene and biological monitoring.

Environmental development has focused on process waste minimization and product recycling [4]. SCI has demonstrated feasibility on a waste treatment process which removes greater than 95% of cadmium from low concentration liquid wastes. SCI has also demonstrated product recycling by shipping modules to a raw material supplier for reintroduction into the smelting process.

SUMMARY, CONCLUSIONS, AND NEXT STEPS

SCI feels that thin film CdTe has provided a technology vehicle by which significant advantages can be obtained over current PV technologies. Key conclusions are:

1) SCI has produced a 60 Watt, 8.4% efficient module the most powerful thin film polycrystalline module on record.
2) SCI has defined and demonstrated cost effective thin-film CdTe modules capable of achieving installed costs of less than $3.00/Watt.
3) Module durability has been demonstrated.
4) The process for making these modules has been proven on a 100kW pilot line.
5) No technology barriers have been identified including safety, health, and environmental issues.

During the next two years, SCI will focus on commercialization of this product. These efforts will include:

1) SCI will expand its current pilot line to 200kW while installing a multi-megawatt manufacturing facility.
2) SCI will increase module testing efforts including product certification.
3) Several test arrays are planned for installation including a 100kW grid-connected field in Toledo, Ohio and 0.5kW array at Southwest Technology Development Instiitute in Las Cruces, New Mexico.
4) SCI will test its recycling strategy in several established markets.

ACKNOWLEDGEMENTS

The author is pleased to acknowledge the efforts of the SCI development team (see Table 6). The extraordinary capability and dedication of this team was essential in achieving the results described in this paper. The work reported here was supported in part by NREL under subcontracts No. ZAI-4-11294-02 and No. ZR-1-11059-1.

Table 6. SCI Development Team.

Karl Bihn	James Brown	Steve Cox
Larry Crosser	Linda Deutsch	Terry Firsdon
Matt Flis	Jim Foote	Rick Harju
Steve Kaake	John Kern	George Khouri
Sabrina Maynard	Harold McMaster	Kevin Miller
Bob Nicholson	Rick Powell	Nick Reiter
Geoff Rich	Dan Sandwisch	Rick Sasala
Mike Steele	Rick Yocum	Teddy Zhou
Peter Meyers	Jim Spillson	Joe Hanak
Bob Maltby	Gary Dorer	

REFERENCES

[1] T. Zhou, et. al., "Vapor Chloride Treatment of Polycrystalline CdTe/CdS Films", *First World Conference on Photovoltaic Energy Conversion*, 1994.

[2] "Interim Qualification Tests and Procedures for Terrestrial Photovoltaic Thin-Film Flat-Plate Modules", SERI/TR 213 3624, 1990.

[3] J.H. Wohlgemuth, "Testing for Module Warranties", *Proceedings of the Photovoltaic Performance and Reliability Workshop*, NREL/CP-410-6033, pp. 200-205, 1993.

[4] R. Sasala, et. al., "Environmentally Responsible Production, Use, and Disposition of Cd Bearing PV Modules", *First World Conference on Photovoltaic Energy Conversion*, 1994.

SIMULATION CALCULATION OF SOLAR ARRAY OUTPUT CONNECTING TWO SUB-ARRAYS WITH DIFFERENT AZIMUTH ANGLES

Hideyuki Takakura, Kousaku Koujitani, *Kholid Akhmad, *Frank Belley,
**Akio Kitamura, and *Yoshihiro Hamakawa
Faculty of Engineering, Toyama Prefectural University, *Faculty of Engineering Science, Osaka University,
**Technical Research Center, Kansai Electric Power Inc.,

ABSTRACT

A systematic investigation on PV array output consisting of sub-arrays installed on plural inclined surfaces has been done based on simulation calculation for the purpose of roof top solar cell array installation. Attention is especially paid to the amount of the outputs and differences between the connection of the modules (parallel or series).

In the case of parallel connection mode, there exist no severe problems if protected by blocking diode and output energy is always larger than that of series connection mode. In summer season, output energy from the array in which all sub-array is faced due south, is almost equal to that from an array in which a part of array facing due south and rest facing due west and even that from facing east and west combination. In the case of series connection of sub-arrays having the same tilt angle but different azimuth angle with bypass diode, maximum output voltage jumps several times in a day depending on the short circuit current from each sub-array.

INTRODUCTION

In order to obtain a maximum output from a solar cell array, it is fundamentally required that insolation to the array should be maximized designing the tilt and azimuth angles appropriately. At the same time, the matching between daily output pattern and daily load requirement pattern result more efficient utilization of solar energy. On the other hand, in some cases such as installing on a roof top of individual house, both the tilt and azimuth angles cannot be arbitrarily selected but determined its structure. Moreover solar cell array should be fixed on plural roof tops having different tilt and/or azimuth angles for getting adequate output. Obviously such arrays have different daily output pattern. Photovoltaic power generation system consisting of such different output is considered to have many problems, especially selecting operating voltage.

To find solutions of the issues mentioned above and to seek the other possibility to array installation condition for the specific application purposes, such as a roof top installation of individual house, the simulation calculation of output performances of PV sub-arrays installed with different tilt and azimuth angles connecting in series or parallel are performed and studied especially on maximum output conditions. Series connection of PV modules results in increased system voltage. However, the current output from serially-connected array is limited by the value of the least current output module. If solar cell array is constructed from sub-arrays installed on different inclined surface and is connected in series, mismatching of photocurrent occurs among the modules and the output current of array will be reduced drastically. This effect is quantitatively discussed from the simulation calculation results.

CALCULATION METHOD

Insolation on an arbitrary inclined surface was calculated utilizing the isotropic sky model by Liu and Jordan[1] and is given by

$$I_p = I_d \cos\beta + I_{hs}\frac{(1+\cos\theta)}{2} + \rho I_g \frac{(1-\cos\theta)}{2} \quad (1)$$

$$\cos\beta = \cos\theta \sin h + \sin\theta \cos h \cos(A-\phi)$$

where the first term represents the direct radiation component, the second is the diffused radiation and the third is the ground-reflected expression.Accuracy of this model was examined comparing calculated insolation values to the measured insolation data[2]. Insolation intensity of an arbitrary surface can be calculated from a direct normal radiation, a horizontal global radiation, and the height and the altitude of the sun. In this work, these data are derived from the data experimentally measured every 15 minutes from 4:15a.m. to 8:00p.m. during 1986 to 1987 at Osaka(long.135°34', 34°43'N.L.), Japan[2]. Albedo is neglected in this calculation but it has some effects especially when the height of the sun is low.

CH3365-4/94/0000-0840 $4.00 © 1994 IEEE
First WCPEC; Dec. 5-9, 1994; Hawaii

Table 1. Output performance of polysilicon solar cell module at the standard condition.

Number of cell	I_{SC} [A]	V_{OC} [V]	P_{max} [W]
36	21.2	3.25	45.5
α [%/°C/cell]	β [mV/°C/cell]	N value	R_{series} [Ω]
0.05	−2.3	1.1	0.3

Output performance of a solar cell module on an arbitrary inclined surface is calculated assuming silicon solar cell module, whose short circuit current is proportional to the insolation intensity and maximum output power is derived. Cell temperature rise due to the insolation and its effects to the short circuit current and open circuit voltage are also considered. Output performance at the standard condition are shown in Table 1, in which α is the temperature coefficient of the short circuit current(I_{SC}) and β is that of the open circuit voltage(V_{OC}). Bypass diode is connected in parallel to the module if the modules are connected in series.

RESULTS AND DISCUSSION

Output power from the solar cell array consisting of the two modules installed with the tilt angle of 30° but with different azimuth angle is calculated. Both parallel or series connection cases are examined. Figure 1 shows the output parameters of the array which consists of the module facing due south and the module facing due west (indicating south-west in this paper for simplicity) calculated on the insolation data of September 23rd, 1986 (clear day).

In parallel connection case (see Fig.1(a)), the current at maximum output power, I_{max}, is simply the sum of I_{max} of two modules and voltage at maximum output power, V_{max}, is almost constant although it has a little falling because of the rising of cell temperature by insolation as well as the atmospheric temperature. While in series connection case (see Fig.1(b)), V_{max} and I_{max} changes several times in a day drastically. This is due to the current mismatching between two modules. The power versus voltage characteristic from two-module solar array on different inclined surfaces in series connection have two peaks as shown in Fig.2 when the short circuit current generated from each module is different with each other. If the short circuit current ratio between two modules is larger than 2, maximum output power point is found at lower voltage side. This instability may have complicated effects to an operation of an inverter with maximum power tracking circuit.

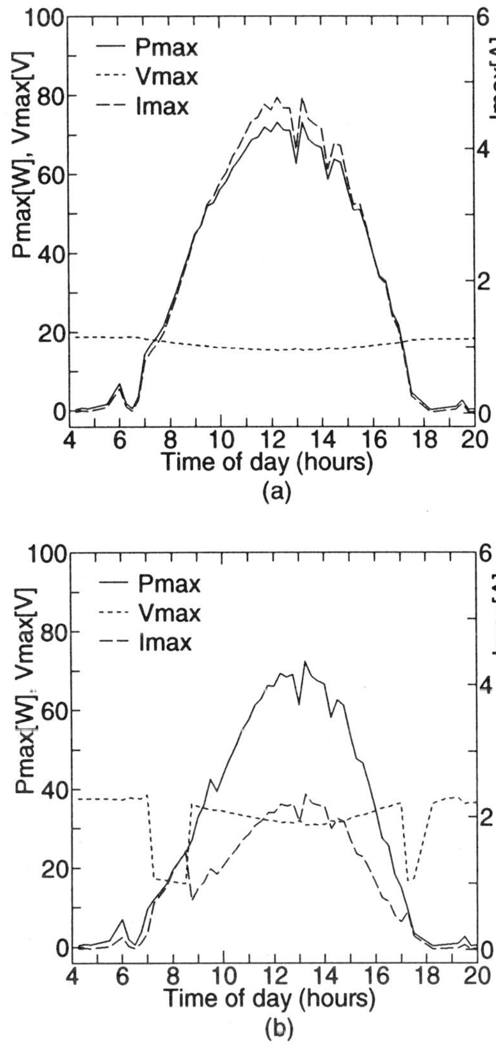

Fig. 1. Output parameters of the array which consists of the module facing due south and one facing due west calculated on the insolation data of September 23rd, 1986 (a)Parallel connection (b)Series connection

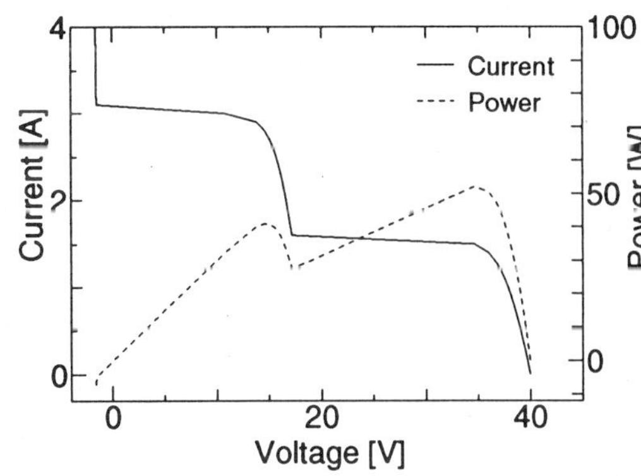

Fig. 2. Output parameters from two-module solar array on different inclined surfaces in series connection.

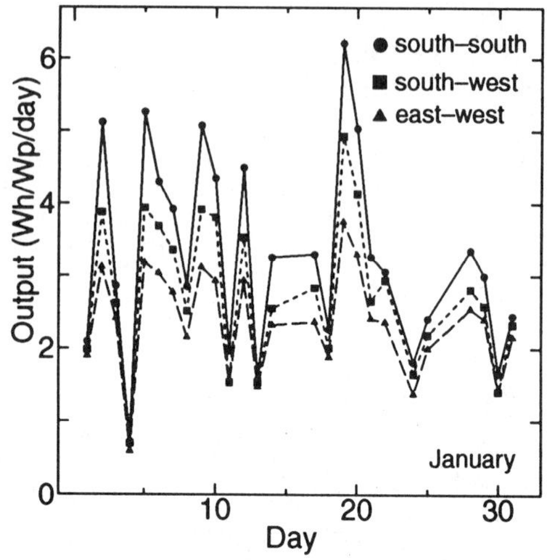

Fig. 3. Monthly variation of daily output energy in the case of parallel connection in January.

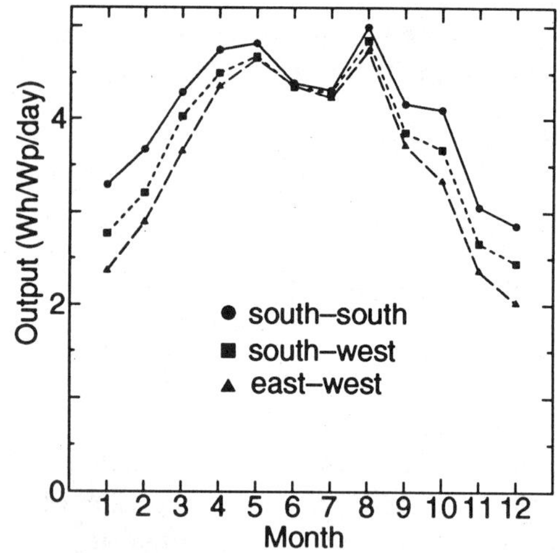

Fig. 5. Monthly-averaged daily output calculated on parallel connection.

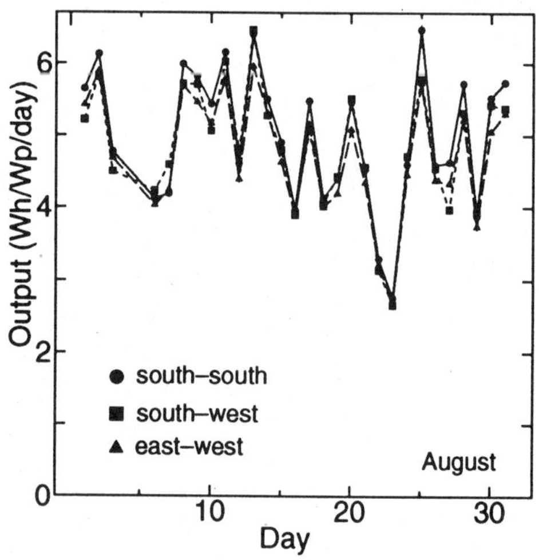

Fig. 4. Monthly variation of daily output energy in the case of parallel connection in August.

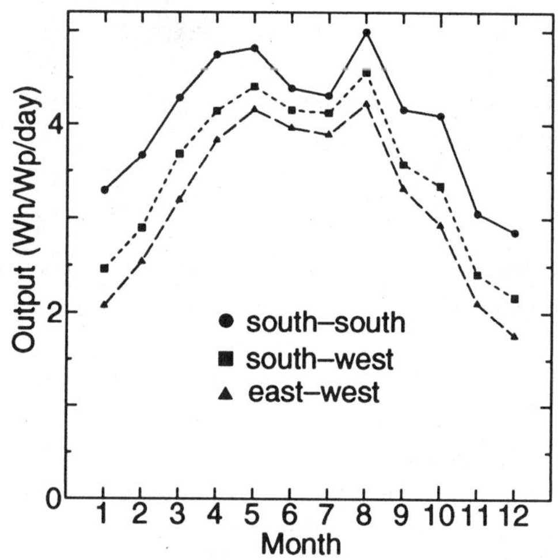

Fig. 6. Monthly-averaged daily output calculated on series connection.

In order to estimate the output electric energy from these array, calculation of daily amount is performed throughout the year. Figures 3 and 4 show the monthly variation of daily output energy in the case of parallel connection in which the output energy is divided by the maximum output power, P_{max}, at the standard condition. As can be seen in Fig.3, on a clear day in January, the differences of the daily output energy among the arrays occur. On a cloudy day, these differences becomes small due to the isotropic radiation of the sky. In August(see Fig.4), daily output is almost the same disregarding the azimuth angles. This can be explained by the average height of the sun, i.e., in summer it is relatively high but in winter it is low.

Figures 5 and 6 show the monthly-averaged daily output calculated on both parallel and series and connection cases. Fig.5 is the result of parallel connection. As shown in Fig.4, the difference of averaged daily output is almost the same in summer season. In other words, almost the same amount of the output energy from the south-south array is obtained from the east-west array. As the height of the sun becomes low, the ratios of the output against the south-south array decrease. Figure 7 shows the ratios of monthly output energy from the south-west array and the east-west array against that from the south-south array. For example, in January, the ratio of output energy from the south-west array against the south-south array is around 84% and 72% from the east-west array

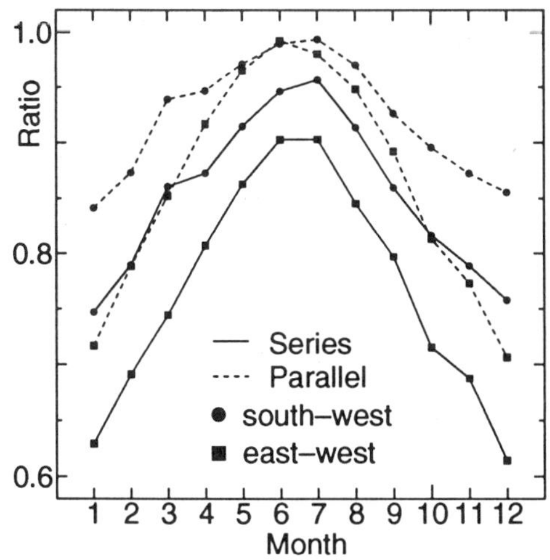

Fig. 7. Ratios of monthly output energy from the south-west and the east-west array against the south-south array.

Table 2. Ratios of output energy throughout the year from the south-west array and the east-west array against the south-south array.

azimuth angles	parallel connection	series connection
south-west	0.93	0.86
east-west	0.88	0.78

in parallel connection.

In the case of series connection, as indicated in Fig.6, large discrepancies are seen as compared with the parallel connection even in summer season because of the current mismatch loss.

Table 2 shows the ratios of output energy throughout the year from the south-west array and the east-west array against the south-south array. As can be seen, the ratios of output energy from the south-west array and the east-west array against the value from the south-south array are about 90% in parallel connection throughout the year.

SUMMARY

Output electric energy from PV array consisting of sub-arrays installed on different inclined surface is systematically studied based on simulation calculation. Maximum output power is calculated at every 15 minutes throughout the year. The insolation data are derived from the data experimentally measured at Osaka, Japan. Attention is especially paid to the amount of the outputs and differences between the connection of the modules (parallel or series).

In the case of series connection of sub-arrays with bypass diode having the same tilt angle but different azimuth angle, maximum output voltage jumps several times in a day depending on the short circuit current from each module. This instability may have complicated effects to an operation of an inverter with maximum power tracking circuit.

In the case of parallel connection mode, there exist no severe problems if protected by blocking diode and output energy is always larger than that of series connection mode. On a clear day in January, the differences of the daily output energy among the arrays occur. On a cloudy day, these differences becomes small due to the isotropic radiation of the sky. In August, daily output is almost the same disregarding the azimuth angles. As compared with the output from the south-south array, output energy in summer season is almost equal to that of the south-west and the east-west combination, but decreases in winter season.

The ratios of output energy from the south-west array and the east-west array against the value from the south-south array are about 90% in parallel connection throughout the year.

REFERENCE

[1] R. M. Klucher: *Conf. Record 13th IEEE Photovoltaic Spec. Conf.*, (Washington D. C., 1978) p.847.

[2] H. Takakura, Y. Kita, M. Murozono and Y. Hamakawa: *Tech. Digests 3rd Intern. PVSEC Japan*, (Tokyo, 1987) p.359.

CALCULATED EFFECT OF CENTRAL DC/AC POWER CONVERSION FROM PV GENERATORS WITH DIFFERENT ORIENTATIONS

T.C.J. van der Weiden, A.J. Kil
Ecofys Consultancy and Research
P.O. Box 8408, NL-3503 RK Utrecht, The Netherlands

ABSTRACT

The energy production of two PV/inverter configurations for a grid connected PV system consisting of roof and façade integrated subarrays with five different orientations and tilt angles was predicted by computer simulations. These calculations show that the total DC energy production of the system is practically not affected if the subarrays have one central inverter and all operate at the same MPP. Application of one central inverter for such a system geometry may lead to a reduction of the specific kWh costs of approximately 4% due to lower specific costs and higher efficiency of large inverters and because the central inverter can be undersized to a larger extent than the 5 smaller inverters.

GENERAL

Integration of grid connected PV systems in the built environment is becoming increasingly important for large scale application of PV. The advantages of integration in the built environment are numerous: the electricity distribution grid can be used to supply the PV power to and PV modules can be mounted on roofs of buildings so that costs for support structures can be avoided and no extra land surface is needed. It is obvious that integration will lead to cost reduction which is absolutely essential for PV to become a substantial source of renewable energy. Making use of buildings as a support structure for PV modules has however a disadvantage as well: in the design of the PV-system account must be taken of the constraints imposed by the architecture of the building.

In Amsterdam, The Netherlands, a project is being realized by the utility company Energiebedrijf Amsterdam in which a 250 kWp PV system will be erected on the roofs of 70 new one-family homes. This PV system will be the largest building integrated system world-wide. Ecofys is involved in this project as a consultant and will monitor the system. The 70 homes in question will be realized on two buildings. The PV modules will be mounted on the roofs of these buildings and on the façade of one of these buildings. Although the architect designed the roofs and façade in close co-operation with the designers of the PV system, the design was not fully optimized with respect to PV energy production, since other interests had to be taken into account as well. Chosen tilt angles and orientations deviate from the optimum for The Netherlands (tilt angle: 36° and orientation: 5° west of the south). In total 6 PV surfaces or subarrays can be distinguished. The characteristics of the subarrays are listed in table 1.

array	tilt	orient.	PV power	irradiation
	(°)	(°)	(kWp)	(kWh/year)
1	25	90	60	55860
2	35	0	65	72930
3	25	-90	60	57180
4	90	-17	15	11865
5	20	28	50	53700
total			250	251535

Table 1. Characteristics of the subarrays

For this geometry a comparison was made between two different configurations of DC/AC inverters. One concerns the application of a single central inverter (with one maximum power point (MPP) tracker) to which all subarrays are connected; the other concerns using a separate inverter for each subarray. Each configuration has its specific advantages. The single inverter has a larger output power and therefore in general lower specific costs (i.e. costs per kW inverter power). Another advantage of the single inverter is its higher efficiency. Finally, as the system costs are minimized by undersizing the inverter power with respect to the PV power [1], the central inverter can be undersized to a larger extent, since the different subarrays do not produce maximum power simultaneously.

On the other hand, a single inverter may lead to a lower energy output: variations in irradiation levels and cell temperatures amongst subarrays lead to a different MPP voltage for each subarray while the inverter input voltage can in principle not be optimized for each subarray separately.

CALCULATIONS

In the design phase of the project in Amsterdam a choice had to be made between one single or a number of smaller inverters. To that end calculations of the energy production for both configurations were carried out using the simulation program INSEL developed by the University of Oldenburg, Germany [2]. The program calculated the DC output of the PV generators for the two different configurations. The simulations are performed using hourly global irradiation and ambient temperature 'test reference year' data (1971-1980) for De Bilt, The Netherlands. The global horizontal irradiation data were converted to irradiation data in the various subarray planes using the Perez model. Furthermore MPP tracking was simulated on an hourly

CH3365-4/94/0000-0844 $4.00 © 1994 IEEE First WCPEC; Dec. 5–9, 1994; Hawaii

basis. The cell temperature rise was assumed to vary linearly with irradiation. The temperature coefficient used was 0.04 ºC/(W/m²), and was derived from measurements on roof integrated systems. Spectral losses, reflection losses, losses due to mismatch of the module characteristics and ohmic losses in the cabling are not incorporated in the model. Therefore the absolute output values of the output energy are overestimated. However, comparing the calculation results of the two configurations gives insight in the difference in tracking losses of the two configurations.

subarray	DC energy production (kWh/year)	inverter/PV power ratio	inverter power (kW)
1	45031	0.715	43
2	59672	0.81	53
3	46027	0.728	43
4	9538	0.65	10
5	43824	0.78	39
total 1..5	204092		188
1 inverter	203742	0.706	177

Table 2. Calculation results for the 5-inverter and the 1-inverter configuration: total DC energy; inverter undersizing factor and corresponding inverter size.

In table 2. the results of the calculations are summarized. For all subarrays, the DC energy produced in one reference year is shown. Dividing these figures by the irradiation figures in table 1. results in a performance ratio of 0.81 for each subarray, which is quite high and confirms the incompleteness of the model used. The most surprising conclusion that can be drawn from the figures in table 2. is that the total energy production of the 1-inverter configuration is merely 0.2% less than that of the 'optimized' system with 5 inverters. Thus from an energetic point of view there is no reason to install five inverters instead of one.

Due to its lower power and its lower specific costs, application of one large inverter would reduce the specific system costs by about 3% under the assumption of Dfl. 2/W for the 5 inverters, Dfl. 1.5/W for the large inverter and Dfl. 18/Wp for PV and other expenses. The lower power stems from the extent to which each of the inverters can be undersized. As a criterion for this undersizing, the total input energy at powers above the nominal inverter power was chosen to be 2% of the total DC energy. Only a small fraction of this energy is lost if the inverter is well regulated.

Together with the higher efficiency of the large inverter (93%) compared to that of the smaller (92%), the total reduction of the specific energy costs in the 1-inverter configuration is approximately 4% compared to the costs in the 5-inverter system.

The result of the calculations is that the loss of energy production and corresponding increase in the specific energy costs due to the non-ideal orientation of the system as a whole (11% yearly) can partially be compensated (4%) by the application of one large inverter for all subarrays.

FUTURE ACTIVITIES

Due to the absence of reliable field data concerning central MPP tracking of differently oriented PV arrays and due to the size of the project, a configuration with more inverters will be realized. Monitoring of this system will, amongst others, concentrate on the verification of the calculated results.

CONCLUSIONS

Calculations show that PV subarrays with a different MPP behavior due to different orientations can be connected to one inverter with a negligible loss of energy production. Moreover, a central inverter allows a larger degree of undersizing than a separate inverter for each subarray with a different orientation. Together with a higher inverter efficiency and a lower price per kW inverter power a system cost reduction of about 4% can be accomplished.

Acknowledgment

This work was supported by the EU (DGXVII), the Netherlands PV R&D&D program (Netherlands Agency for Energy and the Environment - NOVEM), the city of Amsterdam, Energiebedrijf Amsterdam, R&S Renewable Energy Systems and BP Solar.

REFERENCES

[1] T.C.J. van der Weiden, A.J.N. Schoen, "Optimized design and technical potential of grid connected PV systems on office buildings", Proceedings of the 11th E.C. Photovoltaic Solar Energy Conference (1992), pp 1451-1454

[2] J. Schumacher, "Digitale Simulation regenerativer elektrischer Energieversorgungssysteme", thesis University of Oldenburg (1990).

Fault finding in a 1 MW photovoltaic plant by reflectometry

L. Schirone, F.P. Califano
Terza Università di Roma, Via Segre, 2 - 00146 Roma - Italy - FAX +39 6 5561373

U. Moschella, U. Rocca,
ANIT, Via Pacinotti, 20 - 16151 Genova - Italy - FAX +39 10 413012

ABSTRACT

Time Domain Reflectometry has been used for diagnostic monitoring of a large size photovoltaic (PV) plant in operating conditions. By analysing the waveforms obtained when a step-voltage excitation is propagated down the electrical line connecting the PV generators to the inverter, we were able to detect, identify and localise the most common fault conditions, such as breaks of the circuit, insulation defects, wiring anomalies. Best compromise between analysis time and precision was achieved by testing groups of three paralleled strings (panels): in this way we were able to detect in a short time whether a fault was present, while its position was separately determined with a good precision by repeating the test on the single strings. The whole 1 MW PV plant was tested in a couple of days. Finally, we discuss some economical aspects of this fault finding technique.

INTRODUCTION

Fault finding is a major problem in the assembling and the maintenance of a large scale photovoltaic (PV) field, due to the time requested for manual searching. Time Domain Reflectometry (TDR) has proven to be an effective approach to determine both the type and the position of a fault in many kinds of electrical transmission lines and we already demonstrated its application in a single photovoltaic string [1].

We report the results of the tests performed in operating conditions on the 1 MW PV plant ELIO1, located in Vasto, Italy: typical fault conditions are discussed, by illustrating the corresponding registered waveforms. Some economical aspects of this fault finding technique are discussed in conclusion.

FUNDAMENTALS OF TIME DOMAIN REFLECTOMETRY

Reflectometric systems can be regarded as closed loop radar's: a step-voltage excitation propagates down the electrical line under test, while the voltage is monitored by an oscilloscope. The signal propagates down the transmission line consisting of the PV modules and the series-connecting wires and is partially reflected when a faulty module is reached, producing in the waveform a delayed replica of input pulse. Measured delay allows us to calculate the distance travelled by the signal and so provides an indication of the fault position [1]. The type of a fault is revealed by the sign of the reflected signal [2]: being the input step negative in our system, in the case of an open/short circuit a decreasing/increasing ramp adds to the voltage measured in normal conditions (see Fig. 1).

During the first microsecond of the waveforms in Fig. 1 voltage evolution depends on the connection details between PV field and the measurement point: bending of the wires, proximity to other conductors, etc.. After this time, a small slope's modification is caused by crossing sectioning switches. The sign of waveform's slope changes corresponding to the reflected signal produced when the pulse enters photovoltaic modules and meets with an abrupt modification of propagation conditions, due to the lower characteristic impedance of the PV modules [1]. In normal operating conditions, the next change in slope is caused by discontinuity of propagation characteristics at the end of PV string and occurs after a delay Δt_{max} depending on the length of electrical path inside the modules. In the presence of a fault, a reflected wave is produced before Δt_{max}, after a delay depending on fault position.

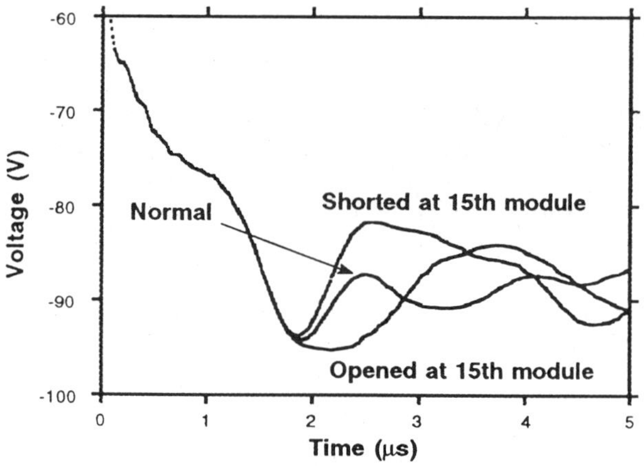

Fig. 1. Typical waveforms recorded when TDR is performed on a string in the presence of a fault on 15th module: in the case of an open/short circuit a decreasing/increasing ramp adds to the voltage measured in normal condition. The string was about 150 meters far from observation point.

CH3365-4/94/0000-0846 $4.00 © 1994 IEEE

First WCPEC; Dec. 5–9, 1994; Hawaii

EXPERIMENTAL SET-UP

The PV plant wiring scheme is based on strings consisting of 28 series-connected 47 W_{pk} modules. The strings are paralleled in 3-elements groups (panels) and 4 panels are grouped in a field. The 64 fields are separately connected to one of two inverters which interface the plant to power grid.

TDR set-up is shown in Fig. 2. A pulsed voltage generator is applied at the inverter side of the field-connecting wires and drives string's anode by negative pulses with 70 V amplitude, 40 ns rise time, 5 µs duration and 1000 Hz repetition rate. The generator (output impedance $Z_O = 50\ \Omega$) is interfaced to the two conductor line (characteristic impedance $Z_e \approx 300\ \Omega$) by an impedance-matching resistor ($R_M = 270\ \Omega$) and by a blocking capacitor (100 nF).

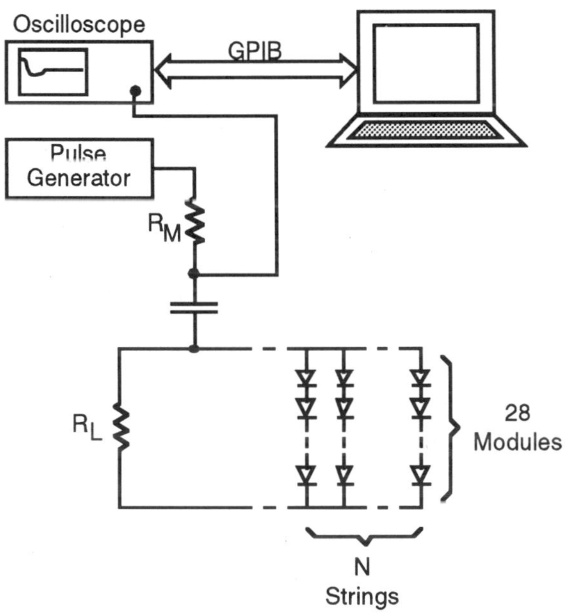

Fig. 2. Experimental set-up.

In the same point a storage oscilloscope records the voltage signal by means of a high-impedance probe. A standard instrumentation link (GPIB) interfaces the oscilloscope to a computer, allowing us to perform numerical analysis of the measured waveforms.

RESULTS

Fig. 1 shows typical waveforms for a string located about 150 meters far from the point of observation. The lowest curve reveals a break in the circuit after 15th module. Before repairing the fault, the same point was deliberately short-circuited to ground, obtaining the waveform shown for comparison in the figure. Also it is reported the waveform obtained after reparation, when the string was normally operating.

The recorded waveforms could also give information on other wiring anomalies, such as the connection of a module in inverted polarity, as shown in Fig. 3. The effect is rather small, and only could be observed if TDR is performed on a loaded string.

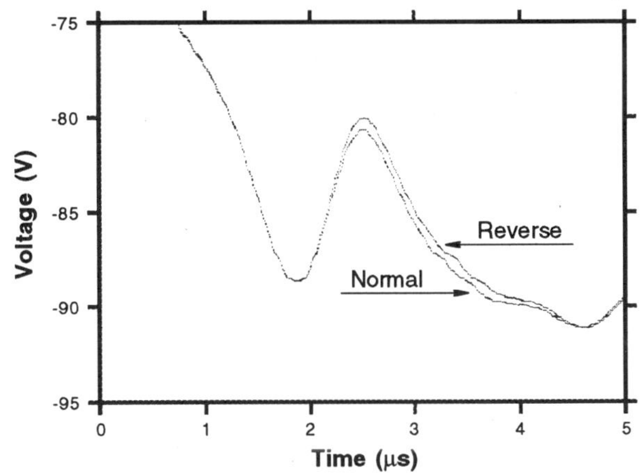

Fig. 3. Waveforms recorded when TDR is performed on a string containing a module connected with inverted polarity.

Figures 4 and 5 show the modifications occurring in voltage waveforms obtained for a string when an (ideal) fault is moved along the string: in the various fault positions, both the delay of the reflected pulse and its amplitude are changed. Delay effect is more evident when the fault occurs in the first 15 modules, while amplitude effect is more evident when it occurs deep inside the string.

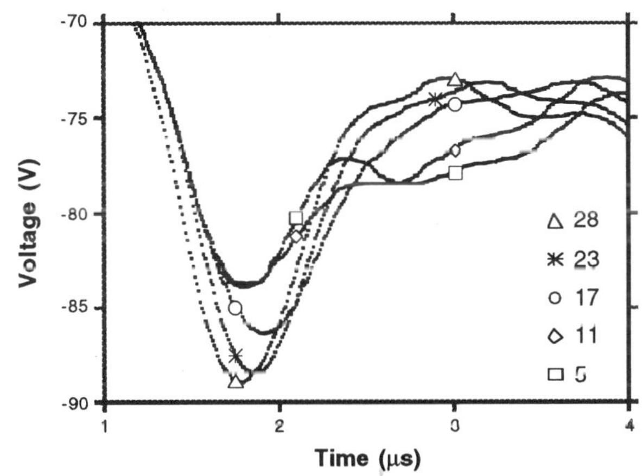

Fig. 4. Waveforms recorded when TDR is performed on a string in the presence of an open circuit at different positions.

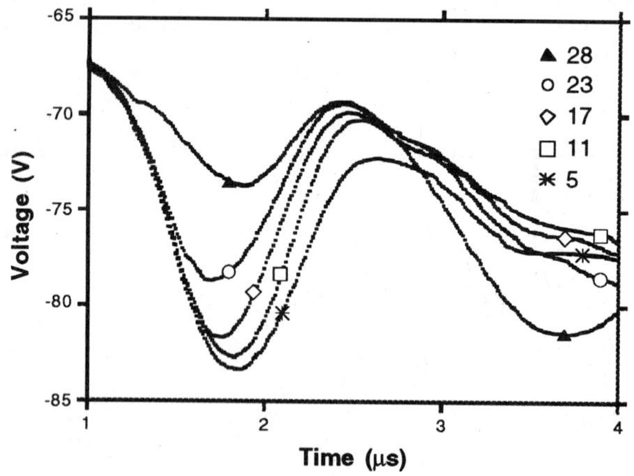

Fig. 5. Waveforms recorded when TDR is performed on a string in the presence of a short circuit to ground at different positions.

The attenuation taking place when the pulse is propagating in the string is influenced by environmental conditions, like temperature or irradiation level [1]. Fig. 6 shows the modifications occurring in the waveforms for a string under an irradiation level ranging from 200 to 800 W/m^2. The short circuited, normally operating and opened strings were nearly located at the same distance from the observation point. Minor differences in the waveforms arise from connection details.

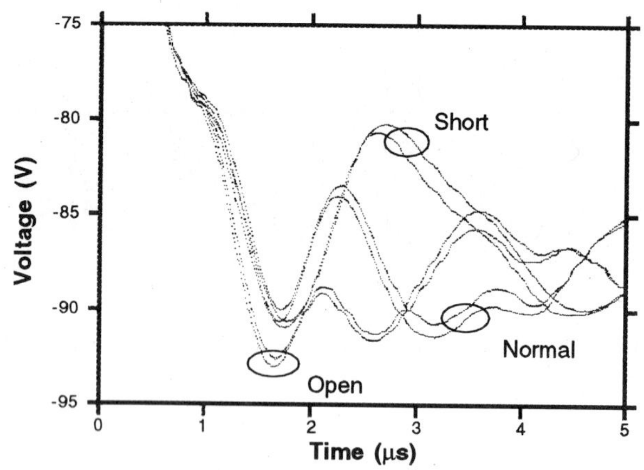

Fig. 6. Irradiation level effect on the waveforms observed for strings under different operating conditions (short circuit to ground, normal, open circuit): upper/lower set of curves were recorded under 200/800 W/m^2.

Former discussion reports some factors affecting waveforms shape and illustrate why fault finding could be based on comparison approach better than on analysis of single waveforms: any anomalous feature in a string can be easily detected by comparison with the waveforms obtained for other strings under the same experimental conditions (see Fig. 7).

Improved sensitivity achieved by comparison method allows to detect a fault even if TDR is performed on several paralleled strings. Whenever fault detection occurs, TDR is repeated on single strings, determining fault position with a good precision. In order to reduce the time required to scan the whole plant, it is important to keep as high as possible the number of paralleled strings. On the other hand, when TDR is performed on a

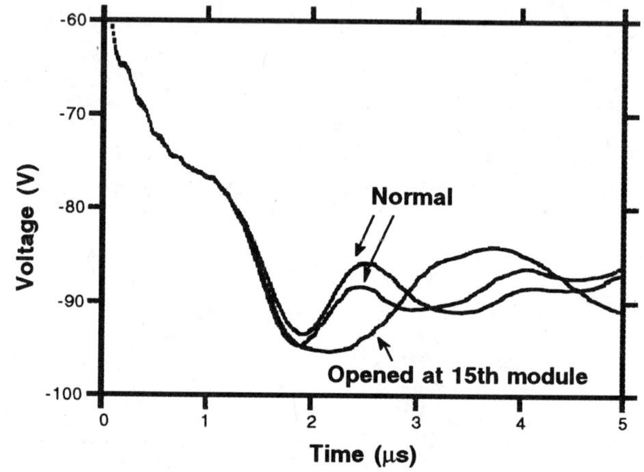

Fig. 7a. Waveforms recorded when TDR is performed on the strings in a panel in the presence of an open circuit at 15th module.

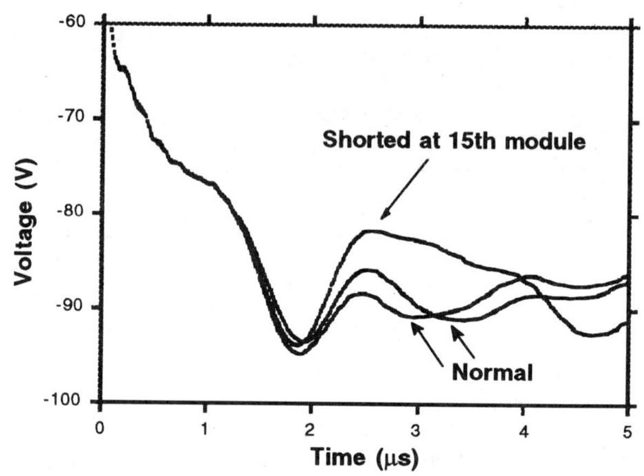

Fig. 7b. Waveforms recorded when TDR is performed on the strings in a panel in the presence of a short circuit to ground at 15th module.

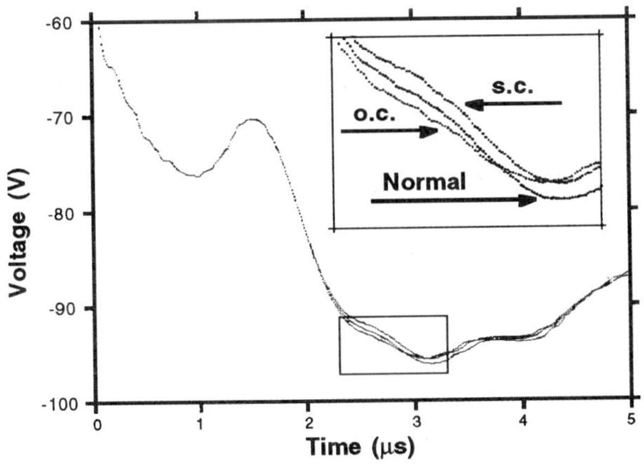

Fig. 8. Waveforms recorded when TDR is performed on a field consisting of 12 paralleled strings in the presence of an open circuit, a short circuit to ground and in normally operating conditions.

field (12 paralleled strings), in the different operating conditions (normal, short circuit to ground, and open circuit) the differences observed in resulting waveforms are too little to allow pratical fault detection (see Fig. 8).

Best compromise between speed and sensitivity was obtained by performing TDR on groups of three paralleled strings, corresponding to the panels, and comparing each other the waveforms for the four panels in each field (see Fig. 9).

We are recording data under several environmental conditions, in order to synthesise general criteria allowing us to perform direct fault detection, without comparison among waveforms.

CONCLUSIONS

TDR has proven effective for fault finding in large size PV plants: after detection of breaks in the circuit, insulation defects and other wiring anomalies, they could be localised in a few minutes. The whole 1 MW PV plant was tested in two days, revealing five breaks in modules bus bars and one short circuit to ground in a switch board. During the test one of two inverters was switched off, in order to keep interference level as low as possible. Lost power and man power have a cost, which can be balanced in a few months by retrieved productivity, amounting to about 10 kW and corresponding to 1% of plant's target power. On the other hand, the cost of TDR system is neglegible, as the instruments are normally available in our laboratories.

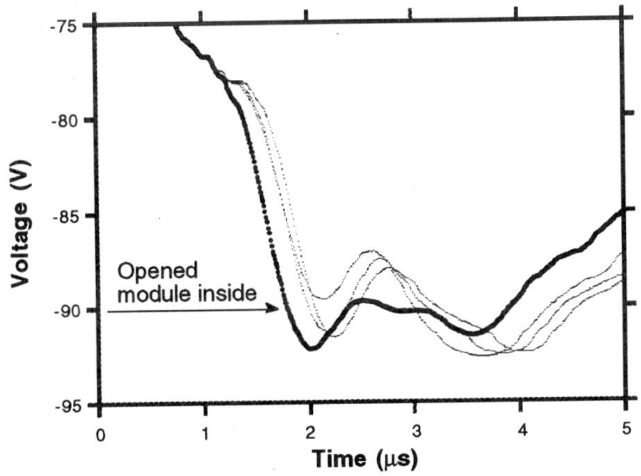

Fig. 9a. Waveforms recorded when TDR is performed on the four panels in the same field in the presence of an open circuit.

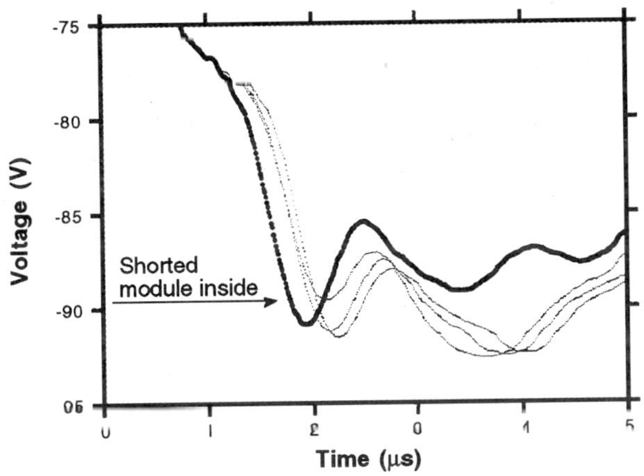

Fig. 9b. Waveforms recorded when TDR is performed on the four panels in the same field in the presence of a short circuit to ground (each panel consists of three strings).

REFERENCES

[1] L. Schirone, F.P. Califano, M. Pastena, *Progress in Photovoltaics*, **2**, 1994, pp. 35-44

[2] S. Ramo, J.R. Winnery and T. Van Duzer, *Fields and waves in Communication Electronics*, Wiley, New York, 1965

TECHNICAL EVALUATION OF A DUAL-JUNCTION SAME-BAND-GAP AMORPHOUS SILICON PHOTOVOLTAIC SYSTEM AT NREL

Troy Strand, Laxmi Mrig, Robert Hansen, and Keith Emery
National Renewable Energy Laboratory
1617 Cole Blvd., Golden, Colorado 80401, USA

ABSTRACT

On December 7, 1992, a 1.8-kW_{ac} utility-interconnect photovoltaic (PV) system using amorphous silicon modules was brought on-line at the National Renewable Energy Laboratory's photovoltaic test site. This system was deployed to conduct an in-situ technical evaluation of the PV array (in a high voltage configuration) and system performance and reliability in a utility-interconnect application. The system is unique due to the installation of construction-grade insulation on the back of each PV module. This use of insulation is an attempt to levelize the annual array power output by elevating the operating temperature of the modules. This paper presents array and system performance data. Emphasis is placed on quantifying the effects of individual losses as well as seasonal changes on PV array and system performance.

INTRODUCTION

A 1.8-kW_{ac} utility-interconnect photovoltaic (PV) system was brought on-line for technical evaluation at the National Renewable Energy Laboratory's (NREL) photovoltaic test site. The system is located at 39.7° N latitude, 105° W longitude, 1782 m elevation. The array is fixed at a 40° tilt angle and is aligned true south. The PV array consists of six monopoles. The positive subarray comprises three monopoles and the negative subarray comprises the remaining three. Each monopole has 17 United Solar Systems Corp. (USSC) UPM-880 PV modules in series for a total of 102 modules. The modules are amorphous silicon (a-Si), with a dual-junction, same-band-gap structure and composition. The array output is fed to an Omnion Series 2200 static power converter rated at 2 kW_{ac}. The output of the power converter is fed to the local utility power grid via NREL's power distribution network. The system has been under evaluation since December 1992, with results being presented in previous publications [1]. Table 1 gives the array ratings at standard test conditions (STC): 1000 W/m^2, 25°C, AM1.5 global spectrum.

The system was originally deployed as six monopoles of 16 UPM-880 PV modules in series totaling 96 modules. The ac output on March 24, 1993, was determined to be 1.65 kW_{ac} at 1000 W/m^2 and 25°C back-of-module temperature. This output indicated that the system, as configured, did not meet the estimated system ac output of 1.8 kW_{ac}. Six additional modules were then added to the PV array on April 19, 1993 (one module per monopole), for a new total of 102 modules. This addition was an attempt to elevate the system's ac power output. The system was determined to be operating at 1.82 kW_{ac} on August 24, 1993, and at 1.81 kW_{ac} on August 17, 1994, at 1000 W/m^2, air mass 0.9 (corrected for pressure), and 25°C back-of-module temperature. As a consequence of adding the six additional modules, ac power and array peak power measured prior to April 20, 1993, have been multiplied by the ratio of 102/96 (1.0625) for this paper.

The Photovoltaic system was deployed to conduct an in-situ technical evaluation of the PV array and system performance and reliability in a utility-interconnect application. The system is unique due to the installation of construction-grade insulation on the back of each PV module. This use of insulation is an attempt to levelize the annual array power output by elevating the operating temperature of the modules. This paper presents array and system performance data. Emphasis is placed on quantifying the effects of individual losses as well as seasonal changes on PV array and system performance.

Table 1. Photovoltaic array ratings at STC

V_{OC}	I_{SC}	V_{MAX}	I_{MAX}	P_{MAX}	Aperture Area
±374 V	±5.4 A	±265 V	±4.2 A	2244 W	37.59 m^2

EXPERIMENTAL PROCEDURE

The data acquisition system (DAS) is centered around a Campbell Scientific CR10 data logger. Data are sampled every 2.25 s (changed to 5 s on July 28, 1994) and stored as hourly averages. Data collected include positive and negative dc currents and voltages; plane-of-array (POA) irradiance; ac current, voltage, and power; and back-of-module temperature. Meteorological data are provided by the Reference Meteorological and Irradiance Station (RMIS) located at the test site. The spectral and spatial optical properties of the atmosphere are measured and monitored by the Atmospheric Optical Calibration System (AOCS).

Current versus voltage (I-V) traces are acquired monthly (weather permitting), at POA irradiance levels between 900 and 1100 W/m^2 (near solar noon), via a portable I-V curve tracer. The spectrum at the time of the I-V trace is either measured by a LiCor spectroradiometer or is calculated based on data from RMIS or AOCS. The spectral corrections used in this paper are based on the calculated spectrum. Prior to deployment, all modules had baseline I-V traces acquired both by a Spire 240A solar simulator at STC and outdoors under prevailing conditions at NREL.

RESULTS

Figure 1 shows ac power (normalized to 1000 W/m^2) versus back-of-module temperature and time. The data set for this chart was restricted to POA irradiance levels greater than 750 W/m^2. The resulting data were then graphed and fit with sixth-order polynomial trend lines. Referencing these trend lines in the figure, a direct correlation among power, back-of-module temperature, and season is evident. This correlation is the well-documented seasonal change in performance exhibited by amorphous silicon PV modules [2, 3]. The decrease in ac power output from the summer of 1993 high to the winter of 1993 low was approximately 7%. The increase (rebound) in ac power output from the winter low of 1993 to the summer high of 1994 was approximately 6%.

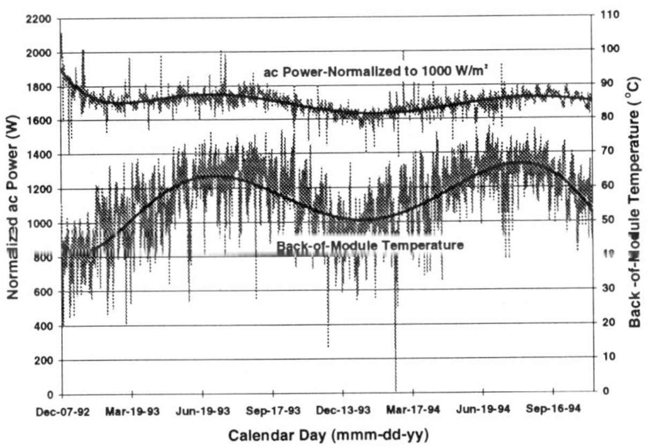

Figure 1. AC power normalized to 1000 W/m^2 versus back-of-module temperature and time

I-V traces of the PV array have been acquired monthly since system installation. Figure 2 presents the measured array peak power normalized to 1000 W/m^2 versus air mass, back-of-module temperature, and time. The array peak power on June 10, 1993 (summer of 1993 high), was 1902 W. The array peak power on January 12, 1994 (winter of 1993 low), was 1744 W. The array peak power on August 15, 1994 (summer of 1994 high), was 1933 W. From these values we see a decrease of ≈8.3% in array output from the summer of 1993 to the winter of 1993. Furthermore, an increase (rebound) of ≈10.8% in array output from the winter of 1993 to the summer of 1994 is also observed.

Given the seasonal changes in PV array and system performance, the focus was turned to quantifying the effects of temperature, spectrum, annealing, and light-induced degradation. In this regard, all known array losses and influences on performance were either measured or calculated. The losses identified were module mismatch, system dc losses (i.e., wire, blocking diodes, switchgear, and protective fuses), and soiling. The influences on performance identified were changes in spectrum, temperature, and insolation.

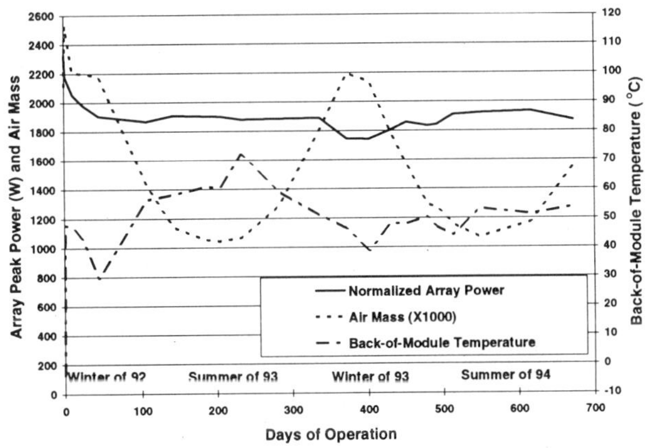

Figure 2. Array Peak power (normalized to 1000 W/m^2) versus air mass, back-of-module temperature, and time

The loss due to module mismatch was calculated based on the baseline I-V curve traces acquired via the Spire 240A solar simulator and on each module's location in the PV array. Module mismatch was based upon the current-limiting module (minimum I_{MAX}) and the summation of V_{MAX} for each monopole. In determining the mismatch for each monopole, the product of the summation of V_{MAX} and the minimum I_{MAX} was divided by the summation of P_{MAX}. The mismatch calculated for all monopoles was then averaged. Module mismatch for the individual monopoles was calculated to range from 2% to 5%. The module mismatch loss for the PV array was thus calculated to be approximately 3.6%.

System dc losses were based on voltage drops measured at approximately 1000W/m^2 and the calculated I^2R losses for all dc wiring. The voltage drops across each blocking diode, array fuses, and dc switch gear were measured. These voltage drops were then multiplied by the maximum power current (normalized to 1000 W/m^2) observed at the time of the I-V trace measurement and the results summed. The calculated resistance of the wire was multiplied by the square of the measured maximum power current (normalized to 1000 W/m^2). The system dc losses calculated in this manner were less than 10 W (0.3 to 0.4 %) at 1000 W/m^2 for this PV array.

To quantify the soiling losses, I-V curve traces of one monopole before cleaning (dirty) and after cleaning (clean) were acquired and compared at 1000 W/m^2. This method involved taking a "dirty" trace, cleaning the monopole, then immediately taking the "clean" trace. Soiling losses were measured at 3.2% on March 23, 1993, and 3.0% on October 28, 1994.

Using the modeled spectrum for the site at time of test, spectral corrections (based on the limiting cell of the dual-junction module: top, bottom, or matched) were calculated. To compare measured powers with respect to a specific reference spectrum (i.e., ASTM E892-87 global reference spectrum), the data must be corrected for the spectral mismatch error [4]. The spectral mismatch error was computed for each of the selected days using measured spectral response data from dual-junction same-band-gap amorphous silicon cells and modules of the same kind as deployed in the system. The spectral response of the reference device (pyranometer) is constant versus wavelength. The measured spectrum was problematic because the spectral irradiance was measured for only a few days and over a limited (300-1100-nm) spectral range. To solve this problem, the spectral irradiance was modeled using a semiempirical spectral model [5]. The model requires as inputs the POA tilt angle, air temperature, barometric pressure, measurement time, total irradiance on a horizontal surface, direct normal irradiance, the diffuse irradiance on a horizontal surface, and, preferably, the POA irradiance. All of these quantities were available. Figure 3 shows a comparison of the modeled spectrum with the spectral irradiance measured at NREL via a LiCor LI-1800 spectroradiometer on January 20, 1993. The spectral corrections (from summer to winter) obtained using this method ranged from 1.6% to 8.6% at the NREL PV test site. It should be noted that these spectral corrections are for current only and the fill factor will be influenced by other spectral effects.

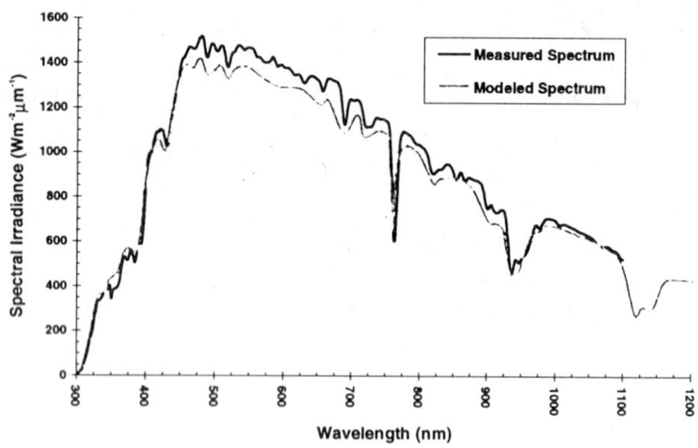

Figure 3. Measured spectrum versus modeled spectrum

Peak power was corrected for temperature based on a coefficient of -0.1% per °C. The temperature coefficient used for correcting peak power was that supplied by the manufacturer (USSC). For this analysis, it was assumed to be linear over the temperature range observed and independent of the spectrum. These corrections were then applied to the prevailing array peak power as measured by a portable I-V curve tracer.

The prevailing array peak power (normalized to 1000 W/m^2), changes in performance, and the previously listed losses were compared to the summation of rated module peak power at STC (102 modules x 22 W=2244 W). Figure 4 presents the normalized array peak power, changes in performance, and losses as percent of the array rating at STC. Table 2 lists the data used in Fig. 4 in numeric format. The first few data points given in Table 2 are elevated due to the fact that the early light-induced degradation has not been fully completed. Consequently, the first six data points were not included in the results of this paper. After correcting the normalized peak power for temperature, spectrum, soiling, module mismatch, and dc system losses, the seasonal change in performance was still observed (see Fig. 4). The corrected array peak power decreased by 7.5% from the summer high on June 10, 1993, to the winter low on January 12, 1994. This decrease in array peak power, even after applying corrections, may be attributed to light-induced degradation. The corrected array peak power rebounded by 7.7% from the winter low on January 12, 1994, to the summer high on June 8, 1994. This rebound in corrected array peak power is attributed to thermally-induced annealing which appears to occur during the warm summer months.

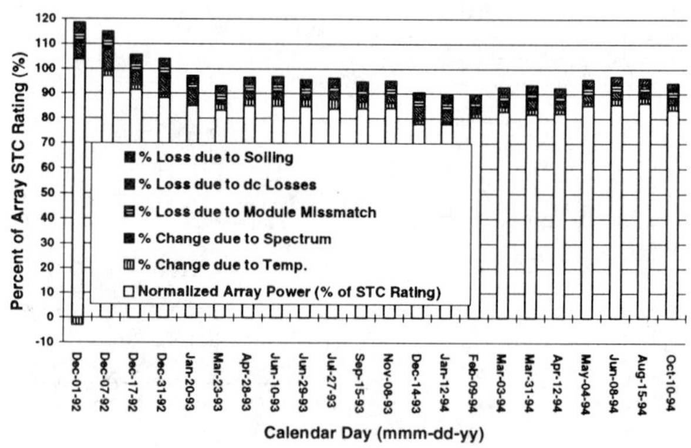

Figure 4. Normalized Peak Power and changes/losses compared to Array Peak Power at STC

Figure 4 shows that the peak powers measured on January 12, 1994, and June 8, 1994 (after applying corrections for the identified losses and changes in performance), were lower than the STC rating for array peak power by 10.8% and 3.1%, respectively. This difference from the STC rating is attributed to measurement error (bias error), light-induced degradation, and annealing.

CONCLUSIONS

A 1.8-kW_{ac} a-Si dual-junction utility-interconnect PV system is operational and under evaluation at NREL's PV test site. The PV array and balance-of-system are performing as designed.

Losses and changes in array performance were identified and quantified. System losses (dc only) were found to be less than 10 W for all data points. Losses due to soiling were measured at approximately 3%. The loss due to module mismatch was calculated to be approximately 3.6%. The change in performance due to temperature, from winter to summer, was found to range between -3% and +4%. Changes in performance due to spectrum from summer to winter were found to be between 1.6% and 8.6%. After correcting for these losses and changes, the measured (prevailing) array peak power was brought within approximately 11% of the STC rating for winter operation and 3% for summer operation.

The array peak power was found to degrade by 7.5% from summer to winter of 1993 and rebound (anneal) by 7.7% from the winter of 1993 to the summer of 1994. The lowest uncorrected array peak power, normalized to 1000 W/m^2 for the winter of 1993, was 77.7% of the 2244 W array peak power rating. The highest uncorrected array peak power, normalized to 1000 W/m^2 for the summer of 1994, was 86.2% of the 2244 W array peak power rating.

As other mechanisms for system losses/degradation may affect the PV array, further research into the influences on array/system performance is being conducted.

Table 2. Array peak power (normalized to 1000 W/m^2), changes in performance, and losses compared to the STC rating for array peak power

Date	Limitting Cell	Spectral Corr.	Prevailing Array Power (% of STC Rating)	Normalized Array Power (% of STC Rating)	% Change due to Temp.	% Change due to Spectrum	% Loss due to Module Missmatch	% Loss due to dc Losses	% Loss due to Soiling	Corrected Array Power (% of STC Rating)
Dec-01-92	TOP	0.933	108.6	103.8	-3.0	7.5	3.7	0.3	3.1	115.4
Dec-07-92	TOP	0.914	97.5	96.9	2.2	9.1	3.5	0.3	2.9	115.0
Dec-17-92	TOP	0.939	85.9	91.4	2.0	5.9	3.3	0.3	2.7	105.8
Dec-31-92	TOP	0.915	83.1	88.2	1.6	8.2	3.2	0.3	2.6	104.1
Jan-20-93	TOP	0.933	86.2	84.9	0.4	6.1	3.1	0.3	2.5	97.3
Mar-23-93	BOTTOM	0.984	85.2	83.2	2.6	1.4	3.0	0.3	2.5	93.0
Apr-28-93	BOTTOM	0.968	85.2	85.0	2.8	2.8	3.1	0.3	2.5	96.5
Jun-10-93	BOTTOM	0.966	81.9	84.8	3.0	3.0	3.1	0.3	2.5	96.7
Jun-29-93	BOTTOM	0.976	82.4	84.6	3.0	2.1	3.0	0.3	2.5	95.7
Jul-27-93	BOTTOM	0.970	83.4	83.7	4.0	2.6	3.0	0.3	2.5	96.1
Sep-15-93	BOTTOM	0.975	89.0	84.0	2.9	2.2	3.0	0.3	2.5	94.8
Nov-08-93	MATCHED	0.967	81.1	84.1	2.2	2.9	3.0	0.4	2.5	95.1
Dec-14-93	TOP	0.933	76.4	77.8	1.7	5.6	2.8	0.3	2.3	90.5
Jan-12-94	TOP	0.939	75.9	77.7	1.0	5.0	2.8	0.3	2.3	89.2
Feb-09-94	MATCHED	0.976	81.3	80.3	1.8	2.0	2.9	0.3	2.4	89.7
Mar-03-94	BOTTOM	0.975	81.8	82.7	1.9	2.1	3.0	0.3	2.5	92.5
Mar-31-94	BOTTOM	0.953	85.7	81.7	2.1	4.0	2.9	0.3	2.5	93.5
Apr-12-94	BOTTOM	[illegible]	[illegible]	[illegible]	1.9	2.7	3.0	0.3	2.5	92.3
May-04-94	BOTTOM	0.966	81.7	85.1	1.6	3.0	3.1	0.3	2.6	95.7
Jun-08-94	BOTTOM	0.967	79.8	85.6	2.4	2.9	3.1	0.3	2.6	96.9
Aug-15-94	BOTTOM	0.980	79.8	86.2	2.3	1.8	3.1	0.4	2.6	96.2
Oct-10-94	BOTTOM	0.969	82.7	83.4	2.4	2.7	3.0	0.3	2.5	94.3

ACKNOWLEDGMENTS

The authors thank R. DeBlasio for his support and leadership of the Photovoltaic Module and System Performance and Engineering Project. We also thank B. Kroposki, D. Myers, J. Burdick, S. Rummel, L. Ottoson, and Y. Caiyem for their technical support. This work was supported by the United States Department of Energy under contract number DE-AC36-83CH10093.

REFERENCES

[1] T. Strand, L. Mrig, and R. Hansen, "Technical Evaluation of Photovoltaic Systems at NREL," PV Performance and Reliability Workshop, Lakewood, CO, September, 1994.

[2] R.G. Pratt and J. Burdick, "Performance of a 4 kW Amorphous-Silicon Alloy Photovoltaic Array at Oakland Community College, Auburn Hills, Michigan," *Twentieth IEEE PVSC*, Las Vegas, NV, pp. 1272-1277, 1988.

[3] B. Kroposki and L. Mrig, "Thin Film Module Stability Testing at NREL," PV Performance and Reliability Workshop, Lakewood, CO, September, 1994.

[4] K.A. Emery, C.R. Osterwald, T.W. Cannon, D.R. Myers, J. Burdick, T. Glatfelter, W. Czubatyj, and J. Yang, "Methods for Measuring Solar Cell Efficiency Independent of Reference Cell or Light Source," *Eighteenth IEEE PVSC*, Las Vegas, NV, pp. 253-274, 1985

[5] S. Nann and C. Riordan, "Solar Spectral Irradiance under Clear and Cloudy Skies: Measurements and a Semiemperical Model," *Jour. Appl. Meteorology*, **30**, pp. 447-462, 1991.

PERFORMANCE OF A 268Wp STAND-ALONE PV SYSTEM TEST FACILITY

K. Godfrey Adanu
Energy Research and Development Laboratory
Department of Physics, University of Ghana, Legon-Accra, Ghana

ABSTRACT

This paper reports on studies carried out on a 268Wp stand-alone PV test facility installed at the University of Ghana. Parameters measured include solar irradiance, ambient temperature, cell temperature, array voltage, battery voltage, and energy output from array. An automatic data acquisition system was used to record the data. The purpose of this study was to determine the effect on performance of a PV array due to seasonal climatic changes: for example, effect due to dust during the period when dust-laden winds blow from the Sahara desert to most sub-saharan West African countries. The reduced intensity of the solar radiation as would be expected reduced the total array output during the harmattan and the rainy seasons. No visible changes were observed in the PV modules themselves over the period of study, but one of the junction boxes attached to the modules had become severely corroded.

INTRODUCTION

It is now generally realized that countries within the tropics are potentially the largest beneficiaries of solar energy technologies, especially photovoltaic electricity generating systems for rural electricity supply. It is therefore important that these countries become acquainted early with the application of these technologies.

The 268Wp PV installation at the University of Ghana was aimed among other things to serve as a means to introduce this technology and also to serve as a demonstration unit. As a demonstration unit the system was exhibited at an international 3-week science and technology fair held in Accra in 1986 where its novelty drew a lot attention and enthusiasm. Many students from the University and secondary schools have since continued to visit the installation on the campus and learned about electricity generation from sunshine. The system is also being used in particular to investigate the performance of PV systems in the climate prevalent in Ghana. Like that in most sub-saharan West African countries the climate of Ghana consists of a dry dust-laden harmattan season (from about November to February), a dry hot and sunny period (March to May) a wet cloudy season (end of May to August), and a variably wet/dry season (September to October). There are variations of course; but these seem to be the general pattern. In the northern regions of Ghana for example there is only one distinct rainy season which peaks around August, a long dry and dusty harmattan period, and a still much longer dry sunny and hot period.

This study carried out over the period between 1987 and 1991 was meant to investigate the variation of the performance of the PV system over the various seasons. It had been anticipated that the results of such a study would help in more accurately sizing PV systems in Ghana and also in the sub-region generally.

CH3365-4/94/0000-0854 $4.00 © 1994 IEEE

CONSTRUCTIONAL INFORMATION

The system consists of an array of four sc-Si modules each of size 128.2 x 64.2 x 4.2 (cm) with a total module area of 0.823m^2, an aperture area of 0.785m^2 and a total overall cell area of 0.565m^2. Each of the modules has 72 sc-Si cells of diameter 100mm connected in series to provide an output voltage of 48volts. The terminals of each module has a three-contact connector. The nominal operating voltage was stated as 32 volts. Visually one can identify several bubbles in the encapsulant in two of the modules, but these did not seem to adversely affect the performance of the modules.

Four 12-volt automobile batteries connected in series served as storage battery bank. These were connected via a charge control unit and an alarm control unit. Three pairs of 24-volt 45W DC bulbs connected in parallel across the battery bank served as load to dissipate energy stored in the battery. The array was mounted on a wooden platform inclined at an angle of 7 degrees to the horizontal. The latitude of the station is 5.5 degrees N and the slope was needed only so as to drain off rain water.

DATA ACQUISITION AND ANALYSIS

The data collected were recorded on a bubble cassette inside and automatic data acquisition system. Quantities which were not directly measured were calculated from the measured values inside a signal processing unit. The following parameters were recorded: incident global irradiance in array-plane, ambient temperature, cell temperature, energy output from array to battery, energy from array to load, array voltage, energy to battery, energy from battery and energy to load.

The bubble cassette with recorded data was mailed to EURATOM, the EC Joint Research Centre at Ispra in Italy for analysis. In a few cases the cassette got lost in the post; considering the fact that each cassette carries data for about seven months every cassette lost was the loss of substantial valuable data. Consequently there is no full year with data for all twelve months. The results for the different months and seasons have therefore been pieced together to obtain the trend in the various seasons, however being aware that no two or three years have completely identical climatic patterns, particularly with regard to the onset and intensity of the harmattan and the rainy seasons, nor of the length and intensity of the solar radiation during the sunny months.

RESULTS

Evidence of the enormous effect of dust particles in the air during the harmattan is shown in reduced visibility, see Fig.1. The contrast is striking in the two pictures taken from about the same location on the University Avenue. The effect on the cleanness of the surface of the modules is shown in Fig.2. The dust was allowed to accumulate over a period of two days before being partly wiped off. This suggests strongly that there is a need for regular cleaning of PV modules especially during the harmattan period and this must be brought home forcefully to the end-users to ensure that they derive maximum benefit from their installations even during the harmattan period. The dust in the air which reduces the solar irradiance is beyond the reach of the PV user but to allow reduced array output due to accumulated dust should be prevented.

In Fig.3 the cumulative frequency of in-plane irradiance is shown for the month of January 1988 at the peak of the harmattan and also for the month of March 1988 when the sun was bright

Fig. 1 The weather at noon for a day during the harmattan (top) and for a day during a non-harmattan period.

Fig. 2 Surface of a module that has been partly wiped after two days of dust accumulation.

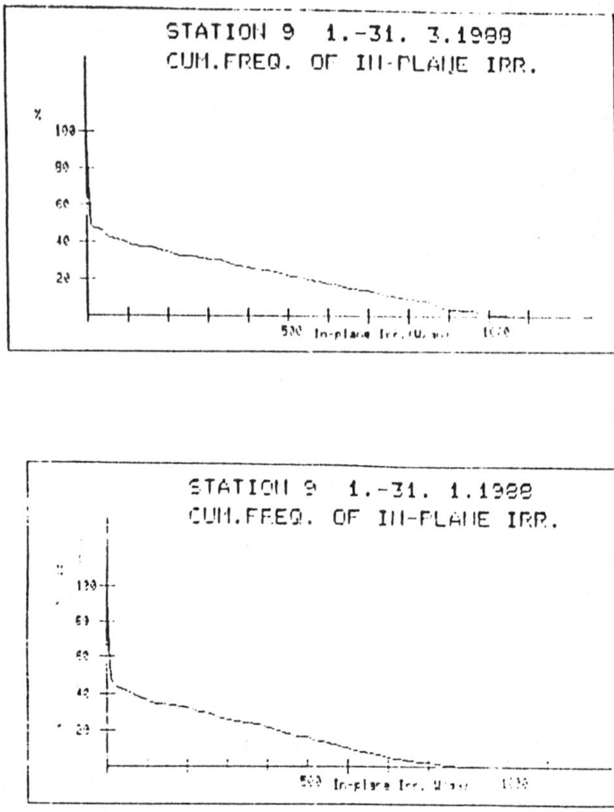

Fig. 3 Cummulative frequency of in-plane irradiance during a month at the peak of the harmattan and during a non-harmattan period.

and the sky clear. These two curves show a maximum solar irradiance of 1.05 Wm^2 for March 1988 as against 0.85 Wm^2 during January 1988.

In fig.4, ambient temperature and cell temperature have been plotted as a function of time of day. Cell temperature is determined principally by solar radiation which is not converted into electricity; this occurs at high solar radiation intensity levels.

The energy output from the PV array and the solar irradiance have been plotted as a function of time of day, see Fig.5. The dip in the array

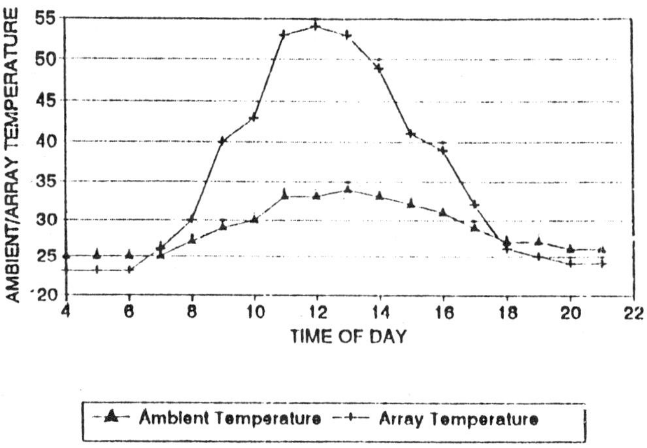

Fig. 4 Ambient temperature and array temperature as functions of time of day.

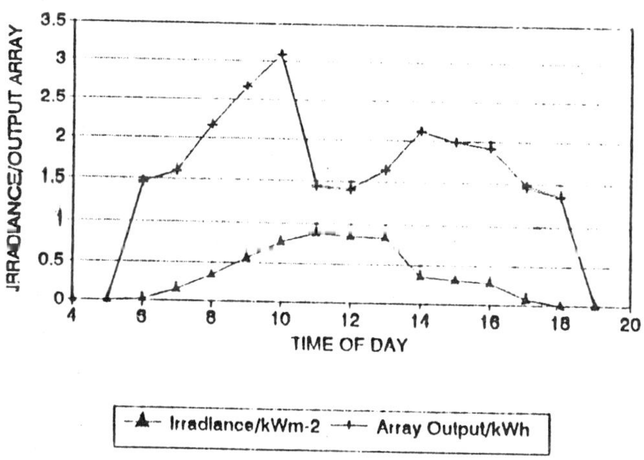

Fig. 5 In-plane irradiance and energy output from array as function of time of day.

output arround 1300h may be attributed to the effect of high cell temperature. This dip corresponds roughly to the time of day when the cell temperature is high as shown in Fig.4. It is generally known that high cell temperatures reduce the efficiency of energy conversion of the cells, and this result is as would be expected.

DISCUSSION AND CONCLUSIONS

The effect of dust particles in the harmattan sky generally reduces the solar irradiance and therefore the energy output from the PV array, not dissimilar to the effect of clouds on array output. Initially it was speculated that the small size of the dust particles (of the order of 2 micron) might noticeably change the spectral distribution in the solar radiation that might affect the response of solar cells. Obviously this is an investigation that will have to be carried out under controlled conditions.

Data obtained from this study have however given information on the effect of irradiance in array output, daily array output, daylight maximum cell temperature, and the cummulative frequency of array-plane irradiance. Such information can be used for more reliable sizing of PV electricity generating systems.

ACKNOWLEDGEMENTS

The author wishes to acknowledge the kind gesture of ESTI at EURATOM; the EC Joint Research Centre at Ispra, Italy in providing the solar modules and the other equipment for this study at the University of Ghana, and also for undertaking the analysis of the data from the data acquisition system.

A COMPARISON OF PHOTOVOLTAIC MODULE PERFORMANCE EVALUATION METHODOLOGIES FOR ENERGY RATINGS

Benjamin Kroposki, Keith Emery, Daryl Myers, and Laxmi Mrig
National Renewable Energy Laboratory
1617 Cole Blvd., Golden, CO 80401, USA

ABSTRACT

The rating of photovoltaic (PV) modules has always been a controversial topic in the PV community. Currently, there is no industry standard methodology to evaluate PV modules for energy production. This issue must be discussed and resolved for the benefit of system planners, utilities, and other consumers. Several methodologies are available to rate a module's peak power, but do any accurately predict energy output for flat-plate modules? This paper analyzes the energy performance of PV modules using six different energy calculation techniques and compares the results to the measured amount of energy produced. The results indicate which methods are the most effective for predicting energy output in Golden, Colorado, under prevailing meteorological conditions.

INTRODUCTION

The rating of photovoltaic (PV) modules can be classified in two ways: as a power rating or as an energy rating. A power rating is an instantaneous measurement with respect to a fixed set of reference conditions or a value that is based on a statistical analysis (e.g., regression). An energy rating is based on the amount of energy a module produces over a certain amount of time under specific conditions. Both of these rating methods have good applications. A power rating is an excellent way to compare modules under specific reference conditions. An energy rating gives a good comparison of module performance under actual operating conditions. Currently, there is no industry standard methodology to evaluate PV modules for energy production. This issue must be discussed and resolved for the benefit of system planners, utilities, and consumers. This paper compares different methods that are used to predict the energy output of PV modules and presents a comparison of results based on measurements obtained on various PV modules in Golden, Colorado.

There are basically two methods of rating the amount of peak power that modules produce. One is by calculating the instantaneous peak power from I-V curves at specific conditions, such as Standard Reporting Conditions (SRC). The second method to rate power is to use regression analysis on long-term data of module power generation. These data can be used to form models of how the module operates at different meteorological and irradiance values. Rated power is then computed at a specific reference condition. PVUSA Test Conditions (PTC) and Nominal Terrestrial Environment (NTE) are examples of methods employing regression analysis. The reference criteria for these three methods are listed in Table 1.

Table 1. Reference criteria of module power rating test methods [1-4]

Method	Irradiance (W/m^2)	Temperature (°C)	Spectrum	Wind Speed (m/s)
SRC	1000	25 (module)	AM1.5	-
PTC	1000	20 (ambient)	-	1
NTE	800	20 (ambient)	-	1

Once the amount of power a module produces at certain reference conditions is known, an energy rating can be determined. This is done by multiplying the instantaneous power by a selected period of time. For this paper, the time period is determined by the amount of peak sun-hours. One peak sun-hour is defined as the equivalent number of hours when the irradiance equals 1000 W/m^2 (i.e. 5 hours of irradiance at 200 W/m^2 equals 1 peak sun-hour). Multiplying the rated power (W) by the number of peak sun-hours (h) gives an approximate energy estimate in watt-hours (Wh). This method does not take into account changing temperature and spectral influences on the module, or the fact that the module may operate differently under different irradiance conditions.

Another method for determining a module energy rating is to use a "standard day" for the reference conditions and sum the energy generated over that standard day. This technique is used in the ENRA [4] and AM/PM [5] energy rating methods. This approach presents a standard method for comparing modules but may not give an accurate picture of the real operating conditions. Finally, the realistic reporting conditions (RRC) [6] method calculates the energy produced over a selected period of time usually monthly or annually, and takes into account actual or typical weather conditions. Typical weather conditions for a site can be found in typical meteorological year (TMY) data files, the Solar Radiation Data Manual for Flat-Plate and Concentrating Collectors [7] developed by the National Renewable Energy Laboratory (NREL), or the Solar and Meteorological surface observation network 1961-1990, available from the National Climatic Data Center [8].

This paper analyzes the energy performance of a set of modules using six different energy calculation techniques and compares the results to the measured amount of energy produced. The different energy rating methods that will be compared are described below.

1. Energy at SRC (ESRC): The module's peak power is determined at SRC and then the power is multiplied by the

First WCPEC; Dec. 5–9, 1994; Hawaii

number of sun-hours during a specified time period to determine energy output. The module is rated at 25°C; however, variations in temperature that the module may experience throughout the time period are not taken into account. This assumption will lead to significant discrepancies in energy output for modules that have relatively large temperature coefficients. Energy calculated by this method will typically be higher than the measured value.

2. Energy at PTC (EPTC): This method determines a power rating based on a regression analysis over a 30-day period, at PTC. Data are collected for at least 30 days, with a cumulative total of at least 10 kWh/m^2 above 1000 W/m^2 of global plane-of-array irradiance. The rating is based on regression analysis of quality checked data using only irradiance of 500 W/m^2 or greater. The power is calculated using the following equation [4]:

$$\mathbf{P} = \mathbf{A_1}*E + \mathbf{A_2}*E^2 + \mathbf{A_3}*E*T_{amb} + \mathbf{A_4}*E*WS \quad (1)$$

where, **P** = output power, Wdc
E = irradiance, W/m^2
T_{amb} = ambient air temperature, °C
WS = wind speed at 10 m above grade, m/s
$\mathbf{A_n}$ = regression coefficients

The PTC-rated power is then multiplied by the number of sun-hours during a specified time period to determine the energy rating. For this study, no wind speed restrictions were used.

3. Energy at NTE (ENTE): This method is similar to the EPTC except that the regression analysis is done at the Nominal Terrestrial Environment (NTE)[3]. The power rating is computed at 800 W/m^2, 20°C ambient temperature, and wind speed of 1 m/s using the PTC power model. Then the NTE-rated power is multiplied by the number of sun-hours during a specified time period to determine an energy rating.

4. ENRA (ENergy RAtings): This method also uses a regression analysis to determine the power output for each module using the following three parameter response function [5].

$$\mathbf{P} = \mathbf{A_1}*E + \mathbf{A_2}*E^2 + \mathbf{A_3}*E*\ln(E) \quad (2)$$

where, **P** = output power, Wdc
E = irradiance, W/m^2
$\mathbf{A_n}$ = regression coefficients

Only data above 500 W/m^2 are used in computing the regression coefficients. The response function is then integrated over a reference day. The reference day is defined by the following function:

$$E(t) = \sum_{n=0}^{4} \mathbf{B_n} t^n [W/m^2] \quad (3)$$

$\mathbf{B_0}$ = 2.602676742e3
$\mathbf{B_1}$ = -1.399250090e3
$\mathbf{B_2}$ = 2.43511524e2
$\mathbf{B_3}$ = -1.5395308e1
$\mathbf{B_4}$ = 3.19483e-1
t = time in hours (from t=6 to t=18)

This function has a maximum of irradiance of 900 W/m^2 at 12:00, and its integral has a value of 6.82 kWh/m^2, or 6.82 sun-hours. This model does not take into account temperature or spectral changes over the reference time period.

5. AM/PM: A model of power is obtained using derived temperature and irradiance coefficients for each module. The power equation is defined below:

$$\mathbf{P} = \mathbf{P_{irr}}*(1+10^{-6}*T_{coef}*(T_{mod} - 25°C)) \quad (4)$$

where, **P** = output power, Wdc
$\mathbf{P_{irr}}$ = calculated power based on irradiance
T_{coef} = max. power temp. coefficient in ppm/°C normalized at 25°C
T_{mod} = module temperature °C

Calculated power based on irradiance is derived from the following formula:

$$\mathbf{P_{irr}} = m*I_{sc} + b = m * E_{tot}/(1000 * I_{src}) + b \quad (5)$$

where, I_{sc} = short circuit current
E_{tot} = total irradiance, W/m^2
I_{src} = I_{sc} at 25°C and 1000 W/m^2

This function assumes that I_{sc} is linear with total irradiance and temperature. The module temperature is computed from the following equation [9]:

$$\Delta T = T_{mod} - T_{air} = 0.314*E_{tot}(1-\eta) / (1 + 0.4 * V^{1/2}) \quad (6)$$

where, T_{air} = air temperature, °F
E_{tot} = total irradiance, W/ft^2
η = efficiency = $\mathbf{P_{irr}}$/(Area*E_{tot})
V = wind velocity, ft/sec

A better fit to the measured-versus-calculated module temperature was obtained when 0.15 was used instead of the 0.314 suggested by the reference. The energy over a standard day is then computed by multiplying the calculated power by the amount of time during a standard day. An "AM/PM" standard day is defined as a function of insolation, temperature, and air mass variations over a 24-hour period. For this experiment, only irradiance and temperature were defined. The equations below for the irradiance and temperature models were derived from a polynomial curve fit to the figures in the reference [5].

$$E(t) = \sum_{n=0}^{4} \mathbf{B_n} t^n [W/m^2] \quad (7)$$

$\mathbf{B_0}$ = 2.965956e3
$\mathbf{B_1}$ = -1.484969e3
$\mathbf{B_2}$ = 2.457535e2
$\mathbf{B_3}$ = -1.532984e1
$\mathbf{B_4}$ = 3.195195e-1
t = time in hours (from t=6 to t=18)

This function has an integral value of 5.35 kWh/m^2, or 5.35 sun-hours.

$$Air\_Temp(t) = \sum_{n=0}^{6} \mathbf{B_n} t^n [W/m^2] \quad (8)$$

B_0 = -1.276272e1
B_1 = 2.454010e1
B_2 = -7.683963
B_3 = 1.115173
B_4 = -8.009432e-2
B_5= 2.801611e-3
B_6 = -3.848843e-5
t = time in hours (from t=6 to t=18)

6. RRC: A model for module power is obtained though laboratory characterization. The same power model that was used for the AM/PM method was used for this methodology. The energy is then summed over a specific time period using realistic meteorological conditions. For this paper, hourly averages of actual measured meteorological data were used to calculate the amount of energy that each module produced. This should lead to accurate predictions of energy, if the power models for each module are valid.

EXPERIMENTAL

A set of modules at NREL has been deployed outdoors and loaded at the peak power point. The modules' voltage and current are measured every 5 s and averaged hourly. This provides the measured energy produced by each module in watt-hour(Wh). A set of seven modules has been selected to represent major types of terrestrial flat-plate PV modules. This set includes single-crystalline silicon (mono-Si); multicrystalline silicon (multi-Si); single-, dual-, and triple-junction amorphous silicon (a-Si, a-Si/a-Si); and a-Si/a-Si/a-Si); copper indium diselenide (CIS); and cadmium telluride (CdTe) modules. Each module's I-V curve is also taken every half hour and stored. These data are used, along with meteorological and radiometric data, to compute regression coefficients for various module performance evaluation models. The meteorological and radiometric data at the site are collected at the Reference Meteorological and Irradiance Station (RMIS). These data include ambient temperature, plane-of-array (POA) irradiance, direct normal irradiance, and wind speed. These data are used to compute power equations for each module using regression analysis over a period of 30 days. The 30-day period that was used for this study was the month of September, 1994.

Module temperature and irradiance coefficients for the AM/PM and RRC energy ratings methods were developed using a Spire 240 solar simulator. Irradiance measurements were taken between 0.1 and 1.2 suns, with the temperature held to 25±0.5°C. Temperature measurements were taken between 10°C and 50°C, with the irradiance held at 1000 W/m^2.

The modules' energy output is compared with the amount predicted by using six different energy rating methods over a clear day, a cloudy day, and a month. The clear day was September 8, 1994. This day had a total sun-hours of 7.65 sun-hours and an average daily temperature of 22°C. The cloudy day was September 30, 1994. This day had a total of 2.21 sun-hours and an average daily temperature of 15°C. The month used was September, 1994, which had a total of 171 sun-hours and an average temperature of 18°C.

Table 2 lists the modules used in this study along with their manufacturers' rated power. Some modules do not contain manufacturer ratings because they are research modules. The table also includes the modules' rated power measured at SRC, and calculated power ratings at PTC and NTE. The table also contains the nominal operating cell temperature (NOCT) for each module for the month of September. The regression analysis for both the PTC and NTE consisted of the 30 days of the month of September. Only data with irradiance above 500 W/m^2 were used. All power ratings are normalized to 1000 W/m^2.

Table 2. Power rating of modules and module NOCT

Module Type	Manf. (W)	SRC (W)	NTE (W)	PTC (W)	NOCT (°C)
a-Si	5	3.27	3.1	3.1	40.7
a-Si/a-Si	22	19.6	18.0	18.9	33.2
a-Si/a-Si/a-Si	-	25.1	23.6	24.7	31.7
mono-Si	75	67.2	60.7	60.8	30.9
multi-Si	60	53.9	49.0	49.2	29.6
CIS	-	31.4	24.6	25.8	33.1
CdTe	-	53.6	41.4	42.9	36.9

RESULTS AND DISCUSSION

Tables 3 and 4 show a comparison of measured energy output (Wh) versus different rating methods for September 8, 1994. This is considered a good solar day in Golden, Colorado.

Table 3. Measured energy (Wh) versus percent difference (%) for a clear day

Module Type	Meas.	ESRC	ENTE	EPTC
a-Si	23.6	5.6	5.2	0.4
a-Si/a-Si	141.1	5.8	1.7	2.4
a-Si/a-Si/a-Si	186.3	2.9	1.1	1.4
mono-Si	454.0	11.8	6.2	2.4
multi-Si	371.5	9.9	5.0	1.3
CIS	187.8	21.8	4.2	4.9
CdTe	329.3	19.7	0.2	-0.3

Table 3 shows that, for a clear day, the ESRC method over-predicts energy outputfor all module types. The ESRC values for a-Si are much closer to the measured values because these modules are not as temperature sensitive as crystalline silicon or other thin-films. The ENTE and EPTC methods predict module energy output within 6% for most technologies on clear days. The regression analysis works well because the models for the technologies are stable, and predict well, especially on clear days.

Table 4. Measured energy (Wh) versus percent difference (%) for a clear day

Module Type	Meas.	ENRA	AM/PM	RRC
a-Si	23.6	-11.3	-30.4	6.3
a-Si/a-Si	141.1	-11.9	-40.0	3.2
a-Si/a-Si/a-Si	186.3	-12.8	-46.6	0.7
mono-Si	454.0	7.4	-33.0	0.9
multi-Si	371.5	-1.0	-35.2	0.1
CIS	187.8	-7.3	-24.1	9.6
CdTe	329.3	-6.8	-15.1	14.4

Table 4 shows that, for a clear day, the ENRA and AM/PM methods under-predict the energy produced for all modules except one. This is because the ENRA and AM/PM method base their energy output on a standard day where the total irradiance was less than that of the clear day. The RRC method was extremely close to the measured value except for the a-Si, CIS and CdTe modules. This is because the models for these technologies may not be accurate (i.e., more scatter or bias in data).

Tables 5 and 6 show a comparison of measured energy output (Wh) versus different rating methods for September 8, 1994. This is considered a cloudy day for the month of September in Golden, Colorado.

Table 5. Measured energy (Wh) versus percent difference (%) for a cloudy day

Module Type	Meas.	ESRC	ENTE	EPTC
a-Si	7.1	1.4	1.4	-4.4
a-Si/a-Si	43.8	-1.4	-5.8	-5.0
a-Si/a-Si/a-Si	56.6	-2.2	-4.0	-3.9
mono-Si	141.6	4.6	-1.4	-5.4
multi-Si	115.2	3.3	-2.0	-6.0
CIS	56.7	18.2	-0.2	0.5
CdTe	101.7	14.1	-6.7	-7.3

Table 5 shows that, for a cloudy day, the ESRC method predicts energy within 3% for a-Si and within 5% for mono- and multi-Si modules. This is due to the lower operating temperatures under cloudy sky conditions. For CIS and CdTe modules, the SRC method overestimates energy output up to 18%. The ENTE and EPTC methods predicts module energy output within 8% for most technologies on cloudy days.

Table 6. Measured energy (Wh) versus percent difference (%) for a cloudy day

Module Type	Meas.	ENRA	AM/PM	RRC
a-Si	7.1	66.5	60.8	15.5
a-Si/a-Si	43.8	65.3	56.5	-12.9
a-Si/a-Si/a-Si	56.6	65.7	55.5	-21.7
mono-Si	141.6	71.1	58.5	-0.3
multi-Si	115.2	68.7	58.1	-3.6
CIS	56.7	67.6	62.5	-0.2
CdTe	101.7	67.0	64.4	18.5

Table 6 shows that, for a cloudy day, the ENRA and AM/PM methods over-predict the energy produced for all modules by at least 55%. This is because the ENRA and AM/PM method base their energy output on a standard day where the total irradiance was more than that of the cloudy day. The RRC method had a lot of variation in predicting energy output. This method's estimation is very close for mono-Si and CIS. This is because the models for these technologies are accurate at low irradiance levels.

Tables 7 and 8 show a comparison of measured energy output (kWh) versus different rating methods for September, 1994.

Table 7. Measured energy (kWh) versus percent difference (%) for the month of September 1994

Module Type	Meas.	ESRC	ENTE	EPTC
a-Si	0.53	4.7	4.1	-0.6
a-Si/a-Si	3.15	5.7	1.6	2.4
a-Si/a-Si/a-Si	4.13	3.5	1.8	2.1
mono-Si	10.25	10.7	5.0	1.2
multi-Si	8.41	8.6	3.7	-0.1
CIS	4.22	21.2	3.5	4.2
CdTe	7.34	19.7	0.3	-0.2

Table 7 shows that, for the month of September, the ESRC method predicts energy within 6% for a-Si and within 11% for mono- and multi-Si modules. For CIS and CdTe modules, the ESRC method overestimates energy output by 22%. The ENTE and EPTC methods can predict module energy output within 5% for all technologies for the month of September.

Table 8. Measured energy (kWh) versus percent difference (%) for the month of September 1994

Module Type	Meas.	ENRA	AM/PM	RRC
a-Si	0.53	16.5	0.4	7.5
a-Si/a-Si	3.15	16.7	-4.2	2.1
a-Si/a-Si/a-Si	4.13	16.7	-8.4	-1.2
mono-Si	10.25	30.3	-0.1	3.4
multi-Si	8.41	23.8	-2.0	1.7
CIS	4.22	19.7	7.0	10.5
CdTe	7.34	20.6	14.4	17.7

Table 8 shows that, for the month of September 1994, the ENRA method over-estimates the energy produced for all module types. This is because the standard day that this method uses is considered a clear day, and this method applies no corrections for temperature. The AM/PM method come considerably closer to predicting a monthly energy output because its standard day is a well-defined average day, and it incorporates temperature effects on modules. The RRC method predicts energy very well for technologies that have a well-defined model for performance, but did not predict as well for technologies where the models for their performance are not well known.

CONCLUSIONS

The ESRC methodology does not accurately predict module energy output for modules that are temperature-sensitive when the modules operate at elevated temperatures. This is because there are no temperature corrections in the power or energy ratings. The ESRC method is a good predictor of energy only when the modules' operating conditions are close to the standard reporting conditions.

Methodologies that use regression analyses like EPTC and ENTE can accurately predict energy output if the regression and prediction are computed for the same time periods. The equations that are developed with regression analyses may not be valid if the module's response changes over conditions that were not used in the regression. An example would be using models developed from summer data, which might not accurately predict winter energy output. Scatter in the data used to derive the models will lead to less-accurate predictions.

Standard day techniques, such as the ENRA and AM/PM, work well for long-term averages if the standard day is representative of an average day in a particular area. These methods do not do well in predicting energy for days or periods when the conditions are not close to the standard day. The energy totals for these methods would be closer to the actual values if they were corrected back to the actual amount of irradiance on a daily basis.

The RRC method works well at predicting energy for modules if the power models for the technology are accurate and stable. This paper showed that the RRC method works well when using actual meteorological and irradiance data, but these results will vary if one were to use typical monthly or yearly data.

FUTURE WORK

The Photovoltaic Module and Systems Performance and Engineering Project at NREL will continue to study the issue of module energy ratings. Work needs to be conducted on developing better models that predict how modules perform under a variety of meteorological conditions. This will lead to more accurate energy predictions. Also, a decision must be made regarding calculating energy based on standard or typical conditions or actual conditions.

ACKNOWLEDGEMENTS

The authors thank Richard DeBlasio for his support of energy ratings work in this project. The authors also thank Jerry Anderson, Yehoshua Caiyam, Steve Rummel, and Larry Ottoson for their help with module characterization and temperature coefficient development, and Troy Strand for stimulating technical discussions. This work was supported by the U.S. Department of Energy under contract No. DE-AC36-83CH10093.

REFERENCES

[1] "Standard Tables for Terrestrial Solar Spectral Irradiance at Air Mass 1.5 for a 37° Tilted Surface," ASTM E 892.

[2] S. Nann and K. Emery, "A Numerical Analysis of PV-Rating Methods," *22nd IEEE PVSC*, Oct. 1991.

[3] "Standard Test Methods Electrical Performance of Nonconcentrator Terrestrial Photovoltaic Modules and Arrays Using Reference Cells," ASTM E 1036.

[4] *PVUSA Technical Specifications*, 19271-EMT-3, Aug. 1991.

[5] E. Gianoli-Rossi and K. Krebs, "Energy Rating of PV Modules by Outdoor Response Analysis," *8th E.C. PV Solar Energy Conference*. Florence, Italy, 1988.

[6] C.F. Gay, J.E. Rumberg, and J.H. Wilson, "AM/PM - All-Day Module Performance Measurements," *16th IEEE PVSC*, San Diego, CA. 1982.

[7] K. Bücher, "Do We Need Site-Dependent and Climate-Dependent Module Rating?," *23rd IEEE PVSC*, Louisville, KY, 1993.

[8] National Climatic Data Center, *Solar and Meteorological Surface Observation Network 1961-1990*, Ver. 1.0, September 1993.

[9] E.M. Sparrow and K.K. Tien, "Forced Convection Heat Transfer at an Inclined and Yawed Square Plate-Application to Solar Collectors," J. Heat Transfer, Vol. 99, pp. 507-512, 1977.

MEASUREMENT TECHNIQUES FOR THE 3RD PEP ROUND ROBIN

R.F.M. Roes*, W.J. Zaaiman, H.A. Ossenbrink
European Commission DG XII, Joint Research Centre - Ispra, Italy
European Solar Test Installation (ESTI)

* Grantholder Enschede Polytechnic, Enschede, The Netherlands.

ABSTRACT

This paper describes the measurement techniques used for the calibration of the Newer Technology reference cells for the 3rd PEP Round Robin, held from 1993 to 1995. The current-voltage characteristics of the devices under test are not only measured close to Standard Test Conditions (STC), but also at a number of different temperatures, allowing the temperature coefficients α and β as well as the curve correction factor κ to be established. Linearity measurements, with respect to the short-circuit current I_{sc} and the open-circuit voltage V_{oc}, are performed and series resistance of cells is determined. An extrapolation procedure is used to correct the output parameters to Standard Test Conditions.

INTRODUCTION

Initiated by the Photovoltaic Solar Energy Project (PEP) of the Technology, Growth and Employment Working Group of the Summit, a third international intercomparison (PEP'93 Round Robin Calibration) on the calibration of reference solar cells is carried out from 1993 to 1995 where NREL (USA) serves as Operating Agent.

OBJECTIVES

The main objective of this Round Robin is the establishment of a World Photovoltaic Scale (WPVS) through comparison of traceable reference cells. The *Newer Technology* sample set are used to determine the possible measurement problems in the characterisation and calibration with special cells as GaAs, $CuInSe_2$, CdTe, a-Si tandem- and bi-cells as well as GaAs concentrator cells.

MEASUREMENT TECHNIQUES

Current - Voltage Characteristics

The current-voltage (I-V) characteristics of the Round Robin cells are measured with two different types of Large Area Pulsed Solar Simulators (LAPSS), close to Standard Test Conditions (AirMass 1.5 Global, 25°C, 1000 $W.m^{-2}$). The shape of the SpectroLab and Pasan LAPSS light pulses is shown in Figure 1a and 1b. In case of the Spectrolab LAPSS, the I-V curve is traced using the voltage ramp of an electronic load. The linear voltage ramp scans the photovoltaic device from I_{sc} to V_{oc} in 1.5 to 2 ms. The Pasan LAPSS uses an electronic load that imposes a current sink on the device under test to measure the I-V characteristics. The scan time for a complete I-V curve is also 1.5 ms.
The software creates the I-V curve and directly corrects the data to an average irradiance level of 1000 $W.m^{-2}$, giving the calibrated values of I_{sc}, V_{oc} and fill factor.

Simulator Flash-Decay Calibration

In addition to the I-V curve calibration described above, the short-circuit current I_{sc} or the open circuit voltage V_{oc} signal together with the irradiance signal from the reference cell are measured throughout the duration of the Pasan simulator flash using the soft decay of the light pulse [4]. This technique results in electrical characteristics over a wide range of irradiance levels and reduces for possible time effects during the relatively short I-V characteristic measurements. The values of I_{sc} and V_{oc}, at 25°C and 1000 $W.m^{-2}$, are determined using linear least-squares fitting. Besides this, the diode ideality factor as well as the saturation current of the cell are calculated from the flash data, giving information about the linearity of the device.

Measurement of Temperature Coefficients

The temperature coefficients α and β, for I_{sc} and V_{oc} respectively, are established by repeating the flash of the Pasan LAPSS at different cell temperatures. In the measurement hardware set-up, the cell is installed in a temperature controlled cabinet with a glass door to permit the cell to be illuminated. For each individual cell, a correction factor for the irradiance attenuation due to the glass, is applied to the measured I_{sc} and V_{oc} data. As during the determination of α and β, full I-V curves are measured, a similar correction procedure allows the curve correction factor κ to be determined.

CH3365-4/94/0000-0863 $4.00 © 1994 IEEE   First WCPEC; Dec. 5-9, 1994; Hawaii

Spectral Response Measurement

The spectral responsitivity of all the Round Robin cells was measured to correct for spectral mismatch that occurs whenever calibration is performed under non-standard spectral irradiance [1]. At ESTI, a relative spectral response method is used in which the short-circuit current of the cell under test is compared with that produced by a calibrated reference cell, when both are illuminated by the same monochromatic light [5]. As light source, the Pasan flash light is used in combination with 17 interference bandpass filters (18 nm bandwidth) that are installed directly in front of the lamp. By changing the filters, 17 measurements are made in the range from 350 to 1100 nm, after which the response curve is obtained by interpolating the data. The spectral mismatch factor is then calculated in accordance with the ASTM standard test method for primary reference cell calibration [2].

Extrapolation Procedure for Non-STC Measurements

When indoor calibrations have not been made at Standard Test Conditions, which is usually the case with artificial light sources like solar simulators, the performance of the cell at STC is obtained by applying an extrapolation procedure. Current values are first adjusted by temperature correction term (α), followed by an irradiance correction term and finally corrected for spectral mismatch. Voltages are similarly adjusted for temperature (β), followed by a series resistance correction term, which accounts for the deviation in voltage drop that occurs due to a non-STC current-flow through the internal resistance of the cell, at non-STC irradiance levels. For this, the series resistance of the cells was determined by tracing the I-V curve at three different irradiance levels, as described by Wolf and Rauschenbach [3], at 25°C. Finally, the voltage is corrected by an irradiance correction term, which directly follows from the diode coefficient (determined in the flash-decay calibration). The described extrapolation procedure was used to determine the curve correction factor κ at Standard Test Conditions.

RESULTS

Current - Voltage Characteristics

Differences in the light pulse between the two Large Area Pulsed Solar Simulators are showed in Figures 1a and 1b. The I-V characteristic on the same sample using the two types of Simulators and two different types of electronic load curves is shown in Figure 2.

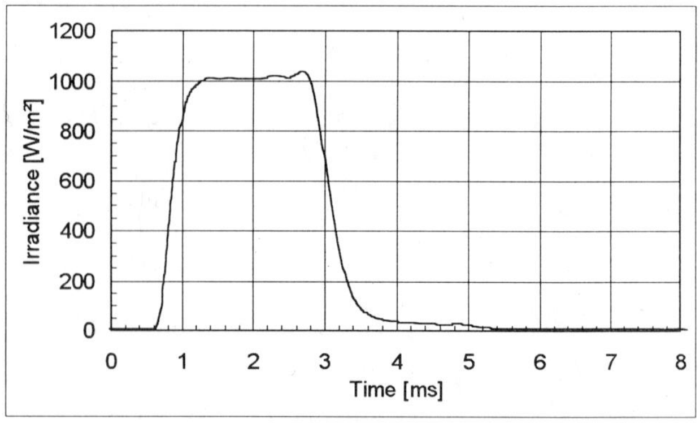

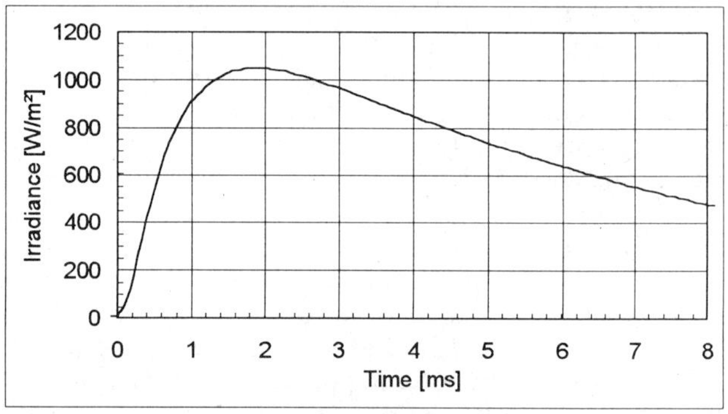

Figure 1a (top) and 1b (bottom) Light intensity curves from SpectroLab (Fig.1a) and Pasan (Fig.1b) Large Area Pulsed Solar Simulators

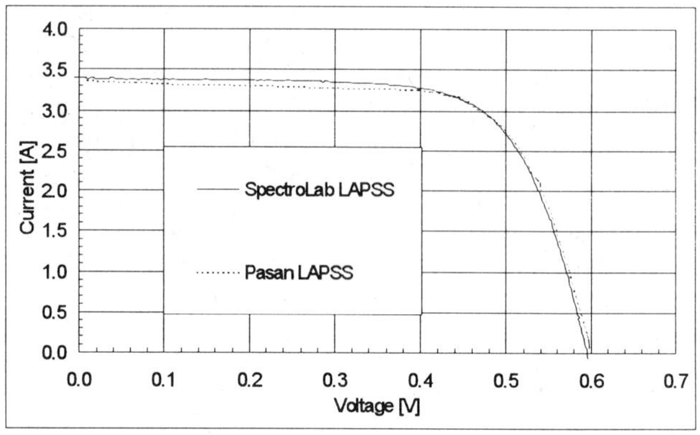

Figure 2 I-V Characteristic results with the two types of Simulators with their dedicated electronic loads.

Simulator Flash-Decay Calibration

With the present equipment two cells can be measured during the I_{sc} and V_{oc} Calibration procedure. One sample is shunted with a precision shunt (voltage drop across the sample less than 5% of the V_{oc} value) and the other sample is measured in V_{oc}. Using the same flash two measurements can be recorded at the same time. A typical measurement result of the raw data is displayed in Figure 3. The irradiance level is normalised to 1.

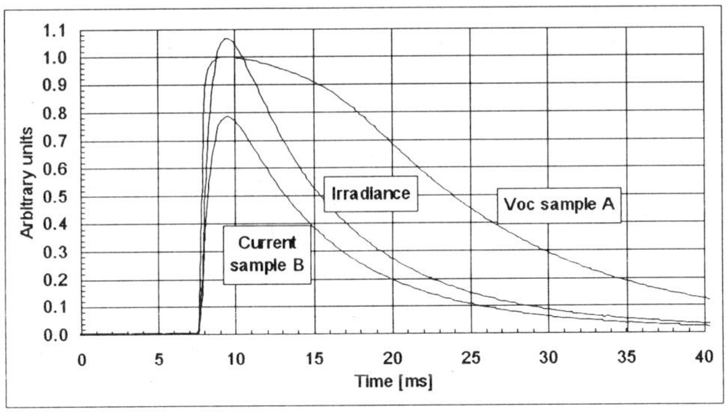

Figure 3 Raw data of two different samples connected respectively in V_{oc} and I_{sc} and exposed to the same light pulse.

From all the data triplets (Irradiance, Voltage and Current) V_{oc} and Isc vs. Irradiance were extracted. Figure 4a shows the result of V_{oc} vs. $\ln(Irr/Irr_{1000})$. The slope of each curve is the value of the diode ideality factor. Figure 4b shows the I_{sc} value vs. Irradiance. For clarity not all samples are displayed in this figure.

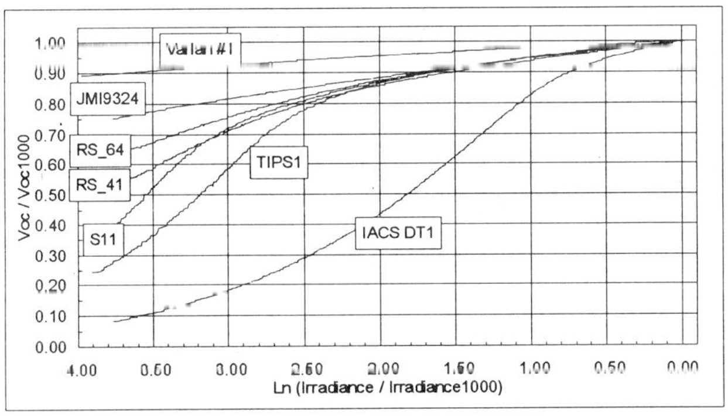

Figure 4a V_{oc} vs. $\ln(Irr/Irr_{1000})$ of different types of *Newer Technology* samples.

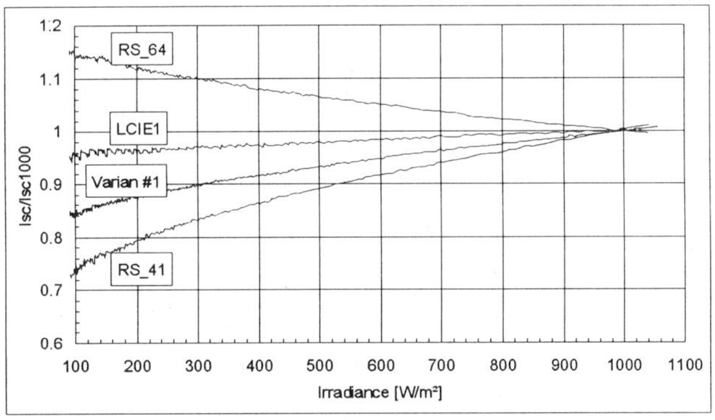

Figure 4b I_{sc} vs. Irradiance. No spectral mismatch correction applied .

Measurement of Temperature Coefficients

Figure 5 shows the temperature dependency of one CdTe sample of the NT series.

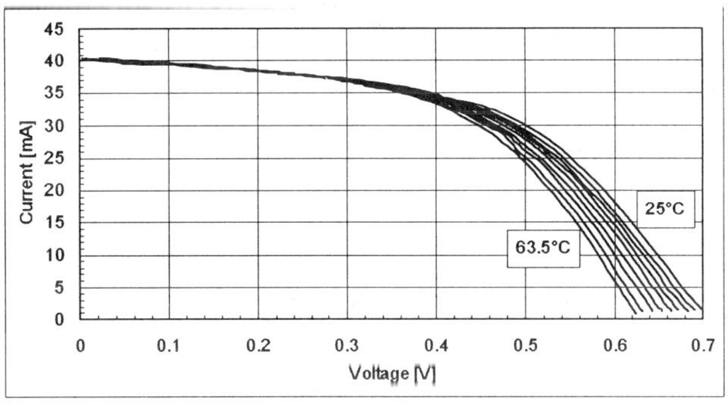

Figure 5 CdTe I-V curves measured at 800 $W.m^{-2}$ at temperatures: 25, 30, 35, 41.1, 45.3, 50.6, 55, 60 and 63.5 °C.

After the determination of α and β, and correcting all the I-V pairs to 25°C and 1000 $W\ m^{-2}$ (including the correction for the series resistance) the transposed curves did coincide in I_{sc} and V_{oc} but not in Fill factor. Figure 6 shows the variation of κ with temperature for a CdTe cell.

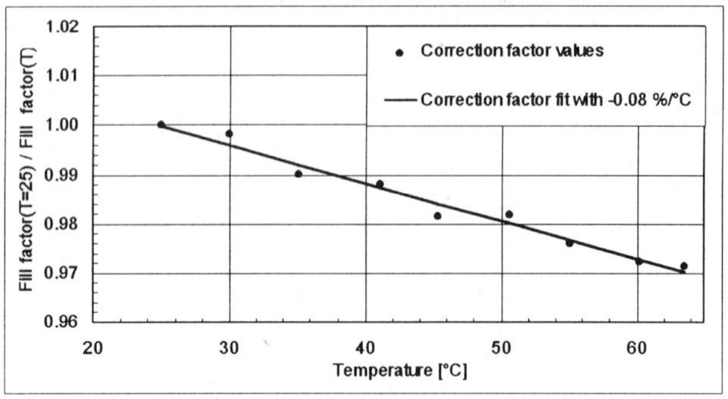

Figure 6 Correction factor vs. temperature for CdTe sample.

A linear decrease of κ with increasing temperature is observed. In contrast with CdTe, a positive temperature dependency of κ was observed for Si cells.

Spectral Response Measurement

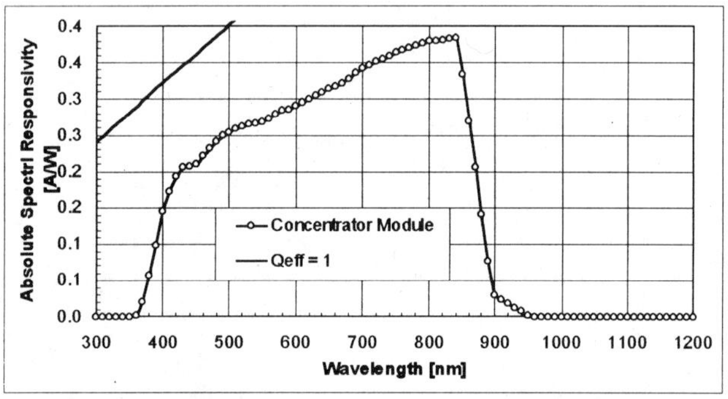

Figure 7 Spectral responsivity of the complete Varian Concentrator Module (cell and lens).

CONCLUSIONS

The results of the *Newer Technology* Samples were sent to NREL and are not reported here. All the current-voltage characteristics of the Round Robin cells were measured with two types of Large Area Pulsed Solar Simulators. In addition, I_{sc} and V_{oc} calibration was performed using the decay of a simulator flash, also giving information about the linearity of the device. Temperature coefficients as well as the curve correction κ of a cell were established by measuring whole I-V curves at different temperatures. Spectral Response was measured with the Pasan LAPSS [5]. On some samples the series resistance of the cells have been determined. During the measurements on the *Newer Technology* Cells difficulties arose in measuring the concentrator cells in front of our flash lamps. These cells were aligned with a small laser.

REFERENCES

[1] "Standard Procedures for Terrestrial Photovoltaic Measurements",European Commission, CEC 101, Issue 2, EUR 7078 EN, 1981.

2] "Standard Methods for Measuring the Spectral Response of Photovoltaic cells", American Society for Testing and Materials, ASTM Standard E 1021, 1984.

[3] "Series Resistance Effects on Solar Cell Measurements", Advanced Energy Conversion, Vol. 3, pp. 455-479, April-June 1963.

[4] Lowcost calibration method for photovoltaic (PV) solar devices. C. Helmke Theses July 1992.

[5] R. van Steenwinkel, *Measurements of Spectral Responsivities of cells and modules*; Proc. 7th European PVSEC Sevilla, Reidel (1987), **325**

WIDE-VIEW-ANGLE ABSOLUTE CAVITY RADIOMETER FOR SOLAR SIMULATOR USE

F. Nagamine, R. Shimokawa*, T. Abe, M. Suzuki, H. Ikeda** and Y. Miyake**
Japan Quality Assurance Organization(JQA),
Solar Cell Test Research Lab.
21-25, Kinuta 1-chome, Setagaya-ku, Tokyo 157, Japan
*Electrotechnical Laboratory(ETL)
1-4, Umezono 1-chome, Tsukuba-shi, Ibaraki-ken 305, Japan
**EKO Instruments Trading Co., Ltd. (EKO)
1-21, Hatagaya 1-chome, Shibuya-ku, Tokyo 151, Japan

ABSTRACT

We have developed a wide-view-angle absolute cavity radiometer which can be used under the light of solar simulator as well as the direct sunlight for the comparison between the Solar Simulator Method(based on the ETL spectral irradiance scale) and the Global Sunlight Method(based on the WRR sacle) for the calibration of reference solar cells.

Simultaneous measurements of direct sunlight were performed by this radiometer and a WRR absolute cavity radiometer PMO-6. Also, the irradiance measured by this radiometer was compared with the integrated irradiance by the spectroradiometer calibrated with ETL irradiance scale under the light of solar simulator.

Consequently, We found that the irradiance based on the WRR scale measured by the developed radiometer was about 1.0% lower than the integrated irradiance based on the ETL spectral irradiance scale.

INTRODUCTION

For the calibration of reference solar cells, Solar Simulator Method [1] based on ETL spectral irradiance scale [2] has been employed in JQA (former JMI), whilo WRR(World Radiometeric Reference)scale is required internationally by IEC for the measurement of irradiance onto the solar cells. Global Sunlight Method [3] has also been carried out by JQA based on WRR scale for the comparison with the solar simulator method.

This method requires a pyranometer [4] calibrated with WRR absolute cavity radiometer and a spectroradiometer for the spectral correction.

We compared the calibration values of the reference solar cells by the indoor calibration method(the Solar Simulator Method) and the outdoor calibration method(the Global Sunlight Method)by using the pyranometer, and found that the difference was about 1.5～2.0%. So, we have reached an idea that a more accurate WRR radiometer than the pyranometer should be developed and used for more accurate comparison and for the supply of reference solar cells based on the WRR scale with better than 1.0% accuracy.

We have finally developed a wide-view-angle absolute cavity radiometer which can be suitably used under the direct sunlight and the light of solar simulator. The uncertainty of this radiometer has been investigated and found to be better than 0.5%.

In this paper, the construction, principle, accuracy and the experimental results of the developed wide-view-angle absolute cavity radiometer are described.

EXPERIMENTAL

Figure 1 shows the construction of the developed absolute cavity radiometer ACR-1(EKO MS-220)having similar construction to currently available WRR absolute cavity radiometer, such as PMO-6. Two special features are introduced to receive light from wide view angle :
1)removable collimation tube ⑧ having a full view angle(or F.O.V.) of 5 degrees to measure direct sunlight is mounted on the front aperture ① having a full view angle of 40 degrees,
2)thermal guard ⑤ surrounding the cavity ④ is placed and its temperature is controlled to be the same with that of the cavity. This idea is introduced for the first time in the developed radiometer.

CH3365-4/94/0000-0867 $4.00 © 1994 IEEE

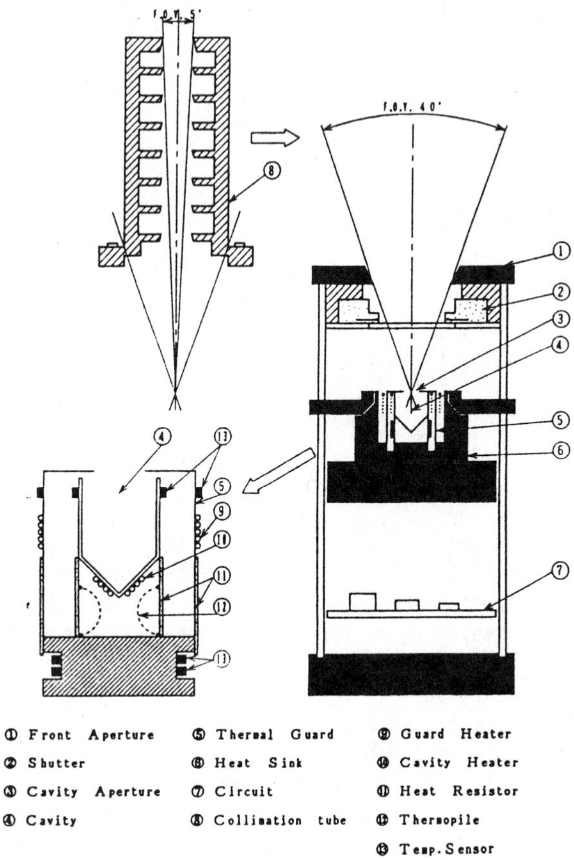

Fig. 1. Construction of Wide-View-Angle Absolute Cavity Radiometer ACR-1

As shown in Figure 2, the cavity ④ is heated up by the incidint light which raises the temperature difference between hot and cold junctions of the thermopile ⑫. The output of the thermopile is directly proportional to the incident irradiance. On the other hand, the temperature difference is also raised due to the supplied heating electric power. Accordingly, the correlation is given in equation(1).

$$I = \frac{1}{A\ \varepsilon}\ P\ \frac{\Delta T}{\Delta T'}\ C$$

where,

- I : irradiance of incident light
- A : aperture area
- ε : absorptivity
- P : electric power
- ΔT : temperature difference due to incident light
- $\Delta T'$: temperature difference due to electric power
- C : correction factor

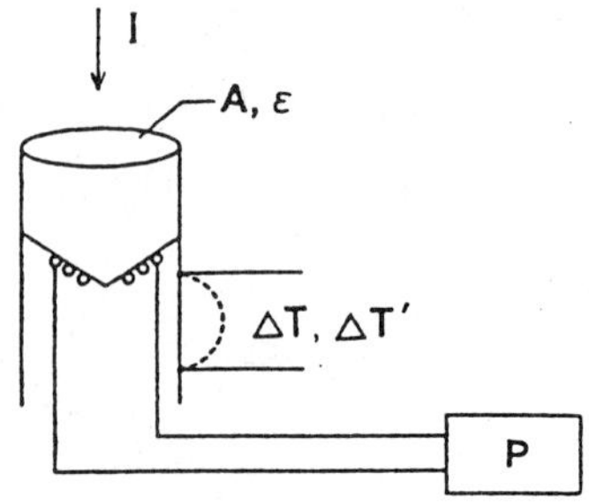

Fig. 2. Measurement Principle of ACR-1

The characterization which is important for the radiometer before the routine use was made with the procedures shown in the next section.

For the comparison of calibration values by the indoor and outdoor calibration methods, the procedures shown in Figure 3 are used. ACR-1 measures the irradiance of the direct sunlight with a full view angle of 5 degrees when the collimation tube is mounted, and this irradiance is compared with that measured by the WRR absolute cavity radiometer PMO-6. Also, ACR-1 measures the irradiance of the Xenon lamp solar simulator with a full view angle of 40 degrees when the collimation tube is removed, and the irradiance is compared with the integrated irradiance of the spectral irradiance within the wavelength region from 250 to 2600nm measured by the spectroradiometer calibrated with ETL spectral irradiance scale.

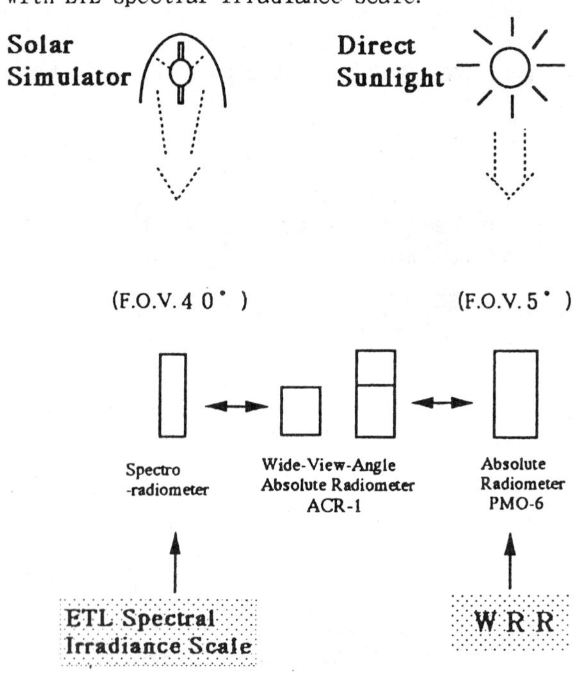

Fig. 3. Comparison Between ETL Spectral Irradiance Scale and WRR Scale by Using Wide-View-Angle Absolute Radiometer, ACR-1

CHARACTERIZATION of RADIOMETER ACR-1

It is necessary to determine the aperture area A, the absorptivity ε and the correction factor C shown in the equation (1) to measure the absolute irradiance by ACR-1. The following are the determining procedures and the results.

(a) Aperture area A

The diameter of the aperture area was measured in four portions and averaged by using a precise projector type scale and the area was calculated.

(b) Absorptivity ε

The cavity was irradiated with a laser beam of 632nm, and the absorptivity ε was determined by measuring the reflectivity.

(c) Correction factor C

The correction factor C is composed of several perameters. These were individually determined using the method developed by ourselves which partially included the method applied to PMO-6 [5].

We determined the cosine error Ea, scattering effect Ed, spectral transmittance Es, non-equivalency Eu and lead heating El with correction factors as shown in Table 1. As the total factor, 0.99929 was obtained for the direct sunlight measurement and also for solar simulator measurement.

Moreover, we investigated the uncertainties which were originated in the power measurement and in each characterization procedure in order to measure absolute irradiance. As the result, the total uncertainty was 0.495%. The details of characterization will be reported in future.

Table 1. The Values and Correction Factors of ACR-1

Item	Symbol	Values and C.Factors	
		D.Sunlight	S.Simulator
Absorptivity	ε	0.99917	0.9917
Area (m²)	A	5.0215×10^{-5}	5.0215×10^{-5}
cosine error	Ea	——	1.00100
scattering effect	Ed	0.9996	0.9986
spectral transmittance	Es	1.00119	1.00119
non-equivalency	Eu	0.9992	0.9992
lead heating	El	0.9993	0.9993
Total Factor	C	0.99929	0.99929

RESULTS and DISCUSSION

Figure 4 shows the results of simultaneous measurements of direct sunlight on a fairly good sunny day by ACR-1 with collimation tube mounted and by WRR absolute cavity radiometer PMO-6 having accuracy of 0.25%. The data by both radiometers seemed to be in good agreement. The deviation of irradiance measured by ACR-1 to PMO-6 was shown in Figure 5. The irradiance data measured by ACR-1 were 0.54% lower in average deviation than those by PMO-6 except for some points. We have repeated the measurement many times, and found that the irradiance measured by ACR-1 was about 0.5% lower than that by PMO-6.

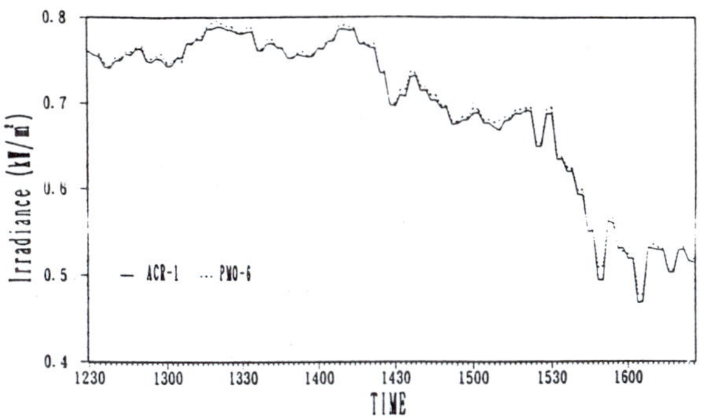

Fig. 4. Irradiance Data Measured at the Same Time by ACR-1 and PMO-6

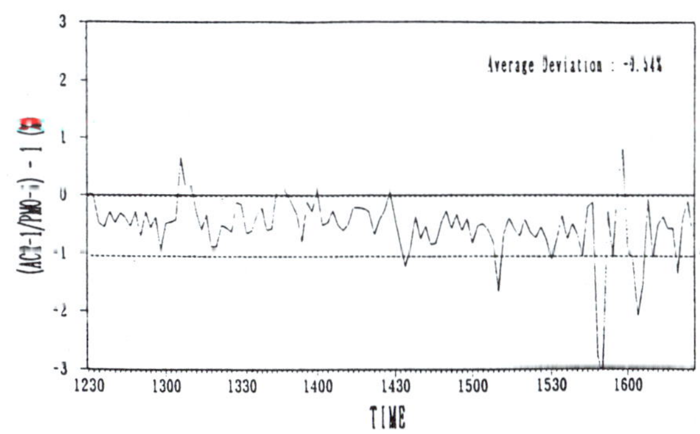

Fig. 5. Deviation of Irradiance Measured by ACR-1 to PMO-6

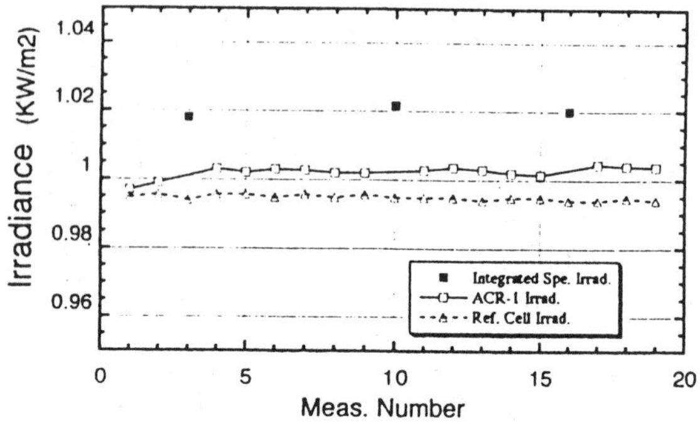

Fig. 6. Comparison of Irradiance by ACR-1 and Integrated Irradiance by Spectroradiometer

Figure 6 shows the comparison of irradiance data measured by ACR-1 without mounting the collimation tube and the integrated irradiance of the spectral irradiance from 250 to 2600nm measured by the spectroradiometer with ETL spectral irradiance scale under the stable light of the solar simulator. The irradiance data measured by a c-Si reference solar cell were also added in the figure so that the stability of the light of the solar simulator could be observed during the measurement period.

The first two points measured by ACR-1 seemed to be a little lower than the other points due to the initial warming-up stage of the radiometer. The irradiance measured by stabilized ACR-1 was approximately 1.5% lower than the integrated irradiance by the spectroradiometer. We have repeated the measurement and found that the difference between the irradiance measured by ACR-1 and the integrated irradiance was almost the same.

Judging from the results shown above, if the irradiance measured by ACR-1 is increased 0.5% so as to be in coincidence with the scale of PMO-6, the irradiance based on the WRR scale measured by ACR-1 was about 1.0% lower than the integrated irradiance by the spectroradiometer based on the ETL spectral irradiance scale.

The uncertainty of the calibration values of reference solar cells will be reported when further experimental works are completed.

CONCLUSION

We have developed a wide-view-angle absolute cavity radiometer whose uncertainty was found to be better than 0.5%. This radiometr ACR-1 has been compared with the WRR absolute cavity radiometer PMO-6 under the direct sunlight and found that the irradiance measured by ACR-1 was about 0.5% lower than that by PMO-6.
The irradiance measured by ACR-1 has also been compared with the integrated irradiance measured by the spectroradiometer calibrated with the ETL spectral irradiance scale under the light of solar simulator, and found that the irradiance measured by ACR-1 was about 1.5% lower than the integrated irradiance. Therefore, the irradiance based on the WRR scale measured by ACR-1 was about 1.0% lower than the integrated irradiance by the spectroradiometer based on the ETL spectral irradiance scale.
We will supply accurately calibrated reference solar cells based on WRR scale by using the developed wide-view-angle absolute cavity radiometer ACR-1 with further investigation.

ACKNOWLEDGEMENT

The present work was carried out under the sponsorship of the Agency of Industrial Science and Technology(AIST) of MITI through the Japan Electrical Manufacturers' Association(JEMA) with the kind advices from the Standardization Commitee for New Type Solar Cells organized by Optoelectronic Industry and Technology Development Association(OITDA).
Authors thank to Mr. I. Saito of ETL for his kind arrangement of calibrating the standard spectral irradiance lamp.

REFERENCE

[1]R. Shimokawa, F. Nagamine et al., "Japanese Indoor Calibration Method for the Reference Solar Cell and Comparison with the Outdoor Calibration", Japanese J. of Appl. Phys. 26 (1987), pp. 86-91.
[2]Walker J. H. et. al., "Results of a CCPR Intercomparison of Spectral Irradiance Measurements by National Laboratories" J. Res. Natl. Inst. Stand. Technol. 96 (1991) 647.
[3]CEI/IEC Publication 904-2 (1989), "Photovoltaic Devices-Part 2:Requirements for Reference Solar Cells"
[4]F. Nagamine, R. Shimokawa et al., "Calibration of Pyranometers for the Photovoltaic Device Field", Japanese J. of Appl. Phys. 29 (1990), pp. 516-521.
[5]R. W. Brusa and C. Frohlich, "The PMO-6 Radiometer and its Characterization", Pub. No. 563A, Sept. 1980, World Radiation Centre.

ULTIMATE TRANSISTOR ELECTRONIC LOAD FOR ELECTRICAL PERFORMANCE MEASUREMENT OF PHOTOVOLTAIC DEVICES USING PULSED SOLAR SIMULATORS

E.G. Mantingh*, W. Zaaiman, H.A.Ossenbrink
European Commission DG XII, Joint Research Centre - Ispra, Italy
European Solar Test Installation (ESTI)

* Grantholder Enschede Polytechnic, Enschede, The Netherlands.

ABSTRACT

This paper presents the development of the Ultimate Electronic Load (UTEL). The load is equipped with an active forward coupling and is adjustable for every single-pulse solar simulator .

INTRODUCTION

The manufacture of Photovoltaic modules requires electrical performance measurement of the finished product at Standard Test Conditions (25ºC, 1000 W/m^2) [2]. With certain types of devices, the electronic loads can function incorrectly. The active feedback of the SpectroLab LAPSS load can cause oscillations with high-current and low voltage devices (e.g. high efficiency cells). Inappropriate setting of the current sink in the Pasan LAPSS load can result in a poor distribution of measurement points along the I-V curve.
The UTEL was developed to measure solar cells and modules up to 200 W. It has an active forward coupling, to avoid oscillations, and provide a good distribution of measurement points. Because of these advantages the UTEL has a better repeatability for the present range of PV-Device technology. Due to a unique trigger and delay circuit, the UTEL is also capable of measuring IV-curves at low irradiance, and is adjustable for every single-pulse solar simulator.

OBJECTIVE

The development of an electronic load without active feedback and capable of measuring Photovoltaic devices (cells and modules) on both Large Area Pulsed Solar Simulators. The principle of operation [1] of the load is shown in Figure 1. Forward coupling is achieved by regulating the collector-base voltage.

DESCRIPTION OF INSTRUMENT

The electronic load part (see Figure 1) consists of a power transistor connected in series with the Photovoltaic Device and a battery. An adjustable voltage source, U_{shift}, is introduced between the base and collector of the transistor.

The circuit exploits the I_e (U_{be}) characteristic of the transistor as an electronic load. The I_e (U_{be}) characteristic is determined by I_b (U_{be}) characteristic. The emitter current, I_e , is the sum of the base and collector current.

$$I_e = I_b + I_c = (1 + \beta)\ I_b \qquad (1)$$

where I_e is the current generated by the illuminated PV device.

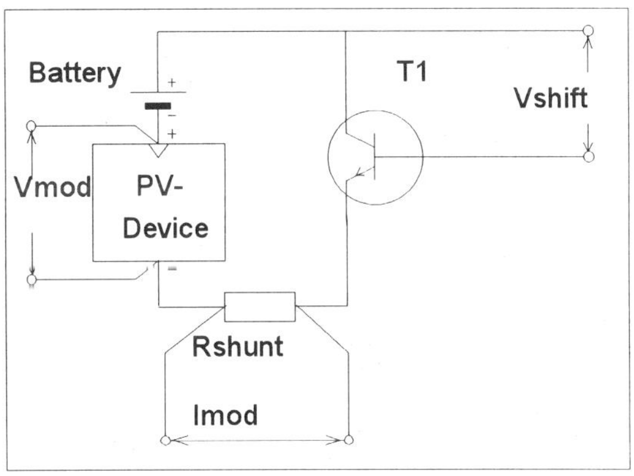

Figure 1 Load principle of the UTEL

The I_e (U_{be}) characteristic is an exponential characteristic, and it is shifted across the photovoltaic output characteristic by means of the adjustable voltage source U_{shift} according to

$$U_{pv} = U_{be} - U_{shift} \qquad (2)$$

where U_{pv} is the voltage across the solar device.
The maximum current which the voltage source has to supply, is the base current I_b. The operating point of the solar cell is given by the point of intersection of the load characteristic and the photovoltaic output characteristic.
Figure 2 reveals that the angle of intersection between the exponential load characteristic and the photovoltaic output characteristic is large enough to ensure ripple-

 First WCPEC; Dec. 5–9, 1994; Hawaii

insensitive measurement of the photovoltaic output characteristic close to both short-circuit point, and open-circuit point. To ensure a near-optimal angle of intersection at the short-circuit point the load transistor should show high conductance in the I_e (U_{be}) characteristic for high currents.

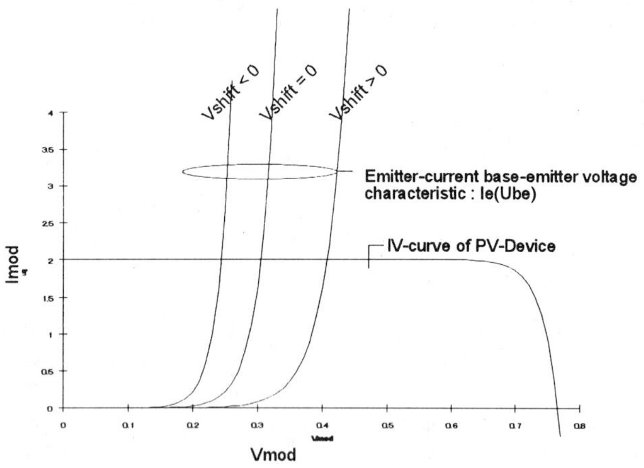

Figure 2 Shifting the I-V characteristic of the power transistor by changing the U_{shift} voltage

In Figure 3 the electronic circuit of the UTEL is shown. The transistor T1, a power Darlington transistor, is controlled by a forward coupling, by a capacitor across base and collector. The capacitor generates a voltage ramp dependent on its capacity, the short circuit current, and the open circuit voltage of the photovoltaic device. The capacitor must be changed for different PV devices. At the beginning of the measurement, transistor T2 is open so that power transistor T1 keeps the PV device in I_{sc}. If the flash occurs, the trigger unit closes T2, and the capacitor is charged by the current through the base-emitter of T1. Because of this base-emitter current a collector-emitter current will flow. While the capacitor is charged the module current will decrease, because with increase of the capacitor voltage T1 passes less current. The I-V measurement finishes when the capacitor is fully charged to that voltage at which no more current can flow. The capacitor is discharged through T3 in series with a resistor. This discharge can be realised at any time before the next measurement. This V_{oc} status of the device allows the measurement of V_{oc} versus irradiance during the decay of the flash (Pasan LAPSS). A battery is inserted to compensate the cable losses and the voltage drop over the transistor (U_{ce}) and the shunt. If the voltage of the battery is high enough, the starting point of the IV-curve will be shifted over the U_{pv} = 0 axis, and no negative U_{shift} has to be used.

The forward coupling in the measurement system effectively eliminates any possible oscillations. A trigger signal, made from a small 9 cell PV module, starts the delay generator. The delay can be adjusted to any desired irradiance level during the flash. This feature allows I-V characteristic measurements at low irradiances.

RESULTS

The UTEL is constructed in a 19" rack subdivided into several units (see Photo 1). The electrical connections between the units are located on the back side and care has been taken to avoid long connections in the power loop between the different units (Input connector, battery, power transistor and shunt). Two small power supplies are used; one for the electronic delay generator and one to charge the NiCd battery.

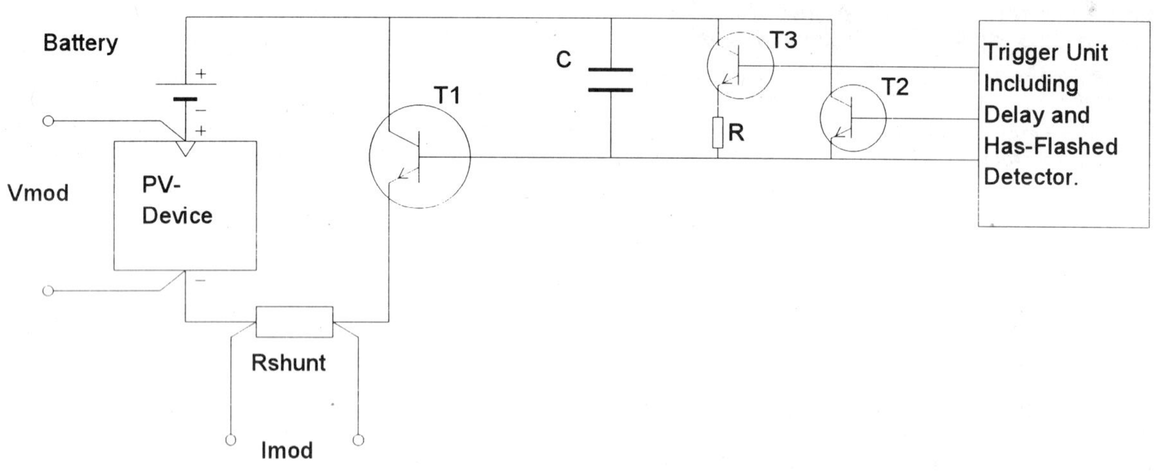

Figure 3 Electronic schematic of the UTEL load

This battery is charged whenever the UTEL is connected to the mains but as soon as the unit is switched on the battery charging power supply is disconnected. The power rating for the present load is limited to 200 W with a maximum voltage of 120 V and a maximum current of 20A. These ranges should be wide enough to cover the present PV module range. Figures 4a,b and 5a,b show the improvements of the measured data between the factory loads (SpectroLab and Pasan) and the UTEL.

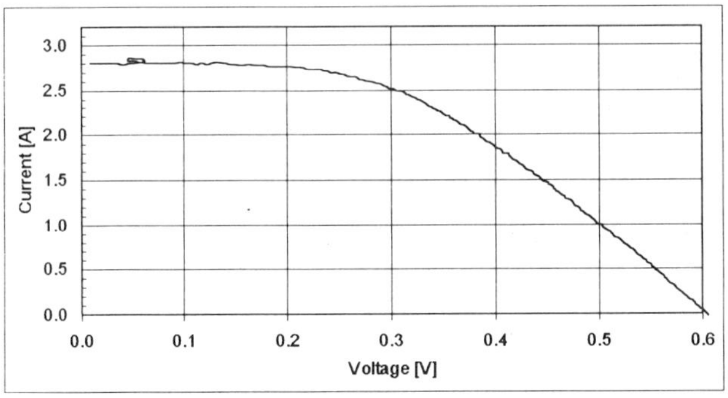

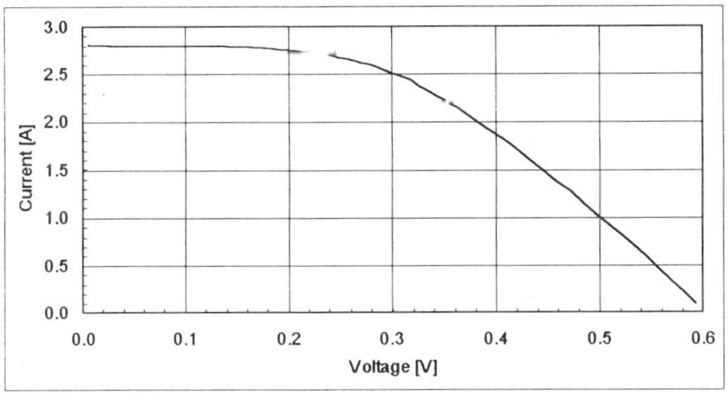

Figure 4a (top), 4b (bottom) I-V curves measured with respectively with the SpectroLab load (**4a**) and UTEL (**4b**)

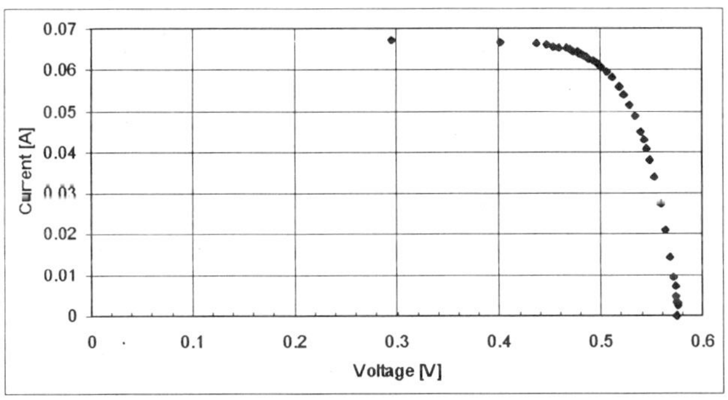

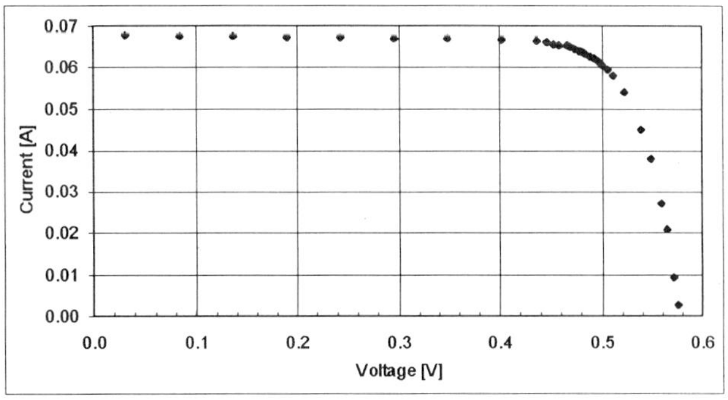

Figure 5a (bottom left), 5b (top) I-V curves measured with respectively with the Pasan load (**5a**) and UTEL (**5b**)

CONCLUSIONS

From measurements during the PEP93 Round Robin on various cells and calibration measurements on Reference modules at ESTI, this electronic load works satisfactorily. Problems such as oscillations (SpectroLab load) or a bad point distributions (Pasan load) were not found using this new electronic load, which results in a better reproducibility of I-V characteristic measurements at ESTI

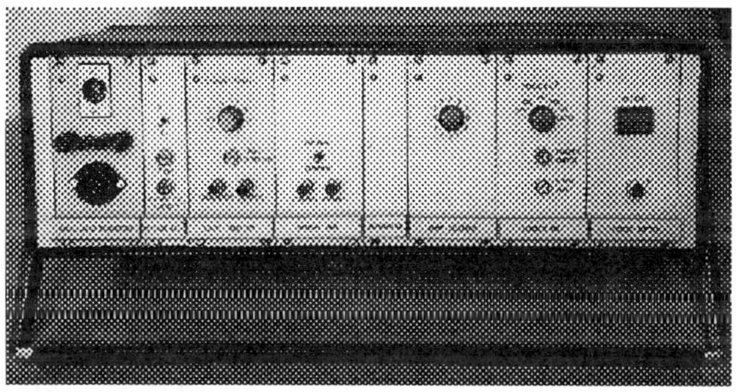

Photo 1: The electronic load UTEL built at ESTI.

REFERENCES

[1] Böhm et al., *Measurement system for large-area solar cells*; Proc. 7th European PVSEC Sevilla, Reidel (1987)

[2] IEC Publication 904-1,1987, *Measurement of Photovoltaic Current-Voltage Characteristics.*

SPECTRAL RESPONSE CALIBRATION FACILITY FOR PHOTOVOLTAIC CELLS

R. Budde*, W. Zaaiman, H.A. Ossenbrink
European Commission DG XII, Joint Research Centre - Ispra, Italy
European Solar Test Installation (ESTI)

* Grantholder Enschede Polytechnic, Enschede, The Netherlands

ABSTRACT

This paper describes a secondary calibration for determining the absolute spectral responsivity of photovoltaic cells. The test facility uses the reference cell method for determination of the relative spectral responsivity. Using the known absolute spectral responsivity data of the reference cell, the absolute spectral responsivity of the test cell is calculated. The whole measurement is controlled by a PC.

INTRODUCTION

The spectral responsivity of a photovoltaic cell is required to interpret laboratory measurements on devices and is used for technical calculations. The knowledge of absolute spectral responsivity allows the precise calibration of the short circuit current. Present, the spectral responsivity of photovoltaic devices is measured at ESTI using the Pasan Large Area Pulsed Solar Simulator (LAPSS). This equipment was developed to perform measurements on solar modules having dimensions up to 1.5 x 1.7 m² [1]. To perform high precision measurements on small photovoltaic cells, of approximately 2 x 2 cm², it is more convenient to use a facility optimized for small-area cells.

OBJECTIVE

The main reason for developing this facility was to determine the spectral responsivity with better wavelength resolution than currently measured with the Pasan LAPSS. The bandwidth of the monochromatic light source should not exceed 20 nm for an absolute spectral responsivity measurement to avoid errors in the short and long wavelength conversion limits of the cell. Care must be taken to ensure that scattered light or higher order light effects are negligible. The light source should be capable of providing a spatial uniformity of ± 2.5% over the area of the test plane [2].

MAIN DESIGN

To illuminate two cells on the test plane simultaneously, a high intensity light source consisting of a 1500 Ws (electrical energy) flash lamp mounted on top of a cabinet, is used. The blackened interior of the cabinet is fitted with two diaphragms to reduce scattered light. By these means, the influence of any background light on the cells inside the cabin is eliminated (see Figure 1).

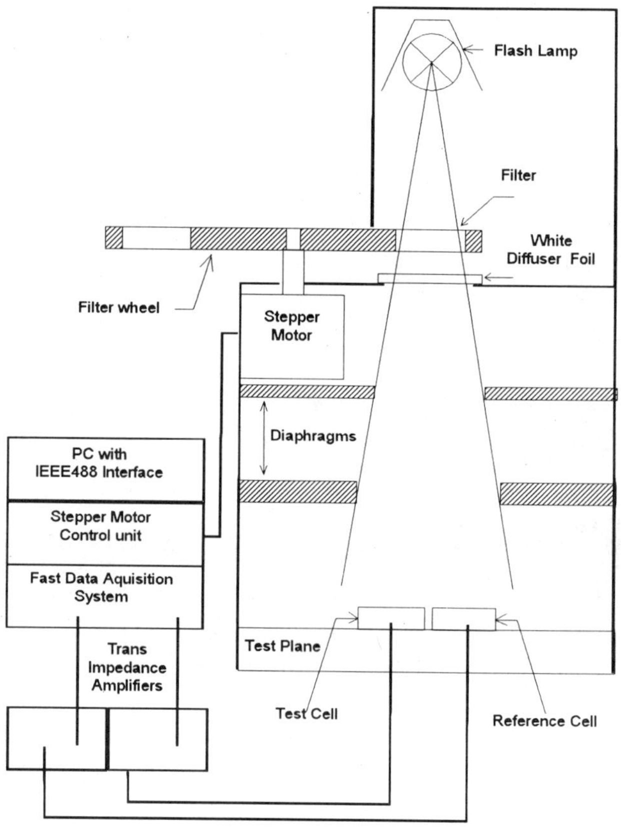

Figure 1. Hardware setup of the facility

A white diffuser foil is placed below the filter wheel to increase the uniformity in the test plane. The measurement of the spectral responsivity requires illumination at different wavelengths. Therefore a filter wheel driven by a stepper motor is located illuminated by the flash lamp. The filter wheel contains 16 filters with different transmission wavelengths in 100 nm

CH3365-4/94/0000-0874 $4.00 © 1994 IEEE

First WCPEC; Dec. 5-9, 1994; Hawaii

increments. A better wavelength resolution can be achieved by using a second filter wheel having intermediate transmission wavelengths.
A digital sampling two channel high speed voltmeter reads out voltages proportional to the short circuit currents of the two illuminated cells. A PC controls the filter wheel positioning, executes the measurement procedure and calculates the results.

MEASUREMENT RESULTS

The spectral transmission of the white diffusing foil and the filters was measured with a PERKIN-ELMER Lambda 9 UV/VIS/NIR Spectrophotometer. The maximum acceptable decrease relative to the maximum irradiance is, within the measurement plane, 2.5% [1]. The uniformity distribution was first measured on the test plane without white diffuser foil (see Figure 2.).

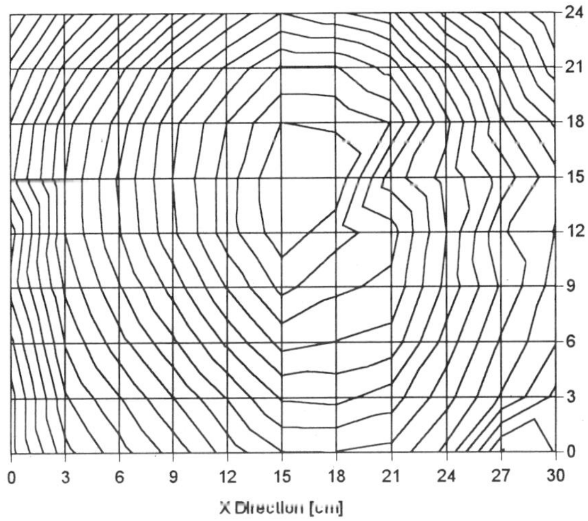

Figure 2 Uniformity at the test plane without diffuser foil. Each line describes a 2.5% decrease in irradiance.

The resulting test area was approximately 15 cm² and too small to perform a spectral responsivity measurement on two reference cells. A white diffuser foil was inserted (see Figure 1). The spectral transmission of the foil (see Figure 3) extends the useful calibration area on the test plane (see Figure 4) but decreases the intensity especially at wavelengths below 400 nm. This decrease of spectral intensity at lower wavelengths due to the transmission of the foil, the spectral intensity of the flash lamp used and the spectral responsivity of photovoltaic devices, increases measurement errors as the generated current from both photovoltaic devices drops below a certain limit.

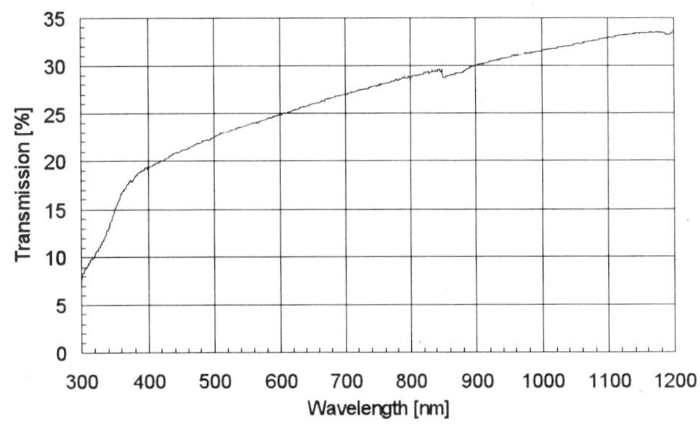

Figure 3 Spectral Transmission of the white diffuser foil.

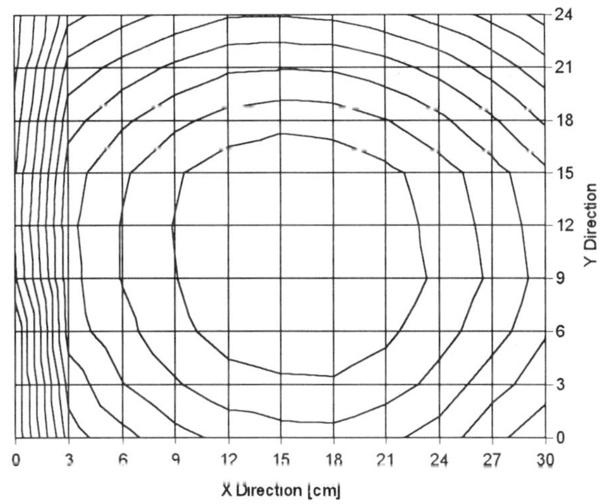

Figure 4 Uniformity at the test plane with diffuser foil. Each line describes a 2.5% decrease in irradiance.

The subsequent calculation of the absolute spectral responsivity is done according to the Differential Spectral Responsivity Method without linearity check [3]. The ratio of the current signals is calculated over a period of 2 ms after the flash pulse had its peak intensity. Multiplying the ratio values with the absolute spectral responsivity of the reference device gives the spectral responsivity of the device under test.
Figure 5 shows the absolute spectral responsivity of a test cell and its factory values using this facility.

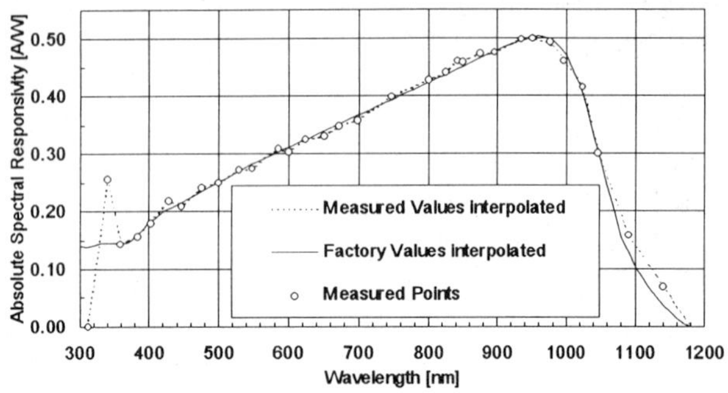

Figure 5 Absolute spectral responsivity of a test cell compared with its factory values.

CONCLUSION

The area, in which the reference cell and the test cell are illuminated with a uniformity within 2.5%, is 100 cm². The calibration facility runs automatically (besides changing between filter wheel one and two) and will in future be used as the preferred precision cell spectral responsivity facility. Measurements in the range below 384 nm are not reliable. However these errors contribute less than 0.5% to the total current at AM1.5 Global (see Figure 6).

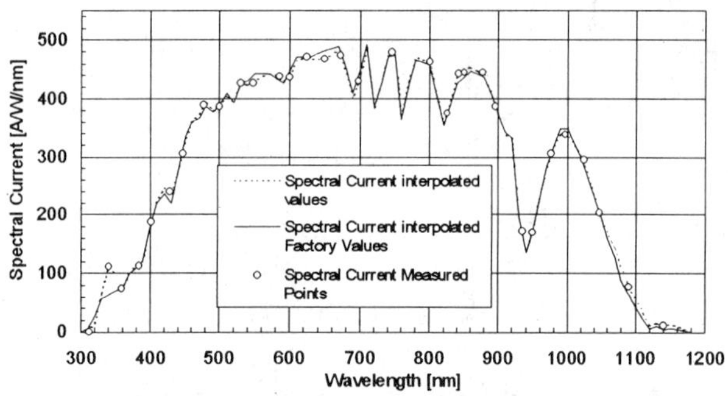

Figure 6 Spectral current comparison between measured and factory values.

The facility does not yet support the use of bias light. This option could later be added to illuminate the photovoltaic cells at irradiance approximately equal to normal end use operating conditions for the cells. Major recommendations to improve the facility, in particular at wavelengths below 384 nm, are to apply a diffuser with higher irradiance transmission, and a data acquisition system with a better resolution at input voltages below 50 mV (one of the Transient Recorders used by ESTI).

REFERENCES

[1] R. van Steenwinkel, *Measurements of Spectral Responsivities of cells and modules*; Proc. 7th European PVSEC Sevilla, Reidel (1987), **325**

[2] ASTM Designation: E1021-84, *Standard Methods for Measuring the Spectral Response of Photovoltaic Cells*

[3] PTB Bericht: *The Result of the PEP87 Round Robin Calibration*, PTB-Opt-31, ISBN 3-89429-067-6

SPECTRACUBE: A FAST SPECTRORADIOMETER FOR PULSED SIMULATOR SPECTRAL MISMATCH CORRECTIONS

W.J. Zaaiman, H. A. Ossenbrink, M. Mertens
European Commission DG XII, Joint Research Centre - Ispra, Italy
European Solar Test Installation (ESTI)

ABSTRACT

This paper describes the principle of the SpectraCube. The instrument consists of an array of independent detector modules. Each module has an optical band pass filter, apertures and a silicon photodiode as detector. During outdoor calibration at ESTI, a commercial spectroradiometer is used to determine the mismatch correction factor but no instrument is available to measure the spectral distribution of the laboratory Large Area Pulsed Solar Simulators as the duration of the light pulse (2 ms) is by far to short for commercial spectroradiometers.

INTRODUCTION

The instrument developed at ESTI and here described consists of an array of 64 independent detector modules. Each module has an optical band pass filter (bandwidth of 10 to 100 nm depending on the filter wavelength), separate apertures and a silicon photodiode as detector. The analogue signal from each of the 64 detectors is fed into two fast data acquisition PC boards. As the two PC boards are operating simultaneously, 2*32 channels are scanned with 200 kS/s, resulting in a full scan of 160 μs. At this scan rate 12 spectral intensity curves can be stored during the 2 ms duration of the light pulse of our Pulsed Solar Simulators.

OBJECTIVE

Spectral radiometers measure the spectral irradiance of a light source. ESTI needs these instruments for the calibration of photovoltaic devices. All data measured have to refer to a particular, standardised spectral irradiance distribution (Air Mass 1.5 Global, IEC 904-3 [1]), which cannot be reproduced well enough by artificial light. One can correct for Air Mass 1.5 Global as long as the spectral response of the device under test an its reference as well the actual spectral irradiance are known [3]. If during the electrical performance measurements the spectral irradiance is known one could directly correct to Air Mass 1.5 Global.

MECHANICAL PART

Each detector module is made from black anodised Aluminium (see figure 1) and consists of 5 sections separated by apertures. The aperture (opening) angle of the detector system is 2.98° and the slope angle is 1.04°. The aperture angle is determined by the ratio of distance between receiver and field stop to radius of receiver. This ratio should be greater than 15 which imply an opening angle of <= 4°. The slope angle should have an angle between 1 and 2°. These are the recommendations of the International Radiation Commission [2].

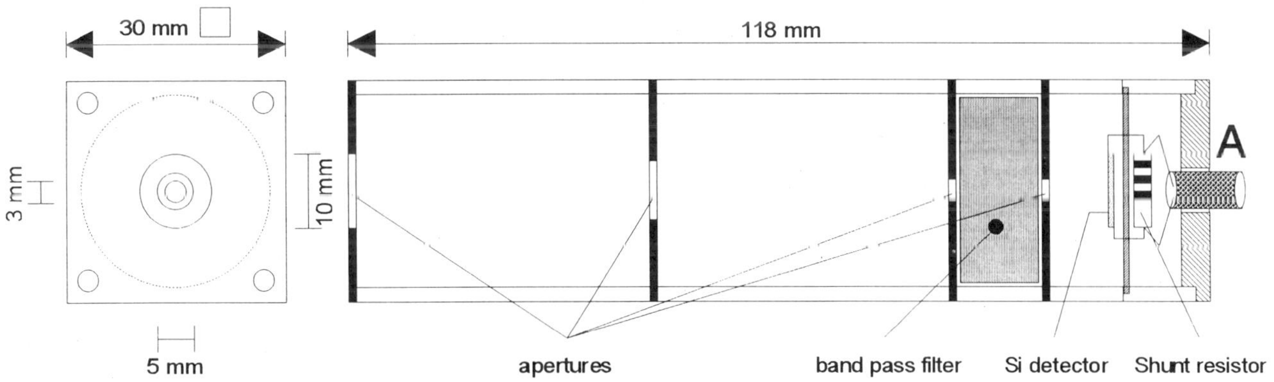

Figure 1: Mechanical construction of one detector module. From point A the irradiance signal is fed to one of the 64 inputs connectors of the A/D boards.

CH3365-4/94/0000-0877 $4.00 © 1994 IEEE

First WCPEC; Dec. 5–9, 1994; Hawaii

The present detector system covers the 400 to 1100 nm range in steps of 10..15 nm. Detector modules covering the 214 to 400 nm range (10..15nm steps); the 1100 to 1200 nm range (50 nm steps) and the 1200 to 2500 nm range (100 nm steps) are under construction. This wide range of detectors allows selective detector systems to be matched to ranges of interest.

On each detector system the wavelength of the incorporated filter is stated. One detector system weights 160 gram including the bandpass filter.

ELECTRICAL PART

The detectors are optimised for the wavelength of the bandpass filters, but are for the majority off-the-shelf silicon photodiodes. The Si photodiode operates in current mode. The shunt is matched to each detector to deliver between 80 and 95 mV when illuminated by one of our Pulsed Solar Simulators. In contrast to the voltage operating mode, the signal is much smaller but almost linear over a wide range of photon flux. With this *normalised* input voltage level, the two 12 bit A/D boards can operate with a fixed gain and reach the desired speed of 200 kSamples/s.

RESULTS

The calibration of the fully assembled SpectraCube is done outdoor against a commercial spectroradiometer (GER) or for separate detector modules a mechanical support was made to allow indoor calibration against a Standard Lamp. During the measurements with the Pulsed Solar Simulators the system is triggered by a *Has Flashed Cell*. From that moment both boards are starting to record the data until their internal memory is filled. The internal memory of each board can contain 512 data points (this are 16 scans of 32 channels). During the indoor flash measurements both boards operates at their full speed (200 kS/s) without wait periods between the scans. For outdoor spectral measurements a repetition rate can be set to perform periodical scans. This feature allows unattended operation if the SpectraCube system is mounted on a tracker. The software controlling the two boards is written in Visual Basic for Windows. A typical elaborated result of the captured data is shown in figure 2. Variations in intensity can be observed at 583 nm and in the range between 680 and 750 nm. There the intensity increases during the duration of the flash pulse. Spectral Mismatch can now be calculated in coincidence with the data points measured at I_{sc}.

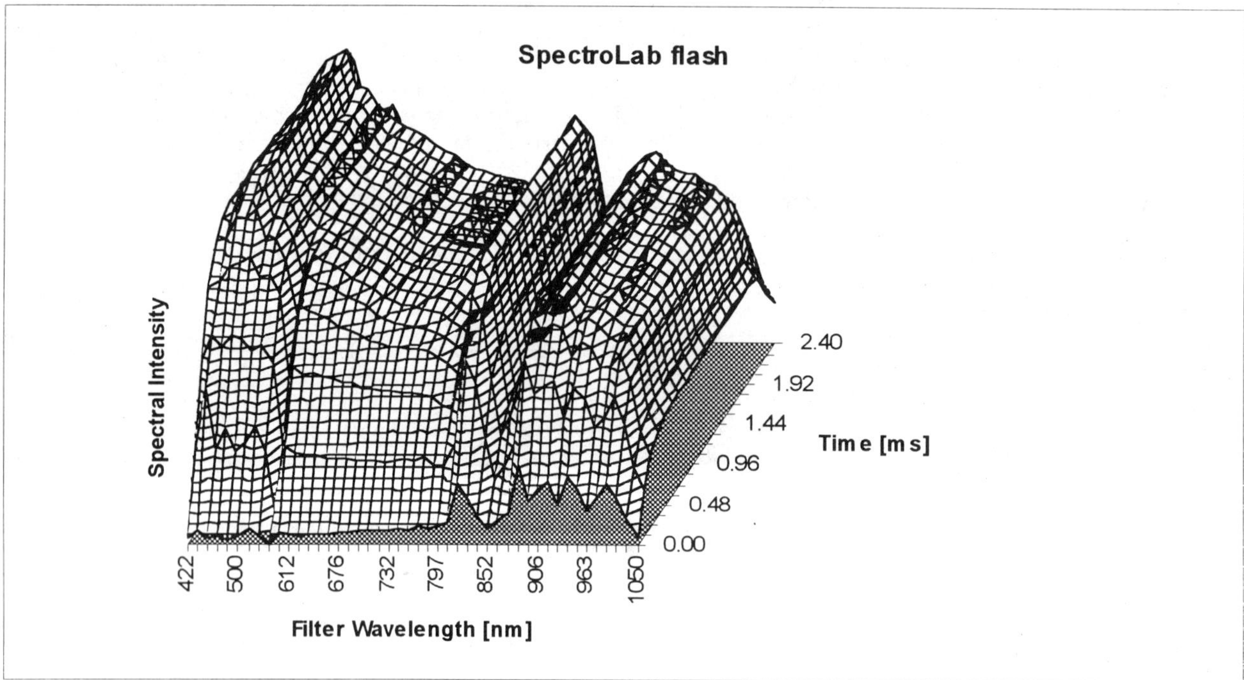

Figure 2: Spectral intensity variation during the flash period of the SpectroLAb LAPSS.

CONCLUSION

This new type of fast spectroradiometer allows real-time correction to Air Mass 1.5 Global during the electrical characterisation of photovoltaic devices (if the spectral responsivity of the device under test and its reference is known [3]). During the I-V measurement a permanent control of the flash-bulb degradation is possible, which improves the time of usage of the flash-bulb. Due to the modularity of the detector modules the system can be adopted for other ranges of interest. The instrument is sufficiently compact to be portable.

REFERENCES

[1] IEC Publication 904-3, *Measurement principles for terrestrial photovoltaic (PV) solar devices with reference spectral irradiance data*; CEI/IEC 904-3: 1989, Genève, Suisse

[2] Kinsell L. Coulson, *Solar and Terrestrial Radiation*, ACADEMIC PRESS 1975

[3]] ASTM 973-83, *Standard Practise for the Determination of the Spectral Mismatch Parameter Between a Photovoltaic Device and a Photovoltaic Reference Cell.*

Qualification Testing of Thin-Film and Crystalline Photovoltaic Modules

Joseph Burdick, Jim Pruett, and Elvira Beck

National Renewable Energy Laboratory
Golden, Colorado USA

ABSTRACT

The results of the first complete module qualification testing performed on the photovoltaic (PV) modules of several different manufacturers and a variety of technologies are presented here. Both thin-film and crystalline PV modules are being evaluated. The National Renewable Energy Laboratory (NREL) now has the capability to perform the tests in the proposed IEEE PV module qualification test sequence. Each test performed will be discussed. In addition, we have developed new test methods and improved techniques as well as engineering solutions to problems encountered with some of the present test procedures. This work not only demonstrates NREL's continuing committment to work and interact with PV manufacturers to improve their modules from both a performance and a reliability standpoint, but also shows the ongoing effort to validate ASTM test methods and IEEE qualification testing procedures. Finally, this research establishes a basis from which further studies of the correlation between indoor, accelerated testing and outdoor, real-world exposure can be made, with the hope of eventually predicting long-term PV module service lifetimes in the field.

INTRODUCTION

The need for continued improvement in both the performance and long-term reliability of photovoltaic (PV) modules for use in field-installed applications is self-evident. Along this line, then, the appropriate use of various test methods and procedures, as well as the development of new testing techniques, is essential in order to improve PV modules, both as "virgin" products directly from the manufacturing line, and as samples that have been exposed to outdoor conditions for many years. The primary series of screening tests employed to do just this are called "module qualification tests": the purpose of these tests is to help determine whether the module *design*, as fabricated -- i.e., the PV material, the interconnects and the environmental packaging -- will exhibit good long-term performance and reliability in the field. There have been many years of history establishing the testing sequence. Starting with the JPL Block V [1], the UL Standard 1703 [2], and the SERI Interim Qualification Tests and Procedures [3], this work continues with the proposed PV module qualification test sequence currently under development in the IEEE standards committee [4]. In the work presented here, modules from several different PV manufacturers and/or technologies are being tested and evaluated by running them through the complete Module Qualification Test (MQT) sequence. Batch #1 (currently underway) involves commercial and prototype modules from eight PV manufacturers; Batch #2 involves testing modules from five additional manufacturers. The data are shared with the individual manufacturers prior to publication or disclosure due to the confidential nature of the testing and results.

DISCUSSION AND RESULTS

The PV Module Qualification Test sequence under proposal and evaluation in the IEEE standards committee [4] is shown in Fig. 1. Each test performed at NREL, along with some of the associated results, is discussed below. We have not only developed or improved some of the test methods, but we have also worked to obtain engineering solutions to issues or problems encountered with some of the current test procedures. Our effort includes steps to comply with environmental, safety, and health (ES&H) policies, where these have become necessary. Examples of these improvements and how they have been implemented at NREL are presented below in the sections that detail the individual tests performed.

The various photovoltaic modules being tested in the MQT program are shown in Table I. These modules include off-the-shelf commercial PV modules as well as state-of-the-art R&D prototypes and deliverables for NREL contracts. Because some of the data are proprietary, the manufacturers' identities must necessarily remain confidential. Some of the manufacturers have not yet supplied NREL with modules for the MQT, while others have sent only a few modules for outdoor deployment at the present time. The "standard" MQT procedure, described below, requires a minimum of nine modules, eight of which go through the various test sequences (A-D), with the ninth module held as a control.

CH3365-4/94/0000-0880 $4.00 © 1994 IEEE First WCPEC; Dec. 5–9, 1994; Hawaii

Initial Tests and Inspections

Visual Inspection / Electrical Performance
Ground Continuity
Electrical Isolation (Dry Hi-Pot)
Wet Insulation-Resistance (Wet Megger)
Electrical Isolation (Wet Hi-Pot)
Visual Inspection / Electrical Performance

Sequence A	**Sequence B**	**Sequence C**	**Sequence D**
Thermal Cycling (200 Cycles)	UV/Thermal	Damp Heat (1000 Hrs)	Outdoor Exposure
Vis. Insp. / Elec. Perf.	*Vis. Insp. / Elec. Perf.*	Elec. Isolation (Dry Hi-Pot)	*Vis. Insp. / Elec. Perf.*
Surface Cut	Thermal Cycling (50 Cycles)	*Vis.Insp. / Elec. Perf.*	Non-Intrusive Hot Spot
	Vis. Insp. / Elec. Perf.	Hail Impact	*Vis. Insp. / Elec. Perf.*
	Humidity-Freeze (10 Cycles)	Mech. Loading (Static)	Bypass-Diode Thermal
	Elec. Isolation (Dry Hi-Pot)		
	Vis. Insp. / Elec. Perf.		
	Termination Robustness		
	Twist / Mech. Loading (Dynamic)		
	Vis. Insp. / Elec. Perf.		
	Surface-Cut Susceptibility		

Final Tests and Inspections

Visual Inspection / Electrical Performance
Electrical Isolation (Dry Hi-Pot)
Wet Insulation-Resistance (Wet Megger)
Electrical Isolation (Wet Hi-Pot)
Ground Continuity
Visual Inspection / Electrical Performance

Fig. 1. Photovoltaic module qualification test sequence (proposed, IEEE [4])

Table I PV modules undergoing NREL's module qualification test (MQT) sequence

Manufacturer (Batch Number)	Structure/ Composition	# of modules completing each test sequence: Initial	Seq. A	Seq. B	Seq. C	Seq. D	Final
A (#1)	a-Si	8	2	U	2		
B (#1)	a-Si	8	2	U	2	U	
C (#1)	a-Si	13	3	U	4		
D (#1)	a-Si/a-Si	8	2	U	2	U	
E (#2)	a-Si/a-Si/a-Si:Ge					U	
F (#1)	CIS	4	1	U	1	U	
G (#2)	CdTe						
H (#2)	CdTe						
I (#1)	mono x-Si	8	3	U	2	U	
J (#1)	multi x-Si	9	2	U	3	U	
K (#1)	Spheral Si	9	U	U	U		
L (#2)	Si-Film						
M (#2)	Ribbon Si						

(U = Currently Under Test)

Visual Inspection Test

The incoming photovoltaic modules are individually inspected for nicks, scratches, cuts, delaminations, and visible blemishes in the semiconductor material, metallization or encapsulant, as well as structural defects or cracks in the frame, junction box, backsheet, etc. These imperfections are noted not only for aesthetic reasons, but also as possible precursors to failures that may occur during the testing. This inspection is also performed each time an electrical performance (I-V) test is taken throughout the testing sequence (see Fig. 1).

Electrical Performance Tests

The PV modules receive an initial and final electrical performance (I-V) test under both controlled, indoor conditions (1000 W/m^2, AM1.5 Global spectrum, 25°C) using a calibrated Spire 240A solar simulator, and under prevailing, outdoor conditions (clear sky, >850 W/m^2) using a standardized outdoor measurement system at the NREL outdoor PV test site. The intermediate I-V tests, taken throughout the test sequence, as indicated in italics in Fig. 1, are performed under the indoor (solar simulator) conditions only.

Ground Continuity Test

The modules with metal frames and, therefore, a common grounding point, undergo the ground continuity test. Those modules with plastic frames, or no frames at all, are not required to undergo this test. Each frame portion (normally each of the four sides of the module frame) is supplied with twice the short-circuit current (I_{sc}) of the module between it and the module grounding point. The voltage between these two points is measured, and a resistance is obtained. This resistance value must be less than 0.1 Ω. *All modules tested passed this test.*

However, one point of concern that became apparent during testing is that, in many of the cases, the corner screws in the frame needed to be securely tightened, or loosened and re-tightened, in order for the module to pass the test. That is, the ground was not very "continuous" to all portions of the frame until the tightening operation was performed. Because many of these modules are commercially available and have been obtained directly from the distributor, they are usually installed in the field as received. Without any additional changes to the modules -- such as tightening the screws -- *many of them would not have passed this test.* This is a concern because these modules are going directly into field-installed applications, many of which are high-voltage PV arrays, in which (one part of) the module frame is used as the grounding point. Without a truly continuous ground, this could lead to a potentially hazardous situation in which one portion of the frame is at earth ground, while another portion is assumed to be at ground, but is really floating at some higher potential. System installers and field personnel such as testers and inspectors need to be aware of this condition.

Dry/Wet Hi-Pot Tests; Wet Insulation-Resistance Test

The dry hi-pot test is performed by shorting the module leads together and passing a high voltage from the shorted leads to the module grounding point, which is either the metal frame or a conductive foil wrapped around the module (in the case of plastic or frameless modules). The voltage applied is twice the maximum system voltage plus 1000 Volts. In most cases, this turns out to be either 2200 V (2 x 600 + 1000) or 1100 V (2 x 50 + 1000). The leakage current is measured and should not exceed 50 μA. The apparatus used to perform this test is called a "Hi-Pot" test unit; hence, the test name. From our results, we have seen that *several of the modules failed this test.* In most of these cases, visible and audible arcing were observed between the frame and the module backsheet. This still constitutes a failure, even if the leakage current limit is not exceeded.

The wet hi-pot test has been a point of much discussion in ASTM standards committee meetings over the past few years. The disagreement has been over whether to use the spray/trough test or the full-immersion test. The immersion test is a much more stringent test, and many manufacturers already use it. Recently, it has been decided that this is the preferred test method over the spray test, as the immersed module is fully wetted, while the sprayed module does not always get fully wet, even when a surfactant solution is used. In our case, the module is placed face down in a large plastic tray containing water with a conductivity of at least 286 μmhos. The back of the module is filled with water up to the bottom edge of the junction box. Once again, the high potential is applied between the shorted module leads and the ground, as described above, and the leakage current is measured. *Several modules, both thin-film and crystalline Si, failed this test.*

The wet insulation-resistance, or wet megger, test was also performed during the initial test sequence. *Several modules (both thin-film and crystalline Si) that passed the wet megger test subsequently failed the wet hi-pot test.* This is an important result, as the IEEE standards committee developing the qualification testing document [4] had recently voted to eliminate the wet hi-pot test, believing it to be both redundant and too dangerous (due to the high voltage combined with the wet conditions). However, these results -- modules passing the wet megger and then failing the wet hi-pot -- show that the wet hi-pot test is still needed to expose severe failures in some modules that only occur under wet, high-voltage conditions. The IEEE committee has subsequently voted to reinstate the wet hi-pot test into the MQT sequence.

Thermal and Humidity-Freeze Cycling Tests

Once the modules completed the initial test sequence and had another I-V curve taken, they were separated into different groups to undergo the four test sequences (A-D) in the MQT procedure. The modules undergoing Sequence A (see Fig. 1) have completed >200 thermal cycles (-40°C to +90°C), while those in Sequence B will undergo 50 thermal cycles to be followed by 10 humidity-freeze cycles (-40°C to 85°C at 85% R.H.) after completing the UV/thermal test (discussed below). These tests tend to be very good indicators of how well the module will perform in the field under long-term exposure to extreme outdoor temperature and moisture conditions. *All Sequence A modules passed* after undergoing >200 thermal cycles: that is, none lost more than 10% of their initial P_{max}.

UV/Thermal Test

The ultraviolet (UV) / thermal weathering test, which is the first test in Sequence B, is under development at NREL at the present time. Proposals have been received from vendors for both indoor and outdoor accelerated-weathering equipment. These will expose modules to the combined effects (weathering) of solar radiation (including UV), temperature, and moisture under accelerated conditions.

In the ASTM standard for solar radiation weathering of PV modules, the requirement is for a test-to-failure. However, because this may take years to occur, even under accelerated conditions, an annex in the test standard gives details for a procedure with a shorter period of exposure, to be used in cases such as a module qualification, or screening, type of testing. This procedure requires a certain amount of total UV exposure, in addition to thermal and moisture conditions. For artificial testing, this standard requires the use of *filtered xenon-arc* lamps.

In the IEEE qualification test document [4], however, the UV/thermal test requires *UV fluorescent* lamps. This latter test has been constructed at NREL and preliminary testing is now underway.

The final goal is to establish test conditions and procedures, including the appropriate UV, visible, and infrared (IR) radiation exposure on the modules, as well as the "proper" level of both the temperature and humidity.

Dynamic and Static Mechanical Loading Tests

In mid-1993, NREL procured a Mechanical Flex Tester, capable of performing the dynamic mechanical loading test. The apparatus was installed and thoroughly tested, and is presently being fitted with flat pressure transducers to record the pressure on the face of the modules undergoing the flex test. The purpose of the dynamic loading test is to simulate wind or other changing mechanical loads that a module might experience in the field. The requirement for this test is for a uniform load of 1440 Pa (30 lbs/ft^2) to be applied alternately to each face of a flat-plate PV module. This cyclic pressure loading is performed for 10,000 cycles at a rate not to exceed 20 cycles/min. NREL's Mechanical Flex Tester is presently operational and will be utilized to perform this test on the modules that are undergoing Sequence B (see Fig. 1).

The static loading test simulates exposure to static mechanical loads, such as snow, which may cause a PV module to bend or undergo stress for extended periods of time. The test requirement is for a uniform mechanical load of 2400 Pa (50 lbs/ft^2) to be applied to the front face and then rear face of the module, each for 30 mins; this is repeated three times. We use stacks of photocopier paper -- evenly distributed over the module surface -- for this test

Hail Impact Test

In early 1992, NREL procured a Hail Impact Test System (appropriately named, H.I.T.S.), which has been used many times in the succeeding years to test prototype and commercial PV modules for their resistance to impact from propelled hail balls. The actual test requirement is for a minimum of 10 hail balls (made from 2.54-cm (1") diameter ice spheres) to be projected onto the face of the photovoltaic module at 23.2 m/sec (~52 mph). The module face must not crack or shatter. NREL's apparatus uses a pneumatic projectile system with air-pressure controls to propel the ice ball, as well as a velocity-measuring unit consisting of two light beams through which the ball passes on its way to the target. The impact sites on the module are usually chosen for their sensitivity to high-velocity impact, such as the glass/frame interface, the module corners, and the areas over the bus bars or junction box.

Surface-Cut Test: Non-Intrusive Hot-Spot Test

The surface-cut test is performed with a sharp edge drawn across the surface of a module to determine its susceptibility to being cut. Four passes were performed on both the front and back surfaces of each module. This was followed by a hi-pot test to determine whether the attempt at cutting the front and back of the module will lead to moisture intrusion, shorting, or other types of electrical failure. NREL's surface-cut apparatus has been designed in order to meet the test requirements as set forth in the UL Standard 1703 [2]. *All modules tested passed this test.*

The *intrusive* hot-spot test, designed to determine whether shadowing or other such phenomena may generate hot spots in a PV module due to reverse-bias heating of some of the interconnected solar cells, is a very painstaking test requiring a specially-fabricated module that can only be used for this one particular test. We have developed a *non-intrusive* method for performing this test using a large-area light source and a thermal/IR video camera. We have performed some exploratory tests on several different types of PV modules and the preliminary indications are that this technique works quite well for observing hot spots in isolated cells within PV modules.

Damp Heat Test

The damp heat test (+85°C, 85% R.H. for 1000 hours) is a very harsh exposure test, used to determine a module's resistance to both high temperature and extreme moisture for extended periods of time. This is the first test in Sequence C (see Fig. 1) and is performed in the same environmental chambers that are used for the thermal and humidity-freeze cycling tests of Sequences A and B. *A few modules (thin-film only) failed this test*, having lost >10% of their P_{max}.

Outdoor Exposure Test

The outdoor exposure test has been added to the proposed IEEE qualification test sequence as the first test in Sequence D (see Fig. 1). The requirement is to expose the modules to a total irradiance equivalent to ~30 days in the field. However, because this is a real-time, real-world test, not performed under accelerated conditions, but rather under natural, outdoor conditions, the longer this test can be performed, the more realistic the results will be. Hence, this portion of the testing was begun in September, 1993, with more modules added in March, 1994. The modules are connected to a multiplexing module I-V curve tracer, which continuously loads the modules at their maximum power point (P_{max}) as well as obtaining an I-V curve of each module every 30 min. These data are also used for other NREL indoor/outdoor studies.

Final Electrical Performance Test

As mentioned earlier, throughout the MQT sequence I-V tests are performed to determine if the modules have failed a particular test or sequence of tests. The final electrical performance (I-V) test, performed under both indoor and outdoor conditions, is used to determine the ultimate success or failure of the PV modules undergoing the complete test sequence. Each of the modules being tested must show <10% loss in power (P_{max}) from the initial I-V test, and must also meet the manufacturer's rated power specifications, in order for that particular module *design* to pass the entire module qualification test. Further, if the manufacturer changes the fundamental design of its module, then the entire MQT sequence must be re-performed, starting with a fresh batch of nine modules fabricated with the new design.

CONCLUSION

NREL now has the capability of performing the tests in the proposed IEEE PV module qualification test (MQT) sequence. In order to demonstrate this, as well as to continue our work with PV manufacturers to help them improve their long-term performance and module reliability, we are performing the complete MQT sequence on a large group of modules from a variety of PV manufacturers and technologies. Batch #1 involves products and prototypes from eight such manufacturers, while Batch #2 includes modules from five additional companies. This testing program is currently underway, with many of the tests already completed, and with the entire sequence of tests to be concluded over the next several months.

ACKNOWLEDGEMENTS

The authors would like to acknowledge the support, input and encouragement of NREL colleagues R. DeBlasio, R. Hulstrom and L. Mrig; the testing and measurements performed by S. Rummel, Y. Caiyem and L. Ottoson; and the productive interaction and cooperation with PV manufacturers and test facilities. Funding is under DOE contract DE-AC36-83CH-10093.

REFERENCES

1. "Block V Solar Cell Module Design and Test Specification for Intermediate Load Applications - - 1981", JPL-5101-161, Pasadena, CA, 1981.
2. "Standard for Safety: Flat-Plate Photovoltaic Modules and Panels", UL 1703, 1986.
3. "Interim Qualification Tests and Procedures for Terrestrial Photovoltaic Thin-Film Flat-Plate Modules", SERI/TR-213-3624, Golden, CO, 1990.
4. "Recommended Practice: Qualification of Photovoltaic Modules", PAR 1262, IEEE SCC 21, 1994.

PHOTOVOLTAIC MODULE CERTIFICATION AND LABORATORY ACCREDITATION CRITERIA DEVELOPMENT

Carl R. Osterwald, National Renewable Energy Laboratory
Robert Hammond, Arizona State University
Gene Zerlaut, SC-International
Robert D'Aiello, Consultant

ABSTRACT

This paper presents an overview of a model product certification and test laboratory accreditation program for photovoltaic (PV) modules that was recently developed by the National Renewable Energy Laboratory and Arizona State University. The specific objective of this project was to produce a document that details the equipment, facilities, quality assurance procedures, and technical expertise an accredited laboratory needs for performance and qualification testing of PV modules, along with the specific tests needed for a module design to be certified. Development of the document was done in conjunction with a criteria development committee consisting of representatives from 30 U.S. PV manufacturers, end users, standards and codes organizations, and testing laboratories. The intent is to lay the groundwork for a future U.S. PV certification and accreditation program that will be beneficial to the PV industry as a whole.

INTRODUCTION

A third-party certification program for PV products may help alleviate the current situation in the PV industry: buyers and end-users of PV modules must either accept manufacturers specifications for the performance and reliability of their products; find an independent laboratory capable of performing the necessary testing; or, as a last resort, attempt to perform the testing themselves. Unfortunately, the majority of buyers and end-users have little or no knowledge of PV testing, a situation that compounds the problem.

Conferring upon a product the status of certified, and the offer for sale by a manufacturer of certified product, creates an inducement to the purchaser or consumer, to rely on the quality of the product, and to rely on the claims of the manufacturer. This reliance creates a chain of potential liability that encompasses the certification body, the testing laboratory(s), the laboratory accreditation organization, and the manufacturer. In point of fact, the existence of these programs have the effect of transferring some of the risk from the manufacturer to entities in the overall certification process.

Once announced, or advertised, the conferring of accreditation on a laboratory creates an inducement to manufacturers to place reliance on the act of accreditation, to thereby rely upon the laboratory's status as an accredited laboratory, and to employ the services of that laboratory. Because of the chain of liability inherent to the laboratory accreditation process, a considerable body of deliberation in all phases of the accreditation process has been developed internationally during the past few years.

Product Certification Options

There are three options that are used for product certification programs: certification body assessment (option 1), independent assessment (option 2), and testing laboratory certification (option 3). In option 1, a certification body assesses the test laboratory(s) as well as certifies products. Option 2 is similar to option 1, except that the test laboratory assessment is performed by an independent assessment agency, such the American Association of Laboratory Assessors (A2LA) or the National Voluntary Laboratory Assessment Program (NVLAP). The assessment agency uses the requirements of the certification body to perform laboratory assessments. For option 3, a single test laboratory also certifies product. An example of this scenario is Underwriters Laboratories (UL) in the United States (the UL label is a well-known and recognized certification for product safety). Figure 1 illustrates these options. The criteria for the model PV certification program recommends using independent assessment. This option frees the certification body from the additional burden of laboratory assessment, and is advantageous for liability reasons.

Testing Laboratory Options

Several laboratory options are also available for product testing in a certification program: single full-service laboratory, multiple full-service laboratories, and multiple partial-service laboratories. There are advantages and drawbacks for each of these. A single full-service laboratory is the easiest to accredit and monitor, and systematic errors are not randomized among multiple laboratories. On the other hand, it is noncompetitive and thus testing costs are higher, and the single laboratory can become technically complacent. Having multiple full-service laboratories available solves the non-competition disadvantage, but increases the complexity needed for administration of the certification program. Also, all errors are randomized among laboratories, which may result in differences. Testing costs may actually be higher for this option because the testing volume, which may be limited, is distributed among the multiple laboratories. The multiple partial-service laboratories option may actually be required in the initial stages of a

First WCPEC; Dec. 5-9, 1994; Hawaii

certification program if no single laboratory has the ability to perform all the certification testing. Its major drawback is the need to transport test samples between the participating laboratories, which increases costs and risks in-transit damage and loss. Because of the present lack of third-party testing services in the PV industry, it was not possible to recommend one of these options at the present time. This question will have to be answered after the certification program is initiated.

Certification Body

The primary purpose of the certification body is to operate and manage the certification program, including the certification of manufacturers' products. It can be an existing certification body, a new body created by a PV trade association, or a new body organized by a PV industry segment. For legal reasons, the certification body must be incorporated. It would be governed by a board of directors that includes persons with interests from, for example, industry participants such as manufacturers, the public sector, and PV technical representatives. The functions of the body are handled by committees of persons from these same interests. An executive director manages the certification program on a day-to-day basis and interfaces with the operational committees as an ex officio member.

A committee on test methods and standards is responsible for adopting test procedures needed for certification. In this capacity it maintains official liaisons with the domestic committees active in PV standards development and maintenance. It must continuously update the test procedures to conform to the relevant standards, as well as additional requirements as they arise.

The laboratory committee adopts, develops, and modifies the testing laboratory accreditation requirements. It reviews, and accepts or rejects, assessment reports submitted by laboratory assessors or the accreditation body. From these reports, it selects one or more laboratories based on need, full or partial capability versus needs, testing fees, and ability to meet scheduling requirements. The committee develops and maintains a between-assessment surveillance and reviews the program for accredited laboratories.

A product certification and rating committee reviews and evaluates results of qualification and electrical per-

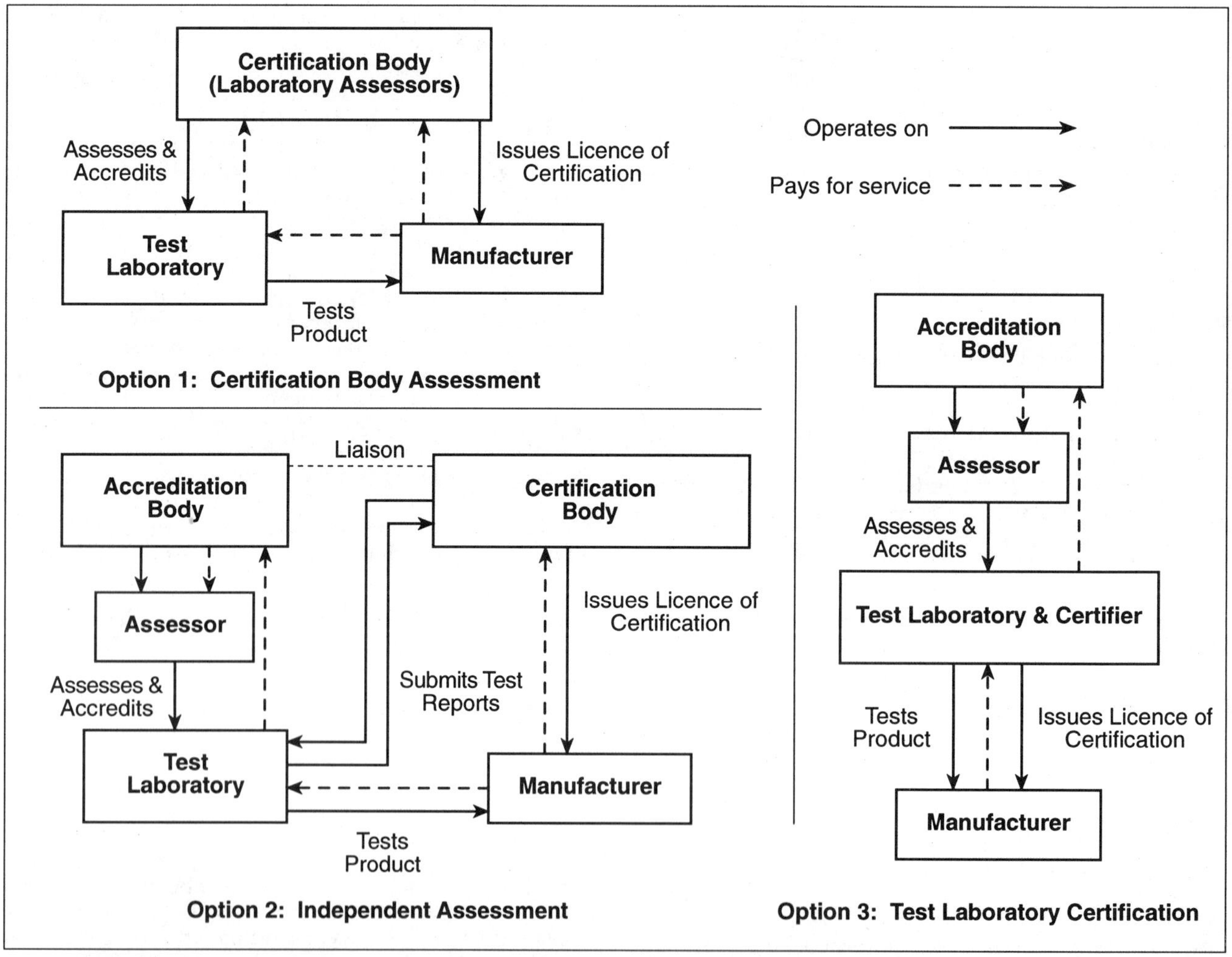

Fig. 1. The three options used for third-party product certification programs.

formance tests submitted by the laboratory. After evaluating any discrepancies and requesting any needed additional information, it renders a decision on the suitability of an applicant's product for certification. If the product is deemed suitable, it requests the executive director to issue a license to allow the manufacturer to use the certification mark or label.

Several other committees handle functions such as manufacturers quality-system and licensing compliance, and auditing and financial concerns. An appeals board handles appeals of test and audit results, technical disputes involving test laboratories, disputes between participants and the certification body, complaints from participants or users about other certified product, and complaints about misuse of certification labeling.

ACCREDITATION/CERTIFICATION CRITERIA DOCUMENT

The criteria document has three separate sections, referred to as PV–1, PV–2, and PV–3, that specify different aspects of the document [18]. PV–1 is titled "Criteria for a Model Quality System for Laboratories Engaged in Testing Photovoltaic Modules," PV–2 is "Model for a Third-Party Certification and Labeling Program for Photovoltaic Modules," and PV–3 is "Testing Requirements for a Certification and Labeling Program for Photovoltaic Modules: Test Standards, Testing Methodologies, Instrumentation and Facilities." These documents are based on relevant ISO/IEC (International Standards Organization and International Electrotechnical Commission) standards, as well as IEEE (Institute of Electrical and Electronic Engineers), ANSI (American National Standards Institute), and ASTM (American Institute for Testing and Materials) standards [1-17]. The model is therefore consistent with accepted practices of national and international accreditation and certification programs.

Although any potential PV certification body is free to use and modify them in any way, the three sections have been written in a fashion that allows their use directly in the by-laws of such an organization. The objective is that such a body will have the necessary information to make intelligent and informed decisions with respect to the balance among costs, benefits, and protection from unnecessary risk.

PV–1

The objective of PV–1 is to set forth the minimum quality standards with which a competent laboratory shall comply and operate under to be accredited for testing PV modules. PV–1 is intended as a model for the development of a laboratory's quality system, for maintaining the quality of services, and for aiding the assessment of conformance toward accreditation. It also establishes the general qualifications of its management and testing personnel, the external and on-the-job training requirements, and the minimum requirements for calibration of instrumentation and traceability of instrumentation and test data, test report requirements, and handling of complaints.

An accredited laboratory must be organized in such a way that confidence in its independence of judgement and integrity is maintained at all times. It must, by means of a quality system, define and document its policies and objectives for good laboratory practice and the quality of testing services. A quality manual specifies and documents the responsibility, authority, and interrelation of all personnel who manage, perform, or verify work affecting the quality of testing. This includes procedures for the calibration, verification, and maintenance of instrumentation and test equipment, and procedures for handling test specimens.

PV–2

The module certification and labeling program criteria (PV–2) details the requirements that a certification body should use for authorizing a module manufacturer to use the certification program's certificate of conformity, or the program's certification mark. The certification mark is used as a label which indicates the module design has passed the testing required by section PV–3 (both performance and qualification) in an independent laboratory accredited to perform the testing. PV–2 details the responsibilities and structure of a certification body, as discussed above. It also specifies the quality system needed by a manufacturer, and how to determine that the manufacturer has the necessary equipment, staff, and facilities required for production of a quality module design, and therefore for participation in the certification program. Procedures for initial testing and periodic re-testing of a manufacturer's module design are specified, along with selection of samples for testing by the accredited laboratory. Details of the procedures for licensing, surveillance of continued conformity, suspension, withdrawal, or cancellation of the license, and liability issues are defined. Finally, the certification body must have procedures for revising section PV–3 under circumstances such as improvements in test methods or the necessity of additional test methods.

PV–3

PV–3 establishes the testing and reporting requirements that a PV qualification laboratory must perform in support of a PV module product certification and labeling program. It also lists the facilities, equipment, and personnel an accredited laboratory needs to perform the certification testing. The testing consists of an initial baseline performance value test, followed by a comprehensive qualification sequence that is intended for evaluating module performance, safety, and susceptibility to known failure mechanisms. Following the qualification sequence, the output power of the modules tested must be within ±10% of the baseline performance value.

Module qualification is tested according to IEEE 1262 [6]. A minimum of nine modules are subjected to the testing, which is intended to discover potential degradation of module performance resulting from environmental weathering, mechanical loading, corrosion, and shadowing. Acceptable results from these tests should provide reasonable assurance that modules manufactured by the same processes as the tested modules will perform reliably in the field after installation. In addition to performance, modules are also characterized with visual inspections, electrical insulation tests, and a ground-path continuity test. Modules are subjected to environmental stresses such as humidity-freeze cycling, thermal cycling, ultraviolet and outdoor exposure, damp heat, ice-ball impact, hot-spot endurance, and static mechanical loading.

Module performance is measured according to ASTM E 1036 [3] at Standard Test Conditions (STC), namely 1000 W/m^2, AM 1.5 global reference spectrum [2], and 25°C module temperature. Reference cells used for the performance measurements must be calibrated according to ASTM E 1362 [5] by at least two independent laboratories. The light source for the performance measurements may be natural sunlight or a solar simulator, but the light source must meet the requirement for a Class A solar simulator [4]. The module manufacturer's performance measurements, submitted with the modules to be tested, must be within ±5% of the results of the initial baseline performance.

CONCLUSION

A document has been developed that can serve as the starting point for a possible U.S. product certification program for PV modules using accredited third-party testing. The document is based on existing standards for certification programs. Details of the quality system needed by accredited testing laboratories, the functions and resposibilities of the certification body, and the testing necessary for module certification, are specified.

ACKNOWLEDGMENT

This work was supported by the U.S. Department of Energy under contract No. DE-AC36-83CH10093.

REFERENCES

[1] "General criteria for evaluating testing laboratories," ASTM standard E 548, ASTM, Philadelphia, PA, USA.
[2] "Tables for terrestrial solar spectral irradiance at air mass 1.5 for a 37° tilted surface," ASTM standard E 892, ASTM, Philadelphia, PA, USA.
[3] "Test methods for electrical performance of nonconcen-trator terrestrial photovoltaic modules and arrays using reference cells," ASTM standard E 1036, ASTM, Philadelphia, PA, USA.
[4] "Specification for solar simulation for terrestrial photovoltaic testing," ASTM standard E 927, ASTM, Philadelphia, PA, USA.
[5] "Test method for calibration of nonconcentrator photovoltaic secondary reference cells," ASTM standard E 1362, ASTM, Philadelphia, PA, USA.
[6] "Recommended practice for qualification of photovoltaic (PV) modules," IEEE standard 1262, IEEE, Piscataway, NJ, USA (in approval process).
[7] "Certification – third party certification programs," ANSI standard Z34.1, ANSI, New York, NY, USA.
[8] "Methods for indicating conformity with standards for third-party certification systems," ISO/IEC guide 23, ISO, Geneva, Switzerland.
[9] "General requirements for the accreditation of calibration and testing laboratories," ISO/IEC guide 25, ISO, Geneva, Switzerland.
[10] "Guidelines for corrective action to be taken by a certification body in the event of misuse of its mark of conformity," ISO/IEC guide 27, ISO, Geneva, Switzerland.
[11] "General rules for a model third-party certification system for products," ISO/IEC guide 28, ISO, Geneva, Switzerland.
[12] "General requirements for the acceptance of testing laboratories," ISO/IEC guide 38, ISO, Geneva, Switzerland.
[13] "Testing laboratory accreditation systems — general recommendations for operation," ISO/IEC guide 55, ISO, Geneva, Switzerland.
[14] "An approach to the review by a certification body of its own internal quality system," ISO/IEC guide 56, ISO, Geneva, Switzerland.
[15] "Quality management and quality assurance standards — guidelines for selection and use," ISO standard 9000, ISO, Geneva, Switzerland.
[16] "Quality systems — model for quality assurance in design/development, production, installation and servicing," ISO standard 9001, ISO, Geneva, Switzerland.
[17] "Quality management and quality system elements — guidelines," ISO standard 9004, ISO, Geneva, Switzerland.
[18] "Photovoltaic Module Certification and Laboratory Accreditation Criteria," NREL technical report, available from the National Technical Information Service, Springfield, VA, USA (in press).

RELIABILITY TESTING OF PV MODULES

John H. Wohlgemuth
Solarex Corporation, Frederick, MD 21701

ABSTRACT

The reliability of PV modules is critical to their acceptance in the market place. The success that PV has achieved is in large part due to the proven reliability of crystalline silicon modules. This paper will discuss the development of reliability testing of PV modules, explaining how accelerated stress tests are developed and what they can tell about the product. Those tests usually included in qualification test sequences will be reviewed in terms of the stress or failure mechanism being tested, what lifetime or reliability predictions can be made from the results and how the tests can be extended to provide additional reliability information. Several new tests, the UV and Bypass Diode Tests, will be presented. Limitations to extending qualification and reliability testing to new technology modules will also be discussed.

DEVELOPMENT OF RELIABILITY TESTING

The first step in module testing is outdoor exposure, since this is where the modules have to work. However, outdoor exposure takes too long and varies from location to location. Therefore, standardized accelerated tests are developed. In designing accelerated environmental tests, extreme care must be taken to insure that the test relates to both acceleration of the environment that the product is to be exposed to as well as to the mechanisms by which the environmental stress causes failure in the product. Manufacturers use the accelerated tests as quick feedback in the development of more reliable modules. By designing modules to pass the accelerated tests, production modules may no longer exhibited the early field failures caused by the failure mechanisms that the accelerated tests were designed to stress.

The next step is field deployment of modules that successfully pass the accelerated environmental tests. Once again, field failures are likely to occur, but this time the modules will probably work for a much longer time. These new failures have to be analyzed, failure mechanisms identified and new or modified accelerated environmental tests developed.

In the late 1970's and 1980's this procedure was applied to crystalline silicon modules through the LSA Program at JPL. [1-4] A sequence of accelerated tests that included thermal cycling, humidity-freeze, hot spot endurance, hail, mechanical load, and dry high pot was defined as the Block V Qualification Test Sequence. [5] One of the results of this program was the commercial availability of highly reliable crystalline silicon modules.

RELIABILITY TESTS VERSUS QUALIFICATION TESTS

In the DOE Block procurements of PV modules, JPL wished to assure a minimum degree of module reliability and to establish a uniform set of minimum requirements that all modules had to meet. To meet these needs, JPL established the series of module qualification specifications (Blocks I through V). These JPL qualifications, as well as subsequent module qualification sequences like CEC 501, 502 and 503, SERI IQT and IEC-1215, are well defined sets of accelerated environmental tests developed out of a module reliability testing program. Every module tested to these sequences is expected to pass, providing evidence that the module type can withstand the level of stress defined in the test and therefore is expected to have some minimum lifetime in normal operation.

Reliability tests are used to develop a scientific understanding of how the module performs under various environmental stresses. Reliability tests go beyond the ranges used in qualification tests, for example running the tests for longer duration or by applying higher stress levels. Another useful approach to reliability testing is to use combinations of stress tests and to experiment with applying different stresses at the same time. Often these tests are continued until failure in an attempt to understand what failures can occur and therefore what changes in module construction or design will ultimately improve module reliability and lifetime.

Having well accepted qualification test sequences can results in two problems:

1. Everyone starts believing that the qualification tests tell them everything they need to know about their modules reliability and so don't perform additional reliability tests.
2. The same qualification tests are applied to all PV technologies without determining if the test

CH3365-4/94/0000-0889 $4.00 © 1994 IEEE First WCPEC; Dec. 5–9, 1994; Hawaii

sequences are really relevant for anything but the module type they were developed for, crystalline silicon in this case.

We have suffered through these fallacies, with Carissa Solar [6-8] demonstrating the first point and the bar graph corrosion of early a-Si modules [9] demonstrating the second.

ACCELERATED TEST RESULTS

For crystalline silicon modules there is over 15 years of outdoor performance data to correlate with accelerated test results. Each of the accelerated tests in the qualification test sequence is designed to evaluate the modules performance against a known module failure mechanism. Some of these tests are discussed below:

Thermal Cycling

Daily changes in module temperature produce stresses on the interconnects between cells and on the bonds between the interconnects and the solar cells. Early PV modules suffered open circuit failures, especially when deployed in desert environments where the temperature extremes were large. The thermal cycling test was developed to simulate this failure mechanism. To accelerate the process, the cycle time was reduced from daily to 4 to 6 hours with a maximum temperature change of 100° C/hour and the extremes increased to -40° C to +90° C for Block I-V and SERI-IQT and -40° C to +85° C for CEC-503 and IEC-1215. The lower extreme (-40° C) was selected because it represented the typical worst case for terrestrial environments and it was the lowest temperature easily achieved by a standard refrigeration unit. The upper extreme (+90° C or +85° C) was selected because flat plate modules are unlikely to experience higher temperatures and because higher temperatures would likely damage many of the plastics used in PV module construction.

The Block II-IV tests required only 50 thermal cycles. However, modules that successfully passed the 50 cycle test failed after 2 to 4 years of exposure in hot dry climates. In Block V the number of cycles increased to 200. Many of the early block modules were built using substrate materials such as aluminum and fiber board [4]. The large thermal expansion coefficients of these materials resulted in large stresses on the interconnects and their bonds. Interconnect failure was a common occurrence. These designs could not pass the 200 cycle test.

By Block IV all of the modules were designed with glass superstrates. The low thermal expansion coefficient of the glass minimizes stress on the system. In the glass superstrate structure, the largest thermal stresses are due to the mismatch of thermal expansion coefficient between the interconnect (usually copper) and the cell (silicon). Actual failure of the interconnect itself is rare. Failures in this design are usually associated with fatigue of the interconnect to cell bond. This is most often a workmanship issue rather than a design issue. Modules that pass the 200 cycle test have now been in the field for close to 10 years, with very low failure rates. [10]

Because of the limits on the temperature extremes and on the rate of temperature change discussed above, thermal cycling tests should only be extended by increasing the number of thermal cycles. At Solarex we have carried thermal cycling tests out to 1000 cycles to study long term module reliability. Successful completion of 1000 cycles was a key factor in extending the module warranty to 20 years. [11]

Damp Heat and Humidity Freeze

The damp heat test is designed to evaluate the ability of the module to withstand long term penetration of moisture. Early PV modules with titanium-silver metallization were susceptible to corrosion in humid environments. The Block I test involved exposure at 70° C and 90% relative humidity for 68 hours. In subsequent Block procurements it was replaced by a cycling test. Blocks II and III called for 5 cycles between 40° C, 90% RH and 23° C, while Block IV increased the upper temperature to 54° C. In Block V the test was changed to a humidity freeze test with 10 cycles between 85° C, 85% RH and -40° C. Subsequent work, however, has shown that the damp heat and humidity freeze tests actually test for different failure mechanisms. This was recognized in the preparation of the never released Block VI specification. [4] Subsequent qualification tests (IEC-1215 and CEC-503) have incorporated both a damp heat and humidity freeze test.

Long term exposure to damp heat is testing the ability of the module metallization system to withstand moisture induced corrosion. The failure of PVB encapsulated modules with silver paste contact cells was directly related to this phenomena, that can be duplicated in the laboratory using the damp heat test. [12] The condition of 85° C and 85% relative humidity, was selected to accelerate the phenomena, but at a humidity level that can usually be controlled without resulting in condensation of liquid water in the chamber. Otth and Ross [13] estimated that 200 hours at 85° C, 85% RH represented 20 years of exposure in Miami, Florida. Ricaud and Petersen [14] calculated a range of probable acceleration factors and concluded that a 20 year field exposure could be simulated with up to 400 hours of exposure at 85° C, 85% RH. So it appears that the 1000 hour qualification test included in IEC-1215 and CEC-503 is more than adequate to test module reliability for a 30 year life.

In the humidity freeze test moisture is pumped into the encapsulant during the high temperature, high humidity part of the cycle. This moisture is then frozen during the low temperature portion of the cycle. This is a test of the strength of the encapsulation system, particularly the adhesion of the various layers. Inadequate priming of the glass, inadequate curing of the encapsulant or the wrong selection of backsheet adhesive are all conditions that can result in field failures that will be observed in this test.

A good encapsulation system does not appear to suffer any damage during the humidity freeze test. There is no evidence to indicate that running more cycles of the humidity freeze test will provide any additional information about the long term reliability of the module. However, there is evidence that extending the time for moisture to enter the encapsulation system may increase the severity of the test. We may actually be able to develop a better test by combining the damp heat and humidity freeze tests, that is a 1000 hour exposure to 85° C, 85% RH with a freeze cycle every 100 hours. Experimental work is continuing in this area.

Mechanical Load (Static and Dynamic)

Mechanical load tests are designed to determine the modules ability to withstand wind, snow or ice loads. The static test in IEC-1215 requires the application of 2400 Pa (~30 lb/ft^2) for 1 hour on each side for 2 cycles. The dynamic test in Block V and SERI-IQT require 10,000 cycles of front and back loading at 2400 Pa at a cycle rate not to exceed 20 cycles per minute. The load of 2400 Pa represents a wind load of 130 km/hr. IEC-1215 recommends increasing the load to 5400 Pa to the front of the module, if the module is to be qualified for heavy accumulations of snow and ice.

Hot Spot Endurance

Hot spot heating occurs in a PV module when the current carrying capability of a particular cell or cells is lower than the operating current of the cell string. This condition results in the affected cell(s) going into reverse bias and absorbing power equal to the product of the reverse voltage and the string current. This condition can be caused by transient events such as local shadowing or soiling or by permanent degradation of the cell(s) such as cell cracking or interconnect failure.

Both an intrusive (Block V and SERI-IQT) and a non-intrusive (CEC-503 and IEC-1215) hot spot test have been utilized. Both of these test identify module designs that will clearly suffer from hot spot failures in the field. However, neither test is really adequate to test modules for use in high voltage systems. Indeed many module types can pass the hot spot tests without bypass diode protection, but can suffer hot spot problems in high voltage systems. In the non-intrusive method the module can not maintain the appropriate current level when cells are shadowed. In the intrusive method the stress level is suppose to be determined based on the expected system configuration. However, this is often unknown by the module manufacturer performing the test. What is needed is a non-intrusive test in which the module current is maintained by an external source at the expected operating current while each of the cells in the module is shadowed.

NEW TESTS

From the long term field results, two new failures mechanisms have been identified from which new accelerated tests are being developed.

UV Test

The discoloration of EVA in the modules at Carrisa Plains was the first indication that UV and high temperature caused module degradation. [6-8] Subsequent field data indicated that the discoloration was observable in hot climates under one sun operating conditions.[15] Laboratory tests showed that while temperature accelerated the process, UV was necessary for the discoloration to occur.[15] The major issue with establishing an accelerated UV test is the spectrum of the light source. While there are many sources of UV, most produce a significant amount of their light at wavelengths shorter and therefore more damaging than the UV spectrum experienced in outdoor exposure. Three sources have been identified that can match the outdoor UV spectrum, xenon, UV fluorescent and concentrated sunlight. All three of these light sources have been utilized to test plastic, paints and PV module materials. Xenon is extremely useful for testing small samples at high acceleration rates, but is not very well suited for long term exposure of large PV modules. Outdoor exposure using Fresnel lenses is also very useful for comparing the UV tolerance of various materials, but because of the variations in weather, it is not recommended for UV testing of modules. Use of fluorescent UV lamps for UV testing has been documented in standards in ASTM G 53-88 and ISO 4892-1981. Banks of these fluorescent lamps can be assembled to illuminate modules of almost any size. Use of fluorescent UV-B lamps has been proposed for incorporation in IEC-1215 and in the draft IEEE module qualification test document.

Bypass Diode Test

A significant number of bypass diode field failures have been reported. Improper diode selection or mounting can result in overheating of the diode with a resultant fire safety issue. The PV community is now working on the development of a qualification test for the bypass diode. The bypass diode should not operate at temperatures above the manufacturer's safety rating under worst case module operating conditions. The proposed worst case conditions are defined as 1.25 times short circuit current at a module temperature consistent with an ambient of 40° C and full sunlight heating.

SYNERGISTIC EFFECTS

In module qualification testing the accelerated tests are run one after the other in series or parallel. There is no attempt to determine what happens if one or more stresses are applied at the same time. Module reliability testing should include combinations of stresses. For example run thermal cycling with the modules passing current equivalent to peak power current or under illumination producing a significant fraction of this current level. Another example is to run the damp heat test with the module biased at several hundred volts above a grounded frame. Such a combination of tests resulted in serious failures for modules with PVB encapsulant, but made little difference when the encapsulant was EVA, duplicating the field performance observed with these 2 materials.[12] Combination of tests can be selected from our knowledge of PV operating conditions or from analysis of observed field failures.

NON-CRYSTALLINE SILICON PV TECHNOLOGIES

For each new PV module technology, reliability testing requires a baseline of outdoor exposure testing to identify failure mechanisms for which accelerated tests can be developed. Until this field work has been completed, there is no basis for believing that passing a crystalline silicon qualification test sequence like Block V or IEC-1215 means anything for a new technology module. At this stage of development these accelerated tests should be considered reliability not qualification tests. New technologies will likely suffer from failure mechanisms that are not even tested for in the crystalline test sequence. In addition, we must be careful about interpreting failures of new module technologies in the tests designed for crystalline silicon. For example a thin film module might fail the thermal cycle test because the cool-down cycle is too fast. Since modules are unlikely to experience a temperature decrease of 100°C/hour during normal operation, one should seriously question whether the module or process should be redesigned to pass the test. However, an estimate should be made of the maximum temperature drop that the module may experience during operation so that the module can be subjected to a more realistic temperature cycle test.

SUMMARY

Reliability testing should be an ongoing program for all PV module technologies. For the PV modules manufactured today we still can not answer the questions:

- How long will a module last?
- What mechanism(s) ultimately limits module lifetime?

Continued module reliability testing combined with a comprehensive field testing program may eventually give us answers to these questions and provide us with methods of extending module lifetime even further.

REFERENCES

[1] A. R. Hoffman and R. G. Ross, "Environmental Qualification Testing of Terrestrial Solar Cell Modules", *Thirteenth IEEE PVSC*, 1978, p. 835.

[2] A. R. Hoffman, J. S. Griffith and R. G. Ross, "Qualification Testing of Flat-Plate Photovoltaic Modules", *IEEE Transactions on Reliability*, Vol. R31, No. 3, 1982, p. 252.

[3] JPL Publication 86-31, "Flat Plate Solar Array Project Final Report", Volume VI: Engineering Science and Reliability, 1986.

[4] M. I. Smokler, D. H. Otth and R. G. Ross, "The Block Program Approach to Photovoltaic Module Development", *Eighteenth IEEE PVSC*, 1985, p. 1150.

[5] JPL Publication 510-161, "Block V Solar Cell Module Design and Test Specification for Intermediate Load Applications", 1981.

[6] D. D. Sumner, C. M. Whitaker and L. E. Schlueter, "Carrisa Plains Photovoltaic Power Plant 1984-1987 Performance", *Twentieth IEEE PVSC*, 1988, p. 1289.

[7] H. J. Wenger, C. Jennings and I. Iannucci, "Carrisa Plains Power Plant Performance", *Twenty First IEEE PVSC*, 1990, p. 844.

[8] J. H. Wohlgemuth and R. C. Petersen, "Reliability of EVA Modules", *Twenty Third IEEE PVSC*, 1993, p. 1090.

[9] D. A. Fagnan, R. V. D'Aiello and J. Mongon, "Solarex/Philadelphia Electric Amorphous Silicon PV Test Site", *Nineteenth IEEE PVSC*, 1987, p. 1508.

[10] A. L. Rosenthal, M. G. Thomas and S. J. Durand, "A Ten Year Review of Performance of Photovoltaic Systems", *Twenty Third IEEE PVSC*, 1993, p. 1289.

[11] J. H. Wohlgemuth, "Testing for Module Warranties", *Photovoltaic Performance and Reliability Workshop*, NREL/CP-410-6033, 1993, p. 200.

[12] G. R. Mon, J. Orehotsky, R. G. Ross and G. Whitla, "Predicting Electrochemical Breakdown in Terrestrial Photovoltaic Modules", *Seventeenth IEEE PVSC*, 1984, p. 1682.

[13] D. Otth and R. G. Ross, "Assessing Photovoltaic Module Degradation and Lifetime from Long Term Environmental Testing", *Proceedings of 30th Institute of Environmental Science Annual Meeting*, 1984, p. 121.

[14] A. M. Ricaud and R. C. Petersen, "Environmental Stress Tests: Acceleration of Corrosion", *Eighth E.C. PVSEC*, 1988, p. 549.

[15] W. Holley, S. Agro, J. Galica, L. Thoma, R. White and R. Yorgensen, "Advanced Development of PV Encapsulants", Annual Subcontract Report, NREL/TP-411-7352, 1994.

INVESTIGATION INTO THE CAUSES OF BROWNING IN EVA ENCAPSULATED FLAT PLATE PV MODULES

William H. Holley, Jr., Susan C. Agro, James P. Galica, Lynne A. Thoma, Robert S. Yorgensen
Springborn Laboratories, Inc.
Dr. Myer Ezrin, Dr. Peter Klemchuk, Gary Lavigne
UCONN Institute of Materials Science
Holly Thomas
National Renewable Energy Laboratory

ABSTRACT

The problem of browning in a number of EVA encapsulated flat plate photovoltaic modules has led to the questioning of EVA as a suitable material for such applications. By isolating the variables that could possibly lead to EVA browning, such as module construction, types of glass superstrates, additives, and processing conditions, we have been able to determine those significant specific variables that seem to have the most influence on discoloration. When "standard-cure" EVA-based laminates were exposed to accelerated UV aging, measurable yellowing of those laminates was evident after only one to two weeks exposure, and visual discoloration was observed after four to six weeks. Some samples yellowed quickly and some not at all, and there were significant differences in the rates of discoloration between "standard-cure" and "fast-cure" EVA. This paper looks at the results of these studies, especially focusing on the effect of additives in the EVA on the rate of yellowing, and discusses how preliminary results can be used to reformulate EVA encapsulants.

INTRODUCTION

Over the last few years we have received several reports of the ethylene vinyl acetate (EVA) encapsulant in fielded flat plate photovoltaic modules yellowing or browning. A literature search and field surveys revealed that this discoloration was not the bane of any single manufacturer, but has been experienced in fielded modules made by all major U.S. manufacturers. Also, we found that this discoloration problem is not limited to EVA encapsulant sheet made by any one supplier.

But what was apparent from our review was that the degree of discoloration appeared to loosely correlate with the degree of average daily direct normal solar radiation, and the module's approximate maximum operating temperature. To this end, field-aged EVA encapsulated modules from an affected site in Carrisa Plains, California, (both mirror enhanced and non-mirror enhanced arrays) were autopsied and analyzed for vinyl acetate content, unsaturation, and concentration of additive levels by Thermogravimetric Analysis (TGA), Fourier Transform Infrared Analysis (FTIR), Gas Chromatography/Mass Spectrometry (GC/MS), and Gas Chromatography/Flame Ionization Detection (GC/FID). These test results were then compared to results produced by in-house developed, UV-aged laminate samples made with either "fast-cure" or "standard-cure" EVA, and using one of four different low iron glasses.

LABORATORY METHODOLOGY

For comparative laboratory accelerated UV-aging studies, coupon-sized laminates, measuring 2.70 x 2.75 inches, were used. The majority of these laminates were glass/glass so as to facilitate visual, colorimetric, and spectrographic measurements. All were prepared by vacuum lamination on laboratory-scale equipment using the commercial time/temperature/vacuum schedule recommended for "standard-cure" EVA with Lupersol 101. Then during cure, a data logger and multiple thermocouples were used to verify the temperature profile of all samples cured in the laminator.

Given our interest in achieving maximum UV-B transmission (i.e., 285 to 330 nm--that wavelength region suspected of being responsible for EVA discoloration/degradation), a low iron glass superstrate was selected. This decision was further reinforced by the fact that low iron glass was the superstrate most commonly used in those field modules which had shown discoloration. Specific low iron glasses tested included AFG's Solite (samples from the mid-1980s), PPG's Starphire, AFG's Solatex II (containing less than 4 percent cerium-oxide by weight), and PPG's Airphire (also containing cerium-oxide).

During testing, all coupon laminates were exposed to an irradiance of 0.55 watts/square meter @ 340 nm--roughly the equivalent of two suns--and a temperature of 100 degrees C for a minimum of 17 weeks or until significant degradation/discoloration had occurred. Exposure was in a Ci35A Atlas xenon-arc Weather-Ometer with quartz/borosilicate glass filters. The nominal lower end UV cut-off was 285 nm, and all samples were exposed in duplicate. Also, most of the samples were rotated from top to bottom and between

CH3365-4/94/0000-0893 $4.00 © 1994 IEEE

First WCPEC; Dec. 5–9, 1994; Hawaii

racks on a weekly basis so as to normalize any minor Temperature/Irradiance (T/I) differences.

Laboratory staff monitored all samples for any color/chemical changes during exposure. Non-destructive tests performed included Yellowness Index per ASTM D-1925 and percent light transmission (%T) between 250 and 900 nm via UV-Visible Spectroscopy (UV-VIS). Destructive testing, meanwhile, concentrated on additive concentration analysis and vinyl acetate content analysis. Then at the conclusion of the accelerated UV exposure testing, selected samples were forwarded to the University of Connecticut's Institute of Materials Science (UCONN/IMS) for analysis.

RESULTS

Use of "Fast-Cure" EVA Reduces Discoloration

In an extensive survey of flat plate photovoltaic modules which had experienced EVA browning, it was found that "standard-cure" EVA was used in all instances. "Standard-cure" EVA contains Lupersol 101 peroxide as the crosslinker. Conversely, there were no reported cases of EVA browning with "fast-cure" 15295 EVA. This "fast-cure" EVA is crosslinked with Lupersol TBEC peroxide.

Laboratory accelerated UV aging studies mimicked these results. "Fast-cure" EVA yellowed at a much slower rate than the "standard-cure" A9918 (see Fig. 1). After 20 weeks in the Weather-Ometer, samples laminated with A9918 had a color index of 48 compared to only 21 for laminates made with "fast-cure" 15295.

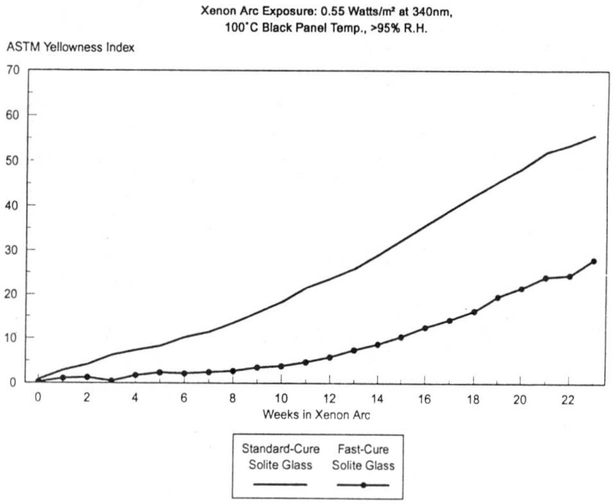

Fig. 1. Accelerated UV aging of "standard-cure" versus "fast-cure" EVA encapsulants.

Further analytical studies have since confirmed that the Lupersol 101 peroxide contained in the "standard-cure" A9918 grade is a major contributor to photothermally induced color development.

Glass With Cerium-Oxide Reduces Discoloration

Laminates of A9918 EVA which used either Solatex II or Airphire glass as the superstrate showed greatly reduced discoloration (see Fig. 2). After 20 weeks in the Weather-Ometer, these samples exhibited a color index of only 10 to 11, compared to Solite glass control samples which had a color index of 48. Yellowness visibility in the control samples was also similar to that of the fielded modules without mirror enhancement.

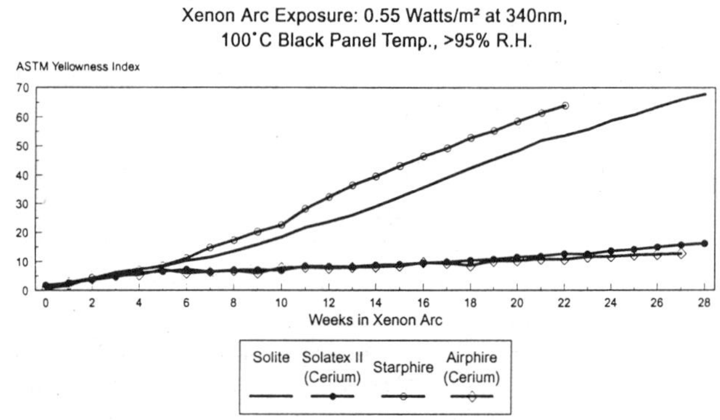

Fig. 2. Effects of glass composition on accelerated UV aging of "standard-cure" A 9918P EVA.

Figure 3 (see Fig. 3) highlights the differences in light transmission over the UV region of the spectrum for these glasses. The spectral irradiance of terrestrial sunlight is included as a reference point. As can be seen, the amount of available energy sacrificed because of selection of a cerium-oxide containing glass is only in the 300 to 350 nm region--an area of the spectrum where solar cells, at best, operate at only 50 percent of their normal efficiency.

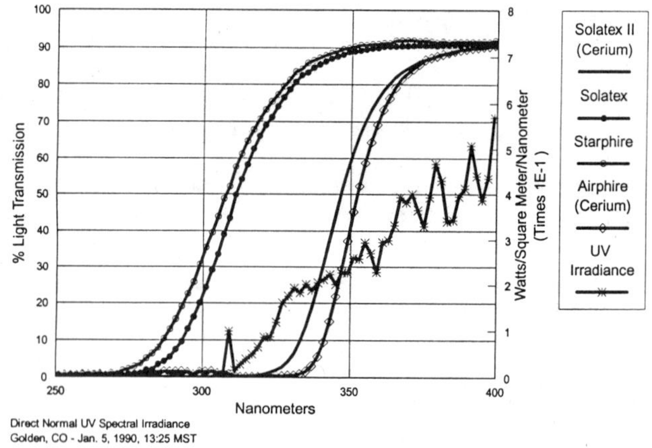

Fig. 3. Light transmission of various glass superstrates versus direct normal UV spectral irradiance.

"Fast-Cure" EVA With Cerium-Oxide Shows No Discoloration

After a full 21 weeks in the Weather-Ometer, laminates prepared with 15295 EVA and either Airphire or Solatex II superstrate had color indices of only 4.4 and 3.3--a yellowness level which is not discernible to the human eye. In contrast, samples with the Solite superstrate which were subjected to the same 21 weeks of accelerated weathering had a color index of 24.0 (see Fig. 4).

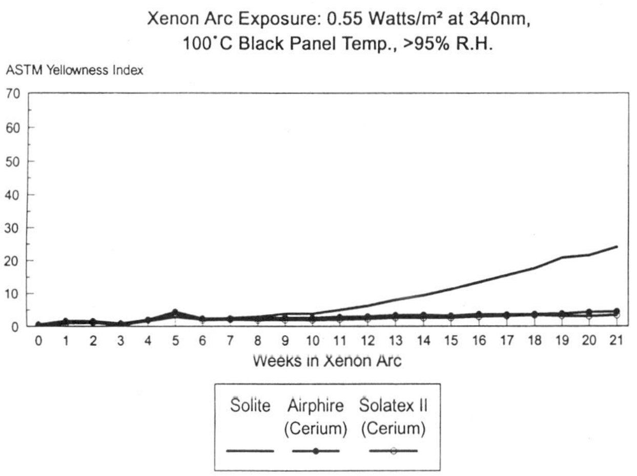

Fig. 4. Effects of glass composition on accelerated aging of "fast-cure" 15295P EVA.

No Loss of Acetate Group in Aged EVA

Carrying testing a step further, virgin EVA pellets of Elvax 3185 and fully formulated "standard-cure" EVA encapsulant were analyzed for vinyl acetate content, both before and after field aging. Analysis was by complete thermolysis of the acetate group using TGA.

In all cases, vinyl acetate content in the EVA resins were equivalent and within experimental error standards for TGA. Also, similar tests performed on browned EVA samples taken from the field-aged, mirror enhanced modules of Carrisa Plains, CA, showed no evidence of photothermolysis of the vinyl acetate groups (see Table 1).

Sample ID	Sample Description	Color	% VA corrected for ash residue
Elvax 150	Raw Material as rec'd.	None	32.6
A9918P	Uncured Sheet	None	33.2
28948-9	Cured, Unaged Glass Laminate	None	34.9
Panel #2	Full X Section Outside or Cell Near Edge	None	34.5
Panel #2	Top of Dark Area (next to glass)	Brown	34.6
Panel #2	Bottom of Dark Area (next to cell)	Clear	35.5
Panel #3	Colorless X Section	Clear	35.1
Panel #3	Dark X Section	Brown	33.9
Panel #3	Top of Dark X Section	Brown	35.5

Table 1. Percent vinyl acetate content by TGA.

Non-Colored EVA in Fielded Modules Shows Oxidation

Finally, FTIR analysis was performed on both non-colored and browned EVA areas of fielded modules. First, the browned Carrisa Plains samples showed no significant double-bond absorbances in areas of the spectrum where one would expect them. This functionality would have been prevalent if the browning was due to conjugated unsaturation in the EVA backbone. But FTIR analysis of the clear EVA areas in fielded modules--namely areas between cells and above cracks in cells--showed strong absorbance at 3550 cm^{-1}. This finding is ascribed to hydroxyl groups resulting from oxidation of the encapsulant (see Fig. 5). Further testing of the browned EVA samples at 3550 cm^{-1} showed much weaker double-bond absorbance in those samples taken from areas over the cells.

Thus, oxidation appears to occur only in those areas where the EVA had access to atmospheric oxygen.

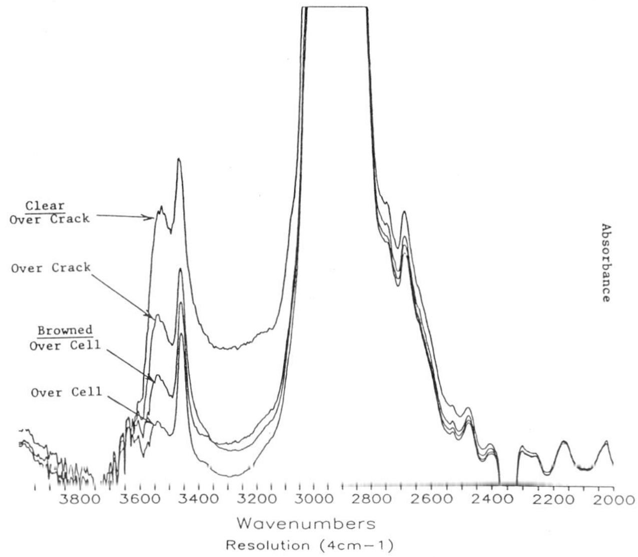

Fig. 5. Infrared spectroscopy of browned versus clear EVA.

CONCLUSIONS

Evidence to date discounts the idea that the browning of EVA in fully EVA encapsulated flat plate photovoltaic modules is based on conjugated unsaturation in the EVA polymer backbone, as a result of the photothermolysis of acetate groups. Instead, the discoloration appears to result from the interaction of additives in the presence of temperature and UV--with a combination of Lupersol 101 peroxide and other additives proving particularly troublesome. Thus, we conclude that the use of EVA as a base polymer for flat plate solar cells is a sound choice, and find EVA a very strong candidate for the development of an encapsulant with improved outdoor exposure stability.

Additionally, glass superstrates containing cerium-oxide are effective in minimizing EVA encapsulant discoloration as they screen out a majority of the received UV-B radiation. This leads to the conclusion that the use of an alternate crosslinking agent, such as Lupersol TBEC, and cerium-oxide-containing superstrates will virtually eliminate the problem of EVA discoloration over a module's normal life span.

NOTE: This work was made possible by the U.S. Department of Energy through subcontract ZAG-3-11219-02-105661 to Springborn Laboratories, Inc. from the National Renewable Energy Laboratory (NREL), Golden, Colorado. These studies are part of NREL's on-going Photovoltaic Manufacturing Technology (PVMaT) program, under Phase 3A, Shared Process Issues. The Technical Monitor on this program is Holly Thomas of NREL.

FACTORS THAT AFFECT THE EVA ENCAPSULANT DISCOLORATION RATE UPON ACCELERATED EXPOSURE

F. J. PERN

Measurements and Characterization Branch
National Renewable Energy Laboratory
1617 Cole Blvd., Golden, CO 80401, USA

ABSTRACT

Several factors that may affect the net discoloration rate of the ethylene-vinyl acetate (EVA) copolymer encapsulants used in crystalline-Si photovoltaic (c-Si PV) modules upon accelerated exposure have been investigated by employing UV visible spectrophotometry, spectrocolorimetry, and fluorescence analysis. A number of laminated films, including the two typical EVA formulations, A9918 and 15295, were studied. The results indicate that the rate of EVA discoloration is affected by the (1) curing agent and curing conditions; (2) presence and concentration of curing-generated, UV-excitable chromophores; (3) UV light intensity; (4) loss rate of the UV absorber, Cyasorb UV 531; (5) lamination; (6) film thickness; and (7) photobleaching rate due to the diffusion of air into the laminated films. In general, the loss rate of the UV absorber and the rate of discoloration from light yellow to brown follow a sigmoidal pattern. A reasonable correlation for net changes in transmittance at 420 nm, yellowness index, and fluorescence peak area (or intensity) ratio is obtained as the extent of EVA discoloration progressed.

INTRODUCTION

Browning of the ethylene vinyl acetate (EVA) encapsulant used in crystalline-Si photovoltaic (c-Si PV) modules can reduce the efficiency because of a decreased light transmittance. One of the worst cases is the browning of EVA in the c-Si PV modules deployed at Carrisa Plains, California, that use a V-trough-mirror configuration. Consecutive decreases in the annual power output have been well documented since 1986 [1,2]. Recently, browning of the EVA has also developed in the c-Si modules with a similar mirror-enhanced configuration deployed in the Negev Desert, Israel [3]. Degradation mechanisms of the EVA encapsulant have been investigated and discussed [4,5]. We have previously conducted post-mortem analyses of field-degraded, yellow to dark brown EVA to understand the chemical and physical changes in the material; laboratory-simulated degradation to characterize the degradation mechanisms; and accelerated exposure of laminated solar cells to quantify the effect of EVA discoloration on cell efficiency [6-8]. In this work, three non-invasive, spectrophotometric, analytical methods--optical transmission, colorimetry, and fluorescence analysis--were employed to investigate the factors that may affect the rate of EVA discoloration in glass-laminated samples that were subjected to accelerated exposure.

EXPERIMENTAL

A number of film samples of various formulations including the two typical (*unprimed*) EVA formulations, A9918 and 15295 [9,10], were prepared. The films were laminated at ~115°C for ~10 min between cleaned glass slides of either 2.5 cm x 2.5 cm x 0.16 cm (1" x 1" x 1/16") quartz or 5 cm x 5 cm x 0.32 cm (2" x 2" x 1/8") borosilicate by using a manually controlled, hydraulic minipress with two heated platens. Aluminum spacers were employed to control the laminated film thickness under applied hydraulic pressure. Curing was conducted by heating for ~40 min at ~145°C for the EVA A9918 using the curing peroxide Lupersol™ 101 (Elf Atochem) and ~12 min at ~145°C for the EVA 15295 using Lupersol™ TBEC [9,10]. The laminated films were exposed to either a concentrated 1-kW Xe light (Oriel) or a 1-kW Xe, enhanced-UV solar simulator (Oriel) filtered by a 0.32 cm (1/8")-thick borosilicate slide to remove UV light below ~280 nm. The irradiance of the light integrated from 300 nm to 400 nm is ~17.6 sun (relative to the global normal August solar spectrum measured in Golden, Colorado) for the concentrated 1-kW Xe light measured with an OL752 spectroradiometer at the test plane and ~4.8 sun for the solar simulator with the borosilicate filter. The exposed samples were analyzed periodically for their transmittance with a Hewlett-Packard Model 8452A UV-Vis spectrophotometer, color indices with a HunterLab UltraScan™ colorimeter, and fluorescence characteristics with a SPEX Model FL112 Fluorolog II fluorescence spectrophotometer.

RESULTS AND DISCUSSION

Effects of Curing and Formulation on EVA Discoloration

The curing step at 145°C introduces two effects on the cured EVA films besides the purposed cross-linking. First, the

CH3365-4/94/0000-0897 $4.00 © 1994 IEEE

First WCPEC; Dec. 5–9, 1994; Hawaii

calculated residual concentration of Lupersol 101 in EVA A9918 is ~40% after 50 min, which is in marked contrast to the nearly complete consumption of Lupersol TBEC in ~20 min because of a lower activation energy and a shorter half-life of the latter [11]. Second, the slow-cured EVA A9918 produces a much greater concentration of UV-excitable chromophores than the fast-cured EVA 15295, as characterized by fluorescence analysis [11,12]. This may result in an incomplete shielding of these chromophores by the UV absorber, Cyasorb UV 531 (referred to as Cyasorb hereafter), from being excited by UV light [12] so that photo-initiated Norrish degradation reactions may occur [13]. The consequence of possessing higher concentrations of residual peroxide and UV-excitable chromophores is a significantly reduced photothermal stability of the cured EVA A9918 than that of EVA 15295 as reported recently [11].

The EVA discoloration is accompanied by photodecomposition of the UV absorber, Cyasorb, as illustrated in Fig. 1a for a laminated and cured EVA A9918 film exposed to the concentrated 1-kW Xe light. The increase in transmittance below 350 nm results from a decreased concentration of Cyasorb. The decrease in the transmittance at wavelengths >360 nm is accompanied by the appearance of a light yellow color that gradually darkens to a light brown as the transmittance in the visible range continues to decrease. The corresponding net changes in color indices are shown in Fig. 1b for the yellowing film. A simplified Adam-Nickerson chromaticity system or CIELAB is used here [14]. When curve-fitted, the YI (or b*) curve exhibits a sigmoidal pattern during the course of color development from light yellow to light brown followed by a slight linear slope. In effect, the increase in the yellowness index (YI) and loss in transmittance at 420 nm are mirror images. The discoloration is attributed to the formation of a mixture of conjugated polyenes and polyenic a,β-unsaturated carbonyls [15]. As expected, when the light intensity is increased, the discoloration occurs earlier and the rate is faster (not shown).

The photodegradation rate can be easily differentiated by monitoring the fluorescence emission (FE) characteristics of both of the exposed EVA laminates. Figure 2 shows the time-dependence of emission peak intensities at 420, 500, and 550 nm (excited at 350 nm) that changes upon photo-initiated structural changes of the chromophores. Three common degradation mechanisms are also indicated in the figure. Results from a recent study show that, when present together, Lupersol 101 and Cyasorb UV 531 induced the yellowing. When used separately, neither component induced noticeable yellowing in the laminates. Furthermore, the cured films discolored more than the uncured films. The details will be published elsewhere [16]. In addition, a thicker laminate made with two layers of EVA A9918 film discolored slower initially but darkened more later than the laminate made of single layer EVA [11]. The net change of total transmittance can be calculated by integrating from 280 nm to 820 nm and subtracting from the original total transmittance. Initially, because of the gradual photodepletion of Cyasorb, a net gain in total transmittance is obtained. Subsequently, a net loss is obtained as the yellow-browning increases and exceeds a gain of YI= ~8-10 [11]. No discoloration is observed for the laminated films that do not contain the curing-generated chromophores such as Elvax 150™ (Du Pont,

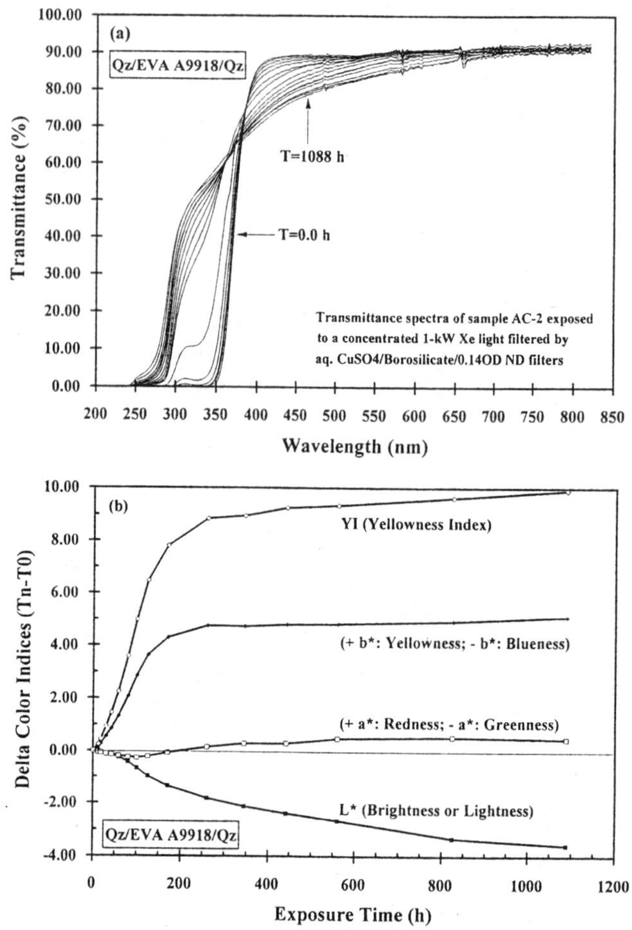

Fig. 1. (a) Transmittance spectra from 0 to 1088 h for an EVA A9918 sample that was laminated between two 2.5 cm x 2.5 cm x 0.16 cm quartz slides, cured, and exposed to a concentrated 1-kW Xe light at 46°C±2°. (b) A plot of yellowness index (YI), a*, b*, and L* measured for the same sample as a function of exposure time. See text for an explanation of the color indices.

the copolymer without any stabilizers or curing agent). The presence of Cu(II) ions in the Elvax 150 films does not greatly promote (as originally expected and verified by fluorescence analysis) the generation of the polyenic chromophores responsible for the EVA discoloration, although the laminates developed an increasingly reddish yellow color upon exposure. More details on the metal ion effects will also be published later [16].

Effects of Delamination and Photobleaching

Delamination of the EVA occurred on two borosilicate laminates after ~1250 h exposure to the enhanced UV light and resulted in a rapid decrease in the yellowness index (ΔYI= ~2-5 within ~300 h) [11]. Eventually, the delamination could be seen visually. The decrease in YI apparently results from photobleaching reactions as air (O_2) diffuses more readily to the yellow-browned EVA as the films gradually delaminate from the glass surfaces. On the other hand, air diffusion into the perimeter of the EVA laminates also accounts for the clear bands (~3-5 mm wide) around the exposed specimens. If the EVA films are only press-sandwiched (not laminated) between glass plates, no discoloration has ever been observed, even for samples exposed to the enhanced UV light for more than 4000 h at ~56°C±3°. Furthermore, EVA laminated with gas-permeable films of Tefzel™ (Du Pont) as superstrates do not discolor when exposed to the enhanced UV light [15], wherein the original chromophores (after curing) are effectively destroyed as verified by fluorescence analysis. The photobleaching reactions cause an oxidative breakdown of the initially existing chromophores or polyenic chromophores in the presence of air, resulting in a color reversal that can be competing with the discoloration of EVA [12].

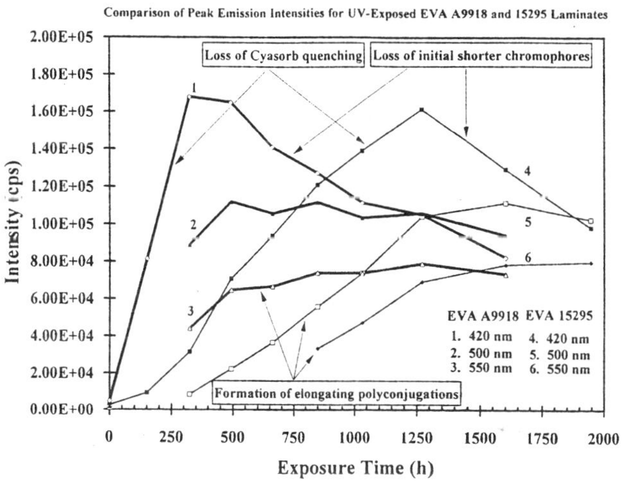

Fig. 2. A plot of fluorescence emission (FE) peak intensities at 420 nm, 500 nm, and 550 nm for films of EVA A9918 and 15295 laminated between two 5 cm x 5 cm x 0.32 cm borosilicate slides, cured, and exposed to the solar simulator's enhanced UV light at 44°C±2°. Three key mechanisms involved in the EVA degradation and discoloration are indicated in the figure.

Correlation of The Spectrophotometric Results

An important objective in this study is to establish a good correlation for the results acquired from the three spectrophotometric measurements in transmittance, color indices (primarily yellowness index), and fluorescence emission (peak intensity or area ratio). For the EVA 15295 laminate, all four curves correlate quite well, as shown in Fig. 3a, after applying some multiplying factors. For EVA A9918 (Fig. 3b), curves for the transmittance changes at 420 nm and YI correlate well with the same multiplying factors, but they show a deviation in the trend from the curves of the emission peak area and peak intensity ratios at T > ~1000 h, which probably results from the delamination problem. Better correlations probably are obtainable if primed EVA films are used.

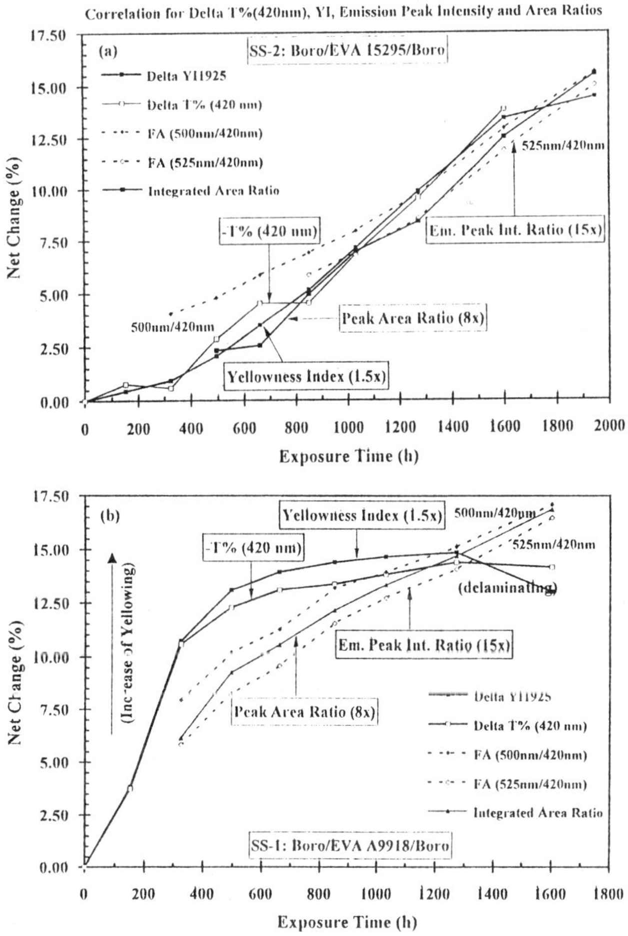

Fig. 3. Correlation plots of net percentage changes in transmittance (T% at 420 nm), yellowness index (YI), and FE peak intensity and area ratios for an (a) EVA 15295 and (b) EVA A9918. Multiplying factors (as indicated) are applied to examine the goodness of fitting of the correlation. The FE peak intensity ratios are for the ratio of an evolving peak in the 500-550 nm range for the polyenic chromophores from EVA degradation to the initial peak at 420 nm. The FE peak area ratios are for the ratio of peak area integrated between 451 nm and 600 nm to the initial peak area integrated between 360 nm and 450 nm. No peak deconvolution for the emission curves was available.

CONCLUSIONS

Several factors that affect the net discoloration rate of EVA have been identified. The EVA formulation, curing agent and treatment, UV light intensity, hermetic lamination, and film thickness are discoloring factors, while delamination, rate of air diffusion, and rate of photobleaching are color reversing factors. The net discoloration rate of EVA is a result of these opposite, competing factors. A reasonably good correlation has been obtained among the transmittance loss, yellowness index increase, and fluorescence peak area (or intensity) ratio changes that should be useful in quantitatively correlating the extent of EVA discoloration to efficiency loss in PV modules in our future studies.

ACKNOWLEDGEMENT

The author thanks S.H. Glick for assistance and D. Trudell and T. Cannon for the spectroradiometric measurements of Xe arc lamps. The author also thank A.W. Czanderna, the team leader of the PV Module Materials and Component Durability task, and R. DeBlasio, the project manager of the PV Module and System Performance and Engineering Project for their support. This work was performed at NREL under support by the U.S. Department of Energy Contract Number DE-AC36-83CH10093.

REFERENCES

[1]. Gay, C. F., and Berman, E., **Chemtech**, March, 1990, 182-6.

[2]. Wenger, H. J., Schaefer, J., Rothenthal, A., Hammond, R., and Schlueter, L., **Proc. of 22nd IEEE PVSC**, IEEE, New York, NY, 1991, pp. 586-591.

[3]. Berman, D., Biryakov, S., and Faiman, D., **Proc. of 12th EC Photovoltaic Solar Energy Conference**, In Press.

[4]. Pern, F. J., and Czanderna, A. W., **Proc. of 11th PV AR&D Review Meeting**, Denver, CO, May 13-15, 1992. American Institute of Physics, Conf. Proc. No. 268, New York, 1992, pp. 445-452.

[5]. Czanderna, A. W., **Proc. of PV Performance and Reliability Workshop**, ed., L. Mrig, NREL/CP-410-6033, Sept. 8-10, 1993, pp. 311-357.

[6]. Pern, F. J., **Proc. of Photovoltaic Module Reliability Workshop**, ed., L. Mrig, SERI/CP-4079, Oct. 25-26, 1990, Lakewood, CO, pp. 279-299.

[7]. Pern, F. J., and Czanderna, A. W., **Solar Energy Materials and Solar Cells, 25**, 1992, 3-23.

[8]. Pern F. J., **Proc. PV Performance and Reliability Workshop**, ed., L. Mrig, SERI/CP-411-5184, Sept. 16-18, Golden, CO, 1992, pp. 326-44.

[9]. Cuddihy, E., Coulbert, C., Gupta, A., and Liang, R., "**Flat-Plate Soalr Array Project Final Report, Vol. VII-Module Encapsulation**," JPL publication 86-31, October 1986. (DOE/JPL-1012-125)

[10]. Willis, P. B., "**Investigation of Matrials and Process for Solar Cell Encapsulation**," Final Report of JPL Contract No. 954527, S/L Project 6072.1 by the Springborn Lab. to JPL, JPL Publication, DOE/JPL-954527-86/29, 1986.

[11]. Pern, F. J., **Proc. of PV Performance and Reliability Workshop**, ed. L. Mrig, NREL/CP-411-7414, Sept. 21-23, 1994, Lakewood, CO, pp. 329-347.

[12]. Pern, F. J., and Glick, S. H., in Noufi R., and Ullal H., ed., **AIP Conference Proceedings for the 12th NREL PV Program Review Meeting**, no. 306, American Institute of Physics, NY, 1994. pp. 573-585. (NREL/TP-412-5996).

[13]. Gugumus, F., in "**Mechanisms of Polymer Degradation and Stabilization**," ed., G. Scott, Chap. 6, Elsevier Applied Science, New York, 1990, pp. 169-210.

[14]. Blakey, R. R., "**Measuring Colour**," a technical publication by Tioxide Group PLC, England, 1990 (registration no.: 249759).

[15]. Pern, F. J., **Polym. Deg. and Stab., 41**, 1993, 125-139. (NREL/TP-213-4605).

[16]. Glick, S. G., and Pern, F. J., to be published (in preparation).

A MORE ELEMENTAL EXAMINATION OF FACTORS GOVERNING PV MODULE ENVIRONMENTAL STABILITY

Theresa L. Jester, Don Aldrich, Jean Hummel, and Robert D. Wieting
Siemens Solar Industries, P.O. Box 6032 Camarillo, CA. 93012

James P. Galica and Lynne Thoma
Springborn Laboratories, Inc. One Springborn Center Enfield, CT. 06082

ABSTRACT

Photovoltaic module reliability studies are essential in understanding both the technical and commercial design lives of various photovoltaic modules.

To date, photovoltaic module reliability studies have primarily focused on retention of initial power when exposed to various environmental stresses. While power of a pv module is one reliability measure, the fundamental parameters affecting photovoltaic module stability are typically not measured for baseline analysis. This study examines these questions in context of fundamental parameters, in an attempt to gain insight at a more elemental level than possible using power measurements.

Reliability parameters must be identified, and analyzed in a systematic and statistical method. One of these fundamental parameters in pv module reliability is the adhesion of the encapsulant EVA to various module surfaces.

The analysis of surface adhesion, both in its initial state, and as it changes with various exposures is studied here, primarily focusing on adhesion of EVA to glass, cell surfaces and the interconnect ribbon. Of particular interest is the affect of manufacturing processes and surface cleanliness on this bond strength.

Siemens Solar Industries, with Springborn Laboratories have begun analyzing adhesion of EVA using a systematic and statistical method of sample coupons measured at multiple points, and through numerous exposures. Specifically, several EVA types were compared for adhesion characteristics. Measurements of EVA adhesion and how it is affected by surface contamination due to residual solder flux have also been specifically quantified. The summary analysis looked at EVA bonds as a function of environmental exposures.

The results of this analysis shows that initial strength of EVA bonds are quite high, these are diminished by both surface contamination effects and environmental exposures. The ultimate bond strength and the ratio to the initial value are variable dependant upon the exposure.

INTRODUCTION

A framework for measuring EVA adhesion has been developed by designing sample coupons to allow quantitative assessment to be made using peel strength measurements of the adhesion of EVA encapsulant film to the various materials embedded in PV modules. These include cell surfaces of different kinds, interconnect ribbons, and glass surfaces. The experimental coupon was designed to accommodate three peel tabs per surface, and is intended to be used in multiple repetitions to obtain statistically significant results. The coupon designed (see Fig 1.) had to be capable of exposure in many types of environmental chambers, and be small enough to allow for multiple units to be fabricated at a time in a vacuum laminator.

Each bondline test specimen set was prepared in triplicate with three 180° "T-Peel" strength measurements made per specimen for a total of nine data points per condition. These peel strength data points were averaged and checked for statistical significance using an Analysis of Variance (ANOVA) technique.

Experiments have been performed to generate baseline adhesive strength data on various grades of

CH3365-4/94/0000-0901 $4.00 © 1994 IEEE First WCPEC; Dec. 5–9, 1994; Hawaii

EVA encapsulant bonded to each critical surface contained in PV modules. Some EVA formulas have substantially more silane primers incorporated into their recipes. There is concern that higher primer concentrations may cause the PV module to be more moisture sensitive due to excess silane at the bondline which can in fact reduce bond strength rather than promote adhesion. The recommended silane primer treatment is to provide a monolithic molecular layer of silane at the bondline to achieve maximum bond strength.

EVA ADHESION

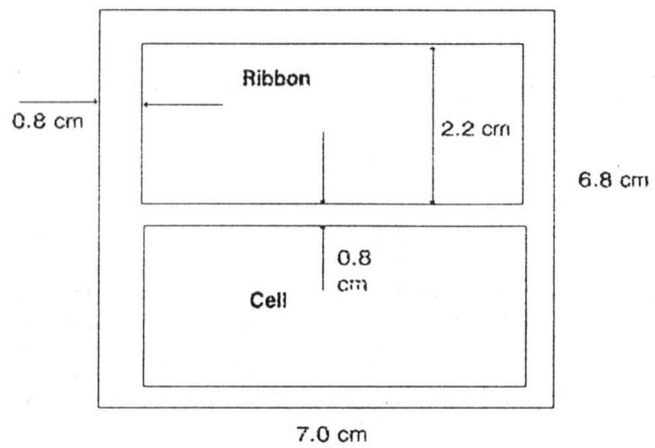

Fig. 1. EVA test coupons.

The objective of the first phase of the experiment was to evaluate the differences in adhesion characteristics of three commonly utilized primed EVA encapsulant. Four critical bondlines were evaluated.

These four critical bondlines were measure in a standard ASTM environment of 50% RH+/- 5% and 23 °C+/-2 °C.

Results of this experiment showed that the average peel strength of all formulas of EVA were high, in excess of 30 pounds/inch on the glass and ribbon surfaces, and approximately 20 pounds/inch on the cell front and cell back. There was no significant dependance on the amount of silane primer present (see Fig. 2), though there was some variation in value due to the surface being adhered to.

The significance of this data is to set a baseline level of adhesion for any process or product change, and for process and product monitoring in manufacturing of photovoltaic modules.

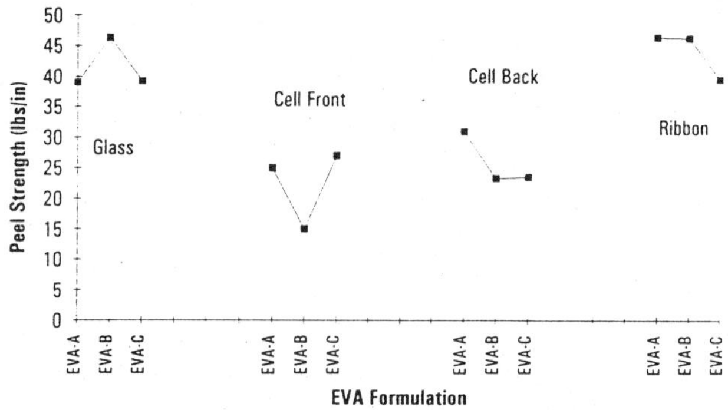

Fig. 2. EVA Adhesion as a function of formulation

MANUFACTURING EFFECTS

Manufacturing processes to produce photovoltaic modules require some amount of monitoring and quality checking on a routine sampling basis. Once a module design is finalized, tested and approved for production, a new set of reliability concerns surface. The manufacturing process is on a path of continuous improvement requiring a certain amount of verification each time a change is made. The fundamental understanding of adhesion mechanisms and a quantified baseline must be checked before, during and after implementation of any change.

The environmental imperative to eliminate chlorofluorocarbons from manufacturing processes has posed significant challenges for soldering operations in stringing silicon solar cells. The removal of freon from the manufacturing sequence has been a significant step in continuously improving manufacturing processes, and has primarily been replaced with new soldering materials such as low-residue or "no clean" solder pastes and fluxes.

The introduction of these no clean pastes have included the standard qualification testing requirements, however it has raised new questions about the potential impact of these foreign materials on the long term stability of a pv module.

A series of experiments has been performed measuring specifically the adhesion of EVA to cell front surfaces, and ribbon surfaces as a function of flux concentration (see Figs. 3 and 4). As a reference, each test had a control sample of materials cleaned with freon. The freon cleaned samples had the highest initial peel strength. A decay in the adhesion occurs as the concentration of flux is increased. The practical implication of this result could mean an extra step for cleaning strings in a manufacturing process. The adhesion ranges for "contaminated" flux surfaces were 20-30 pounds/inch on the cell surface, and 10-20 pounds/inch on the ribbon surface.

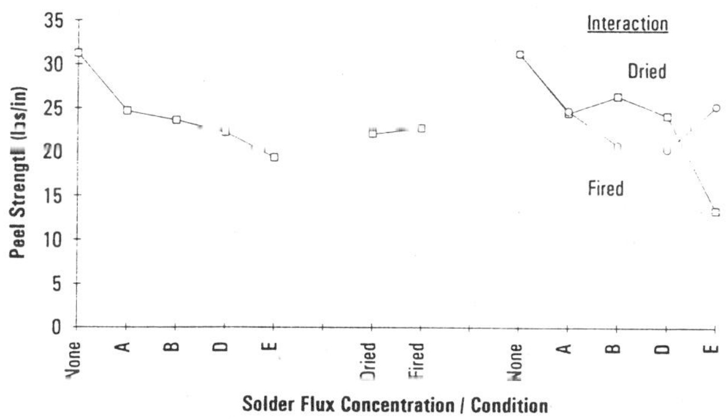

Fig. 3. EVA Adhesion to cell surface with various flux concentrations

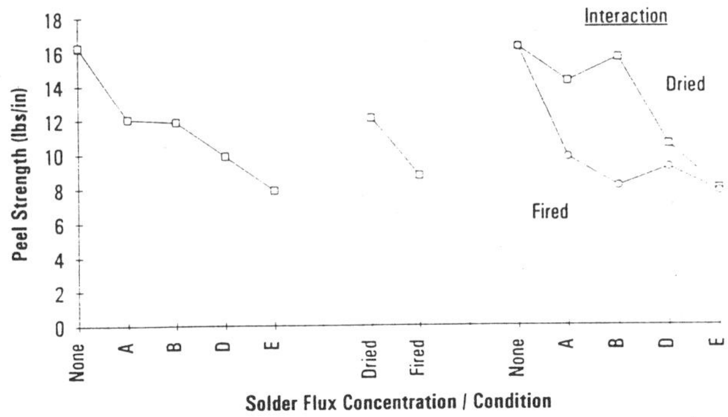

Fig. 4. EVA Adhesion to ribbon surface as a function of flux concentration

ENVIRONMENTAL EFFECTS

Environmental effects on photovoltaic modules can be severe, especially humidity effects.

This phase of the study focused on these effects, with various EVA formulations exposed to three different environmental stresses. The first environmental exposure was 500 hours in arid conditions of 0% RH and 85 °C. The effect on adhesion peel strength was minimal when compared to base line adhesion measurements (see Fig. 5). Typical adhesion values after thermal exposure retained 90% of the baseline value.

The second environmental stress condition was 10 days of humidity freeze cycling from +85 °C to -40 °C at a humidity level of 85% RH. The effect on adhesion peel strength was a dramatic lowering of values when compared to the base line measurements. (see Fig. 6). Typical values after humidity freeze exposure are 50% or less than the base line initial values.

The final environmental exposure was 500 hours at 85% RH and 85°C. This also had an adverse affect on adhesion to each surface when compared to baseline measurements (see Fig. 7). Again, typical values were less than 50% of the baseline measurements.

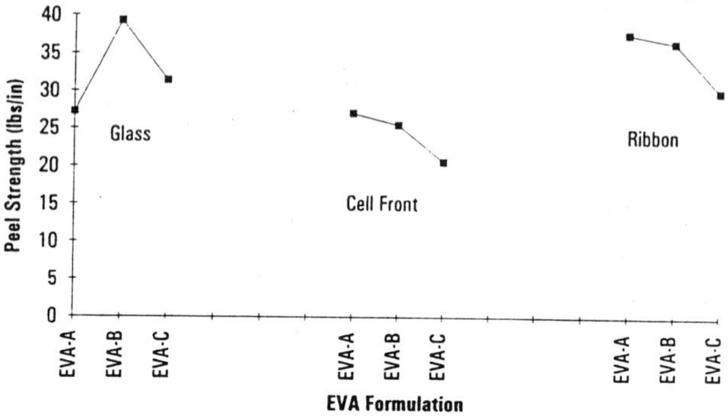

Fig. 5. EVA adhesion after thermal exposure

None of the environmental exposure tests showed a significant dependance on the type of EVA used. The type of environmental exposure was a much more dominant factor.

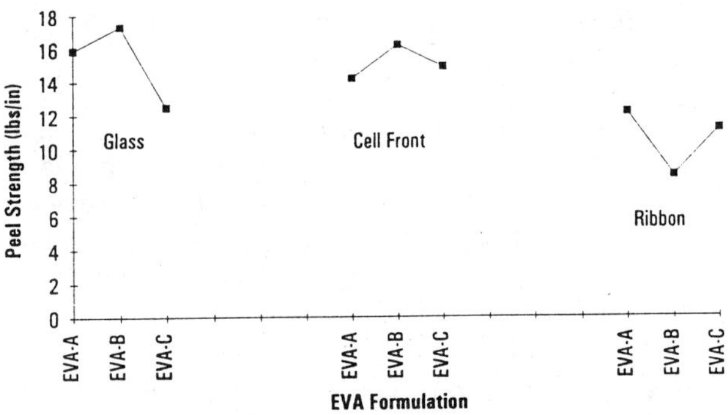

Fig 6. EVA adhesion after humidity/freeze cycling

SUMMARY

EVA adhesion is a fundamental measurement used to understand pv module stability. A systematic measurement and statistical analysis technique is required in order to draw have significant conclusions.

The technique employed in this study is both systematic and statistical, and has shown that the EVA formulation with respect to the amount of silane primer present is not a strong driver in the adhesion peel strengths to various module surfaces.

Contamination of the surfaces in a pv module, such as those left with "no clean" solder fluxes on the surface of the cell and ribbon, have a dramatic affect on adhesion. The lower the concentration of flux on a surface, the better the EVA bond strength is. This has the practical implication of cleaning soldered cells for optimum adhesion characteristics.

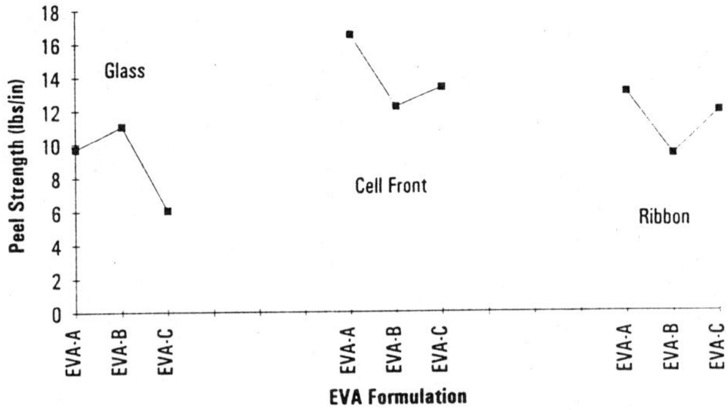

Fig. 7. EVA adhesion after damp heat exposure

The most significant effect on EVA adhesion to module surfaces comes with environmental exposure. While thermal affects are minimal, significantly lower peel strengths were obtained after both humidity freeze and damp heat exposures. Humidity has a large impact on the adhesion values to each module surface.

The summary of this study and its impact lies in the insight it provides at an elemental level of module reliability. The higher the adhesion, generally, the higher the reliability. The understanding and quantification of this adhesion allows for a benchmark for comparison of various product and process changes. The systematic measurement of this adhesion allows for a better understanding of both basic reliability and a monitoring parameter for process changes.

THE EFFECT OF ANNEALING AND OXIDATION ON APCVD TiO_x FILM AND ITS IMPACT ON THE PROCESS OF SILICON SOLAR CELLS

by

David C. Wong and Art Waugh
BTU International, North Billerica, MA

Bob Yui and Paul Sharrock
Texas Instruments, Inc., Dallas, TX

ABSTRACT

Properties of TiO_x film, prepared by Atmospheric Pressure Chemical Vapor Deposition (APCVD) as an Anti-Reflective (AR) coating on silicon solar cells have been extensively investigated. The effect of film densification, after annealing at temperatures as low as 430° C is the transition from amorphous to crystalline. This effect is found to be crucial to the large scale manufacturing of solar cells because of its cost effectiveness and simpler processing requirements. Results of heat treatment and the change in the morphology will be presented. The impact of annealing on Spheral Solar™ cell performance will also be discussed.

INTRODUCTION

Of the 21 megawatts global PV module shipment in 1993, single and poly-crystalline silicon solar cells have accounted for almost 65% and 31%, respectively, of the total [1]. Because of the production volume, any process technology or cell design that could improve the conversion efficiency by a fraction would have a direct impact on the cost. One of the most economical and direct methods to increase a photovoltaic (PV) silicon solar cell output is by anti-reflective coating the surface. A polished silicon surface reflects more than 33% of the incident light because of its high refractive index (R.I. ~ 3.8).

To reduce the reflection, a transparent film of an intermediate refractive index, is required to step between the silicon and the encapsulating material [2, 3]. Among the many AR films available, TiO_x is the best choice for terrestrial PV application for several reasons. It has an appropriate range of refractive index (1.9 to 2.3). The film can be processed at (relative) low cost with high throughput using Atmospheric Chemical Vapor Deposition (APCVD) - in this case, a low temperature process using an in-line conveyor belt furnace [4, 5].

Typical processing costs, in a one megawatt production line, is less than $0.02 per watt [4] (dependent upon cell efficiency). This method has a return of a 45% increase in current output on a polished cell, and up to 14% for a textured silicon cell. An AR coating on polycrystalline cells is even more important because at least 1/3 of the total grains are oriented at the <111> plane that is non-texturable [6]. Without an AR coating, 33% of the incident light on the <111> plane would be reflected. Together with reflection from the other planes, a total loss of 15% in Isc is not unusual.

Another valuable characteristic of TiO_x is the annealing effect even at low temperature. This characteristic enables the TiO_x film to fit into various solar cell processing to be discussed.

EXPERIMENTAL

Polished 5" silicon wafers and Spheral Solar cells were processed in a BTU APCVD belt furnace for the TiO_x deposition.

For the study of the heat treatment effect related to silicon cell processing, TiO_x is prepared respectively at a lower deposition temperature of about 230 °C and at the higher deposition temperature of about 265 °C, a range that would neither cause impurity diffusion nor redistribution of surface dopant concentration at the PN junction. The adhesion of the lower deposition temperature is poor enough to be removable by a light tissue paper touch. In order to limit the number of variables, film thicknesses were preferred to be near 750 A, a quarter wavelength of 6500 A spectrum for minimum reflection on the silicon surface. However, the thickness of the lower deposition temperature film is deliberately adjusted to over 900 A with the anticipation of film densification. The control of film thickness is achieved by adjusting the belt speed, the ratio of chemical flows, and exhaust flow rates.

The typical effect of the post-deposition annealing, as measured in the Texas Instruments Spheral

CH3365-4/94/0000-0905 $4.00 © 1994 IEEE First WCPEC; Dec. 5-9, 1994; Hawaii

Solar™ pilot line, was used a starting point for this work. A wider range of annealing temperatures was next investigated. The index of refraction and film thickness of each wafer, as deposited, are measured with an (Gaertner Scientific Model L116B) ellipsometer. Each wafer is cleaved into quadrants for the annealing experiment at 300 °C, 440 °C, and 870 °C, respectively, in air ambient. The high temperature annealing at 870 °C is only for 20 seconds to simulate a contact sintering process. After the heat treatment, the index of refraction and the corresponding film thickness of each sample was re-measured.

The film surfaces of test wafers from the TI Spheral Solar™ pilot line are analyzed by TEM at 35kx magnification. Film cross-sections are revealed by TEM at 250kx and the structural morphology is studied by x-ray diffraction.

The effect of TiO_x on Spheral Solar™ cell is investigated by comparing the cell performance before and after deposition and annealing.

DISCUSSION

Films as deposited at 230 °C have a low index of refraction at 1.95 versus the crystalline TiO_x (rutile) value of 2.35 (see Table 1).

Heat Treatment	None	300 °C	440 °C	870 °C
Duration time	n/a	30 min.	30 min.	20 sec.
Refractive Index	1.95	2.1	2.15	2.06
--> increment%	n/a	7.8%	10.3%	5.6%
Thickness (A)	941	758	695	820
--> contraction %	0	19.5%	26.1%	12.9%

Table 1. The Effect of Heat Treatment of TiO_x Films deposited on silicon wafers at 230 °C.

Heat treatment at temperature as low as 300 °C for 30 minutes improves the index by 7.8% with almost 20% in thickness contraction. The increase in refractive index and contraction in thickness are even more obvious at the higher annealing temperature of 440 °C for the same time.

Results suggest that the film after heat treatment has gone through a phase transition from amorphous-like to crystalline. The phenomenon is in agreement to Ottermann's observation [7].

Films deposited at 265 °C, however, are more crystalline-like with less free space (i.e., lower in porosity) that limits the rearrangements of atoms by annealing as indicated in Table 2. The film contraction is confirmed by, test wafers that were processed on the TI Spheral Solar™ pilot line, the TEM cross-section analysis before and after 430 °C annealing (see Fig. 1).

Grain structures, after annealing, are shown by TEM at 35kx magnification (see Fig. 2) and confirmed by x-ray diffraction, which indicated an anatase structure (see Fig. 3).

Heat Treatment	None	300 °C	440 °C	870 °C
Duration time	n/a	30 min.	30 min.	20 sec.
Refractive Index	2.15	2.24	2.25	2.18
--> increment%	n/a	4.2%	4.2%	1.4%
Thickness (A)	753	672	677	724
--> contraction %	0	10.8%	10.1%	3.8%

Table 2. The Effect of Heat Treatment of TiO_x Films deposited on silicon wafers at 265 °C.

Figure 1a. Cross-section TEM of as-deposited wafer.

Figure 1b. Cross-section TEM of post-annealed wafer.

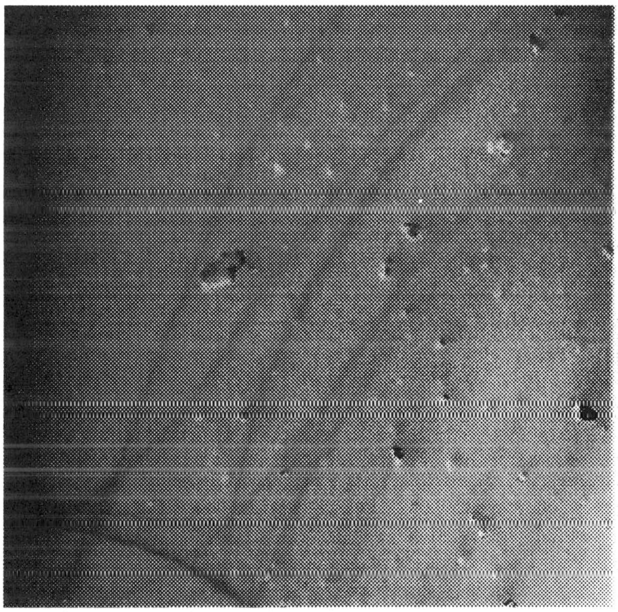

Figure 2a. Top shot TEM of as-deposited wafer.

Figure 2b. Top shot TEM of post-annealed wafer.

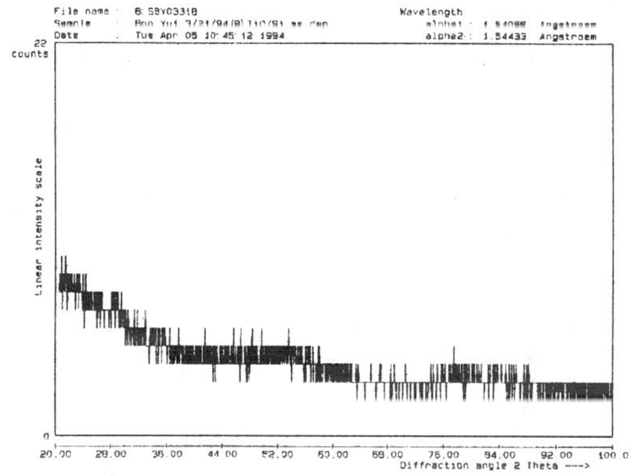

Figure 3a. X-ray diffraction chart for as-deposited wafer

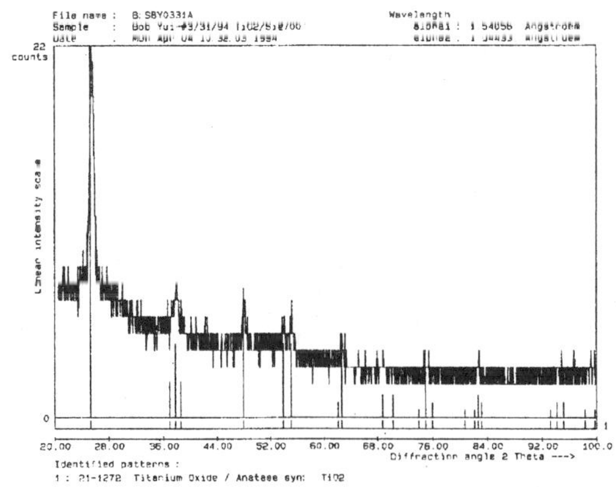

Figure 3b. X-ray diffraction chart for post-annealed wafer.

Heat treatment of films at 870 °C for 20 seconds is of interest to cell processing. The selected temperature and duration time are to simulate silver paste contact sintering for single-crystal silicon solar cells. The data suggests that the film quality would be improved even during the pulsed firing, especially on low temperature-processed TiO_x films of high porosity. Such films are favorable for the "firing-through" process in which the TiO_x film is deposited at the lower temperature followed by screen printing and sintering. A final step would be a 440 °C densification, a temperature below oxygen donor formation [8], through a belt furnace in an air atmosphere. Besides the improvement in film quality, the densification procedure would also stress relieve the silicon contact interface. The fire-through process is preferable because ribbon strings can be soldered directly onto the bus without cleaning.

For Spheral Solar™ cells, no firing-through process is needed. [9] The AR coating is typically deposited at a high temperature about 265 °C followed by a 430 °C heat treatment to further densify the film and stress-relieve the silicon-aluminum bond. As shown in Table 3, the increase in Jsc is due to the antireflection of the TiO_x film and the improvement in fill factor is obviously because of the annealing effect.

	Voc (volts)	Jsc (mA/cm^2)	Fill Factor %
no TiO_x deposition	0.541	19.8	67.5
no TiO_x dep. & post-annealed	0.544	20.2	69.1
TiO_x deposition	0.550	23.0	69.9
TiO_x dep. & post-annealed	0.554	24.3	70.7

Table 3. Spheral Solar™ cell results

Note: There , typically, is a 10-15% boost in Jsc after encapsulation.

CONCLUSION

The versatile property of APCVD TiO_x film as a function of deposition temperature and heat treatment has been studied. It is shown that its low recrystallization temperature from amorphous state to crystalline makes it suitable for a "firing-through" metallization process. By heat treating the film in air, the index of refraction would increase to a better matching of the module lamination material. The porosity of TiO_x film that is permeable for oxygen diffusion to form silicon dioxide on the silicon surface might have a passivation effect for improving the Voc.

ACKNOWLEDGEMENTS

The authors gratefully acknowledge the assistance of John Nicholson and Tracy Whippen of BTU International, and Lorraine Bentle, Mike Coviello, Jim Florence, and Winfred Robinson of Texas Instruments.

REFERENCES

[1] Photovolatic Insider's Report, Vol. XIII, **No. 10**, p. 6, 1994.

[2] E.Y. Wang et al., Conference Record, **10th** IEEE Photovoltaic Specialist Confernece, 1973, p. 188.

[3] P.A. Iles, J. Vac. Sci. Technol., **14(**1977) 1100.

[4] D.C. Wong et al., International SAMPRE Electronics Conference Series, **Vol. 5**, 1991, p. 476.

[5] N.M. Gralenski, SPIE, Optical Coatings for Energy Efficiency and Solar Applications, 1982, p. 44.

[6] A.W. Smith et al., **21st** IEEE Photovolatic Proceedings, 1990, p. 426.

[7] C. Ottermann et al., MRS, **Vol. 308**, 1993, p. 60.

[8] S.C. Baber et al., J. of Electrochemical Society, Vol. 130, **No. 7**, 1983, p. 1551.

[9] J. Levine, et al., **22nd** IEEE Photovoltaics Specialists Conference, 1991, p. 1045.

RECENT TEST RESULTS OF 6 COMMERCIAL INVERTERS FOR GRID CONNECTED PV SYSTEMS

E.J. Wildenbeest, S.W.H. de Haan, N.C. van der Borg

Netherlands Energy Research Foundation ECN
ECN Renewable Energy
P.O. Box 1, NL-1755 ZG Petten
The Netherlands

ABSTRACT

Test results of 6 commercial inverters for grid connected pv systems in the power range up to 5 kW are given. The considered aspects include: energy efficiency, Power Point Tracking efficiency, start-up power, power quality, conformity to EMC standards, acoustic noise and susceptibility to electromagnetic noise. The results are obtained with the PV test facility of ECN.
The results show that in the last few years considerable progress has been made concerning the performance of grid connected inverters which are available at the European market. In general the efficiency is over 88% over a wide power range, with a peak efficiency of about 92%. Most inverters comply with standards concerning power factor, harmonics, EMC and acoustic noise.

1. INTRODUCTION

For designers of grid connected pv systems and for manufacturers of inverters for grid connected pv systems it is important to have access to data concerning the performance of these inverters. For plant designers this information is required to be able to make the proper choice for a specific application; for manufacturers it is important to see in what aspects their product has to be improved to be competitive.
In the last two years 12 inverters have been tested by ECN. The test are carried out in a laboratory at well defined conditions concerning pv array characteristics and grid characteristics.

2. TEST FACILITY

The test facility, shown in fig 1, consists of a computer controlled pv array simulator, line impedance simulation networks, a power failure simulator PFS, a burst generator and computer controlled measurement equipment. The pv simulator can simulate both the static and the dynamic behaviour of PV generators up to 10 kW (from 20A @480V to 80A @120V). Via data files the available dc power can be varied reflecting normal daily insolation patterns or changes for instance caused by the weather. To enable EMC tests the system is equipped with Line Impedance Simulation Networks (LISN) on both dc and ac side. The ac side LISN is a standard LISN as prescribed in VDE0877. The dc side LISN is developed by ECN on bases of measurements of actual pv generators.
The Power Failure Simulator (PFS) enables investigation of the behaviour of inverters in response to power failure, grid dips, TIF signals, voltage and frequency disturbances and so on. Results obtained with the PFS are not reported here.

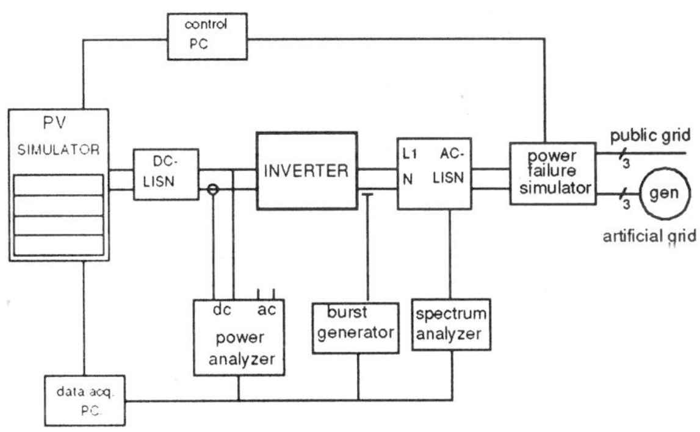

Fig. 1 *Block diagram of test facility at ECN*

2.1 Evaluated characteristics

For this report the following characteristics were examined:

- energy efficiency
- stand by power loss
- available dc power when inverter starts generation of ac power
- MPPT efficiency
- power factor
- line current harmonics
- total harmonic distortion
- conformity to EMC standard VDE0871
- acoustic noise
- susceptibility to burst

CH3365-4/94/0000-0909 $4.00 © 1994 IEEE

First WCPEC; Dec. 5–9, 1994; Hawaii

The measurements of harmonic currents are carried out with the set up as shown in figure 1. The measurements of current harmonics might slightly be affected by voltage harmonics being present in the public grid. For that reason an uncertainty of a few percent should be accounted for in the harmonic measurement results. Note that the AC LISN does not affect these harmonics.

2.2 Evaluated inverters

The following inverters were tested:

Solar Diamant	Egir 010-st	1.75	kW
UFE Wuseltronic	NEG1600 (rev. version)	1.6	kW
SMA	PV-WR 5000	5	kW
Sosnowski	Solwex 2065	2	kW
Mastervolt	Sunmaster 1800 (revised)	1.5	kW
ASP	Topclass GRID II	2.2	kW

During the tests several inverters broke down and were replaced by the manufacturers. Because of repeated problems, the tests with the Egir could not be completed. In the meantime the EGIR 010 has been replced by the EGIR 019. For other inverters modifications were carried out during the testing period. As far as possible all tests have been performed with the latest versions.

3.TEST RESULTS

The most important results of the testing are summarized in table 1.

The energy efficiency of all six inverters is shown in figure 2. The figure shows that in general all inverters have an efficiency of over 80% for $P > 0.1\ P_{nom}$. All inverters have an efficiency of about 90% at 0,7 P_{nom}. In comparison to inverters of the first generation the efficiency of the inverters is quite improved, in particular in the partial load region. The Sunmaster even has an efficiency above 90% at 0,1 P_{nom}. At nominal power the highest efficiency is reached by the SMA5000.

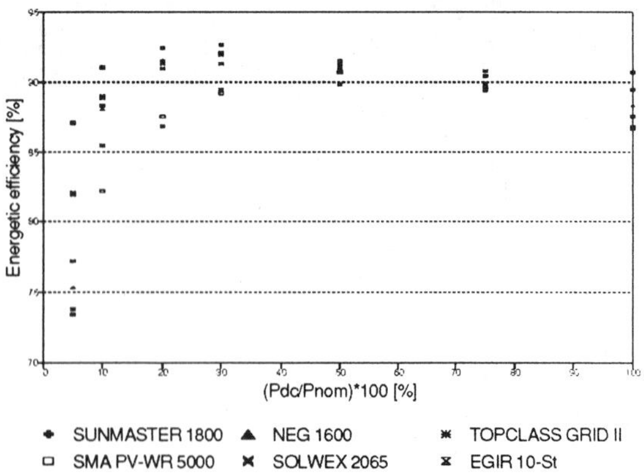

Fig. 2 *Energy efficiency as a function of power*

Equally important as the energy efficiency is the MPPT efficiency, which reflects the ability of the inverter to operate the pv generator in the Maximum Power Point (MPP). The MPPT efficiency is defined as the ratio of dc power extracted from the array to the available dc power in the MPP (P_{dc}/P_{MPP}). For $P > 0.2\ P_{nom}$ all inverters have an MPPT efficiency over 99%, with the Topclass slightly falling behind. In the power region below $0.2P_{nom}$ there is an important difference. At $0.1P_{nom}$ the Sunmaster, NEG1600 and SMA5000 are still above 98%, whilst the Topclass and Solwex are falling behind (92.7% resp. 96.9%). The MPPT efficiency of the Egir could not be evaluated. At very low insolation levels the accuracy of the evaluation of the MPPT efficiency might be affected by the VI characteristic of the simulator.

Both harmonic distortion and power factor are of importance for the grid. The Total Harmonic Distortion (THD) is shown in table 1. Although the THD of the EGIR and SMA is high (>20%), the individual harmonics are not exceeding the limit as given in standard EN60555 very serious (see fig.3). At the specific ECN grid they were exceeding the limit for several harmonics with more than 40%. It should be noted that these measurements might be affected slightly by the grid properties at ECN.

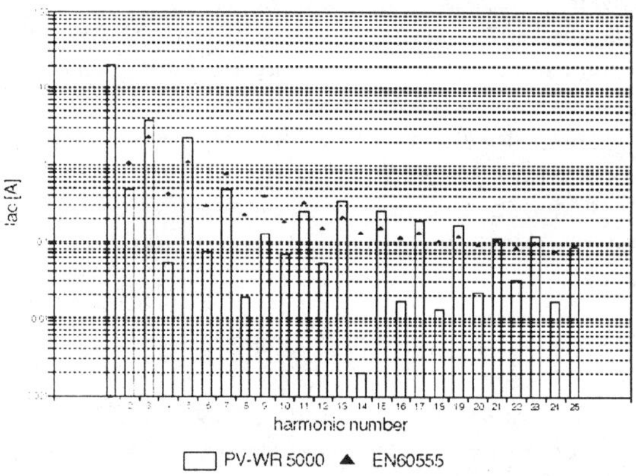

Fig. 3 *Harmonic distortion of the SMA 5000 at nominal power*

With respect to power factor the inverters behave quite different. Some inverters have an excellent power factor in excess of 0,9 over a wide power range (see fig. 4). The power factor of the EGIR10-st is below 0.5 over the complete power range.Because of the switching process inverters produce electromagnetic noise, which may cause interference. It has been determined whether the conducted emission produced by the inverters comply with standard VDE0871 (more or less equivalent to EN 5011. For several reasons the EMC tests have not been completed for all inverters; the Egir010-st broke down during tests, whilst the Topclass GRID II tests have to be confirmed yet. All tested inverters comply with limit line A. Some inverters (Sunmaster, NEG1600), almost meet the more severe limit line B .

Table 1: Summary of measurement results

Inverter type	Sunmaster 1800	NEG 1600	Topclass GRID II	SMA PV-WR 5000	Egir 10-St	Solwex 2065
Pdc nominal	1500	1600	2500	5000	1000	2000
Pac at zero insolation	0.6	3.1	6.3	15.2	-[1]	11.9[2]
Pdc when inverter stops generating AC power	2.13	17	18	262	-[1]	25.2
Energetic eff. [%] at 0.1 Pnom 0.2 0.3 0.5 0.75 1.0	 91.0 92.4 92.6 91.5 89.9 87.5	 87.5 91.7 92.5 92.8 91.6 89.6	 88.3 90.9 91.3 91.2 90.5 89.4	 82.1 87.5 89.2 90.7 90.4 90.7	 85.4 86.8 89.4 89.8 86.4 90.7	 88.9 91.4 92.0 91.0 89.3 86.7
MPPT efficency [%] at 0.1 Pnom 0.2 0.3 0.5 0.75 1.0	 98.1 98.4 99.4 99.8 99.8 99.7	 98.4 98.4 99.3 99.7 99.8 99.4	 92.7 97.9 98.7 98.6 99.0 99.2	 98.9 99.1 99.0 98.3 99.7 99.6	-[1]	 96.9 99.3 99.2 99.5 - 98.5
Power factor [%] at 0.1 Pnom 0.2 0.3 0.5 0.75 1.0	 0.65 0.81 0.89 0.94 0.95 0.94	 0.92 0.97 0.98 0.99 1 1	 0.93 0.98 0.99 0.99 0.99 0.99	 0.55 0.67 0.77 0.90 0.95 0.98	 0.33[3] 0.32 0.39 0.42 0.41 0.48	 0.87 0.95 0.97 0.99 0.99 0.99
THD [%]	10.0	4.4	10.8	20.9	29.1	3.8
Harmonic currents conform EN60555	Yes	Yes	Yes	No[4]	No[1]	Yes
EMC conformity VDE0871 (A,B curve)	A	B	-[5]	A	-[1]	A
Acoustic noise at 1 m (dB(A))	47.6	31.6	42.2	49.6	-[1]	31.0
Burst tests acc. to EN1000-4 Level	2	2	2	2	-[1]	2

1 test not completed because of inverter failure

2 switches to 0.8W after 60 s

3 manufacturer supposes that the low power factor is caused by non-functioning of compensation control

4 close to prescribed limits; results might be affected by local grid conditions

5 test not confirmed yet

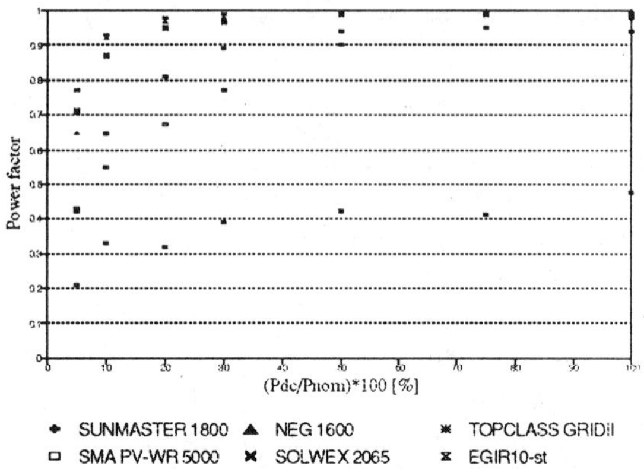

Fig. 4 *Power factor as a function of power*

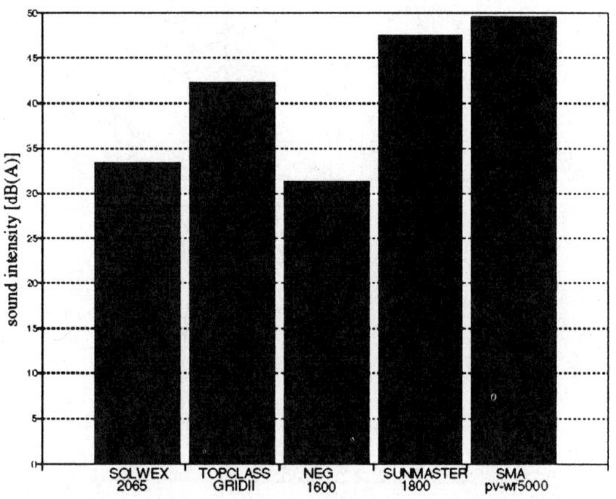

Fig. 5 *Sound intensity in dB(A) at nominal power*

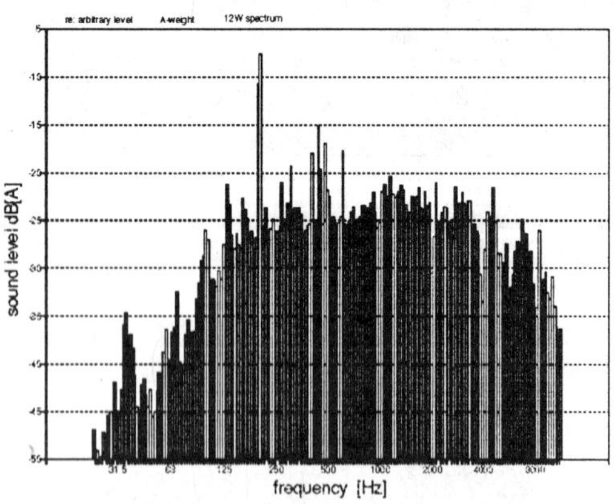

Fig. 6 *Sound pressure spectrum of Sunmaster 1800. At 200 Hz a prominent tone is present*

The acoustic noise produced by inverters is of importance when inverters are installed near living rooms. The produced sound intensity produced by the inverters is between 35 and 50 dB(A) (see fig. 5). In addition to the - sound intensity also the tonality is of importance. An inverter with a low sound intensity level may cause inconvenience if all acoustic energy is concentrated in one tone. Some inverters prove to produce sound with a higher tonality level than other inverters. Figure 6 shows, as an example, the tonality spectrum of the Sunmaster 1800.

For high reliablity during operation it is important that inverters are not very susceptible to electromagnetic noise from the environment. All inverters have been subjected to a burst test. All inverters operate unaffected at burst level 2 of standard EN1000-4. At higher levels the operation is affected during the burst, whilst some inverters switch off and have to be reset manually. The test results are reported more extensively in [3].

4. CONCLUSIONS

The results show that in the last few years considerable progress has been made concerning the performance of grid connected inverters which are available at the European market. In general the efficiency is over 90% over a wide power range. Almost all inverters comply with standards concerning power factor, harmonics (EN 60555), EMC (VDE 0871, EN1000-4) and acoustic noise.

5. REFERENCES

[1] S.W.H. de Haan, E.J. Wildenbeest, A.T. Veltman, Testing of inverters for grid connected pv systems, ECN-C--93-091, 1993.

[2] N.J.C.M. van der Borg, A.T. Veltman, An experimental comparison of seven commercial PV inverters, ECN-C--93-092, 1993

[3] E.J. Wildenbeest, N.J.C.M. van der Borg, S.W.H. de Haan, Comparison of six inverters, ECN draft report ECN-C--94-050, 1994. (draft report; final version in preparation)

HIGH-PERFORMANCE INVERTERS FOR
GRID-CONNECTED PV APPLICATIONS

F.J. ter Heide, Mastervolt; Lemelerbergweg 23, NL-1101 AJ Amsterdam, The Netherlands.
J. Wanner, Power Research Electronics; P.O. Box 10, NL-4849 ZG Dorst, The Netherlands.

Introduction

In a few years time Mastervolt has become one of the leading European manufacturers of stand-alone inverters for marine and mobile applications. The application of photovoltaics in this area is well on its way.
Based on this experience the development of grid-connected inverters evolved naturally within the context of the effort for development of grid-connected photovoltaic systems in the Netherlands.
This work was supported by NOVEM under the NOZ-PV program.

Sunmaster 1800

Application

A 1.8 kW inverter (93 % max. efficiency) was developed for application in roof-integrated and free-standing PV systems on buildings (figure 1). This inverter is meant for installation inside the building or in a waterproof cabinet. It has extensive options for metering of input and output power.
In a comparative test of inverters available in Europe performed by ECN (ref. 1) the inverter proved to be one of the best regarding efficiency and the best regarding noise level and EMC.
The major area of application will be on one-family houses that are to be built in the Netherlands over the next decade.

Technical description

Unlike most inverters of this size and application the Sunmaster 1800 is a three-stage instead of a two-stage inverter (figure 2).
The inverter consists of a double parallel PWM step-down converter at 20 kHz, a 20 kHz square wave converter driving an HF transformer and an IGBT unwrapper. The rectified-sine waveform is generated by the two step-down converters. These are phase-shifted by 180 degrees, thereby relieving stress on the input capacitors.
The rather low input voltage (45-90 V) is transformed to the mains level by the square wave converter. This unregulated converter operates in zero-voltage switching mode.
By splitting the PWM function and the voltage transformation

CH3365-4/94/0000-0913 $4.00 © 1994 IEEE First WCPEC; Dec. 5–9, 1994; Hawaii

function over two independently optimised converters a high overall efficiency is reached. Because component stress is minimised, low-budget components can be used. In this way a successful trade-off between high efficiency and complexity can be reached.
MPP-tracking is performed by ripple phase detection between input signals of voltage and power.

Sunmaster 130

Application
After successful development of the Sunmaster 1800, an inverter was designed for application in AC-modules, a new and exiting development in grid-connected PV (ref. 2).
This inverter is essentially a PCB that can be integrated into a connection box on the backside of a 1 m^2 PV module. This combination forms a small grid-connected PV unit that can be retrofitted on existing buildings. Multiple units can be AC interconnected, covering the range below the area of application of the Sunmaster 1800.
The aggravated temperature conditions on the backside of the PV module have been taken into account in the design of this inverter. Although operating conditions are rather fierce, no concessions had to be made towards efficiency (93 % max.), performance or reliability (figure 3).

Technical description
The design of the Sunmaster 130 is stripped down to essentials (figure 4). The basic design of the square wave converter of the Sunmaster 1800 is used to transform the input voltage (20-40 V) to the mains level. The electrical energy is fed into the grid by a 50 Hz PWM bridge converter and appropriate filtering.
MPP-tracking is performed by a simplified strategy implemented in analog electronics.

Conclusion

A firm commitment to photovoltaic energy was made by developing high-performance grid-connected inverters for a wide range of forseeable PV systems. Further involvement in the development of both grid-connected and autonomous photovoltaic applications can be expected.

Ref.1: An experimental comparison of the characteristics of seven commercial PV-inverters, N.J.C.M. van der Borg e.a.; ECN, March 94.
Ref.2: Design specification of AC-modules, C.P.M. Dunselman e.a.; Ecofys, November 93.

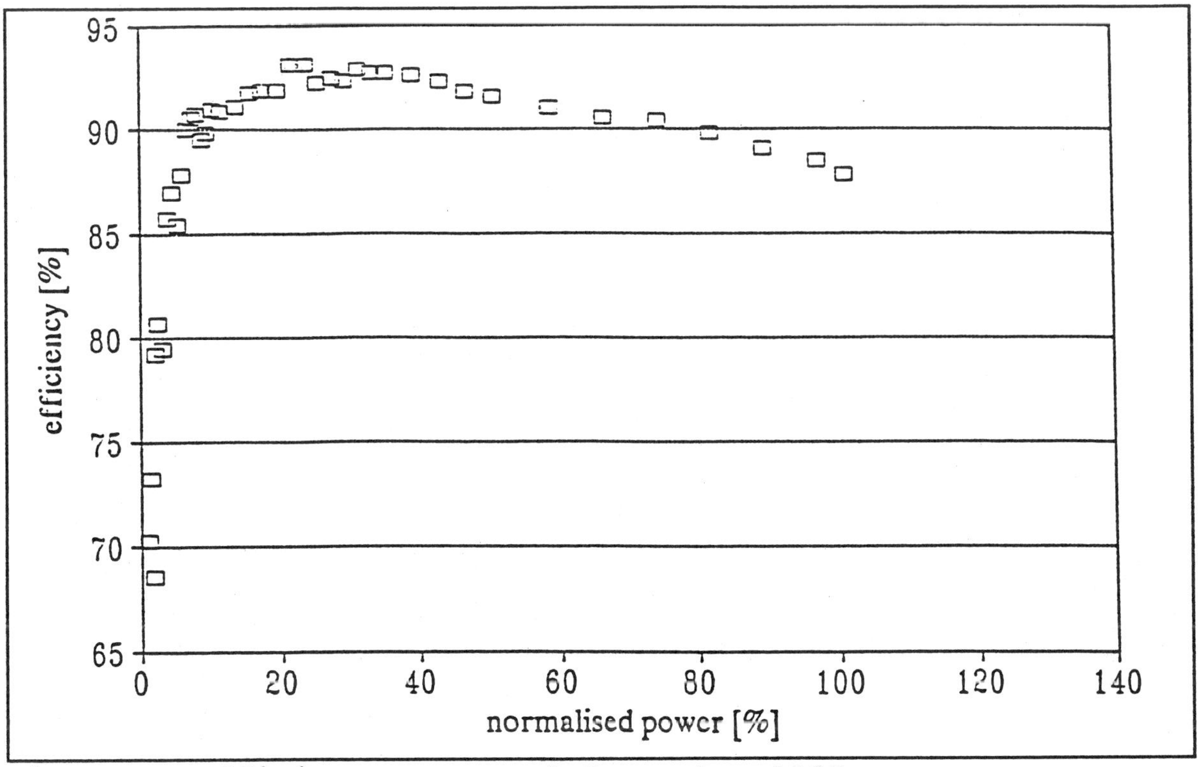

Figure 1: Efficiency curve of Sunmaster 1800.

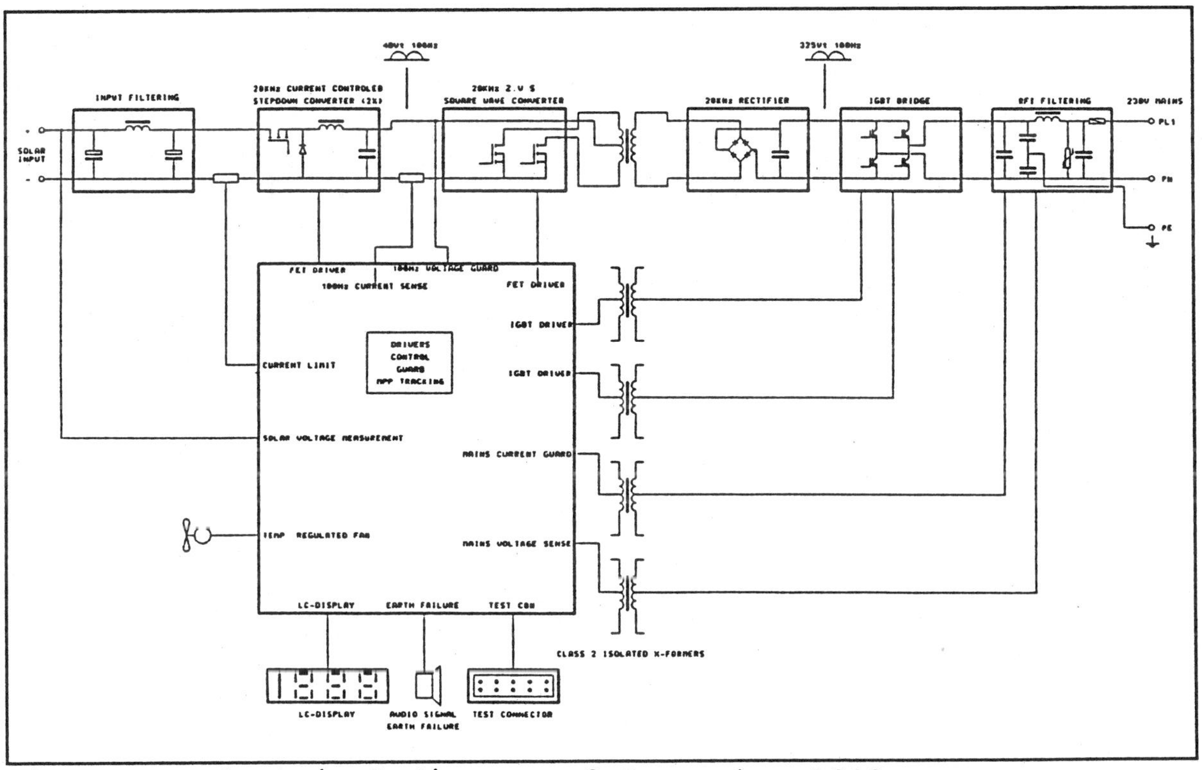

Figure 2: Electrical diagram of Sunmaster 1800.

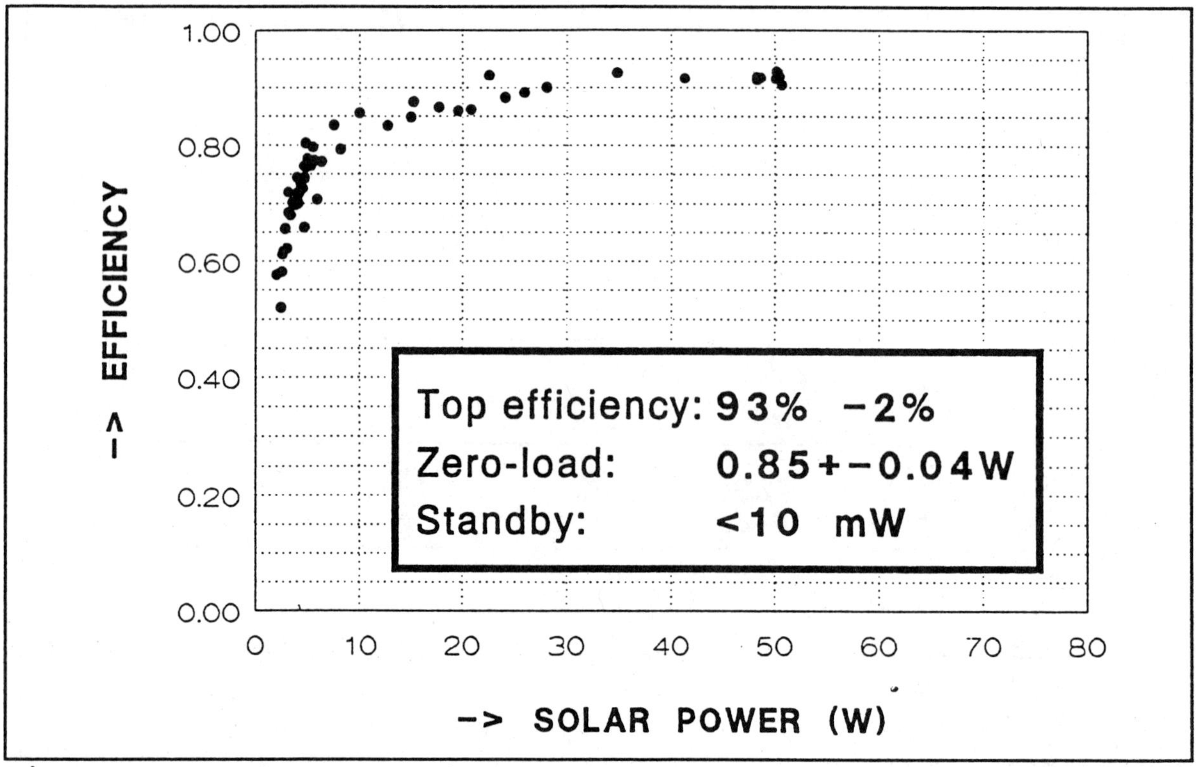

Figure 3: Performance of Sunmaster 130.

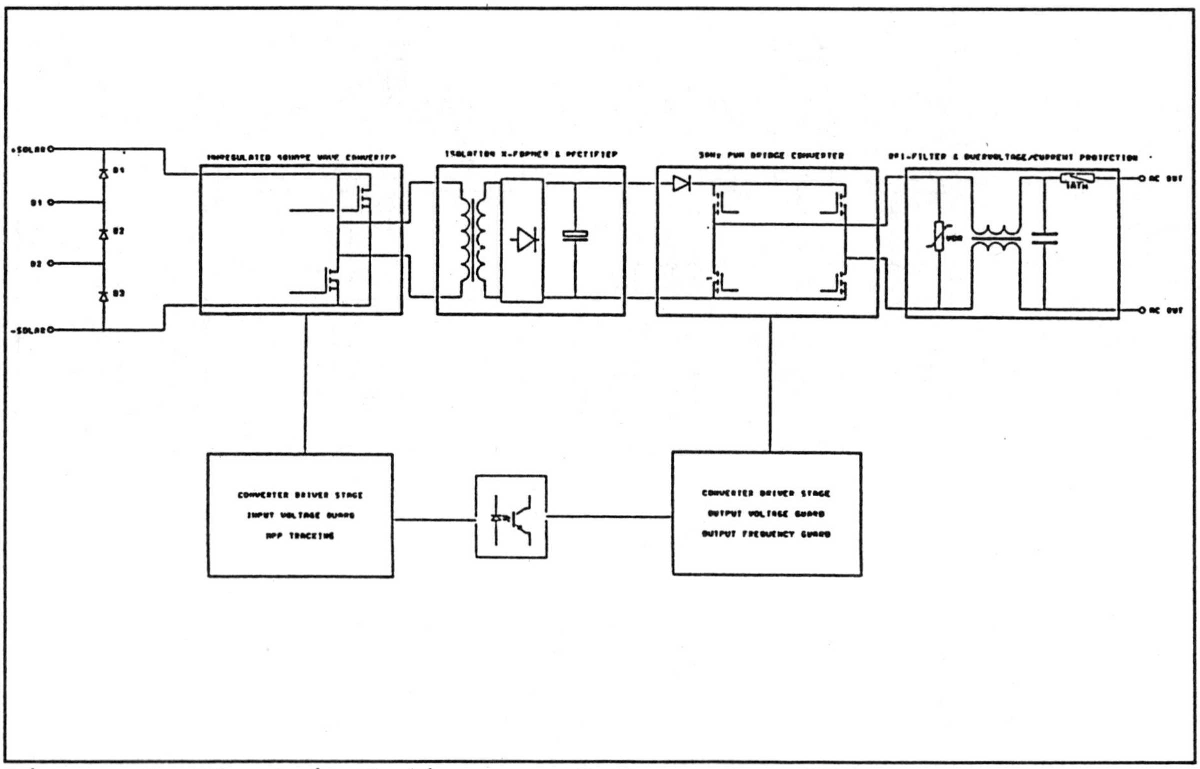

Figure 4: Electrical diagram of Sunmaster 130.

A LIGHTWEIGHT INVERTER FOR OFF-GRID AND GRID-CONNECTED SYSTEMS

Fabio Barone
Centre for Photovoltaic Devices and Systems,
University of New South Wales, Sydney, Australia.

ABSTRACT

This paper presents a novel bi-directional power processing topology particularly suitable as an inverter for both grid and off-grid alternative energy systems. All semiconductor devices are soft-switched, thereby allowing higher switching frequencies, improving efficiency, and reducing electromagnetic interference (EMI). The topology may utilise a small, light-weight, and inexpensive high-frequency transformer. Experimental results of a 1kW bi-directional dc-dc converter are presented.

1. INTRODUCTION

Synthesis of sinusoidal voltage from a dc source has traditionally been done by push-pull or full-bridge inverter topologies. As switching frequencies increase and PWM schemes evolve, the quality of the output waveform has improved dramatically; harmonic content levels as low as 0.1% have been reported [1]. However, most inverters experience high switching losses and must use a bulky low-frequency transformer to provide the necessary isolation and voltage translation.

The bulky transformer may be replaced by a small high-frequency transformer if the inverter is preceded by a high-frequency dc-dc converter step-up stage [2]. Such topologies, however, are uni-directional only, and bi-directional power flow is desirable for both grid and off-grid systems. Grid authorities can employ bi-directional inverters for load levelling and static VAR compensation. Off-grid systems with a diesel alternator may re-charge batteries during diesel operation.

Switching losses may be eliminated by using resonant soft-switching techniques, which have been applied to dc-dc converters [3,4,5], voltage-fed series resonant inverters [6,7], and also to current-fed parallel resonant inverters [8,9]. However the semiconductor devices are subjected to much higher current and/or voltage stress than hard-switched non-resonant topologies, thus negating any saving in switching loss. Further, the resonant tank may require a reactive power rating of over twice the rated output power of the inverter, and can become a significant source of losses [8].

This paper presents a bi-directional ac-link topology and control scheme whereby the semiconductor devices are soft-switched without using resonance. An inexpensive high-frequency transformer is used which improves efficiency and dramatically reduces size and weight. Being capable of bi-directional power flow, it can change between rectifying and inverting modes instantly. It can be controlled to draw or inject sinusoidal current from the mains electrical supply without causing harmonic pollution. Circuit operation and results of a 1kW two-quadrant output prototype inverter are presented.

2. CIRCUIT OPERATION.

The proposed inverter topology consists of two full-bridge inverters on each side of a high frequency link, as shown in fig. 1. The link as drawn includes a high-frequency transformer, although this is not necessary for circuit operation. A zero-voltage-switched voltage-sourced-inverter (ZVS-VSI) is on the primary side, while a zero-current-switched current-sourced-inverter (ZCS-CSI) is on the secondary side. As drawn, the primary side is connected to the input power source, V_g, while the secondary side is connected to the output load. The terms "source" and "load" are somewhat misleading since it will be shown that the topology is fully bi-directional in terms of power flow. The output filter comprises an inductor and capacitor, which presents a current-source to the ZCS-CSI.

The transformer is shown with its magnetising inductance, since the magnetising current plays an important role in circuit operation under light load conditions. As it stands, the inverter is capable of bi-polar output voltage, but only uni-polar output current; a second ZCS-CSI is required for complete four-quadrant operation. It can be shown that multiple ZCS-CSI modules may be connected to the high frequency link and be controlled independently of each other, so that three-phase output may be synthesised. The concept may be extended to any number of ZCS-CSIs and ZVS-VSIs connected via a high-frequency link to form a high-frequency power system. For the purposes of illustration, the simplest configuration will be considered. The following description refers to the state diagrams in fig. 1, and assumes the output filter is replaced by a constant current source for the duration of one switching cycle.

CH3365-4/94/0000-0917 $4.00 © 1994 IEEE First WCPEC; Dec. 5–9, 1994; Hawaii

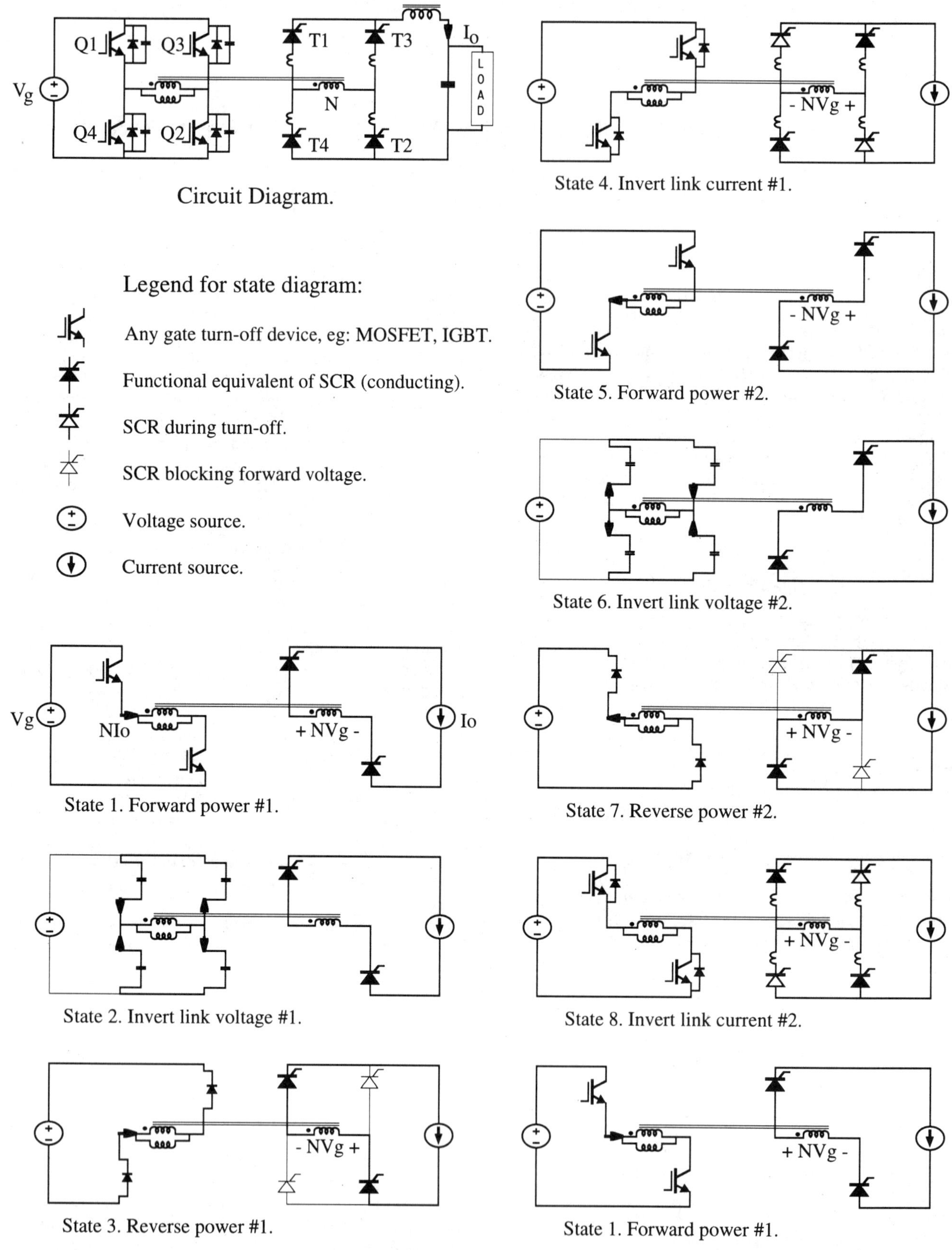

Fig. 1. Inverter circuit diagram and switching states.

State 1. Forward Power # 1.
Q1 and Q2 are both on, which applies a positive voltage to the link via the transformer. On the secondary side both T1 and T2 are on, and they direct the output current in a positive direction with respect to the link. From the convention used for polarities, the link is transferring power from the voltage source to the current source, which we shall denote as positive power flow. Notice that T3 and T4 are reverse biased, and cannot be turned on.

State 2. Invert link voltage #1.
Under command from the control circuit, both Q1 and Q2 are now turned off. The primary current, being driven by the switches on the secondary side, must continue to flow, and so it diverts into the lossless snubber capacitors. The direction of primary current has the effect of increasing the voltages across the outgoing switches Q1 and Q2, but decreasing the voltages across the incoming switches Q3 and Q4. Q1 and Q2 experience some turn off loss, but this is easily controlled by the snubber capacitors. Since it is driven by a constant current, the voltage changes are linear, and will always be driven to completion, ie.: there are no conditions on achieving ZVS, except that some current be present. At no load, the transition is driven by the magnetising current. This state should only occur briefly, and does not take part in energy transfer across the link since all the link current flows in the snubber capacitors, and no current is drawn from the voltage source. The link voltage changes polarity with controlled dv/dt, given by:

$$dv/dt = NI_O / C \qquad (1)$$

where I_O is the load current, N is the secondary-to-primary turns ratio, and C is the value of each snubber capacitor.

State 3. Reverse Power #1.
The voltage transition is now completed, and the anti-parallel diodes of Q3 and Q4 are on, clamping the link voltage and facilitating ZVS for Q3 and Q4. The link voltage has now changed polarity, but the link current has not changed direction. Hence the power is now flowing from the current source to the voltage source, ie: reverse power. Since the capacitors no longer conduct current, they are not shown. Notice now that T3 and T4 are forward biased, and can be turned on. Conversely, Q3 and Q4 cannot be turned off, since their anti-parallel diodes are conducting.

State 4. Invert link current #1.
The secondary side SCRs T3 and T4 are now gated on by the control circuit. The gating of T3 applies a reverse voltage across the combination of T1 and its snubber inductor, while T4 reverse biases T2 and its snubber. Thus current ramps down in T1 and T2, but ramps up in T3 and T4, at a rate determined by the link voltage appearing at the secondary of the transformer and the snubber inductors, L:

$$di/dt = NV_g / L \qquad (2)$$

This means that T1 & T2 turn off losslessly at zero current, and T3 & T4 are turned on with di/dt snubbing to limit turn-on losses. Since the link voltage remains constant, these transitions also are linear and are guaranteed to be completed. The link current changes polarity with controlled rate of change, similar to the primary side transition case. During this brief transition, all the link voltage appears across the snubber inductors, leaving zero volts across the current source, so no power transfer is taking place. In practice, the leakage inductance of the transformer will contribute to any discrete snubber inductors used, and may itself be sufficient.

State 5. Forward Power #2.
Now both the link voltage and the link current have changed polarity with respect to directions shown in state 1. Hence the direction of power flow is again forward, ie.: from voltage source to current source. The duration of this state is determined by the control circuit. Again note that both T1 and T2 cannot be turned on since they are reverse biased, however Q3 and Q4 may be turned off since they are conducting forward current.

State 6. Invert link voltage #2.
Now both Q3 and Q4 are turned off, leaving the circuit in the same state as for state 2 but with link current reversed. The link current continues to flow in the lossless snubber capacitors while the link voltage changes polarity at a controlled rate, until the anti-parallel diodes of Q1 and Q2 clamp the link voltage.

State 7. Reverse Power #2.
Now the link current is flowing into the positive side of the voltage source, so power is flowing from the current source into the voltage source, ie. reverse power. Q1 and Q2 cannot be turned off, but T1 and T2 are forward biased and can be gated on. The duration of this state is determined by the control circuit.

State 8. Invert link current #2.
Now T1 and T2 are gated on, causing a transition of the secondary side similar to that described in state 4. This time T3 and T4 are turned off at zero current, while T1 and T2 turn on with di/dt snubbing. The link voltage stays constant, but the link current changes polarity at a controlled rate.

After the completion of state 8, the circuit reverts to state 1, and the cycle of operation repeats.

2.1. Load current freewheeling.
With the full-bridge ZCS-CSI it is possible to implement a state whereby the load current freewheels through the ZCS-CSI only, without flowing into the primary side. This leads to a reduction in conduction loss, greater flexibility in control, and reduced ratings of filter components. This

is implemented by firing only T3 in state 4, and delaying the firing of T4. Similarly for state 8, T1 may be fired before T2.

2.2. Control of transformer magnetisation.

The full-bridge ZVS-VSI is capable of clamping the link voltage to zero, thus forcing a zero power flow state, and may be used to control the magnetic flux in the transformer. This is an important feature in off-grid systems where battery voltages can fluctuate by up to 30%. This is implemented by turning off Q1 before Q2 in state 2, and turning off Q4 before Q3 in state 6.

3. EXPERIMENTAL RESULTS

A 1kW prototype switching at 20kHz full cycle has been constructed [10]. Instead of using SCRs in the ZCS-CSI, IGBTs with series connected fast diodes were employed, along with turn-off detection circuitry. Parallel MOSFETs were used in the ZVS-VSI. At present, only one ZCS-CSI stage has been built, which allows only uni-directional output current. Another ZCS-CSI stage may be added to facilitate four-quadrant sine-wave output. Control at this stage is simple fixed duty-cycle; feedback control has not yet been implemented.

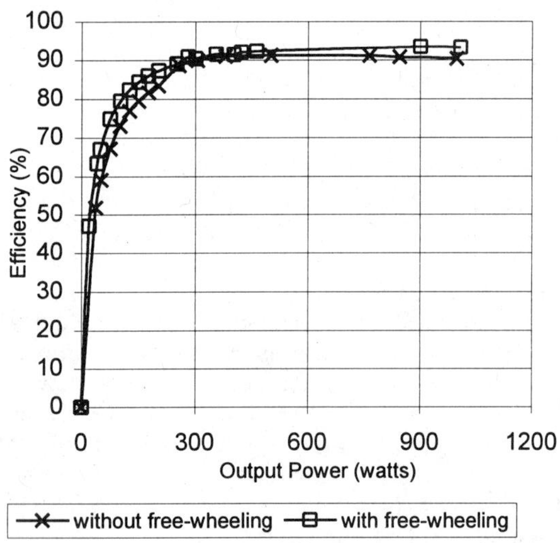

Fig. 2. Efficiency vs output power.

The prototype was tested by generating dc output with varying loads and duty-cycles up to a maximum output of 1kW at 400V dc, from a 50V dc supply. The input and output powers were measured simultaneously using two Voltech PM1000 AC Power Analysers. From the resulting plot in fig. 2 it can be seen that the peak efficiency for operation without free-wheeling mode reached 91% for output powers greater than about 300W. When free-wheeling mode was used the efficiency curve improved, particularly at light load, reaching a peak of over 93%. The no-load power consumption of the inverter, an important consideration in off-grid systems, was measured at 12.4W for both modes of operation.

4. CONCLUSION

A high-frequency ac-link topology has been described and demonstrated. The proposed control scheme provides soft-switching for all devices without incurring an overstress penalty. An efficient laboratory prototype was successfully implemented. A modification of the control scheme allowed a free-wheeling mode of operation, resulting in improved efficiency over the entire load range. The features of high-efficiency, light-weight, and bi-directional power flow make it an ideal candidate for battery-fed grid and off-grid applications.

5. REFERENCES

[1] Legowski, S., Trzynadlowski, A.M., "Power-MOSFET, Hypersonic Inverter with High-Quality Output Current.", *IEEE Applied Power Electronics Conference Record*, 1990, pp. 3-7.

[2] Gosbell, V.J., Sanders, G.J., Basell, M.C., "A High Efficiency Inverter for Remote Homestead Use", *Jol. of Electrical and Electronics Engineering, Australia - IE Aust. & IREE Aust.*, Vol. **5**, No. 1, March 1985, pp. 49-53.

[3] Liu, K.H., Oruganti, R., Lee, F.C., "Resonant Switches-Topologies and Characteristics.", *IEEE Power Electronics Specialists Conference Record*, 1985, pp. 106-116.

[4] Steigerwald, R.L., "A Comparison of Half-Bridge Resonant Converter Topologies.", *IEEE Applied Power Electronics Conference Record*, 1987, pp. 135-144.

[5] Patterson, O.D., Divan, D.M., "Pseudo-Resonant Full Bridge DC/DC Converter." *IEEE Power Electronics Specialists Conference Record*, 1987, pp. 424-430.

[6] Ranganathan, V.T., Ziogas, P.D., Stefanovic, V.R., "A DC-AC Power Conversion Technique Using Twin Resonant High-Frequency Links.", *IEEE trans. Industry Applications*, vol. **IA-19**, no. 3, May/June 1983, pp. 393-400.

[7] Chung, Y.H., Shin, B.S., Cho, G.H., "Bilateral Series Resonant Inverter for High Frequency Link UPS.", *IEEE Power Electronics Specialists Conference Record*, 1989, pp. 83-90.

[8] Espelage, P.M., Bose, B.K., "High-Frequency Link Power Conversion.". *IEEE trans. Industry Applications*, Vol. **IA-13**, no. 5, Sept/Oct 1977.

[9] Sood, P.K., Lipo, T.A., "Power Conversion Distribution System Using a High-Frequency AC Link.", *IEEE trans. Industry Applications*, Vol. **24**, no. 2, Mar/Apr 1988, pp.

[10] Barone, F., Spooner, E.D., Daly, K.C., "A High-Efficiency, Bi-Directional, Lightweight Inverter/Charger.", *Proc. Australasian Universities Power Engineering Conference*, October 1993, pp. 202-209.

CHARACTERISTICS AND PERFORMANCE OF A PV-SYSTEM CONSISTING OF 20 AC-MODULES

L.E. de Graaf, T.C.J. van der Weiden
Ecofys Research and Consultancy
P.O. Box 8408, NL-3503 RK Utrecht, The Netherlands
tel: +31 30 913400, fax: +31 30 913401

ABSTRACT

AC-modules are PV-panels with an inverter on the rear of the module. These modules supply AC-power. PV-systems consisting of AC-modules have several advantages compared with conventional PV-systems. One of the advantages is the possibility to realise small grid connected PV-systems, because each AC-module is an independent unit. Other important advantages are ease of engineering and reduced energy losses caused by mismatch and partial shading. On the roof of the Ecofys building in Utrecht (NL), a pilot system consisting of 20 polycrystalline AC-modules was realized. In this paper the first encouraging results of the monitoring of this pilot system are presented: a system efficiency of over 11% and a performance ratio of over 0.8.

INTRODUCTION

AC-modules are PV-panels with an integrated inverter. Ecofys has developed these modules in cooperation with the PV manufacturer R&S Renewable Energy Systems and the inverter manufacturer Mastervolt. The first prototype has been presented at the 12th EPVSEC in Amsterdam, April 1994 [1]. On the flat roof of the Ecofys building, a pilot system of 20 prototype AC-modules has been realized in close collaboration with the mentioned manufacturers and the utility company REMU. In this paper, advantages and possible problems of AC-modules are described. Furthermore, the first operating results of the pilot system are presented.

ADVANTAGES AND POSSIBLE DISADVANTAGES OF AC-MODULES

Less energy losses

In an AC-module system, energy losses are smaller than in a conventional grid connected PV-system as a result of the following features:

- No mismatch in strings, because the AC-modules are connected in parallel.
- Less resistance losses in cables and connections.
- Less shadow losses, because each independent unit covers a relatively small area.

Other advantages

Besides less energy losses, AC-modules have several other advantages. In [1] all these, and even more, advantages have been described more comprehensively. Here are mentioned:

- Increased safety during installation.
- Reduced susceptibility for lightning.
- Decreased installation costs.
- Possibility to realise small grid connected PV-systems, because each AC-module is an independent unit.

Possible disadvantages of AC-modules

Possible disadvantages of AC-modules are the following:

- Increased zero-load dissipation compared to conventional PV-systems.
- Increased thermal stress on the inverter, because the inverter is mounted on the back of the module.

PILOT SYSTEM

Characteristics AC-modules

One AC module consists of a R&S polycrystalline module and a Mastervolt inverter of 130 Wp. The dimensions of the module are 1.28 x 0.64 m^2. Each module is made up of 72 solar cells and is about 100 Wp, so the pilot system is 2 kWp.

Characteristics pilot system

The 20 prototype AC-modules are stand off mounted on the flat roof of the Ecofys building. The rear of three of the modules is thermally insulated, to imitate the operating conditions of a roof integrated system. This insulation has been carried out very thoroughly: at the rear no ventilation is possible at all, so a worst case

First WCPEC; Dec. 5-9, 1994; Hawaii

situation is simulated. The orientation of the PV-system is south, the inclination is 30°.

FIRST OPERATING RESULTS OF THE SYSTEM

The monitoring of the AC-modules is in operation since August 1994, and will be carried out till the summer of 1995. The sample time is 20 seconds. For six modules, 5-minute averages are computed, for the other modules hourly averages are stored. In this section the following aspects will be discussed: the system efficiency, the performance ratio of the AC-modules, the thermal behaviour of the modules, the thermal load of the inverters and the power quality.

System efficiency

In Table 1 the system efficiency of the AC-modules is shown. The system efficiency is defined as:

$$\frac{E_{fi} * 100\%}{E_{ipc} * A} \qquad (1)$$

In which:

E_{fi} = AC-energy supplied by the inverter [kWh];
E_{ipc} = irradiation energy density [kWh/m^2], the reference cell data have been corrected for the cell temperature;
A = the total cell area of the system [m^2] (0.72 m^2 per module).

The efficiency of the stand-off mounted as well as of the thermally insulated modules is higher than 11%, which is very high for a grid connected PV-system with polycrystalline cells. The only exception is the efficiency of the thermally insulated modules in the month of August. The efficiency in this month is lower, due to the higher module temperatures. August 1994 was a rather warm month in the Netherlands. For August, two system efficiencies are tabulated for the thermally insulated modules: 9.4 and 10.9%. During this month, the insulated modules failed sometimes. The efficiency of 9.4% is the real system efficiency, including all the monitoring data. If the time the insulated modules failed is not taken into account, a system efficiency of 10.9% is found.

Table 1: System efficiency of the AC-modules

Month 1994	System efficiency AC-modules	
	stand-off mounted	thermally insulated
August	11.4 %	9.4/10.9 %
September	11.5 %	11.1 %
October	11.8 %	11.3 %

Performance ratio

The performance ratio (PR) is defined by JRC Ispra as the ratio between the Final Yield Y_f and the Reference Yield Y_r [2]. For grid connected PV-systems, PR can be defined easier as [3]:

$$PR = \frac{E_{fi}}{P_{pv,stc}} * \frac{G_{stc}}{E_{ipc}} \qquad (2)$$

In which:

$P_{pv,stc}$ = installed PV-power [kW$_p$];
G_{stc} = 1 kW$_p$/m^2.

Thus, PR is the ratio between the total amount of energy supplied by the inverter divided by the installed PV-power and the irradiation energy density. In contrast to the system efficiency, the PR is independent of the solar cell type. The PR's of systems using different cell types can be compared with each other. In Table 2 the monthly performance ratios for the AC-modules measured in August till October 1994 are shown. The PR of the stand-off mounted AC-modules ranges from 0.82 to 0.85. The PR of the thermally insulated modules is, as expected, a little bit lower: between 0.68/0.79 and 0.81. The performance ratio of the insulated modules is lower as a result of temperature effects on the modules. In August again two PR's for the insulated modules are given as a result of the failure of the insulated AC-modules during some time. The value 0.68 is the real value. The value 0.79 is the PR which is found if the time during which the AC-modules failed is not taken into account.

In comparison with measured PR's of other PV-systems, the PR of the AC-modules is high. A few examples are given. In the 1000-roof programme in Germany, the PR for grid connected systems varies between 0.64 and 0.71 [4]. In the municipality of

Heerhugowaard in the Netherlands, grid connected PV-systems have been integrated in the roofs of ten dwellings. In 1993 the PR of these PV-systems was 0.60-0.73, depending on the specific configuration of the PV-system. These PR's have been derived from monitoring results in [5]. In the village of Lekkerkerk, also in the Netherlands, a PV-system is integrated in the roof of a house using solar tiles. From the monitoring data of the last half year, the annual PR of this system is expected to be 0.68 [3].

Table 2: Performance Ratio of the AC-modules

Month 1994	PR of AC-modules	
	stand-off mounted	thermally insulated
August	0.82	0.68/0.79
September	0.83	0.80
October	0.85	0.81

Thermal behaviour of the AC-modules

The temperature of three modules is monitored with temperature sensors (two thermally insulated modules and one stand-off mounted module). The performance of a module is depending on the module temperature. As expected, the temperature of the thermally insulated modules is higher than of the stand-off mounted modules, due to a worse heat exchange between module and ambient. In case of the insulated modules, a maximum module temperature as high as 84°C has been measured.

The relation between ambient temperature and module temperature can be described with the formula:

$$T_{module} - T_{ambient} - k \ast G + T_0 \qquad (3)$$

In which G is the irradiation and k a regression coefficient.

For stand-off mounted systems, the coefficient k is usually taken as 0.03 °C/(W/m^2), see [6]. Table 3 shows the calculated monthly values of the coefficient k for the AC-modules. The calculated values for the stand-off mounted AC-modules are 0.020 to 0.024, significantly lower than 0.03 °C/(W/m^2). The values for the insulated modules vary from 0.041 to 0.048. These values are similar to the values found for the roof integrated PV-system in Lekkerkerk (0.042 to 0.047) [3].

Table 3: Computed monthly coefficients k for the AC-modules

Month 1994	k for AC-modules [°C/(W/m^2)]	
	stand-off mounted	thermally insulated
August	0.023	0.048
September	0.020	0.041
October	0.024	0.047

Thermal load of the inverters

In the inverter boxes on the rear of the modules, temperature sensors were fitted. The thermal load of the inverters is an important part of the monitoring research project. One of the possible causes of problems of AC-modules is the temperature in the inverter box, which can reach high values due to the fact that the inverters are mounted directly on the back of the modules. Up to now, the highest measured temperature (5-minute average) in the insulated boxes was about 68°C in the month of August. In the boxes of the stand-off mounted modules, the maximum measured temperature was 58°C.

The inverters have been designed to have a projected operational lifetime of 40,000 h at 85°C or 10,000 h at 105°C. A maximum temperature of 68°C was expected not to cause problems. Nevertheless, in August 1994, the insulated modules sometimes were shut off temporarily, leading to some loss of energy. The monitoring data show falling out of the insulated inverters at high irradiation levels and corresponding high temperatures in the inverter boxes. This falling out was the cause of the low system efficiency and PR of the insulated modules in August, as described in the paragraphs 4.1 and 4.2. Additional research on this phenomenon is needed for further development of the AC-module.

Power quality

Measurements of harmonic currents were carried out. The harmonic currents do not exceed the limits from table 2 of the standard IEC 555.

FURTHER ACTIVITIES

Future activities are the further technical development of the AC-module. Special attention will be paid to the thermal load and the reliability of the inverters. Another aspect to which attention will be paid is the safety certification of AC-modules.

CONCLUSION

The concept of the AC-modules is demonstrated to be very promising. For stand-off mounted AC-modules in the pilot system, an average system efficiency of 11.5% has been shown. For thoroughly thermally insulated modules an average system efficiency of 11% has been measured. The average performance ratio of the AC-modules is 0.80 for the insulated modules and even 0.83 for the stand-off mounted modules. Further development will be needed to enable a well-considered market introduction.

ACKNOWLEDGEMENT

The project is supported by the Netherlands Agency for Energy and the Environment (NOVEM) and the regional utility company REMU.

REFERENCES

[1] C.P.M. Dunselman, T.C.J. van der Weiden *et al.*, *Feasibility and Development of PV-modules with integrated inverter: AC-modules*, 12th EPVSEC 1994, Amsterdam.

[2] G. Blaesser, K. Krebs, *Guidelines for the Assessment of Photovoltaic Plants; Document B, Analysis and Presentation of Monitoring Data (Issue 3)*, CEC Joint Research Centre, Ispra, June 1992.

[3] J.H. Boumans, T.C.J. van der Weiden, *Monitoring van een 3 kW PV-daksysteem in Lekkerkerk, Tussenrapportage, Concept*, Ecofys, October 1994 (in Dutch).

[4] H. Riess *et al.*, *Performance and Reliability of the Photovoltaic Demonstration Plants in the German Measurement and Documentation (MuD) Programme*, 12th EPVSEC 1994, Amsterdam.

[5] N.J.C.M. van der Borg, *Analyse meetresultaten 1993 PV-woningen te Heerhugowaard*, ECN, number report ECN-C-94-028, May 1994 (in Dutch)

[6] M.R. Starr *et al.*, *Guidelines for the Assessment of Photovoltaic Plants; Document C, Initial and Periodic Tests on PV Plants (Issue 1)*, CEC Joint Research Centre, Ispra, November 1988

TEST RESULTS OF A 130W AC MODULE;
a modular solar ac power station

S.W.H. de Haan[1] H. Oldenkamp[2] E.J. Wildenbeest[1]

[1] Netherlands Energy Research Foundation
ECN Renewable Energy
P.O. Box 1
NL-1755 ZG Petten, The Netherlands

[2] OKE-Services
Nieuwstraat 29
NL-5611 DA Eindhoven
The Netherlands

ABSTRACT

An ac module is a pv module with integrated dc to ac converter which generates grid conform ac power. In this paper the ideas behind this portable and expandable ac power systems are highlighted.
The block diagram, the principle of operation and test results of a 130W inverter are shown. The measurements show a power factor exceeding 0.99 at full load, whilst at 12W partial load the power factor is better than 0.89. The harmonic distortion decreases from 8% at 30W to 6% at 50W and slightly decreases further at higher power levels. Previous measurements [1] showed that an efficiency over 90% over a wide operation range is obtained.
It is estimated that the cost price of the "touch safe" inverter (dimensions 100*100*20 mm), when produced in large quantities is between 0.5 and 1 $/Wp. The price performance ratio is obtained by application of resonant switching techniques enabling low component size and costs.

1. INTRODUCTION

Traditional pv systems require complex dc cabling systems giving rise to a variety of problems (see fig. 1).
In particular grid connected PV systems with a central inverter have shown significant problems with respect to high dc-voltage levels, safety, cable losses, risk of dc-arcs, fire hazard and protection. Most of these problems can be overcome by expensive cabling and installation systems, which increases the costs at system level. In addition components like the inverter, switch gear, cabling have to be selected in accordance with system size. This requires an individual design of each system and it is an obstacle to expanding the system at a later stage.
The integration of PV module and inverter into so called "AC modules" (also referred to as MIC, Module Integrated Converter), offers interesting possibilities to overcome these problems.

Fig. 1 *PV system with central inverter*

These AC modules generate floating ac power at module level. The AC module does justice to the inherent modularity of PV: a PV system can be expanded freely without a redesign at dc level; the AC modules can be connected simply at 230 V ac level with standard well developed and low priced ac installation techniques.

2. FEATURES OF MODULAR AC POWER STATIONS

For an ac-module the following list of interesting features can be given:

CH3365-4/94/0000-0925 $4.00 © 1994 IEEE First WCPEC; Dec. 5-9, 1994; Hawaii

- the system is highly modular allowing easy system expansion with units of about 100W;
- for the electrical interconnection system standard 230V ac installation material can be used, thus reducing costs of installation material and system design;
- the conduction losses and cable costs are low because of the relatively high ac voltage;
- the low minimum system size lowers the threshold for application;
- as each module is operated in the MPP, the mismatch losses at system level are low;
- because of the absence of string diodes the associates losses are prevented;
- due to the compact dc system lay out, lightning induced surge voltages are low;
- the number of required inverters per kW_p installed PV power is about ten times larger than the number of traditional inverters. Therefore the production volume required for mass production is more easily reached.

Inverters for ac modules basically include all features of central inverters. To enable low costs per unit and per Wp the principle of operation is based on resonant techniques allowing very high switching frequencies whilst the control is implemented with a micro controller to reduce the component count.

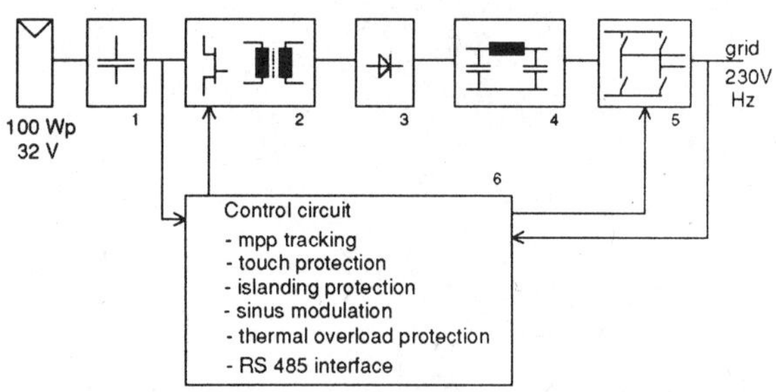

Fig. 2 *Block diagram of integrated inverter*

The inverter consists of the following main stages (see fig.2):

1 a capacitor. With a 50 Hz grid connected the ac power has a large 100 Hz ripple component. The capacitor, which acts as an energy buffer, should be sufficiently large to prevent the ripple to penetrate in the array current.

2 a high-frequency, resonant switching stage with included high-frequency transformer. This zero current switching (ZCS) stage generates a sequence of current pulses at a high repetition rate (350 kHz). The amplitude of the sequence is modulated such that a sinusoidal envelope is obtained.

3 a high frequency rectifier, which converts the bipolar sequence of pulses to a unipolar sequence of pulses.

4 a high frequency filter which eliminates the high frequency components, so that a rectified sine is obtained.

5 a bridge inverter which switches at 50 Hz and unfolds the rectified sine to a bipolar sine.

- a filter (not shown) to block noise which is generated by the inverter and which protects the inverter for spikes coming from the grid.

6 a control circuit. The control circuit includes all main control and protection functions of the inverter. These functions include: sinus modulation, MPPT tracking, islanding protection. In addition an RS 485 interface is included to enable monitoring and inspection of the status of a large number of inverters (up to 128) via a common two-wire signal connection. Apart from the modulator and the interface, all principal functions are incorporated in a micro controller.

A photograph of one of the prototypes is shown in fig. 3.

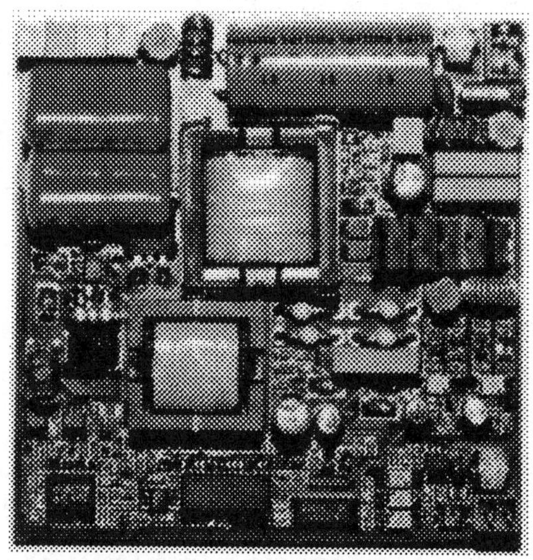

Fig. 3 *Photograph of inverter including power stage and control*

The circuit is touch safe as it needs some grid power to be able to start operation. Without a grid being available the inverter won't start.

3. TEST RESULTS

Starting

After the inverter is connected to the grid it takes about 20 s before the inverter starts delivering power to the grid. During this time very little power is extracted from the grid to charge a capacitor which delivers power to the micro controller. When the capacitor is charged the micro controller goes to stand-by mode. In stand-by mode the micro controller performs some checks (grid voltage, grid frequency, dc voltage). If all checks are ok and when sufficient dc power is available the inverter actually starts with the conversion of dc to ac power. In stand by mode the power consumption is lower than 5 mW.

Islanding detection
During operation the grid voltage and grid frequency are checked continuously to detect loss of grid. In case of loss of grid the inverter is shut down within 100 ms to prevent island operation, which might cause danger to personel. The test revealed that the inverter is shut down if the grid voltage or frequency exceeds one of the following limits:

$U_{grid} < 190V$ or $U_{grid} > 240V$

$f_{grid} < 49$ Hz or $f_{grid} > 51$ Hz

These limits are accepted by many utility companies for detection of islanding.

Power quality
The power quality of the inverter follows from figs.4 to 6. Fig. 4 shows the current waveform of the inverter.

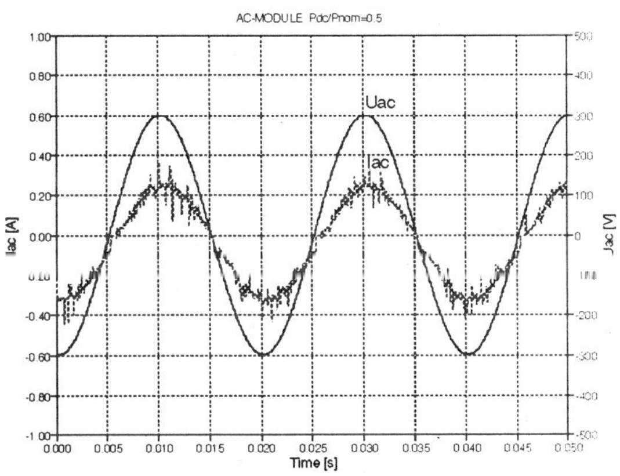

Fig. 4 *Waveform of current to the grid*

In comparison to other pv inverters the current waveform is quite good, although it shows some high fequency noise. The noise in this prototype is expected to be removed by a simple high frequency filter. The fair power quality is confirmed by figs. 5 and 6. Fig. 5 shows the power factor as a function of power. The power factor is better then 0.97 for 20W and above. Above half power the power factor is better then 0.98. Fig 6 gives the total harmonic distortion of the line current as a function of power. The harmonic distortion decreases with increasing power. Above 50 W the distortion is below 6%.
Fig. 7 shows individual line harmonics at 50 W output power. It is seen that all harmonics are below 5% of the fundamental. In the same figure the limits according to standard EN60555 are shown. Based on these results it is expected that parallel connection of a large number of these inverters will not cause any problem concerning harmonics or power factor.

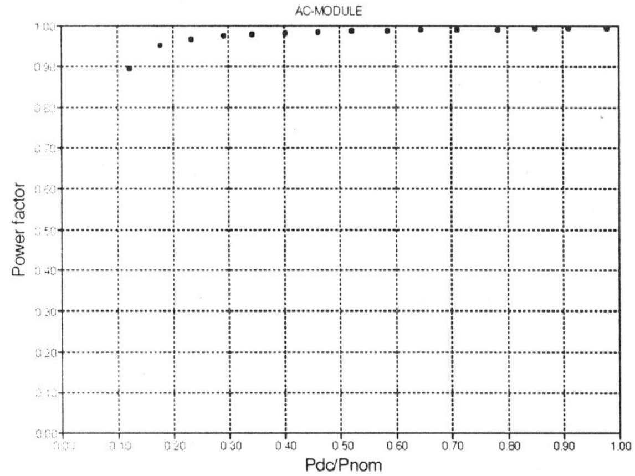

Fig. 5 *Power factor as a function of normalized power*

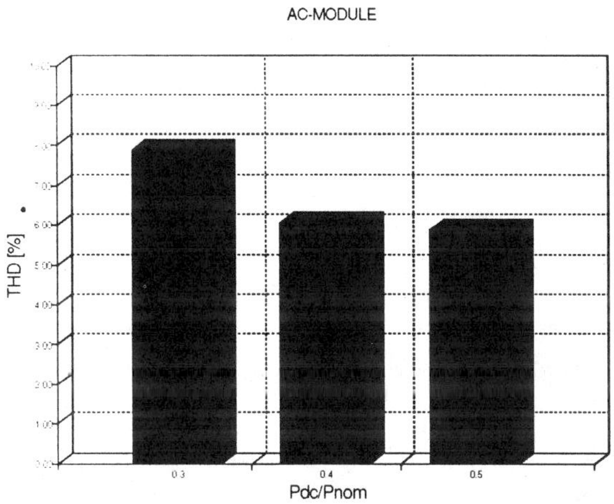

Fig. 6 *Total harmonic distortion as a function of normalized power*

From fig. 7 it also follows that at least 50 ac modules can be paralleled without exceeding the limit. Because of slight differences in the control circuits of the inverters it is likely that not all harmonics will have the same phase, which means that even more modules can be parallelled. The total harmonic distortion caused by a large number of ac modules will be much less than the distortion caused by traditional inverters.

Efficiency
Results of efficiency measurements of one of the prototypes have been reported before [1]. These results showed that an overall efficiency of over 90% is feasible. A detailed experimental analysis of one of the prototypes confirmed the previous measurements.

In the latest half year the development effort was directed to software development and waveform synthesis.

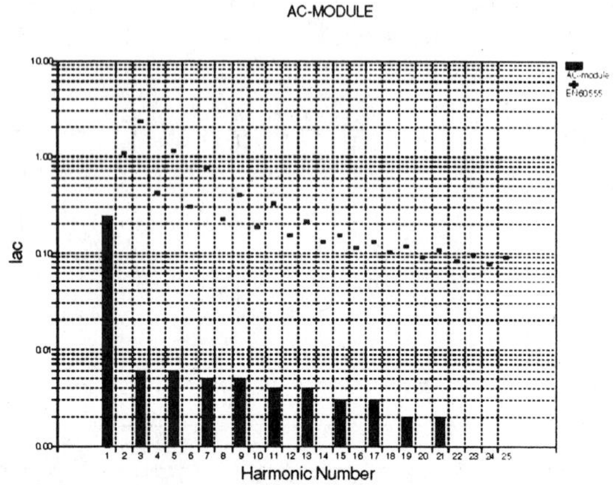

Fig. 7 *Generated harmonics at 60% of nominal power. The piramids reflect limits according to standard EN 60555*

Encapsulation

As these inverters have to operate in a harsh environment (temperatures from -20C to 80C; relative humidity up to 100%), a research programme is going on to develop a method to encapsulate the inverter reliable.

Monitoring

The OKE4 inverter includes an RS485 interface to enable monitoring of the AC modules. Up to 128 inverters can be monitored by a common two-wire signal connection. The interface is electrically insulated from the microcontroller; the information is transmitted via an internal optical link. Via the interface amongst other the following data can be obtained:

- the total amount of energy delivered to the grid since installation
- the current to the grid
- the grid voltage
- the internal temperature

To prevent damage, in case the signal cables are accidentaly connected to the grid, the interface is equipped with components for overvoltage protection.

6. CONCLUSIONS

Because of the simplicity and modularity ac modules (fig. 8) offer a key to wide spread introduction of pv system.
Recent and previous test results show that the prototype inverters meets requirements concerning islanding detection, efficiency abd power quality. Both the excellent power quality figures and monitoring feature prevents disadvantages which else migth occur when many ac modules are connected in parallel.
Research to improve the reliablity of the ac modules, including encapsulation is in progress.

Fig. 8 *PV system with integrated inverters*

7. REFERENCES

[1] S.W.H. de Haan, H. Oldenkamp, C.F.A. Frumau, W. Bonim, 'Development of a 100W resonant inverter for ac modules', 12th Eur. Photovoltaic Solar Energy Conf., Amsterdam, 1994

[2] W. Kleinkauf , J, Sachau, H. Hempel, 'Modulare Energieaufbereitung und Anlagen technik - Ansatze zur Gestaltung PV-gerechter Systemtechnik'. Forschungsverbund Sonnenenergie-Photovoltaik 2, 1992.

[3] C.P.M. Dunselman, T.C.J. van der Weiden, R.J.C. van Zolingen, F.J. ter Heide, S.W.H. de Haan, 'Ontwerpspecificatie AC modules' (in Dutch), Ecofys report E265, 1993.

MODULE ORIENTATED PHOTOVOLTAIC INVERTERS A COMPARISON OF DIFFERENT CIRCUITS

Günter Keller, Thomas Krieger, Michael Viotto
ISET · Institut für Solare Energieversorgungstechnik, Verein an der Universität Gesamthochschule Kassel e.V.
Königstor 59, D-34119 Kassel, Germany, Tel.: ++49(0)561 7294224
Uwe Krengel
IEE · Prof. Dr.-Ing. W. Kleinkauf, Universität Gh-Kassel, D-34109 Kassel, Germany, Tel.: ++49(0)561 8046516

ABSTRACT

An analysis of the PV-modules available on the European market leads to the operating conditions of the module orientated inverters. The efficiency at rated power and the costs have been calculated for a selection of nine power electronic circuits out of the 30 topologies under consideration. The inverter was simulated by a computer model to develop and compare different current control techniques. The results flew in the development of an inverter with a rated power of 700 W (MoSR 700). The results of measurements taken from this type and other inverters are presented.

INTRODUCTION

Photovoltaic technology will become more widely applicable if, in addition to the cost of the modules, also the cost of the balance-of-system is considerably reduced and the reliability is being increased. Using a consequent system structure, modularisation and standardisation of the components of PV-plants, the photovoltaic energy conversion can be more efficient under technical and economical aspects. Important components for supply of AC-consumers or for grid coupled PV-units are small power DC/AC-converters with high or additional requirements.

CONDITIONS

Compared to other energies the photovoltaic energy conversion has a low density and the power is proportional to the area. The basic element is the solarcell with a temperature-dependent voltage-level (ca. 0.5 V) and an area- and irradiation-proportional current. Protected in a construction of glas many cells will be combined to modules from a few watts up to 100 W or more.

Nowadays, in conventional PV-plants a number of modules are connected in series to reach a defined voltage-level. To size up the power some of this arrays are connected parallel at a DC-line. A main DC-line will transport the energy to one or more centralized PV-inverters to supply AC-consumers or feed the energy into public grids. Important disadvantages of this technology are e.g.

- the danger of arcing and thus fire,
- high safety demands on DC-lines,
- high expenditure of special DC-installation,
- MPP adjustable only for the whole generator,
- spezified engineering for each plant,
- enlargements in the power are complicated.

A modular structure of power conditioning can avoid or reduce many of these disadvantages.

Module integrated inverters pick up the basic modular structure of photovoltaic most consistently. The whole DC-technology out of the module will be avoided and each module works in its individual MPP. One gets the smallest complet constituent for simple confguration of PV-plants in fine power-steps. Two years ago, ISET and IEE developed in cooperation a laboratory model with a nominal power of 100 W. The experiences of this project shows that under economic conditions a seriesproduct has to be realized above a nominal power of 150 W.

The compromise of technical advantages und economical aspects are module orientated inverters which are dimensioned for a small group of solar modules.

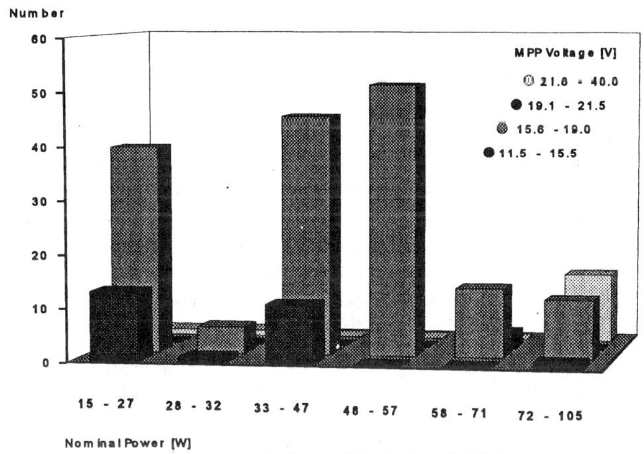

Fig. 1: Frequency of Classified MPP-Voltages as a Function of Nominal Power

In order to develop module orientated inverters one has to define a nominal power and input-voltage range to get a good exploitation of solar energy with the possibility to connect many different modules available on the market to it. An enquiry shows the frequency of the classified MPP-voltages as a function of the nominal power (see Fig. 1).

In most module types the solar cells are connected in series. One can explain the wide spectrum of power in the standard range of 16 - 18 V with various areas and efficiencies of the solar cells. Many applications between 1 and 5 kW use modules from 48 up to 57 W (2.5 - 3.5 A), but also in the range of 33 to 47 W (1.8 - 2.5 A) and 58 to 71 W (3.5 - 4.5 A).

INVERTER-TOPOLOGIES

During the development of a module orientated inverter with a rated power of 300 W 30 different circuits - divided into four categories - have been investigated (s. Tab. 1). The first 29 topologies consist of a serial circuit of an input DC-converter with HF-transformer (10 different circuits) and the suitable inverter (3 different circuits) for grid parallel operation [2]. The kind of the intermediate voltage is used to se-

CH3365-4/94/0000-0929 $4.00 © 1994 IEEE

parate these circuits in the categories DC- (I), Sine-magnitude-modulated (II) und AC-intermediate-voltage (III). The constant energy transformation at the input stage circuits of category I results in low stress for the input power semiconductors and a small size of the transformer. Switching losses due to the modulation of the output stage are the disadvantage of these circuits.

cat./no.	Iprim/Ie	PT_{max}/PT_{nenn}	U_{sek}/U_{netz}	volume
I / 1	4.1	2.00	1.5	+
I / 2	2.0	1.00	1.5	++
I / 3	2.0	1.00	1.5	++
I / 4	4.1	2.00	1.5	+
I / 5	4.1	1.00	1.5	++
I / 6	4.1	1.00	1.5	++
I / 7	2.0	1.00	1.5	++
I / 8	2.0	1.00	1.5	++
I / 9	2.0	0.00	1.5	++
I / 10	2.0	2*0.5	1.5	+
II / 11	10.7	4.00	1.7	0
II / 12	5.3	2.00	1.7	+
II / 13	5.3	2.00	1.7	+
II / 14	10.7	4.00	1.7	0
II / 15	10.5	2.00	1.7	+
II / 16	10.5	2.00	1.7	+
II / 17	5.3	2.00	1.7	+
II / 18	5.3	2.00	1.7	+
II / 19	2.6	2*1	1.7	0
III / 20	10.7	4.00	14.2	0
III / 21	5.3	4.00	7.1	0
III / 22	5.3	2.00	14.2	+
III / 23	10.7	2.00	7.1	+
III / 24	10.5	4.00	14.2	0
III / 25	10.5	4.00	7.1	0
III / 26	5.3	2.00	14.2	+
III / 27	5.3	2.00	7.1	+
III / 28	2.6	2.00	14.2	+
III / 29	5.3	2.00	7.1	+
IV / 30	5.3	2 (50Hz)	-------	--

Tab. 1: Key Properties of the Investigated Circuits

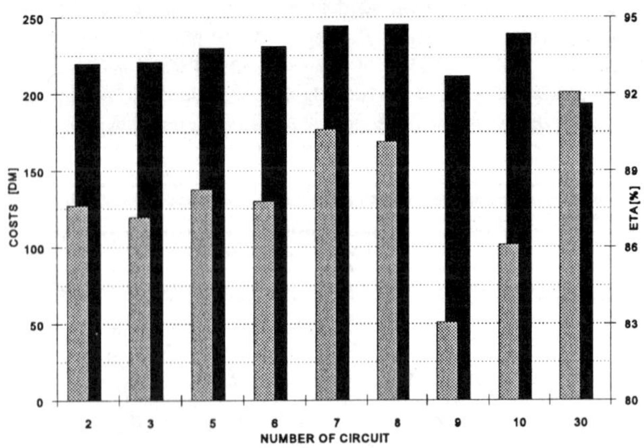

Fig. 2: Efficiencies (left bar) and Costs of Components (right bar) of Selected Circuits

Circuits or categories II and III exhibit a power fluctuation at the input DC-converter leading to higher stress for the relevant semiconductors and a bigger HF-transformer. As an advantage there are almost no switching losses in the output-inverter. Categorie II is identified by the use of a rectifier and a filter to smooth the voltage at the inter-level. In the third category bidirectional power semiconductors are used in the output inverter. Circuit no. 30 (cat. 4) contains only one converter in the powerstage with a 50 Hz -transformer adjusting the voltage level.

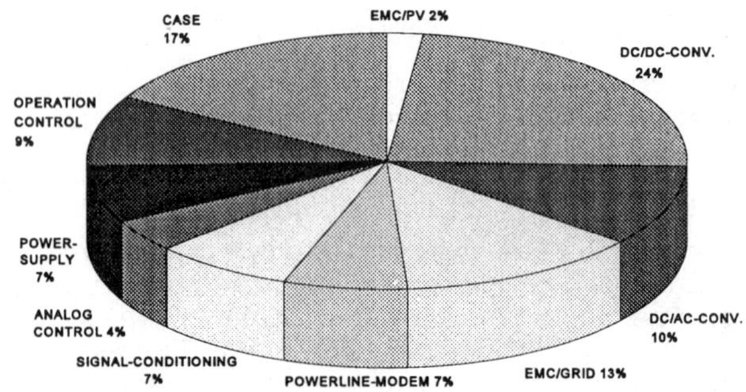

Fig. 3: Distribution of the Component Costs for Circuit 7

Because of the features given in table 1 nine circuits have been selected for a closer investigation. The efficiencies at rated power and the cost of components have been calculated for these topologies (s. Fig. 2).
The small size and other features of circuit no. 7 (double ended full-bridge at the input stage, rectifier, filter, double ended full-bridge at the output) lead to a draft for an inverter based on this topology and a detailled calculation of the cost of components. A laboratory-prototype showed a maximum efficiency of η = 90 %. Figure 3 gives the breakdown of costs. The measured efficiency and the costs lead to a new assessment of the development targets. Small size was traded in for higher efficiency and lower cost. Circuit no. 30 was realized with an increased rated power according to the specific target price. A laboratory-prototype build to this specifications operated in parallel to the grid with an efficiency fairly above 92 %.

CURRENT CONTROL

A low distortion factor of the current fed into the grid is required to operate a multitude of small inverters in parallel. Compliance with the norms and regulations in question requieres an effective current feedback control. Different techniques have been evaluated using circuit simulation. In all cases a direct control of the grid current was inpossible due to delay of the EMC-filter. Control quantity is therefore the current into the filter.

Analog Control

The simple clocked bang-bang controllers which where investigated first did not lead to satisfying results. The analog controller with PWM simulated next showed a much better grid current (s. fig. 4). The control deviation is proportional to the grid voltage and therefore has no negative influences on the distortion factor.

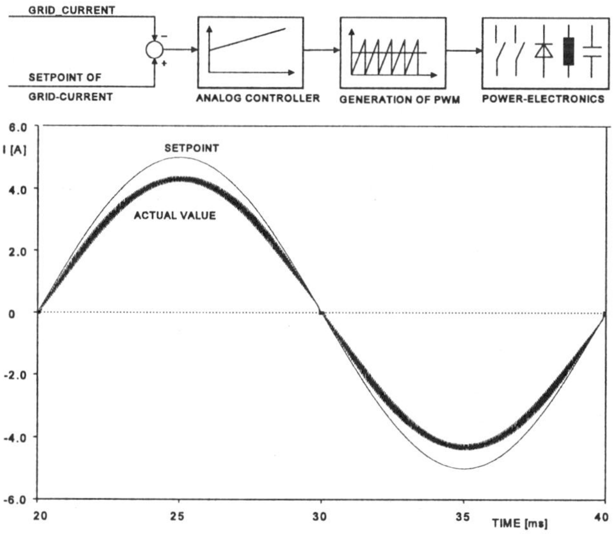

Fig. 4: Analog Controller with Simulated Grid Current

Digital Control

General operation control of the inverter and the MPP-traking can easily be done by a standard 8-Bit-microcontroller. Using this controller for the current control also reduces the chipcount resulting in lower costs. Additional advantages are the flexibility of a digital controller and the possibilty of easy modifications by software. Figure 5 displays a block diagramm of the controller. Setpoint and actual value are sampled and AD-converted. The setpoint of the current is derived from the grid voltage. These values are used to calculate the actual pulse width which is transfered to the power electronic by microcontroller-integrated counters. Simple 8-bit-microcontrollers set the following hardware-limits:

- resolution of the AD-converter: 8 bit
- sampling frequency: 5 kHz

The model used for the construction of the controller considers the discretization in time and value of setpoint and

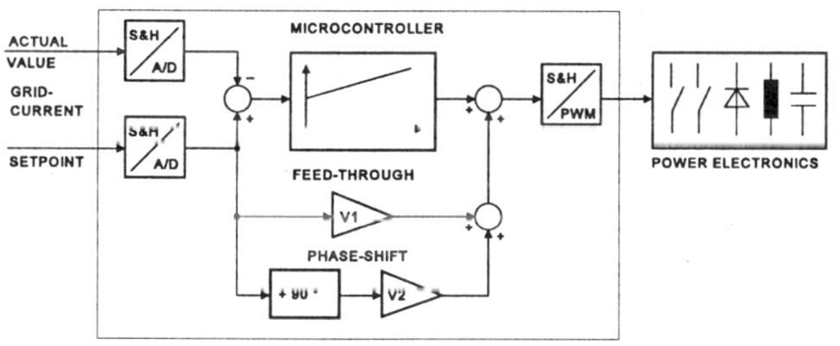

Fig. 5: Block Diagram of the Digital Controller

actual value of the grid current as well as of the pulse width. Different sample times of the input values due to the multiplexing analog port of the prozessor and the calculation

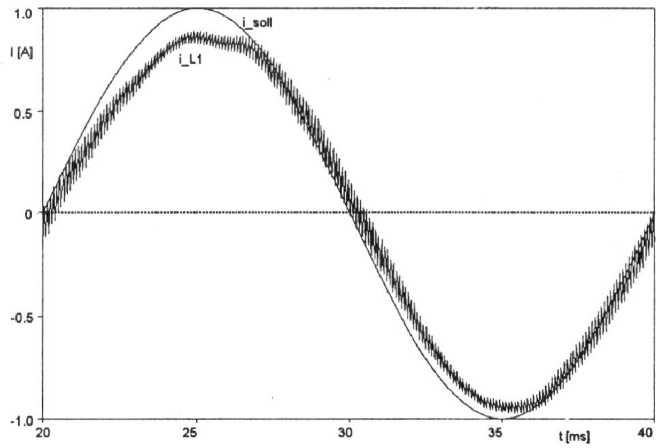

Fig 6: Performance of the Digital Controller

time of the prozessor are taken into account. The delay caused by the digital controller of 200 µs leads to a distinct reduction of the stable loop gain. A direct feed through of the grid voltage on the PWM is required to reduce the control deviation. Due to the dead time of the digital controller and the properties of the EMC-filter the grid current has a phase deviation compared to the grid voltage. This deviation is diminished with a internally generated function with a phase lag of 90° to the grid voltage which is fed additionally on the PWM. The amplitude of this internal signal steers the phase of the current fed into the grid. Figure 6 shows the result of a simulation performed with a digital controller of this type.

Results of Simulation

Circuit simulations demonstrated that even with a low priced 8-Bit-microcontroller a current control with the required quality can be realized. Measurements taken in our laboratory confirm these results. Adjusting the model with the measured data should lead to further improvements of the controllor:

- balancing device tolerances in the EMC-Filter with an adaptive phase correction,
- generation of the setpoint of the grid current via a PLL and a sine table stored in the microcontroller,
- algorithm to increase performance with distorted grid voltages.

MEASUREMENTS

Four inverters were measured at our laboratories:

Type a: Weeber, WE3004-100

27-50 V, 100 W, serial circuit of a half-bridge converter with HF-transformer and a full-bridge to the grid, no galvanic isolation. The DC-converter operates with analog current control whereas the bridge connected to the grid changes the polarity of the half waves. The inverter is prepared for remote diagnostics.

Type b: Mastervolt, Sunmaster 130
27-50 V, 110 W, serial circuit of an uncontrolled ZVS-DC-converter with HF- transformer and a 50 Hz-FET-full-bridge, galvanic isolation

Type c: ZSW, DMI-1, prototype
27-50 V, 100 W, serial circuit of a double ended full-bridge converter and a 50 Hz-transformer, analog current control, galvanic isolation, series production prepared by Dorfmüller

Type d: SMA, MoSR 700, laboratory model
75-250 V (range is covered with multiple taps of the transformer), 700 W, type of circuit like c) but with digital current control. Series product will contain a powerline modem and two independant islanding detectors. No additional components are required to operate in parallel to the grid.

Devices a, c, d contain a microcontroller for operation control. The inverters had various features (e.g. overvoltage protection, different forms of islanding detection). A isolating relay were only in devices c) and d).
The efficiency η, the distortion factor k and the power factor λ have been measured. Figures 7-9 show the results over normalized AC power. The η-plots contain in addition to the data points the polynominal interpolation curves. The maximum efficiency of the inverters were measured as:

WE3004-100:	85,2 % at 49,9 W
Sunmaster 130:	94,5 % at 63,6 W
ZSW:	89,8 % at 37,7 W
MoSR 700):	93,5 % at 306 W

The sinusodial currents of the three self commutated inverters show the expected low distortions and a high power factor. The output current follows the grid voltage and reaches distortion factors as low as 4 % with a distortion factor of the grid voltage of 2,5 % and goes further down to 1,8 % with a grid distortion of 0,5 %. In contrast to this device b) contains no current control and shows - in spite of the use of transistors in the output bridge - the behaviour of a line commutated inverter. This results in a high distortion factor and a power factor that shows a strong dependance on the input voltage.

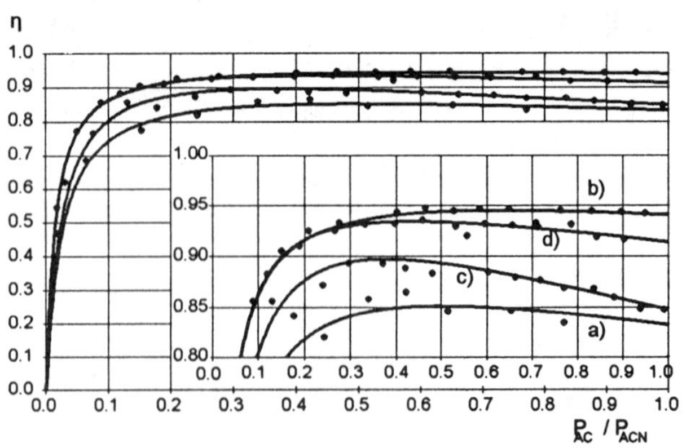

Fig. 7: efficiencies over normalized AC Power

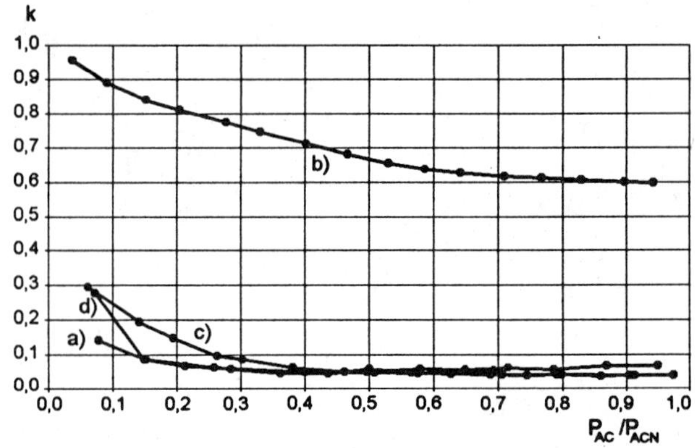

Fig 8: distortion factors over normalized AC Power

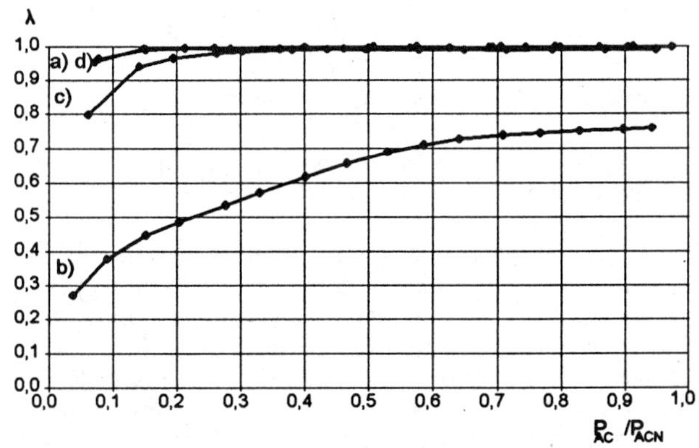

Fig 9: power factors over normalized AC power

CONCLUSION

The variety of available power electronic components facilitates the realisation of inverters with very small rated power. Due to components which are independant on the output power energetical and economical aspects set a lower power limit. It depends on with the requirements and features of the module orientated inverter.

References

W.Kleinkauf, J.Sachau, H.Hempel,
Developments in Inverters for Photovoltaic System;
11. Photovoltaic Solar Energy Conference, Montreux 1992

R.Geipel, B.Gruß, U.Krengel, T.Krieger, J Sachau, M.Viotto, Begleitende Untersuchung zur Entwicklung eines Modul-orientierten Stromrichters; BMFT-Abschlußbericht, ISET, Kassel 1994

EVALUATION OF THE BATTERIES AND CHARGE CONTROLLERS IN SMALL STAND-ALONE PHOTOVOLTAIC SYSTEMS

Joseph R. Woodworth, Michael G. Thomas, John W. Stevens: Sandia National Laboratories
Steven R. Harrington: K-Tech Corporation at Sandia National Laboratories
James P. Dunlop, M. Ramu Swamy, Leighton Demetrius: Florida Solar Energy Center

ABSTRACT

In this paper, we report the results of long-term tests on 14 separate small stand-alone PV systems. These tests were carried out at two separate laboratories in Albuquerque, New Mexico and in Cape Canaveral, Florida. The systems were heavily instrumented to help determine what effects voltage regulation set points have on system water loss, battery capacity lifetime and system reliability. We find that under the proper conditions, the flooded lead-acid batteries in these PV systems can achieve cycle-lifetimes comparable to manufacturer's expectations for conventional battery applications. We also find that both the overcharge and overdischarge protection provided by the system charge controllers were necessary for proper operation of these systems. Details of the data and a number of conclusions for system designers are presented.

INTRODUCTION

Early loss of battery capacity is one of the most common problems restricting the use of small stand-alone PV systems, both in developed countries and in third world applications. [1,2,3,4] A better understanding of how charge controller settings and algorithms affect specific types of batteries would allow users to improve PV system reliability and achieve more predictable battery lifetimes.

In this paper we report the results of two separate long-term tests of flooded lead-acid batteries and charge controllers in fourteen separate PV systems. Previous papers [5,6,7] have reported shorter term experiments involving batteries and charge controllers and given an interim report on the tests reported in this paper. The tests reported here, which were designed to evaluate what parameters affect a battery's state-of-charge, lifetime, and water loss, were performed at Sandia National Laboratories (Sandia) in Albuquerque, New Mexico and at the Florida Solar Energy Center (FSEC) in Cape Canaveral, Florida. Seven complete stand-alone PV systems were tested at each location. The tests lasted slightly less than two years at each location, with the systems operating at Sandia from November 1990 to October 1992 and at FSEC from November 1991 to August 1993.

SYSTEM DESCRIPTION

Each stand-alone PV system consisted of a photovoltaic array, a battery, a charge controller and a resistive load. A simplified schematic of a typical system is shown in Figure 1.

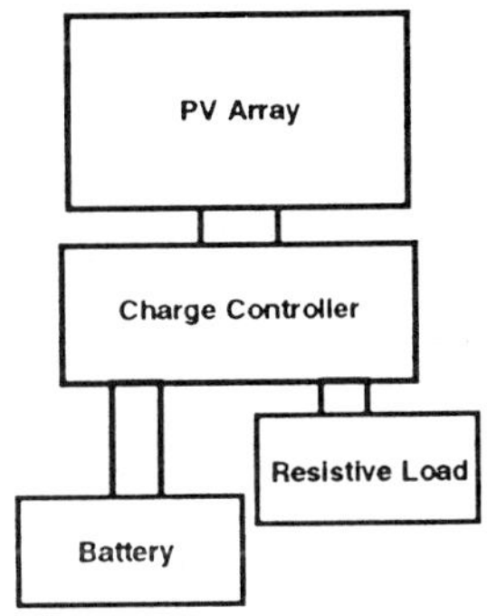

Figure 1: Schematic of typical PV system

Batteries

The batteries in all of these tests were the Trojan Battery Company's "Pacer" 30XH batteries.[8] These 12-volt batteries have a rated capacity of 105 amp-hrs at a 20-hr discharge rate and 110 amp-hrs at a 100-hr discharge rate. They have lead-antimony plates (6% antimony in positive plates, 2.75% antimony in negative plates). The 30XH batteries were chosen for these experiments because they resembled deep-cycle batteries in many respects, but have relatively thin plates and hence would be expected to approach end-of-life during the course of the two-year-long experiments. When subjected to a daily depth of discharge of 15%, as was the case in the Sandia experiments, Trojan suggests that these batteries will have a cycle life of approximately 900 cycles — i.e. after 900 cycles, the battery's capacity will have been reduced to ~50% of its original value, with complete failure of the battery expected to occur relatively rapidly thereafter.

In these tests, we refer both to battery "capacity" and battery "state-of-charge." In this paper, these terms are defined as follows.

1. ***Battery capacity***: The number of amp-hours that can be withdrawn from a battery at a specified discharge rate, before the battery drops to a specified low voltage **after** the battery has been charged by a power supply that provides both bulk and finishing charge cycles. At Sandia, battery capacity measurements involved two steps. First, the battery was charged by a power supply at 7 amps to 14.4 volts and then held at 14.4 volts for 30 hours. Second, the battery was discharged at a 5-amp rate until the battery voltage dropped to 10.75

First WCPEC; Dec. 5–9, 1994; Hawaii

volts. The amp-hrs withdrawn from the battery in the second half of the test were defined as the battery capacity.

2. ***Battery state-of-charge***: The number of amp-hours that can be withdrawn from a battery **at the time of a particular test**, when the battery is discharged at a specified rate to a specified low voltage. At Sandia, the batteries were discharged at a 5-amp rate to 10.75 volts. Note that there is no restriction in the state-of-charge measurements on how the battery was treated prior to being discharged. Battery capacity can be measured only after carefully charging the battery. Battery state-of-charge can be measured anytime. In this paper, battery state-of-charge is expressed as a fraction of the batteries' initial capacity.

Loads

Both experiments used fixed resistors as the loads for these PV systems. At Sandia, the loads drew 1.0 Amps of current from the system 24 hours a day for a total daily load of 24 amp-hrs. At FSEC, the loads were connected to the system only at night, drawing 2.75 amperes for 8 hours producing a total daily load of 22.3 amp-hrs. The FSEC loads approximated the loads found in PV lighting systems. The Sandia loads approximated the loads found in communications systems. This difference in load timing caused important differences in the system operation, which will be discussed in the results section.

PV Arrays

At FSEC, each system was powered by two Solarex MSX 60 modules in parallel, with a combined maximum-power output of 15.9 V, 6.26 A and 100 watts at standard test conditions (25° C, 1000 W/m^2 solar irradiance). At Sandia, each system was powered by four Solarex 5-volt, 60-watt modules wired in series with cell shading over parallel strings to reduce the current output. These systems had relatively long wiring runs between the arrays and controllers, further reducing the array output. As a result, the arrays at Sandia provided a maximum power output measured at the charge controller connections of 15 volts, 6.3 amps and 90 watts at standard test conditions.

Charge Controllers

Charge controllers with a variety of operating algorithms were purchased from seven different manufacturers:

Bobier Electronics[9]
BOSS[10]
Heliotrope General[11]
Integrated Power Corporation[12]
Polar Power[13]
Specialty Concepts Inc.[14]
SunAmp[15]

Since the charge controllers at Sandia and FSEC were purchased a year apart, they were not identical at the two locations. However, the controllers in systems 1 through 5 at Sandia and FSEC were from the same manufacturers (i.e. the controller in system #1 was from the same manufacturer at Sandia and FSEC, the controller in system #2 was from the same manufacturer at Sandia and FSEC . . . etc.). In addition, systems 2 through 5 had settings close enough that they can be used to directly compare data from the two sites.

The controllers used in these experiments had four set points:

- **"Vr"** regulation voltage - voltage at which the controller disconnects the array from the battery to protect the battery from overcharging.
- **"Vrr"** regulation reconnect voltage - voltage at which the controller reconnects the array to the battery to begin full current charging again.
- **"LVD"** low voltage disconnect - voltage at which the controller disconnects the battery from the load to prevent over-discharging of the battery.
- **"LVR"** low voltage reconnect - voltage at which the controller reconnects the battery to the load.

Controllers 3 and 7 at Sandia did not initially have low-voltage disconnects. After these batteries were drained to ~1.5 V in January of 1991 during a period of cloudy weather, external low-voltage disconnects were added.

Two basic types of charge controller algorithms were used in this experiment.

1. **"Off-on" controllers** applied either the full array current or no current to the battery depending upon the battery voltage. The difference between Vr and Vrr in these controllers varied from 0.1 V to 1.7 V. The controllers with large Vr - Vrr differences (0.8V or greater) cycled off and on during regulation with a frequency of 0.1 cycles/sec or less. The two controllers with small Vr - Vrr differences (0.1 to 0.3 V) were called by the manufacturers "constant voltage" (CV) and "pulse width modulated" (PWM). However, observation of these controllers' operating characteristics on an oscilloscope indicate that they are off-on controllers operating at high frequency (~400 cycles/sec) with the exact frequency being a function of the battery state-of-charge, charge rate and the controller setpoints.

2. **A current limiting controller** (#7) at FSEC applied full array current up to the battery's regulation voltage and then limited the charging current to 1 ampere.

Tables I and II list the charge controller set points for the tests at Sandia and FSEC respectively. Calibration of the charge controller set points was accomplished prior to the experiment by the techniques described in Bower, Dunlop and Maytrott.[5] However, the Vr, Vrr and LVD values at FSEC

Table I: Settings for SNL Phase 1 Charge Controllers

System #	Algorithm	Temperature Compensation	Values at charge controller terminals (volts at 25 C) Vr	Vrr	LVR	LVD
1	off-on	no	14.25	12.9	12.3	10.8
2	off-on	internal to CC	14.54	13.45	12.44	11.83
3	off-on PWM	no	14.4	14.1	12.3	10.8
4	off-on CV	yes-at battery	14.4	14.3	13.4	11.8
5	off-on	no	14.36	13.45	13.15	11.48
6	off-on	yes-at battery	14.4	12.66	13.63	11.60
7	off-on	yes-but temperature probe in CC rm.	14.35	13.15	12.3	10.8

Table II: Settings for FSEC Phase 2 Charge Controllers

System #	Algorithm	Temperature compensation	Values at charge controller terminals (volts at 25 C) Vr	Vrr	LVR	LVD	Values at battery (25 C) Vr
1	off-on 2-step	at battery for 1st 9 months, none after 9 months	>15.0* 14.28/13.8*	~13.8 13.4	12.4	11.0	14.7* 14.0/13.54*
2	off-on	internal to CC	14.57	13.46	12.1	11.6	14.21
3	off on PWM	yes- at battery	14.5	14.1	12.8	11.2	14.4
4	off-on CV	internal to CC	14.4	14.4	12.8	11.5	14.35
5	off/on	yes - at battery	14.7	13.8	12.7	11.45	14.45
6	off/on	none	~14.5** 14.00**	~13.0** 13.72**	12.3- 12.5	11.7	~14.2** 14.58**
7	current limiting	external at battery	14.4** 14.8**		12.7	11.1	14.1** 14.45**

* Controller was changed during 7/92. 1st line in table gives initial set point, second line gives final set point. After 7/92, controller initially charged battery to upper Vr set point each day, then regulated between lower Vr set point and Vrr.
** Controller settings were changed in 5/92. 1st line gives initial settings, 2nd line gives final settings.

were also determined from actual system operating voltages. In some cases there was a difference of ~0.2 V in the set point value as determined by the two techniques. In these cases, we used the value obtained from the system operating voltages.

When the array was providing its full output current, resistance in the wiring and fuses and the somewhat smaller resistance of the shunt used to measure battery current, caused a ~0.2 V voltage drop between the charge controller "battery" terminals and the actual terminals on the battery at Sandia and a 0.3 V to 0.4 V drop at FSEC. This effectively lowered the Vr and raised the LVD of the system.

Controller # 1 at FSEC was a two-step off-on controller. It had a higher Vr setpoint for the first regulation cycle of each day than for all successive regulation cycles. Initially, however, this controller malfunctioned, rarely went into regulation at all and continued to bulk charge the battery regardless of the battery voltage. This controller was replaced 9 months into the test (July 1992) with a similar controller from the same manufacturer. The two-step regulation did function on this second controller.

Set point changes were also made to systems 6 and 7 at FSEC in May 1992. The top values for Vr and Vrr in Table II are the original settings and the lower values are the final settings.

TEST PROCEDURES

Battery Initialization

To be sure that the batteries at each location had equal capacities at the beginning of the tests, the batteries were charged and discharged several times, recording the ampere hours in each charge or discharge cycle. For all of the discharge tests (both in battery initialization and state-of-charge measurements), the batteries were discharged in two-hour increments, with the batteries being allowed to rest for one hour between discharge periods. Thus, during discharge, the batteries were repeatedly discharged for two hours, allowed to rest for one hour and then discharged for two more hours until the battery voltage reached the lower cutoff voltage during the first 15 minutes of a two-hour cycle. The off-on cycling in the discharge tests was designed to measure battery internal resistance as a function of state-of-charge by measuring battery voltage under load and in open-circuit conditions at a number of different states of charge. For tests designed only to measure state-of-charge, a much simpler discharge at a constant current to a fixed low-voltage cutoff would have been adequate. When all of the battery capacities were within 2% of each other at each location, the batteries were declared ready for the long-term tests.

At Sandia, batteries were put through nine initial charge/discharge cycles. The Sandia batteries were charged at 7 amps to 14.4 volts and then held at 14.4 volts for 30 hours. The Sandia batteries were discharged at 5 amps to a cutoff voltage of 10.75 volts. At the beginning of the initialization tests, the Sandia battery capacities varied from 76 to 86 amp-hrs. At the end of the initialization tests, all the batteries had capacities between 92 and 94 amp-hrs, with an average of 93 amp hrs.

At FSEC, the batteries were put through five initial charge/discharge cycles. The FSEC batteries were also charged at 7 amps to 14.4 volts and then held at this voltage for 30 hours. The FSEC batteries were discharged at 5 amps to a cutoff voltage of 10.7 volts. At the beginning of the initialization tests, the FSEC battery capacities varied from 94 to 99 amp-hrs. At the end of the initialization tests, all the FSEC batteries had capacities between 97 and 100 amp-hrs, with an average of 98.5 amp-hrs.

State-of-Charge Measurements

At Sandia, the following procedure was used to determine battery state-of-charge:

1. At the end of five sunny days, the batteries were isolated from the PV systems.
2. The next morning, battery open-circuit voltage was measured and specific gravity measurements were taken using a refractometer.
3. The discharge test load was configured for a 5-amp discharge current, and the discharging began, in its 2-hours-on, one-hour-off cycle until the battery reached 10.75 volts in the first 15 minutes of a discharge period.
4. The batteries were then connected in series, recharged until the average battery voltage was 14.4 volts and held at this voltage level for 10 hours.
5. Battery water was added and documented.
6. Batteries were returned to the PV systems.

At FSEC, the following procedure was used to determine battery state-of-charge:

1. The systems were normally allowed to run with the loads disconnected for three to five days prior to testing.
2. The PV arrays were disconnected the evening before the state-of-charge test.
3. The next morning, specific gravities were measured with a refractometer and the loads were configured to provide a 5-amp discharge rate, with the system cycling two hours on and one hour off until it reached 10.7 V in the first 15 minutes of a two-hour discharge cycle.
4. The batteries were recharged in two groups, with up to 4 batteries in series, at a 7-amp rate until the average battery voltage in each group was 14.4 volts. The batteries were then held at this voltage level for approximately 20 hours.
5. Battery water was added and documented
6. The batteries were then returned to the systems.

Note that the battery recharges described here probably did not provide complete equalization charging. Trojan Battery Corporation informs us that in order to equalize these batteries, it is necessary to hold them at a charge voltage of 15.4 V for several hours.[16]

The three-to-five days that the FSEC systems ran without the loads prior to state-of-charge testing probably yielded results that were somewhat higher than the normal system state-of-charge. This will be discussed further in the next section.

RESULTS

Figure 2 shows a sampling of the raw data from system number two at FSEC for three sunny days in July of 1992. This figure shows six-minute averages of battery voltage (measured at the controller terminals), battery current, current from the PV array, load current, tilt irradiance (solar irradiance at the plane of the PV array), and battery temperature.

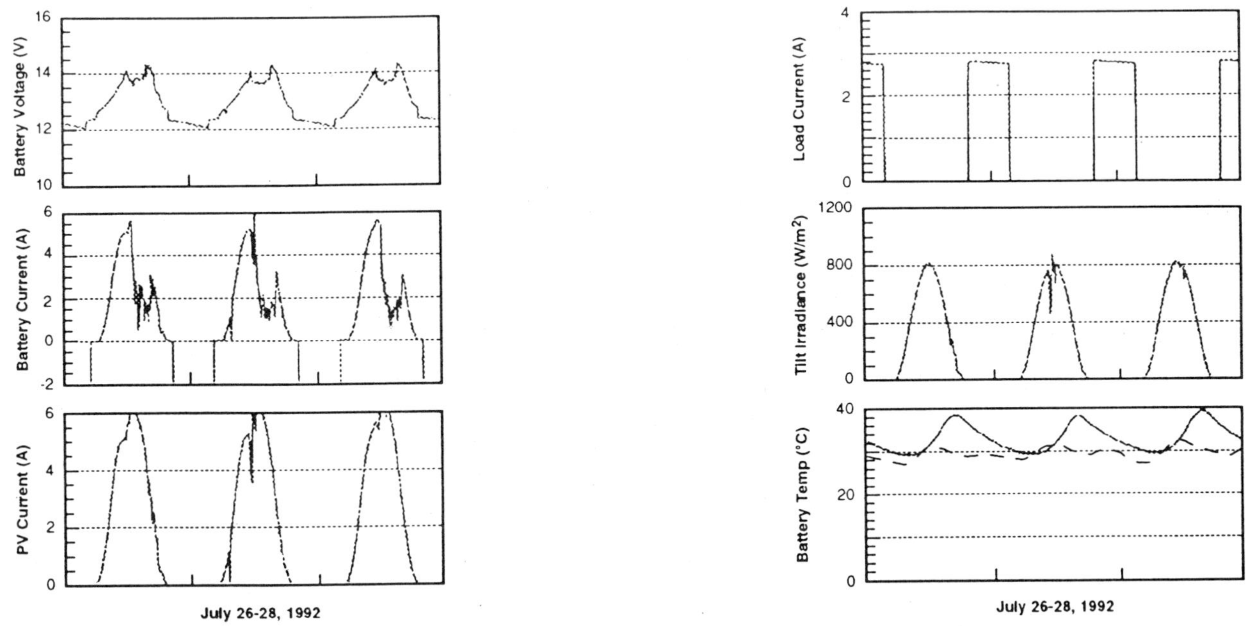

Figure 2: Raw data from system tests at FSEC; 6-minute averages of PV, battery & load current, battery voltage, solar irradiance at the plane of the array and battery temperature. Battery room temperature is also shown as a dashed curve under the curve of battery temperature.

Battery voltage, battery current and PV current all rise with the tilt irradiance in the mornings as the battery receives bulk charging to replace the energy dissipated by the load the previous evening. About noon, the charge controller goes into its regulation cycle. While the off-on cycling during regulation is too fast to be resolved in this six-minute average data, the data still displays a number of interesting trends. The average battery voltage drops as soon as regulation starts and does not rise again until the system quits regulating near sunset. Since the controller on this system is a "shunt" controller, which short-circuits the array during regulation, the battery current drops and the array current rises to its short-circuit current value during regulation. In the evening, the battery voltage first drops sharply from ~12.8 V to ~12.4 V as the load is connected and then drops slowly to ~12.1 V during the 8 hours that the load is on. When the load is disconnected, the battery voltage recovers to an equilibrium value of ~12.4 V and remains there until sunrise.

Battery temperature is influenced by two factors: the temperature in the battery room and the fact that charging batteries is an exothermic process. In the three days shown in Figure 2, temperature in this east-facing battery room begins rising at about 7 a.m. or one hour after sunrise, peaks at ~32 °C about 1 p.m. and then slowly decays to an overnight low of ~28° C. Thus, the batteries at FSEC are above room temperature for all but one or two hours in the early afternoon. During the night, the battery temperature slowly drops toward room temperature, reaching an overnight low of ~30 °C about 8 a.m. Battery temperature then begins rising as the battery is charged, peaking at ~39 °C at sunset. Note that the battery temperature continues to rise sharply while the system is in regulation, despite the fact that the average current entering the battery has been reduced more than a factor of 2. The data in Figure 2 represent three completely sunny days. On days that are partly or completely cloudy, the battery temperature is much closer to room temperature.

Figure 3 shows plots of battery voltage measured at the battery terminals and battery current for system number 2 at FSEC on July 23, 1992. This data was taken with a ~10-second time resolution, fast enough to observe the off-on cycling as this system starts regulation. The full-on, full-off

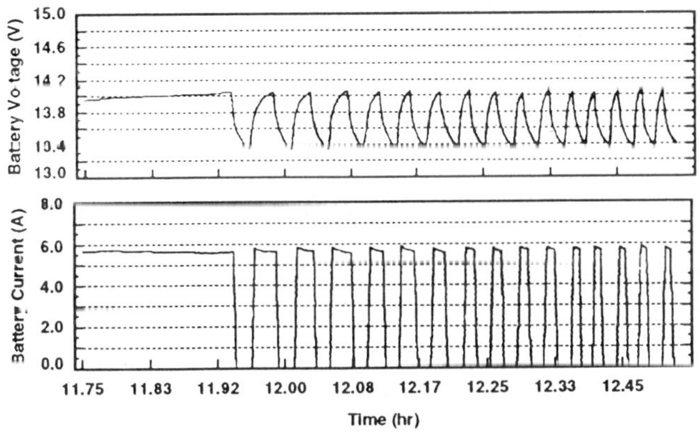

Figure 3: Battery voltage (measured at battery terminals) and battery current on system #2 at FSEC as the system begins regulating on a sunny day in July 1992.

nature of this system during regulation is clearly evident in the battery current. The regulation reconnect voltage (Vrr) of 13.4 is also evident from the voltage trace. The peak battery voltage shown in Figure 3 is gradually rising toward its maximum voltage of 14.26 V, which will be achieved as the regulation cycle occurs with higher frequency later in the day.

Figure 4 shows the measured solar irradiance on the plane of the arrays at Sandia and FSEC. This graph, which shows monthly averages of irradiance, indicates that it is considerably sunnier at Sandia, in Albuquerque, NM than it is at FSEC, in Cape Canaveral, FL. Photovoltaic system designers normally size a system by looking at the ratio of array energy supplied to load energy consumed in the worst month of an average or "typical meteorological" year. PVCAD software [17] was used to predict the array to load ratios of these systems. The typical meteorological year data in PVCAD indicated that the Sandia systems should have a worst month array to load ratio of 1.40 to 1 and the FSEC systems should have a ratio of 1.19 to 1. Recent "typical meteorological year" data published by the National Renewable Energy Laboratory [18] suggest that the Sandia systems should have a worst month array-to-load ratio of 1.31 to 1 and the FSEC systems should have a ratio of 1.15 to 1.

For the batteries, PV arrays and loads in these experiments, we would initially expect the batteries to see an average daily depth of discharge of 22 to 24 amp-hrs, equivalent to 20 to 22 % of the batteries' rated capacity, at FSEC and Sandia respectively. For the FSEC systems this is correct. For the Sandia systems, however, the situation is more complex, since the loads at Sandia drew current 24 hours a day. Some of the current generated by the Sandia arrays during the day went directly to the load, bypassing the battery and reducing the average daily depth of discharge. PVCAD predicts that this effect will lower the daily depth of discharge at Sandia to ~15% of the batteries' rated capacity. Analysis of the time the Sandia systems spent in bulk charging and in regulation in 1991 supports the PVCAD prediction. Thus, while the FSEC and Sandia systems had nominally identical PV arrays, batteries and daily load amp-hours, the difference in load timing caused significantly shallower daily depth of discharge cycles in the Sandia batteries than in the FSEC batteries.

Table III shows the total amount of water that was added to each of the systems at Sandia and FSEC over the course of the 23 months of testing. Note that the average water usage at FSEC is about 60% greater than the average water use at Sandia.

Figure 5 shows weekly averages of daily maximum and minimum battery temperatures at Sandia and FSEC. Weekly averages of daily maximum outside ambient temperatures are also shown at Sandia for comparison. The Sandia data is for 1991, the FSEC data is for 1992. While the Sandia temperature data is incomplete, the battery and ambient temperature data taken together strongly suggest that for the majority of the summer, battery temperatures at FSEC were significantly higher than at Sandia.

Figure 6 shows outside ambient temperature, battery temperature and charge controller temperature for system #4 at Sandia during the summer of 1991. This charge controller limited the current during its regulation cycle by pulsing the current on and off at high frequency (~400 cycles/sec). The battery temperature, which peaks roughly at sundown, rarely exceeds the ambient temperature by more than 5°C. The charge controller, however, often rose ~30° C above the battery temperature due to heat the controller generated internally. Because of this, controllers that have external temperature probes measuring battery temperature directly appear preferable to controllers which use controller temperature to approximate battery temperature, particularly

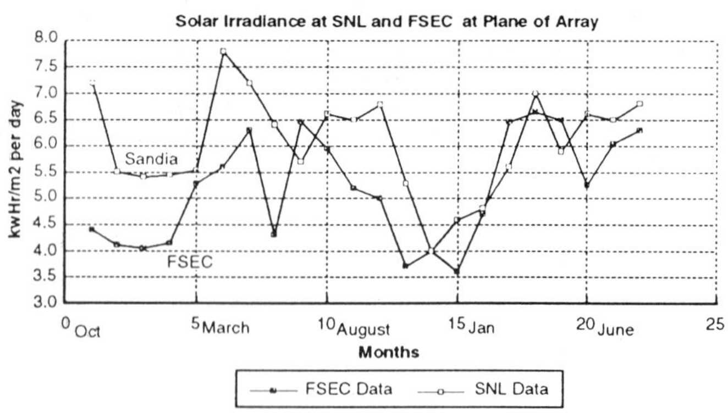

Figure 4: Solar irradiance at the plane of the arrays at Sandia and FSEC. Note that there is consistently more solar energy available at Sandia.

Table III: Total Water addition to batteries during 23 months of testing.

System	Water Loss (grams) Sandia	FSEC
1	2,268	5,726
2	2,778	4,309
3	4,054	6,038
4	4,365	5,669
5	2,835	4,366
6	2,807	5,273
7	3,090	4,139

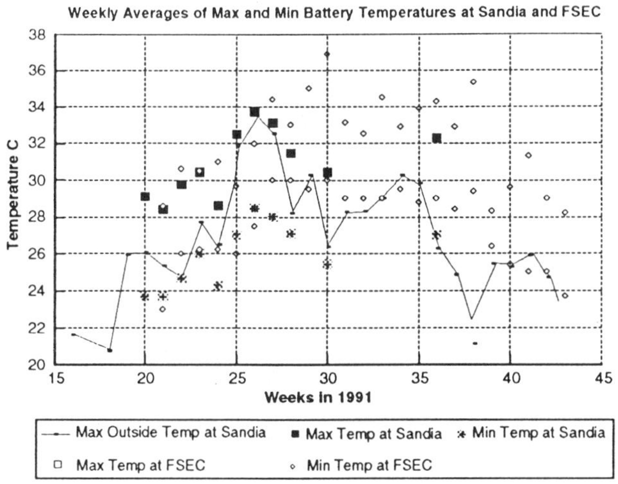

Figure 5: Weekly averages of daily maximum and minimum battery temperatures at Sandia and FSEC along with weekly averages of maximum daily outside ambient temperatures at Sandia. Sandia data is for 1991 and FSEC data is for 1992. Note that during July and August, the batteries are significantly hotter at FSEC than at Sandia.

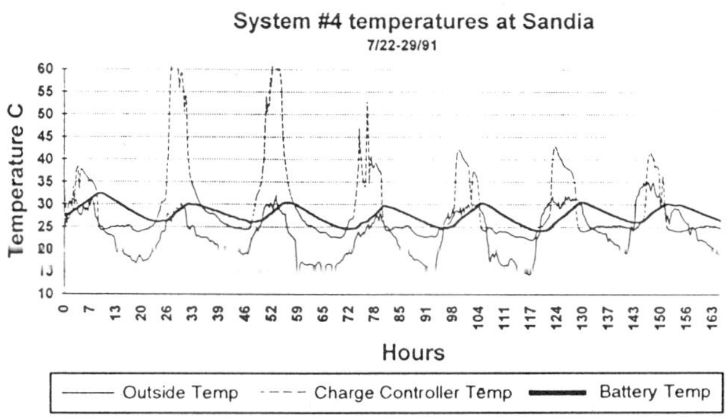

Figure 6: Battery, charge controller and outside ambient temperatures for system #4 at Sandia. Note that the charge controller temperature often rises 30 °C above ambient temperatures due to heat generated inside the controller.

for controllers that operate at high frequency or in a linear or resistive current limiting mode.

Figures 7 and 8 are bar charts showing the state-of-charge of the batteries in the Sandia and FSEC systems as a fraction of the initial capacity of each battery. As discussed earlier, these measurements were made by actually discharging the batteries. Since the temperature of the batteries fluctuated significantly from summer to winter during the tests, we have corrected the state-of-charge measurements to 70° F using the data given by Perez.[19] The box under each figure

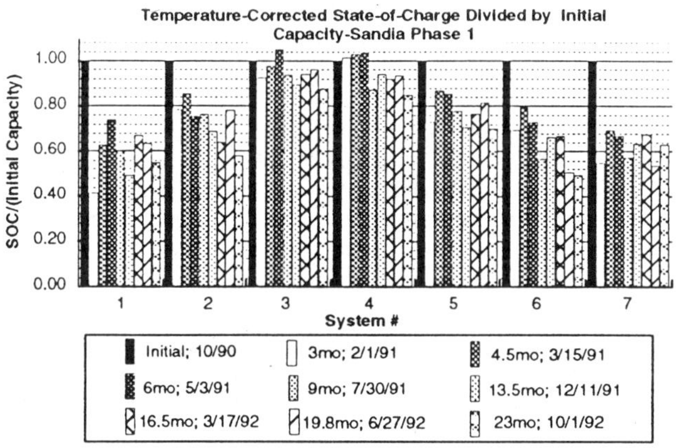

Figure 7: State-of-charge of batteries at Sandia as a fraction of their initial capacity. All state-of-charge measurements are corrected to 21 °C. Note that some of the Sandia batteries remain at or above 85% state-of-charge throughout the tests.

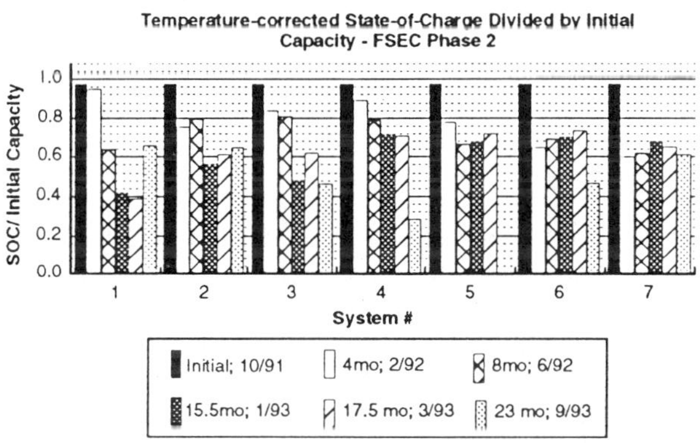

Figure 8: State-of-charge of batteries at FSEC as a fraction of initial capacity corrected to 21 °C as in Figure 7. Note that the state-of-charge of the batteries has decayed to ~70% or less by the end of the second winter.

indicates when the state-of-charge measurements were made. Recall that systems 2 through 5 had controllers from the same manufacturers at both locations.

As mentioned previously, the systems at FSEC were normally allowed to run for three to five days with the load disconnected before the state-of-charge tests. This undoubtedly raised the system above its normal state-of-charge. We can estimate the size of this effect by comparing three battery state-of-charge measurements at FSEC. The first measurement was taken four months into the test after the systems had spent three days without the load. In the second measurement, which was taken eight months into the test, the array and load were disconnected at the same time the evening before the test; The third measurement was also taken at eight months, after a power supply recharge in which the batteries were bulk charged to 14.4 V and then held at 14.4 V for 30 hours (this third set of data is not shown in Fig. 7) . Surprisingly, the solar irradiances at four months (February) and eight months (June) were about the same at FSEC, allowing a meaningful comparison between the two different periods. We conclude that systems 2 through 5 have increased their average state-of-charge by only about 6% during the three days without a load on the system.

ANALYSIS

Are charge controllers really necessary?

The systems described in this paper have battery capacities equal to ~ 4 days of load requirements and are typical of many PV systems used for residential or outdoor lighting applications. The function of the charge controllers in these systems is to protect the batteries from overcharge or overdischarge. There are PV designs which operate satisfactorily without charge controllers. As an example, when load requirements are much smaller than in the systems described here, it is often economical to omit the charge controller entirely and use a battery with a capacity of ~30 days of load requirements instead. The U. S. Coast Guard has used this type of design in many navigational beacons. While the batteries in these navigational systems do suffer from the effects of overcharging and overdischarging, the relatively large battery capacity allows these systems to continue to operate reliably.[20]

Two systems in this test with malfunctioning or incomplete charge controllers demonstrate why controllers are needed for the types of PV systems tested here. System #1 at FSEC initially had a malfunctioning controller that rarely regulated at all and which, on most clear days, continued to try to bulk charge the battery until sunset. This led to maximum battery voltages of 15.0 to 15.3 V on many sunny days, which is excessive for this type of battery. As a result, the battery in system # 1 had very high water loss (see Figure 16) and in the post-mortem inspection, the positive grids of this battery were found to be severely corroded. On the other hand, system # 3 at Sandia initially had no low-voltage disconnect. As a result, the battery in this system was drained to ~1.5 V during a cloudy period in January 1991. At 1.5 V, the controller could not operate properly and disconnected the array from the battery, locking the system into a non-functioning state. Manual intervention was required to restart the system. An external low-voltage disconnect was added to the system after this incident. Note that this problem, which occurred due to the absence of a low voltage disconnect, occurred on the system which otherwise maintained the highest state-of-charge of any of the fourteen systems in this test.

Set Points

The differing set points of the systems at Sandia allow us to look at the effect of the set points on the system's state-of-charge. Figure 9 plots the average system state-of-charge over the 23 months of the test as a function of Vr. Vr values

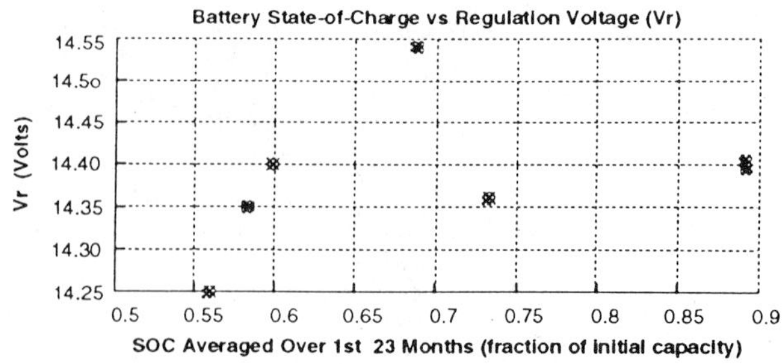

Figure 9: Average state-of-charge of the batteries at Sandia over the 23-month test as a function of regulation voltage (Vr) of the charge controller. Note the absence of any obvious correlation.

at Sandia had been set as close as possible to 14.4 V, therefore we could only examine the effect of Vr over a limited range. Over this limited range, there does not appear to be any correlation between Vr and state-of-charge. Figure 10 shows a plot of the average system state-of-charge as a function of the value of Vrr, the regulation reconnect voltage. This data shows a strong correlation, with a correlation coefficient (R)[21] of 0.95. Analysis of the correlation coefficient and the number of data points indicates that the probability (P) [21] that this apparent correlation is accidental is less than 0.1%. Figure 11 shows a plot of total water loss at Sandia over the 23 months of the experiment as a function of Vrr. Again, we see a significant correlation with R = 0.88 and P = 1%. Plots of water loss versus Vr show no obvious correlation for the Sandia data. Finally, Figure 12 shows that system state-of-charge and water loss are also correlated with each other (R = 0.89, P = less than 1%).

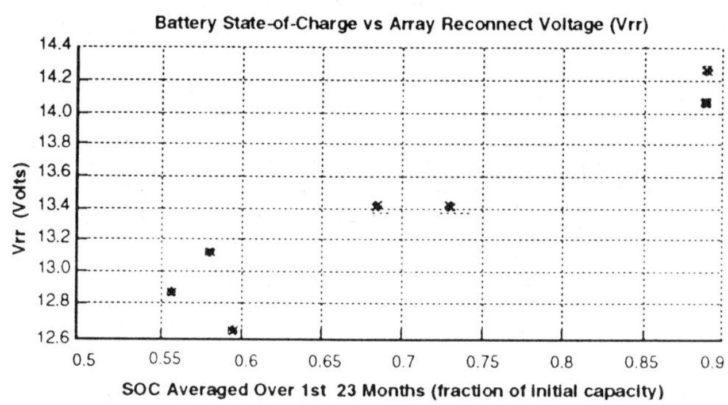

Figure 10: Average state-of-charge of the batteries at Sandia over the 23-month test as a function of regulation reconnect voltage (Vrr). Note that state-of-charge is strongly correlated to Vrr.

Both state-of-charge and water loss are strongly dependent on regulation reconnect voltage, but only weakly dependent on regulation voltage over the range of the Vr and Vrr data shown. It appears that the number of times a system cycles off and on during a day in regulation has a much stronger impact on battery state-of-charge than the maximum voltage

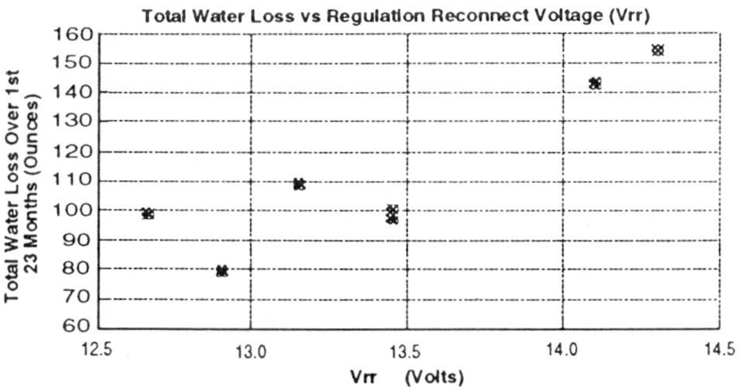

Figure 11: Total water loss in the Sandia PV systems as a function of regulation reconnect voltage. These two variables are also correlated.

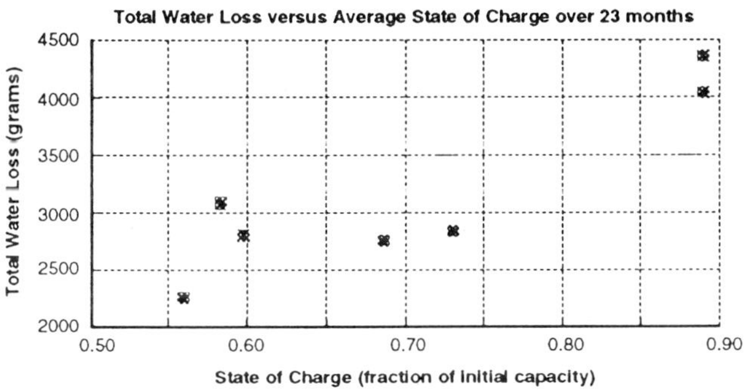

Figure 12: Total water loss in the Sandia PV systems as a function of average state of charge over the 23 months of testing.

reached in any one cycle. Beyond certain limits, it is clear that the value of Vr will become important. As an example of this, the controller in system #1 at FSEC, which was malfunctioning during the first nine months of the test, regularly charged its battery to 15.0 to 15.3 V, causing the high water loss and excessive grid corrosion described in the previous section.

During the second winter at FSEC, which had unusually low solar irradiance, most of the systems had their batteries drained to the low-voltage disconnect point 20 or more times. As a result, the value of the low voltage disconnect (LVD) had an impact on the overall health of the battery during this winter and the summer that followed. Figure 13 shows a plot of state-of-charge for five of the off-on controllers averaged over the winter of 1992 and the summer that followed. The sixth off-on controller at FSEC is not included because its system suffered a load failure during this period. Figure 13 demonstrates a clear relationship between the LVD and the long-term battery state-of-charge for these systems (R = 0.98, P is less than 1%). Thus, relatively high LVD values appear to protect the long-term health of the battery, but at the expense of decreased load availability.

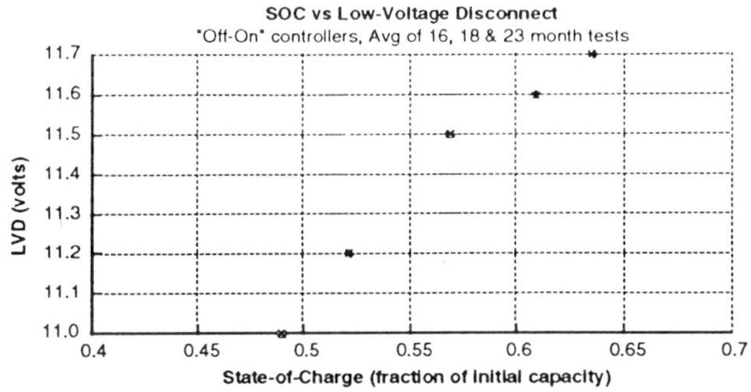

Figure 13: Average state-of-charge for five of the off-on controllers at FSEC as a function of the controller's low voltage disconnect setting (LVD). For systems that spend a significant amount of time in a deficit situation, increasing the value of the low-voltage disconnect improves the long-term health of the battery.

Solar Irradiance Levels

The differing climate at the two test locations caused very different behavior in the PV systems. As stated earlier, the NREL solar irradiance data suggests that the PV systems should have had worst-month array-to-load ratios of 1.31 to 1 at Sandia and 1.15 to 1 at FSEC. Weather patterns are notoriously variable from year to year however and the NREL irradiance data shows worst-month irradiance at Albuquerque varying by ~ ± 25% about the mean over the years from 1960 to 1991. Measured array-to-load ratios averaged over the 3 worst months of the second winter at each site were 1.21 to 1 at Sandia and only 1.06 to 1 at FSEC. Since energy losses in the batteries, wiring, fuses, and charge controllers are expected to be more than 6%, this indicates that the FSEC systems were in an energy deficit situation for about 3 months during the second winter.

Figure 14 compares state-of-charge measurements for systems 3 & 4 at Sandia and FSEC. These systems had the same controllers with nominally identical set points at the two laboratories. It is important to remember, however, that the average daily depth of discharge was 15% at Sandia and 20% at FSEC. In Figure 14, the Sandia state-of-charge data stays above 90% state-of-charge for most of the test, ending at about 85% state-of-charge at 23 months. The FSEC state-of-charge data drops steadily to final values of 28% and 40% in the same time frame. Remembering that the FSEC state-of-charge measurements are probably somewhat higher than the true system state-of-charge makes this difference even larger. Since neither set of batteries had exceeded the manufacturer's expected cycle life in this data, it is probable that the difference in system performance seen in Figure 14 is due to the difference in the array-to-load ratio at Sandia and FSEC.

It is useful to look at how the state-of-charge varied on a daily basis for the FSEC systems. During the second winter,

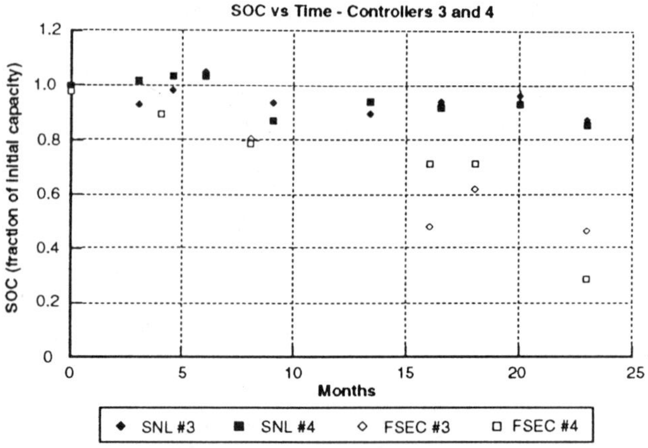

Figure 14: State-of-charge versus time for two sets of nominally identical systems at Sandia and FSEC. These are the systems with the highest Vrr settings.

the average of the state of charge measurements for systems 2 - 7 at FSEC was 64% (We exclude system #1 because its set points were too low). Remember that the true FSEC state-of-charge levels may be ~6% lower than the values given here. Thus, given the ~22 amp-hr nightly load, the systems at FSEC were cycling between 64% and 41% state-of-charge on a daily basis. This type of cycling between two relatively low states-of-charge will lead to stratification of the battery electrolyte and sulfation of the battery plates, both of which will shorten the useful life of the battery.

From the relatively high state-of-charge measurements at Sandia and the relatively low measurements at FSEC, we conclude that the 1.31 to 1 array-to-load ratio at Sandia was adequate for proper operation of the PV systems with a 24-hour load. The 1.15 to 1 ratio at FSEC was inadequate for systems with nighttime loads As a caution, we note that the systems at Sandia rarely saw more than three cloudy days in a row. In areas characterized by long (1 - 2 weeks or more) periods of unbroken cloudy weather, it may be necessary to increase the array-to-load levels above those used in the Sandia systems.

Battery Cycle Life

Since early failure of batteries has been a common problem in PV systems, it is useful to compare the battery lifetime seen in these experiments to the lifetime expected by the battery manufacturer. The Trojan Battery Corporation expects the 30XH batteries used in these experiments to have a useful cycle life of ~900 cycles when they are cycled at 15 % depth-of-discharge. At 900 cycles, Trojan would expect its "average battery" to have declined to 50% of its original capacity and would expect it to fail completely soon thereafter. During the course of these experiments, the batteries at Sandia went through about 690 daily cycles to 15% depth of discharge in addition to the 17 cycles to 100% depth-of-discharge, during the initialization and state-of-charge tests. Since the lifetime for 100% depth of discharge cycling is only about one-sixth as large as for 15% cycling, the 690 cycles at 15% and 17 tests to 100% were roughly equivalent to 800 cycles at 15% depth of discharge, or eight-ninths of the expected life of the battery. At the end of this cycling, the five batteries on the Sandia systems with the highest Vrr values, systems 2,3,4,5 & 7, remained above 50% state-of-charge. Systems 3 and 4, which had the highest Vrr values, remained above 80% state-of-charge. Thus, it appears that under the proper conditions, flooded lead-acid batteries can achieve lifetimes in stand-alone PV systems comparable to manufacturer's expectations for more conventional battery uses. It is important to note that the batteries at Sandia were rarely at temperatures higher than 35° C and that they were recharged about 4 times a year by a power supply that charged the batteries to 14.4V and then held them at that voltage for 10 hours.

Water Loss

Battery water loss is a common cause of PV system failure. For most systems, the dominant cause of water loss is the over-voltage gassing or electrolysis reaction. [16,19] This reaction, which is described by a number of authors,[19,22,23] is a strong positive function of temperature, with hot batteries producing more gas at a given voltage and charging current than cold batteries. The amount of water lost by over-voltage gassing (in the absence of any recombination) and the current driving the gassing reaction are related by: [23]

$$H_2O = 0.0056\, I\, T \qquad (1)$$

where:
- H_2O = grams of water converted to gas per cell
- I = gassing current passing through the cell in amperes
- T = time in minutes

In the light of these facts, we can examine the Sandia and FSEC water loss data shown in Figures 15 and 16 and in Table III. Several things are apparent from the figures. First,

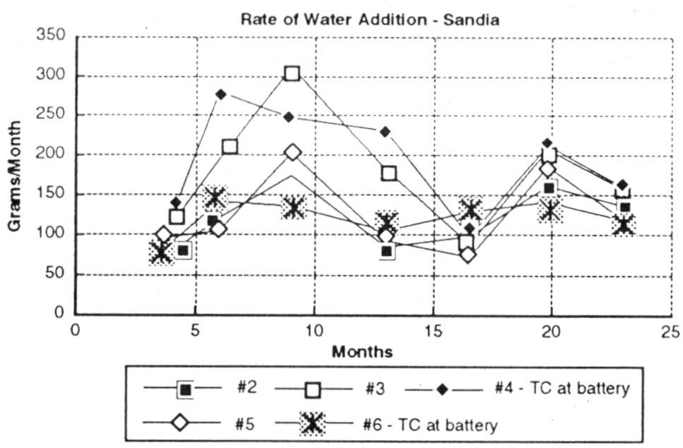

Figure 15: Rate of water addition versus time for some batteries at Sandia.

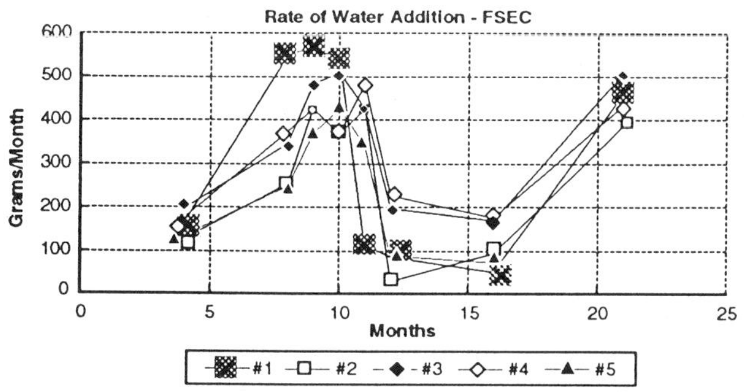

Figure 16: Rate of water addition versus time for some batteries at FSEC. Note the high initial losses of system #1 which malfunctioned and normally continued to bulk charge the battery regardless of battery voltage. The water loss in system #1 dropped dramatically after the controller was replaced 10 months into the test.

summer water loss is a factor of 2-to-3 higher than winter loss for most of the systems. Second, summer water loss is about a factor of 2 higher at FSEC than at Sandia. This causes the overall water loss at FSEC to be about 60% higher than at Sandia. In the summer at FSEC, when water losses were the highest, typical batteries lost ~440 g of water per month. To account for this loss through the over-voltage gassing reaction would require that 6.5 amp-hrs per day was consumed by gassing in the batteries. Since 6.5 amp-hrs is a small fraction of the total energy available from the PV array per day, we feel these results are consistent with overvoltage gassing being the dominant battery water loss mechanism in these tests. The fact that summer water losses were higher than winter water losses is consistent with the strong temperature dependence of the over-voltage gassing reaction. The fact that summer water losses were higher at FSEC than Sandia is also consistent with the temperature dependence of the gassing reaction and the fact that summer temperatures were higher at FSEC than at Sandia (Figure 5).

There is some evidence from the Sandia data that temperature compensation of the controller set points does smooth out the summer-to-winter water loss cycle. In Figure 15, system 6 at Sandia had temperature compensation based on the battery temperature with a coefficient of approximately -30 mV/°C or -5 mV/°C per cell. This system has a much smaller summer-to-winter variation than the other systems. System 4 at Sandia, which was kept at a much higher state of-charge, also had temperature compensation with a coefficient also of about -30mV/°C. We can compare system 4 to system 3 at Sandia, which had a similar state-of-charge, but did not have temperature compensation. During the first year, system 4 does lose more water in cool weather and less water in hot weather than system 3. However, this effect is not apparent in the second year. The effects of the very high voltages (~15.0 - 15.3V) attained on system #1 at FSEC are apparent in Figure 16. System one has very high water use until the charge controller was replaced during the 10th month, after which the water loss drops sharply.

Specific Gravity Measurements

In these experiments, specific gravity measurements were taken just before measuring battery state-of-charge. In order to minimize the impact of local gradients in the electrolyte, these specific gravity measurements were taken after the batteries had been disconnected from the PV array and load for at least 14 hours but less than 20 hours. Figure 17 shows a plot of battery state-of-charge as a function of the average

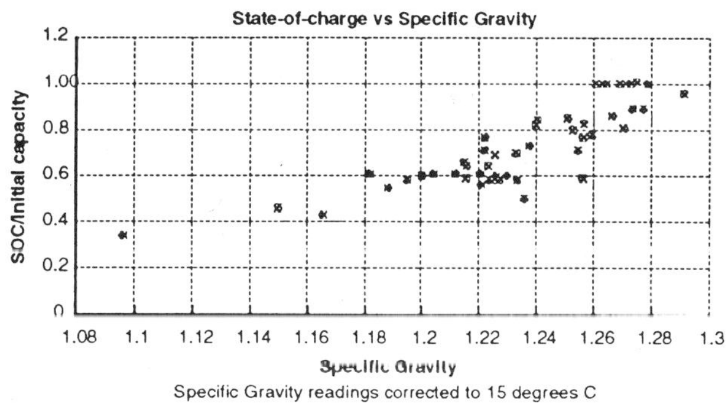

Figure 17: Battery state-of-charge versus average specific gravity of the six cells in each battery. While the variables are clearly correlated, there is enough scatter in the data to prevent specific gravity from being used as an accurate gauge of state-of-charge.

specific gravity in the six cells in each battery for the systems at Sandia. These specific gravity measurements are corrected to 25° C. Note that these tests were performed on nominally identical batteries that were all purchased from the manufacturer at the same time. Although there is clearly a correlation between state-of-charge and specific gravity (R = 0.84, P = less than 0.1%), there is considerable scatter in the data. Nearly identical values of specific gravity in Figure 17 correspond to state of charge measurements that differ by almost 30%. This variation, which may be due to the fact that these measurements were taken using the electrolyte from the top surface of batteries which had significant electrolyte stratification, makes specific gravity measurements only a rough measurement of battery state-of-charge when performed on aging batteries in field situations.

CONCLUSIONS

Conclusions for controller and system design are:

1. Battery state-of-charge was strongly correlated with the value of the regulation reconnect voltage (Vrr). Over the limited range of regulation voltages tested at Sandia (14.25 V to 14.54 V) battery state-of-charge was only

weakly related to the value of the regulation voltage. It appears that the number of times a system cycles off and on in regulation, which is determined by the regulation reconnect voltage, is more important than the maximum voltage reached in any one cycle. At some point beyond the narrow range of regulation voltage settings investigated here, it is clear that regulation voltage settings will become more important.

2. The data taken at Sandia demonstrate that Trojan's 30XH flooded lead-acid battery with lead-antimony grids, when operating in a PV system, can achieve cycle-lifetimes which are comparable to manufacturer's expections for conventional battery applications. It is important to note that the batteries which met manufacturer's expectations were maintained at a high state-of-charge, rarely exceeded temperatures of 35° C and that about four times a year these batteries were recharged by a power supply that charged the batteries to 14.4 V and then held them at that voltage for 10 hours.

3. The systems tested here had battery capacities equal to ~ 4 days load requirements and array-to-load ratios ranging from 1.15-to-1 to 1.31-to-1. For these systems, both the overcharge protection and the overdischarge protection provided by the charge controllers were required for reliable system operation and acceptable battery lifetime.

4. The array-energy-input to load-energy-requirement ratio ("array-to-load ratio") in the worst month of a "typical meteorological year" is an important parameter affecting the overall health of a PV system. The data in these tests indicate that with a load operating only at night, the array-to-load ratio of 1.15 to 1 at FSEC was inadequate for proper operation of the PV systems. The array-to-load ratio of 1.31 to 1 at Sandia was adequate when operated with a 24-hour-per day load.

5. For PV systems that spend a significant amount of time in an energy-deficit situation, increasing the value of the low-voltage cutoff significantly improves the long-term health of the battery, at the expense of decreasing the load availability.

6. While there was clearly a relationship between battery state-of-charge and specific gravity of the battery electrolyte, the scatter in the data is too large for specific gravity measurements to be used as an accurate measurement of battery state-of-charge when performed on aging batteries in field situations. Field measurements of specific gravity still have considerable value in diagnosing failed or marginal cells.

7. Charge controllers which operate at high frequency or in a current limiting mode can generate heat internally during the regulation cycle and in some cases this can raise the temperature of the charge controller 30° C above ambient temperatures. Therefore, external temperature compensation probes that measure battery temperature directly are preferable to temperature compensation probes located inside the charge controller.

Conclusions for battery water loss:

8. For the flooded lead-acid batteries tested here, battery water loss is directly related to the battery state-of-charge. In general, the higher the state-of-charge, the higher the water loss. Allowance for this water loss needs to be included in system design or maintenance plans.

9. Battery water loss varied by a factor of 2 to 3 from summer to winter. The amount of battery water loss seen in these experiments can be accounted for by the over-voltage gassing reaction and variations in the battery temperatures. The fact that the summer water losses at FSEC were more than a factor of two higher than the losses at Sandia is consistent with the fact that the batteries were hotter at FSEC than Sandia in the summer and the fact that the overvoltage gassing rate increases with increasing temperature.

10. Temperature compensation of the charge controller set points can reduce the size of the summer-to-winter water loss cycling, but did not reduce overall system water loss for the systems tested here. It should be noted however, that temperature compensation is also expected to reduce grid corrosion in hot weather and help maintain a high state-of-charge in cold weather. Our data does not give any information on these latter two factors.

ACKNOWLEDGMENTS

Many people contributed to these experiments. We would like to acknowledge a number of helpful conversations with the charge controller manufacturers, with Jim Drizos of the Trojan Battery Corporation and with Garth Corey, Rudy Jungst, and Jim Freese of Sandia's Storage Batteries Department. We also wish to acknowledge that this paper drew heavily on internal reports at Sandia[24] and FSEC.[25] This work was supported by the U. S. Department of Energy under contract DE-AC04-94ALB50009.

REFERENCES

1. Lisa Shepperd and Elizabeth H. Richards, **Solar Photovoltaics for Development Applications**, Florida Solar Energy Center and Sandia National Laboratories, 1993, Available from PV Design Assistance Center Sandia National Laboratories, Mail Stop 0752, Sandia National Laboratories, POB 5800, Albuquerque, NM 87185-0752, (505) 844-3698.
2. Michael G. Thomas, Harold N. Post, Anne VanArsdall, **Photovoltaics Now - Photovoltaic Systems for Government Agencies**, Sandia National Laboratories SAND88-3149, Revised February 1994, Available from

PV Design Assistance Center at Sandia.

3. Herbert Wade, **Photovoltaic Rural Electrification Projects in the Pacific Region and Lessons Learned**, South Pacific Institute for Renewable Energy, BP 11530, Mahina, Tahiti, French Polynesia. This report was prepared for the World Bank as part of its 1991 Pacific Regional Energy Assessment.
4. Jorge M. Huacuz, Roberto Flores, Jaime Agredano, Gonzalo Munguia, **Field Performance of Lead-Acid Batteries in Photovoltaic Rural Electrification Kits**, Instituto de Investigations Electricas, Division Fuentes de Energia, Departmento de Fuentes No-Convencionales de Energia, Apartado Postal 475. Cuernavaca, Mor. 62000, Mexico, May 1994.
5. Ward I. Bower, James P. Dunlop, Craig W. Maytrott, **Performance of Battery Charge Controllers: An Interim Test Report**, Proceedings of the 21st IEEE PV Specialists Conference, 1990, p. 1122, IEEE publication #0160-8371/90/000.
6. James P. Dunlop, Ward Bower, Steve Harrington, **Performance of Battery Charge Controllers: First Year Test Report**, Proceedings of the 22nd IEEE PV Specialists Conference, 1991, p. 640, IEEE publication #CH2953-8/91/000.
7. John Stevens, Jay Kratochvil, Steve Harrington, **Field Investigation of the Relationship Between Battery Size and PV system Performance**, Proceedings of the 23rd IEEE PV Specialists Conference, May 1993, p. 1163, publication #93CH3283-9
8. Trojan Battery Company, 12380 Clark St, Santa Fe Springs, CA 90670; (800) 423-6569
9. Bobier Electronics Inc, 512 37th St. POB 1545, Parkersburg, West Virginia, 26101, (304) 485-7150.
10. BOSS has been purchased by Photocomm Inc, 7681 E. Gray Rd. Scottsdale, Arizona, 85260, (602) 948-8003.
11. Heliotrope General, 3733 Kenora Dr. Spring Valley CA, 91977, (800) 552-8838
12. Integrated Power Corporation: 7524 Standish Place, Rockville, Md 20855; (301) 294-9133
13. Polar Power: 2808 Oregon Ct. Bldg K-4, Torrance, CA 90503; (310) 320-3514
14. Specialty Concepts Inc, 8954 Mason Ave, Chatsworth CA 91311, (818) 998-5238
15. SunAmp Power: 1902 Country Club Dr, No. 8, Mesa AZ 85201; (602) 833-1550.
16. Jim Drizos, Trojan Battery Corporation, private communication.
17. PVCAD, Version 2.10C, available from Photovoltaic Resources International, 1902 N. Country Club Dr. #6 Mesa, AZ 85201, (602) 834-7778
18. **Solar Radiation Data Manual for Flat Plate and Concentrating Collectors,** National Renewable Energy Laboratory, 1617 Cole Blvd, Golden, Colorado, NREL/TP-463-3393, 1994.
19. Richard A. Perez, **The Complete Battery Book**, TAB Books Inc. Blue Ridge Summit, PA, 1985, pps. 22 and 36.
20. V. Alminauskas, **Performance Evaluation of Lead Acid Batteries for Use with Solar Panels**, Proceedings of the 23rd IEEE PV Specialists Conference, May 1993, p. 1258, IEEE publication #93 CH3283-9
21. Philip R. Bevington, **Data Reduction and Error Analysis for the Physical Sciences**, McGraw-Hill Book Co, New York, 1969, pps. 119 & 312.
22. David Linden, **Handbook of Batteries and Fuel Cells**, McGraw Hill, New York City, 1984, ISBN 0-07-037874-6. Out of Print - but available from Independent Battery Manufacturer's Assn, 100 Larchwood Dr, Largo, FL 34640-2811, (813) 586-1408.
23. Elsevier Sequoia and S. A. Lausanne, **Lead Acid Batteries, A Reference and Data Book**, Indian Lead Zinc Information Centre, New Deli, 1977, p. 116.
24. Steven R. Harrington, **PV Battery/Charge Controller Tests at Sandia National Laboratories - Phase One**, issued in draft form 4/19/94.
25. James P. Dunlop, **Evaluation of Battery Charge Controllers - Phase 1 and 2 Summary**, FSEC-CR-516-92, September 1992 .

DEVELOPMENT OF INTERMITTENT REDOX FLOW BATTERY FOR PV SYSTEM

Izumi Tsuda, Kosuke Kurokawa, Ken Nozaki
Electrotechnical Laboratory

ABSTRACT

Redox flow battery has been developed as a storage device for photovoltaic systems. The pump loss is the greatest problem for redox flow battery under the low current condition. An intermittent flow redox battery has been developed for the reduction of the pump loss. The experimental results of this battery show that the efficiency under the intermittent pump operation increases higher than the continuous pump operation. Moreover, inert gas bubble technology has been introduced to improve the performance under the high current condition. It is clear from the experiments that this technology increases the efficiencies. The simulation results of these technologies are coincident with experimental results. It is shown by the simulation that they can improve the Faradic and energy efficiencies of a number of stacks in series.

INTRODUCTION

A storage technology is one of the key-technologies putting PV systems to practical use. Ordinally, a lead acid battery is used as the storage system in PV systems. It seems that the lead acid battery is not congruous with PV systems as follows. First, it is necessary that the lead acid battery is sometimes made a full-charged condition, and the terminal voltage of the lead acid battery rises at this condition. These facts cause the battery to decrease energy efficiency. Secondly, the state of charge (SOC) of the battery decides the terminal voltage. Sometimes, the battery subsystem is directly connected with PV array. Then the operating voltage of the system is mostly different from the maximum output point of the PV array. This effect reduces the system efficiency.

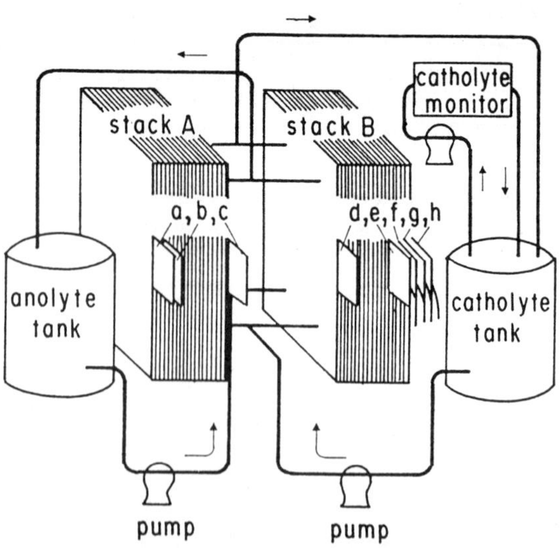

Fig. 1. Construction of redox flow battery for PV system

Redox flow battery (RFB) is one of the advanced type batteries. The RFB for photovoltaic (PV) systems has been developed[1][2][3][4]. Fig. 1 shows the construction of the RFB. This battery is composed of stacks, tanks, pumps, positive and negative electrolytes and so on. Electric energy is charged by the oxidation of the positive electrolyte and the reduction of the negative electrolyte, and discharged by the reverse reactions. The pumps make the positive and the negative electrolytes to circulate through the stacks and the tanks. As shown in Fig. 2, the stack is piled with cells and the electrolytes are ordinally supplied in a parallel to each cell within the stacks. The advantages of the RFB are as follows.

(1) It is easy to select the power and capacity of the storage independently.
(2) The terminal voltage of the battery can be regulated by tap changing. As the result, the maximum output of PV array can be traced by this regulation.
(3) The self-discharge in the electrolyte tanks is negligible.
(4) It is easy to measure the SOC of the battery and so on.

It seems for these advantages that this battery is suitable for the stand-alone PV systems. It is expected that the application of the RFB for the PV system improves the system performance and reduces the system cost.

The preproto- and proto-type of the continuous flow battery were developed for stand-alone PV systems. The

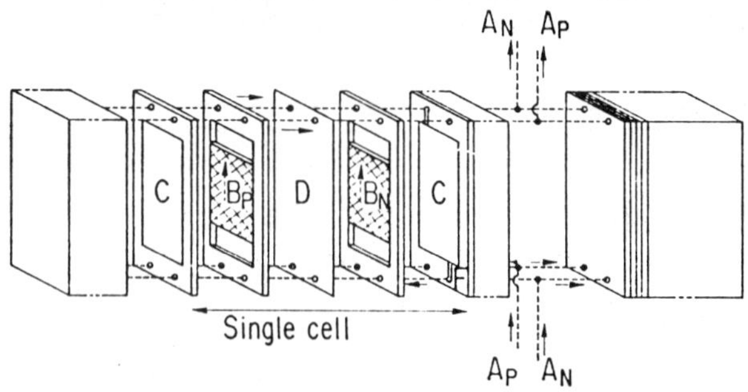

Fig. 2. Composition and electrolyte flow system of RFB

A_P:positive electrolyte flow, A_N:negative electrolyte flow, B_P:positive electrode, B_N:negative electrode, C:bipolar plate, D:cation exchange membrane, E:manifold, F:slit

CH3365-4/94/0000-0946 $4.00 © 1994 IEEE

First WCPEC; Dec. 5–9, 1994; Hawaii

simulation of the proto-type clarifies the advantages of the tap changing effect[4]. On the other hands, the continuous pump operation decreases the power efficiency under the low charge-discharge current condition. The reason is that the shunt current through the electrolyte in the pipe between the stacks decreases the efficiency of the RFB. Therefore, the study of the intermittent flow redox battery (IFRB) started. It is expected by the introduction of the IFRB to improve the efficiency under the low charge-discharge current condition. Moreover, the study of the electrolyte flow with a inert gas bubble has been started. The inert gas bubble is expected to introduce the increasing electric resistance of the electrolyte stream in the piping without increasing fluid pressure.

The experimental results show that the using of these technologies improves the Faradic and energy efficiencies. These experimental results are compared with the simulation results. The simulation results are coincident with the experimental results. The simulation results show that these technologies introduce the efficiency increase of the RFB under the condition that the many stacks are electrically connected in series.

EXPERIMENT OF IFRB

Experimental method

Table 1 shows the specification of the IFRB. The IFRB is composed of stacks as shown in Fig.3, and each stack is piled with 10 or 12 cells. If the pumps were stopped, the positive and the negative electrolytes remaining in the piping among stacks are collected to the tanks due to the potential energy. The experiments are carried out in three modes. The first mode is a continuous flow operation mode. The second is no electrolyte flow mode. The third is an intermittent flow mode. The SOC in the tanks of the IFRB is estimated from the open circuit voltage (OCV) of the mini cell in the first and the third modes. The experiments were carried out from 30% SOC to 70% SOC in these modes. The SOC of the cell is not able to measure in the second mode. The experiments of this mode were carried out in the region of the terminal voltage corresponding to the OCV between 10% and 90% SOC.

The intermittent operation of the IFRB is explained as follows. It is impossible to measure the SOC of the electrolyte in the stacks under the pump stopping condition. As the result, the electrolytes are circulated during the preset time when the terminal voltage of the IFRB is over the upper boundary value or under the lower boundary value. The upper boundary value is the same as the OCV equivalent to the 90% SOC, and the lower boundary value is the OCV of the 10% SOC respectively. The pumps are stopped when the charging or discharging current of the IFRB is zero. The preset time of the pump operation was decided 120sec from preparatory experiment.

Table 1. Specification of IFRB

rated current	10A
number of stacks	4
number of cells	
No.1	10
No.2	12
No.3	12
No.4	12
Diameter of pipe between stacks	5mm
electrode area	$432cm^2$
electrode thickness	3mm
positive electrolyte	1.5M $FeCl_2$ 4M HCl
negative electrolyte	1.5M $CrCl_3$ 4M HCl

The flow rate of the positive electrolyte was 10.2litre/min, and the negative was 11.6litre/min as a whole. The pressures of the outlet of the pumps are

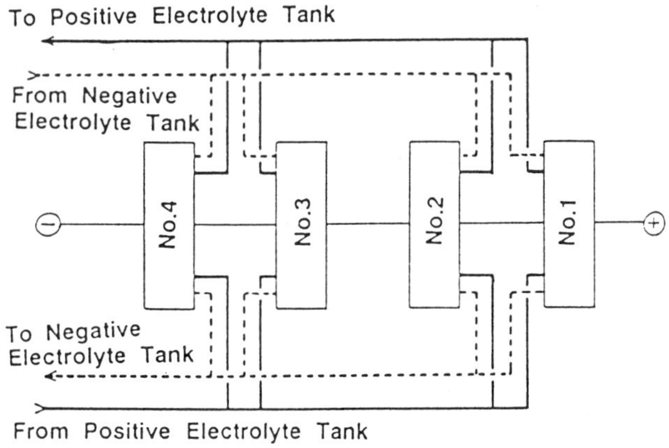

Fig. 3. Construction of IFRB

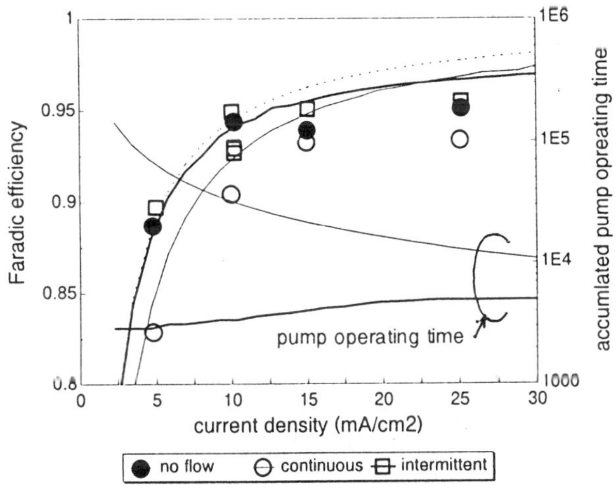

(a)Faradic efficiency

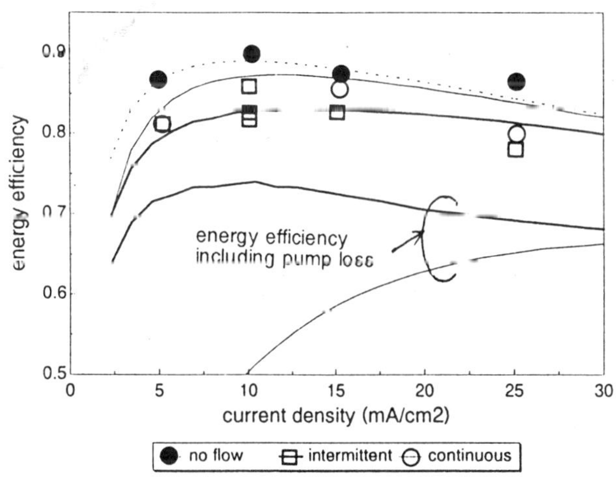

(b)energy efficiency

Fig. 4. Experimental and simulation results of IFRB

1.0kg/cm^2.

Experimental results of IFRB

The first mode is the conventional mode of the RFB. The IFRB of the performance and the irregularity among the cells were measured. The irregularity among the cells was not observed by measuring the voltage of each cell. The experiment of the second mode was carried out to measure of the performance of the stacks.

The experimental results are shown in Fig.4. Faradic efficiency of the conventional mode is the worst because of leak current through piping between the stacks. The Faradic efficiency of the second mode is higher than others because the operating SOC region was more narrower than other case. The reason of the efficiency decrease in the low charge-discharge current region was the leak current through the electrolyte in the manifold of the stack under the no flow mode. The energy efficiency of the intermittent mode is less than other cases without considering the pump power. That reason is as follows. The average terminal voltage of the IFRB is higher than other cases under the charging condition and lower than others under discharging condition. As a result, the voltage and the energy efficiencies decreased.

The cell resistance under the circulating electrolyte condition changes from 0.006Ω to 0.007Ω. The cell resistance values under the no flow condition are calculated from the difference of the OCV of mini cell and the terminal voltage per cell when electrolytes were immediately circulated. The variation of the cell resistance under the no flow condition is from 0.007Ω to 0.008Ω.

Experiment of inert gas bubble effect

In order to suppress the shunt current through the piping between stacks, the inert gas bubble is flushed through the piping with the electrolytes. At first, it was clear by using the equipment shown in Fig.5 that the inert gas bubble had the effect on the electric resistance of the electrolyte in the piping. Two outside electrodes were acted as current flow electrodes, and two inside electrodes were worked as the detecting electrode of the voltage. The diameter of the teflon pipe was 5mm. The resistance of the electrolyte without the inert gas bubble was about 500Ω. The resistance of the electrolyte with the inert gas bubble as shown in Fig.5 was over 1MΩ under the no flow condition. The value of the resistance of the flowing condition showed the decrease of about 6.5kΩ.

This technology was introduced to the IFRB, and the experiments were carried out under the intermittent and the continuous flow conditions. The inert gas bubble effect increases efficiencies slightly as shown in Table 2. These results were appropriate in the point of this size, and they demonstrated shown that the inert gas bubble mixed flow technology is available for the RFB.

SIMULATION

Simulation model

The simulation was carried out by using the stack model[4]. The leak current of the piping among the stacks is calculated on the geometrical dimension of the piping. In the no flow mode simulation the terminal voltage is used as the indicator of the current direction change in the same way of the experiments. The value of the SOC is 30% at the start of the simulation. The effective SOC of the IFRB system is used as indicator for the changing the direction of the charging current at the continuous and intermittent flow modes. The effective SOC is the weighted average of the SOC in the tanks and the stacks. The efficiencies are evaluated on the region of the effective SOC between 30% and 70%. The value of the cell resistance is assumed constant 0.0075Ω under the no flow condition and 0.0065Ω. The resistivity of the electrolytes is assumed constant 2Ω∘cm. The volumes of the electrolyte in the tanks are assumed 30litre for each other.

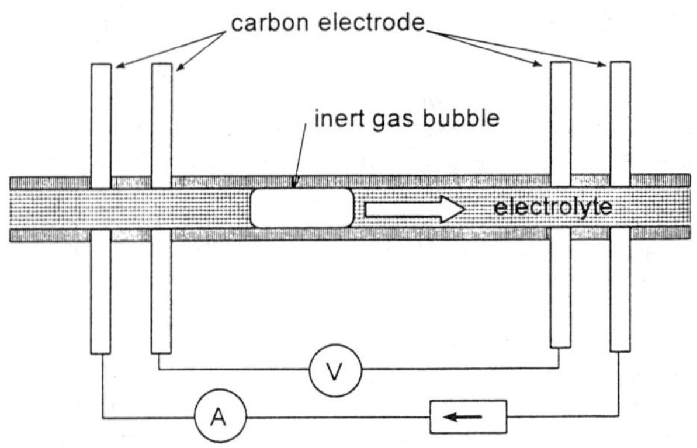

Fig. 5. Measuring equipment of inert gas bubble effect

Comparison with Experimental Results

The curves in the Fig.4 show the simulation results. These results are coincident with the experimental results. The simulation results of Faradic efficiency of the intermittent flow mode is almost the same value of the non-flow mode at low current condition, and is gotten nearer to same value of the continuous flow mode as increasing the charging current (shown in Fig.4 (a)).

The evaluation of the energy efficiency without the pump power shows that the intermittent flow is the worst as shown in Fig. 4 (b). Fig. 4 (a) shows that the long time pump operation is needed at the continuous mode under the low current condition. Therefore, the evaluation of the energy efficiency including pump power is introduced. The realized hydraulic power from experiments is estimated about 20W for each electrolytes from the multiplication of the flow rate and the pump head pressure. It is assumed that the pump efficiency is 50%. The total pump power the both positive and negative electrolytes are estimated 80W. The efficiency Including this effect shows that the continu-

Table 2. Experimental results of gas bubble effect

mode	current(A)	efficiency(%)			
		Faradic		energy*	
with gas bubble	20.03	97.6	(96.8)	83.7	(86.3)
	15.02	89.9	(95.9)	80.7	(87.9)
without gas bubble	20.01	94.4	(95.9)	81.7	(85.1)
	15.01	86.5	(94.7)	77.6	(87.1)

():simulation results, *:without pump power

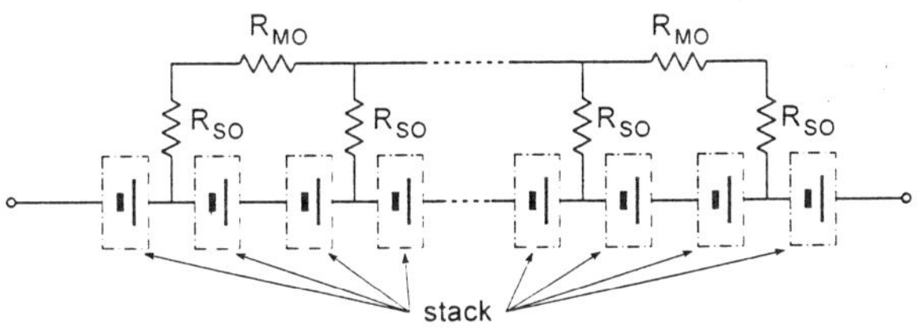

Fig. 6. Equivalent circuit of many stacks construction
R_{SO}:resistance of piping supplied electrolyte
R_{MO}:resistance of external manifold

ous mode is the worst.

Simulation results of the many stacks

By using the simulation program of the preceeding paragraph many stacks simulations introducing these technologies were carried out. The resistance of the piping becomes infinity when the pump is stopped. The inert gas bubble increases the resistance when the pump is operating. It is expected that these technologies increase the efficiencies.

The many stacks are electrically connected in

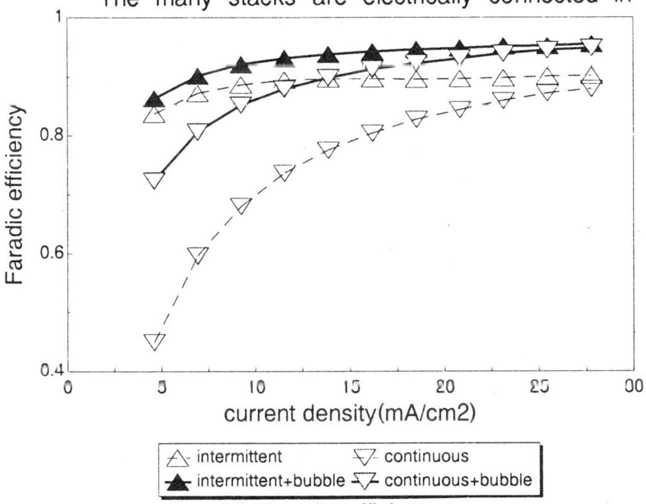

(a) Faradic efficiency

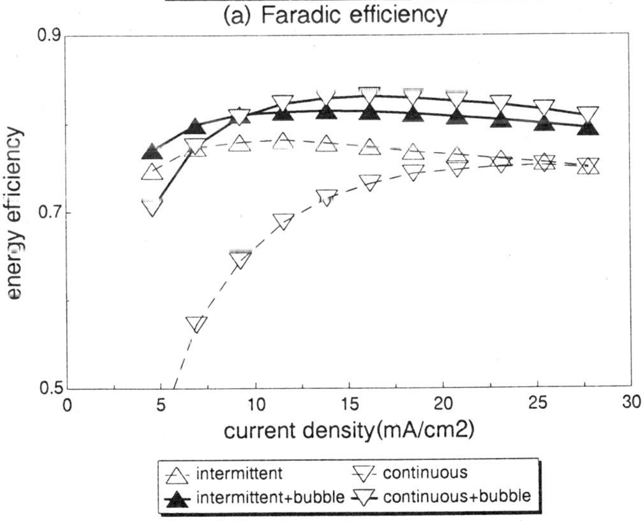

(b) energy efficiency including pump loss
Fig. 7. Simulation results of 20 stacks series connection

series. The electrolyte must be supplied to each stacks, and the current loss through the piping must be made minimum. As the result, the lines supplying electrolytes are prepared so that the electrolytes are supplied to the both side of stacks as shown in Fig. 3. The equivalent circuit of the construction of many stacks is shown in Fig. 6. The number of stacks is 20. Each stack is piled with 12 cells of the same dimension as the previous experiments. The piping is the same dimension as the experiment. The electrolyte tank capacity is 600*l* each other. Fig.7 (a) and (b) show the simulation results. The Faradic and energy efficiency increase compared with the continuous flow mode. The energy efficiency is evaluated including pump power. Especially, the intermittent flow operation steeply increases the efficiency compared with the continuous flowing operation. The characteristics showing the increase of efficiencies under the low current region are suitable for the storage system of the PV system because the effective efficiency increases.

CONCLUSION

The following became obvious from these simulation results and the experimental results.
(1) The current efficiency increases in the non-flow mode.
(2) Otherwise, the energy efficiency decreases under the large current density in non-flow mode.
(3) In the low current density the efficiency of the IFRB at the intermittent mode increases in comparison with the continuous flow mode.
(4) It is clarified from the simulation results that the intermittent flow operation is specially effective at the composition of the many stacks.

In near future, the experiment of the IFRB connected with the PV array will be carried out. In addition, the annual simulation of the IFRB system connected with PV array will be made.

REFERENCE

[1] K. Kurokawa, et. al., "A photovoltaic power generating system provided with redox flow battery", Advances in Solar Energy Technology, 1987, pp. 403-407.
[2] I. Tsuda, et. al., "Development of Prototype Redox Flow Battery for Photovoltaic Power System", 4th International PVSEC Proceeding, 1989, pp. 991-996.
[3] I. Tsuda, et. al., "Development of Redox Flow Battery for Photovoltaic Power Generating Systems", Technical Digest 5th International PVSEC, 1990, pp.675-678.
[4] I. Tsuda, et al., "Development of Redox Flow Battery for PV Power Generating System", New Energy Systems and Conversions, 1993, pp387-392.

DEVELOPMENT OF VANADIUM REDOX FLOW BATTERY FOR PHOTOVOLTAIC GENERATION SYSTEM

Akira Shibata, Mitsubishi Chemical Corporation
Kanji Sato & Masato Nakajima, Kashima-kita Electric Power Corporation

ABSTRACT

Photovoltaic power generation system (PV) requires a battery for night and rainy day. A redox flow battery has advantage over a lead acid one on this application for the capability of deep discharge and needlessness of equalized charge.

The authors have developed the high performance vanadium redox flow battery for this purpose and inexpensive production technology of electrolyte which occupies the majority in the battery cost by chemical reduction from boiler plant by-product.

The 2 kW (10 kWh) battery, the minimum unit for practical size battery (50 kW × 50h), achieved 1.2 kW/cm^2-electrode area at the 100 mA/cm^2 current density.

INTRODUCTION

The PV is recently appearing before footlights world over by the reason such as the environment preservation of the Earth, countermeasure of fossil fuel exhaustion, etc. But it requires the storage battery against night and rainy day. As one of candidate, the lead acid battery seems applicable to this purpose. However, it has the significant disadvantages such as the requisite of equalized charge to each cell, difficulty of deep discharge and short life.

Fe-Cr redox flow battery using Fe^{2+}/Fe^{3+} and Cr^{2+}/Cr^{3+}, reported by L. Thaller[1], Lewis Research Center of NASA in 1975 has been studied for PV application, because it has no bias on the charging condition in each cell and it is possible to choose the operating voltage in conformity with the most effective point of solar cell by change of tap. But the energy density of this battery is low and the decline of cell capacity occurs by mingling with cathode and anode electrolyte. Moreover, as it has the essential shortcoming such as cell performance deterioration by the unbalance of discharge state in cathode and anode due to self-discharge by side reaction of anode electrolyte, the study for PV application has been interrupted.

On the other hand, the vanadium redox flow battery which has been begun to study by Skyllas-Kazacos et al.[2] [3] [4] is composed of electrolyte tanks, pumps for electrolyte circulation and stack which is the assembly of cells. The configuration is shown by Fig. 1.

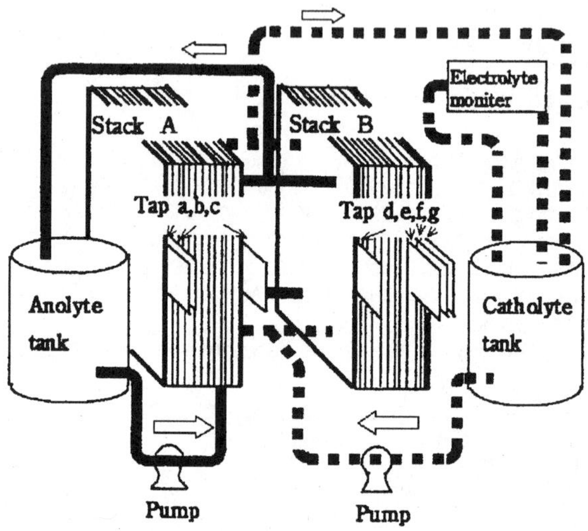

Fig. 1 Schematic Diagram for Vanadium Redox Flow Battery

The reaction of this battery is given by the following equation.

$$\text{Cathode:}\quad VO_2^{+} + e^{-} \underset{\text{Charge}}{\overset{\text{Discharge}}{\rightleftarrows}} VO^{2+}$$

$$\text{Anode:}\quad V^{2+} \underset{\text{Charge}}{\overset{\text{Discharge}}{\rightleftarrows}} V^{3+} + e^{-}$$

Though this vanadium battery has higher energy density than Fe-Cr battery, because the open voltage is high (1.4V) and there is not cell performance deterioration, as the same vanadium compound is used in both of cathod and anode electrolyte. But a weak point is that vanadium electrolyte is expensive.

The authors succeeded to develop the production technology of inexpensive electrolyte by chemical reduction from vanadium compound on the market and also recovered from the soot of boiler plant. This vanadium recovery process[5] [6] [7] [8] is currently commercialized by Kashima-kita Electric Power Corp., subsidiary of Mitsubishi Chemical Corp.

From these circumstances, we began the consignment development work of commercial technology of vanadium battery for photovoltaic system by the request of NEDO (New Energy and Industrial Technology Development Organization), a corporation having a special status of MITI (Ministry of International Trade and Industry) in 1993. The target date of technology establishment is set in 1997.

CH3365-4/94/0000-0950 $4.00 © 1994 IEEE First WCPEC; Dec. 5-9, 1994; Hawaii

This report describes the recent evaluation result on operating 2 kW × 5h (10 kWh) prototype battery and the study result on commercializing electrolyte production techology.

EXPERIMENTAL RESULT

Vanadium battery system

As it is considered that 50 kW × 50h (2,500 kWh) size battery, taking consideration into lack of sunlight for a week due to night and rain, is commercially suitable for the application such as village, hotel, etc. in resort and detached area, 2 kW battery was determined as the prototype and five hour rate was adopted for increasing the cycle number of charge and discharge, because the electricity storage capacity is adjustable by electrolyte volume.

High current density requires such structure as make much quantities of electrolyte circulate, necessary for chemical reaction in the cell. To determine the required quantity of electrolyte, necessary for satisfying the demanded current density, the "flow factor" was examined by using small cell with 20 cm^2 electrode area. The flow factor is a ratio of electric current and electrolyte quantity and given by the following definition.

Flow Factor = X/Y (per cell)
X: Supply quantity of electrolyte, converted to quantity of electricity (Coulomb/second)
Y: Current density (Ampere)

It is necessary to maintain about flow factor 20 from the result shown by Fig. 2.

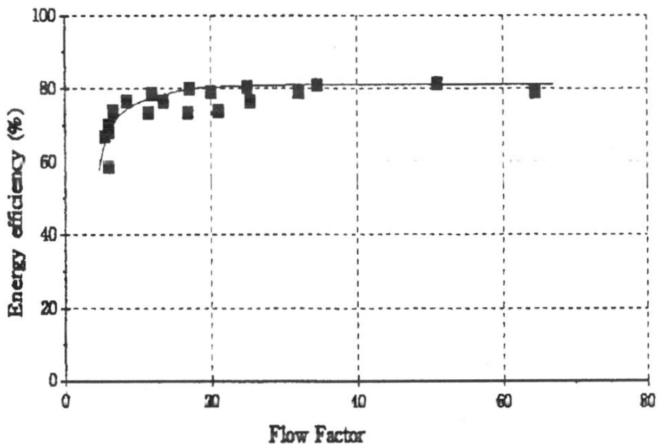

Fig. 2 FLow Factor vs. Cell Efficienty

The special cell was designed and made to circulate the required quantity of electrolyte and decrease shunt current. The piping system and flow meter was designed to minimize the pressure drop. The anionic polysulfone ion exchange membrane, made by Asahi Glass Co. was used and carbon felt XF308, made by Toyobo Co. was used for electrode. The stack consists of 30 cells with 600 cm^2 electrode area and can generate 2.2 to 2.6 kW, because each cell generates about 73W at 100 mA/cm^2 current density and about 88W at 120 mA/cm^2. 250 liter electrolyte, composed of 2 mole vanadium and 2 mole sulfuric acid is filled in the both tanks of cathode and anode to achieve exceeding 10 kWh discharge capability. The picture of 2 kW is shown by Fig. 3.

Fig. 3 Photograph of 2 kW Vanadium Battery

Operation result of battery

The performance evaluation test was carried out by the battery, complied 30 cells with 600 cm^2 electrode area. DCP060-100, made by Takasago Manufacturing Co. was used as power source unit for charge and discharge. The charge and discharge test was performed at the constant current. It stably showed 2.3 kW output, 80% pawer efficiency and 10 kWh electric enery at 100 mA/cm^2 electric density as shown by Fig. 4.

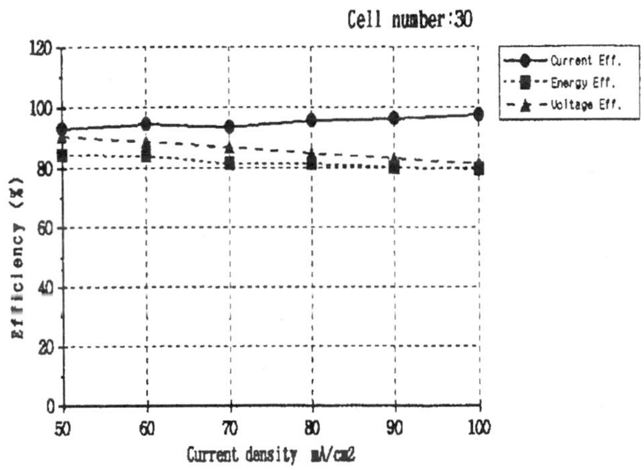

Fig. 4 Performance of 2 kW Vanadium Battery

This is four times as much as the performance of Fe-Cr battery in terms of both output and energy density and this battery is under the test of long term continuous operation.

Manufacture of vanadium electrolyte

Manufacture of V^{4+} electrolyte by SO_2

Kashima-kita Electric Power Corp. is recovering 200 T/Y ammonium metavanadate from 5,000 T/Y boiler soot by the process, shown by Fig. 5 and the electrolyte which generates 25,000 kWh could be obtained from this recovered vanadate.

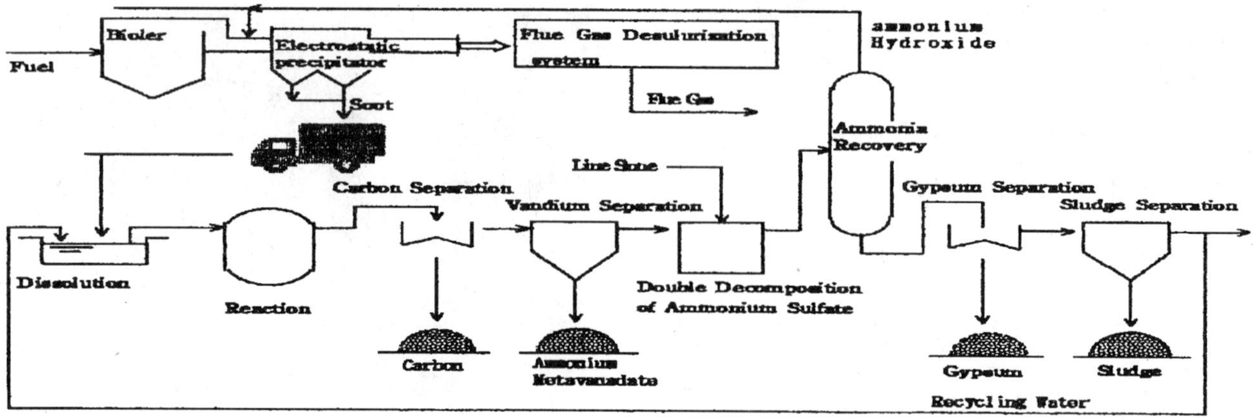

Fig. 5 Schematic Diagram for KMG Process

The authors tried to make vanadium electrolyte at low cost by using this recovered vanadate and refined sulfurous acid gas, obtained from the flue gas desulfurization unit. The table 1 indicates the impurities of recovered vanadate. As it contains much silicate and ammonia compared to high quality vanadium pentoxide (V_2O_5) which is used for making electrolyte at present, it is inevitable to purify, though such impurities are less than highvelt V_2O_5.

Table 1 Impurity of Vanadium Compound

[wt%]

Impurities	Na	K	Si	Fe	Ca	Mg	Ni	NH_4
Highvelt V_2O_5	.119	.026	.010	.050	.015	.003	.003	—
Recovered NH_4VO_3	.003	.002	.142	.007	.004	.008	.008	15.4
High quality V_2O_5	.012	.002	.008	.007	.014	.003	.003	—

The purification method is as follows (see Fig. 6). At first, the recovered vanadate is dissolved by hot water and the insoluble impurities are eliminated by filtration. Then pH of the solution is lowered up till 1.8 by addition of sulfuric acid to form polyvanadate and it is recovered by filtration. Almost all silicate and ammonia are left in the solution and eliminated under these processing. Though silicate is removed over 95% by this treatment, about 6% ammonia remains. Therefore, the purified V_2O_5 is finally obtained by calcination at 550°C to remove the remained ammonia.

The tetravalent vanadium (V^{4+}) electrolyte was made by adding the above purified V_2O_5 to 3 molar sulfuric acid, heated at 60°C and blowing sulfurous acid (SO_2) gas. The surplus SO_2 gas was eliminated by nitrogen gas bubbling. Though the table 2 shows a comparison of impurity, contained in the above V^{4+} electrolyte and the other electrolyte which is made from High quality V_2O_5 and Highvelt V_2O_5 by the same procedure, it is clear that our V^{4+} electrolyte purity is superior.

Table 2 Analytical Results of Various Electrolyte

Electrolyte materials	Impurities in electrolyte (mg/l)							
	Na	K	Si	Fe	Ca	Mg	Ni	NH_4
High quality V_2O_5	23	14	1	17	7	2	8	—
Highvelt V_2O_5	227	172	12	70	28	18	9	—
Recovered NH_4VO_3	6	13	9.6	10	8	5	24	—

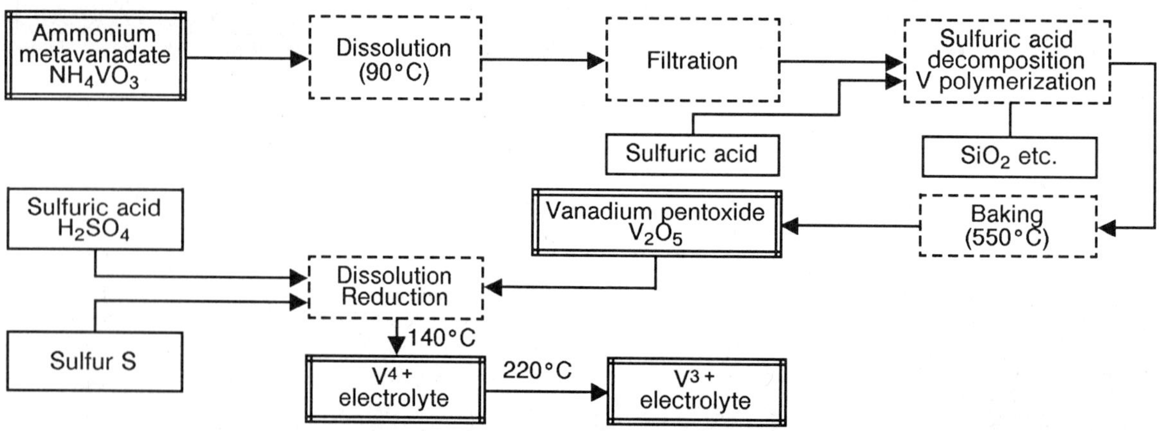

Fig. 6 Electrolyte Manufacturing Method

Manufacture of V^{3+} and V^{4+} electrolyte by sulfur reduction

As cathode electrolyte is V^{4+} and anode electrolyte is trivalent vanadium (V^{3+}) ion at discharge state, it is necessary to make V^{3+} electrolyte besides V^{4+} electrolyte. Though the elecrolytic reduction method is conventional, the authors tried to develop the sulfur reduction technology for reducing manufacturing cost. The characteristic of this method is to successively obtain V^{4+} electrolyte at 140°C and V^{3+} electrolyte at 220°C. It is obvious that the high purity electrolyte is obtained by this method from table 3 which shows the analytical data of electrolyte.

Table 3 Analytical Results of Electrolyte made from Recovery Vanadium

Unit: Wt ppm

Impurities	Umpurified V^{4+} electrolyte	Purified V^{4+} electrolyte	Sulfur reduced V^{4+} electrolyte	Sulfur reduced V^{3+} electrolyte
Na	6	23	2	2
K	13	14	0.2	0.2
Fe	10	17	0.2	0.2
Ca	8	7	8	6
Mg	5	2	19	10
Si	170	1	25	25
Ni	24	8	9	8
SO_4	—	—	—	—
Al	—	—	—	—
P	—	—	—	—

Though the manufacturing cost of this process is under studying in detail by installing the pilot plant, having 1 m^3/D capacity, it is expected that such cost could be one third (1/3) to one fifth (1/5) of the conventional process.

CONCLUSION

2 kW × 5h (10 kWh) redox flow battery was made to realize high performance and energy density of backup battery for photovoltaic power generation system. As the result of test operation, it showed four time as much as the performance of Fe-Cr redox flow battery in terms of output and energy density. It means to make possible to slash the battery manufacturing and operating cost.

Furthermore, it was attempted utilizing the by-product vanadate and sulfuric acid of power plant as raw material to reduce the electrolyte cost which occupies major portion of battery cost. The vanadium is recovered as ammonium metavanadate from boiler soot and sulfuric acid is made from sulfurous acid gas, recovered by flue gas desulfurization unit.

Moreover, it has been developed such process as easily eliminate impurities by the wet method and manufacture V^{3+} electrolyte and V^{4+} electrolyte at low cost from sulfurous acid gas or sulfur. It is expected to reduce the manufacturing cost of electrolyte one third (1/3) to one fifth (1/5) of conventional method.

[1] L.H Thaller, Electrically Rechargeable Redox Flow, *United States Patent* **No. 3,996,064**

[2] E. Sum and M. Skyllas-Kazacos, *J. of Power Sources*, **15, 179** (1985)

[3] E. Sum, M. Rychcik and M. Skyllas-Kazacos, *J. of Power Sources*, **1945** (1987)

[4] M. Rychcik and M. Skyllas-Kazacos, "Evaluation of Electrode Materials for Vanadium Redox Cell", *J. of Power Source*, **1945** (1987)

[5] N. Kaneko, T. Akahoshi, A. Sakuma, H. Ohyama and M. Iijima, "Recovery of Valuables from Oil Fly Ash" *Kagakukougaku* **Vol. 56**, No. 6 (1992)

[6] T. Akahoshi, M. Tsuruta, A. Sakuma, M. Iijima, K. Sato and N. Kaneko, *JPC*-**1771850**

[7] N. Kaneko, T. Hasegawa, A. Chida, S. Ogawa and S. Saito, *JPC*-**1802326**

[8] T. Akahoshi, N. Kaneko, A. Sakuma and T. Sugiyama, *JPC*-**1802014**

MATERIAL AND ENVIRONMENTAL IMPLICATIONS OF ULTRA HIGH EFFICIENCY SOLAR CELLS

A.E. Baumann, K.M. Hynes and R. Hill
Newcastle Photovoltaics Applications Centre, University of Northumbria, Newcastle upon Tyne, UK

ABSTRACT

If electricity generation by photovoltaics (PV) is to expand to a scale comparable with conventional power generation, then PV production on the scale of Gigawatts will be required. For many power production applications high efficiency is important and it is likely that multi-junction devices will be used in these applications. A target of the Japanese PV Programme is to construct flat plate modules of multi-junction cells composed of III-V and Si materials with efficiencies of 40% by the year 2010 [1]. This paper discusses the implications of the production of 40% efficient cells at the rate of 1 million/m^2 per year, i.e. a production rate of 400 MW/year. It will analyse the range of environmental and health impacts which might arise from a production facility of this scale and discuss the choice of flat plate vs concentrator systems.

INTRODUCTION

The development of multi-junction flat-plate and/or concentrator cells based on III-V compounds would allow the establishment of a stable solar energy supply structure in urban and rural areas and reduce power generating costs to levels that are competitive with costs of electricity provided from the grid. Power generation from PV systems would not only be a sustainable way of generating electricity but it would also contribute to reducing greenhouse gas emissions and thus helping to combat climate change.

However, before large-scale commercialisation of these multi-junction cells begins one has to make sure that neither the environment nor public and occupational health and safety are at risk from these new PV devices. Semiconductor materials used in newly developed PV technologies are very often untested in respect to their potential environmental and health hazards. It is therefore of great importance that the health and safety hazards of selected processes and materials are identified so that appropriate control and mitigation strategies can be incorporated into the entire production process and other up- and downstream processes.

This paper will consider concentrator 3-junction cells based on III-V compounds and Silicon (Si) [Aluminium Gallium Arsenide (AlGaAs)/Gallium Arsenide (GaAs)/Si, Gallium Phosphide (GaP)/Gallium Arsenide Phosphide (GaAsP)/Si and Indium Gallium Phosphide(InGaP)/InGaP/Si], produced by Metal-Organic Chemical Vapour Deposition (MOCVD) technology. It will also discuss the material requirements for the production of 40% efficient cells at the rate of 1 million m^2/year, i.e. a production rate of 400 MW/year.

This paper will identify the potential hazards to both occupational and public health that might arise from the development of III-V devices based on the MOCVD technique. As MOCVD involves the use of hydride gases and metal-organic materials which are highly toxic and pyrophoric, the environmental and health impacts associated with these source gases and materials will be analysed.

CELL CHARACTERISATION AND PROCESS DESCRIPTION

Due to their wide bandgap for photo absorption multi-junction PV cells are very promising for the production of ultra high efficiency cells. The candidate materials for ultra high efficiency cells considered in this paper as well as the theoretical efficiencies of these 3-junction cells are shown in Table 1. Although the utilisation of multi-junction cells could greatly improve the conversion efficiency due to the wide-band photo response, the design of 3-junction super high efficiency cells is rather more complex than single-junction cells.

The primary technique used in laboratories for the growth of binary (GaAs, GaP) and ternary (AlGaAs, InGaP) compounds on Si substrates is MOCVD. The main advantage of MOCVD is that the entire cascade structure can be grown in a continuous fashion and that there is uniformity over large areas, accurate and reproducible grading of material compositions, sharp interfaces and thin epitaxial layers [2]. The growth of III-V compounds on Si substrates has been attractive for the application of MOCVD

Cell Structure	Top Cell	Middle Cell	Bottom Cell	Terminal	Calc. Efficiency (1 Sun)
AlGaAs/GaAs/Si	Al(0.3)Ga(0.7)As	GaAs	Si	6	40.3%
GaP/GaAsP/Si	GaP	GaAsP	Si	6	41.0%
InGaP/InGaP/Si	In(0.4)Ga(0.6)P	In(0.7)Ga(0.3)P	Si	6	39.9%

Table 1 Theoretical efficiencies of potentially suitable 3-junction cells [1]

CH3365-4/94/0000-0954 $4.00 © 1994 IEEE

technology "because of the low cost, high thermal conductivity and high strength of Si, with the possibility of complex circuits in monolithic configurations" [3]. However, there are two major drawbacks in obtaining high-quality III-V materials on Si. First, the existence of a high interface defect density (10^8 m^2) due to a lattice mismatch and thermal expansion coefficient differences and second, crack generation also due to thermal expansion coefficient differences. In order to achieve high efficiency III-V on Si the interface defect density has to be reduced by at least a factor of ten. Various methods such as thermal cycle annealing and insertion of SLSs (Strained Layer Superlattices) have been investigated and shown to be successful in improving cell efficiency [1].

The MOCVD epitaxial growth technique used for the production of ultra-high efficiency cells composed of III-V compounds is based on the pyrolysis reaction between a metal-organic material of the III group (e.g. Trimethyl-aluminium) and a group V hydride gas (e.g. arsine). As the alkyls and metal hydrides utilised in the MOCVD process have the potential of being moderately to extremely hazardous to the environment and human health they will be investigated in more detail in section 4 of this paper.

RESOURCE REQUIREMENTS FOR LARGE-SCALE PRODUCTION

If ultra-high efficiency cells based on III-V compounds are to be produced and applied on a large scale (GW/year) problems may arise with the raw material supply. The availability of some of the raw material resources could become severely limited, leading to an increase in costs which again could affect the production economics. Raw material prices are an important determinant in the production cost of PV cells and an increase in raw material prices could seriously jeopardise large scale commercialisation. However, as feedstock requirements are dependent on several factors such as production level, raw material utilisation, layer thickness, overall yield of the production process and solar cell efficiency, changes in the PV cell production technologies will reflect a change in demand for III-V compounds. Table 2 indicates the amount of feedstock required for the various layers of an ultra high efficiency PV cell. By using Si as a substrate material instead of GaAs or Germanium (Ge) wafers, there are no resource constraints to be expected in the future. Furthermore, existing production capacity of some of the more scarce resources such as Ga is estimated to be three times higher than the actual production level [4].

As described in section 2, the materials used during MOCVD are hydride gases and metal-organic materials not elemental materials of the Group IIIA and Group VA of the Periodic Table. For the production of the binary compounds GaAs and GaP, Trimethylgallium (TMGa), Arsine (AsH_3) and Phosphine (PH_3) are used. The ternary III-V alloys AlGaAs, InGaP use Trimethylaluminium (TMAl), TMGa, AsH_3,Trimethylindium (TMIn) and PH_3 respectively.

Table 3 summarises the current estimated world production of the raw materials Si, Al, Ga, In, As and P, their

Si substrate	thickness	appr. 200 μm
AlGaAs/GaAs/Si	GaAs total thickness	appr. 5 μm
	AlAs total thickness	appr. 1 μm
GaP/GaAs/Si	GaP total thickness	appr. 5 μm
	GaAs total thickness	appr. 3 μm
InGaP/InGaP/Si	GaP total thickness	appr. 4 μm
	InP total thickness	appr. 4 μm

Table 2 Assumed thickness of substrate and layers of 3-junction cells

concentrations in earth rocks, their price and the quantities incorporated in 40% efficient flat plate and concentrator cells. It is clear that annual production of 0.4 GWp of flat plate module would incur resource pressures for Ga and In, although the world production of Ga could increase if needed. The use of concentrator cells would not only ease resource constraints but would also minimise the quantity of effluents released during the various production processes and present a cost benefit as the associated control costs for effluent treatment (e.g. wet scrubbers, gas detection systems, high efficiency particulate vacuum cleaners etc.) and disposal would be reduced significantly.

ENVIRONMENTAL AND HEALTH ASPECTS

The environmental and health hazards from the production of multi-junction thin film photovoltaic cells based on III-V compounds in research & development and commercial scale manufacturing facilities mainly arise from the use of hydride gases and alkyl metals in the MOCVD process. The MOCVD technique requires relatively large amounts of metal-organic compounds such as TMAl, TMGa and TMIn as well as metal hydrides such as AsH_3 and PH_3 which are pyrophoric and moderately to extremely hazardous materials.

Potential Environmental and Health Impacts

The major dangers to the environment and occupational and public health lie in the use and storage of arsine and phosphine. During MOCVD, these gases may be accidentally released from a leaking gas cylinder in a gas cabinet or with reaction by-products. As both gases are highly toxic, their accidental release during transport, processing, storage or effluent treatment could pose a serious health risk to workers at the plant or the population living in the vicinity of the plant.

Arsine, the hydride gas of arsenic, is a highly toxic gas with a TLV (Threshold Limit Value) of 0.05 ppm and an OES (Occupational Exposure Limit) of 0.2 mg/m^3. It is particularly poisonous by inhalation and symptoms develop within a few hours of exposure. Arsine poisoning can lead to headache, dizziness and nausea and it can affect the

Cell	Materials	Tonnes incorporated in flat plate 40% cells	Tonnes incorporated in concentrator 40% cells	Estimated World Production t/yr (1992)	Concentration in Earth Rocks µg/g	Raw Material Price
AlGaAs/GaAs/Si	Si	1086	2.17	5.0 x 10^8	277,200	1.00 $/kg
	Al	1	0.002	1.4 x 10^10	81,300	1.20 $/kg
	Ga	12.8	0.03	45	15	170.00 $/kg
	As	16.56	0.03	50,000	5	0.8 $/lb
GaP/GaAsP/Si	Si	1086	2.17			
	Ga	21.68	0.04			
	As	8.24	0.02			
	P	6,36	0.01	0.5 x 10^6	1,200	2000.00 $/t
InGaP/InGaP/Si	Si	1086	2.17			
	Ga	11.44	0.02			
	In	15.08	0.03	160	0.1	150.00 $/kg
	P	9.16	0.02			

Table 3 Quantities of feedstock material for 0.4 GW p.a. production

central nervous system. Acute exposure to arsine results in haemolysis, the destruction of the red blood cell's capacity for carrying oxygen. Severe haemolysis may lead to kidney failure and in some cases to death. Arsine is thought to be a human carcinogen and suspected mutagen [5]. Arsine is also used in MOCVD growth of GaAs, AlAs, AlGaAs, GaAsP and other related ternary and quaternary alloys.

Phosphine, the hydride gas of phosphorus, is a pyrophoric, highly toxic and poisonous gas with acute toxic effects on the gastrointestinal and central nervous systems. Exposure to phosphine can result in pulmonary edema, haemorrhage, cardiac failure and brain damage. In workers, pulmonary edema has caused both sudden and delayed death. Chronic exposures can cause bronchitis, neuralgic symptoms and liver dysfunction.

Because of the highly toxic properties of these hydride gases chemical manufactures are developing arsine substitutes such as dimethylarsenic and diethylarsine and alkyl phosphine derivatives such as tertiary-butyl phosphine or isobutylphosphine for the MOCVD process which are significantly less toxic than the source gases. This replacement would significantly reduce the hazards associated with the production of multi-junction thin film cells based on III-V compounds. The substitution of arsine and phosphine would also be beneficial in terms of costs by saving expenditure for exhaust gas treatment systems and disposal of toxic gases.

The MOCVD process also involves relatively large amounts of metal-organic compounds such as MTAl, MTGa and MTIn. Very little is known about the toxicity of these materials and their risks to the environment and occupational and public health. Tables 4 and 5 indicate and summarise the data on the potential hazards associated with these hydride gases and metal-organic compounds. Current knowledge about the toxicology of III-V compounds is limited to the availability of data on GaAs. Research has shown that GaAs is soluble in biological pH (7.4), releasing arsenic to the bloodstream [6]. Further toxicological tests seem to confirm the *in vivo* solubility of GaAs as intratracheal instillation of GaAs particulates in rats led to a significant increase in concentrations of arsenic in the blood [7]. Although GaAs did not cause pulmonary edema it had effects on the lung tissue, increasing the cell types indicative to pneumo-toxic response. The intratracheal instillation of smaller GaAs particulates caused pulmonary lesions and supported the in vivo solubility of GaAs [6]. Given the results of these tests, it has to be assumed that the exposure to GaAs particulates causes systemic arsenic intoxication and acute inflammatory reactions in the lung tissue. Many arsenic compounds are also known to be carcinogenic. To what extent the data on the toxicity of GaAs can be transferred to other arsenic containing III-V compounds is difficult to say. As many of the compounds have similar crystal structures and bond types there is the possibility that AlAs, AlGaAs and GaAsP release arsenic in the same way.

MOCVD is an increasingly important technology to produce multi-junction high efficiency PV cells. In order to reduce the potential hazards to occupational and public health it is crucial to develop appropriate control methods and safety systems, such as proper storage of flammable and toxic materials, minimising the stock of toxic and combustible materials at the plant and good housekeeping. Reactor design to improve operating safety, gas detection and gas supply shut-down systems as well as appropriate

Materials	Long-Term Exposure Limits	Potential Environmental and Health Hazards	Carcinogen Mutagen Teratogen
Al as TMAL	TLV: 10 TLV: 2 (AL)	Pyrophoric, reacts exposively with water Alkyl metals cause severe burns on skin	no evidence
AS as AsH3	OES:0.2 TLV:0.05ppm	Flammable, moderately explosive Emits highly toxic fumes of arsenic Poison by inhalation Attaches to haemoglobin, results in haemolysis, can lead to kidney failure/death	Human carcinogen Suspected mutagen
Ga as TMGa	No TLV est.	Pyrophoric, releasing Ga2O3 and CH4 Ga remains in tissue for long periods of time when injected intramuscular	no evidence
In as TMIn	OES:0.1 TLV:0.1 (In)	Flammable Poison by subcutaneous route, affects liver, heart, kidneys and blood Inhalation of In compounds highly toxic	Suspected teratogen
P as PH3	OES:0.1 TLV:0.3pp	Pyrophoric Highly toxic and poisonous,with acute effects on gastrointestinal and nervous systems. Severe pulmonary irritant (pulmonary edema)	no evidence
Si	TLV:10 (of total dust)	Elemental Si flammable Harmful when inhaled as dust, can cause silicosis	no evidence

Table 4 Hazards from materials

effluent treatment are further measures to minimise the risk of exposure to toxic materials and adverse health impacts in workers and the public.

Given the toxic and pyrophoric properties of the III-V materials used for the production of ultra high efficient multi-junction cells at a production facility of the scale of 0.4 GW p.a., it is obvious that concentrator systems would pose lower potential hazards in production than the same output of flat plate systems.

Compounds	Long-Term Exposure Limits	Potential Environmental and Health Hazards	Carcinogen Mutagen Teratogen
AlAs	prob. as GaAs	Emits toxic arsenic fumes	no evidence
GaAs	OES:0.01 (As) MEL:0.2 (As)	Releases deadly poisonous arsine when Inreaction with steam,acids and acid fumes Soluble at biological pH (7.4),releasing arsenic into bloodstream Highly toxic when inhaled	Confirmed carcinogen
GaP	n.a.	Highly flammable Highly toxic, affecting nervous system	no evidence
InAs	n.a.	Highly toxic by inhalation and parenteral administration	no evidence
InP	OES:0.1	May evolve toxic fumes in fire Harmful if inhaled as dust In- compounds are highly toxic by subcutaneous injection	no evidence

Table 5 Hazards from compounds

DISCUSSION

The production of III-V compound semiconductor cells on Si will contribute to a significant improvement in the efficiency of PV cells. Although the calculations for cell efficiencies of 40% by the Japanese have so far been only theoretical, there is the prospect to meet these predictions in the future. The target quoted for the Japanese PV programme of 40% flat plate module efficiency by 2010 would be a major advance in PV technology. Large scale production engineering could possibly lead to commercially viable modules with efficiencies in the 40% region. However, the large scale production of modules with III-V multi-junction cells would give rise to a range of environmental impacts and production constraints due to the toxic nature of the materials and their precursors. It would seem from this analysis that these constraints may be sufficiently serious that the large scale use of III-V multi-junction cells should be largely confined to concentrator applications. Concentrator cells utilise less material than flat plate cells, thus release fewer effluents into the environment and minimise the risk to workers and the general public of being exposed to hazardous materials. Although research into alternative materials may eliminate the use of toxic source gases in the future, there are nevertheless other safety issues to be followed to ensure that neither occupational nor public health are at risk.

Appropriate monitoring and control technologies have been developed and are used in the semiconductor industry. It will be essential for the safety of workers and the general public that existing safety procedures are complied with. As there is very little information on the toxicology of III-V compounds with the exception of GaAs further research is needed to investigate the potential health hazards of other III-V materials.

REFERENCES

1. M. Yamaguchi and T. Goto, ATAS Issue 8, United Nations, New York, Autumn 1992.
2. J.C. Lee and P.D. Moskowitz, Potential Health and Safety Hazards Associated with the Production of Gallium Arsenide Alloy Based Multi-Junction Photovoltaic Cells, Brookhaven National Laboratory, Upton, NY, September 1985.
3. K.J. Linden, MOCVD: Explaining the Range of Photonic Materials, Photonics Spectra, Issue February 1991
4. E.A. Alsema, R.F. Cuelenaer and W.C. Turkenburg, Cost Perspectives of GaAs Thin-Film Solar Cells, 10th EC Photovoltiacs Solar Energy Conference, Lisbon, 1991.
5. P.D. Moskowitz, An Overview of Environmental Health and Safety Issues in the Photovoltaic Industry, Brookhaven National Laboratory, Upton, NY, 1992.
6. R.M. Content, Control Methods for Metals and Metalloids in III-V Materials Vapour-Phase Epitaxy, in ACGIH, Hazard Assessment and Control Technology in Semiconductor Manufacturing, Chelsea, USA, 1989.
7. D.R. Webb, S.E. Wilson and D.E. Carter, Pulmonary Clearance and Toxicity of Respirable Gallium Arsenide Particulates Intratracheally Instilled in Rats, in Am. Ind. Hyg. Assoc. J. 48:660, 1987.

ACKNOWLEDGEMENT

This work is supported by the Commission of the European Union under the Contracts JOU2-CT92-0243 and JOU2-CT92-0106.

AN ASSESSMENT OF THE ENVIRONMENTAL IMPACTS OF THIN FILM CADMIUM TELLURIDE MODULES BASED ON LIFE CYCLE ANALYSIS

K.M. Hynes, A.E. Baumann and R. Hill
Newcastle Photovoltaics Applications Centre, University of Northumbria, Newcastle upon Tyne, UK

ABSTRACT

Life cycle analysis has identified the production and decommissioning/disposal of thin film CdTe modules as the stages which have potentially the most severe environmental impacts. This paper investigates these stages with respect to materials, energy input and possible environmental and health implications

INTRODUCTION

If photovoltaics is to penetrate the energy market, the technology has to be suitable for large scale production. Once the technical and economic feasibility have been proved, criteria such as environmental impacts have to be considered. Life Cycle Analysis can be used to assess these environmental impacts. A life cycle analysis on cadmium telluride thin film modules, previously undertaken by this group [1], has shown that the module production and disposal/decommissioning stages have potentially the most severe environmental impacts. It is the intention of this paper to investigate these two stages in detail with respect to materials, energy input, disposal of waste and any possible environmental and health implications. This type of analysis is essential if CdTe modules are to be produced on a large scale without counterbalancing environmental costs.

LIFE CYCLE ANALYSIS

Life cycle analysis applied in the energy sector allows the quantification and comparison of the multiple impacts which energy systems exert on their environment, through all stages of their existence ("cradle to grave"). The assessment of the entire life cycle allows the evaluation of not only the environmental and health impacts but also of energy input and raw materials. The life cycle of CdTe modules can be broken down into four main areas: mining/refining, production of the module, operation and decommissioning/disposal. With each of these stages there are associated environmental and health impacts, common to each area, which include atmospheric and water pollution, impacts on natural ecosystems, accidents, land use, visual intrusion, resource depletion, noise and public and occupational health impacts. Details of such environmental and health impacts at the four stages of the life cycle have been published in [1]. Once the basic life cycle has been determined, its modular nature allows for various steps within the process stages to be substituted with other processing techniques so that a comparison can be made of the various technologies which may be used for production of CdTe modules.

THE CdTe MODULE

For the purposes of this work, a 10MW per annum production plant has been assumed. The plant would thus produce 100,000m^2 of modules per year assuming 10% module efficiency. It is assumed that PV production plant of 10 - 50MW will be necessary if thin film modules are to form a substantial part of the power production market into the next century. Several production techniques are under investigation for use in the manufacture of thin film CdTe modules and this is typical for most thin film PV technologies. Both back and front wall cell configurations are being developed using a variety of contacting and window materials, and it is unlikely that only one technology will be used in production of the absorber and window layers. Table 1 shows examples of the range of techniques and materials which may be used to produce thin film CdTe modules. The techniques use varying amounts of materials within the structure of the module and different feedstock materials are also used.

This paper will consider two production technologies for CdTe thin film modules which use considerably different techniques for the deposition of the window and absorber layers. The first (type 1) is based on the non-vacuum techniques of electrodeposition and chemical bath deposition for absorber and window layers and the second (type 2) uses thermal evaporation for these layers and the CdTe would be evaporated from the compound. The other parts of the modules are assumed to have the same structure as, for this study, it is the absorber and window layer deposition technologies which are being compared. However, the contact materials assumed here are typical of those being developed by various groups. Both types of module have the superstrate (frontwall) configuration which seems to be the most common within the CdTe community. The substrate and superstrate are of 3mm float glass. A thin (120 nm) sputtered layer of tin oxide forms the transparent front contact. The back contact is assumed to be a very thin sputtered layer of copper followed by a thicker (1μm) layer of molybdenum. In type 1, the CdTe layer produced by

CH3365-4/94/0000-0958 $4.00 © 1994 IEEE

PROCESS STEP	MATERIALS	TECHNIQUES
FRONT CONTACT	Aluminium Indium tin oxide Tin oxide Zinc oxide	Sputtering MOCVD Spraying
WINDOW LAYER	Cadmium sulphide Zinc oxide	Chemical bath deposition Thermal evaporation ALE Screen printing Sputtering Electrodeposition Closed space sublimation Spraying
ABSORBER LAYER	Cadmium telluride	Electrodeposition ALE Thermal evaporation Closed space sublimation Screen printing MOCVD Spraying
BACK CONTACT	Copper Nickel Phosphorus Molybdenum Gold Indium tin oxide Zinc telluride Carbon doped graphite	Sputtering Thermal evaporation Electrodeposition

Table 1 Materials and technques used for CdTe module production

electrodeposition would be required to be about 1.5 µm thick and the CdS layer, 0.2 µm. The CdTe layer deposited by thermal evaporation in type 2 would need to be about 5 µm thick and the CdS layer about 1.7 µm thick.

ENERGY ANALYSIS

When considering new energy supply technologies, it is important to consider the energy requirement for their production, operation and decommissioning as they need to be net energy producers over their lifetimes. Moreover, it is assumed that, in the near term at least, the energy used in their production would be supplied by conventional, high environmental impact technologies.

Energy analysis is a technique that is widely used for determining energy flows in industrial processes. It allows the calculation of the required energy input for unit output of product. The work reported here considers the energy requirement for the production of the CdTe PV modules using the two processing technologies detailed earlier. This work conforms to IFIAS Level 3 which includes process energy, material input and an allocation for the energy content of the capital equipment required [2].

As with life cycle analysis, the modular nature of the production process allows each process stage to be analysed in turn. The energy requirement for each process stage consists of three main factors, materials energy, machine energy and process energy in kWh (thermal)/m². The materials energy is the amount of energy consumed in manufacturing the materials which appear in the finished product. The machine energy is the amount of energy consumed in the manufacture of the production equipment used in the step. The process energy is the amount of energy used to perform the process step and may also include materials and/or process gases which do not appear in the finished product. These three factors are summed for each process step, eg. deposition of the back contact, and then the total energy requirement is the total of all the process stages. From these energy requirements it is possible to assess and compare the various production techniques and evaluate energy payback times and energy ratios. The energy payback time is the time taken for the module to produce the energy used to manufacture it and the energy ratio is the ratio of the energy produced by the module over its entire lifetime to the energy required to produce the module. Energy ratios of greater than 10 are

Module Type	Process Step	Material Energy	Machine Energy	Process Energy	Total
1	Front Contact	0.01	0.95	2.74	3.7
	Window Layer	0.04	1.45	21.76	23.25
	Absorber Layer	1.56	11.24	56.71	69.51
	Back Contact	1.44	1.11	11.29	13.84
	Glass and Other Processing Stages	164.31	0.3	0.8	165.41
	Total	167.36	15.04	93.29	275.7
2	Front Contact	0.54	0.9	2.62	4.06
	Window Layer	0.87	7.19	63.92	71.97
	Absorber Layer	4.43	3.3	66.38	74.11
	Back Contact	1.98	1.11	11.29	14.37
	Glass and Other Processing Stages	164.31	0.3	0.8	165.41
	Total	172.12	12.8	144.99	329.92

Table 2 Energy requirements, (kWh(th)/m²), for the production of CdTe modules (types 1 and 2)

Module Type	Location	Payback time (in months)			Energy Ratio		
		10% module	15% module	20% module	10% module	15% module	20% module
1	South West USA	5	4	2	40	60	80
	Southern Europe	7	5	4	24	36	48
	Northern Europe	11	7	6	16	24	32
2	South West USA	6	4	2	33	50	66
	Southern Europe	10	6	5	20	30	40
	Northern Europe	13	10	7	13	20	26

Table 3 Energy payback times and energy ratios (15 year lifetime assumed for energy ratios)

usually necessary for a technology to be economically viable.

Table 2 gives the energy requirements for module types 1 and 2 and the resulting energy payback times and energy ratios are given in Table 3. The results indicate that production of CdTe thin film modules has a low energy requirement. For type 1 modules, the glass and other processing stages takes up 60% of the energy requirement and, for type 2, this percentage is 50%. As would be expected, the energy requirement for thermal evaporation of the window layer is much greater than that for the chemical deposition method, which uses low cost machinery and only very thin layers are required. With thermal evaporation, a thicker layer of CdS is needed and the equipment is much more expensive. However, thermal evaporation techniques usually give a higher deposition rate than the wet chemical methods.

When the energy requirement of the absorber layer is investigated it can be seen that the two techniques give very similar results. This is due to quite different factors. With electrodeposition, low cost equipment is utilised, but deposition rates are fairly slow (0.06µm/minute) with low deposition power requirements and thin films. However, with thermal evaporation the films need to be thicker and deposition rates are usually about 10 times faster, but the power requirements and high cost of the machinery tend to increase the energy requirement. Both technologies produce very short energy payback times and excellent energy ratios which indicate that, even for operation in Northern Europe with 10% efficiency, CdTe modules would seem to have very encouraging prospects for large scale production.

DECOMMISSIONING AND DISPOSAL

This section covers not only the decommissioning and disposal of the CdTe modules at the end of their working life, but also the disposal of waste generated during the production process. Such waste products will differ depending upon the production technique being used as the various production techniques use a range of starting materials.

Manufacturing Waste

This usually comes in one of three forms [3].

1. Solids including CdS, CdTe and various non-stoichiometric mixtures containing Cd and Te. These deposits may need to be scraped off the walls of deposition chambers or other pieces of apparatus, or may include sand if equipment has been cleaned by sand blasting. This could also include cleaning cloths coated with acids and dust collected on particle filters. This type of waste would be typical with thermal evaporation techniques.

2. Liquid wastes including water and alcohols containing Cd compounds. In the electrodeposition of CdTe from an aqueous bath, the solutions are either acidic or basic, containing dissolved Cd salts and TeO_2. These components are continuously added to the solution during electrodeposition as they are continuously being depleted. If the deposition bath becomes contaminated, disposal and recycling of Te and Cd would be needed.

3. Modules rejected during manufacturing, testing etc.

Most waste materials from the CdTe industry will require special handling, however some can be treated very easily. Glass or tin oxide coated glass may have been broken before processing began and it can be disposed of without any special procedures. There are four main types of waste streams generated during the production of CdTe modules. Some modules are rejected during manufacture and these can be recycled using the same types of pyrometallurgical processes currently used in non-ferrous smelting, or they could be used as feedstock in glass recycling. Waste streams such as chemical bath solutions and deposition system residues would contain materials in different forms to the initial raw materials. Recycling of such waste products would depend on their specific nature. Some waste will be in its original form and purity eg. sputtering targets and these can usually be returned to the company from which they were bought. The fourth category is mixed waste which would include liquid waste from electrodeposition baths etc. Some recycling companies accept this type of waste and the solid material eg. Cd could be obtained and sold to Cd producers as feedstock.

Disposal of Spent Modules

When PV modules have reached the end of their useful life, there are two options for safe disposal, recycling or disposal in landfill sites. With present low levels of production of CdTe modules, disposal in landfill sites is an acceptable alternative to recycling as there is no strategy in place for the total recycling of all the materials within the module. However, despite the fact that the cadmium compounds are in an insoluble form and have a low bioavailability, leaching of cadmium from landfill sites has to be closely monitored and possibly prohibited if this form of disposal is used. At present, some metals industries use

CdTe modules only as electronic scrap and they are recycled for the glass. The main conclusions of the 1992 Workshop on Recycling of Cadmium and Selenium from Photovoltaic Modules and Manufacturing Wastes [3] were that there was sufficient capacity in smelters to accommodate all wastes produced by CdTe manufacturers for the next ten to twenty years. The cost of recycling to the PV industry was not expected to be a burden to commercialisation. Various recycling options exist ranging from small on-site plants to large off-site smelters. Even those wastes defined as hazardous could be handled by existing recycling technologies as they are generated and routinely handled in larger quantities by the non-ferrous metals industry.

HEALTH IMPACTS

Table 4 shows the main feedstock materials used to produce CdTe thin film PV modules and the materials which appear in the module. It is important to consider all these materials as some are present during production and others are present in the module during its operation and eventual disposal and all materials may appear in wastes generated. This table indicates that properties of Cd, Te and their compounds can differ significantly. However, both Cd and CdTe are suspected carcinogens and CdS and $CdSO_4$ have been found to cause cancer in laboratory animals. There is evidence to show that CdS, $CdSO_4$ and CdTe may all cause adverse mutagenic and teratogenic effects, but no evidence of this has been found for Cd. All materials in the table are harmful by ingestion and many are irritating to the eyes and skin. It should be recognised that many of these materials may be harmful to human health, and that care must be taken throughout the life cycle of the CdTe modules as some or all of these materials are present at all stages of the life cycle. An impact pathway analysis and detailed study of the toxicity and health impacts of these materials has been undertaken by this group and are published elsewhere [4].

Material	Potential Environmental and Health Hazards	Carcinogenicity	Mutagenicity / Teratogenicity
Cadmium	Spontaneously flammable in air. Prolonged inhalation of fumes or dust can produce systemic effects. May be harmful if ingested.	Suspected carcinogen	No evidence
Cadmium Sulphide	May evolved toxic fumes in fire. Harmful by ingestion, inhalation and skin contact. Irritating to eyes. Can cause increased salivation, choking, vomiting, stomach pains and diarrhoea. Prolonged exposure to dust can cause lung and kidney damage	Found to cause cancer in laboratory animals	May cause adverse mutagenic and teratogenic effects
Cadmium Sulphate	May evolve toxic fumes in fire. Harmful by ingestion, inhalation and skin contact. Irritating to eyes. Can cause increased salivation, choking, vomiting, stomach pains and diarrhoea. Prolonged exposure to dust can cause lung and kidney damage.	Found to cause cancer in laboratory animals	May cause adverse mutagenic and teratogenic effects
Cadmium Telluride	May evolve toxic fumes in fire. Harmful by ingestion, inhalation and skin contact. Irritating to eyes. Can cause increased salivation, choking, vomiting, stomach pains and diarrhoea. Prolonged exposure to dust can cause lung and kidney damage.	Suspected carcinogen	May cause adverse mutagenic and teratogenic effects
Tellurium (powder)	Combustible. May evolve toxic fumes in fire. Harmful by ingestion and inhalation. Irritant to skin and eyes.	No evidence	Evidence of reproductive effects
Tellurium Dioxide	May evolve toxic fumes in fire. Harmful by ingestion and inhalation. Irritant to skin and eyes.	No evidence	Evidence of teratogenic effects

Table 4 Health impacts of the main materials associated with thin film CdTe PV module production

DISCUSSION

CdTe thin film modules can be produced using a variety of techniques and materials. Most of the production techniques are in commercial use and it should be technically possible to recycle most, if not all, waste products using similar technologies to those of the metals industries and recycling companies. At low production levels, disposal of spent modules in landfill sites is an alternative, but as production volume increases this becomes less favourable.

Energy analysis has shown that the energy requirement for the production of these modules is low, leading to energy payback times from a few months to just over a year. Energy ratios also prove to be very favourable.

Many materials present during the life cycle of CdTe modules can be hazardous to human health and the environment. Sensible strategies should be implemented to allow for the presence of such materials.

REFERENCES

1. K.M. Hynes, A.E. Baumann and R. Hill, "Life Cycle Analysis of PV Modules Based on Cadmium Telluride", *12th EC PVSEC*, 1994, pp.309-312
2. I. Boustead and G.F. Hancock, "*Handbook of Industrial Energy Analysis*", Ellis Horwood, Chichester, UK, 1979, p.184
3. *Recycling of Cadmium and Selenium from Photovoltaic Modules and Manufacturing Wastes: A Workshop Report*, Eds. P.D. Moskowitz and K. Zweibel, 1992, Brookhaven National Laboratory, New York.
4. A.E. Baumann, K.M. Hynes and R. Hill, "An Investigation of Cadmium Telluride Thin Film PV Modules by Impact Pathway Analysis", to be published in *Renewable Energy*.

ACKNOWLEDGEMENT

This work is supported by the Commission o the European Union under Contracts JOU2-CT92-0243 and JOU2-CT92-0106.

STRATEGIES FOR RECYCLING CdTe PHOTOVOLTAIC MODULES

Chris Eberspacher and Charles F. Gay
UNISUN, Newbury Park, CA 91320 USA

Paul D. Moskowitz
Brookhaven National Laboratory, Upton, NY 11973 USA

ABSTRACT

Recycling end-of-life cadmium telluride (CdTe) photovoltaic (PV) modules may enhance the competitive advantage of CdTe PV in the marketplace, but the experiences of industries with comparable Environmental, Health and Safety (EH&S) challenges suggest that collection and recycling costs can impose significant economic burdens. Customer cooperation and pending changes to U.S. Federal law may improve recycling economics.

INTRODUCTION

PV technology is undergoing a transition, evolving from the traditional crystalline silicon to a new generation of low-cost thin-films. One of the most promising thin film PV technologies is CdTe. Large-area CdTe modules have been fabricated with efficiencies of ca. 10%, comparable to crystalline silicon modules commercially available today [1]. In full-scale production CdTe PV is projected to be very competitive in both performance and price. Groups in the United States, Japan and Europe are actively working to commercialize CdTe PV technology.

This paper focuses on issues related to the recycling of end-of-life CdTe PV modules. Recycling is one element of a broader study of the Environmental, Health and Safety (E,H&S) issues related to CdTe materials and PV products. The broader study examined issues of public perception and customer preference in the context of analyzing the economics and logistics related to the field use and end-of-life recycling of CdTe PV products. Existing E,H&S data was reviewed, U.S. Federal laws were analyzed, prospective vendors and customers for CdTe PV were surveyed, and public acceptance and recycling strategies of industries facing comparable E,H&S issues were evaluated.

Comparatively little E,H&S data exist for CdTe materials in general and CdTe PV modules in particular. Most existing statutes regulate CdTe based on the better-known hazards of elemental cadmium (Cd) and tellurium (Te). CdTe PV products may face challenges in the marketplace related to customers' apprehensions about CdTe based in a large part on the lack of differentiation between CdTe and Cd + Te. Additional E,H&S data are needed on CdTe-based materials, and programs are essential to better educate customers and the public on the nature of these materials.

Everyday comparisons are useful in orienting the customer to the relative risks and benefits of CdTe PV in comparison with other energy sources and other well-accepted products. For example, a prototypical 2 kWp CdTe PV residential array might contain ca. 150 g Cd, and an average U.S. household might at any given time have eight each size C and size AA rechargeable NiCd batteries, containing in total 60-90 g Cd. Thus, an average household may already have consumer goods containing amounts of Cd comparable to the Cd in a residential CdTe PV array [2]. The relative risks of CdTe modules can be reduced by an effective collection and recycling program aimed at diverting end-of-life CdTe PV modules from uncontrolled landfills, and such a recycling program would likely be more manageable for the ca. $10,000 - 20,000 PV array than for the $30 - 60 battery set.

U.S. LAW

The key law affecting CdTe PV product use and end-of-life handling in the U.S. is the Resource Conservation and Recovery Act (RCRA). RCRA regulates the disposal of all solid wastes and sets strict guidelines for record keeping, labeling, packaging, transportation, siting, inspections, training, and emergency planning for the generation, transport, storage and disposal of hazardous wastes.

A key differentiation in the complexity of handling a waste material under RCRA is the determination of whether the waste is "hazardous" and must be handled according to strict guidelines, or whether a waste is non-hazardous and can be disposed of simply as common trash. Hazardous wastes are determined either by definition (i.e. "listed" in the RCRA text) or by testing. CdTe is not "listed" as hazardous, so CdTe-containing products are tested using a series of tests, including the Toxicity Characteristic Leaching Procedure (TCLP) test designed to predict risks of groundwater contamination by leaching . In preliminary tests on a very limited sample set, experimental CdTe modules failed the TCLP tests [6]. More recently, prototype CdTe modules passed the TCLP test [4,5].

First W PEC: Dec. 5-9, 1994; Hawaii

If modules "pass" the TCLP, then by U.S. Environmental Protection Agency (EPA) toxicity standards end-of-life modules are non-hazardous trash and require no special handling. This greatly simplifies disposal. Hence passing TCLP is a key to profitability for any commercial product and must be a primary design goal for PV modules. However, passing TCLP does not in itself end consideration of collection and/or recycling programs. State and local laws may impose additional requirements, and long-term business planning may dictate pro-active programs beyond those required by present law. For example, rechargeable NiCd batteries used in U.S. households currently are exempted from being handled as hazardous wastes because RCRA exempts all household wastes. In spite of this exemption and at considerable expense, the NiCd battery industry has actively encouraged laws facilitating used NiCd battery collection and recycling in order to reduce the migration of Cd into municipal landfills via used NiCd's and in turn the likelihood that regulations will be promulgated restricting NiCd use.

Conversely, if modules "fail" TCLP then they are defined under RCRA as a hazardous waste at end-of-life. This is not necessarily a severe problem for the module manufacturer in that RCRA disposal regulations apply to the generator of the waste, which in most cases will be the module user, not the module manufacturer. However, if CdTe PV module users are faced with high module disposal costs, then CdTe module sales may suffer; so CdTe PV module manufacturers may choose to set up and co-fund the proper disposal of end-of-life modules that fail TCLP. Classification as a hazardous waste significantly increases disposal costs, whether one chooses disposal in a controlled landfill or reclaim through various means.

Thus, either for public relations, regulatory requirement, or commercial necessity, CdTe module manufacturers and vendors might consider a collection and recycling program for end-of-life CdTe modules.

STRATEGIES OF COMPARABLE INDUSTRIES

Numerous existing industries face E,H&S issues comparable to those facing CdTe PV, and the public acceptance and recycling strategies of those industries are useful guideposts for how the CdTe PV industry might proceed.

Since the vast majority (99.5^{+} wt%) of an unframed CdTe PV module is glass, CdTe PV modules might be recycled through existing glass recycling pathways. However, the high cleanliness specifications of sheet glass manufacturers effectively preclude the recycling of CdTe PV modules as part of the sheet glass recycling stream. Container glass has lower cleanliness standards, but the economics of container glass recycling is generally negative, driven primarily by litter reduction and landfill space conservation, and the addition of CdTe into the beverage container glass supply poses complications.

Pb-acid batteries are widely recycled because the reclaim value of the Pb metal, and to a lessor extent the plastic casings, is sufficient to offset collection and electrolyte treatment costs. Today over 90 % of Pb-acid automotive batteries sold in the U.S. are eventually recycled [6], due largely to laws mandating a "reverse distribution chain" in which battery retailers charge customers a deposit on the sale of new batteries and collect used batteries for recycling. Retailers are in turn paid up to ca. 50 ¢/battery by intermediary collection entities and/or primary recyclers.

End-of-life Pb-acid batteries are defined as "hazardous" under RCRA and would be costly to recycle if not for specific RCRA exclusions intended to encourage battery recycling by easing the costs and complexity of collecting and transporting used Pb-acid batteries. The EPA granted these exclusions in view of the well-established and highly-effective Pb-acid battery recycling system and in order to further encourage recycling and materials reuse. This suggests an opportunity for the PV industry to mitigate collection and recycling costs by setting up an effective PV module recycling program and seeking regulatory relief to improve the economics of such a program. This also underscores the importance of cooperative collective action by CdTe PV module manufacturers to establish workable pro-active recycling and educational programs.

NiCd batteries are widely used in rechargeable applications, and sales of NiCd batteries are increasing as sales of portable electrical devices increase. NiCd batteries "fail" the TCLP test and thus are considered "hazardous" under RCRA rules. Manufacturers of large industrial NiCd's ease the disposal burdens for their large commercial customers by providing collection and recycling programs, often at little or no cost to the customer. As a result, upwards of 80% of industrial NiCd's are currently recycled [7].

The most common technique for recycling NiCd's is pyrometallurgical processing in which the battery casing is broken open, the battery pieces are heated to vaporize the Cd, and the Ni and Fe support grids are smelted. The largest U.S. recycler of NiCd's is INMETCO, a secondary metal smelter with a primary interest in smelting scrap for resale as Fe-Ni stainless steel feedstock. Cd is a "tramp" metal that is processed at an economic penalty. The reclaim values of the Ni and Fe do not fully offset the overall processing costs, and the battery user is typically charged 30 - 40 ¢/lb for recycling. European and Asian NiCd recyclers generally complete the material life cycle by recycling Cd for reuse in new NiCd's, but total reclaim value is again typically less than total NiCd processing costs.

The mechanics and economics of industrial NiCd battery recycling reveals two important lessons for potential CdTe PV module recycling. First, at the prevailing prices of Cd and at the low concentrations of Cd in CdTe PV modules (e.g. ca. 0.05 wt% versus ca. 15 wt% in NiCd batteries) it is not economic for NiCd recyclers to recover Cd from CdTe PV modules. Second, in the absence of secondary components of high reclaim value (e.g. Ni and Fe in NiCd's), CdTe PV module recycling will be at some cost to the end-of-life module owner.

The majority of NiCd batteries sold in the U.S. are smaller consumer batteries, and the majority of such batteries are not at present recycled. RCRA allows such unregulated disposal by private citizens under its broad exemption for all "household" wastes. The EPA has recognized that large volumes of widely disperse small individual quantity wastes like consumer NiCd batteries account for a disproportionate fraction of the hazardous materials entering the uncontrolled solid waste stream. The EPA calls this "universal waste" and intends to amend RCRA with the Universal Waste Rule (UWR). The UWR is directed at facilitating feasible and effective collection and recycling programs by easing key RCRA requirements. Should the UWR be issued, the NiCd battery industry will incorporate and fund an independent entity to collect and recycle consumer NiCd batteries at no cost to the user [7]. Battery collection will be encouraged by setting up retailer collection programs that provide convenient store drop-off points for the user and compensation payments to the retailer.

The corollaries for the CdTe PV industry are clear. First, public relations and regulatory issues can justify collection and recycling programs even if such programs are not economic on their own. Second, the economics of a PV recycling program in the U.S. would be greatly improved if PV modules were included under the UWR exemptions.

PV INDUSTRY AND CUSTOMER SURVEY

A survey focusing on commercialization issues collected quantitative data on the knowledge, perceptions, preferences and values of prospective purveyors and users of CdTe PV modules. Survey results show that many module users are concerned about materials content, especially about Cd content; but that products containing regulated materials (e.g Cd in CdTe) would be purchased if the price and performance were competitive and if the modules are conveniently recycled. Most responders now use products containing regulated materials because those products are reliable and cost-effective, and half of the responders indicated that their customers typically do not know or care about the materials content in products.

Most responders indicated that they would voluntarily participate in a module recycling program, and the balance indicated that they and/or their customers would participate if they were compensated for handling costs and/or if their customers received some incentives. Responders indicated that they and/or their customers would be more likely to purchase modules that could be conveniently recycled, but that they would be less likely to purchase modules that by law must be recycled or handled as hazardous waste. For example, responders indicated that a module that is comparable to standard crystalline silicon modules in all respects except that it must by law be recycled or disposed of as hazardous waste will likely command market prices of 5 - 30 % less than crystalline silicon modules, with recycling being less disadvantageous (i.e. 5 - 15 %) than hazardous waste disposal (i.e. 10 - 30 %). The non-U.S. responders gave the strongest answers, predicting that future local laws would require module manufacturers to take back end-of-life modules and that modules requiring recycling or hazardous waste disposal could not be sold unless a large incentive was offered to adequately motivate users to return end-of-life modules. Although all responders indicated a willingness to participate in module recycling and many indicated that module recycling was a positive influence on their choice of modules, few responders indicated that ease of module recycling would command a higher module selling price.

The survey results underscore the differences between customer segments and the importance of educational and marketing materials being carefully targeted to the different customer groups. For example, module and system vendors were the most concerned about materials content and the most likely to specifically identify Cd content as a primary concern, while utilities were the least likely to cite E,H&S concerns. This suggests that utilities might be good target customers for the market entry of CdTe PV modules in so much as utilities would be technically sophisticated customers with no strong pre-conceived aversions to handling products containing regulated materials.

CdTe PV RECYCLING

Real-world recycling is a dynamic trade-off between costs and benefits and is a strong function of the details of the recycling program and of the regulatory and market environment in which the program operates.

A near-term route for recycling CdTe PV modules is the existing NiCd battery recyclers. While it is likely that NiCd recyclers will charge the module user for recycling, the simplicity of using the existing fully-permitted NiCd recycling chain facilitates CdTe PV recycling in the nascent stages of CdTe PV manufacturing and marketing.

An alternative recycling route is primary metals smelters. A prototypical glass/glass module structure with 0.15 μm CdS and 2 μm CdTe layers has only 0.05 wt% each Cd and Te, well below the typical metals concentrations in the Zn, Cu and Pb processing byproducts from which Cd and Te are generally recovered. At the low concentrations present in a PV module, processing costs would almost assuredly exceed metals value. Furthermore, though non-ferrous smelters purchase silicas as fluxing agents for the smelting process, the low-cost soda lime glass used in most PV modules has less free silica than the raw silica sources typically used, and U.S. smelters often use silicas containing trace amounts of precious metals to improve the economics of the overall smelting operation. Thus, it is likely that primary metals smelters will charge the module user and/or manufacturer for recycling

Another alternative recycling route is secondary metals smelters and cement kilns. Small specialty smelters are being developed for use in treating hazardous materials and might be useable to process CdTe PV modules. Cement kilns now processing a wide range of hazardous wastes at costs significantly below incineration costs might be used to process end-of-life CdTe modules [8].

Hazardous waste landfill disposal of end-of-life CdTe modules by the user is a likely scenario if CdTe modules fail TCLP tests. Module manufacturers are not at present required to take any economic or legal responsibility for end-of-life modules, but laws exist in some locales requiring battery manufacturers to accept returned end-of-life batteries, and such laws may expand in scope to include PV modules. If the user is not a private citizen exempted under RCRA's household waste exemptions, the module user is faced with landfill costs, taxes, and future liability. Landfill costs in California average $400 per 55-gallon drum of dry solid hazardous waste. Assuming 8% efficient glass/glass CdTe modules ground up to better use drum volume, this translates to $0.14/Wp disposal costs. Hazardous waste generator taxes are ca. $200/ton for landfill disposal in California, adding ca. $0.03/Wp. Landfill disposal may incur future liabilities, e.g. clean up expenses.

Hazardous waste incineration by the user largely eliminates future liability, but incineration incurs higher per drum costs and incurs waste generator various taxes. For example, in California solid hazardous wastes can be incinerated at a cost of ca. $600/drum, thus incineration fees alone would be ca. $0.20/Wp.

Hazardous waste collection and recycling, whether by the user, the manufacturer, or some separate recycling entity, incurs various direct and indirect costs. An incentive will likely be necessary to encourage users to return end-of-life modules to centralized collection locations, whether via reverse flow through the retail chain or directly to some specific module collection mechanism. Auto battery collection incentives are $2-10 on a ca. $60 battery; one could reasonably expect a similar amount for a 30 - 40 Wp module selling at $1.50 2/Wp. Assuming a $2/module incentive yields ca. $0.07/Wp in incentive payments. Assuming a centrally located recycling site, one could reasonably expect $500/ton, or $0.08/Wp in shipping and handling costs for bulk, solid, non-radioactive, non-volatile hazardous waste. Actual module recycling would likely come at some cost to the party disposing of the module. NiCd battery recycling charges range from $0.20-0.70/lb, and costs would be higher without the Ni reclaim value. Assuming module recycling charges equal to the lowest present costs for battery recycling, one estimates $0.20/lb, or $0.07/Wp. Summing collection incentives, transport, and recycling costs, an estimate of ca. $0.22/Wp is obtained.

Non-hazardous waste collection and recycling would benefit from the simplified handling of non-hazardous wastes. An incentive would be required to encourage module users to return end-of-life modules to collection points if the user was not otherwise faced with substantial hazardous waste disposal cost. A $2/module incentive is plausible. Shipping and handling costs for non-hazardous wastes are much less than for comparable amounts of hazardous wastes, but end-of-life modules would likely be shipped back in small shipments through transport means similar to those used to send out new modules. Assuming typical ground transport, one reasonably estimates $0.25/lb, or $0.09/Wp. If non-hazardous modules could be reclaimed for their glass value at $5/ton, then one estimates a gain of $0.01/Wp. Summing collection incentives, transport costs, and reclaim value, an estimate of ca. $0.15/Wp is obtained.

It is important to realize that these cost estimates are based on conservative assumptions about the difficulty of collecting and recycling CdTe modules. If the scenarios accurately reflect the disposal options for end-of-life CdTe modules, then CdTe PV technology faces important challenges. It is possible, however, that the desire of modules users to purchase recyclable products and their willingness to voluntarily participate in recycling programs will provide much more positive economics than the scenarios quantified here.

ACKNOWLEDGMENTS

This work was funded by Brookhaven National Laboratory under Contract No. 725008.

REFERENCES

[1] J.M. Woodcock et al., "Thin Film CdTe Photovoltaic Cells", *12th EC PVSEC*, 1994, p. 948.

[2] Ernst & Young, "Household Battery Waste Management Study", Report # P402-92-001, California Integrated Waste Management Board, 1992.

[3] K. Zweibel, "Perspectives and Opportunities in Polycrystalline Thin-Film Photovoltaic Technologies", *23rd IEEE PVSC*, 1993, p. 379.

[4] M. H. Patterson et al., "Health, Safety,and Environmental Aspects of the Use of Cadmium Compounds in Thin Film PV Modules", *Int. PVSEC-7*, 1993, p. 521.

[5] M. H. Patterson et al., "Health, Safety,and Environmental Aspects of the Production and Use of CdTe Thin Film Photovoltaic Modules", *12th EC PVSEC*, 1994, p. 951.

[6] Battery Council International brochure, December 1991, and personal communications in 1994.

[7] Personal communication with Portable Rechargeable Battery Association.

[8] I. Kim, *Chemical Engineering*, April 1994, p. 41.

Amorphous Silicon Alloy-Based Roof Integrated Photovoltaic Systems

P.Nath, C.Vogeli, A.Singh and J. Call
United Solar Systems Corp.
1100 West Maple Road
Troy Michigan 48084

ABSTRACT

A roll-to-roll process is used to deposit tandem amorphous silicon alloy solar cell onto thin (0.005") stainless steel substrate. Using this solar cell material, we have designed and fabricated a photovoltaic (PV) module which can be integrated into building roofs. The module is fabricated by laminating the large area amorphous silicon on stainless steel solar cell material onto a 0.03" thick coated galvanized steel support plate. The module is then formed in such a way to allow installation as a batten and seam roofing system. This paper describes the fabrication and installation details of such PV systems.

INTRODUCTION

Architects designing buildings where economy, beauty and durability are important considerations, prefer architectural sheet metal for batten seam, standing seam, mansard and facia. More often than not, metal roofing and siding systems blend well with all type of masonry building materials and have a history of low maintenance and easy installation. In the photovoltaic industry it is being widely recognized that integration of photovoltaic modules into buildings can effectively decrease the cost of PV systems. At United Solar Systems Corp., a roll-to-roll process is used to deposit tandem amorphous silicon alloy solar cell onto thin (0.005") stainless steel substrate. By laminating this large amorphous silicon on stainless solar cell material onto a 0.03" thick galvalume architectural steel sheet, we have designed and fabricated a PV module which can be integrated into buildings. The module is constructed in such a way to allow installation as a conventional batten and seam roof. The fabrication and installation details of this novel integrated PV roofing system are described.

MODULE DESIGN FEATURES

The module fabrication process has already been described in details elsewhere [1]., but a brief description follows. Amorphous silicon based tandem solar cell material is made using roll-to-roll deposition onto thin (0.005") 2500 foot long rolls of stainless steel. This generic solar cell material can be cut into appropriate sizes, assembled, interconnected and finished as a PV building element. Finished PV modules are shown in Fig. 1. The illustrated module is approximately 1'x 4' and is rated at 22 watts. Key design features of this module are given below.

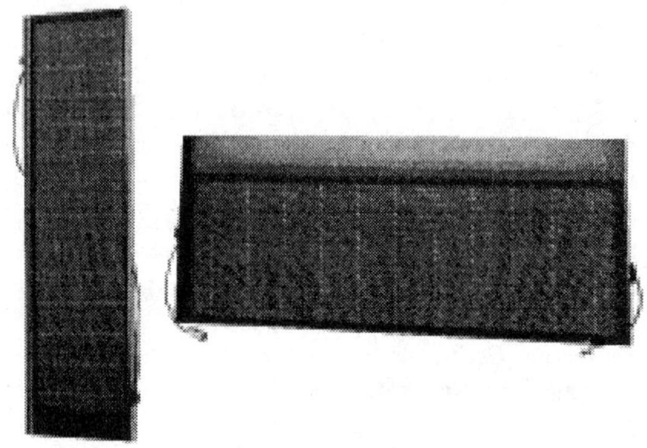

FIG. 1 1'X4' Roofing Module of a Vertical and horizontal design

Top Cover

The top outer cover of these modules is a uniaxially oriented fluoropolymer film. This film is laminated to the cell surface using a multilayer composite based on ethylene vinyl acetate. Ingredients included in the EVA formulation provide for long term weatherability[2]. The proprietary multilayer front cover provides a high cut strength. The cut strength exceeds the limit set forth in the U.L. 1703 document[3]. During lamination the front surface of the modules is embossed resulting in a diffuse surface appearance.

Current Collection

Efficient current collection is accomplished by incorporating an aesthetically pleasing grid structure. If architectural requirements demand a one color surface, it can be achieved by adjusting the thickness of the transparent conducting layer and disguising the grid lines. Prior to applying current collection grids, the PV material is treated by a defect isolation technique [4] which electrically isolates shunted regions of the cell material.

CH3365-4/94/0000-0966 $4.00 © 1994 IEEE

First WCPEC: Dec. 5-9, 1994: Hawaii

Interconnections

The module is assembled by electrically interconnecting the desired number of individual strips of appropriate size. Contact redundancy and flat chip bypass diodes are included to provide for reliable operation even under partial shadowed sunlight conditions.

Backing Plate

To provide stiffness, formability as a building element and easy installation, the large area module assembly is laminated to 0.03" thick coated galvalume steel backing plate. This plate is the same as utilized in various metal roofing applications.

Module Finishing

Depending upon the roof design, the large area PV module is formed. The two edges of the module are upturned allowing batten and seam roof installation. The bottom and top edges are formed in such a way to allow strong mechanical attachment to a roofing deck, as well as secure water tight overlap between modules

Module Termination

Two potted J-Boxes with plug in connectors are mounted onto the upturned edges of the module. This feature allows for an easy plug in water tight electrical connection during installation.

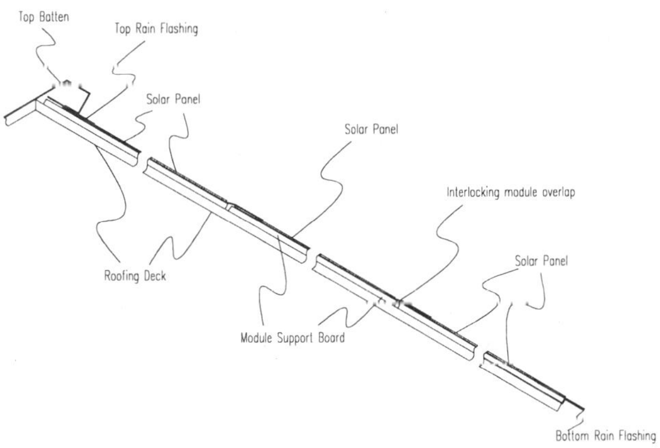

Fig.2 Schematic of installation

INSTALLATION DETAILS

Since the PV module is formed in the shape of a conventional batten and seam roofing sheet metal element, it can be installed like a standard batten and seam roof. This is illustrated in Fig. 2. The modules are installed over a solid substrate such as construction grade plywood and an underlayment of 30 # roofing felt. Non penetrating roofing clips are used to fasten the modules to the roofing deck as illustrated in Fig.3. Hidden screws under the overlap ensure mechanical strength and water tightness of the overlap joint. The overlap crossection of the installation is shown schematically in Fig. 4. After the mechanical installation, modules are interconnected by the plug in connectors. The seams in-between the modules are covered with snap on battens. The edges of the roof are finished with standard edge flashings. For enhanced appearance, the battens and edge flashing are color coordinated. Some examples of the finished roofs are shown in Fig. 5 and Fig. 6.

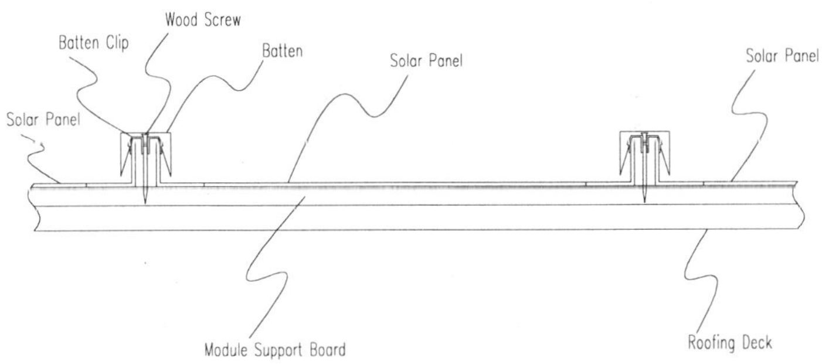

Fig. 3 Installation Details of Roofing Module (Crossection)

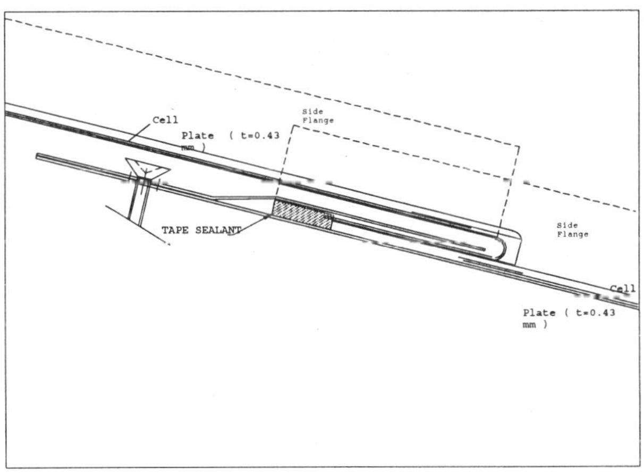

Fig.4 Schematic veiw of the overlap section.

Fig. 5 Roofing Installation using the Vertical Design Roofing Module.

Fig.6 Roofing Installation using horizontal design roofing module. Batten pitch is 4 ft.

CONCLUSIONS

An integral amorphous silicon alloy-based photovoltaic roofing element has been designed and fabricated. These modules are easily installed using conventional batten and seam roofing installation methods and hardware. This integrated PV roofing approach results in a substantial lowering of the total installed cost.

REFERENCES

1. P. Nath et al, proceedings of the 3rd international Photovoltaic Science and Engineering Conference, Tokyo (1987) p.395
2. Contract Report, DOE/ JPL- 954527-81/19, Springborn Laboratories, Enfield, CN, 1981
3. Standard for Flat-Plate Photovoltaic Modules and Panels" ,Underwriters's Laboratories, Northbrook, Il, 1986
4. P. Nath et al. , Appl. Phys. Letters 53 (11), p.986, 1988

PHOTOVOLTAIC MODULES INTEGRATED WITH A METAL CURTAIN WALL

M. Yoshino, N. Nakada, T. Mori
K. Yamagishi, S. Yoshida
YKK Corp.
200 Yoshida, Kurobe Toyama 938 Japan
Tel.: +81-765-54-8333 Fax: +81-765-54-8338

Y. Higashi, K. Shirasawa
KYOCERA Corp.
1166-6 Hebimizo-cho, Yohkaichi,
Shiga, 527, Japan
Tel.: +81-748-55-1211 Fax: +81-748-55-4635

ABSTRACT

An integrated photovoltaic system for buildings has many advantages. To realize building integration of photovoltaics, we have initially designed a PV module integrated with a metal curtain wall. PV modules are installed as spandrel panels and consist of long and slender PV sub-modules. These sub-modules have an encapsulated structure consisting of: glass, EVA, solar cells, EVA, aluminum base plate. We will also present initial PV performance data from the experimental wall. In this wall, almost the same maximum P_{max} of 64 W/m^2 was obtained and the module temperature was approximately 10 ℃ lower compared with conventional superstrate-type PV modules which have 1.3 times the solar cells of this module. Moreover, aesthetic requirements for this module are discussed in this paper.

INTRODUCTION

Recently, photovoltaic (PV) generation systems have been highly expected to be a clean source of energy. Moreover, from the view point of environmental and energy-supply problems, the promotion of the wide use of PV modules is earnestly hoped for. An important subject of this is the reduction of their costs. In densely populated cities and countries, the architectural integration of photovoltaics into buildings will be advantageous because unoccupied land for PV generation systems is limited, and there are a lot of sunny building facades available for PV installation. This architectural integration allows reduction of module costs and installation work expenses by including them in the materials and construction work costs. Furthermore, this will be much more aesthetically pleasing for the building facades.

From the aforementioned point of view, we have developed PV modules integrated with a metal curtain wall that can be applied to building facades as a wall construction material.

In this paper, we will report the initial results of this development.

SCHEMATIC DESIGN

In this development, we integrate PV modules with the metal curtain wall which is constructed by a mullionsystem. (see Fig. 1) Generally, glass, metal, and stone panels are installed to the spandrel unit and are then fitted to building

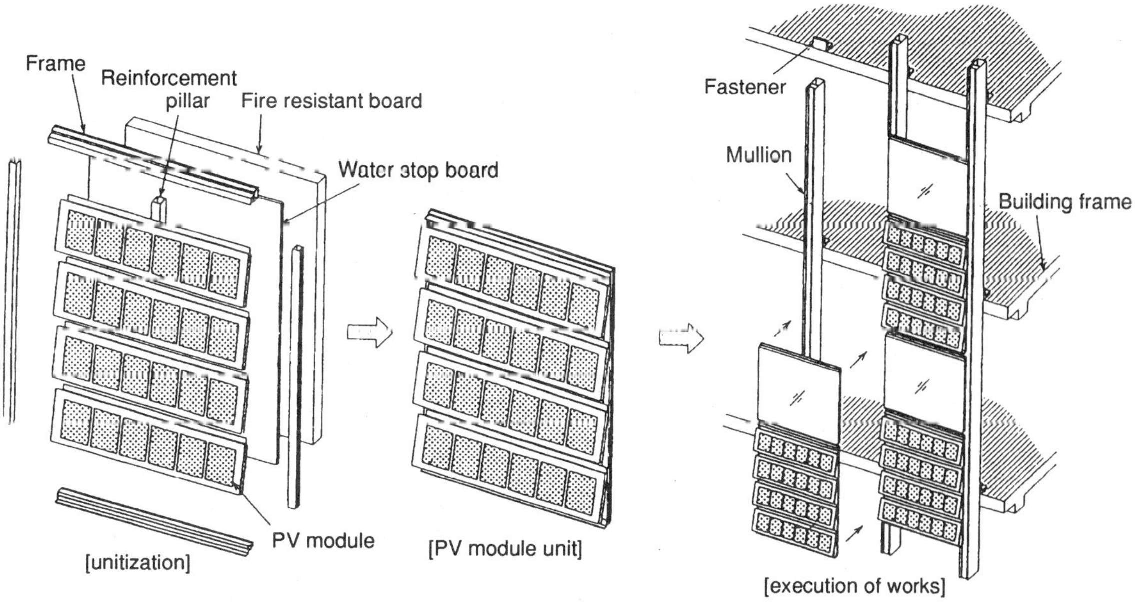

Fig. 1. PV module unit and its principles of construction.

CH3365-4/94/0000-0969 $4.00 © 1994 IEEE

First WCPEC; Dec. 5–9, 1994; Hawaii

facades. As shown in Fig. 1, PV modules are installed as a spandrel unit of the curtain wall. This spandrel unit consists of a PV module, frame, a water stop board, a fire resistant board and a reinforcement pillar.

Furthermore, PV modules consist of long and slender PV sub-modules. These PV sub-modules have an encapsulated structure consisting of: glass, EVA, solar cells, EVA, aluminum base plate (where the aluminum base plate has radiative fins) (see Fig. 2).

These components are prefabricated in a factory. Therefore, the product quality is stable and the construction period can be shortened.

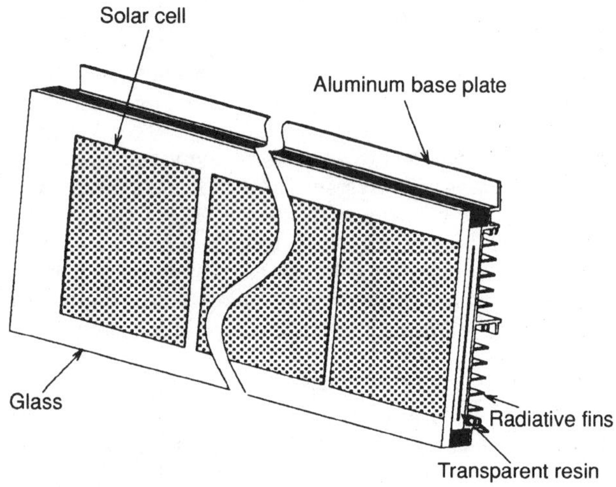

Fig. 2. The structure of PV sub-module.

This module has the following significant features compared with conventional PV modules:

- Various areas of the photovoltaic panel can be constructed by combining suitable length with the suitable number of steps of the modules.
- Modules can be repaired by only changing the damaged modules, so that the replacement cost is lower and the operation is easier.
- Modules can be inclined separately in the limited space of the curtain wall frame.
- We can expect improvements in the durability of photovoltaic modules and reduction of the module temperature by using an aluminum base plate with radiative fins.

PHOTOVOLTAIC PERFORMANCE

Experimental Wall

In order to evaluate PV performance, we have constructed two experimental walls facing south. One is for the optimization of inclination and interval of PV sub-modules (see Fig. 3, 4), and another is for the evaluation of aspects and the measurement of PV module temperatures. (see Fig. 5) On these walls, seventeen PV modules were fitted and their conditions are shown in Table 1, 2. These modules include two conventional superstrate-type modules facing south with fixed angles of 90° and 83° for comparison with the modules of this development. Also, to measure the module temperature, a thermocouple sensor was attached on the back side of the PV sub-module of each PV module.

Fig. 3. Experimental wall for the optimization of inclination and interval of PV sub-modules. (Wall 1)

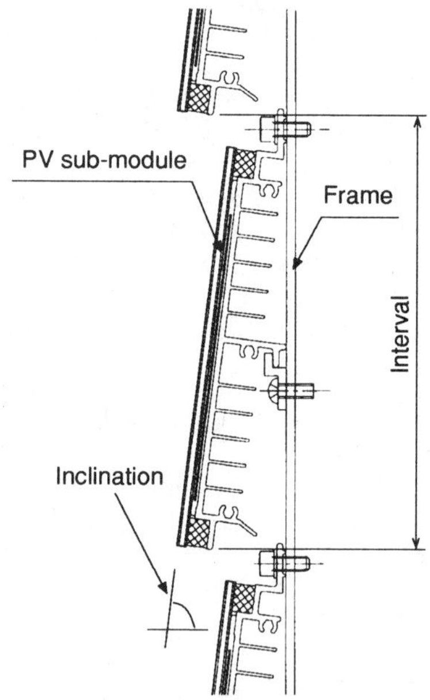

Fig. 4. Cross view of PV module.

Fig. 5. Experimental wall for the evaluation of aspects and the measurement of PV module temperature. (Wall 2)

Table 1. Module conditions of Wall 1

Fixed Angle	Interval
90°	150 mm
83° ,76° ,69°	150 mm, 200 mm
90°	superstrate-type

Table 2. Module conditions of Wall 2.

Fixed Angle	Interval	Radiative Fins
83°	150 mm	non, 4, 6, 10
83°	superstrate-type	

Monitoring System

The electrical parameters (short circuit current, open circuit voltage, fill factor, power output etc.) are measured by an I-V curve tracer EKO MP 123 every 5 minutes and recorded to the hard disk of a personal computer. The meteorological data (global and diffuse irradiance, irradiance on a 83° inclined surface, ambient temperature, wind speed and direction) and the module temperature are measured with proper sensors (pyranometer, thermocouple) by autonomous Datalogger ETO CADAC 100 at a sampling rate of 10 seconds. These data are recorded to the disk as 10-minute average values.

Experimental Data

The measurements have been performed since August first, 1994. Fig. 6 shows the dependance of P_{max} per square meter on the fixed angle of PV sub-modules and the curve of P_{max} of superstrate-type module with fixed angle of 90° for one sunny clear day in September 1994. As shown in the figure, a maximum P_{max} of approximately 64 W/m^2 is obtained at a fixed angle of both 76° and 69° . If the lower sub-module was not shaded by the upper sub-module, the maximum P_{max} was obtained at a fixed angle of 69° . In the same figure, compared with the superstrate-type module, almost the same maximum P_{max} is obtained by the fixed angle of both 76° and 69° , although superstrate-type module has approximately 1.3 times the solar cells of these modules. Therefore, these results indicate that the optimum fixed angle of PV sub-modules is existent. To clarify this point, we have simulated the energy output of the module based on the data obtained from the experimental wall and will describe the initial result about this later.

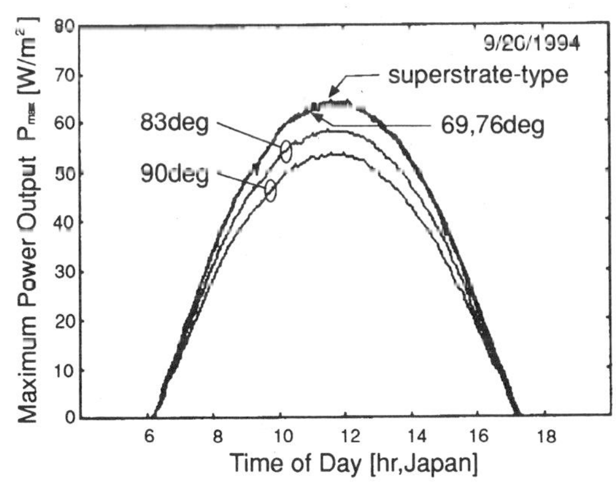

Fig. 6. Dependance of P_{max} on the fixed angle of PV sub-modules and the curve of P_{max} of superstrate-type module. (sub-module interval: 150 mm)

Concerning the reduction of the module temperature, we got distinctive results using an aluminum base plate as a component of the sub-module.

Fig. 7 shows the transitions of the temperature of the sub-modules of this study and the superstrate-type module on the same day of September 26, 1994. We have embedded the thermocouple sensor directly on the surface of the solar cell of both type modules.

From the results of this measurement, the temperature of the sub-module of this study was lower than that of superstrate-type module by approximately 10 ℃ from 9:00 am to noon. It is considered that this difference of 10 ℃ was due to the total effect of high thermal conductivity of the aluminum base plate and the rear space for ventilation between sub-modules and the water stop board.

On the other hand, the difference of temperature between the cell surface and the module rear surface was less than 2.5 ℃ throughout the day. Therefore, the temperature measured at the module rear surface can stand for the module temperature as a whole, including the solar cells.

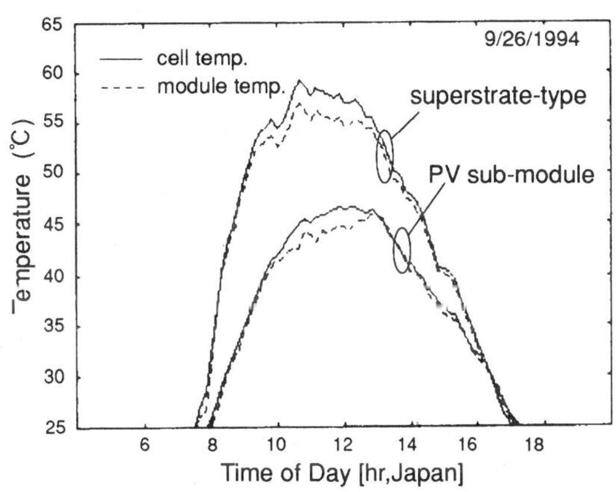

Fig. 7. Transitions of the temperature of the sub-modules and the superstrate-type module.

Computer Simulation

Accurate irradiance data on each module is necessary to simulate the maximum power output. Since we have only measured the global and diffuse irradiance on a horizontal surface and global irradiance on an 83 degree-inclined plane, we first estimate the global irradiance on various surface inclinations from the global and diffuse irradiance data on a horizontal surface, applying the Perez diffuse irradiance model [1] and isotropic ground-reflection model. If shading occurs on the sub-module, the rate of its area is calculated and applied to estimate the reduction of the insolation of direct beam component and diffuse component from circumsolar regions.

The maximum power output is computed with irradiance

on a PV module and module temperature represents the cell temperature. To simulate it for the PV module with a certain fixed angle, we used the module temperature of that with close fixed angle.

The simulation accuracy is illustrated in Fig. 8 as an example of a clear day. The discrepancy between measurement and simulation is seen in the evening due to the inaccurate global irradiance data because of the effect of a shadow which is caused by a nearby building.

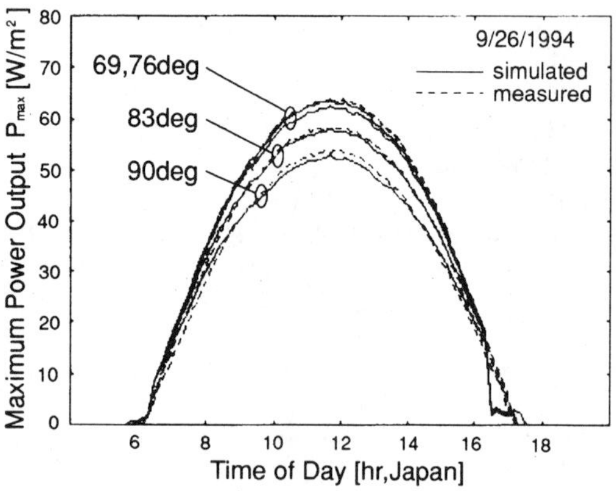

Fig. 8. Comparison of measured and simulated maximum power output. (sub-module interval: 150 mm)

Fig. 9 shows the daily energy output against fixed angle of a PV module on 5 clear days. In the figure, open symbols show the simulated data and closed symbols show the measured data. These curves are fitted through simulated data sets by the least squares method. Various points of view must be considered as reasons for the inconsistency between measured and simulated data in this figure besides the above mentioned Fig. 8. Hence, we will continue to study to improve the exactness of this simulation.

Fig. 9 clearly shows that the daily energy output of each fixed angle is increasing and the maxima of the curves are changing according to the movement of the suns path northward. It means an optimized fixed angle for PV generation varies depending on the season. Therefore, at least, annual data sets are needed for optimization. We will continue to study about it.

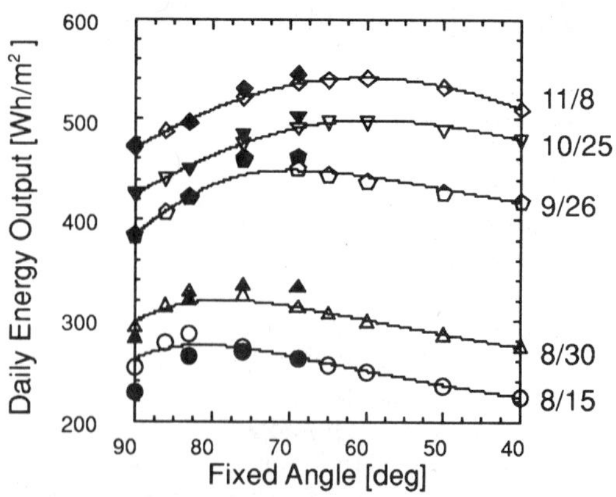

Fig. 9. Daily energy output against fixed angle of a PV module on 5 clear days.

AESTHETICS

Aesthetics is very important to apply PV modules as building facades. It may be one of the most important key factors for the promotion of wide spread PV facades.

As mentioned before, the PV modules developed in this study disposes inclined PV sub-modules, therefore special designs are needed for its pleasing visual and aesthetic aspects as an ornamental panel. Important viewpoints for the design may be described as follows;

1) Coloring of module
2) Aesthetic harmony with other material (glass, metal, stone, etc.) in the spandrel areas.
3) Design for spaces between sub-modules.

In regard to the third item, the inclination and interval of PV sub-modules affect their aesthetics. Therefore, the optimum condition for PV generation might not necessarily be the best disposition of PV sub-modules. In this study, we will propose good designs to create aesthetically pleasing PV modules adding high performance to PV generation.

CONCLUSION

In this development, we have designed PV modules integrated with a metal curtain wall. PV modules are installed as spandrel panels and consist of long and slender PV sub-modules. These sub-modules have an encapsulated structure consisting of: glass, EVA, solar cells, EVA, aluminum base plate. Also we have constructed an experimental wall to measure PV performance of the module. In this wall, almost the same maximum P_{max} of 64 W/m^2 was obtained and the module temperature was approximately 10 ℃ lower compared with conventional PV modules which have 1.3 times the solar cells of this module.

Further development of this module is now underway.

ACKNOWLEDGEMENT

This work was supported by the New Energy and Industrial Technology Development Organization (NEDO) as a part of the New Sunshine Program under the Ministry of International Trade and Industry (MITI). The authors are thanksful to YKK Architectural Products Corp. for the preliminary design of the curtain wall unit.

REFERENCES

[1] R. Perez and R. Seals, "A New Simplified Version of the Perez Diffuse Irradiance Model for Tilted Surfaces", *Solar Energy*, **39**, 1987, pp. 221-231.

Development of Photovoltaic Modules Integrated with Roofing Materials (Heat Insulated Roof Panel)

Y.NITTA, T.HATUKAIWA, T.YAMAWAKI, Y.MATUMURA, S.MIZUKAMI
KANEKA CORPORATION

ABSTRACT

We have started to develop low cost photovoltaic modules integrated with roofing materials for wooden houses. We made a concept of the design for the modules using amorphous silicon solar cells and produced test modules that consist of untempered surface glass, solar cells, waterproof sheet, heat insulating materials and base frames. We have primarily tested the distributed pressure resistance as a building component. When applying a load from the front surface side of the modules, a 3.6 mm deflection at the center of the specimen under 300 kg/m2 load was observed, which is equivalent to a snowfall of 1.2 meters. As a result, we have finally confirmed that modules have enough structural strength to be used as a roof panel. We also tested the impact resistance of untempered surface glass by the testing method in JIS3212. In this test, cracks could not be seen from a height of 75 cm.

INTRODUCTION

From a global environmental perspective, alternative energy sources are recently becoming increasingly important. In particular, photovoltaic power is enormously attractive and offers great expectations. Several incentive plans for the diffusion of PV systems have been recently introduced by the Ministry of International Trade and Industry in Japan. However, the cost of photovoltaics is decidedly more expensive compared with conventional power generation methods at present. In order to promote the installation of PV systems, it is important to reduce the cost of the photovoltaic modules.

House rooftops provide good space for PV systems and the cost of the roofing panels with PV modules is at least partially offset by saving the roofing materials that PV modules replace. In addition, by directly installing the modules to the frame of the house a reduction of roofing hours can be expected. Therefore, the module integrated with roofing materials is extremely effective for low cost and utilization of dead space in a house. This module is also integrated with heat insulating materials to satisfy the important demand for energy conservation in today's homes. The temperature of the module in this structure increases under sunlight. The increase in temperature of the module using crystal silicon solar cell usually affects the cell performance. On the other hand, the output power of the module using amorphous silicon solar cell decreases by a stable state under sunlight for a long time and recovers to its initial one by the thermal annealing at optimum temperature. And the dependence of the peformance on temperature for amorphous silicon solar cell is smaller than that for crystal silicon solar cells. Therefore, the thermal annealing effect can be expected under sunlight in the amorphous silicon solar cells with insulating materials.

Roofing tiles with solar cells have already been reported on. In these cases, however, they are mounted on the rooftop by conventional methods, so it is difficult to reduce the cost for installation. Thus, we propose a new type module with high durability, which can be easily installed as a roof panel.

We have already achieved a stabilized efficiency of 8.5% for an area of 10 square cm on two stacked amorphous silicon solar cells [1]. In the near future, we plan to fabricate a large scale amorphous silicon solar cell for practical photovoltaic use, and this cell will be introduced to the new type modules for realizing a high efficiency and low cost module.

We started our development program for the roof panel with photovoltaic function for wooden houses in the last quarter of FY1993 and have made a basic concept of photovoltaic modules integrated with roofing materials. At this point, we have constructed and primarily examined these modules to conform that they have enough strength as roof panels and to identify problems in assembling the modules. In this report, we describe the results in detail.

STRUCTURE AND MANUFACTURING

In order to see the final modules and to investigate unconfirmed problems in manufacturing, we produced test modules (4.50m x 0.455m) after creating a design. The structure of this module is shown in Figure 1. This module consists of surface glass, solar cells, waterproof sheet, heat insulating materials and base frame. Untempered glass with a thickness of 4mm is used as the surface glass, and the base frame is made of a laminated veneer lumbar(LVL) in considering the dimensional stability of the modules over a long time. The cross section of the LVL frame measures 90mm by 40mm for main frame and 35mm by 40mm for subsidiary frame. The thickness of the insulating material, which is made of polystyrene form, is 55mm for better energy conservation in the houses. The waterproof sheet of a mainly butyl based adhesion is inserted between the solar cell and the insulating material to be watertight. This waterproof sheet is spread over the side of module

CH3365-4/94/0000-0973 $4.00 © 1994 IEEE

First WCPEC; Dec. 5–9, 1994; Hawaii

as shown in Figure 1. If the modules are installed, they can be closely attached to each other at the sides with this adhesive sheet. By placing a cover frame above the attached part the unit becomes watertight.

The manufacturing process is shown in Figure 2. The frame-work of the module is constructed in ladder structure, and the insulating material is attached to it with adhesive tape. After covering it with waterproof sheet, the submodules are mounted on the sheet. Finally, the active power generation area of this module is almost 65% of the total area and the maximum error observed on the assembled modules was 3 mm diagonally, within the specified accuracy. As a result, there were no serious problems in the manufacturing process.

A schematic diagram for installing these modules is shown in Figure 3. The modules are transported by truck, and lifted by crane. They are directly installed to the frame of a house which is expected to reduce the required roofing hours.

EXPERIMENTAL AND RESULTS

The distribution pressure test of this module was carried out using the equipment as shown in Figure 4. This equipment is made of a steel frame without bottom and the required load can be apply to the surface of the

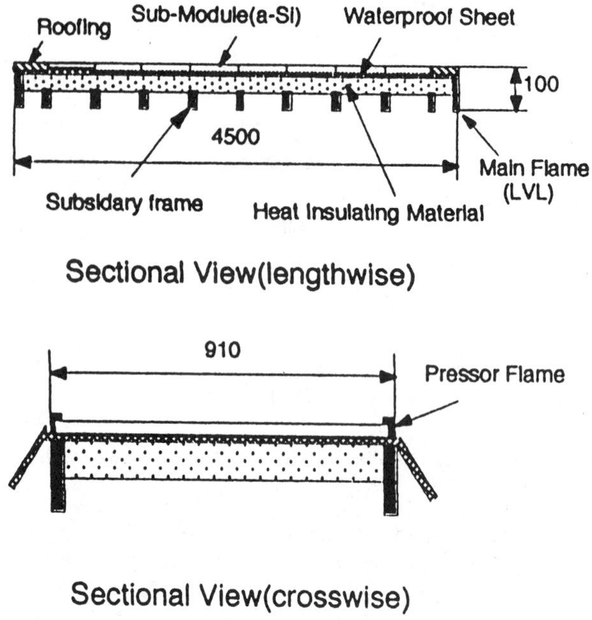

Fig.1. Sectional view of the module.

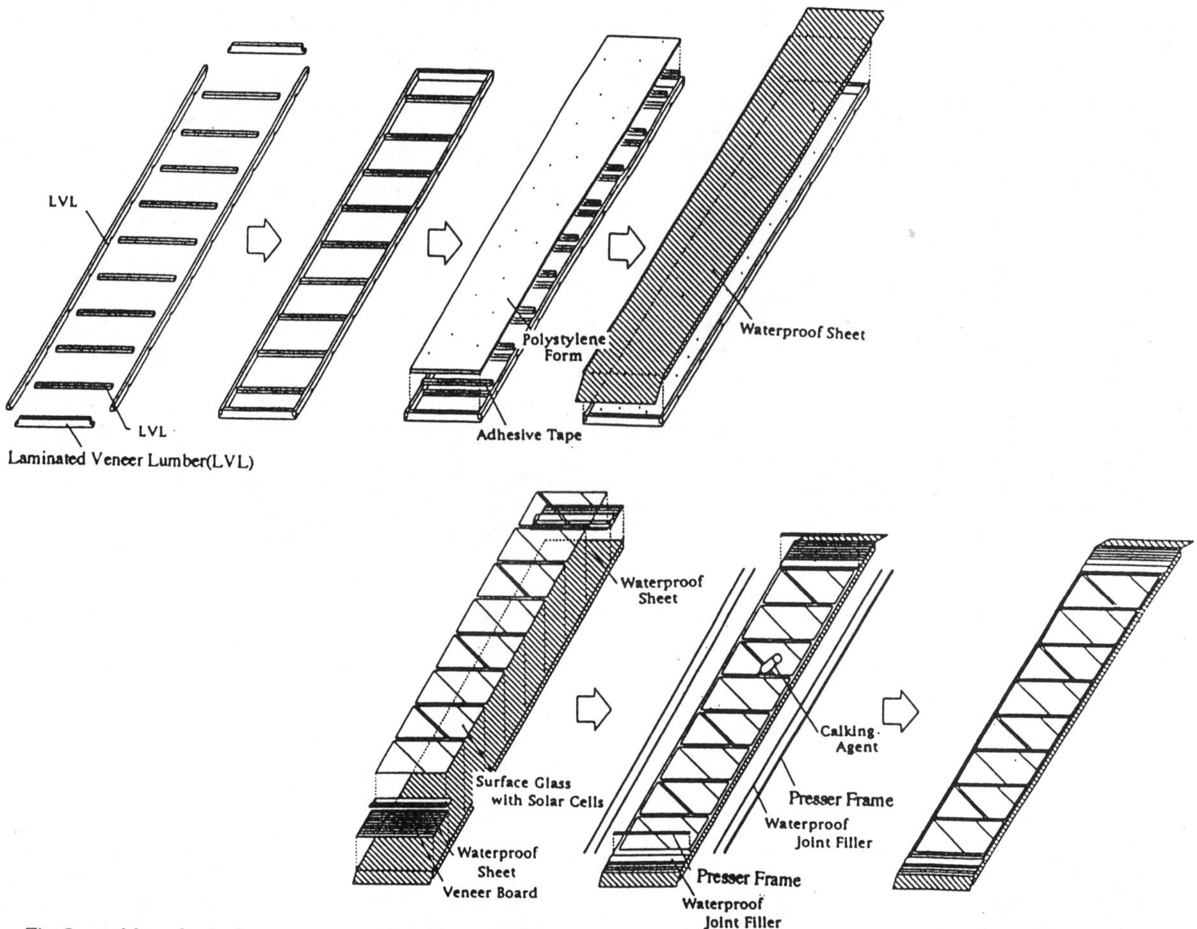

Fig.2 Manufacturing process for the module integrated with roofing materials.

test piece with water. The relationship of load-deflection is shown in Figure 5. When applying a load from the front surface side of the modules, a 3.6 mm deflection at

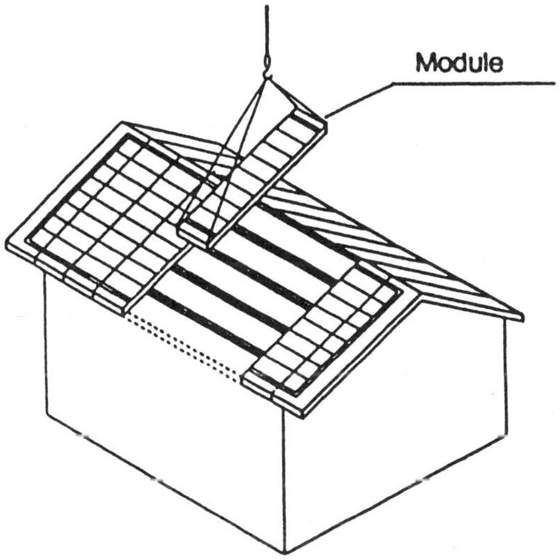

Fig.3. Schematic diagram for installing.

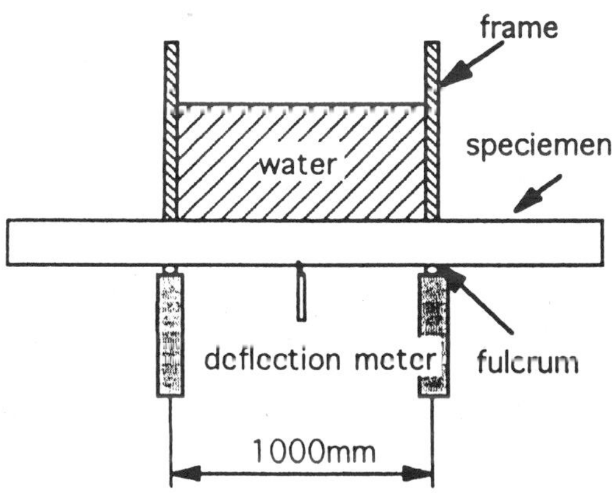

Fig.4. Equipment for the distribution pressure test.

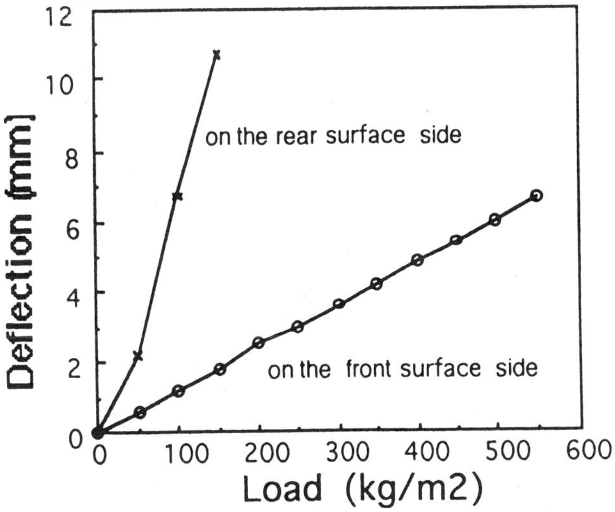

Fig.5. Relation of load-deflection on the module.

the center of the test piece under 300 kg/m2 load, which is equivalent to a snowfall of 1.2 meters, was observed, and it has been confirmed that test modules have the same structural strength as a roof panel. Furthermore, we could increase the load by 550 kg/m2, equivalent to a snowfall of 2.2 meters, without breakage. A deflection of 6.7 mm under 100kg/m2 load and a breaking load of 450kg/m2 was observed, when applying a load from the rear surface side. To conclude, we have confirmed that the modules have enough structural strength.

An increase of the deflection under the load from the rear side is larger than that under the load from the front side. The subsidiary frame is effective protection against the load from front surface side but not against the load from the rear surface side because it is mounted under the insulating material. It is necessary to improve the structure of the modules for stronger protection against the load from the rear side.

We tested the impact resistance of the surface glass by the testing method in JISR3212. We dropped a 227g steel ball with a diameter of 38 mm on the surface glass from a required height. In this test, cracks could not be observed from a height of 75 cm. We have found that heat insulating material under the untempered surface glass also acts as a shock absorber. We have further confirmed that pieces of broken glass by excessive impact do not scatter.

Finally, a roofing test was carried out using the test modules. As a result, we have confirmed that there is no deflection of the modules in lifting them by a crane, and no change occurs in the surface glass and solar cells during roofing tests. Furthermore, it was found that the test modules had enough strength to allow walking on them. A photograph of roofing is shown in Figure 6.

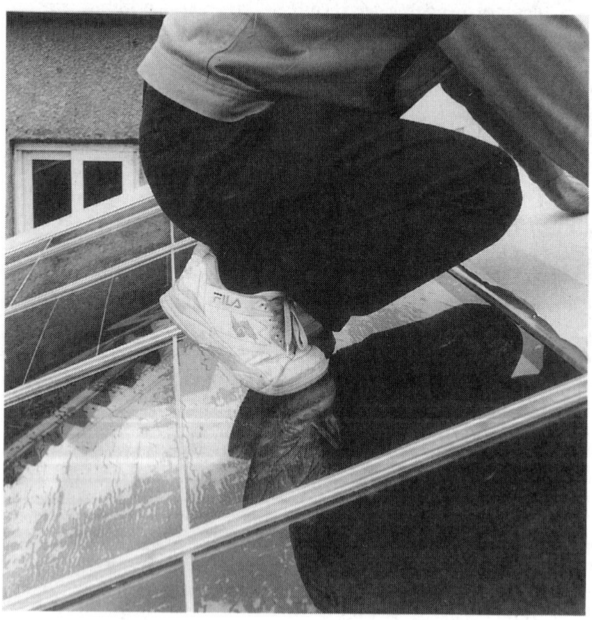

Fig.6. Photograph of roofing the modules integrated with roofing materials.

The test modules have been installed for examination of water tightness in March 1994. Up to this point, there is no problem.

CONCLUSION

We have designed and produced test modules. When applying the load from the front surface side of the modules, a 3.6 mm deflection at the center of the specimen under 300 kg/m2 load, which is equivalent to a snowfall of 1.2 meters, was observed. In conclusion, we have confirmed that modules have enough structural strength as a roof panel. We tested the impact resistance of untempered surface glass that is used in these modules. We dropped a 227g steel ball with a diameter of 38 mm from a required height. In this test, cracks could not be observed from a height of 75 cm.

It was found that these modules had enough strength to allow walking on them during installation of the modules.

Finally, the panels have proved to be watertight since their installation in March 1994.

ACKNOWLEDGMENT

This work was supported by New Energy and Industrial Technology Development Organization as a part of New Sunshine Program.

REFERENCES

1) K.Hayashi et al. "NewType of Large-Area a-Si Module Produced using a Polymer Encapsulation Method",1st WCPEC,1994, to be published

PV MODULES, USING COLOR SOLAR CELLS, DESIGNED FOR BUILDING

N. Ishikawa, H. Yamashita, S. Goda, I. Hide
Chitose Research Laboratory, Daido Hoxan Inc.
T. Yoshida, K. Iwai, N. Sagara
Kajima Technical Research Institute, Kajima Corporation
K. Tazawa, M. Komiyama, H. Nishi, T. Harada
Photovoltaic Business Division, Showa Shell Sekiyu K.K.

ABSTRACT

The designs and structures of a photovoltaic module to be installed in a curtain wall were examined with the object of sharply reducing PV power generation costs. As for the design, solar cells of different colors were produced and an opinion survey of designers and other construction-related persons was conducted. Renderings based on color solar cells were prepared using computer graphics. In general, the output of cells decreases for colors other than shades of dark blue. However, there is a good possibility that greater importance will be put on design, including color and surface condition, than on output. In this case, the market share may expand and, as a result, the cost may drop. As for the structure, various materials that can be used for a building-material-integrated-module were investigated and methods to install PV modules into building materials were examined. Moreover, experimental module samples fitted with stainless steel sheets and aluminum-sash frames were made.

INTRODUCTION

When PV modules are incorporated into the walls or the roofs of buildings, a cost reduction effect is expected, because part of the PV module cost and construction cost, such as the base-frame cost, can be included in the building-material cost and the building-construction cost.

There are some requirements to satisfy quality as building materials. It is a matter of course that durability, fire-resistance, etc. are important factors. In addition, we considered the design of modules. Design is one of the most important points when a building is constructed.

We developed for high-rise buildings a PV module using crystalline silicon solar cells incorporated into the curtain-wall. Curtain-walls are used in a particularly large number of high-rise buildings, therefore it is considered that there is enough of a potential market. This study seeks to produce PV modules incorporated into building materials which have not only sufficient quality but also good designs as building materials.

This paper mainly describes the experimental production of color solar cells, examination of design and the structure of a building-material-integrated module, and experimental production of module samples.

COLOR SOLOR CELLS

Solar cell samples in various colors (brown, yellow, green, etc.) were produced by changing the refractive index and the thickness of TiO_2 anti-reflection (AR) coating. At that time, the relationship between the colors and the thickness was examined by theoretical calculation[1], and the result was applied to make actual color solar cells. As a typical example, Figure 1 shows the calculated value of spectral reflectivity when the cell surface is green. It can be seen that the peak of reflection is around 500nm (green). The refractive index in this case was 2.2 and the film thickness was 240nm. By making this AR film on a polycrystalline silicon substrate, a greenish surface could be obtained and green solar cells could be produced. Figure 2 shows experimental production of polycrystalline silicon color solar cells. Table 1 shows cell performance of color solar cells produced experimentally. The fabrication process of the color solar cells has not been optimized, but it proved that the energy conversion efficiency fell about 15% at a maximum reduction rate, or about 20% for gray, compared with dark blue solar cells.

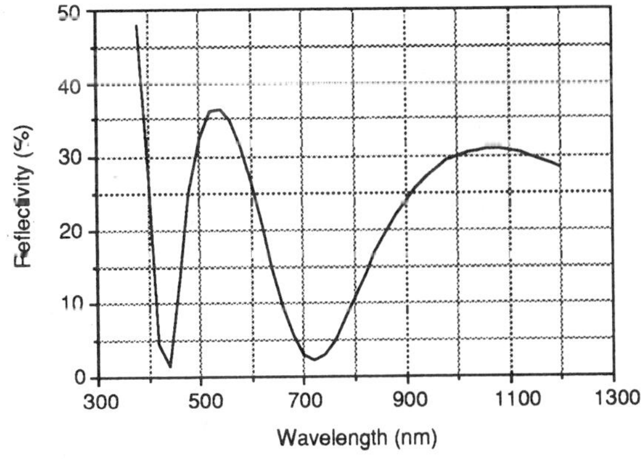

Fig.1. Calculated value of spectral reflectivity (refractive index 2.2, coating thickness 240nm)

CH3365-4/94/0000-0977 $4.00 © 1994 IEEE
First WCPEC; Dec. 5-9, 1994; Hawaii

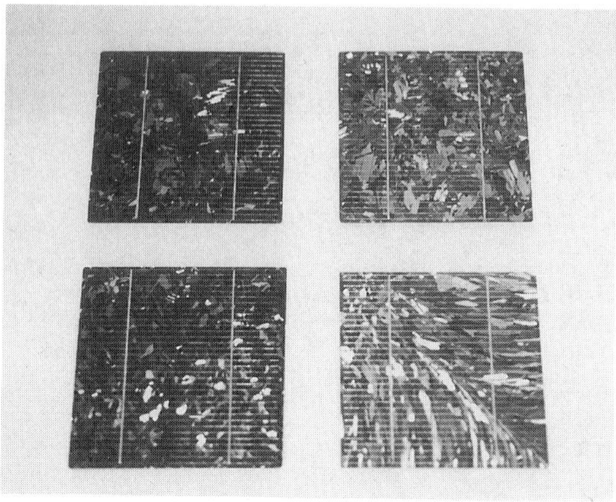

Fig.2. Experimental production of color solar cells

Table 1 Cell perfomance of color solar cells

AM1.5, 100mW/cm², 25℃, 100mm□

Color	Jsc (mA/cm²)	Voc (mV)	FF	Eff (%)
darak bule	31.7	608	0.733	14.1
bule	30.2	594	0.727	13.1
yellow	27.2	598	0.723	11.8
green	26.9	596	0.733	11.8
brown	29.4	599	0.743	13.1
gray	25.4	591	0.752	11.3

EXAMINATION OF DESIGN

Using various substrates and surface conditions of solar cells, such as polycrystalline silicon, single-crystalline silicon and ones which were texture etched and mirror polished, color solar cell samples (Fig.3) were produced, and an opinion survey of designers and other construction-related persons was conducted. The results could be summed up as follows:

- After directivity is taken into consideration, there is a great possibility that they can be used for high-rise and mid-rise buildings as well as for other buildings with wide walls. In densely built-up urban areas, however, they can be used only on facade walls.
- From a distance the modules resemble glass curtain walls, as the module surface is made of glass. On the other hand, seen from a short distance, they look different in texture including design texture, according to different colors or crystals of solar cells.
- Geometrical arrangement of color solar cells widens the possibility of new designs.
- Besides the conventional dark blue, gray (non-AR-coated colors, the original color of crystalline silicon), brown and yellow also made good impressions.
- Tint and surface conditions can yield such texture as that of cut stone or of metal.

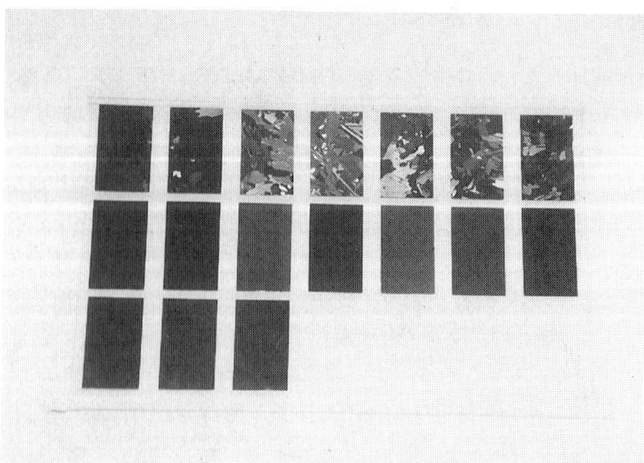

Fig.3. Color solar cell samples

Fig.4. CG in the case of fitting green solar cells into an office building

Moreover, to examine the possibility of using building-material-integrated modules for building walls in the future, designs based on those color solar cell samples were prepared using computer graphics (CG). Figure 4 shows the CG for an office building fitted with green solar cells. The CG rendering of a distant view shows cell electrodes and interconnectors to be hardly noticeable, thus the modules give little incompatibility to building facades.

EXAMINATION OF THE STRUCTURE OF THE MODULE

Figure 5 shows the structure of a building-material-integrated module. It is an example of a PV module which integrates a clear surface material like glass or resin on the base material by using encapsulating material such as ethylene vinyl acetate (EVA).

As for surface material, spectral transmittance was measured, namely of reinforced white sheet glass which is generally used for modules, blue sheet glass, acrylic board which is reputed to excel in transmissiveness, and polycarbonate which excels in strength. Transmittance was highest for acrylic board (5-mm thick), then decreased in order from reinforced white sheet glass (3.2mm) to polycarbonate (3mm) to blue sheet glass (3mm). Transmittance was particularly low in blue sheet glass. This is caused by the impurities in it. By producing modules using blue sheet glass and reinforced white sheet glass and comparing the output, it was found that the output of modules with blue sheet glass were inferior to those with white sheet glass by 5 to 7%. On the other hand, for ordinary glass curtain walls, boards of 10mm thick or more are needed, considering the degree of strength required to resist winds. In that case, however, it is expected that transmissiveness will decrease and the

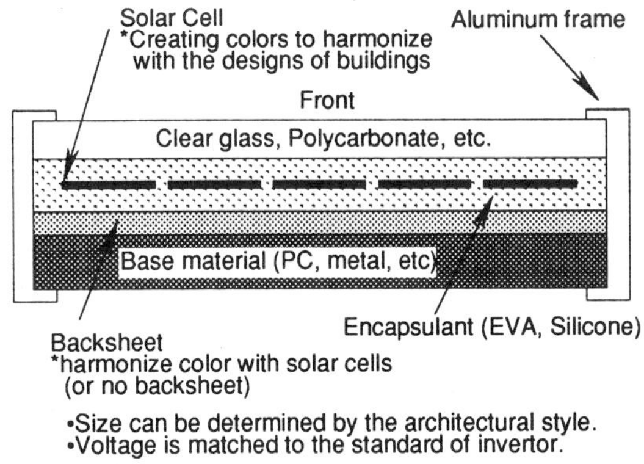

Fig.5. The structure of a building-material-integrated module

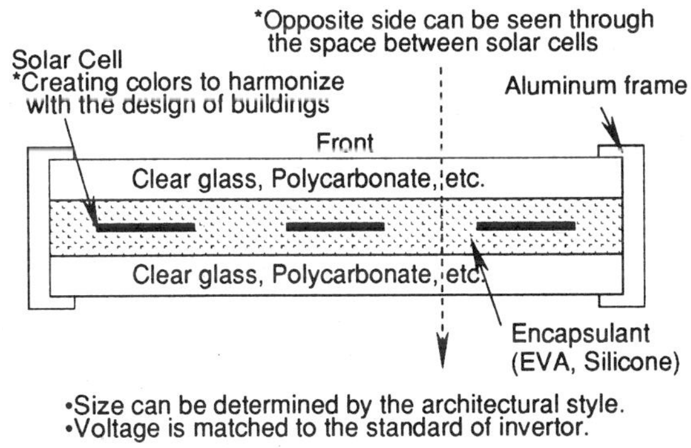

Fig.7. Building-material-integrated module for "top light"

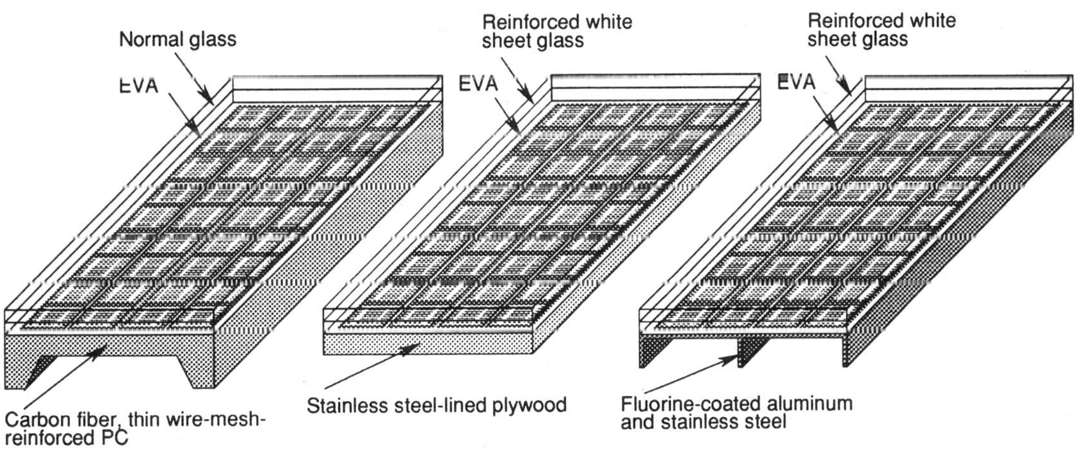

Fig.6. Various materials which can be used as a base

output of the PV module will drop. Moreover, when it is attached to building walls, that is, vertically, it is estimated that the angle of light incidence will increase as will the reflection coefficient. These points should be considered in estimating potential power output.

As for base materials, the precast concrete (PC) and metal substrate which has been used in high-rise building in the past was researched, and examination was made of carbon fiber, thin wire-mesh-reinforced PC, stainless steel-lined plywood, fluorine-coated aluminum and stainless steel, as shown in Fig.6. Examination also was made of a material called "top light" which has cells spaced out between two sheets of glass and functions as a window, as shown in Fig.7.

EXPERIMENTAL PRODUCTION OF MODULE SAMPLES

In order to carry out future market research and preparation of design more efficiently, samples of building-material-integrated modules were produced. Figure 8 shows the material in which a stainless steel sheet is fitted as a base into a PV module. It can be used as a material

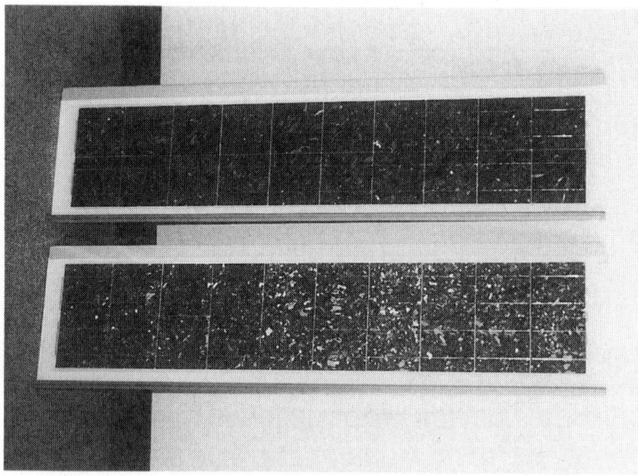

Fig.8. Module samples with stainless-steel base

Fig.9. Module samples with aluminum-sash frame

for roofs and walls. Figure 9 shows a sample of a PV modules fitted with an aluminum-sash frame. Its structure is almost the same as that of an ordinary glass curtain wall. 8mm-thick glass was used for that sample. Close examination of the appearance of these samples demonstrates that the color of back sheets seen between the cells might influence the design of curtain walls. Therefore, it is necessary to evaluate several color combinations, or in some cases, to consider the shape of interconnectors or electrodes.

CONCLUSION

We considered the design of a PV module as a building material and produced solar cell samples of various colors, substrates and surface conditions. Renderings were also prepared using CG. As a result of an opinion survey using those samples, it was confirmed that PV modules using color solar cells have great potential as building materials. In general, the output of cells was seen to decrease for colors other than shades of dark blue. However, there is a good possibility that greater importance will be put on design, including color and surface condition, than on output. In this case, the market share may expand and, as a result, the cost may drop. We also considered the actual structure of the PV module incorporated into building materials and produced some samples.

The designs and structures of building-material-integrated modules will be closely examined further The method of installation, construction and maintenance will also be examined. In addition, we will conduct demonstration tests by installing such modules on a building wall. Furthermore, to estimate potential power output, the amount of solar radiation falling on buildings' walls, and the effects of glass thickness and of angles of sunlight incidence on module output will be examined.

ACKNOELEGMENT

This work was supported by the New Energy and Industrial Technology Development Organization as a part of the New Sunshine Program under the Ministry of International Trade and Industry of Japan.

REFERENCES

[1] Joseph H. Apfel, "Graphics in Optical Coating Design", *APPLIED OPTICS*, **11**, 1972, pp. 1303-1322.

SITING, INSTALLATION, AND PERFORMANCE ISSUES FOR COMMERCIAL ROOF-MOUNTED PV SYSTEMS

J. Berdner and C. Whitaker
Endecon Engineering

H. Wenger
Pacific Energy Group

C. Jennings
Pacific Gas and Electric Company

E. Kern and M. Russell
Ascension Technology

ABSTRACT

Pacific Gas and Electric Co. is one of eleven utilities participating in the first phase of the U.S. Environmental Protection Agency's (EPA) Photovoltaic Demand Side Management (PV-DSM) Program coordinated by Ascension Technology Inc. (ATI). [1] An extensive effort was made by PG&E to install the nominal 10-kWac EPA Photovoltaic (PV) system on a customer's building. Problems arose related to permitting, code compliance, roofing materials, and solar access. As a result, the design was modified slightly and the system was installed on a PG&E-owned building in December 1993.

Viability of roof-mounted PV will require that the system installation be fully code compliant, straightforward and routine. This paper identifies issues and suggests solutions regarding the installation of utility-grade PV systems on customer premises.

INTRODUCTION

PG&E's primary objective in participating in the EPA PV-DSM program was to demonstrate the technical viability of reducing customer electrical loads with roof-mounted PV systems. These types of PV-DSM systems could be another customer option within the context of a traditional demand-side management program. The utility would encourage the installation of PV-DSM systems in areas where large T&D-related capital expenditures were planned, thereby maximizing utility T&D benefits while providing customer demand-side reduction benefits. [2]

The siting and installation process of PG&E's PV-DSM system illuminated several unforeseen hurdles that should be addressed to facilitate wide-spread utility integration. This paper discusses the lessons learned in the siting and installation process.

SYSTEM DESCRIPTION

The system array consists of 252 Siemens Solar M55J single crystal silicon modules rated at 13.35 kWdc at STC[1]. The modules are pre-panelized and wired with quick disconnect bayonet type connectors. The panels are mounted on ATI's innovative ballasted roof jack mounting structure, shown in Figure 1, which is designed to eliminate the requirement for roof penetrations. The combination of the pre-wired panels and the roof jack system greatly reduce installation time compared to more conventional methods where the array is mechanically attached to the roof.

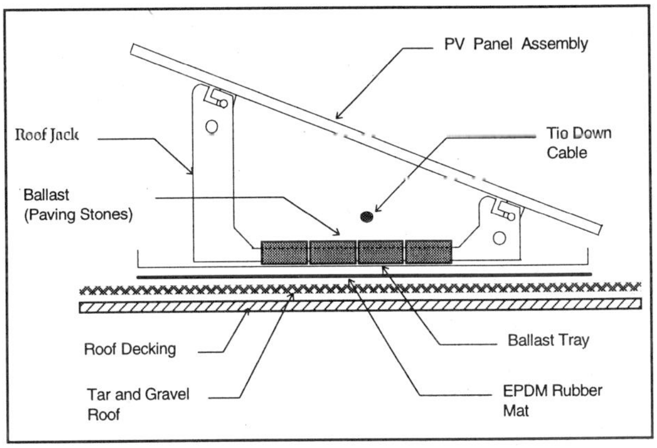

Figure 1 - Ballasted Structure System

The modules are mounted at a 15 degree tilt angle and are oriented 10 degrees east of south. The arrays are connected to three Omnion series 2200 single phase inverters each rated at 4 kWac at 120 Vac. The total system output is approximately 10 kWac @ PTC[2].

The system performance is monitored using a rotating shadow band pyronometer and Campbell datalogger supplied by ATI. The total building load is monitored by a separate Campbell datalogger located at the main service meter.

SITE SELECTION

There were essentially three primary criteria for siting the PV-DSM demonstration system: (1) an area that is approaching T&D capacity constraints and where upgrades are planned (this maximizes utility value); (2) a strong correlation between PV output and building load (this maximizes customer value); and (3) site specific considerations such as available roof area, obstructions,

1 STC = 1000 W/m2, 25° C cell temp, 1.5 AM

2 PTC = 1000 W/m^2, 20° C ambient, 1 m/s wind speed.

CH3365-4/94/0000-0981 $4.00 © 1994 IEEE

First WCPEC; Dec. 5–9, 1994; Hawaii

building orientation (solar access), roof type, and the ability of the roof to support additional dead-loading.
Prior to the final site selection, the Livermore-Pleasanton T&D planning area (30 miles southeast of San Francisco) was identified as meeting the first siting criteria. To meet the second criteria and simplify selection, the search was narrowed to low rise commercial office space, with flat roofs. Low-rise commercial office building loads tend to be highly correlated with solar output in PG&E's service area.

The third siting criteria proved to be the most difficult and uncovered several issues that serve as the focus of this paper.

Area Requirements and Solar Access

The roof area required for the 10-kW array is approximately 2,100 square-feet. Additionally, the space to the south, west, and east of the array must be free of obstructions which could shadow the array. The most common obstructions are HVAC equipment placed in the center of the roof to minimize ducting runs. Other obstructions include ventilation fans, skylights, vent stacks, and parapet walls. Although the roof-top PV system, with its low tilt angle, is designed to maximize summer energy production when shadowing is not typically an issue, a solar estimating tool, such as the solar pathfinder, is essential to evaluate the extent of shadowing due to nearby obstructions.

To maximize the value of the energy to the customer, the array should be oriented to match the building load profile. For low-rise commercial office space in the PG&E service territory, the peak building load typically occurs in the mid to late afternoon. The optimum array orientation is therefore between true south to 30 degrees west of true south.

Roof Types

Roof construction methods vary regionally within the United States. In the eastern half of the country membrane roofs with stone ballast are common, and these roofs are normally designed for significant dead (snow) loads. ATI's ballasted mounting system is well-suited to this type of construction.

In the western part of the country, built up, modified bitumen with mineral cap and tar with impregnated gravel roofs are the most common. While it is possible to use the ballasted system with the built up and modified bitumen roofs, there is still some disagreement between roofing manufactures over the system's impact on roof warranty. The ballasted system is not well suited to tar with impregnated gravel as it is nearly impossible to remove all of the gravel form the tar before installation. Gravel trapped under the pan could be pressed down through the tar creating a possibility for leakage. In addition, the pans become quite warm during the hot summer months and would likely adhere to the exposed tar surface complicating roof maintenance or re-roofing when required.

Until the warranty issues have been resolved for built up and bitumen roofs, a strong preference should be given to membrane roofs when evaluating potential sites suitable for a ballasted PV array design.

Significant snow loading is also relatively uncommon and western roofs are not normally designed for large dead loads. In some areas the roof construction practices would not support the additional dead loading imposed by the ballasted structure. The ability of the roof to support the additional dead loading must be determined early in the site selection process. This typically requires some limited, site-specific engineering for several candidate sites.

PERMITTING AND CODE COMPLIANCE

All rooftop-mounted installations are subject to local and national building codes. Several candidate buildings were evaluated during the siting process and permit requirements were investigated for the two most promising sites. Local building inspectors were invited to preliminary site meetings and were provided with a preliminary design package supplied by ATI.

In both cases, the local building inspectors required additional site-specific engineering drawings and calculations to be included with the permit application. In addition to a basic PV system design package, the following site-specific documents were required to as part of the permitting process.

- Roof plan showing array location.
- Structural details of the roof structure including attachment points.
- Wet-stamped structural engineering drawings and calculations for the additional dead load, wind load, and seismic load imposed on the building by the PV system.
- Single line electrical drawings showing system wiring and identifying the point of interconnection.
- Elevation views of the arrays for planning (visual impact) purposes.
- Manufacturer's data sheets for the major system components or product certifications.

Two significant code compliance issues were uncovered during the permit application process.

The first was Article 90 of the National Electric Code (NEC) and the requirement for listed items [3]. To facilitate a timely installation and avoid factory testing the decision was made to relocate the system from a customer-owned to a utility-owned facility.

The second code compliance issue concerns Chapter 1630 of the 1994 Uniform Building Code (UBC) [4]. This issue remains unresolved and calls into question the appropriateness of the ballasted structure approach in many areas.

National Electric Code

The NEC is widely accepted as a minimum safety code for electrical installations. It has been adopted and is considered law in 40 states. In some cases, state and local codes will also impose additional requirements that augment the NEC.

The applicability of the NEC is described in Article 90-2. In general, the Code covers all installations of electrical conductors and equipment within or on public and private buildings or other structures. Generation, transmission, and distribution equipment that is owned and operated by an electric utility is normally exempted from NEC requirements. Although the PG&E PV-DSM system is utility-owned generation equipment, the code compliance issue is complicated by the fact that the equipment was to be installed on a customer's facility.

While it is possible to design and install systems that would be Code exempt, clearly future systems should be designed to meet NEC requirements. The NEC is adopted within PG&E's service territory, and several different electrical inspectors required that the PV system be NEC compliant. Code compliant systems minimize liability, simplify installation, and maximize the number of potential installations.

Article 90-6 of the NEC requires that all equipment covered by the code must by examined for safety by a qualified testing organization. Qualified testing organizations are nationally recognized testing laboratories such as Underwriter's Laboratory (UL), Edison Testing Laboratory (ETL) and others. Once examined, the products become listed or labeled by that laboratory.

During PG&E's permit application process, the building inspector correctly noted that the inverters were unlisted and requested that they be tested in either the formal testing or a field listing procedure. At that time ATI, with assistance from Sandia National Laboratories, was in the process of submitting the inverter and all of the other electrical system elements to UL for testing. Unfortunately, the UL listing process is quite lengthy and the project schedule did not permit the delay. To meet the project timeline the decision was made to install the system on a utility-owned research facility thereby avoiding the requirements for listed products. Ultimately, UL listing was obtained for all of the system components and ATI's system is now one of the only fully NEC compliant roof-top PV packages available. Use of unlisted items is no longer an issue for new installations of the ATI system but this experience highlights the need for listed components in other systems.

Uniform Building Code - Seismic Loading

Chapter 1630.1 of the 1994 edition of the Uniform Building Code covers lateral forces on non-structural components or equipment that is supported by structures. This chapter stipulates that friction resulting from gravity loads can not be used to provide resistance to seismic forces. The prohibition against use of friction in seismic load calculations effectively requires the system to be mechanically attached to the roof.

The total lateral seismic forces are described in Chapter 1630.2 and shown by equation (1) below:

$$Fp = Z * Ip * Cp * Wp \qquad (1)$$

where:

Fp = total lateral force due to seismic loading.

Z = seismic zone factor. (Z ranges from 0.075 in seismic zone 1 to 0.40 in seismic zone 4. Most of California's urban areas are in seismic zone 4.)

Ip = importance factor. (Ip is 1.0 for most structures and 1.5 for essential or hazardous facilities.)

Cp = horizontal force factor. (Cp is 0.75 for electrical or mechanical equipment.)

Wp = weight of the individual component.

The weight of the ATI 10 kW, 108 m^2, PV array is 8,775 lb. plus an additional 11,250 lb. for ballast (or approximately 1,000 lb/kW). The lateral force due to seismic loading for this 10 kW array mounted on a typical building in a seismic zone 4 area is shown in equation (2) below:

$$Fp = (0.40)*(1.0)*(0.75)*(Wp)$$

$$= (0.30)\ (Wp) = 6{,}008 \text{ lb. (array with ballast)} \qquad (2)$$

To meet the UBC requirement for attachment, PG&E civil engineers designed a tie down cable, shown in Figure 1, which was added to the ATI system. The tie down cable was run down the length of each row of the array and anchored to a custom fabricated steel stantion at each end. The stantions are welded to the underlying roof supporting structure and project up through the roof surface to the cable attachment point. The array is configured in five rows requiring a total of ten roof penetrations.

The wind loading calculations outlined in the next section, show that the seismic loads for the ballasted array in a zone 4 area are similar in magnitude to the lateral forces due to wind loading. Since the attachments are required to meet seismic loading, it may be possible to utilize the same attachments to resist the wind load forces. The ballast could then be eliminated and the dead loading on the roof could be significantly reduced. Additional study should be completed to develop an unballasted structure which minimizes the number of roof penetrations.

Wind Loading

In their preliminary design package ATI shows that the array ballast is needed to resist the sliding rather than uplift forces generated by wind loading. The sliding (or lateral) forces due to wind loading are given as 207 lb. per panel or 7,452 lb. for a 10 kW, 108 m^2, array. As noted above, the sliding force is similar in magnitude to the lateral forces due to seismic loading found in equation (2).

The wind loading calculations supplied in ATI's design package are based on an ASCE method [5]. Additional calculations were performed using the methods in Chapter 16 of the 1994 UBC. In both cases, the ballast requirement was determined by the sliding rather than overturning forces. As can be seen in Table 1, there is a significant variance between the two methods. If the UBC method were used the ballast requirement, and the resultant dead loading on the roof would increase dramatically. The additional loading would have eliminated both of the candidate sites where the site specific engineering calculations were completed. Further study is required to develop standardized wind loading methodology which applies specifically to rooftop mounted PV systems.

Table 1 - Ballast Requirements for Wind Loading

	ASCE		UBC	
	Panel	Array	Panel	Array
Ballast (lb.)	254	9,144	581	20,912
Total (lb.)	469	16,884	796	28,656
Roof Load (psf)	14.6	14.6	24.7	24.7

PV System Performance

The PG&E PV-DSM demonstration system is located on a flat-roofed two-story building with occupancy and load patterns typical of commercial office space. A successful PV-DSM system (and program) will have: (1) output that is highly correlated with building loads and utility T&D capacity requirements, particularly during peak demand; (2) high capacity factor to maximize customer value; and (3) minimum down-time.

From the customer's perspective, a PV-DSM system can provide energy and demand charge savings. Figure 2 presents measured building load profiles (scaled down by a factor of five for illustration purposes.) These load profiles correspond to the days when customer demand charges are calculated. The PV-DSM system consistently shaves building peak loads, yielding measured peak 15-minute demand reductions of 6, 8, 7, 8, 7, and 5 kW for the six PG&E-defined peak months of May through October.

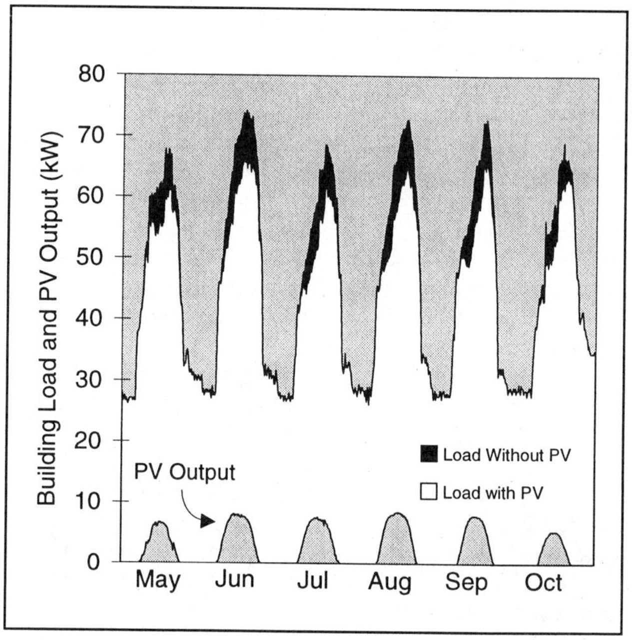

Figure 2. Peak load shaving on demand charge days

If the PV-DSM system was installed on a PG&E customer's premises, electric bill savings would be approximately $2,000 in the first year, of which 15 percent would come from demand charge savings. The remaining 85 percent would come from about 19 MWh of energy savings each year. Since energy production drives utility bill savings, peak demand clipping with battery systems is not recommended. Battery storage for customers on time-of-use billing is, however, worth considering.

From the utility's perspective, a PV-DSM program can provide traditional energy and capacity savings, and non-traditional benefits such as the deferral of capital expenditures on T&D equipment.

Figure 3 presents the capacity and load of the distribution feeder that serves the PV-DSM building, on the peak load day (August 15, 1994) PV-DSM output is scaled by 500 times to represent a 5 MW PV-DSM program. Since feeder capacity declines as a function of ambient temperature and solar irradiance, there is a good match with PV output. The 5 MW of PV-DSM rated capacity provides an effective feeder capacity boost of 2.9 MW. This demonstrates the ability of PV-DSM to potentially defer T&D equipment upgrades such as distribution feeders.

From both the customer and utility perspectives, PV-DSM systems must have high availability and require little maintenance. PG&E's PV-DSM system has encountered only one O&M problem: one of the three inverters was shipped back to the factory for repairs which took about two weeks. System availability has been 100 percent during daylight hours: there has always been power production capability. The system is performing as designed and, with the exception of the single inverter repair, has required no maintenance during the first 10 months of operation.

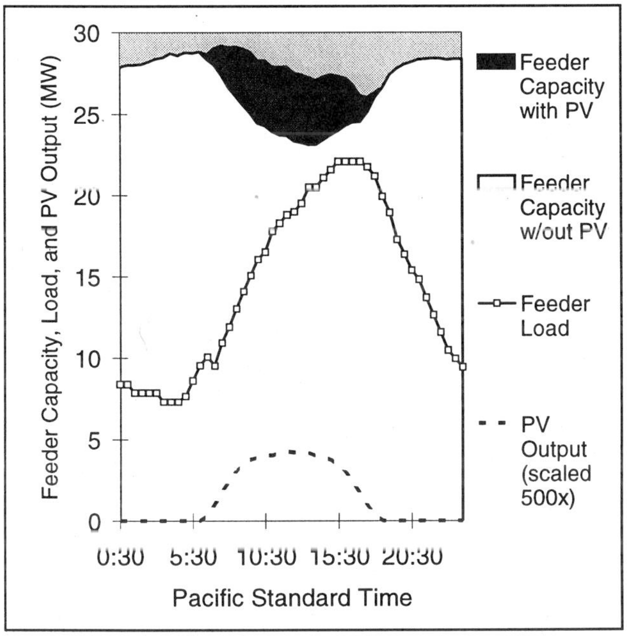

Figure 3. Effect of PVDSM on feeder capacity.

Other Issues and Recommendations

Although ATI's ballasted structure may not be appropriate for all installations, many aspects of the system are significant improvements over traditional mounting methods. Additional effort should be devoted to developing a low cost mounting method that will incorporate these improvements while meeting the UBC's seismic requirement for mechanical attachment to the roof.

ATI's system now meets the NEC requirement for listed items eliminating this as an issue for future installations of the ATI system. Other system integrators and inverter manufacturers are beginning to recognize the necessity of listed equipment and listed equipment should become more widely available in the near future.

The regional variations in code requirements continue to make code compliance an issue. The methods used for calculation of wind loading on roof mounted PV arrays yield substantially differing results. In general however, it appears that structures designed for wind loading will meet the seismic loading requirements in all seismic zones. Uniform wind loading methods should be developed to allow standardization of a low cost structure design.

The integration of the PV system to the existing building wiring could be simplified if the inverter output voltage was 277/480V rather than 120/208V. 277/480V circuits, used for HVAC loads, are more common on the rooftops of commercial buildings. A 277/480V PV system could be easily interconnected to these circuits with a minimum of modification to the existing building wiring. In addition, the inverters should be designed with outdoor rated enclosures and temperature ratings which would allow them to be mounted directly on the rooftop near the existing air conditioning equipment.

The system performance is excellent and, with only one exception, has required no maintenance. The correlation with building and feeder load is very high demonstrating peak demand savings, high capacity factor, and ability of a PV-DSM program to defer T&D capital upgrades.

---

[1] Kern, E.C., and R. Spiegel, "Photovoltaic Demand-Side Management Cost-Shared Demonstrations", Solar Energy Forum, Washington, D.C., 1993.

[2] Wenger, H., Hoff T., and R. Perez, "Photovoltaics as a Demand Side Management Option: Benefits of a Utility-Customer Partnership", World Energy Engineering Congress, Atlanta, GA, October 1992.

[3] "National Electric Code ", 1993 edition, National Fire Protection Association

[4] "Uniform Building Code", 1994 Edition, International Conference of Building Officials

[5] "Wind Forces on Structures", American Society of Civil Engineers, paper #3269, 1969.

A NEW FLEXIBLE A-SI PV MODULE AND ITS APPLICATION TO ROOFTOP PV SYSTEMS

Yukimi Ichikawa, Takuro Ihara, Toshio Hama, Yujiro Watanuki, Shinji Kato, Hiromitu Ota and Hiroshi Sakai
Fuji Electric Corporate Research and Development, Ltd.
2-2-1, Nagasaka, Yokosuka, Kanagawa 240-01, Japan

Norikazu Sakai, Junichi Kurihara and Osamu Ishikawa
Misawa Homes Institute of Research and Development Co., Ltd.
1-1-19, Takaido-nishi, Suginami-ku, Tokyo 168, Japan

ABSTRACT

A novel photovoltaic(PV) module for roof top systems, solar roofing, was proposed. Solar roofing is a flexible amorphous silicon(a-Si) PV sheet having the function of roofing. Tempered glass is used as roof covering material. Technical items for roof top systems using solar roofing were discussed. Preliminary studies on module component materials showed feasibility of solar roofing. We designed a construction technology for tempered glass covered roofs. Effects of shadows on PV module upon its performance were also analyzed.

INTRODUCTION

It has been considered in Japan that application of PV systems to roof top of private houses will open a way to a big market for PV industries. Cost of PV systems is one of key factors for their application to private houses. It is important for installation of solar cells on the roof to take account of appearance of roof, easiness of roof construction method and saving of building materials. Thus it is necessary to develop low cost PV modules that are integrated with construction materials. As such specially designed PV module, we proposed a new PV module named as solar roofing, a flexible a-Si PV sheet having the function of roofing. Its basic structure is that flexible a-Si solar cells with plastic film substrates are sandwiched by a damp-proof cover sheet and a waterproof rear sheet. Solar roofing is rather light in weight and easy in treatment than other PV modules using mono- and poly-crystalline silicon solar cells and a-Si solar cells with glass substrates. In this article, we will discuss a structure of solar roofing and a roof construction method using this PV module.

SOLAR ROOFING

Concept of Solar Roofing

The conventional roofing is a waterproof sheet put upon a roof board. Roof cover materials such as tiles and slates are fixed on it. When we apply PV modules as roof cover materials, we must account of the installation and wiring methods for PV modules on the roof top. This suggested that carpenters could not apply their skills to PV roof and this might be an obstacle for diffusion of roof top PV systems. It is necessary to develop a new PV product so that the conventional roof construction techniques can be used. For this purpose, we proposed a solar roofing,

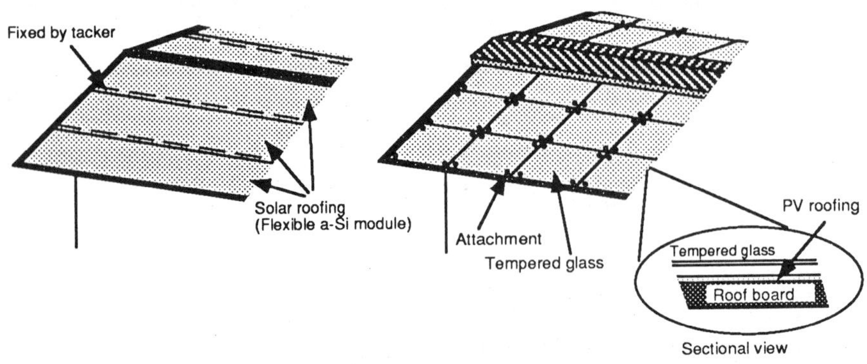

Fig. 1. Roof structure with solar roofing.

CH3365-4/94/0000-0986 $4.00 © 1994 IEEE

First WCPEC; Dec. 5–9, 1994; Hawaii

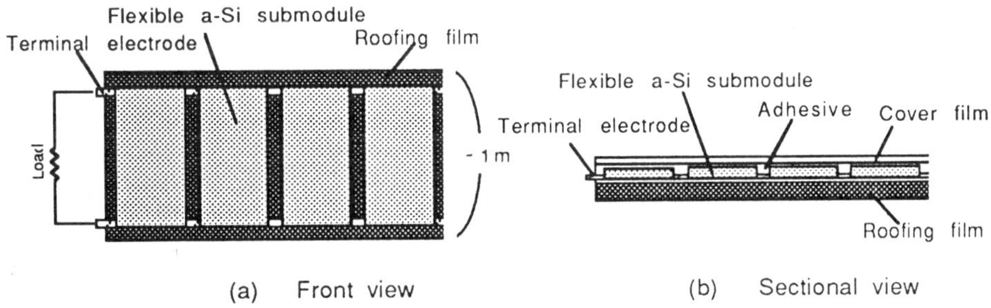

Fig. 2. Structure of solar roofing.

a flexible a-Si PV sheet having the function of roofing.

Figure 1 shows a basic idea of solar roofing. Because its installation method on a roof is designed to be the same as the conventional roofing sheet, we can fix solar roofing on the roof board by tackers or nails(Fig.1(a)). Tempered glass is put on it as a roof covering material(Fig.1(b)).

Structure of Solar Roofing

A basic structure of solar roofing is shown in Fig. 2. Flexible a-Si submodules(solar tiles) connected in parallel are sandwiched by a cover film and a roofing film. The roofing film at rear side guarantees the same waterproofs as the conventional roofing sheet. The cover film is damp-proof to suppress the degradation of photovoltaic properties of solar roofing by the effect of moisture from the surface side. Because each solar tile has a series-connected structure on a large area substrate to generate enough out-put voltage, we can pick up required voltage for an inverter of this roof top PV system from each solar roofing. We develop a novel series-connected structure called as SCAF(Series Connection through Apertures on Film) for the solar tiles with plastic film substrates[1].

Material Research for Solar Roofing

We carried out mechanical tests and environment test on component materials of solar roofing. Solar roofing has to be so flexible as to stick to the roof board closely. Solar tile is flexible. We studied bending properties of component materials such as cover film, adhesive and roofing film, to estimate their flexibility. Figure 3 shows

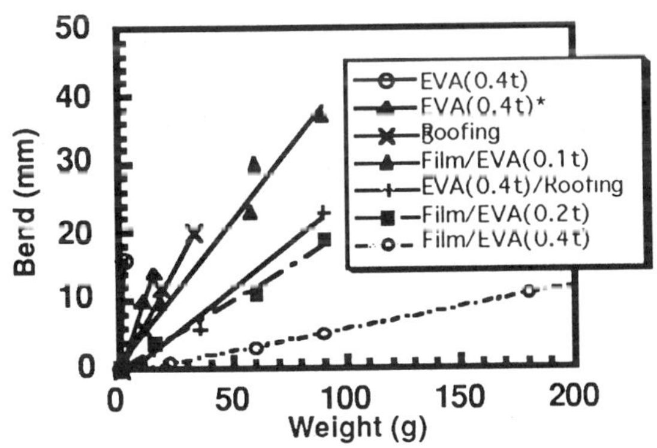

Fig. 3 Results of bending test for various materials.

the results of an experiment by a one-edge fixed beam method. As a reference for the flexibility of the conventional roofing, we used rubberized asphalt roofing. EVA(Ethylene vinyl acetate) is widely used as adhesive for PV modules. Bending property of EVA depended on its thickness. EVA with a thickness of about 1.4mm showed similar bending as the asphalt roofing. Bending of EVA decreased when combined with a damp-proof film. It was found that EVA less than 0.2mm thick with the damp-proof film showed similar flexibility as the combination of EVA and the roofing. From the experiment on damp-proof films combined with EVA, it was found that CPVF(Clear polyvinyl fluoride) was rather flexible than other films such as PAR (Polyarylate), PPS(Polyphenylene sulfide) and so on.

We carried out a thermal cycle test as a preliminary environment experiment. We used a sample module in which a flexible a-Si cell was sandwiched between CPVF

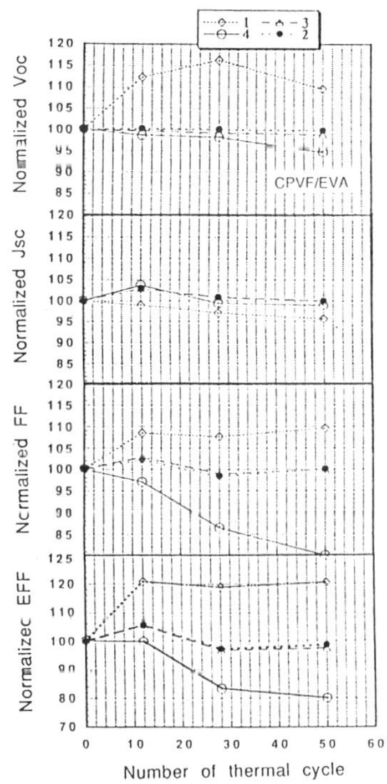

Fig. 4 Change of I-V characteristics of sample modules with CPVF and EVA.

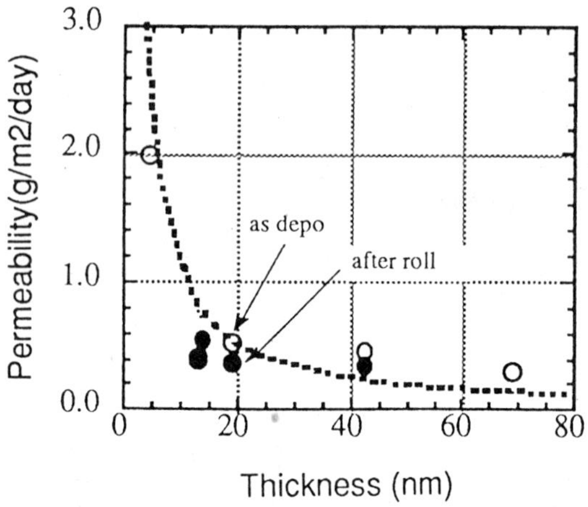

Fig. 5 Moisture permeability of SiOx prepared on PES films by rf sputtering method.

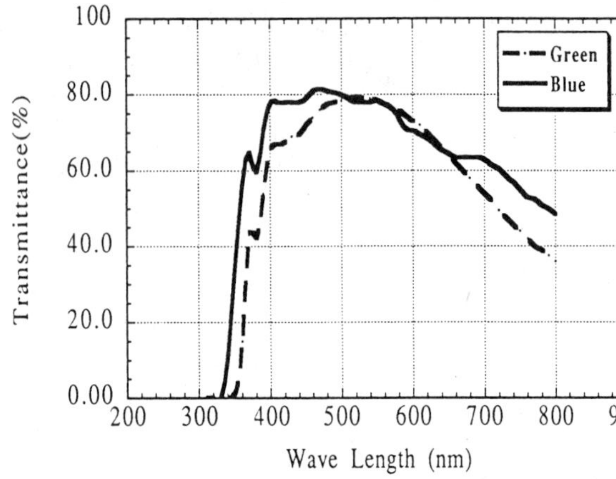

Fig. 6 Transmittance for green- and blue-colored glasses.

film as a cover film and PVF film as a rear film. EVA was used as an adhesive agent. Results of thermal cycle test of four sample modules are shown in Fig. 4. On the average, I-V characteristics for sample modules were sustained after 50 cycles of tests(-20°C - 80°C in 6 hours per cycle). This suggested that the proposed structure of solar roofing is feasible for practical use.

It is important to protect diffusion of moisture into solar tiles. It is well known that silicon oxides or silicon nitrides are efficient materials as a barrier to moisture. For this purpose, we studied the properties of silicon oxide(SiOx) films deposited by rf sputtering method. The condition of sputtering is the following: rf discharge power density of 1.23 - 3.70W/cm^2, Ar gas pressure of 5×10^{-3} - 2×10^{-2}torr, no substrate heating and targets of SiO_2 and SiO. Figure 5 shows a permeability of moisture for SiO_x films deposited on PES(Polyether sulfone) films. The refractive indices of the SiO_x films were 1.4- 1.5 and the absorption coefficient in the range of visible light was small. White circles show data of films as deposited and black circles show data after a roll test in which we coiled a film on a pipe with a diameter of 18mm. As shown in the figure, we obtained a permeability less than 0.5g/m^2/day for SiOx films with thickness larger than 20nm. A degradation of the permeability was not seen in this roll test. The results suggested that SiOx films deposited by rf sputtering method were useful as surface coating materials to improve damp-proofs of solar roofing or solar tiles.

DESIGN OF ROOFTOP STRUCTURE

For users, shapes and colors of the roof are the most important factors as well as floor planing in their house plan. Solar roofing, roofing having the function of photovoltaics, enables us to design the roof by various kinds of roof cover materials. Our basic plan is that the construction of the PV roof with solar roofing will be as similar as possible to the conventional roof, available to any kinds of roofs and superior to the conventional one. In the concrete, (1)the design and execution of the roof will be similar to the conventional roof, (2)an increasing freedom of design with a combination of roof cover materials, (3)a user as well as a designer can consider solar cells separately from the house, (4)it is applicable to reroof a house.

Table 1 Technical subjects for roof construction method

Plans	Technical subjects
Design combined with roof materials	·Design of glass ·Transmittance of frosted glass and colored glass and their effects on photovoltaic performance of PV roofing
Frameless.glass	·Point-supporting of glass ·Thermal condition (temperature and humidity) in a ventilation layer ·Waterproof structure for the roof
Applicability to any types of roofs	·Cutting solar roofing anywhere to fit the size of the roof ·Effect of shadows by the roof constituents
No wiring work on the roof (wiring and troublesome works inside the house)	·Wiring of solar roofing

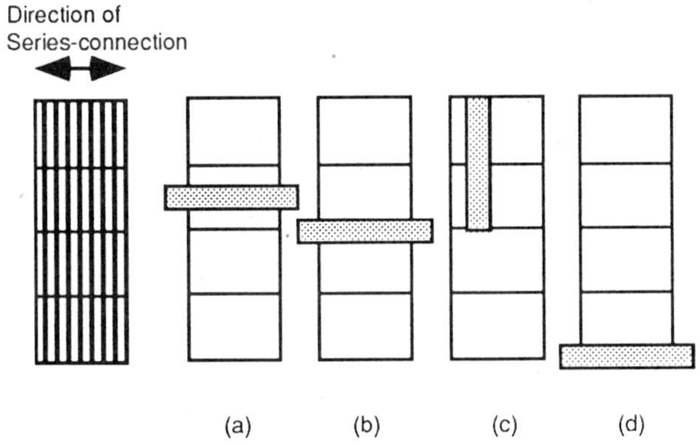

Fig. 7 Four types of shadows on a test module.

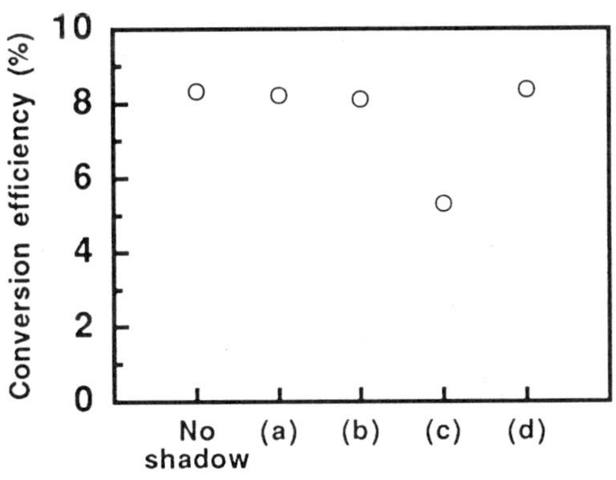

Fig. 8 Conversion efficiencies of a test module with four types of shadows.

In accordance with the above plan, we extracted technical subjects for the roof construction method as shown in Table 1. Design of tempered glass, colors and types, is important to enrich design for the roof. Figure 6 shows results of transmittance measurement on green- and blue-colored glasses. From these spectra, we calculated the decrease in the output power of solar roofing to be about 30% for both glasses. Even so, we consider the improvement in appearance has much advantages for solar roofing to be used as construction material

As tempered glass without frame is used on solar roofing, we designed a point-supporting method of glass as shown in Fig. 1(b) Supporting fixtures for tempered glass are set at edges of solar roofing. Transparent gaskets are put between the glasses to ensure waterproofs of the rooftop. A space between solar roofing and tempered glass is for ventilation to cool solar tiles.

There is ever possibility of shadows caused by parts of the roof. We studied the effects of the shadow on the conversion efficiency of PV modules. In this experiment, we used a PV module in which the direction of series connection in the submodule is orthogonal to that of parallel connections of submodules as similar as solar roofing. Figure 7 shows four types of shadows on PV modules and Fig. 8 shows the conversion efficiencies in each cases. When a shadow was formed along the direction of series connection in submodules as in Fig. 7(a), (b) and (d), decrease in the conversion efficiency of the PV module was small. In this case, the output voltage of the module was not influenced with the shadow. On the other hand, the conversion efficiency dropped largely for the case of the shadow formed as in Fig. 7(c). For this case, the output voltage of the module decreased significantly by the effect of the shadow. As all solar tiles are connected in parallel in a solar roofing and all solar roofing is connected in parallel, the effect of shadows formed along beams of the roof would be very small. This is one of caharacteristic features of solar roofing.

SUMMARY

A new PV module for rooftop systems, named as solar roofing, was proposed. Solar roofing was a flexible a-Si PV module having the function of the conventional roofing. We carried out material research on module components for solar roofing. It was found that a combination of EVA of thickness less than 0.2mm with a damp-proof film of CPVF showed enough mechanical flexibility for roofing. A preliminary heat cycle test on the sample modules showed the feasibility of the solar roofing for practical use. We also showed that SiO_x film formed by rf sputtering method was useful for surface coating against diffusion of moisture into solar tiles.

Our basic plan is that the construction of the PV roof with solar roofing will be as similar as possible to the conventional roof and available to any kinds of roofs. From this plan, we extracted and discussed technical subjects such as design of tempered glass, a point-supporting system for tempered glass, ventilation space between solar roofing and cover glass and effects of shadows on solar roofing.

ACKNOWLEDGEMENT

This work was supported by the New Energy and Industrial Technology Development Organization as a part of the New Sunshine Program under the Ministry of International Trade and Industry.

REFERENCES

[1] Y. Ichikawa et al., "A New Structure a-Si Solar Cells with Plastic Film Substrates", *First WCPEC*, 1994 (This conference).

LIGHTWEIGHT FLEXIBLE ROOFTOP PV MODULE

M. Izu, H. C. Ovshinsky, K. Whelan, L. Fatalski, and S.R. Ovshinsky
Energy Conversion Devices., Inc.
1675 West Maple Road
Troy, MI 48084

T. Glatfelter, K. Younan, K. Hoffman, A. Banerjee, J. Yang, and S. Guha
United Solar Systems Corp.
1100 West Maple Road, Troy, MI 48084

ABSTRACT

Energy Conversion Devices, Inc. (ECD) and United Solar Systems Corp. (United Solar) are developing lightweight, flexible photovoltaic modules that can replace conventional roofing materials and be economically and aesthetically integrated into residential and commercial buildings. The modules will be fabricated from high-efficiency multi-junction a-Si alloy solar cells developed by ECD and United Solar. These cells are produced on thin, flexible, stainless steel substrates. Two types of products, 1 ft by 10 ft overlapping PV shingles and 1.3 ft by 20 ft PV roof panels are being developed by United Solar and ECD, respectively. United Solar's shingle type design uses a roof mounting procedures similar to those used with conventional asphalt shingles, while ECD's PV panel uses mounting procedures conforming to metal roof systems.Thus, they can be installed on roof sheathings, replacing ordinary shingles or metal roofing panels, on a standard wood roof construction.

INTRODUCTION

During the past fifteen years, ECD has made important progress in the development of materials, device designs and manufacturing processes required for the continued advancement of photovoltaic technology. Among these accomplishments, ECD has pioneered, and continues to develop, two key proprietary technologies with significant potential for achieving the cost goals necessary for widespread growth of the photovoltaic market: (1) a low-cost, roll-to-roll continuous substrate thin-film solar cell manufacturing process; and (2) a high-efficiency, multiple-junction, spectrum-splitting thin-film amorphous-silicon alloy device structure [1-10].

Commercial production of multiple-junction a-Si alloy modules has been underway at ECD and its American joint venture company, United Solar, for a number of years using ECD's proprietary roll-to-roll process and numerous advantages of this technology have been demonstrated. These advantages include relatively low semiconductor material cost, relatively low process cost, and the productoin of lightweight, rugged and flexible solar modules with reduced installed PV system costs.

ECD and United Solar are now developing lightweight flexible PV modules that can replace conventional shingles or other roofing materials and be economically and aesthetically integrated into residential and commercial buildings. The modules will be fabricated from high-efficiency, multiple-junction a-Si alloy solar cells developed by ECD and United Solar. These cells are produced on thin, flexible stainless steel substrates and encapsulated with polymer materials.

SCOPE AND METHOD OF APPROACH

Two types of products, 1 ft by 10 ft overlapping PV shingles and 1.3 by 20 ft PV roof panels are being developed by United Solar and ECD, respectively. United Solar's shingle type design utilizes a mounting procedure similar to that for conventional asphalt shingles, while ECD's PV panel design utilizes a mounting procedure similar to that for metal roofing systems. Both types of rooftop PV modules are lightweight and flexible. Consequently they can be installed on roof sheathings, replacing ordinary shingles or metal roofing panels, on a standard wood roof construction. The product design features and installation methods have been refined through discussions and reviews with building industry experts including National Association of Home Builders (NAHB) Research Center, Solar Design Associates, and Minoru Yamasaki Associates.

UNITED SOLAR'S SHINGLE MODULE

United Solar's roofing module is designed to emulate the conventional asphalt shingle in form, structure, function, and installation. This module's form is illustrated conceptually in Fig. 1. The module is meant to be used as the roofing material; additional shingles are not required to be underneath. We anticipate that the shingle module can be installed at the same time as a building is roofed with standard asphalt shingles, using similar techniques and achieving aesthetically appealing integration between the PV and asphalt shingles. Thus, the

CH3365-4/94/0000-0990 $4.00 © 1994 IEEE

First WCPEC; Dec. 5-9, 1994; Hawaii

module will become the roofing material, as well as supplying PV power. The installation procedures envisioned for this module have two distinct advantages: (1) a roofer can physically install the PV shingle using essentially identical procedures to that for conventional asphalt (roofing brads), and (2) the electrical interconnection of the PV modules may be performed as a completely independent task, thereby separating trades during installation. These two advantages combined offer the opportunity of introducing PV into the market without the need to develop a new trade.

A shingle roofing module offers considerable economic advantages when compared to other applications of photovoltaics since a mechanical support structure is not needed for the shingle module. The structure is present at no additional cost in the form of the existing roof deck. The need for module framing is also eliminated for the same reason. United Solar's thin, flexible PV material combined with module construction technology developed in this work can produce environmentally protected modules without the need for edge protection and/or framing.

Module Design Considerations

The new module's geometry is nominally 12 inches by 132 inches, the same width but longer than a conventional asphalt shingle. The module length can be somewhat arbitrary, but is dictated by factors such as weight, ease of handling, electrical connections, and integration with asphalt shingles. As is the case of conventional shingles, the PV shingles will also be overlapped in such a manner as to expose only 5 inches when installed.

A shingle module is composed of several PV sub-panels integrated by laminating the elements together. The sub-panels consist of various thin-film coatings deposited onto thin sheets of stainless steel. The materials used in the module are illustrated in Fig. 2. Environmental protection is provided by ethylene vinyl acetate (EVA) and Tefzel. A polyester-based material is also used on the back surface. The shingle also includes interconnecting wires and current bypass diodes (not shown). Electrical inter-module connections will be made below the sheaves of the roof. There will be one roof penetration for each module via the use of a sealed insert. The result will be a flexible PV module that bears a close resemblance to an asphalt shingle. The PV shingles will protect the roof in the same manner as standard shingles, that is by providing water run off. As indicated above, a clear separation of the building trades is envisioned for the installation of these modules. The roofer would tack down the shingles with customary roofing brads. Later, an electrician would subsequently complete the electrical connections in the attic spaces beneath the roof.

ECD'S METAL ROOF-TYPE MODULE

ECD has designed an aesthetically and mechanically integrated rooftop system using standard metal roofing materials and installation techniques already available in the metal roofing industry. Fig. 3 shows a module schematic. Fig. 4 shows cross-sections of construction components.

As illustrated in Fig. 3, the module size is approximately 1.3 ft by 20 ft. Depending on the application, the width may be varied from 1 to 2 ft and the length may be varied from 10 to 22 ft. Typically, the module consists of seventeen series-connected stainless steel strip cells encapsulated with EVA/Tefzel onto an approximately 1.3 ft by 20 ft standard metal roof panel. Both edges of the module will be rolled into 3/4 inch standing seams after encapsulation so that it will accept the standard metal roof mounting hardware. The module will be installed vertically on the slope of the roof. The length of the module can be adjusted so that it can fit the slope of the roof. The installation procedure is summarized in Table 1. The metal roof-type PV installation does not have horizontal seams. This maximizes watertight integrity. The vertical seams of the metal roof utilize standard metal roof watertight construction. This design does not require any holes on the roof for wiring. All the wiring is made on top of the roof in the ridge vent. These design features will result in ease of installation and reliability of the system, as well as low installed cost. The rooftop PV modules will be mechanically and aesthetically integrated with the adjacent standard metal roof as is seen in Fig. 4. The features and benefits of ECD's d1.3 ft by 20 ft metal rooftop module are summarized in Table 2.

ECD has designed and constructed a large-area 2.5 ft by 22 ft laminator to fabricate its 1.3 ft by 20 ft metal rooftop module.

Table 1. 1.3 FT X 20 FT Metal Roof Module Basic Installation

- → **Prepare Roof Using Standard Roofing Construction**
- → **Locate And Mount First 1.3' X 20' Module And Standard Ridge Vent Bracket With 3 Fasteners At Top Of Module**
- → **Attach Standard Clips Along Sides Of Standing Seam**
- → **Add Standard Metal Roofing, Ends And Flashing To Complete Roof Installation**
- → **Electrically Connect Modules, Accessible From The Top With Ridge Vent Removed Or From The Attic Space**

Table 2. 1.3 FT X 20 FT Metal Roof Module Features

- → **Mechanically And Aesthetically Integrates With Existing Metal Roofing Systems**
- → **Leakproof And Durability Features Of The Best Metal Roofing Systems**
- → **Minimum Wiring Connections -- Wiring Can Be Totally Above The Roof**
- → **Installed By Metal Roofers Using Same Techniques As Metal Roofing**
- → **Installs On Any Roof**
- → **Metal Roof Market Has Grown 270% In 5 Years**

ACKNOWLEDGMENT

The project is supported by DOE/NREL's PV Bonus Program, Building opportunities in the U.S. for

Photovoltaics under subcontract number, DE-FC02-93CH10571. Members of the project team include NAHB Research Center, Detroit Edison, Arizona Public Service Company, Bechtel Corporation, Solar Design Associates, Minoru Yamasaki Associates and others.

The authors wish to express thanks to Joe Weihagen of NAHB Research Center and Steve Strong of Solar Design Associates for their helpful discussions and collaboration.

REFERENCES

[1] M. Izu and S. R. Ovshinsky, Production of Tandem Amorphous Silicon Alloy Cells in a Continuous Roll-To-Roll Process, SPIE Proc. 407, 42 (1983).

[2] M. Izu and S. R. Ovshinsky, Roll-To-Roll Plasma Deposition Machine for the Production of Tandem Amorphous Silicon Alloy Solar Cells, Thin Solid Films 119 55 (1984).

[3] H. Morimoto and M. Izu, Amorphous Silicon Solar Cells Production in a Roll-To-Roll Plasma CVD Process, JARECT 16 (1984); Amorphous Semiconductor Technology & Devices, North Holland Publishing Company, Edited by Y. Hamakawa, 212 (1984).

[4] S. R Ovshinsky, Roll-To-Roll Mass Production Process for Amorphous Silicon Solar Cell Fabrication, Proc. International PVSEC-1, 577 (1988).

[5] High efficiency Multiple-Junction Solar Cells using Amorphous Silicon and Amorphous Silicon-Germanium Alloys, J. Yang, R. Ross, T. Glatfelter, R. Mohr, G. Hammond, C. Bernotaitis, E. Chen, J. Burdick, M. Hopson and S. Guha, Proc. 20th IEEE PV Spec. Conf. 241 (1988).

[6] J. Yang, R. Ross, T. Glatfelter, R. Mohr, and S, Guha, Amorphous Silicon-Germanium Alloy Solar Cells with using Amorphous Silicon and Amorphous Silicon-Germanium Alloys, Proc. 20th IEEE PV Spec. Conf. 241 (1988).Profiled Bandgaps, MRS Symposium Proc. Vol. 149, 435 (1989).

[7] M. Izu, X. Deng, A. Krisko, K. Whelan, R. Young, H. C. Ovshinsky, K. L. Narasimhan, and S. R. Ovshinsky, Manufacturing of Triple-Junction 4 ft^2 a-Si Alloy PV Modules, Proc. of 23rd IEEE PVSC, 919, (1993).

[8] M. Izu, S. R. Ovshinsky, X. Deng, A. Krisko, H. C. Ovshinsky, K. L. Narasimhan, and R. Young, Continuous Roll-to-Roll Amorphous Silicon Photovoltaic Manufacturing Technology, AIP Conf. Proc. 306, 12th NREL PV Program Rev., 198, (1993).

[9] S. Guha, J. Yang, A. Banerjee, T. Glatfelter, K. Hoffman, S. R. Ovshinsky, M. Izu, H. C. Ovshinsky, and X. Deng, Amorphous Silicon Alloy Photovoltaic Technology - From R&D to Production. Proc. of MRS Spring meeting (1994).

[10] X. Deng, M. Izu, K. L. Narasimhan and S. R. Ovshinsky, Stability Test of 4 ft^2 Triple-Junction a-Si Alloy PV Production Modules, Proc. of MRS Spring meeting (1994).

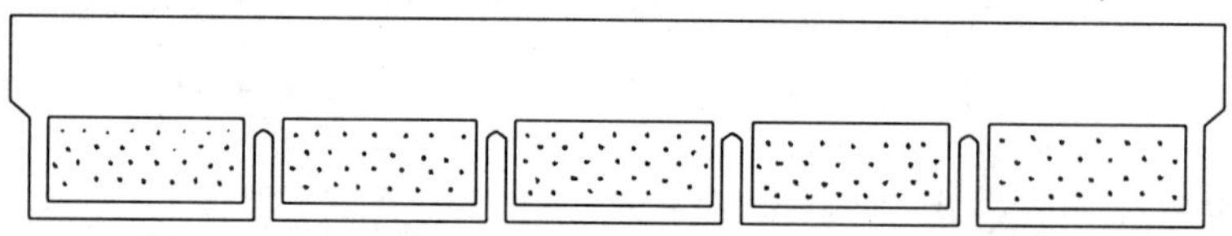

Figure 1 Schematic of shingle module footprint. The shaded area represents the exposed PV.

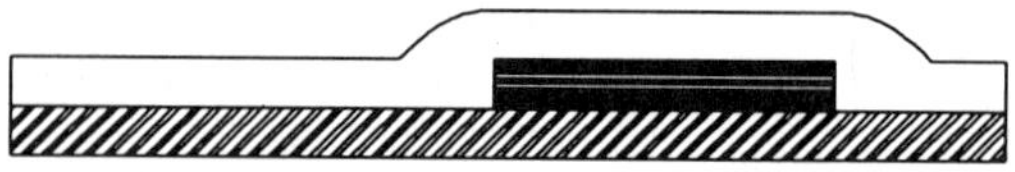

Figure 2 Shingle module materials stack-up

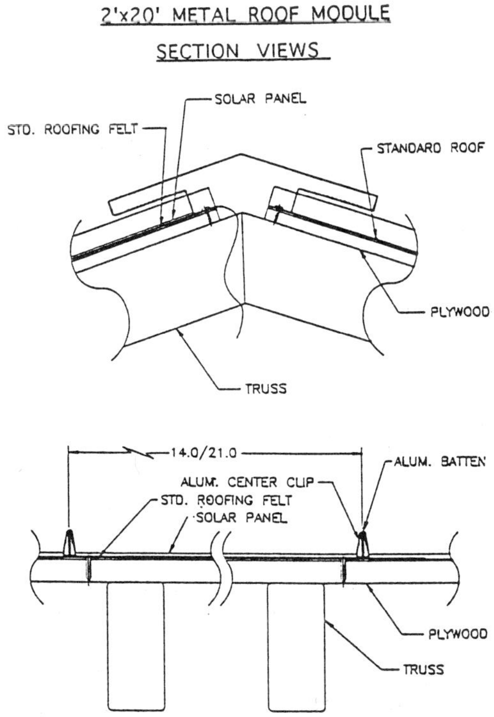

Figure 4 Metal Roof Module Section Views

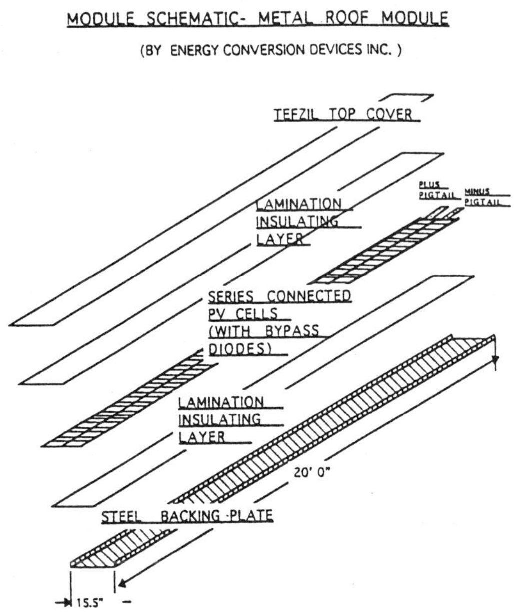

Figure 3 Module Schematic Metal Roof Module

Figure 5 A Model House with ECD's Rooftop Modules

EFFECTS OF INHOMOGENEOUS IRRADIATION DISTRIBUTION ON A PV ARRAY IN AN URBAN ENVIRONMENT

Anne Kovach
System Simulation Group
Fraunhofer Institute for Solar Energy Systems,
79100 Freiburg, Germany

ABSTRACT

The effect of inhomogeneous irradiation on building-integrated photovoltaic arrays is studied with the help of the simulation program, PVPACK, developed in this work. Particular attention is paid to the effect of partial shading and reflection on photovoltaic (PV) arrays from the surrounding environment. By incorporating the raytracing program, RADIANCE, the irradiation distribution on the building surface can be calculated with a high resolution. Simulation results show that a poorly planned lay-out design of a PV array on a building can lead to large energy losses. Such a simulation tool is a helpful aid for successful incorporation of PV in the building sector.

Introduction

In an urban area the integration of photovoltaics into a building facade or roof provides electrical energy for the building in a space-saving and aesthetic manner. The success of a building-integrated PV system is dependent in part, on a well-planned and optimized integration strategy. Both the aesthetics and the energetic performance of a building-integrated PV system must be considered in order to achieve a high acceptance level within the community.

Due to the electrical connections in a PV array, the decrease in energy output is a nonlinear function of the decrease in irradiation due to shading. Even small shadows on the order of a solar cell (100 cm^2) can cause a noticeable reduction in the energy output. In urban surroundings, shading obstacles such as trees, overhangs, telephone poles, and protruding parts of the building, as well as reflection occurring from neighboring windows or white walls near the PV array, can lead to inhomogeneous incident irradiation. Such inhomogeneous irradiation can cause electrical mismatching and hot spots leading to energy output losses and temperature increases. The overall energy efficiency of the PV array is reduced. Reflection, on the other hand, can lead to a higher annual energy output.

As a working tool, a simulation procedure was developed to analyse the effect of inhomogeneous irradiation on a PV array over time. In particular, a raytracing technique based on the program RADIANCE [1] was used to calculate the irradiance distribution over the PV array with a high resolution. The incident irradiation on the PV array and the resulting hourly energy output are calculated. In this paper the technique used to calculate the hourly irradiation distribution on the PV array is briefly described, and simulation results giving the annual energy output of building-integrated photovoltaic arrays are discussed for different integration strategies. With this technique especially the influence of small shadows on the energy output can be quantified.

Calculation of Irradiation Distribution

Due to the sensitivity of a photovoltaic array even to small shadows, an irradiation grid having a high resolution where shading occurs is an important criterion in evaluating the performance of shaded PV arrays. Due to

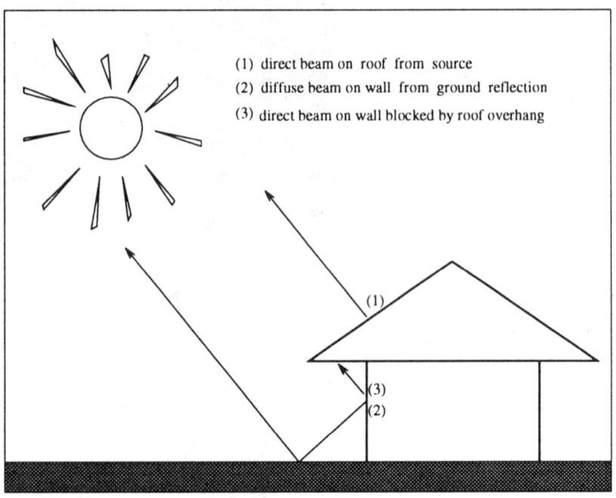

Figure 1: Image-oriented raytracing technique illustrated on a house.

its precision in calculating both shading and reflection effects on a building, the program RADIANCE, a daylighting program based on a technique called image-oriented

CH3365-4/94/0000-0994 $4.00 © 1994 IEEE

raytracing was chosen as the basis for calculating the hourly irradiation distribution on buildings for this work. This technique is illustrated in Figure 1. Solar rays incident on a house are traced back to their source. The points of incidence are designated in a grid defined by the program user.

Since RADIANCE is a daylighting program, several modifications were made to apply it for use with photovoltaic systems, which make use of a larger range of the spectrum. First, all calculations were performed assuming a gray world. Neither the spectral selectivity of solar cells nor of the surroundings was considered. Second, two calculations are carried out for each time step, one using a diffuse isotropic distribution and the other using a sky distribution with direct irradiation. Therefore, shading of diffuse and direct irradiation could be calculated separately. Third, the irradiance values calculated in the RADIANCE scene are used as relative values with respect to the diffuse and direct irradiation at an unshaded point. These relative values are then multiplied by the direct and diffuse irradiation from the site weather data for the corresponding sun position. Ultimately a grid containing the different irradiation values distributed over the building surface is calculated for each daylight hour over the year. For the example in this paper, the grid resolution was 5cm in the shaded portions of the PV array. These realistic irradiation values are used as input for the photovoltaic simulator which calculates the annual PV energy output for arrays experiencing inhomogeneous irradiation in hourly time steps.

Effective Irradiation

A useful measure of the effective use of the available irradiation incident on a building is defined as the effective irradiation, Φ', in this work. This factor represents the irradiation incident on the array as a weighted function of the collective irradiations on the shaded and unshaded portions of the array.

Φ' is calculated as shown in Equation 1.

$$\Phi' = \frac{A_{un}}{A_T}\Phi_{un} + \sum_{i=1}^{n}\frac{A^i_{sh}}{A_T}\Phi^i_{sh} \qquad (1)$$

In this equation A_T is the total array area and there are n different irradiations on n differently shaded portions of the array. Φ_{un} and A_{un} are the irradiation and area of the unshaded part of array, and Φ^i_{shad} and A^i_{sh} are the irradiation and the area respectively on the ith shaded part of the array.

For example, at 1000 W/m^2 full sun 75 % of the PV array area remains unshaded while 25 % of the PV array area is shaded and has an incident irradiation of 400 W/m^2. The resulting effective irradiation Φ' for this time period is calculated to be 850 W/m^2 from Equation (1). In this example n=1; there is only one shaded portion of the array.

Depending on the electrical connections of the PV modules or cells in the shaded area, different energy outputs can be generated for the same Φ'. Therefore, by comparing the energy output to the effective irradiation, different PV building-integration strategies can be considered and assessed, when partial array shading is inevitable for a given building.

Simulation Results

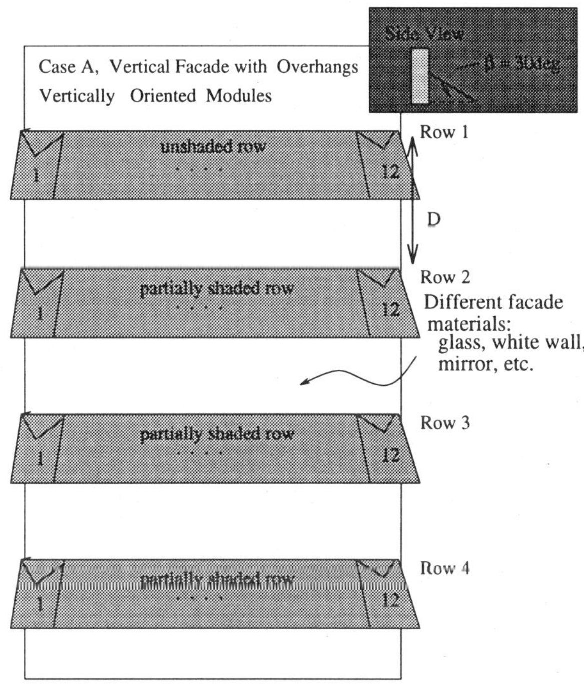

Figure 2: Schematic of PV-shading elements mounted on a south-facing vertical facade.

Using PVPACK the annual energy output was calculated for a 2.16 kWp PV array integrated onto four overhangs on a south-facing facade in Freiburg, Germany (48 °N) as shown in Figure 2. A series string containing 12 modules was mounted on each overhang. These four strings were connected in parallel. A parameter study was carried out investigating the different integration strategies. The tilt of the overhangs, β, was varied between 0 and 60° and the distance between the overhangs, D, was varied between 1 and 3m in this analysis. The overhang width, projecting from the facade, is 1m. Direct and diffuse shading effects from the overhangs mounted above, diffuse shading from the vertical facade, as well as reflection effects incident on the PV array from the vertical facade were considered in these simulations.

Shading losses

Simulation results giving the percentage of energy lost annually due to shading are shown in Figure 3 for varying

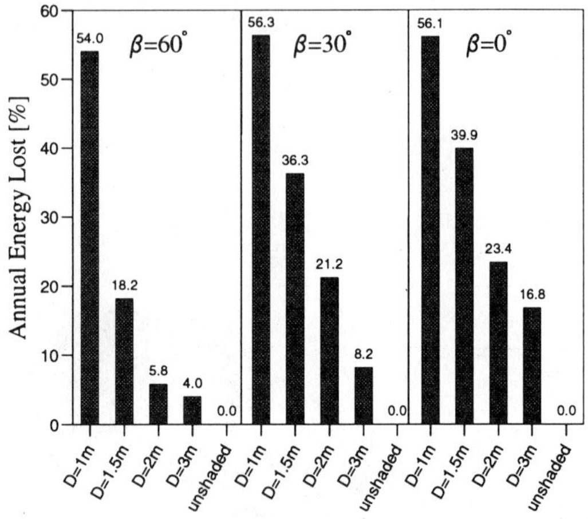

Figure 3: Percentage of annual energy lost due to shading for varying D at different β.

D at $\beta = 60°$, $30°$ and $0°$ respectively. The percentages are calculated with respect to the unshaded case for each β. The results indicate that for each D, the shading losses increase as β approaches the horizontal (i.e, $\beta = 0°$) as expected. For $\beta = 60°$, the lowest annual shading losses were calculated. For $\beta = 0°$, shading of both the diffuse and direct irradiation on the lower overhangs results in the largest energy losses for each D.

Contour graphs

When evaluating different PV integration strategies, not only the shading losses, but more importantly, the total annual energy output incorporating the shading losses should be considered.

In this study the annual energy output for each different integration strategy is represented as a fraction of the annual energy output produced by this array at the optimum integration strategy for the site. For the simulation results presented in this paper, the optimum integration strategy for Freiburg, Germany (48N) was taken to be an unshaded plane, facing south and tilted at $30°$. In Table 1 the annual energy output is given as a percentage of the maximum attainable for this array, based on Freiburg weather data. In the last column, the annual shading losses, corresponding to Fig. 3, are given. These losses are calculated with respect to the unshaded case for each β.

These simulation results are shown as a function of the varying parameters β and D in the contour graph in Figure 4. Here the annual energy output of the PV array

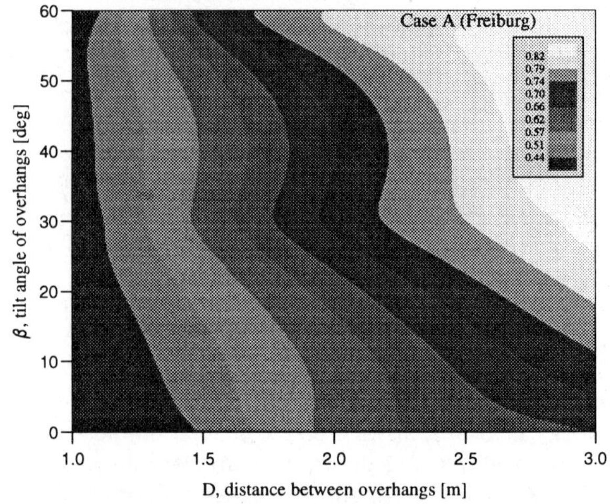

Figure 4: Annual PV energy output as fraction of maximum attainable

for each integration strategy is represented as a fraction of the maximum attainable output energy from this PV array at the site.

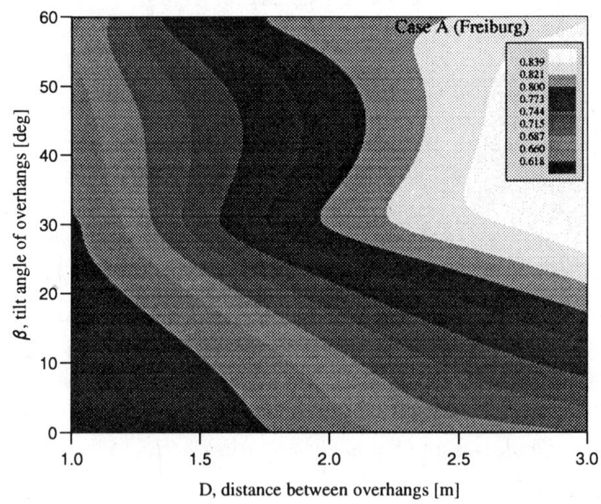

Figure 5: Annual effective irradiation as fraction of maximum attainable

Simulations show that the maximum annual energy output for this integration strategy is approximately 83 % of the maximum attainable for this same array in Freiburg. This occurs for D=3m when $\beta = 30°$ or $60°$. Although an array tilt of $\beta = 30°$ has been shown to be the most optimal for Freiburg [2], for this integration strategy overhangs tilted at $\beta = 60°$ result in a higher annual energy at each D for $1.5 < D < 3$m due to shading losses.

For comparison with Figure 4, the corresponding effective irradiation Φ' for each integration strategy is graphed in Figure 5. By comparing Figures 4 and 5, one sees that the energy output varies between approximately 40 % and 83 % while the effective irradiation only varies between approximately 60 % and 84 %, thus indicating the nonlinear relationship between these two quantities.

Tilt Angle β [°]	Overhang Interval D [m]	Energy Output %	Shading Losses %
60°	1.0	38.4	54.0
60°	1.5	68.3	18.2
60°	2.0	78.0	5.8
60°	3.0	83.3	4.0
60°	unshaded	83.5	0.0
30°	1.0	39.7	56.3
30°	1.5	58.0	36.3
30°	2.0	71.3	21.2
30°	3.0	83.51	8.2
30°	unshaded	90.95	0.0
0°	1.0	33.0	56.1
0°	1.5	44.0	39.9
0°	2.0	57.6	23.4
0°	3.0	65.7	16.8
0°	unshaded	75.2	0.0

Table 1: Annual energy output and shading losses.

Different Facade Materials

The above simulations were carried out assuming a white building facade. In the results presented in this section, different facade materials are considered. Simulations were carried out for an unshaded case for overhangs tilted at $\beta = 30°$ to determine what effect these materials have on the annual PV array energy output. A dark-, glass-covered and mirror-covered facade were simulated for the integration strategy shown in Figure 2. The results are given in Table 2.

Facade Material	Annual Energy Difference
white facade	00.00 %
dark facade	-4.35 %
windows	+9.40 %
mirrors	+24.00 %

Table 2: Difference in annual energy for various facade materials

Using a white facade as a reference case, results show that the annual energy output is reduced by 4.35 % when the PV array is mounted on overhangs on a dark facade. A facade covered with windows increases the annual output by 9.4 %, and for a mirror-covered building facade, reflections from the mirrors increase the annual energy output by 24 %. These simulations were carried out for a south-facing building in Freiburg, Germany, with overhangs tilted at 30°.

These results will vary depending on the site location, weather data, the building orientation and the tilt angle of the overhangs. Methods to increase the annual energy output of a building-integrated PV array using different facade materials promises to be an interesting area for future research.

Conclusions

The results in this paper illustrate the helpfulness of a simulation program such as PVPACK in determining the optimum building-integration strategy under complex situations in an urban environment. In particular, the use of a raytracing technique to determine the irradiation distribution for building-integrated PV arrays was proven to be a powerful method. Subtle influences of diffuse shading and first and secondary reflections can be quantified. Complex shading geometries can be simulated with little effort.

The importance of a well-planned and optimized layout strategy for building-integrated PV arrays is substantiated. The total annual energy output, including shading losses, is an important consideration in planning an optimal integration strategy.

Simulations presented here show that for severe cases over 50 % of the annual energy can be lost due to shading. In contrast, reflection effects from highly reflecting surfaces such as mirrors can increase the annual energy output by up to 24 % for the integration strategy considered in this paper.

Acknowledgements

Many thanks to the co-workers at FhG-ISE for their support and assistance throughout this project. This work was carried out within the framework of the International Energy Agency (IEA) Task 16 program, Photovoltaics in Buildings.

References

[1] G. Ward, **RADIANCE 2.4**, Synthetic Imaging System, Lawrence Berkeley Laboratory, Berkeley, CA, USA 1993.

[2] M. Jantsch et.al. "The Effect of Tilt Angle and Voltage Conditions on PV System Performance." *11th European PV Solar Energy Conf.* Montreux 1993.

THE IMPORTANCE OF HYBRID PV-BUILDING INTEGRATION

M. Posnansky, S. Gnos
ATLANTIS ENERGY LTD, 3012 Bern
Tel.: 031 300 32 20 Fax: 031 300 32 30

S. Coonen
ATLANTIS ENERGY INC, GRASS VALLEY, CA 95945
Tel.: 916 274 27 43 Fax: 916 274 25 64

ABSTRACT

An extensive utilization of photovoltaics for future electricity generation and for hybrid generation of electricity and thermal energy is possible, when PV-panels are designed to become a part of the building envelope itself. Large areas are available, since roofs and façades are perfectly suited for solar energy conversion. ATLANTIS ENERGY LTD has developed special PV-generators which fulfill at the same time the functions and requirements of conventional building elements. Through the successful realization of important large sized PV-façades and PV-roof power plants for different types of buildings it has been further demonstrated:

- the pleasing architectural appearance of PV-installations
- the techno-economical viability of the cogeneration of electric and thermal energy
- the improved and favorable energy pay-back time for the photovoltaic system

In the context of different R&D projects funded by the Swiss government to implement a series of typical building integrated photovoltaic systems, ATLANTIS ENERGY LTD was entrusted to design and build various hybrid building integrated PV-power plants:

- The PV-building integration of the factory building 'Aerni'. This PV-power plant consists of a 9 kW_{el} vertical PV-façade and of a hybrid 53 kW_{el} PV-system integrated into the skylights. Through the backside cooling of the PV-elements with air, approximately 115 kW of additional thermal power is recuperated for heating the building. The thermal energy is partially stored in the concrete slabs of the basement and during the summer period in an earthstorage. The systems have been operating successfully since the start up in July 1991[1].
- The multifunctional PV-façade 'Scheidegger' with an installed peak power of 18 kW_{el}. The PV-building elements substitute conventional building elements and cogenerate electrical and thermal energy in the central part; the installed thermal power is 12 kW. This system has been working successefully without interruption.
- The hybrid PV-shingle roof 'Brig' with an installed peak power of 15 kW_{el} and 30 kW_{th}. On sunny days during the winter and the transitional periods, the produced thermal energy heats the shower rooms of the building. This system has also been working successefully since the start up in March 93.
- The 8,4 kW_{el}, 12 kW_{th} hybrid PV-shingle roof 'Rigi': This PV-roof has been working continuously since the start up. The measurements of different temperatures (of PV-elements / air) prove the optimal thermal energy collection by natural convection.

1. IMPORTANT ASPECTS OF THE PV-BUILDING INTEGRATION

The realization of the PV-projects described under chapter 3 became possible by using a new generation of PV-elements (PV-building elements) developed by ATLANTIS ENERGY LTD. Through the combination of 'energy conversion' and 'building protection functions', important synergy effects result:

- appealing architectural design of the PV-power systems (architects and builders become motivated)
- reduction of investment costs: by substituting conventional sometimes rather costly façade/roof/ shading elements and compared to the present PV-installations by saving frames, supports, foundations, installation costs etc.

CH3365-4/94/0000-0998 $4.00 © 1994 IEEE First WCPEC; Dec. 5-9, 1994; Hawaii

- improved energy pay-back time by saving the energy for manufacturing above items (conventional building elements do not recuperate their energy investments !)
- simultaneous conversion of solar radiation into both electricity and thermal energy more than doubles the overall collection efficiency. By an appropriate cooling of PV-façades and PV-roofs (thereby increasing the electrical output), low temperature heat for heating or warm water production is obtained.

2. MAIN ASPECT OF THE HYBRID PV-BUILDING INTEGRATION

An average photovoltaic-element (monocristalline cells: 15% efficiency) working at the temperature of 65 °C converts about 11% into electrical energy. Approximately 10% of the irradiation is reflected. The remaining 79 % of the irradiation is converted in the PV-element into thermal energy (heat) .The PV-element emits this heat by convection and infrared radiation. An optimized building integrated hybrid PV-power plant is able to collect a considerable part of this thermal energy in the form of warm air or warm water. Measurements show that building integrated hybrid PV-power plants reach thermal efficiencies between 30 to 45 % depending of the way of energy collection (natural convection, forced convection, etc.). The total electro-thermal efficiency of hybrid PV-plants reaches 40 to 55%. The cogeneration of electrical and thermal energy results in a reduction of the energy pay back time and in a reduction of the energy costs. The following descriptions of realized examples show the importance of hybrid PV-building integrations.

3. EXAMPLES OF REALIZED HYBRID PV-PLANTS

3.1 THE HYBRID PV-SHED ROOF 'AERNI FENSTER, ARISDORF'

The factory 'Aerni Fenster' is equipped with a 53 kW_{el} / 115 kW_{th} hybrid PV-power system incorporated into the skylights of the roof (Fig. 8, 9). In addition to the electrical energy, thermal energy (hot air) is generated by blowing ambient air through a gap between the PV-building element and the thermal insulation of the building. This cools the cells, which results in the increase of electrical efficiency of the system. Thermal efficiencies of 25 to 35% (depending on weather conditions) have been expected. Fig. 1 shows the expected thermal efficiencies of the hybrid skylights:

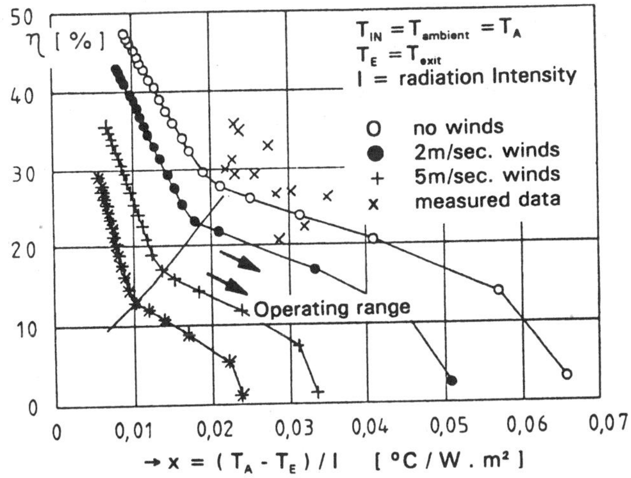

Fig. 1: Expected thermal efficiencies

In the period between January 93 to August 93 the measurements confirm the expected electrical energy production. The measured thermal efficiency of the system was between 32 and 45 % (higher than expected), the measured thermal peak power was 218 kW. Fig. 2 shows the expected and the measured monthly thermal energy production.

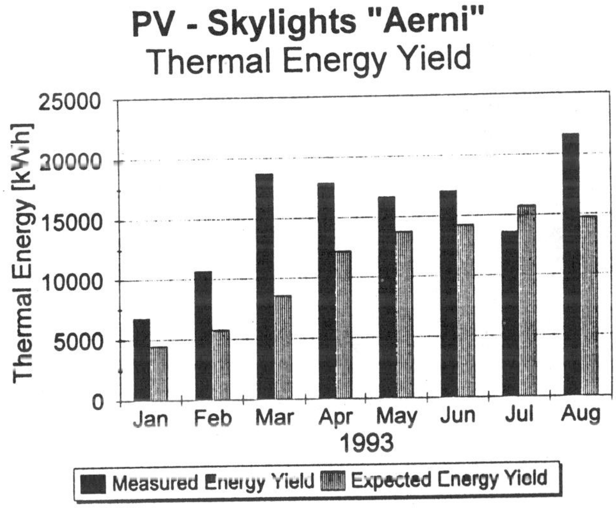

Fig. 2: The expected and the measured thermal energy yield of the hybrid skylight system.

During the heating period the warm air is injected into the factory hall if the heat is needed, or into the basement for short term energy storage in the walls and floors made of concrete. In summer, the thermal energy is stored in an underground storage for retrieval in the winter by a heat pump driven by a cogenerating diesel-plant. The energy concept of the factory is shown in Fig. 10. In the year 1993 the hybrid system covered 70 % of the electro-thermal energy requirements of the factory.

3.2 THE OPTIMIZED MULTIFUNCTIONAL PV-FAÇADE 'SCHEIDEGGER, KIRCHBERG'

The multifunctional PV-façade 'Scheidegger' (Fig. 15) has an installed electrical power of 18 kW and a thermal power of 12 kW (hybrid system only in the central part). The realization of this special PV-façade became possible by using the PV-building elements "SYSTEM ATLANTIS ENERGY". The PV-elements fulfill the following functions at the same time:

- production of electrical energy
- production of thermal energy (hot air)
- shadowing of the offices
- weather protection of the building
- appealing architecture

The thermal energy is collected between the PV-building elements and the insulation of the building by an air flow created by natural convection. Fig. 3 depicts schematically how above aims were attained.

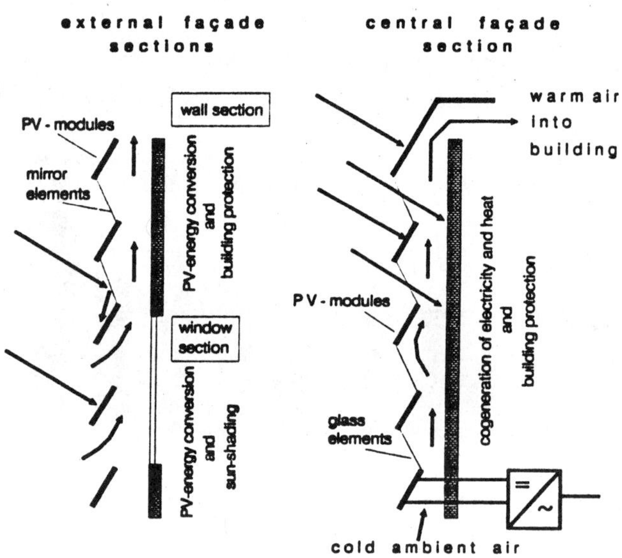

Fig. 3: Functional diagram of the multifunctional PV-façade.

The substitution of conventional façade elements and in particular of the shading shutters contributes to an important reduction of the investment costs.
The measurements over two years confirm the expected performances. The maximal measured electro-thermal efficiency is about 40%. The performance of this multifunctional PV-façade is shown in Fig. 4.

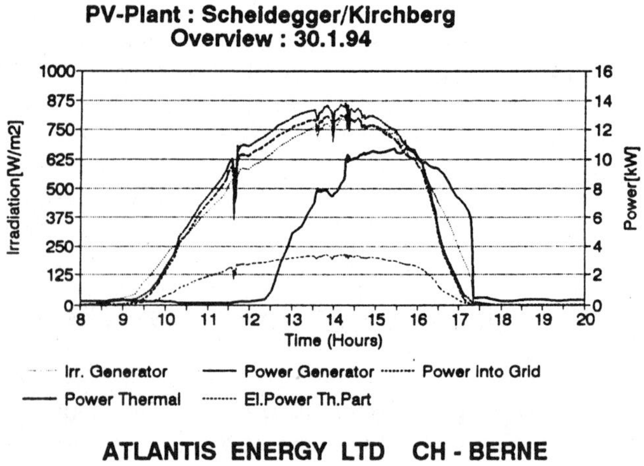

Fig. 4 : trend graphic of January 30th 1994

3.3 THE HYBRID SHINGLE ROOFS BRIG / RIGI

ATLANTIS ENERGY LTD developed also PV-shingle roofs for the combined production of electrical and thermal energy. The simultaneous production of thermal energy is realized by cooling the PV-shingles by natural convection (Fig. 5). By the substitution of conventional shingles and the combined generation of electrical and thermal energy, important cost reductions result.

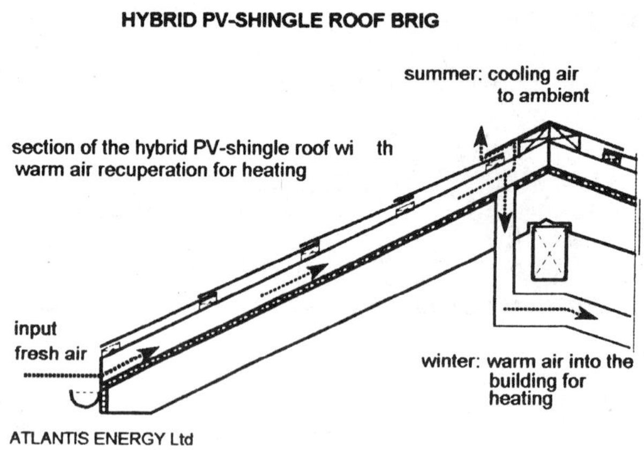

Fig. 5: section of a hybrid PV-shingle roof

The PV-shingle roof 'Brig' (Fig.11. 12) has an installed electrical power of 15 kW, a thermal power of 30 kW and has been in operation since March 93. The thermal energy is used for the direct heating of the shower rooms during the winter and the transitional periods.

The hybrid 8,4 kW_{el}, 12 kW_{th} 'Rigi' shingle roof (Fig. 13, 14) has been operating since October 93. Several temperature sensors in the PV-shingles and in the channel between the shingles and the insulation of the building give sound information about the air flow created by natural convection. The aim of the measurements is to optimize the channel geometry for optimal energy production (= optimal cell cooling). Fig. 6 shows the position of the measure points in the roof:

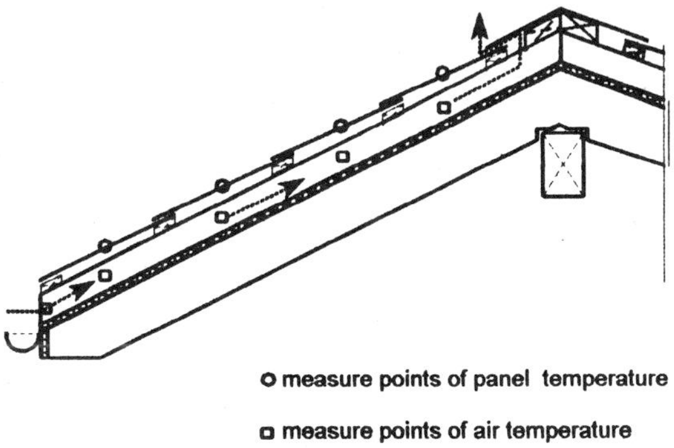

Fig. 6: measure of temperatures

The aim of the analysis of the measurements is to optimize the channel geometry for optimal thermal energy production for the design of future PV-shingle roofs.
The comparison of our simulation and the result of the measurements show, that the optimal channel geometry is nearly reached. Fig. 7 shows the expected and the measured temperatures in the roof of 27 March 94 at an irradiation of 989 W/m².

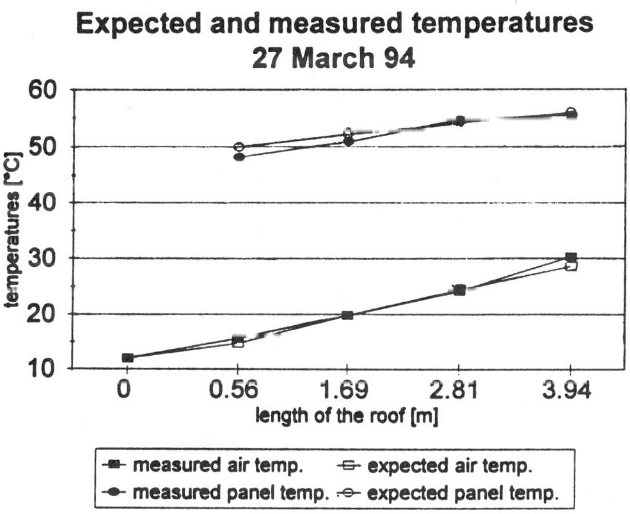

Fig. 7: expected and measured temperatures of 27 March 94

On the basis of these measurements the thermal power of this roof is more than three times as high as the electrical power. As result, the maximum over all efficiency reaches 45 %. With the hybrid technology, the total yearly energy output is two to three times that of a conventional PV-installation, depending upon thermal energy use. It is clear, that an important cost reduction results through a multifunctional (specially hybrid) use of the PV-elements. The aim for future realisations of hybrid PV-roofs is to optimize the efficiency of the use of the produced thermal energy.

4. CONCLUSIONS

The technology of hybrid PV-building integration with multifunctional PV-building elements is mature, shows clear architectural and techno-economic advantages and thus paves the way for a large scale implementation of hybrid energy generation. Various successfully realized hybrid PV-roofs and PV-façades are milestones in this respect and they demonstrate, that the combined conversion of solar radiation into electrical and thermal energy in façades and roofs is of great interest where heating or warm water is required.

5. REFERENCES

[1] Examples of realized PV-roof and PV-façade power plants, M. Posnansky, H. Hochreutener, S. Gnos; PV-Solar Energy Conf., 1992 Montreux
[2] 10 kW roof with large sized PV-shingles for a commercial building, J-M. Cottier, 1992 Montreux
[3] Introduction to a new factory building with PV-façades and PV-skylight that cover 70 % of the total energy requirements, M. Posnansky, S. Gnos, Solar Energy in Architecture and Urban Planning, 1993 Florence
[4] 15 kW Schindeldach mit gleichzeitiger Warmluftnutzung, Th. Hostettler, S. Gnos, CISBAT 93, Lausanne
[5] Examples of realized hybrid PV-power plants with specially conceived PV-modules for building integration, M. Posnansky, S. Gnos, International solar energy conference San Francisco, 1994

Fig. 8: Overview of the building integrated PV-power plant Aerni

Fig. 9: Close up of the hybrid PV-skylight

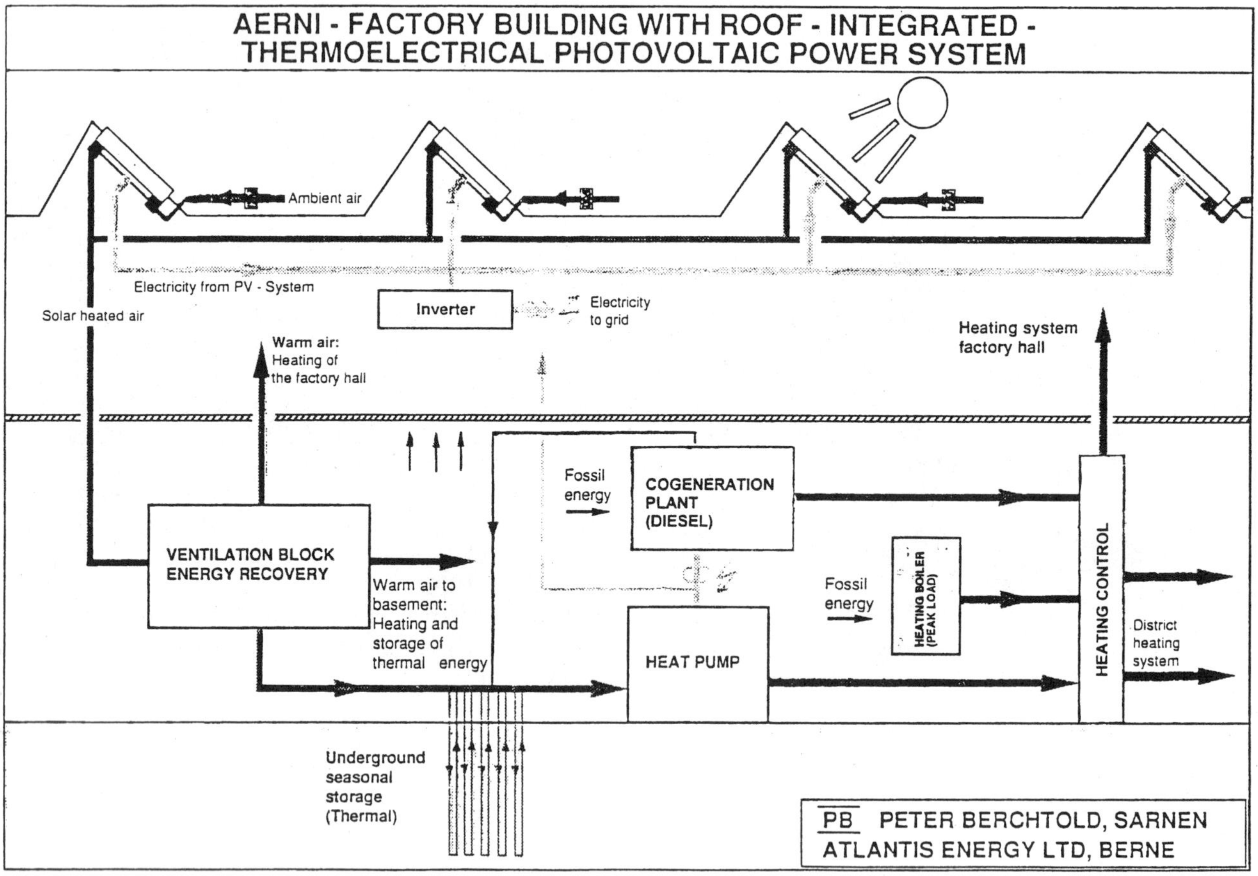

Fig. 10: energy concept of the Aerni factory

Fig. 11: Overview of the hybrid PV-shingle roof Brig

Fig. 12: Close up of the shingle roof Brig

Fig. 13: Overview of the hybrid PV-shingle roof Rigi

Fig. 14: Close up of the shingle roof Rigi

Fig. 15: the multifunctional PV-façade Scheidegger

Optimizing Roof-Integrated Photovoltaics: A Case Study of the PowerGuard™ Roofing Tile

Thomas L. Dinwoodie, AIA, President
POWERLIGHT CORPORATION
2550 Benvenue, Berkeley, CA USA 94704
510.525.9495 tel 510.525.7392 fax

Daniel S. Shugar, P.E., Vice-President
ADVANCED PHOTOVOLTAIC SYSTEMS, INC.
150 Tehama Court, San Bruno, CA 94066
415.871.2275 tel 415.871.2279 fax

ABSTRACT

This paper describes the development and implementation of a building-integrated photovoltaic (PV) roofing tile system that provides grid-connected electric power to buildings. The PV roofing tile system, together with waterproof membrane, insulation, and electrical interconnection, is called *PowerGuard™*. PowerGuard™ is one of the first PV *roofing systems* for flat and moderately-sloped commercial buildings that replaces conventional roof materials without requiring membrane penetrations and mechanical fastening to building structures. When evaluated as a PV system, the building integration *reduces* the cost of a PowerGuard™ system by 14% to 26% rather than *incurring* a structural mounting cost of 18% to 22% to conventionally fasten the system.

Data are reported from a 3.0 kW PowerGuard™ prototype operating in Folsom, California. PowerLight Corporation supplied the system using large-area amorphous silicon modules manufactured by Advanced Photovoltaic Systems, Inc. Performance data indicates the system is exceeding contractual requirements. Sensitivity analysis, based upon performance, installed costs, and supplier data, indicates 1) a marginal economic advantage to tilting the PV array; 2) a marginal economic impact of increased PV efficiency; and 3) economies-of-scale which make PowerGuard™ systems economical today for commercial customers in sunny areas who pay high electricity rates.

INTRODUCTION

PowerLight Corporation and Advanced Photovoltaic Systems, Inc. (APS) are engaged in a systematic effort to drive the costs of installed photovoltaics into a range competitive with grid-connected power. The development of PowerGuard™ provides a case study for cost effective grid-connected PV roofing, designed with stringent allowable-cost criteria.

PowerGuard™ is a photovoltaic roofing assembly for new and re-roofing applications in the commercial/industrial building sectors. PowerGuard™ is in the latter stage of a three-year development and field testing program by PowerLight Corporation (see Fig. 1). Commercial installations of PowerGuard™ will range in size from 10 kW to 2 MW. Annual U.S. markets for building-integrated photovoltaic (BIPV) roofing are estimated to be 400 MW/yr, with over 30 MW/yr in California alone.

The potential for building integration to enhance PV economics is illustrated in Figure 2. The upper curve represents projected price reductions for conventional rack-mounted PV through the year 2004. This plot is based on 100 kWac size PV systems using PowerLight and APS design information, market data from the SMUD 1994 PV Project Solicitation, component supplier cost data and prior analysis of historical installed systems [1]. The lower curve represents the projected net installed price of a hypothetical BIPV system. It is important to note the term

Fig. 1
3 kW POWERGUARD™ PROTOTYPE

First WCPEC; Dec. 5–9, 1994; Hawaii

"net" installed price on the left vertical axis. The *gross* installed cost for BIPV can be higher than for conventional PV, since the PV balance-of-system must be specially designed to function as a building construction material. However, after recieving credits for displaced construction materials and labor, the *net* installed price of BIPV will be less than for conventional PV if the application has been well conceived.

Referring to Figure 2, any BIPV approach will have a characteristic curve corresponding to net installed price on the left vertical axis, market potential on the right vertical axis, and timetable to cost-effectiveness and market penetration on the horizontal axis. The figure illustrates a BIPV product accelerating market adoption of commercial systems priced at $3.50/Wac up to 4 years.

Fig. 2
PHOTOVOLTAIC COMMERCIAL ROOFTOP SYSTEM COSTS:
Conventional Mounting vs: Building-Integrated

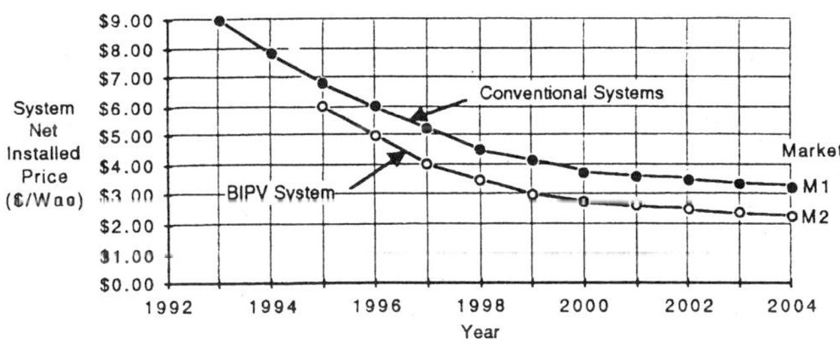

POWERGUARD™ DESIGN OPTIMIZATION

Design Criteria

PowerLight established product design criteria after careful investigation of numerous market variables, including potential synergistic value as a construction material; application size and repeatability; the value of displaced electricity; marketing costs; acceptance by construction trades; decision-criteria on the part of building owners, contractors and architects; financing mechanisms; and requisite investor rates of return. In addition, several functional criteria were analyzed, including PV weight and strength characteristics, size compatibility with standard construction dimensions, compatibility with conventional construction methods and assemblies, aesthetics, environmental impact, design life, safety, and other factors. A comprehensive review of such considerations for BIPV is reported by Paradis et. al. [2].

Design Features

Consideration of the above criteria helped shape the PowerGuard™ design effort. PowerGuard™ consists of photovoltaic tiles situated atop rigid insulation which rests on a waterproof roofing membrane without membrane penetration or mechanical fastening to the building structure. The weight of the PV panel together with edge to edge connection between tiles work to create a significant safety factor against wind uplift. A modular electrical raceway is integral to the assembly. PowerGuard™ features are summarized as a PV and roof construction assembly:

PowerGuard™ as a PV System:

- Generates on-peak power ranging from 10 kW to 2 MW for customer demand-side load reduction
- Eliminates the need for a PV support structure
- Installs quickly and easily, with no roof penetrations
- Displaces roofing components valued at $2-$5/sf
- Is reusable if the building is disassembled
- Includes a 12-year warranty on power production

PowerGuard™ as a Roofing System:

- Installs as roof membrane, insulation and ballast
- Is appropriate for both new and re-roof commercial applications, with slopes from zero to 2:12
- Can be installed on built-up and single ply membranes
- Is free of CFC's / Energy payback less than 6 months
- 30-year assembly life is compatible with the life of PV
- Provides DSM benefits of improved building thermal performance
- Significantly extends the life of otherwise unprotected roofing membranes
- Is less than half the weight of aggregate ballasted roofs (4 lbs/sf vs. 10 lb/sf)
- Flexibly installs around roof penetrations and HVAC
- Eliminates building interior condensation - with dew point above membrane

PowerGuard™ shows added value when evaluated on the basis of potential grid-support benefits when located in areas of high value on a utility's distribution system [3]. The economic feasibility of targeted grid support systems is further enhanced through customer ownership [4].

PowerLight has patents awarded and pending on PowerGuard™ and related BIPV products.

POWERGUARD™ PROTOTYPE DEMONSTRATION

A 3 kW PowerGuard™ prototype was designed and installed in the summer of 1994 at the new Operations Center of the Department of Energy Western Area Power Administration (DOE/WAPA) in Folsom, CA (Fig. 1). The project was funded by the Sacramento Municipal Utilities District (SMUD). Total on-site construction was completed in three days. The system consists of 72 PowerGuard™ tiles arranged in 12 strings, 6 modules per string, with power conversion using an Omnion 4 kW series 2200 inverter. Each PowerGuard™ tile incorporates one EP50 PV module from APS. The APS module has been extensively tested in a large grid-connected utility system and has been shown to have excellent stability and long-term product reliability [5,6].

The PowerGuard™ system interconnects with WAPA's electrical breaker panel located in the building mechanical room. The facility acts as a demand-side management (DSM) system, off-setting building energy consumption during periods of high-value pricing. Preliminary testing shows system output exceeding the contracted requirement of 3 kWdc, as shown in the output profile for September 1, 1994 (Fig. 3). Estimated annual energy production is 4930 kWh/yr (18.4% capacity factor). In addition, PowerGuard™ operates as a 'cool' roof, due to proprietary design features which thermally regulate the assembly. The temperature of the roof membrane beneath the PowerGuard™ modules is plotted in comparison with

Fig. 3
POWERGUARD SYSTEM DC OUTPUT
Folsom, CA 9/1/94

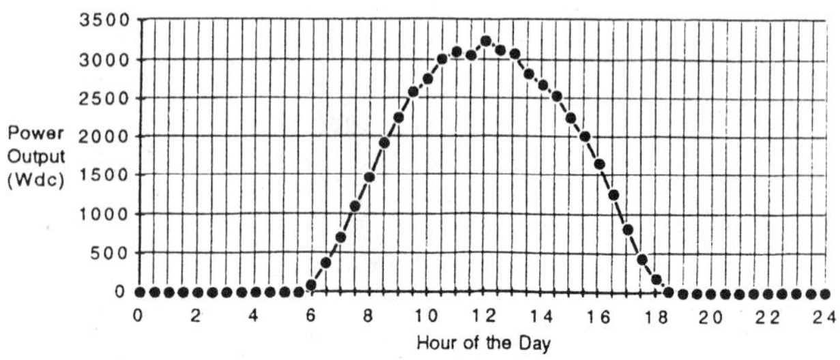

Fig. 4
POWERGUARD SYSTEM TEMPERATURE PROFILE
Folsom, CA 10/6/94

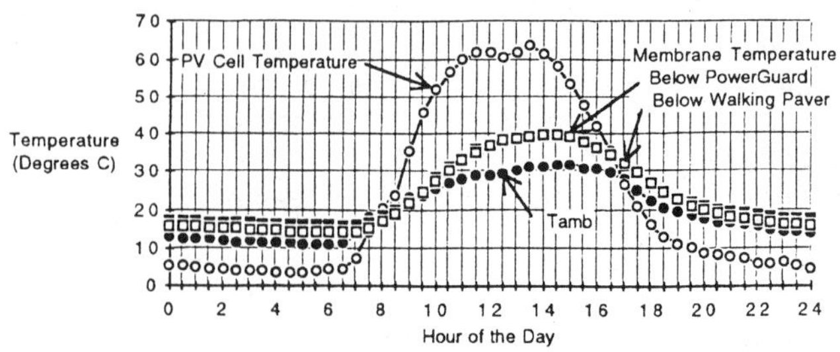

membrane temperature beneath the walking paver (Fig. 4). The latter consists of 2" EPS insulation (R10) covered with a 3/8" layer of latex modified cement, which reflects solar insolation due to its light color. The temperature profiles are indistinguishable. Such cool temperatures indicate PowerGuard™ provides passive DSM benefits through reduction of building thermal load.

The PowerGuard™ system design life is 30 years, or double the life of conventional roofing assemblies. A typical commercial installation, averaging 300 kW and occupying a 120,000 sf rooftop, is expected to produce over 2,500 megawatt-hours of electricity, helping to defer the need for new fossil fuel power plants. Such an installation would avoid over 1300 tons of coal or over 2 million gallons of oil, and would avoid the generation of over 23,000 tons of global warming gases (CO2), over 160 tons of acid rain (SOx), and over 90 tons of smog (NOx). Photovoltaics emit no atmospheric pollution or solid waste, are silent in operation, and consume no water. The APS thin-film modules have an energy payback of under 6 months, while manufactured in an emission free, closed process using safe, abundant and recyclable materials. By integrating into buildings, the PowerGuard™ system conserves conventional building material resources, and can defer or eliminate the need for new electric lines and substations, preserving natural habitats and avoiding additional electromagnetic fields.

POWERGUARD™ SENSITIVITY ANALYSIS

Sensitivity studies have been conducted to examine the relative cost/benefit advantages to increasing array tilt angle, increasing PV cell efficiency, declining PV price, and increasing project size. *Relative benefit/cost ratio* is used as an index to evaluate the economic impact of various independent variables relevant in PowerGuard™ design. The base case system consists of a 250 kW PV installation with flat orientation, 4.5% module edge-to-edge efficiency, and installed costs of \$3.50/Wp. A \$4.00/sf building materials credit was used. PV output was modeled using PVFORM with Sunnyvale, CA insolation data (37.5^o latitude), coupled with Pacific Gas & Electric Company's E19 time-of-use commercial rates.

Sensitivity to a PV Tilt Angle

Figure 5 plots the relative cost-effectiveness of tilting the array from horizontal. The plot shows the relative cost/benefit ratio for a system tilted at 30 degrees for structure costs varying from zero to \$.80/Wp. When the structure costs are below \$.28/Wp, payback improves, while costs higher than \$.28/Wp increase payback period.

The analysis suggests that for locations 40° latitude or less, it is sub-optimal to invest in a support structure simply to achieve an array tilt, unless the support structure is very low cost. This analysis takes account of the *revenue value* of array tilt, not simply energy value. Whereas annual energy output between the two arrays differs by approximately 10%, that difference declines to 7% when energy output is coupled with time-of-use rates for this summer peaking utility. In fact, the monthly output during the summer for the flat array exceeds the output of the tilted array.

Sensitivity to PV Efficiency

A more surprising result concerned the relative value of PV efficiency. Relative cost/benefit verses cell efficiency was examined under three values for displaced building materials (Fig. 6). The analysis uses area-related building material values of \$6.00/sf (high), \$4.00/sf (medium), and \$2.00/sf (low). Where building credit is low, PV efficiency is important. However, the relative cost/benefit ratio increases only 25% when cell efficiency increases to 20% from the 4.5% base case. Where the building credit is higher than the base case, there is a *negative* value to increasing cell efficiency. This is because the building credit, based on dollars per unit area, drives system economics. Higher efficiency PV displaces a smaller area of construction materials per project kW, resulting in a smaller credit relative to project kW capacity.

Sensitivity to PV System Price

The results of varying PV system price relative to the base case at \$3.50/Wac was also evaluated (Fig. 7). The graph indicates a low sensitivity of cost/benefit ratio to system costs. This is due to the fact the systems are financed, and discounted cash flows which capture depreciation and interest deductions are used.

Sensitivity to Project Scale

Although PV is a modular technology, not all costs scale linearly, such as marketing expenses, design fees and construction set-up. An analysis compared relative cost/benefit ratio to project scale, with 250 kW representing the base case (Fig. 8). Our analysis indicates that below 100 kW, cost/benefit ratio exceeds 120% of base case and rises steeply, whereas projects exceeding 400 kW provide diminishing economies. Conversely, significant savings for 250 kW and larger projects enable

Fig. 5
SENSITIVITY TO TILT ANGLE

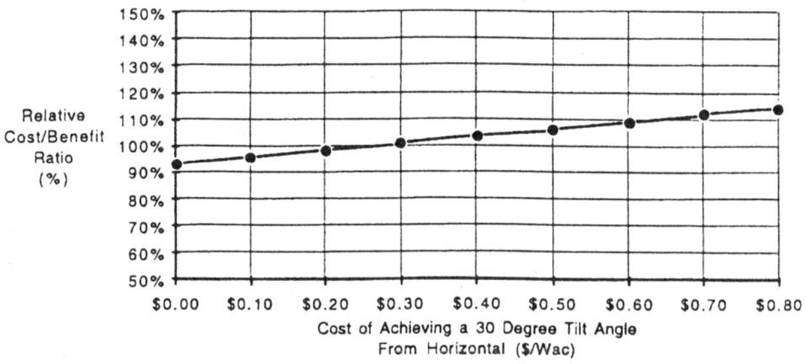

Fig. 7
SENSITIVITY TO NET PV SYSTEM PRICE

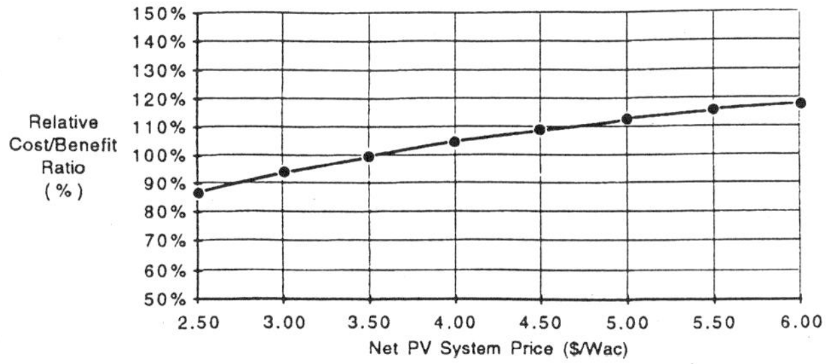

Fig. 6
SENSITIVITY TO PV MODULE EFFICIENCY

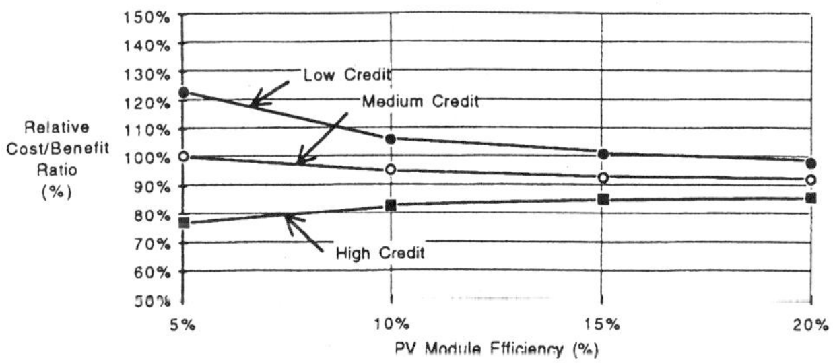

Fig. 8
SENSITIVITY TO PROJECT SIZE

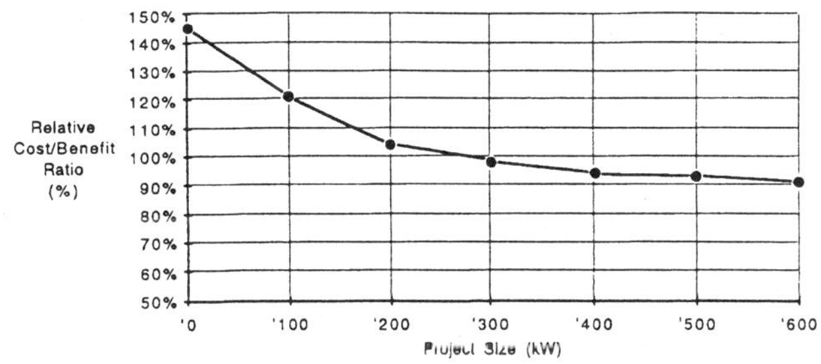

PowerGuard™ systems to be cost-effective today for profitable commercial/industrial customers in locations less than 40° latitude in sunny areas with relatively high utility rates. For such customers, depreciation and other tax benefits have value to reducing tax liability, and play an important role in enhancing system cost-effectiveness [4].

CONCLUSIONS

1. Building integration offers potential advantages over conventional PV systems by eliminating costly BOS components and by enhancing the value of the PV.
2. For BIPV grid-connected products in regions of high solar insolation, high time-of-use utility rates, and below 40° latitude:
   1. A horizontal orientation suffers little disadvantage relative to a tilt-array, and is preferred where the cost of tilting the array exceeds $.28/Wp;
   2. Lower efficiency PV can offer payback advantages where the value of displaced construction materials is high;
   3. Significant economies-of-scale will enable larger PowerGuard™ systems to be economical in 1995 for preferred commercial customers in sunny areas paying high rates for summer peak electricity.
3. Installation and testing of the initial PowerGuard™ prototype indicates:
   1. PowerGuard™ installed quickly and is compatible as a roofing assembly;
   2. The system power output has exceeded its design and contract rating;
   2. The product provides DSM benefits of improved building thermal performance.

ACKNOWLEDGMENTS

The authors acknowledge the assistance of the companies who supported the recent PowerGuard™ demonstration: the Sacramento Municipal Utility District (SMUD), Sacramento, CA (Dave Collier, Don Osborn, Bob Wichert, Larry Wittrup); the Western Area Power Administration (Jim Feider, Guy Nelson, Krishna Shah); Endecon, Inc., San Ramon, CA (Chuck Whitaker); and Western Single Ply (Dave Pope, Jim Tanner, John Tanner).

REFERENCES

[1] Shugar, D.S., et. al. "Comparison of Selected Economic Factors for Large Ground-Mounted Photovoltaic Systems with Roof-Mounted Photovoltaic Systems in Switzerland and the USA", *Proceedings of the 11th European PV Solar Energy Conference*, Oct. 1992.

[2] Paradis, A., Shugar, D.S., "Photovoltaic Building Materials", *Solar Today*, May/June 1994, p. 34-37, published by the American Solar Energy Society.

[3] Shugar, D. S., "Photovoltaics in the Utility Distribution Systems. The Evaluation of System and Distributed Benefits", *IEEE Photovoltaics Specialists Conference Proceedings*, p. 836-844, May 1990.

[4] Wenger, H.J., et. al., "Photovoltaics as a Demand-Side Management Option: Benefits of a Utility-Customer Partnership", *World Energy Congress*, Atlanta, GA, 1992.

[5] Matlin, R., Shugar, D.S., "Design and Two-Year Operational Performance of a 400 kWac Amorphous Silicon Solar Photovoltaic Power Plant at PVUSA", *Proceedings of the IEA Task VI Workshop*, Paestrum, Italy, July, 1994.

[6] Shugar, D.S., "Applications of Amorphous Silicon Photovoltaics: Myths and Facts", *First World Conference on Photovoltaic Energy Conversion*, Waikoloa, HI, 1994.

A NEW TYPE OF PHOTOVOLTAIC SHINGLE

N. Okuda, T. Yagiura, M. Morizane, N. Nakamura, M. Ohnishi and S. Nakano
New Materials Research Center, Sanyo Electric Co., Ltd.
1-1 Dainichi Higashimachi, Moriguchi, Osaka 570, Japan
T. Ito, S. Omoto and Y. Yamashita
Housing Materials R&D Dept., Kubota Corporation
2-1 Takamatsu, Kosei-cho, Koga-gun, Shiga 520-32, Japan
H. Ito and T. Fujiwara
H&E System Dept. Technology Development Lab., Kubota Corporation
1-1-1 Hama, Amagasaki, Hyogo 661, Japan

ABSTRACT

We have proposed a new type of solar cells unified with roofing materials. The photovoltaic (PV) shingle offers many features, such as low cost, simple construction and maintenance, good design, and fire resistance, compared with previous modules. The PV shingle was confirmed to have no major problems as a roofing material by environmental, waterproofing and installation tests.

INTRODUCTION

Recently in Japan, PV modules have been gradually installed on the roofs of ordinary houses[1]. But these conventional PV modules have various problems, such as fewer design, high total cost of construction, complex mounting and replacement procedures and so on. This factor is an obstacle to promoting the PV systems in the market. Up to the present, various kinds of PV modules for houses were studied[2]. In this study, a new type of PV shingle is proposed. The basic specifications and the results of various tests are discussed.

BASIC SPECIFICATIONS

We studied module design concepts from the standpoint of basic performance, installation and design.We then proposed a module unified with roofing materials that was built using a frameless structure, and that incorporated a colored metal plate on the rear side as well as reinforced glass on the front side of the module (see Fig.1) and the main features are as follows.

1) The module is standardized at 910 mm wide by 331 mm high in order to conform to module dimensions for residential construction in Japan.
2) A metal plate placed on the rear side of the module (Back Metal; BM structure) ensures that the module is fire resistant.
3) The frameless structure is introduced in order to improve design characteristics and to reduce module cost.
4) The Just Fit Connection (JFC) structure in which overlaps of each module are very small is adopted. By this structure, individual modules can be easily replaced and the module cost can be reduced.
5) A simple installation procedure using special lock mounting that hooks and mounts the module not only

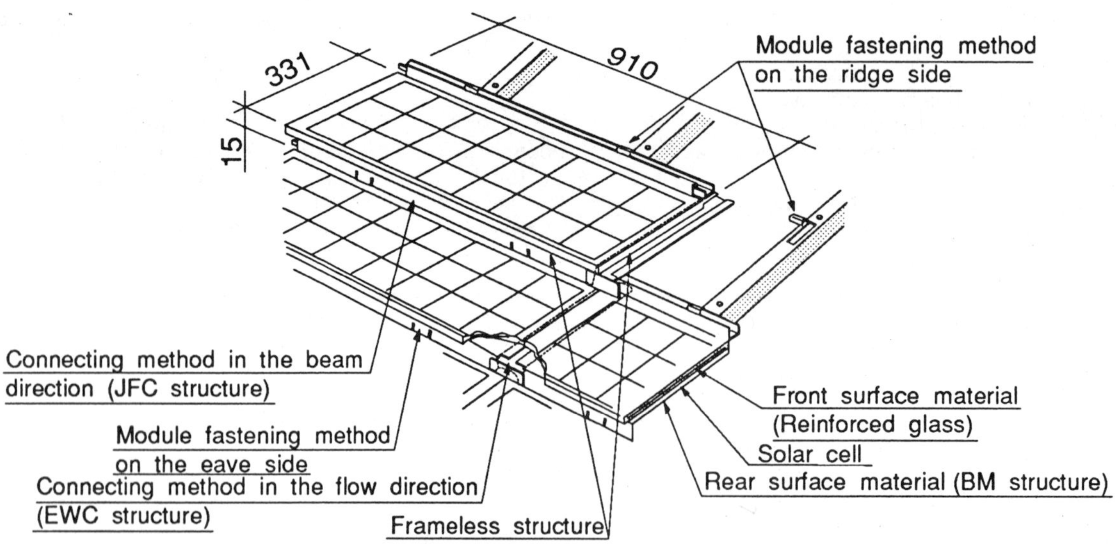

Fig.1. Basic specifications for modules integrated with roofing materials

CH3365-4/94/0000-1008 $4.00 © 1994 IEEE

First WCPEC; Dec. 5-9, 1994; Hawaii

assures easy installation even by conventional contractors, but also reduces installation costs.

6) A mated system that allows modules to be mounted and removed by a single person without special tools increases installation efficiency and reduced installation costs.
7) The ventilation flues of varying heights for cooling the module temperature are used to install PV modules.

STUDIES OF MODULE FABRICATION TECHNOLOGIES

In studies on fabricating modules unified with roofing materials, we studied module materials and structures as well as their reliability.

Studies on module materials and structures

1) Long-term durability was an important consideration in selecting module materials, and as a result, the basic specifications we set included reinforced glass for the front plate material, EVA resin for the cell sealant, and a steel plate with colored aluminum alloy plating for the rear plate.
2) In order to facilitate module attachment and removal by a single person, we set basic specifications for the terminal box that included the connecting method, size, and installation position.

Reliability evaluations

Reliability evaluations based on JIS standards were conducted on modules prepared by cutting the raised section of some of the metal plates in accordance with the basic specifications (see Fig.2). The details and the results of the evaluations are listed in Table 1.

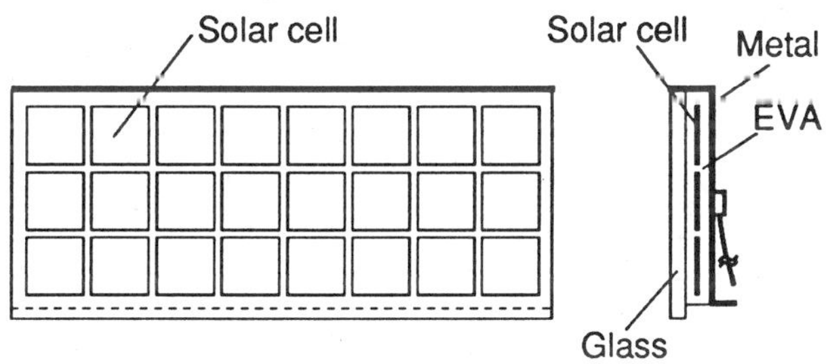

Fig.2. Test sample for reliability evaluations

Table 1. Test details and results

Item	Details	Test results	
		Output power (After test/initial)	Withstand voltage
Saltwater spray test	According to JIS C 8917-8918	100.3%	Passed
Moisture-resistance test	Ditto	100.4%	Passed

As far as the reliability of modules fabricated with the basic specifications is concerned, the results above indicate no basic problems in the output power and withstand voltage.

STUDIES OF INSTALLATION PROCEDURES

The procedure for mounting PV modules installed on roofs must be studied and evaluated from the standpoint of building materials as well as from the standpoint of solar cells.The installation procedure has a tremendous impact on cost, and affects basic performance in terms of roof positioning. Here we also studied design characteristics with a focus on waterproofing performance, basic installation procedures and the industrialization of the modules and cooling efficiency by ventilation/cooling structures.

Confirmation of waterproofing characteristics

Simulated modules fabricated in accordance with the basic specifications were mounted to a test frame under conditions approaching that of an actual installation, and were then exposed to wind blasts and water spray to measure the amount of leakage around the rear surface of the module. The testing apparatus is shown in Fig.3. The results of the tests are listed in Table 2. From the above results, the waterproofing characteristics of the modules were confirmed to have no major problems compared with conventional roofing tiles.

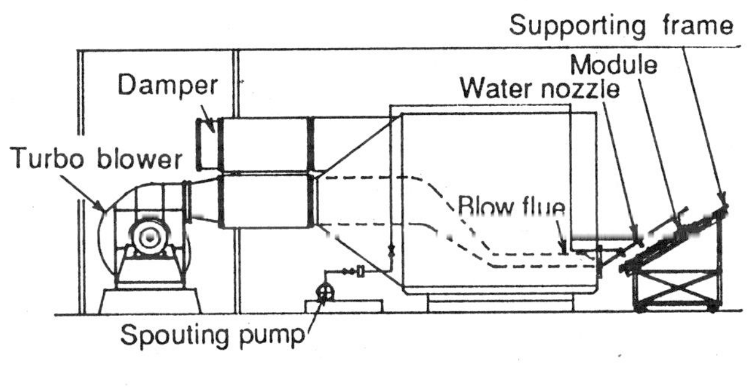

Fig.3. Illustration of waterproof testing apparatus

Table 2. Amount of leakage

Wind speed (m/s)	Amount of leakage (cc/m · min)				
	Test item: Module	Claytile(*2)		colonial(*2)	
	Gradient: 6/10	4/10	10/10	3/10	10/10
10	—(*3)	7.5	6.1	—(*3)	—(*3)
15	0.0(*4)	24.5	11.0	—(*3)	—(*3)
20	4.5	241.0	62.0	0.6	0.0
25	30.0	305.0	239.0	7.8	0.4

Amount of water spray is 5l/m/min (*1)
(*1) Rainfall equivalent to 165mm/h
(*2) Test results from the conprehensive Technological Development Project under the Ministry of Construction
(*3) No leaks
(*4) Water permeated, but could not be measured

Fig.4. Cutout model after installation test

Confirmation of installation procedures

Cutout models using simulated modules fabricated in accordance with the basic specifications were used to confirm the installation procedures. Figure 4 shows the cutout model after installation test. The results are as follows.

1) We confirmed module installation both from the upstream and the downstream side, but we found that mounting from the downstream to the upstream side was ideal from the standpoint of installation efficiency and safety.
2) We found that engaging the top and bottom fittings of the module and waterproofing drains took a significant amount of time.
3) We tested a connecting method by which a single lead wire was attached to the terminal box first before mounting a subsequent module, and found essentially no problems with mounting or removal. We also found that the mounting position of the terminal box as well as the length of the lead wire had a significant impact on installation efficiency.

The results above indicated the direction we needed to go with installation procedures as well as problem areas and future tasks related to module mounting and removal.

Demonstration tests for designs

We studied color schemes for the solar cell color and the color of the metal plate on the rear side of the module structure using cutout samples, and performed demonstration tests by installing the samples in cutout models under array conditions approximating actual conditions. In these tests, polycrystalline silicon cells are used. The results are as follows.

1) From demonstration tests using cutout models, we determined that the colors for the metal plate that matched the solar cell color, and that made the vertical joints less conspicuous were navy blue, dark brown and dark gray.
2) From the standpoint of building materials, we found that the module unified with roofing materials compared favorably with general roofing materials.

Studies on ventilation/cooling structures

Because module temperature affects the power generating characteristics, we demonstrated the cooling effect on the rear surface of the metal plate on the solar cell module using ventilation flues of varying heights on a cutout model (see Fig.5). The results are as follows.

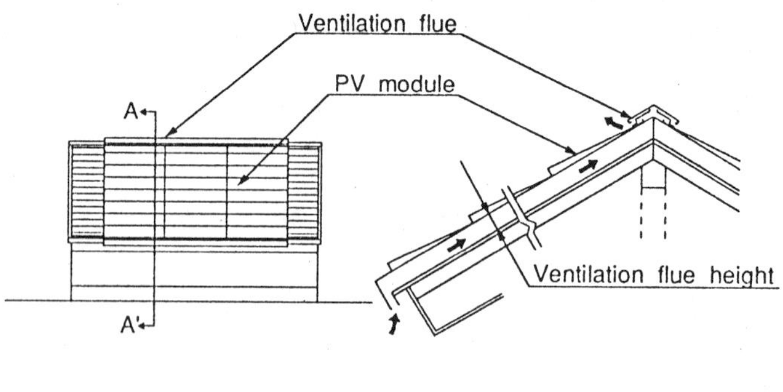

Fig.5. Cutout model for cooling effect test

1) The height of the ventilation flue was set at 45 mm, and the temperature of the rear surface on the metal plate was measured with the ventilation holes open and closed. We found that the temperature rose considerably when the ventilation holes were closed. That temperature was about 5~6 degrees higher than that of open case as shown in Fig.6.
2) When we varied ventilation flue height to 45, 75, 105 and 135 mm in other tests, we found that the temperature tended to decrease as the height of the ventilation flue increased, although the difference was not that significant as shown in Fig.7.

The results above have provided direction for ventilation flue height and the efficiency of structures installed on the ventilation layer.

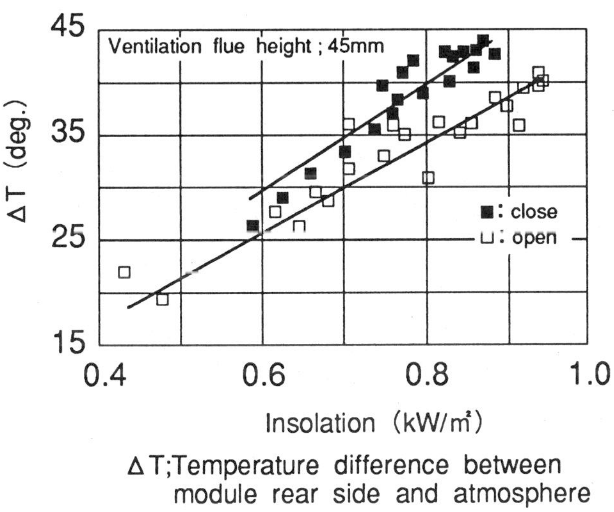

Fig.6. Affection of ventilation flue to module temperature (■; Close. □; Open, Flue height; 45mm)

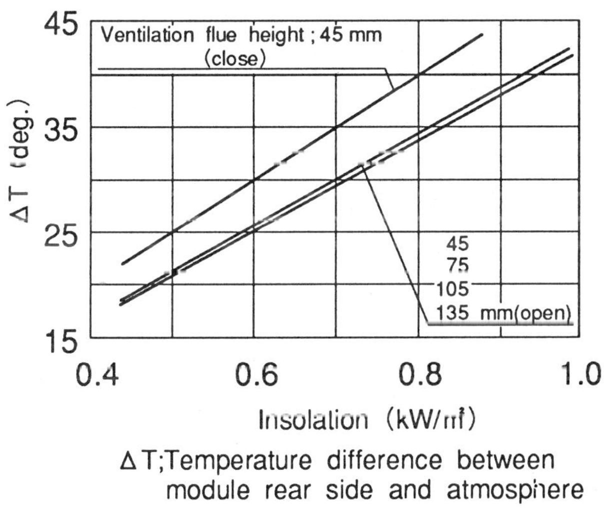

Fig.7. Affection of ventilation flue height to module temperature

CONCLUSIONS

We have proposed a new type of solar cells unified with roofing materials. The PV shingle offers many features, such as low cost, simple construction and maintenance, good design, and fire resistance, compared with previous modules. The PV shingle was confirmed as follows.

1) In regard to the basic specifications, Back Metal (BM) structure for the rear surface material, Just Fit Connection (JFC) structure for the connection method, frameless structure etc are adopted.
2) The results of reliability evaluations based on JIS standards indicate no basic problems in the output power and withstand voltage.
3) From the results of waterproofing tests, the waterproofing characteristics of the modules are confirmed to have no major problems compared with conventional roofing tiles.
4) The results of installation tests indicate the direction we need to go with installation procedures.
5) By studying the ventilation/cooling structure, the direction for ventilation flue height and the efficiency of structures installed on the ventilation layer is obtained.

ACKNOWLEDGEMENT

The work was supported by the New Energy and Industrial Technology Development Organization (NEDO) as a part of the New Sunshine Program under the Ministry of International Trade and Industry.

REFERENCES

[1] N. Tsujino et al., " Residential Photovoltaic Power Generating System Connected to Utility Line" , *Technical Digest of International PVSEC-7.* Nagoya, 1993, pp.345-346
[2] T. Matsuoka et al., " A Solar Cell Unified with a Building Materials" , *Technical Digest of International PVSEC-3.* Tokyo, 1987, pp.565-568

"CAPTHEL", A 66% EFFICIENT HYBRID SOLAR MODULE AND THE "ECOTHEL" CO-GENERATION SOLAR SYSTEM.

Alain Ricaud and Pierre Roubeau, Cythelia .
36 Allée de chanteraine 91 190 Gif sur Yvette. France. Tel/fax: 33 1 60 12 42 92
3 Place de la Gare de Lozère, 91 120 Palaiseau. Tel.33 1 60 10 36 29/fax.33 1 60 10 43 19

ABSTRACT

This study relates to the performance tests of a **66% efficient hybrid solar module "Capthel"**, and to the economic study of the combined generation of solar electricity and hot air ("Ecothel" solar co-generation).

We demonstrate that a-Si:H PV modules are best suited for co-generation because they behave as if they had a positive temperature coefficient. We report experimental results on the first prototype and describe further improvements in the design of the thermal collector. We then study the potential savings and return on investment of an "Ecothel" installation connected to the grid of a private home using electric space heating like most of the modern houses in France and a commercial building with air-conditionning.

Introduction

Abondant litterature has reviewed the cost situation and analysed the cost projection of PV electricity both for industrialised and developing countries applications [1]. All come to the conclusion that, except for specific remote applications (signalization, pumping, telecom), PV electricity is far away from economic competitiveness (presently 60 to 120 US cents per kWh) with conventional sources of energy (12 US cents /kWh) and wind-mills power generation .

However, the authors are convinced that, for economic and psychologic reasons, the industrialized countries which are at the origin of this new technology have to put up with PV within their own territories, prior to transfering it to the massive potential market of developing countries. As a consequence, it is imperative to find the ways and means of making PV economic in our industrial environment and if possible in the most proximate and general usage of energy : **private homes and commercial buildings**.

In a first step toward that direction, with the recent trend of the **building integrated** PV technology, where the area related balance of system cost is considerably reduced, it has be shown that the cost of PV electricity can be squeezed down to 40 US cents/kWh [2] .

To further demonstrate the economic viability of solar energy, we are testing a new concept where the hybrid solar modules **"Capthel"**, are installed underneath the upper panes of domestic verandas, and optimized to **generate simultaneously electrical and thermal energy** (See Fig 1, the "Ecothel" concept).

General context.

So far, PV manufacturers have been reluctant to combine electrical and hot air generation, especially with modules integrated into building architecture, for several reasons: PV module industry has not yet reached the point where custom size modules can be manufactured cost effectively; the most commonly available technology also giving the best PV efficiency performances, the crystalline modules, have a negative temperature coefficient of - 0.4 % / °C and as a consequence their PV performance is reduced by 10 to 15% when they run hot (50 to 65 °C); the roofing industry, the architects, the space-heating installers and the PV distributors, usualy have very little to share. They don't speak the same language, and they consider having enough specific problems to keep focusing on, within their own domains of competence.

Economic context.

If we combine the fonctions of photovoltaïc conversion (with a commercial efficiency ranging from 6% at 240 US $\$/m^2$, up to 12% at 560 $\$/m^2$) with the fonction of hot air generation (with efficiency varying from 30% for an added cost of 60$\$/m^2$, up to 60% for an improved version with added cost of 120 $\$/m^2$), one may transform two pieces of equipment into one monolithic module with a global efficiency of co-generation ranging from 36% up to 72%. With the same m^2 of panel it is now possible to generate 60W of electricity and 600 W of thermal energy. The hybrid module "Capthel" provides a triple advantage:

- it saves the absorber cost.
- it saves installation costs.
- it saves space .

Amorphous silicon modules.

Recent progresses in the industrial production of large area amorphous silicon modules have changed the economic deal. It is now possible to buy a large,

CH3365-4/94/0000-1012 $4.00 © 1994 IEEE First WCPEC; Dec. 5–9, 1994; Hawaii

robust photovoltaïc glass-pane (or plastic encapsulated stainless steel laminate) for 250 US $ / m² [3]. Calculations, confirmed by long run experimentations have shown that a-Si:H PV modules light induced relative degradation stabilizes within the range of 25-30% for single, 10-15% for stack [4] and less than 10 % for triple junctions configurations. In addition it has been shown that **stabilized a-Si:H PV modules behave as if they had a positive temperature coefficient of + 0.2 %/°C.** In other words, running them hot, partially anneals the Staebler-Wronski degradation. This assumption remains valid up to their initial efficiency since it results from the competition of two physical phenomena: a positive annealing effect on one side (estimated to be + 0.3% to + 0.4 % /°C for single junction devices with a junction temperature ranging from 50 to 100 °C), and the natural negative effect of the temperature on the gap of the material, normally resulting in a negative temperature coefficient (- 0.1% to -0.3 % /°C in the same temperature range) depending on the deposition process. In addition to their natural price advantage, the net result of this effect is to position single junction a-Si:H modules as preferred candidates for hybrid collectors

"Capthel" construction.

The "Capthel" hybrid solar modules will utilize existing commercial photovoltaïc modules such as the a-Si:H Unisolar RM 1225 of USSC. The first prototype is based on the Unisolar RM 1225 acting as the thermal absorber. One single pane of standard 6 mm glass - officialy approved by building industry standards - with a 4 cm air gap on front, provides external protection, thermal insulation and accomplishes the green-house effect; although it reduces the incident irradiance by 10-12 %, the electrical performance of the module remains acceptable thanks to the thermal annealing of the light-induced degradation. The air collector is made of aluminum tubes with a 1cm² square section. Back insulation is provided through a 4cm air gap and emissivity is reduced by means of back aluminum foils

Performances and improvements.

Through a selection of 4 parameters, we report thermal performances that have been measured during the summer 1994. We compare them with the therotical performances of a perfectly well insulated one glass-pane solar collector having the same geometry. In a hot air solar collector, the most important element is the heat exchanger. A modified version (*) - CAPTHEL 02, patent pending - is under construction : the heat transfer from the absorber to the air has been dramatically improved, by a totally different surface exchange geometry and the conduction losses from the edges have been reduced, resulting in the following parameters .

Parameter:	Theoretical	CAPTHEL01	CAPTHEL02
Static constant (W/m².°K^4)	$\sigma_0 = 3.610^{-8}$	$\sigma_1 = 4.610^{-8}$	$\sigma_2 = 4.110^{-8}$ (*)
Max temperature T air at 1000W/m²	T = 151 °C	T = 114 °C	T = 133 °C
Cut-off point: with To-Ti= 40 °C	E=170W/m²	E=350 W/m²	E= 215 W/m²
Flow F (m^3/h.m²) Efficiency R (%) with To-Ti = 40°C at 800 W/m².	F = 50 R = 69 %	F= 39 R = 55 %	F= 45 R = 63 %

Fig. 2 illustrates the cut-off point and gives the air flow and the thermal efficiency as functions of irradiance for a temperature elevation of 40 °C.

Applications

Domestic space heating through a rock thermal storage tank.

Summer conditionning of commercial buildings (verandas hot air extraction).

Solar grain dryer in the outback.

Economic evaluation of the "ECOTHEL" concept in the french context.

Fig 3. summarizes the cost of generated electricity and heat throughout the year for a private home and a commercial building. If we compare the home owner's energy bill - especially the one that use electric space-heating, like most of the modern houses in France - to the actual depreciation of the "Ecothel" system, it is clearly shown that the cost of **domestic solar energy comes rather close to competitiveness.**

References

[1] "Photovoltaïc commercial modules, which product for what market ?" Alain Ricaud, Cythelia, Proc. 12th EC PVSEC, April 1994, Amsterdam, pp.7-14.

[2] "Alternative approaches for evaluating the cost effectiveness of building integrated PV systems", P. D. Lung and al, Proc. 11th ECPVSEC, Oct.1992, Montreux, p. 1459.

[3] "Large area amorphous silicon based PV elements for buildings", J. Macneil and A. Paradis, Proc.11th ECPVSEC, Oct.1992, Montreux, p. 1432.

[4] "Stability and reliability of amorphous silicon PIN-PIN encapsulated modules", J.Dutta and all, Solems, Proc. 11th ECPVSEC, Oct.1992, Montreux, p. 545.

Fig. 1: THE "ECOTHEL" CONCEPT: INTEGRATED SOLAR SYSTEM BLOCK DIAGRAM

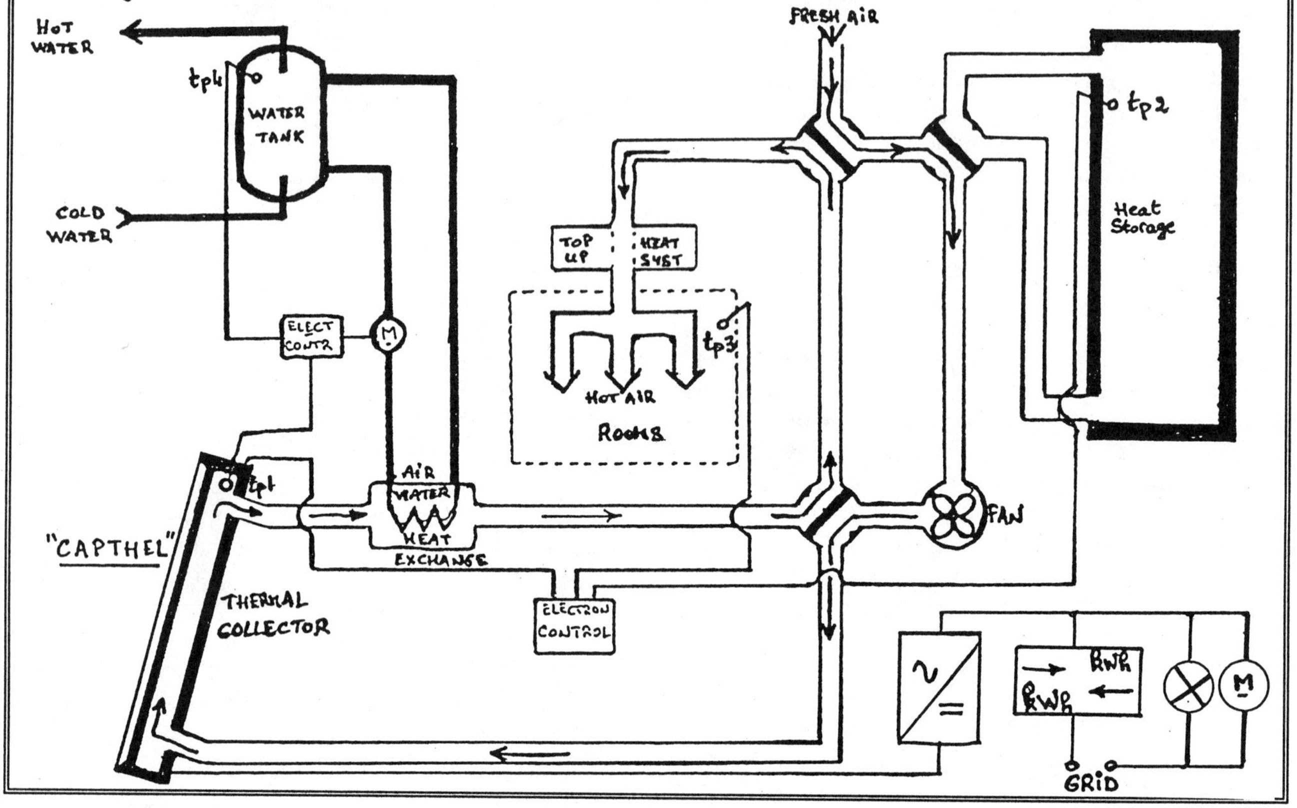

Fig.2: Air flow and thermal efficiency as a function of irradiance for To-Ti=40 °C
("Capthel 01"measured values; "Capthel 02" calculated values)

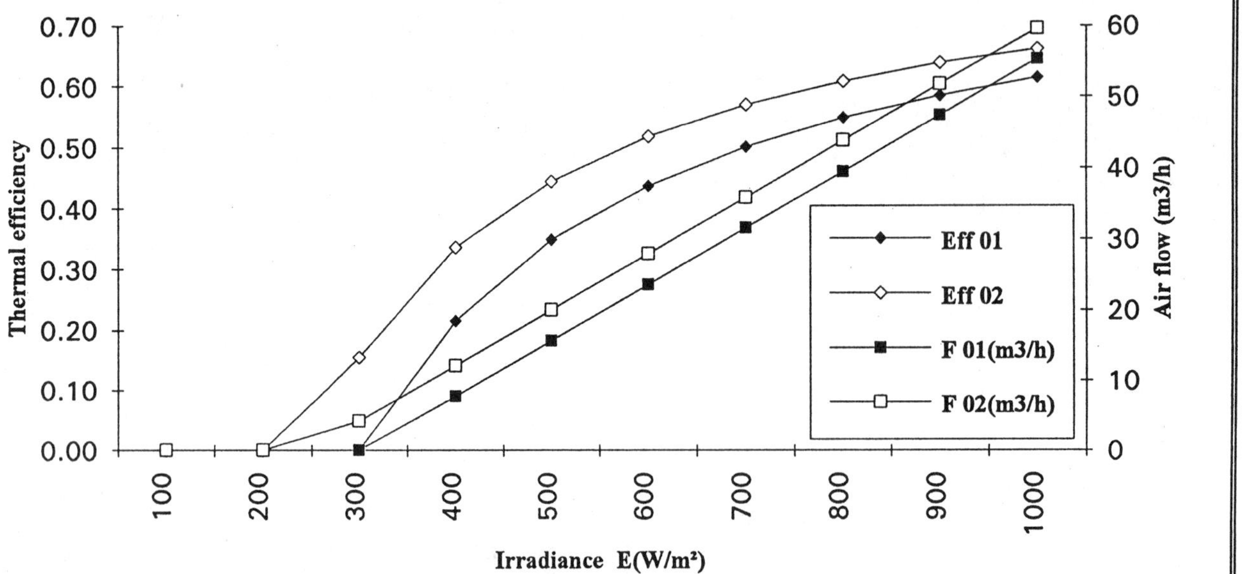

Ecothel concept with Capthel (1)	coef	Private home (2)			Commercial building (3)		
on façade (2) or on roof (3)		**summer**	**winter**	**average**	**summer**	**winter**	**average**
Building volume (m3):surface(...m²)*h(...m)	2.5			250			5000
Multi-functional modules array (m²)	0.2			20			200
PV system capacity at ...% eff (kWp)	6			1.2			12
TH system capacity at ...% eff (kWp)	60			12			120
Total peak power installation (kWp)				13.2			132
PV modules cost at ... $/m² (US $)	250			5000			50000
Inverter, cabling , connection to the grid.				2500			20500
Total cost of electric generator (US$)				**7500**			**70500**
Thermal generator adaptation at ... $/m²	120			2400			24000
Thermal system fans and regul (1000+...*m²)	20			1400			5000
Total cost of thermal generator (US$)				**3800**			**29000**
Façade adaptation at ... $/m²	50			1000			10000
Façade installation at (400 + ...*m²)	50			1400			10400
Avoided cost of bdg material and installation	80			-2000			-16400
Net cost of the façade function				400			4000
"ECOTHEL"Generator total cost (US$)				**11700**			**103500**
Depreciation time n (years)				20			20
Interest rate t (%)				5			5
Actualization coeff. Ka (%)				8.02			8.02
Maintenance cost coeff. Km(% invest)				2			2
Solar irradiance in the modules plane E(kWh/kWp/a)		640	760	700	1400	700	1050
Productivity coeff Kp (%)		85	75	80	85	75	82
Generated electricity (kWh)		652.8	684	668	14280	6300	10290
Generated heat (kWh)		6528	6840	6684	142800	63000	102900
Cost of generated electricity ($/kWh.el)		**1.15**	**1.10**	**1.12**	**0.49**	**1.12**	**0.69**
Cost of generated heat ($/kWh. th)		**0.06**	**0.06**	**0.06**	**0.02**	**0.05**	**0.03**
Cost of generated energy ($/kWh)		**0.16**	**0.16**	**0.16**	**0.07**	**0.15**	**0.09**
Electricity consumption (kWh/year)	12.5	1000	1500	1250	30000	15000	25000
PV share of electricity demand		65%	46%	53%	48%	42%	41%
Selling price of electricity to the grid (US$/kWh)				**0.04**			**0.04**
Cost of electricity from the grid (US $/kWh)				**0.12**			**0.12**
Heating consumption (kWh/year)	50	5000	12500	8750	50000	125000	87500
TH share of thermal energy demand		131%	55%	76%	286%	50%	118%
Average cost of electricity		0.79	0.57	0.66	0.30	0.54	0.34
Average cost of thermal energy		0.06	0.08	0.08	0.02	0.08	0.06
Average cost of total energy demand		**0.18**	**0.14**	**0.15**	**0.12**	**0.13**	**0.13**
Generator life time				20			20
Overall savings in life time (US $)		6783	9029	15812	77136	83160	160296
Savings % of actualized generator cost				**84%**			**97%**
Return on investment (years)				**24**			**21**
Building price (US k$)				100			1500
Generator cost % of building price				**12%**			**7%**

Fig 3: Economic evaluation of the "ECOTHEL" concept in the french context

[1] Double glazing with "Capthel"amorphous silicon modules connected to the grid; overnight storage ; no interseasonal heat storag

[2] Heating comprises domestic hot water; modules tilt angle 90°, optimized for winter.

[3] Air conditionning in building during summer; modules tilt angle 45°, optimized for summer.

PV-MODULES AT THE FACADE OF AN INNER COURTYARD - MEASUREMENTS AND SIMULATION

H.G. Groehn, H. Barthels
Institute of Energy Process Engineering (IEV),Research Center Jülich (KFA),D-52425 Jülich

ABSTRACT

For a PV generator installed at the facade of an inner courtyard the irradiance losses due to shadowing had to be assessed and the PV modules were to be connected in such a way that the energy loss added up during the year is minimized. Besides the direct radiation losses, the buildings surrounding the courtyard also cause a permanent reduction of the view factor, i.e. the sky section from which the modules receive light, which is different for every module of the facade. The influence of the diminished view factor on diffuse radiation under different irradiance conditions was calculated and validated by measurements. Profiles of the shadow, two different PV-module connection and the influence of a different number of bypass diodes on energy conversion are investigated.

SCOPE OF PROBLEMS

A more extensive spread of photovoltaic electricity production also has to consider the facades of existing buildings for the installation of PV generators. More often than is the case for PV modules on roofs, the problem of partial shadowing will be of major importance. The level of energy losses caused by shadowing will be illustrated in the following by the example of a 12.2 kWp solar facade already constructed.

Of course, in contrast to a freely arranged module surface, the shadow cast by the buildings surrounding the courtyard must be considered as a main parameter in calculating the photovoltaically convertable energy. It is thus relatively easy to derive direct radiation losses for the individual solar cells of the PV generator. In addition, however, the buildings surrounding the courtyard also reduce the sky section from which the solar facade can be diffusely irradiated. Finally, the diurnally and seasonally different irradiance on adjacent modules raises the question of the criteria according to which the PV modules should be connected in strings.

IRRADIANCE LOSSES DUE TO SHADOWING

The time and duration of shadowing is different for each point of the solar facade due to its geometrical arrangement with respect to the buildings surrounding the courtyard. Fig. 1 shows the annual shadow time in relation to the time of possible direct irradiation. This does not include the time during which the whole facade cannot be directly irradiated due to its orientation to the SE. The coordinates in Fig. 1 are the length and height of the facade. Moreover, the module distribution on the facade is plotted. As expected, maximum shadow times of more than 40 % occur for the lower corners at the SW and NE edge of the facade, whereas the shadow time is < 10 % for a large number of modules in the

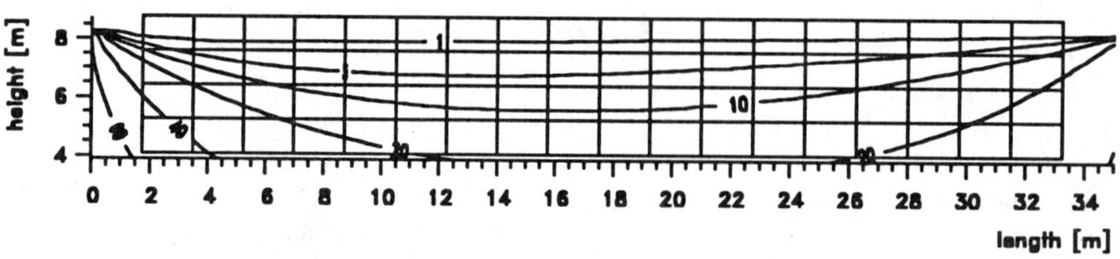

Fig. 1 Percantage of shadow of the annually possible direct irradiation time of the SE-facade of the inner courtyard

The installation of a 43 kWp PV system at the Central Library of the Research Centre Jülich as part of an autonomous solar electric energy supply of the building /1/ also included the vertical facade of the inner courtyard of the library oriented to the southeast for solar electricity generation. The 35 m long wall is covered with 18 PV modules arranged horizontally over a length of 31.5 m, while the vertical extension of the PV field is 4.8 m with four modules arranged above each other. The module surface is 1.75 m x 1.17 m. The module arrangement is shown approximately true to scale in Fig. 1. The inner courtyard is surrounded by buildings with a height only 0.46 m lower than the upper edge of the module field. The depth of the courtyard is 28 m. The PV modules are arranged in the form of a curtain facade approx. 0.4 m in front of the building wall so that the rear of the modules is also ventilated.

central and upper region of the facade.

The permanently existing reduction in diffuse radiation due to the buildings surrounding the courtyard will be quantified in the following. The sky section from which a solar cell is irradiated is to be described by its view factor, EZ. This is the projection of the solid angle of the radiance-emitting plane into the radiance-receiving plane relative to the projection of the whole sky hemisphere /3/. The view factor is thus EZ = 1 for a horizontally arranged plane and EZ = 0.5 for a vertical plane which only "views" half the sky. The reduction ΔEZ of the view factor due to the buildings surrounding the courtyard is different for each solar cell of the PV field. It is greatest for the cells at the lower edge of the field next to the adjacent buildings, where it amounts to about ΔEZ = 0.22, averaged over a PV module with 150 cells. This means that with EZ = 0.5 - 0.22 this module only receives diffuse

CH3365-4/94/0000-1016 $4.00 © 1994 IEEE

First WCPEC; Dec. 5-9, 1994; Hawaii

radiation from approx. 60 % of the sky region from which a PV module not shielded by adjacent buildings is irradiated. The size of the radiation-reflecting surfaces increases inversely proportional to the reduction of the view factor.

An assessment of the influence of a diminished view factor on photovoltaic energy conversion is difficult due to the fact that diffuse radiation is not isotropic, but is rather a complicated function of turbidity, solar altitude and also of the azimuth of the sun and solar cell in the case of inclined surfaces /2/. An attempt was therefore made to empirically evaluate the influence of the diminished view factor on the level of diffuse radiation.

In order to determine the relation between diffuse radiation, Dcl, on a freestanding wall and that on a point of the facade with diminished view factor, Dclv, the signals of two CM11 pyranometers from Kipp & Zonen were used, one of them positioned at the upper edge of the SE facade, and thus undisturbed by neighbouring buildings (EZ=0.5), to measure the global radiation, Gcl, and the second installed at the lower edge of the module field to measure the diminished global radiation, Gclv. At the same time, an Eppley pyrheliometer was used to measure the direct radiation, BcNo, in order to determine the direct radiation component, Bcl, on the facade. The view factor for the lower pyranometer was calculated as EZ=0.379, reduced by the fraction $\Delta EZ=0.121$ of the adjacent walls. Pyranometer data were first evaluated for two selected days, i.e. a cloudless day with anisotropic diffuse radiation and a cloudy day with a diffuse radiation distribution assumed to be isotropic. Fig. 2 and 3 show the global radiation values, Gcl and Gclv, measured for the two days. Over the largest portion of the day, the pyranometer arranged at the lower edge of the PV field shows a 10 .. 15 %

Fig.2-5:Comparison of global irradiation measurements at the upper (Gcl) and lower (Gclv) edge of the module field. Calculation of Gclv with equation (4)

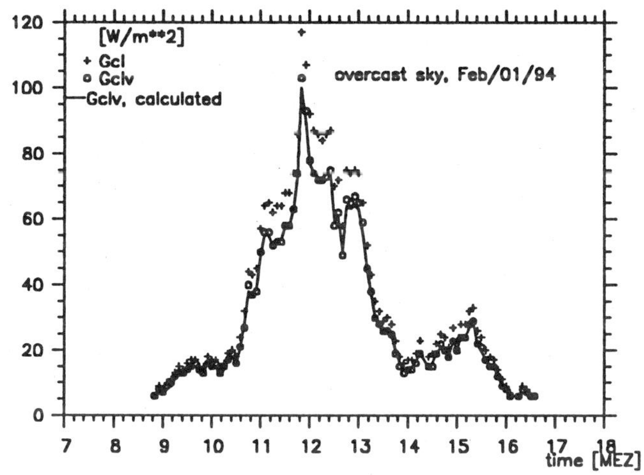

Fig. 2: Cloudy day, /Feb/02/94.
Position of Gclv: Length 8.9m, Height 3.9m

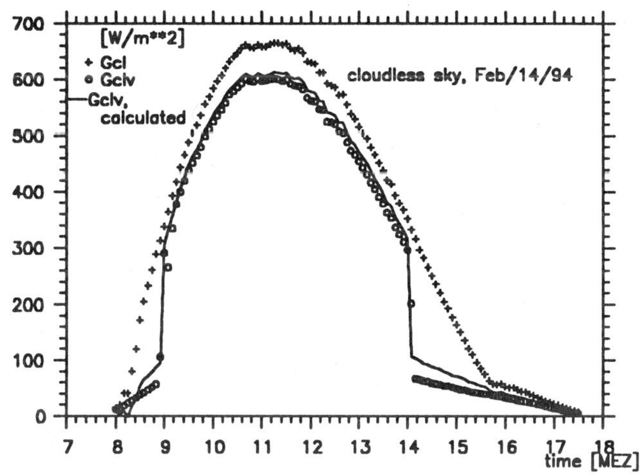

Fig. 3: Cloudless day, /Feb/14/94.
Position of Gclv: Length 8.9m, Height 3.9m

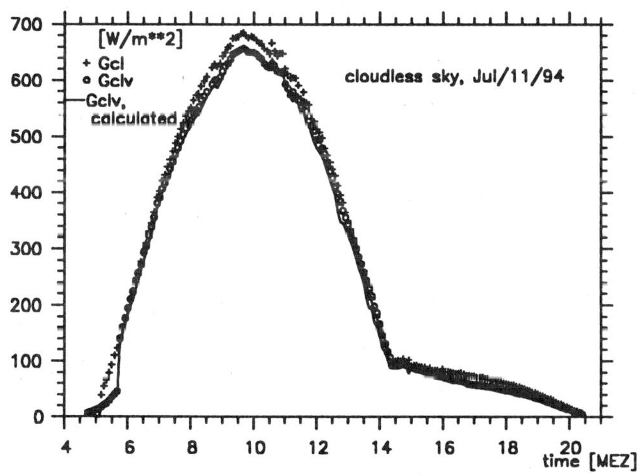

Fig. 4: Cloudless day, /Jul/11/94
Position of Gclv: Length 8.9m, Height 3.9m

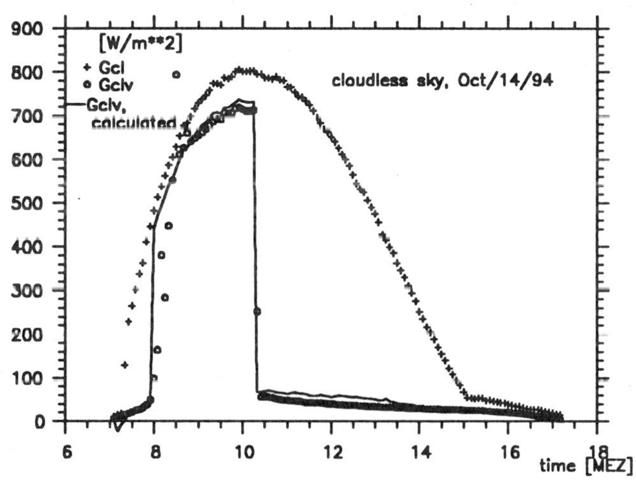

Fig. 5: Cloudless day, /Oct/14/94.
Position of Gclv: Length 2.1m, Height 3.9m/94.

smaller value than the instrument installed at the upper edge. For the cloudless day, a steep Gclv increase is observed at 9.00 a.m. and an equally pronounced decrease of the signal at about 2.00 p.m. At these times of the day, the building shadow leaves and reaches the pyranometer arranged at the lower edge of the SE facade.

For the analytical simulation of Dclv, the diffuse radiation fraction directly reaching the pyranometer, DcSkl, and the fraction reflected from the ground and walls, DcGrl, were corrected in a first step in the ratio of the view factors calculated for the upper and lower edges of the PV field.

$$Dclv=(0.5-\Delta EZ)/0.5*DcSkl+(0.5+\Delta EZ)/0.5*DcGrl \quad (1)$$

The relations

$$Dcl=DcSkl+DcGrl, \quad DcGrl=AL*0.5*Gc, \quad Gcl=Dcl+Bcl \quad (2)$$

give

$$Gcl=(0.5-\Delta EZ)/0.5*Dcl+2*\Delta EZ*AL*Gc+Bcl \quad (3)$$

possible to determine the diurnal sums of global radiation at the lower edge of the PV field from the measurement of global radiation at the upper edge for the days shown in Fig. 2 and 3 with deviations of less than 1 %. Measurements on 11.07.94, i.e. with higher solar altitude, Fig. 4, and with the pyranometer measuring the Gclv positioned at the lower SW edge of the PV field involving a reduction of the view factor by $\Delta EZ=0.219$, Fig. 5, showed, however, that the term given in eq. (3) does not describe the albedos with sufficient accuracy for the whole year and for all points of the wall. At elevated solar altitude, as shown in Fig. 4, the albedo is absolutely and relatively greater than according to eq. (3), so that the difference between Gcl and Gclv is smaller than, for example, in Fig. 3. A more precise description of the albedos requires a separate consideration of the reflection of irradiance on the individual facades surrounding the courtyard and on the ground. Such a calculation was performed, but the computational effort was very high and not justified for a simulation of the irradiance on the whole solar facade. A weighting of the data measured for global radiation, Gclv, at different times of the year and different points on the wall led to the result that, in the case of a cloudless sky, the albedo increase due to the buildings surrounding the courtyard should not be included, but the relevant

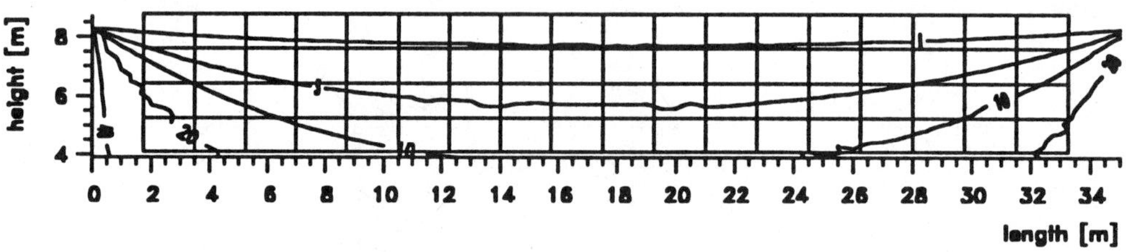

Fig. 6 Annual global radiation losses[%] at the SE-facade due to adjacent buildings

Tab.1 Annually solar irradiation on the SE inner courtyard facade	module field (72 modules)	entire facade (80 modules)
irradiance on a freestandig facade [kWh]	98480	109421
irradiance on the inner courtyard facade [kWh]	92872	101196
direct beem losses [%]	1.6	2.5
global irradiance losses [%]	5.7	7.5
shadow time [%]	11.2	13.6
average view factor	0.472	0.428

For the cloudy day with approximately isotropic diffuse radiation, a very precise description of Gclv was achieved with the selected approach for the lower edge of the PV-module field. An albedo of AL=0.2 was selected for the determination of DcGrl. A smaller or greater value, tested with the example of the cloudy day, led to less good agreement between measurement and simulation.

The selected correction did not prove to be sufficient for the cloudless day with anisotropic diffuse radiation. A numerical simulation of the global radiation Gclv was achieved by assuming in analogy to /4/ that 60 % of the diffuse irradiation would directionally come from the environment of the sun with a view factor of EZ=0.29 /5/. With the equations specified in /5/ it was

term in eq. (3) should be retained for the cloudy sky. This is achieved by introducing the ratio of measured direct irradiant, Bcl, to the direct radiation to be theoretically expected according to /2/, Bcl_0. The global radiation, Gclv, onto an arbitrary point of the facade can thus be determined from radiation measurements at a location not disturbed by adjacent buildings as a function of the view factor according to the following equation:

$$Gclv=(0.5-\Delta EZ)/0.5*Dcl + 2*\Delta EZ*AL*Gc*(1-Bcl/Bcl_0) + Bcl \quad (4)$$

The global radiation Gclv determined with this equation is plotted as an additional curve in Fig. 2-5. The maximum deviation between calculation and measurement for the diurnal sums of irradiance energy is 8 %. The accuracy of the simulation of Gclv should thus be sufficient for determining the photovoltaic energy conversion of the modules of the SE facade. The irradiance losses to be expected during the year were estimated by means of eq. (4). Average radiation values were used for direct, diffuse and global radiation (Bml, Dml, Gml) which cover the average sunshine duration and were determined in /2/ for the Jülich site from several years of weather observation. The calculation result is shown in Fig. 6. The annual global radiation losses are much lower than to be expected in view of the shadow times in Fig. 1. The global radiation losses only amount to about 1 % for the upper module row and are only > 20 % for the two modules at the lower edge of the module field. The results are further broken down in the following table, Tab. 1. Column 2 of Tab. 1 contains the data for the solar facade already constructed, and column 3 shows the values for the case where the entire SE facade would be covered with PV modules. The global radiation losses amount to 5.7 % in comparison to a freestanding facade without shadow. It is remarkable to note that the direct radiation loss caused by shadowing only contributes about 1/3 to the global radiation loss. A comparison of the irradiance energies in columns 2 and 3 suggests that, in the event of the whole facade being covered with modules, the global radiation loss would amount to approx. 24 % for the eight additional modules at the facade edges. In contrast, the irradiance losses of the eight edge modules only amount to 11 % for the 72-module field constructed.

INFLUECE OF SHADOWING ON PHOTOVOLTAIC

In order to determine the electric energy gain from the PV generator, it is necessary to know the current-voltage characteristic of the solar cells and module connection used. The solar cell characteristic was determined on the basis of manufacturer specifications and from field measurements at the complete PV generator. At the SE facade, 9000 monocrystalline solar cells of 0.1 x 0.1 m^2 each are arranged, half of them in modules of 100 cells and half in 150-cell modules. A maximum of 20 cells are provided with a bypass diode for reasons of module safety. Four to six modules with a total of 600 cells are connected in series and the whole module field consists of 15 strings connected in parallel. The calculation of the module connection includes losses due to bypass and string diodes as well as the line losses to the MPP controllers at a distance of approx. 200 m. The input of a different diffuse radiation for virtually each module or submodule due to different view factors also means the determination of an individual characteristic for each module or submodule. In addition to the radiation values according to /2/, the input of module temperatures is also required. They were preliminarily determined from monthly air temperature averages for every hour of the day according to computing rules for freestanding modules /6/.

The PV-module connection was optimized so that modules with similar global radiation losses were combined in a string. The energy production of this connection was compared with that of a non-optimized design combining adjacent modules in a string. Moreover, it was checked to what extent bypass diodes can help reduce shadowing losses. The calculation results are summarized in Tab. 2. In order to determine the annual summation values of converted energy, calculations were carried out for every fifth day of the year at minimum intervals of five minutes. A verification with shorter intervals did not show any significant deviations.

Tab.2 Annual losses of electrical energy at the SE inner courtyard facade due to the surrounding buildings	[%]
freestanding facade, modules unconnected	100
SE inner courtyard facade, modules unconnected, 1 bypass diode per module	93.2
SE inner courtyard facade, optimized module connection, 1 bypass diode per module	89.9
SE inner courtyard facade, optimized module connection, 1 bypass diode for 20 cells each	90.4
SE inner courtyard facade, module connection not optimized, 1 bypass diode per module	88.7

In Tab. 2 the energy production of a freestanding facade not impaired by adjacent buildings is put equal to 100 %. Adding up the energy input of all modules of the shaded inner courtyard facade, without taking account of their connection, gives a production loss of 6.8 % compared to the unshaded facade. This higher loss in comparison to the global radiation loss of 5.7 % from Tab. 1 is due to the fact that a module or submodule, once it is partially shaded, is only considered to be exposed to diffuse radiation. The selected optimized connection gives connection losses of 3.5 % assuming only one bypass diode per module. On the assumption of five bypass diodes per module, i.e. a module design very close to that actually performed, the connection losses decrease to 3.0 %. Adding up the losses due to shadowing of the facade and module connection reduces energy production by 9.6 % in comparison to a freestanding facade. The last line in Tab. 2 contains the annual energy production for a module connection combining, as far as possible, adjacent modules to strings with 600 cells each. This gives an output loss of approx. 1.4 % compared to the optimized connection.

SUMMARY

For the calculation of irradiance on a PV generator mounted at the vertical facade of an inner courtyard, it is also necessary to include the reduction in diffuse radiation due to the buildings surrounding the courtyard, which reduce the sky section from which the facade can be irradiated, in addition to the reduction in direct radiation due to shadowing. The view factor as a measure of the radiance-emitting sky section is different for each module of the facade. Diffuse radiation is proportional to the view factor in the case of isotropic distribution, and the anisotropy of diffuse radiation must be taken into account by means of a correction in the case of a clear sky, for which a preliminary approach is given here. The reduction in annual global radiation input is approx. 5.7 % for the SE inner courtyard facade under consideration, the loss of electrical energy, adding losses due to the model connection to the global radiation losses, amounts to approx. 9.6 %. The module connection was optimized so that, as far as possible, modules with approximately the same annual irradiance were connected in series.

REFERENCES

/1/ H. Barthels, W.A. Brocke, H.G. Groehn, P. Ritzenhoff
Realized Design of an Autonomous Solar Energy Supply, ISES Solar World Congress, Budapest 1993, Proceedings Vol. 2, pp. 431-436

/2/ P. Ritzenhoff
Erstellung eines Modells zur Simulation der Solarstrahlung auf beliebig orientierte Flächen und deren Trennung in Diffus- und Direktanteil
Bericht des Forschungszentrums Jülich, JÜL-2600, Jülich 1992

/3/ E.R.G. Eckert
Wärme- und Stoffaustausch, 3rd edition, 1966, Springer-Verlag

/4/ J.K. Page (editor)
Prediction of Solar Radiation on Inclined Surfaces
Solar Energy R&D in the European Community, Series F, Volume 3
D. Reidel Publishing Company, Dordrecht 1986

/5/ H.G. Groehn, H. Barthels
PV-Module an der Fassade eines Innenhofes - Messung und Simulation, 9. Intern. Sonnenforum, Stuttgart, 1994, Proceedings, pp. 662-669

/6/ H. Karl
Photovoltaische Systeme - Grundlagen zur Dimensionierung und zum Betrieb,
Dissertation, Stuttart 1984

DEVELOPMENT, TESTING AND FIRST INSTALLATIONS OF AN INTEGRATED SOLAR ROOF SYSTEM

Bruce M. Cross
Energy Equipment Testing Service, University of Wales, Cardiff, UK

ABSTRACT

There are some aspects of PV systems which must be carefully considered in the detail design stage. These are a consequence of the nature of the PV conversion (i.e. that the efficiency reduces with increasing temperature), of the requirements of the building envelope (i.e. weatherability) and safety requirements (i.e. high voltage DC systems).

This paper reports on a program to design, test and commercialise a system within these constraints. The concept was tested in full scale prototype in the indoor solar simulator at the University in Cardiff, both for operational performance and for durability and weatherability. A first installation has just been completed in Oxford UK.

DESIGN CONSIDERATIONS

PV system

The PV system operates at higher efficiency the lower is the operating temperature. The ambient temperature is outside the designers control, but the loss of heat to the ambient is greatly affected by the detailed design. In a frame mounted installation the modules are cooled by free air flow over both front and rear faces. This generally limits the operating temperature to around 45°K above ambient. In a building integrated installation the module (or laminate) is only cooled by free air on the front face, the flow of air behind depending on the method of attachment to the building structure. Thus the design should incorporate air flow behind the module to encourage heat transfer into the air, and inlet and outlet positioned to maximise stack effect in order to promote maximum flow.

Building Envelope

The requirement for a roof is to keep out rain, retain heat and to resist wind and snow loads. In addition there are many regulatory considerations. It is generally accepted that a roof will have both a primary and secondary waterproof system. The PV modules (or laminates) may form the primary waterproof layer, or indeed may act as only a rainscreen with both the waterproof layers beneath.

Safety

The use of high voltage DC systems (up to 300V) is not normal practice in buildings and requires careful separation from AC mains wiring. On the roof, components are operating in outdoor conditions, and must be tested and certified for use in such manner.

PRELIMINARY WORK

A DTI/ETSU funded project was completed in 1993 which built and tested variants of full scale roof systems. The testing was performed in all operating conditions (See Fig. 1) and the final report identified an approved system for demonstration in a real installation. (See Reference)

The systems were tested for electrical performance, peak temperature, water ingress under maximum rainfall, condensation removal, and thermal shock. The test roof was fitted with modules from various manufacturers, in each case in a vertical configuration (up the roof) with air inlet at the eaves and exit at the ridge. It was found that the air flowing in the 'duct' created behind the modules increased in temperature as it rose up the roof, leading to higher temperatures for the modules closer to the ridge (Fig. 2). This puts a limit (estimated at 5m) on the eaves to ridge length that can be used without intermediate venting (a complex design) whilst maintaining a reasonable operating temperature.

Rainfall was simulated using an array of spray nozzles to produce worst case conditions. The system is designed with internal drains to allow for the various installation errors that occur on site. These coped with ingress even when cap strips were removed. Secondary protection is provided by a high performance sarking as the lower internal surface of the rear ventilation duct.

FIRST INSTALLATIONS

Utilising the experience gained during prototype testing, a set of aluminium extrusions was defined that would fulfill the same functions and be easy to assemble with a minimum of site work. It was also

CH3365-4/94/0000-1020 $4.00 © 1994 IEEE First WCPEC; Dec. 5–9, 1994; Hawaii

to be suitable for a broad range of other functions, such as use with roof-lights, gable windows and solar hot water panels. The system was to be self supporting over reasonable spans, to allow simple fixing of

Fig 1 Prototype test in Solar Simulator - Cardiff

inner cladding, and to be applied in the widest possible way. Two installations are in progress (November 1994) and two others are planned for 1995.

The main vertical members are attached through the sarking into counterbattens on a standard insulated and boarded roof. Cross-rails are then attached using pre-drilled holes. Each module is pre-wired, framed and opaque coated in the factory and is positioned on the roof. The cable is laid through clips attached to the cross-rails up to the ridge where it passes through a conduit into the roof space. Thus, there are no connections that need access from the roof. This increases the cable lengths but improves overall reliability of the system, reducing maintenance costs. Cap strips provide the clamping force and may be varied in profile and colour to suit the architecture of the building.

Fig 2. Module Temperatures - Trial Roof

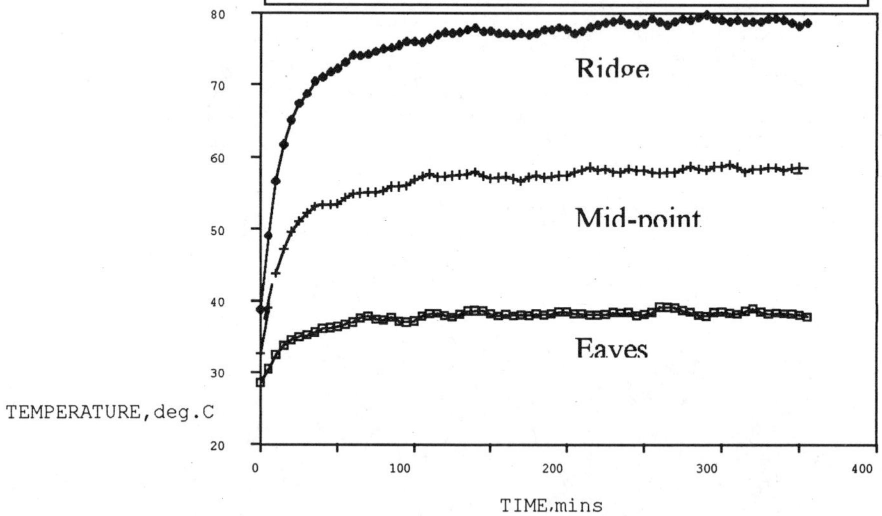

Eaves details have been developed that allow the water to run off at two levels (from module face and inner gulley/sarking) and allow air to freely enter the cooling duct whilst providing a vermin screen. Ridge details are simpler in that run-off is not a consideration. An abutment vent has also been designed for 'lean-to' applications.

An important part of the system is its ability to link with standard roofing systems where not all of the roof is constructed for PV use. The interface with both slated and tiled roofs has been demonstrated in the test roof program. A dummy glass panel has also been used as infill where non standard spacings cannot be avoided and where angled cuts are needed at valleys. These panels may be laminated with a plastic interlayer or screen printed rear layer to match the colour of the PV modules. Other flashings are available to link with roof lights and similar components. A solar domestic hot water heating absorber has been designed to integrate with the system, presenting a uniform external appearance for all devices. A representative solar DHW system has been installed on the first PV roof in Oxford.

ECONOMICS

The cost of the integration system in production is comparable with patent glazing systems. Thus for installations when the PV laminates replace a glazed roof, the excess cost is only due to PV and electrical components, the mounting and installation being equivalent to the non-PV option.

For opaque roof systems the avoided cost is typically the outer covering and battening costs. In this instance, the avoided cost represents only a small part of the integration system cost (20-30%) making the total system cost (£/Wp) higher, (See Table 1). Costs are estimated for moderate scale implementation (1000 roofs/yr). It will be seen that the avoided costs reduce the system cost by one-third for a 'high tech' roofing system.

COMMERCIAL EXPLOITATION

The system has been designed and developed by the Energy Equipment Testing Service at UWCC in collaboration with Inscape Architects (Bristol) and CAT (Machynlleth). It is presently being utilised in two installations with two more planned for 1995. The first installation at Oxford (UK) for Dr. Roaf (Oxford - Brookes School of Architecture) was completed at the end of November and a view of the installation is shown in Fig 3. It should be noted that this first roof demonstrates the full potential of the system since it encorporates 4kWe of PV modules, $5m^2$ of solar domestic hot water panels, and two opening rooflights. The installation will be monitored under an ETSU program to provide data on all PV cladding installations in the UK.

A second installation on an enclosed walkway for an office development is in progress in central England. A third installation on a solar conference centre in mid-Wales will take place during 1995. The site is presently under preparation. The fourth project, an integrated garage roof for a wind/solar electric car program is scheduled for late 1995.

Discussions for licencing its use, replication and manufacture overseas are underway.

REFERENCES

Feasibility and Design Study of a Photovoltaic Building Demonstration
ETSU S/P2/00158/REP
Energy Technology Support Unit, Harwell, Oxfordshire UK.

Table 1. Costs

Roof Type	Avoided Cost $\$/m^2$	PV System cost $/Wp
None	0	10
Tiled	40	9.7
Slated	120	9.5
Patent Glazed	225	8.7
Advanced	600	6.6

Fig 3. PV roof under construction at Oxford - UK.

TEST FACILITIES AND SAFETY REGULATIONS FOR ROOFTOP MOUNTED AND GRIDCONNECTED PV SYSTEMS

J.H. Boumans, A.J.N. Schoen
Ecofys Research and Consultancy, P.O. Box 8408, NL-3503 RK Utrecht, The Netherlands

S.A.M. Verhoeven
KEMA Transmission & Distribution, P.O. Box 9035, NL-6800 ET Arnhem, The Netherlands

ABSTRACT

In order to reach large scale implementation, a PV roof has to comply with building and electrical codes. In the Netherlands, two R&D activities are running in this field: a test facility for PV roofs and the development of protection and safety regulations for building integrated PV systems.
In the test facility for PV roofs, the following aspects are considered: safety, durability, fire resistance, energy efficiency and building physics. Two PV roof integration systems have been tested, results indicating that the PV building quality of these systems can be guaranteed for more than 5 years.
Concerning the electrical safety, a dedicated test installation has been realized by KEMA. In this test facility for electrical safety and protection, faults can be applied to examine the electrical system behaviour. The results of these measurements will lead to guidelines how to design and protect a safe low-cost grid connected PV system.

1. GENERAL

Quality testing and certification of photovoltaic systems is essential, in order to reach large scale application. This is specifically true for the application of building integrated PV (BIPV), since quality issues are of great importance in the building sector. All building components must fulfil a well-defined level of performance, specified in the building codes. These levels of performance include water tightness, fire resistance, noise protection and durability. Quality requirements for houses, constructed under the "GIW-specifications" (guaranteeing high quality) are even more severe than the building codes.
Concerning electrical issues, the quality standard of building integrated photovoltaics must be high. No hazardous situation is allowable. Damage, caused by malfunctioning of a building mounted PV system can be very severe in comparison to free standing PV systems. Fire in an array field is not desirable, but the effects will be limited, whereas fire on a roof may lead to high costs and casualties, which is not acceptable. Damage, caused by a stroke of lighting on an insufficiently protected, building mounted system can be substantial, whereas a stroke of lighting on a free standing system might cause no harm.

In order to reach large scale application of PV systems, both the building and electrical quality has to meet general accepted standards. Only then the building and electrical industry as well as the utilities will accept PV as a safe and reliable technology. Without this acceptance, large scale introduction of PV's will not be possible.
For this reason, a number of testing and certification activities have been started in the Netherlands, focussing on both the building quality as well as the electrical quality and safety. The aim is not only to test and certify PV systems according to national standards, but also to use the results of the tests for the development of more cost-effective solutions.

2. BUILDING QUALITY OF PV SYSTEMS

By order of NOVEM, a test facility for PV roofs has been realized at BDA Buro Dakadvies BV (see Fig. 1). The set-up of the test facility is based on a preliminary study [1].
The test facility enables testing of PV systems, integrated in pitched roofs. In the facility, both the support structure and the angle of inclination can be varied. Watertightness and durability tests can be carried out in the facility.
In addition, the national building code requires approval of the following aspects: safety, fire resistance, sound transmission, energy efficiency and building physics. This

Fig. 1. The BDA test facility during temperature cycling (as part of the durability testing)

approval may be reached by testing (fire resistance) or by design inspection by a certified institute (safety, sound transmission)

CH3365-4/94/0000-1024 $4.00 © 1994 IEEE

First WCPEC; Dec. 5–9, 1994; Hawaii

Recently, two different PV roof systems have been tested and inspected in accordance to the building code [2,3]. Tests are based on national building regulations and standards. It must be noted, however, that no PV-specific building codes are available. Only general codes can be applied. Specific codes for other building elements can be translated for the application of PV.
For this reason, specific quality aspects, brought forward in a steering committee, have been included in these tests. The main national research institutes in the field of roof constructions (BDA, TNO, Bouwcentrum Advies) are represented in this committee. The steering committee also reviews the results of tests and design inspections.

The PV roof

The construction of both PV roof systems consists of a frame work of aluminum profiles combined with frameless modules (laminates). The profiles originate from greenhouse technology, where they have been applied successfully for many years. Vertical profiles consist of an aluminium lower part and a plastic cornice. Horizontally an aluminium H-profile is used. Basically the same mounting system has been used in several Dutch demonstration projects, though minor changes have been made in order to improve mounting on the roof.

Fig. 2. Mounting PV laminates in the profile system

The PV laminates and aluminium profiles are mounted on a conventional roof construction (omitting regularly used tiles). Between the PV laminate surface and the roof construction, a ventilation gap is included (see Fig. 2).
The first PV roof consisted of only PV laminates (in a 'portrait' configuration), whereas the second roof included both a solar collector and PV laminates (in a 'landscape' configuration). In the second system, special attention had to be paid to the connection details between the two solar systems. Moreover, the 'landscape' configuration of the roof was expected to have consequences for the construction safety, requiring additional assessment.

Watertightness

The test on watertightness is carried out in compliance with the NEN 2778 standard 'Moisture in buildings. Method of assessment'. The raintightness of PV-modules is tested according the comparable test described in the NEN 7044 standard 'Test method for concrete tiles'. In this test, it is stated that water penetration of the weather skin may not exceed 2%.
The test consists of spraying the roof at a certain static pressure on the weather skin. The level of this pressure depends on the actual location of the system, defined by the average wind climate of the location, the height of the building and the wind roughness of the surroundings. A positive result in the test is achieved if the roof surface underneath the PV laminates is visually dry.
The watertightness of the PV roof is tested twice (before and after the durability test - see next section).

Durability

The scope of this test is to examine the influence of frequent temperature cycling on the construction. No changes may occur which might influence the normal functioning of the construction. This so-called "thermo-shock testing" consists of heating the construction with infra red lights and cooling the surface with cold water. During the heating of the construction, the temperature will rise to 80 C. After heating, the temperature of the construction will in five minutes fall to a level of 15 C. This is repeated for 500 times which stands for an average number of temperature cycles during a 10 years period.

Construction safety

The safety of the construction is tested using the so called "airbox method", in compliance with the UEATc-guidelines and the NEN 6707 standard. This is a test on the mounting system, whereby a piece of the structure is loaded with a dynamic windpressure until it collapses. The number of cycles and the maximum allowable windpressure are a measure of the lifetime of the mounting system and the maximum circumstances under which the mounting system can be applied.

Moisture regulations

The surface of a PV roof can be seen as a watertight envelope. The ventilation depends on the gaps at the lower and upper side of the PV roof. Due to the heating up during day and cooling down during the night (below ambient temperatures), condensation at the rear of the PV modules can be expected. Source of moisture is the open air but also vapour from the inside of the building. With the aid of the simulation model MATCH it has been determined that the latter needs special attention, even if there is sufficient ventilation of the PV roof. Condensation problems can be solved by making the construction between the roof and the house as airtight as possible.

Noise protection

Considering noise protection and sound transfer, there is a large difference between a PV roof and a conventional tile roof, especially if the PV system lies over more than one house. The PV roof is a stiff construction built up of solid materials; a tile roof will absorb much of the noise. Concerning noise transmission, an extra problem can be expected at the wall between two houses. A study has made clear that noise problems can be prevented by disconnecting the construction at the side wall.

Fire resistance

In order to assess the fire resistance of the PV-roof, two experiments have been done in compliance with the NEN 6063 standard. In this test the effect of fire falling on the roof is examined. A basket with burning woodwool is placed on the roof and left burning for at least 20 minutes (see Fig. 3), after which a visual inspection takes place.

Fig. 3. Testing the fire resistance

There are two conditions which have to be met: the dimensions of the burnt area on the roofing surface may not increase, and no fire penetration may have occurred. In the first experiment, a detail of a PV-roof with a vapour open foil underneath the PV-laminates has been used. In the second test, PUR insulation underneath was PV surface used. In both cases, the PV-roof stands the test.

Results

Two roof mounting systems have been tested according to the test descriptions mentioned before. The steering committee, in which the main national research institutes in the field of roof constructions participate, has evaluated the results of these tests and assessments.
Concerning *watertightness*, both systems withstand the tests, though improvements of the original designs had to be made. These improvements in general dealt with the interconnections of different roof areas (e.g. between the PV and solar thermal area).
According to the results of the tests, no problems with *durability* can be expected. A life time of more than 10 years seems to be feasible.
Results of the *constructional safety* tests indicate that the PV roof design in question is applicable for every regular location in the Netherlands.
In the *Fire resistance* tests, no damage occurred.
Concerning *noise protection*, specific measures have to be taken in order to avoid too much sound transmittance between connecting houses, via the PV roof.
Moisture-regulation: special attention during construction of the houses is required. The roof construction must be as airtight as possible, and examination of the actual work on site is required.

Conclusions

Two different PV roof systems have been tested and inspected in accordance to the national building codes and standards. Overall. it is concluded that no major problems are to be expected in the application of these mounting systems. This overall conclusion has been drawn up by a steering committee in which the main research institutes in the field of roof constructions participate. This conclusion is shared by the 'GIW'-institute. It can thus be expected that the tested PV technologies will be acceptable for the building sector.

3. ELECTRICAL SAFETY AND PROTECTION

PV systems connected to the electricity network have to comply with electrical safety and protection regulations. These regulations prevent damage, both direct (e.g. electrical shock) and indirect (e.g. fire hazard) to either the PV system, the electricity grid and persons.
IEC 364 gives regulations how to design an electrical system with respect to electrical safety. A PV system generates direct current and has a very low short circuit capability. The countermeasures given in IEC 364 are therefore not always adequate.
A PV-system generates electricity and needs a special protection to prevent power injection from the PV system into the electricity network when this network is distorted. A PV system is thus comparable with an individually owned co-generation unit, which produces heat and power. Regulations for protection of the grid at the interconnection point of co-generation units are however not useable. PV systems have electronic inverters instead of rotating machines for co-generation units.
It will now be clear that special attention has to be given to safety and grid interconnection aspects of grid connected PV's. In a research project conducted jointly by KEMA Transmission and Distribution, R&S Renewable Energy Systems and ECN Energy Research Foundation, attention is given to the following topics.

Safety aspects

The DC voltage level defines the insulation level of the PV systems. Voltages above 120 Vdc have to comply to insulation class II, as class I (full metal enclosure) is not applicable for PV systems. PV's have no short circuit capacity and can therefore not be protected by fuses. Blocking diodes are used, sometimes in combination with fuses. Research focuses on the design of an adequate and cost effective DC-side protection of the PV system.

Grid interconnection

Protective devices have to supervise the status of the electricity network to define whether power may be injected in the grid. If some distortion, e.g. islanding, over/under-voltage or over/under-frequency is detected, the inverter must be disconnected from the grid. Research focuses on the limits and settings of these protections.

Grounding and over-voltage protection

Lightning strokes may cause considerable damage to the PV system. Lightning induced voltages [4] can reach high values, damaging the inverter. The research focuses on an effective grounding and ground cable routing. By using this Maxwell-approach (Electro-Magnetic Compatibility), over-voltage devices may be omitted.

4. THE KEMA TEST INSTALLATION

In order to define the best possible way of protecting the PV system, a dedicated test installation was built. This outdoor installation consists of 36 modules of poly-crystalline solar cells which can be easily varied in different array configurations (DC voltages with and without centre grounding). he modules are supported by aluminium profiles, mounted on a wooden structure. In this way, a realistic rooftop mounting system is simulated. As several array configurations are possible, several inverters are available with different operating principles (PWM, thyristor controlled).
In the PV system different types of faults can be applied. The system response is measured dynamically. The following categories of faults can be applied:
1. *PV systems faults*, such as single and multiple grounding, short circuit, partial and full shading, cable disconnection and malfunctioning of diodes.
2. *Network faults*, including voltage dips, phase jumps, frequency variation and voltage bursts.
3. *Lightning faults*: induced voltages and currents on the PV system by an indirect or direct stroke.

Results

At present, a number of experiments has been carried out. The first results show that the actual voltage and current distribution in an array with two ground faults can easily be calculated using the open circuit or MPP voltage and current of every module (see Fig. 4).
In this figure, the voltage Ummp is the maximum power point voltage of a module. Im is the module current at the insolation level present. From the figure, the voltage and current distribution can easily be calculated. When comparing the results with actual measurements, a deviation of several percent is found.
For the lightning induced voltage measurements the test installation will be transported into KEMA's high voltages laboratory, where actual lightning currents are generated and injected into the ground structure of the PV system. These measurements will result in different-mode and common-mode induced voltages and currents from which an appropriate grounding structure can be defined.

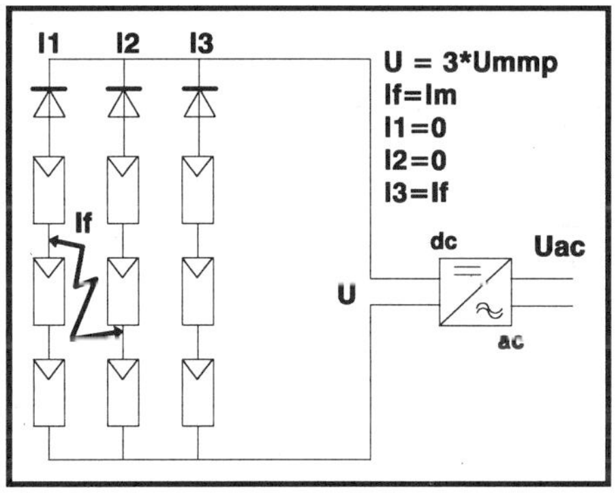

Fig. 4. Voltage and current distribution is an array with two ground faults

Conclusions

A test facility in which faults can be applied is most essential for gaining knowledge about the behaviour of a grid connected PV system. Faults that have to be applied are in the PV system itself, at the AC-grid interconnection and external influences such as lightning strokes. The results of all these measurements will contribute to the cost efective and safe electrical design of a PV system.

LITERATURE

[1] J.H. Boumans, A.J.N. Schoen; 'Requirements and constraints for a BIPV test facility'; Ecofys, August 1993

[2] J.H. Boumans, A.J.N. Schoen; 'Assessment of a solar roof construction Nieuwland' (in Dutch) Ecofys, November 1994

[3] J.H. Boumans, A.J.N. Schoen, 'Realisation and application of a PV roof test facility', (in Dutch) Ecofys Dec 1994

[4] S.A.M. Verhoeven, 'Short circuit behaviour and lighting induced voltages in solar modules', Twelftfh European Photovltaics Solar Energy Conference, April 1994, Amsterdam, The Netherlands

PERFORMANCE OF MODULAR GRID CONNECTED PV SYSTEMS WITH UNDERSIZED INVERTERS IN PORTUGAL AND THE NETHERLANDS

A.J. Kil, T.C.J. van der Weiden
Ecofys Consultancy and Research
P.O. Box 8408, NL-3503 RK Utrecht, The Netherlands

ABSTRACT

Preliminary results on the design and performance of three multi-array grid connected PV systems on flat roofs are presented. From the results it can be concluded that inverters can at least be undersized to 65% of the PV peak power without any significant loss of energy production. These results confirm the findings from model calculations carried out as a part of the design process.

INTRODUCTION

Grid connected PV systems in the built environment will be of crucial importance for PV to become a substantial source of electricity production in Europe. Further cost reduction is however essential for large scale introduction of such systems. In addition, it is important for utility companies to build up experience with the realization and operation of grid connected PV systems. Both aspects are included in the project presented in this paper.

Ecofys has realized three modular grid connected PV systems with a total peak power of 30 kW on flat roofs of office buildings. Two systems have been placed on buildings owned by European utility companies, *sc.* Electricidade de Portugal (Setúbal, Portugal, 38°N) and Openbaar Nutsbedrijf Schiedam (ONS, Schiedam, The Netherlands, 52°N) and the third one on a building of the Energy Research Foundation ECN (Petten, The Netherlands). The project is carried out in co-operation with these three partners.

In this paper the design of the systems will be discussed. Aspects concerning the optimization of the system design are presented. In addition some preliminary results from the monitoring program will be shown.

SYSTEM DESIGN

Each of the three systems has a PV power of 10 kWp and consists of a field of parallel subarrays facing south with tilt angles of 15° (Portugal) or 30° (The Netherlands). The PV modules are mounted on an open construction. Each system is divided in 5 PV/inverter units. Each PV/inverter unit coincides with at least one subarray and consists of semi-crystalline PV modules with a total peak power of about 2 kW and an inverter with a nominal output power of 1.5 kW.

Undersizing of inverters

One of the innovative aspects in this project is the undersizing of the inverters, *i.e.* applying an inverter with a power that is substantially smaller than the nominal peak power of the PV generator. Undersizing of inverters has already been identified several years ago by Ecofys as a possibility to reduce the costs of a PV system with a minimum loss of energy production. Several publications are now available on this issue, based on calculations [1,2,3]. However, field experience is still very limited.

The inverter applied in this project can be undersized since it is capable of automatically avoiding higher temperatures or currents by limiting the input power. Furthermore, the inverter is capable of using excess DC-power for forced cooling. The inverter has a low self consumption: it starts supplying power to the grid at an input power of 4 W, which is only 0.27% of its nominal power. The inverter has a maximum efficiency of 93%. These technical features are essential for a flexible system design. This flexibility is a great advantage if a PV system is to be installed on a flat roof, especially in the case of an existing building with a limited roof area.

Pre-design considerations

To obtain field experience with inverter undersizing the ratio between nominal inverter power and PV peak power was varied for the various units. In the system design phase the optimum PV power ratio r (defined as the nominal inverter power divided by the nominal PV peak power of a unit) was determined using the simulation program INSEL (developed by the University of Oldenburg, Germany) [4]. Using hourly irradiation data (south, tilt 35°, corrected for spectral and reflection losses) for one location in the Netherlands and the efficiency curve of the applied inverter the specific energy output was calculated as well as the price of 1 kWh of PV-energy. We assumed a price of NLG 2/W for the inverter and of NLG 18/Wp for the PV modules and other system components; the life span of the PV system was assumed to be 20 years. Firstly we calculated the specific output and the price per kWh on the basis of a model in which temperature effects, mismatch losses, ohmic losses and losses at low irradiation levels were not included. Interest rates were not taken into account. This calculation resulted in a minimum kWh-price at r=0.75. However, the kWh-price appeared to be not very sensitive for variations around the

CH3365-4/94/0000-1028 $4.00 © 1994 IEEE   First WCPEC; Dec. 5-9, 1994; Hawaii

minimum: in the range $0.67<r<0.83$ the kWh-price is constant within 1%, which implies a large flexibility with respect to system design. Secondly we performed similar calculations using a model in which the loss mechanisms mentioned above were included. The cell temperature coefficient was taken to be 0.03 °C/(W/m²), the yearly cable losses and the yearly mismatch losses were both assumed to be 2%. The results of both sets of calculations are plotted in Figs. 1 and 2.

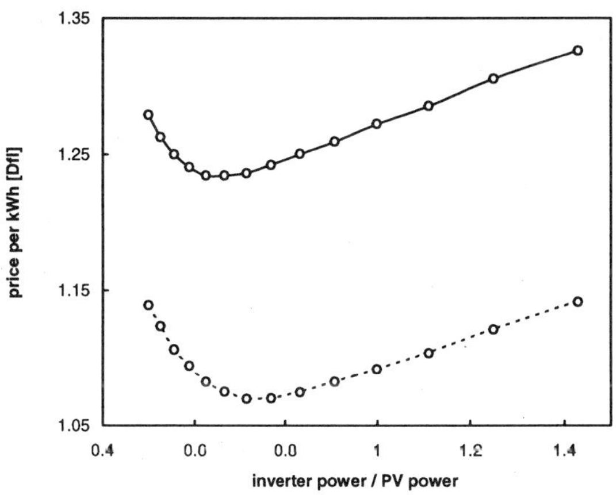

Fig. 1 Calculated price per generated kWh as a function of inverter undersizing factor for the 'no-losses' model (dashed line) and the 'losses' model (full line).

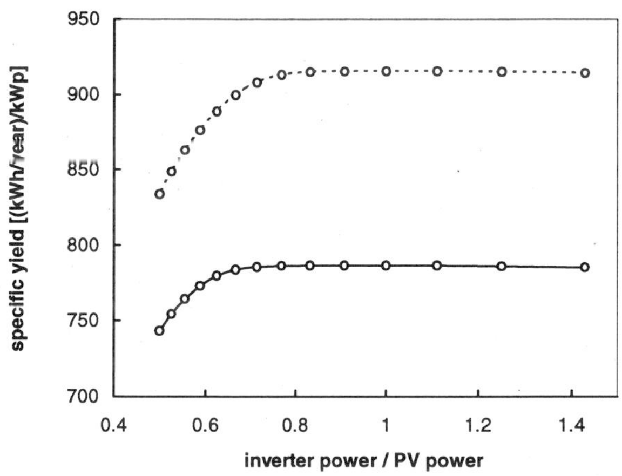

Fig. 2 Calculated specific energy output as a function of inverter undersizing factor for the 'no-losses' model (dashed lino) and tho 'loccoc' modol (full lino).

Fig. 1 shows the dependence of the kWh-price on the power ratio r for both sets. In Fig. 2 the specific energy output as a function of r is shown. Quite obviously, the kWh-price is higher in the model including losses, since the energy output is lower. Most striking is the shift of the minimum kWh-price from r=0.75 to r=0.65, with a variation of less than 1% within the range $0.57<r<0.77$. This shift can be attributed mainly to the reduction of the PV output at high cell temperatures occurring at high irradiation levels. Cable (ohmic) losses, which become prominent at high currents (and thus at high irradiation levels), are to a lesser extent responsible for the shift of the minimum. The specific energy output of the 'losses' model shows a maximum at r=0.95 (see Fig. 2). However, also the specific output depends quite weakly on r, which is reflected in an energy loss of only 0.5% at r=0.65. At r=0.77 the energy loss is a negligible 0.05%. The actual unit PV powers were chosen in accordance with the results of the 'no-losses' model: the power ratio r of the units was varied between r=0.65 and r=0.85. This was established by varying the number of PV strings per unit (from 9 to 11 parallel strings of 4 PV modules in series).

Finally we remark that undersizing is of course not only advantageous with respect to the inverter power but also with respect to other system components like the cabling especially if the current is limited by the inverter.

Variation of the inter-array distance

The costs per Wp are expected to decrease with increasing PV power installed on one flat roof. Therefore, the distance between the subarrays should be as small as possible. On the other hand, the energy output per installed Wp will decrease with the inter-array distance due to shading. Therefore a trade-off has to be made between a high specific energy output and low system costs. Little field experience is available on the effect of inter-array distance on system output. A yearly loss of 2.7% was calculated at a roof cover ratio of 65%. In the system installed at ONS different inter-array distances (from 1.36 to 4.50 m) were applied between the various subarrays. In this way the effect of partial shading of subarrays can be studied. This aspect is again important to the flexible and cost-effective realization of PV systems on roofs of buildings since space on roofs is a limited.

OPERATING EXPERIENCE

The monitoring program started in August 1994. Since then (preliminary) results on the performance of two of the systems have been obtained. Monitoring of the third system has only recently started. The results concern the performance of the systems in general and the influence on the performance of the inverter/PV power ratio and partial shading in particular.

Inverter undersizing

In Fig. 3 the performance ratio PR of the 5 subarrays of the system installed at ECN is presented. The relative PR is plotted as a function of inverter/PV power ratio r on the base of data of the month of August. The lowest PR value is defined as 100%. Data corrected for shadow losses are plotted as well. The corrected PR shows practically no dependence on r. This means that in The Netherlands undersizing down to r=0.65 does not lead to a significant loss of energy production. Thus, the inverters of this system should have been undersized to a larger extent in order to be able to determine the optimum inverter/PV power ratio applicable in practice. The preliminary conclusion is that the inverter power of grid connected systems in The Netherlands can at least be undersized down to r=0.65 without affecting the performance. More data are however necessary to corroborate this statement.

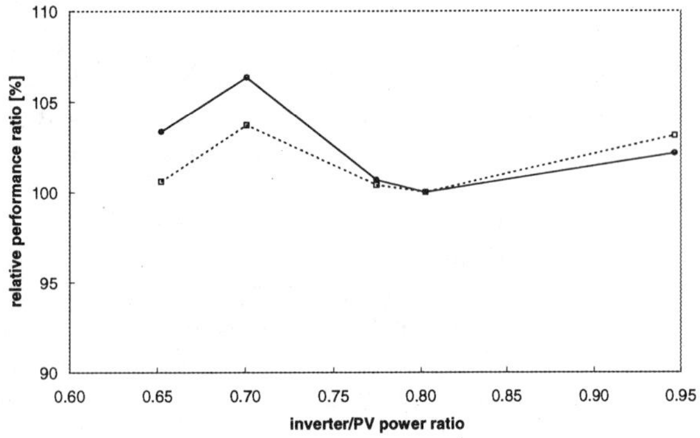

Fig. 3 Performance the 5 ECN subarrays as a function of the degree of undersizing. The lowest *PR*-value is set to 100%. The dashed curve is corrected for shadow losses.

A similar conclusion applies to the system in Portugal: the subarray with the highest PV power only shows a small energy loss, which indicates that under Portuguese climatological conditions the inverter power of a similar PV system can be restricted to 67% of the PV power. This is illustrated in Fig. 4, where for the 3 unshaded subarrays the number of generated kWh per kWp installed PV power has been plotted versus PV power. The figures were calculated from the data of August 1994. In this month losses due to inverter undersizing are most likely to occur, as irradiation levels are maximum.

It must be kept in mind that in months with lower irradiation levels, the units with lower r-values can be expected to have higher *PR*'s, leading to compensation of the losses during summer. Data for a one year period will be used to determine the overall performance of the different units.

Partial shading

The influence of partial shading of the arrays is qualitatively presented in Fig. 5. In this figure the ratio of the energy output of an unshaded and that of a partially shaded subarray has been plotted as a function of the hour of the day. Data from a number of days in summer have been used. We assumed that at noon both subarrays were unshaded, resulting in a normalized ratio of 1 at noon. The effect of shading is evident in the rising ratio in the morning and evening. At this stage the data set is too limited to draw quantitative conclusions on the effect of the inter-array distance and resulting shading on the performance of the system on a yearly basis.

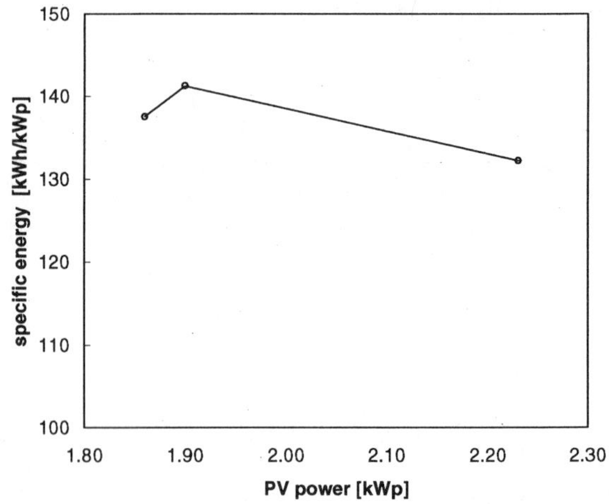

Fig. 4 PV energy per kWp versus PV power of three unshaded subarrays of the EDP system (data from August 1994).

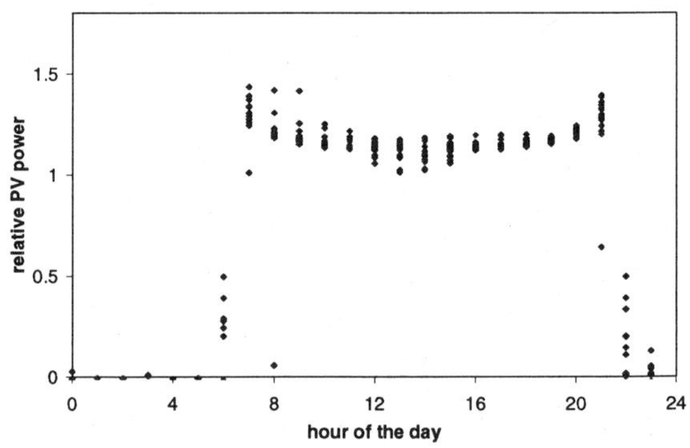

Fig. 5 Power ratio of an unshaded and a shaded subarray of the ECN system as a function of the time of day for a number of days in the summer.

CONCLUSIONS

Calculations have shown that undersizing the inverter power with respect to the PV power is a very effective way of minimizing the specific costs of a grid connected PV system without performance reduction. This was confirmed by the analysis of the first monitoring data: the inverter/PV power ratio can be as low as 0.65 for The Netherlands and 0.67 for Portugal. Theory and practice indicate that even smaller ratios may be applied. The effect of shading on the array performance has been clearly identified. These preliminary conclusions will be given a firmer ground by future measurements.

Acknowledgment

This project is supported by the EU (DGXVII), the Netherlands PV R&D&D program (Netherlands Agency for Energy and the Environment - NOVEM), R&S Renewable Energy Systems, Mastervolt, utility company Electricidade de Portugal, utility company Openbaar Nutsbedrijf Schiedam and Energy Research Foundation ECN.

REFERENCES

[1] T.C.J. van der Weiden and A.J.N. Schoen, "Optimized design and technical potential of grid connected PV systems on office buildings", Proceedings of the 11th E.C. Photovoltaic Solar Energy Conference (1992), pp. 1451-1454

[2] L. Keller and P. Affolter, "Optimizing the panel area of a photovoltaic system, in relation to the static inverter", Proceedings of the 11th E.C. Photovoltaic Solar Energy Conference (1992), pp. 1159-1162

[3] M. Jantsch et al., "Results of the concerted action on power conditioning and control", Proceedings of the 11th E.C. Photovoltaic Solar Energy Conference (1992), pp. 1589-1592

[4] J. Schumacher, "Digitale Simulation regenerativer elektrischer Energieversorgungssysteme", thesis University of Oldenburg (1990).

A PV-House Experience in Japan

Katsuaki Sato, Hirofumi Ida* and Osamu Ishikawa*
Faculty of Technology, Tokyo University of Agriculture and Technology, Koganei, Tokyo 184, Japan
**Misawa Homes Institute of Research & Development, Suginami, Tokyo 168, Japan*

ABSTRACT

A PV-house with a 3 kW solar-cells panel roof was constructed and has been used for ordinary family life of one of the authors. The PV output is interconnected to the utility power network with a reverse current flow. Average power output of the solar panel exceeded 13 kWh per day for sunny days and 6 kWh for cloudy days in May. Time-sequential data of power consumption and supply have been recorded. The PV generation system is found to be capable of generating 300 kWh per month in average and supplying as large as 100-200 kWh per month to the utility power network. This system has been proved to be effective in reducing the peak power demand in the hot summer days in Japan.

1. Introduction

Utility scale interactive dispersed PV system for ordinary consumers' residences is expected to be effective for reduction of peak-power demands on daytime in hot seasons[1], as well as for reduction of delivery loss. For this purpose provisions to maintain the reliability and quality of power system has been required. On March 1993, a new guideline has been established in Japan for grid-interconnection of small scale PV generation system with reversed power flow[2]. (See the Appendix) Simplified interconnection relay is recommended. In addition, low cost integrated PV panels have been developed for roof use. These situation prompted one of the authors (K.S.) who is a researcher of solar cells materials to make his mind to build a house with the PV-roof.

He had his house built under the specification of "Eco-Energy House" which had been proposed by Misawa Homes Institute in preceding paper[3].

The house was built under the concept of reducing the energy consumption and loss as much as possible by adopting an architecture with high heat insulation and air-tightness.

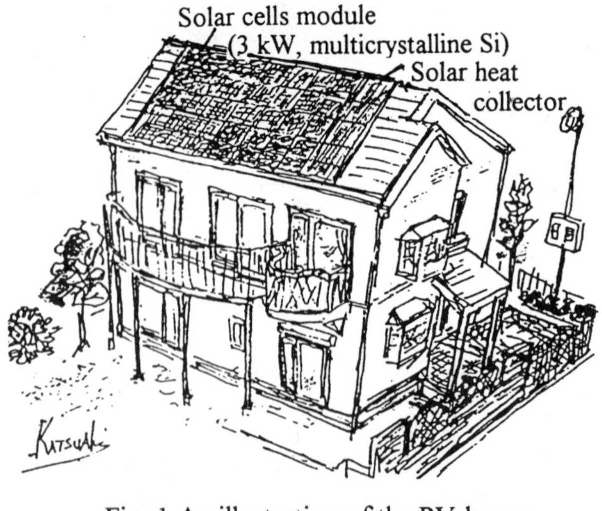

Fig. 1 An illustration of the PV-house for ordinary life use.

2. Specification

The newly built house (illustrated in Fig. 1), which was completed on March, 1994, is equipped with the solar power generation system consisting of roof-integrated 3 kW PV panels, an 3 kW inverter and the inter-connection relay which follows the *new guideline*, as well as solar heat collecting system to warm utility pipe water before supplying to hot-water supply system which utilizes the low-cost night-time power. The total system is illustrated in Fig. 2. Total of 40 multicrystalline silicon PV modules, the area of one module being 90cm x 90cm, cover about 70% of the south-side inclination of the roof. The inclination angle of the roof is 26.5°. The inverter used has an efficiency of 90% in average. It has an ability of automatic shutoff, when the PV-power generated becomes less than that required to operate the apparatus.

A heat pump system described in ref. 3 is adopted for the air conditioning of this house. Owing to the low thermal conduction of the Eco-Energy House system (the measured heat loss coefficient being as small as 1.19 kcal/m^{2}hr °C) the power required to maintain0 the air conditioning has proved to be reduced by 85 % of that in ordinary wooden residences.

CH3365-4/94/0000-1032 $4.00 © 1994 IEEE

First WCPEC; Dec. 5-9, 1994; Hawaii

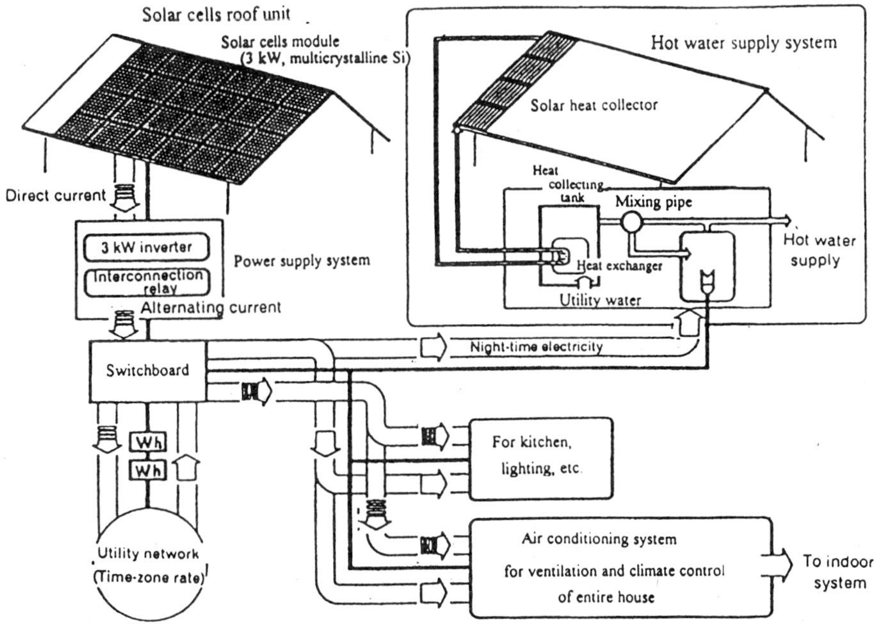

Fig. 2 Schematic illustration of the PV-house with the concept of "Eco-Energy House"

Instead of city-gas cookers usually employed in ordinary Japanese home kitchens electrical heaters are used in order to reduce the CO_2 exhaust.

To save power consumption of lighting, ball-type bulb-shaped fluorescent lamps, the power consumption of which are 1/3 of incandescent lamps of the same brightness, have been employed for lighting of all rooms. Use of fluorescent lamps often leads to reduction of power factor due to inductors, which becomes a problem for grid-interconnection. However, such problem has been solved by use of bulb-shaped fluorescent lamps with built-in inverters.

3. Measurement and Results

The output power from the PV-module/ inverter system and the power consumption/ supply at the interconnection point are monitored using power transducers(PT) inserted into appropriate positions of the wiring. The output voltages of PTs are monitored by a personal computer system.

Fig. 3(a) shows as a typical example a time-sequential profile of the power supply/ consumption (white square) and the PV-output (black square) of the house for the entire day of May 19 (clear). Peak power generated by PV array was 2.24 kW at 11:20. The angle of incidence of the solar radiation at the time of culmination is nearly perpendicular to the surface of the PV modules in this season of the year. It is thus elucidated that the real energy production of the system is 75 % of the nominal power output of solar module. This value is quite reasonable if one takes into account the real radiation and inverter efficiency (90%). Total power generated for time period when solar radiation was available amouted to 17.38 kWh for the day. Total power consumption during the same time period was 6.44 kWh, resulting in the reverse power flow of 10.93 kWh. The total power purchased from the network amounted to 23.55 kWh for the night-time (23:00-7:00) and 4.60 kWh for the day-time (7:00-23:00). The strong jump of power consumption during the night-time seen in Fig. 3(a) was mainly due to the hot-water supply. The time duration for which the hot-water heater is switched on depends on the solar radiation of the preceding day since the water supplied is preheated by the solar-heat collector. Compare it with the data of August 14 shown in Fig. 3(b), in which heater is switched on as short as 90 min.

Total payment for purchase was as small as 313 yen for the entire day since the rate at night-time cost 1/4 of the day-time rate, whereas total sales amounted to 340 yen, resulting in 27 yen surplus.

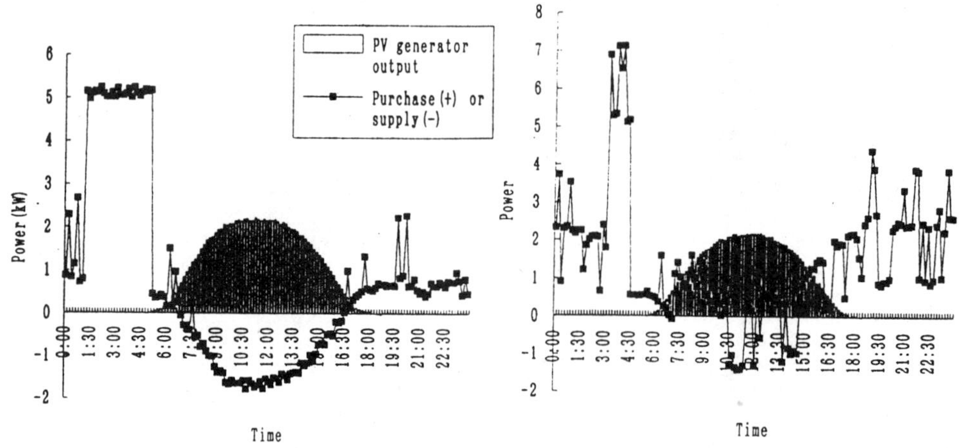

(a) (b)

Fig.3 Time-sequential data of PV-generator output (vertical bars) and Power purchased from or supplied to the utility network.
(a) May 19, without airconditioning, (b)August, 14, with airconditioning

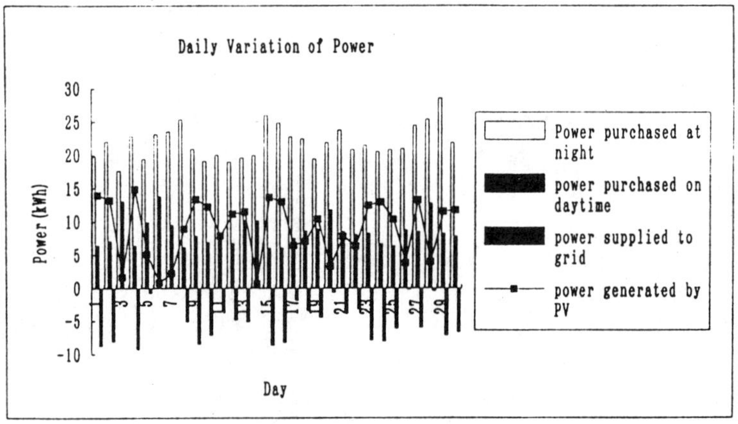

Fig. 4 Daily Variation of Power purchased from and supplied to the network. The power produced by the PV generation system is also plotted.

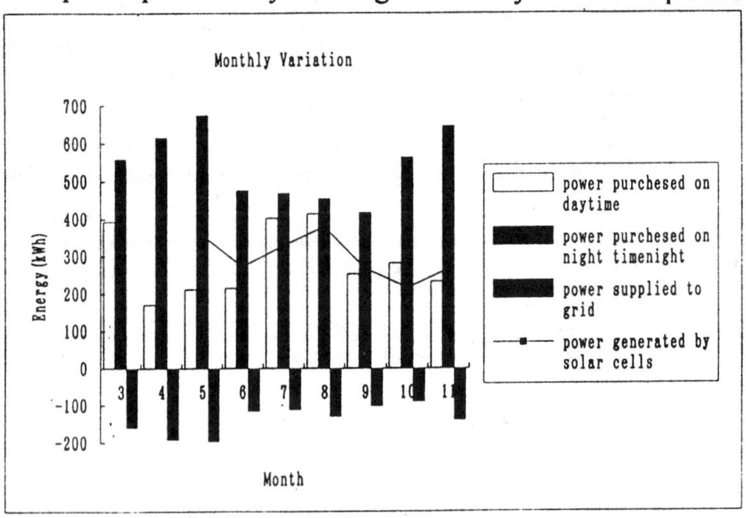

Fig.5 Monthly variatin of the power purchased from and supplied to the network. The power produced by the PV generation system is also plotted.

Fig. 4 illustrates daily variation of the power purchased from and supplied to the grid, as well as power generated by solar cells system in November. The figure shows that average power output of the PV generator exceeded 13 kWh per day for sunny days and 5 kWh for cloudy days. Average of reverse power flow to the network amounted to 4.6 kWh in average for november (8.2 kWh for sunny days and 3 kWh for cloudy days).

Fig. 5 shows monthly variation of the power purchased from and supplied to the network as well as the power produced by the PV generator. Average of 300 kWh was found to be generated per month by the PV system. The energy supplied to the utility grid amounted to 100-200 kWh per month. PV power generated is large on the month when the daytime consumption is large.

4. Conclusion

An experience of living in the PV-house which was constructed under the concept of Eco-Energy House is described. The PV-house proved to be capable of supplying considerable amount of power to the network maintaining comfortable family life, showing effectiveness of PV-system for peak demand reduction.

References

1. H. Kobayashi: Tech. Digest International PVSEC-7, Nagoya Japan 1993, pp.367-370.
2. E. Hashimoto: Tech. Digest International PVSEC-7, Nagoya Japan 1993, pp. 375-377.
3. H. Ida: Tech. Digest International PVSEC-7, Nagoya Japan 1993, pp. 343-344.

Appendix

Guidline for grid-interconnected operation of small PV generation system (for low voltage users)

Items	Requirements
Capacity of solar PV generation system	Capacity of PV gen. system which can be interconnected to low vtg. power distribution line is limitted to 50 kW
Power factor (when interconnected with allowable reverse flow operation)	The power factor at the receiving point of consumer should be 85% or higher. leading power facter not allowed
Harmonics	Should not exceed 5% (total current distortion) 3% (harmonic distortion of each order)
Protective coordination	Overvoltage, low voltage, low frequency relay should be installed. (Incorporation in the Inverter is allowed.) Prevention of islanding is required.
Voltage fluctuation	If the voltage is likely to exceed the rage 101±6V, 202±20V by the reverse flow, adequate measures should be taken to adjust the vtg. by the consumer.
Cost reduction and Simplification	The grid interconnection device should be desirably simplified and should be as compact as possible.

Comments: The grid-interconnected PV generators should have the interconnec-tion device incorporated in the inverter. Isolation transformer may be omitted if automatic cut-off circuit works when DC component is detected.

A NEW PERFORMANCE INDEX FOR PV SYSTEM ANALYSIS

T. Townsend and C. Whitaker
Endecon Engineering
3160 Crow Canyon Rd., Suite 240
San Ramon, CA 94583

B. Farmer
Pacific Gas and Electric Co.
3400 Crow Canyon Rd.
San Ramon, CA 94583

H. Wenger
Pacific Energy Group
32 Valla Court
Walnut Creek, CA 94596

ABSTRACT

This paper describes a term called the Performance Index (P.I.) for analyzing photovoltaic (PV) system performance. The P.I. is the dimensionless ratio of actual to expected generation over any time interval.

PV system performance is often described by applying terms widely used for conventional power plants. However, there are drawbacks to using some of the most common quantities, such as capacity factor, efficiency, and availability. For PV systems, these quantities exhibit wider seasonal variations and smaller magnitudes than for conventional power plants. Consequently, PV systems are often judged to be operating poorly, even in the absence of failures.

At PVUSA, the P.I. is now being used for both real-time power plant monitoring and for longer-term analyses. A description of the calculation procedure and sample results are included.

INTRODUCTION

Answering a simple question regarding PV system performance such as "How well is it working?" proves to be more difficult than it appears. The capacity factor is probably the most useful statistic to answer such a question for many types of power plants, but not for PV. Because of the time-dependent nature of the fuel (solar) resource, normal daily and seasonal variations in solar radiation result in low capacity factors and can mask losses due to component failures or soiling.

Other common performance measures can be calculated, such as energy, average and peak power as a percentage of rated power,[1] efficiency, availability, and equivalent forced outage rate, but with similar drawbacks. Taken together, the additional measures can yield fairly comprehensive assessments, but they also make for cumbersome and potentially misleading interpretations of PV system performance.

For example, a PV system with a wintertime peak equal to 60% of its rated output, operating at 8% efficiency, 15% capacity factor, and 35% availability may have nothing wrong with it, while similar values in the summer would suggest problems exist. Those familiar with conventional plants might mistakenly conclude the PV system had been operating poorly during the winter and might also correctly conclude the summertime performance was poor, but possibly for the wrong reasons.

This confusion prompts another extreme to occur - a tendency to report oversimplified on/off status answers that say little or nothing about how well the system is operating. Both the detailed and oversimplified approaches are inadequate, suggesting the need for a capacity factor-like statistic that doesn't penalize a PV system for externally driven departures from its rated power.

The PVUSA project is now monitoring fourteen PV systems installed at seven sites around the U.S. over the past six years. The installed systems range from 15 kW to 500 kW, with a combined capacity of 1,600 kW. A mix of amorphous, polycrystalline, and single-crystal silicon flat-plate and concentrator technologies are in place, mounted on various fixed tilt and tracking structures [1]. The diversity of locations, sizes, and technology types at PVUSA reinforces the need for a statistic that permits meaningful comparisons by normalizing for weather and plant size/type differences. The P.I. incorporates several desirable characteristics: It is dimensionless, ranges from 0 to 100%, relies on commonly measured quantities for PV systems, applies to real-time and longer-term analyses, normalizes for rated capacity, and independently adjusts for irradiance, temperature, and as many other power loss or gain mechanisms as the user can quantify.

PERFORMANCE INDEX DERIVATION

The P.I. shown in Eqn. (1) is defined as the ratio of actual to expected generation over any time interval [2]. (The following discussion will center on instantaneous power-based P.I. calculations, but an energy P.I. may be calculated by integrating over time.) The actual

[1] Rated power under a set of reference test conditions. Manufacturers' standard test conditions (STC) of 1,000 W/m^2 plane of array irradiance, air mass 1.5, and 25 or 28 C cell temperature have been widely used. As these conditions rarely occur outdoors, PVUSA has developed a more representative set of test conditions (PTC), defined as 1,000 W/m^2 plane of array irrad. (850 W/m^2 direct normal irrad. for concentrators), 20 C air temperature, and 1 m/s wind speed at 10 m height. PTC ratings tend to be about 10% lower than STC ratings.

CH3365-4/94/0000-1036 $4.00 © 1994 IEEE

First WCPEC; Dec. 5–9, 1994; Hawaii

generation is easily measured with a revenue meter or power transducer, but the expected generation is more difficult to calculate.

$$P.I. = \frac{\text{Actual Generation}}{\text{Expected Generation}} \quad (1)$$

Calculating expected generation requires adjusting for effects that drive power up or down from its rated value.

$$\text{Exp. Gen.} = \text{Rated Power x Adjustments} \quad (2)$$

Although a proposed calculation procedure will be derived from Eqn. (2) with recommended adjustments for several key effects, rigid criteria for adjusting the rated power are not defined here. As a general guideline, adjustments should be made for those effects considered uncontrollable (weather-related) or irreversible (design or material-related). While there may be some value in defining a uniform calculation procedure, the statistic may be more useful if the calculation is flexible enough to permit some effects to be adjusted for on an application-specific basis.

Power and energy losses or gains may be attributed to the following effects.

- Irradiance (greater or less than reference)
- Module temperature (greater or less than rated)
- Degradation (long-term rating declines)
- Soiling (dirt, bird droppings, snow)
- Balance of system (BOS) losses, such as:
  - Tracker physical limits (reduced irrad.)
  - Tracker non-ideal performance (reduced irrad.)
  - Shading (mismatch, reduced irrad.)
    Field of view (reduced irrad.)
  - PCU (standby loads, off-peak power tracking, and low-load reduced efficiency)
  - Transformer (no-load losses)
- Reflection (add'l loss at high incidence angles)
- Solar spectrum (some PVs are sensitive to changes)
- Annealing (degradation can be partly reversed for some PVs under sustained elevated temperatures)
- Outages (full, partial; scheduled, unscheduled)

For effects such as unplanned partial or full outages, the P.I. is useful for determining their impact on actual generation (no adjustment to the expected generation). Allowing for other effects becomes more discretionary. For example, it is debatable whether reduced actual generation due to long-term degradation, soiling, and tracker losses should be accompanied by corresponding reductions in expected generation. Allowing for such effects increases the P.I.

Some effects are difficult or impractical to quantify, like spectral or annealing effects. Only a few effects tend to be at least as significant as the principal uncertainty of at least ±3% associated with most irradiance measurements. The minor effects can therefore be neglected or considered as being lumped with other effects.

As a practical expression, the performance index may be defined as:

$$P.I. = \frac{\text{Actual Power}}{\text{Rated Power} \bullet IA \bullet TA \bullet DA \bullet SA \bullet BOSA} \quad (3)$$

where adjustment factors are:

IA = Irradiance Adjustment = actual / rated irrad. (integrated, this can be considered an "insolation factor," analogous to the traditional capacity factor)
TA = Temperature Adjustment = $(1 + \beta \bullet (T_{mod} - T_{mod,PTC}))$
  β = temperature coefficient, 1/C
  T_{mod}, $T_{mod,PTC}$ = actual and rated module temperatures, C
DA = Degradation Adjustment – Present rating / Initial rating
SA = Soiling Adjustment = actual (soiled) power / rated (cleaned) power
BOSA = 1 - (fraction BOS losses)

A term for scheduled outages is not shown here. For real-time monitoring, the P.I. is undefined (and unnecessary) when a system is down for a scheduled outage. When used for longer-term analyses (e.g., monthly), scheduled outages are accounted for by subtracting the insolation received during the outage from the IA term's integrated actual insolation. Integrated, the equation becomes:

$$P.I. = \frac{\text{Actual Energy} \bullet \text{Rating Irrad}}{\text{Rat. Pwr.} \bullet \text{Act. Ins.} \bullet TA \bullet DA \bullet SA \bullet BOSA} \quad (4)$$

Regarding the IA and TA terms: the irradiance and temperature effects are essentially uncontrollable, and the expected power should be adjusted so that the P.I. does not penalize or reward a system as a result of variations in irradiance or temperature. For the integrated Eqn. (4), the correct term to use for the module temperature is the power-weighted-average module temperature. Power-weighting minimizes the effect of cooler temperatures that tend to occur at low power and properly weights the warmer module temperatures characteristic of midday periods.

$$\overline{T_{mod}} = \frac{\sum(\text{Power} \bullet T_{mod})}{\text{Energy}} \quad (5)$$

Regarding the DA term: At PVUSA, many of the systems have degraded (some by 20%) after several years to the point that their performance looks poor relative to their initial rating. While mindful of the change, PVUSA's operators have found it more useful to adjust the rated power by the ratio of the present/initial rating so that the real-time P.I. can readily identify new system problems.

Regarding the SA term: Due to surrounding agriculture and dry summers, soiling losses can be significant at PVUSA's Davis site. Power losses as high as 20% have been measured on systems that have gone unwashed for several months. Losses resulting from electing not to wash an array should reduce the P.I. Conversely, by including soiling losses the P.I. can be used to identify the degree to which system performance has changed due to other mechanisms.

Regarding the BOSA term: There are a number of losses lumped together in this term. Most are small, dynamic, and difficult to quantify independently. (Doing so generally requires special lab and field tests and additional uncommon measurements.) As they tend to be proportionately more important early and late in the day at low irradiance levels, the effects are often neglected, particularly for real-time diagnostics that focus on midday performance. Combined, however, their effects can be significant: Two of PVUSA's single-axis tracking systems lose 5% to 15% of their annual generation, largely due to non-ideal tracking, and another two-axis tracking system's annual energy is reduced by 10% due to shading losses (note that shading and other BOS-related losses aren't represented in the system's rated power; one reason why actual long-term energy production tends to fall short of expectations).

The reflective losses at high incidence angles, solar spectrum, and annealing effects are generally ignored, although with additional instruments (such as spectroradiometers) and special tests (such as flash tests to plot incidence angle effects and study annealing) their effects could be estimated.

Calculations similar to the P.I. have been proposed. Roy [3] proposed the Operating Reliability Factor, where the capacity factor is divided by an "insolation coefficient;" with the insolation coefficient calculated in a manner analogous to the capacity factor. By dropping the temperature, degradation, soiling, and BOS terms, the P.I. equation reduces to the Operating Reliability Factor. By ignoring the temperature effect, though, the Operating Reliability Factor is subject to significant temperature-related seasonal variations of $\pm15\%$, large enough to mask many failure-related reductions in power.

Independent of Roy's work, PVUSA has been using the same calculation for the past few years, with various labels: PV effectiveness, energy-weighted availability, and insolation-weighted capacity factor [1,4].

PERFORMANCE INDEX APPLICATIONS

PVUSA began using the P.I. in 1994 for two applications: real-time monitoring and longer-term performance analyses.

At its Davis site, PVUSA uses a computer display screen to show real-time values for numerous weather quantities, utility and PV system status signals, and PV system electrical parameters. Even with a day-to-day familiarity with the generation patterns exhibited by eight systems ranging from 15 to over 400 kW, the plant's operators have always found it cumbersome to verify correct system performance under prevailing weather conditions. Smaller failures have sometimes gone unnoticed for days because of the overwhelming number of measurements.

A P.I. software calculation was added earlier this year for each PV system, along with a color-coded P.I. display and an audible alarm to alert operators in the event of a system shutdown. At a glance, operators can now see whether each system is operating properly.

The P.I. calculation includes adjustments for irradiance, module temperature, and degradation, but does not adjust for soiling, BOS, or other losses. A green indicator is displayed as long as the P.I. is above 90%. Below 90%, a yellow indicator is used, which generally points to a component failure or heavy soiling. At irradiances below 250 W/m^2, a more generous P.I. threshold of 50% is used to indicate satisfactory (green) performance; without a dynamic BOS loss adjustment included in the P.I. calculation, systems would rarely meet the 90% threshold at low irradiance. A red indicator and accompanying alarm are used to indicate abnormal system shutdowns, while a dark blue indicator is used at night to indicate a normal shutdown.

The other principal use of the P.I. has been for longer-term analyses. Figure 1 illustrates the 500-kW single-axis tracking Kerman PV plant's P.I. for June 1994. Although the system's capacity factor was just 40.5%, its P.I. was 92%. Most of the "lost" generation was simply a result of the sun being unavailable at the rating irradiance of 1,000 W/m^2 for 24 hours each day. In this example, the degree to which soiling reduced the overall generation was unknown and so it was lumped with the failure-related reductions in generation. A BOS loss of 5% was included based on comparisons between the insolation measured on the actual tracking structure compared to that available on an optimally-tracked structure [5].

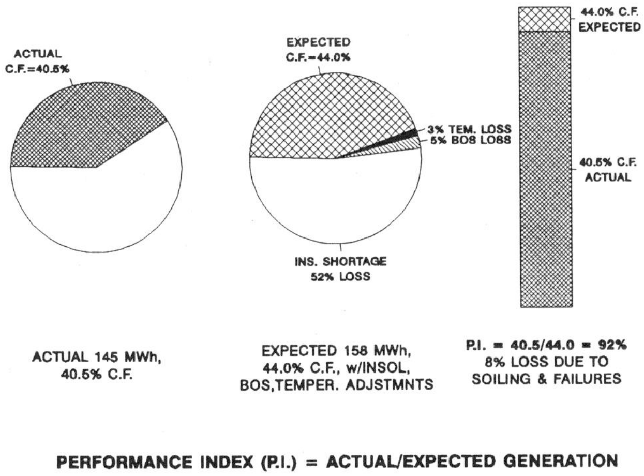

Fig. 1. June 1994 500-kW Kerman PV Plant Performance Index

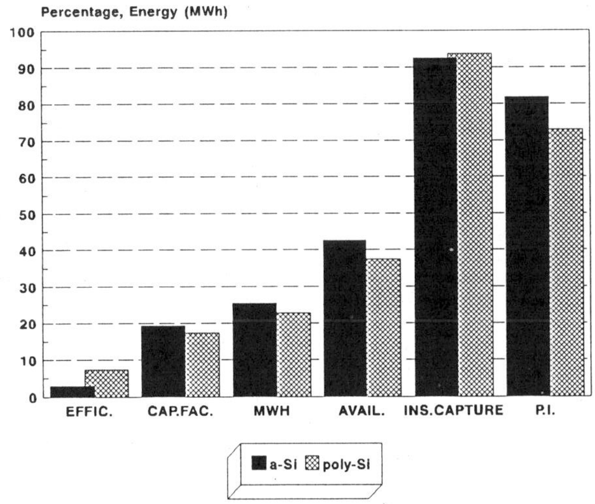

Fig. 2. Annual performance for two 15-kW systems

Figure 2 uses the P.I. and several other quantities to compare the annual performance of two systems with 15 kW ratings. One system consists of amorphous silicon modules and the other consists of polycrystalline silicon modules. The annual efficiency and insolation capture[2] were higher for the poly-Si system, yet the a-Si system produced more energy (therefore, its capacity factor was correspondingly higher) and it was available during a greater percentage of the year. The availability is calculated as the percentage of hours (24-hr. basis) during which net power was supplied to the utility by the PV system. The P.I. shows the a-Si system had the better year, a conclusion that, in this special case of identically sized systems, could have also been reached by comparing the energy produced. Note that while the capacity factor comparison illustrates the same conclusion as the P.I., the capacity factor does not distinguish between changes in weather or system performance.

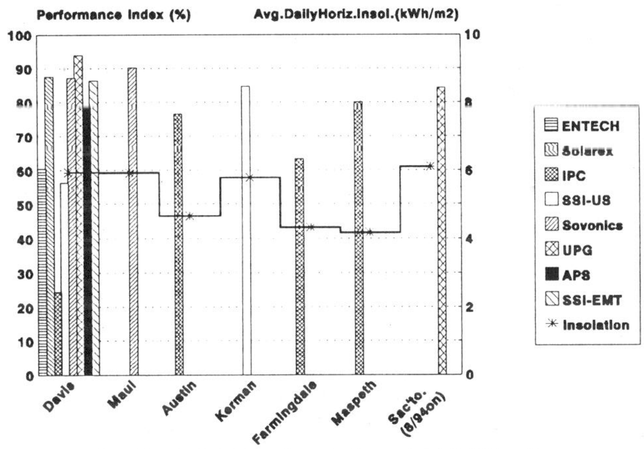

Fig. 3. 1994 P.I., insolation at PVUSA sites

Figure 3 uses the P.I. to compare the cumulative 1994 performance of 14 systems ranging from 15 kW to 500 kW at 7 sites around the U.S. The average global horizontal insolation is also shown for each site. In this case, the P.I. provides a readily interpreted comparison among systems of vastly different sizes, orientations, and locations. The only adjustment factors used here were for insolation, temperature, and degradation. Through September, only two of the systems have maintained a P.I. of 90% or better. Six others have produced at least 80% of their expected generation, while the remaining six systems have suffered more extensive downtime due to various failures and operated at about 20% to 80% P.I.'s.

---

[2] Insolation capture is the % of insolation received during hours when the system was operating, compared to the total insolation received over a given time interval.

REFERENCES

[1] PVUSA Project Team, "1993 PVUSA Progress Report," PG&E R&D Report 007.5-93.8, 1994.

[2] Townsend, T.U., "Performance Index Applications at PVUSA," NREL PV Performance and Reliability Workshop, Lakewood, CO, September 1994.

[3] Roy, A.S., "Comparative Performance of Various Solar-Electric Facilities," *Proc. of the 1989 Annual Conference*, ASES, Lakewood, CO, June 1989.

[4] Townsend, T.U., "Emerging Photovoltaic Module Technologies at PVUSA: A Five-Year Performance Assessment," *Proc. of the 1994 Annual Conference*, ASES, San Jose, CA, June 1994.

[5] Hoff, T., and Wenger, H.J., "The Value of Distributed Generation: The PVUSA Grid-Support Project Serving Kerman Substation," PG&E R&D Report 007.5-94.5, January 1994.

GRID-TIED PV SYSTEM MODELLING: HOW AND WHY

Miles C. Russell
Ascension Technology, Inc.
P.O. Box 314, Lincoln Center, Massachusetts 01773 USA

ABSTRACT

A critical part of Ascension Technology's operation and monitoring of numerous grid-tied photovoltaic (PV) systems for the U.S. Environmental Protection Agency's PV Demand-Side-Management Demonstration Projects is to insure that these PV system are operating properly and that problems can be diagnosed by scrutiny of the data being retrieved from these systems. The objective of the work reported here was to develop accurate, flexible PV array and inverter models, and apply them to help understand the performance of the PV systems and address issues which are important to inverter designers/manufacturers, systems designers and system operators.

Graphical display of measured PV system performance, along with modelled performance has proven useful for remotely diagnosing a wide range of PV system phenomena and quantifying their impact: partial array shading, partial array snow cover, PV array max-power voltage lying outside the inverter tracking window (both an array sizing concern, as well as an inverter design issue) and inverter power limiting. Modelling was also key to discovering and understanding an inverter control problem.

THE PV SYSTEMS

Ascension Technology is under contract to the U.S. Environmental Protection Agency (EPA) to investigate the environmental and demand-side-management (DSM) impacts of grid-tied PV systems. During 1993, for the first EPA PV-DSM Project, we worked with 10 electric utilities to install individual and grouped 4-kW roof-mounted PV systems in 17 sites around the country. The second EPA project involves 11 electric utilities and 11 additional sites where 18-kW PV systems (groups of three independent 6-kW systems) are being roof-mounted during 1994. All of these PV installations are equipped with instrumentation, providing the data which supports our studies for the EPA.

Table I below identifies the components in the PV systems for EPA projects 1 and 2. Note that these systems are utility-interactive and therefore do not contain batteries. The majority of the 28 PV arrays are mounted on flat-roof commercial buildings; four of the EPA 1 systems are installed on residential rooftops. The arrays are fixed in tilt and azimuth.

Table 1
PV System Components

	EPA 1	EPA 2
Modules	Siemens Solar M55	ASE Americas GP-50-DG
Array Configuration	28 series x 3 parallel	10 series x 3 parallel
Array rating (STC)	4.452 kW dc	8.55 kW dc
Tilt angle	15 degrees fixed	25 degrees fixed
Inverter	Omnion 4kW	Omnion 6kW

THE DATABASE

Many efforts to model PV systems and components have been undertaken in the past, yet no existing software tool provided precisely what was required for the objectives we have with our PV system network. In addition, accurate algorithms for translating PV current and voltage with changes in irradiance and temperature are still being evolved. Finally, obtaining the necessary characterization information for PV modules has also been difficult in the past. As a result, we have worked with Sandia National Laboratory to obtain module characterization and models and we have developed custom software to work with the data being retrieved and archived from these PV installations. Table 2 shows the parameters being monitored at each of the EPA PV-DSM Project PV systems.

In addition, a few of the PV systems have been equipped with extra instrumentation to monitor PV array current and voltage. This dc-side information allows greater scrutiny of the operation of the inverters and has proven useful for the development of models.

The data acquisition systems utilize Campbell Scientific CR10 dataloggers and Ascension Technology's Rotating Shadowband Pyranometers.

CH3365-4/94/0000-1040 $4.00 © 1994 IEEE

First WCPEC; Dec. 5-9, 1994; Hawaii

Table 2
PV Systems Instrumentation

Variable	Sensor
AC kWh produced by the PV system	pulse-initiating kWh meter
AC kWh used in the host building	pulse-initiating kWh meter
Plane-of-array irradiance	LiCor 200SZ pyranometer
Direct normal irradiance	Rotating Shadowband Pyranometer
Global horizontal irradiance	Rotating Shadowband Pyranometer
Horizontal diffuse irradiance	Rotating Shadowband Pyranometer
Ambient air temperature	Thermistor housed in gilled radiation shield

Data are retrieved each night and records stored as 15-minute average values for the parameters in Table 2. Each workday morning the data passes through a screening program and a single page is printed, summarizing all sites. Greater scrutiny of performance is accomplished with our custom graphical analysis program, which contains PV system models.

The simulation begins with models of the PV modules and power conditioner. PV array temperature is estimated from POA irradiance and ambient temperature; array dc power output is computed from array temperature, POA irradiance and the array model (which includes IV curve translation equations and coefficients for the specific PV module); ac power is estimated from array dc power and the inverter model. The level of sophistication of these models varies, but the models are designed to utilize only the parameters which are normally available from these sites.

THE MODELS

Sandia National Laboratories has provided accurate measurement of PV module characteristics for our modelling purposes (with the permission of the module manufacturers). The Sandia PV Systems Evaluation Laboratory suggests PV models of the following forms. Note that constants (C_1, C_2 et. al.) in the equations are determined from curve fitting measured data and will differ for each modelled parameter.

$$I_{sc} = e\hat{}[C_1+C_2\{\ln(POA)-C_3\}] + C_4(POA/1000)(T_c-50) \quad (1)$$

$$I_{mp} = e\hat{}[C_1+C_2\{\ln(POA)-C_3\}] + C_4(POA/1000)(T_c-50) \quad (2)$$

$$V_{oc} = C_1+C_2 * \ln(POA/1000) - C_3(T_c-25) \quad (3)$$

$$V_{mp} = C_1-C_2 * \ln(POA/1000) - C_3(T_c-25) \quad (4)$$

$$T_c = Tair + C_1 + C_2(POA), \quad (5)$$

T_c (cell temperature) and Tair (ambient air temperature) are in degrees C, POA (plane-of-array irradiance) is in Watts per square meter, Voc (open-circuit voltage) and Vmp (max-power voltage) are in volts dc, and Isc (short-circuit current) and Imp (max-power current) are in amps dc. PV array dc power output is computed as the product of Imp and Vmp.

These equations, with the specific constants derived for Siemens Solar M55 and ASE Americas GP-50-DG PV modules are used in our simulation software. In addition, we model the efficiency of the power conditioner as a function of the operating power level, expressed as its "percent full load (Pfl)" -- array dc power divided by the inverter dc rating (computed as the inverter ac rating divided by inverter full-load efficiency). Equations of the following form are used:

Linear region (Pfl at or below 10%)

$$Eff = C_1 * Pfl - C_2 \quad (6)$$

Cubic region (Pfl greater than 10%)

$$Eff = C_1 + C_2 * Pfl + C_3 * Pfl^2 + C_4 * Pfl^3 \quad (7)$$

GRAPHICAL DATA ANALYSIS

When these models are used with the meteorological database from Ascension Technology's growing network of solar resource monitoring stations, PV system performance can be predicted and system design tradeoffs made for sites around the country. Graphical display of PV system performance, along with modelled performance has proven useful for remotely diagnosing a wide range of PV system phenomena and quantifying their impact. Examples are cited below.

Array-to-Inverter Power Matching

Observe in Table 1 that the PV array output for the Project 1 PV systems (8.55 kWdc each, or 25.6 kWdc total) exceeds the rating of the inverter (6 kWac each, or 18 kWac total). Operationally this results in "clipping" of the PV array output power when the inverter reaches its power limit. Figure 1 shows an example of a single day's performance for a PV system located in Lawton, OK. The figure shows the measured ac power output of the system, along with the simulated output. The system reaches its nominal ac peak output of 18 kWac and maintains this power level for several hours around solar noon. In order to continue operating at its limit, the inverter moves the dc votlage above the PV array's maximum power point, until a point is found where array current, voltage and power are all within acceptable range.

The simulated ac power output, also shown in the figure, is not restricted by a power limit. That is, the simulated output shows that if the inverter were able to accommodate it, the system could produce higher power during the period of power limiting behavior. The obvious question is: how much energy is lost over the year due to this somewhat-oversized PV array? With our PV system models and meteorological database we are able to answer that question, as shown in Figures 2 and 3.

PV array performance was simulated for an entire year, using meteorological data for Tempe, AZ and Waltham, MA.

The meteorological data are 15-minute averages of irradiance and ambient temperature. At each time-step the maximum power of the PV array was computed. The data were then sorted to produce the power duration curves shown in Figures 2 and 3. Note that a lower limit of 100 W/m^2 was imposed on the POA irradiance for the simulations, since our models are not tested at extremely low irradiance.

The area beneath these curves is the annual dc energy which could be produced by the system; for Tempe, AZ this is 18,350 kWh dc and for Waltham, MA 13,670 kWh dc. The area under the portion of the curves at dc power levels of 6.6 kW (the array dc power when the inverter ac output is 6kW, its limit) and higher is the energy which is "lost" when the inverter limits power (see Fig. 1). In Tempe, AZ, the lost energy is 439 kWh dc, or 2.4% of the annual potential total; in Waltham, MA the lost energy is 305 kWh, or 2.2% of the potential total.

An Inverter Control Problem

On a routine basis the data retrieved from the EPA PV systems is summarized in tabular form. Anomalous performance is flagged. To scrutinize further, the data are displayed with Ascension Technology's custom graphing software. The simulated performance is displayed along with the measured performance; in this way, discrepancies between actual and expected performance are clearly identified.

Figure 4 shows an example of an inverter logic problem, which was identified and diagnosed by Ascension Technology's data analysis procedures. Compare the simulated and measured PV system ac power output. Note that the PV system began operating then levelled off unexpectedly at a low power level of about 750 W ac; output did not rise to expected levels until approximately 1:30 pm. Performance of this type was observed on several other days also. The inverter manufacturer was notified and the phenomenon turned out to be a control problem, which was corrected by a simple software modification.

Note that the simulated PV system ac power aligns reasonably well with the measured values during the afternoon, when inverter behavior is normal. This gives us some confidence in the accuracy of the models being used.

Inverter Maximum-Power-Point Tracking

In a similar situation, where the daily data summary had flagged low system energy production, relative to our expectations based on simulation, we identified an apparent maximum-power-tracking problem. This problem arose at a site which has dc instrumentation, in addition to the standard irradiance and ac transducers.

Figure 5 shows the measured and simulated PV array maximum power voltage for a Project 1 site at Edwards Air Force Base, CA. As shown, the simulated array max-power voltage varies over the day, dropping just below 200 Volts at mid-day when the PV array temperature is highest; the measured array voltage locks up at a level of approximately 220 Volts, contrary to manufacturer's specifications. This behavior has been observed on numerous occasions. At this time we are working with the inverter manufacturer to understand the cause of this behavior.

Future Plans

Our modelling abilities will continue to be refined, using the database of PV system performance measurements for model validation. We intend to adopt an I-V curve based PV array model, to allow better investigation of array-inverter interface design issues.

Battery storage capacity will be modelled and simulations of grid-tied PV systems performed to estimate the potential benefits of shifting PV output for better demand-side-management benefit. This work is being conducted as part of the studies for the EPA PV-DSM Projects and related activities being conducted with Delmarva Power and Light under the National Renewable Energy Laboratory's PV BONUS program.

Final Comments

PV system models have proven to be invaluable for troubleshooting hardware problems and advancing our understanding of the PV system performance issues. Insuring that the EPA PV systems are operating properly -- diagnosing problems using our database -- is one of the highest priorities of the project and a major challenge of managing these numerous widespread PV installations. Key to meeting this challenge is the ability to model the PV systems accurately.

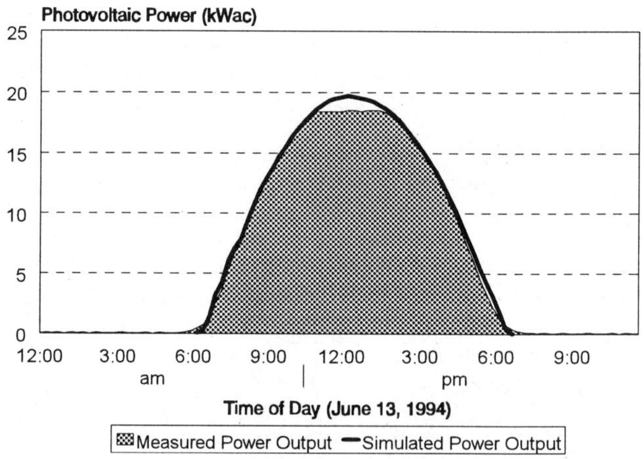

Figure 1 Simulated and measured PV system ac output, showing inverter power limiting

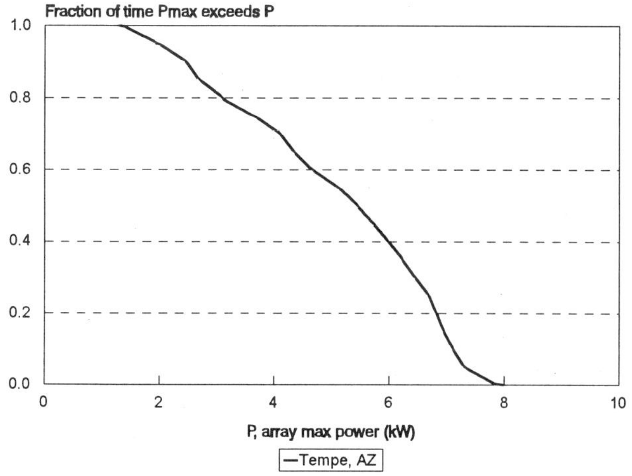

Figure 2 Max-power duration curve, Tempe AZ, for a simulated EPA2 PV array, south-facing, 25-degree tilt

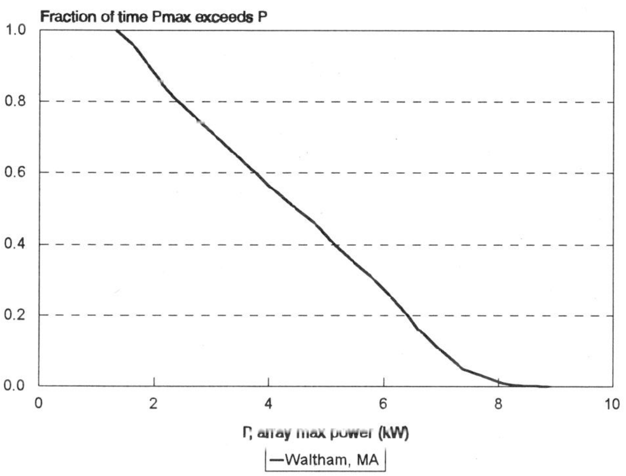

Figure 3 Max-power duration curve, Waltham, MA, for a simulated EPA2 PV array, south-facing, 25-degree tilt

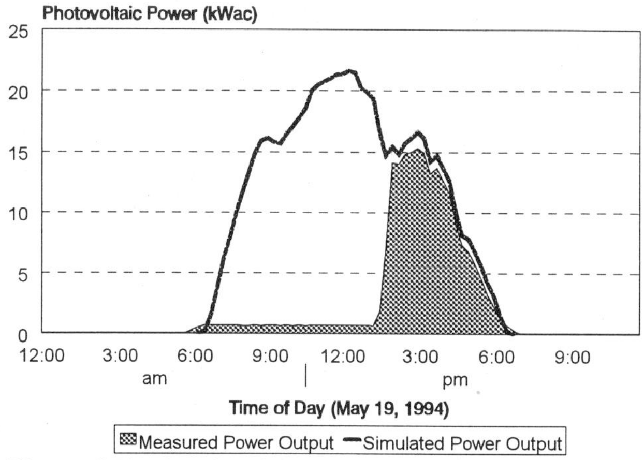

Figure 4 Simulated and measured PV system ac output for an Project 2 system with an inverter maximum power tracking problem

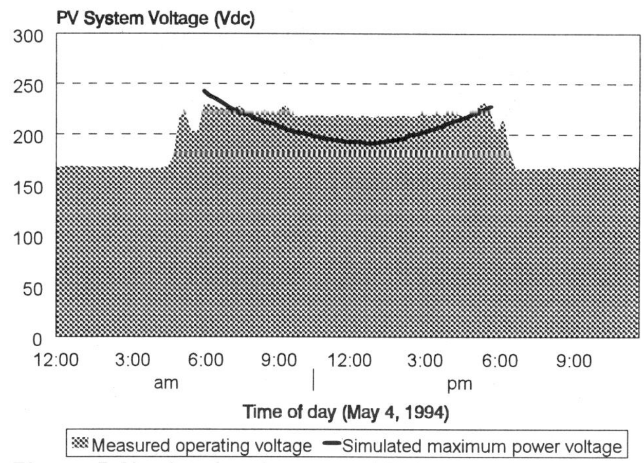

Figure 5 Simulated and measured PV system ac output for an EPA Project 1 system with inverter voltage control unable to track the maximum power point below 220 volts.

PHOTOVOLTAICS FOR DEMAND-SIDE MANAGEMENT UTILITY MARKETS: A UTILITY/CUSTOMER PARTNERSHIP APPROACH

John Byrne,[1] Steven Letendre,[1] Ralph Nigro,[2] Chandra Govindarajalu,[1] William Wallace,[3] Young-Doo Wang[1]

[1]Center for Energy and Environmental Policy
University of Delaware
[2]Delmarva Power
[3]National Renewable Energy Laboratory

ABSTRACT

Photovoltaic (PV) systems located at customer sites can be used to meet utility needs for demand-side management (DSM) applications. PV-DSM can also represent a high-value intermediate market for PV in the utility sector. Maximum value for PV in DSM applications can be achieved by incorporating a dispatching capability to PV systems (through the addition of storage). This enables utilities to evaluate PV systems as a peak-shaving technology. To date, peak-shaving has been the higher value DSM application for U.S. utilities. Our analysis of the value of dispatchable PV-DSM systems indicates that small-scale, customer-sited systems are approaching competitive cost levels in several regions of the U.S. that have favorable load matching and peak demand pricing characteristics. This paper presents the results for PV-DSM systems located within the service territories of five case study utilities.

INTRODUCTION

Utility system peak loads tend to coincide with long, hot sunny days during the summer when high solar insolation is also available. Several recent studies have found that PV output matches closely the daily peaks for summer demand [1]. As a result, PV technologies could deliver high-value power to a utility system in a technically and economically comparable way to DSM peak management programs already in use in the U.S. utility sector. In evaluating the future direction of PV technologies, the U.S. Department of Energy identifies utility peak power reduction technologies as the market segment where PV can be most competitive in the near term.

To date, PV has been investigated largely as an energy supply option. Not surprisingly, in a recent survey of utility managers and state regulators conducted by the Center for Energy and Environmental Policy (CEEP), University of Delaware, DSM ranked as the most preferred option for meeting future load growth [2]. In that same survey, utility managers and state regulators ranked PV well below DSM, and even after next-generation nuclear power, in meeting capacity needs through the end of the century. This low ranking of PV is based primarily on a supply-side analysis of the technology in which PV's high capital cost precludes it from competing with other supply-side options. As a consequence, utility investment in PV remains modest.

Recently, several utilities, including Pacific Gas and Electric and Niagara Mohawk, have expressed interest in investigating in PV as a DSM option [3,4]. This paper focuses on research being conducted by CEEP in conjunction with Delmarva Power (DP&L), a Mid-Atlantic utility servicing most of Delaware and parts of Virginia and Maryland, which indicates that PV-DSM is closer to commercial viability than previously thought. The results of an economic analysis of a PV-DSM system for five case study utilities suggest that PV, when deployed in a DSM role, is approaching cost-effectiveness, especially when a unique utility/customer investment partnership is established for the purpose of purchasing PV-DSM systems.

THE CONCEPT OF PV-DSM

The concept of PV-DSM under investigation by CEEP and DP&L differs from ones explored elsewhere. Under our design, PV technology is deployed in an integrated system design that incorporates some form of energy storage. Figure 1 illustrates this type of system diagrammatically. Most other PV-DSM applications attempt to maximize the energy provided by the system to the customer (i.e., maximum energy savings). In contrast, our application seeks to maximize the power that can be made available for peak management purposes (i.e., maximum peak demand reduction). During the five to ten peak demand days, our system could be dispatched to achieve utility peak-load reduction as needed. On non-peak days, our PV-

 First WCPEC; Dec. 5–9, 1994; Hawaii

DSM system could provide value to the customer in the form of shaved building load and reduced energy needs. Such an application could be especially valuable to commercial customers because the rates they pay usually include demand charges. These charges often constitute a far greater portion of the commercial customer's bill than energy charges.

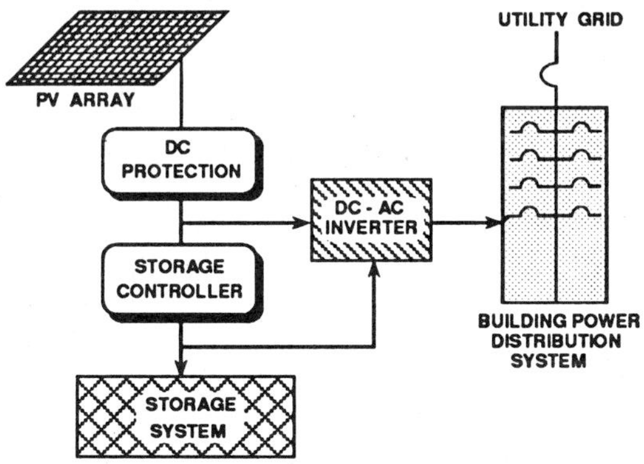

Source: Center for Energy and Environmental Policy, University of Delaware

Figure 1 Block Diagram of a Dispatchable PV Peak Shaving System with Storage

A peak-shaving PV-DSM system offers a utility the capacity to dispatch the technology for load-reduction purposes. Conventional load-reducing DSM programs, such as direct load control (DLC), interrupt the normal operation of electric equipment and, thus, reduce the service this equipment provides to the customer. Dispatchable PV-DSM has the comparative advantage of not interrupting the service provided by the electric equipment. Furthermore, the level of load reduction that can be credited to the system can exceed the rated capacity of the PV array. During the morning hours of low demand, a PV-DSM system can use its storage component to accumulate energy value. Alternatively, the storage system can be charged using off-peak grid electricity. The batteries in our system are sized according to the amount of energy produced by the array prior to dispatch. The maximum of the twelve months determines the amount of storage included in the PV-DSM system. When the dispatch order is given during the utility's peak demand hours, the output of the array can be combined with the stored energy of the system to achieve a higher credited capacity than the PV array's rated capacity.

An alternative to PV-DSM that can provide a dispatchable peak-shaving capacity is a battery-only system. Such a system would utilize off-peak, base-load generating units to charge a bank of batteries. The stored energy (minus round trip losses) would then be available for peak load dispatch. A dispatchable PV system is conceptually a better option than the battery-only system because: (1) the fuel costs for a PV unit are zero; (2) battery systems must purchase energy and a portion of that energy is wasted due to round trip losses; (3) the PV array, as well as its battery unit, supply energy at the time of dispatch and, therefore, the size of the battery bank is considerably smaller for this application than for the battery-only option; and (4) potential cost reductions for PV are much greater than for utility scale battery systems. In fact, our analysis illustrates that the present value revenue requirements (PVRR) for both battery-only systems and PV-DSM systems are approximately equal for intermediate demand charge scenarios ($13/kW and 3.6¢/kWh) and PV-DSM have a lower PVRR under a high demand scenario ($16.5/kW and 6.0¢/kWh) [5].

ECONOMIC ANALYSIS OF PV-DSM FOR FIVE CASE STUDY UTILITIES

To conduct an economic analysis of PV-DSM, CEEP, in conjunction with DP&L, has developed a spreadsheet model to estimate the benefits and costs associated with both non-dispatchable and dispatchable PV-DSM systems. This model was utilized to assess the economic viability of investments in both non-dispatchable and dispatchable peak shaving PV-DSM systems for five case study utilities. The five utilities under investigation are diverse in terms of their geographic location, size, and rate structures. The participating utilities are Niagara Mohawk Power Corporation (NMPC), Austin Municipal Utility (AUSTIN), Sacramento Municipal Utility District (SMUD), Delmarva Power & Light (DP&L), and an East-Coast Urban Utility (ECUU). Table 1 provides some basic characteristics of each utility.

Table 1
Selected Utility Characteristics

	NMPC*	DP&L	ECUU	SMUD	AUSTIN
Peak Demand Charges ($/kW)	11	13	43	9	13
Peak Energy Charge (¢/kWh)	5.7	4.0	6.8	5.9	2.8
Peak Demand (MW)	6,159	2,736	N/A	2,145	1,615

Note: * Winter-peaking; all others are summer-peaking.
Source: Center for Energy and Environmental Policy, University of Delaware

The performance of a 10 kW PV-DSM system with a 25 year life was assessed for the above utilities. Both non-dispatchable and dispatchable system designs were analyzed. Regional Typical Meteorological Year (TMY) data provided insolation values and temperatures for the utility service territories to estimate the AC output of a 10 kW PV array. System load data for each utility were compared to the PV array output to estimate potential load-reduction impacts from the operation of a PV-DSM system. A four-hour dispatch requirement was assumed and dispatch hours were set based on the utility's system load peak. To provide a modest credited capacity based on the coincidence between peak demand and the availability of the solar resource, one hour of buffer storage was included for non-dispatchable systems. Table 2 lists the credited capacities for the systems analyzed (equal to the coincident peak demand reduction averaged over the utility's three peak months from the operation of the PV-DSM system). An installed array cost of $7,500 per kW was assumed, including PV modules and all balance of system components. In addition, battery storage costs of $200 per kWh were used.

Table 2
Credited Capacity for a 10kW PV-DSM System (kW)

	NMPC*	DP&L	ECUU	SMUD	AUSTIN
Non-Dispatchable	0.7	4.3	4.4	3.2	4.9
Dispatchable	4.5	14.9	12.2	17.2	15.3

Note: * Winter-peaking; all others are summer-peaking.
Source: Center for Energy and Environmental Policy, University of Delaware

Two ownership options were analyzed: utility-owned and customer-owned systems. Table 3 provides the benefit-cost ratios for these ownership options. The results presented in Table 3 suggest that the economics of PV-DSM are slightly more favorable for customer-owners. This is due largely to the tax treatment of capital investments afforded customers. However, investment in PV-DSM remains uneconomical under either ownership option in all utility service territories. Our results are consistent with analyses performed by PG&E, which suggest that, currently, the cost of PV systems exceed their value as DSM applications for both utilities and customers when each undertakes the necessary investment alone [3].

Importantly, in all utility service territories, the potential benefits of dispatchable PV-DSM are considerably greater than those of a non-dispatchable (i.e., non-storage) system. When compared to the costs of each system, dispatchable options in all territories are closer to commercial viability. This is demonstrative proof of the importance of deploying PV in a dispatchable DSM mode over the more typical applications that emphasize maximization of energy output. The increased costs of a dispatchable

Table 3
PV-DSM Investment Options:
Benefit/Cost Ratio Comparisons

	NMPC*	DP&L	ECUU	SMUD	AUSTIN
Non-Dispatchable					
Utility Owned	0.40	0.48	0.51	N/A	N/A
Customer Owned	0.61	0.63	0.74	0.62	0.63
Dispatchable					
Utility Owned	0.41	0.66	0.52	N/A	N/A
Customer Owned	0.69	0.76	0.98	0.73	0.75

Note: * Winter-peaking; all others are summer-peaking.
N/A: These are municipal utilities and, therefore, the methods for estimating avoided costs and carrying charges are different from those employed by investor-owned utilities. Currently, methods are being refined to allow for meaningful comparisons.
Source: Center for Energy and Environmental Policy, University of Delaware

system are more than offset to the customer by increased demand savings and to the utility by avoided capacity costs.

UTILITY/CUSTOMER PARTNERSHIP

To speed the penetration of PV into the DSM market, an innovative utility-customer partnership may be needed. We propose a Green Investment Fund (GIF) that has customers and utilities contributing to a fund for the purpose of purchasing a PV-DSM system much in the same manner that "green pricing" joins both parties in a common effort to make advance purchases of technologies that are currently not cost-effective to the individual parties. Both strategies are intended to encourage early sales of new technology in the hope that such sales will stimulate more rapid price reductions by renewable technology manufacturers.

Unlike green pricing, a GIF does not require premium rates. Instead, customers are sought who would be willing to forego the bulk of potential bill savings that would accrue to customer-owners of PV-DSM systems and to invest, as well, a high percentage of the tax savings available to them. The commitments of bill and tax savings may be attractive to commercial customers who have an interest in promoting environmental values and/or technologies.

It is not likely that such savings would be realized because (in most cases, at least) customer-owned systems are not currently cost-effective. And, if the utility partner can assume the initial capital cost (which would then be repaid through tax and bill savings), a customer's participation in a GIF partnership would involve no additional out-of-pocket expenses. The customer might retain a portion of the bill savings for allowing the PV-DSM system to be installed on their premises. Although PV-DSM would have to compete with other DSM options, the environmentally friendly nature of this investment may offer special advantages to customers.

Under a GIF, the utility's contribution is based on its avoided cost. For a capacity-constrained utility, a useful proxy for the value of peaking capacity is the avoided cost of a gas-fired combustion turbine with the capacity factor set to zero. This is reasonable assuming that utility system peaks occur infrequently, and that the utility can take credit for the dispatchable capacity of the system. Double counting of customer and utility benefits is avoided if, during the time the system is dispatched by the utility for peak shaving, no energy or demand savings are calculated. Certain utility peak management programs utilizing customer-sited engine-generation operate in a similar fashion.

Table 4
PV-DSM Investment Options:
Benefit/Cost Ratio Comparisons
(Dispatchable System)

	NMPC*	DP&L	FCUU	SMUD	AUSTIN
Utility Owned	0.41	0.66	0.52	N/A	N/A
Customer Owned	0.69	0.76	0.98	0.73	0.75
Green Investment Fund	0.74	0.93	1.18	N/A	N/A

Note: * Winter peaking; all others are summer-peaking.
N/A: Because the methods for estimating avoided costs and carrying charges has not been finalized for these municipal utilities, a GIF scenario could not be run.
Source: Center for Energy and Environmental Policy, University of Delaware

Under the assumption that a GIF receives the required regulatory sanction, we calculated the net present value benefits and costs of PV-DSM in the five case utilities' service territories. Table 4 compares the benefit-cost ratios for three investment options: customer-owned, utility-owned, and GIF. For the purpose of these calculations, we assumed that the customer contributed 80 percent of the potential bill savings and 100 percent of the tax savings available from the operation of a dispatchable PV-DSM system. The utility was assumed to contribute to the Fund the avoided capacity costs (at zero capacity factor) based on the credited capacity of the PV-DSM system operating in each service territory. And, it was assumed that the utility would manage the Fund and would be responsible for O&M. With these assumptions met, a GIF can be an effective vehicle for the development of PV-DSM.

CONCLUSIONS

The use of PV as a dispatchable peak shaving technology appears to provide higher value in utility DSM applications compared to supply-side or non-dispatchable DSM applications of PV. Our analysis indicates that PV-DSM is closer to commercial viability than previously thought and may be currently cost-effective for certain utilities with above-average commercial rates. An innovative utility/customer partnership is proposed in which benefits are pooled for the purpose of purchasing PV-DSM systems. Under this arrangement, the opportunity for PV to play a role in the utility DSM market is enhanced.

REFERENCES

[1] R. Perez, R. Seals and R. Stewart. 1993. "Assessing the Load Matching Capability of Photovoltaics for US Utilities Based Upon Satellite-Derived Insolation Data." *Proceedings of the 23rd IEEE Photovoltaic Specialists Conference*. Louisville, Kentucky.

[2] J. Byrne, Y.D. Wang and S. Hoffman. 1992. *Photovoltaic Technology as a Demand-Side Tool: A National Survey of Utility and State Commission Views*. Newark, Delaware: Center for Energy and Environmental Policy, University of Delaware.

[3] Tom Hoff and Howard J. Wenger. 1992. *Photovoltaics as a Demand-Side Management Option, Phase I: Concept Development*. Pacific Gas & Electric, Research and Development, Report 007.5-92.4.

[4] Lynne S. Hogeland. 1993. *Analysis of Photovoltaics As a Viable DSM Measure*. Niagara Mohawk Power Corporation.

[5] J. Byrne, Y.D. Wang, R. Nigro and W. Bottenberg. 1993. "Commercial Building Demand-Side Management Tools: Requirements for Dispatchable Photovoltaic Systems." *Proceedings of the 23rd IEEE Photovoltaic Specialists Conference*. Louisville, Kentucky.

AVOIDED COST BENEFITS OF PV ON DIESEL-ELECTRIC GRIDS

Eric Usher, Sylvain Martel
Energy Diversification Research Laboratory
CANMET, Department of Natural Resources Canada
1615 Montée Ste-Julie, Varennes, P.Q., J3X 1S6, Canada

ABSTRACT

EDRL has been assessing the effects on plant operation and avoided cost benefits of using PV systems on diesel-electric grids in Canada's northern communities.

Modelling results to date show that the major impact of PV on a diesel grid is the energy credit associated with reductions in fuel consumption. In addition, although PV has little effect on runtime, it reduces generator starts which can be a benefit with those which are not kept pre-heated when on standby.

This paper will discuss the results of this analysis and briefly present a 3 kW_p grid-connected PV system which is being installed in the remote community of Iqaluit. This work is part of the *PV for the North* program.

INTRODUCTION

Although the Northwest Territories (NWT) abound with mineral wealth, viable forms of energy supply are limited. A majority of the electricity is produced via diesel generation, which is costly both financially and environmentally and new clean and reliable energy options are therefore in high demand. PV is one technology which holds potential in the north, both on community diesel-electric grids and in remote off-grid applications.

Canada has a commitment to sustainable energy development and towards reducing its reliance on fossil fuels. Towards this end, CANMET's *Energy Diversification Research Laboratory* (EDRL) and the *Science Institute of the Northwest Territories* (SINT) have launched *PV for the North*, a five-year photovoltaic research, development and technology transfer program. This program will ensure that a portion of future electricity requirements in the NWT will be met with PV, a clean energy supply technology.

Although the cost of installing PV generating stations in the north can be estimated based on similar installations elsewhere, little has been done previously to evaluate the actual value of PV produced energy on a diesel grid. A study has therefore been initiated to assess the avoided cost benefits of using PV systems on remote community diesel-electric grids. In this study, these avoided cost benefits are being evaluated with an approach used both with other "intermittent" energy sources, such as wind, and with Demand-Side Management (DSM) options.

DIESEL-ELECTRIC SUPPLY IN THE NWT

Only two resources are used to produce electricity on a significant scale in the North: diesel fuel, which accounts for 59% of production, and water, which accounts for the remaining 41%.

The *Northwest Territories Power Corporation* (NWTPC) maintains diesel generating stations in 55 NWT communities. The diesel plants on these remote grids usually include three to four generators which operate in parallel as required by the load. Different operating strategies are employed in an effort to optimize fuel conversion efficiency through maximizing generator load fraction. To respond to load increases quickly, the off-line generators are always kept pre-heated on standby.

The overview of NWTPC operations for the fiscal year 1992/93, summarized in table 1 [1], highlights several aspects of remote stations which are in constrast with centralized plants in southern Canada. As compared with the larger, centralized utilities, where reserve margins on generating capacity is typically 25%, the cumulative reserve margin at NWT plants is 74%. T&D losses are minimal since production is close to the point of use in these small communities; distributed benefits of alternate energy sources will, therefore, never be substantial. Due to the remoteness of sites, sufficient backup capacity is required in each community to maintain reliability. However, because of the low capital cost of diesel plants, maintaining sufficient capacity is not the primary expense on remote diesel-grids. Rather, the cost of diesel produced electricity is related more to fuel consumption, therefore energy, than capacity costs. Depending on location and method of transport, diesel fuel costs across the NWT range from \$0.53/l to \$1.27/l [2] (\$1CDN = \$0.75USD).

Table 2 presents typical electricity production costs and rates for the remote community of Cambridge Bay. Actual costs and rates vary considerably between communities: this distribution of rates at the 55 communities is shown in Table 3.

CH3365-4/94/0000-1048 \$4.00 © 1994 IEEE

Table 1. NWTPC Diesel Operations

NWTPC 1992/93	
Sales	154 GWh
Generation	172 GWh
Purchases	8 GWh
Line-Losses	11.8 GWh
Peak Demand	35 MW
Installed Capacity	135.4 MW
Capacity Factor	15.2%
Load Factor	55.2%

There is a standard capacity charge of \$4.92/kW for NWTPC's commercial customers, however, this is usually a small fraction of their total energy bill [3].

The avoided cost value of an energy supply or DSM option on a diesel grid is considerably different than on a central grid. The avoided cost of an option is comprised of two components,

Table 2. Cambrige Bay Electricity Rates

Residential	First 300 kWh/month	\$0.23/kWh
	Additional	\$0.41/kWh
Commercial		\$0.43/kWh
Government		\$0.45/kWh

namely energy and power. The energy component includes factors such as fuel, O&M and administrative costs. The power component reflects the cost of meeting a unit increase in demand at the time of the utility's annual peak demand, i.e. the cost of installing new capacity to meet the increase. The two components are therefore independent and should be accounted

Table 3. Rate Breakdown by Customer Class in 55 Remote Communities

\$/kWh	Res.	Com.	Gov.
< \$0,25	14	7	1
\$0,25 - 0,50	35	38	18
\$0,50 - 0,75	1	6	19
\$0,75 - 1,00	0	1	8
> \$1,00	1	0	1

for separately. The largest cost savings on central grids usually comes from offsetting power capacity (capacity credit), whereas the largest cost savings on diesel grids should come from simply offsetting kWh production (energy credit). Therefore, although sources such as PV which peak in summer hold little value for a winter peaking central grid utility, they do hold value for a winter peaking diesel grid because of the high year-round energy (kWh) cost, and the low, in relative terms, capacity (kW) cost. In other words, offsetting energy production during the non-peaking summer period is as valuable in avoided cost benefits as offsetting energy during the peaking winter period.

Future annual growth on NWTPC diesel grids is forecast at 2% to the year 2003, involving an additional 19 MW of new capacity expansion to the 55 generating stations [1]. On the customer side, major changes are taking place; rate subsidization is being reduced as a large proportion of NWT government assests are sold off. Customers are therefore looking more towards DSM options and independent power production.

METHODOLOGY

For this study, an assessment of PV at various levels of penetration was undertaken, based on the savings derived from offsetting the load applied to the diesel generating station. Using measured climatic data, the annual performance of a PV system was modelled for the communities of interest using the WATSUN-PV simulation program. To model the effects that the PV produced energy would have on diesel plant operation, the Atlantic Wind Test Site Wind-Diesel simulation program was upgraded to accept the PV energy figures output from WATSUN-PV.

PRELIMINARY RESULTS

Diesel Plant Operation

The diesel generating station at Cambridge Bay (pop. 1006), where the peak load is 1200 kW, consists of four Catapillar generators sized 720 kW, 720 kW, 725 kW and 950 kW. With increasing load, the capacity steps are 725, 950 and 1675.

The results of simulation runs for Cambridge Bay are summarized in Table 4. The work incorporated hourly simulation models of the effects of PV generated energy on the diesel grid. The parameters relevant to PV system operation are: energy produced, energy dumped and the optimal Penetration Ratio (PR), which is the installed PV capacity divided by peak load. The parameters of relevance in diesel plant operation are: fuel consumption, generator efficiency, run-time and start/stop rate.

Although it is seasonal, the Northwest Territories do not lack for sunshine. In fact, they receive about the same insolation annually as most regions of Canada, and even compares with many parts of the United States. On a latitude tilted surface, Cambridge Bay, at 69°N, has average insolation of 3.67 kWh/day

Table 4. Cambridge Bay Simulation Results

ANNUAL RESULTS		PV ARRAY (kWp)		
		400	800	1200
Energy Produced	(MWh)	427	855	1282
Energy Dumped	(MWh)	3	47	229
	(kWh/kWp)	9	60	192
PV Fraction of Load	(%)	7.1	13.6	17.7
Fuel Savings	(l/kWp)	273	253	216
Gen. Starts Reduction	(%)	4.6	13.2	16.0

Fig. 2: PV-Diesel Plant Interaction

(1.2 MW Cambridge Bay Grid)

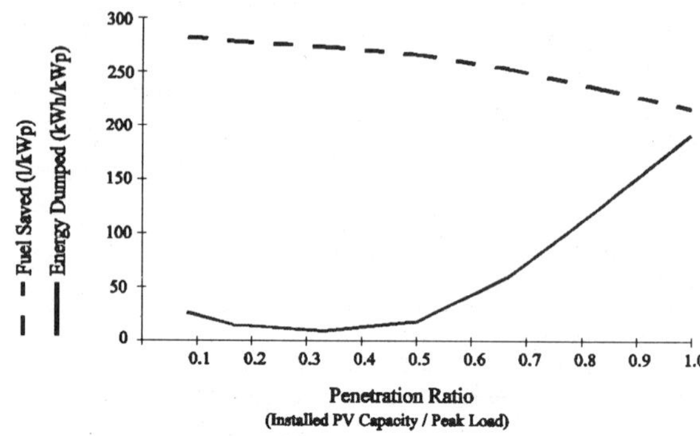

With this solar resource, the energy which would be delivered to the grid annually by a PV system at Cambridge Bay is predicted to be 1.1 MWh per installed kW_p. The load profile for the community and predicted contributions of 800 kW_p and 1200 kW_p PV systems are shown graphically in Fig. 1. Table 4 summarizes the interaction of these PV systems on the Cambridge Bay grid. The dumping of energy, when PV system output exceeds the load, is not a problem at low penetration rates: at 0.5 PR, less than 2% of electricity produced is wasted. However, as shown in Fig. 2, marginal dumped energy increases sharply above 0.5 PR. This affects fuel consumption, also shown in Fig. 2. Reductions in fuel consumption are presented as litres saved annually per kW_p of installed PV. This reduction is 273 l/kW_p at a PR of 0.33 dropping to 216 l/kW_p at full penetration. The additional input power from the PV system lowers generator efficiency marginally; at full penetration, plant efficiency is reduced by 1.7% from 3.45 kWh/l to 3.39 kWh/l.

The capacity factor for the 800 kW_p PV system is 12.2%, close to the ave. 15.2% capacity factor of generators at NWT diesel plants.

Beyond reductions in fuel consumption, an "intermittent" source such as PV or wind can also reduce diesel generator maintenance and replacement costs if it reduces runtime or the frequency of cold-starts.

At penetration ratios up to 1.0, the PV system does not significantly reduce generator runtime. However, by displacing load it does reduce the frequency of generator starts. At a PR of 1.0, generator starts are reduced by 16%.

Although Cambrige Bay's seasonal load profile does not coincide with solar availability, on a daily basis the load and PV output are closely matched. Fig. 3 plots the load profile and output of a 800 kWp PV system (PR = 0.67) for the first week of April. This solar-load correlation tends to reduce the start frequency of the generators. This is a valuable benefit for plants which cold-start their generators, although this is not the case at NWTPC plants where all generators are pre-heated.

These results are typical for diesel grids in the NWT although there is a sensitivity to the choice of diesel plant operating strategy, and the seasonal and daily load profiles. Seasonal variations in load affect the quantity of energy dumped from the PV array.

Fig. 1: Annual Load and PV Output Profile

Cambridge Bay

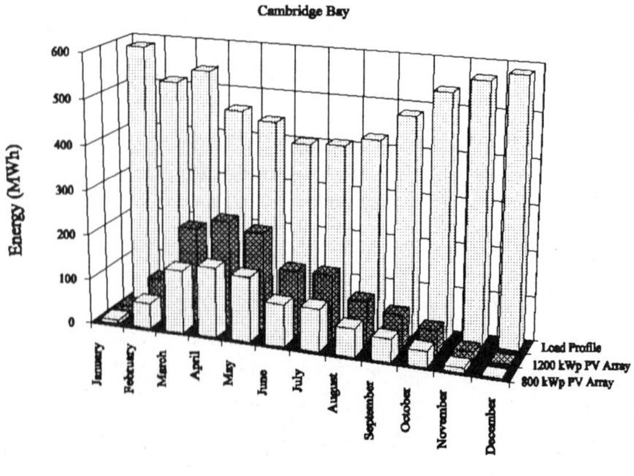

Fig. 3: Daily Load and PV Output Profile

Cambridge Bay

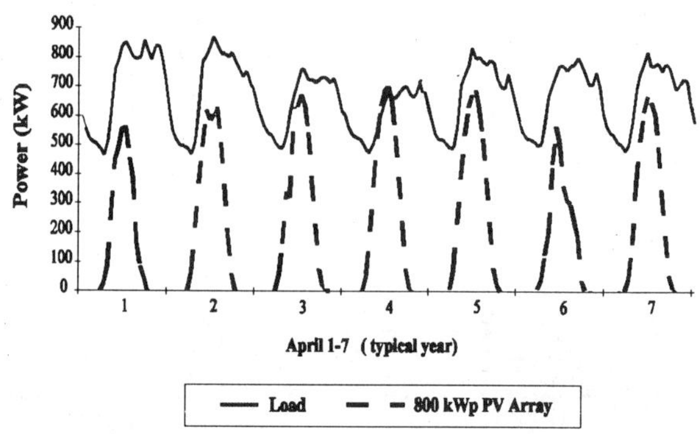

Avoided Cost Benefits

To quantify the value of an energy supply alternative to an existing technology, cumulative life-cycle avoided costs must be determined. The calculation of the total avoided cost benefit consists first in determining the annual fuel benefit as follows:

$$AB_f = FS * (FC + LC) \quad (1)$$

where, AB_f = annual fuel benefit ($\$/W_p$)
FS = fuel saved (l/W_p)
FC = fuel cost ($\$/l$)
LC = lube cost ($\$/l$)

If daily insolation profiles match the load, a PV system can provide additional benefits to generators which are cold-started. Each cold-start is considered equivalent to 25 hours of runtime in marginal O&M and capital replacement costs. The annual benefit of reductions in start frequency is therefore:

$$AB_s = SR * HE * (OC + CRC) \quad (2)$$

where, AB_s = annual cold-start red. benefit ($\$/W_p$)
SR = cold-start reduction (W_p^{-1})
HE = cold-start hours equivalency (h)
OC = overhaul cost per h runtime ($\$/h$)
CRC = capital repl. cost per h runtime ($\$/h$)

To determine the life-cycle avoided cost benefit of a PV system to a diesel grid, the present value of the annual benefits must be summed over the life of the system and brought back to present day values considering the time value of money. To account for these factors, NWTPC discounts annual benefits in any given year to 1995 'real terms' discount factors which are based on an annual discount rate of 7.53. The avoided cost benefit is therefore:

$$ACB = \Sigma\ ((AB_f + AB_s)/(1+r)^n) \quad (3)$$

where, ACB = avoided cost benefits ($\$/W_p$)
r = annual discount rate (%)
n = PV system life (yrs)

A grid-connected PV installation's service life has been estimated at 30 years. Analyses were therefore undertaken for the years 1995 to 2024. The results of an analysis for a potential 400 kWp system in Cambridge Bay is presented in Table 5. Figures are given for a variety of discount and fuel escalation rates for the case of fuel benefits alone, and with cold-start reduction benefits.

DIESEL GRID-CONNECTED PV SYSTEM

As part of the *PV for the North* program, a 3.5 kW_p PV system is to be installed on Arctic College in Iqaluit. Once installed, this system will be monitored and the data used to determine the effects of PV produced incremental load offsets on diesel generator operation and therefore the value of installing PV on remote community diesel grids in the NWT.

Table 5. Avoided Cost Benefits of PV on Diesel Grid

Benefit	Disc. Rate	Fuel Esc.	$\$/W_p$
Fuel Only	7.5%	3.5% (infl.)	2.87
	7.5 %	5.0%	3.31
	3.0 %	3.5% (infl.)	4.62
Fuel and Cold-Starts Reduction*	7.5 %	3.5% (infl.)	4.40
	7.5 %	5.0%	5.09
	3.0%	3.5% (infl.)	7.11
400 kW_p PV, 1200 kW Diesel Grid FC = \$0.80, LC = \$0.03 * O&M = \$7.50/h, Capital Cost = \$9.50/h			

The avoided-cost benefits analysis discussed herein will be compared with data retrieved from this, the first NWT grid-connected system, to produce a general economic impact analysis of using PV on remote community grids in the NWT.

CONCLUSIONS

The principal avoided cost benefits of using PV on diesel-electric grids are reduced fuel consumption and, to a lesser extent, reduced generator start frequency. Depending on the choice of discount and fuel escalation rates, and whether the generator is cold-started, results to date predict benefits ranging from \$2.87/W_p to \$7.11/W_p. PV does not reduce generator run-time significantly. Due to closely matched insolation and diesel plant load during the summer period, PV can be implemented to penetration ratios of 0.67 without dumping substantial energy.

ACKNOWLEDGEMENTS

The authors would like to thank David Gray from NWTPC and Real Reid from Hydro Quebec for their data and information and Gregory Leng from EDRL for his input.

REFERENCES

[1] Load Forecast, Northwest Territories Power Corp., Planning and Technical Services, June (1993).

[2] Martel, S., Technico-economic Market Study of Photovoltaic Applications in the Northwest Territories, EDRL, CANMET, Natural Resources Canada, June (1994).

[3] NWTPC Rates Schedule, April 1, (1992).

[4] Reid, R., Saulnier, B., Potentiel Eolien-Diesel pour les reseaux non-relies; Aspects technico-economiques, Institut de Recherche de l'Hydro Quebec, March, (1987).

A NEW EXPERT SYSTEM BASED CONTROL TOOL FOR POWER SYSTEMS WITH LARGE INTEGRATION OF PVs & WIND POWER PLANTS.

E. Nogaret, G. Stavrakakis, J.C. Bonin, G. Kariniotakis - ARMINES, FRANCE.
M. Papadopoulos, N. Hatziargyriou, S. Papathanassiou - NTUA, GREECE.
J. Gatopoulos, E. Karagounis - AMBER S.A., GREECE.
J. Halliday, G. Dutton - RAL, UK.
J. Pecas-Lopes - INESC, PORTUGAL.
A. Androutsos, P. Pligoropoulos - PPC, GREECE.

ABSTRACT

In the present paper an advanced control system for the optimal operation and management of medium size power systems with large penetration from renewable power sources is presented. This control system is aimed to assist the power system operators in their managing task, by proposing them optimal actions for the economic and secure management of the power system. The control system is composed of various modules performing functions like short-term forecasting of the load and of the renewable sources, economic dispatch of the power units and on-line dynamic security assessment. The pilot control system is being installed and evaluated in the Greek island of Lemnos.

INTRODUCTION

A considerable number of PV and wind power plants are installed or planned to be installed for electricity production in islands of the European Union and in developing countries, where favourable conditions of renewable sources often exist.

In these cases the objective is to maximise the use of renewable energies in order to increase the resulting benefits like fuel savings. However, solar and wind penetration are limited by the repercussions of the solar and wind power variability on the existing power station, which usually includes diesel units.

In order to achieve a larger penetration of renewable sources, without deterioration of the quality of service offered to customers, the development of new control tools is necessary, that will assist the power system operators.

In the present paper the essential elements of an advanced control system for the optimal operation and management of medium size power systems with a large penetration from renewable power sources are presented.

This control system is developed within a JOULE-II project funded by the European Union and the Greek utility PPC and is going to be implemented in the Greek Island of Lemnos.

The developed control system is aimed to assist the operator of a medium size autonomous power system by proposing optimal dispatching scenarios for the various power units, as well as various actions needed to avoid dangerous situations which might result from a poor prediction of load or weather or unforeseen disturbances.

The control system minimises the production costs through an on-line optimal scheduling of the power units which takes into account the technical constraints of the diesel-units as well as short-term forecasts of the load and the renewable resources.

The power system security is maximised through an on-line security assessment module, which supervises the scheduling of the power units and enables the power system states to be capable of withstanding any sudden change in the load or in the production of the renewable sources, without exceeding acceptable limits for frequency excursions and voltage deviations in the grid.

Special attention is given to the monitoring of the security margin defined by the frequency deviation during emergency situations, such as sudden power variations from the renewable units.

An advanced software, having real-time performance and a sophisticated man machine interface, has been developed. The various modules, that perform the control system functions, have been integrated in the software and validated with real data obtained by the existing SCADA system in Lemnos.

The prototype control system will be installed in the island of Lemnos before the end of 1994, where its performance will be evaluated and guidelines for further use in other islands will be prepared.

DESCRIPTION OF THE CONTROL SYSTEM

The developed control system is designed to produce optimal scenarios, both from the economic and security point of view, for the power system operation. For this purpose, various functions have been developed and integrated in the control system structure, which is shown in Fig. 1, and they are performed periodically as described below :

i) Initially, a data-base is continuously updated by a data acquisition system (SCADA). The data concern wind speed and solar radiation, power output for the conventional and renewable units, load consumption levels, temperature, etc. The data-base checks for errors, filters erroneous data and translates the incoming information to be accessed by the control software. The data-base incorporates also the necessary comprehensive information about the entire system - e.g. the various machine parameters.

ii) Using these data and other fixed parameters, the expected load and renewable units power are forecasted for a period equal to the planning period.

CH3365-4/94/0000-1052 $4.00 © 1994 IEEE

First WCPEC; Dec. 5-9, 1994; Hawaii

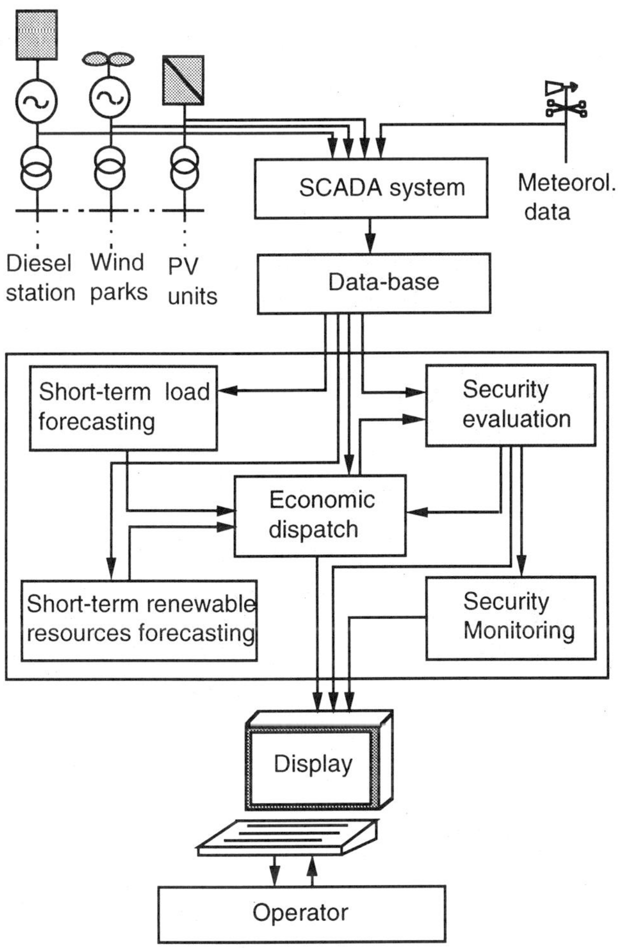

Fig. 1. The control system architecture.

iii) Taking into account the forecasted load and renewable units expected power, a "Logistic" program schedules the operation of the system for the next time period (e.g. two hours, taking decisions every 10 minutes) so that maximum fuel saving is achieved, without deterioration of the quality of service to the consumers of the power system.

iv) The quality of service concerns mainly loss of load events due to a possible deficit between generation and demand, as well as dynamic instabilities and important frequency deviations due to sudden wind speed or solar radiation fluctuations. Both are related to the diesel spinning reserve provided during the scheduling of the operation of the system.

Two different modules perform the security assessment of the power system. The first one checks if the proposed by the logistic module operation scenarios might lead to dynamically unsafe situations. If yes, these scenarios are rejected and the output of the logistic module is revised. The second module provides a real-time quantitative evaluation of the system security for its current operating state, expressed by its frequency deviations if a preselected power disturbance from the renewable sources occurs.

The above functions have been incorporated either as independent modules or in the form of rules within an expert system (ES) shell. The modular structure of the developed system permits the testing of various different methods that perform the same function, e.g. various forecasting or operational planning methods.

An advanced man-machine interface has been designed so that the control system is ergonomic and friendly to the user. The operators can easily communicate with the control system in order to obtain data on the power system behaviour in graphical form.

Special attention has been given to the presentation of the proposed schedules, the predicted load and wind power profiles, the various actions that are proposed in critical situations, recording of historical information and finally on monitoring the power system security.

A multi-tasking operating system - Windows NT - and an Expert System (ES) shell - NEXPERT - provide the real-time performance required by the project. The ES shell is used to incorporate the various security rules. The type of computer chosen for the implementation is a PC 486. In the following sections a more detailed description of the various functions is given.

ECONOMIC DISPATCH-"LOGISTIC"- MODULE

The economic dispatch or "logistic" module has the following two roles:

i) to simulate the operation of the power station on time scales of one minute or greater, to simulate the effect of various alternative control strategies and so assess their effectiveness.
(Dynamic stability over time scales of a few seconds is considered outside the role of the model).

ii) to form the basis of the real-time control system.

To fulfil the first role, the performances of logistic models developed by RAL and NTUA/PPC have been compared for a range of operating conditions [1].

To fulfil the second role - to form the basis of a real time control system - it is required to produce a load dispatch algorithm and to take decisions on whether to stop/start diesel or renewable units. The NTUA load dispatch algorithm has been extracted from the corresponding logistic model and incorporated into the advisory controller for the power station.

The two Logistic Models (LM) developed by NTUA/PPC and RAL are based on time series analysis and utilise a time step in the order of one-minute to include the load and renewable sources fluctuations, as well as their influence on the system performance and a higher time step, in the order of ten minutes, to determine the necessary decisions. The main components of each one of the two LMs are:

The Autonomous Diesel Station model. It simulates a number of diesel units working in parallel. Emphasis is given in the detailed representation of the start and stop procedures. The individual technical characteristics of each unit, e.g. characteristic fuel consumption curve, are taken into account.

The Renewable Power Plants model. The model

simulates arbitrary number of wind parks and PV plants. It considers that all wind turbines within a park and PV units within a plant operate under the same wind or solar conditions respectively.

The Load Management and Control System. This system monitors all the operational parameters of the diesel generators and the renewable units, as well as environmental data. It incorporates the rules for optimal operation of the power system and performs the decision making for the starting and stopping of the diesel units and the renewable units.

The LM calculates, for each diesel unit, the energy production, the hours of operation, the fuel consumption and the number of starts-stops. It calculates also the energy produced by the renewable units and the renewable energy spilt due to the load management commands. Finally, the model records the loss of loads events as well as the relevant lost energy.

A serie of validation studies have been carried out for both the NTUA and the RAL models for the case-study of Lemnos. Both models have been used to simulate the power system operation and their output has been compared with measurements of the actual operation of the power system of Lemnos [1].

Issues on the control algorithm

The economic dispatch is realised by the NTUA LM, which is based on two different time steps; one-minute (basic) and ten-minute (decision-forecasting) steps.

One-minute time step. The logistic model performs the following operations on an one-minute basis :

- It collects all operational data of the power system.
- It calculates the diesel station load and checks the operation limits of the diesel units.
  - if the lower acceptable operation limit of the diesel units is violated some wind turbines are stopped
  - if a diesel unit is overloaded, a part of the load is rejected by a load management command.

Ten-minute time step. The following actions are performed on a ten-minute basis :

- Communication with the forecasting modules in order to obtain the load and renewable power sources profiles for a period of two or three hours ahead with a time step of ten-minutes.
- Decision making on which diesel generator is going to be started or stopped by taking into account the following factors :
  - a priority list of the available diesel units (running or not),
  - the forecasted load and renewable power sources profiles for the planning period under consideration as well as factors expressing the uncertainty of the forecasts,
  - the required spinning reserve of the diesel units as a function of the load and the wind and solar margins,
  - the cost optimisation of the diesel station operation for the planning period,

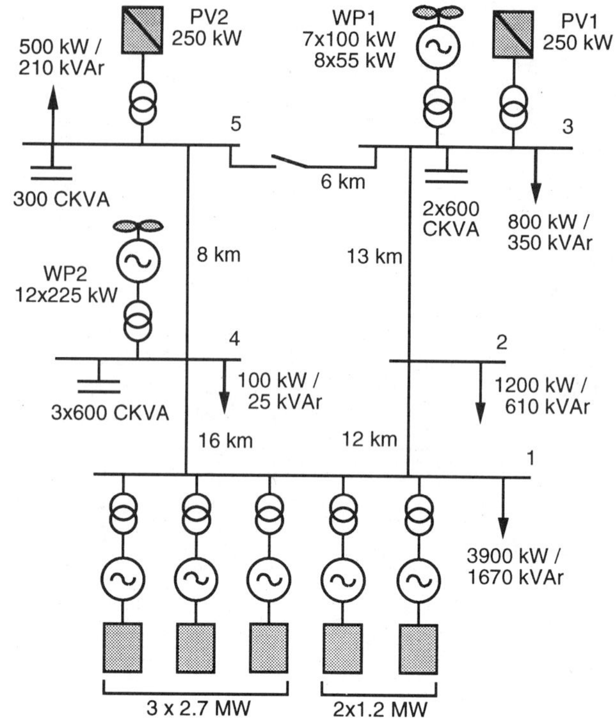

Fig. 2. One-line diagram of the power system model of the island of Lemnos.

  - the security assessment of the system in order to determine if the dispatching scenario examined is dynamically safe for the power system.

FORECASTING MODULES

The objective of the two forecasting modules is to provide an estimation for the load and the renewable power sources profiles for the next planning period (e.g. 2 hours) with a time step equal to the time step used by the logistic module to take decisions (e.g. 10 min).

The principal method used for load forecasting is the Winter method, which is based on exponential smoothing and takes into account seasonal variations.

Concerning wind speed and solar radiation forecasting, the "persistent" method has been used, which assumes that the power from the renewable units during the scheduling period will remain at a constant value equal to an average of past values. It was found that for the case of Lemnos the optimal persistent predictors are obtained by using the average of two or three past values (20 to 30 min) according to the forecast time step ahead.

The efficiency of artificial neural networks has been also examined in wind power forecasting. According to the results obtained using data sets from the island of Lemnos, an improvement in the order of 15% over persistent method can be achieved in the forecasting accuracy [2].

The effectiveness of different wind speed, solar radiation and load forecasting techniques can be proven off-line by the use of the logistic model. By this way, it can be shown their likely value in reducing operational costs before final implementation on-line.

ON-LINE DYNAMIC SECURITY ASSESSMENT MODULE

In the case of autonomous power systems with a large penetration from renewable power sources, large frequency excursions and dynamically unstable situations may occur, due to fast variations of the renewable production, like in the case of very high winds that may result the loss of wind generator production.

Such disturbances, occurring in a time scale of a few seconds, are not foreseen in the logistic modelling and the derivation of the operating plans of the system. A Dynamic Security Assessment (DSA) module is therefore of primary importance, in order to guard the system against pre-selected severe wind or solar radiation disturbance.

The DSA is provided in two ways, each one serving a different objective:

First, during the planning of the power system operation by the economic dispatch module, the dynamic security of each one of the proposed scenarios is evaluated by the DSA module and the scenarios that might result in large frequency excursions or loss of synchronism in case of severe renewable power disturbances are rejected.

Second, a real-time quantitative evaluation of the system security for its current operating state is provided, expressed by its frequency deviations if a preselected wind power disturbance occurs.

In order to obtain from the DSA module fast, credible and easily interpreted results, Artificial Intelligence techniques were used for the elaboration of efficient security classifiers. More specifically:

i) The Decision Trees (DT) method is used for the security assessment of the proposed by the dispatching module operating plans. Using suitable DTs, the proposed operating state is classified as safe or unsafe with respect to preselected disturbances. If it is found unsafe, the dispatching algorithm is instructed to revise its output.

   The DTs have been implemented in the form of rules within an ES-shell. By this approach easy modifications in the case of changes in the power system configuration e.g. installation of new wind parks are possible [3].

ii) Artificial Neural Networks were implemented for the real-time evaluation of the system security, providing an emulation of the expected transient frequency excursion if one of the preselected disturbances took place. In this way, the operators of the system are notified for its security and preventive control actions can be taken, if it is considered necessary [3].

The expected transient frequency excursions that the system can experience were selected, from a practical and operational point of view, as a dynamic security indice, because in several networks, wind park disconnection and/or load shedding are triggered by frequency, according to the relay settings.

EVALUATION OF THE CONTROL SYSTEM

The evaluation of the control system performance includes two stages. The first stage includes the testing of the various modules as well as of the entire software by using data files obtained by the SCADA system of Lemnos.

The second evaluation phase will be performed on-site after the control system installation and the training of the operators. During this phase, which is planned to start by the end of 1994 and is going to last three months, the final tuning of the control system will take place.

The control system performances and its use by the operators will be evaluated on-site. The response of the control system during emergency situations will be of particular interest.

It is anticipated that in certain situations the operators may be inclined to countermand certain decisions of the controller. Such occasions will be recorded and simulations will be carried out using the logistic model in order to demonstrate the consequences of such an action.

CONCLUSIONS

In the present paper an advanced control system for medium sized grids that performs the management of the power units (diesels, wind turbines, PV plants, etc.) has been presented. This control system maximises the benefits coming from the renewable energy sources (in terms of fuel and capacity credits), while maintaining a high level of security to the grid.

Various models that perform functions like optimal scheduling, forecasting, security assessment, etc. have been integrated in a modular way within the developed software. The control system provides a communication interface with a data base fed by a SCADA system and an advanced man-machine interface for an efficient communication with the power system operators.

The developed control system has been tested with measurements from the Greek island of Lemnos. The installation phase is about to start in the island of Lemnos within 1994, where its operation will be evaluated on-site.

The modular development of the control system permits further adaptations for diffusion in other islands than Lemnos; for this purpose guidelines are being prepared presenting the necessary adaptations.

REFERENCES

[1] Dutton, A.G., Halliday, J.A., Androutsos, A., Papadopoulos, M., "A comparison of two logistic models for load and wind power scheduling in an island power station", *EWEC '94*, Thessaloniki, Greece, October 1994.

[2] G. Kariniotakis, G. Stavrakakis, E. Nogaret, "Wind-speed and wind-power short-term forecasting using artificial neural networks", *4th Interim Report of the JOULE-II project JOU2-CT92-0053*, September 1994.

[3] N.D. Hatziargyriou, S.A. Papathanassiou, J.A. Pecas-Lopes, J.N. Fidalgo, V. Van Acker, "Fast Dynamic Security Assessment of Autonomous Power Systems with a Large Penetration from Wind - The Lemnos Study Case", *EWEC'94*, Thessaloniki, Greece, October 1994.

DYNAMIC STABILITY ANALYSIS OF PV-INJECTED POWER INTO A PARALLEL AC-DC POWER SYSTEM.

By

Fadia M.A. Ghali Hisham A. El-Khashab
M. Said Abdel-Motaleb.
Electronics Research Institute, Sh. El-Tahrir, Dokki, Cairo, Egypt

Keywords: Power Systems, Photovoltaic, Stability.

ABSTRACT

The present paper introduces an approach to solve the problem of damping the electro-mechanical oscillations of conventional plant-photovoltaic generator hybrid system. The photovoltaic generator is coupled directly to the DC-link of AC-DC parallel connection to a large power system. The intermittent nature of the photovoltaic generator imposes an unpredictable characterization of negative load which affects the power system dynamic stability. The control process is actuated through a proportional-integral-derivative (PID) power system stabilizer (PSS) and PID-rectifier current regulator (RCR)

INTRODUCTION

There has been increasing interest in intermittent generation such as solar and wind power interconnected with utility systems due to environmental, political and institutional reasons. Intermittent generation is quite different from conventional power plants. Its output is highly unpredictable and somewhat uncontrollable compared with the steady output of conventional generating units. There has been also rising interest in plant response as a factor in generation planning due to the anticipated operating difficulties during the morning load pickup period. Some systems have had to make compromises in economic unit commitment and dispatch because of insufficient response rate from the large fossil and nuclear units pushed into cycling duty. These difficulties will be aggravated by future penetration of intermittent generation into the utility systems.

The intermittent nature of the unconventional generation resources such as solar and wind could add a significant variance to the thermal load of a power system. Utilities have traditionally planned for the normal load fluctuation and the sudden loss of the largest generating units for secure operation . With increasing penetration of intermittent generation, the utilities are forced to change their planning and operating requirements such as load - following and spinning reserve capacity[1].

Over the past decade, an intensive R&D activity has been directed towards the development of efficient ,reliable and cost effective photovoltaic cells , as well as the design of interface and balance-of-system devices which are required for the distribution, storage and consumption of the electric power. At an early stage of the PV development program, it was recognized that, in addition to stand-alone applications , probably the most attractive long term utilization of PV arrays is their operation in a utility interactive mode . Although institutional and economic considerations favor such as interactive operation , many technical issues must be resolved before PV's become a viable and acceptable alternative to utility generation practices. Research problems that are currently being addressed include protection and safety issues, power quality , harmonic current generation and

First WCPEC; Dec. 5-9, 1994; Hawaii

PV-array depends on the intensity of insulation which varies according to time of the day and year, cloud density and wind velocity. So, the output power generated by one PV array may show a sharp change in characteristics. However, it could be assumed that the output fluctuations in all PV -arries constituting a PV-farm will be more smoother , because the insolation at all locations does not change in the same manner and not simultaneously.

The logic diagram of the system dynamics is shown in Fig.4 describing the state of the power network, the generator swing equation, PID-PSS and PID-RCR controllers relations. The behavior of the transmission lines and the system load disturbances is represented by the load flow equations, while the swing equations describe the operations of the conventional units. The feedback loop branches into two subloops unified at the angle controller. The first subloop is PID-PSS needed to generate the control signal to modulate the excitation field of the conventional unit. The second subloop having the same input signal is the PID-RCR controlling the DC power link.

MATHEMATICAL REPRESENTATION

The control policy of the PV-conventional generator hybrid system in utility interactive operation mode can be mathematically represented via appropriate transfer functions. Such a representation of the overall system dynamics is shown in Fig. 5

In general the total power generated by different energy sources must equate the load demand and network losses, mathematically expressed as :

$$P_G + P_{PV} - P_D$$

generation side = demand side

However, for hybrid systems connected on a pool system, as the case studied in this paper, the PV-generation can be considered as negative load (is on the demand side) and moreover the variations in its output as the source of disturbances in the overall system which must be compensated by the

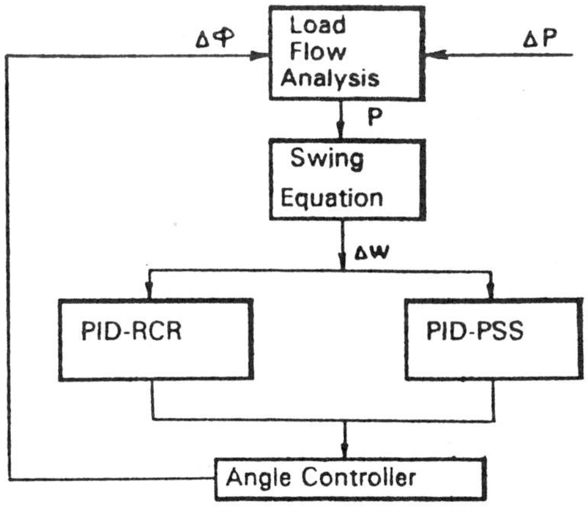

Fig. 4: Logic diagram of the system dynamics

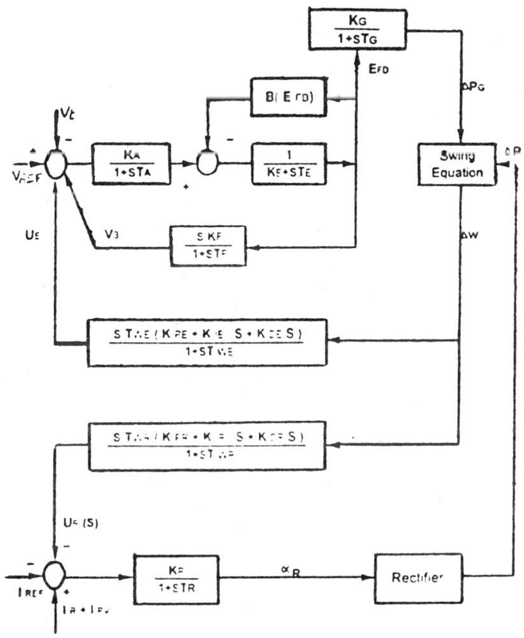

Fig. 5 : Transfer Function Representation Of The Control Scheme

control systems operating on different controllable components of the system, e.g. excitation field of the generator and the rectifier. The system's state equations can be expressed as follows:

$$\Delta PT - \Delta PG = h\,\Delta\dot{\delta} + D\,\Delta\ddot{\delta}$$

where $h = W_K / \pi f_n$, f_n nominal frequency and

propagation , control and islanding of dispersed grid-connected sources [2].

On the other hand, the fast power modulation capacity of a HVDC link has been utilized to improve the damping of electromechanical mode oscillation in a parallel AC-DC power system. Based on model control theory, the design of PSS and RCR was performed requiring only one control input signal for the PID scheme. The PID controller is by far the most widely employed control in the industry and is very simple for practical implementation . The PID-PSS and PID-RCR, which are free from the problems associated with communication links, are capable of controlling both the electromechanical mode and the exciter mode[3-5].

SYSTEM REPRESENTATION

A synchronous machine representing a conventional source of energy, is connected to a large power system through a parallel AC-DC transmission lines. The power system is acting as a pool system and schematically shown as an infinite bus-bar. As shown in Fig.1 , the AC transmission line is represented by an equivalent Π-section. will left sided charging capacity combined with the filter capacitor at the rectifier . On the other hand, the DC-transmission line is demonstrated by. an equivalent T-section with lumped charging capacitance at the mid-point of the DC-link dividing the series impedance into two parts. The photovoltaic generator is mannupilating the facility of existing DC bus-bar and is connected directly to it without the need of excess interface power conditioning subsystem.

The synchronous generator is equipped with the IEEE Type I excitation system which is given in Fig.2 including PID-PSS . The DC-link is connected to current rectifier and inverter at the generator and power system infinite buses respectively for the purpose of power conversion.

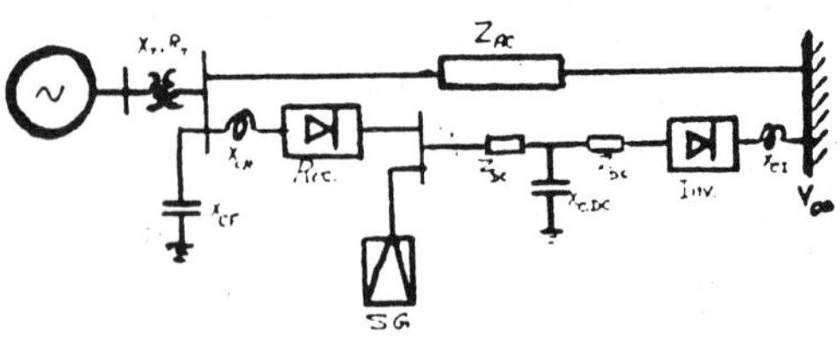

Fig. 1: Single line diagram of the system studied

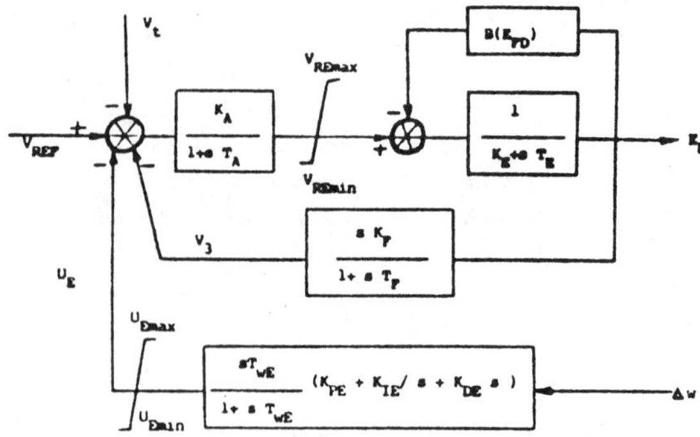

Fig. 2: Block diagram of IEEE type I excitation system

DYNAMIC STABILITY ALGORITHM

Fig. 3 shows an overview of the methodology adopted for controlling a conventional plant-PV generator hybrid system in an utility interactive operation mode . Monitoring of the variation in system frequency (Δ f) and utility grid bus voltage ΔV_l for the purpose of controlling the phase difference $\Delta\Phi$ between the inverter and grid buses and correspondingly the real power flow is the approach suggested for coping with system load changes. In this study load changes are defined as the variations in PV-generator output power. However, because of the variations of solar irradiance occurring in the minute-to-minute time farme, PV systems cannot be a part of the utilities generation schedule. Therefore, in the presence of PV-systems ,operational questions must be answered considering PV-systems as ever changing negative loads. The maximum power generated by

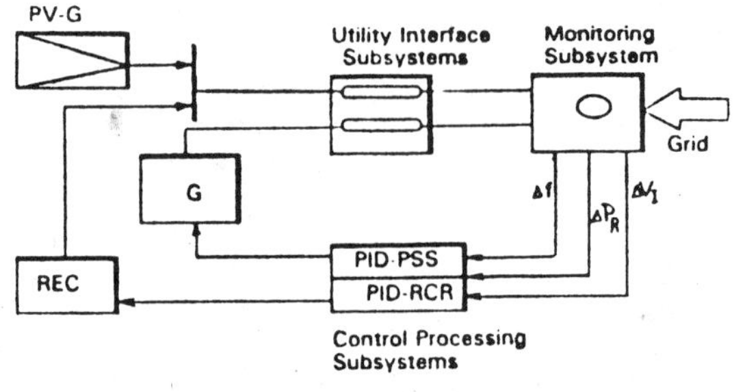

Fig. 3: Proposed feedback configuration for stability study

W_K is the kinetic energy of the generator rotating part,

D damping factor due to the machine damper windings.

Considering the state variables

$$X_1 = \Delta\delta \quad , \quad X_2 = 2\pi\Delta f = \Delta w$$

a state variable formulation of the swing dynamics is expressed as

$$\dot{X}_1 = X_2$$

$$\dot{X}_2 = (1/h)(\Delta P_T - \Delta P_G - DX_2)$$

$$\dot{X}_3 = (K_R/T_R)(X_5 + X_4 - X_6) - (1/T_R)X_3$$

$$X_3 = \Delta\alpha_R \cdot X_{3min} < X_3 < X_{3max} \cdot$$

$$X_4 = \Delta I_{PV}$$

$$X_5 = \Delta I_R \quad . \; X_{5\,min} \cdot < X_5 < X_{5max} \cdot$$

$$X_6 = \Delta U_R$$

The sufixes R , I refer to rectifier and inverter respectively.

Steady state conditions are established using the load flow algorithm. A change in PV- generation initiates the post fault condition, which is a new condition for the load flow equations. The load flow algorithm is used to calculate the changes in grid bus voltage and power generated by the conventional plant. The swing equation is excessed producing the output change in frequency Δf. Through the control components PID-PSS and PID-RCR, the required change in angle $\Delta\Phi$ is calculated and processed achieving a new stable operating point.

CONCLUSIONS

The paper contributes towards fulfilling the need of presenting a comprehensive approach to the problem of integrating PV-system in existing HVDC link. The basic problem in operating PV plant arises from the fact that randomly varying sources cannot be dispatched in the same manner as conventional plants. Besides, adding a PV plant on an existing transmission and distribution system may cause additional security problems. The likelihood of a PV plant generating peak power during a system's peak demand condition is rather high, mostly in summer peaking utilities. Under such conditions, transformers and transmission lines may already be close to being heavily loaded and additional generation from the PV plant can easily overload those components . Unless these operational concerns can be effectively addressed, any long term planning results would be unacceptable.

Since the PID controllers are very simple for practical implementation, they have been employed to form the control scheme of the combined conventional plant-PV-source. As evident from preliminarily computer results, the developed control scheme can exhibit satisfactory dynamic responses under PV-variable generation.

REFERENCES

[1] Stephen T. Lee, Zia A. Yamayee, " Load Flowing and Spinning Reserve Penalties For Intermittent Generation" , IEEE Trans. on PAS, Vol. PAS-100, No. 3, March 1981.

[2] G. Vachtsevanos, H. Kang " Simulation Studies of Islanded Behavior of Grid-Connected Photovoltaic Systems", IEEE Trans. on Energy Conversion, Vol.4, No. 2, June 1989.

[3] Yuan-Yih Hsu, Li Wang, " Damping of a Parallel AC DC Power System Using PID Power System Stabilizers and Rectifier Current Regulators", IEEE Trans. on Energy Conversion, Vol .3, No. 3, September 1988.

[4] G.J. Vachtsevanon, K.G. Kalaitznleis, " A Methodology for Dynamic Utility Interactive Operation of Dispersed Storage and Generation Devices", IEEE Trans. on Power Systems, Vol. PWRS-2, No.1, February 1987.

[5] ibid, " On the Control and Stability of Grid Connented Photovoltaic Sources"., IEEE Trans. on Energy Conversion, Vol. EC-2 , No. 4, December 1987.

A NOVEL HIGH-CONCENTRATION PV TECHNOLOGY FOR COST COMPETITIVE UTILITY BULK POWER GENERATION

Vahan Garboushian, Sewang Yoon, Gerald Turner
AMONIX, Inc., 3425 Fujita Street, Torrance, CA 90505

Allen Gunn, David Fair
Scientific Engineering, Inc., 5960 E. Shirley Lane, Montgomery, AL 36117

ABSTRACT

This paper describes an innovative design for a nominal 20 kW, integrated, high-concentration (260X) photovoltaic (IHCPV) system which has been developed for cost-effective, utility-scale bulk power generation. This technology recently set a new world record for efficiency : 20.3% under STC (18.5%, >20 kW at PVUSA operating conditions). High-concentration PV systems offer several advantages for low cost power generation: (1) cost reduction through the optimum utilization of silicon, (2) higher conversion cell (hence system) efficiency at concentration vs. one-sun, and (3) inherently higher capacity factor in high direct normal insolation areas because of its built-in tracking. Previously little progress has been made in deploying HCPV for large-scale electricity generation because of: (1) the lack of a stable, high performance, high-concentration solar cell, and (2) the high cost associated with the PV modules, structure, tracking system, and ancillary equipment. With the arrival of a stable high performance cell developed by AMONIX, high-concentration PV systems can now be realized. A novel integrated system concept greatly reduces the costs associated with system hardware and labor by; (1) integrating the load bearing structure and the Fresnel lens/receiver plate elements eliminating the need for separate modules, and (2) use of a manufacturing-worthy receiver plate which makes use of "circuit-board" construction techniques. A full-scale 20 kilowatt IHCPV system has been deployed, and test results which validate the system design are reported. The IHCPV system development is complete and only volume production, **not technical breakthroughs**, is needed to meet the cost goals of <$2.00/watt at multi-megawatt levels.

INTRODUCTION

This paper describes a uniquely structured, integrated, high-concentration (260X) photovoltaic (IHCPV) system which has been developed for cost-effective, utility-scale applications. A 20 kW array, based on PVUSA operating conditions, was deployed at Arizona Public Service's Solar Test and Research (STAR) facility in Tempe, AZ in October, 1994 (see Figure 1).

High-concentration PV systems offer several distinct advantages for low cost power generation: (1) cost reduction through the optimum utilization of silicon (concentrator systems reduce the amount of costly silicon required to generate a given amount of electricity by an amount approximately equal to the concentration ratio), (2) higher conversion cell efficiency at concentration vs. one-sun, and (3) inherently higher capacity factor (compared to fixed-tilt, flat-plate systems) in high direct normal insolation (DNI) areas because of its built-in tracking. Despite these inherent cost reducing elements, high-concentration PV systems have not previously emerged for large-scale utility use because of: (1) the lack of a **stable**, **high performance**, **commercially available** high-concentration solar cell, and (2) the high costs associated with the PV modules, structure, tracking system, and ancillary equipment.

Figure 1: 20 kW IHCPV System

The development of the AMONIX HCPV solar cell has paved the way for commercial deployment of HCPV systems [1]. This cell has achieved a world record for commercially manufactured concentration cells: >26% efficiency at 850 DNI, 25^0 C. AMONIX has used the unique approach of utilizing commercial Integrated Circuit foundries for cell production rather than expending the time and capital required to build specialized facilities around this unique technology. This approach results in immediate access to volume cell production capacity along with predictable costs.

CH3365-4/94/0000-1060 $4.00 © 1994 IEEE
First WCPEC; Dec. 5-9, 1994; Hawaii

With the arrival of a stable high performance cell, high-concentration PV systems can now be realized. However to reduce costs, considerable effort has been applied so that all of the savings resulting from greatly reduced silicon usage does not get lost in the cost for the structure, tracker, and ancillary equipment required in concentrating systems. The result is the Integrated High-Concentration PV (IHCPV) array. The new, innovative (patented) IHCPV system concept eliminates much of the costly hardware used in earlier high-concentration designs. This has been accomplished by the simplification of the array structure which (1) eliminated earlier separate "box"-type modules mounted on structure assemblies (Figure 2), and substitutes an integrated design which combines both the load-bearing structure and the Fresnel lens/receiver plate elements thus eliminating the need for separate modules (Figure 3), and (2) a novel, manufacturing-worthy receiver plate which makes use of "circuit-board" construction techniques eliminating costly and labor intensive cell packaging and interconnects. Figure 2 shows a first generation HCPV system and Figure 3 shows the vastly simplified IHCPV design.

The IHCPV system completed electrical and mechanical design qualification to the Department of Energy's Sandia National Laboratories specification in the Fall of 1993. This 2,000 watt testbed system subsequently set a new world record for electrical PV energy conversion efficiency at greater than 20%.

The first full-scale 20,000 watt system has been deployed at Arizona Public Service's STAR facility (see Figure 1). Performance of this system is discussed below.

INTEGRATED ARRAY DESCRIPTION

The AMONIX IHCPV system utilizes a Fresnel lens of 260X geometric concentration ratio to focus sunlight onto back-junction, point-contact photovoltaic cells specifically designed to handle the high solar flux. The cells must be precisely positioned relative to the lens and the assembly must track the sun accurately in order to utilize the direct normal insolation. Tracking tolerance is improved by adding secondary optical elements to redirect any light that might otherwise miss the cell. The conventional method for constructing concentrator PV systems has been to build weather-tight modules (boxes) which support Fresnel lens parquets, individually mounted PV cells, secondary optical elements and associated electrical circuitry; to mount these modules onto a two-axis tracking structure; and, to electrically connect the modules to form an array (see Figure 2). Since concentrators usually have long focal distances, modules are typically bulky and can be difficult to handle and expensive to ship. Installation of modules onto trackers requires an alignment procedure to account for structural imperfections and insure that all modules are aimed in the same direction.

The basic concept of an "integrated array" is to integrate the module housing or module structure into the tracker structure thereby eliminating redundant structural components. The integrated structure was designed by Scientific Engineering, Inc. who is teamed with AMONIX for the commercialization of IHCPV technology. As shown by Figure 3, Fresnel lens parquets attach directly to the top of the tracker structure and PV "panels" or "circuit boards" attach directly to the back of the structure. The concept requires that components be precisely fabricated and assembled in order to meet optical alignment requirements and that the structure have sufficient stiffness to maintain module alignment under all operating conditions. The concept also requires that water sealing be incorporated into the design to prevent electrical shorting.

The lens parquets and PV panels are the same physical size, based upon a 4x6 matrix of cells. The acrylic lens parquet is made using automated compression molding technology. The lenses are manufactured by Fresnel Optics, Inc., another member of AMONIX's commercialization team.

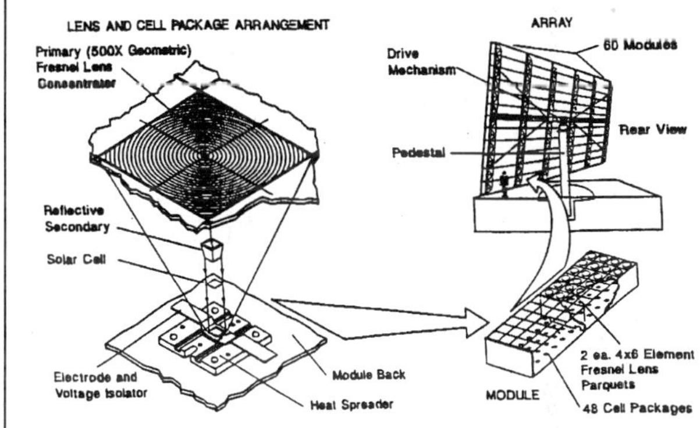

Figure 2: Early HCPV System Design (with modules)

The substrate for the PV panel as illustrated in Figure 3 consists of a three part thermally conductive laminate, with the front surface being electrically conductive and electrically isolated from the back surface. The cells, reflective secondary optical elements, blocking diodes, and inter-panel wiring terminations are all soldered directly to the top surface of the laminate thereby eliminating cell packages. The electrically conductive PV panel surface is also configured to match the mounting pattern of the AMONIX HCPV cell as well as provide the necessary circuitry for series and parallel cell connections (i.e.-no cell to cell wiring). The precise placement of components on the laminate and the subsequent soldering operation can be accomplished with robotics and other common automated manufacturing techniques. Shipments of lens parquets and PV panels are much more compact than assembled module boxes.

A 20 kW IHCPV array consists of 168 Fresnel lens parquets and PV receiver plates mounted to a tracking structure. Each Fresnel parquet has an aperture area of 0.7587 m^2 for a total array aperture area of 127.46 m^2. The overall tracking structure is 159 m^2 and is attached to a center pedestal via a two-axis mechanical drive. The tracker is designed to operate in winds up to 35 MPH and to survive winds up to 90 MPH in a stowed position.

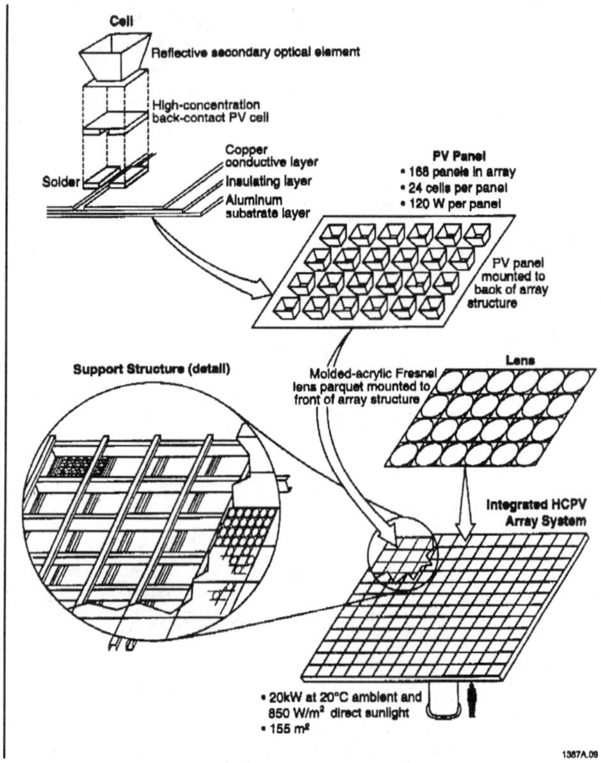

Figure 3: The Integrated High-Concentration PV Concept

The integrated tracking structure (designed and installed by Scientific Engineering, Inc.) is 13.81 m by 11.53 m and consists of a matrix of identical structural steel "cees" and "hats", two "backbone" trusses and a drive interface structure. The cees and hats are fabricated using automated roll-forming and numerical punching technology. Roll-forming allows cost effective customization of the cross-sectional shapes of the structural members such that members can be designed to fit within the available "non-optical areas" thereby keeping the gross array area as small as possible and allowing for optimization in terms of member strength and weight. Fastening holes will ultimately be numerically punched in all members in order to maintain the required precision for alignment. This structural concept eliminates material waste, minimizes fabrication labor and field assembly labor, eliminates the need for welding or painting since galvanized material can be used, and, due to the precision of fabrication, eliminates the need for manual alignment adjustments in the field.

The low cost two-axis utilizes a planocentric ring drive for azimuth rotation with a gear ratio of 31,102:1 and a jackscrew arrangement with a gear ratio of 31,640:1 for elevation rotation. The drive can rotate the array in azimuth plus or minus 180 degrees from south, and 90 degrees in elevation. Tracking control is accomplished with a time based, open loop control system. A microprocessor computes the position of the sun and directs the tracker accordingly. The tracker keeps track of its current position by use of a Hall-Effect encoder on the motor shaft of the azimuth and elevation drives. By counting motor revolutions from a known reference position and given the drive parameters, the control system can align the tracker with the sun to within 0.05 degree accuracy.

Electrically, the 168 Fresnel lens parquets and mating PV receiver plates are subdivided into three subarrays. Each subarray consist of two parallel strings of 28 receiver plates in-series for a nominal rating of 32 amperes at 210 V_{DC}. Each subarray is connected to one phase of a three phase, grid interactive inverter which has a 480 VAC, 60 Hz, three phase delta output. Each of the three phases of the inverter are controlled independently. Once connected to the utility grid, each control circuit allows sufficient current to be delivered to the utility to bring the array to its maximum power point.

SYSTEM PERFORMANCE

Flatness measurements were made on the structure to determine the amount of gravity sag and to validate finite element model predictions. The results of these measurements indicate that misalignment due to gravity sag is less than 0.1 degree and that the model predictions were reasonable.

Two sets of tracking error monitors are installed on the tracker and are used to measure tracking accuracy. The results of these measurements indicate that the system is capable of tracking the sun to within 0.1 degree accuracy throughout the day.

Each PV receiver plate was individually tested at the factory and ranked according to peak power performance. The plates were positioned in the array so as to minimize power losses due to mismatch. Mismatch losses for the overall array are less than 1%. Wiring loses within the array and between the array and inverter are less than 1%

Thermocouples are used to measure various temperatures during the operation of the array, including ambient air temperature. In order to measure variations in temperature, thermocouples were placed on one plate on the lower part of the array and one plate on the upper part of the array. For each plate, one thermocouple was placed immediately behind a cell, one thermocouple was placed immediately between two cells and one was placed in the air just above the inside of each plate. The results of the temperature data suggest that the top plates operate about 5° C higher than the bottom plates.

The electrical performance of each subarray is monitored by measuring current and voltage simultaneously with direct normal insolation. Readings are recorded every minute and averaged over a ten minute period. Power from each subarray is then computed and summed for the total array. Given the aperture area and the above measurements, operating efficiency is then computed. Figure 4 depicts actual data of output power (DC) versus DNI under variable temperature conditions. Note that the array has exceeded the design goal of 20,000 watts.

SUMMARY

A novel, cost-effective, high-concentration PV system has been designed and a 20 kW array has been deployed. Test results validate the high performance capability of the IHCPV system. A new world record for system efficiency has been achieved: 20.3% under STC (18.5% at PVUSA operating conditions) [2]. This performance, combined with the cost-effective integrated design make IHCPV a leading contender for low-cost utility-scale bulk power generation. **No technical breakthroughs (only volume production and deployment) are needed to reduce the cost to <$2.00/W at multi-megawatt levels.**

ACKNOWLEDGMENTS

The authors wish to acknowledge the sponsorship of the Electric Power Research Institute for the development of the IHCPV technology, and its sustained commitment towards the commercialization of high-concentration PV. The efforts of Dr. Edgar DeMeo and Mr. Frank Dostalek significantly contributed to this project.

The authors wish to recognize Arizona Public Service for their leadership in the commercial deployment of the IHCPV technology. In particular, we wish to acknowledge the support of Dr. Peter Johnston, Mr. Herb Hayden, and Mr. Peter Eckert.

REFERENCES

[1] S. Yoon, V. Garboushian, "Commercialization of High-Concentration Back-Junction Point-Contact Photovoltaic Cells; Successful Pilot Production", *Proc. of 11th E.C. Photovoltaic Solar Energy Conference*, 1992, pp. 257-260

[2] **Standard Test Conditions**
1000 W/m^2 direct normal irradiance, 25^0 C **cell temperature**, and 3 m/s wind speed (at 10 meters above grade)
PVUSA Test Conditions for Concentrators
850 W/m^2 direct normal irradiance, 20^0 C **ambient temperature**, and 1 m/s wind speed (at 10 meters above grade)

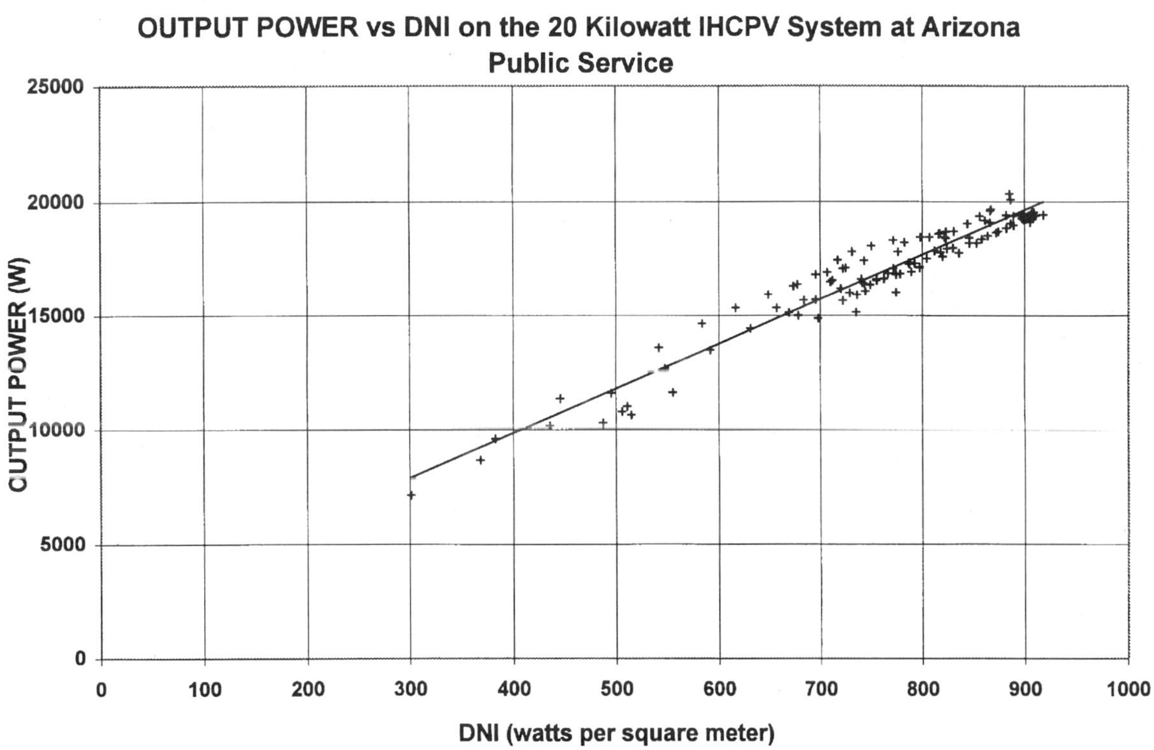

Figure 4: Data from AMONIX 20 kW IHCPV System at Arizona Public Service

A UTILITY-CONNECTED RESIDENTIAL PV SYSTEM ADAPTED A NOVEL SINGLE-PHASE COMPOSITE PWM VOLTAGE SOURCE INVERTER

Sakutaro Nonaka

Faculty of Engineering in Kyushu, Kinki University
11-6, Kayanomori, Iizuka-shi, Fukuoka, 820 Japan.

ABSTRACT

A Utility-interactive residential PV system adapted a novel single-phase composite PWM voltage source inverter (VSI) is proposed. The proposed system is equipped with PV array, battery storage, and a single-phase VSI, which has a new circuit configuration and PWM method.

The VSI circuit consists of the normal single-phase bridge circuit and an additional arm. The two auxiliary self-turn-off devices of the arm avail to adopt a composite PWM control, which contributes to reduce the ripple in the AC output current. Also, the VSI has a *LC* series resonance circuit tuned to twice the utility frequency connected in parallel with the smoothing capacitor. The series resonance circuit absorbs the double-frequency AC components included in the DC pulsated current. Consequently, the smoothing capacitor is drastically reduced .

It was found with the computer simulation and the experiment that the waveform of AC output current shows an ideal sine wave.

INTRODUCTION

Interconnection systems with a sinusoidal output PWM current source inverter (CSI) between photovoltaic arrays and the utility line have reported by the author[1]. It was clarified that a utility interactive PV system with a CSI has advantages as follows.

(1) The PWM CSI makes it possible to supply a sinusoidal current with unity power factor to the utility.

(2) There is no need for any feedback control to connect the PV generation system with the utility line in the CSI system.

(3) Short-circuit failure causes no excessive current in the CSI system because the DC current never exceeds the PV short-circuit current.

Research on the interconnection system of the residential PV power generation has already entered the stage of practical test. However, in these research, almost all of inverters under consideration are VSIs and no examples of CSIs are seen except the author's reports.

In this paper, a novel single-phase composite PWM VSI is described. This VSI circuit configuration was invented by Nonaka [2][3] in duality of the single-phase composite PWM CSI [1].The proposed system is equipped with PV array, battery storage, and a single-phase VSI, which has a new circuit configuration and PWM method.

The VSI provides a sine wave output current for domestic loads and the utility when the PV array's output exceeds the load demand.

The VSI output current is controlled by both of the firing angle α and the modulation index *MI*. The operation mode of VSI is easily changed from inverter action to rectifier action with changing the firing angle α from plus to minus. When α is minus, the VSI acts as a battery charger.

Any shortfall between load demand and PV array output is automatically met by power from the utility grid. There is no need of a regulator between PV array and battery by proper design of the battery's volt corresponded to the output voltage of PV array. The circuit configuration of the utility-interactive PV system adopted the proposed VSI is very simple.

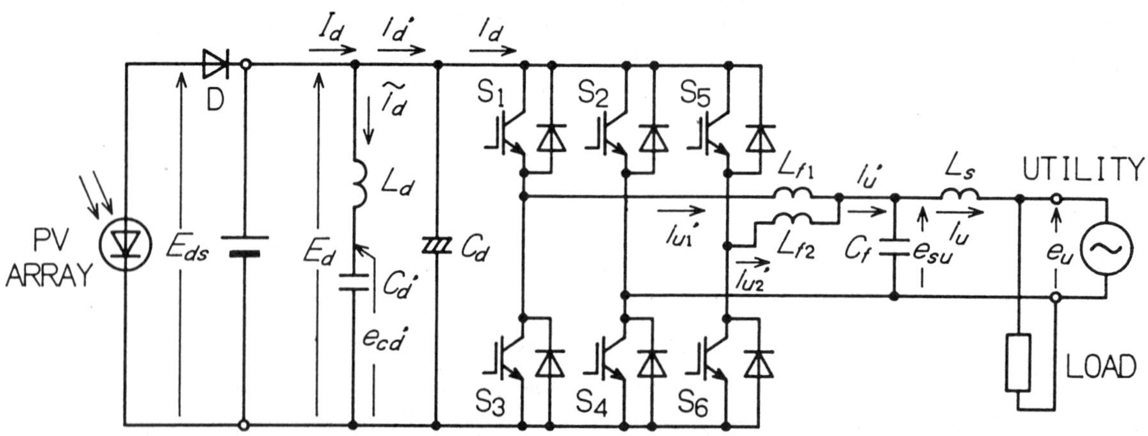

Fig.1. Utility-interactive residential PV system with battery storage employed a novel single-phase composite PWM VSI.

CH3365-4/94/0000-1064 $4.00 © 1994 IEEE

First WCPEC; Dec. 5–9, 1994; Hawaii

CIRCUIT CONFIGURATION

Figure1 shows a novel single-phase composite PWM VSI circuit which composed an important part of the utility-interactive PV system.

In the VSI circuit, four main IGBTs S_1~S_4 with antiparallel diodes, respectively, compose the usual single-phase bridge circuit.

The additional arm consists of IGBT's S_5 and S_6 with antiparallel diodes. The AC filter reactor is divided L_{f1} and L_{f2}, which flow a-half of output current in parallel, respectively. L_s indicates the interconnected reactor between PV system and utility. L_d and C_d' indicate the double-frequency series resonance circuit.

When the parameters L_d and C_d' of the resonance circuit are satisfied with the following relation:

$$2\omega L_d = \frac{1}{2\omega C_d'} \quad (1)$$

The impedance of the series resonance circuit becomes zero for the double-frequency AC component $\tilde{i_d}$ included the DC current i_d'.

Its availability has been pointed out at 1975 by K.Heumann [4].

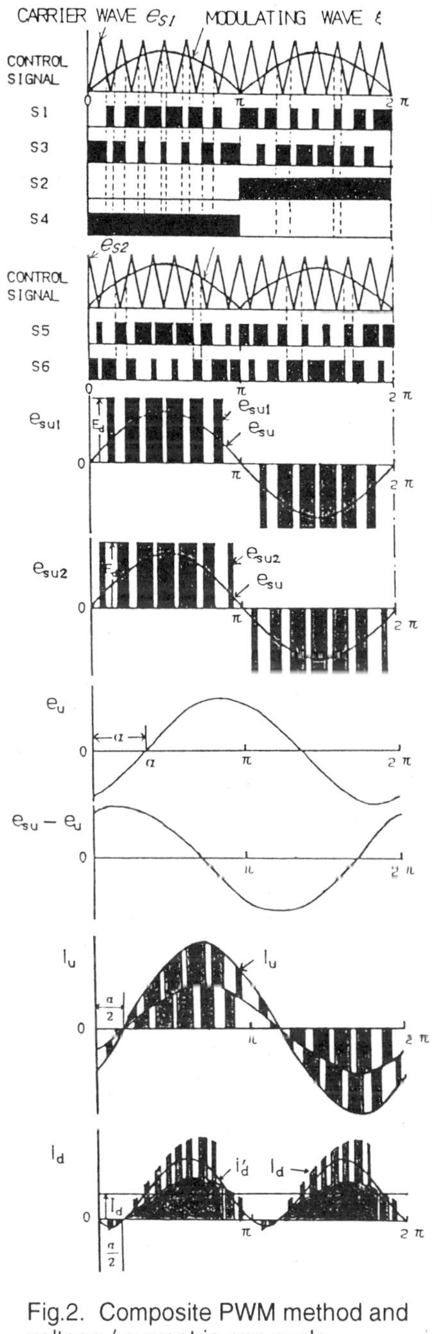

Fig.2. Composite PWM method and voltage / current in one cycle.

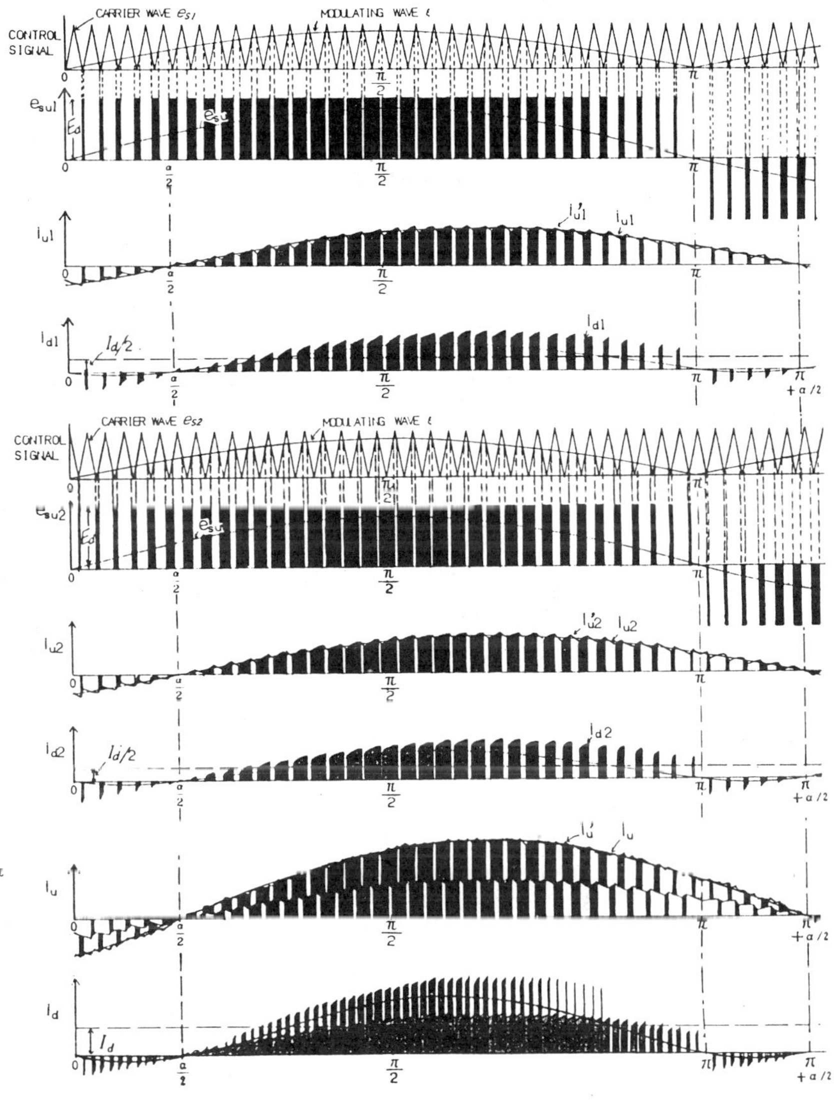

Fig.3. Detail of composite PWM method and voltage / current at fc = 4.2kHz and f=60Hz.

Here, the effective value of the resonance current $\widetilde{I_d}$ is determined by

$$\widetilde{I_d} = MI \cdot I_u / 2 \qquad (2)$$

and $\widetilde{I_d}$ is independent of values of L_d and C_d'. The effective voltage E_{cd}' across the capacitor C_d' becomes

$$E_{cd}' = \frac{\widetilde{I_d}}{2\omega C_d'} = 2\omega L_d \widetilde{I_d} \qquad (3)$$

The values of L_d and C_d' affect the voltage E_{cd}'. As far as E_{cd}' is not over extremely a normal voltage, there is freedom in selection of combination of L_d and C_d'.

It is a solution measure to select the one which can be implemented in smallest size and lightest weight overall with $L_d C_d'$ combination.

The smoothing capacitor C_d operates only as a by-pass capacitor of the ripple current due to PWM switching, and is small capacitance enough to absorb the ripple. Therefore, the capacitance C_d can be greatly reduced in comparison with the traditional one. Also there is need C_d in parallel with the battery storage which has an inner electrical resistance.

COMPOSITE PWM METHOD

Figure 2 illustrates the composite PWM method, which is accomplished by means of two types of PWM patterns produced by using a sinusoidal modulating wave ξ and two triangular carrier waves e_{s1} and e_{s2}.

For easy understanding of the PWM method, in this figure, the pulse number in half cycle is displayed by 7 (carrier frequency f_c =840Hz at f =60Hz) which is less than actual one. In main IGBTs S_1 and S_3 on the same arm, ON/OFF are repeated alternatively by the PWM patterns determined at the intersecting point of the triangular carrier wave e_{s1} and the sinusoidal modulating wave ξ.

In main IGBTs S_4 and S_2 on the other arm, ON/OFF are repeated alternatively by each half-cycle.

On the other hand, in auxiliary IGBTs S_5 and S_6, ON/OFF are repeated alternatively at the intersecting point of the reverse-phase triangular carrier wave e_{s2} and ξ. It is necessary to provide a dead time between gate singles for the IGBTs on the same arm, for short-circuit prevention in the VSI. Figure 2 shows the voltages and currents at each part of the circuit indicated by symbols in Fig.1. The e_{su1} shows a sinusoidal pulse train with amplitude E_d caused by ON state both S_1 and S_4 or S_3 and S_2. The e_{su2} shows the same sinusoidal PWM and amplitude but different pulse train caused by ON state both S_5 and S_4 or S_6 and S_2. However, the output voltage e_{su} appeared across the filter capacitor C_f shows the same sinusoidal waveform because the ripples of e_{su1} and e_{su2} are filtered through L_{f1} and L_{f2}, respectively.

Figure 3 shows the detail of composite PWM method and the waveforms of voltage/current at f_c=4.2kHz and f=60Hz.

In the VSI, it is necessary for performing the control to compare the inverter output voltage e_{su} and the utility voltage e_u through the voltage drop across the interconnected reactor L_s.

Figure 4 shows control methods of AC output current i_u by the phasor diagram. Fig.4(a) gives the firing angle α control and fixed MI at $E_{su}=E_u$.

Fig.4(b) shows α and MI control to keep power factor at unity.

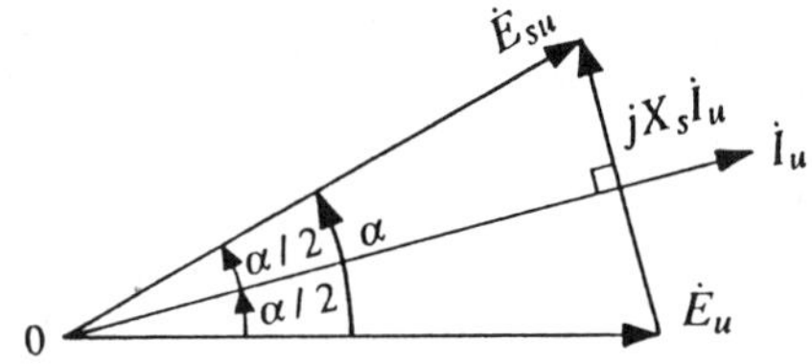

(a) α control and fixed MI at $E_{su} = E_u$.

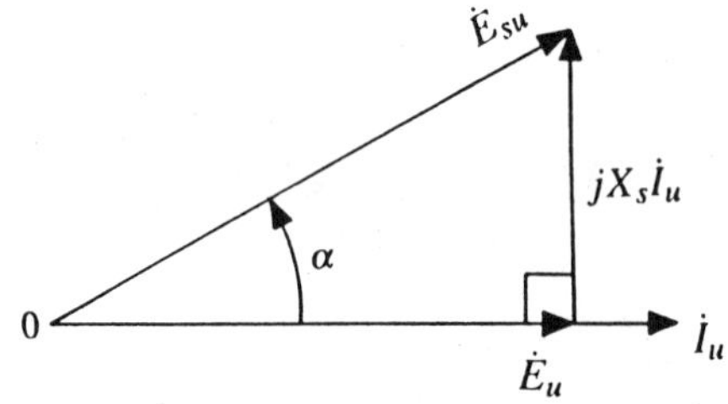

(b) α and MI control to keep power factor at utity.

Fig.4. Control method of AC.

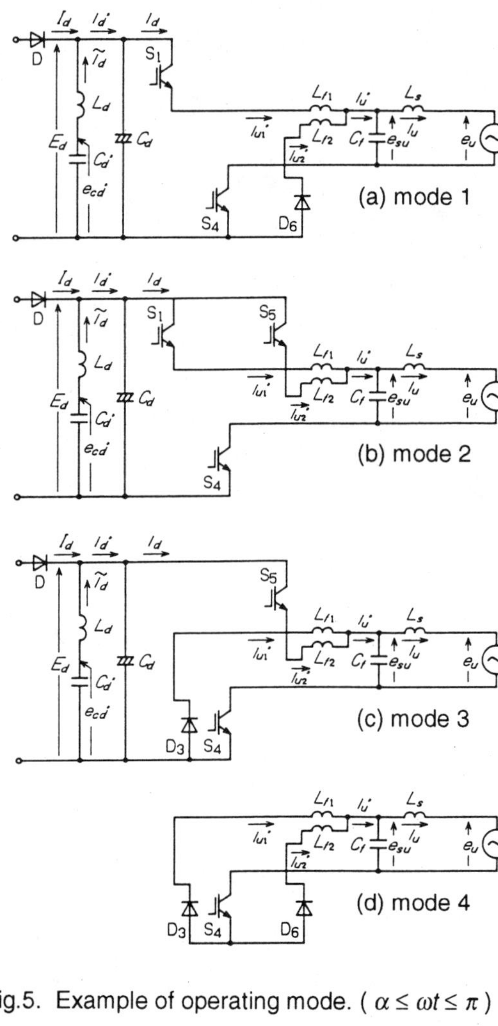

Fig.5. Example of operating mode. ($\alpha \leq \omega t \leq \pi$)

Figures 2 and 3 show the AC current control by the method of Fig.4(a).

The output current i_u' is divided i_{u1}' through L_{f1} and i_{u2}' through L_{f2}. Each current varies sinusoidally with repetition of increase at the power supply mode and decrease at the circulating current mode according to the individual PWM pattern.

The ripple of the sinusoidal output current i_u' which equals to the sum of two currents i_{u1}' and i_{u2}' is smaller than those of two currents.

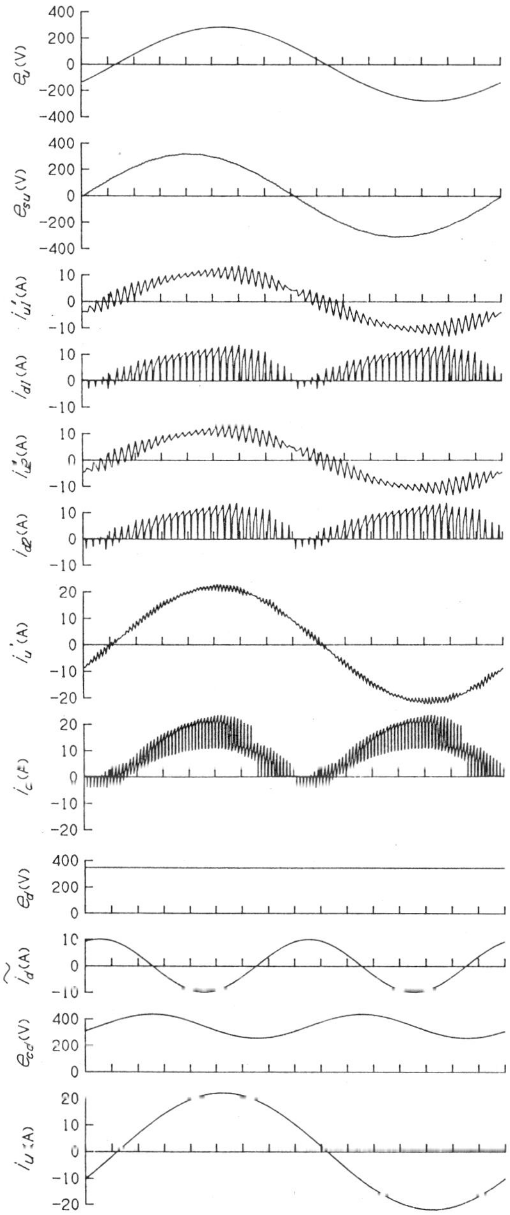

$L_{f1} = L_{f2} = 3$mH, $L_s = 17.7$mH,
$L_d = 12$mH, $C_d' = 147\mu$F.
($p = 3.12$kW, $E_{su} = 221.6$V, $I_u = 15.6$A, $E_u = 200$V)

Fig.6. Computed waveforms operated at unity power-factor. ($\alpha = 30^\circ$, $MI = 0.924$)

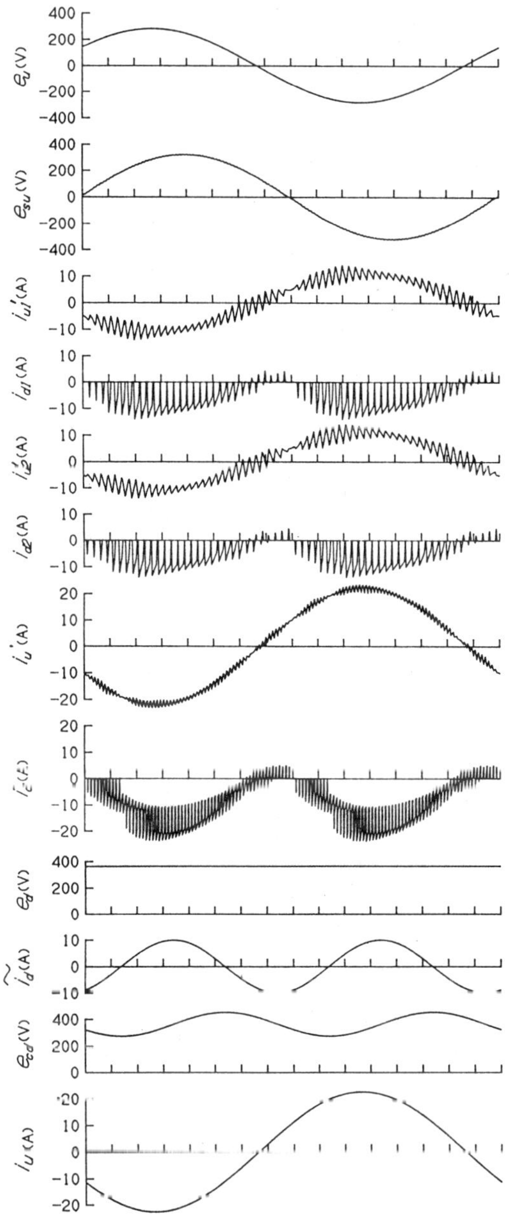

$L_{f1} = L_{f2} = 3$mH, $L_s = 17.7$mH,
$L_d = 12$mH, $C_d' = 147\mu$F.
($p = -3.20$kW, $E_{su} = 227.0$V, $I_u = 16.0$A, $E_u = 200$V)

Fig.7. Computed waveforms at battery charging. ($\alpha = -30^\circ$, $MI = 0.901$, $p.f. = 1.00$)

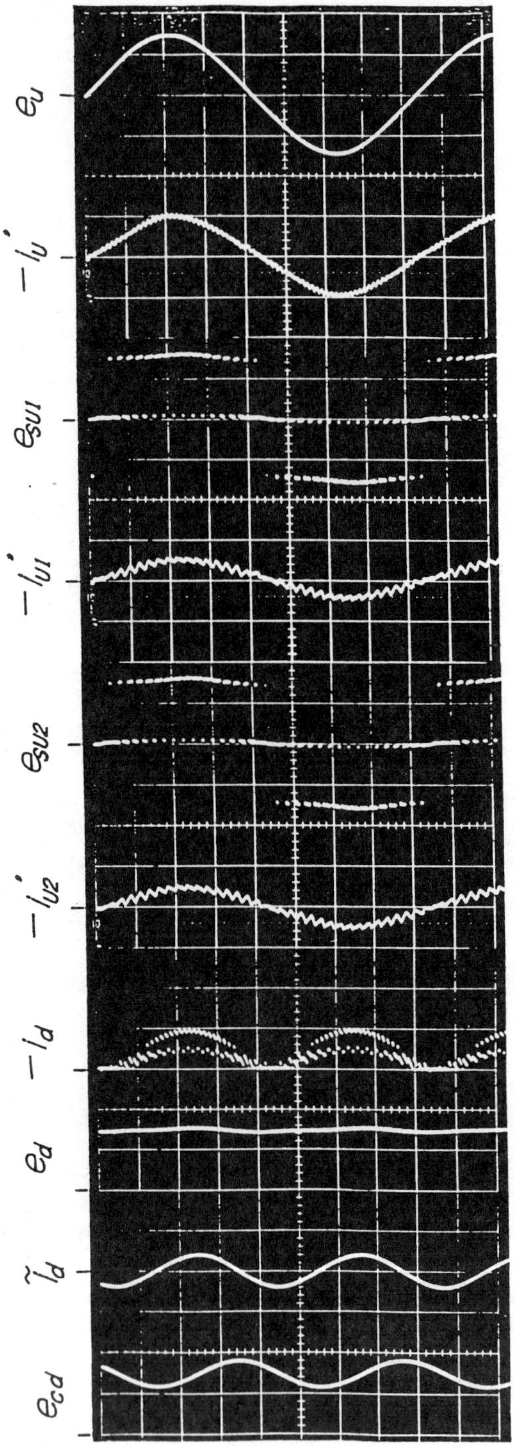

$L_{f1} = L_{f2} = 13.1\text{mH}$, $L_s = 0$, $L_d = 13.3\text{mH}$,
$C_d' = 14.7\mu\text{F}$, $C_d = 100\mu\text{F}$, C_f was removed.
$f = 60\text{Hz}$, $f_c = 3.96\text{kHz}$, $E_u = 50\text{V}$, $R_d = 50\Omega$.

Fig.8. Measured waveforms at rectifier operation with resistance load instead of battery.
($\alpha = -7^\circ$, $MI = 0.9$, $p.f. = 1.0$)

As the ripple of current i_u' is filtered by C_f, the waveform of the output current i_u becomes perfectly ideal sine wave.

On the other hand, both DC current i_{d1} and i_{d2} are pulse trains which their amplitudes are modulated sinusoidally. As the ripple of DC current i_d which is sum of i_{d1} and i_{d2} is filtered by C_d, the current i_d' will become a pulsating current at twice of the utility frequency. The AC component $\tilde{i}_d$ included i_d' is by-passed through the series resonance circuit $L_d C_d'$. Therefore, the input current I_d from the battery storage and PV array to the VSI becomes a flat DC current.

Figure 5 shows example of the operating modes in the period $\alpha \leq \omega t \leq \pi$, that is, $e_{su} > 0$, $e_u > 0$, $i_u > 0$.

WAVEFORMS

The rating of the VSI model was assumed 2.5kW, $E_u = 200\text{V}$, $I_u = 13\text{A}$, $f = 60\text{Hz}$, $p.f. \geq 0.96$, $MI = 0.8$, $E_d = 354\text{V}$, $I_d = 7.8\text{A}$, $C_d = 100\mu\text{F}$, $C_f = 10\mu\text{F}$ and $f_c = 3.6\text{kHz}$.

Figure 6 shows the computed waveforms operated at unity power factor when $\alpha = 30^\circ$, and $MI = 0.924$. Figure 7 gives the computed waveforms at battery charging when $\alpha = -30^\circ$ and $MI = 0.901$. Figure 8 exhibits the measured waveforms at rectifier operation with resistance load instead of battery when $\alpha = -7^\circ$, $MI = 0.9$ and unity power factor [5].

CONCLUSIONS

It has been verified that the results of theoretical analysis agree well with the experimental results. And the feasibility of the proposed utility-interactive PV system is confirmed.

ACKNOWLEDGEMENT

The author is indebted to Director S. Nakamura of the Research Laboratory, Kyushu Electric Power Co., Inc. for this work. The author wishes to thank Dr. K. Yamasaki of the same Reserch Laboratory for his useful discussions on this work.

REFERENCE

[1] S.Nonaka, "A suitable single-phase PWM current source inverter for utility connected residential PV system", *Solar Energy Materials and Solar Cells* 35, 1994, pp.437–444

[2] S.Nonaka, "Voltage Source Type Composite PWM Single-Phase Inverter and Converter Circuit in Duality of Current Source Type", presented at *Kyushu Branch Annual Meeting IEE Japan*, no.418, 1993.

[3] S.Nonaka, "Single-Phase Sinusoidal Wave PWM Voltage Source Inverter", presented at the *Meeting on Semiconductor Power Conversion of IEE in Japan* SPC-93-75, Oct.1993, pp.9-21.

[4] K.Heumann, "Grundlagen der Leistungselektronik", B.G.Teubner Stuttgart, 1975, p.138.

[5] S.Nonaka & Y.Neva, "Single-Phase Composite PWM Voltage Source Converter", *Conference Record of the IEEE Ind. Appl. Soc. Annu. Meeting*, Denver, Oct.1994, pp.761-767.

Analysis of Bidirectional PWM Converter for Application of Residential Solar Air Conditioning System

Hong - Sung Kim Gyu - Ha Choe

Dept. of Electrical Eng.
Kon-Kuk University
Seoul Korea

Gwon - Jong Yu JinSoo Song

Photovoltaic Research Team
Korea Institute of Energy Research
Taejon Korea

ABSTRACT

In the conventional solar air conditioning system the diode rectifier is used to build up DC link voltage from ac utility source. The diode rectifier is simple and cheap but the reverse of power flow cannot be made. It can derate the utilization of solar cell and also there have occurred the problems of low power factor and plentiful harmonics at the ac input side. Hence in addition to peak power cutting in summer, some advantages can be obtained by adopting the PWM converter to the solar air conditioner. As a result, obtained are the characteristics of the PWM converter such as low distorted current waveform, unity power factor. Also the proposed system is verified by examining the dynamics to step load change and also power reversal testing.

1. INTRODUCTION

Much power demand would be required now at the power system because of the use of various home appliances such as air conditioner, refrigerator, etc. The present power demand of the utility is faced to weaken the allowable reserve margin in Korea because most large power loads such as air conditioner are used in summer especially at midday for one day. The peak power demand of the utility usually occurs at 2 or 3 o'clock in the afternoon.[1] This time almost coincides with the time when the amount of the sunshine becomes the highest, and thus the maximum solar power could be obtained. This urgent problem could be solved by the use of Photovoltaic (PV) systems which have the merits of green energy source without pollution, and free choices in power generation scale, and the use of dead space such as rooftop.

For the application of the PV systems to the residential air conditioner the whole system used is composed of diode rectifier, 3-phase inverterand motor. The inverter circuit used to the air conditioning systems could obtain the effect of the energy saving because the speed of the compressor motor can be controlled according to the operating condition. Hence inverter control can improve the whole system efficiency to be higher than the on-off control in the conventional system.

In the inverter-mounted air-conditioning systems, however, the diode rectifiers used to convert the ac input to the dc output is very simple but the power flow can not be made bidirectionally, and also there have occurred the problems of low power factor and plentiful harmonics at the ac input side[2-6].

The PWM converter can be a remedy for those problem that also can make power flow bidirectional. Therefore for the purposes of peak power cutting and power flow control the application of the PWM converter to solar air conditioning system is proposed in this paper and the operating characteristics of the PWM converter are investigated about steady state and transient state characteristics. The PWM converter performs current control and voltage regulation. The two-level PWM hysteresis method is used for the current control and PI controller for the voltage regulation.

The system is studied through digital simulation by ACSL(advanced continuous simulation language) tool and also the experimental results are obtained and then compared to the computed results for verification

2. SYSTEM DESCRIPTION

Fig.1 represents the whole solar air conditioning system with PWM converter. And the system consists of solar cell array, step-up chopper and air conditioner including PWM converter, inverter and 3-phase induction motor for compressor drive. Also Fig.2 shows the overall control block diagram for PWM converter

2-1 Power conversion circuit

(1) Step-up chopper

Step-up chopper is used to control dc power from solar array to the dc link side of the PWM inverter, by which

CH3365-4/94/0000-1069 $4.00 © 1994 IEEE First WCPEC; Dec. 5–9, 1994; Hawaii

the solar cell voltage becomes boosted according to the power flow.

Suppose that the load is constant, or that the dc power transferred to 3-phase induction motor through inverter becomes fixed. As the power amount from the step-up chopper becomes increased, so the power from the utility could be decreased. At this instant the step-up chopper makes the solar cell output power be maximized by controlling the duty ratio according to the environmental situations.

(2) PWM inverter and induction motor

In the inverter-type air conditioners 3-phase induction motor is usually adopted for compressor drive while single-phase motor is used together with on-off relay switch in the conventional ones.

In on-off control the motor speed cannot be controlled precisely, but the accurate speed control can be made by PWM inverter with VVVF (variable-voltage variable-frequency) control. Also the input energy is reduced, and the efficiency is improved, and hence the whole control characteristics would be made better.

(3) PWM converter

PWM converter controls the current to be supplied from the utility bidirectionally, and its magnitude should be adjusted according to the amount from the solar cell output power for fixed load. At every operating instant the solar cell must be always operated at maximum power point by the step-up chopper.

The PWM converter has the function of bidirectional power control, of which direction can be decided by the phase angle between the voltage and current at the utility side. The utility supplies power to the load when the phase angle is zero, and the power flow reverses for the angle of 180 °.

For given load power and utility voltage with constant magnitude, the instantaneous current to be controlled by PWM converter can be expressed as follows:

$$i_{St}(t) = \frac{P_L(t) - P_{SC}(t)}{v_S(t)} \tag{1}$$

where $P_{SC}(t)$: solar cell output power,
and $v_S(t)$: instantaneous utility voltage.

It can be seen that the utility current is expressed as the form of the power difference between load power and solar output.

2-2 System controller

Fig.2 shows the overall control block diagram for PWM converter and step-up chopper. The controller is composed of both the step-up chopper control loop and PWM converter control loop. The constant voltage controller regulates the duty ratio of step-up chopper by comparing detected solar array voltage with reference voltage. Also PWM converter has both functions of current control and DC link voltage regulation. In the latter DC link voltage reguation loop the actual voltage through LPF is compared to the reference and then the error of (2) is sent to PI controller, of which output actuates via limiter the former current controller. Thus the PWM signal obtained is sent to the PWM converter and hence the ac side current could be controlled as the current reference expressed in (3).

The hysteresis current controller is adopted by which the actual current is to be controlled between the preset lower and upper bands.

$$e(t) = V_{ref} - v_C(t) \tag{2}$$

$$i^*(t) = [Kp \cdot e(t) + Ki \cdot \int e(t)dt] \cdot \sin \omega t \tag{3}$$

The Butterworth filter is used as LPF (Low Pass Filter) which has characteristic of the constant gain in the pass band.

3. RESULTS

The proposed system is studied by digital simulation using ACSL tool and also experimental investigation is performed for verification. For simplicity it is assumed that the switch is ideal and also dead time is neglected. The utility voltage is 220[V], and the DC link voltage 320[V], and the switching frequency is set to 10[kHz]. The solar power is designed to be 1[kW] for economical viewpoint, which is a half of the full load power rating.

3-1 Decision of intermediate inductance

The intermediate inductor between the utility voltage and dc link voltage is required because of instantaneous waveform difference. This inductor should be designed suitably for better current regulation. The decision of the inductance is made by evaluating voltage THD(Vss in Fig.1).

Fig.3 shows the characteristics of the voltage THD for the variance of inductance under the loading condition of 100[%] to 50[%]. The vertical axis is the THD, and the horizontal axis the inductance. It can be seen that THD is effected by the inductance only, and not the loading conditions. From these results the inductance can be recommended as about 4.5[mH] in order to maintain the THD value to be lower than 5[%].

3-2 Power factor

The variation of input power factor at every loading

condition factor can be obtained as shown in Fig.4. The results are shown in two cases: One is for rectifying mode and the other for inverting mode.

In the rectifying mode the input power factor becomes increased to be higher for larger loads, and the same tendency can be found for larger inductance. In best case the power factor is maintained to be about 0.98. Also as the difference power between load and solar array is sent to the utility side in the inverting mode, so the zero of loading condition means that all sloar power is sent to the load side. Thus the power factor increases with inductance, and decreases with the loading conditions.

3-3 Experimental results

Fig.5 shows the waveforms of voltage and current at the utility side at the PWM converter. It can be seen that the controlled current becomes nearly in phase with utility voltage, and hence the power factor becomes above 0.9.

The power reversal from the rectifying to the inverting operation occurs when load reduces lower. Fig.6 shows the power reversal pattern when the load reduces from 100[%] to 0[%]. The transient state appears during about 1 cycle, and the current flows entirely from solar array to the utility side rather than load without changing the magnitude.

The dynamic response of utility current is given in Fig.7 when load decreases from 100[%] to 50[%] abruptly and return again to 100[%]. In both cases a new steady-state reaches with 4 cycles of utility voltage, and hence relatively good dynamic response can be shown by the proposed method.

4. CONCLUSION

The solar air conditioning system with PWM converter is proposed and then the several characteristics are investigated by both digital simulation and experiment.

The PWM converter has two main functions of current control and bidirectional power control. For current control the hysteresis controller is used, and also the step-up chopper is adopted for power flow from solar array. The revealing characteristics of the PWM converter can be summarized as low distorted current, high power factor, and high utilization of solar energy by reversible power control.

REFERENCES

[1] G.J.Yu et al.,"Solar Air Conditioner for Electricity Peak Cut", KIEE Summer Meeting Conf.Record, 1992, pp.1045-1047.(Korean)

[2] J.T.Boys,"Current - forced single phase reversible rectifier",IEE PROC-B, Vol.136, No.5,September, 1989, pp.205-211.

[3] A.B.Kulkami et al., "Transient Test on a Voltage - Regulated Controlled-Current PWM Converter",IEEE Trans .on IE,Vol.34, No.3,August 1987, pp.319-324.

[4] P.D Ziogas, "An Active Power Factor Correction Technique for Three-Phase Diode Rectifiers", IEEE PESC Record, 1986, pp.58-66.

[5] B.T.Ooi et al.,"A Single - Phase Controlled Current PWM Rectifier",IEEE Trans. on PE,Vol.3, NO.4, October 1988, pp.453-459.

[6] S.Manias," Novel Full Bridge Semicontrolled Switch Mode Rectifiers", IEE PROC-B, Vol. 138, No.5, 1991, pp. 252-256.

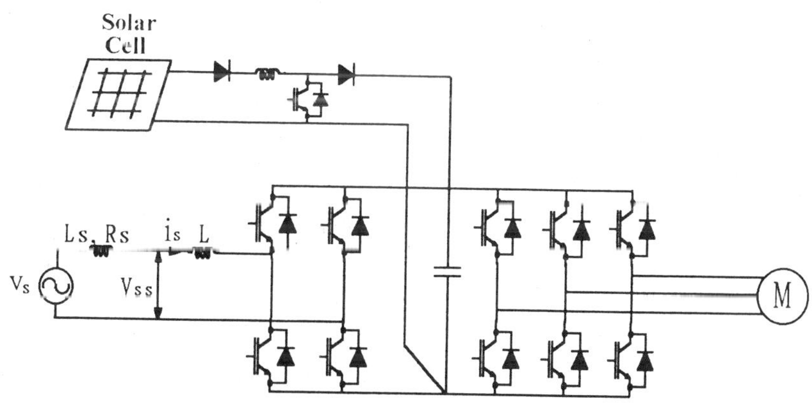

Fig.1. Power conversion circuit

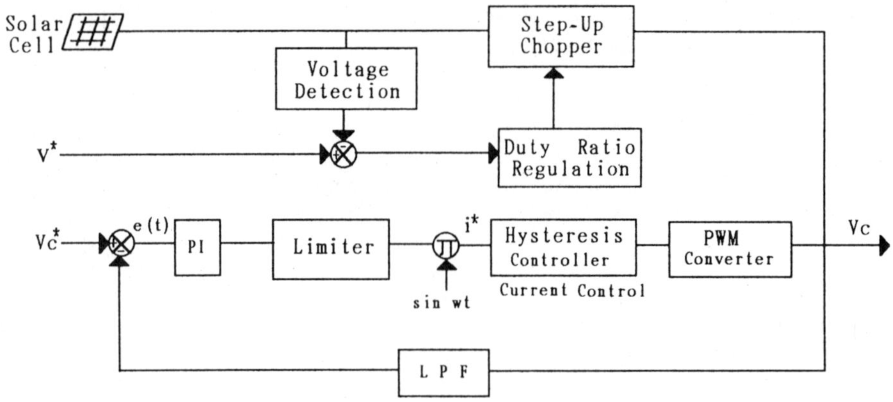

Fig.2. Control block diagram

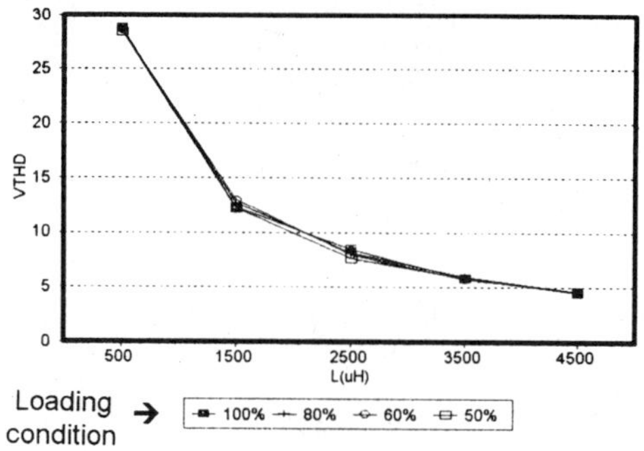

Fig.3. Voltage(Vss) THD

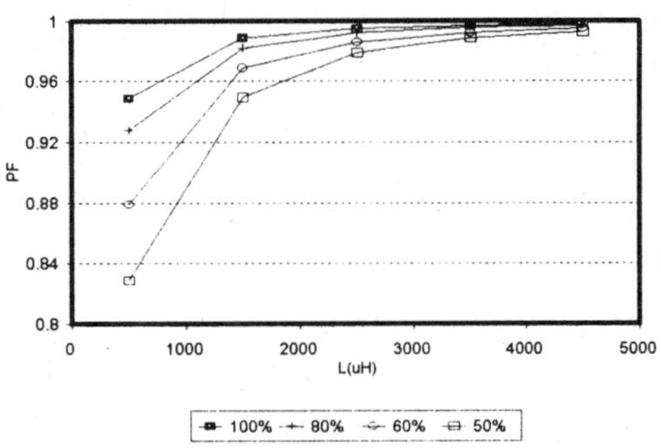

(a) Rectifying mode

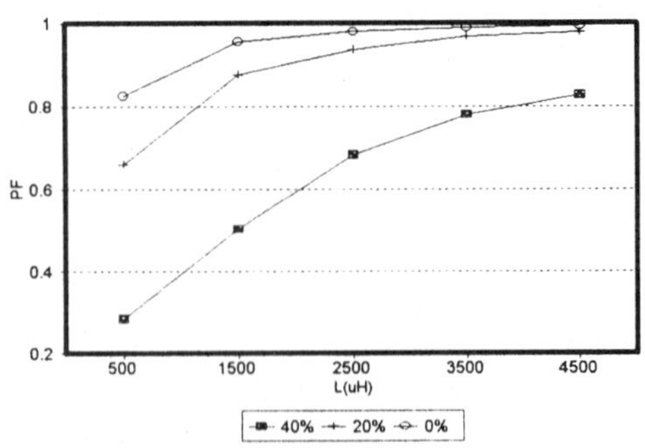

(b) Inverting mode

Fig.4. Power factor

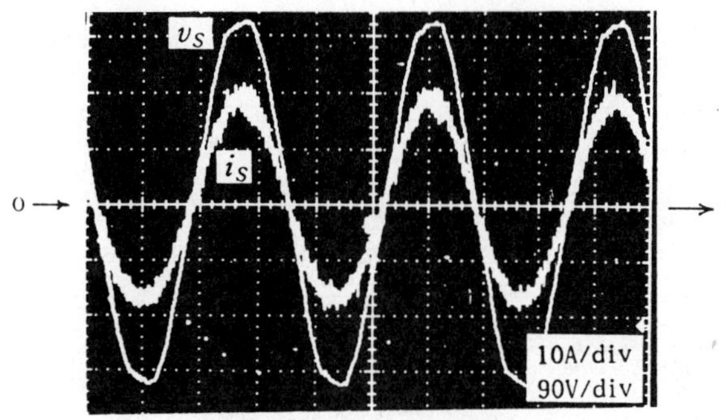

Fig.5. Current waveform

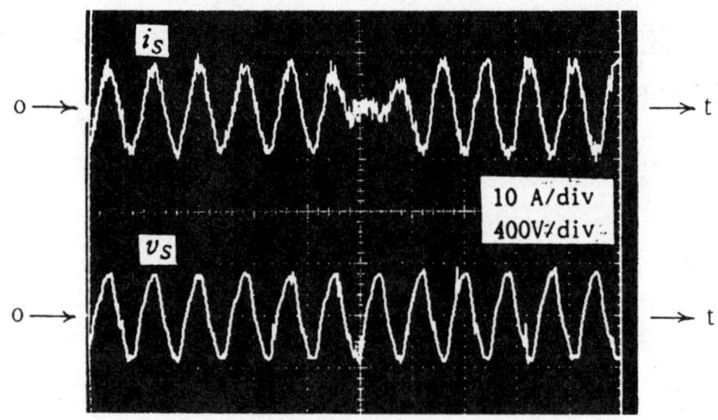

Fig.6. Power reversal test

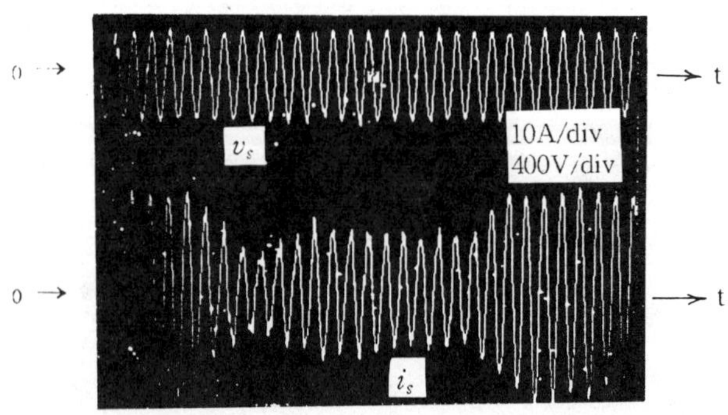

Fig.7. Load variation test

DESIGN, DEVELOPMENT AND INSTALLATION OF 100KW UTILITY AND GRID CONNECTED SOLAR PV POWER PLANTS FOR RURAL APPLICATIONS-AN INDIAN EXPERIENCE

Brig.M.R.Narayanan, D.V.Gupta, R.C.Gupta, R.S.Gupta
Central Electronics Limited, Sahibabad,U.P.-201010,India

ABSTRACT: Two 100KW Solar Photovoltaic Power Plants have been installed in the State of Uttar Pradesh (UP), India, by Central Electronics Limited (CEL) for the Non-Conventional Energy Development Agency (NEDA) of UP. In the absence of prior experience on such large capacity Solar PV Power Plants, an experimental pilot PV power plant of 25 KW capacity was installed in CEL to gain hands-on experience and the lessons learnt were made use of while installing the two power plants. This paper briefly describes the features of the two power plants, the developmental approach adopted based on "Building Block Philosophy" with 25 KW System as the basic unit with the attendant advantages. It includes the indigenous design and development effort made for grid connected operation and most importantly the special design features incorporated to ensure a very high degree of safety and protection so necessary in the rural areas with predominantly non-literate users. The paper is concluded with some important lessons learnt from both the technical and logistics point of view for guiding installation of similar such plants in the remote rural areas in India and other developing countries in the future.

INTRODUCTION : In early 1991, to gain experience in setting up of large Stand Alone and Grid Interactive Solar PV power plants, the Non-Conventional Energy Development Agency (NEDA) of Uttar Pradesh (UP), the largest state in India, took a lead role in the conceptualisation and setting up of two identical 100 KWp Power Plants in the State. The Ministry of Non-Conventional Energy Sources (MNES), Government of India, was closely associated in the pioneering effort from the very beginning providing both technical and financial support. Central Electronics Limited (CEL), a Public Sector Company, was entrusted by NEDA in mid 1992 the task of design, indigenous development, installation and commissioning on a turn-key basis of both the PV Power Plants. One of these is located in village Kalyanpur, in the District Aligarh and the other at village Sarai Sadi, in the District Mau. Each of the two Power Plants provide electricity in the "Stand Alone Mode" to a cluster of villages spread over an approximate area of six square kilometers and cater to a load of approximately 600 domestic lights (CFL-9), 60 street lights (CFL-11) and 15 water pumps each of 3 HP capacity. In addition, both the power plants have the facility to connect part of the surplus power to the grid in the Grid interactive Mode.

In the absence of any previous experience in installing large capacity PV power plants with DC operating voltage of the order of 450 to 500 Volts and the fact that the Power Conditioning Unit (PCU) with Grid Interactive Facility was being developed for the first time in the country suiting the Indian Grid conditions, CEL decided, in mid 1992, to set up a 25 KWp pilot PV power plant at its premises to gain hands-on experience and also validate the indigenous development approach. The lessons learnt were made use of while installing the two power plants.

DEVELOPMENTAL APPROACH : In view of the wide fluctuation in the voltage (±15%) and frequency (±2Hz) in the Indian grids , a PCU of 25 KW capacity was indigeuously developed. To gain hands on experience and expertise in installing larger capacity power plants with DC operating voltage of the order of 450-500V and to test and evaluate the indigenously developed PCU for grid interface, a 25 KW Power Plant was installed in CEL premises during mid 1992. The building block philosophy using modular approach has the following advantages:
Flexibility : The basic units of 25 KW capacity can be suitably combined to provide upto 50,75 and 100 KW capacity depending upon arrangement of sub-arrays connecting them. This permits configuring the PV array as required to meet any load and transferring the remaining generated power to the electric grid, thus optimally utilizing generated power.
Redundancy : The use of single PCU, for village load has

CH3365-4/94/0000-1073 $4.00 © 1994 IEEE First WCPEC; Dec. 5-9, 1994; Hawaii

serious disadvantage of total disruption in case of failure of the PCU. On the other hand, two or three PCUs (of 25 KW capacity) paralleled to cater for the total village load has the advantage that in case any one of them fails or is shut-down for maintenance, power to the village will not be affected. Such redundancy automatically ensures the availaibility of uninterrupted power from the plant to the end-users.

Reliability : Since the basic unit handles only 25 KW power, the solid state devices used are of lower ratings and thus, enhances the reliability of such devices.

Efficiency : The inverters have better efficiency when operated at 50% of the load or above and maximum efficiency at near full load conditions. Thus a 25 KW PCU will be more efficient than a 100 KW PCU for smaller load.

SALIENT FEATURES : Both the 100 KW PV Power plants consist of two independent parts.

Stand Alone mode : 75 KW
Grid Interface mode : 25 KW

STAND ALONE MODE : The stand alone mode has been designed to provide electricity for the domestic lights, street lights and water pumps for irrigation. The lighting and pump loads have been distributed in a cluster of hamlets spread over an area of around 6 Sq.Km. A schematic diagram of the stand alone mode is depicted at Fig.1. A photograph of the PV array is given at Fig.2. Broadly it consists of the following sub-systems/ Balance Of Systems (BOS).

Fig.2 A view of PV Array

a) A PV array of 75 KW comprises of 72 panels of 1.05 KW each. Each panel has 30 PV Modules all connected in series to provide a panel DC output of 1.05 KW at Vm:480 V. The interconnection of 24 such panels in parallel arranged over two rows as in Fig.3 forms a sub array of 25 KW and three numbers of such sub arrays in parallel forms 75 KW as at Fig.1. The DC output from each of the three sub-arrays is routed through necessary junction boxes and isolator switches as shown in Fig.3. to Direct Current Distribution Board (DCDB) where from it is fed to the PCUs.

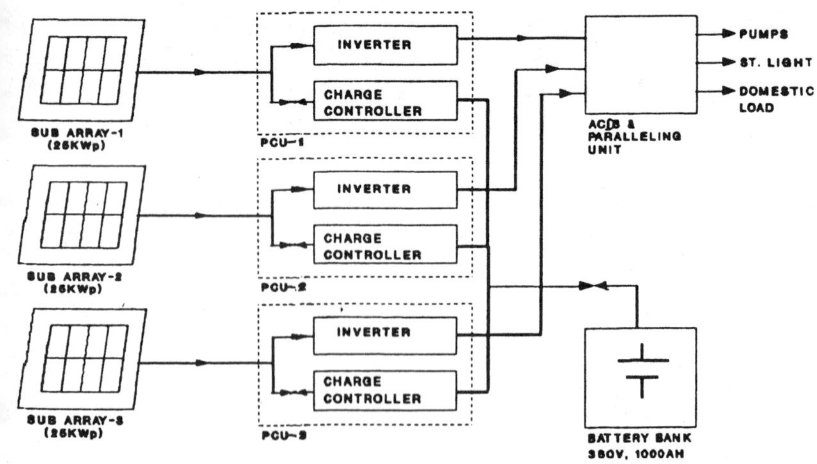

Fig.1 A Schematic diagram of PV Array

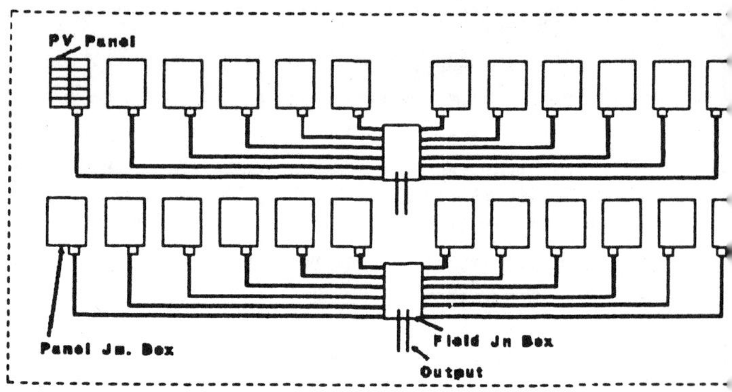

Fig.3 PV Sub Array of 25 KW.

Power Conditioning Unit (PCU) : The PCUs used in the stand alone mode have

i) A charge controller for protection of the the battery bank from over charge/deep discharge.

ii) An inverter to convert DC output of the PV array to AC power meeting all the safety. wave shape and transient behaviour requirements laid down by the Indian Electricity Authority.

iii) Provision for control and protection of components within the PCU as well as certain equipments outside the PCU.

All the three charge controllers are connected in parallel to charge the battery bank of 360 volts, 2000AH. The inverter inverts the DC power into AC, 415V, 50Hz, 3 phase. The output of all the three PCUs is synchronised and is fed to an AC Distribution Board. (ACDB), from where the power is distributed to different loads i.e. domestic lights, street lights and the water pumps through an overhead LT transmission line. as at Fig.1.

Battery Bank : The battery bank is of 360 V, 2000AH capacity using tubular positive plate, lead acid cells of 2V each which are housed in transparent SAN containers. A photograph of the battery bank is shown in Fig.4.

Fig.4. Battery Bank, 360V, 2000AH.

Electric Load : (a) The plant in the stand-alone mode caters to the following electrical loads:

i) 600 Nos Domestic Lights of compact fluorescent lamp (CFL) of 9 W each.

ii) 60 Nos Street Lights, (CFL) of 11 W each.

iii) 8 Nos. of Surface centrifugal monoblock Water Pumps of 5 HP capacity each for one of the plants and,

iv) 15 Nos. Submersible Centrifugal Water Pumps of 3 HP each for the second plant.

GRID INTERACTIVE MODE : The PCU used for grid interface is to directly convert the input power to A.C, three phase. It does not require the charge controller but a facility to regulate the amount of power extracted from the array (Maximum Power Point Tracker, MPPT) is required. The DC output from one of the Sub-arrays (25 KW) is fed to PCU-4. The three phase 415V, 50Hz, output from PCU-4 is interfaced with grid through appropriate power switching and protection devices as at Fig.5. Provision of an indigenously developed data logger has been made to record the specified PV array parameters and the energy transferred to the grid.

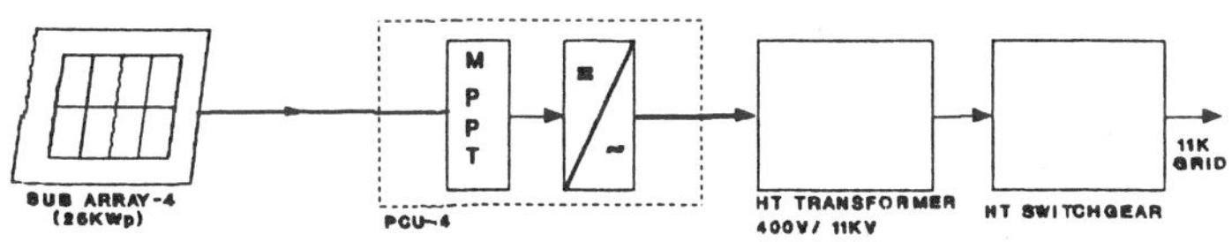

Fig.5. A Schematic diagram of Grid interfaced plant.

Key Parameters addressed in the design and development of the PCU

i) High efficiency charge controller and inverter based on PWM technique have been used. The output voltage and frequency are maintained constant over a wide band of load variations and input voltage variation. Total harmonic distortion in the output sine-wave form is maintained within 5%.

ii) The inverter is capable of handling surge

currents of the order of 150% of its rated capacity for a duration of one minute for operation of 3 phase induction motors.

iii) The inverter is designed to protect itself from output overload and short cirucit conditions. It will trip in case of any short circuit or overload conditions.

iv) It accepts a wide window of DC voltages viz 330 V to 600 V.

v) The PCU starting and shut-off procedure needs to be very well defined and understood. Initiation of starting procedure must be contingent upon the availability of the PV and grid power. Connection of de-energised PCU or a PCU out of phase with grid power can result in reverse flow of power which can lead to components damage.

vi) The PCU must not be attempted for interconnection with a de-energised power grid. It can lead to damage to the PCU and safety of the staff working on grid line is jeopardised. The paralleling/interfacing of the PCU with the grid may be utilized only when the grid power and the minimum required solar power are available. Also the interfacing must be carried out automatically when the synchronous conditions are met.

SAFETY CONSIDERATIONS : Human safety is the most important consideration in the PV Plant installation, operation and maintenance. Therefore, all possible safety aspects have been incorporated in the plant. Some of the important aspects are

i) Floating ground of the battery bank.

ii) All non-current carrying matallic parts in the PV array are grounded.

iii) Fool proof grounding scheme for the PV array, PCU and the transmission line.

iv) ON/OFF switching procedures for grid interfacing to avoid reverse flow of power.

v) Suitable junction boxes to isolate the PCU from the PV array.

vi) Use of MOVs to protect the array from high voltage transients.

vii) Use of MCCBs to isolate the Battery bank from PCU/PV array.

LESSONS LEARNT FOR FUTURE GUIDANCE : Some of the important lessons learnt during execution, operation and maintenance of the PV Plants are summarised below

i) Behaviour of the leakage current to ground in PV panels and sub-arrays.

ii) Insulation requirement at various points in the PCU and Junction boxes under high voltage.

iii) The module up-keep at high voltages and their behaviour over a long period of time and their maintenance.

iv) The behaviour of high voltage battery bank over a period of time and the problems associated with its safe installation, operation and maintanance.

v) The requirement of isolation control and protection switch gears when interfaced with the grid.

vi) The problems associated with the transmission and distribution line in rural areas with respect to safety and efficient use of power.

Acknowledgement : The authors are grateful to Non Conventional Energy Development Agency, U.P.(NEDA) and Ministry of Non Conventional Energy Sources, New Delhi for providing an opportunity to install, operate and maintain both the PV plants.

Anti-Islanding Protection Using a Twin-Peak Band-Pass Filter in Interconnected PV Systems, and Substantiating Evaluations

T.Ishida[1], R.Hagihara[1], M.Yugo[2], Y.Makino[2],
M.Maekawa[2], A.Takeoka[1], R.Suzuki[2] and S.Nakano[1]

[1]New Materials Research Center, SANYO Electric Co., Ltd.
[2]Mechatronics Research Center, SANYO Electric Co., Ltd.
1-1 Dainichi-Higashi, Moriguchi, Osaka 570, Japan

ABSTRACT

We have successfully developed and industrialized a new frequency-shift anti-islanding protection method using a twin-peak band-pass filter (BPF) for grid-interconnected photovoltaic (PV) systems. In this method, the power conditioner has a twin-peak BPF in a current feed back loop in place of the normal BPF. Our new method works perfectly for various kinds of loads such as resistance, inductive and capacitive loads connecterd to the PV system.

Furthermore, because there are no mis-detections, the system enables the most effective generation of electric energy from solar cells. A power conditioner equipped with this protection was officially certified as suitable for grid-interconnection.

INTRODUCTION

The importance of anti-islanding protection is increasing together with the spread of PV systems and the increase of PV systems connected to the utility grid. Islanding is a phenomenon in which the PV power station continues to operate during a power failure as long as the output of the power station is balanced with the power consumption of the load. When the islanding phenomenon occurs, the distribution line of the grid is charged even though transmission is stopped at the upper side of the grid. It has been pointed out that this causes the danger of electric shock to wiring maintenance personnel. Therefore, technology to prevent islanding is necessary to assure maintenance personnel's safety.

The objective of this study is to develop an effective anti-islanding protection method and to evaluate its properties.

THE MECHANISM OF THE TWIN-PEAK BPF METHOD

A view of the power conditioner equipped with our new anti-islanding protection is shown in Fig. 1 (left: the inverter, right: the transformer). Figure 2 shows the configuration of a PV system with a twin-peak BPF. The inverter is a voltage source instantaneous current control pulse width modulation (PWM) type. The inverter receives the sine wave of voltage signal $\boldsymbol{V}$ through the potential transformer (PT) and BPF. Output current control signal $\boldsymbol{I_S}$ is proportional to the amplitude of $\boldsymbol{V}$ and is given as an instantaneous pulse, so the inverter keeps the power factor (PF) at 1.0.

Inverters normally have a single-peak BPF (ex. with a center frequency of 60 Hz). On the other hand, our new inverter has a twin-peak BPF in place of the single-peak BPF. This is a feature of our inverter from the view point of electric circuitry. A twin-peak BPF is constructed from two band-pass filters with different center frequencies, such as 58 Hz and 62 Hz. The gain of the BPF has its

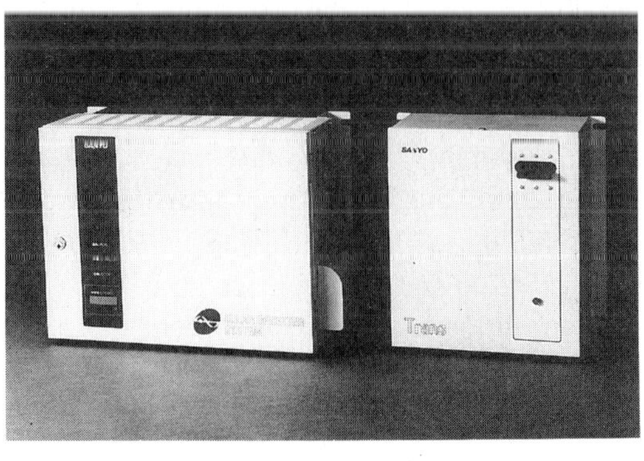

Fig. 1 A view of the power conditioner

CH3365-4/94/0000-1077 $4.00 © 1994 IEEE

First WCPEC; Dec. 5–9, 1994; Hawaii

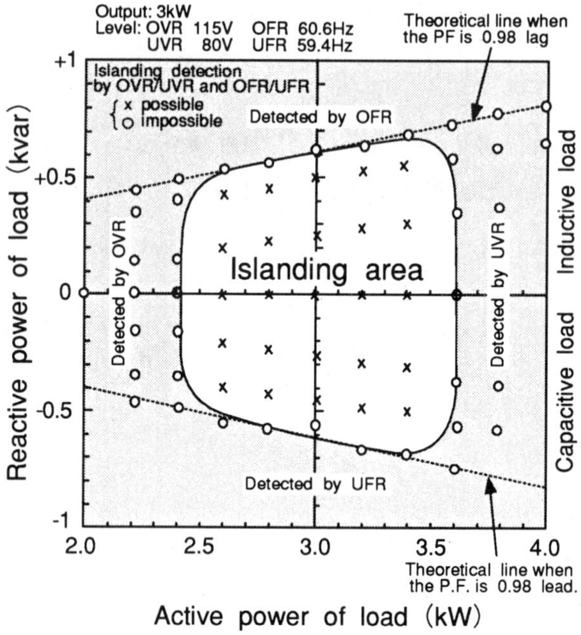

Fig. 5 Islanding area based on the characteristics of the load

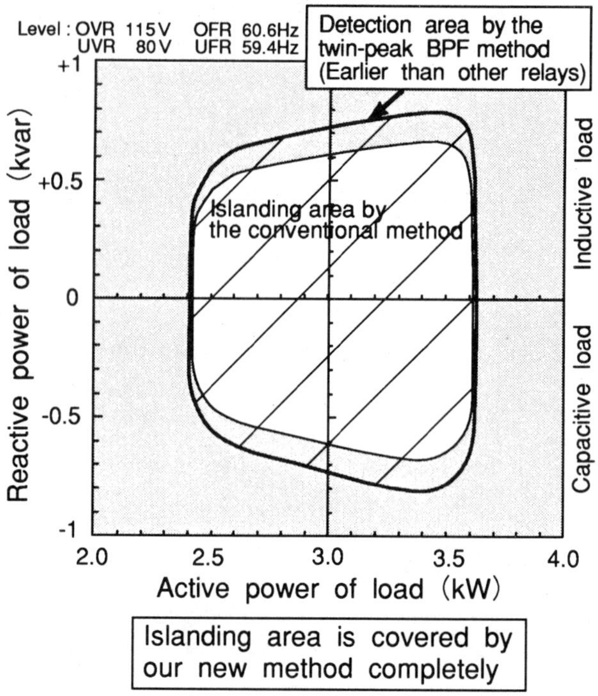

Fig. 6 Detection areas by the twin-peak BPF method

caused by a set of conventional protection relays such as an overvoltage relay (OVR), an undervoltage relay (UVR), an over frequency relay (OFR) and an under frequency relay (UFR).

Figure 4 shows the relationship between the islanding area and the output power of the inverter. The islanding phenomenon is caused not only by a perfectly balanced condition between the inverter and the load but also by an unbalanced condition within ±20% of the output power. The islanding area is increased in proportion to the output power of the inverter. So we can know the condition of the entire output range by investigating at the rated output (ex. 3 kW).

Figure 5 shows the islanding area based on the characteristics of the load at rated output. The horizontal axis indicates the value of the resistance load, and the vertical axis indicates the reactive power of the load. Generally speaking, when the load is larger than the output of the inverter and the PF is kept at 1.0, the output voltage is decreased after power failure. When the load is smaller than the output, the output voltage is increased after power failure. Thus, under an unbalanced condition of over ±20% of the resistance load, the OVR/UVR detects the islanding and stops the operation.

When an inductive load is connected, the frequency of the output is decreased after power failure. When a capacitive load is connected, the frequency is increased vice versa. When the PF is over 0.98, islanding can be detected by an OFR/UFR with the detecting levels

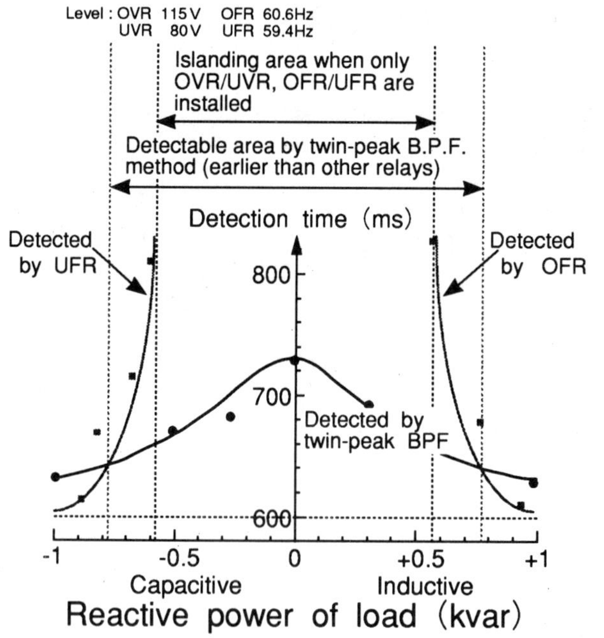

Fig. 7 Relationship between the reactive power of the load and the detection time

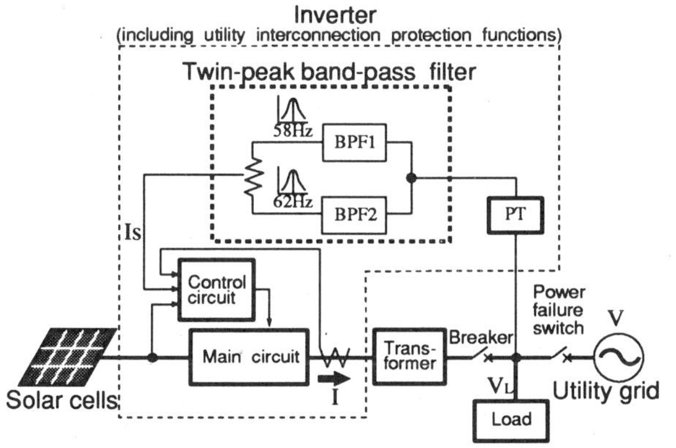

Fig. 2 Configuration of the system with a twin-peak BPF

extreme value at 58 Hz, 60 Hz and 62 Hz as shown in Fig. 3. The phase is also zero at 58 Hz, 60 Hz and 62 Hz as designed.

During normal grid-interconnection operation, the voltage of the load V_L is equal to the voltage of the utility V. So output current I is controlled to be the same phase as current control signal I_S which is received from the voltage waveforms of the utility V. As a result, output current I and utility voltage V become the same phase by feed back loop control.

During power failure, the utility voltage V is lost and replaced by the load voltage V_L as a reference signal. The phase of V_L is unchanged, because the load current I flows to the LCR circuit of the load continuously. The current control signal I_S is determined by V_L through the twin-peak BPF. The frequency changes slightly to that which is decided by the load at the time of power failure. With this change in frequency, the phase of I_S changes according to the phase-characteristics of the BPF transfer function (from V_L to I_S), as shown in Fig. 3.

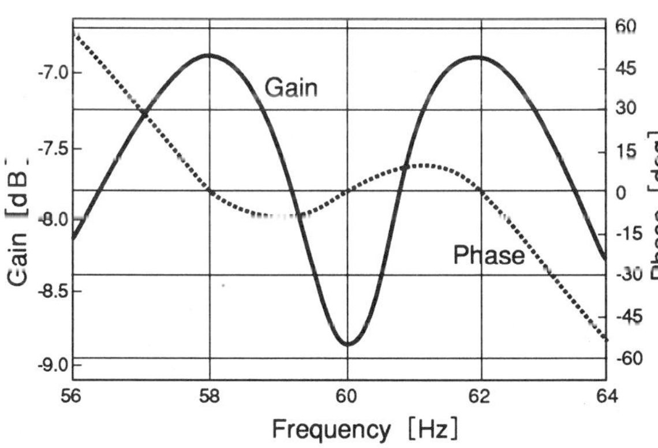

Fig. 3 Characteristics of the twin-peak BPF

When the load is a little inductive and I_s leads to I, the output current feed back loop control system increases the frequency of I while controlling I to have the same phase as Is. Thus, the system continues to increase the frequency until the system reaches the phase (nearly equal to zero) which is determined by the load. Therefore, by detecting the over condition of the frequency, the inverter can detect islanding and stop the operation.When the load is a little capacitive, the frequency drops vice versa.

The larger the gradient of the phase against frequency becomes, the easier the frequency fluctuates at power failure. The gradient of the phase against frequency is decided by the selectivity of the BPF, so the islanding detection time can be adjusted by changing the selectivity of the BPF.

THE EFFECT OF THE TWIN PEAK BPF METHOD

(1)The detection area

To prove the effect of anti-islanding protection using a twin-peak BPF, we first investigated the islanding area

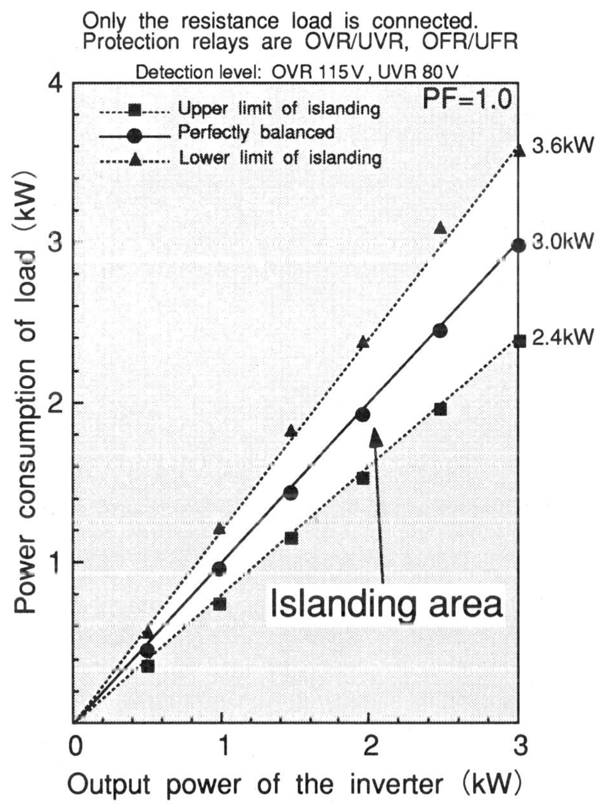

Fig. 4 Relationship between the islanding area and the output power of the inverter

shown in Fig. 5 (with the OFR at 60.6 Hz, and the UFR at 50.4 Hz and so on). In Fig. 5, the dotted line shows the theoretical line which indicates the PF is 0.98. The trapezoidal area enclosed by the dotted lines and the ± 20% of the active power lines show the islanding area in which the conventional protection (OVR/UVR, OFR/UFR)cannot detect islanding.

Our new anti-islanding protection method using a twin-peak band-pass filter detects islanding under any condition. As shown in Fig. 6, it was experimentally proved that the twin-peak BPF method completely covers islanding areas in which conventional protection relays could not detect.

Figure 7 shows the relationship between the detection time and the reactive power of the load at rated output. With only OFR/UFR, the detection time increases greatly when the reactive power comes to zero (the PF comes to 1.0). In the center area in which the reactive power is under 0.6 kvar, the detection time becomes almost infinite. This means that detection is impossible by the conventional method. On the other hand, the twin-peak BPF method can detect islanding within 730 ms in that area.

In the outer area in which the reactive power exceeds 0.8 kvar, the detection time of OFR/UFR is shorter than that of the twin-peak BPF. So OFR/UFR works earlier.

(2)The operating waveforms

As we have discussed above, the detection of islanding is most difficult when the output power is balanced completely at rated output (maximum output)

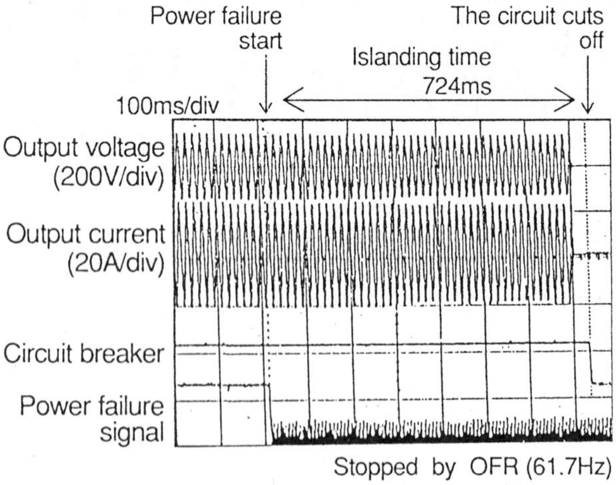

Fig. 8 Operating waveforms of anti-islanding protection

and the power factor is 1.0. Figure 8 shows the operating waveforms of anti-islanding protection under such a condition, in which the pure resistance load was connected (the PF was 1.0) and the system was operated at rated output. In this experiment, power failure was caused by an external switch. To show the start point of power failure, the power failure signal was received from an auxiliary switch linked with the power failure switch. As shown in Fig. 8, the anti-islanding protection was able to detect the power failure, stop the operation, and cut off the circuit breaker of the PV system within 724 ms. The output frequency just before stopping was 61.7 Hz. So the over frequency relay equipped with the inverter worked.

Under other conditions, such as capacitive loads, inductive loads, and with rotating machines connected, the twin-peak BPF method worked completely in the same way. Furthermore, this method has no mis-detection. Generally, other methods such as the 3rd harmonics voltage detection method, have some mis-detection.

In July 1994, Japan Electrical Testing Laboratory Inc. certified a power conditioner equipped with this anti-islanding protection as suitable for grid-interconnection.

CONCLUSION

We have developed a new, perfect anti-islanding protection method using a twin-peak band-pass filter (a kind of frequency-shift method). We have also proved that this method is able to detect islanding completely when various loads were connected to the PV system. The structure of the twin-peak BPF circuit is simple, and there is no mis-detection.

REFERENCES

[1] Ding Lu, Y. Makino, Y. Kuwano, "Twin Peak Band-Pass-Filter as Protection for Islanding of Inverter Connected to Utility Line" , *The Transactions of the Institute of Electrical Engineers of Japan*, vol. **114-B, No. 4**, April, 1994, p. 425.

[2]R. Hagihara, T. Ishida, et al. "Anti-islanding Protection Using Twin-Peak Band-Pass Filter for PV Power Generating Systems Connected to the Utility Grid", *The 1994 National Convention Record of Industry Applications Society of I.E.E. Japan*, 1994, p. 217.

A DEVELOPMENT OF COMPACT AND RELIABLE PROTECTIVE CONTROL UNIT FOR GRID CONNECTED SMALL RESIDENTIAL PV SYSTEMS

Kiyoshi TAKIGAWA Hiromu KOBAYASHI

Central Research Institute of Electric Power Industry (CRIEPI)
Iwato Kita 2–11–1, Komae–shi, Tokyo 201 Japan

ABSTRACT

In the future, a serious problem maybe arise on the maintenance of power quality, stability and safety of utility grid, if a large number of low quality private generators are left without any adequate measures taken.

Especially for the safety problem, it is the most important factor for utilities in Japan to check the reliability of islanding protection. Because the phenomenon will directly affect to the safety of personnel and facility of utility.

In order to solve the problem, a reliable and compact protective control unit for residential application was developed by CRIEPI.

It was confirmed that the unit shows excellent reliability for the islanding protection, on the demonstration test at Rokko Island.

BASIC REQUIREMENTS

A typical small grid connected system will be composed from PV array, PCS and protective control unit as shown in Fig.1.

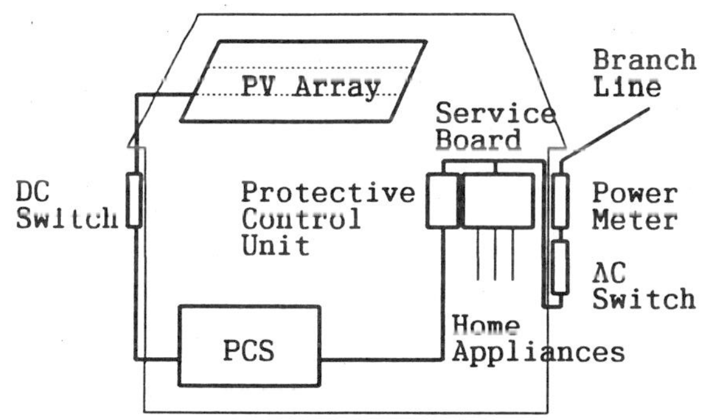

Fig.1 An Example of Configuration of Small Residential PV System.

In this configuration, the protective control unit should play the following roles as safety interface to utility grid.

(a)Function to detect internal failures such as grounding or short circuit of private generator, and to stop and cut off the generator and interrupt fault current.

(b)Function to detect abnormality of utility grid on voltage or frequency including islanding phenomenon, and to stop and cut off the private generator.

(c)Function to protect the private generator from high voltage such as lightning surge.

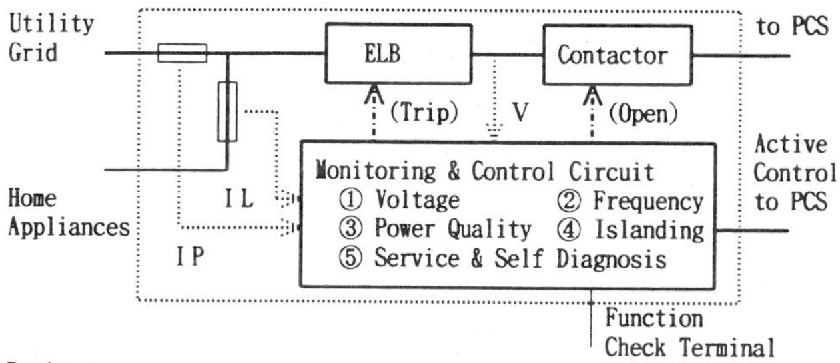

Ratings

1) Circuit Type	:	Single Phase, 2 Wire		
2) Rating Power	:		2	kW
3) Rating Voltage	:	AC	100	V
4) Rating Current	:	AC	20	A
5) Breaking Current	:	AC	2.5	kA max.

Protective Function

1) Over Current	:	AC	30	A	(within 0.1 Sec.) *1
2) Leakage Current	:	AC	15	mA	(within 0.1 Sec.) *1
3) Power Factor	:		95	%	(0.5 ~ 1.0 Sec.) *1
4) Current Harmonics	:	THD	5	%	(0.5 ~ 1.0 Sec.) *1
5) Voltage Fault	:	AC	101±15	V	(0.5 ~ 1.0 Sec.) *2
6) Frequency Fault	:		60±2	Hz	(0.5 ~ 1.0 Sec.) *2
7) Islanding Protect	:				(0.5 ~ 1.0 Sec.) *2
a Passive Method		Δθ	9	Deg.	
		Δh	0.5	%	
b. Active Method		Δf	0.1	%	(5 Deg. Lag)

Note) *1: ELB Trip *2: Contactor Ooen.

Service Function

1) Status Display	:	Line, Generator and Control Circuit Status
2) Service Display	:	Load, Generator and Net Power on Deals

Fig.2 Outline of Protective Control Unit for Small Residential Application. (Islanding Protection: Passive + Active)

First WCPEC; Dec. 5–9, 1994; Hawaii

Studies on protective control carried out to date have employed the method to unitize the all protective functions into PCS control, at the view point of compactness and cost reduction.

However, as a matter of fact, it is so hard to test the accuracy of protective functions in that situation, because they are interspersed with various control functions of PCS.

To avoid the problem, all protective functions required for interconnection should be better to remove from PCU, and operate independently as a protective control unit.

If only the unit was standardized, utility companies can easily inspect the validity of protective functions without confusing.

DEMONSTRATION MODEL

The circuit configuration is very neat and simple as shown in Fig.2. It is composed from ELB, contactor and control electronics.

The control electronics is fully digitalized using a digital signal processor (DSP).

(1)Basic protection

Abnormal current caused by the fault of private generator is cut off by ELB. Meanwhile, when an abnormal event such as interruption takes place on utility grid, contactor is released to remove the generator from utility grid.

The control electronics covers delicate and a wide range of faults on generator and utility grid to meet the the official guide line in Japan.

In particular, it has monitoring functions to detect abnormality on power factor and harmonic current of private generator, and to remove one which has poor performance to avoid degradation of power quality of utility grid.

(2)Islanding protection

It is important to assure the reliability of islanding protection, because the phenomenon will directly affect to the safety of personnel and facility of utility as illustrated in Fig.3.

To avoid such dangerous situation, various

Fig.3 A Situation of Dangerous Islanding Phenomenon.

Table 1 Outline of Various Islanding Prevention Methods Recommended by Official Guideline in Japan.

Measures	Summary of Description
1.Measure with PCS Type	
(1)Employment of Current Regulated Type of PCS	* Giving no ability for voltage regulation.
2.Passive Measures	
(1)Monitoring of Line Voltage & Frequency	* Detection of deviation from regulated boundary of line voltage and frequency.
(2)Monitoring of Line Voltage Phase	* Detection of steep cahnge on phase of line voltage.
(3)Monitoring of Line Voltage Harmonics	* Detection of steep change on harmonics of line voltage.
(4)Detection of steep cgange of frequency	* Detection of steep change on line frequency.
3.Active Measures	
(1)Frequency Shift	* Detection of variation on frequency caused by active frequency shift of PCS.
(2)Output Power Rocking	
a.Effective Power Rocking	* Detection of variation on frequency or voltage caused by active rocking of effective power.
b.Reactive Power Rocking	* Detection of variation on frequency or current caused by active rocking of reactive power.
(3)Rocking of Auxiliary Load	* Detection of variation on voltage or absorbed current to utility line caused by active rocking of auxiliary load.

means of protection are recommended by official guideline in Japan as shown in Table 1, and each passive and active method must be conbined to assure the reliability of protection.

In this unit, passive protection is composed from detections of steep and tiny change of phase and 3rd harmonic of line voltage. And frequency shift is employed for active protection.

The detection of phase change is so effective, becuse frequency change of PCS is very sensitive especially to unbalance of reactive power on islanding condition as shown in Fig.4

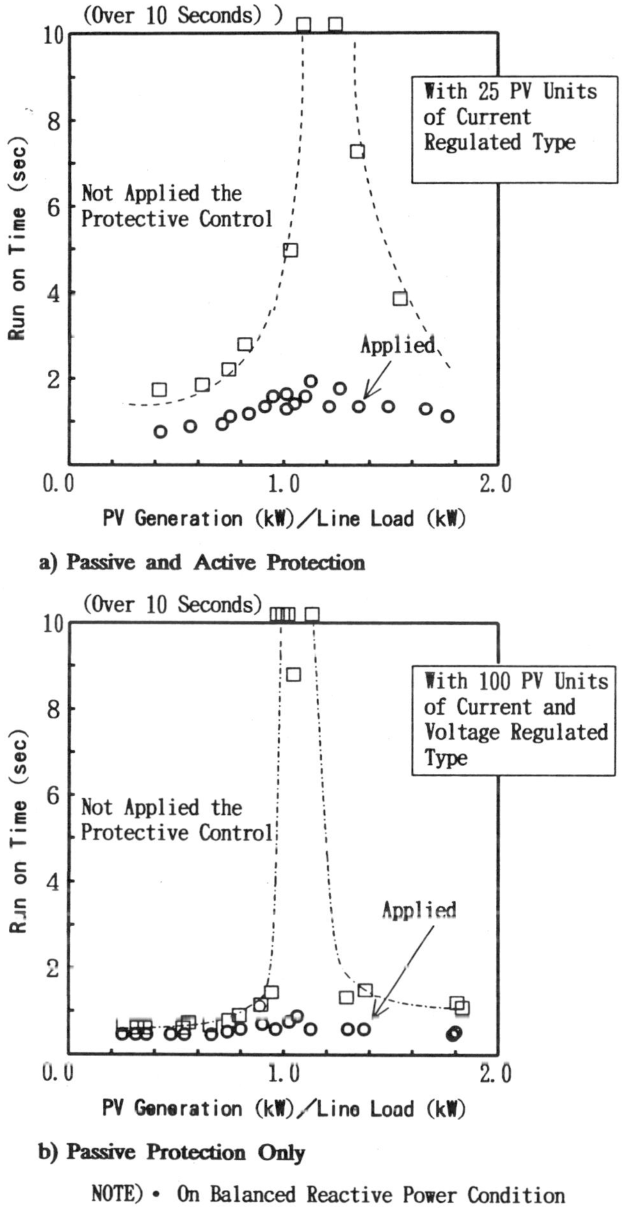

a) Passive and Active Protection

b) Passive Protection Only

Fig.5 An Example of Effect on Islanding Protection (Rokko Island)

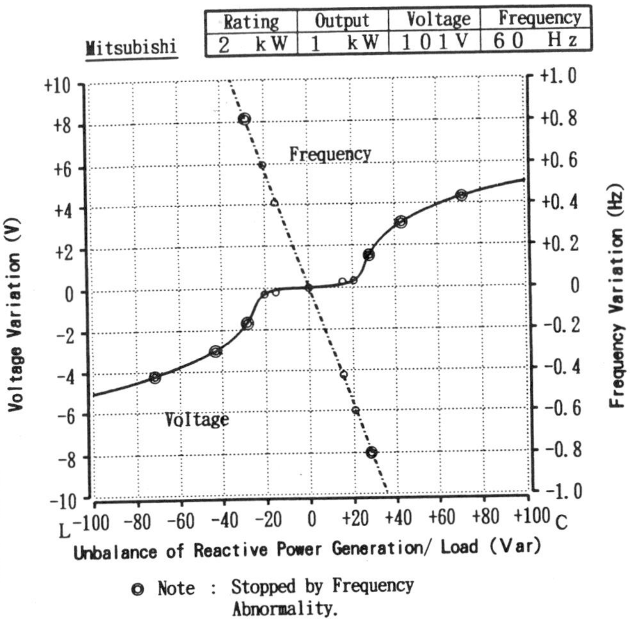

Fig.4 Variation of Output Voltage and Frequency of PCS on Unbalanced Reactive Power Condition. (PCS Type : Voltage Regulated)

In a completely blanced condition, the weak point of phase detection is made up by harmonic detection and active frequency shift.

(3)Automatic fault diagnosis

A function of automatic fault diagnosis is incorporated to assure the operating reliability. When the unit went to disorder itself, ELB is quickly tripped to avoid the irregular operation.

(4)Service indication

It indicates the status including utility grid, generator and protective control unit itself, and displays various measurements such as generating power, load power, net power on deal and each cumulative values.

Reliability of islanding protection

A result of demonstration test conducted at Rokko Island is as shown in Fig.5 and Fig.6. The Islanding phenomenon easily appear if the unit is unit is not applied. However, it was confirmed that the phenomenon is perfectly suppressed within 1 to 2 sec. by use of the unit.

Estimated sales price

As a result of cost analysis, it was cleared that the price on a lot of 1,000 to 10,000 units is in a level of ¥40,000. It is only a tenth of the actual market price in Japan.

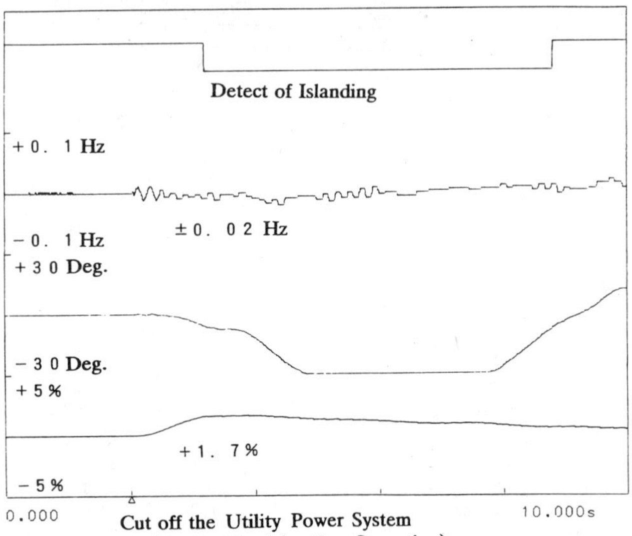

Fig.6–a An Example of Operation on Islanding Protection (Only Passive Method)

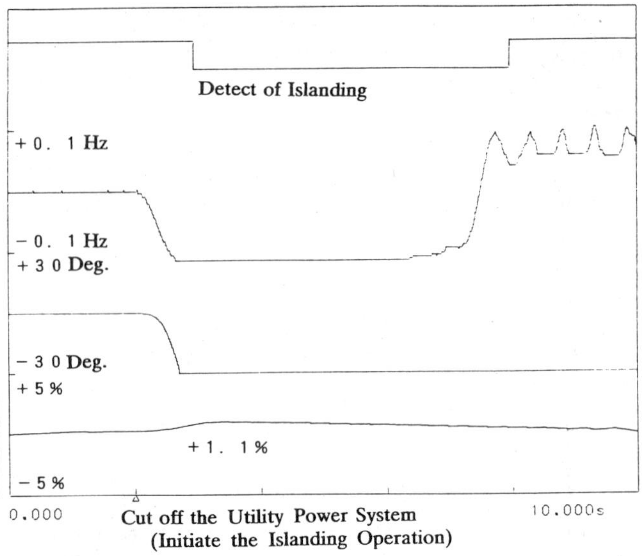

Fig.6–b An Example of Operation on Islanding Protection (Passive + Active)

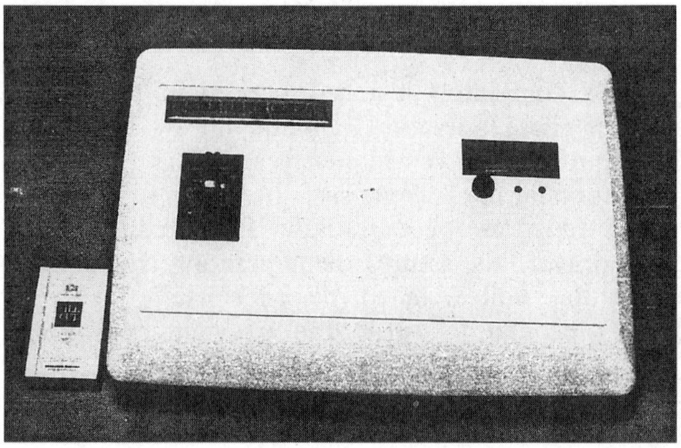

Fig.7–a An Appearance of Simplified Model (Passive Islanding Protection Only)

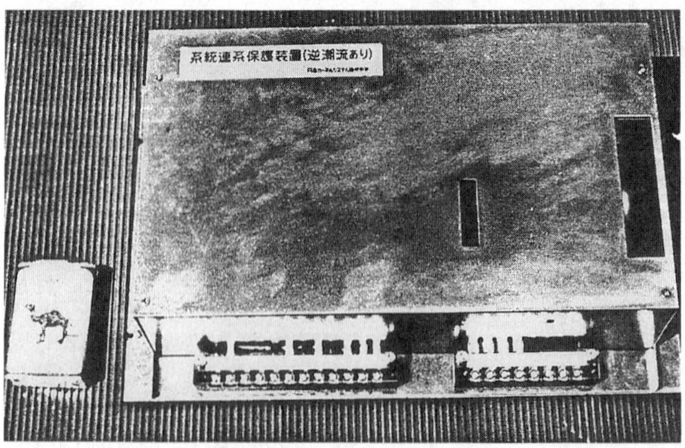

Fig.7–b An Appearance of Professional Model (Passive + Active Protection)

Table 2 Estimated Market Price of the Unit for Small Residential Application.

Component	10 units	100 units	1.000 units	10.000units
(1)PC Board	4.350	3.000	2.250	2.250
(2)Electronics Parts	12.654	12.400	12.153	11.909
(3)Parts Mounting	6.500	4.500	2.500	1.750
(4)Discrete Parts	15.120	13.113	12.553	12.429
①ELB	(6.200)	(5.150)	(4.950)	(4.950)
②Contactor	(730)	(690)	(690)	(685)
③Transformer	(3.400)	(3.060)	(2.910)	(2.849)
④Small PT/CT	(1.640)	(1.220)	(1.100)	(1.100)
⑤Small Parts	(3.150)	(2.993)	(2.903)	(2.845)
(5)Wiring	3.000	1.500	1.300	1.200
(6)Casing	17.000	10.000	4.000	3.500
(7)Test & Adj.	5.000	3.000	2.500	2.000
Sub Total	63.624	47.513	37.256	35.038
General Expense	31.812	14.254	9.314	7.008
Marketing Price	¥ 95.436	¥ 61.767	¥ 46.570	¥ 42.046

CONCLUSION

A pretty protective control unit was developed and it was confirmed that the unit shows excellent reliability for the islanding protection.

The simplified model only installed passive protection, it has also sufficient reliability for small residential application.

Finally, the authors would like to appreciate generous cooperation of the Japan Kernel Systems Inc. for development of demonstration model.

A part of applied technology abobe is a result of NEW SUNSHINE PROJECT.

STATISTICAL EVALUATION OF OPTIMUM ISLANDING PREVENTING METHOD FOR UTILITY INTERACTIVE SMALL SCALE DISPERSED PV SYSTEMS

Hiromu Kobayashi, Kiyoshi Takigawa
Central Research Institute of Electric Power Industry

ABSTRACT

The optimum islanding preventing method for the small scale dispersed PV system were evaluated, based on a simulation analysis concerning the occurrence probability of islanding phenomenon in actual operation. As the results, it is sufficient to adopt only one developed mesure when the PV penetration ratio to distribution line capacity is less than 20% to 30%. It is necessary to adopt combinatation method consisted of 2 or more developed measures when the penetration ratio is more 40%. The harmonic monitoring method is effective in case of using it in combination method.

INTRODUNTION

It is expected that utility interactive small scale dispersed PV system will be widely diffused in Japan. Islanding phenomenon of these PV system in the event of distribution line stoppage at line accident and maintenance is one of the most serious problem on the utility interconnection. Many types of new islanding preventing measures for the PV systems were designed and developed in Rokko Island PV test project and other projects[1][2]. Main measures are shown in Tabel 1. However the optimum method of the designed islanding preventing measures was not cleared yet. In this study, the optimum islanding preventing method for the small scale dispersed PV system considering reliable safety and system cost reduction were evaluated and cleared, based on a simulation analysis concerning the occurrence probability of islanding phenomenon in actual operation.

PROCEDURES OF SIMULATION ANALYSIS

Following procedures were taken for the simulation to clear the islanding occurrence probability, which is the ratio of the number of islanding to the number of distribution line stoppage, and the number of islanding of each islanding preventing measure developed to date.

i) Calculation of failed condition of islanding detection

– To determine the relationship between the power output and loads at which detection of islanding phenomenon is failed. Dynamic characteristics simulation with EMTP model, which was verified by the demonstration results of Rokko Island 200kW PV facility[3], was mainly employed.

Table 1 Islanding preventing measures and their detecting conditions

Passive measures
*Abnormal voltage and frequency detection measure
  – +/−10% of reference voltage
  – +/−1% of reference frequency
*Phase monitoring measure
  – +/−3 degree (instantaneous)
*Harmonic monitoring measure
  – 1% (increasing value)
*Frequency change rate monitoring measure
  – +/− 0.2%/sec

Active measures
*Frequency bias measure
  – Detection of frequency change
*Output power variation measure
  – Detection of frequency divergence

ii) Calculation of the power output profile of PV power generation at each location

– To calculate the overall power output profile of PV power generation unit throughout a year at each location using the solar irradiance profile data with the PV penetration rate, which is the ratio of overall PV capacity to distribution line capacity, taken as a parameter.

iii) Calculation of occurrence probability of islanding phenomenon

– To calculate the probability of appearance of the ratio of PV power generation to load(= power generation/load), at which the islanding phenomenon cannot be detected in the time zone when faults and power interruption for maintenance work are possible to occur for each location, using the load profile data, the overall power output of PV determined in ii), and the power generation/load condition for failed detection of islanding phenomenon as determined in i).

iv) Calculation of the number of islanding phenomenon in a year

– To calculate the number of islanding phenomenon occurred in the whole country in a year using the numbers of faults and power interruption by

CH3365-4/94/0000-1085 $4.00 © 1994 IEEE
First WCPEC; Dec. 5–9, 1994; Hawaii

service work in the whole country in a year and the country wide average of the probability of islanding occurrence at which the islanding phenomenon cannot be detected at each location as determined in iii).

CONDITIONS FOR THE FAILURE TO DETECT ISLANDING

Simulation analysis was conducted to establish conditions for failure to detect the islanding phenomenon with measures to prevent the islanding phenomenon developed to date shown in Table 1 and combination of these measures. Figs.1 to 4 show results of these calculations. It was found from the above that the region of failed detection of each measure was reduced with relatively high significance from the case that the voltage and frequency monitoring measure was employed. The region of failed detection by a combined measure is essentially the superimposed region of regions of failed detection by each of detecting measures. A simulation calculation was separately conducted, however, for the measure to monitor frequency change rate + an active measure was employed, since the measure detects the differential with time of the frequency which is changed by the active measure. Fig.4 shows results of the calculation. It was found from it that the region of failed detection was much reduced for all of combined measures.

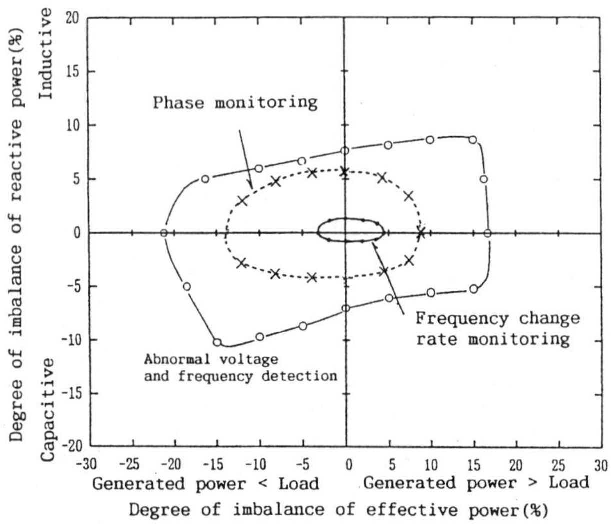

Fig.1 Regions of failed detection of islanding (Passive measures)

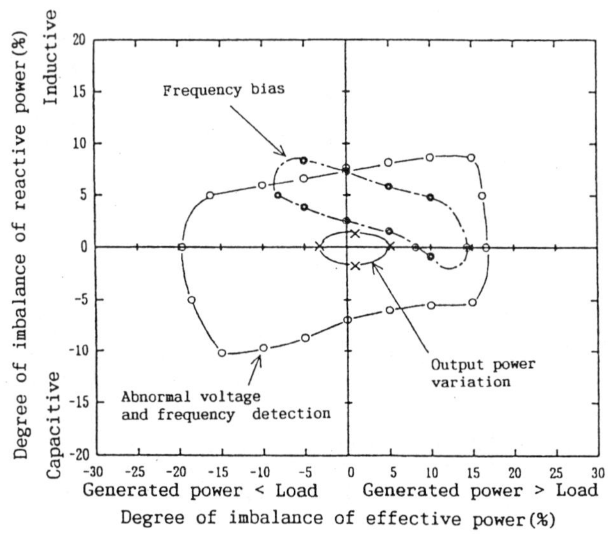

Fig.3 Regions of failed detection of islanding (Active measures)

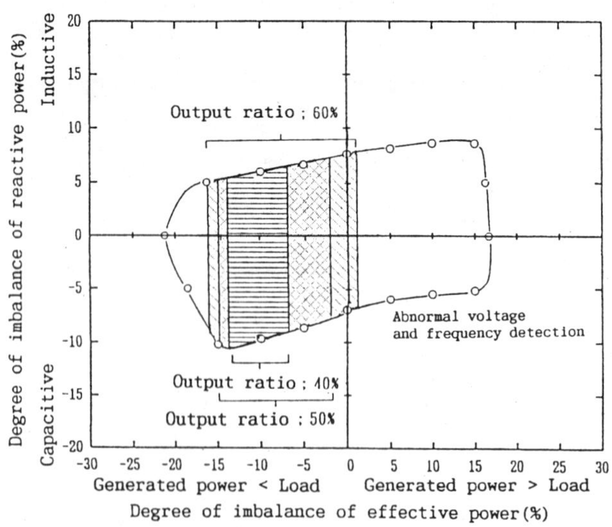

Fig.2 Regions of failed detection of islanding (Harmonic monitoring)

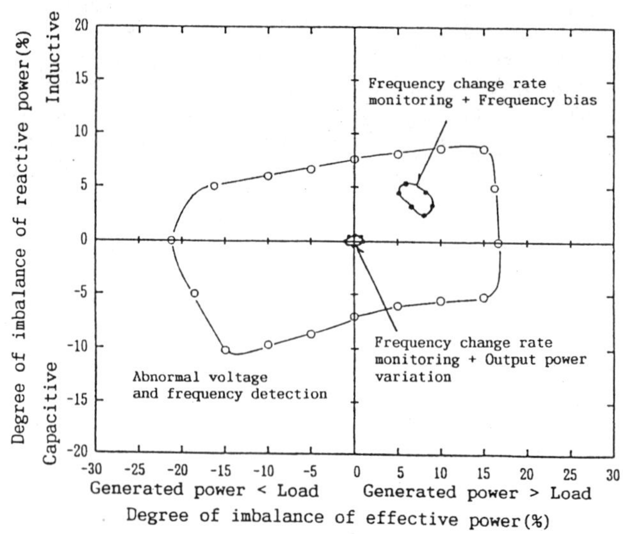

Fig.4 Regions of failed detection of islanding (Combined measures)

PROBABILITY AND THE NUMBER OF ISLANDING PHENOMENON

Setting conditions in the calculation are shown as follows: The data of load profile of the country–wide distribution lines is assumed to follow the standard pattern in the residential area which was uniformly determined by the sampling survey. PV power generation systems are all assumed to have no storage battery and accompany the reverse power flow. Uniform rate of penetration was also assumed. Ten typical cities were chosen to represent the entire country from Hokkaido to Kyushu.

Fig.5 shows an example of result obtained when the interruption of distribution line takes place randomly without depending on the meteorological condition. This is an average over the country. Table 3 gives the number of occurrences of islanding phenomenon over the country in a year, which is calculated based on the values of probability of occurrence of the islanding phenomenon such as above. From these results, No islanding occur with all measures when the rate of penetration is lower than 20% on the distribution line, giving few probability for the islanding phenomenon to take place. It was found that the probability with the measure to monitor abnormal frequency only increases rapidly when the rate of penetration rate exceeds about 30% and reaches about 6% or higher at maximum, whereas with the active or passive measure added, the number of occerrences is reduced to about 0 when the rate of penetration is lower than 30% and 1/2 – 1/10 with the passive measure and 1/3 – 1/10 with the active measure when the penetration rate

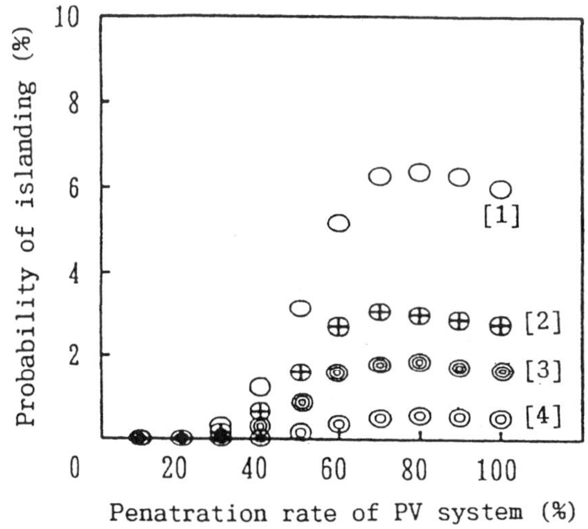

(a) Each of passive measures

Remarks
[1] Abnormal voltage and frequency detection
[2] Phase monitoring
[3] Harmonic monitoring
[4] Frequency change rate monitoring

Fig.5 Probability of occurrence of islanding phenomenon
[Note1]Result when the interruption of power distribution line takes place randomly without dependence on the weather
[Note2]Average over 10 locations in main cities in Japan

Table 2 Total numbers of islanding occurred in a year in Japan

Measures \ Penetration rate (%)	20	30	40	50	60	70	80	90	100
*Abnormal voltage and frequency detection	0	82	3200	11530	19004	23086	23704	22946	21617
*Phase monitoring	0	14	2092	6032	9870	11214	11147	10623	9955
*Harmonic monitoring	0	0	1163	3286	5917	6792	6960	6550	6138
*Frequency change rate monitoring	0	0	232	766	1469	1762	1765	1699	1576
*Output power variation	0	0	164	657	1223	1543	1544	1532	1397
*Frequency bias	0	0	847	3128	7251	9001	9360	9030	8558
*Frequency change rate monitoring +Output power variation	0	0	54	165	302	373	373	360	346
*Frequency change rate monitoring +Frequency bias	0	0	55	383	823	1156	1254	1242	1176
*Harmonic monitoring +Output power variation	0	0	0	0	14	69	152	138	138
*Harmonic monitoring +Frequency bias	0	0	0	0	178	369	727	1096	1096
*Frequency change rate monitoring +Harmonic monitoring	0	0	0	0	14	69	152	138	138

Note) Penetration rate means the retio of overall PV capacity to distribution line capacity is higher than it. With a combined measure, it was found that the islanding phenomenon did not take place until the rate of penetration reaches about 40%. In the condition at which the islanding phenomenon takes place, the number of occurrence with the combined frequency change rate monitoring + active measure was further reduced to less than 1/4 of that with the active measure alone. The number of occurrence is maintained to almost equal to 0 even when the penetration rate reaches 60% with the measure to monitor harmonic + active measure or the measure to monitor harmonic + frequency change rate monitoring. This is attained by the fact that the region of failed detection of the measure to monitor the harmonic overlaps at small area with the region of failed detection of other measure, giving a high effect of mutual complements.

OPTIMUM ISLANDING PREVENTING METHOD

Followings were cleared concerning the optimum islanding preventing method under the condition of restricting the occurrence probability of islanding.

It is sufficient to adopt only one developed mesure that is passive or active measure in addition to conventional utility abnormal voltage and frequency detector when the penetration ratio of PV system to distribution line capacity is less than 20% to 30%.

It is necessary to adopt combinatation method consisted of 2 or more developed measures in addition to the conventional detector when the penetration ratio is more 40%. Especialy the harmonic monitoring method is very effective in case of using it in combination method.

Combination method of 2 passive measures is also effective equal to the combination method of active and passive measures.

CONCLUSIONS

The optimum islanding preventing method for the small scale dispersed PV system were evaluated, based on a simulation analysis concerning the occurrence probability of islanding phenomenon in actual operation. Followings were cleared: It is sufficient to adopt only one developed mesure when the PV penetration ratio is less than 20% to 30%. It is necessary to adopt combinatation method consisted of 2 or more developed measures when the penetration ratio is more 40%. The harmonic monitoring method is very effective in case of using it in combination method. The result of the evaluation will be utilized in future revision of the guideline on utility interconnection.

ACKNOWLEDGEMENT

This study was conducted under the research contract with the New Energy and Industrial Technology Development Organization(NEDO).

REFERENCES

[1] H.Kobayashi et al.," Method for Preventing Islanding Phenomenon on Utility Grid with a Number of Small Scale PV Systems" ,**Twenty Second IEEE PVPS**, 1991, pp.695–700.

[2] K.Takigawa et al.," Development of Compact and Reliable Protective Control Unit for Grid Connected Small Residential Systems" , **7th International Photo–voltaic Science and Engineering Conference**, 1993, pp.491–492.

[3] K.Takigawa and A.Kitamura," State of the Art on Utility Interconnection Technology Development" , **5th International Photovoltaic Science and Engineering Conference**, 1990, pp.669–674.

THE CONTROL SYSTEM OF ENEL's 3.3 MWp PV PLANT

V.Arcidiacono, S.Corsi, L.Lambri
Centro Ricerca di Automatica
ENEL S.p.A. ITALY

ABSTRACT

The paper describes the control system of the Enel's first PV plant for bulk power production, consisting of ten subfields of 330 KWp each, built at Serre (Salerno).

Each subfield is provided with an autonomous line-commutated and high efficiency DC/AC converter having an innovative control system, based on microprocessor technology and touch-screen monitor, which operates by a real-time MPPT control function. The regulators of the converters also accomplish the task of peripheral interfaces for the centralized command and supervisory system, which collects data and sends commands by means of fiber optic radial LAN.

The centralized management system has a wealth of functions for the detailed display of the plant's operational state, the computing and monitoring of plant performances, the historical data collection, and for a wide range of other services. Through it is also possible to control the plant remotely using concise commands and signals, and the full remote monitoring via modem of the plant current state and acquired data.

The paper goes into detail on the innovative control functions of the peripherical regulators, on the centralized control system, and on the overall system field test results.

INTRODUCTION

Over the last ten years Enel has acquired considerable experience in defining, designing and experimenting control criteria for tracking the maximum power point of PV generators

Since 1982 two original and effective MPPT control criteria have been proposed by Enel at the IEEE PV Specialist Conference [1], one of them, named "Auto-Oscillation Criterion" is based on a non-linear closed-loop control law which determines a "limiting cycle" around the PV generator maximum power point. The second, named "Forced-Oscillation Criterion" operates in closed-loop through a small-triangular oscillation impressed around the operating point. The small oscillatory variation allows the slope measurement of the PV generator power-voltage curve. These control algorithms were applied on special high efficiency converters, with very good field test results.

The Vulcano 80kWp experimental PV plant [2] and the Adrano PV generators testing area [3] made use from the beginning of MPPT converters designed by Enel, which showed high reliability and high PV energy conversion efficiency.

Nowadays the microprocessor technology allows more sophisticated and high performances converter controllers with acceptable costs. Among the automatic control functions the Maximum Power Point Tracking is compulsory while interest is growing in the sweeping of the PV generator electrical characteristics, the converter day-start-up and the night shut-down, the real-time computing of the converter efficiency, etc..

In addition a large amount of plant performance data can be acquired and stored, whilst an advanced effective and useful operator interface obtained by monitoring functional synoptic schemes and tables with a large amount of measurements, signals and parameters, can be obtained. The converter controller can also make it possible to exchange data, via Local Area Network (LAN), with the power station Control Room, or via telecommunications with a remote Control Room.

In a PV plant with more then one PV generators and DC/AC converters, the centralized Control Room can nowadays be designed by a computerized Supervisory Control And Data Acquisition (SCADA) system with a high level of automation for the centralized control functions, the monitoring of plant performances, the failures analysis to assist and to facilitate the maintenance activities, the storing in archives of historical data, the comparison of different PV generator performances or PV energy conversion system performances, etc. The

Fig. 1. View of the 3.3 MW-Peak SERRE PV Plant

CH3365-4/94/0000-1089 $4.00 © 1994 IEEE First WCPEC; Dec. 5-9, 1994; Hawaii

centralized control functions can be very innovative when they produce: a detailed friendly guide for the maintenance operator recognizing automatically and in real-time the PV modules degradation and failures, monitor plant historical data and performance curves, record significant transients, count the failures number for each important plant component, show in detail the chronological events, allow the telecontrol, etc.

The design of the main machine interface can be based on full graphic technology to realize coloured synoptic schemes of the plant and tables for an easy and friendly plant operation by monitoring and also for an effective demonstration of the plant performances and control functions during visits.

The LAN to permit the communications among the centralized master station and the remote terminal units can use fiber optic cables with a lot of advantages such as:

- total immunity to the electromagnetic interference;
- an available bandwidth vastly wider than that given by metallic cable;
- immunity to cross talk between cables;
- immunity to lightning and electrical storms;
- long life expectancy in underground and underwater installation.

The cost of the described SCADA system is justified when a detailed plant performances characterization, a large automation, a high level of autodiagnosis and monitoring, a remote telecontrol and an easy maintenance, are requested.

SHORT DESCRIPTION OF THE PLANT

The SERRE PV plant [4] entered into operation in September 1994. It consists of ten electrically independent subfields of 330 kWp each (see Fig. 1), connected to a medium voltage ring terminating at the main control board.

Each subfield is provided with an autonomous DC/AC converter controlled by a MPPT microprocessor based apparatus. The converter of each subfield is connected through a 550 KVA transformer and by a medium voltage switch to a buried cable, which forms a 20 kV ring where the interconnection with the distribution grid is achieved.

The photovoltaic modules are made of multi-crystalline silicon, except for one subfield made of mono-crystalline silicon. Nine of the ten subfields have a fixed metal supporting structure with an inclination angle of 20 degrees.

One subfield has been differently designed with mobile supporting structure, rotating from East to West on a horizontal North-South axis. The plant fills 6 hectares with about 46000 PV modules (36 or 72 cells for each module).

GENERALITIES ON PLANT CONTROL SYSTEM

The power station control system consists of ten peripheral controllers, one for each of the DC/AC 550 kVA converters which are housed at the subfields' cabins, and one centralized supervisory and data acquisition system installed at the central building Control Room (see Fig. 2).

Each peripheral control system allows the full and autonomous management of the corresponding converter; it also acquires all the measurements and operating state signals from the PV subfield and the DC/AC converter and sends them to the centralized command and supervisory system.

The converters' peripheral controllers can be operated from the plant Control Room by an Ethernet, fiber optic LAN.

The centralized SCADA system manages the communications from and towards the periphery, acquires all the plant operation data, processes the data for real-time monitoring and plant historical performances storage, automatically commands the start-up and the shut-down of the plant converters.

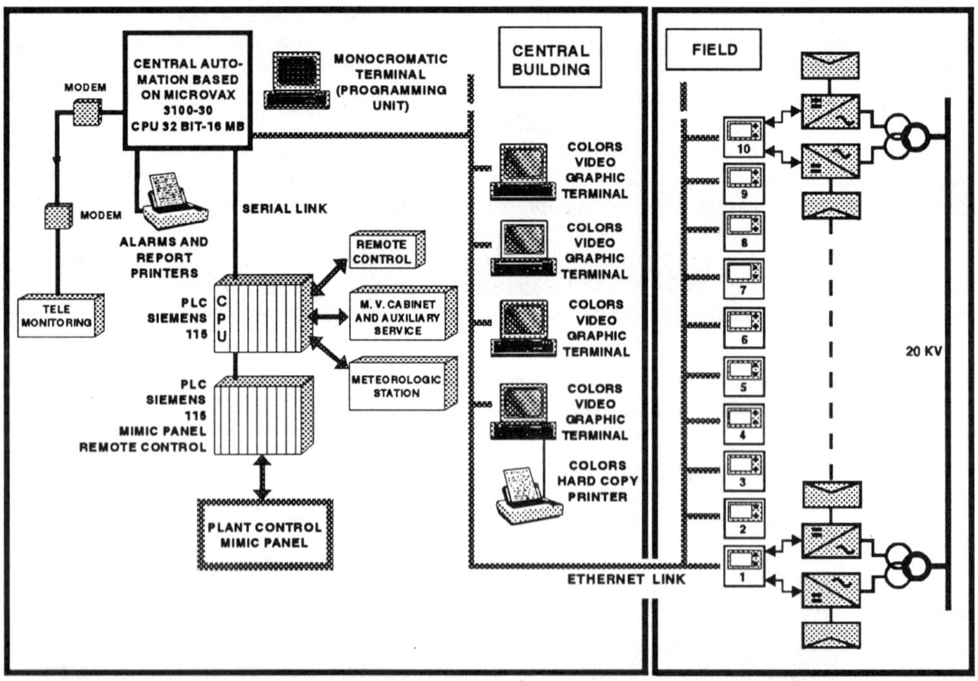

Fig.2. The plant control system: the supervision and data acquisition system at the central building and the controllers of the converters on the field.

In addition the SCADA system makes it possible to watch from a remote site, via modem, the plant operative state and performances. The plant remote control is achieved through adequate interfaces with Enel's telecommunications network.

THE SUBFIELDS DC/AC CONVERTERS AND THEIR CONTROL SYSTEMS

The Subfield Converter

Each of the ten PV subfields delivers power to a 20kV MV grid through a special line-commutated inverter (manufacturer: SEIRA, Italy) characterized by high efficiency and MPPT control law. The inverter consists of two three-phase thyristor bridges, galvanically insulated and connected on the low voltage side to the two star-triangle secondaries of the step-up transformer. On the DC side of each bridge an LC filter is located which has the task of reducing to negligible values the high frequency current and voltage ripples seen by the PV generator. This makes it possible to obtain an effective real-time tracking of the PV generator maximum power point by using small amplitude variation at lower frequency (2 Hz) of the operating point (see the following for a more detailed description of the MPPT control criterion)

The consequent small amplitude oscillation around the maximum power point makes it possible to obtain a high conversion efficiency.

With the chosen electrical scheme of the subfield converter, it is possible to operate one of the two sections autonomously, even when the other is under maintenance.

Fig. 3 shows on the back the converter power electronic board and on the front the microprocessor-based converter control apparatus.

The converter efficiency (including the losses of the bridges, transformer, DC side LC filter, interconnecting cables and converter auxiliary services) is 97% at nominal power, while it becomes progressively lower reducing the power output. At 25% of the nominal power the converter efficiency is close to 94%.

Fig.3. The 550 kVA subfield DC/AC converter and its control unit with touch screen monitor

The Converter Control System

Fig. 3 shows the converter's innovative control system based on microprocessor technology and touch-screen monitor, which provides the closed-loop MPPT function for each section of the PV subfield when the operating conditions vary (irradiance, junction temperature, cleaning of the panels, etc).

The adopted MPPT control criterion, defined and developed by Enel [1] is named "Forced Oscillation Criterion". It is briefly described by the block diagram in Fig. 4 which achieves the measurement of the PV generator power-voltage characteristic slope. The slope is measured by superimposing a small-amplitude triangular oscillation on the PV generator operating voltage.

The measurements of the PV generator powers corresponding to the positive and negative semiwaves of the triangular oscillation are sent to the integrator input with the positive and negative sign respectively.

At the zero crossing and positive slope of the triangular oscillation the output of the integrator is sampled and the memorized value represents the PV generator new operating point for the next period. In this way, at every period the change (positive or negative) of the sample value in comparison with the value of the previous period is proportional to the difference between the areas subtended by the power-voltage characteristic corresponding to the positive and negative triangular semiwaves.

Due to the integrator and the shape of the power-voltage curve, the operating point corresponds to the generator maximum power when the equilibrium point, at which the two areas are equal, is reached.

In addition to the MPPT function the converter control apparatus is able to sweep at every time, by manual or automatic command, the electrical characteristics of the PV subfield sections from no-load to short-circuit operating conditions.

The other functions are related to the automation of the control manoeuvres and sequences, data acquisition from the PV subfield, data exchange with the centralized supervisory and control system, remote control.

The touch-screen interface makes it possible to select

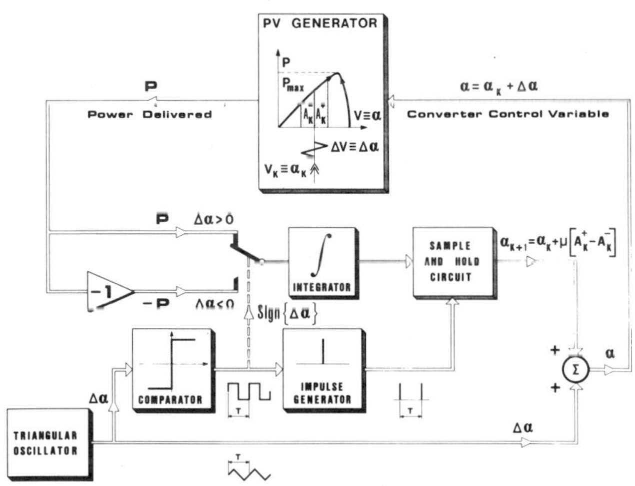

Fig.4. Block diagram of the MPPT Forced-Oscillation Criterion

and operate commands by icons, to choose animated synoptic schemes (see Fig. 5), to monitor the PV generator curves acquired by the sweeping of the electrical characteristics, to compare different curves, to monitor in real-time meaningful measurements, state signals and alarms, etc.

THE CENTRALIZED CONTROL, SUPERVISORY AND DATA ACQUISITION SYSTEM

As the SERRE power station is an experimental and demonstration plant, it has a particularly innovative centralized control, supervisory and data acquisition system called SCGS (manufacturer: C.Gavazzi, Italy), with a wealth of functions for the detailed display of the plant's operational state and a well probed characterization of the plant's performances.

SCGS Hardware, Lay-out and Communication Network

As regards the practical implementation (see Fig.2), SCGS consists of:

- a MicroVax workstation 3100-30, with RAM memory of 40 Mbytes, hard disk of 425 Mbytes and tape streamer with a capacity of 95 Mbytes;
- four 19" graphical colour terminals, with a 1152x900 video resolution, equipped with keyboard and mouse for the man-machine interface;
- a wall-type synoptic panel, for the display of the plant electrical scheme, measurements and the operating conditions of the main components;
- a PLC for acquisition of the signals coming from the auxiliary system and the meteo station;
- an Ethernet fiber-optic LAN for the interconnection, with a TCP/IP protocol, between the inverter control systems and the MicroVax workstation configured as the communication master.

SCGS is generally fed by UPS; in case of inverter failure or excessive lowering of the 110 Vdc battery voltage the SCGS feeding is automatically switched from the inverter to the auxiliary-services 220 Vac bus-bar.

SCGS communicates with the peripheral converter controllers acquiring the converter's operating conditions and the subfields' daily performances curves (voltage-current characteristics). Furthermore, by means of the PLC, SCGS acquires the operating conditions of the plant switchgears, protections, auxiliary services and the meteorologic data.

SCGS sends commands to the main 20 kV switches and the converter control systems. The total number of data exchanged by SCGS is over one thousand;. two hundred of them are measurements. All the logic and analog signals are updated every second.

The most important information for the plant operation are shown on the monitors and on the conventional synoptic panel, (see Figs. 6,7) and a synthesis of them is sent to the telecontrol system.

Data are elaborated in order to compute the mean, the integral, the maximum and the minimum values of the measurements every 15', hour, day, month, year. The repetition of the main failures/alarms is also counted.

The results of these processes are monitored by tables and memorized on the hard-disk and tape. Plant performance curves, as mainly concern the PV conversion efficiency, are computed by SCGS and monitored.

The plant events are managed by SCGS according to ISA-1 standard, drawing the operator's attention when they are not suited to the standard or exceed the normal operation limits.

Fig.6. The plant Control Room

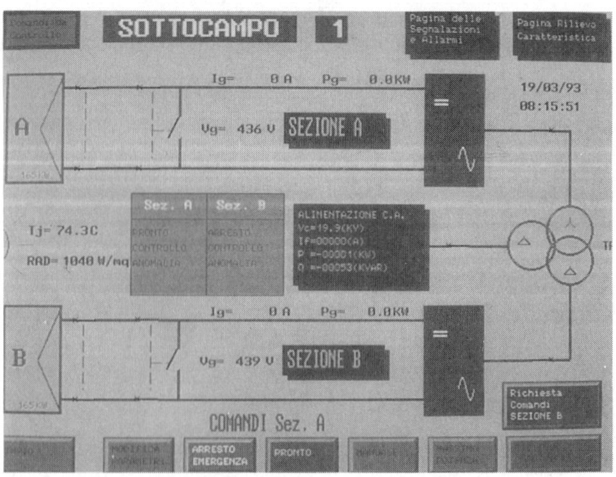

Fig.5. A subfield synoptic conversion scheme shown by the converter control system

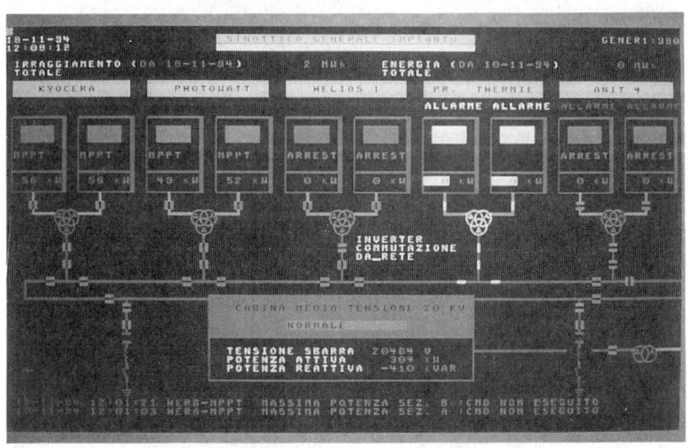

Fig.7. A plant synoptic electrical scheme shown by a SCGS monitor

SCGS Control Function

The SCGS control functions can be subdivided into managing, monitoring and diagnostics functions.

Managing functions:
- congruency check for the commands sent to the plant by the graphic interface;
- automatic daily start-up and night shut-down of the converters on the basis of irradiance measurement;
- automatic switch-on/off of the 20 kV power factor correction filters;
- daily automatic sweeping of all the subfields current-voltage characteristics, for automatic comparisons with reference curves and for an automatic check of the PV modules performance lowering or failures.

Monitoring and Diagnostics functions:
- monitoring of the voltage-current subfields' curves, for those already memorized in the hard-disk or demanded at the present time by the operator, with their overlapping and zoom;
- comparison among the measurements of the subfields' irradiances for an automatic check of failures in their precision transducers;
- comparison between the power measurements at the DC and AC sides for each converter, for an automatic check of possible failures in the active power measurement circuits;
- auto-diagnostics of the correct operation of SCGS HW, LAN and communications;
- check of the tape filling and pertinent signals.

EXPERIMENTAL RESULTS

Some experimental results obtained during the commissioning tests and the subsequent preliminary partial operation of the plant are shown in the following figures.

Fig.8 represents the voltage profile at one of the two PV subfield sections during a cloudy day in Autumn. The voltage variations are determined by the MPPT control law of the converter controller. It is possible to notice that the low frequency variations of the PV generator operating points has a range of about 180 V. Furthermore there are some high frequency variations having a peak to peak amplitude of about 60 V.

Fig.9 shows the inverter starting manoeuvre from no-load to MPPT operating condition.

The voltage of the PV generator operating point ramps from no-load to a given threshold, and consequently the power increases reaching the maximum value and after it decreases to the value corresponding to the mentioned threshold.

After that the MPPT control law is activated and the operating point tracks and reaches the maximum power in a few seconds.

CONCLUSIONS

SERRE, the first Enel's multimegawatt PV plant for bulk power production, has been a good opportunity for the application of the wide and deep experiences on PV generator control criteria and on PV plant supervision and data acquisition acquired by Enel during many PV applications in the last fifteen years.

The converter MPPT controller, microprocessor-based with touch-screen monitor and a lot of innovative control functions, is nowadays one of the most advanced and effective converter control solution in the world which makes it possible a very high conversion efficiency.

The centralized SCADA system with fiber optic LAN represents a modern, immune from electromagnetic noises, and very useful opportunity for a detailed characterization of the PV subfields performances and of the recurrent failures and for a high level automation of the plant.

The first experimental results confirm the validity and the relevant quality of all the control solutions and operation facilities designed for the plant control system.

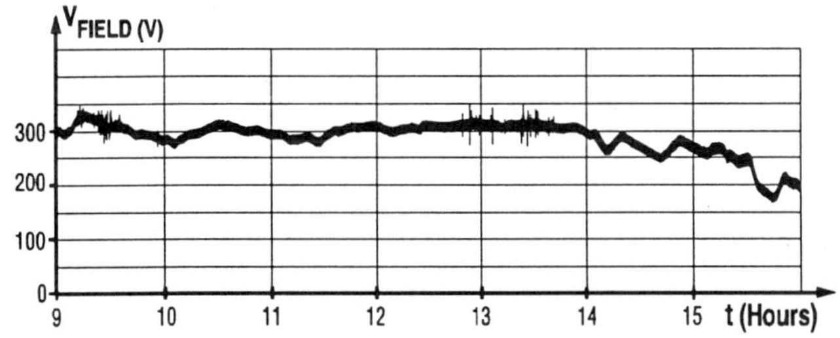

Fig.8. MPPT voltage profile during a cloudy day

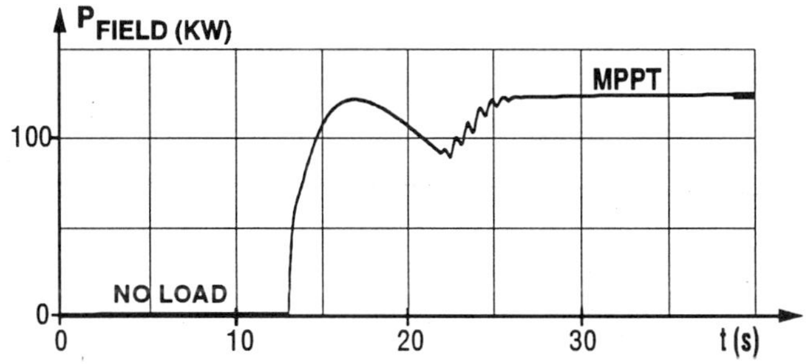

Fig.9. Inverter starting from no-load to MPPT

REFERENCES

[1] V.Arcidiacono, S.Corsi, L.Lambri "Maximum Power Point Tracker for Photovoltaic Power Plants" 16th IEEE Photovoltaic Specialist Conference, S. Diego (CA), September/October 1982.

[2] A.Previ, et al. "The Vulcano Project" International Journal of Solar Energy, 1985, Vol. 3.

[3] G.Chimento, S.Corsi, S.Guastalla, A.Iliceto "L'impianto fotovoltaico sperimentale di Adrano" - ISES, Giornata di Studio su "Gli impianti fotovoltaici connessi alla rete", 14 Maggio 1991, Atti a cura della Sezione Italiana ISES.

[4] A.Previ, A.Iliceto, G.Belli, V.Arcidiacono, S.Corsi, L.Lambri "The 3.3 MW - Peak Photovoltaic Power Station at SERRE" First WECPEC,, WAIKOLOA HAWAII, USA, December 5-9, 1994.

OPTIMIZATION OF SETTING ARRAYS CONSIDERING THE INFLUENCE OF FRONT ARRAY'S SHADOW

Y.Ishisaka*, K.Harada*, Y.Ishihara*, T.Todaka*, H.Yokouchi** and M.Yukawa**
*Dept. of Electrical Engineering, Doshisha Univ.
**Hamamatsu Solar Energy Laboratory, JQA INSTITUTE

ABSTRACT

This paper presents optimization of setting arrays considering the influence of front array's shadow. The total power per year and the total power per year per array are used as optimum condition. The following parameters are used for setting arrays, tilt angle θ, distance to front arrays d. By way of example , Tokyo (Long. 139.77 [deg.] E, Lat. 35.68 [deg.] N) is selected as simulating point, and the following is confirmed.
The relationship between θ and d for the maximum total power per year on array field to have a constant area is shown. For example, modules are laid closely on an array field, i.e. θ=0° and d=0, and the maximum total power per year is got in Tokyo. The relationship θ and d for the total power per year per array is shown.

INTRODUCTION

Since an environmental problem of the earth was paid attention, it has been required strongly to use natural energies. The photovoltaic power generation system, that is one of them, is a power generation system that has no environmental pollution and no limit of energy sourse. This system has the advantage of being noiseless and maintenance free. The system has been researched and developed positively.

" Catching irradiance " is the most important point on the essential nature of photovoltaic system. Therefore, solar arrays setting point, tilt angle and so on are examined from various view points. It is well known that front array's shadow influence the power which is generated by the system. But the influence has not been analyzed in detail.

The relationship between tilt angle θ and distance to front arrays d is examined for total power per year and total power per year per array considering the influence of front array's shadow, and this paper presents optimization of setting arrays considering the influence of front array's shadow.

THE EVALUATION WAY OF FRONT SOLAR ARRAY'S SHADOW[1]

The influence of front array's shadow to object array's power is evaluated by following ways. An solar array is composed of a few solar modules which are connected series or parallel. If a part of series connecting module is shadowed, whole of series connecting module is treated as shadowed and the area is given diffuse irradiance. When whole of series connecting module is sunned, the area is given both direct and diffuse irradiance. Further, the generating power is calculated from irradiance which series connecting module gets.

The photovoltaic current-voltage characteristics is used for calculation of the power which solar module generates.

P_f is the power which arrays generate when whole of arrays are sunned. P_s is the power when a part of arrays is shadowed. The power rate K_p is given by the following equation.

$$K_p = P_s / P_f \qquad (1)$$

THE DETERMINATION WAY OF ARRAY SETTING

In case that an array size is decided 30[m] × 3[m], if number of setting arrays in widthways is varied, K_p has little change. Because the influence of setting arrays number in widthways is very little and is negligible, K_p is calculated as number of setting array in widthways is 1.

If a size of array which is set, width of array field W, length L, tilt angle θ and distance to front arrays d are decided, number of arrays that is set on the field is calculated.

Using the power rate K_p, total power per year on the field is calculated considering the influence of front array's shadow. The setting of arrays to give the maximum total power per year on the field is determined from this results. Further, total power per year per array is calculated from total power per year and setting arrays number. The setting of arrays to give the maximum total power per year per array on the field is determined from this result.

First WCPEC; Dec. 5–9, 1994; Hawaii

SIMULATING MODEL

Figure 1 shows the size of array that is simulated in this paper. All the dimensions are normalized by the length of the panel (3[m]) as 1. The standard weather data for HASP (Heating, Air-conditioning and Sanitary engineering Program) is used as irradiance and temperature data. It has temperature, horizontal diffuse irradiance and normal direct irradiance at the each interval of an hour. The simulating term is a year. Tokyo (Long. 139.77 [deg.] E, Lat. 35.68 [deg.] N) is selected as simulating point. The size of array field is 100 × 100. The load is constant (8.4[Ω]) that an array gives maximum power on the standard test condition (Irradiance is 1.0[kW/m^2], Cell temperature is 25[℃]).

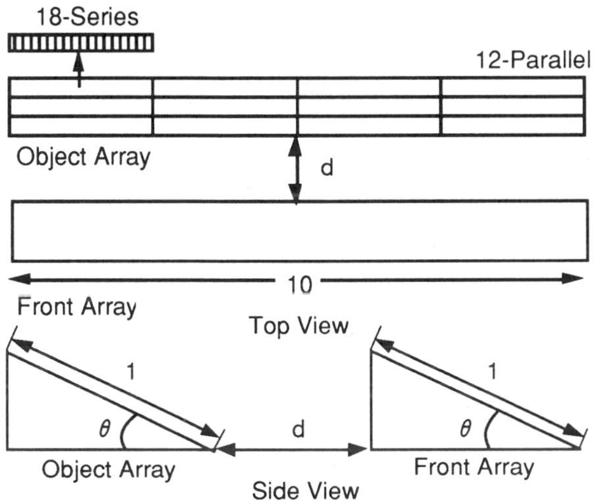

Fig.1 Simulating Model

The calculation of generated power

The current-voltage characteristics of a monocrystalline silicon solar module is given by the following equations in this simulation.

$$I = I_{ph} - I_0 \left\{ \exp \frac{q(V + I \times R_s)}{nkT} - 1 \right\} - \frac{V + I \times R_s}{R_{sh}} \quad (2)$$

$$I_{ph} = I_{pho} \{1 + 5.1029 \times 10^{-4} (T_C - 25)\} \times \{1.03\, G_A - 0.03 (1 - \exp(-8\, G_A))\} \quad (3)$$

$$I_0 = I_{00} \exp(0.09672 (T_C - 25)) \quad (4)$$

$$R_s = R_{s0} \{1 + 3.3717 \times 10^{-3} \times (T_C - 25) + 9.7058 \times 10^{-5} \times (T_C - 25)^2\} \quad (5)$$

$$R_{sh} = R_{sho} / \{ 1 + 5.7987 \times 10^{-3} (T_C - 25) + 1.6129 \times 10^{-4} \times (T_C - 25)^2 \} \quad (6)$$

V : Output Voltage [V], I : Output Current [A]
R_s : Series Resistance [Ω], R_{sh} : Parallel Resistance [Ω]
I_{ph} : Photoelectric Current [A], I_0 : Saturation Current [A]
R_{s0}, R_{sh0}, I_{ph0} and I_{00} : Each values on the standard test condition of R_s, R_{sh}, I_{ph} and I_0.
q : Electron Charge [c], k: Boltzmann Constant [J/K]
n : Diode Performance Index, E_g : Energy gap [eV]
T : Solar Cell Absolute Temperature [K]
T_C : Solar Cell Temperature [℃]
G_A : Incident Irradiance [kW/m^2]

The parameters of monocrystalline photovoltaic module are shown in Table 1. The module name is GL136N that made by Showa Shell Sekiyu.

Table 1 Constants of Photovoltaic Module

R_{S0} [Ω]	0.213
R_{sh0} [Ω]	242.2
I_{ph0} [A]	3.140
I_{00} [A]	6.695×10^{-7}
q	1.6022×10^{-19}
k [J/K]	1.3807×10^{-23}
n	1.5
E_g	1.13

The results calculated by eqs. (2)~(6) are verified under the conditions in Fig.2. The object array sets at 3-Series×2-Parallel modules, θ=0° and d=0. The front array sets at θ=30° on the south of the object array. The power is measured under the constant load (R=25[Ω]). Figure 3 shows comparison between the calculated results and measured ones. Figure 3 shows that the calculated results are very close to the measured ones. The rate of the measured results to the calculated ones is 96% on total power in Nov. 19, '94.

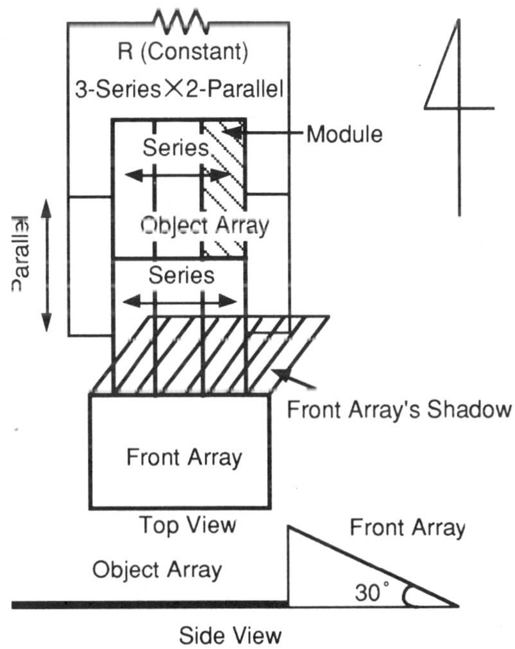

Fig.2 Experimental Model

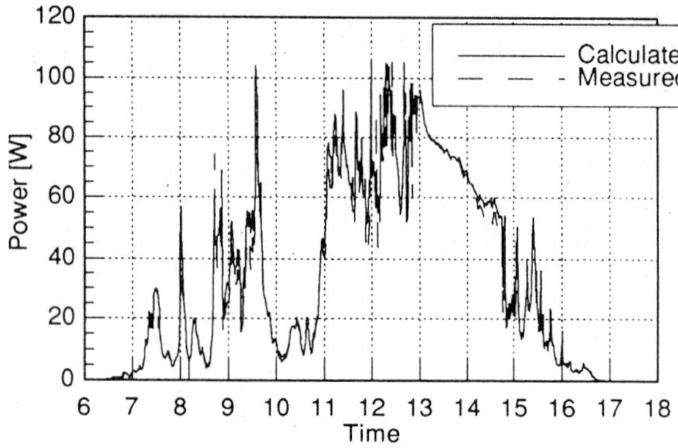

Fig.3 Comparison between Calculated Results with Measured Ones

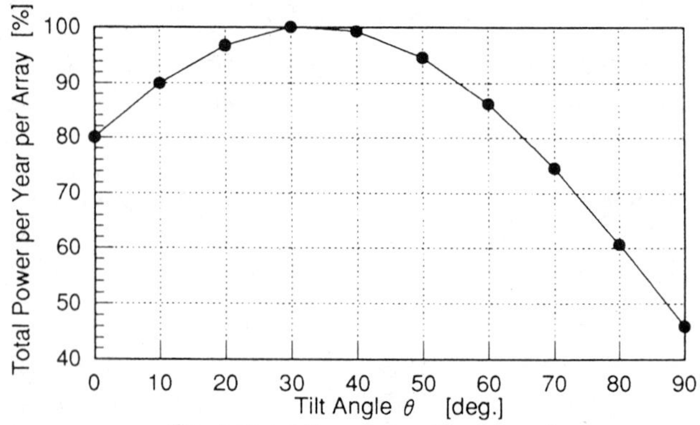

Fig.4 Total Power per Year per Array

SIMULATING RESULTS

The maximum total power per year

The total power per year is shown in Fig.4, when d=3 and θ is varied from 0° to 90° at the each of 10° in Tokyo. The d=3 is the minimum distance which is not influenced by the front array's shadow. The values on each θ are shown as the value of θ=30° being standard. Figure 4 shows the maximum power per year is got at θ=30° in Tokyo.

Then, θ is varied from 0° to 90° at the each of 10° and d is varied from 0 to 3 with regard to each θ. Then, K_p is shown in Fig.5, and the number of arrays which can be set on the array field is shown in Fig.6. The value of each θ and d is shown as the value of θ=30° and d=3 is standard.

The relationship between θ and d that can give the maximum total power per year on the array field is examined from Fig.4 to Fig.6, and shown in Fig.7. The value of each θ and d is shown as the value of θ=30° and d=3 being standard. Figure 7 shows that modules are laid closely on an array field, i.e. θ=0° and d=0, to give the maximum total power per year in Tokyo.

But when arrays are setting practically, it is necessary to have a constant distance to maintain the arrays. When d is decided, θ that can give the maximum total power per year on the array field is determined as shown Fig.7.

The maximum total power per year per array

The way of setting arrays to give the maximum total power per year is determined as shown in Fig.7, when the arrays setting conditions are given. But arrays are not always to operate most efficiently in this way.

The total power per year per is calculated here. The setting of arrays that can be operated most efficiently is decided from that.

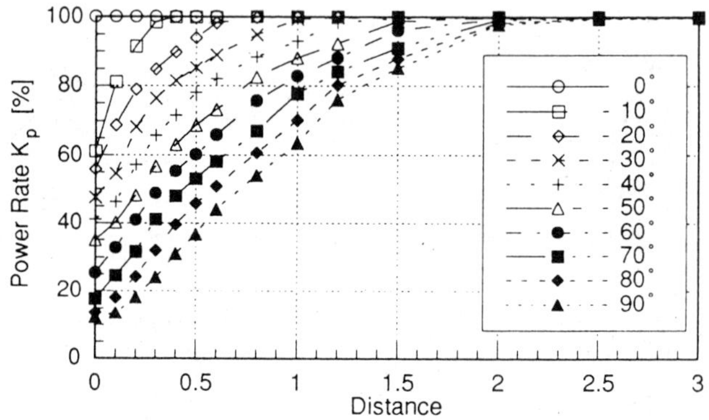

Fig.5 Power Rate K_p

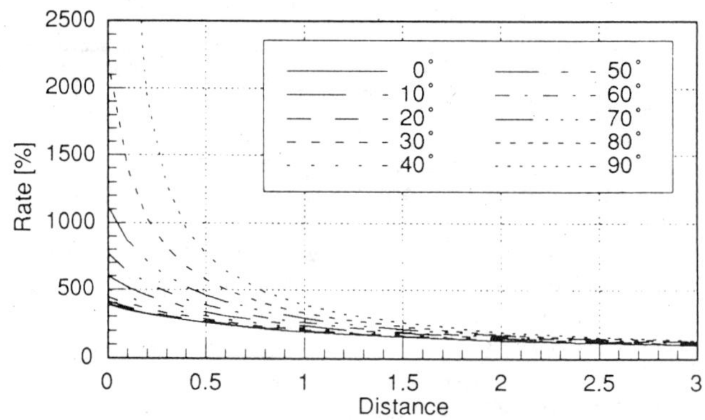

Fig.6 Number of Setting Arrays

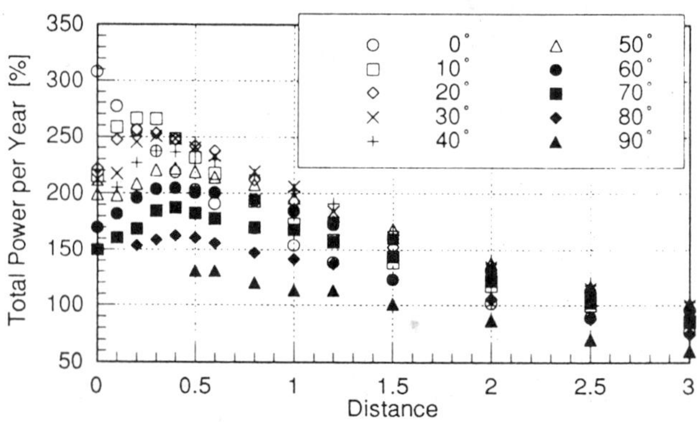

Fig. 7 Total Power per Year

The relationship between θ and d for the total power per year per array is shown irFig.8 as the value of θ=30° and d=3 being standard. When d is given, θ that the total power per year per array is maximum is determined from Fig.8. Then, the setting of arrays that can be operated most efficiently is determined.

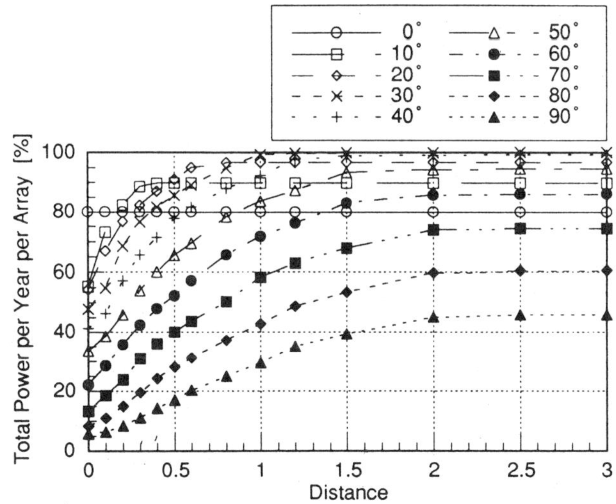

Fig.8 Total Power per Year per Array

The influence of diffuse irradiation

This paper presents the simulating results in Tokyo. These results are influenced by irradiation data. The influence of diffuse irradiation on setting arrays is described here.

The irradiation and rate of diffuse irradiation per month at θ=30° in Tokyo is shown in Fig.9 for reference. The total power per year is shown in Fig.10, when diffuse irradiation is varied from 0% to 100% at the each interval of 20% at θ=30°. Figure 10 shows that total power per year is small as rate of diffuse irradiation is low and the influence of front array's shadow is greater. K_p is about 100% at θ=30° and $d \geqq 1$, and the change is little at the range.

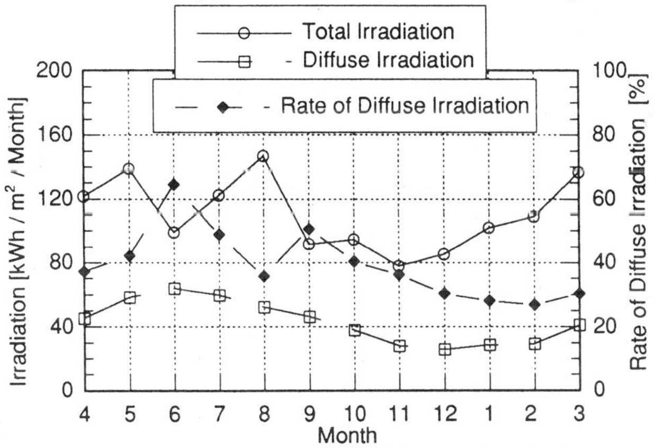

Fig.9 Irradiation and Rate of Diffuse Irradiation

CONCLUSIONS

This paper presents optimization of setting arrays considering the influence of front array's shadow. The following is confirmed.

(1) The relationship between θ and d that give the maximum total power per year on a constant array field is shown.

(2) To give the maximum total power per year in Tokyo, modules are laid closely on an array field, i.e. θ=0° and d=0.

(3) The relationship between θ and d for the maximum total power per year per array is shown.

At present, as the efficiency of an photovoltaic cell is low, it is important that the photovoltaic power generation system is operated efficiently. This method can confirm the effect of front array's shadow which has not been analyzed in detail, so that the proposed method is effective.

This method can apply everywhere in the world if irradiance and temperature data at an object place are given.

The total power per year and the total power per year per array are selected as the optimum conditions in this paper, it is necessary to consider the cost of modules, lands, and arrays construction as more optimum conditions to derive more useful result.

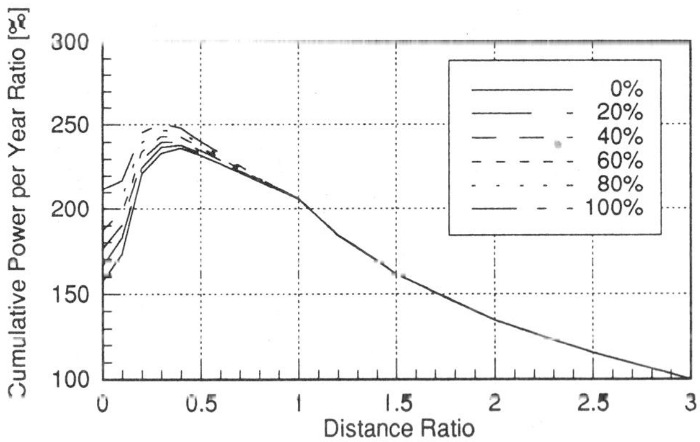

Fig.10 The Influence of Diffuse Irradiation

REFERENCES

[1] Y. Ishisaka et al. , " The Calculation Method of Coefficient to Revise Shadow Effects ", *National Convention Record I.E.E. Japan*, **12**, 1994, pp. 223

[2] Japan Solar Energy Society, *Solar Energy Utilization Handbook*, 1985, pp. 71

A MATRIX FOR INDEXING KNOWLEDGE ABOUT UTILITY PHOTOVOLTAIC PROJECTS

W.P. McNaughton
Cornice Engineering, Inc.

J.E. Bigger
Electric Power Research Institute

ABSTRACT

A means of indexing the state of knowledge in the use of photovoltaic (PV) technologies for utility applications has been developed. It was organized around the thesis that generic "situations" can be defined that will help utilities to obtain and interpret the results from past and current PV projects, and to plan for future projects. To date, both "major" applications and smaller (sometimes termed "cost-effective" or "early") applications have been defined and placed into the situation matrix format. The goal is to provide a tool to U.S. utilities and other organizations that will allow them to determine existing sources of information about specific PV applications and to track incoming results from the large number of envisioned new PV demonstration projects.

PROJECT OBJECTIVES

The objectives of the work reported here were:

1. To characterize potential applications for photovoltaics by the U.S. utility industry.

2. To develop a comprehensive list of questions that utilities must address in assessing the use of PV to meet their needs and those of their customers in any application.

3. To determine which questions have been answered for each application type, and organize that information in a manner such that information transfer programs could make those answers available to interested utilities.

4. To develop matrices of "situations" to show the available knowledge and, by default, illustrate the unanswered questions; such a tool to be made available for use in the planning, organizing and reporting of large-scale PV demonstration projects.

TECHNICAL APPROACH

Definition of Applications and Issues

Twenty major utility applications for PV and thirty-one smaller applications have been defined to date.

Subsequent to (and iterative with) the definition of applications, lists of critical questions about the use of PV by utilities were prepared. This was done in conjunction with key utility personnel, as well as technology and policy experts. Questions were combined into groups; each group then was tagged as an "issue". Table I shows the resulting list of issues that were defined.

Table I
List of Issues and Categories

BASIC UNDERSTANDING
- Basics of Application
- Costs
- Design (Prototype)
- Installation
- Operation and Monitoring
- Performance
- Safety

TECHNICAL PRECURSORS
- Design (Standardized)
- Education and Training
- Interconnection and Utility System Issues
- Inspection, Maintenance and Repair
- Reliability and Lifetime
- Site Choice and Constraints

INTERNAL AND EXTERNAL INFLUENCES
- Benefits and Value
- Codes and Standards
- Financing and Ownership
- Infrastructure
- Marketing and Implementation
- Regulatory and Legal
- Vendors and Procurement

CH3365-4/94/0000-1098 $4.00 © 1994 IEEE First WCPEC; Dec. 5–9, 1994; Hawaii

Developing the Situations Matrix

Following the definition of applications and of major issues, a key organizing table ("Situation Matrix") was then prepared for each of the twenty major applications. The issues which had been identified became the vertical axis in each situations matrix.

A review of past, current/on-going, and future (near-term) photovoltaic projects was completed. Applicable projects were classified into one (or occasionally two) appropriate situations. The issues that have been covered by that project were determined from available open literature, EPRI reports, and personal contacts with key utility and non-utility personnel. This enabled the applicable boxes in each Situation Matrix to be completed.

An example of the result is shown in Table 2. The key to abbreviations used in this Table (each designating a project) is shown in Table 3, which also provides the project name, location, installation date, and references where additional information about that project can be found. **Only references that can be generally obtained by U.S. utilities have been included and only issue coverage that could be confirmed from those references is included in the situations matrices.**

Judging When an Issue Was Covered

An issue was "covered" by a particular project if the issue had been specifically (and thoroughly) explored in available literature to such an extent that another utility can get answers to their questions about that issue without additional information source (such as the personnel who installed the original application).

The last line in each situation matrix (Table 2) was made for projects which fit into a particular situations matrix, but for which either a) we did not find coverage of particular issues in the literature (past projects) or b) it has not yet been determined (by the project team or by the utility) what issues will be covered in a manner that is available to other utilities (future or planned projects). In the case of past projects, issue coverage may be found by additional literature searching and the appropriate boxes in the situations matrix can be completed. For planned and future projects, as the results from those projects become widely available, those issue boxes can also be filled in.

For those projects where issue coverage could be determined there was a subjective evaluation of how well a "covered" issue was "covered". Plain text in the matrix indicates that the topic has been addressed, **bold text** indicates that this was a significant project effort and it is reflected by excellent coverage of the issue in the cited references. *Italic text* is used mostly for recently installed or planned projects to indicate where issues are likely to be covered by future references. No judgments about whether they will be covered in detail or not (that is, there is no italic, bold text), as much of that will depend upon how the project develops.

Note that every project reviewed certainly investigated more issues than are shown as "covered". Two obvious examples will illustrate the point. Very few projects are shown in the situations matrices as having covered the issues of **Safety** or of **Regulatory/Legal** when in fact many projects will have addressed these very important topics in-house. The determining factor used here was whether another utility would be provided with answers about Safety or Regulatory/Legal questions for a similar type project just by using the available open-literature information. If the answer was yes, then the issue is shown as "covered" by that project.

Finally, in judging "coverage", errors of omission were determined to be preferable to errors of commission in our compilation. That is, unless we could find specific answers **in the references we reviewed (**or in limited personal contacts), then we did not "credit" that project with having addressed that issue. The available literature is extensive and it was not within the limits of this project to review more than a fraction of it. It should be easier to expand the "coverage" provided by a particular project with the addition of other reference materials than to delete coverage that had been assumed.

Information Access by Issue

There are some very good references about issues that, because they were not tied to a particular project, are not easily found by reviewing the appropriate situations matrix. The compilation of some of these "issue-specific" references will eventually offer a second path for the interested reader to get to applicable information (either situation-specific projects or issue-specific references).

How Projects were Chosen for Inclusion

The potential data base of PV projects is large. As an example, a survey conducted for EPRI in 1988 [1] found 219 grid-connected PV systems in the United States. A 1993 EPRI survey of non-grid connected ("cost effective") applications [2] identified more than 1850 systems installed. The non-utility experience base for several applications is even larger.

Clearly, to include every installation of PV in this analysis would defeat many of the objectives outlined above. For one or more reasons, many of the potential candidates for inclusion were not used for the development of the situations matrices. A primary requirement was that projects had to have extensive involvement of a U.S. utility. These criteria eliminated from this compilation some good experience that should not be overlooked by utilities, such as that for the U.S. Department of Defense (more than 1,000 uses of PV installed and more than 21,000 potentially cost-effective applications identified [3], the U.S. Coast Guard (more than 10,000 PV-powered navigational aids installed) and the extensive experience base in off-grid, privately owned residential systems (for example, more than 1,200 solar-powered off-grid residences had been identified through 1987 by the Arizona Solar Energy Office [4]).

It was judged, however, that of most relevance were those projects specifically installed to meet the needs of a utility or by a utility to meet the needs of one of its customers. Clearly, the experience base of these other agencies and groups could be added to the existing format is desired.

The projects eventually included met one or more of the following criteria:

a).All known grid-connected PV projects since 1980 of size > 10 kW.

b).Other known large, utility-related systems since 1980

c).Other utility-related systems well documented in the literature (these may or may not be larger than 10 kW)

d).Systems (or studies) that provided some unique coverage of a key issue of concern to interested utilities.

Eliminated from consideration for the situations matrices were demonstrations that were primarily test sites (those which were set up specifically to test one or more technologies side-by-side), and the international experience base.

The process outlined above was iterative and highly interactive. Additions to and deletions from the lists of questions, issues, applications, and situations matrices were performed throughout the development and review effort. New and comprehensive projects continue to be put into place; it is expected that on-going refinement of the situations matrices will continue as the knowledge base increases.

CONCLUDING REMARKS

This work is intended to be a first cut at such questions as: For what issues is the needed information available to help make decisions (positive or negative) about PV use? What technical and non-technical questions have been addressed for each potential use of PV by a U.S. utility? What are the unanswered questions? If answers are known, where is more effort needed to provide technology transfer so that information gets into the hands of decision makers?

The resulting Situations Matrices allow individual utilities contemplating PV demonstrations or pre-commercial installations to determine what information is available, and what critical questions and issues need to be addressed to prepare for large-scale integration of these systems in their service area. Other types of organizations such as the Department of Energy, EPRI, and the Utility Photovoltaic Group, are finding early results from the Situations Matrices to be useful as a guide to planning significant technology demonstration and pre-commercial efforts, insuring that critical precursor questions and issues are addressed, and that information is made available as large-scale adoption of PV technologies occurs.

REFERENCES

[1] K. Smith, *Survey of U.S. Line-Connect Photovoltaic Systems,* Special Report GS-6306, Electric Power Research Institute, Palo Alto, CA, March, 1989.

[2] C.D. Eastwood, *Survey of Cost-Effective Photovoltaic Applications at US. Electric Utilities*, Final Report T-102648, Research Project 3258, Electric Power Research Institute, Palo Alto, CA, August, 1993.

[3] J. Ryan, and R. Sellers, "Overcoming Institutional Barriers", *Solar Today*, March/April, 1992, pp. 18-20.

[4] P. Soderberg, P., "Living Beyond the Top of the World", *Solar Today,* September/October, 1992, pp. 11-13.

ACKNOWLEDGMENTS

This work was funded by the Electric Power Research Institute under RP1667-03, which support is gratefully acknowledged. A number of organizations and individuals provided detailed input about specific PV demonstration projects. These included: Tim Ball (Solar Engineering), Steve Durand (Southwest Technology Development Institute), Mark Kapner (New York Power Authority), Thomas Lepley (Arizona Public Service Company), Mike McGoey (Solar Engineering), Gary McNaughton (La Plata Electric Association, Inc.), Donald Osborn (Sacramento Municipal Utility District), Brian Parry (Western Area Power Administration), John Prescott (Idaho Power Company), Chuck Rhodes (Public Service Company of Colorado), Larry Shirley (North Carolina Solar Center), and Howard Wenger (Pacific Energy Group).

This project was a collaborative effort that utilized some of the leading utility authorities in photovoltaic applications in the United States. Review continues to proceed at the time of writing of this paper. Contributions during early drafts were made by: Edgar DeMeo (EPRI), Steve Hester (Consultant), Peter Mark Jansson (Atlantic Electric), Gary McNaughton (La Plata Electric Cooperative), and Howard Wenger (Pacific Energy Group).

Table 2
Example Situations Matrix

APPLICATION: II - Commercial/industrial, PV penetration moderate compared to load

DISTINGUISHING CHARACTERISTICS: Grid-connected on customer side of the meter; sited at commercial customer site; no storage; PV is a moderate (10 to 30%) percentage of total building load.

ISSUE/ENVIRONMENT	Southwest	Northwest	Midwest	Southeast	Northeast
BASIC UNDERSTANDING					
Basics of Application	*EPA2-1*		*EPA2-11*		
Costs	LEA, 3M, *EPA2-1*		*EPA2-11*		PEP
Design (Prototype)			OMNI	BER	BEV
Installation	*EPA2-1*		*EPA2-11*		
Operation and Monitoring	LEA, *EPA2-1*		*EPA2-11*		PEP
Performance	**3M,** *EPA2-1*		OMNI, *EPA2-11*	BER	BEV
Safety					
TECHNICAL PRECURSORS					
Design (Standardized)	*EPA2-1*		*EPA2-11*		
Education and Training					
Inspection, Maint., and Repair	LEA, **3M,** *EPA2-1*		*EPA2-11*		PEP
Interconnection/System	*EPA2-1*		*EPA2-11*		
Reliability and Lifetime	3M, *EPA2-1*		OMNI, *EPA2-11*	BER	BEV
Site Choice and Constraints	3M				
INTERNAL/EXT. INFLUENCES					
Benefits and Value	*EPA2-1*		*EPA2-11*		
Codes and Standards					
Financing and Ownership					
Infrastructure					
Marketing and Implementation					
Regulatory and Legal					
Vendors and Procurement					
Projects for which issues have not yet been determined					

Key:
Abbreviations in boxes refer to past, on-going or planned projects; see Section 3 for summary descriptions.
Bold means **significant effort** to address this issue.
Italics means project is newly installed or in planning and this *issue is expected* to be covered.

Table 3
Sample Key for Situations Matrix

Abbrev.	Project Name, Location (Installation Date) (Size) **Reference(s)**
BER	Tudor Hill Undersea Laboratory, Bermuda (1983) (56 kW) **Durand, 1993** (BER, BEV, LEA, OMNI, PEP, 3M) **Southwest, 1983** (BER)
BEV	Beverly High School, Beverly, Massachusetts (1981) (99 kW) **Durand, 1993** **Southwest, 1985b** (BEV)
EPA 2-1	EPA Project 2, Number 1, Boyd Deep Canyon Research Center, Anza, California (1994) (18 kW)
EPA 2-11	EPA Project 2,, Number 11, Area Office and Dispatch Center, Lawton, Oklahoma (1994) (18 kW) **Ascension, 1993** (EPA 2-x projects)
LEA	Lovington Shopping Center, Lovington, New Mexico (1982) (104 kW) **Durand, 1993** **Lynette, 1989** (LEA, PEP)
OMNI	Oklahoma Center for Science and Arts (Omniplex), Oklahoma City, Oklahoma, (1982) (105/140 kW) **Durand, 1993** **Southwest, 1987b** (OMNI)
PEP	Georgetown University International Cultural Center, Washington, D.C. (1984) (300 kW) **Durand, 1993** **Lynette, 1989** (LEA, PEP)
3M	3M Austin Center, City of Austin, Texas (1990) (273 kW) **Durand, 1993** **Hoffner, 1991** (3M) **Ramakumar, 1993** (3M) **Rosenthal, 1992** (3M)

Other References: **Kiss, 1993; SDA, 1992; Young, 1992**

Ascension, 1993. Ascension Technology, Inc., Meeting Notes, EPA Project Review, Raleigh, North Carolina, October 27, 1993

Durand, 1993. Durand, S. and D. Bowling, *Field Experience with Photovoltaic Systems: Ten-Year Assessment*, Final Report TR-102138, Research Project 1607-06, Electric Power Research Institute, Palo Alto, CA, March, 1993.

Hoffner, 1991. Hoffner, J.E., "Performance of Austin's 300-kilowatt Photovoltaic Power Plants", in *1991 Solar World Congress*, Proceedings of the Biennial Congress of the International Solar Energy Society, held August 19-23, 1991 in Denver, CO., Pergammon Press, New York, 1991, pp. 119-124.

Kiss, 1993. Kiss Cathcart Anders Architects, P.C., *Building-Integrated Photovoltaics*, A Study Prepared for the National Renewable Energy Laboratory, January, 1993.

Lynette, 1989. Lynette, R. and K. Conover, *Photovoltaic Operation and Maintenance Evaluation*, Final Report GS-6625, Research Project 1607-5, Electric Power Research Institute, Palo Alto, CA, December, 1989.

Ramakumar, 1993. Ramakumar, R. and J. Bigger, "Photovoltaic Systems", *Proceedings of the IEEE,* Vol. 81, No. 3, March, 1993, pp. 365-377.

Rosenthal, 1992. Rosenthal, A.L., *Photovoltaic System Performance Assessment for 1990*, Final Report TR-101678, Research Project 1607-06, Electric Power Research Institute, Palo Alto, CA, November, 1992.

SDA, 1992. Solar Design Associates, *An Overview of Worldwide Development Activity in Building-Integrated Photovoltaics*, NREL Contract AG-2-12096-1, Final Report, National Renewable Energy Laboratory, Golden, CO, December, 1992.

Southwest, 1983. Southwest Technology Development Institute, *Bermuda Grid-Connected PV System: Tudor Hill Undersea Laboratory,* Results of Testing Performed October 17-20, 1983..

Southwest, 1985b. Southwest Technology Development Institute, *Beverly High School Photovoltaic System in Beverly, Massachusetts,* Results of Testing Performed April 25-27, 1985.

Southwest, 1987b. Southwest Technology Development Institute, *Photovoltaic Power System for the Oklahoma Center for Science & Arts (Omniplex),* Results of Testing Performed January 14 & 15, 1987.

Young, 1992. Young, W. and K. Collier, *Evaluation of Roof-Integrated PV Module Designs and Systems*, Florida Solar Energy Center Project FSEC-CR-496-92, submitted to National Renewable Energy Laboratory, Golden, CO, July 15, 1992.

SILICON AND GAAS/GE CONCENTRATOR POWER PLANTS: A COMPARISON OF COST OF ENERGY PRODUCED

R. A. Whisnant, J. A. Hutchby, M. L. Timmons, R. Venkatasubramanian, J. S. Hills
Research Triangle Institute, Research Triangle Park, NC 27709

ABSTRACT

Until now virtually all of the development of terrestrial concentrator PV power plants has utilized silicon cell technology. If silicon concentrator technology achieves its predicted potential performance and cost [1], it would be very close to commercial viability as defined by DOE goals of 12 ¢/kWh. Also until now, GaAs cell technology has not been a viable alternative for these power plants, largely due to the cost of cells grown on GaAs substrates. The ability to grow high-efficiency GaAs solar cells on germanium substrates [2,3] presents the strong possibility of reducing their cost of energy produced. The cost of large-grain, optical-grade, polycrystalline germanium substrates is potentially 15 times less than single-crystal GaAs substrates [4], and cell conversion efficiencies approaching 35% are likely for tandem-junction GaAs cells. This analysis shows that for comparable 50 MW Fresnel lens plants, the 30-year, levelized cost of energy (COE) from GaAs/Ge cells can be $0.118/kWh compared to $0.140/kWh for a 27.4% Si back-contact cell.

INTRODUCTION

The question to be examined here is whether the lower cost Ge substrates available for GaAs cells, combined with expected higher GaAs cell efficiencies translate into a lower cost of energy produced by a utility-scale PV power plant when compared to the best Si cells [5]. The analysis estimates the cost of both single and tandem-junction GaAs cells and modules, as well as the plant BOS costs, and translates those costs into an estimate of total capital requirement for a PV generating plant. The 30-year, levelized cost of energy for plants incorporating each cell considered is used as a measure of comparison

REFERENCE PV SYSTEM DESIGN

A previous study [1,6] for the Electric Power Research Institute (EPRI) provided a detailed design of a 50 MW Fresnel lens power plant using advanced, back-contact, Si cells. The cost to manufacture the cells and modules at a 100 MW/year rate was modeled by the Research Triangle Institute using its STAMPP model. The PV generating plant was defined and its total cost, including the PV modules, was estimated by the prime contractor for the study, Bechtel Group. Annual power production at Carissa Plains, CA was modeled and the 30-year, levelized cost of energy produced was determined. Some basic plant design details for the reference study are provided in Table 1.

Table 1. Reference Si-Cell Plant Design Summary [6]

Cell	Advanced Back-Contact
Concentration Ratio	500:1
Efficiencies	
Cell (@ 25°C)	27.4%
Module (@ oper. temp.)	20.7%
Operating Temp.	79°C
Dimensions	
Cell (active)	0.315 in × 0.315 in
Module (overall)	30 in × 90 in
Array	37 ft × 32 ft
5 MW Subfields	1673 ft × 1073 ft
Plant Land Area	700 acres
Quantities	
Cells per Module	48
Modules per Array	60
Arrays per Subfield	339
Subfields per Plant	10
Design Power Point	50 MW (AC)
Annual Energy Output	140.1 GWh (AC)

GaAs CELL AND MODULE COST ANALYSIS

Approach

The approach in this present analysis was to retain the reference power plant design, module design, and BOS per-unit costs, and to substitute several configurations of GaAs cells in order to determine their impact on the cost of energy produced. Since GaAs cell efficiency was one the major parameters to be examined, the number of PV modules, arrays, and total BOS requirements were scaled to keep the total plant capacity at 50 MW in every case. The number of modules required for 50 MW capacity was determined for each efficiency considered and used to define annual module production levels. GaAs cell manufacturing processes were substituted in the STAMPP model for the Si cell

CH3365-4/94/0000-1103 $4.00 © 1994 IEEE

First WCPEC; Dec. 5-9, 1994; Hawaii

process used in the EPRI study. New module costs for a Fresnel module incorporating GaAs cells were then computed and inserted into the same calculation for Total Capital Requirement used in the prior study (with the exception that an inflation factor was applied to put all results in 1994 dollars).

GaAs Cell Definition

Both the single- and the tandem-junction GaAs cells considered in this analysis are defined by Table 2. Manufacturing processes for a total of six cells were modeled: each of the two cells in Table 2 on three substrates—single-crystal GaAs, single-crystal Ge, and polycrystalline Ge. All wafers were taken as 4 inches in diameter. Cell size was taken to be the same as the silicon cells in the reference study, 1 cm × 1 cm overall with an 0.8 cm × 0.8 cm active area. These dimensions result in 59 cells per wafer. Cells are fabricated by OMCVD at a deposition rate of 0.075 μm /min for all layers. The polycrystalline Ge substrates used in this analysis are large-grain (≅1 cm), optical-grade, 4" diameter wafers. The analysis assumes that similar efficiencies can be obtained on both single- and poly-crystalline Ge substrates.

Table 2. GaAs Cell Structures and Growth Times

Layer	Material	Thickness (μm)	Time (min) @ 0.075 μm/min
Cap	GaAs	0.2	2.7
Window	AlGaAs	0.03	0.4
Emitter*	GaInP	0.3	4.0
Base*	GaInP	0.6	8.0
BSF*	GaInP	0.1	1.33
Tunnel Junction*	GaAs	0.015	0.2
Tunnel Junction*	GaAs	0.015	0.2
Emitter	GaAs	0.5	6.67
Base	GaAs	3.75	50.0
BSF	AlGaAs	0.1	1.33
Buffer: GaAs Wafer	GaAs	0.75	10.0
Buffer: Ge Wafer	GaAs	5.5	73.3
Single-Junction			
Total—GaAs Wafer			71.1
Total—Ge Wafer			134.4
Tandem-Junction			
Total—GaAs Wafer			84.83
Total—Ge Wafer			148.33

*Tandem-Junction Cell Only

Since the GaAs cells have front side metallization and the Si cells do not, it is assumed that the refractory secondary optical element (incorporated in both Si and GaAs systems) has a prismatic bottom surface to reduce interference by the metal gridlines. The cost associated with this assumption is relatively very small.

Modeling

The cost to produce Fresnel modules with GaAs cells was modeled with RTI's STAMPP software code. Process steps were defined for manufacture of modules starting with wafers and ending with a finished module packed for shipping. A work station to accomplish each step was defined including cost and life of capital equipment and facilities, materials consumption, labor requirements, and performance measures such as throughput rate, yield, and availability. A required, or breakeven, price for modules produced at a 100 MW/year level was computed for a business whose costs included the direct manufacturing cost, indirect costs of management and supervision, taxes, debt, and a fixed rate of return on equity capital. The cell fabrication steps were specific to the GaAs cells; the subsequent module fabrication steps were the same as those incorporated in the reference analysis for Si modules.

An overall plant efficiency was determined that took into account thermal derating of the cells, optical losses, shadowing, and mismatch and electrical losses. Annual energy production at Carissa Plains was simulated in the reference study [6].

Cell/Module Cost Results

Table 3 gives calculated required prices for modules and their associated cells as a function of efficiency. The

Table 3. GaAs Cell and Module Price

Cell and Wafer	Cell Efficiency	$/Module	$/Cell	$/48 Cells
SJ/GA-SC	25	741.14	9.58	459.94
	30	741.54	9.59	460.50
SJ/Ge-SC	25	447.01	3.26	156.52
	30	450.40	3.31	159.11
SJ/Ge-PC	25	404.19	2.32	111.28
	30	407.58	2.37	113.79
TJ/GA-SC	30	745.96	9.68	464.68
	35	748.56	9.73	466.99
TJ/Ge-SC	30	456.33	3.43	164.51
	35	456.31	3.44	164.9
TJ/Ge-PC	30	413.50	2.48	119.11
	35	413.49	2.49	119.51

efficiency was used to determine the annual production volume corresponding to 100 MW/year. At these high volumes, volume (or efficiency) has almost no effect on price. The substrate is the major determinant of module price. In comparison, the price of the Si-cell module was $328.42, and the cell price was $0.98/each.

Comparisons to established semiconductor products for the calculated prices for cells or modules are not easily obtained. A recent article [7] suggests that DRAM chip costs are currently at $3.90/cm². The cells in Table 3 are 1 cm² and costs using Ge wafers range from about $2.50 to $3.40 per cell.

The module prices in Table 3 are primarily driven by materials costs, which vary between 65% and 75% of the total price. For example, the contribution of various price components for the tandem-junction cell on poly-Ge are:

- Direct Materials 69.5%
- Direct Labor 4.7%
- Manufacturing Overhead 15.7%
- Administrative Cost 1.1%
- Taxes 3.0%
- Return on Equity 5.9%

Key Input Materials Costs

Although more than 50 materials were accounted for in the analysis, 4 of those are the basic drivers in the materials cost, as shown in Table 4. The module box is a reinforced polystyrene injection-molding that is the basic structural member of the module. The heat sink is the larger of two aluminum stampings that form a single-cell heat sink. The Fresnel lens is a four-lens parquet. GaAs wafer cost is more than an order of magnitude higher than the cost of Ge polycrystalline wafers, and GaAs wafers account for almost 60% of the total materials cost. For the poly-Ge and Si substrate modules, the wafer costs are a smaller percentage of the total than either of the other 3 materials shown.

Table 4. Module Materials Cost Drivers (GaAs Tandem-Junction Cells)

	Substrate (GaAs & Ge = 4", Si = 6")			
	GaAs	Single-xtal Ge	Poly-xtal Ge	Si
Wafer Cost/ea.	$360.46	$64.37	$23.62	$29.94
Material	Percentage of Total Materials Cost			
Wafer	59.8	20.4	8.6	5.3
Module Box ($3.59/ea.)	6.9	13.2	15.1	17.4
Heat Sink ($0.52/ea.)	4.0	7.7	8.8	10.1
Fresnel Lens ($4.07/ft²)	12.2	23.3	26.8	14.3

Plant Total Capital Requirement

Table 5 illustrates the cost elements included in the Total Capital Requirement calculation for each plant. Costs for the reference system using Si cells and for the tandem-junction GaAs/poly-Ge system are given. The capital requirements for PV modules and area-related BOS costs reflect the smaller number of GaAs modules required, although the module cost of $413.49 is higher than the $328.42 for Si modules. Other costs in Table 5 are consistent with the EPRI Technical Assessment Guide methodology as defined in the reference study [6].

Table 5. Plant Total Capital Requirements (1994$)

	Si*	GaAs**
	$/W	$/W
Building and Site Improvements	0.030	0.024
Array Structure	0.853	0.622
Photovoltaic Modules	1.806	1.582
Power Conditioning Unit	0.133	0.133
Balance of System	0.070	0.057
Master Control System	0.003	0.003
Total Field Cost	2.894	2.421
Engineering and Const. Mgmt.	0.116	0.097
Owner's Costs	0.120	0.101
Contingencies	0.314	0.262
Total Plant Cost	3.444	2.881
Escalation (Mixed Yr. Dollars)	-0.066	-0.055
Total Cash Expended (Mixed Yr. $)	3.379	2.826
AFDC (Mixed Year Dollars)	0.151	0.126
Total Plant Investment	3.530	2.953
Preproduction Costs	0.071	0.059
Inventory Capital	0.017	0.014
Land	0.004	0.003
Total Capital Requirement	3.622	3.029

*Silicon Back-Contact (27.4%)
**Tandem Junction GaAs/poly-Ge (35%)

COST OF ENERGY PRODUCED

The 30-year, levelized cost of energy (COE) produced by a 50 MW plant at Carissa Plains, CA is shown in Figure 1 as a function of cell efficiency for 6 different GaAs cell/substrate choices, plus the Si back-contact cell. For either single-junction or tandem-junction cells, the effect of using Ge substrates instead of GaAs substrates is greater than the effect of cell efficiency over the efficiency range examined. Polycrystalline Ge wafers, while costing only about one-third as much as single-crystal wafers, result in a relatively small additional COE reduction. The implicit assumption in the COE calculations is that the same cell efficiencies can be achieved with either single-crystal or polycrystalline wafers. That may not be borne out in practice, and, given the small COE advantage of the polycrystalline material, the single crystal material might be the better choice.

For a given cell efficiency and substrate, the tandem-junction GaAs cells have a very slightly higher COE than the single-junction cells. For example, at 30% efficiency the SJ-Ge/SC configuration has a COE of $0.142/kWh, while the cost for the TJ-Ge/SC configuration is $0.143/kWh. This small difference results from the fact that the cell growth time for the tandem cell is only about 10% more than for the single-junction, owing largely to the long growth time for the buffer layer that is used in both cells (see Table 2). However, it would be expected that single-junction efficiencies may be limited to below 30%, whereas the tandem-junction may achieve 35%. The higher tandem-junction efficiency results in a COE of $0.124/kWh for Ge/SC substrate. A 35% tandem-junction cell on Ge/PC substrate results in a further small reduction in COE to $0.118/kWh.

REFERENCES

1. W. Stolte, et al., Design, Performance and Cost of Energy from High Concentration and Flat-Plate Utility-Scale PV Systems, *Proceedings of 23rd IEEE Photovoltaic Specialists Conference - 1993,* pp. 1292-97.
2. R. Venkatasubramanian, et al., 15.8%-Efficient (1-Sun, AM 1.5G) GaAs Solar Cell on Optical-Grade Polycrystalline Ge Substrate, *Proceedings of the 23rd IEEE Photovoltaic Specialists Conference - 1993,* pp. 691-93.
3. C.H. Cheng, et al., Improved MOCVD Technology in High Throughput High Efficiency GaAs/Ge Solar Cell Manufacture. *Proceedings of the 22nd IEEE Photovoltaic Specialists Conference - 1992,* pp. 393-95.
4. Personal communication from Eagle-Picher Industries.
5. R.A. Sinton, et al, (1990), "Developments in Module Ready Si Backside-Contact Solar Cells", *Proceedings of the 21st IEEE Photovoltaic Specialists Conference - 1990,* Volume I, page 302.
6. Engineering and Economic Evaluation of Central Station PV Power Plants, EPRI TR-101255, December, 1992, Electric Power Research Institute, Palo Alto, CA.
7. J. Robertson, SIA Technology Roadmap Hits the.Streets, *Electronic News*, March 1, 1993, v39, n1952.

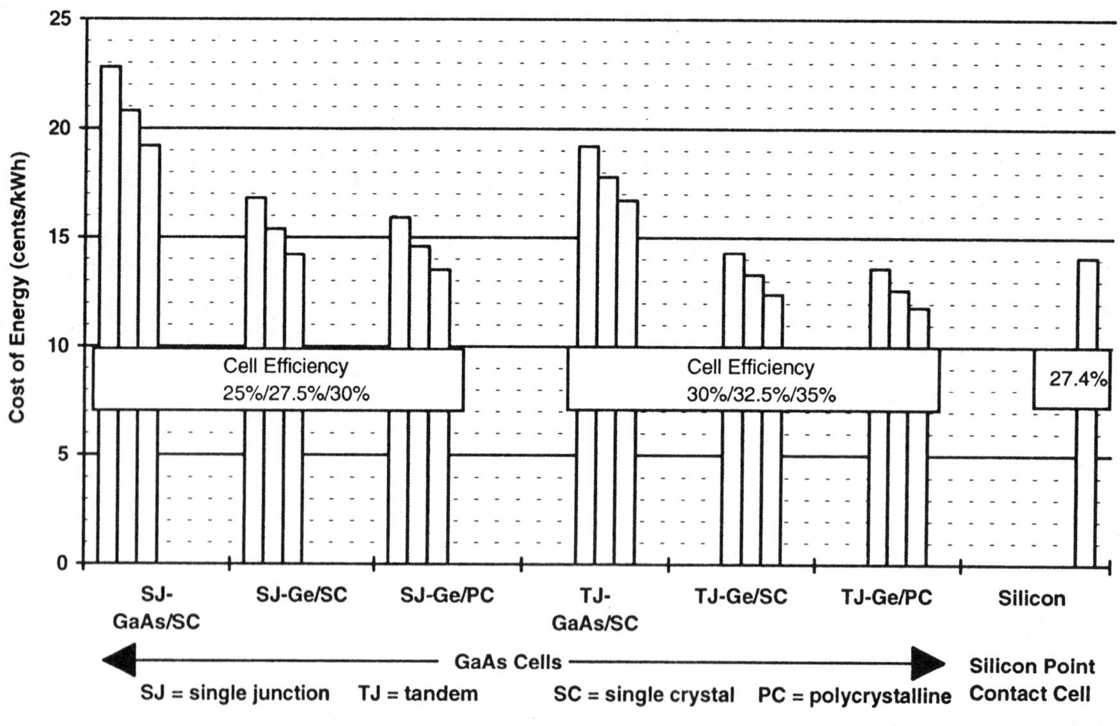

Figure 1. Cost of Energy Produced by 50 MW Fresnel Lens Plant

Design and Cost Optimization of Utility-Grade Off-Grid Power Systems

J. Berdner and C. Whitaker
Endecon Engineering
San Ramon, CA

H. Wenger
Pacific Energy Group
Walnut Creek, CA

C. Jennings
Pacific Gas and Electric Co.
San Ramon, CA

ABSTRACT

Utilities are beginning to provide off-grid electric service to customers in lieu of extending uneconomic electric distribution lines. To a large extent, the success of a utility off-grid service program is dependent on the utility's ability to provide power systems with excellent reliability and power quality at competitive prices.

In December 1992, Pacific Gas and Electric Company (PG&E) tested the provision of off-grid service by procuring a first-generation system which was installed at a customer's residence in April 1993. This pilot was a technical success, demonstrating that a utility can provide reliable, high-quality off-grid power. The project did, however, highlight the need for large cost reductions necessary for significant market penetration. This paper presents a second generation design which attempts to address these issues.

INTRODUCTION

PG&E began to seriously consider the feasibility of an off-grid service option in 1991, and in 1992 focus groups were held to determine customer requirements and expectations. A prototype Off-Grid Power System (OGPS) was procured in late 1992. The OGPS was instrumented and tested at PG&E's R&D facility in San Ramon, CA prior to installation at a customer's residence in Santa Cruz, CA. The system has operated continuously for the last 19 months with virtually 100% availability and high customer satisfaction. While system performance has been very good, the initial cost was too high for widespread use in remote residential applications.

PROTOTYPE SYSTEM DESIGN

The major elements of the OGPS, described in Table 1, are a PV array, an engine generator, a battery bank, a DC to AC inverter, and an enclosure. The prototype system was designed to provide 5 kWh per day to a traditional 240 Vac split-phase service. The PV array was designed to provide approximately 60% of the annual energy requirement (solar fraction) with a targeted maximum generator run time of 500 hours.

The OGPS was originally supplied with a 2.6 kW modified square wave inverter. In September 1993, the inverter was replaced with a Beta test version of a Trace Engineering 4.0 kW sine wave. The power quality of the new sine wave inverter is superior to that of the modified square wave unit. Trace now offers a commercially available, ETL-Listed, sine wave inverter which meets IEEE 519 power quality guidelines for voltage distortion. In addition to the improved power quality and higher power rating, the new inverter incorporates many of the control functions that had been included in a custom controller.

Table 1 - System Configuration

Component	Type	Size
PV array	Solarex MSX-60	1,080 W, 24 Vdc
Battery	GNB 3-75A27	1,015 Ahr, 24 Vdc
Generator	Onan 5 CCK	5.0 kW
Inverter (1)	Trace U2624 SB	2.6 kW
Inverter (2)	Trace SW 4024	4 kW

The valve regulated lead acid (VRLA) battery bank was sized to provide 12 kWh of storage with a maximum 50% depth of discharge. The VRLA design, although higher cost, eliminated the requirement for battery maintenance.

One of the original requirements for the OGPS was to create a self contained unit with the solar array mounted to the OGPS enclosure. Additionally, as a proof-of-concept prototype, it was determined that all system components should be easily accessible for maintenance and data collection. As a result, the first generation OGPS was not a lowest cost design but rather a design to establish the operational characteristics of hybrid power systems. Future units will be designed as least cost solutions meeting the power quality and reliability requirements established by the prototype system.

PROTOTYPE SYSTEM COST

An RFP was issued and the prototype OGPS was purchased for $40,000 from Applied Power Corporation (formerly Solar Engineering) of Lacey, WA.

During the preliminary design phase for the prototype unit, maintenance cost was identified as a significant cost driver. To avoid battery maintenance, a valve regulated lead acid (VRLA) battery was selected. Because the batteries used in the prototype system were maintenance free, all of the normal operation and maintenance (O&M) costs were associated with the engine generator. In order

CH3365-4/94/0000-1107 $4.00 © 1994 IEEE

First WCPEC; Dec. 5–9, 1994; Hawaii

to extend the generator's service interval, an automatic oil makeup system was added. Originally it was hoped that scheduled maintenance could be limited to a single visit per year. A single site visit per year was assumed to be the minimum required for general system maintenance independent of generator run time. At the original design load of 5 kWhac per day with a 60% solar fraction and 40% generator fraction this would result in a generator service interval of 500 hours.

The actual customer load has been less than the original 5 kWh per day design load. The net effect of the reduced load is a higher solar fraction and lower generator run time. The total generator run time to date, including testing prior to installation, is approximately 240 hours. After approximately 200 hours of run time (18 months of service), the generator operation degraded to the point that the inverter could no longer synchronize to generator output for battery charging. After routine servicing, i.e. tune up, the inverter was able to synchronize to the generator and battery charging was once again possible.

This experience suggests that a 500-hour service interval may not be practical and a minimum of two scheduled maintenance visits per year may be required to reach the 500 hour per year target. At 500 hours of run time per year the levelized cost for generator O&M - based on quotations from a local generator service company, and a fuel cost of $1.25 per hour - is approximately $1,000 per year. Using the economic assumptions outlined in Table 2 results in a present value of $6,900 for O&M of the generator used in the prototype system.

DESIGN AND COST OPTIMIZATION

All design changes were examined using present value analysis to determine the overall impact on energy cost. The system sizing and cost analysis are based on a 5 kWhac per day in the Santa Cruz, CA area (i.e. coastal San Francisco Bay Area weather patterns). Results may be significantly different for other load levels or climatic regions. The economic assumptions used in the present value analysis are presented in Table 2 below:

Table 2 - Economic Assumptions

Item	Value
Term	10 Years
Discount Rate	10 %
Inflation Rate	3% per year
Fuel Cost	$1.25 per hour

The design of the second generation OGPSII was largely based on the hardware used successfully in the prototype system. Costs for the OGPSII design presented here are based on quotations from vendors for a 10 system procurement. Each of the cost categories shown in Table 3 were examined independently for potential reductions. The capital cost was reduced by $17,210 from $40,000 to $22,790. Approximately 83% of the cost savings can be achieved through changes in the first five items.

Table 3 - Cost Savings

Description	Prototype Cost	OGPS II Quotes	Savings % of Total
Enclosure	7,105	2,500	26.7%
Storage Battery	6,972	4,000	17.4%
Design, Engineering	3,000	300	15.7%
Generator	5,580	3,200	13.8%
Metering, DAS	1,554	0	9.0%
PV Array	7,413	6,500	5.3%
Assembly, Testing	3,000	2,000	5.8%
Controls	2,121	1,395	4.2%
Misc. Wiring etc.	860	500	2.1%
Inverter	2,395	2,395	0.0%
Other misc. hardware	0	0	0.0%
TOTAL	40,000	22,790	100.0%

The largest cost savings were obtained by down-sizing the enclosure. The dimensions of the enclosure have been reduced from 8'x8'x12' to 4'x4'x8'. As previously noted, the prototype enclosure was designed to be an integral part of the array support structure, for array shipping, and for easy access to all system components for data collection purposes. The second generation design includes a smaller enclosure which is not an integral part of the array support structure and can not be used for shipping the array. With a smaller enclosure, the array must be mounted independently rather than directly on the enclosure. The fixed-angle pole-top structure chosen to support the array reduces the amount of site preparation and allows the array to be mounted above ground level for safety purposes. Wooden utility poles were selected because electric utilities posses' the necessary installation experience and equipment and have inexpensive sources for both new and surplus poles. An additional advantage of pole-mounted structures is that they simplify adjustment of array size to match load or local weather patterns with no other changes in system configuration.

The VRLA battery used in the prototype system is a high cost design but does not require maintenance, or periodic equalization. Although these are significant advantages, it should be possible to achieve an extended watering interval with lower cost flooded batteries and recombining vent caps. Battery servicing could then be incorporated with generator servicing at minimal additional cost. The battery selected for the second generation design is a tubular positive plate flooded cell with recombining vent caps. Makeup water required by the batteries would be added during the normal generator maintenance visits. Manufacturer's literature suggests the cycle life of both the VRLA and flooded batteries should be equal, i.e. 1200 cycles to an 80% depth of discharge. In the prototype

OGPS design, the cycle depth is limited to a maximum of 50% and the projected battery service life is ten years.

Design and engineering is considered a non-recurring expense (NRE). Per-unit costs can be reduced through design standardization and multi-unit procurements. The NRE for the prototype unit has been spread across 10 units for OGPSII, reducing the cost to approximately $300 per unit.

Generator costs and the associated O&M expenses are a significant driver. The trade off between generator run time and array size, or solar fraction (SF), is the basis for the present value analysis. A reference system configuration was established using a commercially available PV sizing program. The reference PV array (SF=100%) used in the prototype OGPS is rated at 2,160 Watts at Standard Test Conditions.

Generator run time was estimated by scaling the PV array size in order to examine total system cost versus solar fraction Monthly generator run time and the present value of the O&M costs are calculated for each configuration and used to develop monthly costs. The monthly service charge is how much the utility would have to charge to recover capital and maintenance costs over the specified period of time.

Figure 1 shows the monthly costs and generator run time for three systems: using a commercial-grade generator, an RV-grade generator, and for a system with lower cost modules and a 20-year recovery period. In addition to the standard PV/generator hybrid systems, two extreme cases are shown for each set of assumptions: a 110% PV-only system and a no continuous operation (24 hr/day) generator with no PV or batteries. As can be seen, the lowest overall hybrid system cost (service charge plus fuel) occurs at a 70% solar fraction. Depending on assumptions, this system results in a monthly service charge of between $256 and $416 plus a $7 fuel cost. The analysis also shows that for a 5 kWh per day load at this location, the 110% PV system with no generator provides the lowest costs. Halving module price (from $5.25/Watt to $2.63/Watt) and doubling the recovery period results in a 39% reduction in monthly fee.

For a remote home site, however, the load is variable and the generator provides an alternate source of power during unplanned high load periods. For a utility-owned and maintained system, the availability of the power system is of critical importance. Given the variable nature of the load and insolation patterns, a generator is probably preferred.

These estimates are for a utility-financed OGPS using the economic assumptions shown in Table 2. The generator O&M assumption used in the analysis were based on quotations from a local generator service company and actual costs incurred during the first 19 months of prototype system operation. O&M assumptions for the Commercial and RV grade generators are presented in Table 4 and 5.

The prototype OGPS system was designed as a test bed and required significant additional metering and Data Acquisition System (DAS) transducers. Costs associated

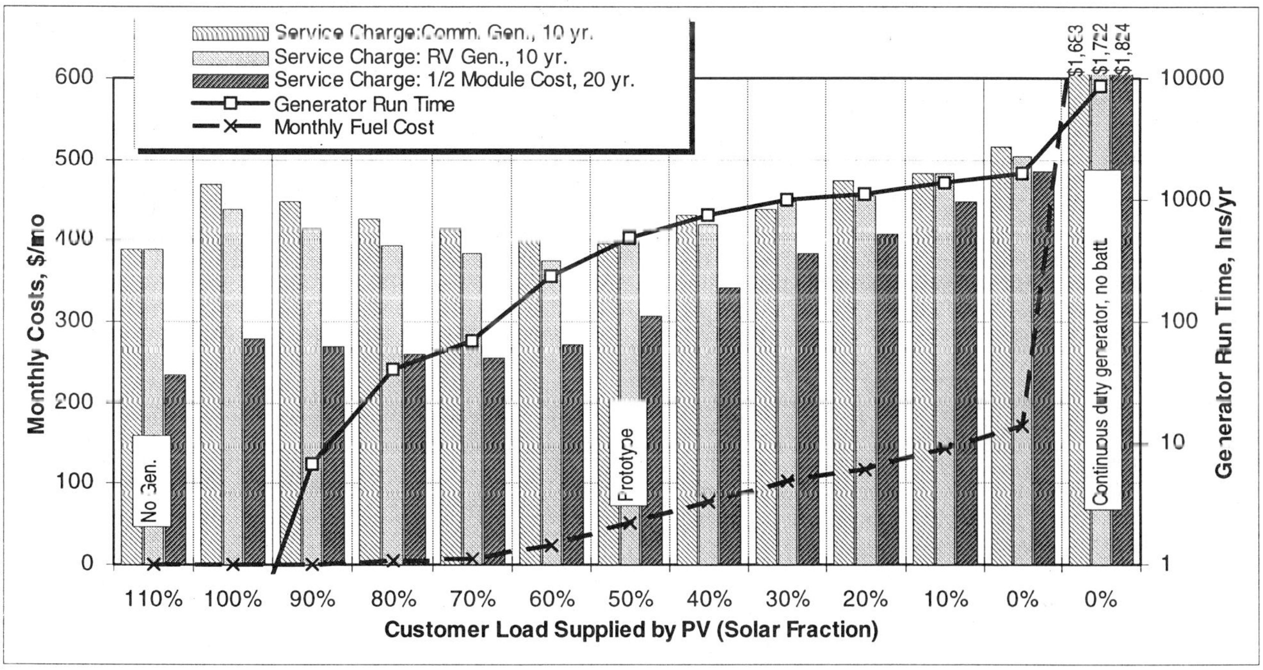

Figure 1 - Monthly Cost Comparison

Table 4 - Commercial Generator O&M Assumptions

Service	Interval (hrs)	Cost ($)
Basic tune-up	250	125
Intermediate w/ valve adjust	1000	181
Rebuild	2500	1118
Replacement	5000	5280

Table 5 - RV Generator O&M Assumptions

Service	Interval (hrs)	Cost ($)
Basic tune-up	250	125
Intermediate w/ valve adjust	750	181
Rebuild	1500	1118
Replacement	3000	3130

with metering and DAS were eliminated in the second generation design.

The remaining 16% of cost savings resulted from higher volume procurement, improved fabrication techniques, and continued reductions in PV module prices.

Customer Satisfaction and Market Penetration

The customer load is probably the key OGPS design criteria, since it drives system cost and, ultimately, customer satisfaction and market penetration. Ideally, the OGPS design is matched to customer load. If customer load exceeds the design load, then excessive generator run time occurs. Excessive generator run time will reduce customer satisfaction because of generator noise and increased costs from fuel usage and a penalty the utility may levy because of increased generator maintenance requirements. If customer load is significantly less than the design load, then the customer is essentially paying for too much service (since the customer is charged a fixed monthly fee) and the utility may be artificially constraining market penetration.

From 1989 to 1992, PG&E conducted a number of surveys, studies and focus groups to investigate the feasibility of the Service Without the Wire concept [1] Two key issues to emerge from this work concern the customer's "willingness to pay" and the ultimate ownership of the power system. The maximum acceptable monthly service charge ranged from $100 to $250 per month to achieve reasonably high penetration of the residential market. Residential customers also had a strong preference for a lease to own type program. Commercial customers were more willing to accept higher monthly service charges and did not show the strong preference for a lease to own program.

The results presented thus far are based on an OGPS design load of 5 kWh/day, which requires a utility charge of approximately $300 to $400 per month, depending on the duration of the service contract. On the other hand, an OGPS with a design load of 3 kWh/day and no back-up generator would cost between $200 and $300 per month. The challenge is to balance customer willingness to pay against service expectations. A utility that has a selection of perhaps a handful of standard OGPS designs with a range of costs and capabilities may be best able to serve the market .

CONCLUSIONS AND RECOMMENDATIONS

The prototype Off Grid Power System has demonstrated that stand-alone systems can achieve the power quality and reliability goals required by utilities. The design changes proposed for the second generation OGPS can significantly reduce overall system cost. and still meet the power quality and reliability established by the prototype system. Field testing of the second generation design should be conducted in order to validate the costs of the proposed design changes. A key to achieving these cost reductions is larger procurements in order to take advantage of quantity discounts and minimize one-time engineering charges.

Operation and maintenance costs are a key design parameter. The principle item requiring periodic maintenance is the engine generator. Maintenance costs are minimized by reducing generator run time. The prototype design specified a 60% solar fraction and a generator service interval of 500 hours. The experience gained with the prototype system indicates that a 500 hour service interval may not be practical for generators of this size.

The present value analysis shows that the lowest overall system costs occur with a 70% solar fraction. The minima is fairly broad, however, and by increasing the solar fraction to 70% or 80%, generator run time can be dramatically reduced with minimal impact on service charge. A higher solar fraction reduces generator run time and permits use of lower cost generators which can further reduce overall system cost.

To reconcile residential customer service expectations and willingness-to-pay with monthly service charges, system cost must be dramatically reduced for utility ownership to be viable. Commercial customer's willingness-to-pay is higher and may represent a viable market for the second generation OGPS. The number of commercial customers is low, however, and therefore utilities should work together to standardize designs for bulk purchases in order to reach the volume necessary for further cost reductions.

[1] Wenger, H., J. Berdner, and C. Jennings, "Service Without the Wire: Developing a new service concept at PG&E", Solar '94 Conference, San Jose, CA, June 1994.

INTEGRATING PHOTOVOLTAICS INTO UTILITY DISTRIBUTION SYSTEMS

H.W. Zaininger
Zaininger Engineering Company, Inc.
San Jose, CA

P.R. Barnes
Oak Ridge National Laboratory
Oak Ridge, TN

ABSTRACT

Electric utility distribution system impacts associated with the integration of distributed photovoltaic (PV) energy sources vary from site to site and utility to utility. The objective of this paper is to examine several utility- and site-specific conditions which may affect economic viability of distributed PV applications to utility systems. Assessment methodology compatible with technical and economic assessment techniques employed by utility engineers and planners is employed to determine PV benefits for seven different utility systems. The seven case studies are performed using utility system characteristics and assumptions obtained from appropriate utility personnel. The resulting site-specific distributed PV benefits increase nonsite-specific generation system benefits available to central station PV plants as much as 46%, for one utility located in the Southwest.

INTRODUCTION

PV is an environmentally beneficial electricity producing renewable energy technology which can be considered as an electric power generation resource when planning electric utility systems. If PV is employed in large central generation stations like conventional generation, nonsite-specific utility generation system economic benefits, such as energy and capacity displacement will be obtained. These nonsite-specific utility generation system economic benefits can be very large. However, delivering this central station PV power to electric utility customers requires increased transmission system costs to deliver the power to the utility load centers, and increased distribution system costs to deliver the power to the end user within the load centers.

Small, distributed PV installed in electric utility distribution systems can obtain local site-specific economic benefits which are not available to large central station PV, such as the deferral of new transmission and distribution (T&D) line and transformer additions, loss reduction, improved reliability and power quality. Electric utility distribution system impacts and resulting distributed economic benefits associated with the integration of distributed PV energy sources vary from site to site and utility to utility based on the local solar insolation characteristics, PV penetration level, whether battery or other energy storage systems are applied, distribution system characteristics, design standards, voltage levels, load density, reliability, and power quality.

Small, distributed PV energy sources installed on utility distribution systems at load centers will also produce nonsite-specific utility generation system economic benefits like large central station PV, such as energy and capacity displacement benefits, in addition to the local site-specific distribution system benefits. Although generation system benefits are not site-specific, they are utility-specific, and will vary significantly among utilities in different regions of the United States. In addition, transmission system benefits, environmental benefits and other benefits may apply.

This paper summarizes the results of a research project [1] [2], in which case studies for the following seven utilities are evaluated to examine a range of expected utility- and site-specific conditions throughout the United States that may effect the economic viability of distributed PV energy sources.

UTILITY	STATE
Southern California Edison	California
Public Service Co. of New Mexico	New Mexico
Green Mountain Power Co.	Vermont
Georgia Power Co.	Georgia
Florida Power & Light Co.	Florida
Lenoir City Utilities Board	Tennessee
Orcas Power & Light Co.	Washington

ASSESSMENT METHODOLOGY

One major objective of this research project was to develop an assessment methodology and produce results that are credible and acceptable to electric utility distribution engineers and system planners. Thus, the benefits are determined from a utility perspective, and the assessment methodology is compatible with the technical and economic assessment techniques employed by utility engineers and planners. The assessment methodology uses a bottom-up approach and provides flexibility for considering PV or other renewable resources at different levels of aggregation on a utility system. The resulting site-specific distributed utility benefits and nonsite-specific generation utility benefits are then combined, and the total benefits compared with the capital and operating costs of the PV system.

CH3365-4/94/0000-1111 $4.00 © 1994 IEEE

First WCPEC; Dec. 5–9, 1994; Hawaii

The primary economic assessment approach [3] [4] [5] used in this research project consists of performing a benefit-cost assessment using present worth of revenue requirements (PWRR) engineering economic analysis, providing for suitable economic and financial parameters corresponding to different investor-owned and public utility perspectives and assumptions. The PWRR approach, which is compatible with the system planning techniques generally employed by both private and public utilities worldwide, consists of calculating the relative annual revenue required to support the alternative utility system expansion plans with and without PV throughout the study period. The benefit-cost calculations for six of the seven utilities were performed using this PWRR economic approach. The calculations for the seventh, Lenoir City Utilities Board, used simplified payback economic methodology, which was compatible with the utility's economic assessment approach.

The best available data were collected from each of the seven utilities during a site visit. Appropriate distribution design standards and associated line and transformer per-unit material and labor cost estimates were obtained, along with representative circuit layouts. The members of the project team familiarized themselves with each utility's distribution planning and design philosophy, reliability and power quality criteria. Energy and capacity costs, and economic assessment procedures and assumptions also were obtained. Transmission costs and applicable environmental externalities were obtained from the utility when available.

A planning study was then performed on rural, suburban, and urban distribution primary and secondary system scenarios for each utility. The studies assumed installation within a typical 5-year distribution planning horizon; the study period was 30 years. First, a base-case distribution system expansion without PV was performed using the cost and distribution planning and design information collected from each utility. Then a similar expansion of the distribution system scenarios was performed, integrating PV applications for an appropriate range of penetration.

The resulting site-specific distribution system benefits were then combined with other utility benefits including energy displacement value, generation capacity value, bulk transmission capacity and loss benefits, and various environmental and other applicable benefits.

PV PERFORMANCE

This paper considers utility interconnected PV applications. Thus energy storage, which increases PV capital investment, is not a requirement for PV systems, and was considered only when the increased cost can be justified by larger utility system benefits.

PV costs and performance strongly depend on the amount of energy generated by the PV plant. The major factor in determining the amount of energy that can be generated by a PV plant is the solar insolation available at the site in question. Typical meteorological year (TMY) hourly insolation based on SOLMET and ERSATZ data [6] [7] were used with Sandia's PVFORM program [8] to determine PV performance for the individual utility case studies.

PV performance varied significantly for the different utilities. For example, per-kW PV performance assumptions for the(Southern California Edison) SCE case study were based on SOLMET data for Daggett, which is located in Southern California. This site is representative of expected solar insolation in SCE's inland valley service areas. Expected annual capacity factor for a 10-m^2 PV system rated at 1 kW, assuming fixed orientation at the site latitude of 34.9^0 is about 25%. The fixed-orientation PV system capacity factor is about 25%, and the capacity factor of the two-axis tracking PV system is about 33.8%. Monthly performance for the two PV system configurations is presented in Fig. 1.

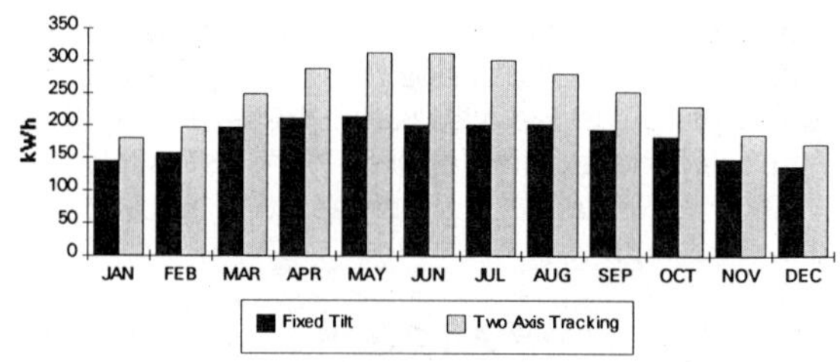

Fig. 1. SCE PV system monthly performance per kW.

Figure 2 shows the hourly performance for the fixed-orientation 1-kW PV system for the peak solar insolation day and average solar insolation day in April and in June. On April peak days when the solar insolation is high, the PV system can attain 1 kW output; on the average day, the PV system can attain over 0.9 kW output. In June, which represents the SCE summer peak period, the PV system does not attain full output.

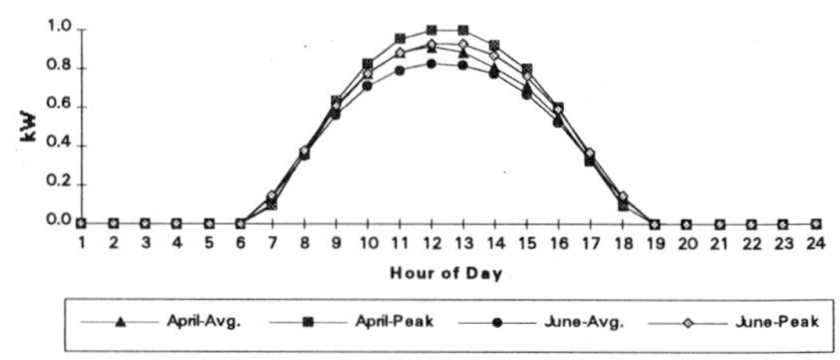

Fig. 2. SCE fixed-orientation PV daily output per kW

Figure 3 shows the corresponding hourly performance for the two-axis tracking 1 kW PV system for June and April. Two-axis tracking provides significantly better PV performance, but at a higher PV capital investment.

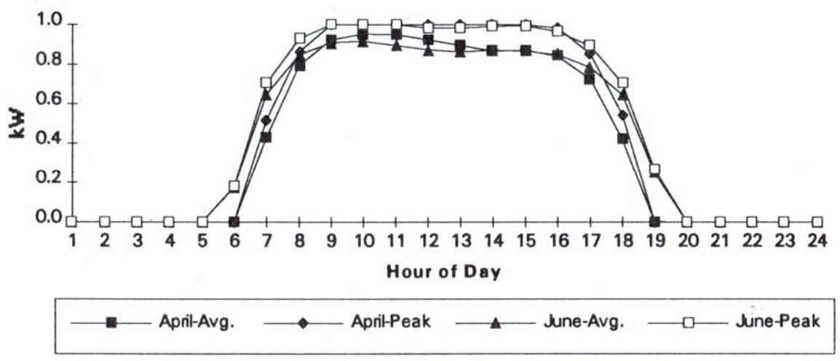

Fig. 3. SCE two-axis tracking PV daily output per kW.

To illustrate a less attractive example, per-kW PV performance assumptions for the Florida Power & Light (FP&L) case study were based on SOLMET data for Miami. This site is representative of expected solar insolation in the FP&L service area. In this case the fixed-orientation PV system capacity factor is about 18.8%, and the capacity factor of the two-axis tracking PV system is about 23.9%.

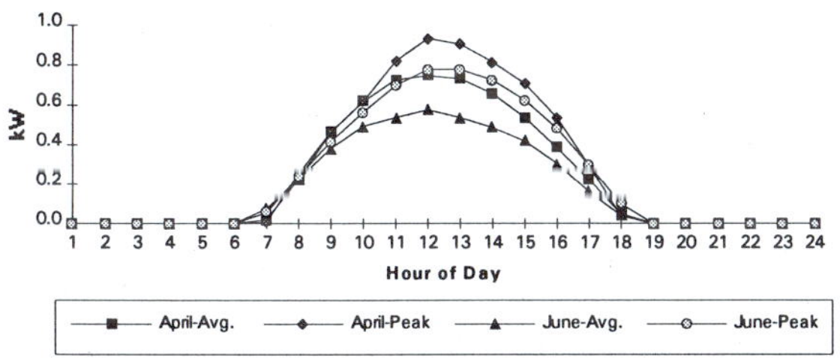

Fig. 4. FP&L fixed-orientation PV daily output per kW.

Figures 4 and 5 show the hourly performance for the fixed-orientation and two-axis tracking 1-kW PV system for the peak solar insolation day and average solar insolation day in April and June for FP&L. In April and June, during high solar insolation periods, the PV can provide fairly high levels of kW output. However, the average monthly output in both June and April is relatively low compared with SCE. Again two-axis tracking provides somewhat better PV performance. However, the average monthly performance of the two-axis tracking PV system is still low compared with SCE. It seems clear from this information that for many days, PV output will not correlate with system daily peak loads during the summer peak season. At FP&L even with two-axis tracking, it is likely that on many days PV MW output will not correlate with daily peak loads during the summer peak season.

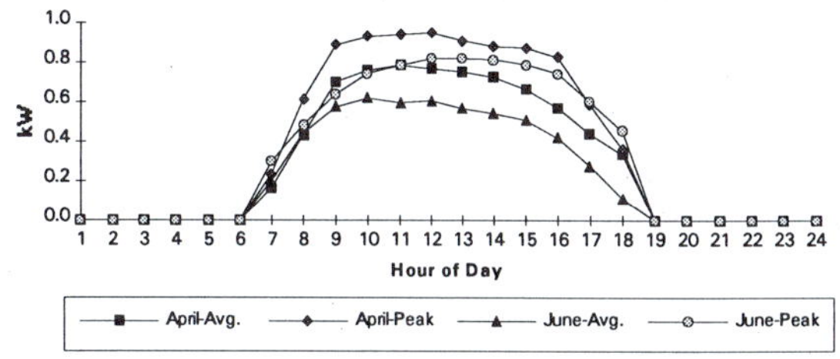

Fig. 5. FP&L two-axis tracking PV daily output per kW.

SUMMARY OF RESULTS

Representative benefits for selected fixed-orientation applications in the seven case studies are shown in Fig. 6. Tracking PV cases resulted in additional benefits. However, the additional investment required for tracking PV appeared to exceed the additional benefits in the applications assessed in the case studies [2].

The value of the distributed T&D utility benefits was found to be utility- and site-specific for the seven utilities evaluated in this study. For PV applications, the distributed utility benefits increased the total benefits $23/kW–$823/kW (4-46%) above the corresponding nonsite-specific benefits applicable to large central station PV applications. The largest increases were in the southwestern U.S., where daily peak loads correlate well with PV electric generation. For regions where the correlation between PV output and peak load is poor, the distributed utility benefits were much smaller. Note that battery or other storage may significantly increase the T&D benefits, at additional cost.

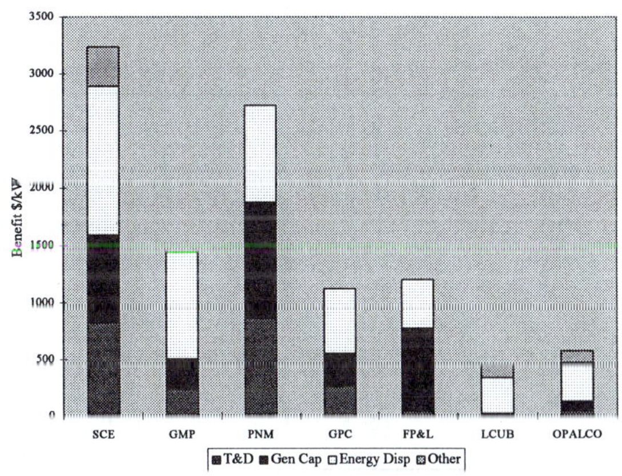

Fig. 6. Representative benefits for seven utilities

Deferred distribution facilities are more likely to be realized with PV systems located in the Southwest. Commercial loads, particularly office buildings, were found to correlate well with PV output. PV applications could utilize the large roof areas associated with these commercial loads for PV electric generation. Large T&D benefits also were obtained from mixed residential-commercial distribution system applications. In these mixed distribution systems, PV could be installed on residential roofs and still obtain large T&D benefits.

The value of voltage and var control power quality benefits were generally small because of the low-cost conventional methods that were currently used by the utilities. In fact, a small penalty occurs when line-commutated inverters are employed. Enhanced reliability of a distribution circuit may be a benefit in some very special cases. In our case studies, utilities did not give credit for a reliability enhancement benefit.

Additional capacity credit due to dispersing the PV throughout the distribution system varied significantly among the utilities. The highest values occurred for PV applications in the southwest U.S. (SCE and PNM).

The energy displacement value was a direct function of the solar insolation. Both capacity value and energy displacement value can be enhanced by the appropriate application of battery storage.

Other benefits include environmental externalities and Federal incentives resulting from the 1992 Energy Policy Act (EPACT) [9]. The consideration of externalities and set-asides in the economic assessment of new generation is mandated in some states by the PUCs. Environmental benefits were included in the economic analysis only when they were part of the utility's economic evaluation criteria for resource planning. EPACT PV production incentives benefits were applied to public utilities only, as the tax credits do not directly apply to private utilities.

CONCLUSIONS AND OBSERVATIONS

Some conclusions and observations resulting from these case studies are as follows;

- Site-specific T&D benefits can significantly increase the economic benefits of small, distributed PV connected to electric utility distribution systems, compared to large central station PV applications.

- PV output has excellent time-of-day correlation with SCE & PNM summer peak day commercial load shapes. Battery storage is not expected to be needed to obtain large T&D benefits.

- PV output has poor correlation with GPC and FP&L summer peak day load shapes, as peak days are likely to occur on cloudy days as well as sunny days. Battery storage is sure to be needed to back up the PV to attain large T&D benefits.

- PV output has poor correlation with GMP, LCUB, and OPALCO winter peak day load shapes. Battery storage is sure to be needed to back up the PV to attain large T&D benefits.

ACKNOWLEDGMENTS

Research for this paper was sponsored by the United States Department of Energy under Contract DE-AC050-840R21400 with Oak Ridge National Laboratory managed by Martin Marietta Energy Systems, Inc.

REFERENCES

[1] P.R. Barnes et al, *The Integration of Renewable Energy Sources into Electric Power Distribution Systems,* Vol. I, National Assessment, ORNL-6775/VI, June, 1994.

[2] H.W. Zaininger et al, *The Integration of Renewable Energy Sources into Electric Power Distribution Systems,* Vol. II, Utility Case Assessments, ORNL-6775/V2, June, 1994.

[3] E.L. Grant, and G.W. Ireson,. *Principles of Engineering Economy,* Ronald Press Company, New York, 1964.

[4] P.H. Jaynes *Profitability and Economic Choice,* The Iowa State University Press, Ames, Iowa, 1968.

[5] Electric Power Research Institute, *TAG™ Technical Assessment Guide Volume 1*, Palo Alto, Calif., 1989.

[6] C.L. Knapp, *Isolation Data Manua*l, Solar Energy Research Institute, Golden, Colorado, 1980.

[7] C.L. Knapp and T.L. Stoffel, *Direct Normal Solar Radiation Data Manual,* Solar Energy Research Institute, Golden, Colorado, 1982.

[8] D.F. Menicucci and J.P. Fernandez, *User's Manual for PVFORM: A Photovoltaic System Simulation Program for Stand-ALone and Grid-Interactive Applications,* Sandia National Laboratories, Albuquerque, N.M., 1988.

[9] U.S. House of Representatives, *Conference Report: Energy Policy Act of 1992 to Accomopany H.R. 776,* U.S. Government Printing Office, Washington D.C., October 26, 1990.

LUMINESCENT CONCENTRATOR MODULE OF A PRACTICAL SIZE

Koichi Sakuta, Shinji Sawata and Mitsumori Tanimoto
Electrotechnical Laboratory

ABSTRACT

A prototype of a luminescent concentrator module of a practical size of 960mm x 420mm was designed, fabricated and evaluated. A new "bottom-mounting" configuration was introduced. Single crystalline silicon solar cells were mounted at the center of the bottom surface of a luminescent plate made of PMMA containing a perylene-derivative dye. Enhancement of the short-circuit current by a factor of 2.23 was achieved. Another advantage in performance of the module under the partial shadowing conditions was demonstrated.

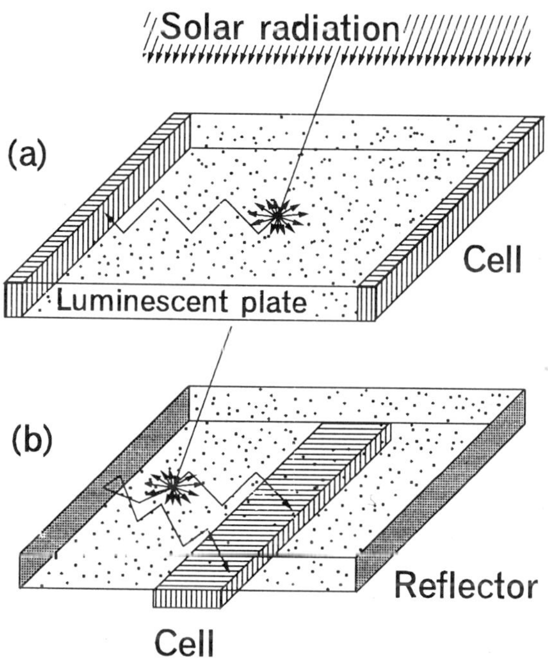

Fig. 1 LCM with (a) an edge-mounting, and (b) a bottom-mounting configuration.

INTRODUCTION

A novel concept of a luminescent solar concentrator (LSC) has attracted a lot of attention in the photovoltaic and other solar-related research communities for almost two decades since the first proposal by Wever and Lambe [1]. An LSC can concentrate not only the direct solar radiation but also the diffuse sky radiation without any tracking device. Due to mainly the poor photo-stability of the organic fluorescent dyes, the research activity in this field almost ceased several years ago. The development of a new class of perylene dyes with excellent photo-stability and high quantum efficiency renewed interest for this concept [2]. We have studied the relation between the collection efficiency and the concentration ratio in LSC experimentally and analytically [3-4].

Although various experimental LSC devises have been tested, there are very few studies on LSC devises of a practical dimension. To demonstrate the usefulness of the LCM concept for the current PV technologies, we built and evaluated a prototype of a luminescent concentrator module (LCM) of a practical size (420mm x 960mm), using commercially available solar cells. The enhancement of the short-circuit current by a factor of over 2 was demonstrated.

NEW CONFIGURATION

The most straightforward way of coupling solar cells to the luminescent plate is to mount the solar cells to the edge surfaces of the plate, which requires solar cells with very small width. On the other hand, solar cells with the size of 100mm x 100mm are most easily available now. We introduced, therefore, a different configuration of "bottom-mounting" for LCMs of a relatively low concentration (Fig. 1).

In the bottom-mounting configuration, a large part of the concentrated light traveling over the cells is expected to be absorbed by the cells if the cell width is sufficiently large compared to the plate thickness. This new configuration has the following advantages over the ordinary "edge-mounting" concept,

- easy assembly and better protection of the cells
- additional direct input to the cells.

The disadvantages of the configuration are;

- reduced concentration ratio
- incomplete coupling to the cells.

Geometrical Coupling Efficiency

The geometrical coupling efficiency of the bottom-mounting configuration can be analytically calculated in approximation. In Fig. 2, beams with the angle $-\theta_d < \theta < \theta_u$ fail to hit the solar cell. Assuming an isotropic distribution of the position and angle of the outgoing lights at the edge of the cell, whose angle is limited to $|\theta| \leq \theta_c$, the complement of the critical angle, the geometrical coupling efficiency Y_c is expressed as follows:

CH3365-4/94/0000-1115 $4.00 © 1994 IEEE First WCPEC: Dec. 5-9, 1994: Hawaii

$$\theta_u = \tan^{-1}\left(\frac{2-x}{W\sec\phi}\right) \quad (1)$$

$$\theta_d = \tan^{-1}\left(\frac{x}{W\sec\phi}\right) \quad (2)$$

$$\theta_c = \cos^{-1}\left(\frac{1}{n}\right) \quad (3)$$

$$Y_c = 1 - \frac{\int_0^1 \int_{-\frac{\pi}{2}}^{\frac{\pi}{2}} \int_{-\theta_d}^{\theta_u} \cos\theta \, d\theta \, d\phi \, dx}{\int_0^1 \int_{-\frac{\pi}{2}}^{\frac{\pi}{2}} \int_{-\theta_c}^{\theta_c} \cos\theta \, d\theta \, d\phi \, dx} \quad (4)$$

$$= 1 - \frac{n}{2\pi\sqrt{n^2-1}}\left\{4\tan^{-1}\left(\frac{2}{W}\right) + W\ln\left(\frac{W^2}{W^2+4}\right)\right\}$$

where n is the refractive index of the plate and W is the relative width of the cell to the thickness of the plate.

The above expression is not correct when W is very small and θ_u exceeds θ_c, that is:

$$W < 2\cot\theta_c = 2/\sqrt{n^2-1} \approx 1.8 \text{ (when } n = 1.5) \quad (5)$$

Figure 3 shows the W-dependence of the coupling efficiency Y_c when n = 1.5. Y_c increases with W, and $Y_c \approx 0.87$ at $W = 20/3$ which is for the case of the prototype mentioned later.

In the above calculation, the output beams which eventually hit the cell after several times of travels by multiple reflections at the edges are ignored. Therefore, a practical efficiency is expected to be slightly better.

EXPERIMENTAL

Luminescent Plate

A new class of perylene-derivative fluorescent dyes, developed by BASF group in Germany, have demonstrated an excellent photo-stability and a fluorescence quantum efficiency of almost unity [2]. A 3mm-thick PMMA (polymethyl methacrylate) plate (SHINKOLITE® #995, Mitsubishi Rayon Co., LTD.) containing one of these dyes was used as a luminescent concentrator plate. The plate absorbs 400 - 600nm radiation and emits fluorescent light in the 600 - 700nm region, where a silicon solar cell has a relatively good response. Figure 4 shows the absorption/emission spectra of the plate.

This plate has excellent stability under the outdoor conditions. No significant degradation in absorption and fluorescence has been observed after 3 years of the outdoor exposure test in our laboratory.

Solar Cell

Rectangular cells, 100mm x 20mm in size, were cut from commercially available single crystalline silicon solar cells (SHARP Corp.), 100mm x 100mm in size.

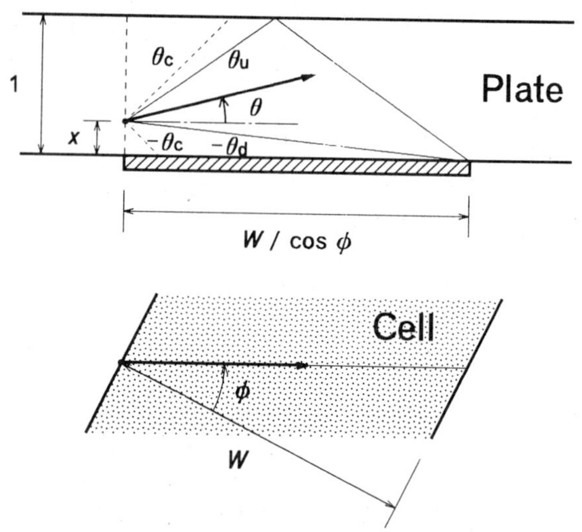

Fig. 2 Notation and geometry used in computing the geometrical coupling efficiency of the bottom-mounting configuration.

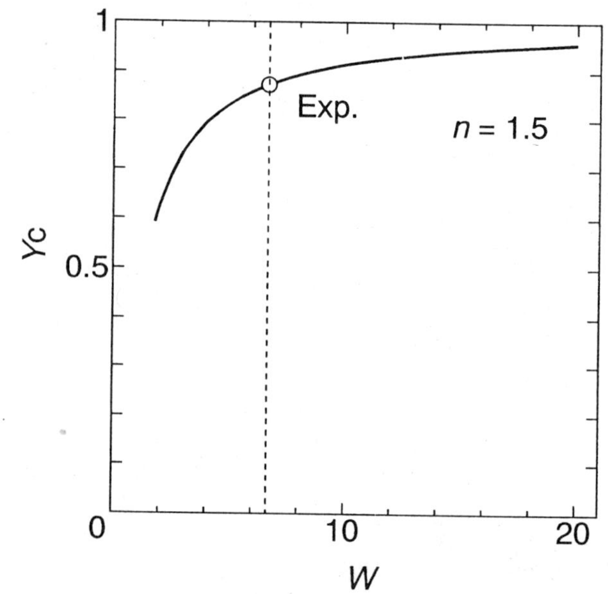

Fig. 3 Geometrical coupling efficiency as a function of the relative width of the cell.

The conversion efficiency of the cell is 15.5%, and the spectral response of the cell is shown in Fig. 5, together with the reference solar spectrum and the emission spectrum of the plate. Nine cells were interconnected in series to make a cell string.

Module

The schematic of the prototype module is shown in Fig. 6. The cell string, placed on an aluminum pipe support, was mounted at the center of the bottom surface of a 960mm x 420mm rectangular luminescent plate. Silicone gel was used as an optical coupler/adhesive.

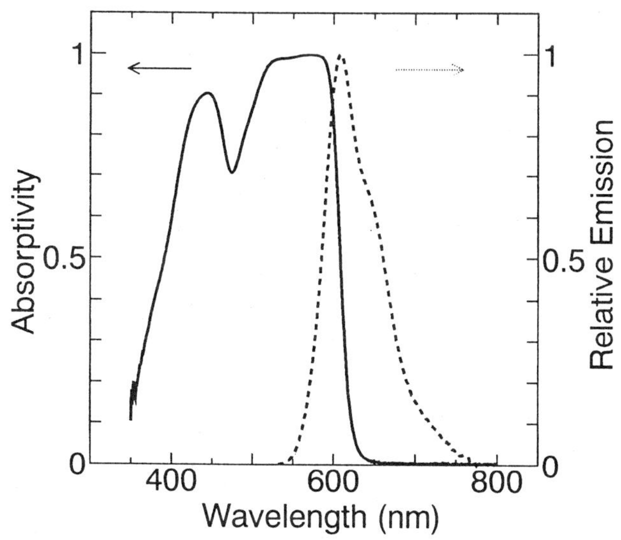

Fig. 4 Absorption and emission spectra of the plate.

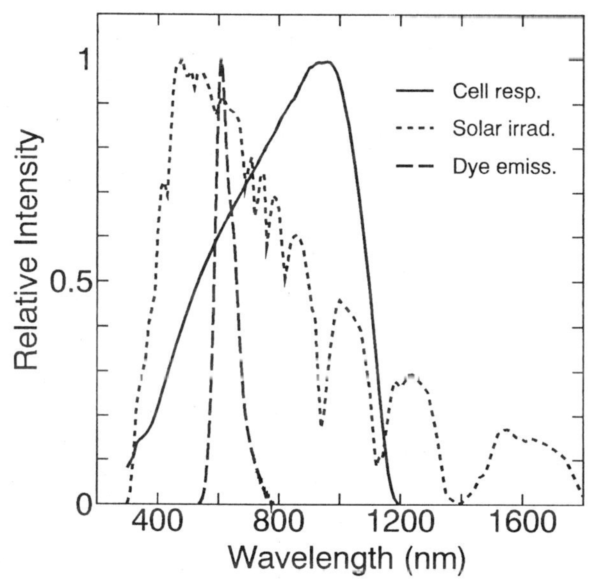

Fig. 5 Spectral response of the cell, reference solar spectrum and emission spectrum of the plate.

The four edge surfaces of the luminescent plate were covered with either silvered plastic films (specular reflectance = 0.95) or diffuse reflective white paint (KODAK White Reflectance Coating, diffuse reflectance = 0.98) to reflect the light back to the solar cells.

Measurements

The module was tested under the standard test condition using a large-area, long-pulse solar simulator at the Solar Cell Test Research Laboratory of the Japan

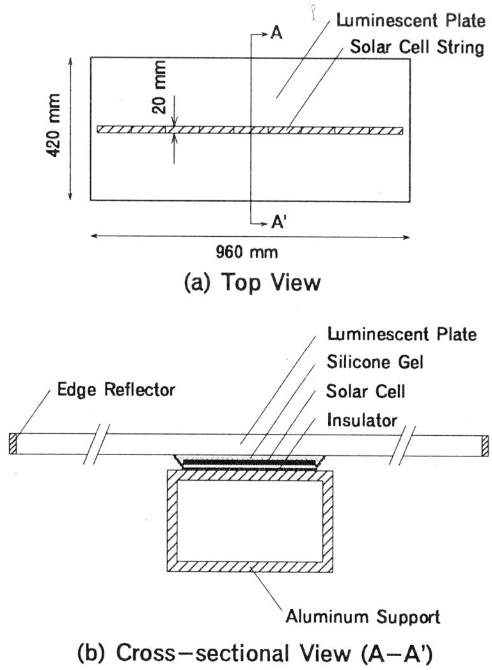

Fig. 6 Schematic of the prototype luminescent concentrator module.

Quality Assurance Organization (JQA). A module with a clear PMMA cover was used as a reference. Either a black-painted PMMA plate or a white-painted one was used as a back reflector.

RESULTS

Figure 7 shows the I-V characteristics of the prototype LCM (with white reflector on the edge) and the reference module. The enhancement of the short-circuit current, which is a direct indicator of the effective concentration ratio, by a factor of 2.23 (with a white back reflector) and 2.05 (with a black back reflector) was observed. The LCM with silver reflector on the edge gave slightly less enhancement of 1.97 (with a black back reflector).

The contribution of the direct input to the cell was evaluated by measuring the output of the LCM with a mask covering the cell area. The enhancement is found to be 1.34 when the cells were masked and the back reflector was white.

An LCM has another advantage in case of partial shadowing, which causes a severe decrease in the output and sometimes a damage to the cell in a conventional module. Because each cell is collecting light from wide area of the luminescent plate in an LCM, it is less sensitive to the partial shadowing. The output decrease was measured when the central part of the module was shadowed by a strip of cardboard 100mm wide covering the entire cell at the center. It is found that the output of the reference module decreases to almost zero, while the LCM keeps an about 50% output (Fig. 8).

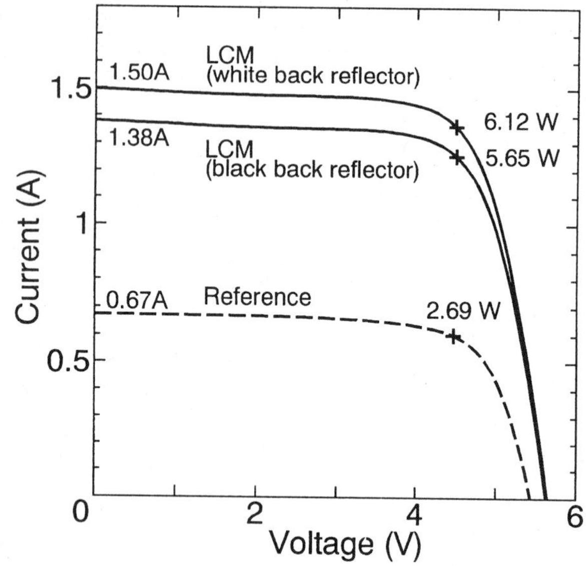

Fig .7 I-V characteristics of the prototype module.

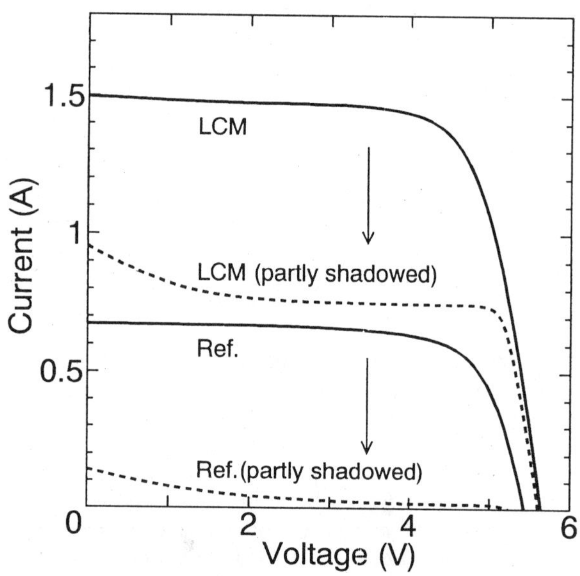

Fig. 8 Effects of partial shadowing.

DISCUSSION

The geometrical concentration ratio of the prototype LCM is 420/20 = 21. Using the efficiency for the direct input Y_{dir}, the efficiency for the non-direct input $Y_{plate,b}$ (with black back reflector) and $Y_{plate,w}$ (with white back reflector), the above results are expressed as follows:

$$2.05 = Y_{dir} + 20Y_{plate,b} \quad (6)$$

$$2.23 = Y_{dir} + 20Y_{plate,w} \quad (7)$$

$$1.34 = 20Y_{plate,w} \quad (8)$$

From these relations, it is found that $Y_{dir} = 0.89$, $Y_{plate,b} = 0.058$ and $Y_{plate,w} = 0.067$.

$Y_{plate,b}$ is expressed as follows:

$$Y_{plate,b} = Y_{abs} \cdot Y_{trans} \cdot Y_c \cdot K \quad (9)$$

where Y_{abs} is the absorption efficiency, Y_{trans} is the transmission efficiency, and K is the enhancement factor of the cell efficiency due to the change of the input spectrum. From the known values of the characteristics of the plate and the cell, it is calculated that Y_{abs} is 0.29 and K is 1.25. Y_c is 0.87 according to the equation (4). Following the simulation [4], it is estimated that Y_{trans} is 0.28. The expected value for $Y_{plate,b}$ is, therefore, .

$$Y_{plate,b}\,(\text{expected}) = 0.29 \times 0.28 \times 0.87 \times 1.25 = 0.088$$

The dominant factor of the reduced $Y_{plate,b}$ is thought to be the imperfect reflection at the edge, which reduces Y_{trans}.

A very unique feature of LSCs, which has been rarely drawn attention [5], is the ability to accept radiation from both sides. Because there is some radiation hitting the backside of a module in any situation, these "bifacial" modules can get extra input compared with conventional modules. Bifacial modules are also useful where the radiation is coming from various directions.

Besides the significant reduction in the per-Watt-cost, LCM has the potential to extend the application of photovoltaics to various fields such as see-through PV windows and PV arcades.

ACKNOWLEDGMENTS

We wish to express our gratitude to Mr. Nagamine and his group of the Japan Quality Assurance Organization for their help in the measurement of the module using the large-area solar simulator.

REFERENCES

[1] W.H.Weber, and J.Lambe, "Luminescent Greenhouse Collector for Solar Radiation", *Appl. Opt.*, **15**, 1976, pp.2299-2300.

[2] G.Seybold, and G.Wagenblast, "New Perylene and Violanthrone Dyestuffs for Fluorescent Collectors", *Dyes and Pigments*, **11**, 1989, pp.303-317.

[3] K.Sakuta, S.Sawata, and M.Tanimoto, "Concentration Ratio and Collection Efficiency of a Luminescent Solar Concentrator", *Proc. JSES/JWEA Joint Conf.*, 1992, pp.91-94.

[4] K.Sakuta, S.Sawata, and M.Tanimoto, "Monte Carlo Analysis of Luminescent Solar Concentrators", *Proc. JSES/JWEA Joint Conf.*, 1993, pp.25-28.

[5] G.Lifante, et al., "Luminescent Solar Concentrators as Bifacial Captors", *Solar Cells*, **8**, 1983, pp.355-360.

AN OPTIMUM DESIGN AND PROPERTIES OF A STATIC CONCENTRATOR WITH A NON-IMAGING LENS

K.Yoshioka, M.Kobayashi, A.Suzuki, K.Endoh, N. Ohe and T.Saitoh
Tokyo A & T University
Koganei, Tokyo 184, JAPAN

ABSTRACT

An optimum design of a lens-type static concentrator was carried out using a ray-tracing program. The optimum concentrator was determined by calculation of annual integrated power for several different concentrators and also flat-plate ones. Using an average sunlight data for ten years, a maximum average output ratio of the concentrator to the flat-plate modules at the same cell area condition was 1.7 at a half-acceptance angle of 25°. Also, the best half-acceptance angle, was calculated to be 25° by taking account shadowing effect. In experiments under a solar simulator, it was shown that the best half-acceptance angle was obtained at 25°. This suggests that an occupied area of costly crystalline Si cells in the static concentrator can be reduced to 60%.

INTRODUCTION

For the past several years, conversion efficiencies of crystalline Si solar cells have been greatly improved, but energy production cost by PV power systems is relatively high compared with conventional power systems. One of the solutions toward the cost reduction is to reduce occupied area of solar cells in PV modules by using a concentrator. Among concentrator systems, high-concentration tracking systems have mainly been developed so far. The high concentrator systems are not suitable for Japanese weather conditions. In Japan, a subsidy system for private houses was implemented in 1994 by the Ministry of International Trade and Industry to promote PV system diffusion. From the backgrounds above, the demands of PV systems for private houses will probably become greater in the future. To realize cost-effective modules for the PV systems, we proposed a non-tracking type concentrator with a new non-imaging lens for a static concentrator (1,2). The lens didn't have so excellent optical properties as the conventional Compound Parabolic Concentrator's (CPC's), but the height of the concentrator lens was about half of that for CPC's.

In this paper, more detailed study is carried out to investigate an optimum design of the non-imaging lens by a ray-tracing method. Firstly, optical concentration is calculated as a function of two angular parameters for the concentrator and flat module. Based on the results, annual integrated power of the static concentrator and flat-plate one are calculated using the average sunlight data at Tokyo and Kumamoto in Japan. In order to confirm the calculated results, the experiment is also carried out. Moreover, a practical design taking account of a shadowing effect by a neighbor lens is also carried out for an assembled module.

OUTLINES OF DESIGN AND FEATURES OF THE LENS

Fig.1 shows the lens for a static concentrator which we proposed before. As seen from the figure, the lens is a 2 dimensional-type concentrator with a shape like a stick.

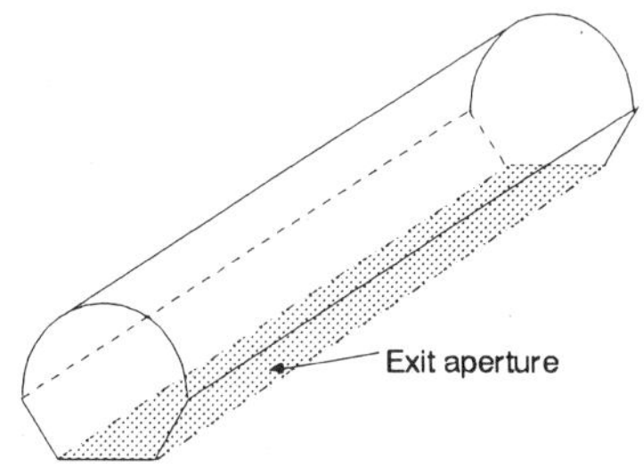

Fig.1 Appearance of the lens mentioned in this paper

The lens, as well as conventional reflective-type static concentrators, can be designed by the method called Edge Ray Principle(3). A concentrator designed by this method has characteristics that over 90% of sunlight entering an entry aperture within a half-acceptance angle reaches an exit aperture and there are not any focal points on the exit aperture. Therefore, a non-imaging concentrator has nearly constant optical concentration within a half-acceptance angle. In order to confirm this fact, the non-imaging lens which we proposed before had also been studied. And, it was found our lens also has much the same characteristics as other non-imaging concentrators.

Geometrical concentration ratio for the non-imaging concentrator is inversely proportional to half-acceptance incidence angle. In the case of this lens, the relation between the geometrical concentration ratio and the half-acceptance angle is given in the following equation.

$$C = n / \sin \theta_{max} , \qquad (1)$$

where n is the refractive index of the lens material and θ_{max} the half-acceptance angle. Equation (1) shows the geometrical concentration ratio of the non-tracking concentrator is a few magnification at most. In fact, the elevation of the sun at noon varies by ±23.4° throughout a year. Also, the elevation varies with time even in the same day. Therefore, optimum acceptance angle of the lens cannot be determined by only optical concentration ratio for sunlight coming from right above the lens top. Thus, the optimum half-acceptance angle of the lens is found by the integration of output power per unit solar cell area.

OPTICAL PROPERTIES OF THE MODULE BY CALCULATION

For the non-tracking concentrators, it is important to know the optical concentration at each incidence angle. Fig.2 shows the celestial sphere on the equator. Also as for areas except on the equator, we can use this celestial sphere by considering the model slanted at the same angle as the latitude. As shown in the figure, a location of the sun is designated with two angular parameters. In this paper, the first one is defined as $\alpha 0$, which corresponds to declination. The second one is defined as $\beta 0$, which corresponds to time angle.

In contrast to $\alpha 0$ and $\beta 0$ defined based on a location of

CH3365-4/94/0000-1119 $4.00 © 1994 IEEE First WCPEC; Dec. 5–9, 1994; Hawaii

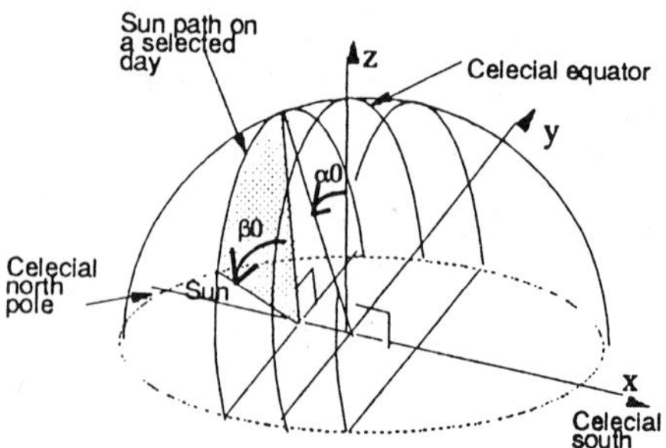

Fig. 2 Definition of the angles, α0 and β0 to describe a location of the sun

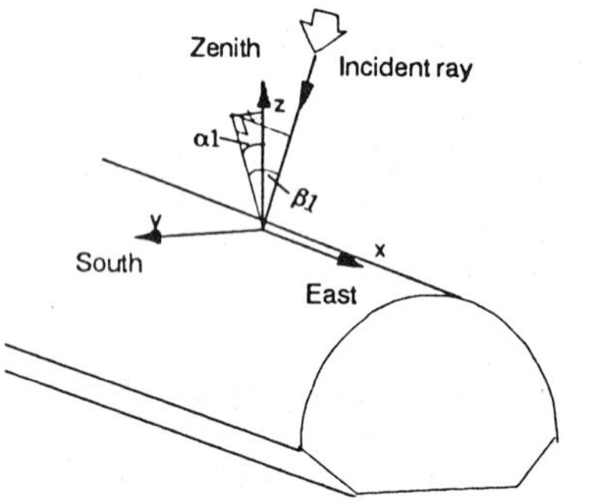

Fig. 3 Definition of the angles, α1 and β1 to describe the direction of a ray entering the lens

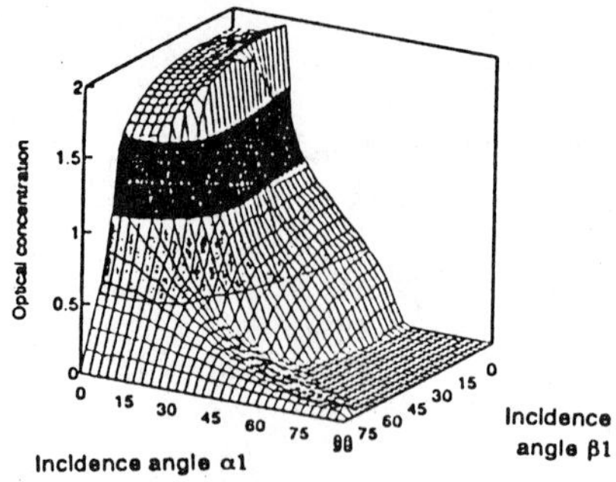

Fig. 4 Relation between optical concentration and incidence angle for the static concentrator designed at a half acceptance angle of 20°

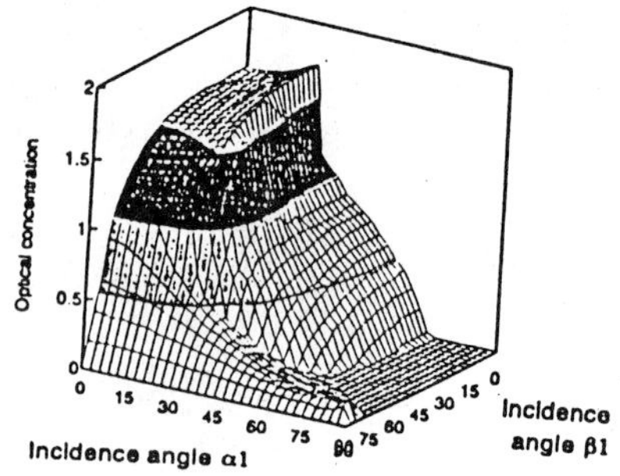

Fig. 5 Relation between optical concentration and incidence angle for the static concentrator designed at a half acceptance angle of 25°

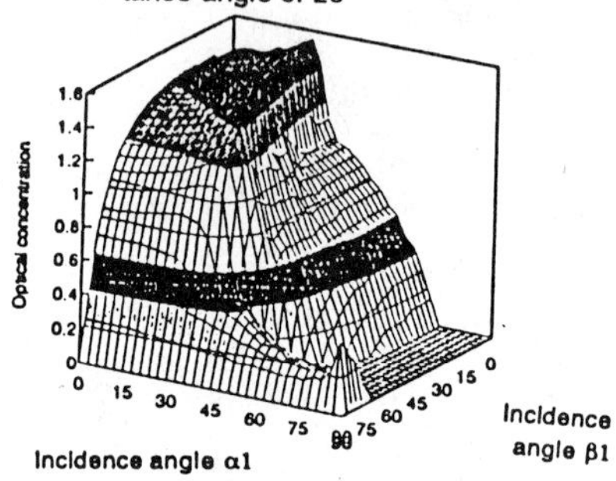

Fig. 6 Relation between optical concentration and incidence angle for the static concentrator designed at a half acceptance angle of 30°

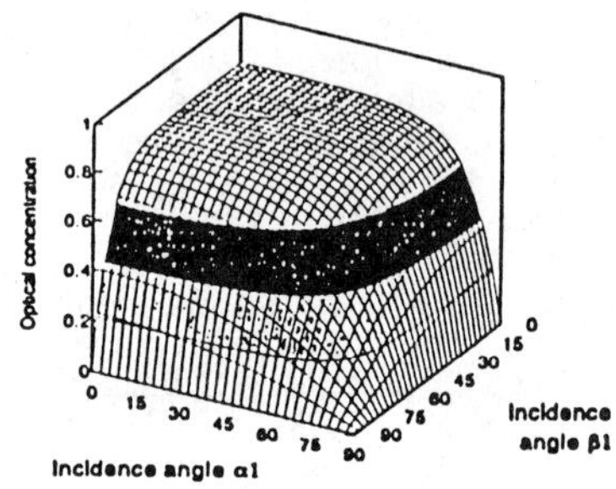

Fig. 7 Relation between optical concentration and incidence angles for the flat plate module

the sun, as shown in Fig.3, based on ray's incidence angle to the lens, α1 and β1 were defined for the sake of convenience in the calculation. Under supposition that the concentrators are aimed at the celestial equator, the transformation from (α0, β0) to (α1, β1) is easily given in Equation (2) using the matrix.

$$\begin{pmatrix} \cos\alpha0\sin\beta0 \\ \sin\alpha0\cos\beta0 \\ \cos\alpha0 \end{pmatrix} = \begin{pmatrix} \sin\beta1 \\ \cos\beta1\sin\alpha1 \\ \cos\beta1\cos\alpha1 \end{pmatrix} \quad (2)$$

With the ray-tracing method, optical concentration ratios for the flat-plate and the concentrator models were calculated at different half-acceptance angle as a function of the incidence angle, (α1, β1). Fig.4,5 and 6 show optical concentration distribution for the concentrators designed at half acceptance angles of 20°, 25° and 30° respectively. Optical concentration at each incidence angle was calculated by the density ratio between the photon energy entering the entry aperture of the lens and absorbed into a Si solar cell put on the exit aperture. As seen from Fig.4, 5 and 6, the smaller the half acceptance angle of the lens becomes, the higher optical concentration becomes in smaller angle regions. Fig.7 shows optical concentration distribution for the flat-plate model. In this case, optical concentration is equal to the ratio between the photon energy density entering the flat acrylic cover and that absorbed into a Si solar cell attached to it.

OPTIMIZATION THROUGH CALCULATION BASED ON ANNUAL INTEGRATED POWER

Based on the results calculated in the previous section, annual integrated power for the each concentrator and flat-plate models was carried out at each half-acceptance angle. A location of the sun at each time can be easily calculated using almanac data. As for direct-normal and diffuse solar radiation on a horizontal plane and air temperature, the data for Heating Air-Conditioning and Sanitary Engineers Program (HASP data) were used. The HASP data contain several constituents such as temperature, humidity, solar radiation, cloud amount and wind velocity. Each constituent contains data every 1h, which are average values of measurement data for the last ten years. Since we had HASP data at Tokyo and Kumamoto in Japan, the annual output power in these two cities could be predicted. The HASP data clearly show that global solar radiation in Tokyo includes a little more diffuse solar radiation than in Kumamoto at any times. In this calculation, it was supposed that the models were directed at the celestial equator in both Kumamoto and Tokyo. An amount of diffuse solar radiation on the surface parallel to the models was obtained by excluding that behind the plane put the modules on under assumption that luminance is constant on all the points on celestial sphere. In this simulation, a temperature rise and the effect of it were taken into account. The efficiency of a solar cell was calculated at a condition that the efficiency decreased by 5% per rise of 1℃ based on that at 25℃ .

From calculation taking account of the conditions above, results such as Fig.8 were obtained. The figure shows output ratio per unit models area of the concentrator to the flat plate types as a function of half-acceptance angle. The output ratio doesn't reach 1.0 at any half-acceptance angles. The peak value is 0.96 at 50°. This is mainly due to optical loss near the both sides of the refractive surface. Also, especially for the region which half acceptance angle is small, the average output ratio is relatively low. This suggests that the concentrator model especially with small half-acceptance angle can generate power only within extremely limited season and time.

Secondly, annual average-output ratio per unit solar cell area between concentrator and flat-plate models were calculated as a function of half-acceptance angle. From Fig.9, it is clearly seen that the output ratio has a maximum value of 1.7 at a half-acceptance angle of 25°. Needless to say, this suggests that an occupied area of solar cells in the model can be reduced to 60% by using the lens with the half-acceptance angle of 25°. In this study, under weather conditions in Tokyo and Kumamoto the optimization of the concentrator model has been performed. But, in a co-author paper(4), it has been reported that with regard to 2 dimensional CPC an optimum half-acceptance angle is 25°

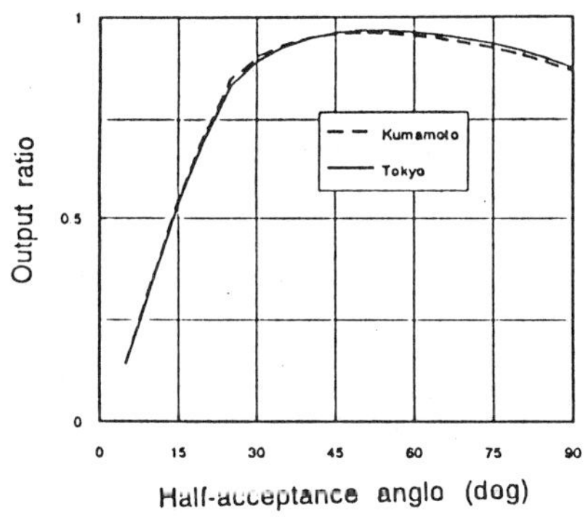

Fig. 8 Output ratio per unit module area of concentrator to flat-plate modules as a fuction of half-acceptance angle.

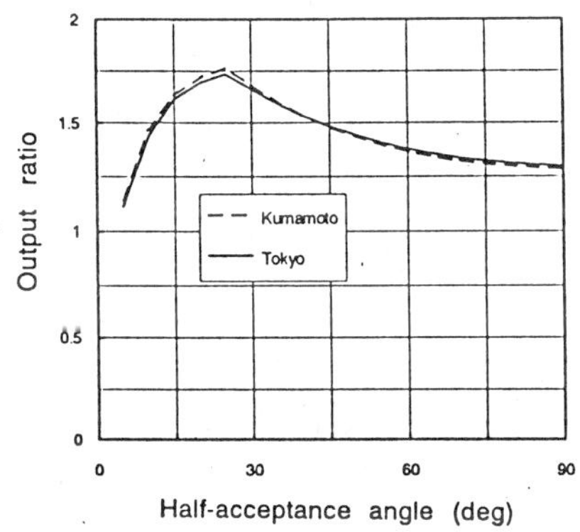

Fig. 9 Output ratio the concentrator to flat-plate models for the same cell area as a function of half-acceptance angle

without regard to the ratio of diffuse solar radiation to global solar radiation. Therefore, also in our concentrator model, much the same tendency would be obtained.

OPTIMIZATION THROUGH THE ADDITION OF THE SHADOWING EFFECT

In the previous section, we discussed about the optimization of the concentrator without considering effects by neighbor lenses. However, in a practical assembled module shown in Fig.10, changes in properties cased by neighbor lenses should occur. One is reflective rays at the surface of neighbor lenses. But, that is small problem since the reflective rays enter another lens at angles far beyond a half-acceptance angle. The other is shadowing effect. The effect is an important problem especially for direct solar radiation coming at a large incidence angle. Fig.11 shows an example of the shadowing effect for the concentrator modules with a half-acceptance angle of 25° at $\alpha 1=25°$ and $\beta 1=0°$. As seen from the figure, the neighbor lens intercept some of rays coming at a large incidence angle. In the same manner as the previous section, annual average output ratios per unit solar cell area between the concentrator and flat-plate models were calculated as a function of half-acceptance angle. As shown in Fig.11, tendency of both the properties of the concentrator modules with and without considering the shadowing effect is much the same. Also from these calculated results, it was found that an op-

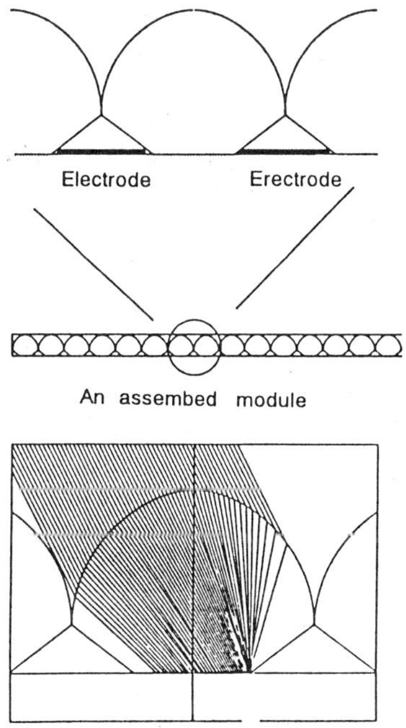

Fig. 10 A shodowing effect by a neighbor lens for the concentrator module

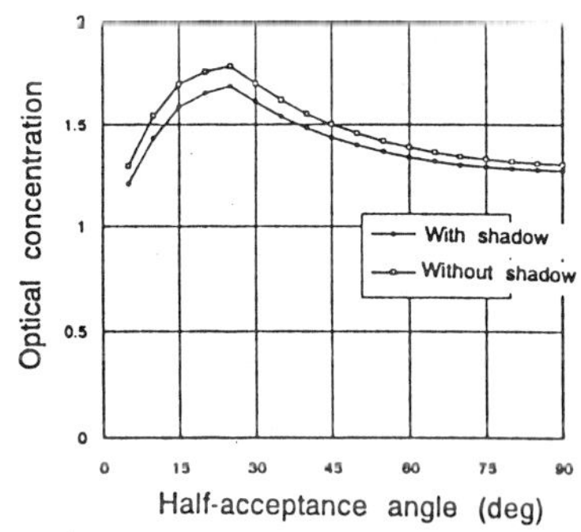

Fig. 11 The effect of shadow by a neighbor lens for optical concentration v.s. half-acceptance angle relation

timum half-acceptance angle was 25° and in this case the loss by shadowing effect was only 5%.

MEASURED PROPERTIES UNDER SOLAR SIMULATOR

In order to confirm the calculated results described above, experiments were carried out under a solar simulator. As for the experimental models, four acrylic lenses designed at a half acceptance angle of 20°, 25°, 30° and 35°, were prepared. Exit aperture of all the lenses was 20cm in length and 2cm in width. Each lens was attached with a Si solar cell of 2cm × 3cm to middle of that using silicone resin. Also, a flat-plate model was prepared in order to compare with the concentrator models. The flat-plate model was fabricated by attaching the Si solar cell to 2mm thick acrylic plate. Before fabricating each model, I-V measurements of the solar cell used in each model were performed in advance at various angles under a solar simulator. In these experiments, $\alpha 1$ and $\beta 1$, which have already been defined, were used as parameters for designating a ray incidence angle. For fixed $\alpha 1$ of 0°, 6°, 12°, 18° and 23°, each measurement was carried out at varying $\beta 1$ from 0° to 80°. Measurement Conditions with regard to the light source were 1.5 in AM and 100mw/cm^2 in intensity. After fabricating each model, I-V measurements for those were performed in the same way. Optical concentration ratios, the ratio between Isc of a solar cell with and without the lens, were calculated from the measured results. Fig.12 and Fig.13 show relations of optical concentration ratio vs. time angle. In Fig.12, optical concentration ratio at declination of 0, i.e., on vernal and autumnal equinoxes, was shown as a function of time angle. It is seen that each concentrator model has optical concentrator ratios over 1.0 for ±5h before and after the meridian since time angle of 15° is equal to 1 hour, and that at smaller half-acceptance angle, better performance is obtained. On the other hand, the relation at declination of 23°, i.e. on summer and winter solstices, is shown in Fig.13 as a function of time angle. From this figure, it was clearly seen that the concentrator model with a half-acceptance angle of 20° is not good in performance compared with the others. With regard to the mod-

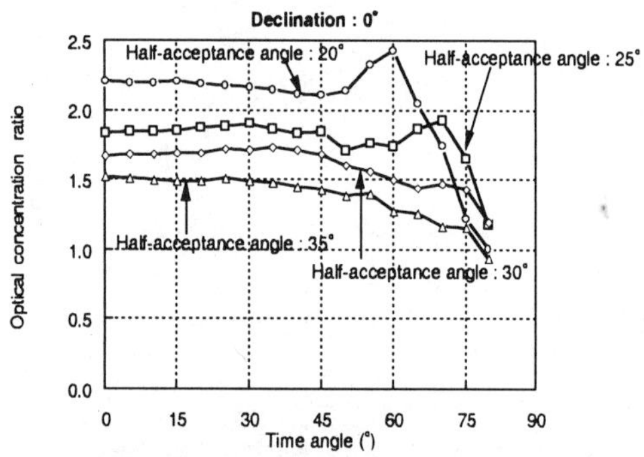

Fig. 12 Optical concentration ratio for each concentrator model vs. time angle relation obtained from a experiment carried out under supposition of declination of 0°

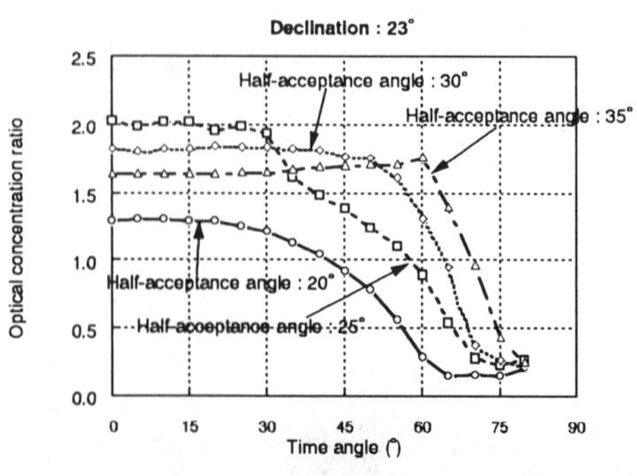

Fig. 13 Optical concentration ratio for each concentrator model vs. time angle relation in a experiment carried out under supposition of declination of 23°

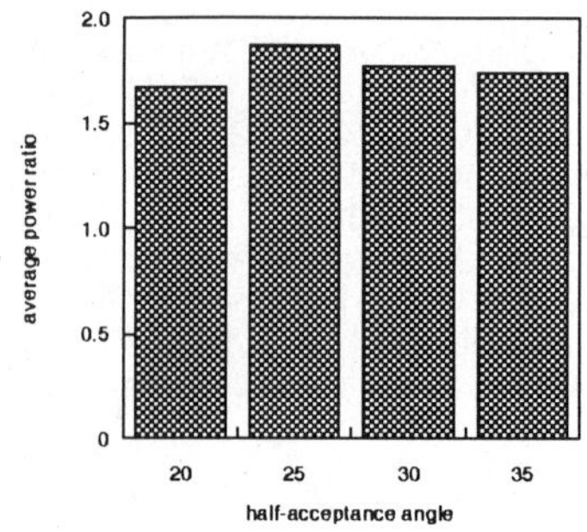

Fig. 14 Average power ratio of a solar cell between each concentrator and flat-plate models obtained from integration of the measured data

els with half acceptance angle of 25°, 30° and 35°, it is difficult to judge which is better from only this figure. So, average power ratio of a solar cell between each concentrator and flat-plate models was calculated by integration of the data obtained from the measurements. As shown in Fig.14, the best half-acceptance angle was determined to be 25° for the refractive concentrator model.

In these experiments, an effect of light intensity distribution on a solar cell was not studied in detail. We would like to clarify the effect by measuring the performance outdoors in the future tasks.

CONCLUSION

An optimum design on PV modules using a lens-type static concentrator was carried out by ray-tracing method. From comparison of the concentrator with flat-plate models through the integration of power for solar cells with the same size, the best half-acceptance incidence angle was determined to be 25°. And from calculation by considering a shadowing effect by neighbor lenses, it was found that the decrease in the output power due to the shadowing was only 5%.

Also from the experiments under a solar simulator, it was found that the best half-acceptance angle for the concentrator module was also 25° and average output ratio to the flat-plate one was 1.85.

ACKNOWLEDGMENTS

This work was supported by the Photovoltaic Power Generation Technology Research Association as a part of New Sunshine Project of the Ministry of International Trade and Industry, for which we would like to thank whom it may concern.

REFERENCES

(1) A. Suzuki et al., Design and Optical Features of a Non-Imaging Lens and its Application to Solar Energy Systems, Proc. of the First International Conference on New Energy Systems and Conversion, Yokohama, June, 1993.

(2) K. Yoshioka et al., Design and Properties of a Refractive Static Concentrator Module, Solar Energy materials & Solar Cells, Vol.34 p125-131, 1994

(3) A. Luque, Solar Cells and Optics for Photovoltaic Concentration, Adam Hilger, Bristol and Philadelphia, 1989

(4) A. Suzuki and S.Kobayashi, Yearly Distributed Insolation Model and Optimum Design of a Two Dimensional Compound Parabolic Concentrator, Solar Energy, to be published.

FLAT HIGH CONCENTRATION DEVICES

J.C. Miñano, J.C. González and I. Zanesco
Instituto de Energía Solar - ETSI Telecomunicación - UPM
E28040 - Madrid - Spain - FAX (34 1)5446341, E-mail minano@ies-def.upm.es

ABSTRACT

During last five years new concentrators achieving the theoretical maximum acceptance-angle-concentration product have appeared. Besides this, some of these concentrators are very compact (concentrator depth/aperture diameter smaller than 1/3). The feasibility of these concentrators for photovoltaic applications is studied. It is concluded that these concentrators may be useful for high concentration cells (irradiance for maximum efficiency greater than 800 suns) if these cells have a small size (diameter smaller than 5 mm). The concentrators may provide for this case an acceptance angle of $\approx \pm 2.7$ degrees with concentration factor around 1000x and a concentrator depth 10 times the cell diameter. Instantaneous direct insolation and ambient temperature measurements of Madrid and a thermal model of the heat sink is used to calculate the annual electric energy output with which different concentration factors are compared. Concentration of 1000x is close to the one giving the maximum annual electrical output.

INTRODUCTION

Among the new concentrators referred above, two of them may have interest for the photovoltaic community: the RXI concentrator[1] and the TIR lens[2]. We are going to focus our analysis in the RXI concentrator with rotational symmetry. As said before this is a very compact concentrator: its depth is about 1/3 of its diameter. Moreover, the acceptance angle ($\pm\alpha$) is the theoretical maximum corresponding to its concentration factor C_g, ($C_g = n^2/\sin^2\alpha$, where n is the refractive index of the material filling the concentrator, $n \approx 1.5$). The total transmission of the RXI, without considering absorption or reflection losses, is $\simeq 97\%$, i.e., using perfect reflectors, antireflection coatings and non absorbing dielectric media, 97% of the incoming radiation within the angle $\pm\alpha$ would be sent to the cell. This means that optical efficiencies around 80% may be achievable with practical optical materials for incidence angles smaller than α. Concentrators with (loseless) total transmissions close to 92% were already known (large focal length parabolic mirrors or Fresnel lenses together with secondary concentrators of the CEC type). What is new of the RXI is its compact size and its simplicity.

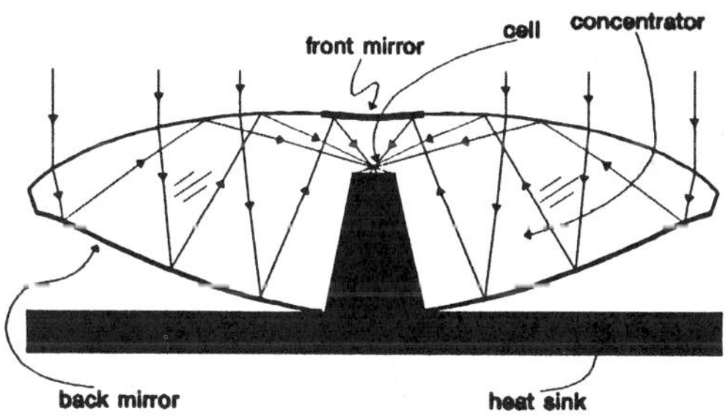

Figure 1. RXI concentrator cross-section. Rays from impinging the entry aperture within ± 2.7 degrees are sent to the cell after a refraction at the upper surface, a reflexion at the lower surface and a total internal reflexion at the upper surface. Concentration factor 1000x.

Let us consider more deeply these characteristics (compactness and high concentration-acceptance angle) to know if these concentrator may have interest for the photovoltaic community:

(1) Compactness is interesting itself not only because aesthetic aspects but also because it is related with the cost of housing. According to Chamberlin[3] housing is the second most costly item of a concentrator module (other items are cells, cell assemblies, optics and other).

(2) High concentration (let's say $C_g > 1000$) with acceptable acceptances angles ($\alpha > \pm 1$ degree) is interesting because the cell cost savings related with concentration. In this way the cell cost (which is the most costly item of a concentrator module according to the Chamberlin's study) may be reduced, although not so simply as reducing the cell area. It has also to be taken into account that cell efficiencies may be very small for high irradiance levels, not only because the losses associated with the series resistance but also because of

CH3365-4/94/0000-1123 $4.00 © 1994 IEEE

the operating cell temperature.

Reaching the concentration-acceptance angle upper bound means, for instance, that an acceptance angle α of ± 1 degree (which may be considered as an acceptable acceptance angle for solar concentration applications) may be achieved with concentration factors as high as $C_g \simeq 7400$ (refractive index of the concentrator dielectric $n = 1.5$). Such concentration factor will produce an irradiance on the cell $\simeq$ 4750 suns at 800 W/m^2 of direct irradiance on the entry aperture (optical efficiency $\simeq$ 80%). Under such high irradiance, photoconversion efficiencies are small. That is why it is usually accepted that photovoltaic concentrators don't need to get the concentration upper bound. Nevertheless, 800 W/m^2 is close to the peak value of direct irradiance and may be not a representative irradiance at the entry aperture; so, even though 4750 suns is too high irradiance for photovoltaic applications, we cannot conclude that such high concentration factors are useless.

An important aspect to be considered under high irradiance concentrations is the thermal resistance from cell to ambient. Temperature drop from the cell to the ambient is smaller under constant irradiance at the aperture if we reduce proportionally the whole device (concentrator, cell and heat sink).

To analyze the feasibility of high concentration devices for photovoltaic applications, we have calculated the annual electricity production under the climatological conditions of Madrid using a simplified model for the heat sink. The purpose of the analysis is to search out if the combined effect of a small cell (and thus of a good thermal resistance from cell to ambient) and the variations of the actual direct irradiance may permit the use of high concentration devices.

The thermal model of the heat sink is the subject of the next section. Cell model used is shown in Section 3 and last section is devoted to the results.

HEAT SINK THERMAL MODEL

Figure 2 shows the structure of the heat sink studied in this work. The cell is assumed to be bonded to an aluminum truncated cone, which is bonded to an aluminum disk plate. Heat conduction through the truncated cone is approximated by the following equation:

$$T_{cell}-T_a=\frac{1}{\sigma\sqrt{A_c\pi}\sin(\alpha)}\left(1-\frac{1}{\sin(\alpha)h+\sqrt{\frac{A_c}{\pi}}}\sqrt{\frac{A_c}{\pi}}\right)Ph$$

where P_h is the heat power, A_c is the cell area, h is the height of the truncated cone, T_{cell} is the cell temperature and T_a is the temperature of the truncated cone surface in contact with the plate, and $\sigma = 2.1$ W cm^{-1} K^{-1} is the thermal conductivity of aluminum and α is the angle between the axis and the wall of the cone. Heat conduction through the truncated cone has been approximated as if the truncated cone had two spheric surfaces (at the top and at the bottom) centred at the cone vertex [4]. No heat conduction through the dielectric is considered.

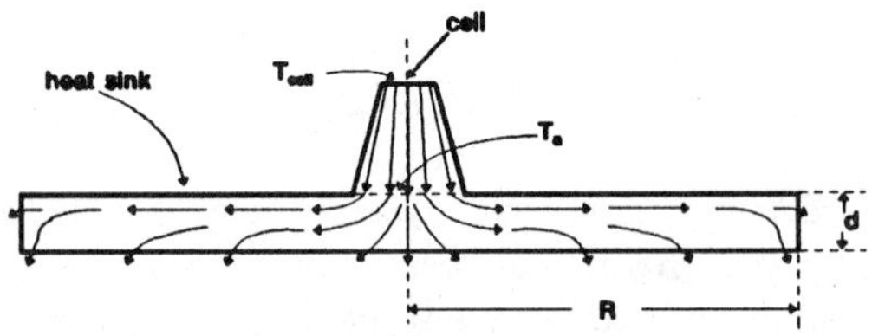

Figure 2. Heat sink is formed by an aluminum truncated cone and a plate whose area is the same as the concentrators entry aperture.

The heat transport through the plate to the ambient is given by the following equation [5].

$$\frac{\partial T^2}{\partial r^2}+\frac{1}{r}\frac{\partial T}{\partial r}+\frac{(h_{conv}+h_{rad})}{d\sigma}T=0$$

where $h_{con} = 5\ 10^{-4}$ Wcm^{-2}K^{-1}. The dependence of h_{rad} with temperature is taken from reference[4] .

The disk plate has always the same area as the concentrators entry aperture. The truncated cone shape differs depending on the cell size and concentration. See Figure 3 shows the heat sink evolves when concentration is varied, keeping a constant cell size or when the cell size varies keeping constant the concentration factor.

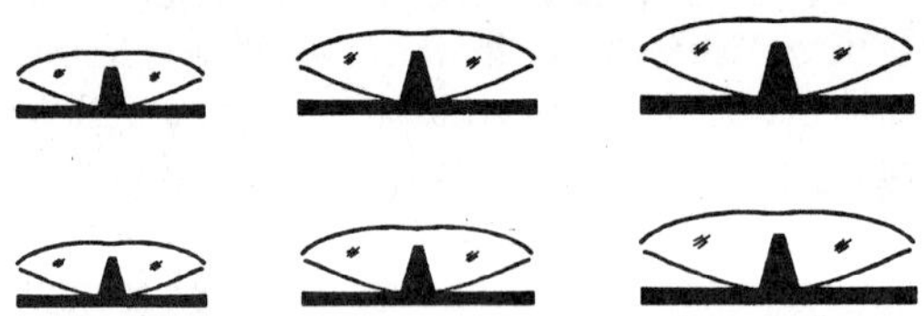

Figure 3. Heat sinks and concentrators used for the comparison. First row are three cases with the same cell size and different concentration, last row are three concentrators with the same concentration factor but different cell sizes.

CELL MODEL, TEMPERATURE AND INSOLATION DATA

A simulated GaAs cell is used for the

calculations. Cell characteristics are shown in Figures 4 and 5. Data have been taken from reference [6]. Cell efficiency versus temperature and concentration have been calculated using the model of reference [4].

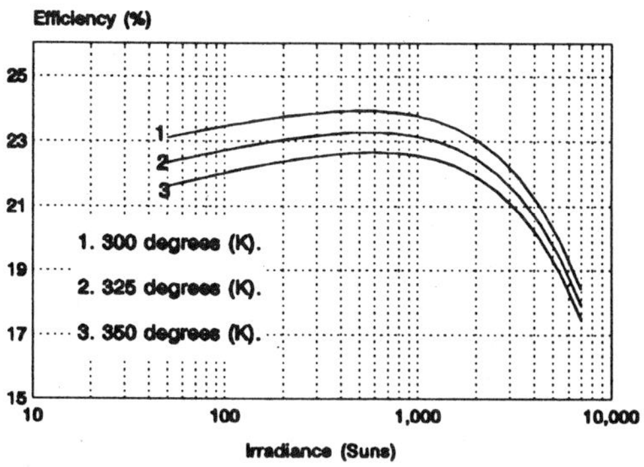

Figure 4. Efficiency versus concentration for the GaAs cell used in the calculations.

Temperature and insolation data have been measured from September 1993 to August 1994 at the IES facilities in Madrid[8]. These data provide direct insolation and ambient temperature for each minute of this period. Averaged hourly insolation data was rejected because averaged values give optimum concentration factors greater than actual values: for instance, if the hourly averaged direct irradiance is 0.5 suns and the irradiance of maximum efficiency is 800 suns, then it

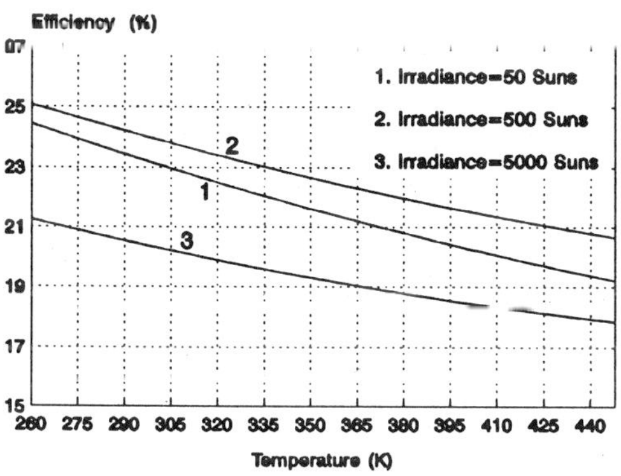

Figure 5. Efficiency versus temperature for the GaAs cell used for the calculations.

may be erroneously concluded that the best concentration factor for this case is 1600x. This would be true if during this hour the direct irradiance is constant. Nevertheless, the result would be very different if, for instance, the direct irradiance has been 1 sun during 1/2 hour and 0 during the other 1/2 hour. Instantaneous direct irradiance obtained by correlations from hourly averaged values also assume a smooth variation.

RESULTS AND DISCUSSION

Figure 6 shows the annual electricity production for three different cell sizes. As expected, the smaller the cell size, the greater the electricity output. This is due to its lower thermal resistance. It is worthy to note that annual electricity output is kept approximately constant for a large range of concentration factors (let us say from 400x to 1500x). Nevertheless, this range of concentrations is smaller than the one obtained by calculations based on hourly averaged values of direct irradiance which extended up to 3000x.

The following conclusions can be taken from the results of Figure 6:

(a) Concentration factors around 1000x are suitable for photovoltaic concentration if the temperature drop from cell to ambient is kept at reasonable levels .

(b) To keep this temperature drop at reasonable levels it is necessary to use small cells (cell diameters smaller than 5 mm).

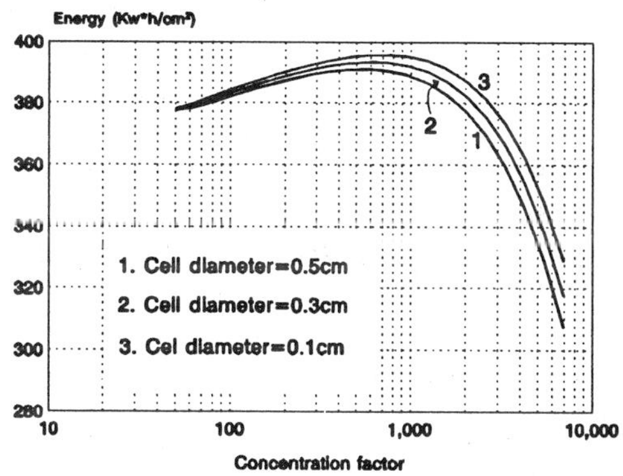

Figure 6. Annual electricity production versus concentration factor for three different cell sizes. Optical efficiency of the concentrator has been taken equal to 1. To get actual values of the electric energy this should be multiplied by the optical efficiency.

To take advantage from such high concentrations it is necessary to use high efficiency cells. Fortunately these cells use to be small, which is necessary to get good cell temperatures. Additionally, the small size of the cells is also implies a small dielectric volume of the concentrator. Note that the volume of the concentrator relative to its entry aperture is proportional to the cell diameter if the concentration factor is fixed. The same happens with the ratio of the heat sink volume to the concentrator entry aperture. That is one of the reasons why the temperature of the cell is smaller when using small cells.

From the point of view of concentrator design there is no problem in getting acceptance angles greater than ± 2 degrees with levels of concentration of 1000x (using RXI type concentrators). Additionally these concentrators present another benefit : compactness.

Heat sink for these cells seems to be qualitatively similar to the ones derived for power semiconductor devices. In our opinion, all the arguments given above enforces the option of small high concentration photovoltaic modules.

REFERENCES

1. J.C. Miñano, J.C. González, P. Benítez, "New nonimaging designs: the RX and the RXI concentrators", in *Nonimaging Optics: Maximum-Efficiency Light Transfer II,* Roland Winston, Robert L. Holman, Editors, Proc. SPIE 2016, 120- 127 (1993).
2. W.A. Parkyn, P.L. Gleckman, D.G. Pelka "Converging TIR lens for nonimaging concentration of light from compact incoherent sources", in *Nonimaging Optics: Maximum-Efficiency Light Transfer II,* Roland Winston, Robert L. Holman, Editors, Proc. SPIE 2016, 78-86 (1993).
3. J.L. Chamberlin, "The costs of photovoltaic concentrator modules" 20th IEEE Photovoltaic Specialists Conference. Las Vegas 1988 pp 1353-1356.
4. G. Sala "Cooling of solar cells" in *Solar Cells and Optics for Photovoltaic Concentration,* A. Luque ed., Adam Hilger, Bristol 1989.
5. V.P. Isachenko, V.A. Osipova, A.S.Sukomel. *"Heat transfer"*. Ed. MIR, 1987.
6. C.Algora, A.Marti, J.C.Maroto, G.L.Araujo and P.A.Davis. *"GaAs concentrator Solar Cell for operating inside light confining cavity"*. 11th E.C. Photovoltaic Solar energy conference.
7. M. Macagnan. *Caracterización de la radiación solar para aplicaciones fotovoltaicas autónomas.* Doctor Thesis, E.T.S.I. Telecomunicación, UPM, Madrid, 1993.

ACKNOWLEDGEMENTS

Juan C. Miñano and Juan C. González acknowledge the support provided by Consejeria de Educación de la Comunidad de Madrid (Project number 052/92). I. Zanesco acknowledges the support provided by CAPES (Coordenação de Aperfeiçoamento de Pessoal de Nivel Superior), a Brazilian sponsoring agency.

DESIGN AND DRAWING SUPPORT SYSTEM FOR PHOTOVOLTAIC ARRAY STRUCTURE

T. Hirasawa K. Iida S. Yamaguchi
Kandenko Co., Ltd.

N. Asayama Y. Naito
Nasu Denki Tekko Co., Ltd.

ABSTRACT

This paper describes a design and drawing support system for a photovoltaic (PV) array structure. The operator inputs data (e.g. structure type, tilt angle, load conditions, etc.) into the system, which computes stress on each element of structure and outputs the calculated results. If the results are within the tolerance limit, a skeleton drawing of the structure is produced. The weight list of materials is also outputted on paper. When the results are beyond the limit, the data is modified by the operator and re-computed until the satisfactory results are shown.

INTRODUCTION

Cost reduction is the most important issue pertaining to the spread of PV systems. Therefore, the cost reduction of BOS as well as that of the solar cell are strongly desired. Owing to standardization and mass production, the costs of the solar cell and the power conditioner can be gradually lowered year by year. However, it is difficult to lower the cost of PV array structures as they must be manufactured and installed in accordance with various conditions. Aiming at the cost reduction of PV array structures, we have now developed a design and drawing support system for the photovoltaic array structure. The cost reduction of the PV array structures can be realized by repeatedly performing simulations using this system so as to enable design of the most durable and economic PV array structures. Our system is intended for those who possess sufficient knowledge of solar cells and electricity but are inexperienced in matters related to structural design. It consists of three independent systems and 43 original programs, for example, data converters, system linkers, system controllers and so on. The three independent systems are a data base (DB) system for data entry, a finite element method (FEM) system for structural analysis and a computer-aided design (CAD) system for drawing.

FUNCTIONS

This system consists of four functions, namely, INPUT, FEM, CAD and OUTPUT. All functions and works will be selected from a layer-structured menu by using a cursor (See Fig.1). When one of the functions which comprise a

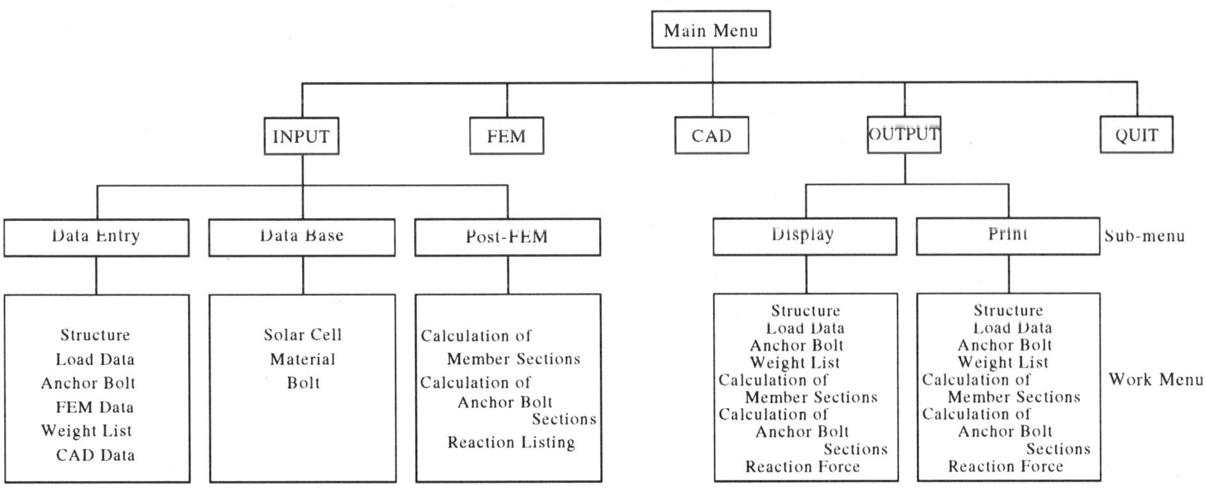

Fig.1 Layer-structured Menu

CH3365-4/94/0000-1127 $4.00 © 1994 IEEE
First WCPEC; Dec. 5-9, 1994; Hawaii

main menu, i.e., INPUT, is chosen, a sub-menu of three subsidiary functions — Data Entry, Data Base, Post-FEM — is displayed on the screen. Furthermore, a work menu is displayed when a subsidiary function is chosen from the sub-menu. The contents of the displayed work menu are as follows; structure, load data, anchor bolt, FEM data, weight list, CAD data in the Data Entry; solar cell, material, bolt in the Data Base; calculation of member sections, calculation of anchor bolt secitons, reaction listing in the Post-FEM. OUTPUT is positioned on two subsidiary functions, Display and Print. Several data and lists can be outputted by these subsidiary functions.

Data Entry

After the structure is selected in the work menu, the specifications of the structure are entered directly from the keyboard to the data entry format on the screen. The type of structure is selected among sixteen(16) different structural types in a sub-file. As to the structure dimensions and tilt angle, the numerical values are input. Regarding members and bolts, the specifications of each member, i.e., pillar, front horizontal member, etc., are selected from the material and bolt data base. In a similar manner, the PV module is selected from the solar cell data base. As to the quantity and layout of a unit of PV modules, the numerical values are input. With regard to load data in the work menu, wind force coefficient, standard velocity pressure, applied standards, expected snow depth, earthquake coefficient, analyzed structure, loading conditions and safety factors are input. Regarding the analyzed structure, a truss frame or rigid frame is selected. For the loading conditions, a load such as a dead load or wind load and the combination of such loads are selected from the sub-file. The maximum loading conditions which can be input are ten (10) cases, with all of the cases able to be analyzed. In regard to anchor bolt in the work menu, the number and dimensions of bolts to be used are input. For FEM data, a data file is made by converting structure data into FEM data. A solid model can be constructed by applying the plane shape obtained from the structure data to each surface of the model. It is also possible to create an asymmetric solid model. When converting data, abnormal structures or materials are detected and an error message is displayed. In such cases, the data should be corrected and converted again. In regard to the weight list, a structure weight file is created on the basis of thestructure data. Regarding CAD data, a data file is created by converting the structure data into the CAD data.

Data Base

The manufacturer's name, model name, dimensions and unit weight are input into the data base of the solar cell. The names of angles and H sections, quality of materials, sectional area, unit weight, geometrical moment of inertia, section modulus and radius of gyration are input into the data base of the material. Names, the quality of materials, diameters, sectional area and diameters of bolt holes are input into the data base of the bolt. For each of the above data bases, additions to and corrections of data can be freely performed.

Post-FEM

With regard to the members, a file of calculation of member sections is created by use of a file compiled from FEM analysis. The number of materials having insufficient

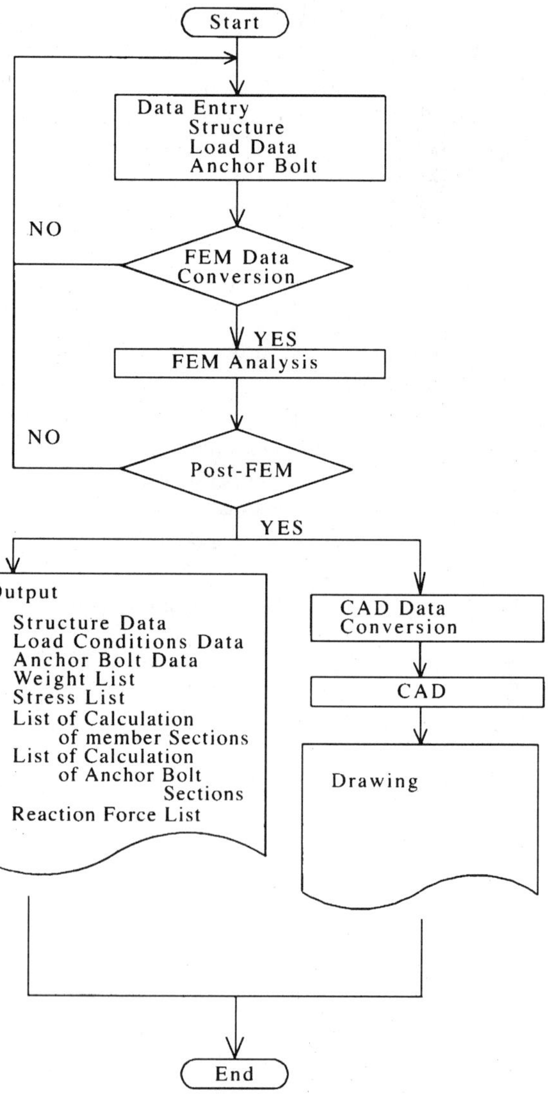

Fig.2 Operation Flow

strength is then displayed by scanning the data of the file of the calculation of member sections. As for the anchor bolts, the file of the calculation of anchor bolt sections is made in the same manner, with indication displayed as to whether or not the strength of the anchor bolts is sufficient. Regarding the reaction listing, a list of the foundational reaction forces is made in accordance with each load condition.

Output

As for output, the structure data file, the load conditions data file, the anchor bolt data file, the weight list file of the structure, the list file of the calculation of member sections and anchor bolt sections and the list file of the foundational reaction forces are displayed on the screen and printed out on paper.

OPERATION AND DATA FLOW

Lists and drawings are drawn up by normal operation (See Fig.2). Since on-line manuals are built into the system, the operator can refer to the system in case he or she does not understand the operation or technical terms. As the system stores the latest data at all times, the operator can easily obtain the lightest structure having sufficient strength by simply correcting a portion of the structure data and load data and by repeating calculation of structural analysis. At this time, the operator can confirm the results of the corrected data on the screen, with only the final results output on paper. Fig.3 shows data flow between the data bases and the programs. Fig.4 shows an example of an output plan.

CONCLUSIONS

It took less than ten (10) minutes to output a drawing after input of data by normal operation. When we applied our system to the array structure installed previously, we could reduce the weight of the structure by thirty (30) percent under the same design conditions. We intend to expand the types of structure and the contents of data bases so as to upgrade system usability.

REFERENCES

[1]"Design Standards for Steel Structure", Architectural Institute of Japan, Tokyo, Japan, May 1973

[2]"Design Standards on Structures for Transmissions", The Institute of Electrical Engineers of Japan, Tokyo, Japan, November 1979

[3]Hiroshi Yamakawa, "Calculating Dynamics and CAE Series No.9 Optimum Design" , Baifukan Co., Ltd., Tokyo, Japan, April 1993

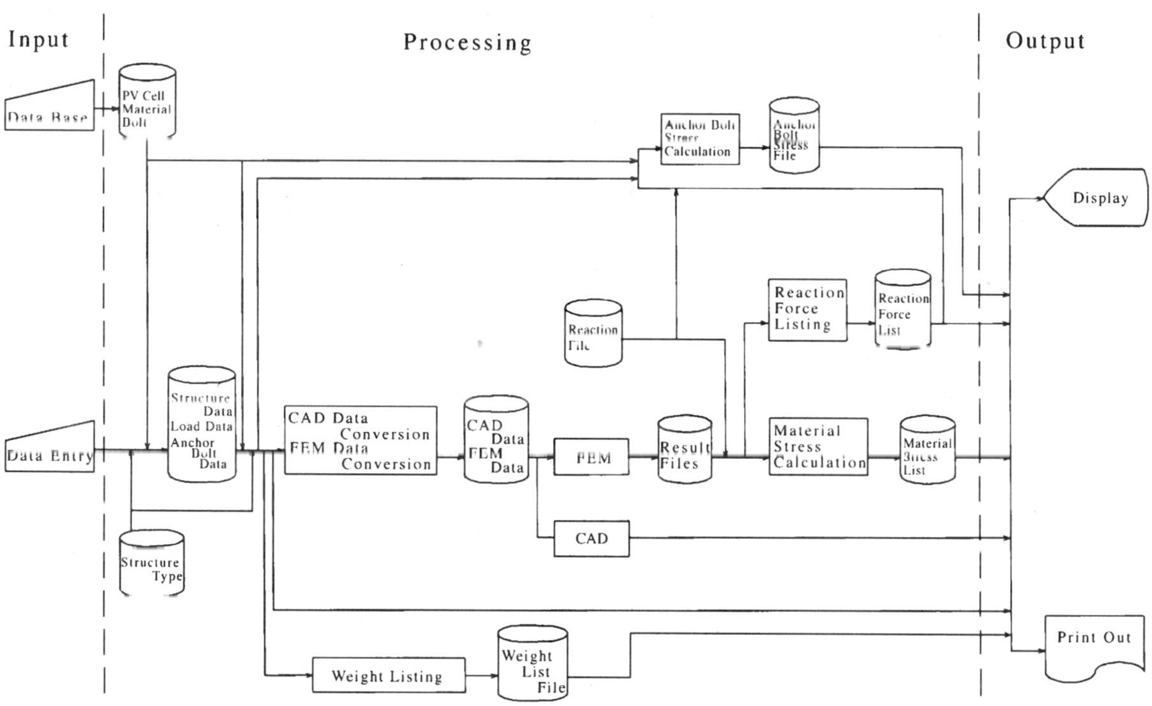

Fig.3 Data Flow

[4]Donald A. Norman, "The Psychology of Everyday Things" , Basic Books Inc., New York, USA, 1988
[5]"Microsoft C Professional Development System Install Guide First Edition", Microsoft Corp., USA, June 1991
[6]"dbMAGIC User's Manual Third Edition", WACOM Co., Ltd., Tokyo, Japan, September 1992
[7]"ANSYS-PC/LINEAR 3 Rev4.4A Verification Manual", Swason Analysis Systems, Inc., Houston, Pennsylvania, USA, April 1990
[8]"AUTOCAD User Reference First Edition", Autodesk Inc., California, USA, November 1990

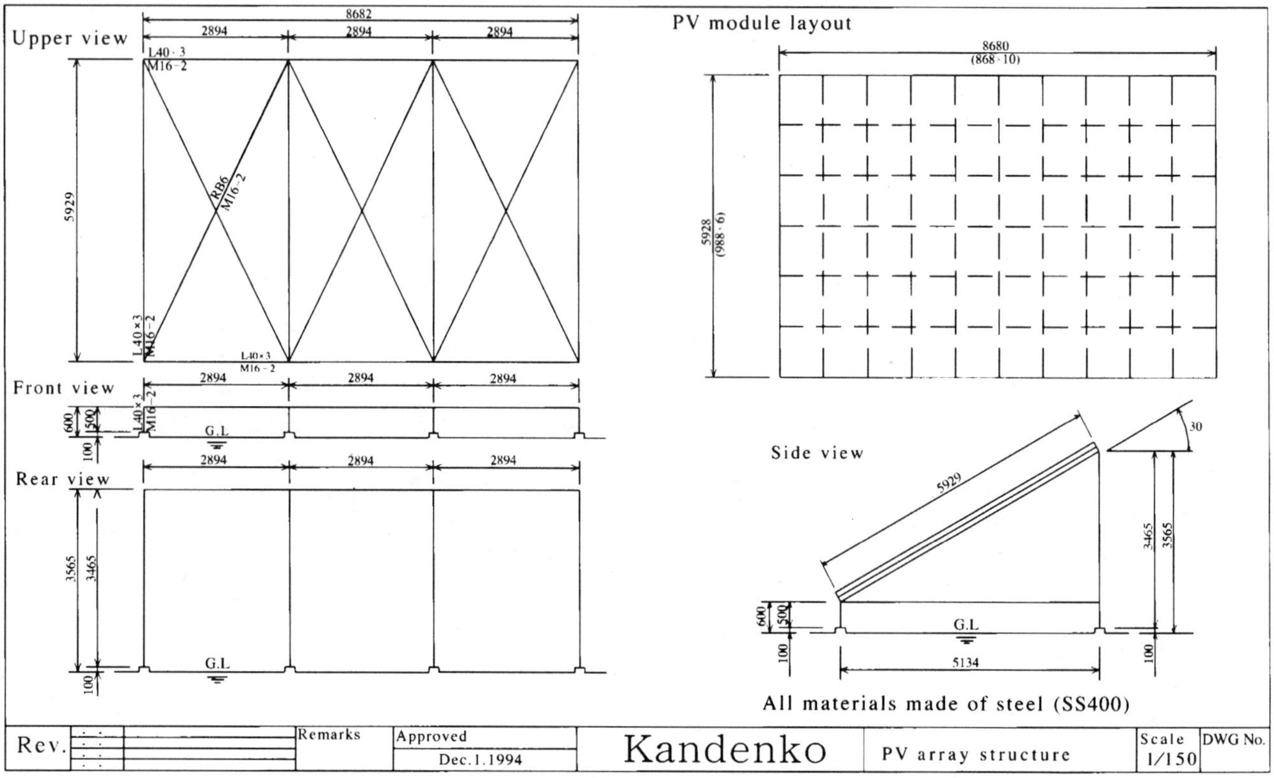

Fig.4 Example of Output Plan

TESTING THE NEW SOLAR TRACKER WITH SHAPE MEMORY ALLOY ACTORS

V.Poulek

Poulek Solar International, Kastanova 1481, CZ-250 01 Brandys n.L., Prague-East District, Czech Republic, Fax: +42 202 804 017

INTRODUCTION

A new low-cost (95 USD/m^2) single axis passive solar tracker based on a shape memory alloy (SMA) actuators has been tested.

The SMA actuator can be easily deformed at relatively low temperatures (by the tracker actuators below 70°C). It returns back to its original shape when heated above transformation temperature, producing mechanical work. During thermal cycles the SMA actuator works as a heat engine. The efficiency of the SMA actuators (~2%) is approximately two orders of magnitude higher than that of bimetallic ones (which could be theoretically also used).

Solar radiation energy focused by the lens is used to heat one of two SMA actuators, made from a longitudinally cut SMA tube, to cause their differential deformation. One end of both actuators is connected to a stator while the another end is connected to a rotor. The deformation of actuators rotate the apparatus until a focus of the lens is directed on a mirror between the ends of SMA actuators. Then the tracker and the solar collectors are directed properly to the Sun (see Fig. 1). Any further change in the focus position will cause heating and contraction of the one of both actuators which will move the rotor with the mirror into the focus.

The system is completely solid state (no gases or fluids), extremely simple (it has only 5 movable parts) and requires no outside control or energy source to operate.

EVALUATION OF THE NEW TRACKER

Shape memory alloys are well developed now [1] and their applications are quickly growing. The tracker application fits very well to SMA actuators because slow motion and low cooling rates are needed. NiTi, CuZnAl and CuAlNi shape memory alloy actuators were tested in our trackers.

To avoid overheating of the alloy in the focus of the lens and for roughly uniform distribution of thermal energy into a volume of the actuator high thermal conductivity copper base shape memory alloy CuAlNi has been selected for the solar tracker. The overheating is also the reason why linear focus cylindrical lens was selected for the tracker instead of point focus spherical lens.

The tracker is designed to last as long as solar arrays. Because of extremely simple solid state design and sealed case an estimated lifetime of the apparatus should be 10-20 years (~7000 tracking cycles). A fatigue life of the SMA actuators is usually at least 100.000 cycles. No fatigue of actuators has been observed in accelerated life test after 10.000 cycles. Long term life tests have to be performed to verify exact lifetime of actuators.

Theoretically, oscillation between two actuators could occur, however, no such effect has been observed during laboratory tests because the CuZnAl and CuAlNi shape memory alloys have highest damping capacity of all known materials (SDC=85% and 30% respectively).

The SMA actuators are placed in a cylindrical case. The case of the tracker is 200mm long and it has diameter of 130mm. Hermetically sealed case protects the apparatus against the dust and humidity. The sealed case was designed and tested in accordance with IP00 standard.

During the night and/or cloudy weather the tracker always slowly (max. 10°/hour) returns to equilibrium position (Fig. 1.a). Because of symmetrical design the tracker can start to follow the sun not only at the morning but also at the late afternoon (anywhere in the range ±60° around equilibrium position)

A torque of the new tracker has been investigated.

CH3365-4/94/0000-1131 $4.00 © 1994 IEEE First WCPEC; Dec. 5–9, 1994; Hawaii

Usual working stress of the selected Cu base SMA alloy is 75-100 MPa at memory effect (deformation) of 2%. The SMA actuators with diameter of 30 mm have torque 0.3 Nm per centimeter of the tube length. Increasing linear dimension of both the actuator tubes and the cylindrical concentrating lens continuous sca ling of the torque from 0 up to at least 30 Nm is possible. Second possibility is to increase diameter of the actuator tubes. The basic actuator diameter is 30mm. Actuator with diameter of 60 mm is evaluated at present. The torque of the tracker with these actuators could be scaled to at least 100 Nm.

The area related torque of the tracker has been adjusted to
1 Nm/m^2. More powerful (i.e. more expensive) SMA actuators (5 Nm/m^2 or more) can be used if necessary.

Because of symmetrical and balanced mounting of the solar panels torque necessary to rotate the system is very low (<<1Nm) and depends on the quality of the bearings. Symmetrical design also enables to decrease wind (or snow) induced torque, however, it can't be eliminated completely. Highest wind speed measured at the test site until now was 85 km/h.

Self-locking transmission solves the problem. Using e.g. worm gearbox the rotation of the tracker can be transmitted to the solar panels but reverse way is not possible. The system is therefore protected against windstorms. We have developed new simple gear-less self-locking transmissions for the tracker.

Influence of the self-locking transmission on performance of the solar tracker has been investigated.

Efficiency of the self-locking transmission has been only about 45% and, therefore, 2 times more powerful actuators have to be used for given torque. On the other hand without wind protection, especially in windy climates, up to 10 times more powerful actuators have to be used to prevent any wind induced deviation out of proper direction and fluttering of solar arrays. The self-locking tracker can, therefore, use less powerful actuators than unprotected one.

The self-locking protection works also at night and cloudy weather when passive trackers have no torque. By trackers without self-locking transmission (with shock dampers only) the wind flow is often more powerful than actuator of the tracker and the wind can rotate solar arrays out of proper direction. It decreases collectible energy. Due to self-locking transmission the wind induced rotation is possible only in direction of tracking, rotation out of tracking direction is locked.

The solar tracker was designed for tracking angle up to 140°. Measured tracking angle has been up to 124° according to weather. Although this tracking angle is smaller than expected, increasing tracking angle above 120° there is no further substantial surplus of the collectible energy [2].

The new tracker was tested in polar axis tracking mode, however, horizontal and/or vertical axis tracking modes are also possible. Tracking accuracy of the tracker with simple lens, which has been approximately ±5°, is good for flat plate solar collectors.

Using achromatic lens, more powerful actuators and high tolerance mechanics tracking accuracy could be up to ±1°. Such accuracy is acceptable even for medium concentration (~20) linear concentrators (parabolic trough or Fresnel). Solar tracker with achromatic lens was not yet tested to verify above mentioned estimated accuracy.

Some trackers without automatic morning orientation are oriented to the west until the sun is high enough above the horizon to heat tracking elements. A substantial portion of the morning solar radiation is therefore lost before orientation is achieved. Tracker with automatic morning orientation solves the problem.

An equilibrium orientation of the new solar tracker at night, which is usually half the way between east and west (see Fig.1a), was adjusted slightly (20°) to the east. Because the "sensor" lens of the tracker is able to collect solar radiation from the angle 120°(±60°), the actuator of the tracker is illuminated immediately after sunrise and the solar tracker is at the morning reoriented directly to the east in a few minutes. Automatic morning orientation is, therefore, included in the design of the new tracker.

CONCLUSION

During laboratory tests and short term field tests the new tracker worked very well. Long term field tests are on the way at present.

REFERENCES

1. T. W. Duerig, D. Stockel and A. Keeley in "Engineering aspects of shape memory alloys", (T. Duerig ed.), Butterworth, London, (1990) 181-194.
2. J. M. Gordon and H.J. Wenger, Solar Energy, Vol.46, No 4. (1991) 211-217.

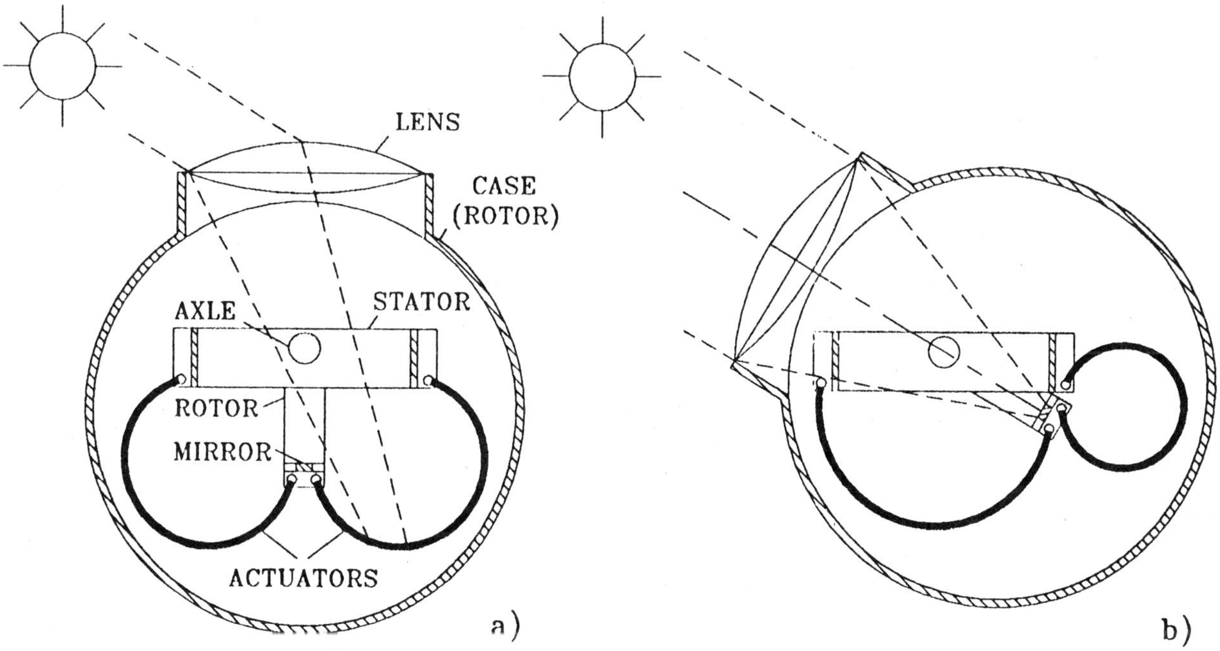

Fig. 1. Solar tracker before (a) and after (b) morning orientation

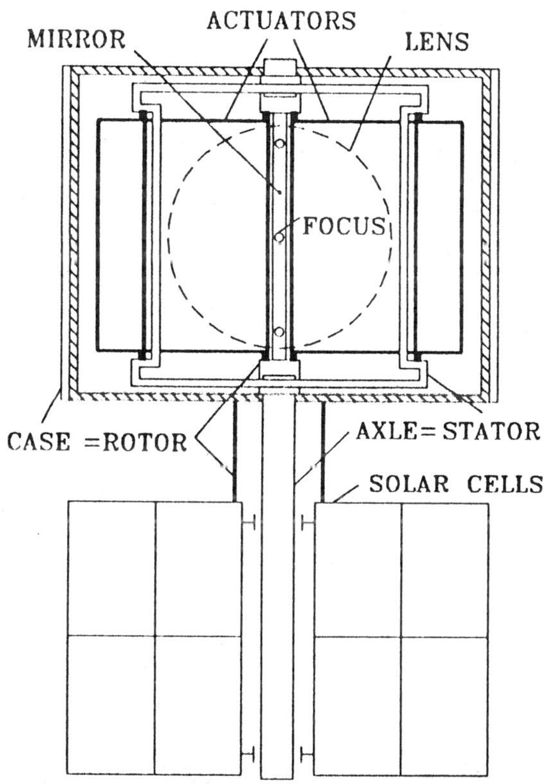

Fig. 2. Plan view of the solar tracker from Fig. 1a

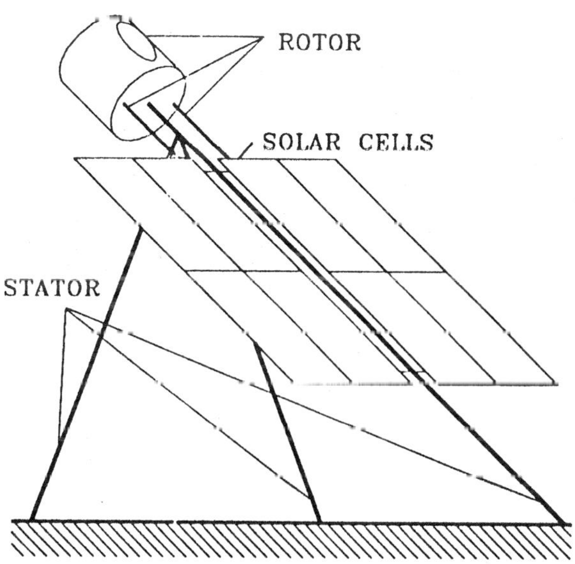

Fig. 3. Solar tracker with a pair of flat plate PV solar arrays

LOS SABINOS PV RURAL ELECTRIFICATION PROJECT: RESULTS AFTER 30 MONTHS OF OPERATION

Jorge M. Huacuz, Jaime Agredano, Gonzálo Munguia, Roberto Flores
Instituto de Investigaciones Eléctricas (IIE), Mexico

ABSTRACT

Los Sabinos, a small rural community in the mountains of Mexico has been electrified by means of small distributed PV systems. The systems were designed and built using for the firts time the Technical Specification for Photovoltaic Rural Electrification Systems originally developed at IIE. Since 1992 the systems have been monitored. In this paper, the results after two and a half years of operation are presented. The main problems found are pointed out and the lessons learned are discussed.

INTRODUCTION

In the late 1980's a hydroelectric dam was built in the Mexican state of Nayarit, some 1000 km west of Mexico City. The area flooded included some small towns which had to be relocated to higher land, Los Sabinos was among them. The electrification project was carried out with photovoltaic (PV) systems, since the utility distribution lines are 70 km away across very rough terrain.

The New Los Sabinos is located on the banks of the Santiago river, at 21 50' North latitude, 104 45' West longitude and 220 meters above the sea level. The nearest city to Los Sabinos is Tepic, the State Capital. There are no roads to Los Sabinos so the only way to get there at the beginning of the electrification program was by walking for around 20 hours or by aircraft. Since the summer of 1993 people can travel by boat across the artificial lake formed by the dam.

In the old Los Sabinos the population did not have electricity; they used candles and kerosene for lighting, and dry cells for radio listening. The use of such energy sources represented a large expense plus the difficulty of getting them. Hence, the use of small PV systems seemed to be a suitable alternative for the electrification project

CLIMATOLOGY OF THE SITE

A Li_Cor Minimum Data Set Recorder was installed in 1992 to measure some climatic and solarimetric variables. The system was taking data for almost three years. Los Sabinos has a hot dry climate almost all the year around. The rainy season goes from June trough September with a yearly precipitation of about 100 mm of rain. Mean temperature during the year is 25 ºC. The daily mean irrandiance is 5.5 kWh/m^2.

CHARACTERISTICS OF SYSTEMS

The PV systems installed at Los Sabinos were designed to meet some basic criteria such as: Reliability (ensuring the energy supply even, during periods of adverse climatic conditions), safety, maturity (technologies that have passed the experimental stage), easy to operate and maintain, sustainable (locally controlled and managed) and reproducible (potential for application in other rural communities). [1}

System's voltage was chosen to be 12 Volts DC, with a maximum voltage drop of less than 3% in each circuit. Autonomy of the systems was designed to be five days. Lead-acid car type batteries were used in all installations for reasons of availability. Table 1 shows the characteristics of the systems installed. [2]

Table 1. CHARACTERISTICS OF SYSTEMS INSTALLED

Installation	Lamps Qty/W	Tapeplayer Qty/W	TV Qty/W	Module Qty/W	Batteries Qty/Ah	Total Systems
Home	4/9	1/10	----	1/48	2/78	20
School	17/9	-----	1/50	2/48	2/100	1
Community hall	2/9 4/13	-----	1/50	4/48	4/100	1
Church	12/9	1/10	----	2/48	2/100	1
Medical dispensary	4/9	----	----	1/48	2/78	1
Street Lights	2/13	----	----	1/48	2/78	8
Basket-ball court Light	1/13	----	----	1/48	2/78	4

All installations have a charge controller (cc) of the shunt type to protect the battery. The cc can handle up to 15 amps. from the PV module and up to 20 amps. to the load. High voltage disconnect/reconnect and Low voltage disconnect/reconnect set points are 14.2/13.0 and 11.6/13.4 Volts

CH3365-4/94/0000-1134 $4.00 © 1994 IEEE

respectively. Lamps are of the compact fluorescent type. Later on four more systems were added as new people settled in town. In the Figure 1 shows a photograph of a typical installation in Los Sabinos.

Figure 1. Photograph of a typical installation

PROJECT MONITORING

PV systems are being monitored through field surveys The first one took place in October 1991 just before the contractor finished the installation of the systems. Since then 4 more inspections have been performed. The last one in May 1994. Evaluation of the project covers three main points. Physical and operative status of PV systems; degree of satisfaction of the users; and important aspects of sustainability, including the amount and effectiveness of information and training received by the user upon system installation, and the local availability of spare parts and maintenance services. While checking the physical and operative status of the installations , instantaneous values of the PV module current and voltage are recorded. Batteries are inspected searching for abnormal conditions like the presence of sulphate at the terminals, loose connectors and electrolyte leaks; the battery state of charge is determined from readings of the specific gravity in each cell. Lamps are checked for operational failures, such as flickering, low light intensity and burned out tubes. Information from the field surveys is processed by means of the computer program ESFER, [3] developed at the Electrical Research Institute of Mexico to evaluate the performance of PV rural electrification systems. ESFER is user friendly and delivers quick and easy to understand graphic results.

MAIN RESULTS

As of this writing most PV systems are three year old. Figure 2 shows the rating of all domestic, and communal systems as computed with ESFER.

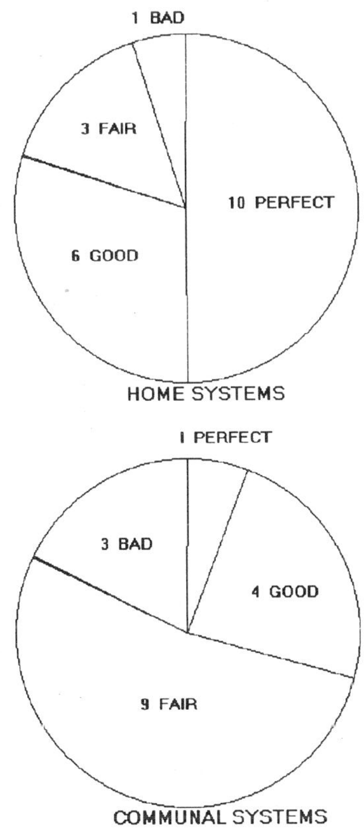

Figure 2. Rating of home and communal PV systems in Los Sabinos

The following comments apply in each case.

Domestic systems: Currently there are 20 domestic systems installed; 16 since 1991 and the rest since the end of 1992. The original design called for one 12 V, 120 Ah car type battery per system. Instead, two 12 V 78 Ah car type batteries connected in parallel were installed. As of May, 1994, 87.5% of the batteries originally installed were working. The rest had already been replaced. Two Charge controllers have been replaced thus far. Only four burned lamps have been reported; the rest are working well. Figure 3 shows the values of specific gravity (SG) measured in the batteries. Average SG values from all inspections were found at or above 1.25 equivalent to 90% state of charge. It can be noticed that in August 1992, average SG values were higher by the higher radiation levels during the summer time. Although average SG values remained within a narrow margin over time, minimum values decreased dramatically in 1993 and 1994, meaning that some batteries were performing poorly after two years of operation. Figure 4 shows the type and frequency of battery problems found incidence. High electrolyte level have been a recurring problem, followed sulphated battery terminals. It should be noticed that since February 1993 , systems maintenance was the sole responsibility of the community.

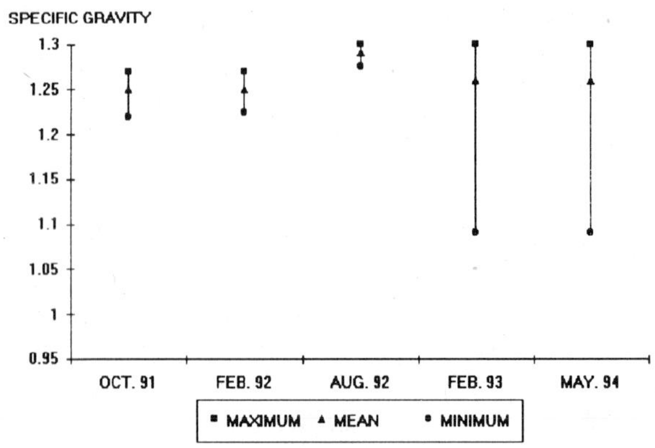

Figure 3. Specific gravity records in the batteries of home systems

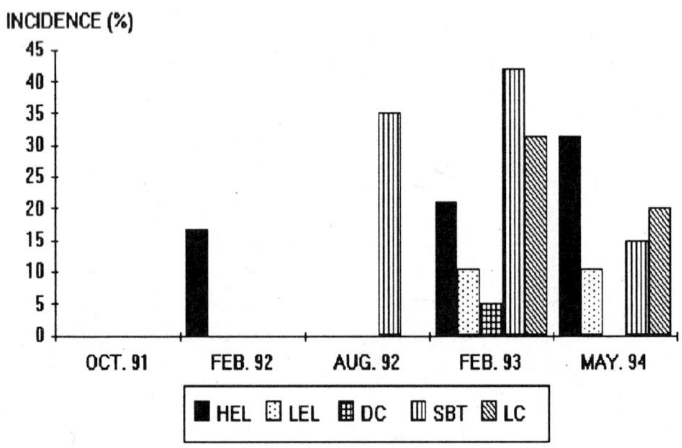

Key: **HEL** High Electrolyte Level **LEL** Low Electrolyte Level **DC** Dry Cells **SBT** Sulphated Battery Terminals **LC** Loose Connectors

Figure 4. Type and frequency of battery problems in domestic systems

Communal systems. The following installations are included in this category: church , schools, community hall, medical dispensary, and street and basket-ball court lighting. The principal problem in these installations has been with the batteries. . It should be noted that in the August 1992 inspection, the batteries were found with low electrolyte level. In the May 1994 inspection, 40% of the batteries were found already dry. This fact clearly points in the direction of poor maintenance of communal systems.

PROJECT SUSTAINABILITY

A number of factors that pose a threat to project sustainability have been identified. Among them are the poor cohesion among members of the community, which results in poor maintenance of communal systems, and the difficult access to spare parts. Users who experience trouble with their installations, need to travel a long way to buy lamps, fuses or new batteries. The most troubling issue up to now has been the apparent lack of responsibility with respect to the communal installations. The only exception is the PV system installed at the church, which is being maintained by the priest. He is a Catholic minister who appreciate the advantages of having a PV system at the church, since in the past he used to bring along a small gasoline generator set from another town across the mountains. The priest's influential position is being used to convince the population to look after the communal PV installations . Similar messages are been broadcasted trough a radio station (also powered by PV), from a nearby town in the mountains, since other communities that have been also electrified with PV in the region are following the same track.

USER SATISFACTION

In spite of the problems experienced, in general all the users have expressed their satisfaction with the PV systems. Their life styles have changed noticeably. Before the PV electrification program took place, they used kerosene lamps and candles for illumination, at an expense averaging 10 US$ per month. Now aside from not having to spend this amount, people appreciate not having around the smell of burning kerosene or paraffin, plus the added benefit of a safer environment (some people used to fall asleep leaving lamps or candles burning, with the inherent fire risk).

CONCLUSION

Local availability of spare parts is acritical element of sustainability of PV rural electrification proyects. It has been observed that the lack of spare parts induces defective operation and mantinience practices such using pieces of thick wire instead of fuses in the controller . Organization of the community to look after communal PV systems is also important . This proyect has shown that correctly sized, open, lead-acid car type batteries perform well in small PV systems, as long as they are properly maintained . This requeriment demands good information and proper training for the users. The batteries in Los Sabinos PV systems had undergone around 1000 crarge-discharge cycles as of May ,1994. In general people have expressed their satisfaction with the use of PV systems to replace candles and kerosene lamps.

REFERENCES

[1} J.M. Huacuz et al., "PV for Rural Electrification: The Road Ahead", Solar Industry Journal; Second Quarter; pp. 20-29, 1992

[2} J.M. Huacuz et al., "PV System Design for the New Los Sabinos Town", (In Spanish), Proceedings of XVI Reunion of ANES, , pp. 49-53, 1992.

[3} Villasenor et al., "ESFER, Computer Program for Rural Electrification PV Systems Evaluation", (In Spanish), Proceeding of XVIII Reunion of ANES, pp. 381-385, 1994.

FIELD EXPERIENCE AND COMPARISON OF PHOTOVOLTAIC OFF-GRID POWER SYSTEMS AND RETICULATED SUPPLY

I.F. MacGill M.E. Watt R.J. Kaye

Centre for Photovoltaic Devices and Systems
University of New South Wales
Sydney Australia

ABSTRACT

This paper reports the findings of a recent survey of remote area residents concerning their experiences with PV based off-grid systems and grid supply. Continued advances in off-grid systems would seem to make them an increasingly cost effective and 'grid equivalent' alternative for remote areas. While economic studies of the two supply options have been reported, less is known about other important factors for consumers assessing the viability of off-grid systems. Residents in remote areas were therefore surveyed on their choice of off-grid or grid power, and their experiences with this supply. Findings include the importance of supply quality and reliability for those who chose grid. Their experiences, however, suggest that off-grid power is more reliable and of similar quality. Overall, findings support the conclusion that PV based off-grid power systems are a viable alternative to reticulated supply for remote consumers.

INTRODUCTION

Off-grid power systems for residents in remote areas are an important and highly promising market for photovoltaics (PV). The growing commercial pressures on electricity utilities worldwide have focussed attention on the very high costs, and often unprofitable returns, of grid supply to remote and sparsely populated areas. At the same time, there have been important advances in off-grid power systems. The traditional single diesel or petrol generator is now giving way to hybrid power systems. These typically incorporate some combination of renewable energy input from photovoltaics (PV), wind turbines or micro-hydro, a diesel or petrol generator, battery storage and power electronics. Progress in power conversion and controllers, together with the falling costs of photovoltaics, have seen off-grid power systems become increasingly 'grid equivalent' and economically viable alternatives to reticulated supply in remote areas.

Utilities in Australia and the US (eg. [1]) are now examining off-grid supply options for their customers. Considerable work has been done on modelling the 'long run' costs of off-grid power systems according to factors including anticipated electrical loads, system configuration, and solar and wind resources.

There has also been some work on assessing the true 'long run' costs of reticulated supply to remote areas (eg. [2]). Certainly in Australia, broader community concerns have often outweighed commercial considerations.

Less is known however, about the views and experiences of consumers in remote areas with their off-grid or reticulated supply. Operational issues including supply reliability, power quality, convenience and lifestyle are clearly important in consumer assessments of whether off-grid systems are an appropriate alternative to grid connection. Only limited information on the operational performance of off-grid power systems is available from logged data. Similarly, electricity utilities normally have only limited, and often aggregated, data on the performance of their remote grid lines.

This paper outlines some results of a recent study [3] which examined and compared off-grid and reticulated supply options for remote areas across a broad range of operational issues. In particular, the study included a field survey of residents in remote areas of the Australian state of New South Wales (NSW). These consumers reported on their views of, and experiences with, off-grid or reticulated supply.

This paper focuses on off-grid systems that incorporate photovoltaics for at least part of their energy input. As shown later, these make up a large and growing proportion of off-grid systems in NSW.

FIELD SURVEY OF RESIDENTS IN REMOTE AREAS

Survey questionnaires were sent to NSW residents who had, within the last six years, chosen and installed either an off-grid system or grid extension for their domestic power supply. Residents were questioned on issues including:

- important factors in their decision for off-grid or reticulated supply,
- the household appliances that are run off their electricity supply,

First WCPEC; Dec. 5–9, 1994; Hawaii

- where an off-grid power system was chosen, its type, technology and equipment ratings,
- their perception of the overall costs of their supply,
- performance measures of their power supply including supply quality, reliability and convenience, and
- their overall satisfaction with their chosen supply option.

Approximately 700 surveys were sent out with over 300 being returned, a response rate of around 40%.

OFF-GRID AND GRID SUPPLY OPTIONS

A trend towards off-grid power in remote areas is clearly evident in New South Wales. The State Government's Remote Area Power Assistance Scheme (RAPAS) offers financial grants that cover a proportion of the costs of either grid connection, or the purchase of an off-grid power system. There has been a marked shift in consumer preference towards off-grid supply over the life of this scheme. Off-grid systems made up only one third of grant applications when the scheme began in 1987, yet almost 60% of applications in 1993.

The capital cost of grid connection for consumers who chose grid supply has averaged approximately US$17,000. The grid connection quotation given to consumers who eventually chose off-grid systems has averaged US$41,000, while the cost of their off-grid power systems has averaged around US$8,500.

Survey respondents identified the main factors behind their choice of power supply (see Fig. 1). The high capital costs of long grid extensions and 'lifestyle' considerations were driving factors for those who chose off-grid PV based supply. Respondents opting for grid connection were more concerned with supply reliability, power quality and perceived operating costs than were off-grid residents. They placed further importance on avoiding potential problems with off-grid system maintenance and load limitations.

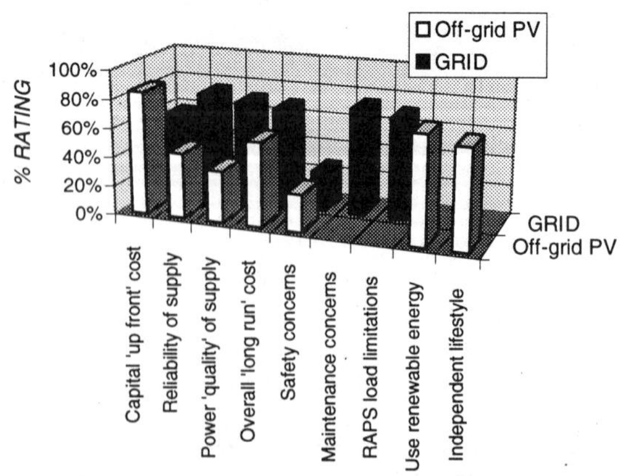

Fig. 1. Consumer rating of the importance of various factors in their choice of power supply.

OFF-GRID POWER SYSTEMS

There have been some important trends in off-grid power system design over the seven years covered by the survey. Hybrid systems combining PV, a diesel or petrol generator, batteries and an inverter made up around 60% of systems installed from 1987 to 1989, yet over 85% of off-grid systems in the period 1992 to 1993 (see Fig. 2).

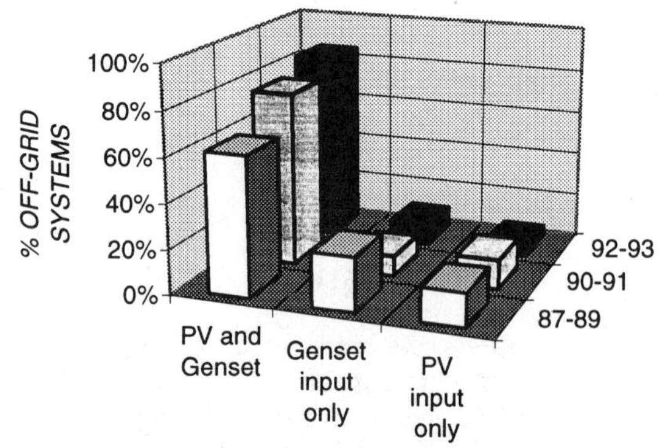

Fig. 2. Trends in off-grid system configurations

The proportion of systems incorporating PV rose from around 75% to 90% over the same period, while the average size of these arrays grew from around 340 watts to 600 watts (see Fig. 3). This trend to larger arrays corresponds with a move towards petrol generators (offering lower capital, though higher operating, costs than diesel), and suggests that system design is moving towards greater reliance on renewable energy input, and lower generator usage.

With power conversion electronics, there is a clear trend towards 'true' sinewave inverters offering superior power quality to square and quasi-sine waveform units.

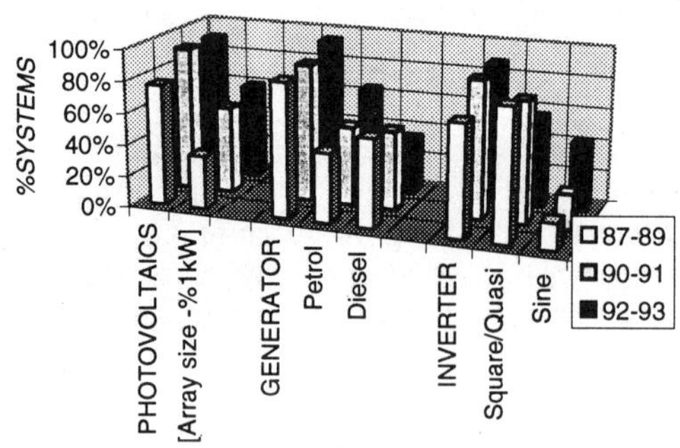

Fig. 3. Trends in off-grid system components.

RESIDENTIAL ELECTRICAL LOADS

The survey has highlighted some marked differences between the electrical loads of consumers with grid power and those with off-grid PV systems (see Fig. 4). Household appliances with high power demands including cooktops, ovens, hot water systems and, to a lesser extent, refrigeration are rarely run from an off-grid electricity supply.

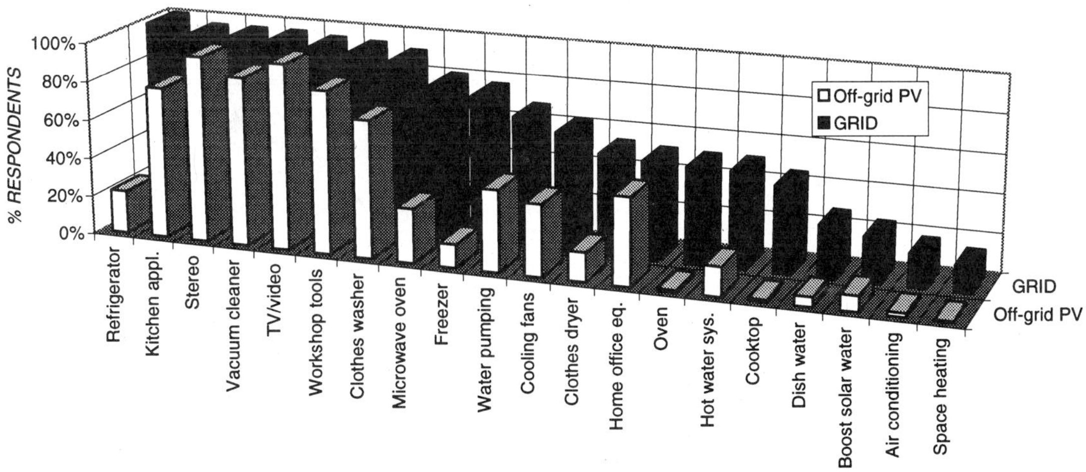

Fig. 4. Household appliances run from consumers' electricity supply.

Owners of these systems will, instead, usually satisfy these needs with other energy options, notably solar thermal, wood and bottled gas. With the exception of refrigeration, a substantial proportion of grid connected dwellings also do not use electricity for these services.

FIELD PERFORMANCE OF PV BASED OFF-GRID SYSTEMS AND GRID SUPPLY

Survey respondents detailed the quality and reliability of their power supply. Across four measures of power quality, grid connected residents reported more problems with lights flickering and momentary blackouts, while those with off-grid PV systems experienced more appliance problems and phone interference (see Fig. 5).
With regard to supply reliability, a considerably greater proportion of consumers with grid connection had experienced supply failure than those with off-grid PV based systems (see Fig. 6).
The survey also asked respondents to rate a range of possible concerns with their electricity supply (see Fig. 7). For those with off-grid PV based systems, important concerns included load restrictions, generator fumes and noise, maintenance and a lack of technical support.

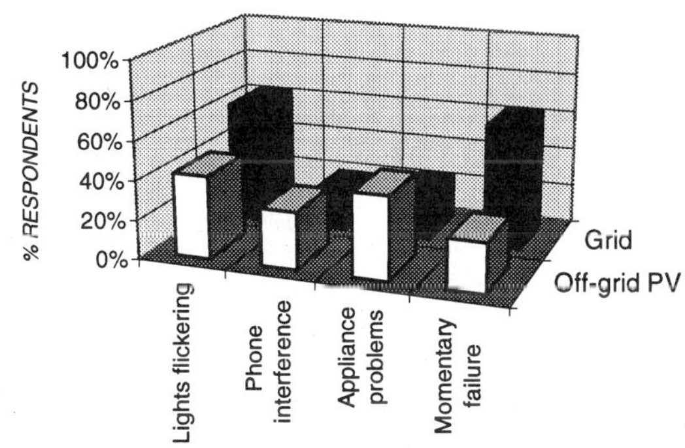

Fig. 5. Respondents reporting problems with the power quality of their electricity supply.

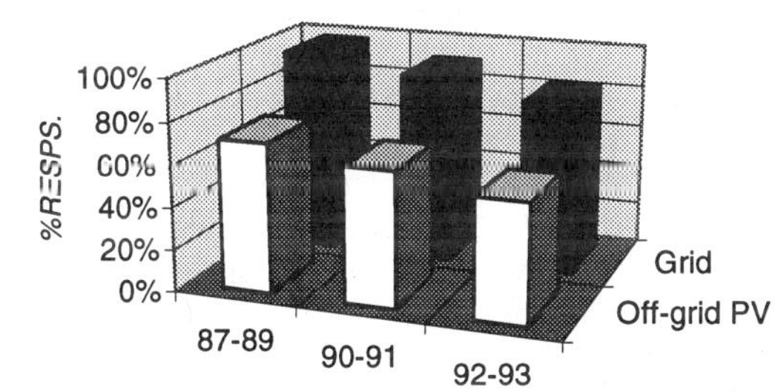

Fig. 6. Respondents reporting supply failure.

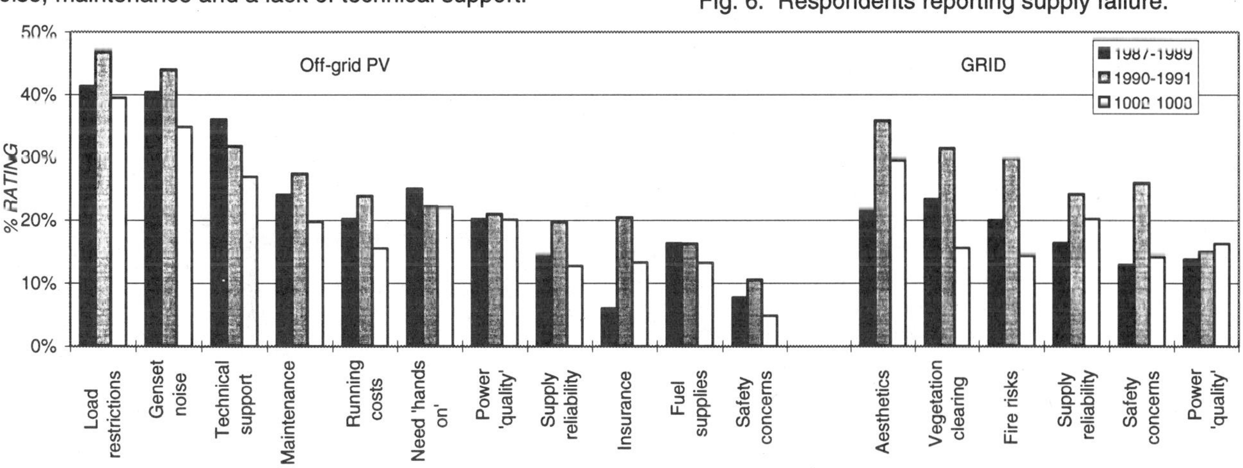

Fig. 7. Respondent ratings of their concerns for a range of potential problems with their grid or off-grid supply.

Overall, grid connected residents had lower rated concerns. The environmental impacts of grid connection were seen as most important, although supply reliability rated as a higher cause of concern than for consumers with off-grid systems.

CONSUMER SATISFACTION

Both respondents with grid connection and those with off-grid PV based power systems reported a high level of satisfaction with their chosen supply option; an overall 78% approval rating for grid and 76% approval rating for PV based off-grid supply. Accordingly, most would choose the same supply option given the choice again (see Fig. 8). Satisfaction with PV based off-grid systems has risen over the last seven years while there has been no equivalent trend with grid connected residents.

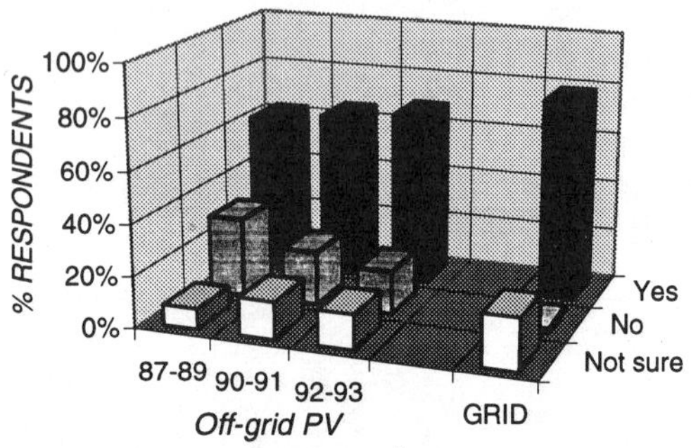

Fig. 8. Respondents who would select the same supply option if given the choice again.

CONCLUSIONS

PV based off-grid power systems would seem to be an increasingly appropriate alternative to grid supply for residents in remote areas. Survey findings reported in this paper highlight some of the important operational issues with off-grid systems and grid extensions as competing options for remote power supply.

Power quality and supply reliability were driving factors for consumers who opted for grid supply. Their major concerns with off-grid systems were maintenance requirements and possible load restrictions.

Field experience has, however, cast new light on these consumer perceptions. Survey findings suggest that off-grid systems provide comparable power quality to the grid. Further, sinewave inverters offer superior power quality than long grid extensions, and are becoming the inverter of choice for off-grid systems.

Off-grid systems are also proving a more reliable power supply option than grid connection. This is not so surprising when one considers the long distances, over often rugged and exposed terrain, that these lines are required to run. Supply failures with off-grid systems can result from equipment failure or an extended period without energy input; for example, a stretch of bad weather. The trend towards advanced hybrid systems combining PV with diesel or petrol generators, together with what often amounts to several days of energy storage, can make off-grid supply a very reliable option.

The potential for load restrictions with off-grid systems arises from both the power quality requirements of some appliances, and their typically lower power capacity than grid connections. The former is becoming less of an issue with the trend towards sinewave inverters. The latter need not be an issue as consumers with off-grid systems tend to use energy efficient appliances, and run household services with high power demands from more appropriate energy sources.

Perceptions of the ongoing costs of their power supply were particularly important to residents who chose grid connection. The true operating costs of off-grid and grid supply are distorted, certainly within Australia, by the high cross subsidies that grid connected consumers in remote areas currently enjoy. In NSW, this has been estimated to average $700 each year for each such consumer [2]. Any move towards more cost reflective operating charges will clearly make grid connection less attractive. The high and growing level of satisfaction with PV based off-grid power systems is even more encouraging in light of these financial inequities.

Many consumers in remote areas, as well as the utilities that serve them, are unaware of the capabilities, performance and potential advantages of PV based off-grid power systems. The shift in consumer preference towards off-grid systems suggests, that in NSW at least, this situation is now changing. Utility interest worldwide is also growing, and is likely to accelerate this trend.

Overall, survey findings support the conclusion that PV based off-grid power systems are now a viable alternative to remote grid supply across the range of economic, technical and broader operational considerations.

ACKNOWLEDGMENTS

The research on which much of this paper is based was funded by the NSW State Energy Research and Development Fund, and the Electricity Council of NSW. The authors would also like to acknowledge the support and participation of the Electricity Association of NSW and four NSW Electricity Distributors: Northern Rivers, Shortland, Western Power and Prospect Electricity.

REFERENCES

[1] L.A. Crowley, "Utility Owned Remote PV Systems", *Solar Today*, March/April, 1993, pp. 29-30.

[2] E.J. Henderson, "Rural Electricity - A Community Service Obligation?", *in Proc. Elec. Supply Eng. Assoc. (ESEA) Regional Conf.*, Australia, 1992, pp. 3.1-3.11.

[3] M.E. Watt and I.F. MacGill, *Opportunities for NSW Electricity Distributors to Supply RAPS Systems*, SERDF project report to NSW Office of Energy, Australia, 1994.

PV BASED SYSTEMS, WITH WIND, DIESEL OR LPG GENSET BACKUP, SUPPLYING SMALL TV REBROADCAST STATIONS IN PORTUGAL

Horácio F. Ramos

ABSTRACT

This paper describes the implementation of a program intended to introduce PV based hybrid power systems to supply electrical power to small size TV rebroadcast stations in Portugal. Reliability is a major concern to this type of application, as well as economical and social constraints, so wind or diesel/LPG genset backup are used. This paper includes a description of the systems behaviour, comparison among these topologies and economical viability data from an users view point.

INTRODUCTION

As direct Television Broadcasting uses UHF and VHF frequency bands, coverage area is very limited by orographic conditions, and on hilly countries there is a need to use several repeaters on high places to feed signals to towns and villages in valleys and low countries. This is the case in Portugal, where most repeaters are located in far and remote locations; being of low radiated power (PAR) they just cover sometimes a few hundred viewers.

Traditionally the cost involved in establishing such infrastructures is very high due mainly to its remoteness. On the other hand, the areas of coverage of these repeaters are usually undeveloped and poor, and TV is considered a mean to help populations grow, both socially and economically. This way broadcasting operators must reduce the investments on these repeaters, in order to interest the TV stations on their move to markets which are not exactly appealing commercially.

One of the major costs is the one related to electrical power supply, due to the remoteness and adverse locations of these sites: the costs of constructing a power line, a transformer station and electrical installation is so high that in most cases projects have to be abandoned. Off-grid PV-based systems become then a very sound and attractive solution, both economical and technical wise.

For increased reliability, in order to reduce initial investment costs (related to not oversizing the PV panels and battery capacity) and to cope with the lack of meteorological data to correctly size the systems, the PV-based hybrid systems are the natural choices, with distinct backup solutions (wind, diesel or LPG gensets) for different power requirements.

Teledifusora de Portugal, SA (TDP), a company that operates 3 nation-wide and 2 regional TV networks in Portugal and recently merged with 2 other companies to create Portugal Telecom, SA, began end of 1992, and is currently running, a program to assess the economical and technical viability of using renewable energy sources to drive its remote low power TV repeaters.

This program has the support of the EC Thermie found (project SE/90/93/PO and SE/166/94/PO/ES) and is being currently implemented together with TUDOR (P) and ATERSA (E), two companies that manufacture and install PV systems. The four different topologies explained above are being tested in 26 sites, 4 of them already in operation, the others to be implemented till the end of 1997.

SYSTEM DESCRIPTION

These systems are intended to run isolated (off-grid), be of very low maintenance (less than 1-visit per year), and no power failures are admitted (100% reliability). The criteria for their design are:

PV stand alone

For "very small power" repeaters (power ratings up to 100W, consumption up to 650 kWh/year) the chosen solution is PV stand alone compact system. This enables the use of all-weather cabinets to shelter the equipment, instead of constructing a full house, thus reducing costs and installation time. Moreover this systems are engineered to be very low maintenance (1 visit per 2 years), with reduced operational costs.

CH3365-4/94/0000-1141 $4.00 © 1994 IEEE First WCPEC; Dec. 5-9, 1994; Hawaii

PV/Wind hybrid

For "small power" repeaters (power ratings up to 300W, consumption up to 3.000 kWh/year) the selected solution is a PV and wind hybrid system, with a 50/50 contribution by each source.

PV/LPG hybrid

Also for "small power" repeaters, and where wind conditions are not adequate, a solution with an electrical generator running with Liquid Propane Gas (LPG) as backup is selected.

PV/diesel hybrid

For higher power ratings and consumption (up to 1.2kW and 8.500 kWh/year) a fourth configuration is selected, with a diesel genset backing up a PV array.

This option was taken mainly due to the fact that in these sites there was already a station running 24h/day with diesel gensets and the intention was to lower operation costs and maintenance cycles, increasing reliability and energy efficiency. The existing gensets are to be overhauled and supply backup power to the PV system.

Components

Topology: the PV array supplies more than 50% of the necessary energy. Each system has PV panels, backup energy source, battery bank, voltage regulator and monitoring equipment. Working voltage is 48Vdc.

Load: in each site (station) the system drives 3 distinct loads (one repeating equipment for each TV network) working at 48Vdc; while in stand-by mode these equipment's need only 10% of their nominal consumption power. In the following discussion peak power and consumption are shown as totals for all loads. Repeaters from different manufacturers and models are to be tested in order to assess their energy efficiency, compatibility with these systems and look for emc related problems.

PV array: the selected PV panels are monocristaline silicium cells of 52Wp power, manufactured by BP (E) or ISOFOTON (E). The supporting structures are in hot galvanised steel, able to withstand winds up to 200 km/h. Special care is given to the architectural integration of panels and the economical/technical advantages of variable tilt PV arrays are to be studied. For compact PV-only systems the panels are located in the same tower used for the antennas, the batteries, controller and transmitting equipment sharing the same cabinet.

Battery bank: the batteries used are maintenance-free, gel type, made by HAGEN (D), which enables them to be installed in the same room with the electronic equipment and in any position, saving space and maintenance costs.

Regulator: the microprocessor-based PV charge controllers are made by ATERSA (E); these units accurately regulate the battery charging according to their working conditions (i.e. temperature, capacity) and start the backup genset when available. For increased reliability the voltage regulators for the PV and the backup are independent, enabling the system to keep working with one of them out of service.

Monitoring: each system has a complete data acquisition and logging system to record internal and external variables (e.g. battery voltage, load current, radiation, air temperature, wind speed and so on). Real-time monitoring will be accomplished in some stations by a satellite-based system (Meteosat), by a private UHF communications system or GSM data communication services. This enables the maintenance services to solve malfunctions without disrupting broadcasting.

Wind generator: the windmill is a 1100 W (at 7m/s) upwind horizontal-axis machine made by LMW (NL) with a three blade rotor of 3m diameter, equipped with a brushless PMG generator, installed either on top of a 36m concrete tower or on a 18m pole. The rectifier and series voltage regulator is installed near the battery bank and charge controller.

LPG genset: this system uses a 1-cylinder 4-stroke 162 cm3 engine driving an ac 2 kVA asynchronous electrical machine, used both as starting motor and generator. The controller starts the LPG system using the main battery bank thus reducing the need to have a separate battery and regulator for start-up. The fuel is stored under pressure in easily replaceable 45kg containers. The quantity of fuel stored is enough to feed the system between routine maintenance visits.

Diesel genset: existing gensets with air-cooled diesel engines driving 5 to 25-kVA AC 3-phase generators supplying a simple rectifier bridge and regulator are to be refurbished and overhauled. Starting motor uses an independent battery, fuel is stored in a large capacity tank and oil level has automatic compensation., which enables maintenance to be performed at much larger intervals (every 500 hours or once a year).

Implementation

This program is currently being implemented; the sites to be constructed in 1996 and 1997 (13) are yet under study.

Site	Type of system	Status
1 - Piodão	PV+wind	completed
2 - Alvoco	PV+wind	completed
3 - Loriga	PV+wind	completed
4 - Alpedrinha	PV+wind	completed
5 - Tourém	PV + LPG	on-going
6 - Pico da Urze	PV + diesel	on-going
7 - Rio d' Onor	PV only	on-going
8 - Barriosa	PV + diesel	1995
9 - Alcaria	PV + LPG	1995

Table 1: the list of sites already in operation and to be installed during 1995

Studies to be performed

This program was designed to enable a complete assessment of all design, technical and economical variables at stake when engineering PV based power systems to telecommunications facilities.

The major innovative aspects in this application are:
- the use of maintenance-free gel batteries;
- mounting wind generators on top of concrete towers;
- using compact small size LPG generators systems to increase reliability of PV power systems;
- the architectural integration of PV panels and structures.

The benefits of using variable inclination PV arrays, the use of low maintenance components to reduce operation costs, and the real-time monitoring to increase reliability, are also aspects to be studied with this program.

PROS AND CONS

Comparing in technical terms the traditional solution of grid connection and this PV-based systems, in any of the topologies decribed above, the advantage goes obviously for the PV systems.

Advantages of these systems are the lower operational costs and their intrinsically safe nature in terms of lightning protection and feed line overvoltage and transients, one of the major causes of malfunctions in telecommunications equipment.

Disadvantages are the lower life expectancy, with the need to replace the battery bank within 15-years of operation and eventually the higher maintainability required by PV-based systems.

The use of hybrid systems instead of only PV gives a greater reliability to the system and the possibility to have an alternative energy source in place to use for emergency and maintenance tasks.

ECONOMICAL ASPECTS

The telecommunications business appears as a very interesting market niche to renewable energy based power systems, namely PV, as their cost is becoming lower than the traditional solution of grid connection.

With the structure of prices of Portugal in 1994, and using the calculation methods explained in [1], it can be figured out the minimum distance from which PV-based systems become cheaper than grid-connection (obviously playback time is null and we are dealing with a sound investment up front):

average yearly energy consumption	distance to LV grid	distance to HV grid	most economical solution
< 650 kWh/year	< 1.000 m	< 100 m	AC grid
	> 1.000 m	> 100 m	PV only
< 3.000 kWh/year	< 1.500 m	< 1.000 m	AC grid
	> 1.500 m	> 1.000 m	PV/wind PV/LPG
< 8.500 kWh/year	< 1.500 m	< 2.300 m	AC grid
	> 1.500 m	> 2.300 m	PV/diesel

Table 2: decision table for selecting the most economical power system

Compared with a PV system, which is of a proven, solid, reliable and maintenance-free construction, grid connection involves constructing a very comprehensive set of infrastructures, like HV power line, transformer station, overvoltage protection, isolating transformer, distribution panels and so on, thus becoming very expensive. Moreover one can always even lower initial investment on renewable energy systems, downsizing the PV array (using more backup produced energy) and upgrade the installation on a later stage to fine tune to the local meteorological conditions.

SOCIAL CONSTRAINTS

In remote locations of less developed countries like Portugal, the population is poor and becoming very old since youngsters are leaving for the big towns; bringing news media and communications facilities to this populations is a tool to invert this situation and avoid desertification of rural areas.

On the other hand major focus is being given nowadays to environmental issues; since these TV repeaters are most of the time in natural parks and other reserved areas, avoiding to construct power lines will for sure help preserving nature thus bringing development in a very balanced way.

Finally in this way the PV technology is promoted within potential users, as they can watch this systems and easily find other uses and applications to this emerging market.

CONCLUSIONS

In Europe the renewable energy markets, namely the PV, are being expanded by an effort of the European Commission and their subsidies strategies (Thermie), enabling manufacturers and users to develop products and applications with reduced financial risks.

The use of hybrid systems in Telecommunications facilities was already tested in various ways [2-3], but never before in such a comprehensive program, as the one described here. We conclude that:

Technically PV-based hybrid systems are definitively more reliable than the traditional solution of grid connection.

Economical wise, even without the possibility of having 49% non-refundable loans from the EC and national Founds, these systems are getting every day more competitive, as their costs lower and the power lines are getting higher.

DEFINITIONS

D	: Germany
E	: Spain
EC	: European Commission
EMC	: Electromagnetic Compatibility
EU	: European Union
HV	: High Voltage
LV	: Low Voltage
LPG	: Liquid Propane Gas
NL	: Netherlands
P	: Portugal
PAR	: Radiated power
PV	: Photovoltaic
TDP	: Teledifusora de Portugal, SA
TV	: Television

REFERENCES

[1] H. Ramos, "Sistemas Fotovoltaicos de Alimentação de Energia Eléctrica a Pequenos Retransmissores de Televisão: Relato duma Experiência", *VII Congresso Ibérico de Energia Solar*, 1994, pp. 405-410

[2] EC Directorate-General for Energy (DG XVII), "THERMIE European Solar Photovoltaic Technology Projects", 1993

[3] Proceedings of the "European Seminar on the Use of Photovoltaic Technologies in Rural and Isolated Areas" held in Freiburg, Germany, 1992

DEVELOPMENT AND OPERATION OF A PHOTOVOLTAIC POWER SYSTEM FOR USE AT REMOTE ANTARCTIC SITES

Michael F. Piszczor, Lisa L. Kohout & Michelle Manzo
NASA Lewis Research Center
Cleveland, OH 44135

Anthony J. Colozza
NYMA
Brook Park, OH 44142

ABSTRACT

A photovoltaic power system, designed and built at the NASA Lewis Research Center, has successfully operated over the past two summer seasons at a remote site in Antarctica, providing utility-type power for a six-person field team. The system was installed at the Lake Hoare site for approximately five weeks during late 1992, put into storage for the Antarctic winter, and then used again during the 1993 season. The photovoltaic power system consists of three silicon photovoltaic sub-arrays delivering a total of 1.5 kWe peak power, three lead-acid gel battery modules supplying 2.4 kWh, and an electrical distribution system which delivers 120 Vac and 12 Vdc to the user. The system worked extremely well in providing quiet, reliable power. The experience gained from early system demonstrations such as this should be beneficial in accelerating the transition toward future PV systems in Antarctica and other similar areas.

INTRODUCTION

In 1991, a small team at the NASA Lewis Research Center began to design and fabricate a photovoltaic system that could provide utility-type power to a six person field team conducting research at Lake Hoare in the Dry Valleys of Antarctica. Previously the camp had been powered by diesel generators, which were noisy and also polluted the pristine environment which was being studied. In an effort to reduce its dependence on diesel fuel, the National Science Foundation (NSF), which sponsored the research team, became interested in the use of alternate energy sources. The move to eventually utilize a solar powered system was motivated by environmental concerns as well as long-term cost issues.

Design, fabrication and operation of the photovoltaic power system at Lake Hoare was initiated as part of the Antarctic Space Analog Program, a joint NASA/NSF program designed to take advantage of the similarities between Antarctica and the harsh environment of space. The Antarctic Space Analog Program was begun in 1990 to unlock the potential of the Antarctic environment for testing hardware and habitats that support human researchers. For NASA, the program provides opportunities to prepare for future missions in remote and hostile environments, while the NSF benefits from the application of NASA-developed technologies and systems that can increase the efficiency, reduce the cost, and minimize the environmental impact of the US Antarctic Program. NASA specifically benefited from the PV power system project by gaining valuable information with regard to system level deployment and operation in a remote location by a minimally trained crew, as well as validation of the initial integration concept.

DESIGN REQUIREMENTS

The unique environment of Antarctica required an unconventional photovoltaic array design. A number of factors needed to be taken into account, such as the severe cold, the remote operational location and the circular path of the sun. From a human factors standpoint, the system had to be easy for the field team to deploy while wearing heavy gloves, which meant minimizing the number of bolted and hardwired connections. Also, the team would have minimal training so the system had to be easy to deploy and operate. Since the entire system could only be transported to the remote site by helicopter, strict volume and mass constraints had to be met.

Power consumption at the site was largely unknown, therefore the system was designed to handle a varied load profile. The loads included the use of personal computers and printers, lab equipment, lighting and a small microwave oven. It was determined that the camp needed a baseline of 0.5 kW from the array and 2.4 kWh of energy storage. Provisions were made to supply both 12 Vdc and 110/120 Vac to accommodate all of the equipment.

CH3365-4/94/0000-1145 $4.00 © 1994 IEEE First WCPEC; Dec. 5–9, 1994; Hawaii

SYSTEM DESCRIPTION

The photovoltaic power system consists of three silicon photovoltaic sub-arrays delivering a total of 1.5 kWe peak power, three lead-acid gel battery modules supplying 2.4 kWh, a field junction stand, an instrument and control rack, and two dc power supplies [1]. Figure 1 shows one of the sub-arrays with the balance of the system components. In order to minimize the number of hardwire connections to be made at the site, the modules were pre-wired prior to shipping and connections were made between modules with plug-type connectors. This design approach not only made for easier installation, but also reduced the risk of electric shock by protecting the field team from exposure to potentially hot wires. The only hardwired connection required during installation was between the power system and the existing ac distribution system at the site. This modular design approach also allowed for portability of the system and ease of relocation from the drop site to the installation point.

The PV array design selected was a flat panel array capable of rotating 360°. The power system was sized to put out the minimal amount of required power based on an array panel positioned at 70° from the horizontal and realigned every 6 hours toward the sun. This design required 24 Siemens Solar Industries M55 solar modules to supply the needed power. These modules were divided into 3 sub-arrays of 8 series-connected modules each, with each sub-array being connected in parallel.

To maximize the ease of installation, the sub-arrays were designed to be deployed and restowed without the use of any materials or tools. In the stowed configuration, the base legs are raised to a vertical position and two wings of two panels each are folded over the four central panels. This provides a compact stowed configuration and provides both protection for the PV cells and eliminates any electrical hazard since light can not reach the panels. (See Fig. 2). During installation, the array legs, which were hinged to the base, would be lowered and a bar along the side of the leg would slide forward, locking the legs in the horizontal position. The base legs could then be secured to the ground with either a ground stake or by weighing down the legs with rocks or sand. The two outer wings which were hinged to the central panels would then be unfolded and the outer edge of each would be secured to the structure with a movable brace connected to the main body of the sub-array. The brace would be attached to the outer panels with the use of a self-locking pin.

The rotation of the array is accomplished by a roller plate between the top and bottom portions of the base. A spring-activated locking pin attached to a foot lever is used to lock the array into position every 90°. There is also a safety pin which could be inserted between the upper and lower portions of the base to secure the base from rotating when the array is exposed to high velocity winds. (The array structure was tested in a wind tunnel with wind speeds up to 60 mph.) The array can be turned up to 360° clockwise and then must be turned back 360° counter clockwise to begin the next cycle. A stop was placed inside the base to prevent the continual rotation of the array. This feature prevented the power wire from coming off the array during rotation of the array.

Batteries are required on the electrical bus to provide stabilization during current surges and also to provide power during shadow periods. The system is designed to provide 2.4 kWh of storage. At room temperature it is capable of delivering 9.6 kWh at a 20 hour discharge rate. However, operational battery capacity was significantly derated because of the extremely low temperatures encountered. The battery storage system consists of 10 sealed lead-acid gel-electrolyte batteries connected in series to provide a nominal 120 Vdc on discharge. The cell capacity is rated at 80 AH at the 20 hour discharge rate at room temperature. Dynasty GC12V80 batteries, manufactured by Johnson Controls, were chosen for the system.

The batteries are configured into three modules: two consisting of three batteries each and one consisting of four batteries. The batteries within each module are connected in series to provide a nominal 36 Vdc and 48 Vdc, respectively. The three modules are further connected in series to achieve the 120 Vdc battery subsystem. Each battery module enclosure is made of aluminum to minimize weight. A layer of styrofoam between the enclosure walls and the batteries provides thermal insulation. Connections between modules are made via Supercon connectors installed through the walls of the enclosures. These connectors were chosen for the inherent safety feature of shrouded contacts so as to minimize the risk of electrical shock.

The field junction stand serves as the electrical junction point for the sub-arrays. At this junction, the individual sub-arrays are electrically connected in parallel and the array output power is routed to the instrument and control rack. For each sub-array, there is an associated non-fused disconnect box mounted on the stand, which isolates the sub-array from the rest of the bus for servicing.

The instrument and control rack provides the main power bus interconnects, circuit protection, battery charge/ discharge control, ac/dc supply connections, and data display and acquistion. Battery state-of-charge is regulated through the use of a battery charge controller, manufactured by Photocomm, Inc. The battery charge controller uses a series regulator to control the battery charge/discharge state. A utility-type ac inverter is also built into the rack. It generates a quasi-sine output, 117 Vac(trms), 60 Hz from the main bus high voltage dc. In order to provide the user with 12 Vdc power, dc-to-dc converters were designed into a power supply package which provides dual outputs for a total of 72 W.

REMOTE SITE OPERATION

The power system was modularized for ease of deployment and operation. The crated units were transported to the remote site by helicopter, with final assembly and placement of the system being done by hand under demanding Antarctic conditions. The system was first installed at a field camp at Lake Hoare in the Dry Valleys of Antarctica in October 1992. (See Fig. 3). The power system supported a team of 6 researchers during the austral summer, providing power to both their living area (a Jamesway hut) and a small laboratory facility. The system was restowed at the end of the five week period, wintered-over at the site and deployed again the following season (October 1993). The photovoltaic power system was officially transferred to NSF in August 1993.

Analysis of the system data showed that the array output voltage approached 180 Vdc, which was higher than expected, most likely due to lower than anticipated ambient temperatures (approximately -22 C). This array output voltage exceeded the rated voltage of a gas discharge tube and metal oxide varistor, causing them to fail. The failure of these devices, which are not necessary for Antarctic applications, did not affect the operation or performance of the system. The battery charge controller was modified to accommodate the higher voltages.

Very favorable comments were received from the remote site team on the simplicity of the design and the reliability of the system. Another important feature was the quiet operation of the system, compared to the diesel generators used before. The significant increase in power also prompted the research team to start planning for the use of additional equipment the following season.

SUMMARY

The photovoltaic power system was installed at the Lake Hoare site for approximately five weeks during late 1992, put into storage for the Antarctic winter, and then used again during the 1993 season. Significantly more power than expected was generated by the system, due to lower average temperatures. Favorable results were received from the field team with regard to the simplicity, reliability and quiet operation of the system. While it is expected that the use of photovoltaics, and other alternate energy systems, will increase in Antarctica, the lessons learned from these early systems should be extremely beneficial in accelerating the transition toward these future systems.

REFERENCES

[1] L.L. Kohout, A. Merolla and A.J. Colozza, "A Solar Photovoltaic Power System for Use in Antarctica", NASA Technical Memorandum 106417, December 1993.

Fig. 1. Antarctic photovoltaic power system showing (left to right) the instrument & control rack, field junction stand, one PV sub-array and the battery modules.

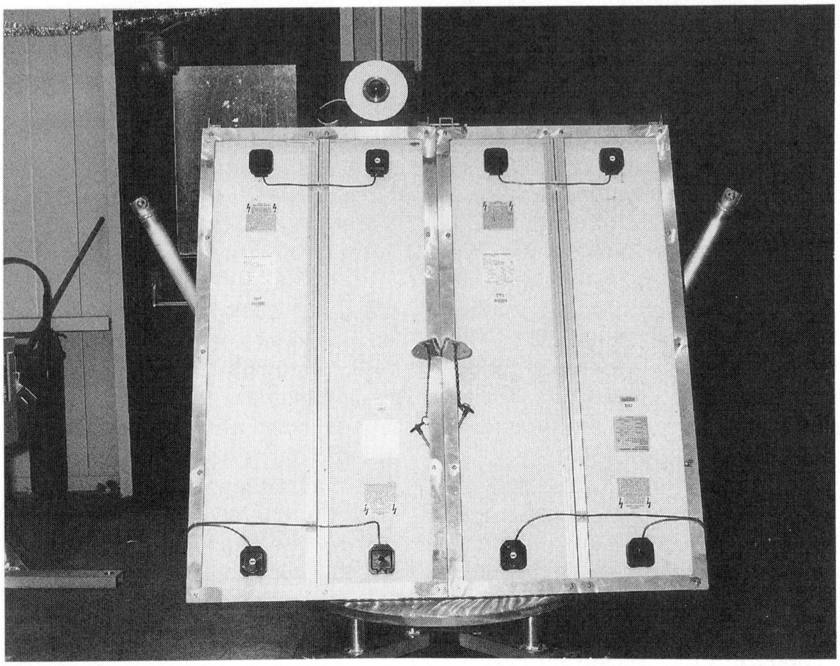

Fig. 2. One photovoltaic sub-array in the stowed configuration (front view).

Fig. 3. Photovoltaic power system in operation at Lake Hoare site in Antarctica.

ANALYSIS OF MARKET DEVELOPMENT FOR PHOTOVOLTAICS

J A Gregory[1], A S Bahaj[2] & R S Stainton[3]

[1] IT Power Ltd, The Warren, Bramshill Road, Eversley, Hants, RG27 0PR, UK,

Tel. +44 1734 730073, Fax. +44 1734 730820

[2]Institute of Cryogenics and Energy Systems, [3] Accounting and Management Studies,

University of Southampton, Highfield, Southampton SO19 5NH, UK,

Tel. +44 1703 592051, Fax. 44-1703-593053

ABSTRACT

The acceptance of any product by the market is dependant upon a multitude of formal and informal mechanisms. These include, amongst others, the technical merit of the product, its presentation, its perceived image and the user feedback to the market place. Further growth or diffusion of the product into the market-place is possible if the image and perception are positive. If the product is difficult to use, and/or lacks the information necessary for implementation and operation, it is unlikely that it will sell well. The result is a negative image and slow market growth.

Today, photovoltaic (PV) systems are technically appropriate for many applications. However, apart from specific niche markets, the overall market penetration for PV is small when compared to other products of similar developmental history.

This paper discusses the market development of PV applications, reviews some of the promoted scenarios and depicts past experiences with PV systems. In particular, it looks at the market development for PV in the telecommunications and health care sectors.

INTRODUCTION

Although PV for terrestrial applications commenced in the mid 1970s, it was not until the mid-1980s that PV markets began to expand in any significant manner, and a real commercial market to develop.

The development of the PV market has variously been compared to the development of the computer, integrated circuits, automobiles etc. But, unlike these markets, the mass market has not yet been established for PV.

The niche applications which expanded first and fastest were for industrial applications such as telecommunications and navigation aids. PV systems were designed to suit the technical requirements and the economic constraints of these customers, and a market developed and matured. Today, there are many large telecommunications installations around the world powered by PV, for instance Australia Telecom has one of the longest experiences with PV and over 8,000 installations powering remote telecommunications [1].

For other applications, however, the market for PV has not grown as significantly as predicted. This is particularly true of the PV vaccine refrigerator market, where PV is considered to be reliable and economic by the World Health Organisation and the PV industry, but the general sentiment in the market-place is that it is still unreliable and too expensive.

PAST PREDICTIONS AND CURRENT SCENARIOS

During the past 15 years, many predictions have been made for market growth, most of which have been over-optimistic and unrealised.

In 1983 Starr & Palz stated that ' assuming that full support is given to the industry and that appropriate financial and fiscal encouragement are provided to purchase the systems ... in 1990 the market [in Europe] could be about 20 MWp per year, rising to between 650 and 1200 MWp per year to the year 2000, implying average growth rates ... of between 45% and 50 % per year.'[2].

In 1993 and 1994 respectively, the European Union developed the *Power for the World* initiative and the World Bank proposed the *Solar Initiative*. Both these programmes aim to significantly increase the deployment of PV, primarily in developing countries. *Power for the World* targets

CH3365-4/94/0000-1149 $4.00 © 1994 IEEE

First WCPEC; Dec. 5–9, 1994; Hawaii

the 1-2 billion people without electricity, particularly the poorest of the poor [3].

The *Solar Initiative* it is looking to facilitate some 50 - 70 MWp of PV for small-scale applications over the next three years. The type of projects identified as a priority are rural electrification, water pumping, rural health clinics, street lighting and school buildings. It also includes other solar technologies and wind power. [4].

PV MARKET DEVELOPMENT

Within the PV industry, it is generally accepted that PV is technically ready and economically-feasible for most market applications. This has been the perception for many years.

The total amount of PV sold in 1993 is officially 61.1 MWp. This represents an increase of just 3.6% over 1992, and is lower than the 64 MWp predicted in 1992, and certainly is a long way behind predictions from the early 1980s.. However, it is expected that the market for PV will continue to grow, though not all sectors.

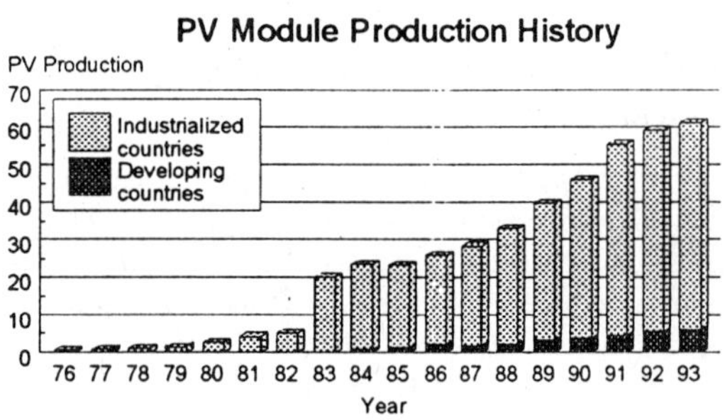

Figure 1

Figure 1 [5] shows the market growth since 1976, including the increasing share of estimated production taken by developing countries. Of all the sectors, the industrial systems markets are where in the 1980s, mainstream business evaluated and bought PV. Today this is a mature market, where PV is considered standard for remote applications by the telecommunications and marine industries.

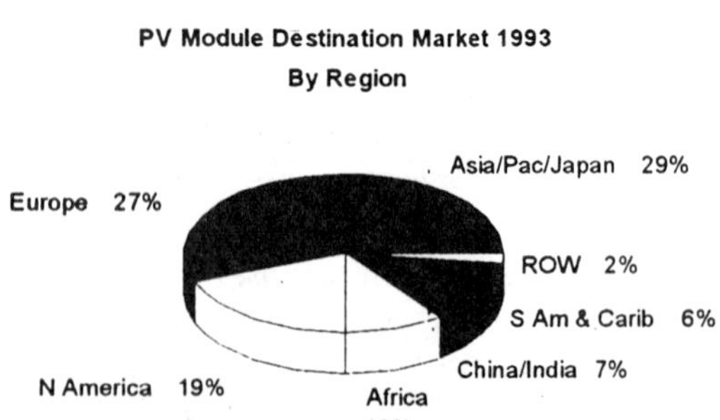

Figure 2

Figure 2 shows the PV module destination market in 1993. Of the markets in Asia/Pacific and Africa, the majority of systems are supplied by companies manufacturing in the developed countries, and international donor agency programmes still make up the bulk of purchasers (Figure 3). Figure 4 shows that world-wide rural systems and remote industrial applications make up the bulk of the market, with remote industrial PV being used for telecommunications and cathodic protection in particular, and rural systems including a variety of very diverse systems, such as pumping, lighting, vaccine storage and village power (100 Wp - 10 kWp).

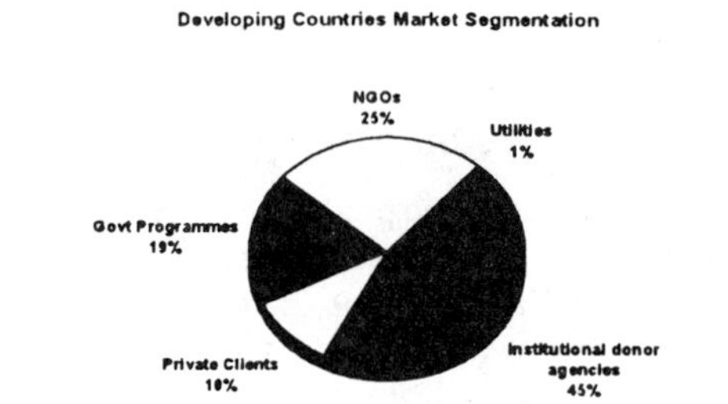

Figure 3

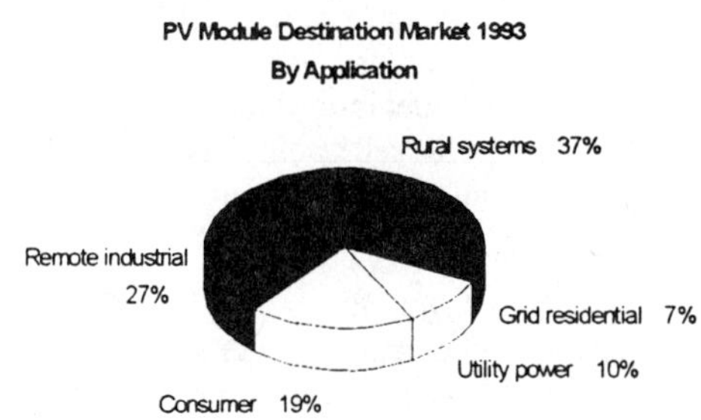

Figure 4

However, there are a number of markets, where the needs are real, and the potential market substantial, though little significant market development has occurred to date. The scenarios of the past have used stand-alone applications particularly in developing countries as their mainstay for market development, and they have not eventuated.

A COMPARISON OF PV MARKETS: TELECOMMUNICATIONS AND VACCINE REFRIGERATION

Telecommunications

The telecommunications sector is divided into three sections. The first is "local access", where typically the distance covered by a single repeater station is 50 - 300 km. These systems are expensive to install as the population density is low compared to equipment costs. The second section is the semi-urban areas, and the third area is urban areas. PV has a significant role in supplying electricity for systems in the first two sectors, and indeed supplies an estimated 50% of electricity requirements for the world telecommunications market [6]. The average size of telecommunications equipment is 100-150 W, and the trend is towards smaller, more efficient systems, with a target of 15 W in the short term and 5 W in the longer term. PV is very well placed to supply these electricity needs. PV telecommunications accounts for around 10 MWp in module sales per annum, being around 10% of world PV shipments. This figure has been fairly constant over the last 2 - 4 years. Increasingly efficient telecommunications equipment means that, while the total number of PV sales to telecoms projects is growing, the number of modules required is decreasing. A second factor affecting the market is that the majority of systems installed today are not for back-bone, country-wide telecommunications systems, but for fill-in systems in rural areas.

The telecommunications sector has a number of very large corporations investing billions of dollars into services and sales, and millions of dollars into increasing the competitiveness of its products. Most telecommunications companies are now either private or becoming increasingly autonomous from government control. PV-powered systems have been installed world-wide, from mountain tops in the Swiss Alps, to remote locations in Africa.

The decision to purchase PV by the telecommunications companies is based on commercial rationale, without concern for the technicalities; it is financed as any other purchase; and the data would suggest that PV has a positive image within this market.

Vaccine refrigeration

There are around 3,000 PV refrigerators installed in developing countries. This application accounts for a very small proportion of the world PV sales annually (about 0.1%), and in 1992 this amounted to only around 60 kWp. The main purchasers for PV vaccine refrigerators are institutional: aid agencies and Non Government Organisations (NGOs), health programmes (national and international), and United Nations agencies such as UNICEF and WHO. Any increase or decrease in these programmes therefore has a significant affect upon the market. For instance, the recent tender for 200 PV refrigerators in Zambia by the Ministry of Health, with funding from the European Union, represents a major increase in the year's market.

Vaccine refrigerators are an important key to successful immunisation programmes in developing countries.. Many of the rural hospitals and clinics do not have grid-electricity or a stable supply of fuel with which to run refrigerators. Almost all have abundant sunshine.

The Expanded Programme on Immunisation (EPI) of the World Health Organisation (WHO) has collaborated with refrigerator manufacturers, PV systems suppliers and development agencies since 1980. It has also introduced standards for PV. Their view for many years has been that PV refrigeration devices work effectively, and are an important solution in most rural areas [7]. Their recommendations at the UNESCO Solar Summit in 1993 were:

"WHO/EPI concludes that the benefits [of PV refrigerators] are sufficiently important to outweigh the marginal differences in the cost of purchase and maintenance, if the burden of cost is shared with other applications sharing the same PV power source."

However, WHO also concluded that the overall lifetime cost is only close to the lifetime cost of conventional fuel alternatives if factors concerning the value of vaccine losses and other assumptions are included in the cost/benefit equation [8].

PV refrigerators are not in use in developed countries, as all hospitals and clinics are linked to the electricity grid. There are some refrigerators powered by PV, but these small numbers are sold into the leisure market (e.g. for camping and caravaning). This market could be expanded, but the product is very expensive, and the return on investment for new, less expensive products is seen by the PV manufacturers as marginal at best.

There are no *fundamental* technical problems which inhibit market growth, as design and manufacturing inadequacies have largely been dealt with through co-operation between WHO and manufacturers, and the specification of WHO-approved equipment in project tender documents. There are problems which require repair and maintenance, such as balance of system components, but these are not insurmountable particularly if there are locally trained technicians and spare parts available. There are many other examples in the field where the equipment has been wrongly specified or undersized (often for short-term commercial reasons), and PV is blamed, propagating the myth that PV does not work.

Another technical factor which may be hindering the large-scale growth of the market is that system optimisation has not yet occurred, due to the small size of the market and inert demand. However, system optimisation did not stop the development of the market for PV within the telecommunications market. Rather it occurred as a result of market expansion..

Poor system performance is often the result of inadequate installation and poor maintenance. Experiences with PV systems in health centres in Zaire, as documented by IT Power, have shown that if health programmes includes provision for training local technicians and users (hospital staff), and setting up a spare parts infrastructure, then systems can operate well for many years [9].

CONCLUSIONS

For PV to become commercial on a large scale and to have mass market appeal for the different sectors, a number of barriers have to be overcome. New programmes which aim to significantly advance the world-wide market for PV must take into account the experiences with PV to date, and look at market developments not just technical advances. For system optimisation to occur, a significant market needs to be available.

There are a number of fundamental differences therefore between the market for PV in the telecommunications industry and the market for PV in health care. Table 1 summarises these differences.

Attribute	Telecoms	Health
Private market	✓	-
Aid, NGO, government market	-	✓
Community service	-	✓
Many users	-	✓
Demonstrations	✓	✓
Major technical problems	-	-
Strong market penetration	✓	-
Good perception by market	✓	-
Range of brands	100+	16*
Range of sizes	✓	✓
In competition with conventional fuels	✓	✓
Need for institutional / political support	-	✓
Advantages of reliability, efficiency, cost-effective **, modular, and little maintenance	✓	✓

Table 1

* *WHO accredited*

** *for locations remote from the grid, and also including social costs for PV refrigeration*

To significantly increase the market, PV must be:

- easy to use and to train users
- easy to install and to train local technicians to carry out any repairs or maintenance
- easy to purchase and finance, and not totally reliant upon one type of customer
- easy to demonstrate and to report on at non-PV conferences
- demonstrated as applicable and not second best
- technically competent and economically competitive

The two initiatives cited earlier, the EU's *Power for the World* and the World Bank *Solar Initiative*, should address these issues.

REFERENCES

[1] D Kuhn, *Telecom Australia's Experience with PV Power Systems*, CADDET Renewable Energy Newsletter, Issue 1, July 1993

[2] Michael R Starr & W Palz, *Photovoltaic Power for Europe*, Solar Energy R&D in the European Community, Series C, Volume 2, 1983.

[3]. Wolfgang Palz, *Power for the World*, International Solar Energy Society Congress *Harmony with Nature,* Budapest, August 1993.

[4] Denis Anderson, and Kulsum Ahmed, *The Case for a Solar Initiative*, Draft, The World Bank, September 1994.

[5] A Derrick, R W Barlow and J A Gregory, *The World Market: 1993 Status Report and Future Prospects*, 12th EC Photovoltaic Solar Energy Conference, Amsterdam, 1994.

[6] Personal communication, telecommunications industry (various sources).

[7] Michel Zaffran, *Solar Refrigeration for the storage of vaccines in the expanded programme on immunisation in developing countries*, 10th EC Photovoltaic Solar Energy Conference, Lisbon, 1991.

[8] Michel Zaffran (co-ordinator), Solar Energy and Health, a strategy for rural health and development, UNESCO Solar Summit, Paris, July 1993.

[9] A Derrick, J M Durand & T J Hart, *Solar powered vaccine refrigerators: results of concerted action on infrastructure support*, 9th EC Photovoltaic Solar Energy Conference, Freiburg, 1989.

PHOTOVOLTAIC POWER WITHOUT BATTERIES FOR CONTINUOUS CATHODIC PROTECTION AND AN ALTERNATE PHOTOVOLTAIC/ULTRACAPACITOR COMBINED POWER SOURCE

Wallace W. Muehl, Sr.
Author/Inventor
Department of the Navy, Coastal Systems Station
Dahlgren Division, Panama City, FL 32407-7001

ABSTRACT - The Coastal Systems Station (COASTSYSTA) designed, installed, and started up on 20 January 1990, a state-of-the-art stand-alone photovoltaic powered impressed current cathodic protection system (PVCPSYS) not requiring any backup power for steel and iron submerged structures. The PVCPSYS, installed on a 775-foot steel sheet piling of a Navy dock bulkhead, provides complete, continuous corrosion protection. The PVCPSYS has been in operation for more than five years, has not required any repair or maintenance, and is environmentally clean. Initial cost savings of the PVCPSYS versus conventional cathodic protection system was $46,000. A second PVCPSYS was installed on another 800-foot bulkhead on 21 May 1993. It is also providing complete corrosion protection without backup power. Performance is well documented. Other potential applications are moth-balled ships, locks, dams, bridges, pipelines, and similar structures. These systems are considered a major advance by Sandia and the Department of Defense (DOD) Photovoltaic Review Committee. An ultracapacitor, a recent hi-tech development that is environmentally clean, will be incorporated in the PVCPSYS when required to enhance the system's capability. A photovoltaic/ultracapacitor (or equivalent) combined power source operating under adverse conditions, and/or to satisfy or meet regulations will assure cathodic protection, including pipelines carrying combustibles or other products that could otherwise create environmental problems. Patents are pending on this PVCPSYS and the photovoltaic/ultracapacitor powered systems.

The objective of the initial project was to successfully demonstrate that renewable energy can efficiently and economically replace, or be used instead of, continuous non-renewable power sources. An opportunity to clearly show that photovoltaic power is practical and reliable was the result of a recommendation to provide cathodic protection to the Naval Diving and Salvage Training Center bulkhead.

The COASTSYSTA in Panama City, Florida, has broken new ground in the application of solar energy for cathodic protection. Photovoltaic arrays without battery backup have been connected to the 775-foot steel sheet piling of a dock bulkhead via a cathodic protection system to prevent corrosion on that steel structure in a salt water environment.

As the name signifies, cathodic protection is the process by which, in the COASTSYSTA impressed current type application, the entire steel sheet piling is transformed into a cathode via a series of anodes mounted in PVC standoff racks, in the water, next to the piling. When direct current (DC) energy is applied to the anodes and sufficient electrical potential is attained by current flow from the anodes via an electrolyte (seawater) to the piling, the corrosion is transferred to the anodes, preventing piling corrosion.

Mr. Wally Muehl, electrical/mechanical engineer at COASTSYSTA, was evaluating power sources to protect the Naval Diving and Salvage Training Center bulkhead when he focused on photovoltaics. Although there were 10 other impressed current cathodic protection systems installed on the docks, all were powered by a continuous power source with the current rectified to DC. Of these 10 systems, 5 were down from 1 to 1-1/2 years and 3 were down for more than 2 years due to rectifier failures and/or the power source secured due to construction. As a result, no corrosion protection was provided. PVCPSYS's would have continued to provide power and corrosion protection and would not have been affected by these type power outages.

The Naval Diving and Salvage Training Center is in a separate location from these docks, and it was determined that power was not readily available and would be expensive to provide rectifiers on the dock due to the dock configuration. Rectifiers would also pose a safety hazard on the

First WCPEC; Dec. 5-9, 1994; Hawaii

dock that is regularly used for diver and salvage training. This bulkhead was 12-years old and other than the initial coating, received no corrosion protection.

Mr. Muehl developed a state-of-the-art solar powered impressed current cathodic protection system requiring no battery backup power for submerged steel and iron type structures. Innovations in design and method of operation permits the photovoltaic arrays to easily provide and maintain complete continuous corrosion protection without the necessity of DC power backup such as batteries. Battery backup power is considered costly and an environmental problem. To date, all impressed current systems require a continuous DC power supply in order to provide cathodic protection.

The COASTSYSTA photovoltaic power system is a fixed-axis system which is suitable for the Panama City latitude of 30°10'N, 85°22'W. The tilt of the adjustable arrays was set at latitude instead of +15° in January 1990, and has not been changed. This is a good indication that other areas with good distribution, but lower insolation levels, would be excellent prospects for a similar type of photovoltaic powered system. For higher latitudes, there are several other options to improve system performance without battery backup. These include one-axis east/west tracking, two-axis north/south, east/west tracking, or simply adding a module or two to meet the additional current requirements. Note: In the Northern Hemisphere, true south should be used in the design, not magnetic south. In most areas, this varies from the magnetic south given by a compass. Magnetic variation could be substantial and can be determined from a isogonic map. This is given in degrees east or west of magnetic south.

Two other problems had to be considered and resolved in order to install an impressed current cathodic protection system. The first problem was ensuring that the steel piling had electrical continuity. Another problem was providing sufficient impression of current "carry over" to overcome a 155-foot section of piling that had to be bypassed, and provide cathodic protection, without anode placement in the area having a water depth of 27 feet where diving takes place. Both problems were overcome in the design.

To facilitate the use of a photovoltaic powered cathodic protection system without battery backup, the steel sheet pilings were provided an initial one-time only preconditioning polarization for a predetermined continuous time period to the extent that these pilings were initially polarized to a relatively high negative potential by a temporary DC power source. The photovoltaic power system was provided with blocking diodes to prevent any possibility of current reversal. It is to be noted that possible evolution of a protective hydrogen film is merely a by-product of the preconditioning polarization at the higher negative potentials. Additionally, depending upon the environment and if higher (more negative) polarized potentials could be maintained other than required to provide basic complete cathodic protection, formation of thicker calcareous deposits having protective value over a period of time could occur. The initial DC power for polarization can be provided by a DC power source such as a portable, motor-driven DC welder or generator.

The COASTSYSTA PVCPSYS tests performed, along with other data obtained, provide a further explanation that the anode/seawater/cathode piling structure acts like a battery, and when allowed to rest the polarity level recovers and is electrochemical in nature. An electrochemical lead/acid battery, for example, can recover charge if allowed to rest after serving a load. The electrochemical reaction reverses slightly when the load is disconnected; however, a capacitor without an external current source cannot recover by simply removing the load. It is believed that the one-time only initial preconditioning polarization (controlled conditions) of the structure embeds single hydrogen atoms in the steel sheet piling that can also migrate and diffuse in the structure. This system delays the decay of the negative potential and permits the photovoltaic arrays to supply sufficient power; thus allowing the system to easily provide complete, continuous cathodic corrosion protection in situations which include cloudy, overcast, rainy and nighttime conditions without the necessity for DC power backup such as batteries. Re: The Hydrogen Program Manager for the National Renewable Energy Laboratory (NREL) has reviewed our PVCPSYS data, and also believes that pre-polarization (controlled conditions) in a similar environment can embed single hydrogen atoms in the mild steel piling that can also migrate and diffuse in the structure.

The foregoing novel method and system of a one-time only preconditioning or prepolarization of the structure prior to energizing the photovoltaic solar array on-line with the system, provides a relatively higher negative potential that has a slow rate of decay. This permits the use of regulated photovoltaic solar energy with excess available power, and without any backup power, to easily provide complete continuous corrosion protection with excellent polarization levels and improving with time. An analogy may be that the steel structure becomes very effectively polarized and remains so by the variable DC charge affect provided by the simple solar array system, much like a piece of steel or iron can become magnetized by the application of a DC electrical current.

The installation, start up, and continuing operation, including underwater inspections, are well documented to date by COASTSYSTA and verified on site, during the day and at night by the U.S. Army Corps of Engineers, Construction Engineering Research Laboratory, Naval Energy Program Office, and members of the DOD Photovoltaic Review Committee. According to the National Weather Service, the average amount of available sunshine for the three weeks prior to visits by these organizations averaged 24 percent.

In 1985, the estimated cost of a conventional cathodic protection system requiring continuous DC power was $75,000.00. In 1990, the estimated cost was $108,000.00. The PVCPSYS cost at contract completion was $61,316.00, complete and ready to use. The initial cost savings by installing a PVCPSYS was in excess of $46,000.00. The DOD Photovoltaic Review Committee and Sandia National Laboratories consider this successful and cost effective system a major advance for the application of photovoltaics.

This system has been in operation over five years and has not required maintenance or adjustment. A patent is pending on the new technology. Other possible applications include mothballed ships, docks, dams, locks, bridges, marinas, offshore structures, and pipelines in various environments.

A photovoltaic power system without any backup power was installed on another 800-foot bulkhead. The two previous 400-foot conventional rectifier powered impressed current cathodic protection systems were modified to allow this conversion. The PVCPSYS successfully started operation on 21 May 1993, without any backup power and is providing complete continuous corrosion protection. A state-of-the-art data collection system is provided via a personal computer that is MS-DOS compatible and located about 1/2 mile away from the site. Along with other capabilities this system will simultaneously monitor, report, analyze, and record the solar energy output in DC volts, DC amps, as well as the DC negative potential of the steel sheet piling.

When required a recent high-tech development ultracapacitor or equivalent that is also environmentally clean will be incorporated into the PVCPSYS to enhance the system's capability. A photovoltaic/ultracapacitor combined power source will assure cathodic protection for submerged or buried steel or iron type structures. It is also effective under adverse conditions and will satisfy environmental regulations in such applications as pipelines carrying combustible materials. Westinghouse Electric Corporation acquired the rights to manufacture the ultracapacitor and has expressed an interest in the combined PVCP/ultracapacitor system.

The ultracapacitor technology developed over the past 12 years by Pinnacle Research Institute, Inc. (PRI), provides both high energy and power. The capacitor breaks all paradigms associated with conventional batteries and capacitors. The ultracapacitor has no conventional dielectric. Unlike batteries there is no chemical reaction; therefore, the ultracapacitor can be discharged and rapidly recharged by photovoltaic power for an indefinite number of times. (Testing by U. S. DoE confirms over 120,000 cycles of operation). Further, it's sponge-like surface gives the plate an actual surface area that is 100,000 times greater than the same surface of the same plate if it were smooth. The higher surface area allows more energy to be stored. The ultracapacitor has 100 times more power per ounce and is significantly smaller than conventional capacitors. Most importantly, a second generation of ultracapacitors has been recently developed that are capable of releasing energy gradually. An ultra-capacitor (if required for the COASTSYSTA PVCPSYS installed in 1993) would weigh about 18 pounds. In the electric powered automobile industry, for example, some initial applications will be for hill climbing and acceleration. PRI is a subcontractor of the Department of Energy Hybrid Electric Vehicle Program.

With reference to this manuscript, it is with pleasure that I acknowledge the "helpful cooperation and information" received from the following personnel:

• Dr. Michael G. Thomas and Dr. Hal Post, Senior Members, Technical Staff, Photovoltaic Research Department, Sandia National Laboratories, Albuquerque, New Mexico.

• Mr. James F. Jenkins, P.E., Corrosion & Metallurgical Engineer, Naval Civil Engineering Laboratory, Port Hueneme, California.

* Mr. L. E. Humble, Photovoltaic Programs, Energy Program Office, Naval Weapons Center, China Lake, California.

* Mr. Roch A. Ducey, Principal Investigator and Ms. Jearaldine I. Northrup, Research Engineer, U.S. Army Construction Engineering Research Laboratory, Champaign, Illinois.

• Mr. Thomas F. Lewicki, P.E., Facilities Corrosion Program Manager, HQ Air Force Civil Engineering Support Agency, Tyndall Air Force Base, Florida.

• Dr. Thomas M. Cawthon, Hydrogen Program Manager and H. Dana Moran, Manager, Research and Technology Applications, National Renewable Energy Laboratory, Golden, Colorado.

• Dr. K.C. Tsai, President, Pinnacle Research Institute, Inc., Los Gatos, California.

• Navy Divers and Dive Locker, Coastal Systems Station, Panama City, Florida.

* Members of the DOD Photovoltaic Review Committee.

MONITORING RESULTS OF PV FOR ELECTRIC PROPULSION IN RECREATIONAL BOATING

G. Loois, F.P.H. Wouters, G.M. Koerts, T.C.J. van der Weiden, ECOFYS Research and Consultancy, P.O.Box 8408, NL-3503 RK Utrecht, The Netherlands, tel. + 31 30 913 400, fax. + 31 30 913 401,

ABSTRACT

In future, grid connected systems will become the most important application of PV in Europe. Until that time a range of applications of autonomous systems needs to be developed to support the introduction of PV. Two projects concerning development and demonstration of PV-systems for powering ships with electric propulsion, aim to open a market with a technical potential of several hundreds of MWp in Europe. A secure energy supply, enhanced comfort (less noise and air pollution) and benevolence for the natural environment are the most significant advantages for introduction.

Eighteen PV/battery-systems for electric propulsion of leisure boats (4 kWp in total) are currently realized and investigated in a pilot project in The Netherlands. The systems in use all run to the satisfaction of the users. Preliminary results of monitoring data on a few systems of these systems will be presented.

A European project comprising about more than 75 PV-powered recreational ships (100 Wp to 2.5 kWp), involving shipbuilders, local authorities, utilities and interest groups has recently started.

INTRODUCTION

In future the main market for PV-systems will consist of grid connected systems. In the Netherlands by 2010 these systems are projected to supply in the order of 0.3% of the national electricity demand. Until that time various economically feasible applications of autonomous systems have to be developed and introduced to increase the confidence in PV and guarantee sufficient turnover for further developments in PV. Ecofys undertakes a variety of activities to stimulate market introduction of autonomous PV applications.

In recreational boating, PV energy supply can play an important role since the use of combustion engines in recreational areas becomes more and more controversial in the eyes of both users and administrators of these areas. It is shown that electrically propelled boats are more energy efficient in terms of primary energy and produce for less pollutants (CO_2, NO_x, SO_2, volatile organic compounds and particles) [1]. The environmentally beneficial characteristics and its carefree application, attract many boat owners. PV energy can be used to propel the boat and for electrical appliances normally on board. This PV application will profit from the general interest for electric propulsion. On the same time PV is expected to improve and extend the applicability of electric propulsion and therefor accelerate the replacement of fuel engines. In two areas in The Netherlands where combustion engines are prohibited, electric propulsion is seen as the only alternative. Several other areas in Europe similar rules are valid or announced to follow soon.

TECHNICAL POTENTIAL

Information on the recreational boating sector in The Netherlands is compiled by organizations such as ANWB, Hiswa and KNWV. These associations represent interest groups on tourist traffic, commerce and general watersports respectively. From these sources [2] a technical potential of 50 MWp is estimated [3], being:

- 22 thousand open boats (motorboats and sailing boats) to be fitted with 60 Wp (total 1.3 MWp) and
- 122 thousand cabin vessels with 400 Wp (49 MWp).

From the typical use of the motor of 50 hours per year a demand for electric propulsion follows of 10.3 GWh.

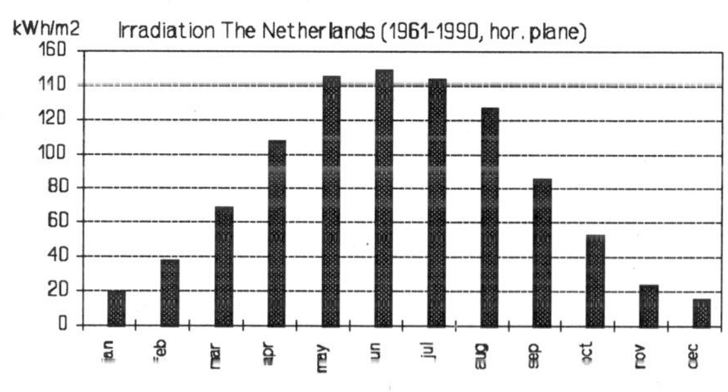

fig.1 Monthly irradiation in The Netherlands

Extrapolation of these figures result in an estimation for a technical potential of several hundreds of megawatts in Europe. Also small canalboats and ferries in touristic areas are important applications, since they professional application they enhance the confidence in the technology.

PV POWER DEMAND FOR PROPULSION

PV-electric propulsion for recreational vessels is best feasible for manoeuvering in marinas and canals. This occurs as a rule only a few times a day for short periods. In many cases only in the weekends. Irradiation however is available throughout the summer season. This means that relatively high demands for short periods can be supplied by relatively small PV-arrays (fig.4). The typical size of applicable electrical motors ranges from 1 to 5 kW (cost 1 US$/Watt). The efficiency of electric traction can be higher than 86% for commercially available products [4]. Within limits PV energy may also be used for electrical appliances for navigation, communication and lighting. This means that the necessary nominal power of the PV array can be limited to about 500 Wp for cabin vessels and to 120 Wp for smaller sailing vessels. Canalboats and cruise ships typically have motors of around 5 kW and a PV-array of 1.5 to 2.5 kWp.

TECHNICAL IMPLEMENTATION

PV-modules
On the sailing vessels slightly flexible MSX-lite™ modules of Solarex are used. On the other vessels, glass modules are used being Dutch RDG 48 modules of R&S and Siemens M55 and M50K-A1 modules.

Seagoing sailing yacht
In fig.2 is shown the electric energy supply system of a seagoing sailing yacht, designed to enable voyages of several weeks at temperate latitudes.

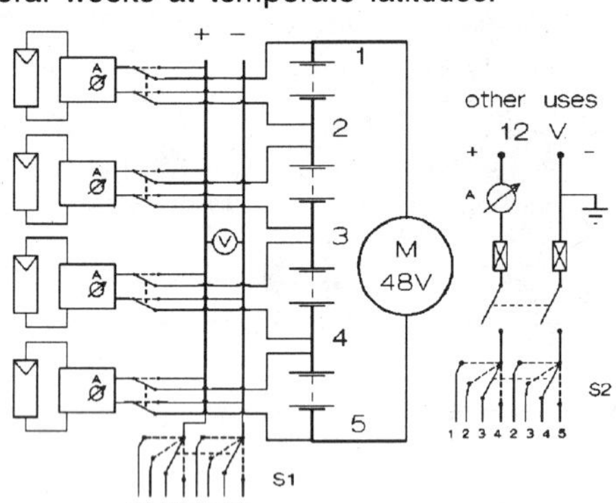

fig.2: Design of the electrical installation aboard a seagoing sailing yacht.

A high reliability is achieved by compartimentalising the installation into 12 Volt sections. The components that make up each section can be switched to other sections if a malfunction would occur. The electric propulsion uses a 48 Volt voltage, while the rest of the equipment for navigation and communication uses 12 Volt.

PV-electric outboard motor
The design shown in fig.3 is much simpler and does not differ much from widespread systems such as drinking water pumps for cattle, beacons and garden houses, systems commonly applied in The Netherlands.

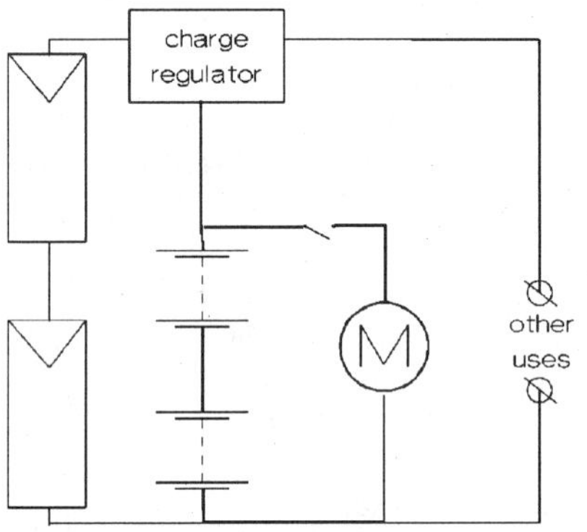

fig.3: Design of the electrical installation of a simple PV/electric outboard motor.

MONITORING

The performance of all systems is monitored. The users record the readings of the Ah-counters and the Voltmeter on a monthly basis or more often. In selected systems the energy flows and ambient circumstances are monitored on an hourly basis by means of a datalogger. Three systems have been monitored succesfully during the summer of 1994. The main measuring campaign for the recreational vessels will take place in 1995. All monitoring is performed according to the guidelines of CEC-JRC.

INSTALLATION AND COMMISSIONING

Most systems were installed in close cooperation with the ship owners. In this process also some training of the owner on system operation and battery use is given. To solve problems concerning the limited available area on board, creative solutions were developed for the positioning of modules, especially for the recreational boats (see pictures). The final installation and commissioning was done by Ecofys.

PILOT PROJECT OPERATING EXPERIENCE

Tne users reported full satisfaction with their systems, in use over the summer period of 1994. Some reported problems with the laminates on the MSX lite modules

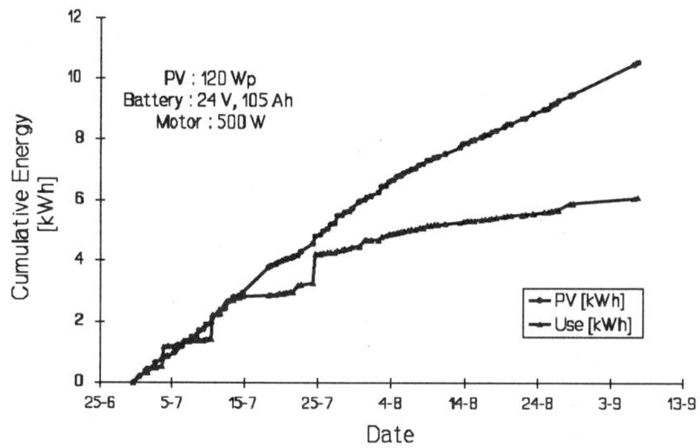

fig.4 Example of production (upper graph) and demand (lower graph) of a PV- powered recreational ship.

In fig.4 is shown a typical PV production and use pattern. The solar energy produced is more than the energy used by the motor. The measurments confirm the limited use of the motor (steps). The performance of the systems is expressed in terms of

Array Yield (Ya) =
energy from the PV-array per installed nominal array power [kWh/day/kWp]

and (neglecting the battery losses)

Performance Ratio (PR) =
array yield / solar irradiation

The array yield varies from 1.2 to 1.5 kWh/day/kWp during the summer season. The performance ratios vary between 0.1 and 0.4. These performance ratios are similar to the performance ratio's of other small autonomous systems in the European Community [5]. Performance ratio itself is not a sufficient measure for the quality of the system. In combination with observations of the situations in which the boats are (no shadowing etc.) it must be concluded that some PV-arrays are overdimensioned. This means that it might be possible to use smaller PV-arrays enhancing the competitiveness of the technology.
In 1995 the quantitative assessment will be carried out for all eighteen ships in the pilot project mainly during the summer season.

FUTURE ACTIVITIES

Experiences from the pilot project will be used to further develop and demonstrate the technology in the European demonstration project 'PV-generators for propulsion of recreational boats in The Netherlands, Germany and Ireland'. This project will lead to more than 75 PV-powered electric boats in the Netherlands, Germany and Ireland. In this project will be cooperated with regular ship builders to demonstrate integration of PV-electric propulsion in boats, aiming at a continuous application of the technology in ship building.

CONCLUSIONS

Autonomous PV-generators are very well applicable for propulsion of recreational vessels, to the full satisfaction of the user.
Array yields lie between 1.2 and 1.5 kWh/kWp/day, the performance ratios lie between 0.1 and 0.4.
Preliminary results show that in most systems no power shortage have occured. The PV-power may be reduced without affecting the performance of the system.
These systems have been realized with commercially available components. Further activities are needed to enable a commercial introduction of the technology.
Feasibility studies showed that the European technical potential for PV-propulsion of ships is in the order of magnitude of hundreds of MWp for these systems.

LITERATURE

[1] R.T.M. Smokers, Energie- en milieuaspecten van elektrisch varen, ECN, 1994 (in Dutch)
[2] Hiswa, Trends en cijfers in de waterrecreatiesector 1991, January 1992 (in Dutch).
[3] T.C.J. van der Weiden, e.a., Voorstudie PV in de recreatieve watersport, April 1992 (in Dutch).
[4] F.P.H. Wouters, e.a. Elektrische aandrijving voor vaartuigen, juni 1994 (in Dutch)
[5] W.B. Gillet, e.a., Review of recent results and experiences from CEC PV Demonstration and Thermie programmes, 11th EPVSEC, November 1992.

ACKNOWLEDGEMENT

These projects are, among others, supported by the European Community, EC-DG XVII and Novem, The Netherlands Agency for Energy and the Environment.

ECOFYS Consultancy and Research, Utrecht, The Netherlands, December 1994

Picture 1 Sailing ship with 500 W motor and 120 Wp installed vertical on the railing

Picture 2 Trimaran with 3.8 kW motor and 420 Wp installed in between and on the hulls.

TRANSPORTABLE STAND-ALONE SOLAR IMMISSION MEASURING STATION WITH OPTIMISED PHOTOVOLTAIC ENERGY SUPPLY *

Rainer Rudischer, Günter Ulbrich
Institute for Air Conditioning and Refrigeration (ILK)
Department of Solar Engineering
Dresden/Germany Fax:+49-351-4081725

ABSTRACT

The damage of forest ecosystems by aerosol wastes and contaminants is increasing. This fact led to an alarming situation in Europe and world-wide, which made it necessary to install measuring stations to investigate causes and observe the ecological immission situation mainly in silvicultures at desirable locations distant from grid connection and urbanisation. The presented measuring station satisfies those requirements by means of an emission-free, self-sufficient and optimised photovoltaic energy supply in combination with an absolute maintenance-free, long-lived battery-storage system. The energy demand was minimised by means of a variable clocked measurement operation and modification of measuring instruments. Further, a so called PMPP- (Programmed MPP) control system, implemented into the overall control system, optimises the energy supply.

PROBLEM

Nowadays it is an undisputed fact that both in Europe and world-wide forest ecosystems are seriously contaminated and damaged by aerosol wastes and contaminants, whereas particularly, the filtering effect of the treetops play an important role. Especially, in Germany immission measuring stations have been installed in many affected areas in order to quantify the contaminant input, particularly SO2, O3 and NO/NOx. A restraining condition with regard to the fixing of the location and the influence on the measured values is the required energy supply of such measuring stations. For desirable locations, where immission concentration values are subjected to substantial variations in dependence on the time of day and the terrain shape and investigation of causes is inevitable, usual commercial immission measuring stations with nominal electric power ≥ 3 kW and battery supply or diesel-electric supply are unsuitable because of the following reasons: Firstly, the locations are mostly distant from urbanisation (i.e. away from electric grids). Secondly, expenditures of maintenance are unacceptable and service life is limited. Thirdly, measured values are falsified by combustion emissions.

* sponsored by the German Federal Ministry of Economy

SOLUTION

To meet the specific requirements, a photovoltaic powered, stand-alone i.e. absolute self-sufficient, transportable, automatic measuring and operating system with minimised energy consumption, no own emissions, data storage and remote control is an adequate solution. Further development emphasises are, apart from a foundation-free and simple installation, the development of an easy-to-handle PV generator, which meets the transport requirements and solutions of problems of upright position safety and operation safety for the unattended operation of the measuring station.

The reduction of the total energy consumption, whereas 98% are required by the own consumptions of the O_3-, SO_2- and $NO/NO_x(NO_2)$-measuring instruments, was carried out in two parallel steps:

1. Reduction of the power input of the immission measuring instruments by means of modification of all energy consumers
2. Installation and testing of a clocked measuring operation with variable frequency

Instrument	nominal AC power	expected DC power	actual DC power
O 341 M (O_3)	**50 W**	**50,0 W**	**43,5 W**
AF 21 M (SO_2)	**120 W**	**72,5 W**	**49,9 W**
AC 31 M (NO/NO_x)	**350 W**	**180,0 W**	**84,5 W**

Table 1 Nominal AC and actual DC power of immission monitoring instruments

Those measures were carried out after the proof, that the measurement data of the used authorised measuring instruments are not influenced and a safe clocked measuring operation without limitations of reliability and service life is possible. After a check of different measuring instruments, it was found out, that the instruments of only one producer were appropriate. In co-operation with this producer the instruments were converted to DC voltage and the own consumption was consequently reduced by means of a common stabilised

CH3365-4/94/0000-1161 $4.00 © 1994 IEEE

power supply for all required voltage levels. Table 1 shows the reached status.

It is possible to install a cycled measuring operation for the application "flue gas damage research", because a reduction of the measured values per day can be realised due to the correlation of the rest period of the trees and the reduced solar radiation during the winter season (November - March). During the vegetation period (April - October) a desired increase of the measured values per day is possible and already realised due to the increased solar radiation. In other words, for this application a very good correlation between needed and available energy is given. Fig. 1 shows the two tested standard measuring cycles up to now. It is possible to extend the measuring times e.g. in case of excessive energy by means of an adapted control mode.

Since satisfying commercial MPP-controllers for monocrystaline Si-PV-generator configurations were not available, a so-called PMPP-(programmed **m**aximum **p**ower **p**oint) method was developed, patented and implemented in the controller. The appropriate MPP of the PV generator characteristics, stored in the controller, is computed and adjusted by an extreme short measurement of short circuit current and open loop voltage, which is carried out continously at intervals of several seconds. After comparing these measured values, depending on temperature and solar radiation flux density, to listed and computer-stored PV-generator characteristics, the correspondent MPP-point is adjusted exactly without search operations. This is accomplished by pulse width modulation and partially applying the load with a frequency in the kHz-range. The PV-modul operating point is stabilised by a capacitor. Fig. 2 shows

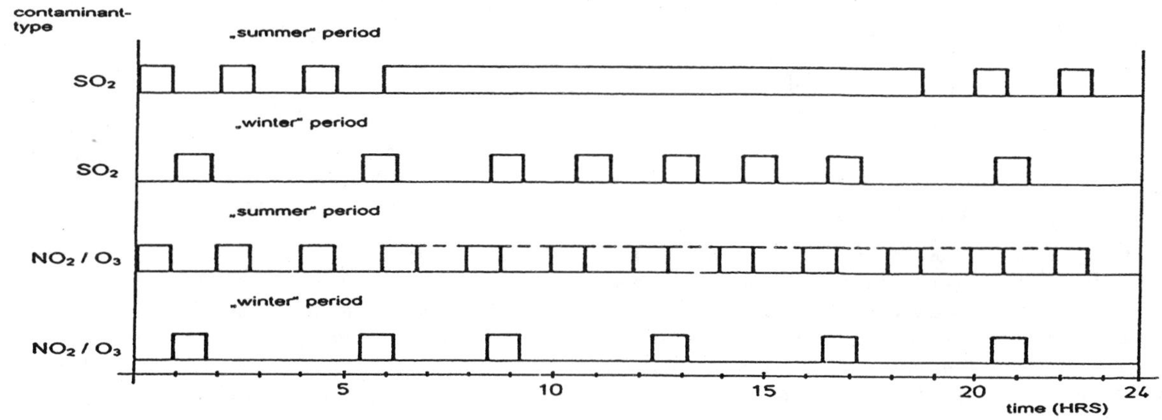

Fig. 1 Daily standard immission measuring cycles during "winter" periode (Nov. - Mar.) and "summer" periode (Apr. - Oct.)

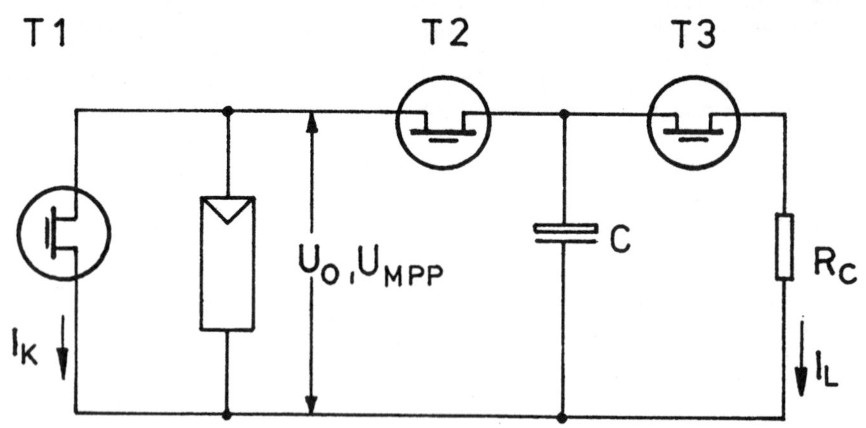

Fig. 2 Control circuit for the PMPP method

the basic control scheme.

The present development and research work is concentrated on the optimization of the MPP-controller and the possible energy efficiency compared to standard MPP-controllers.

Fig. 3 shows the block diagram of the total system.

observation) work without own consumption and inevitable electromechanical switches just need a trigger pulse.

The Dimension of the PV generator and the battery storage were calculated for Central European conditions with max. 16 days of low radiation ($\varnothing E_l$ = 0,3 kWh/m^2/d) and all-year operation. The complete measuring and control technique is installed in an insulated control box with ducted ventilation or heating in winter ($t_{tr} \geq 10°C$). The transmission of the pre-compressed measurement

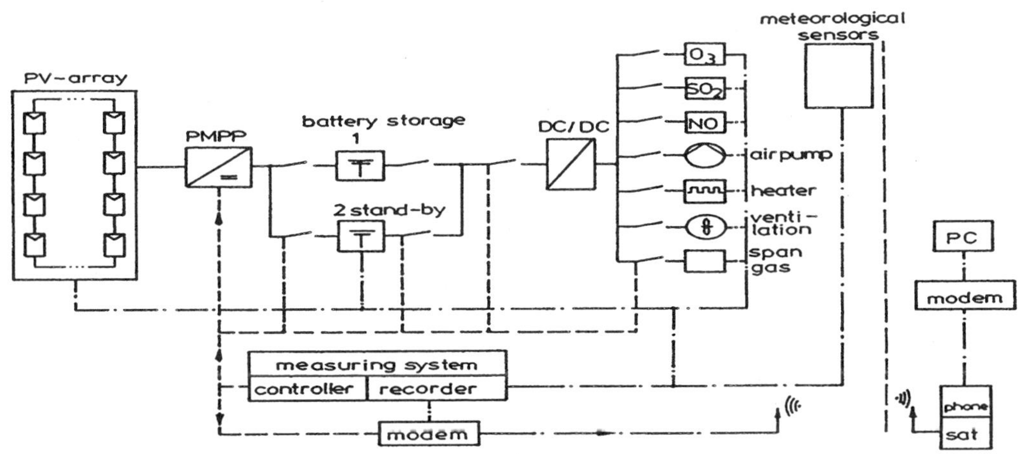

Fig. 3 Block diagram, PV energy supply, controlling and measuring system

The complete control and measuring system is realised in CMOS technology. Sensors (measurement of meteorological data, energy balance, temperature

Fig. 4 Transportation of the Solar Immission Measuring Station with long-distance lorry

data is carried out automatically via radiotelephone to an analyser PC. On-line variations of measurement and control parameters and real-time presentation of the measured values are possible.

The measuring station is operated in co-operation with the ILK and the Technical University Dresden on a forest glade (60x60 m^2) in the Erzgebirge near Dresden since November 1993. The transport test was successful. Fig. 4 shows the transport state and fig. 5 the present location. The station operates without any fault since November 1993.

Technical features:

- PV stand-alone system
- nominal PV generator power 2,1 kWp
- UV fluorenscent SO_2 analyser
- UV absorption O_3 analyser
- Chemiluminiscent NO analyser
- Detectability 1 ppm
- Recording of meteorological data
- Automatic data measuring, recording and system control
- long distance data transmission
- no foundation required

Fig. 5 The Solar Immission Measuring Station in a forest glade located in the Erzgebirge in the south of Dresden

OPTIMISATION OF PHOTOVOLTAIC WATER PUMPS COUPLED WITH AN INTERFACING PULSE WIDTH MODULATED DC/AC INVERTER POWER CONDITIONING DEVICE

Neway Argaw
Tampere University of Technology (TUT)
P. O. Box 692, FIN-33101 Tampere, Finland

ABSTRACT

A proper matching of electromechanical loads to a PV array is a great problem in many PV systems. This is mainly related to the non-linear, solar irradiance and cell temperature-dependent voltage and current characteristic of the PV array generator.

A simple non-linear optimisation technique is used to solve the load matching problem of a PV water pumping system. A number of case studies (at best and worst month solar radiation situations) are presented based on the power consumption of the load. Results show that, an optimum matching factor of 0.74 and 0.55 (at best and worst month respectively) can be achieved using 1.76 kWp M55 type PV array and M402/SP5A-7 type motor/pump with SA1500 dc/ac inverter interfacing device in this works field study area. It is possible to achieve a higher load matching factor by selecting carefully the proper size of the array and the motor/pump.

1. INTRODUCTION

Proper matching of the subsystem to a PV array is essential for maximum utilisation of water pumping systems. The main problems of such systems are related to the non-linear, solar irradiance and cell temperature-dependent voltage and current characteristic of the PV array generator [2]. It is therefore essential to analyse all mathematical relations and models of the subsystem and the PV array using a unified approach.

The load matching factor [2-8], as defined by the ratio of the energy acquired by the motor/pump subsystem to the maximum PV array power produced in one day period, is used as a measure for the quality of load matching. This method is based on maximising the total gross electromechanical energy delivered to a mechanical load for a given solar irradiance and temperature profiles.

This paper attempts to solve the problem of sizing a centrifugal pump coupled with a squirrel cage motor powered by a PV array through a dc/ac inverter.

2. THEORETICAL MODELS

2.1 PV array modelling

The current-voltage (*I-V*) characteristic of a PV array as a function of solar irradiance and cell temperature can be approximated by scaling the characteristics of a single PV cell.

PV cells' current can be ascertained by using the Newton-Raphson method. This technique provides a contraction mapping on the interval of uncertainty of the current domain to cover the true current [6-7, 9-11]. Fig. 1 and 2 shows the equivalent circuit model of a single PV cell and PV module respectively. Fig. 3 illustrates the schematic diagram of the PV water pumping system.

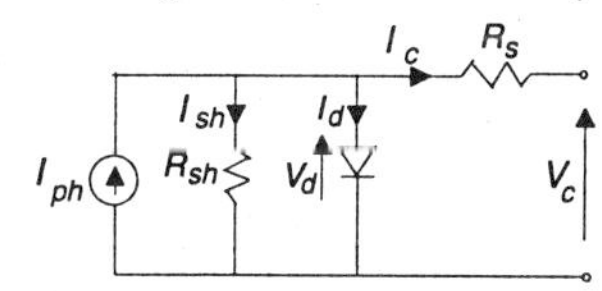

Fig. 1. Equivalent circuit of a PV cell.

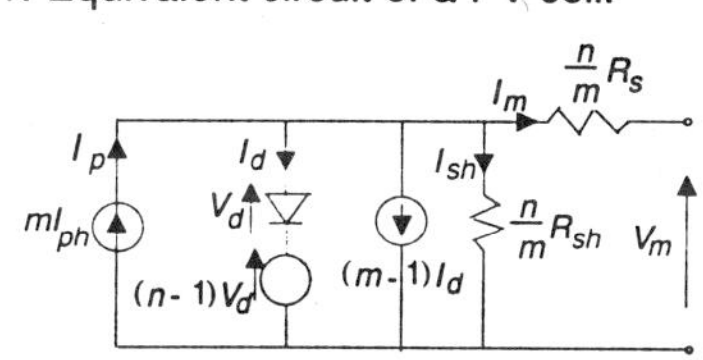

Fig. 2. Equivalent circuit of a PV module.

The operating equation describing the *I-V* characteristic, $I = f(V)$, of a PV cell (silicon cell), derived from Fig. 1, is written as [6-7, 10-15]

$$I_c = I_{ph} - I_d - I_{sh}$$

$$= I_{ph} - I_o\left[\exp\left(\frac{q(V_c + R_s I_c)}{AKT_c}\right) - 1\right] - \frac{(V_c + R_s I_c)}{R_{sh}} \qquad (1)$$

Where,

$$I_{ph} = I_{sc} I / I_N + K_I (T_c - T_r) \qquad (2)$$

For *n* number of series strings with *m* number of cells in parallel, the PV module light generated current is

$$I_p = m I_{ph} \qquad (3)$$

The short circuit current and the open circuit voltage can be related by

$$I_{sc} = I_{ph} - I_o\left[\exp\left(\frac{qR_s I_{sc}}{AkT_c}\right) - 1\right] - \frac{R_s I_{sc}}{R_{sh}} \qquad (4)$$

$$V_{oc} = \frac{AKT_c}{q}\left[\ln\left(\frac{I_{ph} + I_o}{I_o}\right)\right] \qquad (5)$$

Where,

$$I_o = I_r\left(\frac{T_c}{T_r}\right)^3 \exp\left[\left(\frac{qE_g}{BK}\right)\left(\frac{1}{T_r} - \frac{1}{T_c}\right)\right] \qquad (6)$$

The shunt resistor and diode current are expressed by the following relations

First WCPEC; Dec. 5–9, 1994; Hawaii

$$I_{sh} = \frac{m}{nR_{sh}}\left[V_m + \frac{n}{m}R_s I_m\right]$$
$$= \frac{mV_m}{nR_{sh}} + \frac{R_s I_m}{R_{sh}} = \frac{m}{R_{sh}}(V_c + R_s I_c) \quad (7)$$

$$I_d = I_o\left\{\exp\left[\frac{q(mV_m + nR_s I_m)}{mnAKT_c}\right] - 1\right\} \quad (8)$$
$$= I_o\left\{\exp\left[\frac{q(V_c + R_s I_c)}{AKT_c}\right] - 1\right\}$$

The PV module current is then determined from

$$I_m = f(V_m, I_m) = I_p - mI_d - I_{sh} \quad (9)$$

The cell temperature is given by the relation [6-7]:

$$T_c = T_a + \frac{H}{800}(NOCT - 20) \quad (10)$$

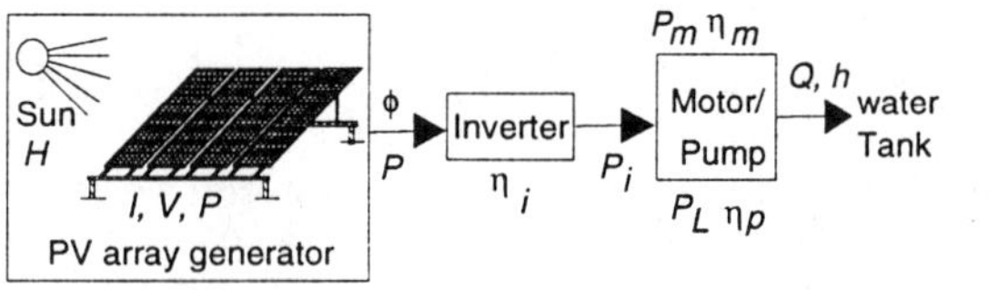

Fig. 3. Schematic diagram of a PV water pumping system.

2.2 DC/AC inverter

The efficiency of a typical inverter varies with its output power. The characteristic of the PV system efficiency in a given day can be very low, for conventional inverters (see Fig. 4), if it is operating at a fraction of its rated load during that day.

A new speed control approach for ac motor drives using programmed pulse-width modulated, PWM switching technique is developed that can give an inverter output voltage close to unity (0.98) [16]. Enjeti et al and many researchers [16-17], identify that PWM switching technique can optimise various objectives, such as minimising voltage harmonics, obtaining minimum losses in the motor, reducing torque pulsation and minimising the generated acoustic noise. The system equations are clearly defined in References [16-18].

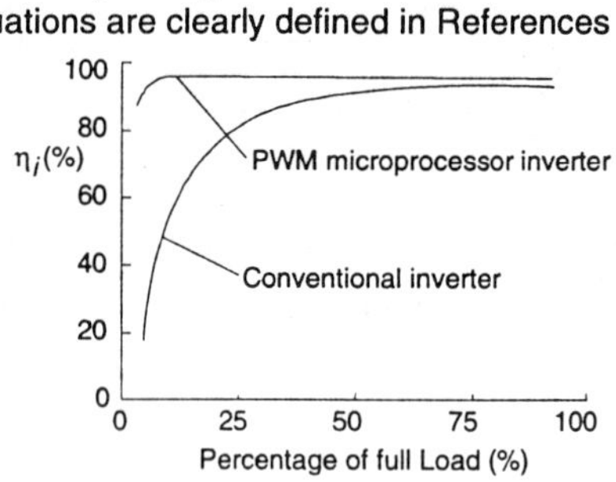

Fig. 4. A typical inverter efficiency curve.

In this work, PWM dc/ac inverter controlled by a microcomputer, made by Grundfos A/S, is used in system analysis. This type of inverter has almost constant efficiency even at 10% load (96%) [1] (see Fig. 4).

2.3 Motor/pump subsystem

The design theory is limited to squirrel cage induction motor coupled with centrifugal pumps.

To produce a starting torque, and subsequently a running torque, it is necessary to have a current flowing through the rotor winding by short-circuiting the winding. Initially the induced emf from the inverter causes a rotor current per phase, I to flow through the short-circuit, producing an ampere conductor distribution that acts with the flux field to produce the starting torque. The sense of this torque is always to cause the rotor to travel in the same direction as the rotating field. The flux field is assumed to revolve clockwise at a speed corresponding to the applied stator frequency and the poles of the stator winding. This speed is called the synchronous speed and is described by [19-20]

$$n_s = \frac{120f}{p} \quad (11)$$

Where f is the stator or source frequency in revolution per second (rps) and p is the number of poles. The relative speed between the rotor and the flux wave or the slip is defined by [19-20]

$$s = \frac{n_s - n_m}{n_s} \quad (12)$$

Where n_m is the operating speed of the motor (or the actual rotor speed) in rpm. The rotor frequency, f_r often called the slip frequency, is defined by $f_r = sf$ [19].

The power transmitted across the air-gap of the motor to the rotor and there converted to mechanical form, absorbed in rotational losses, or dissipated as heat in the rotor windings is, the dissipated three fictitious resistance, R/s of the three phases. This is the real power associated with the equivalent rotor circuit which represents the rotor copper losses in all of each phase and is called the air-gap power. In other words, it is the input power to the rotor and is defined by [19-22]

$P_{ag} = 3\frac{R}{s}I^2$. Or this can be expressed in the form,

$$P_{ag} = 3RI^2 + 3(1-s)\frac{R}{s}I^2 \quad (13)$$

The two terms on the right hand side of the equation, the first is the power dissipated as heat in the rotor winding. The second represents the mechanical power output at the motor coupling plus the rotational losses. The mechanical power output is therefore,

$$P_o = 3(1-s)\frac{R}{s}I^2 - P_{RL} \quad (14)$$

Where P_{RL} is the rotational losses. The efficiency of the motor is then,

$$\eta_m = \frac{P_o}{P_{ag}}, \text{ or; } P_o = 3\eta_m \frac{R}{s}I^2 \quad (15)$$

The maximum terminal voltage applied to the motor will be the rated terminal voltage, and this will give maximum forward speed. Therefore, the motor output torque (in N-m) is

$$T_o = \frac{P_o}{n_m} = \frac{3\eta_m}{n_m}\frac{R}{s}I^2 \quad (16)$$

For the case of centrifugal pump where the load is commonly requiring a first-quadrant drive, the load torque of the centrifugal pump expressed as a function of speed is $T_L = K_L n_m^2$. Where K_L is a constant for the centrifugal pump load. The required mechanical power input to the centrifugal pump is $P_L = T_L n_m$.

The efficiency of the pump is the ratio of hydraulic power divided by the power input to the pump shaft [23]. Since the motor and pump are coupled together, the motor and pump speed are assumed to be the same. Therefore,

$\eta_p = \frac{P_L}{P_o} = \frac{T_L}{T_o}$. Or the torque on the pump is

$$T_L = \eta_p T_o = \frac{3\eta_m \eta_p}{n_m} \frac{R}{s} I^2 \qquad (17)$$

Under steady conditions or moderate acceleration of such low inertia loads, $T_o \cong T_L$, so that the constant K_L will be

$$K_L = \left(3\eta_m / n_m^3\right)\left(\frac{R}{s} I^2\right) \qquad (18)$$

3. SYSTEM ANALYSIS

The two important characteristics for a PV water pumping system are first, there must be enough irradiance to start pumping operation, that is until the irradiance level reaches to the threshold level. The threshold of the PV pumping system depends on the system components. Secondly, the non-linear relationship between the pumping rate and the solar irradiance [6-7].

The load matching factor, as defined in chapter 1, is the ratio of energy acquired by the motor/pump subsystem to the maximum PV array energy in one day period. This can also be defined as the ratio of the actual array output used for water pumping to the array output capability. That is,

$$\phi = \frac{\text{Actual array output used for water pumping}}{\text{Array output capability}} \qquad (19)$$

This parameter is very useful for assessing how much of the PV array's real power production capability is being used.

The load matching factor can be determined by three methods. The first is using theoretical model, the second is from solar irradiance data, and the third is from daily mean hourly PV array power output. Mathematically, from the theoretical model, the load matching factor is expressed by

$$\phi = \frac{P_l}{P} \qquad (20)$$

The array output power capability, P and the power acquired by the motor/pump, P_L are determined from the theoretical models in chapter 2. That is,

$$P = IV = (n_p I_m)(n_s V_m) \qquad (21)$$

$$P_L = \eta_i \eta_m \eta_p P = \rho g h Q \qquad (22)$$

Where, ρ is density of water (kg/m^3), g is acceleration due to gravity (m/s), h is the total pumping head, and Q is the flow rate (m^3/s).

Fig. 5 represents the PV array power output of this work taken from field study, and the 200W line shows the threshold level of the electrical load on the pumping system at 170 W/m^2 solar irradiance. The two curves are the best and the worst month situations. Taking a short time interval (the shaded area of A and B, and C and D for best and worst months respectively), the total array output capability can be obtained by integrating over the given time duration. The actual array output used for water pumping is then the curve above the threshold line. The area below the threshold line is the wasted power due to friction and other losses. Therefore, the load matching factor is the ratio of the area of the curve above the threshold line to the total area of the curve.

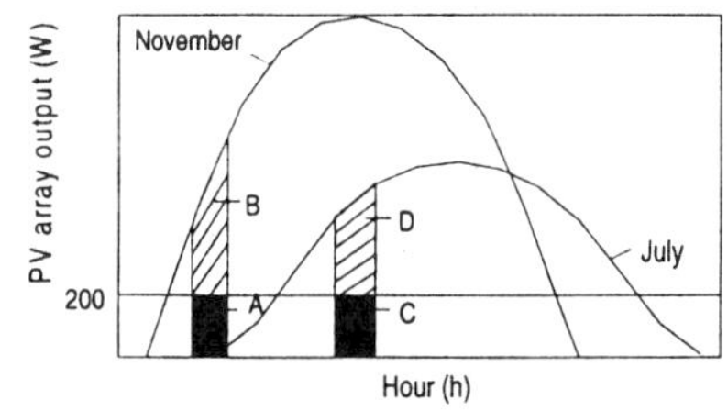

Fig. 5. A typical monthly mean hourly PV array output.

The load matching factor, from daily mean hourly solar irradiance, is determined in the same way as in Fig. 5 (since the utilisability depends on the distribution of solar radiation), and is defined by [24-25]

$$\phi = \sum^{N} (H_N - H_T) \Big/ \sum^{N} H_N \qquad (23)$$

Where, N is the number of daily hour readings, H_T is the critical solar irradiance, or the threshold, to start water pumping, and H_N is the solar irradiance on the tilted surface of the PV array.

4. RESULTS AND DISCUSSION

Optimisation of a PV array water pumping system, using the above three methods mentioned in chapter 3, shows that the utilisability of PV array mainly depends on the solar radiation profile, and the array power output and the load characteristics. From these optimisation techniques, PV arrays can be configured using the required rated power of the motor/pump subsystem and the I-V characteristic of a single cell. Conversely, the optimal size of the motor/pump subsystem can be determined from a known array configuration.

Fig. 6 shows the PV array power output at best and worst month solar radiation situations in this works field study area. The 200 W (at 170 W/m^2) line is the threshold level for the motor/pump.

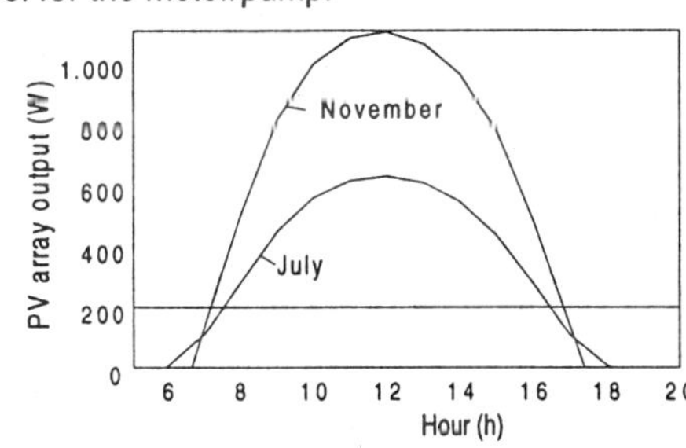

Fig. 6. Monthly mean hourly array power output.

Results show that an optimum matching factor of 0.74 and 0.55 (at best and worst month situations respectively) can be achieved using M55 type (8 series by 4 parallel) PV array and M402/SP5A-7 type motor/pump with SA1500 dc/ac inverter interfacing device.

The load matching factor improves with higher solar radiation and it is possible to achieve a higher load matching factor by carefully selecting the proper sizes of the array and the motor/pump subsystem. The result shows that the utilisability of PV array mainly depends on the solar radiation seasonal variations, and the array and the load characteristics.

***Acknowledgement*:-** This work is financed by the Academy of Finland.

NOMENCLATURE

I_p PV module light-generated current (A)
I_{ph} Light-generated current of a single cell (A)
I_o Reverse saturation current of diode (A)
I_r Reverse saturation current of diode at reference temperature (3.279×10^{-5} A)*
I_d Diode current (A)
q Electric charge (1.60218×10^{-19} C)
E_g Band gap for silicon (1.11 eV)*
R_s Cell series resistance (15 mΩ)*
R_{sh} Cell shunt resistance (1 kΩ)*
I_{sc} Short circuit current at 1 kW/m^2 (at 25°C, I_{sc} = 3.4 A, and at 47°C I_{sc} = 3.37 A)*
V_{oc} Open circuit voltage, at 1 kW/m^2 (at 25°C, V_{oc} = 21.7 V, and at 47°C V_{oc} = 20.1 V)*
I_{max} Maximum current at 1 kW/m^2 (at 25°C, I_{max} = 3.05 A, and at 47°C I_{max} = 3.06 A)
V_{max} Maximum voltage at 1 kW/m^2 (at 25°C, V_{max} = 17.4 V, and at 47°C V_{max} = 15.2 V)
P_{max} Maximum power at 1 kW/m^2 (at 25°C, P_{max} = 53 W, and at 47°C P_{max} = 46.5 W)
K Boltzmann constant (1.3805×10^{-23} N-m/°K)
A, B Diode ideality factor (1.792)*
I_{sh} Shunt resistor current (A)
V_m PV module Voltage (V)
I_m PV module current (A)
I PV array current (A)
V PV array Voltage (V)
V_c PV cell terminal Voltage (V)
I_c PV cell output current (A)
I_m PV module current (A)
T_r Cell reference temperature (298 °K)
T_c Cell temperature (°K)
T_a Ambient temperature (°K)
$NOCT$ Normal operating cell temperature (42 °C)
K_t Short circuit current temperature coefficient (4×10^{-4} A/°K)*
H_N Solar irradiance on tilted surface of array (W/m^2)
H Global solar irradiance (W/m^2)
H_T Threshold (critical) irradiance level (at 170 W/m^2 or 200 W PV array power output)

* For M55 type (n_S = 8 Series by n_P = 4 parallel) PV array coupled with M402/SP5A-7 motor/pump, with SA1500 dc/ac inverter interfacing device.

REFERENCES

[1] Grundfos A/S operation manual.

[2] Y. Kame, and L. Khouzas, A new approach to analyse the matching of direct-coupled electro-mechanical loads and a PV array. *22nd IEEE Phot. Spec. Conf.* Vol. 1, 1991, pp. 733-738.

[3] K. Khouzam, Optimum load Matching in direct-coupled photovoltaic power-Systems application for resistive loads. *IEEE Trans. on Energy conv.* EC-**5** (2), 1990, pp. 265-271.

[4] Y. R. Hsiao, Direct coupling of photovoltaic power source to water pumping system. *Solar Energy* **32** (4), 1984, pp. 489-498.

[5] M. M. Saied, Optimal solar array configuration and DC motor field parameters for maximum annual output mechanical energy. *IEEE Trans. on Energy conv.* EC-**4** (3), 1989, pp. 459-271.

[6] W. B. Lawrance and B. Wichert, A versatile PV module simulation model based on PSI/e. *Solar Energy* **52**(2), 1994, pp. 191-195.

[7] F. Loxsom and P. Durongkaveroj, Estimating the performance of a photovoltaic pumping system. *Solar Energy* **52**(2), 1994, pp. 215-219.

[8] K. Y. Khouzam and P. P. Groumpos, A non-linear Optimisation approach to the load matching of stand-alone PV power systems. *20th IEEE Phot. Spec. Conf.* Vol. 2, 1988, pp. 1298-1303.

[9] L. H. Goldstein and G. R. Case, PVSS-A photovoltaic system simulation program. *Solar Energy* **21**, 1978, pp. 37-43.

[10] F. Lasnier and T. Ang, Photovoltaic engineering handbook. 69-72, 1990, pp. 173.

[11] G. Vachtsevanos and K. Kalaitzakis, A hybrid photovoltaic simulation for utility interactive studies. *IEEE Trans. on Energy conv.* EC-**2** (2), 1987, pp. 227-231.

[12] G. Ambrosone, S. Catalanotti, U. Coscia, and G. Troise, Comparison between power and energy methods of analysis of photovoltaic plants. *Solar Energy* **34** (1), 1985, 1-8.

[13] S. Singer, B. Rozenshtein and S. Surazi, Characterisation of PV array output using a small number of measured parameters. *Solar Energy* **32** (5), 1984, pp. 603-607.

[14] W. Knaupp, Power rating of photovoltaic modules from outdoor measurements. *22nd IEEE Phot. Spec. Conf.* Vol. 1, 1991, pp. 620-624.

[15] Z. M. Salameh and C. Liang, Optimum switching points for array reconfiguration controller. *21st IEEE Phot. Spec. Conf.* Vol. 2, 1990, pp. 971-976.

[16] P. N. Enjeti, P. D. Ziogas, J. F. Lindsay, and M. H. Rashid, A new PWM speed control system for high-performance ac motor drives. *IEEE Industrial electronics*, **37** (2), 1990, pp. 143-151.

[17] K. E. Addoweesh, W. Shepherd, and L. N. Hulley, Induction motor speed control using a microprocessor-Based PWM inverter. *IEEE Industrial electronics*, **36** (4), 1989, pp. 516.

[18] R. Itoh, Stability of a PWM current-source rectifier/inverter fed induction motor drive. *IEE Proceedings-B, Electric power applications*, **137** (6), 1990, pp. 348-354.

[19] V. D. Toro, Electrical machines and power systems, 1985, pp. 172-176.

[20] O. I. Elgerd, Basic electric power engineering, 1977, pp. 369-370.

[21] S. B. Dewan and A. Straughen, Power semi-conductor circuit, 1975, pp. 191-194, 449-450.

[22] S. K. Pillai, A first course on electrical drives, 1982, pp. 58-59.

[23] I. J. Karassik, W. C. Krutzsch, W. H. Fraser, and J. P. Messina, Pump handbook, Vol. 2, 1986, pp. 196.

[24] S. A. Klein and W. A. Beckman, Review of solar radiation utilisability. *Journal of Solar Energy Engineering*. Vol. 106, 1984, pp. 394-402.

[25] D. R. Clark, S. A. Klein and W. A. Beckman, Algorithm for evaluating the hourly radiation utilisability function. *Journal of Solar Energy Engineering*. Vol. 105, 1983, pp. 281-287.

FIELD AND ARMATURE CONTROL FOR SEPARATELY EXCITED DC MOTORS POWERED BY PHOTOVOLTAIC CELLS

Khalid Hassan Hussein, Itsuya Muta, Tsutomu Hoshino, and Manabu Osakada
Department of Electrical & Electronics Engineering
Saga University, 1 Honjo machi, Saga-shi, 840 JAPAN

ABSTRACT

In this paper the authors have investigated, both by simulation and experimentally, the effect of matching the field and armature windings on the performance of a separately excited dc-motor interfaced to two photovoltaic (PV) arrays. Special emphasis has been placed on matching the load line of the field winding which, if directly connected, intersects the *I-V* characteristics of the array at a voltage close to the open circuit voltage resulting in a very small excitation current and hence poor motor performance. Step-up choppers were used as impedance transformers to match the motor windings to the maximum power lines of the PV arrays. The motor was coupled to a torque meter and the step-up choppers were PWM controlled by a personal computer (PC). The importance of matching the field winding was clearly illustrated whereas matching only the armature winding didn't significantly improve the system efficiency.

INTRODUCTION

Powering electromechanical loads by PV cells is one of the most attractive applications of PV energy conversion. This is particularly true for remote rural areas in developing countries where the need for water pumping and vaccines refrigeration is of crucial importance. Utilization of PV energy to drive such mechanical loads has many advantages over the conventional energy sources and has proven to be reliable and cost effective in many sites around the world [1]. For this reason, use of PV cells to drive electrical motors (particularly DC motors) received a lot of interest and attention from many researchers.

Due to the nonlinear current-voltage (*I-V*) characteristics of PV cells, it is an important consideration to match the motor load line as close as possible to the maximum power (P_{max}) line of the PV cells in order to achieve the best utilization of the available PV energy. The coordinates of the maximum power line are defined by the optimum operating voltage (V_{opt}) and current (I_{opt}) which vary according to the atmospheric conditions (solar radiation and ambient temperature).

The performance of different types of DC motors powered by PV cells is extensively studied and reported in the literature [2,3, 4]. Many reports agree on the superior performance of separately exited DC motors (SEM) when powered by PV cells, with regard to matching the motor load line to the maximum power line of the PV cells. However, the effect of the field winding and its array on the performance of the SEM is usually either neglected or assumed to be equivalent to the case of an ordinary DC power supply, i.e. constant rated field current.

In this paper, the actual effect of matching the field winding of a SEM on the motor performance will be investigated and highlighted together with the effect of matching the armature winding. The investigation will be presented by simulating the characteristics of PV cells when interfaced to a SEM, and by the performance of an experimental setup. The experimental setup controls the PV power flow to the motor windings through step-up choppers with the help of a PC which performs the functions of maximum power tracking, as will be illustrated in the following sections.

CHARACTERISTICS OF PV CELLS

Characteristics of PV cells can be simulated by using the following set of equations [5]

$$V = \frac{n_s AkT}{q} \ln\left(1 + \frac{n_p I_s - I}{n_p I_o}\right) \tag{1}$$

where :

- I = PV cells output current
- V = PV cells output voltage
- n_s = number of cells connected in series
- n_p = number of cells connected in parallel
- q = Electron charge
- k = Boltzmann's constant
- A = p-n junction ideality factor
- T = cell temperature in Kelvin
- I_{rs} = cell reverse saturation current

The cell reverse saturation current I_{rs} varies with temperature according to the following equation :

$$I_{rs} = I_{rr}\left[\frac{T}{T_r}\right]^3 \exp\left(\frac{qE_G}{kA}\left[\frac{1}{T_r} - \frac{1}{T}\right]\right) \tag{2}$$

where : T_r is the cell reference temperature, I_{rr} is the reverse saturation current at T_r, and E_G is the band gap

CH3365-4/94/0000-1169 $4.00 © 1994 IEEE

First WCPEC; Dec. 5–9, 1994; Hawaii

energy of the semiconductor used in the cell. The photocurrent I_{Ph} depends on the solar radiation and the cell temperature as follows :

$$I_{Ph} = \left[I_{scr} + k_i(T - T_r)\right]\frac{S}{1000} \tag{3}$$

where : I_{scr} is the cell short circuit current at reference temperature and radiation, k_i is the short circuit current temperature coefficient, and S is the solar radiation in W/m^2 (1 kW/m^2 = 100% solar radiation). The PV array power $P = IV$ can be calculated using equation (1) as follows :

$$P = I\frac{n_s AkT}{q}\ln\left(1 + \frac{n_p I_s - I}{n_p I_o}\right) \tag{4}$$

Most of the parameters in the above equations are characteristic of the particular cells used; in our case, we used Hitachi cells (model LGB2D22V).

In order to facilitate a quick reference to the operating voltage, current, and power of the armature and the field arrays, we adopted a per unit system based on the optimum values V_{opt}, I_{opt}, and P_{max} of each array at 100% solar radiation and 28 °C array temperature [6]. The base values of this per unit system are listed in Table 1 below.

Table 1. Base Values of the Per Unit System.

Parameter	Armature array	Field array
Voltage	45.6 V	22.8 V
Current	6.14 A	3.07 A
Power	280.0 W	70.0 W

Now, by making step variations in the solar radiation S and the cell temperature T in equations (1) ~ (4), the electrical characteristics of the armature and the field arrays can be investigated as a function of the atmospheric conditions.

CHARACTERISTICS OF SEPARATELY EXCITED DC MOTORS

Separately excited DC motors are characterized by the following set of equations [7].

$$V_{arm} = I_{arm} R_{arm} + E \tag{5}$$

where V_{arm}, I_{arm}, and R_{arm} are the armature voltage, current, and resistance respectively. E is the motor back-emf given by :

$$E = k_e \Phi N \tag{6}$$

where k_e is the motor voltage constant, Φ is the motor magnetic flux in Wb and N is the motor speed in rpm.

The field equation is given by :

$$V_f = I_f R_f \tag{7}$$

where V_f, I_f, and R_f are the field voltage, current, and resistance respectively.

The motor electromagnetic torque τ in N-m is given by :

$$\tau = k_t \Phi I_{arm} \tag{8}$$

where k_t is the motor torque constant.

In the above equations, the values of the constants k_e and k_t depend on the system of units used and on the construction of the motor (i.e. type of the armature winding, number of turns in the armature winding, and number of poles in the motor [7]). The value of ϕ depends on the magnetization curve of the motor and it is a function of the field current.

Using equation (5), the input power to the armature winding P_{arm} can be obtained by multiplying both sides of the equation by I_{arm}, hence :

$$P_{arm} = V_{arm} I_{arm} = (I_{arm})^2 R_{arm} + E I_{arm} \tag{9}$$

The term $(I_{arm})^2 R_{arm}$ represents the power loss due to the armature resistance (including the brush contact resistance) and the term $E I_{arm}$ corresponds to the power delivered to the mechanical load. The motor output mechanical power P_m can also be calculated using the following relation :

$$P_m = \left(\frac{2\pi N}{60}\right)\tau = \frac{\pi N}{30}\tau \tag{10}$$

Using the expression for τ from equation (8) and equating P_m to $E I_{arm}$ we get

$$k_t = \frac{30}{\pi} k_e \tag{11}$$

For our analysis purposes, we determined the value of the parameter $k_e \phi$ experimentally as a function of the field current I_f, and it is given by the following equation

$$k_e \phi = 0.003 + 0.404 I_f - 0.950 (I_f)^2 \tag{12}$$

The performance of the SEM when powered by PV arrays can be investigated by simultaneously solving the above set of motor equations with equations (1) ~ (3). The I-V characteristics of the armature and the field windings can be determined by solving equation (5) and equation (7) with equation (1) respectively, for different atmospheric conditions as plotted in Fig. 1.

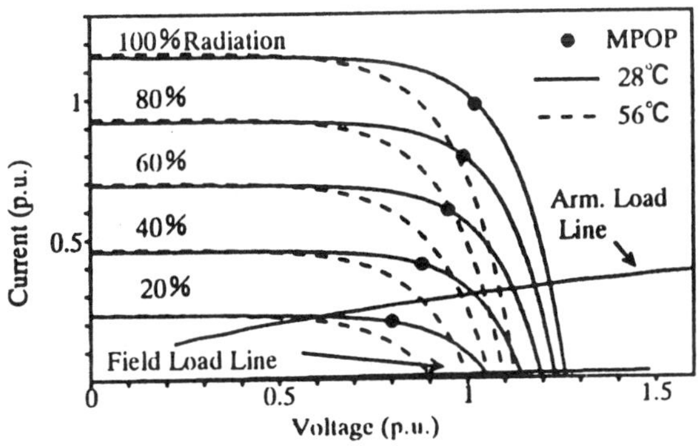

Fig. 1. The I-V characteristics of the field and armature windings when directly connected to their PV arrays.

The operating voltages and currents of the armature and the field arrays are given (in p.u.) by the intersection points shown in Fig. 1. Using the appropriate base values from Table 1, the actual values of these operating points can be calculated.

From Fig. 1 we clearly note that the load line of the field winding, when directly connected to its array, intersects the I-V characteristics of the array at a voltage close to the open circuit voltage. This will result in a very small field current and hence weak motor flux. In other words, the field array operates away from its maximum power operating point (MPOP) and it is poorly utilized because of the large impedance mismatch between its optimum load line and the field load line. The armature load line, on the other hand, is characterized by a small resistance at start-up the thing that enables the armature array to deliver almost its short circuit current to the armature winding. This relatively high current is necessary for the armature to develop the starting torque. After the motor starts running, the increasing back emf will shift the armature load line towards the open circuit voltage of the array. The steady state load line of the armature winding, though better than the field load line, does not coincide with the optimum load line of the armature array.

The effect of the field mismatch (or the small I_f) can be illustrated by simulating the motor characteristics as follows. Solving equations (5) and (6) simultaneously with equation (1) gives the electromechanical plane (or the N-I) characteristics of the SEM for a particular atmospheric conditions as

$$n = \frac{1}{k_e\phi}\left\{\frac{n_s AkT}{q}\ln\left(1 + \frac{n_p I_s - I_{arm}}{n_p I_o}\right) - I_{arm}R_{arm}\right\} \quad (13)$$

where the armature current is substituted for the array current. The value of $k_e\phi$ in the above equation can be calculated by simultaneously solving equation (7) with equation (1) and substituting the resulting value of I_f in equation (12).

Using the relation $I_{arm} = \tau/k_\tau\phi$ from equation (8) and substituting in equation (13) gives the mechanical plane (or, N - τ) characteristics of the SEM for a particular atmospheric conditions as

$$n = \frac{1}{k_e f}\left\{\frac{n_s AkT}{q}\ln\left(1 + \frac{n_p I_s - \tau/(k_t\phi)}{n_p I_o}\right) - \frac{\tau R_{arm}}{k_t\phi}\right\} \quad (14)$$

where the value of $k_t\phi$ can be calculated from equation (11). With the help of equations (2) and (3), the (n-τ) characteristics can be simulated for different atmospheric conditions as shown in Fig. 2.

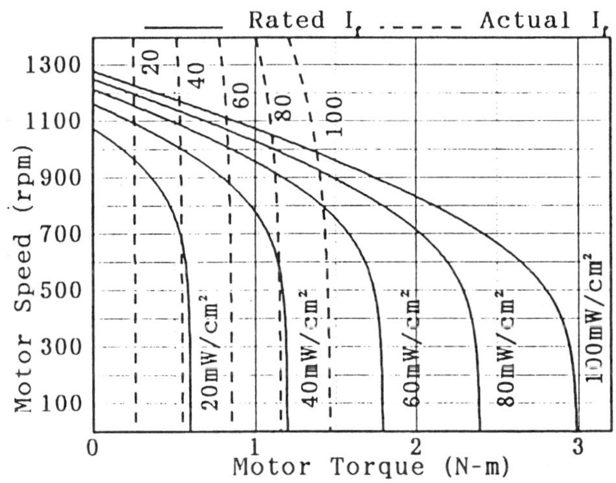

Fig. 2. Simulation of the (N - τ) characteristics.

In order to emphasize the effect of the actual value of the field current (which is supplied from the field array) on the performance of the motor, our simulation results show a comparison between two performances of the motor. The solid lines represent the performance of the motor based on the assumption that the field current is constant (rated value) and the doted lines represent the performance of the motor when the field winding is directly connected to its PV array. From these results, the importance of the field current consideration in PV powered SEM is clearly evident.

EXPERIMENTAL SET-UP

Our experimental set-up consists of a SEM powered by two PV arrays through step-up choppers as shown in Fig. 3. The motor is coupled to a brake-torque-meter with its braking current set to represent a mechanical load within 80% of the peak ratings of the armature array. The step-up choppers are PWM controlled by a personal computer (PC). The PC monitors all the system parameters (including the solar radiation and the array temperature), executes control algorithms responsible for maximum PV power tracking [8], outputs the proper control signals to the PWM circuitry, and generates result files to evaluate the overall performance of the system.

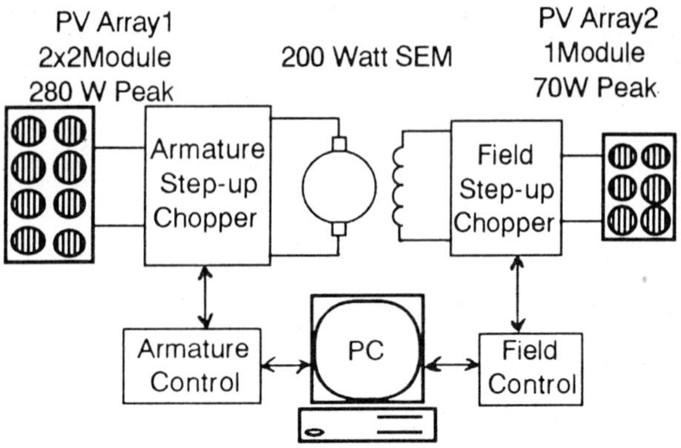

Fig. 3. General block diagram of the experimental set-up.

In actual systems, the major role of the PC can be performed by a maximum power point tracking device.

EXPERIMENTAL RESULTS AND CONCLUSIONS

In order to evaluate the effect of controlling the field and the armature windings of the separately excited dc-motor, four sets of data were recorded under almost the same radiation and temperature conditions : when no control was applied, controlling only the armature winding, controlling only the field winding, and when controlling both windings. The torque-meter braking current was kept constant to ensure the same starting torque condition and torque-speed characteristics for all the tests. Experimental results for two modes of operation are shown graphically in Fig. 4.

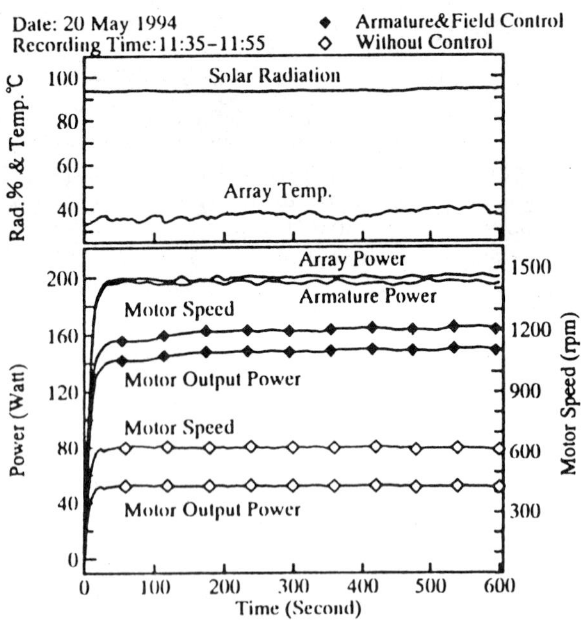

Fig. 4. The effect of matching the field and armature windings on the SEM performance.

The overall system efficiency defined as the ratio between the motor output mechanical power (calculated using the motor torque and speed) to the input power available from the PV array, was calculated for each of the above four sets of data and can be summarized as follows : 29% when no control was applied, 30% when controlling only the armature winding, 72% when controlling only the field winding, and 75% when controlling both windings.

The importance of controlling (or matching) the field circuit is clearly apparent from these results where efficiencies over 70% were achieved. On the other hand, controlling only the armature circuit didn't significantly improve the system efficiency and this is because most of the input power was wasted in the form of heat in the armature winding due to the poor field excitation. This can be further explained by referring to equations (8) and (9) as follows : from equation (8) we note that as the motor flux Φ decreases, due to the small I_f, the armature tends to draw more current in order to meet the load torque . From equation (9), although the power delivered to the mechanical load is proportional to I_{arm} i.e. increases with I_{arm}, the heat loss in R_{arm} increases even further (being proportional to the square of I_{arm}).

REFERENCES

[1] K. Stabes et al. : "Reliability, Cost, and Performance of PV Powered Water Pumping Systems : a Survey for Electric Utilities", *IEEE Trans. on Energy Conversion* Vol.**8**, No. 3, September 1993, pp. 506 - 512.

[2] M. Saied : "Matching of DC Motors to Photovoltaic Generators for Maximum daily Gross Mechanical Energy", *IEEE Trans. on Energy Conversion* ,Vol. **3**, No. 3, September 1988, pp. 465 - 472.

[3] S. Singer et al. : "Starting Characteristics of Direct Current Motors Powered by Solar Cells", *IEEE Trans. on Energy Conversion*, Vol. **8**, March 1993, pp. 47 - 52.

[4] J. Appelbaum : "Starting and Steady State Characteristics of DC Motors Powered by Solar Cell Generators", *IEEE Trans. on Energy Conversion*, Vol. **1**, No. 1, March 1986, pp. 17 - 25.

[5] B. Bose et al. : "Microcomputer Control of a Residential Photovoltaic Power Conditioning System", *IEEE Trans. on Industry Application*, Vol. **21**, No. 5, September 1985, p.p.1182 - 1191.

[6] K. Khouzam et al. : "Optimum Matching of Direct - Coupled Electromechanical Loads to a Photovoltaic Generator" , *IEEE Trans. on Energy Conversion*, Vol. **8**, No. 3, September 1993, pp. 343 - 349.

[7] A. Langsdorf : "*Principles of Direct Current Machines*", sixth Edition, McGraw-Hill Co., 1959.

[8] K. Hussein, et al. : "Modified Algorithms for Photovoltaic Maximum Power Tracking", *Record of 1992 Joint Conference of Electrical and Electronics Engineers in Kyushu - Japan*, October 1992, p. 301.

OPTIMIZING THE NUMBER OF a-Si SOLAR CELLS IN A MODULE FOR LOAD MATCHING

P.K. KONER
Grantholder, TEAM Ispra (Italy)

H. OSSENBRINK
European Solar Test Installation
European Commission DG XII Joint Research Centre
Ispra Establishment

ABSTRACT

The importance of a-Si module design in the optimum DC load matching has been studied. In this context, a directly coupled water pump with the source of a-Si solar cells has been analysed. The comparative matching performances of pumps have been made under two differnt sources of a-Si and x-Si solar cells. The a-Si solar modules powered pumping system has been found as suitable system.

INTRODUCTION

A number of studies have been made on the optimum matching of the DC loads with x-Si (crystalline) solar modules, mainly water pumping system as load. These studies typically focused on the characteristics matching of the load and reconfiguration of solar modules. However, no studies has yet been reported on the optimisation study of DC loads by the possibility of reconfiguration of solar cells in a given size/power module. The reconfiguration of solar cells in a given power module would increase choices towards the reconfiguration of solar modules, but can be realised only in limited resolution. For instance, a typical PV-pump system with 200Wp, powered by four 50 W modules, would only be adjustable by steps of 144, 72, 36 and 18 cells in series, without increasing the cost of manufacturing.

The advantage of the fabrication of a-Si module in industry that the number of series connected solar cells can be altered in a given size/power of module by changing the layout of the laser patterning. Which yields the increasing or decreasing of module voltage. Consequently, the module current varies inversely with the module voltage due to the change of the area of a single solar cell. Therefore, the packing factor of the module does not affected significantly. This could result of the reducing the voltage and current mismatch between source and any DC loads as compared to that for c-Si PV modules. Because the voltage and current are fixed for a given rated c-Si solar module.

The possibility of changing the number of series connected solar cells in a given a-Si solar module can be considered as an advantage of the reconfiguration of solar cells on the module level. This is possible by changing the laser-patterning system used in manufacture of these modules. For example, if the number of series connected solar cells is increased by 10% in a standard a-Si solar module, as a result module current reduces by 10%. This possibility has limits, as the laser scribe causes an inactive zone of appr. 0.7 mm, so a distance of 6 mm of the patterns is considered as a minimum. On the other hand, the laser patterns should be not too wide in order not to increase the series resistance of TCO (Transparent Conductive Oxide) and hence decrease the Fill-Factor of the module. A patterning width of 14 mm can be considered as a maximum with today's TCO materials. Within these limits, the number of cells in series may be varied by factor of two with almost any intermediate values. As water-pumps are usually of given performance and operating parameters (in particular when existing systems should be re-worked with photovoltaic power), the modification of the laser-patterning would be a convenient method to match the module to the pump characteristics, for the benefit of maximum hydraulic coupling efficiency and thus water quantity pumped for an estimated average day of operation.

Another advantage of a-Si solar cell used in directly coupled DC loads is that the power vs. voltage characteristics is wider as compared to that of x-Si solar cell. This gives an opportunity that a load connected to the a-Si solar cells can give better matching efficiency as compared to the same load connected to the c-Si solar cells, as a deviation from the optimum operating voltage would have less influence on the resulting power. Therefore, a study on the matching analysis of a directly coupled pumping load powered by a-Si solar cells has been made by alteration of the number of series and parallel connected solar cells.

CH3365-4/94/0000-1173 $4.00 © 1994 IEEE

First WCPEC; Dec. 5-9, 1994; Hawaii

MODEL CALCULATIONS

The a-Si Solar Cell

The relation of short circuit current (I_{sc}) and iradiance (I_{rad}) of a-Si solar cell has been found from the experimental data as

$$I_{sc} = C_1*Irad + C_2*(I_{rad})^2 \quad (1)$$

where $C_1 = 7.9*10^{-3}$ and $C_2 = 4.62 * 10^{-6}$ (based on a number of APS-modules).

The relation of open circuit current (V_{oc}) with I_{sc} of a-Si solar cell is similar to the that of c-Si solar cell and has been found from the measurements of a-Si modules as

$$V_{oc} = Ln(I_{sc}/I_o+1) \quad (2)$$

where I_o is reverse saturation current = $2.82 * 10^{-5}$ mA/Cm^2

The current(I)-voltage(V) characteristics of a-Si cell can calculated by the following equations (3-6)

$$I_d = I_o*(exp(V_d*q/AkT) - 1) \quad (3)$$

$$I_{rsh} = V_d/(R_{sh}*(1+V_d*C_{rsh})) \quad (4)$$

$$I = I_{sc} - I_d - I_{rsh} \quad (5)$$

$$V= V_d - IR_s \quad (6)$$

where I_d is the diode current, I_{rsh} the shunt current, R_{sh} the shunt resistance (=1100/Irradiance. Kohm.cm^2), Crsh the constant for variation of shunt resistance with voltage(=-1.1 Ohm cm2/W) and R_s the series resistance (= 2.5 Ohm.cm^2). The modelled and experimental current-voltage characteristics of a-Si module (Flagsol) is shown in Fig. [1].

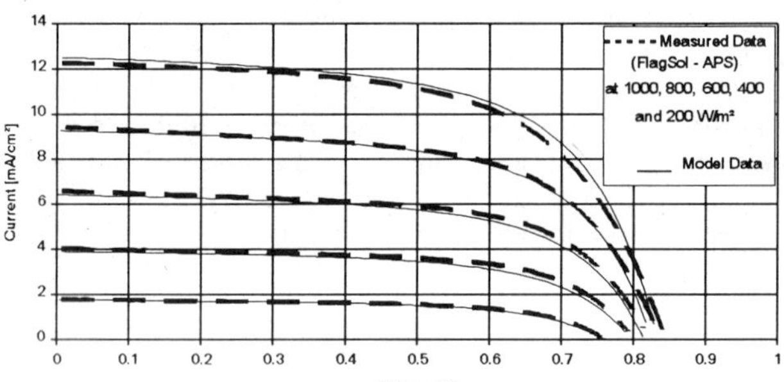

Fig. 1 Modelled IV-characteristics compared to measurement results

MOTOR (DC series) CHARACTERISTICS

$$V^m = MI^m w+I^m R_t \quad (7)$$

$$T = M (I^m)^2 \quad (8)$$

where M is the motor constant, w is speed, R_t is total resistance of armature and field coil and T is the torque of the motor.

PUMP (Centrifugal) CHARACTERISTICS

$$H = K_1*w_o^2 - K_2*(w_o-w)^2 \quad (9)$$

$$Q = K_3*Sqrt(H-H_o) \quad (10)$$

where H is the water head, K_1 & K_2 are constants, wo is the speed of the pump at no-flow condition, Q is the flow of the pump in MKS units, H_o is the water head at no-flow condition and K_3 is a constant.

SOLUTION OF MONOBLOCK MOTOR-PUMP CHARACTERISTICS

$$T = K_4 * w^2 \quad (11)$$

$$V^m = Sqrt(K_4*w/(M*w_o)) *(Mw^2+R_t w_o) \quad (12)$$

$$I^m = Sqrt((K_4/M)*(w*w_o)) \quad (13)$$

$$E_{mp} = qQH/(V^m * I^m) \quad (14)$$

where T is the torque of motor-pump and Emp is the motor-pump efficiency.

SOLUTION OF a-SOLAR CELL-MOTOR-PUMP CHARACTERISTICS

$$I^m = I_{sc} - I_o*(exp(q*(V^m+I*R_s)/AkT)-1) - (V^m+I^m*R_s)/(R_{sh}*(1+C_{rsh}*(V^m+I^m*R_s))) \quad (15)$$

This is a transcendental equation and it has been solved using Newton-Raphson approximation technique

The motor -pump efficiency is maximized under different input voltage of motor and defined as E_{mpm}. Then Hydraulic coupling efficiency is defined as

$$E_{ch} = E_{mp}*E_{ce}/E_{mpm} \quad (16)$$

where E_{ce} is the electrical coupling efficiency of PV powered pump system. The E_{ce} has been calculated as

$$E_{ce} = V^m * I^m/(P^m) \quad (17)$$

where P^m is the maximum power of the PV generator for a certain condition.

Two pumps have been selected for the verification of the present optimum matching analysis, one of which (Pump-1) optimally matched and another one (Pump2) unmatched. The experimentally determined value of different constants of two motor-pump systems have been shown in Table [1]

TABLE 1

Constant	*Pump-1*	*Pump-2*
K_1	$4.74*10^{-5}$	$7.3*10^{-5}$
K_2	$9.7*10^{-3}$	$1.7*10^{-3}$
K_3	0.1	0.76
K_4	$2.83*10^{-7}$	$2.87*10^{-6}$
M	0.22	0.06
R_t	40	2

(All values in MKS unit)

By putting the value of all constants of motor and pump in the theoretical analysis, the required number of solar cells (a-Si and x-Si) have been calculated to achieve 100% hydraulic coupling efficiency of both systems at 800 W/m^2. The solar insolation level of maximum coupling efficiency is assumed here 800 W/m^2 to get the maximum average daily efficiency for a standard day in star dard location. However, it can be altered according to the pattern of solar insolation in a specific location. The equivalent current-voltage characteristics of a-Si and c-SI PV modules and in relation to the characteristics of pump 1 are compared in Fig.[2]. The comparative hydraulic coupling efficiencies of two different pumps under two different sources for a typical day are shown in Fig.[3].

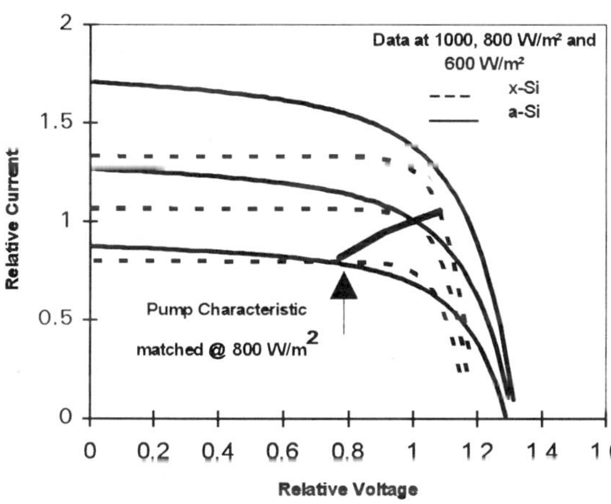

Fig. 2 Pump current-voltage characteristics relative to a-Si and x-Si characteristics.

CONCLUSIONS

The modelled current-voltage characteristics of a-Si module match well with experimental data.

The calculated optimum number of a-Si solar cells (series & parallel) can be achieved in a module of given size and power by changing the width of the laser patterning in the production process, whereas this would not be possible in x-Si PV production without a major modification in the module design.

The relatively broad power vs. voltage characteristics is more tolerant to impedance mismatch between source and DC loads at different irradiance levels, however in this study the daily average hydraulic output of the pump is not improved remarkably when powered by a-Si solar cells because the current-voltage characteristics of the pumps used for this calculation are quite far from the maximum power point characteristics of both types of PV modules.

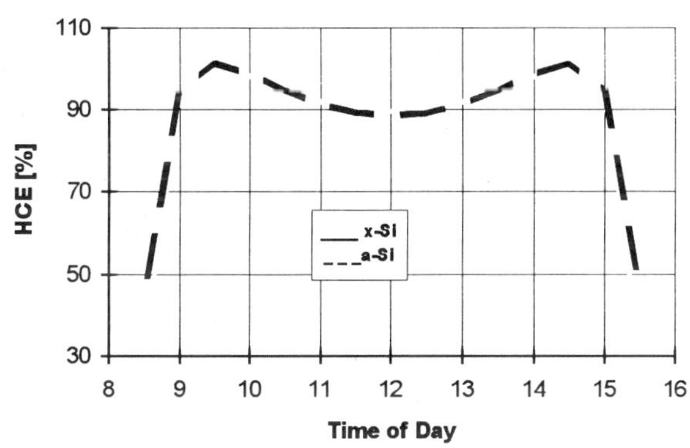

Fig.3 Hydraulic coupling efficiency during a typical day

EFFECTS OF ADVANTAGEOUS PLACING OF PV CELLS AND MODULES IN GENERATING UNITS

K. Reiche, R. Preu, G.Kleiss, K. Bücher
Fraunhofer Institute for Solar Energy Systems (FhG-ISE) Oltmannsstr. 5, D-79100 Freiburg, Germany

ABSTRACT

The effect of advantageous placing when assembling solar cells or modules to PV modules or generating arrays resp. is investigated. The idea is to take advantage of thermal inhomogenities within a generating unit to reduce a possible mismatch between non-identical components when operated at MPP (maximum power point). The cases of series and parallel connecting photovoltaic generating components are studied. We simulated the output of a cell production (or module production resp.) and estimated the effect on generator efficiency. As result we found a possible gain in the energy output of up to 3% for the set-up of a PV array (parallel connected case) while the effect occurring within a module is negligible (series connected case). A simple rule for advantageous module placing within a PV array is formulated.

INTRODUCTION

There have been several studies on the connection of PV components (cells or modules) to generating units (modules or arrays resp.) [1],[2]. However, when the units are operated outdoors, new effects will occur due to thermal inhomogenities inside the generators. These inhomogenities have been observed to be as high as $\Delta T = 8K$ [3],[4] inside modules and substantially higher inside arrays (e.g. $\Delta T = 16K$ [4]). The reasons for the occurring temperature inhomogenities may be summarized in three categories: heat conduction (mounting), free and forced convection [5]. Since temperature has always a considerable effect on the output of photovoltaic generators, we investigated the influence of the thermal inhomogenities on the output of connected units. These investigations are presented for maximum power point operation. Generally, current and voltage at MPP (I_{mpp} and V_{mpp}) for a single component (without connection) will be different for each component. However, when operated at a common MPP, all components will be forced into a working point, which is optimal for the entire generator, but not for the single component. Consequently, there will always be losses in efficiency when connecting (non-identical) components. It is a straightforward idea to reduce this loss that arises due to the inhomogenity in the component's semiconductor properties by taking advantage of the existing inhomogenity in temperature. This means benefiting from the fact, that current and voltage for each component will change with temperature. Placing the components on cold and hot places according to their quality ('good' components have higher current or voltage at MPP whereas 'bad' components perform poorer in I_{MPP} or V_{mpp}) will reduce the effective mismatch in current or voltage. When connected, the components will consequently be forced less far out of their individual MPP. Thus the loss in $\eta^{connected}$ will diminish. It has to be pointed out, that usually the information on the quality of the component is available. First, screening of cells when assembling a PV module (series connected case) has been applied for long time by module producers. We checked the possibility to use the hereby gained information to place the cells in an advantageous way within a PV module. Second, nowadays many producers print the individually measured value of the voltage at maximum power point on the name plate of each module. When assembling modules to a PV array (parallel connection) this turns out to be the most important quantity, as will be shown below.

METHOD

We simulated the electrical characteristics of each cell and module with the 2 diode equation. The I-V curve of the k-th component (cell or module resp.) operating at temperature T^k is:

$$I^k(V^k) = I^k_{photo} - (V^k + I^k R^k_{ser})/R^k_{par} - I^k_{01}\,[\exp(q(V^k + R^k_S I^k)/n^k_1 kT_k) - 1] - I^k_{02}\,[\exp(q(V^k + R^k_S I^k)/n^k_2 kT_k) - 1]. \quad (1)$$

Current and voltage for series and parallel connection are calculated according to Kirchhoff's law. To simulate the spread of the output of a typical cell (or module) production line, we varied the parameters of the above two diode equation (1) by a Gaussian distribution for d/dT I^k_{photo}, I^k_{photo}, n^k_1, n^k_2, R^k_{ser}, R^k_{par}, whereas I^k_{01} and I^k_{02} have been generated with an exponential noise. Tab. 1 summarizes the mean values and the maximum relative deviations of the parameters for a sample of 88 components used in this report. We regard these values as typical for the spread of a production line. Fig. 1 shows the distribution of η_{STC} values for this sample.

First WCPEC; Dec. 5-9, 1994; Hawaii

Results to be presented below have been obtained by calculating the effect of advantageous placing for each possible combination of two components out of this sample of 88 components. For the case of series connection this means, that we estimated the effect by connecting only two cells (instead of e.g. 36 cells) at a time to a 'mini module' with one hot cell and one cold cell. In the case of parallel connection, we simulated the effect by connecting two modules (each one again represented by a 8 parameter set for the two diode equation (1) out of our 88 cell sample) to a 'mini array' with one hot module and one cold module. For both cases, the number of possible different combinations is 88*87/2 = 3828 simulated generating units.

Parameter	mean (units given in left column)	maximum relative deviat. [%]
I_{photo} [A/cm²]	3.11E-02	5.04
I_{01} [A/cm²]	2.18E-12	268.44
I_{02} [A/cm²]	9.18E-08	153.46
R_{ser} [Ωcm²]	1.44E+00	26.03
R_{par} [Ωcm²]	1.31E+03	24.65
$1/I_{photo} dI_{photo}/dT$ [1/K]	5.77E-04	2.65
n_1	1.00	6.28
n_2	2.03	4.84
eta [%]	13.1	10.07

Tab. 1 Means and maximum relative deviations of the simulation of the properties of a production line output.

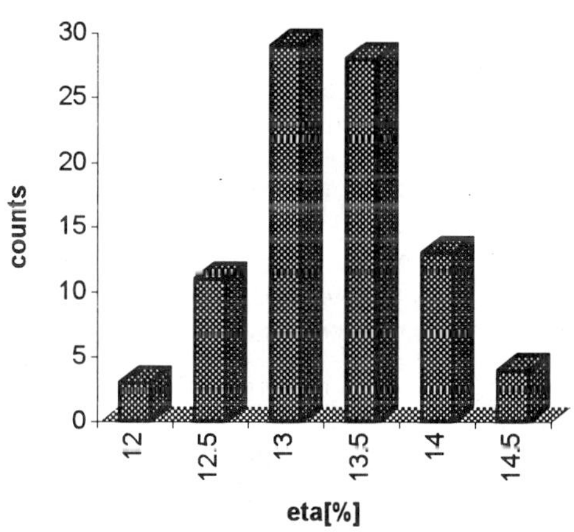

Fig. 1: The η distribution of the 88 cell sample used for the calculations.

All results are presented at standard testing conditions (STC : irradiance G = 1000 W/m², temperature T = 25°C, spectrum E(λ) = AM1.5). Despite the fact, that STC does not deliver a good module rating point [6], we found for all effects to be defined below, that the shifts in efficiency are only weakly dependent on the rating method (point rating with respect to specific G and T or averaging and site-dependent rating). Thus it is sufficient to present the results at one prominent point. We use STC because STC rating is still commonly used.

RESULTS

Series Connection

The effect of exploiting the information acquired during cell screening is investigated. That means the information about the 'quality' of cells is available: 'good cells' have a higher current at MPP (I_{MPP}) than 'bad cells'. The strategy of advantageous placing in the case of series connection is to place a 'good cell' on a cold position and a 'bad cell' on a hot position. Since I_{MPP} is expected to increase with temperature (following the increase in I_{SC}) the mismatch between the two cells will hereby be reduced. In order to compare the expected output of a module production, that does not take advantage of the temperature inhomogenity, but instead assembles classes of 'good cells' to 'good modules' and classes of 'bad cells' to 'bad modules', to the output of a production applying advantageous placing, we define the possible gain in efficiency as follows:

$$\text{effect1} = (\eta_{adv} - \eta_{no\ adv}) / \eta_{adv} \quad (2)$$

where η_{adv} denotes the module efficiency with the just formulated advantageous placing strategy for series connection. $\eta_{no\ adv}$ is the expectation value that would be obtained without considering the effects of cell placing and is calculated as arithmetic mean of the efficiencies one would obtain when placing two 'good cells' on both hot and cold places and placing two 'bad cells' on the two places. As quoted above, the thermal inhomogenity occurring inside one module may reach up to ΔT = 8 K [3],[4]. Simulating the module production by series connecting only two cells (instead of e.g. 36) we assumed for our calculations a lower temperature difference of ΔT = 4 K. The relative deviation for I_{MPP} for our sample of 88 cells was less than 5% corresponding to the maximum relative deviation for I_{photo} as given in Tab.1. The effects of the simulation are shown in Fig. 2. The maximum gain in efficiency for the studied production line is less than 2%, while the mean effect is about 0.1%.

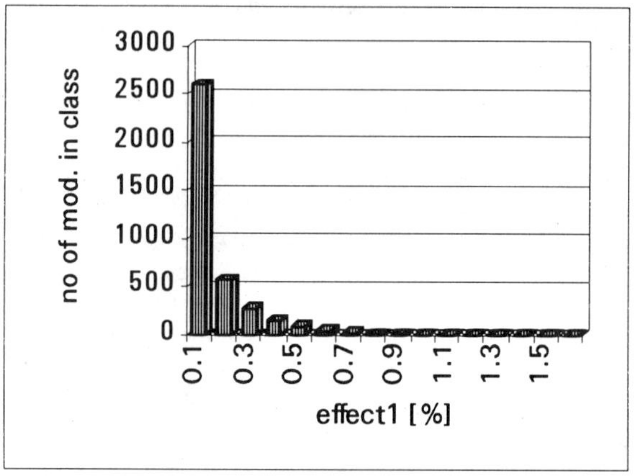

Fig. 2: Distribution of effect1 (see eqn. (2)) as gained from advantageous cell placing according to the thermal inhomogenity inside a module.

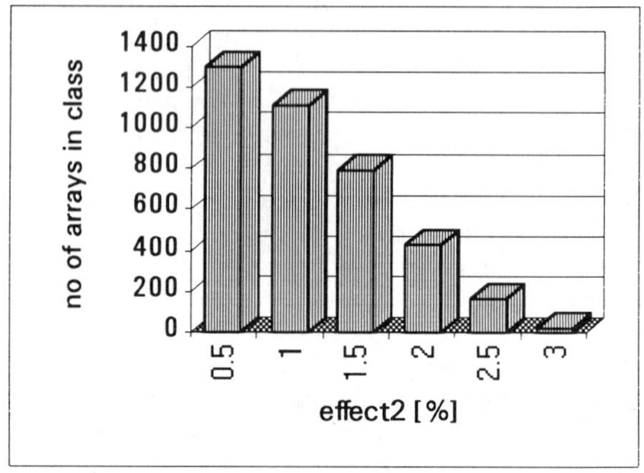

Fig 3: Distribution of effect2 (see eqn. (3)) as gained from advantageous module placing according to thermal inhomogenity inside an array.

Parallel Connection

Again, it is assumed, that information on the quality of the individual modules to be connected to an array is available: 'good modules' have a higher voltage at MPP (V_{MPP}) than 'bad modules'. However, since the sign of the influence of temperature on the prominent quantity is different from the series connection case (V_{MPP} is decreasing with temperature following V_{OC}), a new advantageous placing strategy has to be formulated for the parallel connection case: 'good modules' are to be distributed to warm places within the array and 'bad modules' to cold places. To estimate the difference in performance when assembling the array by advantageous or disadvantageous placing of the modules inside the array, we define the effect slightly different:

$$\text{effect2} = 2\,(\eta_{adv} - \eta_{disadv}) / (\eta_{adv} + \eta_{disadv}) \qquad (3)$$

where η_{adv} denotes the array efficiency with application of the advantageous placing strategy of parallel connection and η_{disadv} denotes the array efficiency when assembled in a disadvantageous way, i.e. contrary to the advantageous placing strategy. We had to define effect2 slightly different, because in this case the aim of the PV user is to form *one* array out of *all* modules purchased. Although temperature inhomogenities of up to $\Delta T = 16K$ [5] and more have been reported in the case of large PV arrays, we assumed only a conservative temperature difference of $\Delta T = 8K$. Again we estimated the possible gain in efficiency by calculating a parallel connection of only two modules. Similar to the case of series connection, the maximum relative difference in V_{MPP} between two arbitrary modules is 5%. As can be seen in Fig. 3, a gain in efficiency of up to 3% is possible, while the mean of effect2 is found to be about 1%.

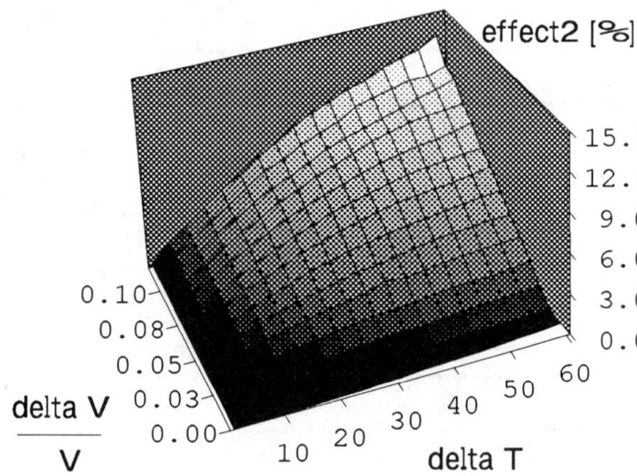

Fig.4: Dependence of the mean effect2[%] on mismatch$_{par.\ con.}$ (4b) and ΔT. We calculated effect2 up to $\Delta T = 60K$ to check a simple ansatz for linear approximation of the effect (6) which seems to hold in this case for the area of interest.

DISCUSSION

In order to interpret the obtained results it is convenient to note once again that the studied effects (effect1 and effect2) measure the capability of diminishing an existing mismatch of individual components. They may be expressed as relative quantities for the two cases studied as:

$$\text{mismatch}_{ser.\ con.} = \Delta I_{MPP} / I_{MPP} \qquad (4a)$$

$$\text{mismatch}_{par.\ con.} = \Delta V_{MPP} / V_{MPP} \qquad (4b)$$

According to their definition, the studied effects may be approximated to be proportional to these mismatches. As a matter of fact the proposed method of advantageous placing will yield no gain for optimally matched components (i.e. for mismatch=0). Further, the process of diminishing these mismatches by advantageous placing is due to thermal shifts of I_{MPP} or V_{MPP} resp. The latter may in good approximation be expressed in terms of normalized temperature coefficients TC of I_{SC} or V_{OC} resp.:

$$TC_{ser.\ con.} \approx 1/\ I_{SC}\ d/dT\ I_{SC} \quad (5a)$$

$$TC_{par.\ con.} \approx 1/\ V_{OC}\ d/dT\ V_{OC} \quad (5b)$$

Finally the effects have been found in good approximation to be proportional to the temperature split ΔT that occurs within the connected unit (4K for the series connected case and 8K for the parallel connected case). Again, no effect can be found for $\Delta T=0$.

Summarizing all individual contributions we expect the effect for our sample to be proportional to the product:

$$\text{effect} \propto \text{mismatch} \times TC \times \Delta T \quad (6)$$

For future work it remains to test, whether this simple linear approximation holds for more general cases, e.g. for samples of real cells/modules instead of our simulated sample. Since the effects are considerable in the case of parallel connection, we investigated a possible over-compensation of the mismatch, in the case of a large temperature split ΔT. As can be seen in Fig. 4 the simple linear approximation (6) with respect to ΔT does not hold for large ΔT. On the other hand, it may be stated, that even for temperature splits as high as $\Delta T=60K$ (which would never occur in realistic installations, but may arise for concentrating installations [7]), a benefit from the advantageous placing method is to be expected if the mismatch is sufficiently large.

When comparing the means of effect1 and effect2 we may use the approximation (6) to explain the observed difference in their orders of magnitude. The mismatches occurring in the samples are in the same order of magnitude and exhibit a similar distribution ($\text{mismatch}_{ser.\ con.} \approx \text{mismatch}_{par.\ con.}$). Further, the split temperatures ΔT differ only by a factor of 2 for the two cases studied. But when regarding the difference in the temperature coefficients TC, the reason for the differences in the observed effects is revealed. Despite the fact, that every single component has of course an individual $TC_{ser.\ con.}$ or $TC_{par.\ con.}$ resp., these temperature coefficients scatter around a mean that is in the typical range of order for crystalline silicon solar cells. $TC_{ser.\ con.} \approx 0.05\%/K$ and $TC_{par.\ con.} \approx -0.3\%/K$ [8]. Thus it is mainly this difference in the temperature coefficients that makes the strategy of advantageous placing more efficient when applied to parallel connected units than in the series connected case.

CONCLUSIONS

Series Connection

In conclusion, it can be stated, that advantageous placing of series connected cells inside a module with respect to the thermal inhomogenity under outdoor conditions is not expected to yield a considerable benefit. The maximum possible gain in efficiency for the studied production line is less than 2%, the mean effect being about 0.1%. However, with respect to temperature measurement for modules operated outdoors, this is good news: The error induced through temperature inhomogenity inside the module may be considerably reduced by measuring at several points and calculating the mean temperature. The distribution of effect1 in Fig. 2 may be interpreted as typical error distribution for the just formulated measurement procedure.

Parallel Connection

Concerning the mounting of modules inside a PV array it is worthwhile placing the modules in an advantageous way by benefiting from the thermal inhomogenities. Without any extra efforts or costs, just by placing the modules with higher V_{mpp} at the hotter places (e.g. the top of the array in the case of free convection), up to 3% on the efficiency may be gained. As individual measurement of each module by the producers is nowadays state-of-the-art, V_{mpp} should always be printed on the module's name plate.

ACKNOWLEDGMENTS

Work reported here is being funded by the German Ministry for Research and Technology BMFT under contract number 0328650F. G.Kleiss gratefully mentions financial support from the German Academic Exchange Service DAAD for enabling research at GENEC, C.E.Cadarache, France.

REFERENCES

[1] R.Janoch, *17th IEEE PVSC, Kissimmee*, pp.558 (1984)

[2] L.L.Buicciarelli, *Solar Energy,* **23** pp.277-288 (1979)

[3] P.C.S. Paes, *'Essais en environnement reel de champs de modules photovoltaiques au silicium multicristallin et au silicium amorphe hydrogene', PhD thesis, Université d'Aix-Marseille III*, (1988)

[4] H. Laukamp, et.al., *12th EC-PVSEC, Amsterdam*, pp.928 (1994)

[5] D.U.Sauer, R.Kaiser, *9. Symposium Photovoltaische Solarenergie, Staffelstein,* pp.485 *(1994),* (in german)

[6] K.Bücher, G.Kleiss, *12th EC-PVSEC, Amsterdam*, pp.299 (1994)

[7] J.L. Watkins, E.L. Burgess, *13th IEEE PVSC, New York*, pp.1061-1066 (1978)

[8] C.R.Osterwald et.al., *19th IEEE PVSC, Las Vegas*, pp.744 (1988)

TRIMODE POWER CONVERTER OPTIMIZES PV, DIESEL AND BATTERY ENERGY SOURCES

George O'Sullivan, Abacus Controls Inc - Life Member IEEE

ABSTRACT

Conservatively, there are over 100,000 localities in the world waiting for the benefits that electricity can provide, and many of these are in climates where sunshine is plentiful. With these locations in mind a prototype 30-kW hybrid system has been assembled at Sandia National Laboratories (SNL) to evaluate the performance, reliability and economics of photovoltaic, diesel and battery energy systems managed by a TRIMODE power converter. In the Power Converter, four IGBT's with an isolation transformer and filter, serve as rectifier and charger to charge the battery from the diesel; as a stand-alone inverter to convert PV and battery energy to ac; and, as a parallel inverter with the diesel-generator. Whenever the diesel is supplying the load, an algorithm assures that the diesel is running at maximum efficiency. Test results are showing the feasibility of the TRIMODE Power Converter in multi-sourced, hybrid energy systems.

INTRODUCTION

Cost effective applications of hybrid systems include remote facility power, remote home and village power, and power for dedicated loads such as communications. The hybrid system market is anticipated to be one of the most important growth opportunities for PV for the rest of the decade.[1]

Hybrid system design is bounded by economics, electrical performance goals and predetermined needs. Reducing diesel fuel consumption and engine run time is an economic necessity in countries where oil importation is a critical drain on financial resources.

POWER CONVERTER

In 1988 a BIMODE Power Converter that was capable of converting power between a battery and an AC power grid in both directions was introduced. Originally developed for Uninterruptible Power Supply (UPS) applications, the BIMODE was soon recognized for its potential in hybrid systems involving diesel engine generators, batteries and renewable energy sources.[2]

The TRIMODE Power Converter adds to the dual capabilities of charging a battery from a diesel engine generator and inverting battery power to an AC grid. It adds a third mode, operation in parallel with the diesel engine generator, thus completing the functional set needed to optimize the power flow in a hybrid system. The term AC grid is used throughout this paper to refer to local AC distribution systems associated with hybrid applications with diesel-generators.

In Figure 1, a block diagram of the TRIMODE Power Converter, power flow is shown with solid lines, control with dashed lines, and the open arrows show the direction of power transfer. The three-phase AC grid can be energized either from the diesel engine generator, the three-phase TRIMODE power processor, or both sources.

The upper solid state relay (SSR) is used to direct the diesel-generator power to the load and battery. It is closed only when inverter and diesel-generator synchronization are confirmed. During these operations, the lower SSR is CLOSED.

With the diesel operating and rated output is sensed by the TRIMODE controller over a few cycles, the TRIMODE will begin to supply supplementary power to the grid through the lower SSR. This controlled SSR acts as the safety device for the battery voltage, diesel voltage, loss of phase lock, and excessive current demand from the TRIMODE. The algorithm that determines how much current to add to the diesel-generator current will be the load current less the diesel generator rated current on a per phase basis.

Operating on a per phase basis has significant advantages in remote village applications since there is a reasonable probability of severe phase imbalance. With the TRIMODE design, the lightly

CH3365-4/94/0000-1180 $4.00 © 1994 IEEE

First WCPEC; Dec. 5-9, 1994; Hawaii

loaded phase will be charging the battery while the heavily loaded phase will be using battery power, thus the load is synthetically transferred from one phase to another. The desirable end result will be load balancing among the phases.

This Phase Load Management algorithm is illustrated in Figure 2. The diesel engine generator power rating is at the origin in the curve. When the generator power rating exceeds the grid demand power, the battery is charged to achieve full generator power. Operation in this mode is to the right of the origin. When the generator power rating is less than the grid demand power, the TRIMODE transfers from battery charging to inverting and adds power to the grid. Operation in this mode is to the left of the origin.

The Power Processor Controller (PPC) keeps record on the status of the array, the battery, and the diesel engine generator and delivers start/stop commands to the diesel; "warmup complete, generator power ready" signal to the SSR; "array power available" signal to the MPT; and, "battery charged" signal to the TRIMODE. Whenever the diesel is supplying power through the SSR, the TRIMODE automatically goes into the charger mode.

Interfacing between the PV array and the battery is a DC to DC Maximum Power Tracker (MPT) whose function is to operate the photovoltaic array at its maximum power point whenever the battery can accept the power, and to operate as a constant voltage DC power supply at the battery float voltage whenever the battery is fully charged.

Considering the rated power of typical village diesel generators and the broad range of operating voltage for both a photovoltaic array and a lead-acid or NiCad battery, the economics of the system cost can be shown to justify the cost of the MPT. The production selling price of a DC to DC MPT is $0.30 per watt, and its efficiency is 97%. The break even point for the justification of the MPT is an increase in output energy of 7% by operating at the maximum power point rather than at the instantaneous demand voltage of the battery. This analysis takes no credit for the on-off controller that is required to assure that the array can not overcharge the battery.

Figure 3 illustrates the performance of an Abacus Controls Inc. MPT on a late cloudy afternoon in Southern California. Figure 3A shows the actual insolation as measured by a pilot cell, and Figure 3B shows the power delivered to the load by the MPT.

TEST PROGRAM

A BIMODE inverter was tested at SNL as part of the TRIMODE development. The laboratory testing

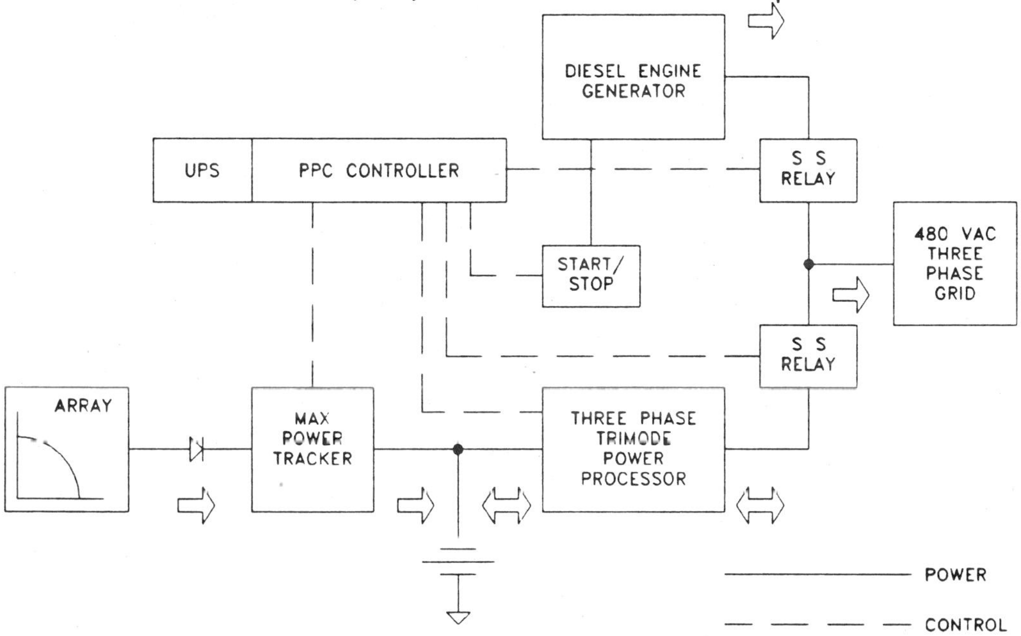

FIGURE 1
TRIMODE POWER CONVERTER

program is designed to improve test guidelines, assist manufacturers by conducting tests too costly for factory testing, and provide acceptance testing. The hybrid test configuration includes a 25-kW polycrystalline PV array, a 300-kWh battery, a 50-kW diesel engine-generator, and resistive, reactive and non-linear loads.

An inverter in a village power system must function with loads imposing a wide range of power factor and distortion typical of non-linear loads, while maintaining efficiency, voltage regulation and low distortion. Figure 4 is a plot of measured data for power factor vs efficiency with the output of each phase maintained at approximately 10 kVA.

A critical feature of the TRIMODE design is its ability to make a bumpless transfer without interruption of power to the load when transferring energy sources. This must be true whether transferring from the diesel to the inverter or from the inverter to the diesel. Figure 5 shows load being transferred from the diesel to the inverter for phases A and B as recorded at Sandia. During diesel operation, to the left in Figure 5, phase A was idle while phase B was battery charging. Both phases transferred without interruption at a negative zero cross.

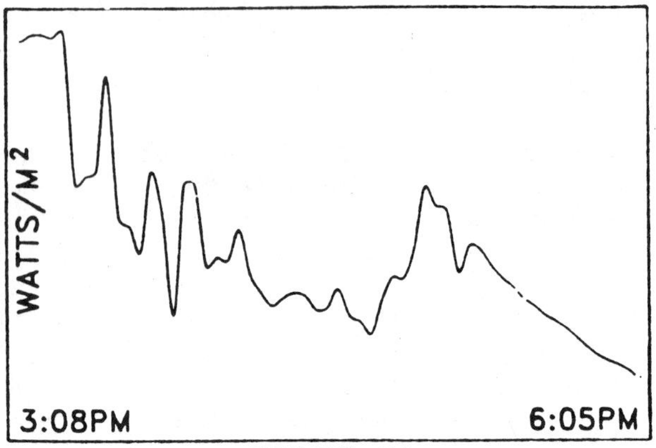

FIGURE 3A

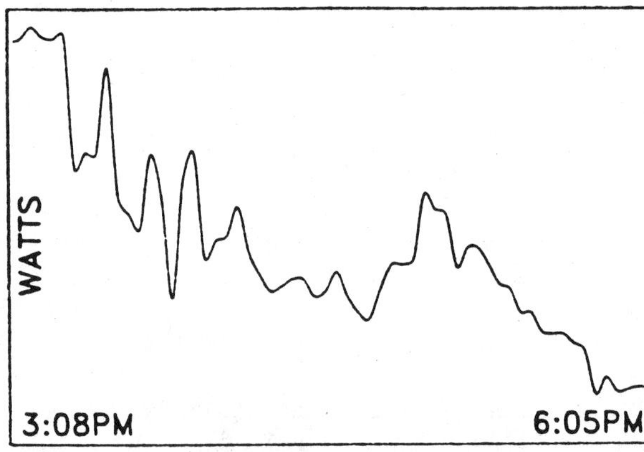

FIGURE 3B

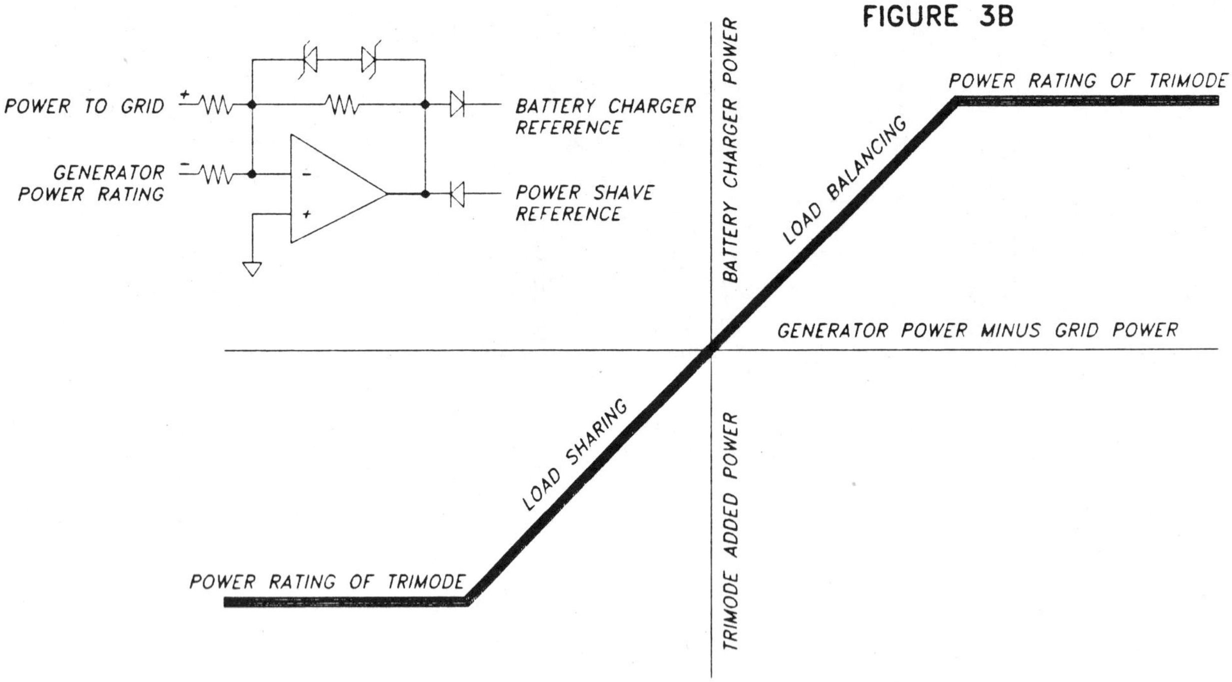

FIGURE 2 PHASE LOAD MANAGEMENT

CONCLUSION

Testing of the Abacus Controls Inc Power Converter with PV, diesel and battery energy sources will soon be completed. To date, test results show that the system is performing to specification.

A much larger system, rated at 300-kW AC, will be installed in central California in 1995. It is expected to prove both the economic and technical advantages of the TRIMODE Power Converter with hybrid energy sources.

REFERENCES

1. Bower, Ward, "Merging Photovoltaic Hardware Development with Hybrid Applications in the U. S. A." *Solar 93*, December 1-5, 1993.

2. Bower, Ward and George O'Sullivan, Bimode Uninterruptible Power Supply Compatibility in Renewable Hybrid Energy Systems," *IEEE AES Magazine*, August 1990.

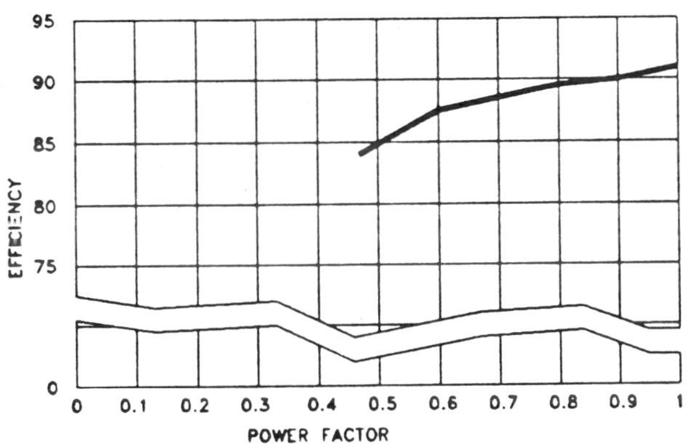

FIGURE 4 INVERTER EFFICIENCY VS POWER FACTOR

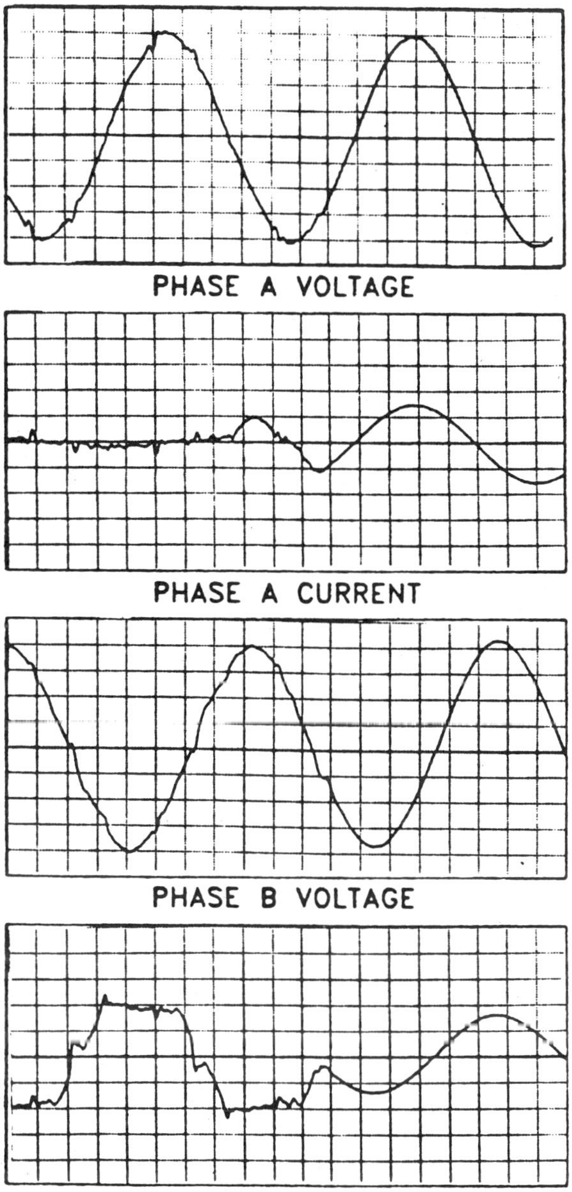

FIGURE 5

Calculating Method of the Capacity of Photovoltaic Array and Battery Considering Power Failure Probability in Stand-Alone PV Systems

A. Ueda*, K. Harada*, Y. Ishihara* , T. Todaka*, M. Yukawa** and H. Yokouchi**
*Doshisha University, **Hamamatsu Solar Energy Laboratory JQA INSTITUTE

ABSTRACT

In this paper the way to decide the most suitable capacity of Photovoltaic arrays and batteries in Stand-Alone Photovoltaic Systems is suggested. It is decided based on the parameter, PFP (Power Failure Probability). The designer can make the nomogram using the suggested method to decide the most suitable capacity of array and battery from the location of the system, the load patterns, and the climate data.

An example of deciding the capacity of arrays and batteries in Tokyo from the calculation and the derived nomogram is presented.

INTRODUCTION

The power generated by the photovoltaic arrays varies constantly. In addition the power required by the load is usually different from the power generated by arrays. Therefore in Stand-Alone Photovoltaic Generating Systems some batteries are needed in which the power left over is stored to satisfy the load when there is insufficient irradiance. But although the batteries account for the considerable part of the system cost, the way to decide the capacity of them is obscure.

The way to decide the most suitable capacity of arrays and batteries based on PFP is suggested in this paper.

PFP DEFINITION

PFP is probability of power failure and it represents the level of confidence that the system will satisfy the load. An PFP of 0 means that the load will always be satisfied, and an PFP of 1 means that the load will never be supplied.

PFP contains LLP(Lack of Load Probability) and PFTP (Power Failure Time Probability). LLP [%] and PFTP [%] are defined as following equations.

$$LLP = \frac{\Sigma L_{ns}}{\Sigma L} \times 100 \quad [\%] \tag{1}$$

$$PFTP = \frac{T_f}{8760} \times 100 \quad [\%] \tag{2}$$

where L [kWh] is the hourly required load, L_{ns} [kWh]$_s$ is the load not satisfied, T_f [hour] is the power failure time. In this paper we calculate LLP practically, and show an example of choosing the suitable capacity of arrays and batteries.

CONDITIONS OF SYSTEM

When the capacity of array and battery are constant, LLP and PFTP are changed by operating the required load and irradiance. In this case the general load patterns are used for the required load. These load patterns are the hourly data, which are made from real load power of residences and factories in seven commercial areas. The load patterns change every month. The load pattern of December is shown in Fig.1 for an example.

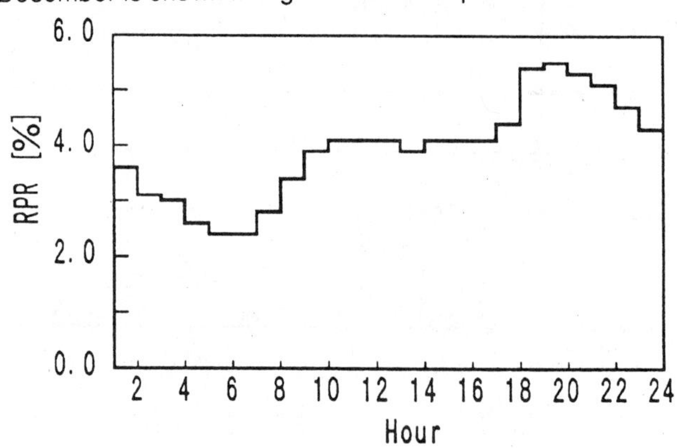

Fig.1 Load pattern of December

Where RPR[%](Required Power Rate) is defined as following equations.

$$RPR = \frac{L}{L_D} \times 100 \quad [\%] \tag{3}$$

RPR expresses the rate of the load to L_D, where L_D [kWh/Day] is the average of required load a day.

Irradiance and temperature data is HASP data (Heating, Air-conditioning and Sanitary engineering Program) , average hourly data of some years.

Voltage - Current characteristic equations of module are given as follows;

First WCPEC; Dec. 5-9, 1994; Hawaii

$$I = I_L - I_O\left[\exp\left\{\frac{q(V+IR_S)}{nkT}\right\}-1\right]-\frac{V+IR_S}{R_P} \quad [V] \tag{4}$$

$$T = T_{emp}+273+0.3E_{tp} \quad [°C] \tag{5}$$

$$I_O = C_O T^3 \exp\left\{-\frac{qE_g}{nkT}\right\} \quad [A] \tag{6}$$

$$I_L = 0.00546E_{tp}\{1+0.001(T-301)\} \quad [A] \tag{7}$$

where symbols in the equations are defined as follows.

- V : output voltage [V]
- I : output current [A]
- R_s : the series resistance [Ω]
- R_p : parallel resistance [Ω]
- I_L : photoelectromotive current [A]
- I_o : saturation current [A]
- q : electron charge [C]
- k : Boltzmann constant [J/K]
- n : junction constant
- T : cell temperature [K]
- C_o : load current temperature coefficient
- E_g : energy gap [eV]
- E_{tp} : module incident irradiance [mW/cm²].

Monocrystalline silicon solar cell (produced by SHOWA SHELL SEKIYU Corporation, GL136/M55) is used for simulation. The specifications are shown in Table 1, and diode parameters are shown in Table 2.

Table 1 Specifications of solar cell

Model	GL136/M55
Power	53.0 [W]
Current at load	3.05 [A]
Voltage at load	17.4 [V]
Short Circuit Curren	3.27 [A]
Open Circuit Voltage	21.8 [V]
Element	103mm cell
Interconnect	36pcs/series
Dimension	1291(H)×328.5(W)×35(D)mm
Weight	6.2kg

Measure conditions : irradiance 1[kW/m²], cell temperature 298[K]

The current and power generated by the photovoltaic arrays can be calculated by Eqs.(4) to (7) and the parameters of Table 2 when the voltage is already known. In the system shown in Fig.2 , the array output voltage equals the voltage of battery, therefore calculation of voltage of battery given by Q(residual capacity of battery)-V(voltage)characteristic curve[1] is needed. When upper limit of charge is chosen 80%, charge characteristics curve is similar to discharge one,

Table 2 Diode parameters of solar cell

R_S[Ω]	0.141
R_P[Ω]	2000
C_O	10
n	1.11
E_g[eV]	1.13

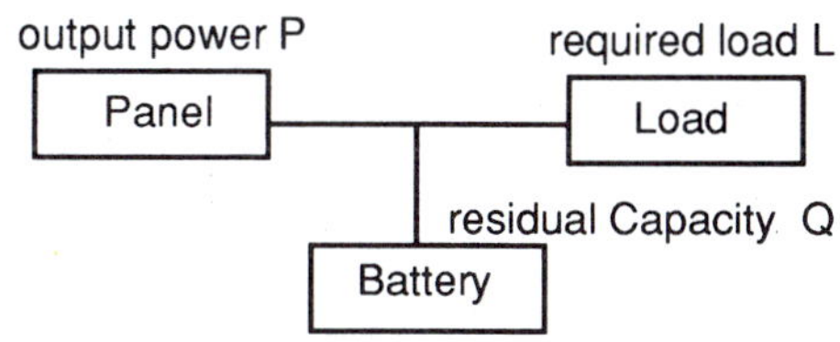

Fig.2 The system of the simulation

then the charge characteristic curve is used in this simulation.

In this simulation, PFP of the system shown in Fig.2 is calculated. And considering the battery life, battery energy is used from 20% to 80% of capacity of battery. In this system, when Q becomes lower than a certain level , L is changed into L' ruled by Table 3.

Table 3 Change rule of L

30% < Q/B ≦ 40%	L' = L×2/3
20% < Q/B ≦ 30%	L' = L×1/3
Q/B ≦ 20%	L' = 0

In the table B[kWh] is capacity of battery. The purpose of this change is the reduction of the state where Q/B is less than 20%.

Then at a time that Q/B is less than 20% the state is named whole power failure, and at a time that 20%<Q/B ≦40%, the state is named partial power failure.

The simulation is calculated at intervals of one hour. But if the result of Q/B calculation is less than 40%, the step is changed to 10 minutes. This reason is that Q increases or decreases rapidly in one hour when Q is low.

CALCULATING METHOD OF PFP

The flowchart of mainroutine and subroutine of PFP calculating program are shown in Figs.3 (a) and (b).

This program calculates PFP from hourly irradiance and temperature.

At first capacity of array and battery are input. Next L is read from required load file in each month. Irradiance and temperature data are read from the file, and P is calculated by Eqs.(4) to (7) and using the Newton-Raphson's method. In the subroutine L is changed according to Table 3, whole power failure and partial P and L. And PFP and PFTP are calculated from lack of load and power failure are judged, and ΔQ is calculated

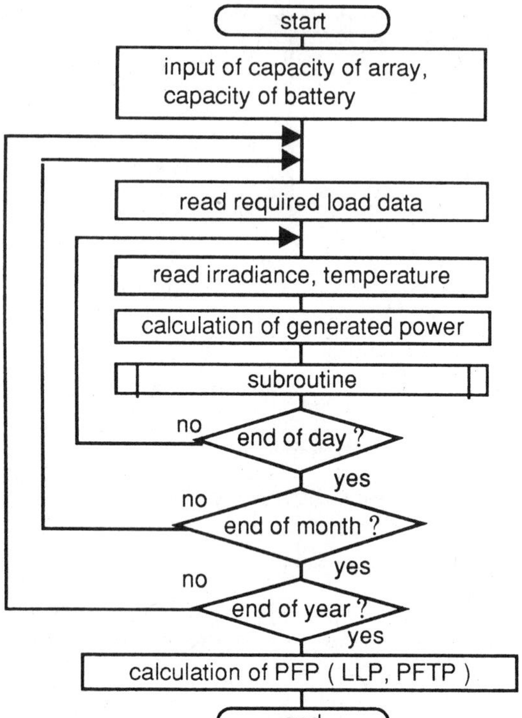

(a) Mainroutine

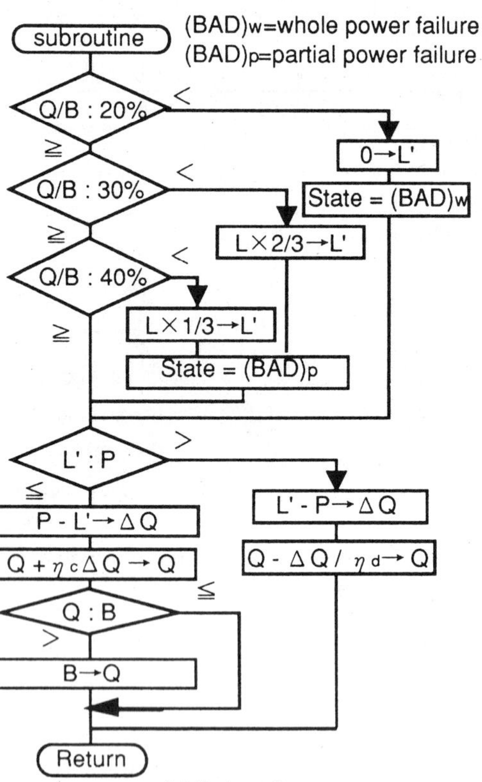

(b) Subroutine

Fig.3 PFP calculating program

comparing power failure time.

Where P [kWh] is the hourly power generated by the photovoltaic arrays, L' [kWh] is the hourly supplied power, ΔQ [kWh] is the residual capacity of battery variation, η_c is charging efficiency, and η_d is discharging efficiency.

CALCULATING RESULT

Because of photovoltaic array capacity decision, EPPR (Effective Photovoltaic Power of Array Rate) defined below is used.

$$EPPR = \frac{\Sigma P}{\Sigma L} \times 100 \quad [\%] \tag{8}$$

where ΣP is the cumulative P per year.

And B_D,that is normalized B[kWh] by L_D is defined as follows;

$$B_D = \frac{B}{L_D} \quad [\text{Day}] \tag{9}$$

In the system shown in Fig.2, when Q is 80% and P is over L, P_S[kWh] (surplus of power) is generated. Thus SPP (Surplus power probability) is defined as follows;

$$SPP = \frac{\Sigma P_s}{\Sigma L} \times 100 \quad [\%] \tag{10}$$

where ΣP_S is the cumulative P_S per year.

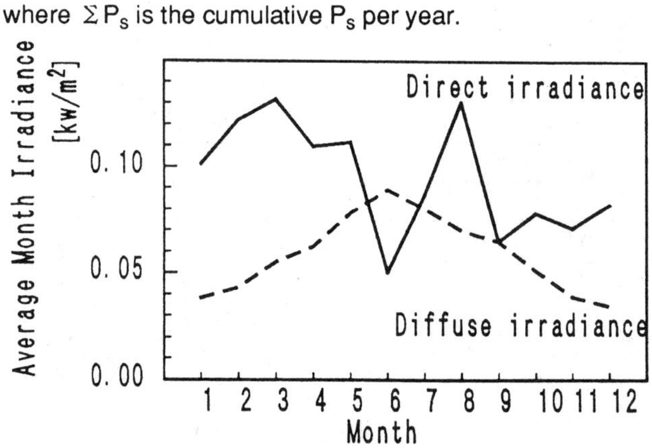

Fig.4 average month total irradiance data of Tokyo (24 hour average, tilt angle = 30°)

LLP and SPP calculated from irradiance data of Tokyo are investigated. Irradiance data of Tokyo is shown in Fig.4. Relation curve between EPPR and LLP, EPPR and SPP are shown in Fig.5, when B_D is 1[Day]. In this figure SPP increases in proportion to EPPR ,while LLP reduces exponentially. It suggests that so many module is needed to LLP equals to 0%.

Figure.6 shows relation between LLP and EPPR when B_D is parameter. In the case EPPR is small, even if B_D increases, LLP doesn't so much change.

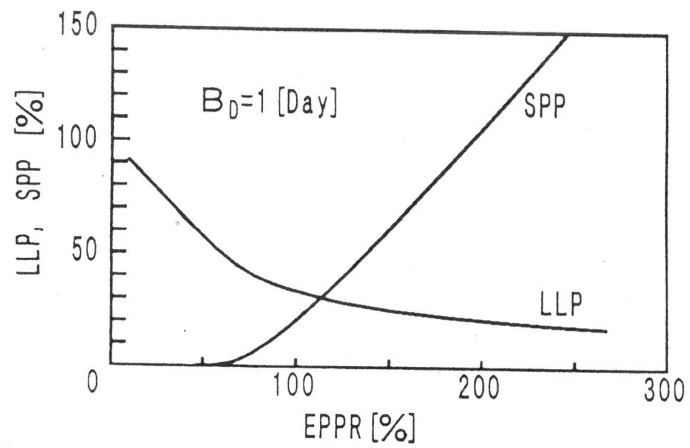

Fig.5 LLP, SPP's characteristics curve (Tokyo)

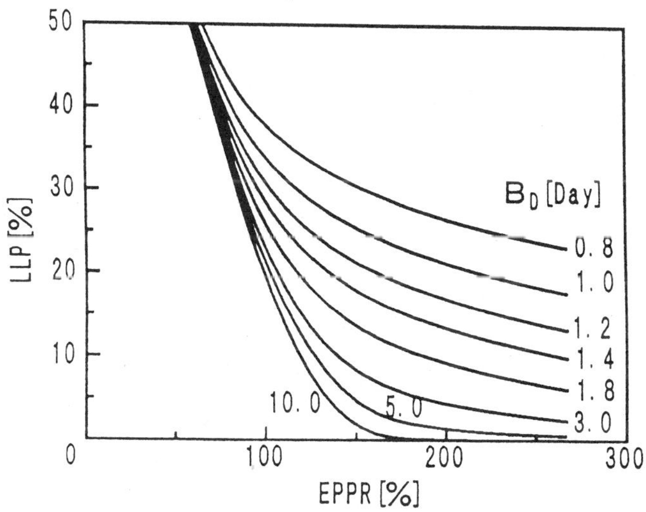

Fig.6 LLP's characteristic curve (Tokyo)
(B_D is parameter)

EPPR - B_D characteristic curve is shown in Fig.7, in the case LLP is parameter. In addition RPPR (Rating Photovoltaic Power of Array Rate) defined as follows is shown in the same figure.

$$RPPR = \frac{P_R}{L_D} \times 100 \quad [\%] \tag{11}$$

where P_R [kWh] is the hourly power generated by the photovoltaic arrays in standard test condition. (irradiance 1 [kW/m^2], cell temperature 298[K])

If the irradiance data and the cumulative required load per year are given, the capacity of arrays and batteries can be decided from Fig.7.

EXAMPLE OF CAPACITY OF ARRAY AND BATTERY DECISION

For example in Tokyo when the cumulative required load per year are 6570[kWh] (therefore the average of required load a day L_D = 6570 / 365 =18[kWh]) and LLP is 5%, and P_R is 10.8[kWh], from Eq.(11) RPPR=10.8/18× 100=60%.

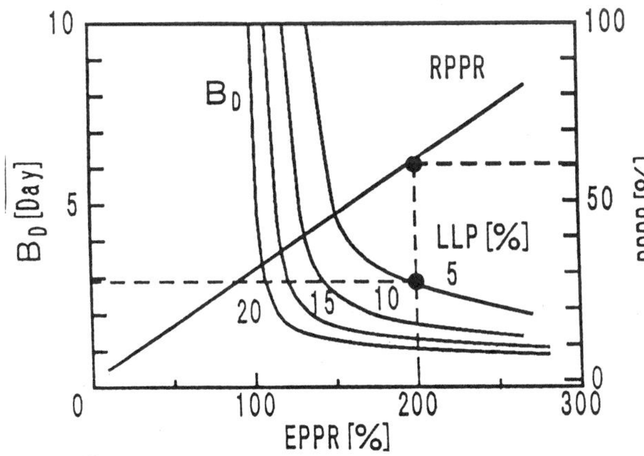

Fig.7 EPPR - B_D characteristic curve (Tokyo)
(LLP is parameter)

From Fig.7 EPPR equals 200%, and because the designer would like that LLP is 5%, B_D=2.95[Day]. From Eq.(9) suitable capacity of batteries B=53.1[kWh].

CONCLUSION

The method of calculation of PFP, and the way to decide the relation of the capacity of arrays and batteries have been presented in the case of Stand-Alone Photovoltaic Systems, when the climate condition, the characteristics of array and the load patterns are given. The designer can decide the most suitable capacity of array and battery with the nomogram (like Fig.7) as a function of PFP.

Furthermore, in the case of utility connected system that be able to reverse power flow, the cumulative power per year gotten from grid connect or the power sent grid connect can be calculated from SPP and LLP.

REFERENCES

[1] I. Tsuda, K.Kurokawa, K.Nozaki, " Influence of I-V Characteristics of PV Cell on Stand Alone PV Power Systems " , T.IEE Japan, Vol. 113-B, No.7, '93, pp. 745-751.

A NEW METHOD FOR OPTIMISING THE OPERATION OF STAND-ALONE PV HYBRID POWER SYSTEMS

M. Hancock H.R. Outhred R. J. Kaye

Centre for Photovoltaic Devices and Systems
University of NSW, Sydney, Australia.

ABSTRACT

This paper describes a new method for optimising the operation of stand-alone hybrid power systems containing some combination of auxiliary generator, PV generation and storage battery. The method provides a fast and accurate solution to the optimisation problem relative to previous methods. The method has been implemented on computer and an example control policy produced by the new method is presented and discussed. A comparison has been made between the new method and a conventional control algorithm under a simulation environment and using a deterministic future. This comparison has identified potential to reduce system operating costs.

INTRODUCTION

Stand-alone power systems, also known as Remote Area Power Supply (RAPS) systems, are increasingly being used as an alternative to the extension of reticulated electricity supply in many areas of the world. A state-of-the-art Hybrid RAPS system, the type considered in this work, is defined here to contain one or more diesel/petrol generators, photovoltaic (PV) or wind generation, battery storage, and associated power conditioning equipment. Typical applications of Hybrid RAPS systems include the supply of electricity to isolated communities or homesteads, and power for remote telecommunications sites.

In developing RAPS systems, the aim is to provide the best system for the lowest overall cost. This is termed optimising RAPS design. A conventional approach to this aim has been trial and error using RAPS systems in the field. An alternative method is to mathematically model RAPS systems so they can be simulated on computer. An extension of this latter approach is to use formal mathematical optimisation techniques to systematically optimise the design of RAPS systems. Systematic optimisation is considered here to have the greatest potential as a method for the design of RAPS systems.

This paper describes the development and analysis of a new method for optimising the operation of Hybrid RAPS systems. Optimisation of operation has been chosen as the focus of this work because it forms an important first step in the process of optimising RAPS system design.

The paper firstly considers previous approaches to optimising RAPS operation and gives reasons for the choice of optimisation algorithm used in this work. A description of the optimisation algorithm and its application to a RAPS system is then given. An example control policy produced by the optimisation algorithm is presented and discussed. Results of a comparison of the control policies of the optimisation algorithm with a conventional RAPS control algorithm are also presented. Areas for further development of the algorithm are identified and conclusions reached.

CHOICE OF OPTIMISATION METHOD

Mathematical optimisation algorithms which have previously been used to optimise RAPS operation include linear programming [1], quadratic programming [2], mixed integer non-linear programming [3], generalised network flow programming [4], discrete dynamic programming [5] and stochastic discrete dynamic programming [6]. Artificial intelligence techniques including knowledge-based systems, neural networks and fuzzy logic can also be considered for improving RAPS operation [7].

For this study we have used a modified form of a mathematical optimisation algorithm combining linear programming and dynamic programming (the LPDP algorithm) described in [8]. There are a number of reasons for this choice. The algorithm is more computationally efficient than non-linear optimisation approaches due to both its linear and dynamic programming nature. Its piece-wise linear nature allows it to more accurately model a RAPS system than would a standard linear optimisation approach. Unlike many other algorithms it can be adapted to incorporate generator on/off states. Unlike discretisation-based algorithms, the results do not suffer from inaccuracies caused by quantisation of state variables. Further, it is possible to extend the algorithm to incorporate modelling of the stochastic (uncertain) nature of the future RAPS operating environment, which is a difficult task for many algorithms.

OPTIMISATION ALGORITHM

Description of the LPDP Algorithm

The RAPS system is modelled using real power flows. The optimisation is performed over a time period (eg 1 day) divided into smaller time intervals (eg 24 x 1hr intervals). For each time interval, using component operating costs, a production cost function is calculated which describes the system operating cost over the time interval as a function of net power flow from the storage. This function must be piece-wise linear, continuous and convex.

First WCPEC; Dec. 5-9, 1994; Hawaii

The LPDP optimisation is then performed using a backward dynamic programming technique by applying a linear programming based optimisation at each time interval, beginning with the final time interval (eg the 24th hour) and ending with the first time interval (eg the 1st hour). This process produces a new function at each time interval, termed the cost-to-go function, which describes the minimum cost of operating the system from the time interval in question into the future as a function of storage level. Once the process is complete, given an initial storage level, the optimal net power flow for each time interval can be simply determined by traversing the cost-to-go functions forwards in time. These net power flows are then used to calculate the corresponding optimal RAPS control actions. The LPDP optimisation process is described in greater detail in [8].

Application to a RAPS System

The optimisation algorithm is applied to a Parallel Hybrid RAPS system. This is because most RAPS systems can be represented as a subset of such a system. The term Parallel describes a system where the inverter can synchronise to the generator so that both supply the load simultaneously. Further, it is assumed that a bi-directional inverter is used. This type of system is modelled as it represents a state-of-the-art Hybrid RAPS system. The following RAPS component models are used in the optimisation:

The battery is modelled as a storage entity with minimum and maximum available capacity levels (kWh), maximum charge and discharge rates (kW) and constant charge and discharge energy efficiency. No battery operating costs are incorporated.

The generator is modelled as a controllable variable power source with minimum and maximum output power and controllable on/off state. Fuel consumption costs are modelled as a linear function of generator output power. The function is described by a no-load running cost and an incremental fuel cost. A start-up cost for the generator is also included in the model.

The PV array is modelled as a variable power source controllable in the range zero to the maximum available PV power for the time interval concerned. No PV operating costs are incorporated. Wind generation can be modelled in a similar way to the PV array.

The load is modelled as a power sink, controllable in the range zero to the maximum load power demand for the time interval concerned. There is a cost associated with not supplying all the load demand, calculated by multiplying the amount of load demand not supplied by an 'incremental loss-of-load cost' parameter.

The bi-directional inverter is modelled by maximum output power limits for charging and inverting, and with constant charging and inverting power efficiencies. There are no operating costs associated with the inverter.

It was necessary to alter the LPDP algorithm in order to incorporate the generator on/off state and generator start-up cost. This resulted in non-convexity in the cost-to-go functions requiring further changes to the algorithm. The altered 'non-convex' LPDP algorithm has been labelled as the NCLPDP algorithm. A mathematical proof has been developed to show that the new algorithm is still optimal[1] [9].

While it is considered possible and desirable to extend the NCLPDP algorithm to incorporate stochastic (uncertain) variation in the future load demand and renewable energy input, this has not yet been incorporated. However, deterministic analysis, which assumes perfect foresight, still provides a useful method of approaching the operation optimisation problem. A deterministic analysis is used in this work.

RESULTS

Validation Against a Discrete Dynamic Programming Algorithm

The NCLPDP algorithm has been implemented on a personal computer using the C programming language. Results have been obtained from this algorithm for a number of RAPS scenarios. The RAPS systems used for the testing were 'Homestead' size (20kWh to 70kWh loads per day) Parallel Hybrid RAPS systems.
The output of the algorithm for these RAPS cases has been compared to the output of a discrete dynamic programming (DDP) algorithm [9], a less computationally efficient (but proven) conventional optimisation algorithm which is considered here to represent the best previous approach to the RAPS operation optimisation problem. The results of the DDP algorithm converge to those of the NCLPDP algorithm as the number of DDP discretisations are increased, showing that the NCLPDP algorithm is a more accurate method. Further, for levels of DDP discretisations which give acceptable accuracy, the computational efficiency of the NCLPDP algorithm is at least an order of magnitude faster.

Analysis of the Algorithm Control Policies

The optimal control policy for one of the RAPS scenarios simulated is displayed in Figure 1. This case is used to highlight a number of intuitively obvious control characteristics that would be expected of the RAPS operation optimisation algorithm. These characteristics are presented below along with a description of how the RAPS scenario displays them. The RAPS system used has been chosen to represent a typical 'Homestead' size Hybrid RAPS system. The load pattern has been chosen in order to allow some of the control characteristics to be more easily displayed. The RAPS parameters are provided in Table 1. The optimisation is performed over 24 one hour time intervals. The characteristics are:

(1) *The control algorithm will act to utilise as much PV energy as possible before using any generator energy.* All of the PV energy is utilised in this example.

(2) *The control algorithm will act to utilise as much generator energy as possible before suffering a loss-of-load.* Loss-of-load is avoided due to generator utilisation.

---

[1] This proof is part of research work by the principal author leading to a Masters of Engineering Degree by research. The assistance of Dr David Clements of the School of Electrical Engineering, UNSW, in the development of this proof is gratefully acknowledged.

Table 1: Parameters for sample RAPS scenario

Daily load demand	53.5kWh	PV array peak capacity	3kW
Nominal battery capacity	42.5kWh	Diesel generator capacity	4kW
Available battery capacity	17kWh	Diesel no-load fuel consumption	0.42 liters/h
Battery charge efficiency	70%	Diesel incremental fuel consumption (C)	0.22 liters/kWh
Battery discharge efficiency	100%	Diesel start-up cost	none
Inverter capacity	3kW	Fuel cost	F $/liter
Inverter efficiency	91%	Incremental loss-of-load cost	$(5*C*F)

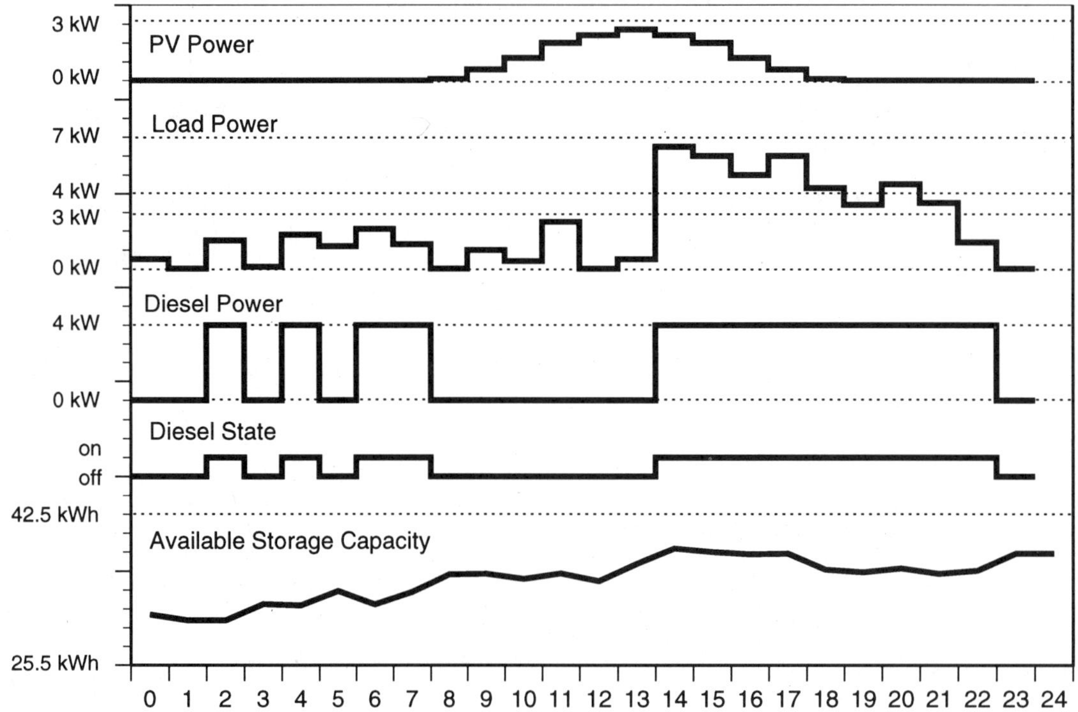

Figure 1: Control actions for the sample RAPS scenario.

(3) *In supplying generator energy, the algorithm will attempt to minimise generator run time (due to the generator no-load fuel consumption) by operating the generator at maximum achievable power output.* It can be observed that for this example the generator is always operated at maximum power output.

(4) *The algorithm will run the generator at times which maximise the amount of net load (load demand - PV power) supplied directly by the generator. This is so that the minimum amount of PV and generator energy is wasted by inverter and storage (efficiency induced) losses.* Diesel on states occur for time intervals 14 to 21 where the loads are greater than the inverter capacity (3kW), however this is because a loss-of-load would occur if the diesel did not turn on. The characteristic under concern is better displayed for the diesel on states at t=2, 4, 6, 7 and 22 which coincide with the largest net loads less than or equal to the inverter capacity. The diesel does not run at t=11 because the net load is lower than at t=2, 4, 6, 7 or 22 due to the PV power. Comparing this control policy against the same RAPS scenario with no PV energy, four extra diesel run hours are required if there is no PV energy. These extra diesel on states also occurred at times of largest net load which are in the first half of the day. This comparison shows that the algorithm will reduce diesel run time early in the day if PV is known to occur later in the day.

(5) *If generator start-up costs are included then the algorithm will attempt to minimise the number of generator off to on transitions.* When diesel start-up costs were increased, the discretionary diesel on states in the first half of the day grouped together to reduce the diesel off to on transitions.

Analysis of control strategies produced by the algorithm for a number of different RAPS scenarios also displayed these characteristics. It is reassuring that the formal optimisation algorithm confirms intuition.

Comparison with a Conventional RAPS Control Algorithm

It is desirable to compare the NCLPDP algorithm against a conventional RAPS control algorithm in order to determine the potential to reduce operating costs. This has been done by comparing the results of the NCLPDP algorithm with those of a RAPS simulation program, RAPSIM [10], which incorporates a conventional RAPS control algorithm

in its simulation process. In order to provide an equivalent RAPS platform on which to compare the two algorithms, the programs have been benchmarked to ensure that the RAPS component models used are the same.
A RAPS system similar to that shown in Table 1 was used for the comparison process. Results have been obtained for six RAPS operating scenarios utilising typical daily load and PV profiles. For the scenarios tested, the diesel generator fuel costs calculated for the NCLPDP control policy were 6% to 11% lower than those for the RAPSIM control policy. This indicates that for the RAPS scenarios used, although the conventional control algorithm operates the system in a reasonably efficient manner there is potential to reduce operating costs further. For the RAPS scenarios considered in the comparison (with average daily load demands of around 30kWh) the cost reductions translated to potential yearly fuel savings of 230 to 340 liters. For a larger system (eg. average daily load of around 150kWh) such as that used for a remote community, the savings would be considerably greater.
It should be noted that these results have been obtained for a small number of RAPS scenarios, using a deterministic future and in a simulation environment. Consequently there is a degree of uncertainty involved in relating these results to savings which might be achievable in the field. Incorporation of a stochastic model in the optimisation algorithm would most likely reduce the size of the predicted fuel savings obtained from the scenarios tested. The effects on fuel savings of using more accurate component models (eg battery model) or applying the algorithms 'in the field' is difficult to predict. Further development and implementation of the algorithm would be required in order to resolve these uncertainties. The identification of potential to reduce operating costs and the expected low costs associated with implementing an operation optimisation algorithm indicate that such further development is desirable.

FURTHER DEVELOPMENT

A number of aspects of the NCLPDP algorithm could be further developed in order to improve the accuracy and applicability of the algorithm:

- Uncertainty in future load demand and PV energy could be incorporated into the algorithm using a technique of stochastic analysis.
- The algorithm could be extended to incorporate more accurate models (particularly for batteries).
- The algorithm could be implemented in a RAPS controller to provide experience with practical issues.

CONCLUSION

A new algorithm for optimising the operation of stand-alone PV Hybrid power systems has been presented. It is fast, accurate and able to incorporate aspects of system operation that many previous methods have failed to address. The algorithm has been validated against a conventional optimisation algorithm and a control policy produced by the new algorithm for a typical stand-alone system has been presented.
The new algorithm has been compared with a conventional control algorithm for a number of stand-alone system scenarios. The comparison has been made in a simulation environment and for a deterministic future. The results of this comparison show there is potential to reduce system operating costs and indicate that further development of the new algorithm is desirable. With further development, the algorithm could be implemented directly in a system controller to optimise system operation.

ACKNOWLEDGEMENT

The support of the Australian Electricity Supply Industry Research Board (AESIRB) in providing financial assistance for this project is gratefully acknowledged.

REFERENCES

[1] S.K. Young, "Integrated Solar Energy System Optimization", *Journal of Solar Energy Engineering*, 1982, **104**(4), pp. 310-316.

[2] G.C. Contaxis, J. Kabouris and J. Chadjivassiliadis, "Optimum Operation of an Autonomous Energy System", *EWEC '86 - European Wind Energy Conference*, 1986, vol.2, pp. 305-310.

[3] S. Swaminathan, et al., "A MINLP Formulation for Power System Scheduling", *SOLAR '93 - Annual Conference of the Australian and New Zealand Solar Energy Society*, 1993, vol.2, pp. 446-451.

[4] M.S. Kandil, S.A. Farghal and A.E. El-Alfy, "Optimum Operation of an Autonomous Energy System Suitable for New Communities in Developing Countries", *Electric Power Systems Research*, 1991, **23**(2), pp. 137-146.

[5] G.C. Contaxis and J. Kabouris, "Short Term Scheduling in a Wind/Diesel Autonomous Energy System", *IEEE Transactions on Power Systems*, 1991, **6**(3), pp. 1161-1167.

[6] A.G. Bakirtzis and E.S. Gavanidou, "Optimum Operation of a Small Autonomous System with Unconventional Energy Sources", *Electric Power Systems Research*, 1992, **23**(2), pp. 93-102.

[7] I.F. MacGill, R.J. Kaye and H.R. Outhred, "Artificial Intelligence Approaches to the Design and Operation of Remote Area Power Supply Systems", *SOLAR '93 - Annual Conference of the Australian and New Zealand Solar Energy Society*, 1993, vol.2, pp. 440-445.

[8] C.H. Bannister and R.J. Kaye, "A Rapid Method for Optimisation of Linear Systems with Storage", *Operations Research*, 1991, **39**(2), pp. 220-232.

[9] M. Hancock, *Optimising the Operation of Remote Area Power Supplies*, Masters Thesis in Progress, School of Electrical Engineering, University of New South Wales, Australia.

[10] D.P. Remmer and M.S. Dymond, "RAPSIM - A Computer Simulation for Remote Area Power Supply Systems.", *SOLAR '91 - Annual Conference of the Australian and New Zealand Solar Energy Society*, 1991, pp. 525-532.

A NEW APPROACH TO OPTIMAL SIZING OF COMPONENTS IN STAND-ALONE PHOTOVOLTAIC POWER SYSTEMS

R J Kaye
School of Electrical Engineering
University of New South Wales, Sydney, Australia

ABSTRACT

The problem of choosing component sizes in a stand-alone photovoltaic power system is expressed as a hierarchical optimisation problem in which system variable costs are calculated by optimising the operation of the system. It is then shown that dual variables which are a by-product of variable cost minimisation can be used to calculate the derivative of variable cost with respect to component sizes. These data can then be used to reduce total system cost in an iterative optimisation algorithm.

INTRODUCTION

This paper presents a new method for determining the size of the key components in a PV power system. The objective is to minimise the system life-cycle costs (purchase, maintenance and operations costs) including inconvenience costs of unsupplied end uses.

Stand-alone PV power systems are becoming increasingly competitive with grid connection for remote consumers. Improvements in stand alone component technology have lead to greater supply reliability and reduced life-cycle costs. Further reductions in life-cycle costs and improvements in user satisfaction can be achieved by optimally matching the sizes of each component to the size and value characteristics of the end uses.

This paper presents a novel approach to optimally sizing system components, based on calculating the derivatives of variable cost with respect to component sizes.

OPTIMISATION OF COMPONENT SIZES

The objective usually pursued in choosing the size stand alone PV power system components such as the battery, the PV panels, the power conditioning equipment and, in some cases, the auxiliary (petrol or diesel) generator is to minimise or to contain life cycle costs. Life cycle costs can be divided into two terms:

- fixed costs which are functions of the sizing decisions and include the cost of purchasing and installing fixed equipment. Fixed costs are not a function of the way in which the system is operated; and
- variable costs which are the running and maintenance costs and the cost of not satisfying demand for electrical energy. This latter can also be expressed as a benefit of satisfying end uses. Variable costs depend on the size of system components and on the way in which the system is operated.

The sizing process will thus require some representation of the functional dependence of variable costs on component sizes. In practice, this can vary from simple rules of thumb derived from experience to complex computer models. While rules of thumb tend to be less sophisticated, computer models require accurate data which may in some cases not be readily available.

Sizing can usually be viewed as an iterative process (see Fig. 1) in which the operations model calculates variable cost information for a sequence of different vectors of component sizes, which are sequentially chosen by the component sizing optimiser in order to reduce variable cost. Algorithms that can be used by the component sizing optimiser include trial and error and linear and non-linear programming.

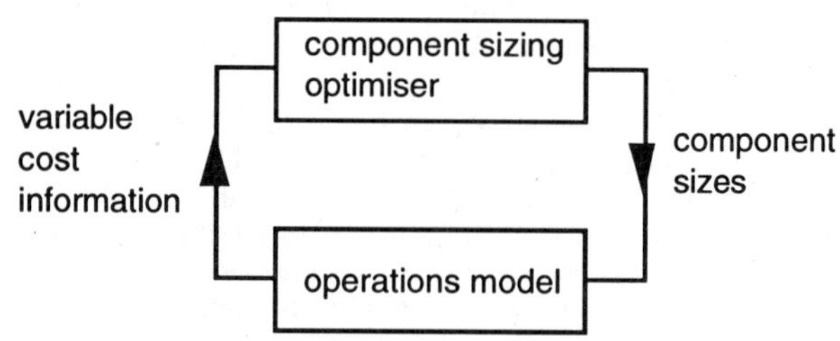

Fig. 1: Conceptual Model of Iterative Component Sizing

Some operations models can produce, in addition to the value of variable cost, other information, such as the derivatives of variable cost with respect to the sizes of each component. These derivatives give a first order indication of the way variable cost changes with small variations in each component. This information can be used in most component sizing optimiser algorithms.

For example, most non-linear programming algorithms require estimates of the derivative of the objective function. In the absence of better information, these are often obtained using finite difference techniques, where the operations model is evaluated for two vectors of component sizes which are identical except for a small difference in one entry. This requires a large number of evaluations of the operations model and can encounter problems of numerical stability.

In the next section, more sophisticated approaches to obtaining derivative information are examined.

OPTIMISATION OF SYSTEM OPERATION

The most detailed representation of variable costs involves both optimisation to determine the operational decisions and simulation to assess how those decisions are implemented. In some situations, the optimisation step is replaced by heuristics. Simpler approaches to variable

CH3365-4/94/0000-1192 $4.00 © 1994 IEEE First WCPEC; Dec. 5-9, 1994; Hawaii

cost modelling exit but will not be discussed here.

Operational decisions at each point in time include the rate of charging or discharging the battery, output from the auxiliary generator and the rate of satisfaction for each end use. The objective in determining these is usually minimisation of variable costs. This is a dynamic optimisation problem since decisions at different times are couple together by energy storage in the battery. Further, in many situations, it is desirable to model the uncertainty in the evolution of certain parameters which affect the operation of the system. For example, available energy from the PV arrays and demand for electrical energy change over time in an unpredictable fashion and are often modelled as stochastic processes. Minimising the variable costs is then a stochastic dynamic optimisation problem.

These types of problems have been solved using a number of different approaches including linear and non-linear programming and dynamic programming. The output of the operations optimisation process is a schedule which expresses operational decisions at each time as a function of the system endogenous state (for example, the amount of energy stored in the battery) and the exogenous state which describes the uncertainty processes. For example, if the random variations in solar radiation are modelled as a Markov process, then the exogenous state might be the present value of the PV array output.

Simulation is then used to calculate the variable cost that results from a particular schedule. In the case where uncertainty in the operations environment is being modelled, Monte-Carlo simulation will be required. Here many runs with different randomly selected environments are performed and the results of all the runs are averaged to obtain an estimator of the expected value of variable cost. In the absence of an explicit operations optimisation step, the operations schedule might be determined heuristically.

In the case where the operation schedule was obtained from heuristic considerations, derivatives of variable cost with respect to component size can sometimes be obtained without recourse to finite difference techniques by using a recently developed infinitesimal perturbation analysis technique [1].

On the other hand when operations optimisation is explicitly performed, it is possible to calculate and use Kuhn-Tucker multipliers to obtain the required derivative [2]. These multipliers, which are often referred to as the dual variables or as the shadow prices, are automatically found in some optimisation algorithms such as linear programming and are easily calculated from the results of others. There is a multiplier associated with each constraint in an optimisation problem, and, under certain smoothness and convexity assumptions, the value of the multiplier is equal to the derivative of the optimised objective function with respect to a relief of that constraint (see [2] for details). Thus in the optimisation of system operation, if the Kuhn-Tucker multipliers can be calculated they can be easily converted into derivatives of the variable cost with respect to component sizes.

A SIMPLE EXAMPLE

In this section, a simple, deterministic model of a PV power system without auxiliary generation is used to demonstrate the use of Kuhn-Tucker to find the derivatives of variable cost with respect to the size of system components. The methodology can be easily generalised to a wide range of more complex models, such as incorporating the effects of uncertainty in solar radiation.

System Model

The system model (see Fig. 2) consists of an energy storage unit, a photovoltaic array and a number of end uses. For this system, sizing is the determination of the peak capacity of the array and the battery storage capacity and its maximum rates of charging and discharge. A simple model of an auxiliary energy source such as a petrol or diesel generator could easily be included. The invertor and other power conditioning and control equipment have not been modelled.

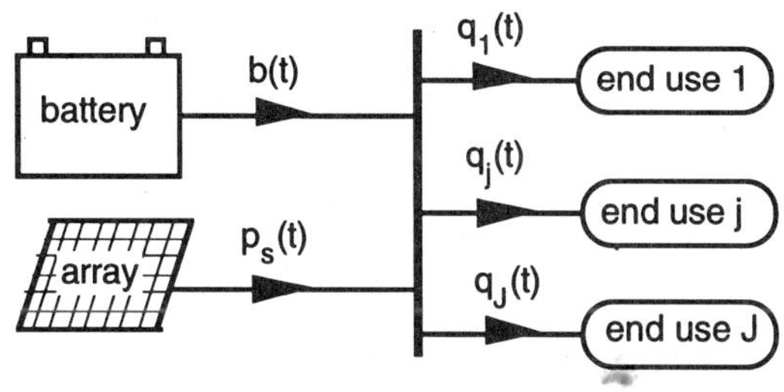

Fig. 2: System model showing power flows.

In the following, time is discretised into intervals of length Δt (hr). Each interval is labelled $t = 1,\ldots,T$ so that the optimisation horizon is $T\ \Delta t$ (hr).

End Uses Model

Each end use is characterised by an incremental value that accrues as the end use is satisfied and by a maximum consumption. Both incremental value and maximum consumption can vary with time.

Suppose that there are J end uses. For each end use $j = 1,\ldots,J$, define

$q_j(t)$ = rate of satisfaction of end use j during time interval t (kW)

$Q_j(t)$ = maximum rate of satisfaction of end use j during interval t (kW)

$v_j(t)$ = incremental value of end use j at during time interval t (¢/kWh)

Here it has been implicitly assumed that value of satisfying an end use can be expressed in monetary terms. In some situations this might involve some approximation.

It follows that for each time t,

$$0 \le q_j(t) \le Q_j(t) \tag{1}$$

and that the total end use satisfaction over the optimisation horizon is

$$V = \sum_{t=0}^{T} \sum_{j=0}^{J} v_j(t)\, q_j(t)\, \Delta t \tag{2}$$

Because each of the other components in the system is assumed to be operated at zero variable cost, the system variable cost will be $-V$. If there were an auxiliary generator, the fuel and variable maintenance costs would be added to $-V$.

End use incremental value data used in this paper are shown in Fig. 2.

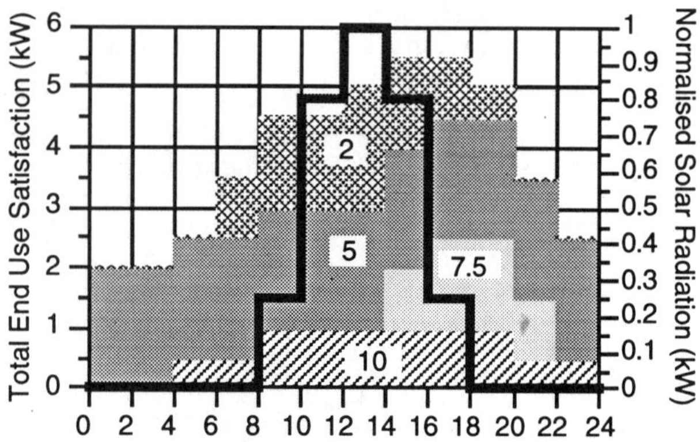

Fig. 2: End use maximum satisfaction and normalised solar radiation as a function of time of day. Each shaded area corresponds to a particular end use, the incremental value of which is shown in ¢/kWh.

Solar Model

The contribution of the photovoltaic array is modelled as a simple injection of power which varies with time. Define

P_s = peak capacity of array (kW)

and for each period t = 1,...,T

$p_s(t)$ = rate of injection of solar energy during time interval t (kW)

It is assumed that there is a function $\sigma(t)$ of time t, which is the normalised radiation, so that, for each t,

$$p_s(t) = P_s\, \sigma(t) \tag{3}$$

The function $\sigma(t)$ (see Fig. 2) is assumed to be a function only of time of day.

The fixed (purchase and installation) cost of the array is assumed to be directly proportional to the capacity of the array. In order to conveniently compare this cost with the variable cost, it is expressed on a per hour basis. That is the total purchase cost is levelised over the expected life of the array. Thus define

χ_s = levelised unit fixed cost of solar array (¢/kW/hr)

More complex models of fixed cost dependency on array size are easily incorporated into the analysis below.

Battery Model

The battery is modelled as a simple energy storage element with charge and discharge limits and a limit on the amount of energy that can be stored. Thus define:

$x(t)$ = quantity of energy stored in battery at the beginning of time interval t (kWh)

x^{max} = maximum energy that can be stored in the battery (kWh)

$b(t)$ = net rate at which energy is discharged from the battery during time interval t (kW)

b^{max} = maximum rate of discharging the battery (kW)

b^{min} = minimum rate of net discharging (kW)

Negative values of b(t) correspond to battery charging and thus b^{min} would normally be negative.

Battery operation is thus modelled as:

$$x(t+1) = x(t) - b(t)\,\Delta t \tag{4}$$

$$b^{min} \le b(t) \le b^{max} \tag{5}$$

$$0 \le x(t+1) \le x^{max} \tag{6}$$

It is assumed that battery life and maintenance requirements are independent of how the battery is operated. This is a gross assumption which can be removed provided an adequate model of the dependence of battery life on utilisation is available.

The sizing parameters for the battery are $(b^{max}, b^{min}, x^{max})$. For simplicity of explanation, only batteries with fixed ratios of x^{max}/b^{max} and b^{min}/b^{max} will be considered here. Thus we assume that there exists fixed ρ_x and ρ_b such that

$$x^{max} = \rho_x\, b^{max} \tag{7}$$

$$b^{min} = \rho_b\, b^{max} \tag{8}$$

In this way, the fixed cost (purchase and installation) can be expressed as a function of the maximum discharge rate, b^{max}. Thus define

χ_b = levelised incremental fixed cost of solar array (¢/kW/hr)

which has been levelised in the fashion described for χ_s above. Note that χ_b is a function of the ratios ρ_x and ρ_b.

Optimisation of Systems Operation

Operations of the components are linked by the energy conservation equation, which can be written as (see Fig. 1):

$$b(t) + p_s(t) = \sum_{j=0}^{J} q_j(t) \tag{9}$$

The minimal variable cost of operating the system, $W(P_s, b^{max}, b^{min}, x^{max})$ is then found by minimising $-V$ (see eqn. (2)) subject to the operating constraints in equations (1), (4), (5) and (6).

In the results presented in this paper, this optimisation problem was solved using the technique described in [3] which combines linear programming and dynamic programming. Examples of the type of operations schedules produced by this optimisation can be found in [4].

Derivatives

One advantage of this operations optimisation technique is that it gives ready access to the Kuhn-Tucker multipliers [2]. For each time interval t, define

$\theta(t)$ = Kuhn-Tucker multiplier for eqn. (9)

$\mu_+^b(t)$ = Kuhn-Tucker multiplier associated with the right hand inequality in eqn. (5)

$\mu_-^b(t)$ = Kuhn-Tucker multiplier associated with the left hand inequality in eqn. (5)

$\mu^x(t)$ = Kuhn-Tucker multiplier associated with the right hand inequality in eqn. (4)

It then follows that

$$\frac{\partial W(P_s, b^{max}, b^{min}, x^{max})}{\partial P_s} = -\sum_{t=1}^{T} \theta(t)\,\sigma(t) \qquad (10)$$

$$\frac{\partial W(P_s, b^{max}, b^{min}, x^{max})}{\partial b^{max}} = -\sum_{t=1}^{T} \mu_+^b(t) \qquad (11)$$

$$\frac{\partial W(P_s, b^{max}, b^{min}, x^{max})}{\partial b^{min}} = \sum_{t=1}^{T} \mu_-^b(t) \qquad (12)$$

$$\frac{\partial W(P_s, b^{max}, b^{min}, x^{max})}{\partial x^{max}} = -\sum_{t=1}^{T} \mu^x(t) \qquad (13)$$

However, since b^{min} and x^{max} are assumed to be related to b^{max} by equations (7) and (8), the total derivative of W with respect to b^{max} is given by

$$\frac{\partial W(P_s, b^{max})}{\partial b^{max}} = -\sum_{t=1}^{T} \left(\mu_+^b(t) - \mu_-^b(t)\,\rho_b + \mu^x(t)\,\rho_x \right) \qquad (14)$$

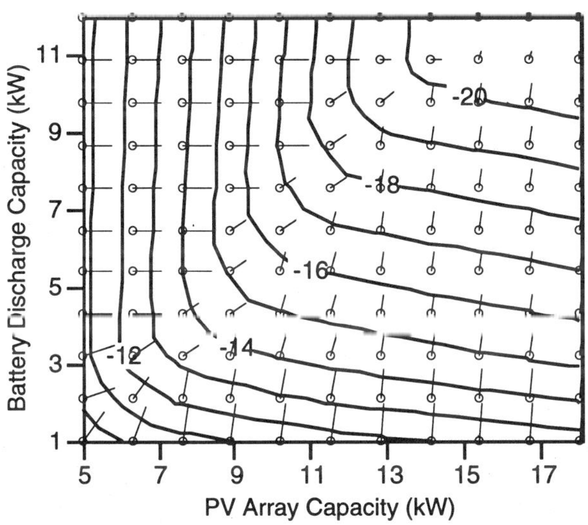

Fig. 3. Contours of constant variable cost (labelled in ¢/hr) for a stand-alone PV power system. The battery storage capacity and maximum charging rate are constrained so that ρ_x = 4 hrs, ρ_b =1 The thin lines indicate the magnitude and direction of the derivative of $-V$.

Fig. 3 illustrates the resulting derivatives of variable cost. These derivatives have been extensively verified using finite differences and were found to be very accurate. In Fig. 3, it can be seen that the derivatives are normal to the contour lines and that their magnitudes decrease with increasing battery and array sizes.

Optimal Sizing

The derivative of the total cost can be found by adding the derivatives of the variable cost (as found in Fig. 3) and the derivative of the fixed cost (i.e. χ_b and χ_s). The results for a grid of array and battery sizes are illustrated in Fig. 4. Note that the gradient vectors are normal to the contour lines.

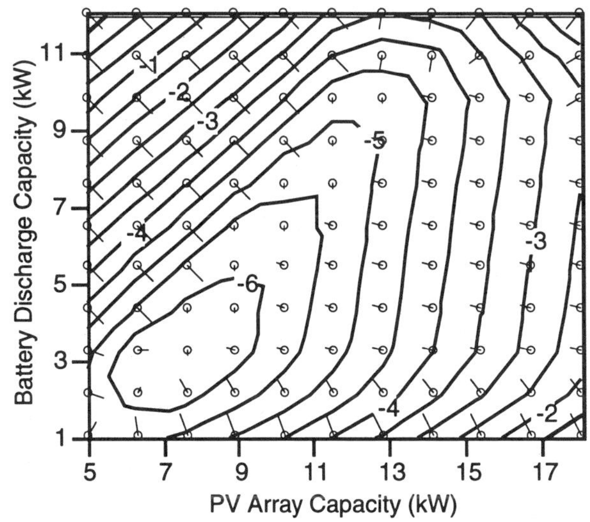

Fig. 4. Contours of constant total (variable + fixed) cost (in ¢/hr). Levelised PV array fixed costs are χ_s = 0.6 ¢/kW/hr and levelised battery fixed costs are χ_b = 0.7 ¢/kW/hr. The battery storage capacity and maximum charging rate are constrained so that ρ_x = 4 hrs, ρ_b =1. The thin lines indicate the magnitude and negative direction of the derivative of total fixed cost.

Optimal sizing of the components would not normally require the construction of Fig. 4. Instead the process would proceed from design point to design point by moving along, for example, the gradient direction. It can be seen that such a process would converge to a point of low total cost.

CONCLUSIONS

The approach described in this paper is easily extended to include stochastic fluctuations in demand for electrical energy and solar radiation by modifying the operations optimisation model and the calculation of derivatives as described in [1].

The work presented in this paper demonstrates that the process of sizing the components of a stand alone PV power system can be improved by extracting additional information from the cost model.

REFERENCES

[1] Y.-C. Ho and X.-R. Cao, *Perturbation analysis of discrete event dynamic systems*, Kluwer, Boston, 1991.

[2] D. G. Luenberger, *Linear and nonlinear programming*, 2nd edn., Addison-Wesley, Reading, Mass, 1984.

[3] C. H. Bannister and R. J. Kaye, "A rapid method for optimisation of linear systems with storage"' *Operations Research*, 1991, **39**(2), pp. 220-232.

[4] R.J. Kaye, "Optimising the value of photovoltaic energy in electric supply systems with storage," *Proc. 12th Photovoltaic Solar Energy Conf.*, 1994.

SIZING PROCEDURE FOR PHOTOVOLTAIC SYSTEMS

Eiichi Endo and Kosuke Kurokawa
Electrotechnical Laboratory
1-1-4 Umezono, Tsukuba-shi, Ibaraki, 305 Japan

ABSTRACT

The purpose of this study is to establish a comprehensive sizing procedure for photovoltaic (PV) systems. To establish it, the knowledge about PV system design is acquired by analyzing existing PV systems which have been designed optimally and interviewing with expert PV engineers. The acquired knowledge has been systematized as a sizing procedure for PV systems. It has following features.
- It covers wide varieties of PV system configurations, including stand-alone systems, hybrid systems and utility-connected systems.
- It is based on an analytical method and could size not only PV arrays but also batteries, inverters and diesel generators.
- It makes PV system design optimal at the practical level on the basis of experiential knowledge.

The proposed sizing procedure has been confirmed that it enables appropriate design of PV systems. It should contribute to wide spread of PV systems in the future.

INTRODUCTION

To use solar energy efficiently and economically, photovoltaic (PV) systems are necessary to be designed in appropriate system configurations and capacities under the conditions of loads, irradiation and usable back-up systems. PV systems have, however, various configurations and wide range of capacities depending on their design conditions. Therefore, the method for their optimal design is very important.

The design methods of PV systems could be classified into several categories and some of their sizing procedures are reported with examples[1,4]. They are, however, applicable to the restricted system configurations and system components.

From this reason, a comprehensive design method for PV systems, which is applicable various system configurations and all system components, is proposed. It consists of three parts as shown in Figure 1. In this paper, the proposed sizing procedure for PV systems is presented in detail.

DESIGN OF SYSTEM CONFIGURATION

Configurations of the PV systems are given based on necessity of PV system components, which are decided on the based of system applications, characteristics of loads and back-up systems. Choice of stand-alone system

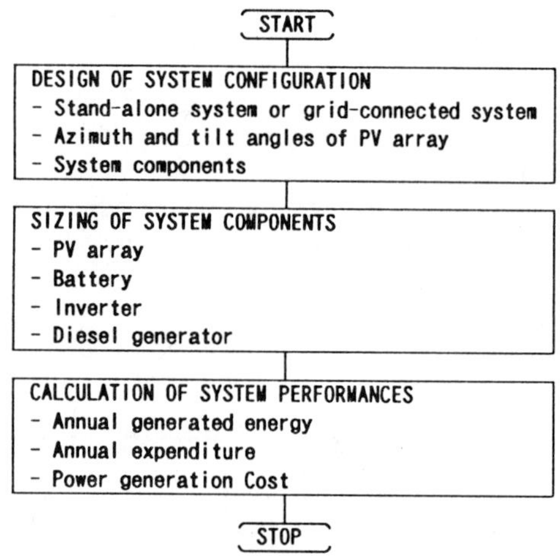

Fig. 1. Outline of the proposed design method for PV systems.

or utility-connected system, connection side of hybrid generators, load shedding, azimuth and tilt angles of PV arrays, and type of inverters should be decided before sizing of the PV system components. These items are decided based on qualitative design conditions.

Stand-alone System or Utility-connected System

If the PV system could use electricity from an existing utility, a utility-connected system is preferable. Otherwise a stand-alone system should be chosen. In case of a utility-connected system, a bidirectional or unidirectional power flow is chosen based on the power flow to the utility is possible or not. In case of a stand-alone system, availability of local energies, diesel generators and load shedding should be decided firstly.

Local energies, such as wind power and small hydro power, should be used, if they are enough and can supplement solar energy. Diesel generators could be used, if fuel is able to supply and no local energy is available nor sufficient and no load could be shed. Load shedding is available, if loads do not require high supply reliability and no diesel generator is used and no local energy is available nor sufficient.

Azimuth and Tilt Angles of PV Array

Azimuth angles of PV arrays are usually taken south in the northern hemisphere. If the PV array will be installed on a inclined roof, its azimuth angle should be set

First WCPEC; Dec. 5–9, 1994; Hawaii

at the azimuth angle of the roof. In case of horizontal roofs or lands must be used efficiently, the azimuth angle must be chosen as the direction of the roofs or the lands. If there are shadows of mountains or buildings, the azimuth angle should be set to avoid them. If the peak-hour of power generation is able to shift at that of loads by adjusting the azimuth angle, the azimuth angle should be set at the peak-hour angle.

Tilt angles of PV arrays without tracking are usually set at the annual optimal tilt angle, which maximize annual irradiation. But if inclined roof is used, the tilt angle should be settled at the tilt angle of the roof. If it is much snowfall in winter, the tilt angle in a stand-alone system must be selected so that snow could be slipped down. If it is not much snow, the tilt angle in a stand-alone system without diesel generators is decided so that its array capacity could be minimized.

System Components

In stand-alone systems, batteries must be used, if the PV system is a stand-alone system and power must be supplied during no sunshine hours, such as cloudy days, rainy days, or nights. Batteries are also necessary, if large or many motors are included in the loads, because motors generate rush currents. In utility-connected systems, batteries are used when high supply reliability is required. Batteries are also used, if the peak of power generation is needed to shift at the peak of loads and the adjustment by the azimuth angle is impossible or insufficient. In case of utility-connected systems with unidirectional power flow, batteries are used when generated energy exceeds energy of loads.

Outputs from PV systems are DC, therefore inverters are necessary if loads are AC or the PV system is a utility-connected system with bidirectional power flow. In case of DC loads, if the PV system is a utility-connected system or a stand-alone system with AC-side-connected hybrid generators, a bidirectional converter is necessary to convert AC to DC.

SIZING OF SYSTEM COMPONENTS

Sizing Method for PV Systems

The sizing methods for PV systems are classified into two categories, that is, simulative methods, for example[3], and analytical methods, for example[1,2,4,6]. The simulation could be applicable when no sizing procedure is established. But it cannot avoid many trials and errors to make the sizing properly. In the preliminary design, it is important to offer an appropriate system easily. Therefore, the analytical methods are chosen in this paper.

The analytical sizing methods could be categorized into three types. They are based on as follows:

- loads and irradiation[4,6],
- available areas[1], and
- LOLP(loss of load probability)[2].

The sizing method based on LOLP is a method that presents combinations of a PV array capacity and a battery capacity which makes LOLP at the desired level analytically. But it is applicable to the PV systems which are consist of only PV arrays and batteries and has few sizing examples.

On the contrary, the sizing method based on available areas is usually applied to utility-connected systems with bidirectional power flow and it gives a PV array capacity under the available areas, such as roofs. The sizing method based on loads and irradiation is usually used in stand-alone systems including hybrid systems and it can size appropriate PV systems evaluating loads, irradiation and all of the system losses.

The LOLP method is applicable to the restricted system configurations and system components. But a combined method of the other two methods could size appropriate PV systems in the various system configurations including both stand-alone systems and utility-connected systems. And it could size not only PV arrays, but also batteries, inverters and diesel generators.

In the PV system sizing, it is important that the sizing method is applicable to various system configurations and all system components. Therefore, the combined sizing method is proposed in this paper.

Sizing Procedure for System Components

The sizing procedure of the proposed method has many sizing examples in the Sunshine Project of Japan and the experiential knowledge, which makes sizing appropriately, is highly accumulated. Therefore, based on a method of knowledge engineering, the knowledge about PV system sizing is acquired by analyzing existing PV systems which have been sized optimally and interviewing with expert PV engineers to complement and update it. The acquired knowledge is systematized as a sizing procedure for PV systems. Figure 2 shows the precise flow and equations of the proposed sizing procedure concerning PV arrays, batteries, inverters and diesel generators.

The proposed sizing procedure for PV arrays could be divided into following cases.

- Utility-connected systems with bidirectional power flow which are sized based on available areas.
- Utility-connected systems with bidirectional power flow which are sized based on loads and irradiation and stand-alone systems with diesel generators.
- Stand-alone systems without diesel generators.

In the first case, PV arrays are sized based on the available areas. In the second case, they are sized so that the annual generated energy equals to the annual load energy. In the third case, they are sized in every month based on loads and irradiation and the maximum PV array capacity in the year is chosen.

Batteries are sized by one of the two procedures. They are selected whether the PV system is a stand-alone system with diesel generators and the surplus of generated energy in the daytime is consumed at night everyday or not. Inverters are sized based on whether the PV system is stand-alone or not.

CALCULATION OF SYSTEM PERFORMANCES

After sizing PV system components, system performances, such as annual generated energy, annual expenditure and power generation cost could be calculated precisely without simulation. The annual generated energy is given by adding annual generated energy from PV arrays and diesel generators. The annual expenditure is consist of several items as follows:

- capital cost, such as equipment cost, land cost and property tax,

- direct cost, such as personnel expenditure, repair cost and land rental cost,
- management cost, and
- fuel cost.

The power generation cost could be calculated by dividing the total expenditure in the lifetime by the total generated energy in the lifetime. Instead of the total generated energy, a total load energy is usable in stand-alone systems. In utility-connected systems, a part of load energy which is supplied by PV arrays plus energy which is sent to the utility is also usable.

CONCLUSIONS

The features of the proposed sizing procedure are summarized as follows.

- It covers various configurations of PV systems, including stand-alone systems, hybrid systems and utility-connected systems.
- It is based on an analytical method and could size not only PV arrays but also batteries, inverters and diesel generators.
- It does not minimize system cost mathematically, but gives optimal design at the practical level on the basis of experiential knowledge.

The proposed sizing procedure has been represented as a knowledge base and incorporated into a personal computer with climate data bases, such as irradiation, temperature and snowfall of the 225 meteorological observatories in Japan. By using the incorporated knowledge bases and the data bases, the Design-Aid Expert System for Photovoltaic Systems has been developed[5].

The effectiveness of the proposed sizing procedure is confirmed through several design examples. The authors convince that it should contribute wide-spread of PV systems in the future.

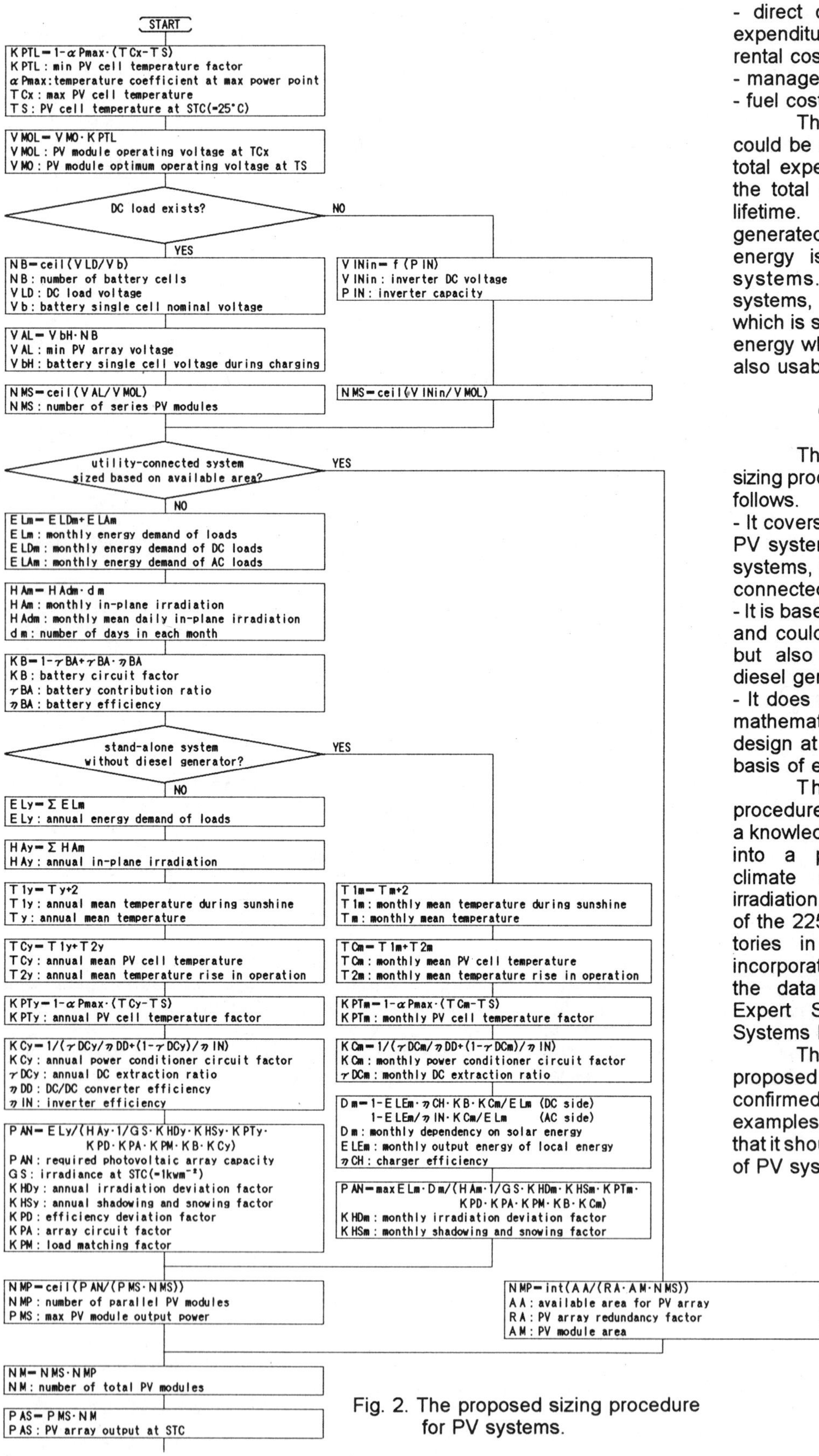

Fig. 2. The proposed sizing procedure for PV systems.

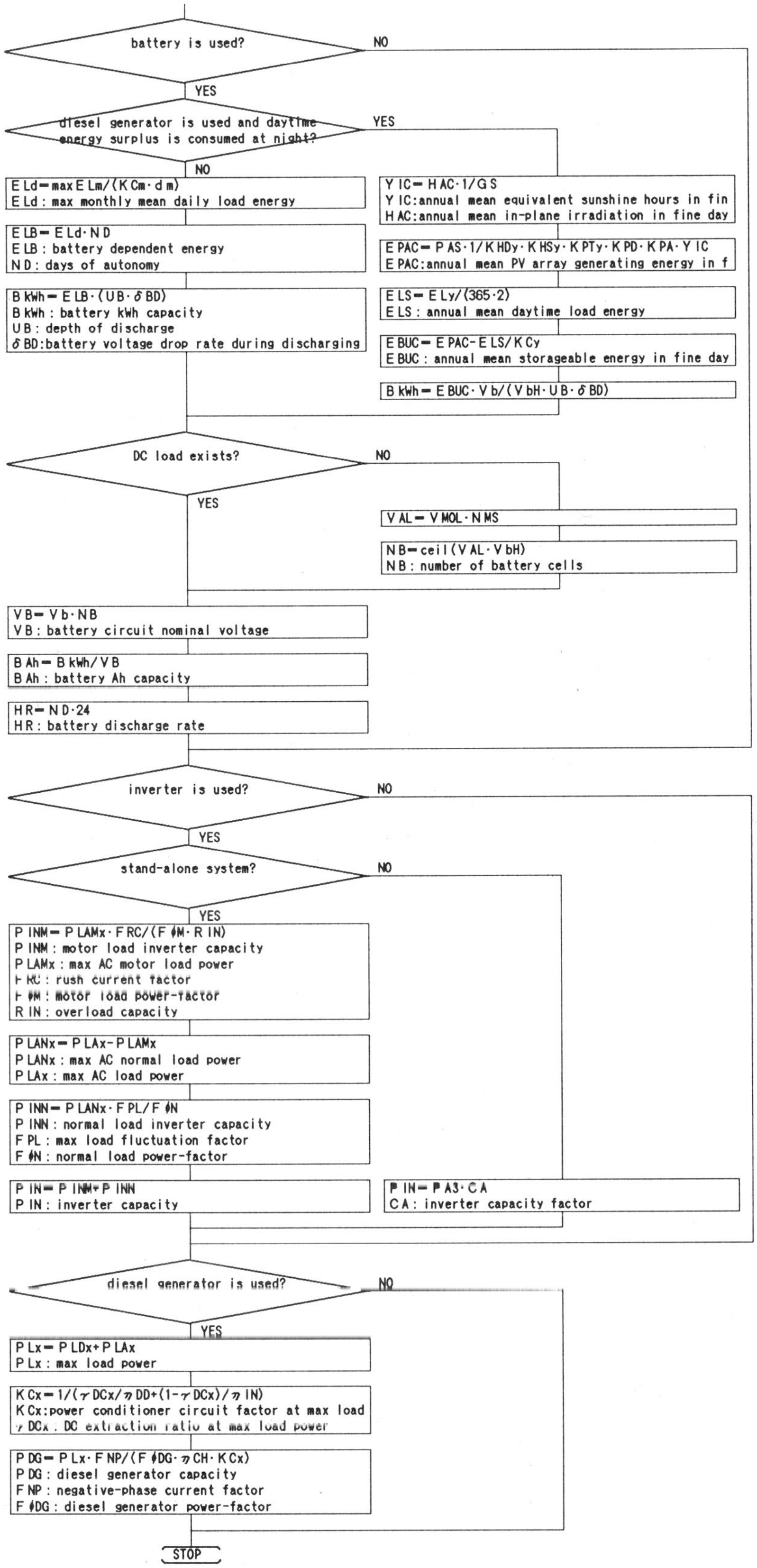

Fig. 2. The proposed sizing procedure for PV systems (continued).

ACKNOWLEDGMENT

The authors would like to thank Mr. Motonobu Yukawa for his kind and helpful comments on the proposed sizing procedure.

REFERENCES

[1] S.J.Strong, "The Solar Electric House: A Design Manual for Home-Scale Photovoltaic Power System", Rodale Press, 1987

[2] J.J.Duffy and S.L.Frye, "An Expert-System-Based PV Sizing Method for Minimum Cost", *Proceedings of the ASES Annual Meeting*, 1989

[3] R.Chapman and J.Fernandez, "SIZEPV: A Simulation Program for Stand-alone Photovoltaic Systems", User's Manual, Sandia Laboratories, 1989

[4] Canada Centre for Mineral and Energy Technology, Energy Mines and Resources Canada, "Photovoltaic Systems Design Manual", 1991

[5] E.Endo, K.Kurokawa and M.Hayashi, "Design-Aid Expert System for Photovoltaic Power Generation System", *Renewable Energy*, Pergamon Press, 1992, **1**, pp.260-264

[6] K.Kurokawa and S.Wakamatsu, "Photovoltaic Systems Design Guidebook", Ohm-sha, 1994 (in Japanese)

DEPLOYMENT OF A DISPATCHABLE PHOTOVOLTAIC SYSTEM: TECHNICAL AND ECONOMIC RESULTS

John Byrne, Young-Doo Wang, Stephen Letendre, Chandrasekar Govindarajalu
Center for Energy and Environmental Policy, University of Delaware, Newark, DE, USA (CEEP)
Ralph Nigro
Delmarva Power and Light Company, Wilmington, DE, USA (DP&L)
William Bottenberg
Delaware Solar Electric Power, Newark, DE, USA

ABSTRACT

This paper discusses the incorporation of PV as a demand-side management (DSM) tool. The valuation of the benefits provided by PV in a DSM role indicates that it is much closer to commercial viability than was thought from economic analyses focusing exclusively on this technology as a supply-side option. However, in order to realize PV's potential, this technology must be deployed in high-value DSM applications; in particular, applications that promise dispatchable peak-shaving capability. Our analysis of the performance of a prototype system installed by Delmarva Power, indicates that small-scale, commercial customer-sited DSM systems incorporating this technology are approaching competitive cost levels

INTRODUCTION

In many parts of the U.S., utility system peak loads tend to coincide with long, hot sunny days during the summer when high solar insolation is available. The high correlation between utility peak load and the rate of solar insolation has prompted recent interest in DSM applications of PV. [1,2]

Ordinarily, a PV-DSM system's peak load reduction capacity is equal to the power it generates at any point in time. However, an approach integrating PV technology with storage offers the possibility for PV to displace a load greater than its power output at peak demand periods. This application of PV-DSM has been developed by the Center for Energy and Environmental Policy (CEEP) and Delmarva Power (a regional utility serving most of Delaware and portions of Maryland and Virginia). Its practical possibility and economic viability are currently being tested in a project underway at Delmarva Power. Analyses to this point indicate that the combined benefits to customers and utilities are approximately 70 percent of current PV system costs. This places the technology much closer to commercialization than previously thought.

THE CONCEPT OF PV-DSM

The concept of PV-DSM under investigation by CEEP and Delmarva Power differs from those explored elsewhere in that PV technology is deployed in an integrated system design which incorporates some form of energy storage. Most PV-DSM applications are based on maximization of the energy provided by the system to the utility or customer. In contrast, our application seeks to maximize the power which can be made available during a utility's peak demand period. In this design, PV-DSM maximizes demand savings. Such an application could be especially valuable to commercial customers because the rates they pay usually include demand charges which often constitute a far greater portion of their bill than energy charges. A PV-DSM system with storage would maximize the benefits to these customers. At the same time, utilities can benefit from the contribution of PV-DSM systems to peak-shaving.

Demand savings are maximized by designing PV-DSM to offer to utilities dispatchable load-reduction capacity. Dispatchability of a PV-DSM system can be achieved by either integrating the solar component with a direct load control device or by incorporating some form of energy storage. [3] Possible forms of storage include batteries and cool storage (e.g., for a system designed to manage air-conditioning loads).

The PV-DSM system analyzed below uses batteries to store the energy produced by the PV system during periods of relatively low demand (early to mid-morning). By sizing the battery bank so that it can be comfortably charged by the morning sun available in a peak demand day (this can be done by sizing on the basis of a "worst case" peak demand day), the system can deliver dependable dispatch capacity to utilities. For example, a system incorporating a 10 kw PV array and 50 kwh of battery storage can consistently provide 16.3 kW of power for a four-hour period in the Mid-Atlantic.

An alternative to PV-DSM which can provide dispatchable peak shaving capacity is a battery only system. Such a system would utilize off-peak, base load generating units to charge a bank of batteries. The stored energy (minus round trip losses) would then be available for peak load dispatch. But a dispatchable PV system is conceptually a better an option than the battery only system because: 1) the fuel costs for a PV unit are zero, and, therefore, investment risks associated with fuel costs are eliminated;

CH3365-4/94/0000-1200 $4.00 © 1994 IEEE First WCPEC; Dec. 5-9, 1994; Hawaii

2) battery systems must purchase energy and a portion of that energy is wasted due to round trip losses; 3) the PV array, as well as its battery unit, supply energy at the time of dispatch, thereby reducing storage requirements and associated hazards and costs; and 4) joint deployment of a PV array and battery storage requires only one inverter per system.

The system depicted in Figure 1 identifies the major components: the PV array, DC interface, battery/inverter module and control module. The system at Delmarva Power was designed for summer peak shaving using the combined output of the PV array and energy in short-term battery storage. Based on data taken during the Summer of 1992 in tests at the University of Delaware and data collected during system testing from July through September 1993, clouds are practically non-existent on the summer days in which Delmarva experiences system peaks. On such days, the output of the PV array is evenly divided between morning and afternoon, peaking at solar noon (approximately 1:00 PM, DST). However, Delmarva's need for capacity, like many other utilities, is skewed towards the afternoon hours. To meet the problem, the array's output during the morning hours was used to charge a battery bank. This stored energy, minus storage losses, was combined with the array's output in the afternoon and fed into the building's distribution grid. The system's capacity is dependent on the amount of energy stored prior to dispatch, and the duration of the dispatch period.

Table 1
Dispatchable PV-DSM System Description

Site: DP&L Northern District General Office, Newark, DE	
Flat roof mounted array, ballast tray configuration	
Fixed flat plate array of Cz silicon based modules	
PV Array DC Rating:	14.5 KW (PVUSA Conditions)
Storage Capacity:	25 kwh @ 4 hr discharge rate
Inverter:	31.25 kw Omnion inverter
System Output:	480 VAC 3 phase

SYSTEM DISPATCH PERFORMANCE

The primary purpose of the Delmarva Power system is to test the practicality of dispatchable PV. The system was dispatched a total of 27 times from July through late September 1993. An example of a "back-to-back" dispatch is shown in Figure 2. This represents the most difficult design condition for the unit, because it requires the batteries to be fully charged in time for the peak-shaving on successive peak days. This ability of the system to be reliably dispatched on successive days is a function of the availability of solar energy for battery charging on system peak days.

The performance of the system during dispatches and charging cycles helped to highlight a few of the problems associated with hardware components (i.e., control system, battery storage capacity, and PV array performance), but also established the soundness of the

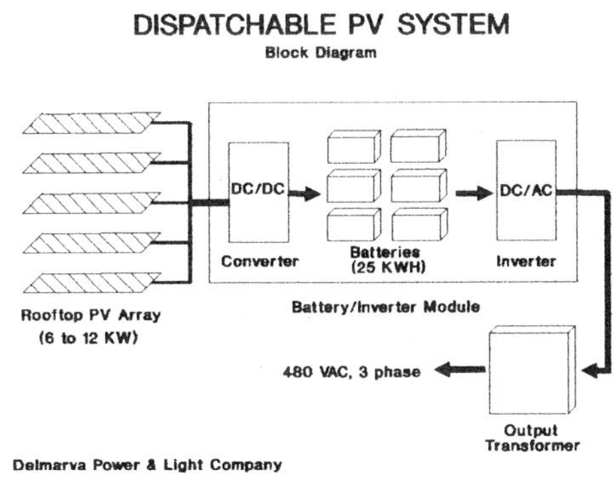

Figure 1

basic concept. This is a major step forward because this type of system was never actually tested before the Summer of 1993. Refinements are now being made for a commercially viable system by summer 1995.

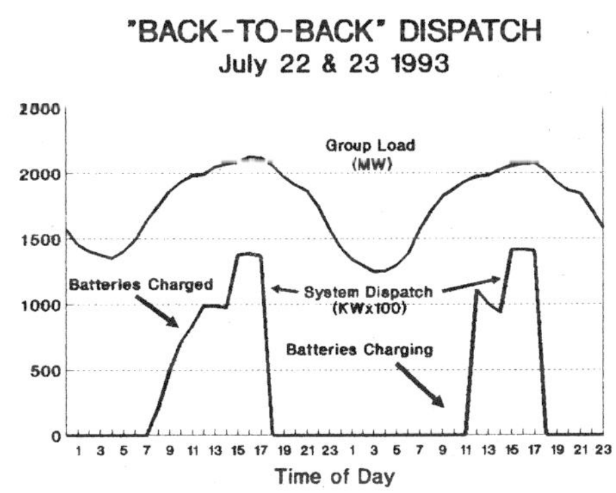

Figure 2

SOLAR RESOURCE AND SYSTEM SIZING ANALYSIS

Solar resource availability on system peak days is an important factor in the operation of the dispatchable PV system. Because the system uses storage, it is not important that the solar resource and load coincide exactly. In Figure 3, the coincidence between Group Load and Global Horizontal Irradiance is shown for the highest hourly loads during the month of July 1993. Although the coincidence is good for about the ten highest load hours, it drops quickly. The main implication of this is that the effective load carrying capability of a PV array without storage in this region is significantly smaller than the nameplate rating for all but a few hours.

Figure 4 shows the coincidence between Total Daily Solar Energy Availability (expressed as kWh/day/sq. meter) and Daily Peak Loads for weekdays from July through September 1993. Because the trends shown are nearly parallel, this figure indicates that, for about the 20 highest summer peak loads, there is a high and relatively predictable level of solar energy available on those days.

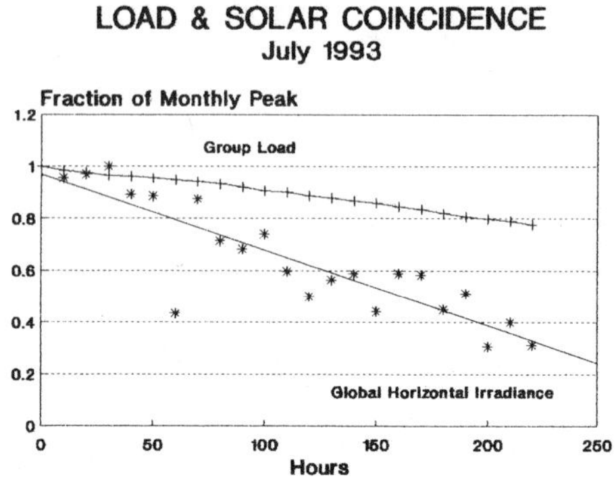

Figure 3

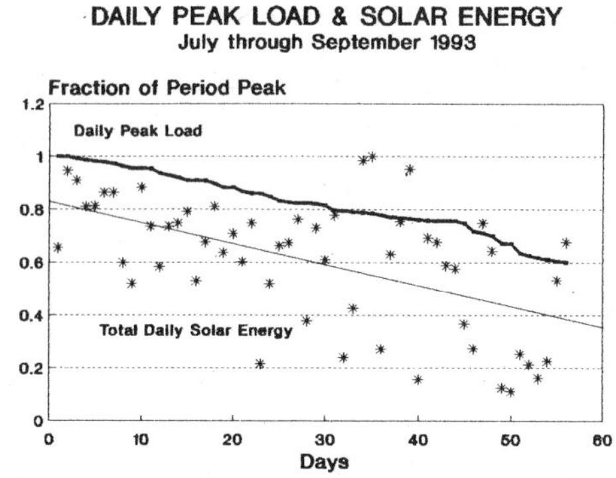

Figure 4

Because the hottest days in the region are also quite humid, the amount of solar energy actually available is somewhat lower than the clearest, cooler, low humidity days. For lower utility loads, the data indicates less predictability. This is because cooler summer days in the mid-Atlantic region (also days in which utility loads are lower) are a mixture of cloudy and very clear sunny days. This figure has a direct impact on the design of dispatchable PV systems because it indicates the relative quantity of solar energy available for collection and storage on peak days.

A combination of PV array and battery storage component performance, resource availability data, and reasonable design assumptions can be used to help size dispatchable PV peak shaving systems. A methodology which accomplishes this would first establish the number of times peak shaving is required during peak months, and then optimize the PV array size for the expected solar resource relative to a given increment of battery storage capacity. This methodology is presently under development at Delmarva Power.

ECONOMIC ANALYSIS OF DISPATCHABLE PV-DSM PEAK SHAVING SYSTEM

The net present values (NPV) of benefits and costs for a 10 kW dispatchable PV peak-shaving system with a 25 year life were calculated using a PV-DSM model developed by CEEP and Delmarva Power, see [2]. With ground source global-horizontal irradiance data supplied by a University of Delaware test facility, it was estimated that the AC output of a 10 kW PV array dispatched daily for four hours during Delmarva's peak demand would typically be credited with a peak-shaving capacity of 16.3 kW during the summer months. Battery storage size was determined by identifying the maximum kwhs generated from 7:30 am until 1:00 pm, using a dispatch period of 2:00 - 6:00 pm for the dispatchable PV-DSM system. The nominal battery storage was 50 kWh.

PV-DSM's benefits and costs were calculated for utility and customer owned systems. Table 2 summarizes the NPVs of benefits and costs for each ownership option. Neither a utility or customer owned system passed the benefit cost test where a value of 1.0 or greater indicates cost effectiveness.

The NPV costs for the utility are made up of three components: capital costs, operating and maintenance (O&M) costs, and carrying charges. Capital costs include $8,500 per kW for the PV array and inverter and $200 per kWh for battery storage. O&M costs include $500 every five years for overhauling the power conditioning system and $150 per kWh every seven years for battery replacement. Delmarva Power current carrying charges

Table 2
Benefit Cost Comparison: 16.3 kW (Credited Capacity) Dispatchable PV-DSM for Delmarva Power

Benefits ($)		Costs ($)	
Utility Owned and Operated			
Avoided Costs	45,690	Capital Costs	95,760
Environmental Benefits	170	Carrying Charges	25,930
Tax Savings	50,340	O&M Costs	24,340
Total	96,200		146,030
Benefit Cost Ratio	0.66		
Customer Owned and Operated			
Energy Savings	9,350	Capital Costs	95,760
Demand Savings	18,260	O&M Costs	15,870
Tax Savings	57,280		
Total	84,890		111,630
Benefit Cost Ratio	0.76		

The utility discount rate: 7.99%, the customer discount rate: 12.0%

were used to cover annual requirements for allowable return, taxes, depreciation, and other fixed overhead costs.

Customer ownership and financing of the PV-DSM system yields a higher benefit cost ratio. Under this ownership option, the customer would retain all bill savings that result from the operation of the PV-DSM unit. Tax benefits include the deductibility of depreciation (on a double declining balance basis) and interest payments and a 10 percent renewable energy tax credit on the purchase of the PV-DSM system. Customer NPV costs include the same capital and O&M costs described above in the utility ownership option. The NPV of O&M costs differ because the customer has a higher discount rate (12%) than the utility's. It is assumed that the customer would not attach carrying charges to this investment. The combined benefits to customers were found to be 76 percent of system costs.

SENSITIVITY ANALYSIS

The above cost-benefit analysis indicates that commercialization of this system is within grasp. Several cost/benefit trials were conducted to assess the sensitivity of the cost/benefit ratios to key assumptions used in the calculations. There are three factors which are the most leveraging. These are array efficiency, system cost and the ownership financing method.

The PV array DC efficiency (at PTC - PVUSA Test Conditions) for the base case was taken as 10.24%. This value was assumed based on measured array values for the Delmarva system. An increase in the array efficiency to 12.24% changes the benefit cost ratio to 0.81. This change assumes that the system (array plus inverter) cost at $8,500/kW is fixed. The increase in benefit cost ratio is due to the increase in the value of the electricity generated for a fixed credited capacity and a reduction in capital cost.

Changing the PV array plus inverter part of the system cost from $8,500/kW to $4,500/kW produces an increase in the benefit cost ratio to 0.87. The change is not as great as might be anticipated because of the effect of tax benefits from capital investment. As the total system cost goes down these benefits decrease commensurately.

The key assumption which influences the benefit cost feasibility for dispatchable PV systems is the basis for ownership of the system. Calculations in Table 2 limit the benefit according to those for the customer or the utility, but not both. If we assume a joint ownership of the system, as described in [3], the utility value for dispatchable peak shaving and the customer value for reduced energy costs can be combined. Such an approach depends on the avoidance of double counting of benefits. This has been addressed in what is elsewhere discussed as a Green Investment Fund (GIF). The benefit cost ratio for a GIF in Delmarva Power's service territory would be 0.92 under the same system used for the calculations in Table 2.

In sum, to speed commercialization, increases in system efficiency, reductions in system cost and joint ownership options need to be pursued. This analysis shows that modest improvements in the cost and performance characteristics of PV-DSM systems as well as expanded ownership options will make these systems cost effective in the Mid-Atlantic region (an area with only moderate solar isolation levels).

CONCLUSIONS

Our analysis suggests that PV-DSM is technically feasible and near commercial viability. In the case of a dispatchable PV-DSM application, the benefits generated by this option equal 76 percent of the current costs for an installed, customer owned system. By reducing costs, and improving the efficiency of major PV system components, as well as employing joint ownership options, dispatchable PV systems could soon emerge as cost-effective DSM options for commercial buildings. The introduction of environmental and fuel risk factors into utility calculations and continued movement toward real-time electricity pricing will all enhance the competitiveness of the PV-DSM application discussed here. The benefits made possible by the application of PV in a DSM role are considerable and place photovoltaic technologies much closer to large market penetration than is conventionally assumed.

ACKNOWLEDGMENTS

This work was partially supported by the National Renewable Energy Laboratory (NREL) under subcontract, XR-2-I 1248-1. We would like to thank William Wallace (NREL) his criticisms and comments. We would also like to acknowledge the contributions of AC Battery Corporation, Ascension Technology Corporation and AstroPower Inc.

REFERENCES

[1] B. Bailey, J. Doty, R. Perez and R. Stewart., "Performance of a Photovoltaic Demand-Side Management System", *Proceedings of the 1991 ISES Solar World Congress*, 1991.

[2] T. Hoff, H. Wenger, D. Keene., "Photovoltaics as a Demand-Side Management Option", *Pacific Gas & Electric, Research and Development , Report 007.5-92.4*, 1994.

[3] J. Byrne, Y-D. Wang, R. Nigro and W. Bottenberg., "Commercial Building Demand-Side Management Tools: Requirements for Dispatchable Photovoltaic Systems", *Proceedings of the 23rd IEEE Photovoltaic Specialists Conference*. Louisville, KY, 1993, pp 1140 - 1145.

[4] J. Byrne, Y-D. Wang, R. Nigro, and S. Letendre., "Photovoltaics in a Demand-Side Management Role", *ACEEE 1994 Summer Study on Energy Efficient Buildings, Proceedings*, 1994, pp 2.43 - 2.48.

BUILDING-INTEGRATED PV—ANALYSIS AND US MARKET POTENTIAL

Lisa Frantzis, Sarah Hill, Peter Teagan and David Friedman
Arthur D. Little, Inc. Cambridge, MA USA

ABSTRACT

Arthur D. Little, Inc., in conjunction with Solar Design Associates, conducted a study for the US Department of Energy (DOE), Office of Building Technologies (OBT) to determine the market potential for building-integrated photovoltaics (BIPV). This study defines BIPV as two types of applications: 1) where the PV modules are an integral part of the building, often serving as the exterior weathering skin, and 2) the PV modules are mounted on the existing building exterior. Both of these systems are fully integrated with the energy usage of the building and have potential for significant market penetration in the US.

BIPV COST REDUCTION OPTIONS

This study initially discussed various cost reduction options for BIPV such as ballasted roof mount, curtainwall applications, AC modules, and other integral systems to evaluate their economic attractiveness in US residential and commercial buildings. Cost data were provided for modules, inverters, wiring, installation, and other indirect costs for the years 1994, 1995, 2000, 2005, and 2010. Cost data were compiled for various types of BIPV systems for both new and retrofit applications. These data were then used as inputs into the ADL PV Market Penetration model to estimate potential national energy savings.

BIPV BUILDING ANALYSIS/ECONOMICS

As part of this study, the impact of BIPV on residential and commercial building load profiles were evaluated and the resulting economic savings that can be achieved by implementing BIPV were calculated. Information on each of the BIPV system's orientation, tilt, and performance characteristics were input into the ADL PV Performance Model, a simulation model developed by Arthur D. Little. The data were then coupled with hourly building load data to simulate the actual impact of the BIPV system on a given building's load profile.

The building runs conducted for retail and office buildings have initially indicated that BIPV without storage or cycling off of AC loads may not reduce commercial building peaks significantly, as their building loads peak longer than coincident PV output.

Commercial Office Building Example

The Southwest orientation/latitude tilt (+/- 10°) is one of the more favorable configurations for BIPV to reduce building demand (kW) for an office building. The model run conducted for the prototypical office building shows the peak demand reduced by 44% of the theoretical maximum peak reduction (i.e. the rated capacity of the PV system) without any battery storage or cycling off of AC loads. Given current utility rate structures for commercial buildings, demand charge savings are minimal. Although individual building loads may peak longer than coincident BIPV output, there is substantial correlation with utility peak load profiles. This analysis is discussed more in the utility benefits section of this paper.

Commercial Curtainwall Example

Office building load profiles with a vertical BIPV curtainwall system resulted in a demand reduction of approximately 25% of the rated capacity of the system without storage and cycling off of AC loads. The annual energy savings from a curtainwall system are 34% less than a BIPV with a latitude tilt, true south orientation. During summer months, which are often utility peak months, the curtainwall BIPV system monthly building electricity savings are ~50% less than a BIPV with a latitude tilt, true south orientation. This loss in energy performance is somewhat compensated for by material credits associated with displacement of Spandrel glass and other curtainwall materials.

Residential Example

The ability of a residential customer to sell BIPV power to the utility at or near the customer cost of power is critical for their economic viability (Fig. 1). Residential building loads tend to be low during the day when BIPV output is greatest. As shown in Fig. 1, areas with shorter cooling seasons such as Albuquerque, need to sell their excess power back to the utility if they do not have storage, in order to be more economical.

BIPV ALLOWABLE SYSTEM PRICE WITHOUT IRP BENEFITS

The BIPV annual energy cost savings were used to determine a maximum allowable system price for BIPV, assuming a five year payback period. Both residential and commercial customers typically use some form of simple payback consideration (e.g. rather than a more formal rate of return or cost of capital type analysis) to assess the economic value of energy saving technologies. Based on field interviews, consumer surveys, and analyses of market experience with a variety of energy saving technologies, Arthur D. Little has developed curves that relate market penetration rates to economic performance from the consumer perspective. These curves show that the threshold payback period (i.e. upper limit) required for significant market penetration is four to five years. This analysis therefore used a five year payback to estimate an allowable system price that customers most typically would be willing to pay for BIPV.

The "allowable BIPV system prices" were determined from the customer perspective. Arthur D.

CH3365-4/94/0000-1204 $4.00 © 1994 IEEE

Little, however, did conduct several scenarios. One of the scenarios used data provided by the State University of New York at Albany (SUNYA) to show BIPV correlation with utility loads. Effective Load Carrying Capability (ELCC) values were derived for nine regions of the US for which the ADL Market Model was developed. ELCC as defined by SUNYA is the effective increase in a utility's installed capacity due to the added resource, at a constant loss of load probability. Using the value of these correlations can potentially increase the allowable system price for BIPV systems. Some preliminary findings using existing utility rate structures in the Northeast and Southwest US indicate:

- Office building allowable system price of between ~$720–1,140/kW (without IRP benefits).
- Office building allowable system price of between $1,135–1,290/kW adjusted for ELCC (without IRP benefits).
- Retail building allowable system price of between ~$430–725/kW (without IRP benefits).
- Retail building allowable system price of between $1,130–1,500/kW adjusted for ELCC (without IRP benefits).

IMPACT OF IRP BENEFITS ON BIPV ECONOMICS

The most favorable BIPV analysis assumes energy savings, potential demand charge savings, and Integrated Resource Planning (IRP) benefits. IRP benefits include transmission and distribution deferred costs, capacity credits, and environmental externality benefits.

This study has shown that IRP benefits, if properly assigned, can increase significantly the allowable BIPV system price for commercial buildings. These IRP benefits are potentially very important to the cost-effectiveness of BIPV systems and can increase the allowable installed system price approximately 50-80% - if benefits gained by the utility for the BIPV systems are passed on to the consumer in some form of rebates or "cash incentives".

A preliminary analysis indicates that the impact of these additional benefits on the allowable system prices using current utility rates are:

- Retail building allowable system prices increase from ~$700/kW (no IRP benefits) to $2,300/kW (with IRP benefits).
- Office building allowable system prices increase from ~$1,200/kW (no IRP benefits) in the Southwest to ~$2,000/kW (with IRP benefits).
- Office building curtainwall applications increase from ~$700/kW (no IRP benefits) to ~1,600/kW (with IRP benefits).
- Single family allowable system prices increase from ~$1,300/kW (no IRP benefits) to ~$1,800/kW (with IRP benefits).

If the industry does not move to encourage utility groups and Public Utility Commissions to value these benefits, the market for BIPV in the US will be more limited based on economics.

IMPACT OF BIPV ON NATIONAL ENERGY SAVINGS

Four ADL PV Market Models were used to assess the potential impact of BIPV on national energy savings for single family, multifamily, and commercial buildings. Separate models for rooftop and curtainwall systems were used to evaluate commercial building applications. The models use existing and forecast data for utility rates, BIPV system performance, installed system pricing with BIPV cost reduction options, and potential IRP benefits to determine consumer payback, consequent market penetration, and the resulting energy savings potential.

As discussed above, the incorporation of IRP benefits results in shorter payback periods and significantly better economics. The value of these potential IRP benefits, for example, in the Pacific Southwest in the year 2005 for rooftop commercial building applications, can reduce the payback period from 16 years with no IRP benefits to 11 years with IRP benefits (Fig. 3).

The addition of IRP benefits, which increases market penetration, results in an annual US market between 600–700MW/yr by the year 2010. If IRP benefits are not assigned, this market size is less than 5MW/yr (based strictly on an economic analysis and not including demonstration programs).

Aggressive module cost reductions can also have a significant impact on the market penetration of BIPV systems. Many projections have been made by module manufacturers that advanced thin-film PV module materials can achieve direct manufacturing costs below $1/Wpac by the year 2005. Adding appropriate profit margins for manufacturers still results in significant BIPV cost reductions. These aggressive module cost reductions, along with the BIPV cost reduction options identified in this study, can further improve BIPV economics. The same payback period discussed above for the year 2005 in the Pacific Southwest would be 13 years without IRP benefits and nine years with IRP benefits (Fig. 4).

Advanced thin-film module materials in combination with BIPV cost reduction options can therefore result in significant annual markets in the 2005–2010 time period. Even a small fraction of building penetration results in relatively large markets because of the magnitude of the building sector. The potential market penetration for the US could reach 600-700MW/yr by 2005 and 1300–1600MW/yr by 2010 <u>if IRP benefits are incorporated</u> along with very aggressive module and installed BIPV cost reduction options. Without IRP benefits, the market penetration will likely be less than 30MW/yr in the US by 2005, based on an economic analysis from the customer perspective.

To achieve these large market penetration values, the industry must promote activity that focusses on installation cost reduction, and provide supporting data to show the BIPV/utility interface benefits. Currently, IRP benefits are not given to BIPV systems. Although the BIPV output may not correlate well with some building load profiles, the utility should acknowledge the BIPV system output correlation with their overall peak load. The PV industry, DOE and Public Utility Commissions (PUCs) need to promote the value of BIPV until the customer receives rebates or cash incentives that more accurately reflect the IRP benefits to the utility.

CONCLUSIONS

- For residential systems, it is essential that BIPV systems be able to sell back to the utility at or near the customer cost of power, rather than at the utility avoided cost. Sell back to the utility at these rates ensures more favorable economics.
- Variations in commercial electric rate structures are as important in defining BIPV economics as variations in insolation.
- The ability to obtain utility IRP benefits, combined with projected BIPV system cost reductions, can potentially result in large BIPV market penetrations from the customers economic perspective.
- The industry, DOE, and the PUCs should work together to properly value BIPV benefits. Proper assignment of capacity credits, T&D avoided and deferred costs, and environmental externalities can accelerate the BIPV market penetration and result in significant energy savings.

Fig. 1. Impact of BIPV Sizing and Buyback Potential on Single Family Electricity Savings

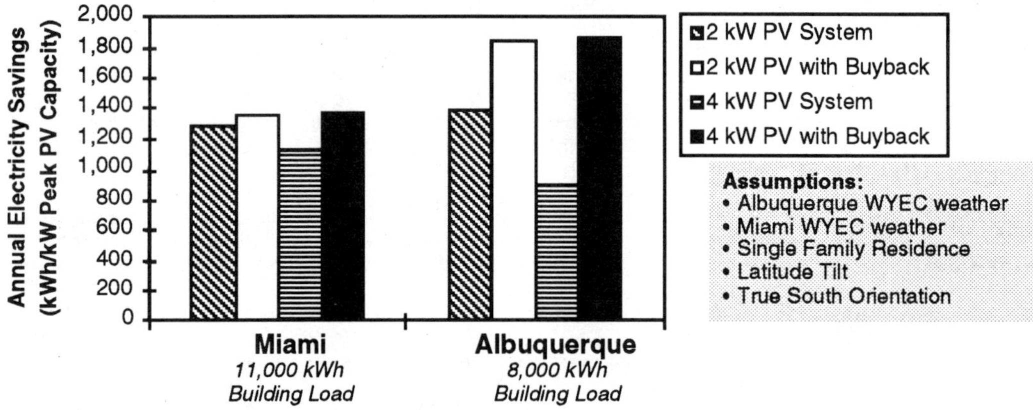

Fig. 2. BIPV ELCC Correlations

Region	Capacity Correlation (%) (ELCC)[1]	Representative Utility
New England (Boston)	50	NEES
Middle Atlantic (New York)	60	LILCO
East North Central (Chicago)	60	NSP
West North Central (Minneapolis)	60	NSP
South Atlantic (Miami)	45	FP&L
East South Central (Birmingham)	55	SES
West South Central (Dallas)	65	TU
Pacific Southwest (Los Angeles)	70	PG&E
Pacific Northwest (Olympia)	20	Relative to all northwest

[1] ELCC—Effective Load Carrying Capability. A statistical measure of effective capacity—the effective increase in a utility's installed capacity due to the added resource, at a constsnt loss of load probability. These percentages assume 2% PV penetration on the considered grid. If 10% PV penetration is assumed, then the above ELCC percentages would be 3–5% less.

Source: Telephone interview with Richard Perez at SUNYA, August 1994.

Fig. 3. Impact of IRP Benefits and ELCC on Payback Periods

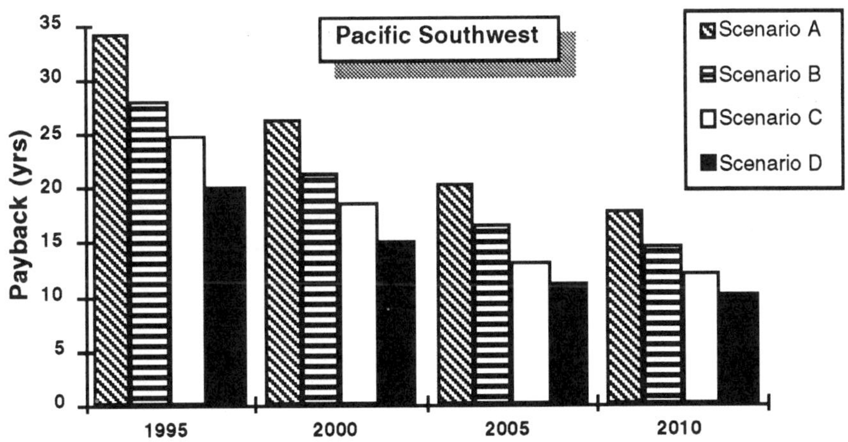

Scenario A: PV with Energy Charge Savings (no additional benefits)
Scenario B: Scenario A with Demand Charge Savings [1]
Scenario C: Scenario B with IRP Benefits (T&D and Capacity Credits included) [2]
Scenario D: Scenario C with Environmental Externalities included [2]

Assumptions: All scenarios represent a 30 kW AC module system at a south-facing, 15-degree tilt and with a ballasted mount on a flat roof. Total system installed costs are, including one-step distribution (15% markup) and a 10% federal tax credit, $7.10/Wpac (1995), $5.30/Wpac (2000), $4.00/Wpac (2005), and $3.50/Wpac (2010).

Fig. 4. Impact of IRP Benefits and ELCC on Payback Periods Using Aggressive Module Pricing

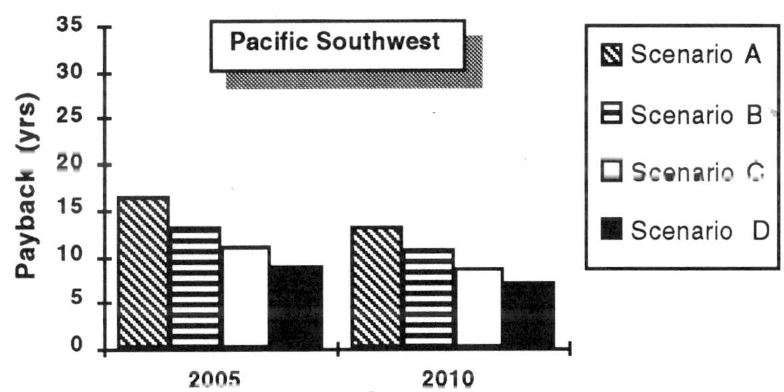

Scenario A: PV with Energy Charge Savings (no additional benefits)
Scenario B: Scenario A with Demand Charge Savings [1]
Scenario C: Scenario B with IRP Benefits (T&D and Capacity Credits included) [2]
Scenario D: Scenario C with Environmental Externalities included [2]

Assumptions: All scenarios represent a 30 kW AC module system at a south-facing, 15-degree tilt and with a ballasted mount on a flat roof. Total system installed costs are, including one-step distribution (15% markup) and a 10% federal tax credit, $3.25/Wpac (2005), and $2.60/Wpac (2010).

Notes for Fig. 3 and 4

1. Assumes a percentage of system capacity is recovered through demand charge savings. This ELCC percentage varies by region.
2. T&D avoided costs and environmental externality benefits assume utilities will pass on "cash incentives" to consumers. To date, monetized values have not been offered for these benefits.

HOLOGRAPHIC SOLAR CONCENTRATOR FOR TERRESTRIAL PHOTOVOLTAICS

Jacques Ludman and Juanita Riccobono
Northeast Photosciences, Inc., Hollis, NH
Nadya Reinhand and Irina Semenova
Ioffe Institute, St. Petersburg, Russia
José Martín, William Tai, Xiao-Li Li, and Geof Syphers
Department of Chemical and Energy Engineering, University of Massachusetts Lowell

ABSTRACT

This paper describes theoretical and experimental work performed on these high efficiency holographic concentrating and spectral splitting photovoltaic systems. The theoretical system efficiency was calculated to be 41% for a GaAs-InGaAs tandem cell when the hologram was recorded at 632.8 nm. A prototype system was used to demonstrate that it maintained a fairly constant efficiency over a period of 15 minutes without a heatsink; whereas the efficiency of a Fresnel system, using the same aperture and cell, dropped a factor of two within 30 seconds when used without a heatsink. Finally, a cost analysis of this system estimates it to generate power for 5.7¢ per kWh when production levels reach 20 MW per year capacity.

INTRODUCTION

It is well known that flat Si solar panels operate at about 10% efficiency. Concentrating solar energy with a concentrating system will increase this efficiency from 10% to approximately 20% [1-2]. It is also well known that splitting the solar spectrum into its component colors brings the efficiency up to approximately 30% [1]. Achieving these high efficiencies has long been a goal of the solar concentrator community.

A holographic device that greatly improves the efficiency of solar energy conversion by both concentrating and spectral splitting, has been developed [3]. This technology uses a single-element hologram to spectrally separate light and focus it perpendicular to the hologram in a thin, concentrated line. Two or more different solar cells are then placed along this line such that each cell absorbs only those wavelengths of the spectrum that it can most efficiently convert to electrical power (see Figure 1).

Because the hologram is a single element performing two functions, reflective losses are lowered. Because the hologram focuses the light into a line, long and narrow solar cells are used. This decreases shading and resistance losses from the contacts on the PV cells since the contacts are shorter and can be made thinner than for normal PV cells. Thus the free electron path is shortened and the recombination rate is decreased despite minimal front-surface contacts.

A major advantage to this technology is the reduced cooling requirements; spectral separation prevents overheating by diffracting unwanted infrared radiation away from the cells, while the spectrally-dispersed side focus replaces the common but difficult-to-cool cascade (i.e., two different solar cells which are grown one on top of the other) design with a tandem (i.e., two different solar cells which are next to each other) design.

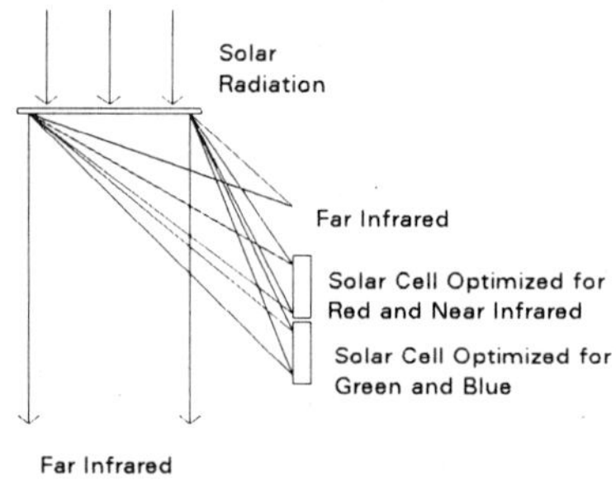

Figure 1. Holographic Solar Concentrator and Spectrum Splitter. One photocell can be optimized for red and near infrared, the other for green and blue. Far infrared, which would contribute only useless heat, is diffracted away from the cells.

CALCULATION OF SYSTEM EFFICIENCY

The diffraction efficiency of holographic optical elements depends on the wavelength of the incident radiation, as well as the spectral sensitivity of the photovoltaic cells which convert the light energy into electrical energy. Thus, the efficiency of the whole system may be evaluated as: solar spectrum (AM1 for earth) * HOE diffraction efficiency as a function of wavelength * solar cell efficiency as a function of wavelength.

The expression for the theoretical efficiency η of the lossless transmission hologram [4] can be modified from $\eta = \eta(\xi)$, where ξ is a parameter depending on the emulsion layer thickness, angle between the recording beams and its change to $\eta = \eta\ (\delta\lambda)$, where $\delta\lambda$ is a change in the wavelength of illumination from the recording wavelength, λ_a. The magnitude of $\delta\lambda$ may vary in a wide range. Thus, η:

$$\eta = \left| -1 \frac{\exp(-\beta\delta\lambda)\,\sin\left(\delta\lambda^2\beta^2 + \upsilon^2\right)^{1/2}}{\left(1+\delta\lambda^2\beta^2/\upsilon^2\right)^{1/2}} \right|^2 \qquad (1)$$

$$\text{where} \quad \upsilon = \frac{\pi n_1 T}{\lambda_a \cos\theta_0} \qquad (2)$$

$$\text{and} \quad \beta = \frac{\xi}{\delta\lambda} = -\frac{1}{\lambda_a} tg\theta_0 \left(\frac{2\pi\bar{n}}{\lambda_a}\right) T \sin\theta_0 \qquad (3)$$

$2\theta_0$ is an angle between two interfering beams in the recording medium. T is the emulsion layer thickness. $\bar{n}$ is the refraction coefficient average value. n_1 is the modulation amplitude of the refraction coefficient. The efficiency will be 100% when the angle of incidence is equal to the Bragg angle and $\sin\nu = 1$, or $\nu = \pi/2$.

The diffraction efficiency of a holographic optical element can then be calculated at different recording (i.e., laser) wavelengths in the form: $\eta = \eta\ (\delta\lambda)$. Calculations were made at λ_a = 441.6 nm; 488 nm; 530 nm; 632.8 nm, using: T=2 microns, $\theta_0 = 10°$, $\bar{n} = 1.52$ (These values for T, θ_0, and $\bar{n}$ were chosen for maximum hologram efficiency).

The spectral sensitivity or conversion efficiency of the solar cells must be taken into account as well. There were two cases of solar cells chosen for these calculations. One was a single cell, Si (bandgap is 1100 nm), and the other was a tandem cell consisting of GaAs (bandgap is 880 nm) and InGaAs (bandgap is 1680 nm).

Thus, the resulting system efficiency is the product of each of these calculations. The most efficient system is the one with a hologram recorded at 632.8 nm (HeNe laser) with the GaAs-InGaAs tandem cell; this combination has an efficiency of 41% for the solar spectrum at AM1. Assuming this efficiency were equal to 1, efficiencies for the other cases would be as follows:

λ_a (nm)	η for Si cell (%)	η for GaAs + InGaAs cell (%)
HeCd laser (441.6)	54	70
Ar laser (488)	63	79
Ar laser (530)	68	88
HeNe laser (632.8)	75	100

EXPERIMENTAL

The output of the hologram concentration system was compared to that of a Fresnel lens concentrator having the same size aperture and using the same photovoltaic cells. The original purpose of the experiment was to directly compare the efficiencies of the two concentrator types using the PV cells available. It was quickly discovered that without heatsinks the higher concentration ratio and inclusion of infrared light by the Fresnel lens heated the cells so quickly that within thirty seconds the efficiency had dropped by a factor of two. The heating effects of both systems were studied by measuring the power output as a function of exposure time (see Tables 1 and 2).

Maximum Power Output Versus Exposure Time

(Test Conditions: 700 W/m^2)

Table 1 Fresnel System Without Heatsink (aperture size: 10.8cm × 14.0cm)

Time	V_{oc}	I_{sc}	P_{max}*	Relative Efficiency
(sec)	(volt)	(mA)	(mW)	(percent)
0	0.95	903.3	559.6	88.5
10	0.77	656.5	327.7	51.8
20	0.73	624.0	296.1	46.8
30	0.71	604.7	278.7	44.0
40	0.69	591.3	266.7	42.1
50	0.68	579.6	256.6	40.5
60	0.67	571.4	249.2	39.3

*The fill factor was assumed to be 0.65 for all cases: $P_{max} = V_{oc} \times I_{sc} \times FF$

Table 2 Hologram System Without Heatsink (aperture size: 9.5cm × 11.4cm)

Time	V_{oc}	I_{sc}	P_{max}	Relative Efficiency
(sec)	(volt)	(mA)	(mW)	(percent)
0	0.77	869.3	435.1	96.0
10	0.76	873.8	431.7	95.2
20	0.76	872.8	431.2	95.2
30	0.78	868.7	440.4	97.2
40	0.76	872.6	431.1	95.2
50	0.77	871.1	436.0	96.2
60	0.80	871.7	453.3	100
1000	0.77	867.2	434.0	95.7

The rapid drop in power shown in Table 1 does not indicate that the Fresnel system is impractical since heatsinks were not used in this comparison and they are an integral part of Fresnel lens systems. Rather, the study shows the ability of the hologram to divert IR and direct the useful light without compromising the overall efficiency.

COST ANALYSIS

Generally, the cost for a renewable energy source (such as solar energy) is in the 10 to 35¢ per kWh range, depending on such variables as location, demand, availability and region. This new technology reduces the cost to around 5.7¢ per kWh. A detailed economic analysis of this technology for earth-based power generation is shown in Table 3 (at end of article). This analysis is based on a modification of the Department of Energy Laboratory (Sandia National Laboratory) Sandia Baseline 3 Photovoltaic Concentrator Module (SBM3), which was projected to be able to produce energy at around 10-11¢ per kWh [5].

This cost analysis shows that it is competitive, not only with Sandia's SBM3 system, but also with other power generation systems, such as nuclear and fossil fuels, when their environmental externality (clean-up) costs are taken into account (see Table 4). For most sources of energy, the cost per kWh is in the 5 to 12¢ per kWh range, depending on the cost of the fuel, as well as the environmental externality cost.

Table 4: **LEVELIZED GENERATING COST BREAKDOWN FOR 1200 MWe CAPACITY POWER PLANTS**

All costs are in ¢/kWh.

Plant/Fuel Type	Capital and O & M Costs	Fuel Cost	Externality Cost	Environmental[4] Externality Cost	Total Cost
Pulverized Coal	2.8	1.7	0.1	3.0	7.6
Gasified Coal	3.0	1.7	0.1	3.0	7.8
Combined-Cycle Gas	1.0	3.2	0	1.0	5.2
Nuclear	3.0	0.7	0.1	3.0	6.8
Flat Plate PV[1]	12	0	0	0	12
Sandia SBM3[2]	10	0	0	0	10
NEPS Holographic PV[3]	5.7	0	0	0	5.7

Source: USCEA [1992] for all costs, except noted as follows:

[1]Source: Scientific American [1994]

[2]Source: Sandia [1989]

[3]Source: NEPS [1994]

CONCLUSION

A holographic solar concentrator can be made which:

- concentrates sunlight,
- splits the spectrum,
- deflects far infrared energy which contributes to heating in solar cells,
- has high efficiency, and
- produces energy at a cost of 5.7¢ per kWh.

This photovoltaic system can be used with a combination of solar cells, although some combinations (GaAs-InGaAs) are more efficient than others. When compared to a Fresnel system, it was found that the holographic system concentrates more total power than the Fresnel system, the holographic system has a larger relative efficiency, and that large heatsinks are not necessary with the holographic system which decreases the bulk and cost of the system as compared to the Fresnel system.

REFERENCES

[1] S.M. Sze, *Physics of Semiconductor Devices, 2cnd Edition*, John Wiley & Sons, NY, 1981.

[2] J.L. Stone, "Photovoltaics: Unlimited Electric Energy from the Sun," *Physics Today*, Sept. 1993, p. 22.

[3] J.E. Ludman, J.L. Sampson, R.A. Bradbury, J.G. Martín, J.R. Riccobono, G. Sliker, E. Rallis, "Photovoltaic Systems Based on Spectrally Selective Holographic Concentrators," *Practical Holography VI, SPIE Proc.*, **1667**, 1992, pp. 182-189.

[4] R.J. Collier, C.B. Burckhardt, L.H. Lin, *Optical Holography*, Academic Press, 1971.

[5] E.H. Richards, "Sandia's Baseline 3 Photovoltaic Concentrator Module," *IEEE Photovoltaic Specialists Conference*, 1993, p. 1292.

ACKNOWLEDGMENTS

This research was supported by the Department of Defense, Ballistic Missile Defense Organization, and the Department of the Air Force, Air Force Systems Command, Wright Laboratory at Wright-Patterson AFB in Ohio. We would also like to thank the Photovoltaic Group at NASA-Lewis in Cleveland, Ohio for helpful discussions.

Table 3 Levelized Energy Cost Calculation for a Holographic Solar Concentrator Power Plant Based on 0.7 sq.m.

DATA INPUT SECTION	Units	'1 to 2 MW/YEAR		'20+ MW/YEAR	
Discount Rate		0.06	0.12	0.06	0.12
Lifetime	years	30	30	30	30
Module Cost	$/.7sq.m	285	285	144	144
BOS area cost	$/.7sq.m	240	240	125	125
BOS power cost	$/kW	500	500	200	200
Average Peak Insulation	kW/.7sq.m	0.8	0.8	0.8	0.8
Cell Efficiency (STC)		33.0%	33.0%	33.0%	33.0%
Cell Efficiency (NOCT)		29.7%	29.7%	29.7%	29.7%
Optical Efficiency		86.0%	86.0%	86.0%	86.0%
Performance Reduction Coefficient		95.0%	95.0%	95.0%	95.0%
PCU Efficiency		96.0%	96.0%	96.0%	96.0%
Tracking Efficiency		95%	95%	95%	95%
Indirect Cost Factor		0.5	0.5	0.3	0.3
Yearly O/M Costs	$/.7sq.m-yr	4	4	2	2
Annual Solar Availability	kW/.7sq.m-yr	2600	2600	2600	2600
Annual Systems Availability		0.95	0.95	0.95	0.95
Ground Cover Ratio		.25	.25	.25	.25
ENERGY COST CALCULATIONS					
Module Efficiency (NOTC)		25.5%	25.5%	25.5%	25.5%
Array Efficiency (DC peak)		24.3%	24.3%	24.3%	24.3%
System Efficiency (Accept)		23.3%	23.3%	23.3%	23.3%
System Efficiency (AC annual)		22.1%	22.1%	22.1%	22.1%
Module Area: 1 kW DC peak	.7sq.m	5.15	5.15	5.15	5.15
Module Cost: 1 kW DC peak	$/kW	1468	1468	742	742
BOS power cost	.7sq.m	97	97	39	39
Total BOS Costs	.7sq.m	337	337	164	164
Total Installed Costs	.7sq.m	933	933	400	400
Total Installed Costs	$/kW(DC pk)	4807	4807	2061	2061
Present Value O & M	.7sq.m	55	32	28	16
Annualized costs	.7sq.m-yr	71.79	119.84	31.07	51.68
Annualized costs		370	617	160	266
Annual System Output		547	547	547	547
Annual System Output		2816	2816	2816	2816
Required Land Area		21	21	21	21
Levelized kWh cost		**$0.1313**	**$0.2192**	**$0.0568**	**$0.0945**

HOLOGRAPHIC SOLAR CONCENTRATOR FOR TERRESTRIAL PHOTOVOLTAICS

Jacques Ludman and Juanita Riccobono
Northeast Photosciences, Inc., Hollis, NH
Nadya Reinhand and Irina Semenova
Ioffe Institute, St. Petersburg, Russia
José Martín, William Tai, Xiao-Li Li, and Geof Syphers
Department of Chemical and Energy Engineering, University of Massachusetts Lowell

ABSTRACT

This paper describes theoretical and experimental work performed on these high efficiency holographic concentrating and spectral splitting photovoltaic systems. The theoretical system efficiency was calculated to be 41% for a GaAs-InGaAs tandem cell when the hologram was recorded at 632.8 nm. A prototype system was used to demonstrate that it maintained a fairly constant efficiency over a period of 15 minutes without a heatsink; whereas the efficiency of a Fresnel system, using the same aperture and cell, dropped a factor of two within 30 seconds when used without a heatsink. Finally, a cost analysis of this system estimates it to generate power for 5.7¢ per kWh when production levels reach 20 MW per year capacity.

INTRODUCTION

It is well known that flat Si solar panels operate at about 10% efficiency. Concentrating solar energy with a concentrating system will increase this efficiency from 10% to approximately 20% [1-2]. It is also well known that splitting the solar spectrum into its component colors brings the efficiency up to approximately 30% [1]. Achieving these high efficiencies has long been a goal of the solar concentrator community.

A holographic device that greatly improves the efficiency of solar energy conversion by both concentrating and spectral splitting, has been developed [3]. This technology uses a single-element hologram to spectrally separate light and focus it perpendicular to the hologram in a thin, concentrated line. Two or more different solar cells are then placed along this line such that each cell absorbs only those wavelengths of the spectrum that it can most efficiently convert to electrical power (see Figure 1).

Because the hologram is a single element performing two functions, reflective losses are lowered. Because the hologram focuses the light into a line, long and narrow solar cells are used. This decreases shading and resistance losses from the contacts on the PV cells since the contacts are shorter and can be made thinner than for normal PV cells. Thus the free electron path is shortened and the recombination rate is decreased despite minimal front-surface contacts.

A major advantage to this technology is the reduced cooling requirements; spectral separation prevents overheating by diffracting unwanted infrared radiation away from the cells, while the spectrally-dispersed side focus replaces the common but difficult-to-cool cascade (i.e., two different solar cells which are grown one on top of the other) design with a tandem (i.e., two different solar cells which are next to each other) design.

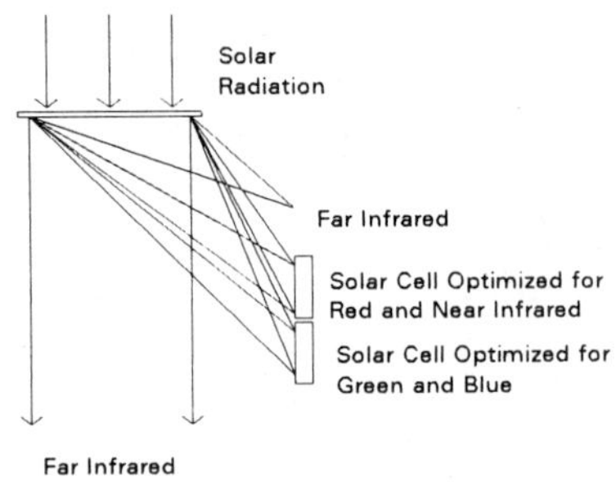

Figure 1. Holographic Solar Concentrator and Spectrum Splitter. One photocell can be optimized for red and near infrared, the other for green and blue. Far infrared, which would contribute only useless heat, is diffracted away from the cells.

CALCULATION OF SYSTEM EFFICIENCY

The diffraction efficiency of holographic optical elements depends on the wavelength of the incident radiation, as well as the spectral sensitivity of the photovoltaic cells which convert the light energy into electrical energy. Thus, the efficiency of the whole system may be evaluated as: solar spectrum (AM1 for earth) * HOE diffraction efficiency as a function of wavelength * solar cell efficiency as a function of wavelength.

The expression for the theoretical efficiency η of the lossless transmission hologram [4] can be modified from $\eta = \eta(\xi)$, where ξ is a parameter depending on the emulsion layer thickness, angle between the recording beams and its change to $\eta = \eta\ (\delta\lambda)$, where $\delta\lambda$ is a change in the wavelength of illumination from the recording wavelength, λ_a. The magnitude of $\delta\lambda$ may vary in a wide range. Thus, η:

$$\eta = \left| -1 \frac{\exp(-\beta\delta\lambda)\sin\left(\delta\lambda^2\beta^2 + \upsilon^2\right)^{1/2}}{\left(1 + \delta\lambda^2\beta^2/\upsilon^2\right)^{1/2}} \right|^2 \quad (1)$$

CH3365-4/94/0000-1212 $4.00 © 1994 IEEE

where $$\upsilon = \frac{\pi n_1 T}{\lambda_a \cos\theta_0} \qquad (2)$$

and $$\beta = \frac{\xi}{\delta\lambda} = -\frac{1}{\lambda_a} tg\theta_0 \left(\frac{2\pi \bar{n}}{\lambda_a}\right) T \sin\theta_0 \qquad (3)$$

$2\theta_0$ is an angle between two interfering beams in the recording medium. T is the emulsion layer thickness. $\bar{n}$ is the refraction coefficient average value. n_1 is the modulation amplitude of the refraction coefficient. The efficiency will be 100% when the angle of incidence is equal to the Bragg angle and $\sin\nu = 1$, or $\nu = \pi/2$.

The diffraction efficiency of a holographic optical element can then be calculated at different recording (i.e., laser) wavelengths in the form: $\eta = \eta(\delta\lambda)$. Calculations were made at λ_a = 441.6 nm; 488 nm; 530 nm; 632.8 nm, using: T=2 microns, $\theta_0 = 10°$, $n = 1.52$ (These values for T, θ_0, and n were chosen for maximum hologram efficiency).

The spectral sensitivity or conversion efficiency of the solar cells must be taken into account as well. There were two cases of solar cells chosen for these calculations. One was a single cell, Si (bandgap is 1100 nm), and the other was a tandem cell consisting of GaAs (bandgap is 880 nm) and InGaAs (bandgap is 1680 nm).

Thus, the resulting system efficiency is the product of each of these calculations. The most efficient system is the one with a hologram recorded at 632.8 nm (HeNe laser) with the GaAs-InGaAs tandem cell; this combination has an efficiency of 41% for the solar spectrum at AM1. Assuming this efficiency were equal to 1, efficiencies for the other cases would be as follows:

λ_a (nm)	η for Si cell (%)	η for GaAs + InGaAs cell (%)
HeCd laser (441.6)	54	70
Ar laser (488)	63	79
Ar laser (530)	68	88
HeNe laser (632.8)	75	100

EXPERIMENTAL

The output of the hologram concentration system was compared to that of a Fresnel lens concentrator having the same size aperture and using the same photovoltaic cells. The original purpose of the experiment was to directly compare the efficiencies of the two concentrator types using the PV cells available. It was quickly discovered that without heatsinks the higher concentration ratio and inclusion of infrared light by the Fresnel lens heated the cells so quickly that within thirty seconds the efficiency had dropped by a factor of two. The heating effects of both systems were studied by measuring the power output as a function of exposure time (see Tables 1 and 2).

Maximum Power Output Versus Exposure Time
(Test Conditions: 700 W/m^2)

Table 1 Fresnel System Without Heatsink (aperture size: 10.8cm × 14.0cm)

Time	V_{oc}	I_{sc}	P_{max}*	Relative Efficiency
(sec)	(volt)	(mA)	(mW)	(percent)
0	0.95	903.3	559.6	88.5
10	0.77	656.5	327.7	51.8
20	0.73	624.0	296.1	46.8
30	0.71	604.7	278.7	44.0
40	0.69	591.3	266.7	42.1
50	0.68	579.6	256.6	40.5
60	0.67	571.4	249.2	39.3

*The fill factor was assumed to be 0.65 for all cases: $P_{max} = V_{oc} \times I_{sc} \times FF$

Table 2 Hologram System Without Heatsink (aperture size: 9.5cm × 11.4cm)

Time	V_{oc}	I_{sc}	P_{max}	Relative Efficiency
(sec)	(volt)	(mA)	(mW)	(percent)
0	0.77	869.3	435.1	96.0
10	0.76	873.8	431.7	95.2
20	0.76	872.8	431.2	95.2
30	0.78	868.7	440.4	97.2
40	0.76	872.6	431.1	95.2
50	0.77	871.1	436.0	96.2
60	0.80	871.7	453.3	100
1000	0.77	867.2	434.0	95.7

The rapid drop in power shown in Table 1 does not indicate that the Fresnel system is impractical since heatsinks were not used in this comparison and they are an integral part of Fresnel lens systems. Rather, the study shows the ability of the hologram to divert IR and direct the useful light without compromising the overall efficiency.

COST ANALYSIS

Generally, the cost for a renewable energy source (such as solar energy) is in the 10 to 35¢ per kWh range, depending on such variables as location, demand, availability and region. This new technology reduces the cost to around 5.7¢ per kWh. A detailed economic analysis of this technology for earth-based power generation is shown in Table 3 (at end of article). This analysis is based on a modification of the Department of Energy Laboratory (Sandia National Laboratory) Sandia Baseline 3 Photovoltaic Concentrator Module (SBM3), which was projected to be able to produce energy at around 10-11¢ per kWh [5].

This cost analysis shows that it is competitive, not only with Sandia's SBM3 system, but also with other power generation systems, such as nuclear and fossil fuels, when their environmental externality (clean-up) costs are taken into account (see Table 4). For most sources of energy, the cost per kWh is in the 5 to 12¢ per kWh range, depending on the cost of the fuel, as well as the environmental externality cost.

Table 4: **LEVELIZED GENERATING COST BREAKDOWN FOR 1200 MWe CAPACITY POWER PLANTS**

All costs are in ¢/kWh.

Plant/Fuel Type	Capital and O & M Costs	Fuel Cost	Externality Cost	Environmental[4] Externality Cost	Total Cost
Pulverized Coal	2.8	1.7	0.1	3.0	7.6
Gasified Coal	3.0	1.7	0.1	3.0	7.8
Combined-Cycle Gas	1.0	3.2	0	1.0	5.2
Nuclear	3.0	0.7	0.1	3.0	6.8
Flat Plate PV[1]	12	0	0	0	12
Sandia SBM3[2]	10	0	0	0	10
NEPS Holographic PV[3]	5.7	0	0	0	5.7

Source: USCEA [1992] for all costs, except noted as follows:

[1]Source: Scientific American [1994]

[2]Source: Sandia [1989]

[3]Source: NEPS [1994]

CONCLUSION

A holographic solar concentrator can be made which:

- concentrates sunlight,
- splits the spectrum,
- deflects far infrared energy which contributes to heating in solar cells,
- has high efficiency, and
- produces energy at a cost of 5.7¢ per kWh.

This photovoltaic system can be used with a combination of solar cells, although some combinations (GaAs-InGaAs) are more efficient than others. When compared to a Fresnel system, it was found that the holographic system concentrates more total power than the Fresnel system, the holographic system has a larger relative efficiency, and that large heatsinks are not necessary with the holographic system which decreases the bulk and cost of the system as compared to the Fresnel system.

REFERENCES

[1] S.M. Sze, *Physics of Semiconductor Devices, 2cnd Edition*, John Wiley & Sons, NY, 1981.

[2] J.L. Stone, "Photovoltaics: Unlimited Electric Energy from the Sun," *Physics Today*, Sept. 1993, p. 22.

[3] J.E. Ludman, J.L. Sampson, R.A. Bradbury, J.G. Martín, J.R. Riccobono, G. Sliker, E. Rallis, "Photovoltaic Systems Based on Spectrally Selective Holographic Concentrators," *Practical Holography VI, SPIE Proc.*, **1667**, 1992, pp. 182-189.

[4] R.J. Collier, C.B. Burckhardt, L.H. Lin, *Optical Holography*, Academic Press, 1971.

[5] E.H. Richards, "Sandia's Baseline 3 Photovoltaic Concentrator Module," *IEEE Photovoltaic Specialists Conference*, 1993, p. 1292.

ACKNOWLEDGMENTS

This research was supported by the Department of Defense, Ballistic Missile Defense Organization, and the Department of the Air Force, Air Force Systems Command, Wright Laboratory at Wright-Patterson AFB in Ohio. We would also like to thank the Photovoltaic Group at NASA-Lewis in Cleveland, Ohio for helpful discussions.

Table 3 Levelized Energy Cost Calculation for a Holographic Solar Concentrator Power Plant Based on 0.7 sq.m.

DATA INPUT SECTION	Units	'1 to 2 MW/YEAR		'20+ MW/YEAR	
Discount Rate		0.06	0.12	0.06	0.12
Lifetime	years	30	30	30	30
Module Cost	$/.7sq.m	285	285	144	144
BOS area cost	$/.7sq.m	240	240	125	125
BOS power cost	$/kW	500	500	200	200
Average Peak Insulation	kW/.7sq.m	0.8	0.8	0.8	0.8
Cell Efficiency (STC)		33.0%	33.0%	33.0%	33.0%
Cell Efficiency (NOCT)		29.7%	29.7%	29.7%	29.7%
Optical Efficiency		86.0%	86.0%	86.0%	86.0%
Performance Reduction Coefficient		95.0%	95.0%	95.0%	95.0%
PCU Efficiency		96.0%	96.0%	96.0%	96.0%
Tracking Efficiency		95%	95%	95%	95%
Indirect Cost Factor		0.5	0.5	0.3	0.3
Yearly O/M Costs	$/.7sq.m-yr	4	4	2	2
Annual Solar Availability	kW/.7sq.m-yr	2600	2600	2600	2600
Annual Systems Availability		0.95	0.95	0.95	0.95
Ground Cover Ratio		.25	.25	.25	.25
ENERGY COST CALCULATIONS					
Module Efficiency (NOTC)		25.5%	25.5%	25.5%	25.5%
Array Efficiency (DC peak)		24.3%	24.3%	24.3%	24.3%
System Efficiency (Accept)		23.3%	23.3%	23.3%	23.3%
System Efficiency (AC annual)		22.1%	22.1%	22.1%	22.1%
Module Area: 1 kW DC peak	.7sq.m	5.15	5.15	5.15	5.15
Module Cost: 1 kW DC peak	$/kW	1468	1468	742	742
BOS power cost	.7sq.m	97	97	39	39
Total BOS Costs	.7sq.m	337	337	164	164
Total Installed Costs	.7sq.m	933	933	400	400
Total Installed Costs	$/kW(DC pk)	4807	4807	2061	2061
Present Value O & M	.7sq.m	55	32	28	16
Annualized costs	.7sq.m-yr	71.79	119.84	31.07	51.68
Annualized costs		370	617	160	266
Annual System Output		547	547	547	547
Annual System Output		2816	2816	2816	2816
Required Land Area		21	21	21	21
Levelized kWh cost		**$0.1313**	**$0.2192**	**$0.0568**	**$0.0945**

Development of a Residential Use, Utility Interactive PV Inverter with Isolation Transformer-less Circuit---development aspects

H. Shinohara, K. Kimoto, T. Itami, T. Ambou, C. Okado,
K. Nakajima, S. Hojo, K. Owada, M. Kuniyoshi, Y. Sato
Toshiba Corporation
1-6, Uchisaiwai Cho, 1-Chome, Chiyoda-ku
Tokyo, 100, JAPAN

Abstract

A 3kW residential use, utility interactive PV inverter with isolation transformer-less circuit has been developed. Aiming at cost effective, compact and highly efficient PV inverter, a prototype inverter was fabricated and tested. In order to improve efficiency, AC reactors and other components were specially designed and evaluated. So far, conversion efficiency of 94.5% was attained. Aimed cost of the inverter is ¥50/Watt on the basis of 3,000 units/month production.

Introduction

In sufficiently electrified areas, it seems feasible to utilize dispersed type PV systems which are connected to the utility glid.[1][2] The research and development of utility-interactive PV inverters has been conducted since 1986.[3]

A PV inverter with isolation transformer-less circuit is now developed from 1993 to 1995. The development is under way after the completion of former development of "high frequency isolation type PV inverter".[4]

These R&D are supported by the New Energy and Industrial Technology Development organization (NEDO) as a part of New Sunshine Project. In this R&D, key issues listed below are pursued.

(1) Effective cost and compactness
(2) High reliability
(3) High efficiency and low distortion of output power

In order to realize a cost effective and compact system, a PV system which allows reverse-directional power flow without battery energy storage devices was selected as a basic concept. A prototype PV inverter was designed and manufactured as a basis for an improved type inverter to be fabricated this year.

In this paper, main features of the inverter with isolation transformer-less circuit and some test results are reported.

Circuit configuration

Main circuit configuration

Main circuit configuration is shown in Fig. 1. Table 1 shows main ratings of the isolation transformer-less PV inverter.

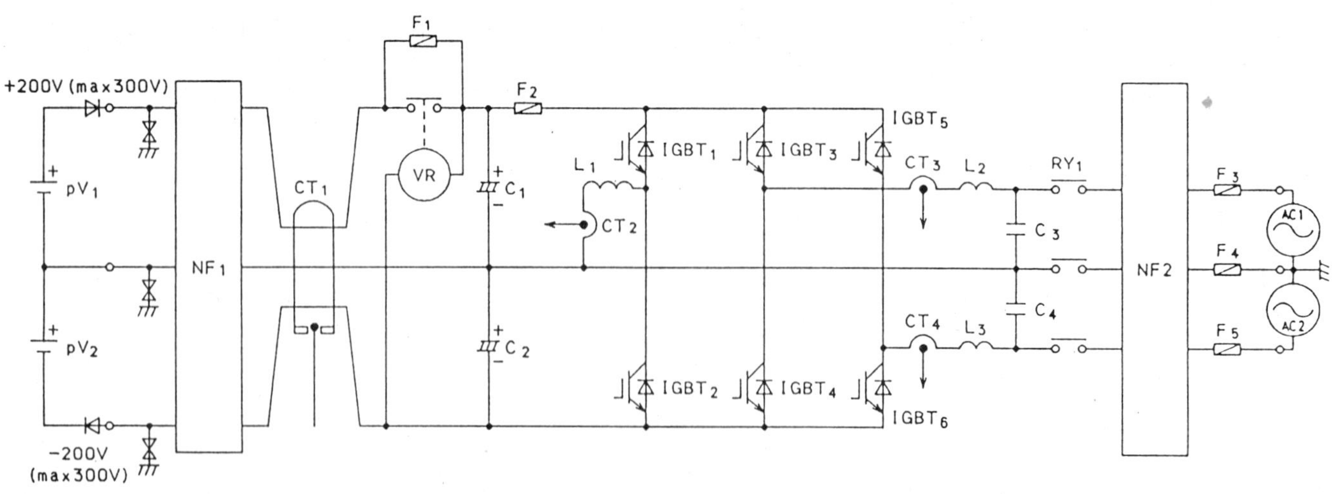

Fig. 1 Main circuit configuration

CH3365-4/94/0000-1216 $4.00 © 1994 IEEE